Encyclopedic Dictionary of Mathematics

Encyclopedic Dictionary of Mathematics

by the
Mathematical Society
of Japan

edited by
Shôkichi Iyanaga
and
Yukiyosi Kawada

translation reviewed by
Kenneth O. May

The MIT Press
Cambridge, Massachusetts,
and London, England

Originally published in 1954 by Iwanami Shoten, Publishers, Tokyo, under the title *Iwanami Sūgaku Ziten*.

Revised and augmented edition, 1960.
Second edition, 1968.
Translated by the Mathematical Society of Japan with the cooperation of the American Mathematical Society.

This book was set in Phototronic Times Roman by Science Typographers, Inc. printed on Warren Old Style by Halliday Lithograph Corp., and bound by Halliday Lithograph Corp. in the United States of America

Publisher's Note. We thank Mr. Stanley Gerr for bringing *Iwanami Sūgaku Ziten* to our attention, for his translation of a portion of the first Japanese edition—which facilitated the appraisal of this work by American mathematicians—and for his continuing interest in seeing an English-language edition reach publication.

The MIT Press

Library of Congress Cataloging in Publication Data

Nihon Sūgakkai.
Encyclopedic dictionary of mathematics.

Translation of Iwanami sūgaku ziten (i.e. jiten)
Bibliography: p.
Includes index.
1. Mathematics—Dictionaries. I. Iyanaga, Shôkichi, 1906- II. Kawada, Yukiyosi, 1916- III. Title.
QA5 N5 1977 510′.3 77-1129
ISBN 0-262-09016-3

Contents

Foreword

The American Mathematical Society welcomes the publication of the *Encyclopedic Dictionary of Mathematics*. For many years we have been fascinated by the publication in Japanese, *Iwanami Sūgaku Ziten*, because we saw that this was an encyclopedia that contained effective and penetrating information about all the fields of advanced mathematical research. We were also frustrated because we could not read Japanese and so we could not really reach out to this expert and effective source of information. We now welcome the fact that the second Japanese edition has been translated into English and we look forward to the fascination which we can now have in getting at this rich mine of information.

Saunders MacLane
President 1973–1974
American Mathematical Society

Preface to the English Edition

The first and second editions of *Iwanami Sūgaku Ziten* (in Japanese) were published, respectively, in April 1954 and June 1968 by Iwanami Shoten, Publishers, Tokyo. Beginning in the late 1950s, a number of unsuccessful attempts were made to arrange for translating the *Sūgaku Ziten* into European languages. Finally an agreement for an English translation was made between the MIT Press and the Mathematical Society of Japan in July 1968. The discussions were carried on first by Professor Kôsaku Yosida, then president of the Mathematical Society of Japan, and later by Yukiyosi Kawada, who succeeded him in April 1968. Throughout these initial negotiations, which lasted from 1966 to 1968, we received the kindest assistance from Dr. Gordon Walker, Executive Director of the American Mathematical Society, and from Professors W. T. Martin and Shizuo Kakutani.

The agreement for the project was shortly followed by the establishment of a committee for the English edition of *Sūgaku Ziten* within the Mathematical Society of Japan, with the following membership: Professors Yasuo Akizuki, Shigeru Furuya, Sin Hitotumatu, Masuo Hukuhara, Isao Imai, Shôkichi Iyanaga, Yukiyosi Kawada, Kunihiko Kodaira, Atuo Komatu, Hirokichi Kudō, Shôji Maehara, Yukio Mimura, Kiyoshi Noshiro, Shigeo Sasaki, Shoji Ura, Nobuo Yoneda, and Kôsaku Yosida. This committee requested the original authors of the articles and other members of the Society to translate the work. A list of translators will be found at the end of this work.

In November 1968, an advisory committee for the project was formed with the following membership: Professor Edwin Hewitt (chairman), Dr. Sydney H. Gould, Professor Shizuo Kakutani, Professor Kenneth O. May, and Professor Isaac Namioka.

As the translating began, we were immediately faced with problems concerning unification of terminology and style, some of which were inherent in the difference between the structures of our two languages; e.g., the Japanese language makes no distinction between singular and plural forms of nouns.

In August 1969, Professor Hewitt kindly arranged a meeting at the University of Washington, Seattle, between the members of the Japanese and American committees, and a representative of the MIT Press. It was agreed during this meeting that the translation should be faithful, with only a minimum number of changes, such as the correction of mathematical errors; whereas the references to each article might be augmented considerably for the convenience of Western readers. Professor Kenneth O. May volunteered to review the whole translated manuscript, and Professors Isaac Namioka and Shizuo Kakutani, who are proficient in both Japanese and English, proposed to read through some of the manuscript of the translated articles. It was also agreed that the Systematic List of Articles should appear in French, German, and Russian, as well as in English.

We owe very much to the American committee: Professor Hewitt organized the whole work, and Professor May revised the entire manuscript and gave us important advice concerning the appendices, according to which we deleted some of the numerical tables which may be easily found in readily accessible Western books. Professor Namioka reviewed a great part of the manuscript, transmitting his views to Professor May, who forwarded them to us with his comments. All of this assistance helped us greatly in making our final decisions. Professor Kakutani gave us very detailed and important advice on the choice of reference works.

We were also assisted concerning English terminology and reference books by the following Japanese mathematicians working in American universities: Tadatoshi Akiba, Professors Kiyosi Itô, Tatsuji Kambayashi, Tosio Kato, Teruhisa Matsusaka, Katsumi Nomizu, Ichiro Satake, Michio Suzuki, and Gaisi Takeuti.

The Mathematical Society of Japan established the following double reviewing system: group A with its twenty subgroups, each headed by the members listed below, reviewed their respective subjects; while group B reviewed the whole manuscript, mainly from the linguistic standpoint.

Group A

1. Foundations of mathematics: Shôji Maehara
2. Set theory: Atuo Komatu and Shôji Maehara
3. Algebra: Akira Hattori, Masayoshi Nagata, and Hideyuki Matsumura
4. Group theory: Shingo Murakami, Mitsuo Sugiura, and Reiji Takahashi
5. Number theory: Yukiyosi Kawada and Tomio Kubota
6. Geometry, 7. Differential geometry: Shôkichi Iyanaga, Shigeo Sasaki, and Kentaro Yano

8. Algebraic geometry: Yasuo Akizuki and Kunihiko Kodaira
9. Topology: Atuo Komatu
10. Real analysis: Sin Hitotumatu, Shunji Kametani, and Shigeki Yano
11. Complex analysis: Kiyoshi Noshiro
12. Functional analysis: Yukio Mimura and Kôsaku Yosida
13. Differential equations: Masuo Hukuhara and Sigeru Mizohata
14. Special functions: Sin Hitotumatu
15. Numerical analysis, 16. Probability theory: Kiyosi Itô
17. Statistics: Hirokichi Kudō
18. Information theory: Tosio Kitagawa and Hirofumi Uzawa
19. Theoretical physics: Isao Imai and Kazuhiko Nishijima
20. History of mathematics: Tamotsu Murata

Group B
Kenichi Iyanaga and Mitsuyo Iyanaga

Professor Sin Hitotumatu also assisted us in translating the titles of Japanese books given in the references, and the explanations attached to the lists of formulas and numerical tables in the appendices. We are also grateful for the generous cooperation offered to us by our colleagues in the Department of Mathematics, Faculty of Science, University of Tokyo: Hiroshi Fujita, Shigeru Furuya, Akio Hattori, Seizô Itô, Nagayoshi Iwahori, Tosihusa Kimura, Kunihiko Kodaira, Hikosaburo Komatsu, Akihiro Nozaki, and Itiro Tamura. We are indebted as well to Professors Walter L. Bailey of the University of Chicago and Yuji Ito of Brown University for many valuable consultations concerning both mathematical and linguistic questions.

In translating the systematic list of articles into French, German, and Russian, we were assisted by Professor Hideya Matsumoto in Paris, Professor Emanuel Sperner in Hamburg, Professor Katsuhiro Chiba in Tokyo, and Professor Arkadiĭ Malcev in Moscow.

We began to send the manuscript to the MIT Press in March 1970 and finished sending it in July 1972. The manuscript was edited there, then sent to Professors May, Namioka, and Kakutani, and finally was sent back to us with their comments and questions. All of the references were carefully checked by Laura Platt.

Iwanami Shoten, Publishers, have always been cooperative with us. In the office of the Mathematical Society of Japan, Yōko Endo, Reiko Nagase, and Chieko Sagawa helped us with their efficient secretarial work.

The fruition of this project was made possible only by the gracious assistance offered to us by many people, including those already mentioned. We should like to express our most sincere gratitude to all those who have helped us so kindly.

Shôkichi Iyanaga, Yukiyosi Kawada
August, 1973

Addition made in January 1976

After the procedures described above, the whole manuscript of this Encyclopedia was sent to the MIT Press in August 1973. It was, however, toward the end of November 1975 that the final decision was made by the MIT Press to send the manuscript to composition in early 1976 in order that the work be published in 1977.

At the same time, we were asked to review and update the manuscript up to the end of February 1976. We are now making our best effort to this effect with the kind help, especially from the linguistic viewpoint, of Dr. E. J. Brody.

In so doing, we have noticed that perhaps too much emphasis has been given to results obtained by Japanese mathematicians and that there are still many things in this book which should be improved.

We hope that ongoing revisions will be carried out in subsequent editions.

Preface to the Second Edition

Seven and a half years have passed since the revised and augmented edition of *Iwanami Sugakū Ziten* was published. The nature and purpose of this book remains the same as described in the preface to the first edition: It is an encyclopedic dictionary with articles of medium length aimed at presenting the whole of mathematics in a lucid system, giving exact definitions of important terms in both pure and applied mathematics, and describing the present state of research in each field, together with historical background and some perspectives for the future. However, mathematical science is in rapid motion, and the "present state of research" changes constantly. The present updated second edition has been published to remedy this situation as far as possible.

The main points of revision are as follows:

(1) On the Articles and the Size of the Dictionary. From the articles of the last edition, we have removed those whose importance has diminished recently (e.g., Geometry of Triangles), while we have added new articles in domains of growing importance (e.g., Categories and Functors; *K*-Theory). Many articles concerning applied mathematics in the first edition were short; in this edition, they have been combined into articles of medium length to save space and to systematize the presentation. The number of articles, 593 in the first edition, has thus diminished to 436. We have made every effort to keep the size of the encyclopedia as it was, but the substantial augmentation of content has necessarily brought about an enlargement of about 30%.

(2) On the Text. When the title of an article has remained the same as in the first edition, we have reviewed the whole text and revised whenever necessary. Especially for the fundamental ideas, we have endeavored to give thorough explanations. In the first edition we gave English, French, and German translations of article titles; in the present edition, we have also added Russian. The bibliographies at the ends of articles have been updated.

(3) On Terminology. In previous editions we endeavored to unify the terminology of the whole encyclopedia so that the reader would have no difficulties with cross-references. Here we have done this once again with the hope of attaining results more perfect than before.

(4) On Appendices. The appendices were designed to supplement the text efficiently. Some overlapping of the appendices with the text found in previous editions has been removed. Also deleted in this edition are elementary formulas in analytic geometry and tables of statistical distributions, which can be easily found in other books. However, we have added some formulas in topology, the theory of probability, and statistics, as well as tables of characters of finite groups, etc.

(5) On the Indexes. Important terms are listed multiply in the indexes to facilitate quick finding, e.g., the term *transcendental singularity* appears under both *transcendental* and *singularity*. Both names in the text and those in the references are included in the Name Index of this edition. The numbers of items in the Index of Mathematical Terms in Japanese and in European Languages, and in the Name Index in this edition are 17740, 10124, and 2438, compared with 8254, 8070, and 1279, respectively, in the previous edition.

The compilation of this edition was organized as follows. In the spring of 1964 we began to select the titles of articles with the aid of the following colleagues: in set theory and foundations of mathematics, Shôji Maehara; in algebra and number theory, Yasuo Akizuki, Yukiyosi Kawada; in differential geometry, theory of Lie groups, and topology, Yozo Matsushima, Atuo Komatu; in analysis, Masuo Hukuhara, Kôsaku Yosida, Shunji Kametani, Sin Hitotumatu; in probability theory, statistics, and mathematics for programming, Kiyosi Itô, Hirokichi Kudō, and Shigeru Furuya; in theoretical physics, Isao Imai; and for the appendices, Sin Hitotumatu. I have participated in compiling the articles on geometry and the history of mathematics. Kawada and Hitotumatu undertook the responsibility of putting the volume in order.

The work of selecting titles was completed in the summer of 1964. We then asked 173 colleagues to contribute articles. The names of these contributors and those of the previous editions are listed elsewhere. To all of them goes our most sincere gratitude.

In editing the manuscript, we were assisted by the following colleagues in addition to those mentioned already: in set theory and foundations of mathematics, Setsuya Seki and Tsurane Iwamura; in algebra and number theory, Masayosi Nagata, Akira Hattori, Hideyuki Matsumura, Ichiro Satake, Tikao Tatuzawa; in geometry, theory of Lie groups, and topology, Singo Murakami, Hideki Ozeki, Noboru Tanaka, Kiiti Morita, Hirosi Toda, Minoru Nakaoka, Masahiro Sugawara, Shôrô Araki; in analysis, Kiyoshi Noshiro, Yûsaku Komatu, Seizo Ito, Hiroshi Fujita, Shige-Toshi Kuroda, Sigeru Mizohata, Masaya

Yamaguti, Tosiya Saito, Tosihusa Kimura, Masahiro Iwano; in probability theory, statistics, and mathematics of programming, Nobuyuki Ikeda, Tadashi Ueno, Masashi Okamoto, Haruki Morimoto, Kei Takeuchi, Goro Ishii, Tokitake Kusama, Hukukane Nikaido, Toshio Kitagawa; and in theoretical physics, Ryogo Kubo, Hironari Miyazawa, Yoshihide Kozai.

After the summer of 1965, we entered into the period of finer technical editing, in which we were assisted by the following colleagues: in algebra, Keijiro Yamazaki, Shin-ichiro Ihara, Takeshi Kondo; in geometry, Tadashi Nagano, Mitsuo Sugiura, Ichiro Tamura, Kiyoshi Katase; in analysis, Nobuyuki Suita, Kotaro Oikawa, Kenkiti Kasahara, Tosinobu Muramatsu, Hikosaburo Komatsu, Setuzô Yosida, Hiroshi Tanaka; and in history, Tamotsu Murata.

We were also assisted by Katsuhiro Chiba in the Russian-language translation of the article titles, and by Osamu Kôta and Kiyoshi Katase in the indexing.

Proof sheets began appearing in the spring of 1966. In proofreading, Kaoru Sekino, Osamu Kôta, Kiyoshi Katase, and Teruo Ushijima helped us, as well as Mrs. Rieko Fujisaki. Miss Yoko Endo worked with all of us throughout the entire revision at the office of the Mathematical Society of Japan. She helped us especially in looking for and checking references and preparing the Name Index.

Yukiyosi Kawada supervised the whole work, succeeding me in the role I had played in the compilation of the first and augmented editions. Sin Hitotumatu collaborated with him throughout, especially on the appendices. The second and the third proof sheets of the text were read by Kawada; the fourth proof by Hitotumatu; the proof sheets of appendices by both Kawada and Hitotumatu.

The Editorial Committee of the Mathematical Society of Japan asked me to write this preface. Having edited the previous editions and realizing full well the difficulty of the task, I would like to express my particular gratitude to Kawada. In the Dictionary Department of Iwanami Shoten, Publishers, Messrs. Hiroshi Horie, Tetsuo Misaka, Shigeki Kobayashi, and Toshio Kouda were very cooperative in their collaboration with us. To them and also to those who typeset and printed this book at Dai-Nippon Printing Co. and Shaken Co. goes our gratitude.

S. Iyanaga
March 1968

Preface to the Revised and Augmented Edition

Six years have passed since the publication of the first edition of this encyclopedia. This revised and augmented edition incorporates the achievement of these years. It contains, together with the correction of errors found in the first edition, some new articles such as Abelian Varieties, Automata, Sheaves, Homological Algebra, Information Theory, and also supplements to articles in the first edition such as Complex Multiplication, Computers, and Manifolds. These additional items render the new edition 93 pages longer than the previous one. Each article has been thoroughly revised, and the indexes have been completely rewritten.

We were assisted by the following colleagues in selecting articles, writing and proofreading: In set theory and foundations of mathematics, Sigekatu Kuroda, Setsuya Seki; in algebra and number theory, Tadao Tannaka, Tsuneo Tamagawa: in real analysis, Shunji Kametani, Kôsaku Yosida; in function theory, Kiyoshi Noshiro, Sin Hitotumatu; in theory of differential and functional equations, Masuo Hukuhara, Masahiro Iwano, Ken Yamanaka; in functional analysis, Kôsaku Yosida, Seizô Itô; in geometry, Shigeo Sasaki, Nagayoshi Iwahori; in topology, Atuo Komatu, Itiro Tamura, Nobuo Yoneda; in theory of probability, Kiyosi Itô, Seizô Itô; in statistics, Toshio Kitagawa, Sigeiti Moriguti, Tatsuo Kawata; in applied mathematics, Sigeiti Moriguti; and in mechanics and theoretical physics, Takahiko Yamanouchi, Isao Imai. The revision and augmentation of the articles concerning the history of mathematics was done by myself. Mlles. Yôko Tao, Eiko Miyagawa, and Mutsuko Nogami worked in the office of the Society.

The names of authors who contributed to the completion of this edition have been added to the original list of contributors.

The project of editing this edition started in the summer of 1958. We acknowledge our deep gratitude to all those who have collaborated with us since that time.

S. Iyanaga
August 1960

Preface to the First Edition

This encyclopedia, *Iwanami Sūgaku Ziten*, was compiled by the Mathematical Society of Japan at the request of the Iwanami Shoten, Publishers, who have hitherto published a series of scientific dictionaries such as *Iwanami Rikagaku-Ziten* (*Iwanami Encyclopedia of Physics and Chemistry*) and *Iwanami Tetugaku-Ziten* (*Iwanami Encyclopedia of Philosophy*). As mentioned in the prefaces to these volumes, the importance of such encyclopedias in clarifying the present state of each science is obvious if we observe the rapid pace of contemporary research. Mathematics is also in rapid motion. As a fundamental part of exact science, it serves as a basis of all science and technology. It also retains its close contact with philosophy. Therefore, the significance of having an encyclopedia of mathematics cannot be overemphasized.

Mathematics has made remarkable progress in the 20th century. As for the situation toward the end of the 19th century, we quote the following passage from the article Mathematics in the 19th Century of this encyclopedia: "Toward the end of [the 19th] century, the subjects of mathematical research became highly differentiated. Branches split into more specialized areas of study, while unexpected relations were found between previously unconnected fields. The situation became so complicated that it was difficult to view mathematics as a whole. It was in these circumstances that in 1898, at the suggestion of Franz Meyer and under the sponsorship of the Academies of Göttingen, Berlin, and Vienna, a project was initiated to compile an encyclopedia of the mathematical sciences. Entitled the *Enzyklopädie der mathematischen Wissenschaften*, it was completed in 20 years. . . ."

One of the characteristics of 20th-century mathematics is the conscious utilization of the axiomatic method, and of general concepts such as sets and mappings, which serve as foundations of different theories. Indeed, mathematics is being reorganized on the basis of topology and algebra. One such example of reorganization is found in Bourbaki's *Eléments de mathématique*; some fifteen volumes of this series have been published since 1939, and more are coming. This encyclopedia, with its limited size, cannot contain proofs for theorems. However, we intend to present a lucid view of the totality of mathematics, including its historical background and future possibilities.

Each article of this encyclopedia is of medium length; it is sufficiently short to permit the reader to find exact definitions of notions, and sufficiently long to contain explanations clarifying how important concepts in the same field are related to each other. The problem of choosing adequate titles required some deliberation. The Systematic List of Articles, classified according to specific fields, shows those we have chosen. The Index of Terms contains detailed references for each notion. The appendices, including formulas and tables, supplement the text, and will be particularly useful for applied mathematicians.

The project of compiling this encyclopedia was proposed in the spring of 1947 by the Steering Committee of the Mathematical Society of Japan. It was promptly adopted, and the selection of articles in specific fields was started by the sectional committees of the Society. After seven years, our encyclopedia is finally appearing. We shall not give a detailed description of how our work proceeded through all these years. We list simply the names of those who assisted us greatly and to whom we should like to express our deep gratitude.

The president of the Society at the start of this project was the late Professor Tadahiko Kubota; but our work has been supported also by Professors Teiji Takagi, Zyoiti Suetuna, and Masatsugu Tsuji as well as by other leading members of the Society.

At the stage of selecting articles, we were assisted by the following colleagues: in history and the foundations of mathematics, Sigekatu Kuroda, Motokiti Kondo; in algebra and number theory, Kenjiro Shoda, Tadasi Nakayama, Masao Sugawara, Yukiyosi Kawada, Kenkichi Iwasawa; in geometry, Kentaro Yano, Asajiro Ichida; in function theory, Kiyoshi Noshiro, Yûsaku Komatu; in the theory of differential and functional equations, Masuo Hukuhara, Shigeru Furuya; in topology, Atuo Komatu, Ryoji Shizuma; in functional analysis, Yukio Mimura, Shizuo Kakutani, Kôsaku Yosida; in the theory of probability and statistics, Tatsuo Kawata, Toshio Kitagawa, Junjiro Ogawa; and in applied mathematics, Ayao Amemiya, Isao Imai, Kunihiko Kodaira, Shigeiti Moriguti.

We asked 190 colleagues to contribute articles, which were collected in 1949. Since then we have spent an unexpectedly long time editing them. Terminology had to be unified throughout the encyclopedia so that the reader would have no trouble with cross-references. Repetitions had to be eliminated and gaps filled. Part of the manuscript thus had to be rewritten a number of times. We have made our utmost effort in this editing work, but we are not completely without apprehension that our result has still left something to be desired. For any shortcomings in the work, I take complete responsibility, as I have acted as the editor-in-chief. Also, since we have rewritten the manuscript, as already mentioned, we have refrained from printing the name of the original author of each article; for this, I must request the understanding of the contributors.

In the stage of editing and proofreading, we were assisted by the following colleagues: Yukio Mimura, Yukiyosi Kawada, Kazuo Matsuzaka, Sin Hitotumatu, Setsuo Fukutomi, Setsuya Seki, Shoji Irie, Shigeo Sasaki, Tatsuo Kawata, Sigekatu Kuroda, Yûsaku Komatu, Ayao Amemiya, Isao Iami, Tosio Kato, Tsurane Iwamura, Morikuni Goto, Kôsaku Yosida, Jirô Tamura, Yasuo Akizuki, Kiyoshi Noshiro, Motosaburo Masuyama, Sigeiti Moriguti, Osamu Kôta, Nobuo Yoneda, Tsuneo Tamagawa, Jun-ichi Hano; and more particularly, in the foundations of mathematics, Sigekatu Kuroda, Tsurane Iwamura; in algebra and number theory, Kazuo Matsuzaka, Yukiyosi Kawada; in algebraic geometry, Yakuo Akizuki; in real analysis, Tatsuo Kawata; in complex analysis, Yûsaku Komatu, Sin Hitotumatu, Jirô Tamura; in functional analysis, Kôsaku Yosida; in topology, Setsuo Fukutomi, Nobuo Yoneda; in the theory of probability and statistics, Motosaburo Masuyama, Sigeiti Moriguti; and in applied mathematics, Ayao Amemia, Isao Imai, Tosio Kato, Sigeiti Moriguti.

The portraits of Abel and Riemann were kindly loaned to us by Torataro Shimomura [these do not appear in the English-language edition—ed].

The formulas in the appendices were compiled by Isao Imai, Sin Hitotumatu, and Sigeiti Moriguti; the Subject Index (in Japanese and European languages) by Osamu Kôta and Mrs. Hiroko Ide; the Name Index and Comments on Journals and Serials by Setsuo Fukutomi. Setsuo Fukutomi has taken an active part in our work ever since 1948 and given much effort to collecting and rewriting the manuscript and to unifying terminology. The editorial staff of Iwanami Shoten, Publishers, has always been cooperative. Without their generous support, this encyclopedia could never have been published.

I should like to express my sincere gratitude to all those who have collaborated with us directly or indirectly.

S. Iyanaga
March 1954

Introduction

The text of this Encyclopedic Dictionary consists of 436 *articles* arranged alphabetically, beginning with 1 Abel and ending with 436 Zeta Functions. Most of these articles are divided into *sections*, indicated by A, B, C, . . . , AA, BB, ... Cross-references to articles, e.g., to the second article, are of the form: (→2 Abelian Groups) or (→2 Abelian Groups A), according as the whole article or a specific section is being referred to. Citations in the indexes are also given in terms of article and section numbers.

Key terms accompanied by their definitions in the text are printed in boldface. All of these terms are found in the Subject Index at the end of the volume.

The sign † means that the term preceded by it can be found in the index. A list of special notations used throughout the work (with explanations of their meanings) appears after the appendices.

A Systematic List of Articles, showing the general structure of the work, follows the text. (French, German, and Russian translations of the systematic list are also given.) The number in parentheses after each article title refers to this systematic classification; e.g., "Abelian Varieties (VIII.5)" means that the article on Abelian varieties is the fifth article in Section VIII of the systematic list.

Books and articles in journals are cited in the text by numbers in brackets: [1], [2], ... At the end of each article there is a section of references in which, for books, the name of the author or authors, title, name of the publisher, year of publication, and the number of the edition are given; for journals, the name of the author, title of the article, name of the journal, and the volume numbers and inclusive page numbers are given in this order. (The names of journals and publishers are abbreviated as indicated in the lists at the end of the work.)

The Cyrillic alphabet is transliterated as follows:

Cyrillic Alphabet			**Transliteration**
А	а	(a)	a
Б	б	(be)	b
В	в	(ve)	v
Г	г	(ge)	g
Д	д	(de)	d
Е	е	(ye)	e
Ё	ё	(yo)	ë
Ж	ж	(zhe)	ž
З	з	(ze)	z
И	и	(i)	i
(Й)	й	(i kratkoe)	ĭ
К	к	(ka)	k
Л	л	(el')	l
М	м	(em)	m
Н	н	(en)	n
О	о	(o)	o
П	п	(pe)	p
Р	р	(er)	r
С	с	(es)	s
Т	т	(te)	t
У	у	(u)	u
Ф	ф	(ef)	f
Х	х	(kha)	h
Ц	ц	(tse)	c
Ч	ч	(che)	č
Ш	ш	(sha)	š
Щ	щ	(shcha)	šč
(Ъ)	ъ	(tvërdiĭ znak)	"
(Ы)	ы	(yery)	y
(Ь)	ь	(myahkiĭ znak)	'
Э	э	(e)	è
Ю	ю	(yu)	ju
Я	я	(ya)	ja

A

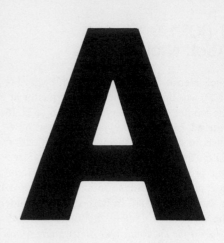

1 (XX.12) Abel, Niels Henrik

Niels Henrik Abel (August 5, 1802–April 6, 1829) was born the son of a poor pastor in the hamlet of Findö in Norway. In 1822, he entered the University of Christiania; however, he studied mathematics almost entirely on his own. Nonetheless he was recognized as a promising student by his senior, Holmboe, and after graduation he studied abroad in Berlin and Paris. In Berlin he came under the favor of A. Crelle, the founder of the *Journal für die Reine und Angewandte Mathematik*, and participated in the founding of this journal. Although he did brilliant work in Paris, he did not gain the fame he deserved. He returned to Norway in May 1827, but, unable to find a job, he was obliged to fight poverty while continuing his research. He died at twenty-six of tuberculosis.

His best-known works are: the result that algebraic equations of order five or above cannot generally be solved algebraically; the result that †Abelian equations can be solved algebraically; the theory of †binomial series and of †elliptic functions; and the introduction of †Abelian functions. His work in both algebra and analysis, written in a style conducive to easy comprehension, reached the highest level of attainment of his time.

References

[1] N. H. Abel, Oeuvres complètes I, II, edited by L. Sylow and S. Lie, Grondahl & Son, Christiania, new edition, 1881.
[2] C. A. Bjerknes, Niels Henrik Abel, Gauthier-Villars, 1885.
[3] F. Klein, Vorlesungen über die Entwicklung der Mathematik im 19. Jahrhundert I, Springer, 1926 (Chelsea, 1956).
[4] T. Takagi, Kinsei sûgakusidan (Japanese; Topics from the history of mathematics of the 19th century), Kawade, 1943 (Kyôritu, 1970).

2 (IV.2) Abelian Groups

A. General Remarks

A †group G is called an **Abelian group** (or **commutative group**) if G satisfies the commutative law $ab = ba$ for all $a, b \in G$. In this article, G always denotes an Abelian group. Every †subgroup of G is a †normal subgroup, and all elements of finite order in G form a subgroup T, for which the †factor group G/T has no elements of finite order except the identity e. T is called the **(maximal) torsion subgroup** of G. If $G = T$, then the Abelian group G is called a **torsion group** (or **periodic group**). On the other hand, if $T = \{e\}$, then G is called **torsion-free**, and if $T \neq G$, $T \neq \{e\}$, then G is called **mixed.** If the order of every element of a torsion group G is a power of a fixed prime number p, then G is called an **Abelian p-group** (or **primary Abelian group**). An Abelian torsion group is the †direct sum of primary Abelian groups. Thus the study of torsion groups is reduced to that of primary Abelian groups.

B. Finite Abelian Groups

The following fundamental theorem on finite Abelian groups was established by L. Kronecker, G. Frobenius, and L. Stickelberger in the 1870s. An Abelian group G of order p^n, where p is a prime number, is a direct product of †cyclic subgroups $Z_1, \ldots, Z_r: G = Z_1 \times \ldots \times Z_r$. If Z_i is of order p^{n_i}, then $n = n_1 + \ldots + n_r$ and we may assume that $n_i \geqslant n_{i+1}$. A direct product decomposition of G, as above, is not unique, but $n_1, \ldots, n_r$ are determined uniquely by G. The system $\{p^{n_1}, \ldots, p^{n_r}\}$ or $\{n_1, \ldots, n_r\}$ is called the system of **invariants** (or **type**) of G, and a system of †generators $\{z_1, \ldots, z_r\}$ of $Z_1, \ldots, Z_r$ is called a **basis** of G. An Abelian group of type $(p, p, \ldots, p)$ is called an **elementary Abelian group.** The decomposition of a finite Abelian group into a †direct sum of subsets (not necessarily of subgroups) was considered by G. Hajós (1942) and applied successfully to a problem of number theory (→ 160 Finite Groups).

C. Finitely Generated Abelian Groups

The theory of †finitely generated Abelian groups, i.e., Abelian groups generated by finite numbers of elements, is as old as that of finite Abelian groups. The direct product of †infinite cyclic groups is called a **free Abelian group**. A finitely generated Abelian group G is the direct product of a finite Abelian group and a free Abelian group. The finite factor is the **torsion subgroup** of G. The free factor of the group G is not necessarily unique; however, the number of infinite cyclic factors of the free factor is uniquely determined and is called the **rank** of G. Two finitely generated Abelian groups are isomorphic if they have isomorphic maximal torsion subgroups and the same rank. This theory can be extended to the theory of †modules over a †principal ideal domain (→ 70 Commutative Rings K).

D. Torsion Groups

The structure of Abelian p-groups is relatively well known, compared with other infinitely generated Abelian torsion groups. In the 1920s, H. Prüfer made the first important contribution to the study of Abelian p-groups, and H. Ulm and L. Zippin completed the theory for countable groups in the 1930s. The uncountable case was first treated by L. Kulikov in the 1940s, but the study of this case is still in progress.

An Abelian p-group $G \neq \{e\}$ is called **divisible** (or **complete**) if for any $a \in G$ there is an element $x \in G$ satisfying $x^p = a$. A divisible group is a †direct sum of **Abelian groups of type** p^∞ (Prüfer). Here a group of type p^∞ is isomorphic to the †multiplicative group of all the p^nth roots of unity ($n = 1, 2, \ldots$) in the complex number field. Let G be any Abelian p-group. The maximal divisible subgroup V of G is a direct product factor of $G: G = V \times R$, where R has no divisible subgroups. An Abelian p-group without a divisible subgroup is called a **reduced Abelian group**.

An element x of an Abelian p-group G is said to have **infinite height** if for any n there is an element $y_n \in G$ satisfying $x = y_n^{p^n}$. The elements of infinite height form a subgroup G^1 of G. If $G^1 = \{e\}$ and G is countable, then G is decomposed uniquely into the direct sum of cyclic groups. This assertion fails if the hypothesis of countability is dropped. By †transfinite induction we can define G^β as follows. If β is an †isolated ordinal number, then $G^\beta = (G^{\beta-1})^1$, and if β is a †limit ordinal number, then $G^\beta = \bigcap_{\alpha<\beta} G^\alpha$. For the least ordinal number τ such that $G^\tau = G^{\tau+1}$, G^τ is the maximal divisible subgroup of G. If G is reduced, then $G^\tau = \{e\}$. We call τ the **type** of an Abelian p-group G. For $\alpha < \tau$, $\bar{G}^\alpha = G^\alpha / G^{\alpha+1}$ is called the **Ulm factor** of G and the sequence $\bar{G}^0, \ldots, \bar{G}^\alpha, \ldots$ $(\alpha < \tau)$ is called the **sequence of Ulm factors** of G. Each Ulm factor $\bar{G}^\alpha$ has no element of infinite height, and if $\alpha < \tau - 1$, then $\bar{G}^\alpha$ has an element of arbitrarily large order. Let τ be a countable ordinal number, and assume that for any ordinal number $\alpha < \tau$ there is given a countable Abelian p-group A_α such that A_α has no element of infinite height, and that for $\alpha \neq \tau - 1$ A_α has an element of arbitrarily large order. Then there is a reduced countable Abelian p-group which is of type τ with a sequence of Ulm factors isomorphic to A_0, $A_1, \ldots, A_\alpha, \ldots$ $(\alpha < \tau)$ (Zippin). Two reduced countable Abelian p-groups A and B are isomorphic if they have the same type τ; and for any $\alpha < \tau$ the Ulm factors $\bar{A}^\alpha$ and $\bar{B}^\alpha$ are isomorphic. The assertion fails if the hypothesis of countability is dropped.

E. Torsion-Free Groups

In Abelian groups, the group operation is often written $a + b$, using the additive notation; and an additively written group, called an **additive group**, is generally assumed to be Abelian. In the rest of this article we consider exclusively additive Abelian groups, of which the additive group **Z** of rational integers is the most primitive example. In such a group the identity element is called the **zero element** and is denoted by 0, the inverse of a is denoted by $-a$, and we write $a + (-b) = a - b$. The direct sum of additive groups A_λ $(\lambda \in \Lambda)$ is called a **free additive group** if each A_λ is isomorphic to **Z**. An additive group G is regarded as a †module over the †ring **Z**, to which the notion of linear independence is applicable (→ 275 Modules). Elements $a_1, \ldots, a_r$ of G are **linearly dependent** if there are integers $n_1, \ldots, n_r$ not all of which are zero such that $n_1 a_1 + \ldots + n_r a_r = 0$. Those that are not linearly dependent are termed **linearly independent**. An infinite set of elements of G is called linearly independent if the elements of any finite subset are linearly independent. If there are N elements of G that are linearly independent, but if any $N+1$ elements of G are linearly dependent, then N is called the **rank** of G. Such a system of N linearly independent elements is called a **maximal independent system**. A torsion-free additive group G is not necessarily free if G is not finitely generated.

The first important work on torsion-free additive groups was done by F. W. Levi (1917). A. G. Kuroš (1937) completed the theory in the case of finite rank. In the general case little is known, and I. Kaplansky, J. Rotman, and others are continuing the investigation.

The additive group **Q** of rational numbers is of rank 1, and conversely any additive group of rank 1 is isomorphic to some subgroup of **Q**. An additive group G is called **divisible** (or **complete**) if for any $a \in G$ and for any integer n there is an element $x_n \in G$ such that $nx_n = a$. A divisible torsion-free additive group is isomorphic to a direct sum of some copies of **Q**. For any torsion-free additive group G there is a divisible torsion-free additive group containing G. A minimal additive group F among these groups is uniquely determined up to isomorphism and has the rational number field $\mathbf{Q}$ as an †operator domain. Let $\mathbf{Q}^{(p)}\{a/b \mid (a,b)=1, p \nmid b\}$ be the ring of p-integers in $\mathbf{Q}$, and let G_p be the smallest $\mathbf{Q}^{(p)}$-subgroup of F containing G. Let $\mathbf{Q}_p$ be the †p-adic number field and $\mathbf{Z}_p$ the ring of †p-adic integers. Extending the operator domain $\mathbf{Q}$ to $\mathbf{Q}_p$ we obtain naturally a $\mathbf{Q}_p$-module F_p from F. Let $\bar{G}_p$ be the natural closure of G_p in F_p. Then $\bar{G}_p$ has $\mathbf{Z}_p$ as an operator domain and thus becomes a $\mathbf{Z}_p$-module. A $\mathbf{Z}_p$-module of rank N is isomorphic to the direct sum of κ_p copies of $\mathbf{Q}_p$ and $N-\kappa_p$ copies of $\mathbf{Z}_p : \bar{G}_p \cong \Sigma_n \mathbf{Q}_p v_n + \Sigma_m \mathbf{Z}_p w_m$ ($n = 1, \ldots, \kappa_p; m = 1, \ldots, N-\kappa_p$). Here κ_p is called the **p-rank** of G. As the invariants of G, Kuroš gives the rank, p-ranks for all primes p, and a certain equivalence class of the sequence of the matrices $\mathfrak{M}_p$. Here p ranges over all primes, and $\mathfrak{M}_p$ is the matrix of coefficients when the elements of a maximal independent system of F are written as linear combinations of (v_n, w_m).

F. General Abelian Groups

An Abelian group is, in general, an †extension of a torsion group by a torsion-free group. A torsion group T is called **bounded** if there is an integer n such that $t^n = 1$ for all $t \in T$. Suppose there is a torsion group T. Then T is a direct summand of an Abelian group G which contains T as its maximal torsion subgroup if and only if T is the direct product of a divisible group and a bounded group (R. Baer and S. Fomin).

G. Characters

A **character** χ of an Abelian group G is a function which assigns to each $a \in G$ a complex number $\chi(a)$ of absolute value 1 and satisfies $\chi(ab) = \chi(a)\chi(b)$ for all $a, b \in G$. The product $\chi = \chi_1\chi_2$ of two characters χ_1 and χ_2 is defined by $\chi(a) = \chi_1(a)\chi_2(a)$, and χ is also a character of G. Thus, all the characters of G form an Abelian group $C(G)$, which is called the **character group** of G. The identity element of the character group is the **identity character** (or **principal character**) χ such that $\chi(a) = 1$ for all $a \in G$. If G is finite, then $G \cong C(G)$. This implies the duality $G = C[C(G)]$. This fact was extended by L. S. Pontrjagin to †locally compact †topological Abelian groups (→ 405 Topological Abelian Groups B–D). For additive Abelian groups with operator domains → 275 Modules.

References

[1] L. Fuchs, Abelian groups, Pergamon, 1960.
[2] I. Kaplansky, Infinite Abelian groups, Univ. of Michigan, 1954.
[3] А. Г. Курош (A. G. Kuroš), Теория групп, Гостехиздат, second edition, 1953;
English translation, The theory of groups I, II, Chelsea, 1960.

3 (VIII.5) Abelian Varieties

A. History

N. H. Abel was the first to consider †algebraic functions as functions of complex variables and to discover double periods of †elliptic functions such as $x = x(u)$, which is the inverse function of an elliptic integral

$$u = \int^x \frac{dx}{\sqrt{f_4(x)}},$$

where $f_4(x)$ is a polynomial in x of degree 4. C. G. Jacobi expressed elliptic functions explicitly as ratios of theta series. Elliptic functions are special cases of †Abelian functions, and it is natural to consider the inverse function of the †hyperelliptic integral, or more generally, the inverse function of the Abelian integral. Actually, by investigating hyperelliptic integrals of the first kind of genus 2,

$$\int^{x_1} \frac{dx_1}{\sqrt{f_6(x_1)}}, \qquad \int^{x_2} \frac{x_2\,dx_2}{\sqrt{f_6(x_2)}},$$

where $f_6(x)$ is a polynomial of degree 6 in x, Jacobi obtained multiple-valued functions with quadruple periods. Furthermore, by considering sums of two integrals

$$u_1 = \int^{x_1} \frac{dx_1}{\sqrt{f_6(x_1)}} + \int^{x_2} \frac{dx_2}{\sqrt{f_6(x_2)}},$$

$$u_2 = \int^{x_1} \frac{x_1\,dx_1}{\sqrt{f_6(x_1)}} + \int^{x_2} \frac{x_2\,dx_2}{\sqrt{f_6(x_2)}},$$

he discovered the remarkable fact that the elementary symmetric functions $s_1 = x_1 + x_2$ and $s_2 = x_1x_2$ of x_1 and x_2 are single-valued functions of u_1 and u_2 with quadruple periods. He also conjectured that these functions s_1 and s_2 might be expressed explicitly in terms of theta series of u_1 and u_2; this conjecture was later confirmed by J. G. Rosenhain and A. Göpel.

In the latter half of the 19th century, efforts to establish a foundation for these results led to the recognition of the importance of Jacobi's inverse problem, which sought to obtain a g-tuple $(P_1, \ldots, P_g)$ of points satisfying the relations $\Sigma_{j=1}^{g} \int^{P_j} \omega_i \equiv u_i$ (mod periods), $i = 1, \ldots, g$, for g linearly independent differentials

$\omega_1, \ldots, \omega_g$ of the first kind on an algebraic curve of genus g and for an arbitrarily given g-tuple $(u_1, \ldots, u_g)$ of complex numbers, and to express each point explicitly as a function of $u_1, \ldots, u_g$. Generally, such a g-tuple $(P_1, \ldots, P_g)$ of points is uniquely determined up to order. Hence the essential part of Jacobi's problem is to express elementary Abelian functions and elementary symmetric functions of the coordinates of these points as functions of $(u_1, \ldots, u_g)$. B. Riemann was the first to succeed in expressing them as rational functions of theta functions.

The theory of functions with multiple periods was further developed by H. Poincaré, G. Frobenius, and E. Picard. In the 20th century, the importance of the theories of Abelian functions and Abelian varieties has become more obvious with the development of the theory of functions of several complex variables and algebraic geometry. Valuable contributions have been made by S. Lefschetz, C.L. Siegel, and others. Recently, A. Weil succeeded in constructing the theory of Abelian varieties from a purely algebraic standpoint [21]. This is important to both algebraic geometry and number theory.

B. Algebraic Theory

When a †group variety is †complete as a variety, the group law is commutative; such a group variety is called an **Abelian variety** (→ 18 Algebraic Varieties). Let B be a subvariety of an Abelian variety A, and assume that B is a subgroup of A as an abstract group. Then B has the structure of an Abelian variety whose law of composition is induced by that of A, and B is called an **Abelian subvariety** of A. More generally, when an algebraic subset $\mathfrak{B}$ is a subgroup of A, then the component B of $\mathfrak{B}$ containing the identity element is an Abelian subvariety, and $\mathfrak{B}$ is a union of B and a finite number of cosets of B. When A is defined over a field k, then any Abelian subvariety of A is defined over a finite †separable extension of k (W. L. Chow's theorem). An Abelian variety A is called **simple** if A has no Abelian subvarieties other than A itself and 0.

Every †rational mapping of an algebraic variety V into an Abelian variety is defined at each simple point of V. This implies that an Abelian variety is the †minimal model among all algebraic varieties that are †birationally equivalent to it.

C. Homomorphisms

A rational mapping of an Abelian variety A into an Abelian variety B is called a **rational homomorphism** (or simply **homomorphism**) if f is a group homomorphism. Let F be a rational mapping of A into B; then F can be uniquely expressed as follows: $F(x) = F_0(x) + F(0)$ $(x \in A)$, where F_0 is a homomorphism and $F(0)$ is the image of the identity element 0 of A. Hence the structure of an Abelian variety is essentially determined by its underlying variety.

When a rational homomorphism f is birational, f is called a **birational isomorphism** (or simply **isomorphism**). It is clear that a rational isomorphism is an abstract isomorphism, but the converse is not necessarily true. Let A, B be two Abelian varieties. We say that A is **isogenous** to B if the dimension of A is equal to that of B and there exists a surjective homomorphism of A onto B, or equivalently, if there exists a surjective homomorphism of A onto B whose kernel is finite. The relation of isogeny is an equivalence relation. Every Abelian variety is isogenous to a product of simple Abelian varieties that are determined uniquely up to isogeny and order.

Let A, B be two Abelian varieties; we denote by $\mathrm{Hom}(A,B)$ the additive group of rational homomorphisms of A into B. When a rational homomorphism λ is surjective, then the **degree** $\nu(\lambda)$ of λ is defined by $\lambda(A) = \nu(\lambda)B$. If λ is an isogeny, then $\nu(\lambda) \neq 0$, and the order of the kernel $\{t \mid t \in A, \lambda t = 0\}$ is at most $\nu(\lambda)$. The additive group $\mathrm{Hom}(A,B)$ is finitely generated. If $A = B$, then $\mathrm{Hom}(A,A)$ has a ring structure; it is called the **ring of endomorphisms** (or **endomorphism ring**) of A and is denoted by $\mathfrak{A}(A)$. The tensor product $\mathfrak{A}_0(A) = \mathfrak{A}(A) \otimes \mathbf{Q}$, where $\mathbf{Q}$ is the field of rational numbers, is an †associative algebra over $\mathbf{Q}$, and its dimension is at most $4n$, where n is the dimension of A. If A is simple, then $\mathfrak{A}_0(A)$ is a †division algebra. More generally, $\mathfrak{A}_0(A)$ is isomorphic to a direct product of some †total matrix algebras over division algebras; thus $\mathfrak{A}_0(A)$ is †semisimple. In particular, if A is 1-dimensional (in other words, if A is an †elliptic curve), the types of $\mathfrak{A}_0(A)$ are well known; when the characteristic $p = 0$, then $\mathfrak{A}_0(A)$ is either the field of rational numbers or an †imaginary quadratic field. When $p > 0$, aside from these two fields, we have a †quaternion algebra over $\mathbf{Q}$ as the possible type of $\mathfrak{A}_0(A)$.

Let k be a finite field with q elements. An algebraic integer is called a **Weil number** for q if every conjugate of it has absolute value $\sqrt{q}$. If A is an Abelian variety defined and simple over k, the qth power endomorphism of $A : x \to x^q$ determines a conjugacy class of Weil numbers for q, as Weil showed (→ 436 Zeta Functions). Moreover, we have the following classification theorem (J. Tate, T.

Honda): There is a one-to-one correspondence between the set of all k-isogeny classes of k-simple Abelian varieties over k and the set of all conjugacy classes of Weil numbers for q. Tate also determined the structure of the division algebra $\mathfrak{A}_0(A)$ over $\mathbf{Q}$, which is described in terms of the decomposition of the qth power endomorphism into prime ideals.

D. Divisors

Let $\mathfrak{G}$ be the additive group of †divisors on an Abelian variety A and $\mathfrak{G}_a$ be the subgroup of divisors that are †algebraically equivalent to 0. Then the factor group $\mathfrak{G}/\mathfrak{G}_a$ has no †torsion part; this implies that for an Abelian variety †numerical equivalence coincides with †algebraic equivalence. We denote this relation by $\equiv$. Given an element a of A, the translation $T_a: A \ni x \rightarrow x + a \in A$ gives a †birational transformation, which is everywhere †biregular, on the underlying variety of A; we denote by X_a the image of a divisor X on A. Then $X \equiv 0$ if and only if X_a is †linearly equivalent to X for each point a of A. Also, there exists an injective homomorphism of $\mathfrak{G}/\mathfrak{G}_a$ into $\mathfrak{A}(A)$. The †Albanese variety of an Abelian variety A is A itself, and the †Picard variety $\hat{A}$ of A is isogenous to A. In particular, for the †Jacobian variety J of an algebraic curve, $\hat{J}$ is isomorphic to J itself. The Picard variety $\hat{\hat{A}}$ of $\hat{A}$ is isomorphic to A (**duality theorem**). Let X be a divisor on A; the mapping $a \rightarrow$the †linear equivalence class of the divisor $X_a - X$, $a \in A$, is a rational homomorphism of A into $\hat{A}$, and we denote it by φ_X. If $\varphi_X = 0$, then $X \equiv 0$, and vice versa. If φ_X is surjective, we say that X is **nondegenerate.** A †positive divisor X is nondegenerate if and only if X is †ample (→ 18 Algebraic Varieties N). There always exist positive nondegenerate divisors on an Abelian variety; therefore an Abelian variety can be embedded in some projective space. For a given divisor X on A, we can find n suitable points $u_1, \ldots, u_n$, where n is the dimension of A, so that the †intersection product $X_{u_1} \cdot \ldots \cdot X_{u_n}$ is defined. We denote by $(X^{(n)})$ the †degree of the zero cycle $X_{u_1} \cdot \ldots \cdot X_{u_n}$. If X is positive nondegenerate, then the dimension $l(X)$ of the †defining module of the †complete linear system determined by X is equal to $(X^{(n)})/n!$ (**Poincaré's theorem**). Furthermore, the degree $\nu(\varphi_X)$ of φ_X, where X is any divisor on A, is given by the formula $\nu(\varphi_X) = (X^{(n)})/n!$ (**Frobenius's theorem**).

E. l-adic Representations

Let A be an Abelian variety of dimension n. For a given prime number l, let $\mathfrak{G}_l(A)$ denote the group of points on A whose order is a power of l. If l is different from the characteristic of the base field of A, then the group $\mathfrak{G}_l(A)$ is isomorphic to the direct product of $2n$ factor groups $\mathbf{Q}_l/\mathbf{Z}_l$, where $\mathbf{Q}_l$ is the field of l-adic numbers and $\mathbf{Z}_l$ is the group of l-adic integers (→ 425 Valuations). We call such an isomorphism the ***l*-adic coordinate system** of $\mathfrak{G}_l(A)$. Now let λ be a rational homomorphism of A into an Abelian variety B of dimension m. Then we can see that λ induces a homomorphism of $\mathfrak{G}_l(A)$ into $\mathfrak{G}_l(B)$. This shows that by placing l-adic coordinate systems on $\mathfrak{G}_l(A)$ and $\mathfrak{G}_l(B)$ respectively, we get a matrix representation $M_l(\lambda)$ of λ with $2m$ rows and $2n$ columns. The representation $\lambda \rightarrow M_l(\lambda)$ is faithful, and $M_l(\lambda)$ is called the ***l*-adic representation** of λ. When $A = B$, then $\lambda \rightarrow M_l(\lambda)$ is a faithful representation of the ring of endomorphisms $\mathfrak{A}(A)$. This representation can be naturally extended to the representation of the algebra $\mathfrak{A}_0(A)$; the characteristic polynomial of the l-adic representation $M_l(\lambda)$ (where λ is an element of $A_0(A)$) is a polynomial with coefficients in $\mathbf{Q}$. Moreover, the polynomial does not depend on the choice of the prime number l. When $\lambda \in \mathfrak{A}(A)$, then $\nu(\lambda)$ is equal to $\det M_l(\lambda)$. The trace of $M_l(\lambda)$ is usually written as $\sigma(\lambda)$.

Let λ be a rational homomorphism of A into B and Y be a divisor on B. Then by the correspondence $\mathrm{cl}(Y) \rightarrow \mathrm{cl}(\lambda^{-1}(Y))$, we obtain a rational homomorphism of $\hat{B}$ into $\hat{A}$, called the **transpose** of λ and denoted by ${}^t\lambda$, where cl means the linear equivalence class (→ 18 Algebraic Varieties M). If X is a nondegenerate divisor on A, then φ_X is surjective; and this implies that there is a rational homomorphism β of $\hat{A}$ onto A such that $\beta \circ \varphi_X = \nu(\varphi_X)\delta$ ($\delta =$ the identity mapping of A). We denote by φ_X^{-1} the element $(1/\nu(\varphi_X))\,\beta$ in $\mathrm{Hom}(\hat{A}, A) \otimes \mathbf{Q}$. The correspondence $\alpha \rightarrow \alpha'$, $\alpha' = \varphi_X^{-1} \circ {}^t\alpha \circ \varphi_X (\alpha \in \mathfrak{A}_0(A))$ is an †involution of $\mathfrak{A}_0(A)$ and is of order 1 or 2. If $\alpha \neq 0$, we note the important theorem: $\sigma(\alpha' \circ \alpha) > 0$ (**Castelnuovo's lemma**). A. Weil was the first to recognize the importance of this theorem in connection with †Riemann's hypothesis on †congruence zeta functions.

F. Differential Forms

Let $\omega = \sum f_{(i)} du_{i_1} \wedge \ldots \wedge du_{i_r}$ be a †differential form on an Abelian variety A of dimension n. Given a point a of A and the translation T_a, we denote by $\omega \circ T_a$ the differential form $\sum (f_{(i)} \circ T_a) d(u_{i_1} \circ T_a) \wedge \ldots \wedge d(u_{i_r} \circ T_a)$, where $\omega \circ T_a$ is determined independently on the expressions of ω. If $\omega \circ T_a = \omega$ for every point a of A, then ω is called an **invariant differential form** on A. The †differential form of the first kind is invariant, and conversely, every invariant differential form is of the first kind. Let K be the †universal domain and $K(A)$ be the †function field of A. The set of the linear differential forms of the first kind on A is a linear space over K of dimension n, and its basis becomes a basis over $K(A)$ of the linear space consisting of all linear differential forms on A. An **invariant derivation** on A is a derivation D in $K(A)$ satisfying $(Df)\circ T_a = D(f\circ T_a)$ for any element f of $K(A)$ and every point a of A. For a linear differential form $\omega = \sum f_i du_i$ and a derivation D, we put $\langle \omega, D\rangle = \sum f_i Du_i$. Then $\langle \omega, D\rangle$ is a bilinear form in ω and D. A derivation D is invariant if and only if $\langle \omega, D\rangle$ is a constant function for every invariant linear differential form ω. Similarly, a linear differential form ω is invariant if and only if $\langle \omega, D\rangle$ is a constant function for every invariant derivation D. The linear space of invariant linear differential forms and that of invariant derivations are dual to each other with respect to the bilinear form $\langle \omega, D\rangle$.

Now consider the case when the characteristic p of the universal domain is positive. The automorphism $a\rightarrow a^p$, $a\in\Omega$, of the universal domain Ω induces a group isomorphism of A; we denote by A^p the image of A and by x^p the image of a point x of A. The image A^p is an Abelian variety, and the group isomorphism $\pi : x\rightarrow x^p$, $x\in A$ is an isogeny of A onto A^p. Let B be another Abelian variety that is isogenous to A, and let λ be an isogeny of A onto B. If there is an isogeny $\mu : B\rightarrow A^p$ such that $\pi = \mu\circ\lambda$, then we say that λ is of **height** 1. The function field $\Omega(B)$ of B can be considered as a subfield of the function field $\Omega(A)$ of A by the mapping λ. If λ is of height 1 and $\nu(\lambda)=p$, there exists an invariant derivation D of $\Omega(A)$ with the constant field $\Omega(B)$, uniquely determined up to constants. Moreover, we can choose D so that $D^p = D$ or $D^p = 0$. In the first case λ is said to be of type (i_1); in the second case it is said to be of type (i_2). An isogeny whose degree is a prime different from the characteristic p is said to be of type (S_1), and a separable isogeny whose degree is p is said to be of type (S_2). Any isogeny can be written as a product of isogenies of these four types.

G. Polarized Abelian Varieties

Let X be a divisor on an Abelian variety A; we denote by $\mathfrak{X}$ the class of divisors X' such that $mX \equiv m'X'$ for suitably chosen positive integers m, m'. When the class $\mathfrak{X}$ contains positive nondegenerate divisors, we say that $\mathfrak{X}$ determines a **polarization** on A, and the couple $(A,\mathfrak{X})$ is called a **polarized Abelian variety**. In particular, if A is a Jacobian variety whose polarization $\mathfrak{X}$ is determined by a theta divisor, we call $(A,\mathfrak{X})$ the **canonically polarized Jacobian variety**. If an endomorphism α of A keeps the polarization invariant, i.e., if the class determined by $\alpha^{-1}(X)$ coincides with the class $\mathfrak{X}$, then α is called an **endomorphism** of the polarized Abelian variety $(A,\mathfrak{X})$. In particular, if α is an automorphism of A, then we say that α is an **automorphism** of $(A,\mathfrak{X})$. The group of all automorphisms of a polarized Abelian variety is finite. In particular, the group of automorphisms of a canonically polarized Jacobian variety is finite. Hence follows the famous theorem concerning the finiteness of the group of automorphisms of an algebraic curve of genus not less than 2.

On the other hand the algebraic equivalence class of a nondegenerate divisor is called an **inhomogeneous polarization**. (The above polarization is then sometimes called a homogeneous polarization.) An inhomogeneous polarization X determines an isogeny $\varphi_X : A\rightarrow\hat{A}$ uniquely. An endomorphism of an inhomogeneously polarized Abelian variety can be defined similarly.

H. Analytic Theory

For the rest of this article we take the complex number field $\mathbf{C}$ as the universal domain, and in this case we can utilize analytic and topological methods. The modern theory of Abelian varieties originated from the study of this situation.

Let $\mathbf{C}^n$ be an n-dimensional vector space over $\mathbf{C}$. In a natural way, the space $\mathbf{C}^n$ becomes a $2n$-dimensional vector space $\mathbf{R}^{2n}$ over the real field $\mathbf{R}$, and the mapping $J: z\rightarrow \sqrt{-1}\, z$, $z\in\mathbf{C}^n$ is an $\mathbf{R}$-linear automorphism of $\mathbf{R}^{2n}$ such that $J^2=-1$. Conversely, if for an even-dimensional $\mathbf{R}$-vector space $\mathbf{R}^{2n}$ such a mapping J is given, then by putting $(a+\sqrt{-1}\, b)x = ax + bJx$ $(x\in\mathbf{R}^{2n};\ a,b\in\mathbf{R})$, we can introduce an n-dimensional complex linear structure into $\mathbf{R}^{2n}$. We then say that J determines a **complex structure** on $\mathbf{R}^{2n}$; we denote by $\mathbf{C}^n = (\mathbf{R}^{2n}, J)$ the space having the complex structure determined by J. Let $\omega_1,\ldots,\omega_{2n}$ be $2n$ $\mathbf{R}$ linearly independent points on $\mathbf{C}^n=(\mathbf{R}^{2n},J)$. Then the subgroup $\mathbf{D}$ of rank $2n$ generated by these points is discrete, and the factor group $\mathbf{T}^n = \mathbf{C}^n/\mathbf{D}$ is a **complex torus** of dimension n. We fix a basis of $\mathbf{C}^n$ and introduce a complex coordinate system on $\mathbf{C}^n$. Utilizing the basis $\omega_1,\ldots,\omega_{2n}$ of $\mathbf{R}^{2n}$, we also introduce a real coordinate system on $\mathbf{R}^{2n}$. We then obtain an $n\times 2n$ matrix $\Omega=(\omega_{ij})$, where the $(\omega_{1i},\ldots,\omega_{ni})$ are the complex coordinates of ω_i; the matrix Ω is called the **period matrix** of $\mathbf{T}^n$. Let $(z_1,\ldots,z_n)$ be the complex coordinates of a point $z \in \mathbf{C}^n$ and $(x_1, \ldots, x_{2n})$ be the real coordinates. Then we have ${}^t(z_1, \ldots, z_n) = \Omega {}^t(x_1, \ldots, x_{2n})$. If we let the same symbol J stand for the representation matrix of the linear transformation J with respect to the basis $\omega_1, \ldots, \omega_{2n}$, then we have $\sqrt{-1}\,\Omega = \Omega J$. The torus $\mathbf{T}^n$ is homeomorphic to the product of $2n$ circles. From this fact, we know that the †Poincaré polynomial of $\mathbf{T}^n$ is given by $(1+x)^{2n}$.

I. Theta Functions

A holomorphic function $f(z)$ on $\mathbf{C}^n = (\mathbf{R}^n, J)$ is called a **theta function** if for every $d \in \mathbf{D}$ we have $f(z+d) = f(z)\exp(\Sigma a_\alpha z_\alpha + c)$ (a_α, $c \in \mathbf{C}$), where the linear form $\Sigma a_\alpha z_\alpha$ and the constant c are both dependent on d. The set of zeros of a theta function f, which we write as (f), determines a positive analytic divisor on $\mathbf{T}^n$. Conversely, for every positive analytic divisor X on $\mathbf{T}^n$, there exists a theta function f such that $(f) = X$. With respect to the real coordinate system $x_1, \ldots, x_{2n}$ determined by the basis $\omega_1, \ldots, \omega_{2n}$, we can find $2n \times 2n$ matrices A, A_0 and a $1 \times 2n$ matrix b, with elements in $\mathbf{C}$, such that the **transformation formula** $f(x+a) = f(x)\exp(2\pi\sqrt{-1}\,({}^t aAx + \frac{1}{2}{}^t aA_0 a + {}^t ba))$ (where $A \equiv A_0 \pmod{\mathbf{Z}}$, ${}^t A_0 = A_0$) holds for every $1 \times 2n$ matrix a whose elements are rational integers. Moreover, if we put $E = A - {}^t A$, then E is an †alternating matrix whose elements are rational integers, and $S = EJ$ is a †positive semidefinite symmetric matrix. Conversely, if there exists such an alternating matrix E we can find a theta function. (There does exist, however, a complex torus on which no theta function exists other than trivial ones, i.e., ones of the form $\exp(\varphi(z))$ where $\varphi(z)$ is a polynomial of degree at most 2.)

A theta function f is called **nondegenerate** if it cannot be a function of $n-1$ complex variables, and f is nondegenerate if and only if the matrix $S = EJ$ is †positive definite. A complex torus has the structure of an Abelian variety if and only if there exists a nondegenerate theta function, i.e., if and only if there exists an alternating matrix E whose elements are rational integers such that EJ is a positive definite symmetric matrix. The latter condition is satisfied if and only if there exists an alternating matrix E whose elements are rational integers such that $\Omega {}^t E^{-1} {}^t\Omega = 0$, $\sqrt{-1}\,\Omega {}^t E^{-1} {}^t\bar{\Omega} > 0$ (positive definite Hermitian matrix). In particular, a period matrix Ω satisfying these conditions is called a **Riemann matrix**, and the rational matrix ${}^t E^{-1}$ is called the **principal matrix** belonging to Ω.

Determining a polarization on an Abelian variety can be reduced to designating a class of principal matrices obtained from a principal matrix by multiplying by positive integers. Let X be a positive divisor on an Abelian variety $\mathbf{T}^n = \mathbf{C}^n/\mathbf{D}$, and let f be a theta function such that $(f) = X$. Then the divisor X is nondegenerate if and only if the theta function f is nondegenerate, and the latter statement holds if and only if the alternating matrix E obtained from f is †invertible. For a given principal matrix ${}^t E^{-1}$, we can choose suitable coordinate systems of $\mathbf{C}^n$ and $\mathbf{D}$ so that $\Omega = (I_n, F)$ and $E = \begin{pmatrix} 0 & \Delta \\ -\Delta & 0 \end{pmatrix}$, where I_n is a unit matrix and Δ is a diagonal matrix whose elements are †elementary divisors of E. In these situations, the imaginary part of ΔF is symmetric and positive definite, i.e., ΔF is a point of the †Siegel upper half-space $\mathfrak{S}_n$. Thus, to the polarized Abelian variety (where the polarization is determined by ${}^t E^{-1}$) there corresponds a point of $\mathfrak{S}_n$. This gives a one-to-one correspondence between the isomorphism classes of Abelian varieties polarized by a principal matrix with given elementary divisors and the points of the factor space $\mathfrak{S}_n/\Gamma_n(\Delta)$, where $\Gamma_n(\Delta)$ is a subgroup that is †commensurable to the Siegel †modular group of degree n and operates on $\mathfrak{S}_n$ discontinuously. $\mathfrak{S}_n/\Gamma_n(\Delta)$ is the coarse †moduli space of Abelian varieties, polarized as above; the projective embedding of $\mathfrak{S}_n/\Gamma_n(\Delta)$ is given by means of †Siegel modular forms.

J. Abelian Functions

The quotient of two theta functions having the same transformation formulas for each element in $\mathbf{D}$ is a periodic function that is everywhere holomorphic; such a function is called an **Abelian function**. Any Abelian function is written as the quotient of two theta functions having the same transformation formulas. The set of Abelian functions with respect to $\mathbf{D}$ is a field, which we call an **Abelian function field**. If $\mathbf{T}^n = \mathbf{C}^n/\mathbf{D}$ has the structure of an Abelian variety, then it coincides with the field of rational functions on $\mathbf{T}^n$; in this sense, the function field of an Abelian variety is sometimes called an Abelian function field.

K. Homomorphisms

Let $\mathbf{C}^{n_i} = (\mathbf{R}^{2n_i}, J_i)$ $(i = 1, 2)$ be complex linear spaces; an $\mathbf{R}$-linear mapping $f: \mathbf{R}^{2n_1} \to \mathbf{R}^{2n_2}$ is $\mathbf{C}$-linear if and only if the relation $f \circ J_1 = J_2 \circ f$ holds. Let $\mathbf{D}_i$ be †lattice groups of $\mathbf{C}^{n_i}$ for $i = 1$, 2. If a $\mathbf{C}$-linear mapping $\Lambda: \mathbf{C}^{n_1} \to \mathbf{C}^{n_2}$ satisfies $\Lambda(\mathbf{D}_1) \subset \mathbf{D}_2$, then Λ induces a complex analytic homomorphism of $\mathbf{T}_1 = \mathbf{C}^{n_1}/\mathbf{D}_1$ to $\mathbf{T}_2 = \mathbf{C}^{n_2}/\mathbf{D}_2$. Conversely, any complex analytic homomorphism of $\mathbf{T}_1$ to $\mathbf{T}_2$ is obtained in this way. Let $\mathbf{T}_1$ and $\mathbf{T}_2$ be Abelian varieties, and let $\Omega_1=(\omega_1^{(1)},\ldots,\omega_{2n_1}^{(1)})$ and $\Omega_2=(\omega_1^{(2)},\ldots,\omega_{2n_2}^{(2)})$ be their Riemann matrices. Then for a homomorphism $\lambda:\mathbf{T}_1\to\mathbf{T}_2$, we can find a representation matrix $W(\lambda)$ with complex coefficients; and with respect to the real coordinate systems $(\omega_1^{(1)},\ldots,\omega_{2n_1}^{(1)})$ and $(\omega_1^{(2)},\ldots,\omega_{2n_2}^{(2)})$, we can find a representation matrix $M(\lambda)$ with coefficients in $\mathbf{Z}$ such that $W(\lambda)\Omega_1=\Omega_2 M(\lambda)$. Conversely, if for a complex matrix W there is a matrix M, with coefficients in $\mathbf{Z}$, satisfying the relation given above, then W gives a homomorphism of $\mathbf{T}_1$ to $\mathbf{T}_2$. The above equation is called **Hurwitz's relation**. The notion of l-adic coordinate system, which is valid for a general characteristic, corresponds to that of the lattice group, and the l-adic representation $M_l(\lambda)$ of λ is the abstraction of the integral representation $M(\lambda)$.

L. Abelian Integrals

Let $\mathfrak{R}$ be a compact †Riemann surface of genus $g\geqslant 1$ (→ 13 Algebraic Functions) and let ω be a sum of †Abelian differentials of the first kind or of the second kind. Then the †period of ω along a cycle γ depends only on the †homology class of γ. The set of all differentials of the first kind forms a complex linear space of dimension g; we denote it by $\mathfrak{D}_0$. Let P be a point and P_0 a fixed point of R; then we denote by $u(P)$ the vector integral $\left(\int_{P_0}^{P}\omega_1,\ldots,\int_{P_0}^{P}\omega_g\right)$, where $(\omega_1,\ldots,\omega_g)$ is a basis of $\mathfrak{D}_0$ and the path from P_0 to P is common to every integral. The correspondence $P\to u(P)$ is not a single-valued mapping; that is, the totality of differences of values of $u(P)$ coincides with the group $\mathbf{D}$ consisting of periods $\left(\int_\gamma\omega_1,\ldots,\int_\gamma\omega_g\right)$, where γ varies over all cycles. Let a set of cycles $\{\gamma_1,\ldots,\gamma_{2g}\}$ be a basis of the homology group, with coefficients in $\mathbf{Z}$; then $2g$ column vectors of the $g\times 2g$ matrix $\Omega=(\omega_{ij})$, $\omega_{ij}=\int_{\gamma_j}\omega_i$, are linearly independent over $\mathbf{R}$. Since the group $\mathbf{D}$ coincides with the set of linear combinations, with coefficients in $\mathbf{Z}$, of column vectors of Ω, this $\mathbf{D}$ is a discrete subgroup of rank $2g$, and the matrix Ω is a period matrix of the complex torus $\mathbf{T}^g=\mathbf{C}^g/\mathbf{D}$ of dimension g.

For a basis of the homology group with coefficients in $\mathbf{Z}$, we take †normal sections $\alpha_1,\ldots,\alpha_g,\alpha_{g+1},\ldots,\alpha_{2g}$ of $\mathfrak{R}$, and let the same symbol Ω stand for the period matrix (ω_{ij}), $\omega_{ij}=\int_{\alpha_j}\omega_i$; then we have

$$\Omega E\,{}^t\Omega=0,\quad \sqrt{-1}\,\Omega E\,{}^t\overline{\Omega}>0$$

(positive definite Hermitian matrix), where

$$E=\begin{pmatrix}0 & I_g\\ -I_g & 0\end{pmatrix}\quad\text{(with } I_g \text{ the unit matrix).}$$

This implies that E is a principal matrix belonging to Ω; we call the equality and inequality just given **Riemann's period relation** and **Riemann's period inequality**, respectively. Furthermore, we can choose a suitable basis of $\mathfrak{D}_0$ so that the period matrix Ω is of the form (I_g,F); and when that is so, the imaginary part of the $g\times g$ matrix F is positive definite and symmetric. We consider the function $\vartheta(u)$, $u=(u_1,\ldots,u_g)$, defined by an infinite series

$$\vartheta(u)=\sum_{\mathbf{m}}\exp\left(2\pi\sqrt{-1}\,(\mathbf{m}{}^tu+\tfrac{1}{2}\mathbf{m}F{}^t\mathbf{m})\right),$$

where the sum is taken over all row vectors $\mathbf{m}=(m_1,\ldots,m_g)$ with coefficients in $\mathbf{Z}$. If u is in a bounded region, then the series for $\vartheta(u)$ is uniformly and absolutely convergent. Hence $\vartheta(u)$ is a holomorphic function of u. This is a theta function corresponding to the principal matrix $E=\begin{pmatrix}0 & I_g\\ -I_g & 0\end{pmatrix}$ and is called the **Riemann theta function**. As $\vartheta(u)$ is nondegenerate, the complex torus $\mathbf{T}^g=\mathbf{C}^g/\mathbf{D}$ has the structure of an Abelian variety. If we regard the Riemann surface $\mathfrak{R}$ as an algebraic curve, this Abelian variety $\mathbf{T}^g$ is precisely the Jacobian variety of the curve $\mathfrak{R}$. We call the zeros of $\vartheta(u)$ the †canonical divisor on $\mathbf{T}^g$.

The correspondence $P\to u(P)$, $P\in\mathfrak{R}$ induces a well-defined mapping φ of $\mathfrak{R}$ onto $\mathbf{T}^g=\mathbf{C}^g/\mathbf{D}$. Moreover, if we set $\varphi(A)=\sum_{j=1}^{r}\varphi(P_j)-\sum_{j=1}^{r}\varphi(Q_j)$ for any divisor $A=P_1\ldots P_r/Q_1\ldots Q_r$ of degree 0, then $\varphi(A)$ is a point on $\mathbf{T}^g=\mathbf{C}^g/\mathbf{D}$, which is represented by the vector $\left(\sum_{j=1}^{r}\int_{Q_j}^{P_j}\omega_1,\ldots,\sum_{j=1}^{r}\int_{Q_j}^{P_j}\omega_g\right)$ of $\mathbf{C}^g$, and the mapping $A\to\varphi(A)$ is a homomorphism of the group $\mathfrak{G}_0$ of divisors of degree 0 onto $\mathbf{T}^g=\mathbf{C}^g/\mathbf{D}$. The kernel of this homomorphism coincides with the group $\mathfrak{G}_l$ of †principal divisors (**Abel's theorem**). Hence a divisor $A=P_1\ldots P_r/Q_1\ldots Q_r$ of degree 0 is a divisor of some function if and only if we have $\sum_{j=1}^{r}\int_{Q_j}^{P_j}\omega_i\equiv 0\ (\mathrm{mod}\,\mathbf{D})$ $(i=1,\ldots,g)$, where the left-hand side is 0 for a suitable path.

Given g fixed points $P_1,\ldots,P_g$ on $\mathfrak{R}$ and given $(u_1,\ldots,u_g)$ as any vector of $\mathbf{C}^g$, the problem of finding g points $Q_1,\ldots,Q_g$ satisfying relations $\sum_{j=1}^{g}\int_{P_j}^{Q_j}\omega_i\equiv u_i\ (\mathrm{mod}\,\mathbf{D})$, $i=1,\ldots,g$, is called **Jacobi's inverse problem**. To solve the problem, we take a divisor A of degree 0 such that the class $\varphi(A)$ (mod $\mathbf{D}$) is represented by $(u_1,\ldots,u_g)$; then, by virtue of the †Riemann-Roch theorem, there exists a divisor $Q_1\ldots Q_g$ satisfying

$$A\equiv Q_1\ldots Q_g/P_1\ldots P_g\ (\mathrm{mod}\,\mathfrak{G}_l).$$

Abel's theorem implies that the set of points $\{Q_1, \ldots, Q_g\}$ is a solution of Jacobi's problem. Moreover, for general $(u_1, \ldots, u_g)$, the solution is unique; i.e., there exists a subvariety $\tilde{\mathfrak{X}}$ of dimension $g-2$ on $\mathbf{C}^g$ such that the solution is unique up to order if and only if $(u_1, \ldots, u_g)$ does not lie on $\tilde{\mathfrak{X}}$. In particular, if every point P_i coincides with the fixed point P_0, that appeared in the definition of $u(P)$, the subvariety on $\mathbf{T}^g = \mathbf{C}^g/\mathbf{D}$ determined by $\tilde{\mathfrak{X}}$ is obtained in the following way: Let $W_1 + \ldots + W_{2g-2}$ be a canonical divisor on $\mathfrak{R}$, and put $c = \varphi(W_1) + \ldots + \varphi(W_{2g-2})$. Then the locus of points $c - \varphi(R_1) - \ldots - \varphi(R_{g-2})$ (where $g-2$ points $R_1, \ldots, R_{g-2}$ are taken independently over all points of $\mathfrak{R}$) is the desired subvariety.

M. Elementary Abelian Functions

Let z be a nonconstant meromorphic function on $\mathfrak{R}$. Then for any $u = (u_1, \ldots, u_g) \in \mathbf{C}^g$ that does not lie on $\tilde{\mathfrak{X}}$, there exist g points $Q_1, \ldots, Q_g$ (uniquely determined up to order) as the solution of Jacobi's problem. Therefore, the elementary symmetric functions

$$s_1(u;z) = \sum_{j=1}^{g} z(Q_j), \quad s_2(u;z) = \sum_{i<j} z(Q_i)z(Q_j), \quad \ldots, \quad s_g(u;z) = \prod_{j=1}^{g} z(Q_j)$$

are well defined if u lies outside the variety $\tilde{\mathfrak{X}}$ of dimension $g-2$. Each function $s_i(u;z)$, regarded as a function of u, can be extended uniquely to an Abelian function in the whole space $\mathbf{C}^g$; the function is denoted by the same symbol $s_i(u;z)$. The Abelian functions $s_1(u;z), \ldots, s_g(u;z)$ are called the **elementary Abelian functions** obtained from z.

Now let K be the field of Abelian functions on $\mathbf{C}^n/\mathbf{D}$ and k the field of meromorphic functions on $\mathfrak{R}$; then the dimension of K over $\mathbf{C}$ is g, and $[K:\mathbf{C}(s_1(u;z), \ldots, s_g(u;z))] = r^g$, where r is the degree of the function z that is given by $[k:\mathbf{C}(z)]$. Moreover, if we take any function w such that $k = \mathbf{C}(z,w)$, then we have

$$K = \mathbf{C}(s_1(u;z), \ldots, s_g(u;z); s_1(u;w), \ldots, s_g(u;w)),$$

where $s_1(u;w), \ldots, s_g(u;w)$ are the elementary Abelian functions obtained from w.

We can write any elementary Abelian function as a rational function of Riemann theta functions; therefore, any Abelian function can be written as a rational function of Riemann theta functions. Furthermore, if u and v are variable vectors, then $s_i(u+v;z)$ can be represented as an algebraic function of $s_1(u;z), \ldots, s_g(u;z), s_1(v;z), \ldots, s_g(v;z)$; i.e., we can choose a suitable polynomial $H_i(Z;X_1, \ldots, X_g; Y_1, \ldots, Y_g)$ with coefficients in $\mathbf{C}$ so that $H_i(s_i(u+v;z); s_1(u;z), \ldots, s_g(u;z); s_1(v;z), \ldots, s_g(v;z)) = 0$. This is the **algebraic addition formula** with respect to the elementary Abelian functions $s_i(u;z)$, $i = 1, \ldots, g$.

As the study of †Abelian integrals of the first kind led us to the theory of Jacobian varieties, †Abelian integrals of the second and the third kind give rise to the theory of †generalized Jacobian varieties (→ 11 Algebraic Curves).

The theory of Abelian varieties has significant applications to number theory, as shown by the following examples: the theory of †unramified Abelian extensions with respect to a function field of several variables defined over a finite field (S. Lang), the theory of heights of points on an Abelian variety (Weil, A. Néron, Tate), and the theory of complex multiplication (→ 75 Complex Multiplication) in the case of higher dimensions (Y. Taniyama, G. Shimura).

N. Some Recent Results

(1) Level Structure, Moduli of Abelian Varieties. Let A be an Abelian variety over k of dimension g and n a positive integer which the characteristic of k does not divide. A **level n structure** on A is defined to be a set of $2g$ points $\sigma_1, \ldots, \sigma_{2g}$ on A which form a basis for the group of points of order n on A.

Let $A(g,d,n;k)$ be the set of triples: (i) an Abelian variety A over k of dimension g, (ii) an inhomogeneous polarization X with $\nu(\varphi_X) = d^2$, and (iii) a level n structure $\sigma_1, \ldots, \sigma_{2g}$ of A, all up to isomorphism. Similarly we can define $A(g,d,n;S)$ for Abelian schemes over a scheme S. The correspondence $S \to A(g,d,n;S)$ defines a functor $\mathfrak{A}(g,d,n)$. D. Mumford has shown that there exists the †coarse moduli scheme $\mathfrak{A}(g,d,n)$ quasi-projective over Spec($\mathbf{Z}$), and that it is even fine if $n \geqslant 3$ [6]. He used the technique of †Hilbert schemes and †stable points (→ 18 Algebraic Varieties). One of the key steps of his proof is to show that for an embedding $\phi: A \to \mathbf{P}^m$ of degree r (i.e., the degree of $\phi(A)$ in $\mathbf{P}^m$) over an algebraically closed field k and a positive integer n such that char(k) does not divide n and $n > \sqrt{(m+1)r}$, the point $(\phi(x_i))_{i=1,\ldots,n^{2g}}$ in $(\mathbf{P}^m)^{n^{2g}}$ is stable with respect to the action of $PGL(m)$, where the x_i are the points of order n on A (with an arbitrary order). Mumford later showed another method of constructing the moduli of polarized Abelian varieties by using algebraic theta constants [7].

(2) Néron Minimal Models, Good and Stable Reduction. Let R be a discrete valuation ring with residue field k and quotient field K. For an Abelian variety A over K, there exists a smooth group scheme $\mathcal{A}$ of finite type over $S = \mathrm{Spec}(R)$, called the **Néron minimal model** of A, such that for every scheme S' smooth over S there is a canonical isomorphism

$$\mathrm{Hom}_S(S', \mathcal{A}) \cong \mathrm{Hom}_K(S'_K, A),$$

where S'_K is the pullback of S' by $\mathrm{Spec}(K) \to \mathrm{Spec}(R)$ (Néron, M. Raynaud). In particular we have $\mathcal{A}_K \cong A$. Denote by A_0 the fiber of $\mathcal{A}$ over the closed point of S.

If $\mathcal{A}$ is proper over S, we say that A has a **good reduction** at R. If the connected component A_0^0 of A_0 containing 0 has no unipotent radical (or equivalently, A_0^0 is an extension of an Abelian variety by an algebraic torus), we say that A has a **stable reduction**. If there is a finite separable extension K' of K with a prolongation R' of R to K' such that $A \times_K K'$ has good (stable) reduction, we say that A has **potential good (stable) reduction** at R. Let K_s be a separable closure of K and $\bar{R}$ a prolongation of R to K_s. For a prime number $l \neq \mathrm{char}(k)$, we have a canonical homomorphism $\rho: \mathrm{Gal}(K_s/K) \to \mathrm{Aut}(\mathfrak{G}_l(A))$, called a monodromy. Then A has potential good reduction if and only if the image of the †inertia group $I(\bar{R})$ by ρ is a finite group (J.-P. Serre and Tate). Every Abelian variety A over K has potential stable reduction at R (**stable reduction theorem**, A. Grothendieck [16]).

(3) Graded Ring of Theta Functions. If f is a theta function with a period system $\mathbf{D}$, f^n is also a theta function with the same period system for every positive n. We denote by S_n the vector space of the theta functions with the period system $\mathbf{D}$ subject to the same transformation law as f^n. If we set $(f) = X$, S_n can be naturally considered as the †defining module of the †complete linear system of X, and the dimension of S_n $(= l(nX))$ is equal to the product of the nonzero diagonal elements of Δ (→ Section I) (Frobenius). For $g \in S_m$ and $h \in S_n$, $gh \in S_{m+n}$, hence $S = \oplus_{n \geqslant 0} S_n$ is a graded ring, which is normal and finitely generated. For $m \geqslant 2$ and $n \geqslant 3$, the product map $S_m \times S_n \to S_{m+n}$ is surjective (D. Mumford, S. Koizumi). If the elementary divisors of E can be divided by an integer $\geqslant 4$, the kernel of the natural graded mapping $\mathbf{S}(S_1) \to S$ (where $\mathbf{S}(S_1)$ denotes the †symmetric algebra over S_1) is generated by the quadratic relations (i.e., the part of degree 2) for sufficiently large degrees (Mumford). Geometrically this means that if $X = (f)$, $f \in S_1$, is nondegenerate, then, with the projective embedding $i: T^n \to \mathbf{P}^N$ defined by the complete linear system of X, $i(T^n)$ is an intersection of quadrics in $\mathbf{P}^N$ containing $i(T^n)$.

Mumford has developed the theory of algebraic theta functions that works also for the positive characteristic case [7] and proved the above results in general.

References

[1] H. Cartan, Séminaire sur la théorie des fonctions de plusieurs variables I-V, 1951–1952, Ecole Norm. Sup. Paris.
[2] F. Conforto, Abelsche Funktionen und algebraische Geometrie, Springer, 1956.
[3] A. Krazer and W. Wirtinger, Abelsche Funktionen und allgemeine Thetafunktionen, Enzykl. Math. Wiss., IIB 7 (1920), 604–873.
[4] J. Igusa, Theta functions, Springer, 1972.
[5] S. Lang, Abelian varieties, Interscience, 1959.
[6] D. Mumford, Geometric invariant theory, Erg. Math., Springer, 1965.
[7] D. Mumford, On the equations defining Abelian varieties I–III, Inventiones Math., I (1966), 287–354; II,III (1967), 75–135, 215–244.
[8] D. Mumford, Varieties defined by quadratic equations, Questioni sulle variete algebraiche, Corsi dal C. I. M. E., Cremona, 1969, 31–94.
[9] D. Mumford, Abelian varieties, Tata Inst. studies in math., Oxford Univ. Press, 1970.
[10] A. Néron, Modèles minimaux des variétés abéliennes sur les corps locaux et globaux, Publ. Math. Inst. HES, no. 21, 1964, p. 5–128.
[11] W. F. Osgood, Lehrbuch der Funktionentheorie II_2, Teubner, 1932.
[12] M. Raynaud, Modèles de Néron, C. R. Acad. Sci. Paris, 262 (1966), 345–347.
[13] B. Riemann, Theorie der Abelschen Funktionen, J. Reine Angew. Math., 54 (1857), 115–155.
[14] J.-P. Serre, Quelques propriétés des variétés abéliennes en caractéristique p, Amer. J. Math., 80 (1958), 715–739.
[15] J.-P. Serre and J. Tate, Good reduction of Abelian varieties, Ann. of Math., (2) 88 (1968), 492–517.
[16] Séminaire de géométrie algébriques (SGA) 7, Groupes de monodromie en géométrie algébrique, Lecture notes in math. 288, 340, Springer, 1972, 1973.
[17] G. Shimura, Moduli and fibre systems of Abelian varieties, Ann. of Math., (2) 83 (1966), 294–338.
[18] C. L. Siegel, Analytic functions of several complex variables, Lecture notes, Institute for Advanced Study, Princeton, 1948–1949.

[19] J. Tate, Classes d'isogénie des variétés abéliennes sur un corps fini (d'après T. Honda), Séminaire Bourbaki, no. 352, Lecture notes in math. 179, Springer 1968–1969.
[20] A. Weil, Théorèmes fondamentaux de la théorie des fonctions thêta, Séminaire Bourbaki, no. 16, 1949, Benjamin.
[21] A. Weil, Variétés abéliennes et courbes algébriques, Actualités Sci. Ind., Hermann, 1948.
[22] A. Weil, Variétés kählériennes, Actualités Sci. Ind., Hermann, 1958.
[23] A. Weil, Généralisation des fonctions abéliennes, J. Math. Pures Appl., (9) 17 (1938), 47–87.

4 (XII.7) Abstract Integrals

A. General Remarks

There are many ways of generalizing the concept of the †Lebesgue integral from various points of view. We distinguish here two main lines along which these generalizations have been carried out. One is to extend the value space of integrands or measures to a †locally convex linear topological vector space (the general integrals in this case are called integrals in linear topological spaces), and the other is to characterize the concept of integrals abstractly in connection with †order relations in function spaces. We shall give explanations of typical concepts of general integrals according to each method.

B. Bochner Integrals in Linear Topological Spaces

A function $x(s)$ defined on a †σ-finite measure space$(S,\mathfrak{S},\mu)$ with values in a †Banach space X is called a **simple function** (or **finite-valued function**) if there exists a division of S into a finite number of mutually disjoint measurable sets $A_1,A_2,\ldots,A_n$ in each of which $x(s)$ takes a constant value c_j. Then $x(s)$ may be written as $\sum_{j=1}^{n}\chi_{A_j}(s)c_j$, where $\chi_{A_j}(s)$ is the †characteristic function of A_j. A function $x(s)$ is **strongly measurable** if it is the strong limit of a sequence of simple functions almost everywhere, that is, $\lim_{n\to\infty}\|x_n(s)-x(s)\|=0$ a.e. If a function $x(s)$ is strongly measurable and $\|x(s)\|$ is †Lebesgue integrable as a real-valued function, then $x(s)$ is said to be **Bochner integrable**. (The usual measurability of $\|x(s)\|$ follows from the strong measurability of $x(s)$.) If, in particular, $x(s)$ is a Bochner integrable simple function $\sum_{j=1}^{n}\chi_{A_j}(s)c_j$, then its Bochner integral is defined by

$$\int_S x(s)\,d\mu=\sum_{j=1}^{n}\mu(A_j)c_j.$$

One can show that for a Bochner integrable function $x(s)$, there exists a sequence of simple functions satisfying the following conditions: (i) $\lim_{n\to\infty}\|x_n(s)-x(s)\|=0$ holds a.e. on S; (ii) $\lim_{n\to\infty}\int_S\|x_n(s)-x(s)\|\,d\mu=0$. (In fact, it follows from the strong measurability of $x(s)$ that there exists a sequence of simple functions $y_n(s)$ which converges strongly to $x(s)$. Let $\{x_n(s)\}$ be a sequence of simple functions defined by

$$x_n(s)=y_n(s)\quad\text{if}\quad\|y_n(s)\|\leqslant 2\|x(s)\|,$$
$$x_n(s)=0\quad\text{if}\quad\|y_n(s)\|>2\|x(s)\|.$$

Then $\{x_n(s)\}$ satisfies conditions (i) and (ii).) Therefore, for such a sequence $\{x_n(s)\}$, $\int_S x_n(s)\,d\mu$ converges in X strongly to an element independently of the approximating sequence $\{x_n(s)\}$. The **Bochner integral** of $x(s)$, $\int_S x(s)\,d\mu$, is defined by

$$\int_S x(s)\,d\mu=\lim_{n\to\infty}\int_S x_n(s)\,d\mu.$$

To distinguish the Bochner integral of $x(s)$ from other kinds of integrals, we shall make use of the notation $(\mathrm{Bn})\int_S x(s)\,d\mu$. A Bochner integrable function on S is Bochner integrable also on every $\mathfrak{S}$-measurable subset in S. Furthermore, the Bochner integral has the basic properties of Lebesgue integrals, such as linearity, complete additivity, and †absolute continuity. †Lebesgue's convergence theorem and †Fubini's theorem, etc., hold also, but the †Radon-Nikodym theorem does not hold. The Bochner integral also has the property that if $x(s)$ is Bochner integrable, then for a †bounded linear operator T from X to another Banach space Y, the function $T(x(s))$ on S with values in Y is Bochner integrable and satisfies:

$$\int_S T(x(s))\,d\mu=T\left\{\int_S x(s)\,d\mu\right\}.$$

If, in particular, S is a finite-dimensional Euclidean space, then the Bochner integral has the property of †strong differentiability [6, 13, 21].

C. Birkhoff Integrals

The Birkhoff integral is defined constructively using decomposition of the domain of an integrand as in the case of the ordinary Lebesgue integral. We shall state the precise definition of this integral with some of its properties. First, we define the concept of unconditional convergence: Let $\{x_j\}$ be an infinite sequence in a Banach space X. Then a series $\sum_{j=1}^{\infty} x_j$ is called **unconditionally convergent** if, for each rearrangement α, the resulting series $\sum_{j=1}^{\infty} x_{\alpha(j)}$ is convergent. If a series $\sum_{j=1}^{\infty} x_j$ is unconditionally convergent, then for each rearrangement α, the sum of the series $\sum_{j=1}^{\infty} x_{\alpha(j)}$ is independent of α. If, in particular, X is the number space, then the concepts of unconditional convergence and absolute convergence coincide. However, in general, an unconditionally convergent series is not necessarily absolutely convergent (in the sense of the norm). Let $x(s)$ be a function defined on a σ-finite measure space $(S, \mathfrak{S}, \mu)$ with values in X. Then $x(s)$ is said to be **summable under a countable decomposition** Δ,

$$\Delta : S = \bigcup_{j=1}^{\infty} A_j, \qquad A_i \cap A_j = \varnothing \qquad (i \neq j),$$

$$\mu(A_j) < +\infty,$$

if $x(s)$ is bounded on each A_j and any series $\sum_{j=1}^{\infty} \mu(A_j) x(s_j)$ $(s_j \in A_j)$ is unconditionally convergent. In this case, the convex closure of the set $\{\sum_{j=1}^{\infty} \mu(A_j) x(s_j) \mid s_j \in A_j\}$ is called the **integral range** of $x(s)$ for Δ and denoted by $J(x, \Delta)$. If for an arbitrary positive number ϵ there exists a countable decomposition Δ_0 of S under which $x(s)$ is summable and the diameter of $J(x, \Delta_0)$ is smaller than ϵ, then $x(s)$ is said to be **Birkhoff integrable**. In this case, the intersection of its integral ranges for all countable decompositions is shown to consist of a single element of X. If $x(s)$ is summable under Δ, then $x(s)$ is also summable under any countable decomposition Δ' finer than Δ (this means that every subset of Δ' is included in a certain subset of Δ) and we have $J(x, \Delta') \subset J(x, \Delta)$. The **Birkhoff integral** of $x(s)$ over S is, by definition, the element $\bigcap_{\Delta} J(x, \Delta)$ which will be denoted by $(\mathrm{Bk})\int_S x(s)\, d\mu$ or simply by $\int_S x(s)\, d\mu$. Birkhoff integrable functions in S are also Birkhoff integrable over any $\mathfrak{S}$-measurable subset in S. The Birkhoff integral as a set function has the properties of complete additivity and absolute continuity and, as an operator from the space of Birkhoff integrable functions, has the property of linearity. However, for this integral, †Fubini's theorem does not hold, and the convergence theorem holds only under some restrictions which are more strict than in the case of the Bochner integral [4]. Bochner integrable functions are Birkhoff integrable, but the converse is not true.

Starting from the construction of the Birkhoff integral, G. Birkhoff and R. S. Phillips defined more general integrals whose integrands take values in a locally convex linear topological space and which include special cases of the Birkhoff integral and the Gel'fand-Pettis integral, which will be described in the following section. (For example, the Gel'fand-Pettis integral of a function $x(s)$ with values in a Banach space X is the Phillips integral of $x(s)$ with values in X considered as a locally convex space endowed with the †weak topology.) To further generalize the Phillips integral, C. E. Rickart has defined an integral for integrands whose values are subsets in a locally convex space and has shown that this integral has a property corresponding to the Radon-Nikodym theorem [20].

D. Gel'fand-Pettis Integrals

The Gel'fand-Pettis integral is more general than that of Birkhoff. Let $x(s)$ be a function defined on a σ-finite measure space $(S, \mathfrak{S}, \mu)$ with values in a Banach space X. Then $x(s)$ is said to be **weakly measurable** if, for each x^* of the †dual space X^* of X, the numerical function $x^*(x(s))$ is $\mathfrak{S}$-measurable. When X is †separable, $x(s)$ is weakly measurable if and only if $x(s)$ is strongly measurable. A weakly measurable function $x(s)$ is said to be **Gel'fand-Pettis integrable** (or **weakly integrable**) on a $\mathfrak{S}$-measurable set A if, for each $x^* \in X^*$, $x^*(x(s))$ is Lebesgue integrable on A and there exists an element x_A of X such that

$$x^*(x_A) = \int_A x^*(x(s))\, d\mu.$$

This element x_A is called the **Gel'fand-Pettis integral** on A, and is denoted by (G-P) $\int_A x(s)\, d\mu$ or simply by $\int_A x(s)\, d\mu$. If $x(s)$ is Gel'fand-Pettis integrable on every $\mathfrak{S}$-measurable set, then it is said to be Gel'fand-Pettis integrable. This integral has properties similar to those of the Birkhoff integral. However, Fubini's theorem and the Radon-Nikodym theorem do not hold [13, 17, 21]. If $x(s)$ is a Birkhoff integrable function, then it is Gel'fand-Pettis integrable, but the converse is not true [18].

We sometimes use the term *weak integrability* in a weaker sense than the one described above. A weakly measurable function $x(s)$ is **weakly integrable** on an $\mathfrak{S}$-measurable set if $x^*(x(s))$ is Lebesgue integrable on A for every $x^* \in X^*$. In this case, it is shown that the functional defined by

$$x_A^{**}(x^*) = \int_A x^*(x(s))\, d\mu, \qquad x^* \in X^*,$$

is a linear continuous functional on X^* [9, 11]. The element $x_A^{**} \in X^{**}$ is, by definition, the **weak integral** of $x(s)$ on A. In another general case where X is a locally convex space we can

$b, a \in A, b \in B\}$. A finite sum of subsets of $\mathbf{N}$ is defined similarly. If $\mathbf{N}$ is the sum $A + \ldots + A$ (r times), then we say that A is a **basis of order r in N.** Let $A(x)$ denote the number of integers in A that do not exceed x. The density of A is defined as $\inf A(x)/x$. If $A(n) \geqslant \alpha n$, $B(n) \geqslant \beta n$ $(0 < \alpha, \beta < 1)$ for all $n \in \mathbf{N}$, then we have $(A+B)(n) \geqslant (\min(1, \alpha + \beta))n$. This result was stated by E. Landau (1937) without proof; A. J. Hinčin had given a proof (1932) for the case $\alpha = \beta$, and H. B. Mann (1942) and E. Artin and P. Scherk (1943) succeeded in proving the statement for the general case. Suppose that the densities of A and B are, respectively, α and 0. If B is a basis of finite order in $\mathbf{N}$, then the density of $A+B$ is greater than that of A (Hinčin and P. Erdös, 1936). Let P be the set of all prime numbers. Though the density of P is 0, the density of $P+P$ is positive (L. G. Šnirel'man, 1930). Hence P is a basis of finite order in $\mathbf{N}$; in other words, there exists a positive integer r such that every natural number can be expressed as a sum of at most r primes. Though the density of the set Q of the kth powers of natural numbers is 0, there exists a positive integer $s(k)$ such that the sum $Q + \ldots + Q$ ($s(k)$ times) is a set of positive density. L. K. Hua (1956) gave a simple proof of this fact, based on Ju. V. Linnik's idea. It follows, therefore, that any natural number n can be expressed as a sum $n = a_1^k + \ldots + a_t^k$ where $a_i \in \mathbf{N}$, $t \geqslant s(k)$. This result had already been shown by D. Hilbert (1909).

An ancient method of finding prime numbers is †Eratosthenes' sieve. V. Brun (1920) devised a new sieve method to express an arbitrary integer n as the sum of two integers $n = a + b$, where the number of prime factors of a and b is as small as possible. This method was improved by H. A. Rademacher (1924), Landau (1931), A. Bustab (1937), and others. Among these the method found by A. Selberg (1952) is notable. Many variations of it appear in papers in *Acta Arithmetica*.

B. Farey Sequences

Let τ be a positive integer. We arrange in increasing order the set of all positive irreducible fractions lying between 0 and 1 whose denominators do not exceed τ. This sequence is called the **Farey sequence** of order τ. For example, the Farey sequence of order 5 consists of

$$\frac{0}{1}, \frac{1}{5}, \frac{1}{4}, \frac{1}{3}, \frac{2}{5}, \frac{1}{2}, \frac{3}{5}, \frac{2}{3}, \frac{3}{4}, \frac{4}{5}, \frac{1}{1}.$$

A necessary and sufficient condition that a fraction a/b be directly followed by a fraction c/d in the Farey sequence of order n is $b + d \geqslant n+1$, $bc - ad = 1$. In this case the fraction $(a+c)/(b+d)$ is called the **mediant** of a/b and c/d. Interpolating the Farey sequence of order n with such mediants $(a+c)/(b+d)$ satisfying $b + d = n+1$, we obtain the Farey sequence of order $n+1$.

Let a/q be a fraction in the Farey sequence of order τ, and a'/q', a''/q'' be adjacent members of a/q in the sequence such that $a'/q' < a/q < a''/q''$. The interval $[(a'+a)/(q'+q),\ (a+a'')/(q+q'')]$ is known as the **Farey arc** surrounding a/q. In particular, if $a/q = 0/1$, then we set $[-1/n, 1/n]$ to be the Farey arc surrounding $0/1$, where $n = [\tau]+1$ ([ ] is †Gauss's symbol). We can thus decompose the interval $[-1/n, 1-1/n]$ into a disjoint union of Farey arcs. If α is contained in the Farey arc surrounding a/q, then $|\alpha - a/q| < 1/q\tau$. Therefore, for a given $\tau \geqslant 1$ and a real α, we can prove the existence of a/q such that

$$(|a|, q) = 1, \quad 0 < q \leqslant \tau, \quad |\alpha - a/q| < 1/q\tau.$$

To estimate the integral of a function of period 1 over the interval [0, 1], we sometimes utilize the decomposition of the interval into a disjoint union of Farey arcs as mentioned above. This method is called the **circle method**, and the subdivision of the interval is known as the **Farey dissection.**

Given a positive number c, a Farey arc around a/q is usually called a **major arc** (or **basic interval**) if q does not exceed the given bound c; otherwise, it is called a **minor arc** (or **supplementary interval**). Usually, the principal part of the previously mentioned integral is derived from the integral over the major arcs, and the residual part is provided by the integral over the minor arcs.

C. Goldbach's Problem

Goldbach's problem is found in letters (1742) he exchanged with L. Euler. In them he stated that every positive integer can be expressed as the sum of primes. More precisely, he conjectured that any even integer not smaller than 6 can be expressed as the sum of two odd primes and that any odd integer not smaller than 9 can be expressed as the sum of three odd primes.

I. M. Vinogradov (1937) proved that every sufficiently large odd integer can be expressed as the sum of three primes. Let N be a sufficiently large odd integer. If we write

$$A(q,N) = \frac{\mu(q)}{\varphi^3(q)} \sum_{\substack{1 \leqslant a \leqslant q \\ (a,q)=1}} \exp\left(-2\pi i \frac{a}{q} N\right),$$

then the series $S(N)=\sum_{q=1}^{\infty}A(q,N)$ is absolutely convergent and is equal to

$$\prod_{p}\left(1+\frac{1}{(p-1)^3}\right)\prod_{p/N}\left(1-\frac{1}{p^2-3p+3}\right).$$

It is known that $S(N)>6/\pi^2$ for all N. If we denote by $r(N)$ the number of solutions of $N=p_1+p_2+p_3$, then $r(N)\sim(N^2/2(\log N)^3)\cdot S(N)$. To prove this, Vinogradov used the circle method. He employed the †prime number theorem for arithmetic progressions to estimate the integrals over the major arcs and devised an ingenious method to estimate the series $\sum_{p<N}\exp(2\pi i\alpha p)$ in the computation of the integrals on the minor arcs. A finite or infinite sum of exponential functions such as this is called a **trigonometric sum**. More generally, we consider trigonometric sums of several variables. Vinogradov provided detailed remarks and calculations [8].

In the case of even integers, the problem is still unsolved, although I. G. van der Corput, T. Estermann, and N. G. Čudakov proved simultaneously (1938) that almost all even integers (i.e., except a set of density 0) can be expressed as the sum of two primes. For these problems, Linnik (1946) and Čudakov (1947) introduced function-theoretic methods. They obtained the density theorem concerning the zeros of L-series, and A. Zulauf (1952) continued along the same lines. These methods had been suggested by G. H. Hardy and J. E. Littlewood (c. 1919), although they had assumed that the †extended Riemann hypothesis held.

D. Polygonal Numbers

Let m be an integer greater than 3, and let $a_1=1,\ a_{n+1}-a_n=(m-2)n+1\ (n=1,2,\dots)$. The sequence $\{a_n\}$ forms the system of **polygonal numbers of order** m. The general term of $\{a_n\}$ is given by $n+\frac{1}{2}(m-2)(n^2-n)\ (n=1,2,\dots)$. Such a_n are said to be **triangular numbers** if $m=3$, **square numbers** if $m=4$, and **pentagonal numbers** if $m=5$.

P. Fermat (1636) stated that every natural number can be expressed as the sum of m polygonal numbers of order m. This conjecture was proved by A. M. Legendre (1798) for $m=3$, by J. L. Lagrange (1772) for $m=4$, and by A. L. Cauchy (1813) for the general case. With regard to Lagrange 's result, Legendre noticed that in order that a positive integer n be expressed as the sum of three squares, it is necessary and sufficient that n is not of the form $4^a(8m+7)$.

Given a positive integer n, the number of integral solutions of the equation $x_1^2+x_2^2+\dots+x_s^2=n$ is denoted by $r_s(n)$. For example, $r_2(5)=8$. The †generating function $\sum_{n=1}^{\infty}r_2(n)n^{-s}$ can be expressed as $\sum_{m,n=-\infty}^{+\infty}(m^2+n^2)^{-s}$, where the term corresponding to $m=n=0$ is omitted; the function is equal to $4\zeta_K(s)$, where $\zeta_K(s)$ is the †Dedekind zeta function of the Gaussian number field $K=\mathbf{Q}(\sqrt{-1}\,)$. The equation $4\zeta_K(s)=4\zeta(s)L(s,\chi)$ (where $\chi(n)=(-4/n)$) leads to

$$r_2(n)=4\sum_{m|n}{}'(-1)^{(m-1)/2},$$

where $\sum'$ means the sum over all odd factors m of n. This result was obtained by C. G. J. Jacobi (1829). He also obtained the following formula:

$$r_4(n)=8\sum_{m|n,4\nmid m}{}'\ m,$$

where $\sum'$ means the sum over all divisors m of n not divisible by 4 (Hardy and E. M. Wright, C. L. Siegel, 1964). Let $q=\exp(2\pi i\tau)$ $(\operatorname{Im}\tau>0)$, and

$$f(q)=1+\sum_{n=1}^{\infty}r_s(n)q^n$$
$$=(1+2q+2q^4+2q^9+\dots)^s.$$

Hardy (1920) considered the variation of $f(q)$ for $q=a\exp(2\pi ih/k)$ (with $0<a<1$) as $a\to 1$; he obtained

$$f(q)\sim\pi^{s/2}\left(\frac{S_{h,k}}{n}\right)^s\log\left(\frac{1}{a}\right)^{-s/2},$$

where $S_{h,k}=\sum_{j=1}^{k}\exp(2\pi i(h/k)j^2)$. Furthermore, he constructed the **singular series**

$$\rho_s(n)=\frac{\pi^{s/2}n^{(s/2)-1}}{\Gamma(s/2)}\sum_{k=1}^{\infty}A_k,\quad A_1=1,$$

$$A_k=k^{-s}\sum_{\substack{1\leqslant h<k\\(h,k)=1}}(S_{h,k})^s\exp\left(-2\pi i\frac{h}{k}n\right),$$

and showed that $r_s(n)=\rho_s(n)$ for $s=5$, 6, 7, and 8. P. T. Bateman (1951) proved the same equation for $s=3$ and 4. If $s\geqslant 5$, then $r_s(n)=\rho_s(n)+O(n^{s/4})$ (Hardy, Littlewood, and S. Ramanujan, c. 1919). The detailed exposition of this result is in the notes of A. Z. Walfisz (1952) and Rademacher [7]. H. D. Kloostermann (1926) and Estermann (1962) studied the equation $ax_1^2+bx_2^2+cx_3^2+dx_4^2=n$, which led to a new field of study concerning the †Kloostermann sum. For instance, estimates such as

$$\left|\sum_{x=1}^{p-1}\exp\left(2\pi i\frac{cx+d/x}{p}\right)\right|\leqslant 2\sqrt{p}$$

are obtained by using the theory of zeta functions in algebraic function fields in one variable (A. Weil, 1948) (→ 436 Zeta Functions).

E. Waring's Problem

The first formulation of **Waring's problem** is found in E. Waring, *Meditationes algebraicae* (1770), in which he discusses the problem of expressing an arbitrary positive integer as the sum of at most nine cubes or as the sum of at most nineteen biquadratics. Hilbert proved (→ Section A) that there exists a positive integer $s(k)$ such that, for any integer N, the equation

$$x_1^k + x_2^k + \ldots + x_s^k = N$$

has a nonnegative integral solution if $s \geqslant s(k)$. We denote by $g(k)$ the least value of $s(k)$, and by $G(k)$ the least value of $s(k)$ for which the equation is solvable with at most finitely many exceptions of N. Research concerning $g(k)$ and $G(k)$ received its initial impetus from the circle method considered by Hardy and Littlewood, and it underwent considerable development in the works of H. Weyl and Vinogradov.

Let $r_s(N)$ be the number of solutions of the above equation. We have then

$$r_s(N) = \int_0^1 \Big(\sum_{x \leqslant N^{1/k}} \exp(2\pi i \alpha x^k) \Big)^s \times \exp(-2\pi i N\alpha)\, d\alpha.$$

If we make the Farey dissection, translating the interval $[0, 1]$ slightly, then the main term of $r_s(N)$ is provided by the integrals over major arcs, and the residual term is derived from the integrals over minor arcs. According to Hua (1959) we have

$$r_s(N) \sim S(N) \frac{\Gamma(1+1/k)^s}{\Gamma(s/k)} N^{s/k-1},$$

provided that $s \geqslant 2k^2(2\log k + \log\log k + c)$. Let p be a prime, and let $M(N, p^l)$ denote the number of solutions of the congruence equation

$$x_1^k + x_2^k + \ldots + x_s^k \equiv N \pmod{p^l}.$$

Then $\lim_{l\to\infty} M(N, p^l)/p^{l(s-1)} = \chi_p(N)$ is not zero, and the infinite product $\Pi_p \chi_p(N) = S(N)$ converges for $s \geqslant 4k$, where $S(N)$ is larger than a positive constant which is determined independently of the choice of N. On the other hand, let

$$S(a,q) = \sum_{x=0}^{q-1} \exp\Big(2\pi i \frac{a}{q} x^n\Big),$$

$$A(q,N) = q^{-s} \sum_{\substack{1 \leqslant a \leqslant q \\ (a,q)=1}} S(a,q)^s \exp\Big(-2\pi i \frac{a}{q} N\Big).$$

Then $\sum_{q=1}^{\infty} A(q,N)$ is absolutely convergent, and the sum is equal to $S(N)$. Next, if we denote by $V(N,\delta)$ the volume of the closed region satisfying $N \leqslant x_1^k + x_2^k + \ldots + x_s^k \leqslant N + \delta$ in the s-dimensional Euclidean space, then $\lim_{\delta\to 0} V(N,\delta)/\delta = \chi_\infty(N)$ exists and is equal to $(\Gamma(1+1/k)^s/\Gamma(s/k))N^{s/k-1}$. Hence we can show that the principal part of $r_s(N)$ is equal to the infinite product $\Pi_p \chi_p(N)$, where p runs over all finite and infinite prime spots in **Q**. This is a generalization of the singular series studied by Hardy.

With regard to $g(k)$, there are studies by L. E. Dickson (1936) and others. It is easy to see that $g(k) \geqslant 2^k + (3/2)^k - 2$ and that $G(k) \geqslant k+1$. It has been shown that $G(3) \leqslant 7$ (Linnik, 1947) and that $G(4) = 16$ (H. Davenport, 1939). More generally, Vinogradov (1959) proved that

$$G(k) \leqslant 2k\log k + 4k\log\log k + 2k\log\log\log k + ck.$$

To prove this, Vinogradov introduced the following integral, which is closely related to the †prime number theorem:

$$I(P) = \int_0^1 \ldots \int_0^1 \Big| \sum_{x \leqslant P} \exp\big(2\pi i(\alpha_1 x + \alpha_2 x^2 + \ldots + \alpha_k x^k)\big) \Big|^{2s} d\alpha_1 \ldots d\alpha_k.$$

Hua (1949) proved that if $s \geqslant \frac{1}{4}k(k+1) + lk$, then

$$I(P) \leqslant (5s)^{5sl} (\log P)^{2l} P^{2s - k(k+1)/2 + \delta},$$

where $\delta = \frac{1}{2}k(k+1)(1-1/k)^{l-1}$. Concerning $I(P)$, another notable approach was made by A. A. Karacuba and N. M. Korobov (1963). Further investigation proved that $I(P) = c_1 c_2 P^{2s-k(k+1)/2} + o(P^{2s-k(k+1)/2})$ if $s \geqslant ck^2 \log k$ (Vinogradov, Hua, 1959). The result is called the **Vinogradov mean value theorem**.

There are many variations and generalizations of this theorem. Vinogradov and Hua (1944) studied the problem of representing an arbitrary N as $N = p_1^k + p_2^k + \ldots + p_s^k$ (with p_i prime). Hua (1937) and others also considered the problem of representing N as $N = f(x_1) + f(x_2) + \ldots + f(x_s)$, where $f(x)$ is a given polynomial. Also, let

$$C(x_1, x_2, \ldots, x_n) = \sum_{i=1}^{n} \sum_{j=1}^{n} \sum_{k=1}^{n} C_{ijk} x_i x_j x_k$$

be a homogeneous polynomial of degree 3 with integral coefficients. Davenport (1957) proved that if $n \geqslant 32$, then the equation $C(x_1, x_2, \ldots, x_n) = 0$ has at least one nontrivial integral solution (the result was later further improved). There are further developments of the theory of representations of integers by forms in many variables by V. A. Tartakovskiĭ, H. Davenport, B. J. Birch, and D. J. Lewis.

Siegel (1922) considered the generalization of Hardy's square sum problem to the case of an algebraic number field. He later (1945) studied the generalized Waring's problem in an algebraic number field K of finite degree: Let I be the principal order of K, and let J_k be the subring of I generated by the kth powers of integers in K. It is easily seen that the index $(I:J_k)$ is finite. Hence our concern regarding $s(k)$ must be restricted to integers contained in J_k. Another question is how to extend the Farey dissection to an algebraic number field. Siegel succeeded in solving these difficulties. His ingenuity is seen in his way of dealing with the minor arcs, which provided a stimulus to the research of T. Mitsui (1960).

A generalization of Goldbach's problem to the case of an algebraic number field was obtained by Mitsui (1960) and O. Körner (1961).

As another extension of the Vinogradov three primes theorem, Mitsui (1971) proved the following theorem: Let K be an algebraic number field of degree n. Let C be the principal ideal class generated by a totally positive number in K, and P be the set of prime ideals of degree 1 contained in C. Let N be a positive integer and $I_s(N)$ be the number of representations of N as the sum of the norms of s prime ideals belonging to P,

$$I_s(N)=\sum_{N=N_{\mathfrak{p}_1}+\ldots+N_{\mathfrak{p}_s}} 1,\qquad \mathfrak{p}_i\in P\quad(1\leqslant i\leqslant s).$$

If N is sufficiently large and $s\geqslant 3$, we have the asymptotic formula

$$I_s(N)=A_sS(N)\frac{N^{s-1}}{(\log N)^s}+O\left(N^{s-1}\frac{\log\log N}{(\log N)^{s+1}}\right),$$

where A_s is a positive constant depending on s and K independent of N, and $S(N)$ denotes the singular series. If $s\equiv N\pmod{2D}$, where D is the discriminant of K, then $S(N)\geqslant c>0$, where c is a constant.

References

[1] R. G. Ayoub, An introduction to the analytic theory of numbers, Amer. Math. Soc. Math. Surveys, 1963.

[2] G. H. Hardy and E. M. Wright, An introduction to the theory of numbers, Clarendon Press, fourth edition, 1965.

[3] L. K. Hua, Additive Primzahltheorie, Teubner, 1959.

[4] L. K. Hua, Die Abschätzung von Exponentialsummen und ihre Anwendung in der Zahlentheorie, Enzykl. Math., Bd. I, 2, Heft 13, Teil I, 1959.

[5] E. G. H. Landau, Über einige neuere Fortschritte der additiven Zahlentheorie, Cambridge Univ. Press, 1937.

[6] H. H. Ostmann, Additive Zahlentheorie I, II, Erg. Math., Springer, 1956.

[7] H. A. Rademacher, Lectures on analytic number theory, Lecture note, Tata Inst. 1954–1955.

[8] И. М. Виноградов (I. M. Vinogradov), Избранные Труды, Изд. АН СССР, 1952.

[9] И. М. Виноградов (I. M. Vinogradov), К вопросу о верхней границе для $G(n)$, Izv. Akad. Nauk SSSR, 23 (1959), 637–642.

[10] T. Mitsui, On the Goldbach problem in an algebraic number field I, II, J. Math. Soc. Japan, 12 (1960), 290–324, 325–372.

6 (XVI.10) Additive Processes

A. General Remarks

The study of sums of †independent random variables has been one of the main topics in modern probability theory (→ 249 Limit Theorems in Probability Theory). In joining the idea of this study to the consideration of †stochastic processes with continuous time parameter, we get the notion of additive processes.

B. Definitions and Fundamental Properties

A real-valued †stochastic process $\{X_t\}_{0\leqslant t<\infty}$, in the rest of this article denoted by $X(t)$ $(0\leqslant t<\infty)$, where for simplicity we assume that $X(0)=0$, is called an **additive process** (or **process with independent increments**) if for any $t_0<t_1<\ldots<t_n$, $X(t_i)-X(t_{i-1})$ $(i=1,2,\ldots,n)$ are †independent. An additive process is essentially the same as a †spatially homogeneous †Markov process (i.e., a Markov process on $\mathbf{R}^1$ that is invariant under translations). When, for any $h>0$ and $t>s$, $X(t+h)-X(s+h)$ and $X(t)-X(s)$ have the same law, i.e., the distribution law of $X(t)-X(s)$ depends only on $t-s$, we call $X(t)$ **temporally homogeneous**. This is essentially the same notion as a temporally and spatially homogeneous Markov process.

Let $X(t)$ be a given additive process. If $f(t)$ is a function of t only, then clearly $Y(t)=X(t)-f(t)$ is also an additive process and we can choose $f(t)$ such that for every $t>0$ and for every sequence $s_n\uparrow t$ $(s_n\downarrow t)$, $Y(s_n)$ converges with probability one. Here, $\lim Y(s_n)$ is independent of a particular choice of s_n, and we denote it by $Y(t-)$ $(Y(t+))$. We shall call such $Y(t)$ a **centered process** and also say that $Y(t)$ is obtained from $X(t)$ by **centering**. This $f(t)$ is given, for example, by the condition $E(\arctan(X(t)-f(t)))=0$.

Let $Y(t)$ be a centered additive process. Then $Y(t-)=Y(t)=Y(t+)$ for all $t>0$, except on an at most countable t-set S, and $t\in S$ is called a **fixed point of discontinuity** of $Y(t)$. Then $Y_1(t)=\lim_{n\to\infty}U_n(t)$ exists with probability 1, where

$$U_n(t)=\sum_{\substack{0<i<n\\ s_i<t}}(Y(s_i+)-Y(s_i-))$$
$$+Y(t)-Y(t-)-C_t^n,$$

C_t^n is a constant determined by $E(\arctan U_n(t))=0$, and $S=\{s_j\}$ $(j=1,2,\ldots)$. $Y_2(t)=Y(t)-Y_1(t)$ is a centered additive process without any fixed point of discontinuity. Furthermore, $Y_1(t)$ and $Y_2(t)$ are independent. Thus, we have a decomposition of $Y(t):Y(t)=Y_1(t)+Y_2(t)$, where $Y_1(t)$ and $Y_2(t)$ are mutually independent additive processes. The structure of $Y_1(t)$ is simple, and it is not worthwhile to study its behavior in more detail. On the other hand, since $Y_2(t)$ is a centered additive process without any fixed point of discontinuity, it is an additive process that is †continuous in probability. Let $\tilde{Y}_2(t)$ be a †separable modification of $Y_2(t)$. Then the discontinuities of almost all sample functions of $\tilde{Y}_2(t)$ are of at most the first kind. If we set $Y_2^*(t)=\tilde{Y}_2(t+)$, $Y_2^*(t)$ is a †modification of $Y_2(t)$, and almost all sample functions of $Y_2^*(t)$ are right continuous and have left-hand limits at every t. In the study of the process $Y_2(t)$, it is always convenient to take such a modification. Thus, we give the following general definition: an additive process is called a **Lévy process** if it is continuous in probability and almost all sample functions are right continuous and have left-hand limits at every $t\in[0,\infty)$ [3, 7].

The notions of additive processes and Lévy processes can also be considered for $\mathbf{R}^N$-valued processes.

C. Additive Processes and Infinitely Divisible Distributions

Let $X(t)$ be a Lévy process and Φ_{st} $(s<t)$ be the †distribution of $X(t)-X(s)$. Then Φ_{st} is an †infinitely divisible distribution (→ 131 Distributions of Random Variables). Conversely, for a given infinitely divisible distribution Φ we can construct an essentially unique temporally homogeneous Lévy process $X(t)$ such that Φ coincides with the distribution of $X(1)$. If $X(t)$ is temporally homogeneous, the †characteristic function $\varphi_{st}(z)=E(e^{iz(X(t)-X(s))})$ of the distribution Φ_{st} is given in the form $\varphi_{st}=\exp((t-s)\psi(z))$; hence the law of the process $X(t)$ is completely determined by the function $\psi(z)$. By the †Lévy-Hinčin canonical form, $\psi(z)$ is written in the form:

$$\psi(z)=imz-\frac{v}{2}z^2$$
$$+\int_{-\infty}^{\infty}\left(e^{izu}-1-\frac{izu}{1+u^2}\right)n(du), \qquad (1)$$

where $m, v\in\mathbf{R}$, $v\geqslant 0$, and $n(du)$ is a nonnegative measure on $\mathbf{R}-\{0\}$ such that $\int_{-\infty}^{\infty}\frac{u^2}{1+u^2}n(du)<\infty$. These m, v, and $n(du)$ are uniquely determined by $\psi(z)$.

D. Basic Additive Processes

Wiener Process. When almost all sample functions of a Lévy process $X(t)$ are continuous, the distribution of $X(t)-X(s)$ is a †normal distribution. If, further, $X(t)$ is temporally homogeneous, $\psi(z)$ has the form $\psi(z)=imz-\frac{1}{2}vz^2$. In particular, if $m=0$ and $v=1$, then $X(t)$ is called a **Wiener process** or **Brownian motion**. This stochastic process was introduced by N. Wiener (1923) as a mathematical model for the random movement of colloidal particles first observed by a British botanist, R. Brown. This is one of the most fundamental and important stochastic processes in modern probability theory (→ 46 Brownian Motion).

Poisson Processes. When almost all sample functions of a Lévy process are increasing step functions with only jumps of size 1, the distribution of $X(t)-X(s)$ is a †Poisson distribution. If, further, $X(t)$ is temporally homogeneous, $\psi(z)$ in (1) has the form $\psi(z)=\lambda(e^{iz}-1)$ $(\lambda>0)$ and $X(t)$ is called a **Poisson process**. Let $X(t)$ be a Poisson process and let $T_0, T_0+T_1, T_0+T_1+T_2,\ldots$ be successive jumping times of a sample function $X(t)$. Then $T_0, T_1, T_2,\ldots$ is a sequence of mutually independent random variables with the common exponential distribution $P(T\in dt)=\lambda e^{-\lambda t}dt$. Conversely, given such a sequence $\{T_n\}$, if we define $X(t)=\inf\{n\,|\,T_0+T_1+\ldots+T_n>t\}$, then $X(t)$ is a Poisson process. Thus, for example, the number of telephone calls at a switchboard is a Poisson process when the intervals between successive calls may be regarded as independent and having a common exponential distribution.

E. The Structure of the General Lévy Process

In this section we restrict ourselves for simplicity to temporally homogeneous Lévy processes. As we noted, the probability law of the process $X(t)$ is determined by the function $\psi(z)$. The Lévy-Hinčin formula (1) in a certain sense shows that ψ is a combination of a Wiener process and Poisson processes. This fact can be seen more clearly from the Lévy-Itô theorem, which states that the sample function of $X(t)$ itself may be expressed as a composite of those of a Wiener process and Poisson processes. The Lévy-Itô theorem actually implies formula (1) and, moreover, clarifies its probabilistic meaning.

The Lévy-Itô theorem may be summarized as follows: Let U be a Borel subset of $\mathbf{R}$ which has a positive distance from the origin, and let $N(t, U)$ be the number of s such that $X(s)-X(s-)\in U, s \leqslant t$. Then $N(t, U)$ is a Poisson process. The expectation $E(N(t, U))$ can be written in the form $tn(U)$, where $n(U)$ defines a nonnegative Borel measure on $\mathbf{R}-\{0\}$; furthermore, it satisfies

$$\int_{-\infty}^{\infty} \frac{u^2}{1+u^2} n(du) < \infty.$$

Next, we set

$$S_\epsilon(t) = \int_{|u|>\epsilon} uN(t, du) = \sum_{\substack{s \leqslant t \\ |X(s)-X(s-)|>\epsilon}} (X(s)-X(s-)).$$

Generally, $S_\epsilon(t)$ diverges as $\epsilon \downarrow 0$. However, with probability one, a centered process

$$\bar{S}_\epsilon(t) = S_\epsilon(t) - t\int_{|u|>\epsilon} \frac{u}{1+u^2} n(du)$$

converges uniformly in t on every finite interval as $\epsilon \downarrow 0$. Furthermore, $X(t) - \lim_{\epsilon\downarrow 0} \bar{S}_\epsilon(t)$ is continuous with probability one. However, a Lévy process $X(t)$ of which almost all sample functions are continuous has the form $mt + \sqrt{v}\, B(t)$, where $m, v \geqslant 0$ are constants and $B(t)$ is a Wiener process. Hence we have

$$X(t) = mt + \sqrt{v}\, B(t) + \lim_{\epsilon\downarrow 0} \int_{|u|>\epsilon} \left(uN(t, du) - \frac{u}{1+u^2} tn(du) \right). \quad (2)$$

Furthermore, we can show that if

$U_1, U_2, \ldots, U_n$

are disjoint, then

$B(t), N(t, U_1), N(t, U_2), \ldots, N(t, U_n)$

are mutually independent Lévy processes. In particular, in (2) the terms are mutually independent. The m, v, and $n(du)$ in (2) correspond, of course, to those in (1). Conversely, given m, v, and $n(du)$, we can construct $B(t)$ and $N(t, U)$ with the above properties, and then (2) defines a Lévy process which corresponds to $\psi(z)$ given by (1). The measure $n(du)$ is called the **Lévy measure** of $X(t)$.

F. Examples of Lévy Processes

Compound Poisson Processes. A temporally homogeneous Lévy process is called a **compound Poisson process** if almost all sample functions are step functions, namely, if $\psi(z)$ in (1) is given by

$$\psi(z) = \int_{\infty}^{\infty} (e^{izu}-1) n(du),$$

$$\lambda = \int_{-\infty}^{\infty} n(du) < \infty.$$

If we set $\Phi(du) = (1/\lambda)n(du)$, then $\Phi(du)$ is a probability distribution on $\mathbf{R}$. Φ is the distribution of the size of jumps when they occur. A compound Poisson process is constructed in the following way: Let $T_0, T_1, T_2, \ldots$; $U_1, U_2, \ldots$ be mutually independent random variables such that

$$P(T_n \in dt) = \lambda e^{-\lambda t} dt, \quad t>0,$$

$$P(U_n \in du) = \Phi(du),$$

and let

$$X(t) = U_1 + U_2 + \ldots + U_n,$$

where

$$n = \inf\{ n \mid T_0 + T_1 + \ldots + T_n > t\}.$$

Then $X(t)$ is a compound Poisson process. Thus the number of jumps of $X(t)$ follows a Poisson process, and the size of each jump obeys the distribution Φ.

Stable Processes. Suppose $X(t)$ is a temporally homogeneous Lévy process such that for each $t>0$, there exists a constant C_t such that the random variables $X(t)$ and $C_t X(1)$ have the same distribution. In such a case, $X(t)$ is called a **stable process**. We have a stable process if and only if the corresponding infinitely divisible distribution is a †stable distribution. The †exponent α $(0<\alpha\leqslant 2)$ of the stable distribution is called the **exponent of the stable process**. When $\alpha=1$, it is called a **Cauchy process**. When $\alpha=2$, it is essentially a Wiener process. $\psi(z)$ in (1) is given as follows:

$$\psi(z) = C_+ \int_0^{\infty} (e^{izu}-1) \frac{du}{u^{1+\alpha}} + C_- \int_{-\infty}^{0} (e^{izu}-1) \frac{du}{|u|^{1+\alpha}}, \quad 0<\alpha<1,$$

$$\psi(z) = C_+ \int_0^{\infty} (e^{izu}-1-izu) \frac{du}{u^{1+\alpha}} + C_- \int_{-\infty}^{0} (e^{izu}-1-izu) \frac{du}{|u|^{1+\alpha}}, \quad 1<\alpha<2,$$

$$\psi(z) = iC_1 z + \frac{C_0}{\pi} \int_{-\infty}^{\infty} \left(e^{izu} - 1 - \frac{izu}{1+u^2} \right) \frac{du}{u^2}, \quad \alpha=1,$$

where C_+, C_-, and C_0 are nonnegative constants and C_1 is a real constant. Hence,

$$\psi(z)=\left(-C_0+i\frac{z}{|z|}C_1\right)|z|^\alpha,$$

where if $0<\alpha<2$ and $\alpha\neq1$,

$$-C_0=(C_++C_-)\Gamma(-\alpha)\cos\frac{\pi}{2}\alpha$$

$$C_1=(C_--C_+)\Gamma(-\alpha)\sin\frac{\pi}{2}\alpha.$$

When $C_1=0$, $X(t)$ is called a **symmetric stable process**. In particular, for the symmetric Cauchy process corresponding to $\psi(z)=-|z|$, we have

$$P(X(t)<x)=\frac{1}{\pi}\int_{-\infty}^{x}\frac{t}{t^2+y^2}dy.$$

Next, when $0<\alpha<1$ and $C_-=0$, almost all sample functions of $X(t)$ are purely discontinuous increasing functions (i.e., sums of positive jumps). In this case, $X(t)$ is called a **one-sided stable process of the exponent** α (or **subordinator of the exponent** α). Now, if $X(t)$ is a symmetric stable process of the exponent β $(0<\beta\leqslant2)$ and $Y(t)$ is a subordinator of the exponent α which is independent of $X(t)$, then $Z(t)=X(Y(t))$ is a symmetric stable process of the exponent $\alpha\beta$. This operation is called a **subordination** and is closely related to the theory of †fractional powers of †infinitesimal generators of semigroups (→ 262 Markov Processes) [2].

A symmetric stable process $X(t)$ is defined in a similar way when $X(t)$ takes values in an N-dimensional space $\mathbf{R}^N$. In particular, if $X(t)$ is a symmetric stable process of the exponent α $(0<\alpha\leqslant2)$ given by

$$E(e^{i(z,X(t))})=e^{-t|z|^\alpha},\quad z\in\mathbf{R}^N,$$

then for every bounded measurable function $f(x)$ with compact support, we have

$$E\left(\int_0^\infty f(x+X(t))\,dt\right)$$

$$=\frac{\Gamma((N-\alpha)/2)}{2^\alpha\pi^{N/2}\Gamma(\alpha/2)}\int_{\mathbf{R}^N}|x-y|^{\alpha-N}f(y)\,dy.$$

The right-hand side is the †Riesz potential of order α (→ 335 Potential Theory). This fact is a generalization of a well-known relation between Brownian motion and Newtonian potential (→ 46 Brownian Motion), and through this relation we can study several properties of sample functions and also compute various quantities related to stable processes [1].

G. Generalizations of Additive Processes

A temporally homogeneous Lévy process is, as we have seen, essentially a temporally homogeneous Markov process on $\mathbf{R}$ which is homogeneous in space (i.e., invariant under translations of the space). Thus, on a homogeneous space where homogeneity in space makes sense, we can generalize the notion of additive processes. Let M be a †homogeneous space with transformation group G. A temporally homogeneous Markov process $X(t)$ is called an **invariant Markov process**, (or **homogeneous Markov process**) if its system of †transition probabilities $\{P(t,x,E)\}$ satisfies $P(t,x,E)=P(t,gx,gE)$ for all $g\in G$. Thus an additive process is exactly an invariant Markov process on $\mathbf{R}^N$ when G is the group of translations. G. A. Hunt determined all invariant Markov processes when M is a †Lie group or a factor space of a Lie group [5].

Let G be a Lie group and $\Lambda=\Lambda(G)$ be the (left-invariant) †Lie algebra of G. Let $G_c=G\cup\{\omega\}$ be a †one-point compactification of G, and C be the set of all continuous functions G_c. We may define $Yf\,(Y\in\Lambda, f\in C)$ as usual by

$$Y(f)=\lim_{t\downarrow0}\frac{R_{\eta(t)}f-f}{t},$$

$$\eta(t)=\exp tY,\quad R_\sigma f(\tau)=f(\tau\alpha),$$

when the limit exists uniformly. Let $C_2=\{f\in C\,|\,Y(Xf)$ exists for every $X,Y\in\Lambda\}$. Let $X_1,X_2,\ldots,X_d$ be a basis of $\Lambda(G)$, and let $x_1,x_2,\ldots,x_d$ be functions in C_2 such that $x_i(e)=0$ and $X_i(x_j)(e)=\delta_{ij}$ $(i,j=1,2,\ldots,d$; e is the unit element of G). Take a neighborhood of e and define a function $\varphi(g)=\sum_{i=1}^d x_i^2(g)$ for g contained in the neighborhood, and extend this function to G_c in such a way that $\varphi\in C_2$ and $\varphi\geqslant k>0$ $(k=\text{constant})$ outside of the neighborhood of e. Then $g\in G$ defines a transformation of G_c by $\tau_g\sigma=g\sigma$, $\tau_g\omega=\omega$, and in this way G_c is supplied with the structure of a †homogeneous space with the transformation group G. Now let $X(t)$ be an invariant Markov process on G_c which is †continuous in probability. Then the †semigroup T_t (which is a †strongly continuous semigroup on C) of the process $X(t)$ is characterized as follows: The domain of the infinitesimal generator A of T_t contains C_2, and for $f\in C_2$

$$Af(\tau)=\sum_{i=1}^d a_i(X_if)(\tau)+\sum_{i,j=1}^d a_{ij}X_i(X_jf)(\tau)$$

$$+\int_{G_c-\{e\}}\left\{f(\tau\sigma)-f(\tau)\right.$$

$$\left.-\sum_{i=1}^d X_if(\tau)\cdot x_i(\sigma)\right\}n(d\sigma),$$

where a_i, a_{ij} are real numbers $(i,j=1,2,\ldots,d)$ such that (a_{ij}) is a symmetric nonnegative definite matrix, and $n(d\sigma)$ is a nonnegative measure on $G_c-\{e\}$ such that $\int_{G_c-\{e\}}\varphi(\sigma)\cdot$

$n(d\sigma) < \infty$. Conversely, given such a_i, a_{ij} and $n(d\sigma)$, there exists one and only one invariant Markov process on G_c whose semigroup is given as above.

A similar result is obtained when M is a factor space of a Lie group by its compact subgroup. Furthermore, for more concrete homogeneous spaces such as spheres or †Lobačevskiĭ spaces (more generally, †symmetric Riemannian spaces) the canonical form of the invariant Markov processes and infinitely divisible laws is obtained by making use of harmonic analysis [4, 11]. For hitting probabilities of additive processes → [12].

References

[1] R. M. Blumenthal, R. K. Getoor, and D. B. Ray, On the distribution of the first hits for the symmetric stable processes, Trans. Amer. Math. Soc., 99 (1961), 540–554.

[2] S. Bochner, Harmonic analysis and the theory of probability, Univ. of California Press, 1955.

[3] J. L. Doob, Stochastic processes, John Wiley, 1953.

[4] U. Grenander, Probabilities on algebraic structures, John Wiley, 1963.

[5] G. A. Hunt, Semi-groups of measures on Lie groups, Trans. Amer. Math. Soc., 81 (1956), 264–293.

[6] K. Itô, On stochastic processes I, Japan. J. Math., 18 (1942), 261–301.

[7] K. Itô, Kakuritu katei (Japanese; Stochastic processes), Iwanami coll. of modern appl. math., 1957.

[8] K. Itô, On stochastic processes, Lecture notes, Tata Inst., 1960.

[9] P. Lévy, Théorie de l'addition des variables aléatoires, Gauthier-Villars, 1937.

[10] P. Lévy, Processus stochastiques et mouvement brownien, Gauthier-Villars, 1948.

[11] R. Gangolli, Isotropic infinitely divisible measures on symmetric spaces, Acta Math., 111 (1964), 213–246.

[12] H. Kesten, Hitting probabilities of single points for processes with stationary independent increments, Mem. Amer. Math. Soc., 93 (1969).

7 (III.21) Adeles and Ideles

A. Introduction

The concept of idele was first introduced by C. Chevalley (*J. Math. Pures Appl.*, (9) 15 (1936); *Ann. of Math.*,(2) 41 (1940)), for †algebraic number fields. Later on, this concept and the allied concept of adele were defined for †simple algebras and also for †algebraic groups over algebraic number fields, and the two concepts became important in the arithmetical theory of these objects. We shall first explain the general concept of restricted direct product, by means of which adeles and ideles will be defined.

B. Restricted Direct Product

Let I be an index set. Suppose we are given, for each $\mathfrak{p} \in I$, a †locally compact group $G_\mathfrak{p}$, and for each $\mathfrak{p}$ except for a given finite set, say $\mathfrak{p}_1, \mathfrak{p}_2, \ldots, \mathfrak{p}_r$, a compact open subgroup $U_\mathfrak{p}$ of $G_\mathfrak{p}$. Let G be the subgroup of the direct product $\prod_{\mathfrak{p} \in I} G_\mathfrak{p}$ consisting of elements $(g_\mathfrak{p})$ whose $G_\mathfrak{p}$-components $g_\mathfrak{p}$ lie in $U_\mathfrak{p}$, except for a finite number of $\mathfrak{p}$. Put $U = \prod_{i=1}^r G_{\mathfrak{p}_i} \times \prod_{\mathfrak{p} \neq \mathfrak{p}_i} U_\mathfrak{p}$. Then U is a locally compact group with respect to the †product topology. The group G can be supplied naturally with a topology with respect to which G is a locally compact group and the quotient space G/U is discrete. The group G together with this topology is called the **restricted direct product** of $\{G_\mathfrak{p}\}$ with respect to $\{U_\mathfrak{p}\}$.

C. Adeles and Ideles

Let k be an †algebraic number field of finite degree and I be the totality of finite and infinite †prime divisors of k. For each $\mathfrak{p} \in I$ we denote by $k_\mathfrak{p}$ and $k_\mathfrak{p}^\times$ the †completion of k with respect to $\mathfrak{p}$ and the multiplicative group of nonzero elements of $k_\mathfrak{p}$, respectively. Furthermore, for each finite prime divisor $\mathfrak{p}$, we denote by $\mathfrak{o}_\mathfrak{p}$ and $\mathfrak{u}_\mathfrak{p}$ the ring of †$\mathfrak{p}$-adic integers of k and the multiplicative group of †units of $\mathfrak{o}_\mathfrak{p}$, respectively.

(1) Since $\mathfrak{o}_\mathfrak{p}$ is a compact open subgroup of $k_\mathfrak{p}$ as an additive group, we can construct the restricted direct product $\mathbf{A}_k$ of $\{k_\mathfrak{p}\}$ with respect to $\{\mathfrak{o}_\mathfrak{p}\}$. Then $\mathbf{A}_k$ is a locally compact ring with respect to the componentwise ring operations. We call $\mathbf{A}_k$ the **adele ring** (or **ring of valuation vectors**) of k, and an element of $\mathbf{A}_k$ an **adele** (or **valuation vector**) of k. The element of the direct product $\prod_\mathfrak{p} k_\mathfrak{p}$ whose $\mathfrak{p}$-component is a fixed element of k for all $\mathfrak{p}$ is an adele. We call such an adele a **principal adele**. Since $\mathfrak{u}_\mathfrak{p}$ is a compact open subgroup of $k_\mathfrak{p}^\times$ for each finite prime $\mathfrak{p}$, we can construct the restricted direct product $\mathbf{J}_k$ of $\{k_\mathfrak{p}^\times\}$ with respect to $\{\mathfrak{u}_\mathfrak{p}\}$. We call $\mathbf{J}_k$ the **idele group** of k and an element of $\mathbf{J}_k$ an **idele** of k. The element of the direct product $\prod k_\mathfrak{p}^\times$ whose $\mathfrak{p}$-component is a fixed element of k for all $\mathfrak{p}$ is an idele. We call such an idele a **principal idele**. Each element $\mathfrak{b}$ of $\mathbf{J}_k$ induces

an automorphism $f_{\mathfrak{b}}$ of the additive group $\mathbf{A}_k$ defined by $f_{\mathfrak{b}}(\mathfrak{a})=\mathfrak{b}\cdot\mathfrak{a}$ ($\mathfrak{a}\in\mathbf{A}_k$). Thus $\mathbf{J}_k$ can be regarded as a subgroup of the automorphism group Aut($\mathbf{A}_k$) of the additive group $\mathbf{A}_k$. The topology of $\mathbf{J}_k$ coincides with the relative topology of $\mathbf{J}_k$ as a subgroup of Aut($\mathbf{A}_k$). We note, however, that the topology of $\mathbf{J}_k$ is different from the relative topology of $\mathbf{J}_k$ as a subspace of $\mathbf{A}_k$, and the former is stronger than the latter. Finally, for a †function field in one variable over a †finite field, the adele ring and the idele group can be defined similarly.

(2) Let $\mathfrak{R}$ be a †normal simple algebra over k and $\mathfrak{O}$ be a †maximal order of $\mathfrak{R}$. For each $\mathfrak{p}\in I$ put $\mathfrak{R}_{\mathfrak{p}}=\mathfrak{R}\otimes_k k_{\mathfrak{p}}$, and for each finite prime divisor $\mathfrak{p}$ put $\mathfrak{O}_{\mathfrak{p}}=\mathfrak{o}_{\mathfrak{p}}\cdot\mathfrak{O}$. Then $\mathfrak{O}_{\mathfrak{p}}$ is a compact open additive subgroup of $\mathfrak{R}_{\mathfrak{p}}$. By the adele ring $\mathbf{A}_{\mathfrak{R}}$ of $\mathfrak{R}$ we mean the restricted direct product of $\{\mathfrak{R}_{\mathfrak{p}}\}$ with respect to $\{\mathfrak{O}_{\mathfrak{p}}\}$. Let $\mathfrak{R}_{\mathfrak{p}}^{\times}$ and $\mathfrak{U}_{\mathfrak{p}}$ be the multiplicative group of nonzero divisors of $\mathfrak{R}_{\mathfrak{p}}$ and the multiplicative group of the units of $\mathfrak{O}_{\mathfrak{p}}$, respectively. ($\mathfrak{U}_{\mathfrak{p}}$ can be defined only if $\mathfrak{p}$ is a finite prime divisor.) By the idele group $\mathbf{J}_{\mathfrak{R}}$ of $\mathfrak{R}$ we mean the restricted direct product of $\{\mathfrak{R}_{\mathfrak{p}}^{\times}\}$ with respect to $\{\mathfrak{U}_{\mathfrak{p}}\}$. The notion of principal adele (or idele) of $\mathfrak{R}$ can be defined similarly, as in (1). The structures, as topological groups, of $\mathbf{A}_{\mathfrak{R}}$ and $\mathbf{J}_{\mathfrak{R}}$ do not depend on the choice of a maximal order $\mathfrak{O}$. The adele ring $\mathbf{A}_k$ and the idele group $\mathbf{J}_k$ described in (1) are special cases of $\mathbf{A}_{\mathfrak{R}}$ and $\mathbf{J}_{\mathfrak{R}}$, respectively.

(3) Let G be a linear †algebraic group defined over k, and let $G_{\mathfrak{p}}$ be the set of $k_{\mathfrak{p}}$-†rational points of the group for each $\mathfrak{p}\in I$. For each finite prime divisor $\mathfrak{p}$, let $U_{\mathfrak{p}}$ be the set of elements α of $G_{\mathfrak{p}}$ such that the coordinates of both α and α^{-1} are $\mathfrak{p}$-adic integers. We can then construct the restricted direct product of $\{G_{\mathfrak{p}}\}$ with respect to $\{U_{\mathfrak{p}}\}$, which is called the **idele group** (or **adele group**) of G.

In the following section we focus on describing the fundamental properties of adeles and ideles of an algebraic number field k. We shall start, however, by observing more generally those adeles and ideles of a normal simple algebra $\mathfrak{R}$ over k. (For the properties of the adele group of algebraic groups → [7]; 15 Algebraic Groups.)

D. The Structures of the Adele Ring and Idele Group

Let $\mathfrak{R}$ be a normal simple algebra over an algebraic number field of finite degree k. We identify the totality of principal adeles of $\mathfrak{R}$ (principal ideles of $\mathfrak{R}$) with $\mathfrak{R}$ ($\mathfrak{R}^{\times}$), and denote it by the same letter $\mathfrak{R}$ ($\mathfrak{R}^{\times}$). Then $\mathfrak{R}$ ($\mathfrak{R}^{\times}$) is a discrete subgroup of $\mathbf{A}_{\mathfrak{R}}$ ($\mathbf{J}_{\mathfrak{R}}$). The quotient group $\mathbf{A}_{\mathfrak{R}}/\mathfrak{R}$ is compact. Denoting by $|\alpha_{\mathfrak{p}}|_{\mathfrak{p}}$ ($\alpha_{\mathfrak{p}}\in k_{\mathfrak{p}}$) and $N_{\mathfrak{p}}(\alpha_{\mathfrak{p}})$ ($\alpha_{\mathfrak{p}}\in\mathfrak{R}_{\mathfrak{p}}$) the †normalized valuation of $k_{\mathfrak{p}}$ and the †reduced norm from $\mathfrak{R}_{\mathfrak{p}}$ to $k_{\mathfrak{p}}$, respectively, we define, for $\mathfrak{a}\in\mathbf{J}_{\mathfrak{R}}$, a positive number: $V(\mathfrak{a})=\prod_{\mathfrak{p}\in I}|N_{\mathfrak{p}}(\alpha_{\mathfrak{p}})|_{\mathfrak{p}}$ where $\mathfrak{a}=(\alpha_{\mathfrak{p}})$. We call $V(\mathfrak{a})$ the **volume** of $\mathfrak{a}$. If $\mathfrak{a}$ is a principal idele, we have $V(\mathfrak{a})=1$ by the †product formula on valuations. Denote by $\mathbf{J}^0_{\mathfrak{R}}$ the set of ideles $\mathfrak{a}$ with $V(\mathfrak{a})=1$ and put $\mathbf{C}^0_{\mathfrak{R}}=\mathbf{J}^0_{\mathfrak{R}}/\mathfrak{R}^{\times}$. Then $\mathbf{C}^0_{\mathfrak{R}}$ has finite volume with respect to the †Haar measure of $\mathbf{J}_{\mathfrak{R}}$. Furthermore, $\mathbf{C}^0_{\mathfrak{R}}$ is compact if and only if $\mathfrak{R}$ is a †division algebra. In particular, $\mathbf{C}^0_k$ is compact. Let $\mathbf{Q}$ be the field of rational numbers. For each rational prime p, we define a character λ_p of the completion $\mathbf{Q}_p$ of $\mathbf{Q}$ with respect to the p-adic topology (by a character of $\mathbf{Q}_p$, we mean a continuous homomorphism from $\mathbf{Q}_p$ to the 1-dimensional torus $\mathbf{R}/\mathbf{Z}$): If $p=p_\infty$ is the infinite prime of $\mathbf{Q}$, then we put $\lambda_{p_\infty}(x)\equiv -x \bmod \mathbf{Z}$ ($x\in\mathbf{Q}_p$). If p is finite, then we let λ_p be the composite of the following three canonical homomorphisms, namely, the one from $\mathbf{Q}_p$ to $\mathbf{Q}_p/\mathbf{Z}_p$, the one from $\mathbf{Q}_p/\mathbf{Z}_p$ to $\mathbf{Q}/\mathbf{Z}$, and the one from $\mathbf{Q}/\mathbf{Z}$ to $\mathbf{R}/\mathbf{Z}$. We define a character $\lambda_{\mathfrak{p}}$ of $\mathfrak{R}_{\mathfrak{p}}$ as follows: $\lambda_{\mathfrak{p}}=\lambda_p\circ\mathrm{tr}(\mathfrak{R}_{\mathfrak{p}}/\mathbf{Q}_p)$, where p is the rational prime divisible by $\mathfrak{p}$ and $\mathrm{tr}(\mathfrak{R}_{\mathfrak{p}}/\mathbf{Q}_p)$ denotes the †reduced trace from $\mathfrak{R}_{\mathfrak{p}}$ to $\mathbf{Q}_p$. For $x,y\in\mathfrak{R}_{\mathfrak{p}}$, put $(x,y)_{\mathfrak{p}}=\exp(2\pi i\lambda_{\mathfrak{p}}(xy))$. Then the additive group $\mathfrak{R}_{\mathfrak{p}}$ is self-dual relative to $(x,y)_{\mathfrak{p}}$. Furthermore, if we put $\langle\mathfrak{a},\mathfrak{b}\rangle=\prod_{\mathfrak{p}}(a_{\mathfrak{p}},b_{\mathfrak{p}})_{\mathfrak{p}}$ for $\mathfrak{a}=(a_{\mathfrak{p}})$ and $\mathfrak{b}=(b_{\mathfrak{p}})\in\mathbf{A}_{\mathfrak{R}}$ then $\mathbf{A}_{\mathfrak{R}}$ is self-dual relative to $\langle\mathfrak{a},\mathfrak{b}\rangle$. The †annihilator of the group of principal adeles with respect to $\langle\mathfrak{a},\mathfrak{b}\rangle$ is $\mathfrak{R}$. Hence it follows from †Pontrjagin's duality theorem that $\mathbf{A}_{\mathfrak{R}}/\mathfrak{R}$ is compact. Henceforth let $\mathfrak{R}=k$. We call the quotient group $\mathbf{C}_k=\mathbf{J}_k/k^{\times}$ (an element of $\mathbf{C}_k$) the **idele class group** of k (an **idele class**). If a character χ of $\mathbf{J}_k$ satisfies the condition $\chi(\alpha)=1$ for all $\alpha\in k$ (i.e., if χ is a character of $\mathbf{C}_k$), we call such a character a **Grössencharakter**. Grössencharakters were introduced by E. Hecke as characters of a certain type of the †ideal group of k (*Math. Z.*, 1 (1918), 5 (1920)), but they are essentially the same as the ones defined above [1]. Let $\mathbf{D}_k$ be the connected component of the identity element of $\mathbf{C}_k$. Then $\mathbf{C}_k/\mathbf{D}_k$ is totally disconnected and compact. Hence a Grössencharakter χ is of finite order if and only if $\chi(\mathbf{D}_k)=1$. We can prove by †class field theory that $\mathbf{C}_k/\mathbf{D}_k$ is canonically isomorphic to the Galois group over k of the maximal Abelian extension of k (→ 62 Class Field Theory). For the structure of $\mathbf{D}_k$, the following fact is known: Let r_1 and r_2 be the number of †real infinite prime divisors and †imaginary infinite prime divisors of k, respectively. Then the dual group of $\mathbf{D}_k$ is isomorphic to

$\mathbf{R}\times\mathbf{Q}^{r_1+r_2-1}\times\mathbf{Z}^{r_2}$, where $\mathbf{R}$ is the additive group of real numbers with the usual topology and $\mathbf{Q}$ ($\mathbf{Z}$) is the additive group of rational numbers (rational integers) with the discrete topology. Let F be a function field in one variable over a finite field $\mathbf{F}_0$. The properties of the adele ring and idele group of F are similar to the properties of $\mathbf{A}_k$ and $\mathbf{J}_k$, while the group $\mathbf{C}_F$ has a simpler structure than $\mathbf{C}_k$. To explain the structure of $\mathbf{C}_F$, let $\tilde{F}$ be the maximal Abelian extension of F, $\tilde{G}$ be the Galois group of $\tilde{F}/F$, and G_F be the subgroup of $\tilde{G}$ consisting of the elements σ such that $\sigma(\alpha)=\alpha^{q^n}$ for all $\alpha\in\overline{\mathbf{F}}_0$ (= the †algebraic closure of $\mathbf{F}_0$), where q is the number of elements of the finite field $\mathbf{F}_0$ and n is a given rational integer. Also, let G_F^0 be the subgroup of G_F consisting of the elements inducing the identity mapping on $\overline{\mathbf{F}}_0$. G_F^0 is a compact group with respect to the †Krull topology. G_F can be naturally supplied with a topology such that the group G_F is a locally compact group and the quotient group G_F/G_F^0 is discrete. Then class field theory implies that G_F is isomorphic to $\mathbf{C}_F$ as a topological group.

The following characterization of the adele ring of a number field or function field in one variable over a finite field is the work of K. Iwasawa (*Ann. of Math.*, 57 (1953)). Let $\mathbf{A}$ be a †semisimple commutative and locally compact topological ring with unity 1. Assume that $\mathbf{A}$ is neither discrete nor compact, and moreover that $\mathbf{A}$ contains a discrete subfield $k\ni 1$ and $\mathbf{A}/k$ is compact. Then k is an algebraic number field or a function field in one variable over a finite field, and $\mathbf{A}$ is isomorphic to the adele ring of k as a topological ring.

E. Ideles and Cohomology

Let K be a Galois extension of finite degree of an algebraic number field k, and $\mathfrak{G}$ be the Galois group of the extension K/k. $\mathfrak{G}$ operates naturally on the idele group $\mathbf{J}_K$ and the idele class group $\mathbf{C}_K$ of K. The structures of the †cohomology groups of $\mathfrak{G}$ with the coefficient groups $\mathbf{J}_k$ and $\mathbf{C}_K$ were investigated by G. Hochschild, T. Nakayama, E. Artin, J. Tate, and others. In particular, we have $H^1(\mathfrak{G},\mathbf{C}_K)=\{0\}$ and $H^2(\mathfrak{G},\mathbf{C}_K)\cong\mathbf{Z}/n\mathbf{Z}$ (cyclic group of order n), where $n=[K:k]$. These facts play an important role in one of the proofs of class field theory (→ [3]; 62 Class Field Theory). Furthermore, A. Weil introduced the so-called **Weil group**, which is a †group extension of a certain type of $\mathbf{C}_K$ by $\mathfrak{G}$. He defined the most general L-functions, which include both †Artin L-functions and †Hecke L-functions with Grössencharakters (→ [2]; 436 Zeta Functions).

F. Fourier Analysis on the Adele Group

†Dedekind zeta functions and Hecke L-functions are †meromorphic on the whole complex plane and satisfy functional equations of certain types. This can be proved by methods of Fourier analysis on the adele group $\mathbf{A}_k$ (Artin, Iwasawa, Tate [1, 8]). For a continuous complex-valued function $\varphi(\mathfrak{a})$ on $\mathbf{A}_k$ satisfying suitable conditions, we define the Fourier transform of $\varphi(\mathfrak{a})$ as follows:

$$\hat{\varphi}(\mathfrak{a})=\int_{A_k}\varphi(\mathfrak{a})\langle\mathfrak{a},\mathfrak{b}\rangle d\mathfrak{b},$$

where $d\mathfrak{b}$ denotes the Haar measure on $\mathbf{A}_k$. By normalizing $d\mathfrak{b}$ suitably and applying †Poisson's summation formula, we get, for each idele $\mathfrak{a}$ of k,

$$\sum_{\alpha\in k}\varphi(\mathfrak{a}\alpha)=V(\mathfrak{a})^{-1}\sum_{\alpha\in k}\hat{\varphi}(\mathfrak{a}^{-1}\alpha).$$

This is called the **Θ-formula**. Consider the following integral on $\mathbf{J}_k$:

$$\xi(s)=\int_{J_k}V(\mathfrak{a})^s\chi(\mathfrak{a})\varphi(\mathfrak{a})\,d^*\mathfrak{a},$$

where $d^*\mathfrak{a}$ denotes the Haar measure on $\mathbf{J}_k$, s is a complex number, and χ is a Grössencharakter of k, namely, a character of $\mathbf{C}_k$. This integral converges if $s>1$, and by using the Θ-formula one can show that $\xi(s)$ is meromorphic on the whole complex plane and satisfies a functional equation of a certain type. When the function φ is of special type, then the above integral can be explicitly expressed as the product of L-functions, Γ-functions, and exponential functions. This method of expressing L-functions by integrals on $\mathbf{J}_k$ and applying the Θ-formula can be applied to investigate †Hey zeta functions and L-functions of various types defined for a simple algebra (→ 436 Zeta Functions) (G. Fujisaki [6]; T. Tamagawa, *Ann. of Math.*, 77 (1963)).

References

[1] J. Tate, Fourier analysis in number fields and Hecke's zeta-functions, Thesis, Princeton, 1950. (Algebraic number theory, edited by J. W. S. Cassels and A. Fröhlich, Academic Press, 1967, ch. 15.)
[2] A. Weil, Sur la théorie du corps de classes, J. Math. Soc. Japan, 3 (1951), 1–35.
[3] E. Artin and J. Tate, Class field theory, Lecture notes, Harvard, 1961 (Benjamin, 1967).
[4] E. Artin, Representatives of the connected component of the idele class group, Proc. Intern. Symp. Alg. Number Theory, Tokyo and Nikko, 1955, p. 51–54.
[5] G. Shimura and Y. Taniyama, Kindaiteki

seisûron (Japanese; Modern number theory), Kyôritu, 1957.
[6] G. Fujisaki, On the zeta-function of the simple algebra over the field of rational numbers, J. Fac. Sci. Univ. Tokyo, 7 (1958), 567–604.
[7] A. Weil, Adeles and algebraic groups, Lecture notes, Institute for Advanced Study, Princeton, 1961.
[8] S. Lang, Algebraic numbers, Addison-Wesley, 1964.
[9] A. Weil, Basic number theory, Springer, 1967.

8 (XIII.14) Adjoint Differential Equations

A. Ordinary Differential Equations

Given a †linear homogeneous ordinary differential equation of the nth order,

$$L[y]\equiv \sum_{k=0}^{n} p_k(x)y^{(n-k)}=0, \tag{1}$$

where $p_k(x)$ $(k=0,1,\ldots,n)$ are analytic functions of a complex variable x, we call the equation

$$M[y]\equiv \sum_{k=0}^{n} (-1)^{n-k}(\bar{p}_k y)^{(n-k)}=0 \tag{2}$$

the **adjoint differential equation** of the given equation (1). (The bar means complex conjugation.) The left-hand side $M[y]$ of (2) is called the **adjoint differential expression** of the left-hand side $L[y]$ of (1). $L[y]$ turns out to be the adjoint expression of $M[y]$. Hence, adjointness is a mutually reciprocal relation between differential expressions. Between mutually adjoint expressions L and M, we have **Lagrange's identity**

$$\bar{z}L[y]-y\overline{M[z]}=dN[y,z]/dx, \tag{3}$$

where $N[y,x]$ denotes the bilinear form

$$N[y,z]\equiv \sum_{k=0}^{n-1}\sum_{h=0}^{n-1-k}(-1)^h(p_k\bar{z})^{(h)}y^{(n-1-k-h)} \tag{4}$$

in $y^{(h)}$ and $\bar{z}^{(h)}$ $(h=0,1,2,\ldots,n-1)$. $N[y,z]$ is called a **bilinear concomitant differential expression**. From this relation it follows that if p independent solutions of $M[y]=0$ are known, it is possible to reduce by p the order of the original equation $L[y]=0$. If $M[y]\equiv L[y]$, the equation $L[y]=0$ is said to be **self-adjoint**.

Now let x be a real variable and p_k be real-valued. If (1) is self-adjoint, its order must be an even number, say, $n=2m$. In this case, $L[u]$ is expressed in the form $L[y]=\sum_{k=0}^{m}(p_k y^{(m-k)})^{(m-k)}$, and (1) becomes the †Euler equation for the variational problem $\delta\int\Sigma p_k(y^{(m-k)})^2dx=0$. Now, if we take n arbitrary functions $z_1,\ldots,z_n$ and determine $g_1,\ldots,g_n$ by the relations

$$\sum g_iz_i=0,\quad \ldots,\quad \sum g_iz_i^{(n-2)}=0,$$
$$\sum g_iz_i^{(n-1)}=(-1)^n p_0^{-1},$$

we have, on account of Lagrange's identity,

$$y=\sum_i c_ig_i-\sum_i g_i\int_a^x z_iL[y]\,dx$$
$$+\sum_i g_i\int_a^x M[z_i]y\,dx,$$

where the c_i are constants. By using this equation, it is possible to reduce the differential equation $L[y]=f$ to an †integral equation of the Volterra type (**Dini's method**). In particular, if we know a certain number of solutions of $M[z]=0$, then we are led to a simpler equation by taking these solutions as the z_i.

Next, let us return to the complex variables and suppose that we are given a system of linear homogeneous differential equations of the first order,

$$y_i'+\sum_{j=1}^{n} p_{ij}(x)y_j=0,\qquad i=1,2,\ldots,n, \tag{5}$$

with n dependent variables y_i $(i=1,2,\ldots,n)$. We call the system

$$z_i'-\sum_{j=1}^{n}\bar{p}_{ji}(x)z_j=0,\qquad i=1,2,\ldots,n, \tag{6}$$

the **adjoint system of differential equations** of (5). It is clear that the adjoint relation is a reciprocal one. For the solutions y_i,z_i of (5) and (6) we have

$$\frac{d}{dx}\left(\sum_{i=1}^{n} y_i\bar{z}_i\right)=0,\quad \text{i.e.,}\quad \sum_{i=1}^{n} y_i\bar{z}_i=\text{constant}.$$

Hence, if p solutions of (6) are known, it is possible to decrease by p the number of unknown variables in the original system (5). The system (5) is self-adjoint if and only if $p_{ij}=-\bar{p}_{ji}$ $(i,j=1,2,\ldots,n)$; we then have $\mathrm{Re}\,p_{ii}=0$.

B. Partial Differential Equations

Given a partial differential equation of the form

$$L[u]\equiv\sum a_p(x)D^pu=0,\quad p=(p_1,\ldots,p_n),$$
$$D^p=\partial^{p_1\ldots p_n}/\partial x_1^{p_1}\ldots\partial x_n^{p_n},$$

we call the equation

$$L^*[u]\equiv\sum(-1)^{p_1+\ldots+p_n}D^p(\bar{a}_p(x)u)=0$$

the **adjoint partial differential equation** of $L[u]=0$. For instance, the adjoint equation of a linear homogeneous partial differential equation of the second order

$$L[u] \equiv \sum p_{jk} u_{jk} + \sum p_j u_j + pu = 0, \qquad p_{jk} = p_{kj} \tag{7}$$

($u_{jk} = \partial^2 u / \partial x_j \partial x_k$, $u_j = \partial u / \partial x_j$), where the p_{jk}, p_j, p are given real-valued functions of $x_1, \ldots, x_n$, is given by

$$M[v] \equiv \sum (p_{jk} v)_{jk} - \sum (p_j v)_j + pv = 0.$$

The relation between L and M is again a reciprocal one. From the identity

$$vL[u] - uM[v] = \sum \partial N_j / \partial x_j,$$

$$N_j = \sum_k \{ p_{jk} u_k v - (p_{jk} v)_k u \} + p_j uv,$$

we have **Green's formula** in the wider sense

$$\int_D (vL[u] - uM[v]) \, dx_1 \ldots dx_n$$

$$= -\sum_j \int_{\partial D} N_j \cos(\nu, x_j) \, dS,$$

where the integrals on the right-hand side are $(n-1)$-dimensional surface integrals taken over the boundary ∂D of a domain D, and ν denotes the inward normal. (7) is self-adjoint if and only if $p_j = \sum_k \partial p_{jk} / \partial x_k$, and this is equivalent to the condition that (7) can be derived from the variational problem $\delta \int \sum p_{jk} u_j u_k \, dx_l \ldots dx_n = 0$.

References

See references to 307 Ordinary Differential Equations and 315 Partial Differential Equations.

9 (VI.16) Affine Geometry

A. Construction of Affine Spaces

An affine space A is constructed as follows: Let V be a †vector space over a †field K, and let A be a nonempty set. For any vector $\mathbf{a} \in V$ and any element p of A, suppose that an addition $p + \mathbf{a} \in A$ is defined as satisfying the following three conditions: (i) $p + \mathbf{0} = p$ ($\mathbf{0}$ being a zero vector); (ii) $(p + \mathbf{a}) + \mathbf{b} = p + (\mathbf{a} + \mathbf{b})$ ($\mathbf{a}, \mathbf{b} \in V$); and (iii) for any $q \in A$ there exists a unique vector $\mathbf{a} \in V$ such that $q = p + \mathbf{a}$. (Condition (i) follows from (ii) and (iii).) Then we call A an **affine space**, V the **standard vector space** of A, and K the **coefficient field** of A. Each element of A is called a **point**.

If we fix an arbitrary point $o \in A$, there is a one-to-one correspondence between A and V given by the mapping sending $p \in A$ to $\mathbf{a} \in V$ such that $p = o + \mathbf{a}$. Such an element $\mathbf{a}$ of V is called a **position vector** of p with the **initial point** o and is denoted by $\overrightarrow{op}$. We say that $r+1$ points p_α $(0 \leqslant \alpha \leqslant r)$ of A are **independent** if r vectors $\mathbf{a}_i = \overrightarrow{p_0 p_i}$ $(1 \leqslant i \leqslant r)$ are linearly independent in V; otherwise, they are said to be **dependent**. This definition of dependence of points p_α is independent of the choice of the initial point among them. If V is of dimension n, we say that A is of **dimension** n, $\dim A = n$; in this case, we sometimes write A^n instead of A and V^n instead of V. The affine space A is of dimension n if and only if the maximum number of independent points in A is $n+1$.

Next, for any vector subspace V^k of V^n and an arbitrary point $p \in A^n$, we put $A_p^k = \{ q \in A^n \mid q = p + \mathbf{x}, \mathbf{x} \in V^k \}$ and call it a **subspace** of A^n. It is an affine space of dimension k. Conversely, every subset of A^n that is an affine space can be expressed in this form. A^1, A^2, and A^{n-1} in A^n are called a **line**, **plane**, and **hyperplane**, respectively. A set that consists of only one point is also considered as a subspace A^0. For subspaces A^r and A^s of A^n, we denote by $A^r \cap A^s$ the **intersection** (i.e., the set-theoretic intersection) of A^r and A^s, and by $A^r \cup A^s$ the join of A^r and A^s (i.e., the intersection of all subspaces that contain both A^r and A^s). Then $A^r \cap A^s$ is the affine space of highest dimension contained in A^r and A^s, and $A^r \cup A^s$ is the affine space of lowest dimension that contains A^r and A^s. If $r+1$ points are given in A^n, there always exists a subspace A^r that contains all of these points. In particular, if the points are independent, then such an A^r is unique. Moreover, if $A^r \cap A^s \neq \varnothing$ ($\varnothing$ is the empty set), then we have $r + s = \dim(A^r \cup A^s) + \dim(A^r \cap A^s)$. This is called the **dimension theorem** (or **intersection theorem**) of affine geometry.

Next suppose that $r+1$ points p_α $(0 \leqslant \alpha \leqslant r)$ in A^n are independent, and put $p_{r+1} = p_0$. Let q_α be an arbitrary point on $p_\alpha \cup p_{\alpha+1}$ that differs from p_α and $p_{\alpha+1}$. If λ^α are elements of K such that $\lambda^\alpha \cdot \overrightarrow{p_\alpha q_\alpha} = \overrightarrow{q_\alpha p_{\alpha+1}}$, then $q_0, \ldots, q_r$ are dependent if and only if $\lambda^0 \lambda^1 \ldots \lambda^r = (-1)^{r+1}$. And if $r \geqslant 2$ and $\sigma_\alpha = q_\alpha \cup p_{\alpha+2} \cup \ldots \cup p_{\alpha-1}$ $(p_{-1} = p_r)$, then $\sigma_0, \ldots, \sigma_r$ have a point in common if and only if $\lambda^0 \lambda^1 \ldots \lambda^r = 1$. The former is called **Menelaus's theorem**, and the latter is called **Ceva's theorem**.

The set $L(A)$ of all subspaces (including $\varnothing$ considered as an affine space of dimension -1) constitutes a †lattice by the inclusion relation. Conversely, assume that a lattice L with the order relation is given and that for each element a of L its dimension $\dim a$ is defined. Furthermore, assume that $\dim a$ takes all values in $\{-1,0,\dots,n\}$, where $n \geqslant 3$. Moreover, suppose that $\alpha \prec \beta$ implies $\dim\alpha \leqslant \dim\beta$ and that the intersection theorem holds. Then L is isomorphic to a lattice composed of all subspaces of an affine space of dimension n over some field K.

B. Parallelism in Affine Spaces

Let A^r and A^s be subspaces of A^n. We say that A^r and A^s are **parallel in the wider sense** if either of the following conditions holds: (i) $A^r \supset A^s$ or $A^s \supset A^r$; or (ii) $A^r \cap A^s = \varnothing$ and $\dim(A^r \cup A^s) \leqslant r+s$. Next, let A^r and B^r be subspaces of A^n of the same dimension. If A^r and B^r coincide, or $A^r \cap B^r = \varnothing$ and $\dim(A^r \cup B^r) = r+1$, then they are said to be **parallel in the narrower sense** (simply **parallel**), and we denote the relation by $A^r /\!/ B^r$. If $r=s=1$, the definitions of parallelism in the narrower sense and wider sense are equivalent. If $r>1$, parallelism in the narrower sense implies parallelism in the wider sense. For two sets (a_α) and (b_α) $(0 \leqslant \alpha \leqslant r)$ of $r+1$ independent points, let V^r and W^r be vector spaces with bases $\overrightarrow{a_0a_i}$ and $\overrightarrow{b_0b_i}$ $(1 \leqslant i \leqslant r)$, respectively. Then $A^r = a_0 \cup \dots \cup a_r$ and $B^r = b_0 \cup \dots \cup b_r$ are parallel if and only if $V^r = W^r$; and for an arbitrary point p, there exists a unique r-dimensional subspace that is parallel to A^r and passes through point p. If A^r and B^s are parallel in the wider sense, then there exist subspaces A^t and B^t $(t \geqslant 1)$ of A^r and B^s that are parallel to each other. Moreover, if neither A^r nor B^s is contained in the other and if t is the largest integer with the property just given, then we have $t = r+s+1-\dim(A^r \cup B^s)$.

Parallelism between subspaces of A^n is an equivalence relation. Specifically, the equivalence class of a 1-dimensional subspace A^1 is called a **point at infinity** and is denoted by A^0_∞. Given a subspace A^r of A^n, the set of points at infinity A^0_∞ represented by lines A^1 contained in A^r is denoted by A^{r-1}_∞; we have $A^r /\!/ B^r$ if and only if $A^{r-1}_\infty = B^{r-1}_\infty$. The set A^{r-1}_∞ can be identified with the equivalence class of A^r with respect to the parallel relation; A^{r-1}_∞ is called a **space at infinity**. In particular, the set A^{n-1}_∞ is called the **hyperplane at infinity**. The set-theoretic sum $A^n \cup A^{n-1}_\infty = \bar{A}^n$ is supplied with the structure of a †projective space; the "points" in $\bar{A}^n$ are elements of $\bar{A}^n$, and the "lines" in $\bar{A}^n$ are $A^1 \cup A^0_\infty$ and A^1_∞.

C. Coordinates of Affine Spaces

If we fix a point o in A^n and a basis $\{\mathbf{e}_1, \dots, \mathbf{e}_n\}$ of the standard vector space V^n, then any point p in A^n is uniquely expressed as

$$p = o + \sum_{i=1}^{n} x^i \cdot \mathbf{e}_i, \qquad x^i \in K. \tag{1}$$

The system $\mathfrak{F} = (o; \mathbf{e}_1, \dots, \mathbf{e}_n)$ is called an **affine frame** (simply the **frame**) of A^n; the point o is called its **origin**, and $\mathbf{e}_i$ is called the ith **unit vector**. The mapping sending p to $(x^1, \dots, x^n)$ gives a †bijection of A^n to K^n; we call $(x^1, \dots, x^n)$ **affine coordinates** of p with respect to $\mathfrak{F}$, and x^i the ith **affine coordinate**. In particular, if K is a topological field (e.g., the real number field $\mathbf{R}$ or the complex number field $\mathbf{C}$), this bijection $A^n \to K^n$ induces a topology of A^n, which can be shown to be independent of the choice of $\mathfrak{F}$. In the rest of this article, by "coordinates" we mean affine coordinates unless otherwise stated. Putting $a_i = o + \mathbf{e}_i$ $(1 \leqslant i \leqslant n)$, we sometimes call $(o; a_1, \dots, a_n)$ an affine frame. Further, putting $l_i = o \cup a_i$, $\pi_i = o \cup a_1 \cup \dots \cup a_{i-1} \cup a_{i+1} \cup \dots \cup a_n$, we call a_i, l_i, and π_i the ith **unit point**, the ith **coordinate axis**, and the ith **coordinate hyperplane**, respectively.

Assume that subspaces A^r and A^s $(r,s>0, r+s=n)$ are not parallel in the wider sense. For a point p of A^n, denote by $A^r(p)$ the subspace that passes through p and is parallel to A^r, and put $q = A^r(p) \cap A^s$. A mapping $\varphi: A^n \to A^s$ defined by $\varphi(p) = q$ is called a **parallel projection** on A^s with respect to A^r. In particular, if $A^r = \pi_i$ and $A^s = l_i$ $(r = n-1, s=1)$, we write $\varphi(p) = p_i$. Then the ith coordinate x^i of p is an element of K such that $\overrightarrow{op_i} = x^i\,\overrightarrow{oa_i}$. Hence such coordinates are also called **parallel coordinates** (or **Cartesian coordinates**).

Suppose that we are given $r+1$ points $b_0, \dots, b_r$ of A^n and $r+1$ elements $\lambda^0, \dots, \lambda^r$ of K such that $\Sigma^r_{\alpha=0}\lambda^\alpha = 1$. We fix a point o of A^n. If a point p in A^n satisfies $\overrightarrow{op} = \Sigma^r_{\alpha=0}\lambda^\alpha\,\overrightarrow{ob_\alpha}$, then $\overrightarrow{b_0p} = \Sigma^r_{i=1}\lambda^i\,\overrightarrow{b_0b_i}$; hence p is contained in the subspace $b_0 \cup \dots \cup b_r$. Conversely, if a point p is contained in the latter subspace, then there exists a system $(\lambda^0, \dots, \lambda^r)$ such that

$$\overrightarrow{op} = \Sigma^r_{\alpha=0}\lambda^\alpha\,\overrightarrow{ob_\alpha} \quad \text{and} \quad \Sigma^r_{\alpha=0}\lambda^\alpha = 1.$$

The system $(\lambda^0, \dots, \lambda^r)$ has a geometric meaning since we have also $\overrightarrow{o'p} = \Sigma^r_{\alpha=0}\lambda^\alpha\,\overrightarrow{o'b_\alpha}$ if we replace the point o by any other point o' of A^n. The elements $\lambda^0, \dots, \lambda^r$ are called **barycentric coordinates** of p with respect to $\{b_0, \dots, b_r\}$. In particular, if $\{b_0, \dots, b_r\}$ are independent, then the barycentric coordinates $(\lambda^0, \dots, \lambda^r)$ are uniquely determined by the point p on $b_0 \cup \dots \cup b_r$. Furthermore, let $(y^1, \dots, y^n)$ be affine coordinates of p with respect to an affine frame $\mathfrak{F}$, and let $(x^1_\alpha, \dots, x^n_\alpha)$ be affine coordinates of b_α $(\alpha = 0, \dots, r)$. Then p belongs to the subspace $A^r =$

$b_0 \cup \ldots \cup b_r$ if and only if $y^i = \Sigma_{\alpha=0}^r \lambda^\alpha x_\alpha^i$ ($i = 1, \ldots, n$). In this case we say that the system of the linear equations $y^i = \Sigma\lambda^\alpha x_\alpha^i$ ($\Sigma\lambda^\alpha = 1$) gives a **parametric representation** of the subspace A^r (by parameters λ^α). Specifically, if $r = n-1$, the solvability of the system of equations $y^i = \Sigma_{\alpha=0}^{n-1} \lambda^\alpha x_\alpha^i$ ($i = 1, \ldots, n$), $1 = \Sigma\lambda^\alpha$ implies the equation $\Sigma_{i=1}^n y^i \mu_i = \mu_0$ for some nontrivial constants $\mu_0, \ldots, \mu_n$. Hence the latter equation represents the hyperplane $\pi = A^{n-1}$. If a point p has barycentric coordinates $\lambda^0 = \lambda^1 = \ldots = \lambda^r = (r+1)^{-1}$ with respect to $\{b_0, \ldots, b_r\}$, it is called the **barycenter** of $b_0, b_1, \ldots, b_r$. The barycenter is uniquely determined by the set $\{b_0, \ldots, b_r\}$ and is denoted by $g(b_0, \ldots, b_r)$. Specifically, the barycenter of two points b_0 and b_1 is called the **midpoint** (or **middle point**) of b_0 and b_1. If we divide $B = \{b_0, \ldots, b_r\}$ into two sets of points and if g_1 and g_2 are barycenters of these two sets of points, respectively, then $g_1 \cup g_2$ passes through the barycenter of B. More generally, a point with barycentric coordinates $(\lambda^0, \ldots, \lambda^r)$ with respect to $\{b_0, \ldots, b_r\}$ is called a barycenter of $b_0, \ldots, b_r$ with **weights** $\lambda^0, \ldots, \lambda^r$.

D. Affine Spaces over Ordered Fields

Suppose the coefficient field K is an †ordered field (e.g., the real number field **R**). Given a hyperplane π of A^n, we take an affine frame for which π is the nth coordinate hyperplane. If we denote coordinates of points with respect to this frame by $(x^1, \ldots, x^n)$, then the equation of π is given by $x^n = 0$. Let A_+^n and A_-^n be sets of points whose nth coordinates are positive and negative, respectively. They are called **half-spaces** of A^n divided by π. The union of π and a half-space is called a **closed half-space**. A half-space of a subspace A^r of A^n (divided by some A^{r-1} on A^r) is called a half-space of dimension r. For a point p of A^n that does not lie on π, the half-space containing p is called the **side** of p with respect to π. In particular, when $n = 1$, let p and q be two points on a line l. The closed side of q with respect to p is called the **(closed) half-line** (or **ray**) from p to q. The intersection of the closed half-lines emanating from p to q and from q to p is called the **segment** joining p and q and is denoted by $\overline{pq}$. Clearly $\overline{pq} = \overline{qp}$. A subset C of A^n is called a **convex set** if the segment joining two arbitrary points of C is contained in C. Each half-space of each dimension is convex. For any family C_τ of convex sets, $\bigcap_\tau C_\tau$ is also convex. Therefore, for any subset D in A^n there exists a minimal convex set that contains D. It is called the **convex closure** (or **convex hull**) of D. The convex closure $C(P)$ of a finite set of points $P = \{p_0, \ldots, p_k\}$ in A^n is called a **convex cell**, and $\dim(p_0 \cup \ldots \cup p_k)$ is called the **dimension** of the convex cell. In particular, when $p_0, \ldots, p_k$ are independent, $C(P)$ is called a k-dimensional **simplex** with **vertices** $p_0, \ldots, p_k$. The 1-dimensional simplex having two distinct points p and q as vertices is the segment $\overline{pq}$, and the vertices p and q are called **ends** of the segment. A point is regarded as a 0-dimensional simplex. Each 2-dimensional or 3-dimensional simplex is called a **triangle** or **tetrahedron**, respectively. A k-dimensional simplex S with vertices $p_0, \ldots, p_k$ is a set of points whose barycentric coordinates λ^α ($0 \leqslant \alpha \leqslant k$, $\Sigma\lambda^\alpha = 1$) with respect to the vertices satisfy $\lambda^\alpha \geqslant 0$. On the other hand, if we put $A^k = p_0 \cup \ldots \cup p_k$ and $\pi_\alpha = p_0 \cup \ldots \cup p_{\alpha-1} \cup p_{\alpha+1} \cup \ldots \cup p_k$, and denote by A_α^k the side of p_α in A^k and by $\overline{A_\alpha^k}$ the closed side of p_α in A^k with respect to π_α, then the simplex S is given by $\bigcap_{\alpha=0}^k \overline{A_\alpha^k}$, and $\bigcap_{\alpha=0}^k A_\alpha^k$ is called an **open simplex**. A^n has the structure of a †topological space in which the set of open n-dimensional simplexes forms a †base of †open sets. In particular, if K is **R**, the topology of A^n thus defined is compatible with the one that is naturally induced by the topology of **R** (→ Section C). With respect to this topology, A^n is a †Hausdorff space. The terms *open* and *closed* as used before for n-dimensional simplexes agree with the corresponding notions of this topology.

A subset of A^n is said to be **bounded** if it is contained in some simplex. A bounded set obtained through a finite process of constructing intersections and unions from a finite number of closed half-spaces is called a **polyhedron**. The points of a convex polyhedron are characterized by several linear inequalities satisfied by their coordinates. A set of points whose coordinates $(x^1, \ldots, x^n)$ satisfy $h^i \leqslant x^i \leqslant k^i$ for $k^i, h^i \in K$ is called a **parallelotope**; it is a polyhedron whose †interior is called an **open parallelotope**. A simplex is a polyhedron, and polyhedra admit †simplicial decompositions. A polyhedron can also be defined as the set-theoretic union of a finite number of simplexes.

Let P be a finite set of points, and let its convex closure $C(P)$ be a convex cell of dimension m. Then we can take a subset Q of P so that $\dim C(Q) = m-1$ and $C(Q)$ is contained in the †boundary of $C(P)$. Such a $C(Q)$ is called a **face** of $C(P)$, and we denote this relation by $C(P) \succ C(Q)$. If $C(P) \succ C(P_1) \succ \ldots \succ C(P_s)$, then $C(P_s)$ is called an $(m-s)$-dimensional face of $C(P)$. A 0-dimensional face is called a **vertex**, and a 1-dimensional face is called an **edge**. Suppose that $C(P) \succ C(Q)$ for $P = \{p_0, \ldots, p_k\}$ and $Q = \{p_{i_0}, \ldots, p_{i_{k-1}}\}$. Then

$F = p_{i_0} \cup \ldots \cup p_{i_{k-1}}$ is a hyperplane of $E = p_0 \cup \ldots \cup p_k$, and $C(P)$ is contained in a closed side of E divided by F. Therefore, if $C(P)$ has $d(m-1)$-dimensional faces, then $C(P)$ is expressed as the intersection of d m-dimensional closed half-spaces. This shows that any convex cell is a polyhedron.

E. Affine Transformations

A mapping $\varphi: A^n \to A^m$ is an **affine mapping** if there is a linear mapping $\bar{\varphi}: V^n \to V^m$ of the standard vector spaces of A^n and A^m such that $\varphi(p+\mathbf{x}) = \varphi(p) + \bar{\varphi}(\mathbf{x})$ holds for any $p \in A^n$ and any $\mathbf{x} \in V^n$. An affine mapping of A^n into itself is called an **affine transformation** (or **affinity**) of A^n. Specifically, a bijective affine transformation is called a **regular** (or **proper**) **affine transformation**. An affine transformation φ of A^n is characterized by each one of the following properties: (i) Let $o \in A^n$ be a fixed point. Then φ is a mapping of A^n onto itself that can be expressed as

$$\varphi(o+\mathbf{x}) = o + \mathbf{a} + f(\mathbf{x}), \tag{2}$$

where $\mathbf{a}$ is a fixed vector of V^n and f is a linear transformation of V^n. (ii) The mapping $\varphi: A^n \to A^n$ is a mapping such that $\overrightarrow{\varphi(a)\varphi(b)} = \lambda \cdot \overrightarrow{\varphi(p)\varphi(q)}$ if $\overrightarrow{ab} = \lambda \cdot \overrightarrow{pq}$ ($\lambda \in K$). Moreover, if the †characteristic of K is not equal to 2, an affine transformation is also characterized as follows: (iii) φ is a mapping that sends lines into lines and preserves the ratio of each pair of parallel segments.

The set $\mathfrak{A}(A^n)$ of all regular affine transformations of A^n constitutes a group that we call the **group of affine transformations**. If the linear mapping f associated with a regular affine transformation φ is the identity mapping, then φ is called a **translation**. The set $\mathfrak{B}(A^n)$ of all translations is a normal subgroup of $\mathfrak{A}(A^n)$ and is called the **group of translations**. The group of translations is isomorphic to V regarded as an additive group. The vector group $\mathfrak{B}(A^n)$ (i.e., an additive group of a linear space) acts †simply transitively on A^n. We see that $\mathfrak{A}(A^n)/\mathfrak{B}(A^n) \cong GL(n,K)$, where $GL(n,K)$ denotes the †general linear group. The set of all regular affine transformations that leave a point o of A^n invariant constitutes a subgroup $\mathfrak{C}(A^n)$ of $\mathfrak{A}(A^n)$; it is called an †isotropy group at o and is isomorphic to $GL(n,K)$. Let $\mathfrak{F} = (o; \mathbf{e}_1, \ldots, \mathbf{e}_n)$ be an affine frame of A^n with origin o, and let φ be a regular affine transformation of A^n given by (2); put $\mathbf{x} = \Sigma x^i \mathbf{e}_i$, $\varphi(o+\mathbf{x}) = o + \Sigma \bar{x}^i \mathbf{e}_i$, $\mathbf{a} = \Sigma a^i \mathbf{e}_i$, and $f(\mathbf{e}_i) = \Sigma a_i^j \mathbf{e}_j$. Then φ is expressed with respect to $\mathfrak{F}$ by the following equation:

$$\bar{x}^i = a^i + \sum_{k=1}^{n} a_k^i x^k, \quad \det(a_j^i) \neq 0, \quad 1 \leq i \leq n. \tag{3}$$

Conversely, a transformation that is given by (3) is a regular affine transformation. Elements of $\mathfrak{B}(A^n)$ and $\mathfrak{C}(A^n)$ are expressed with respect to $\mathfrak{F}$ by

$$\bar{x}^i = x^i + a^i, \qquad 1 \leq i \leq n, \tag{4}$$

and

$$\bar{x}^i = \sum_{k=1}^{n} a_k^i x^k, \qquad \det(a_j^i) \neq 0, \qquad 1 \leq i \leq n, \tag{5}$$

respectively. Hence $\mathfrak{A}(A^n)$ is represented as a †semidirect product group of $\mathfrak{B}(A^n)$ and $\mathfrak{C}(A^n)$. In particular, a regular affine transformation that is represented by $\bar{x}^i = ax^i$ ($1 \leq i \leq n$) for some $a \in K$ ($a \neq 0$) is called a **similarity** (or **homothety**) with the origin o as center.

According to F. Klein, the objects we deal with in affine geometry are the properties (parallelism, barycenters, etc.) that are invariant under regular affine transformations. Subsets S_1 and S_2 of A^n are called **affinely congruent** if there exists a regular affine transformation φ sending S_1 onto S_2. For a fixed k, two k-dimensional simplexes are affinely congruent. Now we fix an affine frame $\mathfrak{F}$ in A^n and denote by x_α^i the coordinates of $n+1$ points p_α ($0 \leq \alpha \leq n$) in A^n. Then the quantity

$$V(p_0, \ldots, p_n) = \frac{1}{n!} \begin{vmatrix} 1 & 1 & \ldots & 1 \\ x_0^1 & x_1^1 & \ldots & x_n^1 \\ \ldots & \ldots & \ldots & \ldots \\ x_0^n & x_1^n & \ldots & x_n^n \end{vmatrix} \tag{6}$$

is called the **volume** with respect to $\mathfrak{F}$ of the n-dimensional simplex with vertices $p_0, \ldots, p_n$. If φ is a regular affine transformation given by (3), we have $\bar{V} = \det(a_j^i) V$. Hence the ratio of volumes of two n-dimensional simplexes is independent of the choice of coordinate systems, and is invariant under regular affine transformations.

A regular affine transformation given by (3) satisfying $\det(a_j^i) = 1$ is called an **equivalent affinity**. The set of all equivalent affinities constitutes a subgroup of $\mathfrak{A}(A^n)$. The geometry belonging to this group is called **affine geometry in the narrower sense**. For instance, the concept of volume is an invariant in affine geometry in the narrower sense.

F. Relation with Projective Geometry

Let $\mathbf{P}^n$ be a projective space over a coefficient field K (→ 340 Projective Geometry). If we fix a hyperplane π_∞ in $\mathbf{P}^n$, then the set of projective transformations that leave π_∞ invariant constitutes a subgroup of the group of †projective transformations of $\mathbf{P}^n$; this subgroup is isomorphic to a group of regular affine transformations. Actually, if we use a †projective frame $[a_0, a_1, \ldots, a_n, u]$ such that $a_1, \ldots, a_n$ are points on π_∞, then each projective transformation leaving π_∞ invariant is expressed by equations of the same form as (3) with respect to the †inhomogeneous projective coordinates. The point set A^n complementary to π_∞ in $\mathbf{P}^n$ is an affine space, and π_∞ coincides with the hyperplane at infinity. Moreover, two distinct lines in $\mathbf{P}^n$ are parallel in A^n if they meet on the hyperplane at infinity. Hence, denoting by $(0, l^1, \ldots, l^n)$ the †homogeneous projective coordinates of the intersection of a line l in A^n and π_∞, we call $(l^1, \ldots, l^n)$ the **direction ratio** of l. A projective transformation leaving each point of π_∞ invariant induces a translation. The †principle of duality that holds in projective geometry does not hold in affine geometry. The †pole of the hyperplane at infinity, with respect to a quadric hypersurface, is called the **center** of the quadric hypersurface. A regular quadric hypersurface is called **central** or **noncentral** according as its center belongs to A^n or is a point at infinity. Quadric hypersurfaces in an affine space are classified in several ways, by taking account of their relations with the hyperplane at infinity (→ 80 Conic Sections, 345 Quadric Surfaces).

References

[1] O. Schreier and E. Sperner, Einführung in die analytische Geometrie und Algebra I, II, Teubner, 1931, 1935; English translation, Introduction to modern algebra and matrix theory, Chelsea, second edition, 1961.

[2] O. Schreier and E. Sperner, Einführung in die analytische Geometrie und Algebra I, II, Vandenhoeck & Ruprecht, 1948, 1951.

[3] S. Iyanaga, Kikagaku zyosetu (Japanese; Introduction to geometry), Iwanami, 1968.

[4] E. Artin, Geometric algebra, Interscience, 1957.

[5] H. Weyl, Raum, Zeit, Materie, Springer, fifth edition, 1923; English translation, Space, time, matter, Dover, 1952.

[6] S. Iyanaga and K. Matsuzaka, Affine geometry and projective geometry, J. Fac. Sci. Univ. Tokyo, 14 (1967), 171–196.

10 (III.1) Algebra

The first concepts concerning "unknowns" in algebra originated in India, whence came also our decimal positional system of numeration. These ideas were introduced to Europe through Arabia in the Renaissance period. F. †Viète systematized them into a symbolic method, called **algebra**, representing numbers by letters. The first problem of algebra was solving equations. Before Viète, G. Cardano and L. Ferrari had solved algebraic equations of degrees 3 and 4; the solution of equations of lower degree had been known from antiquity. The effort to solve equations of higher degree remained unresolved until the middle of the 19th century, when N. H. †Abel and E. †Galois proved the nonexistence of algebraic solutions of such equations. They considered not only individual roots of these equations but also any rational transforms of their roots at the same time, and thus were led to the concept of †fields. They also noticed that the problem of algebraic solution could be characterized by properties of permutation groups of the roots. After the discovery of the Galois group, group theory and group-theoretical considerations maintained the central position in algebra for some time. (→ 177 Galois Theory). They developed into the "abstract algebra" of this century in the general atmosphere of arithmetization and of axiomatization of mathematics. At the turn of the century the monumental textbook in three volumes by H. Weber [1] was considered a standard work on algebra. Then there appeared in 1910 an epoch-making paper [2] by E. Steinitz on the abstract theory of fields.

The main objects of algebra today are †algebraic systems, i.e., sets of elements between which compositions, subject to some laws, are defined, and their †structures. †Groups, †rings, †fields, and †lattices are the most primitive and the most fundamental among these systems. Another fundamental concept of algebra is that of †isomorphism or of †homomorphism. The collection of algebraic systems of a given kind, together with the homomorphisms among them, gives rise to the notion of †category; a functor is a sort of homomorphism between categories (→ 53 Categories and Functors). These notions were first used in †homological algebra, created in the 1940s by methods transferred from topology to algebra; now they are of basic significance to the whole of mathematics.

An important branch of algebra with wide applications is the theory of †vector spaces, or more generally that of †modules over a ring. This branch is called **linear algebra**. Homomorphisms between finitely generated modules can be represented by †matrices. Another branch of algebra, called †representation theory, is concerned with representations of groups or rings by matrices. Today's methods of algebra provide useful and powerful tools for the whole of mathematics, in particular for the theory of numbers and algebraic geometry.

The present development of algebra owes much to the activity of the German school in the late 1920s represented by E. Noether, E. Artin, W. Krull, and B. L. van der Waerden. The book by van der Waerden [4] has had a great impact on mathematics. N. Bourbaki [5] has been influenced by van der Waerden but gives an account of more recent developments, particularly in linear algebra. In Japan, M. Sono, who worked at about the same period as E. Noether, was a forerunner in this field; after him, algebraists of the Kyoto School, Y. Akizuki, M. Nagata, and their followers, did notable research, especially in algebraic geometry. On the other hand, K. Shoda studied with E. Noether toward 1930 in Germany; his school includes such algebraists as T. Nakayama, K. Asano, and G. Azumaya. Finally K. Morita and his disciples have made significant contributions to homological algebra.

References

[1] H. M. Weber, Lehrbuch der Algebra, F. Vieweg, I, 1894; II, 1896; III, 1891.
[2] E. Steinitz, Algebraische Theorie der Körper, J. Reine Angew. Math., 137 (1910), 167–309.
[3] K. Shoda, Tyûsyô daisûgaku (Japanese; Abstract algebra), Iwanami, 1932.
[4] B. L. van der Waerden, Moderne Algebra, Springer, first edition, I, 1930; II, 1931; English translation, Algebra I, II, Ungar, 1970.
[5] N. Bourbaki, Eléments de mathématique II. Algèbre, ch. 1–9, Actualités Sci. Ind., 1144b, 1236b, 1044, 1102b, 1179a, 1261a, 1272a, Hermann, 1958–1964.
[6] G. Birkhoff and S. MacLane, A survey of modern algebra, Macmillan, third edition, 1965.
[7] K. Shoda and K. Asano, Daisûgaku I (Japanese; Algebra I), Iwanami, 1952.
[8] T. Nakayama and G. Azumaya, Daisûgaku II (Japanese; Algebra II—Theory of rings), Iwanami, 1954.

11 (VIII.2) Algebraic Curves

A. General Remarks

An †algebraic variety of dimension 1 is called an **algebraic curve** (for analytic theory → 13 Algebraic Functions). The theory of algebraic curves has two aspects, the geometry of manifolds in projective spaces and the theory of function fields of transcendence degree 1 (→ 3 Abelian Varieties, 18 Algebraic Varieties). The number-theoretic study of algebraic function fields concerns the latter theory (→ 75 Complex Multiplication, 436 Zeta Functions). In this article, the geometric aspect of the theory is emphasized. We denote the †universal domain by **K**.

B. Classical Results on Plane Algebraic Curves

Let $f(X, Y)$ be a polynomial of degree m in two variables X and Y. A point set in an affine two-space defined by $f(X, Y)=0$ is called a **plane algebraic curve** C of **order** m. If we set $F(X, Y, Z)=Z^m f(X/Z, Y/Z)$, the form $F(X, Y, Z)$ defines an algebraic curve of order m in a projective two-space. The curve C is called **irreducible** if $f(X, Y)$ is irreducible. Some results in this section are valid only in the case where the characteristic of **K** is 0.

Let C be a plane curve defined by the equation $f(X, Y)=0$. A point $P=(a, b)$ on C is called an **r-ple point** if $f(X+a, Y+b)$ has no term of degree $<r$ in X and Y. At an r-ple point there are r tangent straight lines (counting multiplicity). An r-ple point with $r>1$ is called a **multiple point** (or **singular point**). A **double point** with distinct tangents is called a **node** or an **ordinary double point**; e.g., the origin for $X^3+Y^3-3XY=0$. An algebraic curve can be transformed birationally into a nonsingular curve by a finite number of †locally quadratic transformations with singular centers. If a locally quadratic transformation with a singular center P transforms P into a simple point, then P is called a **cusp**; e.g., the origin for $Y^2-X^3=0$. A plane curve can be transformed into a plane curve that has only ordinary double points by a finite number of plane †Cremona transformations (†quadratic transformations of the projective plane into itself).

Let C be an irreducible plane curve of order >1 in a projective plane S. The set of tangent straight lines forms a curve C' in the dual space S' of S. The curve C' is called the **dual curve** of C, and the order m' of C' is called the **class** of C. The class m' is equal to the number of tangent lines (counting multiplicity) drawn from a point to C. A simple point P is called a **point of inflection** if the tangent line at P has a contact of order >2. The tangent line at a point of inflection corresponds to a cuspidal point of the dual curve. If $F(X_0, X_1, X_2)=0$ is a form defining the curve C, the curve defined by the equation $\det\left(\frac{\partial^2 F}{\partial X_i \partial X_j}\right)_{i,j=0,1,2}=0$ is called the **Hessian** of C. A simple point P of C is a point of inflection if and only if P is contained in the Hessian of C. If an irreducible curve C of order m has only double points consisting of ν nodes and γ cusps, the effective genus π (→ Section C) of C is given by the formula $\pi=(m-1)(m-2)/2-\nu-\gamma$. The class m' and the number γ' of points of inflection are given by $m'=m(m-1)-2\nu-3\gamma$ and $\gamma'=3m(m-2)-6\nu-8\gamma$, respectively. These formulas are called **Plücker's formulas.**

For example, consider an irreducible nonsingular plane curve C of order 3. The curve C is an elliptic curve ($\pi=1$) of class 6 and has nine points of inflection. A straight line containing two points of inflection also contains the third one. At any point P of C there are four tangent lines; †anharmonic ratios of these lines are determined uniquely by C and are independent of the choice of the point P. Moreover, these anharmonic ratios are birational invariants of C (†absolute invariants of C). An irreducible plane curve of order 3 with a singular point is necessarily a †unicursal curve (→ 96 Curves).

The work of M. Noether [7] classifies algebraic curves in a projective 3-space.

C. Fundamental Notions

Let Γ be an irreducible nonsingular curve. An element of the free Abelian group generated by points of Γ is called a **divisor**. A divisor is written in the form $\mathfrak{a}=\sum n_iP_i$, with $n_i\in Z$. The integer $n=\sum n_i$ is called the **degree** of $\mathfrak{a}$ and is denoted by $\deg\mathfrak{a}$. The expression for a divisor $\mathfrak{a}$ is called reduced if $P_i\neq P_j$ for $i\neq j$. A divisor whose reduced expression has only positive coefficients is called a **positive divisor** (or **integral divisor**), and this is denoted by $\mathfrak{a}\succ 0$. The group of divisors on Γ is denoted by $G(\Gamma)$, and the subgroup consisting of divisors of degree 0 is denoted by $G_0(\Gamma)$. Let P be a point of Γ. The subset of the function field $\mathbf{K}(\Gamma)$ of Γ consisting of functions regular at P forms a valuation ring R_P for a †discrete valuation of $K(\Gamma)$. A prime element t of R_P is called a **local parameter** at P. Let v_P be the †normalized valuation of $\mathbf{K}(\Gamma)$ defined by R_P; the integer $v_P(f)$ is called the **order** of f at P. The point P is a **zero** of f if $v_P(f)>0$; it is a **pole** of f if $v_P(f)<0$. There are only a finite number of poles and zeros of a given function f. The divisor $\sum v_P(f)P$ is called the **divisor of the function** f and is denoted by (f). The set of divisors of functions forms a subgroup G_l of G_0. Any divisor $\mathfrak{a}$ in G_l is called a **principal divisor** (we also say that $\mathfrak{a}$ is †linearly equivalent to zero and write $\mathfrak{a}\sim 0$).

Let $\mathfrak{a}$ be an arbitrary divisor. The set of all positive divisors that are linearly equivalent to $\mathfrak{a}$ forms a **complete linear system** $|\mathfrak{a}|$ determined by $\mathfrak{a}$. We set $L(\mathfrak{a})=\{f\in\mathbf{K}(\Gamma)\mid(f)+\mathfrak{a}\succ 0\}$. Then $L(\mathfrak{a})$ is a finite-dimensional vector space over $\mathbf{K}$, and 1-dimensional subspaces of $L(\mathfrak{a})$ correspond bijectively to the elements of $|\mathfrak{a}|$. We set $l(\mathfrak{a})=\dim_{\mathbf{K}}L(\mathfrak{a})$ and $\dim|\mathfrak{a}|=l(\mathfrak{a})-1$. Then $\dim|\mathfrak{a}|$ is called the **dimension** of $|\mathfrak{a}|$. For any divisor $\mathfrak{a}$, the integer $\deg\mathfrak{a}-\dim|\mathfrak{a}|$ is nonnegative and bounded. The supremum g of such integers is called the **genus** of Γ. The nonnegative integer $i(\mathfrak{a})=g-\deg\mathfrak{a}+\dim|\mathfrak{a}|$ is called the **speciality index** of $\mathfrak{a}$.

Let ω be a †differential form on Γ, P be a point of Γ, and t be a local parameter at P. Then ω can be written in the form $\omega=fdt$. We now set $\nu_P(\omega)=v_P(f)$ and $(\omega)=\sum\nu_P(\omega)P$. Then (ω) is a well-defined divisor, and the class of (ω) in G/G_l is independent of the choice of ω. This divisor class is called the **canonical class**; any divisor in this class is called a **canonical divisor** (or **differential divisor**) and is denoted by $\mathfrak{k}$. We have $l(\mathfrak{k})=g$, $\deg\mathfrak{k}=2g-2$. Given a divisor $\mathfrak{a}$, the index $i(\mathfrak{a})$ is equal to the number of linearly independent differentials ω such that $(\omega)\succ\mathfrak{a}$, $i(\mathfrak{a})=l(\mathfrak{k}-\mathfrak{a})$. The equality $l(\mathfrak{a})=\deg\mathfrak{a}-g+1+i(\mathfrak{a})$ is called the **Riemann-Roch theorem.**

For any irreducible algebraic curve Γ, there exists a birationally equivalent nonsingular curve $\tilde{\Gamma}$ that is unique up to isomorphism. The genus of $\tilde{\Gamma}$ is called the **effective genus** of Γ. A curve whose effective genus is zero is called a **unicursal curve** (or **rational curve**). An **elliptic curve** is a curve whose effective genus is 1.

Let k be a subfield of the universal domain $\mathbf{K}$ such that Γ is defined over k, and denote by $\bar{k}$ the algebraic closure of k in $\mathbf{K}$. Then a Γ-divisor $\mathfrak{p}=\sum n_iP_i$ is called a **prime rational divisor over** k if $\mathfrak{p}$ satisfies the following three conditions: (i) $\mathfrak{p}$ is invariant under any automorphism σ of $\bar{k}/k$; (ii) for any j, there exists an automorphism σ_j of $\bar{k}/k$ such that $P_j=P_1^{\sigma_j}$; (iii) $n_1=\ldots=n_t=[k(P_1):k]_i$. An element in the subgroup of $G(\Gamma)$ generated by prime rational k-divisors is called a k-**rational divisor**. Let $k(\Gamma)$ be the subset of $\mathbf{K}(\Gamma)$ consisting of functions f defined over k. Then $k(\Gamma)$ is a subfield of $\mathbf{K}(\Gamma)$, and $k(\Gamma)\otimes_k\mathbf{K}=\mathbf{K}(\Gamma)$. $k(\Gamma)$ is called the **function field of** Γ **over** k. Let $\mathfrak{p}$ be a prime rational k-divisor, and let P be a point of $\mathfrak{p}$. Then $R_P\cap k(\Gamma)$ is a valuation ring of $k(\Gamma)$ uniquely determined by $\mathfrak{p}$ and independent of the choice of the point P in $\mathfrak{p}$. We call this valuation ring the valuation ring determined by $\mathfrak{p}$.

D. Algebraic Function Fields

Let k be a field, and let K be a finite separable extension of a purely transcendental extension $k(x)$ of k such that k is maximally algebraic in K. Then K is called an **algebraic function field over** k **of dimension** 1 (or **of transcendence degree** 1). The equivalence class of †exponential valuations of K that are trivial over k is called a **prime divisor** of K/k. An element of the free Abelian group generated by prime divisors is called a **divisor** of K/k. The group operation in the divisor group of K/k will usually be denoted multiplicatively. Let R_P be the valuation ring of the prime divisor P, and let M_P be the maximal ideal of R_P. The **degree** deg P of the prime divisor P is defined by $[(R_P/M_P):k]$. If we replace the term a curve Γ by a function field K/k; $\mathbf{K}(\Gamma)$ by K; $\mathbf{K}$ by k; and points on Γ by prime divisors of K/k, we can develop the theory of the function field K/k, which is similar to the theory of nonsingular curves Γ (→ Sections B, C). Thus we define the **genus of the function field** K/k.

Suppose we are given an algebraic function field K/k of dimension 1. An algebraic curve Γ defined over k is called a **model** of K/k if $k(\Gamma)$ and K are k-isomorphic. For any function field of dimension 1, there always exist two elements x and y in K such that $K = k(x,y)$. Let $f(X,Y)$ be an irreducible polynomial such that $f(x,y)=0$. Then the plane curve defined by the equation $f(X,Y)=0$ is a model of K/k. Among the models of K/k there exists a †normal model Γ_0 over k that is unique up to biregular, birational isomorphism (and the uniqueness of the normal model of the function field within the birational equivalence class of varieties holds only for curves). In particular, if k is the complex number field, the normal model Γ_0 is the †Riemann surface of the function field K/k. If Γ_0 has no singular point, the theory of the curve Γ_0 and the theory of the function field K/k are essentially identical. (This occurs, for example, when k is †perfect.) In that case the genus of Γ_0 is equal to that of K/k. In general, the genus of the function field is not less than the genus of the normal model Γ_0, and it is greater than the latter if Γ_0 has a singular point. If the genus of K/k is zero, we can take a plane quadratic curve as a model of K/k. Moreover, K/k has a prime divisor of degree 1 if and only if K is a purely transcendental extension of k. A function field K/k of genus 1 is called an **elliptic function field**. If K has a prime divisor of degree 1, an elliptic function field K has a model of a plane cubic curve. Moreover, if the characteristic of the universal domain is different from 2, we can take as the model Γ_0 the curve defined by an equation of the form $Y^2=4X^3-g_2X-g_3$. This is called **Weierstrass's canonical form**. The number $j=(g_2^3-27g_3^2)^{-1}g_2^3$ $(\neq 0)$ is a birational invariant of Γ_0.

Let k be a field of characteristic $\neq 2$, and let $P_m(X)$ be a polynomial of degree m without multiple roots. A function field K/k of the form

$$Q\left(\frac{k[X,Y]}{Y^2-P_m(X)}\right)$$

is called a **hyperelliptic function field**, where $Q(*)$ denotes the field of quotients of *. Any model of a hyperelliptic function field is called a **hyperelliptic curve**. The genus of K/k is equal to the integral part of $(m-1)/2$. Every function field of genus 2 is a hyperelliptic function field. A hyperelliptic function field is characterized as a quadratic extension of a purely transcendental extension and has an automorphism of order 2. The automorphism group G of the function field K/k of genus $\geqslant 2$ is a finite group and is reduced to $\{1\}$ for a "general" function field of genus $\geqslant 3$.

E. Jacobian Varieties

Let Γ be a nonsingular curve. A †group variety J is called the **Jacobian variety** of Γ if it has the following four properties (we fix an algebraically closed †field k of definition for Γ and J): (i) There exists an isomorphism Φ (of abstract groups) of $G_0(\Gamma)/G_l(\Gamma)$ into J. (ii) Φ is continuous in the following sense: Let $\bar{\mathfrak{a}}$, $\bar{\mathfrak{b}}$ be elements of $G_0(\Gamma)/G_l(\Gamma)$ represented by $\mathfrak{a}$, $\mathfrak{b}$. If $\mathfrak{b}$ is a specialization of $\mathfrak{a}$ over a field $K(\supset k)$, then $\Phi(\bar{\mathfrak{b}})$ is also a specialization of $\Phi(\bar{\mathfrak{a}})$ over K. (iii) If there exists a K-rational divisor in the class $\bar{\mathfrak{a}}$, then the point $\Phi(\bar{\mathfrak{a}})$ is also K-rational. (iv) For any $\xi \in J$, there exists a $k(\xi)$-rational divisor $\mathfrak{a}$ in G_0 such that $\Phi(\mathfrak{a} \bmod G_l)=\xi$. A group variety J satisfying these conditions is necessarily a complete variety, hence an †Abelian variety, and is determined uniquely up to isomorphism. The construction of Jacobian varieties over a field of arbitrary characteristic is due to A. Weil [10] (for analytic construction → 13 Algebraic Functions).

Let P be a †generic point of Γ over k, and let P_0 be a fixed k-rational point. Then $\varphi(P) = \Phi(P-P_0)$ defines a rational mapping of Γ into J, and φ, which is an isomorphism of Γ and its image $\varphi(\Gamma)$, is determined uniquely by Φ up to translation on J. This mapping φ is called the **canonical function on** Γ. The dimension of J is equal to the genus g of Γ. If $P_1, \ldots, P_g$ are independent generic points of Γ over k, then $k(P_1, \ldots, P_g)_s$ is the function field of J over k, where $k(P_1, \ldots, P_g)_s$ is the subfield invariant under the group of $g!$ automorphisms $(P_1, \ldots, P_g) \rightarrow (P_{\alpha_1}, \ldots, P_{\alpha_g})$. The

Jacobian variety of Γ is also the †Picard variety of Γ, and it is equal to the †Albanese variety of Γ (→ 3 Abelian Varieties). Hence for any function f on Γ with values in an Abelian variety A, there exists a unique homomorphism λ of J into A such that $f=\lambda\circ\varphi$ + const. This λ is called the **linear extension** of f.

Let Θ be the set of points on J that can be written as $\varphi(P_1)+\ldots+\varphi(P_g)$. Then Θ is an irreducible subvariety of codimension 1. The divisor Θ is called the **canonical divisor** of J. The Jacobian variety that is polarized by the divisor Θ is called the **canonically polarized Jacobian variety** (→ 3 Abelian Varieties G). If two curves Γ and Γ' are birationally equivalent, the canonically polarized Jacobians of Γ and Γ' are isomorphic. Conversely, if the canonically polarized Jacobian varieties J of Γ and J' of Γ' are isomorphic, then Γ and Γ' are birationally equivalent (**Torelli's theorem**). Let r be any integer such that $1\leqslant r\leqslant g$, and let W_r be the set of points that are written in the form

$$\varphi(P_1)+\ldots+\varphi(P_r)$$

$(W_1=\varphi(\Gamma),\ W_{g-1}=\Theta,\ W_g=J)$. Then we have $\Theta^{(r)}=r!\,W_{g-r}$ (†numerically equivalent) and $(\Theta^{(g)})=g!$, where $\Theta^{(r)}$ is the class of intersections of r copies of Θ. The existence of a divisor Θ is characteristic for Jacobian varieties. Actually, if A is an Abelian variety of dimension n that has an irreducible subvariety X^{n-1} of codimension 1 and a positive 1-cycle C such that $(X^{(n)})=n!$ and $X^{(n-1)}=(n-1)!\,C$, then C is a nonsingular curve, A is the Jacobian variety of C, and X is the canonical divisor [13]. The canonical divisor Θ is defined by a †theta function in the classical case.

Let Γ be a nonsingular curve, and let ω be a differential form on Γ. If the divisor (ω) is $\succ 0$, the ω is called a **differential form of the first kind**. Let Ω be the †sheaf of germs of regular differential forms. A differential form of the first kind is an element of $H^0(\Gamma,\Omega)$, and vice versa. Let $\mathfrak{k}$ be a canonical divisor. Then we have a natural isomorphism $H^0(\Gamma,\Omega)\cong L(\mathfrak{k})$, and the number of linearly independent differential forms of the first kind is equal to the genus g of Γ. The †residue of a differential can be defined as in the classical case. A differential that has nonzero residues is called a **differential of the third kind**. The **residue theorem** $\sum \mathrm{Res}_P\,\omega=0$ holds for any differential ω. The form ω is called a **differential form of the second kind** if for any $P\in\Gamma$ there exists a function f_P such that $\omega-df_P$ is regular at P. The set of differential forms of the second kind forms a linear space G_2 over the universal domain and contains the subspace G_1 consisting of the differential forms of the first kind. The quotient space G_2/G_1 has dimension $2g$ or g according as the characteristic of the universal domain is 0 or not.

When the characteristic p of the universal domain is positive, we have what is called the Cartier operator. Let Γ be a curve defined over a perfect field k, let $L=k(\Gamma)$, and let t be an element of L that is transcendental over k and such that $L/k(t)$ is separable. Then any differential ω of L/k is written uniquely as $\omega=(f_0^p+f_1^pt+\ldots+f_{p-1}^pt^{p-1})\,dt$, where $f_i\in L$. Then the **Cartier operator** C given by $C\omega=f_{p-1}dt$ is well defined and independent of the choice of t and leaves G_1 invariant. Hence given a basis $\omega_1,\ldots,\omega_g$ of G_1, we obtain a matrix (a_{ij}) with coefficient in L by $C\omega_i=\sum a_{ij}\omega_j (1\leqslant i\leqslant g)$. This $g\times g$ matrix A is called the **Hasse-Witt matrix** of Γ. The class of A modulo the transformations of the form $S^{-p}AS$ is a birational invariant of Γ and plays an important role in the theory of unramified cyclic p-extensions of the algebraic function field.

F. Generalized Jacobian Varieties

The notion of linear equivalence of divisors on a nonsingular curve can be extended to a more general situation. Such attempts have been made by M. Noether, F. Severi, and M. Rosenlicht, who succeeded in obtaining such a generalization [8].

Let Γ be an algebraic curve, and let $P_1,\ldots,P_t$ be singular points of Γ. Let $\mathfrak{O}_{P_i}$ be the †local ring of P_i. We set $\mathfrak{O}=\bigcap_{i=1}^{t}\mathfrak{O}_{P_i}$ and $\Gamma'=\Gamma-\{P_1,\ldots,P_t\}$. An element of the free Abelian group $\overline{G}(\Gamma)$ generated by points of Γ' is called a Γ-divisor. Let $\mathfrak{a}$ be a Γ-divisor and set $\overline{L}(\mathfrak{a})=\{f\in\mathfrak{O}\mid(f)+\mathfrak{a}\succ 0\}$. Then $\overline{L}(\mathfrak{a})$ is a finite-dimensional linear space (over the universal domain). The dimension of $\overline{L}(\mathfrak{a})$ is denoted by $\overline{l}(\mathfrak{a})$, and we set $\overline{\dim}|\mathfrak{a}|=\overline{l}(\mathfrak{a})-1$. The upper bound π of $\deg(\mathfrak{a})-\overline{\dim}|\mathfrak{a}|$ is a nonnegative integer and is called the $\mathfrak{O}$-**genus** of Γ. We call $\overline{i}(\mathfrak{a})=\pi-\deg\mathfrak{a}-\overline{\dim}|\mathfrak{a}|$ the $\mathfrak{O}$-**speciality index** of the divisor $\mathfrak{a}$. Let C be a nonsingular curve birationally equivalent to Γ, and let $Q_1,\ldots,Q_s$ be points of C that correspond to singular points of Γ. An $\mathfrak{O}$-**differential** ω is a differential form on C (of $\mathbf{K}(\Gamma)=\mathbf{K}(C)$) such that $\sum_{i=1}^{s}\mathrm{Res}_{Q_i}f\omega=0$ for any $f\in\mathfrak{O}$. Then $\overline{i}(\mathfrak{a})$ is equal to the number of linearly independent $\mathfrak{O}$-differentials ω such that $(\omega)\succ\mathfrak{a}$ in Γ'. The equality $\overline{l}(\mathfrak{a})=\deg\mathfrak{a}-\pi+1+\overline{i}(\mathfrak{a})$ is called the **generalized Riemann-Roch theorem**. An $\mathfrak{O}$-differential ω is called an $\mathfrak{O}$-differential of the first kind if ω is regular everywhere on Γ'. The number of linearly independent $\mathfrak{O}$-differentials of the first kind is equal to the $\mathfrak{O}$-genus π. Let g be the effective genus of Γ, i.e., the genus of C. Then we have the equality $\pi - g = \dim_K(\bar{\mathfrak{D}}/\mathfrak{D}) = \delta$, where $\bar{\mathfrak{D}}$ is the integral closure of $\mathfrak{D}$ in $\mathbf{K}(\Gamma)$. The set of $\mathfrak{D}$-differentials forms an $\mathfrak{D}$-module that is in general not of rank 1. Hence in this case, we do not have the "canonical divisor." Let $\mathfrak{c}$ be the conductor of $\bar{\mathfrak{D}}/\mathfrak{D}$. Then $\mathfrak{c}$ determines a Γ-divisor in a natural way. If we denote the degree of this Γ-divisor by d, we have the inequality $\delta + 1 \leqslant d \leqslant 2\delta$. We have $d = 2\delta$ if and only if the set of $\mathfrak{D}$-differentials forms an $\mathfrak{D}$-module of rank 1. This case occurs, for example, if Γ is a curve on a nonsingular surface or a complete intersection. Two Γ-divisors $\mathfrak{a}$ and $\mathfrak{b}$ are said to be **$\mathfrak{D}$-linearly equivalent** if there exists a unit f of $\mathfrak{D}$ such that $\mathfrak{a} - \mathfrak{b} = (f)$. The set of Γ-divisors that are $\mathfrak{D}$-linearly equivalent to zero forms a subgroup $\bar{G}_l(\Gamma)$ of $\bar{G}(\Gamma)$. There exists a group variety $J_{\mathfrak{D}}$, unique up to isomorphism, that satisfies the four conditions required for Jacobian varieties (→ Section E) with respect to the class group $\bar{G}_0(\Gamma)/\bar{G}_l(\Gamma)$. The variety J_0 is called the **generalized Jacobian variety**. The generalized Jacobian variety is not complete, in general. If J is the Jacobian variety of C, then $J_{\mathfrak{D}}$ is an extension of J by a connected †linear algebraic group $I_{\mathfrak{D}}$. Any Abelian extension of the function field of Γ can be obtained by the †isogenies of the generalized Jacobian variety of Γ [8]. This fact plays an important role in class field theory over algebraic functions (→ 62 Class Field Theory). The theory for nonsingular curves is considered as the special case in which $\mathfrak{D} = \mathbf{K}(\Gamma)$.

Suppose that Γ is situated in a projective space of dimension n. Let $\mathfrak{p}$ be the prime ideal in $k[X_0, X_1, \ldots, X_n]$ defining Γ and $\chi(\mathfrak{p}, m)$ be the number of linearly independent forms of degree m modulo $\mathfrak{p}$. Then $\chi(\mathfrak{p}, m)$ is a polynomial in m for large m. This polynomial is called the **Hilbert polynomial** of $\mathfrak{p}$ (or Γ). Let c be the constant term of the Hilbert polynomial. The number $p_a(\Gamma) = 1 - c$ is called the **arithmetic genus** of Γ and is equal to the $\mathfrak{D}$-genus of Γ. If Γ is a plane curve of order d, then $p_a(\Gamma)$ is given by the formula $(d-1)\cdot(d-2)/2$.

G. Sheaf Theory

Let Γ be an irreducible curve and $\mathfrak{D}_P$ be the local ring of a point P of Γ. Then $\mathcal{O}_\Gamma = \bigcup \mathfrak{D}_P$ is an †algebraic coherent sheaf, which is called the structure sheaf of Γ, and $\dim_{\mathbf{K}} H^1(\Gamma, \mathcal{O}_\Gamma)$ is equal to the arithmetic genus π of Γ. Let $\mathfrak{a}$ be a Γ-divisor, and let $\mathcal{O}_\Gamma(\mathfrak{a})$ be the †sheaf of germs of rational functions f such that $(f) + \mathfrak{a} \succ 0$ and $f \in \mathfrak{D}_Q$ for every singular point Q of Γ (→ 377 Sheaves D). Then $\dim_{\mathbf{K}} H^1(\Gamma, \mathcal{O}_\Gamma(\mathfrak{a}))$ is equal to the speciality index $\bar{i}(\mathfrak{a})$, and $\dim_{\mathbf{K}} H^0(\Gamma, \mathcal{O}_\Gamma(\mathfrak{a}))$ is equal to $\bar{l}(\mathfrak{a})$. When Γ has no singular point, the Riemann-Roch theorem is deduced naturally from †Serre's duality theorem: $H^1(\Gamma, \mathcal{O}_\Gamma(\mathfrak{a})) \cong H^0(\Gamma, \mathcal{O}_\Gamma(\mathfrak{k} - \mathfrak{a}))$.

H. Algebraic Correspondence

Let Γ be a nonsingular curve. A divisor of the product variety $\Gamma \times \Gamma$ is called an **algebraic correspondence** of Γ [9, 10]. Let D_0 be the subgroup consisting of divisors that are linearly equivalent to degenerate divisors $a \times \Gamma + \Gamma \times b$. Then the class group $\mathcal{C}(\Gamma) = G(\Gamma \times \Gamma)/D_0$ is called the **group of classes of algebraic correspondences**. We write $X \equiv 0$ if X is an element of D_0. Let X be an algebraic correspondence, k a field of definition for Γ over which X is rational, and P a generic point of Γ over k. Then $X(P) = \mathrm{pr}_2[X(P \times \Gamma)]$ is rational over $k(P)$. The composite $X_1 \circ X_2$ of two correspondences X_1 and X_2 is defined by $(X_1 \circ X_2)(P) = X_1(X_2(P))$ whenever they have meaning. The composite $X_1 \circ X_2$ determines an element of $\mathcal{C}(\Gamma)$ that depends only on the classes of X_1 and X_2. This multiplication supplies the group $\mathcal{C}(\Gamma)$ with the structure of an associative ring. This ring is called the **correspondence ring** of Γ. The correspondence ring $\mathcal{C}(\Gamma)$ and the ring $\mathcal{A}$ of endomorphisms of the Jacobian variety J are isomorphic, and the isomorphism is given by the following rule: Let ξ be an element of $\mathcal{C}(\Gamma)$, and let X be a divisor in ξ. Let P be a generic point of Γ with reference to k over which X is rational. Let P_0 be a k-rational point of Γ. Then the class of $X(P) - X(P_0)$ modulo $G_l(\Gamma)$ is independent of the choice of a divisor X in the given class. We set $\Psi(P) = \Phi(X(P) - X(P_0))$ and let λ be the linear extension of Ψ. The correspondence $\xi \rightarrow \lambda$ is an isomorphism of $\mathcal{C}(\Gamma)$ and $\mathcal{A}$. Now we set $\mathcal{A}_0 = \mathcal{A} \otimes Q$. Then $\mathcal{A}_0$ contains an automorphism ι of order 2 called an **involution**. Let l be a rational prime different from the characteristic p. Then A has a faithful representation by $2g \times 2g$ matrices with coefficients in l-adic integers. The †trace σ of this representation has the property that $\sigma(\beta \circ \beta^{\iota}) > 0$ if $\beta \neq 0$ (**Castelnuovo's lemma**). $\mathcal{A}_0$ is an algebra of finite rank over Q, and $\mathcal{A}$ is a finitely-generated Abelian group. Based on these results A. Weil proved the Riemann hypothesis for congruent ζ-functions on a nonsingular curve (→ 436 Zeta Functions).

I. Coverings

Let Γ and C be nonsingular curves such that there exists a rational mapping $\pi: \Gamma \to C$. Then there is an injection of the function field $\mathbf{K}(C)$ into $\mathbf{K}(\Gamma)$. If $\mathbf{K}(\Gamma)$ is separably algebraic over $\mathbf{K}(C)$, then Γ is called a **covering (curve)** of C. The integer $[\mathbf{K}(\Gamma):\mathbf{K}(C)]=n$ is called the **degree of covering**. Let P be a point of Γ and let $Q=\pi(P)$. Let t, s be local parameters at P on Γ and at Q on C, respectively. The nonnegative integer $\nu_P(ds/dt)$ is called the **differential index** at P and is denoted by m_P. The index m_P is zero except for a finite number of points. The divisor $\Sigma m_P P$ is called the **branch divisor**. The covering Γ is called an **unramified covering** if the branch divisor is zero. If we denote the branch divisor by $\mathfrak{a}$, we have the formula $2g(\Gamma)-2=n(2g(C)-2)+\deg\mathfrak{a}$, where $g(\Gamma)$ and $g(C)$ are genera of Γ and C, respectively. This is called the **Riemann-Hurwitz formula**. This formula yields at once that a rational curve has no nontrivial unramified covering and that Γ can be an unramified covering of itself if and only if Γ is an elliptic curve.

J. Theory of Moduli

Let S be the class of nonsingular curves of genus g. We define in S an equivalence relation $\sim$ by $\Gamma \sim \Gamma'$ if and only if Γ and Γ′ are isomorphic (over the universal domain). When the residue class $S/\sim$ has the structure of an algebraic variety M_g, M_g is called the **variety of moduli** of genus g. Two curves of genus 0 are isomorphic. Hence M_0 is reduced to one point. When $g=1$, any curve is equivalent to a plane curve with the Weierstrass canonical form $y^2=4x^3-g_2x-g_3$, and the birational invariant j (→ Section D) determines the class uniquely. Thus M_1 is defined over a prime field k_0, and the function field of M_1 is isomorphic to $k_0(j)$. For the case where the universal domain is the complex number field, O. Teichmüller gave a construction for M_g when $g \geqslant 2$ (→ 13 Algebraic Functions). We shall introduce here a construction due to W. L. Baily [12]. Let H_n be the set of symmetric complex matrices Z of order n such that $\mathrm{Im}(Z)>0$ (†positive definite). Let Γ_n be the unitary symplectic group of degree $2n$ (i.e., $\Gamma_n = Sp(n,\mathbf{Z}) \cap Sp(n)$), and let $\gamma=(A_{ij})_{i,j=1,2}$ be an element of Γ_n. The formula $\gamma Z=(A_{11}Z+A_{12})(A_{21}Z+A_{22})^{-1}$ determines the operation of Γ_n on H_n, which is discontinuous. We denote the quotient space $\Gamma_n \backslash H_n$ by V_n. There exists a bijective correspondence between the set of points of V_n and the set of equivalence classes of normally polarized Abelian varieties of dimension n. Let E be a subset of V_n consisting of canonically polarized Jacobian varieties. There exists a compactification $V_n{}^*$ of V_n such that $V_n{}^*$ has the structure of a normal algebraic variety (I. Satake). The closure I of E in $V_n{}^*$ is a $(3n-3)$-dimensional subvariety of $V_n{}^*$, and E is a †Zariski open subset of I. Since the set of canonically polarized Jacobian varieties corresponds bijectively to the set of nonsingular curves, E is precisely the variety of moduli M_n.

D. Mumford treated the problem of moduli as the representability of certain functors. T. Matsusaka constructed the variety of moduli as the quotient space of an algebraic universal family of nonsingular †deformations of algebraic curves. His variety of moduli have the structure of a Q-variety.

References

[1] K. Iwasawa, Daisû kansûron (Japanese; Theory of algebraic functions), Iwanami, 1952.
[2] F. Severi, Vorlesungen über algebraische Geometrie, Teubner, 1921.
[3] H. F. Baker, Principles of geometry, vol. 5. Analytical principles of the theory of curves, Cambridge Univ. Press, 1933.
[4] R. J., Walker, Algebraic curves, Princeton Univ. Press, 1950.
[5] W. Fulton, Algebraic curves, Benjamin, 1969.
[6] A. Seidenberg, Elements of the theory of algebraic curves, Addison-Wesley, 1968.
[7] M. Noether, Zur Grundlegung der Theorie der algebraischen Raumcurven, J. Reine Angew. Math., 93 (1882), 271–318 (Berlin, 1887).
[8] J.-P. Serre, Groupes algébriques et corps de classes, Actualités Sci. Ind., Hermann, 1959.
[9] A. Weil, Sur les courbes algébriques et les variétés qui s'en déduisent, Actualités Sci. Ind., Hermann, 1948.
[10] A. Weil, Variétés abéliennes et courbes algébriques, Actualités Sci. Ind., Hermann, 1948.
[11] A. Weil, Zum Beweis des Torellischen Satzes, Nachr. Akad. Wiss. Göttingen (1957), 33–53.
[12] W. L. Baily, On the moduli of Jacobian varieties, Ann. of Math., (2) 71 (1960), 303–314.
[13] T. Matsusaka, On a characterization of a Jacobian variety, Mem. Coll. Sci. Univ. Kyôto, 32 (1959), 1–19.
[14] D. Mumford, Geometric invariant theory, Erg. Math., Springer, 1965.
[15] T. Matsusaka, Theory of Q-varieties, Publ. Math. Soc. Japan, 1964.

12 (III.6) Algebraic Equations

A. General Remarks

Let $F_1(X_1, \ldots, X_m), \ldots, F_r(X_1, \ldots, X_m)$ be r †polynomials in m variables $X_1, \ldots, X_m$ over a †field k. Then the equations

$$F_1 = 0, \quad \ldots, \quad F_r = 0$$

are called **algebraic equations in** m **unknowns**. When we consider these equations simultaneously, where $r \geqslant 2$, we call them a **system of** r **equations** or **simultaneous equations**. (For $r = 1$, a system of one equation means the single equation $F_1 = 0$.) Coefficients of $F_1, \ldots, F_r$ are called **coefficients** of the system, and the greatest of the degrees of $F_1, \ldots, F_r$ is called the **degree** of the system.

To **solve** a system of equations (henceforth in this article we shall omit the word "algebraic") means to find the common †zero points (in an †algebraically closed field containing k) of elements $F_1, \ldots, F_r$ of the †polynomial ring $k[X_1, \ldots, X_m]$. If there exist no common zero points, the system is said to be **inconsistent**; if there exists a finite number of such points, it is said to be **regular**; and if there are an infinite number of such points, it is called **indeterminate**. The †elimination method allows us to reduce the problem of solving a system of r equations to the case $r = 1$. In particular, any regular system of equations can be reduced to the case $m = r = 1$.

B. Equations in One Unknown

For the above reason, it is important to consider an equation of the form $f(X) = 0$, where

$$f(X) = a_0 X^n + a_1 X^{n-1} + \ldots + a_n, \qquad a_0 \neq 0. \tag{1}$$

This gives the general form of an algebraic equation in one unknown.

According as $f(X)$ is reducible or not in the †polynomial ring $k[X]$, the equation $f(X) = 0$ is called **reducible** or **irreducible** (→ 334 Polynomials). In some †algebraic extension field K of k, $f(X)$ can be factored as follows:

$$f(X) = a_0(X - \alpha_1)(X - \alpha_2)\ldots(X - \alpha_n). \tag{2}$$

$\alpha_1, \ldots, \alpha_n$ are called the **roots** of the equation $f(X) = 0$. Hence, any algebraic equation of degree n has exactly n roots (**Kronecker's theorem**). Now, $(-1)^i a_i / a_0$ is equal to the †elementary symmetric function of degree i of $\alpha_1, \ldots, \alpha_n$. Some of the roots $\alpha_1, \ldots, \alpha_n$ may be identical. If α appears ρ times in $\alpha_1, \ldots, \alpha_n$, we say that α is a ρ-tuple root, and ρ is called the **multiplicity** of the root α. When $\rho = 1$, α is called a **simple root**, and when $\rho \geqslant 2$, α is called a **multiple root**. Let $\beta_1, \ldots, \beta_\nu$ be all the distinct roots among $\alpha_1, \ldots, \alpha_n$, and let ρ_i be the multiplicity of β_i $(i = 1, \ldots, \nu)$. Then

$$f(X) = a_0 (X - \beta_1)^{\rho_1} \ldots (X - \beta_\nu)^{\rho_\nu}, \tag{2'}$$

$$\rho_1 + \ldots + \rho_\nu = n.$$

If $\rho_1, \ldots, \rho_\nu$ are not divisible by the †characteristic of k, the greatest common divisor g of f and

$$f' = n a_0 X^{n-1} + (n-1) a_1 X^{n-2} + \ldots + a_{n-1}$$

is $(X - \beta_1)^{\rho_1 - 1} \ldots (X - \beta_\nu)^{\rho_\nu - 1}$. Thus we can reduce the multiplicity of every root to 1 by dividing f by g. Any irreducible equation over a field of characteristic 0 has no multiple roots. Equation (1) has multiple roots if and only if its †discriminant D is equal to 0 (→ 157 Fields; 177 Galois Theory).

C. Equations of Special Types

In Sections C and D we assume that the characteristic of k is zero.

Binomial Equations. An equation of the type $X^m - a = 0$ is called a **binomial equation**. It is solved by **root extraction**. Let $\sqrt[m]{a}$ (*m*th **root** of a) be one of the roots (if a is a positive real number, $\sqrt[m]{a}$ usually denotes a positive real root). Then $\sqrt[m]{a}$ multiplied by $1, \zeta, \zeta^2, \ldots, \zeta^{m-1}$ are the roots of $X^m - a = 0$, where ζ is a †primitive *m*th root of unity.

Reciprocal Equations. An equation $a_0 X^n + a_1 X^{n-1} + \ldots + a_n = 0$ is called a **reciprocal equation** if $a_0 = a_n, a_1 = a_{n-1}, a_2 = a_{n-2}, \ldots$. A reciprocal equation of an odd degree $n = 2m + 1$ has a root $X = -1$, and dividing the left side by $X + 1$ we get a reciprocal equation of degree $2m$. A reciprocal equation of degree $n = 2m$ is reduced to an equation of degree m in $Y = X + X^{-1}$ and the quadratic equation $X^2 - XY + 1 = 0$.

D. Equations of Lower Degrees

(→ Appendix A, Table 1)

(1) A **linear equation** $a_0 X + a_1 = 0$ has a single root $-a_1/a_0$. (2) the roots of a **quadratic equation** $a_0 X^2 + a_1 X + a_2 = 0$ are given by $\left(-a_1 \pm \sqrt{a_1^2 - 4a_0 a_2}\right)/2a_0$. (3) To solve a **cubic equation** $a_0 X^3 + a_1 X^2 + a_2 X + a_3 = 0$, we set $A_1 = 9a_0 a_1 a_2 - 2a_1^3 - 27a_0^2 a_3$, $A_2 = a_1^2 - 3a_0 a_2$, and solve the quadratic equation $T^2 - A_1 T + A_2 = 0$. Let t_1 and t_2 be the roots of this quadratic equation, and let ω be any cube root of 1. Then $(-a_1 + \omega \sqrt[3]{t_1} + \omega^2 \sqrt[3]{t_2})/3a_0$ is a root of the original cubic equation

(**Cardano's formula**). If we apply this method to a cubic equation $aX^3+bX^2+cX+d=0$ with real coefficients, we need to use complex cube roots even if the roots of the equation are real. In fact, it has been proved that it is not possible to solve this equation within the real numbers in this case; i.e., if the cubic equation is irreducible over the extension $\mathbf{Q}(a,b,c,d)$ of the rational number field $\mathbf{Q}$, and if all of its roots are real, it is impossible to find the roots only by rational operations and with real radicals. This is called the **casus irreducibilis.** (4) A **quartic equation** $a_0X^4+a_1X^3+a_2X^2+a_3X+a_4=0$ can be solved by means of reduction to a cubic equation (L. Ferrari) (→ Appendix A, Table 1). Generally, the procedure of solving an algebraic equation, i.e., finding the roots of a given equation from its coefficients by means of a finite number of rational operations and extractions of radicals, is called a **solution by radicals** (or **algebraic solution**). The †general algebraic equation whose degree is $\geqslant 5$ cannot be solved by radicals (N. H. Abel) (→ 177 Galois Theory).

E. Analytic Theory

In this section, k denotes the field $\mathbf{R}$ of real numbers or the field $\mathbf{C}$ of complex numbers. These cases have been studied for a long time, for practical reasons.

Concerning the case $k=\mathbf{C}$, the field $\mathbf{C}$ is †algebraically closed; i.e., every equation with coefficients from $\mathbf{C}$ has a root in $\mathbf{C}$ (**Gauss's theorem**, called the **fundamental theorem of algebra**). Accordingly, in the field $\mathbf{C}$, we always have equations (2) and (2′).

Let $\alpha_1,\ldots,\alpha_n$ be the roots of equation (1). Then each α_i is a continuous function of coefficients $a_0,a_1,\ldots,a_n$. Concerning the location of roots of $f(X)=0$ and $f'(X)=0$ on the complex plane, we have the following theorems:

(1) Any convex polygon on the complex plane containing the roots of $f(X)=0$ also contains the roots of $f'(X)=0$ (Gauss).

(2) Let C be a rectifiable †Jordan curve not passing through a root of $f(X)=0$. Then the number (C,f) of the roots of $f(X)=0$ lying in the region enclosed by C is equal to $(1/2\pi i)\int_C(f'(z)/f(x))\,dz$, where the multiplicity of the roots is taken into account.

(3) Let C be a Jordan curve on the complex plane. If $|f(z)|>|g(z)|$ at every point z on C, then the equations $f=0$ and $f+g=0$ have the same number of roots (counting multiplicity) within the region enclosed by C (**Rouché's theorem**).

(4) The absolute value of a root of equation (1) is not greater than

$$M=\max(|a_1/a_0|,\ldots,|a_n/a_0|)+1.$$

(5) Let D be the †discriminant of f, and assume $|\alpha_i|\leqslant M$ $(i=1,\ldots,n)$. Then $|\alpha_i-\alpha_j|^2\geqslant D/(2M)^{n(n-1)-2}=E$. Since the value of $|D|$ is known from f and one value of M is given by theorem (4), we have one value of E. If we draw a circle on the complex plane with center at the origin and with radius M, and if we cover it with a net whose meshes have diameters less than $\sqrt{E}\,/2$, then the interior of each mesh contains at most one root of $f=0$.

In the case where $k=\mathbf{R}$, i.e., $f\in\mathbf{R}[X]$, let $\beta_1,\ldots,\beta_\nu$ denote the distinct roots of $f=0$, and recall equation (2′). Suppose that $\beta_1,\ldots,\beta_\lambda\in\mathbf{R}$ and the others $\not\in\mathbf{R}$. Then $\nu-\lambda$ is an even integer 2κ, and we can renumber $\beta_{\lambda+1},\ldots,\beta_\nu$ so that $\bar{\beta}_{\lambda+1}=\beta_{\lambda+\kappa+1},\ldots,\bar{\beta}_{\lambda+\kappa}=\beta_\nu$ ($\bar{\beta}$ denotes the conjugate of β) and $\rho_{\lambda+1}=\rho_{\lambda+\kappa+1},\ \ldots,\ \rho_{\lambda+\kappa}=\rho_\nu$. In this case, $\beta_1,\ldots,\beta_\lambda$ are the **real roots** of equation (2), and the other β's are called the **imaginary roots**.

(6) If $f\in\mathbf{R}[X]$ and $a_0>a_1>\ldots>a_n$, then the absolute value of any root of equation (1) is less than 1 (**Kakeya-Eneström theorem**).

Concerning the real roots of an equation $f=0$, where $f\in\mathbf{R}[X]$, we have the following theorems: Let $N(a,b)$ $(a,b\in\mathbf{R})$ denote the number of real roots in the interval (a,b). Furthermore, let $V(c_1,c_2,\ldots,c_p)$ denote the number of changes of sign in the sequence $c_1,c_2,\ldots,c_p$ of real numbers, which is defined as follows: Suppose that we have the sequence $c_{\nu_1},\ldots,c_{\nu_q}$ after deleting the terms $c_i=0$ from the sequence $c_1,c_2,\ldots,c_p$. Then

$$V(c_1,c_2,\ldots,c_p)=\frac{1}{2}\sum_{j=1}^{q-1}(1-\operatorname{sgn}c_{\nu_j}c_{\nu_{j+1}}).$$

(7) $N(0,\infty)\equiv V(a_0,a_1,\ldots,a_n)\pmod 2$ and $N\leqslant V$ (**Descartes's theorem**).

(8) Let $V(c)=V(f(c),f'(c),\ldots,f^{(n)}(c))$. Then $N(a,b)\equiv V(a)-V(b)\pmod 2$ and $N\leqslant V(a)-V(b)$ (**Fourier's theorem**).

(9) We may assume that $f=0$ has no multiple roots. Construct a finite series $f_0=f,f_1=f',\ldots,f_l$ of polynomials over $\mathbf{R}$ such that $f_{i-1}=f_iq_i-f_{i+1}$ for $i=1,2,\ldots,l-1$ and $f_l\in\mathbf{R}$, by successive application of the †division algorithm. Let $V(c)=V(f_0(c),f_1(c),\ldots,f_l(c))$. Then $N(a,b)=V(a)-V(b)$ (**Sturm's theorem**). By this theorem we can determine the location of real roots as precisely as we wish.

(10) In order that every root α_i of an equation $f=0$ with $a_0>0$ lies on the left side of the imaginary axis, i.e., $\operatorname{Re}\alpha_i<0$, it is necessary and sufficient that in the following matrix the †principal minors composed of the first r rows and first r columns be positive for all

$r=1,2,\ldots,n$ (**Hurwitz's theorem**):

$$\begin{bmatrix} a_1 & a_3 & a_5 & a_7 & \cdots & \cdots \\ a_0 & a_2 & a_4 & a_6 & \cdots & \cdots \\ 0 & a_1 & a_3 & a_5 & \cdots & \cdots \\ 0 & a_0 & a_2 & a_4 & \cdots & \cdots \\ \cdots & \cdots & \cdots & \cdots & \cdots & \cdots \\ 0 & 0 & \cdots & \cdots & a_{n-2} & a_n \end{bmatrix}.$$

Also, for $f \in \mathbf{C}[X]$, various results have been obtained about under what conditions all the roots of $f=0$ lie on one side of a given straight line or inside a given circle (e.g., the unit circle) (→ 296 Numerical Solution of Algebraic Equations).

References

[1] M. Fujiwara, Daisûgaku I (Japanese; Algebra I), Utida-rôkakuho, 1928.
[2] T. Takagi, Daisûgaku kôgi (Japanese; Lectures on algebra), Kyôritu, revised edition, 1965.
[3] M. Moriya, Hôteisiki (Japanese; Algebraic equations), Sibundô, 1964.
[4] L. E. Dickson, Elementary theory of equations, John Wiley, 1914.
[5] J. Dieudonné, La théorie analytique des polynômes, Mémor. Sci. Math., Gauthier-Villars, 1938.
[6] M. Marden, Geometry of polynomials, Amer. Math. Soc. Math. Surveys, second edition, 1966.
[7] W. Specht, Algebraische Gleichungen mit reellen oder komplexen Koeffizienten, Teubner, 1958.

13 (XI.11) Algebraic Functions

A. Definition

An **algebraic function** is a many-valued †analytic function $w=w(z)$ defined by an †irreducible algebraic equation $P(z,w)=0$ with complex coefficients.

B. History and Methods

The theory of algebraic functions evolved from the works of C. F. Gauss, N. H. Abel, and C. G. J. Jacobi on †elliptic functions in the early 19th century. Stimulated by their works, B. Riemann and K. Weierstrass established the foundations of the theory of complex functions and developed the important theory of algebraic functions.

The equation $P(z,w)=0$ defines a curve in the 2-dimensional complex †projective space with inhomogeneous coordinates z,w. Investigations from this point of view were initiated by Riemann, A. Clebsch, and P. Gordan. This approach was followed by A. Brill, M. Noether, and the Italian school (F. Severi, C. Segre, etc.) and has developed into contemporary algebraic geometry (→ 11 Algebraic Curves, 14 Algebraic Geometry).

The set of †function elements $w(z)$ satisfying $P(z,w)=0$ is a †complex manifold $\mathfrak{R}$, a closed (= compact) †Riemann surface, on which z and w are †meromorphic functions. The field $K_{\mathfrak{R}}$ consisting of the meromorphic functions on $\mathfrak{R}$ is an †algebraic function field $\mathbf{C}(z,w)$. Conversely, for any closed Riemann surface $\mathfrak{R}$, the field $K_{\mathfrak{R}}$ is an †algebraic function field in one variable over $\mathbf{C}$, and any pair of functions z and w with $K_{\mathfrak{R}}=\mathbf{C}(z, w)$ has the property that $\mathfrak{R}$ is †conformally equivalent to the Riemann surface determined in the above fashion by the irreducible algebraic equation $P(z, w)=0$ satisfied by z and w. Two Riemann surfaces $\mathfrak{R}_1$, $\mathfrak{R}_2$ determined by the equations $P_1=0$, $P_2=0$ are conformally equivalent if and only if the fields $K_{\mathfrak{R}_1}$ and $K_{\mathfrak{R}_2}$ are $\mathbf{C}$-isomorphic. This condition is equivalent to the existence of a †birational transformation between the algebraic curves $P_1=0$, $P_2=0$. The "analytic method" (the method of studying algebraic functions as functions on Riemann surfaces) is the creation of Riemann. It was extended by F. Klein and D. Hilbert, and later by H. Weyl, who established in a monograph [6] a rigorous foundation of the analytic method for the theory of algebraic functions.

Given an arbitrary algebraic function field K in one variable over $\mathbf{C}$, the set $\mathfrak{R}$ of its †prime divisors with a suitable topology and analytic structure is a closed Riemann surface whose function field $K_{\mathfrak{R}}$ coincides with K. The "algebraic method" (the method of studying algebraic functions as elements of an algebraic function field) was founded by J. W. Dedekind and H. Weber at the end of the 19th century. In the 20th century, the algebraic method has made remarkable progress, owing to the development of abstract algebra. It covers the case of an arbitrary ground field as well as that of more than one variable. The theory of algebraic functions has had considerable influence on the development of number theory because of a basic analogy between the two subjects.

The †universal covering space (surface) $\tilde{\mathfrak{R}}$ of a closed Riemann surface $\mathfrak{R}$ can be mapped conformally onto the Riemann sphere, the plane, or the unit disk (or, equivalently, to the upper half-plane) if the †genus g of $\mathfrak{R}$ is 0, 1, or $\geqslant 2$, respectively. The

†covering transformation group G, consisting of †linear fractional transformations without fixed points in $\tilde{\mathfrak{R}}$, is †properly discontinuous and has a compact †fundamental domain. Conversely, if D is one of the three domains just mentioned and if G is the group just described, then $\mathfrak{R}=D/G$ is a closed Riemann surface such that D and G are its universal covering space and covering transformation group. A meromorphic function on $\mathfrak{R}$ is represented as an †automorphic function on $\tilde{\mathfrak{R}}$ with respect to G. If $g=0$, then $G=\{1\}$, $\tilde{\mathfrak{R}}=\mathfrak{R}$, and $K_{\mathfrak{R}}$ is the field of rational functions. If $g=1$, then $K_{\mathfrak{R}}$ is the field of †elliptic functions. The study of algebraic functions as automorphic functions was initiated by H. Poincaré and Klein. Recently, C. L. Siegel made a remarkable contribution to the investigation of the case of several variables. The theory of automorphic functions is also related to number theory. Works of E. Hecke, M. Eichler, and G. Shimura on this domain are noteworthy (→ 34 Automorphic Functions, 75 Complex Multiplication). In the rest of this article, we shall deal mainly with the analytic method (for the case of two variables → [10]).

C. Abelian Differentials

An **Abelian differential** on a closed Riemann surface $\mathfrak{R}$ is, by definition, a complex †differential form $\omega=a(z)dz$, where $a(z)$ is a meromorphic function. Such a differential is said to be of the **first kind** if $a(z)$ is holomorphic, of the **second kind** if the residue vanishes everywhere, and of the **third kind** otherwise.

The indefinite integral $W(p)=\int_{p_0}^{p}\omega$ of an Abelian differential ω, where p_0 is assumed not to be a pole of ω, is called an **Abelian integral**. It is said to be of the **first**, **second**, or **third kind** if the same holds for ω. If γ is a 1-†cycle on $\mathfrak{R}$, the quantity $\int_\gamma \omega$ is referred to as the **period** of ω with respect to γ. An **elliptic integral** is defined to be an Abelian integral on a closed Riemann surface of genus 1. For example, this is the case if the equation $P(z,w)=0$ defining the surface is of degree 2 with respect to w and of degree 3 or 4 with respect to z. More generally, a closed Riemann surface is called **hyperelliptic** if $P(z,w)$ is of degree 2 with respect to w or, equivalently, if $\mathfrak{R}$ carries a meromorphic function with exactly two poles. An Abelian integral on such a surface is called a **hyperelliptic integral**.

On a closed surface $\mathfrak{R}$, let V_a be the linear space over $\mathbf{C}$ of the Abelian differentials of the first kind. Given a 1-cycle α of $\mathfrak{R}$, there exists a unique $\omega_\alpha\in V_a$ such that $\mathrm{Re}\int_\gamma\omega_\alpha$ is equal to the †intersection number (α,γ) for every 1-cycle γ. This differential is characterized also by the property $\int_{\mathfrak{R}}\omega\wedge *\bar{\omega}_\alpha=2\pi\sqrt{-1}\int_\alpha\omega$ for every $\omega\in V_a$. If $\{\alpha_1,\ldots,\alpha_{2g}\}$ form a basis of the 1-dimensional †homology group with integral coefficients, then $\mathrm{Re}\,\omega_{\alpha_i}$ $(i=1,\ldots,2g)$ form a basis of the linear space V_h over $\mathbf{R}$ of the †harmonic differentials on $\mathfrak{R}$ as well as that of the space $\{\mathrm{Re}\,\omega\,|\,\omega\in V_a\}$. Accordingly,

$$\dim_{\mathbf{C}} V_a=g, \qquad \dim_{\mathbf{R}} V_h=2g.$$

These identities show a close relationship between the topological structure of $\mathfrak{R}$ and the space of the Abelian differentials on $\mathfrak{R}$ (→ 196 Harmonic Integrals).

Let $\alpha_1,\ldots,\alpha_{2g}$ be as above, and let $\omega_1,\ldots,\omega_g$ form a basis of V_a over $\mathbf{C}$. The $g\times 2g$ matrix Ω with $\int_{\alpha_i}\omega_j$ as its (i,j)-component is called a **period matrix**. Corresponding to the change of bases (α) and (ω), it is subject to the transformation into the form $A\Omega M$, where A is a $g\times g$ invertible complex matrix and M is a $2g\times 2g$ integral square matrix with determinant ± 1. Conversely, two Riemann surfaces are conformally equivalent if they possess period matrices transformable to each other in this manner (**Torelli's theorem**). On the complex linear space $\mathbf{C}^g$, consider the subgroup generated by the $2g$ column vectors of a period matrix Ω, which will be denoted by Ω also. Since it is of rank $2g$ and properly discontinuous, a group manifold $\mathbf{C}^g/\Omega$ is obtained. It is determined by $\mathfrak{R}$ uniquely up to analytic isomorphism and is called the **Jacobian variety** of $\mathfrak{R}$. The **generalized Jacobian variety** is introduced in a similar fashion by means of Abelian integrals of the second and third kinds (→ 11 Algebraic Curves).

D. The Riemann-Roch Theorem

In the present context, a 0-†chain with integral coefficients on a Riemann surface $\mathfrak{R}$ is referred to as a **divisor**. A divisor $d=\sum n_ip_i$ $(n_i\in Z, p_i\in\mathfrak{R})$ is an **integral divisor** (or **positive divisor**) if $n_i\geqslant 0$ in the reduced expression; d is a **prime** divisor if it consists of a single point p_1 and $n_1=1$. A divisor of a meromorphic function f or an Abelian differential ω is defined by taking the p_i as the zeros (poles) of f or ω and $n_i(-n_i)$ as the multiplicity of the zero (pole) at p_i. The divisors on $\mathfrak{R}$ constitute an Abelian group $\mathfrak{D}$ in which **principal divisors**, i.e., divisors of meromorphic functions, constitute a subgroup $\mathfrak{P}$. The factor group $\mathfrak{D}/\mathfrak{P}$ is called the **divisor class group**; an element of it is called a **divisor class**. The divisors of Abelian differentials constitute a single divisor class, which is referred to as the **canonical divisor class** (or **differential divisor class**). The **degree** and the **dimension** of a divisor class D are defined as follows, independent of the choice of the representative $d=\sum n_i p_i \in D$: $\deg D = \sum n_i, \dim D = \dim_{\mathbf{C}} \{f \mid f$ is meromorphic, (divisor of f) $+d$ is a positive divisor$\}$. For example, the degree of the principal divisor class is zero.

In terms of these concepts, the **Riemann-Roch theorem** is stated as follows: For a divisor class D on a closed Riemann surface $\mathfrak{R}$ of genus g and for an integer n, we have

$$\dim(D+nW)-\dim(-D-(1-n)W)$$
$$=\deg D+(2n-1)(g-1),$$

where W is the canonical divisor class (→ 11 Algebraic Curves).

This theorem implies the following properties of $\mathfrak{R}$: (i) $\deg W=2g-2$. (ii) The holomorphic invariant forms φdz^2 (i.e., analytic tensors of order 2), referred to as **quadratic differentials**, constitute a linear space over $\mathbf{C}$ of dimension 0 (if $g=0$), 1 (if $g=1$), or of dimension $3g-3$ (if $g \geqslant 2$). (iii) For a point $p \in \mathfrak{R}$, a positive integer m is called a **gap value** if $\mathfrak{R}$ carries no meromorphic function having a pole only at p with multiplicity m. Then if $g=0$, no point has gap values; and if $g \geqslant 1$, every point p has exactly g gap values; in this case, p also has a nongap value $m \leqslant g+1$. A point p is called an **ordinary point** if the gap values at p are $1,2,\ldots,g$; otherwise p is called a **Weierstrass point**. The total number of Weierstrass points is not less than $2g+2$ and not greater than $(g-1)g(g+1)$. (iv) A conformal mapping f of $\mathfrak{R}$ onto itself with the property that a 1-cycle γ is always homologous to $f(\gamma)$ is necessarily the identity transformation. A closed Riemann surface of genus $\geqslant 2$ is known to admit only a finite number of conformal mappings into itself; the total number does not exceed $84(g-1)$.

E. Abel's Theorem

Abel's theorem is stated as follows: A divisor d of degree zero is a principal divisor if and only if it is expressed as $d=\partial\gamma$ by means of a 1-chain γ that has the property that $\int_\gamma \omega = 0$ for every $\omega \in V_a$.

Given a divisor class D of degree zero, consider a 1-chain γ with $\partial\gamma \in D$. For every 1-cycle α there corresponds the quantity

$$\chi_\alpha(D)=\exp\left(2\pi\sqrt{-1}\ \mathrm{Re}\int_\gamma \omega_\alpha\right),$$

independent of the choice of γ. Thus D determines a †character on the 1-dimensional homology group, called the **integral character**. Conversely, every character on the homology group is shown to be the integral character of some D. In terms of this notion, Abel's theorem may be stated as follows: D is the principal divisor class if and only if $\chi_\alpha(D)=1$ for every α. This result shows that the 1-dimensional homology group with integral coefficients and the group of the divisor classes of degree zero (with compact topology) are, with respect to integral characters, mutually dual (in the sense of L. S. Pontrjagin) topological Abelian groups (→ 405 Topological Abelian Groups). For the relationship between Abelian integrals and Jacobian varieties, in particular the †Jacobi inverse problem, †Abelian functions, and †Riemann theta functions, → 3 Abelian Varieties L.

F. Moduli of Riemann Surfaces

Consider the set $\mathbf{M}_g$ consisting of the conformal equivalence classes of closed Riemann surfaces of genus g. In 1859 Riemann stated, without rigorous proof, that $\mathbf{M}_g$ is parametrized by $m(g)$ ($=0$ if $g=0$, $=1$ if $g=1$, $=3g-3$ if $g \geqslant 2$) complex parameters. Later, the introduction of a topology and $m(g)$-dimensional complex structure on $\mathbf{M}_g$ were discussed rigorously in various ways. The following explanation of these methods is due to O. Teichmüller [15, 16], L. V. Ahlfors [11, 12], and L. Bers [13, 14]. For the algebraic-geometric approach → 11 Algebraic Curves.

The trivial case $g=0$ is excluded, since $\mathbf{M}_0$ consists of a single point. Take a closed Riemann surface $\mathfrak{R}_0$ of genus $g \geqslant 1$, and consider the pairs $(\mathfrak{R}, H)$ of closed Riemann surfaces $\mathfrak{R}$ of the same genus g and the †homotopy classes H of orientation-preserving homeomorphisms of $\mathfrak{R}_0$ into $\mathfrak{R}$. Two pairs $(\mathfrak{R}, H)$ and $(\mathfrak{R}', H')$ are defined to be conformally equivalent if the homotopy class $H'H^{-1}$ contains a conformal mapping. The set $\mathbf{T}_g$ consisting of the conformal equivalence classes $\langle \mathfrak{R}, H \rangle$ is called the **Teichmüller space** (with center at $\mathfrak{R}_0$). Let $\mathfrak{H}_g$ be the group of homotopy classes of orientation-preserving homeomorphisms of $\mathfrak{R}_0$ onto itself. $\mathfrak{H}_g$ is a transformation group acting on $\mathbf{T}_g$ in the sense that each $\eta \in \mathfrak{H}_g$ induces the transformation $\langle \mathfrak{R}, H \rangle \to \langle \mathfrak{R}, H\eta \rangle$. It satisfies $\mathbf{T}_g/\mathfrak{H}_g = \mathbf{M}_g$. The set $\mathfrak{I}_g$ of elements of $\mathfrak{H}_g$ fixing every point of $\mathbf{T}_g$ consists only of the unity element if $g \geqslant 3$ and is a normal subgroup of order 2 if $g=1,2$. In the rest of this section we assume that $g \geqslant 2$. The case $g=1$ can be discussed similarly, and the result coincides with the classical one: $\mathbf{T}_1$ can be identified with the upper half-plane and $\mathfrak{H}_1/\mathfrak{I}_1$ is the †modular group.

Denote by $B(\mathfrak{R}_0)$ the set of measurable

invariant forms $\mu \overline{dz}\, dz^{-1}$ with $\|\mu\|_\infty < 1$. For every $\mu \in B(\mathfrak{R}_0)$ there exists a pair $(\mathfrak{R}, H)$ for which some $h \in H$ satisfies $h_{\bar{z}} = \mu_{hz}$ (→ 347 Quasiconformal Mappings). This correspondence determines a surjection $\mu \in B(\mathfrak{R}_0) \to \langle \mathfrak{R}, H \rangle \in \mathbf{T}_g$. Next, if $Q(\mathfrak{R}_0)$ denotes the space of holomorphic quadratic differentials $\varphi\, dz^2$ on $\mathfrak{R}_0$, a mapping $\mu \in B(\mathfrak{R}_0) \to \varphi \in Q(\mathfrak{R}_0)$ is obtained as follows: Consider μ on the universal covering space U (= upper half-plane) of $\mathfrak{R}_0$. Extend it to U^* (= lower half-plane) by setting $\mu = 0$, and let f be a quasiconformal mapping f of the plane into itself satisfying $f_{\bar{z}} = \mu f_z$. Take the †Schwarzian derivation $\psi = \{f, z\}$ of the holomorphic function f in U^*. The desired φ is given by $\varphi(z) = \psi(\bar{z})$ on U. It has been verified that two μ induce the same φ if and only if the same $\langle \mathfrak{R}, H \rangle$ corresponds to μ. Consequently, an injection $\langle \mathfrak{R}, H \rangle \in \mathbf{T}_g \to \varphi \in Q(\mathfrak{R}_0)$ is obtained. Since $Q(\mathfrak{R}_0) = \mathbf{C}^{m(g)}$ by the Riemann-Roch theorem, this injection yields an embedding $\mathbf{T}_g \subset \mathbf{C}^{m(g)}$, where $\mathbf{T}_g$ is shown to be a domain.

As a subdomain of $\mathbf{C}^{m(g)}$, the Teichmüller space is an $m(g)$-dimensional complex analytic manifold. It is a bounded †domain of holomorphy in $\mathbf{C}^{m(g)}$. $\mathfrak{H}_g$ is a properly discontinuous group of analytic transformations, and therefore $\mathbf{M}_g$ is an $m(g)$-dimensional normal †analytic space.

Let $\{\alpha_1, \ldots, \alpha_{2g}\}$ be a 1-dimensional homology basis with integral coefficients in $\mathfrak{R}_0$ such that the intersection numbers are $(\alpha_i, \alpha_j) = (\alpha_{g+i}, \alpha_{g+j}) = 0$, $(\alpha_i, \alpha_{g+j}) = \delta_{ij}$, $i, j = 1, \ldots, g$. Given an arbitrary $\langle \mathfrak{R}, H \rangle \in \mathbf{T}_g$, consider the period matrix Ω of $\mathfrak{R}$ with respect to the homology basis $H\alpha_1, \ldots, H\alpha_{2g}$ and the basis $\omega_1, \ldots, \omega_g$ of V_a with the property that $\int_{H\alpha_i} \omega_j = \delta_{ij}$. Then Ω is a holomorphic function on $\mathbf{T}_g$. Furthermore, the analytic structure of the Teichmüller space introduced previously is the unique one (with respect to the topology defined above) for which the period matrix is holomorphic.

The Teichmüller space carries a naturally defined Kähler metric, which for $g = 1$ coincides with the †Poincaré metric if $\mathbf{T}_1$ is identified with the upper half-plane. The †Ricci curvature, †holomorphic sectional curvature, and †scalar curvature of this Kähler metric are all negative.

By means of the †extremal quasiconformal mappings, it can be verified that $\mathbf{T}_g$ is a complete metric space and admits a homeomorphism (different from the injection previously mentioned) onto $Q(\mathfrak{R}_0)$. The latter implies that the Teichmüller space is topologically equivalent to the unit ball in real $2m(g)$-dimensional space.

Observe that the covering space $\mathbf{T}_g$ of $\mathbf{M}_g$ is obtained by taking into account homotopy classes of homeomorphisms. Another approach is due to W. L. Baily, who considered homology instead of homotopy.

References

[1] P. E. Appell and E. Goursat, Théorie des fonctions algébriques et de leurs intégrales, Gauthier-Villars, second edition, 1929.

[2] R. Dedekind and H. Weber, Theorie der algebraischen Funktionen einer Veränderlichen, J. Reine Angew. Math., 92 (1882), 181–290. (Gesammelte math. Werke, F. Vieweg, 1930, vol. 1, p. 238–350.)

[3] K. Hensel and G. Landsberg, Theorie der algebraischen Funktionen einer Variablen und ihre Anwendung auf algebraische Kurven und Abelsche Integrale, Teubner, 1902.

[4] K. Iwasawa, Daisûkansûron (Japanese; Theory of algebraic functions), Iwanami, 1952.

[5] W. F. Osgood, Lehrbuch der Funktionentheorie II, Teubner, 1932.

[6] H. Weyl, Die Idee der Riemannschen Fläche, Teubner, 1913, third edition, 1955; English translation, The concept of a Riemann surface, Addison-Wesley, 1964.

[7] M. Eichler, Einführung in die Theorie der algebraischen Zahlen und Funktionen, Birkhäuser, 1963.

[8] C. Chevalley, Introduction to the theory of algebraic functions of one variable, Amer. Math. Soc. Math. Surveys, 1951.

For history and references concerning algebraic functions,

[9] A. Brill and M. Noether, Die Entwicklung der Theorie der algebraischen Funktionen, Jber. Deutsch. Math. Verein., 3 (1894), 107–566.

For classical theory of algebraic functions of two variables,

[10] E. Picard and G. Simart, Théorie des fonctions algébriques de deux variables indépendentes, Gauthier-Villars, I, 1897; II, 1906.

For moduli,

[11] L. V. Ahlfors, The complex analytic structure of the space of closed Riemann surfaces, Analytic functions, Princeton Univ. Press, 1960, p. 45–66.

[12] L. V. Ahlfors, Some remarks on Teichmüller's space of Riemann surfaces, Ann. of Math., (2) 74 (1961), 171–191.

[13] L. Bers, Spaces of Riemann surfaces, Proc. Intern. Congr. Math., Edinburgh, 1958; Cambridge Univ. Press, 1960, p. 349–361.

[14] L. Bers, Correction to "Spaces of Riemann surfaces as bounded domains," Bull. Amer. Math. Soc., 67 (1961), 465–466.

[15] O. Teichmüller, Extremale quasikonforme Abbildungen und quadratische Differentiale, Abh. Preuss. Akad. Wiss., 22 (1939).

[16] O. Teichmüller, Bestimmung der extremalen quasikonformen Abbildungen bei geschlossenen orientierten Riemannschen Flächen, Abh. Preuss. Akad. Wiss., 24 (1943).
Also → references to 347 Quasiconformal Mappings, 362 Riemann Surfaces.

14 (VIII.1) Algebraic Geometry

A. Introduction

Algebraic geometry is the branch of mathematics that deals with †algebraic varieties, that is, point sets defined by several algebraic equations in a space of any dimension or those derived from these sets by means of certain constructions (→ 18 Algebraic Varieties). It may also be considered to be a theory of the †field of algebraic functions in several variables in geometric language, and it is related to the problem of characterizing algebraic varieties among general †complex manifolds. It also has an important connection with number theory through the theory of †automorphic functions, †Diophantine equations, etc. Methodologically it is closely related to the theory of †commutative rings and to †homological algebra.

To investigate local properties of algebraic varieties we consider varieties embedded in an †affine space; to study global properties we usually consider varieties contained in †projective spaces. A quantity (or property) that is invariant under †projective transformations, †biregular and †birational transformations, or general birational transformations is called a †projective invariant, a **relative invariant**, or an **absolute invariant**, respectively. The study of projective invariants is a part of projective geometry, whose methods are important in algebraic geometry. The notions of relative invariant and absolute invariant are used, for example, in the classification of algebraic varieties.

We usually assume that the coordinates of each point of the variety belong to a certain fixed †field K. In the classical case, namely, when the field K is the field $\mathbf{C}$ of complex numbers, the algebraic varieties are considered as complex manifolds and are studied by applying the theories of partial differential equations, †theta functions, etc. Topological methods may also be applied. Algebraic geometry originated from such studies, but, for the study of properties such as †rational mappings or †algebraic systems, it became necessary to consider as well the case where the ground field K is not †algebraically closed. Furthermore, to apply this to number theory, it is necessary to establish the theory over the field of any †characteristic p. For this purpose it is necessary to establish a theory for varieties having ground domains as general as possible.

B. History

Analytic geometry began with the study of lines and quadratic curves (surfaces) and later came to include the study of cubic and quartic curves (surfaces), and so on. These subjects originally belonged exclusively to analytic (or projective) geometry. At that time, the study could not have been described by so specific a title as algebraic geometry.

The study of such theories as the construction of an algebraic plane curve by families of curves of lower degree or the †algebraic m-n correspondence on a straight line probably began with research such as that by M. Chasles. The most outstanding event in the history of algebraic geometry was the introduction and development of the theory of algebraic functions (→ 13 Algebraic Functions) by B. Riemann (1857). Before that time the degree of an algebraic curve (surface) was the only quantity known to be a projective invariant of the curve (surface).

With the theory of algebraic functions, Riemann gathered into one family all the curves that can be transformed onto each other by birational transformations. As the basis for his study, Riemann examined birational transformations in place of projective transformations. This idea led to the notion of the so-called †Riemann surface. The †genus of the surface was obtained as the characteristic number of the family of curves. The concept of genus was the first absolute invariant to appear in the history of algebraic geometry. Riemann based his theory on †Abelian integrals using †Dirichlet's principle, under the assumption that any algebraic curve is reducible to one without †singularities.

After Riemann many mathematicians tried to reconstruct the theory more precisely without using transcendental methods. M. Noether attempted this reconstruction by using geometric methods. Using the †Cremona transformation, he confirmed Riemann's assumption for curves: that any algebraic curve on a plane can be transformed by a birational transformation to a curve without singularities except for simple †nodes. He also contributed to making more precise the basic conditions for the †Riemann-Roch theorem, which is considered one of the most important theorems in the field. His results on space curves and surfaces are also noteworthy. J. Plücker defined the concept of genus in geometric terms and introduced the †Plücker coordinates. A. Cayley and A. Brill worked along similar lines. Cayley's idea was developed later by B. L. van der Waerden and W. L. Chow, who introduced the †associate form of an algebraic variety and its †Chow coordinates.

Around 1890 the Italian school of algebraic geometry appeared. Following the tradition established by Noether, they employed algebrogeometric methods and uncovered many new facts concerning algebraic surfaces. Among those who belonged to this school were G. Castelnuovo, F. Enriques, and F. Severi.

In France, H. Poincaré and E. Picard initiated their study of algebraic functions of two complex variables. After them S. Lefschetz investigated the theory of complex algebraic surfaces [4, 5]. The results attained by the Italian and French schools were very suggestive but lacked rigorous foundations.

On the other hand, rigorous number-theoretic theories of algebraic curves appeared in Germany. R. Dedekind and H. Weber developed the theory of algebraic function fields parallel to that of †algebraic number fields. K. Hensel introduced the concept of †p-adic numbers in analogy to †power series expansions of analytic functions. E. Noether constructed an abstract theory of †polynomial ideals from a formal theory by E. Lasker and F. S. Macauley. Under her influence there appeared the arithmetic algebraic geometry (of curves) over an abstract field as developed by F. K. Schmidt and others.

In the higher-dimensional case, van der Waerden attempted to create a more rigorous foundation for algebraic geometry under the influence of Noether's abstract ideal theory (c. 1930) [1]. He introduced the concept of †generic points and †specialization and specifically defined the †multiplicity of intersections of two varieties in a projective space. He succeeded in getting a rigorous proof of **Bezout's theorem**: In n-dimensional projective space, the number of intersections of an r-dimensional algebraic subvariety of degree l with an $(n-r)$-dimensional subvariety of degree m is always lm if they intersect in only a finite number of points.

The problem of intersections was taken up by C. Chevalley and A. Weil in the 1940s. Chevalley developed the ideal theory of †local rings (studied initially by W. Krull); he introduced topological concepts and applied them to the problem of intersections. The theory in this direction was later extended further by P. Samuel, M. Nagata, and J.-P. Serre.

Weil gave foundations of algebraic geometry over an abstract field and reconstructed the theory by introducing geometric language to designate objects of abstract algebra [6]. He thus gave quite a new aspect to the theory and extended H. Hasse's arithmetization of the theory of algebraic functions in one variable to the case of several variables. Reconstructing Severi's theory of algebraic correspondence over abstract fields, he succeeded in proving an analogy of the †Riemann hypothesis on †congruent zeta functions (→ 436 Zeta Functions). He also constructed, purely algebraically, the entire theory of †Abelian varieties independent of characteristic.

Around 1930, O. Zariski gave the foundations of algebraic geometry by applying the generalized †valuation theory that had been introduced by Krull. Zariski clarified especially the properties of birational transformations by using valuation theory. His main theorem states that if a birational mapping is not †regular at a †normal point P (→ 18 Algebraic Varieties), each component of the image of P by the mapping is of dimension $\geqslant 1$.

Zariski also solved the problem of †resolution of singularities in the affirmative in the case of characteristic 0 for varieties of dimension $\leqslant 3$. The affirmative resolution of this problem (which Riemann assumed) says that any algebraic variety in a projective space can be transformed birationally to a projective algebraic variety without singularities. Recently, H. Hironaka gave an affirmative answer for any dimension in the case of characteristic 0.

Along with the achievements in algebraic methods, great developments took place in analytic methods. Unification of the concepts of Riemann surfaces and †Riemannian manifolds led to the concept of †complex analytic manifolds. Furthermore, G. de Rham's theorem on the duality of topologically defined homology and cohomology based on differential forms was proved; also, W. V. D. Hodge's theory of †harmonic integrals was developed. In the case of the complex dimension 1, any compact Riemann surface is derived from a certain projective algebraic curve. However, the situation is not so simple in the case of greater dimensions. Weil's concept of †abstract complete algebraic varieties can be considered as an analog of compact complex manifolds. If a compact complex analytic variety is projective, then it must be an algebraic variety (†Chow's theorem). K. Kodaira proved that a necessary and sufficient condition for a compact complex analytic manifold to be biregularly and birationally equivalent

to the identity group. In view of these theorems, the study of algebraic groups can be reduced, in a sense, to the study of Abelian varieties and linear algebraic groups. For this reason, we henceforth restrict ourselves to linear algebraic groups, which are simply called algebraic groups. (The notion of †generalized Jacobian variety, introduced by M. Rosenlicht [14], is an example of an algebraic group in a general sense; → 11 Algebraic Curves.) Recently the notion of algebraic groups has been further generalized to that of †**group schemes** by A. Grothendieck [27].

C. Lie Algebras

Since an algebraic group G defined over k has no singularities, the †tangent space $\mathfrak{g}$ to G at the identity element e is defined and has the same dimension as G: $\dim \mathfrak{g} = \dim G$. The space $\mathfrak{g}$ can be identified in a natural manner with the space of all left-invariant †derivations of the function field of G_0 and thus has the structure of a Lie algebra defined over k (→ 247 Lie Algebras). We call $\mathfrak{g}$ (the Lie algebra $\mathfrak{g}_k$ over k of all k-rational points in $\mathfrak{g}$) the **Lie algebra** of G (of the k-group G_k). If G is a linear algebraic group contained in $GL(n)$, then $\mathfrak{g}_k$ is a Lie subalgebra of $\mathfrak{gl}(n,k)$ with the Lie product defined by $[x,y] = xy - yx$; a linear Lie algebra corresponding to a linear algebraic group is called an **algebraic Lie algebra**. When the characteristic of k is zero, conditions for a linear Lie algebra to be algebraic can be given in terms of the **replica** [4]. Also, in the case of characteristic zero, for $x \in \mathfrak{gl}(n,k)$, $x \in \mathfrak{g}_k$ if and only if $\exp(tx) \in G$, where t is a variable over k and $\exp(tx)$ is understood as a †formal power series in t (contained in Ω). From this, we can prove, exactly as in the theory of Lie groups (→ 248 Lie Groups), a one-to-one correspondence between k-closed subgroups H of G and algebraic Lie subalgebras $\mathfrak{h}_k$ of $\mathfrak{g}_k$, establishing a complete parallelism of the theories of algebraic groups and Lie algebras [4, 5]. This parallelism breaks down when k has positive characteristic [27]. On the other hand, over a field of characteristic $p > 0$, the notion of **formal groups**, an analog of the local Lie groups was introduced by J. Dieudonné [23, 24].

D. Tori [9]

The group $\mathbf{G}_m = GL(1)$, the multiplicative group of nonzero elements in Ω, is a 1-dimensional connected algebraic group defined over the prime field. In general, an algebraic group G that is isomorphic to the direct product $(\mathbf{G}_m)^n$ is called an **(algebraic) torus**. When a torus G defined over k is isomorphic to $(\mathbf{G}_m)^n$ over an extension K of k, G is called K-**trivial** (or K-**split**), and the field K is called a **splitting field** for G. A torus G defined over k always has a splitting field K which is a finite separable extension of k.

In general, a rational homomorphism χ of an algebraic group G into $\mathbf{G}_m$ is called a **character** of G. If we define the sum of two characters χ_1 and χ_2 of G by $(\chi_1 + \chi_2)(g) = \chi_1(g) \cdot \chi_2(g)$ $(g \in G)$, the totality of characters of G is an additive group, called the **character module** of G and denoted by $X(G)$. Let G be a torus defined over k and $X = X(G)$ its character module, and let K be a splitting field for G that is a finite †Galois extension of k. If a K-isomorphism $G \cong (\mathbf{G}_m)^n$ is given by the correspondence $G \ni g \rightarrow (\chi_1(g), \ldots, \chi_n(g))$, then the χ_i are characters of G, and X is a †free module of rank n generated by $\chi_1, \ldots, \chi_n$. Furthermore, if Γ denotes the Galois group of K/k, then for $\sigma \in \Gamma$ and $\chi \in X$, the conjugate χ^σ is also a character of G, and under this action of Γ, X becomes a right Γ-module. We have complete duality between a torus G and its character module X in the following sense. There exists a one-to-one correspondence between closed subgroups G_1 (defined over k) of G and a (Γ-invariant) submodule X_1 of X such that X/X_1 has no p-†torsion (where p is the characteristic of k), determined by the relation of the annihilators $X_1 = G_1^\perp$, $G_1 = X_1^\perp$, and in this correspondence, the character modules of G_1 and of G/G_1 are canonically identified with X/X_1 and X_1, respectively. Furthermore, let G' be another torus (defined over k and split over K) with the character module X', and suppose that we have a (k-) homomorphism $\varphi: G \rightarrow G'$. Then we can define a ($\Gamma$-) homomorphism ${}^t\varphi: X' \rightarrow X$, called the **dual homomorphism** of φ, by the relation ${}^t\varphi(\chi') = \chi' \circ \varphi$ for $\chi' \in X'$; and conversely, any (Γ-) homomorphism of X' into X is obtained uniquely in this manner. In particular, φ is a (k-) isomorphism if and only if its dual ${}^t\varphi$ is a (Γ-) isomorphism. Since for any free (Γ-) module X of finite rank there always exists a torus G (defined over k and split over K) such that $X(G) \cong X$ (as a special case of the existence theorem of k-forms), the †category of all tori (defined over k and split over K) and that of all free (Γ-) modules of finite rank are mutually dual.

E. Semisimple Elements and Unipotent Elements

A matrix a is called †semisimple if it is diagonalizable, i.e., if the †minimal polynomial of a has only simple roots. A matrix a is called †unipotent if $a-1$ is nilpotent, i.e., if all characteristic roots of a are equal to 1. (When the characteristic of the ground field is zero, the unipotent elements u in $GL(n,k)$ and the nilpotent elements x in $\mathfrak{gl}(n,k)$ are in one-to-one correspondence by the relation $u=\exp x$.) Any nonsingular matrix a can be written uniquely as a product of a nonsingular semisimple matrix a' and a unipotent matrix a'' of the same size which are mutually commutative: $a=a'a''=a''a'$ (†multiplicative Jordan decomposition); a' (a'') is called the **semisimple (unipotent) part** of a and is denoted by a_s (a_u); a_s can be expressed as a polynomial of the matrix a with scalar coefficients. For an element a of a (linear) algebraic group G, the semisimplicity (unipotency) of a does not depend on the matrix representation of G. Moreover, these properties are preserved by homomorphisms of algebraic groups. Also, if $a\in G$, then $a_s, a_u\in G$.

For an algebraic group G, we denote the totality of semisimple (unipotent) elements contained in G by G_s (G_u) and call it the **semisimple (unipotent) part** of G. (Note that G_s and G_u are not necessarily subgroups.) A torus G is then characterized by the property that $G=G_0=G_s$. On the other hand, an algebraic group G such that $G=G_u$ is called **unipotent**. For instance, the additive group of the universal domain, $\mathbf{G}_a \cong \left\{ \begin{pmatrix} 1 & x \\ 0 & 1 \end{pmatrix} \middle| x\in\Omega \right\}$, is a 1-dimensional connected unipotent algebraic group.

F. Solvable Groups and Nilpotent Groups [1, 7, 12]

For two closed normal subgroups H_1, H_2 (defined over k) of an algebraic group G, the †commutator group $[H_1,H_2]$ (in the sense of abstract group theory) is also a closed normal subgroup (defined over k) of G. In view of this fact, an algebraic group G is called **solvable (nilpotent)**, when it is †solvable (†nilpotent) as an abstract group. For example, the totality $T(n)$ of $n\times n$ nonsingular *upper unipotent* matrices, i.e., matrices of the form

$$\begin{bmatrix} * & \cdots & * \\ & \ddots & \vdots \\ 0 & & * \end{bmatrix},$$

is a connected solvable algebraic group. A unipotent algebraic group is always nilpotent.

For any connected solvable algebraic group $G\subset GL(n)$, there always exists an element a in $GL(n)$ such that $a^{-1}Ga\subset T(n)$ (**Lie-Kolchin theorem** [1]). A connected solvable algebraic group G has a †composition series $G=G_0\supset G_1\supset\cdots\supset G_r=\{e\}$ such that each G_i is a connected closed normal subgroup of G and G_{i-1}/G_i is isomorphic to either $\mathbf{G}_m$ or $\mathbf{G}_a$. If G is defined over k, the subgroup G_u is a connected k-closed normal subgroup of G, and for any maximal torus T in G, we have a decomposition into a †semidirect product $G=T\cdot G_u$ (in the sense of algebraic groups, i.e., the natural map $T\times G_u\to G$ is birational). It is known for any algebraic group G defined over k that there always exists a maximal torus defined over k [25, 27]. G is nilpotent if and only if G has a unique maximal torus T; when that is so, $T=G_s$ and T is contained in the †center of G. For a connected solvable algebraic group G defined over k, we can take $a\in GL(n,k)$ such that $a^{-1}Ga\subset T(n)$ (see the Lie-Kolchin theorem) if and only if all characters $\chi\in X(G)$ are defined over k; when this condition is satisfied, G is called **k-solvable**. G_u is then defined over k, and G/G_u is a k-trivial torus.

When the characteristic of k is zero, any commutative unipotent algebraic group (defined over k) is (k-) isomorphic to the direct product $(\mathbf{G}_a)^n$. When k is an algebraically closed field of characteristic $p>0$, any connected commutative unipotent algebraic group defined over k is k-isogenous to a direct product of a certain number of the groups $\mathbf{W}_m$ of †Witt vectors (of length m) (Chevalley-Chow theorem [14]). A 1-dimensional connected unipotent algebraic group defined over a perfect field k is k-isomorphic to $\mathbf{G}_a$.

G. Borel's Theory

Let G be an algebraic group and V an (abstract) algebraic variety (both defined over k). We say that V is a **transformation space** of G (defined over k), or simply G acts on V, if there is given an everywhere regular rational mapping $G\times V\ni(g,v)\to gv\in V$ (defined over k) such that $g_1(g_2v)=(g_1g_2)v$, $ev=v$ ($g_1,g_2\in G$, $v\in V$). When the action of G on V is †transitive, V is called a †homogeneous space of G. For a closed subgroup H of a connected algebraic group G (both defined over k), the quotient space G/H has the natural structure of a homogeneous space of G (defined over k). A. Borel [1] proved the following theorems:

(1) If G is a connected solvable algebraic group and V a complete transformation space of G, then G has at least one fixed point in V. More precisely, in order that a connected algebraic group G defined over k be k-solvable, it is necessary and sufficient that for any complete transformation space V of G defined

over k such that $V_k \neq \varnothing$, G has at least one k-rational fixed point in V [12].

(2) Let G be a connected algebraic group. A maximal connected solvable closed subgroup of G is called a **Borel subgroup** of G. Then (i) all pairs (T, B) formed by a maximal torus T in G and a Borel subgroup B containing it are conjugate to each other with respect to inner automorphisms of G. (ii) For a closed subgroup H of G, the quotient space G/H is complete if and only if H contains a Borel subgroup of G; and, when that is so, G/H is actually a †projective algebraic variety. (For instance, if $G = GL(n)$, $B = T(n)$, then G/B is a so-called †flag manifold.) (iii) The conjugates of B (T) cover the whole group G (G_s). A closed subgroup of G is called **parabolic** if it contains a Borel subgroup of G. A parabolic subgroup H coincides with its own †normalizer $N(H)$; in particular, H is always connected. The notion of parabolic subgroups has significance in the theory of automorphic functions.

When G is a connected algebraic group defined over a perfect field k, proposition (i) can be sharpened as follows: The pairs (A, H) formed by a maximal k-trivial torus A in G and a maximal connected k-solvable subgroup (or k-**Borel subgroup**) H containing it are conjugate to each other with respect to the inner automorphisms defined by elements in G_k. The normalizer $N(H)$ of a k-Borel subgroup H is a minimal k-closed parabolic subgroup of G. When the k-Borel subgroups of G are reduced to the identity group, G is called k-**compact** or k-**anisotropic**. (Otherwise, G is called k-**isotropic**.) For instance, the †orthogonal group $G = SO(n,f)$ of a †quadratic form f of n variables is k-compact if and only if the form f is **anisotropic**, i.e., the homogeneous equation $f = 0$ has no solution other than zero in k. Similar facts hold for other classical groups. When k is a †local field, G is k-compact if and only if G_k is compact as a topological group. In general, a k-compact group is †reductive.

H. The Weyl Group

Let G be a connected algebraic group and Q an arbitrary torus in G. The †centralizer $Z(Q)$ of Q is then connected and coincides with the connected component of the normalizer $N(Q)$. Hence the factor group $W = N(Q)/Z(Q)$ is finite and may be identified with a subgroup of the automorphism group of Q (or of its character module $X(Q)$) in a natural manner. The group W is called the **Weyl group** of G relative to Q. In particular, when $Q = T$ (a maximal torus), the order of W is equal to the number of Borel subgroups containing T. In this case, the centralizer $C = Z(T)$ is called a **Cartan subgroup** of G; it is characterized by the property that C is a (maximal) connected nilpotent closed subgroup of G which coincides with the connected component of its own normalizer $N(C)$. The notions of Borel subgroups, Cartan subgroups, and maximal tori are preserved under rational homomorphisms of algebraic groups.

I. Semisimple Groups and Reductive Groups

In an algebraic group G defined over k, there exists a largest connected solvable closed normal subgroup R, called the **radical** of G. When $R = \{e\}$, G is called **semisimple**. When R is a torus, G is called **reductive**. Semisimplicity and reductiveness are preserved under forming a direct product and taking the image (or inverse image) of an isogeny. For a reductive group G, the †commutator subgroup $D(G)$ is semisimple, and $G = D(G) \cdot R$, $D(G) \cap R =$ finite; in other words, G is isogenous to the direct product of a connected semisimple algebraic group and a torus. In general, if R is the radical of a connected algebraic group G and R_u is the unipotent part of R, then the factor groups G/R, G/R_u are semisimple and reductive, respectively. Furthermore, if the characteristic of the field k is zero, there exists a reductive closed subgroup H of G such that G decomposes into a semidirect product $G = H \cdot R_u$ (**Chevalley decomposition** [5]). (In this case, R and R_u are k-closed, and H can be taken to be k-rational; such an H is unique up to inner automorphisms defined by elements in G_k.) Also in the case of characteristic zero, reductive algebraic groups are characterized by the property that all rational representations are completely reducible. But for the cases where k has the characteristic $p > 0$, this property characterizes tori (M. Nagata).

J. Root Systems [3, 7]

Let G be a connected semisimple algebraic group, T a maximal torus, and $X = X(T)$ its character module. A character $\alpha \in X$ is called a **root** of G relative to T if there exists an isomorphism x_α of $\mathbf{G}_a$ onto its image in G such that

$$t^{-1} x_\alpha(\xi) t = x_\alpha(\alpha(t)\xi) \quad \text{for all} \quad \xi \in \mathbf{G}_a,\ t \in T.$$

For a root α, such an isomorphism x_α is uniquely determined up to a scalar multiplication in $\mathbf{G}_a$; hence we put $P_\alpha = x_\alpha(\mathbf{G}_a)$.

If we denote by $\mathfrak{r}$ the totality of roots (relative to T), $\mathfrak{r}$ satisfies the following conditions, where $E = X \otimes \mathbf{Q}$ and E^* is the †dual space of E with the inner product $\langle\ \rangle$: (i) For each $\alpha \in \mathrm{r}$, there corresponds $\alpha^* \in E^*$ such that $\langle \alpha^*, \alpha \rangle = 2$ and $\langle \alpha^*, \beta \rangle \in \mathbf{Z}$ for all $\beta \in \mathrm{r}$. (ii) If we define a reflection w_α of E by

$$w_\alpha x = x - \langle \alpha^*, x \rangle \alpha \quad \text{for} \quad x \in E,$$

then $w_\alpha \beta \in \mathrm{r}$ for all $\beta \in \mathrm{r}$. (In particular, $w_\alpha \alpha = -\alpha \in \mathrm{r}$.) (iii) If α, $\beta \in \mathrm{r}$ are linearly dependent, then $\beta = \pm \alpha$. (iv) If $\dim E = r$, r contains r linearly independent elements.

In general, a finite subset r in a finite-dimensional vector space E over $\mathbf{Q}$ satisfying conditions (i)–(iv) is called a **root system** in E. For a root system r, the elements α^* of E^* corresponding to $\alpha \in \mathrm{r}$ are uniquely determined by these conditions, and the set $\mathrm{r}^* = \{\alpha^*\}$ is a root system in E^* ($\alpha^* \in \mathrm{r}^*$ is called a **coroot**). Also, the group W of linear transformations of E (E^*) generated by w_α (w_{α^*}) with $\alpha \in \mathrm{r}$ is finite, and is called the **Weyl group** of the root system r. If we identify E^* with E by means of any W-invariant (positive definite) metric on E, then $\alpha^* = (2/\langle \alpha, \alpha \rangle)\alpha$. When r is a root system of a semisimple algebraic group,

$$\langle \alpha^*, \chi \rangle \in \mathbf{Z} \qquad \text{for all} \qquad \alpha \in \mathrm{r}, \quad \chi \in X, \qquad (1)$$

so that X is W-invariant, and the Weyl group of the root system r can be identified with the group $N(T)/Z(T)$ mentioned before. (In general, a maximal torus in a connected reductive algebraic group coincides with its own centralizer, so that the Weyl group W may be identified with $N(T)/T$.)

When a †linear ordering (compatible with the addition) is given in E, we denote by r_+ the set of all positive roots in r. An element $\alpha \in \mathrm{r}_+$ is called a **simple root** if it cannot be written as $\alpha = \alpha' + \alpha''$ with $\alpha', \alpha'' \in \mathrm{r}_+$. If $\Delta = \{\alpha_1, \ldots, \alpha_r\}$ is the totality of (distinct) simple roots in r_+, the elements $\alpha_1, \ldots, \alpha_r$ are linearly independent, and any root $\alpha \in \mathrm{r}$ can be written uniquely in the form $\alpha = \pm \sum_{i=1}^r m_i \alpha_i$, with $m_i \in \mathbf{Z}$, $m_i \geqslant 0$. In general, a subset Δ of r having this property is called a **fundamental system**; a fundamental system is always obtained in the manner explained from a linear ordering on E. For a fundamental system Δ, the cone Λ_Δ in E^* defined as the set of x in E^* satisfying the inequalities $\langle \alpha_i, x \rangle > 0$ $(1 \leqslant i \leqslant r)$ is called a **Weyl chamber**. If we denote by L_α the hyperplane defined by the linear equation $\langle \alpha, x \rangle = 0$ for a root α, then $E^* - \bigcup_{\alpha \in \mathrm{r}} L_\alpha = \bigcup_\Delta \Lambda_\Delta$, and W acts †simply transitively on the set of all Weyl chambers $\{\Lambda_\Delta\}$. The Weyl group W is generated by r reflections w_{α_i} $(1 \leqslant i \leqslant r)$.

In a semisimple algebraic group G, Borel subgroups B containing a (fixed) maximal torus T are in one-to-one correspondence with the fundamental systems Δ (or r_+) relative to T by the relation $B_u = \coprod_{\alpha \in \mathrm{r}_+} P_\alpha$, where $P_\alpha = x_\alpha(\mathbf{G}_a)$. (More precisely, every element in B_u can be written uniquely as a product of the elements in P_α, where the ordering of the P_α is taken arbitrarily.)

K. Bruhat Decompositions

If we take a representative s_w of $w \in W$ in $N(T)$, there is a decomposition $G = \bigcup_{w \in W} B s_w B$ (disjoint union). Furthermore, if for $w \in W$ we put $N_w = \coprod_{\alpha \in \mathrm{r}_+ \cap w\mathrm{r}_+} P_\alpha$, and in particular $N = N_e = B_u$, and denote by w_0 the unique element in W such that $w_0 \Delta = -\Delta$, then the element in $B s_w B$ can be written uniquely as a product of elements in N_{ww_0}, $s_w T$, N. Hence we have

$$G = \bigcup_{w \in W} N_{ww_0} \cdot s_w T \cdot N,$$

which is called a **Bruhat decomposition** of G. In particular, if we put $N' = s_{w_0}{}^{-1} N s_{w_0}$ (which is the unipotent part of the Borel subgroup corresponding to $-\Delta$), then $N'TN$ is a Zariski open set in G, and the natural map $N' \times T \times N \rightarrow G$ is birational. This implies that the function field of G is rational (i.e., †purely transcendental over Ω).

L. Structure of Semisimple Groups

A subset r_1 of a root system r is called a **closed subsystem** if $\mathrm{r}_{1\mathbf{z}} \cap \mathrm{r} = \mathrm{r}_1$, where $\mathrm{r}_{1\mathbf{z}}$ denotes the submodule of X generated by r_1. A closed subsystem satisfies conditions (i), (ii), and (iii) of a root system. For a closed subsystem r_1 of a root system r of a semisimple algebraic group G, the subgroup $G(\mathrm{r}_1)$ of G generated by the P_α ($\alpha \in \mathrm{r}_1$) is a semisimple closed subgroup with a maximal torus $T_1 = (G(\mathrm{r}_1) \cap T)_0$, of which the root system relative to T_1 coincides with the restriction of r_1 on T_1 and the coroot system can be identified with $\mathrm{r}_1^* = \{\alpha^* \mid \alpha \in \mathrm{r}_1\}$. The subgroup $G(\mathrm{r}_1)$ is normal if and only if $\mathrm{r} - \mathrm{r}_1$ is also a closed subsystem; when this is so, $G = G(\mathrm{r}_1) \cdot G(\mathrm{r} - \mathrm{r}_1)$, $G(\mathrm{r}_1) \cap G(\mathrm{r} - \mathrm{r}_1) = \text{finite}$. All connected closed normal subgroups of G are obtained in this manner. In order that G be **simple** (sometimes called **absolutely simple** or **almost simple**) as an algebraic group (i.e., without proper connected normal subgroups), it is necessary and sufficient that r is **irreducible** (i.e., r cannot be decomposed into a disjoint union of two proper closed subsystems). In general, a root system r can be decomposed uniquely into the disjoint union $\mathrm{r} = \mathrm{r}_1 \cup \ldots \cup \mathrm{r}_s$ of irreducible closed subsystems r_i such that $\mathrm{r}_1 \cup \ldots \cup \mathrm{r}_i$ $(1 \leqslant i \leqslant s)$ are also closed subsystems, and correspondingly, G is isogenous to the direct product $G_1 \times \ldots \times G_s$ of (absolutely) simple algebraic groups

$G_i = G(\mathfrak{r}_i)$. (G is actually a direct product if it is simply connected or an adjoint group.) The subgroups G_i are determined uniquely and only by G.

M. k-Forms [28]

Let K be an extension of k and G_1 an algebraic group defined over K. An algebraic group G defined over k is called a **k-form** of G_1 if there is a K-isomorphism f of G onto G_1. Suppose further that K/k is finite separable, and for every Galois automorphism σ of $\bar{k}/k$, put $\varphi_\sigma = f^\sigma \circ f^{-1}$. Then φ_σ is an isomorphism of G_1 onto G_1^σ, and the φ_σ satisfy the relation $\varphi_\sigma^\tau \circ \varphi_\tau = \varphi_{\sigma\tau}$. Conversely, given a collection of isomorphisms $\{\varphi_\sigma\}$ satisfying these conditions, there always exists a k-form G (with a K-isomorphism f onto G_1 such that $\varphi_\sigma = f^\sigma \circ f^{-1}$), which is unique up to k-isomorphism (Weil). In particular, if K/k is a finite Galois extension with Galois group Γ, then $\{\varphi_\sigma\}$ is a (continuous) †1-cocycle of Γ in $\mathrm{Aut}_K(G_1)$ (the group of all K-automorphisms of G_1), and by the above correspondence the k-isomorphism classes of k-forms G are in one-to-one correspondence with the (continuous) 1-†cohomology classes of the cocycle $\{\varphi_\sigma\}$ (in the cohomology set $H^1(\Gamma, \mathrm{Aut}_K(G_1))$).

To a given finite separable extension K/k of degree d and an algebraic group G_1 defined over K of dimension n, we can associate a certain algebraic group $\mathfrak{R}_{K/k}(G_1)$ defined over k of dimension dn, which is obtained from G_1 by restricting the ground field [20]. A more precise definition is as follows. Let $\{\sigma_1, \sigma_2, \ldots, \sigma_d\}$ ($\sigma_1 = 1$) be a set of automorphisms of $\bar{k}/k$ such that $\sigma_i|K$ $(1 \leqslant i \leqslant d)$ are all distinct. Then one can find a k-form $\tilde{G}$ of $\tilde{G}_1 = \prod_{i=1}^d G_1^{\sigma_i}$ with an isomorphism $\tilde{f}: \tilde{G} \rightarrow \tilde{G}_1$ such that $\tilde{\varphi}_\sigma = \tilde{f}^\sigma \circ \tilde{f}^{-1}$ is given by $\tilde{\varphi}_\sigma((x_i)) = (x_{i^\sigma})$, where i^σ is defined by the relation $(\sigma_i\sigma)|K = \sigma_{i^\sigma}|K$. If we denote by p_1 the canonical projection of $\tilde{G}_1$ onto its first component G_1 and put $p = p_1 \circ \tilde{f}$, then the pair $(\tilde{G}, p)$ is uniquely characterized (up to k-isomorphism) by the following universality property: If $\tilde{G}'$ is any algebraic group defined over k and φ is a K-morphism of $\tilde{G}'$ into G_1, then there exists a (uniquely determined) k-morphism $\tilde{\varphi}$ of $\tilde{G}'$ into $\tilde{G}$ such that $\varphi = p \circ \tilde{\varphi}$. The group $\tilde{G}$ (together with p) is denoted by $\mathfrak{R}_{K/k}(G_1)$. For the group of rational points, $\tilde{G}_k = G_{1K}$. When the algebraic group G_1 has some additional structure (such as that of †vector space, †algebra, etc.), then $\mathfrak{R}_{K/k}(G_1)$ automatically has the same kind of additional structure.

N. Chevalley's Fundamental Theorems

Let G, G' be connected semisimple algebraic groups, and let $T(T')$ be a maximal torus in $G(G')$, $X(X')$ its character module, $\mathfrak{r}$ ($\mathfrak{r}'$) a root system of G (G') relative to T (T'), etc. If we have an †isogeny φ of G onto G' such that $\varphi(T) = T'$, then there is a bijection $\alpha \rightarrow \alpha'$ of $\mathfrak{r}$ onto $\mathfrak{r}'$ such that $\psi(\alpha') = q_\alpha \alpha$, where ψ is the dual homomorphism of $\varphi|T$ and q_α is a positive integer, which equals 1 if the characteristic is zero and is a power of p if the characteristic is $p > 0$. Conversely, any injective homomorphism $\psi: X' \rightarrow X$ satisfying this condition (with respect to a certain bijection $\mathfrak{r} \rightarrow \mathfrak{r}'$ and q_α) comes from an isogeny $\varphi: G \rightarrow G'$ in the manner already stated. In particular, φ is an isomorphism if and only if ψ is an isomorphism such that $\psi(\mathfrak{r}') = \mathfrak{r}$ (i.e., $q_\alpha = 1$ for all $\alpha \in \mathfrak{r}$) [7]. The isomorphism class of G is thus completely determined by the pair $(X, \mathfrak{r})$, so that we sometimes write $G = G(X, \mathfrak{r})$. A connected semisimple algebraic group G defined over k is called of **Chevalley type** over k (or **k-split**) if there exists a k-trivial maximal torus T in G. If, in the above theorem, G and G' are of Chevalley type over k and T and T' are k-trivial, then the theorem remains true if we replace isogeny by k-isogeny. In particular, the k-isomorphism class of a connected semisimple algebraic group of Chevalley type over k is completely determined by $(X, \mathfrak{r})$. Chevalley also showed that, for any pair $(X, \mathfrak{r})$ satisfying condition (1) above, there always exists a connected semisimple algebraic group $G(X, \mathfrak{r})$ of Chevalley type defined over the prime field. Therefore, since the classification of semisimple algebraic groups of Chevalley type is reduced essentially to that of root systems $(X, \mathfrak{r})$, it turns out that, over any ground field k, there exist as many connected simple algebraic groups of Chevalley type as connected simple complex Lie groups (→ 248 Lie Groups; Appendix A, Table 5.I).

For a given semisimple algebraic group $G = G(X, \mathfrak{r})$ defined over k, put $X_0 = \mathfrak{r}_{\mathbf{Z}}$ (= the submodule of X generated by $\mathfrak{r}$), $X^0 = \{x \in E \mid \langle \alpha^*, x \rangle \in \mathbf{Z}$ for all $\alpha \in \mathfrak{r}\}$. Then we have natural isogenies $G(X^0, \mathfrak{r}) \rightarrow G(X, \mathfrak{r}) \rightarrow G(X_0, \mathfrak{r})$ (with $q_\alpha = 1$), all of which can be taken to be defined over k. The group $G(X^0, \mathfrak{r})$ ($G(X_0, \mathfrak{r})$) is called the **simply connected group** (the **adjoint group**) isogenous to G. When the characteristic of k is zero, these isogenies (which are already known in the classical theory of complex Lie groups) are essentially the only possible isogenies among the semisimple algebraic groups. But when the characteristic is $p > 0$, there are, as well as these, the Frobenius homomorphism (with $q_\alpha = p$) and the following "singular" isogenies (for which $q_\alpha = 1$ or p depending on α): $B_n \rightleftarrows C_n$, $F_4 \rightarrow F_4$ $(p = 2)$, $G_2 \rightarrow G_2$ $(p = 3)$. In particular, when k is a finite field, taking the set of fixed points of the singular k-isogenies, we obtain the simple finite groups of M. Suzuki and R. Ree [15] (→ 160 Finite Groups).

O. Classification Theory

A connected semisimple algebraic group G defined over k is called k- **(almost) simple** if there is no proper connected closed normal subgroup of G defined over k. (When G is k-simple and k-isotropic, the factor group $D(G_k)$/center is an abstract simple group except for a few special cases [16].) For a k-simple algebraic group G, let G_1 be any one of its absolutely simple components, and let k_1 be the smallest field of definition for G_1 containing k. Then k_1/k is a finite separable extension, and G is k-isogenous to $\mathfrak{R}_{k_1/k}(G_1)$. Hence the problem of classifying all k-simple groups (up to isogeny) is equivalent to that of finding all k_1-forms of simple groups of Chevalley type. This latter problem can be reduced, in principle, to the classification of compact k_1-forms and that of certain diagrams (i.e., †Dynkin diagrams along with an action of the Galois group) [13, 15] (→ Appendix A, Table 5.I). For instance, when k is a finite field (or, more generally, a field of dimension $\leqslant 1$ [28]), there is no compact k-simple group; hence, using a simple classification theory of the diagrams, we can show that the only absolutely simple algebraic groups G defined over k are either of Chevalley type or of the types introduced by R. Steinberg (denoted by ${}^2A_n$, ${}^2D_n$, ${}^3D_4$, ${}^6D_4$, ${}^2E_6$). Connected semisimple algebraic groups composed of the groups of these types are characterized by the property that they have a Borel subgroup defined over k. Such groups are called of **Steinberg type** over k (or k-**quasi-split**). Absolutely simple algebraic groups over a †$\mathfrak{p}$-adic field are classified by M. Kneser [26, 30]). When the characteristic of k is not equal to 2, the classification of simple groups of classical type (except for the type D_4) is known to be equivalent to that of semisimple †associative algebras with †involution [19]. A similar relation also holds between some of the exceptional simple groups and †Cayley algebras or †Jordan algebras (H. Hijikata, T. A. Springer, J. Tits).

Following is a list of absolutely simple algebraic groups of classical type.

I. k-forms of $SL(n)$ ($n \geqslant 2$).

I.1. $G_{1k} = SL(m,\mathfrak{K}) = \{g \in M_m(\mathfrak{K}) \mid N(g) = 1\}$, where $\mathfrak{K}$ is a †central division algebra over k with $(\mathfrak{K}:k) = r^2$, $n = mr$, and N denotes the †reduced norm in $M_m(\mathfrak{K})$.

I.2. $G_{1k} = SU(m,\mathfrak{K},f) = \{g \in SL(m,\mathfrak{K}) \mid f(gx,gy) = f(x,y) \text{ for } x, y \in \mathfrak{K}^m\}$, where $\mathfrak{K}$ is a central division algebra over a quadratic extension k' of k with an involution ι of the second kind (which means that $\{\xi \in k' \mid \xi^\iota = \xi\} = k$), $(K:k') = r^2$, $n = mr$, and f is a (nondegenerate) †Hermitian form of m variables over $\mathfrak{K}$ with respect to the involution ι.

II. k-forms of $SO(n)$ ($n \geqslant 3$, $n \neq 4$), $Sp(n)$ (n even, $n \geqslant 2$).

$G_{1k} = SU(n,\mathfrak{K},f)$, where $\mathfrak{K}$ is a central division algebra over k with an involution ι of the first kind (i.e., such that $\{\xi \in k \mid \xi^\iota = \xi\} = k$), $(\mathfrak{K}:k) = r^2$, $n = mr$, and f is a nondegenerate ε-Hermitian form of m variables over $\mathfrak{K}$ with respect to the involution ι. In this case, $\dim\{\xi \in K \mid \xi^\iota = \xi\} = r(r+\varepsilon_0)/2$ with $\varepsilon_0 = \pm 1$, and G_1 is a k-form of SO or Sp according as $\varepsilon\varepsilon_0 = 1$ or -1. ($SO(8)$ may have other k-forms coming from the so-called triality.)

When k is a local field or an algebraic number field, the only central division algebra with an involution of the first kind is a †quaternion algebra (and, if ι is the "canonical involution," then $\varepsilon_0 = -1$).

P. Algebraic Groups over an Algebraic Number Field

Let G be a connected algebraic group defined over an algebraic number field k of finite degree. Let $\{v\}$ be the totality of †prime divisors (i.e., equivalence classes of valuations) of k. Taking the †restricted direct product of a family of locally compact topological groups $\{G_{k_v}\}$, we obtain a locally compact topological group G_A, called the **adele group** of G [20] (→ 7 Adeles and Ideles). In particular, when $G = \mathbf{G}_m$, the adele group $I = (\mathbf{G}_m)_A$ is exactly the †idele group introduced by Chevalley in class field theory. If we identify $x \in G_k$ with an adele whose components are all equal to x, G_k becomes a discrete subgroup of G_A.

Concerning the finiteness property of G_A/G_k, the following results have been obtained [2, 8]: A character $\chi \in X_k(G)$ (= the module of all k-rational characters of G) gives rise to a (continuous) homomorphism $\chi_A: G_A \to I = (\mathbf{G}_m)_A$. Put $G_A^0 = \{g \in G_A \mid |\chi_A(g)| = 1 \text{ for all } \chi \in X_k(G)\}$, where $|\ |$ is the standard norm in I. Then G_A^0 is †unimodular, and the quotient space G_A^0/G_k is of finite volume with respect to the (unique) invariant measure on it. G_A^0/G_k (G_A/G_k) is compact if and only if the semisimple part G/R (the reductive part G/R_u) of G is k-compact. From the arithmetic point of view, it is important to determine explicitly the volume of G_A^0/G_k

with respect to the invariant measure normalized in a certain manner; such a volume is called the **Tamagawa number** of G and is usually denoted by $\tau(G)$ [10,20]. For instance, Siegel's formulas on the volume of the fundamental domain of the unit group of a quadratic form f over k are essentially equivalent to a theorem on the Tamagawa number stating that $\tau(SO(f))=2$.

Let $\mathfrak{o}$ be the †ring of integers in k and L an $\mathfrak{o}$-lattice in the vector space on which G is acting. We can define in a natural manner an action of G_A on the set of all $\mathfrak{o}$-lattices; then the orbit $G_A L$ ($G_k L$) of L with respect to G_A (G_k) is called the **genus (class)** of L. The †stability subgroup $G_{A,L}$ of L in G_A is open, and the double coset space $G_{A,L}\backslash G_A/G_k$ is finite (finiteness of the class number). Moreover, let $\{v_1,\ldots,v_r\}$ be the totality of †infinite prime divisors of k, and put $G_\infty=\prod_{i=1}^r G_{k_{v_i}}$. Then G_∞ is a Lie group, and the canonical projection G_L on G_∞ of $G_{A,L}\subset G_A$ is a discrete subgroup of finite type. (In general, (discrete) subgroups of G_∞ which are †commensurable with G_L are called **arithmetic subgroups**.) As in the adele case, $\chi\in X_k(G)$ gives rise to a (continuous) homomorphism $\chi_\infty: G_\infty\to(\mathbf{R}^\times)^r$, and if $G_\infty^0=\{g\in G_\infty \mid |\chi_\infty(g)|=1\}$, then the quotient space G_∞^0/G_L is of finite volume. Moreover, G_∞^0/G_L (G_∞/G_L) is compact if and only if G_A^0/G_k (G_A/G_k) is compact.

In addition to these, the †approximation theorem and the †Hasse principle are also extended to (classical, or general) algebraic groups (M. Eichler, M. Kneser, G. Shimura, Hijikata, Springer; → [25]).

References

[1] A. Borel, Groupes linéaires algébriques, Ann. of Math., (2) 64 (1956), 20–82.
[2] A. Borel and Harish-Chandra, Arithmetic subgroups of algebraic groups, Ann. of Math., (2) 75 (1962), 485–535.
[3] A. Borel and J. Tits, Groupes réductifs, Publ. Math. Inst. HES, no. 27,1965, p.55–151.
[4] C. Chevalley, Théorie des groupes de Lie II, Actualités Sci. Ind., Hermann, 1951.
[5] C. Chevalley, Théorie des groupes de Lie III, Actualités Sci. Ind., Hermann, 1954.
[6] C. Chevalley, Sur certains groupes simples, Tôhoku Math. J., (2) 7 (1955), 14–66.
[7] C. Chevalley, Classification des groupes de Lie algébriques, Séminaire C. Chevalley, Secrétariat Mathématique, 1956–1958.
[8] G. D. Mostow and T. Tamagawa, On the compactness of arithmetically defined homogeneous spaces, Ann. of Math., (2) 76 (1962), 446–463.
[9] T. Ono, Arithmetic of algebraic tori, Ann. of Math., (2) 74 (1961), 101–139.
[10] T. Ono, On the Tamagawa number of algebraic tori, Ann. of Math., (2) 78 (1963), 47–73.
[11] M. Rosenlicht, Some basic theorems on algebraic groups, Amer. J. Math., 78 (1956), 401–443.
[12] M. Rosenlicht, Some rationality questions on algebraic groups, Ann. Mat. Pura Appl., 43 (1957), 25–50.
[13] I. Satake, On the theory of reductive algebraic groups over a perfect field, J. Math. Soc. Japan, 15 (1963), 210–235.
[14] J.-P. Serre, Groupes algébriques et corps de classes, Actualités Sci. Ind., Hermann, 1959.
[15] J. Tits, Groupes simples et géométries associées, Proc. Intern. Congr. Math., 1962, Stockholm, p. 197–221.
[16] J. Tits, Algebraic and abstract simple groups, Ann. of Math., (2) 80 (1964), 313–329.
[17] A. Weil, On algebraic groups of transformations, Amer. J. Math., 77 (1955), 355–391.
[18] A. Weil, On algebraic groups and homogeneous spaces, Amer. J. Math., 77 (1955), 493–512.
[19] A. Weil, Algebras with involutions and the classical groups, J. Indian Math. Soc., 24 (1960), 589–623.
[20] A. Weil, Adeles and algebraic groups, Lecture notes, Institute for Advanced Study, Princeton, 1961.
[21] Colloque sur la théorie des groupes algébriques, C. B. R. M., Brussels, 1962.
[22] N. Iwahori, Theory of Lie algebras and Chevalley groups I, II (Japanese; mimeographed note), Seminar notes series, Univ. of Tokyo, 1956.
[23] J. Dieudonné, Hyperalgèbres et groupes de Lie formels, Séminaire Sophus Lie, 1955–1956, Ecole Norm. Sup., 1957.
[24] Ю. И. Манин (Ju. I. Manin), Теория коммутативных формальных групп над полями конечной характеристики, Uspehi Mat. Nauk, 18, no. 6 (1963), 3–90; English translation, The theory of commutative formal groups over fields of finite characteristic, Russian Math. Surveys, 18, no. 6 (1963).
[25] Algebraic groups and discontinuous subgroups, Amer. Math. Soc., Proc. Symposia in Pure Math. IX, 1956.
[26] M. Kneser, Galois-Kohomologie halbeinfacher albebraischer Gruppen über $\mathfrak{p}$-adischen Körpern I, II, Math. Z., 88 (1965), 40–47; 89 (1965), 270–272.
[27] M. Demazure and A. Grothendieck, Schémas en groupes, Séminaire de géometrie algébrique, 1963–1964, Publ. Math. Inst. HES, 1964. (Lecture notes in math. 153, Springer,

1970.)
[28] J.-P. Serre, Cohomologie galoisienne, Lecture notes in math. 5, Springer, 1964.
[29] G. B. Seligman, Modular Lie algebras, Springer, 1967.
[30] A. Borel, Linear algebraic groups, Benjamin, 1969.
[31] F. Bruhat and J. Tits, Groupes réductifs sur un corps local I. Données radicielles valuées no. 41, Publ. Math. Inst. HES, 1972, p. 5–252.

16 (V.13) Algebraic Number Fields

A. Introduction

A complex number that satisfies an algebraic equation with rational integral coefficients is said to be an **algebraic number**. If the coefficient of the term of highest degree of the equation is 1, this algebraic number is said to be an **algebraic integer**. The set A of all algebraic numbers is a field which is the †algebraic closure of the rational number field $\mathbf{Q}$ in the complex number field $\mathbf{C}$. The set I of all algebraic integers is an †integral domain which contains the integral domain $\mathbf{Z}$ of all the rational integers. The †field of quotients of I is A.

B. Principal Order

An extension field k of $\mathbf{Q}$ of finite degree (which we shall always suppose to be contained in $\mathbf{C}$) is said to be an **algebraic number field** of finite degree, and k is a subfield of A. The intersection $\mathfrak{o}=k\cap I$ is an integral domain whose field of quotients is k; $\mathfrak{o}$ is called the **principal order** of k. (More generally, a subring R of $\mathfrak{o}$ containing 1 is said to be an **order** of k if the field of quotients of R is k. The set $\mathfrak{f}$ of all elements γ of $\mathfrak{o}$ such that $\gamma\mathfrak{o}\subset R$ is an **ideal** of $\mathfrak{o}$; in addition $\mathfrak{f}$ is called the **conductor** of R.) Let n be the degree of k over $\mathbf{Q}$. Then the additive group of the principal order $\mathfrak{o}$ of k is a †free Abelian group of †rank n. A †basis $(\omega_1,\dots,\omega_n)$ of $\mathfrak{o}$ as a free Abelian group (or $\mathbf{Z}$-module) is said to be a **minimal basis** of $\mathfrak{o}$ (or of k). Let $\omega_j^{(i)}$ $(i=1,\dots,n)$ be †conjugate elements of ω_j over $\mathbf{Q}$, and let $\Delta=|\omega_j^{(i)}|$ be the determinant whose (i,j) entry is $\omega_j^{(i)}$. Then $D_k=\Delta^2$ is a rational integer which is independent of the choice of a minimal basis of $\mathfrak{o}$. D_k is called the **discriminant** of k. If $k\neq\mathbf{Q}$ then $|D_k|>1$ **(Minkowski's theorem,** 1891). For any given rational integer m there are only a finite number of algebraic number fields whose discriminants equal m (C. Hermite and H. Minkowski, 1896). The proof of these theorems depends on the methods of geometry of numbers (→ 187 Geometry of Numbers).

C. Ideals of the Principal Order

An †ideal $\mathfrak{a}$ of the principal order $\mathfrak{o}$ is said to be an **integral ideal** of k. In particular, a prime ideal of $\mathfrak{o}$ is called simply a prime ideal of k. The domain $\mathfrak{o}$ is not necessarily a †principal ideal ring but is always a †Dedekind domain. That is, every ideal $\mathfrak{a}$ of $\mathfrak{o}$ is uniquely expressed (up to the order of the factors) as a finite product of powers of prime ideals of $\mathfrak{o}$. This theorem is called the **fundamental theorem of the principal order** $\mathfrak{o}$.

The quotient ring $\mathfrak{o}/\mathfrak{a}$ of $\mathfrak{o}$ by an ideal $\mathfrak{a}$ $(\neq 0)$ of $\mathfrak{o}$ is a finite ring. The number of elements of $\mathfrak{o}/\mathfrak{a}$ is called the **absolute norm** of $\mathfrak{a}$ and is denoted by $N(\mathfrak{a})$. We have $N(\mathfrak{a}\mathfrak{b})=N(\mathfrak{a})N(\mathfrak{b})$. Every prime ideal $\mathfrak{p}$ of $\mathfrak{o}$ is a †maximal ideal of $\mathfrak{o}$, and $\mathfrak{o}/\mathfrak{p}$ is a finite field. Let the †characteristic of $\mathfrak{o}/\mathfrak{p}$ be p, where p is a prime number. Then $\mathfrak{o}/\mathfrak{p}$ is a finite extension of the †prime field $\mathbf{Z}/p\mathbf{Z}$. Let the degree of $\mathfrak{o}/\mathfrak{p}$ over $\mathbf{Z}/p\mathbf{Z}$ be f. Then $N(\mathfrak{p})=p^f$, and f is said to be the **degree** of the prime ideal $\mathfrak{p}$.

Let s be a complex variable. The (complex-valued) function

$$\zeta_k(s)=\sum_{\mathfrak{a}}1/(N(\mathfrak{a}))^s=\prod_{\mathfrak{p}}1/\left(1-(N(\mathfrak{p}))^{-s}\right)$$

of $s\in\mathbf{C}$ is called the **Dedekind zeta function** of k (R. Dedekind, 1871). Here the summation extends over all ideals $\mathfrak{a}$ of $\mathfrak{o}$, and the product extends over all prime ideals $\mathfrak{p}$ of $\mathfrak{o}$. This series converges absolutely for $\mathrm{Re}\,s>1$, and the function $\zeta_k(s)$ has a single-valued †analytic continuation to a †meromorphic function on the whole complex plane (→436 Zeta Functions).

D. Units

An algebraic integer ϵ of k is said to be a **unit** of k if ϵ^{-1} is also an algebraic integer. Hence ϵ is a unit of k if and only if the †principal ideal (ϵ) is $\mathfrak{o}$. The set E_k of all units of k forms an Abelian group under multiplication, which is called the **unit group** of k. The set of all elements of E_k of finite order coincides with the set of all the roots of unity contained in k and forms a cyclic group of a finite order w. Let n be the degree of k over $\mathbf{Q}$. Then for each element $\alpha\in k$ there are n conjugate elements $\alpha^{(i)}$ over $\mathbf{Q}$. Let $\alpha^{(i)}$ $(i=1,\dots,r_1)$ be real for any $\alpha\in k$, and let $\alpha^{(r_1+j)}$ and $\alpha^{(r_1+r_2+j)}$ $(j=1,\dots,r_2)$ be complex conjugate for any $\alpha\in k$. Then we have $n=r_1+2r_2$. The unit group E_k of k is the direct product of a cyclic group of order w and the free Abelian multiplicative group of †rank $r=r_1+r_2-1$. This theorem is called **Dirichlet's unit theorem** (1846). A basis $(\epsilon_1,\ldots,\epsilon_r)$ of this free group is called a system of **fundamental units** of k.

Let $l_\alpha^{(i)}=\log|\alpha^{(i)}|$ $(i=1,\ldots,r_1)$, $l^{(j)}\alpha=2\log|\alpha^{(j)}|$ $(j=r_1+1,\ldots,r_1+r_2)$ for $\alpha\in k$. For r elements $\eta_1,\ldots,\eta_r$ of E_k,

$$R[\eta_1,\ldots,\eta_r]=\begin{vmatrix} l^{(1)}\eta_1 & l^{(1)}\eta_2 & \cdots & l^{(1)}\eta_r \\ l^{(2)}\eta_1 & l^{(2)}\eta_2 & \cdots & l^{(2)}\eta_r \\ \cdots & \cdots & \cdots & \cdots \\ l^{(r)}\eta_1 & l^{(r)}\eta_2 & \cdots & l^{(r)}\eta_r \end{vmatrix}$$

is called the **regulator** of $(\eta_1,\ldots,\eta_r)$ (Dedekind). In order for $\eta_1,\ldots,\eta_r$ to be multiplicatively independent, it is necessary and sufficient that $R[\eta_1,\ldots,\eta_r]\neq 0$. The absolute value of $R[\eta_1,\ldots,\eta_r]$ takes the minimum value R for fundamental units $(\epsilon_1,\ldots,\epsilon_r)$. $R=|R[\epsilon_1,\ldots,\epsilon_r]|$ is independent of the choice of fundamental units $(\epsilon_1,\ldots,\epsilon_r)$ of k. R is called the **regulator** of k. In general, $|R[\eta_1,\ldots,\eta_r]|/R$ is equal to the index $[E_k:H]$ of the group H generated by the roots of unity in k and $\eta_1,\ldots,\eta_r$. H. W. Leopoldt conjectured that units in k, which are multiplicatively independent over $\mathbf{Z}$, remain multiplicatively independent over $\mathbf{Z}_p$ (the ring of p-adic integers) when they are considered as elements of the tensor product $k\otimes\mathbf{Q}_p$ over $\mathbf{Q}$. This conjecture was affirmatively proved in some special cases by J. Ax (*Illinois J. Math.*, 9 (1965)) and others.

If $k/\mathbf{Q}$ is a †Galois extension, there exists a unit ϵ of k such that the conjugates of ϵ over $\mathbf{Q}$ contain r multiplicatively independent units (**Minkowski's theorem**).

E. Ideal Classes

An $\mathfrak{o}$-module contained in k (i.e., $\mathfrak{o}\mathfrak{a}\subset\mathfrak{a}$) such that $\alpha\mathfrak{a}\subset\mathfrak{o}$ holds for some element α $(\neq 0)$ of k is said to be a **fractional ideal** of k. For two fractional ideals $\mathfrak{a}$, $\mathfrak{b}$ of k the "product" $\mathfrak{a}\mathfrak{b}$ defined by $\{\Sigma\alpha_i\beta_i$ (finite sum) $|\ \alpha_i\in\mathfrak{a},\beta_i\in\mathfrak{b}\}$ is also a fractional ideal. Thus the set of the fractional ideals of k forms a multiplicative commutative semigroup. For a fractional ideal $\mathfrak{a}$ the set $\mathfrak{a}^{-1}=\{\alpha\in k\mid\alpha\mathfrak{a}\subset\mathfrak{o}\}$ is also a fractional ideal of k, and we have $\mathfrak{a}\mathfrak{a}^{-1}=\mathfrak{o}$. Thus the set of all nonzero fractional ideals of k forms an Abelian group $\mathfrak{J}_k$ under multiplication with $\mathfrak{o}$ as identity. Each fractional ideal $\mathfrak{a}$ $(\neq 0)$ is uniquely expressed as a finite product of powers of prime ideals, if we admit negative powers. Namely, $\mathfrak{J}_k$ is a free Abelian multiplicative group with the set of all prime ideals as basis. Given fractional ideals $\mathfrak{a}$ and $\mathfrak{b}$, we say that $\mathfrak{a}$ is **divisible** by $\mathfrak{b}$ if $\mathfrak{a}\subset\mathfrak{b}$; in this case, we call $\mathfrak{b}$ a **divisor** of $\mathfrak{a}$ and $\mathfrak{a}$ a **multiple** of $\mathfrak{b}$. Also, $\mathfrak{a}\subset\mathfrak{b}$ if and only if there exists an integral ideal $\mathfrak{c}$ such that $\mathfrak{a}=\mathfrak{b}\mathfrak{c}$. Given fractional ideals $\mathfrak{a}=\prod\mathfrak{p}_i^{e_i}$ and $\mathfrak{b}=\prod\mathfrak{q}_j^{f_j}$ $(e_i\neq 0, f_j\neq 0)$, we say that $\mathfrak{a}$ and $\mathfrak{b}$ are **relatively prime** if $\{\mathfrak{p}_i\}$ and $\{\mathfrak{q}_j\}$ are disjoint. Usually a fractional ideal of k is simply called an ideal of k.

For an element α $(\neq 0)$ of k, $(\alpha)=\alpha\mathfrak{o}$ is a (fractional) ideal of k, and (α) is said to be a **principal ideal** of k. The set P_k of all principal ideals (α) $(\alpha\in k, \alpha\neq 0)$ is a subgroup of $\mathfrak{J}_k$. Since $(\alpha)=\mathfrak{o}$ is equivalent to $\alpha\in E_k$, we have $P_k\cong k/E_k$, where k^* is the multiplicative group of all nonzero elements of k.

Each coset of $\mathfrak{J}_k$ modulo P_k is called an **ideal class** of k, and the group $\mathfrak{C}_k=\mathfrak{J}_k/P_k$ is called the **ideal class group** of k. Each ideal class contains an integral ideal $\mathfrak{a}$ with $N(\mathfrak{a})\leqslant\sqrt{|D_k|}$ (more precisely, with $N(\mathfrak{a})\leqslant(4/\pi)^{r_2}(n!/n^n)\sqrt{|D_k|}$). From this it follows that $\mathfrak{C}_k$ is a finite Abelian group. The order h of $\mathfrak{C}_k$ is called the **class number** of k. For the calculation of the class number the †residue at the pole $s=1$ of the Dedekind zeta function is used. Namely,

$$\lim_{s\to 1+0}(s-1)\zeta_k(s)=gh,$$

$$g=2^{r_1+r_2}\pi^{r_2}R_k/w_k\sqrt{|D_k|}\ ,$$

where R_k is the regulator of k and w_k is the number of roots of unity in k (Dedekind, 1877). This formula is used, in particular, for the computation of the class numbers of †quadratic fields and †cyclotomic fields (→ 343 Quadratic Fields). The class numbers of cubic and quartic (real) †cyclic fields over $\mathbf{Q}$ were computed by H. Hasse in the case where the †conductor of $k/\mathbf{Q}$ is less than 100 (*Abh. Deutsch. Akad. Wiss. Berlin*, 2 (1948)). In general, let the degree $n=[k:\mathbf{Q}]$ be fixed and let $|D_k|\to\infty$. Then

$$\lim\left(\log(h_kR_k)/\log\sqrt{|D_k|}\ \right)=1.$$

(This formula was proved for $n=2$ by C. L. Siegel, 1935 and for general n by R. Brauer, *Amer. J. Math.*, 69 (1947).)

F. Valuations

All the †Archimedean and †non-Archimedean valuations of an algebraic number field k can be obtained as follows (→ 425 Valuations):

Archimedean Valuations. Let $n=[k:\mathbf{Q}]$, and let n conjugates of $\alpha\in k$ be $\alpha^{(1)},\ldots,\alpha^{(n)}$ such that $\alpha^{(i)}$ $(i=1,\ldots,r_1)$ is real, and $\alpha^{(r_1+j)}$ and $\alpha^{(r_1+r_2+j)}$ $(j=1,\ldots,r_2)$ are complex conjugate. We write $|\alpha|_j=|\alpha^{(j)}|$ $(j=1,\ldots,r_1+r_2)$; these are Archimedean valuations of k which are not mutually equivalent. The equivalence classes of these valuations are denoted by

$\mathfrak{p}_\infty^{(1)}, \ldots, \mathfrak{p}_\infty^{(r_1+r_2)}$, respectively, and are called the †infinite prime divisors of k. The first r_1 infinite prime divisors are called †real and the rest r_2 are called †imaginary (or complex). The valuations of k defined by

$$|\alpha|_{\mathfrak{p}_\infty^{(j)}} = |\alpha|_j, \quad j=1,\ldots,r_1,$$
$$= |\alpha|_j^2, \quad j=r_1+1,\ldots,r_1+r_2,$$

are called †normal valuations of k. Here $|\ |_{\mathfrak{p}_\infty^{(j)}}$ $(j=r_1+1,\ldots,r_1+r_2)$ are valuations in the wider sense. If $r_1=n$ we call k a **totally real field**, and if $r_1=0$ we call k a **totally imaginary field**.

Non-Archimedean Valuations. Let $\mathfrak{p}$ be a prime ideal of k and α an element of k. Let $(\alpha)=\mathfrak{p}^a\mathfrak{b}$, where $\mathfrak{p}$ and $\mathfrak{b}$ are relatively prime. Put $\nu_\mathfrak{p}(\alpha)=a$. Then for any constant ρ $(0<\rho<1)$,

$$|\alpha|_\mathfrak{p} = \rho^{\nu_\mathfrak{p}(\alpha)}$$

is a non-Archimedean valuation of k. This valuation of k is called the †$\mathfrak{p}$-adic valuation of k; $\mathfrak{p}$-adic valuations for different prime ideals are mutually inequivalent. The valuation $|\alpha|_\mathfrak{p}$ with $\rho=(N(\mathfrak{p}))^{-1}$ is called a †normal valuation of k. The equivalence class of valuations containing $|\ |_\mathfrak{p}$ is denoted by the same letter $\mathfrak{p}$ and is called a †finite prime divisor of k.

A formal finite product of powers of finite or infinite prime divisors $\mathfrak{m}^*=\prod\mathfrak{p}_i^{e_i}$ is called a **divisor** of k. If all $e_i \geqslant 0$, then $\mathfrak{m}^*$ is called an **integral divisor** of k. Given divisors $\mathfrak{m}^*=\prod\mathfrak{p}_i^{e_i}$ and $\mathfrak{n}^*=\prod\mathfrak{p}_i^{f_i}$, we write $\mathfrak{m}^*|\mathfrak{n}^*$ if $e_i \leqslant f_i$ $(i=1,2,\ldots)$.

Any valuation of k is equivalent to one of the valuations defined previously (A. Ostrowski, 1918; E. Artin, 1932). For any element $\alpha(\neq 0)$ of k the †product formula $\prod_\mathfrak{p}|\alpha|_\mathfrak{p}=1$ holds where $\mathfrak{p}$ runs over all finite and infinite prime divisors of k and $|\ |_\mathfrak{p}$ are the normal valuations of k. Conversely, let k be a field, and let $V=\{|\ |_\mathfrak{p}\}$ be a set of inequivalent valuations of k such that (i) for any $\alpha\in k$ $(\alpha\neq 0)$ $|\alpha|_\mathfrak{p}\neq 1$ holds only for a finite number of $\mathfrak{p}$ in V; (ii) the product formula $\prod_\mathfrak{p}|\alpha|_\mathfrak{p}=1$ $(\alpha\in k, \alpha\neq 0)$ holds; and (iii) there is at least one non-Archimedean valuation in V. Then k is an algebraic number field and V is the set of all the prime divisors of k (Artin and G. Whaples, *Bull. Amer. Math. Soc.*, 51 (1945)).

G. Ideal Classes in the Narrow Sense

For α, $\beta\in k$, the expression $\alpha\equiv\beta \pmod{\mathfrak{p}_\infty^{(i)}}$ means $\alpha^{(i)}\beta^{(i)}>0$ for a real infinite prime divisor $\mathfrak{p}_\infty^{(i)}$ and $\alpha^{(i)}\beta^{(i)}\neq 0$ for an imaginary infinite prime divisor $\mathfrak{p}_\infty^{(i)}$. We call an element $\alpha\in k$ **totally positive** if all real conjugates $\alpha^{(i)}$ $(i=1,\ldots,r_1)$ are positive. In the notation just given, this means $\alpha\equiv 1 \pmod{\mathfrak{p}^{(i)}}$ $(i=1,\ldots,r_1)$. The set of all principal ideals (α) generated by totally positive elements $\alpha\in k$ is a multiplicative subgroup P_k^+ of P_k. Each coset of $\mathfrak{J}_k$ modulo P_k^+ is called an **ideal class of k in the narrow sense**. Let E_k^+ be the group of all totally positive units of k. Then we have $(\mathfrak{J}_k:P_k^+)=h2^{r_1}/(E_k:E_k^+)$.

H. Multiplicative Congruence

Let $\mathfrak{m}$ be an integral ideal of k, and let $k^*(\mathfrak{m})$ be the multiplicative group of all elements α in k such that (α) is relatively prime to $\mathfrak{m}$. Any element $\alpha\in k^*(\mathfrak{m})$ can be expressed in the form $\alpha=\beta/\gamma$ such that $\beta,\gamma\in\mathfrak{o}$ and (β), (γ) are relatively prime to $\mathfrak{m}$.

Consider an integral divisor $\mathfrak{m}^*=\mathfrak{m}\prod\mathfrak{p}_\infty^{(i)}$ which is a formal product of $\mathfrak{m}$ and infinite prime divisors $\mathfrak{p}_\infty^{(i)}$ of k. We call $\mathfrak{m}$ the finite part of $\mathfrak{m}^*$. Given an element $\alpha\in k^*(\mathfrak{m})$ and elements $\beta,\gamma\in k^*(\mathfrak{m})\cap\mathfrak{o}$ such that $\alpha=\beta/\gamma$, we set $\alpha\equiv 1\ (\mathrm{mod}^\times\mathfrak{m}^*)$ if $\beta\equiv\gamma \pmod{\mathfrak{m}}$, and $\alpha\equiv 1 \pmod{\mathfrak{p}_\infty^{(i)}}$. The set of all α in $k^*(\mathfrak{m})$ such that $\alpha\equiv 1\ (\mathrm{mod}^\times\mathfrak{m}^*)$ forms a multiplicative group. We write $\alpha\equiv\beta\ (\mathrm{mod}^\times\mathfrak{m}^*)$ for α, $\beta\in k$ if $\alpha/\beta\in k^*(\mathfrak{m})$ and $\alpha/\beta\equiv 1\ (\mathrm{mod}^\times\mathfrak{m}^*)$. This congruence is called the **multiplicative congruence**. In the following discussion we shall write $\mathrm{mod}\,\mathfrak{m}^*$ for $\mathrm{mod}^\times\mathfrak{m}^*$.

We denote by $\mathfrak{J}_k(\mathfrak{m})$ the group of all ideals of k that are relatively prime to an integral ideal $\mathfrak{m}$, and by $S(\mathfrak{m}^*)$ the group of all principal ideals (α) such that $\alpha\in k^*(\mathfrak{m})$, $\alpha\equiv 1 \pmod{\mathfrak{m}^*}$; $S(\mathfrak{m}^*)$ is known as the **ray** modulo $\mathfrak{m}^*$. Any subgroup H of $\mathfrak{J}_k(\mathfrak{m})$ which contains $S(\mathfrak{m}^*)$ is called an **ideal group** modulo $\mathfrak{m}^*$, and the factor group $\mathfrak{J}_k(\mathfrak{m})/H$ is called a **group of congruence classes** of ideals modulo $\mathfrak{m}^*$.

If $\mathfrak{n}^*|\mathfrak{m}^*$ for integral divisors $\mathfrak{m}^*$ and $\mathfrak{n}^*$ of k, then $\mathfrak{J}_k(\mathfrak{m})\subset\mathfrak{J}_k(\mathfrak{n})$ and $S(\mathfrak{m}^*)\subset S(\mathfrak{n}^*)$. If H is an ideal group modulo $\mathfrak{n}^*$, then $\Phi(H)=H\cap\mathfrak{J}_k(\mathfrak{m})$ is an ideal group modulo $\mathfrak{m}^*$, and we have $\mathfrak{J}_k(\mathfrak{n})/H\cong\mathfrak{J}_k(\mathfrak{m})/\Phi(H)$. For any given ideal group H_0 modulo $\mathfrak{m}^*$ there is a smallest integral divisor $\mathfrak{f}^*$ such that $\mathfrak{f}^*|\mathfrak{m}^*$, and there exists an ideal group H modulo $\mathfrak{f}^*$ with $\Phi(H)=H_0$ (i.e., if there is an ideal group H' modulo $\mathfrak{n}^*$ with $\Phi(H')=H_0$, then $\mathfrak{f}^*|\mathfrak{n}^*$). We call $\mathfrak{f}^*$ the **conductor** of the ideal group H. The notion of multiplicative congruence is used in †class field theory and in the theory of †norm-residue symbols.

I. Ideal Theory for Relative Extensions

If an algebraic number field K has a subfield k, we say that K/k is a **relative algebraic**

number field. Let $\mathfrak{O}$ be the principal order of K. For a (fractional) ideal $\mathfrak{a}$ of k, $\mathfrak{O}\mathfrak{a}$ is an ideal of K. We write $\mathfrak{O}\mathfrak{a}=E(\mathfrak{a})$ and call $E(\mathfrak{a})$ the **extension** of $\mathfrak{a}$ to K. For ideals $\mathfrak{a}$, $\mathfrak{b}$ of k, we have $E(\mathfrak{a}\mathfrak{b})=E(\mathfrak{a})E(\mathfrak{b})$ and $E(\mathfrak{a})\cap k=\mathfrak{a}$.

Let $\Psi_i: K\to C$ be k-isomorphisms ($i=1,\ldots,n$) where $n=[K:k]$. We write $K^{(i)}=\Psi_i(K)$ and $A^{(i)}=\Psi_i(A)$ for $A\in K$. For an ideal $\mathfrak{A}$ of K, $\mathfrak{A}^{(i)}=\{A^{(i)}\mid A\in\mathfrak{A}\}$ is an ideal of $K^{(i)}$, and $\mathfrak{A}^{(i)}$ is called the **conjugate ideal** of $\mathfrak{A}$ in $K^{(i)}$. Let L be the composite field of $K^{(1)},\ldots,K^{(n)}$. Then the ideal generated by $\mathfrak{A}^{(1)}\ldots\mathfrak{A}^{(n)}$ in L is the extension of an ideal $\mathfrak{a}$ of k. We write $\mathfrak{a}=N_{K/k}(\mathfrak{A})$ and call $\mathfrak{a}$ the **relative norm** of $\mathfrak{A}$ over k. We have $N_{K/k}(\mathfrak{A}\mathfrak{B})=N_{K/k}(\mathfrak{A})N_{K/k}(\mathfrak{B})$ and $N_{K/k}(E(\mathfrak{a}))=\mathfrak{a}^n$ (for an ideal $\mathfrak{a}$ of k). In particular, for $k=\mathbf{Q}$, $N_{K/\mathbf{Q}}(\mathfrak{A})=(N(\mathfrak{A}))$.

Let $\mathfrak{p}$ be a prime ideal of k. Then $E(\mathfrak{p})=\mathfrak{P}_1^{e_1}\mathfrak{P}_2^{e_2}\ldots\mathfrak{P}_g^{e_g}$ in $\mathfrak{O}$, where $\mathfrak{P}_1,\ldots,\mathfrak{P}_g$ are prime ideals of K. Let f_i be the degree of the finite field $\mathfrak{O}/\mathfrak{P}_i$ over $\mathfrak{o}/\mathfrak{p}$. Then $N_{K/k}(\mathfrak{P}_i)=\mathfrak{p}^{f_i}$; f_i is called the **relative degree** of $\mathfrak{P}_i$ over k, and e_i is called the **(relative) ramification index** of $\mathfrak{P}_i$ over k. We have the relation $n=\sum_{i=1}^{g}e_i f_i$ between these numbers. If $e_1=\ldots=e_g=1$, the prime ideal $\mathfrak{p}$ is said to be **unramified** for K/k. Otherwise, $\mathfrak{p}$ is said to be **ramified** for K/k. If every prime ideal of k is unramified for K/k, we call K/k an **unramified extension**. (For an infinite prime divisor $\mathfrak{p}_\infty$ of k we write $\mathfrak{p}_\infty=\prod_{i=1}^{g}\mathfrak{P}_\infty^{(i)e_i}$ if the Archimedean valuation $|\ |_{\mathfrak{p}_\infty}$ of k can be extended to g Archimedean valuations $|\ |_{\mathfrak{P}_\infty^{(i)}}$ ($i=1,\ldots,g$) of K, where $e_i=2$ if $\mathfrak{P}_\infty^{(i)}$ is imaginary and $\mathfrak{p}_\infty$ is real, $e_i=1$ otherwise.)

J. Relative Differents and Relative Discriminants

Let K/k be a relative algebraic number field and $\mathfrak{o}$, $\mathfrak{O}$ be the principal orders of k, K respectively. Put $\mathfrak{M}=\{A\in K\mid \mathrm{Tr}_{K/k}(A\mathfrak{O})\subset\mathfrak{o}\}$, where $\mathrm{Tr}_{K/k}$ is the †trace (→ 157 Fields G). Then $\mathfrak{M}$ is a (fractional) ideal of K and $\mathfrak{M}^{-1}=\mathfrak{D}_{K/k}$ is an integral ideal of K; $\mathfrak{D}_{K/k}$ is called the **relative different** of K over k. When $k=\mathbf{Q}$, $\mathfrak{D}_{K/\mathbf{Q}}$ is simply called the **different** of K. For $L\supset K\supset k$, we have the **chain theorem**: $\mathfrak{D}_{L/k}=\mathfrak{D}_{L/K}\mathfrak{D}_{K/k}$.

Let the conjugates of $A\in K$ over k be $A^{(1)},\ldots,A^{(n)}$, and assume that $A^{(1)}=A$. Put $\delta_{K/k}(A)=\prod_{i=2}^{n}(A-A^{(i)})$ for $A\in K$. If $A\in\mathfrak{O}$, then $\delta_{K/k}(A)\in\mathfrak{D}_{K/k}$. $\mathfrak{D}_{K/k}$ is generated by $\{\delta_{K/k}(A)\mid A\in\mathfrak{O}\}$. The integral ideal $\mathfrak{E}^{(i)}$ generated by $\{A-A^{(i)}\mid A\in\mathfrak{O}\}$ in the field $L=K^{(1)}K^{(2)}\ldots K^{(n)}$ was called *Element* by D. Hilbert. We also have $\mathfrak{D}_{K/k}=\mathfrak{E}^{(2)}\mathfrak{E}^{(3)}\ldots\mathfrak{E}^{(n)}$. The integral ideal $\mathfrak{d}_{K/k}=N_{K/k}(\mathfrak{D}_{K/k})$ of k is called the **relative discriminant** of K/k. If $k=\mathbf{Q}$, $\mathfrak{d}_{K/\mathbf{Q}}=D_K$.

For the relative different $\mathfrak{D}_{K/k}$ to be divisible by a prime ideal $\mathfrak{P}$ of K, it is necessary and sufficient that $E(\mathfrak{p})=\mathfrak{P}^e\mathfrak{P}_2^{e_2}\ldots\mathfrak{P}_g^{e_g}$ with $e>1$, where $\mathfrak{p}=\mathfrak{P}\cap k$ (**Dedekind's discriminant theorem**, 1882). Hence a prime ideal $\mathfrak{p}$ of k is ramified for K/k if and only if $\mathfrak{p}$ divides the relative discriminant $\mathfrak{d}_{K/k}$; there are thus only a finite number of prime ideals of k which ramify for K/k. In particular, K/k is unramified if and only if $\mathfrak{d}_{K/k}=\mathfrak{o}$.

K. Arithmetic of Galois Extensions

Let K/k be a relative algebraic number field such that K is a †Galois extension of k of degree n, and let G be the †Galois group of K/k. Let $\mathfrak{o}$, $\mathfrak{O}$ be the principal order of k, K respectively. The conjugate ideals of an ideal $\mathfrak{A}$ of K are given by $\mathfrak{A}^\sigma=\{A^\sigma\mid A\in\mathfrak{A}\}$ ($\sigma\in G$). If $N_{K/k}(\mathfrak{A})=\mathfrak{a}$, then $E(\mathfrak{a})=\prod_{\sigma\in G}\mathfrak{A}^\sigma$. For a prime ideal $\mathfrak{p}$ of K, $E(\mathfrak{p})=(\mathfrak{P}_1\mathfrak{P}_2\ldots\mathfrak{P}_g)^e$, where $N_{K/k}(\mathfrak{P}_i)=\mathfrak{p}^f$ ($i=1,\ldots,g$), $n=efg$, and $\mathfrak{P}_1,\ldots,\mathfrak{P}_g$ are mutually conjugate prime ideals of K over k.

Hilbert (1894) developed the decomposition theory of a prime ideal $\mathfrak{p}$ of k for a Galois extension K/k in terms of the Galois group G as follows: Let $\mathfrak{P}$ be a prime ideal of $\mathfrak{O}$. Then

$$Z=\{\sigma\in G\mid \mathfrak{P}^\sigma=\mathfrak{P}\}$$

is a subgroup of the Galois group G of K/k. Z is called the **decomposition group** of $\mathfrak{P}$ over k. Let $G=\bigcup_i Z\tau_i$ be the left coset decomposition of G. Then $\mathfrak{P}_i=\mathfrak{P}^{\tau_i}$ ($i=1,\ldots,g$) are all the conjugate ideals of $\mathfrak{P}$ over k.

The subgroup

$$T=\{\sigma\in Z\mid A^\sigma\equiv A\ (\mathrm{mod}\ \mathfrak{P}), A\in\mathfrak{O}\}$$

of the decomposition group Z is normal, and T is called the **inertia group** of $\mathfrak{P}$ over k. The quotient group Z/T is a cyclic group of order f (the relative degree of $\mathfrak{P}$). There exists an element σ of Z such that

$$A^\sigma\equiv A^{N(\mathfrak{p})}\ (\mathrm{mod}\ \mathfrak{P}),\quad A\in\mathfrak{O},$$

and σ is uniquely determined $\mathrm{mod}\,T$; σT generates the cyclic group Z/T. This σ is called the **Frobenius substitution** (or **Frobenius automorphism**) of $\mathfrak{P}$ over k. For $m=1,2,\ldots$,

$$V^{(m)}=\{\sigma\in Z\mid A^\sigma\equiv A\ (\mathrm{mod}\ P^{m+1}), A\in\mathfrak{O}\}$$

are normal subgroups of Z; the group $V^{(m)}$ is called the mth **ramification group** of $\mathfrak{P}$ over k. Let

$$\begin{aligned}V^{(0)}&=\ldots=V^{(v_1)}\supsetneqq V^{(v_1+1)}=\ldots\\&=V^{(v_2)}\supsetneqq\ldots\supsetneqq V^{(v_{r-1}+1)}=\ldots\\&=V^{(v_r)}\supsetneqq V^{(v_r+1)}=1.\end{aligned}$$

Let $V_\rho=V^{(v_\rho+1)}$ ($\rho=0,1,\ldots,r$), where $v_0=-1$. In particular, $V_0=V^{(0)}=T$. The integers $v_1, v_2,\ldots$ are called the **ramification numbers** of $\mathfrak{P}$. The group $T/V^{(1)}$ is isomorphic to a subgroup of the multiplicative group of the finite field $\mathfrak{O}/\mathfrak{P}$. Hence $T/V^{(1)}$ is a cyclic group whose order e_0 is a divisor of $N(\mathfrak{P})-1$. The group $V^{(m)}/V^{(m+1)}$ $(m\geqslant 1)$ is isomorphic to a subgroup of the additive group of the finite field $\mathfrak{O}/\mathfrak{P}$. Hence $V^{(m)}/V^{(m+1)}$ is an Abelian group of type $(p,p,\ldots,p)$ whose order divides $N(\mathfrak{P})$. From $e=|T|=(T:V^{(1)})|V^{(1)}|$ it follows that $e=e_0p^{\mu}$, $(e_0,p)=1$. Here $|G|$ denotes the order of a finite group G. Hence the decomposition group of $\mathfrak{P}$ is a †solvable group. The relation between the ramification numbers for K/k and those for an intermediate Galois extension F/k was completely determined by J. Herbrand (*J. Math. Pures Appl.*, 10 (1931)) [2, 13].

Let $\mathfrak{P}^d$ be the $\mathfrak{P}$-component of the relative different $\mathfrak{D}_{K/k}$ of a Galois extension K/k. Then

$$d=\sum_{\rho=0}^{r-1}(v_{\rho+1}-v_{\rho})(|V_{\rho}|-1)=\sum_{i=0}^{v_r}(|V^{(i)}|-1).$$

In particular, $d=0$ if $T=1$, and $d=e-1$ if $V^{(1)}=1$.

Let k_Z, k_T, and $k_{V^{(m)}}$ be the intermediate fields which correspond to the subgroups Z, T, and $V^{(m)}$, respectively, in the sense of †Galois theory; the fields k_Z, k_T, and $k_{V^{(m)}}$ are called the **decomposition field**, the **inertia field**, and the mth **ramification field** of $\mathfrak{P}$, respectively. Let $\mathfrak{P}$ be a prime ideal of K containing $\mathfrak{p}$, and let $\mathfrak{p}_Z$ and $\mathfrak{p}_T$ be prime ideals in k_Z and k_T such that $\mathfrak{p}_Z=\mathfrak{P}\cap k_Z$ and $\mathfrak{p}_T=\mathfrak{P}\cap k_T$. Then we have $E(\mathfrak{p})=\mathfrak{p}_Z\mathfrak{p}_Z^{(2)}\ldots\mathfrak{p}_Z^{(g)}$ for k_Z/k; $E(\mathfrak{p}_Z)=\mathfrak{p}_T$ and $N_{k_T/k_Z}(\mathfrak{p}_T)=\mathfrak{p}_Z^f$ for k_T/k_Z; and $E(\mathfrak{p}_T)=\mathfrak{P}^e$ for K/k_T.

If a prime ideal $\mathfrak{p}$ of k is unramified for a Galois extension K/k then we have $E(\mathfrak{p})=\mathfrak{P}_1\mathfrak{P}_2\ldots\mathfrak{P}_g$, $\mathfrak{P}_i=\mathfrak{P}^{\tau_i}$ $(i=1,\ldots,g)$, $N_{K/k}(\mathfrak{P}_i)=\mathfrak{p}^f$, and $n=fg$. The Frobenius automorphism $\sigma_i: A^{\sigma_i}\equiv A^{N(\mathfrak{p})}\pmod{\mathfrak{P}_i}$ $(A\in\mathfrak{O})$ for the prime ideal $\mathfrak{P}_i$ is uniquely determined, and its order is f. Since $\mathfrak{P}_i=\mathfrak{P}^{\tau_i}$, we have $\sigma_i=\tau_i^{-1}\sigma_1\tau_i$. Hence $\sigma_1,\ldots,\sigma_g$ belong to the same †conjugate class of G. In particular, if G is an Abelian group, then $\sigma_1=\ldots=\sigma_g$ and

$$A^{\sigma_1}\equiv A^{N(\mathfrak{p})}\pmod{\mathfrak{p}},\quad A\in\mathfrak{O}.$$

We then write

$$\sigma_1=\left(\frac{K/k}{\mathfrak{p}}\right)\quad(\in G)$$

and call this symbol the **Artin symbol** for $\mathfrak{p}$ for the Abelian extension K/k. For an ideal $\mathfrak{a}=\prod\mathfrak{p}^e$ of k that is relatively prime to the relative discriminant of K/k, we define

$$\left(\frac{K/k}{\mathfrak{a}}\right)=\prod\left(\frac{K/k}{\mathfrak{p}}\right)^e\quad(\in G).$$

Evidently, we have

$$\left(\frac{K/k}{\mathfrak{a}\mathfrak{b}}\right)=\left(\frac{K/k}{\mathfrak{a}}\right)\left(\frac{K/k}{\mathfrak{b}}\right).$$

The arithmetic of quadratic fields (→ 343 Quadratic Fields) and the arithmetic of cyclotomic fields (→ Section L) have developed since the 19th century.

L. Arithmetic of Cyclotomic Fields

A complex number ζ whose mth power is 1 but whose m'th power is not 1 for $m'<m$ is called an mth **primitive root of unity**. There are $\varphi(m)$ primitive roots of unity: $\exp(2\pi ir/m)$ $((r,m)=1)$, where φ is †Euler's function. These $\varphi(m)$ primitive roots of unity are the roots of an irreducible polynomial over $\mathbf{Q}$ of degree $\varphi(m)$:

$$F_m(X)=\prod_{d|m}(X^{m/d}-1)^{\mu(d)},$$

where μ is the †Möbius function. The coefficient of the highest term of $F_m(X)$ is 1, and the other coefficients are all rational integers. $F_m(X)$ is called a **cyclotomic polynomial**. An example is

$$F_{12}(X)=(X^{12}-1)(X^2-1)/(X^6-1)(X^4-1)$$
$$=X^4-X^2+1.$$

The algebraic number field $K_m=\mathbf{Q}(\zeta_m)$ $(\zeta_m=\exp 2\pi i/m)$ obtained by adjoining an mth primitive root of unity to $\mathbf{Q}$ is a Galois extension over $\mathbf{Q}$ of degree $\varphi(m)$ whose Galois group G is isomorphic to the multiplicative Abelian group of †reduced residue classes of $\mathbf{Z}$ modulo m: $G=\{\sigma_r\mid\sigma_r(\zeta_m)=\zeta_m^r,(r,m)=1\}$. K_m is called the mth **cyclotomic field**. Cyclotomic fields are †Abelian extensions of $\mathbf{Q}$. Conversely, every Abelian extension of $\mathbf{Q}$ is a subfield of a cyclotomic field (**Kronecker's theorem**, 1853, 1877).

We can choose $(1,\zeta_m,\zeta_m^2,\ldots,\zeta_m^{\varphi(m)-1})$ as a minimal basis of K_m. Let $m=l_1^{h_1}l_2^{h_2}\ldots l_t^{h_t}$ be the decompositions of m in powers of prime numbers $l_1,\ldots,l_t$. Put $K^{(i)}=K_{l_i^{h_i}}$. Then K_m is the composite field $K_m=K^{(1)}K^{(2)}\ldots K^{(t)}$. The different of K_m is given by $\mathfrak{D}_{K_m/\mathbf{Q}}=\mathfrak{D}_{K^{(1)}/\mathbf{Q}}\mathfrak{D}_{K^{(2)}/\mathbf{Q}}\ldots\mathfrak{D}_{K^{(t)}/\mathbf{Q}}$, and the discriminant of K_m is given by $D_{K_m}=D_{K^{(1)}}^{n_1}\ldots D_{K^{(t)}}^{n_t}$ $(n_i=\varphi(m)/\varphi(l_i^{h_i}))$. If $m=l^h$, the discriminant of K_m is $D_{K_m}=\epsilon l^a$ $(a=l^{h-1}(hl-h-1))$, where $\epsilon=-1$ if $l^h=4$ or $l\equiv 3\pmod 4$, and $\epsilon=1$ otherwise. Hence the discriminant of K_m is $D_{K_m}=(\sqrt{-1}\,m/\prod_{p|m}p^{1/(p-1)})^{\varphi(m)}$. Suppose that either $(2,m)=1$ or $4|m$. Then a prime number p ramifies for $K_m/\mathbf{Q}$ if and only if p divides m. In particular, if $m=l^h$ $(m>2)$, then only l ramifies for $K_m/\mathbf{Q}$ and $(l)=\mathfrak{l}^{\varphi(m)}$, $N(\mathfrak{l})=l$ ($\mathfrak{l}$ is explicitly given by $\mathfrak{l}=(1-\zeta_m)$). For $l\neq 2$ the ramification numbers for l are $v_i=l^i-1$ $(i=1,2,\ldots)$, and the ramification fields are $K_l, K_{l^2},\ldots$. For $l=2$ the ramification numbers are 1, 3, 7,... and the ramification fields are $\mathbf{Q}$, K_4, $K_8,\ldots$.

In $K_m/\mathbf{Q}$ a prime number p ($p\nmid m$) is decomposed as $(p)=\mathfrak{p}_1\ldots\mathfrak{p}_g$, $N(\mathfrak{p}_i)=p^f(i=1,\ldots,g)$ and $fg=\varphi(m)$. Here the degree f of $\mathfrak{p}_i$ is determined as the minimal positive integer f such that $p^f\equiv 1 \pmod m$. Hence the decomposition law of a prime number p in $K_m/\mathbf{Q}$ is determined by its residue class $\bmod m$. This is a primitive form of class field theory (→ 62 Class Field Theory).

The class number of the cyclotomic field K_m can be calculated by Dedekind's formula (→ E; see also Hilbert [6], T. Takagi [1]). Here we shall give the result for $m=l$ (a prime number). Let r be a †primitive root modulo l. For $\zeta=\exp 2\pi i/l$, we put

$$\epsilon=\epsilon(\zeta)=\left(\frac{1-\zeta^r}{1-\zeta}\frac{1-\zeta^{-r}}{1-\zeta^{-1}}\right)^{1/2}.$$

Then ϵ is a unit in K_l. Define an element σ of the Galois group of $K_l/\mathbf{Q}$ by $\zeta^\sigma=\zeta^r$, and put $\epsilon_i=\epsilon^{\sigma^i}$ $(i=0,1,\ldots)$. Then $\epsilon_0,\epsilon_1,\ldots,\epsilon_{\rho-1}$ $(\rho=(l-3)/2)$ are multiplicatively independent units. That is, the regulator $R[\epsilon_0,\epsilon_1,\ldots,\epsilon_{\rho-1}]=E\neq 0$. The units $\epsilon_0,\epsilon_1,\ldots,\epsilon_{\rho-1}$ are called **circular units**. The class number h of K_l is the product of two factors, $h=h_1h_2$. Here h_2 is the class number of the real subfield $K_l'=\mathbf{Q}(\zeta+\zeta^{-1})$. E. E. Kummer called h_1 the **first factor** and h_2 the **second factor** of the class number h. Let $\chi_1,\chi_2,\ldots,\chi_{l-1}$ be the multiplicative characters of the †reduced residue classes of $\mathbf{Z}$ modulo l, and let χ_j $(j=1,\ldots,\rho+1)$ be the characters among them such that $\chi_j(-1)=-1$. Then

$$h_1=\frac{(-1)^{\rho+1}}{(2l)^\rho}\prod_{i=1}^{\rho+1}\left(\sum_{j=1}^{l-1}j\chi_i(j)\right),\qquad h_2=\frac{|E|}{R_0}$$

(E. Kummer, 1850). Here R_0 is the regulator of K_l'. Since circular units belong to K_l', the class number h_2 of K_l' is equal to the index of the subgroup generated by $\pm 1,\epsilon_0,\ldots,\epsilon_{\rho-1}$ in the group of units of $K_{l'}$. The class number h of K_l is equal to 1 for $l\leqslant 19$ and it has been conjectured that there exist no more fields K_l with $h=1$. K. Uchida solved this conjecture by proving that the first factor $h_1>1$ for $l>19$ (*Tôhoku Math. J.*, 23 (1971)).

According to Kummer, l divides h if and only if l divides h_1. Since h_1 can be computed explicitly, we can readily determine whether $l|h$ or not (→ 154 Fermat's Problem; Appendix B, Table 4.III). Let the l-component of the class number of K_m for $m=l^{n+1}$ be denoted by l^{e_n}. It has been proved that for sufficiently large n, $e_n=\lambda n+\mu l^n+\nu$ for some integral constants λ, μ, ν (K. Iwasawa, *Bull. Amer. Math. Soc.*, 65 (1959)). It was verified that $\mu=0$ at least for $l\leqslant 4001$ (by Iwasawa and C. C. Sims, *J. Math. Soc. Japan*, 18 (1966)) and for $l<30{,}000$ by W. Johnson (*Math. Comp.*, 29 (1975)). In particular, $e_n>0$ if and only if the class number of K_l is divisible by l (Ph. Furtwängler, 1911).

Since any quadratic field is a subfield of a cyclotomic field (by a †Gaussian sum formula we have $\mathbf{Q}(\sqrt{m})\subset\mathbf{Q}(\zeta_{|d|})$, where d is the discriminant of $\mathbf{Q}(\sqrt{m})$), the computation of the class number of quadratic fields and the proof of the law of reciprocity for the †Legendre symbol follow from the arithmetic of cyclotomic fields.

M. Arithmetic of Kummer Extensions

Assume that an algebraic number field k contains an nth primitive root of unity. Then a †Kummer extension $K=k(\sqrt[n]{\mu})$ $(\mu\in k)$ is a †cyclic extension of k. Assume that $[K:k]=n$. In order that a prime ideal $\mathfrak{p}$ of k ramify for K/k, it is necessary that $\mathfrak{p}$ divides (n) or (μ). If $\mathfrak{p}\nmid(n)$ and $\nu_{\mathfrak{p}}(\mu)\not\equiv 0 \pmod n$, then $\mathfrak{p}$ ramifies for K/k. A prime ideal $\mathfrak{p}$ which is relatively prime to (μ) has the decomposition $E(\mathfrak{p})=\mathfrak{P}_1\ldots\mathfrak{P}_n$ in K if and only if the equation $\mu\equiv\xi^n \pmod{\mathfrak{p}^m}$ is satisfied by some $\xi\in\mathfrak{o}$ for any positive integer m. In particular, if $\mathfrak{p}\nmid(n)$ and $\nu_{\mathfrak{p}}(\mu)=0$, we have $E(\mathfrak{p})=\mathfrak{P}_1\ldots\mathfrak{P}_n$ if and only if $\mu=\xi^n \pmod{\mathfrak{p}}$ is solvable in $\mathfrak{o}$.

If for an element μ of $\mathfrak{o}$

$$\mu\equiv\xi^n \pmod{\mathfrak{p}}$$

is solvable by some $\xi\in\mathfrak{o}$, μ is said to be a **residue of the nth power** modulo $\mathfrak{p}$. Assume that $\mathfrak{p}\nmid(n)$ and $\nu_{\mathfrak{p}}(\mu)=0$. Let f be the minimal positive integer such that μ^f is a residue of the nth power modulo $\mathfrak{p}$. Then $\mathfrak{p}$ is decomposed in K as $E(\mathfrak{p})=\mathfrak{P}_1\ldots\mathfrak{P}_g$, and $N_{K/k}(\mathfrak{P}_i)=\mathfrak{p}^f$ $(i=1,\ldots,g)$.

N. Power-Residue Symbol

Let $\zeta_n=\exp(2\pi i/n)\in k$, and let $\mathfrak{p}$ be relatively prime to (n) and (α) $(\alpha\in k)$. Then for some r, we have

$$\alpha^{(N(\mathfrak{p})-1)/n}\equiv\zeta_n^r \pmod{\mathfrak{p}},$$

and we write

$$\zeta_n^r=\left(\frac{\alpha}{\mathfrak{p}}\right)_n.$$

This symbol is called the nth **power-residue symbol** (Kummer). Generalizing this definition, we can define the nth power-residue symbol $(\alpha/\mathfrak{b})_n$ for an ideal $\mathfrak{b}$ of k which is relatively prime to the relative discriminant of $k(\sqrt[n]{\alpha})/k$ by using the †Artin symbol

$((K/k)/\mathfrak{b})$:

$$(\sqrt[n]{\alpha})^{\sigma} = \left(\frac{\alpha}{\mathfrak{b}}\right)_n \sqrt[n]{\alpha}, \qquad \sigma = \left(\frac{K/k}{\mathfrak{b}}\right).$$

This symbol satisfies

$$\left(\frac{\alpha}{\mathfrak{b}_1\mathfrak{b}_2}\right)_n = \left(\frac{\alpha}{\mathfrak{b}_1}\right)_n \left(\frac{\alpha}{\mathfrak{b}_2}\right)_n,$$

$$\left(\frac{\alpha_1\alpha_2}{\mathfrak{b}}\right)_n = \left(\frac{\alpha_1}{\mathfrak{b}}\right)_n \left(\frac{\alpha_2}{\mathfrak{b}}\right)_n$$

if all the symbols are well defined. In particular, α is a residue of the nth power modulo $\mathfrak{p}$ if and only if $(\alpha/\mathfrak{p})_n = 1$. This symbol coincides with the †quadratic residue symbol for $m=2$, $k=\mathbf{Q}$, and $p \neq 2$.

O. Law of Reciprocity for the Power-Residue Symbol

Several formulas concerning the power-residue symbol are known which are similar to that for the quadratic residue symbol (F. G. M. Eisenstein, Kummer, Furtwängler, Takagi, Artin, Hasse). These can be proved by means of Artin's †general law of reciprocity in class field theory (→ 62 Class Field Theory).

There are many formulas concerning the reciprocity of the power-residue symbol. One of them is as follows: Let $n = l$ be a prime number. Let $\alpha, \beta \in k$ and assume that (i) α is totally positive; (ii) $\nu_{\mathfrak{p}}(\alpha) \equiv 0 \pmod{l}$ if $\nu_{\mathfrak{p}}(\beta) \not\equiv 0 \pmod{l}$, and $\nu_{\mathfrak{p}}(\beta) \equiv 0 \pmod{l}$ if $\nu_{\mathfrak{p}}(\alpha) \not\equiv 0 \pmod{l}$, for any prime ideal $\mathfrak{p}$; and (iii) $\alpha \equiv 1 \pmod{l}$ and $\beta \equiv 1 \pmod{(1-\zeta_l)}$. Then

$$\left(\frac{\alpha}{\beta}\right)_l \left(\frac{\beta}{\alpha}\right)_l^{-1} = \zeta_l^a,$$

$$a = \mathrm{Tr}_{k/\mathbf{Q}}\left(\frac{\alpha-1}{l}\,\frac{\beta-1}{1-\zeta_l}\right)$$

(**law of reciprocity**, Hasse, 1924). This result is a generalization of the formula of Eisenstein (1850). If α ($\in k$) is totally positive and $\alpha \equiv 1 \pmod{l}$, then

$$\left(\frac{\zeta_l}{\alpha}\right)_l = \zeta_l^b, \qquad b = \mathrm{Tr}_{k/\mathbf{Q}}\left(\frac{\alpha-1}{l}\right).$$

If $\alpha \equiv 1 \pmod{l(1-\zeta_l)}$, then

$$\left(\frac{l}{\alpha}\right)_l = \zeta_l^c, \qquad c = \mathrm{Tr}_{k/\mathbf{Q}}\left(\frac{\alpha-1}{l(1-\zeta_l)}\right);$$

$$\left(\frac{1-\zeta_l}{\alpha}\right)_l = \zeta_l^d, \qquad d = -\mathrm{Tr}_{k/\mathbf{Q}}\left(\frac{\alpha-1}{l(1-\zeta_l)}\right)$$

(**complementary law of reciprocity**, Hasse, 1924).

P. Norm Residue

Let $\mathfrak{m}$ be an integral divisor of k such that $\mathfrak{m} = \Pi_i \mathfrak{p}_i^{e_i} \Pi_j \mathfrak{p}_\infty^{(j)}$ ($e_i > 0$) with finite prime divisors $\{\mathfrak{p}_i\}$ and infinite prime divisors $\{\mathfrak{p}_\infty^{(j)}\}$, and let β be an element of k which is relatively prime to $\mathfrak{m}$. For a relative algebraic number field K/k and an element B of K, we set $\beta \equiv N_{K/k}(B) \pmod{\mathfrak{m}}$ if the following two conditions are satisfied: (i) $\beta \equiv N_{K/k}(B)$ $(\mathrm{mod}\, \mathfrak{p}_i^{e_i})$ for every finite $\mathfrak{p}_i$ and (ii) $\beta^{(j)} > 0$ for every infinite prime $\mathfrak{p}_\infty^{(j)}$ such that $\mathfrak{p}^{(j)}$ is real and its extension to K is imaginary. $\beta \in k$ is then said to be a **norm residue** modulo $\mathfrak{m}$ for K/k if there exists a number B of K such that $\beta \equiv N_{K/k} \pmod{\mathfrak{m}}$.

Let $\mathfrak{p}$ be a prime ideal of k. If β is a norm residue modulo $\mathfrak{p}^c$ for a sufficiently large c, then β is a norm residue modulo $\mathfrak{p}^e$ for any $e > c$. Let c be the smallest such integer ($c \geqslant 0$). Then the ideal $\mathfrak{f}_{\mathfrak{p}} = \mathfrak{p}^c$ is said to be the **$\mathfrak{p}$-conductor** of norm residue for K/k. If $\mathfrak{p}$ is unramified for K/k, then $c=0$; i.e., every $\beta \in k$ which is relatively prime to $\mathfrak{p}$ is a norm residue modulo $\mathfrak{p}^e$ for any $e > 0$. For a ramified $\mathfrak{p}$ put $f_{\mathfrak{p}} = \mathfrak{p}^c$. For a Galois extension K/k, c is not greater than

$$\sum_{\rho=0}^{r-1} \frac{|V_\rho|}{|V_0|}(v_{\rho+1} - v_\rho) = \sum_{i=0}^{v_r} \frac{|V^{(i)}|}{|V_0|}.$$

In particular, for an Abelian extension K/k this value is an integer and is equal to c (Hasse, *J. Fac. Sci. Univ. Tokyo*, 1934). For example, the l-conductor of the cyclotomic field $K_{l^h}/\mathbf{Q}$ is l^h. We define the $\mathfrak{p}_\infty$-conductor for K/k for an infinite prime divisor $\mathfrak{p}_\infty$ of k by $\mathfrak{f}_{\mathfrak{p}_\infty} = \mathfrak{p}_\infty$ if $\mathfrak{p}_\infty$ is real and its extension to K is imaginary, and $\mathfrak{f}_{\mathfrak{p}_\infty} = 1$ otherwise (→ 257 Local Fields F).

Q. Norm Residue Symbol

For an Abelian extension K/k, the positive divisor

$$\mathfrak{f} = \prod_{\mathfrak{p}} \mathfrak{f}_{\mathfrak{p}}$$

(where $\mathfrak{p}$ runs over all finite and infinite prime divisors of k) is called the **conductor** of K/k (→ 62 Class Field Theory). For $\alpha \in k$ ($\alpha \neq 0$) take α_0 such that $\alpha/\alpha_0 \equiv 1 \pmod{\mathfrak{f}_{\mathfrak{p}}}$ and $\alpha_0 \equiv 1$ $(\mathrm{mod}\, \mathfrak{f}\mathfrak{f}_{\mathfrak{p}}^{-1})$, and put $(\alpha_0) = \mathfrak{p}^a\mathfrak{b}$ with $\mathfrak{b}$ relatively prime to $\mathfrak{p}$. Then $\mathfrak{b}$ is relatively prime to the relative discriminant $\mathfrak{d}_{K/k}$. We define a new symbol by

$$\left(\frac{\alpha, K/k}{\mathfrak{p}}\right) = \left(\frac{K/k}{\mathfrak{b}}\right) \quad (\in G),$$

where $((K/k)/\mathfrak{b})$ is the Artin symbol. This value is independent of the choice of the auxiliary element α_0. The new symbol is called the **norm-residue symbol** (Hasse, *J. Reine Angew. Math.*, 162 (1930)). In particular, for an infinite real prime divisor $\mathfrak{p}_\infty^{(j)}$ of k whose extension $\mathfrak{P}_\infty^{(j)}$ for K is imaginary, we have

$$\left(\frac{\alpha, K/k}{\mathfrak{p}_\infty^{(j)}}\right)=1 \text{ or } =\sigma$$

according to whether the conjugate $\alpha^{(j)}$ is positive or negative, where σ is the automorphism of K/k induced from the complex conjugation of the completion **C** of K with respect to $\mathfrak{P}^{(j)}$.

The norm-residue symbol has the following properties:

(1) $\left(\frac{\alpha\alpha', K/k}{\mathfrak{p}}\right)=\left(\frac{\alpha, K/k}{\mathfrak{p}}\right)\left(\frac{\alpha', K/k}{\mathfrak{p}}\right)$;

(2) if $\mathfrak{p}$ is unramified for K/k, then

$$\left(\frac{\alpha, K/k}{\mathfrak{p}}\right)=\left(\frac{K/k}{\mathfrak{p}}\right)^{-\nu_{\mathfrak{p}}(\alpha)};$$

(3) in order that α be a norm residue modulo $\mathfrak{f}_{\mathfrak{p}}$ for K/k, it is necessary and sufficient that

$$\left(\frac{\alpha, K/k}{\mathfrak{p}}\right)=1;$$

(4) the **product formula** for the norm-residue symbol (Hasse) is

$$\prod_{\mathfrak{p}}\left(\frac{\alpha, K/k}{\mathfrak{p}}\right)=1,$$

where $\mathfrak{p}$ runs over all finite and infinite prime divisors of k; and (5) if the domain of the variable α is the whole k $(\neq 0)$, or the set of all α such that (α) is relatively prime to $\mathfrak{p}$, or the set of all α such that $\alpha\equiv 1 \pmod{\mathfrak{p}^m}$ $(u_\rho+1\leqslant m\leqslant u_{\rho+1})$, then the range of $((\alpha, K/k)/\mathfrak{p})$ is the decomposition group Z of $\mathfrak{p}$, the inertia group T of $\mathfrak{p}$, or the ramification group V_ρ of $\mathfrak{p}$, respectively (Hasse, S. Iyanaga, 1933) (→ 257 Local Fields F).

R. Hilbert Norm-Residue Symbol

The symbol which was first introduced by Hilbert for quadratic fields can be defined in a general algebraic number field k containing an nth primitive root ζ_n of unity. Let α, $\beta\in k$ $(\alpha\neq 0,\ \beta\neq 0)$, and let $\mathfrak{p}$ be a prime divisor. Then the following nth root of unity $((\alpha,\beta)/\mathfrak{p})_n$ is defined by using the norm-residue symbol:

$$\left(\frac{\alpha,\beta}{\mathfrak{p}}\right)_n\sqrt[n]{\beta}=\left(\sqrt[n]{\beta}\right)^{\sigma},$$

$$\sigma=\left(\frac{\alpha, k(\sqrt[n]{\beta})/k}{\mathfrak{p}}\right).$$

This symbol $((\alpha,\beta)/\mathfrak{p})_n$ is called the **Hilbert norm-residue symbol**. For α, $\beta,\ldots\in k$, we have

(1) $\left(\frac{\alpha\alpha',\beta}{\mathfrak{p}}\right)_n=\left(\frac{\alpha,\beta}{\mathfrak{p}}\right)_n\left(\frac{\alpha',\beta}{\mathfrak{p}}\right)_n$;

(2) the **law of symmetry**:

$$\left(\frac{\alpha,\beta}{\mathfrak{p}}\right)_n=\left(\frac{\beta,\alpha}{\mathfrak{p}}\right)_n^{-1};$$

and (3) the **product formula** for the Hilbert norm-residue symbol:

$$\prod_{\mathfrak{p}}\left(\frac{\alpha,\beta}{\mathfrak{p}}\right)_n=1$$

(Hilbert, Furtwängler, Takagi, Artin, Hasse). For detailed properties concerning the norm-residue symbol, power-residue symbol, and Hilbert norm-residue symbol and for references for them, see Hasse [8].

S. Density Theorem

Let M be a set of prime ideals of k. If

$$\lim_{s\to 1+0}\sum_{\mathfrak{p}\in M}\frac{1}{(N(\mathfrak{p}))^s}\bigg/\log\frac{1}{s-1}$$

exists, its value is said to be the **density** of M. The density of the set of all prime ideals of k is 1. Let H be an ideal group modulo an integral divisor $\mathfrak{m}$. Then the density of the set of all prime ideals contained in each coset of $\mathfrak{J}(\mathfrak{m})$ modulo H is $1/(\mathfrak{J}(\mathfrak{m}):H)$. In particular, let H be the ray $S(\mathfrak{m})$. Then this result implies that each coset of $\mathfrak{J}(\mathfrak{m})$ modulo $S(\mathfrak{m})$ contains infinitely many prime ideals (a generalization to algebraic number fields of the †prime number theorem for arithmetic progression).

Let K/k be a Galois extension, C be a conjugate class of the Galois group G of K/k, and $M(C)$ be the set of all prime ideals $\mathfrak{p}$ of k such that the †Frobenius automorphism of each prime factor $\mathfrak{P}_i$ of $\mathfrak{p}$ in K belongs to C. Then the density of $M(C)$ is $|C|/|G|$ (**Čebotarev's density theorem**, *Math. Ann.*, 95 (1926)).

Each element σ of the Galois group G of K/k can be expressed by the permutation z of the conjugate fields $K^{(1)},\ldots,K^{(n)}$ of K over k. Let z be expressed as the product of r †cycles of length $f_1,\ldots,f_r$. Hence $n=f_1+\ldots+f_r$. Let $C(f_1,\ldots,f_r)$ be the set of all such z in G, and let $M(f_1,\ldots,f_r)$ be the set of all prime ideals $\mathfrak{p}$ of k such that $\mathfrak{p}$ is decomposed in K/k as the product of r prime ideals of K with relative degree $f_1,\ldots,f_r$. Then the density of $M(f_1,\ldots,f_r)$ is $|C(f_1,\ldots,f_r)|/|G|$ (Artin, *Math. Ann.*, 89 (1923)).

T. Relation with the Arithmetic of Local Fields

It is quite useful to investigate the relation between the arithmetic of algebraic number fields and that of local fields. For example, let a prime ideal $\mathfrak{p}$ of an algebraic number field k be decomposed as $E(\mathfrak{p})=\mathfrak{P}_1^{e_1}\ldots\mathfrak{P}_g^{e_g}$, $N_{K/k}(\mathfrak{P}_i)=\mathfrak{p}^{f_i}$ $(i=1,\ldots,g)$ in an extension K of k. Let $K_{\mathfrak{P}}$ and $k_{\mathfrak{p}}$ be the completion of K and k with respect to $\mathfrak{P}$-adic and $\mathfrak{p}$-adic valuations, respectively. Then we have $[K_{\mathfrak{P}_i}:k_{\mathfrak{p}}]=e_i f_i$ and $K\otimes_k k_{\mathfrak{p}}\cong K_{\mathfrak{P}_1}+\ldots+K_{\mathfrak{P}_g}$ (direct sum). The relative different $\mathfrak{D}_{K/k}$ is expressed as (the $\mathfrak{p}$-component of $\mathfrak{D}_{K/k})=\Pi_{i=1}\mathfrak{D}_{K_{\mathfrak{P}_i}/k_{\mathfrak{p}}}$. For a Galois extension K/k the $\mathfrak{p}$-conductor $\mathfrak{f}_{\mathfrak{p}}=\mathfrak{p}^c$ for the norm-residue and the conductor $\mathfrak{p}^c$ of local extension $K_{\mathfrak{P}}/k_{\mathfrak{p}}$ have the same exponent c. For a local field $K_{\mathfrak{P}}/k_{\mathfrak{p}}$, each norm-residue modulo $\mathfrak{f}_{\mathfrak{p}}$ is a norm of an element of $K_{\mathfrak{P}}$, and precise results concerning the norm-residue are obtained. Hence, these results can be applied immediately to a global field K/k (→ 257 Local Fields).

We can also apply the method of the idele group of an algebraic number field k, and thus we can prove results concerning the ideal class group, unit group, and zeta function of k (→ 7 Adeles and Ideles).

U. History of the Arithmetic of Algebraic Number Fields

It was C. F. Gauss (1832) who first generalized the notion of integers to algebraic number fields in considering the elements of $\mathbf{Z}[\sqrt{-1}\,]$ now called **Gaussian integers** ($\mathbf{Z}[\sqrt{-1}\,]$ is the principal order of $\mathbf{Q}(\sqrt{-1}\,)$). After investigations by G. L. Dirichlet and Kummer, the notion of ideals was introduced by Dedekind (1871) [4]. L. Kronecker gave another foundation for arithmetic of algebraic number fields (1882) [5]. Dirichlet proved the unit theorem and, introducing the analytic method to number theory, gave the class number formula of quadratic fields (→ 343 Quadratic Fields). H. Minkowski first applied the theory of lattice points to number theory (→ 187 Geometry of Numbers), and K. Hensel introduced the $\mathfrak{p}$-adic method (→ 257 Local Fields). Hilbert (1897) [7] and Hasse (1926, 1927, 1930) [8] summarized the main results on the arithmetic of algebraic number fields known at that time. In particular, Hilbert's report was centered around the arithmetic of Galois extensions, and Hasse's around the class field theory obtained by T. Takagi and E. Artin (→ 62 Class Field Theory). Since c. 1950, the notions of ideles and adeles were introduced and cohomology-theoretic methods were applied to number theory with success (→ 7 Adeles and Ideles).

References

[1] T. Takagi, Daisûteki seisûron (Japanese; Algebraic theory of numbers), Iwanami, 1948.
[2] S. Kuroda and T. Kubota, Seisûron (Japanese;Theory of numbers), Asakura, 1963.
[3] K. F. Gauss, Theoria residuorum biquadraticorum, I (1825), II (1831), Werke 2, Göttingen, 1863, p. 65–148.
[4] R. Dedekind, Dirichlet's Vorlesungen über Zahlentheorie, fourth edition, 1894, Supplement XI (Gesammelte mathematische Werke 3, Braunschweig, 1932) (Chelsea, 1969).
[5] L. Kronecker, Grundzüge einer arithmetischen Theorie der algebraischen Grösse, J. Reine Angew. Math., 92 (1882), 1–122 (Werke 2, Teubner, 1897, p. 237–387) (Chelsea, 1968).
[6] D. Hilbert, Die Theorie der algebraischen Zahlkörper, Jber. Deutsch. Math. Verein. 4, (1897), 175–546, (Gesammelte Abhandlungen, I, Springer, 1932, p. 63–363) (Chelsea, 1967).
[7] E. Hecke, Vorlesungen über die Theorie der algebraischen Zahlen, Akademische Verlag, 1923 (Chelsea, 1970).
[8] H. Hasse, Bericht über neuere Untersuchungen und Probleme aus der Theorie der algebraischen Zahlkörper, Jber. Deutsch. Math. Verein., I, 35 (1926), 1–55; Ia, 36 (1927), 233–311; II (1930) (Physica Verlag, 1965).
[9] H. Weyl, Algebraic theory of numbers, Ann. Math. Studies, Princeton, 1940.
[10] H. Hasse, Zahlentheorie, Akademie-Verlag, second edition, 1963.
[11] E. Weiss, Algebraic number theory, McGraw-Hill, 1963.
[12] S. Lang, Algebraic numbers, Addison-Wesley, 1964.
[13] E. Artin, Algebraic numbers and algebraic function, Lecture notes, Princeton, 1950–1951 (Gordon and Breach, 1967).
[14] S. Iyanaga (ed.), Sûron (Japanese; Number theory), Iwanami, 1969; English translation, Theory of numbers, North-Holland, 1975.
[15] A. Weil, Basic number theory, Springer, 1967.
[16] J. W. S. Cassels and A. Fröhlich (eds.), Algebraic number theory, Academic Press, 1967.

17 (VIII.3) Algebraic Surfaces

A. General Remark

An †algebraic variety of dimension 2 is called an algebraic surface.

B. History

The history of algebraic surfaces originated

with the study of algebraic functions of two variables. In the case of algebraic functions of one variable, the introduction of †Riemann surfaces attached to such functions played an essential role in the development of the theory. The study of algebraic functions of two variables led naturally to the consideration of the surfaces defined by a suitable polynomial equation. H. Poincaré and E. Picard are among those who studied the homological structure of the surface defined by the equation $P(x,y,z)=0$. The theory of †Abelian integrals (Picard integrals) is one of the consequences of such topological investigations. S. Lefschetz obtained further results in this direction.

M. Noether and Italian geometers such as F. Enriques, G. Castelnuovo, and F. Severi studied algebrogeometric properties of algebraic surfaces. In particular, Italian geometers recognized the importance of irregularity and deeply investigated its geometric meaning. In the early 20th century they succeeded in constructing the great edifice of the theory of algebraic surfaces. Though some of their results lack rigorous proofs, efforts to erect a foundation for those results have led to the recent developments in algebraic geometry. A significant contribution to the modernization of the theory was made by O. Zariski and K. Kodaira.

The resolution of singularities of an algebraic surface is one of the most fundamental problems in the field. When the †universal domain is the complex number field, function-theoretic methods were used by Italian geometers and R. J. Walker (*Ann. of Math.*, 36 (1935)). Zariski introduced the †valuation-theoretic method to deal with the problem when the characteristic of the universal domain is zero. S. Abhyankar (1966) succeeded in resolving the case of positive characteristics. In this article, every algebraic surface is assumed to be nonsingular and situated in a projective space. The †universal domain will be denoted by **K**, and the †function field of an algebraic surface F will be denoted by $\mathbf{K}(F)$.

C. Divisors and Linear Systems

Let F be a nonsingular algebraic surface, Σ be a †linear system of divisors on F, and f_0, $f_1, \ldots, f_n$ be a basis of the †defining module for Σ over K. Let $\varphi(\Sigma)$ be the mapping of F into a projective n-space that sends a general point P on F to a point $Q=(f_0(P), f_1(P), \ldots, f_n(P))$. $\varphi(\Sigma)$ is a †rational mapping of F into the projective space. Let H be a hyperplane in the ambient space of $\varphi(\Sigma)(F)=F'$. Then the †variable component of Σ is given by $\varphi(\Sigma)^{-1}(H)$. If $\dim F'=2$, the generic member of the linear system (except the fixed component) is irreducible. If $\dim F'=1$, the generic member is a composite of members belonging to an irreducible †algebraic system of dimension 1. This is one of **Bertini's theorems**. When $\dim \Sigma \geqslant 1$, the degree of variable points of the intersection $C \cdot C'$ of two variable components C and C' is called the **effective degree** of Σ. Let C, C' be two arbitrary divisors on F. The **Kronecker index** $I(C \cdot C')$ is defined as the †degree of the intersection $D \cdot D'$, where D, D' are divisors such that D and D' are †linearly equivalent to C and C', respectively, and $D \cdot D'$ is well defined. $I(C \cdot C) = (C^2)$ is called the **virtual degree** of C. If D is a †nondegenerate (or ample) divisor of F, we have $(D^2)>0$ and $I(D \cdot E)>0$ for any positive divisor E. These properties characterize a nondegenerate divisor [14]. Let Σ be an irreducible linear system of dimension r $(\geqslant 1)$, and let C be a generic component of Σ. Let C' be a member of Σ different from C. Then the set of C-divisors $C \cdot C'$ forms a linear system of dimension $r-1$ on C. This is called the **trace** of Σ on C and will be denoted by $\mathrm{Tr}_C \Sigma$. The trace is, in general, not complete. The integer $\dim |\mathrm{Tr}_C \Sigma| - \dim \mathrm{Tr}_C \Sigma = \delta(\Sigma)$ is called the **deficiency** of Σ. The deficiency of the complete linear system $|D|$ is denoted by $\delta(D)$.

Let x be a point of F, and let $\mathcal{O}_x$ be the local ring of x. Then $\mathcal{O}_F = \bigcup_{x \in F} \mathcal{O}_x$ is an †algebraic coherent sheaf, called the †structure sheaf of F. Let D be a divisor on F. The sheaf of germs of rational functions f such that $(f)+D \succ 0$ is denoted by $\mathcal{O}_F(D)$. Then $H^0(F, \mathcal{O}_F(D))$ is a defining module for the complete linear system $|D|$. $D \succ 0$ if and only if $\mathcal{O}_F(-D)$ is a sheaf of $\mathcal{O}_F$-ideals. The quotient sheaf $\mathcal{O}_F / \mathcal{O}_F(-D)$ will be denoted by $\mathcal{O}_D$. If D is a †prime divisor, then $\mathcal{O}_D$ is the structure sheaf of the algebraic curve D (→ 11 Algebraic Curves). Let $\mathcal{F}$ be a sheaf on F. We set

$$\chi(F, \mathcal{F}) = \Sigma_{q=0}^{2}(-1)^q \dim H^q(F, \mathcal{F}).$$

$\chi(F, \mathcal{O}_F)$ will be denoted simply by $\chi(F)$. We call $p_a(F)=\chi(F)-1$ the **arithmetic genus** of the algebraic surface F. Sometimes $\chi(F)$ will be referred to as the arithmetic genus of F. We set $\chi_F(D)=\chi(F)-\chi(F, \mathcal{O}_F(-D))$. The integer $p_a(D)=1-\chi_F(D)$ is, by definition, the **arithmetic genus of the divisor** D. If D is a prime divisor, then $p_a(D)$ coincides with the arithmetic genus of the algebraic curve D (→ 11 Algebraic Curves). We have the modular property $\chi(D_1+D_2)+I(D_1 \cdot D_2) = \chi(D_1)+\chi(D_2)$. In particular, we have $p_a(-D) = (D^2)+2-p_a(D)$.

D. The Riemann-Roch Theorem

The set of †differential forms of degree 2 on an algebraic surface F is a 1-dimensional vector space over $\mathbf{K}(F)$. Hence the divisors of differential forms of degree 2 are mutually linearly equivalent. This class is called the **canonical class**, and its divisor is called a **canonical divisor**. In this article we shall denote a canonical divisor by K. Let D be a divisor. Then $\Omega^i(D)$ denotes the sheaf of germs of differential forms ω of degree i such that $(\omega) \succ -D$ $(i=0,1,2)$ $(\Omega^2(0)=\Omega^2, \Omega^0=\mathcal{O}_F)$. †Serre's duality theorem asserts that

$$H^p(F,\Omega^q(D)) \cong H^{2-p}(F,\Omega^{2-q}(-D)).$$

Since $\Omega^2(-D)$ is canonically isomorphic to $\mathcal{O}_F(K-D)$, we have

$$H^p(F,\mathcal{O}_F(D)) \cong H^{2-p}(F,\mathcal{O}_F(K-D)).$$

These isomorphisms and $\chi_F(-D)=\chi(F)-\chi(F,\mathcal{O}_F(D))$ immediately yield the equality

$$l(D)+l(K-D)=p_a(F)+p_a(-D)+h^1(D),$$

where $h^1(D)=\dim H^1(F,\mathcal{O}_F(D))$ is called the **superabundance** of D. We call $l(K-D)=i(D)$ the speciality index. The inequality

$$\dim|D| \geqslant (D^2)-p_a(D)+p_a(F)+1-i(D)$$

is called the **Riemann-Roch theorem**. $|K+D|$ is called the **adjoint system** of $|D|$. We have $p_a(D)-1=I(D\cdot(K+D))/2$. This last equality holds for any divisor D. Thus the **Riemann-Roch inequality** can also be written in the form

$$\dim|D| \geqslant p_a(F)+I(D\cdot(D-K))/2-i(D).$$

There are other formulations due to Kodaira, F. Hirzebruch, and others. Hirzebruch expressed $\chi(F,\mathcal{O}_F(D))$ as a polynomial of the †cohomology class of D and †Chern classes of F (→ 361 Riemann-Roch Theorems). The formula of M. Noether, $12(p_a+1)=(K^2)+c_2$, is a special case of the Riemann-Roch-Hirzebruch theorem, where c_2 is the Euler number of F that represents the second Chern class of F.

E. Invariants of Algebraic Surfaces

There are many invariants besides arithmetic genus, discussed earlier. We set $h^{q,p}=\dim_K H^p(F,\Omega^q)$. Then $h^{2,0}$ is equal to the number of linearly independent holomorphic 2-forms; it is called the **geometric genus** and is usually denoted by p_g. Since $h^{0,1}$ gives the maximum among the deficiencies $\delta(\Sigma)$ of linear systems on F, it is called the **maximal deficiency** of F. For a divisor C such that $h^1(C)=0$, we have $\delta(C)=h^{0,1}$. The number $h^{0,1}$ was formerly called the **irregularity** of F, because $h^{0,1}$ was considered a correction term in the equality $p_a(F)=p_g-h^{0,1}$. The study of higher-dimensional varieties showed, however, that it was unnatural to regard $h^{0,1}$ as a correction term. At present, by **irregularity** we mean the dimension of the Picard variety of F (→ 18 Algebraic Varieties), and we denote this number by q.

When F is defined over the complex number field, we have $h^{p,q}=h^{q,p}$. In particular, $q=h^{0,1}=h^{1,0}$. This number is equal to the number of linearly independent †Abelian simple integrals of the first kind; it is also equal to one-half the first Betti number of F. In cases with positive characteristic, these equalities do not hold in general. J.-P. Serre gave an example of an algebraic surface F such that $h^{0,1} \neq h^{1,0}$, and J. Igusa gave an example such that $q<h^{0,1}=h^{1,0}$. Let K be a canonical divisor of F. The number $P_i=l(iK)$ is called the ***i*-genus**, and P_i $(i=2,3,\ldots)$ are generally called **plurigenera**. If $P_n=0$ and $d|n$, then P_d is also zero. The numbers $p_a(F)=h^{2,0}-h^{0,1}$, $p_g(F)=h^{2,0}$, $h^{1,0}$, $h^{0,1}$, P_i $(i=2,3,\ldots)$ $(P_1=p_g)$ are †absolute invariants of F; i.e., they take the same values for any nonsingular surface F' that is birationally equivalent to F. However, $h^{1,1}$ is not an absolute invariant. Plurigenera $\{P_i\}$ play important roles in regard to the surface with $p_g=0$. For instance, a **rational surface** F has the properties $q=0$, $P_n=0$ $(n>0)$. Conversely, any nonsingular surface with $q=P_2=0$ is a rational surface (Castelnuovo-Kodaira-Zariski). A **ruled surface** F has the property $P_n=0$ $(n>0)$, and $P_{12}=0$ (hence $P_2=P_4=P_6=0$) characterizes a ruled surface (Enriques-Kodaira-Šafarevič). As an application of the characterization of rational surfaces, Zariski gave an algebrogeometric proof for **Castelnuovo's theorem**: Let k be an algebraically closed field, and let L be a subfield of a rational function field $k(x,y)$ of two variables over k such that $L \supset k$ and $\dim_k L=2$. Then if $k(x,y)$ is separable over L, L is also a rational function field over k.

Another kind of invariant can be seen in $p^{(1)}$, which is called a **linear genus**. If F has a †minimal model F_0, then $p^{(1)}$ is equal to the arithmetic genus of a canonical divisor of F_0; i.e., $p^{(1)}=(K^2)+1$.

F. Characteristic Linear Systems of Algebraic Families

One of the central problems considered by the Italian school was to prove that the irregularity q is equal to the maximal deficiency $h^{0,1}$. For that purpose, Severi introduced the notion of characteristic linear systems of algebraic families. Let Σ be an irreducible algebraic family of positive divisors on F such that a generic member C of Σ is an irreducible nonsingular curve, and let r be the dimension of Σ. Let Σ_1 be a 1-dimensional subfamily of Σ containing C as a simple member, and let C' be a generic member of Σ_1. Then the specialization of $C' \cdot C$ over the specialization $C' \to C$ is a well-defined C-divisor of degree $n = I(C \cdot C')$. The set of C-divisors thus obtained is called the **characteristic set**. The characteristic set forms an $(r-1)$-dimensional linear system and contains $\mathrm{Tr}_C|C|$ as a subfamily. This linear system is called the **characteristic linear system** of Σ. For any algebraic family of dimension r, we have $r \leqslant \dim|C| + q$. In particular, there exists an algebraic family Σ that contains ∞^q linear systems and such that for a generic curve C we have $h^1(C) = 0$. For such an algebraic family, we have the equality $r = \dim|C| + q$; hence the inequality $q \leqslant h^{0,1}$ follows. Moreover, if the characteristic linear system is complete, we have $q = h^{0,1}$. The proof of the completeness of characteristic linear systems given by Severi is valid only in some special cases (e.g., the case $p_g = 0$). For a complex algebraic manifold, a rigorous proof was later given [11]. When the characteristic is positive, the completeness does not hold in general (Igusa); however, for the surface with $p_g = 0$, the completeness holds (Y. Nakai). According to a recent investigation by D. Mumford, the completeness holds if and only if the Picard scheme of F is reduced [8].

G. Birational Transformations of Algebraic Surfaces

Let $F, F', \ldots$ be mutually equivalent nonsingular surfaces, and let $T: F \to F'$ be a birational transformation. If T is everywhere regular on F, we say that F **dominates** F' and denote this by $F \geqslant F'$. Moreover, if T is not biregular, we write $F > F'$, and we call T^{-1} an **antiregular transformation**. If $F > F'$, we have $h^{1,1}(F) > h^{1,1}(F')$. If there is no F' such that $F > F'$, we say that F is a **relatively minimal model**. If we always have $F' \geqslant F$ for every birationally equivalent nonsingular surface F', F is called a **minimal model**. For any nonsingular surface, there always exists a relatively minimal model, but a minimal model does not always exist. A necessary and sufficient condition for a surface F to have a minimal model is that F is not a ruled surface (Castelnuovo-Enriques-Zariski).

Let F be an algebraic surface and P be a point of F. The linear system composed of intersections of F with a quadratic hypersurface passing through P defines a **locally quadratic transformation** of F. This is an antiregular transformation with P as a unique fundamental point. If we denote by E the †total transform of P, then we have $(E^2) = -1$ and $p_a(E) = 0$. Any antiregular transformation is represented as a composite of a finite number of locally quadratic transformations. In the birational transformation $T: F \to F'$, the total transform E of a simple point $P' \in F'$ by T^{-1} is called an **exceptional divisor (curve)**. Moreover, if T is regular along E, it is said to be **of the first kind**, otherwise **of the second kind**. The surface F is a relatively minimal model if and only if F has no exceptional curve of the first kind; it is a minimal model if and only if it has no exceptional curve at all.

H. Examples of Algebraic Surfaces

An algebraic surface in the projective space $\mathbf{P}^3$ defined by a homogeneous polynomial of degree m is called an algebraic surface of **order** m. For a nonsingular surface of order m in $\mathbf{P}^3$, we have $p_a = p_g = (m-1)(m-2)(m-3)/6$. Quadratic and cubic surfaces are rational surfaces. A quadratic surface is biregularly equivalent to the product $\mathbf{P}^1 \times \mathbf{P}^1$ of two projective straight lines. There are 27 straight lines on a cubic surface, and they are all exceptional curves of the first kind. Suppose we are given 6 points in general position on a projective plane $\mathbf{P}^2$ and consider the linear system Σ consisting of cubic curves passing through these 6 points. Then $\dim \Sigma = 3$, and $\varphi(\Sigma)$ $(\mathbf{P}^2)$ is a cubic surface in $\mathbf{P}^3$. Altogether the transforms of 6 points, †proper transforms of 15 lines connecting every two points, and proper transforms of 6 conics passing through every 5 points yield 27 lines. Every nonsingular surface of order $\geqslant 4$ in $\mathbf{P}^3$ has a positive geometric genus. They are neither rational nor ruled, and all are minimal models. The canonical class of a quartic surface is zero, and every plurigenus P_i $(i = 1, 2, \ldots)$ is equal to 1. A quartic surface has very interesting properties. For example, the automorphism group of a quartic surface can be an infinite discrete group. A compact complex analytic surface is called a $K3$ **surface** if its first Betti number and its first †Chern class c_1 vanish. Every $K3$ surface is a deformation of a quartic surface. Thus an algebraic $K3$ surface is characterized by $q = 0, p_g = P_2 = p^{(1)} = 1$. The set of $K3$ surfaces forms a family of 20 dimensions, and algebraic $K3$ surfaces are divided into countably many 19-dimensional families.

A quartic surface in $\mathbf{P}^3$ has at most 16 double points. A quartic surface with 16 double points is called a **Kummer surface**, which is the quotient variety of an †Abelian variety of dimension 2 by the automorphism $u \to -u$ of order 2. A minimal nonsingular surface that is birationally equivalent to a Kummer surface is a $K3$ surface.

Among the many other topics dealing with algebraic surfaces are classification of surfaces by birational transformations, classification of relatively minimal models of rational surfaces or ruled surfaces, series of equivalence, and the theory of †moduli.

References

[1] E. Picard and G. Simart, Théorie des fonctions algébriques de deux variables indépendentes, I, II, Gauthier-Villars, 1897, 1906.
[2] E. Pascal, Repetorium der höheren Mathematik, II_2, Teubner, second edition, 1922.
[3] H. W. E. Jung, Algebraische Flächen, Helwing, 1925.
[4] H. F. Baker, Principles of geometry VI. Introduction to the theory of algebraic surfaces and higher loci, Cambridge Univ. Press, 1933.
[5] J. G. Semple and L. Roth, Introduction to algebraic geometry, Clarendon Press, 1949.
[6] F. Enriques, Le superficie algebriche, Zanichelli, 1949.
[7] M. Baldassarri, Algebraic varieties, Erg. Math., Springer, 1956.
[8] D. Mumford, Lectures on curves on an algebraic surface, Ann. Math. Studies, Princeton Univ. Press, 1966.
[9] И. Р. Шафаревич (I. R. Šafarevič), Алгебраические поверхности, Trudy Mat. Inst. Steklov., 75 (1965); English translation, Algebraic surfaces, Amer. Math. Soc., 1967.
[10] O. Zariski, Algebraic surfaces, Erg. Math., Springer, second edition, 1971.
[11] K. Kodaira and D. C. Spencer, A theorem on completeness of characteristic systems of complete continuous systems, Amer. J. Math., 81 (1959), 477–500.
[12] K. Kodaira, On compact complex analytic surfaces I, II, III, Ann. of Math., (2) 71 (1960), 111–152; (2) 77 (1963), 563–626; (2) 78 (1963), 1–40. (Collected works, III, Iwanami and Princeton Univ. Press, 1975.)
[13] M. Nagata, On rational surfaces I, II, Mem. Coll. Sci. Univ. Kyôto (A), 32 (1960), 351–370; 33 (1961), 271–293.
[14] Y. Nakai, Non-degenerate divisors on an algebraic surface, J. Sci. Hiroshima Univ., (A), 24 (1960), 1–6.
[15] O. Zariski, Introduction to the problem of minimal models in the theory of algebraic surfaces, Publ. Math. Soc. Japan, 1958.

18 (VIII.4) Algebraic Varieties

A. Affine Algebraic Varieties and Projective Algebraic Varieties

Fix a field k. A subset of the n-dimensional †affine space k^n over k is called an **affine algebraic variety** (or simply **affine variety**) if it can be expressed as the set of the common zeros of a (finite or infinite) set of polynomials $F_i(X_1, \ldots, X_n)$ with coefficients in k. Similarly, a subset of the n-dimensional †projective space $\mathbf{P}^n(k)$ over k that can be expressed as the set of the common zeros of a set of homogeneous polynomials $G_j(Y_0, \ldots, Y_n)$ is called a **projective algebraic variety** (or simply **projective variety**). In this section, **variety** will mean either an affine or projective variety (the meaning of **subvariety** is obvious). These varieties are the forerunners of the modern, more general versions of algebraic varieties, which we will discuss later.

When V is an affine variety in k^n, the set of the polynomials in $k[X] = k[X_1, \ldots, X_n]$ that vanish at every point of V form an ideal $I(V)$ of $k[X]$. The residue class ring $A_V = k[X]/I(V)$ is called the **coordinate ring** (or **affine ring**) of V. We can regard A_V as the ring of k-valued functions on V that can be expressed as polynomials of the coordinates of k^n. When V is a projective variety, the †homogeneous ideal generated by the homogeneous polynomials in $k[Y] = k[Y_0, \ldots, Y_n]$ that vanish at every point of V is denoted by $I(V)$, and the ring $A_V = k[Y]/I(V)$ is called the **homogeneous coordinate ring** of V.

A variety V is said to be **reducible** or **irreducible** according as it is the union of two proper subvarieties or not. A maximal irreducible subvariety of V is called an **irreducible component** of V. Any variety can be written uniquely as the union of a finite number of irreducible components. A variety V is irreducible if and only if $I(V)$ is a †prime ideal. When that is the case, the field of quotients of A_V (when V is affine) or the subfield of the field of quotients of A_V consisting of the homogeneous elements of degree 0 (when V is projective) is called the **function field** of V and is denoted by $k(V)$. Elements of $k(V)$ are called **rational functions** (or simply **functions) on the variety** V. The field $k(V)$ is †finitely generated over k. When $k(V)$ is a †purely transcendental extension of k, V is called a **rational algebraic variety** over k. The transcendence degree of $k(V)$ over k is called the **dimension** of V. When V is reducible, the maximum of the dimensions of its irreducible components is called its dimension. If W is a subvariety of an irreducible variety V, then $\dim V - \dim W$ is called the **codimension** of W on V. A subvariety of pure codimension 1 of an affine or projective space can be defined by a single equation and is called a **hypersurface**. If the ideal $I(V)$ of a variety V of dimension r in a projective space $\mathbf{P}^n(k)$ is generated by $n-r$ homogeneous polynomials, then V is called a **complete intersection**. Compared with general varieties, complete intersections have some simpler properties. On the other hand, many important varieties are not complete intersections, e.g., †Abelian varieties of dimension $\geqslant 2$.

The intersections and finite unions of subvarieties on a variety V are also subvarieties. Thus the subvarieties can be taken as the †system of closed sets of a topology on V (→ 409 Topology), which is called the **Zariski topology of the variety** V. When k is the field of complex numbers, V can be viewed as an †analytic space, and the topology of V as such (the "usual" topology) is much stronger than the Zariski topology. For the rest of this article, varieties will be considered as having Zariski topologies unless stated otherwise. Terms such as **Zariski open, Zariski closed**, and **Zariski dense** are used to mean open, closed, or dense in a Zariski topology. Suppose a condition (P) concerning the points of an irreducible variety V (concerning the elements of a set M parametrized by the points of V) is satisfied in a nonempty Zariski open set of V. Then we say that the condition (P) holds at **almost all points of the variety** V (almost all elements of the set M).

Let U and V be affine varieties in k^n and k^m, respectively. Then the product set $U \times V$ is an affine variety in k^{n+m} and is called the **product algebraic variety** (or simply the **product**) of U and V. Note that the Zariski topology on $U \times V$ is stronger than the product of the topologies of U and V. When k is algebraically closed, then $U \times V$ is irreducible if U and V are irreducible.

Suppose that k is †algebraically closed. Let $\mathfrak{P}$ be a †prime ideal of $k[X]=k[X_1,\ldots,X_n]$, and let V be the affine variety in k^n defined as the zero points of $\mathfrak{P}$. Then $I(V)=\mathfrak{P}$ (†Hilbert zero point theorem) (→ 364 Rings of Polynomials). Therefore, there exists a one-to-one correspondence between the set of prime ideals of $k[X]$ and the set of irreducible varieties in k^n. In particular, the †maximal ideals correspond to the points of k^n. Similarly, there exists a one-to-one correspondence between the set of homogeneous prime ideals of $k[Y]$ other than $\Sigma_{i=0}^n Y_i k[Y]$ and the set of irreducible subvarieties in $\mathbf{P}^n(k)$.

When we deal with nonlinear algebraic equations, we cannot expect a simple, clear-cut theory without assuming that k is algebraically closed. Hence we take an algebraically closed field K containing k and regard a variety V in k^n as a subset of the variety V_K in K^n defined by the same equations. From now on, we suppose that k is algebraically closed. If the ideal $I(V)$ of $k[X]$ or $k[Y]$ determined by a variety V is generated by polynomials with coefficients in a subfield k' of k, we say that V is **defined over** k' or that k' is a **field of definition** for V. Any variety has the smallest field of definition, which is finitely generated over the prime field. In the theory of A. Weil, we fix an algebraically closed field K that has an infinite transcendence degree over the prime field. This K is called the **universal domain**. A point of V is called a **k'-rational point** of V if all of its coordinates belong to a subfield k' of K.

B. Generic Points and Specializations

Let K_1, K_2 be two extension fields of a field L, and let $(x)\in K_1^n$, $(y)\in K_2^n$. We say (y) is a **specialization** of the point (x) over L (notation: $(x)\underset{L}{\rightarrow}(y)$) if all polynomials $f(X)\in L[X_1,\ldots,X_n]$ satisfying $f(x)=0$ also satisfy $f(y)=0$; in other words, if there exists a homomorphism of L-algebras $L[x_1,\ldots,x_n]\rightarrow L[y_1,\ldots,y_n]$ mapping x_i to y_i. Let K be the universal domain, V an irreducible variety in K^n, and k' $(\subset K)$ a field of definition for V having a finite transcendence degree over the prime field. Then there exists a point (x) of V such that all points of V are specializations of (x) over k'. Such a point (x) (in general not uniquely determined) is called a **generic point** of V and k'. The ring $k'[x]$ is isomorphic to $k'[X]/I(V)\cap k'[X]$ over k'. (Some authors use the term *generic point* to mean *almost all points* as defined earlier.)

C. Local Rings

Let V be an affine variety and let W be an irreducible subvariety of V. Let $\mathfrak{P}_W$ be the subset of A_V consisting of the elements that vanish identically on W. Then $\mathfrak{P}_W$ is a prime ideal of A_V. The ring of quotients of A_V with respect to $\mathfrak{P}_W$ is denoted by $\mathfrak{O}_{V,W}$ or by $\mathfrak{O}_W$ and is called the **local ring** of W on V (or of V at W). Suppose for simplicity that V is irreducible. Then $\mathfrak{O}_W$ is the subring $\{f/g \mid f,g\in A_V, g\notin\mathfrak{P}_W\}$ of $k(V)$, and the †residue field of $\mathfrak{O}_W$ modulo the maximal ideal $\mathfrak{P}_W\mathfrak{O}_W$ can be identified with $k(W)$. When a function φ on V $(\varphi\in k(V))$ belongs to $\mathfrak{O}_W$, it is said to be **regular** at W. For a given function $\varphi\in k(V)$, the set of the points of V where φ

is regular is Zariski open. In the case of a projective variety, the local ring $\mathfrak{O}_{V,W}$ is defined as the subring of the ring of quotients of A_V with respect to $\mathfrak{P}_W$, consisting of the homogeneous elements of degree 0.

A mapping from an open set U of a variety V to k that is regular at every point of U is called a **regular function** on U. The ring of the regular functions on U is denoted by A_U. By assigning A_U to each open set U, we can define a †sheaf of rings $\mathcal{O}_V$ on V, of which the †stalk $\mathcal{O}_{V,x}$ at a point $x \in V$ is the local ring $\mathfrak{O}_{V,x}$. The sheaf $\mathcal{O}_V$ is called the **sheaf of germs of regular functions** on V (or the **structure sheaf** of V) (→ 377 Sheaves).

D. General Definition

Consider a pair $(V, \mathcal{O})$ of a topological space V and a sheaf $\mathcal{O}$ of germs of mappings from V to k. If V has a finite open covering (U_i) such that each $(U_i, \mathcal{O} | U_i)$ is isomorphic to an affine variety V_i (in the sense that there exists a homeomorphism from U_i to V_i that transforms $\mathcal{O} | U_i$ to the structure sheaf of V_i), the pair $(V, \mathcal{O})$ is called a **prealgebraic variety** over k, and $\mathcal{O}$ is called its **structure sheaf**. Usually $(V, \mathcal{O})$ is denoted simply by V.

A **regular mapping** between prealgebraic varieties is defined as a continuous mapping $g: V \to V'$ satisfying $\varphi \circ g \in \mathcal{O}_{V,x}$ for any $x \in V$ and $\varphi \in \mathcal{O}_{V',g(x)}$. Furthermore, if g is a homeomorphism and g^{-1} is also regular, then g is called a **biregular mapping**. The Cartesian product $X \times Y$ of prealgebraic varieties X and Y is locally a product of affine varieties. Therefore, $X \times Y$ has the structure of a prealgebraic variety. A prealgebraic variety X is called an **algebraic variety** if the image of the diagonal mapping $X \to X \times X$ is closed in the Zariski topology of the product variety $X \times X$ ("separation condition"). (This definition is due to J.-P. Serre.) The separation condition corresponds to †Hausdorff's separation axiom. If W is a locally closed subset (i.e., the intersection of an open set and a closed set) of an algebraic variety V, then it becomes an algebraic variety in a natural manner (the germs of regular functions at $P \in W$ are taken to be the germs of functions induced on W by the functions in $\mathcal{O}_{V,P}$). Locally closed subvarieties of k^n or $\mathbf{P}^n(k)$ are called **quasi-affine** or **quasiprojective algebraic varieties**, respectively. Definitions of irreducibility and local rings for general algebraic varieties are given in the same manner as before. In this article, algebraic varieties will often be referred to simply as varieties.

The notion of an irreducible algebraic variety was developed from that of **abstract algebraic variety** (or simply **abstract variety**) defined by Weil. To define the latter, he took a finite number of irreducible affine varieties over a universal domain and patched up those varieties (or their open subsets) by means of biregular mappings [35].

E. Schemes

The set of prime ideals $(\neq (1))$ of a commutative ring A with unity element 1 is denoted by Spec(A) and is called the **spectrum** of A. For any subset $\mathfrak{a}$ of A, we denote by $V(\mathfrak{a})$ the set of the prime ideals containing $\mathfrak{a}$. We define a topology on Spec(A) in which the closed sets are $V(\mathfrak{a})$. This, again, is called the **Zariski topology** of Spec(A). For an element f of A, the open set $D(f) = \mathrm{Spec}(A) - V(f)$ is called an elementary open set. The elementary open sets form a base of open sets in the Zariski topology of Spec(A). The set of closed points is nothing but the set of maximal ideals of A. Assigning to each point $\mathfrak{P}$ of Spec(A) the †ring of quotients $A_{\mathfrak{P}}$, we obtain a sheaf of rings $\tilde{A}$ on Spec(A). We have the equality $\Gamma(D(f), \tilde{A}) = A_f$, where A_f is the †ring of quotients by the multiplicative system $\{f^n | n \geqslant 0\}$. In particular, we have $\Gamma(\mathrm{Spec}(A), \tilde{A}) = A$. Regarded as a †local-ringed space with $\tilde{A}$ as the structure sheaf, Spec(A) is called an **affine scheme**.

A local-ringed space X which is locally isomorphic to an affine scheme is called a **scheme**. A **morphism of schemes** is, by definition, a †morphism between them as local-ringed spaces. Thus, we obtain a †category whose objects are schemes. We denote it by (Sch). Giving a morphism $f: X \to \mathrm{Spec}(A)$ is equivalent to giving a ring homomorphism $\Gamma(f): A \to \Gamma(X, \mathcal{O}_X)$. Hence the category of affine schemes (which is a †full subcategory of (Sch)) is contravariantly equivalent to the category of commutative rings. If there is given a morphism of schemes $f: X \to S$, X is said to be an **S-scheme** or a **scheme over** S, and f is called the structure morphism and S the base scheme. For two S-schemes $f: X \to S$, $g: Y \to S$, a morphism of S-schemes is defined to be a morphism of schemes $h: X \to Y$ with $f = g \circ h$. Thus we obtain the category of S-schemes denoted by (Sch/S). Spec($\mathbf{Z}$) is the unique final object in (Sch), hence (Sch) is nothing but (Sch/Spec($\mathbf{Z}$)).

The fiber product always exists in (Sch). In fact in the case of affine S-schemes $X = \mathrm{Spec}(B)$ and $Y = \mathrm{Spec}(C)$ with $S = \mathrm{Spec}(A)$ we have $X \times_S Y = \mathrm{Spec}(B \otimes_A C)$, and in the general case we construct $X \times_S Y$ by patching together fiber products of affine schemes.

A morphism $f: X \to S$ is called **separated** if the image of the diagonal morphism $\Delta_{X/S}: X \to X \times_S X$ is closed. We say also that

X is separated over S or X is a **separated** S**-scheme**. A scheme X is said to be **separated** if it is separated over Spec($\mathbf{Z}$). All affine schemes are separated.

When K is a field, Spec(K) is a space having only one point and equipped with K as the stalk of the structure sheaf. For a point x of a scheme X, denote by $k(x)$ the residue field of $\mathcal{O}_{X,x}$. For $f \in \mathcal{O}_{X,x}$ we call the residue class of f in $k(x)$ the value of f at x, denoted by $f(x)$. We have a natural morphism i_x: Spec$(k(x)) \to X$ whose image is $\{x\}$. More generally, we call a morphism i of a spectrum Spec(K) of a field K to X a point of X with values in K. Such a point is determined by a point x in X and an embedding of $k(x)$ in K. A point of X with values in an algebraically closed field is called a **geometric point**. For a morphism $f: X \to S$ and a point s in S, the fiber product $X \times_S$ Spec$(k(s))$ is called the **fiber** of f over s and denoted by $f^{-1}(s)$. For a geometric point Spec$(K) \to S$, $X \times_S$ Spec(K) is called a **geometric fiber**.

A scheme X is called **reduced** if the local ring at each point of X has no †nilpotent elements. A scheme is said to be **irreducible** if its underlying topological space is not a union of two proper closed subsets. A scheme is called **integral** if it is reduced and irreducible. Every local ring of an integral scheme is an †integral domain. If a scheme X has an affine open covering $\{U_i = \text{Spec}(A_i)\}$ such that every A_i is a †Noetherian ring, X is said to be **locally Noetherian**. A locally Noetherian scheme is called **Noetherian** if its underlying topological space is †(quasi-)compact.

A morphism $f: X \to Y = \text{Spec}(A)$ is said to be **locally of finite type (of finite type)** if X has an open affine covering (a finite open affine covering) $\{U_i = \text{Spec}(A_i)\}$ such that each A_i is a finitely generated A-algebra. A general morphism $f: X \to Y$ is said to be **locally of finite type (of finite type)** if there is an open affine covering $\{V_i\}$ of Y such that every restriction of $f: f^{-1}(V_i) \to V_i$ is locally of finite type (of finite type). If $f: X \to Y$ is (locally) of finite type we say that X is (locally) of finite type over Y.

A scheme of finite type over a field K (i.e., over Spec(K)) is called an **algebraic scheme** over K. There is a †natural equivalence of categories between the category of reduced separated algebraic K-schemes (as a full subcategory of (Sch$/K$)) and the category of algebraic varieties over K (defined in Section D) equipped with regular mappings as morphisms. Hence we identify these categories from now on. Occasionally, algebraic variety means irreducible reduced algebraic scheme. Nonalgebraic schemes are also important as tools for the study of algebraic varieties. For example, for a point x in a scheme X there is a canonical monomorphism j_x : Spec$(\mathcal{O}_{X,x}) \to X$ by which the unique closed point of Spec$(\mathcal{O}_{X,x})$ is mapped to x. If for two algebraic K-schemes X, Y and for two points $x \in X, y \in Y$ there is a K-isomorphism $\mathcal{O}_{X,x} \cong \mathcal{O}_{Y,y}$, then suitable neighborhoods of x and y are isomorphic over K.

Many concepts concerning varieties, e.g., dimension, generic points, specialization, can be naturally extended to the case of schemes by virtue of commutative ring theory.

A morphism of schemes $f: X \to Y$ is called **proper** if it satisfies the following two conditions: (1) f is separated and of finite type, (2) for every scheme T and for every morphism $T \to Y$, the morphism $X \times_Y T \to T$ obtained from f by the "change of base" is a closed mapping. We also say that X is a proper Y-scheme or X is proper over Y. A proper algebraic K-scheme is called **complete**. A projective variety is complete, while an affine variety over K is complete only when it is of dimension zero. Every algebraic variety can be embedded in a complete variety (M. Nagata).

A morphism of schemes $f: X \to Y$ is called **affine** if every inverse image by f of an open affine subset of Y is again an affine scheme.

A morphism of schemes $f: X \to Y$ is called **finite** if it is of finite type and there is an affine open covering $\{U_i = \text{Spec}(A_i)\}$ of Y such that $f^{-1}(U_i) = \text{Spec}(B_i)$ where B_i is †integral over A_i. For a locally Noetherian scheme Y and a morphism of schemes $f: X \to Y$ the following three conditions are equivalent: (i) f is finite; (ii) f is affine and proper; (iii) f is proper and every fiber of f is a finite set. For a finite surjective morphism of Noetherian schemes $f: X \to Y$, X is an affine scheme if and only if Y is an affine scheme.

A morphism of schemes $f: X \to Y$ is said to be **flat** if for each point $x \in X$, $\mathcal{O}_{X,x}$ is a †flat $\mathcal{O}_{Y,f(x)}$-module. If, moreover, f is surjective, then f is called **faithfully flat**. Assume that $g: Y' \to Y$ is a faithfully flat morphism of finite type of locally Noetherian schemes and $f: X \to Y$ is a morphism of schemes. Then for many important properties of morphisms f has these properties if and only if the pullback $f_{Y'}: X \times_Y Y' \to Y'$ has the same properties (theory of descent).

F. Simple Points and Singular Points

Let V be a variety over an algebraically closed field k. We say that a point P of V is **simple** or that V is **nonsingular** or **smooth** at P if the local ring $\mathfrak{O}_P$ is a †regular local ring. Since the problem is local, we may assume that V

is an affine variety in k^n. Then the simplicity of P on V is equivalent to the following condition: P is contained in only one irreducible component of V, and if that component has dimension r there exist $n-r$ polynomials $F_i(X)$ in $I(V)$ such that rank $(\partial F_i/\partial X_j)_{(X)=P} = n-r$. A point of V that is not simple is called a **singular point** or a **multiple point**. The set of singular points on V (called the **singular locus** of V) is a proper closed subset of V. A variety with no multiple points is called **smooth** or **nonsingular**.

This notion can be made relative. A morphism $f: X \to Y$ of a locally Noetherian scheme is called **smooth** if f is flat and locally of finite type and all the geometric fibers of f are nonsingular. In the case of an affine morphism $f: X = \mathrm{Spec}(R[X_1, \ldots, X_{r+s}]/(f_1, \ldots, f_s)) \to Y = \mathrm{Spec}(R)$ of relative dimension r (by which we mean the dimension of the general fiber) with a Noetherian ring R, the smoothness of f amounts to a condition that rank $((\partial f_i/\partial X_j)(x)) = s$ at each point x of X.

When for a point P of a variety V the local ring $\mathfrak{O}_P$ is †normal, P is called a **normal** point. A simple point is normal. The set of normal points is a nonempty open subset of V. An irreducible variety whose points are all normal is called a **normal algebraic variety** (or simply **normal variety**). The singular locus of a normal variety has codimension $\geqslant 2$. For an irreducible variety V, there exists a pair (V', f) of a normal variety V' and a birational finite morphism $f: V' \to V$; V' is unique up to isomorphisms and is called the **derived normal model** or **normalization** of V.

Simplicity and normality for V at a subvariety W are defined in the same way as at a point by using the local ring $\mathfrak{O}_{V,W}$.

For a morphism $f: X \to Y$ of locally Noetherian schemes, locally of finite type, the following three conditions are equivalent: (i) f is smooth and every fiber of f is a discrete set; (ii) f is flat and every geometric fiber over $\mathrm{Spec}(K)$ of f is a union of spectra of fields isomorphic to K; (iii) f is flat and every fiber of f over $y \in Y$ is a union of spectra of fields that are finite †separable extensions of $k(y)$. These conditions are local with respect to X. If a morphism f satisfies these equivalent conditions, we say that f is **étale** or X is étale over Y. A morphism

$$f: X = \mathrm{Spec}(R[X_1, \ldots, X_n]/(f_1, \ldots, f_n)) \to Y = \mathrm{Spec}(R)$$

is étale if and only if $\det((\partial f_i/\partial X_j)(x)) \neq 0$ for all $x \in X$. Hence étale morphisms correspond to local isomorphisms in the analytic category. For a surjective étale morphism $f: X \to Y$ many important geometric properties (reduced, integral, normal, nonsingular, etc.) hold on X if and only if they hold on Y (theory of descent).

G. Dimension Theorems

Let V be an irreducible variety, and let U and W be irreducible subvarieties of V. Then any irreducible component of $U \cap W$ that is simple on V has dimension $\geqslant \dim U + \dim W - \dim V$. When the equality holds, the component is called a **proper component** of the intersection $U \cap W$. If every component of $U \cap W$ that is simple on V is proper, we say that U and W **properly intersect** on V. Any two subvarieties U and W of $\mathbf{P}^n(k)$ with $\dim U + \dim W \geqslant n$ intersect each other. When V is an irreducible r-dimensional variety in $\mathbf{P}^n(k)$, the number of points of intersection $V \cap L$ of V with an $(n-r)$-dimensional linear variety L is independent of the choice of L as long as L is in a "general position." This number is called the **degree** of V and is denoted by $\deg(V)$.

H. Group Varieties

An algebraic variety G is called an †algebraic group if it has a group structure and if the mapping $G \times G \to G$ sending (x,y) to xy^{-1} is a regular mapping. Every algebraic group is quasiprojective (Chow). If G is irreducible then it is also called a **group variety**; a complete group variety is called an †Abelian variety (→ 15 Algebraic Groups, 3 Abelian Varieties). A scheme G over another scheme S equipped with morphisms over S: $G \times_S G \to G$, $G \to G$, and $S \to G$, called multiplication, inverse, and unit section, respectively, which satisfy the relations corresponding to the usual axioms of group, is called a **group scheme** (over S). As a point set, G is not a group, while, for any scheme T over S the set $G(T) = \mathrm{Hom}_S(T, G)$ of the morphisms from T to G is a group (→ 53 Categories and Functors). Consider an algebraic group scheme G over $S = \mathrm{Spec}(k)$. If the characteristic of k is zero, then G is necessarily reduced, so an algebraic group scheme over k is essentially the same as an algebraic group; if k has characteristic p, there exist algebraic group schemes over k that are not reduced.

I. Rational Mappings

Let $f: V \to V'$ be a regular mapping of varieties. If V is not complete, the image $f(V)$ is not always closed; the closure of $f(V)$ (in V') is called the **closed image** of V. The image $f(V)$ contains an open dense subset of the closed image.

Let V and W be irreducible varieties. A closed subset T of $V \times W$ is called an **algebraic correspondence** of V and W. We say that points $P \in V$ and $Q \in W$ correspond to each other by T if $(P, Q) \in T$. If T is irreducible and the closed image of the projection $T \rightarrow V$ coincides with V, then the function field $k(V)$ can be identified with a subfield of $k(T)$; if we have $k(V) = k(T)$ with this identification, then T is called a **rational mapping** from V to W. Moreover, if the same conditions are satisfied for W, then T is called a **birational mapping** (of **birational correspondence** or **birational transformation**), and in this case we have $k(V) = k(W)$. If regular mappings are identified with their graphs T, they can be considered a special kind of rational mapping. If T is a rational mapping from V to W and W_1 is the closed image of T in W, then $k(W_1)$ can be regarded as a subfield of $k(T)$ $= k(V)$. If $k(V)$ is †separably generated (†purely inseparable) over $k(W_1)$, then T is said to be **separable (purely inseparable)**.

Let T be a rational mapping from V to W, and let V' and W' be irreducible subvarieties of V and W, respectively. If there exists an irreducible subvariety T' of T whose projections have the closed images V' and W', then we say V' and W' correspond to each other by T. The union of irreducible subvarieties of W that correspond to V' by T is a closed subset of W; it is called the **proper transform** of V' by T and is denoted by $T[V']$. Note that $V' \supset V''$ does not imply $T[V'] \supset T[V'']$. The set of points of W that correspond to the points of V' is called the **total transform** of V' by T and is denoted by $T\{V'\}$. Identifying $k(T)$ with $k(V)$, we have $\mathfrak{O}_{T,T'} \supset \mathfrak{O}_{V,V'}$ in general; if the equality holds, we say that T is **regular** (or **defined) along** V'. In that case, W' is the unique irreducible subvariety of W corresponding to V' by T. If V' is simple and of codimension 1 in V, then T is always regular along V'. The set U of points of V at which T is regular is a nonempty open subset, and the restriction of T to U defines a regular mapping from U to W. A rational mapping can be defined as the closure of the graph of a regular mapping defined on an open subset of V.

J. Birational Correspondences

The study of a rational mapping can be reduced to that of the inverse correspondence of a birational correspondence $T \rightarrow V$. **Zariski's main theorem**: let $S: X \rightarrow Y$ be a birational correspondence, and assume that the inverse correspondence $S^{-1}: Y \rightarrow X$ is regular, and that X is normal along an irreducible subvariety X'. If there exists an irreducible component Y' of $S[X']$ with $\dim X' \geqslant \dim Y'$, then S is regular along X'. It follows from the above main theorem that, if $T: V \rightarrow W$ is a rational mapping and if P is a normal point of V such that $T[P]$ contains an isolated point, then T is regular at P.

For a birational mapping $T: V \rightarrow W$ between complete irreducible varieties, a subvariety V' of V is said to be **fundamental** when $\dim T[V'] > \dim V'$. When V' is a point (curve) we say V' is a **fundamental point (fundamental curve)** with respect to T. The most classical example of a birational correspondence with fundamental points is the **quadratic transformation** T of a projective plane onto itself given by $(x_0 : x_1 : x_2) \rightarrow (x_1x_2 : x_2x_0 : x_0x_1)$. In this case we have $T^2 =$ identity, and the points $P_1 = (1:0:0)$, $P_2 = (0:1:0)$, and $P_3 = (0:0:1)$ are the fundamental points with respect to T; P_i corresponds to the line $x_i = 0$.

A birational correspondence from a projective plane (or generally $\mathbf{P}^n(k)$) to itself is called a **Cremona transformation.**

Let V be a complete nonsingular variety over k. It is called a **relatively minimal model** (over k) if every birational regular mapping $T: V \rightarrow V'$ of V to a complete nonsingular V' over k is an isomorphism. It is called **minimal (absolutely minimal)** if every birational (rational) mapping $T: V' \rightarrow V$ of nonsingular V' to V over k is regular. All complete nonsingular curves and Abelian varieties are absolutely minimal. For every algebraic surface S there is a relatively minimal model birationally equivalent to S (→ 17 Algebraic Surfaces).

Rational Varieties. An irreducible algebraic variety V over k whose function field is purely transcendental over k is called a †rational variety. A complete smooth surface S over an algebraically closed field is rational if and only if $P_2(S) = q(S) = 0$ (the Castelnuovo-Zariski criterion, → 17 Algebraic surfaces E). For higher-dimensional rational varieties we have no good criterion so far [75].

If the function field of V has a finite algebraic extension which is purely transcendental over k, then V is called **unirational**. A unirational curve is in fact rational. More generally, if the function field $k(C)$ of a curve C over k is contained in a field finitely generated and purely transcendental over k, then C is rational (**Lüroth's theorem**). A unirational surface over an algebraically closed field of characteristic zero is rational by virtue of the above criterion, but in the case of positive characteristic there are unirational surfaces which are not rational. There are nonrational unirational threefolds even of characteristic zero; for example, all smooth cubic hypersurfaces in $\mathbf{P}^4$ (C. H. Clemens and P. A. Griffiths [49], J. P. Murre [68]) and some smooth quartic hypersurfaces in $\mathbf{P}^4$ (V. A. Iskovskii and Ju. I. Manin [58]).

K. Monoidal Transformations

Let V be an irreducible variety and $\mathfrak{I}$ be a sheaf of ideals of $\mathcal{O}_V$. For any affine open set U of V, the $\mathfrak{I}|U$ is determined by an ideal $\mathfrak{a}$ of the coordinate ring A of U. Let $a_0, a_1, \ldots, a_m$ be a system of generators of $\mathfrak{a}$, and let U' be the graph of a rational mapping from U to $\mathbf{P}^m(k)$ such that the points $P \in U$ and $(a_0(P):a_1(P):\ldots:a_m(P)) \in \mathbf{P}^m(k)$ correspond to each other by U'. Then U' is uniquely determined (up to biregular isomorphisms) by U and $\mathfrak{a}$ only. Suppose V has a covering by affine open sets U_i. We obtain U_i' over U_i as before. By patching them together, we get a birational transformation $T: V \to V'$, which is unique up to biregular isomorphisms. This T is called the **monoidal transformation (monoidal dilatation** or **blowing-up)** of V by the ideal sheaf $\mathfrak{I}$. The mapping $T^{-1}: V' \to V$ is regular, and if W denotes the †support of the sheaf $\mathcal{O}_V/\mathfrak{I}$, T is regular in $V - W$, and $T\{W\}$ is of codimension 1 in V'. In particular, if W is a subvariety of V and if $\mathfrak{I}$ is the sheaf of germs of functions vanishing on W, then T is called the monoidal transformation of V with **center** W. Moreover, if every point of the subvariety W is simple on V as well as on W, we have $T[W] = T\{W\}$; $T[W]$ has the structure of an algebraic †fiber bundle, of which the base space is W, the fiber is the $(\dim V - \dim W - 1)$-dimensional projective space, and the structure group is a projective transformation group. A monoidal transformation having a point as center is called a **locally quadratic transformation**.

L. Resolution of Singularities

Given an arbitrary irreducible variety V, we have the problem of finding out a nonsingular projective variety V' birationally equivalent to V. This is called the problem of **resolution of singularities**. In the case of characteristic zero, this problem was solved by O. Zariski (1944) for dimension $\leqslant 3$ and by H. Hironaka (1964) [15] for any dimension. In the case of characteristic p, S. Abhyankar solved the 2-dimensional case (1956) and the 3-dimensional case (1966). Hironaka's theorem of resolution of singularities is as follows. Let K be any field of characteristic 0, and V a reduced algebraic scheme over K. Then there exists a finite sequence $V_r \to V_{r-1} \to \ldots \to V_1 \to V_0 = V$ of K-morphisms of algebraic K-schemes satisfying the following three conditions: (1) V_r has no singular points. (2) $V_{i+1} \to V_i$ is the inverse of a monoidal transformation having a closed subscheme D_i of V_i as its center. (3) The points of D_i are singular on V_i but simple on D_i, and V_i is normally flat along D_i. (We say that a scheme V is **normally flat** along a subscheme W if for the sheaf of ideals $\mathfrak{I}$ defined by W, the quotient $\mathfrak{I}_x^p/\mathfrak{I}_x^{p+1}$ is †flat over $\mathcal{O}_{W,x} = \mathcal{O}_{V,x}/\mathfrak{I}_x$ for all points x of W and for all integers $p \geqslant 0$.)

M. Cycles and Divisors

Let V be an irreducible variety. We denote by $\mathfrak{B}_r$ the set of r-dimensional irreducible subvarieties of V that are simple on V (i.e., are not contained in the singular locus of V), and by $\mathfrak{Z}_r(V)$ the free Abelian group with basis $\mathfrak{B}_r$. Elements of $\mathfrak{Z}_r(V)$ are called **cycles** of dimension r (or r-cycles) on V. Let A and B be r-cycles; $A = \Sigma n_i A_i$, $B = \Sigma m_i A_i$ ($A_i \in \mathfrak{B}_r$, $A_i \neq A_j$ if $i \neq j$). If $n_i \geqslant m_i$ for all i, then we write $A \geqslant B$. If $A \geqslant 0$, then A is said to be a **positive cycle**. For a 0-cycle $A = \Sigma n_i P_i$ the integer $\deg(A) = \Sigma n_i$ is called the **degree** of A.

A cycle of codimension 1 is called a **divisor**. A divisor $\geqslant 0$ is usually called **effective** instead of positive. If V is of dimension d and if $W \in \mathfrak{B}_{d-1}$, the local ring $\mathcal{O}_{V,W}$ is a †discrete valuation ring. The †normalized valuation defined by it is denoted by $v_W(\ )$. For a function $f \in k(V)$, we say that W is a **zero** of order n if $v_W(f) = n > 0$, and that W is a **pole** of order $-n$ if $v_W(f) = n < 0$. Any function $f \in k(V)$, other than the constant 0, has at most a finite number of zeros and poles. We denote by $(f)_0$ the sum $\Sigma v_W(f) W$ extended over all the zeros W of f, and put $(f^{-1})_0 = (f)_\infty$ and $(f)_0 - (f)_\infty = (f)$. We call $(f)_0$, $(f)_\infty$, and (f) the **zero divisor**, the **pole divisor**, and the **divisor** of f, respectively. The divisor (f) is equal to $\Sigma v_W(f) W$ where the summation is taken over $W \in \mathfrak{B}_{d-1}$, and we have $(fg) = (f) + (g)$. When V is complete and the singular locus of V has codimension > 1, then f is constant if and only if $(f)_\infty = 0$ (or $(f)_0 = 0$). Let D_1 and D_2 be divisors; if there exists a function $f (\neq 0) \in k(V)$ such that $D_1 - D_2 = (f)$, then D_1 and D_2 are called **linearly equivalent** to each other and we write $D_1 \sim D_2$. The **linear equivalence class** containing a divisor D is denoted by $\mathrm{cl}(D)$. A divisor which is linearly equivalent to 0 on a neighborhood of each point of V is called a **Cartier divisor** (some authors call a Cartier divisor simply a divisor). If V is smooth, then any divisor is a Cartier divisor. If a divisor D can be written as $D = (f)$ on an open set U, then the function f is called a **local equation** of D on U. Let $T: V' \to V$ be a rational mapping from a normal variety V' to V, let D be a Cartier divisor on V and assume that the closed image of T is not contained in D. Since T defines a regular mapping $\varphi: V'-W' \rightarrow V$, where W' is a suitable subvariety of V' of codimension >1, we get a Cartier divisor on $V'-W'$ by composing the local equations of D with φ. Taking the closure of this divisor in V', we obtain a divisor on V', which is denoted by $T^{-1}(D)$.

N. Divisors and Linear Systems

Let V be a complete irreducible variety, $f_0, f_1, \ldots, f_n$ be elements of the function field $k(V)$ of V, and D be a divisor on V satisfying $(f_i)+D \geqslant 0$ for each i. Then the set Σ of the divisors of the form $(\Sigma a_i f_i)+D$, where the a_i are elements of k and not all zero, is called a **linear system**. The linear space $kf_0+kf_1+\ldots+kf_n$ is called a **defining module** of Σ. The divisors in Σ are positive and are linearly equivalent to each other; if every positive divisor that is linearly equivalent to a member of Σ belongs to Σ, then Σ is said to be a **complete linear system**. For any linear system Σ, there exists a unique complete linear system containing it, which is denoted by $|\Sigma|$. The maximal positive divisor D_0 that is contained in all divisors of Σ is called the **fixed component** of Σ, and for each $D \in \Sigma$, we call $D-D_0$ the **variable component** of D (or of Σ). A point P of V is called a **base point** of a linear system Σ if P is on each variable component of Σ. A linear system Σ is called **irreducible** if it has no fixed component and if its generic member is irreducible; otherwise it is called **reducible**. The dimension of a defining module of a linear system Σ is denoted by $l(\Sigma)$; we call $l(\Sigma)-1$ the **dimension** of Σ and denote it by $\dim \Sigma$. A linear system of dimension 1 is called a **linear pencil**.

A defining module of a linear system Σ is determined uniquely up to k-isomorphisms. Let L be a defining module, and let $f_0, f_1, \ldots, f_n$ be a linearly independent basis of L over k. If we associate to each point P of V the point $Q=(f_0(P), f_1(P), \ldots, f_n(P))$ of the n-dimensional projective space, then we obtain a rational mapping $\Phi(\Sigma)$ from V to another variety V'. Outside the base points of Σ, the rational mapping $\Phi(\Sigma)$ is regular; and the base points are the fundamental points of $\Phi(\Sigma)$. We say that $\Phi(\Sigma)$ is the rational mapping defined by the linear system Σ. When $\Phi(\Sigma)$ is a biregular transformation, Σ is said to be **ample** (or **very ample**). For a divisor D, the set of positive divisors that are linearly equivalent to D (if they exist) is a linear system, which is called the **complete linear system defined by** D and is denoted by $|D|$. We usually write $l(D)$ instead of $l(|D|)$. If $|D|$ is ample, we say that D is an **ample** (or **very ample**) **divisor**; if mD is ample for a sufficiently large positive integer m, we say that D is a **nondegenerate** (or **ample**) **divisor**.

O. Differential Forms

Let V be an n-dimensional irreducible variety, and let $\mathfrak{K}=k(V)$ be its function field. We denote by $\mathfrak{D}^*$ the set of **derivations** of $\mathfrak{K}$ over k, i.e., the k-linear mappings $D: \mathfrak{K} \rightarrow \mathfrak{K}$ satisfying $D(fg)=D(f)g+fD(g)$. Then $\mathfrak{D}^*$ is an n-dimensional linear space over $\mathfrak{K}$. Let $\mathfrak{D}$ denote the †dual space of $\mathfrak{D}^*$ over $\mathfrak{K}$. For each $f \in \mathfrak{K}$, let df be an element of $\mathfrak{D}$ defined by $\langle df, D\rangle = D(f)$ $(D \in \mathfrak{D}^*)$. Let $x_1, x_2, \ldots, x_n$ be a separating transcendence basis of $\mathfrak{K}$ over k, in the sense that $x_1, \ldots, x_n$ are algebraically independent over k and $\mathfrak{K}$ is a †separable algebraic extension over $k(x_1, \ldots, x_n)$. (Such a basis exists under the weaker hypothesis that k is †perfect.) Then $dx_1, \ldots, dx_n$ form a basis of $\mathfrak{D}$ over $\mathfrak{K}$. The homogeneous elements of degree r of the †Grassmann algebra of $\mathfrak{D}$ over $\mathfrak{K}$ are called **differential forms of degree** r on V (or of the function field $\mathfrak{K}$). The set of the differential forms of degree n is a 1-dimensional linear space over $\mathfrak{K}$ spanned by $dx_1 \wedge dx_2 \wedge \ldots \wedge dx_n$.

A set of n functions $f_1, \ldots, f_n$ in $\mathfrak{K}$ is called a system of **local uniformizing coordinates** on an open set U of V if $f_1-f_1(P), \ldots, f_n-f_n(P)$ is a †regular system of parameters of the local ring $\mathfrak{O}_P$ for each $P \in U$. In that case, $f_1, \ldots, f_n$ is also a separating transcendence basis of $\mathfrak{K}$. If P is a simple point of V, then there exists a system of local uniformizing coordinates on a suitable neighborhood of P. Let ω be a differential form of degree r on V, and write $\omega=\Sigma_{i_1<\ldots<i_r} \varphi_{(i)} df_{i_1} \wedge \ldots \wedge df_{i_r}$ by means of the differentials df_i of local uniformizing coordinates around P. If the coefficients $\varphi_{(i)}$ are regular at P, then ω is said to be **regular** at P.

When V is a complete variety without singular points, a differential form that is everywhere regular on V is called a **differential form of the first kind**; the differential forms of the first kind are determined by the function field $\mathfrak{K}$ and are independent of the choice of the model V. The number of linearly independent differential forms of the first kind, of degree n, is denoted by p_g and is called the **geometric genus of** V.

Let V be a complete variety, W an irreducible subvariety of V of codimension 1, and P a point of W that is simple on V. Choose a system of local uniformizing coordinates (f_i). Given a differential form ω on V, we write it as a "polynomial" in the df_i, and denote by $v_W(\omega)$ the minimum of the values of the coefficients for the †valuation $v_W(\ )$. The number $v_W(\omega)$ is determined by ω and W, and it is independent of the choice of P and of the local uniformizing coordinates. Then ω defines a divisor $(\omega)=\Sigma_W v_W(\omega)W$ on V, which is called the **divisor of a differential form** ω. The divisor of a differential form of degree n $(=\dim V)$ is called a **canonical divisor** and is usually denoted by K. The canonical divisors form a linear equivalence class of divisors.

Albanese Variety, Picard Variety. Let V be a variety. Then we can construct a couple (A,f) consisting of an †Abelian variety called the **Albanese variety** of V and a rational mapping $f: V \to A$ (called a canonical mapping) such that: (i) the image of f generates A, i.e., the sum of f with itself n times, $F: V^n \to A$, is generically surjective for sufficiently large n; (ii) for every rational mapping $g: V \to B$ of V into an Abelian variety B, there exist a homomorphism $h: A \to B$ and a point $b \in B$ such that $g = h \cdot f + b$. The Albanese variety is uniquely determined up to isomorphisms and f is determined up to translations.

In the case of $k=\mathbf{C}$, if V is a complete nonsingular variety and if q is the dimension of the linear space of differential 1-forms of the first kind on V, then the first †Betti number B_1 is equal to $2q$. Let $\omega_1, \ldots, \omega_q$ be a basis of the linear space and let $\gamma_1, \ldots, \gamma_{2q}$ be a basis of the first Betti group. Put $\alpha_{ji} = \int_{\gamma_j} \omega_i$ and $\alpha_j = (\alpha_{j1}, \ldots, \alpha_{jq})$. Then the period vectors α_j $(1 \leqslant j \leqslant 2q)$ are linearly independent over $\mathbf{R}$ in $\mathbf{C}^q$. If Γ denotes the discrete subgroup of $\mathbf{C}^q$ generated by the α_j, then the quotient group $\mathbf{C}^q/\Gamma$ is the Albanese variety of V. The canonical mapping is given by the mapping $P \to (\int_Q^P \omega_1, \ldots, \int_Q^P \omega_q) \pmod{\Gamma}$ where P is a variable point on V and Q is a fixed point on V (→ 232 Kähler Manifolds).

Replacing the phrase rational mapping by regular mapping in the definition, we can define the **strict Albanese variety** of V and prove its existence. It is a quotient Abelian variety of the Albanese variety of V. If V is nonsingular, both coincide by virtue of the †absolute minimality of an Abelian variety.

Let V be a complete normal variety, U the set of the simple points of V, and D a divisor on V. Then D is said to be **algebraically equivalent to** 0 if there exist a nonsingular curve C, a divisor Γ on $U \times C$, and two points P and Q on C such that D can be written as $D = \varphi_P^{-1}(\Gamma) - \varphi_Q^{-1}(\Gamma)$, where φ_P and φ_Q are the morphisms $\varphi_P: U \to U \times P \to U \times C$ and $\varphi_Q: U \to U \times Q \to U \times C$. We denote by $\mathfrak{G}(V)$, $\mathfrak{G}_a(V)$, and $\mathfrak{G}_l(V)$ the set of all divisors on V, the set of divisors that are algebraically equivalent to 0, and the set of divisors that are linearly equivalent to 0, respectively. We can introduce a canonical structure of an Abelian variety into $\mathfrak{G}_a(V)/\mathfrak{G}_l(V)$, which is called the **Picard variety** of V. The dimension q of the Picard variety is called the **number of irregularity** of V; if $q=0$ we say that V is **regular**.

The Albanese variety and the Picard variety of V are †isogeneous to each other, and each one is the Picard variety of the other. If V is a curve, they are isomorphic and called the **Jacobian variety** (→ 13 Algebraic Curves).

Using Cartier divisors instead of divisors we get an analogous theory to construct another kind of Picard variety which turns out to be isomorphic to the Picard variety of the strict Albanese variety of V [47]. The group of the linear equivalence classes of Cartier divisors can be identified with $H^1(V, \mathcal{O}_V^*)$ where $\mathcal{O}_V^*$ is the sheaf of multiplicative groups of the invertible elements in $\mathcal{O}_V$. From this point of view, we can generalize the theory of Picard variety to the case of schemes also. The theory thus obtained is called the theory of **Picard schemes** [67].

Neron-Severi Group. Let V be a complete normal variety. Denote by $\mathfrak{G}(V)$, $\mathfrak{G}_n(V)$, and $\mathfrak{G}_a(V)$ the group of divisors, group of divisors numerically equivalent to zero (→ Q), and group of divisors algebraically equivalent to zero, respectively. The quotient group $NS(V) = \mathfrak{G}(V)/\mathfrak{G}_a(V)$ is finitely generated [60, 71] and is called the **Neron-Severi group** of V. We call the rank of $NS(V)$ the **Picard number** of V and denote it by $\rho(V)$. In the case of a nonsingular projective variety over $k=\mathbf{C}$ we have an inequality $\rho(V) \leqslant h^{1,1}(V)$ $(=\dim_{\mathbf{C}} H^1(V, \Omega_V^1))$ and the **Lefschetz number** $B_2(V) - \rho(V)$ is a birational invariant (where $B_2(V)$ is the second Betti number of V) [4]. For the positive characteristic case, however, the above inequality does not hold in general [27].

The torsion part of $NS(V)$ is $\mathfrak{G}_n(V)/\mathfrak{G}_a(V)$ (T. Matsusaka). The last fact cannot be generalized for higher codimensional cycles [52].

P. Cohomology Theory

Let $(X, \mathcal{O})$ be a ringed space. An $\mathcal{O}$-Module (i.e., a sheaf of $\mathcal{O}$-modules) F is said to be **quasicoherent** if for each point x of X there exist a neighborhood U of x and an †exact sequence $M \to N \to F_{|U} \to 0$ where M and N are free $\mathcal{O}_{|U}$-Modules. An $\mathcal{O}$-Module F is said to be **of finite type** if F is locally generated by a finite number of sections over $\mathcal{O}$; F is **of finite presentation** if, locally, there exists an exact sequence $\mathcal{O}^p \to \mathcal{O}^q \to F \to 0$ where p and q are positive integers (they need not be globally constant); F is **coherent** if (i) F is of finite type and (ii) the kernel of any homomorphism $\mathcal{O}^n_{|U} \to F_{|U}$ (where n is an arbitrary positive integer, and U is an open set) is of finite type. Obviously, if F is coherent, then F is of finite presentation, which implies that F is quasicoherent and of finite type. In the category of $\mathcal{O}$-Modules, the full subcategory of coherent sheaves is closed under almost all operations of sheaves. If $\mathcal{O}$ itself is coherent as an $\mathcal{O}$-Module, $\mathcal{O}$ is said to be a **coherent sheaf of rings**. In this case every $\mathcal{O}$-Module of finite presentation is coherent.

The structure sheaf of a locally Noetherian scheme is a coherent sheaf of rings. On a locally Noetherian scheme X, every quasicoherent sub-$\mathcal{O}_X$-Module or quotient $\mathcal{O}_X$-Module of a coherent $\mathcal{O}_X$-Module is coherent. A coherent $\mathcal{O}_V$-Module on an algebraic variety V is called a **coherent algebraic sheaf**.

Let $X = \mathrm{Spec}(A)$ be an affine scheme. Then every quasicoherent $\mathcal{O}_X$-Module F on X is generated by its global sections. The correspondence $F \to \Gamma(X,F)$ defines an equivalence beeween the category of quasicoherent sheaves on X and the category of A-modules; if A is Noetherian, then the coherent sheaves and the finite A-modules correspond to each other under this equivalence.

Let X be a separated scheme, and $\mathfrak{U} = \{U_\lambda\}$ an affine open covering of X. For each quasicoherent $\mathcal{O}_X$-Module F, the cohomology group $H^q(V,F)$ is canonically isomorphic to the †Čech cohomology $H^q(\mathfrak{U},F)$ (→ 377 Sheaves). If X is of dimension d, then $H^q(X, F) = 0$ for every sheaf F of Abelian groups on X and $q > d$.

For a scheme X we define the **cohomological dimension** cd(X) to be the largest integer q such that $H^q(X,F) \neq 0$ for a quasicoherent $\mathcal{O}_X$-Module F on X [14]. The cohomological dimension cd(X) does not exceed the dimension of X. If X is an affine scheme, then cd$(X) = 0$. The converse is true under the assumption that X is Noetherian (Serre's criterion). For an algebraic scheme X of dimension n, cd$(X) = n$ if and only if X is complete (S. L. Kleiman).

Let $f: X \to Y$ be a proper morphism of Noetherian schemes. Then for every coherent $\mathcal{O}_X$-Module F and for every $q \geqslant 0$, $R^q f_*(F)$ (→ 202 Homological Algebra) is also coherent. In the special case of $Y = \mathrm{Spec}(k)$ with a field k this means that for an algebraic coherent sheaf F on a complete variety X the cohomology group $H^q(X,F)$ is a finite-dimensional vector space over k.

Let V be a variety over k and let F be a locally free $\mathcal{O}_V$-Module of rank n (i.e., an $\mathcal{O}_V$-Module which is locally isomorphic to $\mathcal{O}_V^n$). If we take an open covering $\{U_i\}$ of V and isomorphisms $\varphi_i : F_{|U_i} \simeq \mathcal{O}^n_{V|U_i}$, then $\varphi_i \circ \varphi_j^{-1}$ defines a regular mapping $g_{ij} : U_i \cap U_j \to GL(n,k)$, which is called the coordinate transformation of F. If we construct a †vector bundle B on V by the same coordinate transformations g_{ij}, then F can be regarded as the sheaf $\mathcal{O}(B)$ of germs of sections of B. By means of the canonical homomorphism $GL(n,k) \to PGL(n-1,k)$ we can construct a projective bundle $\mathbf{P}(F)$ on V (which is said to be associated with F). (Note that in [12], $\mathbf{P}(F)$ is defined to be a projective bundle with coordinate transformations ${}^t g_{ij}^{-1}$, i.e., associated with the dual of E in our sense.) This procedure of associating $\mathbf{P}(F)$ with a locally free $\mathcal{O}_V$-Module F can be generalized to arbitrary schemes.

A closed (locally closed) S-subscheme $f: X \to S$ of $p: \mathbf{P}(E) \to S$ is called a **projective scheme (quasiprojective scheme) over** S, or f is said to be a **projective morphism (quasiprojective morphism)**. A projective morphism is proper. A reduced projective scheme over a field k is nothing but a projective variety over k. We can develop the theory of projective schemes by means of †graded rings in a way similar to affine schemes.

A locally free $\mathcal{O}_V$-Module of rank 1 is called an **invertible sheaf**. Invertible sheaves correspond to complex line bundles in the classical case. Let $P = \mathbf{P}^N(k)$ be a projective space, $(y_0, y_1, \ldots, y_N)$ a system of homogeneous coordinates of P, and U_i the open subset of P defined by $y_i \neq 0$. Denote by $\mathcal{O}(n)$ the invertible sheaf on P defined by the coordinate transformation $g_{ij} = (y_j/y_i)^n$. More generally, let $p: P = \mathbf{P}(F) \to S$ be the projective bundle associated with a locally free $\mathcal{O}_S$-Module F of rank $N+1$ on a scheme S. Then there is an invertible sheaf $\mathcal{O}(n) = \mathcal{O}(1)^{\otimes n}$ on P with the properties: (i) for each $s \in S$ its restriction to the fiber $p^{-1}(s) = \mathbf{P}^N(k(s))$ is $\mathcal{O}(n)$ defined above; (ii) $p_*(\mathcal{O}(n)) = \mathbf{S}^n(F)$ for $n > 0$ where $\mathbf{S}^n(F)$ denotes the nth symmetric product of F. The invertible sheaf $\mathcal{O}(1)$ is called the **tautological line bundle** on P. (Note that $\mathcal{O}(1)$ in the sense of [12] is $\mathcal{O}(-1)$ in our sense, but since the definition of $\mathbf{P}(E)$ is also different, the above property (ii) holds without modification.)

For a quasiprojective S-scheme $f: X \to \mathbf{P}(E) \to S$, the restriction of $\mathcal{O}(1)$ to X is denoted by $\mathcal{O}_X(1)$ (or simply $\mathcal{O}(1)$). An invertible sheaf L on X is called **very ample over** S if there exist a locally free $\mathcal{O}_S$-Module of finite type E on S and an S-immersion $i: X \to \mathbf{P}(E)$ such that $\mathcal{O}_X(1) = L$; L is called **relatively ample** over S or S-ample if $L^{\otimes n}$ is very ample over S for a certain $n > 0$ (→ Section N). When

S is an affine scheme, an ample (very ample) sheaf over S is simply called **ample (very ample)**. There is the following cohomological criterion of ampleness (generalized **Serre's theorem**).

Let Y be a Noetherian scheme, $f: X \rightarrow Y$ a proper morphism, and L an invertible $\mathcal{O}_X$-Module. Then the following four conditions are equivalent: (i) L is f-ample; (ii) for each coherent $\mathcal{O}_X$-Module F there is an integer N such that $R^q f_*(F \otimes L^{\otimes n}) = 0$ for all $n \geqslant N$ and $q > 0$; (iii) for each coherent sheaf of ideals $\mathcal{I}$ of $\mathcal{O}_X$ there is an integer N such that $R^1 f_*(\mathcal{I} \otimes L^{\otimes n}) = 0$ for all $n \geqslant N$. They imply the condition (iv): for each coherent $\mathcal{O}_X$-Module F there is an integer N such that the canonical homomorphisms $f^* f_*(F \otimes L^{\otimes n}) \rightarrow F \otimes L^{\otimes n}$ are surjective for all $n \geqslant N$.

Let V be a projective variety over k with a very ample sheaf $\mathcal{O}(1)$ on V and put $\mathcal{O}(n) = \mathcal{O}(1)^{\otimes n}$. Denote by $F(n)$ the tensor product $F \otimes \mathcal{O}(n)$ for each $\mathcal{O}_V$-Module F; then $F(n+m) = F(n) \otimes \mathcal{O}(m)$. If the dimension of V is r, we put $\chi(F) = \sum_{q=0}^{r} (-1)^q \dim H^q(V, F)$ for each coherent sheaf F. Then, for a given exact sequence $0 \rightarrow F' \rightarrow F \rightarrow F'' \rightarrow 0$, we have $\chi(F) = \chi(F') + \chi(F'')$. Moreover, $\chi(F(n))$ is a polynomial in n whose degree is equal to the dimension of the support of F (i.e., the set of the points x with $F_x \neq 0$). This polynomial is called the **Hilbert characteristic function** of F. By (ii) of Serre's theorem, we have $\chi(F(n)) = \dim H^0(V, F(n))$ for sufficiently large n. In the case of $F = \mathcal{O}_V$, the value $\chi(\mathcal{O}(n))$ for sufficiently large n coincides with the dimension of the k-submodule of the homogeneous coordinate ring of V, consisting of the homogeneous elements of degree n.

In general, for a complete variety V of dimension r, we put

$$\chi(V) = \chi(\mathcal{O}_V) = \sum_{q=0}^{r} (-1)^q h^{0,q}$$

($h^{0,q} = \dim H^q(V, \mathcal{O}_V)$) and call it the **arithmetic genus** of V. Classically, the number $p_a(V)$ defined by $p_a(V) = (-1)^r(\chi(V) - 1) = h^{0,r} - h^{0,r-1} + \ldots \pm h^{0,1}$ was called the arithmetic genus of V, instead of $\chi(V)$. When V is a nonsingular curve, $p_a(V)$ is the usual †genus. If V is a projective variety, the constant term of the Hilbert characteristic function of $\mathcal{O}_V$ is $\chi(V)$ and the coefficient of its highest term is $(\deg V)/r!$.

Let V be a normal variety, and D a divisor on V. If, for each point $x \in V$, we denote by $L(D)_x$ the set of the functions $f \in k(V)$ that satisfy $(f) + D \geqslant 0$ on some neighborhood of x, we obtain a coherent algebraic sheaf $L(D)$, and we have $\dim H^0(V, L(D)) = l(D)$. If, moreover, V is complete, we put $\chi_V(D) = \chi(V) - \chi(L(-D))$ and call it the **virtual arithmetic genus** of D. Classically, the number $p_a(D) = (-1)^{r-1}(\chi_V(D) - 1)$ was called by that name. When D is effective and has no multiple components, $\chi_V(D)$ coincides with the arithmetic genus of D regarded as a variety. In general, $\chi_V(D)$ stays invariant if D is replaced by a divisor that is algebraically equivalent to D.

If D is a Cartier divisor, $L(D)$ is an invertible sheaf. For two Cartier divisors D_1, D_2, $L(D_1 + D_2) = L(D_1) \otimes L(D_2)$, and D_1 and D_2 are linearly equivalent if and only if $L(D_1) \cong L(D_2)$. Let V be a complete variety. For an irreducible subvariety W of dimension s we define the intersection number $(D^s \cdot W)$ by the property:

$$\chi\left(\mathcal{O}_W \otimes L(D)^{\otimes n}\right) = ((D^s \cdot W)/s!)n^s + \text{lower terms in } n.$$

Then we have the following **Nakai-Moishezon criterion** of ampleness. Let V be a complete variety over an algebraically closed field and D a Cartier divisor on V. Then D is ample if and only if for every irreducible subvariety W of arbitrary dimension $s > 0$, $(D^s \cdot W) > 0$.

Let V be an irreducible nonsingular variety and let Ω^p denote the sheaf of germs of regular differential forms of degree p ($\Omega^0 = \mathcal{O}_V$). If V is complete, then we denote $\dim H^q(V, \Omega^p)$ by $h^{p,q}$.

Serre's duality theorem: Let V be a nonsingular complete variety of dimension r, B an algebraic vector bundle over V, and B^* the dual vector bundle of B. Denote by $\mathfrak{B}$ and $\mathfrak{B}^*$ the sheaves of germs of sections of B and B^*, respectively. Then (i) $H^r(V, \Omega^r)$ is canonically isomorphic to k, and (ii) $H^q(V, \mathfrak{B})$ and $H^{r-q}(V, \mathfrak{B}^* \otimes \Omega^r)$ are dual to each other as linear spaces by means of the cup product of the above spaces with $H^r(V, \Omega^r) \cong k$. In particular, $H^q(V, \Omega^p)$ is dual to $H^{r-q}(V, \Omega^{r-p})$; hence we have $h^{p,q} = h^{r-p,r-q}$.

This theorem was extensively generalized by Grothendieck in the category of schemes [13].

When the field k is of characteristic 0, we furthermore have $h^{p,q} = h^{q,p}$ by complex conjugation (→ 232 Kähler Manifolds C); but in characteristic p, there are examples for which this symmetry does not hold. In general, $h^{p,q}$ is a †relative invariant but not an †absolute invariant; however, as $h^{p,0}$ is the dimension of the linear space of the differential forms of the first kind with degree p, $h^{p,0}$ is an absolute invariant. Hence $h^{0,p}$ is also an absolute invariant in the case of characteristic 0. When the characteristic is positive, the absolute invariance of $h^{0,p}$ has been proved only for the case $\dim V \leqslant 2$.

if γ_i is a loop based at 0 going once (counterclockwise) around t_i, we have, for each $x \in H^{n-1}(W, \mathbf{Q})$, $\varphi_p(\gamma_i)(x) = x \pm \langle x, \delta_i \rangle \delta_i$, where $\langle\ \rangle$ is the intersection pairing of $H^{n-1}(W, \mathbf{Q})$. $\varphi_p(\gamma_i)$ is called a **Picard-Lefschetz transformation**. The main results due to Lefschetz are restated as follows. (1) (**Weak Lefschetz theorem**). The natural homomorphism $H^{2n-i}(V, \mathbf{Q}) \to H^{2n-i-2}(W, \mathbf{Q})$ is an isomorphism for $0 \leqslant i \leqslant n-2$ and is an injection for $i = n-1$, or equivalently $H_i(V, W, \mathbf{Q}) = 0$ for $0 \leqslant i \leqslant n-1$. (2) (**Strong Lefschetz theorem**). Let ξ be the cohomology class of $H^2(V, \mathbf{Q})$ corresponding to the hyperplane section W and let $L: H^*(V, \mathbf{Q}) \to H^{*+2}(V, \mathbf{Q})$ be the homomorphism defined by the cup product with ξ. Then for each $i \leqslant n$, L^{n-i}: $H^i(V, \mathbf{Q}) \to H^{2n-i}(V, \mathbf{Q})$ is an isomorphism. The weak Lefschetz theorem is true for a cohomology with integral coefficients. In fact, $V - W$ has the homotopy type of a real n-dimensional finite CW complex, and $\pi_r(V, W) = 0$ for $r < n$ [41]. The strong Lefschetz theorem is equivalent to the statement that $H^{n-1}(W, \mathbf{Q})$ is the direct sum of the vector space spanned by the vanishing cycles δ_i and the vector space spanned by the invariant cycles (i.e., $\varphi_{n-1}(\gamma_i)x = x$, $i = 1, \ldots, d$). Lefschetz's original proof of this statement is incomplete, and no direct topological proof is known. The transcendental proof of (2) is given by the theory of harmonic integrals. A version of Lefschetz pencils is a proper morphism $f: X \to D = \{z \mid |z| < \varepsilon\}$ of a complex manifold X onto a disk D such that $f^* = f|f^{-1}(D^*)$, $D^* = D - \{0\}$, is of maximal rank at every point of $f^{-1}(D^*)$. Fix a point $s \in D^*$. $\pi_1(D^*, s)$ operates on $H^k(W, \mathbf{Z})$, $W = f^{-1}(s)$, and we have a representation $\varphi_k: \pi_1(D^*, s) \to GL(H^k(W, \mathbf{Z}))$. For a loop γ based at s and going once around 0 the Picard-Lefschetz transformation $\varphi_k(\gamma)$ is essentially unipotent (i.e., for a certain integer m, $\varphi_k(\gamma)^m$ is unipotent).

For a nonsingular projective variety defined over a field k with characteristic $p > 0$, the above Lefschetz theorems hold for an l-adic cohomology ($l \neq p$) [12 (SGA 7), 51]. Using the theory of finite étale coverings of an algebraic variety defined over a field k, we can define the **algebraic fundamental group** and the **algebraic homotopy groups**, which are profinite completions of the topological fundamental group and the topological homotopy groups, respectively, where $k = \mathbf{C}$ [12 (SGA 1), 44, 72]. Let (X, x) be a germ of a complex space with isolated singular point x. (X, x) is always algebraizable, i.e., the completion $\hat{\mathcal{O}}_{X,x}$ of the analytic local ring $\mathcal{O}_{X,x}$ is isomorphic to the completion of the local ring of a closed point of an algebraic variety defined over $\mathbf{C}$ [42]. Embed (X, x) in $(\mathbf{C}^m, 0)$. Let $S_\varepsilon(D)$ be a sphere (an open ball) of radius ε with center 0 in $\mathbf{C}^m$. If ε is sufficiently small, $K = X \cap S_\varepsilon$ is a compact oriented differentiable manifold. For $\dim X = 2$, K is homeomorphic to a 2-sphere S^2 if and only if X is smooth at x [65]. Let (X, x) be a hypersurface singularity with defining equation $f = 0$. $\varphi(z) = f(z)/|f(z)|$ defines a differentiable fiber bundle $\varphi: S_\varepsilon - K \to S^1$, called the **Milnor fibration**. Each fiber $F_\theta = \varphi^{-1}(e^{i\theta})$ is parallelizable and has the homotopy type of a bouquet $S^n \vee \ldots \vee S^n$ of μ n-spheres, and K is $(n-2)$-connected, where $n = \dim X$. Moreover, $f^{-1}(c) \cap D_\varepsilon$ is diffeomorphic to F for sufficiently small $|c| \neq 0$ [64]. The number μ is called the **Milnor number** of (X, x). If $\dim X = 1$, (S_ε, K) is a link in S^3; moreover, if K is connected, (S_ε, K) is an iterated torus knot (K. Brauner [82]). Let $(X, 0)$ be defined by the equation $f(z) = z_1^{a_1} + z_2^{a_2} + \ldots + z_{n+1}^{a_{n+1}}$, where $a_i \geqslant 2$, $i = 1, 2, \ldots, n+1$. Put $\Sigma(a_1, \ldots, a_{n+1}) = X \cap S_\varepsilon$ for a small ε. Then $\Sigma(a_1, \ldots, a_{n+1})$ is homeomorphic to S^{2n-1} if and only if $\Delta(1) = \amalg(1 - \omega_1\omega_2 \ldots \omega_{n+1}) = \pm 1$, where each ω_j ranges over all a_jth roots of unity other than 1. For any odd integer $p \geqslant 7$ and for any exotic sphere Σ^p which bounds a parallelizable manifold there are integers $a_i \geqslant 2$, $i = 1, \ldots, n+1 = (p+3)/2$, such that Σ^p is diffeomorphic to $\Sigma(a_1, \ldots, a_{n+1})$ [46]. For example, the manifolds $\Sigma(2, 2, 2, 3, 6k-1)$, $k = 1, \ldots, 28$, constitute 28 different exotic 7-spheres.

V. Hodge Theory

Let $H_{\mathbf{R}}$ be a finite-dimensional real vector space containing a lattice $H_{\mathbf{Z}}$ and let $H = H_{\mathbf{R}} \otimes_{\mathbf{R}} \mathbf{C}$ be its complexification. A **Hodge structure** of weight m on H (or $H_{\mathbf{R}}$) is, by definition, a direct sum decomposition $H = \oplus_{p+q=m} H^{p,q}$, $\overline{H^{p,q}} \cong H^{q,p}$ where $H^{p,q}$ is a complex vector subspace and the overbar denotes complex conjugation. If H and H' carry Hodge structures of weight m and m', respectively, then $H \otimes H'$, $\mathrm{Hom}_{\mathbf{C}}(H, H')$, $\wedge^p H$, and H^* carry Hodge structures of weight $m + m'$, $m' - m$, pm, and $-m$, respectively. For a Hodge structure H of weight m, $F^pH = \oplus_{k \geqslant p} H^{k, m-k}$, $p = 0, 1, \ldots, m$, induces a decreasing filtration. Let H be a Hodge structure of weight m and let Q be a bilinear form on H. If the following three conditions are satisfied, the Hodge structure H is said to be polarized by Q. (i) Q is defined over $\mathbf{Q}$ and is symmetric (skew-symmetric) if m is even (odd). (ii) $Q(H^{p,q}, \overline{H^{p',q'}}) = 0$ unless $p = p'$, $q = q'$. (iii) $(\sqrt{-1})^{p-q} Q(v, \bar{v}) > 0$ for nonzero $v \in H^{p,q}$. Let V be a compact Kähler manifold. Then $H = H^n(V, \mathbf{C})$ carries the

Hodge structure induced by the type (p,q)-decomposition (→ 232 Kähler manifolds). This is also the case if V is a compact complex manifold which is the image of a holomorphic mapping from a compact Kähler manifold of the same dimension. Moreover, if V is projective, the Hodge-Riemann bilinear relations define a natural polarization on the subspace P of H consisting of all primitive cohomology classes.

Let $V(W)$ be a smooth algebraic variety defined over $\mathbf{C}$ (complex manifold) and let $\varphi: V \to W$ be a projective smooth morphism with connected fibers. Then $\mathbf{H} = R^m\varphi_*\mathbf{C}$ is a flat vector bundle on W with the flat connection ∇. ∇ is often called the **Gauss-Manin connection** and it can be defined algebraically if W is also algebraic.

For each fiber $V_s = \varphi^{-1}(s)$, $s \in W$, the filtration $F^pH^m(V_s, \mathbf{C}) = \oplus_{k \geqslant p} H^{k,m-k}(V_s)$ induces a complex subbundle $\mathbf{F}^p$ and the connection ∇ has the property $\nabla(\mathcal{O}(\mathbf{F}^p)) \subset \mathcal{O}(\mathbf{F}^{p-1} \otimes \mathbf{T}^*)$ where $\mathbf{T}$ is the tangent bundle of W. Moreover, if W is algebraic, ∇ is a differential equation with regular singular points on $\overline{W}$ where $\overline{W}$ is a smooth compactification of W such that $\overline{W} - W$ is a divisor with normal crossings. If we consider the subbundle $\mathbf{P}$ of $\mathbf{H}$ consisting of all primitive cohomology classes, the polarization on each fiber induces a Hermitian pseudometric on $\mathbf{P}$. Curvatures of bundles $\mathbf{P} \cap \mathbf{F}^p$ have been studied by Griffiths [53]. There exists a classifying space D for polarized Hodge structures and there exists a holomorphic mapping of the universal covering $\tilde{W}$ of W into D, usually called a period mapping. D may not be a bounded symmetric domain but has several interesting properties [9, 54]. In some cases D/Γ with a suitable discrete subgroup Γ is the moduli space of polarized algebraic varieties (e.g., curves, Abelian varieties).

P. Deligne [50] has generalized Hodge theory to arbitrary algebraic varieties (more generally schemes of finite type over $\mathbf{C}$). The simplest case is the Hodge theory of a smooth noncomplete algebraic variety X. By Nagata's embedding theorem there exists a complete algebraic variety $\overline{X}$ such that $Y = \overline{X} - X$ is a subvariety. By virtue of Hironaka's resolution theorem we can assume that $\overline{X}$ is nonsingular and that Y is a divisor with normal crossings. Let $\Omega^1_{\overline{X}}\langle Y\rangle$ be a sheaf of germs of meromorphic 1-forms with logarithmic pole along Y, i.e., locally written as $\sum_{i=1}^k a_i(x)(dx_i/x_i) + \sum_{j=k+1}^n a_j(x)\,dx_j$, where $(x_1, \ldots, x_n)$ is a system of local coordinates with center $p \in Y$ in X such that $x_1 \cdots x_k = 0$ is a local equation of Y and $a_i(x), a_j(x)$ are holomorphic at p. Using the complex $\{\Omega^p_{\overline{X}}\langle Y\rangle = \bigwedge^p \Omega^1_{\overline{X}}\langle Y\rangle, d\}$, with a suitable filtration, Deligne has shown that $H = H^m(X, \mathbf{C})$ carries a mixed Hodge structure and this structure is independent of the choice of $\overline{X}$. The mixed Hodge structure on H consists of two finite filtrations, i.e., $0 \subset \ldots \subset W_{n-1} \subset W_n \subset \ldots \subset H$, the weight filtration which is defined over $\mathbf{Q}$, and $0 \subset \ldots \subset F^p \subset F^{p+1} \subset \ldots \subset H$, the Hodge filtration such that F^p induces on W_n/W_{n-1} a Hodge structure of weight n. As a corollary he has shown that a meromorphic p-form on $\overline{X}$ with logarithmic pole along Y (i.e., a section of $\Omega^p_{\overline{X}}\langle Y\rangle$) is d-closed on X, and $\omega = 0$ if and only if $\omega_{|X}$ is zero in $H^p(X, \mathbf{C})$. An important application of the theory of this mixed Hodge structure on $H^m(X, \mathbf{C})$ is the following. Let V and W be smooth algebraic varieties and let $\varphi: V \to W$ be a smooth projective morphism. If $\overline{V}$ is a smooth compactification of V, the canonical homomorphism $H^m(\overline{V}, \mathbf{Q}) \to H^0(W, R^m\varphi_*\mathbf{Q})$ is surjective. Fix a point $s \in W$. $\pi_1(W, s)$ operates on $H^m(V_s, \mathbf{Q})$. Then this action is semisimple.

W. Deformations, Moduli, Algebraic Spaces

In this section for simplicity the field k is assumed to be algebraically closed. Let X be an algebraic scheme over k. A **deformation of X over a connected scheme** S over k with base point s_0 consists of the following data: (1) A morphism $p: \mathfrak{X} \to S$ that is flat and of finite type. If $\mathfrak{X}$ is complete, p is also proper. (2) A closed point $s_0 \in S$ such that the fiber $\mathfrak{X} \times_S k(s_0)$ is isomorphic to X. For any closed point $s \in S$, the fiber $X_s = \mathfrak{X} \times_S k(s)$ is called a **deformation** of X. If X is smooth and complete, we assume further that p is smooth. Similarly we can define a deformation of a polarized algebraic manifold, an embedding deformation of X in an algebraic scheme Y over k, a deformation of an affine scheme with isolated singular points, and a deformation of vector bundles on a fixed algebraic scheme over k. The theory has two aspects: local theory and global theory.

Let $(R, \mathfrak{m})$ be a complete Noetherian local ring such that $R/\mathfrak{m} = k$. Set $R_n = R/\mathfrak{m}^n$. A formal deformation X_R of X is a sequence $\{X_n\}$ such that (i) X_n is a deformation of X over $\mathrm{Spec}(R_n)$ and (ii) there is a compatible sequence of isomorphisms $X_n \otimes_{R_n} R_{n-1} \cong X_{n-1}$ for any n. Let (FLA/k) be the category of finite-dimensional commutative local k-algebras. The local theory of deformation is the study of the covariant functor F of (FLA/k) to (Set), where, for $A \in (\mathrm{FLA}/k)$, $F(A)$ is the set of isomorphism classes of deformations of X over $\mathrm{Spec}(A)$. The functor is in general neither representable nor prorepresentable (i.e., there exists a formal deformation X_R of X such that $F(A) = \mathrm{Hom}_{k\text{-alg}}(R, A)$). But, under reasonably mild conditions on F, F has the hull R. That is, there is a formal deformation X_R of X and a natural transformation $j: G \to F$ where $G(A) = \mathrm{Hom}_{k\text{-alg}}(R, A)$, such that j is formally smooth (i.e., for any surjection $A' \to A$ in (FLA/k), $G(A') \to G(A) \otimes_{F(A)} F(A')$ is surjective). The formal deformation X_R is called a versal deformation of X. The hull R is unique up to noncanonical isomorphism. The deformation functor F has the hull R if X is (i) a complete algebraic scheme over k, (ii) an affine scheme with isolated singular points, (iii) a polarized algebraic variety over k, or (iv) a vector bundle on a complete algebraic scheme over k. If there exists a deformation $\pi: \mathfrak{X} \to S$ of X over a scheme S with base point s_0 over k such that $R = \hat{\mathcal{O}}_{S,s_0}, X_n \cong \mathfrak{X} \otimes R_n$, the formal deformation $\{\pi_n : X_n \to \mathrm{Spec}(R_n)\}$ is called algebraizable. Algebraizability of the versal deformation has been studied by M. Artin. Since the assumption that S is a scheme is rather restrictive, Artin has introduced the notion of algebraic spaces and has considered algebraizability in the category of algebraic spaces [2, 43]. For a complete algebraic variety, the versal deformation is not necessarily algebraizable and we need to consider deformations of polarized algebraic varieties. The versal deformation of an affine variety with an isolated singularity is algebraizable in the category of algebraic spaces. For the global theory of deformations, we need the projectivity assumption, and the theory is essentially reduced to the theory of Hilbert schemes (or Chow varieties). The problem of moduli is considered as the study of the set M of all isomorphism classes of deformations of X. Usually we consider the moduli of polarized varieties. In many cases, the moduli space can be obtained as the quotient space of a certain (locally closed) subset H of a Hilbert scheme by the following equivalence relation: $s \sim s' \in H$ if and only if $X_s \simeq X_{s'}$ as polarized varieties, where $\pi: \mathfrak{X} \to H$, $\pi^{-1}(s) = X_s$ (Matsusaka [63]). This equivalence relation is often induced by an action of a reductive algebraic group G. Suppose that a reductive algebraic group G operates on an algebraic k-scheme Z. A G-invariant morphism $f: Z \to Y$ (i.e., for the trivial action of G on Y, f is a G-equivariant morphism) is called a **geometric quotient** if (1) f is a surjective affine morphism and $f_*(\mathcal{O}_Z)^G = \mathcal{O}_Y$, (2) if X is a G-stable closed subset of Z, then $f(X)$ is closed in Y, and (3) for $x_1, x_2 \in Z$, $f(x_1) = f(x_2)$ if and only if the G-orbits of x_1 and x_2 are the same. Let G be a reductive algebraic group, $\chi: G \to \mathrm{Aut}(V)$ a rational representation on a finite-dimensional vector space over k and $v_0 \neq 0$ a G-invariant point. Then there exists a G-invariant homogeneous polynomial F of degree $\geqslant 1$ on V such that $F(v_0) \neq 0$ (Haboush [56]). This implies that if a reductive group G operates on an algebraic k-affine scheme Spec(A), then the invariant ring A^G is a finitely generated k-algebra and, moreover, if any G-orbit in Spec(A) is closed, the natural morphism $\mathrm{Spec}(A) \to \mathrm{Spec}(A^G)$ is a geometric quotient [69]. For a quasiprojective scheme Z over k with an action of a reductive group G we need a notion of stable points [66]. The subset of stable points of Z consists of all geometric points of a G-stable open subscheme Z^s of Z, and there exists a geometric quotient $f: Z^s \to Y$ where Y is quasiprojective (Mumford [66], Seshadri [79], Haboush [56]). In this way Mumford has shown the existence of coarse moduli schemes of nonsingular complete irreducible algebraic curves and polarized Abelian varieties. But, in general, analysis of stable points is very difficult and it is desirable to extend the category of schemes so that it becomes easier to obtain a quotient. Matsusaka has introduced the notion of a Q-variety [62]. M. Artin has introduced a notion of an algebraic space which is a special case of a Q-variety. An **algebraic space** X consists of an affine scheme U and a closed subscheme $R \subset U \times U$ such that (1) R is an equivalence relation, and (2) the projections $p_i: R \to U$ $(i = 1, 2)$ are étale. (These are often written as $R \rightrightarrows U \to X$.) A morphism $g: V \to X$ of an affine scheme V to an algebraic space X consists of a closed subscheme $W \subset U \times V$ such that (1) the projection $W \to V$ is étale and surjective, and (2) the two closed subschemes $R \times_U W$, $W \times_V W$ of $U \times U \times V$ are equal. Let $S \rightrightarrows V \to Y$ be an algebraic space. Then $\mathrm{Hom}(Y, X)$ is defined as the kernel of $\mathrm{Hom}(V, X) \rightrightarrows \mathrm{Hom}(S, X)$. If Y is an affine scheme, this definition of $\mathrm{Hom}(Y, X)$ is equivalent to the previous definition by virtue of the étale descent. Thus, algebraic spaces form a category which contains the category of schemes. We can define the structure sheaf of an algebraic space and construct a cohomology theory. Many important notions and theorems for schemes can be generalized to those for algebraic spaces. Every algebraic space has a dense open subset that is an affine scheme. A group algebraic space is a group scheme. Suppose $k = \mathbf{C}$. If an algebraic group G operates on an algebraic k-scheme properly with a finite stabilizer group, the quotient space exists as an algebraic space. In this way Popp has shown the existence of moduli spaces of algebraic surfaces of general type as algebraic spaces (→ 17 Algebraic Surfaces; also [73]). Moreover, every separated algebraic space X of finite type over $\mathbf{C}$ carries a natural structure X^{an} of an analytic space. If there exists a proper modification morphism $f: X^{an} \to Y$ of a separated algebraic space X onto an analytic space Y, then Y carries a structure of an algebraic space and f becomes a morphism of algebraic spaces (Artin [2]). For any algebraically closed field k, Artin has introduced the notion of formal algebraic space and formal contraction, and obtained results similar to those for algebraic spaces. An irreducible compact complex space X whose algebraic dimension of X is equal to $\dim X$ is called a **Moishezon space**. As a corollary of the above theorem, any Moishezon space X carries a structure M of an algebraic space such that $X \cong M^{an}$.

X. Formal Schemes

Let A be a ring which we assume to be Noetherian, for simplicity, and I an ideal of A. Taking $\{I^n\}_{n>0}$ as a fundamental system of neighborhoods of 0, we can introduce a structure of a topological ring into A called **I-adic topology**. The †completion of A with I-adic topology is isomorphic to the projective limit $\hat{A} = \varprojlim_{n>0} A/I^n$ (here A/I^n are regarded as discrete topological rings) and called the **completion of A along I**. If A is Noetherian, then $\hat{A}$ is again Noetherian. There is a canonical continuous homomorphism $i: A \to \hat{A}$ whose kernel comprises the zero divisors a with $a-1 \in I$ (intersection theorem of Krull; → 70 Commutative Rings). If i is an isomorphism, we say A is **complete** with respect to I. The topology of $\hat{A}$ is the $\hat{I}$-adic topology where $\hat{I} = i(I)\hat{A}$ and $\hat{A}$ is complete with respect to $\hat{I}$.

Take a Noetherian ring A complete with respect to I which we consider as an I-adic topological ring by identifying its completion along I with A. On $\mathfrak{X} = V(I) \subset \mathrm{Spec}(A)$ we can define a sheaf of topological rings $\mathcal{O}_{\mathfrak{X}}$ by $\Gamma(\mathfrak{D}(f), \mathcal{O}_X) = \varprojlim_{n>0} A_f / I^n A_f$ for $\mathfrak{D}(f) = D(f) \cap X$ with $f \in A$. We call $(\mathfrak{X}, \mathcal{O}_{\mathfrak{X}})$ the **formal spectrum** of A and write $\mathrm{Spf}(A)$. I is called a **defining ideal** of $\mathrm{Spf}(A)$. A (locally Noetherian) **formal scheme** is by definition a topological local ringed space which is locally isomorphic to a formal spectrum (of a Noetherian ring). If we define morphisms between two formal schemes by those in the category of topological local ringed spaces, the formal schemes form a category.

For two formal spectra $\mathrm{Spf}(A)$ and $\mathrm{Spf}(B)$ with defining ideals I and J, respectively, the direct product $\mathrm{Spf}(A) \times \mathrm{Spf}(B)$ in the category of formal schemes is the formal spectrum of the completion of $A \otimes B$ along $I \otimes B + A \otimes J$. Similarly we can construct a fiber product of formal schemes. A formal scheme $\mathfrak{X}$ is called **separated** if the image of the diagonal morphism $\Delta_{\mathfrak{X}}: \mathfrak{X} \to \mathfrak{X} \times \mathfrak{X}$ is closed (→ E).

For a ring A with an ideal I, the formal spectrum $\mathrm{Spf}(\hat{A})$ (with a defining ideal $\hat{I}$) is called the **completion of** $\mathrm{Spec}(A)$ **along** $V(I)$. Similarly for a scheme X and a closed subscheme X' we can define the **completion $X_{|X'}$ of X along X'**. Every completion of a separated scheme is separated. For a coherent sheaf F on X one can define its **completion $F_{|X'}$ along X'**, which is again coherent under the assumption that X is locally Noetherian.

Thus we can develop a theory of formal schemes in a way similar to that of schemes, which we call **"formal geometry"** (for the more general definitions and further discussions see [11, 17]). Roughly speaking, a function on $X_{|X'}$ is a formal Taylor series with respect to the direction normal to X' whose coefficients are regular functions on X'. The method of formal completion enables us to introduce "analytic" or "infinitesimal" methods in algebraic geometry. Among many important theorems, we state here the following two theorems. (1) **The fundamental theorem of proper mapping**: let $f: X \to Y$ be a proper morphism of locally Noetherian schemes, Y' a closed subscheme of Y, and $X' = X \times_Y Y'$ the inverse image of Y'. Denote the respective completions of X and Y along X' and Y' by $\hat{X}$ and $\hat{Y}$, respectively. We have the induced proper morphism of formal schemes $\hat{f}: \hat{X} \to \hat{Y}$. Then we have canonical isomorphisms, $(R^n f_*(F))_{|Y'} \cong R^n \hat{f}_*(F_{|X'})$, $n \geqslant 0$, for every coherent $\mathcal{O}_X$-Module F on X. This theorem can be applied to prove **Zariski's connectedness theorem**: for a proper morphism $f: X \to Y$ of locally Noetherian schemes with $f_*(\mathcal{O}_X) = \mathcal{O}_Y$, every fiber $f^{-1}(y)$ of f is connected and nonempty for $y \in Y$. (2) We use the same notation as in (1) and assume, moreover, that $Y = \mathrm{Spec}(A)$ for a Noetherian ring A, complete with respect to an ideal I, and $Y' = V(I)$. Then the correspondence $F \to F_{|X'}$ gives an equivalence between the category of coherent $\mathcal{O}_X$-Modules with proper support over Y and the category of coherent $\mathcal{O}_{\hat{X}}$-Modules with proper support over $\hat{Y}$. This theorem plays an important role in the theory of †deformations of algebraic varieties.

References

[1] S. S. Abhyankar, Resolution of singularities of embedded algebraic surfaces, Academic Press, 1966.
[2] M. Artin, Algebraization of formal moduli I, Global analysis, Univ. of Tokyo Press and Princeton Univ. Press (1969), 21–71; II, Ann. of Math., (2) 91 (1970), 83–135.
[3] M. F. Atiyah, Vector bundles over an

elliptic curve, Proc. London Math. Soc., (3) 7 (1957), 414–452.
[4] W. V. D. Hodge and M. F. Atiyah, Integrals of the second kind on an algebraic variety, Ann. of Math., (2) 62 (1955), 5–91.
[5] Séminaire H. Cartan and C. Chevalley, Géométrie algébrique, Ecole Norm. Sup., 1955–1956.
[6] C. Chevalley, Fondements de la géométrie algébrique, Secrétariat Mathématique, 1958.
[7] Séminaire C. Chevalley, Anneaux de Chow et applications, Ecole Norm. Sup., 1958.
[8] J. Dieudonné, Fondements de la géométrie algébrique moderne, Advances in Math., 3 (1969), 322–413.
[9] P. A. Griffiths, Periods of integrals on algebraic manifolds; summary of main results and discussion of open problems, Bull. Amer. Math. Soc., 76 (1970), 228–296.
[10] A. Grothendieck, Fondements de la géométrie algébrique, Inst. H. Poincaré, Univ. Paris, 1962.
[11] A. Grothendieck, Eléments de géométrie algébriques, Publ. Math. Inst. HES, I. Le langage des schémas, no. 4, 1960 (A. Grothendieck and J. A. Dieudonné, Springer, 1971); II. Etude globale élémentaire de quelques classes de morphismes, no. 8, 1961; III. Etude cohomologique des faisceaux cohérents, no. 11, 1961, no. 17, 1963; IV. Etude locale des schémas et des morphismes des schémas, no. 20, 1964, no. 24, 1965, no. 28, 1966, no. 32, 1967.
[12] Cited as SGA. A. Grothendieck and others, Séminaire de géométrie algébrique: SGA 1, Revêtements étales et groupe fondamental, Lecture notes in math. 224, Springer, 1971; SGA 2, Cohomologie locale des faisceaux et théorèmes de Lefschetz locaux et globaux, North-Holland, 1962; SGA 3 (with M. Demazure), Schémas en groupes I, II, III, Lecture notes in math. 151, 152, 153, Springer, 1970; SGA 4 (with M. Artin and J. L. Verdier), Théorie des topos et cohomologie étale des schémas, Lecture notes in math. 269, 270, 305, Springer, 1972–1973; SGA 5, Cohomologie l-adique et fonctions L; SGA 6, Théorie des intersections et théorème de Riemann-Roch, Lecture notes in math. 225 Springer, 1971; SGA 7, Groupes de monodromie en géométrie algébrique, pt. I, Lecture notes in math. 288; pt. II (by P. Deligne and N. Katz), 340, Springer, 1972–1973.
[13] R. Hartshorne, Residues and duality, Lecture notes in math. 20, Springer, 1966.
[14] R. Hartshorne, Ample subvarieties of algebraic varieties, Lecture notes in math. 156, Springer, 1970.
[15] H. Hironaka, Resolution of singularities of an algebraic variety over a field of characteristic zero, Ann. of Math., (2) 79 (1964), 109–326.
[16] H. Hironaka, Smoothing of algebraic cycles of small dimensions, Amer. J. Math., 90 (1968), 1–54.
[17] H. Hironaka and H. Matsumura, Formal functions and formal embeddings, J. Math. Soc. Japan, 20 (1968), 52–82.
[18] J. Igusa, On Picard varieties attached to algebraic varieties, Amer. J. Math., 74 (1952), 1–22.
[19] J. Igusa, Arithmetic variety of moduli for genus two, Ann. of Math., (2) 72 (1960), 612–649.
[20] K. Kodaira, Some results in the transcendental theory of algebraic varieties, Ann. of Math., (2) 59.(1954), 86–134.
[21] K. Kodaira, On Kähler varieties of restricted type, Ann. of Math., (2) 60 (1954), 28–48.
[22] K. Kodaira, On compact analytic surfaces I, II, Ann. of Math., (2) 71 (1960), 111–152; 77 (1973), 563–626.
[23] K. Kodaira, On the structure of compact complex analytic surfaces I, II, III, IV, Amer. J. Math., 86 (1964), 751–798; 88 (1966), 682–721; 90 (1968), 53–83, 1048–1066.
[24] T. Matsusaka, Polarized varieties, fields of moduli and generalized Kummer varieties of polarized Abelian varieties, Amer. J. Math., 80 (1958), 45–82.
[25] T. Matsusaka, Algebraic deformations of polarized varieties, Nagoya Math. J., 31 (1968), 185–245.
[26] Б. Г. Мойшезон (B. G. Moishezon), Критерий проекивности полных алгебраических абстрактных многообразий, Izv. Akad. Nauk SSSR, 28 (1964), 179–224; English translation, A criterion for projectivity of complete algebraic abstract varieties, Amer. Math. Soc. Transl., (2) 63 (1967), 1–50.
[27] D. Mumford, Pathologies of modular algebraic surfaces, Amer. J. Math., 83 (1961), 339–342.
[28] D. Mumford, Further pathologies in algebraic geometry, Amer. J. Math., 84 (1962), 642–648.
[29] D. Mumford, Pathologies III, Amer. J. Math., 89 (1967), 94–104.
[30] M. Nagata, A general theory of algebraic geometry over Dedekind domains I, II, III, Amer. J. Math., 78 (1956), 76–116; 80 (1958), 382–420; 81 (1959), 401–435.
[31] M. Nagata, Imbedding of an abstract variety in a complete variety, J. Math Kyoto Univ., 2 (1962), 1–10.
[32] J.-P. Serre, Faisceaux algébriques cohérents, Ann. of Math., (2) 61 (1955), 197–278.
[33] F. Severi, Serie, sistemi, d'equivalenza corrispondenze algebrische sulle varietà algebrische I, Cremona, 1942.

[34] F. Severi, Geometria dei sistemi algebrichi sopra una superficie e sopra una varietà algebrica II, III, Cremona, 1958, 1959.
[35] A. Weil, Foundations of algebraic geometry, Amer. Math. Soc. Colloq. Publ., second edition, 1962.
[36] A. Weil, On Picard varieties, Amer. J. Math., 74 (1952), 865–894.
[37] O. Zariski, Foundations of a general theory of birational correspondences, Trans. Amer. Math. Soc., 53 (1943), 490–542.
[38] O. Zariski, Theory and applications of holomorphic functions on algebraic varieties over arbitrary ground fields, Mem. Amer. Math. Soc., 5 (1951).
[39] O. Zariski, Complete linear systems on normal varieties and a generalization of a lemma of Enriques-Severi, Ann. of Math., (2) 55 (1952), 552–592.
[40] O. Zariski, Algebraic sheaf theory, Bull. Amer. Math. Soc., 62 (1956), 117–141.
[41] A. Andreotti, T. Frankel, The second Lefschetz theorem on hyperplane sections, Global Analysis, Univ. of Tokyo Press and Princeton Univ. Press, 1969.
[42] M. Artin, Algebraic approximation of structures over complete local rings, Publ. Math. Inst. HES, no. 36, 1969, p. 23–58.
[43] M. Artin, Théorèmes de représentabilité pour les espaces algébriques, Les Presses de l'Université de Montréal, 1973.
[44] M. Artin and B. Mazur, Etale Homotopy, Lecture notes in math. 100, Springer, 1969.
[45] M. Artin and D. Mumford, Some elementary examples of unirational varieties which are not rational, Proc. London Math. Soc., (3) 25 (1972), 75–95.
[46] E. Brieskorn, Beispiele zur Differentialtopologie von Singularitäten, Inventiones Math., 2 (1966), 1–14.
[47] C. Chevalley, Sur la théorie de la variété de Picard, Amer. J. Math., 82 (1960), 435–490.
[48] W.-L. Chow, On the projective embedding of homogeneous varieties, Algebraic geometry and topology, Princeton Univ. Press, 1957, p. 122–128.
[49] C. H. Clemens and P. A. Griffiths, The intermediate Jacobian of the cubic threefold, Ann. of Math. (2) 95 (1972), 281–356.
[50] P. Deligne, Théorie de Hodge: I, Actes Congr. Intern. Math., 1970, Nice, Gauthier-Villars; II, III, Publ. Math. Inst. HES, no. 40, 1971, p. 5–57; no. 44, 1974.
[51] P. Deligne, La conjecture de Weil I, Publ. Math. Inst. HES, no. 43, 1974, p. 273–307.
[52] P. A. Griffiths, On the periods of certain rational integrals I,II, Ann. of Math., (2) 90 (1969), 460–495, 498–541.
[53] P. A. Griffiths, Periods of integrals on algebraic manifolds III, Publ. Math. Inst. HES, no. 38, 1970, p. 125–180.
[54] P. A. Griffiths and W. Schmid, Recent developments in Hodge theory: a discussion of techniques and results. Discrete subgroups of Lie groups and application to moduli, Oxford Univ. Press, 1975, p. 31–127.
[55] A. Grothendieck, On the de Rham cohomology of algebraic varieties, Publ. Math. Inst. HES, no. 29, 1966, p. 351–359.
[56] W. J. Haboush, Reductive groups are geometrically reductive, Ann. of Math., (2) 102 (1975), 67–183.
[57] R. Hartshorne, Introduction to algebraic geometry, Prentice-Hall, 1976.
[58] V. A. Iskovskiĭ and Ju. I. Manin, Three dimensional quartics and counterexamples to the Lüroth problem, Mat. Sb., 86 (1971), 140–166; English translation, Math. USSR-Sb., 15 (1971), 141–166.
[59] N. Katz, Nilpotent connections and the monodromy theorem; applications of a result of Turittin, Publ. Math. Inst. HES, no. 39, 1971, p. 175–232.
[60] S. Lang and A. Néron, Rational points of Abelian varieties over function fields, Amer. J. Math., 81 (1959), 95–118.
[61] S. Łojasiewicz, Triangulation of semi-analytic sets, Ann. Scuola Norm. Pisa, 18 (1964), 449–474.
[62] T. Matsusaka, Theory of Q-varieties, Publ. Math. Soc. Japan, no. 8, Tokyo, 1965.
[63] T. Matsusaka, Polarized varieties with a given Hilbert polynomial, Amer. J. Math., 94 (1972), 1027–1077.
[64] J. Milnor, Singular points of complex hypersurfaces, Ann. Math. Studies, Princeton Univ. Press, 1968.
[65] D. Mumford, The topology of normal singularities of an algebraic surface and a criterion for simplicity, Publ. Math. Inst. HES, no. 9, 1961.
[66] D. Mumford, Geometric invariant theory, Springer, 1965.
[67] D. Mumford, Lectures on curves on an algebraic surface, Ann. Math. Studies, Princeton Univ. Press, 1966.
[68] J. P. Murre, Reduction of the proof of the non-rationality of Mumford, Compositio Math., 27 (1973), 63–82.
[69] M. Nagata, Note on orbit spaces, Osaka Math. J., 14 (1962), 21–34.
[70] Y. Nakai, A criterion of an ample sheaf on a projective scheme, Amer. J. Math., 85 (1963), 14–26.
[71] A. Néron, Problèmes arithmétiques et géométriques attachés à la notion du rang d'une courbe algébrique, Bull. Soc. Math. France, 80 (1952), 101–166.
[72] H. Popp, Fundamentalgruppen algebraischer Mannigfaltigkeiten, Lecture notes

in math. 176, Springer, 1970.
[73] H. Popp, On moduli of algebraic varieties II, Compositio Math., 28 (1974), 51–81.
[74] A. A. Roitman, Rational equivalence of zero cycles, Mat. Sb., 89 (1972), 569–585; English translation, Math. USSR-Sb., 18 (1972), 571–588.
[75] L. Roth, Algebraic threefolds, Springer, 1955.
[76] P. Samuel, Relations d'équivalence en géométrie algébrique, Proc. Intern. Congr., 1958, Edinburgh, Cambridge Univ. Press, p. 470–487.
[77] M. Schlessinger, Functors on Artin rings, Trans. Amer. Math. Soc., 130 (1968), 208–222.
[78] J.-P. Serre, Exemple de variété projectives conjuguées non homéomorphes, C. R. Acad. Sci. Paris, 258 (1964), 4194–4196.
[79] C. S. Sechadri, Quotient space modulo reductive algebraic groups, Ann. of Math., (2) 95 (1972), 511–556.
[80] I. R. Shafarevich, Basic algebraic geometry, Springer, 1974.
[81] A. N. Tyurin, Uspehi Mat. Nauk; English translation, Five lectures on three-dimensional varieties, Russian Math. Surveys, 27, no. 5 (1972), 1–53.
[82] O. Zariski, On the topology of algebroid singularities, Amer. J. Math., 54 (1932), 453–465.
[83] O. Zariski, Studies in equisingularity I, II, III, Amer. J. Math., 87 (1965), 507–536; 87 (1965), 972–1006; 90 (1968), 961–1023.

19 (XI.14) Algebroidal Functions

A. General Remarks

If an analytic function f satisfies an †irreducible algebraic equation

$$A_0(z)f^k + A_1(z)f^{k-1} + \ldots + A_k(z) = 0 \qquad (1)$$

with single-valued †meromorphic functions $A_j(z)$ in a domain G in the complex z-plane, then f is called a k-**valued algebroidal function** in G. With no loss of generality, we may assume that there is no common zero among the $A_j(z)$ and that all the $A_j(z)$ are †holomorphic in G. When $k=1$, the solution of (1) is a single-valued meromorphic function in G. If all the $A_j(z)$ are polynomials, then f is an †algebraic function. Thus algebroidal functions can be regarded as extensions of single-valued to multiple-valued functions and also as extensions of algebraic to †transcendental functions. Since (1) is irreducible, its discriminant $D(z)$ does not vanish identically. For all the points a satisfying $D(a) \neq 0$, $A_0(a) \neq 0$, $a \in G$, equation (1) determines k holomorphic function elements $f_1(z), \ldots, f_k(z)$ in a suitable neighborhood of a that determine the analytic function f. They can be †prolonged analytically in G in the wider sense. At any point satisfying $A_0(z)=0$, at least one element has a pole; and at any point satisfying $D(z)=0$, there may appear ramified elements. Therefore, an algebroidal function can be defined as a finitely multiple-valued analytic function in G with the exception of poles and †algebraic branch points. Every algebroidal function $f(z)$ determines a †Riemann surface, which may be considered a †covering surface $\mathfrak{Z}$ of G. This surface $\mathfrak{Z}$ is a k-sheeted †covering surface over G with no singular point except for algebraic branch points. Also, f reduces to a single-valued meromorphic function on $\mathfrak{Z}$, and all the function elements over a point z are different. A k-valued algebroidal function can also be characterized by this property.

These two (equivalent) definitions of algebroidal functions give rise to two distinct methods of studying these functions. In adopting the first definition, we can make use of results in the single-valued case, as did G. Rémoundos and G. Valiron. When the latter definition is adopted, we can use several methods that are also applicable in the single-valued case, as did H. Selberg and E. Ullrich.

B. The Maximum Principle

Almost all the known results for algebroidal functions are extensions of those on single-valued functions, except for a few particular results. The theory of †entire functions was extended first to single-valued functions and then to meromorphic algebroidal functions in $|z|<\infty$. The †maximum principle, one of the basic principles, holds on the Riemann surface $\mathfrak{Z}$. The following relation holds among the $A_j(z)$ and $|f_\nu(z)|$. Assume that (1) has the form

$$f^k + A_1(z)f^{k-1} + \ldots + A_k(z) = 0. \qquad (1')$$

Then $\log(1+A(z))/\log(1+F(z))$ is bounded, where $A(z)=\max|A_j(z)|$ and $F(z)=\max|f_\nu(z)|$.

An algebroidal function that has no pole in $|z|<\infty$ is called an **entire algebroidal function**. The successive derivatives of an entire algebroidal function may have poles at every branch point, which is a departure from the single-valued integral case. An algebroidal function defined by (1) or (1′) with entire functions $A_j(z)$ and zero-free $A_0(z)$ is entire. Therefore, the †order, type, and class of an entire algebroidal function coincide with those of the largest $A_j(z)$ (→ 413 Transcendental Entire Functions).

For an entire algebroidal function of order less than $1/2$, $F(z)$ tends to infinity uniformly along a sequence of concentric circles $|z|=r_n$ $(r_n\to\infty)$. However, it can be shown that not all the branches of $f(z)$ necessarily tend to infinity on the Riemann surface $\mathfrak{Z}$; in this sense, †Wiman's theorem does not remain true.

C. Picard's Theorem and its Extension

Rémoundos first extended †Picard's theorem and †Borel's theorem to an algebroidal function. Every k-valued transcendental algebroidal function in the finite plane takes on every value infinitely often with at most $2k$ exceptional values. There are examples where $2k$ values are actually omitted. Hence the theorem is the best one possible. Furthermore, the †convergence exponent of $f(z)-w=0$ coincides with the †order of f except for at most $2k$ values, for which the convergence exponents of f are less than the order of f (E. Borel). There are at most $2k$ polynomials $P(z)$ for which $f(z)-P(z)=0$ has at most a finite number of roots (Borel).

Selberg was the first to extend the †Nevanlinna theory of meromorphic functions to algebroidal functions (→ 271 Meromorphic Functions). Almost simultaneously, Valiron obtained the same results starting from the coefficients of (1); then Ullrich improved the results by considering the effect of branch points of $\mathfrak{Z}$.

Let $\mathfrak{Z}_r$ be the part of $\mathfrak{Z}$ over $|z|<r$, let $n(r,w)$ be the number of roots of $f(z)-w=0$ in $\mathfrak{Z}_r$, and let $n(r,\mathfrak{Z})$ be the number of branch points of $\mathfrak{Z}_r$. With these notations we write

$$N(r,w)=\frac{1}{k}\int_0^r(n(t,w)-n(0,w))\frac{dt}{t}+\frac{n(0,w)}{k}\log r,$$

$$m(r,w)=\frac{1}{2k\pi}\int_{|z|=r}\log^+\frac{1}{|f(re^{i\varphi})-w|}d\varphi,\quad w\neq\infty,$$

$$T(r,w)=m(r,w)+N(r,w).$$

Let $T(r,f)$ be the †logarithmic integral of the spherical area of the image of $\mathfrak{Z}_r$ under $w=f(z)$,

$$T(r,f)=\frac{1}{k}\int_0^r\frac{dt}{t}\iint_{\mathfrak{Z}_t}\frac{|f'(te^{i\varphi})|^2}{(1+|f(te^{i\varphi})|^2)^2}t\,dt\,d\varphi,$$

and let $N(r,\mathfrak{Z})$ be the logarithmic integral of $n(r,\mathfrak{Z})$. Then we have $T(r,w)=T(r,f)+O(1)$ and the **ramification theorem**:

$$N(r,\mathfrak{Z})<(2k-2)T(r,f)+O(1).$$

Also, $T(r,f)=O(\log r)$ holds if and only if f is algebraic. Let $A(z)$ be the maximum of $|A_j(z)|$, and let

$$\mu(r)=\frac{1}{2k\pi}\int_0^{2\pi}\log A(re^{i\varphi})d\varphi.$$

Then $T(r,f)=\mu(r)+O(1)$. As the second fundamental theorem we have

$$\sum_{\nu=1}^{q}N(r,w_\nu)>(q-2)T(r,f)-N(r,\mathfrak{Z})+\sum_{\nu=1}^{q}N_1(r,w_\nu)+O(\log rT)$$

$$>(q-2k)T(r,f)+\sum_{\nu=1}^{q}N_1(r,w_\nu)+O(\log rT),$$

where $N_1(r,w)$ is the logarithmic integral of $n_1(r,w)$ which is the sum of the †multiplicity minus one of all the roots of $f(x)-w=0$ in $\mathfrak{Z}_r$. Furthermore, the **deficiency**, **ramification index** of $f(z)$, and **ramification index** of the surface $\mathfrak{Z}$ are defined by

$$\delta(w)=1-\limsup N(r,w)/T(r,f)=\liminf m(r,w)/T(r,f),$$

$$\mathfrak{D}(w)=\liminf N_1(r,w)/T(r,f),$$

$$\xi=\liminf N(r,\mathfrak{Z})/T(r,f).$$

With these notations, we have

$$\sum\delta(w_\nu)+\sum\mathfrak{D}(w_\nu)\leqslant 2+\xi\leqslant 2k.$$

These results contain the Picard theorem and the Borel theorem. Furthermore, by considering the effect of branch points, the Ahlfors theory of covering surfaces can be extended to algebroidal functions (Y. Tumura). By using this result †Bloch's theorem can be obtained very simply.

D. Asymptotic Values and Other Results

In the single-valued case, Valiron, L. Ahlfors, and others studied the †Borel direction, the number of †asymptotic values, etc. However, almost none of the corresponding results hold for the algebroidal case as shown by several counterexamples.

There is no relationship between the order of an entire algebroidal function and the number of its finite †asymptotic values, which is quite different from the single-valued case. Furthermore, it is possible to have an infinite number of asymptotic values even if the order is equal to zero. If an algebroidal function f satisfies $\liminf T(r,f)/(\log r)^2<+\infty$, then it has at most k asymptotic values (Valiron-

Tumura). The Ahlfors theorem, which is concerned with the number of †direct transcendental singular points of the inverse function and the order of a meromorphic function in the single-valued case, has not yet been extended to the algebroidal case.

It is difficult to define the †Borel direction for an algebroidal function because of the appearance of branch points. A. Rauch proved that if $\int^{\infty} T(r,f)/r^{\rho+1}dr=\infty$, then there is a sector with an angle of at least π/ρ in which $L(\varphi)=\int^{\infty}\log^{+}F(re^{i\varphi})/r^{\rho+1}dr$ diverges. Apart from the theory of distribution of values, Selberg obtained some conditions under which the inverse functions of †Abelian integrals of a special kind reduce to algebroidal functions.

References

[1] G. Rémoundos, Extensions aux fonctions algébroïdes multiformes du théorème de M. Picard et de ses généralisations, Gauthier-Villars, 1927.
[2] H. Selberg, Algebroide Funktionen und Umkehrfunktionen Abelscher Integrale, Avhandl. Norske Videnskaps-Akad., Oslo, 8 (1934), 1–72.
[3] Y. Tumura, Theory of algebroidal functions (Japanese), Sûbutu kaisi (Bull. Physico-Math. Soc. of Japan), 15 (1941), 77–96.

20 (X.25) Almost Periodic Functions

A. History

The theory of almost periodic functions was originated by H. Bohr in 1924 as a result of his study of †Dirichlet series. The theory provides a method of studying a wide class of trigonometric series (→ 167 Fourier Series) of general type. Further generalizations were made by N. Wiener, V. V. Stepanov, A. S. Besikovič, S. Bochner, and others. H. Weyl, J. Von Neumann, and others clarified the relations between this theory and †representation of groups, specifically, the relations between almost periodic functions in a †topological group and representation theory of a †compact group.

B. Almost Periodic Functions in the Sense of Bohr

Let $f(x)$ be a complex-valued continuous function defined for all real values of x. A number τ is called a **translation number** of $f(x)$ belonging to $\epsilon>0$ if

$$\sup_{-\infty<x<\infty}|f(x+\tau)-f(x)|\leqslant\epsilon.$$

If for any $\epsilon>0$ there exists a number $l(\epsilon)>0$ such that any interval of length $l(\epsilon)$ contains a translation number of f belonging to ϵ, then $f(x)$ is called **almost periodic in the sense of Bohr**. We denote by **B** the set of all almost periodic functions in the sense of Bohr.

If $f(x)$ is †periodic with a period p, then $f\in\mathbf{B}$, because each number $l\geqslant p$ plays the role of $l(\epsilon)$ for any $\epsilon>0$. Any $f\in\mathbf{B}$ is bounded and uniformly continuous. A necessary and sufficient condition for a bounded continuous function on $(-\infty,\infty)$ to belong to **B** is that for any given sequence $\{h_n\}$ of real numbers, there exists a subsequence $\{h_{n_\nu}\}$ such that the sequence of functions $\{f(x+h_{n_\nu})\}$ is uniformly convergent in $(-\infty,\infty)$; i.e., the set $\{f(x+h)\mid h\in(-\infty,\infty)\}$ is †totally bounded with respect to the uniform norm $\|f\|=\sup|f(x)|$ in the space of bounded continuous functions in $(-\infty,\infty)$.

If $f(x)\in\mathbf{B}$, then $f(-x)$, $\overline{f(x)}$, $\alpha f(x)$ (where α is a complex number), and $f(x+h)$ (where h is a real number) $\in\mathbf{B}$. If $f(x)$, $g(x)$ $\in\mathbf{B}$, then $f(x)\pm g(x)$ and $f(x)g(x)\in\mathbf{B}$. If $f_n(x)\in\mathbf{B}$ and $\{f_n(x)\}$ converges uniformly to $f(x)$, then $f(x)\in\mathbf{B}$. For any real number λ, $\exp i\lambda x$ (where i is the †imaginary unit) is continuous and periodic. Hence the polynomial function $\sum_{n=1}^{m}\alpha_n\exp i\lambda_n x\in\mathbf{B}$. Moreover, if the latter function converges uniformly to $\sum\alpha_n\exp i\lambda_n x$ as m tends to ∞, then the limit function is also an element of **B**. The polynomial $\sum_{n=1}^{m}\alpha_n\exp i\lambda_n x$ and the series $\sum_{n=1}^{\infty}\alpha_n\exp i\lambda_n x$ are called a **generalized trigonometric polynomial** and a **generalized trigonometric series**, respectively.

For any $f\in\mathbf{B}$, its **mean** exists:

$$M[f]=\lim_{T\to\infty}\frac{1}{T}\int_a^{a+T}f(x)\,dx.$$

The convergence of the right-hand formula is uniform in $a\in(-\infty,\infty)$, and the limit is independent of the choice of a. Thus $M[f]$ is a †linear functional defined on **B**. Since $M[\exp i\lambda x]=1$ for $\lambda=0$ and $=0$ for $\lambda\neq0$, the family $\{\exp i\lambda x\mid-\infty<\lambda<\infty\}$ is an †orthonormal system with respect to the †inner product $(f,g)=M[f(x)\overline{g(x)}]$ defined on **B**. Let $\alpha(\lambda)=M[f(x)\exp(-i\lambda x)]$ for any $f\in\mathbf{B}$; then there exist countably many values of λ for which $\alpha(\lambda)$ differs from zero. Denote these values of λ by $\lambda_1,\lambda_2,\ldots$, and write $\alpha(\lambda_n)=\alpha_n$. We call the numbers $\alpha_1,\alpha_2,\ldots,\alpha_n,\ldots$ **Fourier coefficients** of $f(x)$. The formal series $\sum_{n=1}^{\infty}\alpha_n\exp i\lambda_n x$ is called the **Fourier series** of $f(x)$. Moreover, the **Parseval equality** $M[|f(x)|^2]=\sum_{n=1}^{\infty}|\alpha_n|^2$ is valid for any $f\in\mathbf{B}$.

For every periodic function, these definitions coincide with ordinary Fourier coefficients and the Fourier series (→ 167 Fourier Series). Any almost periodic function in the sense of Bohr is uniquely determined by its Fourier coefficients; i.e., if two almost periodic functions have the same Fourier series, then they are identical. For any $f \in \mathbf{B}$, its Fourier series does not always converge uniformly, but $f(x) \in \mathbf{B}$ can be approximated uniformly by a sequence of trigonometric polynomials. Hence the almost periodic functions in the sense of Bohr are also called **uniformly almost periodic functions**.

C. Generalizations of Almost Periodic Functions

Let $C(-\infty, \infty)$ be the space (→ 173 Function Spaces) of all bounded continuous functions on $(-\infty, \infty)$ with distance $\rho(f,g) = \sup_{-\infty<x<\infty}|f(x)-g(x)|$. Then, a uniformly almost periodic function is a limit of a sequence of trigonometric polynomials with respect to this distance. Generally, let ρ be a †distance function introduced in a function space (whose elements are not necessarily continuous in $(-\infty, \infty)$). Then the closure of generalized trigonometric polynomials with respect to the distance ρ is called the **almost periodic functions with respect to** ρ. For example, for $p \geqslant 1$, we set

$$D_{S^p}[f,g] = \sup_{-\infty<a<\infty}\left\{\int_a^{a+1}|f(x)-g(x)|^p dx\right\}^{1/p},$$

$$D_{\overline{W}^p}[e,g] = \lim_{l\to\infty}\sup_{-\infty<a<\infty}\left\{\frac{1}{l}\int_a^{a+l}|f(x)-g(x)|^p dx\right\}^{1/p}.$$

These are distance functions. The properties of the corresponding almost periodic functions and their relations to other classes of almost periodic functions have been studied by Besikovič [1].

D. Analytic Almost Periodic Functions

Let D be a strip domain, $a < \operatorname{Re} z < b$, defined in a complex plane. For any †holomorphic function $f(z)$ in D and $\epsilon > 0$, a real number τ is called a **translation number** of f belonging to $\epsilon > 0$ if $\sup_{z\in D}|f(z+i\tau)-f(z)| \leqslant \epsilon$. If for any $\epsilon>0$ there exists a number $l(\epsilon)>0$ such that any interval of length $l(\epsilon)$ contains a translation number of f belonging to ϵ, then $f(z)$ is called an **analytic almost periodic function** in D. We denote the set of all analytic almost periodic functions in $D = \{a < \operatorname{Re} z < b\}$ by $\mathbf{A}(a,b)$. If we fix an x in $a<x<b$, then $g(y) = f(x+iy)$ for any $f(z) \in \mathbf{A}(a,b)$ belongs to $\mathbf{B}$.

For any $f \in \mathbf{A}(a,b)$ there corresponds a †Dirichlet series $\sum_{n=1}^{\infty}\alpha_n \exp\lambda_n z$ such that two analytic almost periodic functions are identical if the corresponding Dirichlet series are identical. Here the coefficients

$$\alpha_n = M_y[f(x+iy)\exp(-i\lambda_n y)]$$

are determined independently of x ($a<x<b$), and Parseval's equality

$$M_y[|f(x+iy)|^2] = \sum_{n=1}^{\infty}|\alpha_n|^2 \exp 2\lambda_n x$$

holds (Bohr [3]). If the series

$$\sum \alpha_n \exp\lambda_n x \exp i\lambda_n y$$

at $x=a$ and $x=b$ represent the Fourier series of $f_a(y)$ and $f_b(y) \in \mathbf{B}$, respectively, then there exists $f \in \mathbf{A}(a,b)$ such that $f(z)$ is continuous on $\bar{D}$ and

$$f(a+iy) = f_a(y), \quad f(b+iy) = f_b(y).$$

The behavior of $f \in \mathbf{A}(a,b)$ at the boundary or exterior points of the domain $D = \{a < \operatorname{Re} z < b\}$ has also been investigated Besikovič [1].

E. Almost Periodic Functions on Groups

Von Neumann defined almost periodic functions on any group, generalizing the characterization of uniformly almost periodic functions on $(-\infty,\infty)$. Let $B(G)$ be a set of all complex-valued, bounded functions on a group G. Then $B(G)$ is a metric space with the †distance $\rho(f,g) = \sup_{x\in G}|f(x)-g(x)|$. If for any $f \in B(G)$ the set $A_f = \{f_{a,b}(x) = f(axb) \mid a,b \in G\}$ is totally bounded in the metric space $B(G)$, we call f an **almost periodic function on the group** G. This condition is equivalent to the total boundedness of $B_f = \{f_a(x) = f(xa) \mid a \in G\}$ or $C_f = \{{}_af(x) = f(ax) \mid a \in G\}$. We denote the set of almost periodic functions on G by $\mathfrak{A}(G)$.

For $f(x)$, $g(x) \in \mathfrak{A}(G)$, the linear combinations $a \cdot f(x) + b \cdot g(x)$ ($a, b \in \mathbf{C}$) and the product $f(x)g(x)$ are both contained in $\mathfrak{A}(G)$. If $f_n \in \mathfrak{A}(G)$ and $\{f_n\}$ converges to f uniformly on G, then $f \in \mathfrak{A}(G)$. If $f \in \mathfrak{A}(G)$, then $f_{a,b}$, f_a, ${}_af \in \mathfrak{A}(G)$ also. Hence $\mathfrak{A}(G)$ is a closed subalgebra invariant under two-sided translation in the †Banach algebra $B(G)$. For any $f \in \mathfrak{A}(G)$ there exists only one number $M[f]$ in the closure (with respect to the distance ρ in $B(G)$) of

$$A_f' = \left\{\sum_{i=1}^{n} c_i f(a_i x b_i) \mid c_i > 0, \sum c_i = 1, a_i, b_i \in G\right\}$$

(= the least closed †convex set including A_f).

We call $M[f]$ the **mean** of f on G. The mapping $f \to M[f]$ is a linear functional on $\mathfrak{A}(G)$, and we have $M[f] \geqslant 0$ if $f \geqslant 0$.

F. Relation to Bounded Representation

Suppose that we are given a finite-dimensional matrix representation $D(x) = (d_{ij}(x))$ of a group G. Then the following three conditions are equivalent: (i) All the $d_{ij}(x)$ are bounded on G. (ii) All the $d_{ij}(x)$ are almost periodic in G. (iii) The representation D is †equivalent to a representation by unitary matrices. The inner product $(f,g) = M[f(x)\overline{g(x)}]$ provides the algebra $\mathfrak{A}(G)$ with the structure of a †pre- Hilbert space. Let $H(G)$ be the Hilbert space that is the completion of $\mathfrak{A}(G)$. If we select $D^{\lambda}(x) = (d_{ij}^{\lambda}(x))$ from each L, where L is an equivalence class of bounded irreducible representations of G, and if n_{λ} is the order of D^{λ}, then $\{(1/\sqrt{n_{\lambda}})d_{ij}^{\lambda}(x) | 1 \leqslant i,j \leqslant n_{\lambda}, \lambda \in L\}$ is a †complete orthonormal system in the Hilbert space $H(G)$. Any $f(x) \in \mathfrak{A}(G)$ can be approximated uniformly in G by a finite linear combination of the $d_{ij}(x)$.

G. Almost Periodic Functions on Topological Groups

When G is a †separated topological group, we denote the set of all continuous functions on G contained in $\mathfrak{A}(G)$ by $\mathfrak{A}_*(G)$. The statements of the theorems in the previous section concerning $\mathfrak{A}(G)$ and the representation D remain valid if we replace $\mathfrak{A}(G)$ by $\mathfrak{A}_*(G)$ and replace D by a continuous representation of G. In particular, if G is the additive group of real numbers $\mathbf{R}$, then $\mathfrak{A}_*(\mathbf{R})$ is exactly $\mathbf{B}$.

H. Relation to Compact Groups

Every continuous function on a compact group G is almost periodic; i.e., $\mathfrak{A}_*(G) = C(G)$. The mean value $M[f]$ of $f \in \mathfrak{A}_*(G)$ is identical to $\int_G f(x)dx$, where the †Haar measure dx is normalized such that $\int_G dx = 1$. In this case, the theory of bounded representations already discussed is the Peter-Weyl theory (→ 71 Compact Groups).

In general, let G be a separated topological group. There exists a continuous homomorphism φ of G onto a compact group $K = K(G)$ with the following two properties: (i) For any compact group K' and a continuous homomorphism $\varphi' : G \to K'$, there exists a continuous homomorphism $\psi : K \to K'$ such that $\varphi' = \psi \circ \varphi$. (ii) Such a pair $K = (K, \varphi)$ is unique up to isomorphism. K is called the **Bohr compactification** of G, and φ is called the canonical mapping. In particular, suppose that G is a locally compact Abelian group and G^* is its †character group. We denote by G' the group G^* with discrete topology. Let K be the character group of G', and let φ^* be the identity mapping $G' \to G^*$ and φ be its conjugate mapping $G \to K$, which is a continuous homomorphism. Then K is the Bohr compactification of G with the canonical mapping φ. A necessary and sufficient condition for f on G to be continuous almost periodic is the existence of a continuous function f' on K such that $f = f' \circ \varphi$. If this condition is satisfied, then the mean $M[f]$ is identical to $\int_K f'(x)dx$. For any finite-dimensional continuous unitary representation D' of K, $D = D' \circ \varphi$ is a finite-dimensional continuous unitary representation of G, and vice versa. Hence there exists a canonical isomorphism (determined by $D = D' \circ \varphi$) between the equivalence classes of finite-dimensional unitary representations of a separated topological group G and the equivalence classes of finite-dimensional unitary representations of its Bohr compactification K. The †kernel of the canonical mapping $\varphi : G \to K$ is identical to the intersection of all kernels of finite-dimensional continuous unitary representations of G.

I. Maximally Almost Periodic Groups

Let G be a topological group. If, for each pair a, b of distinct elements of G, there exists a continuous almost periodic function f on G such that $f(a) \neq f(b)$, then G is called a **maximally almost periodic group**. This is the case if and only if G has sufficiently many finite-dimensional unitary representations. For a connected locally compact group G, the following six conditions are equivalent: (1) G is a maximally almost periodic group. (2) There is a one-to-one continuous homomorphism from G into a compact group. (3) G is the direct product of a compact group and a vector group $\mathbf{R}^n$. (4) G is the †projective limit of †Lie groups that are locally isomorphic to compact groups. (5) The quotient group G/Z is compact, where Z is the center of G. (6) The system of all neighborhoods that are invariant under the †inner automorphism constitutes a basis for the neighborhood system of the unit [7].

Moreover, any discrete free group is maximally almost periodic. If there is no continuous almost periodic function except constant functions, the topological group is called **minimally almost periodic**. Any noncompact connected †simple Lie group is minimally almost periodic.

References

[1] A. S. Besicovitch (Besikovič), Almost periodic functions, Cambridge Univ. Press, 1932.
[2] S. Bochner, Berträge zur Theorie der fastperiodischen Funktionen I, II, Math. Ann., 96 (1926), 119–147, 383–409.
[3] H. Bohr, Fastperiodischen Funktionen, Erg. Math., Springer, 1932.
[4] V. Stepanov, Über einige Verallgemeinerungen der fastperiodischen Funktionen, Math. Ann., 95 (1926) 473–498.
[5] N. Wiener, On the representation of functions by trigonometrical integrals, Math. Z., 24 (1926), 575–616.
[6] J. Von Neumann, Almost periodic functions in a group I, Trans. Amer. Math. Soc., 36 (1934), 445–492.
[7] J. Dixmier, Les C^*-algèbres et leurs représentations, Gauthier-Villars, 1964.
[8] T. Tannaka, Isôgunron (Japanese; Topological groups), Iwanami, 1948.
[9] W. Maak, Fastperiodische Funktionen, Springer, 1950.

21 (XV.18) Analog Computers

A. General Remarks

An **analog computer** is a physical system by means of which solutions of mathematical equations are obtained by measuring appropriate quantities in the system. A primitive example is the **slide rule**, whose operation is based on the additive law $z = x + y$, where x and y are the lengths of two consecutive segments of the scales and z is the total length of these segments. Other examples are the **planimeter** for measuring the area bounded by a closed curve and the **harmonic analyzer** for obtaining the †Fourier expansion of a periodic function. In a broad sense, a **pantograph** for copying plans and a †nomogram are also considered analog computers. Modern analog computers first drew public attention c. 1940, when analog computers constructed for antiaircraft gunfire carried out the desired tasks instantaneously. About the same time, V. Bush constructed a **differential analyzer** that solved numerically complicated differential equations. While many computers used to be composed of mechanical devices, **electronic analog computers** have recently made remarkable progress. In an electronic analog computer, numbers are represented by electrical quantities such as voltage and amperage. Its arithmetic unit is composed of circuit elements such as resistances, capacitances, and transformers, together with amplifiers such as vacuum tubes and transistors. Today most analog computers are electronic.

Electronic analog computers are used mainly for solving differential equations. They consist of linear operators (integrator, adder, multiplier, etc.) and nonlinear operators for multiplying, dividing, or for generating functions. These operators are connected freely with patch cords to solve each particular problem. The results are shown and recorded by Braun tubes, pen oscillographs, XY-recorders, etc. For instance, the differential equation

$$\frac{d^2x}{dt^2} + a\frac{dx}{dt} + bx^2 = f(t)$$

can be solved by the circuit shown in Fig. 1.

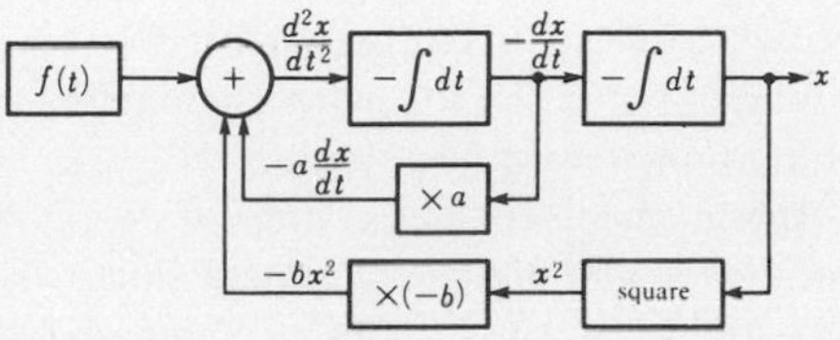

Fig. 1

Linear operators are mostly circuits having negative feedback. They are composed of linear circuit elements and operational amplifiers, that is, amplifiers with sufficiently high gain. The characteristics of such circuits depend mainly on those of circuit components around the feedback loops. An **integrator** is obtained by putting on the feedback line a capacitance which differentiates the feedback current (Fig. 2). A **multiplier** is an operational amplifier having feedback impedance. An **adder** (or **summer**) is also an amplifier which has feedback impedance and several input impedances.

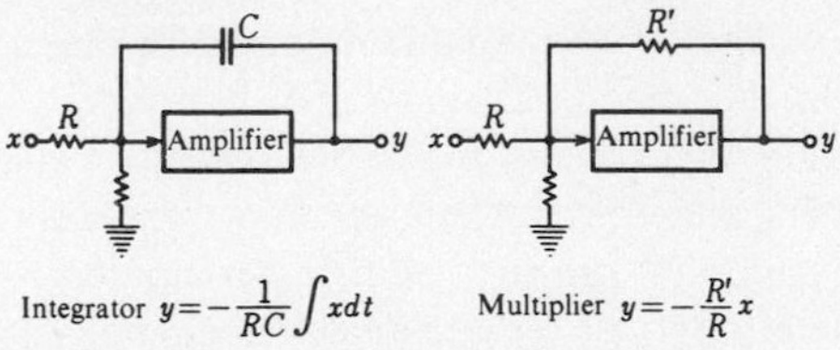

Fig. 2

Electronic analog computers are divided into two classes: high-speed and low-speed computers. A high-speed analog computer repeats fixed processes very quickly, produces a solution in less than one second, and displays it on a Braun tube. A low-speed computer produces a solution much more slowly and gives it in graphical form on paper. An analog computer is called a **direct analog**

computer when its operation is based on the direct analogy between the behavior of the physical system utilized in the computation and the original system about which the problem has been formulated. Another type of analog computer is the **function analog computer**, which solves equations by reductions of a certain type.

B. Analog and Digital Computers

In comparison with digital computers, analog computers have the following merits. They are economical for solving certain restricted types of problems since their construction is rather simple. Besides, the solution can be obtained immediately from the given data. On the other hand, they are deficient in accuracy; to obtain results with errors of order less than 10^{-3} is extremely expensive. The area of applicability of any given analog computer is restricted. To compensate for these shortcomings, the following computers are utilized on certain occasions.

(1) **Hybrid computers**, i.e., combined systems of analog and digital computers. For example, the main part of such a computer may be an analog computer, to which an auxiliary digital computer is attached. In this case the digital computer undertakes miscellaneous services such as change of patch cords, generation of initial values and input signals, auxiliary computations, and tabulation of outputs. Thus the universality of the digital computer helps the main analog computer. There are also hybrid computers whose main bodies are digital computers to which analog computers are attached. These analog computers undertake particular jobs such as the calculation of implicit functions. In this case the high-speed property of the analog computer is fully utilized.

(2) **Digital differential analyzer**. This type of digital computer is used for special purposes and has the same function as the mechanical differential analyzer. It has, however, a high speed comparable to analog computers, and does not lose accuracy, which is an important property of digital computers. Therefore it can replace analog computers unless extremely high speed is required. It is also advantageous to connect it to universal digital computers.

One application of computers is in the automatic control of machinery. In such a case the necessary computation is usually specialized and stability of the computer is strictly required, while the requirements for precision are not so severe. High speed is sometimes required. Analog computers are often used for these purposes.

In ordinary cases when one gives data, receives results, and utilizes them, digital computers are preferable since their speed has increased and programming has become easy on account of the development of simulation languages and translators. Today analog computers are less important than they used to be.

References

[1] H. Yamashita (ed.), Densi keisanki–Anarogu keisanki hen (Japanese; Electronic computers, pt. I. Analog computers), Ohmsya, 1959.

[2] H. Okuno, M. Goto, M. Takeuchi, and H. Wada (eds.), Zyôhô syori hando-bukku (Japanese; Handbook of data-processing), Kôrin-syoten, 1965.

[3] G. A. Korn and T. M. Korn, Electronic analog computers, McGraw-Hill, 1952; second edition, 1956.

[4] G. W. Smith and R. C. Wood, Principles of analog computation, McGraw-Hill, 1959.

22 (X.1) Analysis

The origins of analysis can be traced back to the time when Eudoxus and Archimedes devised the so-called method of exhaustion for calculating the area of a plane figure or the volume of a solid. Their objects were restricted, however, to particular figures or solids. In the 16th and 17th centuries, F. Viète, J. Kepler, and B. Cavalieri again took up this problem. In the 17th century, the problem of drawing a tangent to a given curve was studied by R. Descartes, P. de Fermat, E. Pascal, and J. Wallis. Fermat, particularly, treated this problem in view of its application to finding the maxima and minima of certain functions.

In 1684, G. Leibniz introduced the symbols dx and dy in treating the same problem. He proved that dy/dx represents the slope of the tangent of the curve at a given point. This led him to establish a new operation whose aim was to find the value of dy/dx. In 1686, he established a method he called the inverse method of tangent; this is what we now call integral calculus. He also introduced the notation $\int$.

Newton also performed this inverse operation while laying the foundations of dynamics. He developed what is now called differential and integral calculus under the name of the "method of fluxions." But neither Leibniz nor

Newton was able to formulate rigorously the fundamental concepts of the methods; so they were attacked severely by many contemporary scholars. In England the new calculus was developed by B. Taylor and C. MacLaurin. The contribution of the former in 1715 and that of the latter in 1745 are particularly notable. On the continent, Leibniz's symbolic method was developed by mathematicians of the Bernoulli family, F. A. de l'Hôpital, G. Fagnano, and others. It enabled them to solve a number of scientific problems of that epoch and to pose new kinds of problems.

One of these problems, treated by J. le Rond d'Alembert in relation to the vibration of a chord, concerns the †partial differential equation for $y=y(t,x)$:

$$\partial^2 y/\partial t^2 = a^2 \partial^2 y/\partial x^2, \tag{1}$$

with the boundary conditions $y=0$ for $x=0$ and $x=l$. He obtained the solution $y=f(at+x)-f(at-x)$, where f is an arbitrary function with the period $2l$. In 1753, D. Bernoulli showed that solutions of the equation (1) are given by functions of the form

$$y=\frac{a_0}{2}+\sum_{k=1}^{\infty}\left(a_k\cos\frac{k\pi x}{l}+b_k\sin\frac{k\pi x}{l}\right).$$

These two kinds of solutions gave rise to the question of whether an arbitrary function can be expressed by a †trigonometric series. This problem was studied by A. C. Clairaut, J. L. Lagrange, and L. Euler. In 1807, J. Fourier, in treating a problem on the conduction of heat, showed that an arbitrary function with the period 2π can be expressed by

$$y=\frac{a_0}{2}+\sum_{k=1}^{\infty}(a_k\cos kx+b_k\sin kx), \tag{2}$$

where the coefficients a_k and b_k are given by

$$a_k=\frac{1}{\pi}\int_{-\pi}^{\pi}f(x)\cos kx\,dx,$$

$$b_k=\frac{1}{\pi}\int_{-\pi}^{\pi}f(x)\sin kx\,dx. \tag{3}$$

This series is now called the †Fourier series. It was as late as 1820 that A. L. Cauchy noted that if one treats a series, one must examine its convergence. Fourier's investigation precedes Cauchy's observation, so he did not consider the problem of whether the sum of the series (2) with the coefficients given by (3) actually expressed $f(x)$.

In the 19th century, the concept of †functions, which had been taken in the sense of "analytic expressions," came to be defined by the correspondence relation. Cauchy clarified the ideas of †limit and †continuity, †differentiability and †integrability. He showed that a function that is continuous in a closed interval is integrable in that interval. But his proof was not rigorous, as he lacked the notion of †uniform continuity. In his paper on the trigonometric series, B. Riemann considered the integrability of functions that might be discontinuous and introduced the concept of what we now call the †Riemann integral.

The theory of †sets, initiated by G. Cantor in his paper of 1874, revolutionized analysis. R. Baire, E. Borel, and H. Lebesgue established a theory based on set theory. Baire made a classification of discontinuous functions. Generalizing his results, Lebesgue gave a definition of analytic expressions, thus clarifying terms that had been used vaguely since the time of Euler. Lebesgue also tried to define the concepts of the integral of a function, the length of a curve, and the area of a surface from the most general viewpoint. In generalizing the notion of †measure introduced by Borel, he established the theory of †Lebesgue measure with which he laid the foundations of the theory of †Lebesgue integrals. The introduction of this theory gave a new turn to the theory of the Fourier series.

The study of functions of a complex variable was originated by Cauchy in the first half of the 19th century. He began his research by introducing the notion of "monogenic functions", which are now called †regular functions (or †holomorphic functions). He established †Cauchy's integral theorem and integral formulas for these functions, and deduced from these theorems the †residue theorem for functions with †poles. Making use of this formula, Cauchy proved that a function that is holomorphic at the point a can be expanded in a power series of the form $\sum_{k=0}^{\infty}a_k(z-a)^k$ in the neighborhood of this point.

Riemann considered a complex variable w as a function of another complex variable z when dw/dz is independent of the value of the differential dz. This amounts to the same thing as a "monogenic function" of Cauchy. In his research on †Abelian functions, Riemann introduced the †Riemann surface. This important idea was basic to the progress of analysis and †topology in the 20th century.

K. Weierstrass, who was a contemporary of Riemann, developed the theory of functions of a complex variable from a purely analytic viewpoint. He called a power series $\sum_{k=0}^{\infty}a_k(z-a)^k$ of $z-a$, representing a holomorphic function in the interior of its †circle of convergence, an element of a function. An aggregate of such elements that are derived from one of them by means of †analytic continuations, along all curves having the point a as the initial point, was named by him an †analytic function.

Weierstrass also initiated the study of functions of several complex variables. His ideas were developed by H. Poincaré, P. Cousin, and E. Picard, who tried to extend the theory of functions of one complex variable to that of many variables. Cousin posed in 1859 the problem of constructing a function of several complex variables that has assigned points as zeros and poles with given natures. This problem was not completely solved by Cousin himself; it was successfully pursued by A. Weil and solved by K. Oka. Another problem concerning the †domain of holomorphy of functions of several complex variables was investigated by H. Cartan, P. Thullen, and K. Stein. It developed into the theory of †analytic space, which was studied by H. Grauert, R. Remmert, and Stein.

The study of the †calculus of variations began in the 18th century, and was motivated by the following consideration: Differential calculus gives a general means of finding extreme values of a given function. Likewise, a general method is sought for finding a function that makes the given †functional an extremal. For example, the problem of finding the function $y(x)$ such that the plane curve $y=y(x)$ passing through two given points (a,A) and (b,B) of the plane makes the functional $\int_a^b F(x,y,y')\,dx$ have an extreme value is a case in this category. Euler showed that $y(x)$ must satisfy the differential equation $dF_{y'}/dx-F_y=0$ (1744). Through Lagrange, W. R. Hamilton, and others, the calculus of variations was developed into a theory that is also useful in the fields of †rational mechanics and †quantum mechanics. From research on the continuity or differentiability of the functional with respect to y developed the idea of considering a function as a "point" in a †function space. The branch of analysis that treats functions as elements of function spaces and utilizes the methods of algebra and topology is called **functional analysis**.

Another source of functional analysis is the theory of †integral equations, first introduced in a problem treated by N. H. Abel and generalized later by V. Volterra in the form $A(x)\varphi(x)-\int_a^x K(x,y)\varphi(y)\,dy=f(x)$, where $A(x), K(x,y)$, and $f(x)$ are given and $\varphi(x)$ is the unknown function. E. I. Fredholm established a theory of integral equations of another form: $A(x)\varphi(x)-\int_a^b K(x,y)\varphi(y)\,dy=f(x)$. D. Hilbert introduced the function spaces l_2 and L_2 to study the †eigenvalue problem of integral equations of the Fredholm type with †symmetric kernels. J. von Neumann later established †spectral theory in †abstract Hilbert spaces. He also applied it to lay mathematical foundations of quantum dynamics (1929). Furthermore, S. Banach created a theory of †linear operators in †Banach spaces, which include Hilbert spaces as special cases (1932). This theory was further generalized to that of †linear topological spaces. The theory of †distributions, systematically developed by L. Schwartz (1945) utilizing the theory of linear topological spaces, is an important part of functional analysis and has proved a major factor in the enormous recent progress in the general theory of †partial differential equations.

References

[1] M. B. Cantor, Vorlesungen über Geschichte der Mathematik, Teubner, 1894–1908. Also → references to 265 Mathematics in the 17th Century, 266 Mathematics in the 18th Century, and 267 Mathematics in the 19th Century.

[2] N. Bourbaki, Eléments de mathématique, Fonctions d'une variable réele, Integration, Espace vectoriels topologiques, Théorie spectrales, Actualités Sci. Ind., Hermann, 1955–1969.

[3] C. Carathéodory, Vorlesungen über reelle Funktionen, second edition, Teubner, 1927 (Chelsea, 1968).

[4] H. Cartan, 1. Calcul différentiel, 2. Formes différentielles, Hermann, 1967.

[5] G. Choquet, Lectures on analysis I, II, Benjamin, 1969.

[6] R. Courant and D. Hilbert, Methods of mathematical physics I, II, Interscience, 1953,1962.

[7] R. Courant, Differential- und Integralrechnung I, II, Springer, third edition, 1955; English translation, Differential and integral calculus, Nordemann, 1938.

[8] C. J. de La Vallée-Poussin, Cours d'analyse infinitésimale I, II, Gauthier-Villars, seventh edition, 1930.

[9] J. Dieudonné, Foundations of modern analysis, Academic Press: Treatise on analysis I, second edition, 1969; II, 1970.

[10] J. Dieudonné, Calcul infinitesimal, Hermann, 1968.

[11] M. Fujiwara, Sûgaku kaiseki I; Bibunsekibun gaku (Japanese; Mathematical analysis, pt. 1. Differential and integral calculus), Uchida-rôkakuho, I, 1934; II, 1939.

[12] E. Goursat, Cours d'analyse mathématique: I. Dérivées et différentielles, intégrales définies, développements en séries, applications géométriques; II. Théorie des fonctions analytiques, équations différentielles, équations aux dérivées partielles du premier ordre; III. Intégrales infiniment voisines, équations aux dérivés partielles du second ordre, équations intégrales, calcul des variations,

Gauthier-Villars, fifth edition, 1933–1956.
[13] G. H. Hardy, A course of pure mathematics, Cambridge Univ. Press, seventh edition, 1938.
[14] D. Hilbert, Grundzüge einer allgemeinen Theorie der linearen Integralgleichungen, Teubner, second edition, 1924.
[15] C. Jordan, Cours d'analyse de l'école polytechnique: I. Calcul différentiel; II. Calcul intégral; III. Calcul intégral (équations différentielles), Gauthier-Villars, third edition, I, 1909; II, 1913; III, 1915.
[16] H. L. Lebesgue, Leçons sur l'intégration et la recherche des fonctions primitives, Gauthier-Villars, second edition, 1928.
[17] É. Picard, Traité d'analyse, Gauthier-Villars, I, 1922; II, 1926; III, 1928.
[18] G. Pólya and G. Szegö, Aufgaben und Lehrsätze aus der Analysis I, II, Springer, second edition, 1954.
[19] W. Rudin, Real and complex analysis, McGraw-Hill, 1966.
[20] L. Schwartz, Analyse mathématique I, II, Hermann, 1967.
[20A] L. Schwartz, Analyse—Topologie générale et analyse fonctionelle, Hermann, 1971.
[21] L. Schwartz, Théorie des distributions, Actualites Sci. Ind., Hermann, 1950, second edition, 1966.
[22] В. И. Смирнов (V. I. Smirnov), Курс высшей матемагики, физматгиз, seventeenth edition, 1961; English translation, A course of higher mathematics: I. Elementary calculus; II. Advanced calculus; III-1. Linear algebra; III-2. Complex variables, special functions; IV. Integral equations and partial differential equations; V. Integration and functional analysis, Addison-Wesley, 1964.
[23] T. Takagi, Kaiseki gairon (Japanese; A course of analysis), Iwanami, third edition, 1961.
Also → references to 109 Differential Calculus, 111 Differential Equations, and 200 Holomorphic Functions.

23 (XIX.6) Analytical Dynamics

A. General Remarks

Mechanics as originally formulated by I. Newton was geometric in nature, but later L. Euler, J. L. Lagrange, and others developed the analytical method of treating mechanics that is now called **analytical dynamics**. Lagrange introduced **generalized coordinates** q_j ($j=1,2,\ldots,f$, where f is the number of degrees of freedom of the system considered), which uniquely represent the configuration of the dynamical system, and derived **Lagrange's equations of motion**:

$$\frac{d}{dt}\left(\frac{\partial \mathfrak{L}}{\partial \dot{q}_j}\right)-\frac{\partial \mathfrak{L}}{\partial q_j}=0, \quad j=1,2,\ldots,f,$$

where $\dot{q}_j=dq_j/dt$, and $\mathfrak{L}=T-U$ ($T=$kinetic energy, $U=$potential energy) is a function of q_j and $\dot{q}_j$ called the **Lagrangian function**. Later, W. R. Hamilton introduced

$$p_j=\partial T/\partial \dot{q}_j,$$

$$H=\sum p_j\dot{q}_j-\mathfrak{L}=H(p_1,\ldots,p_f;q_1,\ldots,q_f)$$

and transformed the equations to **Hamilton's canonical equations**:

$$\frac{dq_j}{dt}=\frac{\partial H}{\partial p_j}, \quad \frac{dp_j}{dt}=-\frac{\partial H}{\partial q_j}, \quad j=1,2,\ldots,f.$$

Here p_j is the **generalized momentum** conjugate to q_j, and q_j, p_j are called **canonical variables**. If the functions representing the configuration of the dynamical system in terms of q_j do not explicitly contain the time t, the **Hamiltonian function** (or **Hamiltonian**) H coincides with the total energy of the system $T+U$.

B. Canonical Transformations

The transformation $(p,q)\to(P,Q)$ under which canonical equations preserve their form is called a **canonical transformation**. It is given by

$$p_j=\frac{\partial W}{\partial q_j}, \quad P_j=-\frac{\partial W}{\partial Q_j}, \quad K=H+\frac{\partial W}{\partial t},$$

where $W=W(q_1,\ldots,q_f;Q_1,\ldots,Q_f)$ and K is the Hamiltonian of the transformed system. The set of canonical transformations forms a group, called a **group of canonical transformations**. An infinitesimal transformation is given by

$$dp_j=-\epsilon\frac{\partial S}{\partial q_j}, \quad dq_j=\epsilon\frac{\partial S}{\partial p_j},$$

where ϵ is an infinitesimal constant. Here S is an arbitrary function and is said to be the **generating function** of the infinitesimal transformation. Canonical equations can be interpreted to mean that the variations of p and q during the time interval $\epsilon=dt$ are the infinitesimal canonical transformations whose generating function is $H(p,q,t)$.

The variation of an arbitrary function $F(p,q)$ under an infinitesimal transformation is given by

$$dF=\epsilon(F,S),$$

where

$$(u,v)=\sum_j\left(\frac{\partial u}{\partial q_j}\frac{\partial v}{\partial p_j}-\frac{\partial u}{\partial p_j}\frac{\partial v}{\partial q_j}\right)$$

$$=\sum_j\frac{\partial(u,v)}{\partial(q_j,p_j)}$$

is †Poisson's bracket. Therefore, the time rate of change of a dynamical quantity $F(p,q)$ can be written as

$$dF/dt=(F,H).$$

Thus the function $F(p,q)$ that satisfies $(F,H)=0$ is an †integral of the canonical equations.

If a canonical transformation $(p,q)\to(P,Q)$ such that $P_j=\alpha_j, Q_j=\beta_j$ are constant is found, the motion of the system can be determined by

$$p_j=\frac{\partial W}{\partial q_j},\qquad \beta_j=\frac{\partial W}{\partial \alpha_j},$$

where W is the †complete solution of the **Hamilton-Jacobi differential equation**:

$$\frac{\partial W}{\partial t}+H\left(\frac{\partial W}{\partial q_1},\ldots,\frac{\partial W}{\partial q_f};\quad q_1,\ldots,q_f,t\right)=0$$

(→ 84 Contact Transformations).

References

[1] E. T. Whittaker, A treatise on the analytical dynamics of particles and rigid bodies, Cambridge Univ. Press, fourth edition, 1937.
[2] T. Yamanouchi, Ippan rikigaku (Japanese; General dynamics), Iwanami, 1941; revised edition, 1957.
[3] K. Husimi, Gendai buturigaku o manabu tameno koten rikigaku (Japanese; Classical mechanics for the study of modern physics), Iwanami, 1964.
[4] G. D. Birkhoff, Dynamical systems, Amer. Math. Soc. Colloq. Publ., 1927.
[5] M. Born, Vorlesungen über Atommechanik, Springer, 1925.

24 (XI.2) Analytic Functions

A. General Remarks

A real-valued function $f(t)$ of a real variable t is said to be **analytic** at $t=t_0$ if it can be represented by a †power series in $t-t_0$ in a neighborhood of t_0 in $\mathbf{R}$. If $f(t)$ is defined on an open set of $\mathbf{R}$ at every point of which it is analytic, then $f(t)$ is called an **analytic function**, or, more precisely, a **real analytic function**.

Analogously, a complex-valued function $f(z)$ of a complex variable z defined on a †domain D of the complex plane $\mathbf{C}$ is said to be **analytic** at $z=z_0$ $(\in D)$ if it can be represented by a power series in $z-z_0$ in a neighborhood of z_0 in $\mathbf{C}$, and $f(z)$ is an **analytic function** in D if it is analytic at every point of D. In this article, we are concerned with analytic functions in this sense. To distinguish them from the real case, they are also called **complex analytic functions**. A complex analytic function $f(z)$ defined on D is †differentiable in D; therefore, it is †holomorphic in D. The converse is also true. Thus the term "analytic function" is synonymous with "holomorphic function" insofar as it concerns a complex function (i.e., a complex-valued function of a complex variable) on a domain, but in the theory of functions it takes on an additional meaning that will be explained later.

B. Analytic Continuations

Let $f(z)$ be a holomorphic function in a domain G of the complex plane $\mathbf{C}$ and G^* be a domain containing G as a proper subset. If there exists a function $F(z)$ holomorphic in G^* that coincides with $f(z)$ in G, then $F(z)$ is called an **analytic continuation** (or **analytic prolongation**) of $f(z)$ from G to G^*. If two functions f and g are holomorphic in D and $f(z)=g(z)$ on a subset E that has an †accumulation point in D, then f is identically equal to g in D (**theorem of identity**) since the †zeros of holomorphic functions must be isolated. So for given G, G^*, and $f(z)$, an analytic continuation $F(z)$ is uniquely determined if it exists.

The function $f_1(z)$ defined by the power series $P(z;a)=\sum_{n=0}^{\infty}a_n(z-a)^n$ with the radius of convergence $r_1>0$ is holomorphic in the domain $D_1: |z-a|<r_1$, and at a point b of D_1 it can be expanded into a power series $P(z;b)=\sum_{n=0}^{\infty}b_n(z-b)^n$ with the radius of convergence r_2 $(\geqslant r_1-|b-a|)$. If $r_2>r_1-|b-a|$, the domain $D_2: |z-b|<r_2$ is not entirely contained in D_1. Let $f_2(z)$ be the function defined in D_2 by $P(z;b)$. Then the function $F(z)$ that is equal to $f_1(z)$ in D_1 and to $f_2(z)$ in D_2 is an analytic continuation of $f_1(z)$ from D_1 to $D_1\cup D_2$ (a **direct analytic continuation** by power series).

We have the following classical theorems about analytic continuations:

Let D_1 and D_2 be two disjoint domains, and suppose that their respective boundaries C_1 and C_2 are †rectifiable simple closed curves and that the intersection of C_1 and C_2 contains an open arc Γ. If two holomorphic functions defined in D_1 and D_2, respectively, have finite common †boundary values at every point of Γ, then there exists an analytic continuation $F(z)$ of $f_1(z)$ and $f_2(z)$ to $D_1 \cup \Gamma \cup D_2$ (**Painlevé's theorem**). We sometimes call $f_2(z)$ a continuation of $f_1(z)$ beyond Γ. If Γ is not rectifiable, the continuation beyond Γ does not exist, in general.

Let $f(z)$ be holomorphic in a †Jordan domain D lying in the half-plane on one side of the real axis and containing an open interval I of the real axis in its boundary. If $f(z)$ has finite real boundary values at every point of I, then it can be continued analytically beyond I to the other side of the real axis; there the continued function is given by $\overline{f(\bar{z})}$ (**Schwarz's principle of reflection**). This theorem can be generalized to the case where the real interval is replaced by an †analytic curve.

A **harmonic continuation** of †harmonic functions is defined analogously to analytic continuation. Let D be a Jordan domain lying in the half-plane on one side of the real axis and having an open interval I on the real axis as a part of its boundary. If $u(z)$ is harmonic in D and has the boundary values 0 at every point of I, then $u(z)$ has a harmonic continuation beyond I.

C. Analytic Functions in the Sense of Weierstrass

Let a be a point of the z-sphere and t the †local canonical parameter at a; i.e., $t=z-a$ if $a \neq \infty$, and $t=z^{-1}$ if $a=\infty$. If a power series $P(z;a)=\sum_{n=0}^{\infty} c_n t^n$ has a positive radius of convergence, we call $P(z;a)$ a **function element** with center a on the z-sphere, after K. Weierstrass. $P(z;a)=\sum_{n=0}^{\infty} c_n (z-a)^n$ if $a \neq \infty$, and $P(z;a)=\sum_{n=0}^{\infty} c_n z^{-n}$ if $a=\infty$. They represent a holomorphic function in $|z-a|<r_a$ or in $r_a{}^{-1}<|z|\leqslant\infty$, respectively. If b is a point in the circle of convergence of the function element $P(z;a)$, by the †Taylor expansion of $P(z;a)$ at $z=b$, we obtain the power series $P(z;b)$ in $z-b$, which is a direct analytic continuation of $P(z;a)$. Let a and b be two points on a †Riemann sphere, and let $C: z=z(s)$ $(0\leqslant s\leqslant 1, z(0)=a, z(1)=b)$ be a curve joining a and b. We say that $P(z;a)$ is **analytically continuable** along C and that we obtain $P(z;b)$ at the end point b by the analytic continuation of $P(z;a)$ along C, if the following two conditions are satisfied: (i) To every $s\in[0,1]$ there corresponds a function element $P(z;z(s))$ with center $z(s)$; (ii) for every $s_0\in[0,1]$, we can take a suitable subarc $z=z(s)$ $(|s-s_0|\leqslant\epsilon, \epsilon>0)$ of C contained in the circle of convergence of $P(z;z(s_0))$ such that every function element $P(z;z(s))$ with $|s-s_0|\leqslant\epsilon$ is a direct analytic continuation of $P(z;z(s_0))$. When $P(z;a)$ and the curve C are given, the analytic continuation along C is uniquely determined (**uniqueness theorem of the analytic continuation**).

Given a function element $P(z;a)$ with center a, the set of all function elements obtained by every possible analytic continuation along every curve starting from a is called an **analytic function in the sense of Weierstrass** determined by $P(z;a)$. In this definition, we may restrict the curves to polygonal lines. An analytic function in this sense is completely determined by a single arbitrary function element belonging to it, so two analytic functions are identically equal if they have a common function element.

A †germ of a holomorphic function is identical to a function element, and the set of all germs has the natural structure of a †sheaf $\mathcal{O}$. In the terminology of sheaves, an analytic function is a connected component of $\mathcal{O}$, and an **analytic continuation along a curve** C is a continuous curve Γ in $\mathcal{O}$ whose projection is C.

D. Values and Branches of Analytic Functions

The value of an analytic function at a point b is, by definition, the value at b of its function elements with center b (whose existence is assumed; there may be several such elements). An analytic function is, in general, a multiple-valued function because analytic continuations along different curves with the same end points may lead to different function elements. For a given analytic function $f(z)$, if the maximal number of its function elements with the same center is n, we say it is **n-valued**, and if $n\geqslant 2$ we say it is **multiple-valued** (or **many-valued**). The number of function elements of $f(z)$ with the same center is at most †countably infinite, so the value of $f(z)$ at a point is a countable set (**Poincaré-Volterra theorem**). By introducing a †Riemann surface instead of the complex plane as the domain of definition of an analytic function, we may regard multiple-valued analytic functions as single-valued functions defined on a suitable Riemann surface (→ 362 Riemann Surfaces).

Let $f(z)$ be an analytic function and $P(z;a)$ be a function element belonging to $f(z)$, where a is a point of a domain D. The set of all function elements obtained from $P(z;a)$ by every possible analytic continuation along all curves in D is called a **branch** of $f(z)$ in D determined by $P(z;a)$. When D coincides with the whole complex plane, the branch of $f(z)$ in D is the function $f(z)$ itself. A function holomorphic in a domain D can be expanded in a power series with any point of D as its center, and the set of these power series (function elements) constitutes a branch of an analytic function.

Analytic continuations of a function element along two †homotopic curves, if they are possible, lead to the same result (**monodromy theorem**). In particular, if D is †simply connected and if analytic continuations of $P(z;a)$ are possible along all curves in D starting from a, then the branch of $f(z)$ in D determined by $P(z;a)$ is single-valued.

E. Invariance Theorem of Analytic Relations

Suppose the following four conditions hold: (1) $F(z,w)$ is a holomorphic function of two variables for $z \in \Delta_1$ and $w \in \Delta_2$, where Δ_1, Δ_2 are domains in the complex plane. (2) A curve $C: z = z(s)$ $(0 \leqslant s \leqslant 1, z(0) = a, z(1) = b)$ and two sets of function elements $P(z;z(s))$ and $Q(z;z(s))$ defined for every s $(0 \leqslant s \leqslant 1)$ are given. (3) $P(z;a)$ and $Q(z;a)$ can be continued analytically along C using $P(z;z(s))$ and $Q(z;z(s))$, respectively. (4) There exists a positive number $R(s)$ for every s $(0 \leqslant s \leqslant 1)$ such that, if $|z - z(s)| < R(s)$, the values of $P(z;z(s))$ and $Q(z;z(s))$ belong to Δ_1 and Δ_2, respectively. Under these conditions, if $F(P(z;a), Q(z;a)) = 0$ holds for $|z-a| < R(0)$, then $F(P(z;b), Q(z;b)) = 0$ holds for $|z-b| < R(1)$. In other words, an analytic relation between function elements belonging to two analytic functions that holds in a neighborhood of the starting point of a curve C is conserved for function elements with center at the terminal point b of C. This is called the **invariance theorem of analytic relations**. The same statement is valid for relations among more than two analytic functions and their derivatives (differential equations).

F. Inverse Functions

Suppose that $P(z;a)$ $(a \neq \infty)$ belongs to an analytic function $f(z)$ and $P'(a;a) \neq 0$. We consider the inverse function of $P(z;a)$ in a neighborhood of a and let $\mathfrak{P}(w;\alpha)$ $(\alpha = P(a;a))$ be its expansion as the power series in $w - \alpha$. We call $\mathfrak{P}(w;\alpha)$ the **inverse function element** (or simply **inverse element**) of $P(z;a)$ and the analytic function determined by $\mathfrak{P}(w;\alpha)$ the **inverse analytic function** (or simply **inverse function**) of $f(z)$. The inverse function is completely determined by $f(z)$ and is independent of the choice of $P(z;a)$. For example, analytic functions represented by $\sqrt{w}$ or $\log w$ are defined as the inverse function of z^2 or e^z, respectively.

G. Singularities of Analytic Functions

Hereafter, when we speak of a curve $C: z = z(s)$ $(0 \leqslant s \leqslant 1)$, it is always supposed that C is a curve in the complex plane starting at a and ending at ω. Let K_r be the open disk $|z - \omega| < r$; we denote by C_r the connected component of $C \cap K_r$ that contains ω. If analytic continuations of $P(z;a)$ are possible along any subarc of C with a terminal point arbitrarily near ω but impossible along the whole C, we say that the analytic continuation of $P(z;a)$ along C defines a **singularity** Ω of the coordinate ω, and that Ω lies over ω. For example, if $P(z;a)$ has a finite radius of convergence, for a suitable point ω on the circumference of the circle of convergence the analytic continuation of $P(z;a)$ along the radius $a\omega$ defines a singularity over ω. Now take a point z_r on C_r and denote be $F_r(z)$ the branch of an analytic function determined by $P(z;z_r)$ in K_r. Let Ω be a singularity determined by C and $P(z;a)$, and suppose that we are given another singularity Ω^* over ω determined by C^* and $P(z;a^*)$. If they define the same branch $F_r(z)$ for every K_r, by definition, we put $\Omega = \Omega^*$. Thus $F_r(z)$ defines an †unramified covering surface W_r of the disk K_r, and it is single-valued on W_r.

Singularities are classified according to the geometric structure of W_r and the value distribution of $F_r(z)$ on it. First, if W_r has no †relative boundary over $0 < |z - \omega| < r$ for a suitable r, then Ω is called an **isolated singularity** of the analytic function. In this case, the number k of points of W_r lying over a point z in $K_r - \{\omega\}$ is constant. If $k = \infty$, W_r has a †logarithmic branch point over ω, and Ω is called a **logarithmic singularity**. If $k < \infty$, $F_r(z)$ can be represented as a single-valued holomorphic function in $0 < |t| < r^{1/k}$ by putting $z = \omega + t^k$. In this case, if we introduce an additional point P_0 corresponding to $z = \omega$, then $W_r \cup \{P_0\}$ has only an †algebraic branch point over ω. Now, taking into account the value of $m = F_r(z)$, we call Ω an **algebraic singularity** if $\lim m$ exists. In this case, we have $F_r(z) = \sum_{n=l}^{\infty} c_n t^n$, and if we admit analytic continuations in the wider sense (which is defined later), $P(z;a)$ is analytically continuable along the whole C.

H. The Natural Boundary

Given a domain D and an analytic function holomorphic in D, if all boundary points of D are singularities of $f(z)$ and $f(z)$ is not continuable to the exterior of D, the boundary of D is called the **natural boundary** of $f(z)$. This phenomenon was first discovered for

†elliptic modular functions. Many results are known about power series for which the circumference of the circle of convergence is its natural boundary (→ 336 Power Series). For any given domain D in $\mathbf{C}$, there exists an analytic function whose natural boundary is the boundary of D. The original proof of this fact, given by Weierstrass, contained a defect that was corrected by J. Besse [5].

I. Analytic Continuation in the Wider Sense

Let two †Laurent series (with parameter t) $z=P(t)=\sum_{n=k}^{\infty}a_nt^n$ and $w=Q(t)=\sum_{n=l}^{\infty}b_nt^n$ (k and l are integers, and $a_kb_l\neq 0$) converge in $0<|t|<r$, and let $(P(t_1), Q(t_1))\neq(P(t_2), Q(t_2))$ if $t_1\neq t_2$; then we say that the pair (P,Q) defines a **function element in the wider sense**. If a change of parameter $\tau=r_1t+r_2t^2+\ldots$ ($r_1\neq 0$ and the radius of convergence >0) gives $P(t)=\Pi(\tau)$, $Q(t)=K(\tau)$, we say that (Π,K) and (P,Q) define the same function element. By a suitable choice of parameter, any function element can be given the form $z=t^k+a$ (or $z=t^{-k}$), $w=\sum_{n=l}^{\infty}b_nt^n$, and the elimination of t gives the representation of w as a †Puiseux series of z. So if $k=1$ and $l\geqslant 0$, it reduces to a holomorphic function element. In the case where $k=1$, with $l<0$ not excluded, the above element is called a **rational element**. If $k>1$ it is called a **ramified element**, and if $l<0$ it is called a **polar element**.

If P', Q' are the direct analytic continuations of P and Q at t_0 $(0<|t_0|<r)$, i.e., their Taylor expansions at t_0, the function element (P',Q') is called a direct analytic continuation of (P,Q), which is also considered its own direct analytic continuation. For a fixed r, the set of all direct continuations of (P,Q) thus obtained is called an **analytic neighborhood** of (P,Q), and these neighborhoods define a topology in the set of all function elements. A curve in this topological space is called an **analytic continuation in the wider sense**, and a †connected component of this space is called an **analytic function in the wider sense**. An analytic function in the wider sense is a set of function elements in the wider sense, but it may also be regarded as a function $w=f(z)$ (with an independent variable z and a dependent variable w) defined by each function element $p(z,w): z=P(t)$, $w=Q(t)$. Given an analytic continuation in the wider sense

$$p(s)=p(z,w;s);$$

$$z=z(s)+t^{k(s)},\quad w=\sum_{n=l(s)}^{\infty}c_n(s)t^n,\quad 0\leqslant s\leqslant 1,$$

it is sometimes called an analytic continuation along the curve $C: z=z(s)$ $(0\leqslant s\leqslant 1)$ in the complex plane. If all $p(s)$ are holomorphic function elements, this coincides with the analytic continuation along C in the original sense, but if this is not the case, $p(0)$ and C do not necessarily determine $p(1)$ uniquely. Actually, an analytic function in the wider sense is just an analytic function in the original sense with at most a countable number of ramified or polar elements added.

J. Singularities of Analytic Functions in the Wider Sense

Suppose the following three conditions hold: (1) For every point on C except ω, that is, for $z(s)$ $(0\leqslant s<1)$, a function element in the wider sense $p(z,w;s)$ is given. (2) For every λ (<1), $p(z,w;s)$ $(0\leqslant s\leqslant\lambda)$ constitutes an analytic continuation in the wider sense. (3) It is impossible to find a function element $p(z,w;1)$ such that $p(z,w;s)$ $(0\leqslant s\leqslant 1)$ is an analytic continuation in the wider sense. When these three conditions are satisfied, we say that $p(z,w;s)$ $(0\leqslant s\leqslant 1)$ defines a **transcendental singularity** Ω with ω as its coordinate. The method of determining a branch $w=F_r(z)$ in an open disk with center ω is completely parallel to the case of holomorphic analytic functions. Because of the appearance of function elements in the wider sense in $F_r(z)$, the covering surface W_r of K_r defined by $F_r(z)$ may have algebraic branch points. We call Ω an **isolated singularity** of an analytic function in the wider sense if W_r is unramified and has no relative boundary over $K_r-\{\omega\}$ for a suitable choice of r. In particular, if W_r has a logarithmic branch point over ω, Ω is called a **logarithmic singularity**. If W_r has no point over ω for suitable r, Ω is called a **direct transcendental singularity**; otherwise, it is called an **indirect transcendental singularity**. All isolated singularities are direct singularities. Taking into account the value of $w=F_r(z)$, if the †cluster set of F_r at $\Omega: S_\Omega=\bigcap_{r>0}\overline{\{F_r(z)\}}$ consists of only one point, it is an **ordinary singularity**; if not, it is an **essential singularity**. The inverse function of a single-valued meromorphic function $z=\varphi(w)$ in $|w|<+\infty$ has no essential singularity. For example, the inverse function of $z=w\sin w$ has a direct singularity over $z=\infty$ that is not logarithmic, and the inverse function of $z=(\sin w)/w$ has an indirect singularity over $z=0$.

K. History

A function of a complex variable is **monogenic** in the sense of A. L. Cauchy if it is differentiable at every point of its domain of definition. It was B. Riemann who succeeded in developing Cauchy's concept. Riemann considered an analytic function as a function defined on a †Riemann surface, that is, a 1-dimensional complex analytic manifold. On the other hand, Weierstrass constructed the theory of analytic functions starting from power series. When we speak of single-valued functions defined in a domain of the complex plane, the monogenic functions of Cauchy and the analytic functions of Weierstrass are identical. Although the analytic functions are very special functions, the study of complex analytic functions is usually called the theory of functions of a complex variable, or simply the theory of functions.

By considering the following point set C, which is more general than a domain, E. Borel showed that a monogenic function on C is not necessarily holomorphic in the ordinary sense. Take a countable dense subset in a subdomain D' of a domain D and a double sequence of positive numbers $\{r_n^{(h)}\}$. Put $S_n^{(h)}=\{z \mid |z-z_n|<r_n^{(h)}\}$ and $C^{(h)}=D-\bigcup_{n=1}^{\infty}S_n^{(h)}$. By a suitable choice of $r_n^{(h)}$, we may suppose that the $C^{(h)}$ are connected and monotone increasing with respect to h. Put $C=\bigcup_{h=1}^{\infty}C^{(h)}$. A function defined in C is by definition **monogenic** if it is differentiable in $C^{(h)}$ for every h. For such a monogenic function, Cauchy's †integral formula in a generalized form holds, and the function is infinitely differentiable. If $f(z)$ and $g(z)$ are monogenic in C and coincide on a curve in C, then they are identical in C. Let D be the set $\{z \mid 0<\operatorname{Re}z<1, 0<\operatorname{Im}z<1\}$ and $\{z_n\}$ be all rational points in D ($z_n=(p+iq)/m$). For a natural number h, we define $C^{(h)}$ to be the set D minus the union of open disks with radius $\exp(-e^{m^2})/h$ and center $(p+iq)/m$. The function

$$f(z)=\sum_{m=1}^{\infty}\sum_{p=0}^{m}\sum_{q=0}^{m}\frac{\exp(-e^{m^4})}{z-(p+iq)/m}$$

is monogenic in C in the above-mentioned sense, but not holomorphic in C. The study of these functions developed into the theory of †quasi-analytic functions.

The concept of †analytic functions of several complex variables can also be defined analogously to the case of one variable. Then nonuniformizable singularities appear that lead to a generalization of the concept of †manifolds (→ 27 Analytic Spaces).

References

[1] Y. Yosida, Kansûron (Japanese; Theory of functions), Iwanami, 1938; revised edition, 1965.
[2] K. Noshiro, Kaiseki setuzoku nyûmon (Japanese; Introduction to analytic continuation), Kyôritu, 1964.
[3] K. Noshiro, Kindai kansûron (Japanese; Modern theory of functions), Iwanami, 1954.
[4] H. Weyl, Die Idee der Riemannschen Fläche, Teubner, 1913; third edition, 1955; English translation, The concept of a Riemann surface, Addison-Wesley, 1964.
[5] J. Besse, Sur le domaine d'existence d'une fonction analytique, Comment. Math. Helv., 10 (1937), 302–305.
Also → references to 200 Holomorphic Functions. For Borel's theory,
[6] E. Borel, Leçons sur les fonctions monogènes uniformes d'une variable complexe, Gauthier-Villars, 1917.
For the history and concepts of analytic functions,
[7] G. Julia, Essai sur le développement de la théorie des fonctions de variables complexes, Gauthier-Villars, 1933.

25 (XI.19) Analytic Functions of Several Complex Variables

A. Holomorphic Functions

As in the case of †holomorphic functions of one complex variable, the definition of holomorphic functions can be given in two ways: the first definition utilizes differentiability, following the approach of B. Riemann; and the second method utilizes the notion of power series expansion as developed by K. Weierstrass. In this article we use $\mathbf{N}=\{0,1,2,\ldots\}$.

B. Power Series

Let z be an n-tuple of complex variables $z_1,\ldots,z_n$, and $c=(c_1,\ldots,c_n)$ a point of $\mathbf{C}^n$. A family P of monomials $a_k(z-c)^k=a_{k_1,\ldots,k_n}(z_1-c_1)^{k_1}\ldots(z_n-c_n)^{k_n}$ ($k=(k_1,\ldots,k_n)\in\mathbf{N}^n$), where $a_k\in\mathbf{C}$, is called a **power series** with center c and coefficients a_k. If, for a bijection φ of $\mathbf{N}$ onto $\mathbf{N}^n$, the simple series $\sum_{p\in\mathbf{N}}|a_{\varphi(p)}(z-c)^{\varphi(p)}|$ is convergent at $z=z^0$, we say that P is **absolutely convergent** at z^0. Its sum at z^0, denoted by $\sum a_k(z^0-c)^k$, is defined as the sum $\sum a_{\varphi(p)}(z^0-c)^{\varphi(p)}$, which is independent of the choice of φ. If the family P is uniformly bounded at z^0, then P is absolutely convergent at every point of the open **polydisk** $S=\{z \mid |z_j-c_j|<|z_j^0-c_j|, j=1,\ldots,n\}$. Furthermore, in this case P converges absolutely and uniformly on every compact set in S (N. H. Abel).

The **convergence domain** of a power series P is the set D of points z^0 such that P is absolutely convergent at every point in a neighborhood of z^0. The interior of the set B of points at which the family P is uniformly bounded is equal to D. A **(complete) Reinhardt domain** with center c is a domain D in $\mathbf{C}^n$ such that whenever D contains z_0, the domain D also contains the torus $\{z \,|\, |z_j - c_j| = |z_j{}^0 - c_j|,\ j = 1, \ldots, n\}$ (the closed polydisk $\{z \,|\, |z_j - c_j| \leqslant |z_j^0 - c_j|, j = 1, \ldots, n\}$). If the convergence domain D of the power series P is not empty, it is a complete Reinhardt domain and is also **logarithmically convex**; that is, the set $D - \bigcup_j \{z | z_j = c_j\}$ is mapped onto a convex domain in $\mathbf{R}^n$ by the mapping $z_j \rightarrow \log|z_j - c_j|$ $(j = 1, \ldots, n)$. The set $\tilde{D}$ of points at which P is absolutely convergent is, in general, greater than D, and it is possible that $\tilde{D}$ contains exterior points of D. A **thorn** of D is the set of exterior points of D contained in $\tilde{D}$ that are located on the planes $\{z | z_j = c_j\}$ $(j = 1, \ldots, n)$. An n-tuple $r \in \mathbf{R}^n_+$ is called a set of **associated convergence radii** if P is absolutely convergent at every point of $\{z \,|\, |z_j - c_j| < r_j, j = 1, \ldots, n\}$ but not of $\{z \,|\, |z_j - c_j| > r_j, j = 1, \ldots, n\}$. An n-tuple of associated convergence radii may not be uniquely determined, but it satisfies

$$\limsup_{|k| \rightarrow +\infty} \left(|a_k| r^k\right)^{1/|k|} = 1, \quad |k| = k_1 + \ldots + k_n$$

(E. Lemaire).

Let f be a complex-valued function defined in a neighborhood of $z^0 \in \mathbf{C}^n$. If there exists a convergent power series P with center z^0 such that at every point of a neighborhood of z^0 the value of f and the sum of P coincide, then f is called **analytic** at z^0 in the sense of Weierstrass, and P is the **Taylor expansion** of f at z^0.

C. Differentiability

Let f be a complex-valued function defined in a neighborhood of $z^0 \in \mathbf{C}^n$. If in a neighborhood of z^0 we have

$$f(z) - f(z_0^0) = \alpha_1(z_1 - z_1^0) + \ldots + \alpha_n(z_n - z_n^0) + \epsilon,$$

with $\alpha_1, \ldots, \alpha_n \in \mathbf{C}$ and

$$\lim_{z \rightarrow z^0} \epsilon / (|z_1 - z_1^0| + \ldots + |z_n - z_n^0|) = 0, \tag{1}$$

then we say that f is **(totally) differentiable** at z^0. The function f is then continuous at z^0, and the partial derivatives $\partial f / \partial z_j$ $(j = 1, \ldots, n)$ exist. Furthermore, the Cauchy-Riemann differential equations $\partial f / \partial \bar{z}_j = 0$ $(j = 1, \ldots, n)$ hold, where $\partial f / \partial z_j = (1/2)(\partial f / \partial x_j - i \partial f / \partial y_j)$ and $\partial f / \partial \bar{z}_j = (1/2)(\partial f / \partial x_j + i \partial f / \partial y_j)$ with $z_j = x_j + i y_j$. We say that f is **holomorphic** at z^0 in the sense of Riemann if f is differentiable at every point in a neighborhood of z^0. Analyticity in the sense of Weierstrass is equivalent to holomorphy in the sense of Riemann. Furthermore, if the partial derivatives $\partial f / \partial z_j$ $(j = 1, \ldots, n)$ exist at every point in a neighborhood of z^0, then f is, without assuming continuity, proved to be holomorphic. Thus the holomorphy of f in each variable z_j implies the holomorphy of f in $z = (z_1, \ldots, z_n)$ (**Hartogs's theorem of holomorphy**, 1906).

A complex-valued function in a domain $G \subset \mathbf{C}^n$ is called **holomorphic** in G if it is holomorphic at every point of G. Let $H(G)$ be the ring of holomorphic functions in G. For $f = u + iv \in H(G)$, u and v satisfy in G the differential equations $\partial^2 u(z, \bar{z}) / \partial z_j \partial \bar{z}_k = 0$; that is,

$$\frac{\partial^2 u}{\partial x_j \partial x_k} + \frac{\partial^2 u}{\partial y_j \partial y_k} = 0, \tag{2}$$

$$\frac{\partial^2 u}{\partial x_j \partial y_k} - \frac{\partial^2 u}{\partial x_k \partial y_j} = 0, \qquad j, k = 1, \ldots, n.$$

A †distribution $T \in \mathcal{D}'(G)$ is called **pluriharmonic** in G if it satisfies (2) in G. Then T is harmonic and hence a real analytic function.

Let G_j be a domain in the z_j-plane with piecewise smooth boundary C_j. If $f \in H(G)$ $(G = \prod_{j=1}^n G_j)$ is continuous on $\bar{G}$, then

$$\frac{1}{(2\pi i)^n} \int_{C_1 \times \ldots \times C_n} \frac{f(\zeta)}{(\zeta_1 - z_1) \ldots (\zeta_n - z_n)} d\zeta_1 \wedge \ldots \wedge d\zeta_n = \begin{cases} f(z), & z \in G, \\ 0, & z \notin \bar{G} \end{cases} \tag{3}$$

(**Cauchy's integral representation**). Thus if $n \geqslant 2$, then f is determined by its values only on the proper subset $C = C_1 \times \ldots \times C_n$ of ∂G, which is called the **skeleton** (or **determining set**) of G. For (pluriharmonic) functions of several complex variables, the boundary value problem, not necessarily solvable in its classical form, is not so effective as the Dirichlet problem in one complex variable.

As in the case of one variable, the †Laurent expansion is valid for every holomorphic function in a domain of the form $G = \prod_j G_j$, where the G_j are circular annuli $\subset \mathbf{C}$. Suppose that we are given $f_1 \in H(G_1)$ and $f_2 \in H(G_2)$, where G_1, G_2 are domains in $\mathbf{C}^n$ such that $G_1 \cap G_2$ is nonempty and connected. If $f_1 = f_2$ on $\{z \,|\, |z_j - z_j{}^0| < r_j, y = y^0, j = 1, \ldots, n\}$, where $z^0 = x^0 + i y^0 \in G_1 \cap G_2$, then there exists a unique $f \in H(G_1 \cup G_2)$ such that $f | G_1 = f_1$ and $f | G_2 = f_2$ (**theorem of identity**). Thus †analytic continuation proceeds as in the case of one variable. Similarly, some fundamental theorems in one variable, such as †Liouville's theorem on †entire functions and the †maximum principle, hold also for several variables. However, there are some properties that reveal the differences between the cases of one and several variables. For instance, the set of zeros of a holomorphic function (→ 27 Analytic Spaces B) has no isolated point for $n \geqslant 2$. The investigation of these remarkable differences is one of the purposes of the theory of analytic functions of several complex variables.

D. Šilov Boundaries

While the maximum principle holds for a holomorphic function in a domain G, the set of points where the maximum is attained may be a proper subset S of ∂G. For instance, if G is the product of annuli as before, then the skeleton of G can be taken as S. In connection with the theory of †normed rings, G. E. Šilov proved that there exists a unique smallest member S_0 (called the **Šilov boundary** of G) in the family of closed subsets S such that $\sup\{|f(z)| \mid z \in S\} = \sup\{|f(z)| \mid z \in G\}$ for every $f \in H(G)$ continuous on $\bar{G}$. The structure of S_0 is investigated in detail together with the pseudoconvexity of G connected with it. Applying †Perron's method for the Dirichlet problem to †plurisubharmonic functions and the Šilov boundary, H.-J. Bremermann solved one type of boundary problem.

E. Local Theory

Let f and g be two functions in neighborhoods of $S \subset \mathbf{C}^n$. If $f = g$ in a neighborhood of S, then f and g are called **equivalent** with respect to S. The germ of f on S, denoted by f_S, is the equivalence class of f. A **germ of a holomorphic function** on S is the germ on S of a holomorphic function defined on a neighborhood of S, and $H(S)$ denotes the ring of germs of holomorphic functions on S. Given a point O in $\mathbf{C}^n$, $H(O) = H(\{O\})$ is isomorphic to the ring H_n of convergent power series at O, i.e., the power series that are absolutely convergent in some neighborhoods of O. For every nonzero function $f \in H(O)$, there exists a system of coordinates $(z_1, \ldots, z_n)$ centered at O such that $f(0, \ldots, 0, z_n) \neq 0$ for every z_n in a neighborhood of $z_n = 0$. In a neighborhood of O, f is then equal to the product of an invertible element of H_n and a **distinguished pseudopolynomial**

$$P(z_n) = z_n^p + a_1(z_1, \ldots, z_{n-1}) z_n^{p-1} + \ldots + a_p(z_1, \ldots, z_{n-1}) \in H_{n-1}[z_n],$$

with $a_1(0, \ldots, 0) = \ldots = a_p(0, \ldots, 0) = 0$, and $P(z_n)$ is uniquely determined by f and the coordinates $z_1, \ldots, z_n$ (**Weierstrass's preparation theorem**). It follows from this that H_n is an n-dimensional †regular local ring. Considering $H(O)$ as the †inductive limit of †locally convex rings $H(U)$, where U ranges over a base for a neighborhood system of O, H. Cartan proved the preparation theorem in a more precise form in which the association $f \rightarrow a_j$ is continuous with respect to the supremum norm. Upon careful consideration of this situation, K. Oka proved a theorem of fundamental importance: the †sheaf $\mathcal{O}_{\mathbf{C}^n}$ defined by $\mathcal{O}_{\mathbf{C}^n, z} = H(z)$ $(z \in \mathbf{C}^n)$ is †coherent.

F. Domains of Holomorphy

Given a domain $G \subset \mathbf{C}^n$ for $n \geqslant 2$, it may be that there exists a domain G' strictly greater than G such that all the functions that are holomorphic in G extend to holomorphic functions in G' (**analytic continuation**). For instance, let $S = S' \times \sigma$, where S' and σ are open polydisks in $(z_1, \ldots, z_{n-1})$-space and z_n-space, respectively, and let $T \subset \mathbf{C}^n$ be an open set. If there exists an open set $U (\neq \varnothing) \subset S'$ such that $(U \times \sigma) \cup (S' \times \partial\sigma) \subset T$ and if $S \cap T$ is connected, then all the functions that are holomorphic in S extend uniquely to holomorphic functions in $S \cup T$ (**Hartogs's continuation theorem**). In particular, if A is an †analytic set in a domain $G \subset \mathbf{C}^n$ with $\dim A \leqslant n-2$, then all the functions that are holomorphic in $G - A$ extend uniquely to holomorphic functions in G. Furthermore, if A is an analytic set in G with $A \neq G$, then every $f \in H(G-A)$ that is locally bounded at the points of A extends uniquely to a holomorphic function in G (**Riemann's continuation theorem** for $n \geqslant 2$). The domain $\tilde{G}_f$ of holomorphy for f is defined to be the maximal domain to which f may be continued analytically. A domain G is called a **domain of holomorphy** if $G = \tilde{G}_f$ for some $f \in H(G)$. However, $\tilde{G}_f$ is, in general, not a subdomain of $\mathbf{C}^n$. $\tilde{G}_f$ is, generally, a manifold spread over $\mathbf{C}^n$; i.e., $\tilde{G}_f$ is a connected n-dimensional complex analytic manifold with a holomorphic mapping $\varphi: \tilde{G}_f \rightarrow \mathbf{C}^n$ of maximum Jacobian rank (φ is then an open mapping). The same is true for the common existence domain of functions in a family $H(G)$. The common existence domain $\tilde{G}$ of all the functions in $H(G)$ is called the **envelope of holomorphy**. A **holomorphically complete domain** is a domain G such that $G = \tilde{G}$. These notions carry over to the case where G is a manifold spread over $\mathbf{C}^n$. The (general) Levi problem of determining the conditions for a given domain to be holomorphically complete is fundamental to the theory of analytic functions of several complex variables (→ I). In connection with this problem, various notions of pseudoconvexity of holomorphically complete domains are defined.

G. Pseudoconvexity

An upper semicontinuous real-valued function u $(-\infty \leqslant u < +\infty)$ in a domain $G \subset \mathbf{C}^n$ is said to be **plurisubharmonic** if for every $z^0 \in G$ and every $a \in \mathbf{C}^n$ the function $u(z^0 + ta)$ of t is †subharmonic (including the constant $-\infty$) in all the connected components of $\{t \mid z^0 + ta \in G\}$. A domain G is said to be **pseudoconvex** (or ***d*-pseudoconvex**) if $u = -\log d_G$ is plurisubharmonic in G, where $d_G(z)$ is the distance of $z \in G$ from ∂G with respect to any norm in $\mathbf{C}^n$. Every connected component of the interior of the intersection of a family of pseudoconvex domains is pseudoconvex, and the union of an increasing sequence of pseudoconvex domains is pseudoconvex. Suppose that we are given a domain G and a function u of class C^2 in a neighborhood of $\bar{G}$ such that $G = \{z \mid u(z) < 0\}$ and, for some $\epsilon > 0$, $\Sigma_{jk}(\partial^2 u / \partial z_j \partial \bar{z}_k) a_j \bar{a}_k \geqslant \epsilon |a|^z$ for every $a \in \mathbf{C}^n$. Then the domain G is said to be **strongly pseudoconvex**. Strong pseudoconvexity implies pseudoconvexity. Every pseudoconvex domain is exhausted by an increasing sequence of strongly pseudoconvex domains. An open set $P \subset \mathbf{C}^n$ is called an **analytic polyhedron** if $P = \{z \| \chi_\alpha(z)| < 1, \alpha = 1, \ldots, N\}$, $\chi_\alpha \in H(\bar{P})$ $(\alpha = 1, \ldots, N)$. Then every connected component of P is pseudoconvex. A **Weil domain** is a connected and bounded analytic polyhedron P defined by $\chi_\alpha (\alpha = 1, \ldots, N)$ with $N \geqslant n$, such that for every k $(1 \leqslant k \leqslant n)$ the intersection of the hypersurfaces $|\chi_{\alpha_i}(z)| = 1$ $(1 \leqslant i \leqslant k)$ is of dimension $\leqslant 2n - k$.

H. Holomorphic Convexity

A domain $G \subset \mathbf{C}^n$ is called **holomorphically convex** if for every compact set $K \subset G$, $\hat{K} = \bigcap_{f \in H(G)} \{z \mid |f(z)| \leqslant \sup_{\omega \in K} |f(\omega)|\}$ (the **holomorphic hull** of K) is a compact set contained in G. (For a domain G contained in an †analytic set we can similarly define holomorphic convexity of G.) Every connected component of the intersection of a family of holomorphically convex domains is holomorphically convex, and a holomorphically convex domain is exhausted by an increasing sequence of Weil domains. Holomorphic completeness implies holomorphic convexity. The converse is true for domains in $\mathbf{C}^n$. If G is holomorphically convex, then for every point ζ of ∂G there exists an $f \in H(G)$ such that f is not locally bounded at ζ (H. Cartan and P. Thullen, 1932). Hence a holomorphically convex domain is a domain of holomorphy. Thus a domain in $\mathbf{C}^n$ is holomorphically convex if and only if it is a domain of holomorphy. (The same is true for unramified covering domains over $\mathbf{C}^n$ (Oka, 1953).) The union of an increasing sequence of domains of holomorphy is a domain of holomorphy (H. Behnke and K. Stein, 1938). Suppose that we are given a domain G and domains S_α, T_α $(\alpha = 1, 2, \ldots)$ such that $\overline{S_\alpha \cup T_\alpha} \subset G$ and $\sup_{T_\alpha} |f| = \sup_{S_\alpha \cup T_\alpha} |f|$ for every $f \in H(G)$. Suppose also that $S_0 = \lim S_\alpha$ is bounded. We say that the **continuity principle** holds in G if $\bar{T}_0 \subset G$ $(T_0 = \lim T_\alpha)$ implies $\bar{S}_0 \subset G$. The continuity principle holds in a domain of holomorphy (**Hartogs's theorem of continuity**). Hence, if G is a simply connected bounded domain $\subset \mathbf{C}^n$ $(n \geqslant 2)$ with connected boundary ∂G, then every function holomorphic in a neighborhood of ∂G extends to a holomorphic function in G (**Hartogs-Osgood theorem**). In particular, for $n \geqslant 2$, the set of singular points of a holomorphic function has no isolated point. A domain is pseudoconvex if the continuity principle holds there. Hence a domain of holomorphy is pseudoconvex.

I. The Levi Problem

Let G be a domain in $\mathbf{C}^n$ and $z^0 \in \partial G$. If there exists an open neighborhood U of z^0 such that every connected component of $G \cap U$ is a domain of holomorphy, then G is called **Cartan pseudoconvex** at z^0. On the other hand, if every 1-dimensional analytic set that has z^0 as an ordinary point contains points not belonging to $G \cup \{z^0\}$ in the neighborhoods of z^0, then G is called **Levi pseudoconvex** at z^0. Furthermore, G is called **locally Cartan (Levi) pseudoconvex** if G is Cartan (Levi) pseudoconvex at every point of ∂G. Every domain of holomorphy is locally Cartan pseudoconvex. If G is pseudoconvex and there exists a neighborhood U of z^0 such that $G \cap U = \{z \mid \varphi(z) < 0\}$, where $\varphi \in C^1(U)$, then G is Levi pseudoconvex at z^0.

The (proper) **Levi problem** of whether every pseudoconvex domain is a domain of holomorphy was proposed by E. E. Levi (1911). After unsuccessful efforts by various mathematicians to solve the problem, it was affirmatively solved by Oka (1942 for $n = 2$ and 1953 for manifolds spread over $\mathbf{C}^n$ for $n \geqslant 2$), H. Bremermann, and F. Norguet. A fundamental step in the solution is Oka's **gluing theorem**: Let G be a bounded domain $\subset \mathbf{C}^n$. If every connected component of $G_1 = \{z \mid x_1 > a\} \cap G$

and $G_2 = \{z \mid x_1 < b\} \cap G\,(a<b)$ is a domain of holomorphy, then G is a domain of holomorphy. Indeed, by virtue of the Behnke-Stein theorem and the fact that every pseudoconvex domain is the union of an increasing sequence of bounded locally Cartan pseudoconvex domains, it suffices to solve the Levi problem in the case of a bounded locally Cartan pseudoconvex domain. The Levi problem in this case is solved by the gluing theorem. Various integral representations of holomorphic functions are known besides the Cauchy representation. The Bergmann-Weil integral representation in a Weil domain was used as an important means of solving the Levi problem.

J. Holomorphic Mappings

The notion of holomorphic function with values in a †quasicomplete †locally convex complex vector space E was introduced only recently. The classical theory described before has been generalized, to some extent, to this case. In this way, many applications of the theory have been discovered. An E-valued function in a domain $G \subset \mathbf{C}^n$ is holomorphic if and only if the mapping $u \circ f: G \rightarrow \mathbf{C}$ is holomorphic for every continuous linear form u on E. By this theorem, most problems concerning E-valued holomorphic functions can be reduced to those of ordinary holomorphic functions. Note that the vector space $H(G)$ of ordinary holomorphic functions in G is a †Fréchet space. The spaces $\mathbf{C}^p$ and complex †Banach spaces belong to the above category of E. A $\mathbf{C}^p$-valued holomorphic function in a domain $G \subset \mathbf{C}^n$ is called a **holomorphic mapping** of G into $\mathbf{C}^p$. An isomorphism in the category of domains $G \subset \mathbf{C}^n$ and holomorphic mappings is called an **analytic isomorphism** (or **biholomorphic mapping**). With every domain $G \subset \mathbf{C}^n$ is associated the sheaf $\mathcal{O}_G$ of germs of holomorphic functions over G. Thus we have the notion of a †ringed space $(G, \mathcal{O}_G)$. A complex analytic manifold may be defined as a (Hausdorff) ringed space that is locally isomorphic to some $(G, \mathcal{O}_G)$.

A **meromorphic function** in G is a function that is locally the quotient of two holomorphic functions with denominator $\neq 0$. It may be more strictly defined as a meromorphic mapping of G into $\mathbf{P}_1(\mathbf{C})$ (→ 27 Analytic Spaces D).

K. The Cousin Problems

The Cousin problems are those of constructing meromorphic functions with given zeros or poles. In terms of sheaves the problems are stated as follows: Let $\mathcal{K}_G$ be the sheaf of germs of meromorphic functions over a domain $G \subset \mathbf{C}^n$. The **first Cousin problem** asks whether the mapping $\Gamma(G, \mathcal{K}_G) \rightarrow \Gamma(G, \mathcal{P}_G)$ induced by the exact sequence $0 \rightarrow \mathcal{O}_G \rightarrow \mathcal{K}_G \rightarrow \mathcal{P}_G \rightarrow 0$ ($\mathcal{P}_G = \mathcal{K}_G / \mathcal{O}_G$) is surjective, where $\Gamma(G, \mathcal{F})$ is the module of †sections of $\mathcal{F}$ over G (→ 377 Sheaves C). Let $\mathcal{K}_G{}^*$ be the sheaf of multiplicative groups of germs of meromorphic functions not identically 0 and $\mathcal{O}_G{}^*$ be the subsheaf of $\mathcal{K}_G{}^*$ formed by germs of holomorphic functions. The **second Cousin problem** asks whether the mapping $\Gamma(G, \mathcal{K}_G{}^*) \rightarrow P(G, \mathcal{D}_G)$ ($\mathcal{D}_G = \mathcal{K}_G{}^* / \mathcal{O}_G{}^*$) is surjective.

P. Cousin (1895) solved the first problem for $G = \mathbf{C}^n$ or $\prod_{j=1}^n G_j$ and the second problem for $G = \mathbf{C}^n$. Oka (1935) proved that the first problem is solvable in every domain of holomorphy. In solving the second problem in a domain of holomorphy, Oka established the notion of †fiber bundles and proved that the problem for any domain is reduced to holomorphic triviality of a holomorphic principal fiber bundle over the domain, and that holomorphic triviality is equivalent to topological triviality when the domain is of holomorphy (**Oka's principle**). Using the solutions of the Cousin problems, Oka proved his gluing theorem, described in Section I.

L. Stein Manifolds

Abstracting from certain important properties of a domain of holomorphy, Stein introduced the following category of complex analytic manifolds $(X, \mathcal{O}_X)$: (1) X is paracompact (i.e., each connected component of X has a countable open base). (2) Functions in $\Gamma(X, \mathcal{O}_X)$ separate the points of X. (3) For every point $x \in X$ there exists a system of local coordinates around x that is formed by functions in $\Gamma(X, \mathcal{O}_X)$. (4) X is holomorphically convex. $(X, \mathcal{O}_X)$ is then called a **Stein manifold**. It was later discovered by H. Grauert that conditions (2) and (4) imply (1) and (3).

Applying the theory of †cohomology with coefficients in sheaves, H. Cartan and J.-P. Serre obtained the †**fundamental theorems A** and **B** on Stein manifolds (→ 74 Complex Manifolds). Conversely, for a complex analytic manifold X, if for every †coherent analytic sheaf $\mathcal{I}$ of ideals defined by a 0-dimensional analytic set in X (i.e., a discrete subset of X), $H^1(X, \mathcal{I}) = 0$, then X is a Stein manifold. Furthermore, if $\Gamma(X, \mathcal{O}_X) = \Gamma(Y, \mathcal{O}_Y)$ for a Stein manifold Y (as in the case where $X \subset \mathbf{C}^n$), then the fundamental theorem A for every coherent sheaf of ideals implies that X is a Stein manifold (I. Wakabayashi).

By the fundamental theorems, most results

on domains of holomorphy hold unchanged for Stein manifolds. For instance, the first Cousin problem is always solvable. The second Cousin problem is solvable if and only if $H^2(X,\mathbf{Z})=0$. An n-dimensional Stein manifold can be realized as a domain of holomorphy spread over $\mathbf{C}^n$. Furthermore, some theorems on differentiable manifolds have analogues on Stein manifolds. For instance, the cohomology groups of the complex of holomorphic differential forms over a Stein manifold X are isomorphic to the cohomology groups $H^*(X,\mathbf{C})$ (**analogue of †de Rham's theorem**). Every n-dimensional Stein manifold X is realized as a closed complex analytic submanifold in $\mathbf{C}^{2n+1}$; that is, there exists an injective †proper holomorphic mapping $f: X \to \mathbf{C}^{2n+1}$ with $df \neq 0$. Consider all the holomorphic †principal fiber bundles over a Stein manifold X whose fibers are isomorphic to a complex Lie group G. The analytic isomorphism classes of the bundles and the elements in $H^1(X,G^a)$ (where G^a is the sheaf of germs of holomorphic mappings of X into G) are in one-to-one correspondence. The same is true for the topological isomorphism classes of the bundles and the elements in $H^1(X,G^c)$ (where G^c is the sheaf of germs of continuous mappings of X into G). The mapping $H^1(X,G^a) \to H^1(X,G^c)$ induced by the canonical injection $G^a \to G^c$ is bijective (Grauert). Every relatively compact domain in a complex analytic manifold is holomorphically convex if it is strongly pseudoconvex (Grauert). Hence such a domain is a Stein manifold. It follows from this that every real analytic manifold with countable base for open sets is realized as a closed real analytic submanifold of some $\mathbf{R}^n$.

M. Continuation of Analytic Sets

The application of the theory of cohomology with coefficients in sheaves is not restricted to problems concerning Stein manifolds. Given $G_0=\{z \,|\, |z_j|<1, 1 \leqslant j \leqslant n\}$ $(n \geqslant 3)$, $G_1=\{z \,|\, |z_1|<\epsilon, |z_j|<1, 2 \leqslant j \leqslant n\}$, and $G^{(m)}=G_1 \cup (G_0-\{z \,|\, z_2=\ldots=z_m=0\})$ $(\zeta \leqslant m \leqslant n)$, we have $H^p(G^{(m)}, \mathcal{O}_{G^{(m)}})=0$ $(1 \leqslant p \leqslant m-2)$ (**Scheja's theorem**). Let $\mathfrak{F}$ be a coherent analytic sheaf over a domain $G \subset \mathbf{C}^n$. If, for every point z of an analytic set $A \subsetneqq G$, $\mathfrak{F}_z = \{0\}$ or $p \leqslant n-\dim_z A-2-\mathrm{hd}_z \mathfrak{F}$ (where $\mathrm{hd}_z \mathfrak{F}$ is the †homological dimension of the $\mathcal{O}_{G,z}$-module $\mathfrak{F}_z$), then it follows from Scheja's theorem that the mapping $H^p(G,\mathfrak{F}) \to H^p(G-A,\mathfrak{F})$ induced by the canonical injection $G-A \to G$ is bijective. This generalizes Riemann's continuation theorem for holomorphic functions, which corresponds to the case $p=0$.

As with the continuation of holomorphic functions, we may consider the continuation of analytic sets. Let A be an analytic set in a domain $G \subset \mathbf{C}^n$ and S an analytic set in $G-A$. A point $z \in G$ is said to be **regular (essentially singular)** with respect to S if the closure $\bar{S}$ of S in G is (is not) analytic at z. If $\dim A < \dim_z S$ for every point $z \in S$, then $\bar{S}$ is analytic in G. If S is purely d-dimensional and $\bar{S}$ is analytic at a point of an irreducible component A' of A, then $\bar{S}$ is analytic at every point of A' that is not located in any other irreducible component of A. Furthermore, if $\dim A \leqslant \dim S$ and S is purely d-dimensional, then the following hold: (1) The set E of essential singularities of S is, if not empty, a purely d-dimensional analytic set in G, formed by irreducible components of A. (2) If every irreducible component of A contains points of E not located in any other irreducible component of A, then $A \subset E$, and A is, if not empty, purely d-dimensional. (3) If every d-dimensional irreducible component of A contains points that are regular with respect to S, then $\bar{S}$ is a purely d-dimensional analytic set in G (Thullen, Remmert, and Stein). By these results it is possible to give a proof for †Chow's theorem that every analytic set in $\mathbf{P}_n(\mathbf{C})$ is algebraic.

The continuation of holomorphic functions is related to the continuation of their graphs. W. Rothstein investigated the continuation of analytic sets to obtain the following analogue to Hartogs's theorem of continuity: If $G=G_1 \cup G_2$, $G_1=\{z \,|\, |z_1|<1/2, \Sigma_{j=2}^n |z_j|^2<1\}$, $G_2=\{z \,|\, |z_1|<1, 1/2<\Sigma_{j=2}^n |z_j|^2<1\}$, and $\tilde{G}=\{z \,|\, |z_1|<1, \Sigma_{j=2}^n |z_j|^2<1\}$ (the envelope of holomorphy of G) with $n \geqslant 3$, then every purely $(n-1)$-dimensional analytic set A in G extends to an analytic set in $\tilde{G}$; that is, there exists a purely $(n-1)$-dimensional analytic set $\tilde{A}$ in $\tilde{G}$ such that $A=\tilde{A} \cap G$. Recently, K. Kasahara and H. Fujimoto generalized this theorem to the case of analytic spaces.

N. History

In connection with †Abelian functions, analytic functions of several complex variables have been studied sporadically since the time of Riemann and Weierstrass (H. Poincaré, Cousin). A series of investigations by Hartogs (*Math. Ann.*, 62 (1906), etc.) that revealed the distinctive properties of several complex variables initiated a new epoch in complex analysis. Levi (1910–1911) generalized Hartogs's results to the case of meromorphic functions, introduced the notion of pseudoconvexity, and proposed the so-called Levi problem. After a lapse of time, many contributions to this new area of complex analysis have been made since 1920. The study by K. Reinhardt (1921) of analytic automorphisms was further developed by C. Carathéodory and Behnke. The †kernel function introduced by S. Bochner and S. Bergmann (1922) produced many remarkable results. In contrast with †Picard's theorem in one variable, P. Fatou found a holomorphic mapping $f: \mathbf{C}^2 \rightarrow \mathbf{C}^2$ with nonvanishing Jacobian such that the image $f(\mathbf{C}^2)$ has an exterior point.

The theory of analytic functions of several complex variables has flourished since 1926. Behnke and Thullen in Münster, together with G. Julia and H. Cartan in Paris, were the most active investigators. The results on †normal families of analytic functions of several complex variables (Julia, 1926), the uniqueness theorem of holomorphic mappings (H. Cartan, 1930), and a characterization of a domain of holomorphy by holomorphic convexity (Cartan and Thullen, 1932) are their most remarkable achievements. Behnke and Thullen [2] systematized the results obtained since the discovery of the theory by providing a complete bibliography of articles up to 1934.

The three major unsolved problems at that time—those of Cousin, Levi, and the approximation of holomorphic functions—were intensively studied by Oka in (1936), who has given complete solutions [8]. The investigation of ideals of holomorphic functions by H. Cartan (1944), together with that of ideals with indetermined domains by Oka, has developed into the theory of coherent analytic sheaves. The notion of analytic spaces, first introduced by Behnke and Stein (1951), led to the recent flourishing of the theory of analytic functions of several complex variables. The theory of cohomology with coefficients in sheaves has been effectively applied by H. Cartan and Serre (1951–1952). The introduction of the notion of Stein manifolds (1951) came at the same time. Grauert's deep investigations since 1955, together with those of Stein and Remmert, have contributed greatly to the development of the theory of analytic spaces. In the 1960s, active investigations took place also in the United States [6]. The theory of automorphic functions of several complex variables has been developed by C. L. Siegel, I. Satake, and others in connection with the theory of numbers. The process of analytic continuation has also been successfully applied to the theory of elementary particles in physics.

References

[1] W. F. Osgood, Lehrbuch der Funktionentheorie I, II, Teubner, 1924; revised edition, 1929 (Chelsea, 1965).
[2] H. Behnke and P. Thullen, Theorie der Funktionen mehrerer komplexer Veränderlichen, Erg. Math., Springer, 1934; second edition, 1970.
[3] S. Bochner and W. T. Martin, Several complex variables, Princeton, 1948.
[4] S. Hitotumatu, Tahensû kaiseki kansûron (Japanese; Theory of analytic functions of several complex variables), Baihûkan, 1960.
[5] Б. А. Фукс (B. A. Fuks), Введение в теорию аналитиских функций многих комплексных переменных физматгиз, 1962; English translation, Special chapters in the theory of analytic functions of several complex variables, Amer. Math. Soc. Transl. of Math. Monographs, 1965.
[6] R. C. Gunning and H. Rossi, Analytic functions of several complex variables, Prentice-Hall, 1965.
[7] E. Sakai, Tahensû kansûron (Japanese; Theory of functions of several complex variables), Kyôritu, 1966.
[8] K. Oka, Sur les fonctions analytiques de plusieurs variables (collected papers), Iwanami, 1961.
[9] L. Hörmander, An introduction to complex analysis in several variables, Van Nostrand, 1966.
[10] L. Bers, Introduction to several complex variables, Courant Institute, 1964.
[11] P. Lelong, Fonctions plurisousharmoniques et formes différentielles positives, Gordon and Breach, 1968.
[12] L. Nachbin, Holomorphic functions, domains of holomorphy and local properties, North-Holland, 1970.
For applications to the theory of elementary particles,
[13] В. С. Владимиров (V. S. Vladimirov), Методы теории функций многих комплексных переменных, Наука, 1964; English translation, Methods of the theory of functions of several complex variables, MIT Press, 1966.
Also → references to 27 Analytic Spaces.

26 (I.13) Analytic Sets

A. General Remarks

The notion of analytic sets was first defined by N. N. Luzin and M. Ja. Souslin in 1916, and it was extended to that of projective sets by operations such as complementation and projection (Luzin, 1924). Most mathematicians, including Luzin and W. Sierpiński, who worked in this field, were in agreement with †French empiricism (or †semi-intuitionism), which defended the standpoint of R. Baire, E. Borel, H. Lebesgue, and others. An object is said to be **effectively given** if it can be uniquely, individually, and unambiguously determined in finite terms so that anyone can reach the same object by following the defining procedure. Semi-intuitionists claim that only effectively given objects have mathematical existence, and they do not recognize as a mathematical object something that needs the axiom of choice for its definition. From this point of view, †Borel sets were "well-defined" sets to which classical analysis had to be restricted. Thus the question was raised whether it is possible to extend the class of Borel sets to a wider class of sets with the same certainty. Lebesgue defined a function not belonging to any class of †Baire functions by using the totality of †ordinals of the second class. (Later, this method was systematically developed as the †theory of sieves by Luzin.) However, it did not satisfy Borel as being effective. Can we, then, extend the Borel sets without any use of ordinals of the second class? The discovery of analytic sets gave an affirmative answer.

In this article, we treat a space (denoted by $X, Y, \ldots$) that is †homeomorphic to a †separable, †complete †metric space and its subspace. Denote by $\mathfrak{N}$ the **space of irrational numbers** (a metric space consisting of the irrational numbers $\in \mathbf{R}$ with the metric $|x-y|$ of x and y). The following properties of a subset S of a space X are equivalent: (i) S is a continuous image of $\mathfrak{N}$; (ii) S is a continuous image of a Borel set in X; (iii) S is the projection of a closed set in a product space $X \times \mathfrak{N}$; (iv) S is the projection of a Borel set in $X \times Y$. We call a set satisfying one of these properties an **analytic set** or an **A set** (in X). The complement of an analytic set is called a **complementary analytic set** (or simply **co-analytic set**) or a **CA set**.

B. The Operation A and Sieves

When to each $(n_1, \ldots, n_k)$ of finite sequences of natural numbers there corresponds a unique element $E(n_1, \ldots, n_k)$ of a family F of sets, this correspondence $\{E(n_1, \ldots, n_k)\}$ is called a **schema of Souslin** (or **system of Souslin**) consisting of sets in F. Denoting an infinite sequence of natural numbers by $\{n_i\}$, the set given by $\bigcup_{\{n_i\}} \bigcap_k E(n_1, \ldots, n_k)$ is called the **kernel** of a system of Souslin, and the operation of taking the kernel is called the **operation** A.

Let $\mathbf{Q}$ be the set of all rational numbers between 0 and 1 and F be a family of sets. Take a family $\{C_r\}_{r \in \mathbf{Q}}$ of sets belonging to F with the index set $\mathbf{Q}$ (or more geometrically, a subset $C = \bigcup_{r \in \mathbf{Q}} C_r \times (r)$ of $X \times \mathbf{Q}$ when F is a family of subsets of a space X), and call it a **sieve** consisting of sets in F. Denoting by $\{r_i\}$ a (strictly) monotone decreasing sequence of elements of $\mathbf{Q}$, we call the set $\bigcup_{\{r_i\}} \bigcap_k C_{r_k}$ (namely, the set of all x such that $C^{(x)} = \{r \mid (x,r) \in C\}$ is not well-ordered by the order $\leqslant$ of rational numbers) the set obtained by a sieve C or the **sieved set** obtained by C. If the family F is closed with respect to countable intersection, then the family of all sets obtained by sieves consisting of sets in F is identical to the family of all sets obtained by applying the operation A to F. When F consists of all the closed sets in a given space, this is the family of all analytic sets. In particular, it is sufficient to take the family of closed intervals as F when X is the space of real numbers. Note that we can define sieves and sieved sets more generally by using the space of real numbers $\mathbf{R}$ instead of the set $\mathbf{Q}$ of rationals.

C. Properties of Analytic Sets

It is evident from the definition that every †Borel set is analytic. If a Borel set is uncountable, then it is the union of a countable set and a one-to-one continuous image of $\mathfrak{N}$. The analyticity of sets is invariant under countable unions, intersections, and Cartesian products and the operation A and †Borel-measurable transformations. An uncountable analytic set contains a †perfect subset (Souslin). Therefore, the possible cardinality of an analytic set is at most countable or is that of the continuum. Every analytic set enjoys the †Baire property, and in Euclidean space, every analytic set is †Lebesgue measurable (Luzin, Sierpiński). If a set E in the Euclidean plane is analytic (coanalytic), then $\Gamma(E)$ is also analytic (coanalytic), where $\Gamma(E)$ is the set of all x such that the section $E^{(x)}$ of E that is parallel to the y-axis has a positive measure (Kondô-Tugué). Every analytic (coanalytic) set E can be decomposed into $\aleph_1$ Borel sets. This decomposition is called a decomposition of E into **constituents**. An analytic (coanalytic) set is a Borel set if and only if it is decomposable into a countable number of constituents (Luzin, Sierpiński). In a space with the cardinality of the continuum, there exist analytic sets that are not Borelian. For example, in the space $C([0,1])$ of continuous functions in the interval $[0,1]$ (→ 173 Function Spaces) the set of all differentiable functions is coanalytic but not Borelian (S. Mazurkiewicz).

The following theorems are especially important in analytic set theory. **Luzin's first**

principle (the first separation theorem): For every pair of disjoint analytic sets A_1, A_2, there exists a Borel set B such that $A_1 \subset B$ and $B \cap A_2 = \varnothing$. An immediate corollary of Luzin's first principle is **Souslin's theorem**: If both A and $X-A$ are analytic, then A is a Borel set. **Luzin's second principle** (the second separation theorem): For every pair of analytic sets A and B, there exist complementary analytic sets C and D such that $A-B \subset C$, $B-A \subset D$, and $C \cap D = \varnothing$. A one-to-one continuous image of a Borel set is Borelian (Souslin). More generally, for a given B-measurable function f defined on a Borel set B, the set A ($\subset f(B)$) of all points y whose inverse images $f^{-1}(y)$ are singletons is a complementary analytic set (**Luzin's unicity theorem**). In this theorem, we can replace "a singleton" by "an F_σ-set" (V. Ja. Arsenin, K. Kunugui). Therefore, if a set is the image of a Borel set by a continuous function such that the inverse image of each point is an F_σ set, then it is a Borel set.

D. Generalization to Projective Sets

A **projective set of class** n is inductively defined as follows: (i) the Borel sets are the projective sets of class 0; (ii) the projective sets of class $2n+1$ are the continuous images of the sets of class $2n$; (iii) the projective sets of class $2n$ are the complements of the sets of class $2n-1$.

The projective sets of class 1 are exactly the analytic sets, and those of class 2 are the complementary analytic sets. The following are fundamental properties of projective sets. Denote by L_n the family of the projective sets of class n. Then (1) $L_{2n} \subset L_{2n+k}$ and $L_{2n+1} \subset L_{2n+2+k}$ $(k=1,2,\dots)$; (2) the property of being a set of class n is invariant under countable unions, intersections, and Cartesian products and homeomorphisms; (3) a continuous image of a projective set of class $2n+1$ is of the same class; (4) the projection on X of a set of class $2n+1$ in $X \times Y$ is a set of class $2n+1$ in X; (5) the family of the projective sets of class $2n+1$ in a space X is the family of the projections of all sets of class $2n$ in $X \times X$ (or $X \times \mathfrak{N}$); (6) the kernel of a system of Souslin consisting of sets of class n is a projective set of the same class, where $n \neq 0$, 2.

We frequently call a projective set of class $2n-1$ a P_n **set**, and that of class $2n$ a C_n **set**. A B_n **set** is a set that is both P_n and C_n. The respective families of those sets are also denoted by P_n, C_n, and B_n. In general, for a family $\mathfrak{F}$ of sets in a space X, we denote by $C\mathfrak{F}$ the family of the complements $X-E$ of all sets E in $\mathfrak{F}$. We write $\mathrm{Sep}_{\mathrm{I}}\mathfrak{F}$ and $\mathrm{Sep}_{\mathrm{II}}\mathfrak{F}$ for the propositions obtained by substituting "set in $\mathfrak{F}$," "set in $\mathfrak{F}$ and in $C\mathfrak{F}$," and "set of $C\mathfrak{F}$" for "analytic set," "Borel set," and "coanalytic set," respectively, in Luzin's first principle and Luzin's second principle, respectively. Then $\mathrm{Sep}_{\mathrm{I}} C_2$, $\mathrm{Sep}_{\mathrm{II}} C_2$ hold (P. S. Novikov). If we assume †the axiom of constructibility $V=L$, then $\mathrm{Sep}_{\mathrm{I}}(C_n)$, $\mathrm{Sep}_{\mathrm{II}}(C_n)$ hold for $n \geqslant 3$ (J. W. Addison). Under the assumption that every †game on projective sets is strictly determined, it follows that $\mathrm{Sep}_{\mathrm{I}} P_n$, $\mathrm{Sep}_{\mathrm{II}} P_n$ ($\mathrm{Sep}_I C_n$, $\mathrm{Sep}_{\mathrm{II}} C_n$) hold when n is odd (even) (A. Martin, J. W. Addison and Y. N. Maschovakis).

E. Universal sets

A set U in $\mathfrak{N} \times X$ is called the **universal set** for the projective sets of class n in X if for any projective set P of class n in X, there exists $z_0 \in \mathfrak{N}$ such that $P = \{x \mid (z_0, x) \in U\}$. Concerning universal sets, we have the following result: for every $n > 0$, there exists a universal set for the projective sets of class n in X that is of the same class in $\mathfrak{N} \times X$. Hence in a space with the cardinality of the continuum, there exists a projective set of class $n+1$ which is not of class n.

F. Uniformization Principle

The uniformization problem arose during investigations of implicit functions. For a set E in a space $X \times Y$, **uniformization of** E is the finding of a subset V of E such that

$$\forall x(\exists y((x,y) \in E) \Leftrightarrow \exists! y((x,y) \in V)),$$

where $\exists! y$ is the †quantifier which means "there exists exactly one y." A Borel set can be uniformized by choosing a suitable coanalytic set (Luzin). Any coanalytic set is uniformizable by choosing a coanalytic set, and a P_2 set is uniformizable by choosing a P_2 set (Kondô). Recently, the proof of this was simplified by Addison, Y. Sampei, and Y. Suzuki, and a more elegant one was given by J. R. Schoenfield. The uniformization of an analytic set is, in general, not to be found among analytic or coanalytic sets. There is a conjecture that any analytic set is uniformizable by specifying an A_ρ set (difference of two analytic sets) (Tugué). Assuming that $V=L$, the uniformization of a P_n set is determined by specifying a P_n set, and that of a C_n set by specifying a $C_{n\rho}$ set. On the other hand, if an axiom system of set theory (e.g., ZF; → 35 Axiomatic Set Theory) is consistent, then it is still consistent even if we add to it the following proposition: There exists a C_2 set whose uniformization is impossible by any choice of a definable set in the system (P. J. Cohen, A. Lévy).

G. Kleene's Hierarchy and Effectiveness

First, projective set theory in any space is reducible to the theory in the space of irrational numbers. Second, if we introduce a †weak topology in the set $\mathbf{N}^{\mathbf{N}}$ of †number-theoretic functions α with one argument, the resulting topological space $\mathbf{N}^{\mathbf{N}}$ is homeomorphic to the space $\mathfrak{N}$ of irrational numbers. Third, any subset B of $\mathbf{N}^{\mathbf{N}}$ is open and closed in this topology if and only if there exists a function ξ $(\in \mathbf{N}^{\mathbf{N}})$ and a predicate $A^{\xi}(\alpha)$ that is †general recursive in ξ such that $\alpha \in B \Leftrightarrow A^{\xi}(\alpha)$. Fourth, logical operations such as $\neg$, $\vee$, $\wedge$, $\exists x$ (where x is a variable ranging over the natural numbers), and $\exists \alpha$ exactly correspond to the operations (on sets) complementation, union, intersection, countable union, and projection, respectively. On the basis of these facts, projective set theory is regarded as the theory of the †$\mathbf{N}^{\mathbf{N}}$ analytic hierarchy of Kleene. Here, the following example is remarkable: We can construct a Σ_1^1 set which is universal for the analytic sets (namely, a $\Sigma_1^1[\mathbf{N}^{\mathbf{N}}]$ set) in the space of irrational numbers (→ 197 Hierarchies).

The connection between projective set theory and logic has been discussed by C. Kuratowski and A. Tarski. From their point of view, semi-intuitionists such as Borel conceive the set of natural numbers to be precisely clear in itself and also the continuum to be immediately recognizable by our geometric intuition. In their argument rational numbers do not play such an important role. They take, a priori, the set of irrational numbers as the fundamental domain, and intervals with rational extremities as the simplest sets of points among the subsets of the fundamental domain as the starting point of their argument. Here, the fundamental domain or each interval is not conceived as a totality of its elements, but recognized as a "uniform extent." In contrast with this, singletons and individual irrational numbers are not so simple. For this reason Borel introduced the notion of calculable numbers to study definable real numbers. Following Luzin, we say that a **calculable number** is a constructible real number in the sense that we can give it by an arithmetical approximation as precisely as we want. Now, this notion is nearly identical to the notion of an †effectively calculable real number given by A. Church or A. M. Turing.

In the mathematics of semi-intuitionism, the word "effective" has played an especially important role. Although these mathematicians have always agreed not to accept the †axiom of choice, the exact meaning of "effective" has differed slightly among different members of their group, or in different stages of the development of the theory. Such differences mainly arose in connection with the question: How can we tell whether given entities are finitary or individual? One way to guess the original intention held by Borel and others when they used the term "effective" is to replace the term by "recursive."

It is one thing to "effectively give" an object A satisfying a given property P, but quite another to "effectively prove" that such an A satisfies the property P. If a theorem establishes both, it may be said to be extremely valuable. However, concerning projective sets of class $\geqslant 2$, there are many cases in which we have no effective proofs. This was why Luzin did not quite agree to recognize the mathematical validity of projective sets of class $\geqslant 2$. The problems of finding the cardinality of a complementary analytic set and of showing the measurability of a projective set of class $\geqslant 3$ remain extremely difficult. Under the assumption that $V = L$, the recent development of axiomatic set theory has yielded the following propositions (K. Gödel, Novikov): (i) There exists an uncountable complementary analytic set that does not contain any perfect subset. (ii) There exists a nonmeasurable B_2 set. Recently it was shown that (i), (ii), and other propositions are all independent of the axioms of set theory (Cohen, R. M. Solovay) (→ 35 Axiomatic Set Theory).

References

[1] M. Kondo, Kaiseki syûgôron (Japanese; Theory of analytic sets), series of math. lectures, Osaka Univ., Iwanami, 1938.
[2] C. Kuratowski, Topologie I, Warsaw, revised edition, 1948.
[3] A. A. Liapunov, E. A. Stschegolkow, and V. J. Arsenin, Arbeiten zur deskriptiven Mengenlehre, Deutscher Verlag der Wissenschaften, 1955.
[4] N. N. Lusin (Luzin), Sur les ensembles analytiques, Fund. Math., 10 (1927), 1–95.
[5] N. N. Lusin (Luzin), Leçons sur les ensembles analytiques et leurs applications, Gauthier-Villars, 1930.
[6] W. Sierpiński, Les ensembles projectifs et analytiques, Mémor. Sci. Math., Gauthier-Villars, 1950.

27 (XI.20) Analytic Spaces

A. General Remarks

An †analytic function of a complex variable has as its natural domain of definition a †Riemann surface, i.e., a 1-dimensional complex †analytic manifold. In the case of several complex variables, the set of zeros of an analytic function, the quotient space of a domain by a †properly discontinuous group of analytic automorphisms, the existence domain of an †algebroidal function, etc., are, strictly speaking, not necessarily complex analytic manifolds. It is necessary to consider a more general category of complex analytic manifolds with singularities. The recent theory of analytic functions of several complex variables has been developed over analytic spaces defined on the models of those examples.

B. Analytic Sets

We say that a subset A of a complex analytic manifold G is an **analytic set** in G if it is a closed subset and each point of A has a neighborhood U such that $U \cap A$ is the set of common zeros of a finite number of holomorphic functions in U. Specifically, if A is locally the set of zeros of a single holomorphic function that does not vanish identically, then A is called **principal**. Two subsets S_1 and S_2 of G are called equivalent at $z^0 \in G$ if there exists a neighborhood U of z^0 such that $S_1 \cap U = S_2 \cap U$. By this equivalence relation, every subset S of G defines its germ S_{z^0} at z^0. A **germ of an analytic set** at z^0 is the germ at z^0 of an analytic set in a neighborhood of z^0. Each germ A_0 of an analytic set at $z^0 = 0 \in G$ is associated with an ideal $I(A_0) = \{f \mid f \in H(0), f|A_0 = 0\}$ in the ring $H(0)$ of germs of holomorphic functions at 0. We call A_0 **reducible** if A_0 is the union of two germs of analytic sets A_0' and A_0'' with $A_0' \neq A_0$, $A_0'' \neq A_0$; otherwise, A_0 is called **irreducible**. An analytic set A is called **irreducible at** 0 if the germ at 0 of A is irreducible. Properties of A_0 and $I(A_0)$ correspond to each other. Thus A_0 is irreducible if and only if $I(A_0)$ is prime.

As the ring $H(0)$ is †Noetherian, in a neighborhood of every point z^0 an analytic set A is represented as the union of a finite number of analytic sets A_i that are irreducible at z^0. These A_i are essentially unique. If an analytic set A is irreducible at z^0, then there exists a system of local coordinates $(z_1, \ldots, z_n)$ centered at z^0 and a pair of natural numbers $d \leqslant n$ and k such that, in a neighborhood of z^0, A is a k-sheeted **ramified covering space** with covering mapping $\varphi : (z_1, \ldots, z_n) \to (z_1, \ldots, z_d)$; i.e., for an analytic set R in a neighborhood of $0 \in \mathbf{C}^d$, $\varphi : A - \varphi^{-1}(R) \to \mathbf{C}^d - R$ is, in a neighborhood of z^0, a k-sheeted covering mapping, where $A - \varphi^{-1}(R)$ is a connected d-dimensional complex analytic manifold in the neighborhood. The coordinates of the k points of $A - \varphi^{-1}(R)$ over each point $(z_1, \ldots, z_d)$ are holomorphic in these variables. The number d is the **(local) dimension** of A at z^0 and is denoted by $\dim_{z^0} A$. From this local representation, we obtain **Rückert's zero-point theorem**: Every prime ideal $\mathfrak{P}$ in $H(0)$ is equal to $I(A_0)$, with A_0 an irreducible germ of an analytic set at 0. In this case the dimension at 0 of an analytic set A that defines A_0 is equal to the †Krull dimension of the †local ring $H(0)/\mathfrak{P}$. The theory of local rings is very important in the study of germs of analytic sets. The dimension of a general analytic set A at z^0 is defined by $\dim_{z^0} A = \sup_i \dim_{z^0} A_i$, where $A = \cup_i A_i$ in a neighborhood of z^0, with the A_i irreducible at z^0. If $\dim_{z^0} A_i$ is equal to d for all i, then A is called **purely d-dimensional at** z^0. The **(global) dimension** of A is defined by $\dim A = \sup_{z \in A} \dim_z A$. A **purely d-dimensional analytic set** is defined to be an analytic set that is purely d-dimensional at every one of its points.

A point z^0 of an analytic set A is called **ordinary (regular** or **simple)** if A has the structure of a complex analytic submanifold in a neighborhood of z^0. The set A' of ordinary points of A is dense and open in A. The set $A^* = A - A'$ of **singular** (not ordinary) points is an analytic set in G. If A is purely d-dimensional, then A' is a d-dimensional complex analytic manifold and A^* is an analytic set of dimension $\leqslant d-1$.

Let A be an analytic set of dimension d in G, and B a purely d'-dimensional analytic set in $G - A$ with $d' > d$. Then the closure $\overline{B}$ of B in G is a purely d'-dimensional analytic set (**Remmert-Stein continuation theorem**).

For every analytic set A in G, the †analytic sheaf $\mathfrak{I}(A)$ of germs of holomorphic functions over G that vanish on A is †coherent (H. Cartan). We call $\mathfrak{I}(A)$ the sheaf of ideals defined by an analytic set A. $\mathcal{O}_A = (\mathcal{O}_G / \mathfrak{I}(A))|A$ is a coherent sheaf of rings over A. $\mathcal{O}_A$ is called the sheaf of germs of holomorphic functions on an analytic set A.

C. Analytic Spaces

A †ringed space $(X, \mathcal{O}_X)$ with Hausdorff base space X is called an **analytic space** if for every point $x \in X$, there exists an open neighborhood U of x such that the ringed space $(U, \mathcal{O}_X | U)$ is isomorphic to a ringed space $(A, \mathcal{O}_A)$, where A is an analytic set in an open set G of some $\mathbf{C}^n$. The structure sheaf $\mathcal{O}_X$ is then called a **sheaf of germs of holomorphic functions**. The notion of holomorphic mapping from one open set in $\mathbf{C}^n$ into another is generalized to the case of mappings from one analytic set into another. An analytic set Y in an analytic space X and the sheaf $\mathcal{O}_Y$ of germs of holomorphic functions on Y are defined as in the case where X is a complex manifold. The ringed space $(Y, \mathcal{O}_Y)$ is an analytic space and is called an **analytic subspace** of X. For an analytic space X, the notions of $\dim_x X$, $\dim X$, irreducibility, and pure dimensionality are defined as for an analytic set $A \subset G \subset \mathbf{C}^n$. Every analytic space X is the union of a locally finite family of irreducible analytic subspaces X_i called the **irreducible components** of X.

Let $\varphi: X \rightarrow Y$ be a holomorphic mapping of an analytic space X into another, Y. Its **rank at** $x \in X$ is defined by $r_\varphi(x) = \dim_x X - \dim_x \varphi^{-1}(\varphi(x))$. The number $r_\varphi = \sup_{x \in X} r_\varphi(x)$ is called the **rank** of φ. The **set of degeneracy** E_φ of φ is the set of points $x \in X$ such that $r_{\varphi|X'}(x) < r_{\varphi|X'}$ for an irreducible component X' of X through x. The mapping φ is **nondegenerate** if $E_\varphi = \varnothing$. For any $k \in \mathbf{N}$, $\{x \in X | r_\varphi(x) \leqslant k\}$ is an analytic set (R. Remmert). In particular, E_φ is analytic. For a holomorphic mapping $\varphi: X \rightarrow Y$, the inverse image of an analytic set in Y is an analytic set in X. However, the image of an analytic set is not necessarily analytic. If φ is proper, then the image $\varphi(X')$ of an analytic set X' in X is an analytic set in Y of dimension $r_{\varphi|X'}$ and is irreducible if X' is irreducible (Remmert's theorem).

D. Modifications and Resolution of Singularities

Let M be a subset of an analytic space X. If, for every point $x \in X$, there exists an open neighborhood U of x and an analytic set M^* in U, containing $U \cap M$ such that $U - M^*$ is dense in U, then M is called **analytically thin**. Let $\varphi: X \rightarrow Y$ be a holomorphic mapping. Suppose that there exist two analytically thin sets $M \subset X$, $N \subset Y$ such that φ induces an isomorphism between $X - M$ and $Y - N$. Then X is called a **holomorphic modification** of Y. If furthermore φ is †proper, then X is called a **proper modification** of Y. A **monoidal transformation** of an analytic space X with respect to a coherent sheaf of ideals $\mathcal{I}$ is defined as in the case of a complex manifold. It is a proper modification $f: X^* \rightarrow X$ such that $f^* \mathcal{I}$ is locally principal, and depends on the analytic set $Y = \operatorname{supp}(\mathcal{O}_X / \mathcal{I})$ rather than on $\mathcal{I}$. It is often called the **blowing-up** of X with center Y.

H. Hironaka [7] proved that, if X is an analytic space which is countable at infinity (i.e., a countable union of compact sets), then there is a proper modification $\pi: X' \rightarrow X$ with X' smooth (i.e., free from singular points). Moreover, over any relatively compact open set U of X, π is the product of a finite sequence of blowing-ups $\pi_i: X_i' \rightarrow X_{i-1}'$ $(X_0' = X)$, with smooth centers Y_{i-1}' along which X_{i-1}' is †normally flat. This deep result enables one to derive properties of analytic spaces from those of complex manifolds.

Let X and Y be two analytic spaces and G an analytic set in $X \times Y$. If the canonical projection $\pi: G \rightarrow X$ is a holomorphic (or proper) modification, then we say that a meromorphic mapping (a proper meromorphic mapping) μ of X into Y is defined. The set G is then called the **graph** of μ. A holomorphic mapping $\varphi: X \rightarrow Y$ may be viewed as a proper meromorphic mapping. Let $\mu: X \rightarrow Y$ be a proper meromorphic mapping. Then $\mu(x)$ (the projection of $\pi^{-1}(x)$ into Y) is a nonempty analytic set in Y for every point $x \in X$. Moreover, there exists an analytic set N, with $X - N$ dense in X, such that μ maps $X - N$ into Y holomorphically. The smallest set N with this property is called the **set of points of indeterminacy** or the **singularity set** of μ. A meromorphic mapping $f: X \rightarrow \mathbf{P}^1(\mathbf{C})$ is called a **meromorphic function** on X if none of the irreducible components of X is mapped to $\{\infty\}$ by f. The set $f^{-1}(0) = \pi((X \times \{0\}) \cap G)$ is called the set of zero points of X, and the set $f^{-1}(\infty)$ is called the set of poles. These are analytic sets in X. Let $f_1, \ldots, f_k$ be meromorphic functions on X. Then, by a suitable proper modification of X, one can eliminate the points of indeterminacy of the meromorphic mapping $f: X \rightarrow (\mathbf{P}^1(\mathbf{C}))^k$ defined by $x \rightarrow (f_1(x), \ldots, f_k(x))$, i.e., one can modify f to be holomorphic. The ring of meromorphic functions on X is invariant under proper modifications of X. If X is irreducible and compact, then the field of meromorphic functions on X is a simple algebraic extension of the field of rational functions of $k(\leqslant \dim X)$ variables **(Chow's theorem)**.

Let $(X, \mathcal{O}_X)$ be an analytic space. A point $x \in X$ is called **normal** for X if $\mathcal{O}_{X,x}$ is a normal local ring. The set of nonnormal points for X is an analytically thin analytic set in X (K. Oka). Every ordinary point of X is normal. We call X **normal** if every one of its points is normal. Every nondegenerate holomorphic mapping of an irreducible X into an irreducible and normal Y is an open mapping if its rank is equal to the dimension of Y (Remmert). For every analytic space X, there exists a proper modification $\nu:\tilde{X}\rightarrow X$ with $\tilde{X}$ normal such that ν is nondegenerate and $\nu|\tilde{X}-\nu^{-1}(S)$ is an isomorphism, where S is the set of singular points. Such a proper modification of X is unique up to isomorphisms. We call $\tilde{X}$ a **normalization** of X with normalizing mapping ν.

Let $\varphi:X\rightarrow Y$ be a holomorphic modification. Suppose that Y is normal at $\varphi(x^0)$ ($x^0\in X$). If the set $\varphi^{-1}(\varphi(x^0))$ contains an isolated point, then $\varphi^{-1}(\varphi(x^0))=x^0$, and φ is an isomorphism in a neighborhood of x^0 (an analogue of †Zariski's main theorem). In particular, if $\varphi:X\rightarrow Y$ is a holomorphic mapping and Y is normal, then φ maps $X-E_\varphi$ isomorphically onto the dense open set $\varphi(X-E_\varphi)$. Furthermore, if $\varphi:X\rightarrow Y$ is injective and X and Y are irreducible, normal, and n-dimensional, then $\varphi(X)$ is an open set in Y and $\varphi^{-1}:\varphi(X)\rightarrow Y$ is holomorphic.

E. Analytic Spaces in the Sense of Behnke and Stein

Let $\varphi:\tilde{G}\rightarrow G$ be a proper continuous mapping of a connected locally compact space $\tilde{G}$ onto a domain $G\subset\mathbf{C}^n$. The triple $\mathfrak{G}=(\tilde{G},\varphi,G)$ is an **analytic covering space** over G if the following conditions are satisfied: (i) $\varphi^{-1}(z^0)$ is a finite set for every point $z^0\in G$. (ii) There exists an analytic set $A\subset G$ of dimension $\leqslant n-1$ such that $\varphi|\tilde{G}-\varphi^{-1}(A)$ is a local homeomorphism and every point of $\varphi^{-1}(A)$ has a fundamental system of neighborhoods U such that both U and $U-\varphi^{-1}(A)\neq\varnothing$ are connected. As $\mathfrak{G}$ is unramified over $G-A$, the number of points in $\varphi^{-1}(z^0)$ is constant for $z^0\in G-A$ and is called the **number of sheets** of $\mathfrak{G}$. A point $\tilde{z}\in\tilde{G}$ is called a **ramification point** of $\mathfrak{G}$ if the restriction of φ to any neighborhood of $\tilde{z}$ is not a homeomorphism. Denote by B the set of ramification points of $\mathfrak{G}$. Then $\varphi(B)\subset A$ is an analytic set of dimension $n-1$. Let f be a continuous complex-valued function in an open set D in $\tilde{G}$. We call f **holomorphic** in D if for every point $\tilde{z}^0\in D-B$ and for every open neighborhood V of $z^0=\varphi(\tilde{z}^0)$ over which φ is a homeomorphism, $f\circ\varphi^{-1}$ is holomorphic in V. Denote by $\mathcal{O}_{\tilde{G}}$ the sheaf of germs of holomorphic functions over $\tilde{G}$. Then $(\tilde{G},\mathcal{O}_{\tilde{G}})$ is a ringed space. An **analytic space in the sense of Behnke and Stein** is a Hausdorff ringed space $(X,\mathcal{O}_X)$ that is locally isomorphic to a ringed space of the form $(\tilde{G},\mathcal{O}_{\tilde{G}})$. Riemann's theorem on removable singularities holds for analytic spaces. Every normal analytic space is an analytic space in the sense of Behnke and Stein. An analytic covering space $\mathfrak{G}=(\tilde{G},\varphi,G)$ is a **C-covering space** (covering space in the sense of Cartan) if for every point $z^0\in G$ there exists an open neighborhood V of z^0 and a holomorphic function g in $U=\varphi^{-1}(V)$ which may be expressed as a monic polynomial of degree k whose coefficients are holomorphic functions on V, where k is the number of sheets of $\mathfrak{G}$. A **C-analytic space** is an analytic space in the sense of Behnke and Stein that is locally isomorphic to a C-covering space. The category of C-analytic spaces coincides with that of normal analytic spaces. According to H. Grauert and Remmert [1], every analytic covering space is a C-analytic covering space. Therefore, every analytic space in the sense of Behnke and Stein is a normal analytic space.

Let R be an equivalence relation in an analytic space X. Given a subset A of X, denote by $R[A]$ the set of points of X which are R-equivalent to points of A. We call R **proper** if $R[K]$ is compact for every compact set K in X. Let φ be a proper holomorphic mapping of an analytic space X into another Y. For $x,x'\in X$, let $x\equiv x'(R)$ be defined by $\varphi(x)=\varphi(x')$. The equivalence relation R is then proper. We consider the quotient space X/R and the canonical projection $P:X\rightarrow X/R$. With each open set U in the quotient space X/R we may associate the ring of holomorphic functions in $p^{-1}(U)$ that are constant on $p^{-1}(\bar{x})$ for every $\bar{x}\in U$.

This leads to a ringed space $(X/R,\mathcal{O}_X/R)$, which is proved to be an analytic space by **Grauert's theorem** [5]: All the direct images of a coherent analytic sheaf over X by a proper holomorphic mapping $\varphi:X\rightarrow Y$ are coherent. For every proper equivalence relation R in X, the ringed space $(X/R,\mathcal{O}_X/R)$ is an analytic space if and only if for every point $\bar{x}\in X/R$ there exists an open neighborhood V of $\bar{x}$ such that functions in $\Gamma(V,\mathcal{O}_X/R)$ separate the points of V (H. Cartan).

F. Stein Spaces

For an analytic space $(X,\mathcal{O}_X)$ we have the following conditions: (i) Functions in $\Gamma(X,\mathcal{O}_X)$ separate the points of X. (ii) X is K-**complete**; i.e., for every point $x\in X$ there exist a finite number of $f_i\in\Gamma(X,\mathcal{O}_X)$ ($i=1,\ldots,k$) such that the holomorphic mapping $f=(f_i):X\rightarrow\mathbf{C}^k$ is nondegenerate at X. (iii) Every compact analytic set in X is a finite set. Condition (i) implies (ii), and (ii) implies (iii). If an irreducible analytic space X is K-complete, then X is the countable union of compact sets (Grauert). In fact, if $n = \dim X$, there exist functions $f_i \in \Gamma(X, \mathcal{O}_X)$ $(i = 1, \ldots, n)$ such that the holomorphic mapping $f = (f_i): X \rightarrow \mathbf{C}^n$ is nondegenerate. The notion of holomorphic convexity (→ 25 Analytic Functions of Several Complex Variables F) is carried over to analytic spaces. For a holomorphically convex analytic space, conditions (i), (ii), and (iii) are equivalent (Grauert). A **Stein space** (or **holomorphically complete space**) is a holomorphically convex analytic space that satisfies one of the conditions (i), (ii), or (iii). In a holomorphically convex analytic space $(X, \mathcal{O}_X)$, let R be the equivalence relation defined by $\Gamma(X, \mathcal{O}_X)$; i.e., for $x, x' \in X, x \equiv x'(R)$ if and only if $f(x) = f(x')$ for every $f \in \Gamma(X, \mathcal{O}_X)$. Then R is proper. The analytic space $(X/R, \mathcal{O}_X/R)$ is a Stein space. A Stein space is a generalization of the notion of a †Stein manifold. Fundamental theorems A and B on Stein manifolds hold verbatim for Stein spaces (→ 25 Analytic Functions of Several Complex Variables L). Therefore, the main properties of Stein manifolds are inherited by Stein spaces. Let $\varphi: X \rightarrow Y$ be a holomorphic mapping of an analytic space X into another, Y. If for every $x \in X$ all the connected components of the fibers $\varphi^{-1}(\varphi(x))$ are compact, then the equivalence relation R' defined by those components (i.e., for x and x' in X, $x \equiv x'(R')$ if and only if x and x' belong to the same component of $\varphi^{-1}(\varphi(x))$) is proper, and the ringed space $(X/R', \mathcal{O}_X/R')$ is an analytic space. In particular, if X is a holomorphically convex irreducible analytic space and R is the equivalence relation defined by $\Gamma(X, \mathcal{O}_X)$, then all the fibers of the canonical projection $p: X \rightarrow X/R$ are connected.

G. Other Topics

The †Levi problem in its various forms is also investigated for the case of analytic spaces. In such investigations, deeper results on coherent analytic sheaves and pseudoconvexity are necessary. The notions of holomorphic †vector fields and †differential forms are defined on analytic spaces and are useful in the investigation of analytic spaces themselves.

The notion of analytic space can be generalized as follows (Grauert [5]). A ringed space $(X, \mathcal{O}_X)$ is a **general analytic space** if it is locally isomorphic to a ringed space $(A, \mathcal{H}_A)$, where A is an analytic set in a domain $G \subset \mathbf{C}^n$, and $\mathcal{H}_A = (\mathcal{O}_G/\mathcal{J})|_A$ for some coherent analytic subsheaf $\mathcal{J}$ of $\mathcal{I}(A)$ such that $\operatorname{Supp}(\mathcal{O}_G/\mathcal{J}) = A$. On the other hand, there is an extension to the infinite dimensional case; i.e., one begins by defining analytic sets in an open subset of a complex Banach space, and arrives at the notion of a **Banach analytic space**. These generalizations are useful in the theory of deformation of complex analytic structures.

References

[1] H. Grauert and R. Remmert, Komplexe Räume, Math. Ann., 136 (1958), 245–318.
[2] S. S. Abhyankar, Local analytic geometry, Academic Press, 1964.
[3] Séminaire H. Cartan, Ecole Norm. Sup., 1953–1954, 1957–1958, 1960–1961.
[4] M. Hervé, Several complex variables, local theory, Tata Inst. Studies in Math., Oxford Univ. Press, 1963.
[5] H. Grauert, Ein Theorem der analytischen Garbentheorie und die Modulräume komplexer Strukturen, Publ. Math. Inst. HES, no. 5, 1960.
[6] L. Bers, Introduction to several complex variables, Courant Institute, 1963.
[7] R. Narasimhan, Introduction to the theory of analytic spaces, Lecture notes in math. 25, Springer, 1966.
[8] H. Hironaka, Desingularization of complex-analytic varieties, Act. Congr. Intern. Math., Nice, 1972, Gauthier-Villars, vol. 2, 627–632.

28 (XX.1) Ancient Mathematics

A. General Remarks

To determine the beginning of the history of mathematics, one must define the term "mathematics." Only speculation based on the observation of primitive peoples today can be made on any development of the number concept among prehistoric peoples. The prehistoric period ended in Egypt and Mesopotamia c. 3000 B.C., and a little later in the valleys of the large rivers in India and China.

Since the basis of the civilizations in the river valleys of the ancient world was agriculture, the administrators first had to control watering systems through irrigation, drainage, pumping, and canalization; second, they had to measure land and harvests for tax collection; and third, they had to establish a calendar by observation of the heavenly bodies. All these tasks demanded some knowledge of mathematics.

Additional knowledge of mathematics was

certainly needed for construction of the palaces and tombs. We know something of the development of mathematical knowledge during these ages from some recovered artifacts, but they are not sufficient. There remains a possibility for the discovery of new finds which will bring about a basic change in our knowledge of the history of mathematics during this period.

B. Mathematics In Egypt

The main sources for our understanding of the history of mathematics in Egypt are the Moscow papyrus and the more important Rhind papyrus, both discovered in the 19th century. The Greeks place the origin of their mathematics in Egypt, but it seems that Egyptian mathematics was limited to practical mathematics. The Egyptians had a decimal numeration system, but the place value was not clear; they used fractions, which they always decomposed into the sums of unit fractions (i.e., fractions with 1 as numerator); they solved the problems of everyday arithmetic that were reducible to linear equations; they computed approximate areas and volumes of some figures for the purpose of measurement of farmland or granaries and for construction work; they had exact formulas for the computation of areas of triangles and of trapezoids; and they used $(16/9)^2 = 3.1605\ldots$ as the value of π; but no trace has been found to prove the existence of demonstrative mathematics in ancient Egypt.

C. Mathematics in Mesopotamia

Sources abound for the study of Mesopotamian mathematics, and these studies may very well increase in the future. The Mesopotamians kept exact records of astronomical observations for long periods of time. Their more advanced mathematics was not limited to practical use, as was that of the Egyptians. They used a sexagesimal system of numeration with place value, and also used sexagesimal fractions; however, they lacked a cipher to denote zero until the 4th century B.C., and neither did they have a symbol corresponding to our decimal point, so that exact place value had to be determined from the context of each expression. They had a multiplication table and tables of inverses, squares, and cubes of numbers, and they used these tables to solve equations, even some simple equations of the third degree, as well as simultaneous equations of the second degree for two unknowns. They had accurate solutions for quadratic equations (expressed in words); they discarded negative roots, but they adopted both positive roots when two existed. They studied integral solutions of $a^2+b^2=c^2$ (the largest of their solutions were 12,709, 13,500, and 18,541) and approximate computation of quadratic roots, which suggests some relation to Greek mathematics. We have evidence that some of the geometric algebra in Euclid's *Elements* can be traced to Mesopotamian algebra. Some historians also affirm that the concept of demonstration in Greek mathematics originated with the Mesopotamians, but this theory lacks sufficient proof.

By the 7th century, the Mayas in Central America also possessed a numeration system, with the base 20. As far as we know, the Mesopotamians and the Mayas were the earliest people to possess numeration systems with place value (→ 60 Chinese Mathematics, 190 Greek Mathematics, 213 Indian Mathematics).

References

[1] O. Neugebauer, Vorlesungen über Geschichte der antiken mathematischen Wissenschaften I, Springer, 1934.
[2] O. Neugebauer, Mathematische Keilschrift-Texte, Springer, 1935.
[3] A. B. Chace, The Rhind mathematical papyrus II, Oberlin, 1929.
[4] B. L. van der Waerden, Science awakening, Noordhoff, 1954.
[5] O. Neugebauer, The exact sciences in antiquity, Brown Univ. Press, second edition, 1957.
[6] K. Vogel, Vorgriechische Mathematik I. Vorgeschichte und Ägypten, II. Die Mathematik der Babylonier, Hermann Schroedel, 1958–1959.

29 (XX.4) Arab Mathematics

The role of the Arabs in cultural history was partly that of cultural transmitter. Between the 7th and 13th centuries, they established a religious empire that extended from India to Spain; later it was divided into the eastern and western empires. The caliphs of these empires encouraged research in the sciences, so the capitals Bagdad and Cordova became centers of culture where scholars from different countries gathered.

Arabia is sometimes called the stepfather of European culture. During the 13th century,

Alphonso X (1252–1284) invited Islamic and Hebrew scholars to the Spanish court to translate their writings on algebra, medicine, and astronomy into Spanish. This accomplishment earned him the title of Alphonso the Wise.

The first contact between Greek and Indian mathematics took place in Bagdad under Caliph Al-Mansūr (754–775); Euclid's *Elements* was introduced by way of the Byzantine Empire, while Brahmagupta's *Brahmasphuta-siddhânta* came directly from India. Many mathematical texts found in the Eastern Roman Empire and Syria, including some Greek works, were translated into Arabic. Though it is difficult to discern essential scientific advances in Arabian works, the diffusion of these translations was instrumental in the development of European mathematics.

The Arabs did not use written numerals until Mohammed's time (570–632). Signs representing numbers had been introduced into Arabia when its influence encompassed Egypt and Greece. Indian numerals were imported with Brahmagupta's book and became our present **Arabic numerals** after a series of modifications.

Among all the branches of Arab mathematics, algebra was the most advanced. It started with Alkwarizmi's (820) *Al Gebr W'al Muquabala*, the origin of the word "algebra." It was the first mathematical book written in Arabic. Its content was essentially a variety of methods of solving algebraic equations. *Al gebr* means "transposition of negative terms on one side of the equation to the other side and changing their signs." and *al muquabala* means "simplification of the equation by gathering similar terms." For example, quadratic equations can be brought by these methods into one of the following three types: $x^2=px+q$; $x^2+q=px$; $x^2+px=q$, where p, q are positive numbers. The Arabs expressed the rule of solving these equations verbally. They apparently knew that quadratic equations have two roots, but they adopted only positive roots; when the equation had two positive roots they adopted the smaller root. The proof was given geometrically. It is possible that they learned geometric proofs from the Greeks.

For the Arabs, geometry was secondary to algebra. They did not appreciate proof as seen in Euclid's *Elements*. The book on conic sections by Apollonius was also translated into Arabic, but no essential progress was made in this area. The only remarkable contribution was that of Omar Khayyám, author of *The Rubáiyát,* who applied conic sections in solving the cubic equation $x^3+bx=a$.

In trigonometry, Al Battani (c. 858–929) left a notable contribution. He studied *The Almagest*, the Arabic translation of Ptolemy's astronomical work. He added nothing outstanding to plane trigonometry, but obtained such formulas as $\cos a=\cos b\cos c+\sin b\cdot\sin c\cos A$ for spherical triangles, which were not mentioned in *The Almagest*.

References

[1] M. B. Cantor, Vorlesungen über Geschichte der Mathematik I, Teubner, third edition, 1907.
[2] G. Sarton, Introduction to the history of science I. From Homer to Omar Khayyam, Carnegie Institute of Washington, 1927.

30 (V.18) Arithmetic of Associative Algebras

A. General Theory

Let $\mathfrak{g}$ be a †Dedekind domain (i.e., an integral domain in which every ideal is uniquely decomposed into a product of prime ideals), let F be the field of quotients of $\mathfrak{g}$, and let A be a †separable algebra of finite degree over F. A **$\mathfrak{g}$-lattice** $\mathfrak{a}$ of A is a $\mathfrak{g}$-submodule of A that is finitely generated over $\mathfrak{g}$ and satisfies $A=F\mathfrak{a}$. If a subring $\mathfrak{o}$ of A is a $\mathfrak{g}$-lattice containing $\mathfrak{g}$, then $\mathfrak{o}$ is called an **order.** A **maximal order** is an order that is not contained in any other order. A maximal order always exists although it may not be unique; in particular, if A is commutative, it has only one maximal order. If we put $\mathfrak{o}_l=\{x\in A\,|\,x\mathfrak{a}\subset\mathfrak{a}\}$ ($\mathfrak{o}_r=\{x\in A\,|\,\mathfrak{a}x\subset\mathfrak{a}\}$) for a $\mathfrak{g}$-lattice $\mathfrak{a}$ of A, then $\mathfrak{o}_l$ ($\mathfrak{o}_r$) is an order of A and is called the **left (right) order** of $\mathfrak{a}$. If the left order of $\mathfrak{a}$ is maximal, then so is the right order; the converse is also true. If $\mathfrak{o}_l$ and $\mathfrak{o}_r$ are maximal, we call $\mathfrak{a}$ a **normal $\mathfrak{g}$-lattice**. To describe the same situation we say that $\mathfrak{a}$ is a **left $\mathfrak{o}_l$-ideal** or a **right $\mathfrak{o}_r$-ideal.** If $\mathfrak{o}_l=\mathfrak{o}_r=\mathfrak{o}$, then we say that $\mathfrak{a}$ is a **two-sided $\mathfrak{o}$-ideal**. A lattice $\mathfrak{a}$ with $\mathfrak{a}\subset\mathfrak{o}_l$ (or equivalently with $\mathfrak{a}\subset\mathfrak{o}_r$) is called an **integral $\mathfrak{g}$-lattice** or **integral left (right) ideal**. The product $\mathfrak{a}\mathfrak{b}$ of two normal $\mathfrak{g}$-lattices $\mathfrak{a}$, $\mathfrak{b}$ is called a **proper product** if the right order of $\mathfrak{a}$ coincides with the left order of $\mathfrak{b}$. The proper product defines the structure of a †groupoid on the set of all normal $\mathfrak{g}$-lattices of A. In particular, the inverse $\mathfrak{a}^{-1}$ of a normal $\mathfrak{g}$-lattice $\mathfrak{a}$ satisfying $\mathfrak{a}\mathfrak{a}^{-1}=\mathfrak{o}_l$, $\mathfrak{a}^{-1}\mathfrak{a}=\mathfrak{o}_r$ is given by $\mathfrak{a}^{-1}=\{x\in A\,|\,x\mathfrak{a}\subset\mathfrak{o}_r\}=\{x\in A\,|\,\mathfrak{a}x\subset\mathfrak{o}_l\}$. For a fixed maximal order $\mathfrak{o}$, a maximal **integral two-sided $\mathfrak{o}$-ideal** $\mathfrak{p}$ different from $\mathfrak{o}$ is called a **prime ideal** of $\mathfrak{o}$. If $\mathfrak{p}$ is prime, then $\mathfrak{o}/\mathfrak{p}$ is the matrix algebra of degree κ over a division algebra, and κ is called the **capacity** of the prime ideal $\mathfrak{p}$. The set of all two-sided $\mathfrak{o}$-ideals forms a multiplicative group, of which prime ideals are independent generators.

B. Maximal Orders of a Simple Ring

In the rest of this article, A is a †simple ring, F is the †center of A, and $\mathfrak{o}$ is a maximal order of A. The prime ideals $\mathfrak{q}$ of $\mathfrak{o}$ and the prime ideals $\mathfrak{p}$ of $\mathfrak{g}$ are in one-to-one correspondence by the relation $\mathfrak{q}\cap\mathfrak{g}=\mathfrak{p}$; $\mathfrak{o}/\mathfrak{q}$ is a simple algebra over $\mathfrak{g}/\mathfrak{p}$, and $\mathfrak{q}^e=\mathfrak{p}\mathfrak{o}$ for some natural number e. The **different** $\mathfrak{d}$ of $\mathfrak{o}$ is defined by $\mathfrak{d}^{-1}=\{x\in A \mid \mathrm{Tr}(x\mathfrak{o})\subset\mathfrak{g}\}$. (Tr is the †reduced trace from A to F; $\mathfrak{d}$ is an integral two-sided $\mathfrak{o}$-ideal, and is divisible by $\mathfrak{q}^{e-1}$ if $\mathfrak{q}^e=\mathfrak{p}\mathfrak{o}$.) For $\mathfrak{q}$ to divide $\mathfrak{d}$, it is necessary and sufficient that either $e>1$ or $\mathfrak{o}/\mathfrak{q}$ is not separable over $\mathfrak{g}/\mathfrak{p}$. In particular, if A is a †total matrix algebra over F, then $\mathfrak{d}=\mathfrak{o}$, $\mathfrak{q}=\mathfrak{p}\mathfrak{o}$. The ideal of $\mathfrak{g}$ generated by the †reduced norms (to F) of the elements in a normal $\mathfrak{g}$-lattice $\mathfrak{a}$ of A is denoted by $N_{A/F}(\mathfrak{a})$. If $\mathfrak{a}\mathfrak{b}$ is a proper product, then we have $N_{A/F}(\mathfrak{a}\mathfrak{b})=N_{A/F}(\mathfrak{a})N_{A/F}(\mathfrak{b})$, where $N_{A/F}(\mathfrak{d})$ does not depend on the choice of $\mathfrak{o}$, and is called the **discriminant** of A. If $[A:F]=n^2$ and $\mathfrak{q}^e=\mathfrak{p}\mathfrak{o}$, then $N_{A/F}(\mathfrak{q})=\mathfrak{p}^f$, $ef=n$.

C. Simple Rings over a Local Field

Let F be a field that is complete with respect to a †discrete valuation whose field of residue classes is finite. Let $\mathfrak{g}$ be the †valuation ring of F, $\mathfrak{p}$ be the maximal ideal of $\mathfrak{g}$, and A be a simple algebra with F as its center. If A is a †division algebra and $[A:F]=n^2$, then A has only one maximal order $\mathfrak{o}$, and $\mathfrak{o}$ has only one prime ideal $\mathfrak{q}$ such that $\mathfrak{q}^n=\mathfrak{p}\mathfrak{o}$; $\mathfrak{o}/\mathfrak{q}$ is an extension of degree n of $\mathfrak{g}/\mathfrak{p}$. If A is not necessarily a division algebra, the relation $\xi\mathfrak{o}\xi^{-1}=\mathfrak{o}'$ with an element ξ of A holds between two maximal orders $\mathfrak{o}$ and $\mathfrak{o}'$ of A. Furthermore, for any left (right) $\mathfrak{o}$-ideal $\mathfrak{a}$, there exists an element α such that $\mathfrak{a}=\mathfrak{o}\alpha$ ($\mathfrak{a}=\alpha\mathfrak{o}$). By using the notation of †cyclic algebra, we can express A in the form (K_n, σ,π^r). Here K_n is the unramified extension of degree n over F, σ is the †Frobenius substitution of K_n/F, π is a prime element of F, and $0\leqslant r<n$. The element of $\mathbf{Q}/\mathbf{Z}$ determined by $r/n\ (\mathrm{mod}\,\mathbf{Z})$ is denoted by $\{A\}$ and is called the †Hasse invariant of the algebra class containing A. The mapping $A\to\{A\}$ gives an isomorphism of the †Brauer group of F onto the additive group $\mathbf{Q}/\mathbf{Z}$. If M is an extension of F of finite degree, then $\{A^M\}=[M:F]\{A\}$ holds for the algebra A^M obtained from A by scalar extension (→ 31 Associative Algebras).

D. Simple Rings over an Algebraic Number Field

Let F be an algebraic number field of finite degree, and let A be a simple algebra with center F. Then A is a cyclic algebra and is isomorphic to a total matrix algebra over a division algebra D. The order n of the algebra class of A over F is determined by $n^2=[D:F]$ (H. Hasse, R. Brauer, and E. Noether).

Denote by $F_\mathfrak{p}$ the completion of F with respect to a †prime divisor $\mathfrak{p}$ of F, and let $A_\mathfrak{p}$ be the algebra obtained from A by the scalar extension $F_\mathfrak{p}$ over F. For a finite prime divisor $\mathfrak{p}$, the meaning of $\{A_\mathfrak{p}\}$ is as before; for an infinite prime divisor $\mathfrak{p}$, put $\{A_\mathfrak{p}\}=0$ or $1/2$ $(\mathrm{mod}\,\mathbf{Z})$ according as $A_\mathfrak{p}$ is a total matrix algebra over $F_\mathfrak{p}$ or not. Furthermore, define the subgroup $J_\mathfrak{p}$ of $\mathbf{Q}/\mathbf{Z}$ by

$$J_\mathfrak{p}=\mathbf{Q}/\mathbf{Z}, \qquad \mathfrak{p}: \text{a finite prime divisor,}$$
$$=\{0,1/2\ (\mathrm{mod}\,\mathbf{Z})\}, \qquad \mathfrak{p}: \text{a real infinite prime divisor,}$$
$$=\{0\}, \qquad \mathfrak{p}: \text{a complex infinite prime divisor.}$$

Now let J be the subgroup of the direct product $\Pi_\mathfrak{p} J_\mathfrak{p}$ consisting of all elements of the form $(\alpha_\mathfrak{p})$ $(\alpha_\mathfrak{p}\in J_\mathfrak{p})$ such that $\alpha_\mathfrak{p}=0$ except for a finite number of prime divisors and $\Sigma_\mathfrak{p}\alpha_\mathfrak{p}=0$. Then $A\to(\{A_\mathfrak{p}\})$ gives rise to an isomorphism of the Brauer group over F onto J (Hasse). Each $\{A_\mathfrak{p}\}$ is called the †$\mathfrak{p}$-invariant of A. In particular, A is a total matrix algebra over F if and only if $A_\mathfrak{p}$ is a total matrix algebra over $F_\mathfrak{p}$ for all $\mathfrak{p}$ (→ 31 Associative Algebras). These theorems are closely related to †class field theory.

If $\mathfrak{o}$ is a maximal order of A and $\mathfrak{a}$, $\mathfrak{b}$ are left $\mathfrak{o}$-ideals, then $\mathfrak{a}\xi=\mathfrak{b}$ with an element ξ of A defines an equivalence relation between $\mathfrak{a}$, $\mathfrak{b}$. The number of equivalence classes of left ideals with respect to this equivalence is called the **class number** of A; it is independent of the choice of $\mathfrak{o}$ and is equal to the class number defined by using right ideals. The product of all real infinite prime divisors $\mathfrak{p}$ with $\{A_\mathfrak{p}\}=1/2$ is denoted by P_∞. If P_∞ is the product of all infinite prime divisors and $[A:F]=4$, then A is called a **totally definite quaternion algebra**.

If $\mathfrak{o}$ is a maximal ideal of A and A is not a totally definite quaternion algebra, then $\mathfrak{a}\to N_{A/F}(\mathfrak{a})$ gives a one-to-one correspondence between the classes of left $\mathfrak{o}$-ideals and the congruence classes of ideals of F modulo

P_∞ (→ 16 Algebraic Number Fields H) (Eichler's theorem).

In particular, if A is a total matrix algebra over F, then the class number of A is equal to the class number of F. The class number of a totally definite quaternion algebra was determined by M. Eichler by using the zeta function (→ Section F) of A [5].

Let $\mathfrak{o}$ be a maximal order of A, $\mathfrak{a}$ be an integral two-sided $\mathfrak{o}$-ideal, b be an integer of F, and ξ be an element of $\mathfrak{o}$. Furthermore, assume $b \equiv 1 \pmod{P_\infty}$, $N_{A/F}(\xi) \equiv b \pmod{\mathfrak{a} \cap F}$ (†multiplicative congruence). Then there exists an element β of $\mathfrak{o}$ such that $N_{A/F}(\beta) = b$, $\beta \equiv \xi \pmod{\mathfrak{a}}$ ($N_{A/F}$ is the reduced norm), provided that A is not a totally definite quaternion algebra [6]. This theorem, which is called **Eichler's approximation theorem**, is widely applicable; e.g., it yields the previous theorem on the class number and can be generalized to the case of semisimple †algebraic groups (→ 15 Algebraic Groups).

E. Algebras over a Function Field

The Hasse-Brauer-Noether and Hasse theorems also hold for †normal simple algebras over a †field of algebraic functions of one variable over a finite field. On the other hand, a normal simple algebra over a field K of algebraic functions of one variable over an algebraically closed field is a total matrix algebra over K (**Tsen's theorem**).

F. Other Notions

†Adeles and †ideles for a simple algebra A over an algebraic number field of finite degree can be introduced as in the case of number fields (→ 7 Adeles and Ideles). If $N(\mathfrak{a})$ stands for the number of elements in $\mathfrak{o}/\mathfrak{a}$, where $\mathfrak{o}$ is a maximal order of A and $\mathfrak{a}$ is an integral left $\mathfrak{o}$-ideal, then the **zeta function of the simple algebra** A, called the **Hey zeta function**, is defined by $\zeta_A(s) = \sum N(\mathfrak{a})^{-s}$ (the sum over all integral left $\mathfrak{o}$-ideals). This function has properties similar to Dedekind zeta functions (→ 436 Zeta Functions). Let $\mathfrak{p}$ be an infinite prime divisor of the center F of A, and let $G_\mathfrak{p}$ be the group of elements in $A_\mathfrak{p}$ with the reduced norm 1. Put $G = \prod_\mathfrak{p} G_\mathfrak{p}$ (the product over all infinite prime divisors of F). Then the group Γ of units with the reduced norm 1 in $\mathfrak{o}$ is naturally regarded as a subgroup of G. To be more precise, Γ is a discrete subgroup of G, the volume of G/Γ is finite with respect to an invariant measure, and G/Γ is compact if and only if A is a division algebra (→ 126 Discontinuous Groups). This result can be viewed as a special case of more general facts about semisimple algebraic groups (→ 15 Algebraic Groups). If K is a †maximal compact subgroup of G, then Γ gives rise to a †discontinuous group operating on the homogeneous space G/K, and we obtain †automorphic forms with respect to Γ. The case where A is a †quaternion algebra has been studied extensively (→ 34 Automorphic Functions).

References

[1] K. Asano, Kanron oyobi idearuron (Japanese; Theory of rings and ideals), Kyôritu, 1949.

[2] C. Chevalley, L'arithmétique dans les algèbres de matrices, Actualités Sci. Ind., Hermann, 1936.

[3] M. Deuring, Algebren, Erg. Math., Springer, second edition, 1968.

[4] M. Eichler, Über die Idealklassenzahl hyperkomplexer Systeme, Math. Z., 43 (1938), 481–494.

[5] M. Eichler, Über die Idealklassenzahl total definiter Quaternionen Algebren, Math. Z., 43 (1938), 102–109.

[6] M. Eichler, Allgemeine Kongruenzklasseneinteilungen der Ideale einfacher Algebren über algebraischen Zahlkörpern und ihre L-Reihen, J. Reine Angew. Math., 179 (1938), 227–251.

[7] N. Jacobson, The theory of rings, Amer. Math. Soc. Math. Surveys, 1943.

[8] C. L. Siegel, Discontinuous groups, Ann. of Math., (2) 44 (1943), 674–689. (Gesammelte Abhandlungen, Springer, 1966, vol. 2, p. 390–405.)

[9] A. Weil, Adeles and algebraic groups, Lecture notes, Institute for Advanced Study, Princeton, 1961.

31 (III. 12) Associative Algebras

A. Fundamental Concepts

Let K be a commutative ring with unity element 1 (→ 363 Rings A), and let A be a ring which is a †unitary K-module (→ 275 Modules). Such a ring A is called an **associative algebra over** K (or simply **algebra over** K) if it satisfies the condition $\lambda(ab) = (\lambda a)b = a(\lambda b)$ ($\lambda \in K$; $a, b \in A$). An (associative) algebra A over K is often written A/K, and K is called the **coefficient ring** (or **ground ring**) of the algebra $A = A/K$. In particular, if K is a †field, then it is called the **coefficient field** (or **ground field**) of A. Algebras over fields have been studied in detail. Notions such as **zero algebra, unitary algebra, commutative algebra, (semi) simple algebra, and division algebra** are replicas of the respective ones for rings (→ 363 Rings). Considering both structures as rings and as K-modules, homomorphisms and isomorphisms are defined in a natural manner, and are called **algebra homomorphism** and **algebra isomorphism**, respectively. In this connection, **subalgebra, quotient algebra** (or **residue class algebra**), and **direct product** of algebras are also defined as in the case of rings.

An †**ideal** of an algebra A is defined as an ideal of the ring A, which is at the same time a submodule of the module A over the coefficient ring. The †radical of an algebra A considered as a ring is then an ideal of A in this sense. In fact, the existence of a unity in an algebra A implies that any ideal of A considered as a ring is necessarily an ideal of the algebra A.

In the rest of this article, we assume that all rings have a unity element and that all homomorphisms are unitary. Hence, when we consider subalgebras of an algebra A, we require that they share the unity element with A. If e is the unity element of an algebra A over K, then the mapping $\lambda \rightarrow \lambda e = \lambda'$ ($\lambda \in K$) is a homomorphism $K \rightarrow A$ whose image Ke is contained in the †center of A, and the scalar multiplication λa is equal to the ring multiplication $\lambda' a$ ($\lambda \in K, a \in A$). Conversely, given a homomorphism of K into a ring A whose image is contained in the center of A, we can regard A as an algebra over K in an obvious way. Hence, we are given an algebra A over K if and only if there exists a pair (A, ρ) of a ring A and a homomorphism $\rho: K \rightarrow A$ whose image is contained in the center of A. There exists a uniquely determined (unitary) homomorphism of the ring $\mathbf{Z}$ of rational integers into any ring; hence any ring can be regarded as an algebra over $\mathbf{Z}$. If the coefficient ring K of a nonzero algebra A is a field, then K can be regarded as a subfield contained in the center of A, and the unity element 1 of K coincides with the unity element of A.

Let A and B be algebras over K. Then the †tensor product $A \otimes_K B$ of K-modules is an algebra over K under the multiplication $(a \otimes b)(a' \otimes b') = aa' \otimes bb'$ ($a, a' \in A, b, b' \in B$). This algebra is called the **tensor product** of algebras A and B. Moreover, the mapping $a \rightarrow a \otimes 1$ (resp. $b \rightarrow 1 \otimes b$) ($a \in A, b \in B$) gives an algebra homomorphism $A \rightarrow A \otimes_K B$ (resp. $B \rightarrow A \otimes_X B$), which is called the **canonical homomorphism**. In particular, if A is a commutative algebra, then $A \otimes_K B$ can be regarded as an algebra over A by this canonical homomorphism; and in this case $A \otimes_K B$ is called the algebra obtained by **extension of the coefficient ring** of B to A (or a **scalar extension** of B by A), and is often denoted by B^A. The algebra $K \otimes_K B$ is canonically isomorphic to B. Furthermore, $(A \otimes_K B) \otimes_K C$ and $A \otimes_K (B \otimes_K C)$ are canonically isomorphic, and are written $A \otimes B \otimes C$. Let A and B be commutative algebras over K. Then, for any commutative algebra C and homomorphisms $\alpha: A \rightarrow C$, $\beta: B \rightarrow C$, there exists one and only one homomorphism $\gamma: A \otimes_K B \rightarrow C$ such that $\alpha(a) = \gamma(a \otimes 1)$, $\beta(b) = \gamma(1 \otimes b)$ ($a \in A, b \in B$). This property characterizes the tensor product $A \otimes_K B$ of commutative algebras A and B. In this sense, $A \otimes_K B$ is sometimes called the **coproduct** of A and B (→ 53 Categories and Functors E).

B. Examples of Associative Algebras

As we mentioned, any ring can be regarded as an algebra over the ring $\mathbf{Z}$ of rational integers. But it is often useful to deal with algebras over "larger" or more "efficient" coefficient rings. For instance, we have many rings which are algebras over a commutative ring K, such as the †ring of polynomials, the †ring of formal power series in n variables with coefficients in K, the †endomorphism ring of a K-module, and the †full matrix ring of degree n over K (→ 363 Rings C). There are other important classes of algebras, such as (semi) group algebras, Hecke algebras, and crossed-product algebras, which will be explained later. These algebras are defined by a canonical basis connected directly with a (semi) group structure. On the other hand, the †tensor algebra and the †exterior algebra of linear spaces and the †Clifford algebra associated with a given quadratic form are also important (→ 64 Clifford Algebras, 256 Linear Spaces).

The most frequently used example of a division algebra is the **quaternion field H** (often called **Hamilton's quaternion algebra**, W. R. Hamilton, 1858). This is a four-dimensional linear space over the real number field $\mathbf{R}$ with basis $\{1, i, j, k\}$, with the following laws of multiplication: 1 is a unity element, $i^2 = j^2 = k^2 = -1$, $ij = -ji = k$, $jk = -kj = i$, and $ki = -ik = j$. An element of $\mathbf{H}$ is called a **quaternion**. The only finite-dimensional division algebras over the real number field $\mathbf{R}$ are the real number field $\mathbf{R}$, the complex number field $\mathbf{C}$, and the quaternion field $\mathbf{H}$.

C. Group Algebras and Hecke Algebras

Let K be a commutative ring and $K^{(G)}$ be the

direct sum $\Sigma_{s\in G}K_s$ of modules K_s, where each K_s is isomorphic to K with a group G as the index set (→ 275 Modules F). The elements of $K^{(G)}$ are the families $(\lambda_s)_{s\in G}$ of elements of K whose components are all zero except for a finite number of them. Let $\{u_s\}_{s\in G}$ be the canonical basis of $K^{(G)}$, namely u_s is an element of $K^{(G)}$ of which the sth component is 1 and the others are 0. The module $K^{(G)}$ has the structure of an algebra over K, where the law of multiplication is determined by $u_s u_t = u_{st} (s,t\in G)$. This algebra is called the **group algebra** of G over K. The product $\lambda * \mu$ of elements $\lambda=(\lambda_s)$ and $\mu=(\mu_s)$ of $K^{(G)}$ is then given by

$$(\lambda*\mu)_s=\sum_{s=rl}\lambda_r\mu_l,\quad s\in G. \tag{1}$$

Each basis element u_s is often identified with the group element s, and in this manner, the group G is regarded as a basis of $K^{(G)}$, which is usually written KG or $K[G]$.

In this definition, the group G may be replaced by a semigroup G, and then the algebra $K^{(G)}$ is called a **semigroup algebra**. As an example, let $\mathbf{N}$ be the additive semigroup of all nonnegative rational integers and $\mathbf{N}^n$ be the direct product of n copies of $\mathbf{N}$. If we denote the elements $(i_1,\ldots,i_n)$ of $\mathbf{N}^n$ by $X_1^{i_1}\ldots X_n^{i_n}$ and use multiplication instead of addition, then the semigroup algebra of $\mathbf{N}^n$ over K is exactly the †ring of polynomials $K[X_1,\ldots,X_n]$. On the other hand, if G is a semigroup, then even when it is infinite, it may occur that for any $s\in G$, there exists only a finite number of pairs (r,l) of elements of G such that $s=rl$. In this case, formula (1) also defines the law of multiplication on the †Cartesian product K^G. This algebra is called a **large semigroup algebra** and contains $K^{(G)}$ as a subalgebra. In particular, the large semigroup algebra of $\mathbf{N}^n$ is exactly the †ring of formal power series $K[[X_1,\ldots,x_n]]$.

Let H be a subgroup of a group G, and assume that the index of $H\cap sHs^{-1}$ in H is finite for any $s\in G$. This assumption is equivalent to the condition that any double coset of G by H is a union of a finite number of left as well as right cosets. Let $H\backslash G$, G/H, and $H\backslash G/H$ be the set of all †right cosets, †left cosets, and †double cosets, respectively. Then each element of the direct sum $K^{(H\backslash G)}$, $K^{(G/H)}$, or $K^{(H\backslash G/H)}$ can be regarded as a function defined on G taking a constant value on each right, left, or double coset, respectively. Conversely, any function defined on G can be regarded as an element of $K^{(H\backslash G)}$, $K^{(G/H)}$, or $K^{(H\backslash G/H)}$ if it takes a constant value on each left, right, or double coset, respectively, and if it vanishes everywhere except on a finite number of cosets of respective type. Under such identification, let λ_s, λ_l, λ_r, and λ_t denote the values of a function $\lambda: G\to K$ on $s\in G$, $l\in H\backslash G$, $r\in G/H$, and $t\in H\backslash G/H$, respectively. For any λ, $\mu\in K^{(H\backslash G/H)}$ we define a function $\lambda*\mu$ by

$$(\lambda*\mu)_s=\sum_{s\in rl}\lambda_r\mu_l,\quad s\in G, \tag{2}$$

where the right-hand side is the sum taken over all pairs (r,l) such that $s\in rl$, $r\in G/H$, $l\in H\backslash G$. It can be shown that this sum is a finite sum and that $\lambda*\mu\in K^{(H\backslash G/H)}$. Hence, the module $K^{(H\backslash G/H)}$ has the structure of an algebra over K, which is called the **Hecke algebra** of (G,H) over K and often denoted by $\mathcal{H}_K(G,H)$. If $H=\{e\}$, the $\mathcal{H}_K(G,H)$ is exactly the group algebra of G. In general, $\mathcal{H}_K(G,H)$ can be regarded as an algebra obtained by extending the coefficient ring of $\mathcal{H}_{\mathbf{Z}}(G,H)$ to K. Furthermore, $K^{(H\backslash G)}$ ($K^{(G/H)}$) can be regarded as a right (left) module over the group algebra $K^{(G)}$, and the endomorphism ring $\mathcal{E}_{K^{(G)}}(K^{(H\backslash G)})$ (resp. $\mathcal{E}_{K^{(G)}}(K^{(G/H)})$) is canonically isomorphic (anti-isomorphic) to the Hecke algebra $K^{(H\backslash G/H)}$.

D. General Crossed Product

Let G be a group that operates on a commutative ring L, and denote the operation by $(s,\lambda)\to s(\lambda)$ $(s\in G,\lambda\in L)$; thus, for any $s\in G$ the mapping $\lambda\to s(\lambda)$ $(\lambda\in L)$ is an automorphism of L satisfying $s(t(\lambda))=st(\lambda)$ $(s,t\in G)$. For any λ, $\mu\in L^{(G)}$, we define the product $\lambda*\mu\in L^{(G)}$ by

$$(\lambda*\mu)_s=\sum_{s=rl}\lambda_r r(\mu_l)f(r,l),\quad s\in G, \tag{3}$$

where $\{f(r,l)\}_{r,l\in G}$ is a given family of elements of L. If this family satisfies the equations

$$f(s,r)f(sr,l)=s(f(r,l))f(s,rl),\quad s,r,l\in G,$$

then $L^{(G)}$ forms a ring. In terms of the canonical basis $\{u_s\}_{s\in G}$, this ring structure is defined by the formulas $u_r u_l=f(r,l)u_{rl}$, $u_s\lambda=s(\lambda)u_s$ $(\lambda\in L)$. If K is the subring of L consisting of all $\lambda\in L$ such that $s(\lambda)=\lambda$ $(s\in G)$, then the ring $L^{(G)}$ is an algebra over K, called the **crossed product** of L and G with respect to the given operation and the given **factor set** f of G. In a narrower sense of the term, we consider only the case when L is a field, G is a finite group, $f(r,l)\neq 0$, and G operates on L faithfully. In this case, G can be identified with the †Galois group of a finite †Galois extension L/K, and so the crossed product is written $(L/K,f)$. This is a †central simple algebra over K (→ Sections E, F).

If the operation of G in the above-mentioned general crossed product is trivial, namely, $L=K$, the crossed product of K and G is called an **algebra extension** of G over K with respect to f. Usually, we assume that K is a field and $f(r,l)\neq 0$. If $f(r,l)=1$, the algebra extension is the group algebra. If G is a finite group whose order is prime to the †characteristic of K, then any algebra extension of G over K (in particular, the group algebra) is always †semisimple and †separable. If G is a finite Abelian group and f is a factor set of G, then a bihomomorphism $\varphi: G\times G\rightarrow K^*$ (a mapping which is a homomorphism in each variable) is defined by $\varphi(s,t)=f(s,t)\cdot f(t,s)^{-1}$. An algebra extension of G over K with respect to f is a central simple algebra if and only if φ is nondegenerate. In particular, if G is a direct product of n copies of a group of order 2, then by choosing n elements $s_1, s_2, \ldots, s_n$ of G suitably, any element of G can be uniquely expressed as $s_a s_b \ldots s_z$ where $\{a,b,\ldots,z\}$ is a subset of $\{1,\ldots,n\}$ in the natural order. Hence, if we take $u_{(a,\ldots,z)}=u_a u_b \ldots u_z$ as a basis of an algebra extension, then any factor set f is determined by $f(s_i,s_j)=\lambda_{ij}$, where $\lambda_{ij}=1$ $(i<j)$, $\lambda_{ij}=\pm 1 (i>j)$, and λ_{ii} are arbitrary. When $\lambda_{ij}=-1 (i>j)$, the corresponding algebra extension is a †Clifford algebra. Furthermore if, $\lambda_{ii}=0$, then it is a †Grassmann algebra. If $\lambda_{ii}\neq 0$ and n is even, then it is a central simple algebra whenever the characteristic of K is not 2 (→ 64 Clifford Algebras).

Now let K be the real number field $\mathbf{R}$ and A_n be the Clifford algebra with $\lambda_{ii}=-1$. Then A_2 is the quaternion field. Elements of A_4 are called **sedenions**, and are important in †spinor theory and †Dirac's equation. In general, let K be an arbitrary field whose characteristic is not 2 and put $n=2$, $\lambda_{12}=1$, $\lambda_{21}=-1$, $\lambda_{11}=\lambda\neq 0$, and $\lambda_{22}=\mu\neq 0$, in the previous notation. Then the corresponding central simple algebra Q is called a **(generalized) quaternion algebra**. Thus Q has a basis $\{1,u,v,w\}$ satisfying the following laws: 1 is the unity element, $w=uv=-vu$, $u^2=\lambda$, and $v^2=\mu$ $(\lambda,\mu\in K)$. Any central simple algebra of dimension 4 is isomorphic to a certain quaternion algebra. (In particular, if $K=\mathbf{R}$ and $\lambda=\mu=-1$, then Q coincides with the quaternion field $\mathbf{H}$.) For any element $x=\alpha+\beta u+\gamma v+\delta w$ of Q, the element $\bar{x}=\alpha-\beta u-\gamma v-\delta w$ is said to be **conjugate** to x, and $N(x)=x\bar{x}\in K$ is called the **norm** of x. An element x of Q is invertible in Q if and only if $N(x)\neq 0$.

E. Finite-Dimensional Associative Algebras over a Field

For the rest of this article, we assume that the algebras considered are unitary and finite dimensional over a field K. Then by general properties of left (right) †Artinian rings, any associative algebra A has the following structure: the †radical N of A is the greatest †nilpotent ideal, and the quotient algebra $A/N=\bar{A}$ is †semisimple and decomposed into a direct sum of ideals which are simple algebras

$$\bar{A}=\bar{A}_1+\ldots+\bar{A}_n.$$

Each simple component $\bar{A}_i$ is a full matrix ring of degree r_i over a certain division algebra D_i, and $\bar{A}_i$ is decomposed into a direct sum of r_i minimal left ideals which are mutually $\bar{A}$-isomorphic:

$$\bar{A}_i=\bar{A}\bar{e}_i^{(1)}+\ldots+\bar{A}\bar{e}_i^{(r_i)}, \quad 1\leqslant i\leqslant n,$$

where $\bar{e}_i^{(1)},\ldots,\bar{e}_i^{(r_i)}$ are orthogonal †idempotent elements of $\bar{A}_i$ whose sum is equal to the unity element of $\bar{A}_i$. On the other hand,

$$\bar{A}_i=\bar{e}_i^{(1)}\bar{A}+\ldots+\bar{e}_i^{(r_i)}\bar{A}$$

gives a decomposition of $\bar{A}_i$ into the direct sum of minimal right ideals that are mutually $\bar{A}$-isomorphic. Moreover, we can choose an idempotent element $e_i^{(s)}$ of A from each residue class $\bar{e}_i^{(s)}$ such that $\{e_i^{(s)}\}$ forms a system of orthogonal idempotent elements, whose sum is equal to the unity element 1 of A, and

$$A=\sum_{i=1}^{n}\sum_{s=1}^{r_i}Ae_i^{(s)}=\sum_{i=1}^{n}\sum_{s=1}^{r_i}e_i^{(s)}A$$

gives a decomposition of A into a direct sum of †direct indecomposable left (right) ideals. Here, $Ae_i^{(s)}$ and $Ae_j^{(t)}$ ($e_i^{(s)}A$ and $e_j^{(t)}A$) are A-isomorphic if and only if $i=j$. Conversely, every decomposition of A into indecomposable ideals is obtained as above. The A-submodule $Ne_i^{(s)}$ of $Ae_i^{(s)}$ is the unique maximal proper submodule, and $Ae_i^{(s)}/Ne_i^{(s)}$ and $Ae_j^{(t)}/Ne_j^{(t)}$ are A-isomorphic if and only if $i=j$. Any simple A-module is A-isomorphic to a certain $Ae_i^{(s)}/Ne_i^{(s)}$ (→ Sections H, I).

Any simple algebra A over K is isomorphic to a full matrix ring $M_n(D)$ over a certain division algebra D. This is called **Wedderburn's theorem**. Here n is determined uniquely by A, and D is also determined uniquely by A up to isomorphism. Moreover, the center of A is isomorphic to the center of D. If the center of A coincides with K, then A is called a **central simple algebra** (or **normal simple algebra**) over K. In this case, any isomorphism of two simple subalgebras of A can be extended to an †inner automorphism of A. Let $V(B)$ denote the †commutor of a simple subalgebra B of A. Then $V(B)$ is also a simple subalgebra and $V(V(B))=B$, $\dim A=\dim B\cdot\dim V(B)$. In particular, if B is central over K, then there is a canonical isomorphism $A\cong B\otimes_K V(B)$. If D is a central division algebra, then any maximal commutative subalgebra L of D is a field and satisfies $(\dim L)^2 = \dim D$. Moreover, there are †separable extensions over K among such L. In general, the dimension of a central simple algebra is a square number r^2, where r is called the **degree** of A.

Two central simple algebras are said to be **similar** if they are isomorphic to full matrix rings over the same division algebra. This is an equivalence relation, and each equivalence class is called an **algebra class**. On the other hand, if A is a central simple algebra and B is a simple algebra, then $A \otimes_K B$ is a simple algebra. Moreover, if B is central, then $A \otimes_K B$ is also a central simple algebra. If A and B are similar to A' and B', respectively, then $A \otimes_K B$ is similar to $A' \otimes_K B'$. Hence, the tensor product $\otimes$ defines multiplication in the set $\mathfrak{B}(K)$ of the algebra classes over K. If A is a central simple algebra and A° is an algebra anti-isomorphic to A, then A° is also central simple and $A \otimes_K A^\circ$ is isomorphic to a full matrix ring over K. This shows that $\mathfrak{B}(K)$ forms a group, which is called the **Brauer group**, after R. Brauer, who introduced this concept, or the **algebra class group** over K. For a central simple algebra A, the degree of a division algebra similar to it is called the **Schur index** of A (or of the algebra class of A), and the order of the algebra class of A in the Brauer group is called the **exponent** of A. The Schur index is divisible by the exponent; and conversely, the exponent is divisible by every prime divisor of the Schur index.

F. Extensions of Coefficient Fields

Let L be an †extension field over a field K. Then for any algebra A over K, $L \otimes_K A$ can be regarded as an algebra over L. This algebra is denoted by A^L and is called an **algebra obtained by extending the coefficient field** to L. Let us denote the radical of a ring A by $\mathfrak{R}(A)$. An algebra A over K is called a **separable algebra** if it satisfies $\mathfrak{R}(A^L) = \{0\}$ for any extension field L over K. In the special case when A is an †algebraic extension field over K, A is a separable algebra if and only if every element of A (or every element of a subset which generates A) is †separable over K (→ 157 Fields). A (finite-dimensional) algebra A over K is separable if and only if A is semisimple and the center of every simple component of A is a separable extension over K. If the quotient algebra $A/\mathfrak{R}(A)$ of an algebra A is separable, then there exists a subalgebra S such that $A = S + \mathfrak{R}(A)$ and $S \cap \mathfrak{R}(A) = \{0\}$, and S is uniquely determined up to inner automorphisms (**Wedderburn-Mal'cev theorem**).

In order that an algebra A over K be central simple, it is necessary and sufficient that A^L is simple for any extension field L over K. This latter statement holds if and only if A^L is isomorphic to a full matrix algebra over L for a certain extension field L over K. Such an extension field is a †splitting field of A (→ 358 Representations F). For a central simple algebra A, a (finite) extension field L of degree r over K is a splitting field of A if and only if there exists a central simple algebra B of degree r which is similar to A having a subfield which is K-isomorphic to L. In this case, r is divisible by the Schur index of A. Furthermore, A has a separable splitting field whose extension degree is equal to the Schur index of A. This shows that A has a splitting field which is a finite Galois extension field over K.

Let L be a finite †Galois extension over a field K and G be its †Galois group. If f is a factor set with respect to the operation of G on L^*, then the crossed product $(L/K, f)$ is a central simple algebra over K (→ Section D). The product fg of two factor sets f and g is also a factor set, and the set of all factor sets forms an †Abelian group. Thus, $(L/K, fg)$ is similar to $(L/K, f) \otimes_K (L/K, g)$. On the other hand, factor sets f and g are said to be **associated** with each other if there exists a family $\{\lambda_s\}_{s \in G}$ of elements of L such that

$$f(r,l) = g(r,l) r(\lambda_l) \lambda_{rl}^{-1} \lambda_r, \quad r, l \in G.$$

Hence, f and g are associated with each other if and only if $(L/K, f)$ and $(L/K, g)$ are similar. Therefore, the mapping $f \rightarrow (L/K, f)$ gives a monomorphism of the group $H^2(G, L^*)$ of all associated classes of factor sets (which can be identified with the 2-dimensional †cohomology group of G with coefficients in L^*) into the Brauer group $\mathfrak{B}(K)$ over K. Its image coincides with the subgroup of all algebra classes which have L as a splitting field. In particular, any algebra class is similar to the crossed product of a certain finite Galois extension L and its Galois group G (R. Brauer, E. Noether, A. Albert, K. Shoda, et al.).

G. Cyclic Algebras

Let Z be a †cyclic extension field of degree n over a field K. Then the crossed product of Z and its Galois group G is called a **cyclic algebra** over K. For a fixed generator s of G and any element $\alpha \neq 0$ of K, a factor set $f(s^i, s^j)$ $(0 \leqslant i, j < n)$ can be defined by $f(s^i, s^j) = 1$ $(i+j < n)$ and $f(s^i, s^j) = \alpha$ $(i+j \geqslant n)$. Let (Z, s, α) denote the corresponding crossed product. Then (Z, s, α) and (Z, s, β) are similar if and only if α/β is a norm of a certain element of Z into K. On the other hand, any crossed product of Z and G is similar to a certain (Z, s, α), and the correspondence $\alpha \rightarrow (Z, s, \alpha)$ gives an isomorphism of $K^*/N_{Z/K}(Z^*)$ (the norm class group of Z/K) to the group consisting of the algebra classes over K which have Z as a splitting field. If K is a †𝔭-adic field or a finite †algebraic number field, then any central simple algebra over K is isomorphic to a certain cyclic algebra. In this section, we describe this situation in detail.

Let K be a 𝔭-adic field and q the number of residue classes modulo 𝔭. If A is a central simple algebra over K, then its Schur index coincides with the exponent, which is called simply the **index**. A finite extension field L over K is a splitting field for A if and only if the degree of L is a multiple of the index of A. If n is the degree of A, then the field $W=K(\omega)$ obtained by adjoining to K a †primitive (q^n-1)st root of unity ω is a cyclic (and †unramified) extension of degree n over K, and its Galois group is generated by the automorphism σ of W determined by $\omega^\sigma=\omega^q$, $\sigma|K=$identity. Moreover, we have $A \cong (W, \sigma, \alpha)$ for a certain $\alpha \in K^*$. Let ν be the †exponential 𝔭-adic valuation $v(\alpha)$ of α. Then $\nu/n \pmod{\mathbf{Z}}$ is uniquely determined by the algebra class of A, which is called the **Hasse invariant** of A (or of the algebra class of A). By assigning to each of the algebra classes its Hasse invariant, we get an isomorphism of the Brauer group $\mathfrak{B}(K)$ over K to the group $\mathbf{Q}/\mathbf{Z}$, the additive group of the rational numbers mod $\mathbf{Z}$ (H. Hasse, 1931).

Let K be a finite algebraic number field and A be a central simple algebra over K. Let 𝔭 be a (finite or infinite) †prime divisor of K and $K_\mathfrak{p}$ the †𝔭-adic extension field over K. The algebra $A_\mathfrak{p}$ which is obtained from A by extending the coefficient field to $K_\mathfrak{p}$ is a central simple algebra over $K_\mathfrak{p}$. Except for a finite number of 𝔭, $A_\mathfrak{p}$ is isomorphic to a full matrix ring over $K_\mathfrak{p}$, and A itself is isomorphic to a full matrix ring over K if and only if $A_\mathfrak{p}$ is isomorphic to a full matrix ring over $K_\mathfrak{p}$ for all 𝔭. The index $m_\mathfrak{p}$ of $A_\mathfrak{p}$ is called the **𝔭-index** of A, and the Hasse invariant of $A_\mathfrak{p}$ is called the **𝔭-invariant** of A, which is denoted by $(A/\mathfrak{p})$. If 𝔭 is an †infinite prime divisor, then $m_\mathfrak{p}$ is equal to 1 or 2, and in each case, we define the 𝔭-invariant by setting $(A/\mathfrak{p})=0$ or $1/2 \pmod{\mathbf{Z}}$ correspondingly. The Schur index of A is the L.C.M. of the 𝔭-indices $m_\mathfrak{p}$ for all 𝔭 and coincides with the exponent of A. This is called simply the **index** of A. On the other hand, the 𝔭-invariants satisfy $(A/\mathfrak{p}) \equiv 0 \pmod{\mathbf{Z}}$ except for a finite number of 𝔭, and

$$\sum_{\mathfrak{p}} (A/\mathfrak{p}) \equiv 0 \pmod{\mathbf{Z}}.$$

Conversely, given a rational number $\rho_\mathfrak{p}$ for each 𝔭 such that (i) $\rho_\mathfrak{p} \equiv 0 \pmod{\mathbf{Z}}$ except for a finite number of 𝔭; (ii) $\rho_\mathfrak{p} \equiv 0 \pmod{\mathbf{Z}}$ if 𝔭 is infinite and imaginary, $\rho_\mathfrak{p} \equiv 0$ or $1/2 \pmod{\mathbf{Z}}$ if 𝔭 is infinite and real; (iii) $\Sigma_\mathfrak{p}\rho_\mathfrak{p} \equiv 0 \pmod{\mathbf{Z}}$, then there is a uniquely determined algebra class of central simple algebras A over K such that $(A/\mathfrak{p}) \equiv \rho_\mathfrak{p} \pmod{\mathbf{Z}}$ for each 𝔭. In this way the structure of the Brauer group over a finite algebraic number field is completely determined (Hasse, 1933).

H. Frobenius Algebras

Let A be an algebra over a field K. A is called a **Frobenius algebra** if its †regular representation and †coregular representation (→ 358 Representations E) are similar. Thus A is a Frobenius algebra if the left A-module A and the dual module A^* of the right A-module A are isomorphic as left A-modules. Let

$$A=\sum_{i=1}^{n}\sum_{s=1}^{r_i} Ae_i^{(s)}=\sum_{i=1}^{n}\sum_{s=1}^{r_i} e_i^{(s)}A$$

be direct decompositions of A into indecomposable left (right) ideals (→ Section E). We denote $e_i^{(1)}$ by e_i. Then A is a Frobenius algebra if and only if there exists a permutation π on $1,\ldots,n$ such that (i) $Ae_i \cong (e_{\pi(i)}A)^*$; (ii) $r_i=r_{\pi(i)}$. When there exists a permutation satisfying only condition (i), A is called a **quasi-Frobenius algebra**.

For a subset S of A, the ideals $l(S)=\{a \in A \mid aS=0\}$ and $r(S)=\{a \in A \mid Sa=0\}$ are called the **left annihilator** and **right annihilator** of S, respectively. Then A is a quasi-Frobenius algebra if and only if (iii) $l(r(\mathbf{l}))=\mathbf{l}$ and $r(l(\mathbf{r}))=\mathbf{r}$ for any left ideal $\mathbf{l}$ and any right ideal $\mathbf{r}$. In general, if we are given a left (right) A-module M, we denote the right (left) A-module $\mathrm{Hom}_A(M, A)$ by $\hat{M}$. Then if A is a quasi-Frobenius algebra, there is a canonical isomorphism $\hat{\hat{M}} \cong M$, and the annihilator relation gives a one-to-one and dual correspondence between the set of the submodules of M and the set of submodules of $\hat{M}$ (M. Hall). If a quasi-Frobenius algebra A satisfies (iv) $\dim \mathbf{r}+\dim l(\mathbf{r})=\dim \mathbf{l}+\dim r(\mathbf{l})=\dim A$ for any left ideal $\mathbf{l}$ and any right ideal $\mathbf{r}$, then A is a Frobenius algebra; the converse is also true.

A criterion for an algebra A to be a Frobenius algebra is that there is a linear form $x \rightarrow \lambda(x)$ on A such that if $\lambda(xa)=0$ for all $x \in A$, then $a=0$. Moreover, if λ satisfies $\lambda(xy)=\lambda(yx)$ $(x, y \in A)$, then A is called a

symmetric algebra. For example, semisimple algebras and group algebras are symmetric algebras. If A is a symmetric algebra, then for any left (right) A-module M the right (left) A-modules $M^* = \mathrm{Hom}_K(M, K)$ and $\hat{M} = \mathrm{Hom}_A(M, A)$ are canonically A-isomorphic.

If A is a Frobenius algebra, the radical N of A satisfies $l(N) = r(N)$, and the annihilator of N is a principal left and principal right ideal; the converse is also true. For a two-sided ideal $\mathbf{z}$ of a Frobenius algebra A, the quotient algebra $A/\mathbf{z}$ is a Frobenius algebra if and only if $l(\mathbf{z})$ and $r(\mathbf{z})$ are a principal left and a principal right ideal, respectively. If A is a symmetric algebra, then any two-sided ideal $\mathbf{z}$ satisfies $l(\mathbf{z}) = r(\mathbf{z})$, and $A/\mathbf{z}$ is also a symmetric algebra if and only if $l(\mathbf{z}) = r(\mathbf{z})$ is a principal ideal generated by an element in the center.

Furthermore, for any extension L of the coefficient field K, A^L is a quasi-Frobenius algebra (resp. Frobenius algebra, symmetric algebra) if and only if A is (T. Nakayama, 1939, 1941). The concept of Frobenius algebras has been extended to algebras B over a ring A (F. Kasch, 1954).

I. Uniserial Algebras

In the notation of the preceding section, if each indecomposable left ideal Ae_i (right ideal e_iA) of an algebra A over K has a unique composition series, then A is called a **generalized uniserial algebra**. If an algebra A is decomposed into a direct sum of ideals which are primary rings, then it is called a **uniserial algebra**. Any left module over a generalized uniserial algebra A is decomposed into a direct sum of submodules which are A-homomorphic images of Ae_i: An algebra whose radical N is a principal left and principal right ideal is a generalized uniserial algebra. For an algebra A to be uniserial, it is necessary and sufficient that every two-sided ideal of A is a principal left and principal right ideal. Hence A is uniserial if and only if every quotient algebra of A is a Frobenius algebra. If A^L is uniserial for an extension field L of K, then A itself is uniserial. The converse, however, is not always true. If A^L is uniserial for any extension field L of K, then A is called an **absolutely uniserial algebra**. For A to be absolutely uniserial it is necessary and sufficient that its radical N is a principal ideal generated by an element in the center Z and Z is decomposed into a direct sum of simple extensions of K (i.e., ideals of the form $K[\alpha]$) (K. Asano, G. Köthe, Nakayama, G. Azumaya).

J. Algebraic Algebras

Here, we consider general (not necessarily finite-dimensional) algebras A over a field K. We say that A is an **algebraic algebra** if every element of A is algebraic over K, i.e., every element of A is a root of a certain polynomial with coefficients in K. We say that A satisfies a **polynomial identity** $p(X_1, \ldots, X_n) = 0$ or that A is a ***PI*-algebra** with an identity $p(X_1, \ldots, X_n) = 0$ if there exists a nonzero (noncommutative) polynomial $p(X_1, \ldots, X_n)$ in $X_1, \ldots, X_n$ with coefficients in K such that $p(a_1, \ldots, a_n) = 0$ for all $a_i \in A$. A PI-algebra satisfies an identity which is homogeneous linear in each variable, and also an identity of the form $[x_1, \ldots, x_n]^m = 0$ (where [ ] is the sum $\Sigma \pm x_{i_1} \ldots x_{i_n}$ taken over all permutations of $1, 2, \ldots, n$ and $\pm$ is the sign of the permutation). An algebra is said to be **locally finite** if any finite number of elements of A generate a finite-dimensional subalgebra. For PI-algebras, an affirmative answer was found for Kuroš's problem, which asks whether an algebraic algebra is locally finite if the degree of any element α of A (i.e., $\dim K[\alpha]$) is bounded.

A detailed historical note on the study of algebras can be found at the end of [6].

K. Brauer Group of a Ring

Let R be a commutative ring. An R-algebra is called a **separable algebra** if A is †projective as a two-sided A-module (→ 202 Homological Algebra F). When the base ring is a field, this agrees with the classical notion of separability; and A is separable over R if and only if $A/\mathfrak{m}A$ is separable over the residue field $R/\mathfrak{m}$ for every maximal ideal $\mathfrak{m}$ of R. A **central separable algebra** is also called an **Azumaya algebra**. If P is a finitely generated faithful projective R-module (briefly: **R-progenerator**), the endomorphism ring $\mathrm{End}_R(P)$ is an Azumaya R-algebra. Azumaya algebras A_1 and A_2 are said to be in the same class (similar) if there exist R-progenerators P_1 and P_2 such that $A_1 \otimes \mathrm{End}_R(P_1) \cong A_2 \otimes \mathrm{End}_R(P_2)$. The set of similarity classes forms an Abelian group with respect to $\otimes$. This is called the **Brauer group** $B(R)$ of R (Auslander and Goldman [16]). Every element of $B(R)$ is of finite order [18, 19].

$B(R)$ is a covariant functor from commutative rings to Abelian groups. If R is a field, $B(R)$ coincides with the classically defined one (→ Section E). If R is a †Henselian local ring with the residue field k, the mapping $B(R) \to B(k)$ is an isomorphism [15]. If R is a †regular ring with the quotient field K, $B(R)\to B(K)$ is injective. If further $\dim R \leqslant 2$, we have $B(R)=\bigcap_{\mathfrak{p}} B(R_{\mathfrak{p}})$, $\mathfrak{p}$ running over all primes of height 1 of R, where $R_{\mathfrak{p}}$ is the localization of R at $\mathfrak{p}$ and $B(R)$ and $B(R_{\mathfrak{p}})$ are considered as imbedded in $B(K)$ [16]. As an example, putting these facts together with the structure of $B(K)$ of the algebraic number field K (→ Section G), we have the structure of $B(R)$ of the ring of integers of K: $B(R)=0$ if K is totally imaginary, and $\cong(\mathbf{Z}/2\mathbf{Z})^{r-1}$ if K has $r(>0)$ real infinite places.

A commutative R-algebra S is called a **splitting ring** of A if the S-algebra $S\otimes A$ is isomorphic to $\mathrm{End}_S(P)$ for some S-progenerator P. We denote by $B(S/R)$ the subgroup of $B(R)$ consisting of all algebra classes split by S. Since an Azumaya algebra over a ring need not have a Galois extension as splitting ring, the description of the Brauer groups by means of Galois cohomology ceases to have full generality. Instead we have the following exact sequence of †Amitsur cohomology, assuming S is an R-progenerator (Chase and Rosenberg [17]): $0\to H^1(S/R,U)\to\mathrm{Pic}(R)\to H^0(S/R,\mathrm{Pic})\to H^2(S/R,U)\to B(S/R)\to H^1(S/R,\mathrm{Pic})\to H^3(S/R,U)\to \ldots$, where we denote by $U(T)$ and $\mathrm{Pic}(T)$ of a commutative ring T the unit group and the †Picard group of rank 1 projective modules of T, respectively. The full Brauer group $B(R)$ is mapped monomorphically into $H^2(R,U)=\varinjlim H^2(S/R,U)$, the limit over the †faithfully flat R-algebras S.

Grothendieck and others studied the Brauer groups in more general geometrical context [18].

References

[1] K. Asano, Kanron oyobi idearuron (Japanese; Theory of rings and ideals), Kyôritu, 1949.
[2] G. Azumaya, Tanzyunkan no riron (Japanese; Theory of simple algebras), Kawade, 1951.
[3] K. Shoda, Tyûsyô daisûgaku (Japanese; Abstract algebra), Iwanami, 1932.
[4] K. Shoda, Tagensûron (Japanese; Theory of hypercomplex numbers), Iwanami Coll. of Math., 1935.
[5] T. Nakayama, Kyokusyo ruitairon (Japanese; Local class field theory), Iwanami Coll. of Math., 1935.
[6] T. Nakayama and G. Azumaya, Daisûgaku II (Japanese; Algebra II–Theory of rings), Iwanami, 1954.
[7] A. A. Albert, Structure of algebras, Amer. Math. Soc. Colloq. Publ., 1939.
[8] E. Artin, C. J. Nesbitt, and R. Thrall, Rings with minimum condition, Univ. of Michigan Press, 1944.
[9] M. Deuring, Algebren, Erg. Math., Springer, second edition, 1968 (Chelsea, 1948).
[10] N. Jacobson, The Theory of rings, Amer. Math. Soc. Math. Surveys, 1943.
[11] N. Jacobson, Structure of rings, Amer. Math. Soc. Colloq. Publ., 1956.
[12] B. L. van der Waerden, Algebra II, Springer, 1967.
[13] C. W. Curtis and I. Reiner, Representation theory of finite groups and associative algebras, Interscience, 1962.
[14] J.-P. Serre, Corps locaux, Actualités Sci. Ind., Hermann, 1962.
[15] G. Azumaya, On maximally central algebras, Nagoya Math. J., 2 (1951), 119–150.
[16] M. Auslander and O. Goldman, the Brauer group of a commutative ring, Trans. Amer. Math. Soc., 97 (1960), 367–409.
[17] S. Chase and A. Rosenberg, Amitsur cohomology and the Brauer group, Mem. Amer. Math. Soc., 52 (1965), 34–79.
[18] A. Grothendieck, Le groupe de Brauer, I, II, III, Dix exposés sur la cohomologie des schémas, North-Holland, 1968, p. 46–188.
[19] M.-A. Knus and M. Ojanguren, Théorie de la descente et algèbres d'Azumaya, Lecture notes in math. 389, Springer, 1974.
[20] M. Orzech and C. Small, the Brauer group of commutative rings, Lecture notes in pure and appl. math. 11, Marcel Dekker, 1975.

32 (X. 19) Asymptotic Series

A. Asymptotic Expansions

Let α be a boundary point of an open connected domain D in the complex z-plane or on a Riemann surface. Assume that $\varphi_n(z)$ $(n=0,1,2,\ldots)$ are holomorphic functions of z in D and that for every $n=0, 1, 2, \ldots$, as z tends to α through D, $\varphi_{(n-1)}(z)=o(\varphi_n(z))$ holds. A function $f(z)$ that is holomorphic in $D\cap U$ for some neighborhood U of α will be said to have the **asymptotic expansion** (or be **asymptotically developable** in the form)

$$f(z)\sim a_0\varphi_0(z)+a_1\varphi_1(z)+\ldots+a_n\varphi_n(z)+\ldots \quad (1)$$

as $z\to\alpha$ through D if $f(z)$ satisfies

$$f(z)-a_0\varphi_0(z)-a_1\varphi_1(z)-\ldots-a_n\varphi_n(z) =O(\varphi_{n+1}(z))$$

for any integer $n>0$ as $z\to\alpha$ through D. The

coefficients a_n $(n=0,1,\dots)$ appearing in (1) are uniquely determined. This fact immediately follows from the formulas

$$a_0=\lim_{z\to\alpha} f(z)/\varphi_0(z),\dots,$$
$$a_n=\lim_{z\to\alpha}(f(z)-a_0\varphi_0(z)-\dots-a_{n-1}\varphi_{n-1}(z))/\varphi_n(z).$$

The point α may be the point at infinity.

In most cases, the functions $\varphi_n(z)$ have the form $\varphi(z)^n\psi(z)$ $(n=0,1,2,\dots)$, where $\varphi(z)$ and $\psi(z)$ are holomorphic functions of z in D. For example, suppose that $f(z)=\Gamma(z)$, $D=\{z\,|\,0<|z|<\infty,\ |\arg z|<\pi\}$, and $\alpha=\infty$. Then we have the following asymptotic expansion:

$$\log\Gamma(z)-\{\tfrac{1}{2}\log 2\pi-(z-\tfrac{1}{2})\log z-z\}$$
$$\sim\frac{B_2}{1\cdot 2}\frac{1}{z}-\frac{B_4}{3\cdot 4}\frac{1}{z^3}-\dots$$
$$+(-1)^{n-1}\frac{B_{2n}}{(2n-1)\cdot(2n)}\frac{1}{z^{2n-1}}+\dots, \qquad (2)$$

where B_{2n} is the †Bernoulli number. Hence when z tends to $\alpha=\infty$ through D, we have

$$\Gamma(z)\sim\sqrt{2\pi}\,z^{z-1/2}\,e^{-z} \quad (\text{†Stirling's formula}).$$

If $f(z)$ is holomorphic at $z=\alpha$, its Taylor expansion

$$f(z)=a_0+a_1(z-\alpha)+\dots+a_n(z-\alpha)^n+\dots \qquad (3)$$

can be considered as an asymptotic expansion of $f(z)$ by taking $D=\{z\,|\,0<|z-\alpha|<r\}$, $\varphi_n(z)=(z-\alpha)^n$. Conversely, if $f(z)$ is a holomorphic and single-valued function of z in $0<|z-\alpha|<r$ and, moreover, admits an asymptotic expansion of the form

$$f(z)\sim a_0+a_1(z-\alpha)+\dots+a_n(z-\alpha)^n+\dots, \qquad (4)$$

then α is a removable singular point. Consequently, $f(z)$ admits a Taylor expansion at $z=\alpha$. By virtue of the uniqueness of the coefficients in asymptotic expansions, the power series on the right-hand side of (4) converges uniformly for $|z-\alpha|<r$, and the sum coincides with $f(z)$.

Concerning the uniform convergence of asymptotic expansions, **Carleman's theorem** is well known [4]: If the asymptotic expansion (4) is valid as z tends to α through an arbitrary angular domain in the Riemann surface of $\log(z-\alpha)$ having α as its vertex, then the asymptotic expansion (4) is uniformly convergent.

Assume that both $f(z)$ and $g(z)$ are holomorphic in D and admit asymptotic expansions in power series of $z-\alpha$ as $z\to\alpha$ through D. Then all functions defined by $f(z)+g(z)$, $f(z)-g(z)$, $f(z)g(z)$, and $f(z)/g(z)$ $(g(z)\neq 0,\ z\in D)$ admit asymptotic expansions in power series of $z-\alpha$. Furthermore, functions defined by $\int f(z)dz$ and $f'(z)$ also have asymptotic expansions in power series of $z-\alpha$, which are obtained by termwise integration and termwise differentiation, respectively. (In the case of a real variable, this assertion is generally false.)

Suppose that we are given a sequence $\{a_n\}$ $(n=0,1,\dots)$ and an angular domain D with vertex at α on the Riemann surface of $\log(z-\alpha)$. Then there always exists a function $f(z)$ that is holomorphic in D and admits an asymptotic expansion of the form (4) as $z\to\alpha$ through D. This theorem was first proved by T. Carleman. M. Hukuhara improved the proof (1937) and T. Sato extended the theorem to the case of two variables (1949).

B. Asymptotic Series of Several Variables

From the practical point of view, as in the theory of differential equations (→285 Nonlinear Ordinary Differential Equations (Singular Points)), the case of several variables is often more useful than that of one variable. In the case of several variables, asymptotic expansions may take various forms. For example, let $0=(0,\dots,0)$ be a boundary point of an open connected domain D in the Cartesian product of n complex z_j-planes. If a function $f(z_1,\dots,z_n)$, which is †holomorphic in $D\cap U$ for some neighborhood U of the origin, satisfies, for any positive integer N,

$$f(z_1,\dots,z_n)=\sum_{k_1+\dots+k_n<N} f_{k_1\dots k_n}z_1^{k_1}\dots z_n^{k_n}+O(|z_1|^N+\dots+|z_n|^N) \qquad (5)$$

as $z\to 0$ through D, we say that $f(z_1,\dots,z_n)$ admits the asymptotic expansion

$$f(z_1,\dots,z_n)\sim\sum_{N=1}^{\infty}\sum_{k_1+\dots+k_n=N} f_{k_1\dots k_n}z_1^{k_1}\dots z_n^{k_n}$$

as $z\to 0$, $z\in D$. If, instead of (5), the relation

$$f(z_1,\dots,z_n)=\sum_{j=0}^{N-1} f_j(z_1,\dots,z_{n-1})z_n^j+O(|z_n|^N) \qquad (6)$$

holds uniformly for $(z_1,\dots,z_{n-1})\in D$, where $f_j(z_1,\dots,z_{n-1})$ are holomorphic and bounded functions of their arguments, the function $f(z_1,\dots,z_n)$ is said to admit the asymptotic expansion

$$f(z_1,\dots,z_n)\sim\sum_{j=0}^{\infty} f_j(z_1,\dots,z_{n-1})z^j$$

uniformly for $(z_1,\dots,z_{n-1})$ as $z_n\to 0$, $z\in D$. Moreover, if the coefficients $f_j(z_1,\dots,z_{n-1})$

admit asymptotic expansions of the form (5), a function with an asymptotic expansion of the form (6) always admits an asymptotic expansion of the form (5).

A general theory concerning the four arithmetic operations, differentiation, integration, and existence of implicit functions for asymptotically developable functions was established by Hukuhara (1937).

C. Application to Differential Equations

For a system of linear ordinary differential equations with an irregular singular point at $z=0$, even when there exists a formal power series solution, the power series is generally divergent. Taking the hint given by Stirling's formula for the Γ-function, Poincaré introduced the notion of asymptotic expansions and succeeded in giving an analytic meaning to formal solutions of divergent type. Actual solutions for †difference equations, †difference-differential equations, and †ordinary differential equations of canonical form (including second-order linear †differential equations of confluent type) are represented by asymptotic expansions. A general method of constructing asymptotic expansions of solutions for a system of ordinary differential equations is known as Hukuhara's theory [2]. But this method fails to provide so-called connection formulas of solutions.

When a differential equation admits integral representation for solutions, there is a very powerful method (†method of steepest descent) that enables us to obtain not only asymptotic expansions but also connection formulas of solutions. The elements of this method were established by B. Riemann; P. Debye brought it to fruition (→ 272 Method of Steepest Descent).

References

[1] M. Fujiwara, Zyôbibun hôteisiki ron (Japanese; Theory of ordinary differential equations), Iwanami, 1930.
[2] M. Hukuhara, Zyôbibun hôteisiki (Japanese; Ordinary differential equations), Iwanami, 1950.
[3] N. G. de Bruijn, Asymptotic methods in analysis, North-Holland, 1958.
[4] T. Carleman, Les fonctions quasi-analytiques, Gauthier-Villars, 1926.
[5] A. Erdélyi, Asymptotic expansions, Dover, 1956.
[6] H. Poincaré, Sur les intégrales irrégulières des équations linéaires, Acta Math., 8 (1886), 295–344.
[7] W. R. Wasow, Asymptotic expansions for ordinary differential equations, Interscience, 1965.
[8] E. T. Whittaker and G. N. Watson, A course of modern analysis, Cambridge Univ. Press, fourth edition, 1958.
[9] M. Hukuhara, T. Kimura, and T. Matuda, Equations différentielles ordinaires du premier ordre dans le champ complex, Publ. Math. Soc. of Japan, 1961.
[10] W. B. Ford, Studies on divergent series and summability, 1916, The asymptotic developments of functions defined by Maclaurin series, 1936, Scientific Series, Univ. of Michigan (Chelsea, 1960).

33 (I.16) Automata

An **automaton** (or more precisely a **finite automaton**) is a physical device which assumes one of a finite set of †internal states at each moment t ($t=0,1,2,\ldots$) and at the same time receives an †input symbol chosen from a given finite set U, emits an †output that depends only on the internal state, and assumes the next state determined by the present internal state and the received input symbol. Mathematically, an automaton over a set U of input symbols is a 4-tuple (S, g, s_0, F) consisting of a finite set S of internal states, a function $g: U \times S \rightarrow S$ determining the next state on the basis of the input symbol and the present state, the initial state $s_0 \in S$, and the set of final states $F \subset S$.

A subset of the †free semigroup T generated by U is said to be an **event** (over U). The function g may be extended to the function $f: T \times S \rightarrow S$ that follows.

$$f(e,s)=s; s \in S \text{ and } e \text{ is the identity element of } T,$$

$$f(xu,s)=g(u,f(x,s)); x \in T, u \in T, s \in S.$$

An event E is said to be **representable** by an automaton $A=(S,g,s_0,F)$ if and only if

$$x \in E \Leftrightarrow f(x,s_0) \in F.$$

There are various equivalent statements that an event is representable. One such statement is that the event is regular in the following sense (S. C. Kleene): Consider the operations $\cup$, $\cdot$, and $*$, where $E \cdot F=\{xy \mid x \in E, y \in F\}$, $E * F=\bigcup_{n=0}^{\infty} EF^n$. The smallest class of events that includes the unit set consisting of one element and the empty set and is closed under the operations $\cup$, $\cdot$, $*$ is called the **regular set**. An event belonging to the regular set is a **regular event**.

Automata are often represented by **transition diagrams** as shown in Fig. 1. The letters in the circles standing to the left of the semicolons represent the internal states, and the characters to the right of the semicolons are outputs emitted under the states, while the letters attached to the arrows stand for the inputs.

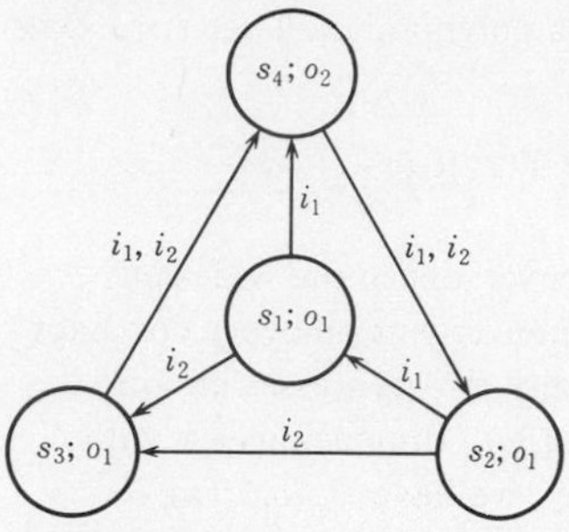

Fig. 1
Transition diagram.

Automata as physical devices include nerve nets and †digital computers. A device for receiving environmental stimuli and converting them into inputs is called a **receptor**, while one for transmitting the output to the environment is an **effector**. An automaton having a receptor and an effector is called a **robot**.

References

[1] C. E. Shannon and J. McCarthy, Automata studies, Ann. Math. Studies, Princeton Univ. Press, 1956.
[2] H. Takahashi, Keisan kikai (Japanese; Computing machines), Iwanami Coll. of Modern Math. Sci., 1958.

34 (IV.17) Automorphic Functions

A. General Remarks

Let X be a (complex) †analytic manifold and Γ a †discontinuous group of (complex) analytic automorphisms of X. A set of †holomorphic functions (without zero) on X, $\{j_\gamma(z)\}$ $(\gamma\in\Gamma)$, is called a **factor of automorphy** if it satisfies the condition $j_{\gamma\gamma'}(z)=j_\gamma(\gamma'(z))j_{\gamma'}(z)$ for all $\gamma, \gamma'\in\Gamma$, $z\in X$. A †meromorphic function f on X is called a **(multiplicative) automorphic function** with respect to the factor of automorphy $\{j_\gamma(z)\}$ if it satisfies the condition $f(\gamma(z))=f(z)j_\gamma(z)$ for all $\gamma\in\Gamma$, $z\in X$. When all $j_\gamma(z)$ are identically equal to 1, i.e., when $f(z)$ is Γ-invariant, f is simply called an **automorphic function with respect to** Γ. When all j_γ are constant (and hence $\gamma\rightarrow j_\gamma$ is a "quasicharacter" of Γ), f is called a **multiplicative function**. If we denote by $J_\gamma(z)$ the functional determinant of the transformation γ, then for an integer m, an automorphic function f with respect to a factor of automorphy of the form $j_\gamma(z)=J_\gamma(z)^{-m}$ is called an **automorphic form of weight** m. (In this case, it is customary to assume that f is holomorphic. Also, in all cases, we usually add a suitable condition on the behavior of f at the "points at infinity," when $\Gamma\backslash X$ is not compact.)

B. The Case of One Variable [2,7,11,12,17]

Except for the cases where Γ is a finite group or its extension by a free Abelian group of rank $\leqslant 2$ (→ 144 Elliptic Functions), it is essentially enough to consider the case where X is the upper half-plane $\mathfrak{H}=\{z\in\mathbf{C}\,|\,\mathrm{Im}\,z>0\}$. In this case, Γ may be obtained as a †discrete subgroup of the †special linear group $G=SL(2,\mathbf{R})$. In the following, we restrict ourselves to the case where Γ is a †Fuchsian group of the first kind (i.e., the case where the †Haar measure $\mu(\Gamma\backslash\mathfrak{H})$ of the quotient space $\Gamma\backslash\mathfrak{H}$ is finite; → 126 Discontinuous Groups). An **automorphic function** (or **Fuchsian function**) f with respect to Γ is, by definition, a meromorphic function f on $\mathfrak{H}$ satisfying the following conditions: (i) f is Γ-invariant, i.e., $f(\gamma z)=f(z)$ for all $\gamma\in\Gamma$; (ii) f is also meromorphic around any †cusp x_0 of Γ, i.e., if φ_0 is a real linear fractional transformation mapping x_0 to ∞ (e.g., $\varphi_0(z)=-1/(z-x_0)$) and if the transformation $z\rightarrow z+h$ $(h>0)$ is a generator of $\varphi_0\Gamma_{x_0}\varphi_0^{-1}$ (where Γ_{x_0} is the †stabilizer of x_0), then $f(\varphi_0^{-1}(z))$ can be expanded into a †Laurent series of $q_h=\exp((2\pi i/h)z)$ (which has only finitely many terms with negative exponent) in a neighborhood $\mathrm{Im}\,z>y_0$ of ∞ (q_h is called the **local parameter** around the cusp x_0). If we denote by $\mathfrak{R}_\Gamma$ the compact †Riemann surface obtained from the quotient space $\Gamma\backslash\mathfrak{H}$ by adjoining a certain (finite) number of "points at infinity" corresponding to the equivalence classes of cusps of Γ, then conditions (i) and (ii) amount to saying that f gives rise in a natural manner to a meromorphic function on $\mathfrak{R}_\Gamma$. Thus the field $\mathfrak{K}_\Gamma$ of all automorphic functions with respect to Γ can be identified with the †algebraic function field belonging to the Riemann surface $\mathfrak{R}_\Gamma$ (→ 13 Algebraic Functions). (While any (nonconstant) automorphic function with respect to a Fuchsian group of the first kind Γ has the real axis as its †natural boundary, an automorphic function with respect to a Fuchsian group of the second kind can always be analytically extended to the lower half-plane through a neighborhood of any "ordinary point" on the real axis.)

Let k be an even integer. A (holomorphic) **automorphic form** (or **Fuchsian form**) f **of weight** $k/2$ or **of dimension** $-k$ with respect to Γ is, by definition, a holomorphic function f on $\mathfrak{H}$ satisfying the following conditions: (i) For every $\sigma=\begin{pmatrix} a & b \\ c & d \end{pmatrix}\in\Gamma, f(\sigma z)=f(z)\cdot(cz+d)^k$. (In other words, $J_\sigma(z)^{-k/2}=(d\sigma(z)/dz)^{-k/2}=(cz+d)^k$ is the factor of automorphy.) (ii) f is also holomorphic around any cusp x_0, i.e., in the above notation, we have an integral "Fourier expansion" $f(\varphi_0^{-1}(z))(d\varphi_0^{-1}(z)/dz)^{k/2}=\sum_{\nu=0}^{\infty}a_\nu q_h^\nu$. If, moreover, the constant term a_0 in this Fourier expansion vanishes at all cusps of Γ, f is called a **cusp form.** In particular, f is a cusp form of dimension -2 if and only if $f(z)dz$ gives rise in a natural manner to a †differential form of the first kind on the Riemann surface $\mathfrak{R}_\Gamma$. (By a slight modification of condition (ii), we can also define an automorphic form of odd dimension $-k$. In case $\begin{pmatrix} -1 & 0 \\ 0 & -1 \end{pmatrix}\in\Gamma$, the condition (i) (for k odd) implies that f is identically equal to zero. So assume that $\begin{pmatrix} -1 & 0 \\ 0 & -1 \end{pmatrix}\not\in\Gamma$. Then cusps x_0 are classified into two categories, according as $\begin{pmatrix} -1 & h \\ 0 & -1 \end{pmatrix}\in\varphi_0\Gamma_{x_0}\varphi_0^{-1}$ or not, and in the first case we should replace the power series in q_h in condition (ii) by $q_h^{1/2}\times$(power series in q_h).)

We denote by $\mathfrak{M}_k(\Gamma)$ (resp. $\mathfrak{S}_k(\Gamma)$) the linear space of all automorphic forms (cusp forms) of dimension $-k$ with respect to Γ. Since clearly the relations $\mathfrak{M}_k\mathfrak{M}_{k'}\subset\mathfrak{M}_{k+k'}$, $\mathfrak{M}_k\mathfrak{S}_{k'}\subset\mathfrak{S}_{k+k'}$ hold, the direct sum $\mathfrak{M}(\Gamma)=\sum_k\mathfrak{M}_k(\Gamma)$ is a (commutative) †graded algebra, of which $\mathfrak{S}(\Gamma)=\sum_k\mathfrak{S}_k(\Gamma)$ is an ideal. It is known that $\mathfrak{M}_k(\Gamma)$ is of finite dimension, and $d_k=\dim\mathfrak{M}_k(\Gamma), d_k^0=\dim\mathfrak{S}_k(\Gamma)$ can be determined (from the †Riemann-Roch theorem) as follows:

$$d_k=0 \quad \text{for} \quad k<0;$$

$$d_0=1,\quad d_0^0=\begin{cases} 1 & \text{for} \quad t=0, \\ 0 & \text{for} \quad t>0; \end{cases}$$

$$d_2=\begin{cases} g & \text{for} \quad t=0, \\ g+t-1 & \text{for} \quad t>0, \end{cases}\qquad d_2^0=g;$$

$$d_k=(k-1)(g-1)+\sum_{i=1}^{s}\left(\frac{k}{2}\left(1-\frac{1}{e_i}\right)\right)+\frac{k}{2}t,$$

$$d_k^0=d_k-t \qquad \text{for} \quad k \text{ even}, \geqslant 4,$$

where s is the number of the equivalence classes of †elliptic fixed points of Γ, e_i ($1\leqslant i\leqslant s$) is the order of the stabilizers of these elliptic points, t is the number of the equivalence classes of cusps of Γ, and g is the †genus of the Riemann surface $\mathfrak{R}_\Gamma$. (When $\begin{pmatrix} -1 & 0 \\ 0 & -1 \end{pmatrix}\not\in\Gamma$, we can obtain a similar formula for k odd except for d_1, d_1^0, which, in general, cannot be determined from the Riemann-Roch theorem.)

One method of constructing automorphic forms is provided by Poincaré series as follows: Let σ_0 be a †linear fractional transformation that maps the unit disk to the upper half-plane (e.g., the †Cayley transformation), and put $\Gamma'=\sigma_0^{-1}\Gamma\sigma_0$; let φ be a function holomorphic on the unit disk including its boundary (e.g., a polynomial). Then for $k\geqslant 4$, the series

$$P_\varphi(z)=\sum_{\sigma'\in\Gamma'}\varphi(\sigma'(z))(d\sigma'(z)/dz)^{k/2}$$

is †uniformly convergent in the wide sense (i.e., uniformly convergent on every compact set) in the unit disk and expresses an automorphic cusp form of dimension $-k$ with respect to Γ', i.e., we have $P_\varphi(\sigma_0^{-1}(z))\cdot(d\sigma_0^{-1}(z)/dz)^{k/2}\in\mathfrak{S}_k(\Gamma)$. Conversely, all elements in $\mathfrak{S}_k(\Gamma)$ ($k\geqslant 4$) can be expressed in this form. A series of this type is called a **theta-Fuchsian series of Poincaré** (or simply **Poincaré series**).

For $k\geqslant 3$, we can define a (positive definite Hermitian) inner product on $\mathfrak{S}_k(\Gamma)$ as follows:

$$(f,g)=\int_F f(z)\,\overline{g(z)}\,y^{k-2}dx\,dy,$$

where $z=x+iy$ and F is a †fundamental region of Γ in $\mathfrak{H}$. This is called the **Petersson metric**. Since this integral converges also for $f,g\in\mathfrak{M}_k(\Gamma)$ if one of them belongs to $\mathfrak{S}_k(\Gamma)$, we can define the orthogonal complement $\mathfrak{E}_k(\Gamma)$ of $\mathfrak{S}_k(\Gamma)$ in $\mathfrak{M}_k(\Gamma)$. As we shall see, $\mathfrak{E}_k(\Gamma)$ is generated by an †Eisenstein series.

Any automorphic function f with respect to Γ can be expressed as a quotient of two automorphic forms $f_1, f_2\in\mathfrak{M}_k(\Gamma)$ for a sufficiently large k. If we put $w=f(z)$, the inverse function $z=f^{-1}(w)$ can be expressed as the quotient of two linearly independent solutions of a †linear differential equation of the form $d^2z/dw^2=\varphi(w)z$, where $\varphi(w)$ is an †algebraic function belonging to the Riemann surface $\mathfrak{R}_\Gamma$.

C. Modular Functions and Modular Forms

[8, 12, 13]

The (elliptic) †modular group

$$\begin{aligned}\Gamma&=\Gamma(1)\\ &=SL(2,\mathbf{Z})\\ &=\left\{\begin{pmatrix} a & b \\ c & d \end{pmatrix}\,\middle|\, a,b,c,d\in\mathbf{Z}, ad-bc=1\right\}\end{aligned}$$

is a Fuchsian group of the first kind acting on the upper half-plane $\mathfrak{H}$. The quotient space $\Gamma\backslash\mathfrak{H}$ can be compactified to a compact Riemann surface $\mathfrak{R}_\Gamma=\Gamma\backslash\mathfrak{H}\cup\{\infty\}$ of genus one by joining one point at infinity corresponding to the cusp ∞, around which we take $q=e^{2\pi iz}$ as a local parameter. The dimension $d_k=$

$\dim \mathfrak{M}_k(\Gamma)$ (k even, $\geqslant 2$) is given as follows:

$$d_k = \begin{cases} [k/12] & \text{for} \quad k \equiv 2 \pmod{12}, \\ [k/12]+1 & \text{for} \quad k \not\equiv 2 \pmod{12}, \end{cases}$$

where [ ] is the †Gauss symbol.

More generally, an automorphic function (automorphic form) with respect to the principal congruence subgroup

$$\Gamma(N) = \left\{ \begin{pmatrix} a & b \\ c & d \end{pmatrix} \equiv \begin{pmatrix} 1 & 0 \\ 0 & 1 \end{pmatrix} (\operatorname{mod} N) \right\}$$

is called a **modular function** (**modular form**) **of level** N. (For the number of cusps of $\Gamma(N)$ and the genus of $\mathfrak{R}_{\Gamma(N)}$ → 126 Discontinuous Groups.) At any cusp of $\Gamma(N)$, the local parameter q_h in (ii) is given by $q_N = \exp(2\pi i z/N)$. For $N, k \geqslant 3$, the dimensions d_k, d_k^0 are given by the general formula of the preceding paragraph (including odd k). For $k \geqslant 3$, we define the (extended) **Eisenstein series** by

$$G_k(z; c_1, c_2, N) = \sum_{m_i \equiv c_i (\operatorname{mod} N)}{}' \frac{1}{(m_1 + m_2 z)^k},$$

where c_1, c_2 are integers such that $(c_1, c_2, N) = 1$ and the symbol Σ' denotes the summation excepting the pair $(m_1, m_2) = (0,0)$. Then $G_k(z; c_1, c_2, N)$ depends only on c_1, c_2 (mod N), and if we take a set $\{(c_1, c_2)\}$ such that $\{(c_1, c_2), (-c_1, -c_2)\}$ forms a complete set of representatives of the (primitive) residue classes (mod N), then the corresponding set of Eisenstein series forms a basis of the space $\mathfrak{E}_k(\Gamma(N))$ $(k \geqslant 3)$, whence we obtain $\dim \mathfrak{E}_k = d_k - d_k^0 = t(N)$ $(k \geqslant 3)$ (for details, including the case $k = 1, 2$, see E. Hecke, *Abh. Math. Sem. Univ. Hamburg*, 5 (1927)).

The Fourier coefficient a_n of the Eisenstein series can be calculated easily, and we get an estimation $a_n = O(n^{k-1+\varepsilon})$ for $k \geqslant 2$, where ε is an arbitrary positive number. On the other hand, for cusp forms, Hecke (*Abh. Math. Sem. Univ. Hamburg*, 5 (1927)) gave $a_n = O(n^{k/2})$. In an attempt to improve this estimate, H. D. Kloosterman was led to consider the sum

$$K(u, v, q) = \sum_{\substack{x (\operatorname{mod} q) \\ (x, q) = 1}} \exp\left(\frac{2\pi i}{q} \left(ux + \frac{v}{x} \right) \right) \qquad (u, v, q \in \mathbf{Z}),$$

called the **Kloosterman sum**, which is also related to the arithmetic of quadratic forms. Using A. Weil's estimate $|K(u, v, p)| \leqslant 2\sqrt{p}$ (where p is an odd prime, $(u,p) = (v,p) = 1$), based on the analog of the Riemann hypothesis, we obtain $a_n = O(n^{k/2 - 1/4 + \varepsilon})$. (If the Ramanujan-Petersson conjecture (→ Section D) is true, this would further be improved to $a_n = O(n^{(k-1)/2 + \varepsilon})$.)

In the case $N = 1$, we obtain the classical Eisenstein series $G_k(z) = G_k(z; 0, 0, 1)$ (k even, $\geqslant 4$), from which we define the modular forms $g_2(z) = 60 G_4(z)$, $g_3(z) = 140\, G_6(z)$, and $\Delta(z) = g_2^3 - 27 g_3^2$ of dimension -4, -6, and -12, respectively. If we denote by $\wp = \wp(u; 1, z)$ the Weierstrass †$\wp$-function with the fundamental period $(1, z)$ we have the relation $\wp'^2 = 4\wp^3 - g_2 \wp - g_3$, and $\Delta(z)$ is the †discriminant of the cubic polynomial appearing in this relation. It is known that every modular form can be expressed uniquely as a polynomial in g_2 and g_3, or in other words, we have $\mathfrak{M}(\Gamma(1)) \cong \mathbf{C}[g_2, g_3]$. The polynomial in g_2, g_3 expressing a modular form is, moreover, **isobaric**, i.e., consists of terms of the form $c g_2{}^{\mu} g_3{}^{\nu}$, $c \in \mathbf{C}$, where $2\mu + 3\nu$ is a constant called the **weight** of the isobaric polynomial. Also, $\Delta(z)$ is a cusp form of the smallest weight, and the ideal of all cusp forms, $\mathfrak{S}(\Gamma(1))$, is a principal ideal in $\mathfrak{M}(\Gamma(1))$ generated by $\Delta(z)$. The Fourier expansion of the Eisenstein series $E_k(z)$ is given as follows:

$$G_k(z) = (2\pi)^k \frac{1}{k!} \left(B_k + (-1)^{k/2} 2k \sum_{\nu=1}^{\infty} \nu^{k-1} \frac{q^\nu}{1 - q^\nu} \right),$$

where $q = e^{2\pi i z}$ and B_k is the †Bernoulli number. The discriminant $\Delta(z)$ is expressed as an infinite product as follows:

$$\Delta(z) = (2\pi)^{12} q \prod_{\nu=1}^{\infty} (1 - q^\nu)^{24}.$$

If we put $J(z) = g_2^3 / \Delta$, the function $J(z)$ is a modular function and gives an †analytic isomorphism of the Riemann surface $\mathfrak{R}_{\Gamma(1)} = \Gamma(1) \backslash \mathfrak{H} \cup \{\infty\}$ onto the Riemann sphere $\mathbf{C} \cup \{\infty\}$ (which maps ζ_3, ζ_4, ∞ to 0, 1, ∞, respectively). Hence the field of modular functions $\mathfrak{K}_{\Gamma(1)}$ is a rational function field $\mathbf{C}(J)$ generated by the function J. The analytic isomorphism class of †complex tori of dimension 1 (= †elliptic curves) $E_{(\omega_1, \omega_2)} = \mathbf{C}/(\mathbf{Z}\omega_1 + \mathbf{Z}\omega_2)$ with the fundamental period (ω_1, ω_2) is uniquely determined by the $\Gamma(1)$-equivalence class of the **modulus** $\tau = \omega_2 / \omega_1$ $(\in \mathfrak{H})$, and hence by the value $J(\tau)$ of the function J. This is the historical origin of the name of †modular function [11, 13].

As an example of a modular function of level 2, we have the **λ-function**:

$$\lambda(z) = \frac{\wp((1+z)/2) - \wp(z/2)}{\wp(1/2) - \wp(z/2)},$$

which gives an analytic isomorphism of the Riemann surface $\mathfrak{R}_{\Gamma(2)} = \Gamma(2) \backslash \mathfrak{H} \cup$ (three points) to the Riemann sphere $\mathbf{C} \cup \{\infty\}$ (mapping 0, 1, ∞ to 1, ∞, 0, respectively; this property is used in a proof of the †Picard theorem). Hence we again have $\mathfrak{K}_{\Gamma(2)} = \mathbf{C}(\lambda)$.

The following relation holds between J and λ:

$$J(z)=\frac{4}{27}\frac{(1-\lambda+\lambda^2)^3}{\lambda^2(1-\lambda)^2}.$$

In general, the field $K_{\Gamma(N)}$ of modular functions of level N is generated by $(g_2/g_3)\cdot \wp((a_1+a_2z)/N;\ 1,z)$ $(a_1, a_2\in\mathbf{Z})$.

D. The Hecke Ring and Its Representation

[5, 9, 10, 14]

In general, let $\tilde{\Gamma}$ be a group, Γ a subgroup of $\tilde{\Gamma}$, and suppose that, for every $\sigma\in\tilde{\Gamma}$, $\sigma\Gamma\sigma^{-1}$ and Γ are †commensurable, which amounts to saying that every double coset $\Gamma\sigma\Gamma$ is a union of a finite number of left or right cosets of Γ. Then in the free module generated by all double cosets $\Gamma\sigma\Gamma$ $(\sigma\in\tilde{\Gamma})$, we can define a bilinear associative product as follows:

$$\Gamma\sigma\Gamma\cdot\Gamma\tau\Gamma=\sum_{\Gamma\rho\Gamma} m(\sigma,\tau;\rho)\Gamma\rho\Gamma,$$

where, writing $\Gamma\sigma\Gamma=\bigcup_i\Gamma\sigma_i$, $\Gamma\tau\Gamma=\bigcup_j\Gamma\tau_j$, we denote by $m(\sigma,\tau;\rho)$ the number of pairs (i,j) such that $\Gamma\sigma_i\tau_j=\Gamma\rho$. The associative ring thus obtained is called the **Hecke ring** (or **Hecke algebra**) and is denoted by $\mathcal{H}(\tilde{\Gamma},\Gamma)$. (If $\tilde{\Gamma}$ is a †topological group and Γ is an open compact subgroup, the Hecke ring $\mathcal{H}(\tilde{\Gamma},\Gamma)$ can also be interpreted as the ring of all $\mathbf{Z}$-valued continuous functions on $\tilde{\Gamma}$ with compact support provided with a †convolution product; → 31 Associative Algebras C.)

As an example, setting $\tilde{\Gamma}=GL^+(2,\mathbf{Q})$, $\Gamma=SL(2,\mathbf{Z})$, we obtain a Hecke ring $\mathcal{H}(\tilde{\Gamma},\Gamma)$. In virtue of the theory of †elementary divisors, a complete set of representatives of $\Gamma\backslash\tilde{\Gamma}/\Gamma$ is given by

$$\left\{\begin{pmatrix}a_1 & 0\\ 0 & a_2\end{pmatrix}\,\middle|\, a_1,a_2\in\mathbf{Q},\ a_1,a_2>0,\ a_1|a_2\right\}.$$

We write $T(a_1,a_2)=\Gamma\begin{pmatrix}a_1 & 0\\ 0 & a_2\end{pmatrix}\Gamma$. From the relation $T(a_1a_2,a_1a_2)T(a_1^{-1},a_2^{-1})=T(a_1,a_2)$ follows the commutativity of $\mathcal{H}(\tilde{\Gamma},\Gamma)$. Furthermore, if we put, for a positive integer n,

$$T_n=T(n)=\sum_{\substack{a_1,a_2\in\mathbf{Z}\\ a_1a_2=n}} T(a_1,a_2),$$

then the following multiplication formula holds:

$$T(n)\cdot T(m)=\sum_{d|n,m} dT(d,d)\cdot T(nm/d^2).$$

We obtain a representation of $\mathcal{H}(\tilde{\Gamma},\Gamma)$ in $\mathfrak{M}_k(\Gamma)$ as follows. For $T=\Gamma\sigma\Gamma=\cup\Gamma\sigma_i\in\mathcal{H}(\tilde{\Gamma},\Gamma)$ and $f\in\mathfrak{M}_k(\Gamma)$ we put

$$(f|T)(z)=\sum_i f(\sigma_i(z))(c_iz+d_i)^{-k}\det(\sigma_i)^l,$$

where $\sigma_i=\begin{pmatrix}a_i & b_i\\ c_i & d_i\end{pmatrix}$ and l is a fixed integer. This representation leaves $\mathfrak{S}_k(\Gamma)$ and $\mathfrak{E}_k(\Gamma)$ invariant. In particular, the representation of T_n (with $l=k-1$), called the **Hecke operator** (Hecke, *Math. Ann.*, 114 (1937)), is †Hermitian with respect to the Petersson metric.

Following Hecke, we associate with every modular form $f(z)=\sum_{n=0}^{\infty}a_nq^n$ $(q=e^{2\pi iz})$ a †Dirichlet series $\varphi(s)=\sum_{n=1}^{\infty}a_nn^{-s}$. Since $a_n=O(n^{k-1+\varepsilon})$ for $f\in\mathfrak{M}_k$ $(k\geqslant 2)$, $\varphi(s)$ is absolutely convergent in the half-plane $\mathrm{Re}\,s>k$. The conditions for f to be a modular form of dimension $-k$ are equivalent to the following conditions for φ: $(s-k)\varphi(s)$ can be extended to an †entire function of finite †genus (actually, of genus 1) and, if we put $R(s)=(2\pi)^{-s}\Gamma(s)\varphi(s)$, $R(s)$ satisfies a functional equation of the form $R(k-s)=(-1)^{k/2}R(s)$. A correspondence between f and φ that satisfies these conditions is one-to-one; in fact we have $a_0=(-1)^{k/2}\mathrm{Res}_{s=k}R(s)$, and the function $g(x)=f(ix)-a_0$ $(x>0)$ and $R(s)$ are related by the †Mellin transforms as follows:

$$R(s)=\int_0^{\infty}g(x)x^{s-1}dx,$$

$$g(x)=\frac{1}{2\pi i}\int_{\mathrm{Re}\,s=\sigma_0}R(s)x^{-s}ds.$$

This correspondence between Dirichlet series and automorphic forms can be generalized further (Hecke, *Math. Ann.*, 112 (1936)).

Now, suppose that a linear subspace $\mathfrak{M}$ of $\mathfrak{M}_k(\Gamma)$ is invariant under all T_n $(n=1,2,\ldots)$, and let $(f_1,\ldots,f_\kappa)$ $(\kappa=\dim\mathfrak{M})$ be a basis of $\mathfrak{M}$. If we denote by $\lambda(n)$ the $\kappa\times\kappa$ matrix representing T_n in this basis, it can be shown that there exist $\kappa\times\kappa$ (complex) matrices $B^{(i)}$ $(1\leqslant i\leqslant\kappa)$ and $\lambda(0)$ such that we have

$$F(z)=\sum_{n=0}^{\infty}q^n\lambda(n)=\sum_{i=1}^{\kappa}f_i(z)B^{(i)}.$$

Similarly we associate with $F(z)$ a matrix-valued Dirichlet series

$$\Phi(s)=\sum_{n=1}^{\infty}n^{-s}\lambda(n).$$

Then from the multiplicative property of the $\lambda(n)$, we obtain the following †Euler product expression of $\Phi(s)$:

$$\Phi(s)=\prod_p\left(I_\kappa-\lambda(p)p^{-s}+p^{\kappa-1-2s}I_\kappa\right)^{-1},$$

where I_κ denotes the unit matrix of degree κ. In particular, when $\dim\mathfrak{M}=1$, i.e., when $\mathfrak{M}=\mathbf{C}f$, where f is an †eigenfunction of all Hecke operators T_n, the corresponding Dirichlet series $\varphi(s)$ has an Euler product of the following form:

$$\varphi(s)=\prod_p\left(1-\lambda_pp^{-s}+p^{k-1-2s}\right)^{-1}.$$

For instance, the Dirichlet series corresponding to the Eisenstein series G_k is $(2(2\pi i)^k(k-1)!)\,\zeta(s)\zeta(s-k+1)$ (where ζ is the †Riemann zeta function), which has a well-known Euler product. Since $\mathfrak{S}_{12}=\mathbf{C}\Delta$, the Dirichlet series corresponding to $\Delta(z)$ also has an Euler product. (This is part of the **Ramanujan conjecture**. Ramanujan also conjectured that the quadratic polynomial $1-\lambda_p X+p^{11}X^2$ appearing in this Euler product has imaginary roots, i.e., $|\lambda_p|<2p^{11/2}$.) Generalizing the Ramanujan conjecture, Petersson conjectured that all eigenvalues of the Hecke operator T_p in $\mathfrak{S}_k$ have absolute value $\leqslant 2p^{(k-1)/2}$ (**Ramanujan-Petersson conjecture**). M. Eichler and G. Shimura (1958) proved this conjecture for $k=2$ (and for almost all p) by reducing it to Weil's result (analog of the Riemann hypothesis) on the †Hasse zeta function of the field of modular functions $\mathfrak{K}_{\Gamma(n)}$. (For a more recent development, see M. Kuga and Shimura, *Ann. of Math.*, 82 (1965).) (→ 436 Zeta Functions)

E. The Case of Many Variables [2,3,21]

Let X be a bounded domain in $\mathbf{C}^N$ and Γ a †discontinuous group of analytic automorphisms of X. The Poincaré series of weight $m\geqslant 2$, defined similarly to the case of the unit disk, converges normally in X and expresses an automorphic form of weight m. If $\Gamma\backslash X$ is compact, let $(f_1,\ldots,f_\kappa)$ be a basis of the space of Poincaré series of weight m. Then for a sufficiently large m (which is a multiple of the order of Γ_x for all $x\in X$), the map $X\ni z\to(f_1(z),\ldots,f_\kappa(z))$ defines in a natural manner a projective embedding of the quotient space $\Gamma\backslash X$ into $P^{\kappa-1}(\mathbf{C})$, which actually gives an analytic isomorphism of $\Gamma\backslash X$ onto a †normal projective algebraic variety. It follows that the field of automorphic functions with respect to Γ is an †algebraic function field of dimension N, and that every automorphic function can be written as the quotient of two Poincaré series of the same weight. In the case where $\Gamma\backslash X$ is not compact, we first discuss some examples.

F. Siegel Modular Functions [3,15,20]

The **Siegel upper half-space $\mathfrak{H}_n$ of degree** n (or **Siegel space of degree** n) is, by definition, the space of all $n\times n$ complex symmetric matrices $Z=X+iY$ with the imaginary part $Y>0$ (positive definite). ($\mathfrak{H}_n$ is analytically equivalent to a †symmetric bounded domain.) The group of all (complex) analytic automorphisms of $\mathfrak{H}_n$ is given by the real (projective) †symplectic group $Sp(n,\mathbf{R})/\{\pm I_{2n}\}$, which acts transitively on $\mathfrak{H}_n$ by $Z\to\sigma(Z)=(AZ+B)/(CZ+D)$ for $\sigma=\begin{pmatrix}A&B\\C&D\end{pmatrix}\in Sp(n,\mathbf{R})$. The subgroup $\Gamma_n=Sp(n,\mathbf{Z})$ consisting of integral matrices, or the corresponding group of linear fractional transformations, is called the **Siegel modular group of degree** n. Γ_n is a †discontinuous group of the first kind acting on $\mathfrak{H}_n$.

A **Siegel modular form** f **of dimension** $-k$ is, by definition, a holomorphic function on $\mathfrak{H}_n$ satisfying the following conditions: (i) for $\sigma=\begin{pmatrix}A&B\\C&D\end{pmatrix}\in\Gamma_n$, $f(\sigma(Z))=f(Z)\det(CZ+D)^k$; (ii) f has an integral Fourier expansion of the form $F(Z)=\sum_{T\geqslant 0}a_T e^{2\pi i\,\mathrm{tr}(TZ)}$, where T runs over all half-integral †positive semidefinite symmetric matrices of degree n. For $n\geqslant 2$, condition (ii) is superfluous (M. Koecher, *Math. Z.*, 59 (1954)). We denote by $\mathfrak{M}_k^{(n)}=\mathfrak{M}_k(\Gamma_n)$ the space of all Siegel modular forms of degree n and dimension $-k$. When we write $\mathfrak{H}_n\ni Z=\begin{pmatrix}Z_1&\mathfrak{z}\\\mathfrak{z}&z\end{pmatrix}$ with $Z_1\in\mathfrak{H}_{n-1}$, $\mathfrak{z}\in\mathbf{C}^{n-1}$, $z\in\mathfrak{H}_1$, and

$$\mathfrak{M}_k^{(n)}\ni f(Z)=\sum_{n=0}^{\infty}a_n(Z_1,\mathfrak{z})e^{2\pi inz},$$

it turns out that $a_0(Z_1,\mathfrak{z})$ depends only on Z_1 and, writing it as $f_1(Z_1)$, we have $f_1\in\mathfrak{M}_k^{(n-1)}$. The mapping $\Phi:f\to f_1$ thus defined is a linear mapping from $\mathfrak{M}_k^{(n)}$ into $\mathfrak{M}_k^{(n-1)}$, which is surjective if k is even and $>2n$ (H. Maass, *Math. Ann.*, 123 (1951)). We denote the kernel of Φ by $\mathfrak{S}_k^{(n)}$ and call $f\in\mathfrak{S}_k^{(n)}$ a **cusp form** (viewing $\mathfrak{H}_{n-1}$ as a cusp of $\mathfrak{H}_n$). For $f\in\mathfrak{M}_k^{(n)}$, the following three conditions are equivalent: (a) $f\in\mathfrak{S}_k^{(n)}$; (b) $f(Z)=\sum_{T>0}a_T e^{2\pi i\,\mathrm{tr}(TZ)}$; (c) $|f(Z)\det(Y)^{k/2}|$ is bounded. We have $\mathfrak{M}_k^{(n)}=\{0\}$ if $k<0$ or if both k and n are odd, and $\mathfrak{M}_0^{(n)}=\mathbf{C}$. In general, $\dim\mathfrak{M}_k^{(n)}$ is finite and $=O(k^{n(n+1)/2})$ $(k\to\infty)$. In particular, for $n=2$ the structure of the graded algebra $\mathfrak{M}(\Gamma_2)=\sum_{k=0}^{\infty}\mathfrak{M}_k^{(2)}$ is determined explicitly. (The even part of this algebra is isomorphic to the polynomial ring of four variables.) Also, $\dim\mathfrak{M}_k^{(2)}$ and the singularities of the quotient variety $\Gamma_2\backslash\mathfrak{H}_2$ are known explicitly (J. I. Igusa, *Amer. J. Math.*, 84 (1962), 86 (1964); U. Christian).

As an example of Siegel modular forms, we have the **Eisenstein-Poincaré series**, defined as follows:

$$E_{S,k}(Z)=\sum_{\sigma\in\Gamma_n(S)\backslash\Gamma_n}e^{2\pi i\,\mathrm{tr}(S\sigma(z))}\det(CZ+D)^{-k},$$

where S is an $n\times n$ rational symmetric matrix $\geqslant 0$, and we put

$$\Gamma_n(S)=\left\{\begin{pmatrix}U&T\,{}^tU^{-1}\\0&{}^tU^{-1}\end{pmatrix}\in\Gamma_n\;\middle|\;{}^tUSU=S,\;e^{2\pi i\,\mathrm{tr}(ST)}=\det(U)^k\right\}.$$

This series is convergent for $k > n + \operatorname{rank} S + 1$ (in particular, for $k > 2n$ if $S > 0$), and the totality of $E_{S,k}$ (those series with $S > 0$) spans $\mathfrak{M}_k^{(n)}$ $(\mathfrak{S}_k^{(n)})$ (Maass, *Math. Ann.*, 123 (1951)). †Theta series defined by integral quadratic forms are examples of Siegel modular forms (with a certain level) and are of great significance in the arithmetic of quadratic forms (C. L. Siegel, *Ann. of Math.*, 36 (1935)).

The quotient space $\Gamma_n\backslash\mathfrak{H}_n$ can be compactified as follows (I. Satake and W. Baily [3]): There exists a positive integer k_0 such that, for any multiple k of k_0, any basis of $\mathfrak{M}_k^{(n)}$ defines in a natural manner a one-to-one biholomorphic projective embedding of $\Gamma_n\backslash\mathfrak{H}_n$, of which the image is a Zariski open set of a normal projective algebraic variety. Since the structure of this projective variety is independent of k, we denote it simply by $\overline{\Gamma_n\backslash\mathfrak{H}_n}$. Then we have $\overline{\Gamma_n\backslash\mathfrak{H}_n} = \bigcup_{r=0}^{n}\Gamma_r\backslash\mathfrak{H}_r$. When $n \geqslant 2$, $\Gamma_n\backslash\mathfrak{H}_n$ is of codimension $\geqslant 2$ in $\overline{\Gamma_n\backslash\mathfrak{H}_n}$, so that (by virtue of †Hartogs's continuation theorem) the conditions at the points at infinity in the definitions of modular functions and forms become superfluous. In fact, for $n \geqslant 2$, if we define a **Siegel modular function of degree** n simply as a Γ_n-invariant meromorphic function on $\mathfrak{H}_n$, then it can automatically be extended to a meromorphic function on the compactification $\overline{\Gamma_n\backslash\mathfrak{H}_n}$, and hence be expressed as the quotient of two modular forms of the same dimension. It also follows that the field of all Siegel modular functions of degree n is an algebraic function field of dimension $n(n+1)/2$. (These results can also be obtained from the †pseudoconcavity of $\Gamma_n\backslash\mathfrak{H}_n$ without using compactification (A. Andreotti and H. Grauert).) For $n=2$, the field of Siegel modular functions is a rational function field.

G. Hilbert Modular Functions

From a suggestion of Hilbert, O. Blumenthal studied a generalization of modular functions of the following type: Let K be a totally real †algebraic number field of finite degree, and let $K^{(1)}, \dots, K^{(n)}$ be the conjugates of K, where $n = [K:\mathbf{Q}]$. ("Totally real" means that all the $K^{(i)}$ are real.) For $\alpha \in K$, we denote by $\alpha^{(i)} \in K^{(i)}$ the ith conjugate of α. Let $\mathfrak{O}$ be the †principal order of K, and consider the group

$$\Gamma_{\mathfrak{O}} = SL(2,\mathfrak{O})$$

$$= \left\{ \sigma = \begin{pmatrix} \alpha & \beta \\ \gamma & \delta \end{pmatrix} \middle| \alpha, \beta, \gamma, \delta \in \mathfrak{O}, \alpha\delta - \beta\gamma = 1 \right\}.$$

We define the action of $\Gamma_{\mathfrak{O}}$ on the n-fold product of the upper half-plane $\mathfrak{H}^n = \{z = (z_1, \dots, z_n) \mid z_i \in \mathfrak{H}\}$ by

$$\sigma(z) = \left(\frac{\alpha^{(1)}z_1 + \beta^{(1)}}{\gamma^{(1)}z_1 + \delta^{(1)}}, \dots, \frac{\alpha^{(n)}z_n + \beta^{(n)}}{\gamma^{(n)}z_n + \delta^{(n)}} \right)$$

for $\sigma = \begin{pmatrix} \alpha & \beta \\ \gamma & \delta \end{pmatrix}$; then $\Gamma_{\mathfrak{O}}$ becomes a discontinuous group of the first kind ($\Gamma_{\mathfrak{O}}$ can also be considered as an †irreducible discrete subgroup of $SL(2,\mathbf{R})^n$). The group $\Gamma_{\mathfrak{O}}$ is called the **Hilbert modular group** of K (in the strict sense). (The definition is sometimes modified by replacing the condition $\alpha\delta - \beta\gamma = 1$ by, say, $\alpha\delta - \beta\gamma =$ (totally positive †unit).) If the †class number of K is h, the quotient space $\Gamma_{\mathfrak{O}}\backslash\mathfrak{H}^n$ can be compactified by adjoining h points at infinity. Therefore, if $n \geqslant 2$, a **Hilbert modular function** can be defined as a meromorphic function on $\mathfrak{H}^n$ invariant under $\Gamma_{\mathfrak{O}}$; and similarly a **Hilbert modular form** f **of dimension** $-k$ as a holomorphic function f on $\mathfrak{H}^n$ such that $(f|\sigma)(z) = f(\sigma(z))\prod_{i=1}^{n}(\gamma^{(i)}z_i + \delta^{(i)})^{-k} = f(z)$ for all $\sigma \in \Gamma_{\mathfrak{O}}$. (In the latter case, f is holomorphic at all cusps, i.e., for every $\sigma \in SL(2,K)$, $f|\sigma$ has an integral Fourier expansion.) For Hilbert modular functions and forms quite similar results as in the case $n=1$ or the case of Siegel modular groups have been obtained (Kloosterman, Maass, K. B. Gundlach, H. Klingen).

H. Further Generalizations

As further generalizations of the notion of modular function, we have Hilbert-Siegel modular functions (I. I. Pjateckiĭ-Šapiro, Baily, Christian [4], where we can find some 150 references), Hermitian modular functions (H. Braun, Klingen), etc. For the most general case, i.e., for an arithmetically defined discontinuous group acting on a †symmetric bounded domain, a unified theory of automorphic functions has been established by recent works of Pjateckiĭ-Šapiro [18, 23], and Baily and A. Borel [1, 24].

On the other hand, exactly as in the classical theory, where the elliptic modular function gave an invariant of 1-dimensional complex tori, generalized modular functions can be viewed as giving an analytic invariant of a certain family of (polarized) †Abelian varieties. From this point of view, a deep number-theoretic (and algebrogeometric) study of automorphic functions (initiated by Hecke and Eichler) has been substantially carried forward by the work of Shimura (see the series of his papers in *Ann. of Math.* starting from vol. 70 (1959); see also [5, 14, 24, 27]).

From the analytic point of view, the theory of automorphic functions is closely connected with the unitary representation of G in the space $L_2(\Gamma\backslash G)$ (or its adelic analogue) [25, 26]. In this respect, the †trace formula of A. Selberg [19], generalizing the †Poisson summation formula, is of fundamental importance; and actually it can be used effectively for calculations of the dimension of the space of automorphic forms and of the trace of Hecke operators (R. P. Langlands; also [24]). When $X = G/K$ is a symmetric domain, we can define, for any representation ρ of K, a (matrix-valued) canonical automorphy factor, by which we define (vector-valued) automorphic forms with respect to a discrete subgroup Γ of G, and under a further condition (say, Γ free, $\Gamma\backslash X$ compact, and the †highest weight of ρ sufficiently large) we obtain a formula for the dimension of the space of such automorphic forms in terms of the †arithmetic genus of $\Gamma\backslash X$ and certain numbers related to the "dual" $X_u = G_u/K$ and the representation ρ [3, 16, 22].

References

[1] W. L. Baily and A. Borel, Compactification of arithmetic quotients of bounded symmetric domains, Ann. of Math., (2) 84 (1966), 442–528.
[2] Séminaire H. Cartan 6, Théorie des fonctions de plusieurs variables, Fonctions automorphes et espaces analytiques, Ecole Norm. Sup., 1953–1954.
[3] Séminaire H. Cartan 10, Fonctions automorphes, Ecole Norm. Sup., 1957–1958.
[4] U. Christian, Zur Theorie der Hilbert-Siegelschen Modulfunktionen, Math. Ann., 152 (1963), 275–341.
[5] M. Eichler, Quadratische Formen und Modulfunktionen, Acta Arith., 4 (1958), 217–239.
[6] M. Eichler, Einführung in die Theorie der algebraischen Zahlen und Funktionen, Birkhäuser, 1963; English translation, Introduction to the theory of algebraic numbers and functions, Academic Press, 1966.
[7] R. Fricke and F. Klein, Vorlesungen über die Theorie der automorphen Funktionen, Teubner, I, 1897; II, 1901; second edition, 1926.
[8] R. C. Gunning, Lectures on modular forms, Ann. Math. Studies, Princeton Univ. Press, 1962.
[9] E. Hecke, Mathematische Werke, Vandenhoeck & Ruprecht, 1959.
[10] E. Hecke, Dirichlet series, modular forms and quadratic forms, Lecture note, Institute for Advanced Study, Princeton, 1938.
[11] K. Iwasawa, Daisû kansûron (Japanese; Theory of algebraic functions), Iwanami, 1952.
[12] Y. Kawada, Theory of automorphic functions in one variable I, II (Japanese; mimeographed notes), seminar notes series, Univ. of Tokyo, 1963–1964.
[13] F. Klein and R. Fricke, Vorlesungen über die Theorie der elliptischen Modulfunktionen, Teubner, I, 1890; II, 1892.
[14] M. Kuga, Fiber varieties over a symmetric space whose fibers are Abelian varieties, Lecture note, Univ. of Chicago, 1963–1964.
[15] H. Maass, Lectures on Siegel's modular functions, Lecture note, Tata Inst., 1954–1955.
[16] Y. Matsushima and S. Murakami, On vector bundle valued harmonic forms and automorphic forms on symmetric Riemannian manifolds, Ann. of Math., (2) 78 (1963), 365–416.
[17] H. Poincaré, Mémoire sur les fonctions fuchsiennes, Acta Math., 1 (1882) 193–294, (Oeuvres, Gauthier-Villars, 1916, vol. 2, p. 167–257.)
[18] И. И. Пятецкий-Шапиро (I. I. Pjateckiĭ-Šapiro), Геометрия классических областей и теория автоморфных функций, физматгиз, 1961.
[19] A. Selberg, Harmonic analysis and discontinuous groups in weakly symmetric Riemannian spaces with applications to Dirichlet series, J. Indian Math. Soc., 20 (1956), 47–87.
[20] C. L. Siegel, Einführung in die Theorie der Modulfunktionen n-ten Grades, Math. Ann., 116 (1939), 617–657. (Gesammelte Abhandlungen, Springer, 1966, vol. 2, p. 97–137.)
[21] C. L. Siegel, Analytic functions of several complex variables, Lecture note, Institute for Advanced Study, Princeton, 1950.
[22] Y. Matsushima and S. Murakami, On certain cohomology groups attached to Hermitian symmetric spaces, Osaka J. Math., 2 (1965), 1–35.
[23] И. И. Пятецкий-Шапиро (I. I. Pjateckiĭ-Šapiro), Арифметические группы в комплексных областях Uspehi Mat. Nauk, 19 (1964), 93–121.
[24] Algebraic groups and discontinuous subgroups, Amer. Math. Soc. Proc. Symp. in Pure Math., IX (1966).
[25] И. Н. Гельфанд, Н. И. Граев, И. И. Пятецкий-Шапиро (I. M. Gel'fand, M. I. Graev, and I. I. Pjateckiĭ-Šapiro), Обобщенные функции, 6, теория Представлений и автоморфные функции, Наука, 1966; English translation, Generalized functions, 6, Representation theory and automorphic functions, Saunders, 1969.
[26] Harish-Chandra, Automorphic forms on semisimple Lie groups, Lecture notes in

math. 62, Springer, 1968.
[27] G. Shimura, Automorphic functions and number theory, Lecture notes in math. 54, Springer, 1968.
[28] G. Shimura, Introduction to the arithmetic theory of automorphic functions, Publ. Math. Soc. Japan, Iwanami Shoten and Princeton, 1971.
[29] H. Jacquet and R. P. Langlands, Automorphic forms on $GL(2)$, Lecture notes in math. 114, Springer, 1970.
[30] J. Lehner, Discontinuous groups and automorphic functions, Amer. Math. Soc. Math. Surveys, 1964.

35 (I.8) Axiomatic Set Theory

A. General Remarks

Axiomatic set theory pursues the goal of reestablishing the essentials of G. Cantor's rather intuitive †set theory by axiomatic constructions consistent with modern theories of the foundations of mathematics.

A system of axioms for set theory was first given by E. Zermelo [16], and was completed by A. Fraenkel [3]. J. Von Neumann [11, 12] expressed it in †symbolic logic, gave a formal generalization, and eliminated ambiguous concepts. P. Bernays and K. Gödel [1, 5] further refined and simplified Von Neumann's formulation. The theories based on the systems before and after the formal generalization are called **Zermelo-Fraenkel set theory** (ZF) and **Bernays-Gödel set theory** (BG), respectively.

B. Zermelo-Fraenkel Set Theory

ZF is a formal system expressed in †first-order predicate logic with the predicate symbol = (equality) and based on the following axioms 1–9 (→ 337 Predicate Logic, 400 Symbolic Logic). These axioms do not contain any predicate symbol other than $\in$, where $x \in y$ is read "x is an element of y." Any formula containing only $\in$ as a predicate symbol is called a **set-theoretic formula**.

Axiom 1:

$$\exists x \forall u \neg (u \in x).$$

This asserts the existence of an **empty set**, i.e., a set having no element. The empty set is denoted by $\varnothing$.

Axiom 2:

$$\forall x \forall y (\forall u (u \in x \rightleftarrows u \in y) \to x = y).$$

This asserts that sets formed by the same elements are equal. It is called the **axiom of extensionality**.

Axiom 3:

$$\forall x \forall y \exists z \forall u (u \in z \rightleftarrows u = x \vee u = y).$$

This asserts the existence for any sets x and y of a set z having x and y as its only elements. This z is called the **unordered pair** of x and y and is denoted by $\{x,y\}$. The set $\{\{x,x\},\{x,y\}\}$ is denoted by (x,y) and is called the **ordered pair** (or simply **pair**) with first element x and second element y.

Axiom 4:

$$\forall x \exists y \forall u (u \in y \rightleftarrows \exists v (u \in v \wedge v \in x)).$$

This asserts the existence for any set x of the **union** (or **sum**) y of all the sets that are elements of x. This y is denoted by $\mathfrak{S}(x)$.

Axiom 5:

$$\forall x \exists y \forall u (u \in y \rightleftarrows \forall v (v \in u \to v \in x)).$$

For any sets s and t, $\forall r (r \in s \to r \in t)$ is denoted by $s \subset t$. This means "s is a **subset** of t." Then Axiom 5 can be expressed by

$$\forall x \exists y \forall u (u \in y \rightleftarrows u \subset x).$$

That is, the axiom asserts the existence for any set x of a set y consisting of all the subsets of x. This y is called the **power set** of x and is denoted by $\mathfrak{P}(x)$.

Axiom 6: For an aribtrary set-theoretic formula $A(u,v)$, we have

$$\forall u \forall v \forall w (A(u,v) \wedge A(u,w) \to v = w)$$
$$\to \forall x \exists y \forall v (v \in y \rightleftarrows \exists u (u \in x \wedge A(u,v))).$$

This states that the image of any set x by any mapping is a set. It was introduced by Fraenkel [3] and is called the **axiom of replacement**. (In fact, it is rather a schema for an infinite number of axioms, for there are an infinite number of $A(u,v)$.)

Axiom 7: For any set-theoretic formula $A(u)$, we have

$$\exists x A(x) \to \exists x \big(A(x) \wedge \neg \exists y (A(y) \wedge y \in x)\big).$$

This is also a schema for an infinite number of axioms. From this axiom, we can deduce that there is no x satisfying $x \in x$. If we assume the following Axiom 8, this is equivalent to the assertion that there is no infinite descending chain $x_1 \ni x_2 \ni x_3 \ni \ldots \ni x_n \ni x_{n+1} \ni \ldots$. This axiom is called the **axiom of regularity**.

Axiom 8:

$$\forall x \Big[\forall u (u \in x \to \exists v (v \in u))$$
$$\wedge \forall u \forall v \big((u \in x \wedge v \in x \wedge \neg u = v)$$
$$\to \neg \exists w (w \in u \wedge w \in v) \big) \to \exists y \big\{ y \subset \mathfrak{S}(x)$$
$$\wedge \forall u (u \in x \to \exists z (z \in u \wedge z \in y$$
$$\wedge \forall w (w \in u \wedge w \in y \to w = z))) \big\} \Big].$$

This asserts that for any family x of nonempty mutually disjoint sets there is a set y that has just one element in common with each member of x. Such a set y is called a **choice set**, and Axiom 8 is called the **axiom of choice**.

Axiom 9:

$$\exists x\Big(\exists u(u\in x)\wedge\forall u\big(u\in x \rightarrow\exists v(v\in x\wedge u\subset v\wedge\neg v=u)\big)\Big).$$

This asserts the existence of an infinite set x and is called the **axiom of infinity**. There are other ways of stating the axiom of infinity. If we assume Axiom 9, Axiom 1 becomes dispensable, because we can deduce it by making use of the axiom of replacement or of the following Axiom 6′.

Zermelo adopted Axiom 6′ instead of the axiom of replacement (Axiom 6):

Axiom 6′: For any set-theoretic formula $A(u)$, we have

$$\forall x\exists y\forall u(u\in y\rightleftarrows u\in x\wedge A(u)).$$

This axiom is called the **axiom of subsets** and asserts the existence of the set y usually denoted by $\{u\,|\,u\in x, A(u)\}$. Axiom 6′ can be deduced from Axiom 6.

The set theory having Axiom 6′ instead of Axiom 6 is called **Zermelo set theory**. It is weaker than ZF. Indeed, the existence of the set ω of all the natural numbers and of $\mathfrak{P}(\omega)$, $\mathfrak{P}(\mathfrak{P}(\omega))$, $\mathfrak{P}(\mathfrak{P}(\mathfrak{P}(\omega)))$,... can be proved in Zermelo set theory, but existence of the set $\{\omega, \mathfrak{P}(\omega), \mathfrak{P}(\mathfrak{P}(\omega)), \mathfrak{P}(\mathfrak{P}(\mathfrak{P}(\omega))), \ldots\}$ cannot be proved. However, we can prove its existence in ZF.

The set theory based on Axioms 1–8 is called **general set theory**. Its †consistency can be reduced to the consistency of the theory of natural numbers, as follows: Let n be any natural number and $n=2^{n_1}+2^{n_2}+\ldots+2^{n_k}$ $(n_1<n_2<\ldots<n_k)$ be its binary expansion. We can make a †model of general set theory in the theory of natural numbers by identifying a set with a natural number and defining "l is an element of n" by "l appears as one of the n_i."

C. Bernays-Gödel Set Theory

The existence of $\{u|A(u)\}$ for an arbitrary set-theoretic formula $A(u)$ cannot be deduced from the axioms of ZF. We call this object a **class** to distinguish it from sets. We introduce a generalized logical system of the first-order predicate logic by adding to the logical system used in ZF class variables, class constants, and inference rules with respect to quantifiers for classes. For any set-theoretic formula $A(u)$ in which no †bound class variable occurs, we adopt

$$\exists X\forall u(u\in X\rightleftarrows A(u))$$

as an axiom, where capital letters X are class variables. The set theory thus obtained is equivalent to **Gödel set theory** [15]. Von Neumann axiomatized set theory by making use of the notion of functions instead of that of classes [11, 12]. In refining this theory and introducing the notion of classes, Bernays and Gödel [1, 5] initiated Bernays-Gödel set theory, BG.

ZF and BG are related as follows: Any formula provable in ZF is provable in BG, and any set-theoretic formula provable in BG and having neither class variable nor class constant is provable in ZF. In this sense, the systems can be regarded as essentially equivalent, but as BG has class variables and class constants, it is more convenient for expressing set-theoretic notions.

Von Neumann defined the following function R by †transfinite induction [13]:

$$R(0)=\varnothing,\qquad R(\alpha)=\bigcup_{\beta<\alpha}\mathfrak{P}(R(\beta)),$$

where α and β are ordinal numbers and $\bigcup_{\beta<\alpha}\mathfrak{P}(R(\beta))$ denotes the set sum of $\mathfrak{P}(R(0)), \mathfrak{P}(R(1)), \ldots, \mathfrak{P}(R(\beta)), \ldots (\beta<\alpha)$. The function R can be defined by a set-theoretic formula, so it exists as a class. Now consider the model $M=M(\alpha)$ for a fixed ordinal number α. We define sets of the model M as elements of $R(\alpha)$, and classes of the model as subsets of $R(\alpha)$. We denote the $\in$ relation of the model by $\in_M$. For classes X and Y of the model, we write $X\in_M Y\rightleftarrows X\in Y$. Then, a necessary and sufficient condition for $R(\alpha)$ to be a model of BG is that α is an †inaccessible ordinal number (→ 306 Ordinal Numbers) [14]. The existence of an inaccessible ordinal number cannot be deduced from the axioms of ZF. There is a series of studies of axiomatization of set theory in which any number of inaccessible ordinal numbers are assumed to exist [6, 7, 8, 15]. When $R(\alpha)$ is a model of BG (ZF), it is called a **natural model** of BG (ZF). Furthermore, if we consider the case where α is the totality of ordinal numbers and denote $R(\alpha)$ by H, H satisfies all the axioms of BG (ZF). As we do not need the axiom of regularity for defining the class H, we see that BG (ZF) is consistent as long as BG (ZF) without the axiom of regularity is consistent [13].

D. Independence of the Continuum Hypothesis and the Axiom of Choice

These axiomatizations of set theory motivated a series of studies from the standpoint of the

†foundations of mathematics on problems that remained unsolved after the appearance of Cantor's primitive set theory. Among these, the problem of the relation between the continuum hypothesis and the axiom of choice was central.

Consistency of the Axiom of Choice and the Continuum Hypothesis. Gödel [5] proved that if ZF without the axiom of choice is consistent, then the system obtained by adding to ZF with the axiom of choice the continuum hypothesis is also consistent. To show this, he constructed a model of ZF satisfying the axiom of choice and the generalized continuum hypothesis as follows. Assume first that M is an arbitrary domain of objects among which the $\in$ relation is defined. By a formula on M, we understand a formula having constants of M as its only constants, having $\in$ as its sole predicate symbol, and further having exclusively variables whose ranges are restricted to M. Let us denote by M' the totality of subsets of M defined by a formula $A(x)$ on M. Now we put $M_0=\{\varnothing\}$, $M_{\alpha+1}=M'_\alpha$, $M_\beta=\bigcup_{\alpha<\beta}M_\alpha$, if β is a †limit ordinal number. We call x **constructible** if $x\in M_\alpha$ for some ordinal number α, assumed to be less than the first inaccessible ordinal number, if any. We denote the totality of constructible sets by L, and the totality of sets of ZF by V. We call the assertion $V=L$, that is, every set is constructible, the **axiom of constructibility**. If we add this axiom to ZF, the axiom of choice and the generalized continuum hypothesis become provable. On the other hand, if we regard elements of L as sets of the model and the original $\in$ relation as the $\in$ relation of the model, we have a model of ZF in which the axiom of constructibility holds.

Independence of the Axiom of Choice and the Continuum Hypothesis. Since the result of Gödel, attempts have been made to prove that the axiom of choice is independent of the other axioms. Fraenkel constructed a model of set theory without satisfying the axiom of choice, starting from a countable number of objects which are not sets [14]. A. Mostowski constructed in ZF a model of set theory having objects that are not sets, and he proved that the model satisfies the axiom: Every set can be †linearly ordered, but does not satisfy the axiom of choice [10]. E. Mendelson constructed a model of set theory that does not satisfy the axiom of choice by making use of an infinite descending chain $a_1\ni a_2\ni a_3\ni\ldots$ [9]. These models, however, do not satisfy all the axioms of ZF or of Zermelo set theory minus the axiom of choice, even though they satisfy most of the axioms. Consequently, they were not sufficient for proving independence.

P. J. Cohen [17] proved the following results in connection with the independence of the axiom of choice, the continuum hypothesis, and $V=L$: If ZF is consistent, each of the following conditions has a model. (1) The axiom of choice as well as the generalized continuum hypothesis holds, but there exists an a satisfying $a\notin L$ and $a\subset\omega$. (2) $\mathfrak{P}(\omega)$ is not well-ordered. (3) The axiom of choice holds, but the continuum hypothesis does not hold. (4) $\mathfrak{P}(\mathfrak{P}(\omega))$ cannot be linearly ordered.

By (1), we see that $V=L$ is independent of the axiom of choice and the generalized continuum hypothesis; (2) shows the independence of the axiom of choice; and (3) shows that the continuum hypothesis is independent of the axiom of choice. In (4), $\mathfrak{P}(\mathfrak{P}(\omega))$ corresponds to the set F of all real-valued functions defined on the interval [0, 1], and (4) shows that the proposition "F can be linearly ordered" is not deducible in ZF without the axiom of choice.

E. Some Recent Results

1. Cardinality and Cofinality (→ 306 Ordinal Numbers). (i) (W. B. Easton [21]) Let $\mathfrak{M}$ be a model of ZFC (the Zermelo-Frankel axioms plus the axiom of choice) in which the GCH (generalized continuum hypothesis) is valid, i.e., $\forall\alpha(2^{\omega_\alpha}=\omega_{\alpha+1})$, and let g be a function from ordinals to ordinals in $\mathfrak{M}$ such that $\forall\alpha,\beta(\alpha<\beta\Rightarrow\omega_{g(\alpha)}\leqslant\omega_{g(\beta)})$ and $\forall\alpha(\omega_\alpha<\mathrm{cf}(\omega_{g(\alpha)}))$. Then there is a Boolean model $\mathfrak{N}$ of ZFC, $\mathfrak{N}\supset\mathfrak{M}$, having the same cofinality and satisfying $2^{\omega_\alpha}=\omega_{g(\alpha)}$ for every regular cardinal. (This means that König's condition $(\mathrm{cf}(2^{\omega_\alpha})>\omega_\alpha)$ is the only restriction on the cardinality of powers of regular ω_α.)

(ii) (J.H. Silver [38]) Suppose

$$\omega<\mathrm{cf}(\alpha)=\mathrm{cf}(\omega_\alpha)<\omega_\alpha.$$

Then for any $\lambda<\mathrm{cf}(\alpha)$

$$\forall v<\alpha(2^{\omega_v}\leqslant\omega_{v+\lambda})\rightarrow 2^{\omega_\alpha}\leqslant\omega_{\alpha+\lambda}.$$

However, the validity of the implication

$$\forall n<\omega(2^{\omega_n}=\omega_{n+1})\rightarrow 2^{\omega_\omega}=\omega_{\omega+1}$$

still remains an open question.

2. Lebesgue Measurability and the Baire Property. As is well known, every Δ^1_1 (Borel) set (and consequently every Σ^1_1 (analytic) set) of real numbers is Lebesgue measurable and has the Baire property.

(i) (K. Gödel) $V=L$ implies the existence of a Δ^1_2 set of real numbers which is neither Lebesgue measurable nor has the Baire property.

Let DC denote the **principle of depending choice:**

$$\forall x \in a \exists y \in a P(x,y) \rightarrow \exists f: \omega \rightarrow a \forall n \in \omega P(f(n), f(n+1)).$$

DC is adequate for the development of the classical notions of measure theory, such as the Jordan decomposition, Radon-Nikodym derivate, etc. Let I denote the hypothesis $\exists \alpha (\mathrm{cf}(\omega_\alpha) = \omega_\alpha \wedge \forall \beta < \alpha (2^{\omega_\beta} < \omega_\alpha))$ **(strongly inaccessible)**.

(ii) (R. Solovay [40]) The consistency of ZFC and I implies that of ZF and either of the following two axioms:

(a) DC plus the hypothesis that every set of real numbers is Lebesgue measurable and has the Baire property.

(b) The axiom of choice plus the hypothesis that every set of real numbers definable by an ω sequence of ordinals is Lebesgue measurable and has the Baire property.

Axiom (a) means that DC is not strong enough to construct a (Lebesgue) nonmeasurable set, while axiom (b) implies that every projective set is Lebesgue measurable and has the Baire property, and hence implies that the set of all constructible real numbers, as a Σ_2^1 set, has Lebesgue measure 0, and is of the first category.

3. Martin's Axiom. Let B be a Boolean algebra. We say that B satisfies ω_α c.c. (chain condition) if the cardinality of every disjoint family of positive elements of B is at most ω_α. Let B^* denote the topological space consisting of all homomorphisms $h: B \rightarrow 2 = \{0, 1\}$ with the open base $U(a) = \{h \mid h(a) = 1\}$ $(a \in B)$; then B^* is a †Baire space. Then **Martin's axiom** (MA) is: Let B be an ω c.c. Boolean algebra and $\alpha < 2^\omega$. Then the intersection of α dense open sets is dense in B^*. Since $\{h \mid \Sigma_\nu h(a_\nu) = h(b)\}$ is dense and open in B^* if $\Sigma_\nu a_\nu = b$, MA means the existence of an $h \in B^*$ preserving any given set of α $(< 2^\omega)$ equations in B. If $2^\omega = \omega_1$, MA merely reduces to the Baire property of B^*. However, if $2^\omega > \omega_1$, then the ω c.c. hypothesis is essential, for there exists a B satisfying ω_1 c.c. such that B^* contains ω_1 dense open sets with empty intersection.

(i) (R. M. Solovay and S. Tennenbaum [42]) The consistency of ZF implies that of ZFC, MA, and $2^\omega > \omega_1$.

(ii) (D. A. Martin and R. M. Solovay [31]) ZFC, MA, and $2^\omega > \omega_1$ imply the following propositions:

(a) $\forall \alpha < 2^\omega (2^\alpha = 2^\omega)$, hence $2^\omega = 2^{\omega_1}$,

(b) the totality of the first category sets of Lebesgue measure zero sets is α-additive for any $\alpha < 2^\omega$;

(c) every Σ_2^1 set of real numbers is Lebesgue measurable and has the Baire property.

4. Souslin's Hypothesis (SH) is: Every dense, linear, order complete set without end points, having at most ω disjoint intervals, is order isomorphic to the continuum of real numbers.

(i) (T. J. Jech, S. Tennenbaum) The consistency of ZF implies that of (a) ZFC, $\neg$ SH, and GCH, as well as (b) ZFC, $\neg$ SH, and $2^\omega > \omega_1$.

(ii) (R. Jensen [26]) The consistency of ZF implies that of ZFC, SH, and $2^\omega = \omega_1$.

(iii) (Solovay-Tennenbaum [42]) ZFC, MA, and $2^\omega > \omega_1$ imply SH.

(iv) (R. Jensen [26]) $V = L$ implies $\neg$ SH.

5. Measurable and Real-Valued Measurable Cardinals. A cardinal $\kappa > \omega$ is said to be **measurable** if there is a measure $\mu: \mathfrak{P}(\kappa) \rightarrow \{0, 1\}$ with (a) $\mu(\kappa) = 1$, (b) $\forall \nu < \kappa (\mu(\{\nu\}) = 0)$ and (c) $\mu(\Sigma_{\nu < \alpha} A_\nu) = \Sigma_{\nu < \alpha} \mu(A_\nu)$; and κ is said to be **real-valued measurable** if there exists a $\mu: \mathfrak{P}(\kappa) \rightarrow [0, 1]$ which satisfies (a), (b) and (c), and is not measurable.

Let MC (RMC) denote the existence of a measurable cardinal (real-valued measurable cardinal).

(i) (S. Ulam [44]) The existence of an ω-additive measure $\mu: \mathfrak{P}(A) \rightarrow [0, 1]$, with $\mu(A) = 1$ and $\forall x \in A (\mu(\{x\}) = 0)$, implies RMC or MC; RMC implies the existence of an extension of Lebesgue measure defined on $\mathfrak{P}([0, 1])$; every real-valued measurable cardinal is $\leqslant 2^\omega$ and weakly inaccessible.

(ii) Every measurable cardinal is strongly (hyper) inaccessible, and (a) $\exists \alpha F(\alpha) \rightarrow \exists \alpha < \omega_1 F(\alpha)$ for any Σ_1^1 formula $F(\alpha)$ on the ordinal numbers, (b) $\exists \alpha F(\alpha) \rightarrow \exists \alpha < \mu_1 F(\alpha)$ for any Π_1^1 formula $F(\alpha)$ and the smallest measurable μ_1. (Many results have been obtained concerning the ordinal magnitude of ω_1 and the measurable cardinals.)

(iii) (R. Solovay [41]) The consistency of ZFC and MC is equivalent to that of ZFC and RMC.

(iv) (Martin and Solovay [31]) ZFC, RMC, and MA are not consistent.

(v) (Lévy and Solovay [30]) The consistency of ZFC and MC implies that of ZFC, MC, MA, and $2^\omega > \omega_2$.

(vi) (Solovay) ZFC and MC imply that every Σ_2^1 set of real numbers is Lebesgue measurable.

(vii) (Martin, Solovay) ZFC, MC, MA, and $2^\omega > \omega_2$ imply that every Σ_3^1 set of real numbers is Lebesgue measurable and has the Baire property.

(viii) (J. H. Silver [38]) The consistency of

ZFC and MC implies that of ZFC and MC as well as the existence of (Lebesgue) nonmeasurable Δ_3^1 sets of real numbers.

References

[1] P. Bernays, A system of axiomatic set theory pt. I, J. Symbolic Logic, 2 (1937), 65–77.
[2] P. J. Cohen, The independence of the continuum hypothesis I, II, Proc. Nat. Acad. Sci. US, 50 (1963), 1143–1148; 51 (1964), 105–110.
[3] A. Fraenkel, Zu den Grundlagen der Cantor-Zermeloschen Mengenlehre, Math. Ann., 86 (1922), 230–237.
[4] A. Fraenkel, Der Begriff 'definit' und die Unabhängigkeit des Auswahlaxioms, S.-B. Preuss. Akad. Wiss., 1922, 253–257.
[5] K. Gödel, The consistency of the axiom of choice and of the generalized continuum-hypothesis with the axioms of set theory, Ann. Math. Studies, Princeton Univ. Press, 1940.
[6] A. Lévy, Axiom schemata of strong infinity in axiomatic set theory, Pacific J. Math., 10 (1960), 223–238.
[7] P. Mahlo, Über lineare transfinite Mengen, Berichte über die Verhandlungen der Königlich Sächsischen Gesellschaft der Wissenschaften zu Leipzig, Math.-Phys. Klasse, 63 (1911), 187–225.
[8] P. Mahlo, Zur Theorie und Anwendung der ρ_0-Zahlen I, II, Berichte über die Verhandlungen der Königlich Sächsischen Gesellschaft der Wissenschaften zu Leipzig, Math.-Phys. Klasse, 64 (1912), 108–112; 65 (1913), 268–282.
[9] E. Mendelson, The independence of a weak axiom of choice, J. Symbolic Logic, 21 (1956), 350–366.
[10] A. Mostowski, Über die Unabhängigkeit des Wohlordnungssatzes vom Ordnungsprinzip, Fund. Math., 32 (1939), 201–252.
[11] J. Von Neumann, Eine Axiomatisierung der Mengenlehre, J. Reine Angew. Math., 154 (1925), 219–240.
[12] J. Von Neumann, Die Axiomatisierung der Mengenlehre, Math. Z., 27 (1928), 669–752.
[13] J. Von Neumann, Über eine Widerspruchsfreiheitsfrage in der axiomatischen Mengenlehre, J. Reine Angew. Math., 160 (1929), 227–241.
[14] J. C. Shepherdson, Inner models for set theory I, II, III, J. Symbolic Logic, 16 (1951), 161–190; 17 (1952), 225–237; 18 (1953), 145–167.
[15] A. Tarski, Über unerreichbare Kardinalzahlen, Fund. Math., 30 (1938), 68–89.
[16] E. Zermelo, Untersuchungen über die Grundlagen der Mengenlehre I, Math. Ann., 65 (1908), 261–281.
[17] P. J. Cohen, Set theory and the continuum hypothesis, Benjamin, 1966.
[18] J. B. Rosser, Simplified independence proofs: Boolean valued models of set theory, Academic Press, 1969.
[19] J. R. Shoenfield, Mathematical logic, Addison-Wesley, 1967.
[20] P. J. Cohen, The independence of the continuum hypothesis I, II, Proc. Nat. Acad. Sci. US, 50 (1963), 1143–1148; 51 (1964), 105–110.
[21] W. B. Easton, Power of regular cardinals, Ann. Math. Logic, 1 (1970), 139–178.
[22] G. Föder, On stationary sets and regressive functions, Acta Sci. Math., 27 (1966), 105–110.
[23] H. Gaifman, Uniform extension operators for models and their applications; Sets, Models and Recursion Theory, North-Holland, (1967), 122–155.
[24] K. Gödel, The consistency of the axiom of choice and of the generalized continuum hypothesis, Proc. Nat. Acad. Sci. US, 24 (1938), 556–557.
[25] T. J. Jech, Non-provability of Souslin's hypothesis, Comm. Math. Univ. Carolinae 8 (1967), 291–305.
[26] R. B. Jensen, The fine structure of the constructible hierarchy, Ann. Math. Logic, 4 (1972), 229–308.
[27] H. J. Keisler and A. Tarski, From accessible to inaccessible cardinals, Fund. Math., 53 (1964), 225–308; corrections: ibid., 57 (1965), 119.
[28] K. Kunen, Inaccessibility properties of cardinals, Ph.D. dissertation, Stanford Univ., Stanford, Calif.
[29] K. Kunen and J. B. Paris, Boolean extensions and measurable cardinals, Ann. Math. Logic, 2 (1971), 359–378.
[30] A. Lévy and R. M. Solovay, Measurable cardinals and the continuum hypothesis, Israel J. Math., 5 (1967), 234–248.
[31] D. A. Martin and R. M. Solovay, Internal Cohen extensions, Ann. Math. Logic, 2 (1970), 143–178.
[32] Y. N. Moschovakis, Determinacy and prewellorderings of the continuum; mathematical logic and foundations of set theory, North-Holland, (1970), p. 24–62.
[33] J. Mycielski, On the axiom of determinateness, Fund. Math., 53 (1964), 205–224.
[34] K. Namba, An axiom of strong infinity and analytic hierarchy of ordinal numbers, Proc. Symp. Pure Math., 13 (1971), 279–319.
[35] K. L. Prikry, Changing measurable into accessible cardinals, Dissertationes Math., 68 (1970).
[36] G. E. Sacks, Forcing with perfect closed

sets, Proc. Symp. Pure Math., 13 (1971), 331–355.
[37] D. S. Scott, Measurable cardinals and constructible sets, Bull. Acad. Pol. Sci., Ser. Sci. Math. Astr. Phys. 7 (1961), 145–149.
[38] J. H. Silver, The consistency of the GCH with the existence of a measurable cardinal, Proc. Symp. Pure Math., 13 (1971), 391–396.
[39] J. H. Silver, Measurable cardinals and Δ_3^1 well-orderings, Ann. of Math., (2) 94 (1971), 414–446.
[40] R. M. Solovay, A model of set theory in which every set of reals is Lebesgue measurable, Ann. of Math., (2) 92 (1970), 1–56.
[41] R. M. Solovay, Real-valued measurable cardinals, Proc. Symp. Pure Math., 13 (1971), 397–428.
[42] R. M. Solovay and S. Tennenbaum, Iterated Cohen extensions and Souslin's Problem, Ann. of Math., (2) 94 (1971), 201–245.
[43] G. Takeuti, Transcendency of cardinals, J. Symb. Logic, 30 (1965), 1–7.
[44] S. Ulam, Zur Masstheorie in der allgemeinen Mengenlehre, Fund. Math., 16 (1930), 140–150.

36 (II.5) Axiom of Choice and Equivalents

A. The Axiom of Choice

In set theory, the following axiom is known as the **axiom of choice**: For any nonempty family $\mathfrak{A}$ of nonempty subsets of a set X, there exists a single-valued function f, called the **choice function**, whose domain is $\mathfrak{A}$, such that for every element A of $\mathfrak{A}$ the value $f(A)$ is a member of A. This axiom is equivalent to each of the following three propositions: (1) If $\{A_\lambda\}_{\lambda\in\Lambda}$ is a family of sets not containing the empty set and with an index set Λ, then the Cartesian product $\Pi_\lambda A_\lambda$ is not the empty set. (2) If a set A is a disjoint union $\bigcup_\lambda A_\lambda$ of a family of subsets $\{A_\lambda\}_{\lambda\in\Lambda}$ which does not contain the empty set, there exists a subset B, called the **choice set**, of A such that every intersection of B and A_λ ($\lambda\in\Lambda$) contains one and only one element. (3) For every mapping f from a set A onto a set B, there is a mapping from B to A such that $f\circ g=1_B$ (identity mapping). Also equivalent to the axiom of choice are the well-ordering theorem and Zorn's lemma, which will be discussed in the following section.

B. The Well-Ordering Theorem

In 1904, E. Zermelo [4] first stated the axiom of choice and used it for his proof of the **well-ordering theorem**, which says that every set can be †well-ordered by an appropriate †ordering. Conversely, the well-ordering theorem implies the axiom of choice. Many important results in set theory can be obtained by using the axiom of choice, for example, that †cardinal numbers are †comparable, or that various definitions of the finiteness or infiniteness of sets are equivalent. Various important theorems outside of set theory, e.g., the existence of bases in a †linear space, †compactness of the direct product of compact †topological spaces (†Tihonov's theorem), the existence of a subset which is not †Lebesgue measurable in Euclidean space, etc., are proved using the axiom of choice. But for those proofs the well-ordering theorem or Zorn's lemma (stated below) are used more often than the axiom of choice.

Using the well-ordering theorem, it can be proved in the following manner that for every field k, any linear space X over k has a basis. Let $\{x_\nu\}_{\nu\in\Lambda}$ be an enumeration of X. By †transfinite induction, we can define the function f from X to $B=\{0,1\}$ such that (i) if $x_0=0, f(x_0)=0$; if $x_0\neq 0, f(x_0)=1$; (ii) for $\nu>0$, if x_ν is expressed as a linear combination of elements of $\{x_\mu \mid \mu<\nu, f(x_\mu)=1\}, f(x_\nu)=0$; otherwise $f(x_\nu)=1$. Then $U=\{x\in X\mid f(x)=1\}$ is a basis of X.

C. Zorn's Lemma

An ordered set X is called an **inductively ordered set** if every †totally ordered subset of X has an †upper bound. A condition C for sets is called a **condition of finite character** if a set X satisfies C if and only if every finite subset of X satisfies C. A condition C for functions is called a **condition of finite character** if C is a condition of finite character for the graph of the function. **Zorn's lemma** [6] can be stated in any one of the following ways, which are all equivalent to the axiom of choice. It is often more convenient to use than the axiom of choice or the well-ordering theorem.

(1) Every inductively ordered set has at least one maximal element.
(2) If every well-ordered subset of an ordered set M has an upper bound, then there is at least one maximal element in M.
(3) Every ordered set M has a well-ordered subset W such that every upper bound of M belongs to W.
(4) For a condition C of finite character for sets, every set X has a maximal (for the relation of the inclusion) subset of X which satisfies C.
(5) Let C be a condition of finite character

for functions from X to Y. Then, in the set of functions which satisfy C, there is a function whose domain is maximal (for the relation of the inclusion).

Using Zorn's lemma (1), we can prove again in the following way any linear space X over a field k has a basis. Let $\mathfrak{A}$ be the set of all nonempty subsets A of X such that arbitrary finite subsets of A are linearly independent over k. $\mathfrak{A}$ is not empty. If we order $\mathfrak{A}$ by the relation of inclusion, then $\mathfrak{A}$ is an inductively ordered set. By Zorn's lemma (1), there is a maximal element U of $\mathfrak{A}$. Since U is maximal, U is a basis of X.

The same theorem is proved as follows using Zorn's lemma (4). Condition C for the subset A of X, that arbitrary finite subsets of A are linearly independent over k, is a condition of finite character. Hence there is a maximal subset U which satisfies C and a basis of X.

Concerning the recent development on axiomatic considerations of the axiom of choice → 35 Axiomatic Set Theory D.

References

[1] T. Nakayama, Syûgô, isô, daisûkei (Japanese; Sets, topology, and algebraic systems), Sibundô, 1949.
[2] S. Iyanaga and K. Kodaira, Gendai sûgaku gaisetu (Japanese; Introduction to modern mathematics) I, Iwanami, 1961.
[3] G. Cantor, Über unendliche, lineare Punktmannigfaltigkeiten, Math. Ann., 21 (1883), 545–591. (Gesammelte Abhandlungen, Springer, 1932.)
[4] E. Zermelo, Beweis, dass jede Menge wohlgeordnet werden kann, Math. Ann., 59 (1904), 514–516.
[5] E. Zermelo, Neuer Beweis für die Möglichkeit einer Wohlordnung, Math. Ann., 65 (1908), 107–128.
[6] M. Zorn, A remark on method in transfinite algebra, Bull. Amer. Math. Soc., 41 (1935), 667–670.
[7] J. W. Tukey, Convergence and uniformity in topology, Ann. Math. Studies, Princeton Univ. Press, 1940.
[8] N. Bourbaki, Eléments de mathématique, I. Théorie des ensembles, ch. 3, Actualités Sci. Ind., 1243, Hermann, second edition, 1966; English translation, Theory of sets, Addison-Wesley, 1968.

37 (I.2) Axiom Systems

A. History

A mathematical theory is based on a specific system of axioms, i.e., a system of hypotheses from which the whole theory is deduced without reliance on other assumptions.

One of the first deductive methods of mathematical reasoning was utilized by Thales, who returned to Greece from Egypt with the knowledge of surveying methods, and who deduced additional results from that empirical knowledge. His method gave impetus to the development of Greek geometry, which flowered with the Pythagorean school and research by members of Plato's Academy. In the course of this development, the deductive method led to the idea of constructing the whole theory upon a system of "absolutely obvious" statements from which the whole theory should be deduced. Euclid systematized Greek geometry in his *Elements* utilizing this idea. His work became the basis of geometry after the Renaissance, and Greek geometry came to be called †Euclidean geometry. In the *Elements* Euclid called the basic obvious statements **common notions** when they are of general nature, and **postulates** when they are specifically geometric. Both were later called **axioms** (or **postulates**).

Among the axioms stated by Euclid, the "fifth postulate" concerning parallels was comparatively longer and more complicated than the other axioms. Many efforts were made to deduce this particular axiom from the other axioms. The failure of these attempts suggested the possibility of establishing a †non-Euclidean geometry, which was actually done by N. I. Lobačevskiĭ and J. Bolyai, who replaced the fifth postulate by its negation and showed that the new system of "axioms" was as valid as the classical one. This development naturally led to a new evaluation of the idea of axioms, and eventually the traditional concept of recognizing the axioms as obvious truths was replaced by the understanding that they are hypotheses for a theory. D. Hilbert [1] established the latter idea as **axiomatization** and claimed that the whole science of mathematics should be built upon a system of axioms. His idea became the foundation of present-day mathematics. Hilbert reorganized classical geometry based upon his idea and published his result in *Foundations of Geometry* (1899).

B. Systems of Axioms

The **system of axioms** of a theory, i.e., the system of basic hypotheses from which we hope to deduce the whole theory, is written in **undefined terms** (or in terms of **undefined concepts**) by means of which all other terms are defined. On the other hand, a given theory is **axiomatized** by specifying such a system of axioms upon which the theory may be reorganized. It should be also noted that a system of axioms determines a †structure (→ 396 Structures).

A system of axioms is considered to be mathematically valid if and only if it is **consistent**. It is also desirable that the axioms in such a system be mutually **independent** (i.e., the negation of any one of the axioms is still consistent with the others). When such a system is not independent, it may be simplified by deleting redundant axioms from it.

When any two models of a system of axioms are isomorphic to each other, we call the system **complete** or **categorical** (→ 396 Structures). For example, the system of axioms (I)–(V) postulated by Hilbert as the foundation for Euclidean geometry is complete (→ 163 Foundations of Geometry), whereas the systems of axioms for the theories of †groups, †rings, or †fields are not complete since there are nonisomorphic groups, etc. Although it is desirable that the systems of axioms postulated for a given theory (e.g., the †theory of real numbers, or Euclidean geometry) be complete, the study of partial systems that may not be complete is also important (→ 164 Foundations of Mathematics).

References

[1] D. Hilbert, Axiomatisches Denken, Math. Ann., 78 (1918), 405–415. (Gesammelte Abhandlungen III, Springer, 1935, p. 146–156.)

[2] N. Bourbaki, Eléments de mathématique, Théorie des ensembles, ch. 1–4 and fascicule de résultats, Actualités Sci. Ind., 1212c, 1243b, 1258a, 1141d, Hermann, 1960, 1967, 1966, 1964.

B

38 (XII.15) Banach Algebras

A. Definition

A real or complex †Banach space R is called a **Banach algebra** (or **normed ring**) if a multiplication law between elements of R is introduced that makes R an †associative algebra and if $\|xy\| \leqslant \|x\| \cdot \|y\|$ is satisfied. The complex case is generally easier to handle, and a real Banach algebra can always be embedded in a complex one isomorphically and isometrically. When a Banach algebra contains a unity element e with respect to the multiplication, we may suppose $\|e\|=1$. If it does not contain a unity element, we can adjoin a unity element to it.

B. Examples

Example 1. The Banach space $C_0(\mathfrak{M})$ consisting of real- or complex-valued continuous functions $x(\xi)$ on $\mathfrak{M}$ which vanish at infinity (i.e., the set $\{\xi \mid |x(\xi)| \geqslant \epsilon\}$ is compact for any $\epsilon > 0$), whose norm is $\|x\| = \sup\{|x(\xi)| \mid \xi \in \mathfrak{M}\}$, and whose multiplication is defined by pointwise multiplication: $xy(\xi) = x(\xi)y(\xi)$.

Example 2. Let X be a Banach space. The set $\mathfrak{B}(X)$ of †bounded linear operators on X forms a Banach algebra if we define addition, multiplication by scalars, multiplication between elements, and the norm in the usual fashion.

C. Spectrum of an Element

We define a new operation between elements by $x \cdot y = x + y - xy$. If $x \cdot y = y \cdot x = 0$, y is called a **quasi-inverse** of x. If a unity element e exists, $e - y$ is an inverse of $e - x$. For a complex number λ such that $|\lambda| > \|x\|$, we see that $\lambda^{-1}x$ possesses a quasi-inverse y given by the strongly convergent series $y = -\sum_{n=1}^{\infty} \lambda^{-n} x^n$. The set of all complex numbers λ such that $\lambda^{-1}x$ does not have a quasi-inverse is called the **spectrum** of x and is denoted by $\mathrm{Sp}(x)$. (When x itself does not have an inverse, in particular, when R does not contain a unity element, we suppose 0 to be in this set.) $\mathrm{Sp}(x)$ is a bounded closed set in the complex plane, and $\sup\{|\lambda| \mid \lambda \in \mathrm{Sp}(x)\} = \lim_{n\to\infty} \|x^n\|^{1/n}$. In Example 2, this spectrum is the spectrum of the operator x, and the problem of determining the behavior of this spectrum constitutes one of the central problems in the theories of Banach and Hilbert spaces.

D. The Gel'fand Representation of a Commutative Banach Algebra

A complex Banach algebra is a field if and only if it coincides with the field **C** of complex numbers (**Gel'fand-Mazur theorem**). Now let R be a commutative Banach algebra and M a †maximal ideal of R. The †quotient algebra R/M is either a ring in which the product of two arbitrary elements is always 0 or a field. In the latter case M is closed in R, R/M is isomorphic to the field of complex numbers, and we call M a **regular maximal ideal**. (For the regularity of an ideal → Section E.) Now for a regular maximal ideal M, the coset $x(M)$ containing an element x of R can be regarded as a complex number by virtue of the isomorphism of R/M and **C**. Then the functional that associates $x(M)$ to x is multiplicative and linear on R, that is, $xy(M) = x(M)y(M)$. Conversely, to each multiplicative linear functional on R, we can associate a regular maximal ideal which relates to this functional in the way specified before. The set $\mathfrak{M}$ of multiplicative linear functionals on R (provided with the functional weak topology called the **Gel'fand topology**) is a locally compact Hausdorff space (compact when R has a unity element), and an element of R is represented as a continuous function on $\mathfrak{M}$ vanishing at infinity. This is the **Gel'fand representation** of R. Concerning this representation, we have (1) $\max\{|x(M)| \mid M \in \mathfrak{M}\} = \lim_{n\to\infty} \|x^n\|^{1/n}$. (When R has a unity element, $\{x(M) \mid M \in \mathfrak{M}\} = \mathrm{Sp}(x)$.) (2) The †kernel of the Gel'fand representation is the set of x such that $\lim_{n\to\infty} \|x^n\|^{1/n} = 0$. An element with this property is called a **generalized nilpotent element**, the set in question is called the **radical** of R, and R is said to be **semisimple** if its radical reduces to $\{0\}$. (3) One important sufficient condition for a semisimple Banach algebra to be representable as a dense subalgebra of $C_0(\mathfrak{M})$ (the algebra of all continuous functions vanishing at infinity) is the following: to any $x \in R$ is associated an $x^* \in R$ such that $x^*(M) = \overline{x(M)}$ for any $M \in \mathfrak{M}$ (an instance of the †Weierstrass-Stone theorem). The norm is not generally preserved through Gel'fand representation (cf. the L_1-algebra of a commutative group mentioned in Section H).

E. Representations of a General Banach Algebra

We understand by a **representation** of a Banach algebra R a law of correspondence which associates with each $x \in R$ a bounded linear operator T_x on a Banach space X that preserves algebraic operations and satisfies $\|T_x\| \leqslant \|x\|$. We call X the **representation space**. A Banach algebra always possesses an isomorphic and isometric representation, but **irreducible representations** are important. A vector subspace (closed or not) Y of X is an invariant subspace if $T_x Y \subset Y$ for any $x \in R$. A representation is algebraically irreducible if the invariant subspaces are trivial, i.e., they are only $\{0\}$ or X. A representation is topologically irreducible if closed invariant subspaces are trivial. The †kernel of an algebraically irreducible representation is called a **primitive ideal**, which may alternatively be defined in the following way: A left ideal J ($\neq \{0\}, R$) is **regular**, by definition, if R contains an element u such that $x - xu \in J$ for any $x \in R$. Such an ideal is always contained in a maximal left ideal, which in turn is necessarily regular and closed. A two-sided ideal I is primitive if it is the set of elements a in R for which $aR \subset J$, where J is some fixed regular maximal left ideal. If R is commutative, a primitive ideal is a regular maximal ideal, and conversely. The intersection of all primitive ideals is the **radical** of R, and when it is $\{0\}$, R is called **semisimple**.

The set of primitive ideals $\mathfrak{J}$ is known as the **structure space** of R, in which the **hull-kernel topology** (or **Jacobson topology**) is introduced. The †closure of a set $\mathfrak{A}$ in $\mathfrak{J}$ is, under this topology, the set of primitive ideals containing the intersection of the ideals in $\mathfrak{A}$. This topology is rather intractable; even in commutative cases, it does not coincide with the Gel'fand topology, in general.

F. Banach Star Algebras

An involution in a Banach algebra R is an operation $x \rightarrow x^*$ that satisfies (1) $(x+y)^* = x^* + y^*$; (2) $(\lambda x)^* = \bar{\lambda} x^*$; (3) $(xy)^* = y^* x^*$; (4) $(x^*)^* = x$. A Banach algebra in which an **involution** is defined is called a **Banach $*$-algebra**. To represent a Banach $*$-algebra we prefer a **$*$-representation**, i.e., a representation $x \rightarrow T_x$ on a Hilbert space such that T_{x^*} is equal to the adjoint T_x^* of T_x for any $x \in R$.

G. C^*-Algebras

An important class of Banach $*$-algebras is the **C^*-algebras**, which consists of Banach $*$-algebras that satisfy $\|x^* x\| = \|x\|^2$ for all elements x. A C^*-algebra is actually a Banach algebra of operators on a Hilbert space (see Example 2) that contains, along with an operator, its adjoint (the Gel'fand-Naĭmark theorem). A C^*-algebra is semisimple, and a commutative C^*-algebra is, by the Gel'fand representation, isomorphic and isometric to $C_0(\mathfrak{M})$ (see Example 1). A topologically irreducible $*$-representation of a C^*-algebra is also algebraically irreducible, and the set of †unitary equivalence classes of these irreducible $*$-representations is called the **dual space**. It becomes a topological space if we introduce the hull-kernel topology inherited from the structure space, but other topologies are also introduced. Moreover, for the study of separable C^*-algebras, Borel structure is a very powerful notion (Mackey's theory [7]). C^*-algebras also have a connection with the important theory of †unitary representations of a topological group (see below). Many works have been published on $*$-representations, dual spaces, etc.

Among various C^*-algebras, CCR algebras (liminal C^*-algebras) are of special interest. They are C^*-algebras whose elements are represented exclusively by †compact operators under irreducible $*$-representations. A slightly more general notion is that of a GCR algebra (I. Kaplansky [8]). These classes of C^*-algebras are quite interesting. For example, the dual space of a C^*-algebra is a †T_0 space (†T_1 space) if and only if the algebra is GCR (CCR) (J. Glimm [9]).

H. Applications to the Theory of Topological Groups

Banach algebras have many applications in different branches of mathematics. Here we mention some that concern topological groups. Let G be a locally compact Hausdorff group (→ 406 Topological Groups) and μ be its †left-invariant measure. The function $\rho(g)$ defined by $\mu(Eg) = \rho(g)\mu(E)$ for any †Borel subset E of G is positive and continuous. By utilizing this function, we may make $L_1(G)$ (with respect to μ) a Banach $*$-algebra by defining

$$xy(g) = \int x(h) y(h^{-1}g) d\mu(h),$$

$$x^*(g) = \overline{x(g^{-1})}\,\rho(g^{-1}).$$

This is the **L_1-algebra** (or **group algebra**). It is not C^* but is semisimple. Considering a unitary representation of G is equivalent to considering a $*$-representation of L_1-algebra. Replace the norm of $x \in L_1(G)$ by $\sup \|T_x\|$, where the supremum is taken with respect to all the $*$-representations. The new norm satisfies the C^* condition, and the completion of $L_1(G)$ with respect to this norm is a C^*-algebra which we call the **C^*-group algebra** of G. The dual of the C^*-group algebra thus defined is called the dual of the group G, and this notion plays an important role in the study of topological groups. Unitary representations of a group G, $*$-representations of the L_1-algebra of G, and $*$-representations of the C^*-group algebra of G are all characterized by positive definite functions on G. A function $p(g)$ is **positive definite**, by definition, if it is measurable on G and

$$\int\int p(g^{-1}h)\overline{x(g)}x(h)d\mu(g)d\mu(h)\geqslant 0$$

is satisfied for any continuous function $x(g)$ with compact support.

The Abelian case. When G is an Abelian group (→ 405 Topological Abelian Groups), a regular maximal ideal M of the L_1-algebra R of G and a character γ of G are in a one-to-one correspondence by the relation

$$x(M)=\int x(g)\overline{\gamma(g)}\,d\mu(g)$$

(the left-hand side is the value of x at M under Gel'fand representation). Moreover, the set of regular maximal ideals of R provided with the Gel'fand topology and the set $\hat{G}$ of characters of G provided with the Pontrjagin topology (the †character group of G) are homeomorphic by this correspondence. Therefore, the Gel'fand transform of an element x of the L_1-algebra R is seen to be a function $\hat{x}(\gamma)$ on $\hat{G}$ defined by the integral on the right-hand side in the above expression, which is properly called the **Fourier transform** of x. Of course, the Fourier transform can be defined for other classes of functions (e.g., for the L_2-space over G, the Fourier transform in the sense of Plancherel), and classical theories of Fourier series, Fourier integrals, and harmonic analysis (→ 194 Harmonic Analysis) are studied from a more extensive point of view. Thus, the statement that the Fourier transform of an element x of the L_1 algebra of G is a continuous function vanishing at infinity (→ Section D), for example, is a version of the classical †Riemann-Lebesgue theorem. **Bochner's theorem** in classical †Fourier analysis is restated thus: A continuous †positive definite function on an Abelian group can be put in the form

$$p(g)=\int\gamma(g)d\rho(\gamma),$$

where ρ is a uniquely determined bounded positive †Radon measure on the character group $\hat{G}$. Developing these theories further, we obtain an alternative proof of the †Pontrjagin duality theorem (H. Cartan and R. Godement). A closed ideal I in the L_1-algebra R determines a set $Z(I)$ in $\hat{G}$ as the set of common zeros of the Fourier transforms of elements of I. We ask whether, conversely, I is characterized by $Z(I)$. This question is the problem of **spectral synthesis**, and many important results have been obtained. The statement that I must coincide with R when $Z(I)$ is empty is a formulation of the **generalized Tauberian theorem** of N. Wiener. A considerable simplification was accomplished by using the theory of Banach algebras, and this was the first fascinating application of the theory (I. M. Gel'fand).

I. Function Algebras

Let X be a †compact Hausdorff space, and let $C(X)(C_R(X))$ be the algebra of all complex- (real-) valued continuous functions on X with the †supremum norm. A closed subalgebra A of $C(X)$ is called a **function algebra** (or **uniform algebra**) on X if it contains the constants and separates the points of X (i.e., for any $x,y\in X$ with $x\neq y$, there exists an $f\in A$ such that $f(x)\neq f(y)$). A typical example of a function algebra is the **disk algebra**, i.e., the algebra of all continuous functions on the unit circle X in the complex plane that can be extended continuously to functions analytic on the open unit disk. Other examples result if we take as X an analytic curve in a †Riemann surface, a †compact Abelian group (e.g., a †torus), and so forth. Also, if we take X to be a domain in the n-dimensional complex space $\mathbf{C}^n$, then the function algebras on X have a close connection with function theory of several complex variables.

Let A be a function algebra on a compact Hausdorff space X, and let F be a closed subset in X. F is called an **antisymmetric set** if $f\in A$ is constant on F whenever f is real on F. G. Šilov inquired whether every function algebra could be expressed in terms of antisymmetric algebras. The problem was completely solved by E. Bishop [12]: There is a partition of X consisting of antisymmetric sets $\{F_\lambda\}$ such that a continuous function f on X is in A whenever $f|F_\lambda\in A|F_\lambda$ for any λ. Bishop's result is an extension of the Stone-Weierstrass theorem.

Let A be the disk algebra and F be a closed subset in the unit circle such that its Lebesgue measure is zero. Then we can prove that $A|F=C(F)$ [28]. For a function algebra A on a compact Hausdorff space X, a closed subset F in X is called an **interpolation set** for A if $A|F=C(F)$. I. Glicksberg [18] characterized interpolation sets for any function algebra and proved various theorems about them.

In the theory of function algebras, maximal algebras play an important role. A is called a **maximal algebra** if $B=C(X)$ or $B=A$ for any function algebra $B\supset A$. The disk algebra is a maximal algebra. This fact is known as Wermer's **maximality theorem** [22]. There are close relationships between abstract function algebras and complex function theory. One of these is the concept of Gleason parts. Let A be a function algebra, and let M_A be the maximal ideal space of A. For $a, b \in M_A$, we write $a \sim b$ if

$$\sup\{|f(a)-f(b)| \mid f \in A, \|f\|<1\}<2.$$

Then the relation $\sim$ is an equivalence relation [17]. Equivalence classes for the relation are called **Gleason parts** for A. Under some conditions a Gleason part has a kind of analyticity. In particular, if A is a Dirichlet algebra as defined below, then any Gleason part P for A either consists of only one point or has the following property: there is a one-to-one continuous mapping τ of the open unit disk D onto P such that $f \circ \tau$ $(f \in A)$ is a holomorphic function on D (Wermer [29]). Here a function algebra A is called a **Dirichlet algebra** if $\operatorname{Re} A$ is dense in $C_R(X)$. The disk algebra is a Dirichlet algebra. A function algebra A is said to be a **logmodular algebra** if $\{\log|f| \mid f \in A^{-1}\}$ is dense in $C_R(X)$, where A^{-1} denotes the set of $f \in A$ with $f^{-1} \in A$. If A is a logmodular algebra, the same result as in the previous theorem can be established.
For $p \in M_A$, we can consider a **representing measure** μ_p for p, i.e., a positive measure μ_p on X such that $f(p)=\int_X f(x)d\mu_p(x)$ for $f \in A$. The representing measure is a generalization of the †Poisson integral representation for analytic functions. Let a, b be any two points in a Gleason part. Then for any representing measure μ_a for a, there is a representing measure μ_b for b such that $\mu_a \leqslant c\mu_b$. Conversely, for a representing measure μ'_b for b, there is a representing measure μ'_a for a such that $\mu'_b \leqslant \mu'_a$ [13]. Also, we can extend the concept of Gleason parts to the case of linear subspaces in $C_R(X)$. In this case, Gleason parts are related to abstract †potential theory because the equivalence relation for Gleason parts has the form of the †Harnack inequality [11]. As applications, there are many problems in connection with function theory; for instance, uniform approximation by rational functions on a compact set in the complex plane [19] and the theory of A-holomorphic functions, which extends the theory of holomorphic functions of several complex variables [26].

In the theory of function algebras, we often use the so-called Rossi's **local maximum modulus principle** [27]: let U be an open set in $M_A - \partial_A$; then the †Šilov boundary of A_U is contained in the topological boundary of U, where A_U denotes the closure of $A|\overline{U}$.

In relation to the theory of function algebras, we can consider the **generalized Hardy classes**. Let A be a function algebra and m be a multiplicative positive measure on A. For $1 \leqslant p < \infty$, the Hardy class $H_p(m)$ is defined as the closure of A in $L_p(m)$, and for $p=\infty$, $H_\infty(m)$ is the weak $*$-closure of A in $L_\infty(m)$. Many theorems that are valid in the classical Hardy classes can be extended to the case of generalized Hardy classes [16, 21, 22, 24, 25]. In the classical H_∞ classes, there was the so-called **Corona problem**: Is the open disk dense in the maximal ideal space of H_∞? This problem was solved affirmatively by L. Carleson [15]. It is equivalent to the following: Let $f_1, f_2, \ldots, f_n$ be functions in H_∞ such that $|f_1|+|f_2|+\ldots+|f_n| \geqslant \varepsilon > 0$. Then there are $g_1, \ldots, g_n \in H_\infty$ such that $f_1 g_1 + \ldots + f_n g_n = 1$. Next, we notice that "the (shift-) invariant subspaces" also are related to function algebras. A closed subspace M in H_2 is an invariant subspace if $zM \subset M$. In fact, we have **Beurling's theorem**: For any nonvoid invariant subspace in H_2, there is an inner function f (i.e., $f \in H_\infty$ and $|f|=1$ almost everywhere) such that $M=fH_2$. There are many theorems for invariant subspaces related to other kinds of spaces, for instance L_2 or spaces of continuous functions [16, 22, 31, 32].

References

[1] C. E. Rickart, General theory of Banach algebras, Van Nostrand, 1960.
[2] M. A. Наймарк (M. A. Naĭmark), Нормированные кольца, гостехиздат, 1956; English translation, Normed rings, Noordhoff, 1964.
[3] L. H. Loomis, An introduction to abstract harmonic analysis, Van Nostrand, 1953.
[4] W. Rudin, Fourier analysis on groups, Interscience, 1962.
[5] K. Yosida, Isôkaiseki (Japanese; Functional analysis), Iwanami, 1951.
[6] Y. Mimura, Isôkaiseki (Japanese; Functional analysis), Kyôritu, 1957.
[7] G. W. Mackey, Borel structure in groups and their duals, Trans. Amer. Math. Soc., 85 (1957), 134–165.
[8] I. Kaplansky, The structure of certain operator algebras, Trans. Amer. Math. Soc., 70 (1951), 219–255.
[9] J. G. Glimm, Type I C^*-algebras, Ann. of Math., (2) 73 (1961), 572–612.
[10] R. F. Arens and I. M. Singer, Function values as boundary integrals, Proc. Amer. Math. Soc., 5 (1954), 735–745.
[11] H. S. Bear and B. Walsh, Integral kernel for one-part function spaces, Pacific J. Math., 23 (1967), 209–215.
[12] E. Bishop, A generalization of the Stone-Weierstrass theorem, Pacific J. Math., 11 (1961), 777–783.

[13] E. Bishop, Representing measures for points in a uniform algebra, Bull. Amer. Math. Soc., 70 (1964), 121–122.
[14] A. Browder, Introduction to function algebras, Benjamin, 1969.
[15] L. Carleson, Interpolations by bounded analytic functions and the corona problem, Ann. of Math., (2) 76 (1962), 542–559.
[16] T. W. Gamelin, Uniform algebras, Prentice-Hall, 1969.
[17] A. M. Gleason, Function algebras, Seminars on analytic functions, II, Institute for Advanced Study, Princeton, 1957, p. 213–226.
[18] I. Glicksberg, Measures orthogonal to algebras and sets of antisymmetry, Trans. Amer. Math. Soc., 105 (1962), 415–435.
[19] I. Glicksberg, Dominant representing measures and rational approximation, Trans. Amer. Math. Soc., 130 (1968), 425–462.
[20] M. Hasumi, Interpolation sets for logmodular Banach algebras, Osaka J. Math., 3 (1966), 303–311.
[21] K. Hoffman, Analytic functions and logmodular Banach algebras, Acta Math., 108 (1962), 271–317.
[22] K. Hoffman, Banach spaces of analytic functions, Prentice-Hall, 1962.
[23] H. Ishikawa, J. Tomiyama, and J. Wada, On the essential set of function algebras, Proc. Japan Acad., 44 (1968), 1000–1002.
[24] G. M. Leibowitz, Lectures on complex function algebras, Scott, Foresman, 1970.
[25] G. Lumer, Algèbres de fonctions et espaces de Hardy, Lecture notes in math. 750, Springer, 1968.
[26] C. Rickart, Holomorphic convexity for general function algebras, Canad. J. Math., 20 (1968), 272–290.
[27] H. Rossi, The local maximum modulus principle, Ann. of Math., (2) 72 (1960), 1–11.
[28] W. Rudin, Boundary values of continuous analytic functions, Proc. Amer. Math. Soc., 7 (1956), 808–811.
[29] J. Wermer, Dirichlet algebras, Duke J. Math., 27 (1960), 373–381.
[30] Function algebras, Proc. of Tulane Univ. Symposium, Scott, Foresman, 1966.
[31] M. Hasumi, (Japanese; On shift-invariant subspaces), Sûgaku, 17 (1966), 214–224.
[32] J. Wada, Norumu kan (Japanese; normed rings), Kyôritu, 1969.
[33] N. Bourbaki, Eléments de mathématique, Théories spectrales, ch. 1, 2, Actualités Sci. Ind., 1332, Hermann, 1967.

39 (XII.3) Banach Spaces

A. General Remarks

The notion of **Banach space** was introduced in analysis in 1922 by S. Banach and N. Wiener independently in order to treat fundamental problems of analysis, such as mapping problems in infinite-dimensional function spaces, by utilizing topological and algebraic methods (→ 173 Function Spaces, 199 Hilbert Spaces, 251 Linear Operators).

B. Definition of Banach Spaces

We associate to each element x of a †linear space X over the real (complex) number field a real number $\|x\|$ satisfying the following conditions: (i) $\|x\| \geqslant 0$ for all x, and $\|x\|=0$ is equivalent to $x=0$; (ii) $\|\alpha x\|=|\alpha|\cdot\|x\|$ for any real (complex) number α; (iii) $\|x+y\| \leqslant \|x\|+\|y\|$. Then $\|x\|$ is called the **norm** of the vector x, and X is called a **normed linear space**. The norm is thus an extension of the notion of the length of a vector in a †Euclidean space. A normed linear space X is a †metric space supplied with the distance $\rho(x,y)=\|x-y\|$, and we write s-$\lim_{n\to\infty}x_n=x$ or simply $x_n\to x$ when $\lim_{n\to\infty}\|x_n-x\|=0$. If this metric space X is †complete, then X is called a **Banach space**.

Examples. †Function spaces C, L_p $(1 \leqslant p < \infty)$, †sequence spaces c, l, m, M, A_p, W_p^l, H_0^l, and BV are all Banach spaces (→ 173 Function Spaces).

In a normed linear space X, we have s-$\lim_{n\to\infty}(\alpha_n x_n+\beta_n y_n)=\alpha x+\beta y$ (where α_n, β_n, α, β are real (complex) numbers and x_n, y_n, $x, y\in X$) if $\alpha_n\to\alpha$, $\beta_n\to\beta$, $x_n\to x$, and $y_n\to y$. This limit equality holds even if we replace (ii) by the weaker (ii′) $\|-x\|=\|x\|$ and $\lim_{n\to\infty}\|\alpha_n x_n-\alpha x\|=0$ if $\alpha_n\to\alpha$ and $\lim_{n\to\infty}\|x_n-x\|=0$. The functional $\|x\|$ satisfying (i), (ii′), and (iii) is called the **quasinorm** of the vector x, and a **quasinormed linear space** X is called a **Fréchet space** if X is complete with respect to the distance $\rho(x,y)=\|x-y\|$. (In Bourbaki's terminology, Fréchet spaces are complete quasinormed linear spaces which are also †locally convex linear topological spaces [5].) The †function space $S(\Omega)$ of real- (complex-) valued †measurable functions $x(t)$ which are almost everywhere finite on a finite measure space Ω with the quasinorm $\|x\|=\int_\Omega|x(t)|(1+|x(t)|)^{-1}d\mu(t)$ is a typical example of a Fréchet space.

A sequence $\{x_n\}$ in a Fréchet space X is called a **basis** (or **base**) for X if to each $x\in X$ there corresponds a unique sequence $\{\alpha_\nu\}$ of real (complex) numbers such that $\lim_{n\to\infty}\|x-\sum_{\nu=1}^{n}\alpha_\nu x_\nu\|=0$. Most separable Fréchet spaces appearing in analysis have bases [10]. However, there are also Banach spaces with no bases [11].

C. Linear Operators and Linear Functionals

Suppose that a linear subspace $D(T)$ of a

linear space X is the (definition) **domain** of a mapping T with values in a linear space X_1 such that $T(\alpha x+\beta y)=\alpha Tx+\beta Ty$. Then T is called a **linear operator**, and in the special case where the **range** $R(T)=\{Tx\in X_1 \mid x\in D(T)\}$ of T is a subset of the real or complex number field, T is called a **linear functional.** If X and X_1 are both quasinormed linear spaces, then T is defined to be continuous if and only if $\text{s-lim}_{n\to\infty}Tx_n=Tx$ whenever $\text{s-lim}_{n\to\infty}x_n=x$. If X and X_1 are both normed linear spaces, then a linear operator T on $D(T)\subset X$ with values in X_1 is continuous if and only if $\sup_{x\in D(T),\|x\|\leqslant 1}\|Tx\|<\infty$. In particular, if $D(T)=X$, the linear operator T is continuous if and only if the set $\{Tx \mid \|x\|\leqslant 1\}$ is bounded. In this case, T is called a **bounded linear operator**, and $\|T\|=\sup_{\|x\|\leqslant 1}\|Tx\|$ is called the **norm** of the operator T. The **scalar multiple, sum,** and **product** of linear operators are defined by $(\alpha T)x=\alpha(Tx)$, $(T+S)x=Tx+Sx$, and $(ST)x=S(Tx)$, respectively. The **identity operator** I in X is defined by $I\cdot x=x$ for all $x\in X$. If the inverse mapping T^{-1} of $x\to Tx$ exists, then it is called the **inverse operator** of T.

D. The Dual Space and the Dual Operator

The totality of continuous linear functionals f defined on a normed linear space X is a Banach space X' under the previously defined linear operations and the norm $\|f\|=\sup_{\|x\|\leqslant 1}|f(x)|$. This X' is called the **dual** (or **conjugate**) **space** of X. In view of useful properties of the †inner product in †Hilbert spaces, it is sometimes convenient to write $\langle x,f\rangle$ for $f(x)$. Let X and Y be normed linear spaces, and let the domain $D(T)$ of a linear operator T with the range $R(T)\subset Y$ be dense in X. Suppose, in this case, that we have a pair (f,g) with $f\in Y'$ and $g\in X'$ satisfying the equation $\langle Tx,f\rangle=\langle x,g\rangle$ identically in $x\in D(T)$. Since in this case g is determined uniquely by f, we may write $g=T'f$. This T' is a linear operator and is called the **dual operator** (or **conjugate operator**) of T. This is an extension of the notion of the †transpose of a matrix in matrix theory. It has been proved that if T is a bounded linear operator then T' is also a bounded linear operator such that $\|T\|=\|T'\|$.

E. The Strong Topology and the Weak Topology

Let X be a normed linear space and X' its dual space. Take a finite number of elements $x_1',x_2',\ldots,x_n'$ from X', and consider the subset of $X:\{x\in X \mid \sup_{1\leqslant i\leqslant n}|\langle x,x_i'\rangle|\leqslant\varepsilon\}$, $\varepsilon>0$. If we take the totality of such subsets of X as a †fundamental system of neighborhoods of 0 of X, then X is a †locally convex linear topological space, denoted by X_w. This topology is called the **weak topology** of X. The convergence of a sequence $\{x_n\}\subset X$ to $x\in X$ with respect to the weak topology of X is denoted by $\text{w-lim}_{n\to\infty}x_n=x$. The original topology of X determined by the norm is then called the **strong topology** of X, and to stress the strong topology we may write X_s in place of the original X. Take a finite number of elements $x_1, x_2, \ldots, x_n$ from X, and consider the subset of $X':\{x'\in X' \mid \sup_{1\leqslant i\leqslant n}|\langle x_i, x'\rangle|\leqslant\varepsilon\}$, $\varepsilon>0$. If we take the totality of such subsets of X' as a fundamental system of neighborhoods of 0 of X', then X' is a †locally convex linear topological space. We shall write this space as X'_{w^*} and call the topology the **weak* topology** of X'. The topology of X' defined by the norm $\|f\|$ is called the **strong topology** of X', and to stress the strong topology we shall write X_s'.

F. The Hahn-Banach Extension Theorem

(1) Let M be a linear subspace of a normed linear space X. Then for any $f_1\in M'$, we can construct an $f\in X'$ such that $f(x)=f_1(x)$ for all $x\in M$ and $\|f\|=\|f_1\|$. (2) For any $x_0\neq 0$ of a normed linear space X, we can construct an $f_0\in X'$ such that $f_0(x_0)=\|x_0\|$ and $\|f_0\|=1$. (3) For any closed linear subspace M of a normed linear space X and a point $x_0\in X-M$, we can construct an $f_0\in X'$ such that $f_0(x_0)>1$, $\|f_0\|\leqslant d^{-1}$ $(d=\sup_{m\in M}\|x_0-m\|)$, and $f_0(x)=0$ for all $x\in M$. These propositions as a whole are the **Hahn-Banach extension theorem**. A more general proposition than (3) is the following, **Mazur's theorem**, which is very useful in applications: (4) Let a closed subset M of a normed linear space X be **convex** (i.e., $x, y\in M$ and $0<\alpha<1$ imply $\alpha x+(1-\alpha)y\in M$) and **balanced** (i.e., $x\in M$ and $|\alpha|\leqslant 1$ imply $\alpha x\in M$). Then for any $x_0\notin M$, we can construct an $f_0\in X'$ such that $f_0(x_0)>1$ and $\sup_{x\in M}|f_0(x)|\leqslant 1$. By (4) we can prove, e.g., that a convex set of a normed linear space is closed in the weak topology if it is closed in the strong topology. We can prove from (1), e.g., the existence of the **generalized limit** $\text{Lim}_{n\to\infty}\zeta_n$ for bounded real sequences $\{\zeta_n\}$ such that $\liminf_{n\to\infty}\zeta_n\leqslant \text{Lim}_{n\to\infty}\zeta_n\leqslant\limsup_{n\to\infty}\zeta_n$ and $\text{Lim}_{n\to\infty}(\alpha\xi_n+\beta\eta_n)=\alpha\,\text{Lim}_{n\to\infty}\xi_n+\beta\,\text{Lim}_{n\to\infty}\eta_n$.

G. Duality in Normed Linear Spaces

An element x_0 of a normed linear space X gives rise to an element x_0'' of $(X_s')'$ determined by $\langle x_0,x'\rangle=\langle x',x_0''\rangle$ for all $x'\in X_s'$. If we write $x_0''=Jx_0$, then J is a linear operator satisfying $\|Jx_0\| = \|x_0\|$ by (2), and so the space X is isomorphic and isometric to a linear subspace of $(X_s')_s'$. In particular, if we can identify X_s with $(X_s')_s'$, we call X a **reflexive** (or **regular**) **Banach space**. A necessary and sufficient condition that the normed linear space X is reflexive is that X is a Banach space such that any bounded sequence $\{x_n\}$ of X contains a subsequence convergent to a point of X in the weak topology of X (**Eberlein-Smul'jan theorem**). A normed linear space is **uniformly convex** if, for any $\varepsilon > 0$, there exists a $\delta > 0$ such that $\|x\| \leqslant 1$, $\|y\| \leqslant 1$, and $\|x-y\| \geqslant \varepsilon$ imply $\|x+y\| \leqslant 2-\delta$. The space L_p with $1 < p < \infty$ is uniformly convex, and any uniformly convex Banach space is reflexive (**Milman's theorem**).

H. The Resonance Theorem

Let $\{T_n\}$ be a sequence of bounded linear operators defined on a Banach space X into a normed linear space Y. The **uniform boundedness theorem (resonance theorem** or **Banach-Steinhaus theorem**) states that $\sup_{n \geqslant 1}\|T_n\| < \infty$ if $\sup_{n\geqslant 1}\|T_n x\| < \infty$ for every $x \in X$. As a corollary, we have $\sup_{n \geqslant 1}\|x_n\| < \infty$ for any weakly convergent sequence of X. Another corollary states that the set $\{x \in X \mid \limsup_{n\to\infty}\|T_n x\| < \infty\}$ either coincides with X or is a subset of X of the †first category. This implies the so-called **principle of condensation of singularities**, which gives a general existence theorem for functions exhibiting various kinds of singularities, for example, a continuous function whose †Fourier expansion diverges at every point of a †perfect set of points of the cardinal number of the continuum. The **Banach-Mazur convergence theorem** states the following: Given a sequence of bounded linear operators $\{T_n\}$ defined on a Banach space X into the Fréchet space $S(\Omega)$ such that $\limsup_{n\to\infty}|(T_n x)(t)| < \infty$ for almost all $t \in \Omega$ whenever x belongs to a fixed subset of X of the second category. Then the set of x for which a finite $\lim_{n\to\infty}(T_n x)(t)$ exists almost everywhere on Ω either coincides with X or is a subset of X of the first category. This theorem plays an important role in the theory of †orthogonal functions and in †ergodic theory.

I. The Closed Graph Theorem

A continuous linear operator defined on a Fréchet space X onto a Fréchet space Y maps open sets of X onto open sets of Y. This is called the **open mapping theorem**. As an application, we can prove the **closed graph theorem**: A linear operator T defined on a Fréchet space X into a Fréchet space Y is continuous if and only if T is a **closed operator**, that is, $\text{s-lim}_{n\to\infty} x_n = x$ and $\text{s-lim}_{n\to\infty} Tx_n = y$ imply $Tx = y$. This theorem plays an important role in a modern treatment of linear partial differential equations.

J. The Closed Range Theorem

Let X and Y be Banach spaces and T a linear closed operator with domain $D(T)$ dense in X and with range $R(T)$ in Y. Under these conditions, the following four propositions are mutually equivalent. (1) $R(T)$ is a closed set in Y. (2) $R(T')$ is a closed set in X'. (3) $R(T) = \{y \in Y \mid \langle y, y^* \rangle = 0$ for all $y^* \in Y'$ such that $T'y^* = 0\}$. (4) $R(T') = \{x^* \in X' \mid \langle x, x^* \rangle = 0$ for all $x \in X$ such that $Tx = 0\}$. These four propositions, as a whole, are called the **closed range theorem**. This theorem implies (5) $R(T) = Y$ if and only if T' has a continuous inverse; and (6) $R(T') = X'$ if and only if T has a continuous inverse.

The Hahn-Banach theorem, the resonance theorem, the open mapping theorem, the closed graph theorem, and the closed range theorem can be extended to various classes of †locally convex linear topological spaces. By virtue of this extension, we are able not only to treat various fundamental problems of analysis from a unified viewpoint but also to develop the theory of functional analysis itself in a new direction (→ 407 Topological Linear Spaces; concerning linear operators on a Banach space → 72 Compact Operators, 135 Eigenvalue Problems, 251 Linear Operators).

K. Interpolation of Banach Spaces

The Riesz-Thorin and Marcinkiewicz-Hunt interpolation theorems (→ 173 Function Spaces) have been generalized to the case of linear operators that map an arbitrary pair of Banach spaces into another.

Let (X_0, X_1) be a pair of Banach spaces that are continuously embedded in a Hausdorff topological linear space. Such a pair is called an **interpolation couple**. Then the spaces $X_0 \cap X_1$ and $X_0 + X_1$ are Banach spaces under the norms

$$\|x\|_{X_0 \cap X_1} = \max\{\|x\|_{X_0}, \|x\|_{X_1}\},$$

$$\|x\|_{X_0+X_1} = \inf\{\|x_0\|_{X_0} + \|x_1\|_{X_1} \mid x = x_0 + x_1\}.$$

A Banach space X is said to be an **intermediate space** for the couple if it satisfies

$$X_0 \cap X_1 \subset X \subset X_0 + X_1$$

with continuous embeddings. An **interpolation method** is a †functor which assigns to each interpolation couple (X_0, X_1) an intermediate space X called the **interpolation space**. There are two important types of interpolation methods, the complex method and the real method, which correspond to the Riesz-Thorin theorem and the Marcinkiewicz-Hunt theorem, respectively.

The **complex method** is due to A. P. Calderón [8], S. G. Krein, and J.-L. Lions.

Let $F(X_0,X_1)$ be the space of all functions $f(\zeta)$, $\zeta=\xi+i\eta$, with values in X_0+X_1 defined in the strip $0\leqslant\xi\leqslant 1$, holomorphic in $0<\xi<1$, continuous and bounded in $0\leqslant\xi\leqslant 1$, and such that $f(i\eta)$ is a continuous and bounded function with values in X_0 and $f(1+i\eta)$ is a continuous and bounded function with values in X_1. $F(X_0,X_1)$ is a Banach space under the norm

$$\|f\|_{F(X_0,X_1)}=\max\left\{\sup\|f(i\eta)\|_{X_0},\ \sup\|f(1+i\eta)\|_{X_1}\right\}.$$

The **complex interpolation space** $[X_0,X_1]_\theta$, $0\leqslant\theta\leqslant 1$, is defined to be the space of values $f(\theta)$ of $f\in F(X_0,X_1)$ with the norm

$$\|x\|_{[X_0,X_1]_\theta}=\inf\left\{\|f\|_{F(X_0,X_1)}\ \middle|\ x=f(\theta)\right\}.$$

$[X_0,X_1]_\theta$ is an intermediate space in which $X_0\cap X_1$ is dense and the following **interpolation theorem** holds: Let (X_0,X_1) and (Y_0,Y_1) be two interpolation couples, and let $T:X_0+X_1\to Y_0+Y_1$ be a linear operator such that

$$\|Tx\|_{Y_i}\leqslant M_i\|x\|_{X_i},\quad x\in X_i,\quad i=0,1.$$

Then T maps $[X_0,X_1]_\theta$ into $[Y_0,Y_1]_\theta$, and

$$\|Tx\|_{[Y_0,Y_1]_\theta}\leqslant M_0^{1-\theta}M_1^{\theta}\|x\|_{[X_0,X_1]_\theta},$$
$$x\in[X_0,X_1]_\theta.$$

Suppose that $X_0\cap X_1$ is dense in $X_{\theta_0}\cap X_{\theta_1}$, where

$$X_{\theta_0}=[X_0,X_1]_{\theta_0},\quad X_{\theta_1}=[X_0,X_1]_{\theta_1},$$
$$0\leqslant\theta_0\leqslant\theta_1\leqslant 1.$$

Then

$$\left[X_{\theta_0},X_{\theta_1}\right]_{\theta'}=[X_0,X_1]_\theta,$$

where $\theta=(1-\theta')\theta_0+\theta'\theta_1$ (**reiteration theorem**).

Suppose that $X_0\cap X_1$ is dense in both X_0 and X_1. Then the dual of $[X_0,X_1]_\theta$ is the space $[X_0',X_1']^\theta$ which is constructed similarly from the interpolation couple (X_0',X_1') in $(X_0\cap X_1)'$ and coincides with $[X_0',X_1']_\theta$ if one of X_0' and X_1' is reflexive (**duality theorem**).

If one of the spaces X_0 and X_1 is reflexive, then so is $[X_0,X_1]_\theta$ for $0<\theta<1$.

Let $1\leqslant p_0\leqslant p_1\leqslant\infty$ and $0<\theta<1$. Then we have

$$\left[L_{p_0}(\Omega),L_{p_1}(\Omega)\right]_\theta=L_p(\Omega),$$

where

$$\frac{1}{p}=\frac{1-\theta}{p_0}+\frac{\theta}{p_1}.$$

This proves that the †Riesz-Thorin theorem is a consequence of the interpolation theorem.

Real interpolation spaces were introduced by Lions as trace spaces and by E. Gagliardo by a different method, both in 1959. These works initiated the modern theory of interpolation of operators. Later, Lions and J. Peetre [9] introduced mean spaces and proved the equivalence of these different definitions.

When X is a Banach space and $1\leqslant p\leqslant\infty$, $L_p^*(X)$ denotes the space of †strongly measurable functions f on $(0,\infty)$ with values in X such that $\|f\|_{L_p^*(X)}<\infty$, where

$$\|f\|_{L_p^*(X)}=\begin{cases}\left(\int_0^\infty\|f(t)\|_X^p\,dt/t\right)^{1/p}, & 1\leqslant p<\infty,\\ \operatorname*{ess\,sup}_{0<t<\infty}\|f(t)\|_X, & p=\infty.\end{cases}$$

Let (X_0,X_1) be an interpolation couple of Banach spaces. The **mean spaces** $(X_0,X_1)_{\theta,p}$ (or $S(p,\theta,X_0;p,\theta-1,X_1)$ in the notation of [9]), $0<\theta<1$ and $1\leqslant p\leqslant\infty$, are defined to be the spaces of means

$$x=\int_0^\infty u(t)\,dt/t$$

in X_0+X_1 when $u(t)$ varies over the space of strongly measurable functions with values in $X_0\cap X_1$ such that

$$\|t^\theta u(t)\|_{L_p^*(X_0)}<\infty,$$
$$\|t^{\theta-1}u(t)\|_{L_p^*(X_1)}<\infty,$$

or equivalently (Peetre) such that

$$\|t^\theta u(t)\|_{L_{p_0}^*(X_0)}<\infty,$$
$$\|t^{\theta-1}u(t)\|_{L_{p_1}^*(X_1)}<\infty,$$

where

$$\frac{1}{p}=\frac{1-\theta}{p_0}+\frac{\theta}{p_1}. \tag{1}$$

Then $(X_0,X_1)_{\theta,p}$ is a Banach space under the norm

$$\|x\|_{(X_0,X_1)_{\theta,p}}{}^S=\inf\left\{\max\left\{\|t^\theta u(t)\|_{L_p^*(X_0)},\|t^{\theta-1}u(t)\|_{L_p^*(X_1)}\right\}\ \middle|\ x=\int u(t)\,dt/t\right\}.$$

An element $x\in X_0+X_1$ belongs to $(X_0,X_1)_{\theta,p}$ if and only if there are functions $v_0(t)$ and $v_1(t)$ on $(0,\infty)$ such that

$$x=v_0(t)+v_1(t)\ \text{a.e. on } t$$

and

$$\|t^\theta v_0(t)\|_{L_{p_0}^*(X_0)}<\infty,$$
$$\|t^{\theta-1}v_1(t)\|_{L_{p_1}^*(X_1)}<\infty,$$

where p_0, p_1, and p satisfy (1). The norm

$\|x\|_{(X_0,X_1)_{\theta,p}}{}^S$ is equivalent to the norm

$\|x\|_{(X_0,X_1)_{\theta,p}}{}^{\underline{S}}$

$$=\inf\Big\{\max\Big\{\|t^{\theta}v_0(t)\|_{L_p^*(X_0)},\ \|t^{\theta-1}v_1(t)\|_{L_p^*(X_1)}\Big\}\Big|\ x=v_0(t)+v_1(t)\text{ for almost all }t\Big\}.$$

When $x\in X_0+X_1$ and $0<t<\infty$, define

$$K(t,x)=\inf\{\|x_0\|_{X_0}+t^{-1}\|x_1\|_{X_1}\,|\,x=x_0+x_1\}$$

($=K(t^{-1},x)$ in the notation of Peetre). It is easily shown that $K(t,x)$ is a continuous function and that $\|x\|_{(X_0,X_1)_{\theta,p}}{}^{\underline{S}}$ is equivalent to

$$\|x\|_{(X_0,X_1)_{\theta,p}}{}^K=\|t^{\theta}K(t,x)\|_{L_p^*}.$$

Now $(X_0,X_1)_{\theta,p}$ is clearly an intermediate space.

Let (X_0,X_1) and (Y_0,Y_1) be two interpolation couples and let $T:X_0+X_1\to Y_0+Y_1$ be a linear operator such that

$$\|Tx\|_{Y_i}\leqslant M_i\|x\|_{X_i},\quad x\in X_i,\quad i=0,1.$$

Then T maps $(X_0,X_1)_{\theta,p}$ into $(Y_0,Y_1)_{\theta,p}$, and

$$\|Tx\|_{(Y_0,Y_1)_{\theta,p}}{}^*\leqslant M_0^{1-\theta}M_1^{\theta}\|x\|_{(X_0,X_1)_{\theta,p}}{}^*,$$

$$x\in(X_0,X_1)_{\theta,p},$$

where $*$ is any one of S, $\underline{S}$, and K (**interpolation theorem**).

If $1\leqslant p_0\leqslant p_1\leqslant\infty$, then $(X_0,X_1)_{\theta,p_0}\subset(X_0,X_1)_{\theta,p_1}$, and the embedding is continuous. If $X_0\supset X_1$ and $\theta_0<\theta_1$, then $(X_0,X_1)_{\theta_0,p}\supset(X_0,X_1)_{\theta_1,q}$ holds for any p and q. Also $X_0\cap X_1$ is dense in $(X_0,X_1)_{\theta,p}$ for $1\leqslant p<\infty$ (**density theorem**).

Suppose that X is an intermediate space. Then $(X_0,X_1)_{\theta,1}\subset X$ holds if and only if

$$\|x\|_X\leqslant C\|x\|_{X_0}^{1-\theta}\|x\|_{X_1},\qquad x\in X_0\cap X_1.$$

Similarly, $X\subset(X_0,X_1)_{\theta,\infty}$ holds if and only if

$$t^{\theta}K(t,x)\leqslant C\|x\|_X,\quad x\in X,$$

or equivalently, if for every $x\in X$ and $0<t<\infty$ there are $x_i\in X_i$ such that $x=x_0+x_1$ and

$$\|x_0\|_{X_0}\leqslant Ct^{-\theta}\|x\|_X,\quad \|x_1\|_{X_1}\leqslant Ct^{1-\theta}\|x\|_X.$$

We say that X is of **class** $K_\theta(X_0,X_1)$ if

$$(X_0,X_1)_{\theta,1}\subset X\subset(X_0,X_1)_{\theta,\infty}.$$

The complex interpolation space $[X_0,X_1]_\theta$ is an example of a Banach space of class $K_\theta(X_0,X_1)$.

Suppose that Y_0 is of class $K_{\theta_0}(X_0,X_1)$ and Y_1 of class $K_{\theta_1}(X_0,X_1)$, with $\theta_0<\theta_1$. Then

$$(Y_0,Y_1)_{\theta',p}=(X_0,X_1)_{\theta,p},$$

where

$$\theta=(1-\theta')\theta_0+\theta'\theta_1,$$

and the two spaces have equivalent norms (reiteration theorem).

Suppose that $X_0\cap X_1$ is dense in both X_0 and X_1 and that $1\leqslant p<\infty$. Then the dual of $(X_0,X_1)_{\theta,p}$ is identified with $(X_0',X_1')_{\theta,p'}$, where $p^{-1}+p'^{-1}=1$ (duality theorem) (Lions, Lions and Peetre).

If one of X_0 and X_1 is reflexive, then so is $(X_0,X_1)_{\theta,p}$ for any $0<\theta<1$ and $1\leqslant p<\infty$ (H. Morimoto).

Let $m>0$ be an integer, $1\leqslant p_i\leqslant\infty$, and α_0 and α_1 real numbers. Denote by $U^m(p_0,\alpha_0,X_0;p_1,\alpha_1,X_1)$ the space of functions $u(t)$ on $(0,\infty)$ with values in X_0+X_1 such that

$$\|t^{\alpha_0}u(t)\|_{L_{p_0}^*(X_0)}<\infty,$$

$$\|t^{\alpha_1+m}u^{(m)}(t)\|_{L_{p_1}^*(X_0)}<\infty,$$

where $u^{(m)}$ is the mth derivative of u in the sense of distributions.

Suppose that $j\geqslant 0$ is an integer and that

$$\alpha_0+j>0,\quad \alpha_1+j<0.$$

Then the **trace** $u^{(j)}(0)$ exists for any $u\in U^m(p_0,\alpha_0,X_0;p_1,\alpha_1,X_1)$. The **trace spaces** $T_j^m(p_0,\alpha_0,X_0;p_1,\alpha_1,X_1)$ ($=T_j^m(p_0,\alpha_0+p_0^{-1},X_0;p_1,\alpha_1+p_1^{-1}-m,X_1)$ in the notation of Lions-Peetre [8]) are defined to be the spaces of the traces $u^{(j)}(0)$ for $u\in U^m(p_0,\alpha_0,X_0;p_1,\alpha_1,X_1)$ with norm

$$\|x\|_{T_j^m(p_0,\alpha_0,X_0;p_1,\alpha_1,X_1)}=\inf\Big\{\max\Big\{\|t^{\alpha_0}u\|_{L_p^*(X_0)},\|t^{\alpha_1+m}u^{(m)}\|_{L_{p_1}^*(X_1)}\Big\}\Big|\ x=u^{(j)}(0)\Big\}.$$

The trace space $T_j^m(p_0,\alpha_0,X_0;p_1,\alpha_1,X_1)$ coincides with the mean space $(X_0,X_1)_{\theta,p}$, where

$$\theta=\frac{\alpha_0+j}{\alpha_0-\alpha_1},\qquad \frac{1}{p}=\frac{1-\theta}{p_0}+\frac{\theta}{p_1},$$

and the two spaces have equivalent norms (**trace theorem**; Lions and Peetre, P. Grisvard).

Let Ω be a †σ-finite measure space. For each measurable function f on Ω, the average function $f^{**}(t)$ (→ 173 Function Spaces) is equal to $K(t,f)$ for the couple $(L_\infty(\Omega),L_1(\Omega))$. Hence, the †Lorentz space $L_{(p,q)}(\Omega)$, $1<p<\infty$ and $1\leqslant q\leqslant\infty$, coincides with the real interpolation space $(L_\infty(\Omega),L_1(\Omega))_{p^{-1},q}$. In particular, the †Marcinkiewicz-Hunt theorem for linear operators follows from the real interpolation theorem (Peetre, Calderón).

Also important are the real interpolation spaces of a Banach space X and the domain $D(A^m)$ of powers of a linear operator A in X equipped with the †graph norm. Case A: A is a closed linear operator such that the resolvent $(\lambda+A)^{-1}$ exists for $0<\lambda<\infty$ and satisfies

$$\|(\lambda+A)^{-1}\|\leqslant M\lambda^{-1};$$

case B: $-A$ generates an †equicontinuous

semigroup T_t of class C^0; case C: $-A$ generates a holomorphic semigroup T_t bounded on a sector $|\arg t| < \omega$ (→ 373 Semigroups of Operators).

Lions and Peetre discussed case B; Grisvard and H. Komatsu, case A; and H. Berens and P. L. Butzer, and Komatsu, case C.

Let $0 < \sigma < m$ with m an integer. Then x belongs to $(X, D(A^m))_{\sigma/m,p}$ if and only if

$\lambda^\sigma\left(A(\lambda+A)^{-1}\right)^m x \in L_p^*(X)$, cases A, B, and C;

$t^{-\sigma}(1-T_t)^m x \in L_p^*(X)$, cases B and C;

$t^{-\sigma+m}A^m T_t x \in L_p^*(X)$, case C.

Also $(X, D(A^m))_{\sigma/m,p}$ does not depend on $m > \sigma$.

Suppose that $0 < k < \sigma$ with k an integer (or a rational number if $D(A)$ is dense). Then $x \in (X, D(A^m))_{\sigma/m,p}$ if and only if $x \in D(A^k)$ and $A^k x \in (X, D(A^m))_{(\sigma-k)/m,p}$.

Berens and Butzer considered also the elements x that satisfy the above conditions with $\sigma = m$.

When $A = -\Delta$ in $L_p(\mathbf{R}^n)$, the above propositions give various equivalent characterizations of the elements in the †fractional Sobolev spaces considered by S. M. Nikolskii, O. V. Besov, and M. H. Taibleson.

Combining this with the interpolation of L_p spaces, we obtain easily the fractional Sobolev embedding theorem (Grisvard, Peetre, A. Yoshikawa).

References

[1] S. Banach, Théorie des opérations linéaires, Warsaw, 1932 (Chelsea, 1963).
[2] E. Hille and R. S. Phillips, Functional analysis and semi-groups, Amer. Math. Soc. Colloq. Publ., 1957.
[3] N. Dunford and J. T. Schwartz, Linear operators I, Interscience, 1958.
[4] K. Yosida, Functional analysis, Springer, 1965.
[5] K. Yosida, Isôkaiseki (Japanese; Functional analysis), Iwanami, 1951.
[6] N. Bourbaki, Eléments de mathématique, Espaces vectoriels topologiques, Actualités Sci. Ind., 1189a, 1229b, 1230a, Hermann, 1966, 1967, 1955.
[7] A. E. Taylor, Introduction to functional analysis, John Wiley, 1958.
[8] A. P. Calderón, Intermediate spaces and interpolation, the complex method, Studia Math., 24 (1964), 113–190.
[9] J.-L. Lions and J. Peetre, Sur une classe d'espaces d'interpolation, Publ. Math. Inst. HES, no. 19, 1964, 5–68.
[10] I. Singer, Bases in Banach spaces I, Springer, 1970.
[11] P. Enflo, A counterexample to the approximation problem in Banach spaces, Acta Math., 130 (1973), 309–317.

40 (XX.13)
Bernoulli Family

The Bernoullis, Protestants who came originally from Holland and settled in Switzerland, were a significant family to the mathematics of the 17th century. In a single century, the family produced eight brilliant mathematicians, all of whom played important roles in the development of calculus.

The brothers Jakob (1654–1705) and Johann (1667–1748) and Daniel (1700–1782), Johann's second son, were especially outstanding. Jakob and Johann were close friends of G. W. †Leibniz, with whom they exchanged the correspondence through which it might be said that calculus developed. Jakob studied the problems of †tautochrone, †brachistochrone, geometry, dynamics, and others, including the †isoperimetric problem. He was the first to change the name *calculus summatoris* to *calculus integralis* (1690). His *Ars conjectandi* was published after his death in 1713; in it are found the rules that made his name prominent in the theory of †probability. He had little guidance, learning mathematics on his own. He was a professor of experimental physics at the University of Basel and later became a professor of mathematics. He taught mathematics to his brother Johann, who succeeded him as professor at the University. Johann's many achievements appeared in such publications of the time as *Acta eruditorum* and *Journal des savants*. In 1701, the beginnings of the †calculus of variations were seen in his solution to the isoperimetric problem. He initiated the term *functio*, the root of the present term *function*, in 1714.

Despite discord between the brothers and also between fathers and sons, the Bernoullis were ardent teachers and brilliant researchers, who instructed not only their sons but also such mathematicians as †Euler. Their achievements were numerous in consolidating the content and form of calculus and also in expanding its application. Daniel was especially outstanding in the theory of probability; he also made contributions in the field of †hydrodynamics and to the †kinetic theory of gases. The eldest Johann's eldest son Nikolaus (1695–1726) achieved distinction as a professor of mathematics in St. Petersburg. Daniel's youngest brother, Johann (1710–1790), succeeded his father, Johann, Sr., as professor at the University of Basel. The son of Johann, Jr., also named Johann (1744–1807), was chairman of mathematics at the Academy of Berlin. A son of the third Johann, another Jakob (1759–1789), was a professor of experimental physics at the University of Basel. Nikolaus (1687–1759), a grandson of the founder Nikolaus (1623–1708) and son of Nikolaus the painter (1662–1716), held Galileo's old chair of mathematics at Padua from 1716 to 1719.

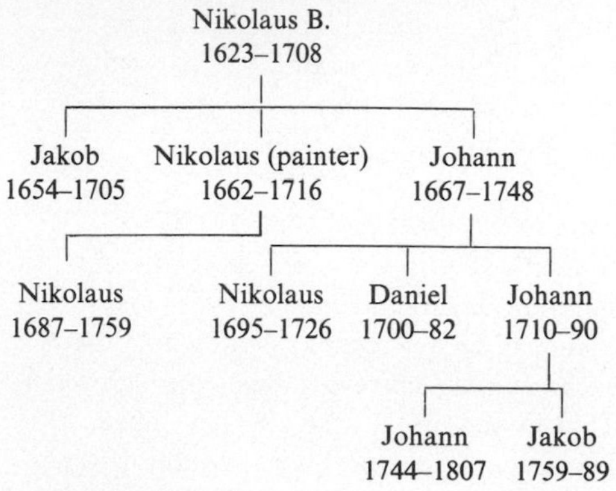

References

[1] M. B. Cantor, Vorlesungen über Geschichte der Mathematik III, Teubner, 1898.

[2] Jacobi Bernoulli, Opera I, II, Cramer, 1744.

[3] Jakob Bernoulli, Die Werke von Jakob Bernoulli, Birkhäuser, 1969.

[4] Jacobi Bernoulli, Ars conjectandi, Bâle, 1713.

[5] Johannis Bernoulli, Opera omnia I–IV, M. M. Bousquet, 1742.

41 (XIV.8) Bessel Functions

A. General Remarks

Bessel functions were first studied in order to solve †Kepler's equation concerning planetary motions and were systematically investigated by F. W. Bessel in 1824. Since then they have appeared in various problems and have become important.

B. Bessel Functions (→ Appendix A, Table 19.III)

Separating variables for the Helmholtz equation $\Delta\Psi + k^2\Psi = 0$ in terms of cylindrical coordinates, we obtain **Bessel's differential equation**

$$\frac{d^2w}{dz^2}+\frac{1}{z}\frac{dw}{dz}+\left(1-\frac{\nu^2}{z^2}\right)w=0 \qquad (1)$$

for the component of the radius vector. The following two linearly independent solutions of (1):

$$H_\nu^{(1)}(z)=\frac{1}{\pi}\int_{L_1} e^{-iz\sin\zeta+i\nu\zeta}d\zeta,$$
$$H_\nu^{(2)}(z)=\frac{1}{\pi}\int_{L_2} e^{-iz\sin\zeta+i\nu\zeta}d\zeta \qquad (2)$$

are called the **Hankel functions of the first and second kind**, respectively, where the contour L_1 of the first integration is a curve from $(-\pi+0)+i\infty$ to $-0-i\infty$, and L_2 is a curve from $+0-i\infty$ to $(\pi-0)+i\infty$. If both z and ν are real, we have

$$\overline{H_\nu^{(1)}(z)}=H_{\bar\nu}^{(2)}(\bar z), \quad \overline{H_\nu^{(2)}(z)}=H_{\bar\nu}^{(1)}(\bar z), \qquad (3)$$

where $\bar z$ is the complex conjugate of z. Hence

$$J_\nu(z)=(H_\nu^{(1)}(z)+H_\nu^{(2)}(z))/2,$$
$$N_\nu(z)\equiv Y_\nu(z)=(H_\nu^{(1)}(z)-H_\nu^{(2)}(z))/2i \qquad (4)$$

are real functions. If both z and ν are complex, the functions $J_\nu(z)$ and $N_\nu(z)$ defined in (4) are also called **Bessel functions** and **Neumann functions**, respectively. The other names for $J_\nu(z)$, $N_\nu(z)$, and $H_\nu(z)$ are **Bessel functions of the first, second, and third kind**, respectively. Each of them satisfies the following recurrence formulas:

$$2\frac{dC_\nu(z)}{dz}=C_{\nu-1}(z)-C_{\nu+1}(z);$$
$$(2\nu/z)C_\nu(z)=C_{\nu-1}(z)+C_{\nu+1}(z). \qquad (5)$$

In general, functions satisfying the simultaneous †differential-difference equations (5) are called **cylindrical functions**. Every cylindrical function $C_\nu(z)$ is represented in the form $C_\nu(z)=a_1(\nu)H_\nu^{(1)}(z)+a_2(\nu)H_\nu^{(2)}(z)$, where $a_1(\nu)$ and $a_2(\nu)$ are arbitrary periodic functions of period 1 with respect to ν.

If $\nu=n$ (an integer) we have

$$J_{-n}(z)=(-1)^nJ_n(z),$$
$$N_{-n}(z)=(-1)^nN_n(z), \qquad (6)$$

which show the linear dependency of J_{-n} and J_n, and N_{-n} and N_n, respectively. If $\nu\neq n$ (an integer), as the fundamental solutions of (1) we may take a pair J_ν and $J_{-\nu}$, or N_ν and $N_{-\nu}$. In (2) if we take as a contour of integration a curve from $(-\pi+0)+i\infty$ to $(\pi-0)+i\infty$, we obtain an integral representation for $J_\nu(z)$, which yields the relations

$$J_\nu(ze^{im\pi})=e^{im\nu\pi}J_\nu(z),$$
$$J_{-\nu}(ze^{im\pi})=e^{-im\nu\pi}J_{-\nu}(z). \qquad (7)$$

If $\nu = n$ (an integer) and $\mathrm{Re}\, z > 0$, we obtain

$$J_n(z) = \frac{1}{2\pi}\int_{-\pi}^{\pi} e^{iz\sin\zeta + in\zeta}\, d\zeta \tag{8}$$

$$= \frac{1}{\pi}\int_0^{\pi} \cos(z\sin\zeta - n\zeta)\, d\zeta, \tag{9}$$

which is called **Bessel's integral**. These representations imply the following expansions by means of †generating functions:

$$e^{iz\sin\zeta} = \sum_{n=-\infty}^{\infty} J_n(z) e^{in\zeta}; \tag{10}$$

$$\cos(z\sin\zeta) = J_0(z) + 2\sum_{n=1}^{\infty} J_{2n}(z)\cos 2n\zeta,$$

$$\sin(z\sin\zeta) = 2\sum_{n=0}^{\infty} J_{2n+1}(z)\sin(2n+1)\zeta. \tag{11}$$

Making a change of variable $u = \exp(-i\zeta)$ in (2), we obtain

$$J_\nu(z) = \frac{1}{2\pi i}\int_L \exp\left(\frac{z}{2}\left(u - \frac{1}{u}\right)\right) u^{-\nu-1}\, du, \tag{12}$$

where L is a contour starting at the point at infinity with the argument $-\pi$, encircling the origin in the positive direction, and tending to the point at infinity with the argument π. From (12) we obtain a power series expansion

$$J_\nu(z) = \left(\frac{z}{2}\right)^{\nu} \sum_{m=0}^{\infty} \frac{(-1)^m}{m!\Gamma(\nu+m+1)}\left(\frac{z}{2}\right)^{2m}, \tag{13}$$

obtained also from (1) by a power series expansion at $z = 0$, which is a †regular singular point of (1). Substituting (13) into

$$N_\nu(z) = (\cos\nu z J_\nu(z) - J_{-\nu}(z))/\sin\nu\pi, \tag{14}$$

we obtain a power series expansion for $N_\nu(z)$. A power series expansion for $N_n(z)$ for an integer n is obtained by taking the limit $\nu \to n$ (→ Appendix A, Table 19). In particular, if $\nu = n + 1/2$ (n is an integer), we have

$$J_{n+1/2}(z) = (-1)^n \frac{(2z)^{n+1/2}}{\sqrt{\pi}} \frac{d^n}{d(z^2)^n}\left(\frac{\sin z}{z}\right),$$

$n = 0, 1, 2, \ldots,$

which is represented by elementary functions and is sometimes called simply the **half Bessel function**. Bessel functions for half-integers have appeared also as radius vector components when the variables in the Helmholtz equation are separated by spherical coordinates. The function

$$j_n(z) = \sqrt{\pi/2z}\, J_{n+1/2}(z)$$

is called the **spherical Bessel function**.

From the differential equations satisfied by $J_\nu(\alpha z)$, we have

$$(\alpha^2 - \beta^2)\int_0^1 z J_\nu(\alpha z) J_\nu(\beta z)\, dz$$

$$= \beta J_\nu(\alpha) J_\nu'(\beta) - \alpha J_\nu'(\alpha) J_\nu(\beta). \tag{15}$$

By letting $\beta \to \alpha$ in (15), we have

$$\int_0^1 z (J_\nu(\alpha z))^2\, dz$$

$$= \frac{1}{2}\left(\left(1 - \frac{\nu^2}{\alpha^2}\right)(J_\nu(\alpha))^2 + (J_\nu'(\alpha))^2\right). \tag{16}$$

If α and β are distinct roots of $J_\nu(z) = 0$, we have from (15)

$$\int_0^1 z J_\nu(\alpha z) J_\nu(\beta z)\, dz = 0, \quad \mathrm{Re}\,\nu > -1. \tag{17}$$

The integral formulas (15), (16), and (17) are called **Lommel's integrals**.

As for the zero points of $J_\nu(z)$, the following facts are well known: $J_\nu(0) = 0$ if $\nu > 0$. $J_\nu(z)$ has no multiple zero points other than $z = 0$. $J_\nu(-\alpha) = 0$ if $J_\nu(\alpha) = 0$. Every zero point of $J_\nu(z)$ is real if $\nu > -1$. Between two adjacent zero points that are positive, there exists one and only one zero point of $J_{\nu-1}(z)$ and $J_{\nu+1}(z)$, respectively. $J_\nu(z)$ has a countably infinite set of zero points on the real axis.

We have the following **addition theorem**:

$$H_n^{(\mu)}(k\rho) e^{in\psi} = \sum_{m=-\infty}^{\infty} J_n(kr_2) H_{n+m}^{(\mu)}(kr_1) e^{im\varphi},$$

$$\mu = 1, 2,$$

where

$$\rho = \sqrt{r_1^2 + r_2^2 - 2r_1 r_2\cos\varphi},$$

$$\rho\cos\psi = r_1 - r_2\cos\varphi, \quad \rho\sin\psi = r_2\sin\varphi.$$

C. Expansion by Bessel Functions

Let $f(r, \varphi)$ be defined for $0 < r < 1$, $-\pi < \varphi < \pi$, and $\alpha_{n,1}, \alpha_{n,2}, \ldots, \alpha_{n,s}, \ldots$ $(0 < \alpha_{n,s} < \alpha_{n,s+1}$, $s = 1, 2, \ldots$ for every $n)$ be zero points of $J_n(x)$ $(n = 0, 1, 2, \ldots)$. Then we have an expansion

$$f(r, \varphi)$$

$$= \sum_{n=0}^{\infty}\sum_{s=1}^{\infty} (a_{n,s}\cos n\varphi + b_{n,s}\sin n\varphi) J_n(\alpha_{n,s} r), \tag{18}$$

which is called the **Fourier-Bessel series**. The coefficients $a_{n,s}$ and $b_{n,s}$ are determined by the properties of †Fourier series and (16) and (17) as follows:

$$\left.\begin{matrix} a_{n,s} \\ b_{n,s} \end{matrix}\right\} = \frac{\varepsilon_n}{\pi (J_{n+1}(\alpha_{n,s}))^2}$$

$$\times \int_0^1 \int_{-\pi}^{\pi} f(r, \varphi)\, J_n(\alpha_{n,s} r) \frac{\cos n\varphi}{\sin n\varphi}\, r\, d\varphi\, dr;$$

$\varepsilon_0 = 1, \quad \varepsilon_1 = \varepsilon_2 = \ldots = 2.$

The integral transformation

$$g(y) = \int_0^{\infty} x f(x) J_n(xy)\, dx \tag{19}$$

is called the **Fourier-Bessel transform**. If $f(x)$ is sufficiently smooth and tends rapidly to zero as $x \to \infty$, the following inversion formula holds:

$$f(x)=\int_0^{\infty} yg(y)J_n(xy)dy. \qquad (20)$$

There are other types of series expansions in terms of Bessel functions as follows: **Dini's series**

$$\sum_{m=1}^{\infty} a_m J_\nu(\lambda_m x)$$

(λ_m is the mth positive root of $xJ_\nu'(x)+HJ_\nu(x)=0$, where H is a real constant); **Kapteyn's series**

$$\sum_{m=1}^{\infty} a_m J_{\nu+m}((\nu+m)x);$$

Schlömilch's series

$$\frac{a_0}{2}+\sum_{m=0}^{\infty} a_m J_0(mx);$$

and the **generalized Schlömilch series**

$$\frac{1}{2}\frac{a_0}{\Gamma(\nu+1)}+\sum_{m=1}^{\infty}\frac{a_m J_\nu(mx)+b_m \mathbf{H}_\nu(mx)}{(mx/2)^\nu},$$

where $\mathbf{H}_\nu(mx)$ is a †Struve function.

D. Asymptotic Expansion

If $|z|$ or $|\nu|$ is sufficiently large, the asymptotic representation for Bessel functions is obtained by applying the †method of the steepest descent for (2). If $|z|>|\nu|$, we have

$$H_\nu^{(1)}(z)\sim\sqrt{\frac{2}{\pi z}}\exp i\left(z-\frac{\pi}{2}\nu-\frac{\pi}{4}\right),$$
$$-\pi<\arg z<2\pi,$$

$$H_\nu^{(2)}(z)\sim\sqrt{\frac{2}{\pi z}}\exp\left(-i\left(z-\frac{\pi}{2}\nu-\frac{\pi}{4}\right)\right),$$
$$-2\pi<\arg z<\pi,$$

$$J_\nu(z)\sim\sqrt{\frac{2}{\pi z}}\cos\left(z-\frac{\pi}{2}\nu-\frac{\pi}{4}\right),$$
$$-\pi<\arg z<\pi,$$

$$N_\nu(z)\sim\sqrt{\frac{2}{\pi z}}\sin\left(z-\frac{\pi}{2}\nu-\frac{\pi}{4}\right).$$

Hence $H_\nu^{(1)}(z)$ tends to zero as $|z|\to\infty$ in the upper half-plane, and becomes large exponentially as $|z|\to\infty$ in the lower half-plane. The results for $H_\nu^{(2)}$ are obtained by interchanging "upper half-plane" with "lower half-plane" in this statement.

If both $|z|$ and $|\nu|$ are sufficiently large, we have the **Debye asymptotic representations**. For example, if $z=\nu\sec\beta$ $(\nu>0,\ \beta>0)$, we have

$$H_\nu^{(1,2)}(\nu\sec\beta)\sim(\pi\nu\tan\beta/2)^{-1/2}$$
$$\times\exp(\pm i(\nu(\tan\beta-\beta)-\pi/4)).$$

If $z=\nu\,\mathrm{sech}\,\alpha$ $(\nu>0,\ \alpha>0)$, we have

$$J_\nu(\nu\,\mathrm{sech}\,\alpha)$$
$$\sim(2\pi\nu\tanh\alpha)^{-1/2}\exp\nu(\tanh\alpha-\alpha),$$
$$N_\nu(\nu\,\mathrm{sech}\,\alpha)$$
$$\sim-(\pi\nu\tanh\alpha/2)^{-1/2}\exp\nu(\alpha-\tanh\alpha).$$

If $|\nu|\sim|z|$, we have

$$H_\nu^{(1,2)}(\nu\sec\beta)\sim\frac{\tan\beta}{\sqrt{3}}$$
$$\times\exp\left(\pm i\left(\frac{\pi}{6}+\nu\left(\tan\beta-\frac{1}{3}\tan^3\beta-\beta\right)\right)\right)$$
$$\times H_{1/3}^{(1,2)}((\nu/3)\tan^3\beta)+0(\nu^{-1}),$$

which is called **Watson's formula.**

E. The Wagner Function

As an application of Bessel functions to the theory of nonstationary aircraft wings, T. Theodorsen introduced the function

$$C(z)=H_1^{(2)}(z)/(H_0^{(2)}(z)+H_1^{(2)}(z))$$

[6], and H. Wagner considered the function

$$k_1(s)=\frac{1}{2\pi i}\int_{\mathrm{Br}} e^{ws}\frac{2C(-iw)}{w}dw$$

[5], where $H_0^{(2)}(z)$, $H_1^{(2)}(z)$ are Hankel functions, and $\int_{\mathrm{Br}}$ means a Bromwich integral giving the inverse Laplace transform. Then $C(z)$ and $k_1(s)$ are called the **Theodorsen function** and **Wagner function**, respectively. The function $k_1(s)$ is equal to the lift coefficient when a 2-dimensional flat wing suddenly moves and proceeds straight forward a distance s holding the angle of incidence at $1/\pi$.

F. Related Miscellaneous Functions

The †modified Bessel functions and †Kelvin functions are obtained from the usual Bessel functions, replacing z by iz and $e^{\pm 3\pi i/4}z$, respectively. The †Struve functions and †Weber functions are related to Bessel functions. †Airy's integral was later shown to be represented by Bessel functions (→ Appendix A, Table 19.IV).

References

[1] G. N. Watson, A treatise on the theory of Bessel functions, Cambridge Univ. Press, 1922.
[2] A. Gray and G. B. Mathews, A treatise on Bessel functions and their application to physics, Macmillan, second edition, 1922.
[3] R. Weyrich, Die Zylinderfunktionen und ihre Anwendungen, Teubner, 1937.
[4] F. Bowman, Introduction to Bessel functions, Longmans-Green, 1938.
[5] H. Wagner, Über die Entstehung des dynamische Auftriebes von Tragflügeln, Z. Angew. Math. Mech., 5 (1925), 17–35.
[6] T. Theodorsen, General theory of aerodynamic instability and the mechanism of

flutter, NASA Tech. Rep. 496 (1935).
Also → references to 172 Functions of Confluent Type, 381 Special Functions.

42 (XVII.14) Biometrics

A. General Remarks

Biometrics deals with mathematical problems in fields of biological sciences such as genetics, epidemiology, and demography.

B. Genetics

Mendel's model in genetics may be understood as follows. At any given locus of any given chromosome there are two genes, one of which is inherited at the time of reproduction from the father, and the other from the mother. Each gene is inherited with probability 1/2 from the pair of genes of each parent, respectively. These genes inherited from ancestors through offspring do not change their properties, except in the case of mutation, which occurs with very small probability. Inherited characteristics, in contrast to those formed by the effects of the environment since birth, are determined by one or more pairs of genes.

Blood type, for example, is determined by a pair of genes. There are three types of genes, and the A, B, and O blood types (phenotypes) are determined by the six possible types (genotypes) of pairs of genes in the following manner: type A: (AA) or (AO); type B: (BB) or (BO); type AB: (AB); and type O: (OO). In this example, the genes A and B are said to be equally dominant over the gene O, because the characteristic O appears only in the pair (OO) whereas it is dominated by A and B in the cases (AO) and (BO).

Let (z_1,z_2), (x_1,x_2), and (y_1,y_2) be the genes at a given locus of an individual, his father, and his mother, respectively. One of the two genes of this individual comes from his father with equal probability and the other from his mother in the same way, and thus $\Pr(z_1=x_i, z_2=y_j)=1/4$ $(i,j=1,2)$. This is the probability-theoretic basis of the **Mendelian principle.** Suppose that there is a consanguineous relationship between the parents, namely, that they have one or more common ancestors, and let (ξ_1,ξ_2) be the genes of a common ancestor. Then $\Pr(z_1=z_2=\xi_1 \text{ or } z_1=z_2=\xi_2)\neq 0$. The **inbreeding coefficient** f is defined as the sum of these probabilities for all the common ancestors or, in other words, the probability that an individual will possess at a given genetic locus two genes identical in their origin. It can be computed by the following recurrence relation: $f=\Sigma(1/2)^{n_1+n_2+1}(1+f_A)$. The summation runs over every loop consisting of two stepwise sequences starting from the father and from the mother and going to the common ancestor who is the only common member in the two sequences. The numbers n_1 and n_2 are the number of steps in the loop ascending to the common ancestor from the father and the mother respectively, and f_A is the inbreeding coefficient of the common ancestor. Different formulas are needed if self-fertilization is involved and for the inbreeding coefficient of sex-linked genes.

A random mating population is a population where the probability of mating is independent of the genotypes of individuals. Consider a population with n types $A_1, \ldots, A_n$ of genes at a specific locus of a chromosome, and let $x_1, \ldots, x_n$ be the gene frequencies in this population. If matings are made at random and there is neither selection nor mutation, the genotype frequencies of A_1A_1, A_1A_2, A_2A_2, etc., are equal to the coefficients x_1^2, $2x_1x_2$, x_2^2, etc., of corresponding terms in the expansion of $(x_1A_1+x_2A_2+\ldots+x_nA_n)^2$. An interesting fact is that when the population has the same number of males and females, the difference in the genotype frequencies in males and females, if any, will disappear in the second generation, and the common frequency will be carried over to subsequent generations as the equilibrium state. This is called the **Hardy-Weinberg law.** One of the factors that cause deviation from this law is consanguinity. To see this, suppose that we are given a population with genes A and a with frequencies p and $1-p$. The gene frequencies of those with the inbreeding coefficient f are given by $AA:(1-f)p^2+fp$; $Aa:2(1-f)p(1-p)$; $aa:(1-f)(1-p)^2+f(1-p)$. This may be proved by considering an infinite population having A and a in proportions p and $(1-p)$, and considering a sampling scheme of drawing two, one each at random with probabilities $(1-f)$ and f, respectively.

Population genetics treats the effects of mutation, linkage, natural selection, consanguinity, immigration, isolation, etc., on genetic properties of a population. The theory of population genetics has been developed by R. A. Fisher, S. Wright, J. B. S. Haldane, M. Kimura, and others.

C. Epidemics

Any infectious disease has some incubation period after infection before symptoms appear. Let us here ignore this incubation period, and assume that an individual starts infecting others as soon as he himself is infected and that the probability of infecting an uninfected individual is proportional to the number of infected individuals in contact with him and also to the length of time of contact. Consider a population with $n+1$ individuals, among whom one was infected initially, and let x be the number of uninfected individuals at time t. Then we have $dx/dt = -\beta x(n+1-x)$, and hence $x = n(n+1)/(n+\exp(n+1)\tau)$, where $\tau = \beta t$. Thus we have $-dx/d\tau = n(n+1)^2\exp(n+1)\tau/(n+\exp(n+1)\tau)^2$, which is called an **epidemic curve**. This shows one of the deterministic approaches. It can be also treated as a problem of †Markov chains, and by solving a differential equation involving a probability-generating function or †moment-generating function, it is possible to obtain the †probability distribution, †mean, and †variance of the number of infected individuals as functions of time. These are topics in the theory of **epidemics**. Because of death, recovery, or quarantine, there may exist those who will neither infect others nor be infected. **General epidemics** treats these variables, i.e., the number of uninfected individuals and the number of those who have been infected and cease to infect others.

D. Population Mathematics

Let P be the population size and t the time. Then the ratio of population increase is given by $(1/P)(dP/dt) = R$. If we assume $R = r$(const), we have the law of geometric progression $P = A\exp(rt)$, while another assumption $R = r(1-P/L)$ with constants r and L leads us to the **logistic curve** $P = L/(1+\exp(-r(t-\beta)))$. **Population mathematics** is concerned with probabilistic treatment of age distribution, birth rate for each age group, distribution of marriage age, change and stability of age distribution with respect to time, etc. A stable population is one whose age distribution does not change with time.

Bioassay includes statistical methods with special reference to biological research.

References

[1] M. Kimura, Syûdan idengaku gairon (Japanese; Introduction to population genetics), Baihûkan, 1960.
[2] N. T. J. Bailey, The mathematical theory of epidemics, Griffin, 1957.
[3] M. S. Bartlett, Stochastic population models in ecology and epidemiology, Methuen, 1960.
[4] D. J. Finney, Statistical method in biological assay, Griffin, 1952.
[5] C. R. Rao, Advanced statistical methods in biometric research, John Wiley, 1952.

43 (II.16)
Boolean Algebra

A. Boolean Algebras

Boolean algebra was introduced by G. Boole to study logical operations (→ 400 Symbolic Logic). It is now included within the more general concept of †lattice or lattice-ordered set (→ 241 Lattices) and appears not only in logic but also very often in analysis in the form of a particular lattice of sets, e.g., the lattice of †measurable sets.

Let L be a given set and suppose that to any pair of its elements x, y there correspond two elements $x\cup y$, $x\cap y$ of L (called **join** and **meet** of x and y, respectively) such that the following laws are valid: (1) $x\cup y = y\cup x$, $x\cap y = y\cap x$ (commutative law); (2) $x\cup(y\cup x) = (x\cup y)\cup z$, $x\cap(y\cap z) = (x\cap y)\cap z$ (associative law); (3) $x\cup(y\cap x) = (x\cup y)\cap x = x$ (absorption law); (4) $x\cup(y\cap z) = (x\cup y)\cap(x\cup z)$, $x\cap(y\cup z) = (x\cap y)\cup(x\cap z)$ (distributive law); $(x,y,z\in L)$. From (1), (2), and (3) it follows further that $x\cup x = x\cap x = x$ (idempotent law). If $x\leqslant y$ is defined to mean $x\cup y = y$, L becomes an †ordered set with respect to the ordering $\leqslant$. Now suppose, moreover, that the following law holds: (5) there exist a least element 0 and a greatest element I, and for any element x there exists an element x' satisfying $x\cup x' = I$, $x\cap x' = 0$ (**law of complementation**). Then L is called a **Boolean algebra** (or †**Boolean lattice**). In this case x' is uniquely determined by x and is called the **complement** of x. The binary operations $(x,y)\to x\cup y$, $x\cap y$ together with the operation $x\to x'$ are called **Boolean operations**. These operations obey **de Morgan's law** $(x\cup y)' = x'\cap y'$, $(x\cap y)' = x'\cup y'$.

B. Generalized Boolean Algebras

Suppose that $a\leqslant b$ holds for two given elements a, b of an ordered set. Then the set of all elements x satisfying $a\leqslant x\leqslant b$ is denoted by $[a,b]$ and is called an **interval**. An interval of a Boolean algebra is also a Boolean algebra with respect to the induced operations $\cup$ and $\cap$, where the least and greatest elements are a and b, respectively, and the complement of x in $[a,b]$ is equal to $a\cup(x'\cap b) = (a\cup x')\cap b$. More generally, if a set L with two operations $\cup$, $\cap$ satisfying (1)–(4) above has a least element 0 and if each interval of L satisfies (5) (i.e., is a Boolean algebra), then L is called a **generalized Boolean algebra**.

C. Boolean Rings

A ring L satisfying the condition $xx=x$ for all $x \in L$ (i.e., all of its elements are †idempotent) is called a **generalized Boolean ring**, and if it has a unity element then it is called a **Boolean ring**. A generalized Boolean ring L satisfies $x+x=0$ for all $x \in L$ and is necessarily a commutative ring. A (generalized) Boolean algebra L becomes a (generalized) Boolean ring if for any elements x, y of L the sum $x+y$ is defined to be the complement of $x \cap y$ in the interval $[0, x \cup y]$, and the product xy is defined to be $x \cap y$. A nonempty subset J of a (generalized) Boolean algebra L is an ideal with respect to the corresponding structure of the ring if and only if $x \cup y \in J$ for $x, y \in J$ and $x \cap y \in J$ for $x \in J, y \in L$. More generally, in any lattice, a nonempty subset that satisfies these conditions is sometimes called an **ideal** of the lattice.

D. Representation of a Boolean Algebra

Any Boolean algebra L is isomorphic to a Boolean lattice of subsets in a set X. If L is of finite †height, then L is isomorphic to the Boolean lattice $\mathfrak{P}(X)$ of all subsets of X. In general X can be taken to be the set of all maximal ideals of L. Let $a \in L$ and let $O(a)=\{\mathfrak{m} \mid \mathfrak{m} \in X, a \notin \mathfrak{m}\}$. The isomorphism is obtained by the mapping $a \rightarrow O(a)$. If we define a topology in X such that $\{O(a) \mid a \in L\}$ is the †open base, then X is a compact, totally disconnected T_1 space and $O(a)$ is characterized as a compact open set in X. Such a space X is called a **Boolean space** (M. H. Stone [3,4]).

In any complete Boolean algebra L, the complete distributive laws hold: $(\sup_I x_i) \cap y = \sup_I (x_i \cap y)$ and its dual. These are equivalent to the stronger relations: $(\sup_I x_i) \cap (\sup_J y_j) = \sup_{IJ}(x_i \cap y_j)$ and its dual. In order that a Boolean algebra L be isomorphic to the Boolean algebra $\mathfrak{P}(X)$ of all subsets of X it is necessary and sufficient that the following strongest complete distributive laws hold: $\inf_I (\sup_{J(i)} x_{ij}) = \sup_F (\inf_I x_{i,\varphi(i)})$ (where F is the set of all functions φ assigning to each $i \in I$ a value $\varphi(i) \in J(i)$) and its dual.

References

[1] P. R. Halmos, Lectures on Boolean algebras, Van Nostrand, 1963.
[2] R. Sikorski, Boolean algebras, Erg. Math., Springer, second edition, 1964.
[3] M. H. Stone, The theory of representations for Boolean algebras, Trans. Amer. Math. Soc., 40 (1936), 37–111.
[4] M. H. Stone, Applications of the theory of Boolean rings to general topology, Trans. Amer. Math. Soc., 41 (1937), 375–481.
[5] G. Boole, Collected logical works I, II, Open Court, 1916.

44 (XI.5) Bounded Functions

A. General Remarks

A complex-valued function defined on a subset E of the complex z-plane is called a **bounded function** defined on E if its range $f(E)$ is bounded, that is, if there exists a positive constant M such that $|f(z)| \leqslant M$ on E. However, when studying the theory of bounded functions, we usually restrict ourselves to the consideration of †analytic or †harmonic functions. On the other hand, the classes of functions $f(z)$ satisfying conditions such as $\operatorname{Re} f(z) > 0$ or $\alpha < \arg f(z) < \beta$ rather than the $|f(z)| \leqslant M$ are studied by a method similar to that applied to the study of bounded functions.

†Schwarz's lemma, †Liouville's theorem, and †Riemann's theorem on the removability of singularities (which will be explained later) are among the classical theorems in the theory of bounded functions.

B. Maximum Principle

When a function $f(z)$ is holomorphic and not constant in a region D of the complex plane, $|f(z)|$ never attains its maximum in the interior of D. In particular, when $f(z)$ is continuous on the closed region $\bar{D} = D \cup \partial D$, the maximum of $|f(z)|$ on D is taken on its boundary ∂D. This fact is called the **maximum (modulus) principle**.

As a direct application of the maximum principle, we can deduce **Schwarz's lemma**: If a holomorphic function $f(z)$ in $|z| < R$ satisfies $|f(z)| \leqslant M$ and $f(0)=0$, we have $|f(z)| \leqslant M|z|/R$ $(|z| < R)$. The equality at z_0, $0 < |z_0| < R$, occurs only for the functions $f(z) = e^{i\lambda} Mz/R$ (where λ is a real constant).

C. Lindelöf's Principle

E. Lindelöf extended Schwarz's lemma and obtained various extensions of the maximum principle, from which he, together with E. Phragmén, deduced several useful theorems on the behavior of a function that is single-valued and holomorphic in a neighborhood of the boundary. We mention some representative theorems:

Let $z=\varphi(\zeta)$ and $w=\psi(\zeta)$ both be †meromorphic and †univalent functions in $|\zeta|<1$ that map $|\zeta|<1$ onto D_z and D_w, respectively. Set $\varphi(0)=z_0$ and $\psi(0)=w_0$. Let $D_z(\rho)$ and $D_w(\rho)$ denote the images of $|\zeta|\leqslant\rho$ $(0<\rho<1)$ under the mappings φ and ψ, respectively. Under these circumstances, if a function $f(z)$ that is holomorphic in D_z satisfies $f(D_z)\subset D_w$ and $f(z_0)=w_0$, then $f(D_z(\rho))\subset D_w(\rho)$. Furthermore, unless $f(z)$ maps D_z onto D_w univalently, $f(D_z(\rho))$ is contained in the interior of $D_w(\rho)$ (an extension of Schwarz's lemma).

Let $f(z)$ be analytic in a bounded region D but not necessarily single-valued. Suppose that $|f(z)|$ is single-valued. Suppose, furthermore, that there is a positive constant M such that, for each boundary point ζ of D, except for a finite number of boundary points and for each $\varepsilon>0$, the inequality $|f(z)|<M+\varepsilon$ holds on the intersection of D with a suitable neighborhood of ζ, and suppose also that each of the exceptional points has a neighborhood such that $f(z)$ is bounded on the intersection of D with this neighborhood. Under these assumptions we have $|f(z)|\leqslant M$. Moreover, if $|f(z_0)|=M$ at a point z_0 of D, then $f(z)$ is a constant (an extension of the maximum principle).

Let $f(z)$ be holomorphic in an angular domain $W: |\arg z|<\alpha\pi/2$. Suppose that there is a constant M such that, for each $\varepsilon>0$, each finite boundary point has a neighborhood such that $|f(z)|<M+\varepsilon$ on the intersection of D with this neighborhood, and that for some positive number $\beta>\alpha$ and for sufficiently large $|z|$ the inequality $|f(z)|<\exp|z|^{1/\beta}$ holds. Under these assumptions we have $|f(z)|\leqslant M$ in D (**Phragmén-Lindelöf theorem**).

Let $f(z)$ be a function that is holomorphic and bounded in a closed angular domain $\overline{W}: \alpha\leqslant\arg z\leqslant\beta$ except for the point at infinity. Suppose that $f(z)\to a$ as $z\to\infty$ along a side of the angle and that $f(z)\to b$ as $z\to\infty$ along the other side of it. Then we have $a=b$ and $f(z)\to a$ uniformly as $z\to\infty$ in W (**Lindelöf's asymptotic value theorem**).

D. Bounded Functions in a Disk

If $f(z)$ is a bounded holomorphic function in the unit disk $|z|<1$, it has a limit at every point z_0 on the circle $C: |z|=1$, except for a set of 1-dimensional measure zero, as z tends to z_0 from within an angle with vertex at z_0 and contained in $|z|<1$ (or along a †Stolz's path at z_0) (**Fatou's theorem**). Under the same assumption, if the boundary value function $f(e^{i\theta})=\lim_{r\to 1-0}f(re^{i\theta})$, which exists by Fatou's theorem, is equal to a constant a for a set of positive measure on the circle C, then $f(z)\equiv a$ in $|z|<1$ (**F. and M. Riesz theorem**). These theorems are valid for meromorphic functions in $|z|<1$ (→ 271 Meromorphic Functions D).

E. Three-Circle Theorem and Related Theorems

Let $f(z)$ be a function that is single-valued, holomorphic, and not identically zero in an annulus $\rho<|z|<R$. Set $M(r)\equiv\max_{|z|=r}|f(z)|$ $(\rho<r<R)$ for $f(z)$. Then $\log M(r)$ is a †convex function of $\log r$ in $\log\rho<\log r<\log R$ (**Hadamard's three-circle theorem**). The same assertion holds for a function $f(z)$ that is not necessarily single-valued, as long as $|f(z)|$ is single-valued. In the case where $f(z)$ is single-valued, a stronger assertion can be obtained (O. Teichmüller). The following theorems are regarded as the analog of Hadamard's three-circle theorem for the respective basic regions:

Set $L(\sigma)\equiv\sup_{-\infty<t<\infty}|f(\sigma+it)|$ $(\alpha<\sigma<\beta)$ for a function that is bounded and regular in a strip $\alpha<\operatorname{Re}z<\beta$. Then $\log L(\sigma)$ is a convex function of σ in $\alpha<\sigma<\beta$ (**Doetsch's three-line theorem**).

Set $l(\sigma)=\limsup_{t\to\infty}|f(\sigma+it)|$ $(\alpha<\sigma<\beta)$ for a function that is holomorphic and bounded in a half-strip $\alpha<\operatorname{Re}z<\beta$, $\operatorname{Im}z>0$. Then $\log l(\sigma)$ is a convex function of σ in $\alpha<\sigma<\beta$ (**Hardy-Littlewood theorem**).

Set

$$I_p(r)=\frac{1}{2\pi}\int_0^{2\pi}|f(re^{i\theta})|^p d\theta$$

for a holomorphic function in a disk $|z|<R$. Then for every $p>0$, $\log I_p(r)$ is an increasing convex function of $\log r$ for $-\infty<\log r<\log R$ (**Hardy's theorem**).

F. Applications of the Maximum Principle

Theorems of the following type are useful for some problems of holomorphic functions:

Let D be a region. Suppose that there exists an arc of angular measure α that is on a circle of radius R centered at a point z_0 of D and not contained in D. Let C denote the intersection of the boundary of D with the disk $|z-z_0|<R$. If $f(z)$ is a single-valued holomorphic function that satisfies $|f(z)|\leqslant M$, and if $\limsup_{z\to\zeta}|f(z)|\leqslant m$ for every $\zeta\in C$, then the inequality $f(z_0)=M^{1-1/n}m^{1/n}$ holds for every positive integer n satisfying $2\pi/n\leqslant\alpha$ (**Lindelöf's theorem**).

Let D be a region bounded by two segments OA, OB both starting from O and making an angle $\pi\alpha$, and a Jordan arc $\widehat{AB}$, and let R be the maximal distance between O and the points on $\widehat{AB}$. Suppose that $f(z)$ is holomorphic in D and that $\limsup|f(z)|$ as $z \to \zeta \in \partial D$ with $z \in D$ is not greater than M for $\zeta \in OA \cup OB$ and m for $\zeta \in \widehat{AB}$. Then we have $|f(z)| \leqslant M^{1-\lambda}m^{\lambda}$ (where $\lambda = (|z|/R)^{1/\alpha})$) at every point on the bisector of the angle $\angle AOB$ in D (**Carlemann's theorem**).

G. Holomorphic Functions with Positive Real Parts

Holomorphic functions with positive real parts are intimately connected with bounded functions. Concerning these functions we have the following classical result, which is equivalent to Schwarz's lemma: If $f(z)$ is holomorphic in $|z|<R$, $\operatorname{Re} f(z) \geqslant 0$ in the same domain, and $f(0)=1$, then $(R-|z|)/(R+|z|) \leqslant \operatorname{Re} f(z) \leqslant (R+|z|)/(R-|z|)$ $(|z|<R)$. The right or left inequality becomes equality for some z_0, $0<|z_0|<R$, only if $f(z) = (Rz_0 \mp z_0 z)/(R \pm z_0 z)$, respectively.

In order to prove various results for the class of functions with positive real part, **Herglotz's integral representation**, which is based on Poisson's integral representation and unique to this class, can be used effectively. It is given by

$$f(z) = \int_0^{2\pi} \frac{e^{i\varphi}+z}{e^{i\varphi}-z}\, d\rho(\varphi), \quad |z|<R,$$

where $\rho(\varphi)$ is monotone increasing (real-valued) with total variation 1 and is determined uniquely up to an additive constant by $f(z)$. An analogous integral representation is introduced for a holomorphic function in an annulus.

The real part $\operatorname{Re} f(z)$ of a holomorphic function $f(z)$ is generally harmonic, and in the case where $f(z) \neq 0$, $\log|f(z)|$ is also harmonic. Conversely, each harmonic function is regarded as the real part of a holomorphic function. By this relation, some properties of holomorphic functions are transferred to corresponding properties of harmonic functions. For instance, the maximum principle for the maximum of $|f(z)|$ is easily translated into that of harmonic functions. It should be remarked that a †conjugate harmonic function of a single-valued harmonic function is, in general, not single-valued in a multiply-connected region. But multiple connectivity has little influence on local properties such as local maxima. The maximum principle holds more generally for †subharmonic functions (for the maximum of the value itself).

H. Coefficient Problems

There are many classical results for partial sums and coefficients of the Taylor expansion of bounded functions in a disk. Let $f(z) = \sum_{n=0}^{\infty} c_n z^n$ be the Taylor expansion of a bounded function in $|z|<1$. Set its partial sum $s_n(z) = \sum_{\nu=0}^{n} c_\nu z^\nu$ $(n=0,1,\ldots)$, and let $t_n(z) = (1/(n+1))\sum_{\nu=0}^{n} s_\nu(z)$ $(n=0,1,\ldots)$, which is the sequence of the arithmetic means of the partial sums (the Fejér sums). Then $|f(z)| \leqslant 1$ in $|z|<1$ if and only if $|t_n(z)| \leqslant 1$ for $|z|=1$ $(n=0,1,\ldots)$ (L. Fejér). Thus the sequences $\{t_n(z)\}$ for bounded functions $f(z)$ are uniformly bounded, whereas the sequences $\{s_n(z)\}$ are not uniformly bounded. Indeed the maximum value of $|s_n(1)|$ is equal to

$$G_n = 1 + \sum_{j=1}^{n} \binom{-1/2}{j}^2 = \sum_{j=0}^{n} \left(\frac{1\cdot 3\cdot \ldots \cdot (2j-1)}{2\cdot 4\cdot \ldots \cdot 2j} \right)^2$$

$(G_n \sim \pi \log n$ as $n \to \infty)$.

The following result is decisive for coefficient problems: Set $h_{\mu\nu} = \sum_{j=0}^{\mu} \bar{c}_{\mu-j} c_{\nu-j}$ $(\mu \leqslant \nu)$, $h_{\nu\mu} = \bar{h}_{\mu\nu}$ for $f(z) = \sum_{n=0}^{\infty} c_n z^n$, let m_n^2 be the maximal eigenvalue (a nonnegative real number) of the Hermitian matrix $(-h_{\mu\nu})_{\mu,\nu=0}^{n}$, and let $m = \lim m_n (\geqslant 0)$.

Let H be a Hermitian form in infinitely many variables given by

$$H = m^2 \sum_{\nu=0}^{\infty} \bar{x}_\nu x_\nu - \sum_{\mu,\nu=0}^{\infty} h_{\mu\nu} \bar{x}_\mu x_\nu = m^2 \sum_{\nu=0}^{\infty} |x_\nu|^2 - \sum_{\mu=0}^{\infty} \left| \sum_{\nu=0}^{\infty} c_\nu x_{\mu+\nu} \right|^2 .$$

Then a necessary and sufficient condition for $|f(z)| \leqslant 1$ when $|z|<1$ is that H be positive semidefinite, i.e., the sequence of the †principal minor determinants of H

$$\Delta\begin{pmatrix} 1 & 2 & \ldots & m \\ 1 & 2 & \ldots & m \end{pmatrix}, \quad m=1,2,\ldots,$$

is all positive or positive for an initial finite number of them and zero for the remainder (I. Schur).

A corresponding result for functions with positive real part in a disk can be stated in a simpler form: A holomorphic function $f(z) = 1/2 + \sum_{n=1}^{\infty} c_n z^n$ in $|z|<1$ satisfies $\operatorname{Re} f(z) \geqslant 0$ if and only if

$$\begin{vmatrix} 1 & c_1 & \ldots & c_n \\ \bar{c}_1 & 1 & \ldots & c_{n-1} \\ & & \ldots & \\ \bar{c}_n & \bar{c}_{n-1} & \ldots & 1 \end{vmatrix} \geqslant 0, \quad n=1,2,\ldots$$

(C. Carathéodory). Furthermore, for $n=$

1,2, ..., when we regard $(c_1, \ldots, c_n)$ as a point of complex n-dimensional Euclidean space, we can determine the domain of existence of this point. This result is generalized for coefficients of the Laurent expansion of a function that is holomorphic and single-valued in an annulus.

Next, there are some results for a function that omits two values: If $f(z)=a_0+a_1z+\ldots$ $(a_1\neq 0)$ is holomorphic in $|z|<R$ and $f(z)\neq 0$, 1 in the same domain $|z|<R$, then there exists a constant $L(a_0,a_1)$, determined only by a_0 and a_1, such that $R\leqslant L(a_0,a_1)$ **(Landau's theorem)**. Under these circumstances, $|f(z)|\leqslant S(a_0,\theta)$ in $|z|\leqslant\theta R$ for $0<\theta<1$, where $S(a_0,\theta)$ is a constant determined only by a_0 and θ **(Schottky's theorem)**. These theorems have applications in value distribution theory.

I. Angular Derivative

Let $f(z)$ be holomorphic in $|z|<1$. If $f(z)\to w_0$ uniformly as $z\to z_0$ along Stolz paths with end point at z_0 and if the limit $\lim_{z\to z_0}((f(z)-w_0)/(z-z_0))\equiv D$ exists, we call D the **angular derivative** of $f(z)$ at z_0. In the case of the half-plane $\mathrm{Re}\,z>0$, the angular derivative at the point z_0 on the imaginary axis is similarly defined. It should be noted that $f(z)-w_0$ is replaced by $1/f(z)$ for $w_0=\infty$ and $1/(z-z_0)$ by z for $z_0=\infty$. In the latter case, a Stolz path is a path contained in an angular domain $|\arg z|\leqslant\alpha$ $(<\pi/2)$ and tending to ∞. The study of angular derivatives was initiated by G. Julia (1920) and J. Wolff (1926) and was further advanced by Carathéodory (1929) and E. Landau and G. Varilon (1929).

A fundamental theorem for angular derivatives can be stated as follows: If a holomorphic function $f(z)$ in $\mathrm{Re}\,z>0$ satisfies $\mathrm{Re}\,f(z)\geqslant 0$, there exists a constant c $(0\leqslant c<+\infty)$ such that $f(z)/z\to c$ and $f'(z)\to c$ uniformly as $z\to\infty$ along every Stolz path. Moreover, the pth derivative of $f(z)$ for an arbitrary positive p, denoted by $D^pf(z)$, has the property that $z^{p-1}D^pf(z)\to c/\Gamma(2-p)$ uniformly. Furthermore, the inequality $\mathrm{Re}\,f(z)\geqslant c\,\mathrm{Re}\,z$ holds everywhere in $\mathrm{Re}\,z>0$. An analogous theorem is valid for the unit disk.

An important problem of the theory of conformal mappings is the study of the conditions under which a function $w=f(z)$ mapping the unit disk (or a half-plane) denoted by D in the z-plane onto a simply-connected region B on the w-plane has a nonzero and finite angular derivative at the boundary point z_0, i.e., the conditions for both conformality and invariance of the ratio of line elements. Carathéodory showed that a sufficient condition is the existence of two circles that are mutually inscribed and circumscribed at the boundary point $w_0=f(z_0)\neq\infty$ of B and that lie inside and outside of B, respectively. Ahlfors later established a necessary and sufficient condition for the existence of the angular derivative by making use of his †distortion for a strip. The angular derivative was used by Wolff in his research on the iteration of conformal mappings.

References

[1] L. Bieberbach, Lehrbuch der Funktionentheorie II, Teubner, 1931 (Johnson Reprint Co., 1969).
[2] C. Carathéodory, Funktionentheorie I, II, Birkhäuser, Basel, 1950; English translation, Theory of functions, Chelsea, I, 1958; II, 1960.
[3] Y. Komatu, Tôkaku syazôron (Japanese; Theory of conformal mappings) I, Kyôritu, 1944.
[4] Y. Komatu, Kansûron, ensyû (Japanese; Theory of functions, exercises on theory of functions), Asakura, 1960.
[5] E. G. H. Landau, Darstellung und Begründung einiger neuerer Ergebnisse der Funktionentheorie, Springer, 1929.
[6] J. E. Littlewood, Lectures on the theory of functions I, Clarendon Press, 1944.
[7] M. Tsuji, Hukuso hensû kansûron (Japanese; Theory of functions of a complex variable), Kyôritu, 1934.
[8] M. Tsuji and Y. Komatu (eds.), Kansûron, ensyû (Japanese; Exercises on the theory of functions), Syôkabô, 1959.

45 (XVI.11) Branching Processes

A. General Remarks

A **branching process** is a mathematical model for random motion of a family of particles each of which is in an isolated process of multiplication and death. Examples of such random motions are population growth, miosis of genes, growth of the numbers of neutrons in an atomic chain reaction, and cascade showers of cosmic rays. The simplest and most fundamental branching process is discussed in the next section.

B. Galton-Watson Branching Processes

Although there is a similar process with continuous parameter $t\in\mathbf{R}$, we consider here only the case of a discrete time parameter

$(t=0,1,2,\ldots)$. Suppose that we are given a family of particles of the same kind. Each member of the family splits into several particles according to a given probability law independently of the other members and its own past history. Let Z_n be the number of particles of the family at a moment (or generation) n; then $\{Z_n\}$ gives rise to a †Markov chain. This is called the **Galton-Watson branching process**.

Let $p_k (k=0,1,2,\ldots)$ be the probability that a particle splits into k particles, and let $f(s)=\sum_{j=0}^{\infty} p_j s^j$. Then the †transition probability p_{ij} of Z_n is given as the coefficient of s^j of $f(s)^i$, since all members of the family split independently. We assume $Z_0=1$ in the rest of this article.

Denote the †generating function $\sum_{k=0}^{\infty} P(Z_n=k)s^k$ $(s\leqslant 1)$ of Z_n by $f_n(s)$. Then we have $f_0(s)=s$, $f_1(s)=f(s)$, $f_{i+j}(s)=f_i(f_j(s))$ $(i,j=0,1,2,\ldots)$, and the †expectation of Z_n is given by $E(Z_n)=m^n$, where $m=f'(1)=E(Z_1)$ is the expectation of Z_1.

If we have $Z_n=0$ for some n, then $Z_{n+1}=Z_{n+2}=\ldots=0$. The probability $q=P(\lim_n Z_n=0)$ is called the **extinction probability**. The case $f_1(s)\equiv s$ is excluded in the following:

$$q=1 \quad \text{for} \quad m\leqslant 1, \qquad 0\leqslant q<1 \quad \text{for} \quad m>1, \tag{1}$$

where q is the nonnegative minimal solution of $f(s)=s$. Moreover, $P(\lim_n Z_n=\infty)=1-q$ when $m>1$, and hence the process $\{Z_n\}$ is †transient.

If we put $W_n=Z_n/m^n$ when $m<\infty$, $\{W_n\}$ gives rise to a †martingale, and the limit $W=\lim_{n\to\infty} W_n$ exists with probability 1. If $m\leqslant 1$, we have $P(W=0)=1$, and if $m>1$, then the †moment-generating function $\varphi(s)=E(\exp(-sW))$ of W satisfies the **Königs-Schröder equation**:

$$\varphi(ms)=f(\varphi(s)), \quad \operatorname{Re} s\geqslant 0. \tag{2}$$

If $m\leqslant 1$, then $P(Z_n=0$ for some $n)=1$, and Yaglom's theorem on the conditional †limit distribution holds: If $m<1$ and $E(Z_1^2)<\infty$, there exist $b_k=\lim_n P(Z_n=k|Z_n\neq 0)$ $(\sum_{k=1}^{\infty} b_k=1)$. The generating function $g(s)=\sum_{k=1}^{\infty} b_k s^k$ satisfies $g(f(s))=mg(s)+1-m$, $(|s|\leqslant 1)$, and hence $g'(1)=\lim_n m^n(1-f_n(0))^{-1}$. If $m=1$ and $f'''(1)<\infty$, $\lim_n P(2Z_n/nf''(1)<u|Z_n\neq 0)=1-\exp(-u)$ $(u\geqslant 0)$.

C. Multitype Galton-Watson Processes

This is a generalization of the simple Galton-Watson process, involving k types of particles $(k\geqslant 2)$, say, $T_1, T_2, \ldots, T_k$. Let $p^l(r_1, r_2, \ldots, r_k)$ be the probability that a particle of type T_l splits into r_m particles of type T_m $(m=1,2,\ldots,k)$, and set $f^l(s_1,s_2,\ldots,s_k)=\sum_r p^l(r_1,r_2,\ldots,r_k)s_1^{r_1}s_2^{r_2}\ldots s_k^{r_k}$. The number of particles $Z_n=(Z_n^1, Z_n^2, \ldots, Z_n^k)$ at a moment n gives rise to a †Markov chain over k-dimensional lattice points, and its transition probability is determined by $p^l(r_1,r_2,\ldots,r_k)$, as in the case of simple Galton-Watson processes [1].

When $m_{ij}=\partial f^i(1,\ldots,1)/\partial s_j<\infty$, the †conditional expectation of Z_{n+m} is given as $E(Z_{m+n}/Z_m=e_m)=e_m M^n$, where M is the matrix (m_{ij}) and $e_m=(e_m^1,\ldots,e_m^k)$, $e_m^i=\delta_{im}$. If every component of M is nonnegative and if there is a whole mumber N such that every component of M^N is positive, the process Z_n is said to be **positively regular**. In this case, M has a positive †eigenvalue λ that is simple, and $|\lambda|>|\mu|$ for all other eigenvalues μ. The eigenvalue λ plays the role of m in the case of the simple Galton-Watson process.

For a positively regular process, (1) is still valid for the extinction probability $q=(q^1,\ldots,q^k)$ $(q^i=P(\lim_n Z_n=0|Z_0=e_i))$, and the transience property holds under a simple condition (where 1 must be understood as $(1,\ldots,1)$, $0\leqslant q<1$ as $0\leqslant q^i<1$ $(i=1,2,\ldots,k)$, and $f(s)=s$ as $f^i(s)=s_i$ $(i=1,2,\ldots,k)$). Moreover, if we assume that $E(Z_1^{i}Z_1^{j}|Z_0=e_r)<\infty$, then $W_n=Z_n/\lambda^n$ converges with probability 1 when $\lambda>1$, and the limit W is a constant multiple of the left †eigenvector $\nu=(\nu_1,\ldots,\nu_k)$ for λ of the matrix M. The moment-generating function of W satisfies a functional equation similar to (2). When $\lambda\leqslant 1$, a theorem analogous to Yaglom's holds.

D. Markov Branching Processes

Let t be a continuous time parameter, $b(t)\Delta t$ be the probability of the occurrence of a branching in a small time interval Δt, and $p_k(t)$ $(k=0,1,2,\ldots)$ be the probability for a particle to split into k particles. When $b(t)$ and $p_k(t)$ are continuous in t, $b(t)>0$, $p_k(t)\geqslant 0$, and $\sum_{k=0}^{\infty} p_k(t)=1$, the branching process is called a **Markov branching process**. The number of particles $\{Z(t)\}$ is a Markov process, and the †transition probability $P_{ik}(\tau,t)$ is the unique solution of †Kolmogorov's forward equation

$$\frac{\partial P_{ik}(\tau,t)}{\partial t}=-kb(t)P_{ik}(\tau,t)+b(t)\sum_{j=1}^{k+1} P_{ij}(\tau,t)jp_{k-j+1}(t),$$

$$P_{ik}(\tau,\tau+0)=1 \quad \text{when } i=k;\ =0 \quad \text{when } i\neq k.$$

When $\sum_{k=0}^{\infty} P_{ik}(\tau,t)<1$, the number of particles attains $+\infty$ in a finite time interval with positive probability. If $\sum_{k=0}^{\infty} kp_k(t)$ converges uniformly in the wide sense in t, then $\sum_{k=0}^{\infty} P_{ik}(\tau,t)=1$.

When $b(t)$ and $p_k(t)$ are constants, $Z(t)$ is said to be a **temporally homogeneous Markov branching process** and yields some results analogous to those obtained for Galton-Watson processes. In particular, concerning the †limit distribution of $Z(t)$ more precise results are obtained.

Some work has also been done on the asymptotic behavior of temporally inhomogeneous cases and multitype Markov branching processes.

E. Age-Dependent Processes

When the probability that a particle splits depends on its age, the branching process is called an **age-dependent branching process**. Each particle has the distribution $G(t)=P(l\leqslant t)$ of the lifetime l ($G(0-)=0, G(0+)<1$, and $G(\infty)=1$), and splits into k particles of age 0 with the probability p_k ($\sum_{k=0}^{\infty}p_k=1$). The number of particles $Z(t)$ at t does not have the †Markov property, in general. Set $F(s,t)=\sum_{k=0}^{\infty}P(Z(t)=k)s^k$ ($|s|\leqslant 1, t\geqslant 0$). Then $F(s,t)$ is the unique solution of

$$F(s,t)=s(1-G(t))+\int_{0-}^{t+}h(F(s,t-u))\,dG(u),$$

$$h(s)=\sum_{k=0}^{\infty}p_k s^k, \qquad (3)$$

where $|F(s,t)|\leqslant 1$. When $G(t)=1-\exp(-bt)$ ($b>0$), $Z(t)$ is a Markov process, and the solution of (3) gives the generating function of the Markov process. When $h'(1)=m<\infty$ and $G(0+)=0$, and if $G(t)$ is not a lattice distribution, the †mean $M(t)=E(Z(t))$ of $Z(t)$ is identically equal to 1 when $m=1$, while $M(t)\sim n_1\exp(\alpha t)$ ($t\to\infty$) when $m\neq 1$. Here $n_1=(m-1)(\alpha m^2\int_0^{\infty}te^{-\alpha t}dG(t))^{-1}$, $m\int_0^{\infty}e^{-\alpha t}dG(t)=1$, and we assume $\int_0^{\infty}t^2e^{-\alpha t}dG(t)<\infty$ if $m<1$. When $m>1$ and $h''(1)<\infty$, and if $G(t)$ is not a lattice distribution, $W_t=(n_1\exp(\alpha t))^{-1}Z(t)$ has the limit $W=\text{l.i.m.}\,W_t$, and $\varphi(s)=E(\exp(-sW))$ satisfies

$$\varphi(s)=\int_0^{\infty}h(\varphi(se^{-\alpha u}))\,dG(u), \qquad \operatorname{Re}s\geqslant 0.$$

Moreover, if we set $dG(t)=g(t)dt$ and assume that $\int_0^{\infty}g(t)^p\,dt<\infty$ ($p>1$), $W=\lim_{t\to\infty}W_t$ exists with probability 1.

For the general theory of branching processes of more general type → [1, 5, 10].

References

[1] T. E. Harris, The theory of branching processes, Springer, 1963.

[2] T. E. Harris, Some mathematical models for branching processes, Proc. 2nd Berkeley Symp. on Math. Stat. and Prob. (1951), 305–328.

[3] Б. А. Севастьянов (B. A. Sevast'janov), теория ветвящихся случайных процессов, Uspehi Mat. Nauk, 6 (1951), no. 6, 47–99.

[4] А. М. Яглом (A. M. Jaglom), Некоторые предельные теоремы теории ветвящихся случайных процессов, Dokl. Akad. Nauk SSSR, (N.S.) 56 (1947), 795–798.

[5] J. E. Moyal, The general theory of stochastic population processes, Acta Math., 108 (1962), 1–31.

[6] В. М. Золотарев (V. M. Zolotarev), Краткие сообщения уточнение ряда теорем теории ветвящихся случайных процессов, Teor. Verojatnost. i Primenen., 2 (1957), 256–266; English translation, More exact statements of several theorems in the theory of branching processes, Theory of Prob. Appl.. 2 (1957), 245–253.

[7] В. П. Цистяков, Н. П. Маркова (V. P. Čistjakov and N. P. Markova), О некогорых теоремах для неоднородных ветвящихся процессов, Dokl. Akad. Nauk SSSR, 147 (1962), 317–320; English translation, On some theorems for inhomogeneous branching processes, Sov. Math., 3 (1962), 1619–1623.

[8] А. А. Савин, В. П. Чистяков (A. A. Savin and V. P. Čistjakov), Некоторые теоремы для ветвящихся процессов с несколькими типами частиц, Teor. Verojatnost. i Primenen. 7 (1962), 95–104; English translation, Some theorems for branching processes with several types of particles, Theory of Prob. Appl., 7 (1962), 93–100.

[9] T. W. Mullikin, Limiting distributions for critical multitype branching processes with discrete time, Trans. Amer. Math. Soc., 106 (1963), 469–494.

[10] K. B. Athreya and P. E. Ney, Branching processes, Springer, 1972.

46 (XVI.8) Brownian Motion

A. General Remarks

An English botanist, R. Brown, noticed that pollen grains floating in water perform peculiarly erratic movements. The physical explanation of this phenomenon is that haphazard impulses are given to the suspended particles by collisions with molecules of the fluid. Let $X(t)$ be the x-coordinate of a particle at time t. Then $X(t)$ is treated as a †random variable, and the distribution of $X(t)-X(s)$ is a †normal distribution $N(0, D|t-s|)$, with †mean 0 and †variance $D|t-s|$, where D is a positive constant. To be more exact, such a family of random variables $\{X(t)\}$ is now considered as the family of random variables determining a †stochastic process. Various aspects of the theory were analyzed by L. Bachelier, A. Einstein, N. Wiener, P. Lévy, and others.

B. Wiener Processes

Let T be the real line $\mathbf{R}^1$ or a subinterval. A †stochastic process $\{X(t)\}_{t\in T}$ defined on a †probability space $\Omega(\mathfrak{B},P)$ is called a **Wiener process** on $\mathbf{R}^d$ if it satisfies the following three conditions: (1) $X(t,\omega)\in\mathbf{R}^d$ $(t\in T, \omega\in\Omega)$. (2) The increments $X(t_j)-X(t_{j-1})$, $j=2,3,\ldots,n$, are †independent random variables for any $t_1<t_2<\ldots<t_n$, $t_j\in T$ $(j=1,2,\ldots,n)$, where n is an arbitrary positive integer. (3) If $X_i(t)$ is the ith component of the vector $X(t)$, then the $\{X_i(t)\}_{t\in T}$ $(1\leqslant i\leqslant d)$ are independent as stochastic processes and every increment $X_i(t)-X_i(s)$ is normally distributed with mean 0 and variance $|t-s|$. A Wiener process is also called a **Brownian motion**. A Wiener process is a †temporary homogeneous additive process. A †separable Wiener process has continuous paths with probability 1. Conversely, if $\{X(t)\}_{t\in T}$ is a temporary homogeneous additive process on $\mathbf{R}^1$ whose †sample function is continuous with probability 1 and the increment $X(t)-X(s)$ has mean 0 and variance $|t-s|$, then it is a Wiener process (→ 395 Stochastic Processes).

Let $\{X_k(\omega)\}$ $(k=0,1,\ldots)$ be a sequence of independent random variables defined on a probability space $\Omega(\mathfrak{B},P)$ such that each X_k has the normal distribution $N(0,1)$. Then the series

$$\frac{t}{\sqrt{\pi}}X_0(\omega)+\sum_{n=0}^{\infty}\sum_{k=2^n}^{2^{n+1}-1}\sqrt{\frac{2}{\pi}}\,\frac{\sin kt}{k}X_k(\omega),$$

$$t\in[0,1]\equiv T,$$

converges uniformly in $t\in T$ with probability 1, and its limit, denoted by $X(t,\omega)$, is a Wiener process.

C. Brownian Motion as a Diffusion Process

In terms of the general framework of †Markov processes, Brownian motion is a typical example of a diffusion process (→ 119 Diffusion Processes). Let $X=\{X_t(\omega),\mathbf{R}^d,P_x\}$ be a continuous Markov process on $\mathbf{R}^d$ (→ 262 Markov Processes) with the †transition probability

$$P(t,x,B)=\int_B(2\pi t)^{-d/2}\exp\Big(-\frac{1}{2t}|x-y|^2\Big)\,dy,$$

$$t>0,\quad x,y\in\mathbf{R}^d,\quad B\in\mathfrak{B}(\mathbf{R}^d),$$

where $\mathfrak{B}(\mathbf{R}^d)$ is the †σ-algebra of all Borel sets in $\mathbf{R}^d$ and $|x|$ denotes the †norm of $x\in\mathbf{R}^d$. For each $x\in\mathbf{R}^d$, the process $\{X_t(\omega),P_x\}$ is a Wiener process in the sense mentioned above. The Markov process X, which is a collection of Wiener processes $\{X_t,P_x\}$ starting at x, is said to be a **d-dimensional Brownian motion**. A d-dimensional Brownian motion possesses the †strong Markov property. Let $\mathfrak{G}$ be the †generator of the †semigroup T_t corresponding to X. A bounded uniformly continuous function f defined on $\mathbf{R}^d$ belongs to the domain of $\mathfrak{G}$ if its partial derivatives $\partial f/\partial x_i$ and $\partial^2 f/\partial x_i\partial x_j$, $i,j=1,2,\ldots,d$, are bounded and uniformly continuous. For such a function f we have $\mathfrak{G}f(x)=(1/2)\Delta f(x)$, where Δ is the †Laplacian operator.

D. Brownian Motions and Potentials

For $\alpha>0$, the function $G_\alpha(x)$ defined by

$$G_\alpha(x)=\frac{1}{2}\int_0^\infty e^{-\alpha t}(2\pi t)^{-d/2}\exp\Big(-\frac{1}{2t}|x|^2\Big)\,dt$$

is said to be the **α-order Green's function**. Since Brownian motion is †nonrecurrent for $d\geqslant 3$, the limit $G_{0+}(x)=\lim_{\alpha\downarrow 0}G_\alpha(x)$, $x\in\mathbf{R}^d$, exists for $d\geqslant 3$, and $G_{0+}(x)$ is equal to $K_0(x)=(\Gamma(d/2-1)/4\pi^{d/2})|x|^{-d+2}$, which is the kernel for the †Newtonian potentials. Brownian motion is †recurrent when $d\leqslant 2$ and $G_{0+}(x)=+\infty$, $x\in\mathbf{R}^d$. In this case, $K_0(x)$ is defined by $K_0(x)=\lim_{\alpha\downarrow 0}(G_\alpha(x)-G_\alpha(x_0))$, and $K_0(x)=(1/2\pi)\log(1/|x|)$, when $d=2$ and $|x_0|=1$. This is the kernel for the †logarithmic potentials. When $d=1$ and $x_0=0$, $K_0(x)=-(1/2)|x|$. Using this relationship, we can express many concepts of classical potential theory in an elegant form in probability language. Denote by σ_A the †hitting time (→ 262 Markov Processes) of the d-dimensional Brownian motion for the set A. For a Green domain (i.e., a domain which is a †Green space) D in $\mathbf{R}^d$ $(d\geqslant 2)$, set $2g^D(t,x,y)\,dy=P_x(X(t,\omega)\in dy,\ \sigma_{\partial D}>t)$, $x,y\in D$. Then the right-hand side of this equation is the †transition probability of the Brownian motion on D with the †absorbing barrier ∂D. Then $G^D(x,y)=\int_0^\infty g^D(t,x,y)\,dt$ is †Green's function of D.

If B is a compact subset of a Green domain

D or an open subset with compact closure $\bar{B} \subset D$, then the hitting probability $p(x) = P_x(\sigma_B < \sigma_{\partial D})$ is the equilibrium potential of B relative to D. A compact subset B in $\mathbf{R}^2$ is of positive †logarithmic capacity if and only if $P_x(\sigma_B < \infty) = 1$ for each $x \in \mathbf{R}^2$.

Given a set A, define $\tilde{\sigma}_A(\omega) = \inf\{t \mid t > 0, x(t, \omega) \in A\}$, where the infimum of the empty set is understood to be $+\infty$. Suppose A is an †analytic subset of $\mathbf{R}^n$. †Blumenthal's 0-1 law implies that $P_x(\tilde{\sigma}_A = 0) = 1$ or 0. The point x is said to be **regular** for A if this probability is 1 and **irregular** for A otherwise. Let B be a compact subset in $\mathbf{R}^d$ $(d \geqslant 2)$. Then †Wiener's test (→ 124 Dirichlet Problem) states that x is regular or irregular for B according to whether the following series diverge or converge:

$$\sum_{k=1}^{\infty} 2^{k(d-2)} C(B_k), \qquad d \geqslant 3,$$

$$\sum_{k=1}^{\infty} k C(B_k), \qquad d = 2,$$

where $C(B_k)$ is the †Newtonian capacity (the logarithmic capacity relative to a bounded domain when $d = 2$) of the set $B_k = \{y \mid 2^{-(k+1)} \leqslant |y - x| < 2^{-k}\} \cap B$. Suppose D is a bounded domain in $\mathbf{R}^d$ $(d \geqslant 2)$. A point $x \in \partial D$ is regular or irregular for $R^d - D$ according as x is regular or irregular in the sense of the †Dirichlet problem for D. Given a continuous boundary function f on ∂D, $u(x) = E_x(f(X_{\tilde{\sigma}_{\partial D}}(\omega)))$ is the solution of the †modified Dirichlet problem for D. Given $x \in D$, the distribution $h(x, B) = P_x(X_{\tilde{\sigma}_{\partial D}}(\omega) \in B)$ $(B \subset \partial D)$ used in the solution $u(x) = \int_{\partial D} f(y) h(x, dy)$ of the modified Dirichlet problem is the †harmonic measure of ∂D as viewed from x.

E. 1-Dimensional Brownian Motion

In his monograph [10], P. Lévy gave a profound description of the fine structure of the individual Brownian path. Let us set

$$\sigma_a = \text{the hitting time to the point } a \text{ in } \mathbf{R}^1,$$

$$m(t, \omega) = \min_{\sigma_0 \leqslant s \leqslant t} X(s, \omega),$$

$$M(t, \omega) = \max_{\sigma_0 \leqslant s \leqslant t} X(s, \omega),$$

$$Y_0(t, \omega) = |X(t, \omega)|,$$

$$Y_1(t, \omega) = \begin{cases} X(t, \omega), & t < \sigma_0(\omega), \\ M(t, \omega) - X(t, \omega), & t \geqslant \sigma_0(\omega), \end{cases}$$

and

$$Y_2(t, \omega) = \begin{cases} X(t, \omega), & t < \sigma_0(\omega), \\ X(t, \omega) - m(t, \omega), & t \geqslant \sigma_0(\omega). \end{cases}$$

Then we have (1) $P_0(M(t) > a) = 2P_0(M(t) > a, X(t) > a) = 2P_0(M(t) > a, X(t) < a) = P_0(|X(t)| > a)$ (**reflection principle** of D. André). (2) The stochastic process $\{\sigma_a, 0 \leqslant a < \infty, P_0\}$ is a †one-sided stable process with exponent $1/2$, that is, it is additive and homogeneous with the law

$$P_0(\sigma_b - \sigma_a \leqslant t) = P_0(\sigma_{b-a} \leqslant t)$$

$$= \int_0^t \frac{b-a}{\sqrt{2\pi s^3}} e^{-(b-a)^2/2s} ds,$$

$$0 \leqslant a < b, \quad t \geqslant 0.$$

(3) Let $\varphi^{-1}(t, \omega)$ be the right continuous inverse function of

$$\varphi(t, \omega) = \int_0^t \chi_{[0,\infty)}(X(s, \omega)) ds,$$

where $\chi_{[0,\infty)}(\cdot)$ is the †indicator function of the interval $[0, \infty)$. Set $Y_3(t, \omega) = X(\varphi^{-1}(t, \omega), \omega)$. Then the four processes $\{Y_i(t, \omega), 0 \leqslant t < \infty, P_x\}$ $(x \in [0, \infty), 0 \leqslant i \leqslant 3)$ on $[0, \infty)$ have the same probability law. Each of them is a diffusion process with transition probability

$$P(t, x, B)$$

$$= \int_B \frac{1}{\sqrt{2\pi t}} (e^{-|x-y|^2/2t} + e^{-|x+y|^2/2t}) dy,$$

$$t > 0, \quad x \in [0, \infty),$$

and is said to be a Brownian motion on $[0, \infty)$ with a †reflecting barrier at the origin. (4) As a consequence of (3), for fixed t $M(t)$, $-m(t)$, and $Y_i(t)$ $(0 \leqslant i \leqslant 3)$ have a common distribution. For example,

$$P_0(M(t) \geqslant a) = P_0(\sigma_a \leqslant t) = 2P_0(X(t) > a)$$

$$= \sqrt{\frac{2}{\pi t}} \int_a^\infty e^{-x^2/2t} dx, \quad a > 0$$

$$P_0(X(t) \in da, M(t) \in db)$$

$$= \sqrt{\frac{2}{\pi t^3}} (2b - a) e^{-(2b-a)^2/2t} da\, db, \quad 0 \leqslant a < b.$$

(5) The diffusion process $\{X(t, \omega), 0 \leqslant t < \sigma_0, P_x\}$ $(x \in (0, \infty))$ obtained from a 1-dimensional Brownian motion by shortening its lifetime is called a Brownian motion on $(0, \infty)$ with an †absorbing barrier at the origin, and its transition probability is given by

$$P(t, x, B)$$

$$= \int_B \frac{1}{\sqrt{2\pi t}} (e^{-|x-y|^2/2t} - e^{-|x+y|^2/2t}) dy,$$

$$t > 0, \quad x \in (0, \infty).$$

The **arcsin law** is valid for many functionals of 1-dimensional Brownian motion. For example,

$$P_0\left(\int_0^t \chi_{[0,\infty)}(X(s, \omega)) ds \leqslant \theta\right)$$

$$= \frac{2}{\pi} \arcsin\sqrt{\frac{\theta}{t}}, \quad 0 < \theta \leqslant t,$$

$P_0(\tau_t(\omega) \leqslant s) = (2/\pi) \arcsin\sqrt{s/t}$, $0 < s \leqslant t$, where $\tau_t(\omega)=\sup\{s \mid X(s,\omega)=0,\ 0\leqslant s\leqslant t\}$.

The visiting set $\mathfrak{Z}(\omega)=\{t \mid X(t,\omega)=0\}$ of a Brownian path is a †totally disconnected set. Its †Lebesgue measure is 0, and the †Hausdorff-Besikovič dimension number of $\mathfrak{Z}(\omega)$ is $1/2$.

The probability law governing the stochastic process obtained through linear interpolation of a 1-dimensional Wiener process is invariant under †projective transformation of the time parameter. This property is called the **projective invariance principle** for the Wiener process. As a special case, it is found that $\{X(t,\omega)/\sqrt{t}\}_{0<t<\infty}$ and $\{X(1/t,\omega)\sqrt{t}\}_{0<t<\infty}$ define the same probability law on $\mathbf{R}^{(0,\infty)}$, where $\{X(t,\omega)\}_{0\leqslant t<\infty}$ is a Wiener process starting at the origin. Hence the properties of the Wiener process in a neighborhood of $t=0$ can be obtained from those in a neighborhood of $t=\infty$, and conversely.

F. d-Dimensional Brownian Motion

Almost all paths of d-dimensional Brownian motion are continuous but are not of †bounded variation on any finite interval. Accordingly they cannot have lengths. A positive, continuous, increasing function φ defined on $[t_0,\infty)$ with $t_0>0$ is said to **belong to the upper (lower) class with respect to local continuity** if $P_0((\inf\{t \mid |X(t,\omega)|>\sqrt{t}\,\varphi(1/t)\})>0)=1$ (0). **Kolmogorov's test** states that φ belongs to the upper class or to the lower class with respect to local continuity according as

$$\int_{t_0}^{\infty} t^{-1}(\varphi(t))^d e^{-\varphi^2(t)/2}\,dt$$

converges or diverges. Consider the **space-time Brownian motion** $\{(-t,X(t,\omega)),P_0\}$. Set $D_\varphi=\{(s,x) \mid -1/t_0\leqslant s\leqslant 0,\ \sqrt{-s}\,\varphi(-1/s)\leqslant x<\infty\}$. Then $0=(0,0)$ is regular or irregular for D_φ according as φ belongs to the lower or upper class with respect to local continuity. Thus Kolmogorov's test is the Wiener test for space-time Brownian motion. For example,

$$\varphi(t)=\big(2\log_{(2)}t+(d+2)\log_{(3)}t+2\log_{(4)}t+\ldots +2\log_{(n-1)}t+(2+\delta)\log_{(n)}t\big)^{1/2}$$

belongs to the upper or lower class with respect to local continuity according as $\delta>0$ or $\delta\leqslant 0$, where $\log_{(n)}t=\log\log_{(n-1)}t$ and $\log_{(1)}t=\log t$. The **law of the iterated logarithm** for 1-dimensional Brownian motion,

$$P_0\left(\limsup_{t\to s}\frac{|X(t,\omega)-X(s,\omega)|}{\sqrt{2|t-s|\log\log 1/|t-s|}}=1\right)=1,$$

is a special case of this example. A positive, continuous decreasing function $\psi(t)$ defined on $[t_0,\infty)$ with $t_0>0$ is said to belong to the upper or lower class according as

$$P_0\Big(\big(\inf\{t \mid |X(t,\omega)|<\sqrt{t}\,\psi(1/t)\}\big)>0\Big)=0 \text{ or } 1.$$

The function ψ belongs to the upper or lower class according as

$$\int_{t_0}^{\infty}\frac{1}{t}(\psi(t))^{d-2}dt=\infty \text{ or } <\infty \quad \text{when } d\geqslant 3,$$

and

$$\int_{t_0}^{\infty}\frac{dt}{t|\log\psi(t)|}=\infty \text{ or} <\infty \quad \text{when } d=2.$$

For example, if $d\geqslant 3$, $\psi(t)=(\log t)^{-(1+\delta)/(d-2)}$ belongs to the upper or lower class according as $\delta\leqslant 0$ or $\delta>0$. If $d=2$ and $\delta>0$, then $\psi(t)=t^{-\delta}$ belongs to the upper class and $\psi(t)=(t\log t)^{-\delta}$ belongs to the lower class.

To describe uniform continuity of a path on the interval [0, 1], take a positive, continuous increasing function ψ defined on $[t_0,\infty)$ with $t_0>0$ and set $\varphi(t)=\sqrt{t}\,\psi(1/t)$. Then ψ is said to **belong to the upper class with respect to uniform continuity** if almost all paths $X(t,\omega)$ $(0\leqslant t\leqslant 1)$ satisfy the †Lipschitz condition relative to φ, that is, for almost all ω there exists an $\varepsilon(\omega)>0$ such that $0<|t-s|<\varepsilon(\omega)$ implies $|X(t,\omega)-X(s,\omega)|<\varphi(|t-s|)$. And ψ is said to **belong to the lower class with respect to uniform continuity** if almost all paths $X(t,\omega)$ $(0\leqslant t\leqslant 1)$ do not satisfy the Lipschitz condition relative to φ. Then ψ belongs to the upper or lower class with respect to uniform continuity according as

$$\int_{t_0}^{\infty}\psi(t)^{d+2}e^{-\psi^2(t)/2}\,dt$$

converges or diverges. For example

$$\psi(t)=\big(2\log t+(d+4)\log_{(2)}t+2\log_{(3)}t+\ldots +2\log_{(n-1)}t+(2+\delta)\log_{(n)}t\big)^{1/2}$$

belongs to the upper or lower class with respect to uniform continuity according as $\delta>0$ or $\delta\leqslant 0$. The following theorem on the uniform continuity of 1-dimensional Brownian motion is a special case of this criterion:

$$P_0\left(\limsup_{\substack{|t-s|\to 0\\ 0\leqslant t,s\leqslant 1}}\frac{|X(t,\omega)-X(s,\omega)|}{\sqrt{2|t-s|\log 1/|t-s|}}=1\right)=1.$$

Now we state some other properties of Brownian paths. Let A be a set of zero †outer capacity in $\mathbf{R}^d$ $(d\geqslant 2)$. Then $P_x(X(t)\in A$ for some $t>0)=0$ for any $x\in\mathbf{R}^d$. Let A be a plane set with positive †inner capacity. Then $P_x(X(t)\in A$ for infinitely many t larger than any given $s>0)=1$ for any $x\in\mathbf{R}^2$. With probability 1, the Lebesgue measure of the set $\{X(t,\omega) \mid 0\leqslant t<\infty\}$ in $\mathbf{R}^d$ is zero for $d\geqslant 2$. This set is everywhere dense in $\mathbf{R}^2$ when $d=2$ but not dense when $d\geqslant 3$. Almost all 2-dimensional Brownian paths have k-fold multiple points for any integer $k \geqslant 2$. In the 3-dimensional case, almost all paths have †double points but cannot have any triple point. In the $d(\geqslant 4)$-dimensional case, almost all paths have no double point. In the $d(\geqslant 3)$- dimensional case, the †Hausdorff measure defined by $t^2 \log\log 1/t$ has the form $\zeta_d t$ for the set $\{X(s,\omega) \mid 0 \leqslant s \leqslant t\}$ with probability 1, where $0 < t \leqslant 1$ and ζ_d is a constant.

G. Additive Functionals

A positive, homogeneous, continuous †additive functional $\varphi(t,\omega)$ of Brownian motion uniquely determines a nonnegative measure de such that

$$\alpha \int_D G^D(x,y) P_\alpha(y) de(y) = 1 - P_\alpha(x)$$

where $G^D(x,y)$ is the Green's function of the bounded Green domain $D \subset \mathbf{R}^d$ and $P_\alpha(x) = E_x(e^{-\alpha\varphi(\sigma_{\partial D})})$. Conversely, under certain conditions, a nonnegative measure determines a positive, homogeneous, continuous additive functional of the Brownian motion (→ 262 Markov Processes). The additive functional of 1-dimensional Brownian motion corresponding to a measure concentrated at $a \in \mathbf{R}^1$ is said to be the **local time** at a. Lévy has studied the structure of the local time. Define the visiting sets $\mathfrak{Z}^+ = \{t \mid Y_0(t,\omega) = 0\}$ and $\mathfrak{Z}^- = \{t \mid Y_1(t,\omega) = 0\}$. Then we have

$$P_0\left(\lim_{\varepsilon\downarrow 0} \sqrt{\frac{\pi\varepsilon}{2}} \times [\text{the number of flat stretches of } M(s,\omega)\,(0 \leqslant s \leqslant t) \text{ of length } \geqslant \varepsilon] = M(t,\omega)\ (t \geqslant 0)\right) = 1.$$

Since the two diffusion processes $X^+ = \{Y_0(t), 0 \leqslant t < \infty, P_x\}$ and $X^- = \{Y_1(t), 0 \leqslant t < \infty, P_x\}$ define the same probability law on $\mathbf{R}^{(0,\infty)}$, there exists a functional $\varphi^+(t,\omega)$ of X^+ corresponding to $M(t,\omega)$ of X^-. The flat stretches of the graph of $\varphi^+(t,\omega)$ are the open intervals $\mathfrak{Z}_n$ $(n \geqslant 1)$ such that $\Sigma_{n\geqslant 1}\mathfrak{Z}_n = [0,\infty) - \mathfrak{Z}^+$. Then we have

$$P_0\left(\lim_{\varepsilon\downarrow 0} \sqrt{\frac{\pi\varepsilon}{2}} \times \left[\text{the number of intervals } \mathfrak{Z}_n \subset [0,t) \text{ of length} \geqslant \varepsilon\right] = \varphi^+(t,\omega)\ (t \geqslant 0)\right) = 1.$$

Furthermore, we have

$$P_0\left(\lim_{\varepsilon\downarrow 0} \sqrt{\frac{\pi}{2\varepsilon}} \times \left[\text{the total length of the intervals } \mathfrak{Z}_n \subset [0,t) \text{ of length } < \varepsilon\right] = \varphi^+(t,\omega)\ (t \geqslant 0)\right) = 1,$$

and

$$P_0\left(\lim_{\varepsilon\downarrow 0} (2\varepsilon)^{-1} \int_0^t \chi_{[0,\infty)}(Y_0(s,\omega))\,ds = \varphi^+(t,\omega)\ (t \geqslant 0)\right) = 1,$$

where $\chi_{[0,\varepsilon)}$ is the indicator function of the interval $[0,\varepsilon)$. Let $d_\varepsilon(t,\omega)$ be the number of times that the reflecting Brownian path $Y_0(s,\omega)$ crosses down from $\varepsilon > 0$ to 0 before time t. Then

$$P_0\left(\lim_{\varepsilon\downarrow 0} \varepsilon d_\varepsilon(t,\omega) = \varphi^+(t,\omega)\ (t \geqslant 0)\right) = 1.$$

The limit

$$\lim_{[x,y]\downarrow a} \frac{1}{2(y-x)} \int_0^t \chi_{[x,y]}(X(s,\omega))\,ds$$

exists and is denoted by $\varphi(t,a,\omega)$. Then we have

$$P_0\left(\limsup_{b-a=\delta\downarrow 0,\, a<b} \frac{|\varphi(t,b,\omega) - \varphi(t,a,\omega)|}{\sqrt{\delta \log 1/\delta}} \leqslant 2\sqrt{\max_{a\in\mathbf{R}^1} \varphi(t,a,\omega)}\right) = 1,$$

and

$$P_0\left(\limsup_{\delta\downarrow 0} \frac{|\varphi(t,\delta,\omega) - \varphi(t,0,\omega)|}{\sqrt{\delta \log\log 1/\delta}} \leqslant 2\sqrt{\varphi(t,0,\omega)}\right) = 1.$$

Let φ be a homogeneous, continuous, but not necessarily positive, additive functional of d-dimensional Brownian motion. Suppose that φ satisfies the condition: for any $C > 0$, $\varepsilon > 0$, and $\delta > 0$ there exists $T > 0$ such that $P_x(|\varphi(t,\omega)| > \varepsilon) < \delta$ for $t \leqslant T$, $|x| \leqslant C$. Then there exist †measurable functions U and V such that V is a †locally square summable real function, U is an $\mathbf{R}^d$-valued function, and

$$\varphi(t,\omega) = V(X(t,\omega)) - V(X(0,\omega)) + \int_0^t (U(X(s,\omega)), dX(s,\omega)),$$

where the last term in the right-hand side means the d-dimensional †stochastic integral (→ 119 Diffusion Processes). In particular, for any function f of class C^2 Itô's formula holds:

$$f(X(t,\omega)) - f(X(0,\omega)) = \int_0^t (\operatorname{grad} f(X(s,\omega)), dX(s,\omega)) + \frac{1}{2}\int_0^t (\Delta f)(X(s,\omega))\,ds.$$

The flow derived from the 1-dimensional Wiener process $\{X_t\}_{-\infty<t<\infty}$ is Kolmogorov's flow. It has †mixing properties of all orders and is †ergodic (→ 386 Stationary

Processes). The †stationary process with independent values at every point corresponding to the †characteristic functional

$$\exp\left(-\frac{1}{2}\int_{-\infty}^{\infty}\varphi(t)^2\,dt\right)$$

on the †Schwartz space $\mathfrak{S}$ defines the same probability law with the stationary process obtained by differentiation of the Wiener process in the †distribution sense.

H. Generalizations of Brownian Motion

In addition to the Brownian motion described above, there are several stochastic processes that are also called Brownian motion. A Gaussian system $\{X(a)\}_{a\in\mathbf{R}^N}$ defined on a probability space $\Omega(\mathfrak{B},P)$ is said to be a **Brownian motion with an N-dimensional time parameter** if (i) $E(X(a))=0$, (ii) $E(X(a)X(b))=\frac{1}{2}(|a|+|b|-|a-b|)$, (iii) $P(X(0)=0)=1$. Let a^* be the †spherical inversion of $a\in\mathbf{R}^N$ with respect to the unit sphere. Set $X^*(a)=|a|X(a^*)$ $(a\neq 0)$ and $X^*(0)=0$. Then $\{X^*(a)\}_{a\in\mathbf{R}^N}$ defines the same probability law with a Brownian motion with an N-dimensional time parameter. This is an analog of the projective invariance principle of Wiener processes. Almost all paths of a Brownian motion with an N-dimensional time parameter are continuous. A positive, continuous, increasing function φ defined on $[t_0,\infty)$ with $t_0>0$ is said to belong to the upper (lower) class with respect to local continuity if the probability that the closure of the set $\{a\,|\,|X(a,\omega)|>\sqrt{|a|}\ \varphi(1/|a|)\}$ contains the origin 0 is equal to 0 (1). Then φ belongs to the upper or lower class with respect to local continuity according as the integral

$$\int_{t_0}^{\infty}\frac{1}{t}(\varphi(t))^{2N-1}e^{-\varphi^2(t)/2}\,dt$$

converges or diverges. For example,

$$\varphi(t)=(2\log_{(2)}t+(2N+1)\log_{(3)}t+2\log_{(4)}t+\ldots+2\log_{(n-1)}t+(2+\delta)\log_{(n)}t)^{1/2}$$

belongs to the upper or lower class with respect to local continuity according as $\delta>0$ or $\delta\leqslant 0$. As a special case, we have

$$P\left(\limsup_{a\to 0}\frac{|X(a,\omega)|}{\sqrt{2|a|\log\log 1/|a|}}=1\right)=1.$$

If almost all paths $X(a,\omega)$ $(|a|\leqslant 1)$ satisfy the Lipschitz condition relative to $\varphi(1/t)\sqrt{t}$, then a positive, continuous, increasing function φ defined on $[t_0,\infty)$ with $t_0>0$ is said to belong to the upper class with respect to uniform continuity. It is said to belong to the lower class with respect to uniform continuity if, with probability 1, these paths do not satisfy the Lipschitz condition relative to $\varphi(1/t)\sqrt{t}$. Then φ belongs to the upper or lower class with respect to uniform continuity according as

$$\int_{t_0}^{\infty}t^{N-1}(\varphi(t))^{4N-1}e^{-\varphi^2(t)/2}\,dt$$

converges or diverges. For example,

$$\varphi(t)=(2N\log t+(4N+1)\log_{(2)}t+2\log_{(3)}t+\ldots+2\log_{(n-1)}t+(2+\delta)\log_{(n)}t)^{1/2}$$

belongs to the upper or lower class with respect to uniform continuity according as $\delta>0$ or $\delta\leqslant 0$. As a special case,

$$P\left(\limsup_{|a-b|\to 0,\,|a|,|b|\leqslant 1}\frac{|X(a,\omega)-X(b,\omega)|}{\sqrt{2N|a-b|\log 1/|a-b|}}=1\right)=1.$$

For general information about Brownian motion with a multidimensional time parameter, see P. Lévy [10] and H. P. McKean [11].

Let $X(t)$ be a Wiener process. L. S. Ornstein and G. E. Uhlenbeck based their investigation of the irregular movements of small particles immersed in a liquid on **Langevin's equation**

$$dU(t)=-\alpha U(t)\,dt+\beta dX(t),$$

where $U(t)$ is the velocity of the particle. The first term on the right-hand side of this †stochastic differential equation (→119 Diffusion Processes) is due to frictional resistance or its analog, which is thought to be proportional to the velocity. The second term represents random external force. The solution of this equation is given by

$$U(t)=\int_{-\infty}^{t}\beta e^{-\alpha(t-u)}\,dX(u).$$

The stochastic process $\{U(t)\}_{-\infty<t<\infty}$ is a stationary Gaussian Markov process with autocorrelation function $\gamma(t)=(\beta^2/2\alpha)e^{-\alpha|t|}$. This process is called **Ornstein-Uhlenbeck Brownian motion**.

For diffusion processes on state spaces that are †Riemannian spaces or †Lie groups, Brownian motion has been defined and is being investigated (→ 6 Additive Processes).

References

[1] K. L. Chung, P. Erdös, and T. Sirao, On the Lipschitz's condition for Brownian motion, J. Math. Soc. Japan, 11 (1959), 263–274.
[2] J. L. Doob, Stochastic processes, John Wiley, 1953.
[3] J. L. Doob, Semi-martingales and subharmonic functions, Trans. Amer. Math. Soc., 77 (1954), 86–121.

[4] Е. Б. Дынкин (E. B. Dynkin), Марковские процессы, физматгиз, 1963; English translation, Markov processes I, II, Springer, 1965.
[5] G. A. Hunt, Some theorems concerning Brownian motion, Trans. Amer. Math. Soc., 81 (1956), 294–319.
[6] K. Itô and H. P. McKean, Jr., Diffusion processes and their sample paths, Springer, 1965.
[7] S. Kakutani, Two-dimensional Brownian motion and harmonic functions, Proc. Imp. Acad. Tokyo, 20 (1944), 706–714.
[8] A. Ja. Khintchin, Asymptotische Gesetze der Wahrscheinlichkeitsrechnung, Erg. Math., Springer, 1933 (Chelsea, 1948).
[9] P. Lévy, Le mouvement brownien plan, Amer. J. Math., 62 (1940), 487–550.
[10] P. Lévy, Processus stochastiques et mouvement brownien, Gauthier-Villars, 1948.
[11] H. P. McKean, Brownian motion with a several-dimensional time, Teor. Verojatnost. i Primenen., 8 (1963), 357–378.
[12] R. E. A. C. Paley and N. Wiener, Fourier transforms in the complex domain, Amer. Math. Soc. Colloq. Publ., 1934.
[13] T. Sirao, On the continuity of Brownian motion with a multidimensional parameter, Nagoya Math. J., 16 (1960), 135–156.
[14] N. Wiener, Differential space, J. Math. and Phys., 2 (1923), 131–174.

C

47 (X.34) Calculus of Variations

A. General Remarks

One of the first objects of differential calculus was to systematize the theory of the extrema of functions of a finite number of independent variables. In the calculus of variations we consider **functionals** (i.e., real-valued or complex-valued functions defined on a functional space $\{u\}$ consisting of functions defined on a certain domain B), originally so named by J. Hadamard. Such a functional is denoted by $J[u|B]$ (or simply $J[u]$). A function u, which is considered an independent variable of the functional, is called an **argument function** (or **admissible function**).

Concrete examples of functionals are the length of a curve $y=f(x)$ and the area of a surface $z=z(x,y)$, which are expressed by

$$L[y]=\int_{x_0}^{x_1}\sqrt{1+(y'(x))^2}\,dx$$

and

$$S[z]=\iint_B\sqrt{1+z_x^2+z_y^2}\,dx\,dy,$$

respectively. Furthermore, consider a curve $y=y(x)$ connecting two given points (x_0,y_0) and (x_1,y_1) with $y_1>y_0$. The time in which a particle slides down without friction from (x_0,y_0) to (x_1,y_1) along this curve under constant gravity acting is the direction of the positive y-axis is expressed by the functional

$$J[y]=k\int_{x_0}^{x_1}\sqrt{(1+y'^2)/(y-y_0)}\,dx,$$

k constant. (1)

Let $F(\ldots)$ be a known real-valued function that depends on a certain number of independent variables and on variable argument functions of them, together with derivatives of the latter functions up to a certain order. Then a typical problem in the calculus of variations is formulated as an extremal problem of a functional that is expressed by the integral with $F(\ldots)$ as the integrand, for example,

$$J[u]=\int_{x_0}^{x_1}F(x,u(x),u'(x),\ldots,u^{(m)}(x))\,dx,$$

$$J[u,v]=\iint_B F(x,y,u(x,y),u_x,u_y,\ldots,$$
$$v(x,y),v_x,v_y,\ldots)\,dx\,dy.$$

For instance, a curve minimizing (1) is the **curve of steepest descent**.

In extremal problems, suitable boundary conditions may be assigned to argument functions. On the other hand, there are **conditional problems in the calculus of variations**. A typical example of this sort is the †isoperimetric problem, i.e., the determination of the curve that bounds a domain with maximal area among all curves on a plane with given length. In general, an extremal problem of a functional under a subsidiary condition that the value of another given functional remain fixed is called a **generalized isoperimetric problem**. In addition to these, there are **Lagrange's problem**, in which a finiteness condition is imposed, and **Hilbert's problem**, in which a condition consisting of differential equations is imposed.

The birth of the calculus of variations was almost simultaneous with that of differential and integral calculus. Johann Bernoulli, Jakob Bernoulli, L. Euler, and others had dealt with several concrete problems of the calculus of variations when in 1760, J. L. Lagrange introduced a general method of dealing with variational problems connected with mechanics. Then an equation bearing the name of Euler or Lagrange was introduced.

B. Euler's Equation

As an example, consider the simplest variational problem

$$J[y]=\int_{x_0}^{x_1}F(x,y(x),y'(x))\,dx=\min. \qquad (2)$$

Let the boundary condition $y(x_0)=y_0,y(x_1)=y_1$ be assigned to the argument function $y(x)$. Consider a family of admissible functions $Y(x;\varepsilon)=y(x)+\varepsilon\eta(x)$, where $\eta(x)$ is any fixed function vanishing at both endpoints and ε is a parameter. If $y(x)$ gives the minimum of $J[y]$, then the function of ε, $J[Y]$, must become minimum for $\varepsilon=0$. The condition $(\partial J[Y]/\partial\varepsilon)_{\varepsilon=0}=0$ is written in the form

$$0=\int_{x_0}^{x_1}(F_y\eta+F_{y'}\eta')\,dx$$
$$=\int_{x_0}^{x_1}\eta\left(F_y-\frac{d}{dx}F_{y'}\right)dx$$

by taking into account the boundary condition. By making use of the arbitrariness of $\eta(x)$, we conclude that

$$0=F_y-\frac{d}{dx}F_{y'}=F_y-F_{y'x}-y'F_{y'y}-y''F_{y'y'} \qquad (3)$$

holds, in view of the following lemma: Let $\varphi(x)$ be continuous in $[x_0,x_1]$ and $\eta(x)$ be a function of class C^p that satisfies $\eta(x_0)=\eta'(x_0)=\ldots=\eta^{(q)}(x_0)=0$, $\eta(x_1)=\eta'(x_1)=\ldots=\eta^{(q)}(x_1)=0$ $(0\leqslant q\leqslant p)$. If $\int_{x_0}^{x_1}\eta(x)\varphi(x)\,dx=0$ holds for any such $\eta(x)$, then $\varphi(x)=0$. (Here class C^p may be replaced by class C^ω while q is supposed finite. If $\eta\in C^\omega$, then $q=\infty$ is admitted.) This is called the **fundamental lemma in the calculus of variations**. We call

(3) the **Euler-Lagrange differential equation** (or **Euler's equation** for the extremal problem). Since this equation is of the second order, $y(x)$ will be determined by means of the boundary condition.

The quantity $[F]_y = F_y - \frac{d}{dx}F_{y'}$ contained in equation (3) is called the **variational derivative** of F with respect to y. Furthermore, $\delta y = \eta d\varepsilon$ and $\delta J = (\partial J[Y]/\partial\varepsilon)_{\varepsilon=0} d\varepsilon$ are called the **first variations** of the argument function y and of the functional $J[y]$, respectively. If $\eta(x)$ is not subject to the condition that it must vanish at the endpoints, then the first variation of J becomes

$$\delta J = \int_{x_0}^{x_1}[F]_y \delta y\, dx + [F_{y'}\delta y]_{x_0}^{x_1}.$$

In comparison with an ordinary extremal problem $f(x_1, \ldots, x_n) = \min$ in differential calculus, $[F]_y$ and δJ correspond to grad$f$ and df, respectively. In general, a solution of the Euler-Lagrange differential equation $[F]_y = 0$ is called a **stationary function** for the variational problem, and its graph is called a **stationary curve**.

For a variational problem involving several argument functions, we have only to write the system of Euler-Lagrange differential equations corresponding to them. For a problem

$$J[y] = \int_{x_0}^{x_1} F(x,y,y',\ldots,y^{(m)})\, dx = \min$$

whose integrand involves derivatives of higher orders of an argument function, the Euler-Lagrange differential equation is

$$[F]_y = \sum_{\mu=0}^{m} (-1)^\mu \frac{d^\mu}{dx^\mu} F_{y^{(\mu)}} = 0.$$

For a problem involving a double integral

$$J[u] = \iint_B F(x,y,u,u_x,y_y)\, dx\, dy = \min,$$

the equation is

$$0 = [F]_u = F_u - \frac{\partial}{\partial x}F_{u_x} - \frac{\partial}{\partial y}F_{u_y}.$$

For a generalized isoperimetric problem, for example $J[y] = \min$, $K[y] = c$, the Euler-Lagrange equation becomes

$$[F + \lambda G]_y = 0$$

with the so-called **Lagrange multiplier** λ, where F and G denote the integrands of J and K, respectively. Two integration constants contained in a general solution of this differential equation of the second order and an undetermined constant λ will be determined, for instance, by the boundary condition $y(x_0) = y_0$, $y(x_1) = y_1$ and a subsidiary condition $K[y] = c$. As an example, for the classical proper isoperimetric problem $F = y$, $G = \sqrt{1+y'^2}$, the equation is $1 - \lambda(y'/\sqrt{1+y'^2})' = 0$, which after integration leads to $(x-\alpha)^2 + (y-\beta)^2 = \lambda^2$ (→ 228 Isoperimetric Problems).

Besides the case of fixed endpoints, there is a boundary condition, for instance, that an endpoint (x_1, y_1) of the argument function $y = y(x)$ must lie on a given curve $T(x,y) = 0$. For the case of such a movable endpoint, the extremal function is subject to the **condition of transversality**,

$$(F - y'F_{y'})T_y - F_{y'}T_x = 0, \qquad x = x_1.$$

C. Sufficient Conditions

A. M. Legendre introduced the notion of the second variation, corresponding to differential quotients of the second order in differential calculus, in order to discuss sufficient conditions. Concerning the simplest problem (2), the inequality $F_{y'y'}(x, y_0(x), y_0'(x)) \geqslant 0$ is necessary in order for $y_0(x)$ to give the minimum. Conversely, the inequality $F_{y'y'} > 0$ and Jacobi's condition (which follows) imply that $y = y_0(x)$ gives a **weak minimum**. Here "weak minimum" means the minimum when a family of admissible functions $\{|y - y_0| < \varepsilon, |y' - y_0'| < \varepsilon\}$ is considered a neighborhood of y_0.

Jacobi's condition: Let u be a solution of a linear ordinary differential equation of the second order,

$$\frac{d}{dx}\left(F_{y'y'}\frac{du}{dx}\right) - \left(F_{yy} - \frac{d}{dx}F_{y'y'}\right)u = 0,$$

$$u(x_0) = 0;$$

then the smallest zero of u that is greater than x_0 (i.e., the **conjugate point** of x_0) is greater than the right endpoint x_1.

K. Weierstrass derived sufficient conditions for a strong minimum by extending the range of admissible functions to $\{|y - y_0| < \varepsilon\}$. Results that were obtained until about this time constitute the content of what is usually called the **classical theory of calculus of variations.**

If for the variational problem (2) there exists a unique curve through every point in a domain on the xy-plane that belongs to a one-parameter family of stationary curves of the functional

$$J[y] = \int_{x_0}^{x_1} F(x,y,y')\, dx,$$

then the domain is called a **field** of stationary curves. Let the parameter value of the curve through a point (x,y) in such a family of stationary curves $y = \varphi(x;\alpha)$ be denoted by $\alpha = \alpha(x,y)$. The slope $p(x,y) = [\varphi'(x;\alpha)]_{\alpha=\alpha(x,y)}$ is called the slope of the field at the point (x,y) or, by regarding x,y as variables, the **slope function** of the field. The value of a curvilinear integral

$$I_C = I[y]$$
$$= \int_C (F(x,y,p) - (y'-p)F_{y'}(x,y,p))\,dx$$

is then determined, and depends only on the two endpoints of the curve C. We call I_C **Hilbert's invariant integral**. In view of the property mentioned above, we may denote the value of the functional J for a function $y = y(x)$ representing a curve C by J_C. Then any admissible curve C that passes through a field embedding a stationary curve C_0 satisfies

$$0 \leqslant \Delta J = J_C - J_{C_0} = \int_C \mathcal{E}(x,y;p,y')\,dx.$$

Here

$$\mathcal{E}(x,y;p,y')$$
$$= F(x,y,y') - F(x,y,p) - (y'-p)F_{y'}(x,y,p)$$

is the $\mathcal{E}$**-function** introduced by Weierstrass. For C_0 to give the minimum of $J[y]$, it suffices that $\mathcal{E} \geqslant 0$ hold for every point (x,y) in the field and every value y' (→ 48 Calculus of Variations in the Large).

D. Optimal Control

Let a system of differential equations

$$dx_i/dt = f_i(x_1, \ldots, x_n; u_1, \ldots, u_k),$$
$$(u_u, \ldots, u_k) \in \Omega; \quad x_i(t_0) = x_i, \quad i = 1, \ldots, n, \tag{4}$$

be given, where $u_1, \ldots, u_k$ are parameters. In general, a problem of **optimal control** is to determine $u_j = u_j(t)$ $(t_0 \leqslant t \leqslant t_1)$ such that the value of a functional

$$J[u] = \int_{t_0}^{t_1} F(x_1, \ldots, x_n; u_1, \ldots, u_k)\,dt$$

becomes minimum, where $x_i(t)$ are the solutions of (4) and are considered functions in $u_1, \ldots, u_k$, and t. Such a problem is a kind of conditional variational problem. But since the existence region of u is restricted, certain conditions in the form of inequalities are imposed, and furthermore u is not necessarily continuous, in many cases the problem cannot be treated within the classical theory of the calculus of variations (→ 88 Control Theory).

E. The Direct Method in the Calculus of Variations

In mathematical physics †variational principles are derived from discussions of formal correspondence between a functional $J[u]$ to be minimized and Euler's equation for $J[u]$. This is certainly one of the important methods in the calculus of variations, but it is also possible to investigate a stationary function u_0 on the basis of its stationary character and independently of Euler's equation. This is called the **direct method** in the calculus of variations. It plays an important role in the theoretical treatment of the existence and uniqueness of solutions, and it is also significant as a technique for approximate or numerical solutions. When a differential equation is given independently of the calculus of variations, it is possible to apply the direct method if a functional whose Euler's equation is the given differential equation can be constructed.

Let D be a bounded domain in m-dimensional space and $f \in L_2(D)$ be a real-valued function. Consider the variational problem of minimizing the functional

$$J[u] = \int_D |\mathrm{grad}\, u|^2\,dx - 2\int_D fu\,dx.$$

Here we suppose the set of admissible functions, denoted by $\bar{A}_J$, to be the Hilbert space obtained by completing the function space $C_0^\infty(D)$ with respect to the norm

$$N(u) = \left(\int_D |\mathrm{grad}\, u|^2\,dx + \int_D u^2\,dx\right)^{1/2}.$$

Utilizing F. Riesz's representation theorem in Hilbert spaces, it can be shown that there exists a minimum value l in $\bar{A}_J$ which is uniquely realized by certain $u_0 \in \bar{A}_J$. Since the function u_0 belongs to $\bar{A}_J$, it can be shown that the boundary condition

$$u|_{\partial D} = 0 \tag{5}$$

is satisfied in a generalized sense. Furthermore, in view of $J[u_0] \leqslant J[u_0 + \varphi]$ being valid for any $\varphi \in C_0^\infty(D)$, it can be verified that the equation

$$-\Delta u = f \tag{6}$$

is satisfied in D in the sense of differentiation of †distributions. In other words, the stationary function u_0 is a solution in the wider sense (a †weak solution) of the classical boundary value problem for †Poisson's equation formulated by (6) and (5). If a function space A_J with $\bar{A}_J \supset A_J \supset C_0^\infty(D)$ is taken as the set of admissible functions, the value $l = J[u_0]$ becomes the greatest lower bound of J in A_J. In this case, if $\{u_n\}_{n=1}^\infty$ is any **minimizing sequence** from A_J, that is, if $u_n \in A_J$, $n = 1, 2, \ldots$; $J[u_n] \to l$ $(n \to \infty)$, then it converges to u_0 in the sense of

$$N(u_n - u_0) \to 0 \quad (n \to \infty). \tag{7}$$

In other words, the solution in the wider sense u_0 of the boundary value problem can be constructed as the limit of a minimizing sequence that consists of sufficiently smooth functions vanishing on the boundary. In the proof of the fact that a solution in the wider sense u_0 coincides with the classical solution of the boundary value problem under an assumption of suitable smoothness for f and ∂D, a standard argument has been established for proving the regularity of a solution in the wider sense.

The technique of obtaining the solution of the boundary value problem as the limit of a minimizing sequence was proposed by B. Riemann concerning the classical †Dirichlet problem and was completed by D. Hilbert. This pioneer work led to the recent treating of boundary value problems by utilizing Hilbert spaces. For †self-adjoint boundary value problems, the method stated for the above example has been generalized almost directly to the cases of higher order and with variable coefficients. By making use of some auxiliary arguments, this technique can be applied extensively to the construction of several kinds of mapping functions in the theory of functions of a complex variable, to solution of †integral equations of the second kind, and also in other fields [3, 4, 7, 8].

The eigenvalue problem, which is formulated by

$$Hu=\lambda u, \qquad u\neq 0, \tag{8}$$

with a †self-adjoint operator H in a Hilbert space, can also be transformed into a variational problem for the †Rayleigh quotient

$$R[u]=(Hu,u)/\|u\|^2 \tag{9}$$

(→ 136 Eigenvalues (Numerical Computation)).

F. Solution of Differential Equations by the Direct Method

In view of the convergence shown in (7), a minimizing sequence may be regarded as an approximating sequence for a solution of the boundary value problem or a stationary function u_0. Let a function

$$u_n=u_n(x;c)=u_n(x;c_1,\ldots,c_n) \tag{10}$$

involving an n-vector $c=(c_1,\ldots,c_n)$ as a parameter be admissible for any c. If $J[u_n(\cdot\,;c)]=F(c)$, obtained by substituting u_n into J, is minimized at $c=c^0$, then $u_n(\cdot\,;c^0)$ is considered as a function that approximates u_0 most precisely within the family $u_n(\cdot\,;c)$. This vector c^0 is obtained, in general, by solving the simultaneous equations

$$\partial J\big[u_n(\cdot\,;c)\big]/\partial c_j=0, \qquad j=1,\ldots,n. \tag{11}$$

The function u_n appearing in (10) is often taken to be a so-called linear admissible function. For instance, in the above example, let $\{\varphi_k\}_{k=1}^{\infty}$ be a system of independent functions complete in $\tilde{A}_J$, and set

$$u_n=c_1\varphi_1+\ldots+c_n\varphi_n. \tag{12}$$

The method that constructs a minimizing sequence $u_1,u_2,\ldots$ by determining the value of c in (12) by (11) is called **Ritz's method**, and φ_k is called a coordinate function in this method. As for the rate of convergence in approximation by Ritz's method as well as estimation of errors, there are several results by the Soviet school in addition to those of E. Trefftz [3, 5]. (Other methods of constructing minimizing sequences are stated in detail in [3]; concerning a connection with †Galerkin's method → 299 Numerical Solution of Partial Differential Equations.)

Ritz's method applied to eigenvalue problems is called the **Rayleigh-Ritz method** [7]. Since a stationary value of the Rayleigh quotient is itself an eigenvalue in this case, the precision of approximation is far better for eigenvalues than for eigenfunctions, so that it is a convenient method of approximate computation of eigenvalues.

References

[1] R. Courant and D. Hilbert, Methods of mathematical physics, Interscience, I, 1953; II, 1962.

[2] P. Funk, Variationsrechnung und ihre Anwendung in Physik und Technik, Springer, 1962.

[3] Л. В. Канторович, В. И. Крылов (L. V. Kantorovič and V. I. Krylov), Приближенные методы высшего анализа, Гостехиздат, 1952.

[4] С. Г. Михлин (S. G. Mihlin), Вариационные методы в математической физике, Гостехиздат 1957; German translation, Variationsmethoden der mathematischen Physik, Akademie-Verlag, 1962; English translation, Variational methods in mathematical physics, Macmillan, 1964.

[5] С. Г. Михлин (S. G. Mihlin), Проблема минимума квадратичного функционала, Гостехиздат, 1952; English translation, The problem of the minimum of a quadratic functional, Holden-Day, 1965.

[6] С. Л. Соболев (S. L. Sobolev), Некоторые применеия функционального анализа в математической физике, Издат. Ленинград. Гос. Унив., 1950; English translation, Applications of functional analysis in mathematical physics, Amer. Math. Soc. Transl. of Math. Monographs, 1963.

[7] K. Terazawa (ed.), Sizen kagakusya no tameno sûgaku gairon (ôyô hen) (Japanese; Introduction to mathematics for natural scientists pt. II. Application), Iwanami, 1960.

[8] K. Yosida and T. Kato, Ôyôsûgaku, ensyû (Japanese; Exercises on applied mathematics), I., Syôkabô, 1961.
[9] O. Bolza, Vorlesungen über Variationsrechnung, Teubner, 1909.
[10] И. М. Гельфанд С. В. Фомин (I. M. Gel'fand and S. V. Fomin), Вариационное исчисление, Физматгиз, 1961; English translation, Calculus of variations, Prentice-Hall, 1963.
[11] Y. Komatu, Henbungaku (Japanese; Calculus of variations), Tôkai Syobô, 1947.
[12] M. Nagumo, Henbungaku (Japanese; Calculus of variations), Asakura, 1951.

48 (VII.8) Calculus of Variations in the Large

A. General Remarks

By a **variational problem** we mean a pair (X,f) of a †topological space X and a real-valued continuous function f on X. Given such a problem (X,f), the **calculus of variations in the large** is concerned with relations between the properties of the space X and those of the function f. As will be shown, some important properties of the space X follow from the properties of f at its critical points. This method is most effectively applied in the following two cases: (i) when X is a †differentiable manifold M and f is a differentiable function on M, and (ii) when X is the †path space Ω on M and f is the energy function E on Ω. In particular, the theory concerning (ii) is based on the theory of †geodesics on †Reimannian manifolds; the methods of the ordinary †calculus of variations can also be utilized (→ 47 Calculus of Variations). The rudiments of the theory concerning both cases can be found in H. Poincaré [1] and G. D. Birkhoff [2]. Later, M. Morse gave a systematic formulation. Recently the Morse theory has been further extended and refined and is now applied to the problems of †differential topology and †differential geometry.

B. Critical Points of Functions on Manifolds

The †tangent vector space of an n-dimensional †differentiable manifold M at a point p will be denoted by T_p. For the rest of this article, by "differentiable" we mean differentiability of class C^∞. Let f be a differentiable real-valued function on M. A point $p \in M$ is called a **critical point** of f if the induced mapping $f_*: T_p \to \mathbf{R}_{f(p)}$ is zero. If we choose a †local coordinate system $(x^1, \ldots, x^n)$ in a neighborhood of p, we have

$$\frac{\partial f}{\partial x^1}(p) = \ldots = \frac{\partial f}{\partial x^n}(p) = 0.$$

In this case the real number $f(p)$ is called a **critical value** of f. If the matrix $(\partial^2 f/\partial x^i \partial x^j(p))$ is invertible, a critical point p is called **nondegenerate**. If this matrix is not invertible, p is called **degenerate**. This definition is independent of the choice of coordinate system around p. The matrix $(\partial^2 f/\partial x^i \partial x^j(p))$ is called the **Hessian** of f at p. The †nullity and **index**, i.e., the number of negative eigenvalues, of the Hessian of f are called the nullity and index of f at p, respectively. The function f attains locally a minimal value at a nondegenerate critical point of index 0 and attains locally a maximal value at a nondegenerate critical point of index n. If p is a nondegenerate critical point of f, then there is a local coordinate system $(y^1, \ldots, y^n)$ in a neighborhood of p with $y^i(p) = 0$ for all i and such that the following identity holds: $f(y) = f(p) - (y^1)^2 - \ldots - (y^\lambda)^2 + (y^{\lambda+1})^2 + \ldots + (y^n)^2$, where λ is the index of f at p.

For constructing manifolds, we now consider the process of †attaching a handle. Let M be a compact manifold with †boundary ∂M. Let D^s be the s-disk, and let $g: (\partial D^s) \times D^{n-s} \to \partial M$ be an †embedding. Then a †manifold $X(M; g; s)$ with a handle attached by g is defined as the quotient set obtained from the disjoint union $M \cup (D^s \times D^{n-s})$ by identifying points in $\partial D^s \times D^{n-s}$ and their images under g and equipped with a natural differentiable structure. Similarly, if $g_i: (\partial D_i^{s_i}) \times D_i^{n-s_i} \to \partial M$ $(i = 1, \ldots, k)$ are embeddings with disjoint images, we can define the †handle body $X(M; g_1, \ldots, g_k; s_1, \ldots, s_k)$ [13].

Let f be a differentiable real-valued function on a manifold M, and let a be a positive number. We put $M^a = f^{-1}(-\infty, a] = \{p \in M \mid f(p) \leqslant a\}$. Then the following fundamental properties on the topology of M and M^a are known (J. Milnor [15]): (i) Let f be a differentiable function on a closed (compact without boundary) manifold with no critical points on $f^{-1}[c-\varepsilon, c+\varepsilon]$ except k nondegenerate ones on $f^{-1}(c)$ with indices $s_1, \ldots, s_k$. Then $M^{c+\varepsilon}$ is †diffeomorphic to $X(M^{c-\varepsilon}; g_1, \ldots, g_k; s_1, \ldots, s_k)$ for suitable embeddings $g_1, \ldots, g_k$. (ii) On every compact manifold M there exists a differentiable function without degenerate critical points. By combining properties (i) and (ii) we see that every compact manifold can be obtained by successively attaching handles to a disk (→ 410 Topology of Differentiable Manifolds D, E).

Concerning the †homotopy types of M and M^a, we have the following two theorems:

(1) Let c be a real number. Suppose that $f^{-1}(c)$ contains k nondegenerate critical points $p_1, \ldots, p_k$ with indices $\lambda_1, \ldots, \lambda_k$, respectively and, $f^{-1}[c-\varepsilon, c+\varepsilon]$ is compact and contains no critical points of f other than $p_1, \ldots, p_k$. Then for all sufficiently small positive numbers ε, the set $M^{c-\varepsilon}$ has the homotopy type of the set $M^{c-\varepsilon} \cup e^{\lambda_1} \cup \ldots \cup e^{\lambda_k}$, where e^{λ_i} are λ_i-cells ($i=1, \ldots, k$).

(2) Let f be a differentiable function on a manifold M with no degenerate critical points, and suppose each M^a is compact. Then M has the homotopy type of a †CW-complex, with one †cell of dimension λ for each critical point of index λ.

Moreover, we have the following theorem: (3) Let f be a differentiable function on a compact manifold M with no degenerate critical points. Let M_λ be the number of critical points of f on M of index λ, and R_λ be a λ-dimensional †Betti number of M. Then we have the following inequalities:

$$M_0 \geqslant R_0,$$
$$M_1 - M_0 \geqslant R_1 - R_0,$$
$$M_i - M_{i-1} + \ldots + (-1)^i M_0 \geqslant R_i - R_{i-1} + \ldots + (-1)^i R_0, \quad 1 < i < n-1,$$
$$M_n - M_{n-1} + \ldots + (-1)^n M_0 = R_n - R_{n-1} + \ldots + (-1)^n R_0.$$

In particular, we have $M_k \geqslant R_k$ for all k. We call these inequalities **Morse inequalities**.

By using these results, S. Smale solved the †Poincaré conjecture in the higher-dimensional case, one of the most remarkable results in differential topology [13] (→ 117 Differential Topology).

C. Geodesics

Let M be a †Riemannian manifold. A curve $\gamma = \gamma(t)$ on M is called a geodesic if the tangent vector field $\gamma'(t) = d\gamma/dt$ on γ is parallel with respect to the †Riemannian connection of M (→ 360 Riemannian Manifolds). A geodesic is characterized as a curve whose sufficiently small subarcs are always minimal paths connecting their endpoints. In differential geometry, geodesics play important roles in both global and local considerations. In a coordinate neighborhood with coordinates $(u^1, \ldots, u^n)$, a geodesic $\gamma = \gamma(t) = (u^1(t), \ldots, u^n(t))$ is given as the solution of the differential equation

$$\frac{d^2u^k}{dt^2} + \sum_{i,j=1}^{n} \left\{ {k \atop ij} \right\} \frac{du^i}{dt} \frac{du^j}{dt} = 0, \quad k = 1, \ldots, n,$$

where $\left\{ {k \atop ij} \right\}$ is †Christoffel's symbol. Hence a geodesic is uniquely determined by its initial point and initial direction. Suppose that the geodesic $\gamma = \gamma(t)$ satisfies the conditions $\gamma(0) = q$ and $\gamma'(0) = v$ and that $\gamma(1)$ exists. If we put $\exp_q v = \gamma(1)$, the geodesic γ can be denoted by $\gamma(t) = \exp_q tv$. The $\exp_q$ diffeomorphically maps a sufficiently small neighborhood of the origin of the tangent vector space T_q onto a neighborhood of q. This mapping is called the **exponential mapping**. If M is †complete, then $\exp_q$ is defined for the whole T_q, and any two points can be joined by a minimal geodesic (H. Hopf and W. Rinow) (→ 360 Riemannian Manifolds C).

For the rest of this article, we will treat only curves that are †piecewise differentiable. Let γ be a curve connecting points p and q in M. If there exists a neighborhood U of γ such that γ is a minimal path in U connecting p and q, we call γ the **relative minimal curve** between p and q. If γ is a relative minimal curve, then γ is a geodesic; however, the converse is not true.

A vector field Y along a geodesic γ is called a **Jacobi field** if it satisfies the **Jacobi differential equation** $Y'' + R(\gamma', Y)\gamma' = 0$, where Y' is the †covariant derivative of Y in the direction of γ' and R is the †curvature tensor of M. A point q is called a **conjugate point** of p along γ if there exists a nonzero Jacobi field along γ that vanishes for p and q. The **multiplicity** of q as a conjugate point is equal to the dimension of the vector space consisting of all Jacobi fields. If the geodesic $\gamma = \widehat{pq}$ contains a conjugate point of p different from q, then γ is not a relative minimal curve between p and q. Conversely, if $\gamma = \widehat{pq}$ contains no conjugate point of p, then γ is a relative minimal curve between p and q. However, a geodesic $\gamma = \widehat{pq}$ is not always a minimal geodesic on M even if it contains no pair of conjugate points. In this case, however, there exists a point p^* on γ such that for any point r in the subarc $\widehat{pp^*}$, the arc $\widehat{pr}$ is a minimal geodesic between p and r. The point p^* on γ farthest from p that has this property is called the **cut point** (or **minimum point**) of p on γ. The following property is well known: Given a compact set $K \subset M$, there exists a number $\delta > 0$ such that any two points of K separated by a distance less than δ can be joined by a unique minimal geodesic. Moreover, such a geodesic depends differentiably on its endpoints. Such a number δ is called the **elementary length** of K.

On the relations between the curvature of M and the distribution of conjugate points and cut points, the following three properties are known [15]:

(1) Suppose that the †sectional curvature of M is everywhere nonpositive. Then no two points of M are conjugate along any geodesic. Furthermore, if M is complete and simply connected, there exists no cut point of an initial point on any geodesic. Hence any two points on M can be joined by a unique geodesic, and M is diffeomorphic to a Euclidean space (J. Hadamard, E. Cartan).

(2) Suppose that the †Ricci curvature of M is everywhere greater than or equal to a fixed positive number k. Then any geodesic whose length is greater than $\pi/\sqrt{k}$ contains cut points and conjugate points of the initial points that are different from the endpoint (S. B. Myers).

(3) Suppose that M is complete and simply connected and that the sectional curvature K satisfies the inequalities $l/4<k\leqslant K\leqslant l$ everywhere where k, l are two positive numbers. Then there exist no cut point and conjugate point of the initial point on any geodesic of length less than $\pi/\sqrt{l}$, while there exist at least one cut point and conjugate point on any geodesic of length $L\geqslant\pi/\sqrt{k}$. Hence any two points separated by a distance $<\pi/\sqrt{l}$ can be joined by a unique minimal geodesic [10]. If M is even-dimensional and the sectional curvature K satisfies the inequalities $0<k\leqslant K\leqslant l$ everywhere, then the same results can be obtained (W. Klingenberg [20]).

Using property (3) we obtain the so-called **sphere theorem** on the topological structures of Riemannian manifolds of positive curvature: If the sectional curvature K of a complete simply connected Riemannian manifold M satisfies the inequalities $c/4<K\leqslant c$ everywhere, then M is homeomorphic to a sphere, where c is a positive constant (H. E. Rauch and [10]).

A geodesic γ with the initial point p in M may have p as its endpoint. Furthermore, if the initial direction of γ coincides with the final direction, then γ is called a **closed geodesic**. Concerning the existence and number of closed geodesics on Riemannian manifolds, we have the following theorems: (i) There exist at least three closed geodesics on a compact Riemannian manifold [12, 21]. (ii) If M is a Riemannian manifold homeomorphic to an n-sphere, there exist at least $2n-s-1$ closed geodesics on M, where s is the integer such that $n=2^k+s$, $0\leqslant s<2^k$. Under the same assumption, there exist at least $n(n+1)/2$ closed geodesics on M, counting their multiplicities (S. L. Al'ber [9], Morse [4]).

Recently, W. Klingenberg obtained similar results on Riemannian manifolds homeomorphic to compact symmetric Riemannian spaces of rank 1 [18, 21] (→ 401 Symmetric Riemannian Spaces). If M is a compact †symmetric Riemannian space of rank 1, all the geodesics on M are closed and have the same length, but the converse is not true. However, if M is a complete simply-connected Riemannian manifold all of whose geodesics are closed and of the same length, then the integral †cohomology ring of M is isomorphic to that of a compact symmetric Riemannian space of rank 1 [7].

D. Variations on the Path Space

Let M be a differentiable manifold, and let p and q be two (not necessarily distinct) points of M. The set of all curves from p to q in M is denoted by $\Omega(M;p,q)$ (or simply Ω). Suppose that a mapping $\omega:[0,1]\rightarrow M$ is an element of Ω. By the tangent vector space of Ω at a curve ω we mean the vector space consisting of all piecewise differentiable vector fields W along ω for which $W(p)=0$ and $W(q)=0$. The notation Ω_ω is used for this vector space. A **variation** of ω (keeping endpoints fixed) is a mapping $\bar{\alpha}:(-\varepsilon,\varepsilon)\rightarrow\Omega$ for some $\varepsilon>0$ such that the following hold: (i) $\bar{\alpha}(0)=\omega$. (ii) There is a subdivision $0=t_0<t_1<\ldots<t_k=1$ of $[0,1]$ such that the mapping $\alpha:(-\varepsilon,\varepsilon)\times[0,1]\rightarrow M$ defined by $\alpha(u,t)=\bar{\alpha}(u)(t)$ is differentiable of class C^∞ on each strip $(-\varepsilon,\varepsilon)\times[t_{i-1},t_i]$ $(i=1,\ldots,k)$. (iii) $\alpha(u,0)=p$, $\alpha(u,1)=q$ for all $u\in(-\varepsilon,\varepsilon)$. More generally, if in the definition of "variation" $(-\varepsilon,\varepsilon)$ is replaced by a neighborhood U of 0 in $\mathbf{R}^n$, then α is called an **n-parameter** variation of ω.

Now suppose that M is a Riemannian manifold. The length of a vector $v\in T_p$ is denoted by $\|v\|$. For $\omega\in\Omega$ we define the **energy** of $\omega=\omega(t)$ from a to b (where $0<a<b<1$) as

$$E_a^b(\omega)=\int_a^b\left\|\frac{d\omega}{dt}\right\|^2dt$$

and write E for E_0^1. The arc length of the curve from a to b is given by

$$L_a^b(\omega)=\int_a^b\left\|\frac{d\omega}{dt}\right\|dt.$$

We have a relation $(L_a^b)^2\leqslant(b-a)E_a^b$.

Let M be a complete Riemannian manifold, and let p,q be points in M separated by a distance d. Then the energy function $E:\Omega(M;p,q)\rightarrow\mathbf{R}$ takes its minimum d^2 precisely on the set of minimal geodesics connecting p and q. Let $\alpha:(-\varepsilon,\varepsilon)\rightarrow\Omega$ be a variation of ω, and let $W_t=(\partial\alpha/\partial u)(0,t)=\alpha_*(\partial/\partial u)(0,t)$ be the associated variation vector field. Furthermore, put $V_t = d\omega/dt$, $A_t = (d\omega/dt)'$ and $\Delta_t V = V_{t+} - V_{t-}$. We get $\Delta_t V = 0$ for all but a finite number of values of t. We then have the **first variation formula**

$$\frac{1}{2}\frac{dE(\alpha(u))}{du}\bigg|_{u=0}$$

$$= -\sum_{t\in(0,1)} \langle W_t, \Delta_t V\rangle - \int_0^1 \langle W_t, A_t\rangle dt,$$

where $\langle\,,\rangle$ is the inner product induced by the Riemannian metric on M. By this formula, we see that the curve ω is a critical point for the function E if and only if ω is a geodesic.

Let $\gamma: [0,1]\to M$ be a geodesic. Given two vector fields $W_1, W_2 \in \Omega_\gamma$, choose a 2-parameter variation $\alpha: U\times[0,1]\to M$, where U is a neighborhood of $(0,0)$ in $\mathbf{R}^2$ such that $\alpha(0,0,t) = \gamma(t), (\partial\alpha/\partial u_1)(0,0,t) = W_1(t), (\partial\alpha/\partial u_2)(0,0,t) = W_2(t)$ $(0\leqslant t\leqslant 1)$. Then the **Hessian** $E_{**}(W_1, W_2)$ is defined as follows:

$$E_{**}(W_1, W_2) = \frac{\partial^2 E(\bar{\alpha}(u_1,u_2))}{\partial u_1 \partial u_2}\bigg|_{(0,0)},$$

where $\bar{\alpha}(u_1,u_2)\in\Omega$ denotes the curve $\bar{\alpha}(u_1, u_2)(t) = \alpha(u_1,u_2,t)$. The **second variation formula** is as follows:

$$\frac{1}{2}E_{**}(W_1, W_2)$$

$$= -\sum_{t\in(0,1)} \langle W_2(t), \Delta_t W_1'(t)\rangle$$

$$- \int_0^1 \langle W_2, W_1'' + R(V, W_1)V\rangle dt,$$

where $V = d\gamma/dt$, $\Delta_t W_1' = W_1'(t+0) - W_1'(t-0)$ $(\Delta_t W_1' = 0$ for all but a finite number of values of t). If γ is a minimal geodesic connecting p and q, then E_{**} is positive semidefinite.

The **index** of the Hessian $E_{**}: \Omega_\gamma\times\Omega_\gamma\to\mathbf{R}$ is defined as the maximum dimension of a subspace of Ω_γ on which E_{**} is negative definite. The **Morse index theorem**, which clarifies the relation between the index of E_{**} and the distribution of the conjugate points on γ, is as follows: The index λ of E_{**} is equal to the number of points $\gamma(t)$ (including their multiplicities) that are conjugate to $\gamma(0)$ along γ $(0<t<1)$. This index is always finite [15].

Given a positive number c, we set $\Omega^c = E^{-1}([0,c])$. We now describe a method to obtain a finite-dimensional manifold that is an approximation to Ω^c. This method is used to study the topological structure of Ω^c. Let M be a connected Riemannian manifold, and let p and q be two (not necessarily distinct) points of M. The set $\Omega = \Omega(M,p,q)$ can be topologized as follows: Let ρ denote the †distance function on M induced by its Riemannian metric. Given $\omega, \omega'\in\Omega$ with arc lengths $s(t), s'(t)$ respectively, we define the distance $d(\omega,\omega')$ as

$$\max_{0<t<1} \rho(\omega(t),\omega'(t)) + \left(\int_0^1 \left(\frac{ds}{dt} - \frac{ds'}{dt}\right)^2\right)^{1/2} dt.$$

The last term is added in order for the energy function $E_a^b(\omega)$ to be a continuous function. This function d supplies the space Ω with the structure of a metric space.

Let $\operatorname{Int}\Omega^c$ denote the interior of $E^{-1}([0,c])$. Choose some subdivision $0 = t_0 < t_1 < \ldots < t_k = 1$ of $[0,1]$. Let $\Omega(t_0, t_1, \ldots, t_k)$ be the subspace of Ω consisting of curves $\omega: [0,1]\to M$ such that the following hold: (i) $\omega(0) = p$ and $\omega(1) = q$. (ii) $\omega|[t_{i-1}, t_i]$ is a minimal geodesic for each $i = 1, \ldots, k$. Now we define the subspaces $\Omega(t_0, t_1, \ldots, t_k)^c = \Omega^c \cap \Omega(t_0, t_1, \ldots, t_k)$ and $\operatorname{Int}\Omega(t_0, t_1, \ldots, t_k)^c = (\operatorname{Int}\Omega^c)\cap\Omega(t_0, t_1, \ldots, t_k)$.

Let M be a complete Riemannian manifold and c be a fixed positive number such that $\Omega^c \neq \varnothing$. Then for a sufficiently fine subdivision $(t_0, t_1, \ldots, t_k)$ of $[0,1]$ the set $\operatorname{Int}\Omega(t_0, t_1, \ldots, t_k)^c$ can be supplied with the structure of a finite-dimensional differentiable manifold in the following way: Choose the subdivision $(t_0, t_1, \ldots, t_k)$ of $[0,1]$ such that the restriction $\omega|[t_{i-1}, t_i]$ of $\omega\in\Omega(t_0, t_1, \ldots, t_k)^c$ is the unique minimal geodesic between its endpoints on which it depends differentiably $(i = 1, \ldots, k)$. The correspondence $\omega\to(\omega(t_1), \ldots, \omega(t_{k-1}))$ defines a homeomorphism between $\operatorname{Int}\Omega(t_0, t_1, \ldots, t_k)^c$ and a certain open subset of the $n(k-1)$-dimensional manifold $M\times M\times \ldots \times M$. The differentiable structure of the latter manifold induces a differentiable structure on $\operatorname{Int}\Omega(t_0, t_1, \ldots, t_k)^c$. To simplify the notation we denote by B the manifold $\operatorname{Int}\Omega(t_0, t_1, \ldots, t_k)^c$. Let $E': B\to\mathbf{R}$ denote the restriction to B of the energy function E. Then the function E' is differentiable. Furthermore, for each $a<c$ the set $B^a = (E')^{-1}[0,a]$ is compact and is a †deformation retract of the corresponding set Ω^a. The critical points of E' coincide precisely with the critical points of E in $\operatorname{Int}\Omega^c$. The index (nullity) of the Hessian E' at each such critical point γ is equal to the index (nullity) of E_{**} at γ (Milnor [15]).

Thus the homotopy type of the infinite-dimensional path space $\operatorname{Int}\Omega^c$ is seen by studying the homotopy type of the finite-dimensional manifold B. As an immediate consequence we have the following basic result on Ω^c: Let M be a complete Riemannian manifold, and let $p, q\in M$ be two points that are not conjugate along any geodesic of length $\leqslant\sqrt{a}$. Then Ω^a has the homotopy type of a finite CW-complex whose cell of dimension λ corresponds to a geodesic in

Ω^a at which E_{**} has index λ (Milnor [15]).

Now we introduce the compact open topology to the space Ω^* of all continuous curves $\omega:[0,1]\rightarrow M$ connecting p and q. The natural mapping $i:\Omega\rightarrow\Omega^*$ is a †homotopy equivalence. It is known that the space Ω^* has the homotopy type of a CW-complex. Since any compact subset of Ω is contained in some Ω^a, we obtain, by passing to the †direct limit, the **fundamental theorem of Morse theory**: Let M be a complete Riemannian manifold, and let $p,q\in M$ be two points that are not conjugate along any geodesic. Then $\Omega(M;p,q)$ has the homotopy type of a countable CW-complex that contains one cell of dimension λ corresponding to each geodesic from p to q of index λ (Milnor [15]).

Let p be a point of M. Only a finite number of points in M are conjugate to p along geodesics. The relation $\pi_i(M)=\pi_{i-1}(\Omega)$ $(i\geqslant 1)$ holds. Using these results, R. Bott obtained remarkable results regarding the periodicity of †stable homotopy groups of †classical groups (→ 205 Homotopy Groups H).

Concerning the number of geodesics, it is known that if M is a complete †noncontractible Riemannian manifold, any two points that are not conjugate can be joined by infinitely many geodesics (J.-P. Serre [6]).

For two disjoint subsets A,B of M, we now consider the set $\Omega_{A,B}$ consisting of the curves $\omega:[0,1]\rightarrow M$, which start at A and end at B. Up to this point, we have treated only those cases where both A and B consist of only one point. The properties of $\Omega_{A,B}$ have been studied for the following cases: (1) A is one point and B is a closed submanifold of M of dimension $d\geqslant 1$; and (2) both A and B are closed submanifolds of M of dimension $d\geqslant 1$ [11, 14]. In the latter case the critical points correspond to the geodesics orthogonal to both closed submanifolds A and B.

Recently, R. S. Palais and S. Smale studied Morse theory on infinite-dimensional manifolds and obtained a fundamental theorem of Morse theory analogous to the one in the finite-dimensional case. They applied the theorem to the study of nonlinear Dirichlet problems [16, 19].

E. Category

Let M be a compact manifold, and let A be any nonempty closed subset of M. The closed set A is said to be of **category** n with respect to M if and only if the following hold: (i) $A=A_1\cup A_2\cup\ldots\cup A_n$, where A_i is a closed subset of M and is contractible to a point $(1\leqslant i\leqslant n)$. (ii) If $A=B_1\cup\ldots\cup B_l$ for closed subsets B_j $(j=1,\ldots,l)$ of M and $l<n$, then some B_i is not contractible to a point. Then the number of critical points of any differentiable function on a compact manifold M is greater than or equal to the category of M (with respect to M) (L. Lusternik and L. Schnirelmann [3], H. Seifert and W. Threlfall [5]).

References

[1] H. Poincaré, Sur les lignes géodésiques des surfaces convexes, Trans. Amer. Math. Soc., 6 (1905), 237–274. (Oeuvres, Gauthier-Villars, 1953, vol. 6, p. 38–85.)

[2] G. D. Birkhoff, Dynamical systems, Amer. Math. Soc. Colloq. Publ., 1927.

[3] L. A. Lusternik and L. Schnirelmann, Méthodes topologiques dans les problèmes variationnels, Actualités Sci. Ind., Hermann, 1934.

[4] M. Morse, Calculus of variations in the large, Amer. Math. Soc. Colloq. Publ., 1934.

[5] H. Seifert and W. Threlfall, Variationsrechnung im Grossen, Teubner, 1938.

[6] J.-P. Serre, Homologie singulière des espace fibrés, applications, Ann. of Math., (2) 54 (1951), 425–505.

[7] R. Bott, On manifolds all of whose geodesics are closed, Ann. of Math., (2) 60 (1954), 375–382.

[8] R. Bott, The stable homotopy groups of the classical groups, Ann. of Math., (2) 70 (1959), 313–337.

[9] С. И. Альбер (S. L. Al'ber), О периодической задаче вариационного исчисления в целом, Uspehi Mat. Nauk (N.S.), 12, no. 4 (1957), 57–124; English translation, On periodicity problems in the calculus of variations in the large, Amer. Math. Soc. Transl., (2) 14 (1960), 107–172.

[10] W. Klingenberg, Über Riemannsche Mannigfaltigkeiten mit positiver Krümmung, Comment. Math. Helv., 35 (1961), 47–54.

[11] W. Ambrose, The index theorem in Riemannian geometry, Ann. of Math., (2) 73 (1961), 49–86.

[12] R. Olivier, Die Existenz geschlossener Geodätischer auf kompakten Mannigfaltigkeiten, Comment. Math. Helv., 35 (1961), 146–152.

[13] S. Smale, Generalized Poincaré's conjecture in dimensions greater than four, Ann. of Math., (2) 74 (1961), 391–406.

[14] L. N. Patterson, On the index theorem, Amer. J. Math., 85 (1963), 271–297.

[15] J. Milnor, Morse theory, Ann. Math. Studies, Princeton Univ. Press, 1963.

[16] R. S. Palais, Morse theory on Hilbert manifolds, Topology, 2 (1963), 299–340.

[17] В. А. Топоногов (V. A. Toponogov),

Оценка длины замкнутой геодезической в компактном римановом пространстве положительной, Dokl. Akad. Nauk SSSR, 154 (1964), 1047–1049; English translation, Estimate of the length of a closed geodesic in a compact Riemann space of positive curvature, Sov. Math. 5 (1964), 251–254.
[18] W. Klingenberg, On the number of closed geodesics on a Riemannian manifold, Bull. Amer. Math. Soc., 70 (1964), 279–282.
[19] K. Shiga, Recent topics in the theory of differentiable manifolds (Japanese), Math. Soc. Japan, mimeographed note, 1964.
[20] W. Klingenberg, Contributions to Riemannian geometry in the large, Ann. of Math., (2) 69 (1959), 654–666.
[21] W. Klingenberg, The theorem of the three closed geodesics, Bull. Amer. Math. Soc., 71 (1965), 601–605.
[22] R. L. Bishop and R. J. Crittenden, Geometry of manifolds, Academic Press, 1964.
[23] R. Hermann, Differential geometry and the calculus of variations, Academic Press, 1968.
[24] M. Morse and S. S. Cairns, Critical point theory in global analysis and differential topology, Academic Press, 1969.
[25] D. Gromoll, W. Klingenberg, and W. Meyer, Riemannsche Geometrie im Grossen, Lecture notes in math. 55, Springer, 1968.

49 (XX.14)
Cantor, Georg

Georg Cantor (March 3, 1845–June 1, 1918), the founder of set theory, was born in St. Petersburg into a Jewish merchant family that settled in Germany in 1856. He studied mathematics, physics, and philosophy in Zürich and at the University of Berlin. After receiving his degree in 1867 in Berlin, he became a lecturer at the University of Halle and served as professor at that university from 1879 to 1905. In 1884, under the strain of opposition to his ideas and his efforts to prove the †continuum hypothesis, he suffered the first of many attacks of depression which continued to hospitalize him from time to time until his death.

The thesis he wrote for his degree concerned the theory of numbers; however, he arrived at set theory from his research concerning the uniqueness of †trigonometric series. In 1874, he introduced for the first time the concept of †cardinal numbers, with which he proved that there were "more" †transcendental numbers than †algebraic numbers. This result caused a sensation in the mathematical world and became the subject of a great deal of controversy. Cantor was troubled by the opposition of †L. Kronecker, but he was supported by †J. W. R. Dedekind and G. Mittag-Leffler. In his note on the history of the theory of †probability, he recalled the period in which the theory was not generally accepted and cried out, "The essence of mathematics lies in its freedom!" In addition to his work on the concept of cardinal numbers, he laid the basis for the concepts of †order types, †transfinite ordinals, and the theory of real numbers by means of †fundamental sequences. He also studied general point sets in Euclidean space and defined the concepts of †accumulation point, †closed set, and †open set. He was a pioneer in †dimension theory, which itself was the origin of the so-called point set theory, which led to the development of general †topology.

References

[1] G. Cantor, Gesammelte Abhandlungen, edited by E. Zermelo and A. Fraenkel, Springer, 1932 (Georg Olms, 1962).
[2] A. Schoenflies, Die Krisis in Cantors mathematischem Schaffen, Acta Math., 50 (1927), 1–23.

50 (X.32)
Capacity

A. General Remarks

The electric capacity of a conductor in the 3-dimensional Euclidean space $\mathbf{R}^3$ is defined as the ratio of a given positive charge on the conductor to the value of the potential on the surface. This definition of capacity is independent of the given charge. The capacity of a set as a mathematical notion was defined first by N. Wiener (1924) and was developed by O. Frostman, C. J. de La Vallée Poussin, and several other French mathematicians in connection with †potential theory.

B. Energy

Let Ω be a †locally compact Hausdorff space and $\Phi(x,y)$ be a †lower semicontinuous function on $\Omega\times\Omega$ such that $-\infty<\Phi\leqslant\infty$. A measure μ will mean a nonnegative †Radon measure with compact †support S_μ. Denote by $\Phi(x,\mu)$ the †potential $\int\Phi(x,y)\,d\mu(y)$ of a measure μ with kernel Φ and by (μ,μ) the

†energy $\iint \Phi\, d\mu\, d\mu$ of μ. Let X be a set in Ω, and denote by $\mathfrak{A}_X$ the class of normalized measures μ (i.e., of measures μ satisfying $\mu(\Omega)=1$) with $S_\mu \subset X$. Let K be a nonempty compact set in Ω. Set $W(K)=\inf(\mu,\mu)$ for $\mu \in \mathfrak{A}_K$, and $W(\varnothing)=\infty$ for the empty set $\varnothing$. For $\Phi(x,y)=1/|x-y|$ in $\Omega=\mathbf{R}^3$, the general solution $u(x)$ of the †Dirichlet problem (†exterior problem) for the boundary function 1 in the unbounded component of $\mathbf{R}^3-K$ is equal to the potential of an †equilibrium mass-distribution. Therefore, if S is a smooth surface surrounding K and normals are drawn outward to S, the integral $-(1/4\pi)\int_S(\partial u/\partial n)\,d\sigma$ of the normal derivative is equal to $1/W(K)$. This is the **capacity** of K as defined by Wiener [19] when K is a closed region. De La Vallée Poussin [18] called the supremum of $\mu(\mathbf{R}^3)$ the **Newtonian capacity** of a bounded †Borel set E, where μ runs through the class of measures μ with $S_\mu \subset E$ whose †Newtonian potentials are not greater than 1 in $\mathbf{R}^3$. If E is compact, the Newtonian capacity coincides with Wiener's capacity. For the †logarithmic potential in $\mathbf{R}^2$, $e^{-W(K)}$ is called the **logarithmic capacity**. When †Green's function $g(z,\infty)$ with the pole at the point at infinity exists in the unbounded component of $\mathbf{R}^2-K$, $\lim_{z\to\infty}(g(z,\infty)-\log|z|)$ is called **Robin's constant** and can be shown to be equal to $W(K)$. (For the relation between Robin's constant and †reduced extremal distance → 152 Extremal Length.) In the case of a general kernel it is difficult to define capacity as above by means of $W(K)$, and hence the value of $W(K)$ itself instead of the capacity of K is often used. When $W(K)=\infty$, we may say that K is of capacity zero. The minimum value of the †Gauss integral $(\mu,\mu)-2\int f d\mu$ is a generalization of $W(K)$, where $\mu \in \mathfrak{A}_K$ and f is an upper semicontinuous function bounded above on K.

C. Minimax Value

Suppose that we are given the kernel Φ as above. For a set $X \subset \Omega$ and a measure μ, set $U(\mu;X)=\sup_{x\in X}\Phi(x,\mu)$ and $V(\mu;X)=\inf_{x\in X}\Phi(x,\mu)$. Next, for $Y\subset\Omega$, set $U(Y)=\inf U(\mu;S_\mu)$, $V(Y)=\sup V(\mu;S_\mu)$, $U(X,Y)=\inf U(\mu;X)$, and $V(X,Y)=\sup V(\mu;X)$, where $\mu\in U_Y$. If $\check{\Phi}(x,y)=\Phi(y,x)$ is taken as a kernel instead of $\Phi(x,y)$, then the notations $\check{W}(K), \check{U}(\mu;X), \check{V}(\mu;X),\ldots$ are used correspondingly. For K the following relations hold:

$$W(K)=\check{W}(K)\leqslant U(K)=\check{U}(K)$$

$$\leqslant\begin{Bmatrix} U(K,K)=\check{V}(K,K)\\ \check{U}(K,K)=V(K,K)\end{Bmatrix}\leqslant\begin{Bmatrix} U(\Omega,K)=\check{V}(K,\Omega)\\ \check{U}(\Omega,K)=V(K,\Omega)\\ V(K)=\check{V}(K).\end{Bmatrix}$$

Examples show that all the inequalities can be strict. The †minimax theorem in the †theory of games plays an important role in the proof of these inequalities [7]. Even if the kernel is symmetric, the inequalities can be strict except for the equality $W(K)=U(K)$. When the kernel is positive, we may define the quantities which correspond to $U(Y)$, $V(Y)$, $U(X,Y)$, $V(X,Y)$ by considering the class of μ with $S_\mu\subset Y$ and $(\mu,\mu)=1$ instead of $\mathfrak{A}_Y$.

D. Transfinite Diameter

As $k\to\infty$,

$$D_k(K)=k^{-1}(k-1)^{-1}\inf_{x_1,\ldots,x_k\in K}\sum_{i\neq j}\Phi(x_i,x_j)$$

decreases and the limit $D(K)$ is equal to $W(K)$. For the logarithmic kernel in $\mathbf{R}^2$, M. Fekete defined $D(K)$ and called $e^{-D(K)}$ the **transfinite diameter** of K (1923). F. Leja and his school in Poland studied relations between transfinite diameter and †conformal mapping. Next, set

$$kR_k(X,Y)=\sup_{x_1,\ldots,x_k\in Y}\inf_{x\in X}\sum_{i=1}^{k}\Phi(x,x_i).$$

Then $R(X,Y)=\lim R_k(X,Y)$ exists as $k\to\infty$, and we have $R(K,Y)=V(K,Y)$.

Fekete introduced $R(K)=R(K,K)$ in $\mathbf{R}^2$ (1923). G. Polya and G. Szegö computed $D(K)$ and $R(K)$ for special K and α-kernel $r^{-\alpha}(\alpha\gtreqless 0)$ in $\mathbf{R}^2$ and $\mathbf{R}^3$ [13]. The equality $D(K)=R(K)$ holds for the logarithmic kernel in $\mathbf{R}^2$ and the Newtonian kernel in $\mathbf{R}^3$. The maximum of the absolute value on K of †Čebyšev's (Tschebyscheff's) polynomial (→ 333 Polynomial Approximations) of order k with respect to K in $\mathbf{R}^2$ is equal to $\exp(-kR_k(K))$.

E. Evans's Theorem

In order that K be of Newtonian capacity zero, it is necessary and sufficient that there exist a measure μ on K such that the Newtonian potential of μ is equal to ∞ at every point of K. This result was proved by G. C. Evans and H. Selberg independently (1935) and is called **Evans's theorem** (or the **Evans-Selberg theorem**). The corresponding theorem in $\mathbf{R}^2$ is often applied in the theory of functions. A similar potential exists in the case of a general kernel if and only if $R(K,K)=\infty$.

F. Nonadditivity of Capacity

Many kinds of capacity satisfy the inequality $\mathrm{cap}(\bigcup_n X_n)\leqslant\sum_n \mathrm{cap}\,X_n$, where a capacity is denoted by cap. Even the Newtonian capacity C is not necessarily additive, but it satisfies $C(K_1 \cup K_2) + C(K_1 \cap K_2) \leqslant C(K_1) + C(K_2)$ (G. Choquet [2]). Choquet [3] proved that X can be divided into mutually disjoint sets X_1 and X_2 such that $C_i(X) = C_i(X_1) = C_i(X_2)$, where $C_i(X)$ is the **Newtonian inner** (or **interior) capacity** defined by $\sup_{K \subset X} C(K)$ if $X \neq \varnothing$ and by 0 if $X = \varnothing$.

G. Relation with Hausdorff Measure

There are many studies of relations between capacity and †Hausdorff measure [1]. Frostman [6] introduced the notion of **capacitary dimension** and observed that it coincides with the Hausdorff dimension. The capacity of product sets has been evaluated from above and below [11]. For compact sets $K \subset \mathbf{R}^n, K' \subset \mathbf{R}^m$, their dimensions α, β, and the dimension γ of $K \times K'$, we have the relation $\alpha + \beta \leqslant \gamma \leqslant \min(m + \alpha, n + \beta)$, where the equalities are attained by general †Cantor sets. There are also works on the evaluation of capacities of general Cantor sets [10, 17]. If K is a continuum of logarithmic capacity 1 in a plane, then its diameter d satisfies $2 \leqslant d \leqslant 4$, and its area A satisfies $A \leqslant \pi$ [8]. Consider the sum $K = \{z_1 + \ldots + z_n \mid z_k \in K_k, 1 \leqslant k \leqslant n\}$ of continua $K_1, \ldots, K_n$ in a plane. The logarithmic capacity of K is strictly greater than the sum of the logarithmic capacities of $K_1, \ldots, K_n$ except in the case where all K_k are convex and similar [16]. By various †symmetrizations the logarithmic capacity decreases in general (→ 228 Isoperimetric Problems; also [14]).

H. Capacitability

The **Newtonian outer** (or **exterior) capacity** $C_e(X)$ is defined by $\inf C_i(G)$, where G ranges over an open set containing X. The inequality $C_i(X) \leqslant C_e(X)$ holds, in general. When the equality holds, X is called **capacitable**. Choquet (1955) [2] proved that all †analytic sets and hence †Borel sets are capacitable but there exists an analytic set whose complement is not capacitable. He himself generalized his result on capacitability in the following way [4]: Let Ω be an abstract space, φ a nondecreasing function defined on the family of all subsets of Ω, and $\mathcal{K}$ some family of subsets of Ω that is closed under the formation of finite unions and countable intersections. Assume that $\varphi(H_n) \downarrow \varphi(H)$ as H_n in $\mathcal{K}$ decreases to H and that $\varphi(X_n) \uparrow \varphi(X)$ as $X_n \uparrow X$. When $\varphi(X)$ is equal to $\sup\{\varphi(H) \mid H \in \mathcal{K}, H \subset X\}$, X is called $(\varphi, \mathcal{K})$-capacitable. Choquet defined $\mathcal{K}$-Suslin sets and showed that they are $(\varphi, \mathcal{K})$-capacitable. M. Kishi [9], Choquet [5], and B. Fuglede [7] investigated capacitability with respect to several kinds of capacity more general than Newtonian capacity. We can discuss capacitability with respect to quantities defined in connection with the †Gauss variational problem.

I. Analytic Capacity

A different kind of capacity $\alpha(K)$, called **analytic capacity**, is defined for K in a plane as follows: Let D be the unbounded component of the complement of K and $\mathfrak{G}$ be the family of holomorphic functions $g(z)$ in D such that $|g(z)| < 1$ and $w = g(z) = b_1/z + b_2/z^2 + \ldots$ outside a compact set. Then $\alpha(K)$ is defined by $\max_{g \in \mathfrak{G}} |b_1|$. In general, $\alpha(K)$ is not greater than the logarithmic capacity $C(K)$. If K is a continuum, then $\alpha(K) = C(K)$ and $\alpha(K)$ is attained by and only by $g(z)$, which maps D onto $|w| < 1$ conformally and $z = \infty$ to $w = 0$.

References

[1] L. Carleson, Selected problems on exceptional sets, Van Nostrand, 1967.
[2] G. Choquet, Theory of capacities, Ann. Inst. Fourier, 5 (1955), 131–295.
[3] G. Choquet, Potentiels sur un ensemble de capacité nulle, Suites de potentiels, C. R. Acad. Sci. Paris, 244 (1957), 1707–1710.
[4] G. Choquet, Forme abstraite du théorème de capacitabilité, Ann. Inst. Fourier, 9 (1959), 83–89.
[5] G. Choquet, Diamètre transfini et comparaison de diverses capacités, Sém. Théorie du Potentiel, 3 (1958–1959), no. 4.
[6] O. Frostman, Potentiel d'équilibre et capacité des ensembles avec quelques applications à la théorie des fonctions, Medd. Lunds Univ. Mat. Sem., 3 (1935).
[7] B. Fuglede, Une application du théorème du minimax à la théorie du potentiel, Colloque Internat. sur la Théorie du Potentiel, Paris, Orsay, 1964, C. N. R. S. exposé no. 8.
[8] G. M. Goluzin, Geometrische Funktionentheorie, Deutsch. Verlag d. Wiss., 1957.
[9] M. Kishi, Capacitability of analytic sets, Nagoya Math. J., 16 (1960), 91–109.
[10] M. Ohtsuka, Capacité d'ensembles de Cantor généralisés, Nagoya Math. J., 11 (1957), 151–160.
[11] M. Ohtsuka, Capacité des ensembles produits, Nagoya Math. J., 12 (1957), 95–130.
[12] M. Ohtsuka, Kansûron tokuron (Japanese; Topics on the theory of functions), Kyôritu, 1957.
[13] G. Pólya and G. Szegö, Über den trans-

finiten Durchmesser (Kapazitätskonstante) von ebenen und räumlichen Punktmengen, J. Reine Angew. Math., 165 (1931), 4–49.
[14] G. Pólya and G. Szegö, Isoperimetric inequalities in mathematical physics, Princeton, 1951.
[15] C. Pommerenke, Über die analytische Kapazität, Arch. Math., 11 (1960), 270–277.
[16] C. Pommerenke, Zwei Bemerkungen zur Kapazität ebener Kontinuen, Arch. Math., 12 (1961), 122–128.
[17] M. Tsuji, Potential theory in modern function theory, Maruzen, 1959.
[18] C. J. de La Vallée Poussin, Extension de la méthode du balayage de Poincaré et problème de Dirichlet, Ann. Inst. H. Poincaré, 2 (1932), 169–232.
[19] N. Wiener, Certain notions in potential theory, J. Math. and Phys., M.I.T., 3 (1924), 24–51.
[20] L. Sario and K. Oikawa, Capacity functions, Springer, 1969.

51 (II.6) Cardinal Numbers

A. Definition

The general concept of cardinal number is an extension of that of natural number (Cantor [1]). When there exists a †one-to-one correspondence whose †domain is a set A and whose †range is a set B, this set B is said to be **equipotent** (or **equipollent**) to A, and this relation is denoted by $A \sim B$. The relation $\sim$ is an †equivalence relation, and each equivalence class under this relation is said to be a **cardinal number**. The class of all sets equipotent to a set A is denoted by $\overline{\overline{A}}$ (or $|A|$) and is said to be the **cardinal number (power, cardinality**, or **potency) of the set** A. When A is a finite set, $\overline{\overline{A}}$ is said to be **finite**, and when A is an infinite set, $\overline{\overline{A}}$ is said to be **infinite** (or **transfinite**). These concepts will be further explained. When the cardinal number of a set A is $\mathfrak{m}$, A is also said to consist of $\mathfrak{m}$ members (or $\mathfrak{m}$ elements). In this sense, 0 and the natural numbers are considered to express finite cardinal numbers. For example, $0=\overline{\overline{\varnothing}}$, $1=\overline{\overline{\{0\}}}$, $2=\overline{\overline{\{0,1\}}}$, etc. Examples of infinite cardinal numbers: A set A which is equipotent to the set $\mathbf{N}$ of all natural numbers is said to be **countably infinite**, and the cardinal number of the set $\mathbf{N}$ is denoted by $\mathfrak{a}$. A set A which is finite or countably infinite is said to be **countable**. The cardinal number of the set of all real numbers is denoted by $\mathfrak{c}$ and is called the **cardinal number of the continuum**. Moreover, the cardinal number of the set of all real-valued functions whose domain is the interval $[0,1]$ is denoted by $\mathfrak{f}$. These three cardinal numbers are known to be distinct. Henceforth in this article, lowercase German letters denote cardinal numbers. For a definition of cardinal numbers using the concept of ordinal numbers → 306 Ordinal Numbers.

B. Ordering of Cardinal Numbers

$\mathfrak{m} \geqslant \mathfrak{n}$ or $\mathfrak{n} \leqslant \mathfrak{m}$ will mean that there exist sets A and B such that $\mathfrak{m}=\overline{\overline{A}}$, $\mathfrak{n}=\overline{\overline{B}}$, and $A \supset B$. $A \neq B$ does not necessarily imply $\mathfrak{m} \neq \mathfrak{n}$. For example, the cardinal number of the set of all positive even numbers is also $\mathfrak{a}$. $\mathfrak{m} \geqslant \mathfrak{n}$ and $\mathfrak{n} \geqslant \mathfrak{m}$ imply $\mathfrak{m}=\mathfrak{n}$ (**Bernšteĭn's theorem**). Since the †reflexive and †transitive laws for the relation $\leqslant$ between cardinal numbers are obvious, the relation is an †ordering relation. The †well-ordering theorem implies that $\geqslant$ is a †total ordering (**comparability theorem for cardinal numbers**). $\mathfrak{m} > \mathfrak{n}$ means that $\mathfrak{m} \geqslant \mathfrak{n}$ and $\mathfrak{m} \neq \mathfrak{n}$. When $\overline{\overline{B}} \leqslant \mathfrak{m}$, B is said to be **at most** $\mathfrak{m}$.

C. Sum, Product, and Power of Cardinal Numbers

For cardinal numbers $\mathfrak{m}$ and $\mathfrak{n}$, choose sets A and B so that $\mathfrak{m}=\overline{\overline{A}}$, $\mathfrak{n}=\overline{\overline{B}}$, and $A \cap B=\varnothing$, and put $\mathfrak{s}=\overline{\overline{A \cup B}}$. Then $\mathfrak{s}$ is uniquely determined by $\mathfrak{m}$ and $\mathfrak{n}$. The $\mathfrak{s}$ is said to be the **sum** of the cardinal numbers $\mathfrak{m}$ and $\mathfrak{n}$ and is denoted by $\mathfrak{m}+\mathfrak{n}$. If the sets A, B are chosen as described above, the cardinal numbers of the †Cartesian product $A \times B$ and of the set of functions A^B are called the **product** of $\mathfrak{m}$ and $\mathfrak{n}$ and the $\mathfrak{n}$th **power** of $\mathfrak{m}$, denoted by $\mathfrak{m}\mathfrak{n}$ and $\mathfrak{m}^{\mathfrak{n}}$, respectively. These operations are also determined by $\mathfrak{m}$ and $\mathfrak{n}$. For these three operations, the following laws are valid: **commutative laws** $\mathfrak{m}+\mathfrak{n}=\mathfrak{n}+\mathfrak{m}$, $\mathfrak{m}\mathfrak{n}=\mathfrak{n}\mathfrak{m}$; **associative laws** $(\mathfrak{m}+\mathfrak{n})+\mathfrak{p}=\mathfrak{m}+(\mathfrak{n}+\mathfrak{p})$, $(\mathfrak{m}\mathfrak{n})\mathfrak{p}=\mathfrak{m}(\mathfrak{n}\mathfrak{p})$; **distributive law** $\mathfrak{p}(\mathfrak{m}+\mathfrak{n})=\mathfrak{p}\mathfrak{m}+\mathfrak{p}\mathfrak{n}$; **exponential laws** $\mathfrak{m}^{\mathfrak{n}+\mathfrak{p}}=\mathfrak{m}^{\mathfrak{n}}\mathfrak{m}^{\mathfrak{p}}$, $\mathfrak{m}^{\mathfrak{n}\mathfrak{p}}=(\mathfrak{m}^{\mathfrak{n}})^{\mathfrak{p}}$, $(\mathfrak{m}\mathfrak{n})^{\mathfrak{p}}=\mathfrak{m}^{\mathfrak{p}}\mathfrak{n}^{\mathfrak{p}}$. In particular, if $\overline{\overline{A}}=\mathfrak{m}$, then $2^{\mathfrak{m}}$ is the cardinal number of the †power set $\mathfrak{P}(A)$ of A.

Addition and multiplication of more than two cardinal numbers can be defined as follows. Let Λ be any set, and suppose that to any element λ of Λ there corresponds a unique cardinal number $\mathfrak{m}_\lambda$. Let M_λ be a set such that $\overline{\overline{M_\lambda}}=\mathfrak{m}_\lambda$, and $M_\lambda \cap M'_\lambda=\varnothing$ for $\lambda \neq \lambda'$. Then the cardinal number of the †direct sum $\Sigma_\lambda M_\lambda$ is said to be the **sum** of all

$\mathfrak{m}_\lambda$ and is denoted by $\Sigma_\lambda \mathfrak{m}_\lambda$. The cardinal number of the Cartesian product $\prod_\lambda M_\lambda$ is said to be the **product** of all $\mathfrak{m}_\lambda (\lambda \in \Lambda)$ and is denoted by $\prod_\lambda \mathfrak{m}_\lambda$. The axiom of choice can be stated as follows: If $\mathfrak{m}_\lambda \neq 0$ for all $\lambda \in \Lambda$, then $\prod_\lambda \mathfrak{m}_\lambda \neq 0$.

D. The Continuum Hypothesis

For $\mathfrak{a}$, $\mathfrak{c}$, and $\mathfrak{f}$ defined as before $\mathfrak{f} = 2^{\mathfrak{c}} > \mathfrak{c} = 2^{\mathfrak{a}} > \mathfrak{a}$. In general, $2^{\mathfrak{m}} > \mathfrak{m}$ holds for any cardinal number $\mathfrak{m}$ (Cantor). The hypothesis which asserts that for any $\mathfrak{m}$, there does not exist an $\mathfrak{n}$ such that $2^{\mathfrak{m}} > \mathfrak{n} > \mathfrak{m}$ is called the **generalized continuum hypothesis**. In particular, this hypothesis restricted to the case where $\mathfrak{m} = \mathfrak{a}$ is called the **continuum hypothesis**. After Cantor stated this hypothesis (*J. Reine Angew. Math*., 84 (1878)), it remained an open question for many years. In particular, Cantor himself repeatedly tried to prove it, and W. Sierpiński pursued various related hypotheses. Finally, the continuum hypothesis and the generalized continuum hypothesis were proved to be independent of the axioms of set theory by K. Gödel (1940) [3] and P. J. Cohen (1963) [4] (→ 35 Axiomatic Set Theory).

E. Cardinality of Ordinal Numbers

Lower-case Greek letters will stand here for †ordinal numbers. The cardinal number of $\{\xi | \xi < \alpha\}$ will be denoted by $\bar{\alpha}$, which is called the **cardinality of the ordinal number** α or the **cardinal number corresponding to** α. When a cardinal number $\mathfrak{m}$ corresponds to some ordinal number, the minimum among ordinal numbers α with $\bar{\alpha} = \mathfrak{m}$ is called the **initial ordinal number** corresponding to $\mathfrak{m}$. An initial ordinal number corresponding to an infinite cardinal number is called a **transfinite initial ordinal number**. There exists a unique correspondence $\beta \to \omega_\beta$ from the class of ordinal numbers onto the class of all transfinite initial ordinal numbers such that $\beta > \gamma$ implies $\omega_\beta > \omega_\gamma$. In particular, $\omega_0 = \omega$, and an ordinal number ξ such that $\xi < \omega_1$ is called a **countable ordinal number**. ω_β is called the βth transfinite initial ordinal number. The cardinality of ω_β is denoted by $\aleph_\beta$ ($\aleph$ is the Hebrew letter **aleph**). In particular, $\mathfrak{a}$ is denoted by $\aleph_0$ (**aleph zero**). We have $\aleph_\beta \geqslant \aleph_\gamma$ if and only if $\beta \geqslant \gamma$, and in this case $\aleph_\beta + \aleph_\gamma = \aleph_\beta$, $\aleph_\beta \aleph_\gamma = \aleph_\beta$. The axiom of choice implies that every infinite cardinal number is an $\aleph_\beta$. Hence, in this case, the continuum hypothesis can be formulated as $2^{\aleph_0} = \aleph_1$, and the generalized continuum hypothesis can be formulated as $2^{\aleph_\beta} = \aleph_{\beta+1}$ for every ordinal number β.

F. Finiteness and Infiniteness

Dedekind [5] defined a set A to be **infinite** if A is equipotent to a proper subset of itself, and to be **finite** otherwise. It is also possible to define finiteness and infiniteness of sets as follows: A set A is finite if there exists a †well-ordering of A such that its †dual ordering is also a well-ordering, and A is infinite otherwise. If a set is finite in the latter sense, then it is also finite in the sense of Dedekind. Under the axiom of choice, these two definitions can be shown to be equivalent.

References

[1] G. Cantor, Beiträge zur Begründung der transfinite Mengenlehre I, Math. Ann., 46 (1895), 471–512. Gesammelte Abhandlungen, Springer, 1932; English translation, Contributions to the founding of the theory of transfinite numbers, Open Court, 1915.
[2] W. Sierpiński, Hypothèse du continu, Warsaw, 1934.
[3] K. Gödel, The consistency of the continuum hypothesis, Ann. Math. Studies, Princeton Univ. Press, 1940.
[4] P. J. Cohen, Set theory and the continuum hypothesis, Benjamin, 1966.
[5] R. Dedekind, Was sind und was sollen die Zahlen? F. Vieweg, 1888. Gesammelte mathematische Werke 3, F. Vieweg, 1932; English translation, Essays on the theory of numbers, Open Court, 1901.
Also → references to 376 Sets.

52 (XX.15)
Cartan, Elie

Elie Cartan (September 4, 1869–June 5, 1951) was born at Dolomieu in the French province of Isère. He entered the Ecole Normale Supérieure in Paris in 1888 and graduated in 1891, having at the same time qualified in the agrégé examination. Beginning his research immediately, he completed his thesis on the structure of continuous transformation groups [2] in 1894 at the age of 25.

Cartan was a professor first at the University of Montpellier, later at the University of Lyon, then the University of Nancy, and finally in 1912 at the University of Paris. He freely used the †moving coordinate system founded by J. G. Darboux and contributed in many areas such as the theory of †Lie groups, the theory of †Pfaffian forms, the theory of †invariant integrals, †topology, †differential geometry (especially the geometry of †connection), and theoretical physics. His doctoral thesis is still an object of interest among young researchers today, and the concept of connection that forms its basis is fundamental in the field of differential geometry. Henri Cartan (1904) is his eldest son.

References

[1] E. Cartan, Oeuvres complètes I–III, Gauthier-Villars, 1952–1955.
[2] E. Cartan, Sur la structure des groupes de transformations finis et continus, Thèse, 1894. (Oeuvres complètes, pt. I, vol. 1, p. 137–287.)
[3] S. S. Chern and C. Chevalley, Elie Cartan and his mathematical work, Bull. Amer. Math. Soc., 58 (1952), 217–250.

53 (II.25) Categories and Functors

A. Categories

Consider the family of all †groups. Given two groups X and Y, denote the set of all †homomorphisms from X to Y by $\mathrm{Hom}(X, Y)$. If X, Y, and Z are groups and if $f: X \to Y$ and $g: Y \to Z$ are homomorphisms, we can compose them to get a homomorphism $g \circ f: X \to Z$.

In general, suppose that we are given, as in this example, (1) a family $\mathfrak{M}$ of **mathematical objects**, and (2) for every pair (X, Y) of objects in $\mathfrak{M}$, a set $\mathrm{Hom}(X, Y)$ whose elements are called **morphisms** from X to Y, and suppose that if $f \in \mathrm{Hom}(X, Y)$ and $g \in \mathrm{Hom}(Y, Z)$, then they determine a morphism $g \circ f \in \mathrm{Hom}(X, Z)$ which is called their **composite**. A morphism $f \in \mathrm{Hom}(X, Y)$ is also written $f: X \to Y$. Suppose further that these morphisms satisfy the following axioms: (1) if $f: X \to Y$, $g: Y \to Z$, and $h: Z \to W$ are morphisms, then $(h \circ g) \circ f = h \circ (g \circ f)$; (2) for each object $X \in \mathfrak{M}$ there exists a morphism $1_X : X \to X$ such that for any $f: X \to Y$ and $g: Z \to X$ we have $f \circ 1_X = f$ and $1_X \circ g = g$; (3) $\mathrm{Hom}(X, Y)$ and $\mathrm{Hom}(X', Y')$ are disjoint unless $X = X'$ and $Y = Y'$. Then we call the whole system (i.e., the family of objects $\mathfrak{M}$, the morphisms, and the composition of morphisms) a **category**. The elements in $\mathfrak{M}$ are called the **objects** of the category.

By axioms (1) and (2), the set $\mathrm{Hom}(X, X)$ is a semigroup (with respect to the composition of morphisms) which has 1_X as the identity element. Hence 1_X is determined uniquely by X. On the other hand, axiom (3) implies that a morphism f determines the objects X and Y such that $f \in \mathrm{Hom}(X, Y)$. From these facts we can give an alternative definition of category using only the morphisms and their composition.

The totality of the objects (morphisms) in a category $\mathcal{C}$ is denoted by **Ob**($\mathcal{C}$) (**Fl**($\mathcal{C}$); the notation Fl comes from the French word *flèche*). The relation $x \in \mathrm{Ob}(\mathcal{C})$ is often abbreviated to $x \in \mathcal{C}$, while $\mathrm{Hom}(X, Y)$ is written $\mathrm{Hom}_{\mathcal{C}}(X, Y)$ if necessary. A **subcategory** of a category $\mathcal{C}$ is a category $\mathcal{C}'$ with $\mathrm{Ob}(\mathcal{C}') \subset \mathrm{Ob}(\mathcal{C})$, such that for $X, Y \in \mathcal{C}'$ we have $\mathrm{Hom}_{\mathcal{C}'}(X, Y) \subset \mathrm{Hom}_{\mathcal{C}}(X, Y)$ and the composition in $\mathcal{C}'$ is the restriction of $\mathcal{C}$ to $\mathcal{C}'$. If $\mathrm{Hom}_{\mathcal{C}'}(X, Y) = \mathrm{Hom}_{\mathcal{C}}(X, Y)$ for all $X, Y \in \mathcal{C}'$, we say that $\mathcal{C}'$ is a **full subcategory** of $\mathcal{C}$.

We define the **product category** $\mathcal{C}_1 \times \mathcal{C}_2$ of two categories in the canonical way, using the pairs of objects and the pairs of morphisms.

B. Examples of Categories

(1) Taking all sets as the objects, all mappings as the morphisms, and the composition of mappings as the composition, we obtain a category called the **category of sets**, denoted by (Sets) (or (Ens) from the French *ensemble*). For the empty set $\varnothing$ we make the convention that $\mathrm{Hom}(\varnothing, Y)$ contains just one element for any Y and that $\mathrm{Hom}(Y, \varnothing)$ is empty if $Y \neq \varnothing$.

(2) As we have seen, taking all groups as the objects and the homomorphisms as the morphisms, we get the **category of groups**, written (Gr). If we limit the objects to †Abelian groups, we get the **category of Abelian groups** (Ab) as a full subcategory of (Gr).

(3) Fix a ring R. The left R-modules and their R-linear mappings define the **category of left R-modules**, which we denote by ${}_R\mathfrak{M}$. The category of right R-modules, $\mathfrak{M}_R$, is defined similarly. When R is †unitary, we usually limit the objects of ${}_R\mathfrak{M}$ and $\mathfrak{M}_R$ to †unitary modules. If R is commutative we can identify ${}_R\mathfrak{M}$ with $\mathfrak{M}_R$. When $R = \mathbf{Z}$ (the ring of rational integers), ${}_R\mathfrak{M}$ can be identified with (Ab). When R is a field, ${}_R\mathfrak{M}$ is also called the **category of linear spaces over R**.

(4) Taking rings as objects and homomorphisms of rings as morphisms, we obtain the **category of rings**. The subcategory consisting of unitary commutative rings and unitary homomorphisms is called the **category of commutative rings** and is denoted by (Rings).

(5) If we take †differentiable manifolds as objects and differentiable mappings as morphisms, we obtain the **category of differentia-**

ble manifolds. Similarly, for †analytic manifolds and analytic mappings we obtain the **category of analytic manifolds**.

(6) Taking topological spaces as objects and continuous mappings as morphisms, we get a category called the **category of topological spaces** and denoted by (Top). On the other hand, if we take the †homotopy classes of continuous mappings as morphisms, and define their composition in the natural way, we obtain another category, which is called the **homotopy category of topological spaces**.

(7) Fix a †preordered set I. Taking the elements of I as the objects and the pairs (x,y) of elements of I with $x \leqslant y$ as the (unique) morphism from x to y, we get a category, in which we define the composite of the morphisms (x,y) and (y,z) to be (x,z).

In examples (1) through (6), the totality of the objects $\mathrm{Ob}(\mathcal{C})$ is not a set, but a †class (→ 376 Sets G; for the logical foundation of category theory, → [3,9]).

C. Diagrams

If a set of arrows $\{A_\alpha\}$ and a set of points $\{B_\beta\}$ are given in such a way that each arrow A_α has a unique initial point and a unique endpoint, then we say that $\{A_\alpha, B_\beta\}$ is a **diagram**. (Usually, we consider the case where each point B_β is the initial point or the endpoint of at least one A_α (Fig. 1).) Let $\mathcal{C}$ be a category and $\{A_\alpha, B_\beta\}$ a diagram. If we associate a morphism f_α in $\mathcal{C}$ with each arrow A_α and an object $Z_\beta \in \mathcal{C}$ with each point B_β so that $f_\alpha \in \mathrm{Hom}(Z_\beta, Z_\gamma)$ whenever A_α has the initial point B_β and the endpoint B_γ, then we say that $\{f_\alpha, Z_\beta\}$ is a **diagram in the category** $\mathcal{C}$ (Fig. 2). Suppose, furthermore, that the following condition is satisfied: For any pair of points B_β and B_γ, and for any sequence of adjacent arrows $A_{\alpha_1}, \ldots, A_{\alpha_m}$ starting at B_β and ending at B_γ (i.e., the initial point of A_{α_1} is B_β, the endpoint of A_{α_i} is the initial point of $A_{\alpha_{i+1}}$ and the endpoint of A_{α_m} is B_γ), the composite $f_{\alpha_m} \circ f_{\alpha_{m-1}} \circ \ldots \circ f_{\alpha_1}$ ($\in \mathrm{Hom}(B_\beta, B_\gamma)$) depends only on B_β and B_γ. Then the diagram in $\mathcal{C}$ is said to be a **commutative diagram**. For example, commutativity of Fig. 2 is equivalent to $f_3 \circ f_1 = f_4 \circ f_2 = f_5$.

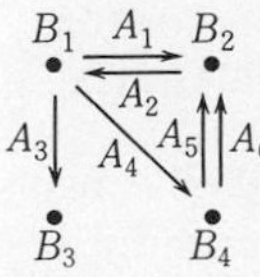

Fig. 1

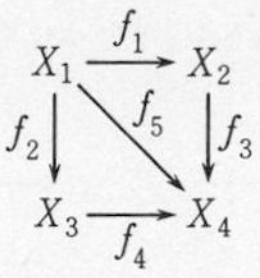

Fig. 2

D. Miscellaneous Definitions

A morphism $f: X \to Y$ in a category $\mathcal{C}$ is called an **isomorphism** (or **equivalence**) if there exists a morphism $g: Y \to X$ such that $f \circ g = 1_Y$, $g \circ f = 1_X$. In this case, g is determined uniquely by f and is itself an isomorphism. We call g the **inverse morphism** of f. Then the inverse of g is f. An isomorphism is sometimes written $f: X \simeq Y$. Two objects X and Y are said to be **isomorphic** if there is an isomorphism $X \to Y$, and then we write $X \cong Y$. The composite of isomorphisms is again an isomorphism. In particular, an isomorphism $X \to X$ is an invertible element of the semigroup $\mathrm{Hom}(X,X)$, and is called an **automorphism** of X. The isomorphisms are the bijections in (Sets), the group isomorphisms in (Gr), the R-isomorphisms in ${}_R\mathfrak{M}$, the ring isomorphisms in the category of rings, the †diffeomorphisms in the categroy of differentiable manifolds, and the homeomorphisms in (Top).

A morphism $f: X \to Y$ is called a **monomorphism** (or **injection**) if for any object Z and for any morphisms $u, v: Z \to X$ $(u \neq v)$ we have $f \circ u \neq f \circ v$. Dually, $f: X \to Y$ is called an **epimorphism** (or **surjection**) if for any $u, v: Y \to Z$ $(u \neq v)$ we have $u \circ f \neq v \circ f$. In the category of sets the monomorphisms and the epimorphisms coincide, respectively, with the injections and the surjections as mappings (→ 376 Sets). A monomorphism which is at the same time an epimorphism is called a **bijection**. An isomorphism is always a bijection, but the converse is false in some categories.

Two monomorphisms $f_1: X_1 \to X$ and $f_2: X_2 \to X$ into the same X are said to be equivalent if there exist $g_1: X_1 \to X_2$ and $g_2: X_2 \to X_1$ such that $f_1 = f_2 \circ g_1$ and $f_2 = f_1 \circ g_2$ (Fig. 3). An equivalence class with respect to this equivalence relation is called a **subobject** of X. Similarly, we define a **quotient object** of X as an equivalence class of epimorphisms from X.

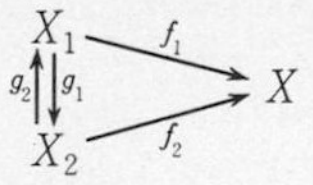

Fig. 3

An object e of a category $\mathcal{C}$ is called a **final object** of $\mathcal{C}$ if for every object Y of $\mathcal{C}$, $\mathrm{Hom}(Y,e)$ contains one and only one element. Dually, an object e' is called an **initial object** (or **cofinal object**) if $\mathrm{Hom}(e',Y)$ contains one and only one element for every $Y \in \mathcal{C}$. If e_1 and e_2 are final objects, then there is a unique isomorphism $e_1 \simeq e_2$, and similarly for initial objects. A set with only one element is the final object in (Sets), and a space with only one point is the final object in (Top). In the category (Gr) (resp. (Ab)), the trivial group $\{1\}$ ($\{0\}$) is the final object and the initial object at the same time. In the category of commutative rings, the zero ring $\{0\}$ is the final object and the ring of rational integers **Z** is the initial object.

E. Product and Coproduct

Let X_1 and X_2 be objects of a category $\mathcal{C}$. We say that a triple (P, p_1, p_2) consisting of an object P and morphisms $p_i: P \to X_i$ $(i=1,2)$ is the **product** (or **direct product**) of X_1 and X_2 if for any pair of morphisms $f_i: X \to X_i$ $(i=1,2)$, there exists a unique morphism $f: X \to P$ with $p_i \circ f = f_i (i=1,2)$ (Fig. 4). If (P', p'_1, p'_2) is another product of X_1 and X_2, then by virtue of this definition there is a unique morphism $f: P \to P'$ such that $p'_i \circ f = p_i$ $(i=1,2)$, and f is an isomorphism. The product is unique in this sense. The product (or any one of the products) of X_1 and X_2 is denoted by $X_1 \times X_2$ or by $X_1 \amalg X_2$.

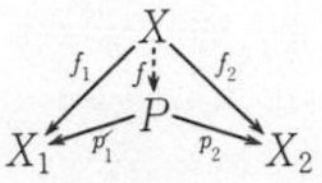

Fig. 4

The product in the categories of sets, of groups, of rings, and of topological spaces coincides with the notion of †direct product in the respective systems. In a general category, the product does not always exist. Suppose the product $X \times X$ exists for an object X; then there is a unique morphism Δ_X: $X \to X \times X$ such that $1_X = p_1 \circ \Delta_X = p_2 \circ \Delta_X$, which is called the **diagonal morphism** of X. Let $f_i: X_i \to X'_i$ $(i=1,2)$ be morphisms and assume that the products $(X_1 \times X_2, p_1, p_2)$, $(X'_1 \times X'_2, p'_1, p'_2)$ exist. Then there is a unique morphism $f: X_1 \times X_2 \to X'_1 \times X'_2$ satisfying $p'_i \circ f = f_i \circ p_i (i=1,2)$. This f is denoted by $f_1 \times f_2$. On the other hand, if $g_i: X \to X_i$ $(i=1,2)$ are given, the unique morphism $g: X \to X_1 \times X_2$ with $p_i \circ g = g_i$ $(i=1,2)$ is denoted by (g_1, g_2). We have $(g_1, g_2) = (g_1 \times g_2) \circ \Delta_X$ if $X \times X$ exists.

The dual notion of product is coproduct. We say that a triple (S, j_1, j_2) of an object S and morphisms $j_i: X_i \to S$ $(i=1,2)$ is the **coproduct** (or **direct sum**) of X_1 and X_2 if for any morphisms $f_i: X_i \to X$ $(i=1,2)$ there exists a unique morphism $f: S \to X$ with $f \circ j_i = f_i$ $(i=1,2)$ (Fig. 5). The coproduct, like the product, is uniquely determined up to canonical isomorphisms. It is denoted by $X_1 + X_2$ or by $X_1 \perp\!\!\!\perp X_2$. The coproduct in (Gr) is the †free product. In (Ab), or more generally in ${}_R\mathfrak{M}$, the product of two objects can be identified with the coproduct (= direct sum) (→ 275 Modules F). The coproduct in the category of commutative rings is the †tensor product over **Z**.

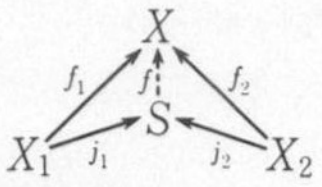

Fig. 5

Product and coproduct can also be defined for a family $\{X_i\}_{i \in I}$ of objects. Namely, the **product** of $\{X_i\}_{i \in I}$ is an object P together with a family of morphisms $p_i: P \to X_i$ $(i \in I)$ having the property that for any family of morphisms $f_i: X \to X_i$ $(i \in I)$, there exists a unique morphism $f: X \to P$ such that $p_i \circ f = f_i$ $(i \in I)$. The product is unique up to canonical isomorphisms, and similarly for the coproduct (→ Sections F and L).

F. Dual Category

In the theory of categories we often encounter the dual treatment of notions and propositions. To be precise, we may define the notion of the **dual category** $\mathcal{C}^\circ$ of a category $\mathcal{C}$ as follows; the objects of $\mathcal{C}^\circ$ are those of $\mathcal{C}$, i.e., $\mathrm{Ob}(\mathcal{C}^\circ) = \mathrm{Ob}(\mathcal{C})$; for any objects X and Y we put $\mathrm{Hom}_{\mathcal{C}^\circ}(X, Y) = \mathrm{Hom}_{\mathcal{C}}(Y, X)$; if $f: X \to Y$ and $g: Y \to Z$ in $\mathcal{C}^\circ$ (i.e., $f: Y \to X$ and $g: Z \to Y$ in $\mathcal{C}$), then the composite $g \circ f$ in $\mathcal{C}^\circ$ is defined to be $f \circ g$ in $\mathcal{C}$. It is clear that $\mathcal{C}^\circ$ then satisfies the axioms of a category. Quite generally, given a proposition concerning objects and morphisms we can construct another proposition by reversing the directions of the morphisms, and we call the latter the dual proposition of the former. The dual proposition of a proposition in $\mathcal{C}$ coincides with a proposition in $\mathcal{C}^\circ$. For instance, a monomorphism (epimorphism) in $\mathcal{C}$ is an epimorphism (monomorphism) in $\mathcal{C}^\circ$, and the final (initial) object in $\mathcal{C}$ is the initial (final) object in $\mathcal{C}^\circ$. The product (coproduct) in $\mathcal{C}$ is the coproduct (product) in $\mathcal{C}^\circ$. Although the notion of the dual category is defined quite formally, it is useful in describing relations between specific categories. The dual category of (Ab), for instance, is equivalent to the category of commutative compact topological groups (†Pontrjagin's duality theorem).

G. Categories over an Object

Fix a category $\mathcal{C}$ and an object $S \in \mathcal{C}$. A pair (X, f) of an object $X \in \mathcal{C}$ and a morphism $f: X \to S$ is called an object over S or an **S-object**, and f is called its **structure morphism**. We often omit f and simply say "an S-object X" if there is no danger of misunderstanding. If (X, f) and (Y, g) are S-objects, a morphism $h: X \to Y$ such that $f = g \circ h$ is called an S-

morphism from (X,f) to (Y,g). The category whose objects are the S-objects and whose morphisms are the S-morphisms is called the **category of S-objects** in $\mathcal{C}$, and is denoted by $\mathcal{C}/S$. It has $(S,1_S)$ as the final object. The product of two S-objects X and Y, taken in $\mathcal{C}/S$, is called the **fiber product** of X and Y **over S** (in $\mathcal{C}$), and is denoted by $X\times_S Y$ or $X\Pi_S Y$. The dual notion of the fiber product is called **fiber sum** (or **amalgamated sum**). Thus for two morphisms $f:S\rightarrow X$ and $g:S\rightarrow Y$, the fiber product of X and Y over S in $\mathcal{C}^\circ/S$ is the fiber sum of X and Y (with respect to S); it is denoted by $X\amalg_S Y$.

Let $\mathcal{C}$ be the category of commutative rings and $K\in\mathcal{C}$. Then the family of K-objects in $\mathcal{C}^\circ$ is precisely the family of commutative K-algebras. The fiber product $A\times_K B$ in $\mathcal{C}^\circ$, i.e., $A\amalg_K B$ in $\mathcal{C}$, is the tensor product $A\otimes_K B$ of algebras.

H. Functors

Let $\mathcal{C}$ and $\mathcal{C}'$ be categories. A **covariant functor** F from $\mathcal{C}$ to $\mathcal{C}'$ is a rule which associates (1) with each object X in $\mathcal{C}$, an object $F(X)$ in $\mathcal{C}'$, and (2) with each morphism $f:X\rightarrow Y$ in $\mathcal{C}$, a morphism $F(f):F(X)\rightarrow F(Y)$ such that $F(g\circ f)=F(g)\circ F(f)$, $F(1_X)=1_{F(X)}$. A **contravariant functor** is defined dually, by modifying this definition to $F(f):F(Y)\rightarrow F(X)$, $F(g\circ f)=F(f)\circ F(g)$. Thus a contravariant functor from $\mathcal{C}$ to $\mathcal{C}'$ is the same as a covariant functor from the dual category $\mathcal{C}^\circ$ to $\mathcal{C}'$ (or from $\mathcal{C}$ to $\mathcal{C}'^\circ$). **Functor** is a general term for both covariant functors and contravariant functors, but some authors use the word exclusively in the sense of a covariant functor. A functor in several variables is defined to be a functor from the product category of the categories in which the variables take their values.

A covariant functor $F:\mathcal{C}\rightarrow\mathcal{C}'$ is said to be **faithful (fully faithful)** if for any $X,Y\in\mathcal{C}$, the mapping $\mathrm{Hom}(X,Y)\rightarrow\mathrm{Hom}(F(x),F(Y))$ induced by F is injective (bijective), and similarly for contravariant functors. A faithful covariant functor $F:\mathcal{C}\rightarrow\mathcal{C}'$ which maps distinct objects of $\mathcal{C}$ to distinct objects of $\mathcal{C}'$ is called an **embedding**, and in this case $\mathcal{C}$ can be identified with a subcategory of $\mathcal{C}'$ by F. A fully faithful covariant functor $F:\mathcal{C}\rightarrow\mathcal{C}'$ is called an **equivalence** (between the categories) if it satisfies the condition that for any object X' of $\mathcal{C}'$, there exists an object X of $\mathcal{C}$ such that $F(X)\cong X'$. In this case we can consider the two categories essentially the same. A contravariant functor from $\mathcal{C}$ to $\mathcal{C}'$ which defines an equivalence from $\mathcal{C}^\circ$ to $\mathcal{C}'$ is called an **antiequivalence**.

I. Examples of Functors

(1) Let $\mathcal{C}$ be the category of groups (or rings). For any $X\in\mathcal{C}$ let $F(X)$ be the underlying set of X (i.e., the set obtained from X by "forgetting" its structure as a group or ring), and for any homomorphism f put $F(f)=f$. Then we get a faithful covariant functor (often called the **forgetful functor**) $F:\mathcal{C}\rightarrow$ (Sets).

(2) Let $\mathcal{C}$ be any category and fix an object X of $\mathcal{C}$. Then we get a covariant functor $h_X:\mathcal{C}\rightarrow$(Sets) as follows: with each $Y\in\mathcal{C}$ we associate the set $\mathrm{Hom}(X,Y)$, and with each morphism $f:Y\rightarrow Y'$ in $\mathcal{C}$ the mapping $f\circ$ (where $f\circ:\mathrm{Hom}(X,Y)\rightarrow\mathrm{Hom}(X,Y')$ is defined by $(f\circ)(g)=f\circ g$). Similarly we define a contravariant functor $h^X:\mathcal{C}\rightarrow$(Sets) by $h^X(Y)=\mathrm{Hom}(Y,X)$ and $h^X(f)=\circ f$.

(3) Let $\rho:A\rightarrow B$ be a homomorphism of rings. With each left A-module M associate the †scalar extension $\rho^*(M)=B\otimes_A M$, and with each A-homomorphism f associate the B-homomorphism $\rho^*(f)=1_B\otimes f$. Then we get a covariant functor $\rho^*:{}_A\mathfrak{M}\rightarrow{}_B\mathfrak{M}$.

(4) Let R be a ring. With each left R-module M associate its dual module $M^*=\mathrm{Hom}_R(M,R)$, and to each R-linear mapping f associate its †dual mapping ${}^tf=\circ f$. Then we get a contravariant functor ${}_R\mathfrak{M}\rightarrow\mathfrak{M}_R$, and similarly for $\mathfrak{M}_R\rightarrow{}_R\mathfrak{M}$.

(5) For each differentiable manifold X let $F(X)$ denote the commutative ring of the differentiable functions on X, and for each differentiable mapping $f:X\rightarrow Y$ let $F(f)$ be the ring homomorphism $\circ f:F(Y)\rightarrow F(X)$. Then F is a faithful contravariant functor.

(6) Fix an Abelian group A. By associating with each topological space X the cohomology group $H(X,A)$ and with each continuous mapping $f:X\rightarrow Y$ the homomorphism $H(Y,A)\rightarrow H(X,A)$ induced by f, we obtain a contravariant functor from (Top) to (Ab).

(7) Fix a topological space X, and let $T(X)$ be the set of the open sets in X. Then $T(X)$ is ordered by inclusion, so it is a category (→ Section B, no. 7). The contravariant functors from $T(X)$ to (Ab) are precisely the †presheaves of Abelian groups over X. We can use any category instead of (Ab) to define a presheaf over X (→ 377 Sheaves).

J. Natural Transformations

Let $\mathcal{C}$ and $\mathcal{C}'$ be categories, and denote by $\mathrm{Hom}(\mathcal{C},\mathcal{C}')$ the collection of all covariant functors $\mathcal{C}\rightarrow\mathcal{C}'$. Let $F,G\in\mathrm{Hom}(\mathcal{C},\mathcal{C}')$. A **natural transformation** (or **functorial morphism**) from F to G is a function which assigns to each object X of $\mathcal{C}$ a morphism

$\varphi(X): F(X) \to G(X)$ in $\mathcal{C}'$ such that for any morphism $f: X \to Y$ in $\mathcal{C}$, the equation $G(f) \circ \varphi(X) = \varphi(Y) \circ F(f)$ holds; in other words, the accompanying diagram is commutative:

$$\begin{array}{ccccc} X & & F(X) & \xrightarrow{\varphi(X)} & G(X) \\ f\downarrow & & F(f)\downarrow & & \downarrow G(f) \\ Y & & F(Y) & \xrightarrow{\varphi(Y)} & G(Y) \end{array}$$

A natural transformation between contravariant functors is defined similarly. For instance, let A and B be Abelian groups and let $H^i(\cdot, A)$ and $H^j(\cdot, B)$ be the contravariant functors of †cohomology viewed as functors (Top)→(Sets). Then the natural transformations between them are the †cohomology operations.

Let $\varphi: F \to G$ be a natural transformation, and suppose that $\varphi(X): F(X) \to G(X)$ is an isomorphism for every $X \in \mathcal{C}$. Then the inverse transformation $G \to F$ of φ exists, and φ is called a **natural equivalence (functorial isomorphism** or **isomorphism)** and is written $\varphi: F \cong G$.

Suppose that $\mathrm{Ob}(\mathcal{C})$ is a set. Then the collection $\mathrm{Hom}(F, G)$ of all natural transformations $F \to G$ is also a set, and hence we can consider $\mathrm{Hom}(\mathcal{C}, \mathcal{C}')$ a category in which the objects are the covariant functors $\mathcal{C} \to \mathcal{C}'$, the morphisms are the natural transformations, and the composition of morphisms is the natural one. Then $\mathrm{Hom}(\mathcal{C}^\circ, \mathcal{C}')$ is the category of contravariant functors from $\mathcal{C}$ to $\mathcal{C}'$. In particular, the category $\mathrm{Hom}(\mathcal{C}^\circ, (\mathrm{Sets}))$ is sometimes denoted by $\hat{\mathcal{C}}$.

Given a category $\mathcal{C}$, a covariant (resp. contravariant) functor $F: \mathcal{C} \to (\mathrm{Sets})$, and an object $X \in \mathcal{C}$, we can define a canonical bijection $\Phi_X: \mathrm{Hom}(h_X, F) \cong F(X)$ (resp. $\mathrm{Hom}(h^X, F) \cong F(X)$) by $\Phi_X(\varphi) = \varphi(X)1_X$. (The functors h_X and h^X were defined in Section I.) The inverse mapping of Φ_X assigns to $\xi \in F(X)$ the natural transformation $\varphi: h_X \to F$ defined by $\varphi(Y)u = F(u)\xi$ $(Y \in \mathcal{C})$. In particular, if we take $F = h_Y$ (h^Y), we obtain a canonical bijection $\mathrm{Hom}(h_X, h_Y) \cong \mathrm{Hom}(Y, X)$ $(\mathrm{Hom}(h^X, h^Y) \cong \mathrm{Hom}(X, Y))$. It follows that there is a fully faithful contravariant (covariant) functor $\mathcal{C} \to \mathrm{Hom}(\mathcal{C}, (\mathrm{Sets}))$ $(\mathcal{C} \to \mathrm{Hom}(\mathcal{C}^\circ, (\mathrm{Sets})) = \hat{\mathcal{C}})$ which associates $h_X (h^X)$ with $X \in \mathcal{C}$.

K. Adjoint Functors

Let $F: \mathcal{C} \to \mathcal{C}'$ and $F': \mathcal{C}' \to \mathcal{C}$ be covariant functors. Suppose that there is a rule which assigns to each pair of objects $M \in \mathcal{C}$ and $M' \in \mathcal{C}'$ a bijective mapping $\theta_{M,M'}$: $\mathrm{Hom}_{\mathcal{C}}(M, F'(M')) \cong \mathrm{Hom}_{\mathcal{C}'}(F(M), M')$ such that for any pair of morphisms $N \to M$ in $\mathcal{C}$ and $M' \to N'$ in $\mathcal{C}'$, the following diagram induced by the morphisms is commutative:

$$\begin{array}{ccc} \mathrm{Hom}_{\mathcal{C}}(M, F'(M')) & \xrightarrow{\theta_{M,M'}} & \mathrm{Hom}_{\mathcal{C}'}(F(M), M') \\ \downarrow & & \downarrow \\ \mathrm{Hom}_{\mathcal{C}}(N, F'(N')) & \xrightarrow{\theta_{N,N'}} & \mathrm{Hom}_{\mathcal{C}'}(F(N), N') \end{array}$$

Then we say that F is a **left adjoint functor** of F' and that F' is a **right adjoint functor** of F. We can regard $\mathrm{Hom}_{\mathcal{C}'}(F(M), M')$ as a functor from $\mathcal{C} \times \mathcal{C}'$ (contravariant in the variable $M \in \mathcal{C}$, covariant in the variable $M' \in \mathcal{C}'$) to the category of sets, and similarly for $\mathrm{Hom}_{\mathcal{C}}(M, F'(M'))$. This commutativity of the diagram means that these two functors are isomorphic (→ Section J).

For instance, let A and B be rings and L a fixed B-A-†bimodule. Let $F: {}_A\mathfrak{M} \to {}_B\mathfrak{M}$ and $F': {}_B\mathfrak{M} \to {}_A\mathfrak{M}$ be the functors defined by

$$F(M) = L \otimes_A M, \quad F'(M') = \mathrm{Hom}_B(L, M'),$$

where the assignment for morphisms is defined in the natural way. Then F is the left adjoint of F' and F' is the right adjoint of F. In particular, let $\rho: A \to B$ be a homomorphism and consider the case $L = B$. Then F is the functor $\rho^*: {}_A\mathrm{M} \to {}_B\mathrm{M}$ and F' is the functor $\rho_*: {}_B\mathfrak{M} \to {}_A\mathfrak{M}$, so that ρ^* is the left adjoint of ρ_* (and ρ_* is the right adjoint of ρ^*) (→ 275 Modules, K, L; for more examples of adjoint functors, see → [11]).

L. Representation of Functors

We begin by discussing an example. Let T be a set, and consider the following problem: Is it possible to find a group X and a mapping $\xi: T \to X$ such that, for any group Y and for any mapping $\eta: T \to Y$, there exists a unique homomorphism $u: X \to Y$ with $u \circ \xi = \eta$? The answer is yes; it is enough to take the †free group X generated by T and the canonical injection $\xi: T \to X$ (Fig. 6). On the other hand, let $F(Y)$ be the set of all mappings $T \to Y$, and for each group homomorphism $f: Y \to Y'$ define the mapping $F(f): F(Y) \to F(Y')$ by $F(f)\eta = f \circ \eta$ $(\eta \in F(Y))$. Then we get a covariant functor from the category $\mathcal{C}$ of groups to the category of sets, $F: \mathcal{C} \to (\mathrm{Sets})$. We can now reformulate the condition on $X \in \mathcal{C}$ and $\xi \in F(X)$ as follows:

For any $Y \in \mathcal{C}$ and for any $\eta \in F(Y)$, there exists a unique morphism $u: X \to Y$ such that $F(u)\xi = \eta$.

Proceeding to the general case, let $\mathcal{C}$ be an arbitrary category and let $F: \mathcal{C} \to (\mathrm{Sets})$ be a

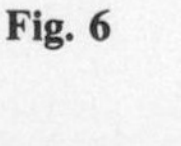

Fig. 6

functor. If there exist an object X of $\mathcal{C}$ and an element ξ of $F(X)$ satisfying the condition just stated (with the modification $u: Y \to X$ in the contravariant case), then we say that the pair (X, ξ) **represents** the functor F, or less specifically, that X represents F, and we call ξ the **canonical element** of $F(X)$. We also say that F is **representable**. The condition stated above is a formulation of the so-called **universal mapping property**. If (X', ξ') also represents F, the unique morphism $u: X \to X'$ (or $X' \to X$) with $F(u)\xi = \xi'$ is necessarily an isomorphism.

When (X, ξ) represents F, the natural transformation $\varphi: h_X \to F$ ($h^X \to F$ in the contravariant case) which corresponds to ξ by the canonical bijection $\Phi_X: \mathrm{Hom}(h_X, F) \rightleftarrows F(X)$ is an isomorphism. Conversely, if there is a functorial isomorphism $\varphi: h_X \to F$ (or $h^X \to F$) for some $X \in \mathcal{C}$, then the object X represents F, with the canonical element of $F(X)$ the element which corresponds to φ by the canonical bijection Φ_X, i.e., $\xi = \varphi(X)1_X$.

We have already seen the example of a free group; here we list a few more examples. (1) Let $\{X_i\}_{i \in I}$ be a family of objects in a category $\mathcal{C}$. For each $Y \in \mathcal{C}$ we put $F(Y) = \prod_{i \in I} \mathrm{Hom}(Y, X_i)$, and for each morphism $f: Y \to Y'$ we define the mapping $F(f): F(Y') \to F(Y)$ by $F(f)(f_i) = (f_i \circ f)$. Then we get a contravariant functor $F: \mathcal{C} \to$(Sets). A pair (X, ξ) which represents F (where $\xi \in F(X) = \prod_{i \in I} \mathrm{Hom}(X, X_i)$) is the product of $\{X_i\}$. Thus, representability of F is equivalent to the existence of the product of $\{X_i\}$, and similarly for the coproduct.

(2) Let R be a ring, M a right R-module, and N a left R-module. For each Abelian group Y let $F(Y)$ denote the set of the R-balanced mappings $M \times N \to Y$ (→ 275 Modules J). Since a homomorphism $f: Y \to Y'$ induces a natural mapping $F(f): F(Y) \to F(Y')$ by composition, we obtain a covariant functor F: (Ab)→(Sets). This functor is representable: the pair consisting of the tensor product $M \otimes_R N$ and the canonical mapping $M \times N \to M \otimes_R N$ represent it.

(3) Let R be a commutative ring and S a subset of R. For each commutative ring Y, let $F(Y)$ denote the set of homomorphisms $R \to Y$ that map the elements of S to invertible elements of Y. As in the preceding example, we obtain a covariant functor F: (Rings)→(Sets). This functor is represented by the †ring of quotients $S^{-1}R$ and the canonical homomorphism $R \to S^{-1}R$.

M. Groups in a Category

Let $\mathcal{C}$ be a category with a final object e, and assume that a finite product always exists in $\mathcal{C}$. If an object $G \in \mathcal{C}$ and morphisms $\alpha: G \times G \to G$, $\beta: G \to G$, $\varepsilon: e \to G$ are given such that the diagrams of Fig. 7 are commutative, then $(G, \alpha, \beta, \varepsilon)$ is called a group in $\mathcal{C}$ (**group object** in $\mathcal{C}$ or $\mathcal{C}$**-group**).

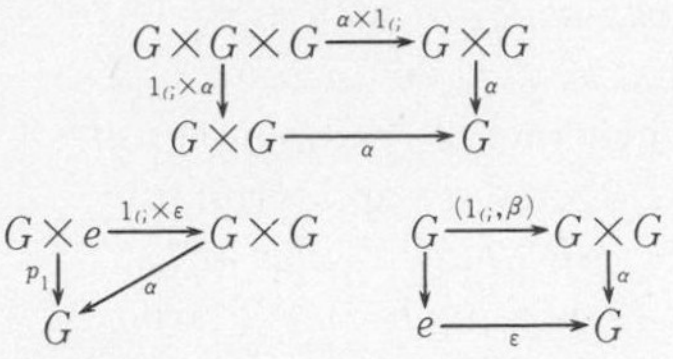

Fig. 7

If $\mathcal{C}$ is the category of sets, then α defines a law of composition in the set G, and the image of e by ε is the identity element and $\beta(x)$ is the inverse of x, so that G is an ordinary group. If $\mathcal{C}$ is the category of topological spaces (analytic manifolds, algebraic varieties, †schemes) then G is a †topological group (†Lie group †algebraic group, †group scheme).

We can also define the $\mathcal{C}$-group by lifting the group concept in (Sets) to the category $\mathcal{C}$ by means of the functor h^X. Namely let G be an object of $\mathcal{C}$, and suppose that for each $Y \in \mathcal{C}$ the set $h^G(Y) = \mathrm{Hom}(Y, G)$ is equipped with a group structure and that for each morphism $f: Y \to Y'$ the induced mapping $h^G(Y') \to h^G(Y)$ is a group homomorphism. In other words, suppose that h^G is a contravariant functor from $\mathcal{C}$ to the category of groups. Then the object G with the additional structure on h^G is called a $\mathcal{C}$-group. This definition is equivalent to the one given above.

N. Additive Categories

A category $\mathcal{C}$ is called an **additive category** if for each pair $X, Y \in \mathcal{C}$, the set of morphisms $\mathrm{Hom}(X, Y)$ has the structure of an additive group such that (1) the composition of morphisms is distributive in both ways: $h \circ (f+g) = h \circ f + h \circ g$, $(f+g) \circ h = f \circ h + g \circ h$; (2) there exists an object $0'$ with $\mathrm{Hom}(0', 0') = \{0\}$; (3) the product (or the coproduct) of any two objects exists. Then the object $0'$ in (2) is a final and initial object, and is called the **zero object**. Both the product and the coproduct of any two objects exist and can be identified. The dual category of an additive category is also an additive category. A functor F from an additive category to another is called an **additive functor** if $F(f+g) = F(f) + F(g)$ holds for morphisms. In an additive category $\mathcal{C}$, $\mathrm{Hom}(X, Y)$ is an additive functor from $\mathcal{C}$ to (Ab) in each variable.

For any ring R, the category of left (or

right) R-modules is an additive category. The following definitions are generalizations of the corresponding concepts in the theory of modules. The **kernel** of a morphism $f: A \to B$ is a pair consisting of an object A' and a monomorphism $i: A' \to A$ with $f \circ i = 0$, such that any morphism $u: X \to A$ with $f \circ u = 0$ is divisible by i (that is, $u = i \circ v$ for some $v: X \to A'$). Dually, the **cokernel** of f is a pair consisting of an object B' and an epimorphism $j: B \to B'$ with $j \circ f = 0$ which divides any morphism $u: B \to X$ with $u \circ f = 0$. We write $A' = \mathrm{Ker} f$, $B' = \mathrm{Coker}\, f$. The kernel of $j: B \to \mathrm{Coker} f$ is called the **image of** f and is denoted by $\mathrm{Im} f$; the cokernel of $i: \mathrm{Ker} f \to A$ is called the **coimage** of f and is denoted by $\mathrm{Coim}\, f$. If all these exist, it follows from the definitions that there is a unique morphism $\mathrm{Coim}\, f \to \mathrm{Im}\, f$ such that the composite of $A \to \mathrm{Coim}\, f \to \mathrm{Im}\, f \to B$ is equal to f.

An additive categroy $\mathcal{C}$ is called an **Abelian category** if it satisfies the following conditions: (1) every morphism has a kernel and a cokernel, (2) for every morphism f, the morphism $\mathrm{Coim}\, f \to \mathrm{Im}\, f$ just mentioned is an isomorphism. The dual category of an Abelian category is also Abelian. The categories of Abelian groups, of R-modules, and of sheaves of $\mathcal{O}$-modules on a †ringed space $(X, \mathcal{O})$ are important examples of Abelian categories. Many propositions which are valid in (Ab) remain valid in any Abelian category. In particular, the notion of an †exact sequence is defined in an Abelian category in the same way as in (Ab), and the fiber product and fiber sum of a finite number of objects always exist in an Abelian category. A functor between Abelian categories which carries exact sequences into exact sequences is called an **exact functor**; (such a functor is automatically additive). If $\mathcal{C}$ is a category of which $\mathrm{Ob}(\mathcal{C})$ is a set, and if $\mathcal{C}$ is an Abelian category, then $\mathrm{Hom}(\mathcal{C}, \mathcal{C}')$ is an Abelian category. Given an Abelian category $\mathcal{C}$ and a subcategory $\mathcal{C}'$ which satisfies certain conditions, one can construct an Abelian category $\mathcal{C}/\mathcal{C}'$ which is called the **quotient category** (Serre's theory of classes of Abelian groups; → [8]).

If $\mathcal{C}$ is an Abelian category of which $\mathrm{Ob}(\mathcal{C})$ is a set, there is an embedding of $\mathcal{C}$ into the category ${}_R\mathfrak{M}$ of modules over some ring R by a fully faithful flat exact covariant functor (**full embedding theorem**, B. Mitchell, *Amer. J. Math.*, 86 (1964)). This remarkable theorem enables us to extend results obtained for modules to the case of Abelian categories.

The notions of category and functor were introduced in [7] and were applied first in topology and then in homological algebra and algebraic geometry (→ 202 Homological Algebra).

References

[1] S. Eilenberg and N. Steenrod, Foundations of algebraic topology, Princeton, 1952.
[2] H. P. Cartan and S. Eilenberg, Homological algebra, Princeton, 1956.
[3] S. MacLane, Homology, Springer, 1963.
[4] P. Freyd, Abelian categories, Harper & Row, 1964.
[5] B. Mitchell, The theory of categories, Academic Press, 1965.
[6] I. Bucur and A. Deleanu, Introduction to the theory of categories and functors, Interscience, 1968.
[7] S. Eilenberg and S. MacLane, General theory of natural equivalences, Trans. Amer. Math. Soc., 58 (1945), 231–294.
[8] A. Grothendieck, Sur quelques points d'algèbre homologique, Tôhoku Math. J., (2) 9 (1957), 119–221.
[9] P. Gabriel, Des catégories abéliennes, Thesis, Paris, 1962.
[10] N. Yoneda, On universality I, II (Japanese), Sûgaku, 13 (1961–1962), 109–112; 14 (1962–1963), 39–43.
[11] S. MacLane, Categorical algebra, Bull. Amer. Math. Soc., 71 (1965), 40–106.

54 (XX.16) Cauchy, Augustin Louis

Augustin Louis Cauchy (August 21, 1789–May 25, 1857) was a French mathematician of the 19th century. He graduated from the Ecole Polytechnique in 1807 and from the Ecole des Ponts et Chaussées in 1810, to become a civil engineer. In 1816, his mathematical works were recognized, and he was appointed a member of the Académie des Sciences while a professor at the Ecole Polytechnique. After the July revolution in 1830, he refused to pledge loyalty to Louis-Philippe and fled to Turin; he later moved to Prague. He returned to France after the revolution of 1848 and became a professor at the University of Paris, where he remained until his death. He was a Catholic and a Royalist all his life.

His scientific contributions were numerous and covered many fields. In algebra, he did pioneer work in †determinants and in the theory of †groups. He also made notable achievements in theoretical physics, optics, and the theory of elasticity. His main field was analysis. He was interested in making analysis rigorous by giving calculus a solid foundation in such works as *Cours d'analyse de l'Ecole Polytechnique* (1821). In his paper "Mémoire sur les intégrales définies prises entre les limites imaginaires" (1825), he proved the main theorem of the theory of functions of a complex variable. Another important work is his proof of the existence theorem for the solutions of †differential equations in the cases of real variables and complex variables.

References

[1] A. Cauchy, Oeuvres complètes I.1–12; II.1–14, Gauthier-Villars, 1882–1958.
[2] F. Klein, Vorlesungen über die Entwicklung der Mathematik im 19. Jahrhundert I, Springer, 1926 (Chelsea, 1956).
[3] T. Takagi, Kinsei sûgakusidan (Japanese; Topics from the history of mathematics of the 19th century), Kawade, 1943.

55 (III.22) Cayley Algebras

Let Q be a †quaternion algebra over a field K of characteristic zero. A **general Cayley algebra** $\mathcal{C}$ is a 2-dimensional Q-†module $Q+Qe$ with the multiplication $(q+re)(s+te) = (qs+\gamma\bar{t}r)+(tq+r\bar{s})e$, where $q,r,s,t\in Q$, γ is a given element in K, and $\bar{t}, \bar{s}$ are the †conjugate quaternions of r, s, respectively. The elements of $\mathcal{C}$ are called **Cayley numbers**; $\mathcal{C}$ is a nonassociative, †alternative algebra of dimension 8 over K (→ 231 Jordan Algebras). The mapping $a=q+re\rightarrow\bar{a}=\bar{q}-re$ is an †antiautomorphism of $\mathcal{C}$. Define two mappings $\mathcal{C}\rightarrow K$ by $N(a)=a\bar{a}=\bar{a}a$ (**norm** of a) and $T(a)=a+\bar{a}$ (**trace** of a). Then every a in $\mathcal{C}$ satisfies the equation $x^2-T(a)x+N(a)=0$. Furthermore, $N(ab)=N(a)N(b)$ for a, b in $\mathcal{C}$. The †quadratic form $N(x)=T(x\bar{x})/2$ characterizes $\mathcal{C}$. In particular, any two (nonassociative) general Cayley algebras over the same field K which are not †alternative fields are isomorphic.

In order for $\mathcal{C}$ to be an alternative field, either of the following two conditions is necessary and sufficient: (i) $N(a)=0$ implies $a=0$; (ii) Q is a noncommutative division algebra and γ cannot be expressed in the form $\sigma^2-\lambda\xi^2-\mu\eta^2+\lambda\mu\zeta^2$ $(\sigma,\xi,\eta,\zeta\in K)$. (For the meaning of λ, μ with respect to Q → 31 Associative Algebras D.) Every alternative field of finite dimension is a general Cayley algebra.

In particular, when Q is the †quaternion field over the real number field with $\lambda=\mu=-1$, the general Cayley algebra over Q with $\gamma=-1$ is called the **Cayley algebra**. When K is an †algebraic number field of finite degree, there are only a finite number of nonisomorphic general Cayley algebras over K.

The Lie algebra $\mathfrak{D}(\mathcal{C})$ of all †derivations of a general Cayley algebra $\mathcal{C}$ is a †simple Lie algebra of type G_2. If K is the real number field, the identity component of the group of all †automorphisms of the Cayley algebra $\mathcal{C}$ is a compact simply connected †simple Lie group of type G_2. The Cayley algebra $\mathcal{C}$ is the unique alternative field over the real number field K. This last fact is important because of the following proposition: In the theory of †non-Desarguesian projective planes, the field which gives rise to the coordinates is an alternative field. Let $\mathfrak{C}P_2$ be the set of all 3×3 †Hermitian matrices A over the Cayley algebra $\mathcal{C}$ such that tr $A=1$, $A^2=A$. Then we can define a structure of a projective plane on $\mathfrak{C}P_2$, which with this structure is called the **Cayley projective plane**. Furthermore, let $\mathfrak{J}$ be the set of all 3×3 Hermitian matrices over $\mathcal{C}$, with a multiplication in $\mathfrak{J}$ defined by $A\cdot B=(1/2)(BA+AB)$. The identity component G of the group of all automorphisms of $\mathfrak{J}$ is a compact simply connected simple Lie group of type F_4. This group G acts on $\mathfrak{C}P_2$ transitively, and $\mathfrak{C}P_2=F_4/\mathrm{spin}(9)$ (→ 248 Lie groups; Appendix A, Table 5.III).

References

[1] L. E. Dickson, Algebren und ihre Zahlentheorie, Orell Füssli, 1927.
[2] M. Zorn, Alternativkörper und quadratische Systeme, Abh. Math. Sem. Univ. Hamburg, 9 (1933), 395–402.
[3] N. Jacobson, Cayley numbers and normal simple Lie algebras of type G, Duke Math. J., 5 (1939), 775–783.

56 (XIX.8) Celestial Mechanics

A. General Remarks

The motions of planets, comets, the Moon, and satellites mostly in our solar system are the main topics in **celestial mechanics**. However, studies in this subject may also include motions of fixed and binary stars in our galaxy, equilibrium figures of celestial bodies, and rotational motions of the Earth and the Moon.

Although celestial mechanics is usually based on †Newtonian mechanics, effects of †general relativity are sometimes taken into account to determine corrections in computations of orbits of celestial bodies. Therefore, the main task of celestial mechanics is to solve differential equations of motion based on Newtonian mechanics. However, since the equations for the problem of n bodies ($n>2$) cannot be solved rigorously (→ 404 Three-Body Problem), appropriate methods are used to obtain approximate solutions of the equations with accuracy comparable to that of observations. Celestial mechanics has developed such methods applicable to actual problems.

The **two-body problem**, which concerns the behavior of two celestial bodies regarded as points exerting mutual interactions, can be reduced to a one-body problem with reference to a central force, since integrals of motion of the center of gravity for the system exist. The †Hamilton-Jacobi equation for the one-body problem is of †separable type and can be solved completely. The orbit for the two-body problem is a †conic with one of its †foci at the center of gravity. The majority of celestial bodies in the solar system actually perform **elliptic motions**. †Kepler's orbital elements for elliptic motion are functions of the integration constants in the solution of the Hamilton-Jacobi equation and are determined by the initial conditions.

B. Perturbations

In studying the †n-body problem, we first solve certain two-body problems and then apply the method of †perturbations, i.e., the **method of variation of constants**, in order to obtain solutions developed as †power series of small parameters. The parameters are ratios of the masses of planets to that of the sun for planetary motions and the ratio of the geocentric lunar distance to the solar distance for lunar motion.

Electronic computers have made it possible to compute planetary coordinates for long intervals of time by solving numerically differential equations of motion including all possible interactions. However, in discussing the stability of the solar system, analytic methods are more effective, particularly the method of obtaining **secular perturbations** by eliminating short-periodic terms by canonical transformations. This is one of the averaging methods of solving differential equations. However, as the solution obtained by the method of perturbation is not always convergent, most important problems related to the stability of motion have not yet been solved rigorously. Secular perturbations for planetary motions can be derived by solving differential equations that are linearized by neglecting cubic powers of orbital eccentricities and inclinations to the ecliptic (→ 303 Orbit Determination), which are small quantities. †Eigenvalues for linear differential equations correspond to mean angular velocities of the perihelion and the ascending node. The equations for the eigenvalues are called **secular equations**.

C. Artificial Satellites

To discuss motions of artificial satellites close to the earth, the latter cannot be regarded as a point or as a sphere but must be assumed to be an oblate spheroid, i.e., an †ellipsoid of revolution. The effects of oblateness on the motion of satellites can be derived as perturbations of the theoretical elliptic motions obtained as the solutions of this two-body problem under the assumption that the earth is spherical. Also, by utilizing a special potential very close to the geopotential, we can find a Hamilton-Jacobi equation of separable type which is solvable. This special potential appears in the problem of two fixed centers with equal masses situated on an imaginary axis. When the geopotential is assumed to be axially symmetric, the equations of motion for the satellites have two degrees of freedom; therefore, there appear two fundamental frequencies related to the special potential. When these two frequencies are equal, the problem is called a **critical inclination problem** and is important from the mathematical point of view. Theories for satellites can be applied to motions of fixed stars in the galaxy.

D. Equilibrium Figures

There has been much discussion about **equilibrium figures** and stabilities of celestial bodies assumed to consist of spinning fluids. The two-body problem with tidal interactions is particularly important; problems concerning the evolution of the Earth-Moon system are special cases of such a problem.

The theory of rotation of the Earth as it is affected by †precession, †nutation, and latitude variations is also a part of celestial mechanics; for this theory, elastic theory and geophysics are applied.

For the n-body problem → 404 Three-Body Problem.

References

[1] H. Poincaré, Les méthodes nouvelles de la méchanique céleste I, II, III, Gauthier-Villars, 1892–1899.

[2] Y. Hagihara, Tentai rikigaku no kiso (Japanese; Foundations of celestial mechanics) I, Kawade, pt. A, 1947; pt. B, 1950.
[3] A. Wintner, The analytical foundation of celestial mechanics, Princeton Univ. Press, 1941.
[4] L. Lichtenstein, Gleichgewichtsfiguren rotierender Flüssigkeiten, Springer, 1933.
[5] C. L. Siegel, Vorlesungen über Himmelsmechanik, Springer, 1956.
[6] F. F. Tisserand, Traité de méchanique céleste 1–4, Gauthier-Villars, 1889–1896.
[7] D. Brouwer and G. M. Clemence, Methods of celestial mechanics, Academic Press, 1961.
[8] Y. Hagihara, Celestial mechanics. I, Dynamical principles and transformation theory, MIT Press, 1970. II, pt. 1 and pt. 2, Perturbation theory, MIT Press, 1972. III, pt. 1 and pt. 2, Differential equations in celestial mechanics, Japan Society for the Promotion of Science, 1974. IV, pt. 1 and pt. 2, Periodic and quasiperiodic solutions, Japan Society for the Promotion of Science, 1975.
[9] C. L. Siegel and J. K. Moser, Lectures on celestial mechanics, Springer, 1971 (revised and enlarged translation of [6]).

57 (III.25) Chain Complexes

A. Graded Modules

Let A be a †ring with unity element and X be a †unitary A-module. If we are given a sequence of A-submodules X_n $(n \in \mathbf{Z})$ such that $X = \sum_{n \in \mathbf{Z}} X_n$ (†direct sum), we call X a **graded A-module** and X_n the **component of degree** n of X. Each element x of a graded A-module X has a unique representation $x = \sum_{n \in \mathbf{Z}} x_n$ $(x_n \in X_n)$; we call x_n the **component** of degree n of x. An A-submodule Y of a graded A-module X is called **homogeneous** if $x \in Y$ implies $x_n \in Y$ $(n \in \mathbf{Z})$. In this case, $Y = \sum_n Y_n$ and the quotient module $X/Y = \sum_n X_n / Y_n$ are graded A-modules, where $Y_n = Y \cap X_n$. Let $X = \sum_n X_n$ and $Y = \sum_n Y_n$ be graded A-modules and $f: X \rightarrow Y$ be an A-homomorphism. If there is a fixed integer p such that $f(X_n) \subset Y_{n+p}$ for any $n \in \mathbf{Z}$, f is called an **A-homomorphism of degree** p. In this case, $\operatorname{Ker} f = \sum_n \operatorname{Ker} f_n$ and $\operatorname{Im} f = \sum_n \operatorname{Im} f_{n-p}$ are homogeneous A-submodules of X and Y, respectively, where $f_n : X_n \rightarrow Y_{n+p}$ is the restriction of f on X_n.

B. Chain Complexes and Homology Modules

By a **chain complex** (X, ∂) over A we mean a graded A-module $X = \sum_n X_n$ together with an A-homomorphism $\partial : X \rightarrow X$ of degree -1 such that $\partial \circ \partial = 0$. Hence a chain complex over A is completely determined by a sequence

$$\ldots \rightarrow X_{n+1} \xrightarrow{\partial_{n+1}} X_n \xrightarrow{\partial_n} X_{n-1} \rightarrow \ldots$$

of A-modules and A-homomorphisms such that $\partial_n \circ \partial_{n+1} = 0$ for all n. We call ∂ the **boundary operator**. For a chain complex (X, ∂), we write $\operatorname{Ker} \partial = Z(X)$, $\operatorname{Ker} \partial_n = Z_n(X)$, $\operatorname{Im} \partial = B(X)$, $\operatorname{Im} \partial_{n+1} = B_n(X)$. Then $Z(X) = \sum_n Z_n(X)$, $B(X) = \sum_n B_n(X)$ are homogeneous submodules of X, called the **module of cycles** and the **module of boundaries**, respectively. $B(X)$ is a homogeneous submodule of $Z(X)$, and the quotient modules $Z(X)/B(X)$, $Z_n(X)/B_n(X)$ are denoted by $H(X)$, $H_n(X)$, respectively. We call $H(X) = \sum_n H_n(X)$ the **homology module** of the chain complex (X, ∂).

If (X, ∂), (Y, ∂') are chain complexes over A, an A-homomorphism $f: X \rightarrow Y$ of degree 0 satisfying $\partial' \circ f = f \circ \partial$ (i.e., $\partial'_n f_n = f_{n-1} \partial_n$) is called a **chain mapping** of X to Y. For a chain mapping f, we have $f(Z_n(X)) \subset Z_n(Y)$, $f(B_n(X)) \subset B_n(Y)$, and hence f induces an A-homomorphism $f_* : H(X) \rightarrow H(Y)$ of degree 0, which is called the **homological mapping** induced by f. We have $(1_X)_* = 1_{H(X)}$, and $(g \circ f)_* = g_* \circ f_*$ for chain mappings $f: X \rightarrow Y$ and $g: Y \rightarrow Z$.

Let $f, g: X \rightarrow Y$ be two chain mappings. If there is an A-homomorphism $D: X \rightarrow Y$ of degree $+1$ such that $f - g = D \circ \partial + \partial' \circ D$, we say that f is **chain homotopic** to g and write $f \simeq g$; D is called a **chain homotopy** of f to g. If f is chain homotopic to g, we have $f_* = g_* : H(X) \rightarrow H(Y)$. For chain complexes X and Y, if there are chain mappings $f: X \rightarrow Y$ and $g: Y \rightarrow X$ such that $f \circ g \simeq 1_Y$ and $g \circ f \simeq 1_X$, we say that X is **chain equivalent** to Y. In this case $f_* : H(X) \rightarrow H(Y)$ is an isomorphism and $g_* : H(Y) \rightarrow H(X)$ is its inverse.

Let (X, ∂) be a chain complex over A and $Y = \sum_n Y_n$ be a homogeneous A-submodule of X such that $\partial Y \subset Y$. Then Y and X/Y are chain complexes over A with the boundary operators induced by ∂. Y is called a **chain subcomplex** of X, and X/Y is called the **quotient chain complex** of X by Y or the **relative chain complex** of $X \bmod Y$. For a chain complex X and its subcomplex Y we have an †exact sequence $0 \rightarrow Y \xrightarrow{i} X \xrightarrow{j} X/Y \rightarrow 0$, where i is the †canonical injection and j the †canonical surjection.

Let (W, ∂'), (X, ∂), (Y, ∂'') be chain complexes over A, and $f: W \rightarrow X$, $g: X \rightarrow Y$ be chain mappings such that $0 \rightarrow W \xrightarrow{f} X \xrightarrow{g} Y \rightarrow 0$ is exact. Then an A-homomorphism $\partial_* : H(Y) \rightarrow H(W)$ of degree -1, called the **connecting**

homomorphism, is defined by $\partial_*(y+B(Y))=f^{-1}\circ\partial\circ g^{-1}(y)+B(W)$ $(y\in Z(Y))$, and we have the **exact sequence of homology**:

$$\ldots\xrightarrow{\partial_*}H_n(W)\xrightarrow{f_*}H_n(X)\xrightarrow{g_*}H_n(Y)$$
$$\xrightarrow{\partial_*}H_{n-1}(W)\xrightarrow{f_*}H_{n-1}(X)\xrightarrow{g_*}H_{n-1}(Y)\xrightarrow{\partial_*}\ldots.$$

For a †commutative diagram

$$\begin{array}{ccccccccc} 0 & \to & W & \to & X & \to & Y & \to & 0 \\ & & \downarrow\varphi & & \downarrow & & \downarrow\psi & & \\ 0 & \to & W' & \to & X' & \to & Y' & \to & 0 \end{array} \qquad (1)$$

consisting of chain complexes and chain mappings in which each row is exact, we have $\partial_*\circ\psi_*=\varphi_*\circ\partial_*:H(Y)\to H(W')$.

For the †inductive limit $\varinjlim X_\lambda$ of chain complexes X_λ over A, we have $H(\varinjlim X_\lambda)=\varinjlim H(X_\lambda)$.

C. Augmented Chain Complexes

A chain complex X is said to be **positive** if $X_n=0$ for all $n<0$. If X is a positive chain complex over A and M is an A-module, then we mean by an **augmentation of** X **over** M an A-homomorphism $\varepsilon:X_0\to M$ such that the composition $X_1\xrightarrow{\partial_1}X_0\xrightarrow{\varepsilon}M$ is trivial: $\varepsilon\circ\partial_1=0$. A positive chain complex X together with an augmentation ε of X over M is called an **augmented chain complex** over M. It is said to be **acyclic** if the sequence

$$\ldots\to X_n\xrightarrow{\partial_n}X_{n-1}\to\ldots\to X_1\xrightarrow{\partial_1}X_0\xrightarrow{\varepsilon}M\to 0$$

is exact, namely, if $H_n(X)=0$ $(n\neq 0)$ and ε induces an A-isomorphism $H_0(X)\cong M$. In this case X is also called a **left resolution** of M. Moreover, if each X_n is a †projective A-module, X is called a **left projective resolution**. For any A-module M, there exists a left projective resolution of M.

Let $\alpha:M\to M'$ be an A-homomorphism of A-modules, and X, X' be augmented chain complexes over M, M' having augmentations ε, ε', respectively. Then a chain mapping $f:X\to X'$ satisfying $\varepsilon'\circ f_0=\alpha\circ\varepsilon$ is called a **chain mapping over** α. If X, X' are left projective resolutions of M, M', respectively, then there exist chain mappings of X to X' over α, and any two such mappings are chain homotopic. In particular, a left projective resolution of an A-module M is uniquely determined up to chain homotopy.

D. Tor

Given a right A-module M and a left A-module N, **Z**-modules $\mathbf{Tor}_n^A(M,N)$ $(n=0,1,2,\ldots)$, called the **torsion product**, are defined as follows: Let

$$Y:\ldots\to Y_n\xrightarrow{\partial_n}Y_{n-1}\to\ldots\xrightarrow{\partial_1}Y_0\xrightarrow{\varepsilon}N\to 0$$

be a projective resolution of N, and consider the chain complex

$$(M\otimes_A Y,1\otimes\partial):\ldots\to M\otimes_A Y_n\xrightarrow{1\otimes\partial_n}M\otimes_A Y_{n-1}\to\ldots\xrightarrow{1\otimes\partial_1}M\otimes_A Y_0\to 0$$

obtained by forming the †tensor product of M and Y. Then we see that the homology module $H_n(M\otimes_A Y)$ is uniquely determined for any choice of left projective resolution of N. We define $\mathrm{Tor}_n^A(M,N)=H_n(M\otimes_A Y)$. In particular, we have $\mathrm{Tor}_0^A(M,N)\cong M\otimes_A N$.

E. Properties of Tor

(1) If M is a †flat A-module, we have $\mathrm{Tor}_n^A(M,N)=0$ $(n=1,2,\ldots)$.

(2) An A-homomorphism $f:M_1\to M_2$ induces a homomorphism $f_*:\mathrm{Tor}_n^A(M_1,N)\to\mathrm{Tor}_n^A(M_2,N)$. We have $(1_M)_*=1$, and $(g\circ f)_*=g_*\circ f_*$ for $f:M_1\to M_2$, $g:M_2\to M_3$.

(3) For an exact sequence $0\to M_1\xrightarrow{f}M_2\xrightarrow{g}M_3\to 0$, we have the following **exact sequence of Tor**:

$$\ldots\xrightarrow{\partial_*}\mathrm{Tor}_n^A(M_1,N)\xrightarrow{f_*}\mathrm{Tor}_n^A(M_2,N)\xrightarrow{g_*}$$
$$\mathrm{Tor}_n^A(M_3,N)\xrightarrow{\partial_*}\mathrm{Tor}_{n-1}^A(M_1,N)\to\ldots$$
$$\to\mathrm{Tor}_1^A(M_3,N)\xrightarrow{\partial_*}M_1\otimes_A N\to M_2\otimes_A N$$
$$\to M_3\otimes_A N\to 0,$$

where ∂_* are the connecting homomorphisms.

(4) For a commutative diagram

$$\begin{array}{ccccccccc} 0 & \to & M_1 & \to & M_2 & \to & M_3 & \to & 0 \\ & & \downarrow\varphi & & \downarrow & & \downarrow\psi & & \\ 0 & \to & M_1' & \to & M_2' & \to & M_3' & \to & 0 \end{array}$$

of A-modules and A-homomorphisms with exact rows, we have $\partial_*\circ\psi_*=\varphi_*\circ\partial_*$.

(5) $\mathrm{Tor}_n^A(\Sigma_\lambda M_\lambda,N)\cong\Sigma_\lambda\mathrm{Tor}_n^A(M_\lambda,N)$.

(6) $\mathrm{Tor}_n^A(\varinjlim M_\lambda,N)\cong\varinjlim\mathrm{Tor}_n^A(M_\lambda,N)$.

On the other hand, take a left projective resolution X of M and consider the chain complex $X\otimes_A N$. Then we have $H_n(X\otimes_A N)\cong\mathrm{Tor}_n^A(M,N)$ for $n=0,1,\ldots$. Therefore properties similar to (1)–(6) hold with respect to the second variable of $N\,\mathrm{Tor}_n^A(M,N)$.

(7) If A° is a ring †anti-isomorphic to A, then $\mathrm{Tor}_n^A(M,N)\cong\mathrm{Tor}_n^{A^\circ}(N,M)$. In particular, if A is commutative, then $\mathrm{Tor}_n^A(M,N)$ is an A-module and we have $\mathrm{Tor}_n^A(M,N)\cong\mathrm{Tor}_n^A(N,M)$.

(8) Let A be a †principal ideal ring. Then $\mathrm{Tor}_n^A(M,N)=0$ $(n=2,3,\ldots)$ and $\mathrm{Tor}_1^A(M,N)$

is also denoted by $\mathbf{M} *_A \mathbf{N}$. For an exact sequence $0 \to M_1 \to M_2 \to M_3 \to 0$, we have the exact sequence $0 \to M_1 *_A N \to M_2 *_A N \to M_3 *_A N \to M_1 \otimes_A N \to M_2 \otimes_A N \to M_3 \otimes_A N \to 0$. In particular, $\mathbf{Z} *_{\mathbf{Z}} N = 0$ and $(\mathbf{Z}/n\mathbf{Z}) *_{\mathbf{Z}} N \cong {}_nN$ $(= \{x \in N \mid nx = 0\})$.

F. Universal Coefficient Theorem for Homology

If (X, ∂) is a chain complex over A and N is a left A-module, then $(X \otimes_A N, \partial \otimes 1)$ is a chain complex. If A is a principal ideal ring and each X_n is a †torsion-free A-module, then we have a formula

$$H_n(X \otimes_A N) \cong H_n(X) \otimes_A N + H_{n-1}(X) *_A N,$$

called the **universal coefficient theorem**.

G. Double Chain Complexes

By a **double chain complex** $(X_{p,q}, \partial', \partial'')$ over A we mean a family of left A-modules $X_{p,q}$ $(p, q \in \mathbf{Z})$ together with A-homomorphisms $\partial'_{p,q}: X_{p,q} \to X_{p-1,q}$ and $\partial''_{p,q}: X_{p,q} \to X_{p,q-1}$ such that $\partial'_{p-1,q} \circ \partial'_{p,q} = \partial''_{p,q-1} \circ \partial''_{p,q} = \partial'_{p,q-1} \circ \partial''_{p,q} + \partial''_{p-1,q} \circ \partial'_{p,q} = 0$. We define the associated chain complex (X_n, ∂) by setting $X_n = \sum_{p+q=n} X_{p,q}$, $\partial_n = \sum_{p+q=n} \partial'_{p,q} + \partial''_{p,q}$. We call ∂ the **total boundary operator**, and ∂', ∂'' the **partial boundary operators**.

Given a chain complex X consisting of right A-modules and a chain complex Y consisting of left A-modules, a double chain complex $(Z_{p,q}, \partial', \partial'')$ is defined by setting $Z_{p,q} = X_p \otimes_A Y_q$, $\partial'_{p,q} = \partial_p \otimes 1$, $\partial''_{p,q} = (-1)^p 1 \otimes \partial_q$, where ∂_p, ∂_q are the boundary operators of X, Y, respectively. It is called the **product double chain complex** of X and Y, and the homology module of its associated chain complex is denoted by $H(X \otimes_A Y)$. With respect to this homology module, the following facts hold. If X is a left projective resolution of a right A-module M and Y is that of a left A-module N, then $H_n(X \otimes_A Y) = \mathrm{Tor}_n^A(M, N)$ (→ 202 Homological Algebra). If A is a principal ideal ring and each X_n is a torsion-free A-module, then we have the formula

$$H_n(X \otimes_A Y) \cong \sum_{p+q=n} H_p(X) \otimes_A H_q(Y) + \sum_{p+q=n-1} H_p(X) *_A H_q(Y),$$

the **Künneth theorem**.

H. Cochain Complexes

By a **cochain complex** (X, d) over A we mean a graded A-module X together with an A-homomorphism $d: X \to X$ of degree $+1$ such that $d \circ d = 0$; d is called the **coboundary operator** or **derivation**. For a cochain complex (X, d), we denote by X^n the component of degree n of X, and by d^n, $X^n \to X^{n+1}$ the restriction of d on X^n. Then a chain complex (Y, ∂) is defined by $Y_n = X^{-n}$ and $\partial_n: Y_n \to Y_{n-1}$ is equal to $d^{-n}: X^{-n} \to X^{-n+1}$.

For a cochain complex (X, d), we write $\mathrm{Ker}\, d^n = Z^n(X)$, $\mathrm{Ker}\, d = Z(X)$ $(Z(X) = \sum Z^n(X))$, $\mathrm{Im}\, d^{n-1} = B^n(X)$, $\mathrm{Im}\, d = B(X)$ $(B(X) = \sum B^n(X))$ and $Z^n(X)/B^n(X) = H^n(X)$, $Z(X)/B(X) = H(X)$ $(H(X) = \sum H^n(X))$. These modules $Z(X)(Z^n(X))$, $B(X)(B^n(X))$, and $H(X)(H^n(X))$ are called the **module of cocycles**, the **module of coboundaries**, and the **cohomology module** of X, respectively. If we consider the associated chain complex (Y, ∂) of (X, d), then $H_{-n}(Y)$ corresponds to $H^n(X)$. In this way, results on chain complexes give results on cochain complexes. Thus the concepts of **cochain mapping, cochain homotopy, cochain equivalence, cochain subcomplex,** and **relative cochain complex** can be defined as in the case of chain complexes in B, and we have corresponding results. In particular, given an exact sequence $0 \to W \xrightarrow{f} X \xrightarrow{g} Y \to 0$ of cochain complexes and cochain mappings, the **connecting homomorphism** $d_*: H^n(Y) \to H^{n+1}(W)$ is defined, and the **exact sequence of cohomology**

$$\ldots \xrightarrow{d_*} H^n(W) \xrightarrow{f_*} H^n(X) \xrightarrow{g_*} H^n(Y)$$
$$\xrightarrow{d_*} H^{n+1}(W) \xrightarrow{f_*} H^{n+1}(X) \xrightarrow{g_*} H^{n+1}(Y) \xrightarrow{d_*} \ldots$$

exists. For a commutative diagram

$$\begin{array}{ccccccccc} 0 \to & W & \to & X & \to & Y & \to 0 \\ & \downarrow\varphi & & \downarrow & & \downarrow\psi & \\ 0 \to & W' & \to & X' & \to & Y' & \to 0 \end{array}$$

of cochain complexes and cochain mappings with exact rows, we have $d_* \circ \psi_* = \varphi_* \circ d_*$.

A cochain complex X is said to be **positive** if $X^n = 0$ for $n < 0$. If X is a positive cochain complex over A and M is an A-module, we mean by an **augmentation** of X over M an A-homomorphism $\varepsilon: M \to X^0$ such that the composition $M \xrightarrow{\varepsilon} X^0 \xrightarrow{d^0} X^1$ is trivial. If the sequence

$$0 \to M \xrightarrow{\varepsilon} X^0 \xrightarrow{d^0} \ldots \to X^n \xrightarrow{d^n} X^{n+1} \to \ldots$$

is exact, X is called a **right resolution** of M. Moreover, if each X^n is an †injective A-module, X is called a **right injective resolution** of M. For any A-module M, there exists a right injective resolution of M, and any two such resolutions are cochain homotopic.

I. Ext

Given left A-modules M and N, **Z**-modules $\mathrm{Ext}_A^n(M,N)$ $(n=0,1,2,\ldots)$ are defined as follows: Let $X:\ldots\to X_n\to X_{n-1}\to\ldots\to X_0\to M\to 0$ be a projective resolution of M, and consider the cochain complex $\mathrm{Hom}_A(X,N)$:

$$0\to\mathrm{Hom}_A(X_0,N)\to\ldots\to\mathrm{Hom}_A(X_{n-1},N)\to\mathrm{Hom}_A(X_n,N)\to\ldots$$

obtained by forming the †module of A-homomorphisms. Then we can show that the cohomology module $H^n(\mathrm{Hom}_A(X,N))$ is uniquely determined for any choice of projective resolution of M. We define $\mathbf{Ext}_A^n(M,N)=H^n(\mathrm{Hom}_A(X,N))$. This can also be defined as the cohomology module $H^n(\mathrm{Hom}_A(M,Y))$ of the cochain complex $\mathrm{Hom}_A(M,Y):0\to\mathrm{Hom}_A(M,Y^0)\to\ldots\to\mathrm{Hom}_A(M,Y^{n-1})\to\mathrm{Hom}_A(M,Y^n)\to\ldots$, where $Y:0\to N\to Y^0\to\ldots\to Y^{n-1}\to Y^n\to\ldots$ is a right injective resolution of N. Furthermore, for a left projective resolution X of M and a right injective resolution Y of N, we see that $\mathrm{Ext}_A^n(M,N)$ is isomorphic to the cohomology module $H^n(\mathrm{Hom}_A(X,Y))$ of the associated cochain complex of the double cochain complex $\mathrm{Hom}_A(X,Y)=(\mathrm{Hom}_A(X_p,Y^q),d',d'')$, where $d'_{p,q}:\mathrm{Hom}_A(X_p,Y^q)\to\mathrm{Hom}_A(X_{p+1},Y^q)$ and $d''_{p,q}:\mathrm{Hom}_A(X_p,Y^q)\to\mathrm{Hom}_A(X_p,Y^{q+1})$ are given by $d'_{p,q}(u)=u\circ\partial_{p+1}$, $d''_{p,q}(u)=(-1)^{p+q+1}d^q\circ u$ $(u\in\mathrm{Hom}_A(X_p,Y^q))$ by using the boundary operator ∂ of X and the coboundary operator d of Y.

J. Properties of Ext

(1)We have $\mathrm{Ext}_A^0(M,N)\cong\mathrm{Hom}_A(M,N)$.

(2) If M is a projective A-module or N is an injective A-module, then $\mathrm{Ext}_A^n(M,N)=0$ $(n=1,2,\ldots)$.

(3) An A-homomorphism $f:M_1\to M_2$ (resp. $f:N_1\to N_2$) induces a homomorphism $f^*:\mathrm{Ext}_A^n(M_2,N)\to\mathrm{Ext}_A^n(M_1,N)$ (resp. $f_*:\mathrm{Ext}_A^n(M,N_1)\to\mathrm{Ext}_A^n(M,N_2)$). We have $1_M^*=1$ and $(g\circ f)^*=f^*\circ g^*$ for $f:M_1\to M_2$, $g:M_2\to M_3$ (resp. $1_{N_*}=1$ and $(g\circ f)_*=g_*\circ f_*$ for $f:N_1\to N_2$, $g:N_2\to N_3$).

(4) For an exact sequence $0\to M_1\to M_2\to M_3\to 0$ (resp. $0\to N_1\to N_2\to N_3\to 0$), we have the **exact sequence of Ext**:

$$0\to\mathrm{Hom}_A(M_3,N)\to\mathrm{Hom}_A(M_2,N)\to\mathrm{Hom}_A(M_1,N)\to\mathrm{Ext}_A^1(M_3,N)\to\mathrm{Ext}_A^1(M_2,N)\to\ldots$$

(resp. $0\to\mathrm{Hom}_A(M,N_1)\to\mathrm{Hom}(M,N_2)\to\mathrm{Hom}(M,N_3)\to\mathrm{Ext}_A^1(M,N_1)\to\mathrm{Ext}_A^1(M,N_2)\to\ldots$).

(5)

$$\mathrm{Ext}_A^n\Big(\sum_\alpha M_\alpha,\ \prod_\beta N_\beta\Big)=\prod_{\alpha,\beta}\mathrm{Ext}_A^n(M_\alpha,N_\beta).$$

(6) If A is a principal ideal ring, then $\mathrm{Ext}_A^n(M,N)=0$ $(n=2,3,\ldots)$, and $\mathrm{Ext}_A^1(M,N)$ is also denoted by $\mathrm{Ext}_A(M,N)$. In particular, $\mathrm{Ext}_z(\mathbf{Z},N)=0$, $\mathrm{Ext}_z(\mathbf{Z}/n\mathbf{Z},N)\cong N/nN$, $\mathrm{Ext}_z(M,\mathbf{Q}/\mathbf{Z})=0$, $\mathrm{Ext}_z(M,\mathbf{Z}/n\mathbf{Z})=\hat{M}/n\hat{M}$, where $\hat{M}=\mathrm{Hom}_z(M,\mathbf{Q}/\mathbf{Z})$.

K. Universal Coefficient Theorem for Cohomology

If X is a chain complex over a principal ideal ring A such that each X_n is a free A-module, then for any A-module N we have the formula

$$H^n(\mathrm{Hom}_A(X,N))\cong\mathrm{Hom}_A(H_n(X),N)+\mathrm{Ext}_A(H_{n-1}(X),N),$$

the **universal coefficient theorem**. This is generalized as follows: Let X be a chain complex and Y a cochain complex, both over a principal ideal ring A. Assume that each X_n is a free A-module or that each Y^n is an injective A-module. Then we have the formula

$$H^n(\mathrm{Hom}_A(X,Y))\cong\sum_{p+q=n}\mathrm{Hom}_A(H_p(X),\ H^q(Y))+\sum_{p+q=n-1}\mathrm{Ext}_A(H_p(X),\ H^q(Y))$$

(→ 202 Homological Algebra; 203 Homology Groups).

References

See references to 202 Homological Algebra, 203 Homology Groups, 275 Modules.

58 (IX.21) Characteristic Classes

A. General Remarks

The theory of characteristic classes arose from the problem of whether or not there exists a †tangent r-frame field on a †differentiable manifold (E. Stiefel [5]) (→ 108 Differentiable Manifolds). The importance of the characteristic class as a fundamental invariant for the †vector bundle structure is now fully recognized (→ 155 Fiber Bundles, 411 Topology of Lie Groups and Homogeneous Spaces, 211 Immersion and Embedding).

B. Stiefel-Whitney Classes

Let $\xi=(E,B,\mathbf{R}^n)$ be an n-dimensional real †vector bundle (called an $\mathbf{R}^n$-bundle) with †paracompact Hausdorff †base space B, fiber

$\mathbf{R}^n$, and †orthogonal group $O(n)$ as the †structure group. Then the element $w_i(\xi)$ of the i-dimensional †cohomology group $H^i(B,\mathbf{Z}_2)$ $(i=1,2,\dots,n)$ of the base space B with coefficient in $\mathbf{Z}_2=\mathbf{Z}/2\mathbf{Z}$, called the i-dimensional (or ith) **Stiefel-Whitney class**, and the element $w(\xi)=1+w_1(\xi)+\dots+w_n(\xi)$ of the †cohomology ring $H^*(B,\mathbf{Z}_2)$, called the **total Stiefel-Whitney class**, are defined as follows. First, we deal with the case $n=1$. We call **infinite-dimensional real projective space** the †inductive limit $\mathbf{P}^\infty(\mathbf{R})=\varinjlim \mathbf{P}^n(\mathbf{R})$ of the finite-dimensional real †projective space $\mathbf{P}^n(\mathbf{R})$. The nontrivial †line bundle γ_1 over the infinite-dimensional real projective space $\mathbf{P}^\infty(\mathbf{R})$ is a universal $\mathbf{R}^1$-bundle (†universal bundle for the orthogonal group $O(1)$); therefore, any $\mathbf{R}^1$-bundle $\xi=(E,B,\mathbf{R}^1)$ is equivalent to an †induced bundle from the universal bundle γ_1 by a †characteristic mapping $f_\xi: B\rightarrow \mathbf{P}^\infty(\mathbf{R}): \xi\equiv f_\xi^*\gamma_1$. We define the 1-dimensional universal Stiefel-Whitney class $w_1(\gamma_1)$ to be the generator of $H^1(\mathbf{P}^\infty(\mathbf{R}),\mathbf{Z}_2)$ and set $w_1(\xi)=f_\xi^* w_1(\gamma_1)$. For general n, we consider the principal $O(n)$-bundle $(P,B,O(n))$ †associated with the given $\mathbf{R}^n$-bundle $\xi=(E,B,\mathbf{R}^n)$. Let Q_n be the subgroup of $O(n)$ consisting of all diagonal matrices. Then the quotient space $Y=P/Q_n$ is the base space of the †principal Q_n-bundle $\eta=(P,P/Q_n,Q_n)$. Let $\rho: Y\rightarrow B=P/O(n)$ be the natural projection. Then the $\mathbf{R}^n$-bundle $\rho^*\xi$ over Y induced by ρ is associated with η and is equivalent to the †Whitney sum of n line bundles (Hirzebruch [4]), $\rho^*\xi\equiv\xi_1+\dots+\xi_n$. Moreover, the homomorphism $\rho^*: H^*(B,\mathbf{Z}_2)\rightarrow H^*(Y,\mathbf{Z}_2)$ is injective (Borel [1]). Therefore, we can uniquely define the total Stiefel-Whitney class $w(\xi)$ of the $\mathbf{R}^n$-bundle by the relation $\rho^* w(\xi)=w(\xi_1)\dots w(\xi_n)$. The Stiefel-Whitney classes defined above are compatible with bundle mappings $f: w(f^*\xi)=f^*w(\xi)$ (→ 155 Fiber bundles L). The Stiefel-Whitney class $w_i(\gamma_n)\in H^i(BO(n),\mathbf{Z}_2)$ $(1\leqslant i\leqslant n)$ of the universal $\mathbf{R}^n$-bundle γ_n over the †classifying space $BO(n)$ is called the i-dimensional **universal Stiefel-Whitney class**. For the Whitney sum of vector bundles, we have $w(\xi\oplus\eta)=w(\xi)\cdot w(\eta)$.

In order that an $\mathbf{R}^n$-bundle ξ be †orientable, namely, for ξ to have an $SO(n)$-structure, it is necessary and sufficient that $w_1(\xi)=0$. For an oriented $\mathbf{R}^n$-bundle ξ, the **Euler-Poincaré class** $X_n(\xi)$ is defined to be the †obstruction class $X_n(\xi)\in H^n(B,\mathbf{Z})$ for constructing a †cross section of the †associated $(n-1)$-sphere bundle. In particular, the Euler-Poincaré class of the universal bundle for $SO(n)$ is called the **universal Euler-Poincaré class**. $X_n(\xi)$ mod 2 is equal to $w_n(\xi)$. If n is odd, we have $2X_n(\xi)=0$.

C. Chern Classes

We consider an n-dimensional complex vector bundle $\omega=(E,B,\mathbf{C}^n)$ (called $\mathbf{C}^n$-bundle in the following) with a paracompact base space B, fiber $\mathbf{C}^n$, and unitary group $U(n)$ as the structure group. The cohomology class $c_i(\omega)\in H^{2i}(B,\mathbf{Z})$ $(i=1,2,\dots,n)$, called the $2i$-dimensional (or ith) **Chern class**, and the **total Chern class** $c(\omega)=1+c_1(\omega)+\dots+c_n(\omega)\in H^*(B,\mathbf{Z})$ are defined as above. Actually, in the case $n=1$, let $\mathbf{C}^\infty=\varinjlim \mathbf{C}^n$ (the inductive limit of the †complex Euclidean spaces $\mathbf{C}^n$), S^∞ be its unit sphere, and $\mathbf{P}^\infty(\mathbf{C})$ be the **infinite-dimensional complex projective space** consisting of all complex lines through the origin O of $\mathbf{C}^\infty$. Then the natural mapping $S^\infty\rightarrow\mathbf{P}^\infty(\mathbf{C})$ defines a universal principal $U(1)$-bundle $(S^\infty,\mathbf{P}^\infty(\mathbf{C}),U(1))$. Let γ_1 be its associated universal $\mathbf{C}^1$-bundle. Then we define the 1-dimensional universal Chern class $c_1(\gamma_1)\in H^2(\mathbf{P}^\infty(\mathbf{C}),\mathbf{Z})$ to be the cohomology class that takes the value -1 on the cycle $S^2(\approx\mathbf{P}^1(\mathbf{C})\subset\mathbf{P}^\infty(\mathbf{C}))$ with the natural orientation. Since a general $\mathbf{C}^1$-bundle $\xi=(E,B,\mathbf{C}^1)$ is induced from γ_1 by a characteristic mapping $f_\xi: B\rightarrow\mathbf{P}^\infty(\mathbf{C})$, we set $c_1(\xi)=f_\xi^* c_1(\gamma_1)$. In the case $n>1$, let $(P,B,U(n))$ be the principal $U(n)$-bundle associated with the given $\mathbf{C}^n$-bundle $\xi=(E,B,\mathbf{C}^n)$. Let T_n be the subgroup of $U(n)$ consisting of all diagonal matrices (which is a †maximal torus of $U(n)$). Then the quotient space $Y=P/T_n$ is the base space of the principal T_n-bundle $\eta=(P,P/T_n,T_n)$. Let $\rho: Y\rightarrow B=P/U(n)$ be the natural projection. Then the $\mathbf{C}^n$-bundle $\rho^*\xi$ over Y is associated with η and is equivalent to the †Whitney sum of n complex line bundles: $\rho^*\xi\equiv\xi_1\oplus\dots\oplus\xi_n$. Moreover, $\rho^*: H^*(B,\mathbf{Z})\rightarrow H^*(Y,\mathbf{Z})$ is a monomorphism (Borel [1], Hirzebruch [4]). Therefore, we can uniquely define the total Chern class $c(\xi)$ of the $\mathbf{C}^n$-bundle ξ by the relation $\rho^*c(\xi)=c(\xi_1)\dots c(\xi_n)$. The Chern classes, as defined above, are compatible with bundle mappings f (i.e., $c(f^*\xi)=f^*c(\xi)$) (→ 155 Fiber Bundles M). The Chern class $c_i(\gamma_n)\in H^{2i}(BU(n),Z)$ $(1\leqslant i\leqslant n)$ of the universal $\mathbf{C}^n$-bundle γ_n over $BU(n)$ is called the $2i$-dimensional **universal Chern class**. Let ξ, η be complex vector bundles over B. Then we have $c(\xi\oplus\eta)=c(\xi)c(\eta)$. By the natural inclusion $U(n)\subset SO(2n)$, we can identify a $\mathbf{C}^n$-bundle ω with an oriented $\mathbf{R}^{2n}$-bundle $\omega_{\mathbf{R}}$. Then we have $c_i(\omega)\bmod 2=w_{2i}(\omega_{\mathbf{R}})$, $w_{2i+1}(\omega_{\mathbf{R}})=0$ $(i=0,1,\dots,n)$, $c_n(\omega)=X_{2n}(\omega_{\mathbf{R}})$.

Examples. Let $(S^{2n+1},\mathbf{P}^n(\mathbf{C}),U(1))$ be the †Hopf bundle, and let γ_1^n be the associated complex line bundle. Then the characteristic mapping of γ_1^n is the natural inclusion $\mathbf{P}^n(\mathbf{C})\rightarrow BU(1)=\mathbf{P}^\infty(\mathbf{C})$, and $c(\gamma_1^n)=1-g_n$, where

$g_n \in H^2(\mathbf{P}^n(\mathbf{C}), \mathbf{Z})$ is the cohomology class dual to the homology class represented by the hyperplane $\mathbf{P}^{n-1}(\mathbf{C})$. On the other hand, for the complex line bundle $\xi_1^n = \{\mathbf{P}^{n-1}(\mathbf{C})\}$ determined by the †divisor (→ 74 Complex Manifolds) $\mathbf{P}^{n-1}(\mathbf{C}) \subset \mathbf{P}^n(\mathbf{C})$, we have $c(\xi_1^n) = 1 + g_n$, and ξ_1^n, γ_1^n are dual to each other. Moreover, the Whitney sum $\tau \oplus \varepsilon_1$ of the tangent bundle $\tau(\mathbf{P}^n(\mathbf{C}))$ of the n-dimensional complex projective space $\mathbf{P}^n(\mathbf{C})$ and the trivial $\mathbf{C}^1$-bundle ε_1 is equivalent to the Whitney sum of $(n+1)$ copies of ξ_1^n. Therefore, we have $c(\tau(\mathbf{P}^n(\mathbf{C}))) = c(\xi_1^n \oplus \xi_1^n \oplus \ldots \oplus \xi_1^n) = (1+g_n)^{n+1}$.

D. Pontrjagin Classes

Utilizing the inclusion map $O(n) \subset U(n)$ of structure groups, we can make a $\mathbf{C}^n$-bundle $\xi_{\mathbf{C}} = \xi \oplus \sqrt{-1}\,\xi$ correspond to an $\mathbf{R}^n$-bundle ξ. We define the $4i$-dimensional **Pontrjagin class** of the $\mathbf{R}^n$-bundle ξ by $p_i(\xi) = (-1)^i c_{2i}(\xi_{\mathbf{C}})$ $(\in H^{4i}(B, \mathbf{Z}))$ $(i = 1, 2, \ldots, [n/2])$ (Hirzebruch [4]). In particular, the Pontrjagin classes of the universal bundle for $O(n)$ are called the **universal Pontrjagin classes**. The **total Pontrjagin classes** $p(\xi) = 1 + p_1(\xi) + \ldots + p_{[n/2]}(\xi)$ can be defined similarly. We have $2c_{2i+1}(\xi_{\mathbf{C}}) = 0$. For $\mathbf{R}^n$-bundles ξ and η, $p(\xi \oplus \eta) - p(\xi)p(\eta)$ has order 2. We have $p_i(\xi) \pmod 2 = (W_{2i}(\xi))^2$ and $p_n(\xi) = (X_{2n}(\xi))^2$ for the oriented $\mathbf{R}^{2n}$-bundle ξ. Moreover, for a complex vector bundle ω, we have (Wu [8])

$$(-1)^k p_k(\omega_{\mathbf{R}}) = \sum_{i=0}^{2k} (-1)^i c_i(\omega) c_{2k-i}(\omega),$$

$$c_0(\omega) = 1.$$

All such classes as defined in Sections B–D are called **characteristic classes**.

E. Other Definitions of Characteristic Classes

Axiomatic Definition. (1) For a $\mathbf{C}^n$-bundle ξ over a paracompact Hausdorff base space B, Chern classes $c_i(\xi) \in H^{2i}(B, \mathbf{Z})$ $(i \geqslant 0)$ are defined, and we have $c_0(\xi) = 1$, $c_i(\xi) = 0$ $(i > n)$. (2) For the total Chern class $c(\xi) = \sum_{i=0}^{\infty} c_i(\xi)$, we have $c(f^*\xi) = f^*c(\xi)$ for each bundle mapping f. (3) For the Whitney sum, we have $c(\xi \oplus \eta) = c(\xi) \cdot c(\eta)$. (4) Normalization condition: For the canonical line bundle ξ_1^n, we have $c(\xi_1^n) = 1 + g_n$ (→ Section C). We can verify the existence and uniqueness of $c_i(\xi)$ satisfying these four conditions, so Chern classes can be defined axiomatically by these conditions (Hirzebruch [4]). We can similarly define Stiefel-Whitney classes axiomatically.

Definition by Obstruction classes. When the base space is a †CW complex, we can define the Chern class of a $\mathbf{C}^n$-bundle ξ over B as follows: Let $V_{n,n-q+1}(\mathbf{C}) = U(n)/I_{n-q+1} \times U(q-1)$ be the †complex Stiefel manifold of all orthonormal $(n-q+1)$-frames in $\mathbf{C}^n$ with Hermitian metric. Then $V_{n,n-q+1}(\mathbf{C})$ is †$(2q-2)$-connected, and its $(2q-1)$-dimensional †homotopy group $\pi_{2q-1}(V_{n,n-q+1}(\mathbf{C})) = \mathbf{Z}$. Let ξ' be the †associated bundle of ξ with fiber $V_{n,n-q+1}(\mathbf{C})$. Then the †primary obstruction class $(\in H^{2q}(B, \mathbf{Z}))$ to constructing a †cross section of ξ' coincides with the Chern class $c_q(\xi)$. Analogously, we can interpret $w_q(\xi)$ for an $\mathbf{R}^n$-bundle ξ as an obstruction class.

Definition by Schubert Cycles. We denote by $\mathbf{C}^k$ the subspace defined by $z_{k+1} = z_{k+2} = \ldots = z_{n+N} = 0$ of the space $\mathbf{C}^{n+N} = \{(z_1, \ldots, z_{n+N}) \mid z_i \in \mathbf{C}, i = 1, \ldots, n+N\}$, and fix the sequence of subspaces $\mathbf{C}^1 \subset \mathbf{C}^2 \subset \ldots \subset \mathbf{C}^{n+N}$. The set of all complex n-planes X through the origin O in $\mathbf{C}^{n+N}$ forms the †complex Grassmann manifold $M_{n+N,n}(\mathbf{C})$. We denote by $E(\gamma_n^N)$ the set of all pairs (X, v), where $X \in M_{n+N,n}(\mathbf{C})$ and v is a vector in X. Then we can define a $2N$-universal complex n-dimensional vector bundle γ_n^N over $M_{n+N,n}(\mathbf{C})$, with projection $(X, v) \to X$. Let $\omega = (\omega(1), \ldots, \omega(n))$ be a sequence of integers satisfying the condition $0 \leqslant \omega(1) \leqslant \ldots \leqslant \omega(n) \leqslant N$. Then the set e_ω of all n-planes $X \subset \mathbf{C}^{n+N}$ through the origin O satisfying $\dim(X \cap \mathbf{C}^{i+\omega(i)}) = i$, $\dim(X \cap \mathbf{C}^{i+\omega(i)-1}) = i-1$, $i = 1, 2, \ldots, n$, forms a real $(2\sum_{i=1}^n \omega(i))$-dimensional open †cell. The set of all these open cells e_ω gives a †cellular subdivision of $M_{n+N,n}(\mathbf{C})$ as a CW complex. The closure $\bar{e}_\omega$ of e_ω is a cellular subcomplex of $M_{n+N,n}(\mathbf{C})$, called a **Schubert variety**. This is a †pseudomanifold with canonical orientation and represents a $(2\sum_{i=1}^n \omega(i))$-dimensional integral cycle, called a **Schubert cycle**. We denote $\bar{e}_\omega$ by $(\omega(1), \ldots, \omega(n))$. All these homology classes form the basis of the homology group $H_*(M_{n+N,n}(\mathbf{C}), \mathbf{Z})$. The cocycle dual to the cycle $(\underbrace{0, \ldots, 0}_{n-q}, \underbrace{1, \ldots, 1}_{q})$ represents the Chern class $c_q(\gamma_n^N) \in H^{2q}(M_{n+N,n}(\mathbf{C}), \mathbf{Z})$. For the real Grassmann manifold $M_{n+N,n}(\mathbf{R})$, we can analogously define the universal Stiefel-Whitney classes.

Thom's Definition. Let ξ be an $\mathbf{R}^n$-bundle over B, B_ξ be its †Thom space, and $U \in H^n(B_\xi, \mathbf{Z}_2)$ be the †fundamental class of B_ξ. Let $j: B \to B_\xi$ be the inclusion induced from the zero cross section and $\varphi: H^k(B, \mathbf{Z}_2) \cong H^{k+n}(B_\xi, \mathbf{Z}_2)$ be the †Thom-Gysin isomorphism. Then we have $j^*U = w_n(\xi)$, $\varphi^{-1}(Sq^i U) = w_i(\xi)$ $(0 \leqslant i \leqslant n)$, where Sq^i is the †Steenrod square (Thom [6]).

Definition by Differential Forms. Let B be a †differentiable manifold and $\xi=(P_\xi, B, U(n))$ be a differentiable principal $U(n)$-bundle over B. Let $\Omega=(\Omega_{i,j})$ be the †curvature form corresponding to the †connection form $\omega=(\omega_{i,j})$, $i,j=1,\ldots,n$, on P_ξ. Then $\Omega_{i,j}$ is a complex-valued 2-form, and $\overline{\Omega}_{i,j}=-\Omega_{j,i}$. For the matrix Ω, we consider the following differential form:

$$\psi=\sum_q \psi_q = det\,|I+(2\pi\sqrt{-1}\,)^{-1}\Omega|,$$

where I is the unit matrix, the multiplication in the determinant is the †exterior product, and ψ_q is the part of degree $2q$ in ψ. Then ψ is defined as a real form independent of the connection ω. We have $d\psi_q=0$, and the cohomology class of $(-1)^q\psi_q$ in $H^{2q}(B,\mathbf{R})$ is the Chern class $c_q(\xi)$ with real coefficients (Borel and Hirzebruch [2], Chern [3]).

Definition by Symmetric Polynomials. (→ 411 Topology of Lie Groups and Homogeneous Spaces.)

F. Characteristic Classes of Manifolds

For a differentiable (complex or almost complex) manifold M, the characteristic classes of its tangent bundle are called **characteristic classes of the manifold** M. We shall denote **Stiefel-Whitney classes, Pontrjagin classes**, and **Chern classes of** M by $w_i(M)$, $p_i(M)$, and $c_i(M)$, respectively. These are invariants of differentiable structures (or complex structures or almost complex structures) of a manifold M if M is a differentiable (or complex or almost complex) manifold. By the **Stiefel-Whitney numbers** of an n-dimensional manifold M, we mean the values of n-dimensional monomials of Stiefel-Whitney classes of M on the fundamental homology class $((w_1(M)^{r_1}w_2(M)^{r_2}\ldots w_n(M)^{r_n})[M]\in\mathbf{Z}_2$, where $r_1+2r_2+\ldots+nr_n=n$, $r_i\geqslant 0$. We can define integer-valued **Pontrjagin numbers** and **Chern numbers** similarly. These numbers are called generally **characteristic numbers** of the given manifold. In particular, $X_n(M)[M]=\chi(M)$ is the †Euler-Poincaré characteristic.

In the case of topological manifolds, we can define characteristic classes in the following sense. Let M be a closed n-dimensional topological manifold and X^n the generator of $H^n(M,\mathbf{Z}_2)$. By defining $X^i(Y^{n-i})=(X^iY^{n-i})\,[M]\in\mathbf{Z}_2$ for $X^i\in H^i(M,\mathbf{Z}_2)$, $Y^{n-i}\in H^{n-i}(M,\mathbf{Z}_2)$, we have an isomorphism $H^i(M,\mathbf{Z}_2)\cong\mathrm{Hom}(H^{n-i}(M,\mathbf{Z}_2),\mathbf{Z}_2)$. The element $u_i\in H^i(M,\mathbf{Z}_2)$, corresponding to the homomorphism $Y^{n-i}\rightarrow Sq^iY^{n-i}[M]$ under this isomorphism is called the **Wu class** of M, where Sq^i is the †Steenrod square. Moreover, we call $w_j=\sum_{i=0}^{j}Sq^{j-i}u_i\in H^j(M,\mathbf{Z}_2)$ the **Stiefel-Whitney class** of the topological manifold M. Then for any †differentiable structure $\mathfrak{D}$, we have $w_j(M,\mathfrak{D})=w_j$. Therefore, Stiefel-Whitney classes of differentiable manifolds are topological invariants (Thom [6], Wu [8]). By contrast, Milnor [9] proved that Pontrjagin classes of differentiable manifolds are not topological invariants.

G. Index Theorem for Differentiable Manifolds

Let M be an oriented closed manifold of dimension $4k$. Putting $f(x,y)=x\cdot y[M]$ for elements x,y of the $2k$-dimensional real cohomology group $H^{2k}(M,\mathbf{R})$, we obtain a bilinear form on $H^{2k}(M,\mathbf{R})$. The †signature of the quadratic form $f(x,x)$ (namely, (number of positive terms)–(number of negative terms)), in its canonical form, is a topological invariant of the manifold M. We call it the **index** of the manifold M and denote it by $\tau(M)$. If the dimension of M is not divisible by 4, we define $\tau(M)=0$. For the product of manifolds we have $\tau(M\times N)=\tau(M)\cdot\tau(N)$, where $\tau(M)$ is an invariant of the †cobordism class of M (Thom [6]).

The index τ of a differentiable manifold gives a homomorphism of the †Thom algebra Ω into the ring $\mathbf{Z}$ of integers. F. Hirzebruch investigated the multiplicative property of τ and gave its expression by means of Pontrjagin numbers. Let P_i be the ith †elementary symmetric function of indeterminates $\beta_1,\ldots,\beta_n$. Then a homogeneous part of the formal power series

$$\prod_{i=1}^{n}\frac{\sqrt{\beta_i}}{\tanh\sqrt{\beta_i}}$$

of $\beta_1,\ldots,\beta_n$ is a symmetric polynomial of $\beta_1,\ldots,\beta_n$, and therefore a polynomial of P_i with rational coefficients. For $k\leqslant n$, we denote the homogeneous part of degree k by $L_k(P_1,\ldots,P_k)$. Specifically, if P_i are the Pontrjagin classes $p_i(M^{4k})$ of a $4k$-dimensional closed differentiable manifold M^{4k}, then $L_k(P_1,\ldots,P_k)$ is a $4k$-dimensional cohomology class of M^{4k}. Then we have the formula

$$\tau(M^{4k})=L_k(P_1,\ldots,P_k)[M^{4k}],$$

called the **index theorem** of differentiable manifolds (or **Hirzebruch index theorem**). For example, $L_1=(1/3)P_1$, $L_2=(1/45)(7P_2-P_1^2)$, and $L_3=(1/945)(62P_3-13P_2P_1+2P_1^3),\ldots$ (→ 410 Topology of Differentiable Manifolds).

H. Combinatorial Pontrjagin Classes

Let K be an oriented n-dimensional compact †homology manifold, and let Σ^r be the boundary of an oriented $(r+1)$-simplex, namely, the †combinatorial r-sphere. Let $f: K \to \Sigma^{n-4i}$ be a †piecewise linear mapping. Then for almost all points y of Σ^{n-4i}, $f^{-1}(y)$ is an oriented $4i$-dimensional compact homology manifold, and its index $\tau(f^{-1}(y))$ is independent of y. We denote this by $\tau(f)$. Then $\tau(f)$ is an invariant of the homotopy class of f. Let σ be the fundamental class of $H^{n-4i}(\Sigma^{n-4i}, \mathbf{Z})$. Then for $n \geqslant 8i+2$, there exists a unique cohomology class $l_i = l_i(K) \in H^{4i}(K, \mathbf{Q})$ such that for any piecewise linear mapping $f: K \to \Sigma^{n-4i}$, we have $(l_i \cdot f^*\sigma)[K] = \tau(f)$. We can remove the restriction $n \geqslant 8i+2$ if we take $K \times \Sigma^m$ for K and define $l_i(K)$ to be $l_i(K \times \Sigma^m)$ for sufficiently large m. If K is a †C^1-triangulation of a differentiable manifold M, $l_i(K)$ coincides with the class $L_i(p_1, \ldots, p_i)$ defined by Hirzebruch (R. Thom [7]; V. Rohlin and A. Švarc), where $p_j = p_j(M)$ is the Pontrjagin class of M. Since the variable P_i can be expressed as a polynomial with rational coefficients of $L_j(P_1, \ldots, P_j), j \leqslant i$, we define the **combinatorial Pontrjagin class** $p_i(K)$ of a homology manifold K as the polynomial of $l_j(K)$ with rational coefficients. Therefore, if K is a C^1-triangulation of a differentiable manifold M, we have $p_i(K) = p_i(M)$ (with integral coefficients). The class $l_i(K)$ and consequently $p_i(K)$ are important combinatorial invariants of K. S. Novikov proved that rational Pontrjagin classes of combinatorial (or smooth) manifolds are topological invariants [10].

I. Gel'fand-Fuks Cohomology

The space $\mathfrak{X}(M)$ consisting of all the smooth vector fields on a smooth manifold M has the structure of a Lie algebra under the bracket operation $[X, Y] = XY - YX$, where the vector fields X and Y are regarded as endomorphisms of the space $C^\infty(M)$ of smooth functions on M, acting as first order differential operators. In $\mathfrak{X}(M)$ we introduce the topology defined by uniform convergence of the components of vector fields and all their partial derivatives on each compact set of M. The Gel'fand-Fuks cohomology of M is just the continuous cohomology of the Lie algebra $\mathfrak{X}(M)$ associated with the trivial representation. Specifically, let $C^0 = \mathbf{R}$, C^p $(p \geqslant 1)$ be the set of all the alternating p-linear continuous mappings φ of $\mathfrak{X}(M) \times \ldots \times \mathfrak{X}(M)$ (p-times) into $\mathbf{R}$. We define the coboundary operator d in C^p by $d\varphi = 0$ $(\varphi \in C^0)$, $d\varphi(X_1, \ldots, X_{p+1}) = \sum_{i<j}(-1)^{i+j}\varphi([X_i, X_j], X_1, \ldots, \check{X}_i, \ldots, \check{X}_j, \ldots, X_{p+1})$ $(\varphi \in C^p, p \geqslant 1)$. Then we have a cochain complex $\oplus\{C^p, d\}$, and the cohomology group $H^*(\mathfrak{X}(M))$ of this complex is called the **Gel'fand-Fuks cohomology group** of M. The exterior multiplication of cochains induces a ring structure in $H^*(\mathfrak{X}(M))$. Gel'fand and Fuks proved that, for any compact oriented manifold M, we have $\dim H^p(\mathfrak{X}(M)) < +\infty$ for all p and $H^p(\mathfrak{X}(M)) = 0$ for $0 \leqslant p \leqslant n$. For example, if M is the circle S^1, then the ring $H^*(\mathfrak{X}(S^1))$ is generated by two generators $\alpha \in H^2$, $\beta \in H^3$, which are explicitly described as cochains in the following way:

$$\alpha\left(f\frac{\partial}{\partial t},\ g\frac{\partial}{\partial t}\right) = \int_{S^1}\begin{vmatrix} f' & f'' \\ g' & g'' \end{vmatrix} dt,$$

$$\beta\left(f\frac{\partial}{\partial t}, g\frac{\partial}{\partial t}, h\frac{\partial}{\partial t}\right) = \int_{S^1}\begin{vmatrix} f & f' & f'' \\ g & g' & g'' \\ h & h' & h'' \end{vmatrix} dt.$$

If all the Pontrjagin classes of a compact oriented manifold M vanish, then $H^*(\mathfrak{X}(M))$ is finitely generated.

Localization of the concept of Gel'fand-Fuks cohomology naturally yields the cohomology of formal vector fields. Here a formal vector field means the expression $\sum f_\mu(x_1, \ldots, x_n)\partial/\partial x_\mu$, f_μ being formal power series in $x_1, \ldots, x_n$. The set of all the formal vector fields forms a Lie algebra $\mathfrak{a}_n$ and the continuous cohomology of $\mathfrak{a}_n$ with respect to the Krull topology is denoted by $H^*(\mathfrak{a}_n)$. Let B_U be the universal classifying space of the group $U(n)$, let $(B_U)_{2n}$ be its $2n$-skeleton and let P_{2n} be the canonical principal $U(n)$-bundle restricted to $(B_U)_{2n}$. Then there is a ring isomorphism $H^*(\mathfrak{a}_n) \cong H^*(P_{2n}; \mathbf{R})$.

An important subcomplex $\oplus\{C^p_\Delta, d\}$ of $\oplus\{C^p, d\}$, the diagonal complex, is defined as follows: $C^p_\Delta = \{\varphi \in C^p \mid \varphi(X_1, \ldots, X_p) = 0$ if $\operatorname{supp} X_1 \cap \ldots \cap \operatorname{supp} X_p = \varnothing\}$. Here, $\operatorname{supp} X_i$ denotes the support of X_i, that is, $\overline{\{x \mid X_i(x) \neq 0\}}$. Let P_M be the principal $U(n)$-bundle associated with the complexified tangent bundle of M $(n = \dim M)$. $U(n)$ acts freely on the product $P_M \times P_{2n}$ and the quotient space $P_M \times P_{2n}/U(n)$ is a fiber bundle over M with fiber P_{2n}. Then, if M is a compact oriented manifold, the cohomology $H^*_\Delta(\mathfrak{X}(M))$ of the diagonal complex is completely determined by the isomorphisms $H^p_\Delta(\mathfrak{X}(M)) \cong H^{p+n}(P_M \times P_{2n}/U(n); \mathbf{R})$ for all p. In particular, if all the Pontrjagin classes of M vanish, then $H^p_\Delta(\mathfrak{X}(M)) = \sum_{i+j=p+n} H^i(M; \mathbf{R}) \otimes H^j(\mathfrak{a}_n)$.

The cohomology theory of $\mathfrak{X}(M)$ in the case where the representation is nontrivial has also been investigated. A typical and important example is the Losik complex which arises from the natural representation of $\mathfrak{X}(M)$ on $C^\infty(M)$. The **Losik complex**

$L(M) = \oplus\{L^p, d\}$ is defined as follows: Let $L^0 = C^\infty(M)$, $L^p = \{\varphi | \varphi$ is an alternating p-linear map of $\mathfrak{X}(M) \times \ldots \times \mathfrak{X}(M)$ to $C^\infty(M)$ with the property $\mathrm{supp}\,\varphi(X_1, \ldots, X_p) \subset \mathrm{supp}\,X_1 \cap \ldots \cap \mathrm{supp}\,X_p\}$ for $p \geqslant 1$ and let $d\varphi(X_1, \ldots, X_{p+1}) = \Sigma_{i<j}(-1)^{i+j}\varphi([X_i, X_j], X_1, \ldots, \check{X}_i, \ldots, \check{X}_j, \ldots, X_{p+1}) + \Sigma_i(-1)^{i+1}X_i \cdot \varphi v(X_1, \ldots, \check{X}_i, \ldots, X_{p+1})$. $L(M)$ canonically includes the de Rham complex $R(M)$ as a subcomplex, because the differential n-forms are characterized as those Losik p-cochains that are p-linear on $\mathfrak{X}(M)$ as a $C^\infty(M)$-module. Let ι be the inclusion map of $R(M)$ to $L(M)$. Then the kernel of the map $H^*(R(M)) \xrightarrow{\iota^*} H^*(L(M))$ is the ideal generated by the Pontrjagin classes p_i $(i \geqslant 1)$ of M.

References

[1] A. Borel, Sur la cohomologie des espaces fibrés principaux et des espaces homogènes de groupes de Lie compacts, Ann. of Math., (2) 57 (1953), 115–207.

[2] A. Borel and F. Hirzebruch, Characteristic classes and homogeneous spaces I, II, III, Amer. J. Math., 80 (1958), 459–538; 81 (1959), 315–382; 82 (1960), 491–504.

[3] S. S. Chern, Characteristic classes of Hermitian manifolds, Ann. of Math., (2) 47 (1946), 58–121.

[4] F. Hirzebruch, Topological methods in algebraic geometry, Springer, third edition, 1966.

[5] E. Stiefel, Richtungsfelder und Fernparallelismus in n-dimensionalen Mannigfaltigkeiten, Comment. Math. Helv. 8 (1936), 3–51.

[6] R. Thom, Espaces fibrés en sphères et carrés de Steenrod, Ann. Sci. Ecole Norm. Sup., (3) 69 (1952), 109–182.

[7] R. Thom, Les classes caractéristiques de Pontrjagin des variétés triangulées, Symposium Internacional de Topologia Algebraica, Mexico, 1958.

[8] W. T. Wu, Sur les classes caractéristiques des structures fibrés sphériques, Actualités Sci. Ind., Hermann, 1952.

[9] J. W. Milnor, Microbundles I, Topology, 3 (1964), Suppl. I., 53–80.

[10] С. П. Новиков (S. P. Novikov), О многообразиях со свободной абелевой фундаментальной группой и их применениях (Классы Понтрягна, гладкости, многомерные узлы), Izv. Akad. Nauk SSSR, 30 (1966), 207–246; English translation, Manifolds with free Abelian fundamental groups and their applications (Pontrjagin classes, smoothness, multidimensional knots), Amer. Math. Soc. Transl., 71 (1968), 1–42.

[11] I. M. Gel'fand, The cohomology of infinite-dimensional Lie algebras; some questions of integral geometry, Actes Congr. Intern. Math., Nice 1970, Gauthier-Villars, vol. 1, p. 95–111.

[12] I. M. Gel'fand and D. B. Fuks, Funkcional analiz priložen., 3 (1969), 4 (1970); English translation, Cohomologies of the Lie algebra of tangent vector fields of a smooth manifold I, II, Functional Analysis Appl., 3 (1969), 194–210; 4 (1970), 110–116.

[13] I. M. Gel'fand and D. B. Fuks, Izv. Akad. Nauk SSSR, 34 (1970); English translation, The cohomology of the Lie algebra of formal vector fields, Math. USSR-Izv. (1970), 322–337.

[14] V. W. Guillemin, Cohomology of vector fields on a manifold, Advances in Math., 10 (1973), 192–220.

[15] M. V. Losik, Funkcional analiz priložen., 4 (1970); English translation, On the cohomologies of infinite-dimensional Lie algebras of vector fields, Functional Analysis Appl., 4 (1970), 127–135.

59 (XVI.3) Characteristic Functions (of Probability Distributions)

A. General Remarks

Consider a †probability measure P defined on a †measurable space $(\mathbf{R}^n, \mathfrak{B}^n)$, where $\mathfrak{B}^n$ is the †σ-algebra of all †Borel sets in $\mathbf{R}^n$. The **characteristic function** of P is the †Fourier transform φ defined by

$$\varphi(z) = \int_{\mathbf{R}^n} e^{i(z,x)} dP(x), \quad z \in \mathbf{R}^n, \tag{1}$$

where (z, x) denotes the †scalar product of z and x $(z, x \in \mathbf{R}^n)$. Let X be an n-dimensional †random variable with †probability distribution F defined on a †probability space $\Omega(\mathfrak{B}, P)$. Then the Fourier transform of F is also called the **characteristic function** of X (→ 131 Distributions (of Random Variables), 339 Probability).

The following properties play a fundamental role in the study of the relationship between probability distributions and characteristic functions: (i) the correspondence defined by (1) between the n-dimensional probability distribution P and its characteristic function φ is one-to-one. (ii) For any $a_p, b_p \in \mathbf{R}$, $a_p <$

$b_p (p=1,2,\ldots,n)$, we have

$$\int_{\mathbf{R}^n} \prod_{p=1}^{n} f(x_p; a_p, b_p) dP(x),$$

$$= \lim_{c\to\infty} \left(\frac{1}{2\pi}\right)^n \int_{-c}^{c} \cdots \int_{-c}^{c} \prod_{p=1}^{n} \frac{e^{-ib_pz_p} - e^{-ia_pz_p}}{-iz_p} \times \varphi(z_1,\ldots,z_n) dz_1,\ldots,dz_n, \quad (2)$$

where $f(x;a,b)$ denotes the **modified indicator function** of $[a,b]$ defined by

$$f(x;a,b) = \begin{cases} 1, & x\in(a,b), \\ 1/2, & x=a \text{ or } b, \\ 0, & x\notin[a,b], \end{cases}$$

and $x=(x_1,\ldots,x_n)\in\mathbf{R}^n$. If an n-dimensional interval $I=[a_1,\ldots,a_n; b_1,\ldots,b_n]$ defined by $a_i \leqslant x_i \leqslant b_i$ $(i=1,2,\ldots,n)$ is an **interval of continuity** for the probability distribution P, i.e., $P(\partial I)=0$ where ∂I denotes the boundary of I, then the left-hand side of (2) is equal to $P(I)$. Equation (2) is called the **inversion formula** for the characteristic function φ.

In general, consider the space $\mathbf{R}^T$ for an arbitrary set T and let $\mathbf{R}_0^T$ be the totality of elements $x\in\mathbf{R}^T$ whose support is a finite subset of T. To every probability distribution P on $\mathbf{R}^T$ corresponds its **characteristic functional** φ, defined by

$$\varphi(z) = \int_{\mathbf{R}^T} e^{i(z,x)} dP(x), \quad z\in\mathbf{R}_0^T.$$

Consider a †nuclear †countably Hilbertian space Φ and denote its †dual space by Φ'. Let P be a probability measure on the σ-algebra $\mathfrak{B}(\Phi')$ generated by the †cylinder sets in Φ'. Then the **characteristic functional** of P is the function φ defined for $\xi\in\Phi$ by

$$\varphi(\xi) = \int_{\Phi'} e^{i(X,\xi)} dP(X), \quad \xi\in\Phi \qquad (3)$$

[2, 10].

The characteristic function φ of an n-dimensional probability distribution has the following properties: (i) For any points $z^{(1)}, \ldots, z^{(p)}$ of the n-dimensional space $\mathbf{R}^n$ and any complex numbers $a_1,\ldots,a_p$, we have

$$\sum_{j,k=1}^{p} \varphi(z^{(j)} - z^{(k)}) a_j \bar{a}_k \geqslant 0.$$

(ii) $\varphi(z^{(k)})$ converges to $\varphi(0)$ as $z^{(k)}\to 0$. (iii) $\varphi(0)=1$. A complex-valued function φ of $z\in\mathbf{R}^n$ is called †positive definite if it satisfies the inequality in (i). Any continuous positive definite function φ on $\mathbf{R}^n$ such that $\varphi(0)=1$ is the characteristic function of an n-dimensional probability distribution (†Bochner's theorem) (→ 194 Harmonic Analysis). A counterpart to Bochner's theorem holds for any positive definite sequence as well (†Herglotz's theorem). Let Φ be a nuclear †countably Hilbertian space. A complex-valued function φ of $\xi\in\Phi$ is called †positive definite if

$$\sum_{j,k=1}^{p} \varphi(\xi^{(j)} - \xi^{(k)}) a_j \bar{a}_k \geqslant 0$$

holds for any points $\xi^{(1)},\ldots,\xi^{(p)}$ of Φ and any complex numbers $a_1,\ldots,a_p$. Then any continuous positive definite function φ on Φ such that $\varphi(0)=1$ is the characteristic function of a probability measure P on the σ-algebra $\mathfrak{B}(\Phi')$ [2]. For general information about criteria that can be used to decide whether a given function is a characteristic function → [9].

The characteristic function is a powerful tool for giving explicit formulas of probability distributions. For characteristic functions of typical probability distributions → Appendix A, Table 22.

B. Convergence of Probability Distributions

Since the condition for the convergence of probability distributions is described in terms of the corresponding characteristic functions, the method of characteristic functions plays an important role in the study of †limit theorems. Let φ_k and φ be the characteristic functions of n-dimensional probability distributions P_k and P, respectively. If P_k converges to P, then φ_k converges to φ as $k\to\infty$ uniformly on every bounded n-dimensional interval. Let P_k be an n-dimensional probability distribution with characteristic function φ_k. If the sequence $\{\varphi_k\}$ converges to the characteristic function φ of an n-dimensional probability distribution P, then the sequence $\{P_k\}$ converges to P. Let φ_k be the characteristic function of an n-dimensional probability distribution P_k. If the sequence $\{\varphi_k\}$ converges pointwise to a limit function φ and the convergence of φ_k is uniform in some neighborhood of the origin, then φ is also the characteristic function of an n-dimensional probability distribution P and the sequence $\{P_k\}$ converges to P [4, 8] (**Lévy's continuity theorem**). (→ 131 Distributions (of Random Variables) F). If P_α is a probability measure on an infinite-dimensional space, e.g., a nuclear countably Hilbertian space, the †tightness of a set $\{P_\alpha | \alpha\in\Lambda\}$ of probability distributions is one of the basic concepts of limit theorems [5, 10, 12].

For any probability distribution concentrated on $[0,\infty)$, the use of †Laplace transforms as a substitute for Fourier transforms provides a powerful tool. The method of †**probability-generating functions** is available for the study of arbitrary probability distribution concentrated on the nonnegative integers [1]. The method of †moment-generating functions is also useful. There are many works on the relation between these functions, probability distributions, and their convergence [1, 11, 13].

References

[1] W. Feller, An introduction to probability theory and its applications I, John Wiley, second edition, 1957.
[2] И. М. Гельфанд, Н. Я. Виленкин (I. M. Gel'fand and N. Ja. Vilenkin), Некоторые применения гармонического анализа, оснащенные гильбертовы пространства, физматгиз, 1961; English translation, Generalized functions IV, Academic Press, 1964.
[3] Б. В. Гнеденко, А. Н. Колмогоров (B. V. Gnedenko and A. N. Kolmogorov), Предельные распределения для сумм независимых случайных величин, Гостехиздат, 1949; English translation, Limit distributions for sums of independent random variables, Addison-Wesley, 1954.
[4] K. Itô, Kakurituron (Japanese; Theory of probability), Iwanami, 1953.
[5] L. LeCam, Convergence in distribution of stochastic processes, Univ. Calif. Publ. Statist., 2 (1957), 207–236.
[6] M. M. Loève, Probability theory, Van Nostrand, third edition, 1963.
[7] L. H. Loomis, An introduction to abstract harmonic analysis, Van Nostrand, 1953.
[8] P. Lévy, Calcul des probabilités, Gauthier-Villars, 1925.
[9] E. Lukacs, Characteristic functions, Hafner, 1960.
[10] Ju. V. Proxorov, The method of characteristic functionals, Proc. 4th Berkeley Symp. on Math. Stat. and Prob., Univ. of California, II (1961), 403–419.
[11] J. A. Shohat and J. D. Tamarkin, The problem of moments, Amer. Math. Soc. Math. Surveys, 1943.
[12] V. S. Varadarajan, Weak convergence of measures on separable metric space, Sankhyā, 19 (1958), 15–22.
[13] D. V. Widder, Laplace transform, Princeton Univ. Press, 1941.
[14] B. Ramachadran, Advanced theory of characteristic functions, Statistical Publishing Society, Calcutta, 1967.

60 (XX.6) Chinese Mathematics

A. Mathematics in the Chao, Han, and Tang Dynasties (3rd Century B.C.–10th Century A.D.)

In ancient China, the art of divination, called *yi*, was used in government administration. This was a kind of calculation that used pieces called *tse*. The book embodying it, called the *I-Ching*, is still popularly used. It shows that "numbers" or mathematics was seriously utilized in China at that time. The multiplication table for numbers up to nine (called the Pythagorean table in the West) was known in China from the legendary period. However, mathematics in the Greek sense, that is, mathematics as a logically systematized science, was unknown in ancient China.

Suanching-Shihshu, or the Ten Books on Arithmetic—namely *Choupi-Suanching, Chiuchang-Suanshu, Haitao-Suanching* (edited by Liu Hui), *Suntzu-Suanching, Wutsao-Suanching, Hsiahouyang-Suanching, Changchiu-Suanching, Wuching-Suanshu, Chiku-Suanching* (edited by Wang Hsiao-Tong), and *Shushu-Chiyi* (edited by Hsu Yue) —came into being between the 2nd century B.C. and the 6th century A.D., from the Chao to the Han eras, with the exception of the *Chiku-Suanching* compiled in the Tang era. These are the only mathematical texts from this early period whose authors and times of publication are known. They were used in the civil service examination for selecting administrators up to the beginning of the Sung era (960 A.D.). The most important among them is *Chinchang-Suanshu*, or the Book of Arithmetic, which contains nine chapters. It treats positive and negative fractions with laws of operations on signed numbers, equations, and the elementary mathematical knowledge of daily life. The *Chiku-Suanching* contains a number of problems reducible to equations of the 3rd and 4th degrees.

There were also two works called *San Tung Shu* (edited by Tong Chuan) and *Chui Shu* (edited by Tsu Chung-Chih), but no copies of them are extant. Later works, one from the Sui era (published in 636) and another from the Tang era, tell us that the latter contained the result $3.1415927 > \pi > 3.1415926$ and the approximate values $355/113$ and $22/7$ for π.

In the 1st century A.D. Buddhism was introduced from India, and paper was invented. However, despite the communication with India, neither the Indian numeration system, written calculation, nor the abacus was at that time widely used in China, although the extraction of square or cube roots was done with calculating rods.

B. Mathematics in the Sung and Yuan Dynasties (10th–14th Centuries)

In the Sung and Yuan periods contact was made with the Arab world. In the 13th century, a mechanical algebra utilizing calculating rods made remarkable progress; this may be attributed to Arab influence. Toward the end of the Sung era appeared the *Shushu-Chiuchang* by Ch'in Chiu-Shao and the *Yiku-Yentan* by Li Chih. The former gives a method like Horner's for approximate solution of equations, and the latter gives the principle of Tienyuan-Shu, i.e., the mechanical algebra of this period. The principle of *Tienyuan-Shu* was further expounded in the *Suanhsueh Chimeng* (1295) and the *Suyuan Yuchien* (1303) by Shih Shih-Chieh, the *Yanghui Suanfa* by Yang Hui, and other works. These were introduced into Japan and influenced the *wasan* of early times. Until recently, no further original mathematical ideas appeared in China.

C. Mathematics after the Ming Era (15th Century)

In this epoch, European renaissance civilization began to influence the Orient. In 1607, Matteo Ricci (1552–1610) translated Books I–VI of Euclid's *Elements* into Chinese with the aid of Hsu Kuang-Chi. In 1592 *Suanfa Tangtsung* by Ch'êng Ta-Wei appeared, which dealt with the use of the abacus. This book had great influence upon *wasan*.

No development was seen in the indigenous mathematics of the Ching era, that is, after the 17th century, but science and technology were imported by Christian missionaries. This brought about calendar reform from the lunar to the solar method. On the other hand, new editions of classical works such as the Ten Books on Arithmetic began to appear in this period. Emperor Kang Hsi-Ti (1655–1722), who was in correspondence with Leibniz, asked Ferdinand Verbiest (renamed Nan Huai Jen in Chinese) to compile *Shuli-Chingwen* (advanced mathematics), whose 53 chapters were completed in 1723. This book dealt with European-style algebra and trigonometry. In the latter half of the 19th century, Alexander Wylie translated a number of Western mathematical books into Chinese, including Books VII–XIII of Euclid's *Elements* and some works on calculus. Many current Chinese and Japanese mathematical terms originated with this translation.

References

[1] Y. Li, History of mathematics in China (Japanese translation), Seikatusya, 1940.
[2] Y. Mikami, Tôzai sûgakusi (Japanese; History of mathematics in Orient and Occident), Kyôritu, 1928.
[3] K. Ogura, Sûgakusi kenkyû (Japanese; Studies in the history of mathematics) I, Iwanami, 1935.
[4] Y. Mikami, The development of mathematics in China and Japan, Teubner, 1913 (Chelsea, 1961).

61 (X.9) C^∞-Functions and Quasi-Analytic Functions

A. General Remarks

An example of a †C^∞-function is a †real analytic function, which is defined to be a function that can be expressed as a power series that converges in a neighborhood of each point of the domain where the function is defined. Many examples, however, show that real analytic functions form a rather small subset of the C^∞-functions. Sometimes C^∞-functions (not real analytic functions) play essential roles in the development of theories of analysis (→ 108 Differentiable Manifolds S). On the other hand, there is a subfamily of C^∞-functions having some remarkable properties in common with the family of real analytic functions. This family is called the **family of quasi-analytic functions**. It has been an important object of study since the beginning of the 20th century. The first part of this article will deal with C^∞-functions, the second part with quasi-analytic functions.

B. C^∞-functions

Let Ω be an open set of the n-dimensional real Euclidean space $\mathbf{R}^n$. A real-valued function $f(x_1, \ldots, x_n)$ defined on Ω is called a **function of class** C^∞ on Ω (or C^∞-**function** on Ω) if $f(x_1, \ldots, x_n)$ is continuously differentiable up to any order. The totality of C^∞-functions defined on Ω is denoted by $C^\infty(\Omega)$. It is an †associative algebra over the real number field $\mathbf{R}$. A continuous function f defined on some closed set F of $\mathbf{R}^n$ is called a C^∞-function on F if there exist an open neighborhood U of F and a $g \in C^\infty(U)$ such that $f = g|F$. This definition is equivalent to the following (H. Whitney [5]): For any multi-index $\alpha = (\alpha_1, \ldots, \alpha_n)$, we can find a continuous function $f^\alpha(x_1, \ldots, x_n)$ on F such that (i) $f^0(x_1, \ldots, x_n) = f(x_1, \ldots, x_n)$ and (ii) for any positive integer r and for every multi-index α with $|\alpha| \leqslant r$,

$$\lim_{y\to x}\left[\frac{1}{\|y-x\|^{r-|\alpha|}}\left|f^{\alpha}(x)-\sum_{|\alpha+\beta|<r} f^{\alpha+\beta}(x)\frac{(x-y)^{\beta}}{\beta!}\right|\right]=0,$$

where $\| \ \|$ denotes the Euclidean norm of $\mathbf{R}^n$.

C. Local Theory of C^∞-Functions

We shall now introduce an equivalence relation $\sim$ in $C^\infty(\mathbf{R}^n)$, defined as follows: $f\sim g \Leftrightarrow f|U=g|U$ for some open neighborhood U of the origin. Let $\mathfrak{E}_n$ denote the quotient set $C^\infty(\mathbf{R}^n)/\sim$, which naturally inherits the structure of an associative algebra from $C^\infty(\mathbf{R}^n)$. An element of $\mathfrak{E}_n$ is called a **germ of a C^∞-function at the origin**. We denote the germ of $f\in C^\infty(\mathbf{R}^n)$ by $\tilde{f}$. To $\tilde{f}\in\mathfrak{E}_n$, we assign the **formal Taylor expansion** $\Sigma(D^\alpha f(0)/\alpha!)x^\alpha$ around 0, where $D^\alpha f$ means $(\partial^{r_1+\ldots+r_n}f)/(\partial^{r_1}x_1\ldots\partial^{r_n}x_n)$ for $\alpha=(r_1,\ldots,r_n)$. This assignment induces a homomorphism τ from $\mathfrak{E}_n$ to the †ring of formal power series $\mathbf{R}[[x_1,\ldots,x_n]]$ of n variables. The homomorphism τ is surjective but not injective. Put $\Lambda_n=\tau^{-1}(0)\subset\mathfrak{E}_n$. A function f whose germ $\tilde{f}$ belongs to Λ_n is called a **flat function**. The function $\varphi(x)$ defined by $\varphi(x)=\exp(-1/x^2)$ when $x\neq0$ and $\varphi(0)=0$ is an example of a flat function on $\mathbf{R}^1$. A close study of the relationship between $\mathfrak{E}_n$ and $\mathbf{R}[[x_1,\ldots,x_n]]$ leads to the **preparation theorem for C^∞-functions**, which can be stated as follows: Let $\tilde{F}(x_1,\ldots,x_n)\in\mathfrak{E}_n$ satisfy $\tilde{F}(0,\ldots,0,x_n)=x_n^p\tilde{g}(x_n)$ ($\tilde{g}\in\mathfrak{E}_1$, $g(0)\neq0$). Then any $\tilde{f}\in\mathfrak{E}_n$ can be expressed as $\tilde{f}=\tilde{F}\tilde{Q}+\tilde{R}$, where $\tilde{Q}\in\mathfrak{E}_n$ and $R=\Sigma_{i=0}^{p-1}r_i(x_1,\ldots,x_{n-1})x_n^i$ with $r_i\in\mathfrak{E}_{n-1}$ (B. Malgrange [3]).

Let $f(x_1,\ldots,x_n)$ be a symmetric function in $(x_1,\ldots,x_n)$ of class C^∞. Then there exists a germ $\tilde{g}\in\mathfrak{E}_n$ such that $\tilde{f}(x_1,\ldots,x_n)=\tilde{g}(\sigma_1,\ldots,\sigma_n)$, where $\sigma_1,\ldots,\sigma_n$ denote elementary symmetric functions with respect to $x_1,\ldots,x_n$ (G. Glaeser, Malgrange). Let $\tilde{f}\in\mathfrak{E}_1$ satisfy $f(x)=f(-x)$. Then there exists a germ $\tilde{g}\in\mathfrak{E}_1$ such that $\tilde{f}(x)=\tilde{g}(x^2)$ (H. Whitney).

D. Global Results

Case of n Variables. $C^\infty(\Omega)$ becomes a †Fréchet space when it is endowed with the topology of uniform convergence on compact sets for all partial derivatives. Let J and J_1 be two closed †ideals of $C^\infty(\Omega)$. Then we have $J=J_1$ if and only if $\tau_x(J)=\tau_x(J_1)$ for each $x\in\Omega$, where τ_x is the mapping from $C^\infty(\Omega)$ to the ring of the formal power series that assigns the formal Taylor series of f around x (Whitney).

Case of One Variable. In the case of one variable, further information can be obtained from various points of view. In the following, f denotes a C^∞-function defined on the unit interval $I=[0,1]$. If f satisfies $f(1)=1, f'(0)=\ldots=f^{(r-1)}(0)=0$, then we have

$$\frac{1}{m_1}+\frac{1}{\sqrt{m_2}}+\ldots+\frac{1}{\sqrt[r]{m_r}}<k,$$

where $m_i=\sup\{|f^{(i)}(x)|\,|\,x\in I\}$ and where k is some constant independent of the choice of f and r (E. Borel). Similar kinds of inequalities were obtained by A. N. Kolmogorov, A. Gorny, and H. Cartan. Let A be an arbitrary countable set of real numbers. If for any $x\in I$ we can find an integer $r(x)$ such that $f^{(r(x))}(x)\in A$, then such a function f is necessarily a polynomial. The interval I can be divided into three disjoint subsets: $S_1^{(f)}, S_2^{(f)}$, and $S_3^{(f)}$. These are characterized as follows: For $x\in S_1^{(f)}$ the formal Taylor series $\tau_x(f)$ of f around x converges to f in some neighborhood of x. For $x\in S_2^{(f)}$, $\tau_x(f)$ diverges. And for $x\in S_3^{(f)}$, $\tau_x(f)$ converges in some neighborhood of x but does not tend to f. Then $S_1^{(f)}$ is an open set and $S_2^{(f)}$ is a G_δ-set, while $S_3^{(f)}$ is an F_σ-set of the †first category. Conversely, let $I=S_1+S_2+S_3$ be any partition of I into an open set S_1, a G_δ-set S_2, and an F_σ-set S_3 of the first category. Then there is some $f\in C^\infty(I)$ with $S_i=S_i^{(f)}$ $(i=1,2,3)$ [4].

E. Relations between C^∞-Functions and Real Analytic Functions

Let $C^\omega(I)$ be the set of real analytic functions on I. Then $C^\omega(I)$ is a subalgebra of $C^\infty(I)$. Applying the above result in the case of $S_1=\varnothing$, we find a function of $f\in C^\infty(I)$ that admits no real analytic function coinciding with f in a subinterval of I. Actually, functions with such a property are distributed densely in $C^\infty(I)$. A necessary and sufficient condition for a function $f\in C^\infty(I)$ to belong to $C^\omega(I)$ is that for a suitable constant k

$$|f^{(n)}(x)|\leqslant k^n n!,\quad x\in I,\qquad n=0,1,2,\ldots$$

is valid (**Pringsheim's theorem**). If $f^{(n)}(x)\geqslant0$ for all $x\in I$ and $n=0,1,2,\ldots$, then $f\in C^\omega(I)$ (S. N. Bernšteĭn). For any open set Ω $(\subset\mathbf{R}^n)$, the set $C^\omega(\Omega)$ of real analytic functions on Ω is dense in $C^\infty(\Omega)$ (**polynomial approximation theorem**). This result is true even when the topology of $C^\infty(\Omega)$ is replaced by a stronger one (Whitney [1]). Let $f\in C^\infty(\Omega)$ and $\varphi\in C^\infty(\Omega)$. Then we can find $g\in C^\infty(\Omega)$ satisfying $f=g\varphi$ if and only if for any $x\in\Omega$, $\tau_x(f)$ is divisible by $\tau_x(\varphi)$ in the ring of formal power series (S. Łojasiewicz, Malgrange [3]).

F. Quasi-Analytic Functions

The investigation of quasi-analytic functions began with the attempt to obtain an intrinsic characterization of analytic functions. Borel defined monogenic functions as functions differentiable on their domains of definition, which may be any subset of the complex plane, not necessarily assumed to be open (→ 24 Analytic Functions). Similar to complex analytic functions, monogenic functions are uniquely determined by their values on any curve. While quasi-analyticity may be defined by such properties, it is customary to approach quasi-analytic functions from another aspect, that is, the behavior of higher derivatives of C$^\infty$-functions.

Generally, a subset B of $C^\infty(I)$ is called the **set of quasi-analytic functions** if the mapping $\tau_x : B \to \mathbf{R}[[x]]$ defined in Section D is injective at each point $x \in I$. The functions belonging to B are called **quasi-analytic**. Here, an important problem is to characterize a set of quasi-analytic functions by specializing the image of $\tau_x(B)$.

Now let $\{M_n\}$ be a sequence of positive numbers. Let $C(M_n)$ be the subset of $C^\infty(I)$ consisting of f such that

$$|f^{(n)}(x)| \leqslant k^n M_n, \quad x \in I, \quad n = 0, 1, 2, \ldots,$$

where $k = k(f)$ is a constant. Then Pringsheim's theorem simply asserts that $C(n!) = C^\omega(I)$.

In 1912, J. Hadamard raised the problem of determining the condition the sequence $\{M_n\}$ should satisfy so that $C(M_n)$ becomes a set of quasi-analytic functions [8]. A. Denjoy showed that if

$$M_n = (n \log^1 n \log^2 n \ldots \log^p n)^n,$$

where

$$\log^1 n = \log n, \ \log^p n = \log(\log^{p-1} n), \qquad p = 2, 3, \ldots,$$

then $C(M_n)$ is a set of quasi-analytic functions [9]. Later he derived the improved condition $\sum M_n^{-1/n} = \infty$. T. Carleman first gave a necessary and sufficient condition for $C(M_n)$ to be a set of quasi-analytic functions, and later A. Ostrowski and T. Bang gave another version of the same condition [10, 11, 12]. The condition states essentially the following: A necessary and sufficient condition that $C(M_n)$ be a family of quasi-analytic functions in the interval (a, b) is given by either (i) $\sum \beta_n^{-1} = +\infty$, where $\beta_n = \inf_{k \geqslant n} M_k^{1/k}$ (Carleman), or (ii) $\int^\infty (\log T(r))/(r^2)\,dr = \infty$, where $T(r) = \sup_{n \geqslant 1}(r^n / M_n)$ (Ostrowski-Bang). S. Mandelbrojt and Bang also gave another condition [12, 13]. (The simplest proof of this theorem is found in [12] or [14], where the proof follows Bang's idea.)

Related to the above theorem, we also have the following: Let $\{M_n\}$ be a sequence of positive numbers with $\Sigma(M_n / M_{n+1}) < \infty$. For $\alpha > 0$ we can find $f \in C(M_n)$ defined on $(-\infty, \infty)$ such that $f(0) > 0$, $f^{(n)}(\pm\alpha) = 0$. Moreover, for $0 < \alpha < \beta$ there exists $f \in C(M_n)$ such that $f(0) > 0$, $f^{(n)}(x) = 0$ ($\alpha \leqslant x \leqslant \beta$, $n = 0, 1, 2, \ldots$) [15].

Suppose that we are given an interval I and increasing sequences $\{\nu_n\}$ and $\{M_n\}$ of positive numbers. Then we have the problem of finding suitable conditions on $\{\nu_n\}$ and $\{M_n\}$ under which the mapping $f \to \{f^{(\nu_n)}(x_0)\}$ gives rise to an injective mapping from $C(M_n)$ to the sequences as above. When $\{\nu_n\}$ and $\{M_n\}$ satisfy the above conditions, then a function belonging to $C(M_n)$ is called **quasi-analytic (ν_n) in the generalized sense**. The study of the inclusion relation between two families $C(M_n)$ and $C(M_n')$ also deserves attention. In [14] the relation between $C(M_n)$ and $C(n!) = C^\omega(I)$ is discussed in detail. There are many open problems concerning the relationship between $C(M_n)$ and $C(M_n')$.

Quasi-analytic functions are closely related to various branches of analysis, in particular the theories of complex analytic functions, Fourier series, Fourier integrals, Dirichlet series, and asymptotic expansions [7, 14, 16].

References

For functions of class C$^\infty$,
[1] R. P. Boas, A primer of real functions, Carus Math. Monographs, John Wiley, 1960.
[2] G. Glaeser, Fonctions composées différentiables, Ann. of Math., (2) 77 (1963), 193–209.
[3] B. Malgrange, Le théorème de préparation en géométrie différentiables, Séminaire H. Cartan, 1962–1963, Exp. 11, Inst. H. Poincaré, Univ. Paris, 1964.
[3A] B. Malgrange, Ideals of differentiable functions, Oxford Univ. Press, 1966.
[4] H. Salzmann and K. Zeller, Singularitäten unendlich oft differenzierbarer Funktionen, Math. Z., 62 (1955), 354–367.
[5] H. Whitney, Analytic extensions of differentiable functions defined in closed sets, Trans. Amer. Math. Soc., 36 (1934), 63–89.
[6] H. Whitney, Differentiable even functions, Duke Math. J., 10 (1943), 159–160.
[7] H. Whitney, On ideals of differentiable functions, Amer. J. Math., 70 (1948), 635–658.
For quasi-analytic functions,
[8] J. Hadamard, Sur la généralisation de la notion de fonction analytique, Bull. Soc.

Math. France, 40 (1912), 28–29.
[9] A. Denjoy, Sur les fonctions quasi-analytiques de variable réelle, C. R. Acad. Sci. Paris, 173 (1921), 1329–1331.
[10] T. Carleman, Les fonctions quasi-analytiques, Gauthier-Villars, 1926.
[11] A. Ostrowski, Über quasianalytische Funktionen und Bestimmtheit asymptotischer Entwickelungen, Acta Math., 53 (1929), 181–266.
[12] T. S. V. Bang, Om quasi-analytiske Funktioner, Thesis, Univ. of Copenhagen, 1946.
[13] S. Mandelbrojt, Analytic functions and classes of infinitely differentiable functions, Rice Institute Pamphlet 29, no. 1, 1942.
[14] S. Mandelbrojt, Séries adhérentes, régularisation des suites, applications, Gauthier-Villars, 1952.
[15] S. Mandelbrojt, Some theorems connected with the theory of infinitely differentiable functions, Duke Math. J., 11 (1944), 341–349.
[16] S. Mandelbrojt, Séries de Fourier et classes quasi-analytiques de fonctions, Gauthier-Villars, 1935.
[17] R. E. A. C. Paley and N. Wiener, Fourier transformations in the complex domain, Amer. Math. Soc. Colloq. Publ., 1934.

62 (V.14) Class Field Theory

A. History

The notion of a class field was first introduced by D. Hilbert (1898). Let k be an †algebraic number field and K a †Galois extension of k. Hilbert called such a field K a class field over k (or K/k was called a class field) if the following property is satisfied: A †prime ideal $\mathfrak{p}$ of k of absolute degree 1 (i.e., a prime ideal whose †absolute norm is a prime number) is decomposed in K as the product of prime ideals of K of absolute degree 1 if and only if $\mathfrak{p}$ is a †principal ideal. (Such a field K is now said to be an **absolute class field** over k in order to distinguish it from a class field later defined more generally by T. Takagi as explained below.) Hilbert conjectured the following theorems (1)–(4) together with the principal ideal theorem (see section D) and proved them in some special cases. (1) For any algebraic number field k there exists one and only one class field K over k. (2) A class field K over k is an †Abelian extension whose †Galois group is isomorphic to the †ideal class group of k. Hence the degree $n=[K:k]$ is equal to the †class number h of k. (3) The †relative different of a class field K/k is the principal order; thus, K/k is an †unramified extension. (4) Let $\mathfrak{p}$ be a prime ideal of k, and let f be the smallest positive integer such that $\mathfrak{p}^f$ is a principal ideal. Then $\mathfrak{p}$ is decomposed in the class field K over k as $\mathfrak{p}=\mathfrak{P}_1\mathfrak{P}_2\ldots\mathfrak{P}_g$, $N_{K/k}(\mathfrak{P}_i)=\mathfrak{p}^f$, $fg=n$.

Hilbert was led to these conjectures from the analogy to the theory of †algebraic functions in one variable. Theorems (1)–(4) were proved by P. Furtwängler (*Math. Ann.*, 63 (1907)), but these results were subsumed under the class field theory of Takagi, who generalized the notion of a class field and proved that every Abelian extension of k is a class field over k (*J. Coll. Sci. Imp. Univ. Tokyo*, (9) 41 (1920)). Since then, the arithmetic of an Abelian extension of k has developed through this theory. In Takagi's paper, L. Kronecker's problem concerning Abelian extensions of an imaginary quadratic field (→ 75 Complex Multiplication) was solved simultaneously. This had also been a long-standing open problem since the 19th century. Actually, class field theory is considered one of the most beautiful theories in mathematics. Later, E. Artin proved the general law of reciprocity (*Abh. Math. Sem. Univ. Hamburg*, 5 (1927)) which put class field theory into its complete form. The original proof by Takagi was rather complicated, and H. Hasse, Artin, J. Herbrand, C. Chevalley, and others tried to simplify it. In particular, Chevalley introduced the notion of †ideles and gave a purely arithmetic proof. On the other hand, attempts are also being made to generalize this theory to non-Abelian extensions. We mention here the results of G. Shimura [19] and Y. Ihara [20].

B. Definition of a Class Field

Let k be an algebraic number field. For the definition of a general class field over k, we need a generalization of the ideal class group of k (→ 16 Algebraic Number Fields H). Let $\mathfrak{m}$ be an †integral divisor of k and let $\mathfrak{J}(\mathfrak{m})$ be the multiplicative group of all †fractional ideals of k which are †relatively prime to $\mathfrak{m}$. For the rest of this article, we mean by an ideal of k a fractional ideal of k. Denote by $S(\mathfrak{m})$ the †ray modulo $\mathfrak{m}$. Let $H(\mathfrak{m})$ be an †ideal group modulo $\mathfrak{m}$, that is, a subgroup of $\mathfrak{J}(\mathfrak{m})$ containing $S(\mathfrak{m})$. A Galois extension K of k is said to be a **class field** over k for the ideal group $H(\mathfrak{m})$ if the following property is satisfied: A prime ideal $\mathfrak{p}$ of k of absolute degree 1 which is relatively prime to $\mathfrak{m}$ is decomposed in K as the product of prime ideals of K of absolute degree 1 if and only if $\mathfrak{p}$ belongs to $H(\mathfrak{m})$. The absolute class field of Hilbert is the case where $\mathfrak{m}=(1)$ and $H(\mathfrak{m})$ is the group of all principal ideals of k.

A class field K/k for an ideal group H is uniquely determined by H (**uniqueness theorem**). The †conductor $\mathfrak{f}$ of H is said to be the **conductor** of the class field for H. The ideal group H corresponding to the class field K/k is determined by K as follows: $H(\mathfrak{f})/S(\mathfrak{f})$ is the set of all cosets C of $\mathfrak{J}(\mathfrak{f})$ modulo $S(\mathfrak{f})$ such that C contains a †relative norm $N_{K/k}(\mathfrak{A})$ of some ideal $\mathfrak{A}$ of K which is relatively prime to $\mathfrak{f}$. In general, let K/k be a Galois extension and $\mathfrak{m}$ be an integral divisor of k. Let $H(\mathfrak{m})$ be the set of all cosets C of $\mathfrak{J}(\mathfrak{m})$ modulo $S(\mathfrak{m})$ such that C contains a relative norm $N_{K/k}(\mathfrak{A})$ of some ideal $\mathfrak{A}$ of K which is relatively prime to $\mathfrak{m}$. Then $H(\mathfrak{m})$ is a multiplicative subgroup of $\mathfrak{J}(\mathfrak{m})$, and the index $h=(\mathfrak{J}(\mathfrak{m}):H(\mathfrak{m}))$ is not greater than the degree $n=[K:k]$. We have $h=n$ if and only if K/k is the class field for H. Hence a class field K over k can be defined as a Galois extension of k such that $h=n$ for a suitable integral divisor $\mathfrak{m}$ of k.

C. Fundamental Theorems in Class Field Theory

1. **Main theorem**: Any Abelian extension K/k is a class field over k for a suitable ideal group H.

2. **Existence theorem**: For any ideal group $H(\mathfrak{m})$ there exists one and only one class field for $H(\mathfrak{m})$.

3. **Composition theorem**: Let K_1 and K_2 be class fields for H_1 and H_2, respectively. Then the composite field K_1K_2 is the class field over k for $H_1 \cap H_2$. Consequently, $K_1 \supset K_2$ if and only if $H_1 \subset H_2$.

4. **Isomorphism theorem**: The Galois group of a class field K/k for $H(\mathfrak{m})$ is isomorphic to $\mathfrak{J}(\mathfrak{m})/H(\mathfrak{m})$. In particular, every class field K over k is an Abelian extension of k.

5. **Decomposition theorem**: Let $\mathfrak{f}$ be the conductor of the class field for H. If $\mathfrak{p}$ is a prime ideal of k relatively prime to $\mathfrak{f}$ and f is the smallest positive integer with $\mathfrak{p}^f \in H$, then $\mathfrak{p}$ is decomposed in K as $\mathfrak{p}=\mathfrak{P}_1\mathfrak{P}_2\ldots\mathfrak{P}_g$, $N_{K/k}(\mathfrak{P}_i)=\mathfrak{p}^f$, $fg=n$.

6. **Conductor-ramification theorem**: Let $\mathfrak{f}$ be the conductor of a class field K/k. Then $\mathfrak{f}$ is not divisible by any prime divisor which is unramified for K/k, and $\mathfrak{f}$ is divisible by every prime divisor which ramifies for K/k. Let $\mathfrak{f}=\prod\mathfrak{f}_\mathfrak{p}$, $\mathfrak{f}_\mathfrak{p}=\mathfrak{p}^c$. Then $\mathfrak{f}_\mathfrak{p}$ coincides with the †$\mathfrak{p}$-conductor of K/k, and the exponent c can be explicitly expressed by the order of the †ramification groups and the †ramification numbers of $\mathfrak{p}$ for K/k (→ 16 Algebraic Number Fields P).

7. Let $\mathfrak{p}$ be a prime ideal of k which ramifies for K/k. Let $H_\mathfrak{p}$ be the ideal group of k such that (i) the conductor of $H_\mathfrak{p}$ is relatively prime to $\mathfrak{p}$ and (ii) $H_\mathfrak{p}$ is the minimal ideal group of k containing H with property (i). Let $n=[K:k]$, $e=(H_\mathfrak{p}:H)$, and $\mathfrak{p}^f \in H_\mathfrak{p}$ where $\mathfrak{p}^d$ $(d<f) \notin H_\mathfrak{p}$. Then $\mathfrak{p}$ is decomposed in K as $\mathfrak{p}=(\mathfrak{P}_1\mathfrak{P}_2\ldots\mathfrak{P}_g)^e$, $N_{K/k}(\mathfrak{P}_i)=\mathfrak{p}^f$, $n=efg$.

8. **Translation theorem**: Let K/k be the class field for an ideal group $H(\mathfrak{m})$, and let Ω be an arbitrary finite extension of k. Then $K\Omega/\Omega$ is the class field for H^*, where H^* is the ideal group of Ω consisting of all ideals $\mathfrak{b}$ of Ω with $N_{\Omega/k}(\mathfrak{b}) \in H(\mathfrak{m})$. In particular, the conductor of $K\Omega/\Omega$ is a divisor of the conductor of K/k.

9. **Artin's general law of reciprocity**: Let K/k be the class field for an ideal group H with the conductor $\mathfrak{f}$. For an ideal $\mathfrak{a}$ of k which is relatively prime to $\mathfrak{f}$ we denote its †Artin symbol by

$$(K/\mathfrak{a})=\left(\frac{K/k}{\mathfrak{a}}\right).$$

Let a mapping Φ from $\mathfrak{J}(\mathfrak{f})$ to the Galois group G of K/k be defined by $\Phi(\mathfrak{a})=(K/\mathfrak{a})$ for $\mathfrak{a} \in \mathfrak{J}(\mathfrak{f})$. Then Φ induces the isomorphism $\mathfrak{J}(\mathfrak{f})/H(\mathfrak{f}) \cong G$. Namely, the isomorphism mentioned in 4. is explicitly given by the Artin symbol. Also, the ideal group $H(\mathfrak{f})$ is characterized as the set of all ideals $\mathfrak{a}$ such that $\mathfrak{a} \in \mathfrak{J}(\mathfrak{f})$ and $(K/\mathfrak{a})=1$. From this theorem we can prove all the known laws of reciprocity for power-residue and norm-residue symbols (→ 16 Algebraic Number Fields O, Q, R).

From the general results of class field theory we can systematically derive all the known theorems concerning the arithmetic of quadratic fields, cyclotomic fields, and Kummer extensions.

D. Principal Ideal Theorem

Let K/k be an absolute class field. Then the extension of any ideal of k to K is a principal ideal of K. This theorem is called the **principal ideal theorem**. It was conjectured by Hilbert, and was formulated by Artin as a theorem of group theory which was proved by Furtwängler (*Abh. Math. Sem. Univ. Hamburg*, 7 (1930)). Later a simple proof was given by S. Iyanaga (*Abh. Math. Sem. Univ. Hamburg*, 10 (1934)). This theorem was also generalized to the following general principal ideal theorem (Iyanaga, *Japan J. Math.*, 7 (1930)): Let K/k be the class field for the ray $S(\mathfrak{f})$ and let $\mathfrak{f}=\mathfrak{F}\mathfrak{D}$ where $\mathfrak{D}$ is the relative different of K/k. Then the extension to K of any ideal of k which is relatively prime to $\mathfrak{f}$ belongs to $S(\mathfrak{F})$. Put $\mathfrak{F}=\prod\mathfrak{P}^v$. Then v is equal to the ramification number v_r+1 (→ 16 Algebraic Number Fields K). For an absolute class field K/k, let the extension of an ideal $\mathfrak{a}$ to K be $(\Theta(\mathfrak{a}))$. Then we can choose $\Theta(\mathfrak{a})\in K$ such that $\Theta(\mathfrak{a})\Theta(\mathfrak{b})^{\sigma(\mathfrak{a})}\Theta(\mathfrak{ab})^{-1}\in k$ where $\sigma(\mathfrak{a})=(K/\mathfrak{a})$ is the Artin symbol for $\mathfrak{a}$ (T. Tannaka, *Ann. of Math.*, 67 (1958)). This result can also be generalized for the class field for $S(\mathfrak{f})$.

E. Theory of Genera

Let K/k be a Galois extension and let $H(\mathfrak{m})$ be an ideal group of k. The set of all ideals $\mathfrak{A}$ of K relatively prime to $\mathfrak{m}$ such that $N_{K/k}(\mathfrak{A})$ belongs to $H(\mathfrak{m})$ forms an ideal group of K. This ideal group is said to be the **principal genus** for H. Each coset of $\mathfrak{J}(\mathfrak{m})$ modulo the principal genus for H is said to be a **genus** for H. In particular, let K/k be a cyclic extension with the conductor $\mathfrak{f}$ and let $H(\mathfrak{f})$ be the ideal group of k generated by $N_{K/k}(A)$ $(A\in K)$ and $S(\mathfrak{f})$. Then the principal genus for $H(\mathfrak{f})$ is the ideal group formed by the ideal classes of K of the form $C^{1-\sigma}$, where σ is a generator of the Galois group of K/k (→ 343 Quadratic Fields F). In general, let K/k be an Abelian extension and let $\mathfrak{f}$ be the conductor of K/k. Then for the ideal $\mathfrak{F}=\prod\mathfrak{P}^{v}$ of K defined in Section D, $N_{K/k}(S(\mathfrak{m}\mathfrak{F}))=S(\mathfrak{m}\mathfrak{f})$ for an arbitrary integral ideal $\mathfrak{m}$ of k. In particular, let K/k be a cyclic extension, and let $H=S(\mathfrak{m}\mathfrak{f})$. Then the principal genus for H is the ideal group consisting of all cosets of $\mathfrak{J}(\mathfrak{m}\mathfrak{F})$ modulo $S(\mathfrak{m}\mathfrak{F})$ of the form $B^{1-\sigma}$ (Herbrand; Iyanaga, *J. Reine Angew. Math.*, 171 (1934)).

F. Class Field Tower Problem and Construction Problem

Furtwängler considered the following problem: Let k be a given algebraic number field, $k=k_0\subset k_1\subset k_2\ldots$ be the sequence of fields such that k_i is the absolute class field over k_{i-1}, and K_∞ be the union of all the k_i. Is K_∞ a finite extension of k? The answer is yes if and only if k_n is of class number 1 for some n. This problem is called the **class field tower problem**. Artin remarked that if for every algebraic number field F of degree n we have the inequality $|D_F|>(\pi/4)^{2r_2}(n^n/n!)^2>(\pi e^2/4)^n/(2\pi n e^{1/6n})$ for the †discriminant D_F, then K_∞/k is always finite [2, p. 46]. Recently, I. R. Šavarevič (1964) solved the class field tower problem negatively; he proved that K_∞/k is infinite if k_i $(i=1,2,\ldots)$ is the maximal unramified Abelian p-extension of k_{i-1} for a fixed prime number p and if the inequality $\gamma\geqslant 3+2\sqrt{\rho+2}$ holds where γ is the minimal number of generators of the p-component of the ideal class group of k and ρ is the rank of the unit group of k. (We call an extension K/k a **p-extension** if the degree $[K:k]$ is a power of a prime number p.) For example, the class field tower K_∞/k is actually infinite if k is an imaginary quadratic field $(\rho=1)$ and $\gamma\geqslant 7$ for $p=2$, for example, $k=Q(\sqrt{-3\cdot5\cdot7\cdot11\cdot13\cdot17\cdot19})$.

Construction problem. Let k be a given algebraic number field and G a finite group. The construction problem asks us whether there exists a Galois extension K/k such that its Galois group $\mathrm{Gal}(K/k)$ is isomorphic to G. If G is Abelian the problem can be solved affirmatively by using class field theory. This problem was also solved affirmatively for p-groups by A. Scholtz and H. Reinhardt in 1937, and for general solvable groups by I. R. Šafarevič in 1954 (*Izv. Akad. Nauk SSSR*, Ser. Mat. 18).

G. Class Field Theory for Algebraic Function Fields and Local Class Field Theory

F. K. Schmidt ([10], 1930) developed an analog of class field theory for Abelian extensions over an algebraic function field in one variable with finite coefficient field. An arithmetic proof was given by M. Moriya (1938). An analog of class field theory for local fields with finite residue-class fields, called **local class field theory** (→ 257 Local Fields) was first developed by Hasse, and later Chevalley gave an algebraic derivation (1933).

H. Cohomology of Groups and Class Field Theory

For the purpose of simplifying the proof of the main theorems in class field theory, the theory of †Galois cohomology was developed by T. Nakayama, G. Hochschild, A. Weil, Artin, J. Tate, and others. In particular, Artin and Tate [11] constructed class field theory on the basis of cohomology theory of finite groups as follows: Let G be a finite group, A a †G-module (or a multiplicative commutative group with the operator domain G), and $\hat{H}^n(G,A)$ the nth †cohomology group $(n=0,\pm1,\pm2,\ldots)$ of G with coefficients in A (→ 202 Homological Algebra I). Then we have $\hat{H}^0(G,A)\cong A^G/N_G(A)$, where A^G is the set of all G-invariant elements in A and $N_G(A)$ is the set of all elements of the form $N_G(a)=\sum_{\sigma\in G}\sigma a$ $(a\in A)$. We can consider $\mathbf{Z}$ a G-module by defining $\sigma n=n$ $(n\in\mathbf{Z}, \sigma\in G)$. Let A, B, C be G-modules such that a G-bilinear mapping $(A,B)\to C$ is defined. Then we can define the †cup product $(\alpha,\beta)\to\alpha\smile\beta$ $(\alpha\in\hat{H}^r(G,A)$, $\beta\in\hat{H}^s(G,B)$, $\alpha\smile\beta\in$

$\hat{H}^{r+s}(G,C)$) for $r, s \in \mathbf{Z}$ with the usual properties. Let A be a G-module and H a subgroup of G. Then the †restriction homomorphism $\mathrm{R}_{G/H}: \hat{H}^n(G,A) \to \hat{H}^n(H,A)$ and the †injection homomorphism $\mathrm{Inj}_{H/G}: \hat{H}^n(H,A) \to \hat{H}^n(G,A)$ are defined for $n \in \mathbf{Z}$. If H is a normal subgroup of G, then the †inflation homomorphism $\mathrm{Inf}_{(G/H)/G}: \hat{H}^n(G/H, A^H) \to \hat{H}^n(G,A)$ can be defined for $n \geqslant 1$ (→ 202 Homological Algebra G).

Let k be an algebraic number field, and let K be a Galois extension of k of degree n with the Galois group $G = G(K/k)$. The multiplicative group $K^\times = K - \{0\}$, the †idele group J_K of K, and the idele class group C_K of K are multiplicative commutative groups with G as their operator domain. The fundamental formulas in Galois cohomology for class field theory are (1) $\hat{H}^1(G, C_K) = 0$ and (2) $\hat{H}^2(G, C_K) \cong \mathbf{Z}/n\mathbf{Z}$.

It is possible to realize the isomorphism of (2) by the **invariant** $\mathrm{inv}_{K/k}: \hat{H}^2(G, C_K) \xrightarrow{\pi} \{(r/n) \pmod{\mathbf{Z}} \mid r = 0, 1, \ldots, n-1\}$ in such a way that the following properties hold, where the **canonical cohomology class** for K/k is the element $\xi_{K/k}$ of $\hat{H}^2(G, C_K)$ such that $\mathrm{inv}_{K/k}\, \xi_{K/k} = (1/n) \pmod{\mathbf{Z}}$:(i) for $k \subset l \subset K$, $G = G(K/k)$, $H = G(K/l)$ the relation $\mathrm{Res}_{G/H} \xi_{K/k} = \xi_{K/l}$ holds; (ii) if l/k is also a Galois extension with $F = G(l/k)$ then we have $\mathrm{Inf}_{F/G} \xi_{l/k} = \xi_{K/k}^m$ $(m = [K:l])$; (iii) for a cyclic extension K/k we have $\mathrm{inv}_{K/k} \xi_{K/k} = \sum_{\mathfrak{p}} \mathrm{inv}_{\mathfrak{p}}(\xi_{K/k})_{\mathfrak{p}} \pmod{\mathbf{Z}}$ where $\mathfrak{p}$ runs over all prime divisors of k and $\mathrm{inv}_{\mathfrak{p}}$ is the invariant in the local theory (→ 257 Local Fields E). By these properties, the canonical cohomology class $\xi_{K/k}$ is uniquely determined. After these preliminaries we can state Tate's theorem, from which the fundamental theorems in class field theory follow.

Tate's theorem. Let K/k be a Galois extension with the Galois group G. Then we have the isomorphism $\Phi_n: \hat{H}^{n-2}(G, \mathbf{Z}) \cong \hat{H}^n(G, C_K)$ $(n = 0, \pm 1, \pm 2, \ldots)$ which is given explicitly by $\Phi_n(\alpha) = \xi_{K/k} \smile \alpha$, where $\xi_{K/k} \in \hat{H}^2(G, C_K)$ is the canonical cohomology class for K/k (*Ann. of Math.* 56 (1952)).

Corollary 1. Since $\hat{H}^{-2}(G, \mathbf{Z}) \cong G/[G,G]$ and $\hat{H}^0(G, C_K) \cong C_k / N_{K/k}(C_K)$, we have the isomorphism $\Phi_0: G/[G,G] \cong C_k / N_{K/k}(C_K)$. Let $f(\tau, \sigma)$ $(\tau, \sigma \in G)$ be a 2-†cocycle belonging to $\xi_{K/k}$. Then by the explicit expression for the cup product we obtain the isomorphism

$$\Phi_0: \sigma \pmod{[G,G]} \to \prod_{\tau \in G} f(\tau, \sigma)^{-1} \pmod{N_{K/k}(C_K)}.$$

This was proved earlier by T. Nakayama and Y. Akizuki (*Math. Ann.*, 112 (1936)).

Corollary 2. For an Abelian extension K/k we have the isomorphism $\Phi_0: G \cong C_k / N_{K/k}(C_K)$. Φ_0^{-1} has the property of being the norm-residue symbol for C_k, and from this isomorphism we can prove immediately Artin's law of reciprocity. Thus we can prove the main theorems in class field theory by the cohomology-theoretic method [11, 15].

We can also see, by generalizing this isomorphism to infinite Abelian extensions, that the Galois group of the maximal Abelian extension of k over the ground field k with †Krull topology is algebraically and topologically isomorphic to C_k / D_k, where D_k is the connected component of the unit element in C_K. The structures of D_k and C_k / D_k were explicitly determined by Artin and T. Kubota, respectively (→ 7 Adeles and Ideles D).

If we assume the fundamental formulas (1) and (2) stated above and several other simple assumptions as axioms for an infinite extension of a fixed ground field, we can develop the results stated in this section purely cohomology-theoretically. Such a system is called a **class formation** (Artin [11]). In addition to the cases of algebraic number fields, algebraic function fields in one variable with finite coefficient fields, and local fields with finite residue-class fields, which we have mentioned already, we also know several other cases for which analogies of class field theory are valid. These analogies can be explained systematically by using class formation theory (Y. Kawada, *Duke Math. J.*, 22 (1955)). Examples are (1) the theory of unramified Abelian extensions of an algebraic function field in one variable with algebraically closed constant field of characteristic 0 (Tate and Kawada, *Amer. J. Math.*, 77 (1955)); (2) the theory of Kummer extensions over a field k such that (i) the characteristic of k is 0, (ii) k contains all the roots of unity, and (iii) for any Galois extension $K/k, N_{K/k}(K) = k$; (3) the theory of Abelian p-extensions of a field of characteristic p (E. Witt, *J. Reine Angew. Math.*, 176 (1963); I. Satake and Kawada, *J. Fac. Sci. Univ. Tokyo*, 7 (1955)); (4) the theory of unramified Abelian p-extensions of an algebraic function field in one variable with algebraically closed constant field of characteristic p (Hasse and Witt, *Monatsh. Math.*, 43 (1936), H. L. Schmid, I. R. Šafarevič, Kawada, T. Tamagawa); and (5) the theory of Abelian extensions of a local field with algebraically closed residue-class fields (G. Whaples, J.-P. Serre, *Bull. Soc. Math. France*, 89 (1961)).

An analogy of class field theory for infinite Abelian extensions was considered by Herbrand, Moriya, M. Mori, and Kawada.

References

[1] T. Takagi, Daisûteki seisûron (Japanese; Algebraic theory of numbers), Iwanami, second edition, 1971.
[2] H. Hasse, Bericht über neuere Untersuchungen und Probleme aus der Theorie der algebraischen Zahlkörper, Jber. Deutsch. Math. Verein, I. 35 (1926), 1–55; Ia. 36 (1927), 231–311; II. (1930) (Physica Verlag, 1965).
[3] H. Hasse, Vorlesungen über Klassenkörpertheorie, Lecture notes, Univ. Marburg, 1932 (Physica Verlag, 1967).
[4] C. Chevalley, Sur la théorie du corps des classes dans les corps finis et les corps locaux, J. Fac. Sci. Univ. Tokyo, 2 (1933), 365–476.
[5] J. Herbrand, Le dévelopement moderne de la théorie des corps algébriques, Gauthier-Villars, 1936.
[6] E. Artin, Algebraic numbers and algebraic functions, Lecture notes, Princeton Univ., 1950–1951 (Gordon and Breach, 1967).
[7] C. Chevalley, La théorie du corps de classes, Ann. of Math., (2) 41 (1940), 394–418.
[8] T. Tannaka, Daisûteki seisûron (Japanese; Algebraic theory of numbers), Kyôritu, 1949.
[9] A. Weil, Sur la théorie du corps de classes, J. Math. Soc. Japan, 3 (1951), 1–35.
[10] F. K. Schmidt, Die Theorie des Klassenkörper über einem Körper algebraischer Funktionen in einer Unbestimmten und mit endlichem Konstantenbereich, S.-B. Phy.-Med. Soz. Erlangen, 62 (1931), 267–284.
[11] E. Artin and J. Tate, Class field theory, Princeton, 1951 (Benjamin, 1967).
[12] S. Iyanaga, Class field theory notes, Univ. of Chicago, 1961.
[13] J.-P. Serre, Groupes algébriques et corps de classes, Actualités Sci. Ind., Hermann, 1959.
[14] J.-P. Serre, Corps locaux, Actualités Sci. Ind., Hermann, 1962.
[15] Y. Kawada, Daisûteki seisûron (Japanese; Theory of algebraic numbers), Kyôritu, 1957.
[16] S. Iyanaga (ed.), Sûron (Japanese; Number theory), Iwanami, 1969; English translation, The theory of numbers, North-Holland, 1975.
[17] A. Weil, Basic number theory, Springer, 1967.
[18] J. W. S. Cassels and A. Fröhlich (eds.), Algebraic number theory, Academic Press, 1967.
[19] G. Shimura, A reciprocity law in non-solvable extensions, J. Reine Angew. Math., 221 (1966), 209–220.
[20] Y. Ihara, Non-Abelian class fields over function fields in special cases, Actes Congr. Intern. Math., 1970, Nice, Gauthier-Villars, 381–389.
[21] Y. Ihara, Some fundamental groups in the arithmetic of algebraic curves over finite fields, Proc. Nat. Acad. Sci. US, 72 (1975), 3281–3284.
[22] Y. Kawada, Class formations, 1969 Number theory institute, Proc. Symposia in Pure Math., Amer. Math. Soc., 20 (1971), 96–114.

63 (IV.5) Classical Groups

A. Introduction

The general linear groups, unitary groups, orthogonal groups, symplectic groups etc. which will be described below are all called **classical groups** (→ 15 Algebraic Groups, 160 Finite Groups, 247 Lie Algebras, 248 Lie Groups).

B. General Linear Groups

Let V be a †linear space of dimension n over a †field K, and let $GL(V)$ denote the set of all †linear mappings of V onto V (hence they are all bijections). Then $GL(V)$ is a group under the composition of mappings. This group is called the **general linear group** (or **full linear group**) on V. Let $e_1, \ldots, e_n$ be a basis of V over K, and let (α_j^i) be the matrix associated with an element A of $GL(V)$: $Ae_i = \sum_j \alpha_i^j e_j$. Then the mapping $A \to (\alpha_j^i)$ is an isomorphism of $GL(V)$ onto the multiplicative group $GL(n,K)$ of all $n \times n$ †invertible matrices over K. We can thus identify the group $GL(V)$ with $GL(n,K)$. $GL(n,K)$ is called the **general linear group of degree** n **over** K. Consider the homomorphism $A \to |A|$ ($|A|$ is the determinant of A) of $GL(V)$ onto the multiplicative group $K^* = K - \{0\}$. Its kernel $SL(V)$ is a normal subgroup of $GL(V)$ and is called the **special linear group** (or **unimodular group**) on V. The subgroup $SL(n,K) = \{A \mid A \in GL(n,K), |A| = 1\}$ of $GL(n,K)$ corresponds to $SL(V)$ under the above isomorphism $GL(V) \cong GL(n,K)$. $SL(n,K)$ is called the **special linear group of degree** n **over** K. Unless $n=2$ and K is the †finite field $\mathbf{F}_2 = GF(2)$, $SL(n,K)$ is the †commutator subgroup of $GL(n,K)$. The †center $\mathfrak{z}$ of $GL(n,K)$ coincides with the set of all scalar matrices αI ($\alpha \in K^*$), and the center $\mathfrak{z}_0$ of $SL(n,K)$ is a finite group given by $\mathfrak{z} \cap SL(n,K) = \{\alpha I \mid \alpha \in K, \alpha^n = 1\}$.

Now let $P(V)$ be the †projective space of

dimension $n-1$ obtained from a linear space V of dimension n. Namely, $P(V)$ is the set of all linear subspaces of dimension 1. Then there exists a natural homomorphism φ of $GL(V)$ into the group of all projective transformations of $P(V)$, and the †kernel of φ coincides with the center $\mathfrak{z}$ of $GL(V)$. Hence $\varphi(GL(V)) \cong GL(V)/\mathfrak{z}$. Both are identified by means of φ. This group is written as $PGL(V)$ and is called the **projective general linear group** on $P(V)$. Similarly, $PGL(n,K) = GL(n,K)/\mathfrak{z}$ is called the **projective general linear group of degree** n **over** K. The quotient group $SL(n,K)/\mathfrak{z}_0$ of $SL(n,K)$ by the center $\mathfrak{z}_0$ is called the **projective special linear group** and is written as $PSL(n,K)$ or $LF(n,K)$ (**linear fractional group**).

The groups $GL(n,K)$, $SL(n,K)$, etc. are also written as $GL_n(K)$, $SL_n(K)$, etc. In particular, when K is the †finite field $\mathbf{F}_q$, these groups are also denoted by $GL(n,q)$, $SL(n,q)$, $PGL(n,q)$, $PSL(n,q)$, $LF(n,q)$.

Simplicity of $PSL(n,k)$. When $n=2$ and $K=\mathbf{F}_2$, $PSL(2,2) \cong \mathfrak{S}_3$ (the †symmetric group of degree 3). When $n=2$ and $K=\mathbf{F}_3$, $PSL(2,3) \cong \mathfrak{A}_4$ (the alternating group of degree 4). Except for these cases, the group $PSL(n,K)$ $(n \geqslant 2)$ is a noncommutative †simple group (→ 160 Finite Groups I).

Suppose that K is the finite field $\mathbf{F}_q$, and let $\alpha(n,q)$, $\beta(n,q)$, $\gamma(n,q)$, $\delta(n,q)$ denote the orders of $GL(n,q)$, $SL(n,q)$, $PGL(n,q)$, $PSL(n,q)$, respectively. Then we have

$$\alpha(n,q) = (q^n - 1)(q^n - q)\dots(q^n - q^{n-1}),$$

$$\beta(n,q) = \gamma(n,q) = \alpha(n,q)/(q-1),$$

$$\delta(n,q) = \gamma(n,q)/d,$$

where $d = (n, q-1)$ (the greatest common divisor of n and $q-1$).

C. Properties as Lie Groups

If the ground field K is the field $\mathbf{R}$ of real numbers (the field $\mathbf{C}$ of complex numbers), the above groups are all †Lie groups (†complex Lie groups). In particular, $SL(n,\mathbf{C})$ is a †simply connected, †simple, and †semisimple complex Lie group of type A_{n-1}, and $PSL(n,\mathbf{C})$ is the †adjoint group of the complex simple Lie algebra of type A_{n-1}.

D. Determination of the Rational Representations of $GL(V)$

In Sections D and E, the field K is assumed to be of characteristic 0. Let ρ be a homomorphism of $GL(V) = GL(n,K)$ into $GL(m,K)$ $(\rho : A = (\alpha_j^i) \to B = (\beta_q^p))$. Then if each β_q^p is a rational function (or polynomial or analytic function) in $(\alpha_1^1, \alpha_2^1, \dots, \alpha_n^n)$ over K, ρ is called a **rational representation** (or **polynomial** or **analytic representation**) of degree m of $GL(V)$. (We suppose that K is $\mathbf{R}$ or $\mathbf{C}$ when we consider analytic functions.) For example, every rational representation of degree 1 can be expressed as $A \to |A|^e$ (e is an integer). In particular, if K is the field $\mathbf{C}$ of complex numbers, every analytic representation of $GL(n,\mathbf{C})$ is a rational representation. Since $GL(n,\mathbf{C})$ is the †complexification of the †unitary group $U(n)$, there exists a one-to-one correspondence between the complex analytic representations of $GL(n,C)$ and the continuous representations of $U(n)$; this correspondence preserves equivalence, irreducibility, †tensor product, and direct sum of the representations (→ 248 Lie Groups). Hence, determining the rational representations of $GL(n,\mathbf{C})$ is equivalent to determining the continuous representations of $U(n)$. Considering the general case, the rational representations of $GL(V) = GL(n,K)$ are all completely reducible. For any rational representation ρ of $GL(V)$, there exists a natural number e such that the representation $\rho' : A \to |A|^e \rho(A)$ is a polynomial representation. Hence in order to determine the rational representations of $GL(V)$, it is sufficient to determine the irreducible polynomial representations of $GL(V)$, which, as described below, can be obtained by decomposing the representations on the †tensor space $V^m = V \otimes \dots \otimes V$ of degree m (m copies of V) $(m = 1, 2, \dots)$. For $A \in GL(V)$, define $D_m(A) \in GL(V^m)$ as the tensor product

$$D_m(A) = A \otimes \dots \otimes A \quad (m \text{ copies of } A).$$

Namely, for $v_1, \dots, v_m \in V$, we have

$$D_m(A)(v_1 \otimes \dots \otimes v_m) = Av_1 \otimes \dots \otimes Av_m.$$

The mapping $GL(V) \ni A \to D_m(A) \in GL(V^m)$ is a polynomial representation of degree n^m of $GL(V)$. Now let $\mathfrak{L}(V^m)$ be the †associative algebra of all linear mappings of V^m into V^m (†total matrix algebra), and let $\mathfrak{A}$ be the subalgebra of $\mathfrak{L}(V^m)$ generated by $\{D_m(A) \mid A \in GL(V)\}$. Next, for an element σ of the symmetric group $\mathfrak{S}_m$ of degree m, define $B_\sigma \in GL(V^m)$ by $B_\sigma(v_1 \otimes \dots \otimes v_m) = v_{\sigma^{-1}(1)} \otimes \dots \otimes v_{\sigma^{-1}(m)}$. Then the mapping $\sigma \to B_\sigma$ is a representation of $\mathfrak{S}_m$ on V^m. Thus we obtain a representation τ of the †group ring $K[\mathfrak{S}_m]$ of $\mathfrak{S}_m$ over K on $V^m : K[\mathfrak{S}_m] \to \mathfrak{L}(V^m)$. Set $\tau(K[\mathfrak{S}_m]) = \mathfrak{B}$. Then $\mathfrak{A}$ and $\mathfrak{B}$ are the †commutors of each other in $\mathfrak{L}(V^m)$, i.e., $\mathfrak{A} = \{X \in \mathfrak{L}(V^m) \mid XB = BX$ (for all $B \in \mathfrak{B}$)$\}$, $\mathfrak{B} = \{X \in \mathfrak{L}(V^m) \mid AX = XA$ (for all $A \in \mathfrak{A}$)$\}$.

Now for a right ideal $\mathfrak{r}$ of $\mathfrak{B}$, let $\mathfrak{r}(V^m)$ be the subspace of V^m composed of all the finite sums of the form ΣBx ($B \in \mathfrak{r}$, $x \in V^m$). Then the following statements hold:

(1) $\mathfrak{r}(V^m)$ is invariant under $\mathfrak{A}$; hence it is a subspace of V^m invariant under $GL(V)$. Conversely, for any subspace U of V^m invariant under $GL(V)$, there exists a unique right ideal $\mathfrak{r}$ of $\mathfrak{B}$ such that $U = \mathfrak{r}(V^m)$.

(2) Let $\mathfrak{r}_1, \mathfrak{r}_2$ be right ideals of $\mathfrak{B}$ and put $U_1 = \mathfrak{r}_1(V^m)$, $U_2 = \mathfrak{r}_2(V^m)$. Then $\mathfrak{r}_1 \cong \mathfrak{r}_2$ (as right $\mathfrak{S}_m$-modules) if and only if $U_1 \cong U_2$ (as representation spaces of $GL(V)$).

(3) The mapping $\mathfrak{r} \to \mathfrak{r}(V^m)$ is a lattice isomorphism of the †lattice of right ideals of $\mathfrak{B}$ onto the lattice of $GL(V)$-invariant subspaces of V^m. Hence if $\mathfrak{r} = \mathfrak{r}_1 + \mathfrak{r}_2$ (direct sum), then $U = U_1 + U_2$ (direct sum). Also, $\mathfrak{r}(V^m)$ gives an irreducible representation of $GL(V)$ if and only if $\mathfrak{r}$ is a minimal right ideal of $\mathfrak{B}$.

Since the algebra $K[\mathfrak{S}_m]$ is a †semisimple algebra, $\mathfrak{B}$ can be considered as a two-sided ideal of $K[\mathfrak{S}_m]$. Hence a minimal right ideal $\mathfrak{r}$ of $\mathfrak{B}$ is also a minimal right ideal of $K[\mathfrak{S}_m]$, and the †idempotent element ε which generates $\mathfrak{r}$ is a †primitive idempotent of $K[\mathfrak{S}_m]$. From the theory of symmetric groups (→358 Representations H) the primitive idempotents of $K[\mathfrak{S}_m]$ are all given (up to isomorphism) by †Young's diagrams $T(f_1, \ldots, f_k)$ ($f_1 \geqslant f_2 \geqslant \ldots \geqslant f_k > 0$, $m = f_1 + \ldots + f_k$). In this setting, we have

(4) Let $\varepsilon = \varepsilon(f_1, \ldots, f_k)$ be the primitive idempotent determined by Young's diagram $T(f_1, \ldots, f_k)$. Then $\varepsilon K[\mathfrak{S}_m] \subset \mathfrak{B}$ if and only if $k \leqslant n$. In this case, put $\varepsilon K[\mathfrak{S}_m] = \mathfrak{r}$, $\mathfrak{r}(V^m) = \varepsilon(V^m) = V^m(T(f_1, \ldots, f_k))$ and denote the irreducible representation of $GL(V)$ on $V^m(T(f_1, \ldots, f_k))$ by $A \to D(A; f_1, \ldots, f_k)$. We call $(f_1, \ldots, f_k)$ the **signature** of this irreducible representation.

(5) The representation $D(A; f_1, \ldots, f_k)$ is an irreducible polynomial representation of $GL(V)$. Furthermore, for any irreducible polynomial representation ρ of $GL(V)$, there exists a unique $D(A; f_1, \ldots, f_k)$ equivalent to ρ. For example, if $k=1$, then $f_1 = m$, $\varepsilon = (m!)^{-1} \Sigma_{\sigma \in \mathfrak{S}_m} \sigma$, and $V^m(T(m))$ is the space of †symmetric tensors of degree m. If $f_1 = \ldots = f_k = 1$, then $k = m$, $\varepsilon = (m!)^{-1} \cdot \Sigma_{\sigma \in \mathfrak{S}_m} (\operatorname{sgn} \sigma)\sigma$, and $V^m(T(1, \ldots, 1))$ is the space of †alternating tensors of degree m.

(6) Let $\chi(A; f_1, \ldots, f_k)$ be the †character of the irreducible representation $D(A; f_1, \ldots, f_k)$. Then

$$\chi(A; f_1, \ldots, f_k) = \begin{vmatrix} \varepsilon_1^{l_1} & \varepsilon_1^{l_2} & \ldots & \varepsilon_1^{l_n} \\ \ldots & \ldots & \ldots & \ldots \\ \varepsilon_n^{l_1} & \varepsilon_n^{l_2} & \ldots & \varepsilon_n^{l_n} \end{vmatrix} \div \begin{vmatrix} \varepsilon_1^{n-1} & \varepsilon_1^{n-2} & \ldots & \varepsilon_1 & 1 \\ \ldots & \ldots & \ldots & \ldots & \ldots \\ \varepsilon_n^{n-1} & \varepsilon_n^{n-2} & \ldots & \varepsilon_n & 1 \end{vmatrix},$$

where $\varepsilon_1, \ldots, \varepsilon_n$ are the †eigenvalues of A and $l_1 = f_1 + (n-1)$, $l_2 = f_2 + (n-2), \ldots, l_n = f_n$ (set $f_{k+1} = \ldots = f_n = 0$). Hence the degree d of $D(A; f_1, \ldots, f_k)$ is expressed as

$$d = D(l_1, \ldots, l_n) / D(n-1, \ldots, 1, 0),$$

where $D(x_1, \ldots, x_n) = \Pi_{i<j}(x_i - x_j)$.

(7) In particular, denote the character of $D(A; m)$ by $p_m = p_m(A)$. Then they satisfy $|I - zA|^{-1} = p_0 + p_1 z + p_2 z^2 + \ldots$ and

$$\chi(A; f_1, \ldots, f_k) = \begin{vmatrix} p_{f_1} & p_{f_1+1} & \ldots & p_{f_1+(n-1)} \\ p_{f_2-1} & p_{f_2} & \ldots & p_{f_2+(n-2)} \\ \ldots & \ldots & \ldots & \ldots \\ p_{f_n-(n-1)} & p_{f_n-(n-2)} & \ldots & p_{f_n} \end{vmatrix},$$

where we put $f_{k+1} = f_{k+2} = \ldots = f_n = 0$, $p_{-1} = p_{-2} = \ldots = 0$. This matrix is simply written as $|p_{l-(n-1)}, \ldots, p_l|$, with the convention that in each row, we set $l_1 = f_1 + (n-1), \ldots, l_{n-1} = f_{n-1} + 1$, $l_n = f_n$.

E. Determination of the Rational Representations of $SL(V)$

The rational representations of $SL(V)$ are completely reducible. By restricting any irreducible representation $D(A; f_1, \ldots, f_n)$ ($f_1 \geqslant f_2 \geqslant \ldots \geqslant f_n \geqslant 0$) of $GL(V)$ to $SL(V)$, we get an irreducible rational representation $\tilde{D}(A; f_1, \ldots, f_n)$ of $SL(V)$. Furthermore, any irreducible rational representation of $SL(V)$ can be obtained in this way. $\tilde{D}(A; f_1, \ldots, f_n)$ and $\tilde{D}(A; f_1', \ldots, f_n')$ are equivalent representations of $SL(V)$ if and only if $f_i - f_{i+1} = f_i' - f_{i+1}'$ ($i = 1, \ldots, n-1$).

F. Unitary Groups

The set $U(n)$ of all $n \times n$ †unitary matrices with complex elements is a group under multiplication (→ 269 Matrices). This group $U(n)$ is called the **unitary group** (or **unitary transformation group**) of degree n. The subset of $U(n)$ consisting of all matrices of determinant 1 is a normal subgroup of $U(n)$. This group is called the **special unitary group** and is denoted by $SU(n)$.

$U(n)$ and $SU(n)$ are subgroups of $GL(n, \mathbf{C})$ and $SL(n, \mathbf{C})$, respectively, and can be obtained from these groups through the †unitary restriction. Hence they are both compact, connected Lie groups; in particular, $SU(1)$ is composed only of the identity and $U(1)$ is the multiplicative group of all complex numbers of absolute value 1. The center $\mathfrak{z}$ of $U(n)$ is the set of all diagonal matrices λI ($\lambda \in \mathbf{C}$, $|\lambda| = 1$), and we have

$$\mathfrak{z} \cong U(1), \qquad \mathfrak{z} \cdot SU(n) = U(n),$$
$$U(n)/SU(n) \cong U(1).$$

Moreover, for $n \geqslant 2$, $SU(n)$ is a simple, semisimple, and simply connected Lie group, which gives one of four infinite series of simple compact Lie groups.

$U(n)/\mathfrak{z}$ is denoted by $PU(n)$ and is called the **projective unitary group**. We have the relations $PU(n) \cong SU(n)/\mathfrak{z} \cap SU(n)$, $\mathfrak{z} \cap SU(n) \cong \mathbf{Z}/n\mathbf{Z}$. Hence $PU(n)$ is locally isomorphic to $SU(n)$.

G. Irreducible Representations of $U(n)$

Restricting the irreducible representation $D(A; f_1, \ldots, f_k)$ of $GL(n, \mathbf{C})$ on $SU(n)$, we obtain a continuous irreducible representation of $SU(n)$, and conversely, all continuous irreducible representations of $SU(n)$ are obtained in this manner. Similarly, any continuous irreducible representations of $U(n)$ are given by $A \rightarrow |A|^e D(A; f_1, \ldots, f_k)$, where e is an integer. Since both $U(n)$ and $SU(n)$ are compact, any continuous representation of these groups can be decomposed into a direct sum of the irreducible representations mentioned above (→ 71 Compact Groups).

The representation theory of $U(n)$ and $SU(n)$ is important as the most typical and concrete example of the representation theory of general compact Lie groups (→ 71 Compact Groups, 247 Lie Algebras, 248 Lie Groups).

H. Unitary Groups over General Fields

A unitary matrix and the unitary group can also be defined over some fields other than the field $\mathbf{C}$ of complex numbers. Namely, let P be a field and K a quadratic extension field of P, and for an element ξ of K, let $\bar{\xi}$ be the †conjugate of ξ over P. Then a matrix of degree n with entries in K is called a unitary matrix of K (relative to P) if it leaves invariant the †Hermitian form $\xi_1\bar{\xi}_1 + \xi_2\bar{\xi}_2 + \ldots + \xi_n\bar{\xi}_n$. The multiplicative group consisting of all unitary matrices is called the **unitary group over** K (relative to P) and is denoted by $U(n, K, P)$; its subgroup consisting of all unitary matrices of determinant 1 is called the **special unitary group over** K and is denoted by $SU(n, K, P)$. The quotient group of $SU(n, K, P)$ by its subgroup consisting of all λI ($\lambda^n = 1, |\lambda| = 1$) is called the **projective special unitary group over** K and is denoted by $PSU(n, K, P)$. In particular, when K and P are the finite fields $\mathbf{F}_{q^2}$ and $\mathbf{F}_q$ ($q = p^m$), $U(n, K, P)$, $SU(n, K, P)$, $PSU(n, K, P)$ are written simply as $U(n, q)$, $SU(n, q)$, $PSU(n, q)$. Then for $n \geqslant 3$, each $PSU(n, q)$ is a noncommutative simple group, except for $PSU(3, 2)$ (→ 160 Finite Groups I).

I. Orthogonal Groups

The set of all †orthogonal matrices of degree n (with real entries) forms a group under multiplication. This group $O(n)$ is called the **orthogonal group** (or **orthogonal transformation group**) of degree n. The subset of $O(n)$ consisting of all orthogonal matrices of determinant 1 forms a normal subgroup of $O(n)$ of index 2. This group $SO(n)$ (also denoted by O_n^+) is called the **rotation group (special orthogonal group** or **proper orthogonal group)** of degree n. Geometrically, $O(n)$ is the set of all orthogonal transformations leaving a point in Euclidean space of dimension n fixed, and $SO(n)$ is composed of all rotations around the point.

Both $O(n)$ and $SO(n)$ are compact Lie groups, and $SO(n)$ coincides with the connected component of $O(n)$ which contains the identity. For $n = 3$ or $n \geqslant 5$, each $SO(n)$ is a simple and semisimple Lie group. Following the theory of Lie algebras, we divide the set of all $SO(n)$ ($n \geqslant 3$ but $n \neq 4$) into two classes according as n is even or odd, and we thus get two of the four infinite series of simple and semisimple compact Lie groups (for $SO(4)$, for example, see [1]).

Although $SO(n)$ ($n \geqslant 3$) is a connected Lie group, it is not simply connected. The simply connected Lie group which is locally isomorphic to $SO(n)$ is called the **spinor group** and is denoted by $Spin(n)$. $SO(n)$ is isomorphic to the quotient group of $Spin(n)$ by a normal subgroup of order 2. Let $\mathfrak{z}$ be the center of $Spin(n)$. Then $\mathfrak{z} \cong \mathbf{Z}/2\mathbf{Z}$ for odd n, $\mathfrak{z} \cong \mathbf{Z}/4\mathbf{Z}$ for $n \equiv 2 \pmod 4$, and $\mathfrak{z} \cong (\mathbf{Z}/2\mathbf{Z}) \oplus (\mathbf{Z}/2\mathbf{Z})$ for $n \equiv 0 \pmod 4$ (→ 13 Clifford Algebras).

The group $O(n, \mathbf{C})$ of all †complex orthogonal matrices is called the **complex orthogonal group**, and the group $SO(n, \mathbf{C})$ of all matrices in $O(n, \mathbf{C})$ of determinant 1 is called the **complex special orthogonal group**. $SO(n, \mathbf{C})$ ($n \geqslant 3, n \neq 4$) is a simple and semisimple complex Lie group.

J. Irreducible Representations of Orthogonal Groups

In the same way as for $GL(n, K)$, the irreducible representations of $O(n)$ can be obtained by decomposing the tensor product $D_m(A) = A \otimes \ldots \otimes A$ of m copies of an orthogonal matrix A using Young's diagram. Namely, consider the Young's diagram $T(f_1, f_2, \ldots, f_k)$ ($f_1 + f_2 \leqslant n$) such that the sum of the lengths of the first column and of the second column is not greater than n, and call it an $O(n)$ diagram. Then to any $O(n)$ diagram $T = T(f_1, \ldots, f_k)$, there corresponds an absolutely irreducible representation $D^0(A;f_1,f_2,\dots,f_k)$, and the $D^0(A;f_1,f_2,\dots,f_k)$ are mutually inequivalent. $D_m(A)$ can be decomposed into the direct sum of those $D^0(A;f_1,f_2,\dots,f_k)$ such that $f=f_1+\dots+f_k$ takes the values $m, m-2, m-4, \dots$. Furthermore, any continuous irreducible representation of $O(n)$ is equivalent to a $D^0(A;f_1,f_2,\dots,f_k)$ obtained from some $O(n)$ diagram $T=T(f_1,f_2,\dots,f_k)$.

In general, two $O(n)$ diagrams T and T' are called **mutually associated diagrams** if the sum of the lengths of their first columns is equal to n and if the lengths of each column other than the first one coincide. In particular, if $T=T(f_1,f_2,\dots,f_k)$ and $2k=n$, then T is said to be self-associated. The set of all $O(n)$ diagrams can be divided into pairs of mutually associated T, T' (and self-associated $T=T'$). Suppose that we are given mutually associated diagrams T and T' and that the length k of the first column of $T=T(f_1,f_2,\dots,f_k)$ is not greater than $n/2$. Then the character $\chi_T(A)$ of $D^0(A;f_1,f_2,\dots,f_k)$ which corresponds to T and the character $\chi_{T'}(A)$ of the irreducible representation corresponding to T' are given by

$$\chi_T(A)=|p_{l-(v-1)}-p_{l-(v+1)}, p_{l-(v-2)}-p_{l-(v+2)}, \dots, p_l-p_{l-2v}|,$$

$$\chi_{T'}(A)=|A|X_T(A), \quad v=[n/2],$$

where p_i and $|p_{l-(v-1)}-p_{l-(v+1)},\dots|$ have the same meaning as in the formula for the characters of irreducible representations of $GL(n,K)$.

The irreducible representations of $SO(n)$ can be immediately obtained from those of $O(n)$. Namely, if $T=T(f_1,f_2,\dots,f_k)$ is not self-associated, $D^0(A;f_1,f_2,\dots,f_k)$ is irreducible as a representation of $SO(n)$, and the representations of $SO(n)$ derived from T and the associated T' coincide. If T is self-associated, $D^0(A;f_1,f_2,\dots,f_k)$ can be decomposed into two irreducible representations of $SO(n)$ of the same degree over the field of complex numbers. Furthermore, the irreducible representations of $SO(n)$ obtained in this way from different pairs of associated diagrams are mutually inequivalent, while any continuous irreducible representation of $SO(n)$ is equivalent to one of these representations. For the representations of $SO(3)$ (the rotation group of degree 3) → 349 Racah Algebra.

Since $SO(n)$ is isomorphic to the quotient group of $Spin(n)$ by a normal subgroup N of order 2, a continuous representation of $Spin(n)$ which is not the identity representation on N can be considered as a double-valued representation of $SO(n)$. This representation is called the **spin representation** and is important in the field of applied mathematics.

The orthogonal group $O(n)$ consists of all $n\times n$ real matrices which leave invariant the quadratic form $\xi_1^2+\dots+\xi_n^2$, while the group of all $n\times n$ real matrices which leave invariant the quadratic form $\xi_1^2+\dots+\xi_r^2-\xi_{r+1}^2-\dots-\xi_n^2$ of †signature $(r,n-r)$ is called the **Lorentz group** of signature $(r,n-r)$. The case for $n=4$ and $r=3$ is used in special relativity (→ 355 Relativity). Let G_0 be the connected component of the identity of the Lorentz group of signature $(3,1)$. Then G_0 is called the **proper Lorentz group**. For $\sigma=(g_{ij})\in G$, we have $|\sigma|=\pm 1$ and $g_{44}\geqslant 1$ or $g_{44}\leqslant -1$. Moreover, we have $G_0=\{\sigma \mid |\sigma|=1, g_{44}\geqslant 1\}$, $G/G_0\cong(\mathbf{Z}/2\mathbf{Z})\oplus(\mathbf{Z}/2\mathbf{Z})$ (†four group), and $G_0\cong SL(2,\mathbf{C})/\{\pm I\}$.

K. Orthogonal Groups over General Fields

Orthogonal groups can also be defined over other general fields than the field of real numbers as follows: Fix a †quadratic form $Q(\xi,\xi)=\sum_{i,j=1}^n\alpha_{ij}\xi_i\xi_j$ $(|\alpha_{ij}|\neq 0)$ over a field K. Then a linear transformation of $\xi_i (i=1,2,\dots,n)$ over K which leaves Q invariant is called an **orthogonal transformation** with respect to Q. The set of all orthogonal transformations forms a group. This group is denoted by $O(n,K,Q)$ or simply $O(Q)$ and is called the **orthogonal (transformation) group over** K **with respect to** Q. In particular, the normal subgroup of all transformations in $O(n,K,Q)$ of determinant 1 is called the **special orthogonal group over** K **with respect to** Q and is denoted by $SO(n,K,Q)$ (or simply $SO(Q)$). $O(n)$ and $SO(n)$ are special cases of $O(n,K,Q)$ and $SO(n,K,Q)$, where K is the field $\mathbf{R}$ of real numbers and $Q(\xi,\xi)$ is the unit quadratic form $\xi_1^2+\xi_2^2+\dots+\xi_n^2$.

Let $\Omega(n,K,Q)$ be the †commutator subgroup of $O(n,K,Q)$. Then this subgroup coincides with the commutator subgroup of $SO(n,K,Q)$. If K is of characteristic $\neq 2$, and if $n\geqslant 5$ and the †index ν of $Q\geqslant 1$, then $\Omega(n,K,Q)/\mathfrak{z}$ ($\mathfrak{z}$ is the center of $\Omega(n,K,Q)$) is a simple group, where $\mathfrak{z}=\{I\}$ or $\mathfrak{z}=\{\pm I\}$ (L. Dickson, J. Dieudonné). Suppose that K is a finite field $\mathbf{F}_q$ (of characteristic $\neq 2$). Then we have $\nu=m$ if $n=2m+1$, and $\nu=m$ or $m-1$ if $n=2m$. Hence $\nu\geqslant 2$ if $n\geqslant 5$. If $\nu=0$ and $K=\mathbf{R}$, we have $\Omega(n,\mathbf{R},Q)=SO(n)$ and, as mentioned before, $SO(n)/\mathfrak{z}$ is simple for $n\geqslant 5$. The same proposition holds also for the case where K is an †algebraic number field (M. Kneser, 1956). If K is of characteristic 2, then $O(n,K,Q)=SO(n,K,Q)$, $\mathfrak{z}=\{I\}$, and $\Omega(n,K,Q)$ is a simple group in many cases (Dieudonné [7]). For the case where K is a finite field (Dickson) → 160 Finite Groups I.

L. Symplectic Groups

Let $\xi_1, \xi_2, \ldots, \xi_{2n}$ and $\eta_1, \eta_2, \ldots, \eta_{2n}$ be two sets of variables, and suppose that the same linear transformation A over a field K acts on them (from the left). If A leaves the †bilinear form $\sum_{i=1}^{n}(\xi_{2i-1}\eta_{2i} - \xi_{2i}\eta_{2i-1})$ invariant, this linear transformation (or the corresponding matrix) A is called a **symplectic transformation (symplectic matrix)** of degree $2n$. The set of all symplectic transformations (or matrices) of degree $2n$ over K forms a group denoted by $Sp(n,K)$ and called the **symplectic group (symplectic transformation group, complex group,** or **Abelian linear group) over** K.

Any matrix in $Sp(n,K)$ is always of determinant 1, and the center $\mathfrak{z}$ of $Sp(n,K)$ consists of I and $-I$. The quotient group $PSp(n,K)$ of $Sp(n,K)$ by $\mathfrak{z}$ is called the **projective symplectic group over** K. Except for the three cases $n=1, K=\mathbf{F}_2$; $n=1, K=\mathbf{F}_3$; and $n=2, K=\mathbf{F}_2$, the group $PS_p(n,K)$ $(n \geqslant 1)$ is always simple.

Properties of symplectic groups as Lie Groups. When K is the field $\mathbf{C}$ of complex numbers or the field $\mathbf{R}$ of real numbers, $Sp(n,K)$ is a Lie group. The intersection of the **complex symplectic group** $Sp(n,\mathbf{C})$ and the unitary group $U(2n)$, namely, the unitary restriction of $Sp(n,\mathbf{C})$, is denoted by $Sp(n)$ and is called the **unitary symplectic group** (or simply **symplectic group**). $Sp(n,\mathbf{C})$ is a simple and semisimple complex Lie group, and both $Sp(n,\mathbf{R})$ and $Sp(n)$ are simple and semisimple Lie groups. Moreover, $Sp(n)$ is compact and simply connected and gives one of four series of simple, semisimple and compact Lie groups (→ 248 Lie Groups).

Let $\mathbf{H}^n$ be the linear space of dimension n over the †quaternion field $\mathbf{H}$. Define the inner product of two elements $x=(x_1, \ldots, x_n)$ and $y=(y_1, \ldots, y_n)$ in $\mathbf{H}^n$ by $(x,y)=x_1\bar{y}_1 + \ldots + x_n\bar{y}_n$ ($\bar{y}_i$ is the †conjugate quaternion of y_i), and consider the group of all linear transformations which leave this inner product invariant. Then this group is isomorphic to $Sp(n)$. $Sp(n)$ is thus compared with the orthogonal group $O(n)$ which leaves invariant the inner product of a linear space over the field $\mathbf{R}$ of real numbers and with the unitary group $U(n)$ which has the same property over the field $\mathbf{C}$ of complex numbers (C. Chevalley [4, ch. 1]).

M. Irreducible Representations of Symplectic Groups

In the same way as for $GL(n,K)$, the representation $D_m(A)=A\otimes \ldots \otimes A$ (tensor product of m copies of A) of $Sp(n,\mathbf{C})$ can be decomposed into irreducible components using Young's diagram. Namely, for any Young's diagram $T=T(f_1,f_2,\ldots,f_k)$ $(k \leqslant n)$ such that the number of k rows is not greater than n, an irreducible representation $D^s(A;f_1,\ldots,f_k)$ of $Sp(n,\mathbf{C})$ is determined. These $D^s(A;f_1,\ldots,f_k)$ are mutually inequivalent, and $D_m(A)$ can be decomposed into the direct sum of representations $D^s(A;f_1,f_2,\ldots,f_k)$ such that $f=f_1+\ldots+f_k$ is equal to any of the values $m, m-2, m-4, \ldots$. The character of $D^s(A;f_1,\ldots,f_k)$ is given by

$$\chi_T(A)=|p_{l-n+1}, p_{l-n+2}+p_{l-n}, \ldots, p_l+p_{l-2n+2}|,$$

where p_i and $|p_{l-n+1}, p_{l-n+2}+p_{l-n}, \ldots|$ have the same meaning as in the formula for the characters of the irreducible representations of $GL(n,K)$.

For the matrices A in $Sp(n)$, $D^s(A;f_1, f_2,\ldots,f_k)$ gives rise to a continuous irreducible representation of $Sp(n)$. Furthermore, any continuous irreducible representation of $Sp(n)$ is equivalent to a representation: $D^s(A;f_1,f_2,\ldots,f_k)$ corresponding to some diagram T.

N. Relations among Various Classical Groups

There are some isomorphisms (homomorphisms) among the classical groups mentioned above. For general fields K → [1,7]. For finite fields K → 160 Finite Groups I. For $K=\mathbf{R}$ or $\mathbf{C}$, the following isomorphisms hold: $SO(3) \cong SU(2)/\{\pm I\}$, $SU(2) \cong Sp(1)$, $SO(5) \cong Sp(2)/\{\pm I\}$, $SO(6) \cong SO(4)/\{\pm I\}$ (→ 247 Lie Algebras, 248 Lie Groups).

O. Classical Groups over Noncommutative Fields

Let V be a right linear space over a noncommutative field K. Then the set of all linear transformations of V forms a group under the multiplication defined by the composition of mappings. This group $GL(V)$ is called the **general linear group** on V. It is isomorphic to the multiplicative group of all $n \times n$ invertible matrices with entries in K. The commutator subgroups $SL(V)$ and $SL(n,K)$ of $GL(V)$ and $GL(n,K)$, respectively, are called the **special linear group** of degree n on V and over K, respectively. Now, suppose that an element A of $GL(V)$ leaves each element of a subspace U of dimension $n-1$ of V fixed. Choose an element x of V which does not belong to U and set $Ax \equiv x\alpha \pmod{U}$; $\alpha \in K$ depends not only on A but on the choice of x. But the conjugate class $\dot{\alpha}=\{\lambda\alpha\lambda^{-1} | \lambda \in K^*\}$ of α in the multiplicative group K^* of K is determined only by A. In particular if $\dot{\alpha}=\{1\}$ and $A \neq I$, then A is called a **transvection**. For a †matrix unit E_{ij}, $B_{ij}(\alpha)=I+\alpha E_{ij}$ is a transvection if $i \neq j$ and $\alpha \neq 0$. $SL(V)$ coincides with the subgroup of $GL(V)$ generated by all transvections. This fact also holds when K is a commutative field, except for the case where $n=2$ and $K=\mathbf{F}_2$. In this case, transvections generate the whole $GL(2,2)$, which is isomorphic to the symmetric group $\mathfrak{S}_3$ of degree 3 and does not coincide with the commutator subgroup. The center $\mathfrak{z}$ of $GL(n,K)$ consists of all scalar matrices corresponding to nonzero elements in the center of K. Let C be the commutator subgroup of the multiplicative group K^* of K. Then for $n \geqslant 2$, $GL(n,K)/SL(n,K)$ is isomorphic to K^*/C. This isomorphism can be obtained by appropriately defining, for $A \in GL(n,K)$, an element $\det A$ of K^*/C which is called the **determinant** of A [8, 11]. The center $\mathfrak{z}_0$ of $SL(n,K)$ is $\{\alpha I \mid \alpha^n \in C\}$. The quotient group $PSL(n,K)=SL(n,K)/\mathfrak{z}_0$ is called the **projective special linear group** of degree n over K. If K is a noncommutative field, then $PSL(n,K)$ $(n \geqslant 2)$ is always a simple group [7, 8].

Next, let K be any field (commutative or noncommutative), and let V be a right linear space of dimension n over K. Consider a Hermitian form $f(x,y)$ (→ 256 Linear Spaces) on V relative to an †involution J of K. If for a fixed element ε in the center of K we have $f(x,y)=\varepsilon f(y,x)$, then f is called an **ε-Hermitian form**. For the rest of this article, f is assumed to be an ε-Hermitian form on V. Let W be a subspace of V. If $f(x,y)=0$ for any $x,y \in W$, then W is called a **totally isotropic subspace**. The largest dimension m of the totally isotropic subspaces of V is called the **index** of f. We always have $2m \leqslant n$. If $f(Ax,Ay)=f(x,y)$ for any $x,y \in V$, then A is called a **unitary transformation** relative to f.

The set $U(n,K,f)$ of all unitary transformations relative to f forms a subgroup of $GL(V)$. This group is called the **unitary group** relative to f. Also, the group $SU(n,K,f)=U(n,K,f) \cap SL(n,K)$ is called the **special unitary group**. When $J=1$ and $\varepsilon=1$, a unitary transformation (unitary group) is called an **orthogonal transformation (orthogonal group)**, and $U(n,K,f)$ is written as $O(n,K,f)$. Also, when $J=1$ and $\varepsilon=-1$, a unitary transformation (unitary group) is called a **symplectic transformation (symplectic group)**, and $U(n,K,f)$ is written as $Sp(n,K)$. In fact, in these cases, for arbitrary choice of f, the corresponding groups are mutually isomorphic.

An ε-Hermitian form f is called an **ε-trace form** if for any $x \in V$, there exists $\alpha \in K$ which satisfies $f(x,x)=\alpha+\varepsilon\alpha^J$. If $J=1$, $\varepsilon=-1$ (hence K is commutative) or $\varepsilon=1$ and K is of characteristic $\neq 2$, then any ε-Hermitian form is an ε-trace form. If f is an ε-trace form, a linear mapping B of any subspace W of V into V such that for any $x,y \in W$, $f(Bx,By)=f(x,y)$ can be extended to an element A of the unitary group $U(n,K,f)$ relative to f **(Witt's theorem)**. In particular, $U(n,K,f)$ acts transitively on the maximal totally isotropic subspaces, and their dimensions are equal to the index m of f. Now, let P be a **Pythagorean ordered field** (an ordered field which contains square roots of any positive element). If $K=P$ and $J=1$, or if $K=P(\sqrt{-1})$, or if the noncommutative field K is a †quaternion algebra over P and J is the operation of †conjugation of K, then for two Hermitian forms f, f', their unitary groups $U(n,K,f)$ and $U(n',K,f')$ are isomorphic if and only if $n=n'$ and the indices of f and f' are equal. In this case, $U(n,K,f)$ can be written as $U(n,m,K)$, where m is the index of f. If the field K is a †quaternion algebra over P and f is an †anti-Hermitian form, there exists an orthogonal basis (e_i) of V such that $f(e_i,e_i)=j$ (quaternion unit), $1 \leqslant i \leqslant n$. Hence, in this case, the unitary group $U(n,K,f)$ relative to f is determined only by n and K.

Suppose that we are given an ε-trace form f over a general field K whose index m is not equal to 0. We exclude the case where $J=1$ and $\varepsilon=1$. Then the unitary group $U(n,K,f)$ contains transvections. Let $T(n,K,f)$ denote the subgroup of $U(n,K,f)$ generated by transvections which are unitary transformations. If $m \geqslant 2$, then $T(n,K,f)$ is the commutator subgroup of $U(n,K,f)$. The center W_n of $T(n,K,f)$ coincides with the intersection of $T(n,K,f)$, and the center $\mathfrak{z}$ of $GL(n,K)$. If $n \geqslant 3$ and K contains more than 25 elements, then the quotient group $T(n,K,f)/W_n$ is a simple group [8]. Also, if K is commutative and $n \geqslant 2$, $m \geqslant 1$, $J \neq 1$, then $T(n,K,f)=SU(n,K,f)$, except for the case where $n=3$, $K=\mathbf{F}_4$.

If K is the field $\mathbf{R}$ of real numbers, the field $\mathbf{C}$ of complex numbers, or the quaternion field $\mathbf{H}$, then $GL(n,K)$, $SL(n,K)$, and $U(n,K,f)$ are all Lie groups. In particular, $SL(n,K)$ and $U(n,K,f)$ are simple Lie groups except for the following three cases: (1) $n=1$, $K=\mathbf{R}$ or $\mathbf{C}$; (2) $n=2$, $K=R$, $J=1$, $\varepsilon=1$; (3) $n=4$, $K=\mathbf{R}$ or $\mathbf{C}$, $J=1$, $\varepsilon=1$, $m=2$. In cases (1) and (2) they are commutative groups, and in case (3) they are locally direct sums of two noncommutative simple groups.

Suppose that $K=\mathbf{H}$. Since $\mathbf{H}$ contains $\mathbf{C}$ as a subfield, a vector space V of dimension n over $\mathbf{H}$ has the structure of a vector space of dimension $2n$ over **C**. From this fact, $GL(n,\mathbf{H})$ can be considered as a subgroup of $GL(2n,\mathbf{C})$ in a natural way.

Each of the complex classical simple groups $G=SL(n,\mathbf{C})$, $SO(n,\mathbf{C})$, $Sp(n,\mathbf{C})$ has the structure of an †algebraic group defined over **R** (→ 15 Algebraic Groups). The **real forms** of G, i.e., the algebraic subgroups of G whose scalar extension to **C** is G, can be realized as $SL(n,K)$, $U(n,K,f)$ corresponding to $K=\mathbf{R}$, **C**, **H**. Namely, a real form of a complex classical group G is conjugate in G to one of the following groups: (i) The real forms of $SL(n,\mathbf{C})$: $SL(n,\mathbf{R})$ (type AI); $SL(k,\mathbf{H})$ only for $n=2k$ (type AII); and the special unitary group $SU(n,m,\mathbf{C})$, $0\leqslant m\leqslant[n/2]$ relative to a Hermitian form f of index m (type AIII). (ii) The real forms of $SO(2n+1,\mathbf{C})$: the proper orthogonal group $SO(2n+1,m,\mathbf{R})$, $0\leqslant m\leqslant n$ relative to a quadratic form of index m on a space of dimension $2n+1$ (type BI). (iii) The real forms of $SO(2n,\mathbf{C})$: $SO(2n,m,R)$, $0\leqslant m\leqslant n$ (type DI); and $U(n,H,f)$ relative to an anti-Hermitian form f on **H** (type DIII). (iv) The real forms of $Sp(n,\mathbf{C})$: $Sp(n,\mathbf{R})$ (type CI); the unitary group $U(2n,m,\mathbf{H})$, $0\leqslant m\leqslant n$ relative to a Hermitian form f of index m on **H** (type CII); and $Sp(n)$ corresponds to the special case $m=0$. The quotient groups of these real forms by their centers can all be realized as the groups of automorphisms of semisimple algebras with involutions J which commute with J (A. Weil [12]).

References

[1] B. L. van der Waerden, Gruppen von linearen Transformationen, Erg. Math., Springer, 1935 (Chelsea, 1948).
[2] H. Weyl, The classical groups, Princeton Univ. Press, 1939, revised edition, 1946.
[3] N. Bourbaki, Eléments de mathématique, Algèbre, ch. 9, Actualités Sci. Ind., 1272a, Hermann, 1959.
[4] C. Chevalley, Theory of Lie groups I, Princeton Univ. Press, 1946.
[5] S. Iyanaga and M. Sugiura, Ôyôsûgakusya no tameno daisûgaku (Japanese; Algebra for applied mathematicians), Iwanami, 1960.
[6] T. Yamanouti, Kaitengun to sono hyôgen (Japanese; Rotation groups and their representations), Iwanami, 1957.
[7] J. Dieudonné, Sur les groupes classiques, Actualités Sci. Ind., Hermann, 1948.
[8] J. Dieudonné, La géométrie des groupes classiques, Springer, 1955.
[9] C. Chevalley, The algebraic theory of spinors, Columbia Univ. Press, 1954.
[10] M. Eichler, Quadratische Formen und orthogonale Gruppen, Springer, 1952.
[11] E. Artin, Geometric algebra, Interscience, 1957.
[12] A. Weil, Algebras with involutions and the classical groups, J. Indian Math. Soc., 24 (1960), 589–623.

64 (III.13) Clifford Algebras

A. Definitions and Basic Properties

Let V be an n-dimensional †linear space over a field K, and let Q be a †quadratic form on V. Denote the †tensor algebra over V by $T(V)$, the tensor multiplication by $\otimes$. Let $I(Q)$ be the two-sided ideal of $T(V)$ generated by the elements $x\otimes x-Q(x)\cdot 1$ $(x\in V)$. The resulting †quotient associative algebra $T(V)/I(Q)$ is then denoted by $C(Q)$ and is called the **Clifford algebra** of the quadratic form Q. The elements of $C(Q)$ are called **Clifford numbers.**

The composite of two canonical mappings $\tau: V\rightarrow T(V)$, $\sigma: T(V)\rightarrow C(Q)$ is a linear injection $\sigma\circ\tau: V\rightarrow C(Q)$. Hence we can regard V as a linear subspace of $C(Q)$ via $\sigma\circ\tau$. Then $C(Q)$ is an associative algebra over K generated by 1 and V. Furthermore, $x^2=Q(x)\cdot 1$ for every x in V.

Indeed, $C(Q)$ is the universal associative algebra with these properties. That is, let A be any associative algebra with a unity element, and let $f: V\rightarrow A$ be a linear mapping such that $f(x)^2=Q(X)\cdot 1$ for every x in V. Then f can be extended uniquely to an algebra homomorphism $\tilde{f}: C(Q)\rightarrow A$ with $\tilde{f}(1)=1$. Furthermore, let Φ be the †symmetric bilinear form associated with $Q: \Phi(x,y)=Q(x+y)-Q(x)-Q(y)$, $x,y\in V$. Then $xy+yx=\Phi(x,y)\cdot 1$ for every x,y in V. $C(Q)$ is of dimension 2^n over K. If $e_1,\dots,e_n$ is a basis of V, then

$$1,\ e_i,\ e_ie_j\ (i<j),\ \dots,\ e_1e_2\dots e_n$$

form a basis of $C(Q)$. In particular, if $\{e_i\}$ is an orthogonal basis relative to Q, we have

$$e_ie_j=-e_je_i,\quad e_i^2=Q(e_i)\cdot 1;\qquad i,j=1,\dots,n,\quad i\neq j. \qquad (1)$$

In this case, $C(Q)$ may be defined as an associative algebra (with a unity element) generated by the $\{e_i\}$ together with the defining relations (1). In particular, for $Q=0$, $C(Q)$ is the †exterior algebra (†Grassmann algebra) over V.

B. The Principal Automorphism and the Principal Antiautomorphism of $C(Q)$

There exists a unique automorphism α of the

algebra $C(Q)$ such that $\alpha(x)=-x$ for every x in V. This automorphism α is called the **principal automorphism** of $C(Q)$, and we have $\alpha^2=1$. Also, there exists a unique antiautomorphism β of the algebra $C(Q)$ such that $\beta(x)=x$ for every x in V. This antiautomorphism β is called the **principal antiautomorphism** of $C(Q)$, and we have $\beta^2=1$.

For the rest of this article we assume that the †discriminant of Q is $\neq 0$. We also assume for the sake of simplicity that the characteristic of K is $\neq 2$. Let $C^+=C^+(Q)=K\cdot 1+V^2+V^4+\ldots$, and $C^-=C^-(Q)=V+V^3+V^5+\ldots$. Then $C(Q)$ is the direct sum of the linear subspaces $C^+(Q)$ and $C^-(Q)$. Furthermore, $C^+C^+\subset C^+$, $C^+C^-\subset C^-$, $C^-C^+\subset C^-$, and $C^-C^-\subset C^+$. Thus, $C(Q)=C^++C^-$ has the structure of a †graded algebra with the index group $\{\pm 1\}$, and C^+ is a subalgebra of $C(Q)$. The elements of $C^+(Q)$, $C^-(Q)$ are called **even elements** and **odd elements**, respectively. We have $\dim C^+(Q)=\dim C^-(Q)=2^{n-1}$.

C. The Structure of $C(Q)$ and $C^+(Q)$

$C(Q)$ and $C^+(Q)$ are both †separable, †semisimple associative algebras over K. Suppose n is even: $n=2r$. Then $C(Q)$ is a †simple algebra with K as its center; the center Z of $C^+(Q)$ is 2-dimensional over K. Let $e_1,\ldots,e_n$ be an orthogonal basis of V. Then 1 and $z=2^re_1\ldots e_n$ form a basis of Z, and we have:

$$z^2=2^{2r}(-1)^rQ(e_1)\ldots Q(e_n)=(-1)^rD,$$

where D is the †discriminant of Φ relative to the basis $\{e_i\}$. Thus if $(-1)^rD$ has a square root in K, $Z\cong K\oplus K$ (direct sum), and so $C^+(Q)$ is decomposed into the direct sum of two simple algebras. If $(-1)^rD$ does not have a square root in K, then Z is a field and $C^+(Q)$ is a simple algebra. In particular, if the †index of Q (i.e., the dimension of a maximal †totally singular subspace of V (→ 344 Quadratic Forms)) is r, $C(Q)$ is isomorphic to the †total matrix algebra of degree 2^r over K, and $C^+(Q)$ is isomorphic to the direct sum of two copies of the total matrix algebra of degree 2^{r-1} over K.

Now suppose that n is odd: $n=2r+1$. Then $C^+(Q)$ is a simple algebra with K as its center. (In particular, if Q is of index r, then $C^+(Q)$ is isomorphic to the total matrix algebra of degree 2^r over K.) The center Z of $C(Q)$ is 2-dimensional over K, and we have $C(Q)\cong Z\otimes_K C^+(Q)$. If $e_1,\ldots,e_n$ is an orthogonal basis of V, then 1 and $z=e_1\ldots e_n$ form a basis of Z. Putting $z'=2^{r+1}z$, we have $z'^2=2(-1)^rD$, where D is the discriminant of Φ relative to $\{e_i\}$. Thus if $2(-1)^rD$ has a square root in K, $C(Q)$ is the direct sum of two 2^{2r}-dimensional simple algebras. If $2(-1)^rD$ has no square root in K, then $C(Q)$ is a simple algebra.

D. The Clifford Group

Let G be the set of all invertible elements s in $C(Q)$ such that $sVs^{-1}=V$. Then G forms a group relative to the multiplication of $C(Q)$. This group G is called the **Clifford group** of the quadratic form Q. The subgroup $G^+=G\cap C^+(Q)$ is called the **special Clifford group**. The linear transformation $\varphi(s):x\to sxs^{-1}$ of V induced by $s\in G$ belongs to the †orthogonal group $O(Q)$ of V relative to Q. Moreover, the mapping $s\to\varphi(s)$ is a homomorphism from G into $O(Q)$. Thus, φ is a †representation of G on V. This representation φ is called the **vector representation** of G. The †kernel of φ consists of invertible elements in the center Z of $C(Q)$. If $x\in G\cap V$, then $Q(x)\neq 0$ and $-\varphi(x)$ is the †reflection mapping of V relative to the hyperplane orthogonal to x. If $n=\dim V$ is odd, $\varphi(G)=\varphi(G^+)=SO(Q)$. If n is even, $\varphi(G)=O(Q)$, $\varphi(G^+)=SO(Q)$.

Exploiting the principal antiautomorphism β of $C(Q)$, we obtain a homomorphism $N:G^+\to K^*$ (the multiplicative group of K) defined by $N(s)=\beta(s)s$ $(s\in G^+)$, and $N(s)$ is called the **spinorial norm** of $s\in G^+$. The normal subgroup of G^+ defined as the kernel of N is denoted by G_0^+ and is called the **reduced Clifford group** (of Q). The subgroup $\varphi(G_0^+)$ of $SO(Q)$ is denoted by $O_0^+(Q)$ and is called the **reduced orthogonal group**.

In particular, in the case where the ground field K is the real number field $\mathbf{R}$, $O_0^+(Q)$ coincides with the †identity component of the †Lorentz group $O(Q)$. Furthermore, if Q is definite, $O_0^+(Q)\cong SO(n)$, so that the identity component $Spin(n)$ of G_0^+ is a †simply connected †covering group of $SO(n)$ via the covering homomorphism φ (with each point in $SO(n)$ covered twice). The group $Spin(n)$ is called the **spinor group** (of degree n).

E. Spin Representations

In this section we assume that the ground field K is the complex number field $\mathbf{C}$ and that $n=\dim V\geqslant 3$. Then we have $O_0^+(Q)\cong SO(n,\mathbf{C})$, so G_0^+ is a simply connected covering group of $SO(n,\mathbf{C})$ via the covering homomorphism φ. In this section we denote G_0^+ by $Spin(n,\mathbf{C})$ and call it the **complex spinor group** (of degree n). $Spin(n,\mathbf{C})$ is the †complexification (→ 248 Lie Groups) of the compact Lie group $Spin(n)$ and is a complex analytic subgroup of the †complex Lie group $C(Q)^*$ consisting of all invertible elements of $C(Q)$. With the bracket operation $[x,y] = xy - yx$, $C(Q)$ becomes the †Lie algebra of $C(Q)^*$. Furthermore, the Lie subalgebra of $C(Q)$ associated with the complex analytic subgroup $Spin(n, \mathbf{C})$ is given by $\sum_{i<j} \mathbf{C} e_i e_j$, where $e_1, \ldots, e_n$ is an orthogonal basis of V. The spin representations of the group $Spin(n, \mathbf{C})$ are defined as follows:

(1) The case where n is odd: $n = 2r+1$. Since $C^+(Q)$ is isomorphic to a total matrix algebra of degree 2^r over $\mathbf{C}$, $C^+(Q)$ has a unique (up to equivalence) †irreducible representation $\tilde{\rho}$, which is of degree 2^r. The restriction of $\tilde{\rho}$ on $Spin(n, \mathbf{C})$ (on $Spin(n)$) defines an irreducible representation ρ of degree 2^r of $Spin(n, \mathbf{C})$ (of $Spin(n)$); ρ is called the **spin representation** of the group $Spin(n, \mathbf{C})$ (of $Spin(n)$). The elements in the representation space of ρ are called **spinors**. Thus, we may say that a spinor is a quantity with 2^r components which obey the transformation law according to the spin representation (→ 384 Spinors). This representation ρ defines a representation of the Lie algebra $\mathfrak{so}(n, C)$ of $Spin(n, \mathbf{C})$ (note that $\mathfrak{so}(n, \mathbf{C})$ is a †complex simple Lie algebra of type B_r). This representation of $\mathfrak{so}(n, \mathbf{C})$ is also called the spin representation of $\mathfrak{so}(n, \mathbf{C})$. Note that ρ is not well defined on $SO(n, \mathbf{C})$ or on $SO(n)$; ρ is of valence 2 on $SO(n, \mathbf{C})$ or on $SO(n)$.

(2) The case where n is even: $n = 2r$. Since $C(Q)$ is isomorphic to a total matrix algebra of degree 2^r over $\mathbf{C}$, $C(Q)$ has a unique (up to equivalence) irreducible representation $\tilde{\rho}$, which is of degree 2^r. The restriction of $\tilde{\rho}$ on $Spin(n, \mathbf{C})$ (on $Spin(n)$) defines a representation ρ of degree 2^r of $Spin(n, \mathbf{C})$ (of $Spin(n)$); ρ is called the spin representation of the group $Spin(n, \mathbf{C})$ (of $Spin(n)$). This representation ρ is, however, not irreducible; ρ is decomposed into the direct sum of two irreducible representations ρ_+ and ρ_-. They are not equivalent to each other, and both are of degree 2^{r-1}. By taking a suitable minimal left ideal L of $C(Q)$ as the representation space of the representation $\tilde{\rho}$, we obtain the representation spaces L^+, L^- of ρ_+, ρ_-, respectively, by putting $L^+ = L \cap C^+(Q)$ and $L^- = L \cap C^-(Q)$. The representation ρ^+ (ρ^-) is called the **even (odd) half-spin representation** of the group $Spin(n, \mathbf{C})$ or of the group $Spin(n)$. The elements in the representation space of ρ^+ (of ρ^-) are called **even (odd) half-spinors**. Again, ρ^+ and ρ^- are not well defined on $SO(n, \mathbf{C})$ or on $SO(n)$. They are of valence 2 on these groups. The representations of the Lie algebra of the Lie group $SO(n, \mathbf{C})$ (note that this Lie algebra is a †complex simple Lie algebra of type D_r) associated with ρ^+, ρ^- are also called **half-spin representations** of this Lie algebra.

References

[1] N. Bourbaki, Eléments de mathématique, Algèbre, ch. 9, Actualités Sci. Ind., 1272a, Hermann, 1959.
[2] C. Chevalley, The algebraic theory of spinors, Columbia University Press, 1954.
[3] M. Eichler, Quadratische Formen und orthogonale Gruppen, Springer, 1952.

65 (XI.10) Cluster Sets

A. Cluster Sets of Functions Meromorphic in an Arbitrary Domain

Let D be an arbitrary †domain in the complex z-plane, Γ its boundary, and E a †totally disconnected closed set contained in Γ. Let $w = f(z)$ be a single-valued †meromorphic function defined in D. Then for each point z_0 in Γ, we can define the following sets related to the mapping $w = f(z)$ in the complex w-plane (or on the complex w-sphere).

The Cluster Set. A value α is called a **cluster value** of $f(z)$ at z_0 if there exists a sequence of points $\{z_n\}$ such that

$$z_n \in D, \quad z_n \to z_0, \quad f(z_n) \to \alpha.$$

The totality $C_D(f, z_0)$ of all the cluster values of $f(z)$ at z_0 is called the **cluster set** of $f(z)$ at z_0 or, more precisely, the **interior cluster set**. It is a nonempty, closed, but not necessarily connected set.

The Boundary Cluster Set. The set of all values α such that there exist a sequence of points $\{\zeta_n\}$ of $\Gamma - \{z_0\}$ (resp. $\Gamma - \{z_0\} - E$) and a sequence of points $\{w_n\}$ in the complex w-plane satisfying

$$\zeta_n \to z_0, \quad w_n \in C_D(f, \zeta_n), \quad w_n \to \alpha$$

is called the **boundary cluster set** of $f(z)$ at z_0 and is denoted by $C_\Gamma(f, z_0)$ (resp. $C_{\Gamma - E}(f, z_0)$). These are closed sets, and

$$C_{\Gamma - E}(f, z_0) \subset C_\Gamma(f, z_0) \subset C_D(f, z_0).$$

If $z_0 \in \Gamma - E$ or z_0 is an isolated point of E, then $C_{\Gamma - E}(f, z_0) = C_\Gamma(f, z_0)$. Furthermore, $C_\Gamma(f, z_0)$ (resp. $C_{\Gamma - E}(f, z_0)$) is empty if and only if z_0 is an isolated boundary point (resp. z_0 is an exterior point of $\Gamma - E$).

Range of Values. The set of values α such that

$$z_n \in D, \quad z_n \to z_0, \quad f(z_n) = \alpha$$

is called the **range of values** of $f(z)$ at z_0 and is denoted by $R_D(f, z_0)$. In other words, $R_D(f, z_0)$ is the set of values α assumed by $f(z)$ infinitely often in any neighborhood of z_0 and is a †G_δ-set.

The Asymptotic Set. Let z_0 be an †accessible boundary point of D. If $f(z)$ converges to a value α as z tends to z_0 along a simple arc in D terminating at z_0, then α is called an **asymptotic value** of f at z_0. The totality $A_D(f, z_0)$ of asymptotic values of f at z_0 is called the **asymptotic set** of f at z_0. If z_0 is an inaccessible boundary point, we let $A_D(f, z_0)$ be the empty set.

B. The Case Where E is of Logarithmic Capacity Zero

Suppose now that E is a subset of Γ of †logarithmic capacity zero, where Γ denotes the boundary of the domain D and $z_0 \in E$. Put

$$\Omega = C_D(f, z_0) - C_{\Gamma - E}(f, z_0).$$

If $\alpha \in \Omega$ is an †exceptional value of $f(z)$ in some neighborhood of z_0, then either α is an asymptotic value of $f(z)$ at z_0 or there exists a sequence of points $\zeta_n \in E$ $(n = 1, 2, \ldots)$ converging to z_0 such that α is an asymptotic value of $f(z)$ at each ζ_n (K. Noshiro, 1937). Furthermore, if z_0 is contained in the closure of $\Gamma - E$, then Ω is an open set (which may be empty), and $\Omega - R_D(f, z_0)$ is at most of logarithmic capacity zero (M. Tsuji, 1943). These are extensions of †Iversen's theorems. If E is contained in a single component Γ_0 of the boundary Γ, z_0 is contained in the closure of $\Gamma - E$, and Ω is nonempty, then $w = f(z)$ assumes every value belonging to each component Ω_n of Ω infinitely often, with two possible exceptions belonging to Ω_n (Noshiro, 1950). In particular, if f is bounded in a neighborhood of z_0, the number of such exceptional values is at most 1. This is still true if each point of E is contained in a boundary component that is a †continuum containing at least two points and Ω is not empty (M. Hervé, 1955). However, this conclusion does not hold if we remove the hypothesis on the set E (K. Matsumoto, 1960). This result is an extension of the Beurling-Kunugi theorem, which deals with the case where E consists of a single point z_0, and an extension of †Picard's theorem on an isolated essential singularity.

C. Cluster Sets of Functions Meromorphic in the Unit Disk

Let D be the unit disk $\{|z| < 1\}$, $z_0 = e^{i\theta_0}$ be a fixed point on the unit circumference Γ, A be an open arc of Γ containing z_0, and E be a set of †linear measure zero such that $z_0 \in E \subset A$. With every $e^{i\theta} \in A - E$ we associate an arbitrary simple arc Λ_θ in D terminating at $e^{i\theta}$ and the **curvilinear cluster set** $C_{\Lambda_\theta}(f, e^{i\theta})$, defined as the set of all values α such that $z_n \in \Lambda_\theta$, $z_n \to e^{i\theta}$, $f(z_n) \to \alpha$. We put

$$C^*_{\Gamma - E}(f, z_0) = \bigcap_{r > 0} M_r,$$

where M_r denotes the closure of the union $C_{\Lambda_\theta}(f, e^{i\theta})$ for all $e^{i\theta}$ in the intersection of $A - E$ with $|z - z_0| < r$.

By using the general boundary cluster set $C^*_{\Gamma - E}(f, z_0)$ instead of the boundary cluster set $C_{\Gamma - E}(f, z_0)$, we obtain results similar to those in the preceding section. Many interesting results have been obtained by F. Bagemihl and W. Seidel, E. F. Collingwood, O. Lehto and K. I. Virtanen, and others concerning the cluster sets of functions meromorphic in the unit disk.

D. The More General Case

The definitions of cluster sets are also available for arbitrary functions for which neither analyticity nor continuity is assumed. If there exist two simple arcs Λ_1 and Λ_2 in the unit disk D terminating at a point $z = e^{i\theta}$ such that

$$C_{\Lambda_1}(f, e^{i\theta}) \cap C_{\Lambda_2}(f, e^{i\theta}) = \varnothing,$$

then $z = e^{i\theta}$ is called an **ambiguous point** of f. Bagemihl proved the following: The set of ambiguous points of an arbitrary complex-valued function defined in the unit disk D is at most countable [2]. Under the same hypothesis, the set of points $e^{i\theta}$ such that $C_D(f, e^{i\theta}) \neq C_\Gamma(f, e^{i\theta})$ is at most countable (Collingwood, 1960). This result shows the importance of introducing the general boundary cluster set $C^*_{\Gamma - E}(f, z_0)$ mentioned before.

E. History

The theory of cluster sets originated from †value distribution theory of analytic functions in the neighborhood of their essential singularities. The first systematic results were those of F. Iversen and W. Gross, obtained about 1920. Subsequent significant contributions were made by Seidel, J. L. Doob, M. L. Cartwright, A. Beurling, and others. Since 1940, some important results have been obtained by K. Kunugi, S. Irie, Y. Tôki, Y. Tumura, S. Kametani, Tsuji, Noshiro, and other Japanese mathematicians. Many results have been extended to †pseudoanalytic functions. As may be seen from the Bagemihl ambiguous point theorem, some properties of cluster sets are not intrinsic to analytic mappings [2, 4]. On the other hand, it seems to be an interesting problem to extend the theory of cluster sets to the case of analytic mappings between open Riemann surfaces.

References

[1] K. Noshiro, Cluster sets, Erg. Math., Springer, 1960.
[2] F. Bagemihl, Curvilinear cluster sets of arbitrary functions, Proc. Nat. Acad. Sci. US, 41 (1955), 379–382.
[3] K. Noshiro, Kindai kansûron (Japanese; Modern theory of functions), Iwanami, 1954.
[4] E. F. Collingwood and A. J. Lohwater, The theory of cluster sets, Cambridge, 1966.

66 (XVIII.11) Coding Theory

In the field of electrical communication, information is often expressed by a sequence of discrete signals. This is done by associating each character with a finite sequence $b_1 \ldots b_n$ of available signals, in accordance with a certain mapping ψ. When q distinct signals are available, such a sequence $b_1 \ldots b_n$ may be regarded as an n-digit number with radix q (in many communication systems, $q=2$). The operation of replacing a character X by the sequence $\psi(X)$ is called **encoding**, and the converse operation is called **decoding**. The sequence $\psi(X)$ is called a **code word** of X, and the image of ψ (a collection of code words) is called a **code**.

The purpose of coding theory is the construction of codes suitable for the transmission of information with a high degree of efficiency. If the signals are transmitted through a noiseless channel, then the only relevant problem is the optimality of the transmission rate (characters/sec). In this connection, information theory is of basic importance, since it provides a theoretical upper bound for the transmission rate in terms of †entropy (→ 216 Information Theory).

When the transmission is subject to error, the automatic detection and correction of errors is of obvious practical importance. A simple method of detecting errors is to enlarge each code word $b_1 \ldots b_n$ by adding another digit b_0 such that the sum $b_0 + \ldots + b_n$ is equal to zero modulo q. This additional digit b_0 is called a **parity digit**. Thus, a single error among these $n+1$ digits can be detected by checking their sum. By adding more digits, more errors can be detected and, in some cases, corrected. However, these additional digits reduce the transmission rate.

Another important problem in this context is the complexity of the required operations, i.e., encoding and decoding plus the detection and correction of errors. Some codes based on probability theory are regarded as virtually useless, since they provide no practical method of decoding.

The brief discussion which follows is primarily concerned with the case of a **binary block code** S, that is, a collection of binary numbers of some fixed length n. Thus, the code S is just a subset of the †Boolean ring $B_n = \{0,1\}^n$. In accordance with the terminology already mentioned above, the elements x of S are known as code words, and the components x_i of x ($x_i = 0$ or $1, i = 1, \ldots, n$) are known as **bits**. The **weight** $w(x)$ of an element $x \in B_n$ is the number of occurrences of the digit 1 in $x = (x_i)$. The **Hamming distance** $d(x,y)$ between the elements x and y of B_n is the total number of unequal bits ($x_i \neq y_i$). Obviously, $d(x,y) = w(x \oplus y)$, where $\oplus$ denotes bitwise addition in the ring B_n. Let d be the minimum distance between distinct code words in S, and let s denote the number of incorrect digits in some received element y. If the corresponding transmitted code word was x, then, obviously, $s = d(x,y)$. If s is positive but less than d, then the presence of an error is thereby detected, since y cannot be another code word in S. If $s \leqslant (d-1)/2$, then the error can be corrected by reading the code word nearest to y as the transmitted word. The minimum distance d of the code S can be increased by expanding each code word x from $(x_1 \ldots x_n)$ to $(x_1 \ldots x_n y_1 \ldots y_k)$. The minimum distance d' of the new code S' ($\subset B_{n+k}$) satisfies the following condition (known as the **Hamming upper bound**): if $d' \geqslant 2t+1$, then

$$\sum_{i=0}^{t} \binom{n+k}{i} (q-1)^i \leqslant q^{(n+k)(1-R)},$$

where q is the number of available signals (here $q=2$) and R is the **information rate** ($= \log_q s'/(n+k)$, where s' is the number of code words in S'). A code attaining this upper bound for some t is said to be **perfect**. Recently, a complete list of perfect codes has been obtained for a prime power q [6].

Many important codes can be defined and analyzed on the basis of various algebraic theories, in particular, the theory of finite fields. If q is a prime power, then the set $V_n = \{0, 1, \ldots, q-1\}^n$ may be regarded as an n-dimensional vector space over $GF(q)$. A vector subspace S of V_n is known as a **linear**

code. If, to each element c of V_n, we associate the polynomial $c(X)=c_1+c_2X+\ldots+c_nX^{n-1}$ over $GF(q)$, then the **cyclic code** S modulo the **generator** $g(X)$ is defined as follows:

$$S=\{c\,|\,c(X)\equiv 0 \pmod{g(X)}\}.$$

In particular, if the dimension n is a factor of q^m-1 for some positive integer m, then any generator $g(X)$ such that

$$g(\beta^t)=g(\beta^{t+1})=\ldots=g(\beta^{t+r-1})=0,$$

for some element β of order n in $GF(q^m)$, determines a so-called **BCH**(Bose-Chaudhuri-Hocquenghem) **code**. Such a code is capable of detecting r errors and correcting $r/2$ errors. The class of BCH codes includes many important types, such as Hamming codes, Reed-Solomon codes, etc..

Other important types include **convolutional codes** for correcting errors in consecutive digits (**burst errors**), **AN codes** for arithmetic circuits and **Goppa codes** (an extension of the BCH class). New codes and new decoding algorithms are still being developed. This field of research is closely linked with information theory, algebra, and various applications of combinatorial analysis, such as †experimental design, etc. [3].

References

[1] J. Wolfowitz, Coding theorems of information theory, Springer, 1961.
[2] R. G. Gallager, Low-density parity-check codes, MIT Press, 1963.
[3] E. R. Berlekamp, Algebraic coding theory, McGraw-Hill, 1968.
[4] S. Lin, An introduction to error-correcting codes, Prentice-Hall, 1970.
[5] W. W. Peterson and E. J. Weldon, Jr., Error-correcting codes, MIT Press, second edition, 1972.
[6] A. Tietäväinen, On the non-existence of perfect codes over finite fields, SIAM J. App. Math., 24 (1973), 88–96.

67 (IX.6) Cohomology Operations

A. General Remarks

The notion of cohomology operations was introduced by Pontrjagin and Steenrod as a result of their study of homotopy classification problems (→ 204 Homotopy). Since then, numerous works have proved the importance of cohomology operations as applied to †homotopy theory, †differential topology, and other branches of topology. In fact, the use of cohomology operations is indispensable in studying problems related to †homotopy groups, †characteristic classes of manifolds, and others.

We denote by the symbol $H^*(X;A)=\Sigma H^n(X;A)$ the †singular cohomology ring of a topological space X with coefficients in a †unitary ring A.

B. Primary Cohomology Operations

A **(primary) cohomology operation** (or simply an operation) φ is a †natural transformation

$$\varphi:\prod_\lambda H^{l_\lambda}(\ ;A_\lambda)\rightarrow\prod_\mu H^{m_\mu}(\ ;B_\mu)$$

between the cohomology functors defined on the †category of topological spaces and continuous mappings. That is, φ is a family of mappings satisfying the following conditions:

(1) For each space X, φ defines a mapping

$$\varphi:\prod_\lambda H^{l_\lambda}(X;A_\lambda)\rightarrow\prod_\mu H^{m_\mu}(X;B_\mu)$$

that is not necessarily additive.

(2) For each mapping $f:X\rightarrow Y$, the commutativity $f^*\circ\varphi=\varphi\circ f^*$ holds in the diagram

$$\begin{array}{ccc}\prod_\lambda H^{l_\lambda}(X;A) & \xrightarrow{\varphi} & \prod_\mu H^{m_\mu}(X;B)\\ \uparrow f^* & & \uparrow f^*\\ \prod_\lambda H^{l_\lambda}(Y;A) & \xrightarrow{\varphi} & \prod_\mu H^{m_\mu}(Y;B).\end{array}$$

We list here two trivial examples.

(I) Addition: Addition of cohomology groups determines an operation $\varphi:H^l(X;A)\times H^l(X;A)\rightarrow H^l(X;A)$.

(II) The †cup product determines an operation

$$\varphi:H^{l_1}(X;A_1)\times H^{l_2}(X;A_2)\rightarrow H^{l_1+l_2}(X;A_1\otimes A_2)$$

defined by

$$\varphi(\alpha,\beta)=\alpha\smile\beta.$$

The **composite** of two cohomology operations is defined in the obvious way. Among cohomology operations the most important ones are operations of one variable. A cohomology operation of type $(A,l;B,m)$ is a natural transformation

$$\varphi:H^l(\ ;A)\rightarrow H^m(\ ;B),$$

where $\mathfrak{O}(A,l;B,m)$ is the Abelian group consisting of all such operations.

Denote by $H^m(A,l;B)$ the mth †cohomology group of an †Eilenberg-MacLane space $K(A,l)$ (→137 Eilenberg-MacLane Complexes), and let $u\in H^l(A,l;A)$ be the †fundamental class of $K(A,l)$. If X is a CW complex, by assigning f^*u to each $f:X\rightarrow K(A,l)$, we obtain a one-to-one correspondence between the set of the homotopy classes $\pi(X,$

$K(A,l)$) and the cohomology group $H^l(X;A)$. Hence by utilizing condition (2), it can be shown that the value of φ on $H^l(X;A)$ is uniquely determined by its operation on $H^l(A,l;A)$. Thus the assignment $\varphi \rightarrow \varphi u$ defines the isomorphism $\mathfrak{O}(A,l;B,m) \cong H^m(A,l;B)$. Let $v \in H^m(B,m;B)$ be the fundamental class of $K(B,m)$. Then an argument similar to the previous one shows that the assignment $g \rightarrow \varphi$, subject to the relation $g^*v = \varphi u$, defines a one-to-one correspondence $\mathfrak{O}(A,l;B,m) \rightarrow \pi(K(A,l), K(B,m))$.

The following four types of operations together with the two above are called elementary operations:

(III) Homomorphisms induced by a coefficient homomorphism: There are homomorphisms $\eta_* : H^l(X;A) \rightarrow H^l(X;B)$ induced by a homomorphism $\eta : A \rightarrow B$.

(IV) Bockstein (cohomology) operation: This operation is the system of homomorphisms $\delta^* : H^l(X;C) \rightarrow H^{l+1}(X;A)$ associated with an †exact coefficient sequence $0 \rightarrow A \rightarrow B \rightarrow C \rightarrow 0$ (→ 57 Chain Complexes). For example, the coefficient sequence $0 \rightarrow \mathbf{Z} \xrightarrow{n} \mathbf{Z} \rightarrow \mathbf{Z}_n \rightarrow 0$ ($\mathbf{Z}_n = \mathbf{Z}/n\mathbf{Z}$) defines a **Bockstein operation** (or **Bockstein homomorphism**), which is usually denoted by $(1/n)\delta$ or Δ_n.

(V) **Steenrod** (or **reduced**) **square operations** $Sq^i (i \geqslant 0) : Sq^i$ is a sequence of operations defined by the following five axioms [2, 5]:

(V1) For each pair of integers $i \geqslant 0$ and $l \geqslant 0$,

$$Sq^i : H^l(X;\mathbf{Z}_2) \rightarrow H^{l+i}(X;\mathbf{Z}_2)$$

is a natural transformation of functors that is a homomorphism.

(V2) $Sq^0 = 1$.

(V3) If $\deg x = i$, then $Sq^i x = x \smile x$ (cup product).

(V4) If $\deg x < i$, then $Sq^i x = 0$.

(V5) **(Cartan formula)**

$$Sq^1(x \smile y) = \sum_{j=0}^{i} Sq^j x \smile Sq^{i-j} y.$$

These five axioms imply the following two formulas:

(V6) Sq^1 is the Bockstein operation β_2 of the coefficient sequence $0 \rightarrow \mathbf{Z}_2 \rightarrow \mathbf{Z}_4 \rightarrow \mathbf{Z}_2 \rightarrow 0$.

(V7) **(Adem relations)** If $0 < i < 2j$, then

$$Sq^i \circ Sq^j = \sum_{k=0}^{[i/2]} \binom{j-1-k}{i-2k} Sq^{i+j-k} \circ Sq^k.$$

The binomial coefficient is taken mod 2.

We can extend the definition of the Sq^i so that they operate on the relative cohomology groups. Now (V1), (V2), and (V5) imply (V8). If $\delta : H^l(Y;\mathbf{Z}_2) \rightarrow H^{l+1}(X,Y;\mathbf{Z}_2)$ is the †coboundary homomorphism, then $\delta \circ Sq^i = Sq^i \circ \delta$.

(V′) **Steenrod** pth **power operations** $\mathcal{P}^i$ ($i \geqslant 0$): Let p be an odd prime. Then $\mathcal{P}^i$ is a sequence of operations defined by the following five axioms [5].

(V′1) For each pair of integers $i \geqslant 0$ and $l \geqslant 0$,

$$\mathcal{P}^i : H^l(X;\mathbf{Z}_p) \rightarrow H^{l+2i(p-1)}(X;\mathbf{Z}_p)$$

is a natural transformation that is a homomorphism.

(V′2) $\mathcal{P}^0 = 1$.

(V′3) If $\deg x = 2k$, then $\mathcal{P}^k x = x^p$.

(V′4) If $\deg x < 2k$, then $\mathcal{P}^k x = 0$.

(V′5) **(Cartan's formula)**

$$\mathcal{P}^i(x \smile y) = \sum_{j=0}^{i} \mathcal{P}^j x \smile \mathcal{P}^{i-j} y.$$

These axioms imply the **Adem relations** for $\mathcal{P}^i$ and the Bockstein homomorphism β_p associated with the coefficient sequence $0 \rightarrow \mathbf{Z}_p \rightarrow \mathbf{Z}_{p^2} \rightarrow \mathbf{Z}_p \rightarrow 0$ (→ Appendix A, Table 6.II).

We can extend the definition of $\mathcal{P}^i$ to the relative cohomology groups too, and we obtain $\delta \circ \mathcal{P}^i = \mathcal{P}^i \circ \delta$.

(VI) **Pontrjagin** pth **power operations** $\mathfrak{P}_p$. Let p be a prime. $\mathfrak{P}_p$ is a family of operations satisfying the following five conditions [3]:

(VI1) For each pair of integers $l \geqslant 0$ and $h \geqslant 1$,

$$\mathfrak{P}_p : H^l(X;\mathbf{Z}_{p^h}) \rightarrow H^{pl}(X;\mathbf{Z}_{p^{h+1}})$$

is a natural transformation.

(VI2) If $\eta : \mathbf{Z}_{p^{h+1}} \rightarrow \mathbf{Z}_{p^h}$ is a homomorphism defined by $\eta(1) = 1$, then we have $\eta_* \circ \mathfrak{P}_p x = x^p$.

(VI3) If $p : \mathbf{Z}_{p^h} \rightarrow \mathbf{Z}_{p^{h+1}}$ is a homomorphism defined by $p(1) = p$, then we have

$$\mathfrak{P}_p(x+y)$$

$$= \mathfrak{P}_p x + \mathfrak{P}_p y + \sum_{i=1}^{p} \left(\binom{p}{i} \Big/ p \right) p_*(x^i \smile y^{p-i}).$$

(VI4) $\mathfrak{P}_p(x \smile y) = \mathfrak{P}_p x \smile \mathfrak{P}_p y$.

(VI5) If $p > 2$ and $\deg x = 2k+1$ (odd), then $\mathfrak{P}_p x = 0$.

Let A, B be finitely generated Abelian groups. Then the computation of $H^*(A,l)$ shows that each element of $\mathfrak{O}(A,l;B,m)$ can be written as the composite of a finite number of the operations of the types (I)–(VI) (H. Cartan).

For example, let u be the fundamental class of $K(\mathbf{Z}_2, 2)$. Then we have $H^2(\mathbf{Z}_2, 2; \mathbf{Z}_4) = Z_2 \cdot 2_* u$, $H^3(\mathbf{Z}_2, 2; \mathbf{Z}_4) = Z_2 \cdot \eta_*(\delta/2) u$, $H^4(\mathbf{Z}_2, 2; \mathbf{Z}_4) = Z_4 \cdot \mathfrak{P}_2 u$, and $H^5(\mathbf{Z}_2, 2; \mathbf{Z}_4) = Z_4 \cdot \eta_* \circ (\delta/4) \circ \mathfrak{P}_2 u \oplus Z_2 \cdot 2_* Sq^2 \circ Sq^1 u$, where $\eta : \mathbf{Z} \rightarrow \mathbf{Z}_4$ and $2 : \mathbf{Z}_2 \rightarrow \mathbf{Z}_4$ are homomorphisms defined by $\eta(1) = 1$ and $2(1) = 2$, respectively, and $\mathbf{Z}_n \cdot a$ is the cyclic group of order n with generator a.

Suppose that we are given sets of integers l_i and m_j. We define the **stable (primary) cohomology operation** φ with respect to these sets of integers l_i and m_j as natural transformations

$$\varphi: \prod_\lambda H^{n+l_\lambda}(\ ;A_\lambda)\to\prod_\mu H^{n+m_\mu}(\ ;B_\mu)$$

satisfying the following condition, for all integers $n\geqslant 0$:

(3) Let $S: H^{l+1}(SX;A)\to H^l(X;A)$ denote the †suspension. Then the commutativity $S\circ\varphi=\varphi\circ S$ holds in the diagram

$$\begin{array}{ccc}\prod_\lambda H^{n+l_\lambda}(X;A) & \xrightarrow{\varphi} & \prod_\mu H^{n+m_\mu}(X;B)\\ \uparrow s & & \uparrow s\\ \prod_\lambda H^{n+l_\lambda+1}(SX;A) & \xrightarrow{\varphi} & \prod_\mu H^{n+m_\mu+1}(SX;B).\end{array}$$

For example, the cohomology operations

$$(-1)^n\beta_p: H^n(X;\mathbf{Z}_p)\to H^{n+1}(X;\mathbf{Z}_p)$$

define a stable cohomology operation β. $Sq^i,\mathfrak{P}^i$ are also examples of stable cohomology operations.

A stable cohomology operation φ of type (A,B) and of degree q is a sequence of cohomology operations of type $(A,n;B,n+q)$ defined for all integers $n\geqslant 0$. $\mathfrak{A}(A,B)_q$ denotes the Abelian group consisting of all stable cohomology operations of type (A,B) and of degree q. In case $A=B$, $\mathfrak{A}(A)=\sum_{q=0}^\infty\mathfrak{A}(A,A)_q$ is a †graded algebra, where multiplication of two operations is given by their composition. Let p be a prime. Then $\mathfrak{A}(\mathbf{Z}_p)$ is called the **Steenrod algebra** mod p and is denoted by $\mathfrak{A}(p)$. $\mathfrak{A}(2)$ is the augmented graded algebra over $\mathbf{Z}_2$ generated by Sq^i subject to the Adem relations. Suppose that we are given a sequence of nonnegative integers $I=(i_1,i_2,\ldots,i_k)$. We call I an **admissible sequence** if $i_{s-1}\geqslant 2i_s$ holds for $2\leqslant s\leqslant k$. We write $Sq^I=Sq^{i_1}\circ Sq^{i_2}\circ\ldots\circ Sq^{i_k}$. If I is an admissible sequence, we say that Sq^I is an **admissible monomial**. The admissible monomials form an additive basis for $\mathfrak{A}(2)$, which has the structure of a †Hopf algebra whose †comultiplication $\psi:\mathfrak{A}(2)\to\mathfrak{A}(2)\otimes\mathfrak{A}(2)$ is given by $\psi Sq^i=\sum_{j=0}^i Sq^j\otimes Sq^{i-j}$. The dual space $\mathfrak{A}(2)^*=\mathrm{Hom}_{\mathbf{Z}_2}(\mathfrak{A}(2),\mathbf{Z}_2)$ gives a Hopf algebra that is the polynomial algebra generated by ξ_i of degree i. The comultiplication $\varphi^*:\mathfrak{A}(2)^*\to\mathfrak{A}(2)^*\otimes\mathfrak{A}(2)^*$ is given by $\varphi^*\xi_i=\sum_{j=0}^i\xi_{i-j}^{2^j}\otimes\xi_j$.

$\mathfrak{A}(p)$ has properties similar to $\mathfrak{A}(2)$ (p is an odd prime). In particular, $\mathfrak{A}(p)$ is a †Hopf algebra generated by $\mathcal{P}^i$ and β subject to the Adem relations with comultiplication ψ given by $\psi\beta=\beta\otimes 1+1\otimes\beta$ and $\psi\mathcal{P}^i=\sum_{j=0}^i\mathcal{P}^j\otimes\mathcal{P}^{i-j}$ [5] (→ Appendix A, Table 6.III).

C. Secondary Cohomology Operations

Here we restrict our attention to a special type of operation treated by Adams [4] that has been proved to be powerful in applications.

For a specified prime p, we write $H^+(X)=\sum_{i=1}^\infty H^i(X;\mathbf{Z}_p)$. We can regard $H^+(X)$ as a †graded left module over $\mathfrak{A}(p)$. Now let C_s $(s=0,1)$ be a pair of †left free modules over $\mathfrak{A}(p)$, and let $d: C_1\to C_0$ be a graded homomorphism over $\mathfrak{A}(p)$. Suppose that we are given an element $z\in C_1$ such that $dz=0$. For future purposes, C_0 and C_1 are assumed to have bases $\{c_{0,\lambda}\}$ and $\{c_{1,\mu}\}$ with $\deg c_{0,\lambda}=l_\lambda$ and $\deg c_{1,\mu}=m_\mu$, respectively, in terms of which d has the representation $dc_{1,\mu}=\sum_\lambda a_{\mu,\lambda}c_{0,\lambda}$. Then z is expressed in the form $z=\sum_\mu b_\mu c_{1,\mu}$.

We say that Φ is a **stable secondary cohomology operation** associated with the pair (d,z) if it satisfies the following four axioms:

(1) Let $D^n(d,X)$ be the module consisting of homomorphisms $\varepsilon: C_0\to H^+(X)$ over $\mathfrak{A}(p)$ of degree n such that $\varepsilon d=0$. Putting $\varepsilon(c_{0,\lambda})=x_\lambda$, we may assume that $D^n(d,X)=\{\varepsilon=\amalg_\lambda x_\lambda\in\amalg_\lambda H^{n+l_\lambda}(X)\mid\sum_\lambda a_{\mu,\lambda}x_\lambda=0\}$. Let $Q^n(z,X)$ be the submodule of $H^+(X)$ consisting of elements of the form $\xi(z)$, where ξ is an $\mathfrak{A}(p)$-homomorphism $: C_1\to H^+(X)$ of degree $n-1$. In other words, $Q^n(z,X)=\sum_\mu b_\mu H^{n+m_\mu-1}(X)$. Then for each $n\geqslant 0$ and each X,Φ is a mapping

$$\Phi: D^n(d,X)\to H^+(X)/Q^n(z,X).$$

(2) For each mapping $f: X\to Y$, the commutativity $f^*\circ\Phi=\Phi\circ f^*$ holds in the diagram

$$\begin{array}{ccc}D^n(d,X) & \xrightarrow{\Phi} & H^+(X)/Q^n(z,X)\\ \uparrow f^* & & \uparrow f^*\\ D^n(d,Y) & \xrightarrow{\Phi} & H^+(Y)/Q^n(z,Y).\end{array}$$

(3) Let $S: H^{n+1}(SX)\to H^n(X)$ denote the suspension. Then the commutativity $S\circ\Phi=\Phi\circ S$ holds in the diagram

$$\begin{array}{ccc}D^n(d,X) & \xrightarrow{\Phi} & H^+(X)/Q^n(z,X)\\ \uparrow s & & \uparrow s\\ D^{n+1}(d,SX) & \xrightarrow{\Phi} & N^+(SX)/Q^{n+1}(z,SX).\end{array}$$

(4) Let $i: Y\to X$ be an injection such that $i^*\circ\varepsilon=0$. That is, $i^*x_\lambda=0$. We can then find homomorphisms (over $\mathfrak{A}(p)$) $\eta: C_0\to H^+(X,Y)$ of degree n and $\zeta: C_1\to H^+(Y)$ of degree $n-1$ such that the following diagram commutes:

$$\begin{array}{ccccccccc}H^+(Y) & \xleftarrow{i^*} & H^+(X) & \xleftarrow{j^*} & H^+(X,Y) & \xleftarrow{\delta^*} & H^+(Y) & \xleftarrow{i^*} & H^+(X)\\ & & \uparrow & & \uparrow\eta & & \uparrow\zeta & & \\ & & \llcorner & \xrightarrow{\varepsilon}\!\!\!\!\!\!\!\! & C_0 & \xleftarrow{d} & C_1 & & \end{array}$$

Then for any such pair (η,ζ) we have $i^*\circ\Phi\varepsilon\equiv\zeta z \bmod i^*Q^n(z,X)$.

For each pair (d,z), there is at least one associated operation Φ. Existence is proved by means of a †Postnikov system. Let Φ and Φ' be two operations associated with the same (d,z). Then they differ by a stable primary operation in the sense that there is an element

$\varphi \in C_0/dC_1$ such that $\Phi'\varepsilon \equiv \Phi\varepsilon + \varphi\varepsilon$ mod $Q(z,X)$, where $\varphi \equiv \sum_\lambda a_\lambda c_{0,\lambda}$ mod dC_1 means $\varphi\varepsilon = \sum_\lambda a_\lambda x_\lambda$.

For example, define the action of $\mathfrak{A}(2)$ on $\mathbf{Z}_2$ by the rule $a \cdot \nu = 0$ if $\deg a > 0$ and $1 \cdot \nu = \nu$ for each $\nu \in \mathbf{Z}_2$. We consider a partial minimal resolution

$$0 \leftarrow \mathbf{Z}_2 \overset{\varepsilon}{\leftarrow} C_0 \overset{d_1}{\leftarrow} C_1 \overset{d_2}{\leftarrow} C_2 \leftarrow \ldots$$

(→ 202 Homological Algebra). First we take $C_0 = \mathfrak{A}(2)$ and define $\varepsilon(1) = 1$. Next we take C_1 to be free over $\mathfrak{A}(2)$ with generators c_i, where $0 \leqslant i$ and $d_1 c_i = Sq^{2^i}$. Furthermore, we take C_2 to be free over $\mathfrak{A}(2)$ with generators $c_{i,j}$ with $0 \leqslant i \leqslant j$, $j \neq i+1$ and

$$d_2(c_{i,j}) = Sq^{2^i} c_j + \sum_{0 \leqslant k < j} b_k c_k$$

with $b_k \in \mathfrak{A}(2)$. Here the $d_1 c_i = Sq^{2^i}$ form a minimal set of generators of $\mathfrak{A}(2)$, and the equations $0 = d_1 d_2 c_{i,j} = Sq^{2^i} \circ Sq^{2^j} + \sum_{0 \leqslant k < j} b_k Sq^{2^k}$ form a minimal set of generators over $\mathfrak{A}(2)$ of the Adem relations. Let $C_1(j)$ be the submodule over $\mathfrak{A}(2)$ generated by c_k with $0 \leqslant k \leqslant j$ in C_1, and $d_1(j)$ be the restriction of d_1 on $C_1(j)$. We write $z_{i,j} = d_2 c_{i,j}$. Let $\Phi_{i,j}$ denote an operation associated with $(d_1(j), z_{i,j})$. Then $\Phi_{i,j}$ is defined on the submodule $D^n_{i,j}(X)$ consisting of elements $x \in H^n(X)$ such that $Sq^{2^k} x = 0$ for $0 \leqslant k \leqslant j$ and takes values in $H^{n+2^i+2^j-1}(X)$ modulo the submodule

$$Q^n_{i,j}(X) = Sq^{2^i} H^{n+2^j-1}(X) + \sum_{0 \leqslant k < j} b_k H^{n+2^k-1}(X).$$

For example, $\Phi_{0,0}: \operatorname{Ker}\beta_2 \to \operatorname{Cok}\beta_2$ is the generalized Bockstein homomorphism defined by $\delta/2^2$ and $\Phi_{1,1}$ is the operation discovered by Adem that appears as the †third obstruction of S^n for $n \geqslant 2$. If $k \geqslant 3$, we have the relation

$$Sq^{2^{k+1}} \equiv \sum_{\substack{0 \leqslant i \leqslant j \leqslant k \\ j \neq i+1}} a_{i,j,k} \Phi_{i,j} \times \bmod \sum_{\substack{0 \leqslant i \leqslant j \leqslant k \\ j \neq i+1}} a_{i,j,k} Q^n_{i,j}(X).$$

These formulas can be applied to prove the nonexistence of an element with †Hopf-invariant 1 in $\pi_{2n-1}(S^n)$ unless $n = 1, 2, 4, 8$ (Adams [4]). Analogous formulas also hold for $\mathfrak{A}(p)$ with $p > 2$. We have no satisfactory theories concerning cohomology operations of orders higher than 2. (For the †extraordinary cohomology operations, → 236 *K*-Theory.)

References

[1] N. E. Steenrod, Products of cocycles and extension of mappings, Ann. of Math., (2) 48 (1947), 290–320.

[2] J.-P. Serre, Cohomologie modulo 2 des complexes d'Eilenberg-MacLane, Comment. Math. Helv., 27 (1953), 198–232.

[3] N. E. Steenrod and P. E. Thomas, Cohomology operations derived from cyclic groups, Comment. Math. Helv., 32 (1957), 129–152.

[4] J. F. Adams, On the non-existence of elements of Hopf invariant one, Ann. of Math., (2) 72 (1960), 20–104.

[5] N. E. Steenrod, Cohomology operations, Ann. Math. Studies, Princeton Univ. Press, 1962.

[6] J. F. Adams, On the structure and applications of the Steenrod algebra, Comment. Math. Helv., 32 (1958), 180–214.

[7] R. E. Mosher and M. C. Tangora, Cohomology operations and applications in homotopy theory, Harper, 1968.

[8] E. H. Spanier, Algebraic topology, McGraw-Hill, 1966.

68 (IX.5) Cohomology Rings

A. General Remarks

Given an Abelian group G and a topological space X, we consider the †homology group $H_*(X;G) = \sum_n H_n(X;G)$ and the †cohomology group $H^*(X;G) = \sum_n H^n(X;G)$ of X with coefficients in G (→ 205 Homology Groups). Although the definition of the latter is "dual" to the former, the cohomology group, unlike the homology group, admits the structure of a †graded ring if G is a †unitary commutative ring. This fact was first discovered by J. W. Alexander, E. Čech, and H. Whitney in the 1930s. In particular, if X is a †manifold, the †Poincaré duality theorem asserts that the product of two cohomology classes is dual to the †intersection of the corresponding homology classes.

B. Cross Products and Slant Products

The following definitions are due to S. Lefschetz and N. E. Steenrod. Let M be a †chain complex over a unitary commutative ring R and let G be an R-module. We denote by $H_*(M:G)$ the homology group of the chain complex $M \otimes G$ and by $H^*(M;G)$ the cohomology group of the †cochain complex $\operatorname{Hom}(M,G)$ (→ 57 Chain Complexes, 203 Homology Groups). Given chain complexes

M_i and modules G_i $(i=1,2)$, a †chain mapping

$$\alpha':(M_1\otimes G_1)\otimes(M_2\otimes G_2)\to(M_1\otimes M_2)\otimes(G_1\otimes G_2)$$

and a †cochain mapping

$$\alpha':\mathrm{Hom}(M_1,G_1)\otimes\mathrm{Hom}(M_2,G_2)\to\mathrm{Hom}(M_1\otimes M_2,G_1\otimes G_2)$$

are defined by

$$\alpha'((c_1\otimes g_1)\otimes(c_2\otimes g_2))=(c_1\otimes c_2)\otimes(g_1\otimes g_2)$$

and

$$\alpha'(u_1\otimes u_2)(c_1\otimes c_2)=u_1(c_1)\otimes u_2(c_2)$$
$$(c_i\in M_i,\ g_i\in G_i,\ u_i\in\mathrm{Hom}(M_i,G_i)),$$

which induce the following homomorphisms (of the †graded modules) of degree 0:

$$\alpha:H_*(M_1;G_1)\otimes H_*(M_2;G_2)\to H_*(M_1\otimes M_2;G_1\otimes G_2),$$

$$\alpha:H^*(M_1;G_1)\otimes H^*(M_2;G_2)\to H^*(M_1\otimes M_2;G_1\otimes G_2).$$

Furthermore, a cochain mapping

$$\beta':\mathrm{Hom}(M_1\otimes M_2,\mathrm{Hom}(G_1,G_2))\to\mathrm{Hom}(M_1\otimes G_1,\mathrm{Hom}(M_2,G_2))$$

and a chain mapping

$$\beta':M_1\otimes M_2\otimes\mathrm{Hom}(G_1,G_2)\to\mathrm{Hom}(\mathrm{Hom}(M_1,G_1),\ M_2\otimes G_2)$$

are defined by

$$((\beta'u)(c_1\otimes g_1))(c_2)=(u(c_1\otimes c_2))\cdot(g_1)$$

and

$$\beta'(c_1\otimes c_2\otimes h)(u_1)=c_2\otimes h(u_1(c_1))$$
$$(u\in\mathrm{Hom}(M_1\otimes M_2,\mathrm{Hom}(G_1,G_2)),\quad h\in\mathrm{Hom}(G_1,G_2)),$$

which induce the following homomorphisms (of the graded modules) of degree 0:

$$\beta:H^*(M_1\otimes M_2;\mathrm{Hom}(G_1,G_2))\to\mathrm{Hom}\big(H_*(M_1;G_1),H^*(M_2;G_2)\big),$$

$$\beta:H_*(M_1\otimes M_2;\mathrm{Hom}(G_1,G_2))\to\mathrm{Hom}\big(H^*(M_1;G_1),H_*(M_2;G_2)\big).$$

Let M_i be the chain complexes defining the (co) homology groups of pairs (X_i,A_i) of topological spaces and their subspaces $(i=1,2)$. We assume that the (co) homology group of the chain complex $M_1\otimes M_2$ is canonically isomorphic to the (co) homology group of the Cartesian product $(X_1,A_1)\times(X_2,A_2)=(X_1\times X_2,A_1\times X_2\cup X_1\times A_2)$ of the pairs (X_1,A_1) and (X_2,A_2). The assumption holds, for example, if the M_i are the †singular chain complexes of the topological spaces X_i or if the M_i are the singular chain complexes of the pairs (X_i,A_i) of †CW complexes and their subcomplexes. Furthermore, suppose that we are given a homomorphism $\eta:G_1\otimes G_2\to G_3$. We consider the following homomorphisms:

$$\times:H_p(X_1,A_1;G_1)\otimes H_q(X_2,A_2;G_2)\to H_{p+q}((X_1,A_1)\times(X_2,A_2);G_3),$$

$$\times:H^p(X_1,A_1;G_1)\otimes H^q(X_2,A_2;G_2)\to H^{p+q}((X_1,A_1)\times(X_2,A_2);G_3),$$

where $\times$ is the composite of α and the homomorphism of (co) homology groups induced by η. For $z_1\in H_p(X_1,A_1;G_1)$ and $z_2\in H_q(X_2,A_2;G_2)$, we define the **cross product** $z_1\times z_2\in H_{p+q}((X_1,A_1)\times(X_2,A_2);G_3)$ to be the image of $z_1\otimes z_2$ under the mapping $\times$. Similarly, we define the **cross product** $\alpha_1\times\alpha_2\in H^{p+q}((X_1,A_1)\times(X_2,A_2);G_3)$ for $\alpha_1\in H^p(X_1,A_1;G_1)$ and $\alpha_2\in H^q(X_2,A_2;G_2)$. Since the homomorphism $\eta:G_1\otimes G_2\to G_3$ can be naturally regarded as a homomorphism of G_1 into $\mathrm{Hom}(G_2,G_3)$, we also have the homomorphisms

$$/:H^n(X_1,A_1)\times(X_2,A_2);G_1)\to\mathrm{Hom}(H_p(X_1,A_1;G_2),H^{n-p}(X_2,A_2;G_3)),$$

$$\backslash:H_n((X_1,A_1)\times(X_2,A_2);G_1)\to\mathrm{Hom}(H^p(X_1,A_1;G_2),H_{n-p}(X_2,A_2;G_3)),$$

which are the composite of the mapping β and the homomorphism of (co) homology groups induced by the mapping

$$G_1\to\mathrm{Hom}(G_2,G_3).$$

Therefore, for

$$w\in H^n((X_1,A_1)\times(X_2,A_2);G_1)$$

and

$$z_1\in H_p(X_1,A_1;G_2)$$

we have an element

$$w/z_1\in H^{n-p}(X_2,A_2;G_3),$$

which is called the **slant product** of w and z_1. Similarly, for

$$z\in H_n((X_1,A_1)\times(X_2,A_2);G_1)$$

and

$$w_1\in H^p(X_1,A_1;G_2),$$

we have the **slant product**

$$z\backslash w_1\in H_{n-p}(X_2,A_2;G_3)$$

of z and w_1.

Cross products and slant products have the following two properties.

(1) Naturality: Cross products and slant products are compatible with the †homomorphisms of (co) homology groups induced by continuous mappings $f_i:(X_i,A_i)\to(Y_i,B_i)$ $(i=1,2)$. For example, the diagram

$$\begin{array}{ccc}
H^p(Y_1,B_1)\otimes H^q(Y_2,B_2) & \xrightarrow{\times} & H^{p+q}((Y_1,B_1)\times(Y_2,B_2))\\
\downarrow f_1^*\otimes f_2^* & & \downarrow (f_1\times f_2)^*\\
H^p(X_1,A_1)\otimes H^q(X_2,A_2) & \xrightarrow{\times} & H^{p+q}((X_1,A_1)\times(X_2,A_2))
\end{array}$$

is commutative.

(2) These products are compatible with the †(co)boundary homomorphisms. For example, the diagrams

$$\begin{array}{ccc} H_n((X_1,A_1)\times X_2) & \xrightarrow{\backslash} & \mathrm{Hom}(H^p(X_1,A_1),H_{n-p}(X_2)) \\ \downarrow \partial_* & & \downarrow \mathrm{Hom}(\delta^*,1) \\ H_{n-1}(A_1\times X_2) & \xrightarrow{\backslash} & \mathrm{Hom}(H^{p-1}(A_1),H_{n-p}(X_2)) \end{array}$$

and

$$\begin{array}{ccc} H_n(X_1\times(X_2,A_2)) & \xrightarrow{\backslash} & \mathrm{Hom}(H^p(X_1),A_{n-p}(X_2,A_2)) \\ \downarrow \partial_* & & \downarrow \mathrm{Hom}(1,(-1)^p\partial_*) \\ H_{n-1}(X_1\times A_2) & \xrightarrow{\backslash} & \mathrm{Hom}(H^p(X_1),H_{n-p-1}(A_2)) \end{array}$$

are commutative.

C. Cup Products and Cap Products

Let $X_1=X_2=X$, and let $d:(X,A_1\cup A_2)\to(X,A_1)\times(X,A_2)$ be the diagonal mapping (i.e., the mapping defined by $d(x)=(x,x)$, $(x\in X)$). Then we have a homomorphism

$$\smile : H^p(X,A_1;G_1)\otimes H^q(X,A_2;G_2) \to H^{p+q}(X,A_1\cup A_2;G_3),$$

which is the composite of the cross product for cohomology and the homomorphism

$$d^*:H^{p+q}((X,A_1)\times(X,A_2);G_3)\to H^{p+q}(X,A_1\cup A_2;G_3).$$

For $w_1\in H^p(X,A_1;G_1)$ and $w_2\in H^q(X,A_2;G_2)$, the **cup product** $w_1\smile w_2\in H^{p+q}(X,A_1\cup A_2;G_3)$ is defined to be the image of $w_1\otimes w_2$ under the homomorphism $\smile$. We also have a homomorphism

$$:\frown : H_n(X,A_1\cup A_2;G_1) \to \mathrm{Hom}(H^p(X,A_1;G_2),H_{n-p}(X,A_2;G_3)),$$

which is the composite of the homomorphism $\backslash$ and the homomorphism $d_*:H_n(X,A_1\cup A_2;G_1)\to H_n((X,A_1)\times(X,A_2);G_1)$. For $z\in H_n(X,A_1\cup A_2;G_1)$ and $w_1\in H^p(X,A_1;G_2)$, we have the element $z\frown w_1\in H_{n-p}(X,A_2;G_3)$ determined by the homomorphism $\frown$; $z\frown w_1$ is called the **cap product** of z and w_1.

The cup product has the following two properties analogous to the corresponding properties of the cross product: (1) Naturality: If $f:(X,A_1,A_2)\to(Y,B_1,B_2)$ is a continuous mapping, then for $w_1\in H^p(Y,B_1)$ and $w_2\in H^q(Y,B_2)$, we have $f^*(w_1\smile w_2)=f^*w_1\smile f^*w_2$ in $H^{p+q}(X,A_1\cup A_2)$. (2) For $w_1\in H^{p-1}(A)$ and $w_2\in H^q(X)$, we have $\delta^*(w_1\smile i^*w_2)=\delta^*w_1\smile w_2$ in $H^{p+q}(X,A)$; for $w_1\in H^p(X)$ and $w_2\in H^{q-1}(A)$, we have $\delta^*(i^*w_1\smile w_2)=(-1)^p w_1\smile\delta^*w_2$ in $H^{p+q}(X,A)$, where $i:A\subset X$ is the inclusion. Given a homomorphism of $G_1\otimes G_2$ to G_3, these properties together with bilinearity characterize the cup product defined on the class of finite CW pairs (a CW pair is a pair of a finite CW complex and its subcomplex). The cap product has the following two properties: (1) Naturality: If $f:(X,A_1,A_2)\to(Y,B_1,B_2)$ is a continuous mapping and $z\in H_n(X,A_1\cup A_2)$ and $w_1\in H^p(Y,B_1)$, then we have $f_*(z\frown f^*w_1)=f_*z\frown w_1$ in $H_{n-p}(Y,B_2)$. (2) For $z\in H_n(X,A)$ and $w\in H^p(A)$, we have $i_*(\partial_*z\frown w)=z\frown\delta^*w$ in $H_{n-p-1}(X)$; for $z\in H_n(X,A)$ and $w\in H^p(X)$, we have $\partial_*(z\frown w)=(-1)^p\partial_*z\frown i^*w$ in $H_{n-p-1}(A)$, where $i:A\subset X$ is the inclusion.

For $z\in H_p(X,A;G_1)$ and $w\in H^p(X,A;G_2)$, we define the **Kronecker index** $\langle z,w\rangle\in G_3$ to be the image of the cap product $z\frown w\in H_0(X;G_3)$ under the homomorphism $H_0(X;G_3)\to G_3$ induced by the †augmentation.

D. Internal Products and External Products

By the definitions in section C we have $w_1\smile w_2=d^*(w_1\times w_2)$ and $z\frown w_1=d_*z\backslash w_1$ for $w_i\in H^*(X,A_i)$ $(i=1,2)$ and $z\in H_*(X,A_1\cup A_2)$. On the other hand, if $p_1:(X_1,A_1)\times X_2\to(X_1,A_1)$ and $p_2:X_1\times(X_2,A_2)\to(X_2,A_2)$ denote the canonical projections, then we have $w_1\times w_2=p_1^*w_1\smile p_2^*w_2$ and $z\backslash w_1=p_{2*}(z\frown p_1^*w_1)$ for $w_i\in H^*(X_i,A_i)$ $(i=1,2)$ and $z\in H_*((X_1,A_1)\times(X_2,A_2))$. Hence the cross product $\times$ and the cup product $\smile$ are defined in terms of each other, and the slant product $\backslash$ and the cap product $\frown$ are defined in terms of each other. In this situation, $\smile$ and $\frown$ are called the **internal product** for $\times$ and $\backslash$, respectively; conversely, $\times$ and $\backslash$ are called the **external product** for $\smile$ and $\frown$, respectively.

The internal products corresponding to the cross product $\times$ for homology and the slant product $/$ cannot generally be defined. However, if X admits a continuous multiplication $\mu:X\times X\to X$, then the corresponding internal products

$$H_p(X;G_1)\otimes H_q(X;G_2)\to H_{p+q}(X;G_3),\quad H^n(X;G_1)\to\mathrm{Hom}(H_p(X;G_2),H^{n-p}(X;G_3))$$

can be defined as the composites $\mu\circ\times$ and $/\circ\mu$, where μ is the homomorphism of (co)homology groups induced by μ (→ 207 Hopf Algebras).

E. Cohomology Rings

Let $G_1=G_2=G_3$ be the ground ring R, and define the homomorphism $G_1\otimes G_2\to G_3$ by the product of R. In this case, if we canonically identify $((X_1,A_1)\times(X_2,A_2))\times(X_3,A_3)$

with $(X_1,A_1)\times((X_2,A_2)\times(X_3,A_3))$ and $(X_1,A_1)\times(X_2,A_2)$ with $(X_2,A_2)\times(X_1,A_1)$, then we have $(w_1\times w_2)\times w_3=w_1\times(w_2\times w_3)$, $w_1\times w_2=(-1)^{pq}w_2\times w_1$, $z\backslash(w_1\times w_2)=(z\backslash w_1)\backslash w_2$ for $w_1\in H^p(X_1,A_1)$, $w_2\in H^q(X_2,A_2)$, $w_3\in H^r(X_3,A_3)$, and $z\in H_n((X_1,A_1)\times(X_2,A_2)\times(X_3,A_3))$. Consequently, we have $(w_1\smile w_2)\smile w_3=w_1\smile(w_2\smile w_3)$, $w_1\smile w_2=(-1)^{pq}w_2\smile w_1$, $z\frown(w_1\frown w_2)=(z\frown w_1)\frown w_2$ for $w_1\in H^p(X,A_1)$, $w_2\in H^q(X,A_2)$, $w_3\in H^r(X,A_3)$, and $z\in H_n(X,A\cup A_2\cup A_3)$. In particular, the cohomology group

$$H^*(X,A;R)=\sum_n H^n(X,A;R),$$

together with the cup product, is an (anti) commutative graded ring over R. The graded ring $H^*(X,A;R)$ is called the **cohomology ring** (or **cohomology algebra**) of (X,A) with coefficients in R. If R has a unity element, the cohomology ring $H^*(X;R)$ of a topological space X also has a unity element. We have $(w_1\times v_1)\smile(w_2\times v_2)=(-1)^{qr}(w_1\smile w_2)\times(v_1\smile v_2)$ for $w_1\in H^p(X,A)$, $w_2\in H^q(X,A)$, $v_1\in H^r(Y,B)$, $v_2\in H^s(Y,B)$. If R is a field, the graded algebras $H^*((X,A)\times(Y,B);R)$ and $H^*(X,A;R)\otimes H^*(Y,B;R)$ are isomorphic.

F. Singular Homology or Cohomology

A basic property utilized in defining cross products, etc., is the fact that the (co) homology group of $(X,A_1)\times(X_2,A_2)$ is canonically isomorphic to the (co)homology group of the tensor product of the chain complexes for (X_1,A_1), (X_2,A_2). In the case of the †singular homology, the isomorphism is induced by the †chain equivalences $\nabla: S(X_1)\otimes S(X_2)\to S(X_1\times X_2)$, $\rho: S(X_1\times X_2)\to S(X_1)\otimes S(X_2)$ such that

$$\nabla(T_1^p\otimes T_2^q)=\sum(-1)^{\varepsilon}(s_{j_q}\ldots s_{j_1}T_1^p\times s_{i_p}\ldots s_{i_1}T_2^q),$$

$$\rho(T_1^n\times T_2^n)=\sum_{i=0}^{n}\partial_{i+1}\ldots\partial_n T_1^n\otimes\partial_0\ldots\partial_{i-1}T_2^n.$$

Here T_1^k is a †singular k-simplex of X_i ($i=1,2$), and $T_1^n\times T_2^n:\Delta^n\to X_1\times X_2$ is the singular n-simplex of $X_1\times X_2$ given by $(T_1^n\times T_2^n)(y)=(T_1^n(y),T_2^n(y))$ $(y\in\Delta^n)$. The s_i and ∂_i are the †degeneracy operator and the †face operator, respectively, and the summation in the first expression runs over all permutations $(i_1,\ldots,i_p,j_1,\ldots,j_q)$ of $(0,1,\ldots,p+q-1)$ such that $i_1<\ldots<i_p$ and $j_1<\ldots<j_q$; $\varepsilon=\varepsilon(i_1,\ldots,i_p,j_1,\ldots,j_q)$ is the sign of the permutation $(i_1,\ldots,i_p,j_1,\ldots,j_q)$. Consequently, given a homomorphism $\eta: G_1\otimes G_2\to G_3$, the cross product for the homology is induced by the chain mapping $\times:(S(X_1)\otimes G_1)\otimes(S(X_2)\otimes G_2)\to S(X_1\times X_2)\otimes G_3$, which is the composite of the chain mapping α' and $\nabla\otimes\eta$. Similarly, the cross product for the cohomology is induced by the cochain mapping $\times$:Hom$(S(X_1),G_1)\otimes$Hom$(S(X_2),G_2)\to$ Hom$(S(X_1\times X_2),G_3)$, which is the composite of the cochain mapping α' and Hom(ρ,η). Thus the cup product for cohomology classes is induced by the cochain mapping $\smile$: Hom$(S(X),G_1)\otimes$Hom$(S(X),G_2)\to$Hom$(S(X),G_3)$, which is the composite of the cochain mapping $\times$ in the case of $X_1=X_2=X$ and the cochain mapping $d^{\#}$: Hom$(S(X\times X),G_3)\to$ Hom$(S(X),G_3)$ induced by the diagonal mapping d. The chain mapping $\times$ defines the cross product $c_1\times c_2$ for chains c_1 and c_2, and the cochain mapping $\times$ defines the cross product $u_1\times u_2$ for cochains u_1 and u_2. Similarly, the cochain mapping $\smile$ defines the cup product $u_1\smile u_2$ for cochains u_1 and u_2. These products are compatible with the (co) chain mappings induced by continuous mappings. For example, we have $f^{\#}(u_1\smile u_2)=f^{\#}u_1\smile f^{\#}u_2$ for a continuous mapping f. For the †(co) boundary operator ∂ (δ), we have the formulas $\partial(c_1\times c_2)=\partial c_1\times c_2+(-1)^p c_1\times\partial c_2$, etc. The slant products $u/c_1, c\backslash u_1$ and the cap product $c\frown u_1$ for chains c_1, c and cochains u, u_1 can be defined similarly, and we have the formulas: $\ldots, f_{\#}c\frown u_1=f_{\#}(c\frown f^{\#}u_1)$, $\ldots,\partial(c\frown u_1)=(-1)^p(\partial c\frown u_1-c\frown\delta u_1)$. For a singular $(p+q)$-simplex T of X, a singular p-cochain u_1, and a singular q-cochain u_2, we have

$$(u_1\smile u_2)(T)=u_1(\partial_{p+1}\ldots\partial_{p+q}T)\cdot u_2(\partial_0\ldots\partial_{p-1}T),$$

$$g_1T\frown u_2=(g_1\cdot u_2(\partial_{p+1}\ldots\partial_{p+q}T))\partial_0\ldots\partial_{p-1}T,$$

where $g_1\cdot g_2=\eta(g_1\otimes g_2)$ $(g_1\in G_1, g_2\in G_2)$. These formulas coincide with the classical definitions of the cup product and cap product. If a triple (X,A_1,A_2) is †proper with respect to the integral singular homology, the cup product $w_1\smile w_2$ for cohomology classes $w_i\in H^*(X,A_i;G:)$ $(i=1,2)$ is well defined. In particular, for each pair (X,A) of a topological space X and its subspace A, we have the cohomology ring $H^*(X,A;R)$.

G. Intersection in Topological Manifolds

Let M^n be a compact connected orientable n-dimensional †topological manifold (→ 259 Manifolds). If $b\in H_n(M^n;\mathbf{Z})$ is a †fundamental homology class of M^n, then an isomorphism $D: H_p(M^n;G)\cong H^{n-p}(M^n;G)$ is defined by $b\frown D(z)=z$ $(z\in H_p(M^n;G))$ in terms of the cap product with respect to the natural homomorphism $\mathbf{Z}\otimes G\to G$ (†Poincaré-Lefschetz duality). For $z_1\in H_p(M^n;G_1)$, $z_2\in H_q(M^n;G_2)$, the **intersection** $z_1\cdot z_2\in H_{p+q-n}(M^n;G_3)$ is defined by $z_1 \cdot z_2 = z_1 \frown Dz_2$ or $D(z_1 \cdot z_2) = Dz_1 \smile Dz_z$. In particular, if $p+q=n$, the image of $z_1 \cdot z_z$ under the augmentation $\varepsilon: H_0(M^n; G) \to G$ is called the **intersection number** of z_1 and z_2 and is denoted by (z_1, z_2). We have $(z_2, z_1) = (-1)^{pq}(z_1, z_2)$, and with respect to the Kronecker index we have $(z_1, z_2) = \langle z_1, Dz_2 \rangle$. Let M^n be a †combinatorial manifold and K be a †simplicial decomposition of M^n. Denote by K' the †barycentric subdivision of K and by K^* the †dual cell complex of K. Then for a p-simplex E^p of K and a q-cell E^q of K^*, the intersection $E^p \cap E^q$ of the sets E^p and E^q is a union of $(p+q-n)$-simplexes of K'. If E^p and E^q are oriented, we consider the algebraic sum of these $(p+q-n)$-simplexes with appropriate orientations. In this manner, for a p-chain c_1 of K and a q-chain c_2 of K^*, we have a $(p+q-n)$-chain $c_1 \cdot c_2$ of K' [4, 5]. If c_1, c_2 represent z_1, z_2, respectively, then $c_1 \cdot c_2$ represents $z_1 \cdot z_2$. In particular, if $p+q=n$, the intersection number (c_1, c_2) can be defined and is equal to (z_1, z_2).

References

[1] R. Godement, Topologie algébrique et théorie des faisceaux, Actualités Sci. Ind., Hermann, 1958.
[2] P. J. Hilton and S. Wylie, Homology theory, Cambridge Univ. Press, 1960.
[3] A. Komatu, M. Nakaoka, and M. Sugawara, Isôkikagaku (Japanese; Topology) I, Iwanami, 1967.
[4] S. Lefschetz, Algebraic topology, Amer. Math. Soc. Colloq. Publ., 1942.
[5] E. H. Spanier, Algebraic topology, McGraw-Hill, 1966.

69 (II.9) Combinatorial Theory

A. Combinatorial Theory

The precise definition of a combinatorial property is not easily given, and the subject matter treated in combinatorial theory is often shared by more than one mathematical discipline. Typical in combinatorial theory is the **enumeration problem**, from which arise various problems such as finding recursive, asymptotic, and congruence relations of solutions. In many cases a problem originally conceived as combinatorial eventually turns out to be number-theoretic, algebraic, analytic, or geometric. For instance, the theory of symmetric forms, considered combinatorial by MacMahon [3], developed into the theory of †representations and †characters of †symmetric groups, owing to Frobenius's discovery of group characters. The theory concerning number of partitions, which was undoubtedly combinatorial at the outset, is now generally recognized as part of number theory. †Graph theory, originally a combinatorial study of 1-dimensional topological †complexes, has developed into an important branch of mathematics. Furthermore, the realization that a complete set of orthogonal Latin squares is equivalent to a finite †projective plane stimulated a new geometric approach. The older title of **combinatorial analysis** is now replaced by the more general "combinatorial theory." When we are given a complex-valued †number-theoretic function a_n, we associate with it the **generating function** of a_n: $f(x) = \sum a_n x^n$. In many important problems the function $f(x)$ of a complex variable x is found to represent a more or less known function, at least a †holomorphic function in a neighborhood of 0, and an analytic study of $f(x)$ provides significant information on the original sequence a_n. This was the background for the name "combinatorial analysis." In some cases, for example when we deal with complicated combinatorial structures, existence theorems are very important, and in such cases an algebraic approach is generally more desirable. Also, when we are chiefly concerned with congruence properties of a sequence, then function theory in †p-adic number fields may be used effectively.

B. The Principle of Inclusion and Exclusion

One of the most powerful techniques applied to an enumeration problem is the **principle of inclusion and exclusion**. Let Ω be a finite set and let $|A|$ denote the number of elements in a subset A of Ω. Writing simply AB in place of $A \cap B$, we have

$$|A_1 \cup A_2 \cup \ldots \cup A_n| = \sum_i |A_i| - \sum_{i<j} |A_i A_j| + \sum_{i<j<k} |A_i A_j A_k| - \ldots + (-1)^{n-1} |A_1 A_2 \ldots A_n|.$$

We may take an arbitrary set Ω and an abstract †measure μ defined over a family of subsets of Ω, under suitable conditions. Thus the †Möbius inversion formula in number theory and the Poincaré formula in probability theory, together with Fréchet's generalization of the latter dealing with the joint probability of dependent and nonexclusive events, belong to this category.

C. Generating Functions

The concept of generating function was found

very early. Let $p(n)$ denote the number of ways of dividing n similar objects into nonempty classes. This is called the **number of partitions** of n. Euler noticed that the following formula is valid for the generating function of $p(n)$:

$$1+\sum_{n=1}^{\infty} p(n)x^n =((1-x)(1-x^2)(1-x^3)\dots)^{-1},$$

whence we obtain $p(n)=p(n-1)+p(n-2)-p(n-5)-p(n-7)+p(n-12)+p(n-15)-\dots$, where the general term contains a †pentagonal number $(3k^2\pm k)/2$ as subtrahend of the argument (→ 323 Partitions of Numbers). Let $B(n)$ denote the number of ways of dividing n completely dissimilar objects into nonempty classes. These are called **Bell numbers**. For this sequence with $B(0)=1$, the formal generating function $\Sigma_{n=0}^{\infty}B(n)x^n$ is not a holomorphic function of a complex variable; the †radius of convergence of the series is 0. However, the **exponential generating function** $g(x)=\Sigma_{n=0}^{\infty}B(n)(n!)^{-1}x^n$ is convergent for all complex numbers x and is equal to e^{e^x-1}. The differential equation $g'=e^xg$ gives rise to a recursive formula

$$B(n+1)=\sum_{k=0}^{n}\binom{n}{k}B(k).$$

There is another kind of generating function for the sequence $a_n:\Sigma a_n n^{-s}$ in the form of †Dirichlet series; it is frequently used in number-theoretic problems.

D. Latin Squares

Let $\Omega=\{a_1,a_2,\dots,a_n\}$ be a set of n symbols. An $n\times n$ square array L of n^2 symbols taken out of Ω is called a **Latin square** over Ω (or Latin square of **order** n) if there is no repetition of symbols in each row and in each column of L. It is convenient to identify the row and the column index sets with Ω. Thus, we may write $z=x\circ y$ if the entry of L in the row x and column y is the symbol z. Then we can describe the Latin square L as a binary operation system $(\Omega,\circ)$, satisfying the two cancellation laws $x\circ y=x'\circ y\rightarrow x=x'$ and $x\circ y=x\circ y'\rightarrow y=y'$. On the other hand, if we take the n^2 triplets xyz formed in this way, then the set of all these triplets forms an †error-correcting code of n^2 words over the alphabetic set Ω with word length 3 and "minimum distance" 2. A Latin square L over Ω is called **reduced** (or in a standard form) if both the row a_1 and and the column a_1 consist of the natural sequence $a_1,a_2,\dots,a_n$. Thus the binary operation system $(\Omega,\circ)$ has an identity element a_1, and hence is a †quasigroup. The number of Latin squares over Ω is $n!(n-1)!$ times the number $L(n)$ of reduced Latin squares of order n. The number $L(n)$ has been calculated for $n\leqslant 8$; $L(1)=L(2)=L(3)=1$, $L(4)=4$, $L(5)=56$, $L(6)=9{,}408$, $L(7)=16{,}942{,}080$, $L(8)=535{,}281{,}401{,}856$, $L(9)=377{,}597{,}570{,}964{,}258{,}816$. Two Latin squares over the same set Ω are called **isomorphic** if one is obtained from the other by a combination of permutations of rows, columns, and the entry alphabets. The number $L^*(n)$ of nonisomorphic Latin squares of order n has been calculated for $n\leqslant 8$; $L^*(1)=L^*(2)=L^*(3)=1$, $L^*(4)=L^*(5)=2$, $L^*(6)=22$, $L^*(7)=563$, $L^*(8)=1{,}676{,}257$. The values of L and L^* for $n=8$ and 9 were calculated by using electronic computers.

E. Orthogonality of Latin Squares

Two Latin squares over Ω are **orthogonal** if every ordered pair (x,y) of elements of Ω appears in the square obtained by superimposing one on the other. If this does occur, the resulting square is called an **Euler square** over Ω. If n is odd or divisible by 4, then it is easy to construct a pair of Latin squares orthogonal to each other, whereas for $n=2$ and for $n=6$ such a pair does not exist. Euler conjectured that there is no pair of two orthogonal Latin squares of order n for $n\equiv 2 \pmod 4$. However, the conjecture was disproved in 1959 by R. C. Bose, S. S. Shrikhande, and E. T. Parker [7]. They proved that $n=2$ and $n=6$ are the sole exceptions, and for any other values of n there is a pair of orthogonal Latin squares.

F. Complete Systems of Orthogonal Latin Squares and Finite Projective Planes

A set of mutually orthogonal Latin squares of order n cannot contain more than $n-1$ Latin squares. If it contains $n-1$ Latin squares, then it is called a **complete system of orthogonal Latin squares** of order n. This is equivalent to a finite †affine plane with exactly n points on each line, and by adjoining $n+1$ points at infinity and a line at infinity, this is equivalent again to a finite †projective plane of order n (in which each line contains exactly $n+1$ points). If n is a power of a prime, then we can construct an analytic projective plane over the †finite field $GF(n)$ of n elements, and thus also a complete system of mutually orthogonal Latin squares of order n. No example of a finite projective plane of order n is known for which n is not a power of a prime. In this connection, R. H. Bruck and H. J. Ryser [9] proved that if $n\equiv 1$ or $\equiv 2 \pmod 4$ and if n contains a prime divisor $q \equiv 3 \pmod 4$ with an odd exponent, then there is no finite projective plane of order n.

G. Combinatorial Configurations

A system D of k-subsets of a given set V of v elements is called a (v,k,b,r,λ)-**configuration** or a **balanced incomplete block design** (BIBD) with the parameter set (v,k,b,r,λ) if D contains exactly b k-subsets (called **blocks**), each element of V is in exactly r blocks, and any two elements of V appear together in exactly λ blocks. Then the conditions $vr = bk$, $\lambda(v-1) = r(k-1)$ are necessary. However, these conditions are not sufficient for the existence of a corresponding configuration. A (v,k,b,r,λ)-configuration is called **symmetric** if $b = v$, $k = r$, and is also called a (v,k,λ)-**configuration**. For these configurations the quantity $n = k - \lambda$ plays an important role. In fact, if there is a (v,k,λ)-configuration and v is even, then n must be a perfect square. The corresponding condition for v odd is that the equation $n = x^2 + (-1)^{(v-1)/2}\lambda y^2$ has a rational solution x, y. This is the **Bruck-Ryser-Chowla theorem** and is based on the fact that the †Minkowski-Hasse invariants of a rational positive definite matrix are invariant under the operations of the rational transformation group of the corresponding rational quadratic forms. It is not known whether this condition is sufficient for the existence of a (v,k,λ)-configuration.

H. Hadamard Matrices

The parameters of a (v,k,λ)-configuration satisfy $n^2 + n + 1 \geqslant v \geqslant 4n - 1$. When $v = n^2 + n + 1$, we have $k = n + 1$, $\lambda = 1$, and the configuration corresponds to the finite projective plane of order n. The lowest case $v = 4n - 1$ corresponds to $k = 2n - 1$, $\lambda = n - 1$ and to a Hadamard matrix of order $4n$. A **Hadamard matrix** of order N is an $N \times N$ matrix with all its entries equal to 1 or -1 and with determinant $N^{N/2}$. Here $N = 1$, 2, or a multiple of 4. Examples of Hadamard matrices of order $\leqslant 264$ have been constructed (→ 106 Design of Experiments).

I. Difference Sets

Assume that D is a (v,k,λ)-configuration over a group V of order v and that all the blocks in D are obtained from one block B of D by a left translation in V. Thus for any element a of V different from the identity, the equation $x^{-1}y = a$ has exactly λ solutions (x,y) in B. Such a set is called a (v,k,λ)-**difference set**. Abelian difference sets have been studied extensively because their properties may be stated in terms of integers of a †cyclotomic number field. In contrast, the study of certain matrices is necessary to describe the structures of a general (v,k,λ)-configuration. All cyclic difference sets with $k \leqslant 100$ have been determined by Baumert [12]. There are 748 possible values of (v,k,λ) satisfying $(v-1)\lambda = k(k-1)$, $2 \leqslant k \leqslant v/2$, of which only 74 correspond to cyclic difference sets.

J. Graphs

A **graph** $G = (V,E)$ consists of a finite set V and a finite set E of unordered pairs of distinct elements of V. The elements of V and E are called **vertices** and **edges** respectively. If $e = \{u,v\}$, we say that u and v are **adjacent**, e **joins** u and v, and that e and $u(v)$ are **incident** with each other. Note that a graph G is a finite 1-dimensional simplicial complex (→ 73 Complexes). Two graphs G and G' are said to be **isomorphic** if they are isomorphic as simplicial complexes. Let $G = (V,E)$ and $G' = (V',E')$ be graphs. A graph $G' = G(V', E')$ is called a **subgraph** of $G = G(V,E)$ if $V' \subset V$, $E' \subset E$; and in particular if $V = V'$, then G' is called a **spanning subgraph** of G. A **pseudograph** is a triple (V,E,Φ) such that V is a finite nonempty set, E is a finite set disjoint from V, and Φ is a mapping from E into the set of all unordered pairs of V. We say that an edge e joins vertices u and v if $\Phi(e) = \{u,v\}$. Adjacency and incidence are also defined for pseudographs in the obvious manner. A **loop** is an edge which joins a vertex of V to itself, and a **multiple edge** is one of two or more edges joining the same pair of vertices which are joined already by another edge. A graph is a pseudograph which contains no loops and no multiple edges. A pseudograph can be converted into a graph by subdividing loops and multiple edges in such a manner that the adjacency of the original pseudograph is preserved. Note, however, that in many books on graph theory, the word graph actually means pseudograph (as defined here). A **directed graph** consists of a finite nonempty set V and a finite set A of the ordered pairs of distinct elements of V. The terms for pseudographs and for directed graphs can be defined in a similar way to those for ordinary graphs.

Connectedness. Let $P = \{v_0, e_1, v_1, e_2, \ldots, e_n, v_n\}$ be an alternating sequence of vertices and edges of a graph G. P is a **path** in G if e_i joins v_{i-1} and v_i for each i and $e_i \neq e_j$ $(i \neq j)$. P is a **closed path** if $v_0 = v_n$. A closed path is called a **circuit** if $v_i \neq v_j$ $(i \neq j, \{i,j\} \neq \{0,n\})$. A graph G is **connected** if any two vertices of G are joined by a path in G. A maximal connected subgraph of G is called a **component** of G. A graph G is said to be n-**connected** if every pair of distinct vertices u and v can be joined by at least n paths which have no common vertices except u and v.

Eulerian Graph and Hamiltonian Graph. A closed path in G is said to be **Eulerian** if all the edges of G appear in P; in that case G is called an **Eulerian graph**. The following proposition is generally regarded as the first theorem in the history of graph theory and was proved by Euler in 1736, when he solved the famous **Königsberg bridge problem**. A connected graph G is Eulerian if and only if the degree of each vertex of G is even, where the **degree** of a vertex v is defined as the number of edges incident with v. A circuit C of G is **Hamiltonian** provided all the vertices of G appear in C, and in that case, G is called a **Hamiltonian graph**. A simple criterion for the Hamiltonian property (analogous to the above-mentioned Euler theorem) has not yet been obtained.

Planar Graph. If $n \geqslant 3$, then any graph G can be realized as a polygonal network in the Euclidean space $\mathbf{R}^n$ in such a manner that no two edges intersect; we then say that G is **embedded** in $\mathbf{R}^n$. A graph that can be embedded in the plane is said to be **planar**. The following characterization of planar graphs is due to Kuratowski. A graph is planar if and only if no subgraph of G is combinatorially equivalent to either of the **Kuratowski graphs**, shown in Fig. 1. Note that graphs G and G' are said to be **combinatorially equivalent** (as simplicial complexes) if G and G' have isomorphic subdivisions (→ 73 Complexes).

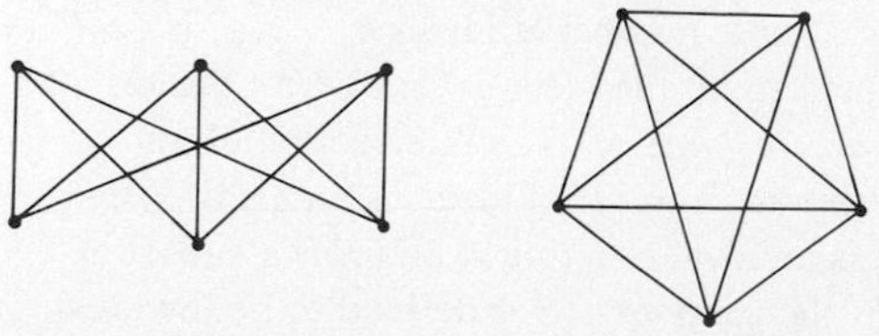

Fig. 1

Coloring Problems. G is said to be n-**colorable** if one of n colors can be assigned to each vertex of G in such a manner that different colors are assigned to any adjacent pair of vertices. Although it is easy to prove that every planar graph is 5-colorable, it is difficult to show that every planar graph is 4-colorable. This problem is, in fact, equivalent to the four-color problem (→165 Four-Color Problem).

Tree. A connected graph G containing no circuit is called a **tree**. Any connected graph contains a spanning subgraph that is a tree; such a subgraph is called a **spanning tree**. Let T be a spanning tree of the graph G. Any edge of G that does not belong to T is called a **chord** of T. Each chord corresponds to a circuit in G, and the set of all the circuits that correspond to some chord of T can be regarded as a basis for the circuits of G (→ 269 Matrices). If $G=(V,E)$ is a graph and K a subset of E, denote by $G-K$ the subgraph $(V,E-K)$ of G. If $G-K$ is not connected K is called a **disconnecting set**. A disconnecting set K is said to be a **cutset** if no proper subset of K is a disconnecting set. Each edge of a spanning tree T of G corresponds to a cutset in G, and the set of all cutsets that correspond to some edges of T can be regarded as a basis for the cutsets of G.

Matrices. Let $\{v_1, \ldots, v_m\}$ and $\{e_1, \ldots, e_n\}$ be the sets of vertices and edges, respectively, of a graph $G=(V,E)$. A graph G is called a **labelled graph** if all the vertices and edges are indexed as in this manner. The **adjacency matrix** $A=(a_{ij})$ of a labelled graph G is an $m \times m$ matrix in which $a_{ij}=1$ if the vertices v_i and v_j are adjacent, and $a_{ij}=0$ otherwise. The **incidence matrix** $B=(b_{ij})$ of a labelled graph G is an $m \times n$ matrix in which $b_{ij}=1$ if v_i is incident with e_j, and $b_{ij}=0$ otherwise. Let $\{C_1, \ldots, C_s\}$ be the set of all circuits in a labelled graph G; then the **circuit matrix** $C=(c_{ij})$ of G is an $s \times n$ matrix in which $c_{ij}=1$ if C_i contains e_j, and $c_{ij}=0$ otherwise. Let $\{K_1, \ldots, K_t\}$ be the set of all cutsets of a labelled graph G; then the **cutset matrix** $K=(k_{ij})$ of G is a $t \times n$ matrix in which $k_{ij}=1$ if K_i contains e_j, and $k_{ij}=0$ otherwise. It is easily seen that $C\,{}^tB \equiv 0 \pmod 2$ where ${}^tB$ denotes the transpose of B. If G is connected, the ranks of B, C and K are $m-1$, $n-m+1$ and $m-1$, respectively. We associate the $m \times m$ matrix $M=(d_{ij})$ with the adjacency matrix defined by $m \times m$ $d_{ij}=-a_{ij}$ if $i \neq j$ and d_{ii} is the degree of v_i. M is uniquely determined by the adjacency matrix. If G is a connected labelled graph, then all the cofactors of M are equal, and the number of spanning trees of G coincides with the common value of these cofactors. This fact is among those demonstrated in the well-known theory of electrical networks due to Kirchhoff.

Applications. Graph theory has useful applications in network flow problems (→ 278

Networks) as well as in the theory of electrical networks; various important parameters of electrical networks can be represented as functions of certain variables corresponding to the respective edges. Applications in the social sciences have also appeared in recent years. In many applications of graph theory, the enumeration of all subgraphs having a certain property often plays an important role. Thus, certain aspects of graph theory are intimately related with many combinatorial problems.

References

[1] M. Hall, Combinatorial theory, Blaisdell, 1967.
[2] H. B. Mann, Addition theorems, Interscience, 1965.
[3] P. A. MacMahon, Combinatorial analysis I, II, Cambridge Univ. Press, 1915, 1916 (Chelsea, 1960).
[4] J. Riordan, An introduction to combinatorial analysis, John Wiley, 1958.
[5] H. J. Ryser, Combinatorial mathematics, John Wiley, 1963.
[6] L. D. Baumert, S. W. Golomb, and M. Hall, Discovery of an Hadamard matrix of order 92, Bull. Amer. Math. Soc., 68 (1962), 237–238.
[7] R. C. Bose and S. S. Shrikhande, On the construction of sets of mutually orthogonal Latin squares and the falsity of a conjecture of Euler, Trans. Amer. Math. Soc., 95 (1960), 191–209.
[8] J. W. Brown, Enumeration of Latin squares with application to order 8, J. Combinat. Theory, 5 (1968), 177–184.
[9] R. H. Bruck and H. J. Ryser, The nonexistence of certain finite projective planes, Canad. J. Math., 1 (1949), 88–93.
[10] S. D. Chowla and H. J. Ryser, Combinatorial problems, Canad. J. Math., 2 (1950), 93–99.
[11] K. Yamamoto, Jacobi sums and difference sets, J. Combinat. Theory, 3 (1967), 146–181.
[12] L. D. Baumert, Cyclic difference sets, Lecture notes in math. 82, Springer, 1971.
[13] W. D. Wallis, A. P. Street, and J. S. Wallis, Combinatorics: Room squares, sum-free sets, Hadamard matrices, Lecture notes in math. 292, Springer, 1972.
[14] J. H. van Lint, Combinatorial theory seminar, Eindhoven Univ. of Tech., Lecture notes in math. 382, Springer, 1974.
[15] C. Berge, The theory of graphs and its applications, John Wiley, 1962.
[16] G. Busacker and T. L. Saaty, Finite graphs and networks, McGraw-Hill, 1965.
[17] F. Harary, Graph theory, Addison-Wesley, 1969.
[18] O. Ore, Theory of graphs, Amer. Math. Soc., 1962.

70 (III.14) Commutative Rings

A. General Remarks

A †ring R (→ 363 Rings) whose multiplication is commutative is called a **commutative ring**. Throughout this article, we mean by a ring a commutative ring with unity element.

B. Ideals

Since our rings are commutative, we need not distinguish right or left ideals from †ideals. A subset $\mathfrak{a}$ of a ring R is an ideal of R if and only if $\mathfrak{a}$ is an R-submodule of R (→ 275 Modules), if and only if $\mathfrak{a}$ is the †kernel of a ring homomorphism from R into some ring (except for the case $\mathfrak{a}=R$). Given an ideal $\mathfrak{a}$ of a ring R, the set of elements which are †nilpotent modulo $\mathfrak{a}$, i.e., $\{x\in R \mid x^n\in\mathfrak{a}\ (\exists n)\}$, is called the **radical** of $\mathfrak{a}$ and is often denoted by $\sqrt{\mathfrak{a}}$. The radical of the zero ideal is called the **radical** of R, or more precisely, the †**nilradical** of R (→ 363 Rings H).

For a subset S of a ring R, the smallest ideal $\mathfrak{a}$ containing S is called the ideal **generated** by S, and S is called a **basis** for $\mathfrak{a}$; if S is a finite set, then S is called a **finite basis**. When $\mathfrak{a}_\lambda$ $(\lambda\in\Lambda)$ are ideals of R, the **sum** of these ideals, denoted by $\Sigma\mathfrak{a}_\lambda$, is defined to be the ideal generated by the union of the ideals $\mathfrak{a}_\lambda$ $(\lambda\in\Lambda)$. If Λ is a finite set, say $\{1,2,\ldots,n\}$, then the sum is denoted also by $\mathfrak{a}_1+\ldots+\mathfrak{a}_n$, and in this case, the sum is also called a finite sum. Note that $\mathfrak{a}_1+\ldots+\mathfrak{a}_n=\{a_1+\ldots+a_n \mid a_i\in\mathfrak{a}_i\}$. The **product** $\mathfrak{a}_1,\ldots,\mathfrak{a}_n$ of a finite number of ideals $\mathfrak{a}_1,\ldots,\mathfrak{a}_n$ is defined to be the ideal generated by the set $\{a_1\ldots a_n \mid a_i\in\mathfrak{a}_i\}$. The intersection of an arbitrary number of ideals is an ideal. When $\mathfrak{a}$ is an ideal of a ring R and S is a subset of R, the **quotient** $\mathfrak{a}:S$ is defined to be the ideal $\{x\in R \mid xS\subset\mathfrak{a}\}$. If $\mathfrak{a},\mathfrak{b},\mathfrak{c},\mathfrak{d}_\lambda$ $(\lambda\in\Lambda)$ are ideals, we have $(\mathfrak{a}:\mathfrak{b}):\mathfrak{c}=\mathfrak{a}:\mathfrak{b}\mathfrak{c}$, $\mathfrak{a}:\Sigma\mathfrak{d}_\lambda=\bigcap_\lambda(\mathfrak{a}:\mathfrak{d}_\lambda)$.

C. Prime Ideals

An ideal $\mathfrak{p}$ of a ring R is called a **prime ideal** if $R/\mathfrak{p}$ is an †integral domain; $\mathfrak{p}$ is a prime ideal if and only if $\mathfrak{p}\neq R$ and also $ab\in\mathfrak{p}$ $(a,b\in R)$ implies $a\in\mathfrak{p}$ or $b\in\mathfrak{p}$ (some literature includes the ring R itself in the set of prime ideals). Let S be a **multiplicatively closed subset** of a ring R, i.e., a nonempty †subsemigroup of R with respect to multiplication. A maximal member among the set of ideals which do not meet S is called a **maximal ideal with respect to** S. Such a member is necessarily a prime ideal; when $S=\{1\}$, it is called a **maximal ideal** of R. An ideal $\mathfrak{m}$ is a maximal ideal of R if and only if $R/\mathfrak{m}$ is a field.

D. The Jacobson Radical

The intersection J of all maximal ideals of a ring R is called the **Jacobson radical** of R; in some cases, this intersection J is called the radical of R (→ 363 Rings H). Let N be an R-submodule of a finite R-module M. If $MJ+N=M$, then $M=N$ (**Krull-Azumaya lemma** or **Nakayama lemma**).

E. Krull Dimension

For a prime ideal $\mathfrak{p}$ of a ring R, the maximum of the lengths n of †descending chains of prime ideals $\mathfrak{p}=\mathfrak{p}_0 \supsetneqq \mathfrak{p}_1 \supsetneqq \cdots \supsetneqq \mathfrak{p}_n$ which begin with $\mathfrak{p}$ (or ∞ if the maximum does not exist) is called the **height** or **rank** of the prime ideal $\mathfrak{p}$. For an ideal $\mathfrak{a}$, the minimum of the heights of prime ideals containing $\mathfrak{a}$ is called the **height** of the ideal $\mathfrak{a}$. The maximum of heights of prime ideals of R is called the **Krull dimension** (or **altitude**) of the ring R. For an ideal $\mathfrak{a}$ of R, the Krull dimension of $R/\mathfrak{a}$ is called the **Krull dimension** (or **depth**) of the ideal $\mathfrak{a}$. (The meanings of the terms *rank*, *dimension*, and *depth* now depend on the writings in which they are found; standardization of definition of these terms is becoming more and more of a necessity.)

F. Primary Ideals

For an ideal $\mathfrak{q}$ of R, if $\mathfrak{q}\neq R$ and every zero divisor of $R/\mathfrak{q}$ is nilpotent, then $\mathfrak{q}$ is called a **primary ideal**. (If R is included in the set of prime ideals, then R is regarded as a primary ideal.) In this case, $\mathfrak{p}=\sqrt{\mathfrak{q}}$ is a prime ideal, and $\mathfrak{q}$ is then said to **belong** to $\mathfrak{p}$ or to be a **$\mathfrak{p}$-primary ideal**. The intersection of a finite number of primary ideals belonging to the same prime ideal $\mathfrak{p}$ is $\mathfrak{p}$-primary. Assume that an ideal $\mathfrak{a}$ is expressed as the intersection $\mathfrak{a}=\mathfrak{q}_1\cap\ldots\cap\mathfrak{q}_n$ of a finite number of primary ideals $\mathfrak{q}_1,\ldots,\mathfrak{q}_n$. If this intersection is **irredundant**, that is, if none of the $\mathfrak{q}_i$ is superfluous in the expression of $\mathfrak{a}$, then the set of $\sqrt{\mathfrak{q}_i}$ ($i=1,\ldots,n$) is uniquely determined by $\mathfrak{a}$. The $\sqrt{\mathfrak{q}_i}$ are called **prime divisors** (or **associated prime ideals**) of the ideal $\mathfrak{a}$; a minimal one among these is called a **minimal** (or **isolated**) **prime divisor** of the ideal $\mathfrak{a}$ and those which are not minimal are called **embedded prime divisors** of the ideal $\mathfrak{a}$. A maximal one among the prime divisors of $\mathfrak{a}$ is called a **maximal prime divisor** of the ideal $\mathfrak{a}$. (For definitions of these concepts in the case where $\mathfrak{a}$ is an arbitrary ideal, see [4].) If, in the expression of $\mathfrak{a}$ as above, n is the smallest occurring in similar expressions, then the expression is called the **shortest representation** of the ideal $\mathfrak{a}$ by primary ideals. In this case, each $\mathfrak{q}_i$ is called a **primary component** of the ideal $\mathfrak{a}$; if $\sqrt{\mathfrak{q}_i}$ is isolated, then $\mathfrak{q}_i$ is called an **isolated primary component** of the ideal $\mathfrak{a}$, and otherwise $\mathfrak{q}_i$ is called an **embedded primary component** of the ideal $\mathfrak{a}$. Isolated primary components are uniquely determined by $\mathfrak{a}$, but embedded primary components are not.

G. Rings of Quotients

Let R be a ring. Then the set U of elements of R which are not zero divisors is multiplicatively closed. In the set $R\times U=\{(r,u)\mid r\in R, u\in U\}$, we define a relation $\equiv$ by $(r,u)\equiv(r',u')\Leftrightarrow ru'=r'u$. Then $\equiv$ is an equivalence relation, and the equivalence class of (r,u) is denoted by r/u. In the set Q of these r/u, addition and multiplication are defined by $r/u+r'/u'=(ru'+r'u)/uu', (r/u)(r'/u')=rr'/uu'$. Then Q becomes a ring, and $r/1$ can be identified with r. Thus Q is a ring containing R, generated by R and inverses of elements of U. This property characterizes Q, which is called the **ring of total quotients** of the ring R. If R is an integral domain, then Q is a field, called the **field of quotients** of the integral domain R. Let S be a multiplicatively closed subset of R such that $0\notin S$, and let $\mathfrak{n}$ be $\{x\in R\mid xs=0\ (\exists s\in S)\}$ and φ be the natural homomorphism $R\rightarrow R/\mathfrak{n}$. Then none of the elements of $\varphi(S)$ is a zero divisor. The subring of the ring of total quotients of $R/\mathfrak{n}$, generated by $R/\mathfrak{n}$ and inverses of elements of $\varphi(S)$, is called the **ring of quotients of the ring R with respect to S**, and is denoted by R_S ($R[S^{-1}]$ or RS^{-1}). When M is an R-module, $M\otimes_R R_S$ is called the **module of quotients of the R-module M with respect to S**. It is significant that R_S is †R-flat. There is a one-to-one correspondence between the set of primary ideals $\mathfrak{q}$ of R which do not meet S and the set of primary ideals $\mathfrak{Q}$ of R_S such that $\mathfrak{q}$ corresponds to $\mathfrak{Q}$ if and only if $\mathfrak{Q}=\mathfrak{q}R_S$ ($\mathfrak{q}=\mathfrak{Q}\cap R$). When $\mathfrak{p}$ is a prime ideal of R, the complement $R-\mathfrak{p}$ is multiplicatively closed, and $R_{R-\mathfrak{p}}$ is called the **local ring of**

$\mathfrak{p}$ or the **ring of quotients of the ring R with respect to the prime ideal** $\mathfrak{p}$, and is denoted by $R_{\mathfrak{p}}$ (→ 281 Noetherian Rings C, D). A ring of quotients is also called a **ring of fractions**.

H. Divisibility

In a ring R, if $a=bc$ $(a,b,c \in R)$, then we say that b is a **divisor** (or **factor**) of a, and that a is a **multiple** of b, or a is **divisible** by b. We denote this by $b|a$. This relation between $a,b \in R$ is called **divisibility relation** in R. If, in this situation, c has its inverse in R, we say that a is an **associate** of b. A factor b of a is called a **proper factor** if b is neither an associate of a nor †invertible. An element which has no proper factor is called an **irreducible element**. A nonzero element which generates a prime ideal is called a **prime element**.

If in an integral domain R every nonzero element is a product of prime elements (up to invertible factors), then we say that the **unique factorization theorem** holds in R, and that R is a **unique factorization domain** (or simply u.f.d.).

Let $A=\{a_1,\ldots,a_n\}$ be a set of nonzero elements of a ring R. A **common divisor** of A is an element which is a factor of a_i. A **common multiple** of A is defined similarly. The **greatest common divisor** (G.C.D.) of A is a common divisor which is a multiple of any common divisor; the **least common multiple** (L.C.M.) of A is a common multiple m which is a factor of any common multiple. Thus, the G.C.D. and L.C.M. exist if R is a u.f.d.

I. Integral Dependence

Let R be a subring of a ring R'' sharing a unity element with R''. An element $a \in R''$ is said to be **integral** (or **integrally dependent**) over R if there are a natural number n and elements c_i of R such that $a^n+c_1a^{n-1}+\ldots+c_n=0$. If every element of a subset S of R'' is integral over R, we say that S is **integral** over R. (When R has no unity element, a similar definition is given under an additional condition that $c_i \in R^i$. An important special case is where R is an ideal. See D. G. Northcott and D. Rees, *Proc. Cambridge Philos. Soc.*, 50 (1954); M. Nagata, *Mem. Coll. Sci. Univ. Kyôto*, 30 (1956).) The set $\tilde{R}$ of elements of R'' which are integral over R is a ring and is called the **integral closure** of R in R''. If $\tilde{R}=R$, then R is said to be **integrally closed in** R''. If R is integrally closed in its ring of total quotients, we say that R is **integrally closed**. An integrally closed integral domain is called a **normal ring**. (In some literature, an integrally closed ring is called a normal ring.) An element $a \in R''$ is called **almost integral** over R if there is an element b of R such that b is not a zero divisor and $a^nb \in R$ for every natural number n. If an element a of the ring of total quotients of R is integral over R, then a is almost integral over R. R is said to be **completely integrally closed** if its ring of total quotients contains no elements which are almost integral over R except the elements of R itself.

J. The Group Theorem

Let Q be the ring of total quotients of a ring R. An R-submodule $\mathfrak{a}$ of Q is called a **fractional ideal** of R if there is a non-zero-divisor c of R such that $c\mathfrak{a} \subset R$. The product of fractional ideals is defined similarly as in the case of products of ideals. The inverse $\mathfrak{a}^{-1}$ of the fractional ideal $\mathfrak{a}$ is defined to be $\{x \in Q \mid x\mathfrak{a} \subset R\}$. If $\mathfrak{a}$ contains an element which is not a zero divisor, then $\mathfrak{a}^{-1}$ is also a fractional ideal. When R is completely integrally closed, we define fractional ideals $\mathfrak{a}$ and $\mathfrak{b}$ to be equivalent if $\mathfrak{a}^{-1}=\mathfrak{b}^{-1}$. This gives rise to an equivalence relation between fractional ideals. The set of equivalence classes of fractional ideals which contain non-zero-divisors forms a group. This result is called the **group theorem**.

An integral domain R is called a **Krull ring** if (i) for every prime ideal $\mathfrak{p}$ of height 1, the ring $R_{\mathfrak{p}}$ is a †discrete valuation ring; (ii) R is the intersection of all of the valuation rings $R_{\mathfrak{p}}$ and (iii) every nonzero element a of R is contained in only a finite number of prime ideals of height 1. In a Krull ring R, for an arbitrary nonzero fractional ideal $\mathfrak{a}$, there is a uniquely determined product of powers of prime ideals of height 1 which is equivalent to $\mathfrak{a}$ in the sense stated above (→ 425 Valuations).

K. Dedekind Domains and Principal Ideal Domains

A ring R is called a **Dedekind domain** if (i) R is a †Noetherian integral domain, (ii) R is a normal ring, and (iii) the Krull dimension of R is 1. For an integral domain R which is not a field, R is a Dedekind domain if and only if the set of all nonzero fractional ideals is a group, if and only if every nonzero ideal of R is expressed as the product of a finite number of prime ideals and such expression is unique up to the order of prime factors. An important example of a Dedekind domain is the ring of all †algebraic integers, i.e., the †principal order of an †algebraic number field of finite degree. In general, if R is a Dedekind domain with field of quotients K and L is a finite algebraic extension of K, then the integral closure of R in L and any ring R' such that $R \subset R' \subsetneqq K$ are Dedekind domains.

An ideal generated by an element is called a **principal ideal**; a fractional ideal generated by an element is called a **principal fractional ideal** (or simply **principal ideal**). In a Dedekind domain, the set P of nonzero principal fractional ideals is a subgroup of the group I of nonzero fractional ideals; I/P is called the **ideal class group** of R, and a member of it is called an **ideal class**. The †order of I/P is called the **class number** of R (→ 16 Algebraic Number Fields). There are many Dedekind domains whose class numbers are infinite.

A ring R is called a **principal ideal ring** if every ideal is principal; furthermore, if R is an integral domain, then R is called a **principal ideal domain**. A principal ideal ring is the direct sum of a finite number of rings of which each direct summand is either a principal ideal domain or a †local ring whose maximal ideal is a principal nilpotent ideal. A principal ideal domain which is not a field is a Dedekind domain and a u.f.d. We consider, for an arbitrary natural number n and a principal ideal ring R, the set $M(n,R)$ of all $n \times n$ matrices over R. Given an element A of $M(n,R)$, there exist elements X, Y in $M(n,R)$ such that (i) X^{-1}, Y^{-1} are in $M(n,R)$, and (ii) denoting by b_{ij} the (i,j)-entry of XAY, we have $b_{11}R \supset b_{22}R \supset \ldots \supset b_{nn}R$ and $b_{ij}=0$ if $i \neq j$. The nonzero members of the set $\{b_{11}, b_{22}, \ldots, b_{nn}\}$ are called the **elementary divisors** of the matrix A. Applying this to a finite module M over the principal ideal ring R, we see that M is the direct sum of $m_1R, \ldots, m_rR$ $(m_i \in M)$ such that, with $\mathfrak{a}_i = \{x \in R \mid m_i x = 0\}$, we have $\mathfrak{a}_1 \subset \mathfrak{a}_2 \subset \ldots \subset \mathfrak{a}_r$ (→ 2 Abelian Groups B; 269 Matrices E).

References

[1] W. Krull, Idealtheorie, Erg. Math., Springer, second edition, 1968.

[2] B. L. van der Waerden, Algebra I, II, Springer, 1966, 1967.

[3] O. Zariski and P. Samuel, Commutative algebra I, II, Van Nostrand, 1958, 1960.

[4] M. Nagata, Local rings, Interscience, 1962.

[5] N. Bourbaki, Eléments de mathématique, Algèbre, ch. 1, 8, Actualités Sci. Ind., Hermann, 1144b, 1964; 1261a, 1958.

[6] N. Bourbaki, Eléments de mathématique, Algèbre commutative, ch. 1–7, Actualités Sci. Ind., Hermann, ch. 1, 2, 1290a, 1961; ch. 3, 4, 1293a, 1967; ch. 5, 6, 1308, 1964; ch. 7, 1314, 1965.

[7] A. Grothendieck, Eléments de géometrie algébrique, Publ. Math. Inst. HES, no. 4, 1960.

[8] D. G. Northcott, Lessons on rings, modules and multiplicities, Cambridge Univ. Press, 1968.

71 (IV.8) Compact Groups

A. Compact Groups

A †topological group G is called a **compact group** if the underlying topological space of G is a †compact Hausdorff space. The †torus group $\mathbf{T}^n = \mathbf{R}^n/\mathbf{Z}^n$ $(n=1,2,\ldots)$ (commutative group), the †orthogonal group $O(n)$, the †unitary group $U(n)$, the †symplectic group $Sp(n)$, and the additive group $\mathbf{Z}_p$ of †p-adic integers are compact groups (→ 63 Classical Groups; for other compact Lie groups, → 248 Lie Groups, 247 Lie Algebras). Let $C(G)$ be the †linear space formed by all complex-valued continuous functions $f, g, h, \ldots$ defined on a compact group G; $C(G)$ is a †Banach space with the norm

$$\|f\| = \sup_{x \in G} |f(x)|.$$

Since a compact group G is †locally compact, there exists a right-invariant †Haar measure on G. Because of the compactness of G, the total measure of G is finite and the measure is also left-invariant. By the condition that the total measure is 1, such measure is uniquely determined. The integral of f in $C(G)$ relative to this measure is called the **mean value** of f. Since for $f, g \in C(G)$, $f(xy^{-1})g(y)$ is continuous in two variables x, y, the †convolution $f \times g(x) = \int f(xy^{-1})g(y)\,dy$ belongs also to $C(G)$. $C(G)$ constitutes a ring under the multiplication defined by the convolution. This ring can be considered as an extension of the notion of †group ring for finite groups; it is called the **group ring** of the compact group G. The function $x \to f(x^{-1})$ will be denoted by f^*, and the inner product of the †function space $L_2(G)$ will be written as (f,g).

B. Representations of Compact Groups

Let $G(E)$ be the group of units of †bounded linear operators on a Banach space E, and suppose that we have a homomorphism U of a topological group G into $G(E)$. The homomorphism U is called a **strongly (weakly) continuous representation** on E of G if, for any $a \in E$, the mapping $x \to U(x)a$ of G into E is continuous with respect to the †strong (†weak) topology on E. In the case where E is a †Hilbert space, a strongly continuous representation U such that every $U(a)$ is a unitary operator is called a †unitary representation. Let U be any strongly continuous representation of a compact group G on a Hilbert space E, and for a,b in E, let $\langle a,b \rangle$ be the mean value of the †inner product $(U(x)a, U(x)b)$ on E. Then U is a unitary representation on the Hilbert space E with the new inner product $\langle a,b \rangle$.

A representation U of G on a Banach space E is said to be **irreducible** if E contains no closed subspace other than $\{0\}$ and E, which is invariant under every $U(x)$ $(x \in G)$. If a weakly continuous representation of a compact group on a Banach space E is irreducible, then E is finite-dimensional. Moreover, any unitary representation U of a compact group on a Hilbert space E can be decomposed into a discrete †direct sum of irreducible representations. Namely, there exists a family $\{E_\alpha\}_{\alpha \in A}$ of irreducible (hence finite-dimensional) invariant subspaces E_α of E which are orthogonal to each other such that $E = \sum_{\alpha \in A} E_\alpha$. In particular, any continuous representation of a compact group on a finite-dimensional space is †completely reducible. In $L_2(G)$ with respect to the †Haar measure on a †locally compact group G, the representation U defined by $(U(x)f)(y) = f(yx)(x,y \in G, f \in L_2(G))$ is a unitary representation of G. This representation U is called the (right) **regular representation** of G. Decomposition of the regular representation of a compact group G into irreducible representations is given by the **Peter-Weyl theory**, described later.

In the rest of this article, the representations under consideration will be (continuous) representations by matrices of finite degree. Let $D_1(x) = (d_{ij}^{(1)}(x))$ and $D_2(x) = (d_{ij}^{(2)}(x))$ be irreducible unitary representations which are not mutually †equivalent. Then from †Schur's lemma follow the orthogonality relations $(d_{ij}^{(1)}, d_{kl}^{(2)}) = 0$ and $(\sqrt{n_1}\, d_{ij}^{(1)}, \sqrt{n_1}\, d_{mn}^{(1)}) = \delta_{im}\delta_{jn}$ (where n_1 is the degree of the representation D_1). From each †class D_α of irreducible representations of G, choose a unitary representative $D_\alpha(x) = (d_{ij}^\alpha(x))$, and let n_α denote its degree. Then the previous fact means that the $\sqrt{n_\alpha}\, d_{ij}^\alpha(x)$ form an †orthonormal system of $L_2(G)$.

Let $h \in C(G)$ and consider the mapping $H: f \to h \times f$ of $C(G)$ to $C(G)$. Then H is a †compact operator in $C(G)$. Since $C(G)$ is contained in $L_2(G)$, we can define the inner product in $C(G)$. By $(h \times f, g) = (f, h^* \times g)$, $h = h^*$ implies $(Hf, g) = (f, Hg)$; that is, H is a †Hermitian operator. For a given f in $C(G)$, there exists h $(= h^*)$ in $C(G)$ such that $h \times f$ is uniformly arbitrarily near f. From the theory of compact Hermitian operators, Hf can be uniformly approximated by linear combinations of the †eigenfunctions of H. Since the †eigenspace of H is finite-dimensional and invariant under $U(a)$, the eigenfunctions of H are linear combinations of a finite number of $\sqrt{n_\alpha}\, d_{ij}^\alpha(x)$. Hence any function f in $C(G)$ can be uniformly approximated by linear combinations of a finite number of $\sqrt{n_\alpha}\, d_{ij}^\alpha(x)$. This fact, like the similar result in †Fourier series, is called the **approximation theorem**. From this, it follows that the orthogonal system $\{\sqrt{n_\alpha}\, d_{ij}^\alpha(x)\}$ is †complete; i.e., if an element of $C(G)$ is orthogonal to each element in this system, then it is 0.

Since $C(G)$ is dense in $L_2(G)$, $\{\sqrt{n_\alpha}\, d_{ij}^\alpha(x)\}$ is a †complete orthonormal system of the Hilbert space $L_2(G)$. Hence for any $f \in L_2(G)$, its "Fourier series" $\sum_\alpha \sum_{ij} c_{ij}^\alpha \sqrt{n_\alpha}\, d_{ij}^\alpha(x)$ (where $c_{ij} = (f, \sqrt{n_\alpha}\, d_{ij}^\alpha)$) converges to f in the mean of order 2 (i.e., by the †metric of $L_2(G)$). In particular, if G is a compact †Lie group and f is sufficiently many times differentiable, then this series converges uniformly to f.

The space V_i^α of dimension n_α spanned by the elements $d_{ij}^\alpha(x)$ $(1 \leqslant j \leqslant n_\alpha)$ in the ith row of the matrix $D_\alpha(x)$ is invariant under the right regular representation U. The representation on V_i^α given by U can only be D_α. Then the fact that $\{\sqrt{n_\alpha}\, d_{ij}^\alpha(x)\}$ is an orthonormal system of $L_2(G)$ means that the regular representation U of a compact group G is decomposable into a discrete direct sum of finite-dimensional irreducible representations. Each irreducible representation D_α is contained in U with multiplicity equal to its degree n_α.

If a function $\varphi(x)$ in $C(G)$ satisfies $\varphi(y^{-1}xy) = \varphi(x)$ for any x,y, then it is called a **class function**. The set $K(G)$ of all continuous class functions coincides with the †center of the group ring $C(G)$. The †character of an irreducible representation of G is a class function, and the set $\{\chi_\alpha(x)\}$ of all characters plays the same role as the orthogonal system $\{\sqrt{n_\alpha}\, d_{ij}^\alpha(x)\}$ in $C(G)$. Namely, $\{\chi_\alpha(x)\}$ is a complete orthonormal system in (the †completion of) $K(G)$, and any class function can be uniformly approximated by linear combinations of a finite number of these characters.

The preceding paragraphs are a brief description of the Peter-Weyl theory. If G is the 1-dimensional †torus group $\mathbf{T}^1 = \mathbf{R}/\mathbf{Z}$, namely, the compact group of real numbers mod 1, then this is actually the theory of Fourier series concerning periodic functions on the line. (For concrete irreducible representations of $O(n)$, $U(n)$, $Sp(n)$, and formulas for characters, → 63 Classical Groups. For representations of compact Lie groups → 248 Lie Groups, 247 Lie Algebras.) The theory of compact groups was completed by F. Peter and H. Weyl (*Math. Ann.*, 97 (1927)), and J. von Neumann's theory concerning almost periodic functions in a group (1934) united the theory of compact groups with H. Bohr's theory of almost periodic functions (→ 20 Almost Periodic Functions).

C. Structure of Compact Groups

Let a be an element of a compact group G different from e. Since the underlying space of a topological group is a †completely regular space, there exists a function in $C(G)$ such that $f(a) \neq f(e)$. Hence there exists a representation $D(x)$ of G such that $D(a)$ is not equal to the unit matrix. This means that any compact group G can be expressed as a †projective limit group of compact Lie groups. Beginning with this fact, Von Neumann (1933) showed that a †locally Euclidean compact group is a Lie group (→ 406 Topological Groups N).

D. Set of Representations

The set $G' = \{D\}$ of representations of G by matrices admits the following operations: (i) $D_1 \otimes D_2$ (†tensor product representation); (ii) $D_1 \oplus D_2 = \begin{pmatrix} D_1 & 0 \\ 0 & D_2 \end{pmatrix}$ (†direct sum representation); (iii) $P^{-1}DP$ (equivalent representation); and (iv) $\overline{D}$ (complex conjugate representation). Let M be a subset of G' such that $\overline{D} \in M$ whenever $D \in M$ and the irreducible components of $D_1 \otimes D_2$ are in M whenever $D_1, D_2 \in M$. Then M is called a **module of representations** of G. There is a one-to-one correspondence between closed normal subgroups H of G and modules M formed by all the representations of G/H.

A representation of G' is a correspondence which assigns to each D a matrix $A(D)$ of the same degree as that of D and preserves the operations of G': $A(D_1 \otimes D_2) = A(D_1) \otimes A(D_2)$, $A(D_1 \oplus D_2) = A(D_1) \oplus A(D_2)$, $A(P^{-1}DP) = P^{-1}A(D)P$, and $A(\overline{D}) = \overline{A(D)}$. Let G'' be the set of all representations of G'. Define the product of $A_1, A_2 \in G''$ by $A_1A_2(D) = A_1(D)A_2(D)$ and a topology on G'' by the †weak topology of the functions $A(D)$ of D. Namely, a typical neighborhood of A_0 is of the form $U(A_0; D_1, \ldots, D_s; \varepsilon) = \{A \mid \|A(D_i) - A_0(D_i)\| < \varepsilon, i = 1, \ldots, s\}$. G'' is a topological group under this multiplication and topology. Then the **Tannaka duality theorem** states that $G'' \cong G$ holds (T. Tannaka, *Tôhoku Math. J.*, 45 (1939)). Let $R(G)$ be the †algebra over the complex number field **C** formed by the set of all linear combinations of a finite number of $d_{ij}^{\alpha}(x)$, and let $\mathrm{Aut}\,R(G)$ be the automorphism group of this algebra $R(G)$. Let G^* be the set of all elements σ in $\mathrm{Aut}\,R(G)$ which commute with every left translation $L(x)$ $((L(x)f)(y) = f(xy))$ and which satisfy $\sigma(\overline{f}) = \overline{\sigma(f)}$. Then G^* is a topological group with respect to the weak topology, and the Tannaka duality theorem means that the correspondence which assigns to each $x \in G$ the right translation $R(x)$ (the restriction of $U(x)$ to $R(G)$) is an isomorphism of G onto G^* as topological groups. In the case where G is a compact Lie group, C. Chevalley restated the Tannaka duality theorem as a theorem giving a relation between compact Lie groups and complex algebraic groups (→ 248 Lie Groups U).

References

See references to 20 Almost Periodic Functions and 406 Topological Groups.

For compact Lie groups → references to 248 Lie Groups.

72 (XII.9) Compact Operators

A. General Remarks

A †linear operator T from a †Banach space X to a Banach space Y is said to be **compact** (or **completely continuous**) if T maps any †bounded set of X to a †relatively compact set of Y. In other words, T is compact if, for any bounded sequence $\{x_n\}$ in X, the sequence $\{Tx_n\}$ in Y contains a †strongly convergent subsequence. A compact operator is necessarily †bounded and hence continuous. A compact operator from X to Y maps any †weakly convergent sequence in X to a strongly convergent sequence in Y. If X is †reflexive, the converse is also true.

In this article the set of all bounded (resp. compact) linear operators from X to Y are denoted by $\mathbf{B}(X, Y)$ (resp. $\mathbf{B}^{(c)}(X, Y)$).

B. Examples of Compact Operators

(1) **Degenerate operators**. An operator $T \in \mathbf{B}(X, Y)$ is said to be degenerate if the †range $R(T)$ of T is finite-dimensional. A degenerate operator is necessarily compact. The †identity operator in X is compact $(Y = X)$ if and only

if X is finite-dimensional. (2) †**Integral operators with continuous kernel**. Let E and F be bounded closed regions in the Euclidean spaces $\mathbf{R}^m$ and $\mathbf{R}^n$, respectively, and let $k = k(t,s)$, $t \in F$, $s \in E$, be a continuous function defined on $F \times E$. Consider the integral operator T with †kernel k, that is, the operator given by

$$(Tx)(t) = \int_E k(t,s)x(s)\,ds, \qquad t \in F.$$

Then T determines a compact operator from the †function space $C(E)$ to $C(F)$. (For the notation for various function spaces, → 173 Function Spaces.) (3) **Integral operators of Hilbert-Schmidt type**. In example (2) let E and F be †Lebesgue measurable sets and let $k \in L_2(F \times E)$. Then the integral operator T determines a compact operator from the †Hilbert space $L_2(E)$ to the Hilbert space $L_2(F)$. (4) Let Ω be a bounded open set in $\mathbf{R}^n$, p a natural number, and $H^p(\Omega)$ $(= W_2^p(\Omega))$ the †Sobolev space of order p (→ 173 Function Spaces). Then the natural injection from $H^p(\Omega)$ into $H^{p-1}(\Omega)$ is compact.

C. Properties of Compact Operators

Any linear combination of compact operators is again compact. If a sequence $\{T_n\}$ of compact operators converges in the norm (i.e., in the †uniform operator topology), then the limit T is compact. Thus, $\mathbf{B}^{(c)}(X, Y)$ is a closed subspace of the Banach space $\mathbf{B}(X, Y)$ with the †operator norm. Any product of a compact operator and a bounded operator is compact. Namely, $A \in \mathbf{B}^{(c)}(X, Y)$, $B \in \mathbf{B}(Y, Z)$, and $C \in \mathbf{B}(Z,X)$ imply $BA \in \mathbf{B}^{(c)}(X, Z)$ and $AC \in \mathbf{B}^{(c)}(Z, Y)$. In particular, $\mathbf{B}^{(c)}(X) = \mathbf{B}^{(c)}(X,X)$ forms a closed †two-sided ideal of $\mathbf{B}(X) = \mathbf{B}(X,X)$. An operator $T \in \mathbf{B}(X, Y)$ is compact if and only if its †dual operator $T' \in \mathbf{B}(Y', X')$ is compact. The range of a compact operator is always †separable. Let X and Y be Hilbert spaces. Then for any $T \in \mathbf{B}^{(c)}(X, Y)$ there exist †orthonormal sets $\{\varphi_n\}$ in X and $\{\psi_n\}$ in Y and a sequence $\{c_n\}$ of nonnegative numbers with $\lim c_n = 0$ such that

$$Tu = \sum_n c_n (u, \varphi_n)\psi_n, \quad u \in X.$$

Consequently, any compact operator between Hilbert spaces X, Y can be approximated by a sequence of degenerate operators. This statement remains true if Y or the dual X' is a Banach space having a †base. However, there are Banach spaces X and Y for which the statement is no longer true [13, 14].

D. The Riesz-Schauder Theorem

Let $T \in \mathbf{B}^{(c)}(X)$ and consider a pair of linear equations

$$u - Tu = f, \qquad f \in X, \tag{1}$$

$$\varphi - T'\varphi = g, \qquad g \in X', \tag{2}$$

where $T' \in \mathbf{B}(X')$ is the dual operator of T in the †dual space X' of X. Put $\mathfrak{M} = \{u \in X \mid u = Tu\}$ and $\mathfrak{M}' = \{\varphi \in X' \mid \varphi = T'\varphi\}$. Then one and only one of the following two cases (i) and (ii) occurs. (i) $\mathfrak{M} = \{0\}$, $\mathfrak{M}' = \{0\}$; for any $f \in X$ equation (1) has a unique solution; and for any $g \in X'$ equation (2) has a unique solution. (ii) $\dim \mathfrak{M} = \dim \mathfrak{M}' = m$, $1 \leqslant m < \infty$; (1) has a solution if and only if f is orthogonal to $\mathfrak{M}'$ (i.e., $\varphi(f) = 0$ for any $\varphi \in \mathfrak{M}'$); and (2) has a solution if and only if g is orthogonal to $\mathfrak{M}$ (i.e., $g(u) = 0$ for any $u \in \mathfrak{M}$). This is called the **Riesz-Schauder theorem**. In particular, when T is an integral operator in a suitable function space, this theorem is also called **Fredholm's alternative theorem** for integral equation (1).

E. Spectra of a Compact Operators

The structure of the †spectrum of a compact operator $T \in \mathbf{B}^{(c)}(X)$ is well-known, owing to the Riesz-Schauder theorem. In particular, the spectrum $\sigma(T)$ of T consists of at most countably many points. Any nonzero point of $\sigma(T)$ is an †eigenvalue of T. When X is infinite-dimensional, 0 always belongs to $\sigma(T)$ but is not necessarily an eigenvalue of T. Each nonzero eigenvalue of T has finite (algebraic) †multiplicity, and hence the †eigenspace $\mathfrak{M}_\lambda(T) = \{u \mid Tu = \lambda y\}$, $\lambda \neq 0$, is finite-dimensional. If λ is an eigenvalue of T, then $\bar{\lambda}$ is an eigenvalue of the †adjoint operator T^* of T with the same (either algebraic or geometric) multiplicity as that of λ.

F. Spectral Representations of Compact Operators

Let T be a compact †normal operator in a Hilbert space H. Then we can find a †complete orthonormal set consisting solely of eigenvectors of T. Namely, for each nonzero eigenvalue λ_j of T, take an orthonormal basis $\{\varphi_k^{(j)}\}$ of the eigenspace associated with λ_j. Rearrange all of $\varphi_k^{(j)}$ into a sequence $\{\varphi_n\}$ and add to it, if 0 is an eigenvalue, a complete orthonormal set of the eigenspace associated with 0. Then we obtain a complete orthonormal set of H as mentioned above. Let μ_n be the eigenvalue associated with φ_n. Then the sequence $\{\mu_n\}$ is precisely an enumeration of nonzero eigenvalues of T with repetitions according to multiplicity. In terms of $\{\varphi_n\}$ and $\{\mu_n\}$, a †spectral representation of T is given as

$$Tu = \sum_{n=1}^{\infty} \mu_n (u, \varphi_n)\varphi_n, \qquad u \in H.$$

G. Classification of Compact Operators

In a Hilbert space there are many subclasses of compact operators that serve to classify compact operators. Fundamental ones among these classes will be discussed below. It is assumed for simplicity that H is a †separable Hilbert space. For any $T \in \mathbf{B}^{(c)}(H)$ the operator $A = (T^* T)^{1/2}$ is a compact nonnegative †self-adjoint operator. Let $\alpha_1 \geqslant \alpha_2 \geqslant \ldots$ be the enumeration in decreasing order of the positive eigenvalues of A, with each repeated according to its multiplicity. The $\alpha_n = \alpha_n(T)$ are sometimes called the **characteristic numbers** of T. For any $p > 0$ the set of all $T \in \mathbf{B}^{(c)}(H)$ such that

$$\|T\|_p = \left(\sum_{n=1}^{\infty} \alpha_n^p \right)^{1/p} < +\infty$$

is denoted by $\mathbf{B}_p(H)$ (or simply $\mathbf{B}_p$). Among these classes, $\mathbf{B}_1$ and $\mathbf{B}_2$ are most important. $\mathbf{B}_1$ is called the **trace class** and $\mathbf{B}_2$ the **Hilbert-Schmidt class**. Correspondingly, $\|T\|_1$ and $\|T\|_2$ are called the **trace norm** and **Hilbert-Schmidt norm** of T, respectively. $\mathbf{B}_1$ is also called the **nuclear class** and any operator $T \in \mathbf{B}_1$ a †**nuclear operator**.

The class $\mathbf{B}_p$ is a two-sided ideal in the Banach algebra $\mathbf{B}$. More precisely, $T \in \mathbf{B}_p$ and $R \in \mathbf{B}$ imply $\|RT\|_p \leqslant \|R\| \, \|T\|_p$ and $\|TR\|_p \leqslant \|R\| \, \|T\|_p$. When $1 \leqslant p < \infty$, the class $\mathbf{B}_p$ becomes a Banach space with the norm $\|T\|_p$. The set of all degenerate operators is dense in any of these Banach spaces.

The norm $\|T\|_p$ $(T \neq 0)$ is a decreasing function of p. Hence $\mathbf{B}_p \subset \mathbf{B}_q$ if $p \leqslant q$. Also, $T \in \mathbf{B}_p$ and $S \in \mathbf{B}_q$ imply $TS \in \mathbf{B}_r$, where $1/r = 1/p + 1/q$. Let $T \in \mathbf{B}_p$ and let $\{\mu_j\}$ be an enumeration of eigenvalues of T with repetitions according to (either geometric or algebraic) †multiplicity. Then

$$\sum_j |\mu_j|^p \leqslant \|T\|_p^p.$$

The Hilbert-Schmidt class. Let $T \in \mathbf{B}_2$. Then for an arbitrary complete orthonormal set $\{u_k\}$, we have

$$\|T\|_2^2 = \sum_{k=1}^{\infty} \|Tu_k\|^2.$$

Then $\mathbf{B}_2$ may be defined as the set of all $T \in \mathbf{B}$ such that the sum on the right-hand side is finite for a certain complete orthonormal set $\{u_k\}$. The space $\mathbf{B}_2$ with the norm $\|T\|_2$ becomes a Hilbert space with the inner product defined by

$$(T, S)_2 = \sum_{k=1}^{\infty} (Tu_k, Su_k),$$

where $\{u_k\}$ is as above.

The trace class. For an operator $T \in \mathbf{B}_1$, the trace $\mathrm{Tr}(T)$ of T is defined as

$$\mathrm{Tr}(T) = \sum_{k=1}^{\infty} (Tu_k, u_k).$$

Here the right-hand side converges absolutely and does not depend on the complete orthonormal set $\{u_k\}$. The trace is a bounded linear functional on the Banach space $\mathbf{B}_1$. The product of two operators of Hilbert-Schmidt type belongs to the trace class, and the converse is also true. For operators of the form $I + T$, $T \in \mathbf{B}_1$, we can define the determinant of $I + T$.

H. Nonlinear Compact Operators

A nonlinear operator defined on a region D in a Banach space X and taking its value in another Banach space Y is said to be **compact** (or **completely continuous**) if F is continuous in D and maps any bounded set in D to a relatively compact set in Y. A †fixed-point theorem which is valid for such an F is often conveniently applied to the existence proofs of solutions of differential equations.

References

[1] S. Banach, Théorie des opérations linéaires, Warsaw, 1932 (Chelsea, 1963).
[2] N. Dunford and J. T. Schwartz, Linear operators I, II, Interscience, 1958, 1963.
[3] A. C. Zaanen, Linear analysis, North-Holland, 1953.
[4] М. А. Красносельский (M. A. Krasnosel'skiĭ), Топологические методы в теорий нелинейных интегральных уравнений, Гостехиздат, 1956.
[5] R. Schatten, A theory of cross spaces, Ann. Math. Studies, Princeton Univ. Press, 1950.
[6] K. Yosida, Functional analysis, Springer, 1965.
[7] И. М. Гельфанд, Н. Я. Виленкин (I. M. Gel'fand and N. Ja. Vilenkin), Обобщенные Функции IV, Физматгиз, 1961; Generalized functions IV. Applications of harmonic analysis, Academic Press, 1964.
[8] И. Ц. Гохберг, М. Г. Крейн (I. C. Gohberg and M. G. Kreĭn), Введение в теорию линейных несамосопяженных операторов, Наука, 1965; English

translation, Introduction to the theory of linear nonselfadjoint operators, Amer. Math Soc. Transl. of Math. Monographs, 1969, vol. 18.
[9] T. Kato, Perturbation theory for linear operators, Springer, 1966.
[10] H. H. Schaefer, Topological vector spaces, Macmillan, 1966.
[11] R. Schatten, Norm ideals of completely continuous operators, Springer, 1960.
[12] H. Helson, Lectures on invariant subspaces, Academic Press, 1964.
[13] A. Grothendieck, Produits tensoriels topologiques et espaces nucléaires, Mem. Amer. Math. Soc., 16 (1955).
[14] P. Enflo, A counterexample to the approximation problem in Banach spaces, Acta Math., 130 (1973), 309–317.

73 (IX.2) Complexes

A. General Remarks

The notion of complexes was introduced by H. Poincaré to study the topology of †manifolds by combinatorial methods (→ 259 Manifolds). Various kinds of complexes have been introduced in the course of the development of topology. These will be dealt with individually in the sections that follow.

B. Euclidean Complexes

A set $\mathfrak{K}$ of †simplexes in a Euclidean space $\mathbf{R}^N$ is called a **Euclidean (simplicial) complex** in $\mathbf{R}^N$ if $\mathfrak{K}$ satisfies the following three conditions: (i) Every face of a simplex belonging to $\mathfrak{K}$ is also an element of $\mathfrak{K}$. (ii) The intersection of two simplexes belonging to $\mathfrak{K}$ is either empty or a face of each of them. (iii) Each point of a simplex belonging to $\mathfrak{K}$ has a neighborhood in $\mathbf{R}^N$ that intersects only a finite number of simplexes belonging to $\mathfrak{K}$. A Euclidean complex is also called a **geometric complex** (or **rectilinear complex**). Each 0-simplex in $\mathfrak{K}$ is called a **vertex** in $\mathfrak{K}$. We define the **dimension of** $\mathfrak{K}$ to be n if $\mathfrak{K}$ contains an n-simplex but no $(n+1)$-simplex, and ∞ if $\mathfrak{K}$ contains n-simplexes for all $n \geqslant 0$.

By a **subcomplex** of a Euclidean complex $\mathfrak{K}$ we mean a Euclidean complex that is a subset of $\mathfrak{K}$. For a Euclidean complex $\mathfrak{K}$, its **r-section** (or **r-skeleton**) is defined to be the subcomplex of $\mathfrak{K}$ consisting of all n-simplexes $(n \leqslant r)$ in $\mathfrak{K}$.

If $\mathfrak{K}$ is a Euclidean complex in $\mathbf{R}^N$, we denote by $|\mathfrak{K}|$ the set of points in $\mathbf{R}^N$ belonging to simplexes in $\mathfrak{K}$. This set $|\mathfrak{K}|$ is called the **Euclidean polyhedron**.

By a **subdivision** $\mathfrak{K}'$ of a Euclidean complex $\mathfrak{K}$ we mean a Euclidean complex such that $|\mathfrak{K}'| = |\mathfrak{K}|$ and each simplex in $\mathfrak{K}'$ is contained in a simplex in $\mathfrak{K}$. Specifically, we can construct a subdivision $\mathfrak{K}'$ of $\mathfrak{K}$ utilizing †barycenters of simplexes in $\mathfrak{K}$; namely, we let $\mathfrak{K}'$ be the set of all r-simplexes whose vertices consist of barycenters of the series $\Delta_0 \subset \Delta_1 \subset \ldots \subset \Delta_r$ of simplexes (Δ_i is an i-simplex) in $\mathfrak{K}$. Then $\mathfrak{K}'$ is a subdivision of $\mathfrak{K}$, called the **barycentric subdivision** of $\mathfrak{K}$ and denoted by $\mathrm{Sd}\,\mathfrak{K}$.

Given a Euclidean complex $\mathfrak{K}$ and a subset A of $|\mathfrak{K}|$, we define the **star** of A in $\mathfrak{K}$ to be the subcomplex of $\mathfrak{K}$ that consists of simplexes $\{\Delta\}$ and their faces such that $\Delta \cap A \neq \varnothing$. Furthermore, we define the **open star** of A in $\mathfrak{K}$ as the union of †open simplexes (the interiors of simplexes) of $\mathfrak{K}$ whose closures intersect A. We denote by $St_{\mathfrak{K}}(A)$ the star of A in $\mathfrak{K}$ and by $O_{\mathfrak{K}}(A)$ the open star of A in $\mathfrak{K}$; then $O_{\mathfrak{K}}(A)$ is an open set whose closure is $|St_{\mathfrak{K}}(A)|$.

The notion of Euclidean simplicial complexes can be generalized to that of **Euclidean cell complexes**; this is done by replacing the term *simplex* by †*convex cell* in the definition of Euclidean simplicial complex. For a Euclidean cell complex $\mathfrak{K}$, the notions of vertex, dimension, subcomplex, and subdivision are defined similarly as in the case of Euclidean simplicial complexes.

C. Simplicial Complexes

Given a Euclidean simplicial complex $\mathfrak{K}$, let K denote the set of all the vertices in $\mathfrak{K}$, and let Σ denote the set consisting of those subsets $\{v_0, \ldots, v_r\}$ of K for which there exist simplexes Δ in $\mathfrak{K}$ such that $\{v_0, \ldots, v_r\}$ coincide with the set of vertices of Δ. Then, we have (1) if $s \in \Sigma$ and $s \supset s'$, $s' \neq \varnothing$, then $s' \in \Sigma$; (2) every set consisting of a single element in K is in Σ, and the empty set is not in Σ. In general, if a pair (K, Σ) of a set K and a set Σ consisting of finite subsets of K satisfy (1) and (2), then the pair (or the set K) is called an **abstract simplicial complex** (or simply **simplicial complex**). If K is a simplicial complex, each element of the set K is called a **vertex** in K, and each set of Σ is called a **simplex** in K. A simplex consisting of $n+1$ vertices is called an **n-simplex**. We say that a simplicial complex K is **finite** if it consists of a finite number of vertices; it is **locally finite** if every vertex of K belongs to only finitely many simplexes in K. We define similarly **countable simplicial complexes** and **locally countable simplicial complexes**. The **dimension** and **r-section** of a simplicial complex are defined as in the case of Euclidean complexes. By a **subcomplex** of a simplicial complex K we mean a simplicial complex K_0 such that each simplex of K_0 is a simplex in K.

If K and L are simplicial complexes, a mapping $\varphi: K \rightarrow L$ is called a **simplicial mapping** of K to L if the following condition is satisfied: If $v_0, v_1, \ldots, v_n$ are vertices of a simplex of K, then $\varphi(v_0), \varphi(v_1), \ldots, \varphi(v_n)$ are vertices of some simplex of L. Two simplicial complexes K and L are said to be **isomorphic** if there exist simplicial mappings $\varphi; K \rightarrow L$, $\psi: L \rightarrow K$ such that $\psi \circ \varphi$ and $\varphi \circ \psi$ are the identity mappings.

Given a simplicial complex K, let $|K|$ denote the set of all functions x from the set of vertices of K to the closed interval $I=[0,1]$ satisfying the following conditions: (i) For any x, the set $\{v \in K \mid x(v) \neq 0\}$ is a simplex of K. (ii) For any x, $\Sigma_{v \in K} x(v)=1$. The value $x(v)$ is called the **barycentric coordinate** of the point $x \in |K|$ with respect to the vertex v. Each vertex v of K is identified with the point of $|K|$ whose barycentric coordinate with respect to the vertex v is 1, and is called a **vertex** in $|K|$. For a simplex $s=\{v_0, v_1, \ldots, v_n\}$ in K, we define $|s|=\{x \in |K| \,\|\, x(v)=0 \ (v \notin s)\}$ and $e(s)=\{x \in |s| \,\|\, x(v_i)>0 \ (i=0,1,\ldots,n)\}$. We call $|s|$ a **simplex** in $|K|$ and $e(s)$ an **open simplex** of $|K|$ or the **interior** of $|s|$. We can define a metric d on $|K|$ by $d(x,y) =(\Sigma_{v \in K}(x(v)-y(v))^2)^{1/2}$. However, $|K|$ is usually supplied with a †stronger topology defined as follows: (1) Each simplex $|s|$ in $|K|$ has the topology given by the metric d. (2) A subset U of $|K|$ is open if and only if $U \cap |s|$ is an open subset of $|s|$ for each simplex s of K. Henceforward, by the topology of $|K|$ we mean the topology just defined, unless otherwise stated. The set $|K|$ with such a topology is called the **polyhedron** of K. The topology of $|K|$ coincides with the above metric topology if and only if K is locally finite. If simplicial complexes K and L are isomorphic, then $|K|$ and $|L|$ are homeomorphic. If K is the simplicial complex defined by a Euclidean simplicial complex $\mathfrak{K}$, then $|K|$ and $|\mathfrak{K}|$ are homeomorphic.

If K and L are simplicial complexes, a mapping $f: |K| \rightarrow |L|$ satisfying the following condition is said to be **linear**: If $s=\{v_0, v_1, \ldots, v_n\}$ is a simplex in K and $x=\lambda_0 v_0+ \ldots +\lambda_n v_n$ ($\lambda_0+\ldots+\lambda_n=1, \lambda_i \geqslant 0$), then $f(v_0), \ldots, f(v_n)$ belong to a simplex in L and $f(x)=\lambda_0 f(v_0)+\ldots+\lambda_n f(v_n)$. The linear mapping determined by a simplicial mapping $\varphi: K \rightarrow L$ is denoted by $|\varphi|: |K| \rightarrow |L|$ and is called a **barycentric mapping**. Sometimes it is also called a **simplicial mapping** and denoted by the same letter φ. For simplicial complexes K and K', if there exists a linear mapping $l: |K'| \rightarrow |K|$, which is a homeomorphism, then we identify $|K|$ and $|K'|$ by l and call K' a **subdivision** of K. The **barycentric subdivision** Sd K of a simplicial complex K is defined as in the case of Euclidean complexes. We also have notions of **star** $St_K(A)$ and **open star** $O_K(A)$ for a simplicial complex K and a subset A of $|K|$. If K and L are simplicial complexes, a mapping $f: |K| \rightarrow |L|$ is called a **piecewise linear mapping** if there exist subdivisions K' and L' of K and L, respectively, such that $f: |K'| \rightarrow |L'|$ is linear.

Given a †covering $\mathfrak{M}=\{M_v\}_{v \in K}$ of a set X, the index set K becomes a simplicial complex if we consider each finite nonempty subset s of K such that $\bigcap_{v \in s} M_v \neq \varnothing$ (empty) to be a simplex. The resulting simplicial complex K is called the **nerve** of the covering $\mathfrak{M}$. Furthermore, if $\mathfrak{M}=\{M_v\}_{v \in K}$ and $\mathfrak{N}=\{N_w\}_{w \in L}$ are coverings of a set X, $\mathfrak{N}$ is a †refinement of the covering $\mathfrak{M}$, and L is the nerve of $\mathfrak{N}$, then a simplicial mapping $\varphi: L \rightarrow K$ is defined by sending each vertex w in L to a vertex v in K such that $N_w \subset M_v$.

Given two disjoint simplicial complexes K and L, a simplicial complex $K * L$, called the **join** of K and L, is defined by the following: (1) the vertices of $K * L$ are the vertices of K and the vertices of L. (2) A nonempty subset of vertices is a simplex of $K * L$ if and only if its subsets in K and L are empty or simplexes there. In particular, the join of a simplicial complex K and a single point is called the **cone** of K.

A simplicial complex K is said to be **ordered** if a †partial ordering is given in the set of vertices in K such that the set of vertices of each simplex is totally ordered. Given ordered simplicial complexes K and L, an ordered simplicial complex $K \times L$, called the **Cartesian product** of K and L, is defined by the following: (1) The vertices in $K \times L$ are pairs (v, w), where v and w are any vertices in K and L, respectively. (2) A set of vertices $(v_0, w_0), \ldots, (v_n, w_n)$ such that $v_0 \leqslant \ldots \leqslant v_n$ and $w_0 \leqslant \ldots \leqslant w_n$ is a simplex in $K \times L$ if $(v_0, \ldots, v_n)$ and $(w_0, \ldots, w_n)$ are simplexes in K and L, respectively; all simplexes in $K \times L$ are obtained in this manner. (3) $(v_1, w_1) \leqslant (v_2, w_2)$ if and only if $v_1 \leqslant v_2$ and $w_1 \leqslant w_2$. Assume that either K or L is locally finite or that both K and L are locally countable. Then the polyhedron $|K \times L|$ is homeomorphic to the product space $|K| \times |L|$ of the topological spaces $|K|$ and $|L|$. The same result holds for the polyhedron $|K * L|$ and the †join $|K| * |L|$.

By a **triangulation** T of a topological space X we mean a pair (K, t) consisting of a simplicial complex K and a homeomorphism $t: |K| \rightarrow X$. A triangulation is also called a **simplicial decomposition**. If $T=(K, t)$ is a triangulation of X, the various concepts defined for K can be transferred to X by means of the mapping t. For example, by a **simplex** of the triangulation T we mean the image of a simplex of $|K|$ under t. We say that a triangulation $T=(K,t)$ is **finite** if K is a finite simplicial complex. If $T=(K,t)$ is a triangulation and (K',l) is a subdivision of K, then $T'=(K',t\circ l)$ is called a **subdivision** of T. If $T_2=(K_1,t_1)$, $T_2=(K_2,t_2)$ are triangulations of topological spaces X_1, X_2, respectively, a mapping $f: X_1\rightarrow X_2$ is called a **simplicial mapping** relative to T_1 and T_2 if $t_2^{-1}\circ f\circ t_1: |K_1|\rightarrow|K_2|$ is a simplicial mapping. The following two problems on triangulations are famous: (1) Under what topological conditions is it possible for a given topological space to be supplied with a triangulation? (2) Given two triangulations T_1, T_2 of a space X, are there subdivisions $T_1'=(K_1',t_1')$, $T_2'=(K_2',t_2')$ of T_1 and T_2, respectively, such that K_1' and K_2' are isomorphic? Concerning the second problem, the conjecture asserting the existence of subdivisions T_1' and T_2' as above is known as the **fundamental conjecture (Hauptvermutung)** in topology. Partial solutions to these problems are known: Every 3-dimensional manifold is triangulizable, and any two of its triangulations admit subdivisions satisfying the condition in (2) (R. H. Bing, E. E. Moise). J. Milnor has shown that the fundamental conjecture is not generally true. (Concerning the fundamental conjecture for combinatorial manifolds → 259 Manifolds.) Any †differentiable manifold is triangulizable, and the fundamental conjecture holds for its †C^r-triangulations ($r\geqslant 1$) (S. S. Cairns, J. H. C. Whitehead).

Let T_1, T_2 be triangulations of topological spaces X_1, X_2, respectively, and let $f: X_1\rightarrow X_2$ be a continuous mapping. Then a simplicial mapping $\varphi: X_1\rightarrow X_2$ relative to T_1 and T_2 is called a **simplicial approximation** to f if, for each $x\in X$, the image $\varphi(x)$ lies on the simplex of T_2 whose interior contains $f(x)$. The following existence theorem is called the **simplicial approximation theorem**: For every continuous mapping $f: X_1\rightarrow X_2$, there exist a subdivision T_1' of T_1 and a simplicial mapping $\varphi: X_1\rightarrow X_2$ relative to T_1' and T_2 that is a simplicial approximation to f. If the triangulation T_1 is finite, then for a sufficiently large n we can choose $\mathrm{Sd}^n T_1$ as the above T_1' (where $\mathrm{Sd}^0 T=T$ and $\mathrm{Sd}^n T=\mathrm{Sd}(\mathrm{Sd}^{n-1}T)$ $(n\geqslant 1)$). If $\varphi: X_1\rightarrow X_2$ is a simplicial approximation to a continuous mapping, then f and φ are †homotopic.

D. Semisimplicial Complexes

By an **ordered simplex** in a simplicial complex K we mean a finite sequence $(v_0,v_1,\ldots,v_n)$ $(n\geqslant 0)$ of vertices in K, contained in the set of vertices of a simplex in K. Let $O(K)_n$ be the set of all ordered simplexes of K of length $n+1$, and define mappings $\partial_i: O(K)_n\rightarrow O(K)_{n-1}$ and $s_i: O(K)_n\rightarrow O(K)_{n+1}$ for $i=0,1,\ldots,n$ by $\partial_i(v_0,\ldots,v_n)=(v_0,\ldots,v_{i-1},v_{i+1},\ldots,v_n)$ and $s_i(v_0,\ldots,v_n)=(v_0,\ldots,v_{i-1},v_i,v_i,v_{i+1},\ldots,v_n)$. Then the following relations hold:

$$\begin{aligned}
&\partial_i\circ\partial_j=\partial_{j-1}\circ\partial_i \quad (i<j),\\
&s_i\circ s_j=s_{j+1}\circ s_i \quad (i\leqslant j),\\
&\partial_i\circ s_j=s_{j-1}\circ\partial_i \quad (i<j),\\
&\partial_i\circ s_j=s_j\circ\partial_{i-1} \quad (i>j+1),\\
&\partial_i\circ s_i=\partial_{i+1}\circ s_i=\text{identity}. \qquad (1)
\end{aligned}$$

Let Δ^n be the n-dimensional simplex in $\mathbf{R}^n$ with vertices $e_0=(0,0,\ldots,0)$, $e_1=(1,0,\ldots,0)$, $\ldots,e_n=(0,\ldots,1)$. By a **singular n-simplex** in a topological space X we mean a continuous mapping $T:\Delta^n\rightarrow X$. Let $S(X)_n$ be the set of all singular n-simplexes in X, and define mappings $\partial_i: S(X)_n\rightarrow S(X)_{n-1}$ and $s_i: S(X)_n\rightarrow S(X)_{n+1}$ for $i=0,1,\ldots,n$ by $\partial_i T(\lambda_0,\ldots,\lambda_n)=T(\lambda_0,\ldots,\lambda_{i-1},0,\lambda_i,\ldots,\lambda_n)$ and $s_i T(\lambda_0,\ldots,\lambda_{n+1})=T(\lambda_0,\ldots,\lambda_{i-1},\lambda_i+\lambda_{i+1},\lambda_{i+1},\ldots,\lambda_{n+1})$, where $(\lambda_0,\ldots,\lambda_n)$ is the point $\Sigma_{i=0}^n\lambda_i e_i$, $\lambda_i\geqslant 0$, $\Sigma_{i=0}^n\lambda_i=1$. Then relation (1) holds between ∂_i and s_i.

Because of the importance of relation (1), which is basic in defining †homology of simplicial complexes and topological spaces (→ 203 Homology Groups), S. Eilenberg and J. A. Zilber gave the following definition: A **semisimplicial complex** K consists of a sequence of sets K_n $(n=0,1,\ldots)$ together with mappings $\partial_i: K_n\rightarrow K_{n-1}$, $s_i: K_n\rightarrow K_{n+1}$ $(i=0,1,\ldots,n)$ satisfying relation (1). An element of K_n is called an **n-simplex** in K, and ∂_i, s_i are called the ith **face operator** and the ith **degeneracy operator**, respectively. A simplex is said to be **degenerate** if it is the image of a simplex under some s_i. A semisimplicial complex is abbreviated as **s.s. complex**. The s.s. complexes $O(K)=\{O(K)_n,\partial_i,s_i\}$ and $S(X)=\{S(X)_n,\partial_i,s_i\}$ are called the **ordered complex** of K and the **singular complex** of X, respectively.

Let K be an s.s. complex, and let L_n be a subset of K_n for $n=0,1,\ldots$. If $\partial_i(L_n)\subset L_{n-1}$ and $s_i(L_n)\subset L_{n+1}$ for each i, then $L=\{L_n,\partial_i|L_n,s_i|L_n\}$ is an s.s. complex, and L is called a **subcomplex** of the s.s. complex K. If A is a subspace of a topological space X, $S(A)$ is a subcomplex of $S(X)$. If K is an ordered simplicial complex, a subcomplex $O'(K)$ of $O(K)$ is obtained by considering the set of all ordered simplexes $(v_0,v_1,\ldots,v_n)$ such that $v_0\leqslant v_1\leqslant\ldots\leqslant v_n$ $(n=0,1,\ldots)$. If K and L are s.s. complexes, a sequence $f=\{f_n\}$ of mappings $f_n: K_n\rightarrow L_n$ defined for each n is called an **s.s. mapping** if $\partial_i \circ f_{n+1} = f_n \circ \partial_i$ and $s_i \circ f_n = f_{n+1} \circ s_i$ $(0 \leqslant i \leqslant n)$. If $f: X \to Y$ is a continuous mapping of topological spaces, then f determines an s.s. mapping $S(f): S(X) \to S(Y)$ by $S(f)(T) = f \circ T$. Two s.s. complexes K and L are said to be **isomorphic** if there is a bijective s.s. mapping of K to L. For two s.s. complexes K and L, we define the **Cartesian product** $K \times L$ to be the s.s. complex given by $(K \times L)_n = K_n \times L_n$, $\partial_i(\sigma, \tau) = (\partial_i \sigma, \partial_i \tau)$, $s_i(\sigma, \tau) = (s_i \sigma, s_i \tau)$ $(\sigma \in K_n, \tau \in L_n)$. If K and L are ordered simplicial complexes, the s.s. complexes $O'(K) \times O'(L)$ and $O'(K \times L)$ are isomorphic. If X and Y are topological spaces, the s.s. complexes $S(X) \times S(Y)$ and $S(X \times Y)$ are isomorphic.

Given an s.s. complex K, we construct a topological space $|K|$ as follows: First, we provide K_n with the †discrete topology and consider the topological space $\bar{K} = \bigcup_{n \geqslant 0} K_n \times \Delta^n$. Next we consider simplicial mappings $\varepsilon^i: \Delta^{n-1} \to \Delta^n$ and $\eta^i: \Delta^{n+1} \to \Delta^n$ defined by $\varepsilon^i(p_j) = p_j$ $(j < i)$, $\varepsilon^i(p_j) = p_{j+1}$ $(j \geqslant i)$ and $\eta^i(p_j) = p_j$ $(j \leqslant i)$, $\eta^i(p_j) = p_{j-1}$ $(j > i)$, where $p_0, \ldots, p_n$ are the vertices of Δ^n. The topological space $|K|$ is defined to be the †quotient space of $\bar{K}$ with respect to an equivalence relation $\sim$ that is defined by the following: $(\partial_i \sigma, y) \sim (\sigma, \varepsilon^i(y))$ $(\sigma \in K_n, y \in \Delta^{n-1})$, $(s_i \sigma, y) \sim (\sigma, \eta^i(y))$ $(\sigma \in K_n, y \in \Delta^{n+1})$, where $i = 0, 1, \ldots, n$. The space $|K|$ is called the **(geometric) realization of the s.s. complex** K (Milnor). Given an s.s. mapping $f: K \to L$, we obtain a continuous mapping $|f|: |K| \to |L|$ defined by $|f|(|\sigma, y|) = |f(\sigma), y|$, where $|\sigma, y|$ is the point in $|K|$ represented by $(\sigma, y) \in \bar{K}$. We call $|f|$ the **realization of the s.s. mapping** f.

The singular complex $S(X)$ of a topological space X has the following property: Given simplexes $\sigma_0, \ldots, \sigma_{k-1}, \sigma_{k+1}, \ldots, \sigma_{n+1} \in K_n$ with $\partial_i \sigma_j = \partial_{j-1} \sigma_i$ $(i < j, i, j \neq k)$, there exists a simplex $\sigma \in K_{n+1}$ with $\partial_i \sigma = \sigma_i$ $(i \neq k)$. An s.s. complex K with this property is called a **Kan complex**. If, in addition, $\partial_i \sigma = \partial_i \sigma'$ $(\sigma, \sigma' \in K_n, i \neq k)$ imply $\partial_k \sigma = \partial_k \sigma'$, we call K a **minimal complex**. For every s.s. complex K, there are minimal subcomplexes M of K that are isomorphic to each other. Moreover, $|M|$ is a †deformation retract of $|K|$. For a Kan complex K, the †homotopy group can be defined combinatorially.

E. Cell Complexes

Let V^n be the †unit n-disk, S^{n-1} be the †unit $(n-1)$-sphere, and X be a Hausdorff space. For a subset e of X, let $\bar{e}$ be the closure of e in X, and let $\dot{e} = \bar{e} - e$. A subset e of the space X is called an ***n*-cell** in X if there is a relative homeomorphism $\varphi: (V^n, S^{n-1}) \to (\bar{e}, \dot{e})$, i.e., a continuous mapping $\varphi: V^n \to \bar{e}$ such that $\varphi(S^{n-1}) \subset \dot{e}$ and $\varphi: V^n - S^{n-1} \to \bar{e} - \dot{e}$ is a homeomorphism. For example, $S^n - \{p\}$ $(p \in S^n)$ is an n-cell. A set $\{e_\lambda \mid \lambda \in \Lambda\}$ of cells in the Hausdorff space X is called a **cellular decomposition** of X if the following three conditions are satisfied: (i) $e_\lambda \cap e_\mu$ is empty if $\lambda \neq \mu$; (ii) $X = \bigcup_{\lambda \in \Lambda} e_\lambda$; (iii) If the dimension of e_λ is $n+1$, then $\dot{e}_\lambda \subset X^n$, where X^n is the union of all the cells e_μ $(\mu \in \Lambda)$ whose dimensions are not greater than n. For example, the n-sphere S^n has a cellular decomposition consisting of a single 0-cell and a single n-cell.

A Hausdorff space X together with its cellular decomposition $\{e_\lambda\}$ is called a **cell complex**, and each e_λ is called a **cell** in the cell complex X. For a cell complex, the notions of **vertex**, ***n*-section**, and **dimension** are defined in the same way as the corresponding notions in Euclidean complexes. Let X be a cell complex and A a topological subspace of X such that the closure of each cell of X intersecting A is contained in A. Then the set of cells e such that $e \cap A \neq \varnothing$ forms a cellular decomposition of A. The set A together with this cellular decomposition is called a **subcomplex** of the cell complex X. A cell complex X with its cells $\{e_\lambda\}$ is said to be **finite** if the number of e_λ is finite. If each point in a cell complex X is an interior point of some finite subcomplex of X, then X is said to be **locally finite**. We define similarly a **countable cell complex** and a **locally countable cell complex**. If the closure of each n-cell of a cell complex X is homeomorphic to $V^n (n = 0, 1, \ldots)$, X is said to be **regular**. If X and Y are cell complexes, a continuous mapping $f: X \to Y$ such that $f(X^n) \subset Y^n$ $(n = 0, 1, \ldots)$ is called a **cellular mapping**. If X and Y are cell complexes, the set of †topological products $e_1 \times e_2$, where e_1, e_2 run over all cells of X, Y, respectively, is a cellular decomposition of the product space $X \times Y$. The resulting cell complex $X \times Y$ is called the **product complex** of the cell complexes X and Y.

A cell complex X is said to be **closure finite** if each cell in X is contained in a finite subcomplex of X; and X is said to have the **weak topology** if a subset $U \subset X$ is open if and only if $U \cap \bar{e}$ is relatively open for each cell e of X. Following J. H. C. Whitehead, we call a cell complex a **CW complex** if it is closure finite and has the weak topology. The cellular decomposition of a CW complex is called a **CW decomposition**. A locally finite cell complex is a CW complex.

Some fundamental properties of CW complexes follow: A CW complex is a †paracompact (hence †normal) space and is †locally contractible. A subcomplex A of a CW complex X is a closed subspace of X, and A itself is a CW complex. A mapping $f: X \rightarrow Y$ of a CW complex X to a topological space Y is continuous if and only if the restriction $f|\bar{e}$ is continuous for each cell e of X. If X and Y are CW complexes and $f: X \rightarrow Y$ is any continuous mapping, then there exists a cellular mapping of X to Y that is homotopic to f (**cellular approximation theorem**). A pair (X,A) consisting of a CW complex X and its subcomplex A has the †homotopy extension property for any topological space. A CW complex has the †covering homotopy property for any †fiber space. The product complex $X \times Y$ of two CW complexes X and Y is not necessarily a CW complex, but it is †homotopically equivalent to a CW complex. If either X or Y is locally finite, or if both X and Y are locally countable, then the product complex $X \times Y$ is a CW complex. For CW complexes X and Y, the †mapping space Y^X is homotopically equivalent to a CW complex. The †covering space of a CW complex has a CW decomposition.

If K is a simplicial complex, the polyhedron $|K|$ is a regular CW complex whose cells are all open simplexes in $|K|$. A simplicial complex K is (locally) finite if and only if the CW complex $|K|$ is (locally) finite. In particular, the Euclidean polyhedron of a Euclidean simplicial (or cell) complex is a locally finite CW complex. A polyhedron $|K|$ generally admits a CW decomposition whose cells are far smaller in number than the simplexes constituting a simplicial decomposition of K. For any CW complex X, there exists a polyhedron $|K|$ that is homotopically equivalent to X. In particular, if X is an n-dimensional finite (countable) CW complex, we can choose as K a simplicial complex that is n-dimensional and finite (countable). The realization $|K|$ of an s.s. complex K is a CW complex whose cells are in one-to-one correspondence with the nondegenerate simplexes in K. For a topological space X, a †weak homotopy equivalence $\rho: |S(X)| \rightarrow X$ is defined by $\rho(|T,y|) = T(y)$ $(T \in S(X)_n, y \in \Delta^n)$. Two CW complexes X and Y are homotopically equivalent if and only if the minimal subcomplexes of $S(X)$ and $S(Y)$ are isomorphic.

References

[1] H. Cartan, Sur la théorie de Kan, Séminaire H. Cartan, Ecole Norm. Sup., 1956–1957.
[2] S. Eilenberg and N. E. Steenrod, Foundations of algebraic topology, Princeton Univ. Press, 1952.
[3] R. Godement, Topologie algébrique et théorie des faisceaux, Actualités Sci. Ind., Hermann, 1958.
[4] P. J. Hilton and S. Wylie, Homology theory, Cambridge Univ. Press, 1960.
[5] Y. Kawada (ed)., Isôkikagaku (Japanese; Topology with exercises), Iwanami, 1965.
[6] A. Komatu, M. Nakaoka, and M. Sugawara, Isôkikagaku (Japanese; Topology) I, Iwanami, 1967.
[7] J. R. Munkres, Elementary differential topology, Princeton Univ. Press, second edition, 1966.
[8] H. Schubert, Topologie, Teubner, 1964.
[9] J. P. May, Simplicial objects in algebraic topology, Van Nostrand, 1967.
[10] E. Curtis, Simplicial homotopy theory, Lecture notes, Aarhus, 1967.
[11] E. H. Spanier, Algebraic topology, McGraw-Hill, 1966.

74 (VII.11) Complex Manifolds

A. Definitions

A Hausdorff topological space X is called a **complex manifold** (or **complex analytic manifold**) of **complex dimension** n if there are given an open covering $\{U_i\}_{i \in I}$ and a family $\{\varphi_i\}_{i \in I}$ of homeomorphisms of U_i onto open sets in the n-dimensional complex affine space $\mathbf{C}^n$ such that in case $U_i \cap U_j \neq \varnothing$, the mapping $\varphi_i \circ \varphi_j^{-1}: \varphi_j(U_i \cap U_j) \rightarrow \varphi_i(U_i \cap U_j)$ is biholomorphic (i.e., $\varphi_i \circ \varphi_j^{-1}$ and its inverse are both †holomorphic functions when expressed in terms of coordinate functions in $\mathbf{C}^n$). We call X the **underlying topological space** of this complex manifold, and we say that an open covering $\{U_i\}_{i \in I}$ and a family $\{\varphi_i\}_{i \in I}$ define a **complex analytic structure** (or simply **complex structure**) on X.

A complex-valued function f defined on an open set U in X is called a **holomorphic function** on U if for any i the function $f \circ \varphi_i^{-1}$ on $\varphi_i(U \cap U_i)$ is holomorphic. When we express the mapping φ_i as $\varphi_i(p) = (z^1(p), \ldots, z^n(p))$ on U_i in terms of the coordinates in $\mathbf{C}^n$, each z^α is a holomorphic function on U_i. We call $(z^1, \ldots, z^n)$ a **holomorphic local coordinate system** on U_i. Given two complex manifolds Y, X, a mapping $\varphi: Y \rightarrow X$ is said to be **holomorphic** if for any open set U in X and any holomorphic function f on U, $f \circ \varphi$ is holomorphic on $\varphi^{-1}(U) \subset Y$. When a mapping $\varphi: Y \rightarrow X$ is bijective and both φ and φ^{-1} are holomorphic, we say Y and X are **isomorphic** by φ as complex manifolds.

As in the case of †differentiable manifolds of class C^∞, we can define concepts such as **complex analytic submanifolds, holomorphic**

tangent vectors, holomorphic vector fields, and **holomorphic differential forms of degree** k (or simply **holomorphic** k**-forms**). **Meromorphic functions** on complex manifolds can also be defined as in the theory of analytic functions of several complex variables (→ 27 Analytic Spaces).

Let X be a complex manifold and p a point of X. Take a holomorphic local coordinate system $(z^1, \ldots, z^n)$ with center p (i.e., $z^\alpha(p)=0$ for all α). A holomorphic function defined on a neighborhood of p can be expressed as a holomorphic function in $(z^1, \ldots, z^n)$, hence as a power series in $(z^1, \ldots, z^n)$ absolutely convergent in a neighborhood of p. If we denote by $\Omega = \Omega(X)$ the †sheaf of germs of holomorphic functions on X, the †stalk Ω_p of Ω at p is isomorphic to the †local ring of convergent power series of n variables $z^1, \ldots, z^n$. At a point p, $(\partial/\partial z^1)_p, \ldots, (\partial/\partial z^n)_p$ form a basis of the holomorphic tangent vector space at p. A holomorphic k-form ω defined on a neighborhood of p can be expressed as $\omega = \sum_{i_1 < \ldots < i_k} f_{i_1 \ldots i_k} dz^{i_1} \wedge \ldots \wedge dz^{i_k}$, where $f_{i_1 \ldots i_k}$ is a holomorphic function for each $(i_1, \ldots, i_k)$.

B. Almost Complex Structures

Let X be a complex manifold, and let $\{U_i, \varphi_i\}_{i \in I}$ be its complex analytic structure, i.e., a covering of X by holomorphic local coordinate systems with $\varphi_i = (z_i^1, \ldots, z_i^n)$. Express z_i^α in the form $z_i^\alpha = x_i^\alpha + \sqrt{-1}\, y_i^\alpha$, where x_i^α and y_i^α are the real and imaginary parts of z_i^α, respectively. Then x_i^α and y_i^α are real-valued functions on the open set U_i of X, and the mapping $\psi_i : U \to \mathbf{R}^{2n}$ defined by $\psi_i(p) = (x_i^1(p), y_i^1(p), \ldots, x_i^n(p), y_i^n(p))$ is a homeomorphism of U_i onto an open set of $\mathbf{R}^{2n}$. This $\{U_i, \psi_i\}_{i \in I}$ defines on X a †differentiable structure of class C^∞ (in fact, a †real analytic structure). Thus a complex manifold of complex dimension n admits canonically a C^∞-structure of real dimension $2n$. For every point p of X there is a real coordinate system on a neighborhood of p, such as $(x^1, y^1, \ldots, x^n, y^n)$, where $(z^1, \ldots, z^n)$ $(z^\alpha = x^\alpha + \sqrt{-1}\, y^\alpha, \alpha = 1, \ldots, n)$ forms a holomorphic coordinate system in X. The real tangent vector space at a point p of X has $\{(\partial/\partial x^1)_p, (\partial/\partial y^1)_p, \ldots, (\partial/\partial x^n)_p, (\partial/\partial y^n)_p\}$ as its basis. Define a linear endomorphism J_p by $(\partial/\partial x^\alpha)_p \to (\partial/\partial y^\alpha)_p$, $(\partial/\partial y^\alpha)_p \to -(\partial/\partial x^\alpha)_p$ $(\alpha = 1, \ldots, n)$; then $J_p^2 = -1$ and J_p does not depend on the choice of holomorphic coordinate system at p. Considering J_p as a tensor of type $(1,1)$, we thus obtain a tensor field J of type $(1,1)$ of class C^∞ on X, which is called the **tensor field of almost complex structure** induced by the complex structure of X.

More generally, when a real differentiable manifold X is provided with a tensor field J of type $(1,1)$ of class C^∞ such that $J^2 = -1$ (considering J as a linear transformation of vector fields), we say that X admits an **almost complex structure** or that X is an **almost complex manifold**. In this case, for contravariant vector fields x and y on X, we define a tensor field S of type $(1,2)$ by $S(x,y) = -[x,y] + [J(x), J(y)] - J([J(x), y]) - J([x, J(y)])$. S is called **Nijenhuis's tensor**. An almost complex structure J is induced by a complex analytic structure if and only if its Nijenhuis's tensor S vanishes identically [3, 10]. A differentiable manifold X of dimension $2n$ admits an almost complex structure if and only if the structure group $GL(2n, \mathbf{R})$ of the bundle of †tangent $2n$-frames of X can be †reduced to $GL(n, \mathbf{C})$. Almost complex manifolds are †orientable.

C. Types of Differential Forms

Let X be a complex manifold, and let $(z^1, \ldots, z^n)$ be a holomorphic local coordinate system in a neighborhood of a point p with $z^\alpha = x^\alpha + \sqrt{-1}\, y^\alpha$ $(\alpha = 1, \ldots, n)$. On the complexified real tangent vector space $T_p(X) \otimes \mathbf{C}$ at p, we define $\partial/\partial z^\alpha$, $\partial/\partial \bar{z}^\alpha$ by

$$(\partial/\partial z^\alpha)_p = (1/2)\{(\partial/\partial x^\alpha)_p - \sqrt{-1}\,(\partial/\partial y^\alpha)_p\},$$

$$(\partial/\partial \bar{z}^\alpha)_p = (1/2)\{(\partial/\partial x^\alpha)_p + \sqrt{-1}\,(\partial/\partial y^\alpha)_p\}.$$

It is easy to see that the operation $\partial/\partial z^\alpha$ on holomorphic functions coincides with that of the holomorphic tangent vector $\partial/\partial z^\alpha$ defined in Section A. A function f is holomorphic at p if and only if $(\partial/\partial \bar{z}^\alpha) f = 0$ $(\alpha = 1, 2, \ldots, n)$. $T_p(X) \otimes \mathbf{C}$ is the direct sum of the subspace spanned by $\{\partial/\partial z^1, \ldots, \partial/\partial z^n\}$ and the subspace spanned by $\{\partial/\partial \bar{z}^1, \ldots, \partial/\partial \bar{z}^n\}$. Moreover, this decomposition is independent of the choice of holomorphic coordinate system. Elements of the two subspaces are respectively called **tangent vectors of type** $(1,0)$ and **tangent vectors of type** $(0,1)$. Similarly, the complexified space of real differentiable 1-forms can be decomposed into the direct sum of two subspaces spanned by $\{dz^1, \ldots, dz^n\}$ and $\{d\bar{z}^1, \ldots, d\bar{z}^n\}$, where $dz^\alpha = dx^\alpha + \sqrt{-1}\, dy^\alpha$ and $d\bar{z}^\alpha = dx^\alpha - \sqrt{-1}\, dy^\alpha$. We say that the elements of the former subspace are of type $(1,0)$ and those of the latter are of type $(0,1)$. Thus the space of differential forms of arbitrary degree can be written as the direct sum of subspaces of type (r,s). Here the subspace of **differential forms of type** (r,s) has as a basis $\{dz^{\alpha_1}\wedge\ldots\wedge dz^{\alpha_r}\wedge d\bar{z}^{\beta_1}\wedge\ldots\wedge d\bar{z}^{\beta_s}\ (1\leqslant\alpha_1<\ldots<\alpha_r\leqslant n, 1\leqslant\beta_1<\ldots<\beta_s\leqslant n)\}$. This decomposition is independent of the choice of a local coordinate system, hence the concept of type can be defined globally on X.

D. d''-Cohomology

In the rest of this article we consider only complex differential forms. For every differential form ω of type (r, s) on X, its †exterior derivative $d\omega$ decomposes into a sum of differential forms of types $(r+1, s)$ and $(r, s+1)$, which we denote by $d'\omega$ and $d''\omega$, respectively. We have $d=d'+d''$, $(d')^2=0$, $(d'')^2=0$, and $d'd''+d''d'=0$. In terms of a local coordinate system, we write

$$d'\omega=\sum_\gamma \frac{\partial f}{\partial z^\gamma}dz^\gamma\wedge dz^{\alpha_1}\wedge\ldots\wedge dz^{\alpha_r}\wedge d\bar{z}^{\beta_1}\wedge\ldots\wedge d\bar{z}^{\beta_s},$$

$$d''\omega=(-1)^r\sum_\gamma \frac{\partial f}{\partial \bar{z}^\gamma}dz^{\alpha_1}\wedge\ldots\wedge dz^{\alpha_r}\wedge d\bar{z}^\gamma\wedge d\bar{z}^{\beta_1}\wedge\ldots\wedge d\bar{z}^{\beta_s},$$

for $\omega=f\,dz^{\alpha_1}\wedge\ldots\wedge dz^{\alpha_r}\wedge d\bar{z}^{\beta_1}\wedge\ldots\wedge d\bar{z}^{\beta_s}$. A differential k-form ω is holomorphic if and only if ω is of type $(k,0)$ and $d''\omega=0$.

For the operator d'' **Dolbeault's lemma** holds: Let ω be a differential form on a neighborhood U of a point p. If $d''\omega=0$, there is a neighborhood V of p contained in U and a differential form θ on V such that $\omega=d''\theta$ on V.

Let $A^{(r,s)}$ and Ω^p be the †sheaf of germs of differential forms of type (r,s) and the sheaf of germs of holomorphic p-forms on X, respectively, and let $\Gamma(X,A^{(r,s)})$ be the set of †sections of $A^{(r,s)}$ on X. $\Gamma(X,A^{(r,s)})$ is the set of differential forms of type (r,s) on X, and $\Sigma_i\Gamma(X,A^{(p,i)})$ forms a †cochain complex with respect to d''. The †cohomology groups of this complex are called the **d''-cohomology groups** or the **Dolbeault cohomology groups**, and the qth cohomology group is denoted by $H^{p,q}(A,d'')$. It follows easily from Dolbeault's lemma that $0\to\Omega^p\to A^{(p,0)}\xrightarrow{d''}A^{(p,1)}\to\ldots$ is an †exact sequence of sheaves. From this we get **Dolbeault's theorem**:

$$H^q(X,\Omega^p)\cong H^{p,q}(A,d''),$$

where the left-hand side is the cohomology group with coefficient sheaf Ω^p.

More generally, for any †complex analytic (holomorphic) vector bundle E on X, we can define the d''-cohomology groups of the differential forms on X with values in E, and they can be shown to be isomorphic to the cohomology groups with coefficient sheaves of germs of holomorphic forms with values in E (→ 196 Harmonic Integrals).

E. Analytic Coherent Sheaves

The structure of a complex manifold X is determined by the †sheaf Ω of germs of holomorphic functions on X, and Ω is a †coherent sheaf (of rings) (**Oka's theorem** [11]). Sheaves of Ω-modules are called **analytic sheaves**, and coherent sheaves of Ω-modules are called **coherent analytic sheaves**. Many properties of X can be expressed in terms of coherent analytic sheaves and their cohomology groups; some examples appear later in this article (also → 361 Riemann-Roch Theorems).

It is important to know whether an analytic sheaf on a complex manifold is coherent. Concerning this question, not only does Oka's theorem apply but also **Cartan's theorem**: The †sheaf of ideals defined by an analytic subset of a complex manifold is coherent [3]. (We say that a subset Y of X is an **analytic subset** if it is a closed subset and each point of Y has a neighborhood U such that $U\cap Y$ is the set of common zeros of a finite number of holomorphic functions on U.) Also relevant is **Grauert's theorem**: If $\pi:X\to Y$ is a †proper holomorphic mapping of complex manifolds (i.e., the inverse image of any compact subset of Y for a holomorphic mapping π is also compact), then for any coherent analytic sheaf F on X its †direct images $\pi_q(F)$ ($q=0,1,2,\ldots,$) are also coherent [4]. (In fact, this theorem holds for analytic spaces; → 27 Analytic Spaces.)

For an analytic coherent sheaf F on a †Stein manifold X, we have the following **fundamental theorems of the Stein manifold**. **Theorem A**: $H^0(X,F)$ generates the stalk F_x (as an Ω_x-module) at every point x of X. **Theorem B**: $H^q(X,F)=0$ for all $q>0$. Conversely, a Stein manifold X is characterized by the following property: For any coherent analytic sheaf F of ideals of Ω, $H^1(X,F)=0$ [11]. If a complex manifold X is compact and F is a coherent analytic sheaf on X, then $H^q(X,F)$ is a complex vector space of finite dimension. If X is an open submanifold of another complex manifold and its closure is compact, then $H^q(X,F)$ is finite-dimensional for some q that depends on various properties (convexity or concavity) of the boundary of X [12, 1].

Let E be a †complex analytic (holomorphic) vector bundle on a complex manifold X of dimension n, and let E^* be the †dual vector bundle of E. Then $H^q(X,\Omega^p(E))$ and $H_*^{n-q}(X,\Omega^{n-p}(E^*))$ (where H_* denotes the cohomology group with compact support) are dual as topological vector spaces under suitable conditions. The duality is given by the integration on X of the exterior products of the differential forms representing the respective elements of the cohomology groups. The duality holds, for example, when $\dim H^q(X, \Omega^p(E)) < \infty$. If X is compact, we need not distinguish H_* from H (**Serre's duality theorem**) [13].

F. Compact Complex Manifolds

On a compact and connected complex manifold X, there are no holomorphic functions except constants (by the †maximum principle of holomorphic functions). The field $K(X)$ of meromorphic functions on X is finitely generated over the complex number field, and its †transcendence degree d does not exceed the complex dimension n of X. For elements of $K(X)$, functional independence and algebraic independence are equivalent [15]. When $n=1$, X is a compact †Riemann surface and the classical theory of algebraic functions shows that $K(X)$ is an †algebraic function field of one variable and X is a †projective algebraic variety. When $n=2$, $d=2$, 1, and 0 can all occur. If $d=2$, X is a projective †algebraic surface (**Chow-Kodaira theorem**). If $d=1$, there exist an †algebraic curve Δ and a holomorphic mapping $\varphi: X \rightarrow \Delta$ such that $K(\Delta)$ is isomorphic to $K(X)$ under φ^*, and $\varphi^{-1}(x)$ is an †elliptic curve for all but a finite number of $x \in \Delta$. K. Kodaira investigated the structure of compact complex surfaces (including the case $d=0$) in detail [5].

On a compact complex manifold X, the free Abelian group generated by the set of irreducible analytic subsets of codimension 1 is called the **divisor group** of X, and an element of it is called a **divisor** of X. For an analytic subset Y of codimension 1, the sheaf of ideals $\mathfrak{J}(Y)$ defined by Y is a sheaf of locally principal ideals of Ω. For a divisor $D = \Sigma a_\alpha Y_\alpha$, the sheaf of locally principal fractional ideals $\mathfrak{J}(D) = \Pi_\alpha \mathfrak{J}(Y_\alpha)^{a_\alpha}$ is called the **sheaf of ideals of** D. The set of nonzero coherent sheaves of locally principal fractional ideals corresponds bijectively to the set of divisors. An element $f \neq 0$ of $K(X)$ generates a sheaf of principal fractional ideals and therefore defines a divisor, which is denoted by (f). The divisor group has an ordering defined by $D = \Sigma a_\alpha Y_\alpha \succ 0$ if and only if all $a_\alpha \geqslant 0$, under which it becomes an †ordered group. For a divisor D, let

$$L(D) = \{ f \in K(X) \mid f \neq 0, \\ (f) + D \succ 0 \} \cup \{0\}.$$

Then $L(D)$ is a $\mathbf{C}$-module of finite dimension. This submodule of $K(X)$ is easy to handle and shows various analytic properties of D. The †Riemann-Roch theorem is used to calculate the dimension of $L(D)$ in terms of other factors (→ 361 Riemann-Roch Theorems). Here is an example of how $L(D)$ shows a property of D: We call two divisors D and D' **linearly equivalent** if there is a $0 \neq F \in K(X)$ such that $(F) = D - D'$. This is an equivalence relation finer than †homological equivalence. If D and D' are linearly equivalent, then $L(D)$ and $L(D')$ are isomorphic by the mapping $L(D) \ni f \rightarrow fF \in L(D')$; therefore, $\dim L(D) = \dim L(D')$. (The latter equation does not follow from the homological equivalence.) A holomorphic vector bundle with fiber $\mathbf{C}$ and structure group $\mathbf{C}^*$ is called a **complex line bundle**. For the sheaf of ideals $\mathfrak{J}(D)$ of a divisor D, we can take a suitable open covering $\{U_j\}$ of X such that for each U_j there is an $R_j \in \Gamma(U_j, \mathfrak{J}(D))$ which generates $\mathfrak{J}(D)_x$ for any $x \in U_j$. Also $g_{jk} = R_j / R_k$ is a holomorphic function nowhere vanishing on $U_j \cap U_k$. With $\{g_{jk}\}$ as the system of †coordinate transformations, we define the **complex line bundle determined by** D and denote it by $[D]$. It is easy to see that $[D]$ is independent of the choice of $\{U_j\}$ or $\{R_j\}$. Moreover, $[D]$ is determined only by the linear equivalence class of D. If we denote by $\Omega([D])$ the sheaf of germs of holomophic sections of $[D]$, the mapping $H^0(X, \Omega([D])) \ni \varphi = \{\varphi_j\} \rightarrow f = \varphi_j / R_j = \varphi_k / R_k \in L(D)$ is an isomorphism of these modules. On an algebraic variety in a projective space, any complex line bundle comes from a divisor (i.e., it can be expressed in the form $[D]$ for some divisor D) [6], but this is not necessarily true on general compact complex manifolds. However, the importance of complex line bundles in the theory of complex manifolds lies in the relation $L(D) \cong H^0(X, \Omega([D]))$, which replaces "things with poles" with "things holomorphic."

Any analytic submanifold X of the projective space $\mathbf{P}^N$ is an algebraic variety (**Chow's theorem**) [2]. Let Y be a general †hyperplane section of X. Then Y is a divisor on X, and the †Chern class of the complex line bundle $[Y]$ corresponds to the canonical Hodge metric on X (→ 232 Kähler Manifolds D). When $[Y]$ is represented by the system of coordinate transformations $\{g_{jk}\}$ with respect to an open covering $\{U_i\}$, we can associate with a coherent analytic sheaf F sheaves $F(n)$ $(n = 0, \pm 1, \pm 2, \ldots)$ as follows. Denoting by F_j the restriction of F to U_j, we glue F_j and F_k together on $U_j \cap U_k$ with the relation $f_j \sim f_k \Leftrightarrow f_j = g_{kj}^n f_k$ (where $f_j \in F_j \subset F$, $f_k \in F_k \subset F$) and obtain a sheaf (denoted by $F(n)$) that is locally isomorphic to F. The following theorems for $F(n)$ hold. For each coherent analytic sheaf F there exists an integer n_0 such that for any $n \geqslant n_0$ the following **fundamental theorems A, B of projective algebraic varieties** hold [12]. **Theorem A**: $\Gamma(X, F(n))$ generates F_x (as an Ω_x-module) for every $x \in X$. **Theorem B**: $H^q(X, F(n)) = 0$ for all $q > 0$. This means that if we permit "poles" on Y of sufficiently high order, then F has sufficiently many sections and the higher cohomology groups vanish.

On a (nonsingular) algebraic variety X in $\mathbf{P}^N$ we have the sheaf Ω of germs of holomorphic functions (the †structure sheaf as a complex manifold) and the sheaf $\mathcal{O}$ of germs of holomorphic rational functions (the structure sheaf as an algebraic variety). Therefore, we have two kinds of coherent sheaves, coherent analytic sheaves and **coherent algebraic sheaves**. In fact, the cohomology theories derived from them are isomorphic. More precisely, for any coherent algebraic sheaf F, $\tilde{F} = F \otimes_{\mathcal{O}} \Omega$ is a coherent analytic sheaf. The correspondence $F \rightarrow \tilde{F}$ gives an equivalence between the †category of coherent algebraic sheaves and that of coherent analytic sheaves, thus giving an isomorphism of their cohomology groups. In other words, as far as the properties that can be expressed by cohomology-theoretic terms of coherent sheaves are concerned, there is no difference between the analytic and algebraic theories of projective algebraic varieties [14].

G. Deformation of Complex Structures

To study the complex analytic structures with which a C^∞-manifold X can be endowed, we consider the so-called **deformation of complex analytic structures** [7]. Let $\mathcal{V}$ and M be C^∞-manifolds, and let ϖ be a C^∞-mapping from $\mathcal{V}$ onto M. Assume that the rank of the Jacobian matrix of ϖ is always equal to $m = \dim M$ and that every point of M has a suitable neighborhood U such that $\varpi^{-1}(U)$ and $X \times U$ are †diffeomorphic and the diffeomorphism maps $V_t = \varpi^{-1}(t)$ $(t \in U)$ onto $X \times t$. Moreover, if there is given a complex analytic structure on each V_t $(t \in M)$ and if $\mathcal{V}$ permits local coordinate systems satisfying the following condition, then we say that $(\mathcal{V}, M, \varpi)$ (or simply $\mathcal{V}$) is a **differentiable family of complex analytic structures** on a differentiable manifold X. The condition on local coordinate systems: There exist an open covering $\{\mathcal{U}_j\}$ of $\mathcal{V}$ and functions $(z_j^1, \ldots, z_j^n, t_j^1, \ldots, t_j^m)$ on $\mathcal{U}_j$ (where the z_j^α are complex-valued and the t_j^λ are real-valued) such that (i) $(\operatorname{Re} z_j^\alpha, \operatorname{Im} z_j^\alpha, t_j^\lambda)$ forms a local coordinate system on $\mathcal{U}_j$ of class C^∞; (ii) (t_j^λ) is the image by ϖ^* of a local coordinate system of M on $\varpi(\mathcal{U}_j)$; and (iii) with t fixed, (z_j^α) forms a holomorphic local coordinate system of the complex manifold V_t.

Moreover, if V and M are complex manifolds, then ϖ is holomorphic, and the coordinates (z_j, t_j) (where the t_j are complex-valued in this case) can be taken as a holomorphic local coordinate system on $\mathcal{V}$ with (t_j) the image of a holomorphic local coordinate system of M. Then we call $(\mathcal{V}, M, \varpi)$ an **analytic family of complex analytic structures**. If M is connected, we say that V_t and $V_{t'}$ are **deformations** of each other for any $t, t' \in M$. We call M the **parameter space** of the family. A differentiable (complex analytic) family ϖ: $\mathcal{V} \rightarrow M$ is said to be **differentiably (complex analytically) trivial** if there is a differentiable (holomorphic) mapping $\Phi: \mathcal{V} \rightarrow V_0 = \varpi^{-1}(0)$ $(0 \in M)$ that maps V_t diffeomorphically (biholomorphically) onto V_0 for every $t \in M$.

The preceding condition for the local coordinate system (z_j, t_j) means that on $\mathcal{U}_j \cap \mathcal{U}_k$ we have $z_j^\alpha = g_{jk}^\alpha(z_k, t_k)$, $t_j^\lambda = h_{jk}^\lambda(t_k)$, where the g_{jk}^α are holomorphic in z_k and the g_{jk}^α and h_{jk}^λ are C^∞-differentiable in all variables (holomorphic in the case of a complex analytic family). For a tangent vector v on M that is expressed on $\varpi(\mathcal{U}_k)$ as $v = \sum_\mu v_k^\mu (\partial / \partial t_k^\mu)$, we write $\theta_{kj}^\alpha = v(g_{jk}^\alpha) = \sum_\mu v_k^\mu (\partial g_{jk}^\alpha / \partial t_k^\mu)$ and $\theta_{jk} = \sum_\alpha \theta_{jk}^\alpha (\partial / \partial z_j^\alpha)$. Then θ_{jk} is a holomorphic vector field on $\mathcal{U}_j \cap \mathcal{U}_k \cap V_t$ for a fixed t. With respect to the covering $\{\mathcal{U}_j \cap V_t\}$ of V_t, $\{\theta_{jk}\}$ forms a cocycle of degree 1 with values in the sheaf Θ_t of germs of holomorphic tangent vector fields on V_t (→ 377 Sheaves F). The cohomology class determined by the cocycle is denoted by $\rho_t(v) \in H^1(V_t, \Theta_t)$ and is called the **infinitesimal deformation** at t. The cohomology class $\rho_t(v)$ is determined only by the values of v at t. In the case of a trivial family, we have $\rho_t(v) = 0$. In fact, $\rho_t(v)$ can be regarded as the measurement of deformation of complex structure of V_t along v.

In the case of 1-dimensional compact complex manifolds, i.e., closed †Riemann surfaces, the details of the deformation are known as the modulus theory of closed Riemann surfaces (→ 13 Algebraic Functions F). In the case of complex manifolds of dimension greater than 1, however, the deformation is usually much more complicated. This is because for some family of holomorphic vector bundles $\mathcal{E} \rightarrow \mathcal{V} \rightarrow M$ (the definition is given later in this section), $\dim H^q(V_t, \Omega(E_t))$ may be discontinuous in t. It is known, however, that $\dim H^q(V_t, \Omega(E_t))$ is upper semicontinuous in t. Here are some general results on deformation: (1) If $H^1(V_0, \Theta_0) = 0$, any family that contains V_0 is locally trivial at V_0 (Frölicher-Nijenhuis). (2) If, for a family $\mathcal{V}$, $\dim H^1(V_t, \Theta_t)$ is independent of t and $\rho_t = 0$ for all $t \in M$, then $\mathcal{V}$ is locally trivial. (3) If

$H^2(V_0, \Theta_0)=0$ and $\dim H^1(V_0, \Theta_0)=m$, then for a sufficiently small neighborhood M of the origin of $\mathbf{C}^n$, there exists an analytic family $\varpi: \mathcal{V} \rightarrow M$ such that $\varpi^{-1}(0)=V_0$, ρ_t is bijective for any $t \in M$, and $\mathcal{V}$ is **analytically complete** at every point of M [8]. Here we say that a family $\mathcal{V} \rightarrow M$ is **differentiably (analytically) complete** at $t \in M$ if for any differentiable (analytic) family $\mathcal{W} \rightarrow N$ such that $W_{s_0}=V_t$ there is a C^∞-differentiable (holomorphic) mapping h from a neighborhood N' of s_0 to M such that $W_s=V_{h(s)}$ $(s \in N')$. (4) If we generalize the notion of deformation to include analytic families with singular analytic sets as parameter spaces, then every compact complex manifold V has an analytic family that contains V and is complete everywhere [9].

We sometimes consider deformation of complex manifolds under some additional conditions. For example, a family of submanifolds of a fixed complex manifold can be treated as we treated differentiable or analytic families. Also, if on a family $\varpi: \mathcal{V} \rightarrow M$ there is a fiber bundle $\mathfrak{B} \rightarrow \mathcal{V}$ whose fiber and structure groups are complex analytic and for any $t \in M$ the fiber $B_t \rightarrow V_t$ is a complex analytic fiber bundle, then we call $\mathfrak{B} \rightarrow \mathcal{V} \rightarrow M$ a **family of fiber bundles**. An example is the †Picard variety $\mathfrak{P}$ of a given †Kähler manifold V, which can be considered as a family of holomorphic line bundles on the trivial family $V \times \mathfrak{P}$ whose parameter space is endowed with a natural complex structure [6].

H. Generalized Monoidal Transformations

Let $\mathfrak{A}$ be a nonzero analytic coherent sheaf of ideals of Ω on a complex manifold X, and let Y be the set of common zeros $\mathfrak{A}$, which is an analytic subset of X. Then X has an open covering such that for each member U of the covering, there are elements $\varphi_1, \ldots, \varphi_m \in \Gamma(U, \mathfrak{A})$ that generate the stalk $\mathfrak{A}_x$ at all $x \in U$. Let W' be the graph of the holomorphic mapping $x \rightarrow (\varphi_1(x): \varphi_2(x): \ldots : \varphi_m(x))$ from $U - U \cap Y$ to $\mathbf{P}^{m-1}$, and denote by W the closure of W' in $U \times \mathbf{P}^{m-1}$. Then W is an †analytic space, possibly with singularities, that does not depend on the choice of the generators $\{\varphi_j\}$ and is determined uniquely by U and $\mathfrak{A}$. Therefore, all W can be glued together to form an analytic space $\tilde{X}$ and to determine a holomorphic mapping $p: \tilde{X} \rightarrow X$. We call $\tilde{X}$ the **generalized monoidal transform** of X with respect to $\mathfrak{A}$. If $\mathfrak{A}$ is the sheaf of ideals defined by Y, the process of making $\tilde{X}$ from X is called the **monoidal transformation** with center Y, and if Y is a point, it is called a **locally quadratic transformation** or a **σ-process**. If Y itself is an analytic submanifold, the transform $\tilde{X}$ is also a manifold. If Y has singularities, $\tilde{X}$ can be very complicated (→ 27 Analytic Spaces).

I. Kodaira Dimension of Complex Manifolds

Let X_1 and X_2 be complex varieties, $\psi: X_1 \rightarrow X_2$ a meromorphic mapping, and $G \subset X_1 \times X_2$ its graph. If the projection $\mathrm{pr}_2: G \rightarrow X_2$ is also a proper modification, ψ is called a bimeromorphic mapping (→ 27 Analytic Spaces). For an n-dimensional compact complex manifold V, we define the **Kodaira dimension** (or **canonical dimension**) $\kappa(V)$ of V as follows (S. Iitaka [27]). Let K be the canonical bundle of V, and $\mathbf{N}_0=\{m>0, m \in \mathbf{Z} \mid P_m(V)=\dim_{\mathbf{C}} H^0(V, \Omega(mK))>0\}$. If $\mathbf{N}_0$ is not empty, let d be the g.c.d. of the integers belonging to $\mathbf{N}_0$. Then there exist a positive integer m_0, positive numbers α, β, and a nonnegative integer κ such that, for $m \geqslant m_0$, the following inequality holds: $\alpha m^\kappa \leqslant P_{md}(V) \leqslant \beta m^\kappa$ for $m \gg 0$. We define $\kappa=\kappa(V)$ in this case. If $\mathbf{N}_0$ is empty, we define $\kappa(V)=-\infty$. $\kappa(V)$ is a bimeromorphic invariant of V and takes one of the following values: $-\infty$, 0, $1, \ldots, n=\dim V$. If $\kappa(V)$ is positive, then there exists a fiber space $f: V^* \rightarrow W$ of complex manifolds such that (1) V^* is bimeromorphically equivalent to V, (2) W is a nonsingular projective variety of dimension $\kappa(V)$, (3) f is a surjective and proper holomorphic mapping, (4) any general fiber $V_w^*=f^{-1}(w)$ $(w \in W)$ is irreducible, and (5) $\kappa(V_w^*)=0$. Moreover, such a fiber space is unique up to bimeromorphic equivalence [27, 34]. Note that the Kodaira dimension is not in general a deformation invariant (I. Nakamura [32]).

J. Analytic Surfaces

In what follows, a surface will mean a 2-dimensional compact complex manifold. For an analytic surface S, an exceptional curve on S and a (relatively) minimal model, etc., are defined with respect to bimeromorphic mappings in analogy with the corresponding concepts for an algebraic surface (→ 17 Algebraic Surfaces). Let $C \subset S$ be an irreducible curve on S. Then there exists a holomorphic mapping φ from S onto another surface S' such that $\varphi(C)$ is a point and such that φ induces the isomorphism $S - C \cong S' - \varphi(C)$ if and only if $C^2=-1$ and C is a nonsingular rational curve (Grauert [23]). S has a minimal model if and only if S is not a ruled surface (Kodaira [17]). The irregularity $q=h^{0,1}$, the geometric genus p_g, i-genus P_i, etc., are also defined in the same way as in the case of algebraic surfaces. Note that, in general, $h^{0,1} \neq h^{1,0}$. The Riemann-Roch theorem and M. Noether's formula are valid also for an analytic surface (Atiyah and Singer; → 361 Riemann-Roch Theorems). We call the transcendence degree of $K(V)$ (the field of meromorphic functions on V) over $\mathbf{C}$ the **algebraic dimension** of S and denote by $a(S)$.

K. Classification of Surfaces

The classification of analytic surfaces by the aid of their numerical invariants was completed by Kodaira, and includes as a special case Enriques's classification of algebraic surfaces [5,17]. By an **elliptic surface** we mean a surface from which there exists a surjective holomorphic mapping φ onto an algebraic curve Δ, such that for a general point p on Δ, $\varphi^{-1}(p)$ is an irreducible nonsingular elliptic curve. If $a(S)=1$ or $\kappa(S)=1$, S has a unique structure as an elliptic surface. The image by φ of the points on S at which φ is not of maximal rank is a finite subset $\{a_1, \ldots, a_r\}$ of Δ. Let t_i be a local coordinate on Δ around a_i with $t_i(a_i)=0$. We call a singular fiber of φ the divisor on S defined by $\{t_i \circ \varphi = 0\}$. The structure and the construction of singular fibers of elliptic surfaces have been completely determined by Kodaira [5]. By an elliptic surface of general type we mean a surface with Kodaira dimension 1. If $\kappa(S)=2$, S is projective algebraic, and is called a **surface of general type**. If $a(S)=0$, then there exist only a finite number of irreducible curves on S. By a **Hopf surface** we mean a surface whose universal covering is $\mathbf{C}^2-(0,0)$. If a surface S is homeomorphic to $S^1 \times S^3$, S is a Hopf surface. Let $b_\nu(S)$ be the νth Betti number of S. If $a(S)=0$, $b_1(S)=1$, $b_2(S)=0$, and S contains a curve, then S is a Hopf surface (Kodaira [17]). By a surface of type VII_0 we mean a minimal surface S with $b_1(S)=1$. M. Inoue constructed two families of new surfaces of type VII_0 with $b_2=0$, which contain no curves (1972). These surfaces have $\mathbf{H} \times \mathbf{C}$ as their universal coverings, where $\mathbf{H}$ is the upper half-plane. He also constructed a class of surfaces of type VII_0 with $b_2>0$, the construction of which is related to the resolution of cusp singularities of Hilbert modular surfaces. By an **Enriques surface** we mean a surface which has a $K3$ surface as its unramified double covering (→ 17 Algebraic Surfaces). Every Enriques surface is an algebraic elliptic surface, with $q=p_g=0$. By a **hyperelliptic surface** we mean an algebraic surface with $q=1$ and $12K \sim 0$. A hyperelliptic surface has an Abelian surface as its finite unramified covering, and is an elliptic fiber bundle over an elliptic curve. The classification of minimal surfaces is given in Table 1. (See [35].) The following relations hold among these invariants: Let b^+ (b^-) be the number of positive (negative) eigenvalues (counted with multiplicities) of the intersection matrix on $H^2(S,\mathbf{R})$ and c_i be the ith Chern class of S. Then

(1) $b^+ - b^- = \frac{1}{3}(c_1^2 - 2c_2)$ (**Hirzebruch signature theorem**);

(2) if b_1 is even, $q = h^{1,0} = \frac{1}{2}b_1$ and $b^+ = 2p_g + 1$;

(3) if b_1 is odd, $q = h^{1,0} + 1 = \frac{1}{2}(b_1+1)$ and $b^+ = 2p_g$.

Table 1. Classification of Minimal Surfaces

κ	p_g	P_{12}	q	b_1	Structure
2		>0			algebraic surface of general type
1					elliptic surface of general type
	1	1	2	4	complex torus
	1	1	2	3	elliptic surface with a trivial canonical bundle
0	0	1	1	2	hyperelliptic surface
	0	1	1	1	elliptic surface belonging to class VII
	1	1	0	0	$K3$ surface
	0	1	0	0	Enriques surface
			0	0	rational surface
$-\infty$	0	0	$\geqslant 1$	$2q$	ruled surface of genus q
			1	1	surface of class VII

Kähler Condition. Every surface with even first Betti number is a deformation of an algebraic surface (Kodaira [5]). Such an S actually admits a Kähler metric except for the case that S is a modification of $K3$ surface with $a(S)=0$ (Miyaoka [31]). Any elliptic surface with odd first Betti number admits an affine structure (Inoue).

Surfaces of General Type. If S is minimal, S is of general type if and only if $P_2(S)>0$ and $c_1^2(S)>0$. We denote by Φ_m the rational mapping defined by the complete linear system $|mK|$. Then for $m \geqslant 5$, Φ_m is birational and, moreover, a morphism if S is minimal. For $m \leqslant 4$, Φ_m is in general not birational. As for the birationality of Φ_m, the following results have been obtained by Kodaira, Bombieri, and Miyaoka [22,30,31]. (1) The image $\Phi_m(S)$ is normal if $m \geqslant 5$. (2) If $c_1^2 \geqslant 2$, Φ_4 is birational. (3) For $m \geqslant 3$, Φ_m is birational except in the following three cases: (i) $c_1^2 = 1$, $p_g = 2$, $m = 3$ and 4, where $\Phi_3(S)$ is a rational ruled surface Σ_2 of degree 4 embedded in $\mathbf{P}^5$ and $\Phi_4(S)$ is a quadric cone in $\mathbf{P}^8$; (ii) $c_1^2 = 2$, $p_g = 3$, $m = 3$, where $\Phi_3(S)$ is $\mathbf{P}^2$ embedded in $\mathbf{P}^9$ by th Veronese embedding; (iii) some surfaces with $c_1^2 = 2$, $p_g = 0$, $n = 3$. (4) If $c_1^2 \geqslant 10$, $p_g \geqslant 6$, Φ_2 is birational if and only if S does not have the structure of a fiber space over a nonsingular curve with a nonsingular curve of genus 2 as generic fiber.

Deformation of Surfaces. Plurigenera of surfaces are deformation invariants. Each class in the above table is closed under deformations [26]. Deformations of elliptic surfaces have been studied in detail. Global deformation of certain surfaces of general type, e.g., quintic surfaces in $\mathbf{P}^3$, have been completely determined [24,25].

References

[1] A. Andreotti and H. Grauet, Théorèmes de finitude pour la cohomologie des espaces complexes, Bull. Soc. Math. France, 90 (1962), 193–259.

[2] W. L. Chow, On compact complex analytic varieties, Amer. J. Math., 71 (1949), 893–914.

[3] A. Fröhlicher, Zur Differentialgeometrie der komplexen Strukturen, Math. Ann., 129 (1955), 50–95.

[4] H. Grauert, Ein Theorem der analytischen Garbentheorie und die Modulräume komplexer Strukturen, Publ. Math. Inst. HES, no. 5, 1960.

[5] K. Kodaira, On compact complex analytic surfaces, Ann. of Math., I, (2) 71 (1960), 111–152; II, (2) 77 (1963), 563–626; III, (2) 78 (1963), 1–40.

[6] K. Kodaira and D. C. Spencer, Groups of complex line bundles over compact Kähler varieties; Divisor class groups on algebraic varieties, Proc. Nat. Acad. Sci. US, 39 (1953), 868–877.

[7] K. Kodaira and D. C. Spencer, On deformations of complex analytic structures, Ann. of Math., I, II, (2) 67 (1958), 328–466; III, (2) 71 (1960), 43–76.

[8] K. Kodaira, L. Nirenberg, and D. C. Spencer, On the existence of deformations of complex analytic structures, Ann. of Math., (2) 68 (1958), 450–459.

[9] M. Kuranishi, On the locally complete families of complex analytic structures, Ann. of Math., (2) 75 (1962), 536–577.

[10] A. Newlander and L. Nirenberg, Complex analytic coordinates in almost complex manifolds, Ann. of Math., (2) 65 (1957), 391–404.

[11] Séminaires H. Cartan, 1951–1952, Ecole Norm. Sup.

[12] Séminaires H. Cartan, 1953–1954, Ecole Norm. Sup.

[13] J.-P. Serre, Un théorème de dualité, Comment. Math. Helv., 29 (1955), 9–26.

[14] J.-P. Serre, Géométrie algébrique et géométrie analytique, Ann. Inst. Fourier, 6 (1955–1956), 1–42.

[15] C. L. Siegel, Meromorphe Funktionen auf kompacten analytischen Mannigfaltigkeiten, Nachr. Akad. Wiss. Göttingen, Math.-Phys. Kl., (1955), 71–77.

[16] F. E. Othmer, Elementarer Beweis des Hauptsatzes über meromorphe Funktionenkörper, Math. Ann., 141 (1960), 99–106.

[17] K. Kodaira, On the structure of compact complex analytic surfaces, Amer. J. Math., I, 86 (1964), 751–798; II, 88 (1966), 682–721; III, IV, 90 (1968), 56–83, 1048–1066.

[18] M. Kuranishi, New proof for the existence of locally complete families of complex structures, Proc. Conference on Complex Analysis, Minneapolis, 1964 (Springer, 1965, p. 142–154).

[19] K. Yano, Differential geometry on complex and almost complex spaces, Macmillan, 1965.

[20] S. S. Chern, Complex manifolds, Lecture notes, Univ. of Chicago, 1955–1956.

[21] S. S. Chern, Complex manifolds without potential theory, Van Nostrand, 1967.

[22] E. Bombieri, Canonical models of surfaces of general type, Publ. Math. Inst. HES, no. 42, 1973, p. 171–219.

[23] H. Grauert, Über Modifikationen und exzeptionelle analytische Mengen, Math. Ann., 146 (1962), 331–368.

[24] E. Horikawa, On deformations of quintic surfaces, Proc. Japan Acad., 49 (1973), 377–379.

[25] E. Horikawa, On the number of moduli of certain algebraic surfaces of general type, J. Fac. Sci. Univ. Tokyo, 22 (1975), 67–78.

[26] S. Iitaka, Deformations of compact complex surfaces: I, in Global analysis, ed. D. C. Spencer and S. Iyanaga, Princeton Univ. Press (1969), p. 267–272; II, III, J. Math. Soc. Japan, 22 (1970), 247–261, and 23 (1971), 692–705.

[27] S. Iitaka, On D-dimensions of algebraic varieties, J. Math. Soc. Japan, 23 (1971), 356–373.

[28] M. Inoue, On surfaces of class VII_0, Inventiones Math., 24 (1974), 269–310.

[29] A. Kas, Deformation of elliptic surfaces, Thesis, Stanford Univ., 1966.
[30] K. Kodaira, Pluricanonical system of algebraic surfaces of general type, J. Math. Soc. Japan, 20 (1968), 170–192.
[31] Y. Miyaoka, Kähler metrics on elliptic surfaces, Proc. Japan Acad., 50 (1974), 533–536.
[32] I. Nakamura, On classification of parallelisable manifolds and small deformations, J. Differential Geometry, 10 (1975), 85–112.
[33] I. R. Šafarevič et al., Algebraic surfaces, Proc. Steklov Inst., Moscow (1965); English translation, Amer. Math. Soc., 1967.
[34] K. Ueno, Classification theory of algebraic varieties and compact complex spaces, Lecture notes in math. 439, Springer, 1975.
[35] K. Uneo, Introduction to classification theory of algebraic varieties and compact complex spaces, classification of algebraic varieties and compact complex manifolds. Proceedings 1974, Lecture notes in math. 412, Springer, 1974, p.288–332.

75 (V.15) Complex Multiplication

A. Classical Theory

If the ratio ω_1/ω_2 of two periods ω_1, ω_2 of an †elliptic function f belongs to an imaginary †quadratic field K, then there exists an †algebraic relation between $f(z)$ and $f(\lambda z)$ for any λ in K, and such an f is said to have **complex multiplication**. This phenomenon for the †sn function with modulus $\sqrt{-1}$ was discovered by C. F. Gauss and was applied to the problem of dividing a †lemniscate into five arcs of equal length. More generally, N. H. Abel showed that the *special dividing equation* of an sn function with complex multiplication is algebraically solvable. From a number-theoretic point of view, L. Kronecker conjectured that every †Abelian extension of an imaginary quadratic number field K is determined by a *transform equation* of an elliptic function with complex multiplication by a number of K (1880). This is an analog of the fact, announced by Kronecker and proved by H. Weber, that every Abelian extension of the rational number field is a subfield of a †cyclotomic field. Kronecker's work was continued by Weber [2], and his conjecture was proved by T. Takagi (1903) for $K=\mathbf{Q}(\sqrt{-1})$, by T. Takenouchi (1916) for $K=\mathbf{Q}(e^{2\pi i/3})$, and by Takagi (1920) for the general case using †class field theory. H. Hasse [5] and M. Sugawara simplified the theory of complex multiplication, and Hasse noticed a relationship between complex multiplication and †congruence zeta functions. Working from Hasse's idea, M. Deuring constructed the theory of complex multiplication purely algebraically and determined †Hasse's zeta function of an elliptic curve with complex multiplication.

In the rest of this article, K always stands for an imaginary quadratic field. Let L be a †lattice group on the complex plane $\mathbf{C}$ generated by ω_1, ω_2, and let z be a complex variable. Define functions $\wp$, g_2, g_3 as follows: $\wp(z,L)=\wp(z;\omega_1,\omega_2)=z^{-2}+\Sigma((z-\omega)^{-2}-\omega^{-2})$, $g_2(L)=g_2(\omega_1,\omega_2)=60\Sigma\omega^{-4}$, $g_3(L)=g_3(\omega_1,\omega_2)=140\Sigma\omega^{-6}$, where Σ means the sum over the elements of L except 0. Let $\wp'$ be the derivative of $\wp$; then $z\rightarrow(1,\wp(z),\wp'(z))$ is a one-to-one correspondence between the points on the complex torus $\mathbf{C}/L$ and those on the †elliptic curve E; $X_0X_2^2=4X_1^3-g_2X_0^2X_1-g_3X_0^3$ in the projective plane. If the quotient ω_1/ω_2 generates K, then the ring of analytic endomorphisms of $\mathbf{C}/L$ (i.e., the ring of endomorphisms of E) is isomorphic to a subring of the †principal order $\mathfrak{o}$ of K. In particular, if the lattice group L is an †ideal of K (for $K\subset\mathbf{C}$; → 343 Quadratic Fields), then the ring of endomorphisms coincides with $\mathfrak{o}$. The function

$$J(\tau)=J(E)=J(L)=2^6 3^3 g_2(L)^3/\Delta(L)$$

($\Delta(L)=g_2(L)^3-27g_3(L)^2$) of $\tau=\omega_1/\omega_2$, $\operatorname{Im}\tau>0$, is a †modular function of level 1, and $J(E)$ is called the **invariant** of the elliptic curve E. If E has a complex multiplication, then $J(E)$ is an algebraic integer. Then the three main theorems of the classical theory of complex multiplication can be stated as follows:

Theorem 1. Let h be the †class number of K, and let $\mathfrak{a}_1,\dots,\mathfrak{a}_h$ be a set of representatives of ideal classes of K. Then $J(\mathfrak{a}_1),\dots,J(\mathfrak{a}_h)$ are exactly the conjugates of $J(\mathfrak{a}_1)$ over K, and $K(J(\mathfrak{a}_1))$ is the maximal unramified Abelian extension (the †absolute class field) of K (→ 62 Class Field Theory).

Next we define the function f by

$$\begin{aligned} f(z;L)&=g_2g_3\Delta^{-1}\cdot\wp(z;L), && K\neq\mathbf{Q}(\sqrt{-1}),\ \mathbf{Q}(e^{2\pi i/3}),\\ &=g_2^2\Delta^{-1}\cdot\wp(z;L)^2, && K=\mathbf{Q}(\sqrt{-1}),\\ &=g_3\Delta^{-1}\cdot\wp(z;L)^3, && K=\mathbf{Q}(e^{2\pi i/3}).\end{aligned}$$

Theorem 2. Let $\mathfrak{o}$ be the †principal order of K, $\mathfrak{m}$ be an integral ideal of K, and $\mathfrak{a}$ be an arbitrary ideal of K. Choose a number ξ of K such that $\mathfrak{m}=\{\lambda\in\mathfrak{o}\mid\lambda\xi\in\mathfrak{a}\}$. Then $K(J(\mathfrak{a}),f(\xi;\mathfrak{a}))$ is the †class field for the †ray mod $\mathfrak{m}$.

The number ξ in this theorem is obtained by $\mathfrak{a}^{-1}\mathfrak{m}(\xi)=\mathfrak{b}$, where $\mathfrak{b}$ is an integral ideal belonging to the ideal class of $\mathfrak{a}^{-1}\mathfrak{m}$. Thus $f(\xi;\mathfrak{a})$ is a "special value" of an elliptic function as well as of a modular function.

The †general law of reciprocity, the †principal ideal theorem, and the ramification in the class field in theorem 2 can also be described in terms of elliptic functions or elliptic curves. In general, the ring of endomorphisms of an elliptic curve defined over a field k of characteristic 0 is either $\mathbf{Z}$ or an order of an imaginary quadratic field. On the other hand, if the characteristic of k is not zero, then the ring of endomorphisms may be an order in a definite †quaternion algebra.

B. Complex Multiplication of an Abelian Variety

Following Kronecker's idea, D. Hilbert posed in his lecture at Paris (1900) the so-called 12th problem: to find an analytic function whose special values generate Abelian extensions over a given algebraic number field (→ 198 Hilbert). E. Hecke constructed unramified Abelian extensions of an imaginary biquadratic field using †Hilbert modular functions [10]. After this, there was no notable development concerning the problem until the theory of complex multiplication was generalized to the case of †Abelian varieties, which was made possible by recent progress in algebraic geometry, in particular A. Weil's geometric theory of Abelian varieties (G. Shimura and Y. Taniyama [11]). The following results have been obtained: Consider a triple $(A,\mathfrak{X},t)$, consisting of an Abelian variety A defined over $\mathbf{C}$, a †polarization $\mathfrak{X}$ of A, and a point t of A. Call two such triples $(A,\mathfrak{X},t)$ and $(A',\mathfrak{X}',t')$ isomorphic if an isomorphism of A onto A' maps $\mathfrak{X}$ onto $\mathfrak{X}'$ and t onto t'. Then there exists one and only one subfield k_0 of $\mathbf{C}$ with the following property: In order that $(A,\mathfrak{X},t)$ and $(A^\sigma,\mathfrak{X}^\sigma,t^\sigma)$ are isomorphic for an automorphism σ of $\mathbf{C}$, it is necessary and sufficient that σ fix all elements of k_0. We call k_0 the **field of moduli** of $(A,\mathfrak{X},t)$. If A is an elliptic curve E and if $t=0$, then $k_0=\mathbf{Q}(J(E))$. In the higher-dimensional case, the field of moduli is generated by special values of a †Siegel modular function.

If F is a totally imaginary number field which is a quadratic extension of a totally real field F_0 of degree n, then there exists a set $\{\varphi_1,\ldots,\varphi_n\}$ of n different isomorphisms of F into $\mathbf{C}$ such that $\bar{\varphi}_i=\varphi_j$ never occurs for $i\neq j$, where the bar denotes complex conjugation. We call $(F;\{\varphi_\lambda\})$ a **CM-type**.

If $\mathfrak{a}$ is an ideal of F, then $L=\{(\varphi_1(\alpha),\ldots,\varphi_n(\alpha))\in\mathbf{C}^n\mid\alpha\in\mathfrak{a}\}$ is a lattice group in the n-dimensional complex linear space $\mathbf{C}^n$, and $\mathbf{C}^n/L$ is analytically isomorphic to an Abelian variety A of dimension n in a complex projective space. The †endomorphism ring $\mathfrak{A}(A)$ of A contains a ring which is isomorphic to the principal order $\mathfrak{o}$ of F. Conversely, every Abelian variety of dimension n such that $\mathfrak{A}(A)$ contains a ring isomorphic to $\mathfrak{o}$ can be constructed in this way.

Let M be a normal extension of $\mathbf{Q}$ containing F. Denote the Galois group of M by G and the subgroup of G corresponding to F by H, and put $S=\bigcup_\lambda\varphi_\lambda H$, where φ_λ stands for a prolongation of the previous φ_λ to G. Then the following three conditions are equivalent: (i) $\mathfrak{A}(A)\cong\mathfrak{o}$; (ii) A is simple; (iii) $H=\{\gamma\in G\mid S\gamma=S\}$. If $H^*=\{\delta\in G\mid\delta S=S\}$, then we can choose $\psi_\mu\in G$ such that $S=\bigcup_\mu H^*\psi_\mu$. If F^* is the subfield of M corresponding to H^*, then $(F^*;\{\psi_\mu\})$ is also a CM-type, and $F^*=\mathbf{Q}(\Sigma_\lambda\varphi_\lambda(\alpha)\mid\alpha\in F)$. Moreover, for an ideal $\mathfrak{x}$ of F^*, $\Pi_\mu\psi_\mu(\mathfrak{x})$ is an ideal of F.

Now let $H_\mathfrak{m}$ be the group of ideals $\mathfrak{x}$ of F^* which are relatively prime to the norm $N(\mathfrak{m})$ of an integral ideal $\mathfrak{m}$ of F^*, and for which there exists a number ξ of F with

$$\prod_\mu\psi_\mu(\mathfrak{x})=(\xi),\quad N(\mathfrak{x})=|\xi|^2,\quad \xi\equiv1\ (\mathrm{mod}^\times\mathfrak{m}).$$

Then $H_\mathfrak{m}$ is an †ideal group modulo $\mathfrak{m}$ in F^*.

Theorem 3. Assume that $\mathfrak{A}(A)\cong\mathfrak{o}$, and let $\mathfrak{X}$ be an arbitrary polarization of A. Denote a point of A with $\mathfrak{m}=\{\lambda\in\mathfrak{o}\mid\lambda t=0\}$ by t, and let $k_\mathfrak{m}$ be the field of moduli of $(A,\mathfrak{X},t)$. Then $k_\mathfrak{m}F^*$ is the †class field over F^* corresponding to $H_\mathfrak{m}$.

The point t in theorem 3 always exists. In particular, if $\mathfrak{m}=\mathfrak{o}$, then $t=0$. In the case where A is an elliptic curve E, F is an imaginary quadratic field and we have $F^*=K$, and theorem 3 coincides essentially with the content of theorems 1 and 2. If $n>1$, $F=F^*$ holds only in special cases.

The theory of complex multiplication of A is closely related to the †Hasse zeta function of A (→ 436 Zeta Functions).

References

[1] L. Kronecker, Zur Theorie der elliptischen Funktionen, Monatber. Königlich Preuss. Akad. Wiss. Berlin, (1881), 1165–1172. (Werke IV, Teubner, 1929, p. 309–318; Chelsea, 1969.)
[2] H. Weber, Lehrbuch der Algebra III, F. Vieweg, second edition, 1908 (Chelsea, 1961).
[3] T. Takagi, Über eine Theorie des relativ-Abelschen Zahlkörpers, J. Coll. Sci. Imp. Univ. Tokyo, 41 (1920), 1–132.

[4] R. Fueter, Vorlesungen über die singulären Moduln und die komplexe Multiplikation der elliptischen Funktionen, Teubner, I, 1924; II, 1927.
[5] H. Hasse, Neue Begründung der komplexen Multiplikation, J. Reine Angew. Math., I, 157 (1926), 115–139; II, 165 (1931), 64–88.
[6] R. Fricke, Lehrbuch der Algebra III, F. Vieweg, 1928.
[7] M. Deuring, Die Typen der Multiplikatorenringe elliptischer Funktionenkörper, Abh. Math. Sem. Univ. Hamburg, 14 (1941), 197–272.
[8] M. Deuring, Die Struktur der elliptischen Funktionenkörper und die Klassenkörper der imaginären quadratischen Zahlkörper, Math. Ann., 124 (1952), 393–426.
[9] M. Deuring, Die Klassenkörper der komplexen Multiplikation, Enzykl. Math., Bd. I_2, Heft 10, Teil II, 23, 1958.
[10] E. Hecke, Höhere Modulfunktionen und ihre anwendung auf die Zahlentheorie, Math. Ann., 71 (1912), 1–37; Über die Konstruktion relativ-Abelscher Zahlkörper durch Modulfunktionen von zwei Variabeln, Math. Ann., 74 (1913), 465–510. (Mathematische Werke, Vandenhoeck & Ruprecht, 1959, p. 21–58, 69–114.)
[11] G. Shimura and Y. Taniyama, Complex multiplication of Abelian varieties and its applications to number theory, Publ. Math. Soc. Japan, 1961.
[12] G. Shimura, Construction of class fields and zeta functions of algebraic curves, Ann. of Math., (2) 85 (1967), 58–159.
[13] A. Borel et al., Seminar on complex multiplication, Lecture notes in math. 21, Springer, 1966.
[14] G. Shimura, Introduction to the arithmetic theory of automorphic functions, Publ. Math. Soc. Japan and Princeton Univ. Press, 1971.

76 (II.12) Complex Numbers

A. Algebraic Properties of Complex Numbers

A **complex number** is an expression of the form $a+ib$ with arbitrary real numbers a and b and the **imaginary unit** i. Writing $\alpha=a+ib$, $\beta=c+id$, we define $\alpha=\beta$ if and only if $a=c$ and $b=d$. As regards algebraic operations with complex numbers, we define $\alpha+\beta=(a+c)+i(b+d)$, $\alpha-\beta=(a-c)+i(b-d)$, $\alpha\beta=(ac-bd)+i(ad+bc)$, and for $\beta\neq0$, i.e., $c^2+d^2\neq0$, $\alpha/\beta=(ac+bd)/(c^2+d^2)+i((ad-bc)/(c^2+d^2))$. Then the addition and multiplication thus defined obey commutative, associative, and distributive laws, and complex numbers form a †commutative field with $0=0+i0$ and $1=1+i0$ as its zero element of addition and identity element of multiplication, respectively. The set of all complex numbers is usually denoted by **C**.

By assigning to each real number a a complex number $a+i0$, algebraic operations on real numbers are carried into those of the corresponding complex numbers. That is to say, the field **R** of all real numbers is mapped isomorphically into the field **C** of all complex numbers. By identifying a with $a+i0$, **R** may be regarded as a subfield of **C**. Also, $0+i1$ will be denoted simply by i. From the previous definition of algebraic operations, it follows that $i^2=-1$. Furthermore, since $\alpha=a+ib=(a+i0)+(b+i0)(0+i1)$, $a+ib$ is not a mere symbolic expression but can also be regarded as the outcome of algebraic operations on a, b, i in **C**. The **real** and **imaginary parts** of a complex number $\alpha=a+ib$ are, by definition, a and b, denoted by $\mathrm{Re}\,\alpha$ and $\mathrm{Im}\,\alpha$, respectively. A complex number which is not a real number is sometimes called an **imaginary number**; in particular, a complex number α with $\mathrm{Re}\,\alpha=0$ is called a **purely imaginary number**. For each complex number $\alpha=a+ib$, we define its **conjugate complex number** as $a-ib$, and denote if by $\bar{\alpha}$. We then have $\overline{\alpha+\beta}=\bar{\alpha}+\bar{\beta}$ and $\overline{\alpha\beta}=\bar{\alpha}\bar{\beta}$. The mapping $\alpha\to\bar{\alpha}$ is an †automorphism of **C** which leaves each element of **R** invariant. Also, the following relations hold: $\mathrm{Re}\,\alpha=(\alpha+\bar{\alpha})/2$ and $\mathrm{Im}\,\alpha=(\alpha-\bar{\alpha})/(2i)$.

Regarded as an overfield of **R**, **C** is an †extension field of **R** of degree 2 obtained by the adjunction of i, which is a root of an irreducible equation $x^2+1=0$. The important algebraic property of **C** is that it is †algebraically closed. Namely, for any polynomial $f(x)$ with coefficients in **C**, the equation $f(x)=0$ possesses at least one root in **C** (†Gauss's fundamental theorem of algebra).

B. Topology of C

The **absolute value** (or **modulus**) of a complex number $\alpha=a+ib$, denoted by $|\alpha|$, is by definition $|\alpha|=\sqrt{a^2+b^2}=\sqrt{\alpha\bar{\alpha}}$. If α is real, then the absolute value of α in the sense of complex numbers is identical to the one in the sense of real numbers. It always holds that $|\alpha|\geqslant0$ and $|\alpha|=0\Leftrightarrow\alpha=0$. It further follows that $|\alpha+\beta|\leqslant|\alpha|+|\beta|$, $|\alpha\beta|=|\alpha||\beta|$, and $|\alpha|=|\bar{\alpha}|$.

For each pair of complex numbers α and β, define $\rho(\alpha,\beta)=|\alpha-\beta|$. Then with $\rho(\alpha,\beta)$

as the †distance function, **C** satisfies the axioms for a †metric space, and in particular, $\lim_{n\to\infty}\alpha_n=\alpha_0 \Leftrightarrow \lim\rho(\alpha_n,\alpha_0)=0 \Leftrightarrow \lim|\alpha_n-\alpha_0|=0 \Leftrightarrow (\lim a_n=a_0$ and $\lim b_n=b_0)$ (where $\alpha_n=a_n+ib_n$, $\alpha_0=a_0+ib_0$). From this it is easily seen, as in the case of the set **R** of all real numbers, that **C** also becomes a †locally compact and †complete metric space.

With respect to this topology, the four operations (except for division by zero) are continuous: $\alpha_n\to\alpha_0$ and $\beta_n\to\beta_0$ imply $\alpha_n+\beta_n\to\alpha_0+\beta_0$; $\alpha_n-\beta_n\to\alpha_0-\beta_0$; $\alpha_n\beta_n\to\alpha_0\beta_0$; and $\alpha_n/\beta_n\to\alpha_0/\beta_0$ (where in the last case we assume $\beta_n\neq0$ and $\beta_0\neq0$). Thus, **C** becomes a topological field. Furthermore, the assignment $\alpha\to\bar\alpha$ gives a continuous mapping $\mathbf{C}\to\mathbf{C}$, a homeomorphic automorphism of **C**.

C. The Complex Plane

If in a plane to which is assigned rectangular coordinate axes, a complex number $\alpha=a+ib$ is represented by a point (a,b), then the plane is called the **complex (number) plane (Gauss-Argand plane** or **Gaussian plane**) (Fig. 1), and the point representing α is called simply the point α. The abscissa and ordinate axes are called the **real** and **imaginary axes**, respectively. A point $\alpha=a+ib$ may be represented by †polar coordinates r, θ with the origin and the real axis as pole and generating line, respectively, where $r=\sqrt{a^2+b^2}$ is the absolute value $|\alpha|$ of α and θ is the **argument** (or **amplitude**), denoted by $\arg\alpha$, of α. The argument of α is uniquely determined $\operatorname{mod} 2\pi$ if $\alpha\neq0$ and is an arbitrary real number if $\alpha=0$.

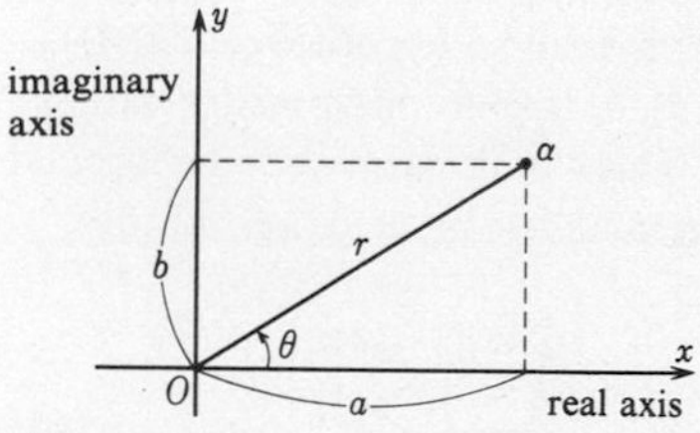

Fig. 1

The absolute value $|\alpha|$ of a complex number α, regarded as a vector from the origin to the point α, is the length of this vector. For complex numbers α and β, to the sum of the vectors α and β corresponds the sum $\alpha+\beta$ of the complex numbers. A complex number α, in terms of its absolute value r and argument θ, is expressed as $\alpha=r(\cos\theta+i\sin\theta)$, which is called the **polar form** of α. For polar forms the following hold: $\bar\alpha=r(\cos\theta-i\sin\theta)=r(\cos(-\theta)+i\sin(-\theta))$; $\alpha^{-1}=\bar\alpha/|\alpha|^2=r^{-1}(\cos(-\theta)+i\sin(-\theta))$ $(\alpha\neq0)$; and $\alpha_1\alpha_2=r_1r_2(\cos(\theta_1+\theta_2)+i\sin(\theta_1+\theta_2))$, where $|\alpha_j|=r_j$ and $\arg\alpha_j=\theta_j$ for $j=1,2$. This last relation when $r_1=r_2=1$ is called **de Moivre's formula.** The nth roots of unity in **C** are given by $\rho_j=\cos2\pi j/n+i\sin2\pi j/n$ $(j=0,1,\ldots,n-1)$ (Fig. 2).

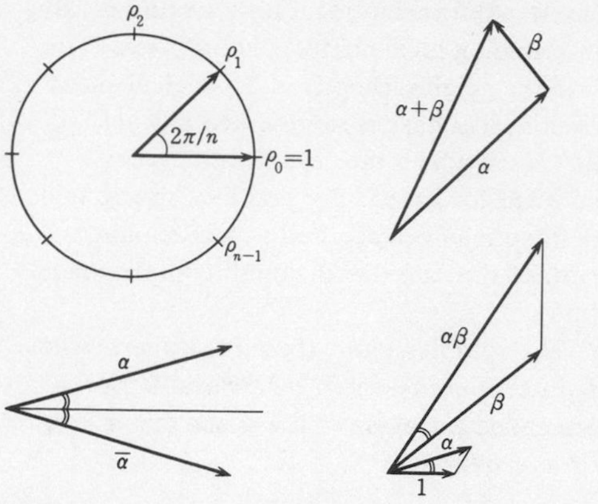

Fig. 2

In the complex plane, the mapping $\alpha\to\bar\alpha$ corresponds to the †reflection of the plane in the real axis, $\alpha\to\alpha+\beta$ to the parallel †translation along a vector β, $\alpha\to\alpha\beta$ $(\beta\neq0)$ to the †rotation with the angle $\arg\beta$ followed by the †homothetic transformation with the center 0 and the ratio constant $|\beta|$, and $\alpha\to\bar\alpha^{-1}$ to the †inversion with respect to the **unit circle** $\{\alpha \mid |\alpha|=1\}$.

The distance $\rho(\alpha,\beta)=|\alpha-\beta|$ between α and β in **C** coincides with their Euclidean distance, provided that α and β are regarded as points in the Euclidean plane, so that the complex plane is †isometric, accordingly homeomorphic, to the Euclidean plane.

D. The Complex Sphere

In the rest of this article, P denotes the complex plane and Σ denotes the sphere of radius 1, with 0, the origin of P, as its center. The points $N(0,0,1)$ and $S(0,0,-1)$ of Σ will be called the **north** and **south pole**, respectively (Fig. 3), where the 1st and 2nd coordinate axes are the real and imaginary axes of P, respectively, and the 3rd coordinate axis is orthogonal to P. A straight line from N through a point z (a complex number) in P intersects Σ at a point $Z=(x_1,x_2,x_3)$ different from N, where $z=(x_1+ix_2)/(1-x_3)$, $x_1=(z+\bar z)/(1+|z|^2)$, $x_2=(z-\bar z)/(i(1+|z|^2))$, and $x_3=(|z|^2-1)/(|z|^2+1)$. The mapping $z\to Z$ is called a **stereographic projection** from N, by means of which P and $\Sigma-\{N\}$ become †conformally equivalent to each other. Consequently z may be represented by a point Z of $\Sigma-\{N\}$, and Σ thus used is called a **complex sphere** (or **Riemann sphere**). Let us adjoin to the complex plane P a new element, denoted by ∞, called the **point at infinity**, which corresponds to the only exceptional point N of Σ. The topology of the complex plane with ∞ may be introduced by the corresponding topology of the Riemann sphere. Indeed, the family of all the sets $\{z \mid |z| > M\} \cup \{\infty\}$ for $M > 0$ forms a †local base around ∞. By introducing local complex coordinates $\zeta = 1/z$ into the neighborhoods of ∞, each element of this local base is represented as $\{\zeta \mid |\zeta| < M^{-1}\}$, in which the convention $\zeta = 0$ is adopted for $z = \infty$. The complex sphere thus defined may be regarded as a †Riemann surface (i.e., a 1-dimensional †complex manifold).

The complex plane (complex sphere) whose points are represented by a variable z or w are called a **z-plane** or a **w-plane** (a **z-sphere** or a **w-sphere**).

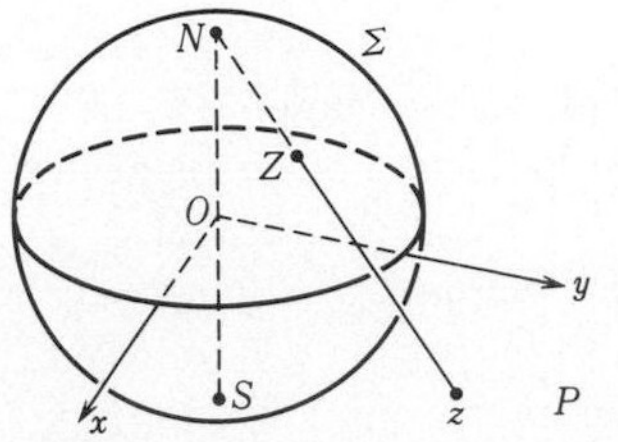

Fig. 3

E. Linear Fractional Functions

Given complex numbers a, b, c, and d with $ad - bc \neq 0$, we define a **linear fractional function** (or simply **linear function**)

$$w = \frac{az+b}{cz+d}. \qquad (1)$$

As a mapping from the z-sphere into the w-sphere, this linear function is called a **Möbius transformation (linear fractional** or simply **linear transformation**). The usual linear transformation, i.e., the one with $c = 0$ in the present case, is sometimes distinguished as an **entire linear transformation**. Since (1) depends only on the proportion $a:b:c:d$, we may assume $ad - bc = 1$ without loss of generality.

(1) is †holomorphic and †univalent on the whole z-sphere with only one exceptional †pole at $-d/c$ (∞ if $c = 0$) of order 1, and the inverse of (1) is also a linear fractional function. The set of all linear transformations forms a †group with composition of transformations as the group operation. One of its subgroups is the †modular group.

Linear transformations carry any circle of the complex plane (or of the Riemann sphere) into a circle of the same plane (or of the same sphere) if we adopt the convention that straight lines are a special kind of circle. (In the case of a Riemann sphere, no such convention is necessary.) Given on a plane a circle with center o and radius r and two points p and p' on a half-line issuing from o satisfying $op \cdot op' = r^2$, the points p and p' are called **symmetric points** (or **reflection points**) with respect to the circle. The transformation $p \rightarrow p'$ is called the **inversion** with respect to this circle. In the complex plane, let z and z' be symmetric points with respect to a circle C. Suppose that by a linear transformation, z, z' and C are carried to points w, w' and a circle D, respectively; then w and w' become symmetric with respect to the circle D (**principle of reflection**). Thus, symmetricity is invariant under linear transformation. Also the †anharmonic ratio of any four points z_1, z_2, z_3, and z_4, $(z_1,z_2;z_3,z_4) = (z_1 - z_3)/(z_1 - z_4):(z_2 - z_3)/(z_2 - z_4)$, is invariant under linear transformations; i.e., $(z_1,z_2;z_3,z_4) = (w_1,w_2;w_3,w_4)$ holds, where w_j is the image of z_j under a linear transformation ($j = 1,2,3,4$).

F. Normal Forms of Linear Transformations

There exist fixed points of the transformation (1) on the Riemann sphere, i.e., points satisfying $z = (az+b)/(cz+d)$. The number of fixed points is 2 or 1, except for the case where $w = z$. If the transformation has two fixed points, they will be denoted here by p and q. The natural convention $p = q$ is adopted if the transformation has one fixed point. If $c = 0$, then p or q is ∞, and furthermore if $c = a - d = 0$, then p and q are both ∞.

For unequal finite p and q, (1) may be rewritten in the following normal form:

$$\frac{w-p}{w-q} = \alpha \frac{z-p}{z-q}, \qquad \alpha = \frac{a-cp}{a-cq} \neq 1,$$

in which, according as $\arg \alpha = 0$, $|\alpha| = 1$, or otherwise, (1) is called a **hyperbolic** (Fig. 4), **elliptic** (Fig. 5), or **loxodromic transformation**, respectively. This classification can be applied also for finite p and infinite q, i.e., to the

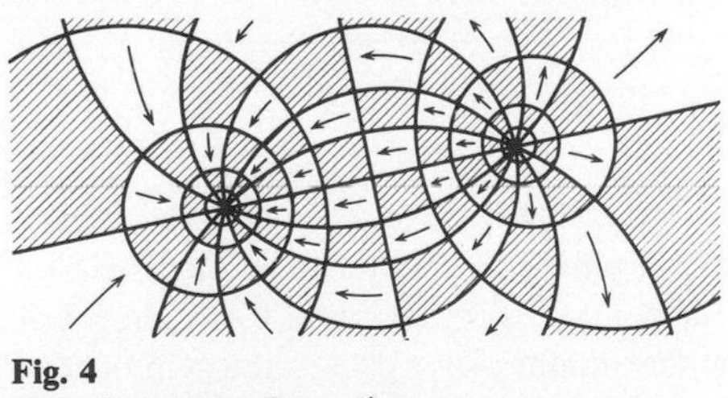

Fig. 4
Hyperbolic transformation.

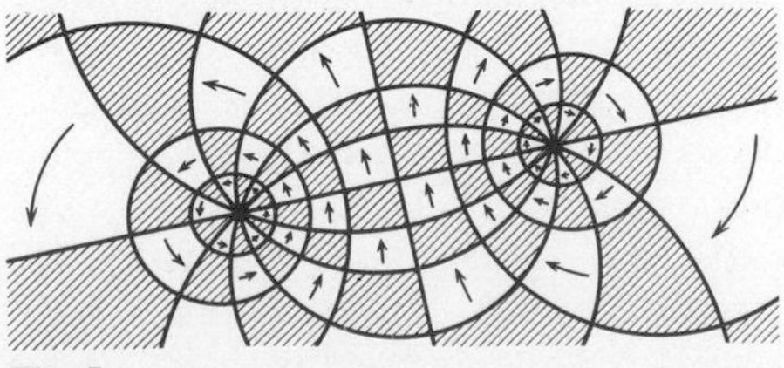

Fig. 5
Elliptic transformation.

transformation $w-p=\alpha(z-p)$. Furthermore, for $p=q\neq\infty$, (1) is rewritten in the following form:

$$\frac{1}{w-p}=\frac{1}{z-p}+\beta,\quad \beta=\frac{c}{a-cp}.$$

In this case (1) is called a **parabolic transformation** (Fig. 6). For $p=q=\infty$, i.e., if $w=z+\beta$, (1) is also called parabolic. We can easily determine to which class (1) belongs by the discriminant $D=(a+d)^2-4$ of the quadratic equation $cz^2-(a-d)z-b=0$ obtained from $z=(az+b)/(cz+d)$ with $ad-bc=1$ by multiplying both sides by $cz+d$. If $a+d$ is real, then according as $D>0$, <0, or $=0$, (1) is hyperbolic, elliptic, or parabolic, respectively, and if $a+d$ is not real, then the transformation is loxodromic.

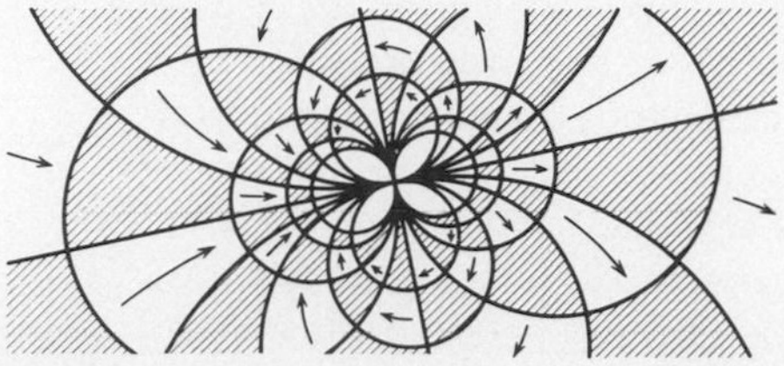

Fig. 6
Parabolic transformation.

Let D and D' be two arbitrary circular disks. Then there always exists a linear transformation which gives a one-to-one †conformal mapping from D onto D'. Conversely, any mapping with this property is given only by linear transformations (provided that the half-plane having a straight line together with the point at infinity as its boundary is regarded as a closed disk), which are uniquely determined by giving three points a, b, c from the boundary of D and as their corresponding points, three arbitrary points a', b', c' from the boundary of D'.

G. The Poincaré Metric

Since conformal mappings from the domain $|z|<1$ onto $|w|<1$ are given by the transformations $w=\varepsilon(z-z_0)/(1-\bar{z}_0z)$ ($|\varepsilon|=1$, $|z_0|<1$) (→ Appendix A, Table 13), for corresponding z and w it holds that

$$|dw|/(1-|w|^2)=|dz|/(1-|z|^2). \qquad (2)$$

$|dz|/(1-|z|^2)$ is called **Poincaré's differential invariant**. With a metric having $ds=|dz|/(1-|z|^2)$ as its †line element, the unit disk $|z|<1$ becomes a †non-Euclidean space in the sense of Lobačevskiĭ, and the metric is called the **Poincaré metric**. Furthermore, since the transformations (2) leave the length of curves invariant, they may be regarded as †motions in this space, where the †geodesic through two points z_1 and z_2 is the circular arc orthogonal to the unit circle. If we denote the intersections of the arc with the unit circle by z_3 and z_4, then the †non-Euclidean distance between two points z_1 and z_2 along the geodesic is given by $(1/2)\log(z_1,z_2;z_3,z_4)$, provided that the points z_4, z_1, z_2, z_3 are arranged on the arc in this order (→ 283 Non-Euclidean Geometry).

References

[1] Y. Kurosu, Hukusosû (Japanese; Complex numbers), Baihûkan, 1959.
[2] Y. Komatu, Hukusosû to sono kansû (Japanese; Complex numbers and functions of a complex variable), Heibonsya, 1950.
[3] T. Matsumoto, Itizi kansû, sono ôyô (Japanese; Linear functions with applications), Huzanbô, 1940.
[4] N. Bourbaki, Eléments de mathématique III. Topologie générale, ch. 8. Nombres complexes, Actualités Sci. Ind., 1235b, Hermann, third edition, 1963; English translation, General topology, pt. 2, Addison-Wesley, 1966.
[5] L. V. Ahlfors, Complex analysis, McGraw-Hill, second edition, 1966.
[6] C. Carathéodory, Funktionentheorie I, Birkhäuser, 1950; English translation, Theory of functions of a complex variable I, Chelsea, 1954.

77 (XV.17) Computers

A. History

Since the beginning of civilization, man has utilized tools for aiding computation. In Japan, bamboo computing rods were used in the 7th century; before the close of the 16th century, the **abacus** was imported from China (→ 230 Japanese Mathematics (Wasan)). The first calculator that worked automatically was the adding machine made by B. Pascal, who invented the "carry transmission mechanism" (1642). This calculator was improved by G. W. Leibniz to execute multiplication and division by repeated addition and subtraction (1671). W. T. Odhner's calculator was a further improvement of these machines. Many calculators manufactured today are equipped with electric motors and other automatic devices; however, these machines will not be described here since they have little to do with modern mathematics. Rather we shall describe automatic digital computers, which have developed rapidly in recent years.

An **automatic computer** is capable of executing automatically a sequence of arithmetic operations according to a given program, while a **desk calculator** can automatically execute by itself only one operation at a time. The automatic computer was conceived by the English mathematician C. Babbage in the 19th century, but mechanical engineering at that time was not advanced enough to allow the construction of such a computer. His idea was first realized by the relay computer, Mark I, of Harvard University c. 1940. In 1947 the first **electronic computer**, ENIAC, appeared, in which vacuum tubes were utilized instead of mechanical components. About 1950, several computers were constructed according to the **stored program** principle proposed by J. Von Neumann. Since then computers have made rapid progress, and we now can be said to live in the "computer age."

Modern electronic computers are fully automatic and thus embody †Turing machines. An automatic computer in general consists of the following five units: the arithmetic unit, memory, control, input, and output. These units are interconnected by wires that exchange information in the course of computation. In the Mark I, the arithmetic unit and memory were composed of relays and cogwheels controlled by electric signals. Several computers were constructed using relays exclusively. Soon afterwards, relays were replaced by electronic elements such as vacuum tubes and transistors, which increased the speed of computation. Together with the development of large-scale memories, the high speed thus achieved caused substantial changes in the capabilities of computers. They can now numerically solve technical problems that had previously been impossible, e.g., some partial differential equations with three independent variables. They are also capable of **nonarithmetic uses**, even for highly intellectual human activities such as translation, information retrieval, theorem proving, etc. It is quite reasonable to call computers universal information-processing machines.

B. Construction of Digital Computers

Information processing in a computer is based on communication among its constituents by electric signals (Fig. 1). In **digital computers**, information is encoded as a sequence of binary numbers 0, 1, whereas continuous values are allowed in †analog computers. The minimum quantity of information in digital computers is therefore a binary digit, called a **bit**. In this article only digital computers are discussed. The binary numbers 0, 1 are represented in practice by two distinct electric signals: two distinct voltages, two distinct phases of alternating current, the existence or nonexistence of a pulse signal, etc. A system of circuits is called **synchronous** when it contains a clock, a generator of periodic pulse signals, which synchronizes the transmission and transformation of information. In an **asynchronous** system, circuits execute each step of information processing independently and advance to the next step after verifying the termination of the preceding step. Most current systems are synchronous.

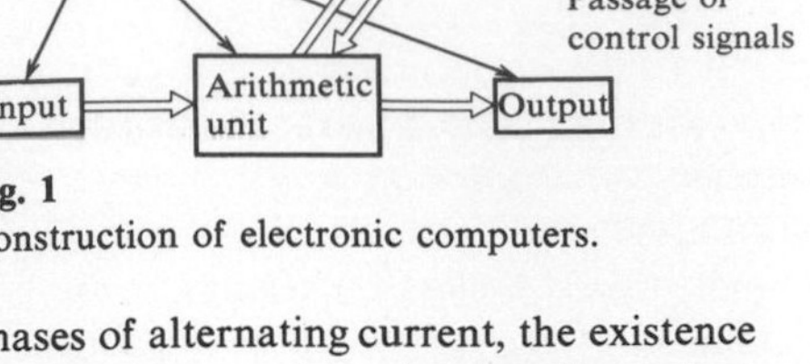

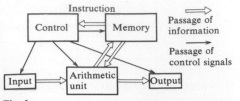

Fig. 1
Construction of electronic computers.

Each unit of a computer is considered as an automaton whose output is determined by its input and internal state. Therefore, it can be composed from logical elements executing basic logical operations such as $\wedge$(*and*), $\vee$ (*or*), $\neg$ (*not*), etc., together with memory elements and amplifiers for supplying energy to signals to be bifurcated. Usually *and* elements and *or* elements are made of diodes, while *not* elements are made of transistors that serve also as amplifiers (Fig. 2). In arithmetic and control units, flip-flop circuits are utilized as memory elements (Fig. 3). The **parametron** is the element that fills by itself the three functions of operating, amplifying, and memorizing. Its activity is based on the parametric-response properties of non-

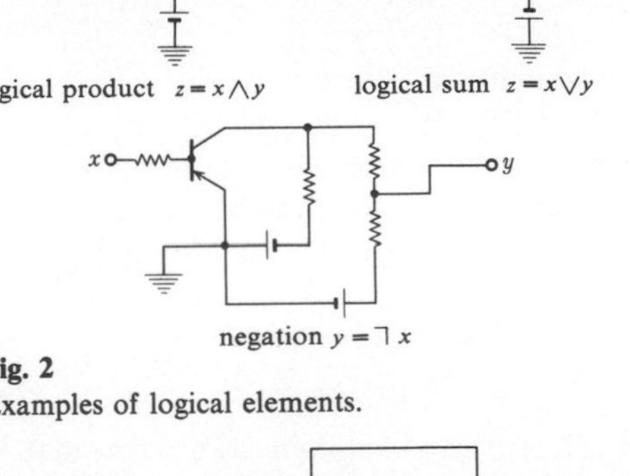

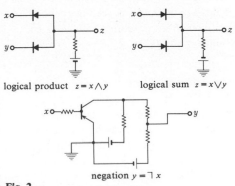

Fig. 2
Examples of logical elements.

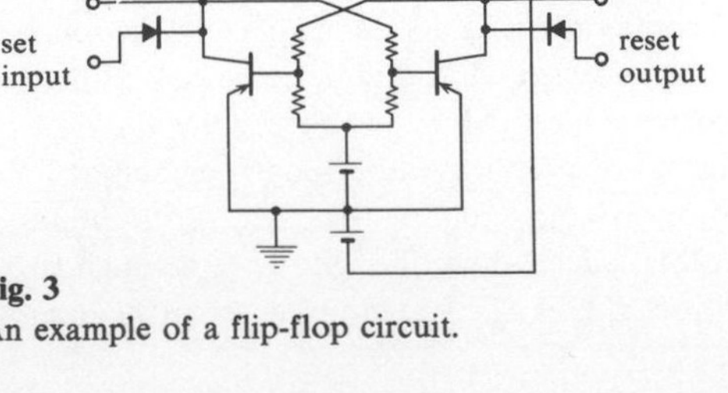

Fig. 3
An example of a flip-flop circuit.

linear magnetic material. Arithmetic and control units are composed of several hundreds or often several thousands of these elements.

The **arithmetic unit** consists mainly of several memory registers and operational circuits associated with them. Each register stores a binary number of n bits (usually $16 \leqslant n \leqslant 64$). An important building block of the operational circuit is the basic adder of 1-bit numbers (Fig. 4). A parallel adder of n-bit numbers can be obtained by connecting n copies of the basic adder. Alternatively, a sequential adder can be composed of a single basic adder which is utilized repeatedly to sum up bit by bit two binary numbers from the lowest bit. Subtraction is usually carried out by adding the complement of the subtrahend. Multiplication is realized by shifting the multiplicand to the left and adding it to the intermediate sum; this addition may be omitted, depending on the relevant bit of the multiplier. In division, we shift the dividend to the left and if possible subtract the divisor from it. By recording at each step whether or not the subtraction is possible, we obtain the quotient. The control of circuits for these operations is also realized by the elements of logical operation and of memory.

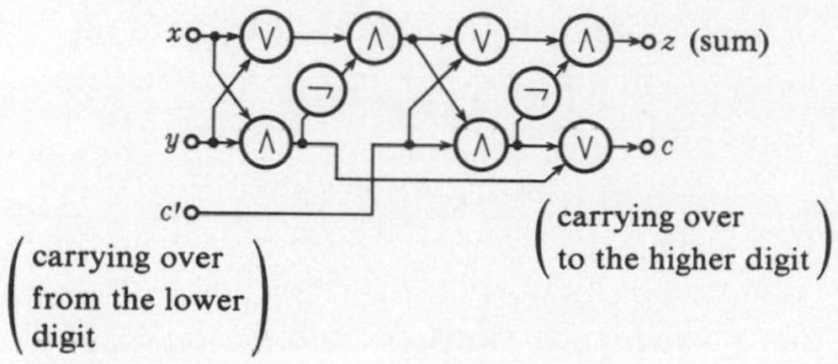

Fig. 4
An example of a binary adder.

The **control unit** repeats the following operations consecutively: (1) takes an instruction from the memory location indicated by the sequential control counter, (2) gets a data word from the memory according to the address part of the instruction, (3) decodes the function part of the instruction and sends control signals to appropriate circuits, and (4) increases the content of the sequential control counter by 1 and, after receiving end signals from the arithmetic unit, returns to step (1). For these purposes the control unit contains a counter, a decoder, an encoder to send control signals, and a register to store the instruction to be executed.

The **memory unit** stores the instructions and given data as well as the necessary data obtained in the course of computation. At present typical memory devices are **magnetic core matrices**, magnetic cores set in array and stitched by conducting wires as a lattice. The magnetic core is a ring of ferrite which records one bit of information by selecting one of two directions of magnetization. Writing and reading information are executed by conducting wires passing through the ring. As auxiliary large-scale memory, there are also magnetic drums, disks, and magnetic tapes, etc. Other elements under investigation are extreme-low-temperature elements, chemical elements, optical elements, etc.

The arithmetic, control, and memory units form the central processor of the computer.

C. Instructions and Programming

In the stored program principle, a computer performs a sequence of calculation according to given instructions. There are various types of instructions, but those most frequently used are **single-address instructions**, each of which contains a single-address part designating an operand. Depending on their function, instructions are classified as follows: arithmetic instructions, store instructions, jump instructions, input and output instructions, etc. An arithmetic instruction executes an arithmetic operation, usually between the contents of a particular register called an **accumulator** and that of the memory register designated by its address part. The result of the operation is left in the accumulator, the heart of the arithmetic unit. A store instruction transfers the contents of the accumulator to the register designated by its address part. A jump instruction specifies by its address part the memory register from which the next instruction should be taken. Without this specification, the computer executes instructions stored in the memory successively according to the order of their addresses. Some jump instructions are called conditional, since the jump to the specified address is executed only when a certain condition is satisfied (for instance, only when the content of the accumulator is negative). Otherwise the computer advances to the next instruction in the ordinary order, that is, the order of the address. By means of such conditional instructions, the computer acquires the ability of judgement and becomes a Turing machine. Input and output instructions specify input-output operations.

A **program** is a finite sequence of instructions arranged suitably for the required computation. Programming, or making a program, is therefore the task of decomposing the required computation into elementary steps each of which corresponds to an instruction.

Every instruction is represented in a computer by a number, a numeric code, which is determined in a definite way by the construction of the control unit. Before starting computation, instructions thus encoded are stored in the memory. In this sense, a program is a sequence of numbers. This sequence is called a **machine-language program**.

A program is usually divided into several blocks, called subprograms or **subroutines**. (Some block may be called the main program according to its function, but this distinction is more or less a matter of convenience.) Some subroutines for common use are made in advance, especially those for frequently required jobs such as evaluating elementary functions and manipulating input-output devices, etc. The system of these ready-made programs is called **software**, in contrast to the **hardware** (i.e., mechanicoelectronic equipment) of the computer. The handiness of a computer depends mainly on its software.

Although programs were formerly written in machine languages, they are now written in certain forms easy to master called **external languages**. A problem-oriented language is an advanced external language in which ordinary arithmetic expressions are available with slight modifications. Programs written in these languages are translated into machine languages by program input routines. The translation of problem-oriented languages is called **automatic programming**, and the translator is called the **compiler**.

Another important translator is the **assembler**, which translates mnemonic codes of instructions (*add* for addition, etc.) into their numeric codes according to a given table. It allows us to utilize symbols for specifying addresses. It also converts decimal numbers into binary, and generates certain segments of the program from rather simple indications. Such an indication, since it is similar to an ordinary instruction, is called a **macroinstruction**. Compilers and assemblers are important constituents of software.

A compiler accepts several macroinstructions. It is moreover equipped with the following facilities: (i) the ability to translate arithmetic expressions into machine language; (ii) the ability to generate linkages to various ready-made subroutines according to certain simple indications; (iii) the ability to automatically allocate programs, subroutines, and data in the memory. Thus it accepts an external form such as:

if x ⩾ 0 **then** printreal (SQRT(x))
        else printstring ('negative').

The first compiler was the **FORTRAN** (FORmula TRANslator) compiler made by IBM (1956) in the United States. Since then many external languages have been designed for various fields. Corresponding translators have been built for every machine. The most widely used problem-oriented languages are FORTRAN (for scientific use), **ALGOL** (ALGOrithmic Language, for scientific use), **COBOL** (COmmon Business Oriented Language, for business use), **LISP** (LISt Processor, for nonarithmetic uses), etc. Interpretative software is utilized for LISP. Today new techniques in the field of numerical analysis are often published in the international language ALGOL.

Large-scale high-speed computers that have recently appeared have made software systems inevitably more complex. There is now software, so-called monitors, which supervises the uninterrupted processing of many programs. Some of them coordinate several assemblers and compilers so that several languages can be mixed in writing a program.

D. Peripheral Devices

In contrast to the central processor, the input and output units are called **peripheral devices (marginal devices)**.

Before starting computation, the computer must have the necessary information (a program) in its memory unit. A program is keyboarded onto cards or tapes with a key punch. These are fed to the card (or tape) reader and converted into electric signals. The reading devices and auxiliary circuits constitute the input unit.

An output device presents the results obtained by the computer. The results are usually printed by teletypewriter or line printer, but on some occasions are entered on cards or tapes, especially when they are to be utilized as input in further computation. Magnetic tape units can be considered as auxiliary input-output devices, since tape reels are removable. The tape units can accept prepared input data as well as record the results, the output.

These devices are operated according to given input-output instructions under the control of the central processor. However, since every input-output device contains mechanical components and is extremely slow in comparison with the central processor, a large-scale computer is often accompanied by satellite computers which undertake the control of input-output devices. Otherwise, the computer may have facilities for interruption (see below) to improve efficiency.

E. Interruption

After the start button of a computer is pushed, the whole process of computation is carried out automatically by a given program and therefore is never interfered with by external occurrences unless it is stopped by means of the stop button of the computer. However, since it is inefficient to let a computer come to an absolute stop, many modern computers have more flexible facilities for **interruption**, i.e., they suspend the performance of a program and transfer the control to another program when they receive particular signals, e.g., an alarm signal indicating erroneous operation, signals from console buttons or from input-output devices indicating an important change of the situation, etc. They also record information on the previously performed program (the location of the instruction under execution, etc.) so as to continue the performance of the program later. Making efficient use of such facilities, a computer and its appropriate software can control a large number of peripheral devices, each of which communicates with the computer independently. Such a processing is called **on-line real time processing** and is used for process control in various plants and in automatic seat reservation systems. It also allows a computer to work concurrently for many users with multiple purposes.

F. Mathematical Linguistics

In accordance with the development of software, programming techniques have gradually accumulated. For instance, we now have an almost satisfactory method of translating arithmetic expressions into machine language. However, we still lack a general theory to cover effectively widespread problems in programming. The design of an adequate metalanguage is still an important problem, if the word "adequate" implies complete description of syntax and semantics of problem-oriented languages. An interesting problem concerning language description is to define and suitably classify grammars with respect to their capabilities of forming a language. Such research is an important branch of **mathematical linguistics**. N. Chomsky has investigated this problem, starting from research on natural languages, and has given a formal definition of grammars and classified them into four types. Many researchers have investigated the relationship among these types and other classes of grammars often defined in terms of the theory of †automata.

References

[1] K. Joh and S. Makinouchi, Keisankikai (Japanese; Computers), Kyôritu, 1953.

[2] H. Yamashita (ed.), Densi keisanki—Dizitaru keisanki hen (Japanese; Electronic computers: pt. II, digital computers), Ohm sya, 1960.

[3] M. Phister, Logical design of digital computers, John Wiley, 1958.

[4] S. Seki and Y. Fujikawa, Densi keisanki nyûmon (Japanese; Introduction to computers), Baihûkan, 1966.

For programming,

[5] 709/7090 FORTRAN programming system, IBM form C 28-6054-2.

[6] P. Naur (ed.), Revised report on the algorithmic language ALGOL 60, Comm. ACM, 6 (1963), 1–17.

[7] Y. Bar-Hillel, Language and information, Addison-Wesley, 1964.

[8] N. Chomsky, Formal properties of grammars, in R. D. Luce, R. R. Bush, and E. Galanter (eds.), Handbook of mathematical psychology II, John Wiley, 1963, p. 323–418.

[9] J. E. Sammet, Programming languages: history and fundamentals, Prentice-Hall, 1969.

78 (VI.19) Conformal Geometry

A. Möbius Geometry

We represent an n-dimensional sphere S^n as the †quadric hypersurface $S^n: x_1^2+x_2^2+\ldots+x_n^2-2x_0x_\infty=0$ in an $(n+1)$-dimensional real †projective space $\mathbf{P}^{n+1}$, where the (x_α) are †homogeneous coordinates in $\mathbf{P}^{n+1}$. We denote by $M(n)$ the group of all †projective transformations of $\mathbf{P}^{n+1}$ that leave S^n invariant. Then the transformation group $M(n)$ acts on S^n. The pair $(S_n, M(n))$ is called the **conformal geometry** or **Möbius geometry**. We call S^n an n-dimensional **conformal space**, a transformation belonging to $M(n)$ a **Möbius transformation**, and $M(n)$ the **Möbius transformation group**. Every point of the projective space $\mathbf{P}^{n+1}$ corresponds to a **hypersphere** on S^n. For example, if a point A lies outside of S^n, then the intersection of S^n and the †polar hyperplane of A with respect to S^n is an $(n-1)$-dimensional sphere S^{n-1}, and the point A corresponds to this **real hypersphere** S^{n-1}. Similarly, if a point A lies on S^n, it corresponds to a **point hypersphere**, and if a point A lies inside of S^n, then A corresponds to an **imaginary hypersphere**. We sometimes identify the point A and the corresponding hypersphere. For any two points $A=(a_\alpha)$ and $B=(b_\alpha)$, we put $AB=a_1b_1+a_2b_2+\ldots+a_nb_n-(a_0b_\infty+a_\infty b_0)$ and call it the **inner product** of the two hyperspheres A and B. The **angle** θ between two intersecting real hyperspheres A and B is defined by $\cos\theta = AB/(\sqrt{A^2}\cdot\sqrt{B^2})$. This angle is invariant under the Möbius transformation.

In the projective space $\mathbf{P}^{n+1}$, we take a †frame $(A_0, A_1, \ldots, A_n, A_\infty)$ that satisfies the following conditions:

$$A_0^2 = A_0A_j = A_iA_\infty = A_\infty^2 = 0, \qquad A_0A_\infty = -1,$$
$$A_iA_j = g_{ij}, \qquad i,j = 1,2,\ldots,n,$$

where (g_{ij}) is a positive definite matrix. We see that A_0 and A_∞ are points on S^n and each A_i is a real hypersphere passing through these two points. Every hypersphere X of S^n can be written as a linear combination of the $A_i : X = u_0A_0 + u_1A_1 + \ldots + u_nA_n + u_\infty A_\infty$. That is, X is represented by †projective coordinates (u_α) with respect to the above frame. We call these homogeneous coordinates (u_α) **$(n+2)$-hyperspherical coordinates** of the hypersphere X. If we use these coordinates, the inner product of two hyperspheres $X = (u_\alpha)$ and $Y = (v_\alpha)$ is given by $XY = \sum_{i,j=1}^{n} g_{ij}u_iv_j - (u_0v_\infty + u_\infty v_0)$.

The Möbius transformation group $M(n)$ is a topological group with two †connected components. If we denote by $M_0(n)$ the maximal connected subgroup of $M(n)$ and by H the subgroup of $M(n)$ that leaves invariant a real hypersphere of S^n, then the set E of all real hyperspheres of S^n can be identified with the †homogeneous space $M_0(n)/H$. The group H also consists of two connected components. If we denote by H_0 the maximal connected subgroup of H, the homogeneous space $\tilde{E} = M_0(n)/H_0$ is a two-fold †covering space of E. For each real hypersphere $A \in E$, an element $\tilde{A} \in \tilde{E}$ over A is called an **oriented real hypersphere**. The Möbius transformation group contains as its subgroups ones that are isomorphic to the †group of congruent transformations of a Euclidean space and ones that are isomorphic to the group of congruent transformations of a non-Euclidean space. Namely the subgroup of $M(n)$ that leaves a point hypersphere invariant is isomorphic to the group generated by congruent transformations and †homotheties of the Euclidean space E^n. The subgroup of †index 2 (the factor group by a †cyclic subgroup of †order 2) of the subgroup of $M(n)$ that leaves invariant a real (imaginary) hypersphere is isomorphic to the group of congruent transformations of n-dimensional hyperbolic (elliptic) †non-Euclidean space.

In the n-dimensional Euclidean space E^n, consider a hypersphere of radius r with center 0. For each point P of E^n, mark a point Q on the †ray OP such that $\overline{OP}\cdot\overline{OQ} = r^2$. We call the point transformation that sends P to Q an **inversion** with respect to the hypersphere. A †symmetry with respect to a hyperplane, considered as an extreme case of inversions, is also called an inversion. We adjoin a point at infinity to the space E^n to construct an n-dimensional sphere S^n. Each inversion can be extended to a transformation of S^n, which we also call an inversion. Then each Möbius transformation is generated by a finite number of inversions. By a Möbius transformation of S^n, each hypersphere is transformed to a hypersphere. Any angle between two curves that intersect at a point of E^n is invariant under Möbius transformations. Conversely, if $n \geqslant 3$, each local transformation of E^n that leaves invariant the angle of each pair of intersecting curves is a †restriction of a Möbius transformation. However, for $n=2$ this is not true in general; any transformation that leaves angles invariant is called a †conformal mapping. Any Möbius transformation $z \to w$ on the †complex sphere $S^2 = \mathbf{C} \cup \{\infty\}$ can be expressed by an equation of the form $w = (\alpha z + \beta)/(\gamma z + \delta)$ or $w = (\alpha\bar{z} + \beta)/(\gamma\bar{z} + \delta)$, where α, β, γ, and δ are complex numbers such that $\alpha\delta - \beta\gamma \neq 0$ and $\bar{z}$ denotes the †complex conjugate of z.

B. Laguerre Geometry

Let Γ be an oriented smooth curve in a Euclidean plane E^2. The tangent line at a point p of Γ is supplied with an orientation that is induced by the orientation of the curve Γ in an obvious manner. The oriented line l thus obtained is called the **oriented tangent line** of Γ at p.

Let S be the set of oriented lines in E^2 and T be a given set of oriented smooth curves in E^2. Consider a bijection γ of the direct product $S \times T$ to itself satisfying the following condition: If l is an oriented tangent line of a curve Γ belonging to T and γ sends (l,Γ) to (l',Γ'), then l' is an oriented tangent of the curve Γ'. The set of such bijections forms a group G. An element γ of G is called an **equilong transformation** if conditions (i) and (ii) below are satisfied. Suppose that we have $l \in S$ for which there exist two curves Γ_1 and Γ_2 in T such that l is a common tangent line of Γ_1 and Γ_2 at p_1 and p_2, respectively. Further suppose that the element $\gamma \in G$ sends (l,Γ_i) to (l',Γ_i') $(i=1,2)$. Then the conditions are
(i) l' is tangent to Γ_i' at points p_i' $(i=1,2)$,
(ii) the distance between p_1 and p_2 is equal to the distance between p_1' and p_2'. The set of equilong transformations forms a subgroup H of G. In particular, if T is the set of oriented circles, the elements of H are called **Laguerre transformations**. In an obvious manner, we can divide the set of oriented circles into two classes, those with "positive"

and those with "negative" orientations. With an oriented circle Γ we associate the pair $\varphi(\Gamma)=(P,r)$, where P is the origin of the circle and r is a real number whose absolute value is equal to the radius of the circle and whose signature coincides with that of the orientation of Γ. An example of a Laguerre transformation, called a **dilatation**, is given by a bijection γ of $S\times T$ to itself satisfying the following condition: Let $\gamma(l,\Gamma)=(l',\Gamma')$; then l' is parallel to l, the distance between l and l' is a given number, and $\varphi(\Gamma')=(P,r+c)$, where $\varphi(\Gamma)=(P,r)$ and c is a given constant. We note here that the action of a Laguerre transformation $\gamma:(l,\Gamma)\rightarrow(l',\Gamma')$ is determined by its action on S (γ acts on S by $\gamma(l)=l'$). Another example of a Laguerre transformation γ, called a **Laguerre inversion**, is determined by means of a given oriented circle O and a line p that is not tangent to O; given an oriented line l, its image l' under the action of γ is determined as follows (here we describe the case where l is not parallel to p). There exists a uniquely determined oriented tangent line g of the circle O parallel to l. Now we have a uniquely determined oriented tangent line g' of O that passes through the point of intersection of the lines g and p. The image l' is the line parallel to g' and passing through the point of intersection of the lines l and p (Fig. 1). Each Laguerre transformation can be written as a product of a finite number of Laguerre inversions. We denote the group of Laguerre transformations by L. The pair (L,S) is, by definition, a model of **Laguerre geometry**. Notions such as Laguerre inversions, dilatations, and transformations can be generalized to cases of higher dimension by utilizing oriented hyperspheres and oriented hyperplanes.

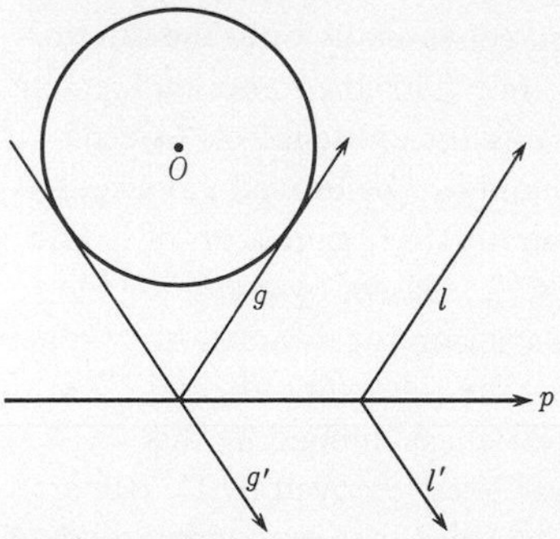

Fig. 1

C. Sphere Geometry

Let S be the set of oriented circles (including point circles and oriented lines) in the Euclidean plane E^2. Two oriented circles C_1 and C_2 are said to be in contact with each other if they have a point p and an oriented tangent line passing through p in common. (An oriented circle C and an oriented line l are in contact with each other if and only if l is an oriented tangent line of C.) In this case, we call the pair (C_1,C_2) a **contact pair**. A bijection γ of S to itself is called a **Lie transformation** if it sends any contact pair to another. An inversion with respect to a circle determines in an obvious manner a Lie transformation, which is also called an inversion. We denote the group of Lie transformations by G. Any element γ of G can be written as the product of a finite number of inversions and Laguerre inversions, and G contains the group of Möbius transformations and the group of Laguerre transformations as subgroups. The pair (G,S) is called a model of **circle geometry**. The notion of circle geometry can be generalized to that of **hypersphere geometry** for the case of higher dimensions. Specifically, when we replace E^2 by E^3 and circles by spheres, we have **sphere geometry**.

Let V be complex 3-dimensional space, and let M, N be the sets of oriented lines and oriented spheres in V, respectively. Then M, N have the structure of 4-dimensional complex manifolds that are homeomorphic to each other. The homeomorphism is given by the **Lie line-sphere transformation** that induces a bijection from the set of pairs of intersecting oriented lines onto the set of pairs of oriented spheres that are in contact with each other.

D. Group-Theoretic Considerations

Here we discuss the preceding three kinds of geometries from the group-theoretic point of view (→ 147 Erlangen Program). Let us consider the quadratic form Q defined by $Q(x)=-x_0^2+x_1^2+x_2^2+x_3^2-x_4^2$ in a real projective space $\mathbf{P}^4$, where $x=(x_0,x_1,x_2,x_3,x_4)$ are homogeneous coordinates. We denote by G the set of all projective transformations of $\mathbf{P}^4$ that leave Q invariant. The group G consists of the set of matrices A of order 5 such that $\det A=1$ and $Q(Ax)=Q(x)$ holds for all x in $\mathbf{P}^4$; we denote by L^3 the set of all points x in $\mathbf{P}^4$ that satisfy $Q(x)=0$. Then G acts transitively on L^3. Hence if we denote by H_a the set of all elements of G that fix a point a in L^3, for example $a=(-1,1,0,0,0)$, we may assume that $L^3=G/H_a$. The circle geometry that belongs to the group of Lie transformations of circles is exactly the geometry of the homogeneous space G/H_a. The group G_b of all transformations of G that leave the hyperplane $x_4=0$ invariant acts transitively on $L^3\cap\{x_4=0\}$ (G_b is isomorphic to $M(2)$). The geometry of the homogeneous space $G_b/H_a\cap G_b$ is plane conformal geometry. Next, the group G_a of all transformations of G that leave invariant $x_0+x_1=0$ coincides with H_a and acts transitively on $L^3 \cap \{x_0+x_1=0\}$. The geometry of the homogeneous space $L^3 \cap \{x_0+x_1=0\}$ (on which G_a acts transitively) is plane Laguerre geometry. In this sense, the circle geometry that belongs to the group of Lie transformations contains the other circle geometries as subgeometries.

We now describe how plane Laguerre geometry (→ Section B) can be realized as the geometry of the space $L^3 \cap \{x_0+x_1=0\}$. Let E^2 be a plane in a Euclidean space E^3. We fix a Cartesian coordinate system (y_0, y_1, y_2) in E^3 so that E^2 is given by $y_0=0$. To each point y of E^3 we can associate an oriented circle in E^2 with center $(0, y_1, y_2)$, radius $|y_0|$, and positive (negative) orientation if y_0 is positive (negative). If y lies on E^2, the corresponding circle is the point circle y itself. Now let us consider the group G_a' of all affine transformations of E^3 whose rotation parts leave the quadratic form $Q'(y) = -y_0^2+y_1^2+y_2^2$ invariant up to scalar factors. (If the rotation part of an affine transformation leaves Q' invariant, it is an isometry with respect to the †metric defined by Q'.) Each element of G_a' induces a transformation of the set of oriented circles in E^2 (including the oriented lines and point circles) onto itself. The mappings of E^3 into L^3 defined by $x_0=(1+Q'(y))/2$, $x_1=(1-Q'(y))/2$, $x_2=y_1$, $x_3=y_2$, $x_4=y_0$ is a one-to-one correspondence of E^3 onto the subset of L^3 such that $x_0+x_1 \neq 0$. This correspondence induces an isomorphism of G_a' onto G_a. In Laguerre geometry, there are no essential distinctions between points and oriented circles on E^2; and the group G_a acts on a 3-dimensional space of oriented circles (including points).

References

[1] W. Blaschke, Vorlesungen über Differentialgeometrie III, Springer, 1929.
[2] T. Kubota, Kaiseki kikagaku (Japanese; Analytic geometry), Utida-rôkakuho, I, 1937; II, 1945.
[3] T. Takasu, Differentialgeometrien in den Kugelräumen I, II, Maruzen, 1938–1939.

79 (XI.15) Conformal Mapping

A. General Remarks

Let a function $w=f(z)$ that maps a †domain D on the †complex z-sphere homeomorphically onto a domain Δ on the complex w-sphere satisfy the following two conditions: (1) Every curve $C_z: z=z(t)$ $(0 \leqslant t \leqslant 1)$ that starts at any point z_0 and possesses a tangent there has an image curve $C_w: w=w(t)=f(z(t))$ $(0 \leqslant t \leqslant 1)$ that also possesses a tangent at the image point $w_0=f(z_0)$. (2) The angle between any two curves $C_z^{(1)}$ and $C_z^{(2)}$ possessing tangents at z_0 is equal to the angle between their image curves $C_w^{(1)}$ and $C_w^{(2)}$, where the direction of the angle is also taken into account. Then the mapping from D onto Δ is said to be **conformal**, and it has been proved that $w=f(z)$ is necessarily a function †analytic in D (D. E. Men'šov, 1931).

Consequently, the theory of conformal mapping is a branch of the theory of analytic functions. That a function $w=f(z)$ maps a domain D conformally onto a domain Δ means that it is a †meromorphic function †univalent in D and its range is Δ. Then $f'(z) \neq 0$ holds at every (finite) point in D, and the ratio of the lengths of the segments between two points $w_0=w(0)$, $w(t)$ on C_w and between two points $z_0=z(0)$, $z(t)$ on C_z tends to a fixed nonvanishing limit $|f'(z_0)|$ as $t \to 0$ independent of the choice of C_z. Hence if z_1 and z_2 lie on $C_z^{(1)}$ and $C_z^{(2)}$, respectively, and w_1 and w_2 are their image points lying on the image curves $C_w^{(1)}$ and $C_w^{(2)}$, respectively, then the two triangles $\triangle z_1 z_0 z_2$ and $\triangle w_1 w_0 w_2$ are nearly similar in the positive sense, provided that z_1 and z_2 are near enough to z_0. This justifies the word "conformal," which means of the same form (→ Appendix A, Table 13).

B. Conformal Mapping onto the Unit Disk

A fundamental theorem in the theory of conformal mapping is **Riemann's mapping theorem**, which states that any †simply connected domain D with at least two boundary points can be mapped conformally onto the interior Δ of the unit circle. This theorem is equivalent to the assertion of the existence of †Green's function of D and can be proved in various ways. B. Riemann (1851) gave a proof, based on an idea of C. F. Gauss, by assuming the existence of a solution for a variational problem minimizing the †Dirichlet integral. The logical incompleteness implied by this assumption was later removed by D. Hilbert and others. The proof that is now regarded as simplest is due to L. Fejér and F. Riesz's method (T. Radó, 1922, 1923), which applies †normal family theory. On the other hand, the **osculating process** due to P. Koebe (1912) is a purely constructive method of proving existence that is also applicable to the case of †multiply connected domains. The mapping function $w=f(z)$ in Riemann's mapping theorem is uniquely determined under the normalization condition $f(z_0)=0$, $\arg f'(z_0)=\theta_0$ at a point z_0 in D, where θ_0 is a given angle.

Let $w=f(z)$ map a simply connected domain D on the z-sphere conformally onto a simply connected domain Δ on the w-sphere. If a sequence $\{z_\nu\}$ in D tends to a boundary point ζ of D, then the corresponding sequence $\{w_\nu\}$ ($w_\nu=f(z_\nu)$) has no †accumulation point in Δ and does not necessarily tend to a boundary point of Δ. Conversely, for two sequences $\{z_\nu\}$ and $\{z'_\nu\}$ tending to the same boundary point ζ, the limits of $\{f(z_\nu)\}$ and $\{f(z'_\nu)\}$ do not necessarily exist and do not necessarily coincide when they exist. If for any sequence $\{z_\nu\}$ tending to ζ, $\{f(z_\nu)\}$ tends to a unique point ω on the boundary of Δ, it is said that $f(z)$ possesses a **boundary value** ω at ζ. To investigate behavior of $\{w_\nu\}$ or of an image curve as z tends to the boundary along a curve in D, the following simplification can be made: Since any simply connected domain with at least two boundary points can be mapped conformally onto a bounded domain by means of an †elementary function, the problem reduces, in view of Riemann's mapping theorem, to the case where D is a bounded simply connected domain and Δ is the unit disk $|w|<1$.

C. Correspondence between Boundaries

Concerning the correspondence between boundaries in the conformal mapping $w=f(z)$ of a bounded simply connected domain D onto the unit disk $|w|<1$, we have the following three theorems:

(1) To any †accessible boundary point z_C of D there corresponds a unique point on $|w|=1$, and to any distinct accessible boundary points z_{C_1} and z_{C_2} of D there correspond distinct points on the unit circumference. Furthermore, the set of all points on $|w|=1$ that correspond to accessible boundary points of D has †angular measure equal to 2π.

(2) There is a one-to-one correspondence between †boundary elements of D and points on $|w|=1$ (C. Carathéodory).

(3) Let $w=f(z)$ map the interior D of a †Jordan curve C conformally onto the unit disk $\Delta:|w|<1$. Then it possesses a boundary value, say $f(\zeta)$, at every point ζ on C that satisfies $|f(\zeta)|=1$. Hence $f(z)$ is continuous on the closed domain $\bar{D}=D\cup C$ and maps $\bar{D}$ bijectively onto the closed disk $\bar{\Delta}:|w|\leqslant 1$. Similarly, the inverse function $z=\varphi(w)$ has an analogous nature and maps $\bar{\Delta}$ bijectively and continuously onto $\bar{D}$; that is, a conformal mapping of the interior D of a Jordan curve onto the unit disk Δ can be extended into a homeomorphism of the closure $\bar{D}$ to $\bar{\Delta}$ (Carathéodory).

In this case, if the Jordan curve C contains a †regular analytic arc Γ, then the mapping function $w=f(z)$ can be prolonged analytically beyond Γ (except at the endpoints of Γ). Hence the mapping $w=f(z)$ is conformal at interior points of Γ.

Problems on the correspondence of angles at the boundary are closely related to †angular derivatives. These problems have been attacked by Carathéodory, S. Warschawski, J. Wolff, and others.

D. Multiply Connected Domains

For multiply connected domains, the term *conformal mapping* is used in a somewhat different sense. In general, if for a †branch of a (single-valued or multiple-valued) analytic function $w=f(z)$ that is meromorphic everywhere in a domain D any two distinct function elements always attain distinct values at their centers, then $w=f(z)$ is said to be **univalent** in D. A (single-valued or multiple-valued) function $w=f(z)$ that is meromorphic and univalent in D whose range is Δ is said to map D conformally onto Δ. In this case the inverse function $z=\varphi(w)$ of $w=f(z)$ is necessarily single-valued in Δ. If $w=f(z)$ is single-valued in D, the situation is the same as in the case of conformal mappings as explained before. In the particular case where D is simply connected, $w=f(z)$ is necessarily single-valued in view of the monodromy theorem. However, $w=f(z)$ may, in general, be multiple-valued in D. To make a distinction, the conformal mapping explained before is said to be a **one-to-one conformal mapping**.

Riemann's mapping theorem in the multiply connected case is stated as follows:

(4) Let D be a multiply connected domain with at least three boundary points. Then there exists a multiple-valued function $w=f(z)$ that is univalent and holomorphic in D, whose range Δ coincides with the unit disk $|w|<1$; that is, D can be mapped conformally onto the unit disk $\Delta:|w|<1$. Furthermore, (a branch of) $w=f(z)$ may be subject to a normalization $f(z_0)=0$, $\arg f'(z_0)=\theta_0$ at any assigned point z_0 in D, and the mapping function is uniquely determined under this condition.

The mapping function $w=f(z)$ in this theorem is necessarily infinitely multiple-valued and has the following properties: If $P(z;z_0)$ is a function element of $w=f(z)$ with an arbitrary fixed point z_0 in D as its center, all function elements at z_0 are expressed by $L_k(P(z;z_0))$ $(k=0,1,\ldots)$, where L_k denotes a †linear transformation by which the unit disk remains invariant. The set of linear transformations L_k $(k=0,1,\ldots)$ forms a group G (called a †Fuchsian group), and the function $z=\varphi(w)$ inverse to $w=f(z)$ is an †automorphic function with respect to the group G. No †elliptic transformation is contained in G. If D possesses an isolated boundary point, G contains †parabolic transformations.

E. Conformal Mapping between Riemann Surfaces

Let F_1 and F_2 be two †Riemann surfaces and a mapping $p_2=\Phi(p_1)$ be given that transforms F_1 onto F_2 bijectively and continuously. For every point p_1^0 on F_1 and its image p_2^0 on F_2, let $z_1=T_1(p_1)$ and $z_2=T_2(p_2)$ be local parameters belonging to them, and a function $z_2=\varphi(z_1)$ be defined by $z_1\to p_1\to p_2\to z_2$. If $z_2=\varphi(z_1)$ maps a neighborhood of $z_1^0=T_1(p_1^0)$ bijectively and conformally onto a neighborhood of $z_2^0=T_2(p_2^0)$, then $p_2=\Phi(p_1)$ is said to map F_1 bijectively and conformally onto F_2. By regarding every function element of a mapping function $w=f(z)$ of the multiply connected domain mentioned before as a point, the †universal covering surface $\tilde{D}$ of the basic surface D is obtained, so that $w=f(z)$ is a single-valued holomorphic and univalent function on $\tilde{D}$. Hence theorem (4) may be restated as follows: Let D be a multiply connected domain with at least three boundary points. Then the universal covering surface $\tilde{D}$ of D can be mapped bijectively and conformally onto the unit disk $|w|<1$. In this case, with respect to accessible boundary points of $\tilde{D}$, a theorem similar to theorem (1) holds. Since $\tilde{D}$ is a simply connected †covering surface of the z-plane, the form of theorem (4) given above may be regarded as a particular case of **Koebe's theorem**, which states that any simply connected open Riemann surface can be mapped conformally onto the unit disk or the whole finite plane.

F. Conformal Mapping of Multiply Connected Domains

It is also important to consider problems concerning one-to-one conformal mapping of a multiply connected domain on the z-sphere onto a suitable multiply connected domain $\mathfrak{D}$ on the w-sphere. The two domains D and $\mathfrak{D}$ are then homeomorphic, but the converse is not true; i.e., there does not necessarily exist a one-to-one conformal mapping between D and $\mathfrak{D}$ even when they are homeomorphic. Now let D and $\mathfrak{D}$ be multiply connected domains on the z- and w-planes, respectively, both possessing at least three boundary points, and be mapped conformally onto the unit disk in accordance with theorem (4). Furthermore, let the Fuchsian groups belonging to D and $\mathfrak{D}$ be denoted by G and $\mathfrak{G}$, respectively. Then in order that D can be mapped one-to-one and conformally onto $\mathfrak{D}$, it is necessary and sufficient that the group $\mathfrak{G}$ is transformed into the group G by a suitable linear transformation.

To a domain of finite †connectivity having only continua for its boundary components, we can associate **conformal invariants** (namely **moduli**) expressed by one real parameter in the doubly connected case and by $3n-6$ real parameters in the $n(>2)$-connected case. A one-to-one conformal mapping is possible only within a class of domains having the same invariants.

While a circular disk is taken as a canonical domain in the simply connected case, an †annulus is often taken as a canonical domain in the doubly connected case. In the latter case, the logarithm of the ratio (>1) of the radii of two concentric boundary circles is usually called the **modulus**. There are various types of $n(\geqslant 2)$-connected canonical domains, for instance, the whole plane, a circular disk or annulus slit along concentric circular arcs or radial segments, a parallel slit plane, etc. The possibility of a one-to-one conformal mapping of a given domain onto a canonical domain of such a type was proved by Hilbert, Koebe, and others in a †potential-theoretic way and by E. Rengel, R. de Possel, H. Grunsky, and others in a purely function-theoretic way. On the other hand, the possibility of one-to-one and conformal mapping onto the whole plane with mutually disjoint circular disks removed was proved by Koebe and later derived by J. Douglas and R. Courant as a particular case of the existence of a solution of †Plateau's problem. L. Bieberbach and Grunsky showed the possibility of mapping an n-connected domain onto an n-sheeted disk. Concerning doubly connected domains, detailed investigations were made by O. Teichmüller, Y. Komatu, and others. For $3\leqslant n<\infty$, an extension of †Schwarz's lemma by L. V. Ahlfors (1947) and, more recently, several important results by means of the †kernel function or Schiffer's variation have been obtained. The case of infinite connectivity is essentially different in some respects from the finite case, and has been discussed by de Possel, H. Grötzsch, and others.

G. Universal Constants

Among various **universal constants** appearing in the theory of conformal mapping, Bloch's constant is an especially famous one. A. Bloch (1924) showed that a covering surface over the w-plane obtained from a mapping $w = F(z) = z + \ldots$ that is one-to-one, conformal, and holomorphic in $|z| < 1$ always contains a †univalent (schlicht) disk whose radius B is a positive number independent of the function F (**Bloch's theorem**). The supremum $\mathfrak{B}$ of such constants B is called **Bloch's constant**. The true value of $\mathfrak{B}$ is yet unknown, but estimations have been given by Ahlfors and Grunsky in the form

$$\frac{\sqrt{3}}{4} < \mathfrak{B} \leqslant \sqrt{\pi}\, 2^{1/4} \frac{\Gamma(1/3)}{\Gamma(1/4)} \left(\frac{\Gamma(11/12)}{\Gamma(1/12)} \right)^{1/2} = 0.4719\ldots.$$

The constant $\mathfrak{A}$ corresponding to the case where the family in Bloch's constant is restricted to univalent functions is known to satisfy $0.5666 < \mathfrak{A} < 0.65647$, while **Landau's constant** $\mathfrak{L}$ corresponding to the case where the image disks are not necessarily univalent satisfies $0.5 \leqslant \mathfrak{L} < 0.54326$ [5].

(For conformal mappings of polygonal domains → 371 Schwarz-Christoffel Transformations. For distortion theorems and coefficient problems → 424 Univalent and Multivalent Functions.)

References

[1] G. Julia, Leçons sur la représentation corforme des aires simplement connexes, Gauthier-Villars, 1931.

[2] G. Julia, Leçons sur la représentation conforme des aires multiplement connexes, Gauthier-Villars, 1934.

[3] C. Carathéodory, Conformal representation, Cambridge Univ. Press, 1932.

[4] S. Bergman, The kernel function and conformal mapping, Amer. Math. Soc. Math. Surveys, 1950.

[5] Y. Komatu, Tôkaku syazôron (Japanese; Theory of conformal mappings), Kyôritu, I, 1944; II, 1949.

[6] Z. Nehari, Conformal mapping, McGraw-Hill, 1952.

[7] Г. М. Голузин (G. M. Golusin), Геометрическая теория функций комплексного переменного, Гостехиздат, 1952; German translation, Geometrische Funktionentheorie, Deutscher Verlag, 1957.

[8] D. Gaier, Konstruktive Methoden der konformen Abbildung, Springer, 1964.

[9] L. Sario and K. Oikawa, Capacity functions, Springer, 1969.

Also → references to 200 Holomorphic Functions and 424 Univalent and Multivalent Functions.

80 (VI.9) Conic Sections

A. General Remarks

Suppose that we are given two straight lines l and m intersecting at V (but not orthogonally) in the 3-dimensional Euclidean space E^3. By rotating the line m around l, we obtain a surface $\mathfrak{F}$. We call this surface $\mathfrak{F}$ a **circular cone** with **vertex** V and **axis** l; a straight line on the surface passing through V is called a **generating line** of $\mathfrak{F}$.

A section C of $\mathfrak{F}$ by a plane π not passing through $V (C = \pi \cap \mathfrak{F})$ is called a **conic section** (or simply a **conic**). This C is a plane curve on the plane π. The point set $\mathfrak{F} - V$ consists of two †connected components $\mathfrak{F}_1$ and $\mathfrak{F}_2$. Let π_i $(i = 1, 2, 3)$ be planes not passing through V. If the conic section $C_1 = \pi_1 \cap \mathfrak{F}$ is †bounded, then C_1 is contained either in $\mathfrak{F}_1$ or in $\mathfrak{F}_2$ and is †connected. We call such a C_1 an **ellipse**. When $C_2 = \pi_2 \cap \mathfrak{F}$ is not bounded but is connected, then π_2 is parallel to one of the generating lines of $\mathfrak{F}$, and C_2 is contained either in $\mathfrak{F}_1$ or in $\mathfrak{F}_2$. We call such a C_2 a **parabola**. When π_3 intersects both of $\mathfrak{F}_1$ and $\mathfrak{F}_2$, then $C_3 = \pi_3 \cap F$ has two connected components and is not bounded. We call such a C_3 a **hyperbola**. These three types exhaust all possible types of conic sections. In particular, if the plane π is perpendicular to the axis l, then $C = \pi \cap \mathfrak{F}$ becomes a circle. Thus a circle is a special kind of ellipse.

B. Foci and Directrices

Let $C = \pi \cap \mathfrak{F}_1$ be an ellipse. The Euclidean space E^3 is divided by π into two †half-spaces E_1^3, E_2^3 (two "sides" of π). If we put $\mathfrak{F}_1 \cap E_1^3 = \mathfrak{F}_{11}$, $\mathfrak{F}_1 \cap E_2^3 = \mathfrak{F}_{12}$, we can construct a sphere S that is contained in E_1^3, tangent to $\mathfrak{F}_{11}$ along a circle K, and tangent to π at a point F. Similarly, we can construct a sphere S' that is in E_2^3, tangent to $\mathfrak{F}_{12}$ along a circle K', and tangent to π at a point F'. We call F, F' the **foci** of the ellipse (Fig. 1).

Let κ, κ' be the planes containing K, K'. Straight lines $d = \kappa \cap \pi$, $d' = \kappa' \cap \pi$ are called **directrices** of C. Unless C is a circle, we have $F \neq F'$, and κ, π (and κ', π) actually intersect; hence d, d' exist. When $C = \pi \cap \mathfrak{F}$ is a parabola or a hyperbola (Figs. 2 and 3), we can similarly define foci (a parabola has only one focus, F, while a hyperbola has two foci, F, F') and directrices (a parabola has only one directrix, d, while a hyperbola has two directrices, d, d').

Let X be a point on the plane π, let $D_F(X)$

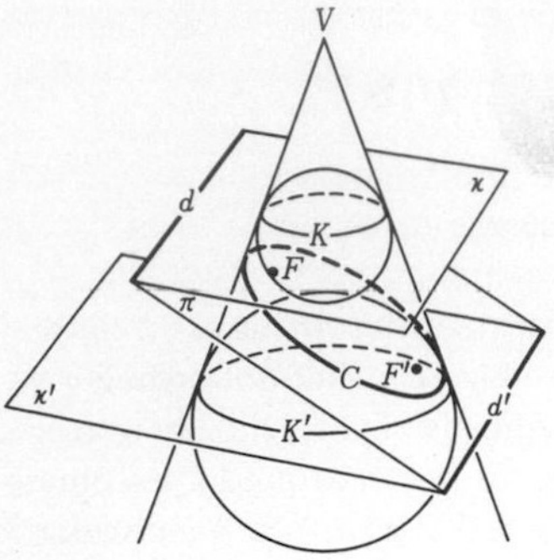

Fig. 1

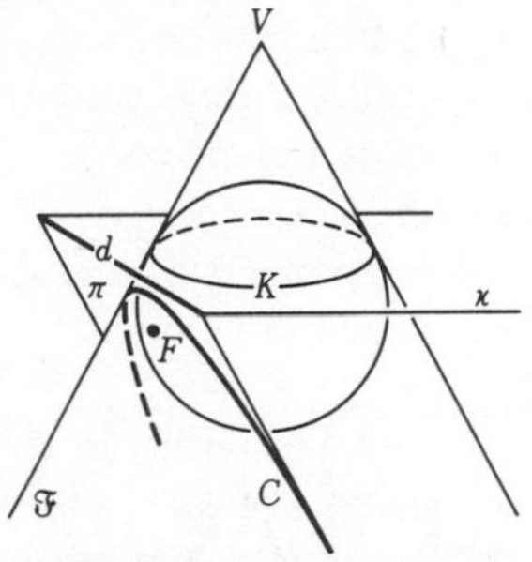

Fig. 2

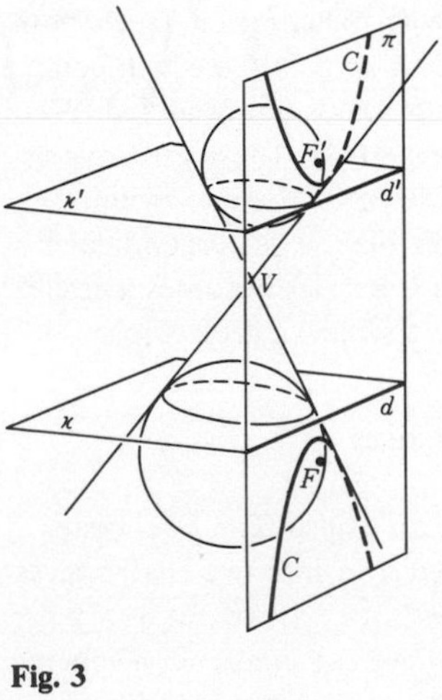

Fig. 3

be the distance between the point X and a focus F, and let $D_d(X)$ be the distance between X and a directrix d. Then the curve C is the locus of the points X satisfying the condition $D_F(X)=e\cdot D_d(X)$, where e is a constant. We call e the **eccentricity** of the conic section C. According as C is an ellipse, a parabola, or a hyperbola, we have $e<1$, $e=1$, or $e>1$. A circle is considered an ellipse whose eccentricity is zero. An ellipse is also characterized as the locus of X such that $FX+F'X=2a$; a hyperbola is the locus of X such that $|FX-F'X|=2a$, where a is a positive constant. When there are two foci, the straight line FF' is perpendicular to directrices d and d'.

C. Canonical Forms of Equations

When C is a hyperbola or an ellipse that is not a circle, C has two foci, F and F'. In this case, the midpoint O of the segment FF' is the center of symmetry of C (when C is a circle, its center O is, of course, the center of symmetry of C). We call O the **center** of C; an ellipse of a hyperbola is called a **central conic.** If we choose a rectangular coordinate system (x,y) having O as the origin and FF' as x-axis, then the equation of C can be expressed in the following form:

$$x^2/a^2 \pm y^2/b^2=1, \qquad a,b>0. \tag{1}$$

According as C is an ellipse or a hyperbola, we take the + or − of the double sign. If C is an ellipse, we have $a>b$. Furthermore, $e=\sqrt{a^2-b^2}/a$ if C is an ellipse and $e=\sqrt{a^2+b^2}/a$ if C is a hyperbola. We also have $F=(ae,0)$ and $F'=(-ae,0)$; the equations of directrices are $x=\pm a/e$.

On the other hand, if C is a parabola, the straight line that is perpendicular to the directrix d and passes through F becomes the axis of symmetry of C. We call this straight line the **axis** of C; the intersection O of the axis and C is called the **vertex** of C. If we choose a rectangular coordinate system (x,y) having O as the origin and having the axis of C as the x-axis, the equation of C can be expressed in the form

$$y^2=4ax, \quad a>0. \tag{2}$$

We call (1) and (2) the **canonical** (or **standard) forms of the equation** of C. We call the associated coordinate system the **canonical coordinate system**. Suppose that C is an ellipse (hence $a>b$). Let A, A' be points of intersection of the x-axis and the ellipse and B, B' be the points of intersection of the y-axis and the ellipse. We call AA' the **major axis** of C and BB' the **minor axis** of C. If C is a hyperbola and (x,y) is the canonical coordinate system, we call the x-axis the **transverse axis** and the y-axis the **conjugate axis.** If C is a central conic, the x- and y-axes of the canonical coordinate system are called the **principal axes**; if C is a parabola, the x-axis is sometimes called the **principal axis.**

D. Properties of Ellipses

An ellipse may be considered the image of a circle under a †parallel projection. Consequently, the section of a circular cylinder by a plane is an ellipse. Also, if we are given a circle C and a fixed diameter D of C, an ellipse is obtained as the locus of the points X lying on lines PM, which are perpendicular to D, with $P \in C$, $M \in D$, satisfying the condition that the ratio $PM:XM$ is constant.

Suppose that we are given two concentric circles having the center at the origin O and with radii a, b. Let P, Q be points of intersection of moving half-lines through O and the two circles. Then the locus of points X of intersection of the ordinates (lines parallel to the y-axis) passing through P and the abscissae (lines parallel to the x-axis) passing through Q is an ellipse (Fig. 4). Suppose that the equation of an ellipse is given by $x^2/a^2 + y^2/b^2 = 1$, with $a > b$. Then the lengths of its major axis and minor axis are $2a$ and $2b$, respectively.

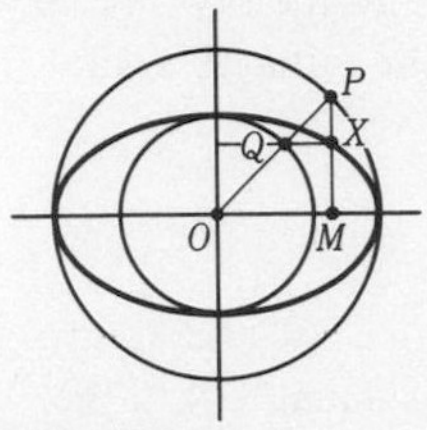

Fig. 4

Given an ellipse and its center O, the circle with center O and diameter equal to the major axis of the ellipse is called the **auxiliary circle** of the ellipse. Given an ellipse C and its focus F, the auxiliary circle of C is the locus of the points X satisfying the condition that the line FX is perpendicular to a tangent line to C passing through X. Suppose that X is a point on an ellipse with foci F, F'. Let TT' be the line tangent to the ellipse at X (X lies between T and T'). Then the angle $\angle TXF'$ is equal to $\angle T'XF$ (Fig. 5). Consequently, the rays starting from one focus of an ellipse and "reflected" by the ellipse converge on the other focus of the ellipse. Also, the product of the distances from two foci of an ellipse to an arbitrary tangent is constant and is equal to b^2.

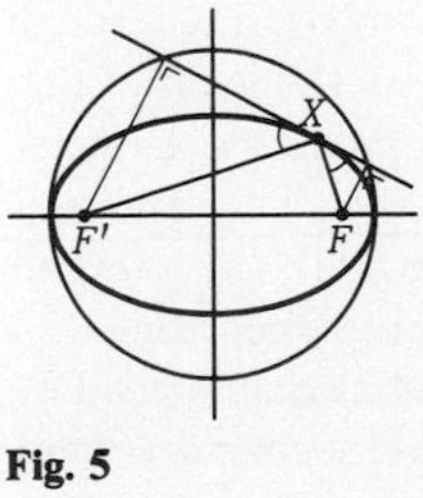

Fig. 5

The ellipse $C: x^2/a^2 + y^2/b^2 = 1$ is expressed parametrically in the form:

$$x = a\cos\theta, \qquad y = b\sin\theta. \tag{3}$$

We call the parameter θ the **eccentric angle** of a point (x,y) on C. Consequently, C is a †Jordan curve and divides the plane into two parts, the inside and the outside. The inside is the set of points (x,y) satisfying $x^2/a^2 + y^2/b^2 < 1$, and the outside is the set of points (x,y) satisfying $x^2/a^2 + y^2/b^2 > 1$. The inside is a †convex set. From a point Q outside C, two tangents to C can be drawn. The locus of points Q such that these two tangents are orthogonal is the circle $x^2 + y^2 = a^2 + b^2$. We call this circle the **director circle**. The area of the "sector" OAX formed by two points $A(a, 0)$, $X(a\cos\theta, b\sin\theta)$ ($\theta > 0$) and the origin O is $ab\theta/2 = (ab/2)\operatorname{Arccos}(x/a)$; the length of the arc $\widehat{AX}$ of the ellipse is represented by the value of the †elliptic integral

$$a\int_0^\theta \sqrt{1 - e^2\cos^2\theta}\, d\theta = aE(\pi/2 - \theta, e).$$

In particular, the area inside an ellipse is equal to πab, and the whole length of the ellipse is $4aE(0,e)$.

With respect to a polar coordinate system (r,θ) having the focus $F(ae, 0)$ as the origin and the ray directed positively along the x-axis as the initial line, the equation of the ellipse C is

$$r = \frac{l}{1 + l\cos\varphi}, \qquad l = \frac{b^2}{a}. \tag{4}$$

Here l is equal to half of the length of the chord that is perpendicular to the major axis and passes through the focus. (This chord is called the **latus rectum** of the ellipse.) Suppose that F is a fixed point and that X is a moving particle attracted toward F by a †central force inversely proportional to the square of the length of FX. Suppose further that X begins with an initial velocity whose direction is tangent to the ellipse C with focus F. Then X always moves on C, and the area velocity described by the radius FX is constant (I. Newton).

E. Properties of Hyperbolas

Two straight lines $x^2/a^2 - y^2/b^2 = 0$, that is, $y/x = \pm b/a$, are †asymptotes of the hyperbola $C: x^2/a^2 - y^2/b^2 = 1$. The hyperbola $C': x^2/a^2 - y^2/b^2 = -1$ is called the **conjugate hyperbola** of C. When $a = b$, the asymptotes are orthogonal to each other, and C, C' are congruent. In this case, we call C a **rectangular hyperbola** (or **equilateral hyperbola**). When we draw parallels to asymptotes from a point X on C, the area of the parallelogram formed by these lines and two asymptotes is constant. (In particular, when C is a rectangular hyperbola, the equation of C becomes $xy = k^2/2$ if we take two asymptotes as coordinate axes. The segment cut off by the asymptotes on the tangent to C at X is divided equally at X. In the case of the hyperbola as well, the product of the distances from two foci to an arbitrary tangent is constant and is equal to b^2, and the angle between two straight lines joining two foci to a point X on C is divided equally by the tangent at X.

A hyperbola C is represented parametrically by

$$x=a\sec\theta, \qquad y=b\tan\theta. \tag{3'}$$

In this case also, we call θ the **eccentric angle** of (x,y). If we use the †hyperbolic functions and the parameter u, the equation of a hyperbola can be written as

$$x=a\cosh u, \qquad y=b\sinh u \tag{3''}$$

instead of (3′). The area of a "sector" OAX formed by two points $A(a,0)$, $X(x,y)$ on the hyperbola and the origin O is, in this case,

$$\frac{abu}{2}=\frac{ab}{2}\operatorname{arc\,cosh}\frac{x}{a}=\frac{ab}{2}\log\frac{x+\sqrt{x^2-a^2}}{a}.$$

The length of the arc $\widehat{AX}$ of the hyperbola is given by the elliptic integral

$$\int_0^x\sqrt{\frac{e^2x^2-a^2}{x^2-a^2}}\,dx.$$

With respect to a polar coordinate system (r,φ) having the focus $F(ae,0)$ as origin and the ray directed positively along the x-axis as the initial line, the equation of the hyperbola C becomes

$$r=\frac{l}{1-e\cos\varphi}, \qquad l=\frac{b^2}{a}, \tag{4'}$$

where l is equal to half the length of the chord passing through the focus and perpendicular to the principal axis. (This chord too is called the **latus rectum**.)

F. Properties of Parabolas

The curve described by a particle attracted by "gravitation" in a fixed direction and affected by no other force is a parabola (G. Galilei). The tangent at a point X on a parabola makes equal angles with the straight line joining F and X and the direction of the principal axis (Fig. 6). Consequently, if the rays starting from the focus of a parabola are "reflected" by the parabola, they all become rays parallel to the principal axis. Let X' be

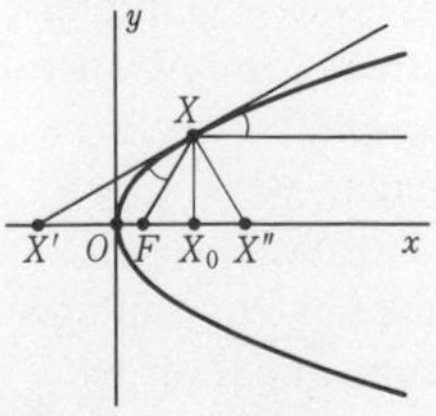

Fig. 6

the point of intersection of the tangent at $X(x_0,y_0)$ and the x-axis, X'' the point of intersection of the normal at X and the x-axis, and X_0 the foot of the perpendicular from X to the x-axis. Then $FX=FX'$, $\triangle FXX'$ is an isosceles triangle, and XX' is divided equally by the y-axis. Consequently, the locus of the foot of a perpendicular from F to a tangent is the y-axis. Also, the length of subtangent $X'X_0=2x_0$, and the length of subnormal $X''X_0=2a=2OF$. Conversely, a curve whose length of subnormal is constant is a parabola. The locus of the midpoints of parallel chords of a parabola is a straight line parallel to the principal axis.

With respect to a polar coordinate system having the focus as origin and the ray directed positively along x-axis as the initial line, the equation of a parabola is

$$r=\frac{l}{1-\cos\varphi}, \qquad l=2a. \tag{4''}$$

The area bounded by a chord BC and an arc $\widehat{BC}$ of a parabola (Fig. 7) is equal to $4/3$ the area of $\triangle ABC$, where A is the point of contact on the tangent of the parabola parallel to BC (Archimedes). Also, the length of the arc $\widehat{OX}$ of parabola (2) is

$$\frac{y_0}{4a}\sqrt{y_0^2+a^2}+a\log\frac{y_0+\sqrt{y_0^2+a^2}}{2a},$$

X having the coordinates (x_0,y_0).

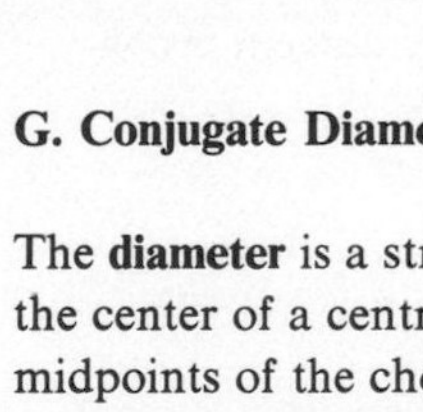

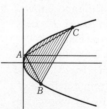

Fig. 7

G. Conjugate Diameters

The **diameter** is a straight line passing through the center of a central conic. The locus of the midpoints of the chords parallel to a diameter d is another diameter d', called **conjugate** to d. Then the diameter conjugate to d' is d. The x-axis and y-axis of a canonical coordinate system form a set of conjugate diameters. Let $2a'$, $2b'$ be the lengths of the segments (sometimes called conjugate diameters) cut off from d,d' by the curve and by ω, the angle between d and d'. Then the following relations hold (as to the double signs $\pm$, we take $+$ in the case of an ellipse and $-$ in the case of a hyperbola): $a'^2\pm b'^2=a^2\pm b^2$, $a'b'\sin\omega=ab$. The product of the slopes of d, d' is equal to $\pm b^2/a^2$. With respect to an oblique coordinate system having d and d' as axes, the equation of the curve is $x^2/a'^2 \pm y^2/b'^2 = 1$.

H. Confocal Conic Sections

The set of ellipses and hyperbolas having two fixed points F, F' as foci is called the **family of confocal central conics** with foci F and F' (Fig. 8). The family of confocal central conics containing the ellipse $x^2/a^2 + y^2/b^2 = 1$ is represented parametrically by

$$x^2\sqrt{a^2+\lambda} + y^2\sqrt{b^2+\lambda} = 1.$$

There exist only one ellipse and only one hyperbola that pass through a point inside each quadrant (for example, a point (x_0, y_0), $x_0 > 0, y_0 > 0$, inside the first quadrant) and belong to the family of curves. The ellipses and hyperbolas of the same family cut each other orthogonally. Thus parameters corresponding to ellipses and hyperbolas belonging to a family of confocal central conics define an orthogonal curvilinear coordinate system, called an †elliptic coordinate system (→ 92 Coordinates).

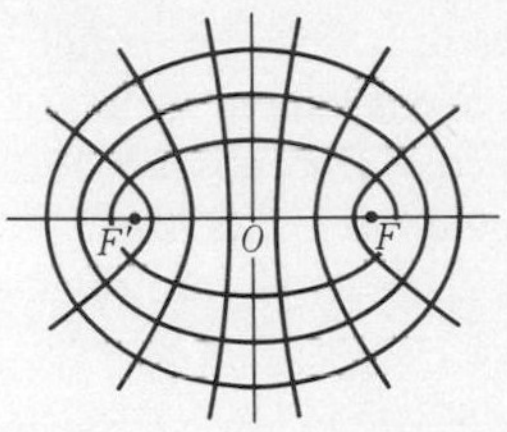

Fig. 8

The set of parabolas having a fixed point F as focus and a straight line passing through F as axis is called a **family of confocal parabolas** (Fig. 9). The family of confocal parabolas containing $y^2 = 4ax$ is the set of curves

$$y^2 = 4(a+\lambda)(x+\lambda).$$

Such families also give rise to orthogonal curvilinear coordinate systems.

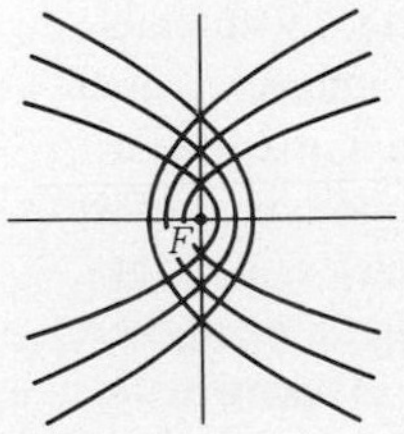

Fig. 9

I. Curves of the Second Order

With respect to a rectangular coordinate system, a curve represented by an equation with real coefficients of the second degree with two variables x,y,

$$ax^2 + 2hxy + by^2 + 2gx + 2fy + c = 0, \qquad (5)$$

where $(a,h,b) \neq (0,0,0)$, is called a **curve of the second order**. A curve of the second order is an empty set, one or two straight lines, or a conic section. For equation (5), we put

$$D_0 = \begin{vmatrix} a & h \\ h & b \end{vmatrix}, \qquad D = \begin{vmatrix} a & h & g \\ h & b & f \\ g & f & c \end{vmatrix} \qquad (6)$$

and call D the **discriminant** of the curve of the second order. If $D_0 \neq 0$, $D \neq 0$, and the curve is not an empty set, then the curve is a central conic. If $D_0 > 0$, then the curve is an ellipse or an empty set. If $D_0 < 0$, then the curve is a hyperbola. If $D_0 = 0$, $D \neq 0$, then the curve is a parabola. If $D = 0$, $D_0 > 0$, then the curve consists of one point. If $D = 0$, $D_0 < 0$, then the curve is two intersecting straight lines. If $D = D_0 = 0$, then the curve is an empty set, one straight line, or two parallel straight lines.

J. Poles and Polars

Let $F(x,y) = ax^2 + 2hxy + by^2 + 2gx + 2fy + c = 0$ be the equation of a conic C and (x_0, y_0) the coordinates of a point P on the plane. A straight line P^* having the equation $(1/2)\cdot(x_0\partial F/\partial x + y_0\partial F/\partial y) = ax_0x + h(x_0y + xy_0) + by_0y + g(x+x_0) + f(y,y_0) + c = 0$ is called the **polar** of P with respect to C (Fig. 10). When the polar of a point P is l, we call P the **pole** of l and denote it by l^*. In general, l^* is uniquely determined by l, and $P^{**} = P$, $l^{**} = l$. If $Q \in P^*$, then $P \in Q^*$. If $P' \in l$, then $l^* \in P'^*$. When a straight line passing through P intersects C at X, Y and intersects P^* at P', then P, P' are †harmonic conjugate with respect to X, Y. In particular, if $P \in P^*$, then $P \in C$, and P^* becomes the tangent of C at P. Given a triangle $\triangle PQR$ on the plane of C, we call the triangle with sides P^*, Q^*, R^* the **polar triangle** of $\triangle PQR$. Let $Q^* \cap R^* = P'$, $R^* \cap P^* = Q'$, and $P^* \cap Q^* = R'$. Then the three straight lines $P \cup P'$, $Q \cup Q'$, $R \cup R'$ meet at a point (M. Chasles). When the polar triangle of $\triangle PQR$ coincides with itself, then $\triangle PQR$ is called a **self-polar triangle**. The polar of a focus is a directrix.

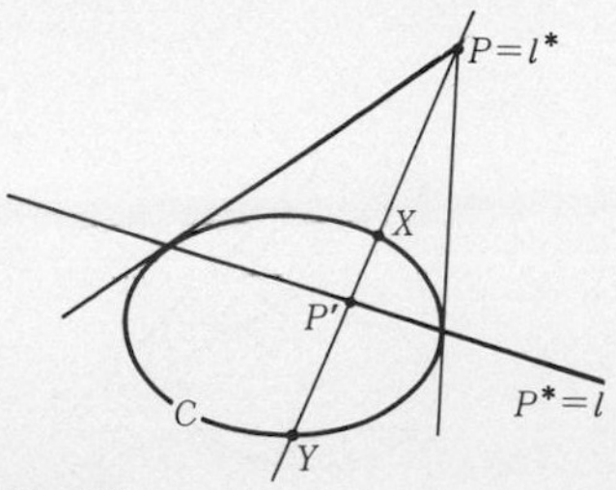

Fig. 10

K. Curves of the Second Class

When the coefficients u, v, w of a straight line $ux+vy+w=0$ satisfy an equation with real coefficients of the second order,

$$Au^2+2Huv+Bv^2+2Guw+2Fvw+Cw^2=0, \tag{5'}$$

where $(A,H,B)\neq(0,0,0)$, the curve enveloped by these straight lines is called a **curve of the second class**. Let Δ be the discriminant defined analogously to D in (6) by using A, H, $B,\ldots$ instead of a, h, $b,\ldots$. A curve of the second class (with $\Delta\neq0$) is essentially the same as a curve of the second order with $D\neq0$. In order that the curve (5′) with $\Delta\neq0$ coincide with the curve (5) with $D\neq0$, it is necessary and sufficient that A, B, C, F, G, H are proportional to the †cofactors of a, b, c, f, g, h in the determinant D given by (6). If $\Delta=0$, then (5′) represents either the empty set, a point (regarded as the set of straight lines passing through the point), or two points.

From a projective point of view, a curve of the second order is defined as a locus of the point of intersection $l\cap l'=X$ of corresponding lines l and l' when two pencils of lines $A(l,m,\ldots)$, $A'(l',m',\ldots)$ passing through two different centers A and A' are in correspondence under a †projective mapping f. (J. Steiner) (Fig. 11). From this it can be proved that three points of intersection of three pairs of opposite sides (AB,DE), (BC,EF), (CD,FA) of a hexagon inscribed in a curve of the second order are on the same straight line (**Pascal's theorem**, Fig. 12). In particular, if the curve of the second order in this theorem consists of two straight lines, the theorem coincides with **Pappus's theorem**, (Fig. 13). We call this straight line l the **Pascal line** of $ABCDEF$. Given a set of six points A, B, C, D, E, F on a curve of the second order, by considering all possible combinations of the points, we get 60 Pascal lines. A configuration consisting of these 60 lines is called **Pascal's configuration**, and has been studied by Steiner, Kirkman, and others. As a †dual to Pascal's theorem, **Brianchon's theorem** holds: Three diagonals of a hexagon with a curve of the second class inscribed meet at a point (Fig. 14).

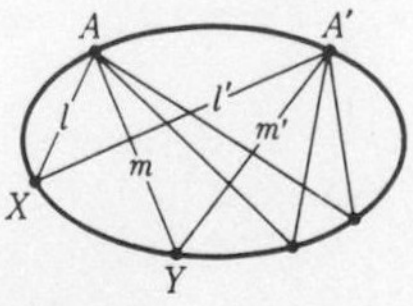

Fig. 11

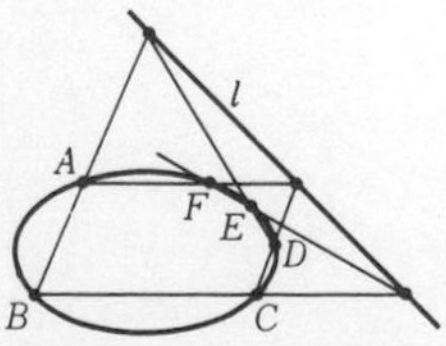

Fig. 12

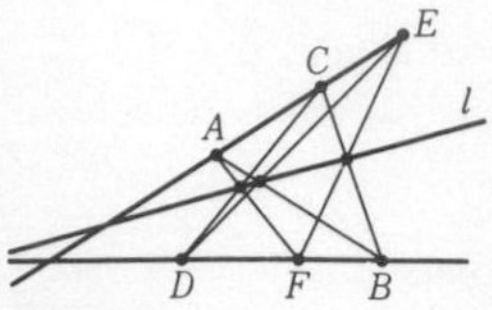

Fig. 13

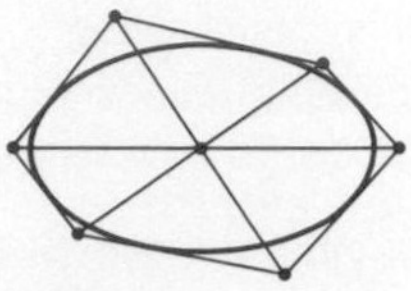

Fig. 14

References

[1] T. Kubota, Kaiseki kikagaku (Japanese; Analytic geometry) I, Utida-rôkakuho, 1937.
[2] G. Salmon, A treatise on conic sections, Longmans, Green and Co., sixth edition, 1879 (Chelsea 1962).
[3] H. F. Baker, Principles of geometry II. Plane geometry, Cambridge Univ. Press, 1922.

81 (II.21) Connectedness

A. Connectedness

A †topological space X is said to be **connected** if there are no proper closed subsets A and B of X such that $A\cap B=\varnothing$ and $A\cup B=X$ (C. Jordan, *Cours d'analyse* I, 1893). A subset S of X is **connected** if S considered as a †subspace of X is connected. If a subset S of X is connected, then the closure $\bar{S}$ is also connected. Let $\{A_\alpha\}$ be a family of connected subsets of X such that $\bar{A}_\alpha\cap A_\beta=A_\alpha\cap\bar{A}_\beta=\varnothing$ never holds for any A_α and A_β. Then the union $\bigcup_\alpha A_\alpha$ is connected. The continuous image of a connected set A is connected. The †product space of a family of connected spaces $\{A_\alpha\}$ is also connected. The union of connected subsets having in common a point p of a topological space X is connected, so that there is a maximal connected subset containing p, called the **connected component** (or simply **component**) of p (F. Hausdorff, 1927). The n-dimensional Euclidean space (for $n\geqslant0$) and n-dimensional sphere S^n (for $n\geqslant1$) are connected, whereas S^0, consisting of two points, is not connected, with each point a component of S^0. A connected open subset of a topological space X is called a **domain** (or **region**) in X. A topological space X is said to be **locally connected** at a point p if for every open set U containing p, there is a connected open set V containing p and contained in U. The space X is locally connected if it is locally connected at each point of X.

B. Arcwise Connectedness

Two points a and b of X are said to be **joined in X by an arc** if there is a continuous function

$f(t)$ on the closed interval $I=[0,1]$ into X such that $f(0)=a$ and $f(1)=b$. A topological space X is **arcwise-connected** if every two points of X are joined in X by an arc. An arcwise-connected space X is connected; however, the converse is not true. The connected spaces mentioned in Section A are all arcwise connected. An **arcwise-connected component** and **locally arcwise connectedness** are defined similarly. The †sinusoid, i.e., the union of the graph of $y=\sin 1/x$ (where x is a nonzero real number) and $\{(0,y) \mid -1 \leqslant y \leqslant 1\}$, is an example of a space that is connected but not locally arcwise connected. Let X and Y be topological spaces. If the †function space $C(X,Y)$ is arcwise connected, then all continuous maps from Y into X are †homotopic to each other, and hence are null-homotopic in X. Let S^n be an n-dimensional sphere $(n \geqslant 0)$. Then X is **n-connected** if $C(S^n, Y)$ is arcwise connected, where 0-connectedness is just arcwise connectedness. A space X is **simply connected** if it is 1-connected and **multiply connected** if it is n-connected, $n \geqslant 2$.

A space X is **contractible** (K. Borsuk, *Fund. Math.*, 24 (1935)) if $C(X,X)$ is arcwise connected, or equivalently, the †identity mapping $e(x)$ defined by $e(x)=x$ is null-homotopic in X. If X is contractible, then it is n-connected for every n. The n-dimensional sphere S^n is not n-connected but is i-connected for $i<n$. The n-†simplex is a contractible topological space. A space X is **locally contractible at a point** p of X if for each neighborhood U of p, there is a neighborhood V of p such that $V \subset U$ and the inclusion mapping of V into U is null-homotopic in U; and X is **locally contractible** if it is locally contractible at every point p of X. Similarly, X is **locally n-connected at a point** p if for each neighborhood U of p, there is a neighborhood V of p such that $V \subset U$ and every continuous map f from S^n into V is null-homotopic in U. X is **locally n-connected** if it is locally n-connected at each of its points. If a space is locally contractible, then it is locally n-connected for each n. †Polyhedra are locally contractible. X is said to be **ω-connected** or **locally ω-connected** if it is n-connected or locally n-connected for each n $(n=0,1,2,\ldots)$. A polyhedron is locally contractible; hence it is locally ω-connected. However, a locally ω-connected space is not necessarily locally contractible.

If a metric space R is †separable and locally connected and if it is a closed subset of a separable metric space X such that the dimension of $X-R$ is finite, then the following theorems hold [4,5,6]. (1) R is a neighborhood retract of X; i.e., there exist an open set $U \supset R$ of X and a continuous mapping f from U onto R such that $f(x)=x$ if $x \in R$. (2) For every continuous mapping f from R into a separable metric space Y, there exist an open set $U \supset R$ of X and a continuous mapping F from U into Y such that $F(x)=f(x)$ for each $x \in R$; i.e., f is extendible over U.

Let X be a locally compact Hausdorff space. Then X is said to be **locally n-homologically connected (locally n-cohomologically connected)** at a point p of X if for each neighborhood U of p, there is a neighborhood V of p such that $V \subset U$ and the n-dimensional †Čech homology group (cohomology group) of U mod V is 0 [3].

C. Continua and Discontinua

A **continuum** is a connected compact metric space consisting of more than one point. A space which does not have a continuum as a subset is a **discontinuum**. A space is said to be **totally disconnected** if the component of each point is the point itself. A subset of the closed interval $I=[0,1]$ consisting of points with coordinates $t=(n_1/3)+(n_2/3^2)+\ldots+(n_i/3^i)+\ldots$, where $n_i=0$ or 2, is homeomorphic to the Cartesian product of countably many Hausdorff spaces consisting of two points 0 and 1. This subset is a compact †perfect set and a discontinuum, called the **Cantor discontinuum** or simply the **Cantor set** or **ternary set** (G. Cantor, *Math. Ann.*, 21 (1883)). Let I_{11}, I_{12} be the closed intervals obtained from I by removing an open interval of length ε_1 lying in the middle of I, and let I_{21}, I_{22}, I_{23}, and I_{24} be the closed intervals obtained from I_{11}, I_{12} by removing open intervals of length ε_2 lying in their middles, (as described in Fig. 1). Inductively, 2^{n+1} closed intervals $I_{n+1,i}$ $(i=1,2,\ldots,2^{n+1})$ are obtained from 2^n closed intervals I_{nj} $(j=1,2,\ldots,2^n)$ by removing open intervals of length ε_{n+1} lying in their middles. Let $I^{(n)}=\bigcup_{i=1}^{2^n} I_{ni}$. Then $C=\bigcap_{n=1}^{\infty} I^{(n)}$ is called the **general Cantor set**, and C is homeomorphic to the ternary set.

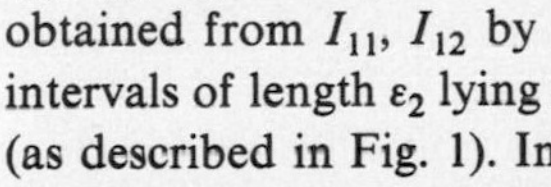

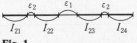

Fig. 1

A metric space X is said to be **well-chained** if for every two points a, b and $\varepsilon>0$ there are points $x_1, x_2, \ldots, x_{n-1}$ such that $d(x_i, x_{i+1})<\varepsilon$ $(x_0=a,\ x_n=b)$. A well-chained compact space is a continuum.

A continuum K is said to be **indecomposable** if there are no proper subcontinua K_1, K_2 such that $K=K_1 \cup K_2$ (L. E. J. Brouwer, *Math. Ann.*, 66 (1910)). Simple examples of indecomposable continua were given by A. Denjoy, *C. R. Acad. Sci.*, Paris, 151 (1910); K. Yoneyama, *Tôhoku Math. J.*, 12 (1917); and B. Knaster, *Fund. Math.*, 3 (1922). Let K be a continuum containing two points a and b. Then K is said to be **irreducible** between a and b if there is no proper subcontinuum of K containing a and b (L. Zoretti, *Ann. Sci. Ecole Norm. Sup.*, 26 (1909)) (→ 96 Curves).

D. The Jordan Curve Theorem

A topological space is called a **Jordan curve** if it is homeomorphic to the circle. The **Jordan curve theorem** states: A Jordan curve J in the plane $\mathbf{R}^2$ separates $\mathbf{R}^2$ into inner and outer regions (C. Jordan, *Cours d'analyse*, 2nd ed., 1893). More precisely, $\mathbf{R}^2 - J$ is the disjoint union $G_1 \cup G_2$ of two regions G_1 and G_2 whose common boundary is J. Let p be a point of J. Then there is a **Jordan arc** (a space homeomorphic to the †segment) with p as an endpoint such that all points of the Jordan arc are contained in G_i ($i=1$ or 2) except for p (A. Schönflies), that is, J is **accessible** from G_i. Conversely, let $J \subset \mathbf{R}^2$, $\mathbf{R}^2 - J = G_1 \cup G_2$, and $G_1 \cap G_2 = \varnothing$, where the G_i are regions such that J is accessible from both G_1 and G_2. Then J is a Jordan curve (Schönflies, 1908). A homeomorphism between a Jordan curve J and the circle C extends to a homeomorphism (more precisely to a †conformal mapping) between a plane containing J and a plane containing the circle C (→ 79 Conformal Mapping). The Jordan curve theorem is generalized as follows: Let S^n be a †topological n-sphere in $\mathbf{R}^{n+1}$ (topological $(n+1)$-sphere S^{n+1}). Then $\mathbf{R}^{n+1} - S^n$ $(S^{n+1} - S^n)$ is $G_1 \cup G_2$, where G_1 and G_2 are regions in $\mathbf{R}^{n+1}$ (S^{n+1}) such that $G_1 \cap G_2 = \varnothing$, S^n is the common boundary of G_1 and G_2, and furthermore S^n is accessible from G_1 and G_2 (**Brouwer theorem**, L. E. J. Brouwer, *Math. Ann.*, 71 (1912)). The **Schönflies problem** asks whether a homeomorphism h from an n-sphere Σ^n onto a topological n-sphere S^n can be extended to a homeomorphism H from an $(n+1)$-sphere Σ^{n+1} containing Σ^n as the equator onto a topological $(n+1)$-sphere S^{n+1} containing S^n. For $n \geqslant 2$ the answer is negative (→ 259 Manifolds F). If the homeomorphism h is †differentiable, then there is a differentiable extension H for $n \neq 3, 6$ (S. Smale, *Ann. of Math.*, 74 (1961)). On the other hand, if the homeomorphism can be extended to an into homeomorphism $h': \Sigma^n \times [-1,1] \rightarrow S^{n+1}$ (i.e., S^n is †bicollared), then there is an extension H of h (**Brown-Mazur theorem**, M. Brown, *Bull. Amer. Math. Soc.*, 66 (1960)).

References

[1] K. Kuratowski, Topologie II, Monograf. Mat. (1950).
[2] S. Lefschetz, Topics in topology, Ann. Math. Studies, Princeton Univ. Press, 1942.
[3] R. L. Wilder, Topology of manifolds, Amer. Math. Soc. Colloq. Publ., 1949.
[4] G. T. Whyburn, Analytic topology, Amer. Math. Soc. Colloq. Publ., 1942.
[5] S. T. Hu, Elements of general topology, Holden-Day, 1964.
[6] O. Hanner, Retraction and extension of mappings of metric and non-metric spaces, Ark. Mat., 2 (1952), 315–360.
[7] K. Borsuk, Theory of retracts, Monograf. Mat., (1967).
Also → references to 408 Topological Spaces.
For the Jordan curve theorem,
[8] P. S. Alexandrov and H. Hopf, Topologie I, Springer-Verlag, 1935.
[9] S. Lefschetz, Introduction to topology, Princeton Univ. Press, 1949.
[10] M. H. A. Newman, Elements of the topology of plane sets of points, Cambridge Univ. Press, second edition, 1951.
[11] S. Iyanaga, Jordan curve theorem (Japanese), Gendai no sûgaku (Modern mathematics) I. Kyôritu, 1947.

82 (VII.3) Connections

A. History

The geometric notion of connections originated with T. Levi-Civita's parallelism (*Rend. Circ. Mat. Palermo*, 42 (1917)) and was later generalized to the notion of connections of differentiable fiber bundles. Notions such as affine connections, Riemannian connections, projective connections, and conformal connections can be described in terms of bundles constructed from the tangent bundles of differentiable manifolds. They are also standard examples of the Cartan connections formulated by E. Cartan and C. Ehresmann.

B. Connections in Principal Bundles

Let $P=(P,\pi,M,G)$ be a differentiable †principal fiber bundle. (For the sake of convenience, we assume that differentiability always means that of class C^∞.) The total space P and the base space M are †differentiable manifolds, and the projection π is a differentiable mapping. The †structure group G is a †Lie group and acts on P from the right as a transformation group. On each fiber, G acts transitively without fixed points. For elements a, x in G, P, we write $R_a(x)=xa$. The mappings induced on †tangent vector spaces by R_a and π will be denoted by the same letters, namely $R_a: T_x(P) \rightarrow T_{xa}(P)$, $\pi: T_x(P) \rightarrow T_{\pi(x)}(M)$. The tangent vector space $T_x(P)$ at each point x of P is mapped by the projection π onto the tangent vector space $T_p(M)$ at the point $p=\pi(x)$ of M. The kernel of this mapping is denoted by $V_x(P)$, and each vector in $V_x(P)$ is said to be **vertical**. The kernel $V_x(P)$ is the totality of elements of $T_x(P)$ that are tangent to the fiber.

C. Connections

We say that a **connection** is given in P if for each point $x \in P$, a subspace Q_x of the tangent space $T_x(P)$ is given such that the following three conditions are satisfied: (i) $T_x(P) = V_x(P) + Q_x$ (direct sum); (ii) $R_a(Q_x) = Q_{xa}$ (Q is invariant under G); and (iii) the mapping $x \rightarrow Q_x$ is differentiable. A vector in Q_x is said to be **horizontal**.

Now suppose that X is an arbitrary †vector field over P. By condition (i), the value X_x of X at each point x of P can be expressed uniquely as $X_x = Y_x + Z_x$, where $Y_x \in V_x(P)$ and $Z_x \in Q_x$. The vector fields Y and Z defined by Y_x and Z_x ($x \in P$) are called the **vertical** and **horizontal components** of X, respectively. Condition (iii) implies that if X is a differentiable vector field, then its horizontal and vertical components are also differentiable vector fields. Let X be a vector field on the base space M. Since π defines an isomorphism of Q_x and $T_p(M)$ ($p=\pi(x)$), we have a unique vector field X^* on P such that (a) $\pi(X^*)=X$ and (b) $X^*_x \in Q_x$. We call X^* the **lift** of X, and it is invariant under G by condition (ii).

Suppose that a connection is given in P. If C is a piecewise differentiable curve in the base space M, we can define a mapping φ that maps the fiber over the initial point p of C onto the fiber over the endpoint q of C as follows: Take an arbitrary point x on the fiber at p. Then we have a unique curve C^*_x in P starting at x such that (a) $\pi(C^*_x) = C$, and (b) each tangent vector to C^*_x is horizontal. (C^*_x is called a **lift** of C that starts at x.) The endpoint y of the curve C^*_x belongs to the fiber over q. We set $\varphi(x)=y$. Because $C^*_{xa} = R_a(C^*_x)$, the mapping φ is commutative with transformations of G. We call this mapping φ the **parallel displacement along the curve** C.

D. Holonomy Groups

Fix a point p in the base space. If C is a closed curve in M starting from p, the parallel displacement along C maps the fiber over p onto itself. So if we fix a point x on the fiber over p, x is transformed by the parallel displacement to a point xa ($a \in G$). Thus each closed curve C starting from p determines an element $a(x,C)$ of G. If C varies over the set of closed curves that start from p, the totality of such elements of G forms a subgroup of G. This subgroup is called the **holonomy group** of the connection defined over P with the reference point x. If M is connected, holonomy groups with different reference points are conjugate. In the above, if we choose as the closed curves C starting from p only those curves that are null-homotopic, the elements $a(x,C)$ form a subgroup of the holonomy group. This is called the **restricted holonomy group**. The holonomy group is a †Lie subgroup of the structure group, and its connected component containing the identity coincides with the restricted holonomy group. Holonomy groups are useful in the study of the behavior of connections.

E. Connection Forms

Let $\mathfrak{g}$ be the Lie algebra (→ 248 Lie Groups) of the structure group of G of a principal fiber bundle $P=(P,\pi,M,G)$. For each A in $\mathfrak{g}$, the 1-parameter subgroup $\exp tA$ ($-\infty < t < \infty$) of G defines a †one-parameter group $R_{\exp tA}$ of transformations on P, and it determines a vector field A^* on P (→ 108 Differentiable Manifolds). Each element of the vector field A^* is vertical at each point x on P, and the A^* ($A \in \mathfrak{g}$) at x generate $V_x(P)$. Moreover, for each element a of G we have $R_a(A^*) = (\mathrm{ad}(a^{-1})A)^*$.

For a connection in P, we define the **connection form** ω on P with values in $\mathfrak{g}$ by the following: (i) $\omega_x(A^*_x) = A$ ($A \in \mathfrak{g}$), and (ii) $\omega_x(X) = 0$ ($X \in Q_x$). The connection form ω thus defined satisfies (iii) $R^*_a(\omega) = \mathrm{ad}(a^{-1})\omega$ ($a \in G$), where $R^*_a(\omega)$ is the †differential form induced by the transformation R_a from the differential form ω. Conversely, given a 1-form ω with values in $\mathfrak{g}$ that satisfies conditions (i) and (ii), we can define a connection in P by defining vectors X such that $\omega(X)=0$ as the horizontal, and its connection form coincides with ω. Thus giving a connection in P is equivalent to giving a connection form in P.

In particular, when a principal fiber bundle P is trivial, i.e., when $P = M \times G$, we can identify the tangent vector space $T_x(P)$ at a point $x=(p,g)$ of P with the direct sum of $T_p(M)$ and $T_g(G)$. If we set $Q_x=T_p(M)$, then Q defines a connection in $P=M\times G$. Such a connection is called **flat**. When a connection can always be expressed as above locally, it is called **locally flat**. Since each principal fiber bundle is locally a product fiber bundle, we see that locally there exists a connection. If the base space M is †paracompact, we can show the existence of connections on P.

F. Extensions and Restrictions of Connections

When a principal fiber bundle $P=(P,\pi,M,G)$ has a †reduced fiber bundle P', we shall consider the relation between the connections of P and of P'. Let G' be a Lie subgroup of G, and let $\mathfrak{g}'$ be its Lie algebra. We shall denote by j both the injection of G' into G and the injection of $\mathfrak{g}'$ into $\mathfrak{g}$. If there exist a differentiable principal fiber bundle $P'=(P',\pi',M,G')$ and a differentiable embedding f of P' into P such that $\pi\circ f=\pi'$ and $f\circ R_a=R_{j(a)}\circ f$ $(a\in G')$ are satisfied, then (P',f) is said to be a reduced fiber bundle of P. Then we have $f_*(A_x^*)=j(A)_{f(x)}^*$ for each $A\in\mathfrak{g}'$ and $x\in P'$.

Suppose that a connection is given in P'; we denote the horizontal space at the point x of P' by Q_x'. At the point $f(x)$ of P, we take $f_*(Q_x')$ as the horizontal space and transform it by right translations of G. Thus we obtain a connection on P. Let ω' and ω be the corresponding connection forms. Then we have $j\circ\omega'=f^*(\omega)$ on P'. Conversely, suppose that we are given a connection in P with the connection form ω. If the induced form $f^*(\omega)$ on P' has values always in $j(\mathfrak{g}')$, we can write $f^*(\omega)=j\circ\omega'$, and ω' defines a connection in P'. In this case the connection in P is called an **extension** of the connection in P', and the connection in P' is called the **restriction** of the connection in P.

G. Curvature Forms

Suppose that a principal fiber bundle $P=(P,\pi,M,G)$ has a connection. Let F be a finite-dimensional vector space and α be a differential form of degree k on P with values in F. We define the **covariant differential** $D\alpha$ of α by

$$(D\alpha)(X_1,\ldots,X_{k+1})=(d\alpha)(hX_1,\ldots,hX_{k+1}),$$

where the X_i are vector fields on P and h denotes the projection to the horizontal component. $D\alpha$ is a differential form of degree $k+1$ on P with values in F.

Let $\rho:G\to GL(F)$ be a †representation of a Lie group G onto F. A differential form α on P with values in F is called a **pseudotensorial form** of type ρ if α satisfies $R_a^*(\alpha)=\rho(a^{-1})\alpha$ $(a\in G)$. In particular, if a pseudotensorial form α satisfies $\iota(A^*)\alpha=0$ for any $A\in\mathfrak{g}$ (→ 108 Differentiable Manifolds Q), it is called a **tensorial form** of type ρ. For each representation ρ of G, we can construct an associated vector bundle E over M with fiber F. A tensorial form of type ρ is identified with a differential form on M with values in E. If α is a pseudotensorial form of type ρ, then $D\alpha$ is a tensorial form of type ρ.

For a connection form ω on P, the covariant differential $D\omega=\Omega$ of ω is called the **curvature form** of the connection. Since ω is a pseudotensorial form of type ad, Ω is a tensorial form of type ad. For the connection form we have the **structure equation** $d\omega=-[\omega,\omega]+\Omega$ [4,6]. Let X and Y be vector fields on M, and let X^* and Y^* be their lifts, respectively. Then we have $\omega([X^*,Y^*])=\Omega(X^*,Y^*)$, which shows that the curvature form Ω for X^*, Y^* gives the vertical component of $[X^*,Y^*]$.

It is known that the following three conditions for a connection are equivalent: (i) The connection is locally flat. (ii) The curvature form vanishes. (iii) The restricted holonomy group is trivial (i.e., the identity group).

The following two theorems are fundamental:

(1) Suppose that a connection is given in a principal fiber bundle $P=(P,\pi,M,G)$. Then the structure group of P can be reduced to the holonomy group [4,6]. In fact, for $x\in P$, let $P(x)$ be the set of points y in P that can be connected to x by a piecewise horizontal curve in P. Then $P(x)$ gives a reduced fiber bundle of P, and the connection in P is an extension of a connection in $P(x)$ [4,6].

(2) The Lie algebra of the holonomy group with a reference point x in P coincides with the vector subspace of $\mathfrak{g}$ spanned by $\{\Omega_y(X,Y)\mid y\in P(x),X,Y\in T_y(P)\}$ [4,6].

The curvature form Ω is used to determine the †characteristic classes of the bundle P [1,2] (→ 58 Characteristic Classes).

In some cases, a connection in the principal fiber bundle induces a connection in an †associated fiber bundle. In particular, when G is $GL(n,\mathbf{R})$ or $GL(n,\mathbf{C})$, we can define a connection in any associated vector bundle. The notion of connections in vector bundles can be defined more algebraically (M. F. Atiyah, *Trans. Amer. Math. Soc.*, 85 (1957)) and can also be defined as a kind of differential operator on vector bundles [8].

H. Affine Connections

Let M be a differentiable manifold of dimension n and P be the †bundle of tangent n-frames over M. Then P has the structure group $GL(n,\mathbf{R})$, and it is the principal bundle associated with the tangent vector bundle of M, which consists of all tangent vectors of M. A connection in the bundle of tangent n-frames is called an **affine connection** (or **linear connection**) on M. An affine connection on M defines (as well as the curvature form Ω) a new form Θ called the torsion form on P, which is given as follows: Let F be an n-dimensional vector space with a fixed basis $(\xi_1,\xi_2,\ldots,\xi_n)$. Since the bundle of tangent n-frames P is the set of all bases (i.e., n-frames) $(e_1,\ldots,e_n)$ in $T_p(M)$ at each point p of M, every point $x=(e_1,\ldots,e_n)$ of P is given as a mapping $\bar{x}$ of F onto $T_p(M)$ $(p=\pi(x))$ defined by $\bar{x}:\xi_i\rightarrow e_i$. We define differential form θ of degree 1 with values in F on P by $\theta_x(X)=\bar{x}^{-1}(\pi_x(X))$ $(X\in T_x(P))$. θ is called a **canonical 1-form** of the bundle of tangent n-frames of the manifold M and has the following property: Any diffeomorphism φ of M onto itself induces a bundle automorphism $\tilde{\varphi}$ of P onto itself, and $\tilde{\varphi}$ preserves θ, that is, $\tilde{\varphi}^*(\theta)=\theta$. Conversely, we can show that any bundle automorphism of P that preserves θ is induced by a diffeomorphism of the base space M.

For an affine connection on M, we define the **torsion form** Θ by $\Theta=D\theta$. Θ is a differential form of degree 2 on P with values in F and satisfies $R_a\Theta=a^{-1}\cdot\Theta$ $(a\in GL(n,\mathbf{R}))$. Furthermore, we have the **structure equation** for Θ, $d\theta=[\omega,\theta]+\Theta$ [2,4,6].

For each element ξ in F, there exists a unique horizontal vector field $B(\xi)$ on P such that $\theta(B(\xi))=\xi$. $B(\xi)$ is called the **basic vector field** corresponding to ξ. At each point $x\in P$, $B(\xi_1)_x,\ldots,B(\xi_n)_x$ form a basis of Q_x. Let $A_1,\ldots,A_m$ $(m=n^2)$ be a basis of $\mathfrak{g}=\mathfrak{gl}(n,\mathbf{R})$. Then at each point $x\in P$, $\{(A_1^*)_x,\ldots,(A_m^*)_x, B(\xi_1)_x,\ldots,B(\xi_n)_x\}$ is a basis of the tangent vector space $T_x(P)$. Thus the bundle of frames P is a †parallelizable manifold. The projection to M of any †integral curve of a basic vector field is a geodesic, which is defined in Section I [4].

An affine connection on M gives a parallel displacement of the tangent vector space of M as follows: Let $C=p_t$ $(0\leqslant t\leqslant 1)$ be a curve in M and $C^*=x_t$ be a lift of C to P. The parallel displacement of the tangent n-frame x_0 at p_0 along the curve C is x_t, and the mapping $\bar{x}_t\circ\bar{x}_0^{-1}:T_{p_0}(M)\rightarrow T_{p_t}(M)$ is called the **parallel displacement** of the tangent space $T_{p_0}(M)$ onto $T_{p_t}(M)$ along C. It is easily seen that the mapping is independent of the choice of lifts.

I. Covariant Differentials

Let $C=\{p_t\}$ $(0\leqslant t\leqslant 1)$ be a differentiable curve in M. If we have a vector Y_t in $T_{p_t}(M)$ for each t and the correspondence $t\rightarrow Y_t$ is differentiable, then $\{Y_t\}$ is called a vector field along the curve C. For $\{Y_t\}$ we set

$$Y_t'=\lim_{h\rightarrow 0}(1/t)(\varphi_{t,h}^{-1}(Y_{t+h})-Y_t),$$

where $\varphi_{t,h}$ is the parallel displacement of $T_{p_t}(M)$ onto $T_{p_{t+h}}(M)$ along the curve C. The vector field $\{Y_t'\}$ along C thus obtained is called the **covariant derivative** of $\{Y_t\}$. $\{Y_t\}$ is parallel along C; that is, $Y_t=\varphi_{0,t}(Y_0)$ if and only if $Y_t'\equiv 0$. In particular, if the tangent vectors to a curve C are parallel along C itself, then C is said to be a **geodesic**.

Let X and Y be vector fields on a manifold M with an affine connection. The **covariant derivative** $\nabla_X Y$ of the vector field Y in the direction of the vector field X is defined as follows: Let p_0 be a point in M, $C=\{p_t\}$ $(-\varepsilon\leqslant t\leqslant\varepsilon)$ be an integral curve of X through p_0, and $\{\varphi_t\}$ be the parallel displacement along C. We set

$$(\nabla_X Y)_{p_0}=\lim_{t\rightarrow 0}(1/t)(\varphi_t^{-1}(Y_{p_t})-Y_{p_0}).$$

Then $\nabla_X Y$ is also a vector field on M.

The mapping $(X,Y)\rightarrow\nabla_X Y$ satisfies the following three conditions: (i) $\nabla_X Y$ is linear with respect to X and Y; (ii) $\nabla_{fX}Y=f\cdot\nabla_X Y$; and (iii) $\nabla_X(fY)=f\cdot\nabla_X Y+(Xf)\cdot Y$, where f is a differentiable function on M. Conversely, if the mapping conditions (i)–(iii) are given, then there exists a unique affine connection on M whose covariant derivative coincides with the given mapping [4,6].

Fix a vector field Y. Then the mapping $X\rightarrow\nabla_X Y$ defines a †tensor field of type $(1,1)$. This tensor field is called the **covariant differential** of Y and is denoted by ∇Y. Now fix a vector field X. Then the mapping $Y\rightarrow\nabla_X Y$ can be naturally extended to tensor fields of arbitrary type, and it commutes with the †contraction of the tensors. For a tensor field K this is denoted by $K\rightarrow\nabla_X K$. Furthermore, the mapping $X\rightarrow\nabla_X K$ is called the **covariant differential** of K and is denoted by ∇K. We call $\nabla_X K$ the **covariant derivative** of K in the direction of X. A tensor field K is invariant under parallel displacements if and only if $\nabla K=0$ (→ 403 Tensor Calculus).

J. Curvature Tensors and Torsion Tensors

For an affine connection on M, the **curvature tensor** R and the **torsion tensor** T are defined by

$$R(X,Y)(Z) = \nabla_X(\nabla_Y Z) - \nabla_Y(\nabla_X Z) - \nabla_{[X,Y]}(Z),$$

$$T(X\cdot Y) = \nabla_X Y - \nabla_Y X - [X,Y],$$

where X, Y, and Z are vector fields on M, and R and T are tensors of types (1,3) and (1,2), respectively. Also, in terms of the curvature form Ω and the torsion form Θ on the bundle of tangent n-frames P over M, they can be defined by

$$R_p(X,Y)(Z) = \bar{x}^{-1}\cdot\Omega_x(X^*,Y^*)\cdot\bar{x}(Z),$$

$$T_p(X,Y) = \bar{x}^{-1}(\Theta_x(X^*,Y^*)),$$

where $\pi(x)=p$, $X, Y \in T_p(M)$, and X^*, Y^* are lifts of X, Y, respectively. The curvature tensor and the torsion tensor satisfy the relations $R(X,Y) = -R(Y,X)$, $T(X,Y) = -T(Y,X)$. Moreover, **Bianchi's identities** hold:

$$\mathfrak{S}(R(X,Y)(Z)) = \mathfrak{S}(T(T(X,Y),Z) + (\nabla_X T)(Y,Z)),$$

and

$$\mathfrak{S}((\nabla_X R)(Y,Z) + R(T(X,Y),Z)) = 0,$$

where $\mathfrak{S}$ denotes the sum of terms that are obtained by cyclic permutations of X, Y, Z [4]. For instance, in the case of a Riemannian connection, to be discussed in Section K, we have $T=0$, and Bianchi's identities reduce to

$$R(X,Y)(Z) + R(Y,Z)(X) + R(Z,X)(Y) = 0,$$

$$(\nabla_X R)(Y,Z) + (\nabla_Y R)(Z,X) + (\nabla_Z R)(X,Y) = 0.$$

We now consider a system of coordinates $(x^1,\dots,x^n)$ in an n-dimensional linear space $M=\mathbf{R}^n$. The vector fields $(X_1,\dots,X_n)$ $(X_i = \partial/\partial x^i)$ form a basis for vector fields on M. If we set $\nabla_{X_i} X_j = 0$, we get an affine connection on $\mathbf{R}^n$. For such a connection we have $R=0, T=0$, and any straight line in $\mathbf{R}^n$ is a geodesic with respect to this connection. The connection is called the **canonical affine connection** on $\mathbf{R}^n$. An affine connection on a manifold M satisfies $R=0$ and $T=0$ if and only if the connection on M is locally isomorphic to the canonical affine connection of $\mathbf{R}^n$.

Let φ be a diffeomorphism of a manifold M with an affine connection onto itself. We call φ an **affine transformation** of M if the induced automorphism $\tilde{\varphi}$ on the bundle of tangent n-frames preserves the connection. In terms of covariant differentials, this condition is equivalent to the condition $\nabla_{\varphi(X)}\varphi(Y) = \varphi(\nabla_X Y)$ for any vector fields X, Y. An affine transformation of the canonical affine connection on $\mathbf{R}^n$ is an ordinary affine transformation. For an affine connection on a manifold M, the set of all affine transformations forms a Lie group and acts on M as a Lie transformation group [4,5].

Let M be a manifold with an affine connection. M is called an affine **locally symmetric space** if $\nabla R = 0$ and $\nabla T = 0$ are satisfied. These conditions are satisfied if and only if at each point p of M, there exist a neighborhood U of p and an affine transformation φ of U such that $\varphi^2 = 1$ and p is the isolated fixed point of φ. If for each point p of M there exists an affine transformation of M such that $\varphi^2 = 1$ and p is an isolated fixed point, then M is called an **affine symmetric space**. A symmetric Riemannian space is a special case of this type (→ 401 Symmetric Riemannian Spaces).

At each point p in a manifold M with an affine connection, we can choose local coordinates $(x^1,\dots,x^n)$ such that $x^i(p)=0$ and the curve $x^i = a^i t$ $(-\delta < t < \delta)$ is a geodesic for each $(a^1,\dots,a^n)$ with $\Sigma(a^i)^2 = 1$. Such local coordinates are called **geodesic coordinates** at p [4]. With respect to geodesic coordinates, we have $(\nabla_{X_i}(X_j))_p = 0$ $(\partial/\partial x^i = X_i)$.

K. Riemannian Connections

When a Riemannian metric g (→ 360 Riemannian Manifolds) is given on a manifold M, it defines a metric on the tangent space $T_p(M)$ at each point p of M, and we may take orthonormal bases in $T_p(M)$. The set P' of all orthonormal bases of tangent spaces is a subset of the bundle P of tangent n-frames of M and forms a subbundle of P; its structure group is the †orthogonal group $O(n)$, and P' gives a reduction of the bundle of tangent n-frames. Conversely, when a reduction of the bundle of frames to $O(n)$ is given, we may define a Riemannian metric on M such that the reduced bundle consists of all orthonormal frames.

For a Riemannian metric g on M, there exists a unique affine connection on M such that (i) $\nabla g = 0$, and (ii) the torsion tensor T vanishes [4]. This connection is called the **Riemannian connection** corresponding to g. The first condition is equivalent to the invariance of the Riemannian metric g under parallel displacement. Thus the affine connection transforms orthonormal bases on M to orthonormal bases and induces a connection in the bundle P'. It is known that the restricted holonomy group of any Riemannian connection is a closed subgroup of $O(n)$ [4]. An affine connection on a manifold M is called a **metric connection** if it preserves a Riemannian metric g on M, i.e., if it satisfies the condition (i).

L. Representations by Local Coordinates

(1) Let $(x^1, \ldots, x^n)$ be a local coordinate system in a manifold M and consider vector fields $X_i = \partial/\partial x_i$. For an affine connection on M, the covariant derivative can be expressed as

$$\nabla_{X_j}(X_k) = \sum_i \Gamma^i_{jk} X_k.$$

The Γ^i_{jk} are called **coefficients of the affine connection** with respect to the local coordinate system $(x^1, \ldots, x^n)$. We denote by $\overline{\Gamma}^i_{jk}$ the coefficients of connection with respect to another local coordinate system $(y^1, \ldots, y^n)$. Then on the intersection of their coordinate neighborhoods, we have

$$\overline{\Gamma}^i_{jk} = \sum_{\alpha,\beta,\gamma} \frac{\partial y^i}{\partial x^\alpha}\left(\frac{\partial x^\beta}{\partial y^j}\frac{\partial x^\gamma}{\partial y^k}\Gamma^\alpha_{\beta\gamma} + \frac{\partial^2 x^\alpha}{\partial y^j \partial y^k}\right).$$

Conversely, if the Γ^i_{jk} are given in each local coordinate system of M and satisfy this relation on each intersection of their coordinate neighborhoods, then there exists a unique affine connection such that the coefficients of the connection are given by Γ^i_{jk}.

(2) The **coefficients of the Riemannian connection** corresponding to a Riemannian metric $g = \sum g_{ij} dx^i dx^j$ on a manifold M is given by

$$\Gamma^i_{jk} = \frac{1}{2}\sum_l g^{il}\left\{\frac{\partial g_{jl}}{\partial x^k} + \frac{\partial g_{lk}}{\partial x^j} - \frac{\partial g_{jk}}{\partial x^l}\right\}.$$

(3) With respect to each local coordinate system $(x^1, \ldots, x^n)$, we express the **torsion tensor** T and the **curvature tensor** R of an affine connection by

$$T = \sum_{ijk} T^i_{jk} dx^j \otimes dx^k \otimes X_i,$$

$$R = \sum_{ijkl} R^i_{jkl} dx^j \otimes dx^k \otimes dx^l \otimes X_i.$$

The components T^i_{jk} and R^i_{jkl} are given by

$$T^i_{jk} = \Gamma^i_{jk} - \Gamma^i_{kj},$$

$$R^i_{jkl} = \left(\partial\Gamma^i_{lj}/\partial x^k - \partial\Gamma^i_{kj}/\partial x^l\right) + \sum_m \left(\Gamma^m_{lj}\Gamma^i_{km} - \Gamma^m_{kj}\Gamma^i_{lm}\right).$$

(4) Let $K = (K^{i_1 \ldots i_r}_{j_1 \ldots j_s})$ be a tensor field of type (r,s). Then the **covariant differential** $\nabla K = (K^{i_1 \ldots i_r}_{j_1 \ldots j_s; k})$ is given by

$$K^{i_1 \ldots i_r}_{j_1 \ldots j_s; k} = \partial K^{i_1 \ldots i_r}_{j_1 \ldots j_s}/\partial x^k + \sum_{\alpha=1}^{r}\left(\sum_l \Gamma^{i_\alpha}_{kl} K^{i_1 \ldots l \ldots i_r}_{j_1 \ldots j_s}\right) - \sum_{\beta=1}^{s}\left(\sum_m \Gamma^m_{kj_\beta} K^{i_1 \ldots i_r}_{j_1 \ldots m \ldots j_s}\right).$$

(5) A curve $x^i = x^i(t)$ is a **geodesic** if and only if

$$\frac{d^2x^i}{dt^2} + \sum_{j,k}\Gamma^i_{jk}\frac{dx^j}{dt}\frac{dx^k}{dt} = 0, \qquad i = 1, 2, \ldots, n$$

(→ 403 Tensor Calculus).

M. Cartan Connections

Let M be a differentiable manifold of dimension n. Consider a homogeneous space $F = G/G'$ of the same dimension n, where G is a Lie group and G' is a closed subgroup of G (→ 201 Homogeneous Spaces). Let $B = (B, M, F, G)$ be a fiber bundle over M with fiber F and structure group G, and $P = (P, M, G)$ be the principal fiber bundle associated with B. Suppose that there exists a cross section f over M to B. Then the structure group of P can be reduced to G'. We denote this reduced fiber bundle by $P' = (P', M, G')$ and the injection of P' into P by j (→ 155 Fiber Bundles).

Suppose that a connection is given in P. Its connection form ω is a differential form of degree 1 on P with values in $\mathfrak{g}$, and the induced form $\omega' = j^*(\omega)$ is also a differential form of degree 1 on P' with values in $\mathfrak{g}$. We call the connection in P a **Cartan connection** on M with the fiber $F = G/G'$ if at each point x of P', ω'_x gives an isomorphism of $T_x(P')$ onto $\mathfrak{g}$ as linear spaces. Such a connection in P is equivalently defined as a 1-form ω' on P' with values in $\mathfrak{g}$ satisfying the following three conditions: (i) $\omega'(A^*) = A$ ($A \in \mathfrak{g}'$ (Lie algebra of G')); (ii) $R^*_a(\omega') = \mathrm{ad}(a^{-1})\omega'$ ($a \in G'$); and (iii) ω'_x gives an isomorphism of $T_x(P')$ onto $\mathfrak{g}$ at each point $x \in P'$. For such ω', we may take a connection form ω in P such that $\omega' = j^*(\omega)$; ω defines a Cartan connection.

N. Soudures

A cross section f over M to B gives a vector bundle $T'(B)$ on M defined as follows: For each point p of M, the projection $B \to M$ defines a mapping $T_{f(p)}(B) \to T_p(M)$. The kernel of this mapping is denoted by $V_{f(p)}(B)$. Then $T'(B) = \bigcup_p V_{f(p)}(B)$ forms a vector bundle over M, and the dimension of its fibers is equal to $n = \dim F$.

A Cartan connection in P gives a bundle isomorphism between $T'(B)$ and the tangent vector bundle $T(M)$ of M as follows: Let x be an arbitrary point in P', and put $p = \pi(x)$. The projection $\pi : P' \to M$ induces an isomorphism of $T_x(P')/V_x(P')$ onto $T_p(M)$. On the other hand, ω'_x gives an isomorphism of $T_x(P')/V_x(P')$ onto $\mathfrak{g}/\mathfrak{g}'$. As a point in P',

x gives a mapping of $F=G/G'$ onto the fiber in B over p and sends the point $\{G'\}$ in F to $f(p)$. By this mapping, $T_0(F)=\mathfrak{g}/\mathfrak{g}'$ is mapped isomorphically onto $V_{f(p)}(B)$. Combining these isomorphisms, we get an isomorphism between $T_p(M)$ and $V_{f(p)}(B)$ that is independent of the choice of $x \in P'$ over p. The set of such isomorphisms for $p \in M$ defines a bundle isomorphism of $T(M)$ and $T'(B)$. If a fiber bundle B over M has an isomorphism such as above through a cross section, then B is said to have a **soudure** [3].

Conversely, if a fiber bundle B over M has a soudure with respect to a cross section f, then there exists a Cartan connection in P such that the soudure given by the connection coincides with the original one [3].

For the tangent vector bundle $T(M)$ of M, the fiber F is an n-dimensional linear space and can be expressed as $F=G/G'$, where G is the †affine transformation group of F and $G'=GL(n,\mathbf{R})$. Then $T(M)$ has the 0-section over M, and there exists a natural soudure. Furthermore, an affine connection on M canonically induces a Cartan connection on M with the fiber $F=G/G'$ [3].

For a Cartan connection on M, we can introduce the notion of **development** of a curve in M into the fiber and also the notion of **completeness** [3].

O. Projective Connections

Let $F_1=G_1/G_1'$ and $F_2=G_2/G_2'$ be homogeneous spaces with $\dim F_1=\dim F_2=n$. Suppose that G_1 is a Lie subgroup of G_2 and G_1' is contained in G_2' by the injection. Then we have a canonical injection $F_1 \rightarrow F_2$ (F_1 is an open subset of F_2 by the assumption).

Suppose that a fiber bundle B_1 with fiber F_1 over M has a cross section f_1. Using f_1, we can construct a bundle B_2 with fiber F_2 over M which also has a cross section f_2. The principal bundle of B_2 is given by extending the structure groups from the principal bundle of B_1. We can show that if B_1 has a soudure with respect to the cross section f_1, then B_2 also has a soudure with respect to the cross section f_2. A Cartan connection in the principal fiber bundle P_1 associated with B_1 that is compatible with a given soudure on B_1 induces a Cartan connection in the principal fiber bundle P_2 associated with B_2, which then induces a soudure on B_2. The latter is called a Cartan connection **induced** from the former.

Let F_1 be an n-dimensional linear space and F_2 be the real †projective space of dimension n. Then the affine transformation group of F_1 can be embedded into the projective transformation group of F_2. Thus the tangent vector bundle of a manifold M induces a fiber bundle over M with the n-dimensional projective space as its fiber. A Cartan connection in this fiber bundle is called a **projective connection** on M. By the argument in this section, every affine connection on M induces a projective connection on M.

Given two affine connections on M, we denote by ∇ and ∇' their corresponding covariant differentials. The two affine connections on M induce the same projective connection on M if and only if there exists a differential form φ of degree 1 on M such that $\nabla'_X Y-\nabla_X Y=\varphi(X)Y+\varphi(Y)X$ for any vector fields X, Y [7]. If a diffeomorphism φ of M preserves the projective connection induced by an affine connection in M, then φ maps geodesics of M into geodesics.

P. Conformal Connections

Let F_1 be an n-dimensional Euclidean space and F_2 an n-dimensional sphere (a †conformal space). We can embed the group of †isometries of F_1 canonically into the group of †conformal transformations of F_2. A Riemannian metric of the tangent vector bundle of a manifold M of dimension n gives a fiber bundle over M with fiber F_2. A Cartan connection in this fiber bundle is called a **conformal connection** on M; a Riemannian connection on M induces a conformal connection on M.

Two Riemannian metrics g_1, g_2 on M induce the same conformal connection on M if and only if there exists a positive function f on M such that $g_2=fg_1$. Thus for a Riemannian manifold M with metric tensor g_1, a diffeomorphism φ of M such that $\varphi^*(g)=fg_1$ leaves invariant the conformal connection induced by g_1. Such a φ is called a **conformal transformation** of M with respect to the given Riemannian metric g_1.

For a Riemannian manifold M with metric tensor g, we define **Weyl's conformal curvature tensor** W by

$$W^i_{jkl}=R^i_{jkl}-\frac{1}{n-2}\left(R_{jk}\delta^i_l-R_{jl}\delta^i_k+g_{jk}R^i_l-g_{jl}R^i_k\right)$$
$$+\frac{R}{(n-1)(n-2)}\left(g_{jk}\delta^i_j-g_{jl}\delta^i_k\right),$$

where the R^i_{jkl} and R_{jk} are components of the curvature tensor and Ricci tensor, respectively, and R is the scalar curvature (→ 360 Riemannian Manifolds, 405 Tensor Calculus). When $\dim M \geqslant 3$, the conformal connection induced by g on M is locally flat if and only if the conformal curvature tensor vanishes [7].

References

[1] H. Cartan, Notions d'algèbre différentielle; application aux groupes de Lie et aux variétés

où opère un groupe de Lie, Colloque de Topologie, Brussels, 1950, p. 15–27.
[2] S. S. Chern, Topics in differential geometry, Lecture notes, Institute for Advanced Study, Princeton, 1951.
[3] C. Ehresmann, Les connexions infinitésimales dans un espace fibré différentiable, Colloque de Topologie, Brussels, 1950, p. 29–55.
[4] S. Kobayashi and K. Nomizu, Foundations of differential geometry, Interscience, I, 1963; II, 1969.
[5] A. Lichnerowicz, Théorie globale des connexions et des groupes d'holonomie, Cremona, 1955.
[6] K. Nomizu, Lie groups and differential geometry, Publ. Math. Soc. Japan, 1956.
[7] K. Yano and S. Bochner, Curvature and Betti numbers, Ann. Math. Studies, Princeton Univ. Press, 1953.
[8] R. S. Palais, Seminar on the Atiyah-Singer index theorem, Ann. Math. Studies, Princeton Univ. Press, 1965.
[9] S. S. Chern, Differentiable manifolds, Lecture notes, Univ. of Chicago, 1959.

83 (I.11) Constructive Ordinal Numbers

To extend the theory of †recursive functions to transfinite ordinal numbers, A. Church and S. C. Kleene considered the set of effectively accessible ordinal numbers and defined the concept of constructive ordinal numbers as explained later in this section [1]. Their work became the basis of fruitful research by Kleene, W. Markwald, C. Spector, and others [2, 3, 4, 5]. A constructive ordinal number was originally introduced as an "expression" in a †formal system utilizing the λ-notation. Since such a system is "effective," we can arithmetize it utilizing †Gödel numbers and assume from the outset that each ordinal number is representable by a natural number. The notations, terminology, and theorems mentioned in this section are mainly those for constructive ordinal numbers of the †second number class.

We call a set of natural numbers satisfying conditions (I) and (II) a **system of notation for ordinal numbers**, and an ordinal number a **constructive ordinal number** when it is representable by a natural number belonging to such a system of notation: (I) No natural number represents two distinct ordinal numbers. (II) There are three †partial recursive functions $K(x)$, $P(x)$, and $Q(x,n)$ defined as follows: (i) for any natural number x representing $X, K(x)$ takes the value 0, 1, or 2 according as X is zero, an †isolated ordinal number, or a †limit ordinal number, respectively; (ii) when X is the ordinal number †immediately after an ordinal number $Y, P(x)$ represents Y for any natural number x representing X; (iii) when X is a limit ordinal number, for any natural number x representing X there exists an increasing sequence $\{Y_n\}$ of ordinal numbers such that $X=\lim_n Y_n$ and $Q(x,n)$ represents Y_n for each natural number n.

The system called S_3 by Kleene is the most useful and convenient among systems of notation for ordinal numbers. Let n_O be a †primitive recursive function of the variable n defined by $0_O=1$, $(n+1)_O=2^{n_O}$. The fundamental notion $a\in O$ and relation $a<_O b$ of the system S_3 are introduced by the following inductive definition: (1) $1\in O$; (2) if $y\in O$, then $2^y\in O$ and $y<_O 2^y$; (3) if a sequence $\{y_n\}$ of natural numbers has the property that for each n, $y_n\in O$ and $y_n<_O y_{n+1}$, and if y is a Gödel number that defines y_n recursively as a function of n_O (i.e., $y_n\simeq\{y\}(n_O)$) (→ 352 Recursive Functions)), then $3\cdot 5^y\in O$, and for each n, $y_n<_O 3\cdot 5^y$; (4) if $x,y,z\in O, x<_O y$, and $y<_O z$, then $x<_O z$; (5) $a\in O, a<_O b$ hold only when they follow from (1)–(4).

Now, for simplicity, we write $a\leqslant_O b$ for $(a<_O b)\vee(a=b)$. The following propositions hold for S_3: (1) If $a<_O b$, then $b\neq 1$; (2) If $a<_O b$, then $a,b\in O$; (3) If $a<_O 2^y$, then $a\leqslant_O y$; (4) If $a<_O 3\cdot 5^y$, then there is a natural number n such that $a\leqslant_O y_n$, where $y_n=\{y\}(n_O)$; (5) If $a\in O$, then $1\leqslant_O a$; (6) If $a\in O$, then for any †number-theoretic function α such that $\alpha(0)=a$ and $\forall n(\alpha(n)\neq 1\rightarrow \alpha(n+1)<_O\alpha(n))$, there is a k such that $\alpha(k)=1$; (7) For each a, $\neg(a<_O a)$; (8) If $c\in O$, $a\leqslant_O c$, and $b\leqslant_O c$, then $a<_O b$ or $a=b$ or $b<_O a$.

Each member a of O **represents** an ordinal number $|a|$ as follows: $|1|=0$; $|2^y|=|y|+1$ for $y\in O$; $|3\cdot 5^y|=\lim_n|y_n|$ for $3\cdot 5^y\in O$, where $y_n=\{y\}(n_O)$. Let b be a member of O. Then $|a|<|b|$ when $a<_O b$; and conversely, for each $\alpha<|b|$, there is a number a such that $|a|=\alpha$ and $a<_O b$. Hence the set $\{a\mid a<_O b\}$ is a †well-ordered set with respect to $<_O$, and its †order type is $|b|$. The least number ξ greater than $|a|$ for every member a of O is the least ordinal number that is not constructive (Church and Kleene denoted it by ω_1). There is a subsystem of S_3 that is well-ordered with respect to $\leqslant_O$ and contains a unique notation for each constructive ordinal number α. For such a subsystem we may take a Π^1_1 set K such that K is a †recursive set in O (S. Feferman and Spector, R. O. Gandy) (→ 197 Hierarchies).

Let $R(x,y)$ be a †predicate on natural numbers. We write $x \leqslant_R y$ for any natural numbers x,y for which $R(x,y)$ holds. We consider only the case where $\leqslant_R$ is a †linear ordering on the set $D_R = \{x \mid \exists y(R(x,y) \vee R(y,x))\}$. If D_R is a well-ordered set with respect to $\leqslant_R$, we denote its †order type by $|R|$. (1) For each (constructive) ordinal number $\alpha < \omega_1$, there is a †general recursive (more strictly, †primitive recursive) predicate R such that $|R| = \alpha$ (Markwald, Spector, Kleene [3]). (2) Conversely, if R is a †hyperarithmetic predicate (e.g., R is general recursive), then $|R| < \omega_1$ (Markwald, Spector). The following theorems are the most fruitful ones in the theory of constructive ordinal numbers, and they fully support the validity of Kleene's idea of †analytic hierarchy. (3) The set O is Π_1^1 (→ 197 Hierarchies), and so is the predicate $a <_O b$. Namely, for O, there is a primitive recursive predicate $R(a,x,\alpha)$ such that

$$a \in O \Leftrightarrow \forall \alpha \exists x R(a,x,\alpha)$$

(Kleene [3]). (4) For each ordinal number $\alpha < \omega_1$, the set $\{a \mid \alpha \geqslant |a|\}$ is a hyperarithmetic set (Spector). (5) O is a †complete set for Π_1^1. That is, for any Π_1^1 set E, there is a primitive recursive function φ such that $a \in E \Leftrightarrow \varphi(a) \in O$ (Kleene [3]). Accordingly, O is not a Σ_1^1 set (→ 197 Hierarchies).

Given a (number-theoretic) predicate Q of one variable (or a set of natural numbers), we can †relativize to Q the notion of constructive ordinal numbers. The least ordinal that is not constructive relative to Q is denoted by ω_1^Q. The relativization to Q of the fundamental notion $a \in O$ and relation $a <_O b$ of the system S_3 of notation are denoted by $a \in O^Q$ and $a <_{O^Q} b$, respectively. Then we can relativize the results of the preceding paragraphs to Q. For example, as the relativization of (3), we have the following: There is a predicate $R^Q(a,x,\alpha)$ which is primitive recursive †uniformly in Q such that

$$a \in O^Q \Leftrightarrow \forall \alpha \exists x R^Q(a,x,\alpha).$$

When Q is hyperarithmetic, we have no generalization of the constructive ordinal numbers by relativizing them to Q, that is, $\omega_1^Q = \omega_1$ holds (Spector). Now by relativizing to O the concept of constructive ordinal numbers, we obtain a proper extension of it ($\omega_1 < \omega_1^O$), and then, performing such extensions successively, we have a (transfinite) sequence $O, O^O, O^{O^O}, \ldots$. On the other hand, we can extend the constructive ordinal numbers to those corresponding to any number class higher than the second number class. There are several extensions done by Church and Kleene, H. C. Wang, D. L. Kreider and H. Rogers, Jr., H. Putnam, A. Kino and G. Takeuti, and others. However, these extensions have not produced as many results as have been obtained for ordinal numbers of the second number class.

References

[1] A. Church and S. C. Kleene, Formal definitions in the theory of ordinal numbers, Fund. Math., 28 (1937), 11–21.

[2] S. C. Kleene, On notation for ordinal numbers, J. Symbolic Logic, 3 (1938), 150–155.

[3] S. C. Kleene, On the forms of the predicates in the theory of constructive ordinals II, Amer. J. Math., 77 (1955), 405–428.

[4] W. Markwald, Zur Theorie der konstruktiven Wohlordnungen, Math. Ann., 127 (1954), 135–149.

[5] C. Spector, Recursive well-orderings, J. Symbolic Logic, 20 (1955), 151–163.

84 (XIII.20) Contact Transformations

A. General Remarks

A transformation of $2n+1$ variables z, x_j, p_j $(j = 1,2,\ldots,n)$,

$$Z = Z(z, x_1, x_2, \ldots, x_n, p_1, p_2, \ldots, p_n),$$
$$X_j = X_j(z, x_1, x_2, \ldots, x_n, p_1, p_2, \ldots, p_n), \quad j = 1,2,\ldots,n,$$
$$P_j = P_j(z, x_1, x_2, \ldots, x_n, p_1, p_2, \ldots, p_n), \quad j = 1,2,\ldots,n, \qquad (1)$$

is called a **contact transformation** in the $(n+1)$-dimensional space $\mathbf{R}^{n+1}$ with the coordinate system $(z, x_1, \ldots, x_n)$ if the †total differential equation

$$dz - p_1 dx_1 - p_2 dx_2 - \ldots - p_n dx_n = 0 \qquad (2)$$

is invariant under the transformation, i.e., if the equality

$$dZ - P_1 dX_1 - P_2 dX_2 - \ldots - P_n dX_n$$
$$= \rho(dz - p_1 dx_1 - p_2 dx_2 - \ldots - p_n dx_n) \qquad (3)$$

holds identically for a suitable nonzero function ρ of z, x_j, p_j. Here we assume that (1) has an inverse transformation. Using **Lagrange's bracket**

$$[f,g] = \sum_{j=1}^{n} \left(\frac{\partial f}{\partial p_j} \left(\frac{dg}{dx_j} \right) - \frac{\partial g}{\partial p_j} \left(\frac{df}{dx_j} \right) \right),$$

$$\left(\frac{df}{dx_j} \right) \equiv \frac{\partial f}{\partial x_j} + p_j \frac{\partial f}{\partial z},$$

we see that (1) is a contact transformation if and only if $[X_j, X_k] = [X_j, Z] = [P_j, P_k] = 0$,

$[P_j, X_k] = \rho\delta_{jk}$, $[P_j, Z] = \rho P_j$, where δ_{jk} is †Kronecker's delta.

From this fact it follows that the composite of two contact transformations and the inverse transformation of a contact transformation are also contact transformations. Since the identity transformation $Z = z$, $X_j = x_j$, $P_j = p_j$ is a contact transformation, the set of all contact transformations forms an infinite-dimensional †topological group. Given a set of scalars $p_1, \ldots, p_n$, a pair consisting of a point (z, x_j) and an n-dimensional hyperplane $z^* - z = \sum_{j=1}^n p_j(x_j^* - x_j)$ in an $(n+1)$-dimensional space is called a **hypersurface element**, and the set of hypersurface elements satisfying (2) is called a **union of hypersurface elements**. Using these concepts, we can state that a transformation (1) of coordinates z, x_j, p_j $(j = 1, 2, \ldots, n)$ is a contact transformation if it transforms each union of hypersurface elements into another one. Consequently, if two n-dimensional hypersurfaces are tangent at a point (z, x_j) in the $(n+1)$-dimensional space, their images under a contact transformation, which are again two n-dimensional hypersurfaces, are tangent at the image point of (z, x_j). The name "contact transformation" is derived from this fact.

For instance, the †correlation with respect to a hypersurface of the second order gives a contact transformation. In fact, from the relation between †poles and †polar lines with respect to the parabola $x^2 + 2y = 0$ in a plane we have **Legendre's transformation** $X = -p$, $Y = xp - y$, $P = -x$ $(\rho = -x)$.

In general, an invertible transformation defined by the three relations $\Omega(x, y, X, Y) = 0$, $\partial\Omega/\partial X + P\partial\Omega/\partial Y = 0$, $\partial\Omega/\partial x + p\partial\Omega/\partial y = 0$ derived from a function $\Omega(x, y, X, Y)$ is a contact transformation. The function Ω is called the **generating function** of this transformation. In this transformation, to each point (x_0, y_0) there corresponds a curve $\Omega(x_0, y_0, X, Y) = 0$. These results are valid also in the case of several variables. For instance, in an $(n+1)$-dimensional space, a transformation $Z = z - x_1p_1 - \ldots - x_\nu p_\nu$; $X_1 = p_1, \ldots, X_\nu = p_\nu$, $X_{\nu+1} = x_{\nu+1}, \ldots, X_n = x_n$; $P_1 = -x_1, \ldots, P_\nu = -x_\nu$, $P_{\nu+1} = p_{\nu+1}, \ldots, P_n = p_n$ represents a contact transformation. Here ν is an integer between 1 and n. In the case $n = 2$, $\nu = 2$, this transformation reduces to a Legendre transformation; and in the case $n = 2$, $\nu = 1$, it is called **Ampère's transformation** (→ Appendix A, Table 15.IV).

B. Canonical Transformations

A transformation of $2n$ variables x_j, p_j $(j = 1, 2, \ldots, n)$

$$X_j = X_j(x_1, x_2, \ldots, x_n, p_1, p_2, \ldots, p_n),$$

$$P_j = P_j(x_1, x_2, \ldots, x_n, p_1, p_2, \ldots, p_n), \quad j = 1, 2, \ldots, n, \tag{4}$$

is called a **canonical transformation** if the †differential form $\sum_{j=1}^n (P_j dX_j - p_j dx_j)$ is †exact in x_j, p_j, i.e., if there exists a function U of x, p such that

$$\sum_{j=1}^n (P_j dX_j - p_j dx_j) = dU. \tag{5}$$

Let (1) denote a contact transformation. If we set

$$z = x_{n+1}, \quad Z = X_{n+1}, \quad P_{n+1} = p_{n+1}/\rho,$$

$$-p_j p_{n+1} = \bar{p}_j, \quad -P_j P_{n+1} = \bar{P}_j, \qquad j = 1, 2, \ldots, n,$$

then (2) becomes

$$\bar{P}_1 dX_1 + \bar{P}_2 dX_2 + \ldots + \bar{P}_n dX_n + P_{n+1} dX_{n+1}$$
$$= \bar{p}_1 dx_1 + \bar{p}_2 dx_2 + \ldots + \bar{p}_n dx_n + p_{n+1} dx_{n+1}.$$

Hence (1) represents a canonical transformation ($U \equiv 0$). Therefore every contact transformation is a canonical transformation.

In canonical transformations Lagrange's bracket becomes the bracket

$$(f, g) = \sum_{j=1}^n \left(\frac{\partial f}{\partial p_j} \frac{\partial g}{\partial x_j} - \frac{\partial g}{\partial p_j} \frac{\partial f}{\partial x_j} \right),$$

which is called **Poisson's bracket**.

If (4) represents a canonical transformation, then we have $(X_j, X_k) = 0$, $(P_j, X_k) = \delta_{jk}$, $(P_j, P_k) = 0$ $(j, k = 1, 2, \ldots, n)$. Conversely, if the X_j $(j = 1, 2, \ldots, n)$ satisfy the relations $(X_j, X_k) = 0$ $(j, k = 1, 2, \ldots, n)$, we can find a function U and uniquely determined functions P_j $(j = 1, 2, \ldots, n)$ for which (5) holds.

C. Applications to the Integration of Differential Equations

Contact transformations have applications to the integration of differential equations since they transform each union of surface elements into another one.

As an example, we shall describe an outline of their application to a partial differential equation of the first order

$$F(x, y, z, p, q) = 0;$$

$$p \equiv \partial z/\partial x, \quad q \equiv \partial z/\partial y. \tag{6}$$

For this purpose, we regard (6) as an equation defining unions of surface elements and transform it into a simpler equation by means of a contact transformation. If the transformed equation can be solved, then the solution of the original equation can be obtained by means of the inverse transformation. Now, let $z = \omega(x, y, a, b)$ be a †complete solution of (6). Then (6) is reduced to $Z - c = 0$ by the transformation generated by the function

$$\Omega \equiv Z - z + \omega(x, y, X, Y) - c = 0,$$

where c is a constant. In this equation the

solution $X=a, Y=b, Z=c, \alpha P+\beta Q=0$ (a,b,c,α,β are constants) plays an important role, and this line element will be called a **characteristic line element**. The characteristic line element satisfies equations that can be transformed by means of the inverse transformation into **Charpit's subsidiary equations** for (6):

$$\frac{dx}{\partial F/\partial p}=\frac{dy}{\partial F/\partial q}=\frac{dz}{p\partial F/\partial p+q\partial F/\partial q}=\frac{-dp}{\partial F/\partial x+p\partial F/\partial z}=\frac{-dq}{\partial F/\partial y+q\partial F/\partial z}. \qquad (7)$$

Consequently, if we have $p=p(x,y,z;a)$, $q=q(x,y,z;a)$ from the solution $G(x,y,z,p,q)=a$ of (7) and $F=0$, then the total differential equation $dz=pdx+qdy$ is †completely integrable, and the †general solution of this equation is a complete solution of (6). Also, if we know two independent solutions $G(x,y,z,p,q)=a$, $H(x,y,z,p,q)=b$ of (7) such that $[G,H]=0$, we can obtain a complete integral of (6) by eliminating p,q among the three equations $F=0$, $G=a$, $H=b$. This method is called the †Lagrange-Charpit method, which is applicable also to the equation $F(z,x_1,\ldots,x_n;p_1,\ldots,p_n)=0$ (→ 317 Partial Differential Equations (Methods of Integration)).

D. Applications to Analytical Dynamics

Consider a partial differential equation

$$F(z,x_1,\ldots,x_n;\partial z/\partial x_1,\ldots,\partial z/\partial x_n)=0. \qquad (8)$$

If we set $W(z,x_1,\ldots,x_n)=0$ and substitute $\partial z/\partial x_j=-(\partial W/\partial x_j)/(\partial W/\partial z)$ into (8), we may solve the equation for $\partial W/\partial z$ and get the †Hamilton-Jacobi equation:

$$\frac{\partial W}{\partial z}+H(z,x_1,\ldots,x_n;\partial W/\partial x_1,\ldots,\partial W/\partial x_n)=0. \qquad (9)$$

Here H is called †Hamiltonian or Hamiltonian function. As usual, we write t,q_j instead of z,x_j and set $\partial W/\partial x_j\equiv p_j$. Charpit's subsidiary equations for (9) then become †Hamilton's canonical equations

$$\frac{dq_j}{dt}=\frac{\partial H}{\partial p_j}=(q_j,H),$$

$$\frac{dp_j}{dt}=-\frac{\partial H}{\partial q_j}=(p_j,H), \qquad (10)$$

where (,) denotes Poisson's bracket (→ 23 Analytical Dynamics). An example of the system (10) is the equations of motion for dynamical systems which are derived from $\delta\int L(q_j,\dot q_j)dt=0$. Here the q_j denote generalized coordinates, $p_j=\partial L/\partial\dot q_j$ ($\dot{}\equiv d/dt$), $H=\Sigma p_j\dot q_j-L$.

If we transform p_j,q_j into P_j,Q_j by means of the canonical transformation $\Sigma(p_jdq_j-P_jdQ_j)=dW$, then (10) becomes $\dot Q_j=\partial K/\partial P_j$, $\dot P_j=-\partial K/\partial Q_j$, $K=H+\partial W/\partial t$, where $p_j=\partial W/\partial q_j$, $P_j=-\partial W/\partial Q_j$. In particular, a transformation that makes $K=0$ is called a **transformation to an equilibrium system**. Such a W is obtained as a solution of (9).

A general infinitesimal transformation is written as $\delta q_j=\varepsilon(q_j,F)$, $\delta p_j=\varepsilon(p_j,F)$ (ε is a small number and F is an arbitrary function of q,p), which entails an infinitesimal variation $\delta A=\varepsilon(A,F)$ on an arbitrary quantity A. Thus, the equations of motion (10) in dynamics can be regarded as defining an infinitesimal transformation.

Furthermore, every mapping in an optical system is a contact transformation, and the quantity corresponding to W is called an **eikonal** (→ 184 Geometric Optics).

References

[1] T. Yoshiye, Syotô ikkai henbibun hôteisiki (Japanese; Elementary theory of partial differential equations of the first order), Syôkabô, 1947.
[2] R. Courant and D. Hilbert, Methods of mathematical physics II, Interscience, 1962.

85 (V.3) Continued Fractions

A. The Notion of Continued Fractions

Let $\{b_n\}$ ($n=0,\ldots,m$) and $\{c_n\}$ ($n=1,\ldots,m$) be finite sequences of elements in a †field F. A fraction of the form

$$b_0+\cfrac{c_1}{b_1+\cfrac{c_2}{b_2+\cdots\cfrac{\cdots}{\cdots+\cfrac{c_m}{b_{m-1}+\cfrac{c_m}{b_m}}}}}$$

is called a **finite continued fraction**. It expresses an element in the field F unless division by 0 occurs in the process of reduction. Symbolically, it is also written in the forms

$$b_0+\frac{c_1}{b_1}+\frac{c_2}{b_2}+\cdots+\frac{c_{m-1}}{b_{m-1}}+\frac{c_m}{b_m},$$

$$b_0+\frac{c_1|}{|b_1}+\frac{c_2|}{|b_2}+\cdots+\frac{c_{m-1}|}{|b_{m-1}}+\frac{c_m|}{|b_m},$$

$$b_0\dotplus\frac{c_1}{b_1}\dotplus\frac{c_2}{b_2}\dotplus\cdots\dotplus\frac{c_{m-1}}{b_{m-1}}\dotplus\frac{c_m}{b_m},$$

$$\left[b_0,\frac{c_1}{b_1},\frac{c_2}{b_2},\ldots,\frac{c_{m-1}}{b_{m-1}},\frac{c_m}{b_m}\right],$$

etc., or more briefly,

$$b_0+\left[\frac{c_n}{b_n}\right]_{n=1}^{m}.$$

If $\{b_n\}$ $(n=0,1,\ldots)$ and $\{c_n\}$ $(n=1,2,\ldots)$ are infinite sequences, the expression

$$b_0+\cfrac{c_1}{b_1+\cfrac{c_2}{b_2+\cdots \; \cdots+\cfrac{c_n}{b_n+\cdots}}}$$

is called an **infinite continued fraction**. By analogy with the finite case, it is expressed by

$$b_0+\frac{c_1}{b_1}+\frac{c_2}{b_2}+\cdots+\frac{c_n}{b_n}+\ldots$$

or by $b_0+\left[\frac{c_n}{b_n}\right]_{n=1}^{\infty}$.

For the infinite continued fraction, the quantity

$$k_n=b_0+\frac{c_1}{b_1}+\cdots+\frac{c_n}{b_n}, \qquad k_0=b_0$$

is called its nth **convergent**, b_0 is called the **initial term**, and b_n and c_n $(n\geqslant 1)$ are called the **partial numerator** and **partial denominator**, respectively. If F is a †topological field (e.g., the real or complex number field) and the sequence $\{k_n\}$ of its elements converges, then the infinite continued fraction is said to **converge**, and the limit is called its **value**.

A finite or infinite continued fraction in which c_n $(n\geqslant 1)$ are all equal to 1, b_0 is a rational integer, and b_n $(n\geqslant 1)$ are all positive rational integers is called a **regular continued fraction**. It is expressed by $[b_0,b_1,\ldots]$. In the following paragraphs we shall mostly discuss regular continued fractions.

For a real number x we mean by $[x]$ the greatest integer not exceeding x. [ ] is called the **Gauss symbol**. Let ω be any given real number, and put

$$\omega=b_0+\frac{1}{\omega_1}, \quad b_0=[\omega];$$

$$\omega_n=b_n+\frac{1}{\omega_{n+1}}, \quad b_n=[\omega_n], \quad n=1,2,\ldots.$$

Then an expansion of ω into a regular continued fraction

$$\omega=b_0+\frac{1}{b_1}+\cdots+\frac{1}{b_n}+\cdots$$

is obtained. If ω is irrational, this expansion is determined uniquely. If ω is rational, the process is interrupted at a finite step ($\omega_m=b_m$), resulting in

$$\omega=b_0+\frac{1}{b_1}+\cdots+\frac{1}{b_m}.$$

An alternative representation of a rational number by a regular continued fraction is given by replacing b_m above by $(b_m-1)+1/1$.

Examples of infinite regular continued fractions are:

$$\frac{e^{2/p}+1}{e^{2/p}-1}=p+\frac{1}{3p}+\cdots+\frac{1}{(2n+1)p}+\cdots,$$

where p is a natural number (J. H. Lambert),

$$e=2+\frac{1}{1}+\frac{1}{2}+\frac{1}{1}+\cdots+$$
$$\frac{1}{1}+\frac{1}{2n}+\frac{1}{1}+\cdots$$

(L. Euler).

B. Convergents

Let the nth convergent of a regular continued fraction be expressed in the form of an irreducible fraction

$$\frac{P_n}{Q_n}=b_0+\frac{1}{b_1}+\cdots+\frac{1}{b_n}, \quad n\geqslant 0,$$

and for convenience put $P_{-2}=0$, $P_{-1}=1$, $Q_{-2}=1$, $Q_{-1}=0$. Then we have the recurrence relations

$$P_n=b_nP_{n-1}+P_{n-2}, \qquad Q_n=b_nQ_{n-1}+Q_{n-2},$$
$$n\geqslant 0,$$

whence follows

$$P_nQ_{n-1}-P_{n-1}Q_n=(-1)^{n+1}, \qquad n\geqslant -1.$$

Any regular continued fraction represents a real number ω which satisfies

$$\omega=(\omega_{n+1}P_n+P_{n-1})/(\omega_{n+1}Q_n+Q_{n-1})$$

in terms of the notation defined in this and the previous section.

In particular, let ω be an irrational number. Then each of the fractions

$$\frac{P_n^{(k)}}{Q_n^{(k)}}=\frac{P_{n-2}+kP_{n-1}}{Q_{n-2}+kQ_{n-1}}, \quad k=1,2,\ldots,b_n-1,$$

which are inserted between two convergents P_{n-2}/Q_{n-2} and $P_n/Q_n=(P_{n-2}+b_nP_{n-1})/(Q_{n-2}+b_nQ_{n-1})$, is called an **intermediate convergent**, while the original convergent P_n/Q_n is called a **principal convergent**.

If a fraction P/Q approximating an irrational number ω satisfies $|\omega-P/Q|<|\omega-p/q|$ for any other fraction p/q with $q\leqslant Q$, then it is said to give the **best approximation**. The fraction giving the best approximation of ω is always a principal or intermediate convergent $P_n^{(k)}/Q_n^{(k)}$ of ω with $k>b_n/2$ or $k=b_n/2$, $Q_n>Q_{n-1}\omega_n$.

The convergents satisfy the relation $P_n/Q_n-P_{n-1}/Q_{n-1}=(-1)^{n-1}Q_nQ_{n-1}$, hence the sequence $\{P_{2n}/Q_{2n}\}$ (resp $\{P_{2n+1}/Q_{2n+1}\}$) is monotonically increasing (decreasing). Approximation by convergents is shown in the relations

$$\left|\omega-\frac{P_n}{Q_n}\right|<\frac{1}{Q_{n+1}Q_n},\quad \left|\omega-\frac{P_n}{Q_n}\right|<\left|\omega-\frac{P_{n-1}}{Q_{n-1}}\right|,$$

$$\omega=\lim_{n\to\infty}\frac{P_n}{Q_n}=b_0+\sum_{n=0}^{\infty}\frac{(-1)^n}{Q_{n+1}Q_n}.$$

There are several results concerning the measure of approximation by convergents. Any ω satisfies $|\omega-P_n/Q_n|<1/(\sqrt{5}\ Q_n^2)$ for infinitely many n, while there exists ω which satisfies $|\omega-P_n/Q_n|<1/(\lambda Q_n^2)$ for only a finite number of n provided that $\lambda>\sqrt{5}$ (A. Hurwitz); at least one of two adjacent convergents satisfies the inequality $|\omega-P_n/Q_n|<1/(2Q_n^2)$ (K. T. Vahlen); at least one of three successive convergents satisfies the inequality $|\omega-P_n/Q_n|<1/(\sqrt{5}\ Q_n^2)$ (E. Borel). There are also results by G. Humbert, M. Fujiwara, H. Shibata, L. R. Ford, and others.

C. Irrational Numbers of the Second Degree

If an infinite regular continued fraction $[b_0,b_1,\ldots]$ satisfies $b_{m+k+\nu}=b_{m+\nu}$ ($\nu=0,1,2,\ldots$), it is called a **recurring continued fraction** and is denoted by the symbol $[b_0,b_1,\ldots,\dot{b}_m,\ldots,\dot{b}_{m+k-1}]$. According as $m=0$ or $m\geqslant 1$, the recurring continued fraction is said to be **pure** or **mixed**, and $[b_m,b_{m+1},\ldots,b_{m+k-1}]$ is called a **block of recurring terms**. In order that the continued fraction of ω be recurring, it is necessary and sufficient that ω is a quadratic irrational number, i.e., a root of $ax^2+bx+c=0$ with rational integral coefficients a, b, c and nonsquare discriminant b^2-4ac (J. L. Lagrange). In order that ω be represented by a purely recurring continued fraction, it is necessary and sufficient that ω is an **irreducible quadratic irrational number**, i.e., it satisfies $\omega>1$ and $0>\omega'>-1$, where ω' is the conjugate root of ω (E. Galois). In order that ω be equal to the square root of a nonsquare rational number, it is necessary and sufficient that its continued fraction is of the form $[b_0,\dot{b}_1,\ldots,b_{k-1},2\dot{b}_0]$ ($b_{k-\nu}=b_\nu$) (A. M. Legendre).

D. Diophantine Equations and Diophantine Approximations

Let $ax-by=1$ ($(a,b)=1$) be a †Diophantine equation of the first degree, and $a/b=[b_0,b_1,\ldots,b_m]=P_m/Q_m$. Since $P_mQ_{m-1}-P_{m-1}Q_m=(-1)^{m-1}$, a solution of the equation is given by $x_0=(-1)^{m-1}Q_m$, $y_0=(-1)^{m-1}P_m$. The general solution is then represented in the form x_0+bt, y_0+at ($t\in\mathbf{Z}$). This method of obtaining a solution is essentially the same as the method which uses the †Euclidean algorithm.

†Pell's equation $x^2-Dy^2=1$ (D is a nonsquare integer) was solved by Lagrange in terms of continued fractions. If the length of the period of $\sqrt{D}$ is k, all positive solutions of Pell's equation are given by $x=P_{2\nu k-1}$, $y=Q_{2\nu k-1}$ ($\nu=1,2,\ldots$), where P_n/Q_n denotes a convergent of the continued fraction of $\sqrt{D}$. Incidentally, $x=P_{(\nu-1)k-1}$, $y=Q_{(2\nu-1)k-1}$ ($\nu=1,2,\ldots$) are positive solutions of $x^2-Dy^2=-1$ provided that k is odd. There are no solutions of $x^2-Dy^2=\pm 1$ other than $x_\nu,y_\nu(\nu=1,2,\ldots)$ given by $(x_1+\sqrt{D}\ y_1)^\nu=x_\nu+\sqrt{D}\ y_\nu$, where x_1,y_1 is the least positive solution. For instance, the least positive solution of $x^2-211y^2=1$ is $x=278{,}354{,}373{,}650$, $y=19{,}162{,}705{,}353$.

Lagrange made further use of continued fractions in order to obtain approximate values of roots of algebraic equations. The method is especially useful for precise computation of neighboring roots.

The theory of continued fractions may be investigated geometrically making use of lattices (→ 187 Geometry of Numbers) (F. Klein, Humbert). For instance, a measure of approximation of P_n/Q_n to ω in Diophantine approximation is represented by the closeness of a lattice point (P_n,Q_n) to the straight line $y=\omega x$ on the plane.

E. Continued Fractions with Variable Terms

There are few results on continued fractions with variable terms. It is noteworthy that from the expansion of $\tan z$ into a continued fraction

$$\tan z=\frac{z}{1}+\frac{-z^2}{3}+\frac{-z^2}{5}+\frac{-z^2}{7}+\cdots$$

(Lambert), the irrationality of π and of $\tan z$ for rational z ($\neq 0$) can be deduced (A. Pringsheim).

Among continued fractions with variable terms, those of the form

$$[a_0,a_nz]_1^\infty=\frac{a_0}{1}+\frac{a_1z}{1}+\cdots+\frac{a_nz}{1}+\cdots$$

are called **normal continued fractions**. Let the convergent of such a continued fraction be $P_n(z)/Q_n(z)$, and for convenience put $P_{-1}(z)=0$, $Q_{-1}(z)=1$. Then we have the recurrence formulas

$$P_n(z)=P_{n-1}(z)+a_nzP_{n-2}(z),$$

$$Q_n(z)=Q_{n-1}(z)+a_nzQ_{n-2}(z),\quad n\geqslant 1.$$

There are the further relations

$$P_n(z)Q_{n-1}(z)-P_{n-1}(z)Q_n(z)=(-1)^nz^n\prod_{\nu=0}^{n}a_\nu,$$

$$[a_0, a_n z]_1^\infty = \sum_{n=0}^{\infty} \frac{(-1)^n z^n}{Q_{n-1}(z) Q_n(z)} \prod_{\nu=0}^{n} a_\nu,$$

where the latter is formal. Let the †power series expansion of the nth convergent of $[a_0, a_n z]_1^\infty$ be

$$\frac{P_n(z)}{Q_n(z)} = \sum_{\nu=0}^{\infty} b_{n\nu} z^\nu, \qquad n \geqslant 1.$$

Then $b_{m\nu} = b_{n\nu}$ $(0 \leqslant \nu \leqslant m \leqslant n)$. If $[a_0; a_n z]_1^\infty$ has a power series expansion about the origin, then

$$[a_0, a_n z]_1^\infty = \sum_{n=0}^{\infty} b_{nn} z^n.$$

If the supremum $\bar{g}$ of $\{|a_n|\}_1^\infty$ is finite, then $[a_0, a_n z]_1^\infty$ converges uniformly for $|z| \leqslant (1/4)\bar{g}$, and hence it represents an †analytic function which is holomorphic in $|z| < (1/4)\bar{g}$.

References

[1] H. Shibata, Renbunsû ron (Japanese; Theory of continued fractions), Iwanami Coll. of Math., 1933.
[2] M. Fujiwara, Daisûgaku I (Japanese; Algebra I), Utida-rôkakuho, 1928.
[3] T. Takagi, Syotô seisûron kôgi (Japanese; Lectures on elementary theory of numbers), Kyôritu, 1931.
[4] Y. Komatu, Kansû to kyokugen (Japanese; Functions and limits), Tôkai, 1949.
[5] A. Pringsheim, Vorlesungen über Zahlen- und Funktionenlehre, Teubner, 1932, p. 926–959.
[6] O. Perron, Die Lehre von den Kettenbrüchen, Teubner, third edition, 1954 (Chelsea, 1950).
For functions defined by continued fractions and applications to moment problems,
[7] H. S. Wall, Analytic theory of continued fractions, Chelsea, 1967.
[8] А. Я. Хинчин (A. Ja. Hinčin), Цепные дроби, Гостехиздат, second edition, 1949; English translation, Continued fractions, Noordhoff, 1963.
[9] А. Н. Хованский (A. N. Hovanskiĭ), Приложение цепных дробей и их обобщений к вопросам приближенного анализа, Гостехиздат, 1956; English translation, The application of continued fractions and their generalizations to problems in approximation theory, Noordhoff, 1963.

86 (X.2) Continuous Functions

A. General Remarks

The notion of continuity is defined for a mapping or a function $f: X \to Y$ from a topological space X to a topological space Y (→ 408 Topological Spaces G). In the present article, however, we shall be concerned mainly with the case where both X and Y are †metric spaces with the distances ρ_X and ρ_Y, respectively. The most usual case is that where $X = \mathbf{R}^n$ (Euclidean space), $Y = \mathbf{R}$ (real numbers).

A function $f: X \to Y$ is said to be **continuous** at a point $x_0 \in X$ if for every positive number ε, we can select a suitable positive number δ (depending on ε and also on x_0) such that $\rho_X(x, x_0) < \delta$ implies $\rho_Y(f(x), f(x_0)) < \varepsilon$. This is equivalent to the condition that $x \to x_0$ implies $f(x) \to f(x_0)$ (→ 89 Convergence). We call f **continuous** (on X) if it is continuous at every point x_0 of X. If for every positive number ε, we can select a suitable positive number δ independent of x and y such that $\rho_X(x, y) < \delta$ implies $\rho_Y(f(x), f(y)) < \varepsilon$ for all $x, y \in X$, we call f **uniformly continuous** on X.

The †supremum $\omega(\delta)$ of $\rho_Y(f(x), f(y))$ for $x, y \in X$ satisfying $\rho_X(x, y) < \delta$ is called the **modulus of continuity** of the function f in X. Uniform continuity means that $\omega(\delta) \to 0$ for $\delta \to 0$.

If $\omega(\delta) \leqslant M\delta^\alpha$ for suitable constants $M, \alpha > 0$, that is, if the inequality $\rho_Y(f(x), f(y)) \leqslant M(\rho_X(x, y))^\alpha$ holds for $x, y \in X$, then f is said to satisfy the **Hölder condition of order** α, also known as the **Lipschitz condition of order** α. If $\alpha = 1$, this condition is called simply the **Lipschitz condition**. A function satisfying one of these conditions is uniformly continuous. The family of functions satisfying the Lipschitz condition of order α is sometimes denoted by **Lip** α.

In general, the composite function $g \circ f: X \to Z$ is continuous if both functions $f: X \to Y$ and $g: Y \to Z$ are continuous. If the ranges of f, g are both the real field $\mathbf{R}$ (or the complex field $\mathbf{C}$, or more generally a †topological field), then $f \pm g, fg$ are continuous if f and g are continuous; and f/g is continuous provided that $g(x) \neq 0$. If $\mathbf{R}$ is the range of both f and g, then $\min(f, g)$ and $\max(f, g)$ are continuous when f and g are continuous. If X is †connected (for example, an interval I in $\mathbf{R}$) and if f is continuous, the image $f(X)$ is again connected.

B. Continuity from One Side

In this section, we always assume that the domain X is an interval I in $\mathbf{R}$ and f is a function from I to a metric space Y. A point x_0 of X is called a **discontinuity (point) of the first kind** of f if both limits $\lim_{x \uparrow x_0} f(x)$ and $\lim_{x \downarrow x_0} f(x)$ exist in Y and are different. Then we say also that f has a **jump** (or **gap**) at x_0. When these two limits exist and have the same value, then f is continuous at x_0.

We say that f has a discontinuity **of at most the first kind** at x_0 if f is continuous at x_0 or if x_0 is a discontinuity of the first kind of f. (Sometimes the phrase "discontinuity of the first kind" is used in the sense of "discontinuity of at most the first kind.") A **discontinuity point** of f (i.e., a point at which f is not continuous) that is not of the first kind is called a **discontinuity point of the second kind**. When $\lim_{x\downarrow x_0} f(x)=f(x_0)$, we call f **right continuous** (or **continuous on the right**) at x_0. In this case, $\lim_{x\uparrow x_0} f(x)$ need not exist. Replacing $x\downarrow x_0$ by $x\uparrow x_0$, we can similarly define the concept of being **left continuous** (or **continuous on the left**). If a function f has a finite number of discontinuity points of the first kind in the interval $[a,b]$ and is continuous at all other points, we call f a **piecewise continuous function** in $[a,b]$.

C. Semicontinuous Functions

In this section, we shall assume that the domain of the functions is a subset E of a metric space X, and that the range is the set of real numbers extended to include $\pm\infty$. Letting x be a point in the closure of E, we denote by $M(x,\delta)$ and $m(x,\delta)$, respectively, the supremum and the infimum of the values of a given function f in the δ-neighborhood of x. We put

$$M(x)=\lim_{\delta\to 0} M(x,\delta), \quad m(x)=\lim_{\delta\to 0} m(x,\delta)$$

and call them the **upper limit function** and **lower limit function** at x, respectively. We have $-\infty\leqslant m(x)\leqslant M(x)\leqslant+\infty$. If $M(x_0)=f(x_0)$ at $x_0\in E$, then f is called **upper semicontinuous** at x_0. If $m(x_0)=f(x_0)$ at $x_0\in E$ (i.e., if $-f$ is upper semicontinuous at x_0), then f is called **lower semicontinuous** at x_0. The function with one of these two properties is said to be **semicontinuous** at x_0.

Either of the following two conditions is necessary and sufficient for the function f to be upper semicontinuous at $x_0\in E$: (1) $f(x_0)=+\infty$, or for every constant λ such that $f(x_0)<\lambda$, there exists a δ-neighborhood such that $M(x_0,\delta)<\lambda$. (2) For every sequence x_n of E converging to x_0 we have $\limsup_{n\to\infty} f(x_n)=f(x_0)$.

A function f is called **upper (lower) semicontinuous in** E if it is upper (lower) semicontinuous at every point $x\in E$. A necessary and sufficient condition for the upper semicontinuity of the function $f(x)$ in E is that $\{x \mid f(x)<\alpha\}$ is a †relative open set in E for every real number α. We can define semicontinuity for functions on a topological space by using this latter property.

A real-valued function $f(x)$ is continuous at $x_0\in E$ if and only if it is upper and lower semicontinuous at x_0 and $f(x_0)$ is finite. A function $f(x)$ is continuous on E if and only if it takes finite real values on E and both $\{x \mid f(x)<\alpha\}$ and $\{x \mid f(x)>\alpha\}$ are relative open sets in E for any real number α. When E is †compact, an upper (lower) semicontinuous function on E actually takes its upper (lower) bound at a point in E. In particular, a continuous function on a compact set E is bounded and assumes its maximum and minimum on E (**Weierstrass's theorem**). Furthermore, if E is connected (e.g., the interval I in **R**), it follows from the connectedness of the image $f(E)$ that if $\alpha,\beta\in f(E)$ and γ lies between α and β, then $\gamma\in f(E)$ (**intermediate-value theorem**).

A real-valued function $f(x)$ on a set E of **R** satisfies the Lipschitz condition if it is †differentiable and the derivative is bounded. Such a function is also †absolutely continuous, continuous, and of †bounded variation. (For the polynomial approximation of real continuous functions → 333 Polynomial Approximation.)

The limit function $f(x)$ of a monotone decreasing sequence of upper semicontinuous functions $f_n(x)$ is also upper semicontinuous. The limit function $f(x)$ of a uniformly converging sequence of continuous functions is continuous. (Regarding the †equicontinuous family of functions → 421 Uniform Convergence.)

D. Baire Functions

The limit function of a pointwise converging sequence of continuous functions defined on a metric space X is not necessarily continuous. R. Baire (*Ann. Mat. Pura Appl.*, (1899)) introduced the notion of Baire functions as follows: He named continuous functions the **functions of class** 0. Then he called a function that is a pointwise limit of a sequence of continuous functions a **function of at most class** 1. A function is said to be of class 1 if it is of at most class 1 and is not of class 0. He similarly defined the notion of **class** n for arbitrary natural number n.

Further a function is called of at most class ω if it is a pointwise limit of a sequence of functions of class n_ν for a sequence of natural numbers n_ν. A function is said to be of **class** ω if it is of at most class ω and is not of class n for any finite number n. In general, using †transfinite induction, we can define the notion of functions of **class** ξ for an arbitrary †ordinal number ξ.

All these functions are called **Baire func-**

tions. Actually, there is no function of class ξ for uncountable ordinal numbers ξ. If X is a †perfect set in Euclidean space, then there is actually a function defined on X of class ξ for an arbitrary countable ordinal number ξ. Hereafter, we shall be concerned with this case only.

If X has the cardinality of the †continuum, then the set of all Baire functions on X has the cardinality of the continuum. On the other hand, the cardinality of all functions on X is actually greater than that of the continuum. Hence there exist functions that are not Baire functions on X. A function is a Baire function if and only if it is †Borel measurable (H. Lebesgue). Therefore, a necessary and sufficient condition for a function f to be a Baire function is that the set $\{x \mid f(x)>\alpha, x\in X\}$ be a Borel set for any real number α (→ 270 Measure C). The limit of a countable sequence of Baire functions is again a Baire function. If $f(x)$ and $g(x)$ are of at most class α on X, then the following functions are also of at most class α: $|f(x)|, f(x)\pm g(x), f(x)\cdot g(x)$, and $f(x)/g(x)$ (provided that $g(x)\neq 0$ on X).

The condition that a function f is of at most class 1 on X is equivalent to either of the following two conditions: (1) For any closed subset F of X, the restriction f^* of f to F has a continuity point in F. (2) For every real number α, the set $\{x \mid f(x)<\alpha, x\in X\}$ is an †F_σ set (Baire). In a †complete metric space, a necessary and sufficient condition for a function f to be of at most class 1 is that the set of continuity points be dense in X.

For example, the **Dirichlet function**, which takes the value 1 at rational points and 0 at irrational points, is expressed as

$$\lim_{\nu\to\infty}\left(\lim_{k\to\infty}(\cos\nu!\,\pi x)^{2k}\right),$$

which is of class 2. A function $f(x,y)$ of two real variables that is continuous in each variable x and y separately is a function of at most class 1.

References

[1] S. Saks, Theory of the integral, Warsaw, 1937.
[2] T. Takagi, Kaiseki gairon (Japanese; A course of analysis), Iwanami, third edition, 1961.
[3] M. Tsuji, Zitu kansûron (Japanese; Theory of real functions), Maki, 1962.
[4] N. Bourbaki, Eléments de mathématique, Topologie générale, ch. 10, Actualités Sci. Ind., 1084b, Hermann, second edition, 1967. For Baire functions,
[5] R. Baire, Leçons sur les fonctions discontinues, Gauthier-Villars, 1905.
[6] C. J. de La Vallée-Poussin, Intégrales de Lebesgue, fonctions d'ensemble, classes de Baire, Gauthier-Villars, second edition, 1934.
[7] W. Rudin, Principles of mathematical analysis, McGraw-Hill, second edition, 1964.

87 (II.17) Continuous Geometry

A. General Remarks

The concept of continuous geometry was introduced by J. Von Neumann as an abstraction of lattice-theoretic properties from a class of †lattices (lattice-ordered sets) which he observed in his research on †operator rings in Hilbert spaces [1]. It can be regarded as an infinite-dimensional generalization of the concept of the lattice of linear subspaces in a projective geometry.

A **continuous geometry** is a †complete and †complemented †modular lattice L (→ 241 Lattices F) which has the following property and its dual (both called **properties of continuity**): For any element a of L and any subset W of L which is †well ordered with respect to the ordering in L, we have $a\cap\sup w=\sup(a\cap w)$ $(w\in W)$. The †center Z of the lattice L is called the **center** of the continuous geometry L, which is said to be **irreducible** when Z has no elements other than the †least element 0 and the †greatest element I; otherwise L is said to be **reducible**. A reducible continuous geometry is isomorphic to a sublattice of a †direct product of irreducible continuous geometries.

On any continuous geometry L, there can be defined a function $d(x)$ whose values belong to a complete lattice-ordered linear space M and which satisfies the following four conditions: (1) $d(x)\geqslant 0$; (2) $d(x)=d(y)$ implies the existence of a common complement of x and y; (3) $d(x\cup y)+d(x\cap y)=d(x)+d(y)$; (4) $\sup d(w)=d(\sup w)$ $(w\in W)$ for any subset W of L which is well ordered with respect to the ordering in L. Such a function $d(x)$ is called a **dimension function** on L. If a group G of †automorphisms of L is given, there can be introduced a generalized dimension function which is invariant under G, with the other conditions slightly weakened (T. Iwamura [3]). Irreducibility of L is equivalent to the property that a real-valued dimension function can be introduced; in this case, the values of $d(x)$ constitute either a finite set or an interval.

B. Representation of Continuous Geometry

A ring R is called a **regular ring** if it has a unity element and if, for any element a of R, there exists an element x in R such that $axa = a$. A continuous geometry L is isomorphic to the lattice (with $\subset$ as its ordering) of †principal left ideals of a regular ring R provided that $d(I) = n \cdot d(x)$ $(n \geqslant 4)$ for some natural number n and some element x of L. The decomposition of R into a †direct sum of ideals corresponds to the decomposition of L into a direct product of lattices. The condition that L is irreducible and finite-dimensional is equivalent to the condition that R is a matrix ring over a †skew field; when these conditions hold and L is considered as a projective geometry, then the coordinates are given by this skew field. In continuous geometries, join and meet are often denoted by the symbols for sum and product, respectively, and sometimes a direct product of continuous geometries is called their direct sum. Sometimes, the requirement that a continuous geometry be complete is weakened to the requirement that it be †conditionally σ-complete.

References

[1] J. Von Neumann, Continuous geometry, Princeton Univ. Press, 1936–1937.
[2] F. Maeda, Renzoku kikagaku (Japanese; Continuous geometry), Iwanami, 1952; German translation, Kontinuierliche Geometrie, Springer, 1958.
[3] T. Iwamura, On continuous geometries 1, II, Japan. J. Math., 19 (1944), 57–71; J. Math. Soc. Japan, 2 (1950), 148–164.
[4] Л. А. Скорняков (L. A. Skornjakov), Дедекиндовы структуры с дополнениями и регулярные, Физматгиз, 1961; English translation, Complemented modular lattices and regular rings, Oliver and Boyd, 1964.

88 (XVIII.15) Control Theory

A. General Remarks

The classical theory of **automatic control** [14] mostly deals with linear control systems whose mathematical structures are described in terms of †ordinary linear differential equations with constant coefficients and whose central problems concern the stability of the system. Applications of the †Laplace transform play an essential role in discussing the response characteristics of the system. The theory of automatic control remained in this stage until the 1930s. Therefore, as far as its mathematical techniques were concerned, there was no novelty for mathematicians, although there were many works that enabled the existing mathematical techniques to be utilized in practical applications. Nevertheless, it is to be noted that even under these circumstances, there were many elaborations of mathematical aspects such as (i) use of †difference-differential equations with constant coefficients in considering the time lag for operations of control; (ii) discussions of transient characteristics of the system; (iii) representation of external disturbances; and (iv) discussion of frequency-response analysis connected with (iii). The whole aspect of mathematical principles adopted in the classical theory of control can be said to belong to the realm of generalized †harmonic analysis and Cauchy series applied to linear translatable functional operators.

Revolutionary technological innovations in electronics and the invention of new automatic control instruments and systems developed after World War II have opened the way to an era of automation or, more adequately, to a cybernetic era. In this connection, current control theory has shown remarkable progress.

First, generalizations ot mathematical formulations have enabled us to take into consideration various real features that we encounter in actual circumstance. Nonlinear control theory is based on the theories of nonlinear vibration, while the theory of sampled-data control systems appeals to phase-space analysis. Discontinuous control systems come from discrete levels of control and discrete transfer of control signals.

Second, irregular inputs, including random noise in particular, naturally lead us to probabilistic formulations of control problems. The modern theory of †stochastic processes has important applications in this area. Statistical criteria for evaluating control characteristics were introduced, and generalized harmonic analysis of irregular motions plays an important role as an analytic technique for attacking these problems.

Last, and most remarkable, a new feature of current control processes is the centralized control system with the computer center as its headquarters. A computer with a large amount of storage for information and equipped with high data-processing capacity yields advanced control features such as those aiming at optimization, adaptation, and learning, which have become crucially important in both control theory and engineering practice.

These new areas have stimulated the development of †cybernetics, the science of control and communication. Control theory is an important contemporary topic. In fact, current control theory has many features, some of which might belong as well to other areas, such as self-organization theory, learning theory, and adaptation theory. It may be concluded that control theory, as well as the three theories just mentioned, may be considered one of the fundamental branches of information science. In fact, current control theory has a vast range of surrounding areas. However, we shall confine ourselves to some important theories that are currently topics of control theory.

B. Deterministic Formulation of the Optimal Control Problem

Let the state of a physical system at a time t be represented by a real n-dimensional vector $x(t)=(x^1(t),x^2(t),\ldots,x^n(t))$. The state of the system is determined by a system of differential equations and initial conditions

$$\frac{dx^i}{dt}=G^i(t,x,u), \qquad x^i(t_0)=x_0^i, \qquad i=1,2,\ldots,n,$$

where $u=(u^1(t),u^2(t),\ldots,u^m(t))$ is called the **control function** (or **control**). It is chosen from a prescribed class of admissible functions and is usually required to satisfy **constraints** $R^j(t,x,u)\geqslant 0, j=1,2,\ldots,r$. The choice is also made so that the system reaches a given terminal state x_1 at t_1. We usually assume that (t,x) varies in a domain $\mathfrak{R}$ of $(n+1)$-dimensional Euclidean space and u varies in a domain $\mathfrak{U}$ of m-dimensional Euclidean space. The problem of **optimal control** is to choose the control $u(t)$ so that a given functional

$$J(u)=g(t_1,x_1)+\int_{t_0}^{t_1}f(t,x,u)\,dt \tag{1}$$

is minimized (or maximized), where g is a given function defined on the set of terminal states, f is a given function defined on $\mathfrak{R}\times\mathfrak{U}$, and the integral is evaluated along the solution (1) corresponding to the choice of $u(t)$. Usually we suppose that the given functions f,G,R are defined and have certain continuity and differentiability properties on $\mathfrak{R}\times\mathfrak{U}$. We also suppose that the terminal states form a p-dimensional manifold $\mathfrak{T}$ in $\mathfrak{R}$, where $0\leqslant p\leqslant n$.

The existence problem associated with the optimal control concerns whether there exists a lower bound for the functional $J(u)$, and if so, whether this lower bound may be attained by some admissible control u^*.

The earliest existence theorems were obtained in connection with the study of **time-optimal problems** for linear systems, that is, problems in which (1) is of the form

$$G^i=\sum_{j=1}^{n}a^{ij}x^j+\sum_{k=1}^{m}b^{ik}u^k+h^i, \quad i=1,2,\ldots,n,$$

and it is required to bring the system from some initial state x_0 at time t_0 to the origin of the x-space in minimum time. The components of the control vector u are assumed to satisfy the constraints $|u^i|\leqslant 1$, $i=1,2,3,\ldots,m$. The result obtained by R. Bellman, I. Glicksberg, and O. Gross [1] and N. N. Krasovskiĭ [2] were generalized by J. P. La Salle [3] for linear systems in which the matrices $A=(a^{ij})$ and $B=(b^{ij})$ and the vector $h=(h^1,h^2,\ldots,h^n)$ are time dependent. La Salle proposed the problem of hitting a moving path $Z(t)$ in the minimum time and showed that if there is a control u that gives rise to a trajectory $x(t)$ that hits the moving target, then there is an optimal control of **bang-bang type**. That is, each component u^j of such an optimal control u takes only the values $+1$ and -1. With further restrictions on A and B, La Salle showed that there exist controls that hit $Z(t)$ and that the optimal controls are unique. A general result for the time-optimal problem with fixed terminal point is due to A. F. Filippov [4], in which the requirement that the differential equation be linear is dropped and the control vector u is permitted to lie in a closed bounded set $\Omega(t,x)$ that can vary with time. E. B. Lee and L. J. Markus obtained a result that includes the existence theorem of La Salle by considering the problem of hitting a compact target set that is moving continuously over a fixed time interval.

C. Pontrjagin's Maximum Principle

Suppose that we are given a system of differential equations

$$dx^i/dt=f^i(x,u), \qquad i=1,2,\ldots,n,$$

for the n-dimensional state vector $x=(x^1,x^2,\ldots,x^n)$ and r-dimensional parameter $u=(u^1,u^2,\ldots,u^r)$. The functions $f^i(x,u)$ and $\partial f^i(x,u)/\partial x^j$ are assumed to be continuous in the Cartesian product $\mathbf{R}^n\times\overline{U}$, where $\overline{U}$ is the closure of U in $\mathbf{R}^r$. A piecewise continuous function defined over the closed interval $[t_0,t_1]$ is said to be an admissible control function if $u(t)\in U$ for every t in $[t_0,t_1]$. Suppose that we are given two points x_0 and x_1 in $\mathbf{R}^n$ and an admissible control function u. Suppose, furthermore, that there exists a solution $x(t)$ of the equation at the beginning of this section that takes the initial value $x(t_0)=x_0$ and the terminal value $x(t_1)=x_1$. We consider here the following functional:

$$J=\int_{t_0}^{t_1} f^0(x(t),u(t))\,dt,$$

where f^0 is a suitable function that is continuous in $\mathbf{R}^n \times \bar{U}$. The control u that minimizes J is called an **optimal control** corresponding to a transition from x_0 to x_1. The corresponding trajectory $x(t)$ is called an **optimal trajectory**. Let x^0 vary according to the law

$$dx^0/dt=f^0(x^1,x^2,\ldots,x^n,u).$$

Introduce the $(n+1)$-dimensional vector $x=(x^0,x^1,x^2,\ldots,x^n)$ and the set of $n+1$ auxiliary functions $(\psi_0,\psi_1,\ldots,\psi_n)$ for which the following system of differential equations holds:

$$\frac{d\psi_i}{dt}=-\sum_{\alpha=0}^{n}\frac{\partial f^\alpha(x,u)}{\partial x^i}\psi_\alpha, \quad i=0,1,2,\ldots,n.$$

We define

$$H(\psi,x,u)\equiv(\psi,f(x,u))\equiv\sum_{\alpha=0}^{n}\psi_\alpha f^\alpha(x,u).$$

Let π be a line in $\mathbf{R}^{n+1}$ that is parallel to the x^0-axis and passes through the point $(0,x_1)$. **Pontrjagin's maximum principle** makes the following assertion:

Let $u(t)$, $t_0 \leqslant t \leqslant t_1$, be an admissible control corresponding to the trajectory $x(t)$ with the initial point x_0 and whose terminal point may be identified with a point on the line π. In order for $u(t)$ and $x(t)$ to be optimal, it is necessary that there exist a nonzero absolutely continuous vector function $\psi(t)=(\psi_0(t), \psi_1(t),\ldots,\psi_n(t))$ corresponding to the functions $u(t)$ and $x(t)$ such that: (i) The function $H(\psi(t),x(t),u)$ of the variable $u\in U$ attains its maximum at the point $u=u(t)$ almost everywhere in the interval $t_0\leqslant t\leqslant t_1$,

$$H(\psi(t),x(t),u(t)) = \max_{u\in U} H(\psi(t),x(t),u) \quad \text{(almost everywhere)};$$

(ii) at the terminal time t_1, the relations

$$\psi_0(t_1)\leqslant 0, \qquad H(\psi(t_1),x(t_1),u(t_1))=0$$

are satisfied.

The maximum principle can be generalized to the cases of nonautonomous differential equations, bounded measurable control functions, and moving terminal points instead of fixed ones.

D. Characterization of Optimal Controls

Assuming that the existence of an optimal control has been established, the next problem is to characterize it. It has been generally recognized that, particularly in the absence of the constraints, the problem of optimal control can be treated as a problem in the †calculus of variations. We need merely write $y'=u$ to reduce the problem to a problem in txy-space. To cope with the constraint, we may introduce a **slack variable** $\xi=(\xi^1,\ldots,\xi^r)$ by means of the differential equations

$$\frac{d(\xi^i)^2}{dt}=R^i(t,x,u), \quad \xi^i(t_0)=0.$$

However, no substantial gains are obtained by this procedure. Another method has been recently developed by Pontrjagin and his colleagues V. G. Boltjanskiĭ and R. V. Gamkrelidze [6, 7]; this method utilizes the maximum principle. Still another approach, heuristic in character, is the †dynamic programming method of Bellman [8, 9, 10]. Each of these two approaches gives us a version of the necessary conditions for optimal control. We give here one version of the necessary conditions for an optimal solution. We assume that the problem is as formulated before. The functions g,f,G,R are of class C^2 on appropriate domains of definition. The following assumptions are made concerning the constraints: (i) If $r>m$, then at most m components of the vector R can vanish at a given point (t,x,u). (ii) At each point (t,x,u), let $\hat{R}_u$ denote the matrix $(\partial R^i/\partial u^j)$, where $j=1,\ldots,m$ and i ranges over the indices such that $R^i(t,x,u)=0$. Then at each point (t,x,u), $\hat{R}_u$ has the maximum rank.

The controls are assumed to be piecewise continuous functions defined on an appropriate fixed interval $t'\leqslant t\leqslant t''$, where $t_0>t'$ and $t''\geqslant t_1$ for all t_1 such that (t_1,x_1) belongs to $\mathfrak{R}$. We introduce a function H of the arguments $(t,x,u,\lambda_0,\lambda)$, where λ_0 is a scalar and λ is a vector $\lambda=(\lambda^1,\ldots,\lambda^n)$. We define

$$H(t,x,u,\lambda_0,\lambda) =\lambda_0 f(t,x,u)+\sum_{i=1}^{n}\lambda^i G^i(t,x,u).$$

L. D. Berkovitz [11] proved the following theorem:

Let u^* be an optimal control in the class of admissible controls, K^* be the corresponding curve, and $x^*(t)$ be the function defining K^* on $[t_0,t_1]$. Then there exist a constant $\lambda_0\geqslant 0$, an n-dimensional vector $\lambda(t)$ defined and continuous on $[t_0,t_1]$, and an r-dimensional vector $\mu(t)\leqslant 0$ defined and continuous on the interval $[t_0,t_1]$ (except perhaps at values of t corresponding to corners of K^* where it possesses unique right-hand and left-hand limits) such that the vector $(\lambda_0,\lambda(t))$ never vanishes and the following conditions are fulfilled:

Condition I. Along K^* the following equations hold:

$$\frac{dx^i}{dt}=\frac{\partial H}{\partial\lambda^i},\quad i=1,\dots,n,$$

$$\frac{d\lambda^i}{dt}=-\frac{\partial H}{\partial x^i}-\sum_{k=1}^{r}\mu^k\frac{\partial R^k}{\partial x^i},\quad i=1,\dots,n,$$

$$\frac{\partial H}{\partial u^i}+\sum_{k=1}^{r}\mu^k\frac{\partial R^k}{\partial u^i}=0,\quad i=1,\dots,m,$$

$$\mu^iR^i=0,\quad i=1,\dots,r.$$

At the endpoint (t_1,x_1^*) of K^*, the transversality condition holds:

$$\lambda_0\frac{\partial g}{\partial\sigma^j}+H\frac{\partial t_1}{\partial\sigma^j}-\sum_{i=1}^{n}\lambda^i\frac{\partial x_1^i}{\partial\sigma^j}=0,\quad j=1,\dots,p,$$

or equivalently:

$$\lambda_0\left(\frac{\partial g}{\partial\sigma^j}+f\frac{\partial t_1}{\partial\sigma^j}\right)+\sum_{i=1}^{n}\lambda^i\left(G^i\frac{\partial t_1}{\partial\sigma^j}-\frac{\partial x_1^i}{\partial\sigma^j}\right)=0,\quad j=1,\dots,p.$$

Along K^* the function H is continuous.

Condition II. For every element (t,x^*,u^*) of K^* and every u such that $u=u(t)$ for some admissible control u,

$$H(t,x^*,u,\lambda_0,\lambda)\geqslant H(t,x^*,u^*,\lambda_0,\lambda).$$

Condition III. Let

$$a^{ij}=\frac{\partial^2\left(H+\sum_{k=1}^{r}\mu^kR^k\right)}{\partial u^i\partial u^j},\quad i,j=1,\dots,m,$$

and let $I(t,x)$ be the subset of indices $i=1,\dots,r$ such that $R^i(t,x,u^*)=0$. Then at each point of K^*,

$$\sum_{i,j=1}^{m}e^ia^{ij}e^j\geqslant 0$$

for all vectors $e=(e^1,\dots,e^m)$ satisfying the system

$$\sum_{j=1}^{m}\frac{\partial R^i}{\partial u^j}e^j=0,\quad i\in I.$$

This theorem can be proved by introducing a slack variable ξ and reducing the control problem to a problem in the calculus of variations. We can apply certain necessary conditions obtained by G. A. Bliss [13] and extended by E. J. McShane [35]; these conditions are then translated back to conditions of the control problem that lead to the proof of the theorem. Condition I comes from the multiplier rule, Condition II comes from the Weierstrass condition, and Condition III and the inequality $\mu\leqslant u$ come from the Clebsch condition. Conditions I and II constitute the so-called maximum principle of Pontrjagin for this problem. In Pontrjagin's formulation, the function H is the negative of the H in the present theorem and $\lambda_0\leqslant 0$. The maximum principle of Pontrjagin has a vast class of applications. Its generalizations to a partial differential equation formulation as well as to a stochastic formulation have been made in recent years.

E. Dynamic Programming and Control Processes

If $W(t,x)$ is the minimum value of $J(u)$ as a function of initial time and position (t,x), then the equation obtained by Bellman by †dynamic programming arguments can be written as follows:

$$W_t(t,x)=-\min_u\left[f(t,x,u)+\sum_{i=1}^{m}W_{x^i}G^i(t,x,u)\right],$$

where $u=u(t)$ ranges over all admissible controls u.

From this equation a form of the †Hamilton-Jacobi equation was deduced by Bellman. He also showed the effectiveness of a dynamic programming approach to more general control processes, including adaptive control processes. His approach is to utilize functional equations that are valid for optimal controls.

F. Some Control Problems in Function Spaces

Since Pontrjagin and his colleagues published their treatise on optimal control [7], considerable effort has been directed toward generalizing their results to problems involving partial differential equations. Some of these problems have been treated in the context of control theory in function spaces. For example, a control problem for an evolution equation can be reduced to an analogous problem for an ordinary differential equation in a function space, provided we regard a spatial differential operator as a functional operator on an appropriate domain. Here we shall briefly summarize recent developments relating to Pontrjagin's maximum principle and controllability in function spaces.

Pontrjagin's Maximum Principle. First, Ju. V. Egorov [15] extended Pontrjagin's principle to ordinary differential equations in Banach spaces. Applying the heuristic approach of dynamic programming, P. K. C. Wang [16] in 1964 and W. L. Brogan [17] in 1965 also obtained a maximum principle for systems governed by partial differential equations. In 1965, A. G. Butkovskiĭ [18] also derived a maximum principle along with many other interesting results on control systems involving partial differential or integral equations.

Also in 1965, A. V. Balakrishnan [19], using the method of convex programming, studied the time-optimal and final-value problems for evolution equations in Banach spaces. Using different methods, A. Friedman [20] subsequently improved upon the results of Balakrishnan. Their results are formally analogous to those previously obtained for ordinary differential equations. Using functional analysis arguments, L. J. Lions [21] solved many problems of control theory for partial differential equations and, in particular, derived certain "variational inequalities" which correspond to the Pontrjagin maximum principle. Lions also derived a series of "unilateral boundary problems" from these inequalities.

Controllability. There are essentially two types of controls which appear in the control theory for partial differential equations, that is, a distributed control and a boundary control. A distributed control is a function distributed over the domain of the appropriate space while a boundary control is a function defined on the boundary of the domain. Butkovskiĭ [18] dealt with the problems of controllability for partial differential equations by reducing them to a moment problem. Similarly, D. L. Russell [22], analyzed the controllability problem for the one-dimensional wave equation by solving a moment problem, using the theory of non-harmonic Fourier series. Controllability of the distributed controls for an evolution equation may be described as follows. Consider the evolution equation

$$\frac{du}{dt} = Au + Bf, \qquad (2)$$

with initial condition $u(0)=0$, in a Hilbert space X, where A is the infinitesimal generator of a strongly continuous semigroup $\{e^{tA}, t \geqslant u\}$ in X and B a bounded operator from a Hilbert space Y into X. We take $C^1([0,\infty), Y)$ as our admissible class of controls. For any control f, the unique solution $u(t) = \int_0^t e^{(t-s)A} Bf(s)\,ds$ of (2) is called a trajectory. We say that (2) is controllable at time T if and only if X is the closure $\overline{R_T}$ of the set

$$R_T = \left\{ \int_0^T e^{(T-s)A} Bf(s)\,ds \mid f \in C^1([0,T], Y) \right\}.$$

Note that (2) is controllable at some finite time if and only if $X = \overline{\cup_{T>0} R_T}$. Obviously, $\overline{R_T} \subset \overline{R_S}$ if $T \leqslant S$. If $\{e^{tA}\}$ is a holomorphic semigroup, then $\overline{R_T} = \overline{R_S}$, and therefore controllability in this case is independent of T. Thus, for a heat equation, in which $A = \Delta$, $\{e^{tA}\}$ is holomorphic, hence controllability is independent of time. However, if a wave equation is written as a system of first-order evolution equations, the controllability does depend upon time (Russell [22]). H. O. Fattorini [23, 24] has studied the relation between the controllability of (2) and the properties of A and B, generalizing results due to La Salle in the case of ordinary differential equations under the assumption that A is a self-adjoint operator which is semibounded from above. Fattorini gave a necessary and sufficient condition to be satisfied by A in order that (1) be controllable for some finite-dimensional Y and for some B. Controllability for evolution equations of higher order has also been investigated by Fattorini [25]; moreover, the same author has studied controllability for boundary control problems by reducing the situation to the case of distributed controls [26].

References

[1] R. Bellman, I. Glicksberg, and O. Gross, On the "bang-bang" control problem, Quart. Appl. Math., 14 (1956), 11–18.
[2] Н. Н. Красовский (N. N. Krasovskiĭ), К Теории оптимального регулирования, Automat. i Telemeh., 18 (1957), 960–970.
[3] J. P. La Salle, The time optimal control problem, Contributions to the theory of nonlinear oscillations V, Ann. Math. Studies, Princeton Univ. Press, 1960, 1–24.
[4] А. Ф. Филиппов (A. F. Filippov), О некоторых вопросах теории оптимального регулирования, Vestnik Moskov. Univ., no. 2 (1959), 25–32.
[5] E. B. Lee and L. Markus, Optimal control for nonlinear processes, Arch. Rational Mech. Anal., 8 (1961), 3 –58.
[6] Р. В. Гамкрелидзе (R. V. Gamkrelidze), Теория оптимальных по быстродействию процессов в линейных системах, Izv. Akad. Nauk SSSR, 22 (1958), 449–474.
[7] Л. С. Понтрягин, В. Г. Болтянский, Р. В. Гамкрелидзе, Е. Ф. Мищенко (L. S. Pontrjagin, V. G. Boltjanskiĭ, R. V. Gamkrelidze, and E. F. Miščenko), Математическая теория оптимальных процессов, физматгиз, 1961; English translation, The mathematical theory of optimal processes, Interscience, 1962.
[8] R. E. Bellman, Dynamic programming, Princeton Univ. Press, 1957.
[9] R. E. Bellman, Adaptive control processes; a guided tour, Princeton Univ. Press, 1961.
[10] R. E. Bellman and S. Dreyfus, Applied dynamic programming, Princeton Univ. Press, 1962.
[11] L. D. Berkovitz, Variational methods in problems of control and programming, J. Math. Anal. Appl., 3 (1961), 145–169.

[12] G. A. Bliss, Lectures on the calculus of variations, Univ. of Chicago Press, 1946.
[13] E. J. McShane, On multipliers for Lagrange problems, Amer. J. Math., 61 (1939), 809–819.
[14] Handbook of control engineering (Japanese), Asakura, 1964.
[15] Ju. V. Egorov, Sufficient conditions for optimality in Banach spaces, Mat. Sb. 64 (1964), 79–101.
[16] P. K. C. Wang, Control of distributed parameter systems, in Advances in control systems 1, ed. by C. T. Leondes, Academic Press (1964), p. 75–172.
[17] W. L. Brogan, Optimal control theory applied to systems described by partial differential equation, Ph.D. thesis, U.C.L.A., 1965.
[18] A. G. Butkovskiĭ, Theory of optimal control of distributed parameter systems (in Russian) Moscow, 1965. English Translation, American Elsevier, 1969.
[19] A. V. Balakrishnan, Optimal control problems in Banach spaces, SIAM J. Control, 3 (1965), 152–180.
[20] A. Friedman, Optimal control in Banach spaces, Math. Anal. Appl., 18 (1967), 35–55.
[21] J. L. Lions, Optimal control of systems governed by partial differential equations (in French), Dunod 1968. English Translation, Springer, 1971.
[22] D. L. Russell, Non harmonic Fourier series in the theory of distributed parameter systems, J. Math. Anal. Appl., 18 (1967), 542–560.
[23] H. O. Fattorini, Some remarks on complete controllability, SIAM J. Control, 4 (1966), 686–693.
[24] H. O. Fattorini, On complete controllability of linear systems, J. Differential Equations, 3 (1967), 391–402.
[25] H. O. Fattorini, Controllability of higher order linear systems, in Mathematical theory of control, ed. by Balakrishnan and Neustadt, Academic Press (1967), p. 301–311.
[26] H. O. Fattorini, Boundary control systems, SIAM J. Control, 6 (1968), 349–385.

89 (II.20) Convergence

A. Introduction

The notion of convergence was first introduced in the real number system to deal with sequences of numbers, sequences of functions, series, and definite integrals (→ 374 Series; 218 Integral Calculus). The notion was then generalized and introduced in †ordered sets, and it is now used in †metric spaces, †topological spaces, and †uniform spaces.

B. Convergence of Sequences of Numbers

A sequence $\{a_n\}$ of numbers is said to be **convergent** to a number a or to **converge** to a, written $\lim_{n\to\infty}a_n = a$ or $a_n \to a$ as $n\to\infty$, if for any positive number ε we can choose a (sufficiently large) natural number n_0 such that for every n larger than n_0 the inequality $|a_n - a| < \varepsilon$ holds. Then a is called the **limit** (or **limit point**) of the sequence $\{a_n\}$. Any sequence has a unique limit whenever a limit exists. A sequence which is not convergent is said to be **divergent** or to **diverge**.

A set A of real numbers is said to be **bounded from above** if there is a real number b such that $a \leqslant b$ for all $a \in A$, **bounded from below** if there is a real number c such that $a \geqslant c$ for all $a \in A$, and **bounded** if it is bounded from above and below. A sequence $\{a_n\}$ of real numbers is said to be **monotonically increasing (monotonically decreasing)**, written $a_n\uparrow (a_n\downarrow)$, if $a_1 \leqslant a_2 \leqslant \ldots \leqslant a_n \leqslant a_{n+1} \leqslant \ldots$ $(a_1 \geqslant a_2 \geqslant \ldots \geqslant a_n \geqslant a_{n+1} \geqslant \ldots)$. A monotonically increasing or decreasing sequence is called a **monotone sequence**.

C. Criteria of Convergence of Sequences of Numbers

Every bounded monotone sequence of real numbers is convergent; its limit is $\sup\{a_n\}$ $(\inf\{a_n\})$ if it is monotonically increasing (decreasing). For any bounded sequence $\{a_n\}$ of real numbers, setting $\alpha_n = \inf\{a_n, a_{n+1}, \ldots\}$ and $\beta_n = \sup\{a_n, a_{n+1}, \ldots\}$, we have $\alpha_n\uparrow$, $\beta_n\downarrow$, and $\alpha_n \leqslant a_n \leqslant \beta_n$. Hence $\lim_{n\to\infty}\alpha_n = \alpha$ $(= \sup\{\alpha_n\})$ and $\lim_{n\to\infty}\beta_n = \beta$ $(= \inf\{\beta_n\})$ exist. α is called the **inferior limit** (or **limit inferior**) of $\{a_n\}$, written $\liminf_{n\to\infty}a_n$ or $\underline{\lim}_{n\to\infty}a_n$, while β is called the **superior limit** (or **limit superior**), written $\limsup_{n\to\infty}a_n$ or $\overline{\lim}_{n\to\infty}a_n$. The limit of a convergent subsequence of a sequence $\{a_n\}$ of numbers is called an **accumulation point** of $\{a_n\}$. For any bounded sequence of real numbers, its superior (inferior) limit is the supremum (infimum) of its accumulation points. Moreover, if β is the superior limit of a sequence $\{a_n\}$, then for any positive number ε, there exist only a finite number of a_n which are greater than $\beta+\varepsilon$, while there exist an infinite number of a_n which are greater than $b-\varepsilon$. The inferior limit of the sequence has a similar property.

Suppose that we are given a sequence $\{a_n\}$ of real numbers and that there exist two sequences $\{u_n\}$ and $\{v_n\}$ such that $u_n \leqslant a_n \leqslant v_n$, $\lim(u_n - v_n) = 0$, $\{u_n\}$ is monotonically increasing, and $\{v_n\}$ is monotonically decreasing. Then $\lim a_n$ exists and is equal to $\lim u_n = \lim v_n$ **(principle of nested intervals)**. In particular, if $\limsup a_n = \liminf a_n$, then $\lim a_n$ exists; the converse also holds.

If $\{a_n\}$ is convergent, then $|a_n - a_m| \to 0$ as $n, m \to \infty$, and vice versa; that is, if for any positive number ε there exists a positive integer n_0 such that $|a_n - a_m| < \varepsilon$ for all $n, m \geqslant n_0$, then $\{a_n\}$ is convergent **(Cauchy's criterion)**.

D. Infinity

For a set A of real numbers, the expression $\sup A = +\infty$ means that A is not bounded from above; $\inf A = -\infty$ means that A is not bounded from below. For a sequence $\{a_n\}$ of real numbers, $\lim a_n = +\infty$ means that for any real number b there exists a positive integer n_0 such that $a_n > b$ for all $n \geqslant n_0$; the notation $\lim a_n = -\infty$ has a similar meaning. The symbols $+\infty$ and $-\infty$ are called **positive** (or **plus**) **infinity** and **negative** (or **minus**) **infinity**, respectively. We say that the limit of $\{a_n\}$ is $+\infty$ $(-\infty)$ if $\lim a_n = +\infty$ $(-\infty)$. In these cases, we customarily say that $\{a_n\}$ **diverges** (or is **divergent**) **to** $+\infty$ $(-\infty)$, or a_n becomes positively (negatively) infinite as $n \to \infty$. We also define $\limsup a_n = +\infty$ ($\liminf a_n = -\infty$) to mean $\sup\{a_n\} = +\infty$ ($\inf\{a_n\} = -\infty$). Among divergent sequences, those whose limit is either $+\infty$ or $-\infty$ are said to be **definitely divergent**, and other sequences are said to be **indefinitely divergent** or to **oscillate**.

We now have the following propositions concerning sequences of numbers: If $\lim a_n = a$ and $\lim b_n = b$, then $\lim(\alpha a_n + \beta b_n) = \alpha a + \beta b$, $\lim(a_n b_n) = ab$, and $\lim(a_n / b_n) = a/b$ (provided that $b_n \neq 0$, $b \neq 0$). For sequences of real numbers, these formulas also hold when a or b is infinity. In those cases we set $\alpha \cdot (\pm\infty) = \pm\infty$ $(\alpha > 0)$, $\alpha \cdot (\pm\infty) = \mp\infty$ $(\alpha < 0)$, $\alpha \pm \infty = \pm\infty$, $\alpha/(\pm\infty) = 0$ for a real number α. The cases $0 \cdot (\pm\infty)$, $+\infty + (-\infty)$, $\pm\infty/(\pm\infty)$ are excluded.

E. Convergence of Sequences of Points in a Topological Space

A sequence $\{a_n\}$ of points in a topological space (→ 408 Topological Spaces) is said to **converge** to a point a if for any †neighborhood U of a there exists a positive integer n_0 such that $a_n \in U$ for all $n \geqslant n_0$. The point a is called a **limit** (or **limit point**) of $\{a_n\}$ and we write $\lim_{n \to \infty} a_n = a$ or $a_n \to a$ as $n \to \infty$.

In particular, the set $\mathbf{R}$ of all real numbers is a topological space in which the set of intervals $(a - \varepsilon, a + \varepsilon)$ for some $\varepsilon > 0$ is a †base for the neighborhood system of a point a, so that the notion of limit in $\mathbf{R}$ explained previously is a special case of the same notion in a topological space. By adding the symbols $+\infty$ and $-\infty$ to $\mathbf{R}$, we obtain the topological space $\overline{\mathbf{R}}$, in which any set containing $\{x \mid x > \alpha, x \in \overline{\mathbf{R}}\}$ ($\{x \mid x < \alpha, x \in \overline{\mathbf{R}}\}$) for some $\alpha \in \mathbf{R}$ is a neighborhood of $+\infty$ $(-\infty)$, where the ordering is defined as $-\infty < \alpha < +\infty$ $(\alpha \in \mathbf{R})$. Then $\lim a_n = +\infty$ $(-\infty)$ is interpreted as convergence in the topological space $\overline{\mathbf{R}}$. The elements of $\overline{\mathbf{R}}$ are called **extended real numbers**.

In the case where the topological space is a †metric space (→ 273 Metric Spaces) with metric ρ, $a_n \to a$ is equivalent to $\rho(a_n, a) \to 0$.

For convergence of sequences of points in a topological space, the following properties (**S**) hold: (**S**) (1). If $a_n = a$ for all n, then $\lim a_n = a$; (2) if $a_n \to a$, then $a_{n_k} \to a$ for any subsequence $\{a_{n_k}\}$; (3) if there is a point a such that any subsequence $\{a_{n_k}\}$ of $\{a_n\}$ has a suitable subsequence converging to a, then $a_n \to a$. In particular, in a †Hausdorff space (e.g., a metric space), it further holds that (**S***) any sequence $\{a_n\}$ has a unique limit $\lim a_n$ whenever a limit exists.

F. Limits of Functions

Let a real-valued function $f(x)$ of a real variable x be defined for $x \neq a$ belonging to a neighborhood of a point a. We say that the **limit** of $f(x)$ is b as x tends to a, and write $\lim_{x \to a} f(x) = b$ or $f(x) \to b$ as $x \to a$, if for any positive number ε there exists a positive number δ such that $0 \neq |x - a| < \delta$ implies $|f(x) - b| < \varepsilon$. Replacing $0 \neq |x - a| < \delta$ by $a < x < a + \delta$ $(a - \delta < x < a)$, we define $f(x) \to b$ as $x \to a + 0$ $(x \to a - 0)$ and say that b is the **limit on the right** (**left**) of $f(x)$ as x tends to a. We define $f(x) \to +\infty$ or $f(x) \to -\infty$ as $x \to a$ analogously to the case of sequences. The expression $f(x) \to b$ as $x \to +\infty$ means that for any positive number ε there is a real number k such that $|f(x) - b| < \varepsilon$ for any $x > k$. There are similar definitions for $x \to -\infty$ and $b = \pm\infty$. When $f(x) \to \pm\infty$ as $x \to a$, we often say that f diverges definitely at a.

In general, for a mapping f from a topological space X into a topological space Y, a point a in X, and a point b in Y, $\lim_{x \to a} f(x) = b$ or $f(x) \to b$ as $x \to a$ means that any neighborhood V of b contains $f(U \cap D - \{a\})$ for some neighborhood U of a, where D is the domain of f. If Y is a Hausdorff space, b is unique (if it exists) for given f and a. This point b is called the **limit** (or **limit value**) of $f(x)$ as $x \to a$.

It is easy to see that this definition of $\lim_{x\to a}f(x)=b$ is a generalization of the cases where the topological spaces are $\bar{\mathbf{R}}$ or $\mathbf{R}$. Let $\mathbf{N}$ be the set of all natural numbers, and let $\bar{\mathbf{N}}=\mathbf{N}\cup\{+\infty\}$ be supplied with the †relative topology as a subspace of $\bar{\mathbf{R}}$. A sequence $\{a_n\}$ of real numbers may be identified with a mapping f from $\bar{\mathbf{N}}$ into $\mathbf{R}$ defined by $f(n)=a_n$. Such a sequence $\{a_n\}$ converges to a if and only if $\lim_{n\to\infty}f(n)=a$.

Suppose, in particular, that f is a mapping from a metric space (X,ρ) into a metric space (Y,σ). Then, $f(x)\to b$ as $x\to a$ means that for any $\varepsilon>0$ there exists a $\delta>0$ such that $\sigma(f(x),b)<\varepsilon$ for all $x\in D$ such that $0<\rho(x,a)<\delta$. Thus $f(x)\to b$ as $x\to a$ if and only if $f(x_n)\to b$ for any $\{x_n\}$ in D with $x_n\to a$. If we set $\rho(z_1,z_2)=|z_1-z_2|$ for complex numbers z_1,z_2, the function ρ supplies the set of all complex numbers $\mathbf{C}$ with a metric and $\mathbf{C}$ becomes isometric to the plane $\mathbf{R}^2$ (→ 76 Complex Numbers). Thus the cases $X=\mathbf{C}$ or $Y=\mathbf{C}$ are particular cases of the above generalization. Furthermore, we introduce the †Riemann sphere $\bar{\mathbf{C}}=\mathbf{C}\cup\{\infty\}$ by adding the †point at infinity ∞ to $\mathbf{C}$. We can define a topology on $\bar{\mathbf{C}}$ such that any set containing $\{\infty\}\cup\{z\,|\,|z|>r\}$ for some positive number r is a neighborhood of ∞. Thus we can define the notions $f(z)\to b$ as $z\to\infty$, $f(x)\to\infty$ as $x\to a$, etc., for a complex-valued function f by considering f as a mapping from the topological space $\bar{\mathbf{C}}$ into itself. Then $f(x)\to\infty$ is equivalent to $1/f(x)\to 0$.

G. Orders of Infinities and Infinitesimals

Let f be a complex-valued function defined on a topological space X and a a fixed point of X. Then f is called an **infinity** (at a) or an **infinitesimal** (at a) if $f(x)\to\infty$ as $x\to a$ or $f(x)\to 0$ as $x\to a$, respectively. Suppose that f and g are infinities and f/g is an infinitesimal. Then f is said to be **of lower order** than g, and g is said to be **of higher order** than f. If both f/g and g/f are bounded, then f is said to be **of the same order** as g. This last relation, written $f\sim g$, is an equivalence relation. An infinity f is said to be **of the nth order** with g if $f\sim g^n$. For two infinitesimals f and g, f is called **of higher order** than g and g **of lower order** than f if f/g is an infinitesimal. For infinitesimals, the terms **of the same order** and **of the nth order** are defined similarly as above. In particular, when $X=\mathbf{C}$ and $a=\infty$, we usually omit the phrase "at ∞." Also, for such a function f, we customarily say that the **order of an infinity (infinitesimal)** is n if $f\sim z^n$ (z^{-n}).

To describe the order of an infinity or an infinitesimal simply, the following notions, due to E. Landau [10], are in common use. Let f and g be two infinities. If $|f(x)/g(x)|$ is bounded as $x\to a$, then f is called **at most of the order of** g as $x\to a$, and we write $f(x)=O(g(x))$ as $x\to a$. Second, if f is of lower order than g, we write $f(x)=o(g(x))$ as $x\to a$. The symbols O, o, indicating the word "order," are called **Landau's symbols**. The notation $f(x)=h(x)+O(g(x))$ means $f(x)-h(x)=O(g(x))$. When we use the symbols O, o we should indicate clearly the phrase "as $x\to a$," which is sometimes omitted when no confusion is to be feared (e.g., for the case of a complex variable with $a=\infty$). These symbols are employed for sequences to describe their behavior as $n\to\infty$.

H. Convergence of Nets (Moore-Smith Convergence) [7, 8]

Let $\mathfrak{A}$ be a (preordered) †directed set. A family of points in a set X with index set $\mathfrak{A}$ (namely, a mapping from $\mathfrak{A}$ to X) is called a **net** in X. A net is denoted by $\{x_\alpha\}_{\alpha\in\mathfrak{A}}$ ($\{x_\alpha\}_{\mathfrak{A}}$ or $\{x_\alpha\}$). A net is called a **universal net** if either $\{\alpha\,|\,x_\alpha\in Y\}$ or $\{\alpha\,|\,x_\alpha\in X-Y\}$ is †cofinal in $\mathfrak{A}$ for any subset Y of X. A net $\{y_\beta\}_{\mathfrak{B}}$ in X is called a **subnet** of $\{x_\alpha\}_{\mathfrak{A}}$ if there exists a mapping $\varphi:\mathfrak{B}\to\mathfrak{A}$ such that (1) $y_\beta=x_{\varphi(\beta)}$ and (2) for any $\alpha_0\in\mathfrak{A}$ there exists a $\beta_0\in\mathfrak{B}$ such that $\beta\geqslant\beta_0$ implies $\varphi(\beta)\geqslant\alpha_0$. In particular, $\{x_\beta\}_{\mathfrak{B}}$ is a subnet of $\{x_\alpha\}_{\mathfrak{A}}$ if $\mathfrak{B}$ is a cofinal directed subset of $\mathfrak{A}$.

For a net $\{x_\alpha\}_{\mathfrak{A}}$ in a topological space X, $\{x_\alpha\}$ is said to **converge** to a point x in X if for any neighborhood U of x there is an α_0 such that $\{x_\alpha\,|\,\alpha\geqslant\alpha_0\}\subset U$. Then a is called a **limit** of the net $\{x_\alpha\}$. We then write $x_\alpha\to x$ $(\alpha\in\mathfrak{A})$ (or simply $x_\alpha\to x$). The convergence of sequences of points is the special case where $\mathfrak{A}=\mathbf{N}$. The notion of convergence using nets was introduced by E. H. Moore and H. E. Smith.

Concerning this convergence, we have the following propositions (**D**): (**D**) (1) If $x_\alpha=x$ for all α, then $x_\alpha\to x$. (2) If $x_\alpha\to x$ and $\{y_\beta\}$ is a subnet of $\{x\}$, then $y_\beta\to x$. (3) If for a net $\{x_\alpha\}$ there is a point x such that any subnet $\{y_\beta\}$ of $\{x_\alpha\}$ has a suitable subnet converging to x, then $x_\alpha\to x$. (4) Suppose that there exist directed sets $\mathfrak{B}_\alpha$ $(\alpha\in\mathfrak{A})$ and (i) $\mathfrak{C}=\mathfrak{A}\times\prod\mathfrak{B}_\alpha$, (ii) $p:\mathfrak{C}\to\mathfrak{A}$ and $p_\alpha:\mathfrak{C}\to\mathfrak{B}_\alpha$ are the projections, and (iii) we are given points in X, $z_\gamma=y_{\alpha\beta}$ for $\gamma\in\mathfrak{C}$ such that $p(\gamma)=\alpha$, $p_\alpha(\gamma)=\beta$. Suppose also that $x_\alpha\to x$ and $y_{\alpha\beta}\to x_\alpha$ $(\beta\in\mathfrak{B}_\alpha)$ for any $\alpha\in\mathfrak{A}$. Then $z_\gamma\to x$ $(\gamma\in\mathfrak{C})$. Furthermore, the space X is a †Hausdorff space if and only if we have the condition (**D***): Any net in X has at most one limit.

A limit of $\{x_\alpha\}$ is denoted by $\lim x_\alpha$ or $\lim_{\alpha\in\mathfrak{A}}x_\alpha$. Then $x_\alpha\to x$ if and only if x is contained in the closure of any subnet $\{y_\beta | \beta \in \mathfrak{B}\}$ of $\{x_\alpha\}$. (We may consider this as a definition of $x_\alpha \to x$.) $\{x_\alpha\}$ is said to **partially converge** to x if x is contained in the closure of $\{x_\alpha | \alpha \geqslant \alpha_0\}$ for all $\alpha_0 \in \mathfrak{A}$.

I. Convergence of Filters [9]

Let X be a set. A set Φ of subsets of X is called a **filter** if the following conditions are satisfied: (i) $X \in \Phi$; (ii) $\varnothing \notin \Phi$ ($\varnothing$ is the empty set); (iii) $A \subset B \subset X$ and $A \in \Phi$ imply $B \in \Phi$; (iv) $A, B \in \Phi$ imply $A \cap B \in \Phi$. Let $\mathfrak{B}$ be a set of subsets of X and Φ be the collection of subsets of X such that each element A of Φ contains a subset belonging to $\mathfrak{B}$. If Φ is a filter, then $\mathfrak{B}$ is called a **filter base** which **generates** Φ. $\mathfrak{B}$ is a filter base if and only if (i) $\varnothing \notin \mathfrak{B}$; (ii) $A, B \in \mathfrak{B}$ implies that there is a $C \in \mathfrak{B}$ with $A \cap B \supset C$. A filter is called an **ultrafilter** (or **maximal filter**) if there exists no filter which contains Φ properly. For any filter there exists an ultrafilter containing it. If $\{\Phi_\lambda\}_\Lambda$ is a family of filters, then the intersection $\bigcap \Phi_\lambda$ is a filter. If $\mathfrak{F}$ is a filter base of Λ, then $\bigcup_{M \in \mathfrak{F}} (\bigcap_{\lambda \in M} \Phi_\lambda)$ is also a filter.

We denote by $\mathfrak{U}(x)$ the †neighborhood system of a point x in a topological space X. A filter Φ in X is said to **converge** to a point a, written $\Phi \to a$, if $\mathfrak{U}(x) \subset \Phi$. A filter base $\mathfrak{B}$ is said to **converge** to a if the filter generated by $\mathfrak{B}$ converges to a.

The convergence of filters just defined has the following fundamental properties **(L)**: (1) for a point a in X the filter $\Phi_a = \{A | a \in A \subset X\}$ converges to a; (2) for two filters Φ and Ψ, $\Phi \to a$ and $\Phi \subset \Psi$ imply $\Psi \to a$; (3) if $\Phi_\lambda \to a$ for all members in a family $\{\Phi_\lambda\}$ of filters, then $\bigcap \Phi_\lambda = \Phi \to a$; (4) suppose that filters $\Phi_y \to y$ are assigned for all points y in a subset Y of X, and that we are given a filter Ψ in X which converges to a, generated by a filter base $\mathfrak{B}$ in Y; then $\bigcup_{B \in \mathfrak{B}} (\bigcap_{y \in B} \Phi_y) \to a$. Furthermore, the space X is Hausdorff if and only if we have the condition **(L*)**: Each filter in X has at most one limit.

J. Relations among Various Definitions of Convergence

Convergence of sequences of points is a special case of that of nets. Properties (1), (2), and (3) of **(D)** imply (1), (2), and (3) of **(S)**, respectively, and **(D*)** implies **(S*)**. Consider a net $\{x_\alpha\}_{\mathfrak{A}}$ in X. Then the set $\{\{x_\alpha | \alpha \in \mathfrak{A}, \alpha \geqslant \alpha_0\} | \alpha_0 \in \mathfrak{A}\}$ of subsets of X is a filter base in X which generates a filter Φ, and $\Phi \to x$ if and only if $x_\alpha \to x$. In this situation, **(L)** implies **(D)**, and **(L*)** implies **(D*)**. Suppose that we are given a function $f: X \to Y$ with the domain D and a point $a \in X$. Let $\mathfrak{U}(a)$ be the neighborhood system of a and assume that, for any $U \in \mathfrak{A}(a)$, $U \cap D - \{a\} \neq \varnothing$. Then the set $\{f(U \cap D - \{a\}) | U \in \mathfrak{U}(z)\}$ is a filter base. Let Φ be the filter generated by it. Then $f(x) \to b$ as $x \to a$ if and only if $\Phi \to b$. Consequently, the various types of convergence described previously can be expressed by means of convergence of filters.

K. Convergence and Topology

In a topological space X, the concept of convergence of nets and that of filters can be defined. Conversely, convergence of nets in X defines a topology of X. In fact, let us assume that we are given a set X and a definition of convergence of filters which satisfies the properties **(L)**. Then convergence of nets that satisfies **(D)** can be introduced as above. If $\bar{A}$ is defined as the set of limits of all nets contained in a subset A of X, then $\bar{A}$ satisfies the axiom of closures (→ 408 Topological Spaces), and a topology can be defined on X. Then we have the following propositions: (i) $a \in \bar{A}$ if and only if there is a net $\{x_\alpha\}$ with $x_\alpha \in A$ converging to a; (ii) $a \in A^i$ if and only if $x_\alpha \to a$ implies $\{x_\alpha\} \cap A \neq \varnothing$; (iii) U is a neighborhood of a if and only if $x_\alpha \to a$ implies $\{x_\alpha\} \cap U \neq \varnothing$. Thus, if X is a topological space, it carries a "new" topology defined by way of convergence of nets. But this "new" topology coincides with the original one. Similarly, starting from convergence of filters (or nets), we can obtain a "new" definition of convergence of filters (or nets), which coincides with the initial one. In conclusion, to define a topology on a space X is the same thing as to define convergence of filters in X or of nets in X.

We shall describe here a few notions on topological spaces in terms of convergence. The fact that a topological space X is compact is equivalent to the fact that every universal net in X converges, and is also equivalent to the fact that every ultrafilter in X converges. A mapping f from a topological space X into a topological space Y is continuous at a if and only if one of the following conditions is satisfied: (1) for any net $\{x_\alpha\}$ in X converging to $a \in X$, we have $f(x_\alpha) \to f(a)$ in Y; (2) for any filter Φ in X converging to $a \in X$, we have $f(\Phi) = \{f(M) | M \in \Phi\} \to f(a)$ in Y; (3) $f(x) \to f(a)$ in Y as $x \to a$ in X (in the sense of the limit of a function at a).

It was M. Fréchet [6] who gave a definition of a topology on a space using the notion of convergence as a foundation. A set is called an **L-space** (or **Fréchet L-space**) if convergence of sequences of points in it is defined so as to satisfy conditions (1) and (2) of (S) and (S*) (1906). Such convergence is called **star convergence** if it also satisfies (3) of (S), and in that case the space is called an **L^*-space**. For any subset A of an L-space X, define $\bar{A}$ as the set of all points a such that $x_n \to a$ for some sequence $\{x_n\}$ contained in A. Then the axioms $\bar{A} \supset A$, $\overline{A \cup B} = \bar{A} \cup \bar{B}$, and $\bar{\varnothing} = \varnothing$ are satisfied, so that X is a †generalized topological space (the axiom $\bar{\bar{A}} = \bar{A}$ is not necessarily satisfied). For a Hausdorff space X with the †first countability axiom and convergence of sequences defined by means of its topology, the closure operation defined before gives the same topology as the initial one.

L. (*o*)-Convergence

A sequence $\{a_n\}$ of elements of an ordered set S is said to be **(*o*)-convergent** to an element a of S if there exist two sequences $\{u_n\}$ and $\{v_n\}$ such that $u_n \leqslant a_n \leqslant v_n$, $u_n \leqslant u_{n+1}$, and $v_n \geqslant v_{n+1}$, and $a = \sup u_n = \inf v_n$. When we write this $a_n \to a$, properties (1) and (2) of (S) and (S*) of convergence of sequences hold. Next, a sequence $\{a_n\}$ is said to be **(*o*)-star convergent** to a if any subsequence of $\{a_n\}$ has a suitable subsequence which converges to a. Then (*o*)-star convergence has the properties (S) and (S*).

For any set X the set $\mathfrak{P}(X)$ of all subsets of X is an ordered set under the inclusion relation. The fact that a sequence $\{A_n\}$ of subsets is (*o*)-convergent to a subset A is equivalent to:

$$A = \bigcap_{m=1}^{\infty} \bigcup_{n=m}^{\infty} A_n = \bigcup_{m=1}^{\infty} \bigcap_{n=m}^{\infty} A_n.$$

The set A is also equal to $\lim A_n$, which is the †limit of a sequence $\{A_n\}$ of subsets.

References

[1] T. Takagi, Kaiseki gairon (Japanese; A course of analysis), Iwanami, third edition, 1961.
[2] S. Hitotumatu, Kaisekigaku zyosetu (Japanese; Elements of analysis) I, Syôkabô, 1962.
[3] G. Birkhoff, Lattice theory, Amer. Math. Soc. Colloq. Publ., revised edition, 1948.
[4] N. Bourbaki, Eléments de mathématique, III. Topologie générale, ch. 1, Actualités Sci. Ind., 1142d, Hermann, fourth edition, 1965; English translation, General topology, Addison-Wesley, 1966.
[5] J. W. Tukey, Convergence and uniformity in topology, Ann. Math. Studies, Princeton Univ. Press, 1940.
[6] M. Fréchet, Sur quelques points du calcul fonctionnel, Rend. Circ. Mat. Palermo, 22 (1906), 1–74.
[7] E. H. Moore and H. L. Smith, A general theory of limits, Amer. J. Math., 44 (1922), 102–121.
[8] G. Birkhoff, Moore-Smith convergence in general topology, Ann. of Math., (2) 38 (1937), 39–56.
[9] H. Cartan, Théorie des filtres, C. R. Acad. Sci. Paris, 205 (1937), 595–598; Filtres et ultrafiltres, ibid., 777–779.
For Landau's symbol,
[10] E. G. H. Landau, Handbuch der Lehre von der Verteilung der Primzahlen, Teubner, 1909 (Chelsea, 1953).

90 (X.4) Convex Functions

A. Convex Functions

A real-valued function $f(x)$ defined on a †convex set D in a linear space over $\mathbf{R}$ is called a **convex function** if for every $x,y \in D$ and $0 \leqslant \lambda \leqslant 1$, we have

$$f(\lambda x + (1-\lambda)y) \leqslant \lambda f(x) + (1-\lambda) f(y). \qquad (1)$$

The function $f(x)$ is called a **strictly convex function** if the sign $\leqslant$ in (1) is replaced by $<$ except when λ is either 0 or 1. If $-\psi(x)$ is convex (strictly convex), the function $\psi(x)$ is called a **concave function (strictly concave function)**. The notion of convex function was introduced by J. L. W. V. Jensen [1] for the domain D an interval on the real line (i.e., $\mathbf{R}$).

Sometimes the condition for a convex function is weakened such that (1) is assumed only for $\lambda = 1/2$. However, if D is a †topological linear space and f is continuous, then the weakened condition implies the original one. Hereafter, we mainly consider the case where D is an interval on the real line. In this case, a convex function $f(x)$ (in the weaker sense) is continuous in the interior of the interval if $f(x)$ is †measurable or bounded from above on a set of positive measure (the latter was proved by A. Ostrowski [2]). In particular, suppose that $f(x)$ is defined in the interval I and is bounded from below. Then either $f(x)$ is continuous or its graph is dense in the set $\{(x,y) \mid x \in I, y \geqslant g(x)\}$, where $g(x)$ is a suitable convex continuous function (Hukuhara [3]). We note here that the original definition of convex function implies the continuity of $f(x)$ in the interior of the interval. In such a case, $f(x)$ always has †right and †left derivatives and satisfies $f'_-(x) \leqslant f'_+(x) \leqslant f'_-(y) \leqslant f'_+(y)$ for $x<y$. Hence it is differentiable except for at most countably many points.

A function $f(x)$ is a continuous convex function in $a \leqslant x \leqslant b$ if and only if it is expressible in the form:

$$f(x)=f(a)+\int_a^x \varphi(t)\,dt,$$

where $\varphi(t)$ is a monotone increasing function. If $f(x)$ is twice differentiable, $f''(x) \geqslant 0$ $(a<x<b)$ is a necessary and sufficient condition for $f(x)$ to be convex in (a,b).

B. Convex Functions and Inequalities

If $f(x)$ is convex (in the original sense), we have, for $a_\nu>0$,

$$f\left(\sum a_\nu x_\nu \Big/ \sum a_\nu\right) \leqslant \sum a_\nu f(x_\nu) \Big/ \sum a_\nu. \qquad (2)$$

Similarly, we have the inequality:

$$f\left(\int \varphi\psi\,dx \Big/ \int \varphi\,dx\right) \leqslant \int \varphi f(\psi)\,dx \Big/ \int \varphi\,dx. \qquad (3)$$

The functions x^a $(a>1$ or $a<0)$, $-x^a$ $(0<a<1)$, $-\log x$, $x\log x$ are strictly convex for $x>0$, and the functions x^{2n} $(n \geqslant 1)$, $\exp x$, $\log(1+e^x)$, $\sqrt{a^2+x^2}$ $(a \neq 0)$ are strictly convex in $-\infty<x<+\infty$. Applying the inequalities (2) or (3) to these functions, we obtain various inequalities, including the inequalities on means (→ 215 Inequalities).

A continuous convex function $f(x)$ over a topological linear space satisfying the relation $f(\alpha x)=\alpha f(x)$ for an arbitrary positive number α is called a **subadditive functional**, often utilized in functional analysis.

C. M. Riesz's Convexity Theorem

Let $x=(\xi_1,\dots,\xi_n)$ be an n-tuple of complex numbers and let $\nu \geqslant 0$. We put $N_\nu(x)=(\sum_{j=1}^n |\xi_j|^{1/\nu})^\nu$ for $\nu>0$ and $N_0(x)=\sup|\xi_j|$. Let (α_{ij}) be an $m\times n$ complex matrix, $x=(\xi_1,\dots,\xi_n)$, $z=(\zeta_1,\dots,\zeta_m)$, $\nu \geqslant 0$, and $\mu \geqslant 0$. We put

$$M(\nu,\mu)=\sup_{N_\nu(x)\leqslant 1,\,N_\mu(z)\leqslant 1}\left|\sum_{i=1}^m\sum_{j=1}^n \alpha_{ij}\zeta_i\xi_j\right|.$$

Then $\log M(\nu,\mu)$ is a convex function of (ν,μ) in the following sense: Let $0<\nu_i\leqslant 1$, $0<\mu_i\leqslant 1$, and $\nu_i+\mu_i \geqslant 1$ $(i=1,2)$. Then $\log M((1-t)\nu_1+t\nu_2,(1-t)\mu_1+t\mu_2)$ is a convex function with respect to t for $0 \leqslant t \leqslant 1$ [4,5]. These results are called **M. Riesz's convexity theorem**. Famous inequalities such as the †Hölder inequality or the †Minkowski inequality follow from this theorem. For example, let T be an †additive operator from the †function space $L_p(\Omega)$ into $L_p(\Omega)$ for all $1 \leqslant p \leqslant \infty$. If T is a continuous operator for $p=1$ and $p=\infty$, and the norm of T is $\leqslant C$ for $p=1$ and $p=\infty$, then T is continuous for all p $(1<p<\infty)$, and its norm is always $\leqslant C$.

References

[1] J. L. W. V. Jensen, Sur les fonctions convexes et les inégalités entre les valeurs moyennes, Acta Math., 30 (1906), 175–193.
[2] A. Ostrowski, Mathematische Miszellen XIV, Über die Funktionalgleichung der Exponentialfunktion und verwandte Funktionalgleichungen, Jber. Deutsch. Math. Verein., 38 (1929), 54–62.
[3] M. Hukuhara, Sur la fonction convexe, Proc. Japan Acad., 30 (1954), 683–685.
[4] M. Riesz, Sur les maxima des formes bilinéaires et sur les fonctionelles linéaires, Acta Math., 49 (1927), 465–497.
[5] G. O. Thorin, An extention of a convexity theorem due to M. Riesz, Medd. Lunds Univ. Mat. Sem., 4 (1939), 1–5.
[6] G. H. Hardy, J. E. Littlewood, and G. Pólya, Inequalities, ch. 3, Cambridge Univ. Press, 1934, revised edition, 1952.
[7] N. Bourbaki, Eléments de mathématique, Espaces vectoriels topologiques, ch. 2, Actualité Sci. Ind., 1189a, Hermann, second edition, 1966.
[8] M. Tsuji, Zitu kansûron (Japanese; Theory of real functions), Maki, 1962, ch. 8.
[9] F. A. Valentine, Convex sets, McGraw-Hill, 1964.

91 (VI.11) Convex Sets

A. General Remarks

A nonempty subset X of the n-dimensional Euclidean space $\mathbf{R}^n$ is called a **convex set** if for any elements x, y in X and any number a such that $0 \leqslant a \leqslant 1$, the element $ax+(1-a)y$ of $\mathbf{R}^n$ is also contained in X. The †interior and the †closure of a convex set are also convex. A point x of a convex set X is called an **extreme point** of X if x cannot be expressed as $(x_1+x_2)/2$ in terms of a pair of distinct points x_1, x_2 in X. A bounded closed convex set is called a **convex body** if it has †interior points. Given an arbitrary nonempty subset X of $\mathbf{R}^n$, the minimum convex set containing X exists, called the **convex hull** of X and denoted by $[X]$. Each point x

of $[X]$ may be expressed as $x=\sum_{i=1}^{n+1}a_ix_i$, where x_i belongs to X and the a_i are nonnegative numbers such that $\sum_{i=1}^{n+1}a_i=1$. When X is a finite set, $[X]$ is called a **convex polyhedron**. If $\dot{X}$ denotes the set of extreme points (also called **vertices**) of a convex polyhedron X, then $X=[\dot{X}]$.

For elements x, y of $\mathbf{R}^n$, denote the inner product by (x,y). Given a nonzero element v of $\mathbf{R}^n$ and a fixed number a, the †hyperplane $H=\{x\,|\,(v,x)=a\}$ divides the space $\mathbf{R}^n$ into two †half-spaces $\{x\,|\,(v,x)\leqslant a\}$ and $\{x\,|\,(v,x)\geqslant a\}$, each of which is a closed convex set. If a convex set X is contained in one of the half-spaces S determined by the hyperplane H and the boundaries of X and H intersect, then we say that H is a **supporting hyperplane** of X and S is a **supporting half-space** of X. A closed convex set X is the intersection of its supporting half-spaces. A boundary point of a convex set X is contained in a supporting hyperplane of X. Given mutually disjoint convex sets X and Y, the **separation theorems** (1) and (2) hold.

(1) If X has inner points, then there exists a nonzero element v of $\mathbf{R}^n$ and a number a such that X is contained in the set $\{x\,|\,(x,v)\geqslant a\}$ and Y is contained in the set $\{x\,|\,(x,v)\leqslant a\}$.

(2) If one of X and Y is bounded, we may replace the signs $\leqslant$ and $\geqslant$ in the statement of (1) by $<$ and $>$, respectively (when the separation of convex sets X and Y is described by strict inequalities, we say that X and Y are **strongly separated**).

As an immediate consequence of the separation theorems, we obtain the following proposition: Suppose that A is an $m\times m$ matrix with real entries. Given any element z in a Euclidean space, we write $z\geqslant 0\ (>0)$ if each component of z is $\geqslant 0\ (>0)$. Furthermore, if ${}^tAy>0$ never holds for an m-dimensional vector $y>0$, then there exists a nonzero n-dimensional vector $x\geqslant 0$ such that $Ax\leqslant 0$ (→ 178 Game Theory).

The definitions given previously for subsets of $\mathbf{R}^n$ may be naturally extended to the case of †real topological linear spaces (see Section G). Also, in the theory of analytic functions of several complex variables, various notions of convexity of the subsets of $\mathbf{C}^n$ are considered (→ 25 Analytic Functions of Several Complex Variables).

B. Helly's Theorem

Suppose that we are given an index set Λ of cardinality greater than $n+1$, and bounded closed convex sets C_λ $(\lambda\in\Lambda)$ in $\mathbf{R}^n$. If any $n+1$ sets of the C_λ have nonempty intersection, then the intersection of all the C_λ is nonempty (**Helly's theorem**).

This theorem has a wide range of applications. For example, we have propositions (1)–(4).

(1) If a convex set X of $\mathbf{R}^n$ is covered by a finite number of half-spaces, then X may be covered by not more than $n+1$ half-spaces among them. (2) Let X and Y be finite subsets of $\mathbf{R}^n$. X and Y are strongly separated by a hyperplane if for an arbitrarily chosen subset S of $X\cup Y$ consisting of at most $n+2$ points, the sets $S\cap X$ and $S\cap Y$ are strongly separated by a hyperplane. (3) If the †diameter of a subset X of $\mathbf{R}^n$ is not greater than 2, then X is contained in a †ball of radius $(2n/(n+1))^{1/2}$. (4) Let X be a convex body in $\mathbf{R}^n$. There exists a point x in X such that $\|x-u\|/\|v-u\|\leqslant n(n+1)$, where u, v are points of intersection of an arbitrary straight line passing through x with the boundary of X, and $\|x\|$ denotes the length $(x,x)^{1/2}$ of x.

Helly's theorem may also be applied to problems of approximation of functions.

C. Ovals

The boundaries of convex bodies in $\mathbf{R}^2$ and $\mathbf{R}^3$ are called **ovals** and **ovaloids**, respectively. An oval E is a †Jordan curve $C(t)$ whose every point $P_0=C(t_0)$ admits left and right "tangents" $l_{P_0}^+$ and $l_{P_0}^-$, where $l_{P_0}^\pm$ is a straight line expressed as the set of points $P(\lambda)$, $\lambda\in\mathbf{R}$ such that $P(\lambda)-P_0=\lambda a^\pm$, with $a^\pm=\lim_{t\to\pm 0}(C(t)-P_0)$. There may exist exceptional points P on E for which left and right tangents do not coincide, but the set of such points is at most countable. Each tangent $l_P^\pm$ shares a point or a segment with E, is called a **supporting line** of the oval, and is also a supporting hyperplane in $\mathbf{R}^2$ of the convex body $[E]$ in the sense of Section A. If we fix an interior point O of a convex body X and take an arbitrary point P different from O, then the boundary E of X admits one and only one supporting line $l(P)$ which is perpendicular to the line OP and meets the half-line OP. Take a rectangular coordinate system with the origin O, and let (x,y) denote the coordinates of P. Then the points (ξ,η) on $l(P)$ satisfy the equation $\xi x+\eta y=H(x,y)$, where $H(x,y)$ is a function determined for all (x,y) in $\mathbf{R}^2$ and satisfying the following conditions: (i) $H(0,0)=0$; (ii) $H(tx,ty)=tH(x,y)$, for $t\geqslant 0$; (iii) $H(x_1+x_2,y_1+y_2)\leqslant H(x_1,y_1)+H(x_2,y_2)$. The function $H(x,y)$ is called the **supporting line function** of E. The magnitude and shape of E are determined by H, and any function satisfying conditions (i)–(iii) is a supporting line function of an oval. An oval E has a finite length $L=L(E)$, and the convex body $[E]$ has a finite area $F=F(E)$. If OP' denotes the half-line with direction opposite to that of OP and l' denotes the supporting line $l(P')$, the distance between the parallel lines l and l' is called the **breadth** of E in the direction PP'. Let $D=D(E)$ and $\Delta=\Delta(E)$ be the maximum and minimum of the breadth of E, respectively. D is the †diameter of E (or $[E]$), and Δ is called the **thickness** of E (or $[E]$). In particular, if $D=\Delta$, then the oval E is called a **curve of constant breadth**. In the following inequalities, equality holds only when E is one of the figures mentioned in parentheses: (1) $L^2 \geqslant 4\pi F$ (circles, J. Steiner (1838), → 228 Isoperimetric Problems); (2) $\pi D^2 \geqslant 4F$ (circles, L. Bieberbach (1915)); (3) $L \leqslant \pi D$ (curves of constant breadth, W. Blaschke (1916)); (4) $F \geqslant \Delta^2/\sqrt{3}$ (regular triangles, J. Pál (1921)). (See T. Kubota, *Tôhoku Sci. Bull.*, I, 12, 13, *Tôhoku Math. J.*, 24, 49.)

D. Linear Combinations of Ovals

Let H_1 and H_2 be supporting line functions of ovals E_1 and E_2, and let t_1 and t_2 be positive numbers. Since the function $t_1H_1+t_2H_2$ satisfies conditions (i)–(iii), given before, it is a supporting line function of an oval $E(t_1,t_2)$. In this case, we may also write $E(t_1,t_2)=t_1E_1+t_2E_2$ and call it a **linear combination** of E_1 and E_2. In particular, the oval $(E_1+E_2)/2$ is called the **mean oval** of E_1 and E_2. In general, there exists a quantity M, called the **mixed area** of E_1 and E_2, such that $F(E(t_1,t_2))=F(E_1)t_1^2+2Mt_1t_2+F(E_2)t_2^2$. M does not depend on the choice of t_1 and t_2, and $M^2 \geqslant F(E_1)F(E_2)$. Here, the equality holds if and only if E_1 and E_2 are homothetic and situated in a position of homothety. Furthermore, if $0 \leqslant t \leqslant 1$, then the square root of $F(E(t,1-t))$ is a †convex function of t (H. Minkowski, *Math. Ann.*, 57 (1903)).

E. Specific Ovals

Suppose that we are given an equilateral triangle ABC. Draw three circles C_1, C_2, and C_3 with centers A, B, and C and radii equal to the length of the sides of ABC. The minor arcs AB, BC, and CA of the circles form an oval which is called a **Reuleaux triangle**. This oval is of constant breadth. Furthermore, given a fixed breadth D, the area $F(E)$ of an oval E of constant breadth D attains its minimum when E is a Reuleaux triangle. A Reuleaux triangle obtained from a triangle ABC revolves freely within the square of side AB and touches each side. In general, an oval which revolves touching the sides of a convex polygon from the inside is called an **inrevolvable** oval. Any such oval revolves inside some regular polygon (M. Fujiwara, S. Kakeya, *Tôhoku Sci. Bull.* I, 4; *Tôhoku Math. J.*, 11).

Various properties of an oval already described may be generalized to the case of a boundary of a convex body in $\mathbf{R}^n$.

F. Convex Cones

A nonempty subset X of $\mathbf{R}^n$ is called a convex cone if for any elements x, y of X and a nonnegative number a, ax and $x+y$ are contained in X. A convex cone is a convex set. Given any nonempty subset X of $\mathbf{R}^n$, the minimum convex cone $K(X)$ containing X exists. Given two convex cones X and Y, a convex cone $X+Y$, called the **sum** of X and Y, is defined as the set of elements $x+y$, where x, y are elements of X, Y. The intersection of convex cones X and Y is also a convex cone. Given a convex cone X, the subset of $\mathbf{R}^n$ consisting of the elements y such that $(x,y) \leqslant 0$ for any element x in X is a convex cone which is called the **dual convex cone** (or **conjugate convex cone**) of X, denoted by X^*. If X is a finite set, $K(X)$ is called a **convex polyhedral cone**. For example, if v is a nonzero vector, then the half-line $(v)=\{x \mid x=av, a \geqslant 0\}$ or the half-space $(v)^*=\{x \mid (v,x) \leqslant 0\}$ is a convex polyhedral cone. A convex polyhedral cone is closed. A convex cone X is a convex polyhedral cone if and only if X is the sum of a finite number of half-lines. Given convex cones X and Y, we have propositions (1)–(3): (1) If $X_1 \subset X_2$, then $X_2^* \subset X_1^*$; (2) $(X_1+X_2)^*=X_1^* \cap X_2^*$; (3) $X_1^*+X_2^* \subset (X_1 \cap X_2)^*$. If X_1 and X_2 are convex polyhedral cones, then $X_1^*+X_2^*=(X_1 \cap X_2)^*$. Generally, $X \subset (X^*)^*=X^{**}$ for a convex cone X. If X is a closed convex cone, then $X=X^{**}$. Namely, the **duality principle** holds for closed convex cones. A linear subspace of $\mathbf{R}^n$ is a convex polyhedral cone. Also, if A is an $m \times n$ real matrix, the subsets $\{x \mid Ax=0, x \geqslant 0\}$ and $\{x \mid Ax \geqslant 0\}$ are convex polyhedral cones. Since the duality principle holds for convex polyhedral cones, we obtain the †Minkowski-Farkas theorem (i.e., if A is an $m \times m$ real matrix and v is an element of $\mathbf{R}^m$, then the equation $Ay=v$ has a solution $y \geqslant 0$ in $\mathbf{R}^m$ if and only if $(v,x) \geqslant 0$ for all $x \in \mathbf{R}^m$ such that ${}^tAx \geqslant 0$). For linear inequalities → 255 Linear Programming.

G. Convex Sets in Function Spaces

The definitions of convex sets and convex cones in $\mathbf{R}^n$ may be naturally extended to the case of any real linear space. Some of their properties may be generalized and applied to the case of function spaces.

(1) Let E be a †locally convex real topological linear space satisfying Hausdorff's †separation axiom. Let A and B be convex sets in E, and assume that B has interior points and $A \cap B$ is empty. Then A and B are separated by a hyperplane. Namely, there exists a nonzero †continuous linear functional f on E such that $\sup f(A) \leqslant \inf f(B)$.

(2) Let E be as in (1), and let C be a convex set in E. If a boundary point x of C admits a nonzero continuous linear functional f such that $f(x) = \sup f(C)$, we call such a point x a **supporting point** of C, and f a **supporting functional** of C. If C has interior points, then any boundary point x of C is a supporting point of C.

(3) Let C be a closed convex set of a †Banach space E. The set of supporting points of C is dense in its boundary.

A convex set C contained in the dual space E^* of a real topological linear space E is called a **regularly convex set** if for any f_0 in E^* not contained in C, there exists a point x_0 in E such that $\sup\{f(x_0) \mid f \in C\} < f_0(x_0)$.

Let E be a real topological linear space satisfying Hausdorff's separation axiom, and let C be a closed convex cone having 0 as its extreme point. Furthermore, assume that $C \cap (-C) = \{0\}$. If we set $x \leqslant y$ when $y - x \in C$, a partial ordering $\leqslant$ is defined in E. For example, if E is $\mathbf{R}^n$, then the **positive orthant** $C = \{x = (x_i) \mid x_i \geqslant 0, i = 1, \ldots, n\}$ satisfies these requirements, and the partial ordering $x \leqslant y$ defined by means of C is equivalent to the relation $x_i \leqslant y_i$ for all i.

Some of the properties of matrices of positive entries or †integral operators whose †kernel functions are positive-valued may be generalized to properties of mappings $f: E \rightarrow E$ such that $f(C) \subset C$.

Also → 255 Linear Programming; for the †Kreĭn-Milman theorem → 407 Topological Linear Spaces.

References

[1] W. Blaschke, Kreis und Kugel, Verlag von Veit, 1916 (Chelsea, 1949).
[2] T. Bonnesen and W. Fenchel, Theorie der konvexen Körper, Erg. Math., Springer, 1934 (Chelsea, 1948).
[3] W. Fenchel, Convex cones, sets and functions, Lecture notes, Princeton Univ. Press, 1953.
[4] Convexity, Amer. Math. Soc. Proc. Symp. in Pure Math. VII (1963).
[5] М. Г. Креин, М. А. Рутман (M. G. Kreĭn and M. A. Rutman), Линейны операторы, оставляющие инвариантным конус в пространстве Банаха Uspehi Mat. Nauk (N.S.), 3, no. 1 (1948), 3–95; English translation, Linear operators leaving invariant a cone in a Banach space, Amer. Math. Soc. Transl., 26 (1950).
[6] T. Kubota, Kinsei kikagaku (Japanese; Modern geometries), Iwanami, 1947.
[7] T. Kubota, Rankeisen oyobi rankeimen ni kansuru bibun kikagaku (Japanese; Differential geometry of ovals and ovaloids), Iwanami Coll. of Math., 1934.
[8] H. Nikaido, Convex structures and economic theory, Academic Press, 1968.
[9] B. Grünbaum, Convex polytopes, Interscience, 1967.
[10] И. М. Яглом, В. Г. Болтянский (I. M. Jaglom and V. G. Boltjanskiĭ), Выпуклые фигуры, Гостехиздат 1951; English translation, Convex figures, Holt, Rinehart and Winston, 1961.

92 (VI.14) Coordinates

A. General Remarks

Suppose that we are given a Euclidean plane E^2 and two lines $X'X$ and $Y'Y$ in E^2 perpendicular to each other. Let O be the point of intersection of $X'X$ and $Y'Y$. We identify each of the straight lines $X'X$ and $Y'Y$ with the set of real numbers $\mathbf{R}$; the point O on each line is identified with zero. Let P be an arbitrary point in E^2. We draw lines PQ, PR parallel to $Y'Y$, $X'X$, where Q, R are in $X'X$, $Y'Y$, respectively. Let x and y be the real numbers corresponding to Q and R. Thus we obtain a mapping sending the point P to the ordered pair (x,y) of real numbers. This mapping gives a one-to-one correspondence between the points P of E^2 and the ordered pairs (x,y) of real numbers in $\mathbf{R}^2$. The numbers x and y are called the **coordinates** of P.

In general, given a set of mathematical objects, if we have machinery that assigns quantitative objects to each element of the set, then such machinery is called a **coordinate system** on the set, and the quantitative objects corresponding to each element are called its **coordinates**. In the previous example, the machinery is called a rectangular coordinate system. Coordinate systems are also useful in expressing quantitative concepts by geometric ones which are intuitively easier to grasp, e.g., diagrams of train schedules and †nomograms. †Map projection, †graphical calculation, †descriptive geometry, etc., may be considered as applications of the concept of coordinate systems.

In many cases, when we introduce a coordinate system in a space, it is determined uniquely by fixing a basic figure in the space. In the case of a rectangular coordinate system on a plane E^2, the basic figure consists of $X'X$ and $Y'Y$, which are called coordinate axes (the point O is called the origin). Sometimes it is convenient to consider real-valued functions f and g on $\mathbf{R}$ and a coordinate system on the plane E^2 determined by the function that sends an arbitrary point P to $(f(x),g(y))$, where (x,y) are the coordinates of P in the rectangular coordinate system. Logarithmic papers, †probability papers, and †stochastic papers (binomial probability papers), etc., are constructed in this way to fit their respective purposes.

In the branches of mathematics there are many varieties of coordinate systems. In this article we deal with frames and coordinates, curvilinear coordinates, and local coordinates.

B. Frames and Coordinates

Suppose that we are given a space M and a †transformation group G acting on M. It is desirable to introduce a coordinate system that best represents the geometric structure of M. Let G_* be a set of figures in M such that G acts †simply transitively on G_*. Each element of G_* is called a **frame**. Utilizing each frame as basic figure, we introduce a coordinate system that is "G-invariant" in the following sense: Let $R \in G_*$, $X \in M$, and $C_R(X)$ be the coordinates of X with R as basic figure. Then the coordinate system is G-invariant if $C_R(X) = C_{gR}(gX)$ for any element g in G. If we have such a coordinate system for each $R \in G_*$, then the expressions of geometric properties of M in terms of the coordinates are independent of the choice of frames.

(1) Projective Coordinates. Let M be an n-dimensional †projective space $\mathbf{P}^n$ over a field K, and let G be the †projective transformation group of $\mathbf{P}^n$. As a frame we can take the system of $n+1$ points $(A_0, A_1, \ldots, A_n)$ in general position. The †homogeneous coordinates of an arbitrary point $X \in \mathbf{P}^n$ are given by the $(n+1)$-tuple $(x_0, x_1, \ldots, x_n)$ satisfying the equation $X = \sum_{j=0}^n x_j A_j$, $x_j \in K$. They are called †projective coordinates. In fact, if $(x_0, x_1, \ldots, x_n) \neq (0, 0, \ldots, 0)$, then $(x_0, x_1, \ldots, x_n)$ and $(\lambda x_0, \lambda x_1, \ldots, \lambda x_n)$ $(\lambda \neq 0)$ represent the same point in $\mathbf{P}^n$. A †hyperplane π of $\mathbf{P}^n$ is expressed as the set of points whose coordinates $(x_0, x_1, \ldots, x_n)$ satisfy a linear homogeneous equation $\sum_{j=0}^n x_j u_j = 0$, $u_j \in K$. Therefore, the hyperplane π is represented by the homogeneous coordinates $(u_0, u_1, \ldots, u_n)$, called †hyperplane coordinates of π.

(2) Affine Coordinates. Let M be an n-dimensional †affine space E^n, and let G be the †group of affine transformations of E^n. As a frame we can take the system $(O; \mathbf{e}_1, \mathbf{e}_2, \ldots, \mathbf{e}_n)$, where O, called the origin, is a point in E^n, and the set of vectors $\{\mathbf{e}_i\}$ is a basis of the †standard vector space of E^n. Then †inhomogeneous coordinates of an arbitrary point $X \in E^n$ are given by the n-tuple $(x_1, x_2, \ldots, x_n)$, where $X = O + \sum_{i=1}^n x_i \mathbf{e}_i$. They are also called †affine coordinates of X in E^n. Sometimes we replace G above by the group of †equivalent affinity and consider the frames $(O; \mathbf{e}_1, \mathbf{e}_2, \ldots, \mathbf{e}_n)$ such that the volume of $[\mathbf{e}_1, \ldots, \mathbf{e}_n] = 1$.

Furthermore, if E^n has the structure of a †Euclidean space, we sometimes replace G by the †group of motions and consider a system of **rectangular coordinates** determined by an †orthogonal frame, that is, a frame $(O; \mathbf{e}_1, \mathbf{e}_2, \ldots, \mathbf{e}_n)$ such that the inner product $(\mathbf{e}_i, \mathbf{e}_j) = \delta_{ij}$, where δ is the †Kronecker delta. By contrast, the general affine coordinate system of a Euclidean space is called a system of **oblique coordinates**. In this case the inner products $(\mathbf{e}_i, \mathbf{e}_j) = g_{ij}$ are invariants of Euclidean geometry, and the distance ρ between two points (x_i) and (y_i) is given by $\rho = (\sum_{i,j=1}^n g_{ij}(y_i - x_i)(y_j - x_j))^{1/2}$. We sometimes consider an oblique coordinate system satisfying $(\mathbf{e}_i, \mathbf{e}_i) = 1$ $(i = 1, \ldots, n)$. In such cases the angle θ_{ij} between two basis vectors $\mathbf{e}_i$ and $\mathbf{e}_j$ is determined by $g_{ij} = \cos\theta_{ij}$.

(3) Barycentric Coordinates. In an n-dimensional affine space E^n, we take $n+1$ linearly independent points $A_0, A_1, \ldots, A_n$ and denote the position vectors from a point O to these points by $\mathbf{a}_0, \mathbf{a}_1, \ldots, \mathbf{a}_n$, respectively. Then for any point $X \in E^n$ there exists a unique set of numbers $(\lambda_0, \lambda_1, \ldots, \lambda_n)$ such that $X = O + \sum_{j=0}^n \lambda_j \mathbf{a}_j$, $\sum_{j=0}^n \lambda_j = 1$. We call these numbers **barycentric coordinates** of X in E^n. They are independent of the choice of the point O.

(4) Plücker Coordinates. Let $V(n,m)$ be the set of all m-dimensional subspaces in an n-dimensional projective space $\mathbf{P}^n$. Then $V(n,m)$ has the structure of a †Grassmann manifold. In order to introduce a coordinate system on $V(n,m)$, we fix a projective coordinate system on $\mathbf{P}^n$. An m-dimensional subspace $\pi \in$

$V(n,m)$ in $\mathbf{P}^n$ is spanned by $m+1$ independent points $B_0, B_1, \ldots, B_m \in \mathbf{P}^n$. We denote projective coordinates of these points by $(b_{0j}), (b_{1j}), \ldots, (b_{mj})$ and construct the following determinants:

$$p_{j_0 j_1 \ldots j_m} = \begin{vmatrix} b_{0j_0} b_{0j_1} \ldots b_{0j_m} \\ \cdots\cdots\cdots \\ b_{mj_0} b_{mj_1} \ldots b_{mj_m} \end{vmatrix}, \quad 0 \leqslant j_0, \ldots, j_m \leqslant n.$$

Then the subspace π can be represented by homogeneous coordinates $(\ldots, p_{j_0 j_1 \ldots j_m} \ldots)$. These coordinates are independent of the choice of $m+1$ points that span π and are called **Plücker coordinates** (or **Grassmann coordinates**) of π in $V(n,m)$. In these coordinates, the $p_{j_0 j_1 \ldots j_m}$ are alternating and satisfy the **Plücker relations**:

$$\sum_{k=0}^{m+1} (-1)^k p_{i_1 i_2 \ldots i_m j_k} p_{j_0 \ldots \hat{j}_k \ldots j_{m+1}} = 0,$$

where $\hat{j}_k$ means that j_k is removed. In particular, when $n=3$ and $m=1$, we have only one Plücker relation $Q: p_{01}p_{23} - p_{02}p_{13} + p_{03}p_{12} = 0$, which is a homogeneous equation of the second degree. In other words, the set $V(3,1)$ of all lines in a 3-dimensional projective space $\mathbf{P}^3$ is realized as a quadric surface Q in a 5-dimensional projective space $\mathbf{P}^5$ that has $(p_{01}, p_{02}, p_{03}, p_{12}, p_{13}, p_{23})$ as projective coordinates. Moreover, when $\mathbf{P}^3$ is a complex projective space, we put

$$p_{01} = \xi_0 + i\xi_3, \quad p_{02} = \xi_1 + i\xi_4, \quad p_{03} = \xi_2 + i\xi_5,$$
$$p_{23} = \xi_0 - i\xi_3, \quad p_{13} = -\xi_1 + i\xi_4, \quad p_{12} = \xi_2 - i\xi_5$$

and obtain a relation

$$\xi_0^2 + \xi_1^2 + \xi_2^2 + \xi_3^2 + \xi_4^2 + \xi_5^2 = 0$$

corresponding to the Plücker relation. Thus every line in $\mathbf{P}^3$ can be represented by homogeneous coordinates $(\xi_0, \xi_1, \ldots, \xi_5)$, which we call **Klein's line coordinates**.

(5) $(n+2)$**-Hyperspherical Coordinates.** Let $(x_0, x_1, \ldots, x_n, x_\infty)$ be projective coordinates in an $(n+1)$-dimensional real projective space $\mathbf{P}^{n+1}$. An n-dimensional †conformal space S^n is realized as a quadric hypersurface S^n in $\mathbf{P}^{n+1}: \sum_{i,j=1}^n g_{ij} x_i x_j - 2x_0 x_\infty = 0$, where (g_{ij}) is a positive definite symmetric matrix. A general point in $\mathbf{P}^{n+1}$ represents a †hypersphere of S^n. That is, a hypersphere represented by a point $X \in \mathbf{P}^{n+1}$ is realized as the intersection of S^n with the †polar hyperplane of X with respect to S^n; according as X lies outside of S^n, on S^n, or inside of S^n, it represents a real hypersphere, a point hypersphere, or an imaginary hypersphere. Therefore, any hypersphere of S^n in $\mathbf{P}^{n+1}$ is expressed by homogeneous coordinates $(x_0, x_1, \ldots, x_n, x_\infty)$, called $(n+2)$**-hyperspherical coordinates**. When $n=2$, they are called **tetracyclic coordinates**, and when $n=3$, **pentaspherical coordinates**. Therefore, if we restrict $(n+2)$-hyperspherical coordinates for points on S^n, then they satisfy the quadratic relation stated before. In the frame $(A_0, A_1, \ldots, A_n, A_\infty)$ of $\mathbf{P}^{n+1}$ which defines the $(n+2)$-hyperspherical coordinates, A_0 and A_∞ are points on S^n, and the other A_i are real hyperspheres passing through the points A_0 and A_∞. It is possible to choose a frame $(A_0, A_1, \ldots, A_n, A_\infty)$ such that the equation for S^n becomes $\sum_{i=1}^n x_i^2 - 2x_0 x_\infty = 0$ (i.e., $g_{ij} = \delta_{ij}$). Among hypersurfaces in S^n, one that is expressed by a homogeneous equation of the second degree with respect to $(n+2)$-hyperspherical coordinates is called a **cyclide**. It is an algebraic surface of the fourth order and is an enveloping surface of the family of hyperspheres that are tangent to n fixed hyperspheres.

(6) Moving Coordinates. When we study the differential geometry of an m-dimensional surface W in a space M on which a transformation group G acts, it is often preferable to take a frame or frames at each point of W and consider a †connection among them. These frames are called **moving frames**, and the set of coordinate systems with respect to moving frames is called a **moving coordinate system** (→ 114 Differential Geometry of Curves and Surfaces).

C. Curvilinear Coordinates

Let $(x_1, x_2, \ldots, x_n)$ be a rectangular coordinate system on an n-dimensional Euclidean space E^n. If $x_i = x_i(u_1, u_2, \ldots, u_n)$, $i = 1, \ldots, n$, are functions of n variables $(u_1, u_2, \ldots, u_n)$ of class C^r $(r \geqslant 1)$ and the †functional determinant $D(x_1, \ldots, x_n)/D(u_1, \ldots, u_n)$ is not equal to zero in some open domain, then $(u_1, u_2, \ldots, u_n)$ are considered local coordinates in E^n. We call them **curvilinear coordinates** of E^n. A hypersurface $u_i = \text{constant}$ (obtained by fixing the value of one of the variables u_i) is called a **coordinate hypersurface**, and a curve $u_j = \text{constant}$ $(j \neq i)$ is called a **coordinate curve**. The line element ds of a Euclidean space E^n is given by

$$ds^2 = \sum_{k=1}^n dx_k^2 = \sum_{i,j=1}^n g_{ij} du_i du_j,$$

$$g_{ij} = \sum_{k=1}^n \frac{\partial x_k}{\partial u_i} \frac{\partial x_k}{\partial u_j}.$$

Thus E^n is equipped with a †Riemannian metric. However, as E^n is †flat, its †curvature tensor satisfies $R^i_{jkl} = 0$. If the metric is diagonal, namely, if $ds^2 = \sum_{i=1}^n g_i du_i^2$, the coordinates are called **orthogonal curvilinear coordi-**

nates. Moreover, if $g_1 = \ldots = g_n$, the coordinates are called **isothermal coordinates**. The metric is diagonal if and only if coordinate curves are mutually perpendicular at the points of intersection. Actually, most curvilinear coordinate systems that are used practically are diagonal. The concept of curvilinear coordinates has been generalized to the case of †differentiable manifolds and is utilized to determine their local coordinates.

(1) Curvilinear Coordinates on Planes or Spaces (→ Appendix A, Table 3). Let (x,y) be rectangular coordinates of a point in a Euclidean plane E^2. The following various coordinate systems on E^2 are also used:

Polar coordinates: (r,θ), where

$$x = r\cos\theta, \qquad y = r\sin\theta.$$

Elliptic coordinates: (λ,μ), where

$$x^2 = (\lambda + a^2)(\mu + a^2)/(a^2 - b^2),$$
$$y^2 = (\lambda + b^2)(\mu + b^2)/(b^2 - a^2),$$
$$a > b > 0, \qquad \lambda > -b^2 > \mu > -a^2.$$

Parabolic coordinates: (α,β), where

$$x = -(\alpha + \beta), \qquad y = \sqrt{-4\alpha\beta}\,, \qquad \alpha > 0 > \beta.$$

Equilateral (or rectangular) hyperbolic coordinates: (u,v), where

$$x = uv, \qquad y = (u^2 - v^2)/2.$$

Bipolar coordinates: (ξ,η), where

$$x = a\sinh\xi/(\cosh\xi + \cos\eta),$$
$$y = a\sin\xi/(\cosh\xi + \cos\eta),$$
$$-\infty < \xi < \infty, \qquad 0 \leqslant \eta \leqslant 2\pi.$$

Next we consider the case of a 3-dimensional Euclidean space E^3 and let (x,y,z) be rectangular coordinates on E^3. The following systems of coordinates on E^3 are sometimes used:

Cylindrical coordinates: (r,θ,z), where

$$x = r\cos\theta, \qquad y = r\sin\theta, \qquad z = z.$$

Spherical coordinates: (r,θ,φ), where

$$x = r\sin\theta\cos\varphi, \quad y = r\sin\theta\sin\varphi, \quad z = r\cos\theta.$$

Ellipsoidal coordinates: (λ,μ,ν), where

$$x^2 = (\lambda + a^2)(\mu + a^2)(\nu + a^2)/(a^2 - b^2)(a^2 - c^2),$$
$$y^2 = (\lambda + b^2)(\mu + b^2)(\nu + b^2)/(b^2 - c^2)(b^2 - a^2),$$
$$z^2 = (\lambda + c^2)(\mu + c^2)(\nu + c^2)/(c^2 - a^2)(c^2 - b^2),$$
$$a > b > c > 0, \qquad \lambda > -c^2 > \mu > -b^2 > \nu > -a^2.$$

These coordinate systems are all called systems of **orthogonal curvilinear coordinates.** Suppose that we are given two rectangular coordinate systems (ξ,η,ζ) and (x,y,z) sharing the same origin. The correlation of the two is given by **Euler's angles** (θ,φ,ψ), where θ, φ, and ψ are the angles between the z-axis and ζ-axis, zx-plane and $z\zeta$-plane, and $\zeta\xi$-plane and ζz-plane, respectively. The Euler angles θ, φ, and ψ are subject to the inequalities $0 \leqslant \theta \leqslant \pi$ and $0 \leqslant \varphi, \psi < 2\pi$. They are often utilized in the dynamics of rigid bodies.

(2) Multipolar Coordinates. Let $P_1, P_2, \ldots, P_m$ be m points in general position in an n-dimensional Euclidean space E^n. If we denote by ρ_i ($\geqslant 0$) the distance between a point X of E^n and P_i, then $(\rho_1, \rho_2, \ldots, \rho_m)$ can be regarded as coordinates of a point X contained in a suitable domain of E^n. They are called **multipolar coordinates**. In particular, if $m=2$ they are called **bipolar coordinates**, and if $m=3$, **tripolar coordinates**. When $m>n$, these coordinates satisfy $m-n$ relations. Next let $\alpha_1, \alpha_2, \ldots, \alpha_m$ be m hyperplanes in general position in E^n. For an arbitrary point X of E^n, we denote by ξ_i the directed distance of X from each hyperplane α_i. The m-tuple $(\xi_1, \xi_2, \ldots, \xi_m)$ provides coordinates of X that are called **multiplanar coordinates** in E^n. When $m>n$, these coordinates satisfy $m-n$ relations. In particular, when $n=2$ and $m=3$, they are called **trilinear coordinates**. In this case, if we denote by S the area of the triangle defined by three lines $\alpha_1, \alpha_2, \alpha_3$, and by a_1, a_2, a_3 the lengths of the three sides of the triangle, then the trilinear coordinates (ξ_1, ξ_2, ξ_3) satisfy a linear relation $a_1\xi_1 + a_2\xi_2 + a_3\xi_3 = 2S$.

(3) Tangential Polar Coordinates. In a Euclidean plane E^2, we take a directed line l_0 passing through a point O. For an arbitrary directed line g, let p be the directed distance between O and g, and let θ be the angle between l_0 and g. Then (p,θ) are called **tangential polar coordinates** (Fig. 1). They are useful for representing tangent lines to curves in E^2. Let C be an †oval in E^2. A line is called a †supporting line of C if its intersection with C consists of a point or a line segment. In this case we take the origin O inside C and consider the coordinates (p,θ) of the supporting lines of C. Then the equation of C can be represented as $p = p(\theta)$, where $p(\theta)$ is a periodic function of period 2π. The coordinates (p,θ) are especially useful when the function

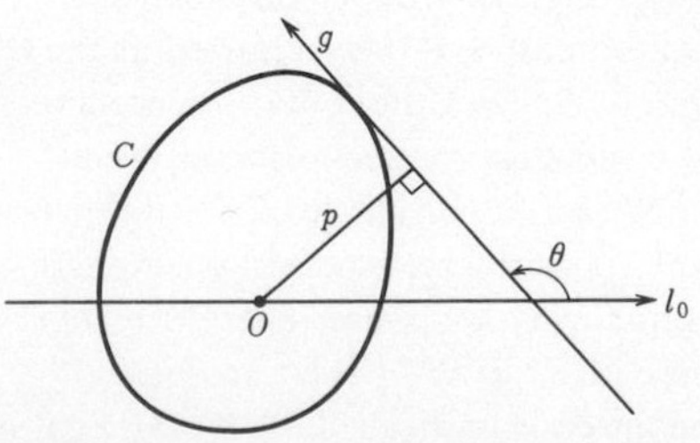

Fig. 1
Tangential polar coordinates.

$p(\theta)$ can be expanded in a †Fourier series. The notion of tangential polar coordinates is also used in the theory of †gears. In the case of Euclidean space E^3, the notion of tangential polar coordinate system can be defined by using tangent planes.

(4) Normal Coordinates. Let M be an n-dimensional †Riemannian manifold, and let $T_A(M)$ be the tangent space of M at a point A. For each tangent vector $\mathbf{v} \in T_A(M)$, we draw a †geodesic through A with the initial direction $\mathbf{v}$ and take a point P on the geodesic such that the distance from A to P is equal to the length of $\mathbf{v}$. Then the correspondence that sends $\mathbf{v}$ to P is a †diffeomorphism of a neighborhood of the zero vector 0 of $T_A(M)$ with a neighborhood of A in M. Therefore the components $(v^1, v^2, \ldots, v^n)$ of $\mathbf{v}$ with respect to a basis of $T_A(M)$ give the coordinates of the points P contained in a suitable neighborhood of A. We call them **normal coordinates** about the point A of M. In these coordinates, each geodesic passing through A is given by equations $v^i = \alpha^i r$ $(i = 1, 2, \ldots, n)$, where the (α^i) are components of the unit vector in the direction of $\mathbf{v}$ and r is the parameter that represents the arc length from A to the point $(v^1, \ldots, v^n)$. In particular, when $n = 2$, we fix a tangent vector $\mathbf{v}_0$ at A and denote the angle between $\mathbf{v}$ and $\mathbf{v}_0$ by θ. Then (r, θ) are coordinates of P called **geodesic polar coordinates**. The notion of normal coordinates can also be defined for †Lie groups or differentiable manifolds with †affine connections.

D. Local Coordinates

Suppose that we have a space M that has a covering by a family of open neighborhoods with coordinate systems. If, for each pair of neighborhoods with nonempty intersection, the coordinate transformation in the intersection satisfies certain specified conditions, then a mathematical structure on M can be defined.

Let E be a †topological space. Suppose that Φ is a family of open sets in E such that the union of any number of open sets in Φ and the intersection of any finite number of open sets in Φ also belong to Φ. A set Γ of †homeomorphisms is called a **pseudogroup of transformations** on E if Γ satisfies the following three conditions: (i) Any homeomorphism $f \in \Gamma$ is defined on an open set $U \in \Phi$, and $f(U) \in \Phi$. (ii) When an open set $U \in \Phi$ is expressed as the union of a family $\{U_i\}$ of open sets $U_i \in \Phi$, a homeomorphism f defined on U belongs to Γ if and only if its restriction to each U_i belongs to Γ. (iii) For any open set $U \in \Phi$, the identity mapping on U belongs to Γ, and if $f, g \in \Gamma$, then the inverse f^{-1} and the composition $g \circ f$, if it exists, belong to Γ.

Let E and M be topological spaces. A homeomorphism $\varphi: U \rightarrow V$ of an open set U in E to an open set V in M is called a **local coordinate system** of M with respect to E. For two local coordinate systems $\varphi_1: U_1 \rightarrow V_1$ and $\varphi_2: U_2 \rightarrow V_2$, the homeomorphism

$$\varphi_2^{-1} \circ \varphi_1 : \varphi_1^{-1}(V_1 \cap V_2) \rightarrow \varphi_2^{-1}(V_1 \cap V_2)$$

is called a **transformation of local coordinates**.

Let Γ be a pseudogroup of transformations on E. A set Σ of local coordinate systems of M with respect to E is said to define a **Γ-structure** on M if Σ satisfies the following two conditions: (1) the totality of the images of local coordinate systems belonging to Σ covers M; (2) if two local coordinate systems φ_1 and φ_2 of Σ have a transformation of local coordinates, it belongs to Γ. Now suppose that two sets Σ and Σ' of local coordinate systems define Γ-structures on M. If the union of Σ and Σ' defines a Γ-structure, then we say that the first two Γ-structures are equivalent. Let Γ be a pseudogroup of †diffeomorphisms, each defined from an open subset of the n-dimensional space $\mathbf{R}^n$ onto another open set. If a Γ-structure is defined on a space M, then M is an n-dimensional †differentiable manifold.

On the other hand, let Γ be a pseudogroup of complex analytic homeomorphisms in an n-dimensional complex number space $\mathbf{C}^n$. If a Γ-structure is defined on a space M, then M is an n-dimensional †complex analytic manifold. **Locally homogeneous spaces**, †foliated manifolds, and †fiber bundles are all equipped with local coordinate systems with suitable Γ-structures.

References

[1] E. Cartan, La méthode du repère mobile, la théorie des groupes continus et les espaces généralisés, Actualités Sci. Ind., Hermann, 1935.
[2] C. Ehresmann, Sur la théorie des espaces fibrés, Colloques Intern. Topologie algébrique, Paris, 1949, C. N. R. S. XII, p. 3–15.
[3] F. Klein, Vorlesungen über höhere Geometrie, Springer, third edition, 1926 (Chelsea, 1957).
[4] B. L. van der Waerden, Einführung in die algebraische Geometrie, Springer, 1939.
[5] O. Veblen and J. H. C. Whitehead, The foundations of differential geometry, Cambridge Univ. Press, 1932.

[6] Wu Wen-Tsün and G. Reeb, Sur les espaces fibrés et les variétés feuilletées, Actualités Sci. Ind., Hermann, 1952.

93 (IX.10) Covering Spaces

A. General Remarks

A continuous mapping $p:\tilde{Y}\rightarrow Y$ of an †arcwise connected topological space $\tilde{Y}$ onto a connected topological space Y is called a **covering mapping** if the following condition (C) is satisfied: (C) Each point of Y has an open neighborhood V such that every †connected component of $p^{-1}(V)$ is mapped homeomorphically onto V by p.

If there is a covering mapping $p:\tilde{Y}\rightarrow Y$, we call $\tilde{Y}$ a **covering space** of Y and $(\tilde{Y},p,Y)$ a **covering**. In particular, for a †differentiable manifold Y, if $\tilde{Y}$ is also a differentiable manifold and p is differentiable, then $\tilde{Y}$ is called a **covering (differentiable) manifold** of Y. (In the theory of †Riemann surfaces, a †covering surface may have some †branch points violating condition (C). Upon removing such points, we obtain a covering space as defined above.)

For each †path $w:I\rightarrow Y$ $(I=[0,1])$ of Y, a path $\tilde{w}:I\rightarrow\tilde{Y}$ with $p\tilde{w}=w$ is uniquely determined by the point $\tilde{w}(1)\in p^{-1}(w(1))$, and a bijection $w_{\#}:p^{-1}(w(1))\approx p^{-1}(w(0))$ is determined by $w_{\#}(\tilde{w}(1))=\tilde{w}(0)$. Thus there exists a one-to-one correspondence between $p^{-1}(y)$ and $p^{-1}(y')$ whenever points $y,y'\in Y$ can be connected by a path and $(\tilde{Y},p,Y,p^{-1}(y_0))$ is a †locally trivial fiber space with discrete fiber $p^{-1}(y_0)$. When the cardinal number of $p^{-1}(y)$ is a finite number n, we call $(\tilde{Y},p,Y)$ an ***n*-fold covering**. In this case, for a †loop $w(I,\dot{I})\rightarrow(Y,y_0)$ with base point $y_0, w_{\#}:p^{-1}(y_0)\approx p^{-1}(y_0)$ is a permutation of the n elements in $p^{-1}(y_0)$, and we obtain a homomorphism of the †fundamental group $\pi_1(Y)=\pi_1(Y,y_0)$ of Y into the †symmetric group $\mathfrak{S}_n$, given by the correspondence $w\rightarrow w_{\#}$. The permutation group $\mathfrak{M}$, which is the image of this homomorphism, is called the **monodromy group** of the n-fold covering.

Two coverings $(\tilde{Y}_i,p_i,Y)$ $(i=1,2)$ are said to be **equivalent** if there is a homeomorphism $\varphi:\tilde{Y}_1\approx\tilde{Y}_2$ with $p_2\circ\varphi=p_1$; such a φ is called an **equivalence**. In particular, a self-equivalence $\varphi:\tilde{Y}\rightarrow\tilde{Y}$ of a covering $(\tilde{Y},p,Y)$ is called a **covering transformation**. The set π of all covering transformations forms a group by the composition of mappings, which is called the **covering transformation group** of $\tilde{Y}$. We call $(\tilde{Y},p,Y)$ a **regular covering** if for each $y\in Y$ and $\tilde{y}_1,\tilde{y}_2\in p^{-1}(y)$, there exists a unique covering transformation that maps $\tilde{y}_1$ to $\tilde{y}_2$. In this case, the †orbit space $\tilde{Y}/\pi$ is homeomorphic to Y, $(\tilde{Y},p,Y,\pi)$ is a †principal bundle, and the monodromy group $\mathfrak{M}$ is isomorphic to π.

For a covering $(\tilde{Y},p,Y)$, we call $\tilde{Y}$ a **covering group** of Y if $\tilde{Y}$ and Y are topological groups and p is a homomorphism. Then $(\tilde{Y},p,Y)$ is a regular covering, and its covering transformation group is isomorphic to $p^{-1}(e)$ (e is the identity element of Y), which is a discrete subgroup lying in the center of $\tilde{Y}$ (→ 406 Topological Groups O).

B. Universal Covering Spaces

For a covering $(\tilde{Y},p,Y)$, we have the following relations of the †homotopy groups: $p_*:\pi_i(\tilde{Y})\rightarrow\pi_i(Y)$ is isomorphic (monomorphic) for $i\geqslant 2$ $(i=1)$, and $\pi_1(Y)/p_*(\pi_1(\tilde{Y}))$ is in one-to-one correspondence with $p^{-1}(y_0)$ $(y_0\in Y)$. If $\tilde{Y}$ is †simply connected, $\tilde{Y}$ is called a **universal covering space** of Y; if in addition $\tilde{Y}$ is a covering group, $\tilde{Y}$ is called a **universal covering group** of Y.

For a †locally arcwise connected space Y, a covering $(\tilde{Y},p,Y)$ is regular if and only if $p_*(\pi_1(\tilde{Y}))$ is a normal subgroup of $\pi_1(Y)$, and a universal covering space of Y is a covering space of any covering space of Y. Moreover, if Y is a topological group, any covering space $\tilde{Y}$ of Y can be given a unique topological group structure with which $\tilde{Y}$ is a covering group of Y.

Let Y be an arcwise connected, locally arcwise connected, and †locally simply connected space. Then the following classification theorem of coverings holds: The set of equivalence classes of coverings of Y is in one-to-one correspondence with the set of conjugate classes of subgroups of the fundamental group $\pi_1(Y)$; in particular, the equivalence class of a covering $(\tilde{Y},p,Y)$ corresponds to the conjugate class of the subgroup $p_*(\pi_1(\tilde{Y}))$. Also, there is a unique universal covering space $\tilde{Y}$ of Y up to homeomorphism. If in addition Y is a topological group, then $\tilde{Y}$ is a unique universal covering group of Y up to isomorphism of topological groups. Such a space $\tilde{Y}$ is obtained as follows: Consider the †path space $\Omega(Y;y_0,Y)$ of all paths in Y starting from a fixed point $y_0\in Y$, and define two paths $w_0,w_1:(I,0)\rightarrow(Y,y_0)$ to be equivalent if and only if there is a †homotopy $w_t:(I,0)\rightarrow(Y,y_0)$ with $w_t(1)=w_0(1)$ $(0\leqslant t\leqslant 1)$. Then we obtain the †identification space $\tilde{Y}$ of $\Omega(Y;y_0,Y)$ by this equivalence relation and the mapping $p:\tilde{Y}\rightarrow Y$ by $p\{w\}=w(1)$; this $\tilde{Y}$ is the universal covering space of Y.

Let $(\tilde{Y},p,Y)$ be a regular covering with the covering transformation group π. Then there is a †locally trivial fiber space $(Y',q,B,\tilde{Y})$ such that Y' has the same †(co) homology groups as Y, and B is an †Eilenberg-MacLane space $K(\pi,1)$. The (co) homology †spectral sequence of this fiber space is called that of the given regular covering $(\tilde{Y},p,Y)$, E_∞ is a bigraded module †associated with a certain filtration of the †singular (co) homology module $H(Y)$, and E_2 is the (co) homology module $H(\pi;H(\tilde{Y}))$ of the group π (π operates on the coefficient module $H(\tilde{Y})$ via the induced homomorphisms of covering transformations; → 156 Fiber Spaces).

For any group π, there is a regular covering (E_π,p,B_π) with the covering transformation group π such that E_π is †contractible; B_π is an Eilenberg-MacLane space $K(\pi,1)$. We can take S^1 (1-sphere) as B_z of the infinite cyclic group $\mathbf{Z}$, and the following infinite lens space as $B_{\mathbf{Z}_k}$ of the finite cyclic group $\mathbf{Z}_k$ (→ 137 Eilenberg-MacLane Complexes).

C. Lens Spaces

Let k be a positive integer and $l_1, \ldots, l_n$ be integers prime to k. Let $S^{2n+1}=\{(z_0,\ldots,z_n)\in \mathbf{C}^{n+1} \mid |z_0|^2+\ldots+|z_n|^2=1\}$ be the unit sphere in the $(n+1)$-dimensional complex linear space $\mathbf{C}^{n+1}$, and define the rotation γ by $\gamma(z_0,z_1,\ldots,z_n)=(z_0\exp 2\pi i/k, z_1\exp 2\pi l_1 i/k, \ldots, z_n \exp 2\pi l_n i/k)$. Then the orbit space $S^{2n+1}/\mathbf{Z}_k=L(k;l_1,\ldots,l_n)$, where $\mathbf{Z}_k=\mathbf{Z}/k\mathbf{Z}$ is interpreted as the cyclic group generated by γ, is called a **lens space**. It is an orientable $(2n+1)$-dimensional †differentiable manifold. Also, the **infinite lens space** $L^\infty(k)=L(k;1,\ldots,1,\ldots)$ is defined by taking $n=\infty$; the infinite sphere S^∞ is a k-fold covering space of $L^\infty(k)$, and $L^\infty(k)=B_{\mathbf{Z}_k}=K(\mathbf{Z}_k,1)$. Its †cohomology ring is given as follows:

(1) For integral coefficients, $H^{2i+1}(L^\infty(k))=0$, $H^{2i}(L^\infty(k))=\mathbf{Z}_k$ $(i>0)$, and the †cup product of generators of degree $2i$ and $2j$ is a generator of degree $2(i+j)$.

(2) Let $k=pk'$ (p is a prime). If $p\neq 2$, or $p=2$ and k' is even, $H^*(L^\infty(k);\mathbf{Z}_p)=\wedge(e_1)\otimes\mathbf{Z}_p[e_2]$. If $p=2$ and k' is odd, $H^*(L^\infty(k);\mathbf{Z}_2)=\mathbf{Z}_2[e_1]$ (e_i is an element of degree i). Here $\wedge$ indicates the †exterior algebra over $\mathbf{Z}_p$, and $\mathbf{Z}_p[\ ]$ the †polynomial ring over $\mathbf{Z}_p$.

If two lens spaces $L(k;l_1,\ldots,l_n)$ and $L(k';l'_1,\ldots,l'_n)$ are homeomorphic, then $k=k'$ and there is an integer m prime to k with

$$l_1\ldots l_n \equiv \pm m^{n+1} l'_1 \ldots l'_n \pmod{k}$$

[4]. For $n=1$, this condition holds if and only if $L(k;l_1)$ and $L(k';l'_1)$ have the same †homotopy type [5]. Also, the condition $k=k'$, $l\equiv\pm l'^{\pm 1} \pmod{k}$ holds if and only if $L(k;l)$ and $L(k';l')$ are homeomorphic. (Sufficiency is shown in [1]; necessity follows from the fact that the †Hauptvermutung is valid for combinatorial 3-manifolds and that the condition holds if the polyhedra $L(k;l)$ and $L(k;l')$ have isomorphic subdivisions [6].)

References

[1] H. Seifert and W. Threlfall, Lehrbuch der Topologie, Teubner, 1934 (Chelsea, 1965).
[2] C. Chevalley, Theory of Lie groups I, Princeton Univ. Press, 1946.
[3] S.-T. Hu, Homotopy theory, Academic Press, 1959.
[4] S. Eilenberg and S. MacLane, Homology of spaces with operators II, Trans. Amer. Math. Soc., 65 (1949), 49–99.
[5] J. H. C. Whitehead, On incidence matrices, nuclei and homotopy types, Ann. of Math., (2) 42 (1941), 1197–1239.
[6] K. Reidemeister, Homotopieringe und Linsenräume, Abh. Math. Sem. Univ. Hamburg, 11 (1935), 102–109.
[7] A. Komatu, M. Nakaoka, and M. Sugawara, Isôkikagaku (Japanese; Topology) I, Iwanami, 1967.
[8] W. S. Massey, Algebraic topology; An introduction, Harcourt, Brace & World, 1967.

94 (IV.16) Crystallographic Groups

A. General Remarks

Let $\mathfrak{G}$ be a †discrete subgroup of the †group of motions $\mathfrak{B}$ in an n-dimensional Euclidean space $\mathbf{R}^n$. If $\mathfrak{G}$ contains exactly n linearly independent translations, then $\mathfrak{G}$ is called an n-dimensional **crystallographic group**. Let $\mathfrak{A}$ be the subgroup of $\mathfrak{G}$ consisting of all the translations in $\mathfrak{G}$. Then $\mathfrak{A}$ is a commutative †normal subgroup generated by n translations, say $\mathfrak{a}_1,\ldots,\mathfrak{a}_n$, and is called the **lattice group** of $\mathfrak{G}$. Take a point $\mathfrak{x}$ of $\mathbf{R}^n$. Then the †$\mathfrak{A}$-orbit of the point $\mathfrak{x}$ is called a **lattice** of $\mathfrak{G}$ (or $\mathfrak{A}$).

For $\mathfrak{A}=\langle\mathfrak{a}_1,\ldots,\mathfrak{a}_n\rangle$, we set $(\mathfrak{a}_i,\mathfrak{a}_j)=a_{ij}$, where $(\mathfrak{a}_i,\mathfrak{a}_j)$ is the †inner product of $\mathfrak{a}_i$ and $\mathfrak{a}_j$. The $n\times n$ matrix $A=(a_{ij})$ is †symmetric, and the †quadratic form $(A\mathfrak{x},\mathfrak{x})=\sum a_{ij}x_ix_j$ is †positive definite. The a_{ij} are called the **lattice constants** of $\mathfrak{A}$.

Two crystallographic groups $\mathfrak{G}_1$ and $\mathfrak{G}_2$, or two lattice groups $\mathfrak{A}_1$ and $\mathfrak{A}_2$, are called **equivalent** (written $\mathfrak{G}_1\sim\mathfrak{G}_2$, $\mathfrak{A}_1\sim\mathfrak{A}_2$) when they are †conjugate in $\mathfrak{B}$. Now let $\mathfrak{G}$ be a crystallographic group. For two points $\mathfrak{x}$, $\mathfrak{y}$ of $\mathbf{R}^n$, denote the translation which brings $\mathfrak{x}$ to $\mathfrak{y}$ by $\tau_{\mathfrak{x},\mathfrak{y}}$. The set $\mathfrak{G}^{\mathfrak{x}}$ of all elements of $\mathfrak{B}$ of the form $\tau_{\mathfrak{x},\sigma\mathfrak{x}}^{-1}\cdot\sigma$, where σ runs over the elements of $\mathfrak{G}$, forms a subgroup of the †stabilizer $\mathfrak{B}_{\mathfrak{x}}$ of $\mathfrak{x}$ in $\mathfrak{B}: \mathfrak{B}_{\mathfrak{x}} = \{\rho \in \mathfrak{B} \mid \rho(\mathfrak{x}) = \mathfrak{x}\}$. Two crystallographic groups $\mathfrak{G}_1$ and $\mathfrak{G}_2$ are said to belong to the same **crystal class** (written $\mathfrak{G}_1 \approx \mathfrak{G}_2$) if $\mathfrak{G}_1^{\mathfrak{x}}$ and $\mathfrak{G}_2^{\mathfrak{x}}$ are conjugate in $\mathfrak{B}_{\mathfrak{x}}$. This definition does not depend on the choice of $\mathfrak{x}$. The relations $\sim, \approx$ are †equivalence relations, and from $\mathfrak{G}_1 \approx \mathfrak{G}_2$ it follows that $\mathfrak{G}_1 \sim \mathfrak{G}_2$. The crystal class to which $\mathfrak{G}$ belongs is sometimes called the **point group** of $\mathfrak{G}$. (In fact, $\mathfrak{G}$ determines $\mathfrak{G}^{\mathfrak{x}}$ as an abstract group that is a subgroup of the †orthogonal group $O(n)$.) For a given dimension n, there are only a finite number of crystal classes (L. Bieberbach).

B. Crystallographic Groups of 3-Dimensional Space

Three-dimensional crystallographic groups, called **space groups**, are used in the application of crystallography. Their lattices are called **space lattices** (or **Bravais lattices**). We usually use symbols $\mathfrak{a}$, $\mathfrak{b}$, $\mathfrak{c}$ instead of $\mathfrak{a}_1$, $\mathfrak{a}_2$, $\mathfrak{a}_3$ to denote a system of generators of the lattice group $\mathfrak{A}$ and set

$$a = |\mathfrak{a}| = \sqrt{a_{11}}\ , b = |\mathfrak{b}| = \sqrt{a_{22}}\ , c = |\mathfrak{c}| = \sqrt{a_{33}}\ ,$$

$$\alpha = \angle(\mathfrak{b},\mathfrak{c}) = \arccos\left(a_{23}/\sqrt{a_{22}a_{33}}\right),$$

$$\beta = \angle(\mathfrak{c},\mathfrak{a}) = \arccos\left(a_{31}/\sqrt{a_{33}a_{11}}\right),$$

$$\gamma = \angle(\mathfrak{a},\mathfrak{b}) = \arccos\left(a_{12}/\sqrt{a_{11}a_{22}}\right).$$

The numbers a, b, c and the angles α, β, γ are often used instead of the a_{ij} $(i,j = 1,2,3)$ and are also called the **lattice constants** of $\mathfrak{A}$. Let a', b', c', α', β', and γ' be the lattice constants of $\mathfrak{A}'$. If $a:b:c = a':b':c'$, $\alpha = \alpha'$, $\beta = \beta'$, and $\gamma = \gamma'$, then $\mathfrak{A} \sim \mathfrak{A}'$.

Space lattices are classified into seven **crystal systems** by these lattice constants as follows (C. Weiss). **Cubic system**: $a:b:c = 1:1:1$, $\alpha = \beta = \gamma = 90°$. **Tetragonal system**: $a:b:c = 1:1:x$ $(x \neq 1)$, $\alpha = \beta = \gamma = 90°$. **Rhombic system**: $a:b:c = 1:x:y$ $(x \neq 1, y \neq 1, x \neq y)$, $\alpha = \beta = \gamma = 90°$. **Monoclinic system**: $a:b:c = 1:x:y$ $(x \neq 1, y \neq 1, x \neq y)$, $\alpha = \gamma = 90°$, $\beta \neq 90°$. **Hexagonal system**: $a:b:c = 1:1:x$, $\alpha = \beta = 90°$, $\gamma = 120°$. **Rhombohedral system**: $a:b:c = 1:1:1$, $\alpha = \beta = \gamma \neq 90°$. **Triclinic system**: otherwise. Two space lattices belonging to different crystal systems are not equivalent.

When a lattice group $\mathfrak{A}$ with generators $\{\mathfrak{a},\mathfrak{b},\mathfrak{c}\}$ is given, the space lattices of the groups $\langle\mathfrak{a},\mathfrak{b},(\mathfrak{b}+\mathfrak{c})/2\rangle$, $\langle(\mathfrak{c}+\mathfrak{a})/2,\mathfrak{b},\mathfrak{c}\rangle$, $\langle\mathfrak{a},(\mathfrak{b}+\mathfrak{c})/2,\mathfrak{c}\rangle$ and $\langle(\mathfrak{a}+\mathfrak{b}+\mathfrak{c})/2,\mathfrak{b},\mathfrak{c}\rangle$ are respectively called the A-, B-, C-, and **I-lattices** determined by $\mathfrak{A}$. These lattices are known as **double lattices**. The first three lattices are known as **base-centered lattices**, and the I-lattice is called the **body-centered lattice**. The lattices of the group $\langle(\mathfrak{c}+\mathfrak{a})/2, (\mathfrak{a}+\mathfrak{b})/2, (\mathfrak{b}+\mathfrak{c})/2\rangle$ are called **F-lattices (face-centered lattices** or **quadruple lattices**) determined by $\mathfrak{A}$. The lattices of $\mathfrak{A}$ itself are called **P-lattices (simple lattices** or **primitive lattices**) (Fig. 1).

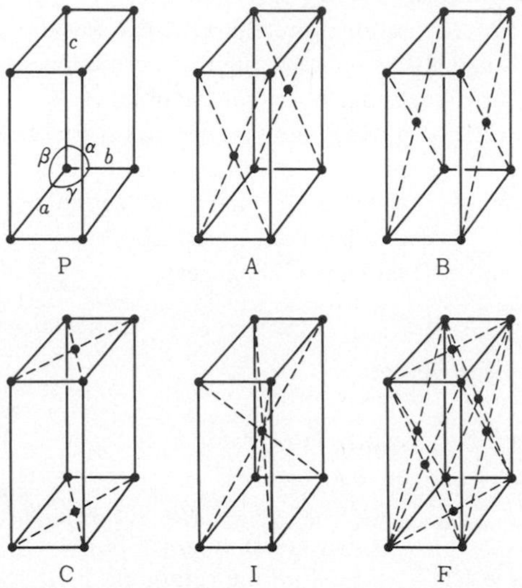

Fig. 1

The following 14 kinds of space lattices are constructed as previously outlined from space lattices belonging to seven crystal systems (A. Bravais): triclinic P; monoclinic P, C (or I); rhombic P, A (or B, C), I, F; tetragonal P, I; cubic P, I, F; hexagonal P; rhombohedral P. If two space lattices are equivalent to each other, they are of the same kind in this classification, but the converse is not always true.

Using the results of G. Eisenstein [2] on ternary quadratic forms, P. Niggli [5] classified space lattices into 42 kinds. It was proved by Eisenstein that, given a lattice L, there exists a unique lattice that is equivalent to L and satisfies the following two conditions. (1) The case $a_{23}, a_{31}, a_{12} > 0$ (i.e., $\alpha, \beta, \gamma < 90°$). The main condition: $a_{11} \leqslant a_{22} \leqslant a_{33}$, $2a_{23} \leqslant a_{22}$, $2a_{31} \leqslant a_{11}$, $2a_{12} \leqslant a_{11}$. The secondary condition: If $a_{11} = a_{22}$, then $a_{23} \leqslant a_{31}$; if $a_{22} = a_{33}$, then $a_{31} \leqslant a_{12}$; if $2a_{23} = a_{22}$, then $a_{12} \leqslant 2a_{31}$; if $2a_{31} = a_{11}$, then $a_{12} \leqslant 2a_{23}$; if $2a_{12} = a_{11}$, then $a_{31} \leqslant 2a_{23}$. (2) The case a_{23}, a_{31}, $a_{12} \leqslant 0$ (i.e., $\alpha, \beta, \gamma \geqslant 90°$). The main condition: $a_{11} \leqslant a_{22} \leqslant a_{33}$, $2|a_{23}| \leqslant a_{22}$, $2|a_{31}| \leqslant a_{11}$, $2|a_{12}| \leqslant a_{11}$, $0 \leqslant a_{11} + a_{22} + 2(a_{23} + a_{31} + a_{12})$. The secondary condition: If $a_{11} = a_{22}$, then $|a_{23}| \leqslant |a_{31}|$; if $a_{22} = a_{33}$, then $|a_{31}| \leqslant |a_{12}|$; if

$2|a_{23}|=a_{22}$, then $a_{12}=0$; if $2|a_{31}|=a_{11}$, then $a_{12}=0$; if $2|a_{12}|=a_{11}$, then $a_{31}=0$; if $a_{11}+a_{22}+2(a_{23}+a_{31}+a_{12})=0$, then $a_{11}\leqslant 2|a_{31}|+|a_{12}|$. Conditions (1) and (2) are called **Eisenstein conditions of reduction**, and a lattice satisfying one of them is said to be **reduced**. By the **reduction** of a given lattice, we mean the determination of the reduced lattice equivalent to the given lattice. For example, there exists a unique reduced lattice equivalent to each of the *P*-, *I*-, and *F*-lattices of the cubic system. These reduced lattices are given as follows (Fig. 2; the thick lines in the diagrams denote $\mathfrak{a}$, $\mathfrak{b}$, $\mathfrak{c}$ of the reduced lattices). *P*: $a_{11}=a_{22}=a_{33}=a^2$, $a_{23}=a_{31}=a_{12}=0$; *I*: $a_{11}=a_{22}=a_{33}=3a^2/4$, $a_{23}=a_{31}=a_{12}=-a^2/4$; *F*: $a_{11}=a_{22}=a_{33}=a^2/2$, $a_{23}=a_{31}=a_{12}=a^2/4$.

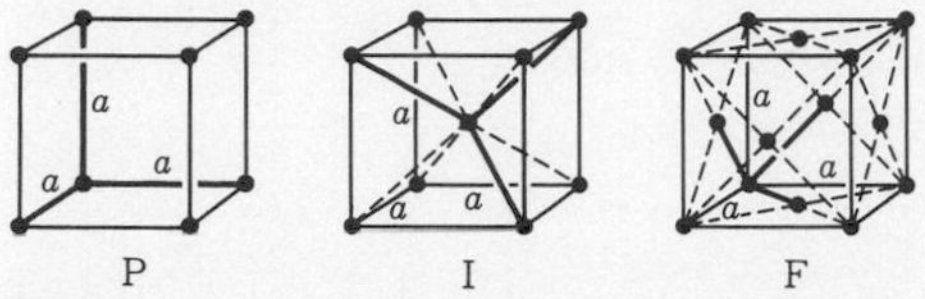

Fig. 2

To reduce a space lattice, we sometimes use geometric intuition, although this "geometric" method is not necessarily simple because the conditions for reduction are complicated. An algebraic method was given by P. Bachmann [4], but this too is quite complicated. A simpler method of reduction was considered by E. Selling [3], and B. Delaunay gave its graphical description [6]. But the reduced lattice was not uniquely determined by these methods. B. W. Jones refined Selling's idea and obtained a method of determining a unique reduced lattice [7].

The method of Selling and Delaunay is as follows: Given lattice constants a_{ij} $(i,j=1,2,3)$ of a lattice group $\mathfrak{A}$, we set $a_{14}=-a_{11}-a_{12}-a_{13}$, $a_{24}=-a_{21}-a_{22}-a_{23}$, $a_{34}=-a_{31}-a_{32}-a_{33}$. If a_{23}, a_{31}, a_{12}, a_{14}, a_{24}, a_{34} are all $\leqslant 0$, then we say that $\mathfrak{A}$ is reduced. Suppose that $\mathfrak{A}$ is not reduced. For example, suppose that $a_{23}>0$. Then we set $a'_{23}=-a_{23}$, $a'_{13}=a_{13}+a_{23}$, $a'_{12}=a_{24}+a_{23}$, $a'_{14}=a_{14}-a_{23}$, $a'_{24}=a_{12}+a_{23}$, $a'_{34}=a_{34}+a_{23}$, and if a'_{23}, a'_{13}, a'_{12}, a'_{14}, a'_{24}, a'_{34} are all $\leqslant 0$, then the lattice group $\mathfrak{A}'$ with the lattice constants $a'_{11}=-a'_{12}-a'_{13}-a'_{14}$, $a'_{22}=-a'_{21}-a'_{23}-a'_{24}$, $a'_{33}=-a'_{31}-a'_{32}-a'_{34}$, a'_{23}, a'_{13}, a'_{12} is reduced and equivalent to $\mathfrak{A}$. If one of a'_{23}, a'_{13}, a'_{12}, a'_{14}, a'_{24}, a'_{34} is positive, then by successive applications of the same process, we can reduce $\mathfrak{A}$. This process is indicated by Delaunay's diagram in Fig. 3.

As in the left-hand side of Fig. 3, we write down the lattice constants of $\mathfrak{A}$ and also a_{14}, a_{24}, a_{34} at the vertices and edges of a tetrahedron; if $a_{23}>0$, then the edge on which a_{23}

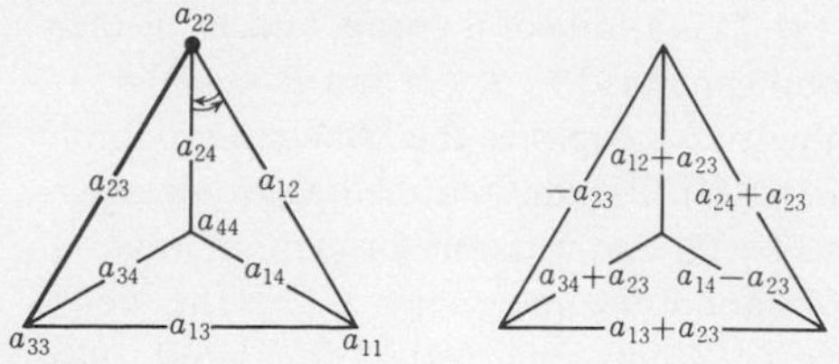

Fig. 3

lies is drawn as a bold line and a solid circle will be added at one of its ends (a_{22} in the figure). In order to pass from left to right in Fig. 3, we first change the sign of the number on the bold line and add the number (a_{23}) to all numbers on the edges except for the edge opposite the bold line. We subtract a_{23} from the number on the opposite edge. Finally we interchange the two numbers on the edges on which the vertex with the solid circle lies as in Fig. 3. Thus we obtain the numbers a'_{23}, a'_{13}, a'_{12}, a'_{14}, a'_{24}, a'_{34}. If we apply this reduction to the *P*-, *I*- and *F*-lattices of the cubic system, we obtain the diagrams in Fig. 4, respectively.

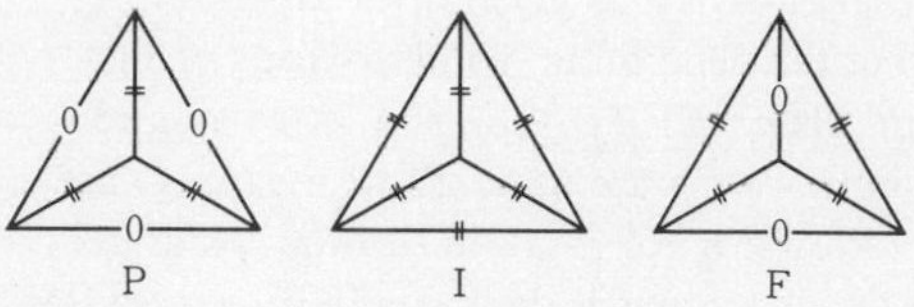

Fig. 4

Delaunay classified space lattices into 24 kinds by this method of reduction.

There exist 32 3-dimensional crystal classes, each of which is generated by at most three elements of the orthogonal group $O(3)$ of degree 3. In 1831, J. F. C. Hessel first proved the existence of these 32 crystal classes, and A. Bravais (1848), A. Gadolin (1871), and P. Curie (1884) systematized the theory. A. Schoenflies (1891) gave its group-theoretic description.

To name these crystal classes, both Schoenflies's notation and the international notation are used. **Schoenflies's notation** consists of the letters C, D, S, T, O, V with one or two subscripts. For subscripts we use the numbers n $(=1,2,3,4,6)$ and the letters i, s, h, v, d. Let (x,y,z) be an orthogonal coordinate system of $\mathbf{R}^3$ such that the x-axis is parallel to $\mathfrak{a}$ and the xy-plane contains $\mathfrak{b}$. We denote rotation through an angle $2\pi/n$ around the z-axis by $R_n(z)$, reflection in the xy-plane (i.e., transformation $(x,y,z)\rightarrow(x,y,-z)$) by $S(z)$, and inversion with respect to the origin (i.e., $(x,y,z)\rightarrow(-x,-y,-z)$) by I. The meanings of these symbols are: C_n is the †cyclic group of order n generated by

$R_n(z)$. D is a †dihedral group, and D_n is the group generated by $R_n(z)$ and $R_2(x)$. $V = D_2$ is the four-group (i.e., the †Abelian group of type $(2,2)$). T is the †tetrahedral group generated by D_2 and rotation through an angle $2\pi/3$ about the line $x=y=z$. O is the †octahedral group generated by $R_4(z)$ and rotation through an angle $2\pi/3$ about the line $x=y=z$. S is used only once in S_4, which is a cyclic group of order 4 generated by $S(z)R_4(z) = IR_4(z)$. The subscript i means inversion and is used in C_2 and C_{3i}, which denote the groups generated by I and $IR_3(z)$, respectively. The subscript s is used only once in C_s and means †reflection. C_s is the group generated by $S(z)$. The subscript h means horizontal; for example, C_{2h} is the group generated by C_2 and $S(z)$. The subscript v means vertical; for example, C_{2v} is the group generated by C_2 and $S(x)$ (or $S(y)$). The subscript d means diagonal and is used in D_{3d}, D_{2d}, and T_d. D_{nd} is the group generated by D_n and reflection in the plane $y/x = \tan\pi/n$, and T_d is the group generated by T and reflection in the plane $y=x$.

In the **international notation**, a crystal class is expressed by its generators. In order to name the generators, combinations of the symbols n $(=1,2,3,4,6)$, $\bar{n}$, a letter m, and a symbol n/m are used. Their meanings are as follows: n is a rotation through an angle $2\pi/n$ whose axis is determined by the crystal system and the position in which the number n appears in the notation; for example, if n appears in the first position, then the axis is the z-axis. Also, $\bar{n}$ is IR, where R is the rotation expressed by n, the notation m is an abbreviation of "mirror" and denotes a reflection and in the first position denotes $S(z)$, but in the second or third position denotes a reflection in a plane perpendicular to the axes determined by a rule described below. Finally, n/m denotes SR, where R is a rotation expressed by n (its angle is $2\pi/n$ and its axis is determined according to the position of n) and S is reflection in a plane perpendicular to the axis of R. The following is the rule concerning the axis of the rotation corresponding to the position of these symbols: The second and third symbols do not appear in the triclinic and monoclinic systems. In the rhombic system, the second and third axes are the x-axis and the y-axis, respectively. In the tetragonal, rhombohedral, and hexagonal systems, the second and third axes are the x-axis and the line $x=y$, $z=0$, respectively. In the cubic system, the second and third axes are the line $x=y=z$ and the line $x=y$, $z=0$, respectively.

With this rule, the 32 crystal classes are as given in Table 1. There exist 320 space

Table 1. 32 Crystal Classes

	Schoenflies's Notation	International Notation
Triclinic system	C_1	1
	C_i	$\bar{1}$
Monoclinic system	C_s	m
	C_2	2
	C_{2h}	2/m
Rhombic system	C_{2v}	2mm
	D_2	222
	D_{2h}	2/m 2/m 2/m
Hexagonal system	C_{3h}	$\bar{6}$
	C_6	6
	C_{6h}	6/m
	D_{3h}	$\bar{6}$2m
	C_{6v}	6mm
	D_6	622
	D_{6h}	6/m 2/m 2/m
Tetragonal system	S_4	$\bar{4}$
	C_4	4
	C_{4h}	4/m
	D_{2d}	$\bar{4}$2m
	C_{4v}	4mm
	D_4	422
	D_{4h}	4/m 2/m 2/m
Rhombohedral system	C_3	3
	C_{3i}	$\bar{3}$
	C_{3v}	3m
	D_3	32
	D_{3d}	$\bar{3}$2/m
Cubic system	T	23
	T_h	2/m$\bar{3}$
	T_d	$\bar{4}$3m
	O	432
	O_h	4/m $\bar{3}$2/m

groups, of which a table is given in [13, 15, 16, 17]. They were enumerated by E. von Fedorov (1885), A. Schoenflies (1891), and W. Barlow (1894) independently under the influence of P. Curie. In the description of space groups by international notation in [13] and [17], the crystal system to which the space group belongs is written first, then the space group is expressed by notations such as P_m, C_m, P_2, P_{2_1}. Here P denotes a simple lattice of its crystal system, C denotes the base-centered lattice, and the subscripts m, 2 mean reflection and rotation by an angle π as in the case of crystal classes. For example, P_2 in the tetragonal system is the space group generated by $\mathfrak{A} = \langle \mathfrak{a}, \mathfrak{b}, \mathfrak{c} \rangle$ and $R_2(z)$. Similarly, P_{2_1} in the tetragonal system is the space group generated by $\mathfrak{A} = \langle \mathfrak{a}, \mathfrak{b}, \mathfrak{c} \rangle$ and the motion composed of the translation $\mathfrak{c}/2$ and $R_2(z)$ (in the tetragonal system the subscript n_k in P_{n_k} generally denotes a motion composed of the translation $k\mathfrak{c}/n$ and $R_n(z)$ (a so-called n-fold screw with pitch k/n)). The meanings of the other notations may be understood similarly.

All 32 crystal classes as abstract groups are of well-known types whose †characters have been calculated.

References

[1] A. Bravais, Abhandlung über die Systeme von regelmässig auf einer Ebene oder Raum

verteilten Punkten, Ostwald's Klassiker exak. Wissenschaften, no. 90, Engelmann, 1897.
[2] G. Eisenstein, Tabelle der reducirten positiven ternären quadratischen Formen, nebst den Resultaten neuer Forschungen über diese Formen, in besonderer Rücksicht auf ihre tabellarische Berechnung, J. Reine Angew. Math., 41 (1851), 141–190.
[3] E. Selling, Des formes quadratiques binaires et ternaires, J. Math. Pures Appl., (3) 3 (1877), 21–60, 153–207.
[4] P. Bachmann, Die Arithmetik der quadratischen Formen, Teubner, 1923, 1925.
[5] P. Niggli, Kristallographische und strukturtheoretische Grundbegriffe, Handb. exp. Physik VII 1, Akademische Verlag., 1928.
[6] B. Delaunay, Neue Darstellung der geometrischen Kristallographie I, Z. Krist., 84 (1932), 109–149.
[7] B. W. Jones, A table of Eisenstein-reduced positive ternary quadratic forms of determinant $\leqslant 200$, Bull. Nat. Res. Council, 97 (1935), 1–51.
[8] B. W. Jones, On Selling's method of reduction for positive ternary quadratic forms, Amer. J. Math., 54 (1932), 14–34.
[9] T. Ito, Seisûron to kessyôgaku (Japanese; Number theory and crystallography), Iwanami Coll. of Math., 1935.
[10] T. Ito, X-ray study on polymorphism, Tokyo, 1950.
[11] A. M. Schoenflies, Kristallsysteme und Kristallstruktur, Teubner, 1891.
[12] H. Hilton, Mathematical crystallography and the theory of groups of movements, Clarendon, 1903.
[13] Internationale Tabellen zur Bestimmung von Kristallstrukturen I, 1935.
[14] J. J. Burckhardt, Die Bewegungsgruppen der Kristallographie, Basel, 1947.
[15] R. W. G. Wyckoff, The analytical expression of the results of the theory of space groups, Carnegie Institute of Washington, second edition, 1930.
[16] P. Niggli, Geometrische Kristallographie des Diskontinuums, Gebrüder Borntraeger, 1919.
[17] International tables for X-ray crystallography I, Birmingham, 1952.
[18] A. Speiser, Die Theorie der Gruppen von endlicher Ordnung, Springer, third edition, 1937.
[19] D. Hilbert and S. Cohn-Vossen, Anschauliche Geometrie, Springer, 1932; English translation, Geometry and the imagination, Chelsea, 1952.
[20] G. Frobenius, Gruppentheoretische Ableitung der 32 Kristallklassen, S.-B. Preuss. Akad. Wiss., (1911), 681–691, (Gesammelte Abhandlungen, Springer, 1968, vol. 3, p. 519–529.)
[21] F. D. Murnaghan, The theory of group representations, Johns Hopkins Press, 1938.
[22] Е. Н. Белова, Н. В. Белов, А. В. Шубников (E. N. Belova, N. V. Belov, and A. V. Šubnikov), О числе и составе абстрактных, отвечающих 32 кристаллографическим классам, Dokl. Akad. Nauk SSSR, 63 (1948), 669–672.

95 (XV.16) Curve Fitting

A. Curve Fitting

The technique of **curve fitting** is utilized to find a simple curve $y=f(x)$ supplying the best possible approximation to values $y_1, y_2, \ldots$ (usually given by observations or experiments) for some discrete values $x_1, x_2, \ldots$ of the independent variable x. This technique is commonly used to obtain empirical formulas from given experimental data and also for functional approximation and smoothing data. The method of †interpolation is used to construct a polynomial passing through all the given points. However, since all experimental data contain errors, we usually construct a curve by methods such as the †method of least squares, in which the curve need not pass through all the points (x_ν, y_ν).

B. Empirical Formulas

An **empirical formula** contrasts with a **theoretical formula**, deduced theoretically for the construction of a function $y=f(x)$ expressing a law of phenomena. An empirical formula is not only utilized to obtain a simple description of experimental results, but also helps in the discovery of laws. Usually, to obtain an empirical formula we first choose a suitable function that contains several parameters called **empirical constants**, and then we specialize the parameters to fit the experimental data. In general, the accuracy of the result is expected to increase with the number of parameters. However, a formula including too many parameters may be complicated or impractical for representing the essential laws of the phenomena. To choose the initial function, it is customary to estimate it by drawing a graph. Sometimes, by observing various known curves, we try to find a curve that seems to be closest to the data. A suitable change of variables is often utilized in order to reduce a curve to a simpler one, e.g., a straight line.

Here we give typical examples of functions along with changes of the variables (in brackets) to transform the graphs into straight lines.

(1) $y=1/(a+bx)$ $[x,1/y]$;

(2) $y=a+bx^2$ $[x^2,y]$;

(3) $y=a+b/x$ $[1/x,y]$;

(4) $y=ax+bx^2$ $[y/x,x]$;

(5) $y=ab^x$ $[x,\log y]$;

(6) $y=ax^b$ $[\log x,\log y]$;

(7) $(x+a)(y+b)=c$ $\left[x-x_1,(x-x_1)/(y-y_1)\right]$;

(8) $y=a\exp(b/(x+c))$ $\left[\log(y/y_1)/(x-x_1),\log(y/y_1)\right]$;

(9) $y=\varphi((x-m)/\sigma)$ $[x,t]$ (where $y=\varphi(t)=(2\pi)^{-1/2}\times\int_{-\infty}^{t}\exp(-t^2/2)\,dt$);

(10) $y=\varphi((\log x-m)/\sigma)$ $[\log x,t]$ (where $y=\varphi(t)$ given in (9)).

Semilogarithmic paper, **logarithmic paper**, and **probability paper** are special graph papers facilitating transformations (1)–(9) in cases (5), (6), and (9), respectively.

Aside from linear formulas, polynomials are most frequently used, sometimes after applying suitable transformations such as (1)–(9). Usually the degree of such polynomials is at most 5. If the discrete arguments of x have equal increments, it is sometimes possible to estimate the necessary degree of such polynomials by observing the †differences Δy, $\Delta^2 y,\ldots$. If $\Delta^n y$ is constant, the degree of such a polynomial is n. Let y_i be given for discrete values x_i with constant differences. If the point $(y_{i+1}/y_i, y_{i+2}/y_i)$ lies on the straight line $\eta=M\xi+B$, then we have the following three propositions: (i) When $M>0$, $B<0$, and $M^2+4B>0$, then $y=ae^{\lambda x}+be^{\mu x}$; (ii) when $M>0$ and $M^2+4B=0$, we have $y=(a+bx)e^{\lambda x}$; (iii) when $M^2+4B<0$, we have $y=e^{\lambda x}(a\cos\omega x+b\sin\omega x)$. For periodic phenomena, it is customary to apply partial sums of †Fourier series $y=a_0+\Sigma(a_n\cos nx+b_n\sin nx)$ (→ 194 Harmonic Analysis). To analyze an S-shaped curve appearing in growth or decay, it is customary to apply the †logistic curve $y=L/(1+e^{-at})$ or $y=A(\tanh\lambda t+1)$.

For determination of empirical constants the following methods may be used: graphing (especially simple when the graph is a straight line); selection of certain points (whose number is equal to the number of the parameters); the mean value method; or the method of least squares. Usually, a theoretical estimation of the laws accompanies these methods of determination.

C. Orthogonal Polynomials

Suppose that we are given the values $y_0,y_1,\ldots,y_{n-1}$ for discrete values $x=0,1,\ldots,n-1$ with equal differences. A polynomial $f(x)$ of degree k ($\leqslant n$) that minimizes the sum of the squares $G_k(n)=\Sigma(y_m-f(m))^2$ is given by

$$f(x)=\sum_{\nu=0}^{k}a_\nu q_\nu(n,x),$$

$$a_\nu=\sum_{m=0}^{n-1}y_m q_\nu(n,m)/S_\nu(n),$$

$$S_\nu(n)=\sum_{m=0}^{n-1}(q_\nu(n,m))^2,$$

and we have $G_k(n)=\Sigma_{m=0}^{n-1}y_m^2-\Sigma_{\nu=0}^{k}a_\nu^2 S_\nu(n)$. Here the function $q_\nu(n,x)$, called a **Čebyšev** **q-function**, is defined by

$$q_\nu(n,x)=\sum_{m=0}^{\nu}\binom{\nu+m}{m}\binom{\nu-n}{\nu-m}\binom{x}{m}.$$

In practical applications, it is better to replace $q_\nu(n,x)$ by the function

$q_\nu^*(n,x)=q_\nu(n,x)/2^{-\nu}\nu!\,M_\nu(n)$,

where $M_\nu(n)$ is the greatest common divisor of

$\binom{\nu+m}{m}\binom{n-1-m}{\nu-m}$ $(m=0,1,\ldots,\nu)$.

The values $q_\nu^*(n,m)$ $(m=0,1,\ldots,n-1)$ are mutually coprime integers. The function $q_\nu^*(n,m)$ is called the **simplest Čebyšev q-function** or **simplest orthogonal polynomial**, and is sometimes denoted by $X_{n,m}(x)$, $\xi'_{n,m}(x)$, or $\varphi_{n,m}(x)$.

When two functions satisfy the condition

$$(f,g)_n=\sum_{m=0}^{n}f(m)g(m)=0,$$

they are called **orthogonal for a finite sum**. If from $1,x,x^2,\ldots,x^\nu$ we make a system orthogonal with respect to this definition, then we have the polynomial

$$P_{\nu,n}(x)=\sum_{m=0}^{\nu}(-1)^m\binom{\nu}{m}\binom{\nu+m}{m}\binom{x}{m}\Big/\binom{n}{m},$$

which has the following connection with q_ν:

$q_\nu(n,x)$
$=((-1)^\nu(n-1)!/2^\nu(n-\nu-1)!)P_{\nu,n-1}(x)$.

The polynomial $P_{\nu,n}(x)$ is called a **Čebyšev orthogonal polynomial** or sometimes simply an **orthogonal polynomial**. The reader should not confuse these with polynomials in an †orthogonal system of functions (→ 312 Orthogonal Functions). The polynomials $P_{\nu,n}(x)$ are orthogonal for the finite sum

$$\sum_{m=0}^{n} P_{\nu,n}(m)P_{\mu,n}(m)=\delta_{\mu\nu}\frac{(\nu+n+1)!(n-\nu)!}{(2\nu+1)(n!)^2}.$$

If $n\to\infty$ and $0\leqslant x\leqslant 1$, the function $P_{\nu,n}(x)$ tends to $P_\nu(1-2x)$, where P_ν is the †Legendre polynomial (→ Appendix A, Table 20. VII).

There are similar investigations on algorithms for approximating polynomials when the arguments are unequal intervals, and on the minimax approximation that minimizes $\max|y_m-f(x_m)|$.

References

[1] M. Masuyama, Zikkenkôsiki no motomekata (Japanese; How to obtain experimental formulas), Takeuti Syoten, 1962.
[2] T. R. Running, Empirical formulas, John Wiley, 1917.
[3] P. G. Guest, Numerical methods of curve fitting, Cambridge Univ. Press, 1961.

96 (VI.22) Curves

A. Introduction

In the beginning of his *Elements*, Euclid gave definitions such as: A line is a length having no width; an end of a line is a point. However, he left notions such as width and length undefined. Thus his definitions were far from satisfactory. Actually, it was only during the latter half of the 19th century that efforts were made to obtain exact definitions of lines and curves. Euclid, among others, distinguished two kinds of curves: **straight lines** and **curves**. Nowadays, however, **lines** in the sense of Euclid are called curves, and a straight line is considered a curve. A first effort to give an exact definition of a curve using analytic methods was made by C. Jordan in his *Cours d'analyse* I (1893).

B. Jordan Arcs and Jordan Curves

Following Jordan, we define a **continuous plane curve** C to be the image of a †continuous mapping sending the interval [0, 1] into the Euclidean plane E^2. Namely, C is the set of points (x,y) in E^2 such that

$$x=f(t)\quad y=g(t),\quad 0\leqslant t\leqslant 1$$

with continuous functions f,g defined on [0, 1]. A continuous curve is also called a **continuous arc**. We call $(f(0),g(0))$ and $(f(1), g(1))$ the **ends** of the arc. Given continuous functions f,g defined on (0, 1), the set $\{(x,y)|x=f(t),y=g(t),0<t<1\}$ is called an **open arc**. More generally, the image of a continuous mapping of [0, 1], (0, 1), [0, 1), or (0, 1] is called an **arc** (or **curve**). Suppose that C is an arc that is the image of an interval I and $P=(x,y)$ is a point on C to which there correspond two elements t_1,t_2 $(t_1<t_2)$ of I such that P is the image of both t_1,t_2. In this case, the point P is called a **multiple point** on C. An arc having no multiple point is called a **simple arc** (or **Jordan arc**).

An arc with one and only one multiple point $P=(f(0),g(0))=(f(1),g(1))$ is called a **Jordan curve** (or **simple closed curve**) (→ 81 Connectedness D). A Jordan curve can be regarded as a topological image in a plane of a circle. Let C be a curve that is the image $\varphi(I)$ of an interval. Then C is said to be of **class C^k (analytic)** if the mapping φ is of †class C^k (†analytic). In general, if S is a topological space, then the image $\varphi(I)$ in S of an interval is called a **curve** in S. In particular, if S has the structure of a differentiable (analytic) manifold, we can define the notion of **curve of class C^k (analytic curve)** in S.

C. Ordinary Curves

A †connected subset of E^2 that is the union of a finite number of simple arcs meeting at a finite number of points is called an **ordinary curve**. An ordinary curve is called a tree if it does not contain a subset that is homeomorphic to a Jordan curve. Let p be a point on an ordinary curve C. The †boundary of a sufficiently small †neighborhood of p meets C at a finite number of points, and this number is independent of the choice of the small neighborhood. We call it the **order** of p in C. A point of order 1 is called an **endpoint** of C, a point of order 2 an **ordinary point**, and a point of order $\geqslant 3$ a **branch point**. If we can represent an ordinary curve C as a continuous curve passing each simple arc of C just once, we say that C is **unicursal** (Fig. 1). A necessary and sufficient condition for C to be unicursal is that the number of points of odd orders in C is less than or equal to 2 (L. Euler).

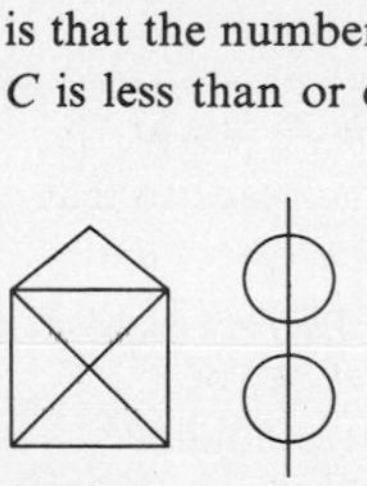

Fig. 1

D. Further Considerations on Definitions

Although the set of ordinary curves as defined above contains various familiar curves, it does not contain a point set defined by $y=\sin 1/x$ in $0<x\leqslant 1$ and $-1\leqslant x<0$, and by $-1\leqslant y\leqslant 1$ at $x=0$ (Fig. 2). This point set is called a **sinusoid**, and it is desirable to obtain a definition of curves wide enough to contain a sinusoid. On the other hand, the notion of continuous curves is, in a sense, too wide, because a curve such as a †Peano curve, which covers a whole square, is among such curves. The notion of simple arcs is too narrow, because even a circle is not a simple arc. As a point set in E^2, a continuous arc is characterized as a †locally connected †continuum and is sometimes called a **Peano continuum** (H. Hahn, S. Mazurkiewicz). On the other hand, A. Schoenflies, inspired by the statement of the †Jordan curve theorem, considered a closed set that divides the plane into two parts, forming the common boundary of both domains, and called it a closed curve. According to this definition, however, a simple curve is not a closed curve. Thus as a general definition of curves it is not appropriate.

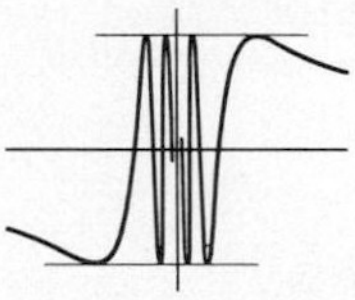

Fig. 2

To give a general notion of curves on a plane (containing sinusoids), we may define a curve as a continuum that is †nowhere dense in E^2 (i.e., a continuum that is a boundary of open sets on the plane) (G. Cantor). Furthermore, to deal with the curves on a topological space, P. S. Uryson and K. Menger defined a **general curve** to be a 1-dimensional continuum (Menger, 1921–1922 [1]). In E^2, the latter notion coincides with the notion of curves defined by Cantor, while in E^3 a general curve is a continuum that does not divide any †domain (→ 81 Connectedness).

E. Universal Curves

Consider a 3-dimensional cube I^3 ($I=[0,1]$). Draw two planes parallel to each face so that the two planes parallel to a face trisect the edges of the cube meeting the planes. Thus I^3 is divided into 3^3 cubes. Let M_1 be the closure of the subset of I^3 that is obtained from I^3 by deleting the cube I_1^3 ($I_1=[1/3,2/3]$) and the 6 cubes having common faces with I_1^3. Then M_1 consists of 20 cubes (Fig. 3). We apply to each of the 20 cubes forming M_1 the same operation that we applied to I^3, and denote by M_2 the union of point sets thus obtained (consisting of 20^2 cubes the length of whose edges equals $1/3^2$). Repeating this process, we obtain a point set M_n consisting of 20^n cubes the length of whose edges equals $1/3^n$. Thus we obtain the series $M_1\supset M_2\supset M_3\supset\ldots$. The set $U=\bigcap_{n=1}^{\infty}M_n$ is a general curve in the sense of Uryson-Menger. Moreover, we can prove that an arbitrary general curve is homeomorphic to a subset of U. Hence we call U the **universal curve**.

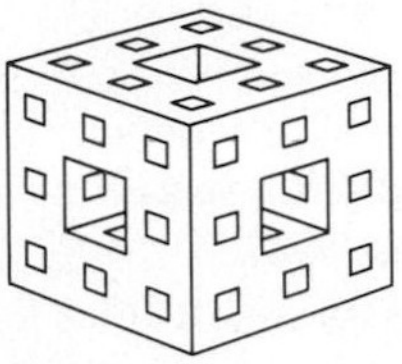

Fig. 3

F. Length of a Curve

In this section, by a curve we mean a continuous curve in a Euclidean space E^n. Let C be a curve in E^n defined by $x_i=f_i(t)$ ($i=1,2,\ldots,n$; $a\leqslant t\leqslant b$; the f_i are real-valued continuous functions defined in $[a,b]$). (We sometimes write this simply as $X=\mathfrak{f}(i)$, where $X=(x_1,\ldots,x_n)$.) We divide $[a,b]$ arbitrarily and denote the dividing points by $a=t_0<t_1<t_2<\ldots<t_r=b$. Let $X_k=\mathfrak{f}(t_k)$, $k=0,\ldots,r$, and let $\overline{X_{k-1}X_k}$ be the length of the straight line segment joining X_{k-1} and X_k. If the length $l=\sum_{k=1}^{r}\overline{X_{k-1}X_k}$ of the broken line $(X_0X_1\ldots X_r)$ (Fig. 4) is bounded for any subdivision of $[a,b]$, C is called a **rectifiable curve**, and the upper limit of l with respect to the subdivisions is called the **length** of C. For C to be rectifiable, it is necessary and sufficient that the f_i ($i=1,2,\ldots,n$) be of †bounded variation (Jordan). Thus if C is a rectifiable curve, then any of the $f_i(t)$ ($i=1,2,\ldots,n$) is almost everywhere differentiable (H. Lebesgue). In particular, if C is of class C^1, then C is rectifi-

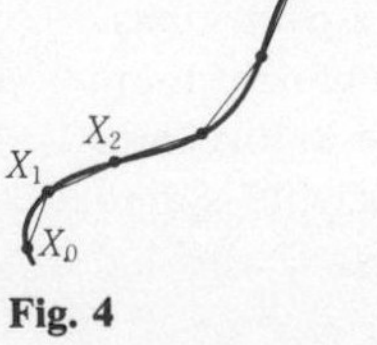

Fig. 4

able, and its length can be represented by

$$s=\int_a^b\left(\sum_{i=1}^{n}\left(\frac{df_i}{dt}\right)^2\right)^{1/2}dt$$

(→ 245 Length and Area).

G. Shapes of Curves

In this section by a curve we mean the image $\varphi(I)$ in a Euclidean space E^n of an interval I (bounded or unbounded), where φ is a continuous mapping. When a curve C of class C^k is given in a Euclidean space E^n, it often becomes necessary to examine the shape of C globally. The determination of the global shape of C from the equation of the curve is called the **curve tracing** of C. The problem has been thoroughly studied in the particular case in which $n=2$ and the equation of C is given by $f(x,y)=0$ in a rectangular coordinate system (by $F(r,\theta)=0$ in a †polar coordinate system), where f (F) is an analytic function. The problems in the case of a rectangular coordinate system are as follows:

Let φ be a single-valued analytic function and I a (bounded or unbounded) interval on the x-axis. If a curve C_0 represented by $y=\varphi(x)$ $(x\in I)$ is a subset of C, then C_0 is called a **branch** of C. According as I is bounded or unbounded, C_0 is said to be a **finite branch** or **infinite branch** of C. When a curve represented by $x=\psi(y)$ $(y\in J)$ (ψ is a single-valued analytic function and J is an interval on the y-axis) is a subset of C, it is also called a branch of C. If it is necessary to distinguish these two branches, we call the former the *x-branch* and the latter the *y-branch*. C consists of an at most denumerable set of branches. If $P(x_0,y_0)\in C$ and $f_y=\partial f/\partial y\neq 0$ at P, then there exists an x-branch containing P; if $f_x=\partial f/\partial x\neq 0$, there exists a y-branch of C. If $\partial f/\partial x=0, \partial f/\partial y=0$ at P, then P is called a **singular point** of C. Points on C that are not singular points are called **ordinary points** of C.

When P is an ordinary point of C, a branch C_0 of C containing P is determined, and the **tangent line** and **normal line** to C at P are the same as those to C_0 and are uniquely determined. The equations of these are $(x-x_0)f_x(x_0,y_0)+(y-y_0)f_y(x_0,y_0)=0$ and $(x-x_0)f_y(x_0,y_0)-(y-y_0)f_x(x_0,y_0)=0$, respectively. If we choose a coordinate system with P as origin and the tangent line and normal line as ξ-axis and η-axis, respectively, then the equation of C with respect to this coordinate system is of the form $\eta=c_2\xi^2+c_3\xi^3+\ldots$ in the neighborhood of P. If we denote by ρ the †curvature of C at P, then $\rho=2c_2=-(f_{xx}f_y^2-2f_{xy}f_xf_y+f_{yy}f_x^2)/(f_x^2+f_y^2)^{3/2}$. When $c_2=\rho=0$, P is a †stationary point. A stationary point of a curve of class C^2 on C is also called a **point of inflection**. When P is not a point of inflection and (ξ,η) are points on C in a neighborhood V of P, the sign of η is definite if V is small enough (Fig. 5). However, if P is a point of inflection and $c_3\neq 0$, then C is of the shape shown in Fig. 6. At a point of inflection, if $c_3=\ldots=c_{\nu-1}=0$, $c_\nu\neq 0$, and ν is even, C is of the shape shown in Fig. 5, and if ν is odd, C is of the shape shown in Fig. 6.

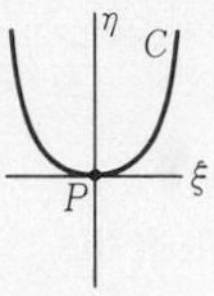

Fig. 5

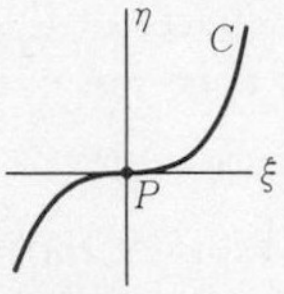

Fig. 6

In a neighborhood of a singular point, C takes various shapes. For example, consider a curve represented by $y^2=x^2(x+a)$, and let P be the origin $(0,0)$. If $a>0$, then there are two branches of C passing through P, and they have different tangents at P (Fig. 7). As in this case, if there are a finite number of different branches passing through P with different tangents, P is called a **node** of C. If $a<0$, then $P\in C$, but there is no other point of C in the neighborhood of P (Fig. 8). Such a point is called an **isolated point** of C. If $a=0$, then there are two branches of C starting from P, and the tangents to these at P are the same (Fig. 9). Such a point is called a **cusp** of C. When C is an †algebraic curve, we can examine the shape of a curve in a neighborhood of a singular point using the †Puiseux series.

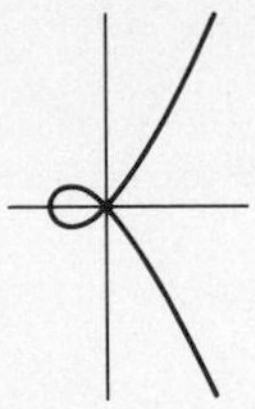
Fig. 7

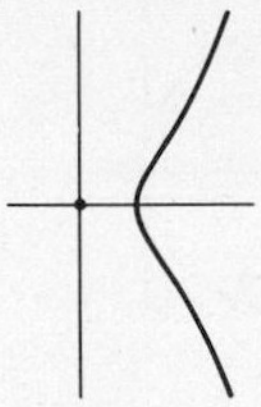
Fig. 8

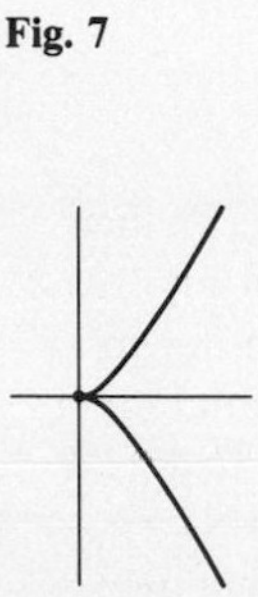
Fig. 9

When C has an infinite branch C_0 (for example, when $C_0: y=\varphi(x)$ $(x \in I)$ is an x-branch of C and $I=[a,\infty)$), if the tangent to C_0 at $P(x_0,y_0)$ $(x_0 \in I)$ has a limiting line for $x_0 \rightarrow \infty$, then the limiting line l is called an **asymptote** of C_0. In this case, the distance from a point $P(x_0,y_0)$ of C_0 to l converges to zero when $x_0 \rightarrow \infty$. An asymptote of an infinite branch of C is also called an asymptote of C.

H. Special Plane Curves

The following are the well-known curves.

Among curves of the third order, those having an equation of the form

$$y^2=f(x)/(x-a) \qquad (1)$$

($f(x)$ is a rational expression of at most the third order in x) are symmetric with respect to the x-axis and have $x=a$ as an asymptote. In particular, if $a>0, f(x)=-x^3$ in (1), then the curve is as shown in Fig. 10 and the origin is a cusp. Let a half-line starting from the origin meet the curve, the circle with diameter $[0,a]$, and the straight line $x=a$ at points X, Y, and A, respectively. This curve is called a **cissoid of Diocles**.

If $a=0, f(x)=c^2(c-x)$ $(c>0)$ in (1), then the curve takes the shape shown in Fig. 11. Let A, C be the points whose coordinates are $(a,0)$, $(c,0)$ $(0<a<c)$, respectively, and let X, Y be the points in the first quadrant at which the straight line parallel to the y-axis and passing through A meets the curve and the circumference with diameter OC, respectively (Fig. 11). Then we have $AX:AY=OC:OA$. This curve is called a **witch of Agnesi**.

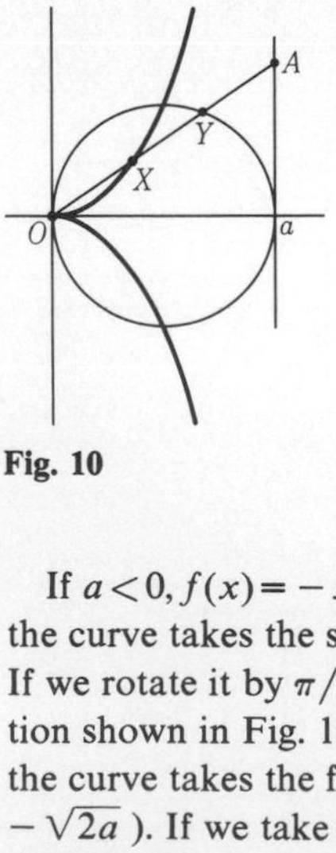

Fig. 10

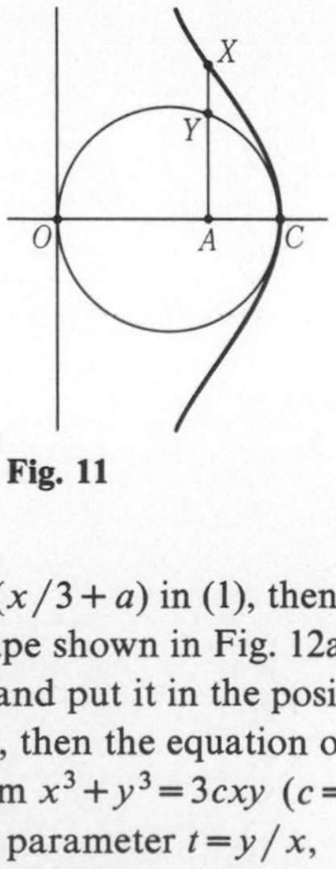

Fig. 11

If $a<0, f(x)=-x^2(x/3+a)$ in (1), then the curve takes the shape shown in Fig. 12a. If we rotate it by $\pi/4$ and put it in the position shown in Fig. 12b, then the equation of the curve takes the form $x^3+y^3=3cxy$ $(c=-\sqrt{2a}\,)$. If we take as parameter $t=y/x$, then we get the parametric representation $x=3ct/(1+t^3), y=3ct^2/(1+t^3)$. This curve is called a **folium cartesii** (or **folium of Descartes**). A curve that has a parametric representation of the form $x=\varphi(t), y=\psi(t)$, where φ, ψ are rational functions, is called an (algebraic) **unicursal curve** (or **rational curve**). Such a curve is an algebraic curve of †genus 0.

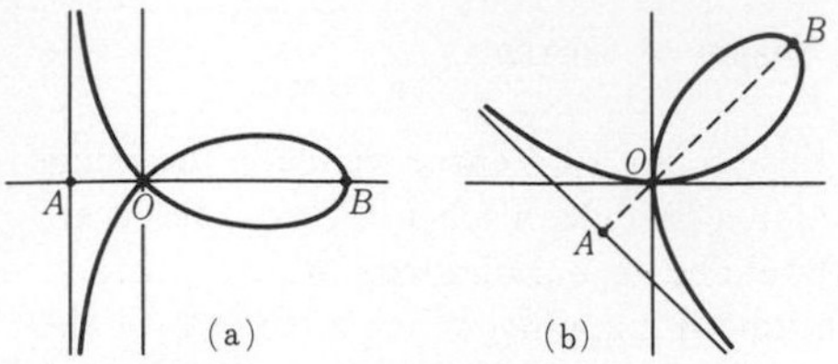

Fig. 12

Let $r=f_1(\theta), r=f_2(\theta), \ldots, r=f_k(\theta)$ be equations of curves $C_1, C_2, \ldots, C_k$ with respect to a polar coordinate system with origin O. A curve C having equation $r=\lambda_1 f_1(\theta)+\ldots+\lambda_k f_k(\theta)$ (the λ_i are constants, usually $+1$ or -1) in the same coordinate system is called a **cissoidal curve** with respect to O. (Fig. 13: $r=-f_1(\theta)+f_2(\theta)$). In Fig. 10, let C_1 be the circumference of the circle with the diameter $[0,a]$, let C_2 be the straight line $x=a$, and put $\lambda_1=-1, \lambda_2=1$. Then we have a cissoid of Diocles. We can regard the folium cartesii as a cissoidal curve obtained from a straight line and an ellipse. When C_1 is a circle with center at O, we call C a **conchoidal curve** of C_2 with respect to O. In particular, when C_2 is a straight line and O is not on C_2, the conchoidal curve is called a **conchoid of Nicomedes**.

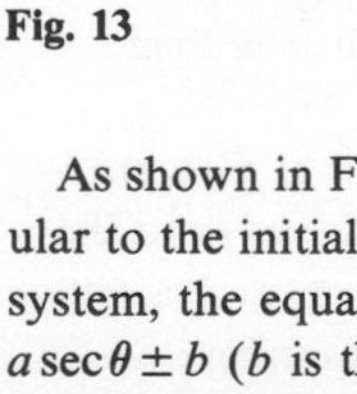

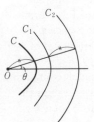

Fig. 13

As shown in Fig. 14, when C_2 is perpendicular to the initial line of the polar coordinate system, the equation of the curve is $r=a\sec\theta \pm b$ (b is the radius of C_1), and the rectangular equation of the curve is $(x-a)^2(x^2+y^2)=b^2x^2$. According as $a>b$, $a=b$, or $a<b$, the curve has a node, cusp, or isolated point, respectively. When C_2 is a circle and O is on C_2, the conchoidal curve of C_2 with respect to O is called a **limaçon** (or **limaçon of Pascal**) (Fig. 15). The equation of a limaçon C with respect to a polar coordinate system having the diameter of a circle passing through O as its initial line is $r=a\cos\theta \pm b$, while the equation of C with respect to a

rectangular coordinate system is $(x^2+y^2-ax)^2=b^2(x^2+y^2)$. In this case, if $a>b$, O is a node of the curve; if $a=b$, O is a cusp. In the case where $a=b$, the curve is called a **cardioid** (the curve shown by a dotted line in Fig. 15).

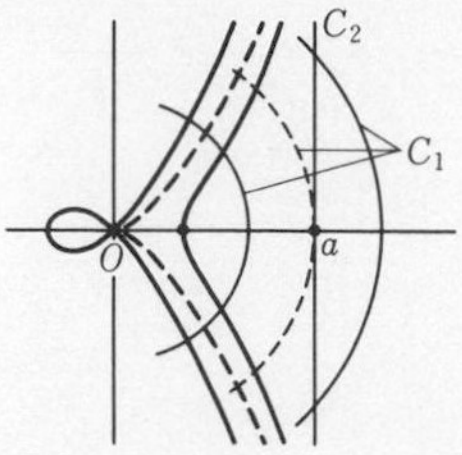

Fig. 14

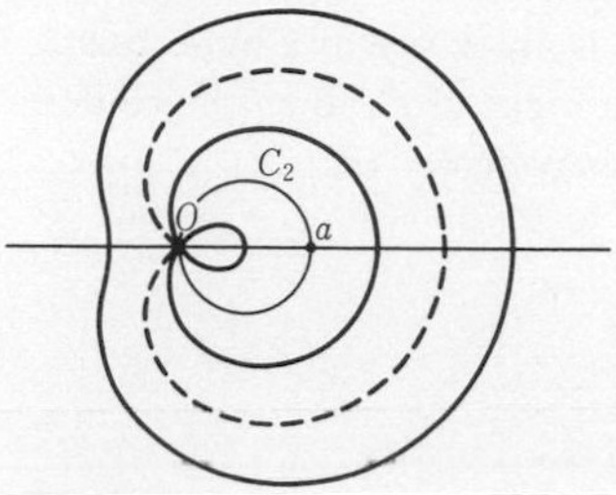

Fig. 15

The locus of a point X having a constant product of its distances from two fixed points A, B is called **Cassini's oval** (Fig. 16). The equation of this curve with respect to the rectangular coordinate system whose origin O is the midpoint of the segment AB and whose x-axis is the straight line AB is $(x^2+y^2)^2-2a^2(x^2-y^2)=k^4-a^4$ (where $AB=2a$, $k^2=AX\cdot BX$). In particular, if $a^2=k^2$, then O is a nodal point of the curve. In this case, the curve (shown by the dotted line in Fig. 16) is called a **lemniscate** (or **Bernoulli's lemniscate**) (Jakob Bernoulli).

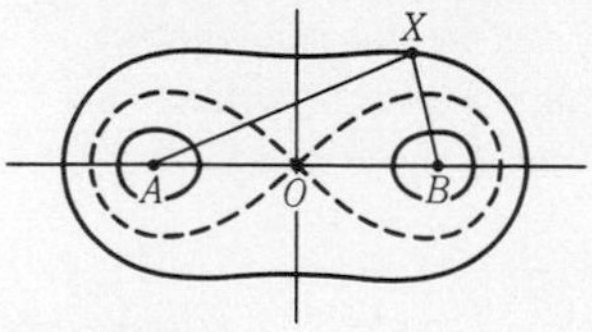

Fig. 16

The locus of the foot of the perpendicular drawn from a fixed point O to the tangent of a fixed curve C at each point of the curve is called the **pedal curve** of C with respect to O. The pedal curve of a rectangular hyperbola with respect to its center is a lemniscate (Fig. 17), and the pedal curve of a circle with respect to a point is a limaçon.

When a curve C' rolls on a fixed curve C without slipping and always tangent to C, the locus Γ of a point X keeping a fixed relation

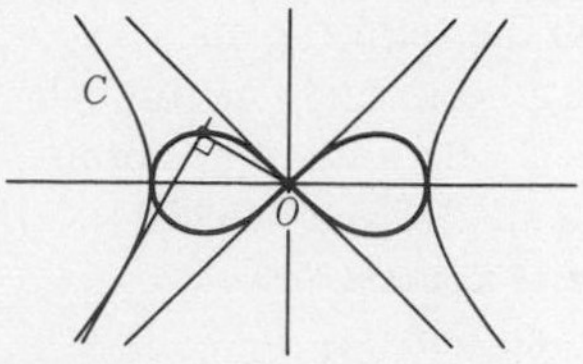

Fig. 17

with the curve C' is called a **roulette** whose **base** is C, **rolling curve** is C', and **pole** is X. In particular, when C is a straight line, C' is a circle, and X is on C', Γ is called a **cycloid** (Fig. 18). When X is not on C', Γ is called a **trochoid** (Fig. 19). A trochoid is represented parametrically by the equations $x=a\theta-b\sin\theta$, $y=a-b\cos\theta$, where the parameter θ is the angle of rotation of C'. When $a=b$, the equation represents a cycloid. The †evolute and †involute of a cycloid are also cycloids (Fig. 20).

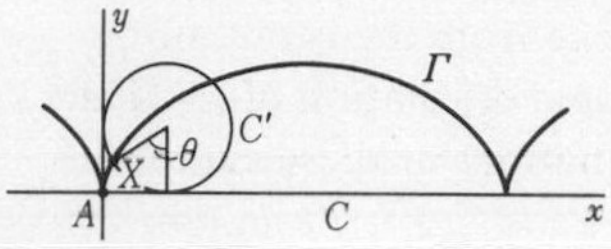

Fig. 18

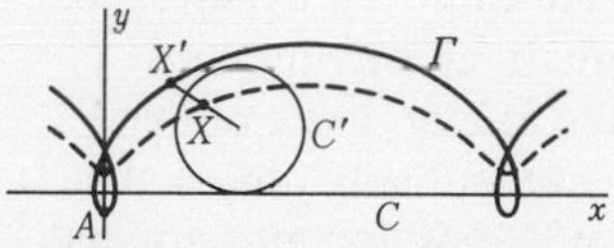

Fig. 19

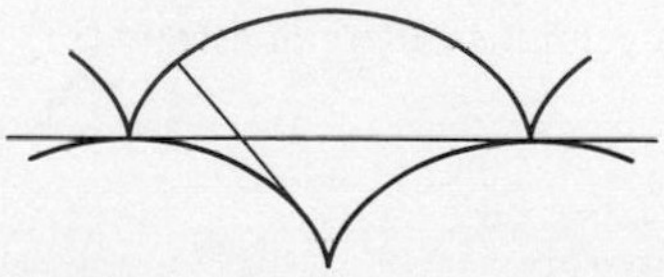

Fig. 20

Suppose that we are in a gravitational field with a given path represented as a cycloid, as is shown in Fig. 21. Assuming that there is no friction, the time necessary for a particle to slide down the path from a point X on the curve to the lowest point C of the curve is independent of the initial position X (C. Huygens). Because of this property, the cycloid is also called a **tautochrone**. Suppose that a particle point starts from a point A in the space and slides down to a lower point B along a curve Γ (without friction) under the effect of a gravitational force. To minimize the elapsed time, we simply take Γ as

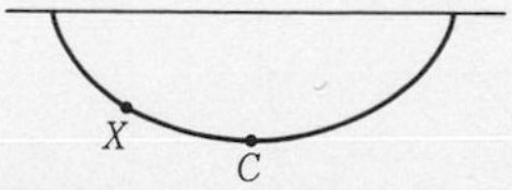

Fig. 21

a cycloid that lies in the vertical plane containing AB and has a horizontal line through A as the base (Fig. 22). Because of this property, the cycloid is called also a **brachistochrone**, i.e., the **line of swiftest descent** (Johann Bernoulli and others).

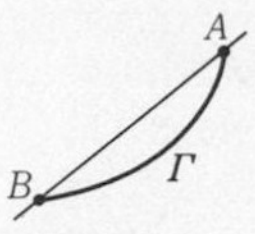

Fig. 22

When the base curve C and the rolling curve C' are both circles and X is on C', we call Γ an **epicycloid** if C, C' are externally tangent (Fig. 23), and a **hypocycloid** if C, C' are internally tangent (Fig. 24). When X is not on C', corresponding to these two cases, we have an **epitrochoid** and a **hypotrochoid**, respectively. Let a, b be radii of C, C', respectively, c the distance from the center of C' to X, and θ the angle of rotation of C'. Then the parametric equations of these curves are $x=(a\pm b)\cos\theta \mp c\cos((a\pm b)/b)\theta$, $y=(a\pm b)\sin\theta - c\sin((a\pm b)/b)\theta$. (Take the upper signs when the curve is an epicycloid and the lower signs when the curve is a hypotrochoid. When $b=c$ the equations are equations of cycloids.) When the ratio $a:b$ is a rational number p/q (p,q are mutually prime), then C' returns to its initial position after rotating q times around C; in this case each Γ becomes an algebraic curve. In particular, when $a=4b=4c$, the hypocycloid is called an **astroid** (Fig. 25). Its equation (with respect to a rectangular coordinate system) is $x^{2/3}+y^{2/3}=a^{2/3}$. The envelope of segments of length a whose endpoints are on the x-axis and y-axis, respectively, is an astroid (Fig. 25). When $a=b=c$ an epicycloid is a **cardioid** (Fig. 26).

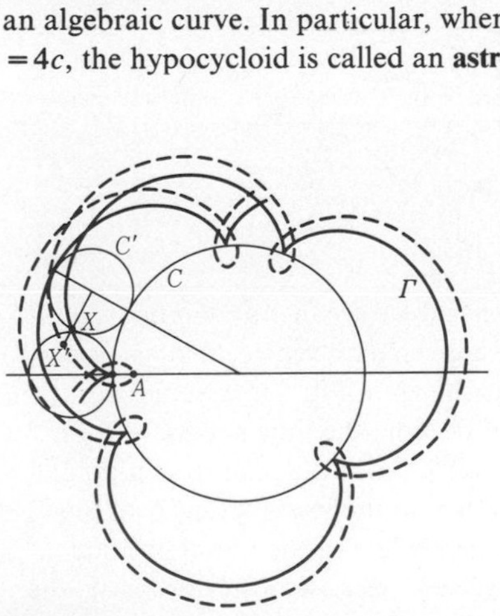

Fig. 23

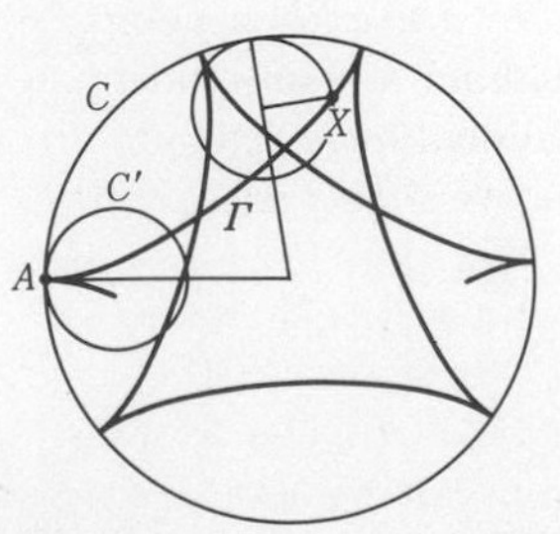

Fig. 24

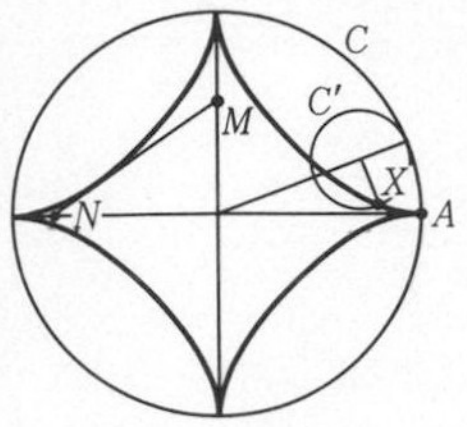

Fig. 25

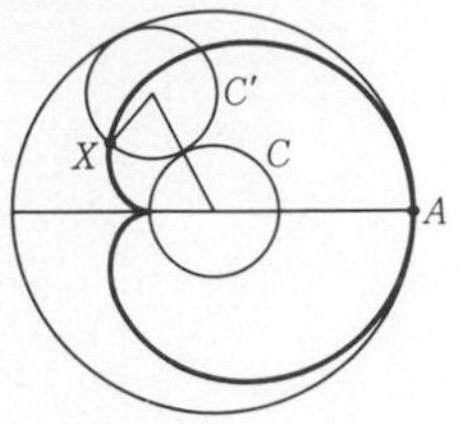

Fig. 26

When the base C is a straight line, the rolling curve C' is an ellipse or a hyperbola, and the pole is a focus of C', then the roulette is called a **Delaunay curve** (Fig. 27).

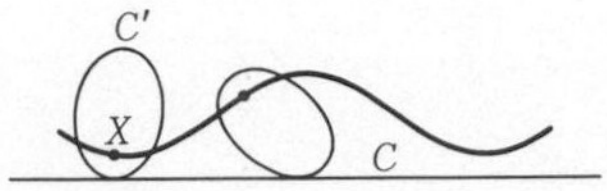

Fig. 27

When C is a straight line, C' is a parabola, and X is the focus of C', the roulette is called a **catenary** (Fig. 28). When we hold two ends of a string of homogeneous density in the gravitational field, the string takes the form of this curve. The equation of the catenary with respect to a rectangular coordinate system is $y=a\cosh x/a=a(e^{x/a}+e^{-x/a})/2$. The involute starting at the point $A(0,a)$ of this curve is called a **tractrix** (Fig. 29). Let Q be the point of intersection of the tangent at P to the tractrix and the x-axis; then the length of PQ is constant and is equal to a. Consequently, when we drag a weight at A by a string of length OA along the x-axis, the curve described by the weight is a tractrix. The parametric equations of the tractrix are $x=a(\log\tan t/2+\cos t)$, $y=a\sin t$.

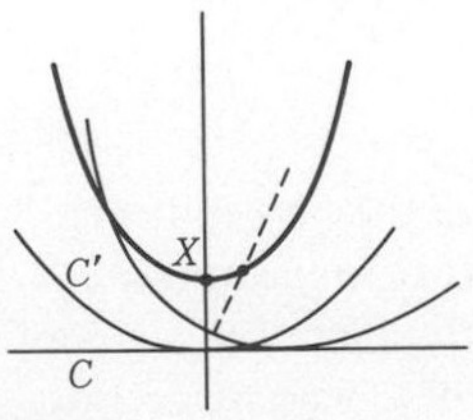

Fig. 28

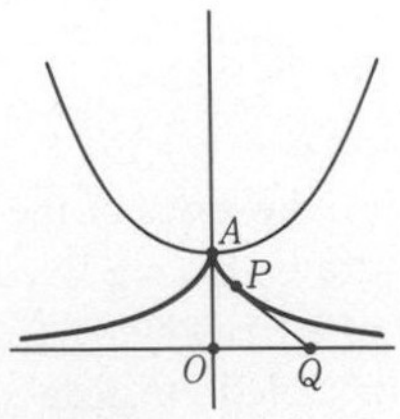

Fig. 29

Suppose that a point Q moves with constant velocity on the x-axis and another point P also moves with constant velocity always toward Q. The locus of the point P is called a **curve of pursuit** (Fig. 30). When the velocity of Q is α times that of P, the equation of the curve of pursuit is $2(x-a)=y^{1-\alpha}/c(1-\alpha)-cy^{1+\alpha}/(1+\alpha)$ if $\alpha\neq 1$ and $2(x-a)=(1/c)\log y-cy^2/2$ if $\alpha=1$. We can consider similar problems when Q moves on a general curve instead of on the x-axis.

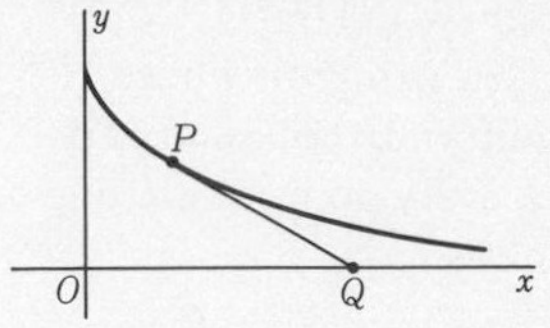

Fig. 30

Many plane curves that are called **spirals** can be expressed by $r=f(\theta)$ (f monotonic) in polar coordinates (r,θ). **Archimedes spiral** is a curve having the equation $r=a\theta$ (Fig. 31). Archimedes found that the area bounded by two straight lines $\theta=\theta_1, \theta=\theta_2$, $(\theta_1<\theta_2)$ and the curve is $a^2(\theta_2^3-\theta_1^3)/6$. A **logarithmic spiral (equiangular spiral** or **Bernoulli spiral)** is a curve having the equation $r=ke^{a\theta}$ (Fig. 32). The angle between the straight line $\theta=$ constant and the tangent to the curve is constant. Johann Bernoulli found that the involute and evolute of this curve are congruent to the original curve. A curve having the equation $r=a/\theta$ is called a **hyperbolic spiral** (or **reciprocal spiral**). If this spiral has the equation $r^2\theta=a$, it is called a **lituus**. These two spirals are shown in Figs. 33 and 34, respectively. Let $\rho=\varphi(s)$ be the †natural equation of a curve. Specifically, a curve having the natural equation $\rho=ks$ (k is a constant) is a logarithmic spiral. A curve having the equation $\rho=a^2/s$ is called a **Cornu spiral** (or **clothoid**; → 172 Functions of Confluent Type). Its parametric representation is

$$x=a\sqrt{\pi}\int_0^t\cos\frac{\pi t^2}{2}\,dt,$$

$$y=a\sqrt{\pi}\int_0^t\sin\frac{\pi t^2}{2}\,dt$$

(†Fresnel integral). M. A. Cornu used this curve in the representation of diffraction in physical optics.

Fig. 31

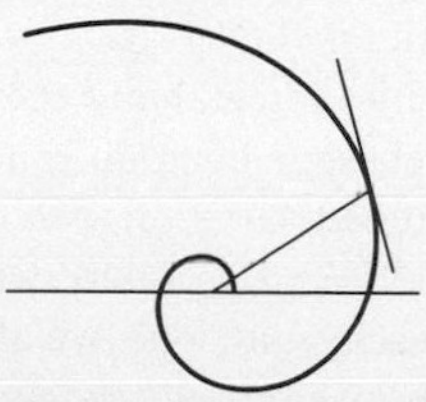

Fig. 32

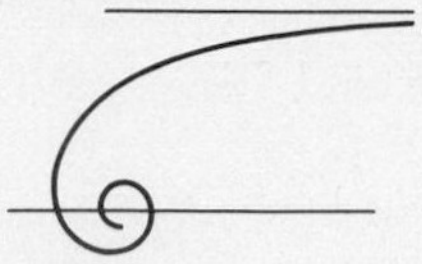

Fig. 33

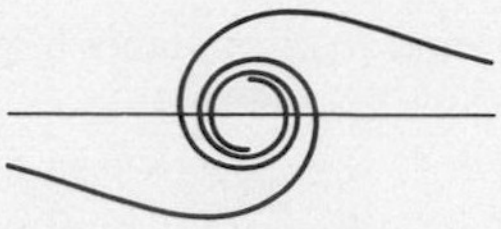

Fig. 34

There are also curves that appear as graphs of †elementary functions. For example, a curve having the equation $y=\sin x$ is called the **sine curve**, and the graphs of equations $y=e^x$ and $y=\log x$ are called the **exponential curve** and the **logarithmic curve**, respectively, although they are congruent. In contrast to algebraic curves, these analytic curves that are not algebraic are called **transcendental curves.**

Regarding the differential geometric properties of plane and space curves → 114 Differential Geometry of Curves and Surfaces; for plane algebraic curves → 11 Algebraic Curves.

I. Envelopes

Let $\mathfrak{f}(s,t)$ be a function of class C^1 of real variables s, t. If we fix $t=t_0$, then $\mathfrak{x}=\mathfrak{f}(s,t_0)$ is the equation of a curve C_{t_0} with a parameter s. If $s(t)$ is a function of t, then $\mathfrak{f}(s(t_0),t_0)$ represents a point P_{t_0} on C_{t_0}. Let E be the locus of P_{t_0} when t_0 moves. If $s(t)$ is a function of class C^1, then E is a curve of class C^1. If E and C_{t_0} are always tangent at each point P_{t_0}, we call E the **envelope** of the family of curves C_t. When $\mathfrak{f}(s,t)$ is given, to find E we need only determine the function $s(t)$. We note that $\partial f/\partial s=\lambda\partial f/\partial t$ is a condition that must be satisfied by the function $s(t)$. When $n=2$ and the equation of C_{t_0} is given in the form $f(x,y,t_0)=0$, the point of intersection of C_{t_0} and $f_t(x,y,t_0)=0$ ($f_t=\partial f/\partial t$) is P_{t_0}. The equation $R(x,y)=0$ obtained by eliminating t from $f(x,y,t)=0$ and $f_t(x,y,t)=0$ is called the **discriminant** of $f(x,y,t)=0$. The set of points (x,y) satisfying the discriminant $R=0$ is the union of E and the locus of the singular points of C_{t_0}.

References

[1] K. Menger, Kurventheorie, Teubner, 1932 (Chelsea, 1967).
[2] R. L. Wilder, Topology of manifolds, Amer. Math. Soc. Colloq. Publ., 1949.
[3] G. T. Whyburn, Analytic topology, Amer. Math. Soc. Colloq. Publ., 1942.
[4] T. Kubota, Bibun kikagaku (Japanese; Differential geometry), Iwanami, 1940.
[5] G. Loria, Spezielle algebraische und transzendente ebene Kurven. Theorie und Geschichte, I, II, Teubner, 1910–1911.

[6] F. G. Teixeira, Traité des courbes spéciales remarquables planes et gauches I–III, Coïmbre, 1908–1915.
[7] T. Saku, Kyokusen no tuiseki (Japanese; Curve tracing), Seibido (Kawade), 1937.

97 (X.11) Curvilinear Integrals and Surface Integrals

A. General Remarks

The integral of a function (or more precisely, a †differential form) along a †curve (†surface) is called a **curvilinear integral (surface integral)** respectively. Because a curvilinear integral is a special case of the Stieltjes integral, we shall first explain this notion, formulated by T. J. Stieltjes (1894) as a generalization of the †Riemann integral. The notion was introduced in connection with Stieltjes's study of †continued fractions, and was the origin of integrals with respect to general measures.

B. The Riemann-Stieltjes Integral

Suppose that $f(x)$, $\alpha(x)$ are real-valued bounded functions defined on $[a,b]$. Take a partition of the interval $a = x_0 < x_1 < x_2 < \ldots < x_{n-1} < x_n = b$ (→ 218 Integral Calculus) and consider the Riemann sum with respect to $\alpha(x)$:

$$\sum_{i=0}^{n-1} f(\xi_i)(\alpha(x_{i+1}) - \alpha(x_i)), \quad \xi_i \in [x_i, x_{i+1}].$$

Suppose that the Riemann sum tends to a fixed number as $\max(x_{i+1} - x_i)$ tends to zero. Then the limit is called the **Riemann-Stieltjes integral** (or simply **Stieltjes integral**) of $f(x)$ with respect to $\alpha(x)$ and is denoted by $\int_a^b f(x)\,d\alpha(x)$. The Riemann integral of $f(x)$ is a special case, where $\alpha(x) = x$.

The Riemann-Stieltjes integral has the elementary properties, such as linearity, of the usual Riemann integral. We also have the following theorem: The integral $\int f(x)\,d\alpha(x)$ exists for every continuous function $f(x)$ if and only if $\alpha(x)$ is of †bounded variation. Hence when we consider the Stieltjes integral of $f(x)$ with respect to $\alpha(x)$, we usually assume that $f(x)$ is continuous and $\alpha(x)$ is of bounded variation. However, the Stieltjes integral may be defined if $f(x)$ is of bounded variation (not necessarily continuous) and $\alpha(x)$ is continuous (not necessarily of bounded variation). If a sequence $f_n(x)$ $(n = 1, 2, \ldots)$ of uniformly bounded continuous functions defined on the interval $[a,b]$ converges to a continuous function $f(x)$ on the interval $[a,b]$, we have

$$\lim_{n\to\infty} \int_a^b f_n(x)\,d\alpha(x) = \int_a^b f(x)\,d\alpha(x),$$

where $\alpha(x)$ is a function of bounded variation. Further, if $\alpha(x)$ and $\alpha_n(x)$ $(n = 1, 2, \ldots)$ are functions of bounded variation whose †total variations are uniformly bounded and $\lim_{n\to\infty} \alpha_n(x) = \alpha(x)$ at every point of continuity of $\alpha(x)$, then we have

$$\lim_{n\to\infty} \int_a^b f(x)\,d\alpha_n(x) = \int_a^b f(x)\,d\alpha(x)$$

for every continuous function $f(x)$ on $[a,b]$ **(Helly's theorem)**.

Let $\alpha(x)$ be a †strictly monotone increasing function, and let $\beta(y)$ be its inverse function. Then we have

$$\int_a^b f(x)\,d\alpha(x) = \int_{\alpha(a)}^{\alpha(b)} f(\beta(y))\,dy, \qquad (1)$$

where the right-hand side is the usual Riemann integral. A function $\alpha(x)$ of bounded variation is represented as the difference of two strictly monotone increasing functions $\alpha_1(x)$ and $\alpha_2(x)$. If we denote by $\beta_i(y)$ the inverse function of $\alpha_i(x)$ $(i = 1, 2)$, we have

$$\int_a^b f(x)\,d\alpha(x) = \int_{\alpha_1(a)}^{\alpha_1(b)} f(\beta_1(y))\,dy - \int_{\alpha_2(a)}^{\alpha_2(b)} f(\beta_2(y))\,dy. \qquad (2)$$

If $\alpha'(x)$ exists and is continuous, we have

$$\int_a^b f(x)\,d\alpha(x) = \int_a^b f(x)\alpha'(x)\,dx. \qquad (3)$$

C. The Lebesgue-Stieltjes Integral

Suppose that $\alpha(x)$ is a monotone increasing function and $I = (x_1, x_2)$. We define an †interval function $U(I) = \alpha(x_2 + 0) - \alpha(x_1 + 0)$. It is nonnegative and countably additive. Hence by utilizing $U(I)$ we can construct the outer measure and also a completely additive measure (→ 270 Measure Theory; 375 Set Functions). The Lebesgue integral with respect to this measure is called the **Lebesgue-Stieltjes integral** (or **Lebesgue-Radon integral**), and is denoted by $\int_a^b f(x)\,d\alpha(x)$. If $\alpha(x)$ is a strictly monotone increasing function and $\beta(y)$ its inverse function, then formula (1) is true if the left-hand side is a Lebesgue-Stieltjes integral and the right-hand side is a Lebesgue integral. If $\alpha(x)$ is a function of bounded variation, decomposing $\alpha(x)$ into the difference of two strictly monotone increasing functions, we also have formula (2). If $\alpha(x)$ is †absolutely continuous, formula (3) is valid, where the right-hand side is a Lebesgue integral.

The Stieltjes integral has the following two properties:

Integration by parts: In the interval $[a,b]$, we have

$$\int_a^b U\,dV + \int_a^b V\,dU = U(b)V(b) - U(a)V(a)$$

if one of $U(x), V(x)$ is continuous and the other is of bounded variation.

Second mean value theorem: If $U(x)$ is monotone increasing and $V(x)$ is continuous, then there exists a ξ in $[a,b]$ such that

$$\int_a^b U\,dV$$
$$= U(a)(V(\xi) - V(a)) + U(b)(V(b) - V(\xi)).$$

D. The Curvilinear Integral

A continuous mapping from an interval $a \leqslant t \leqslant b$ on $\mathbf{R}^1$ into $\mathbf{R}^n$: $\varphi(t) = (\varphi_1(t), \ldots, \varphi_n(t))$ is an oriented curve. Suppose that a function $f(x_1, \ldots, x_n)$ is defined in a neighborhood U of the image C of the mapping $\varphi(t)$. The Stieltjes integral

$$\int_a^b f(\varphi_1(t), \ldots, \varphi_n(t))\,d\varphi_i(t), \quad i = 1, \ldots, n, \tag{4}$$

is called the **curvilinear integral** of the function $f(x_1, \ldots, x_n)$ along the curve C with respect to x_i and is denoted by $\int_C f\,dx_i$. The curve C is called the **contour** (or **path**) **of the integration**, $\varphi(a)$ is called the **initial point** (or **lower end**), and $\varphi(b)$ is called the **terminal point** (or **upper end**) of the integration. Furthermore, the Stieltjes integral of f with respect to the function $\sqrt{(\varphi_1(t))^2 + \ldots + (\varphi_n(t))^2}$ corresponding to the line element is called the **curvilinear integral with respect to the line element** and is denoted by $\int_C f\,ds$. If the integrand in (4) is of bounded variation as a function of t, the curvilinear integral is defined. If C is a †rectifiable curve, the curvilinear integral is defined for an arbitrary continuous function. In the usual case, we are concerned mainly with this sort of situation. For a differential form $\omega = f_1\,dx_1 + \ldots + f_n\,dx_n$ defined on U, the curvilinear integral $\int_C \omega$ is defined by $\sum_{i=1}^n \int_C f_i\,dx_i$.

The curvilinear integral is linear with respect to its integrand. If the terminal point of C_1 is the initial point C_2, we can construct the joint curve $C = C_1 + C_2$, and we have **additivity for the contours**

$$\int_C f\,dx_i = \int_{C_1} f\,dx_i + \int_{C_2} f\,dx_i$$

(a similar formula holds if we replace dx_i by ds). Monotonicity, which asserts that $\int_C f\,dx_i \leqslant \int_C g\,dx_i$ whenever $f \leqslant g$, holds if $\varphi_i(t)$ is monotone increasing, and monotonicity also holds for the curvilinear integral with respect to the line element.

If $n=2$, $\mathbf{R}^2$ may be identified with the †complex plane $\mathbf{C} = \{z = x + iy\}$, and we define $\int_C f(z)\,dz$ by

$$\left\{\int_C u(z)\,dx - \int_C v(z)\,dy\right\}$$
$$+ i\left\{\int_C v(z)\,dx + \int_C u(z)\,dy\right\},$$

where $f(z) = u(z) + iv(z)$. The integral above is called an integral in the complex domain. (For the application of integrals in the complex domain to complex analysis → 200 Holomorphic Functions.)

E. The Surface Integral

By an m-dimensional smooth surface S we mean the image S of a †regular mapping of class C^1 from a domain G in $\mathbf{R}^m$ into $\mathbf{R}^n$ $(m < n)$, $\xi(u) = (\xi_1(u_1, \ldots, u_m), \ldots, \xi_n(u_1, \ldots, u_m))$. Given a continuous function $f(x_1, \ldots, x_n)$ defined in a neighborhood U of S in $\mathbf{R}^n$, the multiple integral

$$\int \cdots \int_G f(\xi_1(u), \ldots, \xi_n(u)) \times \frac{D(x_{i_1}, \ldots, x_{i_m})}{D(u_1, \ldots, u_m)}\,du_1 \ldots du_m,$$
$$\{i_1, \ldots, i_m\} \subset \{1, \ldots, n\}, \tag{5}$$

is called the **surface integral** of f along S with respect to $x_{i_1}, \ldots, x_{i_m}$ and is denoted by $\int_S f\,dx_{i_1} \ldots dx_{i_m}$ or $\int \ldots \int_S f\,dx_{i_1} \ldots dx_{i_m}$. If we replace the Jacobian $D(x_{i_1}, \ldots, x_{i_m})/D(u_1, \ldots, u_m)$ in (5) by the quantity

$$\left(\sum_{i_1 < \ldots < i_m} \left(\frac{D(x_{i_1}, \ldots, x_{i_m})}{D(u_1, \ldots, u_m)}\right)^2\right)^{1/2},$$

which corresponds to the surface element of S, the integral is called a **surface integral with respect to the surface element** and is denoted by $\int_S f\,dS$ or $\int_S f\,d\sigma$. The surface integral of a differential form of degree m in $\mathbf{R}^n$ is similarly defined. In the case $m=1$, the surface integral reduces to a curvilinear integral. As the Stieltjes integral is a generalization of the curvilinear integral, there are several theories that generalize the notion of surface integral without assuming that the mapping $\xi(u)$ is of class C^1.

F. Stokes's Formula

Let S be an m-dimensional smooth surface in $\mathbf{R}^n$ $(m \leqslant n)$ and ∂S be the $(m-1)$-dimensional surface corresponding to the boundary of S. Let ω be a differential form of class C^1 of degree $(m-1)$ and $d\omega$ be its †exterior derivative. Then we have $\int_{\partial S}\omega=\int_S d\omega$, which is called **Stokes's formula** (or the **Green-Stokes formula**). (For the Stokes formula on a general differentiable manifold → 108 Differentiable Manifolds.) As special cases of Stokes's formula, we have the following three classical theorems:

(1) The case of a plane domain: Let D be a bounded domain on the xy-plane bounded by a finite number of smooth curves C with positive directions. If $\omega=P\,dx+Q\,dy$ is a differential form of class C^1 on $\bar{D}$, we have

$$\int_C P\,dx+Q\,dy=\int\int_D\left(\frac{\partial Q}{\partial x}-\frac{\partial P}{\partial y}\right)dx\,dy, \tag{6}$$

since $d\omega=(\partial Q/\partial x-\partial P/\partial y)dx\wedge dy$. This is called **Green's formula** (or **Green's formula on the plane**). Equality (6) is true if P,Q are totally differentiable and the integrand on the right-hand side is continuous (even if the functions $\partial Q/\partial x, \partial P/\partial y$ themselves are not continuous) (E. Goursat).

(2) The case of a domain in a 3-dimensional space: Let D be a bounded domain in xyz-space surrounded by a finite number of smooth surfaces S. For a †vector field $\mathbf{V}=\langle P,Q,R\rangle$ of class C^1 on $\bar{D}$, we put $\omega=P\,dy\wedge dz+Q\,dz\wedge dx+R\,dx\wedge dy$. Then since $d\omega=(\partial P/\partial x+\partial Q/\partial y+\partial R/\partial z)dx\wedge dy\wedge dz$, we have

$$\int\int_S P\,dy\,dz+Q\,dz\,dx+R\,dx\,dy$$
$$=\int\int\int_D \operatorname{div}\mathbf{V}\,dx\,dy\,dz$$
$$=\int\int\int_D\left(\frac{\partial P}{\partial x}+\frac{\partial Q}{\partial y}+\frac{\partial R}{\partial z}\right)dx\,dy\,dz. \tag{7}$$

Equality (7) is called **Gauss's formula (Ostrogradskiĭ's formula** or the **divergence theorem**). The left-hand side of (7) is equal to the surface integral $\int\int_S(\mathbf{V},\mathbf{n})\,d\sigma$, which is the †vector flux through S (where $\mathbf{n}$ means the outer unit normal vector of the surface S).

(3) The case of a bordered surface in a 3-dimensional space: Let $\bar{S}: x=x(u,v), y=y(u,v), z=z(u,v)$ $((u,v)\in\bar{G})$ be a smooth surface in xyz-space, and suppose that the boundary Γ of the domain of the parameters $\bar{G}$ consists of a finite number of smooth curves with positive direction. The boundary C of the surface S is the image curve of Γ. Now let $\mathbf{V}=\langle P,Q,R\rangle$ be a vector field of class C^1 on S, $\mathbf{n}$ be the unit normal vector of S, and $\mathbf{t}$ be the unit tangent vector, and set $\omega=P\,dx+Q\,dy+R\,dz$. Then since $d\omega=(\partial R/\partial y-\partial Q/\partial z)dy\wedge dz+(\partial P/\partial z-\partial R/\partial x)dz\wedge dx+(\partial Q/\partial x-\partial P/\partial y)dx\wedge dy$, we have

$$\int\int_S(\operatorname{rot}\mathbf{V},\mathbf{n})\,d\sigma=\int\int_S\left(\left(\frac{\partial R}{\partial y}-\frac{\partial Q}{\partial z}\right)dy\,dz+\left(\frac{\partial P}{\partial z}-\frac{\partial R}{\partial x}\right)dz\,dx+\left(\frac{\partial Q}{\partial x}-\frac{\partial P}{\partial y}\right)dx\,dy\right)$$
$$=\int_C(P\,dx+Q\,dy+R\,dz)=\int_C(\mathbf{V},\mathbf{t})\,ds. \tag{8}$$

Equality (8) is called **Stokes's formula** (→ Appendix A, Table 3.III).

References

[1] T. Takagi, Kaiseki gairon (Japanese; A course of analysis), Iwanami, third edition, 1961.
[2] M. Fujiwara, Sûgaku kaiseki I; Bibunsekibun gaku (Japanese; Mathematical analysis pt. 1; Differential and integral calculus), II, Utida-rôkakuho, 1939.
[3] S. Hitotumatu, Kaisekigaku zyosetu (Japanese; Elements of analysis) II, Syôkabô, 1963.
For the Stieltjes integral,
[4] H. L. Lebesgue, Leçons sur l'intégration et la recherche des fonctions primitives, Gauthier-Villars, second edition, 1928, ch. II.
[5] S. Saks, Theory of the integral, Warsaw, 1937.
[6] D. V. Widder, The Laplace transform, Princeton Univ. Press, 1941, ch. I.
[7] M. Tsuji, Zitu kansûron (Japanese; Theory of real functions), Maki, 1962.
[8] H. Flanders, Differential forms, with applications to the physical sciences, Academic Press, 1963.
[9] H. K. Nickerson, D. C. Spencer, and N. E. Steenrod, Advanced calculus, Lecture notes, Van Nostrand, 1959.
[10] H. Cartan, Formes différentielles, Hermann, 1967.
For integrals in the complex domain → references to 200 Holomorphic Functions.
For Stokes's formula, etc., → references to 427 Vectors.

98 (XVIII.9) Cybernetics

A. General Remarks

The term **cybernetics** was invented in 1947 by Norbert Wiener to denote a field of science that treats the system of control and communication in animals and machine.

The term was derived from a Greek word *κυβερνήτης*, also the source of the word "governor." The stimulus for establishing such a new area of science came from studies of (i) automatic computation, (ii) automatic control, and (iii) information processing. In fact, these fields had given rise to such technological innovations as the high-speed electronic computer, automatic sights, electrical communication apparatus, and automatic control instruments; and hence laid the basis for the recent developments in electronics and control engineering that were the technological foundation of automation.

The central problem of cybernetics is concerned not with matter or energy but with information, and hence the problem of **artificial intelligence** has been crucially important. There is also an important area in cybernetics, called **biocybernetics**, which is concerned with cybernetic approaches to biological information phenomena. Biocybernetics supplies useful hypotheses in biological research, while on the other hand, experimental knowledge and discoveries in this field stimulate the development of cybernetics.

Cybernetics has made remarkable progress since about 1954. First, there was a new development due to Wiener himself of **nonlinear random theory**, which was utilized to establish the theory of self-organizing systems and learning. The latter theory is deeply connected with organization theory in biocybernetics, and caused the development of a new aspect of cybernetics. Second, Soviet scientists made remarkable progress in two areas: (1) establishing cybernetics as a fundamental science to be compared with mathematics and physics, and (2) cultivating economic planning in cybernetics. Third, there has also been remarkable development in cybernetics in Western Europe regarding fundamental philosophical considerations as well as applications to engineering and humanities.

In spite of these advances, Wiener's original formulation of cybernetics in terms of certain mathematical models utilizing the †statistical mechanics of J. W. Gibbs is still of central importance. Wiener defined cybernetics in the following way: We have two state variables, where one can be adjusted by us, while the other is out of our control. We face the problem of how to realize the most favorable state for us by assigning an optimal value to the adjustable variable on the basis of information on the uncontrollable variable ranging from the past to the present. Cybernetics aims to give a methodology to solve such problems.

B. Mathematical formulation of Cybernetics

(I) We have to deal with the problem of giving a mathematical formulation of time series of uncontrollable variables. Wiener considered the totality of irregular motions corresponding to mathematical models of †Brownian motion (→ 46 Brownian Motion).

In fact, Wiener gave a constructive definition of $X(t,\omega)$ $(0 \leqslant t \leqslant 1, \omega \in \Omega)$ by giving a correspondence between sample functions $X(t)$, where $0 \leqslant t \leqslant 1$ and $-\infty < X < \infty$, and ω in a †probability space Ω, satisfying (1) $X(0,\omega)=0$; (2) for any real numbers t_1 and t_2, $0 < t_1 < t_2 \leqslant 1$, the †random variable $X(t_2,\omega) - X(t_1,\omega)$ is distributed according to the †normal distribution $N(0, t_2 - t_1)$ with mean 0 and variance $t_2 - t_1$; (3) for any real numbers t_1, t_2, t_3, and t_4, $0 < t_1 < t_2 \leqslant t_3 < t_4 \leqslant 1$, the random variables $X(t_2,\omega) - X(t_1,\omega)$ and $X(t_4,\omega) - X(t_3,\omega)$ are mutually †independent.

The process $X(t,\omega)$, $0 \leqslant t \leqslant 1$, $\omega \in \Omega$ is called a **Wiener process**, and can be shown to have the following properties: (i) $X(t,\omega)$ is continuous with probability 1 as a function of t; (ii) $X(t,\omega)$ is nondifferentiable with probability 1 as a function of t. Also, condition (1) can be removed, and the domain of definition can be enlarged to $-\infty < t < \infty$. These two generalizations suggest that $X(t,\omega)$ $(-\infty < t < \infty, \omega \in \Omega)$ coincides with the set of all continuous functions defined on the whole real t-axis and having a specific †probability measure in terms of which almost all members of the family are almost everywhere nondifferentiable functions of t. This family of functions was set up by Wiener as a family of input functions.

There are two essential aspects of Wiener's approach. The first one is the validity of the †ergodic theorem, which implies equality between the phase average and time mean and hence it suffices to estimate only the time mean of sample function in the probability space of Wiener, because for almost all sample functions the sample mean tends to the phase mean as time approaches infinity.

The second characteristic feature is the possibility of executing detailed calculations concerning random variables derived from $X(t,\omega)$, by virtue of properties (2) and (3).

For instance, we have

$$\int_\Omega (X(t_2,\omega) - X(t_1,\omega))^n d\omega$$

$$= \frac{1}{\sqrt{2\pi(t_2 - t_1)}} \int_{-\infty}^{\infty} u^n e^{-u^2/2(t_2 - t_1)} du,$$

$$\int_\Omega X(t_1,\omega) X(t_2,\omega) \ldots X(t_n,\omega) d\omega$$

$$= \begin{cases} 0, & n = 2m+1, \\ \sum \prod_{i=1}^{m} \int_\Omega X(t_{i_1},\omega) X(t_{i_2},\omega) d\omega, & n = 2m, \end{cases}$$

where Σ on the right-hand side runs through all the possible combinations $(1_1 1_2, 2_1 2_2,$

$\ldots, m_1 m_2)$ of m sets of pairs of t_i ($i = 1, 2, \ldots, m$).

For $t \geqslant s$, we have

$$\int_\Omega X(t,\omega) X(s,\omega)\, d\omega = s.$$

Furthermore, we have

$$\int_\Omega \prod_{i=1}^{n} \left(\int_\Omega \varphi_i(t_i)\, dX(t_i,\omega) \right) d\omega$$

$$= \begin{cases} 0, & n = 2m+1, \\ \sum \prod_{i=1}^{m} \int_\Omega \varphi_{i_1}(t) \varphi_{i_2}(t)\, dt, & n = 2m, \end{cases}$$

where Σ is the same as before.

(II) There is a need to design an operator that transforms input functions into output functions.

Operators in the Wiener formulation of cybernetics have two properties. (i) **Translatability**: They are translatable in the sense that they are commutative with translations $T^\tau f(t) = f(t+\tau)$ for any real number τ. (ii) **Realizability**: They are dependent at most on the present and the past history of $X(t_1, \omega)$, that is, on $X(t-\tau, \omega)$ $(0 \leqslant \tau < \infty)$. A third property may be added: (iii) linearity of operators in the fundamental approaches to cybernetics. It is to be noted that one of the characteristic features of the Wiener formulation of cybernetics is that it can be applied to nonlinear problems as well as to linear ones.

(III) There is a need for setting up criteria for determining the meaning of optimality in designing operators. Since the Wiener approach is essentially statistical, it is quite natural to introduce the notions of †unbiasedness, †minimum variance, †autocorrelation functions, and †correlation analysis both for input and output functions. Hence †generalized harmonic analysis was introduced and developed by Wiener himself, and was applied to each individual sample function to get information by virtue of ergodic theorems. Generalized harmonic analysis can be applied to situations where classical harmonic analysis cannot be applied because the functions appearing in the theory do not always satisfy conditions such as †periodicity, †almost periodicity, or absolute convergence to 0 as t tends to $\pm\infty$.

Throughout these mathematical formulations, the characteristic features of Wiener's approches to †prediction theory and to filtering of time series can be easily recognized. For instance, the problem of obtaining a prediction of the future value $X(t+\alpha)$ $(\alpha > 0)$ when we have information about the value $X(s)$ $(-\infty < s \leqslant t)$ is reduced to the problem of finding the †kernel function $K(\tau)$ by means of which a linear translatable realizable operation $\int_0^\infty X(t-\tau)\, dK(\tau)$ is defined so as to satisfy the condition that

$$\lim_{T\to\infty} \frac{1}{2T} \int_{-T}^{T} \left| X(t+\alpha) - \int_0^\infty X(t-\tau)\, dK(\tau) \right|^2 dt = \text{minimum}.$$

The solution is given by an application of generalized harmonic analysis. We define a **spectral density function** $\Phi(\lambda)$ by the relation

$$\varphi(\tau) = \lim_{T\to\infty} \frac{1}{T} \int_{-T}^{-\tau} X(t+\tau) X(t)\, dt$$
$$= \int_{-\infty}^{\infty} e^{i\tau\lambda} \Phi(\lambda)\, d\lambda.$$

Now, because of the assumption that

$$\int_{-\infty}^{\infty} \frac{|\log|\Phi(\lambda)||}{1+\lambda^2}\, d\lambda < \infty$$

we obtain the decomposition representation $\Phi(\lambda) = |\Psi(\lambda)|^2$, where $\Phi(u+iv)$ is shown to have neither singularities nor zeros in the lower half-plane $v < 0$. Then the solution $K(\tau)$ is given by

$$\int_0^\infty e^{-it\lambda}\, dK(t)$$
$$= \frac{1}{2\pi\Psi(\lambda)} \int_0^\infty e^{-i\lambda t}\, dt \int_{-\infty}^{\infty} \Psi(u) e^{iu(t+\alpha)}\, du,$$

which is verified to be useful for various applications.

Nonlinear operators discussed by Wiener have been shown to be useful in verifying the **phenomenon of pulling off**, which has an important application in dealing with self-reproduction and self-organization. It is the idea of Wiener that nonlinear random theory must be a basis for biocybernetics. In the Wiener theory nonlinear operators are expanded in series of orthogonal operators that correspond to †Hermite polynomial expansions with coefficients that constitute a set of functions $\{k_n(\tau_1, \tau_2, \ldots, \tau_n)\}$ $(n = 1, 2, \ldots)$ each of which can be expanded into a series of †Laguerre functions.

References

[1] N. Wiener, Cybernetics; or, control and communication in the animal and the machine, John Wiley, second edition, 1961.

[2] N. Wiener, Nonlinear problems in random theory, MIT Press, 1958.

[3] N. Wiener and J. P. Schadé, Cybernetics of the nervous system, Progress in brain research XVII, Elsevier, 1965.

[4] K. Steinbuch and S. W. Wagner, Neuere Ergebnisse der Kybernetik, R. Oldenbourg, 1964.

[5] S. Beer, Decision and control, John Wiley, 1966.

D

99 (XVIII.14) Data Processing

A. General Remarks

With the recent development of electronic computers, effective systems for data transmission and processing have been created on a large scale, and many results concerning data-processing methods and techniques are being obtained with these systems. In a broad sense, **data processing** includes statistical treatment, †design of experiments, and †operations research, which are described in other articles (→ 106 Design of Experiments, 302 Operations Research).

Research on data processing concerns both individual techniques of a specific kind and systems of processing as a whole, for example the system of programming techniques (→ 77 Computers C). As domains of applications, we have, for example, **information retrieval**, stock management, **program evaluation** and **review techniques**. We denote the set of data by D and assume that D is finite. According as the properties of D, we have various processing problems.

B. The Case Where D Is a Totally Ordered Set

The problems of **order correction** and **table look-up** are fundamental here. Order correction means to put the given elements of D into the order defined for D. Since, historically, sorting machines were used to put punched cards in order, this process is also called **sorting**. The process of arranging several individually well-ordered data packs into one well-ordered pack is called **merging**. This technique is especially important when magnetic tapes are used for memories of computers, and it is often needed in business data processing.

A fundamental data-sorting operation is "comparison" with respect to the order for D. It is not simple to find a way of well-ordering a pack of n elements that requires the least possible number of comparisons. Solutions are known for $n \leqslant 11$ and $n=20, 21$ [1]. Generally, a lower limit ($\log_2 n!$) of the number of comparisons is known. Also, several methods that are convenient for practical use have been devised. We list some methods and the corresponding approximate values of the number T of comparisons for large n: **sorting by merging**, $T \sim n\log_2 n$; **sorting by selection** (repetitions of the operation to choose the minimum element), $T \sim n(n-1)/2$; **sorting by insertion** (adding new elements one by one to the pack already obtained), $T \sim n(n-1)/4$; **sorting by exchanging** (repetitions of comparing two elements and exchanging them if necessary), $T \sim n^2$; and so on. Moreover, when the elements of D are †p-adic numbers of at most s figures, we have methods such as **radix sorting** (sorting in a narrow sense, repetitions of sorting concerning one fixed column), **sorting by address calculation** (first an element $d \in D$ is put into the memory with address d, and after all the elements are put at their addresses they are gathered according to the order of the addresses), or modifications of them. (In the latter two methods, comparison is not necessarily needed.) In practice, the type and the size of memory must be taken into consideration [2, 9, 10].

If there is given a univalent correspondence $f: D \rightarrow D'$ and the correspondence table is stored in the memory in a suitable form, the problem of **table look-up** arises, which requires one to find $f(d)$ for a given $d \in D$. In this process, a fundamental operation is a comparison of $d \in D$ with some $x \in D$ in the table. It is preferable that the average frequency of comparison and the size of the memory used are small. Some possible table organization methods are: (i) to line up the pairs of x and $f(x)$ according to how frequently x occurs; (ii) to line up the pairs $(x, f(x))$ according to the order of x in D. In case (ii), there are methods for finding $f(d)$ such as (iia) to read the table following its order; (iib) to pick the midpoint of the whole table (which is mechanically determined by the system of addressing of the memory), to choose one of the half-intervals by comparing the midpoint to the given d, to pick the midpoint of the candidate interval again, and to repeat this process so that the size of the interval becomes satisfactorily small (binary chopping). Suppose that D has just n elements and that each element appears with equal frequency. Then the averages of the number of comparisons required are $n/2$ in (iia) and $\log_2 n$ in (iib). According as how x is distributed, we use one of the methods (i) or (ii). For example, if the distribution approximately follows the †exponential distribution, then (i) is preferable (for sufficiently large n), whereas if it approximately follows the uniform distribution, then (i) resembles (iia) and is less preferable than (iib). Whichever method we use, it is more convenient for practical applications that all $d \in D$ have the same size. In the general case there is another method that uses the tree representation explained below.

In the case where each $d \in D$ is a natural number, we may store $f(d)$ in the memory at the address d. When this method is employed, table look-up becomes very quick since the location of $f(d)$ is immediately determined by each $d \in D$ and there is no need of comparison. However, unless D is confined to a certain interval, we need certain devices to utilize the memory efficiently [3,9,10].

Techniques concerning these processes are fundamental, for example, in stock management.

C. The Case Where D Is a Partially Ordered Set

If the given finite set D has an order which satisfies the reflexive and transitive laws, addressing is often used as a medium for representing the order in memory. Let us show a simple method for this as an example.

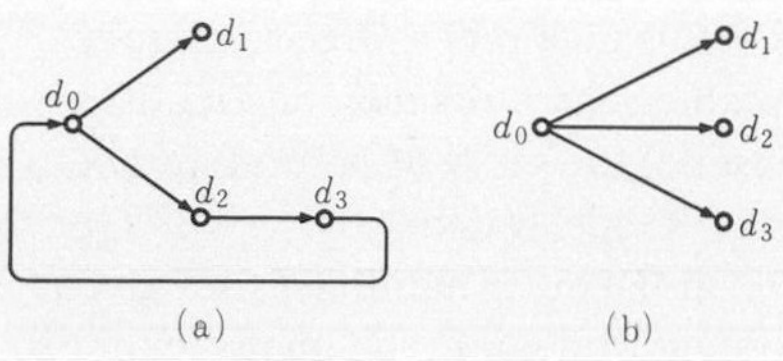

Fig. 1

Fig. 1 shows examples of partially ordered sets. Each circle (called a **node**) represents an element of D. With an element $d \in D$, let us associate a set of three quantities: d, L, and R. Let d^* be the address in memory at which this triple is stored. We suppose that $d^* \neq 0$. The meanings of L and R will be understood by the following example of representations.

(a) (d_0, d_1^*, d_2^*), $(d_1, 0, 0)$, $(d_2, d_3^*, 0)$, $(d_3, d_0^*, 0)$.
(b) (d_0, d_1^*, e_1^*), $(d_1, 0, 0)$, (e_1, d_2^*, d_3^*), $(d_2, 0, 0)$, $(d_3, 0, 0)$.

Here, e_1 is used for convenience, and we may choose any other element not contained in D. Even when many addresses $e_1^*, e_2^*, \ldots$ are needed, it is sufficient to take $e_1 = e_2 = \ldots = e$. There are many other methods of implementation.

A set of data like (a) or (b) in Fig. 1 is called a **tree representation** or **list**. The set of logical formulas is partially ordered, with the order given by rules of inference. The set consisting of a series of inferences forms an ordered subset. Thus, if tree representation can be automatically treated, so will the process of inference. The addition of a new node, the deletion of some nodes, etc., are fundamental operations for dealing with tree representations. There are examples of practical processing systems in which all these operations are realized, and which are universal in a certain sense [4,5]. To prove universality, a result obtained in the foundations of mathematics is used (→ 33 Automata). These systems are widely applicable for symbol manipulation in general (calculation by symbols, inference, language processing). They are also used for program evaluation and review techniques.

D. The Case Where D Is a Semigroup

In dealing with linguistic data (words or sentences, for example), it is often natural to consider a noncommutative †free semigroup D generated by a finite number of generators (the alphabet or vocabulary). In this case, if there exists a natural order for the generators, it determines in D a lexicographic partial ordering. Then the tree representation can be used for representing a dictionary whose entries are elements of D. This method is not so efficient with respect to speed of table look-up and economy of memory used, but it sometimes has the advantage of simplifying the treatment of complicated data.

Sometimes it happens that a partial order compatible with the †group operation has been given in advance for D. In this case, the partial ordering generated by a finite number of †fundamental relations (by the group operations and the transitive law) is especially important. It is fundamental to determine what kind of order relation holds between $d_1, d_2 \in D$. For this problem again, tree representation and the push-down storage system explained in the next section provide useful techniques.

E. Other Cases

In dealing with algebraic formulas or languages with parentheses, data maintenance methods such as tree representation or push-down storage are often convenient as auxiliary memory controlling methods. The characteristic of the **push-down storage** method is that it returns the data in reverse order with respect to the time of acceptance and remittance. This corresponds to the characteristic of parentheses, and in the returning process they are closed in reverse order with respect to the time they were opened. This method is a fundamental technique for increasing the capability of programming languages.

We do not yet have a uniform and systematic theory of data processing. The theory of the abstract computer (→ 33 Automata, 77 Computers) is being completed to some degree. Recently, relations of this theory to computational linguistics have been noticed, and many results are being collected [6, 7, 8].

F. Information Retrieval

Information retrieval is the process of arranging and storing in memory a large quantity of data, and then finding and offering (printing) the needed data automatically. The needed data are designated as those satisfying certain conditions. The problem becomes easy or difficult depending on the way of describing these conditions. If human language can be used with its original meaning, it can be safely said that a kind of man-made brain is achieved. In reality, information retrieval is put to practical use by restricting description or by simplifying the way of grasping the meaning. If a mark is given to each datum, and if the problem is to find a datum with a given mark, then it coincides with the problem of table look-up. If there are some independent marks and the problem is to find all the data that satisfy a certain logical proposition concerning the marks, it will be very difficult. Problems in this field have not yet been mathematically formalized.

References

[1] Z. Kiyasu and N. Ikeno, A theoretical consideration on the classification of information (Japanese), a pamphlet of the Committee on the Theory of Information, 1961.
[2] K. Huti, Sorting (Japanese), Zyôhô syori, Information Processing Society of Japan, 2 (1962), 86–92, 146–151.
[3] K. Ibuki, A searching method using the file without indices (Japanese), Zyôhô syori, Information Processing Society of Japan, 3 (1963), 184–188.
[4] A. K. Scidmore and B. L. Weinberg, Storage and search properties of a tree-organized memory system, Comm. ACM, 6 (1963), 28–31.
[5] J. McCarthy (ed.), Lisp 1.5, Programmer's manual, MIT Press, 1962.
[6] R. D. Luce, R. R. Bush, and E. Galanter, Handbook of mathematical psychology II, John Wiley, 1963.
[7] Y. Bar-Hillel, Language and information, Addison-Wesley, 1964.
[8] J. A. Fodor and J. J. Katz, The structure of language; readings in the philosophy of languages, Prentice-Hall, 1964.
[9] D. E. Knuth, The art of computer programming, vol. 1, ch. 2, Addison-Wesley, 1973.
[10] D. E. Knuth, The art of computer programming, vol. 3, Addison-Wesley, 1973.

100 (I.12) Decision Problem

Suppose that we are given a set S and a proposition $P(x_1, x_2, \ldots, x_n)$ for elements x_i of S. Then we have the **problem of universal validity of** P, which is the problem of finding a general **algorithm** (i.e., a finitary procedure) by which we can discern whether $P(x_1, \ldots, x_n)$ is true for all n-tuples $(x_1, \ldots, x_n)$. The problem of finding an algorithm by which we can discern the validity of $P(x_1, \ldots, x_n)$ for some specifically chosen n-tuples $(x_1, \ldots, x_n)$ is called the **problem of satisfiability of** P. These two problems are customarily called **decision problems**. The problems are such that affirmative solution of one of them implies negative solution of the other.

To give a precise definition of decision problems, let us note that a †free semigroup with countable generators may be identified with a subset of the set $\mathbf{N}$ of natural numbers (by virtue of †Gödel numbering; → 188 Gödel numbers). On the other hand, if $\mathfrak{S}$ is a given †formal system with countably many symbols, the set of all †formulas in $\mathfrak{S}$ is a subset of the free semigroup generated by the symbols in $\mathfrak{S}$. Thus the set of all formulas in $\mathfrak{S}$ is identified with a subset of $\mathbf{N}$. A subset M of $\mathbf{N}$ (or $\mathbf{N} \times \mathbf{N} \times \ldots \times \mathbf{N}$) is **(general) recursive** if its †representing function is general recursive (→ 352 Recursive Functions). By using the concept of recursive function, a precise definition of the decision problem may be given as follows: **The decision problem of** M **is solved affirmatively** if and only if we can obtain †effectively a procedure defining the representing function of M, and the function is general recursive. **The decision problem of** M **is solved negatively** if and only if we can obtain a proof that M is not recursive.

For a set A of formulas in $\mathfrak{S}$, we let $g(A)$ be the set of all Gödel numbers corresponding to the elements of A. Let A' be the set of formulas in A that are †deducible in $\mathfrak{S}$, and let $\tau(A) = g(A')$. The decision problem of the set A of formulas is said to be solved affirmatively (negatively) if the decision problem of $\tau(A)$ is solved affirmatively (negatively). By refining this concept we arrive at the notion of the **degree of (recursive) unsolvability**. Let A and B be subsets of $\mathbf{N}$. The relation "A is recursive in B and B is recursive in A" (→ 352 Recursive Functions) is reflexive, symmetric, and transitive. Hence this relation decomposes the class of all subsets in $\mathbf{N}$ into disjoint nonempty equivalence classes. A and B are defined to have the same degree of unsolvability if they belong to the same equivalence class. Thus the degrees of unsolvability may be identified with the equivalence classes. The degree of recursive sets is **0**.

The relation $\mathbf{a} \leqslant \mathbf{b}$ is defined between the degrees **a** of A and **b** of B to mean "A is recursive in B." Clearly, for any degree **a**, we have $\mathbf{0} \leqslant \mathbf{a}$. The partially ordered system of degrees constitutes an †upper semilattice.

Research on the decision problem has been mostly in areas related to the †first-order predicate calculus L^1 and the formal systems on it. We now list some important results.

(I) Results concerning L^1. The decision problem has been solved negatively for the sets of formulas of the following forms. (Here it is assumed that no function symbols appear and that $\mathfrak{A}$ represents a formula involving no occurrence of $\forall$, $\exists$, or free variables.)

(1) All formulas in L^1 (A. Church, A. M. Turing),
(2) $\exists x_1 \exists x_2 \ldots \exists x_m \forall y_1 \forall y_2 \ldots \forall y_n \ \mathfrak{A}$ (T. Skolem),
(3) $\exists x_1 \exists x_2 \exists x_3 \forall y_1 \forall y_2 \ldots \forall y_n \exists z \ \mathfrak{A}$ (K. Gödel),
(4) $\exists x_1 \exists x_2 \forall y_1 \forall y_2 \ldots \forall y_n \exists z \ \mathfrak{A}$ (L. Kalmar),
(5) $\exists x_1 \exists x_2 \forall y \exists z_1 \exists z_2 \ldots \exists z_n \ \mathfrak{A}$ (J. Pepis),
(6) $\forall x \exists y \forall z \exists u_1 \exists u_2 \ldots \exists u_n \ \mathfrak{A}$ (W. Ackermann),
(7) $\exists x_1 \exists x_2 \exists x_3 \forall y \ \mathfrak{A}$ or $\exists x_1 \exists x_2 \forall y \exists z \ \mathfrak{A}$ (J. Surányi),
(8) $\exists x \forall y_1 \forall y_2 \exists z_1 \exists z_2 \ \mathfrak{A}$ or $\forall x \exists y \forall z \exists u_1 \exists u_2 \ \mathfrak{A}$ (Surányi).

The decision problem has been solved affirmatively for the sets of formulas of the following forms, where it is assumed again that no function symbols appear and that $\mathfrak{A}$ is as above.

(1) All formulas involving variables only on predicates with one argument. (L. Löwenheim, Skolem, H. Behmann),
(2) $\forall x_1 \forall x_2 \ldots \forall x_m \ \mathfrak{A}$ (P. Bernays, M. Schönfinkel, Ackermann),
(3) $\forall x_1 \forall x_2 \ldots \forall x_m \exists y_1 \exists y_2 \ldots \exists y_n \ \mathfrak{A}$ (Bernays, Schönfinkel, Ackermann),
(4) $\forall x_1 \forall x_2 \ldots \forall x_m \exists y_1 \exists y_2 \forall z_1 \forall z_2 \ldots \forall z_n \ \mathfrak{A}$ (Gödel, Kalmar, K. Schutte).

(II) Results concerning formal systems on L^1. Throughout the rest of this article we assume that no †function variables appear. Predicate constants, function constants, and object constants may appear. By the decision problem for a formal system $\mathfrak{L}$ we mean the decision problem for all †closed formulas in $\mathfrak{L}$. Most of the results obtained so far concerning the decision problem for formal systems have been negative. Such results include those for formal systems formalizing natural number theory, the theory of rational integers, the elementary theory of †groups, †rings, †fields, †lattices, and the like, and axiomatic set theory (A. Tarski et al.). The decision problem for a formal system formalizing the elementary theory of †Abelian groups has been solved affirmatively (W. Szmielew). Little is known about the decision problem concerning partial systems of formulas of given formal systems except the following: (1) the decision problem for the set of formulas of the form $\forall x_1 \forall x_2 \ldots \forall x_m \ \mathfrak{A}$ in a formal system (→ 169 Free Groups); (2) the Hilbert-type problem, which is the decision problem for the set of formulas of the form $\exists x_1 \exists x_2 \ldots \exists x_m$ $(t = s)$ in a formal system. In particular, the Hilbert-type problem in a formal system formalizing natural number theory is called Hilbert's tenth problem (→ 198 Hilbert). The latter is the problem of finding an algorithm for deciding whether a †Diophantine equation has an integral solution.

This decision problem was studied by M. Davis, H. Putnam, J. Robinson, and others, and finally Ju. V. Matijasevič solved it negatively by showing that every recursively enumerable relation is Diophantine [9]. (A relation $R(m_1, \ldots, m_j)$ is called **Diophantine** if there is a polynomial $P(x_1, \ldots, x_j, y_1, \ldots, y_k)$ with integer coefficients such that $R(m_1, \ldots, m_j)$ holds if and only if $P(m_1, \ldots, m_j, y_1, \ldots, y_k) = 0$ has a solution for $y_1, \ldots, y_k$ in natural numbers. In addition, some investigations have been made about the †second-order predicate calculus L^2, †intuitionistic logic, etc. [1, 2].

References

[1] W. Ackermann, Solvable cases of the decision problem, North-Holland, 1954.
[2] S. C. Kleene, Introduction to metamathematics, Van Nostrand, 1952.
[3] A. Tarski, Undecidable theories, North-Holland, 1953.
[4] A. Church, An unsolvable problem of elementary number theory, Amer. J. Math., 58 (1936), 345–363.
[5] S. C. Kleene and E. L. Post, The upper semi-lattice of degrees of recursive unsolvability, Ann. of Math., (2) 59 (1954), 379–407.
[6] G. E. Sacks, Degrees of unsolvability, Ann. Math. Studies, Princeton Univ. Press, 1963.
[7] M. Davis, H. Putnam, and J. Robinson, The decision problem for exponential diophantine equations, Ann. of Math., (2) 74 (1961), 425–436.
[8] M. Davis and H. Putnam, Reductions of Hilbert's tenth problem, J. Symbolic Logic, 23 (1958), 183–187.
[9] Ю. В. Матиясевич (Ju. V. Matijasevič), Диофантовость Перечислимых Множеств, Dokl. Akad. Nauk SSSR, 191 (1970), 279–282.
[10] H. Hermes, Enumerability, decidability,

computability, Springer, 1965.
[11] H. Rogers, Theory of recursive functions and effective computability, McGraw-Hill, 1967.
[12] J. R. Schoenfield, Mathematical logic, Addison-Wesley, 1967.

101 (XX.17) Dedekind, Julius Wilhelm Richard

Julius Wilhelm Richard Dedekind (October 6, 1831–February 12, 1916) was born in the city of Braunschweig in central Germany and studied at the University of Göttingen under C. F. †Gauss, who was then in his later years. He received his doctorate at Göttingen with a thesis on the †Euler integral. He was professor of mathematics from 1858 to 1862 at Zürich and from 1863 to 1894 at the Technische Hochschule in Braunschweig. During his early twenties, he wrote works concerning analysis and the theory of probability, but in 1857 he began publishing works on the theory of numbers. He edited †Dirichlet's lectures on number theory (*Vorlesungen über Zahlentheorie*, first edition 1863, fourth edition 1899) and concentrated on research in arithmetic and algebra. The theory of †ideals, of which he was the founder, was originally set out in a supplement (1863) to Dirichlet's *Vorlesungen.*

Dedekind treated subjects ranging from the axiomatic foundations of the theory of ideals to †lattices and †groups as algebraic systems. He was a pioneer of the abstract algebra of the 20th century. Among his notable achievements are the †Dedekind zeta functions of †algebraic number fields, †Dedekind cuts in the theory of real numbers, the algebraic theory of †algebraic functions (of which he was coauthor with H. Weber), and the theory of natural numbers. He was one of the first to support †Cantor's set theory. His theory of natural numbers was founded on the concept of sets and is a predecessor of the later logicist approach. The concept of †recursive functions appeared for the first time in his theory.

References

[1] R. Dedekind, Gesammelte mathematische Werke I–III, Vieweg, 1930–1932 (Chelsea, 1969).
[2] M. Cantor and R. Dedekind, Briefwechsel, Cantor-Dedekind, Actualités Sci. Ind., Hermann, 1937.

102 (IX.12) Degree of Mapping

A. Degree of a Mapping

Let Γ^n and M^n be n-dimensional †simplicial complexes (e.g., †n-spheres) and $|\Gamma^n|$ and $|M^n|$ be †pseudomanifolds provided with orientations determined by their †fundamental cycles ζ^n, z^n. A continuous mapping $f: |\Gamma^n| \rightarrow |M^n|$ induces a homomorphism $f_*: H_n(\Gamma^n, \mathbf{Z}) \rightarrow H_n(M^n, \mathbf{Z})$ (→ 203 Homology Groups). Since $H_n(\Gamma^n, \mathbf{Z})$ and $H_n(M^n, \mathbf{Z})$ are generated by the †homology classes $[\zeta^n]$ and $[z^n]$, there exists an integer d_f such that $f_*([\zeta^n]) = d_f[z^n]$. This integer d_f is called the **degree of the mapping** f.

If a continuous mapping $g: |\Gamma^n| \rightarrow |M^n|$ is †homotopic to f $(f \sim g)$, then we have $d_f = d_g$. If f is homotopic to a constant mapping $(f \sim 0)$, then $d_f = 0$, while if f is a homeomorphism, then $d_f = \pm 1$. In particular, when $|\Gamma^n| = |M^n|$, f is a homeomorphism, and $\zeta^n = z^n$, then f is called an **orientation-preserving mapping** if $d_f = 1$ and an **orientation-reversing mapping** if $d_f = -1$.

Suppose that a continuous mapping $f: |\Gamma^n| \rightarrow |M^n|$ is induced by a †simplicial mapping $\varphi: \Gamma^n \rightarrow M^n$. We set $\zeta^n = \sum_i \sigma_i^n$, $z^n = \sum_j s_j^n$ (σ_i^n, s_j^n are n-simplexes). Let p_j (resp. q_j) be the number of n-simplexes σ_i^n such that $\varphi(\sigma_i^n)$ is equal to s_j^n (resp. $-s_j^n$). Then $p_j - q_j$ is independent of the choice of the index j and is equal to d_f.

Suppose that f, g are continuous mappings from Γ^n to an n-sphere S^n $(n \geqslant 1)$. Then $f \sim g$ if and only if $d_f = d_g$ (**Brouwer's mapping theorem**). Hence we can show that $\pi^n(S^n) \cong \mathbf{Z}$ (→ 204 Homotopy Groups).

B. Local Degree of a Mapping

Suppose that M^n and N^n are n-dimensional oriented manifolds and $f: M^n \rightarrow N^n$ is a continuous mapping. Suppose further that a point p of M^n has a neighborhood U such that $f(p) \neq f(q)$ for any point q contained in $U - \{p\}$. Then f induces a homomorphism $f_*: H_n(U, U - \{p\}) \rightarrow H_n(N, N - \{f(p)\})$ of n-dimensional †local homological groups with integral coefficients that are both isomorphic to $\mathbf{Z}$. If u and v are generators of the groups $H_n(U, U - \{p\})$ and $H_n(N, N - \{f(p)\})$, respectively, then there exists an integer k such that $f_*(u) = kv$. We call this integer k the **local degree of the mapping** f at p. If M^n and N^n are closed pseudomanifolds and $f: M^n \rightarrow N^n$ is a continuous mapping, then there exists a point r of N^n such that the set $f^{-1}(r)$ is a discrete subset $\{p_1, \ldots, p_t\}$ of M^n and each

p_i has a neighborhood U_i satisfying the condition above. If k_i is the local degree of f at p_i, then $d_f = \Sigma k_i$.

C. Linking Coefficients

Given two mutually disjoint smooth closed curves C_1 and C_2 in Euclidean 3-space, a quantity $v(C_1, C_2)$ indicating how closely they are interlinked with each other was given by Gauss as follows: Let C_i be expressed by the parameters $x_i = x_i(t_i)$ $(i=1,2)$, where $x_i(t_i)$ are †continuously differentiable. Then the quantity

$$v(C_1,C_2) = -\frac{\pi}{4}\int_{C_1C_2}\frac{1}{|x_2-x_1|^3}\det\left(x_2-x_1,\ \frac{dx_1}{dt_1},\frac{dx_2}{dt_2}\right)dt_2\,dt_1$$

is an integer called the **linking coefficient** of C_1 and C_2.

More generally, let $|M^n|$ be a †combinatorial manifold and K and K^* its †cellular decompositions such that K^* is dual to K. Let z_1^r and z_2^s $(r+s=n-1)$ be †boundaries belonging to the complex K and K^*. Suppose C^{r+1} is any †chain of K whose boundary is z_1^r. Then the †intersection number $KI(C^{r+1}, z_2^s)$ does not depend on the choice of such a chain C^{r+1}. We set $v(z_1^r, z_2^s) = KI(C^{r+1}, z_2^s)$ and call it the **linking coefficient** of z_1^r and z_2^s. The **linking coefficient** $v(\mathfrak{z}_1^r, \mathfrak{z}_2^s)$ of †singular boundaries $\mathfrak{z}_1^r$, $\mathfrak{z}_2^s$ $(r+s=n-1)$ of $|M^n|$ is similarly defined by considering the approximations z_1^r, z_2^s of $\mathfrak{z}_1^r$, $\mathfrak{z}_2^s$ belonging to a suitable cellular decomposition K and its dual K^*. The coefficient $v(\mathfrak{z}_1^r, \mathfrak{z}_2^s)$ is bilinear with respect to $\mathfrak{z}_1^r$, $\mathfrak{z}_2^s$, and we have $v(\mathfrak{z}_1^r,\mathfrak{z}_2^s) = (-1)^{rs+1}v(\mathfrak{z}_2^s,\mathfrak{z}_1^r)$. In the example shown in the left side of Fig. 1, we have $v(\mathfrak{z}_1^1,\mathfrak{z}_2^1)=1$, while $v(\mathfrak{z}_1^1,\mathfrak{z}_2^1)=2$ for the example shown in the right side of the same figure. In particular, if $\mathfrak{z}_1^r$ is homologous to 0 in $|M^n| - |\mathfrak{z}_2^s|$, then we have $v(\mathfrak{z}_1^r, \mathfrak{z}_2^s)=0$ (Fig. 2). Generally, if $\mathfrak{z}_1^r$ and $\mathfrak{z}_1'^r$ are homologous in $|M^n| - |\mathfrak{z}_2^s|$, then $v(\mathfrak{z}_1^r,\mathfrak{z}_2^s) = v(\mathfrak{z}_1'^r,\mathfrak{z}_2^s)$.

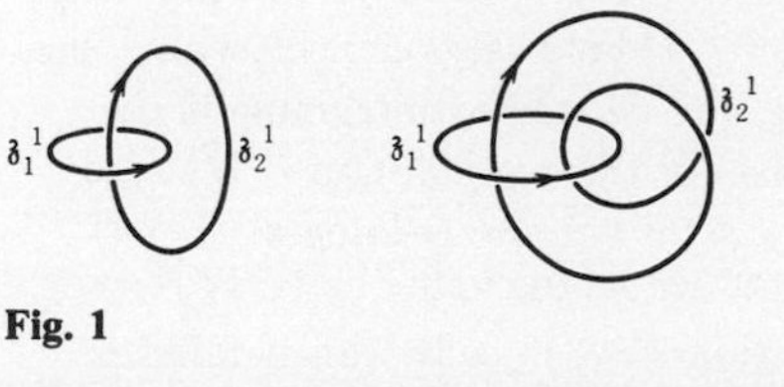

Fig. 1

D. Order of a Point with Respect to a Cycle

Let M^n be an n-dimensional manifold, $\mathfrak{z}^{n-1}$ an $n-1$ dimensional singular boundary of M^n, and o a point of M^n that is not contained in $|\mathfrak{z}^{n-1}|$. We set $\operatorname{ord}(\mathfrak{z}^{n-1}, o) = v(\mathfrak{z}^{n-1}, o)$ and call it the **order of the point** o with respect to $\mathfrak{z}^{n-1}$. For example, when $M^n = \mathbf{R}^2$ and $\mathfrak{z}^1 = \{f(t) \mid 0 \leqslant t \leqslant 1, f(0)=f(1)\}$, where f is a continuous function, the order $\operatorname{ord}(\mathfrak{z}^1, o)$ is equal to the **rotation number** around o of a moving vector $\overrightarrow{of(t)}$ as t changes from 0 to 1. This $\operatorname{ord}(\mathfrak{z}^1, o)$ stays invariant as the point o moves in a †connected component of the complement $\mathbf{R}^2 - |\mathfrak{z}^1|$ (Fig. 3). On the other hand, if $\mathfrak{z}_i^1 = \{f_i(t) \mid 0 \leqslant t \leqslant 1, f_i(0) = f_i(1)\}$ $(i=0,1)$ are closed curves in $\mathbf{R}^2$ and the distance $\rho(f_0(t), f_1(t))$ is smaller than $\rho(f_0(t), o)$ for all t in the interval [0, 1], then we have $\operatorname{ord}(\mathfrak{z}_0^1, o) = \operatorname{ord}(\mathfrak{z}_1^1, o)$ (**Rouché's theorem**).

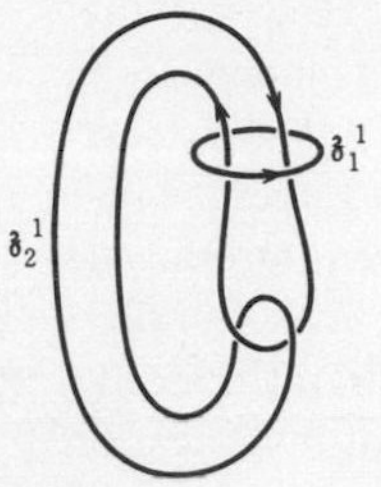

Fig. 2

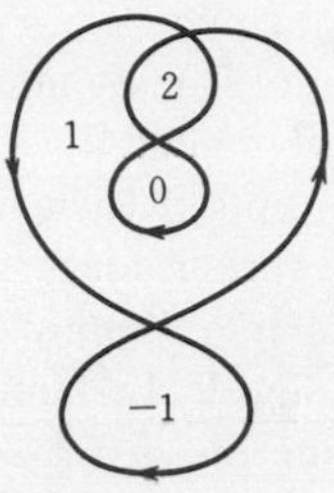

Fig. 3

References

[1] P. S. Aleksandrov and H. Hopf, Topologie I, Springer, 1935 (Chelsea, 1965).
[2] H. Seifert and W. Threlfall, Lehrbuch der Topologie, Teubner, 1934 (Chelsea, 1968).
[3] M. Nagumo, Syazôdo to sonzai teiri (Japanese; The degree of mappings and existence theorems), Kawade, 1948.

103 (X.16) Denjoy Integrals

A. History

For a real-valued function $f(x)$ of a real variable to be †Lebesgue integrable, it is necessary and sufficient that there exists an †absolutely continuous function $F(x)$ such that $F'(x) = f(x)$ at †almost all points x (→ 375 Set Functions). In general, the derivative of a function is not necessarily Lebesgue integrable. A function $f(x)$ is Lebesgue integrable if and only if $|f(x)|$ is integrable. Hence a function that is Riemann integrable in the broader sense is not necessarily Lebesgue integrable (→ 243 Lebesgue Integral). For this reason, it became desirable to extend the concept of Lebesgue integrals. In 1912, A. Denjoy constructively defined a new idea of integrals (the Denjoy integral in the restricted sense; →Section D), which was an extension of both Lebesgue and Riemann integrals. Later, N. N. Lusin provided the descriptive theory of this integral. Independently, and nearly simultaneously, A. J. Hinčin and Denjoy defined a more general integration (the Denjoy integral in the wider sense (1916); →Section D).

In 1914, O. Perron, independent of Denjoy, defined a concept of integrals (Perron integrals) that is equivalent to that of Denjoy integrals in the restricted sense. To establish this concept Perron considered the differential equation $y'=f(x)$ and utilized a method similar to the one used in the proof of the existence theorem for the solution of the differential equation $y'=f(x,y)$. However, the concept of Denjoy integrals is inadequate to treat unbounded functions. Thus to extend the concepts of Riemann and Lebesgue integrals, various ideas have been introduced: For example, Denjoy (1921), J. C. Burkill (1951), and R. D. James (1950) introduced new concepts as byproducts of investigations concerning the coefficients of trigonometric series [2,3]. The A-integral concept devised by K. Kolmogorov (1951) was meant to deal with the problem of the conjugate function of Fourier series [4]. As a certain completion of the space of functionals of step functions, K. Kunugi defined the notion of E. R. integrals (1956), which coincides with that of A-integrals in a special case [5,6].

What has been stated so far deals only with functions of a real variable. Concerning the extension of Denjoy integrals to the case of several variables, research has been done by M. Loomis, S. Kempisty, S. Nakanishi, and others [7,8].

B. Approximate Derivative

If we have $\lim_{h\to 0,k\to 0} m\{E\cap(\xi-h,\xi+k)\}/(h+k)=1$ at a point ξ of a †measurable subset E of the real line, the point ξ is called a **point of density** for E. Almost all the points of E are points of density for E (**Lebesgue's density theorem**). Let E be a measurable set having x_0 as a point of density, and let $F(x)$ be a measurable function on E. If there exists a number l such that for each $\varepsilon>0$, x_0 is a point of density for the set $\{x \mid l-\varepsilon \leqslant (F(x)-F(x_0))/(x-x_0) \leqslant l+\varepsilon, x\in E\}$, then l is called the **approximate derivative** of $F(x)$ at x_0 and is denoted by $\mathrm{AD}\,F(x_0)$. If $\mathrm{AD}\,F(x_0)$ exists, $F(x)$ is said to be **approximately derivable** at x_0. If $F(x)$ is approximately derivable at each point of E, then $F(x)$ is said to be approximately derivable in E. If $F'(x)$ exists at a point x, then $\mathrm{AD}\,F(x)$ exists at x, and we have $\mathrm{AD}\,F(x)=F'(x)$. However, there exists a continuous function $F(x)$ that is approximately derivable at almost all points of an interval and yet not derivable in the ordinary sense at any point of a set of positive measure.

C. Generalized Absolute Continuity

Let E be a set in $\mathbf{R}$, and let $F(x)$ be a real-valued function whose domain contains E. If for each $\varepsilon>0$, there is a $\delta>0$ such that for every sequence $\{[a_n,b_n]\}$ of nonoverlapping intervals whose endpoints belong to E the equality $\Sigma(b_n-a_n)<\delta$ implies $\Sigma|F(b_n)-F(a_n)|<\varepsilon$, then the function $F(x)$ is said to be **absolutely continuous** on E. We denote by AC the set of all functions that are absolutely continuous on E. If $F(x)$ is continuous on E and E is the sum of a countable sequence of sets E_n on each of which $F\in AC$, then $F(x)$ is called a **generalized absolutely continuous function**, and we write $F\in GAC$. If $F\in GAC$, $\mathrm{AD}\,F(x)$ exists almost everywhere.

If, for each $\varepsilon>0$, there is a $\delta>0$ such that for every sequence $\{[a_n,b_n]\}$ of nonoverlapping intervals whose endpoints belong to E the equaltiy $\Sigma(b_n-a_n)<\varepsilon$ implies $\Sigma_n O\{F;[a_n,b_n]\}<\varepsilon$ ($O\{F;[a_n,b_n]\}$ denotes the oscillation of the function $F(x)$ in $[a_n,b_n]$, i.e., the difference between the upper bound and the lower bound of the values assumed by $F(x)$ on $[a_n,b_n]$), then $F(x)$ is said to be **absolutely continuous in the restricted sense** (or **absolutely continuous** $(*)$) on E; and we write $F\in AC(*)$. Just as we defined the notions of generalized absolute continuity and absolute continuity, so we define the notions of **generalized absolute continuity in the restricted sense** and **generalized absolute continuity** $(*)$. Thus $F\in GAC(*)$ means that $F(x)$ is a generalized absolute continuous $(*)$ function on E. If $F\in GAC(*)$, then $F'(x)$ exists almost everywhere.

D. Definitions of Denjoy Integrals

Let $f(x)$ be a real-valued function defined on $I=[a,b]$. If for $f(x)$ there exists a function $F(x)$ that belongs to GAC on I and for which $\mathrm{AD}\,F(x)=f(x)$ holds almost everywhere, then $f(x)$ is said to be **Denjoy integrable in the wider sense** (or **D-integrable**) on I. We call $F(b)-F(a)$ the **definite D-integral** of $f(x)$ over I, and denote the value by $(D)\int_a^b f(x)\,dx$. The function $F(x)$ is called an **indefinite D-integral** of $f(x)$ on I. Similarly, we obtain the definition of the **Denjoy integral in the restricted sense** (or **D(∗)-integral**) by replacing GAC by $GAC(*)$ and $\mathrm{AD}\,F(x)$ by $F'(x)$ in the definition of the D-integral. If a continuous function $F(x)$ satisfies the equality $\mathrm{AD}F(x)=f(x)\neq\pm\infty$ $(F'(x)=f(x)\neq\pm\infty)$ for all except countably many points in I, then $F(x)$ is an indefinite D-integral (D(*)-integral) of $f(x)$. A Lebesgue-integrable function is D(*)-integrable, a D(*)-integrable function is D-integrable, and a D-integrable function that is almost everywhere nonnegative is Lebesgue integrable.

E. Constructive Definition of Integrals

Let S be a functional whose domain $\bigcup_I K(S;I)$ consists of the union of sets $K(S;I)$ of real-valued functions defined on closed intervals $I=[a,b]$. If f belongs to $K(S;I)$, we denote the value $S(f)$ by $S(f;I)$. Such a functional S is called an **integral operator** if the following three conditions are satisfied: (1) If $f\in K(S;I_0)$ and I is an arbitrary interval contained in I_0, then the †restriction f_I on I of f also belongs to $K(S;I)$. Also, $S(f;I)$ is a †continuous additive function of the interval $I\subset I_0$. (2) Let $I_1=[a,b]$, $I_2=[b,c]$, and $I=[a,c]$ $(a<b<c)$. If for a function f defined on I, $f_1\in K(S;I_1)$ and $f_2\in K(S;I_2)$, where $f_1=f_{I_1}$ and $f_2=f_{I_2}$, then $f\in K(S;I)$. (3) If f is identically 0 on I, then $f\in K(S;I)$ and $S(f;I)=0$. For two integral operators S_1 and S_2, we say that S_2 includes S_1 (or S_1 is weaker than S_2) if $K(S_1;I)\subset K(S_2;I)$ for every I and $S_1(f;I)=S_2(f;I)$ for every $f\in K(S_1;I)$. The D-integral (D(*)-integral) is the weakest integral operator containing the Lebesgue integral and satisfying the following two conditions, (C) and (H) (resp. H(*)): (C) **Cauchy's condition**. If, for every function f defined on I_0, we have $f_I\in K(S;I)$ for any $I=[a+\delta,b-\varepsilon]\subsetneq I_0=[a,b]$, and also if the finite limit $\lim_{\delta\to0,\varepsilon\to0}S(f;I)$ exists, then $f\in K(S;I_0)$ and $S(f;I_0)$ coincides with the above limit value. (H) **Harnack's condition**. Let E be a closed subset of I_0, $\{I_k\}$ be a sequence of intervals contiguous to the set consisting of the points of E and the endpoints of I_0, and f be a function on I_0 satisfying the following three conditions: (i) $f_E\in K(S;I_0)$, where $f_E(x)=f(x)$ whenever $x\in E$ and $f_E=0$ otherwise; (ii) $f_k=f_{I_k}\in K(S;I_k)$ for each k; and (iii) $\sum_k|S(f_k;I_k)|<+\infty$ and $\lim_{k\to0}O(S;f_k;I_k)=0$ when the sequence $\{I_k\}$ is infinite. Then it follows that $f\in K(S;I_0)$ and $S(f;I_0)=S(f_E;I_0)+\sum_k S(f_k;I_k)$. (Here $O(S;f_k;I_k)$ denotes the **variation** of $S(f_k)$ on I_k, that is, the upper bound of the numbers $|S(f_J;J)|$, where J denotes any subinterval of I_k.) We obtain condition (H(*)) by replacing condition (iii) in (H) with a more restrictive condition: $\sum_k O(S;f_k;I_k)<+\infty$. The constructive definition of the Denjoy integral in the wider sense (the Denjoy integral in the restricted sense) is obtained by an †induction starting with the Lebesgue integral and using two methods, (C) and (H) (resp. (H(*)), of extension.

F. Perron Integrals

Given a function $f(x)$ defined on an interval $[a,b]$, suppose that $F(x)$ is a function defined on the same interval such that (1) $\underline{F}(x)\geqslant f(x)$; (2) $\underline{F}(x)\neq-\infty$ (resp. (1′) $\overline{F}(x)\leqslant f(x)$; (2′) $\overline{F}(x)\neq+\infty$) at every point x, where $\underline{F}(x)$ (resp. $\overline{F}(x)$) denotes the †lower (upper) derivative of $F(x)$. In this case, $F(x)$ is called a **major (minor) function** of $f(x)$. If for any $\varepsilon>0$ there is a major function $\psi(x)$ and a minor function $\varphi(x)$ of $f(x)$ such that $\psi(b)-\varphi(b)<\varepsilon$, then $f(x)$ is said to be **Perron integrable**. We denote by $(P)\int_a^b f(x)\,dx$ the value $\inf_\psi\{\psi(b)-\psi(a)\}=\sup_\varphi\{\varphi(b)-\varphi(a)\}$.

G. Properties of Integrals

If $\{f_n\}$ is a nondecreasing sequence of functions that are D-integrable on an interval $[a,b]$ and whose D-integrals over $[a,b]$ constitute a sequence bounded from above, then the function $f(x)=\lim_{n\to\infty}f_n(x)$ is itself D-integrable on $[a,b]$, and we have

$$(D)\int_a^b f(x)\,dx=\lim_{n\to\infty}(D)\int_a^b f_n(x)\,dx.$$

If $F(x)$ is a function of †bounded variation and $g(x)$ is a D-integrable function on an interval $[a,b]$, then $F(x)g(x)$ is D-integrable on $[a,b]$; moreover, denoting by $G(x)$ the indefinite D-integral of $g(x)$, the following formula is valid:

$$(D)\int_a^b F(x)g(x)\,dx$$

$$=G(b)F(b)-G(a)F(a)-\int_a^b G(x)\,dF(x),$$

where the integral of the last term is the †Stieltjes integral (**integration by parts**).

If $F(x)$ is a nondecreasing function and $g(x)$ is D-integrable on $[a,b]$, there is a point ξ in $[a,b]$ for which the following formula is valid:

$$(D)\int_a^b g(x)F(x)\,dx=F(a)\cdot(D)\int_a^\xi g(x)\,dx+F(b)\cdot(D)\int_\xi^b g(x)\,dx$$

(**the second mean value theorem**).

The above theorems remain valid if D is replaced by $D(*)$ in the hypotheses and conclusions.

References

[1] S. Saks, Theory of the integral, Warsaw, 1937.
[2] R. D. James, Integrals and summable trigonometric series, Bull. Amer. Math. Soc., 61 (1955), 1–15.
[3] R. Henstock, Theory of integration, Butterworths, 1963.
[4] Ю. С. Очан (Ju. S. Očan), Обобщенный интеграл, Mat. Sb., 28 (70), no. 2 (1951), 293–336.
[5] K. Kunugui, Sur une généralisation de l'intégrale, Fundamental and Applied Aspects of Math., 1 (1959), 1–30.
[6] H. Okano, Sur une généralisation de l'intégral (E. R.) et un théorème général de l'intégration par parties, J. Math. Soc. Japan, 14 (1962), 430–442.
[7] S. Kempisty, Sur les fonctions absolument continue d'intervalle, Fund. Math., 27 (1936), 10–37.
[8] S. Nakanishi (former name S. Enomoto), Sur une totalisation dans les espaces de plusieurs dimensions I, II, Osaka Math. J., 7 (1955), 69–102, 157–178.
[9] S. Izumi, Sekibunron (Japanese; Theory of integrals), Series of math. lec. Osaka Univ., Iwanami, 1937.

104 (XX.18) Descartes, René

René Descartes, (March 31, 1596–February 11, 1650), philosopher, mathematician, and natural scientist, was born in the province of Touraine in France. He became dissatisfied with his studies on scholastic philosophy in the Jesuit Academy in La Flèche, and later, in 1619, while stationed in Ulm during a tour of duty in the army, he underwent a philosophical conversion. He became intrigued with the idea of methodologically unifying the various fields of his interest using mathematics as a model. He returned to Paris in 1621, but moved to Holland in 1628 to concentrate on his work. Sweden's Queen Christina invited him in 1649 to that country, where he died the next year, evidently from a combination of cold and overwork.

Descartes, often considered the founder of modern philosophy, early discarded the traditional theological world view, and stated that all knowledge should be logically recognized only after it has been submitted to rational criticism. This view is the basis of most modern mathematics and physics. In 1637, he published *Géométrie* as an appendix to his *Discours de la méthode*, which also contained his works on optics and meteorology. In it, he promoted F. †Viète's symbolic algebra, which he applied to geometric problems. His idea that algebra could be used as a general method for geometry established him as the founder of †analytic geometry.

References

[1] R. Descartes, Oeuvres I–XII, edited by C. Adam, P. Tannery, and Léopold Cerf, 1897–1910.
[2] H. Lefèbvre, Descartes, Paris, 1947.

105 (VI.18) Descriptive Geometry

A. General Remarks

Descriptive geometry is a discipline developed for practical purposes which deals with exact representation of 3-dimensional figures on a plane. There are seven main methods in common use.

B. Method of Orthogonal Projections

Take two planes π and π_1 intersecting orthogonally along a straight line XY. π, π_1, and XY are called the **horizontal** and **vertical planes** and the **ground line**, respectively. Let $\mathfrak{P}$ be a point in space. The †orthogonal projections P, P_1 of $\mathfrak{P}$ to π, π_1 determine $\mathfrak{P}$ uniquely. Rotating π_1 through a right angle around XY, we superpose it on π. By this operation, called **rabatting**, the point P_1 is brought to a point P' on π. Thus $\mathfrak{P}$ is represented by a pair of points P and P' on π. When the two points P and P' are distinct, the segment joining them is perpendicular to XY. Let $\mathfrak{F}$ be any figure in space, F and F_1 the orthogonal projections of $\mathfrak{F}$ to π and π_1, and F' the figure on π obtained from F_1 by rabatting. The **method of orthogonal projections** utilizes F and F', called the **plan** and the **elevation** of $\mathfrak{F}$, respectively, to represent $\mathfrak{F}$. Sometimes the orthogonal projection of $\mathfrak{F}$ to another plane π_2 perpendicular to π or π_1, or to both, is added. Such a projection is called a **side elevation**. A plane E in space is represented in this method by a pair of lines e and e', called **horizontal** and **vertical traces** of E: e is the intersection of E with π, and e' is the line obtained by rabatting the intersection e_1 of E with π_1. Unless E is parallel to XY, e and e' intersect at a point on XY.

C. Method of Contours

The **method of contours**, when utilized to represent the figure $\mathfrak{F}$, uses the **height (level** or **index)** of relevant points on $\mathfrak{F}$, i.e., their distances from π (measured by some unit length) marked on the plan F. (The height is positive or negative according as the point is "above" or "below" the horizontal plane). A curved surface is represented by its **level curves** or **contour lines**, i.e., the projections of lines on which the level is constant. A plane is represented by a line of the steepest slope, called its **scale of slope**.

D. Method of Oblique Projections

The **method of oblique projections** is a generalization of the method of orthogonal projections. Here general †parallel projections are used instead of orthogonal projections. Generally, the plan F of $\mathfrak{F}$ is the same as in the method of orthogonal projections. However, the elevation F' of $\mathfrak{F}$ is replaced by a (rabatted) parallel projection F'' of $\mathfrak{F}$ to π_1. Let A_qB_q be a parallel projection of a segment AB perpendicular to π_1. Then the ratio $A_qB_q/AB=\mu$ and the angle δ between A_qB_q and the ground line XY are determined by the parallel projection independently of the choice of AB; μ and δ are called the **ratio** and the **angle** of the parallel projection, respectively. The parallel projection with $\mu=1$ and $\delta=45°$ is called the **Cavalieri projection**, and the one with $\mu=0.5$, $\delta=45°$ is called the **cabinet projection**. Both are used in practice.

E. Method of Central Projections or Perspective

Let π_1 be a plane (which may be considered the vertical plane as in Section B) and let S be a fixed point not lying on π_1. Let F be the central projection of a space figure $\mathfrak{F}$ from S to π_1. Since π_1, S, and F do not uniquely determine the original figure $\mathfrak{F}$, some additional data are given in the **method of central projection or perspective** so that we may recover $\mathfrak{F}$ (Figs. 1, 2(a), 2(b)). The point S is called the **point of sight**, and its orthogonal projection V_c on π is called the **visual center**.

Given a straight line g, the intersection G of g with π_1 is called its **trace**, and the intersection G_∞ with the plane π_1 of a line parallel to g passing through S is called the **vanishing point** of g. The line g is represented by the line segment (called the **total perspective** of g) connecting the points G and G_∞. A point P in space is represented by its image p, to which we add the total perspective of a line

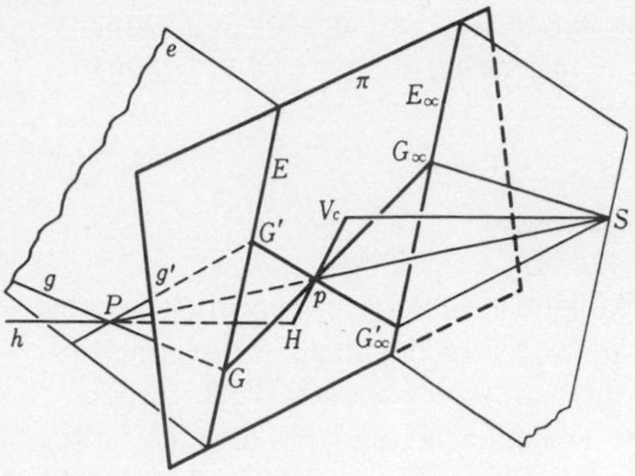

Fig. 1

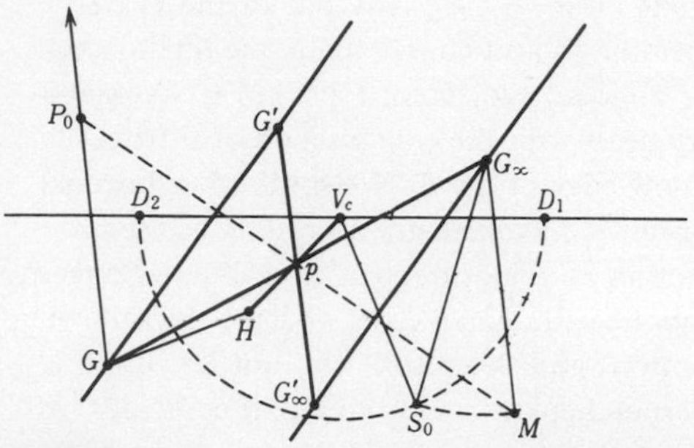

Fig. 2(a)
The line $D_2V_cD_1$ is the horizon, D_1 is the distance point, M is the measuring point of g, the line GP_0 is the measuring line of g, and the lengths of GP_0, V_cS_0, $G_\infty M=G_\infty S_0$ are the actual lengths of GP, SV_c, SG_∞, respectively.

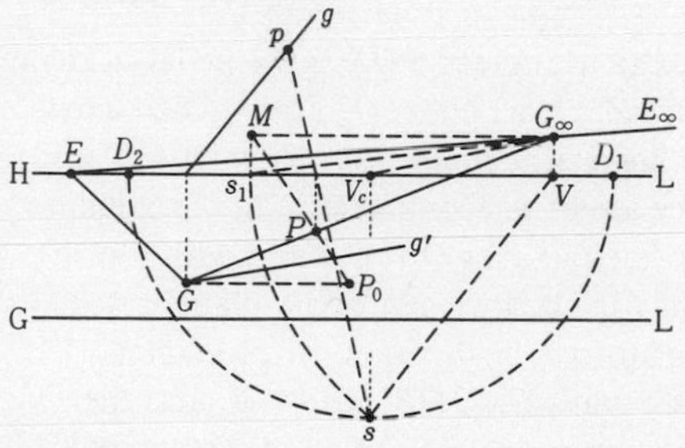

Fig. 2(b)
GL is the ground line, HL is the horizon, s is the projection on the plane of the point of sight S, V_c is its projection on the elevation or the visual center, P is the perspective of p, $G_\infty M$ is the actual length of SG_∞ (i.e., $=G_\infty s_1$), M is a measuring point, and GP_0 is the actual length of Gp.

passing through P. The total perspective HV_c of the line h perpendicular to π_1 passing through P is called the **perpendicular** of the point P. Given a plane e, the trace of all vanishing points of straight lines on e is a line E_∞ which is parallel to E and is called the **vanishing line** of e. The plane e is usually represented by its intersection E with π_1 and the vanishing line E_∞. In practice, the following data are added in order to measure the actual length: the **horizon**, which is the vanishing line of a horizontal plane; the **distance point**, which is the vanishing point of a horizontal line intersecting π_1 at an angle of 45°, i.e., the **diagonal**; the **measuring point** M of g which is a point on π_1 differing from G_∞

by the length $G_\infty V_c$ in an arbitrary direction; and the **measuring line** parallel to $G_\infty M$ in an opposite direction from G.

F. Axonometry

Let $Oxyz$ be a rectangular coordinate system in space and π be the plane intersecting the coordinate axes at points X, Y, Z, respectively. The coordinate planes Oxy, Oyz, Ozx are denoted by π_1, π_2, π_3. Let $\mathfrak{F}$ be a space figure and F_1, F_2, F_3 be its orthogonal projections on π_1, π_2, π_3. Any two of the F_i determine $\mathfrak{F}$, and consequently the (orthogonal or oblique) projection F of $\mathfrak{F}$ to π. **Axonometry** deals with the construction of F from the knowledge of the F_i. We speak of **orthogonal** or **oblique axonometry** according as the projection to π is orthogonal or oblique. Concerning the orthogonal case, we have the following notions and theorems. The foot O^* of the perpendicular from O to π that coincides with the orthocenter of $\triangle XYZ$ is called the **axonometric trace** (Fig. 3). The **contracting ratios** of the coordinate axes defined by $\cos\angle OXO^* = \lambda$, $\cos\angle OYO^* = \mu$, and $\cos\angle OZO^* = \nu$ satsify the relations (**Schlömilch's theorem**):

$$\lambda^2+\mu^2+\nu^2=2, \qquad \lambda^2:\mu^2:\nu^2 = QR:RP:PQ,$$

where the triangle $\triangle PQR$ is the pedal triangle of $\triangle XYZ$. O^*X, O^*Y, O^*Z are called **axes of (orthogonal) axonometry**. Now let OA' and OB' be a pair of †conjugate radii of an ellipse with principal axes OA, OB and with foci at F_1 and F_2. Then we can find a plane π with respect to which the axes of axonometry coincide with OA', OB', and OB, and the contracting ratios are OA'/OA, OB'/OA, and F_1F_2/OA, respectively (**Schwarz's theorem**; Fig. 4). In practice, the cases with the contracting ratios $1:1:1$, $2:1:2$, or $3:1:1$ are mostly used. In the first case, the triangle $\triangle XYZ$ is equilateral; this case is called **isometric axonometry**. When we consider an oblique axonometry, we call the image O^* of O by the oblique projection the **axonometric trace**, and O^*X, O^*Y, O^*Z the **axonometric axes**. For any three line segments O^*A, O^*B, and O^*C of which at most two coincide and at most one is of length 0, there always exists an oblique axonometry whose axes coincide with the given line segments (**Pohlke's theorem**). This theorem supplies the foundation of a method of drawing a picture of a space figure having as "axes" any three line segments O^*A, O^*B, and O^*C satisfying the conditions.

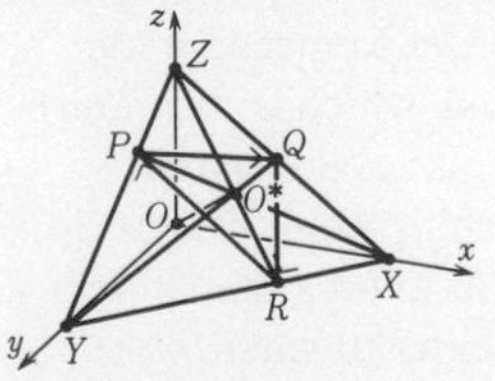

Fig. 3

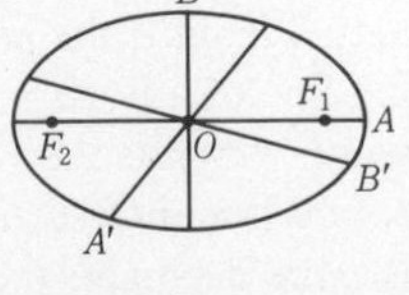

Fig. 4

G. Photogrammetry

Photogrammetry deals with the problem of reconstructing a space figure $\mathfrak{F}$ and the positions of the points of sight from several images by perspective projections from different points of sight. Let e_1 and e_2 be two planes on which the images $\mathfrak{F}'$ and $\mathfrak{F}''$ of the same figure $\mathfrak{F}$ from two different points of sight O_1 and O_2 are given. Utilizing the images of seven points of $\mathfrak{F}$ we can determine the images O_2' and O_1' of the points O_2 and O_1 on e_1 and e_2. O_1 and O_2 are called **principal points** (Fig. 5). Then we have two bundles of lines on e_1 and e_2 with centers at O_2' and O_1', respectively, in projective correspondence. Putting the corresponding points along a line l, we have a 5-parameter family of original figures $\mathfrak{F}$. If three images $\mathfrak{F}'$, $\mathfrak{F}''$, and $\mathfrak{F}'''$ are given, we have $\mathfrak{F}$ with three parameters. From four images, the original figure $\mathfrak{F}$ is determined uniquely up to a scale factor. However, the position of the original center of projection is undetermined in all these cases. When $\mathfrak{F}$ has some symmetries, the construction becomes simpler and is useful in crystallography. Furthermore, if we know the position of O_1 for the image $\mathfrak{F}'$, the **method of Scheinpflug** may be used to simplify the construction.

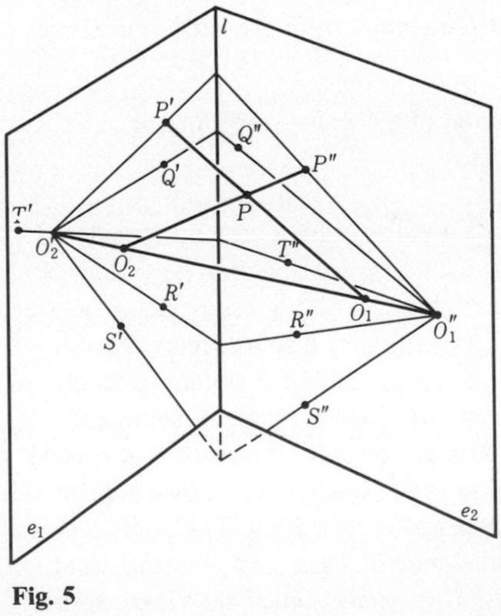

Fig. 5

H. Cyclography

In **cyclography**, we represent points in space by **directed circles**. Take a fixed plane π. A point P is represented by the circle whose center is at the orthogonal projection of P onto π, whose radius is the distance from π

to P, and whose direction (orientation) is positive or negative according as P is "above" or "below" π. (When $P \in \pi$, P is represented by P itself, considered as a circle with radius zero.) The tangent of a directed circle at a point is directed in accordance with the direction of the circle. Two directed circles are said to be in contact if they share a directed tangent. A space figure is represented on π as a set of directed circles. For example, a straight line intersecting π at a point A is represented by a set of directed circles passing through A with the same directed tangent. A plane intersecting π is represented by a set of directed circles with a common directed tangent, and a conic is represented by a set of directed circles that are in contact with a fixed directed circle.

I. History

Descriptive geometry was first developed by G. Monge (1746–1818), and the method of orthogonal projections is often called **Monge's method**. Depending mainly upon this method, Monge showed how to draw intersections of figures and shadows of solid bodies. He also treated the methods of oblique and central projections (*J. Ecole Norm. Sup.* (1795)). Methods in Sections D, E, and F were developed during the 19th century. Axonometry was introduced by Hauck and was completed in a paper by S. Finsterwalder (1899). Photogrammetry was introduced by W. Fiedler (1878) and was completed by E. Müller (c. 1920). Now we have mechanical devices for reconstructing the original figure from a number of photographs. However, the theory of descriptive geometry needs adaptation to the recent development of high-speed computers and picture processing by means of those computers. An important problem related to such an adaptation is the **hidden line problem**, i.e., the problem of obtaining an algorithm to eliminate automatically the lines or parts of lines not visible from a given point of sight.

References

[1] G. Monge, Géométrie descriptive, Paris, 1798 (new edition, Gauthier-Villars, 1922).
[2] S. Finsterwalder, Die geometrischen Grundlagen der Photogrammetrie, Jber. Deutsch. Math. Verein, 6 (1899), 1–41.
[3] G. Loria, Vorlesungen über darstellende Geometrie, Teubner, I, 1907; II, 1913.
[4] J. Hjelmslev, Darstellende Geometrie, Teubner, 1914.
[5] E. L. Ince, A course in descriptive geometry and photogrammetry for the mathematical laboratory, Bell, London, 1915.
[6] K. Rohn and E. Papperitz, Lehrbuch der darstellenden Geometrie, Veit, I, 1893; II, 1896.
[7] E. A. Müller, Lehrbuch der darstellenden Geometrie, Teubner, I, 1908; II, 1916; III, 1927.
[8] D. A. Low, Practical geometry and graphics, Longmans-Green, 1912.
[9] J. F. Dowsett, Advanced constructive geometry, Oxford, 1927.
[10] F. Rehbock, Darstellende Geometrie, Springer, 1957.

106 (XXII.10) Design of Experiments

A. General Remarks

The purpose of the **design of experiments** is (1) to analyze a given †statistical linear model and (2) to determine a "good" statistical linear model. Sometimes this term refers to a statistical method including the †analysis of variances (→ 392 Statistical Linear Models).

Let an n-dimensional †random variable $\mathbf{X}=(X_1, \ldots, X_n)'$ be represented by a linear model

$$\mathbf{X}=A\boldsymbol{\xi}+\mathbf{W}, \tag{1}$$

where A is a given $n\times s$ real matrix, $\boldsymbol{\xi}=(\xi_1,\ldots,\xi_s)'$ is an s-vector, and $\mathbf{W}=(W_1, \ldots, W_n)'$ is a random vector with the †expectation $E(\mathbf{W})=0$. Then $\mathbf{X}$ is called an **observation vector**, W the **error term**, and $\boldsymbol{\xi}$ the **effect** of $\mathbf{X}$. A is called the **design matrix**. In most cases the entries of A are 0 or 1.

According to the properties of the effect $\boldsymbol{\xi}$, the linear models (1) are divided into three classes: (i) The class of **fixed-effect models** for which $\boldsymbol{\xi}$ is a fixed unknown parameter. In this case, the component ξ_i of $\boldsymbol{\xi}$ is called a **fixed effect**, and a linear function $\pi=\mathbf{F}'\boldsymbol{\xi}$ of $\boldsymbol{\xi}$ with a given coefficient vector $\mathbf{F}$ is called a **linear parameter** or **parametric function**. (ii) The class of **random-effect models** for which the components Ξ_i of $\boldsymbol{\xi}$ are random variables. In this case, each component Ξ_i is called a **random effect**, and $\boldsymbol{\xi}$ is denoted by $\boldsymbol{\Xi}$. (iii) The class of **mixed models** for which there are both fixed effects ξ_i and random effects Ξ_j in $\boldsymbol{\xi}$. In this case, the model (1) becomes

$$\mathbf{X}=A_1\boldsymbol{\xi}^1+A_2\boldsymbol{\Xi}^2+\mathbf{W}, \tag{2}$$

where $\boldsymbol{\xi}'=(\xi_1, \ldots, \xi_r)'$ is a fixed-effect vector and $\boldsymbol{\Xi}^2=(\Xi_1^2, \ldots, \Xi_s^2)$ is a random-effect vector. The conditions frequently assumed for the †distribution law of $\mathbf{X}$ are: (a) The errors

W_i $(i=1,\ldots,n)$ are mutually independent and $E(W_1)=\ldots=E(W_n)=0$. (b) The errors W_i $(i=1,\ldots,n)$ have a common unknown †variance σ^2. (c) The errors W_i $(i=1,\ldots,n)$ have the †normal distribution. (d) The random effects Ξ_j $(j=1,\ldots,s)$ are mutually independent and independent of the error term $\mathbf{W}$. (e) The random effects Ξ_j $(j=1,\ldots,s)$ have a common unknown variance σ_1^2. (f) The random effects Ξ_j have the normal distribution.

Let $L(A)$ be a linear subspace of $\mathbf{R}^n$ spanned by the column vectors of A. The linear model

$$\mathbf{X}=B\boldsymbol{\xi}+\mathbf{W} \tag{3}$$

is called a hypothesis on the linear model (1) if $L(B)\subsetneqq L(A)$.

The main issues of the theory of design of experiments are concerned with (I) †statistical inferences, such as †estimation or †testing statistical hypotheses, for models (1), (2), (3), (i), (ii), (iii); (II) determination of the matrix A satisfying certain requirements; (III) construction of a theoretical foundation that can explain the validity of the statistical treatment of the observed data by means of the above models.

B. Block Design

The design of experiments will be described here in terms of so-called **block design**. There are n observation units $\alpha=1,\ldots,n$ called **plots**, and an observation X_α is assigned to each plot α. A **block** is constructed with several plots, and the number of plots in a block is called the **size** of the block, the jth denoted by $k_j, j=1,\ldots,b$, with $\Sigma_j k_j=n$. One of v operations, called **treatments**, is applied to each plot. It is assumed that the observation X_α at the plot α in the jth block under the ith treatment has the following structure:

$$X_\alpha=\xi_i+\eta_j+W_\alpha.$$

The ξ_i, $i=1,\ldots,v$, are called the **treatment effects**, and the $\eta_j, j,\ldots,b$, the **block effects**. It is also assumed that $\Sigma_i\xi_i=0$. In this case, $\mathbf{X}$ is represented in matrix notation by

$$\mathbf{X}=\Phi\boldsymbol{\xi}+\Psi\boldsymbol{\eta}+\mathbf{W}, \tag{4}$$

where $\Phi=(\varphi_{\alpha i})$, $\alpha=1,\ldots,n$, $i=1,\ldots,v$, with

$$\varphi_{\alpha i}=\begin{cases}1 & \text{when the } i\text{th treatment is applied to the plot } \alpha,\\ 0 & \text{otherwise,}\end{cases}$$

and $\Psi=(\psi_{\alpha j})$, $\alpha=1,\ldots,n, j=1,\ldots,b$, with

$$\psi_{\alpha j}=\begin{cases}1 & \text{when the plot } \alpha \text{ belongs to the } j\text{th block,}\\ 0 & \text{otherwise.}\end{cases}$$

Here it will be assumed that $\Sigma_i\varphi_{\alpha i}=1$, $\Sigma_\alpha\varphi_{\alpha i}=r_i\geqslant 1$, $\Sigma r_i=n$, $\Sigma_j\phi_{\alpha j}=1$, and $\Sigma_\alpha\psi_{\alpha j}=k_j\geqslant 1$. We call r_i the **number of replications** of the ith treatment. We set $N=(n_{ij})=\Phi'\Psi$. Then n_{ij} is the number of observations in the jth block to which the ith treatment is applied. The matrix N is called the **incidence matrix** of the block design. Treatments i_0 and i_h are called **connected** if there exists a chain

$$i_0 j_1 i_1 j_2 \ldots i_{h-1} j_h i_h$$

of integers such that $1\leqslant i_p\leqslant v$ $(p=1,2,\ldots,h)$, $1\leqslant j_q\leqslant b$ $(q=1,2,\ldots,h)$, and $n_{i_0 j_1}>0$, $n_{i_1 j_1}>0$, $n_{i_1 j_2}>0,\ldots,n_{i_{h-1}j_h}>0$, $n_{i_h j_h}>0$. The notion of connectedness between two blocks is defined in a similar way. If all treatments and all blocks are mutually connected, then the design is called **connected**. In this case, the rank of the matrix C defined below is $v-1$. If the design is disconnected, then the incidence matrix N can be partitioned into two or more connected parts, e.g.,

$$N=\begin{bmatrix}N_1 & 0\\ 0 & N_2\end{bmatrix}.$$

Thus without loss of generality we may restrict ourselves to the connected case.

C. Estimation Under the Fixed-Effect Model

Conditions (a) and (b) in Section A are assumed here. The normal equation which gives the least square estimate $\hat{\boldsymbol{\xi}}$ and $\hat{\boldsymbol{\eta}}$ of $\boldsymbol{\xi}$ and $\boldsymbol{\eta}$, respectively, is

$$\begin{bmatrix}\Phi'\\ \Psi'\end{bmatrix}[\Phi,\Psi]\begin{bmatrix}\hat{\boldsymbol{\xi}}\\ \hat{\boldsymbol{\eta}}\end{bmatrix}=\begin{bmatrix}\Phi'\\ \Psi'\end{bmatrix}\mathbf{X}. \tag{5}$$

Set $\Phi'\Phi=\mathrm{diag}(r_1,\ldots,r_v)=D_r$, $\Psi'\Psi=\mathrm{diag}(k_1,\ldots,k_b)=D_k$, $C=D_r-ND_k^{-1}N'$, $\mathbf{Q}=(\Phi'-ND_k^{-1}\Psi')\mathbf{X}$, where diag(...) means a diagonal matrix with the diagonal elements.... Then (5) reduces to

$$C\hat{\boldsymbol{\xi}}=\mathbf{Q},\quad \hat{\boldsymbol{\eta}}=D_k^{-1}(\Psi'X-N'\boldsymbol{\xi}). \tag{6}$$

Let L be an orthogonal matrix that transforms the matrix C to a diagonal form; that is, $L'CL=\mathrm{diag}(\rho_1,\ldots,\rho_{v-1},0)=\Lambda$, $\rho_i>0$ for all i. Set $\Lambda^*=\mathrm{diag}(\rho_1^{-1},\ldots,\rho_{v-1}^{-1},0)$, $C^*=L\Lambda^*L'$. Then $\hat{\boldsymbol{\xi}}=C^*Q$ is a particular solution of (6) and $\Sigma\,\hat{\xi}_i=0$. A parametric function $\pi=\mathbf{F}'\boldsymbol{\xi}$ with coefficient vector $\mathbf{F}=(F_1,F_2,\ldots,F_v)'$ is called a **treatment contrast** if the sum $\Sigma_i F_i$ of coefficients vanishes. A treatment contrast $\pi=\mathbf{F}'\boldsymbol{\xi}$ is called a **normalized contrast** if $\mathbf{F}'\mathbf{F}=1$. When a design is connected, any contrast π is estimable, and the †best linear unbiased estimator of π is $\hat{\pi}=\mathbf{F}'\hat{\boldsymbol{\xi}}$. Furthermore, if f_i is the eigenvector of the matrix C with unit length corresponding to an eigenvalue ρ_i and $F=\Sigma_i a_i f_i$, then the variance of estimator $\hat{\pi}$ is given by $\sigma^2\Sigma_i a_i^2/\rho_i$.

D. Test of a Hypothesis $H:\xi_1=\ldots=\xi_v$ in the Fixed-Effect Model

Conditions (a), (b), and (c) of Section A are assumed here. The hypothesis H: $\xi_1=\ldots=\xi_v$ is represented by

$$\mathbf{X}=\Psi\eta+\mathbf{W}. \tag{7}$$

Consider a direct sum decomposition $\mathbf{R}^n=L(\Gamma)+L_\Gamma^\perp(\Psi)+L_\Psi^\perp(\Phi,\Psi)+L_{\Phi\Psi}^\perp$ of $\mathbf{R}^n$, where $L_B^\perp(A)$ and $L_B^\perp$ stand for the orthocomplements of $L(B)$ with respect to $L(A)$ and $\mathbf{R}^n$, respectively, and $\Gamma=(1,1,\ldots,1)'\in\mathbf{R}^n$. The projective operator matrices for the decomposed subspaces $L(\Gamma)$, $L_\Gamma^\perp(\Psi)$, $L_\Psi^\perp(\Phi,\Psi)$, and $L_{\Phi\Psi}^\perp$ are denoted by P_1, P_2, P_3, and P_4, respectively. Then we have

$$P_1=n^{-1}E_{nn},\quad P_2=\Psi D_k^{-1}\Psi'-n^{-1}E_{nn},$$
$$P_3=(I_n-\Psi D_k^{-1}\Psi')\Phi C^*\Phi'(I_n-\Psi D_k^{-1}\Psi'),$$
$$P_4=I_n-P_1-P_2-P_3,$$

where $E_{a,b}$ is an $a\times b$ matrix whose entries are all unity and I_n is the $n\times n$ unit matrix.

The †analysis of variance for the hypothesis (7) in model (4) is given by

$$\mathbf{X}'\mathbf{X}=\mathbf{X}'P_1\mathbf{X}+\mathbf{X}'P_2\mathbf{X}+\mathbf{X}'P_3\mathbf{X}+\mathbf{X}'P_4\mathbf{X}.$$

This is called **intrablock analysis**. A usual test for the hypothesis H is given by a critical region with

$$F=\frac{n-v-b+1}{v-1}\,\frac{\mathbf{X}'P_3\mathbf{X}}{\mathbf{X}'P_4\mathbf{X}}>\text{constant}$$

(→ 392 Statistical Linear Models).

E. Optimal Design

A block design is said to be **optimal** when it minimizes the variance of the estimator $\hat{\pi}$ of a normalized contrast π. Suppose that the number v of treatments, the number b of blocks, and the size k_j of each block, $j=1,\ldots,b$, are given. Under each of the following conditions, the corresponding block design is optimal: (I) $\prod_{i=1}^{v-1}\rho_i$ is maximal, (II) $\min\rho_i$ is maximal, (III) the sum of variances of the estimator of the parameter $\xi_i-\xi_{i'}$ for the pairs (i,i') is minimal.

If $\rho_1=\ldots=\rho_{v-1}=(n-b)/(v-1)$ $(=\rho$, say) and n_{ij} is either 1 or 0, then the design is optimal for each of the optimal condititions (I), (II), and (III). In this case, we have

$$C=\rho(I_v-v^{-1}E_{vv}).$$

Such a design is called a **balanced block design**. When all block sizes k_j equal some number k independent of j, all numbers r_i of replications equal some number r independent of i, and $\lambda_{ii'}=\sum_j n_{ij}n_{i'j}$ (number of times that the treatments i and i' are applied together to the same block) equals some number λ independent of i and i', then the design is balanced. The design is called a **balanced incomplete block design (BIBD)** if these three conditions are fulfilled and $k<v$. Such a BIBD is denoted by (v,k,b,r,λ) and we have the relations $vr=bk$, $\lambda(v-1)=r(k-1)$, $v\leqslant b$, and $r\geqslant k$. The design is said to be **symmetric** when $v=b$. Furthermore, if $v=b$ is even, then $r-\lambda$ must be a perfect square. Necessary conditions for the existence of a BIBD have been obtained. One of these conditions is stated in terms of the Minkowski-Hasse p-invariant $C_p(A)=(-1,-1)_p\prod_{i=1}^n(D_{i'},-D_{i-1})_p$, where

$$(m,n)_p=\left(\frac{m,n}{p}\right)$$

is the †Hilbert norm-residue symbol and D_i is the principal minor of A. No effective condition necessary and sufficient for the existence of a BIBD has been obtained yet.

A method of constructing a BIBD utilizes as blocks the subspaces of the projective space and the affine space over a †finite field. However, this method of constructing a BIBD is not sufficiently powerful. To explain another method of constructing a BIBD, we let G be an additive group of order n and $x_1^{(i)},\ldots,x_m^{(i)}$ be m treatments corresponding to each element $x^{(i)}$ of the group $(i=1,2,\ldots,n)$. The treatment $x_\alpha^{(i)}$ is said to belong to the αth class $(\alpha=1,2,\ldots,m)$, and a pair $(x_\alpha^{(i)},x_\beta^{(j)})$ of treatments is called a **difference** of type $(\alpha,\beta,x^{(p)})$ if $x^{(i)}-x^{(j)}=x^{(p)}$ $(\neq 0)$. We may form t blocks of size k

$$B_1=\{x_{\alpha_1}^{(i_1)},\ldots,x_{\alpha_k}^{(i_k)}\},\quad\ldots,$$
$$B_t=\{x_{\beta_1}^{(j_1)},\ldots,x_{\beta_k}^{(j_k)}\},$$

such that each block B_i contains exactly r treatments belonging to the αth class (for $\alpha=1,\ldots,m$) and among all pairs of treatments in the same block there are λ differences of each type $(\alpha,\beta,x^{(p)})$. Such a set of t blocks is called a **difference set**. The t blocks in a difference set are called its **initial blocks**. Given such a difference set, we can obtain nt blocks by joining elements of G to the elements of each B_s $(s=1,2,\ldots,t)$. These nt blocks form a BIBD $(v=mn,\ k=rm/t,\ b=nt,\ r,\ \lambda)$ (→ 69 Combinatorial Theory).

F. Estimation in a Mixed Model

Consider a block design (4), where every block has the same size and every treatment has the same number of replications. Let $\boldsymbol{\xi}$ be a fixed effect and $\boldsymbol{\eta}$ a random effect denoted by $\mathbf{H}$, and assume that $\mathbf{W}$ satisfies conditions (a) and (b) of section A and that $\mathbf{H}$ satisfies conditions (d) and (e) (where the Ξ_j are replaced by the coordinates H_j of $\mathbf{H}$).

If $E(H_j)=\gamma, j=1,\ldots,b$, then, changing the notation $\mathbf{H}-\Gamma_v\gamma$ to $\mathbf{H}$, (4) is rewritten as

$$\mathbf{X}=\mathbf{\Gamma}\gamma+\Phi\xi+\Psi\mathbf{H}+\mathbf{W} \qquad (8)$$

with $E(\mathbf{H})=\mathbf{0}$. The †normal equation that gives the least square estimate of Ξ is

$$\left(C+\sigma^2(\sigma^2+k\sigma_1^2)^{-1}C_1\right)\xi = \mathbf{Q}+\sigma^2(\sigma^2+k\sigma_1^2)^{-1}\mathbf{Q}_1, \qquad (9)$$

where C and $\mathbf{Q}$ are the same as in (6) and $C_1=ND_k^{-1}N'-rv^{-1}E_{vv}$, $\mathbf{Q}_1=(ND_k^{-1}\Psi'-v^{-1}E_{v1}\Gamma')\mathbf{X}$. Equation (9) cannot be solved unless the ratio $\sigma^2:\sigma_1^2$ is given. When $\sigma^2:\sigma_1^2$ is not known, substituting in $\sigma^2+k\sigma_1^2$ its unbiased estimator given by analysis of variance, the solution of (9) tends to a consistent estimator of ξ as the number of blocks tends to infinity.

G. Estimation in a Random-Effect Model

Let ξ (for simplicity denoted by Ξ) and $\mathbf{H}$ in the model (8) be random effects. Suppose that Ξ, $\mathbf{H}$, and $\mathbf{W}$ are mutually independent, and that the distributions of Ξ, $\mathbf{H}$, and $\mathbf{W}$ are $N(\mathbf{0},\sigma_2^2I_v)$, $N(\mathbf{0},\sigma_1^2I_b)$, and $N(\mathbf{0},\sigma^2I_n)$, respectively. The distribution of $\mathbf{X}$ in (8) contains four parameters γ, σ^2, σ_1^2, σ_2^2. When $k<v$, the †minimal sufficient statistic is generally incomplete, and therefore the optimal estimators of γ, σ^2, σ_1^2, and σ_2^2 cannot be determined. As an example, the minimal sufficient statistic for the random-effect model of a BIBD (v,k,b,r,λ) is

$$(\Sigma X_i, \mathbf{X}'P_{21}\mathbf{X}, \mathbf{X}'P_{22}\mathbf{X}, \mathbf{X}'P_3\mathbf{X}, \mathbf{X}'P_4\mathbf{X}, \mathbf{X}'P_5\mathbf{X}),$$

where

$$P_{21}=k^{-1}(r-\lambda)^{-1}BTB-kr(r-\lambda)^{-1}n^{-1}E_{nn},$$

$$P_{22}=k((k-1)r+\lambda)^{-1} \times r^{-1}(T-k^{-1}BT)(T-k^{-1}TB),$$

$$P_3=k^{-1}B-k^{-1}(r-\lambda)^{-1}BTB + v\lambda(r-\lambda)^{-1}n^{-1}E_{nn},$$

$$P_4=I_n-k^{-1}B-P_{22}, \quad P_5=r^{-1}T-n^{-1}E_{nn},$$

with $T=\Phi\Phi'$ and $B=\Psi\Psi'$. In this case, $E(X'P_4\mathbf{X})=(n-v-b+1)\sigma^2$, $E(\mathbf{X}'P_3\mathbf{X})=(b-v)(\sigma^2+k\sigma_1^2)$. From these equations unbiased estimators of σ^2 and σ_1^2 can be derived, but their optimality is not guaranteed (→ 387 Statistic).

H. Randomization

In any experiment, each plot has its own effect. The blocks are constructed so that this plot effect in each block becomes as homogeneous as possible, although it is impossible to annihilate the effect completely. For this purpose a procedure called **randomization** is adopted. Suppose that we are given k plots in a block and v treatments ($k \leqslant v$). Then randomization is utilized to select a treatment out of v treatments to be allocated to each plot so that the selection is "random". Thus the plot effects are random, and the error term in model (4) can be considered as the sum of a plot effect and an original error. When $k=v$ and $N=E_{vb}$, the design is called a **randomized block design**.

I. Factorial Experiments

Suppose that there are h factors $F_1,\ldots,F_h$ which affect $\mathbf{X}$, and each factor F_i has s_i **levels** $(i=1,\ldots,h)$. It is assumed that $v=s_1\times s_2\times\ldots\times s_h$ treatments are derived by all the combinations of the levels of h factors, and that v treatment effects are represented by the sum of subeffects called **main effects** and **interactions**. Such an experiment is called a **factorial experiment**. Specifically, suppose that we have the case $h=2$, called a **two-way layout**. Let main effects be denoted by $\xi^1=(\xi_1^1,\ldots,\xi_{s_1}^1)'$, $\xi^2=(\xi_1^2,\ldots,\xi_{s_2}^2)'$, and interaction by $\xi^{12}=(\xi_{11}^{12},\xi_{12}^{12},\ldots,\xi_{s_1s_2}^{12})'$, where $\Sigma_i\xi_i^1=0$, $\Sigma_j\xi_j^2=0$, $\Sigma_i\xi_{ij}^{12}=\Sigma_j\xi_j^{12}=0$. When there is no restriction on the number of observations, the observations are replicated t times for each of $v=s_1\times s_2$ combinations. Components X_{ijk} of the observation vector $\mathbf{X}$ are represented by a linear model

$$X_{ijk}=\gamma+\xi_i^1+\xi_j^2+\xi_{ij}^{12}+W_{ijk},$$

$$i=1,\ldots,s_1; \quad j=1,\ldots,s_2; \quad k=1,\ldots,t,$$

or by the vector notation

$$\mathbf{X}=\mathbf{\Gamma}\gamma+A_1\xi^1+A_2\xi^2+A_{12}\xi^{12}+\mathbf{W}.$$

The analysis of variance in this case is given by

$$\mathbf{X}'\mathbf{X}=\sum_i \mathbf{X}'P_i\mathbf{X},$$

where P_i is the projective operator matrix for the subspace derived by a decomposition

$$\mathbf{R}^n=L(\mathbf{\Gamma})+L_\Gamma^\perp(A_1)+L_\Gamma^\perp(A_2)+L_{A_1A_2}^\perp(A_{12}) + L_{A_{12}}^\perp.$$

Denoting by $\bar{X}_{ij\cdot}$ the arithmetic mean of X_{ijk} over the subscript k and using similar notation $\bar{X}_{i\cdot\cdot}$, $\bar{X}_{\cdot j\cdot}$, and $\bar{X}_{\cdot\cdot\cdot}$, we have $\mathbf{X}'P_1\mathbf{X}=$

Table 1 Analysis of Variance

Factor	Sum of Squares	Degrees of Freedom
F_1 main effect	$\mathbf{X}'P_2\mathbf{X}$	s_1-1
F_2 main effect	$\mathbf{X}'P_3\mathbf{X}$	s_2-1
F_1F_2 interaction	$\mathbf{X}'P_4\mathbf{X}$	$(s_1-1)(s_2-1)$
Error term	$\mathbf{X}'P_5\mathbf{X}$	$s_1s_2(t-1)$

$n\overline{X}^2_{...}$, $\mathbf{X}'P_2\mathbf{X}=s_2t\Sigma(\overline{X}_{i..}-\overline{X}_{...})^2$, $\mathbf{X}'P_3\mathbf{X}=s_1t\Sigma(\overline{X}_{.j.}-\overline{X}_{...})^2$, $\mathbf{X}'P_4\mathbf{X}=t\Sigma(\overline{X}_{ij.}+\overline{X}_{...}-\overline{X}_{i..}-\overline{X}_{.j.})^2$, $\mathbf{X}'P_5\mathbf{X}=\Sigma(X_{ijk}-X_{ij.})^2$. The analysis of variance is given in Table 1.

J. Application of Algebras

In the theory of design of experiments, the ideas of association algebra and relationship algebra play an important role. Let A_i be a $v\times v$ symmetric matrix with entries 0 or 1 $(i=0,1,\ldots,m)$. If a set $\{A_i \mid i=0,1,\ldots,m\}$ satisfies the conditions

$$A_0=I_v,\qquad \sum_{i=0}^{m} A_i=E_{vv},$$

and there is a real number p^i_{jk} for every i,j,k such that

$$A_jA_k=\sum_{i=0}^{m} P^i_{jk}A_i,$$

then the A_i are called association matrices. If a component $a^i_{\alpha\beta}$ of A_i is unity, then the treatments α and β are said to be the ith associates. The †algebra $\mathfrak{A}$ over the real number field generated by the matrices $A_0, A_1, \ldots, A_m$ is called an **association algebra**. $\mathfrak{A}$ is commutative, and $A_k\rightarrow\mathfrak{P}_k=(p^i_{jk})$ is the †regular representation of $\mathfrak{A}$. There is a regular matrix $M=(M_{ij})$ that transforms all $\mathfrak{P}_i$ into diagonal matrices simultaneously:

$$M\mathfrak{P}_iM^{-1}=\operatorname{diag}(z_{0i},\ldots,z_{mi}),\quad i=0,1,\ldots,m.$$

$$A_i^{\#}=\Big(\sum_u M_{iu}z_{iu}\Big)^{-1}\sum_j M_{ij}A_j,\quad i=0,1,\ldots,m,$$

are mutually orthogonal idempotent elements of $\mathfrak{A}$. For example, consider the case where $A_0=I_{s_1}\otimes I_{s_2}$, $A_1=(E_{s_1s_1}-I_{s_1})\otimes I_{s_2}$, $A_2=I_{s_1}\otimes(E_{s_2s_2}-I_{s_2})$, $A_3=(E_{s_1s_1}-I_{s_1})\otimes(E_{s_2s_2}-I_{s_2})$ with $m=3$, where $\otimes$ means the †Kronecker product. The algebra $\mathfrak{A}$ generated by A_0, A_1, A_2, A_3 is called a G_3 type association algebra. The association matrices A_0, A_1, A_2, and A_3 correspond to the relationships between the treatments in a two-way layout. The orthogonal idempotent elements in this case are $A_0^{\#}=s_1^{-1}E_{s_1s_1}\otimes s_2^{-1}E_{s_2s_2}$, $A_1^{\#}=(I_{s_1}-s_1^{-1}E_{s_1s_1})\otimes s_2^{-1}E_{s_2s_2}$, $A_2^{\#}=s_1^{-1}E_{s_1s_1}\otimes(I_{s_2}-s_2^{-1}E_{s_2s_2})$, and $A_3^{\#}=(I_{s_1}-s_1^{-1}E_{s_1s_1})\otimes(I_{s_2}-s_2^{-1}E_{s_2s_2})$. For the factorial experiment with h factors $(h\geqslant 2)$, association matrices can be constructed in a similar way. If $h=3$, the number m of association matrices is 7. Many types of association schemes are known, and some of them are classified as group divisible, triangular, or cyclic types.

An experimental design consists of a set of n experimental units which we call plots. Define a relationship R between the plots as a set of ordered pairs (i,j) of plots. A relationship R among a set of n plots can be expressed as a symmetric $n\times n$ matrix (r_{ij}) of 0s and 1s:

$$r_{ij}=\begin{cases}1 & \text{if } i \text{ is related to } j \text{ by the relationship } R,\\ 0 & \text{otherwise,}\end{cases}$$

and this matrix is also denoted by R. If there are k sets of relationships $R_1,\ldots,R_k$ among n plots, the algebra $\mathfrak{R}$ over the real number field generated by the matrices $R_1,\ldots,R_k$ is called the **relationship algebra** of the design. $\mathfrak{R}$ is a †semisimple algebra.

Example (1). An algebra $\mathfrak{R}$ generated by the following matrices over the real number field is called the relationship algebra of the factorial experiment with $h=2$ and t replications. $B_1=A_0\otimes I_t$, $B_2=A_0\otimes(E_{tt}-I_t)$, $B_3=A_1\otimes E_{tt}$, $B_4=A_2\otimes E_{tt}$, $B_5=A_3\otimes E_{tt}$, where A_i, $i=0,1,2,3$, are the association matrices of G_3 type. The orthogonal idempotents that correspond to the two-sided ideal decomposition of $\mathfrak{R}$ are $B_1^{\#}=A_0^{\#}\otimes t^{-1}E_{tt}$, $B_2^{\#}=A_1^{\#}\otimes t^{-1}E_{tt}$, $B_3^{\#}=A_2^{\#}\otimes t^{-1}E_{tt}$, $B_4^{\#}=A_3^{\#}\otimes t^{-1}E_{tt}$, and $B_5^{\#}=A_0^{\#}\otimes(I_t-t^{-1}E_{tt})$. These are the same as the projection P_i in the two-way layout design.

Example (2). Consider a block design with v treatments, each having the same number of replications r, and with b blocks, each having the same size k $(<v)$. Suppose that association matrices A_i, $i=0,1,\ldots,m$, are given, by which the associations among the treatments are defined. Let $\lambda^i_{\alpha\beta}$ be the number of blocks to which the ith associate treatments α and β are applied. The design is called a **partially balanced incomplete block design (PBIBD)** if $\lambda^i_{jl}=\lambda^i\geqslant 0$ independently of j and l. When $m=1$, the design is a BIBD. Let the observation vector be represented by (4). The relationship algebra $\mathfrak{R}$ of a PBIBD is generated by $n\times n$ matrices I_n, E_{nn}, $B=\Psi\Psi'$, $T_i=\Phi A_i\Phi'$, $i=1,2,\ldots,m$, where $NN'=\Sigma\lambda_iA_i=\Sigma\rho_iA_i^{\#}$, with some constants ρ_i $(0\leqslant\rho_i\leqslant rk)$. Write $T_i^{\#}=\Phi A_i^{\#}\Phi'$. According as $\rho_i=rk$, $0<\rho_i<rk$, or $\rho_i=0$, $L(T_i^{\#})$ is said to be **confounded** with the blocks, **partially confounded** with the blocks, or **orthogonal** to the blocks. $\mathfrak{R}$ is noncommutative, †completely irreducible, and isomorphic to the algebra of matrices of the type shown in Fig. 1.

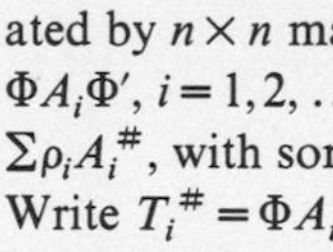

Fig. 1

The analysis of variance of a PBIBD is given by a decomposition of I_n into mutually orthogonal idempotent elements of $\mathfrak{R}$. For a PBIBD, the matrix C in (6) is of the form $C=\Sigma\tau_i A_i^{\#}$, where $\tau_i=r-k^{-1}\rho_i$.

K. Design for Two-Way Elimination of Heterogeneity

Consider a design with v treatments in a $u\times w$ rectangular block. The row effect and the column effect of this block are denoted by $\boldsymbol{\eta}$ and $\boldsymbol{\nu}$. Thus the observation vector $\mathbf{X}$ is of the form

$$\mathbf{X}=\boldsymbol{\Gamma}\gamma+\Phi\xi+\Psi\boldsymbol{\eta}+\Pi\boldsymbol{\nu}+\mathbf{W}, \tag{10}$$

where the definitions of $\boldsymbol{\Gamma}$, Φ, Ψ and Π are similar to those for block designs. Set $L=\Phi'\Pi$, $M=\Phi'\Psi$, $D_r=\Phi'\Phi$, $F=D_r-w^{-1}LL'-u^{-1}MM'+u^{-1}w^{-1}LE_{uu}L'$. The matrix F plays a similar role to that of the matrix C in (6). When the rank of F equals $v-1$, the design is called **connected**. If

$$F=\tau(I_v-v^{-1}E_{vv}), \tag{11}$$

then the design satisfies the optimal conditions (I), (II), and (III) given in section E. When $u=w=v$ and (11) holds, the design is called a **Latin square**. When $u=v$ and (11) holds, the design is called a **Youden square**. When $u>v$ and (11) holds, the design is called a **Shrikhande square**.

If the associations among v treatments are defined in terms of an association scheme, the partially balanced design for **two-way elimination of heterogeneity** can be defined in a way similar to the PBIBD. In this case, the equation $F=\Sigma\tau_i A_i^{\#}$ holds, and if $\mathfrak{A}=\{I_v, E_{v_1v}-I_v\}$, then (11) holds, hence the optimum conditions are fulfilled. The definition of the relationship algebra $\mathfrak{R}$ of a partially balanced design for two-way elimination of heterogeneity is similar to that used for the PBIBD. $\mathfrak{R}$ is isomorphic to the algebra of matrices shown in Fig. 2.

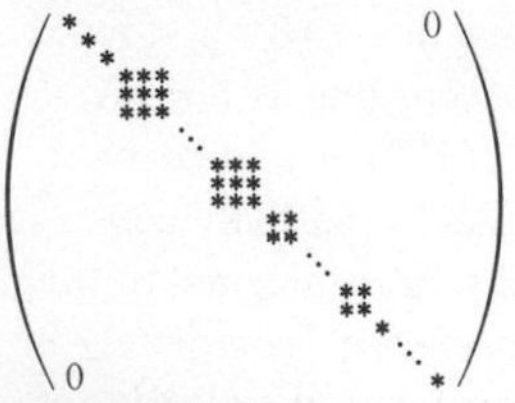

Fig. 2

The analysis of variance of this design is given by a decomposition of I_n into mutually orthogonal idempotent elements of $\mathfrak{R}$.

L. Orthogonal Layouts

Consider an arrangement of v different kinds of letter in an $N\times k$ matrix. If for any set of d columns ($2\leqslant d<k$), every arrangement of d letters from v kinds of letters (the number of possible arrangements is v^d) is repeated λ times in N rows, the arrangement is called an **orthogonal layout** (N,k,v,d) of size N with constraint k, level v, and strength d. When $d=2$ and $\lambda=1$, the orthogonal layouts $(v^2,k,v,2)$ are equivalent to the $k-2$ orthogonal Latin squares. There is a very close relation between the existence of an orthogonal layout and that of a BIBD. For example, the existence of an orthogonal layout $(k^2,k+1,k,2)$ is equivalent to the existence of a BIBD ($v=k^2,k,b=k(k+1)$, $r=k+1,\lambda=1$).

There are still many topics in the theory of experimental design besides the ones mentioned in this article (→ [4] for multiple comparison, [5] for fractional replication, and [7] for response surface).

References

[1] Article on "Experimental design" (Japanese), Gendai tôkeigaku daiziten (Encyclopedia of modern statistics), Tôyôkeizai simpôsya, 1962.
[2] M. Masuyama, Zikken keikaku hô (Japanese; Methods of experimental design), Iwanami Coll. of Modern Math. Sci., 1958.
[3] A. T. James, The relationship algebra of an experimental design, Ann. Math. Statist., 28 (1957), 993–1002.
[4] H. Scheffé, The analysis of variance, John Wiley, 1959.
[5] O. Kempthorne, The design and analysis of experiments, John Wiley, 1952.
[6] W. G. Cochran and G. M. Cox, Experimental designs, John Wiley, 1950.
[7] G. E. P. Box and N. R. Draper, Evolutionary operation, John Wiley, 1968.

107 (III.3) Determinants

A. Definition

The **determinant** of an $n\times n$ †matrix $A=(a_{ik})$ in a †commutative ring R is defined to be the following element of R:

$$\sum(\operatorname{sgn}P)a_{1p_1}a_{2p_2}\dots a_{np_n},$$

where

$$P=\begin{pmatrix}1 & 2 & \dots & n\\ p_1 & p_2 & \dots & p_n\end{pmatrix}$$

is a permutation of the numbers $1, 2, \dots, n$, $\operatorname{sgn}P$ denotes the **sign** of the permutation P (that is, $\operatorname{sgn}P=1$ if P is an even permutation and $\operatorname{sgn}P=-1$ if P is an odd permutation), and the summation extends over all $n!$ permutations of $1, 2, \ldots, n$. The determinant of A is denoted by

$$\begin{vmatrix} a_{11} & a_{12} & \cdots & a_{1n} \\ a_{21} & a_{22} & \cdots & a_{2n} \\ & & \cdots & \\ a_{n1} & a_{n2} & \cdots & a_{nn} \end{vmatrix}.$$

It is written $|a_{ik}|$ or $|A|$ and is also denoted by $\det A$. Usually we suppose that R is the field $\mathbf{R}$ of real numbers or the field $\mathbf{C}$ of complex numbers, but the following theorems are also valid for cases in which R is any commutative ring, unless otherwise stated.

B. Relation with Exterior Algebras

Consider an †exterior algebra (†Grassmann algebra) of a linear space (†free module) of dimension n over R with a basis $(e_1, e_2, \ldots, e_n)$. Set

$e_i' = a_{i1}e_1 + a_{i2}e_2 + \ldots + a_{in}e_n$, where $a_{ij} \in R$.

Then we have $e_1' \wedge e_2' \wedge \ldots \wedge e_n' = |a_{ik}| e_1 \wedge e_2 \wedge \ldots \wedge e_n$. Conversely, we can define the determinant $|a_{ik}|$ by this relation. The properties of determinants can be easily deduced from those of exterior algebras.

C. Fundamental Properties of a Determinant

(1) The determinant of the †transpose ${}^tA$ of a matrix A is equal to the determinant of A. Hence the theorems on determinants, stated for rows, are also valid for columns.

(2) If the elements of one row (column) of a matrix are multiplied by a factor c, the determinant of the matrix is also multiplied by c. If the elements of one row (column) of a matrix are zero, its determinant is equal to zero.

(3) If in a matrix $A = (a_{ik})$, we obtain two matrices A' and A'' by replacing one row, for instance the ith row, by $a_{i1}', \ldots, a_{in}'$ and by $a_{i1} + a_{i1}', \ldots, a_{in} + a_{in}'$, respectively, then $|A''| = |A| + |A'|$. This relation is equally valid for a column.

(4) If we obtain A_Q by a permutation Q on the rows of a matrix A, then $|A_Q| = (\operatorname{sgn} Q)|A|$. In particular, if two rows (columns) of a matrix are interchanged, then the determinant changes sign.

(5) The determinant of a matrix is zero if two rows (or columns) are identical.

(6) The determinant of a matrix is not changed if the elements of any row (column), each multiplied by the same factor, are added to the corresponding elements of another row (column).

(7) Suppose that R has unity element. Let x_{ik} $(i, k = 1, \ldots, n)$ be n^2 variables in R, and denote a function (having its values in R) of these variables by $\varphi(X)$, $X = (x_{ik})$. Assume that $\varphi(X)$ has the following properties: (i) if the elements of one row of X are multiplied by a factor λ, the value of φ is also multiplied by λ; (ii) if we obtain two matrices X' and X'' by replacing one row of X, for instance the ith row, by $x_{i1}', \ldots, x_{in}'$ and by $x_{i1} + x_{i1}', \ldots, x_{in} + x_{in}'$, respectively, then $\varphi(X'') = \varphi(X) + \varphi(X')$; and (iii) if two rows of X are equal, $\varphi(X) = 0$. Then $\varphi(X) = c|X|$ for some constant c (in R).

(8) Suppose now that R is a field K. Assume that a function $\varphi(X)$ (in K) has the following properties: (i) if the elements of one row in X are multiplied by λ, the value of φ is also multiplied by λ; and (ii) the value of φ is not changed if the elements of any row are added to the corresponding elements of another row. Then $\varphi(X) = c|X|$ for some constant c (in K).

D. The Laplace Expansion Theorem

Let $A = (a_{ik})$ be an $n \times n$ matrix. Take r-tuples $(i_1, \ldots, i_r)$ and $(k_1, \ldots, k_r)$, where i_a and k_b belong to $\{1, \ldots, n\}$ and $i_1 < \ldots < i_r$, $k_1 < \ldots < k_r$. Let $(i_{r+1}, \ldots, i_n)$ and $(k_{r+1}, \ldots, k_n)$ be $(n-r)$-tuples such that $i_{r+1} < \ldots < i_n$, $k_{r+1} < \ldots < k_n$ and $\{i_1, \ldots, i_r, i_{r+1}, \ldots, i_n\} = \{k_1, \ldots, k_r, k_{r+1}, \ldots, k_n\} = \{1, \ldots, n\}$. Let

$a_{(i_1, \ldots, i_r)(k_1, \ldots, k_r)}$

be the determinant of an $r \times r$ matrix whose (p,q)-component is the (i_p, k_q)-component of A for each p and q. We call this determinant a **minor** of degree r of the matrix A. (The corresponding submatrix of A is sometimes also called the minor of A.) In particular, if $(i_1, \ldots, i_r) = (k_1, \ldots, k_r)$, then it is called a **principal minor**. Furthermore, we define the **cofactor** of the minor $a_{(i_1, \ldots, i_r)(k_1, \ldots, k_r)}$ of A to be

$\tilde{a}_{(i_1, \ldots, i_r)(k_1, \ldots, k_r)} = (-1)^{\lambda+\mu} a_{(i_{r+1}, \ldots, i_n)(k_{r+1}, \ldots, k_n)}$,

where $\lambda = i_1 + \ldots + i_r$ and $\mu = k_1 + \ldots + k_r$. In the particular case $r = 1$, the cofactor of a_{ik} is $\tilde{a}_{ik} = (-1)^{i+k}\Delta_{ik}$, where Δ_{ik} is the determinant of the $(n-1) \times (n-1)$ matrix obtained from A by eliminating its ith row and kth column. For simplicity, we abbreviate $(i_1, \ldots, i_r)$, $(k_1, \ldots, k_r)$, and $(j_1, \ldots, j_r)$ as (i), (k), and (j), respectively. Then we have

$$\sum_{(j)} a_{(i)(j)} \tilde{a}_{(k)(j)} = \begin{cases} |A| & \text{if } (i) = (k), \\ 0 & \text{if } (i) \neq (k), \end{cases}$$

$$\sum_{(j)} a_{(j)(k)} \tilde{a}_{(j)(i)} = \begin{cases} |A| & \text{if } (i) = (k), \\ 0 & \text{if } (i) \neq (k), \end{cases}$$

where $\sum_{(j)}$ means that the sum is taken over all combinations (j). This is called the **Laplace**

expansion theorem. If a matrix A has the form

$$A=\begin{pmatrix} B & 0 \\ * & C \end{pmatrix} \text{ or } A=\begin{pmatrix} B & * \\ 0 & C \end{pmatrix},$$

and B and C are square matrices, then by this theorem we have $|A|=|B||C|$. If we number the combinations $(i)=(i_1, \dots, i_r)$ and $(k)=(k_1, \dots, k_r)$ appropriately (for instance, in lexicographical order) and regard the numbers assigned to them as row numbers and column numbers, respectively, to form a matrix $(a_{(i)(k)})$, then the Laplace theorem can be expressed as

$$(a_{(i)(k)})(\tilde{a}_{(k)(i)})=(\tilde{a}_{(k)(i)})(a_{(i)(k)})$$

$$=\begin{bmatrix} |A| & & & 0 \\ & \cdot & & \\ & & \cdot & \\ 0 & & & |A| \end{bmatrix}.$$

In the particular case $r=1$, we have

$$\sum_{j=1}^{n} a_{ij}\tilde{a}_{kj}=\begin{cases} |A| & \text{if } i=k, \\ 0 & \text{if } i\neq k, \end{cases}$$

$$\sum_{j=1}^{n} a_{jk}\tilde{a}_{ji}=\begin{cases} |A| & \text{if } i=k, \\ 0 & \text{if } i\neq k. \end{cases}$$

E. Product of Determinants

Let $A=(a_{ik})$ and $B=(b_{ik})$ be two $n\times n$ matrices. For the product $AB=C=(c_{ik})$, where $c_{ik}=\sum_{j=1}^{n} a_{ij}b_{jk}$ $(i,k=1, \dots, n)$, we have $|AB| =|A||B|$. The †inverse matrix A^{-1} exists for an $n\times n$ matrix $A=(a_{ik})$ if and only if $|A|\neq 0$, and then $A^{-1}=(b_{ik})$ with elements $b_{ik}=\tilde{a}_{ki}/|A|$. Moreover, we have $|A^{-1}|=|A|^{-1}$. (In the case where the elements a_{ik} are in the commutative ring R with unity element, A^{-1} exists if and only if $|A|$ is a †regular element of R.)

F. Theorems on Determinants

(1) Let $\tilde{a}_{ik}$ be the cofactor of a_{ik} in the determinant of an $n\times n$ matrix $A=(a_{ik})$. Then the determinant $|\tilde{a}_{ik}|$ is equal to $|A|^{n-1}$. In general,

$$|a_{(i_1, \dots, i_r)(k_1, \dots, k_r)}|=|A|^{\binom{n-1}{r-1}},$$

$$|\tilde{a}_{(i_1, \dots, i_r)(k_1, \dots, k_r)}|=|A|^{\binom{n-1}{r}}.$$

(2) The determinant of a submatrix of the matrix $(\tilde{a}_{ik})$, composed of the i_1th, $\dots, i_r$th rows and k_1th, $\dots, k_r$th columns of $(\tilde{a}_{ik})$ is equal to

$$|A|^{r-1}a_{(i_1, \dots, i_r)(k_1, \dots, k_r)}.$$

(3) Let $\Delta\begin{pmatrix} i_1, \dots, i_r \\ k_1, \dots, k_r \end{pmatrix}$ be the determinant of the $(n-r)\times(n-r)$ matrix obtained from an $n\times n$ matrix A by eliminating its i_1th, $\dots$, i_rth rows and k_1th, $\dots, k_r$th columns. Then

$$|A|\Delta\begin{pmatrix} i & j \\ k & l \end{pmatrix}=\Delta\begin{pmatrix} i \\ k \end{pmatrix}\Delta\begin{pmatrix} j \\ l \end{pmatrix}-\Delta\begin{pmatrix} i \\ l \end{pmatrix}\Delta\begin{pmatrix} j \\ k \end{pmatrix},$$

$$i<j, \quad k<l.$$

(4) **Sylvester's theorem.** Let b_{ik} $(i,k=1, \dots, n-r)$ denote the minor $a_{(1, \dots, r, r+i)(1, \dots, r, r+k)}$ of an $n\times n$ matrix $A=(a_{jl})$. Then

$$|b_{ik}|=|A|\begin{vmatrix} a_{11} & \dots & a_{1r} \\ & \dots & \\ a_{r1} & \dots & a_{rr} \end{vmatrix}^{n-r-1}.$$

(5) Let A be an $n\times m$ matrix and B an $m\times n$ matrix. Then AB is an $n\times n$ matrix. If $n>m$, then $|AB|=0$. If $n\leqslant m$, let $(i)=(i_1, \dots, i_n)$ $(i_1<\dots<i_n)$ be a combination of $1,2, \dots, m$, taken n at a time. Let $A_{(i)}$ be the $n\times n$ matrix composed of the i_1th, $\dots, i_n$th columns of A, and $B_{(i)}$ the $n\times n$ matrix composed of the i_1th, $\dots, i_n$th rows of B. Then $|AB|=\sum_{(i)}|A_{(i)}||B_{(i)}|$, where the summation extends over all possible combinations (i).

(6) Determinant of a †Kronecker product. If A is an $m\times m$ matrix and B is an $n\times n$ matrix, then $|A\otimes B|=|A|^n|B|^m$.

(7) Let H be an $n\times n$ †Hermitian matrix, and let H_k denote the matrix composed of its first k rows and columns. Then H is positive definite if and only if $|H_k|>0$ for all $k=1, \dots, n$.

G. Special Determinants

(1) A determinant of the form

$$\begin{vmatrix} 1 & 1 & \dots & 1 \\ x_1 & x_2 & \dots & x_n \\ x_1^2 & x_2^2 & \dots & x_n^2 \\ & & \dots & \\ x_1^{n-1} & x_2^{n-1} & \dots & x_n^{n-1} \end{vmatrix}$$

is called a **Vandermonde determinant**. It is equal to the †simplest alternating function $\prod_{i>k}(x_i-x_k)$.

(2) The **cyclic determinant** is a determinant of the following form:

$$\begin{vmatrix} x_0 & x_1 & x_2 & \dots & x_{n-1} \\ x_{n-1} & x_0 & x_1 & \dots & x_{n-2} \\ & & & \dots & \\ x_1 & x_2 & x_3 & \dots & x_0 \end{vmatrix}$$

$$=\prod_{i=0}^{n-1}(x_0+\zeta^i x_1+\zeta^{2i}x_2+\dots+\zeta^{(n-1)i}x_{n-1}),$$

where ζ is a †primitive nth root of unity.

(3) Consider the vectors $\alpha_i=(a_{i1}, a_{i2}, \dots, a_{in})$ $(i=1,2, \dots, n)$, and let (α_i, α_j) denote the †inner product of α_i and α_j. Then the following determinant is called the **Gramian** of these

vectors:

$$\begin{vmatrix} (\alpha_1,\alpha_1) & (\alpha_1,\alpha_2) & \dots & (\alpha_1,\alpha_n) \\ (\alpha_2,\alpha_1) & (\alpha_2,\alpha_2) & \dots & (\alpha_2,\alpha_n) \\ & & \dots & \\ (\alpha_n,\alpha_1) & (\alpha_n,\alpha_2) & \dots & (\alpha_n,\alpha_n) \end{vmatrix} = \begin{vmatrix} a_{11} & a_{12} & \dots & a_{1n} \\ a_{21} & a_{22} & \dots & a_{2n} \\ & & \dots & \\ a_{n1} & a_{n2} & \dots & a_{nn} \end{vmatrix}^2.$$

(4) For an †alternating matrix (namely, a square matrix X such that ${}^tX = -X$), we have the following identity:

$$\begin{vmatrix} 0 & x_{12} & x_{13} & \dots & x_{1n} \\ -x_{12} & 0 & x_{23} & \dots & x_{2n} \\ -x_{13} & -x_{23} & 0 & \dots & x_{3n} \\ & & & \dots & \\ -x_{1n} & -x_{2n} & -x_{3n} & \dots & 0 \end{vmatrix} = \begin{cases} P_n(\dots,x_{ij},\dots)^2 & \text{if } n \text{ is even,} \\ 0 & \text{if } n \text{ is odd,} \end{cases}$$

where $P_n(\dots,x_{ij},\dots)$ is a polynomial of variables x_{ij}, which (equipped with appropriate sign) is called the **Pfaffian** of these variables.

References

[1] T. Takagi, Daisûgaku kôgi (Japanese; Lectures on algebra), Kyôritu, revised edition, 1965.
[2] M. Fujiwara, Daisûgaku I (Japanese; Algebra I), Utida-rôkakuho, 1928.
[3] M. Fujiwara, Gyôretu oyobi gyôretusiki (Japanese; Matrices and determinants), Iwanami, revised edition, 1961.
[4] H. Aramata, Gyôretu oyobi gyôretusiki (Japanese; Matrices and determinants), Tôkai, 1947.
[5] K. Asano, Senkei daisûgaku teiyô (Japanese; A manual of linear algebra), Kyôritu, 1948.
[6] H. Tôyama, Gyôreturon (Japanese; Theory of matrices), Kyôritu, 1951.
[7] I. Satake, Gyôretu to gyôretusiki (Japanese; Matrices and determinants), Syôkabô, 1958.
[8] S. Furuya, Gyôretu to gyôretusiki (Japanese; Matrices and determinants), Baihûkan, 1957.
[9] G. W. H. Kowalewski, Einführung in die Determinantentheorie, Walter de Gruyter, second revised edition, 1925 (Chelsea, 1948).
[10] N. Bourbaki, Eléments de mathématique, Algèbre, ch. 3, Actualités Sci. Ind., 1044, Hermann, new edition, 1958.
[11] Ф. Р. Гантмахер (F. R. Gantmaher), Теория Матриц, Гостехиздат 1953; English translation, The theory of matrices, Chelsea, I, II, 1959.
[12] А. Г. Курош (A. G. Kurosh), Курс высшей алгебры, Гостехиздат, 1955; English translation, Lectures on general algebra, Chelsea, 1970.
[13] R. Godement, Cours d'algèbre, Hermann, 1963.
[14] S. MacLane and G. Birkhoff, Algebra, Macmillan, 1967.

108 (VII.2) Differentiable Manifolds

A. Local Coordinates

An n-dimensional †topological manifold M is a †Hausdorff space such that every point has an open neighborhood homeomorphic to an open set of $\mathbf{R}^n$. A pair (U,ψ) consisting of an open set U of M and a homeomorphism ψ of U onto an open set of $\mathbf{R}^n$ is called a **coordinate neighborhood** of M. If we denote by $(x^1(p),\dots,x^n(p))$ $(p\in U)$ the coordinates of the point $\psi(p)$ of $\mathbf{R}^n$, then $x^1,x^2,\dots,x^n$ are real-valued continuous functions defined on U. We call these n functions the **local coordinate system** in the coordinate neighborhood (U,ψ) and the n real numbers $x^1(p)$, ..., $x^n(p)$ the **local coordinates** of the point $p\in U$ (with respect to (U,ψ)).

A set $S=\{(U_\alpha,\psi_\alpha)\}_{\alpha\in A}$ of coordinate neighborhoods is called an **atlas** of M if $\{U_\alpha\}_{\alpha\in A}$ forms an †open covering of M.

B. Differentiable Manifolds

Let $S=\{(U_\alpha,\psi_\alpha)\}_{\alpha\in A}$ be an atlas of an n-dimensional topological manifold M. For each pair of coordinate neighborhoods (U_α,ψ_α) and (U_β,ψ_β) in S such that $U_\alpha\cap U_\beta\neq\varnothing$, $\psi_\beta\circ\psi_\alpha^{-1}$ is a homeomorphism of the open set $\psi_\alpha(U_\alpha\cap U_\beta)$ of $\mathbf{R}^n$ onto the open set $\psi_\beta(U_\alpha\cap U_\beta)$ of $\mathbf{R}^n$. Let $x=(x^1,\dots,x^n)\in\psi_\alpha(U_\alpha\cap U_\beta)$. Then we can write $(\psi_\beta\circ\psi_\alpha^{-1})(x)=(f^1_{\beta\alpha}(x),\dots,f^n_{\beta\alpha}(x))$. If the n real-valued functions $f^1_{\beta\alpha},\dots,f^n_{\beta\alpha}$ defined in $\psi_\alpha(U_\alpha\cap U_\beta)$ are of †class C^r $(1\leqslant r\leqslant\infty)$ (resp. †real analytic) for any α, β in A such that $U_\alpha\cap U_\beta\neq\varnothing$, then we call S an **atlas of class** C^r (resp. C^ω) of M. When an n-dimensional topological manifold M has an atlas S of class C^r $(1\leqslant r\leqslant\omega)$, we call the pair (M,S) an n-dimensional **differentiable manifold of class** C^r (or C^r**-manifold**). A C^ω-manifold is also called a **real analytic manifold**. We call M the **underlying topological space** of (M,S), and we say that S defines a **differentiable structure of class** C^r (or C^r-**structure**) in M.

In particular, a C^ω-structure is called a **real analytic structure**. A C^r-manifold whose underlying topological space is compact (†paracompact) is called a **compact (paracompact)** C^r**-manifold**. A coordinate neighborhood (U,ψ) of M is called a **coordinate neighborhood of class** C^r of (M,S) if the union $S \cup \{(U,\psi)\}$ is also an atlas of class C^r of M. In particular, each coordinate neighborhood of M belonging to S is of class C^r. The set $\tilde{S}$ of all coordinate neighborhoods of class C^r of (M,S) is an atlas of M containing S, and we call $\tilde{S}$ the maximal atlas containing S. Let S and S' be two atlases of class C^r of M. If $\tilde{S}=\tilde{S}'$, then we say that S and S' define the same differentiable structure of class C^r on M and that the differentiable manifolds (M,S) and (M,S') of class C^r are equivalent. In particular, (M,S) and $(M,\tilde{S})$ are equivalent C^r-manifolds. Let S and S' be atlases of class C^r and class C^s, respectively, where $1 \leqslant r < s \leqslant \omega$. Since $s > r$, we may consider S' an atlas of class C^r. If S and S' define the same C^r-structure in M, then we say that the C^s-structure defined by S' is **subordinate** to the C^r-structure defined by S. If M is paracompact, then there exists a C^ω-structure subordinate to a C^r-structure of M (Whitney [13]).

C. Differentiable Manifolds with Boundaries

Let H^n be the half-space of $\mathbf{R}^n$ consisting of all $(x^1,\ldots,x^n)\in\mathbf{R}^n$ such that $x^1 \geqslant 0$. We denote the boundary of H by ∂H. Let U and U' be open sets in H^n, and let $\varphi: U \rightarrow U'$ be a continuous mapping. If there exist open sets W and W' in $\mathbf{R}^n$ containing U and U', respectively, and a mapping $\psi: W \rightarrow W'$ of class C^r that extends φ, we call φ a mapping of U into U' of class C^r. Let M be a Hausdorff topological space. A structure of a C^r-manifold with boundary in M is defined by a set $S=\{(U_\alpha,\psi_\alpha)\}_{\alpha\in A}$, where $\{U_\alpha\}_{\alpha\in A}$ is an open covering of M, and, for each α, ψ_α is a homeomorphism of U_α onto an open set of H^n such that for any $\alpha, \beta \in A$ with $U_\alpha \cap U_\beta \neq \varnothing$, $\psi_\beta \circ \psi_\alpha^{-1}$ is a mapping of class C^r from $\psi_\alpha(U_\alpha \cap U_\beta)$ onto $\psi_\beta(U_\alpha \cap U_\beta)$. The pair (M,S) is called an n-dimensional **differentiable manifold with boundary of class** C^r (or C^r**-manifold with boundary**). Let p be a point of M, and let $p \in U_\alpha$ for some $\alpha \in A$. If $\psi_\alpha(p)$ belongs to the boundary ∂H^n of H^n, we call p a **boundary point** of M. The set ∂M of all boundary points of M forms an $(n-1)$-dimensional C^r-manifold, which we call the **boundary** of M. If we put $U_\alpha' = U_\alpha \cap \partial M$ and denote the restriction of ψ_α to U_α' by ψ_α', then $S'=\{(U_\alpha',\psi_\alpha')\}_{\alpha\in A}$ is an atlas of class C^r of ∂M. If ∂M is empty, then (M,S) is a C^r-manifold. In this sense a C^r-manifold is sometimes called a C^r**-manifold without boundary**.

D. Orientation of a Manifold

Let $S=\{(U_\alpha,\psi_\alpha)\}_{\alpha\in A}$ be an atlas of class C^r in M, and for each α let $\{x_\alpha^1,\ldots,x_\alpha^n\}$ be the local coordinate system in the coordinate neighborhood (U_α,ψ_α). If U_α and U_β intersect, then there exist n real-valued functions F^i $(i=1,\ldots,n)$ defined on $\psi_\alpha(U_\alpha \cap U_\beta)$ such that $x_\beta^i(p)=F^i(x_\alpha^1(p),\ldots,x_\alpha^n(p))$ for $p \in U_\alpha \cap U_\beta$ and $i=1,\ldots,n$. The †Jacobian $D_{\alpha\beta} = D(F^1,\ldots,F^n)/D(x_\alpha^1,\ldots,x_\alpha^n)$ is different from zero at each point $(x_\alpha^1,\ldots,x_\alpha^n)$ of $\psi_\alpha(U_\alpha \cap U_\beta)$. If we can choose an atlas S of M so that, for any α, β such that $U_\alpha \cap U_\beta$ is nonempty, the Jacobian $D_{\alpha\beta}$ is always positive, then we say that the C^r-manifold M is **orientable**, and we call S an **oriented atlas**.

Let $S=\{(U_\alpha,\psi_\alpha)\}_{\alpha\in A}$ and $S'=\{(V_\lambda,\varphi_\lambda)\}_{\lambda\in\Lambda}$ be two oriented atlases of a connected C^r-manifold M. If M is connected, then the sign of the Jacobian $D_{\alpha\lambda}(p)$ of the transformation of local coordinates is independent of the choice of $\alpha\in A$, $\lambda\in\Lambda$, and $p \in U_\alpha \cap V_\lambda$. We say that S and S' define the **same (opposite) orientation** if $D_{\alpha\lambda}$ is always positive (negative). Hence if M is connected, the set of all oriented atlases of class C^r is composed of two subsets such that atlases belonging to one of them have the same orientation while atlases belonging to the other have the opposite orientation. Each of these subsets is called an **orientation** of the connected C^r-manifold M. When we assign to M one of two possible orientations, M is called an **oriented manifold**; the assigned orientation is called its **positive orientation** and the other its **negative orientation**. If $S=\{(U_\alpha,\psi_\alpha)\}_{\alpha\in A}$ belongs to the positive orientation, S and (U_α,ψ_α) are called an atlas and local coordinate system, respectively, compatible with the positive orientation.

E. Differentiable Functions

Let f be a real-valued function defined in a neighborhood of a point p of a C^∞-manifold M. Let (U,ψ) be a coordinate neighborhood of class C^∞ such that $p \in U$. If the function $f \circ \psi^{-1}$ is of class C^r $(1 \leqslant r \leqslant \infty)$ in a neighborhood of the point $\psi(p)$ in $\mathbf{R}^n$, then the function f is called a **function of class** C^r **at** p. This definition is independent of the choice of a coordinate neighborhood of class C^∞. If we denote the local coordinate system in (U,ψ) by $(x^1,\ldots,x^n)$, there exists a function $f(x^1,\ldots,x^n)$ of n variables defined in a neighborhood of $\psi(p)$ in $\mathbf{R}^n$ such that $f(q)=f(x^1(q),\ldots,x^n(q))$ for each point q in the neighborhood of $\psi(p)$. Here we use the same symbol f for the function f defined in a neighborhood of p in M and for the function $f\circ\varphi^{-1}$ defined in the image of the neighborhood by ψ in $\mathbf{R}^n$. The function f is of class C^r at p if and only if $f(x^1,\ldots,x^n)$ is of class C^r in a neighborhood of the point $(x^1(p),\ldots,x^n(p))$ of $\mathbf{R}^n$. A **function of class** C^r (or C^r-**function) in** M is a real-valued function in M that is of class C^r at every point of M.

F. Tangent Vectors

Let M be a C^∞-manifold, and let $\mathfrak{F}(M)$ be the real vector space consisting of all C^∞-functions in M. (For the sake of simplicity, we treat only C^∞-manifolds, and we denote a manifold (M,S) by M.) A **tangent vector** L at a point p of M is a linear mapping $L:\mathfrak{F}(M)\to\mathbf{R}$ such that $L(fg)=L(f)g(p)+f(p)L(g)$ for any f and g in $\mathfrak{F}(M)$. For any two tangent vectors L_1, L_2 and any two real numbers λ_1, λ_2, we define $\lambda_1L_1+\lambda_2L_2$ by $(\lambda_1L_1+\lambda_2L_2)(f)=\lambda_1L_1(f)+\lambda_2L_2(f)$, $f\in\mathfrak{F}(M)$.

Thus tangent vectors at p form a real vector space T_p, which we call the **tangent vector space** (or simply **tangent space**) of M at the point p. The dimension of the tangent vector space T_p equals the dimension of M. The set of all tangent vectors of M forms a †vector bundle (→ 155 Fiber Bundles) over the base space M, called the **tangent vector bundle** (or **tangent bundle**) of M.

By a **tangent r-frame** ($r\leqslant n$) at p we mean an ordered set of r linearly independent tangent vectors at p. The set of all tangent r-frames also forms a fiber bundle over M called the †tangent r-frame bundle (or **bundle of tangent r-frame**).

G. Differential of Function

For a C^∞-function f in M and a point p of M we can define a linear mapping $df_p:T_p\to\mathbf{R}$ by $df_p(L)=L(f)$ for all $L\in T_p$, and we call df_p the **differential of** f **at** p. The totality of differentials at p of C^∞-functions in M forms the †dual vector space of the tangent vector space T_p.

H. Differentiable Mappings

Let φ be a continuous mapping of a C^∞-manifold M into a C^∞-manifold M'. We call φ a **differentiable mapping of class** C^r (or simply C^r-**mapping**) ($1\leqslant r\leqslant\infty$) if the function $f\circ\varphi$ is of class C^r for any C^r-function f on M'. If φ is a homeomorphism of M onto M' and φ and φ^{-1} are both of class C^r, then we call φ a **diffeomorphism of class** C^r. If there exists a diffeomorphism of class C^∞ of a C^∞-manifold M onto a C^∞-manifold M', then M and M' are said to be **diffeomorphic**.

I. Differentials of Differentiable Mappings

Let M and M' be C^∞-manifolds and φ be a C^∞-mapping of M into M'. For a tangent vector L of M at p, a tangent vector L' of M' at $\varphi(p)$ is defined by $L'(g)=L(g\circ\varphi)$, $g\in\mathfrak{F}(M')$. The mapping $L\to L'$ defines a linear mapping $(d\varphi)_p$ of the tangent vector space T_p of M at p into the tangent vector space $T_{\varphi(p)}$ of M' at $\varphi(p)$. The linear mapping $(d\varphi)_p$ is called the **differential of the differentiable mapping** φ **at** p. If $(d\varphi)_p$ is injective, we say the differentiable mapping φ is **regular** at p, and if φ is regular at every point of M, we say that φ is an **immersion** of M into M'. If φ is an injective immersion, then it is an **embedding** of M into M'.

J. Submanifolds

A C^∞-manifold M is said to be a **submanifold** of a C^∞-manifold M' if M is a subset of M' and the identity mapping of M into M' is an embedding. When the topology of M is the one induced from the topology of M', then M is called a **regular submanifold** of M'. A regular submanifold M of M' is called a **closed submanifold** if M is a closed subset of M'. Every n-dimensional paracompact C^∞-manifold is diffeomorphic to a closed submanifold of the Euclidean space E^{2n+1} (**Whitney's theorem**).

Let M be a submanifold of an n-dimensional Euclidean space E^m. We can identify the tangent vector space T_p of M at p with the geometric tangent space of M at p in the Euclidean space E^m. A vector in E^m that is orthogonal to the tangent vector space T_p of M at p is called a **normal vector** to M at p. The set of all vectors normal to M forms a vector bundle over M, which we call the **normal vector bundle** (or **normal bundle**) of M. If M is compact, then the totality of vectors normal to M whose length is smaller than ε (where ε is a sufficiently small positive real number) forms an open neighborhood of M in E^m which we call a **tubular neighborhood** of M.

K. Vector Fields

Let N be a subset of a C^∞-manifold M. By a **vector field** in N we mean a mapping X that assigns to each point p of N a tangent vector

X_p of M at p. We can consider X a †cross section over N of the tangent vector bundle of M. Let X be a vector field in M, and let f be a C^∞-function in M. Then we can define a function Xf in M by $(Xf)(p)=X_pf$. We call X a **vector field of class** C^r if the function Xf is of class C^r for any C^∞-function f in M. Let $(x^1,\ldots,x^n)$ be the local coordinate system in a coordinate neighborhood (U,ψ), and let $(\partial/\partial x^i)_pf=(\partial f/\partial x^i)(p)$ for $p\in U$ and $f\in \mathfrak{F}(M)$. Then $\partial/\partial x^i$ $(i=1,\ldots,n)$ are vector fields in U, and the $(\partial/\partial x^i)_p$ form a basis of T_p at every point $p\in U$. A vector field X in U is written uniquely as $X_p=\sum_i\xi^i(p)(\partial/\partial x^i)_p$ at each point $p\in U$. Then $\xi^1,\ldots,\xi^n$ are real-valued functions defined in U, called the **components** of X with respect to the local coordinate system $(x^1,\ldots,x^n)$. A vector field X in M is of class C^r if and only if its components ξ^i with respect to each coordinate system are functions of class C^r $(0\leqslant r\leqslant\infty)$. Let $(\bar{x}^1,\ldots,\bar{x}^n)$ be another local coordinate system in a neighborhood U of p, and let $(\bar{\xi}^1,\ldots,\bar{\xi}^n)$ be the components of X with respect to $(\bar{x}^1,\ldots,\bar{x}^n)$. Then we have $\bar{\xi}^i(q)=\sum_j(\partial\bar{x}^i/\partial x^j)(q)\xi^j(q)$ at each point $q\in U$.

For the rest of this article we mean by a vector field in M a vector field of class C^∞, and we denote by $\mathfrak{X}(M)$ the set of all vector fields in M. Then $\mathfrak{X}(M)$ is an $\mathfrak{F}(M)$-†module, where $\mathfrak{F}(M)$ denotes the algebra of all C^∞-functions in M. In fact, for $f,g\in\mathfrak{F}(M)$ and $X,Y\in\mathfrak{X}(M)$, we can define a vector field $fX+gY$ by $(fX+gY)_p=f(p)X_p+g(p)Y_p$, and this defines an $\mathfrak{F}(M)$-module structure in $\mathfrak{X}(M)$.

In a coordinate neighborhood (U,ψ), we can write $X=\sum_i\xi^i(\partial/\partial x^i)$. The right-hand side of this equation is sometimes called the **symbol** of the vector field X. A vector field X can also be interpreted as a linear differential operator that acts on $\mathfrak{F}(M)$, called a **first-order differential operator** in M.

Let X and Y be vector fields in M. Then there exists a unique vector field Z in M such that $Zf=X(Yf)-Y(Xf)$ for any C^∞-function f in M. We denote Z by $[X,Y]$ and call it the **Poisson bracket** (or simply **bracket**) of X and Y. If ξ^i and η^i denote the components of X and Y, respectively, in a coordinate neighborhood (U,ψ), then the components ζ^i of $[X,Y]$ are given by $\zeta^i=\sum_k\{\xi^k(\partial\eta^i/\partial x^k)-\eta^k(\partial\xi^i/\partial x^k)\}$. The bracket of two vector fields has the following properties: (i) $[X,Y]f=X(Yf)-Y(Xf)$, (ii) $[fX,gY]=fg[X,Y]+f(Xg)Y-g(Yf)X$, (iii) $[X+Y,Z]=[X,Z]+[Y,Z]$, (iv) $[X,Y]=-[Y,X]$, and (v) $[[X,Y],Z]+[[Y,Z],X]+[[Z,X],Y]=0$ (**Jacobi identity**). These identities show that $\mathfrak{X}(M)$ is a Lie algebra (→ 247 Lie Algebras) over **R**.

If φ is a diffeomorphism of M onto M', then for any vector field X in M we can define a vector field φ_*X in M' by the condition $(\varphi_*X)_p=d\varphi_q(X_q)$, $p=\varphi(q)$. Then φ_* is an isomorphism of the Lie algebra $\mathfrak{X}(M)$ onto the Lie algebra $\mathfrak{X}(M')$.

L. Vector Fields and One-Parameter Groups of Transformations

A **one-parameter group of transformations** of M is a family φ_t $(t\in\mathbf{R})$ of diffeomorphisms satisfying the following two conditions: (i) the mapping of $\mathbf{R}\times M$ into M defined by $(t,p)\rightarrow\varphi_t(p)$ is of class C^∞; and (ii) $\varphi_s\circ\varphi_t=\varphi_{s+t}$ for $s,t\in\mathbf{R}$.

Let φ_t be a one-parameter group of transformations of M. Then we can define a vector field X by $X_pf=\lim_{t\to0}(f(\varphi_t(p))-f(p))/t$, where $p\in M$ and $f\in\mathfrak{F}(M)$. The vector field X thus defined is called the **infinitesimal transformation** of φ_t. We also say that φ_t is generated by X, and sometimes we denote φ_t by the symbol $\exp tX$. In this case, if $(x^1,\ldots,x^n)$ is a local coordinate system, then at each point p of the coordinate neighborhood, we have $X_p=\sum_i(dx^i(\varphi_t(p))/dt)_{t=0}(\partial/\partial x^i)_p$.

If M is compact, then every vector field in M is the infinitesimal transformation of a one-parameter group of transformations; that is, every vector field generates a one-parameter group of transformations. For M noncompact, this is not always true. Nevertheless, for each vector field X we have the following result concerning local properties of X: For each point p of M, there exist a neighborhood U of p, a positive real number ε, and a family φ_t $(|t|<\varepsilon)$ of mappings of U into M satisfying the following three conditions. (1) The mapping of $(-\varepsilon,\varepsilon)\times U$ into M defined by $(t,q)\rightarrow\varphi_t(q)$ is of class C^∞, and for each fixed t, φ_t is a diffeomorphism of U onto an open set $\varphi_t(U)$ of M. (2) If $|s|$, $|t|$, and $|s+t|$ are all smaller than ε and q and $\varphi_t(q)$ both belong to U, then $\varphi_s(\varphi_t(q))=\varphi_{s+t}(q)$. (3) $X(q)f=\lim_{t\to0}(f(\varphi_t(q))-f(q))/t$ for $q\in U$ and $f\in\mathfrak{F}(M)$. We call φ_t the **local one-parameter group of local transformations** around p generated by X.

Let X and Y be vector fields in M, and let φ_t be the local one-parameter group of local transformations around p generated by X. Then $[X,Y]_p=\lim_{t\to0}(Y_p-((\varphi_t)_*Y)_p)/t$, where $(\varphi_t)_*Y$ is a vector field defined as follows: Let U be a neighborhood of the point p where φ_t $(|t|<\varepsilon)$ is defined. Then $(\varphi_t)_*Y$ is the vector field on $\varphi_t(U)$ that is the image of Y under the diffeomorphism φ_t. In particular, if X generates a one-parameter group of transformations of M, then we have $[X,Y]=$

$\lim_{t\to 0}(Y-(\varphi_t)_* Y)/t$ for any vector field Y in M.

M. Tensor Fields

Let $T_s^r(p)$ be the vector space consisting of all r-times contravariant and s-times covariant tensors over the tangent vector space T_p of a C^∞-manifold M, that is,

$$T_s^r(p)=\left(\overset{r}{\otimes}\, T_p\right)\otimes\left(\overset{s}{\otimes}\, T_p^*\right),$$

where T_p^* denotes the dual linear space of T_p (→ 256 Linear Spaces). A **tensor field** (more precisely, **contravariant of order** r **and covariant of order** s, or simply **tensor field of type** (r,s)) on a subset N of M is a mapping K that assigns to each point p of N an element K_p of the vector space $T_s^r(p)$. In particular, if $r=s=0$, K is a real-valued function on N, and we call K a **scalar field**. If $r=1$ and $s=0$, K is a vector field, called a **contravariant vector field**. When $r=0$ and $s=1$, we call K a **covariant vector field** (or **differential form of degree** 1). If $r\neq 0$, $s=0$ or $r=0$, $s\neq 0$, we call K a **contravariant tensor field of order** r or a **covariant tensor field of order** s, respectively. A contravariant or covariant tensor field K is said to be **symmetric (alternating)** if K_p is a symmetric (alternating) tensor at every point p of M.

Let $(x^1,\dots,x^n)$ be the local coordinate system in a coordinate neighborhood (U,ψ). Then at each point p of U, the $(\partial/\partial x^i)_p$ $(i=1,\dots,n)$ form a basis of the tangent vector space T_p, the differentials $(dx^i)_p$ $(i=1,\dots,n)$ form a basis of the dual space T_p^*, and these bases are dual to each other. A tensor field K of type (r,s) defined on M is written at any point p of U in the following form:

$$K_p=\sum K_{j_1\dots j_s}^{i_1\dots i_r}(p)(\partial/\partial x^{i_1})_p\otimes\dots\otimes(\partial/\partial x^{i_r})_p \otimes(dx^{j_1})_p\otimes\dots\otimes(dx^{j_s})_p.$$

The functions $(K_{j_1\dots j_s}^{i_1\dots i_r})_p$ defined in U are called the **components** of the tensor field K of type (r,s) with respect to the local coordinate system $(x^1,\dots,x^n)$. If $\bar{K}_{j_1\dots j_s}^{i_1\dots i_r}$ are the components of K with respect to the local coordinate system $(\bar{x}^1,\dots,\bar{x}^n)$ in another coordinate neighborhood (U',ψ') such that $U\cap U'\neq\varnothing$, then for each $q\in U\cap U'$, the following relations hold:

$$\bar{K}_{j_1\dots j_s}^{i_1\dots i_r}(q)=\sum_{k,l}(\partial\bar{x}^{i_1}/\partial x^{k_1})_q\dots(\partial\bar{x}^{i_r}/\partial x^{k_r})_q \times(\partial x^{l_1}/\partial\bar{x}^{j_1})_q\dots(\partial x^{l_s}/\partial\bar{x}^{j_s})_q K_{l_1\dots l_s}^{k_1\dots k_r}.$$

A tensor field K in M is called a **tensor field of class** C^t $(0\leqslant t\leqslant\infty)$ if the components are functions of class C^t for any coordinate neighborhood of M.

The sum $K+L$ and the tensor product $K\otimes L$ of two tensor fields K and L are defined by the rules $(K+L)_p=K_p+L_p$ and $(K\otimes L)_p=K_p\otimes L_p$, respectively. The contraction of two tensor fields is also defined by taking the contraction pointwise.

Let φ be a diffeomorphism of M into M'. Then the differential $(d\varphi)_q$ is an isomorphism of T_q onto T_p $(p=\varphi(q))$ for each $q\in M$ and hence induces an isomorphism $\tilde{\varphi}_q$ of the vector space $T_s^r(q)$ onto the vector space $T_s^r(p)$ (→ 256 Linear Spaces). For any tensor field K in M we can define a tensor field $\tilde{\varphi}K$ in M' by $(\tilde{\varphi}K)_p=\tilde{\varphi}_q(K_q)$, $p=\varphi(q)$, $q\in M$. Then $\tilde{\varphi}(K+L)=\tilde{\varphi}K+\tilde{\varphi}L$, $\tilde{\varphi}(K\otimes L)=\tilde{\varphi}K\otimes\tilde{\varphi}L$, and the mapping $\tilde{\varphi}$ commutes with contraction.

N. Lie Derivatives of Tensor Fields

Let K be a tensor field and X be a vector field (both of class C^∞) in M. We define a tensor field L_XK by $(L_XK)_p=\lim_{t\to 0}(K_p-(\tilde{\varphi}_tK)_p)/t$, where φ_t denotes the local one-parameter group of local transformations around p generated by X. We call L_XK the **Lie derivative** of K with respect to the vector field X. The operator $L_X: K\to L_XK$ has the following six properties: (i) $L_X(K+K')=L_XK+L_XK'$, (ii) $L_X(K\otimes K')=(L_XK)\otimes K'+K\otimes L_XK'$, (iii) the operator L_X commutes with contraction, (iv) $L_Xf=Xf$ for a scalar field f and $L_XY=[X,Y]$ for a vector field Y, (v) $L_{[X,Y]}=L_XL_Y-L_YL_X$, that is, $L_{[X,Y]}K=L_X(L_YK)-L_Y(L_XK)$, and (vi) K is invariant under φ_t, i.e., $\tilde{\varphi}_tK=K$ for all t, if and only if $L_XK=0$.

O. Covariant Tensor Fields and Multilinear Mappings of the $\mathfrak{F}(M)$-Module $\mathfrak{X}(M)$

Let K be a covariant tensor field of order r in M. We always assume that K is of class C^∞. The value K_p of K at $p\in M$ is an element of the vector space $T_p^*\otimes\dots\otimes T_p^*$ (r times tensor product of T_p^*); hence we may consider K_p an r-linear mapping of T_p into $\mathbf{R}$ (→ 256 Linear Spaces). If $X_1,\dots,X_s$ are vector fields in M, we define a C^∞-function $K(X_1,\dots,X_r)$ by $K(X_1,\dots,X_r)(p)=K_p((X_1)_p,\dots,(X_r)_p)$. Then the mapping that assigns to each r-tuple $(X_1,\dots,X_r)$ of vector fields the C^∞-function $K(X_1,\dots,X_r)$ is an r-linear mapping on the $\mathfrak{F}(M)$-module $\mathfrak{X}(M)$ consisting of all vector fields of class C^∞ in M into $\mathfrak{F}(M)$; that is, $K(X_1,\dots,fX_i+gY_i,\dots,X_r)=fK(X_1,\dots,X_i,\dots,X_r)+gK(X_1,\dots,Y_i,\dots,X_r)$ $(i=1,\dots,r)$ for $f,g\in\mathfrak{F}(M)$. Conversely any r-linear mapping of the $\mathfrak{F}(M)$-module $\mathfrak{X}(M)$ into $\mathfrak{F}(M)$ can be interpreted as a covariant tensor field of order r in M. If the tensor field K is symmetric (alternating), the corresponding r-linear mapping $K(X_1,\ldots,X_r)$ is symmetric (alternating) with respect to $X_1,\ldots,X_r$. For the Lie derivative L_XK of a covariant tensor field of order r of K, we have the following formula: $(L_XK)(X_1,\ldots,X_r)=X(K(X_1,\ldots,X_r))-\sum_{i=1}^{r}K(X_1,\ldots,[X,X_i],\ldots,X_r)$.

P. Riemannian Metrics

A symmetric covariant tensor field g of order 2 and of class C^∞ in M is called a **pseudo-Riemannian metric** if the symmetric bilinear form g_p on the tangent vector space T_p is nondegenerate at each point $p\in M$; and g is called a **Riemannian metric** if g_p is positive definite for all p. If g is a Riemannian metric, the length $\|L\|$ of a tangent vector $L\in T_p$ is defined by $\|L\|^2=g_p(L,L)$. On a paracompact C^∞-manifold there always exists a Riemannian metric. A pair consisting of a differentiable manifold and a Riemannian metric on it is called a **Riemannian manifold** (→ 360 Riemannian Manifolds).

Q. Differential Forms

An alternating covariant tensor field in M of order r and of class $C^t (0\leqslant t\leqslant\infty)$ is also called a **differential form** (or **exterior differential form**) **of degree** r. A differential form of degree 1 is sometimes called a **Pfaffian form**. Let ω be a differential form of degree r. Since each alternating covariant tensor of order r at a point p is an element of $\bigwedge^r T_p^*$, the r-fold †exterior product of T_p^*, the form ω is a mapping that sends each point p of M to an element ω_p of $\bigwedge^r T_p^*$. We can also regard ω as an alternating r-linear mapping of $\mathfrak{X}(M)$ into $\mathfrak{F}(M)$. Let $(x^1,\ldots,x^n)$ be the local coordinate system in a local coordinate neighborhood (U,ψ). Since $(dx^i)_p$ $(i=1,\ldots,n)$ is a basis of T_p^* at each point p of U, we can express ω_p $(p\in U)$ uniquely in the form

$$\omega_p=\sum_{i_1<\ldots<i_r} a_{i_1\ldots i_r}(p)(dx^{i_1})_p\wedge\ldots\wedge(dx^{i_r})_p,$$

where the sum extends over all ordered r-tuples $(i_1,\ldots,i_r)$ of indices such that $1\leqslant i_1<i_2<\ldots<i_r\leqslant n$. For an ordered r-tuple $(i_1,\ldots,i_r)$ of indices with repeated indices we put $a_{i_1\ldots i_r}=0$, and for $(i_1,\ldots,i_r)$ with r distinct indices we put $a_{i_1\ldots i_r}=(\operatorname{sgn}\sigma)a_{j_1\ldots j_r}$, where $(j_1,\ldots,j_r)$ $(j_1<\ldots<j_r)$ is a permutation of $(i_1,\ldots,i_r)$ and $\operatorname{sgn}\sigma$ denotes the sign of the permutation $\sigma: i_k\to j_k$ $(k=1,\ldots,r)$. Then we can write

$$\omega_p=\frac{1}{r!}\sum_{i_1,\ldots,i_r=1}^{n} a_{i_1\ldots i_r}(p)\times(dx^{i_1})_p\wedge\ldots\wedge(dx^{i_r})_p.$$

The functions $a_{i_1\ldots i_r}$ are the components of the tensor field ω, and ω is of class C^t if these components are of class C^t for any coordinate neighborhood. By the **support** (or **carrier**) of a differential form ω we mean the closure of the subset of M consisting of all p such that $\omega_p\neq 0$.

In the rest of this article, differential forms are always of class C^∞, and we denote by $\mathfrak{D}^r(M)$ the real vector space consisting of all differential forms of degree r and of class C^∞. In particular, $\mathfrak{D}^0(M)=\mathfrak{F}(M)$ and $\mathfrak{D}^r(M)=\{0\}$ for $r>n$, $n=\dim M$.

For differential forms we have the following five important operations.

(1) *Exterior product*. Let ω and θ be differential forms of degree r and s, respectively. The **exterior product** $\omega\wedge\theta$ of ω and θ is the differential form of degree $r+s$ defined by $(\omega\wedge\theta)_p=\omega_p\wedge\theta_p$, $p\in M$. Let $X_1,\ldots,X_{r+s}$ be $r+s$ vector fields in M. Then we have

$$(\omega\wedge\theta)(X_1,\ldots,X_{r+s})=\sum\operatorname{sgn}(i;j)\omega(X_{i_1},\ldots,X_{i_r})\theta(X_{j_1},\ldots,X_{j_s}),$$

where the summation runs over all possible partitions of $(1,2,\ldots,r+s)$ such that $i_1<i_2<\ldots<i_r$ and $j_1<j_2<\ldots<j_s$, and $\operatorname{sgn}(i;j)$ means the sign of the permutation $(1,2,\ldots,r+s)\to(i_1,\ldots,i_r,j_1,\ldots,j_s)$. In particular, if $\omega_1,\ldots,\omega_r$ are differential forms of degree 1, then we have $(\omega_1\wedge\ldots\wedge\omega_r)(X_1,\ldots,X_r)=\det(\omega_i(X_j))$.

(2) *Exterior differentiation*. Let ω be a differential form of degree r, and let $\omega=(1/r!)\sum a_{i_1\ldots i_r}dx^{i_1}\wedge\ldots\wedge dx^{i_r}$ in a coordinate neighborhood (U,ψ). Then we can define a differential form $d\omega$ of degree $r+1$ by the condition $d\omega=(1/r!)\sum da_{i_1\ldots i_r}\wedge dx^{i_1}\wedge\ldots\wedge dx^{i_r}$ in U. The differential form $d\omega$ is called the **exterior derivative** (or **exterior differential**) of ω. A differential form ω satisfying the condition $d\omega=0$ is called a **closed differential form**, and a differential form that can be expressed as $\eta=d\omega$ for some ω is called an **exact differential form**. If $a,b\in\mathbf{R}$ and $\omega,\omega'\in\mathfrak{D}^r(M)$, then we have $d(a\omega+b\omega')=ad\omega+bd\omega'$. Therefore, the set $\mathfrak{C}^r(M)$ of all closed differential forms of degree r and the set $\mathfrak{E}^r(M)$ of all exact differential forms of degree r are linear subspaces of $\mathfrak{D}^r(M)$.

For the exterior derivative $d\omega$, we have the following formula:

$$(d\omega)(X_1,\ldots,X_r)=\sum_{i=1}^{r+1}(-1)^{i+1}X_i\big(\omega(X_1,\ldots,\hat{X}_i,\ldots,X_{r+1})\big)+\sum_{i<j}(-1)^{i+j}\omega\big([X_i,X_j],X_1,\ldots,\hat{X}_i,\ldots,\hat{X}_j,\ldots,X_{r+1}\big),$$

where the variables under the sign ˆ are omitted.

(3) *Interior product with vector field.* Let ω be a differential form of degree r and X a vector field. When $r \geqslant 1$, we can define a differential form $\iota(X)\omega$ of degree $r-1$ by the formula $(\iota(X)\omega)(X_1, \ldots, X_{r-1}) = \omega(X, X_1, \ldots, X_{r-1})$ for any $r-1$ vector fields $X_1, \ldots, X_{r-1}$; if $r=0$, we put $\iota(X)\omega = 0$. The differential form $\iota(X)\omega$ is called the **interior product** of ω with X.

(4) *Lie derivative.* The **Lie derivative** $L_X\omega$ of a differential form of degree r with respect to a vector field X is a differential form of the same degree. For any r vector field $X_1, \ldots, X_r$ we have, by definition, $(L_X\omega)(X_1, \ldots, X_r) = X(\omega(X_1, \ldots, X_r)) - \Sigma_{i=1}^r \omega(X_1, \ldots, [X, X_i], \ldots, X_r)$.

(5) Let φ be a C^∞-mapping of M into M', and let $\varphi_p^* : T_{\varphi(p)}^* \rightarrow T_p^*$ be the †transpose (or †dual) of the linear mapping $(d\varphi)_p : T_p \rightarrow T_{\varphi(p)}$ for $p \in M$, i.e., the mapping defined by the condition $((d\varphi)_p L, \alpha) = (L, \varphi_p^* \alpha)$ for each $L \in T_p, \alpha \in T_{\varphi(p)}^*$. We denote the linear mapping of $\bigwedge T_{\varphi(p)}^*$ into $\bigwedge T_p^*$ induced by φ_p^* by φ_p^* also. Let ω be a differential form of degree r in M. Then a differential form $\varphi^*\omega$ in M, the **pullback** by φ of ω, is defined by $(\varphi^*\omega)_p = \varphi_p^* \omega_{\varphi(p)}, p \in M$.

The operations defined previously satisfy the following six important relations: (i) $d^2 = 0$, that is, $d(d\omega) = 0$; (ii) $d(\omega \wedge \theta) = d\omega \wedge \theta + (-1)^r \omega \wedge d\theta$, where ω is of degree r; (iii) $\varphi^*(\omega \wedge \theta) = \varphi^*\omega \wedge \varphi^*\theta$, $\varphi^*(d\omega) = d(\varphi^*\omega)$; (iv) $L_X(\omega \wedge \theta) = (L_X\omega) \wedge \theta + \omega \wedge (L_X\theta)$; (v) $L_X = \iota(X) \cdot d + d \cdot \iota(X)$, $L_X(d\omega) = d(L_X\omega)$, and (vi) $L_{[X,Y]} = L_X \cdot L_Y - L_Y \cdot L_X$, $\iota([X,Y]) = L_X \cdot \iota(Y) - \iota(Y) \cdot L_X$.

R. De Rham Cohomology

Let $\mathfrak{D}(M) = \Sigma_{r=0}^n \mathfrak{D}^r(M)$, where $n = \dim M$. Then $\mathfrak{D}(M)$ is a †cochain complex with †coboundary operator d. We denote by $H^r(\mathfrak{D})$ the r-dimensional cohomology group of this cochain complex, and we call it the r-dimensional **de Rham cohomology group** of the differentiable manifold M. If we denote by $\mathfrak{C}^r(M)$ and $\mathfrak{E}^r(M)$ the subspaces of $\mathfrak{D}^r(M)$ consisting of closed differential forms and exact differential forms, respectively, then $H^r(\mathfrak{D}) = \mathfrak{C}^r(M)/\mathfrak{E}^r(M)$ $(0 \leqslant r \leqslant n)$ by definition. If $\omega \in \mathfrak{C}^i(M)$ and $\theta \in \mathfrak{C}^r(M)$, then $\omega \wedge \theta \in \mathfrak{C}^{i+r}(M)$, and if $\omega \in \mathfrak{C}^i(M)$ and $\theta \in \mathfrak{E}^r(M)$ (or $\omega \in \mathfrak{E}^i(M)$ and $\theta \in \mathfrak{C}^r(M)$), then $\omega \wedge \theta \in \mathfrak{E}^{i+r}(M)$. So if we put $H(\mathfrak{D}) = \Sigma_{r=1}^n H^r(\mathfrak{D})$ (direct sum), we can define a product in $H(\mathfrak{D})$ by $[\omega] \cdot [\theta] = [\omega \wedge \theta]$ for each $[\omega] \in H^i(\mathfrak{D})$, $[\theta] \in H^r(\mathfrak{D})$. With respect to this product, $H(\mathfrak{D})$ forms an algebra over $\mathbf{R}$ called the **de Rham cohomology ring** of M.

S. Partitions of Unity

Let M be a paracompact C^∞-manifold, and let $\{V_i\}_{i \in I}$ be a †locally finite open covering of M such that the closure of V_i is compact for each index i. Then there exists a C^∞-function f_i in M for each i satisfying the following three conditions: (i) $0 \leqslant f_i \leqslant 1$, (ii) the support of f_i is contained in V_i, and (iii) $\Sigma_{i \in I} f_i(x) = 1$ for every $x \in M$. The family of C^∞-functions f_i, $i \in I$, is called a **partition of unity of class** C^∞ subordinate to the open covering $\{V_i\}_{i \in I}$.

T. Integrals of Differential Forms

Integrals Over an Oriented Manifold. Let M be an n-dimensional paracompact oriented C^∞-manifold and ω a differential form of degree n in M with compact support. We can choose a positively oriented atlas $S = \{(U_\alpha, \psi_\alpha)\}_{\alpha \in A}$ such that $\{U_\alpha\}_{\alpha \in A}$ is a locally finite covering, $\overline{U}_\alpha$ is compact for each α, and $\psi_\alpha(U_\alpha)$ is an open cube in $\mathbf{R}^n$ whose faces are parallel to the coordinate hyperplanes. Suppose first that the support of ω is contained in U_α for some index α. Then $(\psi_\alpha^{-1})^*\omega_\alpha$ is a differential form of degree n in the cube $\psi_\alpha(U_\alpha)$, and we can express $(\psi_\alpha^{-1})^*\omega_\alpha$ in the form $a dx^1 \wedge \ldots \wedge dx^n$, where $(x^1, \ldots, x^n)$ are the coordinates in $\mathbf{R}$ such that $x_\alpha^i = x^i \circ \psi_\alpha$ $(i = 1, \ldots, n)$ give a local coordinate system compatible with the orientation of M and a is a C^∞-function with compact support in the cube. Then we define the integral of ω over M by

$$\int_M \omega = \int_{\psi_\alpha(U_\alpha)} a\, dx^1 \ldots dx^n.$$

For the general case let $\{f_\alpha\}_{\alpha \in A}$ be a partition of unity of class C^∞ subordinate to $\{U_\alpha\}_{\alpha \in A}$. Then the support of $f_\alpha\omega$ is contained in U_α, and except for a finite number of the indices α, $f_\alpha\omega$ vanishes identically. Therefore, we may define the integral of ω over M by

$$\int_M \omega = \sum_\alpha \int_M f_\alpha \omega,$$

and we can show that this definition of the integral is independent of the choice of oriented atlas S and of a partition of unity subordinate to S.

Integrals Over a Singular Chain. We fix rectangular coordinates in $\mathbf{R}^r$. Let d_0 be the origin and d_i be the unit point on the ith coordinate axis. Let S^r denote the oriented r-simplex $(d_0, d_1, \ldots, d_r)$ with vertices $d_0, d_1, \ldots, d_r$. When we regard S^r as a point set, we denote it by $|S^r|$. An **oriented singular r-simplex of class**

C^∞ in M is, by definition, a pair (S^r, φ) consisting of S^r and a C^∞-mapping φ of an open neighborhood of $|S^r|$ into M. An element of the free **Z**-module generated by singular r-simplexes of class C^∞ is called an integral **singular r-chain of class** C^∞ in M. We define a real singular r-chain of class C^∞ analogously. Let ω be a differential form of degree r and (S^r, φ) be an oriented singular r-simplex of class C^∞ in M. Then $\varphi^*\omega$ is a differential form of degree r in a neighborhood of $|S^r|$, and we can express $\varphi^*\omega$ in the form $\varphi^*\omega = a\,dx^1 \wedge dx^2 \wedge \ldots \wedge dx^r$. We define the integral of ω over (S^r, φ) by

$$\int_{(S^r,\varphi)} \omega = \int_{|S^r|} a\,dx^1 \ldots dx^r$$

and the integral of ω over a singular r-chain C of class C^∞ by

$$\int_C \omega = \sum_i m_i \int_{(S^r,\varphi_i)} \omega,$$

where $C = \Sigma_i m_i\,(S^r, \varphi_i), m_i \in \mathbf{Z}$ (or $m_i \in \mathbf{R}$). When $r=0$, then ω is a function in M, and S^0 is a point o. In this case we put $\int_C \omega = \Sigma_i m_i \omega(\varphi_i(o))$.

Let $C_r(S, \mathbf{Z})$ $(C_r(S, \mathbf{R}))$ be the **Z**-module (vector space over **R**) of integral (real) singular r-chains of class C^∞ in M, and let $\omega(C)$ be the value of the integral of ω over a chain C. Then ω is a linear function in the vector space $C_r(S, \mathbf{R})$, and hence we can consider ω a **singular r-cochain of class** C^∞.

U. Stokes's Formulas

(1) Let D be a †domain in an n-dimensional C^∞-manifold M, and let ∂D and $\bar{D}$ be the boundary and the closure of D, respectively. Let $S = \{(U_\alpha, \psi_\alpha)\}_{\alpha \in A}$ be an atlas of class C^∞ of M, $U'_\alpha = U_\alpha \cap \bar{D}$, ψ'_α be the restriction of ψ_α to U'_α, and $T = \{(U'_\alpha, \psi'_\alpha)\}_{\alpha \in A}$. If the pair $(\bar{D}, T)$ is a C^∞-manifold with boundary under a suitable choice of S, then the domain D is called a **domain with regular** (or **smooth**) **boundary**. The boundary ∂D of $(\bar{D}, T)$ is then an $(n-1)$-dimensional closed submanifold of M, and if M is orientable, ∂D is also orientable. Now let M be a paracompact and oriented manifold and D be a domain with regular boundary. Let C be a characteristic function of D in M, i.e., a function defined by the condition $C(p) = 1$ for $p \in D$ and $C(p) = 0$ for $p \notin D$. Let θ be a differential form of degree n in M with compact support. We define the integral of θ over D by

$$\int_D \theta = \int_M C \cdot \theta.$$

Let ω be a differential form of degree $n-1$ in M with compact support. We then have **Stokes's formula**:

$$\int_D d\omega = \int_{\partial D} i^*\omega,$$

where i denotes the identity mapping of the submanifold ∂D into M with ∂D having the orientation induced naturally from that of M.

(2) Let C be a singular r-chain of class C^∞ in M, and let ∂C be the boundary of C. Then for any differential form ω of degree $r-1$, we have

$$\int_C d\omega = \int_{\partial C} \omega.$$

This formula is also called **Stokes's formula.**

V. De Rham's Theorem

Let M be a connected paracompact C^∞-manifold. If we consider ω as a singular cochain, we have $(d\omega)(C) = \omega(\partial C)$; by Stokes's formula, this means that the exterior differential $d\omega$ of ω is equal to the coboundary of the singular cochain ω. Let ω and C be a closed differential form of degree r and a singular r-cycle of class C^∞, respectively, and let $[\omega]$ and $[C]$ be the de Rham cohomology class and the singular homology class represented by ω and C. Using Stokes's formula, we can define the inner product $([\omega], [C])$ by

$$([\omega], [C]) = \int_C \omega.$$

Through this inner product, it follows that the de Rham cohomology group $H^r(\mathfrak{D})$ is isomorphic to the rth singular cohomology group $H^r(M, \mathbf{R})$, the dual space of the rth homology group of the complex of real singular chains of class C^∞. Moreover, the de Rham cohomology ring $H(\mathfrak{D})$ is isomorphic to the singular cohomology ring $H^*(M, R)$ **(de Rham's theorem)**.

W. Divergence of a Vector Field

Let M be an n-dimensional oriented C^∞-manifold, and let S be an oriented atlas. Let ω be a differential form of degree n, and let $(x^1, \ldots, x^n)$ be the local coordinate system in a coordinate neighborhood in S. Then we can express ω in the coordinate neighborhood uniquely in the form $\omega = a\,dx^1 \wedge \ldots \wedge dx^n$. If the function a is positive for any coordinate neighborhood in S, we call ω a **volume element** of M. In a paracompact oriented manifold, there always exists a volume element. (We remark that an n-dimensional differentiable manifold M is orientable if and only if there exists an everywhere nonvanishing differential form of degree n.) Let f be a C^∞-function in M with compact support. Then

$f \cdot \omega$ is a differential form of degree n with compact support, and so the integral

$$\int_M f \cdot \omega$$

is defined. We call this integral the **integral of the function f with respect to the volume element ω.**

Let g be a Riemannian metric in M and g_{ij} the components of g with respect to the local coordinate system $(x^1, \ldots, x^n)$ as before. Then we can define a volume element ω in M by putting $\omega = \sqrt{G}\, dx^1 \wedge \ldots \wedge dx^n$, $G = \det(g_{ij})$ in each coordinate neighborhood. The volume element thus defined is called the **volume element associated with the Riemannian metric** g.

Let ω be a volume element and X a vector field in M. Then the Lie derivative $L_X \omega$ is also a differential form of degree n, and we can express $L_X \omega$ in the form $L_X \omega = f_X \cdot \omega$, where f_X is a scalar field, i.e., a function in M. We call f_X the **divergence** of the vector field X with respect to the volume element ω and denote it by $\operatorname{div} X$. If ω is associated with a Riemannian metric, then $\operatorname{div} X$ is called the **divergence** of X with respect to the Riemannian metric.

If M is compact, we have

$$\int_M \operatorname{div} X \cdot \omega = 0$$

for any vector field X. This result is known as **Green's theorem**.

X. Jets

Let M and N be C^∞-manifolds. We define an equivalence relation in the set of all C^∞-mappings of M into N. Let f and g be such mappings and p be a point of M. Choosing local coordinate systems, we write $f(p) = (f_1(x), \ldots, f_n(x))$, $g(p) = (g_1(x), \ldots, g_n(x))$, $x = (x_1, \ldots, x_n)$. We say f and g are equivalent at p if $f(p)$, $g(p)$, and the values at p of all the partial derivatives of f_i and g_i up to the order r (r an integer, $r \geqslant 0$) are equal ($i = 1, \ldots, n$). An equivalence class with respect to this equivalence relation is called a **jet of order** r at p. A jet of order r at p represented by a function f is denoted by $j_p^r f$, and the points p and $f(p)$ are called the **source** and the **target** of the jet $j_p^r f$, respectively. We denote by $J_p^r(M,N)$ the set of all jets of order r with source at p and target in N and let $J^r(M,N) = \bigcup_{p \in M} J_p^r(M,N)$. For any jet j, let $\pi_s(j)$ and $\pi_t(j)$ denote the source and the target of j, respectively. We can introduce the structure of a C^∞-manifold in $J^r(M,N)$ in a natural way such that the projections $\pi_s : J^r(M,N) \to M$ and $\pi_t : J^r(M,N) \to N$ are both of class C^∞ and $J^r(M,N)$ is a fiber bundle over M (N) with projection $\pi_s(\pi_t)$. As examples, we have:

(1) $J_0^1(\mathbf{R}, N)$ is identified with the tangent vector bundle of N.

(2) The set $J^r(\xi)$ of all jets of order r determined by the sections of a vector bundle ξ of class C^∞ is also a vector bundle of class C^∞.

Let $f: M \to N$ and $g: N \to L$ be C^∞-mappings. We define a composition of jets by $j_{f(p)}^r g \cdot j_p^r f = j_p^r (g \circ f)$. A jet $j_p^r f \in J^r(M,N)$ is **invertible** if there exists a mapping $g: N \to M$ such that $j_{f(p)}^r g \cdot j_p^r f = j_p^r(1_M)$, where 1_M denotes the identity mapping of M onto itself. We denote by $I^r(M,N)$ the set of all invertible jets in $J^r(M,N)$, and put $I_p^r(M,N) = I^r(M,N) \cap J_p^r(M,N)$.

(3) Let $G^r(n)$ be the set of all invertible jets in $I^r(\mathbf{R}^n, \mathbf{R}^n)$ whose source and target are the origin of $\mathbf{R}^n$. Then with respect to the composition of jets, $G^r(n)$ forms a Lie group that is an †extension of $G^1(n) = GL(n, \mathbf{R})$ by a simply connected nilpotent Lie group. The projection $G^r(n) \to G^1(n)$ is a special example of the natural projection $J^r(M,N) \to J^s(M,N)$ ($r \geqslant s$), which is defined in general.

(4) We can identify $I_0^1(\mathbf{R}^m, M)$ ($m = \dim M$) with the tangent m-frame bundle over M. More generally, $I_0^r(\mathbf{R}^m, M)$ is a $G^r(m)$-bundle over M.

Y. Foliations

Let M be an n-dimensional C^∞-manifold. A codimension q, C^r-**foliation** of M ($0 \leqslant q \leqslant n, 0 \leqslant r \leqslant \infty$) is a family $\mathfrak{F} = \{L_\alpha ; \alpha \in A\}$ consisting of arcwise connected subsets of M, called **leaves**, with the following properties: (i) $L_\alpha \cap L_{\alpha'} = \varnothing$ if $\alpha \neq \alpha'$; (ii) $\bigcup_{\alpha \in A} L_\alpha = M$; (iii) Every point in M has a local coordinate system (U, ψ) of class C^r such that, for each leaf L_α, the arcwise connected components of $U \cap L_\alpha$ are described by $x^{n-q+1} =$ constant, $\ldots, x^n =$ constant, where $x^1, x^2, \ldots, x^n$ denote the local coordinates in the system (U, ψ). In particular, every leaf of $\mathfrak{F}$ is an $(n-q)$-dimensional submanifold of M.

The totality of solutions of a completely integrable nonsingular system of Pfaffian equations $\omega_i = a_{i1}(x)dx_1 + a_{i2}(x)dx_2 + \ldots + a_{in}(x)dx_n = 0$ ($i = 1, 2, \ldots, q$) forms a codimension q foliation, and the totality of integral curves of a nonsingular vector field of class C^r on M ($r \geqslant 1$) constitutes a codimension $n-1$, C^r-foliation.

Let Q be a q-dimensional C^r-manifold ($q \leqslant n$) and let $f: M \to Q$ be a C^r-**submersion** (i.e., the rank of the differential df of f is q); then, f induces a codimension q, C^r-foliation of M with the arcwise connected components of $f^{-1}(x)$ $(x \in Q)$ as follows:

If a closed C^∞-manifold M admits a codimension 1, C^r-foliation, then the Euler number of M must be zero. In 1944, G. Reeb constructed a codimension 1, C^∞-foliation of the 3-sphere S^3 as follows [20]. Let $f(x)$ be a C^∞-function, defined on the open interval $(-1, \quad 1)$, such that $\lim_{|x|\to 1} f^{(k)}(x) = \infty$ $(k = 0, 1, 2, \ldots)$. The graphs of the equations $y = f(x) + c$ $(-1 < x < 1, c \in \mathbf{R})$, together with the lines $x = c'$ $(|c'| \geqslant 1)$ constitute a codimension 1, C^∞-foliation of $\mathbf{R}^2$. Then, by rotating the strip $\{(x,y) \in \mathbf{R}^2 \mid -1 < x < 1\}$ around the y-axis in $\mathbf{R}^3$, we obtain a codimension 1, C^∞-foliation of the set $\operatorname{Int} D^2 \times \mathbf{R}$, where D^2 denotes the 2-disk. This foliation is invariant under vertical translations, and therefore defines a codimension 1, C^∞-foliation of $\operatorname{Int} D^2 \times S^1$. Since S^3 is the union of two solid tori intersecting in a common toroidal surface, foliations of the interior of each solid torus, constructed in the manner described above, together with the common toroidal surface, define a so-called **Reeb foliation** of S^3. Every open manifold admits a codimension 1, C^∞-foliation (A. Phillips); every closed 3-dimensional manifold admits a codimension 1, C^∞-foliation (S. P. Novikov [19], W. Lickorish, J. Wood); every odd-dimensional sphere admits a codimension 1, C^∞-foliation (I. Tamura, A. Durfee). For further results on the existence of foliations, see Tamura [21] and Thurston [22].

Let $\mathfrak{F}$ be a codimension 1, C^r-foliation of a closed C^∞-manifold M $(r \geqslant 2)$ arising from a completely integrable 1-form ω (i.e., $d\omega \wedge \omega = 0$). Then, if θ is any 1-form such that $d\omega = -\theta \wedge \omega$, the de Rham cohomology class $\Gamma_{\mathfrak{F}} \in H^3(M; \mathbf{R})$ of the closed form $\theta \wedge d\theta$ is an invariant of $\mathfrak{F}$, known as the **Godbillon-Vey invariant** [16].

Two closed, oriented n-dimensional C^∞-manifolds M_0 and M_1 with codimension q, C^r-foliations are said to be **foliated cobordant** if there exists a compact, oriented $(n+1)$-dimensional C^∞-manifold W with boundary $\partial W = M_0 - M_1$ and a codimension q, C^r-foliation of W which is transverse to ∂W and includes the given foliations of M_0 and M_1. The resulting foliated cobordism classes $\{\mathfrak{F}\}$ form a group $\mathfrak{F}\Omega^r_{m,q}$ with respect to disjoint union. The Godbillon-Vey number $\Gamma_{\mathfrak{F}}[M]$ is an invariant of $\mathfrak{F}\Omega^r_{3,1}$ $(r \geqslant 2)$, and Thurston has proved that the homomorphism $\mathfrak{F}\Omega^r_{3,1} \to \mathbf{R}$ defined by $\{\mathfrak{F}\} \to \Gamma_{\mathfrak{F}}[M]$ is surjective. For more details concerning foliations, → [15, 18].

Z. Pseudogroup Structure

Let X be a topological space, and let Γ be a set consisting of homeomorphisms $f: U_f \to V_f$, where U_f, V_f are open subsets of X. We call Γ a **pseudogroup** of topological transformations if Γ satisfies the following four conditions: (i) Γ contains the identity mapping of X onto X; (ii) if $f \in \Gamma$, then the restriction of f onto any open subset U of U_f is also contained in Γ; (iii) if f and g are in Γ and $V_f \subset U_g$, then $g \circ f$ is contained in Γ; and (iv) if $f \in \Gamma$, then $f^{-1}: V_f \to U_f$ is also in Γ.

Following the definition of differentiable manifolds we define a **pseudogroup structure** of M (or, more precisely, a **Γ-structure** of M) as a set A of bijections, with each member α defined on a subset U_α of M onto an open set V_α of X, satisfying the following three conditions: (i) $\bigcup_\alpha U_\alpha = M$; (ii) if $\alpha, \beta \in A$, then $\alpha \circ \beta^{-1} \in \Gamma$, where the domain of definition of $\alpha \circ \beta^{-1}$ is $\beta(U_\alpha \cap U_\beta)$; and (iii) A is the maximal set of bijections that satisfies conditions (i) and (ii). We introduce in M the weakest (coarsest) topology such that every bijection α is a homeomorphism. If Γ' is a pseudogroup of X containing Γ and A' is a Γ'-structure of M such that $A \subset A'$, then we say that A' is subordinate to A. If $X = \mathbf{R}^n$ (or H^n, a half-space of $\mathbf{R}^n$), Γ is the totality of diffeomorphisms of class C^r of open sets of X onto open sets of X, and M is a space with Hausdorff topology, then the Γ-structure is the C^r-structure with or without boundary which we have already defined. When we replace $\mathbf{R}^n$ and H^n with a †Banach space X in the definition above, M endowed with such a Γ-structure is called a **X-manifold of class C^r** or a **C^r-manifold modelled on X** [8]. The notion of X-manifold can be generalized to **X-manifold with boundary** by replacing X with $\lambda^{-1}([0,\infty))$, where λ is a continuous linear mapping of the Banach space onto $\mathbf{R}$. We give three examples of Γ-structures subordinate to Γ', where Γ' is the totality of local transformations of class C^r $(r \geqslant 1)$ in $\mathbf{R}^n$.

(1) When n is even, we identify $\mathbf{R}^n$ with $\mathbf{C}^{n/2}$ and denote the totality of holomorphic transformations of connected open domains by Γ. The Γ-structure in this case is called a **complex structure**.

(2) When n is odd, we define Γ as the totality of transformations of connected open domains in $\mathbf{R}^n$ that leave invariant a Pfaffian form $\sum_{i=1}^m x^i dx^{m+i} + dx^{2m+1}$ $(n = 2m+1)$ up to scalar factors. The Γ-structure in this case is called a **contact structure**.

(3) We consider $\mathbf{R}^n = \mathbf{R}^p \times \mathbf{R}^{n-p}$ and define Γ as the set of all diffeomorphisms $U \to V$ (where U, V are open in $\mathbf{R}^n$) such that each set of the form $U \cap (\mathbf{R}^p \times \{y\})$ is mapped onto a set of the form $V \cap (\mathbf{R}^p \times \{y'\})$. The Γ-structure in this case is called a **foliated structure**.

The problem of determining whether there exists a Γ-structure for given Γ and M involves widely ranging problems of topology and analysis. The classification of Γ with reasonable conditions is another important open problem.

Haefliger has constructed the classifying space $B\Gamma$ for Γ-structures [17].

References

[1] Y. Akizuki, Tyôwa sekibunron (Japanese; Harmonic integrals) I, Iwanami, 1955.
[2] L. Auslander, Differential geometry, Harper, 1967.
[3] R. L. Bishop and S. I. Goldberg, Tensor analysis on manifolds, Macmillan, 1968.
[4] E. Cartan, Les systèmes différentiels extérieures et leurs applications géométriques, Actualités Sci. Ind., Hermann, 1945.
[5] C. Chevalley, Theory of Lie groups I, Princeton Univ. Press, 1946.
[6] G de Rham, Variétés différentiables, Actualités Sci. Ind., Hermann, 1955.
[7] S. Kobayashi and K. Nomizu, Foundations of differential geometry, Interscience, I, 1963; II, 1969.
[8] S. Lang, Introducing to differentiable manifolds, Interscience, 1962.
[9] Y. Matsushima, Tayôtai nyûmon (Japanese; Introduction to the theory of manifolds), Syôkabô, 1965; English translation, Differentiable manifolds, Marcel Dekker, 1972.
[10] J. R. Munkres, Elementary differential topology, Ann. Math. Studies, Princeton Univ. Press, 1963.
[11] K. Nomizu, Lie groups and differential geometry, Publ. Math. Soc. Japan, 1956.
[12] S. Sternberg, Lectures on differential geometry, Prentice-Hall, 1964.
[13] H. Whitney, Differentiable manifolds, Ann. of Math., (2) 37 (1936), 645–680.
[14] H. Whitney, Geometric integration theory, Princeton Univ. Press, 1957.
[15] R. Bott, Lectures on characteristic classes and foliations, Lecture notes in math. 279, Springer, (1972).
[16] C. Godbillon and J. Vey, Un invariant des feuilletages de codimension 1, C. R. Acad. Sci. Paris, 273 (1971), 92–95.
[17] A. Haefliger, Homotopy and integrability, Lecture notes in math., 197 (1971), 133–163, Springer.
[18] H. Lawson, Foliations, Bull. Amer. Math. Soc., 80 (1974), 369–418.
[19] A. Novikov, Topology of foliations, Trudy Mosc. Mat. Ob. 14 (1965), 248–278.
[20] G. Reeb, Sur certains propriétés topologiques des variétés feuilletées, Actualités Sci. Ind., Hermann, 1952.
[21] I. Tamura, Foliations and spinnable structures on manifolds, Ann. Inst. Fourier, 23 (1973), 197–214.
[22] W. Thurston, The theory of foliations of codimension greater than one, Comment. Math. Helv., 49 (1974), 214–231.

109 (X.6) Differential Calculus

A. First-Order Derivatives

Let $y=f(x)$ be a real-valued function of x defined on an interval I in the real line $\mathbf{R}$. If for a fixed $x_0 \in I$, the limit

$$\lim_{\substack{h\to 0\\ x_0+h\in I}} \frac{f(x_0+h)-f(x_0)}{h}$$

exists and is finite, then f is called **differentiable at the point** x_0, and the limit is the **derivative** (**differential coefficient**, or **differential quotient**) of f at the point x_0. If f is differentiable at every point of a set $A \subset I$, then f is said to be **differentiable on** A. The function that assigns the derivative of f at x to $x \in A$ is called the **derivative** (or **derived function**) of $f(x)$, which is denoted by $dy/dx, y', \dot{y}$, $df(x)/dx$, $(d/dx)f(x)$, $f'(x)$, or $D_x f(x)$. The process of determining f' is known as the **differentiation** of f. The derivative of f at the point x_0 is written $f'(x_0)$, $(df/dx)(x_0)$, $D_x f(x_0)$, $[dy/dx]_{x=x_0}$, etc. We say that f is **right (left) differentiable** or **differentiable on the right (left)** if the limit on the right, $\lim_{h\to+0}(f(x_0+h)-f(x_0))/h$ (the limit on the left, $\lim_{h\to+0}(f(x_0-h)-f(x_0))/h$) exists and is finite. This limit is called the **right (left) derivative** or **derivative on the right (left)** and is denoted by $D_x^+f(x_0)$ or $f'_+(x_0)$ ($D_x^-f(x_0)$ or $f'_-(x_0)$). For instance, if f is defined on $I=[a,b)$, then $D_x f(a)$ is identical to $D_x^+f(a)$.

B. Differentials

In the definitions given above, neither dx nor dy in dy/dx has a meaning by itself. In the following, however, we give a definition of dx and dy, using the concept of increment, so that $dy=f'(x)dx$. Let Δy denote the **increment** $f(x+\Delta x)-f(x)$ of f corresponding to the increment Δx of x. Suppose that $f(x)$ is differentiable at x. We set $\Delta y/\Delta x=f'(x)+\varepsilon$. Then we have $\lim_{\Delta x\to 0}\varepsilon=0$. This may be written utilizing †Landau's notation as $\Delta y = f'(x)\Delta x+o(|\Delta x|)$ $(\Delta x \to 0)$; in other words, Δy is the sum of two terms, of which the first, $f'(x)\Delta x$, is proportional to Δx and the second is an †infinitesimal of an order higher than

Δx. Here the principal part $f'(x)\Delta x$ of Δy is called the **differential** of $y=f(x)$ and is denoted by dy. The differential dy thus defined is a function of two independent variables x and Δx. In particular, if $f(x)=x$, from the definition we get $dx=1\cdot\Delta x=\Delta x$. Hence, in general, we have $dy=f'(x)dx$ and $f'(x)=dy/dx$.

With respect to the rectangular coordinates (x,y), the straight line with slope $f'(x_0)$ through a point $(x_0,f(x_0))$ of the graph of $y=f(x)$ is the †tangent line of the graph at the point $(x_0,f(x_0))$. A function is continuous at a point where the function is differentiable, but the converse of this proposition does not hold. In fact, Weierstrass showed that the function defined by the infinite series $\sum_{n=0}^{\infty}a^n\cos b^n\pi x$, where $0<a<1$ and b is an odd integer with $ab>(3/2)\pi+1$, is continuous everywhere and nowhere differentiable on $(-\infty,\infty)$ [2,7].

C. Differentiation

For two differentiable functions f and g defined on the interval I, the following formulas hold: $(\alpha f+\beta g)'=\alpha f'+\beta g'$, where α and β are constants; $(fg)'=f'g+fg'$; and $(f/g)'=(f'g-fg')/g^2$ (at every point where $g\neq 0$). Let $y=f(x)$ be a function of x defined on the interval (a,b) and $x=\varphi(t)$ a function of t defined on (α,β). If $\varphi(t)\in(a,b)$ whenever $t\in(\alpha,\beta)$, then the composite function $y=F(t)=f(\varphi(t))=(f\circ\varphi)(t)$ is well defined. Assume further that f and φ are differentiable on (a,b) and (α,β), respectively. Then the composite function $F(t)=(f\circ\varphi)(t)$ is differentiable on (α,β), and we have the **chain rule**, $F'(t)=f'(x)\varphi'(t)$ $(x=\varphi(t))$, or $dy/dt=(dy/dx)(dx/dt)$. Assume that a function $y=f(x)$ is †strictly increasing or decreasing and differentiable at x_0. If furthermore we have $f'(x)\neq 0$, then the inverse function $x=f^{-1}(y)$ is also differentiable at y_0 $(=f(x_0))$ and satisfies $(dx/dy)_{y=y_0}(dy/dx)_{x=x_0}=1$. However, if $f'(x_0)=0$, then even though $f^{-1}(y)$ is not differentiable at y_0, $\lim_{\Delta y\to 0}(f^{-1}(y_0+\Delta y)-f^{-1}(y_0))/\Delta y$ exists and is $+\infty$ or $-\infty$.

D. Higher-Order Derivatives

If the derivative $f'(x)$ of a function $y=f(x)$ is again differentiable on I, then $(f'(x))'=f''(x)$ is well defined as a function of x on I. In general, if $f^{(n-1)}(x)$ is differentiable on I, then $f(x)$ is called ***n*-times differentiable** on I, and the ***n*th derivative** (or ***n*th derived function**) $f^{(n)}(x)$ of $f(x)$ is defined by $f^{(n)}(x)=(f^{(n-1)}(x))'$ and is also denoted by d^ny/dx^n or $D^{(n)}y$. The nth derivative for $n\geqslant 2$ is called a **higher-order derivative**.

Concerning the nth derivative of the product of two functions, **Leibniz's formula** holds:

$$(fg)^{(n)}=f^{(n)}g+\binom{n}{1}f^{(n-1)}g'+\ldots$$
$$+\binom{n}{k}f^{(n-k)}g^{(k)}+\ldots+fg^{(n)}.$$

Analogous to $dy=y'\Delta x$, which is a function of x and Δx, we may define d^2y in the notation d^2y/dx^2 by $d^2y=d(dy)=d(y'\Delta x)=(y'\Delta x)'\Delta x=y''\Delta x^2$. Since $\Delta x=dx$, it follows from the above that $d^2y=y''dx^2$. Similarly, $d^ny=y^{(n)}dx^n$ and is called the ***n*th differential** (or **differential of *n*th order**) of $f(x)$.

E. The Mean Value Theorem

Let $f(x)$ be a continuous function defined on $[a,b]$, and suppose that for every point x_0 on (a,b) there exists a limit $\lim_{h\to 0}(f(x_0+h)-f(x_0))/h$, which may be infinite. (These conditions are satisfied if $f(x)$ is differentiable on $[a,b]$.) Then there exists a point ξ such that

$$\frac{f(b)-f(a)}{b-a}=f'(\xi),\quad a<\xi<b.$$

This proposition is called the **mean value theorem**. A special case of the theorem under the further condition that $f(a)=f(b)$ is called **Rolle's theorem**. If we put $b-a=h$, $\xi=a+\theta h$, then the conclusion of the theorem may be written as $f(a+h)=f(a)+h\cdot f'(a+\theta h)$ $(0<\theta<1)$.

This theorem implies the following: Let $f(x)$ be a function as in the hypothesis of the mean value theorem, and assume further that $A\leqslant f'(x)\leqslant B$ holds for all x with $a<x<b$. Then $A\leqslant(f(b)-f(a))/(b-a)\leqslant B$. (French mathematicians sometimes call this the "théorème des accroissements finis.") Using the mean value theorem, the following theorem can be proved: If $f(x)$ is continuous on $[a,b]$ and $f'(x)$ exists and is positive on (a,b), then $f(b)>f(a)$. Accordingly, if $f'(x)>0$ at every point x of an interval I, then $f(x)$ is †strictly increasing on that interval. (If $f'(x)<0$ on I, then f is strictly decreasing.) The converse of the previous statement does not always hold ($f(x)=x^3$ is a counterexample, since $f'(0)=0$). Furthermore, from the mean value theorem it follows that if $f'(x)=0$ everywhere in an interval, then $f(x)$ is constant on that interval. Consequently, two functions with the same derivative on an interval differ only by a constant.

Suppose that $f(x)$ is n-times differentiable on an open interval I. For a fixed $a\in I$ and an arbitrary $x \in I$, we put

$$f(x)=f(a)+\frac{f'(a)}{1!}(x-a)+\ldots+\frac{f^{(n-1)}(a)}{(n-1)!}(x-a)^{n-1}+R_n.$$

Then $R_n=f^{(n)}(\xi)(x-a)^n/n!$ for some ξ between a and x. This is called **Taylor's formula**, where R_n is the **remainder** of the nth order given by Lagrange. We also have several other forms for R_n (→ Appendix A, Table 9). If $f^{(n)}(x)$ is continuous at $x=a$, then $\xi \to a$ as $x \to a$, and accordingly, $f^{(n)}(\xi)\to f^{(n)}(a)$. Hence $f(x)=\sum_{k=0}^{n}(f^{(k)}(a)/k!)\,(x-a)^k+o((x-a)^n)$. If $f^{(n)}(x)$ is continuous at $x=a$, then, by Taylor's formula, the value of the polynomial $\sum_{k=0}^{n}(f^{(k)}(a)/k!)(x-a)^k$ may be considered an approximate value of $f(x)$ for x near a. This approximation is called the ***n*th approximation** of $f(x)$, and its error is given by $|R_{n+1}|$. By applying this formula, it is sometimes possible to calculate a limit such as $A=\lim_{x\to a}f(x)/g(x)$, where $f(x)\to 0$ and $g(x)\to 0$ as $x\to a$. For instance, if $f'(x)$ and $g'(x)$ are both continuous at $x=a$ and $g'(a)\neq 0$, then by taking the first approximations of $f(x)$ and $g(x)$ it is easily seen that $A=f'(a)/g'(a)$. A limit of this type is often called a **limit of an indeterminate form** $0/0$. Similarly, we can calculate limits of such indeterminate forms as $0\cdot\infty$ or 0^∞ (for limits of indeterminate forms → [10]).

F. Partial Derivatives

Let $w=f(x,y,\ldots,z)$ be a real-valued function of n independent real variables $x,y,\ldots,z$ defined on a domain G contained in n-dimensional Euclidean space $\mathbf{R}^n$. We obtain a function of a single variable from f by keeping $n-1$ variables (say, $(x,\check{y},\ldots,z)$ except y) fixed. If such a function $\varphi(y)=f(x_0,y,\ldots,z_0)$ is differentiable, that is, if,

$$\varphi'(y_0)=\lim_{\Delta y\to 0}\frac{\varphi(y_0+\Delta y)-\varphi(y_0)}{\Delta y}=\lim_{\Delta y\to 0}\frac{f(x_0,y_0+\Delta y,\ldots,z_0)-f(x_0,y_0,\ldots,z_0)}{\Delta y}$$

exists and is finite, then f is called **partially differentiable** with respect to y at $(x_0,y_0,\ldots,z_0)$, and the derivative is called the **partial derivative** (or **partial differential coefficient**) of $f(x,y,\ldots,z)$ with respect to y at $(x_0,y_0,\ldots,z_0)$. It is denoted by $[\partial w/\partial y]_{x=x_0,\ldots,z=z_0}$, $(\partial/\partial y)f(x_0,y_0,\ldots,z_0)$, $f_y(x_0,y_0,\ldots,z_0)$, or $D_yf(x_0,y_0,\ldots,z_0)$, etc. We usually assume that the point $(x,y,\ldots,z)$ where partial derivatives are considered is an †interior point of the †domain of the function. Since in a space of dimension higher than 1, the †boundary of a domain may be complicated, partial derivatives at boundary points are usually not considered. If a function f possesses a partial derivative with respect to x at every point of an open set G, then f_x is a function on G and is called a **partial derivative** of f with respect to x. The process of determining partial derivatives of f is called the **partial differentiation** of f.

G. Total Differential

Let $w=f(x,y,\ldots,z)$ be a function defined on a domain G, and let $P=(x,y,\ldots,z)$ be an interior point of the domain G of a function $w=f(x,y,\ldots,z)$. Put $\Delta w=f(x_0+\Delta x,y_0+\Delta y,\ldots,z_0+\Delta z)-f(x_0,y_0,\ldots,z_0)$. If there exist constants $\alpha,\beta,\ldots$ such that $\Delta w=\alpha\Delta x+\beta\Delta y+\ldots+\gamma\Delta z+o(\rho)$ $(\rho\to 0)$ where $\rho=\sqrt{\Delta x^2+\Delta y^2+\ldots\Delta z^2}$, then f is called **totally differentiable** (or **differentiable in the sense of Stolz**) at P. In this case, f is partially differentiable at P with respect to each of the variables $x,y,\ldots,z$, and $\alpha=f_x(x_0,y_0,\ldots,z_0)$, $\beta=f_y(x_0,y_0,\ldots,z_0),\ldots,\gamma=f_z(x_0,y_0,\ldots,z_0)$. The principal part of Δw as $\rho\to 0$ is $\alpha\Delta x+\beta\Delta y+\ldots+\gamma\Delta z$, which is called the **total differential** of w at P. If f is totally differentiable at every point of G, then f is said to be totally differentiable on G. The total differential of w is denoted by dw. Since the total differentials of $x,y,\ldots,z$ are $dx=\Delta x$, $dy=\Delta y,\ldots$, $dz=\Delta z$, respectively, we may write $dw=f_xdx+f_ydy+\ldots+f_zdz$; and dw is a function of independent variables $x,y,\ldots,z,dx,dy,\ldots,dz$. The total differentiability of f implies the continuity of f, whereas the partial differentiability of f with respect to each variable does not imply that f is continuous. (Example: Define $f(x,y)=xy/(x^2+y^2)$ for $(x,y)\neq(0,0)$, and $f(0,0)=0$; then the function f is not continuous at $(0,0)$, even though both f_x and f_y exist at $(0,0)$.) The function f is totally differentiable on G if all $f_x,f_y,\ldots,f_z$ exist and are continuous on G, or, more weakly, if all $f_x,f_y,\ldots,f_z$ exist and, with possibly one exception, are continuous. Suppose that $w=f(x,y)$ is totally differentiable at (x,y), and let $\Delta x=\rho\cos\theta$, $\Delta y=\rho\sin\theta$. As $\rho\to 0$, for a fixed θ there exists the limit $\lim_{\rho\to 0}(\Delta z/\rho)=f_x(x,y)\cos\theta+f_y(x,y)\sin\theta$. This limit is called the **directional derivative in the direction** θ at (x,y). The partial derivatives f_x and f_y are special cases of the directional derivative for $\theta=0$ and $\pi/2$, respectively. Suppose that we are given a curve lying in the †interior of the domain of f, and that the curve passes through the point (x,y), where the curve is differentiable. Then the partial derivative of $w=f(x,y)$ in the direction of the normal of the curve at (x,y) is called the **normal derivative** of w at the point (x,y) on the curve and is denoted by $\partial w/\partial n$. Analogous definitions and notations have been introduced for functions of more than two variables.

To see the geometric significance of the total differentiability of $w=f(x,y)$, we consider the graph of the function $w=f(x,y)$ and a point $(a,b,f(a,b))$ on the graph. Then the plane represented by $w-f(a,b)=\alpha(x-a)+\beta(y-b)$ is the †tangent plane to the surface at $(a,b,f(a,b))$ if and only if $\alpha=f_x(a,b)$ and $\beta=f_y(a,b)$. The existence of f_x and f_y depends on the choice of coordinate axes, while the total differentiability of f does not.

H. Higher-Order Partial Derivatives

Suppose that a partial derivative of a function $w=f(x,y,\ldots,z)$ defined on an open set G again admits partial differentiation. The latter partial derivative is called a second-order partial derivative of f. We may similarly define the **nth order partial derivatives**. Higher-order partial derivatives are denoted as follows:

$$\frac{\partial}{\partial x}\left(\frac{\partial w}{\partial x}\right)=\frac{\partial^2 w}{\partial x^2}=f_{xx}(x,y,\ldots,z),$$

$$\frac{\partial}{\partial y}\left(\frac{\partial w}{\partial x}\right)=\frac{\partial^2 w}{\partial x\partial y}=f_{xy}(x,y,\ldots,z),$$

$$\frac{\partial}{\partial x}\left(\frac{\partial^2 w}{\partial x\partial y}\right)=\frac{\partial^3 w}{\partial x\partial y\partial x}=f_{xyx}(x,y,\ldots,z),\ldots.$$

In general, f_{xy} and f_{yx} are not equal. (Peano's example: Let $f(x,y)=xy(x^2-y^2)/(x^2+y^2)$ for $(x,y)\neq(0,0)$ and $f(0,0)=0$. Then $f_{xy}=-1$, $f_{yx}=1$ at $(0,0)$.) However, if both f_{xy} and f_{yx} are continuous on an open set G', then they coincide in G'. Furthermore, if f_x, f_y, and f_{xy} exist in a neighborhood U of a point P belonging to the domain of f and f_{xy} is continuous at P, then f_{yx} exists at P and $f_{xy}=f_{yx}$ (H. A. Schwarz). If f_x and f_y exist in U and are totally differentiable at P, then $f_{xy}=f_{yx}$ at P (W. H. Young). Similarly, if the partial derivatives of order $\geqslant 3$ $f_{\ldots xy\ldots}$ and $f_{\ldots yx\ldots}$ are all continuous, then $f_{\ldots xy\ldots}=f_{\ldots yx\ldots}$. Hence we may change the order of differentiation if all the derivatives concerned are continuous.

I. Composite Functions of Many Variables

Let w be a function of $x,y,\ldots,z$ and let each of $x,y,\ldots,z$ be a function of t. Suppose that the range of $(x(t),y(t),\ldots,z(t))$ is contained in the domain of w. Then w is a function of t. If further w is totally differentiable and x, $y,\ldots,z$ are all differentiable, then w as a function of t is differentiable, and we have

$$\frac{dw}{dt}=\frac{\partial w}{\partial x}\frac{dx}{dt}+\frac{\partial w}{\partial y}\frac{dy}{dt}+\ldots+\frac{\partial w}{\partial z}\frac{dz}{dt}.$$

If partial derivatives of order $\geqslant 2$ are totally differentiable, then d^2w/dt^2, $d^3w/dt^3,\ldots$ are obtained by repeating the above procedure. A similar consideration is valid when x, $y,\ldots,z$ are functions of many variables.

J. Taylor's Formula for Functions of Many Variables

Suppose that $f(x,y)$ is defined on an open domain G, $f(x,y)$ has continuous partial derivatives of orders up to n, and the line segment $(a+(x-a)t,\ b+(y-b)t)$ $(0\leqslant t\leqslant 1)$ is contained in the domain G. Then there exists a number θ $(0<\theta<1)$ such that

$$\begin{aligned}
&f(x,y)\\
&=f(a,b)+\left((x-a)\frac{\partial}{\partial x}+(y-b)\frac{\partial}{\partial y}\right)f(a,b)\\
&\quad+\frac{1}{2!}\left((x-a)\frac{\partial}{\partial x}+(y-b)\frac{\partial}{\partial y}\right)^2 f(a,b)+\ldots\\
&\quad+\frac{1}{(n-1)!}\left((x-a)\frac{\partial}{\partial x}+(y-b)\frac{\partial}{\partial y}\right)^{n-1}\\
&\qquad\times f(a,b)+\frac{1}{n!}\left((x-a)\frac{\partial}{\partial x}+(y-b)\frac{\partial}{\partial y}\right)^{n}\\
&\qquad\times f(a+(x-a)\theta,b+(y-b)\theta),
\end{aligned}$$

where, for instance, the third term $((x-a)\cdot(\partial/\partial x)+(y-b)(\partial/\partial y))^2 f(a,b)$ means $(x-a)^2(\partial^2 f/\partial x^2)(a,b)+2(x-a)(y-b)\cdot(\partial^2 f/\partial x\partial y)(a,b)+(y-b)^2(\partial^2 f/\partial y^2)(a,b)$, with $(\partial^2 f/\partial x^2)(a,b)$, $(\partial^2 f/\partial x\partial y)(a,b)$, and $(\partial^2 f/\partial y^2)(a,b)$ denoting the values of $(\partial^2 f/\partial x^2)$, $(\partial^2 f/\partial x\partial y)$, and $(\partial^2 f/\partial y^2)$ at (a,b), respectively. The displayed formula is called **Taylor's formula** for a function of two variables. A similar formula is valid for a function of n variables $(n\geqslant 3)$. As in the case of functions of one variable, we may derive approximation formulas for f from Taylor's formula.

K. Classes of Functions

If all the partial derivatives of order n of $f(P)$ are continuous on an open set G, then f is said to be a **function of class** C^n (or **n-times continuously differentiable**) on G. The set of all n-times continuously differentiable functions is denoted by $C^n(n=1,2,\ldots)$. A continuous function is of **class** C^0. A **function of class** C^1 is also called a **smooth** function. It is obvious that $C^0\supset C^1\supset C^2\supset\ldots$. A partial derivative of order $r \leqslant s$ of a function belonging to class C^s does not depend on the order of the differentiation. A function belonging to $C^\infty = \bigcap_{r=1}^{\infty} C^r$ is said to be of **class** C^∞ or **infinitely differentiable**. We sometimes say that a function has a certain "nice" property or is "well behaved" if it belongs to some C^r $(r \geqslant 1)$.

Let $w = f(x, y, \ldots, z)$ be a function defined on an open set G of $\mathbf{R}^n$ and $P = (a, b, \ldots, c) \in G$. If

$$f(x,y,\ldots,z) = f(a,b,\ldots,c) + \sum_{r_1=1}^{\infty} \ldots \sum_{r_n=1}^{\infty} \alpha_{r_1 r_2 \ldots r_n} (x-a)^{r_1} (y-b)^{r_2} \ldots (z-c)^{r_n}$$

holds in some open neighborhood U of P, where the right-hand side of the equality is an absolutely convergent series, then f is said to be **real analytic** at P. In this case, f is r-times differentiable at P for any r, and we have

$$\alpha_{r_1 r_2 \ldots r_n} = \frac{r_1! r_2! \ldots r_n!}{(r_1 + r_2 + \ldots + r_n)!} \times \frac{\partial^{r_1 + r_2 + \ldots + r_n} f}{\partial x^{r_1} \partial y^{r_2} \ldots \partial z^{r_n}} (a, b, \ldots, c).$$

If f is real analytic at every point P of the domain G, then f is called a **real analytic function** on G. Sometimes, a real analytic function is called a function of **class** C^ω. A real analytic function belongs to C^∞, but the converse is not true (→ 61 C^∞-Functions and Quasi-Analytic Functions E).

L. Extrema

Let f be a real-valued function defined on a domain G in an n-dimensional Euclidean space $\mathbf{R}^n$ that has the point P_0 in its interior. If there exists a neighborhood U of P_0 such that for every point $P (\neq P_0)$ of U we have $f(P) \geqslant f(P_0)$, then we say that f has a **relative minimum** at P_0, and $f(P_0)$ is a **relative minimum** of f. Replacing $\geqslant$ by $\leqslant$, we obtain the definition of **relative maximum**. Either a relative maximum or relative minimum is called a **relative extremum**.

To find relative extrema, the following facts concerning the sign of the derivative are useful. Suppose that a function of a single variable is differentiable on an interval I. Then we have the following: (1) If f has a relative extremum at an interior point x_0 of I, then $f'(x_0) = 0$. (2) If $f'(x_0) = 0$ and $f'(x)$ changes its sign at x_0 from positive (negative) to negative (positive), then f has a relative maximum (minimum) at x_0. (3) If $f'(x_0) = 0$ and f is twice differentiable on some neighborhood of x_0, then f has a relative maximum or minimum according as $f''(x_0) < 0$ or > 0. If $f''(x_0) = 0$, then nothing definite can be concluded about a relative extremum of f at x_0. In general, if there exists a neighborhood of x_0 in which f is r-times differentiable (r is even) and $f^{(r)}$ is continuous, and if $f'(x_0) = f''(x_0) = \ldots = f^{(r-1)}(x_0) = 0$, $f^{(r)}(x_0) > 0$ (or < 0), then f has a relative minimum (maximum) at x_0. On the other hand, if this condition holds with odd r, then $f(x_0)$ is not a relative extremum. If $f'(x_0) = 0$, then $f(x_0)$ is called a **stationary value** of f.

If a function f of n variables $x, y, \ldots, z$ has a relative extremum at $(x_0, y_0, \ldots, z_0)$, then we have $f_x(x_0, y_0, \ldots, z_0) = 0$, $f_y(x_0, y_0, \ldots, z_0) = 0, \ldots, f_z(x_0, y_0, \ldots, z_0) = 0$, provided that the partial derivatives of f exist. Assume that for a function f of class C^2 of two variables x and y, we have $f_x(x_0, y_0) = 0$ and $f_y(x_0, y_0) = 0$, and let $\delta = f_{xx}(x_0, y_0) f_{yy}(x_0, y_0) - f_{xy}^2(x_0, y_0)$. Then we have the following: (1) If $\delta > 0$, then according as $f_{xx}(x_0, y_0) < 0$ or > 0, f has a relative maximum or minimum at (x_0, y_0). (2) If $\delta < 0$, then f does not have a relative extremum at (x_0, y_0). (3) If $\delta = 0$, then without further information nothing definite can be said about a relative extremum of f at the point.

In general, let $x_1, \ldots, x_n$ be independent variables. If a function f of variables $x_1, \ldots, x_n$ has a relative extremum at a point $P_0 = (x_1^0, \ldots, x_n^0)$, then $f_i = f_{x_i}(P_0) = 0$ $(i = 1, \ldots, n)$, provided that all the partial derivatives of f exist. In general, a point P_0 where these conditions are satisfied is called a **critical point** of f. The value $f(P_0)$ at a critical point is called a **stationary value**. If further f is of class C^2, then consider a †quadratic form of n variables $Q = Q(X_1, \ldots, X_n) = \sum_{i,k} f_{ik} X_i X_k$, where $f_{ik} = f_{x_i x_k}(P_0)$. Suppose that $|f_{ik}| \neq 0$. Then according to whether Q is †positive definite, †negative definite, or †indefinite, f has a relative minimum, relative maximum, or no relative extremum at P_0. If $|f_{ik}| = 0$, then nothing can be said in general. A critical point P of f is said to be **nondegenerate** if $|f_{ik}| \neq 0$ and **degenerate** if $|f_{ik}| = 0$.

We may also apply the method of differentiation of †implicit functions to find relative extrema of functions defined implicitly. Given functions $\varphi_1, \ldots, \varphi_m$ $(m < n)$, the problem of finding a relative extremum of $f(x_1, \ldots, x_n)$ under the condition that $\varphi_1(x_1, \ldots, x_n) = 0, \ldots, \varphi_m(x_1, \ldots, x_n) = 0$ is called the problem of finding a **conditional relative extremum**. This problem may be reduced to the problem of finding a relative extremum of an implicit function. Actually, if the functions $f, \varphi_1, \ldots, \varphi_m$ are of class C^1 and the †Jacobian $\partial(\varphi_1, \ldots, \varphi_m)/\partial(x_{n-m+1}, \ldots, x_n)$ does not vanish in the domain considered, then $y_1 = x_{n-m+1}, \ldots, y_m = x_n$ may be regarded as implicit functions of $x_1, \ldots, x_l$

$(l=n-m)$. Hence we may set $f(x_1,\dots,x_l, y_1,\dots,y_m)=f^*(x_1,\dots,x_l)$. Then f has a relative extremum at $(x_1^0,\dots,x_n^0)$ under the condition $\varphi_1=\dots=\varphi_m=0$ if and only if f^* has a relative extremum at $P_0=(x_1^0,\dots,x_l^0)$. The latter condition implies that all $\partial f^*/\partial x_j$ $(j=1,\dots,l)$ vanish at P_0, which holds if and only if for arbitrary constants $\lambda_1,\dots,\lambda_m$ the function $F(x_1,\dots,x_n)=f+\lambda_1\varphi_1+\dots+\lambda_m\varphi_m$ satisfies $\partial F/\partial x_i=0$ $(i=1,\dots,n)$, and further $\varphi_1=0,\dots,\varphi_m=0$ at $(x_1^0,\dots,x_n^0)$. From this system of equations we may often find the values of $x_1^0,\dots,x_n^0$. This method of finding conditional relative extrema is called **Lagrange's method of indeterminate coefficients**, or the **method of Lagrange multipliers** (→ 212 Implicit Functions; 374 Series G; 218 Integral Calculus H).

References

[1] T. Takagi, Kaiseki gairon (Japanese; A course of analysis), Iwanami, third edition, 1961.
[2] M. Fujiwara, Sûgaku kaiseki I; Bibunsekibun gaku (Japanese; Mathematical analysis, pt. 1. Differential and integral calculus) I, Utida-rôkakuho, 1934.
[3] S. Kametani, Syotô kaisekigaku (Japanese; Elementary analysis), Iwanami, I, 1953; II, 1958.
[4] S. Hitotumatu, Kaisekigaku zyosetu (Japanese; Elements of analysis), Syôkabô, I, 1962; II, 1963.
[5] T. M. Apostol, Mathematical analysis, Addison-Wesley, 1957.
[6] N. Bourbaki, Eléments de mathématique, Fonctions d'une variable réelle, Actualités Sci. Ind., 1074b, 1132a, Hermann, second edition, 1958, 1961.
[7] R. C. Buck, Advanced calculus, McGraw-Hill, second edition, 1965.
[8] R. Courant, Differential and integral calculus I, II, Nordemann, 1938.
[9] G. H. Hardy, A course of pure mathematics, Cambridge Univ. Press, seventh edition, 1938.
[10] E. Hille, Analysis, Blaisdell, I, 1964; II, 1966.
[11] W. Kaplan, Advanced calculus, Addison-Wesley, 1952.
[12] E. G. H. Landau, Einführung in die Differentialrechnung und Integralrechnung, P. Noordhoff, 1934; English translation, Differential and integral calculus, Chelsea, 1965.
[13] J. M. H. Olmsted, Advanced calculus, Appleton-Century-Crofts, 1961.
[14] A. Ostrowski, Vorlesungen über Differential-und Integralrechnung I, II, III, Birkhäuser, second edition, 1960–1961.
[15] M. H. Protter and C. B. Morrey, Modern mathematical analysis, Addison-Wesley, 1964.
[16] W. Rudin, Principles of mathematical analysis, McGraw-Hill, second edition, 1964.
[17] V. I. Smirnov, A course of higher mathematics. I, Elementary calculus; II, Advanced calculus, Addison-Wesley, 1964.
[18] A. E. Taylor, Advanced calculus, Ginn, 1955.

110 (XIII.18) Differential-Difference Equations

A. General Remarks

A differential-difference equation is a functional equation that involves derivatives and differences of unknown functions. As will be shown, this type of equation often appears in time-lag phenomena. Historically, since the first use of differential-difference equations in the problem of a string of finite length by Johann Bernoulli in 1732, many works have been devoted to the investigation of these equations. Summaries of work up to 1960, and from then up to mid 1962 may be found in [1] and [2], respectively, with detailed references. A systematic treatment of †special functions including †gamma functions by means of differential-difference equations has been developed (→ 381 Special Functions).

In general, suppose that the derivative $dx(t)/dt$ of a phenomenon $x(t)$ is a function $dx(t)/dt=f(t,x(t),x(s))$ of t, $x(t)$, and the "history" of the phenomenon $x(s)$, where s is smaller than t and varies in the interval determined by t. Among these phenomena, the simplest satisfy the equation

$$\frac{dx(t)}{dt}=f(t,x(t),x(t-h_1),\dots,x(t-h_m)) \tag{1}$$

($h_1,\dots,h_m$ are positive constants and $h_1<\dots<h_m$), which is called a **differential-difference** (or **difference-differential**) **equation** or **differential equation with retarded arguments** (**time lags**). The same name, differential-difference equation, is given to other types of equations, such as $dx(t)/dt=f(t,x(t),x(t+h_1),\dots,x(t+h_m))$ or $dx(t)/dt=f(t,x(t-h_1),\dots,x(t-h_m),x(t+l_1),\dots,x(t+l_k))$ $(h_1,\dots,h_m,l_1,\dots,l_k>0)$, which include only advanced arguments or both advanced and retarded arguments. The equation

$$\frac{dx(t)}{dt}=f(t,x(t),x(t-h_1),\dots,x(t-h_m),x'(t-h_1),\dots,x'(t-h_m))$$

is said to be of **neutral type**, equation 1 of **retarded type**, and the equation that involves only advanced arguments of **advanced type**. In the definitions in this paragraph, the variable x may represent a vector, but if we are concerned with scalar variables, the equation of neutral type, for example, is given by

$$f(t,u(t),u(t-h_1),\ldots,u(t-h_m),u'(t),u'(t-h_1),\ldots,u'(t-h_m),\ldots,u^{(n)}(t),u^{(n)}(t-h_1),\ldots,u^{(n)}(t-h_m))=0.$$

B. Initial Value Problems

The **initial value problems** for differential-difference equations of retarded type (1) are the problems of obtaining solutions of (1) for $t\geqslant 0$ under the initial conditions $x(t)=\varphi(t)$ $(-h_m\leqslant t<0)$ and $x(0)=x_0$. Let $f(t,x,y_1,\ldots,y_m)$ be continuous for $0\leqslant t\leqslant t_0, |x-x_0|\leqslant a$, $|y_k-x_0|\leqslant a$ $(k=1,\ldots,m)$, and suppose that $|f|\leqslant M$ there. Let $\varphi(t)$ be a given function that is continuous in $-h_m\leqslant t<0$, admits a finite $\lim_{t\to -0}\varphi(t)$ and satisfies $|\varphi(t)-x_0|\leqslant a$. Then there exists a continuous solution of (1) in $0\leqslant t\leqslant \min(t_0,a/M)$ under the initial conditions $x(t)=\varphi(t)$ $(-h_m\leqslant t<0)$ and $x(0)=x_0$. The fixed-point theorem is applicable to the proof of the existence of such solutions; for the existence proof, a method is also used of reducing the problem to the initial value problems of differential equations by decomposing the intervals and then successively connecting the subintervals. This method is effective for neutral type. For the uniqueness of the solution assumptions stronger than the continuity of f are needed. For example, if f satisfies the †Lipschitz condition

$$|f(t,x_1,y_1,\ldots,y_m)-f(t,x_2,z_1,\ldots,z_m)|\leqslant L|x_1-x_2|+L_1|y_1-z_1|+\ldots+L_m|y_m-z_m|$$

$(L_1,\ldots,L_m$ are constant), the uniqueness of solutions can be established by means of the successive approximation method. Also, as in the theory of differential equations, we have criteria of Osgood type, or other criteria such as the one that utilizes †Ljapunov functions to ascertain the uniqueness of solutions. Ljapnuov functions may be applied also to problems of the dependency of solutions on parameters such as $x_0,\varphi(t)$, and $h_1,\ldots,h_m$.

On the other hand, instead of equation (1), if we consider the functions $x_i(t)$ which satisfy the differential-difference inequalities

$$\left|\frac{dx}{dt}-f(t,x(t),x(t-h_1^{(i)}),\ldots,x(t-h_m^{(i)})\right|\leqslant\varepsilon_i(t),\quad i=1,2 \qquad (2)$$

$(x(t)=\varphi_i(t),\ -h_m^{(i)}\leqslant t<0)$, an estimate for the difference $|x_1(t)-x_2(t)|$ can be obtained by making use of the Ljapunov function. If $\varepsilon_i(t)\equiv 0$ in this estimate, the results obtained correspond to those of (1) [3].

C. Linear Differential-Difference Equations

The most general linear differential-difference equation is of the form

$$\sum_{n=0}^{m}A_n(t)x'(t+h_n)+\sum_{n=0}^{m}B_n(t)x(t+h_n)=w(t),\quad (3)$$

where $A_n(t),B_n(t)$ $(n=0,1,\ldots,m)$ are $N\times N$ matrices, $w(t)$ is an N-dimensional vector, h_n $(n=1,\ldots,m)$ are constant, and $0=h_0<h_1<\ldots<h_m$. Corresponding to (3), let the function $K(s,t)$ be an $N\times N$ matrix solution of the **adjoint equation** of (3) such that

$$-\frac{\partial}{\partial s}K(s,t)+\sum_{n=0}^{m}K(s+h_m-h_n,t)B_n(s+h_m-h_n)=0,$$

$t_0<s<t$, under the initial conditions $K(s,t)=0$ for $t<s\leqslant t+h_m, K(s,t)=I$ for $s=t$ (I is the unit matrix). Then the uniquely determined function $K(s,t)$ is called the **kernel function** of (3). In terms of this function, the solution of (3) with the condition $x(t)=0$ $(t_0\leqslant t\leqslant t_0+h_m)$ is represented by

$$x(t+h_m)=\int_{t_0}^{t}K(s,t)w(s)\,dx,\quad t>t_0.$$

In the linear inhomogeneous differential-difference equation of order 1 of neutral type

$$x'(t+h_m)+\sum_{n=0}^{m-1}A_n(t)x'(t+h_n)+\sum_{n=0}^{m}B_n(t)x(t+h_n)=w(t),\quad t>t_0,\quad (4)$$

let S be a set of points written as $t_0+jh_m-\sum_{n=1}^{m-1}i_nh_n$ $(i_1,\ldots,i_{m-1}$ are integers; $j=1,2,\ldots)$ and such that $0\leqslant\sum_{n=1}^{m-1}i_n\leqslant j$, and let S' be a subset of S satisfying $0\leqslant\sum_{n=1}^{m-1}i_n\leqslant j-1$. Let T be a set of points written as $t+h_m-jh_m+\sum_{n=1}^{m-1}i_nh_n$ $(j=1,2,\ldots)$ and such that $0\leqslant\sum_{n=1}^{m-1}i_n\leqslant j$, and let T' be a subset of T satisfying $0\leqslant\sum_{n=1}^{m-1}i_n\leqslant j-1$. Let $A_n(t)$, $A_n'(t)$, $B(t)$ be continuous for $t\geqslant t_0$, and $w(t)$ continuous for $t\geqslant t_0$ except possibly where there exist †discontinuities of the first kind in S. Let $K(s,t)$ be a unique matrix solution of

$$-\frac{\partial K(s,t)}{\partial s}-\sum_{n=0}^{m-1}\frac{\partial}{\partial s}(K(s+h_m-h_n,t)A_n(s+h_m-h_n))+\sum_{n=0}^{m}K(s+h_m-h_n,t)B_n(s+h_m-h_n)=0,$$

$$t_0<s<t,\quad s\notin T$$

under the initial conditions $K(s,t)=0$ for

$t<s\leqslant t+h_m$, $K(s,t)=I$ for $s=t$, satisfying a condition that $K(s,t)+\sum_{n=0}^{m-1}K(s+h_m-h_n,t)A_n(s+h_m-h_n)$ is continuous on $t_0\leqslant s\leqslant t$. Then the solution of (4) under the initial condition $x(t)=0$ $(t_0\leqslant t\leqslant t_0+h_m)$ is represented by

$$x(t+h_m)=\int_{t_0}^{t}K(s,t)w(s)\,ds,\quad t>t_0,$$

and the following relation holds:

$$x'(t+h_m)=w(t)+\int_{t_0}^{t}\frac{\partial}{\partial t}K(s,t)w(s)\,ds$$
$$-\sum_{s\in T'-(t)}(K(s+0,t)-K(s-0,t))w(s),$$
$$t>t_0,\quad t\notin S$$

[4].

D. The Case of Constant Coefficients

In equations (3) and (4), if the coefficients are all constant, then we have

$$\sum_{n=0}^{m}(A_nx'(t-h_n)+B_nx(t-h_n))=f(t),\qquad(5)$$

where $\det A_0\neq 0$, $0=h_0<h_1<\ldots<h_m$, and A_n, B_n are constant matrices. If we put

$$H(s)=\sum_{n=0}^{m}(A_ns+B_n)e^{-h_ns},$$

then the equation $\det H(s)=0$ is called the **characteristic equation** corresponding to (5), and its roots are called the **characteristic roots** corresponding to (5). Let S be a set of points written as $t=\sum_{n=0}^{m}j_nh_n$ (the j_n are integers) and S_1 the intersection of S and $[h_m,\infty)$. If $g(t)\in C^1[0,h_m]$, $f(t)$ is continuous for $[0,\infty)$ except for possible discontinuity of the first kind in S_1, and $\|f\|\leqslant c_1\exp c_2t$ $(c_1>0,c_2>0)$ $(t\to\infty)$, then the continuous solution of (5) satisfying the initial condition $x(t)=g(t)$ $(0\leqslant t\leqslant h_m)$ is represented by the integral

$$x(t)=\int_{c-i\infty}^{c+i\infty}e^{ts}H(s)^{-1}(p(s)+q(s))\,ds,$$
$$t>0$$

for sufficiently large c, where $p(s)$ and $q(s)$ are defined as follows:

$$p(s)=e^{-h_ms}\sum_{n=0}^{m}A_ng(h_m-h_n)$$
$$+\sum_{n=0}^{m}(A_ns+B_n)e^{-h_ms}\int_{0}^{h_m-h_n}g(t)e^{-st}\,dt,$$
$$q(s)=\int_{h_m}^{\infty}f(t)e^{-st}\,dt.$$

If $A_n=0$ $(n=1,\ldots,m)$, $g\in C^0[0,h_m]$, and $f\in C^0[0,\infty)$, then the solution of (5) satisfying the condition $x(t)=g(t)$ $(0\leqslant t\leqslant h_m)$ is represented by

$$x(t)=\int_{c-i\infty}^{c+i\infty}e^{ts}H(s)^{-1}(p_0(s)+q(s))\,ds,$$
$$t>h_m$$

for sufficiently large c, where $q(s)$ is the same as before, and

$$p_0(s)$$
$$=e^{-h_ms}\sum_{n=0}^{m}A_ng(h_m-h_n)$$
$$-\sum_{n=0}^{m}(A_ns+B_n)e^{-h_ns}\int_{h_m-h_n}^{h_m}g(t)e^{-st}\,dt$$
$$=e^{-h_ms}\sum_{n=0}^{m}A_ne^{-h_ns}g(h_m)$$
$$-\sum_{n=0}^{m}e^{-h_ns}\int_{h_m-h_n}^{h_m}(A_ng'(t)+B_ng(t))e^{-st}\,dt.$$

Let S_2 be the intersection of S and $(0,\infty)$ and $K(t)$ be a matrix function satisfying: (i) $K(t)=0$ $(t<0)$; (ii) $K(0)=A_0^{-1}$; (iii) $\sum_{n=0}^{m}A_nK(t-h_n)\in C^0[0,\infty)$; and (iv) $\sum_{n=0}^{m}(A_nK'(t-h_n)+B_nK(t-h_n))=0$ $(t>0,t\notin S_2)$. Then $K(t)$ is called the kernel function of (5). If $g\in C^1[0,h_m]$, $f\in C^0([0,\infty)-S_1)$, and f is continuous except for possible discontinuity of the first kind in S_1, then a continuous solution of (5) satisfying the initial condition $x(t)=g(t)$ $(0\leqslant t\leqslant h_m)$ is represented by

$$x(t)=\sum_{n=0}^{m}K(t-h_n)A_ng(0)$$
$$+\int_{h_m}^{t}K(t-t_1)f(t_1)\,dt_1$$
$$+\sum_{n=0}^{m}\int_{0}^{h_m-h_n}K(t-t_1-h_n)(A_ng'(t_1)$$
$$+B_ng(t_1))\,dt_1,\quad t>0$$

$$=\sum_{n=0}^{m}K(t-h_m-h_n)A_ng(h_m)$$
$$+\int_{h_m}^{t}K(t-t_1)f(t_1)\,dt_1$$
$$-\sum_{n=0}^{m}\int_{h_m-h_n}^{h_m}K(t-t_1-h_n)$$
$$\times(A_ng'(t_1)+B_ng(t_1))\,dt_1,\quad t>h_m.$$

In particular, if $A_n=0$ $(n=1,\ldots,m)$, the latter holds for the case where $g\in C^0[0,h_m]$ and $f\in C^0[0,\infty)$.

If $f(t)\equiv 0$ and $g\in C^1[0,h_m]$ in (5), and if the curves C_l $(l=1,2,\ldots)$ are suitably chosen, the solution is represented by

$$x(t) = \lim_{l\to\infty} \Big(\text{the sum of residues of } H(s)^{-1} p(s) e^{st} \text{ inside } C_l\Big)$$

$$= \lim_{l\to\infty} \sum_{C_l} p_r(t) e^{s_r t}$$

for $t > Nh_m$ (N is the dimension of x), where s_r is a root of the characteristic equation $\det H(s) = 0$, $p_r(t)e^{s_r t}$ the residue at s_r, and $p_r(t)$ the polynomial (vector) with degree less than the multiplicity of s_r. This limit is uniformly convergent for any finite interval $t_0 \leqslant t \leqslant t_0'$ ($t_0 > Nh_m$). If every characteristic root s_r lies in the left half-plane $\mathrm{Re}\, s \leqslant c_1 < \infty$, the limit is uniformly convergent for $t_0 \leqslant t < \infty$.

For the 1-dimensional case, let $u(t)$ be a continuous solution of

$$a_0 u'(t) + b_0 u(t) + b_1 u(t-h) = 0, \quad t > h, \quad a_0 \neq 0, \tag{6}$$

satisfying the initial condition $u(t) = g(t)$ ($0 \leqslant t \leqslant h, g(t) \in C^0[0,h]$). Let $p_r(t)e^{s_r t}$ be the residues of $e^{st}p_0(s)/h(s)$ at the root s_r of the characteristic equation $h(s) \equiv a_0 s + b_0 + b_1 e^{-hs} = 0$, and let c be any number such that no zeros of $h(s)$ lie on the line $\mathrm{Re}\, s = c$. Then there exists a positive constant c_1 independent of t and g such that the following inequality holds:

$$\Big| u(t) - \sum_{\mathrm{Re}\, s_r > c} e^{s_r t} p_r(t) \Big| \leqslant c_1 m_g e^{ct},$$

$$t > h, \quad m_g = \max_{0 \leqslant t \leqslant h} |g(t)|.$$

Hence a necessary and sufficient condition that every continuous solution $u(t)$ of (6) tend to zero as $t \to \infty$ is that every root of the characteristic equation has negative real part.

E. Asymptotic Behavior

Assume that the characteristic equation

$$s + a_0 + b_0 e^{-hs} = 0 \tag{7}$$

corresponding to the equation $v'(t) + a_0 v(t) + b_0 v(t-h) = 0$ admits a unique root λ with largest real part and λ is real and simple. Assume also that $a(t)$ satisfies one of the following two sets of conditions: (1) $\int^\infty |a(t)| \cdot dt < \infty$, or (2) (i) $a(t) \to 0$ ($t \to \infty$), (ii) $a(t) \neq 0$ ($t \geqslant t_0$), (iii) $a'(t) = o(a(t))$ ($t \to \infty$), and (iv) $\int^\infty a(t)^2 dt < \infty$, $\int^\infty |a'(t)| dt < \infty$, $\int^\infty |a''(t)/a(t)| dt < \infty$. Then the equation

$$u'(t) + (a_0 + a(t))u(t) + b_0 u(t-h) = 0 \tag{8}$$

has a solution of the form

$$u(t) = c(1 + o(1)) \exp\Big(\lambda t - c_1 \int_{t_0}^{t} a(r)\, dr\Big), \quad t \to \infty,$$

where c and c_1 are constants and

$$c_1 = (1 - b_0 h e^{-h\lambda})^{-1}.$$

Furthermore, suppose that the **principal root** λ of (7) is real and simple, and that $a(t)$ and $b(t)$ possess †asymptotic expansions

$$a(t) \sim \sum_{n=1}^{\infty} a_n t^{-n}, \quad b(t) \sim \sum_{n=1}^{\infty} b_n t^{-n}, \quad t \to \infty.$$

Suppose that $a'(t)$, $b'(t)$, $a''(t)$, and $b''(t)$ exist and possess asymptotic expansions. Then there exists a solution $u(t)$ of

$$u'(t) + (a_0 + a(t))u(t) + (b_0 + b(t))u(t-h) = 0 \tag{8'}$$

with an asymptotic expansion of the form

$$u(t) \sim e^{\lambda t} t^r \sum_{n=0}^{\infty} u_n t^{-n}.$$

In particular,

$$r = -(a_1 + b_1 e^{-h\lambda})/(1 - b_0 h e^{-h\lambda})$$

[4].

F. Stability Problems

Let $x_0(t)$ be continuous for $t > 0$ and a solution of

$$x'(t) = f(t, x(t), x(t-h)) \tag{9}$$

for $t > h$. If for any $\varepsilon > 0$ and $t_0 \geqslant 0$, there exists a δ such that for every continuous solution $x(t)$ of (9) the inequality

$$\max_{t_0 \leqslant t} |x(t) - x_0(t)| \leqslant \varepsilon$$

is satisfied when

$$\max_{t_0 \leqslant t \leqslant t_0 + h} |x(t) - x_0(t)| \leqslant \delta, \tag{10}$$

then the solution $x_0(t)$ is said to be **stable**. Furthermore, if $x_0(t)$ is stable and for every $t_0 \geqslant 0$ there exists a $\delta = \delta(t_0)$ and

$$\lim_{t\to\infty} |x(t) - x_0(t)| = 0,$$

then $x_0(t)$ is said to be **asymptotically stable**.

In the 1-dimensional (scalar) equation

$$a_0 u'(t) + b_0 u(t) + b_1 u(t-h) = 0, \quad a_0 \neq 0,$$

let $u_0(t)$ be a solution satisfying the initial condition $u(t) = g(t)$ ($0 \leqslant t \leqslant h$). Then if every root of the corresponding characteristic equation $a_0 s + b_0 + b_1 e^{-hs} = 0$ has real part less than $-\lambda_1$ (< 0), the inequality

$$|u_0(t)| \leqslant c_1 m_g e^{-\lambda_1 t}, \quad t \geqslant h, \quad m_g = \max_{0 \leqslant t \leqslant h} |g(t)|$$

holds.

On the stability problem for perturbed equations

$$u'(t) + a(t)u(t) + b(t)u(t-h) = w(t), \quad t > h,$$

there are many theorems similar to the Dini-Hukuhara type, the Poincaré-Ljapunov type, and others in the theory of ordinary differential equations. For the most general equation

(9), when the Ljapunov function may be defined for differential-difference equations, general results on stability and boundedness have been obtained [3,4,7].

G. Differential Equations with Retarded Arguments

The differential-difference equation with nonnegative retarded arguments $\Delta_i(t)$ is, in general, represented by

$$x^{(m_0)}(t)=f\big(t,\ x(t),\ x'(t),\ \ldots,\ x^{(m_0-1)}(t),\ x(t-\Delta_1(t)),\ \ldots,\ x^{(m_1)}(t-\Delta_1(t)),\ \ldots,\ x(t-\Delta_n(t)),\ x'(t-\Delta_n(t)),\ \ldots,\ x^{(m_n)}(t-\Delta_n(t))\big).$$

These equations have been investigated extensively by mathematicians in the USSR and Eastern Europe, and results on stability, boundary value problems, optimal processes, periodic solutions, perturbations, and approximations have been summarized in detail for the case of constant retardations in [1,2,8].

H. Functional Differential Equation

Let $\tau \geqslant 0$ be a real given number and let $C([a,b],\mathbf{R}^n)$ be the Banach space of continuous mappings of the interval $[a,b]$ into a real (or complex) n-dimensional linear space $\mathbf{R}^n$, with the topology of uniform convergence. If $A>0$, $t_0\in\mathbf{R}$ and $x\in C([t_0-\tau,t_0+A],\mathbf{R}^n)$, then, for any $t\in[t_0,t_0+A]$, we define $x_t\in C=C([-\tau,0],\mathbf{R}^n)$ by $x_t(\theta)=x(t+\theta)$ $(-\tau\leqslant\theta\leqslant 0)$. Given a function $f:\mathbf{R}\times C\to\mathbf{R}^n$, the relation

$$x'(t)=f(t,x_t) \tag{11}$$

is called a **functional differential equation**. For a given $t_0\in\mathbf{R}$ and a given $\varphi\in C$ we say that $x(t_0,\varphi)$ is a solution of (11), with initial value φ at t_0, provided there exists an $A>0$ such that $x(t_0,\varphi)$ is a solution of (11) on $[t_0-\tau,t_0+A)$ and $x_{t_0}(t_0,\varphi)=\varphi$. The following existence theorem for equation (11) can be proved by methods analogous to those used in the conventional theory of differential equations [10,12,14,15].

Suppose that D is an open set in $\mathbf{R}\times C$ and $f:D\to\mathbf{R}^n$ is continuous. Then, if $(t_0,\varphi)\in D$, there exists a solution of (1) with initial value φ at t_0. Furthermore, if x is a nonextendable solution of (1) on $[t_0-\tau,b)$, then for any compact set $K\subset D$, there exists a t^* such that $(t,x_t)\notin K$ for $t^*\leqslant t\leqslant b$.

If the function f does not depend explicitly on t, that is, if the equation is of the form

$$x'(t)=f(x_t), \tag{12}$$

then the equation is said to be **autonomous**.

Suppose that $f:C\to\mathbf{R}^n$ is continuous and maps closed bounded sets of C into bounded sets of $\mathbf{R}^n$. Then, assuming that the solution $x(\varphi)$ of (12) through the initial point $(0,\varphi)$ is defined and unique on $[-\tau,\infty)$, it follows that $x(\varphi)$ satisfies the relations

$$x_0(\varphi)=\varphi,\qquad x_{t+s}(\varphi)=x_t(x_s(\varphi))$$

for any $t,s\geqslant 0$ and therefore defines a dynamical system. The three sets defined by

$$\gamma^*(\varphi)=\bigcup_{t\geqslant 0}x_t(\varphi),\qquad \omega(\varphi)=\bigcap_{\tau\geqslant 0}\overline{\bigcup_{t\geqslant\tau}x_t(\varphi)},$$

$$\alpha(\varphi)=\bigcap_{\tau\geqslant 0}\overline{\bigcup_{t\leqslant\tau}x_t(\varphi)}$$

are called the **orbit** through φ, the ω-**limit set** of $\gamma^*(\varphi)$, and the α-**limit set** of $\gamma^*(\varphi)$, respectively. Then, corresponding to well-known facts in the theory of ordinary differential equations, we have the following results [13,14,15].

If there exists a constant $m>0$ and a solution x of (12) such that $|x(t)|<m$ for $t\in[t_0-\tau,\infty)$, then $\gamma^*(x_0)$ is contained in a compact subset of C. Furthermore, if $|x(t)|<m$ for $t\geqslant-\tau$, then $\omega(\gamma^*(x_0))$ is a nonempty, compact connected invariant set, and dist $(x_t,\omega(\gamma^*(x_0)))\to 0$ as $t\to\infty$.

As a special case of (11), we consider a linear functional differential equation

$$x'(t)=L(t,x_t)+f(t), \tag{13}$$

where $f\in LL_1(t_0,\infty)$ (locally $L_1(t_0,\infty)$), and $L(t,\varphi)$ is a linear functional in φ such that there exists an $n\times n$ matrix function $\eta(t,\theta)$, measurable in (t,θ) and of bounded variation in $\theta\in[-\tau,0]$ for each t, and also a function $l(t)\in LL_1(-\tau,\infty)$, satisfying

$$L(t,\varphi)=\int_{-\tau}^{0}(d_\theta\eta(t,\theta))\varphi(\theta),$$

$$|L(t,\varphi)|\leqslant l(t)\|\varphi\|$$

for any $t\in(-\infty,\infty)$ and $\varphi\in C$. Let $x(t_0,\varphi,f)$ be a solution of (13) with initial value φ at t_0. Then, $x(t_0,\varphi,0)$ is uniquely determined and linear in φ, $x(t_0,0,f)$ is linear in f, and moreover the uniqueness implies that

$$x(t_0,\varphi,f)=x(t_0,\varphi,0)+x(t_0,0,f).$$

Furthermore, $x(t_0,\varphi,f)$ can be expressed by the so-called **variation of constants formula**, i.e.,

$$x(t_0,\varphi,f)(t)=x(t_0,\varphi,0)(t)+\int_{t_0}^{t}U(t,s)f(s)\,ds,\quad t\geqslant t_0,$$

with the initial condition $x_{t_0}=\varphi$; here $U(t,s)$ denotes the solution of the equation

$$U(t,s)=\int_s^t L(u,U_u(\cdot,s))\,du+I$$

(a.e. in s for $t\geqslant s$) satisfying $U(t,s)=0$, $s-\tau\leqslant t<s$; as usual, I is the unit $n\times n$ matrix, U

an $n \times n$ matrix, and $U_t(\cdot, s)(\theta) = U(t+\theta, s)$ $(-\tau \leqslant \theta \leqslant 0)$ [14].

Still more specifically, consider a linear autonomous functional differential equation of the form

$$x'(t) = L(x_t), \tag{14}$$

where $L(\varphi)$ is linear in φ. Again, letting $x(\varphi)$ denote the unique solution of (14) with the initial functional value $\varphi \in C$ at $t_0 = 0$, if we define an operator $T(t)$: $C \to C$ by the relation $x_t(\varphi) = T(t)\varphi$, then the family of mappings $\{T(t) \mid t \geqslant 0\}$ forms a strongly continuous semigroup on C. Defining the infinitesimal generator A of $T(t)$ by

$$A\varphi = \lim_{t \to +0} \frac{1}{t}(T(t)\varphi - \varphi), \quad \varphi \in C$$

whenever this limit exists, it follows that $D(A)$, the domain of A, is dense in C, and that $R(A)$, the range of A, lies in C and consists of the functions

$$A\varphi(\theta) = \begin{cases} \dfrac{d}{d\theta}\varphi(\theta), & -\tau \leqslant \theta < 0, \\ \displaystyle\int_{-\tau}^{0} (d\eta(\theta))\varphi(\theta), & \theta = 0, \end{cases}$$

where φ has a continuous derivative on $[-\tau, 0)$ and $\eta(\theta)$ is an $n \times n$ matrix defined on $-\tau \leqslant \theta \leqslant 0$, with elements of bounded variation such that

$$L(\varphi) = \int_{-\tau}^{0} (d\eta(\theta))\varphi(\theta)$$

for any $\varphi \in C$. Furthermore, the relation

$$\frac{d}{dt}T(t)\varphi = T(t)A\varphi = AT(t)\varphi$$

is satisfied for any $\varphi \in D(A)$ [14, 16].

Now consider a system (11) such that $f(t, 0) \equiv 0$, $t \in \mathbf{R}^+$, and such that $f: \mathbf{R}^+ \times C_\rho \to \mathbf{R}^n$ is continuous, where $C_\rho = \{\varphi \in C \mid \|\varphi\| < \rho\}$ and $\mathbf{R}^+ = [0, \infty)$. Then the null solution $x = 0$ is said to be **stable** if for any $\varepsilon > 0$ and $t_0 \geqslant 0$, there exists a $\delta = \delta(\varepsilon, t_0) > 0$ such that the solution $x(t_0, \varphi)$ of (11) satisfies $x_t(t_0, \varphi) \in C_\varepsilon$ for all $\varphi \in C_\delta$ and all $t \geqslant t_0$. The solution $x = 0$ is said to be **asymptotically stable** if it is stable and, for any ε and $t_0 \geqslant 0$, there are $\rho_0 = \rho_0(t_0)$ and $T = T(\varepsilon, t_0)$ such that $\varphi \in C_{\rho_0}$ implies $\|x_t(t_0, \varphi)\| < \varepsilon$ for $t \geqslant t_0 + T(\varepsilon, t_0)$. In the above definitions, if δ is independent of t_0, the solution $x = 0$ is said to be **uniformly stable**, and if ρ_0 and T are independent of t_0, **uniformly asymptotically stable**. If the solution $x = 0$ is not stable, it is said to be **unstable**. Among the various results concerning stability problems [10, 12, 13, 14, 15], we mention the following criterion for the uniform or uniformly asymptotic stability of $x = 0$, which represents an effective application of the second method of Ljapunov [14].

For any continuous function $V: \mathbf{R}^+ \times C_\rho \to \mathbf{R}$, let

$$\dot{V}(t, \varphi) = \overline{\lim_{h \to +0}} \frac{1}{h}\left(V(t+h, x_{t+h}(t, \varphi)) - V(t, \varphi)\right),$$

where $x_{t+h}(t, \varphi)$ is the solution of (1) through (t, φ). Suppose that $f: \mathbf{R}^+ \times C_\rho \to \mathbf{R}^n$ is continuous, that $a(r)$, $b(r)$, and $c(r)$ are continuous for $r \in [0, \rho)$, that $a(r)$ and $b(r)$ are positive and nondecreasing for $r > 0$, that $a(0) = b(0) = 0$, and that $c(r)$ is nonnegative and nondecreasing. Then, if there exists a continuous function $V: \mathbf{R}^+ \times C \to \mathbf{R}$ such that $a(|\varphi(0)|) \leqslant V(t, \varphi) \leqslant b(\|\varphi\|)$, $\dot{V}(t, \varphi) \leqslant -c(|\varphi(0)|)$, the solution $x = 0$ of equation (11) is uniformly stable. If, in addition, $c(r) > 0$ for $r > 0$, then the solution $x = 0$ is uniformly asymptotically stable.

In the analysis of the phenomena in which the past exerts a significant influence upon the future, we frequently encounter differential equations with time lag which, roughly speaking, belong to one of two general types. One is the type of functional differential equation described above, and the other consists of so called **delay-differential equations** of the form

$$x'(t) = f(t, x(\cdot)); \tag{15}$$

here $x(\cdot)$ represents the state of the system from time α to time t, where α is a given constant $\geqslant -\infty$, and f is a given functional defined on the appropriate domain. The general theory for equations of type (15) and various related topics have been discussed in [11], and the relevant stability problems have been studied in [5], using the method of Ljapunov functionals. Clearly type (15) includes **integrodifferential equations**, which form the subject of [17].

References

[1] А. М. Зверкин, Г. А. Каменский, С. Б. Норкин, Л. Э. Эльсгольц (A. M. Zverkin, G. A. Kamenskiĭ, S. B. Norkin, and L. È. Èl'sgol'c), Дифференциальные уравнения с отклоняющимся аргументом, Uspehi Mat. Nauk, 17, no. 2 (1962), 77–164.
[2] А. М. Зверкин, Г. А. Каменский, С. Б. Норкин, Л. Э. Эльсгольц (A. M. Zverkin, G. A. Kamenskiĭ, S. B. Norkin, and L. È. Èl'sgol'c), Дифференциальные уравнения с отклоняющимся аргументом II; Труды семинара по теории дифференциальных уравнеий с отклоняющимся аргументом II, Москва, 1963, p. 3–49.
[3] S. Sugiyama, On the theory of difference-differential equations I, II, III, Waseda Univ. Bull. Sci. Engr. Res. Lab., 26 (1964), 97–111;

27 (1964), 74–84; Mem. School Sci. Engr. Waseda Univ., 28 (1964), 73–84.
[4] R. Bellman and K. L. Cooke, Differential-difference equations, Academic Press, 1963.
[5] R. D. Driver, Existence and stability of solutions of a delay-differential system, Arch. Rational Mech. Anal., 10 (1962), 401–426.
[6] J. K. Hale, Asymptotic behavior of the solutions of differential-difference equations, Proc. Intern. Symp. Non-linear Vibrations II (1961), 409–426.
[7] N. N. Krasovskiĭ, Stability of motion, Stanford Univ. Press, 1963.
[8] Л. Э. Эльсгольц (L. È. Èl'sgol'c), Введение В Теорию дифференциальных уравнений с откдоняющимся аргументом, Наука, 1964.
[9] R. Bellman and K. L. Cooke, Asymptotic behavior of solutions of differential-difference equations, Mem. Amer. Math. Soc., 1959.
[10] T. Yoshizawa, Stability theory by Ljapunov's second method, Publ. Math. Soc. Japan, 1966.
[11] M. N. Oğuztöreli, Time-lag control system, Academic Press, 1966.
[12] A. Halanay, Differential equations, stability, oscillations, time lag, Academic Press, 1966.
[13] J. K. Hale, Sufficient conditions for stability and instability of autonomous functional-differential equations, J. Differential Equations, 1 (1965), 452–482.
[14] J. K. Hale, Functional differential equations, Springer, 1971.
[15] V. Lakshmikantham and S. Leela, Differential and integral inequalities II, Academic Press, 1969.
[16] G. E. Ladas and V. Lakshmikantham, Differential equations in abstract spaces, Academic Press, 1972.
[17] Constantin Corduneanu, Integral equations and stability of feedback systems, Academic Press, 1973.

111 (XIII.1) Differential Equations

A. Ordinary Differential Equations

It was Galileo who found that the acceleration of a falling body is a constant and thence derived his law of a falling body $x(t)=gt^2/2$ as what we would now view as a solution of the differential equation $x''(t)=g$, where $x(t)$ denotes the distance the body has fallen during the time interval t and g is the constant gravitational acceleration. This pioneering work may be regarded as the first example of solution of a differential equation. Also, the †equations of motion, proposed by I. †Newton as the mathematical formulation of the law of motion, including Galileo's law as a special case, were differential equations of the second order. Thus differential equations appeared, simultaneously with differential and integral calculus, as an indispensable tool for the unified and concise expression of the laws of nature. Such laws are generally called **differential laws**.

Newton completely solved the equations of the †two-body problem proposed by himself; G. W. †Leibniz also succeeded in solving many simple differential equations.

In the 18th century, many mathematicians, such as the †Bernoullis, A. C. Clairaut, J. F. Riccati, L. †Euler, and †Lagrange, attacked and solved differential equations of various types independently. In that period, the emphasis was on solution by **quadrature**, that is, by applying a finite number of algebraic operations, transformations of variables, and indefinite integrations to †elementary functions. It was toward the end of the 18th century that new methods, such as integration by infinite series, came to be discussed. A method of †variation of parameters for the solution of linear ordinary differential equations was invented by Lagrange in 1775. Also, at the beginning of the 19th century, C. F. †Gauss developed his study of differential equations satisfied by †hypergeometric series.

The problem of existence of solutions, which supplies a foundation of modern differential equation theory, was first treated by A. L. †Cauchy. His proof of the existence theorem was later improved by R. L. Lipschitz (1869).

Pioneers in function-theoretic treatment of differential equations were C. A. A. Briot and J. C. Bouquet, who investigated the singular points of a function defined by an analytic differential equation. Also, B. †Riemann proposed a new viewpoint which influenced L. Fuchs in his development of the theory of linear ordinary differential equations in a complex domain (1865). Works of A. M. Legendre on †elliptic functions and of H. †Poincaré on †automorphic functions should also be mentioned in this connection.

After the Cauchy-Lipschitz existence theorem for the equation $y'=f(x,y)$ was known, efforts were directed toward weakening the conditions imposed on $f(x,y)$. G. Peano first succeeded in giving a proof under the continuity assumption only (1890), and his results were sharpened by O. Perron (1915).

Regarding the uniqueness of solutions of †initial value problems, there are various results by W. F. Osgood (1898), Perron (1925), and many Japanese mathematicians. The necessary and sufficient condition for uniqueness was successfully formulated in a concise form (→ 310 Ordinary Differential Equations (Initial Value Problems)).

For linear differential equations with periodic coefficients, investigations were carried out by C. Hermite (1877), E. Picard (1881), G. Floquet (1883), G. W. Hill (1886), and others. For instance, solutions such that $y(x+\omega)=\lambda y(x)$ were found to exist, where ω is the period of the coefficients. Analogous results followed in the case of †doubly periodic coefficients.

Techniques of factorization of linear differential equations developed by G. Frobenius (1873) and E. Landau (1920) should also be noted. Picard (1883), J. Drach (1898), and E. Vessiot (1903, 1904) established a remarkable result on the solvability (in the sense of solution by quadrature) of linear differential equations, successfully extending the †Galois theory in this new direction.

The concept of †asymptotic series, which in a sense approximate the solution of differential equations, was introduced by Poincaré (1886) and furthered by M. A. Ljapunov (1892), J. C. C. Kneser (1896), J. Horn (1897), C. E. Love (1914), and others. Poincaré was also the founder of topological methods in differential equation theory, and his ideas were developed extensively by I. Bendixson (1900), Perron (1922, 1923), G. D. Birkhoff, and others (→ 311 Ordinary Differential Equations (Qualitative Theory)).

In 1890, Picard invented an ingenious technique of †successive approximation for the proof of existence theorems, and his technique is now widely used in every field of functional equations. The technique of reducing linear differential equations to linear †integral equations of Volterra type was also developed.

On the †boundary value problems and †eigenvalue problems that appear in many problems of physics, there was extensive research by mathematicians such as J. C. F. Sturm (1836), J. Liouville, L. Tonelli, Picard, M. Bôcher (1898, 1921), Birkhoff (1910, 1911), and others. In this connection the problem arises of expanding a given function by an †orthogonal system of functions obtained as †eigenfunctions of a given boundary value problem. Those problems were brought into unified form by D. †Hilbert (1904) in his theory of †integral equations. Subsequently boundary value problems of ordinary and partial differential equations came to be discussed in this framework.

Finally, it should be mentioned that the †calculus of variations created by Euler and Lagrange gave rise to the study of a certain class of differential equations bearing the name of Euler (→ 47 Calculus of Variations).

B. Partial Differential Equations

The origin of partial differential equations can be traced back to the study of hydrodynamic problems by J. d'Alembert (1744) and Euler. However, perhaps †Lagrange and P. S. †Laplace were the first to investigate the general theory. Subsequently, during the 18th and 19th centuries, G. Monge, A.-M. Ampère, J. F. Pfaff, C. G. †Jacobi, Cauchy, S. †Lie, and many other mathematicians developed it. The fundamental existence theorem for the initial value problem, now called the Cauchy-Kovalevskaja theorem, was proved by S. Kovalevskaja in 1875 (→ 316 Partial Differential Equations (Initial Value Problems)).

Because of their close connection with problems of physics, linear equations of the second order have been a chief object of research. Up to the 19th century, classification into †elliptic, †hyperbolic, and †parabolic types and the study of †boundary and †initial value problems for each of these types constituted the main part of the theory.

In the 20th century, more complicated problems—†nonlinear problems appearing in the study of viscous or compressible fluids, or the study of equations of †mixed type in connection with supersonic flow—have emerged as important topics. Also, the newly developed techniques of functional analysis have brought about remarkable changes. Especially in the study of the †Schrödinger equations of quantum mechanics and of more general †evolution equations, this method has proved to be a powerful tool.

Also, we should not fail to mention that the recent development of electronic computers has made it possible to obtain numerical solutions and to reveal many important facts. †Numerical analysis is now becoming an indispensable part of the theory (→ 315 Partial Differential Equations).

References

[1] E. A. Coddington and N. Levinson, Theory of ordinary differential equations, McGraw-Hill, 1955.
[2] М. А. Наймарк (M. A. Naĭmark), Линейные Дифференциальные Операторы, Гостехиздат, 1954; English translation, Linear differential operators I. Elementary theory of linear differential operators, II. Linear differential operators in Hilbert space, Ungar, 1967.

[3] L. Bieberbach, Theorie der gewöhnlichen Differentialgleichungen, Springer, second edition, 1965.
[4] G. Sansone, Equazioni differenziali nel campo reale I, II, Zanichelli, second edition, 1948, 1949.
[5] P. Hartman, Ordinary differential equations, John Wiley, 1964.
[6] M. Hukuhara, Zyôbibun hôteisiki (Japanese; Ordinary differential equations), Iwanami, 1950.
[7] Л. С. Понтрягин (L. S. Pontrjagin), Обыкновенные Дифференциальные уравнения, Физматгиз, 1961; English translation, Ordinary differential equations, Addison-Wesley, 1962.
[8] E. Goursat, Cours d'analyse mathématique III, Gauthier-Villars, fourth edition, 1927.
[9] R. Courant and D. Hilbert, Methods of mathematical physics, Interscience, I, 1953; II, 1962.
[10] И. Г. Петровский (I. G. Petrovskiĭ), Лекция об уравнениях с частными производными, Гостехиздат, 1950; English translation, Lectures on partial differential equations, Interscience, 1954.
[11] C. Miranda, Partial differential equations of elliptic type, Springer, second edition, 1970.
[12] L. Hörmander, Linear partial differential operators, Springer, 1963.
[13] S. Mizohata, Henbibun hôteisiki ron (Japanese; Theory of partial differential equations), Iwanami, 1965.
[14] K. Yosida, Bibun hôteisiki no kaihô (Japanese; Solutions of differential equations), Iwanami, 1954.
[15] M. Nagumo, Kindaiteki henbibun hôteisiki ron (Japanese; Modern theory of partial differential equations), Kyôritu, 1957.
Also → references to 307 Ordinary Differential Equations, 315 Partial Differential Equations.

112 (VII.1) Differential Geometry

In differential geometry in the classical sense, we use differential calculus to study the properties of figures such as curves and surfaces in Euclidean planes or spaces. Owing to his studies of how to draw tangents to smooth plane curves, P. †Fermat is regarded as a pioneer in this field. Since his time, differential geometry of plane curves, dealing with curvature, †circles of curvature, †evolutes, †envelopes, etc., has been developed as a part of calculus. Also, the field has been expanded to analogous studies of space curves and surfaces, especially of †asymptotic curves, †lines of curvature, †curvatures and †geodesics on surfaces, and †ruled surfaces. C. F. †Gauss founded the theory of surfaces by introducing concepts of the †geometry on surfaces. Thus differential geometry came to occupy a firm position as a branch of mathematics. The influence that differential-geometric investigations of curves and surfaces have exerted upon branches of mathematics, dynamics, physics, and engineering has been profound. For example, E. Beltrami discovered an intimate relation between the geometry on a †pseudo-sphere and †non-Euclidean geometry. The study of †geodesics is a fertile topic deeply related to dynamics, calculus of variations, and topology, on which there is excellent work by J. Hadamard, H. †Poincaré, P. Funk, and G. D. Birkhoff, among others. The study of †minimal surfaces is intimately related to the †theory of functions of a complex variable, †calculus of variations, and topology; K. †Weierstrass, H. A. Schwartz, and J. Douglas are among those who have worked on this subject.

Euclidean geometry is a geometry belonging to F. Klein's Erlangen program (→ 147 Erlangen Program). For other geometries in the sense of F. Klein we may also consider corresponding differential geometries. For instance, in †projective differential geometry we study by means of differential calculus the properties of curves and surfaces that are invariant under projective transformations. This subject was studied by E. J. Wilczynski, G. Fubini, and others; †affine differential geometry and †conformal differential geometry were studied by W. Blaschke and others (→ 113 Differential Geometry in Specific Spaces).

Influenced by Gauss's geometry on a surface, in 1854 B. Riemann introduced †Riemannian geometry (*Göttinger Abh.*, 13 (1968), 1–20; *Werke* 2nd ed., 1892, p. 272–287) (→360 Riemannian Manifolds). Riemannian geometry includes Euclidean and non-Euclidean geometry as special cases, and is important for the great influence it exerted on geometric ideas of the 20th century. Under the influence of the algebraic theory of invariants, Riemannian geometry was then studied as a theory of invariants of quadratic †covariant tensors by E. B. Christoffel, C. G. Ricci, and others. Riemannian geometry attracted wide attention after A. Einstein applied it to the †general theory of relativity in 1916.

In the same year, T. Levi-Civita introduced the notion of †Levi-Civita's parallelism, which contributed greatly to the clarification of geometric properties of Riemannian spaces. Observing parallelism to be an affine-geometric concept, H. Weyl and A. S. Eddington developed a theory of Riemannian spaces "affinely" based on the notion of parallelism without using metrical methods. Such a geometry is called a geometry of an affine connection (→ 82 Connections).

Every straight line in a Euclidean space has the property that all tangents to the line are parallel. In a space with an affine connection, we may define a family of curves called †paths as an analog of straight lines. Such curves are solutions of a system of ordinary differential equations of the second order of a certain type. Coefficients of such differential equations determine a parallelism and hence an affine connection. H. Weyl discovered transformations of coefficients that leave the family of paths invariant as a whole, namely, projective transformations of an affine connection. A geometry that aims to study properties of paths or affine connections that are invariant under these transformations is called a **projective geometry of paths**. Such geometry was studied by L. P. Eisenhart, O. Veblen, and others. The concept of projective connections was an outcome of such studies. Similarly, the concept of conformal connections was developed from the consideration of †conformal transformations of Riemannian spaces.

These geometries cannot in general be regarded as geometries in the sense of Klein. Actually, any one of these geometries generally has no transformations that correspond to †congruent transformations of geometries in the sense of Klein; even if it has such transformations, they do not act transitively on the space. Thus, geometries are naturally divided into two categories, one consisting of geometries in the sense of Klein (based on the group concept) and the other of geometries based on Riemann's idea. Under such circumstances, E. Cartan unified the thoughts of Klein and Riemann from a higher standpoint and constructed his theory of connections.

At each point x of a †manifold we attach a tangent space P_x in the sense of Klein upon which a preassigned topological group G acts. If we can provide the set P of spaces P_x with a suitable structure compatible with the action of G, then the manifold is said to have a connection having G as its structural group (→ 82 Connections B). The space in the sense of Klein with structural group G is the simplest among these manifolds. If G is the group of congruent transformations in Euclidean space, a manifold with connection having G as its structural group is called a **manifold with Euclidean connection**. Among manifolds with Euclidean connection, Riemannian manifolds are characterized as those without †torsion. If we take the group of congruent transformations of projective (conformal) geometry as G, we have manifolds with projective (conformal) connection in the sense of Cartan. Among these, there are remarkable ones called manifolds with normal projective (or conformal) connections, which are essentially the same as the ones studied by Veblen and others. Cartan's idea had a profound influence on modern differential geometry.

A tangent line to a curve C at a point P of C is the limit line of the line PQ, where Q is a point on C approaching P; hence we can define it locally. A concept (or property) such as this which can be defined in an arbitrary small neighborhood of a point of a given figure or a space is called a **local concept** (or **local property**) or a concept (or property) **in the small**. In the early stages of the development of differential geometry, differential calculus was the main tool of study, so most of the results were local. On the other hand, a concept (or property) that is defined in connection with a whole figure or a whole space is called **global** or **in the large**. In modern differential geometry, the study of relations between local and global properties has attracted the interest of mathematicians. This view was emphasized by Blaschke, who worked on the differential geometry of †ovals and †ovaloids. The study of †rigidity of ovaloids by S. Cohn-Vossen belongs in this category, and many works on geodesics and minimal surfaces were done from this standpoint.

From the viewpoint of modern mathematics, the basic concepts on which we construct Riemannian geometry and geometries of connections are global concepts of †differentiable manifolds. However, in Riemann's time the theory of †Lie groups and topology were not yet developed; consequently, Riemannian geometry remained a local theory. In 1925, H. Hopf began to study the relations between local differential-geometric structures and topological structures of Riemannian spaces. Gradually the concept of differentiable manifolds was clarified, the global theory of Lie groups made progress, and topology developed. Contributions such as the proof of the †Gauss-Bonnet formula by S. S. Chern (1944) and the theory of †harmonic integrals by W. V. D. Hodge (1941) made it possible to study Riemannian geometry as a theory of differentiable manifolds with nondegenerate quadratic covariant tensors. Since C. Eheresman's work in 1950, the geometry of connections with structural group G has been studied as a theory of linear differential forms called

†connection forms on a †principal fiber bundle with structural group G over a differentiable manifold M. These geometries are making remarkable progress in connection with other branches, such as the theory of Lie groups, topology, classical differential geometry, the calculus of variations, the theory of differential forms, algebraic geometry, and the theory of functions of several complex variables.

In connection with algebraic geometry and the theory of functions of several complex variables, the theories of †complex manifolds, †complex structures, and †almost complex structures on differentiable manifolds are also studied actively. A connection can also be considered as a kind of structure on a differentiable manifold. Differentiable manifolds are presently objects of research in both differential geometry and topology; in contrast with topology, differential geometry can be considered as a theory of differentiable manifolds with a certain structure, such as a quadratic covariant tensor, a Finsler metric, a complex structure, or a connection. This viewpoint seems to be generally accepted at present.

In Japan, following T. Kubota, A. Kawaguchi, K. Yano, S. Sasaki, and others, many researchers are contributing to the progress of modern differential geometry.

References

[1] L. P. Eisenhart, A treatise on the differential geometry of curves and surfaces, Ginn, 1909.
[2] G. Darboux, Leçons sur la théorie générale des surfaces, Gauthier-Villars, I, second edition, 1914; II, second edition, 1915; III, 1894; IV, 1896.
[3] W. Blaschke, Vorlesungen über Differentialgeometrie, Springer, I, third edition, 1930; II, 1923; III, 1929 (Chelsea, 1967).
[4] L. P. Eisenhart, Riemannian geometry, Princeton Univ. Press, 1926.
[5] E. Cartan, Leçons sur la géométrie des espaces de Riemann, Gauthier-Villars, second edition, 1946.
[6] J. A. Schouten, Ricci-calculus, Springer, second edition, 1954.
[7] O. Veblen and J. H. C. Whitehead, The foundations of differential geometry, Cambridge Tracts, 1932.
[8] S. S. Chern, Topics in differential geometry, Lecture notes, Inst. for Adv. Study, Princeton, 1951.
[9] W. V. D. Hodge, The theory and applications of harmonic integrals, Cambridge Univ. Press, second edition, 1952.
[10] S. Kobayashi and K. Nomizu, Foundations of differential geometry, Interscience, I, 1963; II, 1969.
[11] E. Cartan, Oeuvres complètes I, II, III, Gauthier-Villars, 1952–1955.

113 (VII.10) Differential Geometry in Specific Spaces

A. The Method of Moving Frames

The main theme of this article is the theory of surfaces (i.e., submanifolds) in a differentiable manifold V on which a †Lie transformation group G acts.

If a †Lie group G of dimension r acts †transitively on a space G_* and the †group of stability of any point of G_* consists of the identity element only, then G_* is called the **group manifold** of G, and an element of G_* is called a **frame**. If f_0 is a fixed frame, then the mapping $a \rightarrow af_0 \in G_*$ $(a \in G)$ gives a †diffeomorphism of G to G_*. Let $I(G)$ be the set ω of all †differential forms of degree 1 on G_* that are invariant under transformations of G. Then $I(G)$ is a linear space of dimension r and is the †dual space of the Lie algebra $\mathfrak{g}$ of G. A basis $\{\omega_\lambda \mid 1 \leqslant \lambda \leqslant r\}$ of $I(G)$ is called a set of **relative components** of G. The **equations of structure** hold:

$$d\omega_\lambda = \frac{1}{2} \sum_{\mu,\nu=1}^{r} c_{\lambda\mu\nu}\omega_\mu \wedge \omega_\nu,$$

where $c_{\lambda\mu\nu}$ are †structure constants of the Lie algebra $\mathfrak{g}$.

Let G be a Lie transformation group of a space E. Then a G-invariant submanifold of E on which G acts transitively is called an **orbit**. Each point $y \in E$ determines an orbit containing y. When there exist parameters k_j $(1 \leqslant j \leqslant t)$ such that any G-invariant on E is a function of $k_1, \ldots, k_t$, then these parameters are called the **fundamental invariants** of E. Let H $(\subset G)$ be the group of stability at a point y_0 on an orbit M; then M is identified with the †homogeneous space G/H by the diffeomorphism $\varphi: M \rightarrow G/H$, $\varphi(ay_0) = aH$ $(a \in G)$. Furthermore, a †principal fiber bundle (G_*, M, H, τ) is determined by the projection $\tau: G_* \rightarrow M$, $\tau(af_0) = ay_0$ $(a \in G)$. The †fiber H_y on a point $y \in M$ is a group manifold of H. H_y is called the **family of frames** on y, and an element of H_y is a frame on y. Local coordinates θ_μ $(1 \leqslant \mu \leqslant s)$ of the group H are called the **secondary parameters** and are used to indicate frames in H_y. When H is not connected, let H^0 be the connected component of the identity of H and $\tilde{M}$ be the †covering manifold G/H^0 of M. An element $\tilde{y} \in \tilde{M}$

over $y \in M$ is called an **oriented element**. Now assume that the group H is connected. Then the family of frames H_y on each $y \in M$ is given as an †integral manifold of a †completely integrable system of total differential equations $\pi_i = 0$ $(1 \leqslant i \leqslant r-s, \pi_i \in I(G))$ on the group manifold G_*. Here the π_i are linearly independent and are called the **horizontal components** of M. The π_i are linear combinations of the relative components ω_λ of G, and their coefficients are generally functions of the fundamental invariants k_j. For simplicity, we assume that the relative components $\{\omega_\lambda\}$ are chosen such that the horizontal components π_i and components ω_α $(r-s < \alpha \leqslant r)$ are linearly independent. Then the ω_β $(1 \leqslant \beta \leqslant s)$ are called the **secondary components**. The differentials $d\theta_\mu$ of the secondary parameters are linear combinations of the ω_β. Furthermore, let $\{x_\sigma\}$ $(1 \leqslant \sigma \leqslant n)$ be local coordinates of E; then the differentials dx_σ are linear combinations of the differentials dk_j and the horizontal components π_i.

Let G be a Lie transformation group of a space V. We regard two m-dimensional surfaces W_1 and W_2 passing through $x \in V$ as equivalent if they have a †contact of order p at x. Then an equivalence class of submanifolds is called a **contact element** of order p at the point x. Let E_p be the set of all contact elements of order p at x, where x runs over all the points in V. A contact element of order p naturally determines a contact element of order $p-1$, and we denote this correspondence by $\psi: E_p \rightarrow E_{p-1}$. Thus we obtain the series of correspondences

$$V = E_0 \overset{\psi}{\leftarrow} E_1 \leftarrow \ldots \leftarrow E_{p-1} \overset{\psi}{\leftarrow} E_p \leftarrow \ldots,$$

where a contact element of order 0 is identified with a point of V. Since a transformation on the space V induces a transformation on E_p, G is also a Lie transformation group of E_p, and this transformation commutes with the mapping ψ. The fundamental invariants k_j of E_p are said to be of order p. We use similar terminology (such as frames of order p, etc.) throughout this article. The fundamental invariants k_j of order p $(1 \leqslant j \leqslant t_p)$ may be chosen such that they contain the fundamental invariants k_i $(1 \leqslant i \leqslant t_{p-1})$ of order $p-1$. The additional $t_p - t_{p-1}$ invariants k_α $(t_{p-1} < \alpha \leqslant t_p)$ are called the **invariants of order** p. The family H_y^p $(y \in E_p)$ of frames of order p may be chosen such that H_y^p is contained in the family H_z^{p-1} $(z = \psi y \in E_{p-1})$ of frames of order $p-1$. If necessary, the family H_y^p of frames of order p can be made connected by defining an **orientation** of contact elements of order p. Furthermore, the horizontal components π_j $(1 \leqslant j \leqslant r - s_p)$ of order p may be chosen such that they contain the horizontal components π_i $(1 \leqslant i \leqslant r - s_{p-1})$. The additional $s_{p-1} - s_p$ components π_α $(r - s_{p-1} < \alpha \leqslant r - s_p)$ are called the **principal components of order** p.

Let W be an m-dimensional surface of a space V. The contact element of order p $(\geqslant 0)$ is determined at every point of W and expressed by the family of frames of order p and the values of invariants of orders less than or equal to p. Let $\{u_i\}$ $(1 \leqslant i \leqslant m)$ be local coordinates on W. Then the differentials du_i are given as linear combinations of linearly independent differential forms π_i $(1 \leqslant i \leqslant m)$, where the π_i, called the **basic components** of W, are certain linear combinations of the differentials of the fundamental invariants of V and of the horizontal components of orbits of V. Let $F^p(W)$ be the set of all families of frames of order p; then $F^p(W)$ depends on m parameters u_i and s_p secondary parameters θ_μ of order p. On the space $F^p(W)$, the differentials of invariants of orders less than or equal to $p-1$ and the principal components of orders less than or equal to $p-1$ are linear combinations of the basic components, whose coefficients are functions of the invariants of orders less than or equal to p. The differentials of invariants of order p and the principal components of order p are linear combinations of the basic components:

$$dk_\alpha = h_{\alpha 1}\pi_1 + \ldots + h_{\alpha m}\pi_m, \qquad t_{p-1} < \alpha \leqslant t_p,$$

$$\pi_\alpha = b_{\alpha 1}\pi_1 + \ldots + b_{\alpha m}\pi_m, \quad r - s_{p-1} < \alpha \leqslant r - s_p,$$

where the coefficients $h_{\alpha i}$, $b_{\alpha i}$ are functions of the invariants of orders less than or equal to p and, in general, the secondary parameters θ_μ of order p. These coefficients are called the **coefficients of order** p. Let Γ_p be a subgroup of G preserving a family of frames of order p and D_p be a space whose coordinates are coefficients $(h_{\alpha i}, b_{\alpha i})$ of order p. Then Γ_p acts on D_p as a transformation group. Knowledge of the properties of contact elements of order less than or equal to p can be utilized to obtain information about the invariants of order $p+1$, etc. In fact, if we can choose in the Γ_p-space D_p a subspace C_p that intersects each orbit in D_p at one and only one point, then generally, the secondary parameters of order p associated with the points in C_p correspond to the frames of order p, and the parameters associated with the points in C_p are the invariants of order $p+1$. The restrictions of the coefficients of order p to C_p are functions of the invariants of orders less than or equal to $p+1$; they are independent of the secondary parameters of order p. Thus the frames of order $p+1$ and the invariants of orders less than or equal to $p+1$ determine the contact elements of order p of W and their differentials; generally, the latter can be utilized to determine the contact elements of order $p+1$.

This process of obtaining information of "order $p+1$" utilizing a suitable subspace C_p of D_p is the so-called general **method of moving frames**. However, the surface W may contain points for which the general method does not apply. Actually, there are surfaces W for which the method does not apply for any point in W. Thus various methods of moving frames are necessary to cope with different kinds of surfaces. In the actual application of the method of moving frames, we use certain devices that help to simplify the calculations. In fact, an infinitesimal transformation $(\delta h_{\alpha i}, \delta b_{\alpha i})$ of the group Γ_p acting on the space D_p is expressed as a linear combination of the secondary components of order p; this expression is easily obtained by means of the structural equations of G. The group Γ_{p+1} is a subgroup of Γ_p fixing every point of the subspace C_p, and its infinitesimal transformation is such that $\delta h_{\alpha i}=0$, $\delta b_{\alpha i}=0$. The secondary components of order $p+1$ are immediately obtained from the equations for dk_α and π_α. Furthermore, when $m \geqslant 2$, the condition for the principal components of every order to satisfy the structural equations of G is essential to the problem of the existence of (m-dimensional) surfaces.

As we apply the method of moving frames consecutively to a surface W, we eventually arrive at the order q having the following properties: The families of frames of order $q+1$ coincide with those of order q, and the invariants k_β of order $q+1$ are expressed as functions $\varphi_\beta(k_\alpha)$ of invariants k_α of order less than or equal to q. In this case, the families of frames of orders $q+j$ $(j \geqslant 1)$ are all equal, and the invariants of orders $q+j$ are partial $(j-1)$-derivatives of functions $\varphi_\beta(k_\alpha)$. The family of frames of order q is called the **Frenet frame**. The **differential invariants** on a surface are defined to be differential forms generated by the basic components and the invariants of each order.

Specifically, assume that the group G is an analytic transformation group of V and m-dimensional surfaces W_1, W_2 are analytic. Then there exists an element g of G such that $gW_1 = W_2$ if and only if W_1 and W_2 are of the same kind and have the same relations among the invariants of orders less than or equal to $q+1$. These relations are called the **natural equations** of the surface. The theory of surfaces based on the analysis of the natural equations of surfaces is called **natural geometry**. The **reduction formula** may be obtained by utilizing the Frenet frame; it gives the equation of the surface in the form of power series containing the invariants of each order.

Various results are known concerning the theory of surfaces of the spaces V_1, V_2 sharing the same transformation group G. We also have a theory of **special surfaces** whose invariants satisfy specific functional relations. Furthermore, we have problems concerning the **deformation of a surface** (preserving some differential invariants). Actually, the theory of surfaces of dimension m other than curves and hypersurfaces is generally quite difficult. The method of tensor calculus can be applied to the study of surfaces. The theory of †connections may be considered as an outgrowth of the study of surfaces by means of the method of moving frames and tensor calculus.

B. Projective Differential Geometry

The rudiments of differential geometry subordinated to the †projective transformation group, or **projective differential geometry**, can be found in *Theory of surfaces* by J. G. Darboux. Systematic study of the subject was done by H. G. H. Halphen, Wilczynski, and Fubini. The Fubini theory was enriched substantially by Cartan, E. Čech, E. Bompiani, and J. Kanitani.

In this section we consider a surface S in a 3-dimensional projective space. Let $A(u^1,u^2)$ (u^1,u^2 are parameters on S) be a point of S, and associate with A all the frames $[A,A_1,A_2,A_3]$ $(|A,A_1,A_2,A_3|=1)$, where A_1, A_2, A_3 are points of the tangent plane to S at A. A family of such frames is called the **family of frames of order 1**, and we express its differential by

$$dA_\alpha = \sum_{\beta=0}^{3} \omega_\alpha^\beta A_\beta, \qquad \alpha=0,1,2,3, \qquad A_0=A.$$

The ω_α^β are †Pfaffian forms that depend on two principal parameters determining the origin A and ten secondary parameters determining the frame. We have $\omega_0^0+\omega_1^1+\omega_2^2+\omega_3^3=0$, $\omega_0^3=0$. Furthermore, $\omega^1=\omega_0^1$, $\omega^2=\omega_0^2$ are independent of each other and depend on the principal parameters only. Let z^1, z^2, z^3 be †nonhomogeneous coordinates with respect to a frame of order 1. Then in a neighborhood of the origin, S is expressed by $z^3=\sum_{r=2}^{\infty} f_r$, where the f_r are homogeneous functions of degree r with respect to z^1, z^2.

If we write $f_2=(a_0(z^1)^2+2a_1z^1z^2+a_2(z^2)^2)/2$, then it follows from the equations of structure of the projective transformation group that $\omega_1^3=a_0\omega^1+a_1\omega^2$, $\omega_2^3=a_1\omega^1+a_2\omega^2$. If we put $\varphi_2=a_0(\omega^1)^2+2a_1\omega^1\omega^2+a_2(\omega^2)^2$, then a curve on S defined by $\varphi_2=0$ is called the **asymptotic curve** and its tangent the **asymptotic tangent**. At any point of this curve, the plane tangent to S is in contact of order 2 with this curve, and there are in general two asymptotic curves through any point of S. Equations of the asymptotic tangent at A are given by $z^3=0, f_2=0$. A point of S at which the asymptotic tangents coincide is called a **parabolic point**. If every point of S is parabolic, then S is a †developable surface, and the general theory is not applicable to such a surface.

Among the family of frames of order 1, a frame satisfying $a_0=a_2=0, a_1=1$ is called the **frame of order 2**. For this frame, the straight lines $\overline{AA_1}, \overline{AA_2}$ are asymptotic tangents. With respect to this frame, if $f_3= -(b_0(z^1)^3+3b_1(z^1)^2z^2+3b_2z^1(z^2)^2+b_3(z^2)^3)/3$, then $\omega_1^2=b_0\omega^1+b_1\omega^2$, $-\omega_0^0+\omega_1^1+\omega_2^2-\omega_3^3= 2(b_1\omega^1+b_2\omega^2)$, $\omega_2^1=b_2\omega^1+b_3\omega^2$, and the quadric surface $z^3=z^1z^2-z^3(b_1z^1+b_2z^2+pz^3)$ (with p arbitrary) is called **Darboux's quadric** at A, an especially interesting one among contact quadrics of S. **Darboux's curve** is a curve on S such that Darboux's quadric is in contact of order 3 at any point of it. Its tangent is called **Darboux's tangent** and is given by $z^3=0$, $b_0(z^1)^3+b_3(z^2)^3=0$.

We have $b_0b_3\neq 0$, except in the case of †ruled surfaces. We take special frames of order 2 determined by $b_1=b_2=0, b_0=b_3=1$ and call them the **frames of order 3**. If a frame of order 3 satisfies $f_4=-(c_0(z^1)^4+4c_1(z^1)^3z^2+6(c_2-1)(z^1z^2)^2+4c_3z^1(z^2)^3+c_4(z^2)^4)/12$, then $\omega_0^0-2\omega_1^1+\omega_2^2=c_0\omega^1+c_1\omega^2$, $\omega_3^2-\omega_1^0= c_1\omega^1+c_2\omega^2$, $\omega_3^1-\omega_2^0=c_2\omega^1+c_3\omega^2$, $\omega_0^0+\omega_1^1-2\omega_2^2=c_3\omega^1+c_4\omega^2$. With respect to this family of frames of order 3, $((\omega^1)^3+(\omega^2)^3)/2\omega^1\omega^2$ is an invariant associated with two neighboring points of S, called the **projective line element**. Also, with respect to this frame, two straight lines $\overline{AA_3}, \overline{A_1A_2}$ are polar with respect to Darboux's quadric.

Among the families of frames of order 3, a frame satisfying $c_1=c_2=c_3=0$ is called the **frame of order 4**. With respect to this frame, there exist λ, μ, ν, ρ such that $\omega_1^0=\lambda\omega^1+\mu\omega^2$, $\omega_2^0=\nu\omega^1+\rho\omega^2$, $\omega_3^0=\rho\omega^1+\lambda\omega^2$. Hence if we put $c_0=-3a$, $c_4=-3b$, it follows that

$$(\omega_\alpha^\beta)=\begin{bmatrix}\tau_0^0 & \omega^1 & \omega^2 & 0\\ \omega_1^0 & \tau_1^1 & \omega^1 & \omega^2\\ \omega_2^0 & \omega^2 & -\tau_1^1 & \omega^1\\ \omega_3^0 & \omega_2^0 & \omega_1^0 & -\tau_0^0\end{bmatrix},$$

$\tau_0^0=-(3/2)(a\omega^1+b\omega^2)$, $\tau_1^1=(1/2)(a\omega^1+b\omega^2)$. Thus the frame of order 4 is the Frenet frame and is attached to every point of S. This frame is called the **normal frame**, and the invariants $a, b, \lambda, \mu, \nu, \rho$ are called the **fundamental differential invariants**. The straight lines $\overline{AA_3}$ and $\overline{A_1A_2}$ associated with the normal frame are called **directrices of Wilczynski** of the first and second kind, respectively. With respect to the normal frame, S is expressed by

$$z^3=z^1z^2-((z^1)^3+(z^2)^3)/3+(a(z^1)^4+b(z^2)^4)/4+(z^1z^2)^2/2+\ldots.$$

A necessary and sufficient condition for two surfaces $S, \bar{S}$ to be projectively equivalent is that there are normal frames having the same ω^1, ω^2 and the same six fundamental differential invariants. For six quantities $a, b, \lambda, \mu, \nu, \rho$ to be fundamental differential invariants of a surface, they must satisfy a certain condition of existence [6].

A frame of order 1 such that $\overline{AA_3}$ and $\overline{A_1A_2}$ are polar with respect to Darboux's quadric is called **Darboux's frame**. With respect to this frame also, a theory of surfaces has been established.

Consider a pointwise correspondence between two surfaces $S, \bar{S}$, and denote by $\bar{A}\in\bar{S}$ the point corresponding to $A\in S$. If there exists a projective transformation φ that transforms A into $\bar{A}$ and the image $\varphi(S)$ is in contact of order 2 with $\bar{S}$ at $\bar{A}$, then the pointwise correspondence is called a **projective deformation**. A necessary and sufficient condition for the existence of a projective deformation between two surfaces is that these surfaces have the same projective line element [6]. A ruled surface is projectively deformable only to a ruled surface. Given an arbitrary surface S, it is generally impossible to find a surface that is different from S and projectively deformable to S; some conditions must be satisfied [6, 8].

Let $p^{01}, p^{02}, p^{03}, p^{12}, p^{13}, p^{23}$ be †Plücker coordinates of a straight line in a 3-dimensional projective space P^3. Then we have $p^{01}p^{23}-p^{02}p^{13}+p^{03}p^{12}=0$, and there is a one-to-one correspondence between the ratios of $\{p^{ij}\}$ and straight lines (→ 92 Coordinates B). If the p^{ij} are regarded as homogeneous coordinates of a 5-dimensional projective space P^5, then the previous equation defines a hyperquadric Q in P^5. Thus there is a one-to-one correspondence between points of Q and straight lines in P^3. A curve on Q corresponds to a set of one-parameter families of straight lines, or a ruled surface. Sets of 2-parameter or 3-parameter families of straight lines corresponding to surfaces of 2 or 3 dimensions on Q in P^5 are called **congruences of lines** or **complexes of lines**, respectively. Thus by using a theory of surfaces in P^5, it is possible to establish the theory of congruences and complexes [2, 6, 8], which is an important part of projective differential geometry.

Specifically, if the surface is either a curve or a hypersurface, there are numerous interesting results [2,4,6].

C. Affine Differential Geometry

The theme of general affine differential geometry is the study of differential-geometric properties of a point or set of points in a space that are invariant under the action of the †affine transformation group. **Affine differential geometry** is the study of the properties invariant under the action of the †equivalent affine transformation group, i.e., a subgroup of the affine transformation group formed by elements sending (x_i) to $(\bar{x}_i)$ such that

$$\bar{x}_i = a_i + \sum_{j=1}^{n} a_{ij}x_j, \qquad i=1,\dots,n, \qquad \det(a_{ij})=1.$$

The latter transformation leaves invariant the volume surrounded by an oriented closed hypersurface. The method of moving frames is effective in affine differential geometry.

Let C be a plane curve, and associate with any point $A=A(t)$ of C a family of frames $[A,\mathbf{e}_1,\mathbf{e}_2]$, where the area of the parallelogram determined by the two vectors $\mathbf{e}_1$, $\mathbf{e}_2$ is equal to 1. This frame is called the **frame of order** 0. Its differential is expressed by

$$dA = \sum_{s=1}^{2} \omega^s \mathbf{e}_s, \qquad d\mathbf{e}_r = \sum_{s=1}^{2} \omega_r^s \mathbf{e}_s, \qquad r=1,2,$$

$$\omega_1^1 + \omega_2^2 = 0.$$

The **frames of order 1, 2, and 3** are characterized by $\omega^2=0$; $\omega^2=0$, $\omega_1^2=\omega^1$; and $\omega^2=0$, $\omega_1^2=\omega^1$, $\omega_1^1=0$, respectively. Then a frame of order 3 can be associated with each point of C and coincides with the Frenet frame. We call $\omega^1=d\sigma$ the **affine arc element**; the **affine curvature** κ is defined by $\omega_2^1=-\kappa\, d\sigma$. Then the Frenet formula is given by

$$dA = d\sigma\mathbf{e}_1, \qquad d\mathbf{e}_1 = d\sigma\mathbf{e}_2, \qquad d\mathbf{e}_2 = -\kappa\, d\sigma\mathbf{e}_1.$$

With respect to this frame, C is expressed as

$$y = x^2/2 + \kappa x^4/8 + (d\kappa/d\sigma)x^5/40 + \dots.$$

Further, $d\sigma$ and κ are given analytically by

$$d\sigma = |dA, d^2A|^{1/3}, \qquad \kappa = |d^2A/d\sigma^2, d^3A/d\sigma^3|,$$

where $|M,N| = \det(M,N)$. M, N are column vectors with two entries. We call σ the **affine arc length**. The straight line on which $\mathbf{e}_2$ is situated is called the **affine normal**, the diameter of the parabola osculating C at A. If κ is constant, then C is a conic section. Furthermore, C is an ellipse, hyperbola, or parabola according as the constant κ is positive, negative, or zero. In affine geometry, parabolas play a role similar to that played by straight lines in Euclidean geometry.

There are numerous results concerning the theory of skew curves and surfaces [1]. Concerning the theory of skew curves, results on **affine length, affine curvature, affine torsion, affine principal normals,** and **affine binormals** are similar to those in Euclidean geometry. The affine transformation group is situated between the projective transformation group and the congruent transformation group and hence has properties analogous to theirs. The theory of surfaces has a character similar to that of projective differential geometry [1]. Next, we consider the variation of the affine area of a surface surrounded by a closed skew curve C. We call the extremal surface the **affine minimal surface**. Blaschke and others obtained many results on the global properties of surfaces.

D. Conformal Differential Geometry

Let S^n be a †conformal space of dimension n, and associate with each point $A_0 \in S^n$ a frame $\mathfrak{R}\ [A_0, A_1, \dots, A_n, A_\infty]$ of the †$(n+2)$-hyperspherical coordinates with origin A_0 (→ 78 Conformal Geometry). Then denoting by $A \cdot B$ the †inner product of hyperspheres A, B, we obtain

$$A_\alpha \cdot A_\beta = g_{\alpha\beta}, \qquad \alpha, \beta = 0, 1, \dots, n, \infty,$$

where

$$(g_{\alpha\beta}) = \begin{bmatrix} 0 & 0 & -1 \\ 0 & g_{ij} & 0 \\ -1 & 0 & 0 \end{bmatrix},$$

$$i,j=1,\dots,n, \qquad g_{ij}=g_{ji}.$$

Let z^α be homogeneous coordinates with respect to $\mathfrak{R}$. Then the †Möbius transformation $z \to \bar{z}$ of S^n is characterized by $\bar{z}^\alpha = c_\beta^\alpha z^\beta$, where $g_{\alpha\beta}c_\sigma^\alpha c_\tau^\beta = g_{\sigma\tau}$, $|c_\beta^\alpha| \neq 0$. The differential of the family of the frames is defined by

$$dA_\alpha = \sum_{\beta=0,1}^{n,\infty} \omega_\alpha^\beta A_\beta, \tag{1}$$

where

$$(\omega_\alpha^\beta) = \begin{bmatrix} \omega_0^0 & \omega_0^j & 0 \\ \sum g_{ik}\omega_\infty^k & \omega_i^j & \sum g_{ik}\omega_0^k \\ 0 & \omega_\infty^j & -\omega_0^0 \end{bmatrix},$$

$$\sum \left(g_{ik}\omega_j^k + g_{jk}\omega_i^k \right) = dg_{ij}.$$

There are $(n+1)(n+2)/2$ linearly independent forms among ω, and this is the number of parameters of the Möbius transformation group. The equations of †structure of this group are

$$d\omega_\alpha^\beta = \sum \omega_\alpha^\sigma \wedge \omega_\sigma^\beta. \tag{2}$$

The theme of **conformal differential geometry** is the properties of Pfaffian forms ω_α^β satisfying (1) and (2).

Consider a transformation $c:\Sigma z^\alpha A_\alpha \rightarrow \Sigma z^\alpha(A_\alpha+dA_\alpha)$. (1) If all ω vanish except ω_0^0, then all the circles through A_0, A_∞ are invariant, and any point P is transformed to a neighboring point $\bar{P}$ on the circle, such that the cross ratio $(P,\bar{P};A_0,A_\infty)$ is constant. This transformation is called the **homothety** with centers A_0, A_∞. (2) If all ω vanish except ω_0^j, $\omega_i^\infty=\Sigma g_{ik}\omega_0^k$, then all the circles tangent to a fixed direction at A_∞ are transformed into themselves, and any hypersphere through A_∞ and orthogonal to those circles is transformed into a hypersphere having the same property. This transformation is called the **elation** with center A_∞. (3) If all ω vanish except $\omega_i^0=\Sigma g_{ik}\omega_\infty^k$, ω_∞^j, then the transformation is an elation with center A_0. (4) If all ω vanish except ω_i^j, then the transformation is an infinitesimal rotation with center A_0, with A_∞ regarded as a point at infinity. Thus any infinitesimal Möbius transformation is decomposed into the previous four types of transformation.

To study the theory of curves and hypersurfaces in S^n, we again utilize the Frenet frame chosen from a family of frames associated with A_0. For example, the Frenet formula of a curve in S^3 is given by

$$(\omega_\alpha^\beta)=\begin{pmatrix} 0 & d\sigma & 0 & 0 & 0\\ \kappa d\sigma & 0 & 0 & 0 & d\sigma\\ -d\sigma & 0 & 0 & \tau d\sigma & 0\\ 0 & 0 & -\tau d\sigma & 0 & 0\\ 0 & \kappa d\sigma & -d\sigma & 0 & 0\end{pmatrix}.$$

We call $d\sigma$, κ, and τ the **conformal arc element**, **conformal curvature**, and **conformal torsion**, respectively. There are many results on conformal deformation [5].

Concerning Laguerre differential geometry, we have results dual to those in conformal differential geometry (a point is replaced by a straight line and an angle by a distance between the points of contact of the common tangents of two oriented circles).

E. Contact Manifold

Consider a $(2n+1)$-dimensional differentiable manifold M^{2n+1} with a 1-form η such that $\eta\wedge(d\eta)^n\neq 0$, where $d\eta$ is the exterior derivative of η and $\wedge$ denotes exterior multiplication. (Note that this is true for the 1-form in the left-hand side of eq. (2) in 84 Contact Transformations A.) Such a manifold is called a **contact manifold** with **contact form** η. The structure group of the tangent bundle of a contact manifold M^{2n+1} reduces to $U(n)\times 1$, where $U(n)$ is the unitary group, hence every contact manifold is orientable. Simple but typical examples are given by the unit sphere S^{2n+1} in Euclidean space E^{2n+2} and the tangent sphere bundle of an $(n+1)$- dimensional Riemannian manifold M^{n+1}, both with natural contact forms (S. S. Chern [10]). Every 3-dimensional compact orientable differentiable manifold is a contact manifold (J. Martinet [15]).

Now, a differentiable manifold M^{2n+1} is said to be an **almost contact manifold** if it admits a tensor field φ of type (1, 1), a vector field ξ, and a 1-form η such that

$$\varphi^2 X=-X+\eta(X)\xi, \qquad \eta(\xi)=1, \tag{1}$$

where X is an arbitrary vector field on M^{2n+1}, and the triple (φ,ξ,η) is then called an **almost contact structure**. (1) implies that $\varphi\xi=0$ and $\eta(\varphi X)=0$ (S. Sasaki [15], I). The structure group of the tangent bundle of an almost contact manifold M^{2n+1} reduces to $U(n)\times 1$. Indeed, J. W. Gray [11] took this property as his definition of almost contact structure. For any pair of vector fields X and Y on M^{2n+1}, let

$$\begin{aligned}N(X,Y)=&[X,Y]+\varphi[\varphi X,Y]\\&+\varphi[X,\varphi Y]-[\varphi X,\varphi Y]\\&-\{X\cdot\eta(Y)-Y\cdot\eta(X)\}\xi,\end{aligned}$$

where [,] is the Poisson bracket; then N is a tensor field of type (1,2) over M^{2n+1}, which we call the **torsion tensor** of the almost contact structure (φ,ξ,η). When N vanishes identically on M^{2n+1}, we say that the almost contact structure is **normal**.

An almost contact structure (φ,ξ,η) on M^{2n+1} induces naturally an almost complex structure J on $M^{2n+1}\times R$ (resp. $M^{2n+1}\times S^1$), which reduces to a complex structure if and only if (φ,ξ,η) is normal. A similar statement is also valid for the product space of two almost contact manifolds (A. Morimoto [14]).

If M^{2n+1} is an almost contact manifold with structure tensor (φ,ξ,η), we can find a positive definite Riemann metric g so that $g(\varphi X,\varphi Y)=g(X,Y)-\eta(X)\eta(Y)$ for any pair of vector fields X and Y, and the set of vector fields (φ,ξ,η,g) is then said to be an **almost contact metric structure**.

When M^{2n+1} is a contact manifold with contact form η, there exists a unique vector field ξ which satisfies $d\eta(X,\xi)=0$, $\eta(\xi)=1$ for any vector field X. We can then find a tensor field φ of type (1, 1) and a positive definite metric tensor g so that (i) $d\eta(X,Y)=g(X,\varphi Y)$ is satisfied for any pair of vector fields X and Y and (ii) (φ,ξ,η,g) is an almost contact metric structure. The almost contact metric structure determined in this way by a contact form η is called a **contact metric structure**. A differentiable manifold with normal contact metric structure is called a **normal contact Riemannian manifold**. Brieskorn manifolds are examples of such manifolds. They include, besides the standard sphere S^{2n+1}, all exotic $(2n+1)$-spheres which bound compact oriented parallelizable manifolds. An almost contact manifold is said to be **regular** or nonregular according as the foliation determined by ξ is regular or not. A compact regular contact manifold is a principal circle bundle over a symplectic manifold and it admits a normal contact metric structure if and only if the base manifold is a Hodge manifold (Boothby and Wang [9], Hatakeyama [12]).

Many research papers on the topology and differential geometry of manifolds with the structures defined above have been published by S. Tanno, S. Tachibana, S. I. Goldberg, and others.

References

[1] W. Blaschke, Vorlesungen über Differentialgeometrie, Springer, II, 1923; III, 1929 (Chelsea, 1967).
[2] G. Bol, Projektive Differentialgeometrie I, II, Vandenhoeck & Ruprecht, 1950.
[3] E. Cartan, La théorie des groupes finis et continus, Gauthier-Villars, 1951.
[4] E. Cartan, Leçons sur la théorie des espaces à connexion projective, Gauthier-Villars, 1937.
[5] P. C. Delens, Méthodes et problèmes des géométries différentielles euclidienne et conforme, Gauthier-Villars, 1927.
[6] G. Fubini and E. Čech, Introduction à la géométrie projective différentielle des surfaces, Gauthier-Villars, 1931.
[7] J. Kanitani, Géométrie différentielle projective des hypersurfaces, Mem. Ryojun Coll. of Eng., 1931.
[8] J. Kanitani, Syaei bibun kikagaku (Japanese; Projective differential geometry), Iwanami Coll. of Math., 1933.
[9] W. M. Boothby and H. C. Wang, On contact manifolds. Ann. of Math., (2) 68 (1958), 721–734.
[10] S. S. Chern, Pseudo-groupes continus infinis. Coll. intern. du CNRS, Geom. Diff., Strasbourg (1953), 119–135.
[11] J. W. Gray, Some global properties of contact structure, Ann. of Math., (2) 69 (1959), 421–450.
[12] Y. Hatakeyama, Some notes on differentiable manifolds with almost contact structure, Tôhoku Math. J., (2) 15 (1963), 176–181.
[13] J. Martinet, Formes de contact sur les variétés de dimension 3. Proc. Liverpool Singularities Symposium II (1971), 142–163.
[14] A. Morimoto, On normal almost contact structures, J. Math. Soc. Japan, 15 (1963), 420–436.
[15] S. Sasaki, On differentiable manifolds with certain structures which are closely related to almost contact structure I. Tôhoku Math. J., (2) 12 (1960), 459–476. II (with Y. Hatakeyama), ibid., (2) 13 (1961), 281–294.

114 (VII.5) Differential Geometry of Curves and Surfaces

A. General Remarks

Let f be an †immersion of an m-dimensional †differentiable manifold M of class C^r into an n-dimensional Euclidean space E^n. More precisely, f is a differentiable mapping of class C^r such that the †differential df_p is injective at every point p of M. The image $f(M)$ is not necessarily a submanifold of E^n. The pair (M,f) is called an **immersed submanifold** (or a **surface**) of E^n. When $m=1$, we call it a **curve** of E^n, and when $m=n-1$, a **hypersurface** in E^n. The cases of $n=2$ and $n=3$ have been the main objects of study in differential geometry of curves and surfaces (→ 108 Differentiable Manifolds).

B. Frames in E^n

Every †Euclidean motion in E^n can be expressed as the product of a parallel translation and an †orthogonal transformation that keeps the origin of E^n fixed. The set of all parallel translations is a commutative group that can be identified with $\mathbf{R}^n$. It is a normal subgroup of the group of motions $I(E^n)$ of E^n. So we see that $I(E^n)$ is a †semidirect product of $\mathbf{R}^n$ and the †orthogonal group $O(n)$. The Lie algebra of $I(E^n)$ is the direct sum of $\mathbf{R}^n$ and the Lie algebra $\mathfrak{o}(n)$ of the orthogonal group, where both are regarded as additive groups. Corresponding to this decomposition, we can write the †Maurer-Cartan differential form over $I(E^n)$ as $\omega+\Omega$, where ω belongs to $\mathbf{R}^n$ and Ω to $\mathfrak{o}(n)$. The †structural equation $d(\omega+\Omega)=-(1/2)(\omega+\Omega)\wedge(\omega+\Omega)$ can be divided into the following two parts: $d\omega=\Omega\wedge\omega$; $d\Omega=-(1/2)\Omega\wedge\Omega$. These are known as the **equations of structure** of E^n. By an **orthogonal frame** in E^n we mean an ordered set $(x, \mathbf{e}_1, \ldots, \mathbf{e}_n)$ consisting of a point x and a set of †orthonormal vectors $\mathbf{e}_1, \mathbf{e}_2, \ldots, \mathbf{e}_n$. We denote by $\mathfrak{O}(n)$ the set of all orthogonal frames in E^n. If we denote the translation identified with $x \in \mathbf{R}^n$ by T_x, then there is a one-to-one correspondence $\varphi: I(E^n) \to \mathfrak{O}(n)$ given by $\varphi(T_x A) = (x, A\mathbf{e}_1, \ldots, A\mathbf{e}_n)$ $(A \in O(n))$. We can make $\mathfrak{O}(n)$ into a differentiable manifold so that φ is a †diffeomorphism. We denote the differential forms over $\mathfrak{O}(n)$, which are images of ω and Ω under the †dual mapping of φ^{-1}, by the same letters ω and Ω, respectively. For $\mathfrak{O}(n)$ as a †principal fiber bundle over $\mathbf{R}^n$ with the projection $\pi: \pi(x, \mathbf{e}_1, \ldots, \mathbf{e}_n) = x$ and n vector-valued functions φ_i: $\varphi_i(x, \mathbf{e}_1, \ldots, \mathbf{e}_n) = \mathbf{e}_i$ over $\mathfrak{O}(n)$, we have

$$\omega = \sum_i \omega^i \mathbf{e}_i, \qquad \Omega = \sum_{i<j} \Omega^{ij} E_{ij},$$
$$\omega^i = (d\pi, \varphi_i), \qquad \Omega^{ij} = (d\varphi_i, \varphi_j), \tag{1}$$
$$d\omega^i = \sum_j \Omega^{ij} \wedge \omega^j, \quad d\Omega^{ij} = \sum_k \Omega^{ik} \wedge \Omega^{kj},$$

where $\{E_{ij}\}$ is a basis of $\mathfrak{o}(n)$ defined by $E_{ij}\mathbf{e}_j = \mathbf{e}_i$, $E_{ij}\mathbf{e}_i = -\mathbf{e}_j$, $E_{ij}\mathbf{e}_k = 0$ $(k \neq i, j)$ and $(\ , \ )$ is the scalar product of vector-valued forms induced from the scalar product of E^n. Any diffeomorphism of $\mathfrak{O}(n)$ onto itself preserving ω and Ω must be a Euclidean motion.

C. Theory of Curves

Let (M, f) be an immersion of a 1-dimensional differentiable manifold M into E^n. We identify the tangent space of E^n at each point with E^n itself. Then df_x maps the origin of the tangent space M_x to $f(x)$, and the image $df_x(M_x)$ of M_x by df_x is a straight line passing through $f(x)$ in E^n, called the **tangent line** of $f(M)$ at $f(x)$. By $\mathfrak{O}_f(M)$ we mean the set of all ordered sets $(x, \mathbf{e}_1, \ldots, \mathbf{e}_n)$, where $x \in M$ and $\{\mathbf{e}_i\}$ is an orthonormal basis of E^n such that $\mathbf{e}_1 \in df_x(M_x)$. Then $\mathfrak{O}_f(M)$ can be naturally immersed in $\mathfrak{O}(n)$ by the mapping $\hat{f}$: $\hat{f}(x, \mathbf{e}_1, \ldots, \mathbf{e}_n) = (f(x), \mathbf{e}_1, \ldots, \mathbf{e}_n)$. We can pull back the differential forms ω, Ω, ω^i, Ω^{ij} over $\mathfrak{O}(n)$ to $\mathfrak{O}_f(M)$ by $\hat{f}^*$ and denote them by θ, Θ, θ^i, and Θ^{ij}, respectively; then we have $\theta^i = 0$ $(i > 1)$. Let f_1 and f_2 be two immersions of M into E^n. Then in order that there exist a Euclidean motion α of E^n such that $f_1 = \alpha \circ f_2$, it is necessary and sufficient that there exist a diffeomorphism φ of $\mathfrak{O}_{f_1}(M)$ onto $\mathfrak{O}_{f_2}(M)$ such that $\theta_{f_1} = \varphi^*(\theta_{f_2})$, $\Theta_{f_1} = \varphi^*(\Theta_{f_2})$. Let π_f be the projection of the fiber bundle $\mathfrak{O}_f(M)$, and let φ_i be naturally defined vector-valued functions over $\mathfrak{O}_f(M)$. Then we have $d(f \circ \pi_f) = \theta^1 \varphi_1$, $d\varphi_i = \sum_{j=1}^n \Theta^{ij} \varphi_j$. If we put $ds^2(X) = \|df_x(X)\|^2$ $(X \in M_x)$, then we have $(\theta^1)^2 = \pi_f^*(ds^2)$. For each point $x \in M$ there are two possibilities for the choice of $\mathbf{e}_1$ corresponding to two orientations of the curve. But since $(d\varphi_1, d\varphi_1) = \pi_f^*(\rho^2 ds^2)$, ρ^2 depends only on the point x of M. We call ρ $(\geqslant 0)$ the **absolute curvature**. We now choose an orientation of the curve and then $\mathbf{e}_1$ in accordance with the orientation. Thus we get a submanifold of $\mathfrak{O}_f(M)$, which we again express by the notation $\mathfrak{O}_f(M)$. If we define the form ds by $ds(X) = (df(X), \mathbf{e}_1)$, we have $\theta^1 = \pi_f^*(ds)$, and ds is called the **line element**. Any †local cross section $R: \pi_f \circ R = 1$ of the bundle $\mathfrak{O}_f(M)$ is called a **moving frame**. Putting

$$R^*(\theta^1) = ds, \qquad R^*(\Theta^{ij}) = \rho^{ij} ds,$$

we see that the following equation holds over M:

$$d\mathbf{e}_i = \sum_j \rho^{ij} ds\, \mathbf{e}_j. \tag{2}$$

For two immersions (M, f_1) and (M, f_2), we have $f_1 = \alpha \circ f_2$ (α is a Euclidean motion) if and only if they have the same ds and ρ^{ij} for some moving frames.

D. Frenet's Formulas

In order to study local properties of curves it is sufficient to consider them on †Jordan arcs of class C^r. With respect to orthogonal coordinates $(x^1, \ldots, x^n)$ in E^n, such a curve is represented parametrically by $x^i = f^i(t)$ $(t \in [a, b], \sum (dx^i/dt)^2 > 0)$ or by a vector representation $\mathbf{x} = \mathbf{x}(t)$. If φ is a diffeomorphism of a closed interval $[a', b']$ onto $[a, b]$, then $f \circ \varphi$ and f are representations of the same arc in E^n, and φ is called a **transformation of the parameter**. Any curve of class C^1 is †rectifiable, and its **arc length** is given by $s = \int_a^b (\sum_{i=1}^n (dx^i/dt)^2)^{1/2} dt$. We may choose the arc length s measured from a point on the arc as a parameter, called the **canonical parameter** of the arc. Consider an arc C of class C^n given by the vector representation $\mathbf{x} = \mathbf{x}(s)$, $s \in [a, b]$. We assume that its †Wronskian $|\mathbf{x}'(s), \ldots, \mathbf{x}^{(n)}(s)|$ is not identically zero (we denote by $'$ the derivative with respect to the canonical parameter), which means that the arc C is not contained in a hyperplane in E^n. A point at which the Wronskian vanishes is called a **stationary point**, and we assume that there exists no stationary point on C. By the †Gram-Schmidt orthonormalizing process we obtain an orthonormal basis $\mathbf{e}_1, \ldots, \mathbf{e}_n$ $(|\mathbf{e}_1, \ldots, \mathbf{e}_n| > 0)$ from n vectors $\mathbf{x}'(s), \ldots, \mathbf{x}^{(n)}(s)$ at each point of C. We call the frame thus determined the **Frenet frame**. With respect to the Frenet frame, (2) is rewritten as follows:

$$\begin{aligned} &\mathbf{e}_i'(s) = -\rho_{i-1}(s)\mathbf{e}_{i-1}(s) + \rho_i(s)\mathbf{e}_{i+1}(s), \\ &\qquad\qquad\qquad\qquad i = 1, \ldots, n; \\ &\rho_0(s) = \rho_n(s) = 0; \\ &\rho_j(s) > 0, \qquad\qquad j = 1, \ldots, n-2. \end{aligned} \tag{3}$$

These are called **Frenet's formulas** (or the **Frenet-Serret formulas**). We call $\rho_1, \rho_2, \ldots,$

ρ_{n-2} the first, second, ..., $(n-2)$nd **curvature**, respectively, while we call ρ_{n-1} the **torsion** for $n \geqslant 3$. For a curve in a lower-dimensional subspace $E^m \subset E^n$, we set $\rho_i = 0$ $(i > m)$. The curvatures and the torsion of a straight line are zero. To get Frenet's formulas in these special cases, we fix $\mathbf{e}_i$ $(i > m)$ in the subspace orthogonally complementary to E^m in E^n and proceed as in the general case. Suppose that C_1, C_2 are arcs such that both of their Frenet frames are of class C^1. If there exists a diffeomorphism of C_1 to C_2 that preserves arc length and the ρ_i $(i = 1, \ldots, n-1)$ are equal at corresponding points, the C_1 and C_2 are mapped onto each other by a motion of E^n. This is the **fundamental theorem of the theory of curves**. Given $n-1$ functions of class C^1 $\rho_1(s) \geqslant 0, \ldots, \rho_{n-2}(s) \geqslant 0$ (we assume that the equality signs occur at most at a finite number of points) and $\rho_{n-1}(s)$ for $0 \leqslant s \leqslant L$, there exists an arc that has $\rho_1, \ldots, \rho_{n-2}, \rho_{n-1}$ as its first, ..., $(n-2)$nd curvatures and its torsion, respectively. The equations $\rho_i = \rho_i(s)$ are called the **natural equations** of the curve.

E. Plane Curves

Let $\mathbf{x} = \mathbf{x}(s)$ be a curve of class C^2 in E^2, and $(\mathbf{x}(s), \mathbf{e}_1, \mathbf{e}_2)$ its Frenet frame. The tangent and the normal of this curve at $\mathbf{x}(s)$ have parametric representations $\mathbf{x}(s) + t\mathbf{e}_1$, $\mathbf{x}(s) + t\mathbf{e}_2$, respectively (with parameter t). Frenet's formulas are written as $\mathbf{x}' = \mathbf{e}_1$, $\mathbf{x}'' = \mathbf{e}_1' = \rho \mathbf{e}_2$, and ρ is called the **curvature** of the curve C. The natural equation is given by $\rho = \rho(s)$. If $\rho(s) =$ constant$\neq 0$ along C, C must be a portion of a circle. Another way of defining the curvature is as follows: We take a fixed direction (for example, the positive direction of the x-axis on E^2) and denote by $\theta(s)$ the angle made by the tangent T_s of the curve C at $\mathbf{x}(s)$ with the direction. Then we have $\rho(s) = d\theta / ds$. If $n = 2$, the curvature may take both positive and negative values. Figs. 1 and 2 suggest a geometric meaning of $\rho > 0$ and $\rho < 0$, respectively. The circle with center $\mathbf{x} = \mathbf{x}(s) + (1/\rho)\mathbf{e}_2$ and radius $1/\rho$ has a contact of higher order than any other circle in E^2. We call this circle the **osculation circle** (or **circle of curvature**) at the point $\mathbf{x} = \mathbf{x}(s)$, its center **the center of curvature**, and $1/\rho$ the **radius of curvature**. The locus C' of the center of curvature of a curve C is called an **evolute** of C. Conversely, C is called an **involute** of C', the †envelope of the family of normal lines of C. When a curve is given in terms of its canonical parameter s, the curvature is given by $|\mathbf{x}'(s), \mathbf{x}''(s)|$; when the curve is given by another parameter t as $\mathbf{x} = \mathbf{x}(t)$, the curvature is given by $\rho(t) = |\mathbf{x}'(t), \mathbf{x}''(t)| / |\mathbf{x}'(t)|^3$, where $'$ means d/dt.

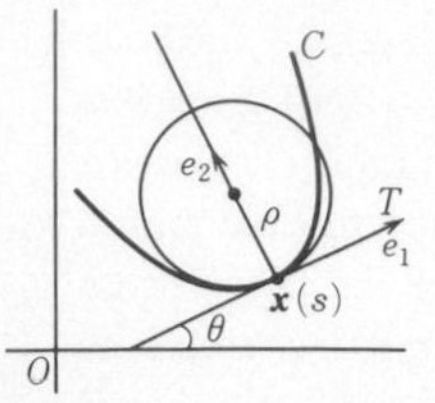

Fig. 1
$\rho > 0$.

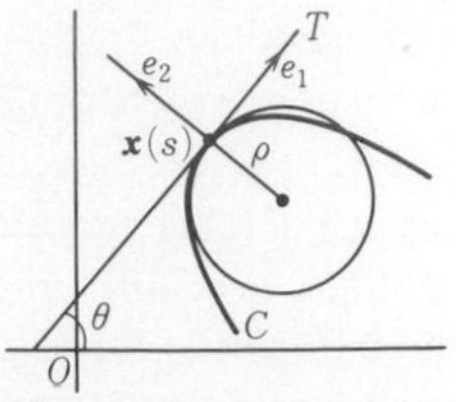

Fig. 2
$\rho < 0$.

The facts we have just stated concern local properties of plane curves. We shall now discuss the global theory of curves, which deals with properties of each curve as a whole. Let $\bar{D}$ be a closed domain consisting of points in the interior and on the boundary of a simple closed curve C. C is called a **closed convex curve** or an **oval** if $\bar{D}$ is †convex in E^2. Among all ovals of given length, the circle has the maximum area. Various generalizations of this theorem have been obtained, and the collection of problems of this kind is called the **isoperimetric problem**. This problem has intimate connections with such fields as †integral geometry. There are also some results concerning the relations between local properties (for example, curvature) and properties of the whole figure. An example is given by the **four-vertex theorem**. A vertex on a curve C is by definition a point where $d\rho/ds = 0$. Then there are at least four vertices on an oval of class C^3. A simple closed curve with $\rho \geqslant 0$ $(\leqslant 0)$ must be convex (→ 91 Convex Sets).

F. Space Curves

Let $\mathbf{x} = \mathbf{x}(s)$ $(s \in [a, b])$ be a curve C of class C^3 in E^3 defined in terms of the canonical parameter s. Let $(\mathbf{x}(s), \mathbf{e}_1, \mathbf{e}_2, \mathbf{e}_3)$ be Frenet frames along C. Then we have the following Frenet formula,

$$\mathbf{e}_1' = \rho_1 \mathbf{e}_2, \quad \mathbf{e}_2' = -\rho_1 \mathbf{e}_1 + \rho_2 \mathbf{e}_3, \quad \mathbf{e}_3' = -\rho_2 \mathbf{e}_2.$$

We call $1/\rho_1$, $1/\rho_2$ the **radius of curvature** and the **radius of torsion**, respectively. The line $\mathbf{x} = \mathbf{x}(s_0) + t\mathbf{e}_1$ is the tangent of C at $\mathbf{x}(s_0)$. The two straight lines through the point $\mathbf{x}(s_0)$ defined by $\bar{\mathbf{x}} = \mathbf{x}(s_0) + t\mathbf{e}_2$ and $\bar{\mathbf{x}} = \mathbf{x}(s_0) + t\mathbf{e}_3$ are called the **principal normal** and the **binormal** of C at $\mathbf{x}(s_0)$, respectively. The three planes through $\mathbf{x}(s_0)$ defined by $\bar{\mathbf{x}} = \mathbf{x}(s_0) + t\mathbf{e}_2 + \bar{t}\mathbf{e}_3$, $\bar{\mathbf{x}} = \mathbf{x}(s_0) + t\mathbf{e}_3 + \bar{t}\mathbf{e}_1$, and $\bar{\mathbf{x}} = \mathbf{x}(s_0) + t\mathbf{e}_1 + \bar{t}\mathbf{e}_2$ are called the **normal plane**, the **rectifying plane**, and the **osculating plane**, respectively. At a point $\mathbf{x}(s_0)$ of a curve $\mathbf{x} = \mathbf{x}(s)$ of class C^ω, we take $\mathbf{e}_1(s_0)$, $\mathbf{e}_2(s_0)$, and $\mathbf{e}_3(s_0)$ as unit vectors of the coordinate axes. Substituting the Frenet formulas into the Taylor expansion of $\mathbf{x}(s)$, we see that the new coordinates $\mathbf{x}_1(s)$,

$\mathbf{x}_2(s)$, $\mathbf{x}_3(s)$ of C are given by

$$x_1=(s-s_0)-(\rho_1(s_0)/6)(s-s_0)^3+\ldots,$$

$$x_2=(\rho_1(s_0)/2)(s-s_0)^2$$
$$+(\rho_1'(s_0)/6)(s-s_0)^3+\ldots,$$

$$x_3=(\rho_1(s_0)\rho_2(s_0)/6)(s-s_0)^3+\ldots.$$

These are called **Bouquet's formulas**. Utilizing these formulas we can see the nature of the curve with given ρ_1 and ρ_2. A curve and its osculating plane at a point on it have contact of order higher than any other plane through that point. The family of osculating planes of C †envelops a †developable surface S and coincides with the locus of tangent lines to C. We call S the **tangent surface** of C, and C the **line of regression** of S. The family of rectifying planes of C also envelops a developable surface called the **rectifying surface**, and C is a †geodesic on this surface. The family of normal planes of C envelops either a cone or a tangent surface of another curve $\bar{C}$. When the natural equation of a space curve has a special form, the shape of the curve is simple. For example, $\rho_1(s)=\text{constant}$, $\rho_2(s)=\text{constant}$ represent a curve, called an **ordinary helix**, on a cylinder which cuts all the generators of the cylinder at a constant angle. More generally, it is known that if $\rho_1/\rho_2=\text{constant}$, the tangent at each point of the curve makes a constant angle with a fixed direction. Such a curve is called a **generalized helix** or a **curve of constant inclination**. Each curve satisfying $a\rho_1+b\rho_2=c$ $(ab\neq 0)$ is called a **Bertrand curve**. For a Bertrand curve there exists another curve $\bar{C}$ and a correspondence of C onto $\bar{C}$ such that they have a common principal normal at corresponding points. Conversely, this property is also a sufficient condition for C to be a Bertrand curve. A **Mannheim curve** is defined analogously as a curve having a correspondence with another curve $\bar{C}$ such that the principal normal of C and the binormal of $\bar{C}$ coincide at corresponding points. When a correspondence of C and $\bar{C}$ has the property that tangents at corresponding points are parallel, then the correspondence is called a **correspondence of Combescure**.

We have mainly stated local properties of space curves. There are also several results about global properties of curves in E^3 analogous to the case of plane curves. For a simple closed curve C of length L, we call $K=\int_0^L\rho_1(s)\,ds$ the **total curvature** of C. Generally we have $K\leqslant 2\pi$, while $K=2\pi$ if and only if C is a closed convex curve lying in a plane (W. Fenchel). We fix an origin O in E^3 and draw a unit tangent vector with initial point O parallel to the unit tangent vector at each point of a space curve C; then the endpoint of this vector traces a curve $\bar{C}$ on the unit sphere with the center O. We call $\bar{C}$ the **spherical indicatrix** of C and the correspondence of C to $\bar{C}$ a **spherical representation**. The total curvature K of a curve C is equal to the length of $\bar{C}$. Consequently, we have $K=\oint_C d\theta$, where θ is the angular deflection of the tangent line along the closed curve C.

G. Theory of Hypersurfaces

Let (M,f) be an immersion of an $(n-1)$-dimensional differentiable manifold M of class C^r into E^n. Then we can define on the hypersurface M a positive definite differential form g of degree 2 induced from the inner product of E^n: $g_x(X,X)=(df_x(X),df_x(X))$, $X\in M_x$. Then M becomes a †Riemannian manifold with †Riemann metric g. We call g the **first fundamental form** of (M,f). By $\mathfrak{O}_f(M)$ we mean the set of all the ordered sets $(x,\mathbf{e}_1,\ldots,\mathbf{e}_n)\in\mathfrak{O}(n)$, where $x\in M$ and $\{\mathbf{e}_i\}$ $(i=1,2,\ldots,n)$ is an orthonormal system of E^n such that $\mathbf{e}_i\in df_x(M_x)$ $(i=1,\ldots,n-1)$. Then $\mathfrak{O}_f(M)$ with natural projection π_f and natural differentiable structure is a †principal fiber bundle over M and has a natural immersion $\hat{f}:\hat{f}(x,\mathbf{e}_1,\ldots,\mathbf{e}_n)=(f(x),\mathbf{e}_1,\ldots,\mathbf{e}_n)$ in the principal fiber bundle $\mathfrak{O}(n)$. We can †pull back the forms on $\mathfrak{O}(n)$ to $\mathfrak{O}_f(M)$ by $\hat{f}^*$ and put $\theta=\hat{f}^*(\omega)$, $\Theta=\hat{f}^*(\Omega)$; then the structural equations of E^n are transformed to $d\theta=\Theta\wedge\theta$, $d\Theta=(-1/2)\Theta\wedge\Theta$. Furthermore, if we put $\theta^i=\hat{f}^*(\omega^i)$, $\Theta^{ij}=\hat{f}^*(\Omega^{ij})$, then $\theta^n=0$ and θ^i, Θ^{ij} $(i,j<n)$ depend only on the first fundamental form of (M,f). Let f_1 and f_2 be two immersions of M into E^n. Then in order that there exist a Euclidean motion α of E^n such that $f_1=\alpha\circ f_2$, it is necessary and sufficient that there exist a diffeomorphism φ of $\mathfrak{O}_{f_1}(M)$ onto $\mathfrak{O}_{f_2}(M)$ such that $\theta_{f_1}=\varphi^*(\theta_{f_2})$ and $\Theta_{f_1}=\varphi^*(\Theta_{f_2})$. Suppose that M is †orientable and oriented. Then the unit vector field normal to $df_x(M_x)$ at every point $x\in M$ in E^n defines a mapping of M into the unit sphere in E^n called the **spherical representation** of M. Regarding the unit normal vector field N of M as a vector-valued function over M, we can define a symmetric product of df and dN by $-(df,dN)(X,Y)=(1/2)[(df(X),dN(Y))+(df(Y),dN(X))]$, called the **second fundamental form** of (M,f).

Two immersions f_1 and f_2 of M that induce the same first and second fundamental forms have a Euclidean motion α such that $f_1=\alpha\circ f_2$; and the converse is also true. This fact is called the **fundamental theorem of the theory of surfaces**.

H. Theory of Surfaces in E^3 (→ Appendix A, Table 4.1)

A surface in E^n is locally expressed by parametric equations $x_i = x_i(u_\alpha)$ $(i=1, \ldots, n; \alpha = 1, \ldots, m)$ or by a single vector equation $\mathbf{x} = \mathbf{x}(u_\alpha)$. We are mainly concerned with the case $n=3$, $m=2$, and we express the surface by a vector representation $\mathbf{x}=\mathbf{x}(u,v)$. The first and second fundamental forms are written as

$$E\,du^2 + 2F\,du\,dv + G\,dv^2,$$
$$P\,du^2 + 2Q\,du\,dv + R\,dv^2.$$

If we use the usual notation of †tensor analysis, then the first and second fundamental forms are also denoted by $g_{\alpha\beta}\,du^\alpha\,du^\beta$ and $H_{\alpha\beta}\,du^\alpha\,du^\beta$, respectively, where u^α $(\alpha=1,2)$ are parameters (with Σ omitted by †Einstein's convention). We call $\{g_{\alpha\beta}\}$, $\{H_{\alpha\beta}\}$ the **first** and **second fundamental quantities**, respectively. At the point $p_0 = x(u_0, v_0)$ on a surface S that corresponds to parameter values (u_0, v_0), the curves expressed by $v=v_0$ and $u=u_0$ are called a **u-curve** and a **v-curve** through p_0, respectively. Let $\mathbf{x}_u$, $\mathbf{x}_v$ denote the tangent vectors $\partial\mathbf{x}/\partial u$, $\partial\mathbf{x}/\partial v$ at p_0 to the u-curve and v-curve, respectively, through the given point p_0 and $\mathbf{N}$ denote the unit vector orthogonal to $\mathbf{x}_u$ and $\mathbf{x}_v$. Then $\mathbf{N}$ is called the **normal vector** of S at p_0 and $(\mathbf{x}_u, \mathbf{x}_v, \mathbf{N})$ the **Gaussian frame** of S at p_0. Although a Gaussian frame is not in general an orthogonal frame, it is intimately related to local parameters. The plane that passes through the point p_0 and is spanned by $\mathbf{x}_u$, $\mathbf{x}_v$ is called the **tangent plane** to S at p_0. The coefficients of the second fundamental form $P(u,v)$, $Q(u,v)$, $R(u,v)$ are expressed by the inner products $P = (-\mathbf{x}_u, \mathbf{N}_u)$, $Q = (-\mathbf{x}_u, \mathbf{N}_v)$, $R = (-\mathbf{x}_v, \mathbf{N}_v)$ $(\mathbf{N}_u = \partial\mathbf{N}/\partial u$, $\mathbf{N}_v = \partial\mathbf{N}/\partial v)$.

Let (X, Y) be the coordinates of a point on the tangent space at p_0 with respect to the Gaussian frame. We call the curve of the second order defined by $PX^2 + 2QXY + RY^2 = \varepsilon$ (ε is a suitable constant) the **Dupin indicatrix**. The point p_0 is called an **elliptic point** or a **hyperbolic point** on S according as the Dupin indicatrix at the point is an ellipse or a hyperbola. If p_0 is an elliptic point, then points near p_0 on the surface lie on one side of the tangent plane at p_0, whereas if p_0 is a hyperbolic point, points near p_0 on the surface lie on both sides of the tangent plane at p_0 (Figs. 3, 4). A hyperbolic point is also called a **saddle point**, since in a neighborhood of the point the surface looks like a saddle. A point that is neither elliptic nor hyperbolic is called a **parabolic point**; at a parabolic point we have $PR - Q^2 = 0$. If at least one of P, Q, R does not vanish at p_0, then there is a neighborhood of p_0 of the surface that lies on one side of the tangent plane at p_0 (Fig. 5). If a vector (X, Y) on the tangent plane of the surface at p_0 satisfies the equation $PX^2 + 2QXY + RY^2 = 0$, then the direction of the vector is called an **asymptotic direction**. If the point p_0 is elliptic, such a direction does not exist; if p_0 is hyperbolic, the direction is an †asymptotic direction of the Dupin indicatrix on the tangent plane at p_0. A curve C on a surface such that the tangent line at each point of the curve coincides with an asymptotic direction of the surface at the point is called an **asymptotic curve**.

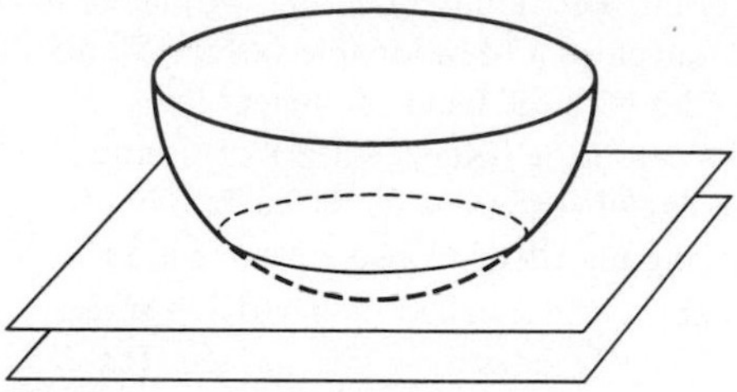

Fig. 3
Elliptic point.

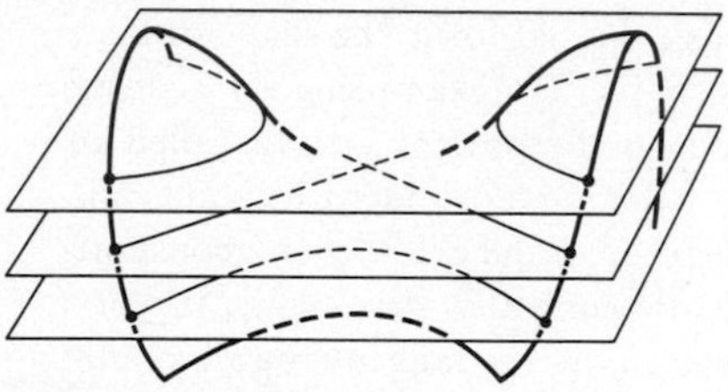

Fig. 4
Hyperbolic point.

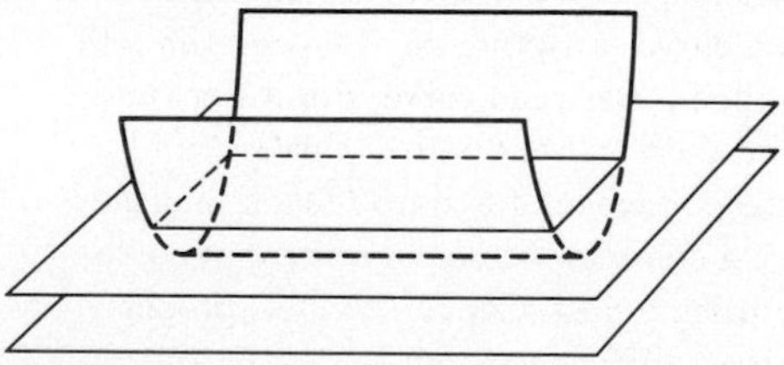

Fig. 5
Parabolic point.

Let $C: \mathbf{x} = \mathbf{x}(u(t), v(t))$ be a curve through p_0 on the surface $\mathbf{x} = \mathbf{x}(u,v)$. Then the curvature ρ of C as a space curve is given by

$$\rho\cos\theta = \frac{P\,du^2 + 2Q\,du\,dv + R\,dv^2}{E\,du^2 + 2F\,du\,dv + G\,dv^2},$$

where $du\,dv$ is the direction of C on the surface at p_0 and θ is the angle between the normal of the surface at p_0 and the principal normal of C at p_0. The center of curvature at a point p_0 of a curve C of class C^2 on a surface of class C^2 is the projection on its osculating plane of the center of curvature of the section C^* (of the surface) cut by the plane determined by the tangent to the curve at p_0 and the normal of the surface at the point (**Meusnier's theorem**). The curvature of the curve C^* at p_0 is called the **normal curvature** of the surface at the point for the tangent direction. Since the normal curvature for a direction at a point is a continuous function of this direction that can be represented as a point on a unit circle, there exist two directions that realize the maximum and minimum of the normal curvature. These directions are given by the equation

$$\begin{vmatrix} E\,du+F\,dv & F\,du+G\,dv \\ P\,du+Q\,dv & Q\,du+R\,dv \end{vmatrix}=0.$$

When this quadratic equation in du/dv has nonzero discriminant, it determines two directions defined by its two roots. These directions are called **principal directions** at the point. A curve C on a surface such that the tangent line at each point of the curve coincides with a principal direction at the point is called a **line of curvature**. When all lines of curvature of a surface are circles, the surface is called a **cyclide of Dupin**. The two normal curvatures corresponding to two principal directions are given by $1/R$, satisfying the following equation of the second order:

$$\left(\frac{1}{R}\right)^2-\frac{ER+GP-2FQ}{EG-F^2}\frac{1}{R}+\frac{PR-Q^2}{EG-F^2}=0.$$

They are called **principal curvatures**, and each of their inverses is called a **radius of principal curvature**. The mean value $H=(\rho_1+\rho_2)/2$ of two principal curvatures $\rho_i=1/R_i$ $(i=1,2)$ is called the **mean curvature** (or **S. Germain's curvature**), and the product $K=\rho_1\rho_2$ is called the **total curvature** (or **Gaussian curvature**). These are given by

$$H=\frac{1}{2}\frac{ER+GP-2FQ}{EG-F^2},\qquad K=\frac{PR-Q^2}{EG-F^2}.$$

A point on a surface is elliptic, hyperbolic, or parabolic according as $K>0$, $K<0$, or $K=0$ at the point. A point where the second fundamental form is proportional to the first fundamental form is called an **umbilical point**, and a point where the second fundamental form vanishes is called a **flat point**. If a surface consists of umbilical points only, the ratio $(P(du)^2+2Q\,du\,dv+R(dv)^2)/(E(du)^2+2F\,du\,dv+G(du)^2)$ is a constant, and the surface is either a sphere or a portion of it. If a surface consists of flat points only, the surface must be either a plane or a portion of it. The mean curvature and the Gaussian curvature of a sphere are constant, and those of a plane are both equal to zero. If we use the spherical representation of a surface stated in Section G, we may give to the Gaussian curvature the following geometric meaning: Let A be the area of the domain enclosed by a closed curve C around a point p_0 on a surface, and let A^* be the area of the domain on the unit sphere enclosed by the curve that is the image of C under the spherical representation of the surface. Then the limit of A^*/A as the closed curve C tends to the point p_0 is equal to K at p_0.

Let us denote by $(g^{\alpha\beta})$ the inverse matrix of the matrix $(g_{\alpha\beta})$ whose elements are coefficients of the first fundamental form $g_{\alpha\beta}\,du^\alpha du^\beta$. We see easily that $g^{11}=G/(EG-F^2)$, $g^{12}=g^{21}=-F/(EG-F^2)$, $g^{22}=E/(EG-F^2)$. We introduce the symbols

$$[\beta\gamma,\alpha]=\frac{1}{2}\left(\frac{\partial g_{\alpha\beta}}{\partial u^\gamma}+\frac{\partial g_{\gamma\alpha}}{\partial u^\beta}-\frac{\partial g_{\beta\gamma}}{\partial u^\alpha}\right),$$

$$\left\{ {\alpha \atop \beta\gamma} \right\}=g^{\alpha\delta}[\beta\gamma,\delta],$$

which are called the **Christoffel symbols** of the first and second kinds, respectively. Suppose that a surface is given by the vector representation $\mathbf{x}=\mathbf{x}(u_1,u_2)$, and put $\mathbf{x}_\alpha=\partial\mathbf{x}/\partial u^\alpha$, $\mathbf{x}_{\alpha\beta}=\partial\mathbf{x}_\alpha/\partial u^\beta$. Then for the derivatives of the Gaussian frame, we obtain

$$\mathbf{x}_{\alpha\beta}=\left\{ {\gamma \atop \alpha\beta} \right\}\mathbf{x}_\gamma+H_{\alpha\beta}\mathbf{N},\qquad \mathbf{N}_\alpha=-g^{\gamma\beta}H_{\beta\alpha}\mathbf{x}_\gamma.$$

We call the former **Gauss's derived equation** and the latter **Weingarten's derived equation**. The integrability conditions of these partial differential equations are

$$R^\delta_{\alpha\beta\gamma}=H_{\alpha\gamma}H^\delta_\beta-H_{\alpha\beta}H^\delta_\gamma,\qquad H^\beta_\gamma=g^{\alpha\beta}H_{\gamma\alpha},$$

$$\frac{\partial H_{\alpha\beta}}{\partial u^\gamma}-\frac{\partial H_{\alpha\gamma}}{\partial u^\beta}+\left\{ {\delta \atop \alpha\beta} \right\}H_{\delta\gamma}-\left\{ {\delta \atop \alpha\gamma} \right\}H_{\delta\beta}=0,$$

where

$$R^\delta_{\alpha\beta\gamma}=\frac{\partial}{\partial u^\beta}\left\{ {\delta \atop \alpha\gamma} \right\}-\frac{\partial}{\partial u^\gamma}\left\{ {\delta \atop \alpha\beta} \right\}+\left\{ {\sigma \atop \alpha\gamma} \right\}\left\{ {\delta \atop \sigma\beta} \right\}-\left\{ {\sigma \atop \alpha\beta} \right\}\left\{ {\delta \atop \sigma\gamma} \right\}$$

are components of the curvature tensor. The former are called the **Gauss equations**, and the latter the **Codazzi-Mainardi equations**. In connection with these equations, **Bonnet's fundamental theorem** states the following: Suppose that a positive definite symmetric matrix $(g_{\alpha\beta})$ and a symmetric matrix $(H_{\alpha\beta})$ are given that are functions of class C^2 and C^1, respectively, defined over a †simply connected domain D in $\mathbf{R}^2$. If they satisfy the Gauss equations and the Codazzi-Mainardi equations, then there exists a surface $\mathbf{x}=\mathbf{x}(u_1,u_2)$ with the given $(g_{\alpha\beta})$ and $(H_{\alpha\beta})$ as coefficients of its first and second fundamental forms, respectively. Such a surface is determined uniquely if, for an arbitrary fixed point (u_1^0,u_2^0) of D, we assign an arbitrary point p_0 and a frame $(\mathbf{x}_1^0,\mathbf{x}_2^0,\mathbf{N}^0)$ at p_0 so that $\mathbf{x}_1^0$, $\mathbf{x}_2^0$ are orthogonal to the unit vector $\mathbf{N}^0$ and

$(\mathbf{x}_\alpha^0, \mathbf{x}_\beta^0) = g_{\alpha\beta}(u_1^0, u_2^0)$ as the Gaussian frame at p_0. The Riemannian geometry on a surface with its first fundamental differential form as Riemannian metric is called **geometry on a surface** (→ 360 Riemannian Manifolds).

A †diffeomorphism between two surfaces preserving arc length is called an **isometric mapping**. The condition of preserving arc length is equivalent to the condition that the first fundamental quantities of the surfaces coincide at each pair of corresponding points, provided that we have introduced parameters on the two surfaces so that corresponding points have the same parameter values. In such a case two surfaces are said to be **isometric**. From the Gauss equation we can see that the total curvature depends only on the first fundamental quantities. So K is a quantity that is preserved under isometric mappings **(Gauss's theorema egregium)**.

A vector field $\lambda^\alpha(t)\mathbf{x}_\alpha$ defined along a curve $u^\alpha = u^\alpha(t)$ on a surface is said to be **parallel in the sense of Levi-Civita** along the curve if its †covariant derivative along the curve vanishes, i.e.,

$$\delta\lambda^\alpha/dt = d\lambda^\alpha/dt + \left\{ \begin{matrix} \alpha \\ \beta\gamma \end{matrix} \right\} \lambda^\beta du^\gamma/dt = 0.$$

The length of a vector belonging to a vector field that is parallel along a curve C is constant along C. The angle of two vectors both belonging to vector fields that are parallel along C is also constant along C. Choose two vector fields $\lambda^\alpha_{(a)}$ parallel along a curve C on a surface that satisfy $g_{\alpha\beta}\lambda^\alpha_{(a)}\lambda^\beta_{(b)} = \delta_{ab}$. Then the tangent vector to C is expressed by $du^\alpha/dt = \lambda^\alpha_{(a)}v^a(t)$. Take a 2-plane and fix an orthogonal coordinate system on it; then the integral curve C of a set of ordinary differential equations $dx^\alpha/dt = C^\alpha_{(a)}v^a(t)$ $(C^\alpha_a = \lambda^\alpha_{(a)}(P_0))$ is called the **development** of C (→ 82 Connections). We denote by ρ the curvature of a curve C of class C^2 on a surface S of class C^2 and by σ the angle between the binormal of C and the normal of S at the same point. Then $\rho_g = \rho\cos\sigma$ is a quantity belonging to geometry on a surface and is called the **geodesic curvature** of the curve at the point. A curve with vanishing geodesic curvature is called a **geodesic**. It satisfies the differential equations

$$\frac{d^2u^\alpha}{ds^2} + \left\{ \begin{matrix} \alpha \\ \beta\gamma \end{matrix} \right\} \frac{du^\beta}{ds}\frac{du^\gamma}{ds} = 0.$$

The development of a geodesic is a straight line (→ 360 Riemannian Manifolds).

Let us consider a simply connected, orientable bounded domain D on a surface such that the boundary of D is a simple closed curve C that consists of a finite number of arcs of class C^2. If we denote by α_i $(i = 1, 2, \ldots, m)$ the external angles at vertices of the curvilinear polygon C (Fig. 6), we have

$$\int_C \rho_g\, ds + \sum_{i=1}^m \alpha_i + \int\int_D K\, d\sigma = 2\pi.$$

This is called the **Gauss-Bonnet formula**. In particular, if all the arcs of C are geodesics, we have

$$\sum_{i=1}^m \alpha_i + \int\int_D K\, d\sigma = 2\pi.$$

This formula implies as special cases the following well-known theorems in Euclidean geometry and spherical trigonometry: (i) The sum of interior angles of a triangle is equal to π. (ii) The area of a spherical triangle is proportional to its spherical excess. The formula also implies the following theorem: On any closed orientable surface we have $\int\int K\, d\sigma = 2\pi\chi$, where χ is the †Euler characteristic of the surface. We call $\int\int K\, d\sigma$ the **integral curvature** (or **total Gaussian curvature**).

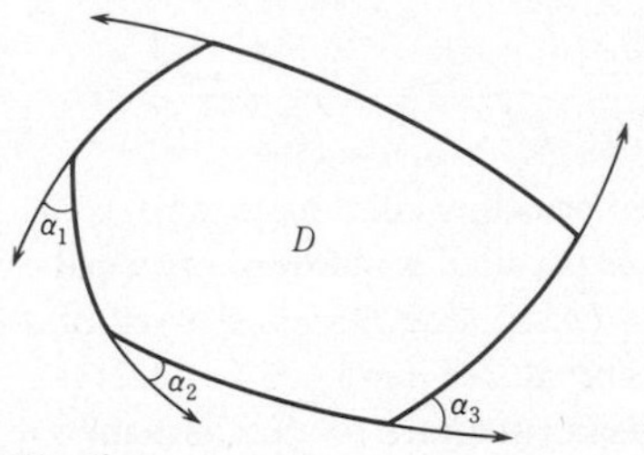

Fig. 6

I. Special Surfaces in E^3

A surface is called a **surface of revolution** if it is generated by a curve C on a plane π when π is rotated around a straight line l in π. Then l and C are called an **axis of rotation** and a **generating curve**, respectively. A surface of revolution having the x_3-axis as the axis of revolution is given by the equations $x_1 = r\cos\theta$, $x_2 = r\sin\theta$, $x_3 = \varphi(r)$; its first fundamental form is $(1+\varphi'^2)dr^2 + r^2 d\theta^2$. The section of a surface of revolution by a half-plane through its axis of revolution is called a **meridian**. According as the meridian is a straight line parallel to the axis of rotation or a straight line intersecting the axis nonorthogonally, the surface of revolution is called a **circular cylinder** or a **circular cone**, respectively. If the meridian is a circle that does not intersect the axis of rotation, it is called a **torus**.

A surface of class C^2 whose mean curvature H vanishes everywhere is called a **minimal**

surface. A surface of class C^1 realizing a relative minimum of areas among all surfaces of class C^1 with a given closed curve as their boundaries is an †analytic surface such that $H=0$. Conversely, a surface of class C^2 with vanishing mean curvature is an analytic surface (→ 330 Plateau's Problem). The equation of a surface of revolution with a catenary as its generating line is given by $x_1^2+x_2^2=a(e^{x_3/a}+e^{-x_3/a})/2$. This surface is called a **catenoid** and is a minimal surface. Conversely, a minimal surface of revolution is necessarily a catenoid. For a surface obtained by rotating a †Delaunay curve around its base line, the mean curvature H is equal to a constant ($\neq 0$). Conversely, a surface of revolution with nonzero constant mean curvature must be such a surface. A surface with constant Gaussian curvature is called a **surface of constant curvature**, and is a 2-dimensional Riemannian space of constant curvature (→ Riemannian Manifolds D). A non-Euclidean plane can be represented locally as a surface of constant curvature (→ 283 Non-Euclidean Geometry E). Two surfaces of the same constant curvature are locally isometric to each other.

Surfaces of revolution of constant curvature are classified. The simplest surface of constant negative curvature is a **pseudosphere**, which is a surface of revolution obtained by rotating a †tractrix $x_1=a\cos\varphi$, $x_3=a\log\tan((\varphi/2)+(\pi/4))-a\sin\varphi$ $(-\pi/2<\varphi<\pi/2)$ around the x_3-axis. A surface generated by a 1-parameter family of straight lines is called a **ruled surface**; a hyperboloid of one sheet, a hyperbolic paraboloid, a circular cylinder, and a circular cone are examples. The first two can be regarded as ruled surfaces in two ways. Each of the straight lines that generate a ruled surface is called a **generating line**. A surface consisting of straight lines parallel to a fixed line and passing through each point of a space curve C is called a **cylindrical surface** with the director curve C. A surface generated by a straight line that connects a certain point o with each point of a curve C is called a **conical surface**. Both a cylindrical surface and a conical surface are ruled surfaces such that $K=0$ everywhere. For ruled surfaces we have $K\leqslant 0$. In particular, a surface such that $H\neq 0$ and $K=0$ everywhere is called a **developable surface**. A developable surface must be either a cylindrical surface, a conical surface, or a tangent surface of a space curve. There exist ruled surfaces that are not developable, for example, hyperboloids of one sheet and hyperbolic paraboloids. A nondevelopable ruled surface is called a **skew surface**. A ruled surface generated by a straight line that moves under a certain rule intersecting a fixed straight line l orthogonally is called a **right conoid**. If we take l as the x_3-axis, the surface is given by the equations $x_1=u\cos v$, $x_2=u\sin v$, $x_3=f(v)$. A surface generated by a curve C (C may be chosen as a plane curve) that moves in the direction of a fixed line l with constant velocity and turns around l with certain constant angular velocity is called a **helicoidal surface**. If we take l as the x_3-axis, the surface is given by the equations $x_1=u\cos v$, $x_2=u\sin v$, $x_3=f(u)+kv$, where k is a constant and $x_3=f(x_1)$ is the equation of C. In particular, if C is a straight line that intersects l orthogonally, then $f(u)=0$, and the surface is called a **right helicoid** (or **ordinary helicoid**). A right conoid is both a ruled surface and a minimal surface. Conversely, a ruled surface that is also a minimal surface is necessarily a right conoid. A helicoidal surface with a tractrix as the curve C is called a **Dini surface** and is a surface of constant negative curvature. On the normal of a surface S there are two points q_i $(i=1,2)$, which are centers of principal curvature at p. The locus of each of these points is a surface called a **center surface** of S. When S is a sphere, two center surfaces degenerate to a point; if S is a surface of revolution, one of the center surfaces degenerates to the axis of revolution and the other is a certain surface of revolution. If S is general, each of the center surfaces is the locus of an edge of regression of the developable surface generated by normals of S along a line of curvature.

When a 1-parameter family of surfaces S_t is given by the equation $F(x_1,x_2,x_3,t)=0$, a surface E that does not belong to this family is called an **enveloping surface** of the family of surfaces $\{S_t\}$ if E is tangent to some S_t at each point of E, that is, if E and S_t have the same tangent plane. The equation of E is obtained by eliminating t from $F(x_1,x_2,x_3,t)=0$ and $(\partial F/\partial t)(x_1,x_2,x_3,t)=0$. In general, if we denote by $\varphi(x_1,x_2,x_3)=0$ the equation obtained by eliminating t from $F=0$ and $\partial F/\partial t=0$, then the surface defined by $\varphi=0$ is either the enveloping surface of $\{S_t\}$ or the locus of singular points of S_t. The intersection C_{t_0} of the enveloping surface E of $\{S_t\}$ and S_{t_0} is a curve defined by $F(x_1,x_2,x_3,t_0)=0$, $(\partial F/\partial t)(x_1,x_2,x_3,t_0)=0$. We call C_{t_0} a **characteristic curve** of $\{S_t\}$. Since $\{C_t\}$ is a family of curves on the enveloping surface E, there may exist an envelope F on E. In such a case, F is called the **line of regression** of $\{S_t\}$. The equation of F is obtained by eliminating t from $F=0$, $\partial F/\partial t=0$, and $\partial^2F/\partial t^2=0$. In particular, the enveloping surface of a family of planes is a developable surface, and their characteristic curves are straight lines. Moreover, the line of regression coincides with the line of regression of the tangent surface.

If there exists a diffeomorphism between two surfaces such that first fundamental forms at each pair of corresponding points are proportional, then the surfaces are said to be in a **conformal correspondence**. In particular, when the proportionality factor is a constant, they are said to be in a **similar** (or **homothetic**) **correspondence**. There exists a local conformal correspondence between any analytic surface and a plane. Namely, if we choose suitable parameters, we can reduce the first fundamental form of any analytic surface to the form $A(\xi,\eta)(d\xi^2+d\eta^2)$. Such parameters are called **isothermal parameters**. From the existence of isothermal parameters we can see that there exists a local conformal correspondence between any two analytic surfaces. The assumption of analyticity in these theorems is not necessary [11]. If there exists a diffeomorphism between two surfaces under which geodesics are mapped to geodesics, then the surfaces are said to be in **geodesic correspondence**. A surface has a locally geodesic correspondence with a plane if and only if it is a surface of constant curvature. If two surfaces are in geodesic correspondence, then with respect to parameters with the same values at corresponding points, we have the relation

$$\overline{\left\{ \begin{matrix} \alpha \\ \beta\gamma \end{matrix} \right\}} = \left\{ \begin{matrix} \alpha \\ \beta\gamma \end{matrix} \right\} + \delta^\alpha_\beta A_\gamma + \delta^\alpha_\gamma A_\beta$$

for coefficients of connections of the two surfaces.

By the †Alexander-Pontrjagin duality theorem, a submanifold M in E^3 that is homeomorphic to S^2 divides E^3 into two domains, and two points belonging to different domains cannot be connected by a broken segment unless the segment meets the surface. Such a manifold M is called a **closed surface**. One of the two domains consists of those points with bounded distance from a point belonging to the domain. Such a domain is called the interior of the closed surface M. If the set M^* consisting of M and its interior is convex in E^3, the surface M is called a **closed convex surface** (or **ovaloid**).

The Gaussian curvature of an ovaloid cannot be negative at any point. Conversely, if a closed surface of class C^4 with an arbitrary †genus embedded in E^3 has constant Gaussian curvature, then it is a sphere (H. Liebmann). A closed surface with $K>0$ must be an ovaloid. Moreover, it is known that on any closed surface there exists at least one point where $K>0$ (J. Hadamard). If there exists no umbilical point and K is strictly positive in a domain on a surface, then the two principal curvatures regarded as continuous functions on the domain cannot take their maximum and minimum values in the domain (D. Hilbert). If on a closed surface with positive Gaussian curvature, there is a point where one of the principal curvatures, k_1, takes its maximum value and the other, k_2, takes its minimum value at the same time, then the closed surface must be a sphere (H. Hopf). A closed surface with $K>0$ and $H=$constant is a sphere (Liebmann). On a compact surface with $K>0$, if there is a relation $k_2=f(k_1)$ between the two principal curvatures k_1, k_2 (where f is a monotone decreasing function), then the closed surface is a sphere (A. D. Alexandrov, S. S. Chern).

When we remove the assumption $K>0$ from these theorems, we have a problem proposed by Hopf which asks whether these theorems hold for closed orientable surfaces. In connection with this problem, Hopf showed that a closed orientable surface of class C^3 such that its †Euler characteristic is zero and mean curvature is constant is a sphere. If there is a certain relation $W(k_1, k_2)=0$ between the two principal curvatures k_1, k_2 $(k_1 \geqslant k_2)$ of a surface, the surface is called a **Weingarten surface** (or ***W*-surface**). When the function W is symmetric with respect to k_1, k_2, the surface is called a **symmetric *W*-surface**. On Hopf's problem, some remarkable results are known for W-surfaces and symmetric W-surfaces.

The Gaussian curvature K is invariant under isometries. Hence a sphere is transformed to a sphere by each isometry. This fact is sometimes described as **rigidity** of a sphere. More generally, if two ovaloids are isometric, then they are congruent (**Cohn-Vossen's theorem**). It is known that if we remove a small circular disk from a sphere, then the remaining portion of the sphere is isometrically deformable [8]. On the existence of closed geodesics on ovaloids, G. D. Birkhoff proved the following theorem: There exist at least three closed geodesics on any ovaloid of class C^3. It is also known that there exist surfaces of revolution that are not spheres but whose geodesics are all closed (→ 48 Calculus of Variations in the Large). On a hyperbolic non-Euclidean compact †space form of genus p $(p \geqslant 2)$ there exists a geodesic whose points are everywhere dense in it (E. Hopf) (for the †ergodicity of flows along geodesics on this surface → 146 Ergodic Theory; 311 Ordinary Differential Equations (Qualitative Theory)).

J. Singular Points of a Surface

Suppose that a neighborhood of a point p_0 of a surface S in E^3 is given by a certain vector-valued function f of class C^r as $\mathfrak{x}=\mathfrak{f}(u,v)$. Then a point p_0 where two vectors $(\partial\mathfrak{f}/\partial u)_{p_0}$, $(\partial\mathfrak{f}/\partial v)_{p_0}$ are linearly independent is called a **regular point**. A point on S that is not regular is called a **singular point**. If for suitable parameters we have $(\partial\mathfrak{f}/\partial u)_{p_0}=0$ but $(\partial\mathfrak{f}/\partial v)_{p_0}$, $(\partial^2\mathfrak{f}/\partial u^2)_{p_0}$, $(\partial^2\mathfrak{f}/\partial u\,\partial v)_{p_0}$ are linearly independent, then such a singular point is called a **semiregular point**. In general, shapes of neighborhoods of singular points are extremely complicated. However, we note the following: (i) by a small deformation of the function f (and its derivatives of orders at most r) we may reduce p_0 to a regular or semiregular point of the deformed surface; (ii) if p_0 is semiregular, we may choose suitable parameters and curvilinear coordinates of class C^r in E^3 near p_0 so that the surface S in the neighborhood of the origin p_0 is expressed by the equations $x_1=u^2$, $x_2=v$, $x_3=uv$ (H. Whitney's theorem [16]). (The higher-dimensional case has also been considered (Whitney [17]).)

K. Gaussian Curvature of Hypersurfaces in E^n

Let (M,f) be a hypersurface in E^n, where M is assumed to be orientable. We regard the spherical representation N as a vector-valued function over M. The quadratic form over M defined by $(dN_x(X), dN_x(X))$ (where $dN_x : M_x \to (S^{n-1})_{N(x)}$ is the †differential of N) is called the **third fundamental form** of (M,f). We denote by $\Lambda^{n-1}(M_x)$ and $\Lambda^{n-1}(S^{n-1}_{N(x)})$, the spaces of †alternating tensors of degree $n-1$ that are constructed on M_x and $(S^{n-1})_{N(x)}$, respectively. Both of these are 1-dimensional vector spaces and have a natural metric induced from the first and third fundamental forms over M, respectively. We give an orientation to M and take natural bases on M_x and $S^{n-1}_{N(x)}$. Then the mapping $\Lambda^{n-1}(M_x)\to\Lambda^{n-1}(S^{n-1}_{N(x)})$ induced from dN_x is represented by a quantity $K(x)$; thus we get a scalar field K, which depends on the orientation of M and is called the **Gaussian curvature** of (M,f). By virtue of $\theta^n=0$, we have $0=d\theta^n=\sum_{\alpha<n}\Theta^{n\alpha}\wedge\theta^\alpha$, and so we get a matrix $(A_{\alpha\beta})$ such that $\Theta^{n\alpha}=A_{\alpha\beta}\theta^\beta$, $A_{\alpha\beta}=A_{\beta\alpha}$. We see that $K=|A_{\alpha\beta}|$. Closed and convex hypersurfaces in E^n are defined as in the case $n=3$.

The following two statements about an immersion (M,f) of a compact orientable manifold of class C^3 are equivalent: (1) $f(M)$ is a convex hypersurface. (2) The †mapping degree of the spherical representation determined by a normal vector field is ± 1, and the Gaussian curvature K is nonpositive or nonnegative (Chern and R. K. Lashof). Therefore, if the Gaussian curvature of an immersion (M,f) of a compact orientable manifold M is positive, then $f(M)$ is a convex hypersurface.

In the case of an m-dimensional submanifold M $(1<m<n-1)$ of E^n, the first fundamental form is defined similarly, but there are $n-m$ forms corresponding to the second fundamental form. The geometric properties of such manifolds are studied by considering them as submanifolds of Riemannian manifolds (→ 360 Riemannian Manifolds).

References

[1] W. Blaschke, Vorlesungen über Differentialgeometrie I, Springer, 1924 (Chelsea, 1967).
[2] A. Duschek and W. Mayer, Lehrbuch der Differentialgeometrie I, Teubner, 1930.
[3] L. P. Eisenhart, An introduction to differential geometry, Princeton Univ. Press, 1940, revised edition, 1947.
[4] K. Yano, Bibun kikagaku (Japanese; Differential geometry), Asakura, 1949.
[5] T. Otsuki, Bibun kikagaku, bibun kikagaku ensyû (Japanese; Differential geometry, Exercises on differential geometry), Asakura, 1961.
[6] T. Kubota, Bibun kikagaku (Japanese; Differential geometry), Iwanami, 1940.
[7] S. Sasaki, Bibun kikagaku (Japanese; Differential geometry), Kyôritu, 1958.
[8] S. Sasaki, Bibun kikagaku—Taiiki teki kôsatu o tyûsin ni (Japanese; Differential geometry—especially from the standpoint of global theory), Sibundô, 1957.
[9] N. Iwahori, Bekutoru kaiseki (Japanese; Vector analysis), Syôkabô, 1960.
[10] E. Cartan, La théorie des groupes finis et continus et la géométrie différentielle traitées par la méthode du repère mobile, Gauthier-Villars, 1937.
[11] L. Bers, Riemann surfaces, Courant Institute of Mathematical Sciences, 1957–1958.
[12] S. Sternberg, Lectures on differential geometry, Prentice-Hall, 1964.
[13] А. Д. Александров (A. D. Aleksandrov), Внутренная геометрия выпуклых поверхностей, Гостехиздат, 1948; German translation, Die innere Geometrie der konvexen Flächen, Akademie-Verlag, 1955.
[14] S. S. Chern, Topics in differential geometry, Lecture notes, Institute for Advanced Study, Princeton, 1951.

[15] H. Hopf, Zur Differential Geometrie geschlossener Flächen in Euklidischen Raum, Convegno Internazionale di Geometria Differenziale, Italy (1953), 45–54 (Cremona, 1954).
[16] H. Whitney, The singularities of a smooth n-manifold in $(2n-1)$-space, Ann. of Math., (2) 45 (1944), 247–293.
[17] H. Whitney, Singularities of mappings of Euclidean spaces, Symposium Internacional de Topologia Algebraica, Universidad Nacional Autonoma de México and UNESCO (1958), 285–301.
[18] T. J. Willmore, An introduction to differential geometry, Clarendon Press, 1959.
[19] N. J. Hicks, Notes on differential geometry, Van Nostrand, 1964.
[20] D. Laugwitz, Differential and Riemannian geometry, Academic Press, 1965.
[21] B. O'Neill, Elementary differential geometry, Academic Press, 1966.
[22] J. J. Stoker, Differential geometry, John Wiley, 1969.

115 (XII.10) Differential Operators

A. Definition

A mapping (or an operator) A of a function space $\mathbf{F}_1$ to a function space $\mathbf{F}_2$ is said to be a **differential operator** if the value $f(x)$ of the image $f=Au$ $(u\in\mathbf{F}_1, f\in\mathbf{F}_2)$ at each point x is determined by the values at x of u and a finite number of its derivatives. If u and f are †distributions, the definition applies with the derivative interpreted in the sense of distributions (→ 130 Distributions (Generalized Functions)). In this article we restrict ourselves to the case of linear differential operators and consider only those of the form:

$$P(x,D)=\sum_{|\alpha|\leqslant m} a_\alpha(x)D^\alpha, \qquad (1)$$

where α denotes n-tuples $(\alpha_1,\alpha_2,\ldots,\alpha_n)$ of nonnegative integers, called **multi-indices**; $|\alpha|$ the **length** of $\alpha: |\alpha|=\alpha_1+\alpha_2+\ldots+\alpha_n$; and D^α the differential operator $D^\alpha=D_1^{\alpha_1}D_2^{\alpha_2}\ldots D_n^{\alpha_n}$, with $D_j=(-i)\partial/\partial x_j$. The coefficient $(-i)$ is sometimes omitted. The coefficients $a_\alpha(x)$ are functions defined on an open set Ω in n-dimensional space. We call $P(x,D)$ an **ordinary differential operator** if the dimension n of Ω is 1 and a **partial differential operator** if $n\geqslant 2$. Ordinary differential operators and partial differential operators behave quite differently in many respects.

We set

$$P(x,\xi)=\sum_{|\alpha|\leqslant m} a_\alpha(x)\xi^\alpha, \qquad \xi^\alpha=\xi_1^{\alpha_1}\ldots\xi_n^{\alpha_n},$$

where $\xi=(\xi_1,\xi_2,\ldots,\xi_n)\in\mathbf{R}^n$ or $\mathbf{C}^n$. The **order** of $P(x,D)$ is the greatest integer $|\alpha|$ for which $a_\alpha(x)\not\equiv 0$. In expression (1) m is assumed to be equal to the order, and in that case

$$P_m(x,D)=\sum_{|\alpha|=m} a_\alpha(x)D^\alpha$$

is called the **principal part** of $P(x,D)$, and the corresponding polynomial $P_m(x,\xi)$ the **characteristic polynomial**.

A differential operator $P(D)$ with constant coefficients is said to be an **elliptic operator** if the characteristic polynomial $P_m(\xi)$ has no real zeros except for $\xi=0$, and a **hypoelliptic operator** if $P(\xi+i\eta)=0$ and $|\xi+i\eta|\to\infty$ imply that $|\eta|\to\infty$. Elliptic operators are hypoelliptic. For example, the Laplacian $\Delta=-(D_1^2+\ldots+D_n^2)$, the Cauchy-Riemann operator $\partial/\partial\bar{z}=(1/2)(\partial/\partial x+i\partial/\partial y)$, and ordinary differential operators are hypoelliptic. The heat operator $iD_{n+1}-\Delta$ is hypoelliptic but not elliptic. For other types of operators, see the appropriate articles (e.g., for hyperbolic operators → 320 Partial Differential Equations of Hyperbolic Type).

A differential operator $P(x,D)$ with variable coefficients is said to be an **elliptic operator** if the characteristic polynomial $P_m(x,\xi)$ has no real zeros except for $\xi=0$ for each $x\in\Omega$.

Differential operators have been investigated for a long time in connection with the linear differential equations

$$P(x,D)u(x)=f(x), \qquad x\in\Omega. \qquad (2)$$

Except for ordinary differential operators, however, it is only recently that general properties of such operators have been studied.

We denote a differential operator with constant coefficients by $P(D)$ and, generally, a linear differential operator with coefficients that are C^∞-functions by $P(x,D)$. However, many of the results in the case of C^∞-coefficients also hold in the case where coefficients are sufficiently differentiable. (For †function spaces $\mathfrak{D}(\Omega)$, $\mathfrak{D}'(\Omega)$, $\mathfrak{S}(\Omega)$, $\mathfrak{S}'(\Omega)$, $C(\Omega)$, $C^\infty(\Omega)$, $C_0^\infty(\Omega)$, $L_p(\Omega)$, etc., → 173 Function Spaces.)

B. Fundamental Solutions

If a differential operator $P(x,D)$ with $\mathfrak{D}(\Omega)$ as its domain has a left inverse F and is expressed as an †integral operator with †kernel distribution (in $\mathfrak{D}'_{x,y}$; → 130 Distributions (Generalized Functions) F), then the kernel is said to be a **fundamental solution** (or **elementary solution**). F is usually a right inverse of the weak extension (→ Section F) of $P(x,D)$ and maps $\mathfrak{D}(\Omega)$ into $\mathfrak{E}(\Omega)$. The image is mapped to the original function by $P(x,D)$. Nevertheless, F is not a genuine right inverse, and hence the fundamental solution is not unique if it exists.

In the case where $P(D)$ is a differential operator with constant coefficients, we call a distribution $E(x)$ a **fundamental solution** if it satisfies

$$P(D)E(x)=\delta(x), \tag{3}$$

where $\delta(x)$ is †Dirac's distribution (δ function). If $E(x)$ is a fundamental solution in this sense, then $F(x,y)=E(x-y)$ is the kernel of a left inverse of $P(D)$ and is a fundamental solution in the sense of the preceding paragraph.

Every differential operator $P(D)$ with constant coefficients has a fundamental solution in the sense of (3) (**Ehrenpreis-Malgrange theorem**; see L. Hörmander [4] for a proof).

General operators $P(x,D)$ with variable coefficients do not necessarily have fundamental solutions. However, if $P(x,D)$ belongs to one of the classical types of operators (elliptic, hyperbolic, or parabolic), then it has a fundamental solution at least locally. (See F. John [8] for elliptic operators, J. Leray [13] for strongly hyperbolic operators, and S. Mizohata [14] and S. D. Eidel'man [9] for parabolic operators.) Leray has generalized John's method to strongly hyperbolic operators in an enormous work [15].

C. Ranges of Differential Operators

Let $P(D)$ be a differential operator with constant coefficients. Then it follows from the Ehrenpreis-Malgrange theorem that $P(D)\mathfrak{D}'(\Omega)\supset\mathfrak{D}(\Omega)$ holds for any open set Ω. However, there are differential operators $P(x,D)$ with variable coefficients such that for any Ω, $P(x,D)\mathfrak{D}'(\Omega)\not\supset\mathfrak{D}(\Omega)$. H. Lewy first devised such an example:

$$P(x,D)=-iD_1+D_2-2(x_1+ix_2)D_3.$$

Let $C_{2m-1}(x,D)$ be the homogeneous part of order $2m-1$ of the commutator

$$P(x,D)\,\overline{P(x,D)}-\overline{P(x,D)}\,P(x,D).$$

Then in order that $P(x,D)\mathfrak{D}'(\Omega)\supset\mathfrak{D}(\Omega)$, it is necessary that

$$P_m(x,\xi)=0 \quad \text{implies} \quad C_{2m}(x,\xi)=0$$

for all $x\in\Omega$, $\xi\in\mathbf{R}^n$ (**Hörmander's theorem** [4]). When $P(x,D)$ is a differential operator which does not satisfy this condition (e.g., Lewy's operator), choose an $f(x)\in\mathfrak{D}(\Omega)$ that is not in $P(x,D)\mathfrak{D}'(\Omega)$. Then the differential equation (2) has no distribution solutions at all. P. Schapira extended this result to the case of †hyperfunctions.

Concerning the ranges of differential operators $P(D)$ with constant coefficients, we have the following detailed results due to L. Ehrenpreis, B. Malgrange [16], and Hörmander [4].

An open set Ω is said to be ***P*-convex** for a differential operator $P(D)$ if for each compact set $K\subset\Omega$ there exists a compact set $K'\subset\Omega$ such that $\varphi\in C_0^\infty(\Omega)$ and $\operatorname{supp}P(-D)\cdot\varphi\subset K$ imply $\operatorname{supp}\varphi\subset K'$. Convex sets are P-convex for any $P(D)$. All open sets are P-convex if and only if $P(D)$ is an elliptic operator.

Theorem: The following conditions are equivalent: (i) Ω is P-convex; (ii) $P(D)\mathfrak{D}'(\Omega)\supset\mathfrak{E}(\Omega)$; (iii) $P(D)\mathfrak{E}(\Omega)=\mathfrak{E}(\Omega)$. Property (iii), the Mittag-Leffler theorem, and the solvability of †Cousin's first problem for the solutions of $P(D)u=0$ are equivalent (Ehrenpreis [17]).

An open set Ω is said to be **strongly *P*-convex** if for each compact set $K\subset\Omega$ there exists a compact set K' such that $\mu\in\mathfrak{E}'(\Omega)$ and $\operatorname{supp}P(-D)\mu\subset K$ imply $\operatorname{supp}\mu\subset K'$; and $\mu\in\mathfrak{E}'(\Omega)$ and $\operatorname{sing\,supp}P(-D)\mu\subset K$ imply $\operatorname{sing\,supp}\mu\subset K'$; where the **singular support** of μ is the closure of the set of all points at which μ is not a C^∞-function. Convex sets are strongly P-convex, and strongly P-convex sets are P-convex.

Theorem: Ω is strongly P-convex if and only if $P(D)\mathfrak{D}'(\Omega)=\mathfrak{D}'(\Omega)$. On the other hand, R. Harvey has shown that every domain Ω is P-convex in the sense of the hyperfunctions of M. Sato, i.e., the equation $P(D)u=f$ always has a hyperfunction solution u on Ω for any hyperfunction f on Ω.

D. Interior Regularity

Next we discuss **interior regularity** for differential operators. Differential operators of order m map functions u of class C^k $(k\geqslant m)$ to functions f of class C^{k-m}, but the converse holds only for ordinary differential operators. As shown by the example $P=D_1D_2$, the solutions u of the homogeneous equation $Pu=0$ may not have any regularity. We have, however, the converse in a weak sense when the class of differential operators is restricted.

The following results are known for differential operators with constant coefficients: (1) The solutions of $P(D)u=0$ are all †real analytic if and only if $P(D)$ is elliptic (**Petrovskiĭ's theorem**) [18]. (2) The solutions of $P(D)u=0$ are all C^∞-functions if and only if $P(D)$ is hypoelliptic (**Hörmander's theorem** [19]). In these theorems, the solutions are required to satisfy the equation only in the sense of distributions. If $P(D)$ is elliptic, the hyperfunction solutions of $P(D)u=0$ are also real analytic (R. Harvey, G. Bengel, H. Komatsu). The result that the distribution solutions of the Laplace equation $\Delta u=0$ are C^{∞}-functions is known as **Weyl's lemma**.

Let $P(x,D)$ be an elliptic differential operator of order m with variable coefficients. A distribution solution u of $P(x,D)u=f$ is $(m+k)$-times $(0 \leqslant k \leqslant \infty)$ locally differentiable in the sense of L_p $(1<p<\infty)$ on the open set on which f is k-times locally differentiable in the sense of L_p (K. O. Friedrichs, P. D. Lax), where f is said to be k-times **locally differentiable** if the derivatives $D^{\alpha}f$ up to order k $(0 \leqslant |\alpha| \leqslant k)$ of f in the sense of distributions (→ 130 Distributions (Generalized Functions)) are measurable functions on Ω, and for each point $x \in \Omega$ there exists a neighborhood Ω_x $(\subset \Omega)$ of x such that $D^{\alpha}f \in L_p(\Omega_x)$. If the coefficients of $P(x,D)$ are real analytic, then u is real analytic on the open set on which f is real analytic (I. G. Petrovskiĭ, C. B. Morrey and L. Nirenberg, P. Schapira, and M. Sato). More precisely, a distribution u is real analytic on an open set Ω if and only if for each compact set K in Ω there exists a constant C such that $\|P^k u\|_K \leqslant C^{k+1}(mk)!$, where $\| \ \|_K$ denotes the L_p-norm on K (H. Komatsu, T. Kotake and M. S. Narasimhan).

Such a simple generalization is not possible for hypoelliptic operators. The solution of $P(x,D)u=0$ is not necessarily of class C^{∞} even when $P(x,D)$ is a hypoelliptic operator for each fixed x. Set

$$\tilde{P}(\xi)^2 = \sum_{|\alpha|>0} |P^{(\alpha)}(\xi)|^2$$

for differential operators $P(D)$ with constant coefficients, where $P^{(\alpha)}(\xi)=\partial^{|\alpha|}P(\xi)/\partial\xi_1^{\alpha_1}\ldots\partial\xi_n^{\alpha_n}$. We say $Q(D)$ is **weaker** than $P(D)$ (or $P(D)$ is **stronger** than $Q(D)$) if

$$\tilde{Q}(\xi)/\tilde{P}(\xi) \leqslant C < \infty, \qquad \xi \in \mathbf{R}^n.$$

$P(D)$ and $Q(D)$ are equally strong (or have equal strength) if $P(D)$ is weaker and stronger than $Q(D)$.

Suppose that for each fixed x, $P(x,D)$ has the same strength as a hypoelliptic operator with constant coefficients independent of x. Then it follows that the solution u of $P(x,D)u=f$ is of class C^{∞} on the open set on which f is of class C^{∞} (Malgrange, Hörmander, Mizohata and Y. Hamada). Furthermore, a distribution u is of class C^{∞} if and only if $P(x,D)^k u$ are functions for any k.

There are two methods of proof for the theorems in the preceding paragraph. One relies on the properties of the fundamental solutions. The other makes use of inequalities called the **Schauder estimates** (or **a priori estimates**), such as

$$\|\nabla^{\sigma}u\|_{L_p(\Omega_\delta)} \leqslant C\left(\|Pu\|_{L_p(\Omega)} + (1+\delta^{-\sigma})\|u\|_{L_p(\Omega)}\right),$$

where Ω_δ is the subset of Ω of all points such that the distance to the boundary is greater than δ. When P is elliptic, we may take $\sigma=m$.

E. Differential Operators in Banach Spaces

We consider differential operators $P(x,D)$ defined on a domain Ω as operators in the function spaces $C(\Omega)$ or $L_p(\Omega)$. Differential operators of order $m \geqslant 1$ are always unbounded operators in a Banach space $X=C(\Omega)$ or $L_p(\Omega)$. Moreover, their domains of definition as operators in X are not generally determined uniquely by the expressions $P(x,D)$ as differential operators.

$P(x,D)$ is a linear operator which maps $C_0^{\infty}(\Omega)$ into X. This operator has a closed extension. The minimal closed extension P_0 is called the **minimal operator** of $P(x,D)$ in X. We have $u \in \mathfrak{D}(P_0)$ and $P_0u=f$ if and only if there exists a sequence $\varphi_n \in C_0^{\infty}(\Omega)$ such that $\varphi_n \to u$, $P(x,D)\varphi_n \to f$. On the other hand, the closed linear operator P_1 whose domain is the set of all $u \in X$ such that $P(x,D)u \in X$ in the sense of distribution is called the **maximal operator** (or **weak extension**) of $P(x,D)$. We have $u \in \mathfrak{D}(P_1)$ and $P_1u=f$ if and only if $\langle u, {}^tP(x,D)\varphi\rangle = \langle f,\varphi\rangle$ for any $\varphi \in C_0^{\infty}(\Omega)$, where ${}^tP(x,D)$ is the **transposed operator**

$${}^tP(x,D)\varphi(x) = \sum_{|\alpha|\leqslant m} (-D)^{\alpha}(a_{\alpha}(x)\varphi(x)).$$

Integration by parts shows that P_1 is an extension of P_0 and that when X is the †dual space of a space Y, the weak extension P_1 in X is the †dual of the minimal operator of ${}^tP(x,D)$ in Y. Let $X=L_p(\Omega)$ $(1<p<\infty)$, Ω be a bounded open set with smooth boundary, and $P(D)$ have constant coefficients. Then P_1 coincides with the smallest closed extension of the operator $P(D)$ having as its domain the set of all $u \in C^{\infty}(\Omega) \cap X$ such that $P(D)u \in X$. The latter closed extension is called the **strong extension**. The difference between the weak and the strong extension is not obvious in the variable coefficient case.

P_0 coincides with P_1 when Ω is the entire space and $P(x,D)$ is an elliptic operator whose coefficients are constants or close to constants (J. Peetre [20]; T. Ikebe and T. Kato [21]). In general, we have $P_0 \neq P_1$. Let $P(x,D)$ be an ordinary differential operator with bounded coefficients such that $|a_m(x)| \geqslant \delta > 0$ and Ω be the bounded interval (a,b). Then the domain of P_1 coincides with the set of all $(m-1)$-times continuously differentiable functions u such that the $(m-1)$st derivative is absolutely continuous and $P(x,D)u \in X$, while the domain of P_0 is the set of all functions u which satisfy in addition the boundary conditions

$$u(a)=u'(a)=\ldots=u^{(m-1)}(a)$$
$$=u(b)=\ldots=u^{(m-1)}(b)=0.$$

(Moreover, $u^{(m)}(a)=u^{(m)}(b)=0$ when $X=C(a,b)$.)

Let $G(P_0)$, $G(P_1)$ $(\subset X\times X)$ be the †graphs of P_0, P_1. Then the quotient space $\mathfrak{B}=G(P_1)/G(P_0)$ is called the **boundary space**, and an element of the dual $\mathfrak{B}'$ of $\mathfrak{B}$, i.e., a continuous linear functional on $G(P_1)$ which vanishes on $G(P_0)$, is called a **boundary value** relative to $P(x,D)$. In the case of ordinary differential operators discussed before, the boundary space is the set of all linear combinations of $u^{(i)}(a)$ and $u^{(j)}(b)$. When $P(x,D)$ is an ordinary differential operator, we can explicitly determine the boundary values even in the case where the interval (a,b) is infinite, the coefficients $a_\alpha(x)$ are not bounded, or $a_m(x)\to 0$ $(x\to a,b)$; and we can show that $\mathfrak{B}$ is finite-dimensional. When $P(x,D)$ is a partial differential operator, $\mathfrak{B}$ is generally of infinite dimension, and the concrete forms of the elements of $\mathfrak{B}$ and $\mathfrak{B}'$ are not known. However, we have some information by M. I. Višic [22] about the boundary values of elliptic operators of the second order. Combining this with the results by J.-L. Lions and E. Magenes [23], we may obtain information for elliptic operators of higher order.

F. Differential Operators with Boundary Conditions

A closed operator between the minimal operator P_0 and the maximal operator P_1 is determined by designating a closed subspace B of the boundary space $\mathfrak{B}$. This operator is called the **operator with the boundary condition** B. Particularly important are boundary conditions expressed in the form

$$Q_i(x,D)u(x)=0, \qquad x\in\partial\Omega, \qquad i=1,\ldots,k, \tag{4}$$

with differential operators $Q_i(x,D)$ $(i=1,\ldots,k)$ defined on the boundary $\partial\Omega$ of Ω.

When $P(x,D)$ is an ordinary differential operator defined on a finite interval and the orders of Q_i are at most $m-1$ (or m), (4) always has a definite meaning. However, in the case of partial differential operators, we need an interpretation of (4), i.e., (4) does not necessarily determine the subspace B of $\mathfrak{B}$ uniquely.

Let P_s be the smallest closed extension of $P(x,D)$ with $\{u\in C^\infty(\bar\Omega)\cap X \mid Q_i(x,D)u(x)=0,\ x\in\partial\Omega;\ P(x,D)u\in X\}$ as its domain. P_s is called the **strong extension** of the differential operator $P(x,D)$ with boundary condition (4).

On the other hand, when Ω, P, and Q_i satisfy suitable conditions, we can define the transposed differential operator ${}^tP(x,D)$ with the transposed boundary operators $R_j(x,D)$ $(j=1,\ldots,l)$. Namely, there are differential operators $R_j(x,D)$ on the boundary such that a necessary and sufficient condition for $u\in C^\infty(\bar\Omega)$ to satisfy (4) and $Pu(x)=f(x)$ is

$$\int_\Omega f(x)\,v(x)\,dx=\int_\Omega u(x)\,{}^tP\,v(x)\,dx \tag{5}$$

for all $v(x)\in C^\infty(\bar\Omega)$ with the boundary condition $R_jv=0$, $x\in\partial\Omega$. Then the operator P_w defined by $P_wu(x)=f(x)$ for the pairs $u(x)$, $f(x)\in X$ satisfying (5) is called the **weak extension** of the differential operator $P(x,D)$ with boundary condition (4). As in the case of operators without boundary conditions, the weak extension is an extension of the strong extension, and generally is the dual of the strong extension in the dual space X' of the transposed differential operator with the transposed boundary condition.

Regularity up to the boundary. The fundamental problems for the differential operator $P(x,D)$ with the homogeneous boundary condition (4) are to determine, for both the strong and the weak extensions, the spaces of solutions of the homogeneous equations $Pu=0$ and their ranges. The problems mostly reduce to determining when the strong and the weak extensions coincide and, including this, also to the problem of regularity on the closed domain $\bar\Omega$ containing the boundary of the solutions u of the equation $P_wu=f$.

This problem was solved by Nirenberg in the case of strongly elliptic operators in $L^2(\Omega)$ with the Dirichlet boundary condition

$$\partial^{j-1}u(x)/\partial n^{j-1}=0, \qquad j=1,2,\ldots,m/2,$$

and generalized later by F. Browder, M. Schechter [24], S. Agmon, Lions, and others in the case of elliptic operators in $L_p(\Omega)$ with a kind of coercive boundary condition (→ Section H).

Consequently, when Ω is bounded and smooth, P_w is equal to P_s for those operators. Write P for P_w. Then the space $N(P)$ of the solutions of $Pu=0$ is a subspace of finite dimension, and the range $R(P)$ is a closed subspace of finite codimension. In particular, it follows that the **index** of P, $\dim N(P)-\operatorname{codim} R(P)$, is finite.

G. Strongly Elliptic Operators

A differential operator $P(x,D)$ is said to be

strongly elliptic if its characteristic polynomial satisfies

$$\operatorname{Re} P_m(x,\xi) \geqslant C|\xi|^m > 0, \quad \xi \neq 0.$$

Many of the elliptic operators, such as the Laplacian, that appear in applications are strongly elliptic. L. Gårding's work [25] on strongly elliptic operators with the Dirichlet condition initiated the recent study of differential operators. His theory is based on the following inequality, called **Gårding's inequality**:

$$\|\nabla^{m/2}u\|_{L_2}^2 \leqslant C\left(\operatorname{Re}\int P\,u\cdot\bar{u}\,dx + \|u\|_{L_2}^2\right),$$

$$u \in C_0^\infty(\Omega).$$

H. Coercive Boundary Conditions

The boundary condition (4) is said to be **coercive** if

$$\|\nabla^m u\| \leqslant C(\|P\,u\| + \|u\|) \qquad (6)$$

holds for any $u \in C^\infty(\bar{\Omega})$ that satisfies (4). In order that a differential operator P have a coercive boundary condition, it is necessary that it be a special type of elliptic operator. In this case, Agmon, N. Aronszajn, Schechter, and others found conditions under which (4) is coercive. Agmon, A. Douglis, and Nirenberg [26] show that the inequality (6) holds in $L_p(\Omega)$ and in the normed spaces of Hölder continuous functions under a suitable condition. The classical boundary conditions $au + b\partial u/\partial n = 0$ ($a \geqslant 0$, $b \geqslant 0$) are coercive for elliptic operators of the second order. However, problems remain when the coefficients of $Q_j(x,D)$ are discontinuous. In order to have coincidence of the strong and weak extensions or regularity up to the boundary, it is not necessary that P be elliptic nor that the boundary condition be coercive. But it is not known to what extent these conditions can be weakened. At present, major contributions are Hörmander's work [27] dealing with operators with constant coefficients and flat boundaries, and works by J. J. Kohn [28], Nirenberg, and Hörmander concerning noncoercive boundary conditions. The latter works are connected with the theory of several complex variables, and have attracted much attention.

I. Self-Adjoint Extension

One of the fundamental problems in the case $X = L_2(\Omega)$ is whether the minimal operator P_0 has a self-adjoint extension. P_0 is symmetric if and only if $P(x,D)$ is **formally self-adjoint**: $P(x,D) = {}^tP(x,D)$. Under this condition the boundary space $\mathfrak{B}$ turns out to be the direct sum of two subspaces $\mathfrak{B}_\pm = \{(x, P_1x) \mid x \in \mathfrak{D}(P_1), P_1x = \pm ix\} + G(P_0)$. The numbers $n_\pm = \dim \mathfrak{B}_\pm$ are called the **deficiency indices** of P_0, and P_0 has a self-adjoint extension if and only if $n_+ = n_-$.

H. Weyl gave a method for computing $n_\pm$ for the **Sturm-Liouville operators**:

$$P(x,D) = -\left(\frac{d}{dx}p(x)\frac{d}{dx}\cdot\right) + q(x),$$

$$x \in (a,b).$$

We say that a (resp. b) is of **limit circle type** if the solutions $u(x)$ of $P(x,D)u(x) + lu(x) = 0$ ($l \in \mathbf{C}$) always belong to L_2 in a neighborhood of a (resp. b) and of **limit point type** if a solution does not belong to L_2. This classification does not depend on the choice of $l \in \mathbf{C}$. (1) If both a and b are of limit point type, then $n_+ = n_- = 0$ and hence P_0 is self-adjoint. (2) If a is of limit circle type and b is of limit point type, then $n_+ = n_- = 1$, and the self-adjoint extensions of P_0 are the operators P_α which are obtained from P_1 by assigning the boundary condition

$$p(a)u'(a)\cos\alpha + u(a)\sin\alpha = 0.$$

The same is true when a and b are interchanged. (3) If both a and b are of limit circle type, then $n_+ = n_- = 2$, and we may impose two boundary conditions to obtain the self-adjoint extensions.

These results have been extended by K. Kodaira [29] and N. Dunford and J. Schwartz [2] to the case of ordinary differential operators of order m. There are formally self-adjoint operators that have no self-adjoint extensions. For example, the operator $-id/dx$ in $L_2(0,\infty)$ has deficiency indices $n_+ = 1 \neq n_- = 0$.

In the case of partial differential operators, it is difficult to determine explicitly all self-adjoint extensions of a given formally self-adjoint operator because the boundary space is complicated. If the boundary condition (4) is formally self-adjoint and coercive, then it follows from the results of Schechter and others that the differential operator with boundary condition (4) is self-adjoint. Furthermore, conditions under which P_0 is self-adjoint or has a self-adjoint extension are known. The following theorem is often used as a condition of the latter type. If a †symmetric operator defined on a dense subspace of a Hilbert space X is positive definite:

$$(Tx,x) \geqslant 0, \qquad x \in \mathfrak{D}(T),$$

then there is a positive definite self-adjoint extension $\tilde{T}$ (Friedrichs's theorem). The self-adjoint extension obtained by this theorem is called the **Friedrichs extension**.

J. Generators of Semigroups

From the point of view of probability theory, W. Feller investigated the extensions of the Laplacian d^2/dx^2 and similar operators that are the generators of positive semigroups. Recently various attempts have been made to generalize his results to the multidimensional case (→ 119 Diffusion Processes; 373 Semigroups of Operators).

P. D. Lax and A. N. Milgram proved that if $P(x,D)$ is a strongly elliptic operator, then $-P(x,D)$ with the Dirichlet condition in $L_2(\Omega)$ is the generator of a semigroup [1].

K. Boundary Value Problems

There are two methods of solving the inhomogeneous boundary value problem

$$P(x,D)u(x)=f(x), \quad x\in\Omega$$

$$Q_i(x,D)u(x)=g_i(x), \quad x\in\partial\Omega, \quad i=1,\dots,k.$$

In the first method, we take a function $v(x)$ that satisfies $Q_i(x,D)v(x)=g_i(x)$ and reduce the problem to the homogeneous one for $u_0=u-v$. In the second method, we consider the pair $P(x,D)$ and $Q_i(x,D)$ as an operator that maps a function u to the pair of functions (Pu,Q_iu) and investigate it directly. The latter method was adopted by Peetre and Hörmander [4].

L. Estimates in Weighted Spaces

J. F. Treves, Hörmander, and H. Kumano-go obtained estimates similar to (6) in L_p-spaces relative to the weighted measure $w_t(x)\,dx$ instead of the usual L_p-spaces, and applied them to the proof of the uniqueness of Cauchy problems for differential equations with variable coefficients (→ 316 Partial Differential Equations (Initial Value Problems)). Hörmander applied similar estimates to the proof of the †fundamental theorems of Stein manifolds.

M. Eigenfunction Expansions

When a self-adjoint operator P in the Hilbert space $L_2(\Omega)$ is a self-adjoint extension of a differential operator $P(x,D)$, the †spectral decomposition of P is concretely expressed by the expansion of functions $u\in L_2(\Omega)$ into eigenfunctions of $P(x,D)$.

If Ω is bounded, $P(x,D)$ is an elliptic operator defined on a neighborhood of $\overline{\Omega}$, and the boundary condition is coercive, then the †spectrum of P is composed solely of eigenvalues, and the eigenvectors of P are eigenfunctions of $P(x,D)$ in the classical sense and are of class C^∞ up to the boundary.

N. The Asymptotic Distribution of Eigenvalues

If $P=-\Delta$, the number $\nu(\lambda)$ of eigenvalues less than λ satisfies the asymptotic relation

$$\nu(\lambda)\sim\frac{\lambda^{n/2}A}{2^{n-1}\pi^{n/2}n\Gamma(n/2)}, \quad \lambda\to\infty$$

regardless of the shape of the domain and the boundary condition, where n is the dimension and A is the volume of Ω [3]. This was first proved by Weyl and extended by T. Carleman, Gårding, and others to the case of operators of higher order with variable coefficients.

O. The Weyl-Stone-Titchmarsh-Kodaira Theory

When Ω is unbounded or the coefficients of $P(x,D)$ have singularities near the boundary of Ω, the spectrum of P may have a continuous part.

Let $P(x,D)$ be an ordinary differential operator on an interval (a,b). Then for each $\lambda\in\mathbf{C}$ the equation $(P(x,D)-\lambda)\varphi(x)=0$ has m linearly independent solutions $\varphi_k(x,\lambda)$ $(k=1,\dots,m)$, and any solution is represented as a linear combination of them. Weyl and M. H. Stone obtained the spectral decomposition of P (of the second order) in the form:

$$u(x)=\sum_{j,k=1}^{m}\int_{-\infty}^{\infty}\varphi_j(x,\lambda)\,d\rho_{jk}(\lambda)\times\int_a^b\overline{\varphi_k(y,\lambda)}\,u(y)\,dy,$$

$$Pu(x)=\sum_{j,k=1}^{m}\int_{-\infty}^{\infty}\lambda\varphi_j(x,\lambda)\,d\rho_{jk}(\lambda)\times\int_a^b\overline{\varphi_k(y,\lambda)}\,u(y)\,dy,$$

where the $\rho_{jk}(\lambda)$ are functions of bounded variation and their variations represent the spectral measure. This formula shows that linear combinations of the $\varphi_j(x,\lambda)$ form generalized eigenfunctions even when λ belongs to the continuous spectrum. Later, E. C. Titchmarsh and Kodaira gave a formula to obtain the density matrix $\rho_{jk}(\lambda)$ and completed the theory (Titchmarsh [6], Kodaira [29], Dunford and Schwartz [2]). This expansion theorem makes it possible to deduce in a unified manner expansion theorems for classical special functions, such as the †Fourier series expansion theorem, the expansions by †Hermite polynomials and †Laguerre polynomials, the †Fourier integral theorem, and various expansions in terms of †Bessel functions [6,7].

The relation between the coefficients of the differential operator and the spectral distribution of P is important in applications and is the subject of many papers [2,5,6].

For example, let $P(x,D) = -d^2/dx^2 + q(x)$ and $\Omega = (-\infty, \infty)$. If $q(x) \to \infty$ as $|x| \to \infty$, then $P(x,D)$ is essentially self-adjoint, the spectrum is entirely composed of the point spectrum, and a detailed estimate of the jth eigenvalue λ_j is also known [6]. If $q(x)$ converges rapidly to 0 as $|x| \to \infty$, then $P(x,D)$ is essentially self-adjoint, and there is only a continuous spectrum for $\lambda > 0$ and eigenvalues for $\lambda < 0$ with at most 0 as accumulation point. If $q(x)$ is a periodic function of x, then $P(x,D)$ is again essentially self-adjoint, and the spectrum consists of a continuous spectrum decomposed in a sequence of nonoverlapping intervals. The converse problem of determining $q(x)$ when the spectral measure is given has been studied by I. M. Gel'fand and B. M. Levitan.

P. Partial Differential Operators

The theory of eigenfunction expansion for partial differential operators with continuous spectra is not completed as it is for ordinary differential operators. What causes difficulties is that the solutions of $(P(x,D) - \lambda)u = 0$, which should be the generalized eigenfunctions, form an infinite-dimensional space, and except for special cases it is impossible to introduce convenient parameters in it. Many proofs are known for a general result that any self-adjoint elliptic operator has an eigenfunction expansion into generalized eigenfunctions of the form

$$u(x) = \int_{-\infty}^{\infty} \sum_{j=1}^{k(\lambda)} \varphi_j(x,\lambda)\, d\rho_j(\lambda) \int_{\Omega} \overline{\varphi_j(x,\lambda)}\, u(y)\, dy$$

[10]. There are few operators, however, for which we know how to construct $\varphi_j(x,\lambda)$ and the measure $d\rho_j(\lambda)$. The Fourier transform gives the expansion for operators with constant coefficients defined on the whole space. By means of a generalized form of the Fourier transform, Ikebe gave an eigenfunction expansion for the Schrödinger operator $-\Delta + q(x)$ in $\mathbf{R}^3$ under the condition that $q(x) \in L_2$ and $O(|x|^{-2-\varepsilon})$ as $|x| \to \infty$ [30]. Y. Shizuta, Mizohata, Lax and R. S. Phillips, N. A. Shenk, Ikebe, D. K. Fadeev, and others proved the same results for similar operators defined on an exterior domain with a bounded set deleted from a higher-dimensional Euclidean space. These theories are closely related to the †scattering theory of the Schrödinger equations $(-id/dt - P)u = 0$ or the wave equations $(d^2/dt^2 + P)u = 0$ associated with those operators.

Many of the problems in †quantum mechanics reduce to finding the spectral distribution of self-adjoint partial differential operators.

Q. Expansion Theorems for Non-Self-Adjoint Operators

A kind of eigenfunction expansion theorem may hold for non-self-adjoint differential operators or for non-Hilbert spaces. (See papers by the Russian school for ordinary differential operators and those by Browder, Agmon, and others for partial differential operators.)

R. Systems of Differential Operators

We have so far dealt with single differential operators which map functions u to functions f. A linear differential operator which maps a p-tuple $(u_1, \ldots, u_p)$ of functions to a q-tuple $(f_1, \ldots, f_q)$ of functions may be written

$$f_i(x) = \sum_{j=1}^{p} P_{ij}(x,D)u_j(x), \quad i = 1, \ldots, q,$$

where $P(x,D) = (P_{ij}(x,D))$ is a matrix of single differential operators. Such a matrix is called a **system of differential operators**. A system $P(x,D)$ is said to be **underdetermined** if $p > q$, **determined** if $p = q$, and **overdetermined** if $p < q$. Many propositions which hold for single operators hold for determined systems under appropriate conditions.

However, there is a fundamental difference between overdetermined (underdetermined) systems and determined systems, as is seen from the theory of several complex variables, which is the theory of a typical overdetermined system $\partial/\partial\bar{z} = (\partial/\partial\bar{z}_i)$. The theory is much more difficult for overdetermined (underdetermined) systems. The general theory of overdetermined and underdetermined systems with constant coefficients has been constructed by Ehrenpreis, Malgrange, V. Palamodov, and Hörmander for C^∞-functions and distributions. It has been extended by H. Komatsu to the case of hyperfunctions.

S. Symmetric Systems of the First Order

Determined systems of the first order are important in applications. Many problems in mathematical physics are formulated in terms of them. Also, single equations of higher order may be reduced to determined systems of the first order by regarding the derivatives as unknown functions. In some cases determined systems of the first order are easier to handle than single operators of higher order. In particular, a system of differential operators

$$P(x,D)=\sum A_i(x)\partial/\partial x_i+B(x)$$

is said to be a **symmetric positive system** if the matrices $A_i(x)$ and $B(x)$ satisfy the following conditions: $A_i^*=A_i$; $B+B^*+\sum\partial A_i/\partial x_i$ is positive semidefinite. Symmetric positive systems have been studied in detail by Friedrichs [31], Phillips [32], C. S. Morawetz, Lax, and others.

T. Pseudodifferential Operators

Pseudodifferential operators are natural extensions of differential operators. The investigation of pseudodifferential operators grew out of the investigation of †singular integral operators. Systematic study was first done by Kohn and Nirenberg [36] and soon followed by the research of Hörmander [37]. Then it became possible to avoid the delicate treatment of singular kernels in singular integral operators. Let $P(x,D_x)$ be a differential operator of the form (1), and let $u(x)$ be a function of class $C_0^\infty(\Omega)$. Then by means of the †Fourier inversion formula, $P(x,D_x)u(x)$ can be represented as

$$P(x,D_x)u(x)$$
$$=\frac{1}{(2\pi)^{n/2}}\int P(x,\xi)\,Fu(\xi)\exp(ix\xi)\,d\xi. \qquad (7)$$

When a polynomial $P(x,\xi)$ is extended to a function of a wider class, the operator $P(x,D_x)$ defined by (7) is called a **pseudodifferential operator** with the **symbol** $P(x,\xi)$. A symbol class is determined in accordance with various purposes, but it is always required that the corresponding operators have essential properties in common with partial differential operators. Hörmander [38] defined a symbol class $S^m_{\rho,\delta}(\Omega)$, $0<\rho$, $0\leqslant\delta$, for any real number m in the following way: Let $P(x,\xi)$ be a C^∞-function defined in $\Omega\times\mathbf{R}^n$. If for any multi-index α, β and any compact set $K\subset\mathbf{R}^n$, there exists a constant $C_{\alpha,\beta,K}$ such that

$$|D_x^\alpha D_\xi^\beta P(x,\xi)|\leqslant C_{\alpha,\beta,K}(1+|\xi|)^{m+\delta|\alpha|-\rho|\beta|},$$
$$x\in K,\quad \xi\in\mathbf{R}^n,$$

then $P(x,\xi)$ is said to be of class $S^m_{\rho,\delta}(\Omega)$, and the operator $P(x,D_x)$ defined by (7) is called a pseudodifferential operator (of order m) of class $S^m_{\rho,\delta}(\Omega)$. When $\Omega=\mathbf{R}^n$ and constants $C_{\alpha,\beta,K}=C_{\alpha,\beta}$ are independent of K, we denote $S^m_{\rho,\delta}(\mathbf{R}^n)$ simply by $S^m_{\rho,\delta}$, and set

$$S^{-\infty}=\bigcap_{-\infty<m<\infty}S^m_{1,0}\left(=\bigcap_{-\infty<m<\infty}S^m_{\rho,\delta}\right),$$

$$S^\infty_{\rho,\delta}=\bigcup_{-\infty<m<\infty}S^m_{\rho,\delta}.$$

Differential operators (1) with coefficients of class $\mathcal{B}$ (→ 173 Function Spaces) belong to $S^m_{1,0}$. The complex power $(1-\Delta)^{z/2}$ of $1-\Delta=1-\sum_{j=1}^n\partial^2/\partial x_j^2$ is defined as a pseudodifferential operator of class $S^{\operatorname{Re}z}_{1,0}$ by the symbol $(1+|\xi|^2)^{z/2}$. Operators of class $S^m_{\rho,\delta}$ are continuous maps of $\mathcal{S}$ into $\mathcal{S}$. Therefore, for any real s, the operator $(1-\Delta)^{s/2}$ can be uniquely extended to be a map of $\mathcal{S}'$ into $\mathcal{S}'$ by the relation

$$\langle(1-\Delta)^{s/2}u,v\rangle=\langle u,(1-\Delta)^{s/2}v\rangle,$$
$$u\in\mathcal{S}',\quad v\in\mathcal{S}.$$

For any $1\leqslant r<\infty$ and real s, the †Sobolev space $H^{s,r}$ is defined by

$$H^{s,r}=\{u\in\mathcal{S}'\,|\,(1-\Delta)^{s/2}u\in L_r(\mathbf{R}^n)\},$$

which is a Banach space provided with the norm $\|u\|_{s,r}=\|(1-\Delta)^{s/2}u\|_{L_r}$. In particular, $H^s=H^{s,2}$ is a Hilbert space with the norm $\|u\|_s=\|u\|_{s,2}$. Set

$$H^{-\infty,r}=\bigcup_{-\infty<s<\infty}H^{s,r},$$

$$H^{\infty,r}=\bigcap_{-\infty<s<\infty}H^{s,r},\qquad H^{-\infty}=H^{-\infty,2},$$

$$H^\infty=H^{\infty,2}.$$

Then $\mathcal{S}'\supset H^{-\infty}\supset\mathcal{E}'$, $H^{-\infty}\supset L_2(\mathbf{R}^n)\supset H^\infty(\subset\mathcal{B})$.

The theory of pseudodifferential operators has been mainly a study of the mappings of $H^{-\infty}$ into $H^{-\infty}$. Here, choosing the Hörmander class $S^m_{\rho,\delta}$ in the case $0\leqslant\delta<\rho\leqslant1$ as a model class, main properties of pseudodifferential operators are listed below (for a symbol class of non-C^∞-functions, see Friedrichs [39]. (1) Pseudolocal property. The operator P of class $S^m_{\rho,\delta}$, in general, does not have the **local property**: $u\in\mathcal{E}'\Rightarrow\operatorname{supp}Pu\subset\operatorname{supp}u$, but has the **pseudolocal property**: $u\in\mathcal{E}'\Rightarrow\operatorname{sing\,supp}Pu\subset\operatorname{sing\,supp}u$ [38]. (2) H^s-boundedness. For $P\in S^m_{\rho,\delta}$ and any real s there exists a constant C_s such that

$$\|Pu\|_s\leqslant C_s\|u\|_{s+m},\qquad u\in H^{s+m}$$

[38,40]. (3) A sharp form of Gårding's inequality. Let $P(x,\xi)=(P_{jk}(x,\xi);j,k=1,\dots,l)$ be a Hermitian symmetric and nonnegative matrix of $P_{j,k}(x,\xi)\in S^m_{\rho,\delta}$. Then there exists a constant C such that

$$\operatorname{Re}(P(x,D_x)u,u)\geqslant-C\|u\|^2_{(m-(\rho-\delta))/2},$$
$$u=(u_1,\dots,u_l)\in H^{m/2}.$$

Here $u\in H^s$ means

$$u_j\in H^s,\quad j=1,\dots,l,\quad \|u\|_s^2=\sum_{j=1}^l\|u_j\|_s^2$$

[40, 41, 42]. (4) Asymptotic expansion formulas. Let $P(x, D_x) \in S^m_{\rho,\delta}$ and $P_j(x, D_x) \in S^{m_j}_{\rho,\delta}$, $j = 1, 2$. Then there exist $P^*(x, D_x) \in S^m_{\rho,\delta}$ and $R(x, D_x) \in S^{m_1+m_2}_{\rho,\delta}$ such that $(P(x, D_x)u, v) = (u, P^*(x, D_x)v)$ for $u, v \in \mathfrak{S}$ and $R(x, D_x) = P_1(x, D_x)P_2(x, D_x)$. Furthermore, if we set

$$P^*_\alpha(x,\xi) = D^\alpha_x (iD_\xi)^\alpha \overline{P(x,\xi)},$$

$$R_\alpha(x,\xi) = (iD_\xi)^\alpha P_1(x,\xi) D^\alpha_x P_2(x,\xi),$$

then for any positive integer N, we have

$$P^*(x,D_x) - \sum_{|\alpha|<N} \frac{1}{\alpha!} P^*_\alpha(x,D_x) \in S^{m-(\rho-\delta)N}_{\rho,\delta},$$

$$R(x,D_x) - \sum_{|\alpha|<N} \frac{1}{\alpha!} R_\alpha(x,D_x) \in S^{m_1+m_2-(\rho-\delta)N}_{\rho,\delta}.$$

Hence the operator class $S^\infty_{\rho,\delta}$ is an algebra in the sense:

$$P \in S^m_{\rho,\delta}, \quad P_j \in S^{m_j}_{\rho,\delta}, \quad j = 1,2,$$
$$\Rightarrow P^* \in S^m_{\rho,\delta},$$
$$P_1 + P_2 \in S^{m_0}_{\rho,\delta} \quad (m_0 = \max(m_1, m_2)),$$
$$P_1P_2 \in S^{m_1+m_2}_{\rho,\delta}.$$

(5) Smoothing operators. The operator class $S^{-\infty}$ is an ideal in the class $S^\infty_{\rho,\delta}$, and elements of $S^{-\infty}$ are **smoothing operators** in the following sense: Let $P_{-\infty} \in S^{-\infty}$. Then for any $1 \leqslant r \leqslant q < \infty$, $P_{-\infty}$ maps $H^{-\infty,r}(\supset \mathcal{E}')$ into $H^{\infty,q}(\subset \mathfrak{B})$, and for any real numbers s_1, s_2, there exists a constant C_{r,q,s_1,s_2} such that $\|P_{-\infty}u\|_{s_2,q} \leqslant C_{r,q,s_1,s_2}\|u\|_{s_1,r}$, $u \in H^{s_1,r}$.

(6) Invariance under coordinate transformations. Assume that $0 \leqslant 1-\rho \leqslant \delta < \rho \leqslant 1$ [38]. Let $x(y) = (x_1(y), \ldots, x_n(y))$ be a C^∞-coordinate transformation from $\mathbf{R}^n_y$ onto $\mathbf{R}^n_x$ such that $\partial x_k(y)/\partial y_j \in \mathfrak{B}$, $j, k = 1, \ldots, n$, and $C^{-1} \leqslant |\det(\partial_y x(y))| \leqslant C$ for a constant $C > 0$, where $\det(\partial_y x(y))$ denotes the determinant of the Jacobian matrix $(\partial_y x(y)) = (\partial x_k(y)/\partial y_j)$. Then for any $P(x, D_x) \in S^m_{\rho,\delta}$ in $\mathbf{R}^n_x$, there exists a $Q(y, D_y) \in S^m_{\rho,\delta}$ in $\mathbf{R}^n_y$ such that

$$Q(y,D_y)w(y) = (P(x,D_x)u)(x(y)), \qquad w(y) \in \mathfrak{S}.$$

This enables us to define pseudodifferential operators on C^∞-manifolds. (7) Parametrix. An operator $E \in S^\infty_{\rho,\delta}$ is called a **left (right) parametrix** for $P \in S^m_{\rho,\delta}$ when $EP - I\,(PE - I)$ is of class $S^{-\infty}$. If E is a left and right parametrix for P, we call it a parametrix for P. For a differential operator P, the existence of a left (right) parametrix is a sufficient condition for P to be hypoelliptic (the equation $Pu = f \in \mathfrak{D}'$ is locally solvable) [38].

Concerning applications of the theory of pseudodifferential operators, there are many works, such as Atiyah and Bott (*Ann. of Math.*, 86 (1967)) on †Lefschetz fixed point formula, Friedrichs and Lax (*Comm. Pure Appl. Math.*, 18 (1965)) on symmetrizable systems, Hörmander [41] on subelliptic operators, Kumano-go (*Comm. Pure Appl. Math.*, 22 (1969)) on uniqueness in the Cauchy problems, and Mizohata and Ohya (*Publ. Res. Inst. Math. Sci.*, (A) 4 (1968)) on weakly hyperbolic equations. It should be noted that Hörmander [43] defined a more extended class of pseudodifferential operators called **Fourier integral operators**, which have no pseudolocal property, and, using such operators, obtained a best possible result for the asymptotic behavior of the spectral function of an elliptic operator. For literature on recent research see [39] and [44].

U. Fourier Integral Operators [50, 52, 57]

A **Fourier integral operator** B: $C_0^\infty(\mathbf{R}^n) \to \mathfrak{D}'(\mathbf{R}^n)$ is a locally finite sum of linear operators of the following type:

$$Af(x) = (2\pi)^{-(n+N)/2} \int_{R^{N+n}} a(x,\theta,y) \times \exp(i\varphi(x,\theta,y)) f(y)\, dy\, d\theta. \qquad (8)$$

Here $a(x,\theta,y)$ is a C^∞-function satisfying the inequality

$$|D^\alpha_x D^\beta_\theta D^\gamma_y a(x,\theta,y)| \leqslant C(1+|\theta|)^{m-\rho|\beta|+(1-\rho)(|\alpha|+|\gamma|)}$$

for some fixed m and ρ, $1 \geqslant \rho > 1/2$, and any triple of multi-indices α, β, γ, and $\varphi(x,\theta,y)$ is a real-valued C^∞-function which is homogeneous of degree 1 in θ for $|\theta| > 1$. The function φ is called the **phase function** and a the **amplitude function**.

Let $C_\varphi = \{(x,\theta,y) \mid d_\theta\varphi(x,\theta,y) = 0, \theta \neq 0\}$ and $W = \{(x,y) \in \mathbf{R}^n \times \mathbf{R}^n \mid \exists\theta \neq 0$ such that $(x,\theta,y) \in C_\varphi\}$. If $d_{x,\theta,y}\varphi(x,\theta,y) \neq 0$ for $\theta \neq 0$, then the kernel distribution $k(x,y)$ of A is of class C^∞ outside W. There have been detailed studies of the case where the $d_{x,\theta,y}(\partial\varphi(x,\theta,y)/\partial\theta_j)$, $j = 1, 2, \ldots, N$, are linearly independent at every point of C_φ. In this case, C_φ is a smooth manifold in $\mathbf{R}^{n+N+n}$ and the mapping Φ: $C_\varphi \ni (x,\theta,y) \to (x,y,\xi,\eta)$, $\xi = d_x\varphi(x,\theta,y)$, $\eta = d_y\varphi(x,\theta,y)$, is an immersion of C_φ to $T^*(\mathbf{R}^n \times \mathbf{R}^n)\backslash 0$, the cotangent bundle of $\mathbf{R}^n \times \mathbf{R}^n$ minus its zero section. The image $\Phi C_\varphi = \Lambda_\varphi$ is a **conic Lagrange manifold**, i.e., the canonical 2-form $\sigma = \sum_j d\xi_j \wedge dx_j - \sum_j d\eta_j \wedge dy_j$ vanishes on Λ_φ and the multiplicative group of positive numbers acts on Λ_φ. Let $\lambda_1, \lambda_2, \ldots, \lambda_{2n}$ be a system of local coordinates in Λ_φ. These, together with $\partial\varphi/\partial\theta_1$, $\partial\varphi/\partial\theta_2, \ldots, \partial\varphi/\partial\theta_N$, constitute a system of local coordinate functions of $\mathbf{R}^{n+N+n}$ in a neighborhood of C_φ. Let J denote the

Jacobian determinant

$$\frac{D\left(\lambda_1,\lambda_2,\ldots\lambda_{2n},\frac{\partial\varphi}{\partial\theta_1},\ldots,\frac{\partial\varphi}{\partial\theta_N}\right)}{D(x,\theta,y)}.$$

The function $a_{\Lambda_\varphi}=\sqrt{J}\, a_{|C_\varphi}\Phi^{-1}$ is called the **symbol** of A. Here $a_{|C_\varphi}$ is the restriction of a to C_φ. The conic Lagrange manifold $\Lambda_\varphi=\Lambda_\varphi(A)$ and the symbol $a_{\Lambda_\varphi}=a_{\Lambda_\varphi}(A)$ essentially determine the singularity of the kernel distribution $k(x,y)$ of the Fourier integral operator A. Conversely, given a conic Lagrange manifold Λ in $T^*(\mathbf{R}^n\times\mathbf{R}^n)\backslash 0$ and a function a_Λ on it, one can construct a Fourier integral operator A such that $\Lambda_\varphi(A)=\Lambda$ and $a_{\Lambda_\varphi}(A)=a_\Lambda$. Those Fourier integral operators whose associated conic Lagrange manifolds are the graphs of homogeneous canonical transformations of $T^*(\mathbf{R}^n)$ are most frequently used in the theory of linear partial differential equations. Let A be a Fourier integral operator such that $\Lambda_\varphi(A)$ is the graph of a homogeneous canonical transformation χ. Then the adjoint of A is a Fourier integral operator such that the associated conic Lagrange manifold is the graph of the inverse transformation χ^{-1}. Let A_1 be another such operator; if $\Lambda_\varphi(A_1)$ is the graph of χ_1, then the composed operator A_1A is also a Fourier integral operator and $\Lambda_\varphi(A_1A)$ is the graph of the composed homogeneous canonical transformation $\chi_1\chi$.

A pseudodifferential operator of class $S^m_{\rho,1-\rho}(\mathbf{R}^n)$ is a particular type of Fourier integral operator. In fact, a Fourier integral operator A is a pseudodifferential operator of class $S^\infty_{\rho,1-\rho}(\mathbf{R}^n)$ if and only if $\Lambda_\varphi(A)$ is the graph of the identity mapping of $T^*(\mathbf{R}^n)$. Hence for any Fourier integral operator A, A^*A and AA^* are pseudodifferential operators.

The following theorem is due to Egorov [53]: Let $P(x,D)$ and $Q(x,D)$ be pseudodifferential operator of class $S^m_{\rho,1-\rho}(\mathbf{R}^n)$ with the symbols $p(x,\xi)$ and $q(x,\xi)$, respectively, and let A be a Fourier integral operator such that the associated conic Lagrange manifold $\Lambda_\varphi(A)$ is the graph of a homogeneous canonical transformation χ of $T^*(\mathbf{R}^n)$. If the equality $P(x,D)A=AQ(x,D)$ holds, then $q(x,\xi)-p(\chi(x,\xi))$ belongs to the class $S^{m+1-2\rho}_{\rho,1-\rho}(\mathbf{R}^n)$.

Assume that $m=1$, $\rho=1$, and that $p_1(x,\xi)$ is a real-valued C^∞-function, homogeneous of degree 1 in ξ for $|\xi|>1$, such that $p(x,\xi)-p_1(x,\xi)\in S^0_{1,0}(\mathbf{R}^n)$ and $d_\xi p_1(x^0,\xi^0)\neq 0$ at (x^0,ξ^0) where $p_1(x^0,\xi^0)=0$. Then one can find a Fourier integral operator A such that the function $q(x,\xi)$ of Egorov's theorem satisfies the relation $q(x,\xi)-\xi_1\in S^0_{1,0}(\mathbf{R}^n)$.

The boundedness of Fourier integral operators in the space $L_2(\mathbf{R}^n)$ (or the spaces $H^s(\mathbf{R}^n)$) has also been studied in several cases. Some sufficient conditions for boundedness can be found in [50, 56, 59].

The theory of Fourier integral operators has its origin in the asymptotic representation of solutions of the wave equation. (see, e.g., [57, 60, 61, 62, and 63]. Eskin [56] used a type of Fourier integral operator in deriving the energy estimates and constructing the fundamental solutions for strict hyperbolic operators. Hörmander [48] introduced the term "Fourier integral operators," and applied these operators to the derivation of highly accurate asymptotic formulas for spectral functions of elliptic operators. Egorov applied his theorem and the corollary stated above to the study of hypoellipticity and local solvability of nondegenerate pseudodifferential operators [54]. Using Egorov's theorem and the same corollary, Nirenberg and Trèves [55] obtained decisive results concerning local solvability of nondegenerate linear partial differential equations; these results were completed by Beals and Fefferman [65]. Hörmander and Duistermaat [50, 51] constructed a general global theory of Fourier integral operators making use of Maslov's theory [57], which was originally published in 1965. By virtue of these researches, the Fourier integral operator has come to be recognized as a powerful tool in the theory of linear partial differential equations. An interesting application of the global theory of Fourier integral operators appears in [58].

The works of Egorov, Nirenberg and Trèves, and Hörmander motivated the theory of hyperfunctions developed by M. Sato and gave rise to the concept of quantized contact transformations, which correspond to Fourier integral operators in the theory of distributions. The above-stated transformation theorem of Egorov has been studied in detail with reference to systems of pseudodifferential equations with analytic coefficients and one unknown function [64].

References

[1] K. Yosida, Functional analysis, Springer, 1965.
[2] N. Dunford and J. T. Schwartz, Linear operators pt. II, Interscience, 1963.
[3] R. Courant and D. Hilbert, Methods of mathematical physics, Interscience, I, 1953; II, 1962.
[4] L. Hörmander, Linear partial differential operators, Springer, 1963.

[5] М. А. Наймарк (M. A. Naĭmark), Линейные Дифференциальные операторы, Гостехиздат, 1954; English translation, Linear differential operators I, II, Ungar, 1967.
[6] E. C. Titchmarch, Eigenfunction expansions associated with second-order differential equations, Claredon Press, I, revised edition, 1962; II, 1958.
[7] K. Yosida, Sekibun hôteisiki ron (Japanese; Theory of integral equations), Iwanami, 1950; English translation, Lectures on differential and integral equations, Interscience, 1960.
[8] F. John, Plane waves and spherical means applied to partial differential equations, Interscience, 1955.
[9] С. Д. Эйдельман (S. D. Èĭdel'man), Параболические системы, Наука, 1964; English translation, Parabolic systems, North-Holland, 1969.
[10] И. М. Гельфанл, Н. Я. Виленкин (I. M. Gel'fand and N. Ja. Vilenkin), Обобщенные Функции IV, Физматгиз, 1961; English translation, Generalized functions IV. Applications of harmonic analysis, Academic Press, 1964.
[11] A special issue on partial differential equations, Sûgaku, 10, no. 4 (1959).
[12] Partial differential equations, Proc. Symposia in Pure Mathematics IV, Amer. Math. Soc., 1961.
[13] J. Leray, Hyperbolic differential equations, Lecture notes, Institute for Advanced Study, Princeton, 1953.
[14] S. Mizohata, Hypoellipticité des équations paraboliques, Bull. Soc. Math. France, 85 (1957), 15–50.
[15] J. Leray, Problème de Cauchy I–V, Bull. Soc. Math. France, 85 (1957), 389–429, 86 (1958), 75–96, 87 (1959), 81–180, 90 (1962), 39–156, 92 (1964), 263–361.
[16] B. Malgrange, Existence et approximation des solutions des équations aux derivées partielles et des équations de convolution, Ann. Inst. Fourier, 6 (1955–1956), 271–355.
[17] L. Ehrenpreis, Sheaves and differential equations, Proc. Amer. Math. Soc., 7 (1956), 1131–1138.
[18] I. G. Petrovskiĭ, Sur l'analyticité des solutions des systèmes d'équations différentielles, Mat. Sb., 5 (47) (1939), 3–70.
[19] L. Hörmander, On the theory of general partial differential operators, Acta Math., 94 (1955), 161–248.
[20] J. Peetre, Théorèmes de régularité pour quelques classes d'opérateurs différentiels, Medd. Lunds Univ. Mat. Sem., 16 (1959), 1–122.
[21] T. Ikebe and T. Kato, Uniqueness of the self-adjoint extension of singular elliptic differential operators, Arch. Rational Mech. Anal., 9 (1962), 77–92.
[22] М. И. Вишик (M. I. Višik), Об общих краевых задачах для эллиптических дифференциальных уравнений, Trudy Moskov. Mat. Obšč., 1 (1952), 187–246; English translation, On general boundary problems for elliptic differential equations, Amer. Math., Soc. Transl., (2) 24 (1963), 107–172.
[23] J.-L. Lions and E. Magenes, Problèmes aux limites non homogènes VI, J. Analyse Math., 11 (1963), 165–188.
[24] M. Schechter, On L^p estimates and regularity I, II, Amer. J. Math., 85 (1963), 1–13; Math. Scand., 13 (1963), 47–69.
[25] L. Gårding, Dirichlet's problem for linear elliptic partial differential equations, Math. Scand., 1 (1953), 55–72.
[26] S. Agmon, A. Douglis, and L. Nirenberg, Estimates near the boundary for solutions of elliptic partial differential equations satisfying general boundary conditions I, II, Comm. Pure Appl. Math., 12 (1959), 623–727; 17 (1964), 35–92.
[27] L. Hörmander, On the regularity of the solutions of boundary problems, Acta Math., 99 (1958), 225–264.
[28] J. J. Kohn, A priori estimates in several complex variables, Bull. Amer. Math. Soc., 70 (1964), 739–745.
[29] K. Kodaira,, On ordinary differential equations of any even order and the corresponding eigenfunction expansions, Amer. J. Math., 72 (1950), 502–544.
[30] T. Ikebe, Eigenfunction expansions associated with the Schroedinger operators and their applications to scattering theory, Arch. Rational Mech. Anal., 5 (1960), 1–34.
[31] K. O. Friedrichs, Symmetric positive linear differential equations, Comm. Pure Appl. Math., 11 (1958), 333–418.
[32] R. S. Phillips, Dissipative operators and parabolic partial differential equations, Comm. Pure Appl. Math., 12 (1959), 249–276.
[33] C. B. Morrey, Some recent developments in the theory of partial differential equations, Bull. Amer. Math. Soc., 68 (1962), 279–297.
[34] L. Schwartz, Some applications of the theory of distributions, Lectures on modern mathematics, John Wiley, 1963, vol. 1, p. 23–58.
[35] L. Nirenberg, Partial differential equations with applications in geometry, Lectures on modern mathematics, John Wiley, 1964, vol. 2, p. 1–41.
[36] J. J. Kohn and L. Nirenberg, An algebra of pseudo-differential operators, Comm. Pure Appl. Math., 18 (1965), 269–305.

[37] L. Hörmander, Pseudo-differential operators, Comm. Pure Appl. Math., 18 (1965), 501–517.
[38] L. Hörmander, Pseudo-differential operators and hypoelliptic equations, Singular Integrals, Proc. Symposia in Pure Mathematics IX, Amer. Math. Soc., 10 (1967), 138–183.
[39] K. O. Friedrichs, Pseudo-differential operators, Lecture notes, Courant Institute, 1968.
[40] H. Kumano-go, Algebras of pseudo-differential operators, J. Fac. Sci. Univ. Tokyo, 17 (1970), 31–50.
[41] L. Hörmander, Pseudo-differential operators and non-elliptic boundary problems, Ann. of Math., (2) 83 (1966), 129–209.
[42] P. D. Lax and L. Nirenberg, On stability for difference schemes, a sharp form of Gårding's inequality, Comm. Pure Appl. Math., 19 (1966), 473–492.
[43] L. Hörmander, The spectral function of an elliptic operator, Acta Math., 121 (1969), 193–218.
[44] Singular Integrals, Proc. Symposia in Pure Mathematics X, Amer. Math. Soc., 1967.
[45] A. Friedman, Generalized functions and partial differential equations, Prentice-Hall, 1963.
[46] L. Ehrenpreis, Fourier analysis in several complex variables, John Wiley, 1970.
[47] В. П. Паламодов (V. P. Palamodov), Линейные дифференциалные операторы с постоянными коэффициентами, Наука 1967; English translation, Linear differential operators with constant coefficients, Springer, 1970.
[48] L. Hörmander, Fourier integral operators I, Acta Math., 127 (1971), 79–183; II (with J. J. Duistermaat), ibid., 128 (1972), 183–269.
[49] L. Nirenberg (ed.,) Pseudo-differential operators, C.I.M.E. (1968), Rome.
[50] L. Hörmander, Fourier integral operators I, Acta Math., 127 (1971), 79–183.
[51] L. Hörmander and J. J. Duistermaat, Fourier integral operators II, Acta Math., 128 (1972), 183–269.
[52] J. J. Duistermaat, Fourier integral operators, Lecture notes, Courant Institute, 1973.
[53] Ju. V. Egorov, On canonical transformation of pseudo-differential operators. Usp. Mat. Nauk, 25 (1969), 235–236.
[54] Ju. V. Egorov, On non-degenerate hypoelliptic pseudo-differential operators. Soviet Math. Dokl., 10 (1969), 697–699.
[55] L. Nirenberg and F. Trèves, On local solvability of linear partial differential equations I, Comm. Pure Appl. Math., 23 (1970), 1–38; II, ibid., 459–510.
[56] G. I. Eskin, The Cauchy problem for hyperbolic systems in convolutions, Math. USSR-Sb., 3 (1967), 243–277.
[57] V. P. Maslov, Théorie des perturbations et méthodes asymptotiques. Gauthier-Villars, 1972.
[58] J. Chazarain, Formule de Poisson pour les variétés riemannienes, Inventiones Math., 24 (1974), 65–82.
[59] D. Fujiwara, On the boundedness of integral transformations with highly oscillatory kernels, Proc. Japan Acad., 51 (1975), 96–99.
[60] P. D. Lax, Asymptotic solutions of oscillatory initial value-problems, Duke Math. J., 24 (1957), 627–646.
[61] D. Ludwig, Uniform asymptotic expansions at a caustic, Comm. Pure Appl. Math., 19 (1966), 215–250.
[62] D. Ludwig, Uniform asymptotic expansions of the field scattered by a convex object at high frequencies, Comm. Pure Appl. Math., 20 (1967), 103–138.
[63] A. Sommerfeld, Optics, Lectures on theoretical physics 4, Academic Press, 1954.
[64] H. Komatsu (ed.), Hyper-functions and pseudo-differential equations, Lecture notes in math. 287, Springer, 1973.
[65] R. Beals and C. Fefferman, On local solvability of linear partial differential equations, Ann. of Math., (2) 97 (1973), 482–498.

116 (III.18) Differential Rings

Let R be a †commutative ring with a unity element 1. If a mapping δ of R into R is such that for any pair x, y of elements of R, (i) $\delta(x+y)=\delta x+\delta y$, and (ii) $\delta(xy)=\delta x\cdot y+x\cdot\delta y$, then δ is called a **derivation** (or **differentiation**) in R. A ring R provided with a finite number of mutually commutative differentiations in R is called a **differential ring**. In this article we consider only the case where R contains a subfield that has the unity element in common with R. In particular, if R is a field, we call it a **differential field**.

In the above definition of differential ring, it is not necessary to mention the †characteristic of the subfield contained in R. However, to make it more effectively applicable in the case of nonzero characteristics, we may define differential rings using *higher differentiation* in place of the differentiation defined above. If a sequence $\delta=\{\delta_\nu\}$ of mappings δ_0, δ_1, $\delta_2, \ldots$ of R into R satisfies conditions (i)–(iv) for any pair x, y of elements of R and any pair λ, μ of nonnegative integers, then δ is called a **higher differentiation** in R: (i) $\delta_\lambda(x+y)=\delta_\lambda x+\delta_\lambda y$; (ii) $\delta_\lambda(xy)=\Sigma\delta_\alpha x\cdot\delta_\beta y$ (the addition is performed over all pairs α, β of nonnegative integers that satisfy $\alpha+\beta=\lambda$); (iii) $\delta_\lambda(\delta_\mu x)=\binom{\lambda+\mu}{\lambda}\delta_{\lambda+\mu}x$; (iv) $\delta_0 x=x$. Two higher differentiations $\delta=\{\delta_\nu\}$ and $\delta'=\{\delta'_\nu\}$ are said to be commutative if and only if δ_λ and δ'_μ commute for all pairs λ, μ of nonnegative integers. Higher differentiations were introduced by H. Hasse (1935) for the study of the field of †algebraic functions of one variable in the case of nonzero characteristics.

These two definitions of differential rings coincide if the characteristic of R is zero. For simplicity, we shall restrict ourselves to that case.

Let $\delta_1, \ldots, \delta_m$ be the differentiations of the differential ring R. If x is an element of R, $\delta_1^{s_1}\delta_2^{s_2}\ldots\delta_m^{s_m}x$ ($s_1, s_2, \ldots, s_m$ are nonnegative integers) is called a **derivative** of x. We call x constant if and only if $\delta_1 x=\ldots=\delta_m x=0$. An †ideal $\mathfrak{a}$ of R with $\delta_i\mathfrak{a}\subset\mathfrak{a}$ $(i=1,2,\ldots,m)$ is called a **differential ideal** of R. If it is a †prime ideal (**semiprime ideal** (i.e., an ideal containing all those elements x that satisfy $x^g\in\mathfrak{a}$ for some natural number g)), then $\mathfrak{a}$ is called a **prime differential ideal (semiprime differential ideal)**. A subring S of R with $\delta_i S\subset S$ can be regarded as a differential ring with respect to the differentiations $\delta_1, \ldots, \delta_m$. We call S a **differential subring** of R and R a **differential extension ring** of S.

Let $X_1, \ldots, X_n$ be elements of a differential extension field of a differential field K with the differentiations $\delta_1, \ldots, \delta_m$, and let $\delta_1^{s_1}\delta_2^{s_2}\ldots\delta_m^{s_m}X_i$ $(s_1\geqslant 0, \ldots, s_m\geqslant 0, 1\leqslant i\leqslant n)$ be †algebraically independent over k. The totality of their polynomials over K, which forms a differential ring, is called the **ring of differential polynomials** of the **differential variables** $X_1, \ldots, X_n$ over K, and is denoted by $K\{X_1, \ldots, X_n\}$. Its elements are called **differential polynomials**. For this ring of differential polynomials we have an analog of †Hilbert's basis theorem in the ring of ordinary polynomials, **Ritt's basis theorem**: If we are given any set $\mathfrak{M}$ of differential polynomials of $X_1, \ldots, X_n$ over K, we can choose a finite number of differential polynomials $P_1, \ldots, P_r$ from $\mathfrak{M}$ such that each element Q of $\mathfrak{M}$ has an integral power Q^g equal to a linear combination of $P_1, \ldots, P_r$ and their derivatives, where the coefficients of the linear combination are elements of $K\{X_1, \ldots, X_n\}$. This theorem implies that in the ring of differential polynomials, every semiprime differential ideal can be expressed as the intersection of a finite number of prime differential ideals; if this expression is †irredundant, it is unique (→ 70 Commutative Rings).

The equation obtained by equating a differential polynomial to zero is called an **algebraic differential equation**. Concerning these equations, we are able to use methods similar to those used in †algebraic geometry in studying the usual algebraic equations. J. Ritt made interesting studies on solutions of algebraic differential equations by such methods, principally in the case when the ground field K consists of †meromorphic functions.

Since that time, basic study of differential rings and fields has been fairly well organized and has developed into theories such as the two following:

(1) **Picard-Vessiot theory**. This is a classical theory of †linear homogeneous differential equations originated by E. Picard and E. Vessiot; it resembles the †Galois theory concerning algebraic equations. The Galois group in this case is a linear group, and its structure characterizes the solution of the differential equation. E. Kolchin introduced the general concept of the **Picard-Vessiot extension field** of a differential field and studied in detail the group of **differential automorphisms** (i.e., the group of all those automorphisms that commute with the differentiations), thus making the classical theory more precise and more general.

(2) **Galois theory of differential fields**. Generalizing the concept of the Picard-Vessiot extension, Kolchin introduced the notion of the **strongly normal extension field** and established the Galois theory for such extensions. In this theory, we see that the Galois group is an †algebraic group relative to a †universal domain over the field of constants of the ground field. Conversely, every algebraic group is the Galois group of a strongly normal extension field. We also see that a strongly normal extension can be decomposed, in a certain sense, into a Picard-Vessiot extension and an Abelian extension (i.e., an extension whose Galois group is an †Abelian variety) (→ Kolchin [2–5], Okugawa [6]).

References

[1] I. Kaplansky, An introduction to differential algebra, Actualités Sci. Ind., Hermann, 1957.

[2] E. R. Kolchin, Algebraic matric groups and the Picard-Vessiot theory of homogeneous ordinary linear differential equations, Ann. of Math., (2) 49 (1948), 1–42.

[3] E. R. Kolchin, Galois theory of differential fields, Amer. J. Math., 75 (1953), 753–824.
[4] E. R. Kolchin, On the Galois theory of differential fields, Amer. J. Math., 77 (1955), 868–894.
[5] E. R. Kolchin, Abelian extensions of differential fields, Amer. J. Math., 82 (1960), 779–790.
[6] K. Okugawa, Basic properties of differential fields of an arbitrary characteristic and the Picard-Vessiot theory, J. Math. Kyoto Univ., 2 (1963), 295–322.
[7] J. F. Ritt, Differential algebra, Amer. Math. Soc. Colloq. Publ., 1950.

117 (IX.23) Differential Topology

Differential topology may be defined as the study of those properties of †differentiable manifolds that are invariant under †diffeomorphisms. Some of the remarkable contributions, such as H. Whitney's embedding theorem [1], S. S. Cairns's triangulation theorem [2], and J. H. C. Whitehead's theorem on regular neighborhoods [3] were made in the 1930s. In the late 1950s, outstanding results were obtained by R. Thom, J. Milnor, S. Smale, M. Kervaire, E. C. Zeeman, and B. Mazur, among others. Differential topology thus became a new, fascinating branch of mathematics (→ 211 Immersions and Embeddings; 259 Manifolds; 410 Topology of Differentiable Manifolds).

The basic objects studied in this field are the topological, combinatorial, and differentiable structures of manifolds and the relationships between combinatorial and differentiable manifolds. Notions from differential geometry such as †connections, †curvature, and †geodesics are not involved here.

The fundamental role in †topology, once played by continuous mappings, is now played by †differentiable mappings. Starting from Whitney's result [1] on the †singularities of mappings of a Euclidean space into another Euclidean space, J. H. C. Whitehead, Smale, Thom, J. C. Moore, and others have made rapid progress in the study of singularities of mappings of a differentiable manifold into another one, †transversal regular mappings, raising the differentiability of mappings, and the like.

Regarding the relationship between combinatorial and differential structures it was shown by Cairns and Whitehead that every differentiable manifold has a C^1-triangulation. On the other hand, according to Kervaire [6], a †combinatorial manifold may have no differentiable structure at all. Moreover, a differentiable manifold may have several essentially distinct differentiable structures, even though they are homeomorphic. Actually, Milnor showed that there exists a differentiable 7-dimensional manifold that is combinatorially equivalent to the 7-dimensional sphere S^7 but not diffeomorphic to S^7. Analogous examples were also given by I. Tamura and N. Shimada.

In the work on differentiable structures of †homotopy spheres, Milnor and Kervaire showed that the †h-cobordism classes of homotopy n-spheres form an Abelian group θ_n under the †connected sum operation (→ 410 Topology of Differentiable Manifolds; Appendix A, Table 6.I). According to Smale [9], θ_n may well be considered as classes of (oriented) differentiable structures of n-spheres for $n \geqslant 5$ (→ 410 Topology of Differentiable Manifolds). Thom then reduced †cobordism classification to a problem of homotopy theory, namely, to a problem concerning †stable homotopy groups of some spaces. Concerning this, there are several important problems, such as those of †parallelizability, †almost parallelizability of manifolds, and †π-manifolds. J. F. Adams obtained results concerning †tangent r-frame bundles by means of †K-theory. Furthermore, Smale and Zeeman have answered the †generalized Poincaré conjecture affirmatively. Thus in the last decade many important results have been obtained in which such notions from algebraic topology as †homotopy groups have been used extensively with success.

Finally, we mention remarkable contributions by Cairns, A. Haefliger, M. Brown, R. H. Bing, and others to the problems of smoothing combinatorial manifolds and the combinatorial, locally flat, and tame †embedding of manifolds.

References

[1] H. Whitney, Differentiable manifolds, Ann. of Math., (2) 37 (1936), 645–680.
[2] S. S. Cairns, Triangulation of the manifold of class one, Bull. Amer. Math. Soc., 41 (1935), 549–552.
[3] J. H. C. Whitehead, On the homotopy type of manifolds, Ann. of Math., (2) 41 (1940), 825–832.
[4] R. Thom, Quelques propriétés globales des variétés différentiables, Comment. Math. Helv., 28 (1954), 17–86.
[5] J. W. Milnor, On manifolds homeomorphic to the 7-sphere, Ann. of Math., (2) 64 (1956), 399–405.

[6] M. A. Kervaire, A manifold which does not admit any differentiable structure, Comment. Math. Helv., 34 (1960), 257–270.
[7] J. W. Milnor, Differential topology, Lecture notes, Princeton, 1958.
[8] J. W. Milnor, Sommes de variétés différentiables et structures différentiables des sphères, Bull. Soc. Math. France, 87 (1959), 439–444.
[9] S. Smale, On the structures of manifolds, Amer. J. Math., 84 (1962), 387–399.
[10] S. Smale, Generalized Poincaré's conjecture in dimensions greater than four, Ann. of Math., (2) 74 (1961), 391–406.
[11] J. R. Stallings, Polyhedral homotopy-spheres, Bull. Amer. Math. Soc., 66 (1960), 485–488.
[12] J. F. Adams, Vector fields on spheres, Ann. of Math., (2) 75 (1962), 603–632.
[13] J. R. Munkres, Obstructions to the smoothing of piecewise differentiable homeomorphisms, Bull. Amer. Math. Soc., 65 (1959), 332–334.
[14] E. C. Zeeman, The generalized Poincaré conjecture, Bull. Amer. Math. Soc., 67 (1961), 270.
[15] A. Haefliger, Plongements differéntiables de variétés dans variétés, Comment. Math. Helv., 36 (1961), 47–82.
[16] S. S. Cairns, The manifold smoothing problem, Bull. Amer. Math. Soc., 67 (1961), 237–238.
[17] J. H. C. Whitehead, Manifolds with transverse fields in Euclidean space, Ann. of Math., (2) 73 (1961), 154–212.
[18] R. H. Bing, Locally tame sets are tame, Ann. of Math., (2) 59 (1954), 145–158.
[19] M. Brown, A proof of the generalized Schoenflies theorem, Bull. Amer. Math. Soc., 66 (1960), 74–76.
[20] B. C. Mazur, On imbeddings of spheres, Bull. Amer. Math. Soc., 65 (1959), 59–65.
[21] A. H. Wallace, Differential topology, Benjamin, 1968.

118 (XIX.20) Diffusion

A. General Remarks

Diffusion is a physical phenomenon that is illustrated by the following example. Suppose that a vessel containing water is separated into two parts by a wall. One part is colored by a certain dye and the other part is not. When the wall is quietly removed without causing flow of water, a process of mixing will occur because dye molecules migrate by irregular thermal motion, so that the whole vessel of water will become uniformly colored after some time. Another typical example of diffusion, which can be directly observed under a microscope, is the †Brownian motion of colloidal particles suspended in liquid.

When a great number of particles are under observation, the diffusion process is described by the "concentration" $C(\mathbf{r},t)$ of particles at a point $\mathbf{r}$ at time t. If a single particle is traced, the process $\mathbf{r}(t)$ representing its motion is considered a †stochastic process, so that the ensemble of such Brownian particles is described by a †probability density function $P(\mathbf{r},t)$ defined as the probability density that the particle is found at position $\mathbf{r}$ at time t. Both $C(\mathbf{r},t)$ and $P(\mathbf{r},t)$ are considered to satisfy an equation of the type

$$\frac{\partial}{\partial t}f(\mathbf{r},t)=\operatorname{div}(D\operatorname{grad}f(\mathbf{r},t)),$$

called the **diffusion equation**. D is the **diffusion constant**, and $j=-D\operatorname{grad}f$ is the particle flow or the probability flow. This relation between j and $\operatorname{grad} f$ is called **Fick's law**. The diffusion constant is determined by the nature of the particle, the interaction of the particle with surrounding molecules, and temperature. For a 1-dimensional diffusion process the diffusion equation takes the form

$$\frac{\partial f(x,t)}{\partial t}=D\frac{\partial^2 f(x,t)}{\partial x^2},$$

a †parabolic partial differential equation. The †fundamental solution of this equation,

$$f(x,t)=\frac{1}{(4\pi Dt)^{1/2}}\exp\left(-\frac{x^2}{4Dt}\right),$$

represents diffusion of particles that are concentrated at the origin at time $t=0$.

Molecular processes of diffusion are generally complex. But two typical idealizations may be considered. One is diffusion in dilute gases: A given molecule will repeatedly collide with other molecules, change the direction of its motion, and follow a zig-zag trajectory. This is a **random walk** or **random flight process**. Let l be the mean free path between two successive collisions and v the velocity of the molecule. Then a simple theory gives $D=lv/3$. The other idealization of diffusion is Brownian motion, in which the Brownian particle receives incessant irregular impulses from surrounding liquid molecules and at the same time experiences a frictional force $-\eta v$ (v is the velocity). If the absolute temperature of the medium is T, the **Einstein relation** $D=kT/\eta$ holds. For the first type of molecular process, the diffusion equation is not exact in the spatial and temporal scales of molecular motion, but becomes asymptotically correct in a large scale in space and in a long scale of time. This corresponds to describing diffusion as a †normal Markov process.

B. Random Walks

K. Pearson presented the following problem in 1905: Suppose that a walker starts from the origin 0 and walks a distance a, then changes the direction at random and repeats this process n times. What is the probability density that he finds himself at the distance $\mathbf{r}$ from the origin? This problem and similar problems are called **random walk problems**. Lord Rayleigh [1] treated the same problem in 1880 when he considered the intensity resultant of a number of sound waves with different random phases.

In one dimension the probability of finding the walker at a distance x from the origin is given by the †binomial distribution,

$$P_n = \frac{n!}{2^n}\left(\frac{n+x}{2}!\,\frac{n-x}{2}!\right)^{-1}, \tag{4}$$

if the length of each step is taken as unity. In two dimensions [2], the required probability density is given by

$$P_n = r\int_0^{\infty} J_1(rt)(J_0(at))^n\,dt.$$

In three dimensions [3], it is given by

$$P_n(r) = \frac{r}{2a^2}\frac{n^{n-2}}{(n-2)!} \times \sum_{s=0}^{k}(-)^s\binom{n}{s}\left(\frac{1}{2}\left(1-\frac{r}{na}\right)-\frac{s}{n}\right)^{n-2}, \tag{5}$$

where

$$\frac{k}{n} \leqslant \frac{1}{2}\left(1-\frac{r}{na}\right) < \frac{k+1}{n}.$$

As long as no †correlation exists between successive walks, these problems deal only with the probability distribution of a sum of †independent random quantities. If there is correlation, the problem is concerned with a probability distribution of a sum of random quantities which make a †Markov process. Generally, we are interested in the asymptotic behavior for large n. The expectation of r^2 is of the order of na^2. The †central limit theorem holds for r that satisfy $r \sim \sqrt{n}\,a \ll na$. For example, expression (4) becomes

$$P_n(x) \sim (1/\sqrt{2\pi n}\,)\exp(-x^2/2n),$$

which is easily obtained by using †Stirling's formula. In three dimensions, the probability density is asymptotically

$$\frac{1}{(2\pi na^2)^{3/2}}\exp\left(-\frac{n(x^2+y^2+z^2)}{2a^2}\right).$$

The theory of random walks is applied to many problems in physics, such as Brownian motion, diffusion of gaseous molecules, composition of waves, and statistical mechanics of polymers and 1-dimensional substances [4].

In some cases the space for a random walker may be limited; for example, he may be confined by walls, or he may not return after passing a given boundary. These conditions generalize the random walk problems. The first passage problem is also related to such problems. These generalizations are important in the general theory of stochastic processes [5].

References

[1] Lord Rayleigh, The theory of sound, Macmillan, second revised edition, 1929, vol. 1, p. 36.
[2] G. N. Watson, A treatise on the theory of Bessel functions, Cambridge Univ. Press, 1922, p. 419.
[3] L. R. G. Treloar, The physics of rubber elasticity, Oxford Univ. Press, 1949, p. 100.
[4] S. Chandrasekhar, Stochastic problems in physics and astronomy, Rev. Mod. Phys., 15 (1943), 2–89.
[5] W. Feller, Introduction to probability theory and its applications, John Wiley, second edition, I, 1957; II, 1966.

119 (XVI.9) Diffusion Processes

A. General Remarks

Let $(\Omega, \mathfrak{B}, P)$ be a †probability space. A †Markov process $\{X_t\}_{0\leqslant t<\infty}$ on a topological space S with †continuous time parameter t is called a **diffusion process** if the †sample function $X_t(\omega)$ is continuous in t with probability 1 (i.e., for almost all $\omega \in \Omega$); here $X_t(\omega)$ may reach the †terminal point (or **death point**) ∂ adjoined to S at some random time $t = \zeta(\omega)$ provided that $X_t(\omega)$ is continuous for $0 \leqslant t < \zeta(\omega)$ and $X_s(\omega) = \partial$ for $s \geqslant \zeta(\omega)$. In the latter case we say that X_t "vanishes" at time $\zeta(\omega)$, meaning intuitively that the particle performing the random motion described by $X_t(\omega)$ disappears at the moment $\zeta(\omega)$. Such a process is said to be **1-dimensional** or **multidimensional** according as S is an interval or a manifold (possibly with boundary) with dimension $\geqslant 2$. Brownian motion is the typical diffusion process (→ 46 Brownian Motion; 118 Diffusion).

Diffusion processes are intimately related to a certain class of †partial differential equations. Let S be the real line. Assume that the †transition probability $P(s,x,t,E)$ $(s<t)$ of the process $\{X_t\}_{0\leqslant t<\infty}$ satisfies

$$1-P(s,x,s+h,(x-\varepsilon,x+\epsilon))=o(h)\quad (h\downarrow 0) \tag{1a}$$

for every $\epsilon>0$ and that the following limits exist:

$$\lim_{h\to 0+}\frac{1}{h}\int_{-\epsilon}^{\epsilon}(y-x)^2P(s,x,s+h,dy)=2a(s,x)>0, \tag{1b}$$

$$\lim_{h\to 0+}\frac{1}{h}\int_{-\epsilon}^{\epsilon}(y-x)P(s,x,s+h,dy)=b(s,x), \tag{1c}$$

$$\lim_{h\to 0+}\frac{1}{h}(P(s,x,s+h,S)-1)=c(s,x)\leqslant 0. \tag{1d}$$

Assume further that the transition probability is †absolutely continuous with respect to Lebesgue measure: $P(s,x,t,(y,y+dy))=p(s,x,t,y)dy$. Then under some suitable additional conditions, $p(s,x,t,y)$ satisfies

$$\frac{\partial p}{\partial s}=-a(s,x)\frac{\partial^2 p}{\partial x^2}-b(s,x)\frac{\partial p}{\partial x}-c(s,x)p,$$
$$p(t-0,x,t,y)=\delta(x-y), \tag{2}$$

and

$$\frac{\partial p}{\partial t}=\frac{\partial^2}{\partial y^2}(a(t,y)p)-\frac{\partial}{\partial y}(b(t,y)p)+c(t,y)p,$$
$$p(s,x,s+0,y)=\delta(y-x), \tag{3}$$

where δ is the †Dirac delta function. The coefficient c vanishes if $\{X_t\}$ is †conservative. Equations (2) and (3) are called **Kolmogorov's backward equation** and **forward equation**, respectively. They are also called the **Fokker-Planck partial differential equations.**

A. N. Kolmogorov [7] derived equations (2) and (3) in 1939, and W. Feller proved that (2) (or (3)) has a unique solution under certain regularity conditions on the coefficients a, b, and c, and that the solution $p(s,x,t,y)$ is nonnegative, has an integral with respect to y that does not exceed 1, and satisfies the †Chapman-Kolmogorov equation

$$\int p(s,x,t,y)p(t,y,u,z)\,dy=p(s,x,u,z).$$

Hence $p(s,x,t,y)$ determines a †Markov process analytically. However, rigorous proof establishing the sufficiency of condition (1) for the Markov process $\{X_t\}$ to be a diffusion process had not appeared by 1950.

In the †temporally homogeneous case, $p(s,x,s+h,y)$ does not depend on s and can be written as $p(h,x,y)$. Then $a(s,x)$, $b(s,x)$, and $c(s,x)$ are also independent of s, and $p(t,x,y)$ satisfies

$$\frac{\partial p}{\partial t}=a(x)\frac{\partial^2 p}{\partial x^2}+b(x)\frac{\partial p}{\partial x}+c(x)p,$$
$$p(0+,x,y)=\delta(x-y). \tag{4}$$

Feller made an intensive study of this case and completely solved the problem of existence and uniqueness of the solution of (4) assuming that $p(t,x,y)$ is nonnegative and that its integral with respect to y does not exceed 1 [3]. In particular, when t varies in $[0,+\infty)$ and S is an interval $[r_1,r_2]$, Feller used the Hille-Yosida theory of †semigroups of operators to determine the conditions that should be satisfied by r_1 and r_2 in order that the differential equation (4) (with the initial condition and his additional assumptions) yields one and only one solution. Feller also introduced the notion of generalized differential operators, which expresses the differential operator in the right-hand side of (4) in the most general form [4]. The probabilistic meaning of his results was clarified by E. B. Dynkin, H. P. McKean, Jr., K. Itô, D. B. Ray, V. A. Volkonskiĭ, and others, and all 1-dimensional diffusion processes with the †strong Markov property have now been completely determined. Multidimensional diffusion processes are much more complicated, however, and many problems remain unsolved.

To give the conditions for a Markov process to be a diffusion process, let us assume that S is a †complete metric space with metric ρ. Let $\{X_t\}$ be a Markov process on S with time parameter t ranging over a finite interval $[t_1,t_2]$ and satisfying

$$\sup P(s,x,t,S-U_\varepsilon(x))=O(h)\quad (h\downarrow 0) \tag{5}$$

for every $\varepsilon>0$, where $U_\varepsilon(x)$ is the ε-neighborhood of x and sup is taken over all $x\in S$, and s, $t\in[t_1,t_2]$ such that $0<t-s<h$. Then $\{X_t\}$ is a diffusion process if and only if

$$\int_{t_1}^{t_2}P(\rho(X_t,X_{t+h})>\varepsilon)\,dt=o(h)\quad (h\downarrow 0)$$

for every $\varepsilon>0$ [9]. This includes the following result due to Dynkin (1952) and J. R. Kinney (1953) as a special case: If in (5) we may replace $O(h)$ with $o(h)$, then $\{X_t\}$ will be a diffusion process. Specifically, when $\{X_t\}$ is a 1-dimensional †strong Markov process, the latter condition is also necessary for the process to be a diffusion process.

Since not much research on temporally nonhomogeneous diffusion processes has been done so far, we restrict our explanation to temporally homogeneous ones. Let $\mathfrak{M}=(X_t,W,P_x|x\in S)$ be a Markov process, where S is a †state space, W is the †path space consisting of all paths $w:[0,+\infty]\to S\cup\{\partial\}$

which are continuous in t for $0 \leqslant t < \zeta(w)$ ($w(t) = \partial$ for $t \geqslant \zeta(w)$ while $w(t) \in S$ for $0 \leqslant t < \zeta(w)$), and P_x is a probability measure on W under the condition that the process departs from x at $t=0$ (→ 262 Markov Processes). We may actually identify W with the †basic space Ω and set $X_t(w) = w(t)$ for $w \in \Omega$. Assume that $\mathfrak{M}$ has the †strong Markov property. By the continuity of paths, the hitting measure $H_E(x, \cdot)$ of a set E from a point $x \notin E$ is concentrated on the boundary of E. Conversely, given a system of hitting measures concentrated on boundaries satisfying certain conditions, there exist (many) corresponding temporally homogeneous diffusion processes. It follows from †Dynkin's formula for †infinitesimal generators (→ 262 Markov Processes) that the infinitesimal generator $\mathfrak{G}$ of a diffusion process has the local property that if u and v belong to the domain of $\mathfrak{G}$ and coincide in a neighborhood of x_0, then $\mathfrak{G}u(x_0) = \mathfrak{G}v(x_0)$.

B. 1-Dimensional Diffusion Processes

Let S be a straight line. A point $x \in S$ is called a **right singular point** if $X_t(w) \geqslant x$ for all $t \in [0, \zeta(w))$ with P_x-probability 1. A **left singular point** is defined analogously, with $\geqslant$ replaced by $\leqslant$. A right and left singular point is called a **trap**, while a right singular point which is not left singular is called a **right shunt** (a **left shunt** is defined analogously). A point is called a **regular point** if it is neither right nor left singular.

The set of all regular points is open. Let (r_1, r_2) be a connected component of this open set. One of the most important results concerning this situation is the proof of the existence of a strictly increasing function $s(x)$ defined on (r_1, r_2) and two measures m and k on (r_1, r_2) such that the infinitesimal generator $\mathfrak{G}$ of $\mathfrak{M}$ is represented as

$$\mathfrak{G}u(x) = \frac{u^+(dx) - u(x)k(dx)}{m(dx)}, \qquad (6)$$

where $u^+(dx)$ is the measure $du^+(x)$ induced by the †right derivative $u^+(x)$ of $u(x)$ with respect to $s(x)$ (i.e., $u^+(x) = \lim_{\Delta x \to +0}\{u(x+\Delta x) - u(x)\}/\{s(x+\Delta x) - s(x)\}$). Equation (6) gives a generalization of second-order †differential operator $au'' + bu' + cu$ ($a > 0, c \leqslant 0$). Here m is positive for nonempty open sets, both m and k are finite for compact sets in (r_1, r_2), and s, m, and k are unique in the following sense: If there are two sets of values of s_i, m_i, and k_i ($i = 1, 2$), then $s_2(x) = cs_1(x) +$ constant, $m_2(dx) = c^{-1}m_1(dx)$, and $k_2(dx) = c^{-1}k_1(dx)$ for some positive constant c. We call s, m, and k, respectively, the **canonical scale, canonical measure** (or **speed measure**), and **killing measure** for $\mathfrak{M}$. They determine the behavior of $w(t)$ belonging to $\mathfrak{M}$ inside the interval (r_1, r_2). Conversely, given any such set of s, m, and k, we can find a 1-dimensional diffusion process $\mathfrak{M}$ such that s, m, and k are, respectively, the canonical scale, canonical measure, and killing measure of $\mathfrak{M}$. If $X_t(w)$ is nonvanishing in (r_1, r_2) with probability 1, the killing measure k is identically zero, and the canonical scale s satisfies the equation

$$P_x(\sigma_{x_2} < \sigma_{x_1}) = \frac{s(x) - s(x_1)}{s(x_2) - s(x_1)}$$

for $x_1 < x < x_2$, where σ_y is the †hitting time required by the path $w(t)$ to attain y.

The motion $w(t)$ belonging to the process $\mathfrak{M}$ and contained in (r_1, r_2) may be constructed from the Brownian motion by means of a topological transformation of the state space (interval) based on s, a †time change based on m, and a †killing based on k. More precisely, we first transform the interval (r_1, r_2) by $x \to s(x)$ into the interval $(s(r_1+0), s(r_2-0))$ so that the diffusion process on this new interval has a canonical scale coincident with x. The speed and killing measures are transformed accordingly. We may, therefore, assume that the canonical scale is x. Let us consider the case $(r_1, r_2) = (-\infty, +\infty)$ for simplicity. Let $\mathfrak{t}(t, x)$ be the †local time of Brownian motion at x. Next, we apply the †time change to the Brownian motion by means of the †additive functional

$$\varphi(t) = \int_{-\infty}^{+\infty} \mathfrak{t}(t, x) m(dx),$$

and finally †kill the latter process by means of the †multiplicative functional

$$\alpha(t) = \exp\left(-\int_{-\infty}^{+\infty} \mathfrak{t}(\varphi^{-1}(t), x) k(dx)\right).$$

Thus we obtain the process $\mathfrak{M}$ in $(-\infty, \infty)$. In particular, if $m(dx) = a(x)^{-1}dx$ and $k(dx) = |c(x)|dx$, we have

$$\varphi(t) = \int_0^t a(X_\tau)^{-1} d\tau,$$

$$\alpha(t) = \exp\left(-\int_0^t |c(X_{\varphi^{-1}(\tau)})| d\tau\right),$$

and $\mathfrak{G}u = au'' + cu$.

At a shunt the infinitesimal generator $\mathfrak{G}$ has a form that is a generalization of the first-order differential operator $bu' + cu$ ($c \leqslant 0$), with $b > 0$ or $b < 0$ according as it is a right shunt or a left shunt. At a trap we have $\mathfrak{G}u(x) = -u(x)/E_x(\zeta)$.

When S is an interval with endpoints r_1 and r_2 and all interior points are regular, the left endpoint r_1 is classified into the following 4 types, according to the behavior of $\mathfrak{M}$ near r_1: Take an arbitrary fixed point $r \in (r_1, r_2)$ and set $n(dx) = m(dx) + k(dx)$ and

$$\alpha = \int_{(r_1, r)} (s(r) - s(x)) n(dx),$$

$$\beta = \int_{(r_1, r)} n((x, r)) s(dx).$$

Then r_1 is a **regular boundary** if $\alpha < \infty$, $\beta < \infty$; an **entrance boundary** if $\alpha < \infty$, $\beta = \infty$; an **exit boundary** if $\alpha = \infty$, $\beta < \infty$; and a **natural boundary** if $\alpha = \infty$, $\beta = \infty$.This classification is independent of the choice of r. A similar classification of r_2 may be established. $X_t(w)$ approaches r_1 in finite time with positive or null probability according as β is finite or infinite. If $\alpha = \infty$, it never happens that $X_t(w)$ starts from r_1 and reaches the interior of the interval S even if $r_1 \in S$, whereas if $\alpha < \infty$ we can construct (adjoining r_1 to S if necessary) a diffusion process that enters the interior from r_1 and whose motion in the interior coincides with that of $X_t(w)$.

If r_1 is a regular boundary for $\mathfrak{M}$ and $r_1 \in S$, then there are various possibilities for the behavior of $X_t(w)$ at r_1. They are expressed by the boundary conditions satisfied by the functions u belonging to the domain of the infinitesimal generator $\mathfrak{G}$. The condition is in general of the form

$$\gamma u(r_1) + \delta \mathfrak{G} u(r_1) + \mu u^+(r_1) = 0,$$

where γ, δ, and μ are constants, $\gamma, \delta \leqslant 0$, $\mu \geqslant 0$, and $|\delta| + \mu > 0$. If $\gamma = \delta = 0$, then r_1 is said to be a **reflecting barrier**. If r_1 is regular for $\mathfrak{M}$ and does not belong to S, then $X_t(w)$ vanishes exactly as $X_t(w)$ reaches r_1, and r_1 is called an **absorbing barrier**. This case corresponds to the boundary condition $u(r_1) = 0$. Whatever the boundary condition may be, $\mathfrak{M}$ is constructed from the Brownian motion with reflecting barrier by topological transformation of the state space, time change, and killing. Here if $\gamma \neq 0$, then killing may occur at r_1; if $\delta \neq 0$, the set of visiting times of r_1 has positive Lebesgue measure; and if $\mu \neq 0$, the trace of the motion may go beyond the point r_1 and reach the interior points of S [2, 3, 4, 6].

If we weaken the assumption of continuity of paths and admit jumps from r_1, the general boundary condition becomes

$$\gamma u(r_1) + \delta \mathfrak{G} u(r_1) + \mu u^+(r_1) + \int_{(r_1, r_2]} (u(x) - u(r_1)) \nu(dx) = 0,$$

where ν is a measure with respect to which $\min(1, s(x) - s(r_1 + 0))$ is integrable.

When $S = (r_1, r_2)$, the transition probability is absolutely continuous with respect to the canonical measure, the density $p(t, x, y)$ has an †eigenfunction expansion

$$p(t, x, y) = \int_{-\infty}^{0} e^{\lambda t} \mathfrak{e}(d\lambda; x, y),$$

and $p(t, x, y)$ is positive, jointly continuous in 3 variables, and symmetric in x and y. A similar result is also known when S is half-open or closed [6].

If $\mathfrak{M}$ is †recurrent, i.e., $P_x(\sigma_y < +\infty) = 1$ for every x and y in S, then there exists a unique (up to a multiplicative constant) †invariant measure for S that is finite for all closed intervals in the interior of S. If $\mathfrak{M}$ is conservative and all the interior points of S are regular, then the canonical measure is an invariant measure, provided that the endpoints are either entrance, natural, or regular reflecting. Local times similar to the Brownian standard local times are defined if the interior points of S are regular.

C. Multidimensional Diffusion Processes

Let the state space S be a domain or the closure of a domain in the n-dimensional Euclidean space $\mathbf{R}^n$. Consider a temporally homogeneous diffusion process $\{X_t\}_{0 \leqslant t < \infty}$ on S for which jumps from the boundary are permitted. Under suitable regularity conditions the infinitesimal generator $\mathfrak{G}$ coincides, for sufficiently smooth functions u in its domain, with the following †elliptic partial differential operator A:

$$A = \sum_{i,j=1}^{n} a^{ij}(x) \frac{\partial^2}{\partial x^i \partial x^j} + \sum_{i=1}^{n} b^i(x) \frac{\partial}{\partial x^i} + c(x), \quad c \leqslant 0; \qquad (7)$$

when S has a boundary, u satisfies a boundary condition of the form

$$\sum_{i,j=1}^{n-1} \alpha^{ij}(x) \frac{\partial^2 u(x)}{\partial x^i \partial x^j} + \sum_{i=1}^{n-1} \beta^i(x) \frac{\partial u(x)}{\partial x^i} + \gamma(x) u(x) + \delta(x) A u(x) + \mu(x) \frac{\partial u(x)}{\partial n} + \int_{y^n > 0} \left(u(y) - u(x) - \chi_U(y) \sum_{i=1}^{n-1} \frac{\partial u(x)}{\partial x^i} \times (y^i - x^i) \right) \nu_x(dy) = 0, \qquad (8)$$

where, for simplicity, we assume that S is the closed half-space defined by $x^n \geqslant 0$, (α^{ij}) is a symmetric nonnegative definite matrix, $\gamma \leqslant 0$, $\delta \leqslant 0$, $\mu \geqslant 0$, $\partial / \partial n$ is the inward directed †conormal derivative associated with a^{ij}, χ_U is the †indicator function of a bounded neighborhood of x, and $\nu_x(dy)$ is a measure that is finite outside of U and with respect to which $\Sigma_{i=1}^{n-1}(y^i-x^i)^2+y^n$ is integrable on U. This condition was discovered by A. D. Ventcel' [11]. Conversely, given an operator A such as (7) and a boundary condition (8) (if S has a boundary), the existence and uniqueness of the corresponding diffusion process are unknown except for special cases. If $S=\mathbf{R}^n$ and A has continuous coefficients, there exists at least one diffusion process corresponding to A.

Suppose that S is the closure of a bounded domain with a sufficiently smooth boundary and A is given with sufficiently smooth coefficients. If the boundary condition is $\gamma u+\mu\partial u/\partial n=0$, $\mu\neq 0$, then there exists a unique diffusion process on S corresponding to this situation. Moreover, if $\gamma=0$, then the process is said to have a **reflecting barrier**. Suppose that a general boundary condition (8) is given. We write the left-hand side of (8) as Lu. Write $u=Hf$ for the solution of $Au=0$ with boundary value f. Under some natural additional conditions, T. Ueno [10] proved that if LH †generates a Markov process on the boundary, then there exists a diffusion process (having possible jumps from the boundary) for A with boundary condition (8). If $\{X_t\}$ has a reflecting barrier, the Markov process on the boundary is, conversely, obtained from $\{X_t\}$ through time change by a nonnegative continuous †additive functional which increases only when the value of X_t is on the boundary.

When we have a diffusion process on S with infinitesimal generator of the form A, we can obtain a probabilistic expression for the solutions of various partial differential equations involving A. Let σ be the hitting time of the boundary of S. Then $Hf(x)$ can be expressed as $E_x(f(X_{\sigma-0}))$. The solution of $Au=-f$ with boundary value 0 is given by $u(x)=E_x(\int_0^\infty f(X_t)\,dt)$, while the solution $u(t,x)$ of $\partial u/\partial t=Au$ with boundary value 0 and initial value $u(0,x)=f$ is $E_x(f(X_t);t<\sigma)$. The first case gives the solution of a †Dirichlet problem; and in this case, the condition for a boundary point to be †regular relative to the Dirichlet problem may also be expressed probabilistically. Furthermore, if $f(X_t)$ is replaced by $f(X_t)\exp(\int_0^t k(X_s)\,ds)$ in the expressions for $u(x)$ and $u(t,x)$, A is replaced by $A+k$ (M. Kac 1951). When $k\leqslant 0$, this replacement gives rise to a killing.

Multidimensional diffusion processes whose movement may proceed in all directions are generalizations of 1-dimensional diffusion processes whose state points are all regular. However, in contrast to the 1-dimensional case, not all such processes can be obtained from the multidimensional Brownian motion by the combination of state space transformation, time change, and killing; in fact, even if we add the drift transformation (to be explained in the next paragraph), there are still multidimensional diffusion processes that cannot be obtained from Brownian motions by these operations.

Let $\mathfrak{M}=(X_t,W,P_x\,|\,x\in\mathbf{R}^n)$ be an n-dimensional Brownian motion, and define a new measure $\tilde{P}_x$ such that

$$\tilde{P}_x(B)=E_x\left(\exp\left(\sum_{i=1}^{n}\int_0^t b^i(X_s)\,dX_s^i-\frac{1}{2}\sum_{i=1}^{n}\int_0^t b^i(X_s)^2\,ds\right);B\right)$$

holds for $B\in\mathbf{B}_t$, i.e., for any event B determined by its history up to the time t, where X_s^i is the ith coordinate of X_s and the integral by dX_s^i is a †stochastic integral. Then $\tilde{\mathfrak{M}}=(X_t,W,\tilde{P}_x\,|\,x\in\mathbf{R}^n)$ is a diffusion process with infinitesimal generator

$$\frac{1}{2}\sum_{i=1}^{n}\frac{\partial^2}{\partial x^{i2}}+\sum_{i=1}^{n}b^i\frac{\partial}{\partial x^i}.$$

This transformation from $\mathfrak{M}$ to $\tilde{\mathfrak{M}}$ is called a **drift**. Drift transformation can be defined for more general diffusion processes as an operation that changes the first-order terms of the infinitesimal generators (G. Maruyama studied this for 1-dimensional processes, I. V. Girsanov, Dynkin, and M. Motoo for multidimensional cases).

When $\sqrt{a}$ and b satisfy the †Lipschitz condition, the diffusion process $\{\tilde{X}_t\}$ with infinitesimal generator $a(x)d^2/dx^2+b(x)d/dx$ is obtained from the 1-dimensional Brownian motion $\{X_t\}$ by solving a **stochastic differential equation**

$$d\tilde{X}_t=\sqrt{2a(\tilde{X}_t)}\;dX_t+b(\tilde{X}_t)\,dt$$

[5]. This fact can be generalized to the multidimensional case, for which the corresponding equation is

$$d\tilde{X}_t^i=\sum_{j=1}^{n}\alpha_j^i(\tilde{X}_t)\,dX_t^j+b^i(\tilde{X}_t)\,dt,$$

where $\{X_t\}$ is an n-dimensional Brownian motion and $\Sigma_{k=1}^n\alpha_k^i\,\alpha_k^j=2a^{ij}$.

Stochastic differential equations are also used in constructing processes with the boundary condition $\Sigma\,\beta^i\partial u/\partial n+\partial u/\partial n=0$, for the cases with dimensions greater than 2, and for general boundary conditions (8) in the 2-dimensional case.

For a conservative diffusion process $\{X_t\}$ on a †differentiable manifold, (7) is replaced by

$$A=\sum_{i,j=1}^{n}\frac{1}{\sqrt{a}}\frac{\partial}{\partial x^i}\left(a^{ij}\sqrt{a}\,\frac{\partial}{\partial x^i}\right)+\sum_{i=1}^{n}b^i\frac{\partial}{\partial x^i},$$

where a^{ij} is a †contravariant tensor of order 2, b^i is a contravariant vector, and a is the determinant of the inverse matrix of (a^{ij}) (the b^i here are different from those in (7)). Under some smoothness conditions, $\{X_t\}$ is determined by A and has a unique (up to a multiplicative constant) invariant measure, say m, which is expressed as

$$\rho(x)\sqrt{a(x)}\ dx^1\ldots dx^n$$

by a function ρ. Suppose that the total measure of m is finite. Then we may assume that the total measure is 1. The diffusion process $\{X_t^*\}$ is said to be †adjoint to $\{X_t\}$ if the infinitesimal generator of X_t^* is of the form

$$\sum_{i,j=1}^{n}\frac{1}{\sqrt{a}}\frac{\partial}{\partial x^i}\left(a^{ij}\sqrt{a}\ \frac{\partial}{\partial x^j}\right)$$
$$-\sum_{i=1}^{n}b^i(x)\frac{\partial}{\partial x^i}+2\sum_{i,j=1}^{n}a^{ij}\frac{\partial\log\rho}{\partial x^i}\frac{\partial}{\partial x^j},$$

and $\{X_t^*\}$ coincides with $\{X_t\}$ if and only if the b^i have a potential, i.e., there is a function f such that $b^i=\sum_{j=1}^{n}a^{ij}\partial f/\partial x^j$. When the state space has a boundary, $\{X_t\}$ has a reflecting barrier if and only if $\{X_t^*\}$ does.

In general, if a diffusion process $\{X_t\}$ is given on a noncompact space S and X_t has no limit points in S as $t\uparrow\zeta$, then some natural compactification should be induced by $\{X_t\}$. The notion of a †Martin boundary for Markov processes is introduced in this connection.

References

[1] Е. Б. Дынкин (E. B. Dynkin), Одномерные непрерывные строго марковские процессы, Teor. Verojatnost. i Primenen., 4 (1959), 3–54; English translation, One-dimensional continuous strong Markov processes, Theory of Prob. Appl., 4 (1959), 1–52.
[2] Е. Б. Дынкин (E. B. Dynkin), Марковские процессы, Физматгиз, 1963; English translation, Markov processes I, II, Springer, 1965.
[3] W. Feller, The parabolic differential equations and the associated semi-groups of transformations, Ann. of Math., (2) 55 (1952), 468–519.
[4] W. Feller, On second order differential operators, Ann. of Math., (2), 61 (1955), 90–105.
[5] K. Itô, On stochastic differential equations, Mem. Amer. Math. Soc., 1951.
[6] K. Itô and H. P. McKean, Jr., Diffusion processes and their sample paths, Springer, 1965.
[6A] H. P. McKean, Jr., Stochastic integrals, Academic Press, 1969.
[7] A. N. Kolmogorov, Über die analytischen Methoden in der Wahrscheinlichkeitsrechnung, Math. Ann., 104 (1931), 415–458.
[8] M. Motoo, Application of additive functionals to the boundary problem of Markov processes, Proc. 5th Berkeley Symp. Math. Stat. Prob. II, pt. 2, Univ. of California Press (1967), 75–110.
[9] Л. В. Серегин (L. V. Seregin), Условия непрерывности вероятностых пропессов, Teor. Verojatnost. i Primenen., 6 (1961), 3–30; English translation, Continuity conditions for stochastic processes, Theory of Prob. Appl., 6 (1961), 1–26.
[10] T. Ueno, The diffusion satisfying Wentzell's boundary conditions and the Markov process on the boundary I, II, Proc. Japan Acad., 36 (1960), 533–538, 625–629.
[11] А. Д. Вентцель (A. D. Ventcel'), О граничных условиях для многомерных диффузионных процессов, Teor. Verojatnost. i Primenen., 4 (1959), 172–185; English translation, On boundary conditions for multidimensional diffusion processes, Theory of Prob. Appl., 4 (1959), 164–177.
[12] A. Friedman, Stochastic differential equations and applications 1, Academic Press, 1975.
[13] A. V. Skorokhod, Studies in the theory of random processes, Addison-Wesley, 1965.

120 (XIX.2) Dimensional Analysis

The system of units for physical quantities is derived from a certain set of †fundamental units. If the fundamental units are denoted by θ, φ, ψ, etc., any other unit (called a **derived unit**) α can always be expressed in the form $\alpha=c\,\theta^l\varphi^m\psi^n\ldots$ $(c,l,m,n,\ldots$ are constants), by definition or by physical laws. The exponents $l,m,n,\ldots$ are called the **dimensions** of α, and the content of the previous statement is expressed as $[\alpha]=[\theta^l\varphi^m\psi^n\ldots]$, which is called the **dimensional formula**. The usual practice is to take as fundamental units length, time, mass, temperature, and energy, which are denoted by L, T, M, θ, and H, respectively. **Dimensional analysis** investigates the relation between physical quantities by use of the π theorem and the law of similitude given below.

A. The π Theorem

If a relationship $f(\alpha,\beta,\ldots)=0$ holds among n physical quantities α, β, ... independently of the choice of fundamental units, the equation $f(\alpha,\beta,\ldots)=0$ can always be transformed into $F(\pi_1,\pi_2,\ldots)=0$, where the π_i are $n-m$ dimensionless quantities (m is the number of fundamental units) of the form $\pi_i=\alpha^{a_i}\beta^{b_i}\ldots$. If we choose the π_i so that $\pi_1=\alpha\beta^{-r_1}\gamma^{-z_1}\ldots$ and π_2, π_3, etc., do not contain α, then $f=0$ implies $\alpha=\beta^{r_0}\gamma^{z_0}\ldots\Phi(\pi_2,\pi_3,\ldots)$, which clearly shows the manner in which the quantity α is related to other quantities β, $\gamma,\ldots$.

B. The Law of Similitude

In general, if two physical systems of the same kind have the same values of the π_i, then the physical states of the systems are similar. If we are given a family of mutually similar systems, it is sufficient to observe a particular one among them (a "model") in order to estimate physical values attached to any one of the given systems.

Consider, for example, the case of the drag D acting on geometrically similar bodies placed in the flow of a viscous incompressible fluid. If v is the velocity, l the representative length of the body, ρ the density of the fluid, and μ the viscosity (which has the dimensional formula $ML^{-1}T^{-1}$), then the π theorem gives $D/\rho v^2 l^2=f(\rho v l/\mu)$. Hence the drag coefficient as given by the left-hand side can be obtained by the experiments performed on a geometrically similar model. The dimensionless quantity $R=vl/\nu$ ($\nu=\mu/\rho$) is called the **Reynolds number**. If the wave resistance due to gravity as well as the effect of compressibility are taken into account, gravitational acceleration g and sound velocity a must be included, so that we have $D/\rho v^2 l^2=f(vl/\nu, v^2/lg, v/a, C_1, C_2,\ldots)$, where $C_1, C_2,\ldots$ are other dimensionless quantities depending on the physical properties of the fluid. $Fr=v^2/lg$ is called the **Froude number**, and $M=v/a$ the **Mach number**.

Next, consider the case of heat transfer between a solid surface and a flow of fluid. Let the area of the solid surface be denoted by S, the heat transferred per unit time by $Q\,(HT^{-1})$, the thermal conductivity of the fluid by $k(HL^{-1}T^{-1}\theta^{-1})$, the specific heat by $C(HM^{-1}\theta^{-1})$, the two representative temperatures by T_0 and T_1 and the representative length by l, where expressions in parentheses represent dimensional formulas. Then we have, as dimensionless quantities, the **Nusselt number** $Nu=Q/(kS(T_1-T_0)/l)$, the **Prandtl number** $Pr=\nu/\kappa$ ($\kappa=k/\rho c$), the **Grashoff number** $Gr=d^3g(T_1-T_0)/\nu^2T_0$, and R, so that from the π theorem we have the relation $Nu=f(R,Pr,Gr,C_1,C_2,\ldots)$. Furthermore, $Pe=vl/\kappa=Pr\,R$ is called the **Péclet number**.

Since the physical relations that hold between a certain number of physical quantities are theoretically obtained by applying mathematical operations to the fundamental equations (†equations of motion), the numerical coefficients appearing in them are in most cases expected to be of unit order of magnitude. Hence, conversely, if the values of the numerical coefficients determined from the experimental data after applying dimensional analysis to the relation between physical quantities are neither too large nor too small, it may be expected that some physical correlation should exist between them.

Reference

[1] A. W. Porter, The method of dimensions, Methuen, 1933.

121 (II.22) Dimension Theory

A. Introduction

Toward the end of the 19th century, G. Cantor discovered that there exists a one-to-one correspondence between the set of points on a line segment and the set of points on a square; and also, G. Peano discovered the existence of a †continuous mapping from the segment onto the square. Soon, the progress of the theory of point-set topology led to the consideration of sets which are more complicated than familiar sets, such as polygons and polyhedra. Thus it became necessary to give a precise definition to dimension, a concept which had previously been used only vaguely. In 1913, L. E. J. Brouwer [9] gave a definition of dimension based on an idea of H. Poincaré. In 1922, the foundations of dimension theory for separable metric spaces were established by K. Menger [11] and P. Uryson [10]. Subsequently, P. S. Aleksandrov and W. Hurewicz greatly contributed to the development of the theory. The foundations of dimension theory for general metric spaces were established independently by M. Katětov and K. Morita. More general theory for †normal spaces has also been investigated; the same results as in metric spaces, however, do not always hold.

B. Definition of Dimension

Let X be a normal space. If any finite open †covering of X has an open covering of †order $\leqslant n+1$ as its refinement (→ 408 Topological Spaces R) (i.e., if for any open sets G_i ($i = 1, \ldots, s$) such that $X = G_1 \cup \ldots \cup G_s$, there exist open sets H_i ($i = 1, \ldots, s$) such that $H_i \subset G_i$, $X = H_1 \cup \ldots \cup H_s$ and any $n+2$ of the H_i have no point in common), then we write $\dim X \leqslant n$. If $\dim X \leqslant n$ and $\dim X \leqslant n-1$ does not hold, then we define X to be **n-dimensional** and write $\dim X = n$. We call $\dim X$ the **covering dimension**, (or **Lebesgue dimension**) of X. This definition is due to H. Lebesgue.

There are other definitions of *dimension* which are given inductively. Let us define $\operatorname{Ind} X = -1$ if X is empty. If for any pair consisting of a closed set F and an open set G with $F \subset G$ in X there exists an open set V such that $F \subset V \subset G$ and $\operatorname{Ind}(\bar{V} - V) \leqslant n-1$, then we define $\operatorname{Ind} X \leqslant n$. Next, we define $\operatorname{ind} \varnothing = -1$. For any point p of X and any neighborhood G of p, suppose that there exists an open neighborhood V of p such that $V \subset G$ and $\operatorname{ind}(\bar{V} - V) = n-1$. Then we define $\operatorname{ind} X \leqslant n$. As before we set $\operatorname{Ind} X = n$ ($\operatorname{ind} X = n$) if $\operatorname{Ind} X \leqslant n$ ($\operatorname{ind} X \leqslant n$) and $\operatorname{Ind} X \leqslant n-1$ ($\operatorname{ind} X \leqslant n-1$) does not hold. (The definition of $\operatorname{ind} X$ is due to Menger.) We call $\operatorname{Ind} X$ ($\operatorname{ind} X$) the **large inductive dimension** of X (the **small inductive dimension** of X).

If $\dim X \leqslant n$ does not hold for any n, then X is called **infinite-dimensional**, written $\dim X = \infty$; we define $\operatorname{Ind} X = \infty$ and $\operatorname{ind} X = \infty$ similarly. These dimensions are invariant under †homeomorphisms.

The set of irrational points in a Euclidean space, the †Cantor discontinuum, and †Baire's zero (-dimensional) spaces are all 0-dimensional. The set of rational points in a †Hilbert space is 1-dimensional.

C. Dimension of Metric Spaces

The following theorems hold for the dimension of metric spaces [13, 15]. Let X and Y be metric spaces. The equality $\dim X = \operatorname{Ind} X$ holds. If $Y \subset X$, then $\dim Y \leqslant \dim X$. If X is a union of a countable number of closed sets F_i ($i = 1, 2, \ldots$), then $\dim X = \max(\dim F_i)$ (**sum theorem for dimension**). The inequality $\dim(X \cup Y) \leqslant \dim X + \dim Y + 1$ holds. If $\dim X = n$, then X is a union of $n+1$ 0-dimensional subsets (**decomposition theorem for dimension**). We have $\dim(X \times Y) \leqslant \dim X + \dim Y$, where $X \neq \varnothing$ (**product theorem for dimension**).

Each of the following is a necessary and sufficient condition for $\dim X \leqslant n$: (i) There exists a subspace A of a Baire zero-dimensional space $B(\tau)$ and a continuous closed mapping f of A onto X such that $f^{-1}(x)$ consists of at most $n+1$ points for each point x of X (K. Morita [16]); (ii) there exists a metric of X which gives the same topology on X such that for any positive number ε, any point x of X, and any $n+2$ points x_i ($i = 1, \ldots, n+2$) at a distance less than ε from the ($\varepsilon/2$)-neighborhood of x, there are at least two points x_i and x_j ($i \neq j$) with distance $< \varepsilon$ (J. Nagata [18]).

Hurewicz's problem asked whether the equality $\dim X = n + m$ ($m > 0$) implies the existence of an m-dimensional space A and a mapping f of A onto X having property (i). It was solved affirmatively for separable metric spaces by J. H. Roberts and for general metric spaces by K. Nagami [19].

If X is the union of a countable number of closed †strongly paracompact subspaces, in particular if X is separable, then $\operatorname{Ind} X = \operatorname{ind} X$ [1, 2, 15]. However, it was shown by P. Roy [20] that this equality does not hold in general.

D. Euclidean Spaces and Dimension

The n-dimensional †Euclidean space $\mathbf{R}^n$ is exactly n-dimensional in the sense mentioned above; thus this concept of dimension agrees with our intuition. The proof of $\dim \mathbf{R} \geqslant n$ comes from **Lebesgue's Theorem**: If each member of a finite closed covering of an n-cube has sufficiently small diameter, then the order of the covering is not less than $n+1$. (The proof of $\dim \mathbf{R}^n \leqslant n$ is easy.) Let X be a subset of $\mathbf{R}^n$ and f a homeomorphism from X onto a subset $f(X)$ of $\mathbf{R}^n$. If x is an interior point of X, then $f(x)$ is an interior point of $f(X)$. Also, if an open set A of $\mathbf{R}^n$ is homeomorphic to a subset B of $\mathbf{R}^n$, then B is open in $\mathbf{R}^n$ (**Brouwer's theorem on the invariance of domain** [8]). This theorem holds for any manifold but not for general separable metric spaces. By the theorem of invariance of domain it can be shown that $\mathbf{R}^m$ and $\mathbf{R}^n$, $m \neq n$, are not homeomorphic (**theorem on invariance of dimension of Euclidean spaces**). Any n-dimensional separable metric space is embedded in a Euclidean space $\mathbf{R}^{2n+1}$, or more precisely, in the subset of $\mathbf{R}^{2n+1}$ consisting of all points x of which at most n coordinates are rational (**Menger-Nöbeling embedding theorem**, G. Nöbeling, *Math. Ann.*, 104 (1930)). Thus, from the topological point of view, any finite-dimensional separable metric space can be identified with a subset of a Euclidean space. Moreover, it is known that any n-dimensional separable metric space is homeomorphic to a subset of some n-dimensional compact metric space.

If F is a bounded closed subset of $\mathbf{R}^m$, then $\dim F \leqslant n$ if and only if for any positive number ε, there exists a continuous mapping f from F into an n-dimensional polyhedron in $\mathbf{R}^m$ such that the distance between x and $f(x)$ is less than ε for each point x of F.

E. Dimension of Normal Spaces

Let X be a normal space. $\operatorname{Ind} X \geqslant \dim X$ and $\operatorname{Ind} X \geqslant \operatorname{ind} X$, but the equalities do not necessarily hold here. The following theorems were obtained by E. Čech, Aleksandrov, C. H. Dowker, E. Hemmingsen, and Morita [1]. If $\dim X \leqslant n$, then any locally finite open covering of X has an open covering of order $\leqslant n+1$ as its refinement; if A is an †F_σ subset of X or A is strongly paracompact, then $\dim A \leqslant \dim X$; if X has a †σ-locally finite closed covering $\{F_\alpha\}$, then $\dim X = \max(\dim F_\alpha)$.

In order that $\dim X \leqslant n$, it is necessary and sufficient that any continuous mapping from a closed subset of X into an n-sphere S^n can be extended continuously to X. If X and Y are †paracompact and X is †locally compact, or if $X \times Y$ is strongly paracompact, then $\dim(X \times Y) \leqslant \dim X + \dim Y$, where $X \neq \varnothing$; if X is a †CW complex, then the equality holds [11]. Katětov proved that $\dim X$ is determined by the ring $C^*(X)$ of the set of bounded real-valued continuous functions on X [13].

F. Homological Dimension

Aleksandrov contributed much to the development of dimension theory in introducing the concept of homological dimension (*Math. Ann.*, 106 (1932)). The **homological dimension** of a compact Hausdorff space X with respect to an Abelian group G is the largest integer n such that the n-dimensional †Čech homology group $\check{H}_n(X,A;G)$ is nonzero for some closed subset A of X. The **cohomological dimension** $D(X;G)$ is defined similarly by using the †Čech cohomology group $\check{H}^n(X,A;G)$. If $\dim X < \infty$, then $\dim X = D(X;\mathbf{Z})$ ($\mathbf{Z}$ is the additive group of integers). The cohomological dimension of X with respect to an arbitrary Abelian group is determined by the cohomological dimension with respect to some specified groups, and the cohomological dimension of the product space $X \times Y$ is expressed in terms of those of X and Y (M. F. Bockstein, [5]). A compact Hausdorff space X has the property that $\dim(X \times Y) = \dim X + \dim Y$ for any compact Hausdorff space Y if and only if $\dim X = D(X;\mathbf{Q}(p))$, where $\mathbf{Q}(p)$ is the additive group of rationals mod 1 of the form m/p^s for any prime number p (V. Boltyanskiĭ, [5]); this result holds when X is paracompact (Y. Kodama [21]).

G. Dimension and Measure

Let X be a separable metric space. Then $\dim X \leqslant n$ if and only if X is homeomorphic to a subset of a Euclidean space $\mathbf{R}^{2n+1}$ whose $(n+1)$-dimensional measure is zero (E. Szpilrajn, *Fund. Math.*, 28 (1937); also → [1,2]).

H. Dimension Type (Fréchet's Definition)

In analogy to the theory of †cardinal numbers in set theory, M. Fréchet (1909) defined the dimension type of topological spaces as follows: Two spaces X and Y are said to have the same **dimension type** if X is homeomorphic to a subset of Y and Y is homeomorphic to a subset of X. K. Kunugui [12] studied the relation between dimension and dimension type.

References

[1] K. Morita, Zigenron (Japanese; Dimension theory), Iwanami, 1950.
[2] W. Hurewicz and H. Wallman, Dimension theory, Princeton Univ. Press, 1941.
[3] J. Nagata, Modern dimension theory, Noordhoff, 1965.
[4] K. Nagami, Dimension theory, Academic Press, 1970.
[5] П. С. Александров (P. S. Aleksandrov), Современное состояние теории размерности, Uspehi Mat. Nauk, (N.S) 6, no. 5 (1951), 43–68; English translation, The present status of the theory of dimension, Amer. Math. Soc. Transl., (2) 1 (1955), 1–26.
[6] L. E. J. Brouwer, Beweis der Invarianz der Dimensionenzahl, Math. Ann., 70 (1911), 161–165.
[7] H. Lebesgue, Sur la non-applicabilité de deux domaines appartenant respectivement à des espaces à n et $n+p$ dimensions, Math. Ann., 70 (1911), 166–168.
[8] L. E. J. Brouwer, Beweis der Invarianz des n-dimensionalen Gebiets, Math. Ann., 71 (1912), 305–313.
[9] L. E. J. Brouwer, Über den natürlichen Dimensionsbegriff, J. Reine Angew. Math., 142 (1913), 146–152.

[10] P. Uryson, Les multiplicités cantoriennes, C. R. Acad. Sci. Paris, 175 (1922), 440–442.
[11] K. Menger, Dimensionstheorie, Teubner, 1928.
[12] K. Kunugui (Kunugi), Sur la théorie du nombre de dimensions, Gauthier-Villars, 1930.
[13] M. Katětov, On the dimension of non-separable spaces I, Czechoslovak Math. J., 2 (1952), 333–368.
[14] K. Morita, On the dimension of product spaces, Amer. J. Math., 75 (1953), 205–223.
[15] K. Morita, Normal families and dimension theory for metric spaces, Math. Ann., 128 (1954), 350–362.
[16] K. Morita, A condition for the metrizability of topological spaces and for n-dimensionality, Sci. Rep. Tokyo Kyoiku Daigaku, 5 (1955), 33–36.
[17] C. H. Dowker, Local dimension of normal spaces, Quart. J. Math., (2) 6 (1955), 101–120.
[18] J. Nagata, Note on dimension theory for metric spaces, Fund. Math., 45 (1958), 143–181.
[19] K. Nagami, Mappings of finite order and dimension theory, Japan. J. Math., 30 (1960), 25–54.
[20] P. Roy, Failure of equivalence of dimension concepts for metric spaces, Bull. Amer. Math. Soc., 68 (1962), 609–613.
[21] Y. Kodama, Note on cohomological dimension for non-compact spaces, J. Math. Soc. Japan, 18 (1966), 343–359.

122 (V.9) Diophantine Equations

A. General Remarks

A **Diophantine equation** is an †algebraic equation whose coefficients lie in the ring **Z** of rational integers and whose solutions are sought in that ring. The name comes from Diophantus, an Alexandrian mathematician of the third century A.D., who proposed many Diophantine problems; but such equations have a very long history, extending back to ancient Egypt, Babylonia, and Greece. As early as the sixth century B.C., Pythagoras is said to have partially solved the equation $x^2+y^2=z^2$ by $x=2n+1, y=2n^2+2n, z=y+1$. A general solution is given by the Pythagorean numbers $x=m^2-n^2, y=2mn, z=m^2+n^2$. †Fermat's problem also concerns a Diophantine equation. The development of the †Diophantine approximation (→ 187 Geometry of Numbers) made it possible to treat not only problems of Diophantine equations over **Z** but also over rings finitely generated over **Z**. In this article, we shall note the principal results on Diophantine equations and also deal with the problem of rational points on †algebraic varieties.

Systematic studies of Diophantine equations over **Z** have been made for the linear equation $\sum_{i=1}^n a_i x_i = a$ $(a_i, a \in \mathbf{Z})$ and for the quadratic equation $ax^2+bxy+cy^2=k$ $(a,b,c,k \in \mathbf{Z})$ in two unknowns. The latter forms a principal topic of C. F. Gauss's *Disquisitiones arithmeticae* and can be regarded as a starting point of modern algebraic number theory. The special quadratic equation $t^2-Du^2=\pm 4$ $(D \in \mathbf{Z})$ is called **Pell's equation.** If $D<0$, then Pell's equation has only a finite number of solutions. If $D>0$, then all solutions t_n, u_n of Pell's equation are given by $\pm((t_1+u_1\sqrt{D})/2)^n=(t_n+u_n\sqrt{D})/2$, provided that the pair t_1, u_1 is a solution with the smallest $t_1+u_1\sqrt{D}>1$. Using continued fractions (→ 85 Continued Fractions), we can determine t_1, u_1 explicitly. A general quadratic Diophantine equation $ax^2+bxy+cy^2=k$ with two unknowns can be solved completely if we use solutions of Pell's equation; this is an application of the arithmetic of quadratic fields (→ 343 Quadratic Fields) [1]. On quadratic Diophantine equations of several unknowns, there are deep studies by C. L. Siegel (→ 344 Quadratic Forms).

B. Rational Integral Solutions of Diophantine Equations

While we have very little knowledge about Diophantine equations of higher degree, some famous results in this area are the following:

(1) **Thue's theorem** (1908): If $f(x)=\sum_{\nu=0}^n a_\nu x^\nu$ $(a_\nu \in \mathbf{Z}, n>2)$ has distinct roots, then the number of rational integral solutions of $\sum_{\nu=0}^n a_\nu x^\nu y^{n-\nu}=a$ $(\mathbf{Z} \ni a \neq 0)$ is finite. This theorem is a direct consequence of Thue's theorem on Diophantine approximation, which says that there are only a finite number of rational numbers p/q $(p,q \in \mathbf{Z}, q>0)$ with $|\alpha-p/q|<1/q^{(n/2)+1}$ for a given algebraic number α of degree n $(n>2)$ [3, p. 122]. K. F. Roth proved that $(n/2)+1$ in this formula can be replaced by $2+\varepsilon$ (ε is an arbitrary positive number independent of n) (*Mathematika*, 2 (1955), 1–20). **Roth's theorem** was generalized to some cases of number fields and function fields (→ 187 Geometry of Numbers) and is effectively applied in the

recent theory of Diophantine equations [2,3].

(2) **Siegel's theorem** (1929): Assume that the equations $f_i(X_1, \ldots, X_n)=0$ $(1 \leq i \leq m)$ determine an algebraic curve with a positive †genus in an †affine space of dimension n. Then the number of rational integral solutions of $f_i(X_1, \ldots, X_n)=0$ $(1 \leq i \leq m)$ is finite. This theorem was generalized by S. Lang in the following form: Let K be a finitely generated field over $\mathbf{Q}$ and I a subring of K that is finitely generated over $\mathbf{Z}$. Furthermore, let C be a nonsingular projective algebraic curve with a positive genus defined over K, and let φ be a rational function on C defined over K. Then there are only a finite number of points P on C with $\varphi(P) \in I$ [2]. The proof of this theorem is based on a generalization of Roth's theorem in the above sense and on the weak Mordell-Weil theorem (→ Section C). A. Robinson and P. Roquette gave another approach to Siegel's theorem from the standpoint of †nonstandard arithmetic (*J. Number Theory* 7 (1975)). The finiteness of solutions for other special types of Diophantine equations is also known [3,4]. On the other hand, a necessary condition for the existence of infinitely many solutions of $f(X, Y)=0$ with rational integral coefficients was given by C. Runge (*J. Reine Angew. Math.*, 100 (1887)) (→ [4]; 154 Fermat's Problem).

As we have seen, many theorems about Diophantine equations are qualitative; quantitative studies have been made only about quadratic forms (→ 344 Quadratic Forms).

C. Rational Solutions

Let V be an †abstract algebraic variety defined over a field k, and let P be a point of V. Then P is called a **rational point** over k of V if the coordinates of the †representative P_α contained in an †affine open set V_α of V are in k (→ 18 Algebraic Varieties D). This definition is independent of the choice of the representative P_α. In particular, if V is a †projective variety, the point P given by the †homogeneous coordinates $(x_0, x_1, \ldots, x_n)$ is rational if and only if $x_i / x_p \in k$ $(0 \leq i \leq n,$ $x_p \neq 0)$. In the following we state main results concerning rational points of algebraic varieties, especially results concerning †Abelian varieties, restricting k to be either an †algebraic number field of finite degree, a †$\mathfrak{p}$-adic number field, or a †finite field.

(1) **Mordell-Weil theorem**: Let A be an Abelian variety of dimension n defined over an algebraic number field k of finite degree. Then the group A_k of all k-rational points on A is finitely generated. This theorem was proved by L. J. Mordell (1922) for the case of $n=1$ and by A. Weil (1928) for the general case [2]. The assertion that A_k / mA_k is a finite group for any rational integer m is called the **weak Mordell-Weil theorem**; this theorem is basic in the proof of the Mordell-Weil theorem and is used in the proof of Siegel's theorem, too. A generalization of the Mordell-Weil theorem is obtained in the case where k is a field (of arbitrary characteristic) finitely generated over the †prime field [2].

If A is defined over a finite algebraic number field k, we have the following conjectures of Birch, Swinnerton-Dyer, and Tate on the rank of A_k. Let $\mathfrak{p}$ be a prime ideal of k at which A has a good reduction, and denote by $A_\mathfrak{p}$ the reduced variety. Let $\pi_\mathfrak{p}^{(1)}, \ldots, \pi_\mathfrak{p}^{(2n)}$ be the eigenvalues of the $N(\mathfrak{p})$th power endomorphism of $A_\mathfrak{p}$ with respect to an l-adic representation (→ 3 Abelian Varieties), and put $L_\mathfrak{p}(s,A)=\prod_{i=1}^{2n}(1-\pi_\mathfrak{p}^{(i)}N(\mathfrak{p})^{-s})^{-1}$. The L-function of A defined by $L(s,A)=\prod' L_\mathfrak{p}(s,A)$, where the product ranges over all good primes, is the principal part of the zeta function of A (→ 436 Zeta Functions). Birch and Swinnerton-Dyer conjectured that if $k=\mathbf{Q}$ and A is of dimension 1, then there exists a constant $C \neq 0$ such that $L(s,A) \sim C(s-1)^g$ as $s \to 1$. Tate generalized this conjecture to any A and k. Moreover, the constant C, appropriately modified by factors corresponding to the bad primes and the infinite primes, is thought to be expressed by certain arithmetic invariants of A [10]. These conjectures are supported by machine computation and some theoretical evidence.

(2) **Lutz-Mattuck theorem**: The group of rational points of an Abelian variety A of dimension n over a †$\mathfrak{p}$-adic number field k contains a subgroup of finite index isomorphic to the direct sum of n copies of the †ring $\mathfrak{o}$ of $\mathfrak{p}$-adic integers in k (E. Lutz, *J. Reine Angew. Math.*, 177 (1937); A. Mattuck, *Ann. of Math*, 62 (1955)).

(3) **Mordell's conjecture**: The number of rational points of an algebraic curve C defined over $\mathbf{Q}$ with genus larger than 1 is finite. This is still an open problem [2]. If C is defined over a function field, the conjecture is true, as was shown by Ju. I. Manin (1963), H. Grauert (1965), P. Samuel (1966), and M. Miwa (1966).

(4) For an algebraic variety defined over a finite field k, B. Dwork proved the following remarkable result: Let N_m be the number of rational points over the extension k_m of degree m over k. Then the function $Z(t)$ defined by $(d/dt)\log Z(t)=\sum_{m=1}^{\infty} N_m t^{m-1}$ is a rational function of t. We call $Z(t)$ the †con-

gruence zeta function of V over k. (For the details of $Z(t)$ → 436 Zeta Functions.)

D. C_i-Fields

Let F be a field, and let $i \geqslant 0$, $d \geqslant 1$ be integers. Let f be a homogeneous polynomial of n variables of degree d with coefficients in F. If the equation $f=0$ has a solution $(x_1, \ldots, x_n) \neq (0, \ldots, 0)$ in F for any f with $n > d^i$, then F is called a **$C_i(d)$-field**. If F is a $C_i(d)$-field for any $d \geqslant 1$, then F is called a **C_i-field**. In order for F to be a C_0-field, it is necessary and sufficient that F be an †algebraically closed field. A C_1-field is sometimes called a **quasi-algebraically closed field**. There exists no noncommutative algebra over a C_1-field F. A finite field is C_1 (C. Chevalley (1936)). If F_0 is algebraically closed, then $F = F_0(X)$ (rational function field of one variable) is a C_1-field (**Tsen's theorem**). A homogeneous polynomial f of $n = d^i$ variables of degree d with coefficients in F such that $f=0$ has no solution in F except $(0, \ldots, 0)$ is called a **normic form** of order i in F. If a C_i-field F_0 has at least one normic form of order i, then (i) $F_0(X_1, \ldots, X_k)$ is a C_{i+k}-field; and (ii) an extension of F_0 of finite degree is a C_i-field. A complete field F with respect to an †exponential valuation is a C_1-field whenever its residue field F_0 is algebraically closed. The field F of power series of one variable over a finite field F_0 is a C_2-field (Lang [7]). Artin conjectured that a p-adic field $\mathbf{Q}_p$ is a C_2-field. It was proved by H. Hasse (1923) that $\mathbf{Q}_p$ is a $C_2(2)$-field and by D. Lewis (1952) that $\mathbf{Q}_p$ is a $C_2(3)$-field. However, G. Terjanian (1966) gave a counterexample to Artin's conjecture; that is, he gave a quartic form of 18 variables with coefficients in $\mathbf{Q}_2$ having only trivial zero in $\mathbf{Q}_2$. J. Ax and S. Kochen (1965) proved that for any integer $d \geqslant 1$ there exists an integer $p_0(d)$ such that $\mathbf{Q}_p$ is a $C_2(d)$-field for $p > p_0(d)$ (→ 274 Model Theory E).

References

[1] T. Takagi, Syotô seisûron kôgi (Japanese; Lectures on elementary theory of numbers), Kyôritu, 1931.
[2] S. Lang, Diophantine geometry, Interscience, 1962.
[3] W. J. LeVeque, Topics in number theory II, Addison-Wesley, 1956.
[4] T. Skolem, Diophantische Gleichungen, Erg. Math., Springer, 1938 (Chelsea, 1950).
[5] J. F. Koksma, Diophantische Approximationen, Erg. Math., Springer, 1936 (Chelsea, 1950).
[6] L. J. Mordell, Diophantine equations, Academic Press, 1969.
[7] S. Lang, On quasi algebraic closure, Ann. of Math., (2) 55 (1952), 373–390.
[8] J. W. S. Cassels, Diophantine equations with special reference to elliptic curves, J. London Math. Soc., 41 (1966), 193–291.
[9] J. Ax and S. Kochen, Diophantine problems over local fields I, II, Amer. J. Math., 87 (1965), 605–648; III, Ann. of Math., (2), 83 (1966), 437–456.
[10] P. Swinnerton-Dyer, The conjectures of Birch and Swinnerton-Dyer, and of Tate, Proc. conference on local fields, Driebergen, 1966 (Springer, 1967).
[11] M. J. Greenberg, Lectures on forms in many variables, Benjamin, 1969.
Also → references to 187 Geometry of Numbers.

123 (XX.19) Dirichlet, Peter Gustav Lejeune

Peter Gustav Lejeune Dirichlet (February 13, 1805–May 5, 1859) was born of a French family in Düren, Germany. From 1822 to 1827 he was in Paris, where he became a friend of J.-B. †Fourier. In 1827, he was appointed lecturer at the University of Breslau; in 1829, lecturer at the University of Berlin; and in 1839, professor at the University of Berlin. In 1855, he was invited as successor to †Gauss to the University of Göttingen, where he spent his last four years as a professor.

His works cover many aspects of mathematics; however, those on number theory, analysis, and potential theory are most famous. He greatly admired Gauss and is said to have kept Gauss's *Disquisitiones arithmeticae* at his side even when traveling.

In number theory, he created the †Dirichlet series and proved that a sequence in arithmetic progression contains infinitely many prime numbers provided that the first term and the common difference are relatively prime. Also, using his "drawer principle," which states that if there are $n+1$ objects in n drawers then at least one drawer contains at least 2 objects, he clarified the structure of †unit groups of †algebraic number fields. In †potential theory he dealt with the †Dirichlet problem concerning the existence of †harmonic functions. He also gave †Dirichlet's condition for the convergence of trigonometric series.

References

[1] P. G. L. Dirichlet's Werke I, II, G. Reimer, 1889, 1897 (Chelsea, 1969).
[2] P. G. L. Dirichlet, Vorlesungen über Zahlentheorie, with supplements by R. Dedekind, F. Vieweg, Braunschweig, fourth edition, 1894 (Chelsea, 1968).
[3] F. Klein, Vorlesungen über die Entwicklung der Mathematik im 19. Jahrhundert I, Springer, 1926 (Chelsea, 1956).
[4] T. Takagi, Kinsei sûgakusidan (Japanese; Topics from the history of mathematics of the 19th century), Kyôritu, 1933 (Kawade, 1942).

124 (X.31) Dirichlet Problem

A. The Classical Dirichlet Problem

Let D be a bounded or unbounded †domain in $\mathbf{R}^n$ ($n \geqslant 2$) with compact boundary S. The classical **Dirichlet problem** is the problem of finding a †harmonic function in D that assumes the values of a prescribed continuous function on S. This problem is also called the †first boundary value problem (→ 195 Harmonic Functions). In this article f always stands for a boundary function given on S. The problem is called an **interior problem** if D is bounded and an **exterior problem** if D is unbounded. In an exterior problem, it is further required that when an †inversion with center at an exterior point P_0 of D is performed on D and a †Kelvin transformation is performed on the solution in D (when the solution exists), the function thus obtained on the inversion image of D be harmonic at P_0 ($n \geqslant 3$). When $n=2$, the solution in D is already regarded as a function on the inversion image of D, which is required to be harmonic at P_0. Thus an interior problem can be transformed to an exterior problem, and vice versa. We now explain the history of the classical problem.

Let D be a bounded domain with boundary S in $\mathbf{R}^3$. G. Green (1828) asserted that if S is sufficiently smooth,

$$u(P) = -\frac{1}{4\pi}\int_S f(Q)\frac{\partial G(P,Q)}{\partial n_Q}\,d\sigma(Q) \qquad (1)$$

is the solution for the Dirichlet problem, where f is prescribed on S, $G(P,Q)$ is †Green's function with the pole at Q in D, n_Q is the outward-drawn normal to S at Q, and $d\sigma$ is the †surface element on S. He took for granted the existence of Green's function from the physics of the problem. Thus his discussion was not quite rigorous. This defect was corrected by A. M. Ljapunov (1898) under a certain condition on S. Denote by u_m the Newtonian potential of a measure with density $m \geqslant 0$ on S. Assume that a continuous function f on S and a positive constant a are given. In 1840, C. F. Gauss investigated the existence of a density $m_f \geqslant 0$ on S of total mass a which satisfies $\int_S (u_{m_f} - 2f) m_f\, d\sigma = \min_m \int_S (u_m - 2f) m\, d\sigma$, where the total mass of m is equal to a. He asserted also that $u_{m_f} - f$ is equal to a constant b on S. If $f \equiv 0$, then u_{m_0} must be equal to a positive constant c on S, and hence $u_{m_f} - bc^{-1}u_{m_0}$ must be a solution of the exterior problem for the boundary function f on S. However, his discussion was incomplete because we cannot always assure the existence of a density that gives a measure minimizing the integral. Moreover, even in the case where D is a ball, there exists a continuous function $f \geqslant 0$ on S such that there is no Newtonian potential that is equal to f on S up to a constant (M. Ohtsuka, 1961). Gauss (1840), W. Thomson (Lord Kelvin) (1847), and G. L. Dirichlet solved the Dirichlet problem by making use of the so-called **Dirichlet principle**, which will be explained in detail in Section F. After K. Weierstrass (1870) pointed out that there is a case where no minimizing function exists, D. Hilbert (1899) gave a rigorous proof of the Dirichlet principle under a certain condition. Meanwhile, C. G. Neumann (1870) solved the Dirichlet problrem rigorously for the first time, although he assumed that D is a convex domain with a smooth boundary. First, he considered the potential $W_1 = (1/2\pi)\cdot \int_S f(\partial r^{-1}/\partial n)d\sigma$ of a double layer in D; then he formed the potential $W_2 = (1/2\pi)\cdot \int_S f_1(\partial r^{-1}/\partial n)d\sigma$ of a double layer with the values f_1 of W_1 on S and defined $W_3, W_4, \ldots$ similarly. The series $W_1 + W_2 + W_3 + W_4 + \ldots$ plus a suitable constant gives a solution to the exterior Dirichlet problem for the boundary function f. In 1887, H. Poincaré also used (1) to solve the Dirichlet problem. He obtained Green's function with the pole at O in a bounded domain D in the following manner: Let D' be the image of D by an inversion with center O, S_0 be a spherical surface surrounding the boundary $\partial D'$ of D', and μ be a uniform measure on S_0 such that the potential of μ is equal to 1 inside S_0. By †sweeping out μ to $\partial D'$, the solution in D' of the exterior problem for the boundary function 1 is obtained. A †Kelvin transformation of this solution yields the solution h in D of the interior problem for the boundary function $1/\overline{OP}$. Now $1/\overline{OP}-h(P)$ is Green's function in D. In 1899 Poincaré used another method (without utilizing (1)) to solve the Dirichlet problem [9]. He observed that it is sufficient to consider the case where f is equal to the restriction to S of a polynomial g, and that g is expressed in D as the sum of the Newtonian potential of the measure τ of general sign with density $-\Delta g/(4\pi)$ and a function that is harmonic in D and continuous on $D\cup S$. If it is possible to sweep out τ to ∂D, then the solution is obtained. He showed that this is in fact so if at every point P of S there exists a cone that is disjoint from D and has its vertex at P. This condition is called **Poincaré's condition.** In 1900, I. Fredholm discussed the Dirichlet problem by reducing it to a problem of †integral equations. A domain D is called a **Dirichlet domain** if any (classical) Dirichlet problem is solvable in D. H. Lebesgue (1912) showed that a solution is obtained by the method of iterative averaging in every Dirichlet domain.

B. The Dirichlet Problem in a General Domain

It had been believed that any classical Dirichlet problem is solvable in every domain until S. Zaremba observed in 1909 that the problem is not solvable for some boundary condition in a punctured ball. In 1913, Lebesgue gave a decisive example in which the domain is homeomorphic to a ball and bounded by a surface sufficiently smooth except at one point. Thus the central interest shifted to finding a harmonic function in D that depends only on a continuous function f given on ∂D and coincides with the classical solution when D is a Dirichlet domain. Extend f to a continuous function in the whole space, and denote it by f_0. Approximate D by an increasing sequence $\{D_n\}$ of Dirichlet domains, and denote by u_n the solution in D_n of the Dirichlet problem for the boundary function f_0. N. Wiener proved in 1924 that u_n converges to a harmonic function that is independent of the choice of extension of f and $\{D_n\}$. The question of where on ∂D the general solution assumes the given boundary values will be treated in Section C. O. D. Kellogg (1928) found a general method that includes Poincaré's method of sweeping out, Schwarz's alternating method, and the result of Wiener. Poincaré's method of sweeping out and Lebesgue's method of iterative averaging both yield Wiener's general solution.

C. Perron's Method

We explain O. Perron's method (1923) in the form improved by M. Brelot (1939). For simplicity we assume that the domain D is bounded in $\mathbf{R}^3$. Given a boundary point M, let U be the family of †subharmonic functions u bounded above and satisfying $\limsup u(P) \leqslant f(M)$ as P tends to M. Define $\underline{H}_f(P)$ as $\sup_u u(P)$, where u runs through U, if this family is not empty; otherwise, set $\underline{H}_f\equiv-\infty$. Call $\underline{H}_f$ a **hypofunction.** Define $\overline{H}_f$ by $-\underline{H}_{-f}$ and call it a **hyperfunction.** If $\underline{H}_f=\overline{H}_f$, the common function is denoted by H_f; if $H_f(p)<\infty$, then H_f is harmonic. This is called a **Perron-Brelot solution (Perron-Wiener-Brelot solution** or simply PWB solution). The method of defining H_f is called **Perron's method** (the Perron-Brelot method or the Perron-Wiener-Brelot method).

Wiener showed in 1923 that the †Daniell-Stone integral may be regarded as a general solution if a †Daniell-Stone integrable function f is given on the boundary of a Dirichlet domain; in 1925, he showed that the same is true for a general domain (not necessarily Dirichlet). He showed also that his solution coincides with the Perron-Brelot solution H_f if f is continuous. Unfortunately, however, from a wrong example he concluded that $\underline{H}_f\neq\overline{H}_f$ can hold even for a simple discontinuous f, and so he lost interest in Perron's method. Brelot (1939) corrected Wiener's erroneous conclusion and proved that the Daniell upper and lower integrals are equal to $\overline{H}_f$ and $\underline{H}_f$, respectively. To any continuous f there corresponds an H_f, and there exists a †Radon measure μ_P satisfying $H_f(P)=\int f d\mu_P$. This measure is called a **harmonic measure** or harmonic measure function. Brelot showed that $\underline{H}_f=\overline{H}_f$ if and only if f is μ_P-integrable for one (or every) P. In particular, if D is a Dirichlet domain and E is a closed set on the boundary ∂D, then the harmonic measure function $\mu_P(E)$ takes the value 1 at an inner point (in the space ∂D) of E and vanishes on $\partial D-E$. We note that μ_P is equal to the measure obtained by sweeping out the unit mass at P to ∂D.

D. Regular Boundary Points

If $H_f(P)\rightarrow f(M)$ as $P\rightarrow M\in\partial D$ for any continuous function f on ∂D, then M is called **regular**. To be a regular point is a local property. A boundary point that is not regular is called **irregular**. The regularity of M is equivalent to the convergence of μ_P as $P\rightarrow M$ to the unit measure at M with respect to the †vague topology. There are many sufficient conditions and necessary conditions for a boundary point to be regular. The existence of a **barrier** is a qualitative condition that is necessary and sufficient for a boundary point to be regular. It was used by Poincaré and named and used effectively by Lebesgue. A barrier is a continuous superharmonic function in D that assumes the boundary value 0 at M and has a positive lower bound outside every ball with center at M. A positive superharmonic function defined in the intersection of D and a neighborhood of M and taking the boundary value 0 may be used as a barrier. A necessary and sufficient condition for a boundary point M to be regular is the existence of a Green's function in D assuming the value 0 at M. This condition was given by G. Bouligand (1925), and it follows from the existence of a barrier. Another necessary and sufficient condition of a quantitative nature was obtained by Wiener. It is equivalent to the requirement that the complement of D be †thin at M. Kellogg conjectured that the set of irregular boundary points is of capacity zero and verified this in $\mathbf{R}^2$ (1928). The conjecture was proved first by G. C. Evans (1933) in $\mathbf{R}^3$, and different proofs were given by F. Vasilesco (1935) and O. Frostman (1935). The conjecture is also true in $\mathbf{R}^n$ for $n \geqslant 4$.

E. The More General Dirichlet Problem

So far, we have been concerned with $\mathbf{R}^n$. More generally, Brelot and G. Choquet [3] obtained the following result in a Green space $\mathfrak{G}$ (→ 195 Harmonic Functions): Consider a metric space that contains $\mathfrak{G}$ and in which $\mathfrak{G}$ is everywhere dense, and denote by Δ the complement of $\mathfrak{G}$ with respect to the space. Let $\{F\}$ be a family of †filters on $\mathfrak{G}$ such that each F converges to a certain point of Δ. Suppose that $u \leqslant 0$ whenever u is subharmonic and bounded above on $\mathfrak{G}$ and $\limsup u \leqslant 0$ along every F. Assume the existence of a barrier v in a neighborhood in $\mathfrak{G}$ of the limit point 0 of every F; that is, v is to be positive superharmonic, to tend to 0 along F, and to have a positive lower bound outside every neighborhood of Q. Under these assumptions, we obtain the PWB solution on $\mathfrak{G}$ as in $\mathbf{R}^3$. There are various examples of Δ and F that satisfy these conditions. In particular, L. Naïm [6] investigated in detail the case where Δ is a †Martin boundary. More generally, it is possible to treat the Dirichlet problem axiomatically (→ 397 Subharmonic Functions).

F. The Dirichlet Principle

Let D be a bounded domain with a sufficiently smooth boundary in $\mathbf{R}^n$, and f be a piecewise C^1-function in D with finite **Dirichlet integral** $\|f\|^2 = \int_D |\operatorname{grad} f|^2 \, d\tau$, where $d\tau$ is the volume element. Suppose that f has continuous boundary values φ on ∂D. The classical Dirichlet principle asserts that the solution of the Dirichlet problem for φ has the smallest Dirichlet integral among the functions that are piecewise of class C^1 in D and assume boundary values φ. In a general domain, H_φ minimizes $\|u-f\|$ among harmonic functions u in D. Brelot [2] discussed the principle for a family of competing functions that are defined in a domain on $\mathfrak{G}$ and whose boundary values may not be defined in the classical manner.

References

[1] M. Brelot, Familles de Perron et problème de Dirichlet, Acta Sci. Math. Szeged., 9 (1939), 133–153.
[2] M. Brelot, Etude et extensions du principe de Dirichlet, Ann. Inst. Fourier, 5 (1955), 371–419.
[3] M. Brelot and G. Choquet, Espaces et lignes de Green, Ann. Inst. Fourier, 3 (1952), 199–263.
[4] R. Courant, Dirichlet's principle, conformal mapping, and minimal surfaces, Interscience, 1950.
[5] O. D. Kellogg, Foundations of potential theory, Springer, 1929.
[6] L. Naïm, Sur le rôle de la frontière de R. S. Martin dans la théorie du potentiel, Ann. Inst. Fourier, 7 (1957), 183–281.
[7] M. Ohtsuka, Kansûron tokuron (Japanese; Topics on the theory of functions), Kyôritu, 1957.
[8] M. Ohtsuka, Dirichlet problem, extremal length and prime ends, Lecture notes, Washington Univ., St. Louis, 1962–1963.
[9] H. Poincaré, Théorie du potentiel Newtonien, Leçons professées à la Sorbonne, Gauthier-Villars, 1899.

125 (XI.4) Dirichlet Series

A. Dirichlet Series

For $z = x + iy$, $\lambda_n > 0$, and $\lambda_n \uparrow +\infty$, the series of the form

$$f(z) = \sum_{n=1}^{\infty} a_n \exp(-\lambda_n z) \tag{1}$$

is called a **Dirichlet series** (more precisely, a **Dirichlet series of the type** $\{\lambda_n\}$). If $\lambda_n = n$, then (1) is a power series with respect to e^{-z}.

If $\lambda_n = \log n$, the series (1) becomes

$$\sum_{n=1}^{\infty} a_n / n^z, \tag{2}$$

which is called an **ordinary Dirichlet series**. If $a_n = 1$, then (2) is the †Riemann zeta function. Series of the form (2) were introduced by P. G. L. Dirichlet in 1839 and utilized in an investigation of the problems of †analytic number theory. Later J. Jensen (1884) and E. Cohen (1894) extended the variable z to complex numbers. The Dirichlet series is not only a useful tool in analytic number theory, but is also investigated as a generalization of power series. The †Laplace transform is the generalization of the Dirichlet series to the integral, and similar formulas often hold for both cases.

B. Convergence Regions

If the series (1) converges at $z = z_0$, then it converges in the half-plane $\operatorname{Re} z > \operatorname{Re} z_0$. Therefore, there is a uniquely determined real number S such that (1) converges in $\operatorname{Re} z > S$ and diverges in $\operatorname{Re} z < S$. If (1) always converges (diverges), we put $S = -\infty$ $(+\infty)$. We call S the **abscissa of convergence** (or **abscissa of simple convergence**). Similarly, there is a uniquely determined real number A such that (1) converges absolutely in $\operatorname{Re} z > A$ and is not absolutely convergent in $\operatorname{Re} z < A$. We call A the **abscissa of absolute convergence**. Furthermore, there is a uniquely determined real number U such that for every $U' > U$, (1) converges uniformly in $\operatorname{Re} z > U'$ and does not converge uniformly in $\operatorname{Re} z > U''$ for every $U'' < U$. The number U is called the **abscissa of uniform convergence**. Among these abscissas we always have the following relations:

$$-\infty \leqslant S \leqslant U \leqslant A \leqslant +\infty,$$

$$A - S \leqslant \limsup_{n \to \infty} \frac{\log n}{\lambda_n}.$$

The latter was proved by Cohen in (1894). The numbers S, A, U are determined from a_n, λ_n by the following formulas:

$$S = \limsup_{x \to \infty} \frac{1}{x} \log \left| \sum_{[x] \leqslant \lambda_n < x} a_n \right|, \tag{3}$$

$$A = \limsup_{x \to \infty} \frac{1}{x} \log \left(\sum_{[x] \leqslant \lambda_n < x} |a_n| \right), \tag{4}$$

$$U = \limsup_{x \to \infty} \frac{1}{x} \log T_x,$$

$$T_x = \sup_{-\infty < y < +\infty} \left| \sum_{[x] \leqslant \lambda_n < x} a_n \exp(-i\lambda_n y) \right|, \tag{5}$$

where [ ] is the †Gauss symbol. Formulas (3) and (4) were proved by T. Kojima (1914), and (5) by M. Kunieda (1916). In particular, when $\lim_{n \to \infty} (\log n)/\lambda_n = 0$, we have the following:

$$S = U = A = \limsup_{n \to \infty} \frac{\log |a_n|}{\lambda_n} \tag{6}$$

(O. Szász, 1922; G. Valiron, 1924).

The series (1) converges uniformly in the angular domain $\{z \mid |\arg(z - z_0)| < \alpha < \pi/2\}$, where the vertex z_0 lies on the line $\operatorname{Re} z_0 = S$. Hence it represents a holomorphic function in the domain $\operatorname{Re} z > S$, but it is possible that there is no singularity on the line $\operatorname{Re} z = S$. For example, if $a_n = (-1)^n$, then the series (2) has $S = 0$, but the sum is an †entire function $(2^{1-z} - 1)\zeta(z)$. Taking the analytic continuation $f(z)$ of the series (1), the infimum R of ρ such that $f(z)$ is holomorphic in $\operatorname{Re} z > \rho$ is called the **abscissa of regularity**. However, it is still possible that there is no singularity on the line $\operatorname{Re} z = R$. We always have $R \leqslant S$, and R is given by the following formula:

$$R = \sup_{-\infty < y < \infty} \limsup_{x \to -\infty} (\log \log^+ |\varphi(x + iy)| + x),$$

$$\varphi(z) = \sum_{n=1}^{\infty} \frac{a_n \exp(-\lambda_n z)}{\Gamma(1 + \lambda_n)}, \tag{7}$$

where $\log^+ a = \max(\log a, 0)$ (C. Tanaka, 1951). The infimum B of ρ such that $f(z)$ is bounded in $\operatorname{Re} z > \rho$ is called the **abscissa of boundedness**. We always have $R \leqslant B \leqslant A$. H. Bohr proved the following three theorems concerning these values: (1) If $\{\lambda_n\}$ are linearly independent over the ring of integers, then $A = B$ (1911). (2) If $(\lambda_{n+1} - \lambda_n)^{-1} = O(\exp e^{\lambda_n \varepsilon})$ for every $\varepsilon > 0$, then $U = B$ (1913). (3) If $\limsup (\log n)/\lambda_n = 0$, then $S = U = A = B$ (1913). In the final case, the values are given by (6).

C. Properties of Functions Given by Dirichlet Series

The coefficients a_n in (1) are given by the function $f(z)$ as follows:

$$\sum_{\nu=1}^{n} a_\nu = \frac{1}{2\pi i} \int_{c - i\infty}^{c + i\infty} f(z) \frac{e^{\omega z}}{z} dz, \tag{8}$$

where $c > \max(S, 0)$, $\lambda_n < \omega < \lambda_{n+1}$, and the integration contour does not pass through $\{\lambda_n\}$. If $\omega = \lambda_n$, then the term a_n in the sum of the left-hand side of (8) is replaced by $a_n/2$ (O. Perron, 1908). Furthermore, if $S < x$, then we have

$$a_n = \lim_{T \to \infty} \frac{1}{T} \int_{\alpha_0}^{\alpha_0 + T} f(x + iy) \times \exp[\lambda_n(x + iy)] dy \tag{9}$$

(J. Hadamard, 1908; C. Tanaka, 1952).

If $x=\operatorname{Re} z>S$, then $f(z)=o(|y|)$ $(|y|\to\infty)$. In order to investigate its behavior more precisely, Bohr introduced

$$\mu(x)=\limsup_{|y|\to+\infty}\frac{\log|f(x+iy)|}{\log|y|}$$

in his thesis (1910) and called it the **order** over $\operatorname{Re} z=x$. The function $\mu(x)$ is nonnegative, monotone decreasing, convex, and continuous with respect to x. Bohr later found that there is a kind of periodicity for the values of $f(z)$ over $\operatorname{Re} z=x_0$; this was the origin of the theory of †almost periodic functions.

As for the zeros of the function $f(z)$, the following theorems are known: If $f(z)$ is not identically zero, it has only a finite number of zeros in $x\geqslant S+\varepsilon$, $e^{-Mx}\leqslant y\leqslant e^{Mx}$ for arbitrary positive numbers ε, M (Perron, 1908). If we denote by $N(T)$ the number of zeros in $x>S+\varepsilon$, $T<y<T+2\delta\log T$, then $\limsup_{T\to\infty}N(T)/(\log T)^2\leqslant\delta/\varepsilon$ (E. Landau, 1927).

There have been many investigations into the connection between the singularities of $f(z)$ and the coefficients a_n. If the a_n are real and positive, the point $z=S$ is always a singularity of $f(z)$. Moreover, if $S=0$, $\operatorname{Re} a_n\geqslant 0$, and $\lim_{n\to\infty}(\cos(\arg a_n))^{1/\lambda_n}=1$, then $z=0$ is a singularity of $f(z)$ (C. Biggeri, 1939). Furthermore, if $\lambda_n/n\to\infty$, $\liminf_{n\to\infty}(\lambda_{n+1}-\lambda_n)>0$, then the line $\operatorname{Re} z=S$ is the †natural boundary of $f(z)$ (F. Carleson and Landau, 1921; A. Ostrowski, 1923). If $S=0$, $\liminf_{n\to\infty}(\lambda_{n+1}-\lambda_n)=q>0$, then there always exist singularities on every interval on the imaginary axis with the length $2\pi/q$ (G. Pólya, 1923). S. Mandelbrojt (1954, 1963) gave some interesting results concerning the relations between the singularities of (1) and the Fourier transform of an entire function.

If $U=-\infty$, the function $f(z)$ is an entire function. Its †order (in the sense of entire function) ρ is given by

$$\rho=\limsup_{x\to-\infty}\frac{\log^+\log^+M(x)}{|x|},$$
$$M(x)=\sup_{-\infty<y<\infty}|f(x+iy)|.$$

There have been many investigations into the †Julia direction of $f(z)$ and related topics by Mandelbrojt, Valiron, and Tanaka.

D. Tauberian Theorems

As in the case of power series, if the series Σa_n converges to s, then $f(+0)=s$ (**Abel's continuity theorem**). The converse is not necessarily true. The converse theorems, with additional conditions on a_n and λ_n, are called **Tauberian theorems**, as in the case of power series. Many theorems are known about this field. The most famous additional conditions are $\lim_{n\to\infty}\lambda_n a_n/(\lambda_n-\lambda_{n-1})=0$ (Landau, 1926), and $a_n=O((\lambda_n-\lambda_{n-1})/\lambda_n)$ (K. Ananda-Rau, 1928). (Regarding the summation of Dirichlet series, especially †Riesz summability, → 398 Summability L).

E. Series Related to Dirichlet Series

A series of the form

$$\sum_{n=1}^{\infty}\frac{n!a_n}{z(z+1)(z+2)\dots(z+n)},$$

$z\neq 0,-1,-2,\dots$

is called a **factorial series** with the coefficients $\{a_n\}$. It converges or diverges simultaneously with the ordinary Dirichlet series $\Sigma a_n/n^z$ except at $z=0$ and negative integers. The series

$$\sum_{n=1}^{\infty}\frac{(z-1)(z-2)\dots(z-n)}{n!}a_n$$
$$=\sum_{n=1}^{\infty}a_n\binom{z-1}{n}$$

is called a **binomial coefficient series**. It converges or diverges simultaneously with the ordinary Dirichlet series $\Sigma(-1)^n a_n/n^z$ except at $z=0$ and positive integers.

References

[1] S. Izumi, Dirikure kyûsûron (Japanese; Theory of Dirichlet series), Iwanami, 1931.
[2] V. Bernstein, Leçons sur les progrès récents de la théorie des séries de Dirichlet, Gauthier-Villars, 1933.
[3] K. Chandrasekharan and S. Minakshisundaram, Typical means, Oxford Univ. Press, Bombay, 1952.
[4] G. H. Hardy and M. Riesz, The general theory of Dirichlet's series, Cambridge, 1915.
[5] S. Mandelbrojt, Séries lacunaires, Actualités Sci. Ind., Hermann, 1936.
[6] S. Mandelbrojt, Séries adhérentes, régularisation des suites, applications, Gauthier-Villars, 1952.
[7] G. Valiron, Théorie générale des séries de Dirichlet, Mémor. Sci. Math., Gauthier-Villars, 1926.

126 (IV.15) Discontinuous Groups

A. Definitions [2,9,13]

Suppose that a group Γ is acting continuously on a †Hausdorff space X, that is, for every

$\gamma \in \Gamma$ and $x \in X$, an element γx of X is assigned in such a way that the mapping $x \to \gamma x$ is a homeomorphism of X onto itself and that we have $\gamma_1(\gamma_2 x) = (\gamma_1\gamma_2)x$, $1x = x$, where 1 is the identity element of Γ. Two points $x, y \in X$ are said to be **Γ-equivalent** if there exists a $\gamma \in \Gamma$ such that $y = \gamma x$. (Γ-equivalence for subsets of X is defined similarly.)

We consider the following conditions of discontinuity of Γ. (i) For every $x \in X$ and any infinite sequence $\{\gamma_i\}$ consisting of distinct elements of Γ, the sequence $\{\gamma_i x\}$ has no †cluster point in X. (ii) For every $x \in X$, there exists a neighborhood U_x such that $\gamma U_x \cap U_x = \varnothing$ for all but finitely many $\gamma \in \Gamma$. (ii′) If $x, y \in X$ are not Γ-equivalent, there exist neighborhoods U_x, U_y of x, y, respectively, such that $\gamma U_x \cap U_y = \varnothing$ for all $\gamma \in \Gamma$. (iii) For any compact subset M of X, $\gamma M \cap M = \varnothing$ for all but finitely many $\gamma \in \Gamma$.

It is easy to see that (ii)⇒(i), (ii)+(ii′)⇒(iii), and if, moreover, X is †locally compact, we also have (iii)⇒(ii), (ii′). When (i) holds, Γ is called a **discontinuous transformation group** of X, and when (ii) holds, Γ is called a **properly discontinuous transformation group**. In particular, when X can be identified with a †homogeneous space G/K of a locally compact group G by a compact subgroup K, the conditions (i), (ii), and (iii) for a subgroup Γ of G are all equivalent, and they are also equivalent to the condition that Γ is a †discrete subgroup of G.

For a discontinuous group Γ acting on X, the †stabilizer $\Gamma_x = \{\gamma \in \Gamma \mid \gamma x = x\}$ of $x \in X$ is always a finite subgroup. When $\Gamma_x = \{1\}$ for all $x \in X$, Γ is said to be **free** (or to **act freely** on X). If $\Gamma_X = \bigcap_{x \in X} \Gamma_x = \{1\}$, Γ is said to act †effectively on X. A point $x \in X$ is called a **fixed point** of Γ if $\Gamma_x \neq \Gamma_X$. In the following, we assume for simplicity that Γ acts effectively on X, unless otherwise specified.

Since Γ-equivalence is clearly an †equivalence relation, we can decompose X into Γ-equivalence classes, or Γ-†orbits. The space of all Γ-orbits, called the **quotient space** of X by Γ, is denoted by $\Gamma\backslash X$. When Γ satisfies the conditions (ii) and (ii′), the space $\Gamma\backslash X$ becomes a †Hausdorff space with respect to the topology of the quotient space. If, moreover, Γ is free, X is an (unramified) †covering space of $\Gamma\backslash X$ with the †covering transformation group Γ. (Conversely, a covering transformation group is always a free, properly discontinuous transformation group.) In general, X may be viewed as a covering space of $\Gamma\backslash X$ with ramifications, and the ramifying points (in X) are nothing but the fixed points of Γ.

B. Fundamental Regions

A complete set of representatives F of $\Gamma\backslash X$ in X (that is, a subset F of X such that $\Gamma F = X$, $\gamma F \cap F = \varnothing$ for $\gamma \in \Gamma, \gamma \neq 1$) is called a **fundamental region** of Γ in X if it further satisfies suitable topological or geometrical requirements. Here we assume that $\bar{F}$, the closure of F, is the closure of its interior F^i. (In this case, $\bar{F}$ or F^i is sometimes called the fundamental region of Γ instead of F itself.) The existence of such a fundamental region is known if Γ satisfies the conditions (ii), (ii′), and the set of fixed points is †nowhere dense in X (R. Baer, F. W. Levi, 1931). A fundamental region F is called **normal** if the set $\{\gamma F\}$ $(\gamma \in \Gamma)$ is locally finite, that is, if, for every $x \in X$, there exists a neighborhood U_x such that $\gamma F \cap U_x = \varnothing$ for all but a finite number of $\gamma \in \Gamma$. If X is †connected and F is normal, then Γ is generated by the set of $\gamma \in \Gamma$ such that $\gamma \bar{F} \cap \bar{F} \neq \varnothing$. Thus, it is useful to know a fundamental region in order to find a set of generators of Γ and a set of †fundamental relations for them. When X has a Γ-invariant †Borel measure μ and Γ is countable, then the measure $\mu(F)$ of F is independent of the choice of F. Hence, it is legitimate to put $\mu(\Gamma\backslash X) = \mu(F)$; Γ is called a **discontinuous group of the first kind** (C. L. Siegel [9]) if Γ is a discontinuous transformation group which has a normal fundamental region F such that $\{\gamma \mid \gamma\bar{F} \cap \bar{F} = \varnothing\}$ is finite and $\mu(F) < \infty$. For instance, if X is locally compact and $\bar{F}$ is compact (⇔$\Gamma\backslash X$: compact), then Γ is of the first kind.

When we are concerned only with the qualitative properties of Γ, it is sometimes convenient to loosen the conditions for a fundamental region and replace it by a **fundamental (open) set** Ω of Γ, that is, an (open) subset Ω of X such that $\Gamma\Omega = X$ and $\gamma\Omega \cap \Omega = \varnothing$ for all but a finite number of $\gamma \in \Gamma$ [1, 3, 14, 18].

C. The Case of a Riemann Surface

Let Γ be a discontinuous group of analytic automorphisms of a †Riemann surface X. In virtue of †uniformization theory, it is enough, in principle, to consider the case where X is †simply connected. Thus we have the following three cases:

(1) $X = \mathbf{C} \cup \{\infty\}$ (†Riemann Sphere). Γ is a finite group. Since Γ can also be considered as a group of motions of the sphere, it is either a cyclic, †dihedral, or †regular polyhedral group [5].

(2) $X = \mathbf{C}$ (Complex Plane). Γ is contained in the group of motions of the plane. The subgroup consisting of all parallel translations contained in Γ is a †free Abelian group of rank $\nu \leqslant 2$. If $\nu = 0$, then Γ is a finite cyclic group. In case $\nu > 0$, Γ consists of the transformations of the following form:

When $\nu = 1$, $\quad z \rightarrow \varepsilon^k z + m\omega \qquad (k, m \in \mathbf{Z})$,

When $\nu = 2$, $\quad z \rightarrow \varepsilon^k z + m_1\omega_1 + m_2\omega_2 \qquad (k, m_i \in \mathbf{Z})$,

where ω, ω_1, ω_2 are nonzero complex numbers with $\operatorname{Im}(\omega_2/\omega_1) > 0$, and $\varepsilon = \pm 1$ in general, except in special cases when $\nu = 2$ and $\omega_2/\omega_1 = \zeta_4$ (resp. ζ_3 or ζ_6, where $\zeta_l = \exp(2\pi i/l)$), in which cases we may put $\varepsilon = \zeta_4$ (resp. ζ_3 or ζ_6). For the fundamental regions corresponding to these values of ε, see Fig. 1. In the cases $\nu = 1$ and 2, the †automorphic functions with respect to Γ are essentially given by exponential functions and elliptic functions, respectively (→ 144 Elliptic Functions).

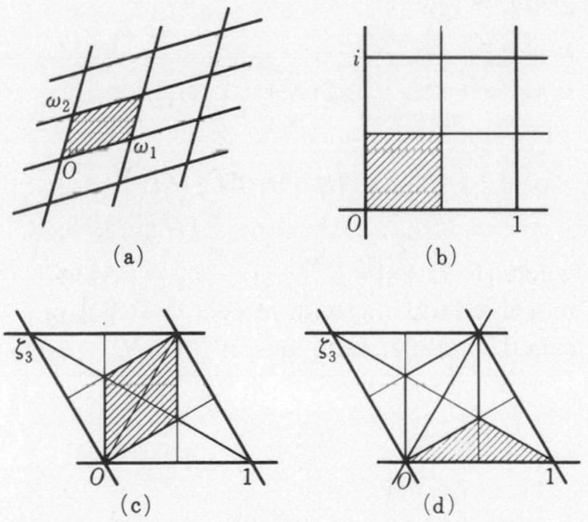

Fig. 1
(a) $\nu = 2$, $\varepsilon = 1$. (b) $\omega_1 = 1$, $\omega_2 = i$, $\varepsilon = i$. (c) $\omega_1 = 1$, $\omega_2 = \zeta_3$, $\varepsilon = \zeta_3$. (d) $\omega_1 = 1$, $\omega_2 = \zeta_3$, $\varepsilon = \zeta_6$.

(3) $X = \{|z| < 1\}$ (Unit Disk) [2,5,8]. By a †Cayley transformation, the unit disk can be transformed to the upper half-plane $\mathfrak{H} = \{z = x + iy \mid y > 0\}$. Any analytic automorphism of $\mathfrak{H}$ is given by a real †linear fractional transformation (Möbius transformation) $z \rightarrow (az + b)(cz + d)^{-1}$ $(a, b, c, d \in \mathbf{R}, ad - bc = 1)$. The totality of real linear fractional transformations act transitively on $\mathfrak{H}$. Hence, $\mathfrak{H}$ can be identified with the †homogeneous space G/K of $G = SL(2, \mathbf{R})$ by $K = SO(2)$ (which is the stabilizer of the point $\sqrt{-1}$). Hence, discontinuous groups Γ of analytic automorphisms of $\mathfrak{H}$ are obtained as discrete subgroups of G. Actually, every element of G defines an analytic automorphism of the whole Riemann sphere, which leaves the real axis $\mathbf{R} \cup \{\infty\}$ invariant. For any $z \in \mathbf{C} \cup \{\infty\}$ and a sequence $\{\gamma_i\}$ consisting of distinct elements of Γ, a cluster point of the sequence $\{\gamma_i z\}$ in $\mathbf{C} \cup \{\infty\}$ is called a **limit point** of Γ. In case there exist only one or two limit points, Γ can easily be transformed to one of the groups given in (2). Otherwise, the set $\mathfrak{L}$ of all limit points of Γ is infinite, and either $\mathfrak{L} = \mathbf{R} \cup \{\infty\}$ or $\mathfrak{L}$ is a †perfect, †nowhere dense subset of $\mathbf{R} \cup \{\infty\}$. When $\mathfrak{L}$ is infinite, Γ is called a **Fuchsoid group**.

Since $\mathfrak{H}$ has a G-invariant †Riemannian metric $ds^2 = y^{-2}(dx^2 + dy^2)$ (called the †Poincaré metric), by which $\mathfrak{H}$ becomes a **hyperbolic plane** (†non-Euclidean plane with negative curvature), we can construct a fundamental region F of Γ which is a normal polygon bounded by geodesics, that is, the arcs of circles orthogonal to the real axis. A set of generators of Γ and the fundamental relations for them are easily obtained by observing the correspondence of the equivalent sides of the fundamental polygon. (Conversely, starting from a normal polygon satisfying a suitable condition, one can construct a discontinuous group Γ having F as a fundamental region. In this manner, we generally obtain a (nontrivial) continuous family of discrete subgroups of G.)

A Fuchsoid group Γ is finitely generated if and only if the fundamental polygon F has a finite number of sides, and in that case Γ is called a **Fuchsian group**. (More generally, a finitely generated discontinuous group of linear fractional transformations acting on a domain in the complex plane is called a **Kleinian group**.) A Fuchsian group Γ is **of the first kind** if and only if $\mathfrak{L} = \mathbf{R} \cup \{\infty\}$; otherwise, it is **of the second kind**. It is also known that a discontinuous group Γ is a Fuchsian group of the first kind if and only if $\mu(\Gamma\backslash\mathfrak{H}) < \infty$ [10]. For a real point $x \in \mathbf{R} \cup \{\infty\}$, we also denote by Γ_x the stabilizer of x (in Γ). The point x is called a **(parabolic) cusp** of Γ if Γ_x is a free cyclic group generated by a †parabolic transformation ($\neq \pm 1$). Cusps of Γ are represented by vertices of the fundamental polygon on the real axis. On the other hand, if a fixed point z of Γ lies in $\mathfrak{H}$, then the stabilizer Γ_z is always a finite cyclic group generated by an †elliptic transformation. Hence, such a point z is also called an **elliptic point** of Γ. For a Fuchsian group Γ of the first kind, let $\{z_1, \ldots, z_s\}$ be a complete set of representatives of the Γ-equivalence classes of the elliptic points of Γ (which can also be chosen from among the vertices of the fundamental polygon), and let e_i be the order of Γ_{z_i}; furthermore, let t be the number of the Γ-equivalence classes of parabolic cusps of Γ. Then the quotient space $\Gamma\backslash\mathfrak{H}$ can be compactified by adjoining t points at infinity, and the resulting space becomes a compact Riemann surface $\mathfrak{R}_\Gamma$ if we define an analytic structure on it in a suitable manner. The area $\mu(\mathfrak{R}_\Gamma)$ measured by the Poincaré metric is given by the following †Gauss-Bonnet formula:

$$\mu(\mathfrak{R}_\Gamma)=\int_F \frac{dx\,dy}{y^2}$$

$$=2\pi\left[2g-2+\sum_{i=1}^{s}\left(1-\frac{1}{e_i}\right)+t\right],$$

where g is the †genus of the Riemann surface $\mathfrak{R}_\Gamma$. It is known that there exists a lower bound $(=\pi/21)$ of $\mu(\mathfrak{R}_\Gamma)$ [10]. Automorphic functions (or Fuchsian functions) with respect to a Fuchsian group Γ, which are essentially the same thing as the algebraic functions on the Riemann surface $\mathfrak{R}_\Gamma$, have been objects of extensive study since Poincaré (1882).

D. Modular Groups [5,6]

The group

$$\Gamma=SL(2,\mathbf{Z})$$

$$=\left\{\begin{pmatrix} a & b \\ c & d\end{pmatrix}\,\middle|\, a,b,c,d,\in\mathbf{Z},\ ad-bc=1\right\}$$

(or the corresponding group of linear fractional transformations) is called the **(elliptic) modular group**. The modular group Γ is a Fuchsian group of the first kind acting on $\mathfrak{H}$, and its fundamental region together with the correspondence of the equivalent sides is shown in Fig. 2. Fig. 3 illustrates the transformations under Γ of the fundamental triangle, where Γ is regarded as acting on the unit disk. From Fig. 2 we obtain the following generators of Γ (mod $\{\pm I_2\}$) ($I_2=\begin{pmatrix} 1 & 0 \\ 0 & 1\end{pmatrix}$) and fundamental relations:

$$\sigma_1=\begin{pmatrix} 1 & 1 \\ 0 & 1\end{pmatrix},\qquad \sigma_2=\begin{pmatrix} 0 & -1 \\ 1 & 0\end{pmatrix},$$

$$\sigma_2^2=(\sigma_2\sigma_1)^3=-I_2.$$

There are two Γ-equivalence classes of elliptic points of Γ, which are represented by $\zeta_4=i$ and ζ_3, with $[\Gamma_i:\{\pm I_2\}]=2$, $[\Gamma_{\zeta_3}:\{\pm I_2\}]=3$, and only one Γ-equivalence class of parabolic cusps which coincides with $\mathbf{Q}\cup\{\infty\}$. The corresponding Riemann surface $\mathfrak{R}_\Gamma$ is analytically equivalent to the Riemann sphere.

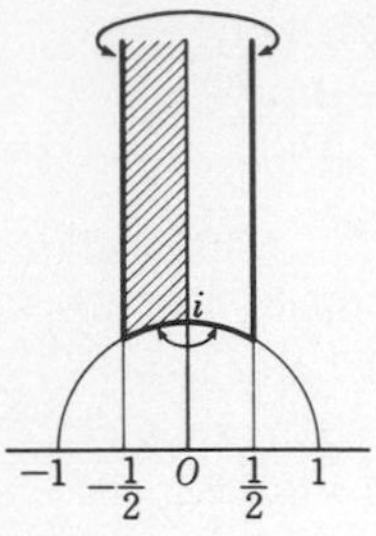

Fig. 2

Fig. 3

For a positive integer N, the totality $\Gamma(N)$ of elements in Γ satisfying the condition $\begin{pmatrix} a & b \\ c & d\end{pmatrix}\equiv\begin{pmatrix} 1 & 0 \\ 0 & 1\end{pmatrix}$ (mod N) forms a normal subgroup of Γ, called a **principal congruence subgroup of level** N. (For the case $N=2$, see Fig. 4.) In general, a subgroup Γ′ of Γ containing $\Gamma(N)$ for some N is called a **congruence subgroup** of Γ. (It is known that there actually exists a subgroup Γ′ of Γ with a finite index, which is not a congruence subgroup.) For $N\geqslant 3$, $-I_2\notin\Gamma(N)$, so that $\Gamma(N)$ is effective. (For $N=1, 2$, we have $\Gamma(N)_{\mathfrak{H}}=\{\pm I_2\}$.) If $N\geqslant 2$, $\Gamma(N)$ has no elliptic point. The number $t(N)$ of the equivalence classes of cusps of $\Gamma(N)$ and the genus $g(N)$ of the corresponding Riemann surface $\mathfrak{R}_{\Gamma(N)}$ are given as follows:

$$t(1)=1,\quad t(2)=3,$$

$$t(N)=(1/2N)[\Gamma:\Gamma(N)]\qquad (N\geqslant 3),$$

$$g(1)=g(2)=0,$$

$$g(N)=1+((N-6)/24N)[\Gamma:\Gamma(N)]\qquad (N\geqslant 3),$$

where $[\Gamma:\Gamma(N)]=N^3\prod_{p|N}(1-1/p^2)$. Automorphic functions with respect to $\Gamma(N)$ are called †modular functions of level N.

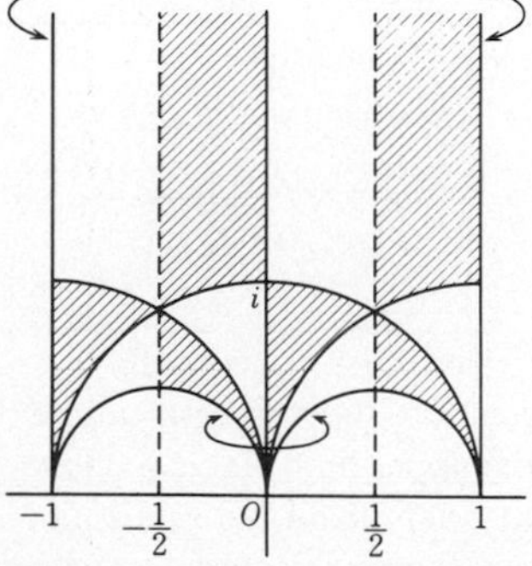

Fig. 4
A fundamental region of Γ(2) which consists of six fundamental regions of Γ(1).

E. The Case of Many Variables

Up to the present time, discontinuous groups Γ and the corresponding automorphic functions have been studied only in the following cases: (2′) $X = \mathbf{C}^n$, $\Gamma \cong \mathbf{Z}^{2n}$ (the free Abelian group of rank $2n$, consisting of parallel translations) [12] (→ 3 Abelian Varieties); (3′) X is a bounded domain in $\mathbf{C}^n$ and Γ is a discontinuous group of analytic automorphisms of X. (In this case, conditions (i), (ii), (iii) are equivalent.)

In the case (3′), the group $\mathfrak{A}(X)$ of all (complex) analytic automorphisms of X, endowed with its natural (†compact-open) topology, becomes a †Lie group, of which Γ is obtained as a discrete subgroup. When $\Gamma \backslash X$ is compact, it is known by the theory of automorphic functions (or by a theorem of Kodaira) that $\Gamma \backslash X$ becomes a †projective variety, which is a †minimal model [2; 12]. In particular, when X is a †symmetric bounded domain, i.e., when X becomes a †symmetric Riemannian space with respect to its †Bergman metric, the connected component G of the identity element of $G' = \mathfrak{A}(X)$ (which incidentally coincides with that of the group $I(X)$ of all †isometries of X) is a †semisimple Lie group of noncompact type (i.e., without compact simple factors), and X can be identified with the homogeneous space of G by a maximal compact subgroup K of G. The theory of discontinuous groups of this type, initiated by Siegel (especially in the case where $X = \mathfrak{H}_n = Sp(n, \mathbf{R})/K$, †Siegel's upper half-space; $\Gamma = Sp(n, \mathbf{Z})$, †Siegel's modular group of degree n), O. Blumenthal, H. Braun, and L. K. Hua, and continued by those in the German school such as M. Koecher, H. Maass, and others, has undergone substantial development in recent years under the influence of the theory of algebraic groups [3, 11, 12, 18] (→ 34 Automorphic Functions).

On the other hand, for a symmetric Riemannian space X of negative curvature, the group of isometries $G = I(X)$ is a †semisimple Lie group of noncompact type with a finite number of connected components and with a finite center, and X can be identified with the homogeneous space of G by a maximal compact subgroup. Therefore the study of discontinuous groups of isometries of X can be reduced to that of discrete subgroups of a Lie group G of this type. A typical example is the case where X is the space of all real †positive definite symmetric matrices of degree n with determinant 1, which can be identified with the quotient space $SL(n, \mathbf{R})/SO(n)$ ($A \in SL(n, \mathbf{R})$ acts on X by $X \ni S \rightarrow {}^tASA$). The unimodular group $\Gamma = SL(n, \mathbf{Z})$ is a discontinuous group of the first kind acting on this space X, and a method of constructing a fundamental region of Γ in X is provided by the **Minkowski reduction theory** [7, 14].

F. Discrete Subgroups of a Semisimple Lie Group

Two subgroups Γ, Γ' of a group G are called **commensurable** if $\Gamma \cap \Gamma'$ is of finite index in both Γ and Γ'. For a real †linear algebraic group $G \subset GL(n, \mathbf{R})$ defined over $\mathbf{Q}$, a subgroup Γ commensurable with $G_{\mathbf{Z}} = G \cap GL(n, \mathbf{Z})$ is called an **arithmetic subgroup** of G (examples: $SL(n, \mathbf{Z})$, $Sp(n, \mathbf{Z})$). An arithmetic subgroup Γ is always discrete, and when G is semisimple, the quotient space $\Gamma \backslash G$ is of finite volume ($\mu(\Gamma \backslash G) < \infty$), with respect to an invariant measure μ. Moreover, $\Gamma \backslash G$ is compact if and only if G is †$\mathbf{Q}$-compact (or †$\mathbf{Q}$-anisotropic), that is, if $G_{\mathbf{Q}}$ or $G_{\mathbf{Z}}$ consists of only †semisimple elements (A. Borel, Harish-Chandra, G. D. Mostow, and T. Tamagawa [1, 14]); the same results remain true if G is †Zariski connected and has no †character defined over $\mathbf{Q}$. The proofs of these facts (and the compactification of the quotient space $\Gamma \backslash X$ for the noncompact case) depend on a construction of fundamental open sets which generalizes the reduction theory of Minkowski and Siegel [3, 7, 11, 18].

For a connected semisimple Lie group G of noncompact type and a discrete subgroup Γ with $\mu(\Gamma \backslash G) < \infty$, the following **density theorem** holds (A. Borel, *Ann. of Math.*, (2) 72 (1960)): (i) For any linear representation ρ of G, the linear closure of $\rho(\Gamma)$ coincides with that of $\rho(G)$; (ii) if G is algebraic, Γ is †Zariski dense in G. Furthermore, suppose G is a direct product of simple groups G_i and the center of G is finite; Γ is called **irreducible** if its projection on any (proper) partial product of $\{G_i\}$ is not discrete. For instance, if G is a †$\mathbf{Q}$-simple algebraic group then $\Gamma = G_{\mathbf{Z}}$ is irreducible. In general, there exists a partition of the set of indices $\{i\}$ such that Γ is commensurable with a direct product of irreducible discrete subgroups of the partial products corresponding to this partition, and these irreducible components are unique up to commensurability.

For a semisimple Lie group G, the problem of counting up all discrete subgroups Γ with $\mu(\Gamma \backslash G) < \infty$ is still open. On the other hand, for a †nilpotent or †solvable Lie group, a general method of constructing discrete subgroups is known; see, for example, M. Saito, *Amer. J. Math.*, 83 (1961). The method of constructing a discrete subgroup Γ of $G = SL(2, \mathbf{R})$ in a geometric manner using the upper half-plane can be generalized to some extent to the construction of discrete subgroups of certain groups using hyperbolic spaces of low dimensions (Vinberg). Except for these few cases, the only examples of Γ known so far are the arithmetic ones.

There are a number of facts which seem to support the conjecture that there should be only very few discrete subgroups of a semisimple Lie group G of higher dimensions. First, the only subgroups of $SL(n,\mathbf{Z})$ $(n \geqslant 3)$, $Sp(n,\mathbf{Z})$ $(n \geqslant 2)$ with finite index are congruence subgroups (H. Bass, M. Lazard, and J.-P. Serre; this result has been generalized to the case of an arbitrary †Chevalley group over an algebraic number field by C. C. Moore and H. Matsumoto). Second, $G_{\mathbf{Z}} = SL(n,\mathbf{Z})$, $Sp(n,\mathbf{Z})$ are maximal in G [18]. Finally, it is known that if a connected semisimple Lie group G with a finite center does not contain a simple factor which is locally isomorphic to $SL(2,\mathbf{R})$, then any discrete subgroup Γ of G with compact quotient $\Gamma \backslash G$ has no nontrivial †deformation (i.e., all deformations are obtained from inner automorphisms of G) (A. Selberg, E. Calabi, and E. Vesentini, A. Weil [15]). This last result amounts to the vanishing of the cohomology group $H^1(\Gamma, X, ad)$, and in this connection an extensive study has been made by Y. Matsushima, S. Murakami, G. Shimura, and K. G. Raguanathan to determine the †Betti numbers of $\Gamma \backslash X$, and more generally the cohomology groups of the type $H(\Gamma, X, \rho)$ with an arbitrary representation ρ of G. These cohomology groups are closely related to automorphic forms with respect to Γ [16, 18].

G. Geometric Discontinuous Groups [4, 13]

The study of discontinuous groups Γ acting on a Euclidean or projective space X as a transformation group of a given structure is a classical problem. All possibilities for such Γ have been counted in low-dimensional cases. For instance, there are 230 kinds of discontinuous groups of †Euclidean motions acting on 3-dimensional Euclidean space without fixed subspaces, which are classified into 32 †crystal classes (A. Schönflies and E. S. Fedorov, 1891–1892; → 94 Crystallographic Groups). All discontinuous groups of a Euclidean space generated by †reflections have also been counted (H. S. M. Coxeter, 1934 [4, 19]).

H. Kleinian Groups

The last decade has seen considerable research on (finitely generated) Kleinian groups. These researches are closely related to the theory of †quasiconformal mappings and the †moduli of Riemann surfaces.

Making use of Eichler cohomology and †potentials, L. V. Ahlfors established his finiteness theorem and L. Bers his area theorem. Bers and B. Maskit investigated the boundaries of †Teichmüller spaces and discovered Kleinian groups with the property that the complement of the set $\mathfrak{L}$ of limit points is connected and simply connected.

Subsequently numerous mathematicians have discussed the classification, deformation, and stability properties of the set $\mathfrak{L}$, uniformization and deformation of Riemann surfaces with or without nodes, and other geometric properties. In their discussion, the theory of quasiconformal mappings plays an important role. Studies of discontinuous groups of motions of the hyperbolic 3-space have also been made.

References

[1] A. Borel and Harish-Chandra, Arithmetic subgroups of algebraic groups, Ann. of Math., (2) 75 (1962), 485–535.

[2] Séminaire H. Cartan 6, Fonctions automorphes et espaces analytiques, Ecole Norm. Sup., 1953–1954.

[3] Séminaire H. Cartan 10, Fonctions automorphes, Ecole Norm. Sup., 1957–1958.

[4] H. S. M. Coxeter and W. O. J. Moser, Generators and relations for discrete groups, Erg. Math., Springer, 1957.

[5] R. Fricke and F. Klein, Vorlesungen über die Theorie der automorphen Funktionen I, Teubner, 1897, second edition, 1926.

[6] F. Klein and R. Fricke, Vorlesungen über die Theorie der elliptischen Modulfunktionen I, Teubner, 1890.

[7] H. Minkowski, Diskontinuitätsbereich für arithmetische Äquivalenz, J. Reine Angew. Math., 129 (1905), 220–274 (Gesammelte Abhandlungen, Teubner, 1911, vol. 2, p. 53–100; Chelsea, 1967).

[8] H. Poincaré, Théorie des groupes fuchsiens, Acta Math., 1 (1882), 1–62 (Oeuvres, Gauthier-Villars, 1916, vol. 2, p. 108–168).

[9] C. L. Siegel, Discontinuous groups, Ann. of Math., (2) 44 (1943), 674–689 (Gesammelte Abhandlungen, Springer, 1966, vol. 2, p. 390–405).

[10] C. L. Siegel, Some remarks on discontinuous groups, Ann. of Math., (2) 46 (1945), 708–718 (Gesammelte Abhandlungen, Springer, 1966, vol. 3, 67–77).

[11] C. L. Siegel, Symplectic geometry, Amer. J. Math., 65 (1943), 1–86 (Gesammelte Abhandlungen, Springer, 1966, vol. 2, p. 274–359; Academic Press, 1964).

[12] C. L. Siegel, Analytic functions of several complex variables, Lecture notes, Institute for Advanced Study, Princeton, 1948–1949.

[13] B. L. van der Waerden, Gruppen von linearen Transformationen, Erg. Math., Springer, 1935 (Chelsea, 1948).
[14] A. Weil, Discontinuous subgroups of classical groups, Lecture notes, Univ. of Chicago, 1958.
[15] A. Weil, On discrete subgroups of Lie groups, Ann. of Math., (2) 72 (1960), 369–384; II, Ann. of Math., (2) 75 (1962), 578–602.
[16] Y. Matsushima and S. Murakami, On vector bundle valued harmonic forms and automorphic forms on symmetric Riemannian manifolds, Ann. of Math., (2) 78 (1963), 365–416.
[17] J. Lehner, Discontinuous groups and automorphic functions, Amer. Math. Soc. Math. Surveys, 1964.
[18] N. Bourbaki, Eléments de mathématique, Groupes et algèbres de Lie, ch. 4, 5, 6, Actualités Sci. Ind., 1377, Hermann, 1968.
[19] Algebraic groups and discontinuous subgroups, Amer. Math. Soc. Proc. Symposia in Pure Math., 1966, vol. 9.
[20] L. Bers and I. Kra, A crash course on Kleinian groups, Lecture notes in math. 400, Springer, 1974.

127 (XIX.29) Dispersion Relations

A. General Remarks

When a physical quantity $M(\omega)$ depending on a variable ω can be expressed in the form

$$M(\omega)=\sum_i \frac{c_i}{\omega_i-\omega}+\int\frac{\lambda(\omega')}{\omega'-\omega}d\omega', \qquad (1)$$

this formula is called a **dispersion relation**. The name comes from the fact that the index of refraction $n(\omega)$ of a dispersive medium can be expressed in this form as a function of the frequency ω. The dispersion relation appears in optics, the theory of nuclear reactions, particle scattering, etc.

If a complex function $f(z)$ is †holomorphic in the upper half of the z-plane and $f\to 0$ as $z\to\infty$, then we have for its real and imaginary parts on the real axis ($\omega,\omega'\in\mathbf{R}$) the following formula involving the †Hilbert transform:

$$\mathrm{Re}f(\omega)=\frac{1}{\pi}\mathrm{p.v.}\int_{-\infty}^{\infty}\frac{\mathrm{Im}f(\omega')}{\omega'-\omega}d\omega'.$$

Here p.v. means the †principal value of the integral, and $\mathrm{Im}f(\omega)$ is allowed to include †δ-functions. The term *dispersion relation* is often used in this restricted form. If $f=O(1)$ as $z\to\infty$, we can consider $g(z)=f(z)/(z-\omega_0+i\varepsilon)$ instead to obtain a dispersion relation of the form

$$\mathrm{Re}f(\omega)-\mathrm{Re}f(\omega_0)=\frac{\omega-\omega_0}{\pi}\mathrm{p.v.}\int\frac{\mathrm{Im}f(\omega')}{(\omega'-\omega)(\omega'-\omega_0)}d\omega'.$$

This procedure is called **subtraction**. If $z=\infty$ is a pole of a finite order of f, a dispersion relation for f can be obtained after a finite number of subtractions.

B. Causality and Dispersion Relations

The dispersion relation for a physical quantity is often linked to **causality**, i.e., the principle that any event is a consequence of a cause. For instance, in the case of scattering of light by a particle, the amplitude of the scattered wave $F(t)$ satisfies, by causality, $F(t)=0$ ($t<0$), where $t=0$ is the instant at which the light hits the particle. The †Fourier transform $f(\omega)$ of $F(t)$ is the boundary value for $\mathrm{Im}z\to 0$ of an analytic function holomorphic at $\mathrm{Im}z>0$ and satisfies a dispersion relation. However, causality and the validity of a dispersion relation are not equivalent; neither of them necessarily follows from the other.

C. Theory of Elementary Particles and Dispersion Relations

The theory of †elementary particles deals with the †probability amplitude T of scattering or creation of particles, which is a function of the energy ω, etc. The probability amplitude $T(\omega)$ is originally defined for real values of $\omega>\omega_0$, but it can be continued analytically for complex values of ω, denoted by z. From the unitarity and symmetry of the S-matrix (→ 379 S-Matrices), it follows that $\mathrm{Im}\,T(\omega)=(1/2)T^*T$ for $\omega>\omega_0$. If we define $\mathrm{Im}\,T(\omega)=0$ for $\omega<\omega_0$, the †principle of reflection implies $T(\bar z)=\overline{T(z)}$. Then starting from $\omega>\omega_0$ and continuing analytically into the upper half-z-plane and then into the lower half-plane after crossing the real axis at $\omega<\omega_0$, we can show that $T(z)$ has a discontinuity of $2i\,\mathrm{Im}\,T(\omega)$ along the real axis for $\omega>\omega_0$. That $T(z)$ is holomorphic except for the cut along the real axis for $\omega>\omega_0$ (called the **unitarity cut**) can sometimes be proved from †field theory. In that case, $T(\omega)$ satisfies a dispersion relation, and T can be determined if its imaginary part is given.

When the masses of particles in question satisfy a certain relation, there are discontinuities in addition to that along the unitarity cut. This phenomenon is called the appearance of **anomalous thresholds**. Without this anomaly, the amplitude T can be calculated from $\operatorname{Im} T$ by a dispersion relation, and $\operatorname{Im} T$ can be calculated from T by unitarity. In this way, the dispersion relation plays the role of the equation of motion in S-matrix theory.

D. Double Dispersion Relations

There are various ways to extend the dispersion relation (1) to a function $f(s,t)$ of two variables s and t, but the following choice is physically important:

$$f(s,t)=\frac{1}{\pi^2}\int_{s_0}^{\infty}\int_{t_0}^{\infty}\frac{\rho_{12}(s',t')}{(s'-s)(t'-t)}ds'\,dt'$$
$$+\frac{1}{\pi^2}\int_{t_0}^{\infty}\int_{u_0}^{\infty}\frac{\rho_{23}(t',u')}{(t'-t)(u'-u)}dt'\,du'$$
$$+\frac{1}{\pi^2}\int_{u_0}^{\infty}\int_{s_0}^{\infty}\frac{\rho_{31}(u',s')}{(u'-u)(s'-s)}du'\,ds',$$

where $u=\text{constant}-s-t$. This form is called the **Mandelstam representation**, and scattering amplitudes are expressed in this form with a suitable choice of variables. This representation and unitarity determine the scattering amplitudes.

For functions of three or more variables, no such representations have been proposed.

References

[1] M. L. Goldberger, Causality conditions and dispersion relations I, Phys. Rev., 99 (1955), 979–985.
[2] G. R. Screaton (ed.), Dispersion relations, Oliver and Boyd, 1961.

128 (V.6) Distribution of Prime Numbers

A. General Remarks

Given a real number x, we denote by $\pi(x)$ the number of primes not exceeding x. A. M. Legendre (1808) obtained empirically the formula $\pi(x)\fallingdotseq x/(\log x-B)$, and C. F. Gauss (1849) obtained the formula:

$$\pi(x)\fallingdotseq\int_2^x\frac{du}{\log u},$$

assuming the average density of primes to be $1/\log x$. The **Bertrand conjecture**, which asserts the existence of at least one prime between x and $2x$, was proved by P. L. Čebyšev (1848), who introduced the functions

$$\theta(x)=\sum_{p\leqslant x}\log p$$

and

$$\psi(x)=\sum_{p^m\leqslant x}\log p$$
$$=\theta(x)+\theta(\sqrt{x})+\theta(\sqrt[3]{x})\ldots.$$

He thereby proved $Ax+O(\sqrt{x})<\theta(x)<\psi(x)<(6/5)Ax+O(\sqrt{x})$, where $A=\log 2^{1/2}3^{1/3}5^{1/5}30^{-1/30}$. G. F. B. Riemann (1858) considered the function $\zeta(s)$ (where $s=\sigma+it$ is a complex variable), expressed by the †Dirichlet series $\sum_{n=1}^{\infty}n^{-s}$, which is convergent for $\sigma>1$. He found relations between the zeros of $\zeta(s)$ (→ 436 Zeta Functions) and $\pi(x)$. F. Mertens (1874) obtained the following formulas:

$$\sum_{p\leqslant x}\frac{\log p}{p}=\log x+O(1),$$
$$\sum_{p\leqslant x}\frac{1}{p}=\log\log x+B+O\left(\frac{1}{\log x}\right),$$
$$\prod_{p\leqslant x}\left(1-\frac{1}{p}\right)=\frac{e^{-c}}{\log x}\left(1+O\left(\frac{1}{\log x}\right)\right).$$

B. Prime Number Theorem

The **prime number theorem**

$$\lim_{x\to\infty}\frac{\pi(x)\log x}{x}=1,\quad\text{or}\quad\pi(x)\sim\frac{x}{\log x},$$

was proved almost simultaneously (1896) by J. Hadamard and C. J. de La Vallée Poussin. Without using the theory of †entire functions, E. Landau (1908) established the following formula:

$$\pi(x)=\operatorname{Li}x+O\left(xe^{-c\sqrt{\log x}}\right),$$

where

$$\operatorname{Li}x=\lim_{\delta\to+0}\left(\int_0^{1-\delta}+\int_{1+\delta}^{x}\right)\frac{du}{\log u}$$

is the †logarithmic integral. It can be shown by partial integration that

$$\operatorname{Li}x=\frac{x}{\log x}+\frac{1!x}{\log^2 x}+\ldots$$
$$+\frac{(k-1)!x}{\log^k x}+O\left(\frac{x}{\log^{k+1}x}\right).$$

For example, by taking $x=10^7$, we get $\pi(x)=664{,}579$, $\operatorname{Li}x=664{,}918$, and $x/\log x=620{,}417$.

If the Dirichlet series $f(s)=\sum_{n=1}^{\infty}a_n n^{-s}$ satisfies the condition $\sum_{n\leqslant x}a_n\sim cx$, then its abscissa of convergence is 1, and we have $\lim_{s\to1+0}(s-1)f(s)=c$. The converse is known as the †Tauberian theorem. If $F(s)=\sum_{n=1}^{\infty}a_n n^{-s}$ $(a_n\geqslant0)$ converges absolutely for

$\sigma>1$ and $F(s)-c/(s-1)$ is analytic for $\sigma\geqslant 1$, then we obtain $\Sigma_{n\leqslant x}a_n\sim cx$ (**Wiener-Ikehara-Landau theorem**, 1932). Specifically, if we put $-\zeta'(s)/\zeta(s)=\Sigma_{n=1}^{\infty}\Lambda(n)n^{-s}$, then the conditions of the theorem are satisfied, and we obtain $\Sigma_{n\leqslant x}\Lambda(n)=\psi(x)\sim x$.

It is easily seen that the prime number theorem is equivalent to $\psi(x)\sim x$ or $\theta(x)\sim x$. The †number theoretic function $\Lambda(n)$ (**Mangoldt's function**) defined by the previous equation satisfies $\Sigma_{d|n}\Lambda(d)=\log n$. It follows from the †Möbius inversion formula that $\Lambda(n)=\Sigma_{d|n}\mu(d)\log(n/d)$. Hence $\Lambda(n)=\log p$ ($n=p^m$) and $=0$ otherwise. Thus we obtain $\psi(x)=\Sigma_{n\leqslant x}\Lambda(n)=O(x)$. When $f(x)=\theta(x)$ or $\psi(x)$, it is easy to show that $\int_2^x(f(t)/t^2)dt=\log x+O(1)$ and $\liminf_{x\to\infty}f(x)/x\leqslant 1\leqslant\limsup_{x\to\infty}f(x)/x$. However, it is not easy to prove $f(x)\sim x$. To do so, introduce the number-theoretic function $M(n)$, which satisfies $\Sigma_{d|n}M(d)=\log^2 n$. As before, we have $M(n)=\Sigma_{d|n}\mu(d)\log^2(n/d)$, hence $M(n)=(2l-1)\log p$ ($n=p^l, l\geqslant 1$), which $=2\log p\log q$ ($n=p^lq^m, l\geqslant 1, m\geqslant 1$) and $=0$ otherwise. Thus we obtain $\Sigma_{n\leqslant x}M(n)=2x\log x+O(x)$. This leads to the well-known Selberg formula (1949),

$$\theta(x)\log x+\sum_{p\leqslant x}\theta(x/p)\log p$$
$$=2x\log x+O(x),$$

which enabled him to prove $\theta(x)\sim x$. Thus he obtained for the first time a proof of the prime number theorem that does not use complex analytic methods. The simple formulas $\Sigma_{n=1}^{\infty}\mu(n)/n=0$ and $\Sigma_{n\leqslant x}\mu(n)=o(x)$, obtained by H. von Mangoldt (1897), were revealed by Landau to have deep meaning concerning the prime number theorem. Let $\pi_r(x)$ denote the number of integers not exceeding x that can be expressed as the product of r distinct primes. In generalizing the prime number theorem, Landau (1911) proved that

$$\pi_r(x)\sim\frac{1}{(r-1)!}\frac{x(\log\log x)^{r-1}}{\log x}.$$

Let us write $\vartheta(x)=\Sigma_{n=1}^{\infty}e^{-\pi xn^2}$. Riemann proved that

$$\Gamma\left(\frac{s}{2}\right)\pi^{-s/2}\zeta(s)$$
$$=\frac{1}{s(s-1)}+\int_1^{\infty}\vartheta(x)(x^{s/2-1}+x^{-1/2-s/2})dx$$

and obtained the well-known functional equation for the zeta function (→ 436 Zeta Functions)

$$\pi^{-s/2}\Gamma(s/2)\zeta(s)$$
$$=\pi^{-1/2+s/2}\Gamma(1/2-s/2)\zeta(1-s).$$

This enables us to extend $\zeta(s)$ as a meromorphic function to the whole complex plane.

Utilizing this extended $\zeta(s)$ and the following result of O. Perron on Dirichlet series, we can estimate $\psi(x)$. Let σ_0 ($\neq\infty$) be the abscissa of convergence of $F(s)=\Sigma_{n=1}^{\infty}f(n)n^{-s}$, and let $a>0$, $a>\sigma_0$, and $x>0$. If

$$\lim_{T\to\infty}\frac{1}{2\pi i}\int_{a-iT}^{a+iT}F(s)\frac{x^s}{s}ds$$

exists, then the limit is equal to $\Sigma'_{n\leqslant x}f(n)$, where Σ' means that in the summation the last term $f(x)$ is replaced by $f(x)/2$ if x is an integer. In many cases, $F(s)$ has a pole at $s=1$, and the principal part of the sum is obtained from the residue at $s=1$, whereas the residual part is given by a certain contour integration. To estimate $\psi(x)$, we use $-\zeta'(s)/\zeta(s)$ as $F(s)$; hence the problem arises of determining the zeros of $\zeta(s)$. Riemann conjectured that all zeros of $\zeta(s)$ in the strip $0\leqslant\sigma\leqslant 1$ must be situated on the vertical line $\sigma=1/2$. If this so-called †Riemann hypothesis (→ 436 Zeta Functions) is true, then it follows that $\pi(x)=\mathrm{Li}\,x+O(\sqrt{x}\log x)$. The ultimate validity of Riemann's hypothesis remains in doubt. Concerning this, the most recent major result is the following formula, obtained by I. M. Vinogradov (1958): $\pi(x)=\mathrm{Li}\,x+O(x\exp(-c\log^{3/5}x/\log\log^{1/3}x))$. Without using Riemann's hypothesis, Littlewood (1918) proved that

$$\limsup_{x\to\infty}\frac{\pi(x)-\mathrm{Li}\,x}{(\sqrt{x}/\log x)\log\log\log x}>0,$$

$$\liminf_{x\to\infty}\frac{\pi(x)-\mathrm{Li}\,x}{(\sqrt{x}/\log x)\log\log\log x}<0.$$

If we denote by $N(T)$ the number of zeros of $\zeta(s)$ in the domain $0<\sigma<1$, $0<t<T$, then we have

$$N(T)=\frac{1}{2\pi}T\log T-\frac{1+\log 2\pi}{2\pi}T+O(\log T).$$

Let $N_0(T)$ denote the number of zeros of $\zeta(s)$ on the interval $\sigma=\frac{1}{2}$, $0<t<T$. Selberg (1942) obtained the following impressive result: $N_0(T)>cT\log T$.

E. C. Titchmarsh (1936) showed that there exist 1041 zeros of $\zeta(s)$ in the domain $0<\sigma<1$, $0<t<1468$ and that all lie on the line $\sigma=\frac{1}{2}$. Computers have provided further results that justify the Riemann hypothesis. N. Levinson proved in 1974 by another method that at least one-third of the zeros of the Riemann zeta function are on the line $\sigma=\frac{1}{2}$. The minimum of the modulus of the imaginary part of the zeros with $\sigma=\frac{1}{2}$ is $t=14.13\ldots$.

C. Twin Primes

Let p_n be the nth prime. We know from the prime number theorem that $p_n\sim n\log n$; more precisely, $p_n = n\log n + n\log\log n + O(n)$. A pair of primes differing only by 2 are called **twin primes**. It is still unknown whether there exist infinitely many twin primes. There exist infinitely many n satisfying $p_{n+1} - p_n < c\log p_n$ (P. Erdös, 1940). Suppose that $\zeta(1/2+it) = O(|t|^c)$. A. E. Ingham (1937) proved that $p_{n+1} - p_n < p_n^{\Theta}$, $\Theta = (1+4c)/(2+4c)+\varepsilon$, by using the following density theorem related to the zeros of $\zeta(s)$: If we denote by $N(\alpha, T)$ the number of zeros of $\zeta(s)$ in the domain $\alpha \leqslant \sigma \leqslant 1$, $0 \leqslant t \leqslant T$ $(1/2 \leqslant \alpha \leqslant 1)$, then there is a positive constant c such that $N(\alpha, T) = O(T^{2(1+c)(1-\alpha)}\log^5 T)$. The **Lindelöf hypothesis** asserts that the constant c can be made arbitrarily small. If the Riemann hypothesis holds, then the Lindelöf hypothesis also holds. It is clear that we can substitute $1/6+\varepsilon$ $(\varepsilon > 0)$ for c, ε being arbitrarily small. Haneke (1963) showed that c can be replaced by $6/37+\varepsilon$. Rankin (1935) proved that

$$p_{n+1} - p_n > c\log p_n \log\log p_n \log\log\log\log p_n \times (\log\log\log p_n)^{-2}$$

holds for infinitely many n. If we denote by $\pi_2(x)$ the number of primes $p \leqslant x$ such that $p+2$ is also a prime, then it has been conjectured that

$$\pi_2(x) \sim C\int_2^x \frac{du}{\log^2 u}$$

as $x \to \infty$, where

$$C = 2\prod_{p>2}\left\{1 - \frac{1}{(p-1)^2}\right\} = 1.32032\cdots$$

The numerical evidence provided by the computation $\pi_2(10^9) = 3424506$ (R. P. Brent, *Math. Comp.*, 28 (1974)) tends to indicate the truth of this conjecture. At present, $76 \cdot 3^{139} \pm 1$ seems to be the largest known pair of twin prime numbers (H. C. Williams and C. R. Zarnke, *Math. Comp.*, 26 (1972)).

D. Prime Numbers in Arithmetic Progressions

Let k be a positive integer, $\chi(n)$ be a residue character modulo k (→ 291 Number-Theoretic Functions), and $L(s,\chi) = \sum_{n=1}^{\infty}\chi(n)n^{-s}$ $(\sigma > 1)$ be the †Dirichlet L-function. The function $L(s,\chi)$ of s defined by this series can be extended to an analytic function in the whole complex plane in the same way as the Riemann zeta function. In particular, when χ is the principal character, then $L(s,\chi)$, thus extended, is a meromorphic function whose only pole is situated at $s=1$ and is simple; otherwise the function $L(s,\chi)$ is holomorphic on C. Using this function $L(s,\chi)$ and in connection with his research concerning the †class numbers of quadratic forms, P. G. L. Dirichlet (1837) proved that there exist infinitely many primes in the arithmetical progression $l, l+k, l+2k, \ldots$, where l is the initial term and k a common difference relatively prime to l. This result is called the **Dirichlet theorem** (or **prime number theorem for arithmetic progressions**).

Suppose that $\mathfrak{a}$ runs over all integral ideals in a †quadratic number field K of discriminant d. Then the †Dedekind zeta function $\zeta_K(s)$ of K is defined by $\sum(N\mathfrak{a})^{-s}$ for $\sigma > 1$. By virtue of the decomposition law of prime ideals (→ 343 Quadratic Fields), we have $\zeta_K(s) = \zeta(s)L(s,\chi)$, where $\chi(n) = (d/n)$ is the †Kronecker symbol. Utilizing $\zeta_K(s)$, we obtain formulas concerning the class number $h(d)$ of the field K. If $d>0$, then $h(d) = (\sqrt{d}/2\log\varepsilon)L(1,\chi)$, where ε is the †fundamental unit of K. On the other hand, if $d<0$, then $h(d) = (w\sqrt{-d}/2\pi)L(1,\chi)$, where w denotes the number of the roots of unity contained in K. It follows that $L(1,\chi) \geqslant 2\log((1+\sqrt{5})/2)/\sqrt{|d|}$. Let χ be a character modulo k induced by a primitive character χ^0. Since we have $L(s,\chi) = L(s,\chi^0)\cdot \prod_{p|k}(1-\chi^0(p)p^{-s})$, it can be shown that $L(1,\chi) \neq 0$ for a real character χ. It is easy to prove that $L(1,\chi) \neq 0$ for a complex character χ. These statements then lead to the Dirichlet theorem. The proof was simplified by H. N. Shapiro (1951). Besides Landau's three proofs for $L(1,\chi) \neq 0$ for real character χ (1908), there are elegant proofs by T. Estermann (1952), Selberg (1949), and others. For a character $\chi \bmod k$, we always have $L(1,\chi) = O(\log k)$, while $L(1,\chi)^{-1} = O(\log k)$ with one possible exception, which may occur only if χ is a real character. Even in this case, we have $L(1,\chi)^{-1} = O(k^{\varepsilon})$ (where $\varepsilon > 0$ is arbitrary, but O depends on ε). This result was obtained by Siegel (1934) from his study concerning class numbers of imaginary quadratic number fields. His proof was simplified by Estermann (1948) and S. D. Chowla (1950). The importance of the prime number theorem for arithmetic progressions was revealed when it was applied to the Goldbach problem. Concerning this problem, the manner in which the remainder term depends on the modulus k became an object of investigation. The Page-Siegel-Walfisz theorem is convenient to use: Denote by $\pi(x;k,l)$ the number of primes not exceeding x and of the form $ky+l$, where $(k,l)=1$. If $x \geqslant \exp(k^{\varepsilon})$ (where $\varepsilon > 0$ is arbitrary), then we have

$$\pi(x;k,l) = \frac{\mathrm{Li}\,x}{\varphi(k)} + O\left(\frac{xe^{-c(\varepsilon)\sqrt{\log x}}}{\varphi(k)}\right).$$

Further research on the distribution of zeros of $L(s,\chi)$ is necessary for the study of $\pi(x; k,l)$ when x takes smaller values. If χ is a nonprincipal real character, then $L(s,\chi)$ may have at most one real zero β_1 around 1. Because of this fact, when x is small we are unable to obtain any formula to indicate the uniform distribution of primes. However, we have the following deep result, obtained by E. Fogels (1962). For a given positive ε, there exist $c_0(\varepsilon)$ and $c(\varepsilon)$ such that $\pi(x;k,l) > c(\varepsilon)x/\varphi(k)k^{\varepsilon}\log x$, provided that $x \geqslant k^{c_0(\varepsilon)}$. On the other hand, Titchmarsh (1930), using the †sieve method, obtained $\pi(x;k,l) = O(x/\varphi(x)\log x)$ for $x \geqslant k^{c_0}$. Fogels's theorem is based on the following theorem by Ju. V. Linnik (1947) and K. Prachar (1957), which is an extension of Page's theorem (1935): We let δ be the function defined by $\delta = 1-\beta_1$ if $L(s,\chi)$ has an exceptional real zero β_1, and $\delta = c_1/\log(k(|t|+2))$ otherwise (where c_1 is a suitably small number). If we denote by β the real part of any zero ($\neq \beta_1$) of $L(s,\chi)$, then the theorem states that

$$\beta \leqslant 1 - \frac{c_1}{\log(k(|t|+2))}\log\left(\frac{c_1 e}{\delta\log(k(|t|+2))}\right),$$

provided that $\delta\log(k(|t|+2)) \leqslant c_1$.

Let s be positive, b_j, z_j $(1 \leqslant j \leqslant s)$ be complex, and l, m be real numbers satisfying $\max|z_j| \geqslant 1$, $l \geqslant s$, and $m \geqslant 0$. Under these conditions P. Turán (1953) obtained the following fundamental theorem:

$$\max_{m \leqslant r \leqslant l+m} |b_1 z_1^r + \ldots + b_s z_s^r|$$

$$\geqslant \left(\frac{1}{8e(l+m)}\right)^l \min_{1 \leqslant j \leqslant s} |b_1 + \ldots + b_j|.$$

This theorem is effective in research on the distribution of zeros of zeta functions. Based on this new method, Turán (1961), S. Knapowski (1962), and Fogels (1965) reached the results cited above.

There is another method of research, called a new sieve method, on the distribution of primes that was introduced by Selberg and Vinogradov. This method was followed by W. B. Jurkat and H. E. Richert (1965). Linnik, A. Rényi (1950), and E. Bombieri (1965) founded still another method, called the **large sieve method**, by which P. T. Batemann, Chowla, and P. Erdös studied the value of $L(1,\chi)$.

E. Sieve Method

Let A be a set of integers, and P a set of primes. For each $p \in P$, let Ω_p be a set of residues mod$p$, and $\omega(p)$ the number of residues belonging to Ω_p. The sieve method is a device for estimating (from above or below) the number of integers n belonging to the set $S(A,P,\Omega_p) = \{n \mid n \in A, n \bmod p \notin \Omega_p \text{ for } p \in P\}$. The combinatorial methods of the Brun, Buchštab, and Richert sieves are interesting and efficient but very complicated. Here, we shall briefly describe the **Selberg sieve**. As an example, denote by $S(x;q,l)$ the number of n satisfying $n \equiv l \bmod q$, $n \leqslant x$, $(n,D)=1$, where q is a prime not exceeding x, $z \leqslant x$, $D = \prod_{p \leqslant z} p$, and $(l,q)=1$; then

$$S(x;q,l) = \sum_{\substack{n \leqslant x \\ n \equiv l \bmod q}} \sum_{d|(n,D)} \mu(d)$$

$$\leqslant \sum_{\substack{n \leqslant x \\ n \equiv l \bmod q}} \left(\sum_{d|(n,D)} \lambda_d\right)^2,$$

where $\lambda_1 = 1$ and λ_d is arbitrary for $d > 1$. Thus, the problem reduces to the optimization of λ_d. Proceeding in this manner, C. Hooly (1967) and Y. Motohashi (1975), using analytic methods, obtained certain deep results relating to the Brun-Titchmarsh theorem.

Let $n_1, n_2, \ldots, n_Z$ be Z natural numbers not exceeding N, and $Z(p,a)$ the number of n_j such that $n_j \equiv a \bmod p$. A. Rényi (1950) proved that

$$\sum_{p < \sqrt[3]{N/12}} p \sum_{a=0}^{p-1} \left\{Z(p,a) - \frac{Z}{p}\right\}^2 \leqslant 2NZ.$$

Set $a_n = 1$ for $n = n_j$ and $a_n = 0$ otherwise, and set $S(\alpha) = \sum_{n \leqslant N} a_n \exp(2\pi i n\alpha)$; then

$$p\sum_{a=1}^{p-1}\left\{Z(p,a) - \frac{Z}{p}\right\}^2 = \sum_{a=1}^{p-1}\left|S\left(\frac{a}{p}\right)\right|^2.$$

In view of this simple fact, E. Bombieri (1965) and P. X. Gallagher (1968) extended the problem and proved that, in general,

$$\sum_{q \leqslant Q} \sum_{\substack{a=1 \\ (a,q)=1}}^{q} \left|\sum_{M < n \leqslant M+N} a_n \exp\left(2\pi i \frac{a}{q} n\right)\right|^2$$

$$\leqslant (N+2Q^2) \sum_{M < n \leqslant M+N} |a_n|^2.$$

Similar results can be obtained for character sums $\sum_{M<n\leqslant M+N} a_n\chi(n)$. In this connection H. L. Montgomery proved that

$$S(\{n; M < n \leqslant M+N\}; P, \Omega_p)$$

$$\leqslant (N+2Q^2)\left\{\sum_{\substack{q \leqslant Q \\ q|P(z)}} \prod_{p|q} \frac{\omega(p)}{p-\omega(p)}\right\}^{-1},$$

$$P(z) = \prod_{\substack{p \in P \\ p \leqslant z}} p.$$

Using these methods, the following estimate was obtained by Bombieri:

$$\sum_{q \leqslant x^{1/2}(\log x)^{-B}} \max_{y \leqslant x} \max_{\substack{l \\ (q,l)=1}} \left|\pi(y,q,l) - \frac{1}{\varphi(q)}\int_2^y \frac{du}{\log n}\right| \ll x(\log x)^{-A}$$

where A is arbitrary and B is a certain function of A (R. C. Vaughan). These methods, known collectively as the **large sieve**, were first directed toward proving the **Rényi theorem** stating that every sufficiently large even integer can be represented as the sum of a prime and an almost prime integer (M. B. Barban). Afterwards, combining Richert's sieve with this large sieve, C. J. Ch'en (1973) obtained the Rényi theorem with an integer having at most two prime factors in place of an almost prime integer. Several applications of Bombieri's theorem have been demonstrated by P. D. T. A. Elliot and H. Halberstam (1966): e.g., the estimation of the number of representations of n as $p+x^2+y^2$ (Hooley, Ju. V. Linnik) and the estimation of $\sum_{p<n} d(n-p)$ (Linnik, B. M. Bredihin).

Let $N(\alpha, T, \chi)$ denote the number of zeros of $L(s,\chi)$ in the rectangle $\alpha \leqslant \sigma \leqslant 1$, $|t| \leqslant T$. Combining the large sieve with new Fourier integral techniques and the Turán power sum method, Gallagher (1970) proved that there exists a positive constant c satisfying

$$\sum_{\chi \bmod Q} N(\alpha, T, \chi) \quad \text{or}$$

$$\sum_{q \leqslant Q} \sum_{\chi \bmod q} N(\alpha, T, \chi) \ll (QT)^{c(1-\alpha)}.$$

Similar results were also obtained by G. Halasz and H. L. Montgomery (1969) using another method which was further exploited by M. Jutilla and M. H. Huxley (1972) and others. In particular, Huxley (1972) deduced that $\pi(x+y)-\pi(x) \sim y(\log x)^{-1}$ if $y \geqslant x^{\frac{7}{12}+\varepsilon}$. Combined with the Deuring-Heilbronn phenomenon, the above zero density theorem not only establishes the Linnik theory, but also yields the following result, due to K. A. Rodoskiĭ, T. Tatuzawa, and I. M. Vinogradov:

$$\pi(x;q,l) = \frac{1}{\varphi(q)} \int_2^x \frac{du}{\log u}$$

$$- E\frac{x_1(l)}{\varphi(q)} \int_2^x \frac{u^{\beta-1}}{\log u} du + O\left(\frac{1}{\varphi(q)} x^{1-\frac{c}{\Delta}}\right)$$

if $x \geqslant \exp(a \log q \log\log q)$, where $E=1$ or 0 according as Siegel's zero β exists or not and

$$\Delta = \operatorname{Max}\left(\log q, (\log x)^{2/5} (\log\log x)^{1/5}\right).$$

Throughout these researches, estimates of the type

$$\sum_{\chi \bmod q} \left| L\left(\frac{1}{2}+it, \chi\right)\right|^4$$

$$\ll \varphi(q)(|t|+2)\log^c\{q(|t|+2)\}$$

and

$$\sum_{\chi \bmod q} \int_{-T}^{T} \left| L\left(\frac{1}{2}+it, \chi\right)\right|^4 dt$$

$$\ll \varphi(q)(T+2)\log^c q(T+2)$$

are of great importance, and have been studied by A. F. Lavrik (1968). Linnik (1961), Huxley (1972), and K. Ramachandra (1975).

F. The Prime Ideal Theorem in Algebraic Number Fields

In an algebraic number field of finite degree, the prime number theorem is replaced by the **prime ideal theorem** (Mitsui, 1956; Fogels, 1962), which is based on the theory of the Hecke †L-function (E. Hecke, 1917; Landau, 1918). Let K be a finite Galois extension over an algebraic number field k of finite degree. Suppose that $\mathfrak{p}$ is a prime ideal of k and is not ramified in K. The †Frobenius automorphism of a prime divisor of $\mathfrak{p}$ in K determines a conjugate class C of the Galois group of K/k. Let $\pi(x;C)$ denote the number of prime ideals in k associated with the class C in the above sense and whose norm does not exceed x. Then we have

$$\pi(x,C) = \frac{h(C)}{(K:k)} \operatorname{Li} x + O\left(xe^{-c\sqrt{\log x}}\right),$$

where $h(C)$ is the number of elements contained in C and c is a positive constant depending on K/k. This is an extension of Čebotarev's theorem (Artin, 1923).

References

[1] R. Ayoub, An introduction to the analytic theory of numbers, Amer. Math. Soc. Math. Surveys, 1963.

[2] H. Bohr and H. Cramér, Die neuere Entwicklung der analytischen Zahlentheorie, Enzykl. Math., II C (8), 1922.

[3] E. Bombieri, Le grand crible dans la théorie analytique des nombres, Société mathématique de France, 1974.

[4] K. Chandrasekharan, Introduction to analytic number theory, Springer, 1968.

[5] P. Erdös, Some recent advances and current problems in number theory, Lectures on modern mathematics III, edited by T. L. Saaty, John Wiley, 1965, p. 196–244.

[6] H. M. Edwards, Riemann's zeta function, Academic Press, 1974.

[7] H. Halberstam and H. E. Richert, Sieve methods, Academic Press, 1974.

[8] G. H. Hardy, Divergent series, Clarendon Press, 1949.

[9] E. Hecke, Vorlesungen über die Theorie der algebraischen Zahlen, Akademische Verlag, Leipzig, 1923 (Chelsea, 1970).

[10] M. N. Huxley, The distribution of prime numbers, Oxford Univ. Press, 1972.

[11] A. E. Ingham, The distribution of prime numbers, Cambridge Univ. Press, 1932.

[12] E. Landau, Handbuch der Lehre von der Verteilung der Primzahlen I, II, Teubner, 1909 (Chelsea, 1969).
[13] E. Landau, Vorlesungen über Zahlentheorie I, II, Hirzel, 1927 (Chelsea).
[14] D. N. Lehmer, List of prime numbers from 1 to 10006721, Carnegie Institution of Washington, 1914.
[15] N. Levinson, More than one third of zeros of Riemann's zeta-function on $\sigma=\frac{1}{2}$, Advances in Math., 13 (1974), 383–486.
[16] H. L. Montgomery, Topics in multiplicative number theory, Lecture notes in math., 227, Springer, 1971.
[17] K. Prachar, Primzahlverteilung, Springer, 1957.
[18] A. Selberg, An elementary proof of the prime number theorem for arithmetic progressions, Canad. J. Math., 2 (1950), 66–78.
[19] Z. Suetuna, Kaise kiteki seisûron (Japanese; Analytic theory of numbers), Iwanami, 1950.
[20] E. C. Titchmarsh, The theory of the Riemann zeta-function, Clarendon Press, 1951.
[21] P. Turán, Eine neue Methode in der Analysis und deren Anwendungen, Akadémiai Kiadó, Budapest, 1953.
[22] И. М. Виноградов (I. M. Vinogradov), Новая Оценка Функции $\zeta(1+it)$, Izv. Akad. Nauk SSSR, 22 (1958), 161–164.

129 (XI.9) Distributions of Values of Functions of a Complex Variable

A. General Remarks

Suppose that we are given a function $f: A \rightarrow B$ and that the variables z, w take on values in A, B, respectively. A **value distribution** of $f(z)$ is a set of points z where $f(z)$ takes on a certain value w (called w-points of $f(z)$). Value distribution theory is usually concerned with the study of value distributions of complex †analytic functions.

B. History

For a †transcendental entire function $f(z)$, every value (including ∞) is a value of the †cluster set of $f(z)$ at the point at infinity (†**Weierstrass's theorem**). This theorem was improved in the following way by E. Picard in 1879: A transcendental entire function $f(z)$ has an infinite number of w-points for any finite value w except for at most one finite value (†Picard's theorem). E. Borel gave a precise form of this theorem, and G. Julia proved the existence of †Julia's directions (→ 413 Transcendental Entire Functions; 271 Meromorphic Functions). After other results in value distribution theory had been obtained by J. Hadamard, G. Valiron, and others, R. Nevanlinna published an important work in 1925 in which he established the so-called **Nevanlinna theory** of meromorphic functions in $|z|<R\leqslant\infty$, unifying results obtained until that time (→ 271 Meromorphic Functions). T. Shimizu and L. V. Ahlfors gave a geometric meaning to the Nevanlinna †characteristic function $T(r)$. Ahlfors established further the theory of covering surfaces by metricotopological methods in 1935, and as its applications obtained the Nevanlinna theory and many other results on meromorphic functions. This theory clarified that the topological meaning of the number 2 of Picard's exceptional values is closely related to the †Euler characteristic -2 of the sphere. H. Selberg established the value distribution theory of †algebroidal functions and gave a precise form of G. Rémoundos's theorem, which corresponds to Picard's theorem in the case of algebroidal functions.

C. Value Distribution Theory on General Domains

The value distributions of meromorphic functions defined in a general domain or an open Riemann surface depend on the function-theoretic "size" of the set of singularities (→ 174 Function-Theoretic Null Sets) or the type of the Riemann surface (→ 362 Riemann Surfaces). For instance, we have this theorem of the Picard type: A single-valued meromorphic function with a set of singularities of †logarithmic capacity zero takes on every value infinitely often in any neighborhood of each singularity except for at most an F_σ-set of values of logarithmic capacity zero (**Hällström-Kametani theorem**). For the study of value distribution at general singularities, it is useful to investigate cluster sets (→ 65 Cluster Sets). In order to generalize the Nevanlinna theory to the case of general domains or Riemann surfaces, we represent their exhaustions by a real parameter r and define the †counting function (→ 271 Meromorphic Functions). G. J. Hällström established the value distribution theory of meromorphic functions defined in the complementary domain D of a compact set E of logarithmic capacity zero by taking $D_r=\{z|v(z)<r\}$ as the exhaustion of D, where $v(z)$ denotes the Evans potential for E, i.e., $v(z)$ is the

potential corresponding to a positive mass distribution μ on E of total mass 1 which tends to $+\infty$ as z tends to any point of E. Thus the number of Picard's exceptional values is not greater than $2+\xi$, where $\xi = \limsup_{r\to\infty} F(r)/T(r)$ with $n(r)$ = the Euler characteristic of D_r and $F(r)=\int_{r_0}^{r} n(r)\,dr$ (Hällström-Tsuji theorem). J. Tamura, L. Sario, and others studied the value distributions of meromorphic functions defined on Riemann surfaces. Sario succeeded in extending the Nevanlinna theory to analytic mappings of a Riemann surface $\mathfrak{R}$ into another Riemann surface $\mathfrak{S}$ by introducing a suitable metric in $\mathfrak{S}$ to define the †proximity function.

In the Nevanlinna theory on general domains, we must sometimes impose conditions that the functions must satisfy in order to obtain a concrete conclusion, for instance, the condition that ξ be finite in the Hällström-Tsuji theorem. But it is also important to determine the domains where some result can be obtained without imposing any additional conditions on the functions. The Hällström-Kametani theorem is an example. Although the set of exceptional values in this theorem cannot be replaced generally by a smaller set than an F_σ-set of logarithmic capacity zero, we have the following theorem: Let E be a †Cantor set with successive ratios $\xi_n = 2l_n/l_{n-1}$, where l_n denotes the length of the segments that remain after repeating n times the process of deleting an open segment from the middle of another segment. Then any single-valued meromorphic function with E as the set of singularities has at most 3 Picard's exceptional values if $\lim_{n\to\infty}\xi_n=0$ and at most 2 if $\xi_{n+1}=o(\xi_n^2)$ (L. Carleson, K. Matsumoto).

Some results show relations between †Nevanlinna's exceptional values and the order or asymptotic values of the functions. S. S. Chern and others attempted to extend the Nevanlinna theory to analytic mappings on manifolds of higher dimensions.

References

[1] K. Noshiro, Kindai kansûron (Japanese; Modern theory of functions), Iwanami, 1954.
[2] A. Dinghas, Vorlesungen über Funktionentheorie, Springer, 1961.
[3] L. Sario, Value distribution under analytic mappings of arbitrary Riemann surfaces, Acta Math., 109 (1963), 1–10.
[4] K. Matsumoto, Existence of perfect Picard sets, Nagoya Math. J., 27 (1966), 213–222.
[5] M. Kurita, A theorem on the value distribution of complex analytic mappings in several complex variables (Japanese), Sûgaku, 16 (1965), 195–202.
[6] W. K. Hayman, Meromorphic functions, Clarendon Press, 1964.
[7] L. Sario and K. Noshiro, Value distribution theory, Van Nostrand, 1966.

130 (XII.6) Distributions (Generalized Functions)

A. History

The advancement of analysis, particularly in the field of partial differential equations and harmonic analysis, necessitated the generalization of the notion of function. For instance, there was knowledge of "functions" such as Dirac's †delta function and †Heaviside's function, which were used by physicists and engineering scientists even though they were not functions in the classical sense. The finite parts of divergent integrals, used by J. Hadamard to investigate the fundamental solutions of wave equations (1932), and the Riemann-Liouville integrals due to M. Riesz (1938) were the first notions leading to the theory of distributions. The rudiments of the idea of distribution, however, may also be found in other earlier works. S. Bochner (1932) and T. Carleman (1944) discussed the Fourier transforms of locally integrable functions on the reals with growth as large as a polynomial. S. L. Sobolev introduced the notion of generalized derivative by means of integration by parts in studying Cauchy problems for hyperbolic equations (1936) and also the notion of generalized solution of differential equations; J. Leray (1934), K. O. Friedrichs (1939), and C. B. Morrey, Jr. (1940) also discussed generalized derivatives. On the other hand, L. Fantappié (1943) investigated analytic functionals that are elements of the dual of the space of analytic functions and applied them to the theory of partial differential equations. Based on a systematic generalization of these investigations, L. Schwartz established the theory of distributions (1945), which not only provided a mathematical foundation for a number of formal methods that had been used in mathematical physics, but also gave new and powerful tools for the theories of differential equations and †Fourier transforms. Furthermore, it has been applied to †representation theory of locally compact groups, the theory of probability, and the theory of manifolds. In particular, in †homology theory on manifolds, the notion of a †current (a differential form whose coefficients are distributions) plays an important role. As will be seen in Section B, distributions are defined as continuous †functionals on a certain function space, and it is essential to select a function space appropriate to the problems concerned. For this reason, I. M. Gel'fand and G. E. Šilov defined various classes of **generalized functions** [12].

B. Definition of Distributions

Let $\varphi(x)$ be a complex-valued function of $x=(x_1,\ldots,x_n)$ defined in the n-dimensional Euclidean space $\mathbf{R}^n$. By the **support** (or **carrier**) of φ, denoted by $\operatorname{supp}\varphi$, we mean the †closure of $\{x \mid \varphi(x)\neq 0\}$. For multi-indices p, i.e., n-tuples $p=(p_1,\ldots,p_n)$ of nonnegative integers, we set $|p|=p_1+\ldots+p_n$. For a function $\varphi(x)$ of class $C^{|p|}$, we write

$$D^p\varphi=\frac{\partial^{|p|}\varphi(x)}{\partial x_1^{p_1}\ldots\partial x_n^{p_n}}. \qquad (1)$$

In particular, $D^{(0,\ldots,0)}\varphi=\varphi$. To indicate the variables x we shall adopt the notation D_x^p.

$\mathfrak{D}_{\mathbf{R}^n}$ denotes the set of all complex-valued functions of class C^∞ defined in $\mathbf{R}^n$ with †compact support, which is a †linear space under the usual addition and scalar multiplication in function spaces. A sequence $\{\varphi_m\}$ in $\mathfrak{D}_{\mathbf{R}^n}$ is said to converge to 0 (the function identically equal to zero) as $m\to\infty$, denoted by $\varphi_m\Rightarrow 0$, if there exists a compact set E such that E contains $\operatorname{supp}\varphi_m$ for every m, and for every p, $\{D^p\varphi_m\}$ converges uniformly to 0 as $m\to\infty$.

We abbreviate $\mathfrak{D}_{\mathbf{R}^n}$ either as $\mathfrak{D}$ or, when we want to indicate the variables x, as $\mathfrak{D}_x$.

A complex-valued †linear functional T defined on $\mathfrak{D}$ is called a **distribution** (or **generalized function in the sense of Schwartz**) if it is continuous on $\mathfrak{D}$, i.e., $\varphi_m\Rightarrow 0$ implies $T(\varphi_m)\to 0$. The set of all distributions is denoted by $\mathfrak{D}'_{\mathbf{R}^n}$ (or $\mathfrak{D}'$). For distributions S and T, the sum $S+T$ and scalar multiple αT are defined by $(S+T)(\varphi)=S(\varphi)+T(\varphi)$ and $(\alpha T)(\varphi)=\alpha T(\varphi)$, respectively, which are also distributions. Hence $\mathfrak{D}'$ is a †linear space.

C. Examples of Distributions

(1) Let $f(x)$ be a †measurable and †locally integrable function. Then a distribution T_f is defined by $T_f(\varphi)=\int\varphi(x)f(x)\,dx$. Here dx is the †Lebesgue measure in $\mathbf{R}^n$, and the domain of integration is $\mathbf{R}^n$ (in fact, $\operatorname{supp}\varphi$). If $T_f=T_g$, then $f(x)=g(x)$ †almost everywhere. Thus we may identify the distribution T_f with the corresponding function f, and sometimes T_f will be denoted simply by f. (2) Let $\mu(M)$ be a complex-valued †completely additive set function on †measurable sets of $\mathbf{R}^n$, and assume that $\mu(M)$ is finite for every compact set M. Then a distribution T_μ is defined by $T_\mu(\varphi)=\int\varphi(x)\,\mu(dx)$. Example (1) is a special case where $\mu(dx)=f(x)\,dx$. If the measure is concentrated at the origin, then $T_\mu(\varphi)=c\varphi(0)$, denoted by $c\delta$, and δ is called **Dirac's distribution**. Sometimes δ is denoted by δ_x, $\delta_{(x)}$, or $\delta(x)$ to indicate that it operates on functions of x. (3) For given p the distribution $\delta^{(p)}$ is defined by $\delta^{(p)}(\varphi)=(-1)^{|p|}D^p\varphi(0)$. $\delta^{(0,\ldots,0)}=\delta$. (4) Let $g(x)$ be a function defined but not integrable on an interval (a,b), and assume that for any positive number ε it is integrable on $(a+\varepsilon,b)$. Moreover, assume that

$$g(x)=\sum_{\nu=1}^{n}A_\nu(x-a)^{-\lambda_\nu}+h(x),$$

where $\operatorname{Re}\lambda_\nu>1$, λ_ν is not an integer, and $h(x)$ is integrable on (a,b). Then

$$\int_{a+\varepsilon}^{b}g(x)\,dx-\sum A_\nu(\lambda_\nu-1)^{-1}\varepsilon^{1-\lambda_\nu}=F(\varepsilon)$$

tends to a finite value as $\varepsilon\to 0$. This limit is called the **finite part** (in French *partie finie*) of the integral $\int_a^b g(x)\,dx$, denoted by $\operatorname{Pf}\int_a^b g(x)\,dx$:

$$\operatorname{Pf}\int_a^b g(x)\,dx$$
$$=-\sum_\nu\frac{A_\nu}{\lambda_\nu-1}(b-a)^{1-\lambda_\nu}+\int_a^b h(x)\,dx.$$

In the same way, for every $\varphi\in\mathfrak{D}_{\mathbf{R}^1}$, $T(\varphi)=\operatorname{Pf}\int_a^b g(x)\varphi(x)\,dx$ is defined, and T is a distribution which will be denoted by $\operatorname{Pf}g$, and which is frequently called a **pseudofunction**. This notion will be extended to the n-dimensional case and used to express the fundamental solutions (→ Section Q; Appendix A, Table 15.V) of †hyperbolic partial differential equations.

D. Supports of Distributions

A distribution T is said to vanish in an open set Ω if $T(\varphi)=0$ for every $\varphi\in\mathfrak{D}$ with support contained in Ω. The union of such open sets is the maximal open set in which T vanishes, and its complement is called the support (or carrier) of the distribution T. The support of the distribution given in example (1) coincides with that of the function f. The support of $\delta^{(p)}$ is the origin. A closed set F is said to be **regular** if for every point a in F we have a neighborhood U of a and constants $\omega\geqslant 0$ and $1\geqslant\alpha>0$ such that every pair of points x and y in $F\cap U$ is connected by a curve contained in F with length less than $\omega|x-y|^\alpha$. If the support F of T is contained in a regular closed set F, then

$$T=\sum_{j=1}^{N}D^{p_j}T_{\mu_j}$$

(the expression is not necessarily unique, and includes the case $N=\infty$), where the μ_j are complex-valued measures with support contained in F. In particular, if the support of a distribution T contains only one point a, then T can be represented uniquely as a finite linear combination

$$T=\sum_{|p|\leqslant m}\alpha_p D^p\delta_{(x-a)}$$

of derivatives of the distribution $\delta_{(x-a)}$ defined by $\delta_{(x-a)}(\varphi)=\varphi(a)$. Here D^pT denotes the distribution derivative (→ Section I).

E. Currents on a Manifold

For simplicity, we consider here an orientable n-dimensional †differentiable manifold X of class C^∞ (→ 108 Differentiable Manifolds). By the support of an †exterior differential form α, we mean the †closure of the set $\{x|\alpha_x\neq 0\}$ (α_x denotes the value of α at x). We denote by $\Phi_k(X)$ the set of all exterior differential forms of degree k with compact support. For a sequence $\{\alpha_m\}$ in $\Phi_k(X)$, the convergence $\alpha_m\Rightarrow 0$ as $m\to\infty$ is defined as follows: There is a compact set E such that $E\supset\operatorname{supp}\varphi_m$, $m=1,2,\ldots$, and for every coordinate neighborhood Q the coefficients of α_m with respect to a local coordinate system in Q and all their derivatives are uniformly convergent to zero in Q. A continuous †linear functional T on $\Phi_k(X)$ (i.e., if $\alpha_m\Rightarrow 0$, then $T(\alpha_m)\to 0$) is called a **current** of degree $n-k$.

F. Examples of Currents

(1) Let β be an exterior differential form of degree $n-k$ with locally integrable coefficients. Then a current T_β is defined by $T_\beta(\alpha)=\beta(\alpha)=\int_X\beta\wedge\alpha$. (2) Let C be a p-dimensional †chain. Then a current T_C is defined by $T_C(\alpha)=C(\alpha)=\int_C\alpha$.

G. Distributions on a Differentiable Manifold X

We denote by $\mathfrak{D}_X$ the set of all functions of class C^∞ with compact support in X: $\mathfrak{D}_X=\Phi_0(X)$. A continuous †linear functional on $\mathfrak{D}_X$, i.e., a current of degree n, is called a distribution on X. $\mathfrak{D}'_x$ denotes the set of all distributions on X. In particular, if the volume element dx is given, a locally integrable function $f(x)$ can be identified with an n-form $f(x)dx$, and $f(x)$ can be considered as a distribution on X in view of the preceding examples.

H. Restrictions of Distributions

Let Ω be an open set in X. Every function $\varphi\in\mathfrak{D}_\Omega$ can be extended to a function $\bar\varphi\in\mathfrak{D}_X$ by setting $\bar\varphi(x)=0$ for $x\notin\Omega$. Thus if $T\in\mathfrak{D}'_X$, then a distribution $S\in\mathfrak{D}'_\Omega$ is defined by $S(\varphi)=T(\bar\varphi)$ for every $\varphi\in\mathfrak{D}_\Omega$, and S is called the **restriction** of T to Ω. Currents are restricted similarly. Two distributions T and S are said to be equal in Ω if their restrictions to Ω are equal. If every point in X has a neighborhood where $T=S$, then $T=S$. In this sense, a distribution is determined completely by its local "values," although the notion of pointwise value, having exact meaning for functions, has no meaning for distributions. Moreover, we can construct a distribution with given local "values" in the following sense: Suppose that an †open covering $\{\Omega_j\}$ of X and a set of distributions $T_j\in\mathfrak{D}'_{\Omega_j}$ are given and that T_j and T_k are equal in $\Omega_j\cap\Omega_k$ for any j and k. Then there exists a unique distribution $T\in\mathfrak{D}'_X$ such that $T=T_j$ in Ω_j for each j. A current has the same properties and can be represented locally as a differential form whose coefficients are distributions.

I. Derivatives of Distributions

In example (1) in Section C, if $f(x)$ is a function of class C^k, then by partial integration $T_{D^pf}(\varphi)=(-1)^{|p|}T_f(D^p\varphi)$ for $|p|\leqslant k$. The right-hand side defines a distribution even if f is not differentiable. In view of this example, we define **derivatives** D^pT of any distribution T by

$$(D^pT)(\varphi)=(-1)^{|p|}T(D^p\varphi),\quad \varphi\in\mathfrak{D}_X. \tag{2}$$

Any distribution is infinitely differentiable. Any locally integrable function is infinitely differentiable in the sense of distributions, and its derivatives D^pT_f are called **distribution derivatives** (or **generalized derivatives**).

For example: (1) $D^p\delta=\delta^{(p)}$; (2) (1-dimensional case) $dx_+/dx=\mathbf{1}$, $d\mathbf{1}/dx=\delta$, where $x_+=\max(x,0)$ and $\mathbf{1}(x)$ is **Heaviside's function**, which is equal to 1 for $x\geqslant 0$ and to 0 for $x<0$.

J. Exterior Derivatives of Currents and Homology Theory

We denote by $d\beta$ the †exterior derivative of a differential form β of degree $n-k-1$. Then

$$T_{d\beta}(\alpha)=\int_X d\beta\wedge\alpha=(-1)^{n-k}\int_X\beta\wedge d\alpha$$
$$=(-1)^{n-k}T_\beta(d\alpha).$$

As suggested by this example, we define the

exterior derivative dT by

$$dT(\alpha)=(-1)^{n-k}T(d\alpha).$$

By †Stokes's formula $\int_C d\alpha=\int_{\partial C}\alpha$, we have $dT_C=(-1)^{n-k-1}T_{\partial C}$ for a k-dimensional †chain C. Thus exterior differential forms and chains may be regarded as special cases of currents, and the notion of the exterior derivative of forms and that of the boundary of chains are given unified meaning as the exterior derivative of currents. This similarity leads to the development of homology theory of currents. A current T is called a **closed** current if $dT=0$, while it is called **homologous to** 0 if there is a current S such that $T=dS$. Utilizing these notions, we can introduce homology group of currents, and we define ∂ by $\partial T=(-1)^{n-k-1}dT$ (when T is of degree $n-k$), so that we may consider currents not only as †chains with respect to ∂ but also as †cochains with respect to d. Therefore, the notion of currents gives a correspondence between the homology group of a manifold and its cohomology group. Furthermore, along with the theory of †intersection numbers, the theory of currents leads to the homology theory of manifolds due to G. de Rham (→ 108 Differentiable Manifolds).

K. Applications of Currents to the Theory of Harmonic Integrals

The †adjoint form $*\alpha$ of an exterior differential form α can be defined in a †Riemannian manifold. Then $(T,\alpha)=T(*\alpha)$ defines the scalar product of a current T and a differential form α of the same degree. The **codeviative** δT of a current T is defined by $(T,d\varphi)=(\delta T,\varphi)$. On a compact C^∞-manifold Ω, every current T can be decomposed into $T=T_1+\Delta T_2$, $\Delta T_1=0$, where the Laplacian Δ is defined to be the operator $d\delta+\delta d$. In this case an extension of Weyl's lemma also holds: A current T is infinitely differentiable if and only if ΔT is infinitely differentiable. Therefore, in the decomposition of T, T_1 is equal to a differential form of class C^∞. The fact that the notion of currents has been extremely fruitful in the theory of †harmonic integrals corresponds to the similar fact that the notion of distributions has been a powerful tool in the study of elliptic partial differential operators.

L. The Topology of $\mathfrak{D}_{\mathbf{R}^n}$ and $\mathfrak{D}'_{\mathbf{R}^n}$

Let $\{a\}=\{a_0,a_1,\dots\}$ be a nondecreasing sequence of positive numbers, and let $\{k\}=\{k_0,k_1,\dots\}$ be a nondecreasing sequence of nonnegative integers. Then we set

$$\rho_{\{a\},\{k\}}(\varphi)=\sup_{j\geqslant 0}\ \sup_{|p|\leqslant k_j}\ \sup_{|x|\geqslant j} a_j|D^p\varphi(x)| \qquad (3)$$

for $\varphi\in\mathfrak{D}$. We define the topology of the space $\mathfrak{D}$ by employing the totality of the †seminorms $\rho_{\{a\},\{k\}}$ as a fundamental system of continuous seminorms. Then $\mathfrak{D}$ is a †locally convex linear topological space, and the convergence $\varphi_m\Rightarrow 0$ is identical to the convergence $\varphi_m\rightarrow 0$ in this topology. In this topology a subset $B\subset\mathfrak{D}$ is †bounded if and only if there exist a compact set $E\subset\mathbf{R}^n$ such that $\operatorname{supp}\varphi\subset E$ for every $\varphi\in B$, and positive numbers M_p for all p such that $\sup|D^p\varphi(x)|\leqslant M_p$ for every $\varphi\in B$.

The topology of the space $\mathfrak{D}'$ is the †strong topology of the †dual of $\mathfrak{D}$: the †topology of uniform convergence on every bounded set in $\mathfrak{D}$. Under these topologies $\mathfrak{D}$ and $\mathfrak{D}'$ are †reflexive linear topological spaces.

M. Multiplication by Functions

Let T be a distribution and $\alpha(x)$ a C^∞-function. Then we can define the product αT by $(\alpha T)(\varphi)=T(\alpha\varphi)$, since the right-hand side is a linear continuous functional on $\mathfrak{D}$ since $\alpha\varphi\in\mathfrak{D}$ for each $\varphi\in\mathfrak{D}$.

N. The Operation of Partial Differential Operators on Distributions

We define the **dual operator** (or **conjugate operator**) $P'(x,D)$ of a partial differential operator $P(x,D)=\sum a_p(x)D^p$ by

$$P'(x,D)\varphi=\sum(-1)^{|p|}D^p\left(a_p(x)\varphi(x)\right), \qquad (4)$$

where the $a_p(x)$ are C^∞-functions. Combining differentiation of distributions, multiplication by functions, and addition, we can apply partial differential operators to distributions, and we have $(P(x,D)T)(\varphi)=T(P'(x,D)\varphi)$.

O. Convergence Theorems

By virtue of the following convergence theorems, various limiting processes concerning distributions may be treated easily. If the limit $\lim T_j(\varphi)=T(\varphi)$ exists for every $\varphi\in\mathfrak{D}$, where $\{T_j\}$ is a sequence of distributions, then $T\in\mathfrak{D}'$ and $\{T_j\}$ is convergent to the distribution T (**convergence theorem**). Moreover, for any p, D^pT_j is convergent to D^pT (**theorem of termwise differentiation**). Any bounded set in $\mathfrak{D}$ is totally bounded. Thus weak convergence of a sequence $\{T_j\}$ implies strong convergence (convergence in the topology of the space $\mathfrak{D}'$) (**strong convergence theorem**).

P. Distributions Depending on a Parameter

Consider distributions T_λ depending on a parameter λ, where λ ranges over the real line, the complex plane, or more generally, an open set in Euclidean space. The convergence theorem and the strong convergence theorem also hold in the case of a continuous parameter λ.

We say that T_λ is **continuous (differentiable) with respect to the parameter** λ if $T_\lambda(\varphi)$ is continuous (differentiable) with respect to λ for any $\varphi\in\mathfrak{D}$. If T_λ is defined and continuous in an interval $[a,b]$, then a distribution $T\in\mathfrak{D}'$ defined by

$$T(\varphi)=\int_a^b T_\lambda(\varphi)d\lambda,\quad \varphi\in\mathfrak{D},$$

is called the **integral of T_λ with respect to the parameter** λ and denoted by $T=\int T_\lambda d\lambda$. Let λ be a real variable and T_λ be differentiable with respect to λ at $\lambda=\lambda_0$. Then

$$\frac{d}{d\lambda}T_\lambda(\varphi)=\lim_{\lambda\to\lambda_0}\left(\frac{T_\lambda-T_{\lambda_0}}{\lambda-\lambda_0}\right)(\varphi)=S(\varphi)\qquad(5)$$

defines a distribution S, called the **derivative of T_λ at $\lambda=\lambda_0$ with respect to the parameter** λ and denoted by $\partial T_\lambda/\partial\lambda$. The formula

$$\frac{\partial}{\partial\lambda}\left(\frac{\partial}{\partial x_j}T_\lambda\right)=\frac{\partial}{\partial x_j}\left(\frac{\partial}{\partial\lambda}T_\lambda\right)\qquad(6)$$

holds. The same facts hold for the case of several real parameters. If T_λ is continuous with respect to λ, then $D_x^pT_\lambda$ is also continuous with respect to λ. In particular, if T_λ is continuous with respect to λ in an interval $[a,b]$, then the formula

$$\int_a^b D_x^pT_\lambda d\lambda=D_x^p\int_a^b T_\lambda d\lambda\qquad(7)$$

holds.

For a complex parameter λ, we say that T_λ is **analytic with respect to** λ if $T_\lambda(\varphi)$ is †analytic with respect to λ for any $\varphi\in\mathfrak{D}$. The fundamental properties of analytic functions are also verified in this case.

Q. Examples of Analytic Continuation of Distributions

For $\mathrm{Re}\lambda>-1$ the **distribution** x_+^λ on $\mathfrak{D}_{\mathbf{R}^1}$ is defined by the locally integrable function x_+^λ, which is analytic with respect to λ and can be analytically continued to the domain $\lambda\neq-1,-2,\ldots$. Namely,

$$x_+^\lambda(\varphi)=\int_0^\infty x^\lambda\left\{\varphi(x)-\varphi(0)-\ldots-\frac{x^{n-1}}{(n-1)!}\varphi^{(n-1)}(0)\right\}dx\qquad(8)$$

for $-n-1<\mathrm{Re}\lambda<-n$. Also, x_+^λ has a simple pole at $\lambda=-k$, and the †residue there is $(-1)^{k-1}\delta_x^{(k-1)}/(k-1)!$ for $k=1,2,\ldots$. The formula

$$\frac{dx_+^\lambda}{dx}=\lambda x_+^{\lambda-1}\qquad(9)$$

holds for $\lambda\neq-1,-2,\ldots$, and $x_+^\lambda/\Gamma(\lambda)$ is an †entire function of λ. We can extend $x_-^\lambda=(-x)_+^\lambda$ $(\mathrm{Re}\lambda>-1)$ analytically to $\lambda\neq-1,-2,\ldots$ in the same manner. Define $(x\pm i0)^\lambda=\lim_{y\to+0}(x\pm iy)^\lambda$. Then $(x\pm i0)^\lambda=x_+^\lambda+\exp(\pm i\lambda\pi)\cdot x_-^\lambda$, where $\mathrm{Re}\lambda>-1$, and $(x\pm i0)^\lambda$ can be analytically continued to the whole complex plane.

For an n-dimensional point x, we write $|x|=\sqrt{x_1^2+\ldots+x_n^2}$. Since $|x|^\lambda$ is locally integrable for $\mathrm{Re}\lambda>-n$, it defines the **distribution** $|x|^\lambda$, which is analytic with respect to λ and can be analytically continued to $\lambda\neq-n,-n-2,\ldots$. Also, $|x|^\lambda$ has a simple pole at $\lambda=-n-2k$ $(k=0,1,2\ldots)$, and the residue there is $\omega_n\Delta^k\delta_x/2^kk!n(n+2)\ldots(n+2k-2)$, where Δ is the †Laplacian and ω_n is the surface area of the unit sphere in $\mathbf{R}^n$. Since $\Delta(|x|^\lambda)=\lambda(\lambda+n-2)|x|^{\lambda-2}$ for $\mathrm{Re}\lambda>2-n$, the uniqueness of the analytic continuation implies that the same formula also holds for $\lambda\neq-n-2k$. The uniqueness of the analytic continuation also shows that

$$\Delta\left(\frac{|x|^\lambda}{\omega_n\Gamma((\lambda+n)/2)}\right)=\frac{2\lambda|x|^{\lambda-2}}{\omega_n\Gamma((\lambda+n-2)/2)}\qquad(10)$$

for any complex λ, since $|x|^\lambda/\Gamma((\lambda+n)/2)$ is entire in λ. In particular, taking $\lambda=2-n$, we have

$$\Delta(|x|^{2-n}/\omega_n)=(2-n)\delta_x.\qquad(10')$$

Operating with this distribution on $\varphi(x+y)$, where $\varphi\in\mathfrak{D}$ and $y\in\mathbf{R}^n$, we have, after the change of variables $x+y\to x$ in the left-hand side,

$$\Delta\int\frac{\varphi(x)}{|x-y|^{n-2}}dx=-(n-2)\omega_n\varphi(y).\qquad(10'')$$

This formula is the †Poisson equation, which is fundamental in †potential theory and can be expressed as an obvious form (10′) in the theory of distributions. Generally, a distribution T is said to be a **fundamental solution** (or **elementary solution**) of a partial differential operator $P(D)$ if $P(D)T=\delta_x$. The formula (10′) means that a fundamental solution of the Laplacian Δ is $|x|^{2-n}/\omega_n(2-n)$. In formulas (10′) and (10″) we always assume that $n>2$. In the same manner, we can also find fundamental solutions of the iterated Laplacian Δ^k (→ Appendix A, Table 15.V).

This idea of "analytic continuation" is due to M. Riesz. The finite parts of divergent integrals, used by Hadamard in the theory of †hyperbolic partial differential equations, may be given an exact meaning by the notion of distributions (→ (4) in Section C), or more specifically, by utilizing the idea of analytic continuation.

R. Plane-Wave Expansion of $|x|^\lambda$

Let ω be a point on the unit sphere in $\mathbf{R}^n$ and λ be a complex number. For $\operatorname{Re}\lambda > -1$, $|x\omega|^\lambda$ ($x\omega = x_1\omega_1 + \ldots + x_n\omega_n$) defines a distribution depending on parameters ω and λ. Since $|x\omega|^\lambda$ is analytic in λ, it can be analytically continued. For $\operatorname{Re}\lambda > -1$,

$$\frac{1}{\pi^{(n-1)/2}\Gamma((\lambda+1)/2)}\int_{|\omega|=1}|x\omega|^\lambda d\omega$$

$$=\frac{2|x|^\lambda}{\Gamma((\lambda+n)/2)} \tag{11}$$

holds, where $d\omega$ is the surface element of the unit sphere $|\omega|=1$. Because of the uniqueness of the analytic continuation, formula (11) also holds for other λ. It is called the **plane-wave expansion of** $|x|^\lambda$ and is used to obtain fundamental solutions of elliptic partial differential operators.

S. Substitution

Let $f=(f_1,\ldots,f_m)$ be a C^∞-mapping from $\mathbf{R}^n$ into $\mathbf{R}^m$, and assume that the †rank of the Jacobian matrix $(\partial f_i/\partial x_j)$ ($i=1,\ldots,m$; $j=1,\ldots,n$) is equal to m in a neighborhood of the †inverse image E of the support of a distribution $S=S_{(y)}\in\mathfrak{D}'_{\mathbf{R}^m}$. In a neighborhood U of E, we can choose $u_1=g_1(x)$, $\ldots, u_{n-m}=g_{n-m}(x)$ so that the transformation $(y,u)=(f(x),g(x))$ has the inverse transformation $x=\psi(y,u)$ of class C^∞. If the support of $\varphi\in\mathfrak{D}_{\mathbf{R}^n}$ is contained in U, then defining $\tilde\varphi$ by

$$\tilde\varphi(y)=\int_{\mathbf{R}^{n-m}}\varphi(\psi(y,u))J(y,u)\,du,$$

we have $\tilde\varphi\in\mathfrak{D}_{\mathbf{R}^m}$, where $J(y,u)$ is the absolute value of the †Jacobian of $\psi(y,u)$, and $\tilde\varphi$ is independent of the choice of $g_1,\ldots,g_{n-m}$. Now we define $T(\varphi)=S(\tilde\varphi)$ if $\operatorname{supp}\varphi\subset U$, and $T(\varphi)=0$ if $\operatorname{supp}\varphi$ does not intersect E. Since a distribution may be determined by its local behavior, we have a distribution $T\in\mathfrak{D}'_{\mathbf{R}^n}$ with support E, which is denoted by $T=S\circ f=S(f)=S_{f(x)}$ and is called the **substituted distribution** of $S_{(y)}$ by $y=f(x)$. The chain rule for derivatives of composites

$$\frac{\partial}{\partial x_j}(S\circ f)=\sum_{i=1}^m\frac{\partial f_i}{\partial x_j}\left(\frac{\partial S}{\partial y_i}\circ f\right) \tag{12}$$

also holds.

For example: (1) If $S=\delta^{(q)}_{(y)}$ ($y\in\mathbf{R}^n$), then E is the surface $f_1(x)=f_2(x)=\ldots=f_n(x)=0$. Assuming that $(f_1,\ldots,f_n)$ satisfies the condition in the previous paragraph, we obtain the **distribution** $\delta^{(q)}(f_1,\ldots,f_n)$ **with support contained in a surface**. From the fact $\delta^{(q)}=D_y^q\delta$ we may write

$$\delta^{(q)}(f_1,\ldots,f_m)=\frac{\partial^{|q|}\delta(f_1,\ldots,f_m)}{(\partial f_1)^{q_1}\ldots(\partial f_m)^{q_m}}$$

(Gel'fand-Šilov notation). (2) For the mapping $f(x)=Ax+b$, where A is an $n\times n$ regular matrix and b is an n-vector, we can define the substituted distribution of S by $f=f(x)$ for any $S\in\mathfrak{D}'_{\mathbf{R}^n}$. For instance, $\delta_{(x-b)}(\varphi)=\varphi(b)$, $\delta_{(x^2-c^2)}=(2c)^{-1}(\delta_{(x-c)}+\delta_{(x+c)})$ ($c>0$, $x\in\mathbf{R}^1$).

T. Direct Product

Let x and y be any points in $\mathbf{R}^n$ and $\mathbf{R}^m$, respectively. We denote $\mathfrak{D}_{\mathbf{R}^n}$ and $\mathfrak{D}_{\mathbf{R}^m}$ by $\mathfrak{D}_x$ and $\mathfrak{D}_y$, respectively. If a bilinear functional $B(\varphi,\psi)$ on $\mathfrak{D}_x\times\mathfrak{D}_y$ is continuous separately in φ and ψ, then there exists a unique distribution $W\in\mathfrak{D}'_{(x,y)}$ such that $B(\varphi,\psi)=W(\varphi(x)\psi(y))$ (**kernel theorem**). Applying this theorem to $B(\varphi,\psi)=T(\varphi)S(\psi)$, where $T\in\mathfrak{D}'_x$, $S\in\mathfrak{D}'_y$, we find that there exists a unique $W\in\mathfrak{D}'_{(x,y)}$ such that $W(\varphi(x)\psi(y))=T(\varphi)S(\psi)$, denoted by $T_{(x)}\times S_{(y)}$ or $S_{(y)}\times T_{(x)}$ and called the **direct product**. By the subscript (x) we indicate that T operates on the space $\mathfrak{D}_x$. Fubini's theorem, $T_{(x)}\times S_{(y)}(\varphi(X,Y))=S_{(y)}(T_{(x)}(\varphi(x,y)))$, holds for any $\varphi\in\mathfrak{D}_{(x,y)}$.

U. Convolution

For distributions S and T, assume that either S or T has a compact support. Then the correspondence

$$\varphi\in\mathfrak{D}\to S_{(x)}T_{(y)}(\varphi(x+y))$$

defines a distribution called the **convolution** of S and T, and denoted by $S*T$. In particular, $T_f*T_g=T_{f*g}$, where $f*g$ is the †convolution of functions f and g, and T_f, T_g, and T_{f*g} denote the distributions corresponding to f, g, and $f*g$, respectively.

For example: $T*\delta=\delta*T=T$, $D^pT*S=T*(D^pS)=D^p(T*S)$.

Thus a solution of the partial differential equation $P(D)T=S$ with constant coefficients is given by the convolution $S*E$ with a fundamental solution E of $P(D)$, whenever the convolution exists.

If $T\in\mathfrak{D}'$, and $\varphi\in\mathfrak{D}$, the convolution $T*\varphi$ is equal to a function f of class C^∞ (or the distribution corresponding to f), and $f(y)=$

$T_{(x)}(\varphi(x-y))$. f is called the **regularization** of T. For distributions S and T, assume that $|f(x)g(y-x)|$ is integrable on $\mathbf{R}^n$ for any regularizations $f=S*\varphi$ and $g=T*\psi$. Then there exists a unique distribution V such that

$$V*(\varphi*\psi)=\int f(x)g(y-x)\,dx.$$

This distribution V is called the **generalized convolution** and is denoted by $S*T$ (C. Chevalley).

V. Distributions Corresponding to Measures

We denote by $\mathcal{C}_{\mathbf{R}^n}=\mathcal{C}$ the linear space of all continuous complex-valued functions defined in $\mathbf{R}^n$ with compact support. A sequence $\{\varphi_m\}$ of elements of $\mathcal{C}$ is said to converge to zero if all supports of φ_m are contained in a fixed compact set and $\{\varphi_m\}$ converges uniformly to zero. Thus we can define a topology in $\mathcal{C}$. By $\mathcal{C}'$ we mean the set of all continuous linear functionals on the space $\mathcal{C}$ with this topology. $\mathcal{C}'\subset\mathfrak{D}'$. If $T\in\mathcal{C}'$, then $T=T_\mu$ ($\rightarrow$ (2) in Section C), where μ is a †completely additive complex-valued set function defined on the bounded Borel sets in $\mathbf{R}^n$. A distribution is called a **positive distribution** if $T(\varphi)\geqslant 0$ for any $\varphi\in\mathfrak{D}$ that has nonnegative values at all points x in $\mathbf{R}^n$. Every positive distribution is equal to a T_μ corresponding to a positive measure μ.

W. Distributions with Compact Support

We denote by $\mathcal{E}_{\mathbf{R}^n}=\mathcal{E}$ the set (linear space) of all complex-valued C^∞-functions on $\mathbf{R}^n$. We define a family of seminorms on $\mathcal{E}$ by

$$\rho_{p,m}(\varphi)=\sup_{|x|\leqslant m}|D^p(x)|, \tag{13}$$

where p ranges over all multi-indices and m ranges over all positive integers. The **topology of the space** $\mathcal{E}$ is the locally convex, linear topology under which $\{\rho_{p,m}\}$ is a fundamental family of continuous seminorms. A sequence $\{\varphi_m\}$ converges to zero in this topology if and only if $\{D^p\varphi_m(x)\}$ converges uniformly on E to zero for any compact set E and any p. By $\mathcal{E}'$ we mean the set of all continuous linear functionals on the space $\mathcal{E}$ with the topology just defined, i.e., the dual of $\mathcal{E}$. The **topology of the space** $\mathcal{E}'$ is the strong topology of the dual space. Under these topologies, $\mathcal{E}$ and $\mathcal{E}'$ are reflexive. For a distribution T with compact support, choosing $\alpha\in\mathfrak{D}$ such that α is identically equal to 1 on the support of T, we define a linear functional S on $\mathcal{E}$ by $S(\varphi)=T(\alpha\varphi)$. Then $S\in\mathcal{E}'$, and S is independent of the choice of α. In this sense, we can consider that S is the extension of T to $\mathcal{E}$, and identify S and T. Thus $\mathcal{E}'$ coincides with the set of all distributions with compact support.

X. The Space $\mathcal{S}$

We write $x^p=x_1^{p_1}\cdot\ldots\cdot x_n^{p_n}$. A function $\varphi(x)$ of class C^∞ is called a †rapidly decreasing function of class C^∞ if

$$\rho_{p,q}(\varphi)=\sup_x|x^pD^q\varphi(x)| \tag{14}$$

is finite for all p and q. We denote by $\mathcal{S}$ the set of all rapidly decreasing functions. The **topology of the space** $\mathcal{S}$ is the locally convex, linear topology under which $\{\rho_{p,q}\}$ is a fundamental family of continuous seminorms. The dual space of $\mathcal{S}$ is denoted by $\mathcal{S}'$. The **topology of the space** $\mathcal{S}'$ is the strong topology of the dual space.

Y. Slowly Increasing Distributions

When a distribution $T\in\mathfrak{D}'$ can be continuously extended to $\mathcal{S}$, its extension is unique. The extension of T exists if any regularization $f=T*\varphi$ of T is a slowly increasing continuous function (i.e., there is a polynomial $P(x)$ such that $|f(x)|\leqslant|P(x)|$), and in this case, T is called a **tempered distribution** (or **slowly increasing distribution**). The set of all distributions of this kind coincides with the space $\mathcal{S}'$.

Z. Fourier Transforms

For $\varphi\in\mathcal{S}$ the integral

$$\mathcal{F}_\varphi(x)=(\sqrt{2\pi}\,)^{-n}\int\varphi(y)\exp(-ixy)\,dy$$

converges, where $xy=x_1y_1+\ldots+x_ny_n$ and $i=\sqrt{-1}$. For $T\in\mathcal{S}'$, the **Fourier transform** $\mathcal{F}T$ is defined by

$$(\mathcal{F}T)(\varphi)=T(\mathcal{F}\varphi),\quad \varphi\in\mathcal{S}. \tag{15}$$

The inverse Fourier transform is defined similarly, except that $-i$ is replaced by i. For example: (1) $\mathcal{F}1=(\sqrt{2\pi}\,)^n\delta$, where 1 is the distribution corresponding to the function 1. (2) $\mathcal{F}(D^pT)=i^{|p|}x^p\mathcal{F}T$.

A function φ of class C^∞ is a †slowly increasing C^∞-function if φ and its derivatives of any order are slowly increasing continuous functions. The space $\mathcal{O}_M$ is the set of all such functions. If $\alpha\in\mathcal{O}_M$ and $T\in\mathcal{S}'$, then $\alpha T\in\mathcal{S}'$, $\mathcal{F}(\alpha T)=\mathcal{F}\alpha*\mathcal{F}T$, and $\mathcal{F}(ix_jT)=-\partial(\mathcal{F}T)/\partial y_j$. Then $\mathcal{F}(\mathcal{O}_M)$, denoted by $\mathcal{O}'_C$, coincides with the set of all distributions T such that any regularization $f=T*\varphi$ is a rapidly decreasing C^∞-function. A member of the space $\mathcal{O}'_C$ is called a **rapidly decreasing distribution**.

AA. Fourier Series and Distributions on Tori

The n-dimensional †torus X is a compact C^∞-manifold, so that $\mathfrak{D}_X = \mathfrak{E}_X$, $\mathfrak{D}'_X = \mathfrak{E}'_X$. The torus X is the quotient space of $\mathbf{R}^n$ with respect to the equivalence relation $x_j \equiv y_j \bmod \mathbf{Z}$ $(j=1,\ldots,n)$. Therefore, the volume element dx of X is defined from that of $\mathbf{R}^n$. Thus for an integrable function f, we can define a distribution T_f by

$$T_f(\varphi) = \int_X f(x)\varphi(x)\,dx, \quad \varphi \in \mathfrak{D}_X.$$

Consider the family of functions $f_p(x) = \exp(2\pi ipx)$, where p ranges over all n-tuples of integers. Then any distribution T on X has the **Fourier series** expansion $T = \sum c_p f_p(x)$. Conversely, if a sequence $\{c_p\}$ is slowly increasing, i.e., $|c_p| \leqslant C(1+|p|^2)^k$ for some k and C, then $\sum c_p f_p$ converges to a distribution T on X.

BB. Structure of Distributions

Any distribution with compact support can be represented as a finite linear combination of derivatives (in the distribution sense) of functions. The restriction of any distribution T to Ω has the same property, where Ω is an open set whose closure is compact. In this sense, the notion of distribution is a generalization of that of function. Furthermore, for any distribution T there exists a sequence $\{\varphi_j\} \subset \mathfrak{D}$ such that $\{\varphi_j\}$ converges to T in the distribution sense.

CC. Justification of Improper Functions

Improper functions such as Dirac's delta functions, which are simply called "functions" in applications, have been given their theoretical foundation by the theory of distribution even though they are not functions in the classical sense.

For example: (1) **Dirac's delta function** $\delta(x)$ is a "function" such that $\delta(0)=\infty$, $\delta(x)=0$ for $x \neq 0$, and it has the following properties: $\delta(-x)=\delta(x)$, $d\mathbf{1}/dx = \delta(x)$, $\delta(cx) = |c|^{-1}\delta(x)$, $\delta(x^2-c^2) = (2|c|)^{-1}\{\delta(x-c)+\delta(x+c)\}$, $x\delta'(x) = -\delta(x)$, $\int_{-\infty}^{\infty}\delta(x)\,dx = 1$, $\int_a^b f(x)\delta(x-c)\,dx = f(c)$ (where $f(x)$ is any continuous function of the interval $[a,b]$, $a<c<b$), $\int_{-\infty}^{\infty}\delta(x-a)\delta(x-b)\,dx = \delta(a-b)$, $\int_a^b f(x)\delta'(x-c)\,dx = -f'(c)$ (where $f(x)$ is any C^1-function on $[a,b]$, $a<c<b$), and $\int_{-\infty}^{\infty}\exp(2\pi icx)\,dx = \delta(c)$. These formulas are valid for Dirac's distribution δ; that is, we may interpret $\delta(\varphi) = \varphi(0) = \int_{-\infty}^{\infty}\delta(x)\varphi(x)\,dx$, where δ in the left-hand side is defined as in Section C.

(2) The **invariant delta function** $D(x,y,z,t)$, used in field theory, has the following properties:

$$D(x,y,z,t) = (2\pi)^{-2}\int\int\int_{\mathbf{R}^3}\sin(\xi x + \eta y + \zeta z - \omega ct)\omega^{-1}\,d\xi\,d\eta\,d\zeta,$$

where $\omega = \sqrt{\xi^2+\eta^2+\zeta^2+\kappa^2}$, c is the speed of light, and $\kappa \geqslant 0$ is a nonnegative constant. $D(x,y,z,t)=0$, $(\partial D/\partial t)_{t=0} = -4\pi c\delta(x,y,z)$. $D(x,y,z,t)$ is invariant with respect to †proper Lorentz transformations. $D(x,y,z,-t) = -D(x,y,z,t)$. $D = r^{-1}\partial F/\partial r$, where $F = J_0(\kappa\sqrt{c^2t^2-r^2})$ $(ct>r)$, $F=0$ $(r>ct>-r)$, $F = -J_0(\kappa\sqrt{c^2t^2-r^2})$ $(ct<-r)$, $r = \sqrt{x^2+y^2+z^2}$, and J_0 is the †Bessel function. $D(x,y,z,t)$ is a "function" whose values equal ∞ on the light cone $ct=r$ and 0 elsewhere. These facts are also formulated in the theory of distributions.

DD. Gel'fand-Šilov Generalized Functions

Following Gel'fand and Šilov, a function space on $\mathbf{R}^n$ or $\mathbf{C}^n$ is called a **fundamental space** (or **test-function space**) if it is a †countably normed space or a countable union of such spaces and its topology is stronger than that of pointwise convergence. The members of the fundamental space are called **fundamental functions** (or **test functions**), and continuous linear functionals on it are called **generalized functions (in the sense of Gel'fand and Šilov)** [12, 13]. Various types of fundamental spaces and generalized functions on them are investigated and applied to many problems concerning partial differential equations. We shall mention some of them.

EE. The Fundamental Spaces $K\{M_k\}$ and $K(a)$

Let $1 \leqslant M_1(x) \leqslant M_2(x) \leqslant \ldots$ be a sequence of real-valued functions defined on a subset S_M of $\mathbf{R}^n$. By $K\{M_k\}$ we mean the linear space of all C^∞-functions φ such that $\varphi(x)=0$ for $x \notin S_M$ and

$$\rho_k(\varphi) = \sup_{|p|\leqslant k}\ \sup_{x\in S_M} M_k(x)|D^p\varphi(x)| \qquad (16)$$

is finite for $k=1,2,\ldots$. The topology of $K\{M_k\}$ is defined in terms of seminorms ρ_k $(k=1,2,\ldots)$. For example: (1) Let S_M be a compact set E in $\mathbf{R}^n$, and let $M_k(x)=1$ $(k=1,2,\ldots)$. Then $K\{M_k\}$ is the set $K(E)$ of all C^∞-functions with support contained in E. In particular when $E = \{x \mid |x_j| \leqslant a_j, j=1,2,\ldots,n\}$, $a=(a_1,\ldots,a_n)$ we write $K(E)=$

$K(a)$. (2) If $S_M = \mathbf{R}^n$ and $M_k(x) = (1+|x|)^k$, then $K\{M_k\} = \mathcal{S}$.

FF. The Fundamental Spaces $Z\{M_k\}$ and $Z(a)$

Let $0 < C(y) \leqslant M_1(z) \leqslant M_2(z) \leqslant \ldots$ be a sequence of real-valued functions on $\mathbf{C}^n$, where $z = x + iy$ and $C(y)$ is a continuous function. The fundamental space $Z\{M_k\}$ consists of all functions $\varphi(x)$ which can be extended into entire functions $\varphi(z)$ on $\mathbf{C}^n$ and for which all the seminorms

$$\rho_k(\varphi) = \sup_{z \in \mathbf{C}^n} M_k(z)|\varphi(z)|, \quad k = 1, 2, \ldots, \tag{17}$$

are finite. The topology is defined in terms of the seminorms ρ_k $(k = 1, 2, \ldots)$. For example, if we take $M_k(z) = (1+|z|)^k \exp(-a|y|)$, where $\exp(-a|y|) = \exp(-a_1|y_1|) \cdot \ldots \cdot \exp(-a_n|y_n|)$ and $a_1 > 0, \ldots, a_n > 0$, we obtain the space $Z(a)$ as a particular case of $Z\{M_k\}$.

GG. Spaces of Type S

Let α, β, A, B, $\bar{A}$, and $\bar{B}$ be n-vectors. By $A \geqslant B$ we mean $A_j \geqslant B_j$ $(j = 1, \ldots, n)$. We use notations $A^p = A_1^{p_1} \ldots A_n^{p_n}$, $p^{p\alpha} = p_1^{p_1\alpha_1} \ldots p_n^{p_n\alpha_n}$. (i) For $\alpha \geqslant 0$, $A > 0$ the space $S_{\alpha,A}$ consists of all C^∞-functions φ such that seminorms

$$\rho_{q,\bar{A}}(\varphi) = \sup_x \sup_p \frac{|x^p D^q \varphi(x)|}{\bar{A}^p p^{p\alpha}} \tag{18}$$

are finite for all $\bar{A} > A$ and q. For example, $S_{0,A} = K(A)$. (ii) For $\beta \geqslant 0$, $B > 0$ the space $S^{\beta,B}$ consists of all C^∞-functions φ such that the seminorms

$$\rho_{p,\bar{B}}(\varphi) = \sup_x \sup_q \frac{|x^p D^q \varphi(x)|}{\bar{B}^q q^{q\beta}} \tag{19}$$

are finite for all $\bar{B} > B$ and q. (iii) For α, $\beta \geqslant 0$ and A, $B > 0$, the space $S_{\alpha,A}^{\beta,B}$ consists of all C^∞-functions such that the seminorms

$$\rho_{\bar{A},\bar{B}}(\varphi) = \sup_x \sup_p \sup_q \frac{|x^p D^q \varphi(x)|}{\bar{A}^p \bar{B}^q p^{p\alpha} q^{q\beta}} \tag{20}$$

are finite for all $\bar{A} > A$ and $\bar{B} > B$. The topologies for these spaces are given in terms of the seminorms (18), (19), and (20), respectively. These spaces are generically called **spaces of type** S.

$\mathcal{F}(S_{\alpha,A}) = S^{\alpha,A}$ $(\alpha > 0)$,

$\mathcal{F}(S^{\beta,B}) = S_{\beta,B}$ $(\beta > 0)$,

$\mathcal{F}(S_{\alpha,A}^{\beta,B}) = S_{\beta,B}^{\alpha,A}$ $(\alpha + \beta > 1)$.

$\mathcal{F}(S_{0,A}) = S^{0,A'}$,

$\mathcal{F}(S^{0,B}) = S_{0,B'}$,

$\mathcal{F}(S_{\alpha,A}^{\beta,B}) = S_{\beta,B'}^{\alpha,A'}$ $(\alpha + \beta = 1)$,

where $A' = A\exp(1/A)$ and $B' = B\exp(1/B)$. $S_{\alpha,A}^{\beta,B} = \{0\}$ for $\alpha + \beta < 1$.

$S_\alpha = \bigcup S_{\alpha,A}$, $S^\beta = \bigcup S^{\beta,B}$, and $S_\alpha^\beta = S_{\alpha,A}^{\beta,B}$, where A and B range over all positive n-vectors.

HH. Entire Functions of Exponential Type

An entire function f is said to be of **exponential type** $\leqslant b$, where $b = (b_1, \ldots, b_n)$, if for any $\varepsilon = (\varepsilon_1, \ldots, \varepsilon_n) > 0$ there is a constant c_ε such that $|f(z)| \leqslant c_\varepsilon \exp(b+\varepsilon)|z|$. If $\varphi \in \mathcal{S}$ can be extended into an entire function of type $\leqslant b$, then $\varphi \in Z(b+\varepsilon)$ for any $\varepsilon > 0$ (†Phragmén-Lindelöf theorem). From $\mathcal{F}(K(b)) = Z(b)$ and $\bigcap_{\varepsilon>0} K(b+\varepsilon) = K(b)$, we have $Z(b) = \bigcap Z(b+\varepsilon)$. Therefore, the set of all rapidly decreasing C^∞-functions that can be extended into entire functions of exponential type $\leqslant b$ coincides with $\mathcal{F}(K(b))$ **(Paley-Wiener theorem)**. An extension of this theorem is the following: A slowly increasing continuous function f can be extended into an entire function of exponential type $\leqslant b$ if and only if the support of its Fourier transform $\mathcal{F}f$ (considered as an element of $\mathcal{S}'$) is contained in the set $\{x \,|\, |x_1| \leqslant b_1, \ldots, |x_n| \leqslant b_n\}$.

By $\mathring{Z}(b)$ we mean the set of all complex functions analytic on $B = \{z \,|\, |z_1| \leqslant b_1, \ldots, |z_n| \leqslant b_n\}$. A sequence $\{\varphi_m\}$ in $\mathring{Z}(b)$ is said to be convergent to zero if there exists a neighborhood Ω of B such that φ_m is defined and analytic on Ω for $m = 1, 2, \ldots$ and $\{\varphi_m\}$ is convergent uniformly to zero in Ω. The set of entire functions of exponential type $\leqslant b$ coincides with the totality of Fourier transforms of linear functionals on $\mathring{Z}(b)$ which are continuous with respect to the above convergence (Gel'fand-Šilov).

II. Nuclearity of Spaces of Generalized Functions

The space $K\{M_k\}$ is †nuclear if the following four conditions are satisfied: (i) $M_k(x) = M_k(|x|)$, where the $M_k(t)$ are monotone increasing, (ii) for any n and k there exist c and p such that $|M_n^{(k)}(t)| \leqslant cM_p(t)$, (iii) for any n there is a p such that $m_{np}(t) = M_n(t)/M_p(t)$ tends to zero as $t \to \infty$, and (iv) $m_{np}(|x|)$ is integrable over $\mathbf{R}^n$. In particular, $\mathcal{S}$ and $K(m)$ are nuclear. Since $K\{M_k\}$ is an †F-space, the space of generalized functions on $K\{M_k\}$ (i.e., the dual of $K\{M_k\}$) is †nuclear if $K\{M_k\}$ is nuclear. Since $\mathcal{D}$ is an †LF-space, that is, the inductive limit of $\{K(m)\}$ $(m = 1, 2, \ldots)$, both $\mathcal{D}$ and $\mathcal{D}'$ are nuclear. The spaces S_α^β are also nuclear and †LF-spaces. Therefore, the $(S_\alpha^\beta)'$ are nuclear. $\mathcal{E}$, $\mathcal{E}'$, $\mathcal{O}_M$, and $\mathcal{O}_C'$ are also nuclear. $\mathcal{E}$, $\mathcal{S}$, $\mathcal{D}$, $\mathcal{E}'$, $\mathcal{S}'$, and

$\mathfrak{D}'$ are †Montel spaces, because they are either F-spaces or LF-spaces or the duals of such spaces (→ 407 Topological Linear Spaces).

JJ. Sato Hyperfunctions

On a real analytic manifold we can define Sato hyperfunctions, which are more general than Schwartz distributions.

We consider $\mathbf{R}^1$ as a subset of $\mathbf{C}^1$. In the set of all functions which are †analytic in some $U-\mathbf{R}^1$, where U is a complex neighborhood of $\mathbf{R}^1$ (i.e., an open set of $\mathbf{C}^1$ containing $\mathbf{R}^1$), we define an †equivalence relation by $F \sim G$ if and only if $F(z)-G(z)$ can be extended into an analytic function in a complex neighborhood of $\mathbf{R}^1$. Then an equivalence class with respect to this relation is called a **Sato hyperfunction**, and the equivalence class containing $F(z)$ is written as $F(x+i0)-F(x-i0)$. This notion may be considered a generalization of that of "boundary values" of analytic functions. The derivative of a hyperfunction $F(x+i0)-F(x-i0)$ is defined as the class containing dF/dz.

These definitions can be extended to the case $\mathbf{R}^n$. By $U_1, U_2, \ldots, U_n$; $V_1, V_2, \ldots, V_n$; and $W_1, W_2, \ldots, W_n$, we mean complex neighborhoods of $\mathbf{R}^1$. Let φ and ψ be analytic in $(U_1-\mathbf{R}^1)\times \ldots \times(U_n-\mathbf{R}^1)$ and $(V_1-\mathbf{R}^1)\times \ldots \times(V_n-\mathbf{R}^1)$, respectively. The equivalence relation $\varphi \sim \psi$ is defined as follows: $\varphi \sim \psi$ if there exist $W_j \subset U_j \cap V_j$ $(j=1,\ldots,n)$ and analytic functions $\chi_j(z)$ in $(W_1-\mathbf{R}^1)\times \ldots \times W_j \times \ldots \times(W_n-\mathbf{R}^1)$ (only the jth factor is equal to W_j) such that $\varphi(z)-\psi(z)=\chi_1(z)+\ldots+\chi_n(z)$. An equivalence class with respect to this relation is called a Sato hyperfunction. We can similarly define hyperfunctions on an open set Ω in $\mathbf{R}^n$.

Using the relative †cohomology with coefficients in the sheaf of analytic functions (→ 377 Sheaves), it is possible to give another, coordinate-free and more natural definition of hyperfunctions. Hyperfunctions can also be locally defined.

Hyperfunctions on an open set cannot be defined as elements of the dual of a topological linear space. However, hyperfunctions with support in a compact set K are identified with analytic functionals with carriers in K, i.e., the space of such hyperfunctions is the dual of the space of local analytic functions defined on a neighborhood of K. Since the sheaf of hyperfunctions is †flabby, we may also define the hyperfunctions as locally finite sums of analytic functionals with compact carriers in $\mathbf{R}^n$.

It is an old idea to represent generalized functions as boundary values of analytic functions. This was actually the starting point of the complex method of Fourier analysis. As we remarked earlier, T. Carleman considered also the case in which the boundary values are not functions. Another example is found in Carleman's proof of the spectral decomposition theorem. Closely related to this are the †dispersion relations appearing in physics. Furthermore, G. Köthe proved that any distribution on a simple closed analytic curve C in $\mathbf{C}$ is the boundary value of an analytic function on $\mathbf{C}-C$. It was M. Sato, however, who proved the local property of hyperfunctions for the first time and extended the theory to the higher-dimensional case.

KK. Ultradistributions

Let M_p be a sequence of positive numbers such that $C(M_p)$ is not quasi-analytic (→ 61 C^∞-Functions and Quasi-Analytic Functions). The spaces

$$\mathfrak{D}_{\{M_p\}}=\{\varphi\in\mathfrak{D} \mid \exists h\ \exists C \text{ such that (21) holds}\},$$

$$\mathfrak{D}_{(M_p)}=\{\varphi\in\mathfrak{D} \mid \forall h>0\ \exists C\ \text{"(21)"}\},$$

where

$$\sup|D^p\varphi(x)| \leqslant Ch^{|p|}M_p, \qquad (21)$$

have natural locally convex topologies. Usually we impose conditions on M_p so that multiplication and differentiation are continuous. Of particular importance is the case of Gevrey sequence $M_p=(p!)^s$ with $s>1$.

The continuous linear functionals on these spaces are called **ultradistributions** (of C. Roumieu type and A. Beurling type, respectively). Distributions are ultradistributions, and ultradistributions are embedded in the hyperfunctions. Ultradistributions have the same local property as distributions, and hyperfunctions and ultradistributions are often useful in the theory of differential equations.

References

[1] L. Schwartz, Théorie des distributions, Hermann, revised edition, 1966.
[2] G. de Rham, Variétés différentiables, Hermann, 1955.
[3] C. Chevalley, Theory of distributions, Columbia Univ. lectures 1950–1952.
[4] L. Schwartz and I. Halperin, Introduction to the theory of distributions, Univ. of Toronto Press, 1952.
[5] T. Iwamura, Tyôkansû (Japanese; Theory of distributions), Iwanami Coll. of Modern Math. Sci., 1958.
[6] K. Yosida, Functional analysis, Springer, 1965.
[7] L. Schwartz, Transformation de Laplace des distributions, Medd. Lunds Univ. Mat. Sem., tome suppl. à M. Riesz (1952), 196–206.

[8] L. Schwartz, Théorie des distributions à valeurs vectorielles 1, Ann. Inst. Fourier, 7 (1957), 1–141.
[9] B. Malgrange, Equations aux dérivées partielles à coefficients constants I, II, C. R. Acad. Sci. Paris, 237 (1953), 1620–1622; 238 (1954), 196–198.
[10] L. Ehrenpreis, Analytic functions and Fourier transform of distributions I, Ann. of Math., (2) 63 (1956), 129–159.
[11] M. Sato, Theory of hyperfunctions I, II, J. Fac. Sci. Univ. Tokyo, 8 (1959), 139–193, 387–437.
[11A] On the theory of hyperfunctions (Japanese), Sûgaku, 10 (1958), 1–27.
[12] И. М. Гельфанд, Г. Е. Шилов (I. M. Gel'fand and G. E. Šilov), Преобразования Фурье быстро растущих функций и вопросы единственности решения задачи Коши, Uspehi Mat. Nauk (N.S.), 9 (61) (1954), 141–148; English translation, Fourier transforms of rapidly increasing functions and questions of uniqueness of the solution of Cauchy's problem, Amer. Math. Soc. Transl., (2) 5 (1957), 221–274.
[13] И. М. Гельфанд, Г. Е. Шилов (I. M. Gel'fand and G. E. Šilov), Обобщенные функции, физматгиз, 1958–1966; English translation, Generalized functions. I. Properties and operations; II. Spaces of fundamental and generalized functions; III. Theory of differential equations; IV (I. M. Gel'fand and N. Ja. Vilenkin). Applications of harmonic analysis; V (I. M. Gel'fand, M. I. Graev, and N. Ja. Vilenkin). Integral geometry and representation theory, Academic Press, 1964, 1968, 1967, 1964, 1966.
[14] С. Л. Соболев (S. L. Sobolev), Об одной теореме функционального анализа, Mat. Sb., 4 (1938), 471–496.
[15] A. Friedman, Generalized functions and partial differential equations, Prentice-Hall, 1963.
[16] L. Ehrenpreis, Fourier analysis in several complex variables, John Wiley, 1970.
[17] В. П. Паламодов (V. P. Palamodov), Линейные дифференциальные операторы постоянными коэффициетами, Наука, 1967; English translation, Linear differential operators with constant coefficients, Springer, 1970.
[18] P. Schapira, Théorie des hyperfonctions, Lecture notes in math. 126, Springer, 1970.
[19] H. Komatsu, Hyperfunctions and partial differential equations with constant coefficients (Japanese), Lecture notes, Univ. of Tokyo, 1968.
[20] M. Sato and M. Kashiwara, On the structure of hyperfunctions (Japanese), Sûgaku no Ayumi, 15 (1970), 9–72.
[21] Hyperfunctions and pseudo-differential equations, Proceedings of a conference at Katata, 1971, Lecture notes in math., Springer, 1973.
[22] M. Sato, T. Kawai, and M. Kashiwara, Microfunctions and pseudo-differential equations, in [21].
[23] J. L. Lions and E. Magenes, Non-homogeneous boundary value problems and applications, vol. 3, Springer, 1973.

131 (XVI.2) Distributions (of Random Variables)

A. Characteristics of Probability Distributions

Given an n-dimensional †random variable $X=(X_1,\ldots,X_n)$ defined on a †probability space $(\Omega,\mathfrak{F},P)$, the set function $\Phi(E)=P(X\in E)$, $E\in\mathfrak{B}^n$, i.e., $\Phi(E)=P(\{\omega\,|\,(X_1(\omega),\ldots,X_n(\omega))\in E\})$, determines a **probability distribution** on the †σ-algebra $\mathfrak{B}^n$ of all n-dimensional †Borel sets (→ 339 Probability). There are several characteristics indicating the properties of probability distributions in one dimension: the **mean** (or **mathematical expectation**) $m=\int_{-\infty}^{\infty}x\,d\Phi(x)$, the **variance** $\sigma^2=\int_{-\infty}^{\infty}|x-m|^2d\Phi(x)$, the **standard deviation** σ, the kth **moment** $\alpha_k=\int_{-\infty}^{\infty}x^k d\Phi(x)$, the kth **absolute moment** $\beta_k=\int_{-\infty}^{\infty}|x|^k d\Phi(x)$, the kth **moment about the mean** $\mu_k=\int_{-\infty}^{\infty}(x-m)^k d\Phi(x)$, etc.

When Φ is the distribution of a random variable X, we write $m=E(X)$, $\sigma^2=E(X-m)^2$, etc. The moments and the moments about the mean are connected by the relation $\mu_r=\sum_{k=0}^{r}\binom{r}{k}\alpha_{r-k}(-m)^k$ $(r=1,2,\ldots)$. When Φ is an n-dimensional distribution, the following quantities are frequently used: the **mean vector**, which is an n-dimensional vector whose ith component is given by $m_i=\int x_i d\Phi(x)$; the **covariance matrix**, which is an $n\times n$ matrix whose (i,j)-element is $\sigma_{ij}=\int(x_i-m_i)(x_j-m_j)d\Phi(x)$; the **moment matrix**, which is an $n\times n$ matrix whose (i,j)-element is $m_{ij}=\int x_i x_j d\Phi(x)$. (The covariance matrix is also called the **variance matrix** or the **variance-covariance matrix**.) The covariance matrix and the moment matrix are †positive definite and symmetric. The quantities listed above are defined only under some integrability conditions.

Given an n-dimensional distribution Φ, we can consider its †characteristic function, defined by

$$\varphi(z)=\int\exp(i(z,x))d\Phi(x)\quad(Z\in\mathbf{R}^n),$$

where (z,x) is the inner product of z and x. Conversely, a characteristic function determines a distribution function uniquely (→ 59 Characteristic Functions (of Probability Distributions)). The **moment-generating function** defined by $f(z)=\int\exp(-(z,x))d\Phi(x)$ $(z\in\mathbf{R}^n)$ does not necessarily exist for all n-dimensional distributions but does exist for a number of useful probability distributions Φ, and then $f(z)$ uniquely determines Φ. Given a 1-dimensional distribution Φ with $\beta_k<+\infty$, we denote by γ_k the coefficient of $(iz)^k/k!$ in the †Maclaurin expansion of $\log\varphi(z)$.

We call γ_k the (kth order) **semi-invariant** of Φ. The moments and semi-invariants are connected by the relations $\gamma_1=\alpha_1$, $\gamma_2=\alpha_2-\alpha_1^2=\sigma^2$, $\gamma_3=\alpha_3-3\alpha_1\alpha_2+2\alpha_1^3$, $\gamma_4=\alpha_4-3\alpha_2^2-4\alpha_1\alpha_3+12\alpha_1^2\alpha_2-6\alpha_1^4,\ldots$.

There exists a one-to-one correspondence between a 1-dimensional distribution Φ and its **(cumulative) distribution function** F defined by $F(x)=\Phi((-\infty,x])$. A distribution function is characterized by the following properties: (1) It is monotone nondecreasing; (2) it is right continuous; (3) $\lim_{x\to-\infty}F(x)=0$ and $\lim_{x\to+\infty}F(x)=1$. Similar statements hold for the multidimensional case.

B. Specific Distributions

Given an n-dimensional distribution, if there exists a positive measure Φ on the point a such that $\Phi(\{a\})>0$, a is called a discontinuity point of Φ. The set D of all discontinuity points of Φ is at most countable. In case $\Phi(D)=1$, a distribution is called a **purely discontinuous distribution**. In particular, if D is a lattice, Φ is called a **lattice distribution**. If the distribution of Φ is a continuous function, Φ is called a **continuous distribution**. By virtue of the †Lebesgue decomposition theorem, every probability distribution can be expressed in the form

$$\Phi=a_1\Phi_1+a_2\Phi_2+a_3\Phi_3,$$
$$a_1,a_2,a_3\geqslant 0,\qquad a_1+a_2+a_3=1,$$

where Φ_1 is purely discontinuous, Φ_2 is †absolutely continuous with respect to †Lebesgue measure, and Φ_3 is continuous and †singular. Let Φ be an absolutely continuous distribution. Then there exists a unique (up to Lebesgue measure zero) measurable nonnegative function $f(x)$ $(-\infty<t<\infty)$ such that $F(x)=\int_{-\infty}^{x}f(t)dt$. This function $f(t)$ is called the **probability density** of Φ.

We now list some frequently used 1-dimensional lattice distributions (for explicit data → Appendix A, Table 22): the **unit distribution** with $\Phi(\{0\})=1$; the **binomial distribution** $Bin(n,p)$ with parameters n and p; the **Poisson distribution** $P(\lambda)$ with parameter λ; the **geometric distribution** $G(p)$ with parameter p; the **hypergeometric distribution** $H(N,n,p)$ with parameters N, n, and p; and the **negative binomial distribution** $NB(m,q)$ with parameters m and q. The following k-dimensional lattice distributions are used frequently: the **multinomial distribution** $M(n,p)$ with parameters n and p; the **multiple hypergeometric distribution**; the **negative multinomial distribution**; etc.

The following 1-dimensional distributions are absolutely continuous: the **normal distribution (or Gaussian distribution)** $N(\mu,\sigma^2)$ with mean μ and variance σ^2 (sometimes $N(0,1)$ is called the **standard normal distribution**); the **Cauchy distribution** $C(\mu,\sigma)$ with parameters μ (†median) and σ; the **uniform distribution** $U(\alpha,\beta)$ on an interval $[\alpha,\beta]$; the **exponential distribution** $e(\sigma)$ with parameter σ; the **gamma distribution** $\Gamma(p,\sigma)$; the †χ^2 distribution $\chi^2(n)$; the **beta distribution** $B(p,q)$; the **F-distribution** $F(m,n)$; the **Z-distribution** $Z(m,n)$; the **t-distribution** $t(n)$; etc. Furthermore, there are several k-dimensional absolutely continuous distributions such as: the **k-dimensional normal distribution** $N(\mu,\Sigma)$ with mean vector $\mu=(\mu_1,\mu_2,\ldots,\mu_k)$ and covariance matrix $\Sigma=(\sigma_{ij})$; the **Dirichlet distribution**; etc.

C. Convolution

Given any two n-dimensional distributions Φ_1, Φ_2, the n-dimensional distribution $\Phi(E)=\int_{\mathbf{R}^n}\chi_E(x+y)\,d\Phi_1(x)d\Phi_2(y)$ is called the **composition** (or **convolution**) of Φ_1 and Φ_2 and is denoted by $\Phi_1*\Phi_2$, where χ_E is the indicator function of the set E. Let X_1 and X_2 be independent random variables with distributions Φ_1 and Φ_2. Then the distribution of X_1+X_2 is $\Phi_1*\Phi_2$. When F_i is the distribution function of Φ_i $(i=1,2)$, the distribution function F_1*F_2 of $\Phi_1*\Phi_2$ is expressed in the form $F_1*F_2(x)=\int_{\mathbf{R}^n}F_1(x-y)dF_2(y)$. If Φ_1 has a density $f_1(x)$, then $\Phi_1*\Phi_2$ has a density $f(x)=\int_{\mathbf{R}^n}f_1(x-y)dF_2(y)$. If $\varphi(z)$ is the characteristic function of the convolution of two probability distributions Φ_1 and Φ_2 with characteristic functions φ_1 and φ_2, then φ is the product of φ_1 and φ_2: $\varphi(z)=\varphi_1(z)\cdot\varphi_2(z)$. Therefore, for every k the kth order semi-invariant of the convolution of two distributions is equal to the sum of their semi-invariants. Suppose that we are given a family of distributions $\Phi=\{\Phi(\alpha_2,\beta_2,\ldots)\}$ indexed with parameters $\alpha,\beta,\ldots$. If for $(\alpha_1,\beta_1,\ldots)$ and $(\alpha_2,\beta_2,\ldots)$ there exists $(\alpha_3,\beta_3,\ldots)$ such that $\Phi(\alpha_1,\beta_1,\ldots)*\Phi(\alpha_2,\beta_2,\ldots)=\Phi(\alpha_3,\beta_3,\ldots)$, then we say that Φ has a **reproducing property**. Some of the distributions listed above have the reproducing property: $P(\lambda_1)*P(\lambda_2)=P(\lambda_1+\lambda_2)$, $Bin(n_1,p)*Bin(n_2,p)=Bin(n_1+$

$n_2,p)$, $NB(m_1,q)*NB(m_2,q)=NB(m_1+m_2,q)$, $N(\mu_1,\sigma_1^2)*N(\mu_2,\sigma_2^2)=N(\mu_1+\mu_2,\sigma_1^2+\sigma_2^2)$, $\Gamma(p_1,\sigma)*\Gamma(p_2,\sigma)=\Gamma(p_1+p_2,\sigma)$, $C(\mu_1,\sigma_1)*C(\mu_2,\sigma_2)=C(\mu_1+\mu_2,\sigma_1+\sigma_2)$, etc. Given a 1-dimensional distribution function $F(x)$,

$$Q_F(l)=\max_{-\infty<x<\infty}(F(x+l)-F(x-l)),\qquad l>0,$$

is called the **maximal concentration function** of F (P. Lévy [14]). Since it satisfies the relation $Q_{F_1*F_2}(l)\leqslant Q_{F_i}(l)$ $(i=1,2)$, we can use it to study the properties of sums of independent random variables. The **mean concentration function** defined by

$$C_F(l)=\frac{1}{2l}\int_{-\infty}^{\infty}(F(x+l)-F(x-l))^2dx$$

is also useful for similar purposes (T. Kawata [10]).

Let $N(m,v)$ be the 1-dimensional normal distribution with mean m and variance v, and let $P(\lambda,a)$ be the distribution obtained through translation by a of the Poisson distribution with parameter λ. If 1-dimensional distributions Φ_k, Ψ_k $(k=1,2)$ exist such that $N(m,v)=\Phi_1*\Phi_2$, $P(\lambda,a)=\Psi_1*\Psi_2$, we have $\Phi_k=N(m_k,v_k)$, $\Psi_k=P(\lambda_k,a_k)$ $(k=1,2)$ for some m_k,v_k,λ_k,a_k $(k=1,2)$. These are known, respectively, as Cramér's theorem and Raikov's theorem. Ju. V. Linnik proved a similar fact (the decomposition theorem) for a more general family with reproducing property by using the theory of analytic functions.

D. Infinitely Divisible Distributions

An n-dimensional probability distribution Φ is called **infinitely divisible** if for every positive interger k, there exists a probability distribution Φ_k such that $\Phi=\Phi_k*\Phi_k*\ldots*\Phi_k$ $(=\Phi_k^{*k})$. The normal and Poisson distributions are infinitely divisible. The probability distribution Φ of an n-dimensional random variable X is infinitely divisible if and only if for every positive integer k we have independent and identically distributed random variables $X_{k1},X_{k2},\ldots,X_{kk}$ such that $X=X_{k1}+X_{k2}+\ldots+X_{kk}$ by extending the basic probability space if necessary. If an n-dimensional distribution Φ satisfies the condition $\int_{|x|>\varepsilon}\Phi(dx)<\varepsilon$, we say that $\Phi\in v(\varepsilon)$. Then Φ is an infinitely divisible distribution if and only if for every $\varepsilon>0$, we can find Φ_1, Φ_2, $\ldots,\Phi_k\in V(\varepsilon)$ such that $\Phi=\Phi_1*\Phi_2*\ldots*\Phi_k$. The characteristic function of a 1-dimensional infinitely divisible distribution can be written in the form

$$\varphi(z)=\exp\left(i\gamma z+\int_{-\infty}^{\infty}A(u,z)\frac{1+u^2}{u^2}dG(a)\right),\tag{1}$$

where γ is a constant, $G(a)$ is a nondecreasing bounded function with $G(-\infty)=0$, $A(u,z)=\exp(iut)-1-izu/(1+u^2)$, and the value of $A(u,z)(1+u^2)/u^2$ at $u=0$ is defined to be $-z^2/2$. Formula (1) is called **Hinčin's canonical form**. For the characteristic function of an infinitely divisible n-dimensional distribution, the canonical form is as follows:

$$\varphi(z)=\exp\left(i(m,z)-\sum_{p,q=1}^{n}c_{pq}z_pz_q+\int_{\mathbf{R}^n}\left(e^{i(z,x)}-1-\frac{i(z,x)}{1+|x|^2}\right)n(dx)\right),$$

$$z=(z_1,\ldots,z_n)\in\mathbf{R}^n,\tag{2}$$

where $m\in\mathbf{R}^n$, (C_{pq}) is a positive semidefinite matrix, and $n(dx)$ is a measure on $\mathbf{R}^n$ such that $n(\{0\})=0$ and

$$\int_{\mathbf{R}^n}\frac{|x|^2}{1+|x|^2}n(dx)<\infty.$$

Formula (2) is called **Lévy's canonical form**. If a 1-dimensional infinitely divisible distribution P satisfies $\int_{\mathbf{R}^1}x^2dP(x)<\infty$, then its characteristic function is given by

$$\varphi(z)=\exp\left(imz+\int_{-\infty}^{\infty}(e^{izu}-1-izu)\frac{1}{u^2}dK(u)\right),\tag{3}$$

where m is a real constant and $K(u)$ is a nondecreasing bounded function such that $K(-\infty)=0$. It is called **Kolmogorov's canonical form** (for infinitely divisible distributions on a homogeneous space → 6 Additive Processes).

Let Φ and Ψ be n-dimensional distributions. If for some $\lambda>0$, $\Psi(E)=\Phi(\lambda E)$ $(\lambda E=\{\lambda\xi\mid\xi\in E\})$ for every set E, we say that Φ and Ψ are equivalent. Let Φ and Ψ be probability distributions with distribution functions F and G and characteristic functions φ and ψ. Then the following three statements are equivalent: (1) Φ and Ψ are equivalent; (2) $G(x)=F(\lambda x)$ for every x; and (3) $\psi(\lambda z)=\varphi(z)$ for every z. We call Φ a **stable distribution** if for every pair of distributions Φ_1, Φ_2 equivalent to Φ, the convolution $\Phi_1*\Phi_2$ is equivalent to Φ. If Φ is stable, every distribution equivalent to Φ is also stable. We can characterize stable distributions in terms of their characteristic functions $\varphi(z)$ as follows: For every pair $\lambda_1,\lambda_2>0$, there exists a $\lambda=\lambda(\lambda_1,\lambda_2)>0$ such that $\varphi(\lambda z)=\varphi(\lambda_1 z)\varphi(\lambda_2 z)$.

We can restate this characterization as follows: Φ is stable if and only if for every pair of independent random variables X_1 and X_2 with identical distribution Φ and for any positive numbers λ_1 and λ_2, there exists a positive number λ such that $(\lambda_1 X_1 + \lambda_2 X_2)/\lambda$ has the distribution Φ. By the definition we see that all stable distributions are infinitely divisible. Putting $\varphi(z) = \exp\psi(z)$, we have $\psi(\lambda z) = \psi(\lambda_1 z) + \psi(\lambda_2 z)$, which implies $\psi(z) = (-c_0 + i(z/|z|)c_1)|z|^\alpha$, where $c_0 \geqslant 0$, $-\infty < c_1 < \infty$, $0 < \alpha \leqslant 2$. The parameter α is called the **exponent** (or **index**) of the stable distribution. The stable distributions with exponent $\alpha = 2$ are the normal distributions, and the stable distributions with exponent $\alpha = 1$ are the Cauchy distributions. We have $\psi(z) = -c_0|z|^\alpha$ for a symmetric stable distribution. For the stable distribution with exponent $1/2$ → Appendix A, Table 22.

Generalizing stable distributions, we can define **quasistable distributions**, which B. V. Gnedenko and A. N. Kolmogorov [5] called stable distributions also. Let F be the distribution function of Φ. A distribution function is said to be quasistable if to every $\lambda_1 > 0$, $\lambda_2 > 0$, and real b_1, b_2 there correspond a positive number λ and a real number b such that we have the relation $F((x-\lambda_1)/b_1) * F((x-\lambda_2)/b_2) = F((x-\lambda)/b)$.

Let $\{X_i\}$ be a sequence of independent random variables with identical distribution Φ. If for suitably chosen constants A_n and B_n, the distribution functions of the sums $B_n^{-1}(\sum_{i=1}^n X_i) - A_n$ converges to a distribution function, the distribution Φ is a quasistable distribution (Lévy). A necessary and sufficient condition in terms of the characteristic function $\varphi(z)$ for a distribution to be quasistable is that its characteristic function satisfies the relation $\varphi(b_1 z)\varphi(b_2 z) = \varphi(bz)e^{i\gamma z}$ $(\gamma = \lambda - \lambda_1 - \lambda_2)$. The characteristic function of a stable distribution has the canonical representation

$$\varphi(z) = \exp\psi(z),$$
$$\psi(z) = imz - c|z|^\alpha(1 + i\beta(z/|z|)\omega(z,\alpha)),$$

where m is a real number, $c \geqslant 0$, $0 < \alpha \leqslant 2$, $|\beta| \leqslant 1$, and $\omega(z,\alpha) = \tan(\pi\alpha/2)$ $(\alpha \neq 1)$, $\omega(z,\alpha) = (2/\pi)\log|z|$ $(\alpha = 1)$.

A quasistable distribution with $\alpha \neq 1$ is obtained from a stable distribution by translation, but quasistable distributions with $\alpha = 1$ are not. **Semistable distributions** are another generalization of stable distributions. A distribution function is called semistable if its characteristic function $\varphi(z)$ satisfies the relation $\psi(qz) = q^\alpha\psi(z)$ for a positive number q $(\neq 1)$, where $\varphi(z) = \exp(\psi(z))$. Also in this case, the general form was obtained by Lévy [14].

E. The Shape of Distributions

Let $F(x)$ be a 1-dimensional distribution. The quantity ζ_p such that $F(\zeta_p - 0) = p \leqslant F(\zeta_p)$ $(0 < p < 1)$ is called the **quantile of order** p of F. In particular, the quantity $\zeta_{1/2}$ is called a **median**. If a 1-dimensional distribution satisfies the relation $1 - F(m+x) = F(m-x)$, it is called a **1-dimensional symmetric distribution function**. In any 1-dimensional symmetric distribution, every moment of odd order about the mean (if it exists) is equal to zero.

The ratio $\gamma_1 = \mu_3/\sigma^3$ is used as a measure of departure from symmetry of a distribution and is called the **coefficient of skewness**. Furthermore, the ratio $\gamma_2 = \mu_3/\sigma^4 - 3$ is called the **coefficient of excess**. For the normal distribution, we have $\gamma_1 = \gamma_2 = 0$. If $\gamma_2 \neq 0$, γ_2 expresses the degree of deviation from the normal distribution.

A distribution function $F(x)$ is called **unimodal** if there exists at most one value $x = a$ such that $F(x)$ is convex for $x < a$ and concave for $x > a$. All 1-dimensional stable distribution functions are unimodal.

F. Convergence of Distributions

The concept of convergence of distributions plays an important role in limit theorems and other fields of probability theory. When Ω is a topological space. we consider **convergence** of probability measures on Ω with regard to the †weak topology introduced in the space of measures on Ω (→ 39 Banach Spaces). Such convergence is called **weak convergence** or **convergence in distribution** in probability theory. For a sequence of n-dimensional distributions Φ_k $(k = 1, 2, \ldots)$ to converge to Φ weakly, each of the following conditions is necessary and sufficient. (1) For every continuous function with compact support, $\lim_{k\to\infty}\int_{\mathbf{R}^n} f(x)\,d\Phi_k(x) = \int_{\mathbf{R}^n} f(x)\,d\Phi(x)$. (2) At every continuity point of the distribution function $F(x_1, \ldots, x_n)$ of Φ, $\lim_{k\to\infty} F_k(x_1, \ldots, x_n) = F(x_1, \ldots, x_n)$ (F_k is a distribution function of Φ_k). (3) For every continuity set E of Φ (namely, a set such that $\Phi(\bar{E} - E^0) = 0$), $\lim_{k\to\infty}\Phi_k(E) = \Phi(E)$. (4) For all open $G \subset \mathbf{R}^n$, $\liminf_{k\to\infty}\Phi_R(G) \geqslant \Phi(G)$. (5) For all closed $F \subset \mathbf{R}^n$, $\limsup_{k\to\infty}\Phi_k(F) \leqslant \Phi(F)$. (6) $\lim_{k\to\infty}\rho(\Phi_k, \Phi) = 0$, where ρ is a metric defined in the following way: Given any n-dimensional distributions Φ_1, Φ_2, we put $\varepsilon_{ij} = \inf\{\varepsilon \mid \Phi_j(F) < \Phi_i(F^\varepsilon)$ for every closed $F\}$ (F^ε is the ε-neighborhood of F) and define $\rho(\Phi_1, \Phi_2) = \max(\varepsilon_{12}, \varepsilon_{21})$. The metric ρ, called the **Lévy distance**, was introduced by Lévy [14] in one dimension and by Ju. V. Prohorov in metric spaces [18]. Each of these conditions except (2) is necessary and sufficient for Φ_n to converge to Φ weakly if Φ_n and Φ are probability measures on a complete separable metric space.

We can give a criterion for the convergence of probability measures in terms of characteristic functions of distributions. Suppose that Φ_n and Φ are probability measures with characteristic functions φ_n and φ. Then Φ_n converges weakly to Φ if and only if for every z, $\lim_n \varphi_n(z) = \varphi(z)$. Let $\Phi_1, \Phi_2, \dots$ and Φ be 1-dimensional distributions. If all absolute moments exist, $\sum_{j=1}^{\infty} \beta_j^{-1/j} = \infty$ for $\beta_j = \int_{-\infty}^{\infty} |x|^j d\Phi(x) < \infty$, and

$$\lim_{k\to\infty} \int_{-\infty}^{\infty} x^j d\Phi_k(x) = \int_{-\infty}^{\infty} x^j d\Phi(x) \qquad (j=0,1,2,\dots),$$

then Φ_k converges to Φ weakly. This condition is sufficient but not necessary.

A family Φ_α $(\alpha \in \Lambda)$ of n-dimensional probability measures is said to be **tight** if for every $\varepsilon > 0$ there exists a compact set $K = K(\varepsilon)$ such that $\Phi_\alpha(K^c) < \varepsilon$ for all $\alpha \in \Lambda$. A family Φ_α $(\alpha \in \Lambda)$ is called **tight** if it is totally bounded with respect to the topology introduced by the Lévy distance.

G. Existence Theorems

Consider a family $\xi_t(\omega)$ ($t \in T$, an arbitrary parameter set) of random variables on a probability measure space $(\Omega, \mathfrak{F}, P)$, with values in a measurable space $(X, \mathfrak{B})$. For arbitrary $t_1, t_2, \dots, t_n \in T$, we put $P(\xi_{t_1}(\omega) \in E_1, \dots, \xi_{t_n}(\omega) \in E_n) = \Psi_{t_1, t_2, \dots, t_n}(E_1, E_2, \dots, E_n)$ $(E_i \in \mathfrak{B})$. Then the family of functions $\Psi_{t_1, t_2, \dots, t_n}(E_1, E_2, \dots, E_n)$ satisfies the following conditions: (1) It induces a probability measure $\Phi_{t_1, t_2, \dots, t_n}(\cdot)$ on the product measurable space such that $\Phi_{t_1, t_2, \dots, t_n}(E_1 \times E_2 \times \dots \times E_n) = \Psi_{t_1, t_2, \dots, t_n}(E_1, E_2, \dots, E_n)$. (2) $\Psi_{t_1, t_2, \dots, t_n}(E_1, E_2, \dots, E_n) = \Psi_{t_{i_1}, t_{i_2} \dots t_{i_n}}(E_{i_1}, E_{i_2}, \dots, E_{i_n})$ for any permutation $(i_1, i_2, \dots, i_n)$ of the numbers $1, 2, \dots, n$. (3) $\Psi_{t_1, t_2, \dots, t_{n-1}, t_n}(E_1, E_2, \dots, E_{n-1}, X) = \Psi_{t_1, \dots, t_{n-1}}(E_1, E_2, \dots, E_{n-1})$ $(n > 1)$. Conditions (2) and (3) are called **consistency conditions**.

Suppose, conversely, that we are given a family of functions $\Psi_{t_1, t_2, \dots, t_n}(E_1, E_2, \dots, E_n)$ $(t_1, t_2, \dots, t_n \in T, E_i \in \mathfrak{B}$ $(i = 1, 2, \dots, n))$ satisfying conditions (1), (2) and (3). Let X^T be the set $\{x | x_t \in X, t \in T\}$ and $\mathfrak{B}^T$ the completely additive class generated by the sets $\{X_t \in E\}$, $E \in \mathfrak{B}$, $t \in T$. Does there exist a probability measure P on the product measure space $(X^T, \mathfrak{B}^T)$ such that $P(x(t_1) \in E_1, x(t_2) \in E_2, \dots, x(t_n) \in E_n) = \Phi_{t_1, t_2, \dots, t_n}(E_1, E_2, \dots, E_n)$? This is Kolmogorov's problem. Kolmogorov gave an affirmative answer to this problem for $X = \mathbf{R}^1$. Generally, it is known that if X is a locally compact Hausdorff space and a countable union of compact sets, the answer is also affirmative. There exist several variants of **Kolmogorov's extension theorem** (S. Bochner [1], E. Nelson [17]), which are useful in constructing stochastic processes (→ 395 Stochastic Processes).

References

[1] S. Bochner, Harmonic analysis and the theory of probability, Univ. of California Press, 1955.
[2] H. Cramér, Mathematical methods of statistics, Princeton Univ. Press, 1946.
[3] J. L. Doob, Stochastic processes, John Wiley, 1953.
[4] И. М. Гельфанл, Н. Я. Виленкин (I. M. Gel'fand and N. Ja. Vilenkin), Некоторые применения гармонического анализа, оснащенные Гильбертовы пространства физматгиз; English translation, Generalized functions IV. Applications of harmonic analysis, Academic Press, 1964.
[5] Б. В. Гнеденко, А. Н. Колмогоров (B. V. Gnedenko and A. N. Kolmogorov), Предельные распределения для сумм независимых случайных величин, Гостехиздат, 1949; English translation, Limit distributions for sums of independent random variables, Addison-Wesley, 1954.
[6] K. Itô, Kakurituron no kiso (Japanese; Foundations of the theory of probability), Iwanami, 1944.
[7] K. Itô, Kakurituron (Japanese; Theory of probability), Iwanami, 1953.
[8] K. Itô, Kakuritu katei (Japanese; Stochastic processes), Iwanami Coll. of Modern Applied Math., 1957.
[9] M. Kac, Probability and related topics in physical sciences, Interscience, 1959.
[10] T. Kawata, Fûrie kaiseki to kakurituron (Japanese; Fourier analysis and the theory of probability), Tyûbunsya, 1946.
[11] Y. Kawada, Kakurituron (Japanese; Theory of probability), Kyôritu, 1953.
[12] A. N. Kolmogorov (Kolmogoroff), Grundbegriffe der Wahrscheinlichkeitsrechnung, Springer, 1933; English translation, Foundations of the theory of probability, Chelsea, 1950.
[13] K. Kunisawa, Kakurituron ni okeru kyokugenteiri (Japanese; Limit theorems of the theory of probability), Tyûbunkan, 1947.
[14] P. Lévy, Théorie de l'addition des variables aléatoires, Gauthier-Villars, 1937.
[15] Ю. В. Линник (Ju. V. Linnik), Разложения вероятностных законов, Ленинград, 1960; English translation, Decomposition of probability distributions, Oliver and Boyd, 1964.
[16] M. M. Loève, Probability theory, Van Nostrand, second edition, 1960.

[17] E. Nelson, Regular probability measures on function space, Ann. of Math., (2) 69 (1959), 630–643.
[18] Ju. V. Prohorov, The method of characteristic functionals, Proc. of 4th Berkeley Symp. Math. Stat. Prob. II, Univ. of California Press, (1961), 403–419.
[19] Ю. В. Прохоров (Ju. V. Prohorov), Сходимость случайных процессов и предельные теоремы теории вероятностей, Teor. Verojatnost. i Primenen., 1 (1956), 177–238; English translation, Convergence of random processes and limit theorems in probability theory, Theory of Prob. Appl., 1 (1956), 155–214.

132 (XVIII.5) Dynamic Programming

A. General Remarks

There are two types of multistage †decision processes. In one of them, an outcome of the whole process is determined at the final stage without any consideration of the outcome for each intermediate stage. The †extensive form of a †game belongs to this type. In the other type, an outcome is assigned at each stage of a multistage decision process. The theory of **dynamic programming**, dealing with this latter type, has been developed by R. Bellman and others since 1950 and is now one of the fundamental branches of mathematical programming theory, along with the theories of linear and nonlinear programming.

There are five features common to most of the fundamental mathematical structures discussed in the current theory of dynamic programming. (1) In each case we have a physical system characterized in any state by a small set of parameters, the **state variables**. (2) For each state of the system we have a choice of a number of **decisions**. (3) The effect of a decision is a transformation of the state variables. (4) The past history of the system is of no importance in determining future actions (†Markov property). (5) The purpose of the process is to maximize some function of the state variables. A **policy** is a rule for making decisions that yields an allowable sequence of decisions; an **optimal policy** is a policy which maximizes a preassigned function of the final state variables. A convenient term for this preassigned function of the final state variables is **criterion function**. One of the characteristic features of Bellman's methodology of dynamic programming is the appeal to the **principle of optimality**: An optimal policy has the property that whatever the initial state and initial decision are, the remaining decisions must constitute an optimal policy with regard to the state resulting from the first decision. Decision processes can be classified according to whether they are deterministic or stochastic and discrete or continuous.

B. Discrete Deterministic Processes

By a deterministic process we mean a process in which the outcome of a decision is uniquely determined by the decision. We assume that the state of the system, apart from time dependence, is described in any stage by an M-dimensional vector $p=(p_1,p_2,\ldots,p_M)$ constrained to lie within some region D. Let $T=\{T_q\}$ (where q runs over a set S which may be finite, enumerable, composed of continua, or a combination of sets of this type) be a set of transformations with the property that $p\in D$ implies that $T_q(p)\in D$ for all $q\in S$, i.e., any transformation T_q carries D into itself. The term "discrete" signifies here that we have a process consisting of a finite or denumerably infinite number of stages. A policy, for the finite process which we consider first, consists of a selection of N transformations in order, $P=(T_1,T_2,\ldots,T_N)$, yielding successively the sequence of states $p_i=T_i(p_{i-1})$ $(i=2,3,\ldots,N)$ with $p_1=T_1(p)$. These transformations are to be chosen to maximize a given function R of the final state p_N. Observe that the maximum value of $R(p_N)$, as determined by an optimal policy, will be a function of the initial vector p and the number N of stages only. Let us then define our basic auxiliary functions $f_N(p)=\max R(p_N)=$ the N-stage return obtained starting from an initial state p and using an optimal policy. This sequence is defined for $N=1,2,\ldots$, and $p\in D$. The essential uses of the principle of optimality can be observed from the following two features. The first is the use of the **embedding principle**. The original process is embedded in a family of similar processes. In place of attempting to determine the characteristics of an optimal policy for an isolated process, we attempt to deduce the common properties of the set of optimal policies possessed by the members of the family. The second feature is the derivation of recurrence relations by which the functional equations connecting the members of the sequence $\{f_k(p)\}$ are established. Assume that we choose some transformation T_q as a result of our first decision, obtaining in this way a new state vector $T_q(p)$. The **maximum return** from the following $k-1$ stages is, by definition, $f_{k-1}(T_q(p))$. It follows that, if we wish to maximize the total k-stage return, q must now be chosen to maximize this $(k-1)$-stage return. The result is the basic recurrence relation $f_k(p)=\max_{q\in S} f_{k-1}(T_q(p))$, for $k \geqslant 2$, with $f_1(p)=\max_{q\in S} R(T_q(p))$. For the case of an unbounded process, the sequence $\{f_k(p)\}$ is replaced by a single function $f(p)$, the total return obtained from using an optimal policy starting from state p, and the recurrence relation is replaced by the functional equation $f(p)=\max_q f(T_q(p))$.

C. Discrete Stochastic Processes

We again consider a discrete process, but one in which the transformations are stochastic rather than deterministic. The initial vector p is transformed into a stochastic vector z with an associated distribution function $dG_q(p,z)$ dependent on p and the choice of q. We assume that z is known after the decision has been made and before the next decision is to be made. We agree to measure the value of a policy in terms of some average value of the function of the final state. Let us call this expected value the **return**. Beginning with the case of a finite process, we define $f_k(p)$ as before. The expected return as a result of the initial choice of T_q is therefore

$$\int_{z\in D} f_{k-1}(z)\,dG_q(p,z).$$

Consequently, the recurrence relation for the sequence $\{f_k(p)\}$ is

$$f_k(p)=\max_{q\in S}\int_{z\in D} f_{k-1}(z)\,dG_q(p,z), \quad k\geqslant 2,$$

with $f_1(p)=\max_{q\in S}\int_{z\in D} R(z)\,dG_q(p,z)$. Considering the unbounded process, we obtain the functional relation

$$f(p)=\max_{q\in S}\int_{z\in D} f(z)\,dG_q(p,z).$$

D. Continuous Deterministic Processes

There are a number of interesting processes that require that decisions be made at each point of a continuum, such as a time interval. The simplest examples of processes of this character are furnished by the †calculus of variations. Let us denote by $f(p;T)$ the return obtained over a time interval $[0,T]$ starting from the initial state p and employing an optimal policy. Although we consider the process as one consisting of choices made at each point t on $[0,T]$, it is better to begin with the concept of choosing policies (functions) over intervals, and then pass to the limit as these intervals shrink to points. The application of the principle of optimality suggests

$$f(p;S+T)=\max_{D[0,S]} f(p_D;T), \tag{1}$$

where the maximum is taken over all allowable decisions made over the interval $[0,S]$. In general, however, in the discussion of processes of continuous type, it is better to use initially the equation $f(p;S+T)=\sup_D f(p_D;T)$, which is usually easy to establish, and then show, under suitable assumptions, that the maximum actually is attained. The limiting form of (1) as $S\to 0$ is a †nonlinear partial differential equation. This is the important form for actual analytic utilization. It is to be remarked that the functional equation just given has some resemblance to the functional equation $f(c,s+t)=f(f(c,s),t)$, which expresses the †semigroup property of the unique solution of the †autonomous differential equation $x(t)=f(c,t)$ with initial value $x(0)=c$.

The theory of dynamic programming gives us the existence and uniqueness of solutions of some of these functional equations. Besides a set of fairly general theorems, there are several interesting results obtained by specific techniques for each individual functional equation. The theory of dynamic programming has opened a new area in the theory of functional equations since the 1950s. The following examples illustrate some features of multiple decision processes discussed in the theory of dynamic programming.

1. Multistage Allocation Processes. We are given a quantity $x>0$ that may be divided into two parts y and $(x-y)$. From y we obtain a return $g(y)$, and from $(x-y)$ a return $h(x-y)$. In so doing, we expend a certain amount of our original resources and are left with a new quantity, $ay+b(x-y)$, $0<a$, $b<1$, with which the process is continued. How do we proceed so as to maximize the total return obtained in a finite or unbounded number of stages? Denote by $f(x)$ the total return obtained using an optimal policy for allocation of resources at each stage, where an unlimited number of operations is permitted. Then we have

$$f(x)=\sup_{0\leqslant y\leqslant x}\big[g(y)+b(x-y) + f(ay+b(x-y))\big].$$

2. Multistage Choice Processes. Suppose that we possess two gold mines A and B, the first of which contains an amount x of gold, while the second contains an amount y. If the only gold-mining machine we have is used in A, there is a probability p_1 that r_1 percent of the gold will be brought up safely, the machine still being usable, and a probability $(1-p_1)$ that the machine will be damaged, will mine no gold, and will be of no further use. Similarly, the mine B has the probabilities p_2, r_2 and $(1-p_2)$ associated with it. How do we proceed in order to maximize the total amount of gold before the machine is defunct? Denote by $f(x,y)$ the expected amount of gold obtained using an optimal sequence of choice. Then we have

$$f(x,y)=\max\begin{bmatrix} p_1\{r_1x+f((1-r_1)x,y)\} \\ p_2\{r_2y+f(x,(1-r_2)y)\}\end{bmatrix}.$$

The optimal policy can be described in the following way. We choose A or B according as $p_1r_1x/(1-p_1)$ is greater or less than $p_2r_2y/(1-p_2)$. We may choose either A or B if equality holds. After an operation according to such a choice, the machine may become defunct and terminate the process. If the machine is usable, then we can apply our policy to a new combination of the amounts of gold in A and B.

E. Markov Decision Processes

Assume that the †transition probabilities p_{ij} depend on a parameter q, which may be a vector, and that at each stage of the process q is to be chosen so as to maximize the probability that the system is in the state S_1. We obtain the nonlinear system

$$y_1(n+1)=\max_q \sum_{j=1}^{N} p_{ij}(q)y_j(n)$$

$$=\sum_{i=1}^{N} y_i(n)p_{i1}(q^*)$$

$$y_i(n+1)=\sum_{j=1}^{N} p_{ij}(q^*)y_j(n), \quad i=2,3,\ldots,N,$$

where $q^*=q^*(n)$ in the remaining $N-1$ equations is one of the values of q that maximize $y_1(n+1)$. There are similar processes that can be considered as continuous analogs of this type of decision process. These are called **Markov decision processes** and were discussed by Bellman. There is, however, another type of Markov decision process in which a reward is given at each stage. For each state i of the system there are k alternatives $1,2,3,\ldots,k$. If we choose the alternative h among these k alternatives, then the transition probabilities p_{ij}^h $(j=1,2,\ldots,n)$ are determined, and a reward r_{ij}^h is associated with each state j. Let us denote by $v_i(n)$ the total reward obtained at the nth stage by appealing to an optimal policy when the initial state is i. Then the principle of optimality in the theory of dynamic programming yields

$$v_i(n+1)=\max_h\left(\sum_{j=1}^{N} p_{ij}^h\left(r_{ij}^h+v_j(n)\right)\right).$$

A policy-iteration method involving a value-determination operation with a policy-improvement routine was given by R. A. Howard. D. H. Blackwell gave the general convergence theorem for Markov decision processes.

F. Dynamic Programming and the Calculus of Variations

In some formulations of multistage decision processes, the dynamic programming approach has certain features in common with the classical †calculus of variations. For instance, suppose that we are concerned with the problem of finding a control function $y(t)$ that minimizes the functional

$$J(y)=\int_0^T g(x(t),y(t))\,dt$$

under the restriction $|y(t)|\leqslant m<\infty$ (m is a constant), and for which

$$dx(t)/dt=h(x(t),y(t)), \qquad x(0)=c.$$

Let us define

$$f(c,T)=\max_{y\in D(0,T)}\int_0^T g(x,y)\,dt,$$

where $D(0,T)$ is the domain of y satisfying the conditions just mentioned. Then the principle of optimality gives

$$f(c,T)$$

$$=\max_{y\in D(0,s)}\left(\int_0^s g(x,y)\,dt+f(x(s),T-s)\right),$$

and then, for $s\to 0$,

$$\frac{\partial f}{\partial T}=\max_{y_0}\left(g(c,y_0)+\frac{\partial f}{\partial c}h(c,y_0)\right),$$

from which we obtain a formula that has an intimate connection with the †Euler differential equation in the calculus of variations. The principles of dynamic programming can be applied in such cases to reformulate classical problems of the calculus of variations to lead to functional equations for the optimal solution.

G. Dynamic Programming and the Maximum Principle

In general, the method of dynamic programming carries a more universal character than the maximum principle in optimal control theory. However, in contrast to the latter, this method does not have a rigorous logical foundation. Recently, in the case of the following optimal problem, V. G. Boltyanskii [6] has presented the justification of the dynamic programming method. We consider the following optimal problem:

Let $f_i(x,u)$ $(i=0,1,\ldots,n)$ be defined for $x \in V \subset \mathbf{R}^n$ and $u \in U \subset \mathbf{R}^r$, where V is an open set, and continuously differentiable on $V \times U$. Suppose that two points x^0 and x^1 are given in V. Among all the piecewise continuous controls $u(t)=(u_1(t),\ldots,u_r(t)) \in U$ which transfer the phase point moving in accordance with

$$\frac{dx_i}{dt}=f_i(x,u(t)), \quad i=1,\ldots,n,$$

from $x^0=x(t_0)$ to $x^1=x(t_1)$, find the control $u(t)$ for which the functional

$$J=\int_{t_0}^{t_1} f_0(x(t),u(t))\,dx$$

takes the smallest value.

A continuous function $\omega(x)=\omega(x_1,\ldots,x_n)$ is called a **Bellman function** relative to a point $a \in V$ if it possesses the following properties: (1) $\omega(a)=0$; (2) there exists a set M (the singular set of $\omega(x)$), which is closed in V and does not contain interior points, such that the function $\omega(x)$ is continuously differentiable on the set $V-M$ and satisfies the condition

$$\sup_{u \in U}\left(\sum_{i=1}^{n}\frac{\partial\omega}{\partial x_i}f_i(x,u)-f_0(x,u)\right)=0,$$

$x \in V-M$.

The following theorem gives a sufficient optimality condition.

Theorem: Assume that for $dx/dt=f(x,u(t))$ given in a region $V \subset \mathbf{R}^n$ there exists a Bellman function $\omega(x)$ relative to the point $a \in V$ with a piecewise smooth singular set. Assume, furthermore, that for any point $x^0 \in V$ there exists a control $u(t)$ which transfers the phase point from $x^0=x(t_0)$ to $a=x(t_1)$ and satisfies the relation

$$\int_{t_0}^{t_1} f_0(x(t),u(t))=-\omega(x^0).$$

Then any such control $u(t)$ is optimal in V.

The characteristic features of the dynamic programming approach can be summarized in the following five points: (1) the advantage of lower dimensionality in comparison with the enumeration approach; (2) the possibility of finding maxima and/or minima of functions defined over restricted domains for which differential calculus may not work well; (3) the availability of numerical solutions in recursive forms; (4) the possibility of formulating certain problems to which classical methods do not apply; and (5) the applicability of the method to most types of problems in mathematical programming, such as †inventory and production control, optimal searching, and some optimal and adaptive control processes.

References

[1] R. E. Bellman, Dynamic programming, Princeton Univ. Press, 1957.
[2] R. E. Bellman, Applied dynamic programming, Princeton Univ. Press, 1962.
[3] R. E. Bellman, Adaptive control processes; a guided tour, Princeton Univ. Press, 1961.
[4] S. M. Roberts, Dynamic programming in chemical engineering and process control, Academic Press, 1964.
[5] G. Leitman (ed.), Optimization techniques, with applications to aerospace systems, Academic Press, 1962.
[6] V. G. Boltyanskii, Sufficient conditions for optimality and the justification of the dynamic programming method, SIAM J. Control, 4 (1966), 326–361.

133 (XIX.5) Dynamics of Rigid Bodies

A. Rigid Bodies

A **rigid body** is defined as a system of particles whose mutual distances are permanently fixed. Like a point particle, it is an ideal concept introduced into mechanics to simplify theoretical treatment. Actual solid bodies can in most cases be regarded as rigid under the action of forces of ordinary magnitude. Since a rigid body can be imagined to be made up of an infinite number of particles, the equations of motion for a system of particles can also be applied to it. Thus the motion of a rigid body can be completely determined by the theorems of momentum and angular momentum.

B. Theorem of Momentum

The **momentum** of a rigid body is defined by

$$\mathbf{Q}=\int_K \frac{d\mathbf{r}}{dt}dm,$$

where dm is the mass of the volume element at a point $\mathbf{r}$ of the rigid body K and $d\mathbf{r}/dt$ is its velocity. If the external forces acting on K are denoted by $\mathbf{F}_i$ $(i=1,2,\ldots)$, we have

$$d\mathbf{Q}/dt=\sum \mathbf{F}_i,$$

which expresses the **theorem of momentum**. If the velocity and acceleration of the **center of gravity (center of mass** or **barycenter)** $\int \mathbf{r}\,dm/\int dm$ of the rigid body are denoted by $\mathbf{V}_G$ and $\mathbf{A}_G$, respectively, and the **mass** $\int dm$ by m, we have $\mathbf{Q}=m\mathbf{V}_G$ and the theorem of momentum becomes $m\,d\mathbf{V}_G/dt=m\mathbf{A}_G=\Sigma\mathbf{F}_i$.

The **angular momentum** of a rigid body K about an arbitrary point $\mathbf{r}_0$ is defined by

$$\mathbf{H}=\int_K(\mathbf{r}-\mathbf{r}_0)\times\frac{d\mathbf{r}}{dt}\,dm$$

($\times$ denotes the †vector product). If $\mathbf{P}_i$ is the position vector of the point at which $\mathbf{F}_i$ acts, we have

$$d\mathbf{H}/dt=\sum(\mathbf{P}_i\times\mathbf{F}_i)=\mathbf{G}.$$

This is called the **theorem of angular momentum**. For the case of a rigid body with one point $\mathbf{r}_0$ fixed, the angular momentum $\mathbf{H}$ and the angular velocity $\boldsymbol{\omega}$ are related by

$$H_x=A\omega_x-F\omega_y-E\omega_z,$$
$$H_y=-F\omega_x+B\omega_y-D\omega_z,$$
$$H_z=-E\omega_x-D\omega_y+C\omega_z,$$

where H_x,H_y,H_z; $\omega_x,\omega_y,\omega_z$ are the components of $\mathbf{H}$ and $\boldsymbol{\omega}$ in the xyz-coordinate system fixed in space with its origin at the fixed point $\mathbf{r}_0$, and

$$A=\int(y^2+z^2)\,dm,\qquad B=\int(z^2+x^2)\,dm,$$
$$C=\int(x^2+y^2)\,dm,\qquad D=\int yz\,dm,$$
$$E=\int zx\,dm,\qquad F=\int xy\,dm,$$

with the integrals taken over the whole rigid body. We call A, B, and C the **moments of inertia** about the x-, y-, and z-axes, respectively, and D, E, F the corresponding **products of inertia**. The rotational motion of a rigid body with one axis fixed is completely determined by the theorem of angular momentum. However, for rotation about a fixed point, it is not very convenient to use, because A, B, C, D, E, and F are generally unknown functions of time.

C. The Ellipsoid of Inertia

The †quadric

$$Ax^2+By^2+Cz^2+2Dyz+2Ezx+2Fxy=1$$

represents an †ellipsoid with its center at the origin, called the **ellipsoid of inertia**. If the principal axes ξ, η, and ζ are taken as coordinate axes, the equation of the ellipsoid of inertia becomes $A\xi^2+B\eta^2+\Gamma\zeta^2=1$, where A, B, Γ are the moments of inertia about the ξ-, η-, ζ-axes and are called the **principal moments of inertia**, while the ξ-, η-, ζ-axes themselves are called the **principal axes of inertia**. If the components of the angular momentum $\mathbf{H}$ and angular velocity in the direction of the principal axes of inertia are denoted by (H_1,H_2,H_3) and $(\omega_1,\omega_2,\omega_3)$, respectively, then $H_1=A\omega_1$, $H_2=B\omega_2$, $H_3=\Gamma\omega_3$. Furthermore, if the ξ-, η-, ζ-components of the resultant moment of the external forces $\mathbf{G}=\Sigma(\mathbf{P}_i\times\mathbf{F}_i)$ are denoted by (G_1,G_2,G_3), then $d\mathbf{H}/dt=\mathbf{G}$ becomes

$$A\,d\omega_1/dt=G_1+(B-\Gamma)\omega_2\omega_3,$$
$$B\,d\omega_2/dt=G_2+(\Gamma-A)\omega_3\omega_1,$$
$$\Gamma\,d\omega_3/dt=G_3+(A-B)\omega_1\omega_2.$$

These are called **Euler's differential equations**.

The study of the motion of a rigid body may be mostly reduced to the study of the motion of its ellipsoid of inertia, since the latter is attached to the rigid body. The method of describing the motion of a rigid body by means of its ellipsoid of inertia is known as **Poinsot's representation**. The motions of two bodies having equal ellipsoids of inertia are the same if the external forces acting on them have equal resultant moments, even if their geometric forms are different.

When a rigid body moves under no constraint, the motion of its center of gravity G can be determined by use of the theorem of momentum. Also, the rotational motion about the center of gravity can be found from $d\mathbf{H}'/dt=\mathbf{G}'$, which is a modification of the theorem of angular momentum. Here $\mathbf{H}'$ and $\mathbf{G}'$ are respectively the angular momentum and moment of external forces about the center of gravity. In this case also, simplification can be achieved by considering the equation in the reference system that coincides with the principal axes of the ellipsoid of inertia with its center at the center of gravity.

References

[1] F. Klein and A. Sommerfeld, Über die Theorie des Kreisels I, II, III, IV, Teubner, 1910.

Also → references to 23 Analytical Dynamics.

E

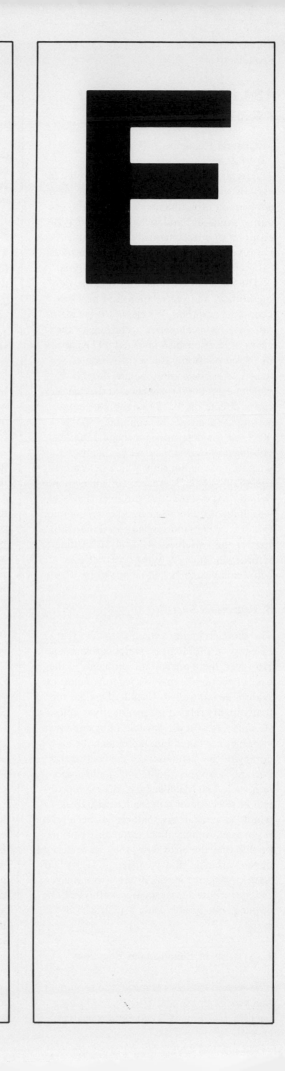

134 (XVII.13) Econometrics

A. General Remarks

The term **econometrics** can be interpreted in various ways. In its widest sense, it means the application of mathematical methods to economic problems, and includes mathematical economics, †mathematical programming, etc. However, here we shall use it to mean statistical methods applied to economic analysis.

The object of econometrics is to provide methods to analyze relationships between economic variables. We classify these methods into four categories according to the types of relationships involved: (1) Analysis of causal relations: If a set of variables $X_1, \ldots, X_n$ affects an economic variable Y, we can estimate the direction and extent of those effects on Y. (2) Analysis of equilibrium: When a set of economic variables $Y_1, \ldots, Y_m$ is determined through a market equilibrium mechanism, we can analyze the structure of relationships which determines the equilibrium. (3) Analysis of correlation: When a set of economic variables is affected simultaneously by some (unknown) common factors, we can analyze the correlation structure of the variables. (4) Analysis of time interdependence: A time process of a set of economic variables can be analyzed.

B. Regression Analysis

The most common technique for the first category of problems is †regression analysis. However, there are specific problems in the case of economic analysis, where a factor can seldom be controlled. Usually there are too many highly related independent variables. In such cases, if all possible independent variables are taken into the model, the accuracy of the †estimators of the coefficients becomes extremely poor. Such a phenomenon, called **multicollinearity**, raises the problem of selection of independent variables, to which no satisfactory solution has been given. Also, assumptions about the error terms may be dubious, the error terms may be correlated, or the variances may be different. If the †variance-covariance matrix of the errors is given, the †generalized least squares method can be applied, but usually such a matrix is not given.

C. Systems of Simultaneous Equations

The second category is specific to economic analysis. Suppose that $\mathbf{Y}=(Y_1, \ldots, Y_G)$ is a vector consisting of G economic variables, among which there exist G relationships that determine the equilibrium levels of the variables. We also suppose that there exist K variables $\mathbf{Z}=(Z_1, \ldots, Z_K)$ that are independent of the economic relations but affect the equilibrium. The variables Y are called **endogenous variables**, and the Z are called **exogenous variables**. If we assume linear relationships among them, we have an expression such as

$$\mathbf{Y}=B\mathbf{Y}+\Gamma\mathbf{Z}+\mathbf{u}, \tag{1}$$

where B and Γ are constant matrices and $\mathbf{u}$ is a vector of **disturbances** or errors. (1) is called the **linear structural equation system** and is regarded as a system of simultaneous equations. By solving the equation formally, we have

$$\mathbf{Y}=\Pi\mathbf{Z}+\mathbf{v}, \tag{2}$$

where $\Pi=(I-B)^{-1}\Gamma$, $\mathbf{v}=(I-B)^{-1}\mathbf{u}$, which is called the **reduced form**. The relation of $\mathbf{Y}$ to $\mathbf{Z}$ is determined through the reduced form (2), and if we have enough data on $\mathbf{Y}$ and $\mathbf{Z}$ we can estimate Π. The problem of **identification** is to know whether we can determine the unknown parameters in B and Γ uniquely from the parameters in the reduced form. A necessary condition for the parameters in one of the equations in (1) to be identified is that the number of unknown parameters (or, since known constants in the system are usually set equal to zero, the number of variables appearing in the equation) is not greater than $K+1$. If it is exactly equal to $K+1$, the equation is said to be **just identified**, and if it is less than $K+1$, the equation is said to be **over identified**.

If all the equations in the system are just identified, for arbitrary Π there exist unique B and Γ that satisfy $\Pi=(I-B)^{-1}\Gamma$. Therefore, if we denote the †least squares estimator of Π by $\hat{\mathbf{\Pi}}$, we can estimate B and Γ from the equation $(I-\hat{\mathbf{B}})\hat{\mathbf{\Pi}}=\hat{\mathbf{\Gamma}}$. This procedure is called the **indirect least squares method** and is equivalent to the †maximum likelihood method if we assume normality for $\mathbf{u}$.

When some of the equations are over-identified, the estimation problem becomes complicated. Three kinds of procedures have been proposed: (1) full system methods; (2) single equation methods; (3) subsystem methods. In full system methods all the parameters are considered simultaneously, and if normality is assumed, the maximum likelihood estimator can be obtained. Since it is usually difficult to compute the maximum likelihood estimator, the simpler, but asymptotically equivalent, **three-stage least squares method** was proposed. The single equation and the subsystem method take into consideration only the information about the parameters in one equation or in a subset of the equations, and estimate the parameters in each equation separately. There is a single equation method, called the **limited information maximum likelihood method**, based on the maximum likelihood approach, and also a **two-stage least squares method**, which first estimates Π by least squares, computes $\hat{\mathbf{Y}} = \hat{\Pi}\mathbf{Z}$, and then applies the least squares method to the model

$$\mathbf{Y} = B\hat{\mathbf{Y}} + \Gamma\mathbf{Z} + \tilde{\mathbf{u}}.$$

These two and also some others are asymptotically equivalent.

Only partial results have been obtained about the small sample properties of these methods, and the relative advantages of them are not yet known.

D. Other Problems

Problems in the third category can be approached by †multivariate analysis techniques. Sometimes †principal component analysis and †canonical correlation analysis have been applied to analyze the variations of a large amount of data. However, the practical meaning of the results obtained is often dubious.

The fourth category is the problem of time-series analysis. The sophisticated theory of stochastic processes has little relevance for economic time series, because usually they do not satisfy such conditions as being stationary or having the †Markov property, etc. But †autoregressive type models are often used. Traditionally, fluctuations of economic time series have been thought to consist of trend, cyclic variation, seasonal variation, and error. Various ad hoc techniques have been used to separate or eliminate such components, but the theoretical treatment of such problems is far from satisfactory.

References

[1] W. C. Hood and T. C. Koopmans (eds.), Studies in econometric method, Cowles Commission Monograph, John Wiley, 14, 1953.

[2] H. Theil, Economic forecasts and policy, North-Holland, 1958.

[3] J. Johnston, Econometric methods, McGraw-Hill, 1963.

135 (XII.11) Eigenvalue Problems

A. General Remarks

Throughout this article, X stands for a †complex linear space and A for a †linear operator in X. Except when X is finite-dimensional, A need not be defined over all X. A linear operator A in X is by definition a linear mapping whose †domain $D(A)$ and †range $R(A)$ are linear subspaces of X. A complex number λ is said to be an **eigenvalue (proper value** or **characteristic value)** of A if there exists an $x \in D(A)$ such that $Ax = \lambda x$, $x \neq 0$. Any such x is called an **eigenvector (eigenelement, proper vector, characteristic vector)** associated with λ. When X is a †function space, the word **eigenfunction** is also used. For an eigenvalue λ of A, the †subspace $M(\lambda)$ of X given by

$$M(\lambda) = M(\lambda; A) = \{x \mid Ax = \lambda x\},$$

i.e., the subspace consisting of 0 and all eigenvectors associated with λ is called the **eigenspace** associated with λ, and the number $m(\lambda) = \dim M(\lambda)$ is called the **geometric multiplicity** of λ. The eigenvalue λ is said to be **(geometrically) simple** or **degenerate** according as $m(\lambda) = 1$ or $m(\lambda) \geqslant 2$. The problem of seeking eigenvalues and eigenvectors is referred to as the **eigenvalue problem**.

When X is a †linear topological space, the notion of eigenvalues leads to a more general object called the spectrum of A. Let λ be a complex number and put $A_\lambda = \lambda I - A$, where I is the †identity operator in X. Furthermore, put $R_\lambda = (A_\lambda)^{-1} = (\lambda I - A)^{-1}$, if the inverse exists. Then the †resolvent set $\rho(A)$ of A is defined to be the set of all λ such that R_λ exists, has domain †dense in X, and is continuous. The †spectrum $\sigma(A)$ of A is, by definition, the complement of $\rho(A)$ in the complex plane, and it is divided into three mutually disjoint sets: the **point spectrum** $\sigma_P(A)$, the **continuous spectrum** $\sigma_C(A)$, and the **residual spectrum** $\sigma_R(A)$. These are defined as follows: $\sigma_P(A) = \{\lambda \mid R_\lambda \text{ does not exist}\} = \{\lambda \mid \lambda \text{ is an eigenvalue of } A\}$; $\sigma_C(A) = \{\lambda \mid R_\lambda \text{ exists and has domain dense in } X, \text{ but is not continuous}\}$; $\sigma_R(A) = \{\lambda \mid R_\lambda \text{exists, but its domain is not dense in } X\}$.

Let X be a †Banach space and $\mathbf{B}(X)$ the set of all †bounded linear operators with domain X. If A is a †closed operator in X, then $\lambda \in \rho(A)$ if and only if $R_\lambda \in \mathbf{B}(X)$. Moreover, $\sigma(A)$ is a closed set. In particular, if $A \in \mathbf{B}(X)$, then $\sigma(A)$ is compact. In this case the **spectral radius** $r_\sigma(A)$ is defined as $r_\sigma(A) = \sup_{\lambda \in \sigma(A)} |\lambda|$. Then $r_\sigma(A) \leqslant \|A^n\|^{1/n}$, $n = 1, 2, \ldots$, and $\|A^n\|^{1/n} \to r_\sigma(A)$, $n \to \infty$.

In many problems of analysis crucial roles have been played by methods involving the spectrum and other related concepts. This branch of analysis is called **spectral analysis**. For an infinite-dimensional X the theory is well developed when X is a †Hilbert space and A is †self-adjoint.

B. Eigenvalue Problems for Matrices

Throughout this section let X be an N-dimensional complex linear space ($N<\infty$) and A a linear operator in X. (We assume that A is defined over all X.) With respect to a fixed basis $(\psi_1, \dots, \psi_N)$ of X, the operator A is represented by an $N \times N$ matrix, also denoted by A. Then the eigenvalues of A coincide with the roots of the †characteristic equation $\det(\lambda I - A)=0$. There are no points of the spectrum other than eigenvalues, that is, $\sigma(A)=\sigma_P(A)$. Hence $\sigma(A)$ consists of at most n points. Let $\lambda \in \sigma(A)$. The multiplicity $\tilde{m}(\lambda)$ of λ as a root of the characteristic equation is called the **(algebraic) multiplicity** of the eigenvalue λ. The sum of $\tilde{m}(\lambda)$ over all the eigenvalues of A is equal to N. The eigenvalue λ is said to be **(algebraically) simple** or **degenerate** according as $\tilde{m}(\lambda)=1$ or $\tilde{m}(\lambda)\geqslant 2$. Let $\nu=1,2,\dots$, and $N_\nu(\lambda)=\{x \mid (\lambda I - A)^\nu x=0\}$. Then $\{N_\nu(\lambda)\}$ forms a nondecreasing chain of subspaces $M(\lambda)=N_1(\lambda)\subset N_2(\lambda)\subset \dots$, which ceases to increase after a finite number of steps. When $\nu \geqslant \tilde{m}(\lambda)$, the space $N_\nu(\lambda)$ is equal to a fixed subspace $\tilde{M}(\lambda)$, sometimes called the **principal subspace** (or **generalized eigenspace**) of A associated with λ. Then $\dim \tilde{M}(\lambda)=\tilde{m}(\lambda)$ and hence $m(\lambda)\leqslant \tilde{m}(\lambda)$. When A is a †normal matrix, $m(\lambda)=\tilde{m}(\lambda)$ and $M(\lambda)=\tilde{M}(\lambda)$. Thus for normal matrices there is no distinction between the geometric and the algebraic multiplicities.

If two matrices A and B are †similar, i.e., if there exists an invertible matrix P such that $B=P^{-1}AP$, then A and B have the same eigenvalues with the same algebraic (or geometric) multiplicities. The same conclusion holds for A and A', where A' is the †transpose of A. For the adjoint matrix A^* we have $\sigma(A^*)=\overline{\sigma(A)}\equiv\{\bar{\lambda} \mid \lambda \in \sigma(A)\}$. For an arbitrary polynomial f the relation $\sigma(f(A))= f(\sigma(A))\equiv\{f(\lambda) \mid \lambda\in\sigma(A)\}$ holds (Frobenius's theorem). These relations can be extended to operators in a Banach space. In particular, $\sigma(f(A))=f(\sigma(A))$ if A is a bounded operator and f is a function holomorphic in a neighborhood of $\sigma(A)$ (for the †spectral mapping theorem → 251 Linear Operators).

In the next four paragraphs, in which the spectral properties of normal or Hermitian matrices will be discussed, we introduce into X the Euclidean †inner product $(\ ,\ )$, regarding X as a space of N-tuples. Let A be an $N \times N$ normal matrix. Then the eigenspaces associated with different eigenvalues of A are mutually orthogonal. Moreover, the eigenspaces of A as a whole span the entire space X. One can therefore choose a †basis of X formed by a †complete orthonormal set of eigenvectors of A. Specifically, there exists a basis $\{\varphi_j \mid j=1,\dots,N\}$ of X such that $A\varphi_j=\mu_j\varphi_j$ and $(\varphi_i,\varphi_j)=\delta_{ij}$, where δ_{ij} is the †Kronecker delta. Moreover, $\mu_1, \dots, \mu_N$ exhaust all the eigenvalues of A. In terms of the basis $\{\varphi_j\}$, an arbitrary $x \in X$ can be expanded as

$$Ax=\sum_{j=1}^{N}\mu_j(x,\varphi_j)\varphi_j=\sum_{\lambda\in\sigma(A)}\lambda P_\lambda x, \tag{1}$$

where P_λ is the orthogonal †projection on the eigenspace associated with the eigenvalue λ.

Of particular importance among normal matrices are Hermitian matrices and unitary matrices. The eigenvalues of a Hermitian matrix are real, and those of a unitary matrix have the absolute value 1.

Solving the eigenvalue problem of a normal matrix A leads immediately to the diagonalization of A. For instance, let U be the $N \times N$ matrix whose jth column is equal to φ_j. Here the basis $\{\varphi_j\}$ is as before, and each φ_j is regarded as a column vector. Then U is unitary, and the transform U^*AU of A by U is the diagonal matrix whose diagonal entries are the μ_j. The problem of transforming a †Hermitian form to its canonical form can also be discussed by means of U. In fact, a Hermitian form $Q=Q(x)$ on X is expressed as $Q(x)=(Ax,x)$ with a Hermitian matrix A. For this A, construct U as before. Then by the transformation $x=Uy$ of the coordinates of X, the form Q is converted to its canonical form $Q=\mu_1|y_1|^2+\dots+\mu_N|y_N|^2$. When X is a real linear space and A is a real symmetric matrix, U is an orthogonal matrix. By means of the orthogonal transformation $x=Uy$, the surface of the second order $Q(x)=1$ in $\mathbf{R}^N$ is converted to the form $\mu_1 y_1^2+\dots+\mu_N y_N^2=1$. The orthogonal transformation $x=Uy$ is called the **transformation to principal axes** of the surface $Q(x)=1$.

The eigenvalue problem of a Hermitian matrix A may be viewed as the variational problem for **Rayleigh's quotient** $R(x)= (Ax,x)/\|x\|^2$, $x\neq 0$. Namely, the smallest eigenvalue μ_1 of A can be obtained as $\mu_1= \min_{x\in X} R(x)$, and any nonzero $x=\varphi_1$ which attains this minimum is an eigenvector associated with μ_1. This is called **Rayleigh's principle**. The second smallest eigenvalue μ_2 is then obtained as the minimum of $R(x)$ in the $(N-1)$-dimensional subspace orthogonal to φ_1. Proceeding similarly, we obtain a nondecreasing sequence $\mu_1\leqslant\mu_2\leqslant\dots\leqslant\mu_N$. The sequence $\{\mu_n\}$ gives the enumeration of all the eigenvalues of A in increasing order with repetitions according to multiplicity. If the maximum of $R(x)$ is used instead of the minimum, the eigenvalues appear in decreasing order: $\mu_N\geqslant\mu_{N-1}\geqslant\dots\geqslant\mu_1$. A more direct characterization of μ_n, $1<n<N$, involving

no previous eigenvectors, is given by

$$\mu_n = \max_{\substack{f_1,\ldots,f_{n-1}\in X\\ f_1,\ldots,f_{n-1}\neq 0}} \left(\min_{\substack{(x,f_j)=0\\ j=1,\ldots,n-1}} R(x) \right).$$

This formula is referred to as the **minimax principle**.

Among nonnormal matrices particular attention has been drawn to matrices whose entries are all nonnegative. Let $A \neq 0$ be such a matrix. Then there exists a positive eigenvalue λ_0 of A such that $|\lambda| \leqslant \lambda_0$ for all eigenvalues λ of A. Furthermore, there is an eigenvector φ_0 associated with λ_0 such that all the components of φ_0 are nonnegative. Theorems of this type are called **Frobenius's theorems** on matrices with nonnegative entries [17].

C. Eigenvalue Problems in Hilbert Spaces

Throughout the rest of the present article except for the last section, X is assumed to be a Hilbert space with inner product (,). Furthermore, most of the discussions will be confined to †normal or †self-adjoint operators. Let A be a normal operator. Then $\sigma(A)$ is nonempty, $\sigma_R(A)$ is empty, eigenspaces associated with different eigenvalues are mutually orthogonal, and if A is bounded, the spectral radius of A coincides with the norm of A. The spectrum of a self-adjoint operator is contained in the real axis, and that of a unitary operator is contained in the unit circle.

The notion of **numerical range** $W(A)$ is important for an operator A in a Hilbert space. It is defined as $W(A)=\{(Ax,x) \mid x\in D(A), \|x\|=1\}$ and is a †convex set in the complex plane. For a self-adjoint operator H, $W(H)$ is contained in the real axis, and the closure of $W(H)$ is the smallest closed (possibly infinite) interval containing $\sigma(H)$.

A fundamental theorem in spectral analysis for self-adjoint (or normal) operators is the spectral theorem, which asserts that a representation such as (1) holds in a generalized form. This will be discussed in detail in Sections D and E.

D. Spectral Measure

Let $\mathfrak{B}$ be a †completely additive class of subsets of a set Ω, that is, $(\Omega, \mathfrak{B})$ is a †measurable space. An operator-valued set function $E=E(\cdot)$ defined on $\mathfrak{B}$ is said to be a (self-adjoint) **spectral measure** if (i) $E(M)$, $M\in\mathfrak{B}$, is an †orthogonal projection in X; (ii) $E(\Omega)=I$; and (iii) E is †countably additive, that is,

$$E\left(\bigcup_{n=1}^{\infty} M_n\right) = \sum_{n=1}^{\infty} E(M_n)$$

(†strong convergence) for a disjoint sequence $\{M_n\}$ of subsets in $\mathfrak{B}$. A spectral measure E satisfies $E(M\cap N)=E(M)E(N)=E(N)E(M)$, $M, N\in\mathfrak{B}$. Spectral measures which are frequently used in spectral analysis are those defined on the family $\mathfrak{B}_r$ ($\mathfrak{B}_c$) of all †Borel sets in the field of real (complex) numbers **R** (**C**). A spectral measure on $\mathfrak{B}_r$ ($\mathfrak{B}_c$) is sometimes referred to as a **real (complex) spectral measure**. For such a spectral measure E the **support** (or the **spectrum**) of E, denoted by $\Lambda(E)$, is defined to be the complement of the largest open set G for which $E(G)=0$. A complex spectral measure such that $\Lambda(E)\subset\mathbf{R}$ may be identified with a real spectral measure.

Let E be a spectral measure on $\mathfrak{B}_r$, and put

$$E_\lambda = E((-\infty,\lambda]), \quad -\infty<\lambda<\infty. \tag{2}$$

Then E_λ satisfies the following relations:

$$E_\lambda E_\mu = E_{\min(\lambda,\mu)}, \qquad \operatorname*{s-lim}_{\lambda\to\mu+0} E_\lambda = E_\mu,$$
$$\operatorname*{s-lim}_{\lambda\to-\infty} E_\lambda = 0, \qquad \operatorname*{s-lim}_{\lambda\to\infty} E_\lambda = I, \tag{3}$$

where s-lim stands for strong convergence. A family $\{E_\lambda\}_{\lambda\in\mathbf{R}}$ of orthogonal projections satisfying the relation (3) is called a **resolution of the identity**. Relation (2) gives a one-to-one correspondence between the resolutions of the identity and the spectral measures on $\mathfrak{B}_r$.

Let E be a spectral measure on $\mathfrak{B}_r$, and let $x, y\in X$. Then the set function $M\to(E(M)x,x)=\|E(M)x\|^2$ is a bounded regular †measure in the ordinary sense, and the set function $M\to(E(M)x,y)$ is a complex-valued regular †completely additive set function. For every complex Borel †measurable function f on **R**, the operator $S(f)$ in X is defined by the relations

$$D(S(f)) = \left\{ x \,\middle|\, \int_{-\infty}^{\infty} |f(\lambda)|^2 (E(d\lambda)x,x) < \infty \right\},$$
$$(S(f)x,y) = \int_{-\infty}^{\infty} f(\lambda)(E(d\lambda)x,y), \tag{4}$$
$$x\in D(S(f)), \quad y\in X.$$

$S(f)$ is a densely defined closed operator and is denoted as $S(f)=\int_{-\infty}^{\infty} f(\lambda)E(d\lambda)$. The correspondence $f\to S(f)$ satisfies formulas of the so-called operation calculus (→ 251 Linear Operators). In particular, $S(\bar{f})=S(f)^*$, and hence $S(f)$ is self-adjoint if f is real-valued. If f is bounded on the support of E, then $S(f)$ is everywhere defined in X and is bounded. $S(f)$ is sometimes called the **spectral integral** of f with respect to E. The operator $S(f)$ can be defined in a similar way for a spectral measure on $\mathfrak{B}_c$ (or for a more general spectral measure).

E. Spectral Theorems

For every self-adjoint operator H in a Hilbert space X, there exists a unique spectral measure E on the family of all real Borel sets $\mathfrak{B}_r$ such that

$$H=\int_{-\infty}^{\infty}\lambda E(d\lambda). \tag{5}$$

In other words, H and E correspond to each other by the relations

$$D(H)=\left\{x \mid \int_{-\infty}^{\infty}\lambda^2(E(d\lambda)x,x)<+\infty\right\},$$

$$(Hx,y)=\int_{-\infty}^{\infty}\lambda(E(d\lambda)x,y), \qquad x\in D(H),\quad y\in X. \tag{6}$$

This is the **spectral theorem** for self-adjoint operators. The support of E is equal to $\sigma(H)$, so that we may write

$$H=\int_{\sigma(H)}\lambda E(d\lambda)=\int_{-\infty}^{\infty}\lambda\chi_{\sigma(H)}(\lambda)E(d\lambda),$$

where χ_M stands for the †characteristic function of M. Formulas (5) and (6) are sometimes called the **spectral resolution** (or **spectral representation**) of H. We call E the spectral measure for H, and the $\{E_\lambda\}$ corresponding to E by formula (2) (or sometimes E itself) the resolution of the identity for H. Let λ be a real number. Then $\lambda\in\sigma_P(H)$ if and only if $E(\{\lambda\})\neq 0$. Also, $\lambda\in\sigma_C(H)$ if and only if $E(\{\lambda\})=0$ and $E(V)\neq 0$ for any neighborhood V of λ. The spectral measure E can be represented in terms of the resolvent $R(\alpha;H)=(\alpha I-H)^{-1}$ of H by the formula

$$E((a,b))=\lim_{\delta\downarrow 0}\lim_{\varepsilon\downarrow 0}\frac{1}{2\pi i}\int_{a+\delta}^{b-\delta}\{R(\mu-\varepsilon i;H)-R(\mu+\varepsilon i;H)\}\,d\mu$$

(strong convergence).

For every normal operator A in X, there exists a unique spectral measure E on the family of all complex Borel sets $\mathfrak{B}_c$ such that

$$A=\int_{\mathbf{C}}zE(dz).$$

This is called the **complex spectral resolution** (or **complex spectral representation**) of A. The support $\Lambda(E)$ is equal to $\sigma(A)$. For a unitary operator U, the support of the associated spectral measure is contained in the unit circle Γ, so that U can be represented as

$$U=\int_{\Gamma}e^{i\theta}F(d\theta) \tag{7}$$

with a spectral measure F defined on Γ. Formula (7) is the spectral resolution of U.

F. Functions of a Self-Adjoint Operator

Let $H=\int_{-\infty}^{\infty}\lambda E(d\lambda)$ be a self-adjoint operator in X. For a complex-valued Borel measurable function f on $\mathbf{R}$, the operator $S(f)$ determined by (4) in reference to the resolution of the identity E associated with H is denoted by $f(H)$:

$$f(H)=\int_{-\infty}^{\infty}f(\lambda)E(d\lambda).$$

The correspondence $f\rightarrow f(H)$ gives an operational calculus for H (→ 251 Linear Operators).

For an arbitrary $a\in X$, we denote by $L_2(a)$ the L_2-space over the measure $\mu_a=\mu_a(\cdot)=(E(\cdot)a,a)$. In other words, $f\in L_2(a)$ if and only if $a\in D(f(H))$. The correspondence $f\leftrightarrow f(H)a$ gives an isometric isomorphism between $L_2(a)$ and the subspace $M(a)=\{f(H)a \mid f\in L_2(a)\}$ of X. (In particular, $M(a)$ is closed.) H is reduced by $M(a)$, and the part of H in $M(a)$ corresponds to the multiplication $f(\lambda)\rightarrow\lambda f(\lambda)$ in $L_2(a)$.

For a given self-adjoint operator H there exists a (not necessarily countable) family $\{a_\theta\}_{\theta\in\Theta}$ of elements a_θ of X such that

$$X=\sum_{\theta\in\Theta}M(a_\theta), \tag{8}$$

where Σ stands for the †direct sum of mutually orthogonal closed subspaces. Consequently, X is represented by the direct sum $\Sigma_{\theta\in\Theta}L_2(a_\theta)$ of L_2-spaces. If $x\in D(H)$ is represented by $\{f_\theta\}_{\theta\in\Theta}$ in this representation, Hx is represented by $\{\lambda f_\theta\}_{\theta\in\Theta}$.

G. Unitary Equivalence and Spectral Multiplicity

In this section X is assumed to be a †separable Hilbert space. Then (8) can be made more precise. Namely, for a self-adjoint operator H, we can find a countable family $\{a_n\}_{n=1}^{\infty}$ of elements of X such that

$$X=\sum_{n=1}^{\infty}M(a_n)\cong\sum_{n=1}^{\infty}L_2(a_n), \tag{9}$$

$$\mu_{a_{n+1}} \text{ is †absolutely continuous with respect to } \mu_{a_n},\quad n=1,2,\ldots. \tag{10}$$

Furthermore, if $\{a_n'\}$ is another family satisfying (9) and (10), then μ_{a_n} and $\mu_{a_n'}$ are absolutely continuous with respect to each other (Hellinger-Hahn theorem).

Two self-adjoint operators H_1 and H_2 are said to be **unitarily equivalent** if there exists a unitary operator U such that $H_2=U^*H_1U$. A criterion for unitary equivalence can be given in terms of the spectral representation given previously. Namely, let $\{a_n^{(i)}\}$, $i=1,2$, be a sequence satisfying (9) and (10) with respect to H_i. Then H_1 and H_2 are unitarily equivalent if and only if $\mu_{a_n^{(1)}}$ and $\mu_{a_n^{(2)}}$ are absolutely continuous with respect to each other for all $n=1,2,\ldots$.

Using a spectral decomposition with properties (9) and (10), we can define the **spectral**

multiplicity $m_\sigma(H)$ of H in the following way. If

$$\mu_{a_k} \neq 0, \quad k=1,\ldots,n, \quad \text{and} \quad \mu_{a_{n+1}}=0, \qquad (11)$$

then $m_\sigma(H)=n$; if there is no such n, then $m_\sigma(H)=\infty$. The spectral multiplicity is invariant under a unitary transformation U: $m_\sigma(H)=m_\sigma(U^*HU)$. A real number λ_0 is said to have multiplicity n (to be of infinite multiplicity) with respect to $\sigma_C(H)$ if (11) holds (if $\mu_{a_k}\neq 0$, $k=1,2,\ldots$) in a neighborhood of λ_0. In this case $\sigma_C(H)$ is said to be degenerate at λ_0 with multiplicity n (with infinite multiplicity). H is said to have a **simple spectrum** if $m_\sigma(H)=1$, or equivalently if there exists an $a\in X$ such that $M(a)=X$. Such an $a\in X$ is called a **generating element** of X with respect to H.

Self-adjoint operators with simple spectra are closely related to Jacobi matrices. Let H be such an operator with a generating element $a\in X$. Take a complete orthonormal set $\{G_n\}_{n=1}^{\infty}$ in $L_2(a)$ such that $G_n=G_n(\lambda)$ is a polynomial of degree $n-1$ and $\lambda G_n(\lambda)\in L_2(a)$. Then $\{g_n\}_{n=1}^{\infty}$, $g_n=G_n(H)a$, is a complete orthonormal set in X. The matrix representation $\{a_{mn}\}$, $a_{mn}=(Hg_n,g_m)$ of H with respect to the basis $\{g_n\}$ has the following properties: (i) $a_{mn}=0$ if $|m-n|\geqslant 2$; (ii) $a_{n,n+1}=a_{n+1,n}\neq 0$; (iii) a_{nn} is real. Any infinite matrix $\{a_{mn}\}$ satisfying (i), (ii), and (iii) is called a **Jacobi matrix**. A Jacobi matrix determines a †symmetric operator whose †deficiency index is either (0,0) or (1,1). Any self-adjoint extension has a simple spectrum. (For more details about Jacobi matrices and their applications → [11].)

H. Eigenvalue Problems in Banach Spaces

Spectral analysis becomes rather involved for general operators in a Banach space as well as for nonnormal operators in a Hilbert space.

For a †compact operator A, the nature of $\sigma(A)$ and the structure of A in the principal subspace associated with a nonzero eigenvalue are well known (→ 72 Compact Operators). However, a full spectral analysis may not be possible without further assumptions.

For a †closed operator with nonempty resolvent set $\rho(A)$, an operational calculus can be developed by means of a function-theoretic method based on the fact that the resolvent $R_\lambda=(\lambda I-A)^{-1}$ is a $\mathbf{B}(X)$-valued holomorphic function of λ in $\rho(A)$. In particular, the spectral mapping theorem holds (→ [13]; 251 Linear Operators).

A general class of operators having associated spectral resolution was introduced by N. Dunford. Let X be a Banach space. An operator $E\in\mathbf{B}(X)$ is called a projection if $E^2=E$. As before we can define a (projection-valued countably additive) spectral measure on $\mathfrak{B}_c$. An operator $A\in\mathbf{B}(X)$ is said to be a **spectral operator** if there exists a spectral measure E on $\mathfrak{B}_c$ satisfying the following properties: (i) $E(M)A=AE(M)$, $M\in\mathfrak{B}_c$; (ii) $\sigma(A_{|E(M)X})\subset\overline{M}$, $M\in\mathfrak{B}_c$, where $A_{|Y}$ is the restriction of A to Y and $\overline{M}$ is the closure of M; (iii) there exists a $k\geqslant 0$ such that $\|E(M)\|\leqslant k$ for all $M\in\mathfrak{B}_c$. E is unique. A spectral operator A is expressed as $A=S+N$, where $S=\int_C zE(dz)$ and N is a †generalized nilpotent. A is said to be a **scalar operator** if $N=0$. Unbounded spectral operators are defined similarly, with (i') $E(M)A\subset AE(M)$ in place of (i). However, for the unbounded spectral operators A we no longer have the decomposition $A=S+N$. (For more details about spectral operators → [18]. For other topics related to the material discussed in this section → 72 Compact Operators; 136 Eigenvalues (Numerical Computation); and 251 Linear Operators).

References

[1] Н. И. Ахиезер, И. М. Глазман (N. I. Ahiezer and I. M. Glazman), Теория линейных операторов в Гильбертовом пространстве, Москва, 1950; English translation, Theory of linear operators in Hilbert space, Ungar, 1961–1963.
[2] R. Courant and D. Hilbert, Methods of mathematical physics I, Interscience, 1953.
[3] N. Dunford and J. T. Schwartz, Linear operators, Interscience, I, 1958; II, 1963.
[4] S. Furuya, Gyôretu to gyôretusiki (Japanese; Matrices and determinants), Baihûkan, 1957.
[5] И. М. Гельфанд, Г. Е. Шилов (I. M. Gel'fand and G. E. Šilov), Обобщенные функции III, Физматгиз, 1958; English translation, Generalized functions III. Theory of differential equations, Academic Press, 1967.
[6] S. H. Gould, Variational methods for eigenvalue problems, Univ. of Toronto Press, 1957.
[7] P. R. Halmos, Introduction to Hilbert space and the theory of spectral multiplicity, Chelsea, second edition, 1957.
[8] T. Kato, Kansû kûkanron (Japanese; Theory of function spaces), Kyôritu, 1957.
[9] J. Von Neumann, Mathematische Grundlagen der Quantenmechanik, Springer, 1932.
[10] I. Satake, Gyôretu to gyôretusiki (Japanese; Matrices and determinants), Syôkabô, 1958.
[11] M. H. Stone, Linear transformations in Hilbert space and their applications to analysis, Amer. Math. Soc. Colloq. Publ., 1932.

[12] B. von Sz.-Nagy, Spektraldarstellung linearer Transformationen des Hilbertschen Raumes, Erg. Math., Springer, 1942.
[13] A. E. Taylor, Introduction to functional analysis, John Wiley, 1958.
[14] K. Yosida, Functional analysis, Springer, 1965.
[15] K. Yosida, Isôkaiseki (Japanese; Functional analysis), Iwanami, 1951.
[16] K. Yosida and T. Kato, Ôyôsûgaku, ensyû (Japanese; Exercises on applied mathematics) I, Syôkabô, 1961.
[17] Ф. Р. Гантмахер (F. R. Gantmaher), Теория матриц, Гостехиздат,1953; English translation, The theory of matrices I, II, Chelsea, 1959.
[18] N. Dunford, A survey of the theory of spectral operators, Bull. Amer. Math. Soc., 64 (1958), 217–274.
[19] T. Kato, Perturbation theory for linear operators, Springer, 1966.
Also → references to 199 Hilbert Spaces; 39 Banach Spaces; 251 Linear Operators; 72 Compact Operators; 269 Matrices; and 136 Eigenvalues (Numerical Computation).

136 (XV.6) Eigenvalues (Numerical Computation)

A. Numerical Computation of Eigenvalues

Numerical computation of the †eigenvalues and †eigenvectors of a matrix is the basic technique for the numerical solution of various eigenvalue problems. Roughly speaking, there are two kinds of methods. One is first to determine the †characteristic polynomial $p(\lambda)=\det(\lambda I-A)$ (where I is the unit matrix) of A (or to give an algorithm to calculate the value of $p(\lambda)$ for an arbitrary λ), then to solve the algebraic equation $p(\lambda)=0$ numerically to obtain the eigenvalues λ_μ ($\mu=1,2,\ldots$) (→ 296 Numerical Solution of Algebraic Equations), and finally to determine the eigenvectors $\mathbf{x}_\mu$ by means of the equations $(\lambda_\mu I-A)\mathbf{x}_\mu=0$. The other is to obtain eigenvalues and eigenvectors directly without resorting to the solution of an algebraic equation. In general, efficient numerical methods are known for real symmetric (or †Hermitian) matrices, but no such methods are known for other matrices. For simplicity we confine ourselves to $n\times n$ matrices $A=(a_{ij})$ $(i,j=1,\ldots,n)$ with real elements. The extension to the case of matrices with complex elements is easy.

B. The Jacobi Method

The **Jacobi method** is an iterative technique for determining all the eigenvalues and eigenvectors of a real symmetric matrix [7]. Before the advent of high-speed computers it was not considered practical, but at present it is one of the most efficient and reliable methods. The following algorithm can be extended to Hermitian matrices by replacing †orthogonal transformations by suitable †unitary transformations.

In outline the method transforms a given matrix $A=(a_{ij})$ $(a_{ji}=a_{ij};\ i,j=1,\ldots,n)$ into a diagonal one by repeated application of 2-dimensional rotations of the reference axes. We first put $A^{(0)}=A$, $U^{(0)}=I$, and compute $A^{(1)}, A^{(2)},\ldots;\ U^{(1)}, U^{(2)},\ldots$ successively as follows:

(1) Select an off-diagonal element of $A^{(l)}=(a_{ij}^{(l)})$ with the maximum absolute value and denote it by $a_{pq}^{(l)}$.

(2) Compute

$$\tan\theta=\frac{2a_{pq}^{(l)}\operatorname{sgn}(a_{pp}^{(l)}-a_{qq}^{(l)})}{|a_{pp}^{(l)}-a_{qq}^{(l)}|+\sqrt{(a_{pp}^{(l)}-a_{qq}^{(l)})^2+4(a_{pq}^{(l)})^2}}$$

(where $\operatorname{sgn}x=1,0,$ or -1 according as $x>0$, $=0$, or <0), $\cos\theta=(1+\tan^2\theta)^{-1/2}$, and $\sin\theta=\tan\theta\cdot\cos\theta$; and form $T^{(l)}=(t_{ij}^{(l)})$, where $t_{pp}^{(l)}=t_{qq}^{(l)}=\cos\theta$, $t_{ii}^{(l)}=1$ for $i\neq p,q$, $-t_{pq}^{(l)}=t_{qp}^{(l)}=\sin\theta$, and $t_{ij}^{(l)}=0$ for all other (i,j).

(3) Determine $A^{(l+1)}$ and $U^{(l+1)}$ by $A^{(l+1)}=T^{(l)\prime}A^{(l)}T^{(l)}$ ($'$ means transposition) and $U^{(l+1)}=U^{(l)}T^{(l)}$. In this process, $T^{(l)}$ represents an orthogonal transformation (rotation) in the plane spanned by the pth and qth coordinate axes such that $a_{pq}^{(l+1)}=a_{qp}^{(l+1)}$ is nullified. If we put $N(B)=\sum_{i,j}b_{ij}^2$ and $M(B)=\sum_{i\neq j}b_{ij}^2$, then $N(B)$ is invariant under an orthogonal transformation, so that $N(A^{(l)})=N(A)$. Furthermore, since $a_{ii}^{(l+1)}=a_{ii}^{(l)}$ $(i\neq p,q)$ and $(a_{pp}^{(l+1)})^2+(a_{qq}^{(l+1)})^2=(a_{pp}^{(l)})^2+(a_{qq}^{(l)})^2+2(a_{pq}^{(l)})^2$, we have $M(A^{(l+1)})=M(A^{(l)})-2(a_{pq}^{(l)})^2$. Since $a_{pq}^{(l)}$ has the maximum absolute value among all the off-diagonal elements, we have $(a_{pq}^{(l)})^2\geqslant M(A^{(l)})/(n^2-n)$. Therefore,

$$\begin{aligned}M(A^{(l+1)})&\leqslant(1-2/(n^2-n))M(A^{(l)})\\&\leqslant(1-2/(n^2-n))^{l+1}M(A)\\&<M(A)\exp(-2(l+1)/(n^2-n)).\end{aligned}$$

It has been proved [13] a fortiori that, after $M(A^{(l)})$ comes down below a certain threshold value, the convergence of the iteration process becomes quadratic, i.e., there is a number c determined by the order n of A and the arrangement of the eigenvalues of A such that $M(A^{(l+n(n-1)/2)})<c(M(A^{(l)}))^2$. Since the set of eigenvalues of $A^{(l)}$ coincides with that of A and the eigenvalues of an

arbitrary symmetric matrix B can be made to correspond one-to-one to its diagonal elements in such a way that the difference between an eigenvalue and the corresponding diagonal element is not greater than $M(B)^{1/2}$, $a_{ii}^{(l)}$ tends to λ_i $(i=1,\ldots,n)$ as l tends to infinity. Moreover, as l tends to infinity, each column vector of $U^{(l)}=(u_{ij}^{(l)})$ tends to the corresponding eigenvector, in the sense that $\sum_{k=1}^{n}a_{ik}u_{kj}^{(l)}-\lambda_j u_j^{(l)}\to 0$.

The number of arithmetic computations required to obtain $A^{(l+1)}$ and $U^{(l+1)}$ from $A^{(l)}$ and $U^{(l)}$ is at most proportional to n, so that for a given ε (>0), the arithmetic required to reduce $\max_i|a_{ii}^{(l)}-\lambda_i|$ below $\varepsilon(M(A))^{1/2}$ is at most proportional to n^3 (because l is at most proportional to n^2). On the other hand, the search for an off-diagonal element of $A^{(l)}$ with the maximum absolute value, if it is done by simply comparing all the elements, will require effort proportional to n^2, so that the amount of work required by the searching process is proportional to n^4. To bypass this searching process, the **cyclic Jacobi method** and the **threshold Jacobi method** have often been used. The former method adopts as $a_{pq}^{(l)}$ the off-diagonal element for which $q>p$ and $l=(p-1)(n-p/2)+(q-p)$, that is, $a_{12}^{(1)}$, $a_{13}^{(2)},\ldots,a_{1,n}^{(n-1)}$, $a_{23}^{(n)}$, $a_{24}^{(n+1)},\ldots$ are adopted in this sequence. The latter method adopts as $a_{pq}^{(l)}$ off-diagonal elements in a sequence similar to the one above so long as they exceed a given threshold value; but if an element is less than that threshold value, then the element next in the sequence is to be a candidate for adoption as $a_{pq}^{(l)}$, where the threshold value is made to decrease gradually as the iteration process proceeds. However, the search for an element with the maximum absolute value can be done more effectively by taking account of the fact that only the elements of the matrix lying in rows p and q and in columns p and q change their values when we transform $A^{(l)}$ into $A^{(l+1)}$. In fact, we may record for each row the value as well as the position of the (off-diagonal) element with the maximum absolute value in that row. By so doing, the effort of searching for an off-diagonal element with the maximum absolute value can be reduced to something proportional to n on the average.

C. The Frame Method

The **Frame method**, proposed by J. S. Frame in 1949, is a rather theoretical method which determines at the same time the characteristic polynomial $p(\lambda)$ of a matrix A and the †**adjugate matrix** (i.e., the transpose of the matrix obtained by replacing each element by its †cofactor) $C(\lambda)$ of $\lambda I-A$ [2]. If we put

$$p(\lambda)=\det(\lambda I-A)$$
$$=\lambda^n+p_1\lambda^{n-1}+\ldots+p_{n-1}\lambda+p_n$$

and

$$C(\lambda)=\lambda^{n-1}C_0+\lambda^{n-2}C_1+\ldots+\lambda C_{n-2}+C_{n-1},$$
$$C_0=I,$$

the p_i and C_i are calculated by the formulas

$$p_i=-\operatorname{tr}AC_{i-1}/i,\qquad C_i=AC_{i-1}+p_iI,\qquad i=1,2,\ldots,n,$$

where, up to rounding errors, $C_n=0$ and $-C_{n-1}/p_n$ equals the inverse of A. The eigenvalues λ_μ $(\mu=1,\ldots,n)$ are obtained as the roots of $p(\lambda)=0$, and if λ_μ is a simple root, the corresponding eigenvector is given by any column vector of $C(\lambda_\mu)$. The amount of computation, except for that required to solve the algebraic equation $p(\lambda)=0$, is proportional to n^4. This method often suffers from serious rounding errors, especially when A is †ill conditioned.

D. The Power Method

The **power method** is suitable for obtaining only the eigenvalue of maximum absolute value [6]. Let us assume that the eigenvalues $\lambda_1,\ldots,\lambda_n$ of A are arranged so that $|\lambda_1|>|\lambda_2|\geqslant|\lambda_3|\geqslant\ldots\geqslant|\lambda_n|$, with λ_1 real, and denote by $\mathbf{y}_1$ the left eigenvector corresponding to λ_1 (which means $\mathbf{y}_1(\lambda_1I-A)=0$). Starting from an arbitrary (real) vector $\mathbf{x}^{(0)}$ such that $(\mathbf{y}_1,\mathbf{x}^{(0)})=0$ and $x_{i_0}^{(0)}=1$ for a prescribed i_0, we compute $\theta^{(0)},\theta^{(1)},\ldots$ and $\mathbf{x}^{(1)},\mathbf{x}^{(2)},\ldots$ by $A\mathbf{x}^{(l)}=\theta^{(l)}\mathbf{x}^{(l+1)}$ $(l=0,1,2,\ldots;\mathbf{x}_{i_0}^{(l+1)}=1)$. Then we have $\lim_{l\to\infty}\theta^{(l)}=\lambda_1$ and $\lim_{l\to\infty}\mathbf{x}^{(l)}=\mathbf{x}_1$ (the eigenvector corresponding to the eigenvalue λ_1). The rate of convergence depends on $|\lambda_1/\lambda_2|$ if the †elementary divisor of A corresponding to λ_1 is linear, but in the case of a nonlinear elementary divisor, the convergence is too slow for practical purposes. If $\lambda_1\neq\lambda_2$ and $|\lambda_1|=|\lambda_2|>|\lambda_3|\geqslant\ldots\geqslant|\lambda_n|$, then the sequences of $\theta^{(l)}$ and $\mathbf{x}^{(l)}$ computed by the formulas above do not converge but in general oscillate. However, from $\theta^{(l)},\theta^{(l+1)}$, $\mathbf{x}^{(l)},\mathbf{x}^{(l+1)},\mathbf{x}^{(l+2)}$ for a sufficiently large l, we can obtain approximate eigenvalues λ_1 and λ_2 as the two roots of the quadratic equation in λ:

$$\begin{vmatrix} x_i^{(l)} & x_j^{(l)} & \theta^{(l)}\theta^{(l+1)} \\ x_i^{(l+1)} & x_j^{(l+1)} & \theta^{(l)}\lambda \\ x_i^{(l+2)} & x_j^{(l+2)} & \lambda^2 \end{vmatrix}=0$$

(i and j arbitrary, $i\neq j$).

The corresponding eigenvectors are given by $\mathbf{x}_1=\lambda_2\mathbf{x}^{(l+1)}-\mathbf{x}^{(l+2)}$ and $\mathbf{x}_2=\lambda_1\mathbf{x}^{(l+1)}-\mathbf{x}^{(l+2)}$.

Complex conjugate pairs of eigenvalues can be dealt with in this manner. This is useful also for the case where $|\lambda_1| \fallingdotseq |\lambda_2|$. The extension to the case of more than two eigenvalues with the same maximum absolute value is obvious.

In order to determine the remaining eigenvalues by the power method we have to combine with it the deflation or transformation of matrices, as mentioned in Sections F and G. The amount of computation depends on the arrangement of the eigenvalues of A and the required accuracy. We note that the multiplication of a matrix by a vector requires an amount of computation proportional to n^2.

E. Improvement of Approximations

If $\mathbf{v}$ $(=\mathbf{x}_\mu + O(\varepsilon))$ is an approximation to the eigenvector $\mathbf{x}_\mu$ corresponding to an eigenvalue λ_μ of a real symmetric matrix A, then the **Rayleigh quotient** $\lambda_R = (\mathbf{v}, A\mathbf{v})/(\mathbf{v},\mathbf{v})$ affords a good approximation to λ_μ. In fact, we have $|\lambda_\mu - \lambda_R| = O(\varepsilon^2)$.

If λ is an eigenvalue of A and $\mathbf{x}$ the corresponding eigenvector, then $P(\lambda)$ is an eigenvalue of $P(A)$ and $\mathbf{x}$ the corresponding eigenvector, where $P(\xi)$ is a polynomial in ξ and ξ^{-1}. This fact may be utilized to transform the magnitudes of eigenvalues to accelerate the convergence of the power method, to separate eigenvalues with the same absolute value, or to obtain intermediate eigenvalues.

Aitken's δ^2-method is efficient in accelerating the convergence of the power method.

F. Deflation

If an eigenvalue λ_μ and the corresponding eigenvector $\mathbf{x}_\mu$ (and also the corresponding left eigenvector ${}^t\mathbf{y}_\mu$ if necessary) of A are known, it is possible to "subtract" them from A to get a problem containing only the remaining eigenvalues. Such a process of reduction, often used in combination with the power method, is called **deflation**. The following are two examples of deflation methods.

(1) Assuming that $\mathbf{x}_\mu$ and $\mathbf{y}_\mu$ are normalized in such a way that $(\mathbf{y}_\mu, \mathbf{x}_\mu) = 1$, form $B = A - \lambda_\mu \mathbf{x}_\mu {}^t\mathbf{y}_\mu$. Then B has the same set of eigenvalues and eigenvectors as A except for λ_μ. The eigenvalue and the eigenvector of B corresponding to λ_μ of A are 0 and $\mathbf{x}_\mu$, respectively. This kind of deflation process can be generalized to the case of nonlinear elementary divisors, but that becomes somewhat complicated.

(2) After normalizing $\mathbf{x}_\mu$ such that its nth component $x_{\mu n}$ is equal to 1, form $B = (b_{ij})$: $b_{ij} = a_{ij} - x_{\mu i} a_{nj}$ $(i,j = 1, \ldots, n-1)$. Then B has the same set of eigenvalues as A except for λ_μ. If $\mathbf{w}_\kappa$ is the eigenvector corresponding to the eigenvalue λ_κ of B, then the corresponding eigenvector $\mathbf{x}_\kappa$ of A is given by $x_{\kappa i} = w_{\kappa i} + d_\kappa x_{\mu i}$ $(i = 1, \ldots, n-1)$, $x_{\kappa n} = d_\kappa$, where d_κ is determined from $\sum_{i=1}^{n-1} a_{ni} w_{\kappa i} = (\lambda_\kappa - \lambda_\mu) d_\kappa$ if $\lambda_\kappa \neq \lambda_\mu$. If $\lambda_\kappa = \lambda_\mu$ and $r \equiv \sum_{i=1}^{n} a_{ni} w_{\kappa i} = 0$, then we may put $d_\kappa = 0$. If $\lambda_\kappa = \lambda_\mu$ and $r \neq 0$, A has a nonlinear elementary divisor for $\lambda_\kappa = \lambda_\mu$, and the $\mathbf{x}_\kappa$ defined by $x_{\kappa i} = w_{\kappa i}/r$ and $x_{\kappa n} = 0$ is a generalized eigenvector of A in the sense that $A\mathbf{x}_\kappa = \lambda_\mu \mathbf{x}_\kappa + \mathbf{x}_\mu$.

G. Transformation of Matrices

There are a number of methods of transforming a given matrix A by means of a suitable similarity transformation $A \rightarrow B = S^{-1}AS$ into another matrix B for which it is easier to solve the eigenvalue problem. The **Givens method** [9] transforms a symmetric matrix A into a **tridiagonal matrix** B (i.e., a matrix such that $b_{ij} = 0$ for $|i-j| \geqslant 2$) by means of an S which is the product of 2-dimensional rotation matrices. The **Householder method** [10] also transforms a symmetric A into a tridiagonal B by means of an orthogonal matrix S of special type, and the **Lanczos method** [8] transforms a general A into a tridiagonal B. To general matrices the following methods are also applicable. (1) The **Danilevskiĭ method** [1] transforms A into its companion matrix B by repeated application of elimination operations. (2) The **Hessenberg method** [3] transforms A into a B such that $b_{ij} = 0$ for $i - j \geqslant 2$ with a triangular S. (3) The **Givens method** [9] transforms A into a B of the same form as in (2) by repeated application of 2-dimensional rotations. All these methods require an amount of computation proportional to n^3 except for the solution of an algebraic equation. In general, special treatment is necessary for the case of multiple eigenvalues. As an example, we explain the Givens method for general matrices. Let $N = (n-1)(n-2)/2$. For $l = 0, 1, \ldots, N-1$, choose $(p,q) = (3,2), (4,2), \ldots, (n,2); (4,3), (5,3), \ldots, (n,3); \ldots; (n-1, n-2), (n, n-2); (n, n-1)$ in this order. Using $T^{(l)}$ of the same form as in the Jacobi method, and setting $\tan\theta = a^{(l)}_{p,q-1}/a^{(l)}_{q,q-1}$ in this case, calculate $A^{(0)} = A$, $U^{(0)} = I$, $A^{(l+1)} = {}^tT^{(l)}A^{(l)}T^{(l)}$, $U^{(l+1)} = U^{(l)}T^{(l)}$ $(l = 0, 1, \ldots, N-1)$ and put $B = A^{(N)}$. Then we have $b_{ij} = 0$ for $i - j \geqslant 2$. We may solve the eigenvalue problem for this simplified B and then retransform the eigenvectors of B thus obtained into those of A by means of $U^{(N)}$. It should be noted that the method of bisection based on †Sturm's theorem is effectively used to solve the characteristic equation of a tridiagonal matrix.

In regard to the numerical computation of eigenvalues, there are many works on the estimation of upper and lower bounds for eigenvalues, the estimation of errors, etc. [3,4,5,12].

References

[1] H. Wayland, Expansion of determinantal equations into polynomial form, Quart. Appl. Math., 2 (1945), 277–306.
[2] P. S. Dwyer, Linear computation, John Wiley, 1951.
[3] R. Zurmühl, Matrizen, Springer, 1950.
[4] L. Collatz, Eigenwertaufgaben mit technischen Anwendungen, Geest & Portig, 1949.
[5] R. von Mises and H. Pollaczek-Geiringer, Praktische Verfahren der Gleichungsauflösung I, II, Z. Angew. Math. Mech., 9 (1929), 58–77, 152–164.
[6] E. Bodewig, Matrix calculus, North-Holland, 1956.
[7] C. G. J. Jacobi, Über ein leichtes Verfahren die in der Theorie der Säcularstörungen vorkommenden Gleichungen numerisch aufzulösen, J. Reine Angew. Math., 30 (1846), 51–94.
[8] C. Lanczos, An iteration method for the solution of the eigenvalue problem of linear differential and integral operators, J. Res. Nat. Bur. Standards, 45 (1950), 255–282.
[9] W. Givens, Computation of plane unitary rotations transforming a general matrix to triangular form, SIAM J. Appl. Math., 6 (1958), 26–50.
[10] J. H. Wilkinson, Householder's method for symmetric matrices, Numer. Math., 4 (1962), 354–361.
[11] P. A. White, The computation of eigenvalues and eigenvectors of a matrix, SIAM J. Appl. Math., 6 (1958), 393–437.
[12] S. H. Crandall, Engineering analysis, McGraw-Hill, 1956.
[13] A. Schönhage, Zur quadratischen Konvergenz des Jacobi-Verfahrens, Numer. Math., 6 (1964), 410–412.
[14] J. H. Wilkinson, The algebraic eigenvalue problem, Clarendon Press, 1965.
[15] T. Kato, Perturbation theory for linear operators, Springer, 1966.

137 (IX.17) Eilenberg-MacLane Complexes

A. General Remarks

Given an integer $n \geqslant 1$ and a group π (Abelian if $n \geqslant 2$), there exists an †arcwise connected topological space X for which the †homotopy groups $\pi_i(X)$ are trivial for $i \neq n$ and $\pi_n(X) \cong \pi$. Such a space is called an **Eilenberg-MacLane space** of type (π, n). Let $\Omega(X; X, *)$ be the †path space over X, and let p_0: $\Omega(X; X, *) \rightarrow X$ be the natural projection. If X is an Eilenberg-MacLane space of type (π, n), then $(\Omega(X; X, *), p_0, X)$ gives rise to a standard †contractible †fiber space whose fiber is an Eilenberg-MacLane space of type $(\pi, n-1)$. Assume that X is an Eilenberg-MacLane space of type (π, n) with Abelian group π. Then X is $(n-1)$-connected; hence the Hurewicz theorem can be utilized to show the existence of an isomorphism $h: \pi_n(X) \cong H_n(X)$, while the universal coefficient theorem can be utilized to show that $H^n(X; \pi) \cong \operatorname{Hom}(H_n(X), \pi)$. Since in this case we have $\pi_n(X) \cong \pi$, the element $h^{-1} \in \operatorname{Hom}(H_n(X), \pi_n(X))$ may be regarded as an element of $\operatorname{Hom}(H_n(X), \pi)$. Now the **fundamental class** of X is defined to be the cohomology class $u \in H^n(X; \pi)$ corresponding to h^{-1}. Let Y be a †CW complex and $\pi(Y; X)$ be the set of †homotopy classes of continuous mappings from Y to X. Then there exists a one-to-one correspondence $\pi(Y; X) \rightarrow H^n(Y; \pi)$ given by the assignment $[f] \rightarrow f^* u$. Let $S(X)$ be the †singular complex of X, and let $M(X)$ be a †minimal complex of $S(X)$. If X is an Eilenberg-MacLane space of type (π, n), $M(X)$ is isomorphic to a certain complex determined uniquely by π and n. This complex is called the **Eilenberg-MacLane complex** of type (π, n) and is denoted by $K(\pi, n)$. The notation $K(\pi, n)$ is also used to mean the space X itself.

B. $K(\pi, n)$

We are mainly concerned with the †category of †s. s. complexes and s. s. mappings, which we call the s. s. category.

Let $\Delta(q)$ be a simplicial complex whose simplexes are all subsets of $\{0, 1, \ldots, q\}$. Let $\varepsilon_i: \Delta(q-1) \rightarrow \Delta(q)$ be the simplicial mapping defined by $\varepsilon_i(j) = j$ $(0 \leqslant j \leqslant i-1)$, $\varepsilon_i(j) = j+1$ $(i \leqslant j \leqslant q-1)$, and let $\eta_i: \Delta(q+1) \rightarrow \Delta(q)$ be the mapping defined by $\eta_i(j) = j$ $(0 \leqslant j \leqslant i)$, $\eta_i(j) = j-1$ $(i+1 \leqslant j \leqslant q+1)$. Now $K(\pi, n)$ is a †Kan complex defined by $K(\pi, n)_q = Z^n(\Delta(q); \pi)$, $\partial_i \sigma = \sigma \circ \varepsilon_i$, $s_i \sigma = \sigma \circ \eta_i$, where $Z^0(\Delta(q); \pi) = \pi$, $Z^1(\Delta(q); \pi)$ is the set of functions defined on the set of pairs (i, j) such that $0 \leqslant i < j \leqslant q$ and satisfying the equality $\sigma(j, k) \cdot \sigma(i, k)^{-1} \cdot \sigma(i, j) = 1$ for $0 \leqslant i < j < k \leqslant q$, and $Z^n(\Delta(q); \pi)$ $(n \geqslant 2)$ is the group of alternating cocycles. If π is Abelian, the structure of the Abelian group $Z^n(\Delta(q); \pi)$ gives $K(\pi, n)$ the structure of an Abelian group in the s. s. category. This structure yields a one-to-one correspondence $K(\pi, n)_q \rightarrow K(\pi, n-1)_{q-1}$

$\times \ldots \times K(\pi,n-1)_0$ for $n \geqslant 1$ and leads to the expression of $\tau \in K(\pi,n)_q$ in the form $\langle \sigma_{q-1}, \ldots, \sigma_0 \rangle$ with $\sigma_i \in K(\pi,n-1)_i$. The **W-construction** of $K(\pi,n-1)$ for $n \geqslant 1$ is a Kan complex $W(\pi,n-1)$ defined by $W(\pi,n-1)_q = K(\pi,n-1)_q \times K(\pi,n)_q$ and $\partial_0(\sigma_q \times \tau_q) = (\partial_0 \sigma_q) \cdot \sigma_{q-1} \times \sigma_0 \tau_q$, $\partial_i(\sigma_q \times \tau_q) = \partial_i \sigma_q \times \partial \tau_q$ for $1 \leqslant i \leqslant q$, $s_i(\sigma_q \times \tau_q) = s_i \sigma_q \times s_i \tau_q$, where $\sigma_q \in K(\pi,n-1)_q$ and $\tau_q = \langle \sigma_{q-1}, \ldots, \sigma_0 \rangle \in K(\pi,n)_q$. Let $p: W(\pi,n-1) \to K(\pi,n)$ be a natural projection. Then $(W(\pi,n-1), p, K(\pi,n))$ plays the role of the †universal bundle for $K(\pi, n-1)$ in the s. s. category in the following sense: Let L be an s. s. complex, and let $f: L \to K(\pi,n)$ be an s. s. mapping. We define $f^{\#}W(\pi,n-1)$ to be the subcomplex of $W(\pi,n-1) \times L$ generated by simplexes $(\sigma_q \times \tau_q) \times \rho_q$ such that $\tau_q = f(\rho_q)$, where $\sigma_q \times \tau_q \in W(\pi,n-1)_q$ and $\rho_q \in L_q$. Let $p: f^{\#}W(\pi,n-1) \to L$ be the natural projection. Then $(f^{\#}W(\pi,n-1), p, L)$ is called the principal fiber bundle induced from $W(\pi,n-1)$ by f. Any principal bundle over L with group $K(\pi,n-1)$ can be expressed as an induced bundle, as before. This property means that $(W(\pi,n-1), p, K(\pi,n))$ is universal. On the other hand, we have an algebraic analog of the universal bundle for the chain group of $K(\pi,n-1)$, called the **bar construction** [1]. Both these concepts were defined by Eilenberg and MacLane in order to determine the structure of the (co)homology of $K(\pi,n)$, which is denoted by $(H^*(\pi,n))H_*(\pi,n)$. This object was later achieved by H. Cartan, who introduced an improved notion called Cartan construction (→ Appendix A, Table 6.III). Let $\pi(L, K(\pi,n))$ be the set of s. s. homotopy classes of s. s. mappings from L to $K(\pi,n)$. If π is Abelian, there exists a one-to-one correspondence $\pi(L, K(\pi,n)) \to H^n(L;\pi)$ given by the assignment $[f] \to f^*u$, where $u \in H^n(\pi,n;\pi)$ is the fundamental class of $K(\pi,n)$. By virtue of this correspondence, $k = f^*u \in H^n(L;\pi)$ determines an induced bundle $f^{\#}W(\pi,n-1)$ uniquely up to equivalence denoted by $K(\pi,n-1) \times_k L$.

Let X be a CW complex. Then the Postnikov system of X is an inverse system (X_n, p_n) $(n = 0,1,2,\ldots)$ consisting of topological spaces X_n, continuous mappings $p_n: X_n \to X_{n-1}$, and a system (X, q_n) $(n = 0,1,2,\ldots)$ consisting of continuous mappings $q_n: X \to X_n$ such that $p_n \circ q_n = q_{n-1}$ satisfies the following three properties: (1) X_0 is one point. (2) (X_n, p_n, X_{n-1}) is a †fiber space induced from a standard contractible fiber space over an Eilenberg-MacLane space of type $(\pi_n(X), n+1)$ by a mapping corresponding to a cohomology class $k^{n+1} \in H^{n+1}(X_{n-1}; \pi_n(X))$. (3) $q_{n*}: \pi_i(X) \to \pi_i(X_n)$ gives an isomorphism for $0 \leqslant i \leqslant n$. These cohomology classes k^{n+1} are called **Eilenberg-Postnikov invariants** (or simply **k-invariants**).

C. The Postnikov Complex

In this section, all spaces are assumed to be †simple (→ 205 Homotopy Groups). Let X be an arcwise connected topological space. Then the minimal complex $M(X)$ can be obtained as the inverse limit of a certain inverse system $(K(n), p(n))$ consisting of Kan complexes $K(n)$ and s. s. mappings $p(n): K(n) \to K(n-1)$ defined by $K(0) = K(0,0)$ and $K(n) = K(\pi_n, n) \times_{k^{n+1}} K(n-1)$ for $n \geqslant 1$, where $\pi_n = \pi_n(X)$ and $k^{n+1} \in H^{n+1}(K(n-1); \pi_n)$. This system is determined uniquely up to s. s. homotopy equivalence by its limit, called the **Postnikov complex** and denoted by $K(\pi_1, k^3, \pi_2, \ldots, k^{n+1}, \pi_n, \ldots)$. As yet we are ignorant of an effective method of computing the cohomology of a Postnikov complex from π_n and k^{n+1}.

D. Symmetric Product

Let X be an arcwise connected topological space. Let X^n be the Cartesian product of n copies of X. Clearly, the symmetric group $\mathfrak{S}_n$ of degree n operates on X^n. The n-fold **symmetric product** SP^nX of X is defined to be the quotient space of X^n under the action of $\mathfrak{S}_n$. If we specify a reference point of X, we have a natural inclusion $SP^{n-1}X \subset SP^nX$ and can consider the inductive limit space $\bigcup_{1 \leqslant n} SP^nX$, denoted by $SP^\infty X$. Then the Dold-Thom theorem shows that $M(SP^\infty X) \cong \prod_{i=1}^{\infty} K(H_i(X), i)$. In particular, we have $M(SP^\infty S^n) \cong K(\mathbf{Z}, n)$ for $n \geqslant 1$. This result can be applied to obtain a direct relationship between the axiomatic definition of †cohomology operations using $K(\pi,n)$ due to Eilenberg and Serre and the constructive definition using the symmetric groups due to Steenrod (A. Dold [6], T. Nakamura, [7]). For a detailed study of the (co)homology of SP^nX, the reader may refer to the works of M. Nakaoka [8].

References

[1] S. Eilenberg and S. MacLane, On the groups $H(\pi,n)$ I–III, Ann. of Math., (2) 58 (1953), 55–106; (2) 60 (1954), 49–139; (2) 60 (1954), 513–557.

[2] J.-P. Serre, Cohomologie modulo 2 des complexes d'Eilenberg-MacLane, Comment. Math. Helv., 27 (1953), 198–232.

[3] Séminaire H. Cartan, Ecole Norm. Sup., 1954–1955.

[4] A. Dold and R. Thom, Quasifaserungen und unendliche symmetrische Produkte, Ann. of Math., (2) 67 (1958), 239–281.
[5] E. H. Spanier, Algebraic topology, McGraw-Hill, 1966.
[6] A. Dold, Über die Steenrodschen Kohomologieoperationen, Ann. of Math., (2) 73 (1961), 258–294.
[7] T. Nakamura, On cohomology operations, Japan. J. Math., 33 (1963), 93–145.
[8] M. Nakaoka, Decomposition theorem for homology groups of symmetric groups, Ann. of Math., (2) 71 (1960), 16–42.

138 (XX.20) Einstein, Albert

Albert Einstein (March 14, 1879–April 18, 1955) was born of Jewish parents in the city of Ulm in southern Germany. He became a Swiss citizen soon after graduating from the Eidgenössische Technische Hochschule of Zürich in 1900. Afterwards, he obtained a position as examiner of patents at Bern, and while at that post, he published his theories on light quanta, †special relativity, and †Brownian motion. After briefly holding professorships at the University of Zürich and the University of Prague, he became a professor at the University of Berlin in 1913. His general theory of relativity was announced in 1916, and in 1921 he won the Nobel Prize in physics for his contributions to theoretical physics. To escape Nazi persecution, he fled to the United States in 1933, and until his retirement in 1945 he was a professor at the Institute for Advanced Study at Princeton. He advised President Roosevelt of the feasibility of constructing the atomic bomb, but after World War II, like others who had been connected with the bomb, he was active in promoting the nuclear disarmament movement and the establishment of a world government.

The theory of relativity raises fundamental epistemological problems concerning time, space, and matter. The results of the general theory were proved in 1919 by observations of the solar eclipse.

Through his latter years, Einstein continued to work on †unified field theory and on the generalization of relativity theory.

References

[1] P. Frank, Einstein, his life and time, Knopf, 1947.
[2] Ainsutain zensyû (Japanese; Collected works of Einstein) I–IV, Kaizôsya, 1922–1924.
[3] A. Einstein, The meaning of relativity, Princeton Univ. Press, fifth edition, 1956.
[4] C. Seelig, Albert Einstein, eine dokumentarische Biographie, Europa Verlag, 1954.

139 (XIX.11) Elasticity

A. General Remarks

Suppose that a solid body is deformed elastically by the action of external forces. We may inquire about the kind of forces and the state of deformation prevailing at each point of the body. The **theory of elasticity** studies this problem, assuming the body to be a continuum and utilizing classical mechanics as a basis. Because the theory is essentially a macroscopic approximation, the limits of its applicability to practical problems should be kept in mind. Moreover, it should be remembered that the proportionality relation between stress and strain mentioned in Section C is not always a valid assumption.

B. Stress

Consider a surface element with normal ν toward the interior of the body. The body is then divided into two parts by the surface. The force exerted across the surface per unit area by one part of the body on the other is called the **stress** on the surface and is represented by the vector $\mathbf{p}_\nu(p_{\nu x},p_{\nu y},p_{\nu z})$. The components of the stress vector normal and parallel to the surface are called the **normal stress** and **tangential stress**, respectively. The latter is also called the **shearing stress**. At any point in the body, $\mathbf{p}_\nu$ is in equilibrium with the three stress vectors $\mathbf{p}_x(p_{xx},p_{xy},p_{xz})$, $\mathbf{p}_y(p_{yx},p_{yy},p_{yz})$, $\mathbf{p}_z(p_{zx},p_{zy},p_{zz})$ on the mutually perpendicular planes passing through that point. Thus:

$$\mathbf{p}_{y_\nu}=\mathbf{p}_x\cos(\nu,x)+\mathbf{p}_y\cos(\nu,y)+\mathbf{p}_z\cos(\nu,z).$$

This can be expressed in the form of a matrix product as

$$\begin{bmatrix} p_{\nu x}\\ p_{\nu y}\\ p_{\nu z}\end{bmatrix}=\begin{bmatrix} p_{xx} & p_{yx} & p_{zx}\\ p_{xy} & p_{yy} & p_{zy}\\ p_{xz} & p_{yz} & p_{zz}\end{bmatrix}\begin{bmatrix}\cos(\nu,x)\\ \cos(\nu,y)\\ \cos(\nu,z)\end{bmatrix}.$$

The nine quantities $(p_{xx},p_{xy},\ldots,p_{zz})$ form a †tensor called the **stress tensor**. Since it can be shown that $p_{yz}=p_{zy}$, $p_{zx}=p_{xz}$, and $p_{xy}=p_{yx}$, the stress tensor is a †symmetric tensor. It is usually represented by

$$\begin{bmatrix} \sigma_x & \tau_z & \tau_y \\ \tau_z & \sigma_y & \tau_x \\ \tau_y & \tau_x & \sigma_z \end{bmatrix},$$

where σ and τ are the normal and tangential stress, respectively. The †divergence of the stress tensor and the body force (including inertial force for the case of a moving body) are in equilibrium, as expressed by the equations of equilibrium. Since it is symmetric, the stress tensor must have at least one set of principal directions, in which all the τ-components vanish and the remaining σ-components are called **principal stresses**. The values of the principal stresses are equal to the roots of the cubic †characteristic equation of the σ-matrix.

C. Strain

If the displacements of two arbitrary neighboring points $\mathbf{x}(x,y,z)$, $\mathbf{x}+\delta\mathbf{x}$ of the body are denoted by $\mathbf{u}(u,v,w)$ and $\mathbf{u}+\delta\mathbf{u}$, the relative displacement $\delta\mathbf{u}(\delta u,\delta v,\delta w)$ is expressed as

$$\delta\mathbf{u}=(\partial\mathbf{u}/\partial\mathbf{x})(\delta\mathbf{x}),$$

if the second- and higher-order terms in $\delta\mathbf{x}$ are neglected. The tensor $(\partial\mathbf{u}/\partial\mathbf{x})$ on the right-hand side can be decomposed into symmetric and †alternating parts:

$$\frac{1}{2}\begin{bmatrix} 2\varepsilon_x & \gamma_z & \gamma_y \\ \gamma_z & 2\varepsilon_y & \gamma_x \\ \gamma_y & \gamma_x & 2\varepsilon_z \end{bmatrix}+\frac{1}{2}\begin{bmatrix} 0 & -\omega_z & \omega_y \\ \omega_z & 0 & -\omega_x \\ -\omega_y & \omega_x & 0 \end{bmatrix},$$

where $\varepsilon_x=\partial u/\partial x, \ldots$; $\gamma_x=\partial w/\partial y+\partial v/\partial z$, $\ldots$; $\omega_x=\partial w/\partial y-\partial v/\partial z, \ldots$. The second part represents a rigid-body rotation (given by the vector $\omega/2$). The first part represents the deformation of the body and is called the **strain tensor**. It is given by the changes in lengths and angles within the body. Thus ε_x, ε_y, ε_z represent the extension of lines, while γ_x, γ_y, γ_z represent the changes of angles between lines originally at right angles. However, in the principal directions of the strain tensor, only changes in length and not angular changes occur. Furthermore, from the relations between the displacement (u,v,w) and strain components $(\varepsilon_x,\ldots,\gamma_x,\ldots)$, six identities (compatibility conditions) connecting the latter can be derived. For studying large deformations, quadratic terms in the displacement must be taken into consideration.

If σ and ε obey the proportionality relation (**Hooke's law**) and the principal axes of the stress and strain are assumed to coincide with each other, then the relations between normal stress and extension in isotropic bodies can be expressed as

$$E\varepsilon_x=\sigma_x-\nu(\sigma_y+\sigma_z), \ldots,$$

where E is Young's modulus and ν is Poisson's ratio. For isotropic bodies, we need only two elastic constants. However, for the sake of convenience, another constant $G=E/2(1+\nu)$ is also used. Thus the relations between shearing stress and strain can be expressed by $\tau_x=G\gamma_x, \ldots$. Since the strain components are expressible as linear forms in the displacements, the stress components can also be expressed as linear forms in u, v, w, and therefore the **equations of equilibrium** can be rewritten as

$$\Delta u+\frac{1}{1-2\nu}\frac{\partial e}{\partial x}+\frac{X}{G}=0, \ldots,$$

where

$$e=\frac{\partial u}{\partial x}+\frac{\partial v}{\partial y}+\frac{\partial w}{\partial z}.$$

For the case of bodies in motion, $X, \ldots$ should be replaced by $X-\rho\ddot{u}, \ldots$, where ρ is the density and $\ddot{u}$ the acceleration.

D. The Energy Principle

When a small variation is given to the displacement, the influence of the external force (together with the inertial force) appears, and hence a change in the amount of energy contained in the body occurs. To the latter change, the heat transfer that may occur simultaneously also contributes. These changes should satisfy the energy equation. Neglecting the fact that deformation is usually not perfectly elastic, the change of internal energy is equal to that of elastic strain energy for the case of slow (isothermal) deformation. For oscillatory (adiabatic) deformation, the temperature change causes a less simple state of affairs. However, in both cases, except for the imperfect elasticity, the strain energy function can be determined, which leads to such important results as the uniqueness of solutions, the minimum energy principle, and the reciprocal theorem concerning work and displacement.

For treating concrete problems such as those concerning beams and plates, we may also use stress functions, instead of working only in terms of displacements. In either case, the usual method involves †boundary value problems of differential equations. It is, however, also possible to use †direct methods in the calculus of variations based on the minimum energy principle.

E. Symbols

Various symbols are used for the stress and strain. For example, σ, τ and ε, γ are in common use in the engineering literature, where the components of $\mathbf{p}_x$ are denoted by τ_{xx}, τ_{xy},

τ_{xz}. In A. E. H. Love's treatise [1], they are written as X_x, Y_x, Z_x, while e_{xx}, e_{yz} are used for ε_x, γ_x, respectively, although the components of the strain tensor are always written as $\gamma_x/2$, $e_{yz}/2$. Also, elastic constants are given various symbols. While E is widely used, G is less common, and Love employs Lamé's symbol μ for G. He also uses the symbol σ for Poisson's ratio, which is denoted by ν in this article to avoid confusion with stress. It may be added that, in the engineering literature, the reciprocal of Poisson's ratio m is frequently used and called the "Poisson number."

References

[1] A. E. H. Love, A treatise on the mathematical theory of elasticity, Cambridge Univ. Press, fourth edition, 1934.

[2] I. S. Sokolnikoff, Mathematical theory of elasticity, McGraw-Hill, second edition, 1956.

140 (XIX.18) Electromagnetism

A. Maxwell's Equations

Mathematical formulation of electromagnetics leads to †initial value and †boundary value problems for Maxwell's equations according to the geometric nature of the medium. **Maxwell's equations** for a vacuum are written in the following form:

$$\varepsilon_0 \partial \mathbf{E}/\partial t = \operatorname{rot}\mathbf{H} - \mathbf{J}_e, \qquad \varepsilon_0 \operatorname{div}\mathbf{E} = \rho_e,$$
$$\mu_0 \partial \mathbf{H}/\partial t = -\operatorname{rot}\mathbf{E} - \mathbf{J}_m, \qquad \mu_0 \operatorname{div}\mathbf{H} = \rho_m, \tag{1}$$

where $\mathbf{E}$ and $\mathbf{H}$ are the electric and magnetic field vectors, ρ_e and ρ_m the electric and magnetic charge densities, $\mathbf{J}_e$ and $\mathbf{J}_m$ the electric and magnetic current densities, ε_0 and μ_0 are constants, and the quantity $1/\sqrt{\varepsilon_0\mu_0} = c$ is the speed of light in a vacuum (2.99797×10^8 m/s). Charge and current densities must satisfy the equations of continuity

$$\partial \rho_e/\partial t + \operatorname{div}\mathbf{J}_e = 0,$$
$$\partial \rho_m/\partial t + \operatorname{div}\mathbf{J}_m = 0. \tag{2}$$

Since it is empirically accepted that $\rho_m = 0$ and $\mathbf{J}_m = 0$ in the actual situation, we henceforward set them equal to zero. This causes an asymmetry between the electric and magnetic quantities. On the other hand, the proposition "$\rho_m = 0$ and $\mathbf{J}_m = 0$" cannot be deduced from the classical theory itself.

In the presence of matter, additional charge and current appear due to the electric and magnetic polarizations $\mathbf{P}$ and $\mathbf{M}$ of the material. Therefore, in this case, it is necessary to make the following substitutions in (1):

$$\rho_e \to \rho = \rho_e - \operatorname{div}\mathbf{P},$$
$$\mathbf{J}_e \to \mathbf{J} = \mathbf{J}_e + \partial \mathbf{P}/\partial t + \operatorname{rot}\mathbf{M},$$
$$\mathbf{H} \to \mathbf{H} + \mathbf{M}. \tag{3}$$

Moreover, if we define the **electric flux density** (or **electric displacement**) $\mathbf{D}$ and the **magnetic flux density** (or **magnetic induction**) $\mathbf{B}$ by

$$\mathbf{D} = \varepsilon_0\mathbf{E} + \mathbf{P}, \qquad \mathbf{B} = \mu_0(\mathbf{H} + \mathbf{M}), \tag{4}$$

then Maxwell's equations (1) are transformed into

$$\partial \mathbf{D}/\partial t = \operatorname{rot}\mathbf{H} - \mathbf{J}_e, \qquad \operatorname{div}\mathbf{D} = \rho_e,$$
$$\partial \mathbf{B}/\partial t = -\operatorname{rot}\mathbf{E}, \qquad \operatorname{div}\mathbf{B} = 0. \tag{5}$$

In the electromagnetic field in a vacuum there is energy with a density

$$u = (\varepsilon_0/2)\mathbf{E}^2 + (\mu_0/2)\mathbf{B}^2 \tag{6}$$

and energy flux with a density expressed by the **Poynting vector**

$$\mathbf{S} = \mathbf{E}\times\mathbf{H}. \tag{7}$$

Between these quantities the following relation holds:

$$\partial u/\partial t + \operatorname{div}\mathbf{S} = 0. \tag{8}$$

An electric charge q moving with velocity $\mathbf{v}$ in an electromagnetic field is subject to the force (**Lorentz force**)

$$\mathbf{F} = q\mathbf{E} + q\mathbf{v}\times\mathbf{B}. \tag{9}$$

This force can be interpreted as being caused by the **Maxwell stress tensor**

$$T_{ik} = (\varepsilon_0/2)(-E_iE_k + 2\delta_{ik}\mathbf{E}^2)$$
$$+ (\mu_0/2)(-H_iH_k + 2\delta_{ik}\mathbf{H}^2). \tag{10}$$

By introducing the **scalar potential** V and the **vector potential** $\mathbf{A}$, we can express the field vectors as follows:

$$\mathbf{B} = \operatorname{rot}\mathbf{A}, \qquad \mathbf{E} = -\operatorname{grad}V - \partial\mathbf{A}/\partial t. \tag{11}$$

Furthermore, if we impose an auxiliary condition (**Lorentz condition**)

$$\varepsilon_0\mu_0 \operatorname{div}\mathbf{A} + \partial V/\partial t = 0, \tag{12}$$

then we obtain from (1) the wave equations

$$\Box V = -\rho_e/\varepsilon_0, \qquad \Box\mathbf{A} = -\mu_0\mathbf{J}_e, \tag{13}$$

where $\Box \equiv \Delta - \varepsilon_0\mu_0\partial^2/\partial t^2$ is called the **d'Alembertian** and is sometimes written as $\diamondsuit^2$. From (13) we conclude that the electromagnetic field can propagate in a vacuum as a wave with speed $c = 1/\sqrt{\varepsilon_0\mu_0}$. In †quantum theory the potentials V and $\mathbf{A}$ are regarded as being more fundamental than $\mathbf{E}$ and $\mathbf{H}$ themselves. However, they are not uniquely determined, in the sense that the **gauge transformation**

$$V \to V + \partial\psi/\partial t, \qquad \mathbf{A} \to \mathbf{A} - \operatorname{grad}\psi \tag{14}$$

with an arbitrary function ψ of the space and

time variables does not affect the fields.

Maxwell's equations are invariant under the †Lorentz transformation. Therefore they can be written in 4-dimensional tensor form. In this formulation V and $\mathbf{A}$ constitute a 4-vector $A_i:(A_x,A_y,A_z,(i/c)V)$, while ρ_e and $\mathbf{J}_e$ constitute another 4-vector J_i: $(J_{ex},J_{ey},J_{ez},-ic\rho_e)$. Furthermore, the components of $\mathbf{E}$ and $\mathbf{B}$ are regarded as those of an †antisymmetric tensor of the second order:

$$f_{ik}:\begin{bmatrix} 0 & B_z & -B_y & iE_x/c \\ -B_z & 0 & B_x & iE_y/c \\ B_y & -B_x & 0 & iE_z/c \\ -iE_x/c & -iE_y/c & -iE_z/c & 0 \end{bmatrix}. \tag{15}$$

Since it has six independent components, the tensor f_{ik} is sometimes called a **6-vector**. With this notation Maxwell's equations (1) are written in the form

$$\sum_k \partial f_{ik}/\partial x_k=\mu_0 J_i,$$

$$\partial f_{ij}/\partial x_k+\partial f_{jk}/\partial x_i+\partial f_{ki}/\partial x_j=0, \tag{16}$$

while equations (11) and (12) become $f_{ik}=\partial A_k/\partial x_i-\partial A_i/\partial x_k$ and $\Sigma\partial A_i/\partial x_i=0$, respectively.

We can regard Maxwell's equations as the †wave equations for †bosons with spin 1 (†photons). The equations of quantum electrodynamics are obtained if we regard the field quantities as quantum-mechanical variables (**q-numbers**) and then perform the †second quantization.

B. Concrete Problems

In solving Maxwell's equations concretely, we usually make additional assumptions for polarizations, electric current, and field vectors

$$\mathbf{P}=\chi_e\mathbf{E},\qquad \mathbf{M}=\chi_m\mathbf{H},\qquad \mathbf{J}_e=\sigma\mathbf{E}. \tag{17}$$

Then we have

$$\mathbf{D}=\varepsilon\mathbf{E},\qquad \mathbf{B}=\mu\mathbf{H}, \tag{18}$$

where $\varepsilon=\varepsilon_0+\chi_e$, $\mu=\mu_0(1+\chi_m)$. Therefore, the equations (5) become identical with (1) if ε_0 and μ_0 are replaced by ε and μ, respectively. (ρ_m and $\mathbf{J}_m$ are set equal to zero.)

Some cases of practical importance are given below.

1. Electrostatics. If the fields are time-independent and there is no electric current, then $\mathbf{E}$ and $\mathbf{H}$ are mutually independent. The static electric field is calculated from the solution of the boundary value problem of the †Poisson equation $\Delta V=-\rho_e/\varepsilon$. Specifically, V takes a constant value in each conductor.

2. Magnetostatics. For zero electric current, the problem of magnetostatics is solved in the same way as in electrostatics. For the case of nonvanishing stationary electric current, the problem is reduced to that of solving

$$\Delta\mathbf{A}=-\mu\mathbf{J}_e,\qquad \operatorname{div}\mathbf{A}=0. \tag{19}$$

3. Quasistationary Electric Circuit. The problem appearing most often in electrical engineering is that of a quasistationary circuit. Its characteristic feature is that the electric currents exist only in the circuit elements (inductors, capacitors, and resistors) and in the lines connecting them. The current (both $\mathbf{J}_e$ and $\partial\mathbf{D}/\partial t$) may be neglected in all other parts of the system. (This could be compared to the situation in dynamics where we consider systems of material points or of rigid bodies having a finite number of degrees of freedom, although every material body is essentially a continuum.) The network system is constructed as a †linear graph with the circuit elements as its branches. Topological †network theory deals with the relation between the structure of the linear graph and the electrical characteristics of the network, whereas function-theoretic network theory deals with the relation between current and voltage at each part of the network. In the latter theory, current and voltage are considered as functions of the frequency of the sinusoidal alternating voltage applied to some point of the network. Both constitute a unique theoretical system in engineering mathematics for designing a network system in accordance with engineering demand (→ 278 Networks; 127 Dispersion Relations).

4. Theory of Electromagnetic Waves. The theory of electromagnetic waves deals with the case where the changes of all field quantities are proportional to $e^{i\omega t}$, and in addition the frequency ω is so large that the term $\partial\mathbf{D}/\partial t$ in (5) is of the same order as or larger than $\mathbf{J}_e$. In such a situation the electromagnetic field behaves like a wave.

Problems of various types arise depending on the geometry of the conducting and dielectric substances, on the type of the energy source, etc. Important problems are: (i) radiation of a wave from a point source into free space; (ii) scattering of a plane wave by small bodies or cylinders; (iii) diffraction of a wave through holes in a conducting plate; (iv) reflection and refraction of a wave at the boundary between different media; (v) wave propagation along a conducting tube (wave guide); and (vi) resonance of the electromagnetic field in a cavity surrounded by a conducting substance. Theoretical treatment similar to that for ordinary networks is possible for microwave circuits consisting of wave guides, cavities, etc.

References

[1] H. Takahasi, Denzikigaku (Japanese; Electromagnetism), Syôkabô, 1961.
[2] J. A. Stratton, Electromagnetic theory, McGraw-Hill, 1941.
[3] B. I. Bleaney, Electricity and magnetism, Clarendon Press, 1957.
[4] E. Kähler, Bemerkungen über die Maxwellschen Gleichungen, Abh. Math. Sem. Univ. Hamburg, 12 (1938), 1–28.
[5] R. Fano, L. J. Chu, and R. B. Adler, Electromagnetic fields, energy and forces, John Wiley, 1960.
[6] E. G. Hallen, Electromagnetic theory, Chapman & Hall, 1962.
[7] Л. Д. Ландау, Е. М. Лифшиц (L. D. Landau and E. M. Lifshitz), Электродинамика сплошных сред, физматгиз, 1959; English translation, Electrodynamics of continuous media, Pergamon, 1960.

141 (X.8) Elementary Functions

A. Definition

A function of a finite number of real or complex variables that is †algebraic, exponential, logarithmic, trigonometric, or inverse trigonometric, or the composite of a finite number of them, is called an **elementary function**. This is the most common type of function in elementary calculus.

J. Liouville (*J. Ecole Polytech.*, 23 (1834); *J. Reine Angew. Math.*, 13 (1835)) defined the elementary functions as follows: An algebraic function of a finite number of complex variables is called an elementary function of class 0. Then e^z and $\log z$ are called elementary functions of class 1. Inductively, we define the notion of elementary functions of class n under the assumption that the notion of elementary functions of class at most $n-1$ has already been defined. Let $g(t)$ and $g_j(w_1, \ldots, w_n)$ $(1 \leqslant j \leqslant m)$ be elementary functions of class at most 1 and $f(z_1, \ldots, z_m)$ be an elementary function of class at most $n-1$. Then the composite functions $g(f(z_1, \ldots, z_m))$ and $f(g_1(w_1, \ldots, w_n), \ldots, g_m(w_1, \ldots, w_n))$ (and only such functions) are called the elementary functions of class at most n. An elementary function of class at most n and not of class at most $n-1$ is called an **elementary function of class** n. A function that is an elementary function of class n for some integer n is called an **elementary function**. In this article, we explain the properties of the most common elementary functions.

B. Exponential and Logarithmic Functions of a Real Variable

Let $a>0$, $a \neq 1$. A function $f(x)$ of a real variable satisfying the functional relation

$$f(x+y)=f(x)f(y), \qquad f(1)=a \tag{1}$$

satisfies $f(n)=a^n$ for positive integers n and $f(-n)=1/a^n$ for negative integers $-n$. In general, $f(n/m)=\sqrt[m]{a^n}$ for every rational number $r=n/m$. If we assume that $f(x)$ is continuous, then there is a unique strictly monotone function $f(x)$ defined in $(-\infty, \infty)$ whose range is $(0, \infty)$. The function $f(x)$ is called the **exponential function with the base** a and is denoted by a^x, read "a to the power x" and also called a **power** of a with **exponent** x. Its inverse function is called the **logarithmic function to the base** a, and is denoted by $\mathbf{log}_a x$. The specific value $\log_a x$ is called the **logarithm** of x to the **base** a. If $g(x)=\log_a x$, we have

$$g(xy)=g(x)+g(y), \qquad g(a)=1. \tag{2}$$

Hence we have $xy=f(g(x)+g(y))$. Therefore, we can compute the product of x and y by adding $g(x)$ and $g(y)$ using a numerical table of the function $g(x)$.

C. Logarithmic Computation

The logarithm to the base 10 is called the **common logarithm**. If two numbers x, y expressed in the decimal system differ only in the position of the decimal point (i.e., $y=x \cdot 10^n$ for an integer n), they share the same fractional parts in their common logarithms. The integral part of the common logarithm is called the **characteristic**, and the fractional part is called the **mantissa**. (We note that in recent years, the word "mantissa" is frequently used for the fractional part a in the †floating point representation $x=a \cdot 10^n$, $10^{-1} \leqslant a<1$, or $1 \leqslant a<10$.) The common logarithms of integers have been computed and published in tables.

D. Derivatives of Exponential and Logarithmic Functions

The function $f(x)=a^x$ is differentiable, and $f'(x)=k_a f(x)$, where k_a is a constant determined by the base a. If we take the base a to be

$$e=\lim_{\nu \to \infty}\left(1+\frac{1}{\nu}\right)^{\nu}=\sum_{\nu=0}^{\infty}\frac{1}{\nu!}=2.71828\ldots,$$

then we have $k_e = 1$. The number $\Sigma(1/\nu!)$ is usually called **Napier's number** and is denoted by e after L. Euler (see his letter to C. Goldbach of 1731; → Appendix B, Table 6). In 1873, C. Hermite proved that e is a transcendental number. We sometimes denote e^x by $\mathbf{exp}\, x$; the term **exponential function** usually means the function $\exp x$. The function e^x is invariant under differentiation, and conversely, a function invariant under differentiation necessarily has the form Ce^x. The logarithm to the base e is called the **Napierian logarithm** (or **natural logarithm**), and we usually denote it by $\mathbf{log}\, x$ without explicitly naming the base e (sometimes it is denoted by $\ln x$). The derivative of $\log x$ is $1/x$, hence we have the integral representation:

$$\log x = \int_1^x \frac{dx}{x}. \tag{3}$$

The constant factor k_a in the derivative of a^x is equal to $\log a$. The graphs of $y = e^x$ and $y = \log x$ are shown in Fig. 1. The functions e^x and $\log(1+x)$ are expanded into the following Taylor series at $x = 0$:

$$e^x = \sum_{\nu=0}^{\infty} \frac{x^\nu}{\nu!}, \tag{4}$$

$$\log(1+x) = \sum_{\nu=1}^{\infty} (-1)^{\nu+1} \frac{x^\nu}{\nu}. \tag{5}$$

The power series in the right-hand side of (4) and (5) are called the **exponential series** and the **logarithmic series**, respectively. The radii of convergence of (4) and (5) are ∞ and 1, respectively.

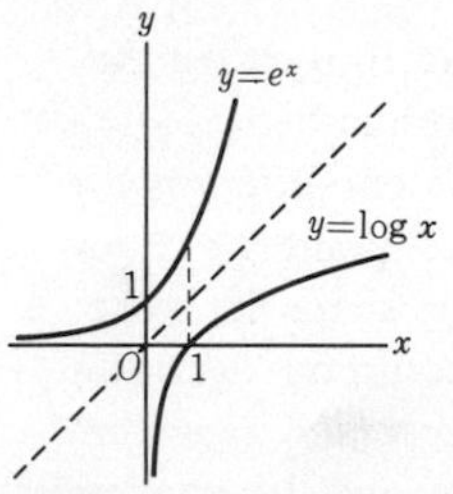

Fig. 1

E. Trigonometric and Inverse Trigonometric Functions of a Real Variable

The trigonometric functions of a real variable x are the functions $\mathbf{sin}\, x$, $\mathbf{cos}\, x$ (→ 417 Trigonometry) and the functions $\mathbf{tan}\, x = \sin x/\cos x$, $\mathbf{cot}\, x = \cos x/\sin x$, $\mathbf{sec}\, x = 1/\cos x$, and $\mathbf{cosec}\, x = 1/\sin x$ derived from $\sin x$ and $\cos x$. The derivatives of $\sin x$ and $\cos x$ are $\cos x$ and $-\sin x$, respectively. They have the following Taylor expansions at $x = 0$:

$$\sin x = \sum_{\nu=0}^{\infty} \frac{(-1)^\nu}{(2\nu+1)!} x^{2\nu+1}, \tag{6}$$

$$\cos x = \sum_{\nu=0}^{\infty} \frac{(-1)^\nu}{(2\nu)!} x^{2\nu}. \tag{7}$$

The radii of convergence of (6) and (7) are both ∞.

The inverse functions of $\sin x$, $\cos x$, and $\tan x$ are the **inverse trigonometric functions** and are denoted by $\mathbf{arc\,sin}\, x$, $\mathbf{arc\,cos}\, x$, and $\mathbf{arc\,tan}\, x$, respectively. (Instead of this notation, $\mathbf{sin}^{-1} x$, $\mathbf{cos}^{-1} x$, and $\mathbf{tan}^{-1} x$ are also used.) These functions are infinitely multiple-valued, as shown in Fig. 2. But if we restrict their ranges within the part shown by solid lines in Fig. 2, they are considered single-valued functions. To be more precise, we restrict the range as follows: $-\pi/2 \leqslant \arcsin x \leqslant \pi/2$, $0 \leqslant \arccos x \leqslant \pi$, $-\pi/2 < \arctan x < \pi/2$.

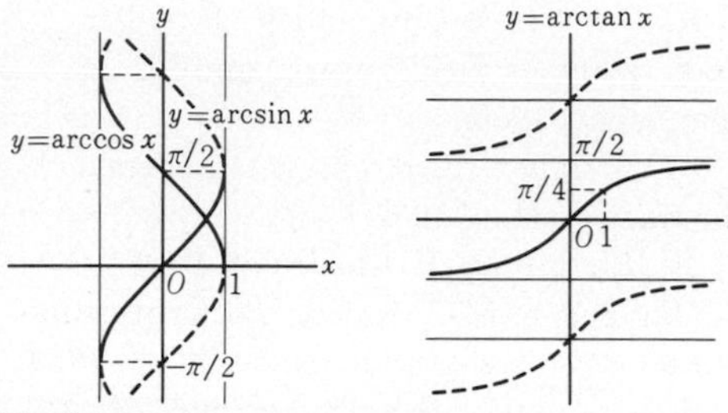

Fig. 2

The functions having these ranges are called the **principal values** and are sometimes denoted by $\mathbf{Arc\,sin}\, x$, $\mathbf{Arc\,cos}\, x$, and $\mathbf{Arc\,tan}\, x$, respectively. The derivatives of these functions are $(1-x^2)^{-1/2}$, $-(1-x^2)^{-1/2}$, $(1+x^2)^{-1}$, respectively (→ Appendix A, Table 9.I; for the Taylor or Laurent expansions of $\tan x$, $\cot x$, $\sec x$, $\operatorname{cosec} x$, $\arcsin x$, $\arccos x$, $\arctan x$, etc., → Appendix A, Table 10.IV).

F. Hyperbolic Functions

Let P be a point on the branch of the hyperbola $x^2 - y^2 = 1$, $x > 0$, and let O be the origin and A the vertex $(1, 0)$ of the hyperbola. Denote by $\theta/2$ the area of the domain surrounded by the line segments OA, OP, and the arc $\widehat{AP}$ of the hyperbola. Then we define the coordinates of P to be $(\cosh\theta, \sinh\theta)$ as functions of θ. We have

$$\mathbf{cosh}\, x = (e^x + e^{-x})/2,$$

$$\mathbf{sinh}\, x = (e^x - e^{-x})/2, \tag{8}$$

called the **hyperbolic cosine** and **hyperbolic sine**, respectively. As in the case of trigonometric functions, we define the **hyperbolic tangent** by $\mathbf{tanh}\, x = \sinh x/\cosh x$, the **hyperbolic cotangent** by $\mathbf{coth}\, x = \cosh x/\sinh x$, the **hyperbolic secant** by $\mathbf{sech}\, x = 1/\cosh x$, and the **hyperbolic cosecant** by $\mathbf{cosech}\, x = 1/\sinh x$. They are called the **hyperbolic functions**. The graphs of $\sinh x$ and $\cosh x$ are shown in Fig. 3. The trigonometric functions are sometimes called **circular functions**.

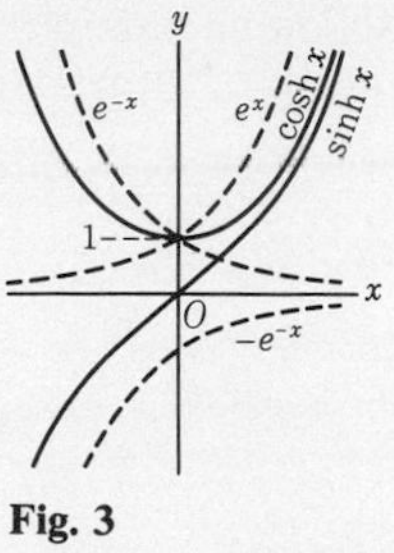

Fig. 3

We now introduce the **Gudermannian** (or **Gudermann function**):

$$\theta = \mathrm{gd}\, u = 2 \arctan e^{u} - \pi/2,$$

$$u = \mathrm{gd}^{-1}\theta = \log[\tan\theta + \sec\theta] = \frac{1}{2}\log\frac{1+\sin\theta}{1-\sin\theta}.$$

Then the hyperbolic functions can be expressed in terms of the trigonometric functions. For example,

$$\sinh u = \tan\theta, \quad \cosh u = \sec\theta, \quad \tanh u = \sin\theta.$$

G. Elementary Functions of a Complex Variable (→ Appendix A, Table 10)

1. Exponential Function. The power series (4) converges for all finite values if we replace x by the complex variable z and gives an †entire function of z with an †essential singularity at the point at infinity. This is the exponential function e^{z} of a complex variable z. It satisfies the **addition formula** (1), $e^{z_1+z_2} = e^{z_1}e^{z_2}$, and it is also the †analytic continuation of the exponential function of a real variable. For a purely imaginary number $z = iy$, we have the **Euler formula**

$$e^{iy} = \cos y + i \sin y. \qquad (9)$$

The function $w = e^{z}$ gives a †conformal mapping from the z-plane to the w-plane, as shown in Fig. 4, which maps the imaginary axis of the z-plane onto the unit circle of the w-plane ($w = u + iv$). For $z = x + iy$ (x,y are real numbers), we have $e^{z} = e^{x}e^{iy}$; hence e^{z} is a †simply periodic function with fundamental period $2\pi i$.

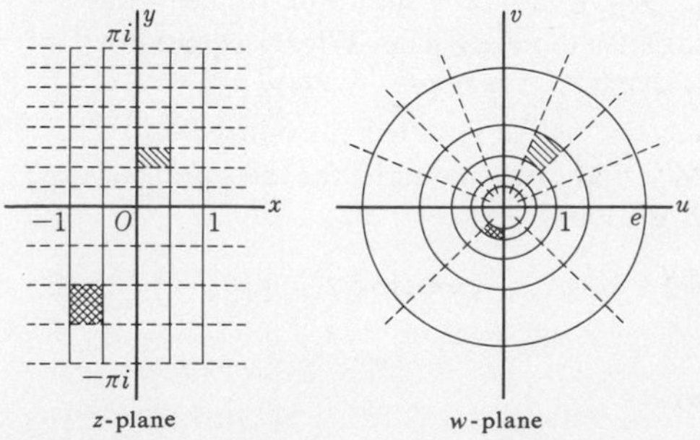

Fig. 4

2. Logarithmic Function. The logarithmic function $\log z$ of a complex variable z is the inverse function of e^{z}. It is an infinitely multiple-valued analytic function that has †logarithmic singularities at $z=0$ and $z=\infty$. All possible values are expressed by $\log z + 2n\pi i$ (n is an arbitrary integer), where we select a suitable value $\log z$. The **principal value** of $\log z$ is usually taken as $\log r + i\theta$, where $z = re^{i\theta}$ ($r = |z|$, θ is the argument of z) and $0 \leqslant \theta < 2\pi$. (Sometimes the range of the argument is taken as $-\pi < \theta \leqslant \pi$.) The principal value of $\log z$ is sometimes denoted by $\mathbf{Log}\, z$. The power series (5) gives one of its †functional elements. The integral representation (3) holds for a complex variable z. The multivalency of $\log z$ results from the selection of a contour of integration; the integral of $1/z$ around the origin is $2\pi i$, which is the increment of $\log z$.

3. Power. The exponential function a^{z} for an arbitrary complex number a is defined to be $\exp(z \log a)$. Similarly, z^{a} is defined to be $\exp(a \log z)$. The function z^{a} is an algebraic function if and only if a is rational. In other cases, the function z^{a} is an elementary function of class 2.

4. Trigonometric Functions, Inverse Trigonometric Functions, Hyperbolic Functions. The trigonometric, inverse trigonometric, and hyperbolic functions of a complex variable are defined by the analytic continuations of the corresponding functions of a real variable. For example, $\sin z$ and $\cos z$ are defined by the power series (6) and (7), respectively. They are entire functions whose zero points are $n\pi$ and $(n-\frac{1}{2})\pi$ (n is an integer), respectively. They are also represented by †Weierstrass's infinite product (→ Appendix A, Table 10.VI).

The functions $\tan z$, $\cot z$, $\sec z$, and $\operatorname{cosec} z$ are †meromorphic functions of z on the complex z-plane, and they are expressed by †Mittag-Leffler partial fractions (→ Appendix A, Tables 10.IV, 10.V). As can be shown from (8) and (9), we have

$$\cos z = (e^{iz} + e^{-iz})/2, \quad \sin z = (e^{iz} - e^{-iz})/2i,$$

$$\cosh z = \cos iz, \quad \sinh z = (\sin iz)/i. \qquad (10)$$

The formula (10) is also called the **Euler formula**. For a complex variable, the trigonometric and hyperbolic functions are composites of exponential functions, the inverse trigonometric functions are composites of logarithmic functions, and all of them are elementary functions of class 1. The definition of elementary functions by Liouville described in Section A refers, of course, to the functions of a complex variable. We remark that the inverse function of an elementary function is not necessarily an elementary function. For example, the inverse function of $y = x - a \sin x$ is not an elementary function (→ 303 Orbit Determination B).

The derivative of an elementary function is also an elementary function. However, the †primitive function of an elementary function is not necessarily an elementary function. The primitive function of a rational function or an algebraic function of †genus 0 is again an elementary function. Similar properties hold for rational functions of trigonometric functions. Liouville carried through a deep investigation of the situation where the integral of an elementary function is also an elementary function.

References

[1] T. Takagi, Kaiseki gairon (Japanese; A course of analysis), Iwanami, third edition, 1961.

[2] N. Bourbaki, Eléments de mathématique, Fonctions d'une variable réele, ch. 3, Fonctions élémentaires, Actualités Sci. Ind., 1074b, Hermann, second edition, 1958.

[3] R. Kurokawa, Theorems of Liouville on the elementary functions I, II, III (Japanese), Sûbutu kaisi (Bull. Physico-Math. Soc. Japan), 1 (1924), 117–146.

Also → references to 109 Differential Calculus.

142 (XIX.32) Elementary Particles

A. General Remarks

The ultimate particles of matter are called **elementary particles.** They are classified into **photons; leptons,** including **electrons** and μ-**particles; mesons;** and **baryons** (Table 1). Each particle is characterized by charge, intrinsic angular momentum (called spin), and other attributes. The **theory of elementary particles** deals with these properties and interactions among particles.

Table 1

	Isomultiplet	I	S	n_B
Baryons	(p, n)	1/2	0	1
	Λ	0	-1	
	$(\Sigma^+,\Sigma^0,\Sigma^-)$	1	-1	
	(Ξ^0,Ξ^-)	1/2	-2	
Mesons	(K^+,K^0)	1/2	1	0
	η	0	0	
	(π^+,π^0,π^-)	1	0	
	$(\bar{K}^0,K^-)$	1/2	-1	

B. Representations by the Lorentz Group

The most fundamental group in the theory of elementary particles is the †inhomogeneous Lorentz group $\bar{G}$. The corresponding †Lie algebra is generated by the following infinitesimal generators M_{kl} and P_k $(k,l=1,2,3,4)$:

$$[M_{kl},M_{mn}]$$
$$=i(g_{lm}M_{kn}+g_{kn}M_{lm}-g_{km}M_{ln}-g_{ln}M_{km}),$$
$$[P_k,P_l]=0,\quad [M_{kl},P_m]=i(g_{lm}P_k-g_{km}P_l),$$

where

$$g_{11}=g_{22}=g_{33}=-g_{44}=-1$$
$$g_{kl}=0,\quad k\neq l,$$

and P_k is the 4-dimensional momentum and M_{23}, M_{31}, and M_{12} are the three components of †angular momentum. The subalgebra generated by P_k corresponds to an invariant Abelian subgroup of $\bar{G}$. The subalgebra generated by M_k corresponds to a simple subalgebra of $\bar{G}$, which is called the †proper Lorentz group G_0. The †Lorentz group G is the group of linear transformations $x_i'=G_{ij}x_j$ that leave $g_{ij}x_ix_j$ invariant, and G_0 is its subgroup consisting of elements satisfying $\det(G_{ij})=1$. The quotient group G/G_0 is an Abelian group of order 4 and of type (2,2): G can be written as $G=G_0\cup SG_0\cup TG_0\cup STG_0$, where S and T are the transformations $(x_i\to -x_i\ (i=1,2,3),\ x_4\to x_4)$ and $(x_i\to x_i\ (i=1,2,3),\ x_4\to -x_4)$, respectively (→ 355 Relativity; 247 Lie Groups).

An elementary particle corresponds to an irreducible representation of the inhomogeneous Lorentz group $\bar{G}$ including S and T. It is specified by the two numbers corresponding to the †Casimir operators $M^2=P_kP^k$ and $W=M_{kl}M^{kl}P_mP^m/2-M_{km}M^{lm}P^kP_l$ (where $P^k=g^{kl}P_l$, $M^{kl}=g^{kk'}g^{ll'}M_{k'l'}$, $g^{ka}g_{ja}=\delta_j^k$). M is the mass, and $W=M^2S^2$ (where S is the spin) in the **center of mass system** defined by $P_1=P_2=P_3=0$. When $M=0$, $\omega_k\omega_l$ $(\omega_k=\varepsilon_{klmn}M_{lm}P_n)$ is used instead of W. Representations of $\bar{G}$ can be obtained by the standard method. Bases of an irreducible representation are obtained by decomposing the †tensor products of Dirac †spinors ψ_α and their conjugate spinors $\bar{\psi}_\alpha$ (→ 346 Quantum Mechanics E; 384 Spinors). A spinor of arbitrary rank satisfies the **Bargmann-Wigner equation** $(P_\mu\gamma^\mu+m)_\alpha^{\alpha'}\psi_{\alpha'\beta\gamma\ldots}^{\delta\varepsilon\ldots}=0$, $(P_\mu\gamma^\mu+m)_\beta^{\beta'}\psi_{\alpha\beta'\gamma\ldots}^{\delta\varepsilon\ldots}=0,\ldots.$ Here γ^μ is the γ-matrix introduced by P. A. M. Dirac. For example, for the spinor of rank 2, we have

$$\Phi_\alpha^\beta=\delta_\alpha^\beta\varphi+\gamma_\alpha^{5\beta}\varphi_5+i(\gamma^\mu\gamma^5)_\alpha^\beta\varphi_{\mu5}$$
$$+\gamma_\alpha^{\mu\beta}\varphi_\mu+\sigma_\alpha^{\mu\nu\beta}\varphi_{\mu\nu}/2,$$

where

$$\gamma^5=\gamma^1\gamma^2\gamma^3\gamma^4,\quad 2i\sigma^{\mu\nu}=\gamma^\mu\gamma^\nu-\gamma^\nu\gamma^\mu,$$

and the equation reads

$$\varphi=0,\quad P_\mu\varphi_5=im\varphi_{\mu5},\quad P_\mu\varphi_{\mu5}=-im\varphi_5,$$
$$P_\mu\varphi_\nu-P_\nu\varphi_\mu=im\varphi_{\mu\nu},\quad P_\nu\varphi_{\nu\mu}=-im\varphi_\mu,$$

where φ_5 corresponds to an elementary particle of spin 0 and φ_μ to spin 1. The Dirac particle has †spin 1/2 and satisfies

$$(P_\mu \gamma^\mu + m)\psi = 0.$$

This is also derived as the †Euler-Lagrange differential equation for the variational problem

$$\delta \mathfrak{L} = \delta \int \bar{\psi} (P_\mu \gamma^\mu + m)\psi \, d^4x,$$

where $P_\mu = i\partial / \partial x_\mu$. The †Lagrangian function $\bar{\psi}(P_\mu \gamma^\mu + m)\psi$ is invariant under the Lorentz transformations.

C. Interactions

Interactions among particles are classified into four categories: **strong interactions**, **electromagnetic interactions**, **weak interactions**, and **gravitational interactions**.

Electromagnetic interactions can be introduced most naturally by requiring the Lagrangian function to be invariant under the local †gauge transformation $\Phi(x) \rightarrow e^{ieQ\lambda(x)}\Phi(x)$, where $\Phi(x)$ is a spinor of arbitrary rank, $\lambda(x)$ is an arbitrary function satisfying $\Box\lambda(x) = 0$, Q is an integer denoting the **charge** of the particle, and ε is the elementary charge. The Lorentz invariance and the positive definiteness of the energy of observable particles lead to the following condition of †quantization

$$\Phi^{\alpha\beta\cdots}_{\zeta\eta\cdots}(x)\Phi^{\zeta'\eta'\cdots}_{\alpha'\beta'\cdots}(x') \pm \Phi^{\zeta'\eta'\cdots}_{\alpha'\beta'\cdots}(x')\Phi^{\alpha\beta\cdots}_{\zeta\eta\cdots}(x)$$
$$= i\delta(x_1 - x_1')\delta(x_2 - x_2')\delta(x_3 - x_3')\delta^\alpha_{\alpha'}\delta^\beta_{\beta'}\ldots,$$

where $x_4 = x_4'$ and the plus sign is taken for a half-odd integer spin and the minus sign for an integer spin. In general, the invariance of the Lagrangian function under a contact phase transformation is called the **internal symmetry**. The theory of electromagnetic interactions is invariant under phase transformations corresponding to the conserved baryon number, lepton number, and hypercharge, in addition to the charge.

Regarding strong interactions, the group of internal symmetry is assumed to be a †simple or †semisimple Lie group including at least $SU(2) \otimes U(1)$ (→ 63 Classical Groups). The theory is expected to be eventually invariant under a very wide group including the inhomogeneous Lorentz group ⊗ the internal symmetry group. Invariance under the $SU(2)$ group is called **charge independence**, and the element I_i of the corresponding Lie algebra satisfying $[I_i, I_j] = 2i\epsilon_{ijk} I_k$, where ϵ_{ijk} is the Levi-Civita symbol of the 3rd rank, is called the **isospin**. The generator of the $U(1)$ group is the **hypercharge** Y, and we have the relation $Q = I_3 + Y/2$. Particles can be classified by the group $SU(2) \otimes U(1)$, and a set of particles belonging to the same irreducible representation is called an **isomultiplet**. Table 1 lists some isomultiplets. **Strangeness** S is defined by $Y = S + n_B$, where n_B is the **baryon number**. Y, Q, I_3, S, and n_B are all additive quantum numbers and can be introduced by means of phase transformations.

D. Models of Elementary Particles

If the internal symmetry is assumed to be a †Lie group of rank 2 or higher, additive quantum numbers besides I_3 are conserved. The **Sakata model** assumes an approximate symmetry under the transformations of $U(3)$ (rank 3), the unitary group of order 3, and its fundamental 3-dimensional representations correspond to the baryon N (p, n, Λ) and the antibaryon $\bar{N}$ ($\bar{\text{p}}$, $\bar{\text{n}}$, $\bar{\Lambda}$). In this model, irreducible decompositions of the tensor product $N\bar{N}$ correspond to mesons, and irreducible decompositions of $NN\bar{N}$ give other baryons. Three conserving quantities correspond to I_3, S, and n_B.

Consider $SU(3)$ (rank 2), the †special unitary group of order 3 that is obtained from $U(3)$ by excluding the phase transformations for n_B. Two conserved quantities are I_3 and Y, and it is possible to assume a correspondence between the 8-dimensional irreducible representation of $SU(3)$ and the 8 baryons and 8 pseudoscalar mesons. In this **octet model**, the observed 8 vector mesons and 10 resonances in the baryon-pseudoscalar meson system (spin 3/2, parity +) are given by the 8-dimensional and 10-dimensional irreducible representations, respectively.

Transition matrices of weak interactions have the following transformation properties: For nonleptonic decays the selection rules $\Delta I = 1/2$ and $\Delta S = \pm 1$ are obeyed, and for semileptonic decays the rules are $\Delta I = 1/2$ and $\Delta S/\Delta Q = 1$ or $\Delta I = 1$ and $\Delta S/\Delta Q = 0$.

References

[1] G. Takeda and H. Miyazawa, Soryûsi buturigaku (Japanese; Physics of elementary particles), Syôkabô, 1965.
[2] M. Taketani, S. Sakata, and S. Nakamura (eds.), Soryûsi no honsitu (Japanese; The nature of elementary particles), Iwanami, 1963.
[3] K. Nishijima, Fundamental particles, Benjamin, 1963.
[4] E. Fermi, Elementary particles, Yale Univ. Press, 1951.
[5] W. Heitler, Quantum theory of radiation, Oxford Univ. Press, second edition, 1944.
[6] R. H. Dalitz, Strange particles and strong interactions, Oxford Univ. Press, 1962.

[7] A. O. Barut, Electrodynamics and classical theory of fields and particles, Macmillan, 1964.
[8] S. Gasiorowicz, Elementary particle physics, John Wiley, 1966.

143 (XIV.9) Ellipsoidal Harmonics

A. Ellipsoidal Coordinates

If $a>b>c$, then for any given $(x,y,z)\in\mathbf{R}^3$, the three roots of the cubic equation in θ

$$F(\theta)=\frac{x^2}{a^2+\theta}+\frac{y^2}{b^2+\theta}+\frac{z^2}{c^2+\theta}-1=0$$

are real and lie in the intervals $\theta>-c^2$, $-c^2>\theta>-b^2$, and $-b^2>\theta>-a^2$. Denoting these three roots by λ, μ, and ν (they are labeled so as to satisfy the inequalities $\lambda>-c^2>\mu>-b^2>\nu>-a^2$), $F(\lambda)=0$, $F(\mu)=0$, and $F(\nu)=0$ represent an ellipsoid, a hyperboloid of one sheet, and a hyperboloid of two sheets, respectively. They are confocal with the ellipsoid

$$\frac{x^2}{a^2}+\frac{y^2}{b^2}+\frac{z^2}{c^2}-1=0,$$

pass through the point (x,y,z), and mutually intersect orthogonally.

The quantities λ, μ, ν are called the **ellipsoidal coordinates** of the point (x,y,z). Rectangular coordinates (x,y,z) are expressed in terms of ellipsoidal coordinates (λ,μ,ν) by the formula

$$x^2=\frac{(a^2+\lambda)(b^2+\lambda)(c^2+\lambda)}{(a^2-b^2)(a^2-c^2)}, \qquad (1)$$

and two others obtained from (1) by cyclic permutations of (a,b,c) and (x,y,z).

B. Ellipsoidal Harmonics

When a †harmonic function ψ of three real variables is constant on the surface $\lambda=$ constant, $\mu=$ constant, or $\nu=$ constant in ellipsoidal coordinates, the function ψ is called an **ellipsoidal harmonic**. A solution of Laplace's equation $\Delta\psi=0$ in the form $\psi=\Lambda(\lambda)M(\mu)N(\nu)$ can be obtained by the method of separation of variables. The equation $\Delta\psi=0$ is written in the form

$$\sum(\mu-\nu)\Delta_\lambda\frac{\partial}{\partial\lambda}\left(\Delta_\lambda\frac{\partial\psi}{\partial\lambda}\right)=0,$$

where the summation is taken over the even permutations of (λ,μ,ν), and

$\Delta_\lambda=\sqrt{(a^2+\lambda)(b^2+\lambda)(c^2+\lambda)}$.

The ordinary differential equation

$$4\Delta_\lambda\frac{d}{d\lambda}\left(\Delta_\lambda\frac{d\Lambda}{d\lambda}\right)=(K\lambda+C)\Lambda \qquad (2)$$

is satisfied by Λ and also by M and N if we replace λ by μ and ν, respectively. Equation (2) is called **Lamé's differential equation**, with K and C the separation constants.

Let $K=n(n+1)$ for $n=0,1,2,\ldots$. Then equation (2), for a suitable value (the eigenvalue) of C, has a solution that is a polynomial in λ or a polynomial multiplied by one, two, or three of $\sqrt{a^2+\lambda}$, $\sqrt{b^2+\lambda}$, and $\sqrt{c^2+\lambda}$.

Among these solutions $2n+1$ are linearly independent. We denote these solutions by $\Lambda=f_n^m(\lambda)$ $(m=1,2,\ldots,2n+1)$. They are essentially equivalent to the Lamé functions, to be defined at the end of this section. To be precise, by setting

$\lambda+(a^2+b^2+c^2)/3=\xi$,

$C=B+n(n+1)(a^2+b^2+c^2)/3$,

$e_1=(b^2+c^2-2a^2)/3,\ldots;\quad e_1+e_2+e_3=0$,

we have

$$\frac{d^2\Lambda}{d\xi^2}+\left(\frac{1/2}{\xi-e_1}-\frac{1/2}{\xi-e_2}-\frac{1/2}{\xi-e_3}\right)\frac{d\Lambda}{d\xi}$$

$$=\frac{n(n+1)\xi+B}{4(\xi-e_1)(\xi-e_2)(\xi-e_3)}\Lambda. \qquad (3)$$

This can also be written in the form

$$\frac{d^2\Lambda}{du^2}=(n(n+1)\wp(u)+B)\Lambda \qquad (4)$$

by the change of variable $\xi=\wp(u)$ with the Weierstrass †$\wp$-function.

The differential equation (3) has $\xi=e_1$, e_2,e_3,∞ as †regular singular points. A solution of (3) that is a polynomial in ξ or a polynomial multiplied by one, two, or three of $\sqrt{\xi-e_1}$, $\sqrt{\xi-e_2}$, and $\sqrt{\xi-e_3}$ is called a **Lamé function of the first kind**.

C. Classification of Lamé Functions

The $2n+1$ linearly independent solutions $f_n^m(\lambda)$ of (2) are classified into the four following families. If n is an even number $2p$, then $p+1$ solutions $f_n^m(\lambda)$ among the $2n+1$ solutions are polynomials in λ of degree p, and the other $3p$ functions are polynomials in λ of degree $p-1$ multiplied by

$\sqrt{(b^2+\lambda)(c^2+\lambda)}$, $\sqrt{(c^2+\lambda)(a^2+\lambda)}$,

or

$\sqrt{(a^2+\lambda)(b^2+\lambda)}$.

Since all these polynomials are products with real factors of degree 1, the solutions belonging to the first family are of the type

$$f_n^m(\lambda)=(\lambda-\theta_1)(\lambda-\theta_2)\ldots(\lambda-\theta_{n/2}), \qquad (5)$$

while solutions of the latter kinds are of the type

$$f_n^m(\lambda)=\left\{\begin{matrix}\sqrt{(b^2+\lambda)(c^2+\lambda)}\\ \sqrt{(c^2+\lambda)(a^2+\lambda)}\\ \sqrt{(a^2+\lambda)(b^2+\lambda)}\end{matrix}\right\}$$
$$\times(\lambda-\theta_1)(\lambda-\theta_2)\dots(\lambda-\theta_{n/2-1}). \quad (6)$$

The functions (5) and (6) are called **Lamé functions of the first species** and **of the third species**, respectively. On the other hand, if n is an odd number $2p+1$, then $3(p+1)$ solutions among the $2n+1$ functions $f_n^m(\lambda)$ are of the type

$$f_n^m(\lambda)=\left\{\begin{matrix}\sqrt{a^2+\lambda}\\ \sqrt{b^2+\lambda}\\ \sqrt{c^2+\lambda}\end{matrix}\right\}$$
$$\times(\lambda-\theta_1)(\lambda-\theta_2)\dots(\lambda-\theta_{(n-1)/2}), \quad (7)$$

and the other p functions are of the type

$$f_n^m(\lambda)=\sqrt{(a^2+\lambda)(b^2+\lambda)(c^2+\lambda)}$$
$$\times(\lambda-\theta_1)(\lambda-\theta_2)\dots(\lambda-\theta_{(n-3)/2}). \quad (8)$$

The functions (7) and (8) are called **Lamé functions of the second species** and **of the fourth species**, respectively. Hence, in either case we have $2n+1$ linearly independent Lamé functions.

When n is even, we obtain an ellipsoidal harmonic

$$\psi_n^m=\prod_{p=1}^{n/2}(\lambda-\theta_p)(\mu-\theta_p)(\nu-\theta_p)$$

by multiplying $f_n^m(\lambda)$, $f_n^m(\mu)$, and $f_n^m(\nu)$ belonging to the first family. Also, in this case, by setting

$$\Theta_p=\frac{x^2}{a^2+\theta_p}+\frac{y^2}{b^2+\theta_p}+\frac{z^2}{c^2+\theta_p}-1$$
$$=\frac{(\lambda-\theta_p)(\mu-\theta_p)(\nu-\theta_p)}{(a^2+\theta_p)(b^2+\theta_p)(c^2+\theta_p)}$$

we have

$$\psi_n^m=\Theta_1\Theta_2\dots\Theta_{n/2} \quad (9)$$

up to constant coefficients. Utilizing Lamé functions of the third species (instead of functions of the first species) and formula (1), we find that

$$\psi_n^m=(yz \text{ or } zx \text{ or } xy)\times\Theta_1\Theta_2\dots\Theta_{n/2-1}. \quad (10)$$

For even n, every ellipsoidal harmonic expressible in terms of polynomials in x, y, z of degree n may be written as a linear combination of the functions (9) and (10), which are called the **ellipsoidal harmonics of the first** and **of the third species**, respectively. Similarly for odd n, the **ellipsoidal harmonics of the second** and **of the fourth species**

$$\psi_n^m=(x \text{ or } y \text{ or } z)\Theta_1\Theta_2\dots\Theta_{(n-1)/2}, \quad (11)$$

$$\psi_n^m=xyz\Theta_1\Theta_2\dots\Theta_{(n-3)/2} \quad (12)$$

are composed of the Lamé functions of the second and fourth species, respectively.

For odd n, these forms are a complete system of ellipsoidal harmonics that are linearly independent and expressible in terms of polynomials in x, y, z of degree n.

The zeros $\xi_1, \xi_2, \dots, \xi_p$ of the Lamé functions are real, and $\xi_i\neq\xi_j$ $(i\neq j)$. They never coincide with any one of e_1, e_2, and e_3. If $e_1>e_2>e_3$, then $\xi_1,\dots,\xi_p$ all lie between e_1 and e_3. If m is an integer such that $0\leqslant m\leqslant p$, we have one and only one Lamé function (with the species given) with m of its zeros lying between e_1 and e_2 and the remaining $p-m$ zeros between e_2 and e_3 (**Stieltjes's theorem**). In this way, a complete system of linearly independent Lamé functions of the specified type may be obtained, since m assumes $p+1$ different values. When the constant B appearing in the differential equation (3) takes specific values so that the equation has Lamé functions of the first kind as its solution, (3) also yields a solution Λ such that $\Lambda\to\xi^{-(n+1)/2}$ as $\xi\to\infty$. This function Λ is called the **Lamé function of the second kind.**

D. Ellipsoids of Revolution (Spheroids)

When the fundamental ellipsoid is a spheroid

$$\frac{x^2+y^2}{a^2}+\frac{z^2}{c^2}=1,$$

it is convenient to use the spheroidal coordinates (ξ,η,φ) given by

$$x=l\sqrt{(\xi^2-1)(1-\eta^2)}\cos\varphi,$$
$$y=l\sqrt{(\xi^2-1)(1-\eta^2)}\sin\varphi, \quad (13)$$
$$z=l\xi\eta, \quad l=\sqrt{c^2-a^2}$$

for $a^2<c^2$ (prolate) and

$$x=l\sqrt{(\xi^2+1)(1-\eta^2)}\cos\varphi,$$
$$y=l\sqrt{(\xi^2+1)(1-\eta^2)}\sin\varphi,$$
$$z=l\xi\eta, \quad l=\sqrt{a^2-c^2} \quad (14)$$

for $a^2>c^2$ (oblate). The solutions of Laplace's equation, which are regular at all finite points, are given by

$$\psi=P_n^m(\xi)P_n^m(\eta)_{\sin}^{\cos}m\varphi$$

in the prolate case, and

$$\psi=P_n^m(i\xi)P_n^m(\eta)_{\sin}^{\cos}m\varphi$$

in the oblate case. Here P_n^m is the †associated Legendre function of the first kind. Solutions which are regular outside a finite ellipsoid can be composed of the †associated Legendre functions of the second kind, $Q_n^m(\xi)$ or $Q_n^m(i\xi)$ instead of $P_n^m(\xi)$ or $P_n^m(i\xi)$, respectively.

E. Spheroidal Wave Functions

Transforming the †Helmholtz equation in prolate spheroidal coordinates (13), we have

$$\frac{1}{(\xi^2-\eta^2)}\left(\frac{\partial}{\partial\xi}(\xi^2-1)\frac{\partial\Psi}{\partial\eta}+\frac{\partial}{\partial\eta}(1-\eta^2)\frac{\partial\Psi}{\partial\eta}\right)$$

$$+\left(\frac{1}{\xi^2-1}+\frac{1}{1-\eta^2}\right)\frac{\partial^2\Psi}{\partial\varphi^2}+\kappa^2\Psi=0,\quad \kappa=kl. \tag{15}$$

By separating variables in the form $\Psi=X(\xi)Y(\eta)^{\cos}_{\sin}m\varphi$, we have the equations:

$$\frac{d}{d\xi}\left((1-\xi^2)\frac{dX}{d\xi}\right)+\left(\lambda-\kappa^2\xi^2-\frac{m^2}{1-\xi^2}\right)X=0, \tag{16a}$$

$$\frac{d}{d\eta}\left((1-\eta^2)\frac{dY}{d\eta}\right)+\left(\lambda-\kappa^2\eta^2-\frac{m^2}{1-\eta^2}\right)Y=0, \tag{16b}$$

which X and Y, respectively, must satisfy. The only difference between equations (16a) and (16b) arises from the fact that the domain of (16a) is given by $1<\xi$ whereas the domain of (16b) is given by $-1<\eta<1$. For the oblate spheroid, utilizing formula (14) we have

$$\frac{1}{(\xi^2+\eta^2)}\left(\frac{\partial}{\partial\xi}(\xi^2+1)\frac{\partial\Psi}{\partial\xi}+\frac{\partial}{\partial\eta}(1-\eta^2)\frac{\partial\Psi}{\partial\eta}\right)$$

$$+\left(\frac{1}{1-\eta^2}-\frac{1}{\xi^2+1}\right)\frac{\partial^2\Psi}{\partial\varphi^2}+\kappa^2\Psi=0. \tag{17}$$

By separating variables as before, $Y(\eta)$ satisfies the same equation as (16b), while $X(\xi)$ satisfies equation (16a) with ξ replaced by $i\xi$. All these equations are of the type

$$\frac{d}{dz}\left((1-z^2)\frac{du}{dz}\right)+\left(\lambda-\kappa^2z^2-\frac{m^2}{1-z^2}\right)u=0. \tag{18}$$

A solution of (18) is known as a **spheroidal wave function**. The equation (18) has ±1 as regular singular points and ∞ as an irregular singular point of class 1. Hence spheroidal wave functions behave like †Legendre functions in the interval $[-1,1]$ and like †Bessel functions in the neighborhood of ∞.

When we write a solution of (18) it is customary to write x instead of z when z is contained in the interval $[-1,1]$. We denote solutions of (18) which are regular on the whole domain $-1\leqslant x\leqslant1$ by $pe_n^m(x)$, and the corresponding eigenvalues by $\lambda_{n,m}$ (assuming the boundary condition stated in this paragraph concerning singularities). In particular, in case $\kappa\to0$, equation (18) reduces to †Legendre's associated differential equation, the eigenvalues of λ become $n(n+1)$ (n is a positive integer), and the corresponding eigenfunctions become the associated Legendre functions of the first kind:

$$P_n^m(x)=(1-x^2)^{m/2}\frac{d^mP_n}{dx^m}. \tag{19}$$

Hence $pe_n^m(x)$ is a solution which tends to a constant multiple of $P_n^m(x)$ as $\kappa\to0$. Using a system of orthogonal functions $P_n^m(x)$, we can expand $pe_n^m(x)$ as:

$$pe_n^m(x)=\sum_{l\geqslant m}A_{n,l}^mP_l^m(x),$$

$$|l-n|=\text{even number}. \tag{20}$$

The coefficients A satisfy a recurrence formula

$$\left(\lambda_{n,m}-l(l+1)+\kappa^2\frac{2l^2+2l-1-2m^2}{(2l-1)(2l+3)}\right)A_{n,l}^m$$

$$-\kappa^2\frac{(l-m-1)(l-m)}{(2l-3)(2l-1)}A_{n,l-2}^m$$

$$-\kappa^2\frac{(l+m+1)(l+m+2)}{(2l+3)(2l+5)}A_{n,l+2}^m=0. \tag{21}$$

The functions $pe_n^m(x)$ and $pe_l^m(x)$ are orthogonal in the domain $-1<x<1$.

Another solution of (18) exists which corresponds to the same eigenvalue $\lambda_{n,m}$, is independent of $pe_n^m(x)$, and has the opposite parity:

$$qe_n^m(x)=\sum_{l\geqslant -m,\,l-n=\text{even}}A_{n,l}^mQ_l^m(x)$$

$$+\sum_{j\geqslant m,\,j-n=\text{odd}}B_{n,j}^mP_j^m(x), \tag{22}$$

where the $A_{n,l}^m$ are the same as in (20) and are determined by the recurrence formula (21), while for $j\geqslant m+2$, the $B_{n,j}^m$ satisfy the recurrence formula

$$\left(\lambda_{n,m}-j(j+1)+\kappa^2\frac{2j^2+2j-1-2m^2}{(2j-1)(2j+3)}\right)B_{n,j}^m$$

$$+\kappa^2\frac{(j-m-1)(j-m)}{(2j-3)(2j-1)}B_{n,j-2}^m$$

$$+\kappa^2\frac{(j+m+1)(j+m+2)}{(2j+3)(2j+5)}B_{n,j+2}^m=0. \tag{23}$$

Since the associated Legendre function of the second kind

$$Q_l^m(x)=(1-x^2)^{m/2}d^mQ_l/dx^m$$

is of the form

$$Q_l^m(x)=P_l^m(x)\log\sqrt{(1+x)/(1-x)}$$
$$+(1-x^2)^{-m/2}\times(\text{a polynomial in }x)$$

for $l\geqslant m$, the $qe_n^m(x)$ have $x=\pm1$ as singular points.

By expressing the solution of equation (18) in integral form we find that $pe_n^m(x)$ satisfies an integral equation

$$i^{n-m}\nu_{n,m}\,pe_n^m(x) = \int_{-1}^{1}(1-x^2)^{m/2}(1-\xi^2)^{m/2}e^{i\kappa x\xi}\,pe_n^m(\xi)\,d\xi, \tag{24}$$

where the coefficient $\nu_{n,m}$ is related to $P_n^m(0)$ or $P_n^{m\prime}(0)$.

In order to extend the domain of definition of $pe_n^m(x)$ and $qe_n^m(x)$ beyond the interval $[-1,1]$, we adopt, in the domain G obtained by deleting the interval $[-1,1]$ from the complex plane, the Heine-Hobson definition of the associated Legendre function,

$$P_n^m(z)=(z^2-1)^{m/2}d^mP_n/dz^m, \tag{25}$$

instead of N. M. Ferrers's definition (19), and construct a solution of (18) in G:

$$pe_n^m(z)=\sum_{l\geq m}A_{n,l}^mP_l^m(z),$$

$$|l-n|=\text{even number}, \tag{26}$$

which is like (20) and again satisfies the integral equation (24). From this we can obtain the expansion formula

$$pe_n^m(z)=\frac{\sqrt{2\pi}\,(z^2-1)^{m/2}}{\nu_{n,m}(\kappa z)^m}\times\sum_{l\geq m}(-1)^{(l-n)/2}\frac{(l+m)!}{(l-m)!}A_{n,l}^m\frac{J_{l+1/2}(\kappa z)}{\kappa z},$$

$$|l-n|=\text{even number}. \tag{27}$$

Multiplying this by a constant, we define

$$je_n^m(z)=\sqrt{\frac{\pi}{2}}\,\frac{(z^2-1)^{m/2}}{z^m}\times\sum_{l\geq m}(-1)^{(l-n)/2}F_{n,l}^m\frac{J_{l+1/2}(\kappa z)}{\sqrt{\kappa z}}\Bigg/\sum_{l\geq m}F_{n,l}^m,$$

$$F_{n,l}^m=\frac{(l+m)!}{(l-m)!}A_{n,l}^m.$$

This expression asymptotically assumes the form

$$je_n^m(z)\sim\sin(\kappa z-n\pi/2)/\kappa z$$

for $|z|\gg 1$. In a similar manner we find a solution

$$ne_n^m(z)=-\sqrt{\frac{\pi}{2}}\,\frac{(z^2-1)^{m/2}}{z^m}\times\sum_{l\geq m}(-1)^{(l-n)/2}F_{n,l}^m\frac{N_{l+1/2}(\kappa z)}{\sqrt{\kappa z}}\Bigg/\sum_{l\geq m}F_{n,l}^m$$

having the asymptotic form

$$ne_n^m(z)\sim\cos(\kappa z-n\pi/2)/\kappa z.$$

Disregarding a constant factor, this coincides with the function defined by (22) with $Q_n^m(z)=(z^2-1)^{m/2}\,d^mQ_n/dz^m$ in place of the associated Legendre function of the second kind.

References

[1] M. J. O. Strutt, Lamésche-Mathieusche- und verwandte Funktionen in Physik und Technik, Erg. Math., Springer, 1932 (Chelsea, 1967).
[2] E. W. Hobson, The theory of spherical and ellipsoidal harmonics, Cambridge Univ. Press, 1931 (Chelsea, 1955).
[3] Y. Hagihara, Kaiten ryûtai no heikôkeizyô ron no kaiko (Japanese; Historical review on the theory of equilibrium figures of a rotating fluid), Iwanami Coll. of Math., 1933.
[4] M. Kotani and H. Hashimoto, Tokusyu kansû (Japanese; Special functions), Iwanami Coll. of Modern Appl. Math., 1958.
[5] C. Flammer, Spheroidal wave functions, Stanford Univ. Press, 1957.
[6] J. A. Stratton, P. M. Morse, L. J. Chu, J. D. C. Little, and F. J. Corbató, Spheroidal wave functions, including tables of separation constants and coefficients, John Wiley, 1956.

144 (XIV.3) Elliptic Functions

A. Elliptic Integrals

Let $\varphi(z)$ be a polynomial in z of degree 3 or 4 with complex coefficients and $R(z,w)$ a rational function in z and w. Then $R(z,\sqrt{\varphi(z)})$ is called an **elliptic irrational function**. An integral of the type $\int R\,dz$ is called an **elliptic integral**. The origin of the name comes from the integral that appeared in calculating the arc length of an ellipse. Any elliptic integral may be expressed by a suitable change of variables as a sum of elementary functions and elliptic integrals of the following three kinds:

$$\int\frac{dz}{\sqrt{(1-z^2)(1-k^2z^2)}},$$

$$\int\sqrt{\frac{1-k^2z^2}{1-z^2}}\,dz,$$

and

$$\int\frac{dz}{(1-a^2z^2)\sqrt{(1-z^2)(1-k^2z^2)}}$$

(→ Appendix A, Table 16.I). These three kinds of integrals are called **elliptic integrals of the first, second,** and **third kind**, respectively, in **Legendre-Jacobi standard form.** This classification corresponds to that of †Abelian integrals. The constant k is called the **modulus** of these elliptic integrals, and a is called the **parameter**. Let the four zeros of $\varphi(z)$ be α_1,

$\alpha_2, \alpha_3, \alpha_4$ (we take one of them as ∞ when $\varphi(z)$ is of degree 3). The †Riemann surface $\mathfrak{R}$ corresponding to the elliptic irrational function has the zeros $\alpha_1, \alpha_2, \alpha_3, \alpha_4$ as †branch points with degree of ramification 1, and is of two sheets and of †genus 1. If the integrand does not have a pole with a †residue, then the integral is multivalued only because the value (called the **periodicity modulus**) of the integral taken along the normal section (basis of the homology group) is not equal to zero.

B. Elliptic Integrals of the First Kind

When R is a function without singularities other than branch points, only the topological structure of R gives rise to the multivaluedness of the integral of R. The standard form is

$$\int_0^z \frac{dz}{\sqrt{(1-z^2)(1-k^2z^2)}} = \int_0^\varphi \frac{d\varphi}{\sqrt{1-k^2\sin^2\varphi}} = w = F(k,\varphi), \qquad (1)$$

where $z=\sin\varphi$. This integral is the inverse function of sn w. The periodicity moduli are $2iK'$ and $4K$, where

$$K=K(k)=\int_0^1 \frac{dz}{\sqrt{(1-2^2)(1-k^2z^2)}} = \int_0^{\pi/2} \frac{d\varphi}{\sqrt{1-k^2\sin^2\varphi}} = F\left(k,\frac{\pi}{2}\right),$$

$$K'=K(k'), \quad k'^2=1-k^2.$$

We call $K(k)$ a **complete elliptic integral of the first kind** and $F(k,\varphi)$ an **incomplete elliptic integral of the first kind** (→ Appendix A, Table 16). Setting

$$\sin\varphi_1 = \frac{(1+k')\sin\varphi\cos\varphi}{\sqrt{1-k^2\sin^2\varphi}}, \quad k_1=\frac{1-k'}{1+k'}, \qquad (2)$$

we have the relation

$$F(k,\varphi)=(1+k_1)F(k_1,\varphi_1)/2,$$

which is called **Landen's transformation**. Since $k_1<k$ when $0<k<1$, this transformation reduces the calculation of elliptic integrals to those with smaller values of k.

C. Elliptic Integrals of the Second Kind

When R has poles with residue zero, its integral has no singularities other than poles. The standard form is

$$F(z)=\int_0^z \sqrt{\frac{1-k^2z^2}{1-z^2}}\,dz = \int_0^\varphi \sqrt{i-k^2\sin^2\varphi}\,d\varphi = E(k,\varphi), \qquad (3)$$

where $z=\sin\varphi$. We have

$$F(z)=\int_0^u \mathrm{dn}^2 u\,du = \frac{\Theta'(u)}{\Theta(u)}+\frac{E}{K}u$$

if we set $z=\mathrm{sn}\,u$ (Section J). Here, Θ is the Jacobian theta function and

$$\Theta(u)=\vartheta_4\left(\frac{u}{2K},\frac{iK'}{K}\right),$$

where ϑ_4 is a theta function to be described in Section I and K, K' are the same as in the case of elliptic integrals of the first kind. The quantity

$$E=\int_0^1 \sqrt{\frac{1-k^2z^2}{1-z^2}}\,dz = \int_0^{\pi/2}\sqrt{1-k^2\sin^2\varphi}\,d\varphi = E\left(k,\frac{\pi}{2}\right)$$

is called a **complete elliptic integral of the second kind.**

D. Elliptic Integrals of the Third Kind

When R has poles with nonzero residue its integral has logarithmic singularities. In this case, residues also contribute to multivaluedness of the integral. The standard form is

$$F(z)=\int_0^z \frac{dz}{(1-a^2z^2)\sqrt{(1-z^2)(1-k^2z^2)}} = \int_0^\varphi \frac{d\varphi}{(1-a^2\sin^2\varphi)\sqrt{1-k^2\sin^2\varphi}},$$

and it is expressed as

$$F(z) = \frac{\mathrm{sn}\,\alpha}{\mathrm{cn}\,\alpha\,\mathrm{dn}\,\alpha}\left(\frac{1}{2}\log\frac{\Theta(u-\alpha)}{\Theta(u+\alpha)}+u\frac{\Theta'(\alpha)}{\Theta(\alpha)}\right)+u$$

if we set $z=\mathrm{sn}\,u$ and $a^2=k^2\mathrm{sn}^2\alpha$ (→ Appendix A, Table 16).

E. Elliptic Functions and Periodic Functions

Historically the elliptic function was first introduced as the inverse function of the elliptic integral. However, since it has been realized that elliptic functions are characterized as functions with double periodicity, it is now customary to define them as doubly periodic functions.

If $f(x)$, defined on a linear space X, satisfies the relation $f(x+\omega)=f(x)$ for some $\omega\in X$ and all $x\in X$, the number ω is called a **period** of $f(x)$, and $f(x)$ with a period other than zero is called a **periodic function**. The set P of all periods of $f(x)$ forms an additive group contained in X. If a basis $\omega_1,\ldots,\omega_n$ of the additive group P exists, its members are called **fundamental periods** of $f(x)$.

Any continuous, nonconstant, periodic function of a real variable has only one positive fundamental period and is called a **simply periodic function.** The †trigonometric functions are typical examples; $\sin x$ and $\cos x$ have the fundamental period 2π; $\tan x$ and $\cot x$ have the fundamental period π (→ 167 Fourier Series).

A single-valued nonconstant †meromorphic function of n complex variables cannot have more than $2n$ fundamental periods that are linearly independent on the real number field. A function of one complex variable with two fundamental periods is called a **doubly periodic function.**

Let ω, ω' be the fundamental periods of a doubly periodic function. For a given number a, the parallelogram with vertices a, $a+\omega$, $a+\omega'$, $a+\omega+\omega'$ is called the **fundamental period parallelogram.** The complex plane is covered with a network of congruent parallelograms, called **period parallelograms**, obtained by translating the fundamental period parallelogram through $m\omega+n\omega'$ ($m,n=0,1,\ldots$).

A doubly periodic function $f(u)$ meromorphic on the complex plane is called an **elliptic function.** For simplicity, we usually denote the fundamental periods of an elliptic function by $2\omega_1$ and $2\omega_3$, and introduce ω_2 defined by the relation $\omega_1+\omega_2+\omega_3=0$. The first, and therefore also higher, derivatives of any elliptic function are elliptic functions with the same periods. The set of all elliptic functions with the same periods forms a †field. The number of poles in a period parallelogram is finite. The sum of the orders of the poles is called the **order** of the elliptic function. An elliptic function with no poles in a period parallelogram is merely a constant (**Liouville's first theorem**). The sum of the residues of an elliptic function at its poles in any period parallelogram is zero (**Liouville's second theorem**). Hence there can be no elliptic function of order 1. An elliptic function of order n assumes any value n times in a period parallelogram (**Liouville's third theorem**). The sum of the zeros minus the sum of the poles is a period (**Liouville's fourth theorem**).

F. Weierstrass's Elliptic Functions

Weierstrass defined

$$\wp(u)=\frac{1}{u^2}+\sum{}'\left(\frac{1}{(u-\Omega)^2}-\frac{1}{\Omega^2}\right)$$

as the simplest kind of elliptic function. Here $\Omega=2m\omega_1+2n\omega_3$, with m, n integers. The summation Σ' extends over all integral values (positive, negative, and zero) of m and n, except for $m=n=0$. $\wp(u)$ is an elliptic function of order 2 with periods $2\omega_1$ and $2\omega_3$, called a **Weierstrass $\wp$-function**. The following functions $\zeta(u)$ and $\sigma(u)$ are called the **Weierstrass zeta** and **sigma functions**, respectively:

$$\zeta(u)=\frac{1}{u}+\sum{}'\left(\frac{1}{u-\Omega}+\frac{u}{\Omega^2}+\frac{1}{\Omega}\right)$$

and

$$\sigma(u)=u\prod{}'\left(\left(1-\frac{u}{\Omega}\right)\exp\left(\frac{u}{\Omega}+\frac{u^2}{\Omega^2}\right)\right).$$

These have quasiperiodicity, expressed by

$$\zeta(u+2\omega_i)=\zeta(u)+2\eta_i, \tag{4}$$

$$\sigma(u+2\omega_i)=-e^{2\eta_i(u+\omega_i)}\sigma(u), \tag{5}$$

$$\eta_1+\eta_2+\eta_3=0, \quad \eta_i=\zeta(\omega_i), \quad i=1,2,3;$$

and they satisfy the relations

$$\wp(u)=-\zeta'(u) \tag{6}$$

and

$$\zeta(u)=\frac{d\log\sigma(u)}{du}=\frac{\sigma'(u)}{\sigma(u)}. \tag{7}$$

The function $\wp(u)$ is an even function of u, and $\zeta(u)$ and $\sigma(u)$ are odd functions of u. By considering the integral $\int\zeta(u)du$ once around the boundary of a fundamental period parallelogram, we have

$$\left.\begin{array}{l}\eta_1\omega_3-\eta_3\omega_1\\ \eta_2\omega_1-\eta_1\omega_2\\ \eta_3\omega_2-\eta_2\omega_3\end{array}\right\}=\pm\frac{\pi}{2}i, \quad \mathrm{Im}\left(\frac{\omega_3}{\omega_1}\right)\gtrless 0, \tag{8}$$

which is called the **Legendre relation.**

The derivative

$$\wp'(u)=-2\sum 1/(u-\Omega)^3$$

of a $\wp$-function is an elliptic function of order 3 and bears the following relation to $\wp(u)$:

$$(\wp'(u))^2=4(\wp(u))^3-g_2\wp(u)-g_3$$

$$=4(\wp(u)-e_1)(\wp(u)-e_2)(\wp(u)-e_3),$$

$$g_2=60\sum{}'1/\Omega^4, \quad g_3=140\sum{}'1/\Omega^6,$$

$$e_i=\wp(\omega_i), \quad i=1,2,3. \tag{9}$$

By differentiating this relation successively, we see that $\wp^{(n)}(u)$ is expressed as a polynomial in $\wp(u)$ if n is an even number and as a product of polynomials in $\wp(u)$ and $\wp'(u)$ if n is an odd number.

In particular, writing $\wp(u)=z$ in (9) we find that the $\wp$-function is the inverse function of the elliptic integral

$$u=\int_\infty^z\frac{dz}{\sqrt{4z^3-g_2z-g_3}}$$

(→ Appendix A, Table 16.IV).

Any elliptic function can be expressed in terms of Weierstrass's functions. Specifically, let the poles of $f(u)$ and their orders be $a_1, a_2, \ldots, a_m$ and $h_1, h_2, \ldots, h_m$, respectively, and let the principal part in the expansion of $f(u)$ near the pole a_k be

$$\sum_{j=1}^{h_k} \frac{A_{kj}}{(u-a_k)^j}, \quad k=1,2,\ldots,m. \tag{10}$$

Then we obtain

$$f(u)=C+\sum_{k=1}^{m}\left(A_{k1}\zeta(u-a_k) + \sum_{j=1}^{h_k}\frac{(-1)^j A_{kj}}{(j-1)!}\wp^{(j-2)}(u-a_k)\right), \tag{11}$$

where C is a constant depending on $f(u)$. This can be reduced to

$$f(u)=A+B\wp'(u)$$

by using the addition theorems (→ Appendix A, Table 16) for $\wp$ and zeta functions, where A and B are rational functions of $\wp(u)$. Therefore, given two elliptic functions with the same periods, after expressing them as rational functions of $\wp$ and $\wp'$ in the above form and eliminating $\wp$ and $\wp'$, we obtain an algebraic equation with constant coefficients. In particular, for any elliptic function $f(u)$ we obtain an †algebraic differential equation of the first order by using this method, with $f'(u)$ an elliptic function with the same periods. Furthermore, the functions $f(u+v)$, $f(u)$, and $f(v)$ satisfy an algebraic equation. Thus for any elliptic function an algebraic addition theorem holds.

G. Elliptic Functions of the Second Kind

As an extension of the definition of elliptic functions, if a†meromorphic function f satisfies the relations

$$f(u+2\omega_1)=\mu_1 f(u), \quad f(u+2\omega_3)=\mu_3 f(u) \tag{12}$$

(μ_1 and μ_3 are constants) with the fundamental periods $2\omega_1$, $2\omega_3$, we call $f(u)$ an **elliptic function of the second kind**. What we have called simply an elliptic function may now be called an **elliptic function of the first kind**. For constants ρ and v, the function

$$f(u)=e^{\rho u}\sigma(u-v)/\sigma(u) \tag{13}$$

is an example of an elliptic function of the second kind. In this case

$$\mu_i=e^{2\rho\omega_i-2v\eta_i}, \quad i=1,3. \tag{14}$$

Furthermore, for given constants μ_1 and μ_3, any elliptic function of the second kind is expressed as the product of an elliptic function of the first kind and the function (13) with ρ and v determined by (14).

H. Elliptic Functions of the Third Kind

If a meromorphic function f satisfies

$$f(u+2\omega_i)=e^{a_i u+b_i}f(u), \quad i=1,3 \tag{15}$$

(a_i and b_i are constants) with periods $2\omega_1$, $2\omega_3$, we call it an **elliptic function of the third kind** (→ 3 Abelian Varieties I).

The Weierstrass sigma function $\sigma(u)$ is an example of an elliptic function of the third kind. The functions σ_1, σ_2, and σ_3, defined by the equations

$$\sigma_i(u)=-\frac{e^{\eta_i u}\sigma(u-\omega_i)}{\sigma(\omega_i)}, \quad i=1,2,3, \tag{16}$$

are also elliptic functions of the third kind. In the case of elliptic functions of the second and third kinds, $2\omega_1$ and $2\omega_3$ are not periods in the strict sense defined earlier, but are conveniently referred to as the periods. The functions σ_i $(i=1,2,3)$ are called **cosigma functions**.

I. The Theta Functions

The **theta functions** are defined by

$$\vartheta_1(v,\tau)=2\sum_{n=0}^{\infty}(-1)^n q^{(n+1/2)^2}\sin(2n+1)\pi v,$$

$$\vartheta_2(v,\tau)=2\sum_{n=0}^{\infty}q^{(n+1/2)^2}\cos(2n+1)\pi v,$$

$$\vartheta_3(v,\tau)=1+2\sum_{n=1}^{\infty}q^{n^2}\cos 2n\pi v,$$

$$\vartheta_4(v,\tau)=1+2\sum_{n=1}^{\infty}(-1)^n q^{n^2}\cos 2n\pi v, \tag{17}$$

where $q=e^{i\pi\tau}$, $\operatorname{Im}\tau>0$. We call (17) the **q-expansion formulas** of the theta functions, and we sometimes write ϑ_0 in place of ϑ_4. A theta function is an elliptic function of the third kind with periods 1 and τ. Any elliptic function may be expressed as a quotient of theta functions (→ Appendix A, Table 16 for specific examples). The q-expansion formula is quite suitable for numerical computation because of its rapid convergence; its terms decrease as the n^2 powers of q when $n\to\infty$.

An elliptic function with the fundamental periods $2\omega_1$ and $2\omega_3$ may also be viewed as having the periods $2\omega_1'=2\omega_3$, $2\omega_3'=-2\omega_1$. Consequently, theta functions formed with the parameter $\tau=\omega_3/\omega_1$ may be expressed in terms of the parameter $\tau'=\omega_3'/\omega_1'=-\omega_1/\omega_3=-1/\tau$, and we have

$$\vartheta_1(v,\tau)=iA\vartheta_1(v/\tau,-1/\tau),$$

$$\vartheta_2(v,\tau)=A\vartheta_4(v/\tau,-1/\tau),$$

$$\vartheta_3(v,\tau)=A\vartheta_3(v/\tau,-1/\tau),$$

$$\vartheta_4(v,\tau)=A\vartheta_2(v/\tau,-1/\tau),$$

$$A=\sqrt{i/\tau}\,\exp(-\pi i v^2/\tau). \tag{18}$$

These are called **Jacobi's imaginary transformations**. If $\text{Im}(-1/\tau)\gg 1$, then $|q| \fallingdotseq 1$, and therefore the series in the q-expansion formula converges slowly. However, by an imaginary transformation we get $\text{Im}(\tau)\gg 1$ and $|q| \fallingdotseq 0$, so that computations become much easier.

Each of the theta functions satisfies the following partial differential equation of the heat-conduction type

$$\partial^2\vartheta(u,\tau)/\partial u^2=4\pi i\,\partial\vartheta(u,\tau)/\partial\tau \qquad (19)$$

(→ Appendix A, Table 16.II).

J. Jacobi's Elliptic Functions

C. G. J. Jacobi defined elliptic integrals as inverse functions of elliptic integrals of the first kind in the Legendre-Jacobi standard form (1). They are in the above notations

$$\operatorname{sn} w=\sqrt{e_1-e_3}\,\frac{\sigma(u)}{\sigma_3(u)}=\frac{\vartheta_3(0)\vartheta_1(v)}{\vartheta_2(0)\vartheta_4(v)}, \qquad (20)$$

$$\operatorname{cn} w=\frac{\sigma_1(u)}{\sigma_3(u)}=\frac{\vartheta_4(0)\vartheta_2(v)}{\vartheta_2(0)\vartheta_4(v)}, \qquad (21)$$

$$\operatorname{dn} w=\frac{\sigma_2(u)}{\sigma_3(u)}=\frac{\vartheta_4(0)\vartheta_3(v)}{\vartheta_3(0)\vartheta_4(v)}, \qquad (22)$$

where $w=\sqrt{e_1-e_3}\,u$ and $v=u/2\omega_1$.

These functions satisfy the relations

$$\operatorname{sn}^2 w+\operatorname{cn}^2 w=1, \quad k^2\operatorname{sn}^2 w+\operatorname{dn}^2 w=1, \qquad (23)$$

where

$$k^2=\frac{e_2-e_3}{e_1-e_3}=\frac{(\vartheta_2(0))^4}{(\vartheta_4(0))^4}. \qquad (24)$$

The constants k and $k'=\sqrt{1-k^2}$ are called the **modulus** and the **complementary modulus**, respectively. Furthermore, the relation $d\operatorname{sn} w/dw=\operatorname{cn} w\operatorname{dn} w$ holds. The function $z=\operatorname{sn} w$ is the inverse function of the elliptic integral (1) (→ Appendix A, Table 16.III).

References

[1] S. Tomotika, Daen kansûron (Japanese; Theory of elliptic functions), Kyôritu, 1958.
[2] T. Takenouchi, Daen kansûron (Japanese; Theory of elliptic functions), Iwanami, 1936.
[3] C. G. J. Jacobi, Fundamenta nova theoriae functionum ellipticarum (1829) (C. G. J. Jacobi's Gesammelte Werke, vol. 1, G. Reimer, 1881, p. 49–239).
[4] G. H. Halphen, Traité des fonctions elliptiques et de leurs applications I, II, III, Gauthier-Villars, 1886–1891.
[5] A. Hurwitz and R. Courant, Vorlesungen über allgemeine Funktionentheorie und elliptische Funktionen, Springer, third edition, 1929.

145 (II.3) Equivalence Relations

A. General Remarks

Suppose that we are given a relation R between elements of a set X such that for any elements x and y of X, either xRy or its negation holds. The relation R is called an **equivalence relation** (on X) if it satisfies the following three conditions: (1) xRx, (2) xRy implies yRx, and (3) xRy and yRz imply xRz. Conditions (1), (2), and (3) are called the **reflexive**, **symmetric**, and **transitive laws**, respectively. Together, they are called the **equivalence properties**. Condition (1) can be replaced by the following: (1′) For each x there exists an x' such that xRx'. The relation "x is equal to y" is an equivalence relation. If xRy means that x and y are in X, then R is also an equivalence relation. An equivalence relation is often denoted by the symbol $\sim$. The relations of congruence and similarity between figures are equivalence relations. If X is the set of integers and $x\equiv y$ means that $x-y$ is even, then the relation $\equiv$ is an equivalence relation.

B. Equivalence Classes and Quotient Sets

Let R be an equivalence relation. "xRy" is read: "x and y are **equivalent**" (or "x is **equivalent to** y"). The subset of X consisting of all elements equivalent to an element a is called the **equivalence class** of a. By (1), (2), and (3), each equivalence class is nonempty, the equivalence class of a contains a, and different equivalence classes do not overlap. Namely, X is decomposed into a †disjoint union of equivalence classes. This †partition is called the **classification** of X with respect to R. For example, the set of integers is classified into the equivalence class of even numbers and that of odd numbers by the relation $\equiv$. Conversely, since the relation "x and y belong to the same member of a partition" is an equivalence relation, we can regard any partition as a classification. An element chosen from an equivalence class is called a **representative** of the equivalence class. In the example we can take 0 and 1 as the representatives of equivalence classes of even and odd numbers, respectively.

X/R denotes the set of equivalence classes of X with respect to R, and is called the **quotient set** of X with respect to R. The mapping $p: X\to X/R$ that carries x in X into the equivalence class of x is called **canonical surjection** (or **projection**). The idea of equivalence relations can be generalized to deal with the case when X is a †class.

C. Stronger and Weaker Equivalence Relations

Let R and S be two equivalence relations on X. If xRy always implies xSy, then we say that R is **stronger** than S, S is **weaker** than R, the classification with respect to R is **finer** than the one with respect to S, or the classification with respect to S is **coarser** than the one with respect to R. The relations "x is equal to y" and "x and y are in X" are the strongest and the weakest equivalence relations, respectively. Any two equivalence relations on X are ordered by their strength, and the set of equivalence relations on X forms a †complete lattice with respect to this ordering.

References

[1] N. Bourbaki, Eléments de mathématique I. Théorie des ensembles, ch. 2, Actualités Sci. Ind., 1212c, Hermann, second edition, 1960; English translation, Theory of sets, Addison-Wesley, 1968.
Also → references to 376 Sets.

146 (XII.14) Ergodic Theory

A. General Remarks

The origin of ergodic theory was the so-called **ergodic hypothesis**, which provided the foundation for classical statistical mechanics as created by L. Boltzmann and J. Gibbs toward the end of the 19th century (→ 393 Statistical Mechanics). Attempts by various mathematicians to give a rigorous proof of the hypothesis resulted in the **recurrence theorem** of H. Poincaré and C. Carathéodory and the **ergodic theorems** of G. D. Birkhoff and J. Von Neumann, which marked the beginnings of ergodic theory as we know it today. As the theory developed it acquired close relationships with other branches of mathematics, for example, the theory of dynamical systems, probability theory, functional analysis, number theory, differential topology, and differential geometry.

The principal object of modern ergodic theory is to study properties of †measurable transformations, particularly transformations with an invariant measure. In most cases, the transformations studied are defined on a **Lebesgue measure space with a finite** (or σ-**finite) measure**. A Lebesgue measure space with a finite measure (σ-finite measure) is a †measure space that is measure theoretically isomorphic to a bounded interval (to the real line) with the usual †Lebesgue measure, possibly together with an at most countable number of atoms. It is known that any separable complete †metric space with a complete regular Borel †probability measure is a Lebesgue measure space with a finite measure. We assume, unless stated otherwise, that the measure space $(X, \mathfrak{B}, m)$ is a Lebesgue measure space. All the subsets of X mentioned are assumed measurable, and a pair of sets or functions that coincide almost everywhere are identified. We also use the abbreviation "a.e." to denote the statement "†almost everywhere."

B. Ergodic Theorems

Let $(X, \mathfrak{B}, m)$ be a σ-finite measure space. A transformation φ defined on X is called **measurable** if for every $B \in \mathfrak{B}$, $\varphi^{-1}(B) \in \mathfrak{B}$. A †bijective transformation φ on X is called **bimeasurable** if both φ and φ^{-1} are measurable. A measurable transformation φ is called **measure-preserving** (or equivalently, the measure m is **invariant** under φ) if $m(\varphi^{-1}(B)) = m(B)$ holds for every B. It is called **nonsingular** if $m(\varphi^{-1}(B)) = 0$ whenever $m(B) = 0$, and **ergodic** if $m((\varphi^{-1}(B) \cup B) - (\varphi^{-1}(B) \cap B)) = 0$ implies either $m(B) = 0$ or $m(X - B) = 0$.

The **mean ergodic theorem** of Von Neumann (1932) states that if φ is a measure-preserving transformation on $(X, \mathfrak{B}, m)$, then for every function f belonging to the †Hilbert space $L_2(X) = L_2(X, \mathfrak{B}, m)$ (→ 173 Function Spaces), the sequence

$$A_n f(x) = \frac{1}{n}\left(\sum_{k=0}^{n-1} f(\varphi^k x)\right)$$

converges in the †norm of $L_2(X)$ as $n \to \infty$ to a function f^* that satisfies $f^*(\varphi x) = f^*(x)$ a.e. The **individual** (or **pointwise**) **ergodic theorem** of Birkhoff (1932) states that for every f belonging to the †function space $L_1(X)$, the sequence $A_n f(x)$ converges a.e. to f^*. From either of these theorems it follows that for any set E satisfying $\varphi^{-1}(E) = E$ and $m(E) < \infty$, the limit function f^* satisfies $\int_E f^* dm = \int_E f dm$. In particular, if $m(X) = 1$ and φ is ergodic, then the limit f^* equals the constant $\int f dm$ a.e. This fact therefore gives a mathematical justification to the ergodic hypothesis, which states that the "time mean" $(\sum_{k=0}^{n-1} f(\varphi^k x))/n$ of what is observable over a sufficiently long time can be replaced by the "phase mean"

$\int f\,dm$. Both Von Neumann's theorem and Birkhoff's theorem were subsequently generalized in various directions by many authors.

1. Mean ergodic theorems are concerned with the †strong convergence of the sequence of averages $A_n = (\sum_{k=0}^{n-1} T^k)/n$ of the iterates of a †bounded linear operator T on some †Banach space. A generalization of Von Neumann's theorem due to F. Riesz, K. Yosida, and S. Kakutani dispenses with the assumption that the linear operator T is induced by a measure-preserving transformation φ and that T acts on the Hilbert space $L_2(X)$. A version of this generalization states that if a linear operator T defined on a Banach space $\mathfrak{X}$ satisfies conditions

(i) $\sup_{n \geqslant 1} \left\| \frac{1}{n} \sum_{k=0}^{n-1} T^k \right\| < \infty$,

(ii) $\lim_{n\to\infty} \frac{1}{n} \| T^n \| = 0$,

then for an element $f \in \mathfrak{X}$ the sequence of averages $A_n f$ converges strongly to an element $f^* \in \mathfrak{X}$ if and only if there exists a subsequence converging †weakly to f^*. From this theorem of Riesz, Yosida, and Kakutani follows the L_p-mean ergodic theorem $(1 < p < \infty)$ for so-called **Markov operators**: If T is a linear operator defined on each of the Banach spaces $L_p(X)$ $(1 \leqslant p \leqslant \infty)$ by means of the formula $Tf(x) = \int f(y) P(x, dy)$, where $P(x, B)$ is the †transition probability of a †Markov process on $(X, \mathfrak{B})$ leaving the measure m invariant (i.e., $\int P(x, B)\,dm = m(B)$ for every $B \in \mathfrak{B}$)(→ 262 Markov Processes), then for every f belonging to $L_p(X)$ $(1 < p < \infty)$ the sequence $A_n f$ converges in the norm of $L_p(X)$ to a limit function f^*.

2. Birkhoff's ergodic theorem has been extended to the following individual ergodic theorem by E. Hopf (1954): If T is a †positive linear operator mapping $L_1(X)$ into $L_1(X)$ and $L_\infty(X)$ into $L_\infty(X)$ with $\|T\|_1 \leqslant 1$ and $\|T\|_\infty \leqslant 1$, then for every f in $L_1(X)$ the sequence $A_n f$ converges a.e. to a limit f^*. If T is a Markov operator, then T maps each $L_p(X)$ into itself and satisfies $\|T\|_p \leqslant 1$ for each p $(1 \leqslant p \leqslant \infty)$, and therefore Hopf's ergodic theorem applies to such T. Special cases of this theorem were proved earlier by J. Doob and by Kakutani. Later, N. Dunford and J. Schwartz showed that the assumption of the positivity of T can be dispensed with in Hopf's theorem. For a positive linear operator T on $L_1(X)$ satisfying $\|T\|_1 \leqslant 1$, R. Chacon and D. Ornstein (1960) proved that the **ratio ergodic theorem** holds: For every pair of functions f and g in $L_1(X)$ with $g \geqslant 0$ a.e., $\lim_{n\to\infty} \sum_{k=0}^{n-1} T^k f(x) / \sum_{k=0}^{n-1} T^k g(x)$ exists and is finite a.e. on the set $\{x \mid \sum_{k=0}^{\infty} T^k g(x) > 0\}$. This theorem extends earlier results of Hopf and W. Hurewicz dealing with special classes of operators arising from measurable transformations. Hopf's ergodic theorem can be deduced from the Chacon-Ornstein theorem, while it is known that there are positive operators T on $L_1(X)$ satisfying $\|T\|_1 \leqslant 1$ for which $\lim_{n\to\infty} A_n f$ fails to exist on a set of positive measure for some $f \in L_1(X)$. This shows that the assumption $\|T\|_\infty \leqslant 1$ is crucial in Hopf's theorem.

3. As was the case in the original proof by Birkhoff of his ergodic theorem, every known proof of an individual ergodic theorem depends crucially on the so-called **maximal ergodic lemma** (or **maximal inequality**). For the case of a positive linear operator T on $L_1(X)$ with $\|T\|_1 \leqslant 1$, Hopf proved the relevant maximal ergodic lemma: If $E(f)$ is the set $\{x \mid \sup_{n \geqslant 1} A_n f(x) > 0\}$ for each f in $L^1(X)$, then $\int_{E(f)} f\,dm \geqslant 0$. Hopf's original proof of this lemma was quite intricate. Even for the special case where T is the operator induced by a measure-preserving transformation as in Birkhoff's theorem, a proof of the maximal ergodic lemma required delicate and involved arguments until A. Garsia (1965) succeeded in giving an extremely simple and elegant proof of Hopf's lemma. A. Brunel extended Hopf's lemma further and used this extension to give a neat proof of the ratio ergodic theorem of Chacon and Ornstein. Recent results by D. Burkholder, E. Stein, S. Sawyer, and others show that the validity of a maximal inequality of some type is necessary as well as sufficient for the almost everywhere convergence of a sequence of functions in many situations, including the case of individual ergodic theorems.

4. Both mean and individual ergodic theorems can be extended without difficulty to a continuous time parameter semigroup $\{T_t \mid t \geqslant 0\}$ of bounded linear operators such that $T_t T_s = T_{t+s}$ $(T_0 = I)$, under a suitable continuity assumption on T_t with respect to t, by replacing the discrete time average $(\sum_{k=0}^{n-1} T^k)/n$ with $(\int_0^t T_s\,ds)/t$. Further extensions to n-parameter semigroups were obtained by N. Wiener and by N. Dunford and A. Zygmund. For mean ergodic theorems, even further extensions were possible to †amenable semigroups of bounded linear operators. For 1-parameter semigroups, the behavior of the mean at zero, $(\int_0^t T_s\,ds)/t$ as $t \downarrow 0$ (**local ergodic theorem**) or $\lambda \int_0^\infty e^{-\lambda s} T_s f\,ds$ as $\lambda \downarrow 0$ or $\lambda \uparrow \infty$ (**Abelian ergodic theorem**), has also been investigated by Wiener, U. Krengel, E. Hille, Yosida, and others. Abelian ergodic theorems are related to properties of the †resolvent of the semigroup $\{T_t\}$ (→ 373 Semigroups of Operators). Other extensions of

individual ergodic theorems and further extensions of the maximal ergodic lemma were obtained by Chacon, M. Akcoglu, and others.

Garsia, Stein, Burkholder, R. Gundy, and others investigated broader questions concerning the almost everywhere convergence of sequences of functions encountered in ergodic theory, probability theory, and the theory of Fourier series and singular integrals, and they obtained a number of significant and far-reaching results [13, 36].

5. The so-called **random ergodic theorem** was introduced first by Von Neumann and S. Ulam in 1945. Their result was subsequently generalized by H. Anzai, Kakutani, A. Beck and J. Schwartz, and S. Tsurumi. One special case of the random ergodic theorem in its generalized version states: Let $T^{(0)}$ and $T^{(1)}$ be positive linear operators both mapping $L^1(X)$ into $L^1(X)$ and $L^\infty(X)$ into $L^\infty(X)$ and satisfying $\|T^{(t)}\|_1 \leqslant 1$, $\|T^{(t)}\|_\infty \leqslant 1$ for $t=0,1$. Consider for each t in the unit interval $[0,1]$ its dyadic expansion $0.t_1t_2t_3\ldots$. Then, for every f in $L^1(X)$, the sequence of averages $(\sum_{k=1}^n T^{(t_k)}T^{(t_{k-1})}\ldots T^{(t_1)}f(x))/n$ converges as $n\to\infty$ for almost all pairs (x,t).

C. Recurrence and Invariant Measures

In this section we assume that the measure space $(X,\mathfrak{B},m)$ is nonatomic. A nonsingular measurable transformation φ defined on $(X,\mathfrak{B},m)$ is called **recurrent (infinitely recurrent)** if for every set B and for almost all $x\in B$, there exists an $n\in\mathbf{Z}^+$ (infinitely many $n\in\mathbf{Z}^+$) such that $\varphi^n(x)\in B$. A set W is called **wandering under** φ if $\varphi^{-n}(W)\cap\varphi^{-k}(W)=\varnothing$ for $n\neq k$. The transformation φ is called **conservative** if no sets of positive measure are wandering under φ, and **incompressible** if $B\supset\varphi^{-1}B$ implies $m(B-\varphi^{-1}B)=0$. The following statements about a nonsingular measurable transformation φ are equivalent: (i) φ is recurrent; (ii) φ is infinitely recurrent; (iii) φ is incompressible; (iv) φ is conservative. An immediate consequence of this is the following **recurrence theorem** of Poincaré (in the form formulated by Carathéodory): A measure-preserving transformation on a finite measure space is infinitely recurrent. In fact, in order for a nonsingular measurable transformation φ to be recurrent (and hence infinitely recurrent) it is sufficient that there exist a finite measure μ invariant under φ and equivalent to (i.e., mutually †absolutely continuous with) the given measure m.

The **invariant measure problem** is one of the basic problems in ergodic theory and is formulated in the following way: Given a nonsingular measurable transformation φ on a σ-finite measure space $(X,\mathfrak{B},m)$, find necessary and sufficient conditions for the existence of a finite (or σ-finite) measure invariant under φ and equivalent to m. The given measure m specifies only the class of equivalent measures among which an invariant measure is to be found. Therefore, we may assume without loss of generality that m is a finite measure. For the remainder of this section, we always mean by an invariant measure the one that is equivalent to m.

The Poincaré recurrence theorem states that φ being recurrent is necessary for the existence of a finite invariant measure. The recurrence of φ is, however, not sufficient, since an ergodic transformation with an infinite but σ-finite invariant measure is recurrent and has no finite invariant measure. A necessary and sufficient condition for the existence of a finite invariant measure was given by a theorem of A. Hajian and Kakutani (1964) which states: φ has a finite invariant measure if and only if φ has no **weakly wandering sets** (a set W is called **weakly wandering under** φ if there exists an infinite subset $\{n_k\}$ of $\mathbf{Z}^+$ such that $\varphi^{-n_k}W\cap\varphi^{-n_j}W=\varnothing$, $k\neq j$). Hajian also proved that a bimeasurable transformation φ has a finite invariant measure if and only if φ is **strongly recurrent** in the following sense: For every set E with $m(E)>0$, there exists a positive integer $k=k(E)$ such that

$$\max_{0\leqslant j\leqslant k} m(\varphi^{n-j}E\cap E)>0$$

for every $n\in\mathbf{Z}$. The existence of a finite invariant measure for φ is closely related to the validity of ergodic theorems, in particular the mean ergodic theorem. For instance, the condition that $\lim_{n\to\infty}(\sum_{k=0}^{n-1} m(\varphi^k E))/n$ exists and is positive for every set E with $m(E)>0$ is necessary and sufficient for the existence of a finite invariant measure for a bimeasurable φ. Actually, it was shown by A. Calderón and Y. Dowker that each of the following conditions alone is necessary and sufficient: (i) $m(E)>0$ implies

$$\liminf_{n\to\infty}\left(\frac{1}{n}\sum_{k=0}^{n-1} m(\varphi^k E)\right)>0;$$

(ii) $m(E)>0$ implies

$$\limsup_{n\to\infty}\left(\frac{1}{n}\sum_{k=0}^{n-1} m(\varphi^k E)\right)>0.$$

These and other related results follow immediately from the theorem of Hajian and Kakutani.

For a nonsingular bimeasurable transformation φ, a pair of sets A and B are said to be **countably equivalent under** φ if there exist countable decompositions $\{A_k | k\in\mathbf{Z}^+\}$ and $\{B_k | k\in\mathbf{Z}^+\}$ for A and B, respectively, and an infinite subset $\{n_k\}$ of $\mathbf{Z}$ such that $\varphi^{n_k}A_k = B_k$ for each k. A and B are said to be **finitely equivalent under** φ if finite decompositions $\{A_k\}$ and $\{B_k\}$ can be chosen. It was proved by Hopf that (i) φ is recurrent if and only if no set of positive measure is finitely equivalent under φ to one of its proper subsets, and (ii) φ has a finite invariant measure if and only if no set of positive measure is countably equivalent under φ to one of its proper subsets. It can be shown that if φ is ergodic, then φ has no σ-finite invariant measure if and only if every pair of sets of positive measure are countably equivalent under φ.

The first example of a transformation admitting no σ-finite invariant measure was constructed by Ornstein in 1960. Since then, a number of simpler examples have been obtained by L. Arnold, Brunel, and others. Arnold also obtained a useful necessary and sufficient condition for the existence of a σ-finite invariant measure. Although examples of transformations admitting no σ-finite invariant measure are still scarce, it was shown by A. Ionescu-Tulcea that in the group of all nonsingular bimeasurable transformations with a suitable metric, those having a σ-finite invariant measure form a subset of the †first category.

In order to construct examples of †factors in the theory of Von Neumann algebras (→ 430 Von Neumann Algebras), Von Neumann considered various ergodic groups of bimeasurable nonsingular transformations on a finite measure space. In this context a group $\mathcal{G} = \{g\}$ of transformations is called ergodic if $m((g^{-1}(B) \cup B) - (g^{-1}(B) \cap B)) = 0$ for every $g \in \mathcal{G}$ implies either $m(B) = 0$ or $m(X - B) = 0$. A measure μ is said to be invariant under the group $\mathcal{G} = \{g\}$ if μ is invariant under every transformation g in $\mathcal{G}$. Von Neumann's construction gives a type II_1 †factor if the group admits a finite invariant measure, a type II_∞ factor if it admits an infinite (but σ-finite) invariant measure, and a type III factor if it has no σ-finite invariant measure. Hajian and K. Itô extended the theorem of Hajian and Kakutani and proved that an arbitrary group $\mathcal{G} = \{g\}$ of nonsingular bimeasurable transformations admits a finite invariant measure if and only if no set of positive measure is weakly wandering under the group $\mathcal{G}$ (a set W is said to be **weakly wandering under a group** $\mathcal{G}$ if there exists an infinite subset $\{g_n \mid n \in \mathbf{Z}^+\}$ of $\mathcal{G}$ such that $g_n(W) \cap g_k(W) = \varnothing$ for $n \neq k$). L. Pukánszky and C. Moore studied a special class of ergodic groups of transformations and determined when these groups admit an invariant measure, finite or σ-finite, or no invariant measures. D. Hill and W. Krieger extended these results further and, among other things, clarified the connection between the work of Moore and the previously mentioned examples of Ornstein, Arnold, and Brunel.

For a bimeasurable nonsingular transformation φ, define the **full group** $[\varphi]$ to be the group of all bimeasurable nonsingular transformations ψ such that, for some $n = n(x)$, $\psi(x) = \varphi^n(x)$ for almost all x. Two transformations φ_1 and φ_2 are said to be **weakly equivalent** if there exists a bimeasurable nonsingular transformation θ such that $\theta[\varphi_1]\theta^{-1} = [\varphi_2]$. H. Dye proved that any pair of ergodic finite measure-preserving transformations are weakly equivalent. Krieger showed, on the other hand, that among ergodic transformations having no σ-finite invariant measures there are uncountably many weakly nonequivalent ones. By using the theory of weak equivalence, Krieger was able to produce type III factors that are not isomorphic to any of the previously known examples. Recent results of Hajian, Itô, and Kakutani show that the structure of orbits and full groups plays a significant role in the problem of σ-finite invariant measure.

A number of results on the invariant measure problem for nonsingular transformations have been extended to the case of nonsingular Markov transition functions (where the problem is formulated in a similar manner) by K. Itô, J. Neveu, S. Foguel, S. Horowitz, and others [10].

D. Examples and Construction of Measure-Preserving Transformations

Examples of measure-preserving transformations appear in many different contexts. We describe some of the important ones.

1. Let G be a †locally compact Abelian group satisfying the second axiom of †countability (→ 406 Topological Groups), $\mathfrak{B}$ a σ-algebra of Borel subsets of G, and m its †Haar measure (normalized if G is compact). Then $(G, \mathfrak{B}, m)$ is a Lebesgue measure space. For a fixed element $g_0 \in G$, define the transformation $\varphi_{g_0}: G \to G$ by $\varphi_{g_0}(g) = g + g_0$. Then φ_{g_0} is a bijective measure-preserving transformation on $(G, \mathfrak{B}, m)$ and is called the **rotation** on G by the element g_0. If G is compact, then the rotation φ_{g_0} is ergodic if and only if the cyclic subgroup generated by the element g_0 is dense in G. If this happens, the element g_0 is called the **topological generator** of G. A group is called **monothetic** if it has a topological generator.

2. If φ is a group †endomorphism of a compact Abelian group G, then φ preserves

the Haar measure m. If φ is a group †automorphism, then it is a bijective measure-preserving transformation on $(G, \mathfrak{B}, m)$. A continuous group automorphism φ induces a group automorphism φ^* of the character group G^*. The measure-preserving transformation φ is ergodic if and only if every character except the identity has an infinite orbit under the induced automorphism φ^*. When the group is the n-dimensional torus $\mathbf{T}^n$, a continuous group automorphism φ is uniquely represented by an $n \times n$ matrix with integer entries and with determinant ± 1. In this case, φ is ergodic if and only if no roots of unity appear among the †eigenvalues of the representing matrix.

3. Let $(Y, \mathfrak{A})$ be a measurable space, let $(Y_n, \mathfrak{A}_n) = (Y, \mathfrak{A})$ for each $n \in \mathbf{Z}$, and define $(Y^*, \mathfrak{A}^*)$ to be the †product measurable space $(\Pi_{n\in\mathbf{Z}} Y_n, \Pi_{n\in\mathbf{Z}} \mathfrak{A}_n)$. The transformation φ defined on $(Y^*, \mathfrak{A}^*)$ by

$$y^* = (\ldots, y_{-1}, y_0, y_1, \ldots) \to \varphi(y^*) = (\ldots, y'_{-1}, y'_0, y'_1, \ldots)$$

with $y'_n = y_{n+1}$ for each n is called the **shift transformation**. Let μ be a probability measure on $(Y, \mathfrak{A})$ such that $(Y, \mathfrak{A}, \mu)$ is a Lebesgue measure space, let $\mu_n = \mu$ for each n, and define μ^* to be the †product measure $\Pi_{n\in\mathbf{Z}} \mu_n$ on $(Y^*, \mathfrak{A}^*)$. Then $(Y^*, \mathfrak{A}^*, \mu^*)$ is a Lebesgue measure space, and the shift transformation φ is a bijective measure-preserving transformation. Considered with the product measure, φ is called a **generalized Bernoulli shift**. When the set Y is at most countable and the measure μ on Y is given by a sequence $\{p_j\}$ of positive numbers with $\Sigma p_j = 1$, φ is called a **Bernoulli shift**. Suppose that $P(y, A)$ is a Markov transition function on $(Y, \mathfrak{A})$ and π is a probability measure invariant under $P(y, A)$. Then we can define the **Markov measure** π^* on the product space $(Y^*, \mathfrak{A}^*)$ by setting

$$\pi^*(E^*) = \int_{A_{i+s}} \cdots \int_{A_{i+1}} \int_{A_i} \pi(dy_0) P(y_0, dy_1) \times P(y_1, dy_2) \ldots P(y_{s-1}, dy_s),$$

for a cylinder set

$$E^* = \Big(\prod_{j<i} Y_j\Big) \times \Big(\prod_{i \leq k \leq i+s} A_k\Big) \times \Big(\prod_{n > i+s} Y_n\Big),$$

and extending it to all of $\mathfrak{A}^*$. The shift transformation φ preserves the Markov measure π^*. Considered as a measure-preserving transformation on $(Y^*, \mathfrak{A}^*, \pi^*)$, φ is called a **Markov shift**.

A generalized Bernoulli shift is always ergodic. A Markov shift is ergodic if and only if the corresponding Markov process is †irreducible, which is the case if and only if the following property is satisfied: For every pair of sets A and B with $\pi(A)\pi(B) > 0$, there exists an $n \in \mathbf{Z}^+$ such that $\int_A P^n(y, B) d\pi > 0$.

There are other measures besides the product measure and the Markov measure that can be defined on the product space $(Y^*, \mathfrak{A}^*)$ and are invariant under the shift transformation φ. For example, if Y is a Borel subset of $\mathbf{R}$ and $\mathfrak{A}$ is the σ-algebra of Borel subsets of Y, then any †stationary stochastic process taking values in Y induces such a measure on $(Y^*, \mathfrak{A}^*)$. When considered with a measure of this type, the shift transformation φ is called the **shift associated with the stationary process**. Properties of the shift associated with a †stationary Gaussian process have been investigated by G. Maruyama, I. Girsanov, H. Totoki, and others. In particular, it is known that the shift is ergodic if and only if the †spectral measure for the †covariance function of the associated Gaussian process is continuous.

4. Other important examples of measure-preserving transformations arise from classical dynamical systems, which will be described in Section F.

5. There are several ways of constructing new measure-preserving transformations from given ones. We describe important cases.

(i) Let φ be a nonsingular, measurable, recurrent transformation (not necessarily measure-preserving) on a σ-finite measure space $(X, \mathfrak{B}, m)$, and let A be a set of positive measure. For $x \in A$, let $n(x) = \min\{n \in \mathbf{Z}^+ \mid \varphi^n(x) \in A\}$. The transformation $\varphi_A : A \to A$ defined by $\varphi_A(x) = \varphi^{n(x)}(x)$ is a nonsingular measurable transformation on the measure space $(A, A \cap \mathfrak{B}, m_A)$, where $m_A(B) = m(A \cap B)/m(A)$, and it is measure-preserving if φ is. We call φ_A the **transformation induced by φ on A**. It is ergodic if φ is ergodic.

(ii) Let φ be a nonsingular measurable transformation on a σ-finite measure space $(X, \mathfrak{B}, m)$, and suppose that $\{A_n\}$ is a countable (possibly finite) partition of X. Define a function $f: X \to \mathbf{Z}^+$ by setting $f(x) = n$ for $x \in A_n$, and let $\tilde{X} = \{(x, j) \mid x \in X, 1 \leq j \leq f(x)\}$ be a subspace of the product measure space $(X \times \mathbf{Z}^+, \mathfrak{B} \times \mathfrak{C}, m \times \mu)$, where $\mathfrak{C}$ is the σ-algebra of all subsets of $\mathbf{Z}^+$, and μ is the measure on $(\mathbf{Z}^+, \mathfrak{C})$ defined by $\mu(\{n\}) = 1$ for each $n \in \mathbf{Z}^+$. The transformation $\tilde{\varphi}$ defined on $\tilde{X}$ by $\tilde{\varphi}(x, j) = (x, j+1)$ if $1 \leq j < f(x)$ and $= (\varphi(x), 1)$ if $j = f(x)$ is a nonsingular measurable transformation on the measure space $(\tilde{X}, \tilde{X} \cap (\mathfrak{B} \times \mathfrak{C}), (m \times \mu)_{\tilde{X}})$, and it is measure-preserving if φ is. We call $\tilde{\varphi}$ the **transformation**

built from φ **with the ceiling function** f. If φ is ergodic and $m(A_n)\to 0$ as $n\to\infty$, then $\tilde{\varphi}$ is also ergodic.

(iii) Suppose that ψ is a measure-preserving transformation on $(X, \mathfrak{B}, m)$ and φ_x is a measure-preserving transformation on $(Y, \mathfrak{A}, \mu)$ for each $x\in X$. Assume that the mapping $(x,y)\to\varphi_x(y)$ is measurable with respect to the σ-algebras $\mathfrak{B}\times\mathfrak{A}$ and $\mathfrak{A}$. The transformation θ defined on the product space $(X\times Y, \mathfrak{B}\times\mathfrak{A}, m\times\mu)$ by $\theta(x,y)=(\psi(x),\varphi_x(y))$ is measure-preserving and is called the **skew product** of ψ and $\{\varphi_x\}$. If $\varphi_x=\varphi$ for all $x\in X$, then we get a **direct product** transformation $\theta(x,y)=(\psi(x),\varphi(y))$.

(iv) A measure-preserving transformation φ on $(Y, \mathfrak{A}, \mu)$ is said to be a **factor transformation** (or a **homomorphic image**) of a measure-preserving transformation ψ on $(X, \mathfrak{B}, m)$ if there exists a measurable transformation η from X onto Y such that $m\circ\eta^{-1}=\mu$ and $\varphi\eta=\eta\psi$. If $\mathfrak{B}'$ is a σ-subalgebra of $\mathfrak{B}$ and ψ leaves $\mathfrak{B}'$ invariant (i.e., $\psi^{-1}\mathfrak{B}'\subset\mathfrak{B}'$), then ψ induces a factor transformation φ on the measure space $(X,\mathfrak{B}',m)$. Conversely, if a measure-preserving transformation φ on $(Y,\mathfrak{A}, \mu)$ is a factor transformation of ψ on $(X,\mathfrak{B}, m)$ via a mapping η, then $\mathfrak{B}'=\eta^{-1}\mathfrak{A}$ is a σ-subalgebra of $\mathfrak{B}$ invariant under ψ.

6. A one-parameter family $\{\varphi_t \mid t\in\mathbf{R}\}$ of bijective measure-preserving transformations on a measure space $(X, \mathfrak{B}, m)$ is called a **flow**. A flow is called **continuous** if the mapping $t\to T_t$ is †weakly continuous where $\{T_t\}$ is the one-parameter family of †unitary operators on $L^2(X)$ induced by the flow $\{\varphi_t\}$. A flow is called **measurable** if the mapping $(t,x)\to\varphi_t(x)$ is a measurable transformation of $\mathbf{R}\times X$ into X. A measurable flow is continuous. A. Vershik and Maruyama proved that for any continuous flow there exists a measurable flow, unique in a specified sense, which is spatially isomorphic (in the sense specified in the next section) to the given flow.

Important examples of flows are given by classical dynamical systems (→ Section F), and by continuous-time stationary stochastic processes.

An important tool in the study of flows is provided by the theorem of W. Ambrose and Kakutani: Every measurable ergodic flow without a fixed point is spatially isomorphic to an S-**flow**. A measurable flow $\{\varphi_t\}$ is called an S-**flow** (**special flow** or **flow built under a function**) if there exist a measure-preserving transformation φ of a measure space $(X, \mathfrak{B}, m)$ and an $\mathbf{R}^+$-valued function f on $(X,\mathfrak{B},m)$ such that each φ_t is a measure-preserving transformation on the subspace $\tilde{X}=\{(x,u)\mid x\in X, 0\leqslant u\leqslant f(x)\}$ of the product measure space $(X\times\mathbf{R}^+, \mathfrak{B}\times\mathfrak{M}, m\times\lambda)$ given by

$$\begin{aligned}\varphi(x,u) &= (x,u+t) \text{ if } -u\leqslant t<-u+f(x),\\ &= \left(\varphi^n(x),\ u+t-\sum_{k=0}^{n-1} f(\varphi^k(x))\right) \text{ if }\\ &\quad -u+\sum_{k=0}^{n-1} f(\varphi^k(x))\leqslant t<-u+\sum_{k=0}^{n} f(\varphi^k(x)),\\ &= \left(\varphi^{-n}(x),\ u+t+\sum_{k=-n}^{-1} f(\varphi^k(x))\right) \text{ if }\\ &\quad -u+\sum_{k=-n}^{-1} f(\varphi^k(x))\leqslant t<-u-\sum_{k=-n+1}^{-1} f(\varphi^k(x)),\end{aligned}$$

$n\geqslant 1$. Here $\mathfrak{M}$ is the σ-algebra of Borel subsets of $\mathbf{R}^+$, and λ is the usual Lebesgue measure.

E. Isomorphism Problems

In this section, we assume that the Lebesgue measure spaces considered are probability spaces. For simplicity, following the common usage among Russian mathematicians, we call a measure-preserving transformation on $(X,\mathfrak{B},m)$ an **endomorphism** and a bijective measure-preserving transformation an **automorphism**.

An automorphism φ_1 on $(X_1, \mathfrak{B}_1, m_1)$ is said to be **spatially isomorphic** (or **metrically isomorphic**) to an automorphism φ_2 on $(X_2, \mathfrak{B}_2, m_2)$ if there exist sets N_1 and N_2 with $m_1(N_1)=m_2(N_2)=0$ and a bijective measurable transformation θ from X_1-N_1 to X_2-N_2 such that $m_2\circ\theta=m_1$ and $\theta\varphi_1=\varphi_2\theta$.

Classification of automorphisms into isomorphism classes constitutes the central problem of modern ergodic theory. Properties of automorphisms that are preserved under spatial isomorphisms are called **isomorphism invariants** (or **metric invariants**). There are several isomorphism invariants that are essential to the study of the isomorphism problem.

1. Spectral Invariants. Two automorphisms φ_1 and φ_2 are said to be **spectrally isomorphic** if the †unitary operators T_1 and T_2 induced by φ_1 and φ_2 on the Hilbert spaces $L_2(X_1)$ and $L_2(X_2)$, respectively, are unitarily equivalent (i.e., there exists an isometric isomorphism V of $L_2(X_1)$ onto $L_2(X_2)$ such that $VT_1=T_2V$). Properties preserved under spectral isomorphisms are called **spectral invariants** (or **spectral properties**). If φ_1 and φ_2 are spatially isomorphic, it is clear that they are spectrally isomorphic; but the converse is not true in general.

(i) The property of φ being ergodic is a

spectral property since φ is ergodic if and only if the number 1 is a simple eigenvalue of the induced unitary operator T. If φ is ergodic, then the set of all eigenvalues of the induced operator T forms a subgroup of the circle group, each eigenvalue is simple, and each †eigenfunction has constant absolute value. If the †spectrum of the induced operator T consists entirely of eigenvalues, φ is said to have **discrete spectrum** (or **pure point spectrum**). A theorem due to Von Neumann and P. Halmos—the first theorem on the question of isomorphism—states that two ergodic automorphisms φ_1 and φ_2 with discrete spectra are spatially isomorphic if and only if they are spectrally isomorphic, which is the case if and only if the induced operators T_1 and T_2 have the same set of eigenvalues. Furthermore, every ergodic automorphism with discrete spectrum is spatially isomorphic to an ergodic rotation on a compact Abelian group.

Analogous results were obtained by L. Abramov for a bigger class of automorphisms, namely, for ergodic automorphisms having so-called **quasidiscrete spectra**.

(ii) An automorphism φ is ergodic if and only if for every pair of sets A, B,

$$\lim_{n\to\infty}\frac{1}{n}\left(\sum_{k=0}^{n-1} m(\varphi^k(A)\cap B)\right)=m(A)m(B).$$

Strengthening this condition, we can define φ to be **weakly mixing** if for every pair of sets A, B,

$$\lim_{n\to\infty}\frac{1}{n}\left(\sum_{k=0}^{n-1}|m(\varphi^k(A)\cap B)-m(A)m(B)|\right)=0;$$

strongly mixing if

$$\lim_{k\to\infty} m(\varphi^k(A)\cap B)=m(A)m(B);$$

and k-**fold mixing** if for arbitrary choice of sets A_j, $j=0,1,\ldots,k$,

$$\lim m(A_0\cap\varphi^{n_1}A_1\cap\varphi^{n_2}A_2\ldots\cap\varphi^{n_k}A_k)$$
$$=m(A_0)m(A_1)\ldots m(A_k),$$

where the limit is taken as $n_1,n_2,\ldots,n_k\to\infty$ in such a way that $n_1<n_2<\ldots<n_k$ and $\min_{1\leqslant j\leqslant k}(n_j-n_{j-1})\to\infty$. The property of an automorphism φ being weakly mixing or strongly mixing is a spectral property. For instance, φ is weakly mixing if and only if the number 1 is a simple eigenvalue and is the only eigenvalue of the induced operator T. It is also known that φ is weakly mixing if and only if the direct product automorphism $\varphi\times\varphi$ is ergodic. The set of all weakly mixing automorphisms forms a dense †G_δ-set in the group of all automorphisms on $(X,\mathfrak{B},m)$ considered with the so-called weak topology (Halmos's theorem). On the other hand, it was shown by V. Rohlin that the set of all strongly mixing automorphisms is a set of first category with respect to the weak topology. However, there are only a few known examples of automorphisms that are weakly mixing but not strongly mixing.

(iii) An automorphism φ is said to have **countable Lebesgue spectrum** if the maximal spectral type of the induced unitary operator T restricted to the †orthocomplement of the subspace of constant functions in $L_2(X)$ is equivalent to the Lebesgue measure and its †multiplicity is countably infinite.

(iv) An automorphism φ is called a K-**automorphism** (or **Kolmogorov automorphism**) if there exists a σ-subalgebra $\mathfrak{B}_0$ of $\mathfrak{B}$ such that (α) $\varphi\mathfrak{B}_0\supset\mathfrak{B}_0$ and $\varphi\mathfrak{B}_0\neq\mathfrak{B}_0$, ($\beta$) $\bigvee_{n\in\mathbf{Z}}\varphi^n\mathfrak{B}_0=\mathfrak{B}$, and ($\gamma$) $\bigwedge_{n\in\mathbf{Z}}\varphi^n\mathfrak{B}_0=\mathfrak{N}$, where $\mathfrak{N}$ is the σ-subalgebra of $\mathfrak{B}$ consisting of null sets and their complements. K-automorphisms are k-fold mixing for all orders k and have countable Lebesgue spectra.

Generalized Bernoulli shifts are all K-automorphisms. An ergodic Markov shift is a K-automorphism if and only if it is strongly mixing, which is the case if and only if the corresponding Markov process is †irreducible and †aperiodic. A continuous group automorphism of a compact Abelian group is a K-automorphism if and only if it is ergodic. In particular, a continuous group automorphism of the n-dimensional torus $\mathbf{T}^n$ is a K-automorphism if and only if no roots of unity appear among the eigenvalues of the representing matrix. Automorphisms arising from classical dynamical systems also provide examples of K-automorphisms. In particular, each automorphism (except the identity) of a †geodesic flow on a surface of negative curvature is a K-automorphism. For the shift transformation φ associated with a stationary Gaussian process, it was shown by Maruyama that φ is (α) weakly mixing if and only if it is ergodic, (β) strongly mixing if and only if the covariance function of the associated Gaussian process tends to 0 as $n\to\infty$, and (γ) a K-automorphism if and only if the †spectral measure of the covariance function is absolutely continuous with respect to the Lebesgue measure.

(v) Examples of automorphisms having various types of spectra have been constructed by a number of authors by using stationary Gaussian processes and the theory of approximation developed by A. Katok and A. Stepin.

(vi) An ergodic automorphism with quasidiscrete spectrum has a mixed spectrum, that is, the spectrum of the induced unitary operator T has a continuous component and eigenvalues in addition to 1. Anzai (1951) constructed a special class of skew product automorphisms having mixed spectra and showed that in this class there are automorphisms that are spectrally isomorphic but not spatially isomorphic. However, the question whether two spatially nonisomorphic automorphisms exist among automorphisms having the same purely continuous spectrum remained unanswered for a long time, until in 1958 Kolmogorov settled it affirmatively by using a new isomorphism invariant called **entropy**.

2. Generators and Entropy. (i) By a **partition** $\xi=\{A_\lambda\}$ of the space X we mean a collection of sets A_λ such that $A_\lambda \cap A_{\lambda'}=\varnothing$ whenever $\lambda\neq\lambda'$ and $\bigcup A_\lambda = X$. We denote by ξ_0 the partition of X into individual points, and by ν the trivial partition $\{X\}$. A partition into a finite (countable) number of sets is called a finite (countable) partition. A partition ξ is said to be finer than another partition ζ (or ζ is coarser than ξ) if for every $A\in\xi$ there is a set $B\in\zeta$ such that $A\subset B$. For a collection $\{\xi_\alpha\}$ of partitions of X, we denote by $\bigvee_\alpha\xi_\alpha$ the coarsest partition that is finer than each ξ_α, and by $\bigwedge_\alpha\xi_\alpha$ the finest partition that is coarser than each ξ_α. If $\xi_k=\{A_{k,n}\}$ is a sequence of countable (or finite) partitions, then $\bigvee_k\xi_k$ is precisely the partition of X into nonempty intersections of the form $\bigcap_k A_{k,n_k}$ with $A_{k,n_k}\in\xi_k$ for each k. With a partition ξ of X we associate a σ-subalgebra $\mathfrak{B}(\xi)$ of $\mathfrak{B}$ which is the σ-algebra of all $\mathfrak{B}$-measurable sets that are a union of elements in ξ. Two partitions ξ and ζ are said to coincide a.e. if $\mathfrak{B}(\xi)=\mathfrak{B}(\zeta)$ a.e. (i.e., for every $A\in\mathfrak{B}(\xi)$ there exists a set B in $\mathfrak{B}(\zeta)$ such that $m(A\cup B - A\cap B)=0$ and conversely).

(ii) Suppose that φ is an endomorphism of $(X,\mathfrak{B},m)$ and ξ a partition of X. By $\varphi^{-1}\xi$ we mean the partition $\{\varphi^{-1}(A)|A\in\xi\}$. If φ is an automorphism, we also define $\varphi\xi=\{\varphi(A)|A\in\xi\}$. A partition ξ is called a **generator** for an endomorphism φ if $\bigvee_{n=0}^{\infty}\varphi^{-n}\xi=\xi$ a.e. If $\bigvee_{n=-\infty}^{\infty}\varphi^{n}\xi=\xi$ a.e., ξ is called a **two-sided generator** for an automorphism φ.

An endomorphism φ is said to be **periodic at a point** $x\in X$ if there exists a positive integer n such that $\varphi^n(x)=x$, and **aperiodic** if the set of points of periodicity has measure zero. If the measure space $(X,\mathfrak{B},m)$ is nonatomic, then every ergodic endomorphism on it is aperiodic. A theorem of Rohlin states that every aperiodic automorphism φ has a countable two-sided generator. This implies that every such φ is spatially isomorphic to the shift transformation on the infinite product space $(Y^*,\mathfrak{C}^*)$ considered with some invariant measure μ^*, where each coordinate space $Y_j=Y$ has at most a countable number of points. Recently Krieger improved this result by showing that if an ergodic automorphism φ has finite entropy, then φ has a finite two-sided generator.

(iii) For a finite or countable partition $\xi=\{A_n\}$, define the **entropy** $H(\xi)$ **of the partition** to be $-\sum_n m(A_n)\log(m(A_n))$ (the logarithms here and below are natural logarithms). We denote by $\mathfrak{Z}$ the set of all partitions ξ with $H(\xi)<\infty$. If $\xi\in\mathfrak{Z}$, then for any endomorphism φ the limit

$$h(\varphi,\xi)=\lim_{n\to\infty}\frac{1}{n}\left(H\left(\bigvee_{k=0}^{n}\varphi^{-k}\xi\right)\right)$$

exists and is finite. The **entropy** $h(\varphi)$ **of the endomorphism** φ is defined to be $\sup\{h(\varphi,\xi)|\xi\in\mathfrak{Z}\}$ and is an isomorphism invariant. Properties of entropy have been investigated extensively since the notion was introduced by Kolmogorov. We cite a few results.

(a) If a partition $\xi\in\mathfrak{Z}$ is a generator for an endomorphism φ or a two-sided generator for an automorphism φ, then $h(\varphi)=h(\varphi,\xi)$ (Sinaĭ's lemma). (b) For every integer n, $h(\varphi^n)=|n|h(\varphi)$. (c) If an automorphism φ is periodic, then $h(\varphi)=0$. (d) If φ_1 is a factor transformation of φ_2, then $h(\varphi_1)\leqslant h(\varphi_2)$. (e) $h(\varphi_1\cdot\varphi_2)=h(\varphi_1)+h(\varphi_2)$. (There is a more complicated formula (due to Rohlin) for the entropy of a skew product automorphism.) (f) If φ is a recurrent automorphism and φ_A is the automorphism induced by φ on a subset A with $m(A)>0$, then $h(\varphi_A)=h(\varphi)/m(A)$. (g) If φ is a Bernoulli shift with probability distribution $\{p_n\}$, then $h(\varphi)=-\sum_n p_n\log p_n$. (h) If φ is a Markov shift based on the Markov transition probability P_{ij} (defined on a countable or finite state space) and an invariant measure π_i, then

$$h(\varphi)=-\sum_i\sum_j\pi_i P_{ij}\log P_{ij}.$$

(i) If φ is ergodic and has a pure point spectrum, or more generally, has a quasidiscrete spectrum, then $h(\varphi)=0$. (j) For an ergodic group automorphism φ on an n-dimensional torus, $h(\varphi)=\sum\log|\lambda|$, where the sum is taken over all eigenvalues λ of modulus >1 of the representing matrix. (k) If an automorphism φ has positive entropy, then in $L_2(X)$ there exists a subspace invariant under the induced unitary operator T such that the spectrum of T restricted to this subspace is countable Lebesgue (Rohlin's theorem). It follows from (k) that automorphisms with †singular spectra or spectra of finite multiplicity must have zero entropy. In proving assertion (g), Kolmogorov established for the first time the fact that there are uncountably many spatially nonisomorphic Bernoulli shifts.

(iv) An automorphism φ is said to have **completely positive entropy** if $h(\varphi,\xi)>0$ for every partition $\xi\neq\nu$. It was shown by Rohlin and Ja. Sinaĭ that an automorphism φ has completely positive entropy if and only if φ is a K-automorphism. M. Pinsker proved that for every automorphism φ there exists a partition, called the **Pinsker partition**, that is invariant under φ and such that the factor transformation of φ with respect to this partition has zero entropy and is the largest among the factor transformations of φ with zero entropy.

(v) Rohlin showed that $h(\varphi)=0$ for an endomorphism φ, if and only if all of its factor transformations are automorphisms. An endomorphism φ is called **exact** if $\bigwedge_{n=0}^{\infty}\varphi^{-n}\xi=\nu$. Rohlin introduced a way to associate to each endomorphism a certain automorphism, called the **natural extension**, which reflects the properties of the endomorphism. For example, an endomorphism and its natural extension are simultaneously ergodic or nonergodic, are mixing of the same order, and have equal entropy. The natural extension of an exact endomorphism is a K-automorphism.

(vi) Automorphisms φ_1 and φ_2 are said to be **weakly isomorphic** if each of them is a factor transformation of the other. Sinaĭ proved (1962) that two strongly mixing Markov shifts with equal entropy are weakly isomorphic. In 1969 Ornstein succeeded in proving the following remarkable result: Two Bernoulli shifts with equal entropy are spatially isomorphic. Partial results in this direction were obtained earlier by L. Meshalkin, J. Blum, and D. Hanson. In the proof of Ornstein's theorem, essential use was made of the following theorem of C. Shannon and B. McMillan, which plays a fundamental role in information theory (→ 216 Information Theory): Suppose that φ is an ergodic endomorphism on $(X, \mathfrak{B}, m)$ and ξ is a partition of X. For a point $x\in X$, let $A_n(x)$ denote the element in the partition $\bigvee_{k=0}^{n-1}\varphi^{-k}\xi$ that contains x. Then for almost all x,

$$\lim_{n\to\infty}\left(-\frac{1}{n}\log m(A_n(x))\right)$$

exists and equals $h(\varphi,\xi)$.

A sequence $\{\xi_n\}$ of partitions is said to be **independent** if for every choice of sets $A_1,\ldots,A_s$ with $A_k\in\xi_{n_k}$, $m(\bigcap_{k=1}^{s}A_k)=\prod_{k=1}^{s}m(A_k)$ whenever $n_1,n_2,\ldots,n_s$ are all distinct. For a fixed $\delta>0$, two partitions ξ and ζ are said to be **δ-independent** if

$$\sum_{A\in\xi,B\in\zeta}|m(A\cap B)-m(A)m(B)|<\delta.$$

A partition ξ is called a **Bernoulli partition** for an automorphism φ if the sequence of partitions $\{\varphi^n\xi | n\in\mathbf{Z}\}$ is independent, and a **weak Bernoulli partition** for φ if for every $\delta>0$ there exists a $k>0$ such that the partitions $\bigvee_{i=k}^{k+n}\varphi^i\xi$ and $\bigvee_{i=-n}^{0}\varphi^i\xi$ are δ-independent for all $n\geqslant 0$. A partition ξ is called a two-sided Bernoulli (weak Bernoulli) generator for an automorphism φ if it is a Bernoulli (weak Bernoulli) partition and is a two-sided generator for φ. An automorphism having a two-sided Bernoulli generator is spatially isomorphic to a generalized Bernoulli shift, which, in turn, is isomorphic to a Bernoulli shift of equal entropy. Ornstein and N. Friedman refined the techniques used in the proof of the theorem of Ornstein and proved that an automorphism having a two-sided weak Bernoulli generator is spatially isomorphic to a Bernoulli shift of equal entropy. Strongly mixing Markov shifts have two-sided weak Bernoulli generators and therefore are spatially isomorphic to Bernoulli shifts. Ornstein was able to prove further that (a) A nontrivial factor transformation of a Bernoulli shift is a Bernoulli shift. (b) There exists a K-automorphism that is not spatially isomorphic to a Bernoulli shift. Consequently, entropy is not a complete isomorphism invariant among K-automorphisms. (c) Every Bernoulli shift can be embedded in a flow. (d) There exists an automorphism that cannot be written as a direct product of a K-automorphism and an automorphism with zero entropy. This example shows that a conjecture made earlier by Pinsker was false.

(vii) Y. Katznelson showed that every ergodic group automorphism on an n-dimensional torus has a two-sided generator satisfying a condition somewhat weaker than the weak Bernoulli condition (he called it the almost weak Bernoulli condition). Using an extension of the theorem of Friedman and Ornstein, we can prove that an automorphism having an almost weak Bernoulli two-sided generator is also spatially isomorphic to a Bernoulli shift. Consequently, an ergodic group automorphism on an n-dimensional torus is spatially isomorphic to a Bernoulli shift with equal entropy. R. Adler and B. Weiss earlier proved by entirely different methods that on a 2-dimensional torus, two ergodic group automorphisms having the same entropy are spatially isomorphic. A partition ξ is called a **Markov partition** for an automorphism φ if for every $A\in\mathbf{B}(\xi)$, $m(A|\bigvee_{n=1}^{\infty}\varphi^{-n}\xi)=m(A|\varphi^{-1}\xi)$ a.e. If φ has a Markov partition that is also a two-sided generator, then φ is spatially isomorphic to a Markov shift. Call a partition a two-sided Markov generator if it is a Markov partition and a two-sided generator. Adler and Weiss showed that every ergodic group automorphism of a 2-dimensional torus has a two-sided Markov generator. Sinaĭ proved later that the same is true for an ergodic group automorphism of an n-dimensional torus if the representing matrix of the group automorphism has no eigenvalue of modulus 1. Combined with the theorem of Friedman and Ornstein, this gives a different proof that such a group automorphism is spatially isomorphic to a Bernoulli shift.

(viii) Examples of exact endomorphisms arise in connection with problems in number theory; for example, †continued fraction expansion and β-expansion. The natural extensions of the exact endomorphisms associated with continued fraction expansion and β-expansion are now known to be spatially isomorphic to Bernoulli shifts.

F. Classical Dynamical Systems

By a **classical dynamical system** we mean a †diffeomorphism or a flow generated by a smooth †vector field on some †differential manifold M^n. Such a system is nonsingular with respect to a measure defined by any †Riemannian metric on M^n. For a fixed Riemannian metric, we call measures **smooth** if they have a smooth density with respect to the measure given by the metric.

1. Among classical dynamical systems, **geodesic flows** have been investigated most extensively. Let $\mathfrak{I}_1(M)$ be the unitary †tangent bundle over the manifold M^n. A point $(x,e)\in\mathfrak{I}_1(M)$ defines a unique †geodesic through x in the direction of e. The geodesic flow on $\mathfrak{I}_1(M)$ is the flow defined by $\varphi_t(x,e)=(x_t,e_t)$, where x_t is the point in M^n reached from x after time t under a motion with unit speed along the geodesic determined by (x,e), and e_t is the unit vector at x_t tangent to the geodesic. The classical †Liouville theorem in this context implies that the measure on $\mathfrak{I}_1(M)$ that is the product of the measure on M^n induced by the metric and the Lebesgue measure on the $(n-1)$-dimensional sphere gives a smooth invariant measure for the geodesic flow. A wide class of systems arising from mechanics can be described as geodesic flows.

Hopf and G. Hedlund proved that if the manifold M^n is compact and has constant negative curvature, then the geodesic flow is strongly mixing. Later, by using the theory of group †representations, I. Gel'fand and S. Fomin proved that the spectrum of a geodesic flow on a compact manifold of constant negative curvature is Lebesgue, and is even countable Lebesgue in the case where the manifold is of dimension 2. F. Mautner and later L. Auslander, L. Green, and F. Hahn extended this algebraic method to flows obtained under the action of some one-parameter subgroup of a †Lie group acting on its †homogeneous space and obtained extensive results for the case of †nilpotent and some †solvable Lie groups.

2. The flow on an n-dimensional torus defined by

$$\varphi_t(x_1,x_2,\ldots,x_n)$$
$$=(x_1+\omega_1 t,x_2+\omega_2 t,\ldots,x_n+\omega_n t)$$

is called a **translational flow**. The numbers $\omega_1,\omega_2,\ldots,\omega_n$ are called **frequencies**. Every orbit of $\{\varphi_t\}$ is dense in the torus if and only if the frequencies are linearly independent over **Z**. The motion under a translational flow with independent frequencies is called a **quasiperiodic** motion. A translational flow for a quasiperiodic motion has discrete spectrum. Under a flow generated by an integrable †Hamiltonian vector field on a manifold M^{2n}, the manifold is decomposed into n-dimensional †submanifolds, each of which is †diffeomorphic to an n-dimensional torus and is invariant under the flow, and the restriction of the flow on each of these submanifolds is isomorphic (in fact, diffeomorphic) to a translational flow (Arnold's theorem). Kolmogorov, V. I. Arnold, and J. Moser investigated a number of conditions under which a small †perturbation of a Hamiltonian function preserves the phase portrait of the flow in a large portion of the manifold so that in this part the decomposition into invariant tori on each of which the motion is quasiperiodic still takes place under the perturbed flow.

3. Sinaĭ obtained a useful criterion for a classical dynamical system to be a K-system. Let M^n be compact, and suppose that $\{\varphi_t\}$ is a flow on M^n defined by a smooth vector field and preserving some smooth measure μ. A one-parameter group $\{\psi_t\}$ of transformations of the space M^n, given by a vector field, is called the flow **transversal** to the flow $\{\varphi_t\}$ if (i) the decomposition of the space M^n into the trajectories of the flow $\{\psi_t\}$ is invariant under $\{\varphi_t\}$; (ii) the limit $\lim_{s\to0}\lim_{t\to0}(W_s(t,x)-t)/ts=\alpha(x)$ exists for the function $W_s(t,x)$, which is defined to be the time length of the segment $\{\varphi_s\psi_u(x)\mid 0\leqslant u\leqslant t\}$ of the trajectory of the flow $\{\psi_t\}$. Sinaĭ's fundamental theorem states that if a flow $\{\varphi_t\}$ is ergodic and has a transversal ergodic flow $\{\psi_t\}$ for which $\int\alpha(x)\,d\mu<0$, then $\{\varphi_t\}$ is a K-flow. If $\alpha(x)<0$, then we can even drop the assumption that $\{\varphi_t\}$ is ergodic. If $\int\alpha(x)\,d\mu>0$, the theorem holds for the flow $\{\varphi_{-t}\}$.

A geodesic flow on a 2-dimensional manifold of constant negative curvature always has a transversal flow, called a **horocycle flow**. The ergodicity of a horocycle flow was proved by Hedlund. It follows from Sinaĭ's fundamental theorem, therefore, that a geodesic flow on a surface of constant negative curvature is a K-flow. Sinaĭ proved even more: A geodesic flow on any surface of negative curvature is a K-flow. There is an extension of the notion of transversal flow to higher dimensions, called **transversal field**. Using this notion Sinaĭ proved that a geodesic flow on a manifold (of any dimension) of constant negative curvature is a K-flow.

4. D. Anosov considered a class of flows and diffeomorphisms satisfying a condition that characterizes unstable motions such as geodesic flows on a manifold of negative curvature. A flow $\{\varphi_t\}$ on a closed connected Riemannian manifold M is called an **Anosov flow** (or **C-flow**) if (i) the velocity vector $d\varphi_t/dt|_{t=0}$ is nonvanishing; (ii) the †tangent space $\mathfrak{T}M_x$ at x splits into a direct sum $\mathfrak{T}M_x = X_x \oplus Y_x \oplus Z_x$, where Z_x is the one-dimensional space generated by the velocity vector at x and $\dim(X_x) = k \neq 0$, $\dim(Y_x) = s \neq 0$ for each x; and (iii) there exist constants a, b and λ (independent of x and t) such that for any positive real number t,

$$\|\tilde{\varphi}_t q\| \geqslant a e^{\lambda t}\|q\|, \quad \|\tilde{\varphi}_{-t} q\| \leqslant b e^{-\lambda t}\|q\| \qquad \text{for} \quad q \in X_x,$$

$$\|\tilde{\varphi}_t q\| \leqslant b e^{-\lambda t}\|q\|, \quad \|\tilde{\varphi}_{-t} q\| \geqslant a e^{\lambda t}\|q\| \qquad \text{for} \quad q \in Y_x,$$

where $\{\tilde{\varphi}_t\}$ is the flow induced on the tangent space by $\{\varphi_t\}$. Conditions analogous to (ii) and (iii) can be formulated also for a diffeomorphism, which defines an **Anosov diffeomorphism** (or **C-diffeomorphism**). Anosov proved that if an Anosov flow has a smooth invariant measure, then it is ergodic and either it has a continuous nonconstant eigenfunction or it is a K-flow. Anosov diffeomorphisms with smooth invariant measures are K-automorphisms. Geodesic flows on manifolds of negative curvature are Anosov diffeomorphisms. All the results mentioned for geodesic flows and their generalizations, therefore, follow from the theorem of Anosov. Anosov proved also that his systems are **structurally stable**. This means that any diffeomorphism sufficiently close to a given one in the C^1-topology is conjugate to the given one via some homeomorphism of the manifold that is close to the identity mapping. Structural stability and other related properties have been investigated in detail for a large class of systems by S. Smale and others [1, 2, 35].

5. An important example of a system that is not an Anosov system because of nonsmoothness has been studied by Sinaĭ: the simplest mechanical model due to Boltzmann and Gibbs of an ideal gas, which is described as a system generated by tiny rigid spherical pellets moving inside a rectangular box and colliding elastically. Sinaĭ succeeded in proving that this is a K-system, thereby giving an affirmative answer to the classical question of the ergodicity of the basic model of statistical mechanics [34].

G. Topological Dynamics

1. Let X be a compact metric space and $\varphi: X \to X$ a †homeomorphism. N. Krylov and N. Bogoljubov showed that there always exists on X a Borel probability measure μ that is invariant under φ. Let $\mathfrak{P}$ be the collection of all Borel probability measures on X, and let $\mathfrak{P}_\varphi$ be the subset of $\mathfrak{P}$ consisting of those invariant under φ. Then $\mathfrak{P}$ and $\mathfrak{P}_\varphi$ are both convex sets compact with respect to the †weak * topology. If $\mathfrak{E}_\varphi$ is the set of all extreme points in $\mathfrak{P}_\varphi$, then by the †Kreĭn-Milman theorem $\mathfrak{E}_\varphi$ is not empty. A measure μ in $\mathfrak{P}_\varphi$ belongs to $\mathfrak{E}_\varphi$ if and only if φ is ergodic with respect to μ. When the set $\mathfrak{E}_\varphi$ consists of a single element, φ is called **uniquely ergodic**; φ is called **minimal** if for every point $x \in X$, the orbit of x under $\varphi = \mathrm{orb}_\varphi(x) = \{\varphi^n(x) \mid n \in \mathbf{Z}\}$ is dense in X. φ is called **strictly ergodic** if it is both minimal and uniquely ergodic. A theorem of J. Oxtoby states that φ is strictly ergodic if and only if for every continuous real-valued function f on X, the sequence of averages $(\sum_{k=0}^{n-1} f(\varphi^k(x)))/n$ converges uniformly to a constant $M(f)$. There are homeomorphisms that are minimal or uniquely ergodic but not strictly ergodic.

2. If for each $i \in \mathbf{Z}$, $X_i = X$ is a compact metric space, the Cartesian product $X^* = \prod_{i \in \mathbf{Z}} X_i$ is compact with respect to the †product topology, and the shift transformation φ on X^* is a homeomorphism. If X consists of a finite number of elements, (X^*, φ) is called a **symbolic dynamical system**. If X has at least two points, (X^*, φ) is not minimal. However, there are closed φ-invariant subsets Ω of X^* such that the system (Ω, φ) is minimal or is strictly ergodic. For a point x^* of X^*, necessary and sufficient conditions are known for the shift dynamical system $(\overline{\mathrm{orb}_\varphi(x^*)}, \varphi)$ to be minimal or strictly ergodic. For example, the Morse sequence x^* generates a minimal dynamical system.

3. R. Jewett proved that any weakly mixing measure-preserving transformation on a Lebesgue space is spatially isomorphic to a uniquely ergodic transformation. Krieger extended the result by showing that if the entropy of an ergodic transformation φ is finite, then φ is spatially isomorphic to a strictly ergodic transformation. Similar results were obtained for flows by K. Jacobs.

4. A homeomorphism φ on a compact metric space X is said to be **equicontinuous** if $\{\varphi^n \mid n \in \mathbf{Z}\}$ is an †equicontinuous family of transformations on X. An equicontinuous homeomorphism is minimal and isometrically isomorphic to a rotation on a compact Abelian group. If $d(x,y) > 0$ implies

$$\inf_{n \in \mathbf{Z}} d(\varphi^n(x), \varphi^n(y)) > 0,$$

φ is called **distal**. An equicontinuous homeomorphism is distal, but the converse is not true in general. A theorem of H. Fürstenberg states that every minimal distal transformation can be obtained by a succession (possibly †transfinite) of isometric extensions of the trivial homeomorphism on the space consisting of a single point. Extensive work has been done on minimal dynamical systems, symbolic dynamical systems, and other related topics by R. Ellis, W. Gottschalk, Hedlund, Fürstenberg, and others [9, 14].

5. A topological analog of the notion of entropy, called **topological entropy**, was introduced by Adler, A. Konheim, and J. McAndrew. This is defined as follows: For every open covering $\mathfrak{A}$ of a compact topological space X, let $N(\mathfrak{A})$ be the number of sets in the minimal subcovering of $\mathfrak{A}$. For open coverings $\mathfrak{A}$ and $\mathfrak{B}$, let $\mathfrak{A}\vee\mathfrak{B}$ be the open covering $\{A\cap B \mid A\in\mathfrak{A}, B\in\mathfrak{B}\}$. For any open covering $\mathfrak{A}$ and a continuous mapping φ on X, the limit $\lim_{n\to\infty}(\log N(\mathfrak{A}\vee\varphi^{-1}\mathfrak{A}\vee\ldots\vee\varphi^{-(n-1)}\mathfrak{A}))/n=h_{\mathrm{top}}(\varphi,\mathfrak{A})$ exists. Topological entropy $h_{\mathrm{top}}(\varphi)$ of the continuous transformation φ is now defined by $h_{\mathrm{top}}(\varphi)=\sup\{h_{\mathrm{top}}(\varphi,\mathfrak{A}) \mid \mathfrak{A}$ an open covering of $X\}$.

L. Goodwyn showed that $h_{\mathrm{top}}(\varphi)\geqslant h_\mu(\varphi)$ for any φ-invariant probability measure μ, where $h_\mu(\varphi)$ is the measure-theoretic entropy of φ regarded as a μ-preserving transformation. T. Goodman went further and succeeded in proving that $h_{\mathrm{top}}(\varphi)=\sup\{h_\mu(\varphi) \mid \mu$ a φ-invariant probability measure$\}$. This result confirms a conjecture made earlier by Adler, Konheim, and McAndrew. For an ergodic group automorphism φ of a compact Abelian group, $h_{\mathrm{top}}(\varphi)$ equals $h_m(\varphi)$, where m is the Haar measure of the group, and the Haar measure is the unique measure having this property (K. Berg and R. Bowen).

6. Fürstenberg introduced an interesting and useful notion called **disjointness** and investigated various consequences of this concept in the study of measure-preserving transformations and dynamical systems [12]. Ellis began an algebraic theory of minimal dynamical systems and succeeded in obtaining systematically a number of results that were previously obtained by diverse methods [9].

H. Miscellany

1. An arbitrary aperiodic automorphism can be approximated by periodic automorphisms. More precisely, if φ is an aperiodic automorphism, then for any positive integer n and $\delta>0$, there exists a periodic automorphism ψ of period n such that $m(\{x \mid \varphi(x)\neq\psi(x)\})<(1/n)+\delta$ (theorem of Halmos and Rohlin). The question as to how quickly this approximation can be carried out has been investigated in detail by Katok, Stepin, and others. The rate of this approximation was shown to have a close relationship with the entropy and spectral properties of the automorphism φ. By utilizing this relationship, various examples of automorphisms with specified spectral properties have been constructed [20].

2. When an arbitrary automorphism φ is given on a Lebesgue measure space $(X,\mathfrak{B},m)$, there exists a unique decomposition $\xi=\{A_\lambda\}$ of X such that (i) each A_λ is invariant under φ and (ii) except for a negligible set (in a specified sense) of A_λ, each A_λ is turned (in a natural manner) into a Lebesgue measure space, and the restriction of φ to A_λ is an ergodic automorphism. This decomposition is called the **ergodic decomposition** of X with respect to φ. There is a corresponding decomposition with respect to a flow. A formula also exists that enables us to compute the entropy $h(\varphi)$ in terms of the entropies of ergodic components of φ.

3. Application of ergodic theory to problems in analytic number theory has been made by several authors. Ergodic or mixing properties of particular measure-preserving transformations that arise in connection with various problems in number theory have been exploited to give answers to these problems. New and more striking applications of ideas of ergodic theory to different types of questions in number theory have been started by Ju. Linnik and by Fürstenberg [12, 23].

References

[1] Ф. В. Аносов, Я. Г. Синай (D. V. Anosov and Ja. G. Sinaĭ), Некоторые гладкие эргодические системы, Uspehi Mat. Nauk, 22 (5) (1967), 107–172; English translation, Some smooth ergodic systems, Russian Math. Surveys, 22 (5) (1967), 103–167.
[2] V. I. Arnold and A. Avez, Ergodic problems of classical mechanics, Benjamin, 1968.
[3] J. Auslander and W. H. Gottschalk (eds.), Topological dynamics, an international symposium, Benjamin, 1968.
[4] L. Auslander, L. Green, F. Hahn, et al., Flows on homogeneous spaces, Ann. Math. Studies, Princeton Univ. Press, 1963.
[5] A. Avez, Ergodic theory of dynamical systems I, II, Lecture notes, Univ. of Minnesota, 1966–1967.
[6] P. Billingsley, Ergodic theory and information, John Wiley, 1965.
[7] N. Dunford and J. T. Schwartz, Linear Operators I, Interscience, 1958.

[8] H. A. Dye, On groups of measure preserving transformations I, II, Amer. J. Math., 81 (1959), 119–159, ibid., 85 (1963), 551–576.
[9] R. Ellis, Lectures on topological dynamics, Benjamin, 1969.
[10] S. R. Foguel, Ergodic theory of Markov processes, Van Nostrand, 1969.
[11] N. A. Friedman, Introduction to ergodic theory, Van Nostrand, 1970.
[12] H. Fürstenberg, Disjointness in ergodic theory, minimal sets and a problem in diophantine approximation, Math. Systems Theory, 1 (1967), 1–49.
[13] A. M. Garsia, Topics in almost everywhere convergence, Markham, 1970.
[14] W. H. Gottschalk and G. A. Hedlund, Topological dynamics, Amer. Math. Soc. Colloq. Publ., 1955.
[15] P. R. Halmos, Lectures on ergodic theory, Publ. Math. Soc. Japan, 1956.
[16] E. Hopf, Ergodentheorie, Springer, 1937.
[17] K. Jacobs, Neuere Methoden und Ergebnisse der Ergodentheorie, Springer, 1960.
[18] K. Jacobs, Lecture notes on ergodic theory I, II, Aarhus Univ., 1962–1963.
[19] S. Kakutani, Ergodic theory, Proc. Intern. Congr. Math., 1950, Cambridge, 128–142.
[20] А. Б. Каток, А. М. Стёрин (A. B. Katok and A. M. Stepin), Аппроксимации в эргодической теории, Uspehi Mat. Nauk, 22 (5) (1967), 81–106; English translation, Approximations in ergodic theory, Russian Math. Surveys, 22 (5) (1967), 77–102.
[21] А. Н. Колмогоров (A. N. Kolmogorov), Оьщая теория динамических систем и классическая механика, Proc. Intern. Congr. Math., 1954, Amsterdam (North-Holland), 1957, p. 315–333; English translation, General theory of dynamical systems and classical mechanics, in R. Abraham, Foundations of mechanics, app. D, Benjamin, 1967, p. 263–279.
[22] W. Krieger, On entropy and generators of measure-preserving transformations, Trans. Amer. Math. Soc., 149 (1970), 453–464.
[23] Ju. V. Linnik, Ergodic properties of algebraic fields, Springer, 1968.
[24] V. V. Nemytskiĭ and V. V. Stepanov, Qualitative theory of differential equations, Princeton Univ. Press, 1960.
[25] D. S. Ornstein, Bernoulli shifts with the same entropy are isomorphic, Advances in Math., 4 (1970), 337–352.
[26] D. S. Ornstein, Some new results in the Kolmogorov-Sinaĭ theory of entropy and ergodic theory, Bull. Amer. Math. Soc., 77 (1971), 878–890.
[27] J. C. Oxtoby, Ergodic sets, Bull. Amer. Math. Soc., 58 (1952), 116–136.
[28] W. Parry, Entropy and generators in ergodic theory, Benjamin, 1969.
[29] В. А. Рохлин (V. A. Rohlin), Избранные вопросы метрической теории динамических систем, Uspehi Mat. Nauk, (N.S.) 4 (2) (1949), 57–128; English translation, Selected topics from the metric theory of dynamical systems, Amer. Math. Soc. Transl., (2) 49 (1966), 171–240.
[30] В. А. Рохлин (V. A. Rohlin), Новый прогресс в теории преобразований с инвариантной мерой, Uspehi Mat. Nauk, (N.S.) 15 (4) (1960), 1–26; English translation, New progress in the theory of transformations with invariant measure, Russian Math. Surveys, 15 (4) (1960), 1–22.
[31] В. А. Рохлин (V. A. Rohlin), Лекции по энтропийной теории преобразовании с инвариантной мерой, Uspehi Mat. Nauk, 22 (5) (1967), 3–56; English translation, Lectures on the entropy theory of measure-preserving transformations, Russian Math. Surveys, 22 (5) (1967), 1–52.
[32] Я. Г. Синай (Ja. G. Sinaĭ), Вероятностные идеи в эргодической теории, Proc. Intern. Congr. Math., Stockholm, 540–559; English translation, Probabilistic ideas in ergodic theory, Amer. Math. Soc. Transl., (2) 31 (1963), 62–84.
[33] Я. Г. Синай (Ja. G. Sinaĭ), Классические динамические системы со счетнократным лебеговским спектром, Akad. Nauk SSSR, 30 (1966), 15–68; English translation, Dynamical systems with countably-multiple Lebesgue spectrum II, Amer. Math. Soc. Transl., (2) 68 (1968), 34–88.
[34] Я. Г. Синай (Ja. G. Sinaĭ), Динамические системы с упругими отражениями, Uspehi Mat. Nauk, 25 (2) (1970), 141–192; English translation, Dynamical systems with elastic reflections, Russian Math. Surveys, 25 (2) (1970), 137–189.
[35] S. Smale, Differential dynamical systems, Bull. Amer. Math. Soc., 73 (1967), 748–817.
[36] E. M. Stein, Topics in harmonic analysis related to the Littlewood-Paley theory, Ann. Math. Studies, Princeton Univ. Press, 1970.
[37] L. Sucheston (ed.), Contributions to ergodic theory and probability, Proc. First Midwestern Conference on Ergodic Theory, Ohio State Univ., 1970, Springer, 1970.
[38] A. M. Vershik and S. A. Yuzvinskiĭ, Dynamical systems with invariant measure, Progress in mathematics VIII, Plenum, 1970, p. 151–215.
[39] F. B. Wright (ed.), Ergodic theory, Proc. Intern. Symposium at Tulane Univ., Academic Press, 1963.
[40] K. Yosida, Functional Analysis, Springer, second edition, 1968.

[41] Proc. Symposium on Topological Dynamics and Ergodic Theory, Univ. of Kentucky, 1971.

147 (VI.13) Erlangen Program

When F. Klein succeeded K. G. C. von Staudt as professor at the Philosophical Faculty of Erlangen University in 1872, he gave an inauguration lecture entitled "Comparative Consideration of Recent Geometric Researches," which later appeared as an article (*Math. Ann.*, 43 (1893) [1]). In it he developed a penetrating idea, now called the Erlangen program, in which he utilized group-theoretic concepts to unify various kinds of geometries that until that time had been considered separately.

The concept of transformation is not new; it was, however, not until the 18th century that the concept of transformation groups was recognized as useful. The theory of †invariants of linear groups and the †Galois theory of algebraic equations attracted attention in the 19th century. In the same century, †projective geometry made remarkable progress, for example, when A. Cayley and E. Laguerre discovered that metrical properties of Euclidean and †non-Euclidean geometries can be interpreted in the language of projective geometry. Cayley proclaimed, "All geometry is projective geometry." After learning geometry under J. Plücker, Klein made the acquaintance of S. Lie. Both men understood the importance of the group concept in mathematics. Lie studied the theory of †continuous transformation groups, and Klein studied discontinuous transformation groups from a geometric standpoint. Klein was thus led to the idea of the Erlangen program, which provided a bird's-eye view of geometry.

Klein's idea can be summarized as follows: A space S is a given set with some geometric structure. Let a transformation group G of S be given. A subset of S, called a **figure**, may have various kinds of properties. The study of the properties that are left invariant under all transformations belonging to G is called the **geometry of the space S subordinate to the group** G. Let this geometry be denoted by (S,G). Two figures of S are said to be **congruent** in (S,G) if one of them is mapped to the other by a transformation of G. The geometry (S,G) is actually the theory of invariants of S under G, with the term **invariants** to be understood in a wider sense; it means both invariant quantities and invariant properties or relations.

Replacing G in (S,G) by a subgroup G' of G, we obtain another geometry (S,G'). A series of subgroups of G gives rise to a series of geometries. For instance, let A be a figure of S. The elements of G leaving A invariant form a subgroup $G(A)$ of G that operates on $S'=S-A$. We thus obtain a geometry $(S',G(A))$ in which A is called an **absolute figure**. In this way, many geometries are obtained from projective geometry. Klein gave numerous examples. It is noteworthy that he mentioned even the groups of †rational and †homeomorphic transformations.

Klein's idea not only synthesized the geometries known at that time, but aso gave a guiding principle for the development of new geometries.

In 1854 Riemann published his epoch-making idea of **Riemannian geometry**. This geometry has a metric, but in general lacks congruence transformations (isometries). (Even if they exist, the group of congruence transformations is generally "smaller" than the corresponding group for the Euclidean geometry of the same dimension.) Thus Riemannian geometry is a geometry that is not included in the framework of the Erlangen program. The importance of Riemannian geometry was acknowledged when it was used by A. Einstein in 1916 as a foundation of his general theory of relativity. H. Weyl, O. Veblen, and J. A. Schouten discovered geometries that are generalizations of affine, projective, and †conformal geometries in the same way as Riemannian geometry is a generalization of Euclidean geometry. It became necessary to establish a theory that reconciled the ideas of Klein and Riemann; E. Cartan succeeded in this by introducing the notion of †connection (→ 82 Connections). However, the Erlangen program, which gave an insight into the essential character of classical geometries, still maintains its role as one of the guiding principles of geometry.

References

[1] F. Klein, Vergleichende Betrachtungen über neuere geometrische Forschungen, Math. Ann., 43 (1893), 63–100 (Gesammelte mathematische Abhandlungen, Springer, 1921, vol. 1, p. 460–497).
[2] G. Fano, Kontinuierliche geometrische Gruppen, Enzykl. Math., Leipzig, 1907–1910, Geometrie III, pt. 1, AB4b, 289–388.

148 (XV.3) Error Analysis

A. General Remarks

The data obtained by observations or measurements in astronomy, geodesy, and other sciences do not usually give exact values of the quantities in question. The **error** is the difference between the approximation and the exact value. The **theory of errors** originated from systematic work with data accompanied by errors. Thus statistical treatment played the main role in the beginning stages (→ 392 Statistical Linear Models). However, due to the recent development of high-speed computers it has become possible to carry out computations on a tremendously large scale. As a result, fine and detailed analysis of errors has become an absolute necessity for modern numerical computations. The analysis of errors in relation to numerical computations has become the center of research in error theory.

B. Errors

One rarely makes a mistake in counting a small number of things; therefore, the exact value of the count can be determined. On the other hand, the exact value in decimals is never obtainable for a continuous quantity, say length, no matter how fine measurements are made, and a large or small stochastic error is thus inevitable in measuring a continuous quantity. A discrete finite quantity is a **digital quantity**, and a continuous quantity is an **analog** quantity. The natures of these two quantities are quite different. The values of a digital quantity are distributed on some discrete set, while the values of an analog quantity are distributed with a continuous probability. The error might occur even for treatment with digital quantities, although checking the results for these quantities is easy. It is preferable to regard digital quantities as being analog quantities if the possible positions are densely distributed.

On the other hand, when an †analog computer is not available, analog quantities receive treatment similar to digital quantities. They are expressed as x times some unit, and x is expanded in the decimal or binary systems. An approximation to such an expansion is obtained by rounding off a numeral at some place, the position depending on the capacity for computation by available methods. There are two ways of rounding off numbers, the **fixed point method** and the **floating point method**. The former specifies the place of digits where rounding off is made, and the latter specifies the number of significant digits.

Classification of Errors. (1) **Errors of input data** are errors that occur when we represent constants such as $1/3$, $\sqrt{2}$, π by finite decimals. (2) **Truncation errors** occur in an approximate expression for the computation formula under consideration. (3) **Rounding-off errors** occur in taking some finite number of digits from the earlier digits in the numerical value at each step. If the computation of an infinite number of digits were actually possible, no errors of this type would appear.

The difference between fixed point and floating point rounding off is that addition and subtraction are appropriate operations for the former, while multiplication and division are appropriate for the latter. In fixed point rounding off, if a number is multiplied many times by numbers less than 1, a so-called **underflow** may occur, and many digits may disappear; a great deal of information can thus be lost. In computation for scientific research that involves frequent multiplication and division, floating point rounding off is preferable. It should be noted that rounding off for addition and subtraction may also cause a critical loss of information. This phenomenon is called **canceling** digits. For instance, in the subtraction $7.6325071 - 7.6318425 = 0.0006646$, where the subtrahend and minuend share several early significant digits, the difference loses those digits. Thus, relative errors may be magnified tremendously, a circumstance that further affects subsequent multiplication and division and finally may cause an extremely inaccurate result. By taking a large number of significant digits, this situation may be avoided to some extent. So-called **high precision computation** shows its effectiveness in such cases.

C. Propagation of Errors

In order to analyze **propagation of errors**, let us assume that all numbers are carried to infinitely many digits so that no rounding-off error occurs. Suppose that we are to evaluate the function $y = f(x_1, \ldots, x_n)$ when $x_1, x_2, \ldots, x_n$ are assigned. Let η be the truncation error of an approximate expression. If an input error δ_i for x_i exists, then the corresponding error for y is

$$\eta + \sum_{i=1}^{n} \frac{\partial f}{\partial x_i} \delta_i.$$

Moreover, suppose that at the final step, we round off to get a result with a finite number of digits, by which an error ε is introduced. Then the final error δ for y is

$$\delta=\varepsilon+\eta+\sum_{i=1}^{n}\frac{\partial f}{\partial x_i}\delta_i.$$

This procedure is performed for each step needed in the computation. If $y=f(x_1,\ldots,x_n)$ is a specified step, then the input error δ_i for that step involves all the errors arising before that step, i.e., δ_i is an **accumulated error.** In this instance, if the error from the previous step slightly affects the present step, and if it tends to diminish as we proceed further, then the whole computation eventually becomes stable. However, if previous errors begin to accumulate more and more, then the result of the computation is very far from the correct solution.

Let us take a reduction formula of a †Bessel function as an example for the latter case:

$$J_{n+1}(x)=(2n/x)J_n(x)-J_{n-1}(x).$$

It is a classic method to compute $J_n(x)$ by this formula, starting with the values of $J_0(x)$ and $J_1(x)$, with x given. By putting $J_{n-1}(x)=y_n$, $J_n(x)=z_n$, the reduction formula may be regarded as a linear transformation of the point $P_n(y_n,z_n)$ in a plane into another point $P_{n+1}(y_{n+1},z_{n+1})$, where

$$y_{n+1}=z_n, \qquad z_{n+1}=-y_n+(2n/x)z_n.$$

The †eigenvalues of this †difference equation are

$$\lambda_1=\frac{n}{x}+\sqrt{\frac{n^2}{x^2}-1}\,,\quad \lambda_2=\frac{n}{x}-\sqrt{\frac{n^2}{x^2}-1}\,,$$

where $\lambda=|\lambda_2|=1$ as long as $n<|x|$, while λ_1 is greater than 1 and increases very rapidly as n increases. Consequently, even the slightest discrepancy in the position of P_n gives rise to a greatly magnified error in the result, and ultimately it becomes meaningless. If all numbers were carried to infinitely many digits, such a phenomenon indeed would not occur. It is preferable that the steps with large discrepancies appear close to the end of the computation. For that purpose, we use a large number of significant digits. Such examples have been discussed [3].

Many studies have been made on the propagation of errors in the numerical solution of ordinary differential equations ([2,4]; → 298 Numerical Solution of Ordinary Differential Equations).

In order to try to prevent rounding-off errors there are interesting devices, such as representing all the numbers by means of integers and utilizing modular arithmetic with a prime modulus [6]. For errors in linear systems → [1,5].

References

[1] J. Von Neumann and H. H. Goldstine, Numerical inverting of matrices of higher order, Bull. Amer. Math. Soc., 53 (1947), 1021–1099.
[2] P. Henrici, Discrete variable methods in ordinary differential equations, John Wiley, 1962.
[3] T. Uno, The problem of error propagation (Japanese), Sûgaku, 15 (1963), 30–40.
[4] M. Iri, A stabilizing device for unstable numerical solutions of ordinary differential equations—Design principle and applications of a "filter" (Japanese), Zyôhô syori, Information Processing Society of Japan, 4 (1963), 249–260.
[5] H. Nagasaka, Error propagation in the solution of some tridiagonal linear equations (Japanese), Zyôhô syori, Information Processing Society of Japan, 5 (1965), 38–44.
[6] H. Takahashi and Y. Ishibashi, A new method for "exact calculation" by a digital-computer, (Japanese), Zyôhô syori, Information Processing Society of Japan, 1 (1960), 78–86.
[7] J. H. Wilkinson, Rounding errors in algebraic processes, Prentice-Hall, 1963.
[8] L. B. Rall (ed.), Error in digital computation I, II, John Wiley, 1965.

149 (VI.3) Euclidean Geometry

A. History

Attempts to construct axiomatically the geometry of ordinary 3-dimensional space were undertaken by the ancient Greeks, culminating in Euclid's *Elements* (→ 190 Greek Mathematics). The **fifth postulate** of Euclid's *Elements* requires that two straight lines in a plane that meet a third line, as shown in Fig. 1, in angles α, β whose sum is less than 180°, have a common point. In the *Elements*, two straight lines in a plane without a common point are said to be **parallel**. It can be proved from other axioms in the *Elements* that if $\alpha+\beta=180°$, the two lines l and l' in Fig. 2 are parallel. Hence given a line l and

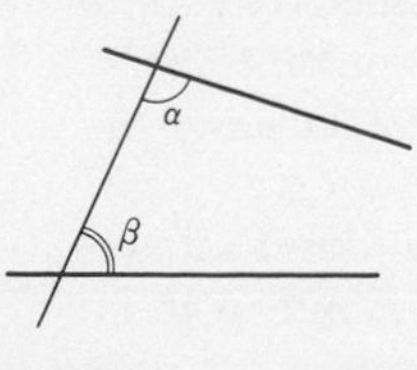

Fig. 1

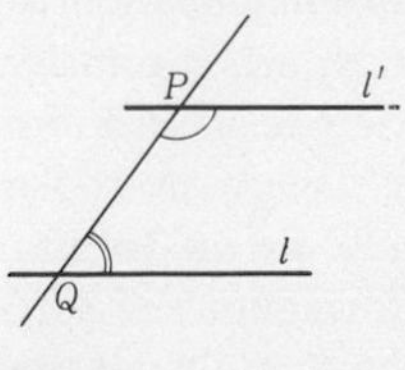

Fig. 2

a point P not lying on l, there exists a line l' passing through P that is parallel to l. The fifth postulate assures the uniqueness of the parallel l' passing through the given point P. For this reason, the fifth postulate is also called the **axiom of parallels**. Utilizing this axiom, we can prove the well-known theorems on parallel lines, the sum of interior angles of triangles, etc. The axiom plays an important role in the proof of the Pythagorean theorem in the *Elements.* The axiom is also called **Euclid's axiom.**

However, Euclid states this axiom in a quite complicated form, and unlike his other axioms, it cannot be verified within a bounded region of the space.

Many mathematicians tried in vain to deduce it from other axioms. Finally the axiom was shown to be independent of other axioms in the *Elements* by the invention of non-Euclidean geometry in the 19th century (→ 283 Non-Euclidean Geometry).

The term *Euclidean geometry* is used in contrast to *non-Euclidean geometry* to refer to the geometry based on Euclid's axiom of parallels as well as on other axioms explicit or implicit in Euclid's *Elements*. It was in the 19th century that a complete system of Euclidean geometry was explicitly formulated (→ 163 Foundations of Geometry). From the standpoint of present-day mathematics, it would be natural to define first the group of motions by the axiom of free mobility due to Helmholtz (→ Section B) and then, following Klein, to define Euclidean geometry as the study of properties of spaces that are invariant under the groups of these motions (→ 147 Erlangen Program).

B. Group of Motions

Let P be an †ordered field and A^n the n-dimensional †affine space over P. Let B^r be an r-dimensional affine subsace of A^n, B^{r-1} an $(r-1)$-dimensional subspace of B^r, B^{r-2} an $(r-2)$-dimensional subspace of B^{r-1}, etc. In the sequence of subspaces $B^r, B^{r-1}, \ldots, B^0$, each $B^k - B^{k-1}$ consists of two †half-spaces $(k=r, r-1, \ldots, 1)$. Let H^k be one of these half-spaces. Then the sequence of half-spaces $H^r, H^{r-1}, \ldots, H^1$ is called an r-dimensional **flag**, denoted by $\mathfrak{H}^r$ $(n \geqslant r \geqslant 1)$, and B^r and H^r are called the **principal space** and the **principal half-space** of $\mathfrak{H}^r$, respectively. If f is a †proper affine transformation of A^n, $f(H^r)$, $f(H^{r-1}), \ldots, f(H^1)$ form an r-dimensional flag $\mathfrak{K}^r$. We write $f(\mathfrak{H}^r)=\mathfrak{K}^r$.

Let $\mathfrak{A}^n$ be the group of all proper affine transformations of A^n. The subgroup $\mathfrak{B}^n$ of $\mathfrak{A}^n$ with the following two properties is called the **group of motions**, and any element of $\mathfrak{B}^n$ is called a **motion** (or **congruent transformation**). (1) Let r be an integer between 1 and n, and let $\mathfrak{H}^r$, $\mathfrak{K}^r$ be any two r-dimensional flags. Then there exists an element f of $\mathfrak{B}^n$ that carries $\mathfrak{H}^r$ to $\mathfrak{K}^r$: $f(\mathfrak{H}^r)=\mathfrak{K}^r$. (2) Let f, g be two elements of $\mathfrak{B}^n$ with $f(\mathfrak{H}^r)=\mathfrak{K}^r$, $g(\mathfrak{H}^r) = \mathfrak{K}^r$, and let p be any point on the principal space of $\mathfrak{H}^r$. Then $f(p)=g(p)$, that is, f, g have the same "effect" on the principal space. In particular, when $r=n$, then $f=g$. That $\mathfrak{A}^n$ possesses a subgroup $\mathfrak{B}^n$ with properties (1) and (2) is called the **axiom of free mobility**.

When $n=1$, it is easy to see that the elements of $\mathfrak{B}^1$ are only those elements f of $\mathfrak{A}^1$ that can be expressed in the form $f(x)=\pm x + a$ $(a \in P)$. When $n \geqslant 2$, P must satisfy the following condition in order that a subgroup $\mathfrak{B}^n$ with properties (1) and (2) exists in $\mathfrak{A}^n$: If $a, b \in P$, then P contains an element x such that $x^2=a^2+b^2$. When this condition is satisfied, the ordered field P is called a **Pythagorean field**. Every †real closed field (e.g., the field **R** of real numbers) is Pythagorean. If $\mathfrak{B}^n$ exists, its uniqueness is assured by (1) and (2). Furthermore, if P contains a square root of every positive element (this condition is satisfied, for example, by **R**), then conditions (1) and (2) are reducible to the case $r=n$ only, i.e., conditions (1) and (2) for other values of r follow from (1) and (2) with $r=n$. Hereafter, we assume the existence of $\mathfrak{B}^n$.

Suppose that we have $A^n \supset B^r \supset B^k$ $(n \geqslant r \geqslant k \geqslant 0)$, and let $\mathfrak{H}^r$ be a flag with the principal space B^r: $\mathfrak{H}^r=(H^r, \ldots, H^k, \ldots, H^1)$. Let $\mathfrak{K}^r$ be another flag with the same principal space B^r: $\mathfrak{K}^r=(K^r, \ldots, K^k, \ldots, K^1)$, where we suppose $H^j=K^j$ for $k \geqslant j \geqslant 1$, whereas for $r \geqslant i \geqslant k+1$, we suppose that H^i and K^i are different half-spaces on B^i divided by B^{i-1}. The flag $\mathfrak{K}^r$ is denoted by $\mathfrak{H}^r_k$. An element f of $\mathfrak{B}^n$ with $f(\mathfrak{H}^r)=\mathfrak{H}^r_k$ is called a **symmetry** (or **reflection**) of B^r with respect to B^k. It leaves every point on B^k invariant, and its effect on B^r is determined only by B^k independently of the choice of half-spaces in $\mathfrak{H}^r$ and $\mathfrak{K}^r$ (subject to the conditions mentioned above). In particular, the symmetry of A^n with respect to a point $A^0=p$ is called a **central symmetry** with respect to the **center** p; and the symmetry of A^n with respect to a hyperplane $A^{n-1}=h$ is called a **hyperplanar symmetry**. They are uniquely determined by p and h, respectively, and are denoted by $S(p)$ and $S(h)$, respectively. If $H(A^n)$ is the set of all hyperplanes of A^n, then $\mathfrak{B}^n$ is generated by $\{S(h) \mid h \in H(A^n)\}$. Furthermore, if p, q are two points of A^n, the composite S_pS_q is a parallel translation by $2\cdot\overrightarrow{pq}$ (Fig. 3). The parallel translations generate a normal subgroup $\mathfrak{T}^n$ of $\mathfrak{B}^n$. For $p, q \in A^n$, the element of $\mathfrak{T}^n$ that carries p to q is denoted by τ_{pq}. The

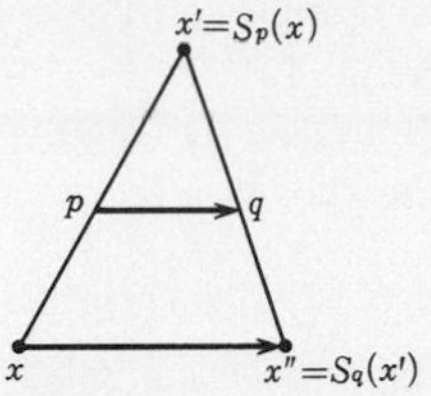

Fig. 3

subgroup of $\mathfrak{B}^n$ that leaves a point p of A^n invariant is denoted by O_p^n. Obviously, we have $O_q^n = \tau_{pq} O_p^n \tau_{pq}^{-1}$. Thus all the O_p^n (for $p \in A^n$) are isomorphic. We call O_p^n the **group of rotations** around p and any element of O_p^n a **rotation** around p. More generally, any element of $\mathfrak{B}^n$ that leaves a subspace A^k of A^n invariant is called a **rotation around the subspace** A^k.

An element of $\mathfrak{B}^n$ that preserves the **orientation** of A^n, i.e., is represented by a †proper affinity with a positive determinant, is called a **proper motion**. Proper motions form a subgroup $\mathfrak{B}_0^n$ of $\mathfrak{B}^n$. Proper rotations are, by definition, rotations belonging to $\mathfrak{B}_0^n$. Sometimes $\mathfrak{B}_0^n$ is called the group of motions; then $\mathfrak{B}^n$ is called the group of motions in the wider sense. In this article, however, we shall continue to use the terminology introduced above.

The study of the properties of A^n invariant under $\mathfrak{B}^n$ is ***n*-dimensional Euclidean geometry**. Since $\mathfrak{A}^n \supset \mathfrak{B}^n$, every proposition in affine geometry (→ 9 Affine Geometry) can be considered a proposition in Euclidean geometry, but there are many propositions that are proper to Euclidean geometry. Sometimes the subgroup of $\mathfrak{A}^n$ generated by $\mathfrak{B}^n$ and the **homotheties** of A^n, i.e., elements of $\mathfrak{A}^n$ represented by †scalar matrices, is called the **group of motions in the wider sense**, and the study of properties of A^n invariant under this group is called n-dimensional **Euclidean geometry in the wider sense**.

C. Length of Segments

Two figures F, F' in A^n are said to be **congruent** if there exists an $f \in \mathfrak{B}^n$ such that $f(F) = F'$. Then we write $F \equiv F'$. The congruence relation is an †equivalence relation. Let $s = \overline{pq}$, $s' = \overline{p'q'}$ be two †segments in A^n. We say that s, s' have equal **length** when $s \equiv s'$. Length is an attribute of the equivalence class of segments. The length of s is denoted by $|s|$. All segments of the form $\overline{pp}$ are congruent, and we define: $|\overline{pp}| = 0$. If we are given a length and a †half-line starting from a point p, we can find a unique point q on it such that $|\overline{pq}| =$ the given length (Fig. 4). Let r be a point on the extension of $\overline{pq}$. The length $|\overline{pr}|$ is then uniquely determined by $|\overline{pq}|$ and $|\overline{qr}|$.

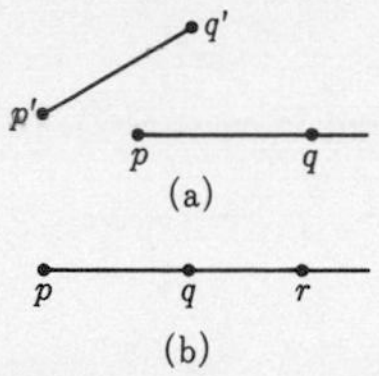

Fig. 4

It is defined as the sum of the lengths: $|\overline{pr}| = |\overline{pq}| + |\overline{qr}|$. With respect to the addition thus defined, lengths of segments in A^n form a †commutative semigroup with the cancellation law, which can be extended to an †Abelian group M with 0 as the identity element (→ 193 Groups).

Let $|s| \neq 0$ and $|s'|$ be any length. On a half-line starting from p, we can find points q, r with $|\overline{pq}| = |s|$, $|\overline{pr}| = |s'|$. Then the element $pr/pq = \lambda \in P$ (→ 9 Affine Geometry) is a positive element of P uniquely determined by $|s|$ and $|s'|$. We call λ the **measure** of $|s'|$ with the **unit** $|s|$ and denote it by $|s'| : |s|$. If P is †Archimedean, λ can be represented by a real number (→ 157 Fields). We have $(|s'| + |s''|) : |s| = (|s'| : |s|) + (|s''| : |s|)$, $(|s''| : |s'|)(|s'| : |s''|) = |s''| : |s|$ (if $|s'| \neq 0$). Thus the mapping $|s'| \rightarrow |s'| : |s|$ sends the additive semigroup of lengths to that of the positive elements of P. This is actually an isomorphism, which can be extended to an isomorphism of M onto the additive group of the field P.

Let $\varphi(X, X', \ldots, X^\alpha)$ be a †homogeneous rational function of a finite number of variables $X, X', \ldots, X^\alpha$. If a relation $\varphi(\lambda, \lambda', \ldots, \lambda^\alpha) = 0$ holds for $\lambda = |s| : |s_0|$, $\lambda' = |s'| : |s_0|$, $\ldots, \lambda^\alpha = |s^\alpha| : |s_0|$, where $|s_0|$ is a length $\neq 0$, then $\varphi(\lambda_1, \lambda_1', \ldots, \lambda_1^\alpha) = 0$ holds also for $\lambda_1 = |s| : |s_1|$, $\lambda_1' = |s'| : |s_1|, \ldots, \lambda_1^\alpha = |s^\alpha| : |s_1|$, where $|s_1|$ is any other length $\neq 0$. Hence, in this case, the expression $\varphi(|s|, |s'|, \ldots, |s^\alpha|) = 0$ is meaningful.

D. Angles and Their Measure

An **angle** $\angle AOB$ is a figure constituted by two half-lines OA, OB starting from the same point O but belonging to different straight lines. The point O is called the **vertex**, and the two half-lines OA, OB are called the **sides** of $\angle AOB$ (Fig. 5). Two congruent angles are said to have the same **measure**, denoted by $|\angle AOB|$ or sometimes simply by α. Let $\mathfrak{H}^2 = (H^2, H^1)$ ($H^1 =$ half-line QR) be a given 2-dimensional flag and the given measure of an angle. Then we can find a unique half-line QP in the given half-plane H^2 such that $|\angle PQR| = \alpha$ (Fig. 6). The angle $\angle PQR$ is said to "belong" to $\mathfrak{H}^2$. Let K^2 be the half-plane separated by the line $P \cup Q$ containing the half-line QR. Then $H^2 \cap K^2$ is called the

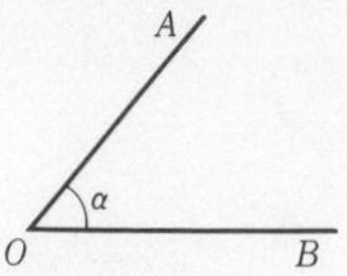

Fig. 5

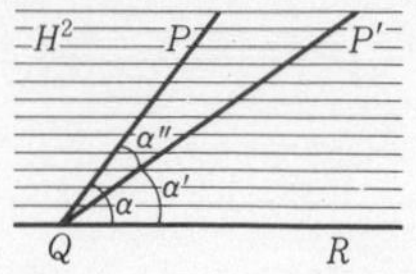

Fig. 6
$\alpha' + \alpha'' = \alpha$.

interior of the angle $\angle PQR$. Let $\angle P'QR$ be another angle belonging to $\mathfrak{H}^2$. If the interior of the latter angle is a subset of the interior of $\angle PQR$ and $QP \neq QP'$, then $|\angle PQR|$ is said to be greater than $|\angle P'QR|$, and we write $|\angle PQR| > |\angle P'QR|$. Actually, it can be shown that $>$ is a relation between the measures of $\angle PQR$ and $\angle P'QR$, and that the set of measures of angles forms a †linearly ordered set with respect to the relation $\geqslant$ defined in the obvious way. When $|\angle PQR| > |\angle P'QR|$, we write $|\angle PQR| = |\angle PQP'| + |\angle P'QR|$. Actually, these are relations between measures of angles. Furthermore, if the measure α of an angle is given, the set of measures of angles $\leqslant \alpha$ forms a linearly ordered set order-isomorphic to a segment and satisfying: (i) If $\beta < \gamma$ then there exists a δ such that $\beta + \delta = \gamma$; (ii) $\beta + \delta = \delta + \beta$; $(\beta_1 + \beta_2) + \beta_3 = \beta_1 + (\beta_2 + \beta_3)$ if all these sums exist; and (iii) $\beta_1 + \delta = \beta_2 + \delta$ implies $\beta_1 = \beta_2$. When P is Archimedean, these properties imply that the measure of angles $\leqslant |\angle PQR|$ can be represented by positive real numbers $\leqslant k$ (k is any given positive number) such that the relations of ordering and addition are preserved.

This one-to-one correspondence between the measures of angles and a subset S of the interval $(O,k]$ of real numbers can be extended to a correspondence between the measures of **general angles** and the subset of $\mathbf{R}$ obtained from S through a translation by nk, $n \in \mathbf{Z}$. When $P = \mathbf{R}$, then we have $S = (O,k]$, and any real number appears as a measure of a general angle. We can choose $\angle PQR$ and the positive number k arbitrarily, but it is customary to choose them as follows. Suppose we are given an angle $\angle AOB$. Let the extensions of the half-lines OA and OB in the opposite directions be OA' and OB', respectively. The angles $\angle AOB$ and $\angle A'OB$ are called **supplementary angles** of each other, and so are $\angle AOB$ and $\angle AOB'$. The angles $\angle AOB$ and $\angle A'OB'$ are called **vertical angles** to each other (Fig. 7). Any angle is congruent to its vertical angle, and an angle that is congruent to its supplementary angle

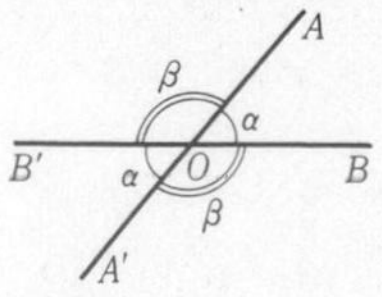

Fig. 7

has a fixed measure. Such an angle (or its measure) is called a **right angle**. In the description of the measurement of angles, we usually consider the case where the special angle $\angle PQR$ is a right angle, and we set $k = \pi/2$. (The existence and uniqueness of the right angle can be proved.) An angle that is greater (smaller) than a right angle is called an **obtuse (acute) angle**. A general angle whose measure is twice (four times) a right angle is called a **straight angle (perigon)**. Sometimes we choose as the "unit angle" $1/90$ of a right angle, which is called a **degree** (hence a right angle $=$ 90 degrees, denoted by $90°$); $1/60$ of a degree is called a **minute** ($1° = 60$ minutes, denoted by $60'$), and $1/60$ of a minute is called a **second** ($1' = 60$ seconds, denoted by $60''$). If, as usual, we put the right angle equal to $\pi/2$, then the unit angle is $(2/\pi)$(right angle). This is called a **radian**, and 1 radian $= 180°/\pi = 57°17'44.806\ldots'' \fallingdotseq 57.3°$.

If a straight line m intersects two straight lines l, l', eight angles $\alpha, \beta, \gamma, \delta, \alpha', \beta', \gamma', \delta'$ appear, as in (Fig. 8). In this figure, α and α', β and β', γ and γ', and δ and δ' are called **corresponding angles**, while α and γ', β and δ', γ and α', and δ and β' are called **alternate angles** to each other. When l and l' are parallel, each of these angles is congruent to its corresponding or alternate angle.

The **Pythagorean theorem** asserts that if a triangle $\triangle ABC$ is given for which $\angle ABC$ is a right angle (Fig. 9), then $|AB|^2 + |BC|^2 = |CA|^2$ (which makes sense since $X^2 + Y^2 - Z^2$ is a homogeneous polynomial).

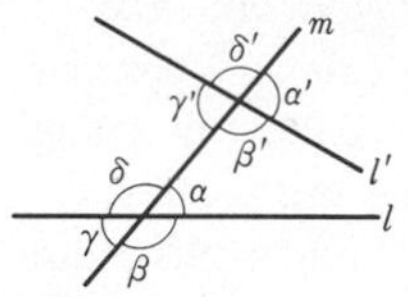

Fig. 8

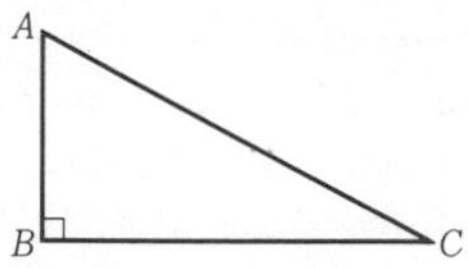

Fig. 9

E. Rectangular Coordinates

When two straight lines l, m intersect, two pairs of vertical angles appear. If one of these angles is a right angle, then all are. Then we say that l and m are **orthogonal** (or **perpendicular**) to each other, and write $l \perp m$. Let l be a line and A^r an r-dimensional subspace of A^n ($1 \leqslant r \leqslant n-1$) intersecting l at a point $O = A^r \cap l$.

If l is orthogonal to all lines on A^r passing through O, then l is said to be orthogonal to A^r, and we write $l \perp A^r$ (Fig. 10). If A^{n-1} is any hyperplane in A^n, then there exists a unique line l through a given point P of A^n that is orthogonal to A^{n-1}; this l is called the **perpendicular** to A^{n-1} through P, and the

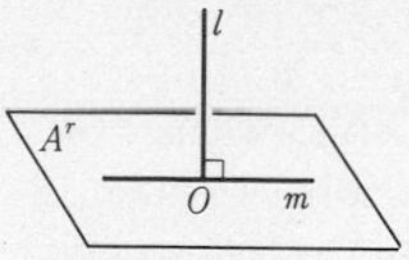

Fig. 10

intersection $l \cap A^{n-1}$ is called the **foot of the perpendicular** through P. When A^{n-1} is given, the mapping from A^n to A^{n-1} assigning to every point P of A^n the foot of the perpendicular through P is called the **orthogonal projection** from A^n to A^{n-1}.

Let $\mathfrak{H}^n = (H^n, H^{n-1}, \ldots, H^1)$ be an n-dimensional flag of A^n and O the initial point of the half-line H^1. Then we can find a point E_i in H^i $(i=1,2,\ldots,n)$ such that $O \cup E_i \perp O \cup E_j$ $(i \neq j,\ i,j=1,2,\ldots,n)$. Moreover, if $|e|$ is any unit of length, then E_i can be chosen uniquely so that $|OE_i| = |e|$ $(i=1,2,\ldots,n)$. Then $O, E_1, \ldots, E_n$ are †independent points in A^n, and we have $A^n = O \cup E_1 \cup \ldots \cup E_n$. Thus we have a †frame $\Sigma = (O; E_1, \ldots, E_n)$ of A^n with O as origin and the E_i as unit points. Such a frame is called an **orthogonal frame**. A coordinate system with this frame, called an **orthogonal coordinate system adapted to** $\mathfrak{H}^n$, is uniquely determined by $\mathfrak{H}^n$. A motion is characterized as an †affinity sending one orthogonal frame onto another or onto itself.

Utilizing an orthogonal coordinate system, the lengths of segments and the measures of angles can be expressed simply. Let $(x_1, \ldots, x_n)$ be the coordinates of X with respect to such a coordinate system. Then the length of the segment $|OX|$ (with $|e|$ as unit) is equal to $(\sum_{i=1}^n x_i^2)^{1/2}$, and when Y is another point, with coordinates $(y_1, \ldots, y_n)$, $O \neq X$, $O \neq Y$, then we have

$$\cos|\angle XOY| = \frac{\sum_{i=1}^n x_i y_i}{\left(\sum_{i=1}^n x_i^2\right)^{1/2} \left(\sum_{i=1}^n y_i\right)^{1/2}}.$$

In particular, we have $O \cup X \perp O \cup Y$ if and only if $\sum_{i=1}^n x_i y_i = 0$.

We may write $\mathbf{x} = \overrightarrow{OX}$ for the †location vector of X. Then the †affinity $A\mathbf{x} + \mathbf{b}$ is a motion if and only if A is an †orthogonal matrix. Thus the †inner product $(\mathbf{x}, \mathbf{y})$ is invariant under motions; it therefore has meaning in Euclidean geometry. If we put $|\mathbf{x}| = (\mathbf{x}, \mathbf{x})^{1/2}$, then the right-hand sides of the formulas for $|OX|$ and $\cos|\angle XOY|$ can be written as $|\mathbf{x}|$ and $(\mathbf{x}, \mathbf{y})/(|\mathbf{x}| \cdot |\mathbf{y}|)$. More generally, we have $|XY| = |\mathbf{y} - \mathbf{x}|$. This is the **Euclidean distance** (or simply **distance**) between X and Y. Then A^n becomes a †metric space with this distance, i.e., a Euclidean space. Historically, the notion of metric spaces was introduced in generalizing Euclidean spaces (→ 273 Metric Spaces).

F. Area and Volume

The subset I^n of A^n consisting of points $(x_1, \ldots, x_n)$ with respect to an orthogonal coordinate system, with $0 \leqslant x_i \leqslant 1$, $i = 1, \ldots, n$, is called an n-dimensional **unit cube**. A functional m that assigns to †polyhedra in the wider sense $P, Q, \ldots$ in A^n nonnegative real numbers $m(P), m(Q), \ldots$ is called an n-dimensional **volume** if it satisfies the following four conditions: (1) $m(\varnothing) = 0$. (2) $m(P \cup Q) + m(P \cap Q) = m(P) + m(Q)$. (3) If P is sent to Q by a translation, then $m(P) = m(Q)$. (4) $m(I^n) = 1$. It has been proved that such a functional is unique and has the property that $P \equiv Q$ implies $m(P) = m(Q)$. Thus the concept of volume can be defined in the framework of Euclidean geometry. More generally, if the affinity $f(\mathbf{x}) = A\mathbf{x} + \mathbf{b}$ sends P onto Q, then $m(Q) = cm(P)$, where c is the absolute value of the determinant $|A|$. If P is covered by a finite number of hyperplanes, then $m(P) = 0$, and if P is a †parallelotope with n independent edges $\mathbf{a}_1, \ldots, \mathbf{a}_n$, then $m(P) = \operatorname{abs}|\mathbf{a}_1, \ldots, \mathbf{a}_n|$, where $|\mathbf{a}_1, \ldots, \mathbf{a}_n|$ is the determinant of the $n \times n$ matrix with $\mathbf{a}_i$ as the ith column vector, and abs x is the absolute value of the real number x. If P is an †n-simplex whose vertices have location vectors $\mathbf{x}_0, \mathbf{x}_1, \ldots, \mathbf{x}_n$, then we have

$$m(P) = \frac{1}{n!} \operatorname{abs} \begin{vmatrix} 1 & 1 & \cdots & 1 \\ \mathbf{x}_0 & \mathbf{x}_1 & \cdots & \mathbf{x}_n \end{vmatrix}.$$

The volume of any polyhedron can be obtained by dividing it into n-simplexes and summing their volumes. If P is an r-dimensional polyhedron in A^n, then the r-dimensional volume of P is obtained by dividing P into r-simplexes and summing their r-dimensional volumes (in the respective r-dimensional Euclidean spaces containing them). In particular, when $r = 1$, we speak of **length** (e.g., the length of a broken line), and when $r = 2$, of **area**. If V is the r-dimensional volume of an r-dimensional parallelotope with r edges $\mathbf{a}_1, \ldots, \mathbf{a}_r$, we have the formula

$$V^2 = \begin{vmatrix} (\mathbf{a}_1, \mathbf{a}_1) & \cdots & (\mathbf{a}_1, \mathbf{a}_r) \\ & \cdots & \\ (\mathbf{a}_r, \mathbf{a}_1) & \cdots & (\mathbf{a}_r, \mathbf{a}_r) \end{vmatrix}.$$

The notion of measure of point sets other than polyhedra is a generalization of the notion of volume of polyhedra (→ 270 Measure Theory).

G. Orthonormalization

Let $O, A_1, \ldots, A_n$ be $n+1$ †independent points in A^n. Then the n vectors $\overrightarrow{OA_i} = \mathbf{a}_i$, $i = 1, \ldots, n$, are independent. The points $O, A_1, \ldots, A_n$ determine an n-dimensional flag $\mathfrak{F}^n$ of A^n as follows. Let H_1 be the half-line OA_1, H_2 be the half-plane on the plane $O \cup A_1 \cup A_2$ separated by the line $O \cup A_1$ in which A_2 lies, ..., H_n be the half-space on A^n separated by the hyperplane $O \cup A_1 \cup \ldots \cup A_{n-1}$ in which A_n lies. Let $\mathbf{b}_1, \ldots, \mathbf{b}_n$ be the unit vectors of the rectangular coordinate system adapted to $\mathfrak{F}^n$. Suppose further that we are given a rectangular coordinate system. Then $\mathbf{b}_1, \ldots, \mathbf{b}_n$ can be obtained from $\mathbf{a}_1, \ldots, \mathbf{a}_n$ by the following procedure, called **orthonormalization** (E. Schmidt): First put $\mathbf{b}_1 = \mathbf{a}_1/|\mathbf{a}_1|$, so that $|\mathbf{b}_1| = 1$. Then $\mathbf{c}_2 = \mathbf{a}_2 - (\mathbf{a}_2, \mathbf{b}_1)\mathbf{b}_1$ satisfies $(\mathbf{b}_1, \mathbf{c}_2) = 0$, $\mathbf{c}_2 \neq 0$. Put $\mathbf{b}_2 = \mathbf{c}_2/|\mathbf{c}_2|$. Then we have $(\mathbf{b}_1, \mathbf{b}_2) = 0$, $|\mathbf{b}_2| = 1$. When $\mathbf{b}_1, \ldots, \mathbf{b}_{i-1}$ are obtained in this way, so that $(\mathbf{b}_j, \mathbf{b}_k) = \delta_{jk}$ for $1 \leqslant j, k \leqslant i-1$, then $\mathbf{c}_i = \mathbf{a}_i - (\mathbf{a}_i, \mathbf{b}_1)\mathbf{b}_1 - \ldots - (\mathbf{a}_i, \mathbf{b}_{i-1})\mathbf{b}_{i-1}$ satisfies $(\mathbf{b}_j, \mathbf{c}_i) = 0$, $\mathbf{c}_i \neq 0$. Hence $\mathbf{b}_i = \mathbf{c}_i/|\mathbf{c}_i|$ added to $\mathbf{b}_1, \ldots, \mathbf{b}_{i-1}$ retains the property $(\mathbf{b}_j, \mathbf{b}_k) = \delta_{jk}$ for $1 \leqslant j, k \leqslant i$, and this procedure can be continued to $i = n$.

Two vectors $\mathbf{u}$, $\mathbf{v}$ are called **orthogonal** (denoted $\mathbf{u} \perp \mathbf{v}$) if $(\mathbf{u}, \mathbf{v}) = 0$, and $\mathbf{u}$ is called **normalized** when $|\mathbf{u}| = 1$. Thus any two of the vectors $\mathbf{b}_1, \ldots, \mathbf{b}_n$ are orthogonal, and each of them is normalized. Between given vectors $\mathbf{a}_1, \ldots, \mathbf{a}_n$ and $\mathbf{b}_1, \ldots, \mathbf{b}_n$ we have the relation $\{\mathbf{a}_1, \ldots, \mathbf{a}_i\}$ (= the linear space generated by $\mathbf{a}_1, \ldots, \mathbf{a}_i$) = $\{\mathbf{b}_1, \ldots, \mathbf{b}_i\}$, $i = 1, \ldots, n$.

Let $\mathfrak{M}_1$, $\mathfrak{M}_2$ be two subspaces of the linear space $\mathfrak{M}$ of the vectors of the Euclidean space A^n. If any element of $\mathfrak{M}_1$ is orthogonal to any element of $\mathfrak{M}_2$, then $\mathfrak{M}_1$ and $\mathfrak{M}_2$ are called **orthogonal** and written $\mathfrak{M}_1 \perp \mathfrak{M}_2$. For any proper subspace $\mathfrak{M}_1$ of $\mathfrak{M}$, it can be shown by the method of orthonormalization that there exists a unique proper subspace $\mathfrak{M}_2$ of $\mathfrak{M}$ such that $\mathfrak{M} = \mathfrak{M}_1 \cup \mathfrak{M}_2$, $\mathfrak{M}_1 \perp \mathfrak{M}_2$. Such a subspace $\mathfrak{M}_2$ is called the **orthocomplement** of $\mathfrak{M}_1$ (with respect to $\mathfrak{M}$). Then $\mathfrak{M}_1 \cap \mathfrak{M}_2 = 0$ follows, and hence $\mathfrak{M} = \mathfrak{M}_1 + \mathfrak{M}_2$. Every element $\mathbf{a}$ of $\mathfrak{M}$ is therefore written uniquely in the form $\mathbf{a}_1 + \mathbf{a}_2$, $\mathbf{a}_1 \in \mathfrak{M}_1$, $\mathbf{a}_2 \in \mathfrak{M}_2$; we call $\mathbf{a}_1$ the $\mathfrak{M}_1$-component of $\mathbf{a}$ and $\mathbf{a}_2$ the **orthogonal component** of $\mathbf{a}$ with respect to $\mathfrak{M}_1$. The mapping from $\mathfrak{M}$ to $\mathfrak{M}_1$ assigning $\mathbf{a}_1$ to $\mathbf{a}$ is called the **orthogonal projection** from $\mathfrak{M}$ to $\mathfrak{M}_1$; it is a linear and †idempotent mapping.

H. Distance between Subspaces

Since the Euclidean space A^n is a metric space, the distance is defined between any two nonempty subsets of A^n (→ 273 Metric Spaces). Let A^r, B^s be two subspaces of dimensions r, s of A^n, and let d be the distance between them. Then it can be shown that there exist points $p \in A^r$, $q \in B^s$ such that $d = \overline{pq}$, and if $p' \in A^r$, $q' \in B^s$ are any other points with $d = \overline{p'q'}$, then $\overrightarrow{pq} = \overrightarrow{p'q'}$. In particular, when $r = 0$ and $s = n-1$ (i.e., when $A^r = p$ is a point and $B^s = B^{n-1}$ is a hyperplane), the distance d can be obtained as follows: If $(\mathbf{a}, \mathbf{x}) = b$ is an equation of B^{n-1} and $\mathbf{p}$ is the location vector of p, then $d = |(\mathbf{a}, \mathbf{p}) - b|/|\mathbf{a}|$. If $|\mathbf{a}| = 1$ in this equation of B^{n-1}, then d is given simply by $|(\mathbf{a}, \mathbf{p}) - b|$. An equation $(\mathbf{a}, \mathbf{x}) = b$ of a hyperplane is said to be in **Hesse's normal form** if $|\mathbf{a}| = 1$.

I. Spheres and Subspaces

The set of points in a Euclidean space lying at a fixed distance r from a given point is called the **sphere** of **radius** r with **center** at the given point. If $\mathbf{p}$ is the location vector of the center of this sphere with respect to a given rectangular coordinate system, then the equation of the sphere is $|\mathbf{x} - \mathbf{p}| = r$ or $(\mathbf{x}, \mathbf{x}) - 2(\mathbf{p}, \mathbf{x}) + (\mathbf{p}, \mathbf{p}) - r^2 = 0$. The set of points lying at equal distances from $k+1$ points with location vectors $\mathbf{p}_0, \mathbf{p}_1, \ldots, \mathbf{p}_k$ $(k \geqslant 1)$ is a linear subspace of the space (which may be $\varnothing$ or the entire space). If these points are independent, then the subspace has dimension $n-k$, where n is the dimension of the entire space. In particular, if these points are vertices of an n-dimensional †simplex, then there is a unique sphere passing through them, called the **circumscribing sphere** of the simplex. In this case, the simplex is said to be **inscribed** in the sphere. If $\mathbf{p}_0, \mathbf{p}_1, \ldots, \mathbf{p}_n$ are location vectors of the vertices of the simplex, then the equation of the circumscribing sphere of the simplex is given by

$$\begin{vmatrix} 1 & 1 & \cdots & 1 & 1 \\ \mathbf{p}_0 & \mathbf{p}_1 & \cdots & \mathbf{p}_n & \mathbf{x} \\ \mathbf{p}_0^2 & \mathbf{p}_1^2 & \cdots & \mathbf{p}_n^2 & \mathbf{x}^2 \end{vmatrix} = 0.$$

When $n = 2$ or 3, there are many classical results concerning the circumscribing circle of a triangle, the circumscribing sphere of a simplex, and other figures related to a triangle or a simplex.

References

[1] H. Terasaka, Syotô kikagaku (Japanese; Elementary geometry), Iwanami, 1952.
[2] S. Iyanaga, Kikagaku zyosetu (Japanese; Introduction to geometry), Iwanami, 1968.
[3] G. D. Birkhoff and R. Beatley, Basic geometry, Scott, Foresman, 1941 (Chelsea, third edition, 1959).
[4] E. E. Moise, Elementary geometry from an advanced standpoint, Addison-Wesley, 1963.
[5] H. Weyl, Mathematische Analyse des

Raumproblems, Springer, 1923 (Chelsea, 1960).
[6] F. Klein, Vorlesungen über höhere Geometrie, Springer, third edition, 1926 (Chelsea, 1957).
[7] J. Dieudonné, Algèbre linéaire et géométrie élémentaire, Hermann, second corrected edition, 1964; English translation, Linear algebra and geometry, Hermann, 1969.

150 (VI.4) Euclidean Spaces

A space satisfying the axioms of Euclidean geometry is called a **Euclidean space**. An †affine space having an n-dimensional Euclidean †inner product space on a real number field $\mathbf{R}$ as †standard vector space is an n-dimensional Euclidean space $\mathbf{E}^n$. In an n-dimensional Euclidean space $\mathbf{E}^n$, we fix an †orthogonal frame $\Sigma=(O,E_1,\ldots,E_n),e_i=\overrightarrow{OE_i}$, $(e_i,e_j)=\delta_{ij}$. The frame Σ determines †rectangular coordinates $(x_1,x_2,\ldots,x_n)$ of each point in $\mathbf{E}^n$. We can thus establish a one-to-one correspondence between $\mathbf{E}^n$ and $\mathbf{R}^n=\{(x_1,\ldots,x_n)\,|\,x_i\in\mathbf{R}\}$. In this sense we identify $\mathbf{E}^n$ and $\mathbf{R}^n$ and usually call $\mathbf{R}^n$ itself a Euclidean space. The 1-dimensional space $\mathbf{R}^1$ is a straight line, and the †Cartesian product of n copies of $\mathbf{R}^1$ is an n-dimensional Euclidean space (or **Cartesian space**). Given points $x=(x_1,x_2,\ldots,x_n)$ and $y=(y_1,y_2,\ldots,y_n)$ in the Euclidean space $\mathbf{R}^n$, the †distance $d(x,y)$ between them is given by

$$\sqrt{(y_1-x_1)^2+\ldots+(y_n-x_n)^2}\,.$$

Thus the distance $d(x,y)$ supplies $\mathbf{R}^n$ with the structure of a †metric space. We call x_i the ith coordinate of the point x, the point $(0,\ldots,0)$ the **origin** of $\mathbf{R}^n$, and the set of points $\{x\,|\,-\infty<x_i<\infty;\ x_j=0, j\neq i\}$ the x_i-**axis** (or ith **coordinate axis**). For an integer m such that $-1\leqslant m\leqslant n$, we define m-dimensional †planes in $\mathbf{R}^n$; a -1-dimensional plane is the empty set, a 0-dimensional plane is a point, and a 1-dimensional plane is a straight line. If we take an orthogonal frame, an m-dimensional plane is represented as an $\mathbf{R}^m$ (→ 149 Euclidean Geometry; 9 Affine Geometry).

As a †topological space, $\mathbf{R}^n$ is †locally compact and †connected. A bounded closed set in $\mathbf{R}^n$ is †compact (**Bolzano-Weierstrass theorem**).

Given a point $a=(a_1,\ldots,a_n)$ in $\mathbf{R}^n$ and a real positive number r, the subset $\{x\,|\,d(x,a)\leqslant r\}$ of $\mathbf{R}^n$ is called an n-dimensional **solid sphere** with **center** a and **radius** r, its †interior $\{x\,|\,d(x,a)<r\}$ an n-dimensional **open sphere**, and its †boundary $\{x\,|\,d(x,a)=r\}$ an $(n-1)$-dimensional **sphere**. In particular, a 2-dimensional solid sphere is called a **circular disk**, its interior an **open circle**, and its boundary a **circumference**. A disk or a circumference is sometimes called simply a **circle**. The family of n-dimensional open spheres with center a gives a base for a neighborhood system of the point a. Suppose that we are given a sphere S and two points x,y on S. The points x,y are called **antipodal points** on the sphere S if there exists a straight line L passing through the center of S such that $S\cap L=\{x,y\}$. The segment (or the length of the segment) whose endpoints are antipodal points is called the **diameter** of the solid sphere (or of the sphere). The notion of †diameter (→ 273 Metric Spaces) of a solid sphere or of a sphere considered as a subset of the metric space $\mathbf{R}^n$ coincides with the notion of diameter of the corresponding set defined above. When $n\geqslant 3$, the intersection of a sphere and a 2-dimensional plane passing through the center of the sphere is called a **great circle** of the sphere. For m such that $1\leqslant m\leqslant n$, we consider an m-dimensional solid sphere or an $(m-1)$-dimensional sphere in an m-dimensional plane $\mathbf{R}^m$. These spheres are also called m-dimensional solid spheres or $(m-1)$-dimensional spheres in $\mathbf{R}^n$.

In particular, the solid sphere of radius 1 having the origin as its center is called the **unit disk** (or **unit cell**), and its boundary is called the **unit sphere**. (In particular, when we deal with the 2-dimensional space $\mathbf{R}^2$, we use the term *circle* instead of *sphere*, as in **unit circle**.) The points $(0,\ldots,0,1)$ and $(0,\ldots,0,-1)$ are called the **north pole** and **south pole** of the unit sphere, respectively. The $(n-2)$-dimensional sphere, which is the intersection of the unit sphere and the hyperplane $x_n=0$, is called the **equator**; the part of the unit sphere that is "above" this hyperplane (i.e., in the half-space $x_n\geqslant 0$) is called the **northern hemisphere**, and the part that is "below" the hyperplane (i.e., in the half-space $x_n\leqslant 0$) the **southern hemisphere**.

Let a_i, b_i be real numbers satisfying $a_i<b_i$ $(i=1,2,\ldots,n)$. The subset $\{x\,|\,a_i<x_i<b_i,\ i=1,2,\ldots,n\}$ of $\mathbf{R}^n$ is called an **open interval** of $\mathbf{R}^n$, and the subset $\{x\,|\,a_i\leqslant x_i\leqslant b_i\}$ a **closed interval**. They are sometimes called **rectangles** (when $n=2$), **rectangular parallelopipeds**, or **boxes**. An open interval is actually an open set of $\mathbf{R}^n$, and a closed interval is a closed set. We can take the set of open intervals as base for a neighborhood system of $\mathbf{R}^n$. In particular, the closed interval $\{x\,|\,0\leqslant x_i\leqslant 1,\ i=1,2,\ldots,n\}$ is called the **unit cube** (or **unit n-cube**) of $\mathbf{R}^n$.

All the †convex closed sets (for example, closed intervals) having interior points in $\mathbf{R}^n$ are homeomorphic to an n-dimensional solid sphere. A topological space I^n that is homeomorphic to an n-dimensional solid sphere is called an n-dimensional **(topological) solid sphere, (topological)** n**-cell**, or n**-element**. A topological space S^{n-1} homeomorphic to an $(n-1)$-dimensional sphere is called an $(n-1)$-dimensional **topological sphere** (or simply $(n-1)$-dimensional sphere. The spaces I^n and S^{n-1} are †orientable †combinatorial manifolds, whose orientations are determined by assigning the generators of the (relative) †homology groups $H_n(I^n, I^n)$ and $H_{n-1}(S^{n-1})$, respectively (both are infinite cyclic groups). Using the †boundary operator $\partial: H_n(I^n, I^n) \rightarrow H_{n-1}(S^{n-1})$, the orientation of I^n or S^{n-1} determines that of the other.

References

See references to 149 Euclidean Geometry.

151 (XX.21) Euler, Leonhard

Leonhard Euler (April 15, 1707–September 18, 1783) was born in Basel, Switzerland. In his mathematical development he was greatly influenced by the Bernoullis (→ 40 Bernoulli Family). He was invited to the St. Petersburg Academy in 1726 and remained there until 1741, when he was invited to Berlin by Frederick the Great (1712–1786). Euler was active at the Berlin Academy until 1766, when he returned to St. Petersburg. Already having lost the sight of his right eye in 1735, he now became blind in his left eye also. This, however, did not impede his research in any way, and he continued to work actively until his death in St. Petersburg.

Euler was the central figure of the mathematical activity of the 18th century. He was interested in all fields of mathematics, but especially in analysis in the style of †Leibniz, which had been passed down through the Bernoullis and was developed by him into a form that led toward the mathematics of the 19th century. Through his work analysis became more easily applicable to the fields of physics and dynamics. He developed calculus one step further, and he dealt formally with complex numbers. He also contributed to such fields as †partial differential equations, the theory of †elliptic functions, and the †calculus of variations. He had, however, little of the concern with rigorous foundations that characterized the 19th century. He was the most prolific mathematician of all time, and his collected works are still incomplete, though some seventy volumes have already been published.

References

[1] L. Euler, Opera omnia, ser. 1, vol. 1–29, ser. 2, vol. 1–30, ser. 3, vol. 1–13, Teubner and O. Füssli, 1911–1967.

152 (XI.17) Extremal Length

A. General Remarks

The notable relation between the lengths of certain families of curves in a plane domain and the area of the domain has long been recognized and utilized in function theory. L. V. Ahlfors and A. Beurling formulated this relation by introducing the notion of extremal length for families of curves [1]. Although there are various definitions of extremal length, they are essentially the same except for one due to J. Hersch [4] and A. Pfluger [3].

The image of an open interval or of a circle under a continuous mapping is called a curve. We say that it is **locally rectifiable** if every †arc of the curve is †rectifiable. Let C be a finite or countable collection of locally rectifiable curves in a plane and ρ, $0 \leqslant \rho \leqslant \infty$, be a †Baire function defined in the plane. Represent C in terms of arc length s (→ 245 Length and Area), and set $\langle C, \rho \rangle = \int_C \rho\, ds$. For a family Γ of finite or countable collections C, ρ is called **admissible** if $\langle C, \rho \rangle \geqslant 1$ for every $C \in \Gamma$. If no $C \in \Gamma$ consists of a finite or countable number of points, then $\rho \equiv \infty$ is always admissible for Γ. Call $\inf \{\int\int \rho^2 dx\, dy\}$, where ρ runs over admissible Baire functions, the **module** of Γ, and denote it by $M(\Gamma)$. The reciprocal $\lambda(\Gamma) = 1/M(\Gamma)$ is called the **extremal length** of Γ. We obtain the same value for $\lambda(\Gamma)$ if we require an admissible ρ to be †lower semicontinuous. If ρ is required to be continuous, then the **extremal length defined by Hersch and Pfluger** is obtained. As is shown in example (1) of Section B, there is a case where the two definitions actually differ.

If an admissible ρ yields $M(\Gamma) = \int\int \rho^2 dx\, dy$, then $\rho|dz|$ is called an extremal metric. Beurling gave a necessary and sufficient condition for a metric to be extremal [6].

We list four properties of extremal length: (1) $\lambda(\Gamma_1) \geqslant \lambda(\Gamma_2)$ if $\Gamma_1 \subset \Gamma_2$. (2) $M(\bigcup_n \Gamma_n) \leqslant \sum_n M(\Gamma_n)$. (3) Let $\{\Gamma_n\}$ and Γ be given. Suppose that there are mutually disjoint measurable sets $\{E_n\}$ such that each $C_n \in \Gamma_n$ is contained in E_n. If each element of $\bigcup_n \Gamma_n$ contains at least one $C \in \Gamma$, then $M(\Gamma) \geqslant \sum_n M(\Gamma_n)$, and hence

$$M\Big(\bigcup_n \Gamma_n\Big) = \sum_n M(\Gamma_n).$$

If each $C \in \Gamma$ contains at least one $C_n \in \Gamma_n$ for every n, then $\lambda(\Gamma) \geqslant \sum_n \lambda(\Gamma_n)$. (4) Let f be an analytic function in a domain Ω and $\{C\}$ be given in Ω. Denote by $f(C)$ the image of C by f. Then $\lambda(\{C\}) \leqslant \lambda(\{f(C)\})$. The equality holds if f is one-to-one. This shows that $\lambda(\{C\})$ is conformally invariant.

B. Extremal Distance

Let Ω be a domain in a plane, $\partial\Omega$ its boundary, and X_1, X_2 sets on $\Omega \cup \partial\Omega$. The extremal length of the family of curves in Ω connecting points of X_1 and points of X_2 is called the **extremal distance** between X_1 and X_2 (relative to Ω) and is denoted by $\lambda_\Omega(X_1, X_2)$.

Example (1) Let $\Omega = \{z \mid |z| < 2\}$, $X_1 = \partial\Omega$, and X_2 be a countable set in $|z| < 1$ such that the set of accumulation points of X_2 coincides with $|z| = 1$. Then $\lambda_\Omega(X_1, X_2) = \infty$, but the extremal distance in the sense of Hersch and Pfluger is equal to $(2\pi)^{-1}\log 2$.

Example (2) In a rectangle with sides a and b, the extremal distance between the sides of length a is b/a.

Example (3) Let Ω be an annulus $r_1 < |z| < r_2$. The extremal distance between the two boundary circles of Ω is equal to $(2\pi)^{-1}\log(r_2/r_1)$. The extremal length of the family of curves in Ω homotopic to the circles is equal to the reciprocal value.

Example (4) Let Ω be a domain in the extended z-plane such that $\infty \in \Omega$. Let $z_0 \in \Omega$ and $\{|z - z_0| = r\} \subset \Omega$, and denote by λ_r the extremal distance between $\{|z - z_0| = r\}$ and a set $X \subset \partial\Omega$ relative to Ω. Then $\lambda_r - (2\pi)^{-1}\log r$ increases with r. We call the limit the **reduced extremal distance** and denote it by $\tilde{\lambda}_\Omega(X, \infty)$. †Robin's constant for †Green's function in Ω with pole at $z = \infty$ is equal to $2\pi\tilde{\lambda}_\Omega(\partial\Omega, \infty)$.

Extremal length is also defined on Riemann surfaces. Some classical conformal invariants can be given in a generalized form in terms of extremal length. The notion of extremal length has applications in various branches of function theory, such as †conformal and †quasiconformal mappings, the Phragmén-Lindelöf theorem, the †coefficient problem, and the †type problem of Riemann surfaces. It is also applied to problems in differential geometry. Extending the notion of extremal length, M. Ohtsuka considered **extremal length with weight**, and B. Fuglede introduced the notion of **generalized module** in higher-dimensional spaces [2]. These notions have useful properties and applications.

References

[1] L. V. Ahlfors and A. Beurling, Conformal invariants and function-theoretic null-sets, Acta Math., 83 (1950), 101–129.

[2] V. Wolontis, Properties of conformal invariants, Amer. J. Math., 74 (1952), 587–606.

[3] A. Pfluger, Extremallängen und Kapazität, Comment. Math. Helv., 29 (1955), 120–131.

[4] J. Hersch, Longueurs extrémales et théorie des fonctions, Comment. Math. Helv., 29 (1955), 301–337.

[5] M. Ohtsuka, Kansûron tokuron (Japanese; Topics on the theory of functions), Kyôritu, 1957.

[6] M. Ohtsuka, Dirichlet problem, extremal length and prime ends, Van Nostrand, 1970.

[7] B. Fuglede, Extremal length and functional completion, Acta Math., 98 (1957), 171–219.

[8] J. A. Jenkins, Univalent functions and conformal mapping, Erg. Math., Springer, 1958.

F

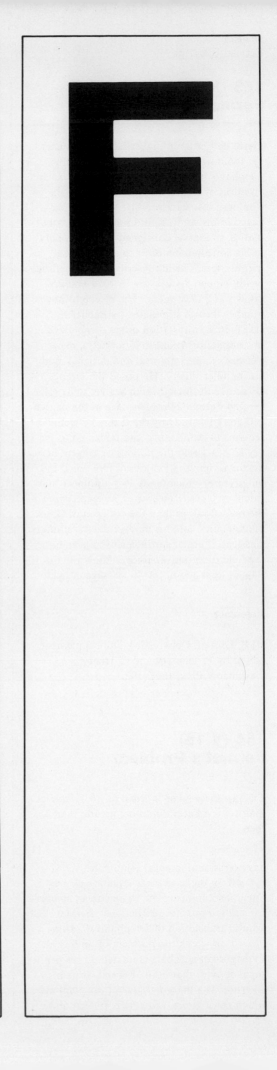

153 (XX.22)
Fermat, Pierre de

Pierre de Fermat (August 20, 1601–January 12, 1665) was born into a family of leather merchants near Toulouse, France. He became an attorney and in 1631 a member of the Toulouse district assembly. When not engaged in his work, he did research in mathematics, so that he consigned his results only to his correspondence or to unpublished manuscripts. The manuscripts were published posthumously by his son in 1679 and are known as *Varia opera*. His research into number theory, stimulated by Bachet's (1581–1638) translation of the *Arithmetika* of Diophantus (published in 1621), made Fermat's name immortal and initiated modern number theory. He posed the famous †Fermat's Problem, which has yet to be solved (→ 154 Fermat's Problem). He began analytic geometry by studying the theory of †conic sections of Apollonius, and utilizing this theory he dealt with the notions of tangent lines, maximal (minimal) values of functions, and quadrature, which made him a pioneer in calculus. Unlike †Descartes, he emphasized the revival rather than the criticism of Greek mathematics, and his mathematics is strongly classical. †Fermat's principle is important in the field of optics, where it is known as the †law of least action for the passage of light.

Reference

[1] P. Fermat, Oeuvres I–V and supplement, edited by P. Tannery and C. Henry, Gauthier-Villars, 1891–1922.

154 (V.16)
Fermat's Problem

The **last theorem of †Fermat** (c. 1637) asserts that if n is a natural number greater than 2, then

$$x^n+y^n=z^n \qquad (1)$$

has no rational integral solution x, y, z with $xyz\neq0$. In the case $n=2$, equation (1) has integral solutions called **Pythagorean numbers** (→ 122 Diophantine Equations). Fermat read a Latin translation of Diophantus' *Arithmetika*, in which the problem of finding all Pythagorean numbers is treated. In his personal copy of that book, Fermat wrote his assertion as a marginal note at the point at which $n=2$ in equation (1) is treated and added the famous words, "I have discovered a truly remarkable proof of this theorem which this margin is too small to contain." It is not known whether Fermat actually had a proof. Fermat's problem asks for a proof or disproof of this conjecture, which itself has not been solved despite centuries of efforts by many mathematicians; but its study has promoted remarkable advances in number theory. In particular, E. E. Kummer's theory of ideal numbers and the development of the theory of †cyclotomic fields were originally conceived in treating Fermat's problem.

In this article, we consider only those integral solutions x, y, z of equation (1) with $xyz\neq0$ that are relatively prime. We also restrict ourselves to the cases $n=l$ (odd prime) and $n=4$, without loss of generality.

For smaller values of n, the nonsolvability of equation (1) was proved long ago, for $n=3$ by L. Euler (1770), and later again by A. M. Legendre; for $n=4$ by Fermat and Euler; for $n=5$ by Legendre (1825); and for $n=7$ by G. Lamé (1839). S. Germain and Legendre found some results on more general cases, but the most remarkable result was obtained by Kummer (*J. Reine Angew. Math.*, 40 (1850), *Abh. Akad. Wiss. Berlin* (1857)).

Let l be an odd prime, ζ a primitive lth root of unity, and h the †class number of the cyclotomic field $\mathbf{Q}(\zeta)$. Then the class number h_2 of the real subfield $\mathbf{Q}(\zeta+\zeta^{-1})$ of $\mathbf{Q}(\zeta)$ divides h. We call $h_1=h/h_2$ and h_2 the †first and †second factors of h, respectively.

(1) If l is †regular, that is, if $(h,l)=1$, then $x^l+y^l=z^l$ has no solution (Kummer, 1850).

There are infinitely many irregular primes [4]; those under 100 are 37, 59, and 67. There are 334 regular primes and 216 irregular primes between 3 and 4001. It is not yet known whether there are infinitely many regular prime numbers, although the beginning part of the sequence of natural numbers contains a larger number of these than the number of irregular prime numbers. The condition $(l,h)=1$ is equivalent to saying that the numerators of †Bernoulli numbers B_{2m} $(m=1,2,\ldots,(l-3)/2)$ are not divisible by l (Kummer, 1850).

Kummer obtained a result on irregular primes (1857) which was improved later as follows. Note that if l is not regular then h_1 is divisible by l (Kummer, 1850) (→ 16 Algebraic Number Fields).

(2) If $(h_2,l)=1$ and the numerators of Bernoulli numbers B_{2ml} $(m=1,2,\ldots,(l-3)/2)$ are not divisible by l^3, then $x^l+y^l=z^l$ has no solution (H. S. Vandiver, *Trans. Amer. Math. Soc.*, 31 (1929)). By computation Vandiver confirmed that $x^l+y^l=z^l$ has no solution for $l<619$. At present, this procedure

has been continued for $l \leqslant 30{,}000$ using computers by the method of D. H. Lehmer, E. Lehmer, and H. S. Vandiver, *Proc. Nat. Acad. Sci. US*, 40 (1954) (W. Johnson, *Math. Comp.*, 29 (1975)).

When the condition $(xyz, l)=1$ or $(xyz, l)=l$ is added, we speak of Case I or Case II, respectively. The following theorems hold for Case I.

(3) If $(h_2, l)=1$, then $x^l+y^l=z^l$ has no solution in Case I (Vandiver, 1934).

(4) If $x^l+y^l=z^l$ has a solution in Case I, then

$$B_{2m}f_{l-2m}(t)\equiv 0 \pmod{l}, \quad m=1,2,\ldots,(l-3)/2 \tag{2}$$

holds for $-t=x/y, y/x, y/z, z/y, x/z$, and z/x, where $f_m(t)=\sum_{r=0}^{l-1} r^{m-1}t^r$, and B_m is the mth Bernoulli number. This is called **Kummer's criterion** (D. Mirimanov, 1905).

A simplification of the above result is

(5a) If $x^l+y^l=z^l$ has a solution in Case I, then

$$(2^{l-1}-1)/l\equiv 0 \pmod{l}$$

(A. Wieferich, *J. Reine Angew. Math.*, 136 (1909)). This result created a sensation at the time of its publication. It was first shown that 1093 and 3511 are the only primes with $l<3700$ for which the above congruence holds; it is presently known that no other l with $l \leqslant 31{,}059{,}000$ satisfies this congruence. The criterion (5a) was gradually improved by Mirimanov (1910, 1911), P. Furtwängler (1912), Vandiver (1914), G. Frobenius (1914), F. Pollaczeck (1917), T. Morishima (1931), and J. B. Rosser (1940, 1941). For example:

(5b) If $x^l+y^l=z^l$ has a solution in Case I, then

$$(m^{l-1}-1)/l\equiv 0 \pmod{l} \tag{3}$$

holds for all m with $2 \leqslant m \leqslant 43$. By means of this result, Rosser (1941) showed for $l<41{,}000{,}000$, and D. H. Lehmer and E. Lehmer (*Bull. Amer. Math. Soc.*, 47 (1941)) showed for $l<253{,}749{,}889$ that $x^l+y^l=z^l$ has no solution in Case I.

We have hitherto been concerned with rational integral solutions of $x^l+y^l=z^l$. We may also consider the problem of proving or disproving that $\alpha^l+\beta^l=\gamma^l$ has no solution α, β, γ with $\alpha\beta\gamma\neq 0$ in the ring of †algebraic integers of $\mathbf{Q}(\zeta)$. Case I means the impossibility of

$$\alpha^l+\beta^l+\gamma^l=0, \quad (\alpha\beta\gamma, l)=1, \tag{4}$$

and Case II means the impossibility of

$$\alpha^l+\beta^l=\varepsilon\lambda^{nl}\gamma^l, \quad (\alpha\beta\gamma, l)=1, \tag{5}$$

where n is a natural number, ε is a †unit in $\mathbf{Q}(\zeta)$, and $\lambda=(1-\zeta)$. We have the following results:

(1*) If $(h, l)=1$, then neither equation (4) nor equation (5) has a solution (Kummer, 1850).

(2*) Under the same conditions as in statement (2), equation (4) has no solution. If we additionally restrict α, β, γ to relatively prime integers of $\mathbf{Q}(\zeta+\zeta^{-1})$ and replace λ by $(1-\zeta)(1-\zeta^{-1})$, then equation (5) also has no solution (Vandiver, 1929).

(3*) If $(h_2, l)=1$, then equation (4) has no solution in the ring of integers in $\mathbf{Q}(\zeta+\zeta^{-1})$ (Morishima, 1934).

(5b*) If equation (4) has solutions α, β, γ in $\mathbf{Q}(\zeta)$, then equation (3) holds for all m with $2 \leqslant m \leqslant 43$ (Morishima, 1934).

For the case where l is sufficiently large, there are results of M. Krasner (*C. R. Acad. Sci. Paris* (1934)) and Morishima (*Proc. Japan Acad.*, 11 (1935)).

Bibliographies are given in Vandiver and Wahlin [1] and Vandiver [2].

References

[1] H. S. Vandiver and G. E. Wahlin, Algebraic numbers II, Bull. Nat. Res. Council, no. 62, 1928.
[2] H. S. Vandiver, Fermat's last theorem, Amer. Math. Monthly, 53 (1946), 555–578.
[3] T. Morishima, Feruma no mondai (Japanese; Fermat's problem), Iwanami Coll. of Math., 1934.

155 (IX.20) Fiber Bundles

A. General Remarks

E. Stiefel [2] introduced certain †diffeomorphism invariants of †differentiable manifolds by considering a field of a finite number of linearly independent vectors attached to each point of a manifold; and H. Whitney [3] obtained the notion of fiber bundles as a compound idea of a manifold and such a field of tangent vectors. S. S. Chern [4] emphasized the global point of view in differential geometry by recognizing the relation between the notion of †connections (due to E. Cartan) and the theory of fiber bundles. The theory of fiber bundles is also applied to various fields of mathematics, for example, the theory of †Lie groups, †homogeneous spaces, †covering spaces, and general vector bundles, vector bundles of class C^r, or analytic vector bundles.

Homological properties of fiber bundles are studied by means of †spectral sequences, and cohomology structures of several homogeneous spaces and several characteristic classes are determined explicitly by means of †cohomology operations. Also, the group $K(X)$, formed by equivalence classes of vector bundles over a finite †CW complex X, is a †generalized cohomology group, treated in †K-theory, in which further development is expected (→ 233 K-Theory).

B. Definition

Let E, B, F be topological spaces, $p: E \to B$ a continuous mapping, and G an †effective left †transformation group of F. If there exist an †open covering $\{U_\alpha\}$ $(\alpha \in \Lambda)$ of B and a homeomorphism $\varphi_\alpha: U_\alpha \times F \approx p^{-1}(U_\alpha)$ for each $\alpha \in \Lambda$ having the following three properties, then the system $(E,p,B,F,G,U_\alpha,\varphi_\alpha)$ is called a **coordinate bundle**: (1) $p\varphi_\alpha(b,y) = b$ $(b \in U_\alpha, y \in F)$. (2) Define $\varphi_{\alpha,b}: F \approx p^{-1}(b)$ $(b \in U_\alpha)$ by $\varphi_{\alpha,b}(y) = \varphi_\alpha(b,y)$; then $g_{\beta\alpha}(b) = \varphi_{\beta,b}^{-1} \circ \varphi_{\alpha,b} \in G$ for $b \in U_\alpha \cap U_\beta$. (3) $g_{\beta\alpha}: U_\alpha \cap U_\beta \to G$ is continuous. We say that this bundle is **equivalent** to a coordinate bundle $(E,p,B,F,G,U'_\mu,\varphi'_\mu)$ if $\bar{g}_{\mu\alpha}(b) = \varphi'^{-1}_{\mu,b} \circ \varphi_{\alpha,b} \in G$ $(b \in U_\alpha \cap U'_\mu)$ and $\bar{g}_{\mu\alpha}: U_\alpha \cap U'_\mu \to G$ is continuous. An equivalence class $\xi = (E,p,B,F,G)$ of coordinate bundles is called a **fiber bundle** (or G-**bundle**), and E is called the **total space** (or **bundle space**), p the **projection**, B the **base space**, F the **fiber**, and G the **bundle group** (or **structure group**). Also, U_α of a coordinate bundle $(E,p,B,F,G,U_\alpha,\varphi_\alpha)$ belonging to the class ξ is called the **coordinate neighborhood**, φ_α the **coordinate function**, and $g_{\beta\alpha}$ the **coordinate transformation** (or **transition function**).

Let $\xi = (E,p,B,F,G)$ and $\xi' = (E',p',B',F,G)$ be two fiber bundles with the same fiber and group. A continuous mapping $\Psi: E \to E'$ is called a **bundle mapping** from ξ to ξ' if the following two conditions are satisfied: (1) There is a continuous mapping $\psi: B \to B'$ with $p' \circ \Psi = \psi \circ p$. (2) $\psi_{\mu\alpha}(b) = \varphi'^{-1}_{\mu,b'} \circ \Psi \circ \varphi_{\alpha,b} \in G$ $(b \in U_\alpha \cap \psi^{-1}(V'_\mu), b' = \psi(b))$, and $\psi_{\mu\alpha}: U_\alpha \cap \psi^{-1}(V'_\mu) \to G$ is continuous, where $\{U_\alpha, \varphi_\alpha\}$ and $\{V'_\mu, \varphi'_\mu\}$ are pairs of coordinate neighborhoods and functions of ξ and ξ', respectively. Moreover, if ψ is a homeomorphism, then Ψ is also a homeomorphism and Ψ^{-1} is a bundle mapping.

Let $\xi = (E,p,B,F,G)$ and $\xi' = (E',p',B,F,G)$ be two fiber bundles with the same base space, fiber, and group. If there is a bundle mapping $\Psi: E \to E'$ such that $\psi: B \to B$ as described before is the identity mapping, then we say that ξ is **equivalent** to ξ' and write $\xi \equiv \xi'$. Take the same coordinate neighborhoods $\{U_\alpha\}$, and let $g_{\beta\alpha}$ and $g'_{\beta\alpha}$ be the coordinate transformations of ξ and ξ', respectively. Then $\xi \equiv \xi'$ if and only if there are continuous mappings $\lambda_\alpha: U_\alpha \to G$ with $g'_{\beta\alpha}(b) = \lambda_\beta(b)\, g_{\beta\alpha}(b)\, \lambda_\alpha(b)^{-1}$ $(b \in U_\alpha \cap U_\beta)$.

For a system $\{g_{\beta\alpha}\}$ of coordinate transformations of a fiber bundle, we have $g_{\gamma\beta}(b)\, g_{\beta\alpha}(b) = g_{\gamma\alpha}(b)$ $(b \in U_\alpha \cap U_\beta \cap U_\gamma)$. Conversely, given a system of $g_{\beta\alpha}: U_\alpha \cap U_\beta \to G$ ($\{U_\alpha\}$ is an open covering of B) satisfying this condition, there is a unique G-bundle (E,p,B,F,G) with $\{g_{\beta\alpha}\}$ as a system of coordinate transformations. Actually, E is the †factor space of $\tilde{E} = \{(b,y,\alpha) \mid b \in U_\alpha\} \subset B \times F \times \Lambda$ obtained by identifying two points (b,y,α), (b',y',β) with $b = b'$, $y' = g_{\beta\alpha}(b) \cdot y$, and p is defined by $p\{(b,y,\alpha)\} = b$, where the index set $\Lambda = \{\alpha\}$ is considered a discrete space.

C. Principal Fiber Bundles

A fiber bundle $\eta = (P,q,B,G,G)$ is called a **principal fiber bundle** (or simply **principal bundle**) if G operates on G by left translations. This is also defined by the following conditions: G is a right †topological transformation group of P, and there exist an open covering $\{U\}$ of B and homeomorphisms $\varphi: U \times G \approx q^{-1}(U)$ with $q\varphi(b,g) = b$, $\varphi(b,g)\cdot g' = \varphi(b,gg')$ $(b \in U; g,g' \in G)$. A bundle mapping $\Psi: P \to P'$ between two principal bundles $\eta = (P,q,B,G)$ and $\eta' = (P',q',B',G)$ is also defined as a continuous mapping Ψ with $\Psi(x \cdot g) = \Psi(x) \cdot g$.

D. Associated Fiber Bundles

Let $\eta = (P,q,B,G)$ be a principal bundle, and let F be a topological space having G as an effective left topological transformation group. Then G is a right topological transformation group of the product space $P \times F$ by $(x,y)\cdot g = (x \cdot g, g^{-1} \cdot y)$ $(x \in P, y \in F, g \in G)$. Consider the orbit space $P \times_G F = (P \times F)/G$, and define the continuous mapping $p: P \times_G F \to B$ by $p\{(x,y)\} = q(x)$. Then $\eta \times_G F = (P \times_G F, p, B, F, G)$ is a fiber bundle, called the **associated fiber bundle** of η with fiber F. On the other hand, η is called an **associated principal bundle** of $\xi = (E,p,B,F,G)$ if $\xi \equiv \eta \times_G F$. A principal bundle η having the same coordinate transformations as ξ is an associated principal bundle of ξ, and two fiber bundles are equivalent if and only if their associated principal bundles are equivalent. Therefore, given a fiber bundle ξ, there exists a principal bundle η such that $\xi = \eta \times_G F$.

E. Examples of Fiber Bundles

(1) Product bundle. $(B \times F, p_1, B, F, G)$, where p_1, the projection of the product space, is called a **product bundle** if there is just one

coordinate neighborhood B and the coordinate function is the identity mapping of $B \times F$. A bundle that is equivalent to a product bundle is called a **trivial bundle**.

(2) A †covering $(\tilde{Y}, p, Y)$ is a fiber bundle whose fiber is the discrete space $p^{-1}(y_0)$ $(y_0 \in Y)$, and the structure group is a factor group of the †fundamental group $\pi_1(Y, y_0)$. In particular, a †regular covering is a principal bundle.

(3) Hopf bundle. Let Λ be the real number field $\mathbf{R}$, the complex number field $\mathbf{C}$, or the quaternion field $\mathbf{H}$, $\lambda = \dim_{\mathbf{R}} \Lambda$, and Λ^{n+1} the $(n+1)$-dimensional linear space over Λ. Identify two points $(z_0, \ldots, z_n)$, $(z'_0, \ldots, z'_n)$ of the subspace $\Lambda^{n+1} - \{0\}$ (0 is the origin) if there is a $z \in \Lambda$ such that $z_i = z'_i z$ $(i = 0, \ldots, n)$. Then we obtain the †factor space $P^n(\Lambda)$, called the n-dimensional **projective space over** Λ. Let S^n_Λ $(= S^{\lambda(n+1)-1}$, the $(\lambda(n+1)-1)$-sphere) be the unit sphere in Λ^{n+1}. Then S^0_Λ is the †topological transformation group of S^n_Λ by the product of Λ, and the †orbit space $S^n_\Lambda / S^0_\Lambda$ is $P^n(\Lambda)$. Furthermore, $(S^n_\Lambda, q, P^n(\Lambda), S^0_\Lambda)$ (q is the projection) is a principal bundle called the **Hopf bundle** (or **Hopf fibering**). These comments are valid also for $n = \infty$. When $n = 1$, $P^1(\Lambda)$ is homeomorphic to S^λ, and the Hopf bundle is $(S^{2\lambda-1}, q, S^\lambda, S^{\lambda-1})$ $(\lambda = 1, 2, 4)$. A Hopf bundle is defined similarly for $\lambda = 8$ using the †Cayley algebra, and the projection $q: S^{2\lambda-1} \to S^\lambda$ $(\lambda = 2, 4, 8)$ is the **Hopf mapping**.

(4) Let G be a topological group, H its closed subgroup, and $r: G \to G/H$, $r(g) = gH$ the natural projection. If there exist a neighborhood U of $r(H) \in G/H$ and a continuous mapping $f: U \to G$ such that $r \circ f$ is the identity mapping, then we say that H has a **local cross section** f in G, and $(G/K, p, G/H, H/K, H/K_0)$ is a fiber bundle for any closed subgroup K of H (where p is the natural projection $gK \to gH$ and K_0 is the largest subgroup of K invariant in H). The associated principal bundle of the latter fiber bundle is $(G/K_0, p, G/H, H/K_0)$. Any closed subgroup H of a †Lie group G has a local cross section in G; hence the above bundles can be obtained.

F. Vector Bundles

A system $\xi = (E, p, B)$ of topological spaces E, B and a continuous mapping $p: E \to B$ is called an n-dimensional real **vector bundle** if the following two conditions are satisfied: (1) $p^{-1}(b)$ is a real vector space for each $b \in B$. (2) There exist an open covering $\{U_\alpha\}$ $(\alpha \in \Lambda)$ of B and a coordinate function $\varphi_\alpha: U_\alpha \times F \approx P^{-1}(U_\alpha)$ for each $\alpha \in \Lambda$, where $F = \mathbf{R}^n$; furthermore, the $\varphi_{\alpha,b}: \mathbf{R}^n \approx p^{-1}(b)$ are isomorphisms of vector spaces. In this case, $g_{\beta\alpha}(b) = \varphi_{\beta,b}^{-1} \circ \varphi_{\alpha,b}: \mathbf{R}^n \approx \mathbf{R}^n$ $(b \in U_\alpha \cap U_\beta)$ is an element of the †general linear group $GL(n, \mathbf{R})$. Hence a vector bundle is a fiber bundle with fiber $\mathbf{R}^n$ and group $GL(n, \mathbf{R})$, and the converse is also true. A 1-dimensional vector bundle is called a **line bundle**. A vector bundle $\xi' = (E', p', B)$ is called a **subbundle** of a vector bundle $\xi = (E, p, B)$ if $E' \subset E$, $p|E' = p'$, and $p'^{-1}(b)$ is a vector subspace of $p^{-1}(b)$ for each $b \in B$.

Let ξ_1 and ξ_2 be two vector bundles of dimension n_1 and n_2 with the same base space B. Let E be the union of the direct sum $p_1^{-1}(b) + p_2^{-1}(b)$ for $b \in B$, and define $p: E \to B$ by $p(p_1^{-1}(b) + p_2^{-1}(b)) = b$. Take the same coordinate neighborhoods U_α for ξ_1 and ξ_2, and define $\varphi_\alpha: U_\alpha \times \mathbf{R}^{n_1+n_2} \to p^{-1}(U_\alpha)$ by $\varphi_\alpha(b, y) = (\varphi^1_{\alpha,b} + \varphi^2_{\alpha,b})(y)$ $(y \in \mathbf{R}^{n_1+n_2} = \mathbf{R}^{n_1} + \mathbf{R}^{n_2})$, where the $\varphi^i_\alpha: U_\alpha \times \mathbf{R}^{n_i} \approx p_i^{-1}(U_\alpha)$ are the coordinate functions of ξ_i. Then E is topologized by taking the family $\{\varphi_\alpha(O)\}$ (O is open in $U_\alpha \times \mathbf{R}^{n_1+n_2}$) as the †open base, and we obtain an (n_1+n_2)-dimensional real vector bundle (E, p, B), denoted by $\xi_1 \oplus \xi_2$ and called the **Whitney sum** of ξ_1 and ξ_2. Similarly, we can define the **tensor product** $\xi_1 \otimes \xi_2$, the **p-fold exterior power** $\wedge^p \xi$ (or **bundle** $\xi^{(p)}$ **of p-vectors**), and $\mathrm{Hom}(\xi_1, \xi_2)$, of dimension $n_1 n_2$, $\binom{n}{p}$, and $n_1 n_2$, respectively, using the tensor product $\mathbf{R}^{n_1} \otimes \mathbf{R}^{n_2} = \mathbf{R}^{n_1 n_2}$, the p-fold †exterior power $\wedge^p \mathbf{R}^n = \mathbf{R}^{\binom{n}{p}}$ (the space $(\mathbf{R}^n)^{(p)}$ of p-vectors in $\mathbf{R}^n$), and $\mathrm{Hom}(\mathbf{R}^{n_1}, \mathbf{R}^{n_2}) = \mathbf{R}^{n_1 n_2}$. (For the last one, we use $\mathrm{Hom}((\varphi^1_{\alpha,b})^{-1}, \varphi^2_{\alpha,b})$ and not $\mathrm{Hom}(\varphi^1_{\alpha,b}, \varphi^2_{\alpha,b})$.) $\mathrm{Hom}(\xi, \varepsilon^1) = \xi^*$ is called the **dual (vector) bundle** of ξ, where ε^1 is the trivial line bundle. If we use coordinate transformations, $\oplus$, $\otimes$, $\wedge^p$, and ξ^* are obtained by the direct sum, †Kronecker product, matrix of †p-minors, and †transpose of matrices, respectively. If ξ_2 is a subbundle of ξ_1, the **quotient bundle** ξ_1/ξ_2 of dimension $n_1 - n_2$ is defined by using $\mathbf{R}^{n_1}/\mathbf{R}^{n_2} = \mathbf{R}^{n_1-n_2}$, and $\xi_2 \oplus (\xi_1/\xi_2)$ is equivalent to ξ_1. These operations preserve the equivalence relation of bundles. Also, $\oplus$ and $\otimes$ are commutative up to equivalence and satisfy the associative and distributive laws. For each ξ having a finite-dimensional †CW complex as base space, there is a ξ' such that $\xi \oplus \xi'$ is trivial.

Using the complex number field $\mathbf{C}$ or the quaternion field $\mathbf{H}$ instead of the real number field $\mathbf{R}$, we can define similarly the **complex vector bundle** or the **quaternion vector bundle** and the operations $\oplus$, $\otimes$, etc.

(5) Tangent bundles, tensor bundles. Let M be an n-dimensional †differentiable manifold of class C^r. Consider the †tangent vector space $T_p(M)$ at $p \in M$, set $T(M) = \bigcup_{p \in M} T_p(M)$, and define $\pi: T(M) \to M$ by $\pi(T_p(M)) = p$. For a †coordinate neighborhood U_p of p with local coordinate system $(x_1, \ldots, x_n)$, each point of $\pi^{-1}(U_p)$ is represented by $\sum_{i=1}^n f_i \dfrac{\partial}{\partial x_i}$,

and $\pi^{-1}(U_p)$ has a coordinate system $(x_1, \ldots, x_n, f_1, \ldots, f_n)$. Hence $T(M)$ is a C^{r-1}-manifold, and $\mathfrak{T}(M) = (T(M), \pi, M, \mathbf{R}^n, GL(n, \mathbf{R}))$ is an n-dimensional real vector bundle. $\mathfrak{T}(M)$ is called the **tangent (vector) bundle**, its dual bundle $\mathfrak{T}^*(M)$ the **cotangent (vector) bundle**, and the tensor product $\mathfrak{T}(M) \otimes \ldots \otimes \mathfrak{T}^*(M) \otimes \ldots$ a **tensor bundle** of M. The line bundle $\wedge^n \mathfrak{T}^*(M)$ is called the **canonical bundle** of M.

For a †complex manifold M, $T(M)$ is a complex manifold and $\mathfrak{T}(M)$ is a complex vector bundle. Therefore, these bundles are defined as complex bundles.

(6) Tangent r-frame bundle. In the preceding example, the space of all †tangent r-frames of M is a bundle space with base space M and group $GL(n, \mathbf{R})$. It is called the **tangent r-frame bundle** of M.

G. The Classification Problem

For a fiber bundle $\xi = (E, p, B, F, G)$ and a continuous mapping $\psi: B' \to B$, consider the subspace $E' = \{(x, b') \in E \times B' \mid p(x) = \psi(b')\}$ of $E \times B'$ and the projections $p': E' \to B'$ and $\Psi: E' \to E$. Then $\psi^{\#}\xi = (E', p', B', F, G)$ is a fiber bundle, and Ψ is a bundle mapping from $\psi^{\#}\xi$ to ξ; $\psi^{\#}\xi$ is called the **induced bundle** of ξ by ψ. Let $\{U_\alpha\}$ and $\{g_{\beta\alpha}\}$ be the systems of coordinate neighborhoods and transformations of ξ. Then $\{\psi^{-1}(U_\alpha)\}$ and $\{g_{\beta\alpha} \circ \psi\}$ are corresponding systems of $\psi^{\#}\xi$. If $\Psi: E' \to E$ is a bundle mapping from ξ' to ξ having $\psi: B' \to B$ as the mapping of base spaces, then $\xi' \equiv \psi^{\#}\xi$. Also, we have $\psi^{\#}\xi_1 \equiv \psi^{\#}\xi_2$ if $\xi_1 \equiv \xi_2$; $(\psi \circ \psi')^{\#}\xi \equiv \psi'^{\#}(\psi^{\#}\xi)$. If ξ is a principal bundle, then $\psi^{\#}\xi$ is also principal, and $\psi^{\#}(\eta \times_G F) \equiv (\psi^{\#}\eta) \times_G F$. For a †paracompact space B', $\psi_1^{\#}\xi \equiv \psi_2^{\#}\xi$ if $\psi_1, \psi_2: B' \to B$ are †homotopic.

For a topological group G, a principal bundle $\xi(n, G) = (E(n, G), p, B(n, G), G)$ is called an **n-universal bundle** if $E(n, G)$ is †n-connected ($n \leqslant \infty$); its base space $B(n, G)$ is called an **n-classifying space** of G. In particular, $\xi(\infty, G) = \xi_G = (E_G, p, B_G, G)$ is called simply a **universal bundle** and B_G a **classifying space** of G. Then we have the **classification theorem**: Let B be a CW complex with $\dim B \leqslant n$; then the set of equivalence classes of principal G-bundles with base space B is in one-to-one correspondence with the †homotopy set $\pi(B; B(n, G))$ of continuous mappings of B into $B(n, G)$. Such a correspondence is given by associating with the induced bundle $\psi^{\#}\xi(n, G)$ a continuous mapping $\psi: B \to B(n, G)$, called the **characteristic mapping** of $\psi^{\#}\xi(n, G)$. Furthermore, if G is an effective left topological transformation group of F, the set of equivalence classes of G-bundles with base space B and fiber F is in one-to-one correspondence with $\pi(B; B(n, G))$. The correspondence is given by associating $\psi^{\#}(\xi(n, G) \times_G F)$ to ψ.

The existence of a universal bundle for any topological group is known, and in particular, a classifying space B_G is a countable CW complex for any countable CW group G (i.e., a topological group that is a countable CW complex such that the mapping $g \to g^{-1}$ of G into G and the product mapping $G \times G \to G$ are both cellular) (J. Milnor [6]). The following examples for Lie groups are also useful. Note that every CW complex B_G of a given G has the same †homotopy type.

H. Examples of Universal Bundles

(1) G is either $O(n)$, $U(n)$, or $Sp(n)$: Let Λ and λ be as in (3) of Section E. According as Λ is $\mathbf{R}$, $\mathbf{C}$, or $\mathbf{H}$, we let $U(n, \Lambda)$ be the †orthogonal group $O(n)$, the †unitary group $U(n)$, or the †symplectic group $Sp(n)$. Then the †Stiefel manifold $V_{m+n,m}(\Lambda) = U(m+n, \Lambda)/I_m \times U(n, \Lambda)$ (I_m is the unit element of $U(m, \Lambda)$) is $(\lambda(n+1)-2)$-connected. Hence the principal bundle $\xi(\lambda(n+1)-2, U(m+n, \Lambda)) = (V_{m+n,m}(\Lambda), M_{m+n,m}(\Lambda), U(m, \Lambda))$ from (4) of Section E is a $(\lambda(n+1)-2)$-universal bundle of $U(m, \Lambda)$, where the base space $M_{m+n,m}(\Lambda) = U(m+n, \Lambda)/U(m, \Lambda) \times U(n, \Lambda)$ is the †Grassmann manifold.

(2) G is either $O(\infty)$, $U(\infty)$, or $Sp(\infty)$. The examples in (1) are valid for $m, n = \infty$. Consider the †inductive limit group $U(\infty, \Lambda) = \bigcup_n U(n, \Lambda)$ under the natural inclusion $U(n, \Lambda) \subset U(n+1, \Lambda)$, and supply the **infinite classical group** $U(\infty, \Lambda)$ with the weak topology (this means that a set O of $U(\infty, \Lambda)$ is open if and only if each $O \cap U(n, \Lambda)$ is open in $U(n, \Lambda)$). Then the **infinite Stiefel manifold** $V_{m+n,m}(\Lambda)$ and the **infinite Grassmann manifold** $M_{m+n,m}(\Lambda)$ ($m = \infty$ or $n = \infty$) are defined as before, and we have

$$M_{\infty,m}(\Lambda) = \bigcup_n M_{m+n,m}(\Lambda),$$

and so on. Furthermore, these manifolds are CW complexes, and $V_{\infty,m}(\Lambda)$ ($m \leqslant \infty$) is ∞-connected. Although $U(\infty, \Lambda)$ is not a Lie group, $U(m, \Lambda) \times U(n, \Lambda)$ has a local cross section in $U(m+n, \Lambda)$ for $m, n \leqslant \infty$ [7]. Therefore, setting $n = \infty$ in (1), $\xi(\infty, U(m, \Lambda))$ is a universal bundle of $U(m, \Lambda)$, and the infinite Grassmann manifold $M_{\infty,m}(\Lambda)$ is a classifying space $B_{U(m,\Lambda)}$. Also, $\xi(\lambda(n+1)-2, U(\infty, \Lambda))$ in (1) is a $(\lambda(n+1)-2)$-universal bundle of $U(\infty, \Lambda)$ ($n \leqslant \infty$).

(3) G is either $SO(m)$ or a general Lie group. For the †rotation group $SO(m)$, we have $\xi(n-1, SO(n)) = (V_{m+n,m}(\mathbf{R}), p, \tilde{M}_{m+n,m}, SO(m))$ and $B_{SO(m)} = \tilde{M}_{\infty,m}$, where $\tilde{M}_{m+n,m} =$

$SO(m+n)/SO(m)\times SO(n)$ is the oriented Grassmann manifold. For any compact Lie group G, we have $\xi(n-1,G)=(V_{m+n,m}(\mathbf{R}), p, O(m+n)/G\times O(n), G)$, where $G\subset O(m)$. For any connected Lie group G, we have $\xi(n,G)=\xi(n,G_1)\times_{G_1}G$, where G_1 is the maximum compact subgroup of G (since G/G_1 is homeomorphic to a Euclidean space, $\xi(n,G)$ †reduces to $\xi(n,G_1)$).

I. Reduction of Fiber Bundles

Let G be a topological group and H its closed subgroup. We say that the structure group of a G-bundle ξ is **reduced** to H if ξ is equivalent to a G-bundle whose coordinate transformations take values in H. For a principal H-bundle $\eta_0=(P,q,B,H)$, the associated H-bundle $\eta_0\times_H G=(P\times_H G,p,B,G)$ with fiber G is defined, where H operates on G by the product of G; it is also a principal G-bundle if we define an operation of G on $P\times_H G$ by $\{(x,g)\}\cdot g'=\{(x,gg')\}$. For a principal G-bundle η, we say that η is **reducible** to an H-bundle if there is a principal H-bundle η_0 with $\eta = \eta_0\times_H G$, and we call η_0 a **reduced bundle** of η. It is easy to see that the group of a G-bundle ξ is reducible to H if and only if the associated principal G-bundle of ξ is reducible to H. Also, if η_0 is a reduced bundle of η, then $\psi^{\#}\eta_0$ is a reduced bundle of $\psi^{\#}\eta$.

Now, assume that H has a local cross section in G and G/H is ∞-connected. Then for an n-universal bundle $\xi(n,H)$ of H, $\xi(n,H)\times_H G$ is an n-universal bundle of G (n-connectedness of $E(n,H)\times_H G$ is shown by the †exact homotopy sequence of †fiber spaces). Therefore, by the classification theorem, the group of any G-bundle is reducible to H, and the equivalence classes of G-bundles are in one-to-one correspondence with those of H-bundles.

(1) A G-bundle is trivial if and only if its group is reducible to e (identity element). A $2n$-dimensional differentiable manifold M of class C^∞ has an †almost complex structure if and only if the group of the tangent bundle $\mathfrak{T}(M)$ is reducible to $GL(n,\mathbf{C})$, i.e., $\mathfrak{T}(M)$ is considered as an n-dimensional complex vector bundle (→ (5) in Section F).

(2) Since $GL(n,\mathbf{R})\approx O(n)\times\mathbf{R}^{n(n+1)/2}$ and $GL(n,\mathbf{C})\approx U(n)\times\mathbf{R}^{n^2}$, n-dimensional real (complex) vector bundles can be considered as $O(n)$ ($U(n)$)-bundles with fiber $\mathbf{R}^n$ ($\mathbf{C}^n$).

J. Homotopy and Homology Theory of Bundles

Since a fiber bundle is a †locally trivial fiber space, the exact sequence and the spectral sequence of fiber spaces (→ 156 Fiber Spaces) are applicable to fiber bundles. For example, the cohomology structures of homogeneous spaces of classical groups have been determined by A. Borel, J.-P. Serre, and others.

(1) Characteristic class. For a classifying space B_G of a topological group G, we have an isomorphism $\pi_n(B_G)\cong\pi_{n-1}(G)$ of homotopy groups and the following classification theorem of fiber bundles over the n-sphere S^n. The set of the equivalence classes of principal G-bundles or G-bundles with fiber F over the base space S^n is in one-to-one correspondence with the set $\pi_{n-1}(G)/\pi_0(G)$ of equivalence classes under the operation of G on $\pi_{n-1}(G)$ given by the inner automorphisms of G; such a correspondence is given by associating with each principal G-bundle $\eta=(P,q,S^n,G)$ the class (called the **characteristic class** of η) containing the image $\Delta(\iota_n)$ of a generator $\iota_n\in\pi_n(S^n)$ by the homomorphism $\Delta:\pi_n(S^n)\cong\pi_n(P,G)\to\pi_{n-1}(G)$. Take U_1 and U_2 (the open sets of S^n such that the last coordinates t_{n+1} are $>-1/2$ and $<1/2$, respectively) as coordinate neighborhoods of η. Then the restriction $T=g_{12}|S^{n-1}$ represents the characteristic class of η, where g_{12}: $U_1\cap U_2\to G$ is the coordinate transformation and S^{n-1} is the equator of S^n.

(2) For the principal bundle $\eta=(SO(n+1),q,S^n,SO(n))$, the mapping $T:S^{n-1}\to SO(n)$ is given by

$$T(t_1,\ldots,t_n)=(I_n-2(t_it_j))\begin{pmatrix} I_{n-1} & 0\\ 0 & -1\end{pmatrix}$$

(I_n is the unit matrix of degree n). Hence the degree of the composite $q'\circ T:S^{n-1}\to S^{n-1}$ (of T and the natural projection $q':SO(n)\to S^{n-1}$) is equal to 0 if n is odd and 2 if n is even. From this fact and the homotopy exact sequence, we have

$$\pi_n(V_{m+n,m}(\mathbf{R}))=\begin{cases}\mathbf{Z} & \text{if } m=1 \text{ or } n \text{ is even},\\ \mathbf{Z}_2=\mathbf{Z}/2\mathbf{Z} & \text{if } m>1 \text{ and } n \text{ is odd}\end{cases}$$

for the real Stiefel manifold $V_{m+n,m}(\mathbf{R})$, which is $(n-1)$-connected.

(3) Sphere bundles. An $O(n+1)$-bundle with fiber S^n is called an n-**sphere bundle**. The set of equivalence classes of n-sphere bundles with base space S^m is in one-to-one correspondence with $\pi_{m-1}(O(n+1))/\pi_0(O(n+1))$. For example, any 1-sphere bundle over S^m ($m\geqslant 3$) and any n-sphere bundle over S^3 is trivial. Every 3-sphere bundle over S^4 is equivalent to one of $\{\xi_{m,n}|m$ an integer, n a positive integer$\}$, where $\xi_{m,n}$ is defined as follows: Let $\rho,\sigma:S^3\to O(4)$ be defined by $\rho(q)q'=qq'q^{-1}$, $\sigma(q)q'=qq'$ (q,q' are †quaternions of norm 1). Then these mappings represent generators of $\pi_3(O(4))\cong\pi_3(S^3\times S^3)\cong\mathbf{Z}+\mathbf{Z}$, and $\xi_{m,n}$ is the 3-sphere bundle over S^4 corre-

sponding to the element $m\{\rho\}+n\{\sigma\}\in \pi_3(O(4))$ (i.e., to the mapping $f_{m,n}:S^3\to O(4)$ defined by $f_{m,n}(q)(q')=q^{m+n}q'q^{-m}$). (Here we use the fact that the operation of the element $r\in O(4)$ $(r(q)=q^{-1})$ is given by $r\rho r^{-1}=\rho$, $r\sigma r^{-1}=\rho\sigma^{-1}$.)

K. Cross Sections

For a fiber bundle $\xi=(E,p,B,F,G)$, a cross section $f: B_0\to E$ over a subspace B_0 $(\subset B)$ is a continuous mapping such that $p\circ f$ is the identity mapping of B_0; a cross section over B is called a **cross section** of ξ. A bundle ξ is trivial if and only if the associated principal bundle of ξ has a cross section. More generally, given a principal bundle $\eta=(P,q,B,G)$ and a closed subgroup H having a local cross section in G, η is reducible to H if and only if the associated bundle $\eta\times_G(G/H)=(P/H,q',B,G/H)$ with fiber G/H has a cross section.

Suppose that the base space B of the fiber bundle $\xi=(E,p,B,F,G)$ is a †polyhedron. We denote the †r-skeleton of B by B^r and consider the problem of extending cross sections $f_r: B^r\to E$ successively for $r=0,1,\ldots$. Clearly, there is a cross section f_0. For each r-simplex σ of B, we have $(p^{-1}(\sigma),p,\sigma,F)\equiv(\sigma\times F,p_1,\sigma,F)$ since σ is †contractible. Hence there is a bundle mapping $\varphi_\sigma:\sigma\times F\approx p^{-1}(\sigma)$ with $p\circ\varphi_\sigma=p_1$. Assume the existence of a cross section $f_{r-1}: B^{r-1}\to E$, and consider the mapping $h_{\dot\sigma}=p_2\circ\varphi_\sigma^{-1}\circ(f_{r-1}|\dot\sigma):\dot\sigma\to F$ $(p_2:\sigma\times F\to F$ is the projection and $\dot\sigma$ is the boundary of $\sigma)$. Then if $h_{\dot\sigma}$ is extendable to $h_\sigma:\sigma\to F$, an extended cross section $f_\sigma:\sigma\to E$ of $f_{r-1}|\dot\sigma$ is defined by $f_\sigma(b)=\varphi_\sigma(b,h_\sigma(b))$ $(b\in\sigma)$, and the extension $f_r: B^r\to E$ of f_{r-1} is defined by $f_r|\sigma=f_\sigma$, $f_r|B^{r-1}=f_{r-1}$. If $\pi_{r-1}(F)=0$, for example, there is an extension h_σ of $h_{\dot\sigma}$ since $(\sigma,\dot\sigma)\approx(V^r,S^{r-1})$, and f_{r-1} is extendable to a cross section f_r.

Now assume that the base space B of a G-bundle $\xi=(E,p,B,F,G)$ is an †arcwise connected polyhedron and F is †$(n-1)$-connected. Then there is a cross section $f: B^n\to E$ constructed by the stepwise method of the previous paragraph. But if $\pi_n(F)\neq 0$, we have an obstruction to extending f over B^{n+1}. Now we explain how to measure this obstruction. Suppose that F is †n-simple. Then for each $(n+1)$-simplex σ of B, the mapping $h_{\dot\sigma}:\dot\sigma\to F$, defined by f as in the previous paragraph, determines a unique element $c(f)(\sigma)$ of the homotopy group $\pi_n(F)$. Hence we have a †cochain $c(f)\in C^{n+1}(B;\pi_n(F))$, and f is extendable to a cross section over B^{n+1} if and only if $c(f)=0$. Thus there is a cross section over B^{n+1} if and only if the set of $c(f)$ for every cross section f over B^n contains the cochain 0; $\{c(f)\}$ is considered as a measure of the obstruction.

Let $w: I\to B$ be a †path. We consider the space $I\times F$ and the canonical projection $p_1: I\times F\to I$. Then there is a bundle mapping $\Omega: I\times F\to E$ with $p\circ\Omega=w\circ p_1$, since $w^\#\xi\equiv(I\times F,p_1,I,F)$; and a homeomorphism $w_\#: F\approx F$ is defined by $w_\#=\varphi_{0,b_0}^{-1}\circ\Omega_0\circ\Omega_1^{-1}\circ\varphi_{1,b_1}$, where $b_\varepsilon=w(\varepsilon)$ $(\varepsilon=0,1)$, $\varphi_\varepsilon: U_\varepsilon\times F\approx p^{-1}(U_\varepsilon)$ is a coordinate function of a coordinate neighborhood $U_\varepsilon\ni b_\varepsilon$, and $\Omega_\varepsilon(=\Omega|\varepsilon\times F): F\approx p^{-1}(b_\varepsilon)$. The homeomorphism $w_\#$ induces an isomorphism $w_\#:\pi_n(F)\cong\pi_n(F)$, and $\pi_n(F)$ forms a †local coefficient on B. Then the cochain $c(f)$ is a †cocycle with the local coefficient $\pi_n(F)$, called the **obstruction cocycle** of f. Furthermore, the set $\{c(f)\}$ for every cross section $f: B^n\to E$ is a cohomology class $c^{n+1}(\xi)\in H^{n+1}(B;\pi_n(F))$ (local coefficient), and $c^{n+1}(\xi)$ is called the **primary obstruction** to the construction of a cross section. There is a cross section over the $(n+1)$-skeleton B^{n+1} if and only if $c^{n+1}(\xi)=0$. The local coefficient $\pi_n(F)$ is trivial if B is †simply connected or, for example, if the structure group G of ξ is connected (in which case ξ is called an **orientable fiber bundle**); when this is true, $c^{n+1}(\xi)$ is an element of $H^{n+1}(B;\pi_n(F))$, where $\pi_n(F)$ is not a local coefficient. Furthermore, if $c^{n+1}(\xi)=0$ and $\pi_i(F)=0$ $(n<i<m)$, then the secondary obstruction $c^{m+1}(\xi)\in H^{m+1}(B;\pi_m(F))$ is defined similarly (→ 300 Obstructions).

L. Stiefel-Whitney Classes

Let $\xi=(E,p,B,F,O(n))$ be an $O(n)$-bundle over an arcwise connected polyhedron B. Consider the Stiefel manifold $V_{n,n-k}=V_{n,n-k}(\mathbf{R})=O(n)/I_{n-k}\times O(k)$, which is $(k-1)$-connected, and the associated bundle $\xi^k=\xi\times_{O(n)}V_{n,n-k}$ with fiber $V_{n,n-k}$. The primary obstruction $W_{k+1}(\xi)=c^{k+1}(\xi^k)\in H^{k+1}(B;\pi_k(V_{n,n-k}))$ $(k=0,1,\ldots,n-1)$ is called the **Stiefel-Whitney class** of ξ. We have $2W_{k+1}(\xi)=0$ unless $k=n-1$ and k is odd. Hence we usually consider $W_{k+1}(\xi)\in H^{k+1}(B;\mathbf{Z}_2)$. ξ is orientable, i.e., the group of ξ is reducible to $SO(n)$, if and only if $W_1(\xi)=0$. The **Stiefel-Whitney classes of an n-dimensional †differentiable manifold M** are defined to be those of the tangent bundle $\mathfrak{T}(M)$. Since the orientability of M coincides with that of $\mathfrak{T}(M)$, M is orientable if and only if $W_1(M)=0$. The condition $W_{k+1}(M)=0$ is necessary for the existence of a continuous field of orthonormal tangent $(n-k)$-frames over M (if $k=n-1$ this condition is also sufficient). Also, $W_n(M)$ is equal to $\chi(M)\mu$, where μ is the †fundamental cohomology class of M and $\chi(M)$ is the †Euler characteristic of M (→ 58 Characteristic Classes B).

M. Chern Classes

For a $U(n)$-bundle $\xi=(E,p,B,F,U(n))$, the primary obstruction $C_{k+1}(\xi)=c^{2k+2}(\xi^k)\in H^{2k+2}(B;\mathbf{Z})$ $(k=0,1,\ldots,n-1)$ of the associated bundle $\xi^k=\xi\times_{U(n)}V_{n,n-k}(\mathbf{C})$ is called the **Chern class** of ξ. If we consider ξ as an $O(2n)$-bundle by $U(n)\subset O(2n)$, then $W_{2k+1}(\xi)=0$ and $W_{2k}(\xi)=C_k(\xi)\ (\mathrm{mod}\ 2)$. The **Chern classes** of a real $2n$-dimensional almost complex manifold are defined to be those of the tangent bundle $\mathfrak{T}(M)$ (→ 58 Characteristic Classes C).

N. Microbundles

A system $x: B \xrightarrow{i} E \xrightarrow{j} B$ of topological spaces E, B and continuous mappings i, j is called an n-dimensional **microbundle** over B if for each $b\in B$, there exist a neighborhood U of b, a neighborhood V of $i(U)$, and a homeomorphism $h: V\approx U\times\mathbf{R}^n$ with $h\circ i\,|\,U=i_1$, $j|V=p_1\circ h$ ($i_1: U\approx U\times 0\subset U\times\mathbf{R}^n$, and $p_1: U\times\mathbf{R}^n\rightarrow U$ is the projection). Let $H_0(n)$ be the topological group of all homeomorphisms of $\mathbf{R}^n$ onto itself fixing the origin with compact-open topology. Then the equivalence classes of n-dimensional microbundles over B are naturally in one-to-one correspondence with the equivalence classes of $H_0(n)$-bundles with base space B and fiber $\mathbf{R}^n$ [10]. The tangent microbundle is defined for any †topological manifold, and J. Milnor [9] used this notion to show that the tangent bundle and the †Pontrjagin classes of a differentiable manifold are not topological invariants.

O. Bundles of Class C^r, Analytic Bundles

A fiber bundle $\xi=(E,p,B,F,G)$ is called a **fiber bundle of class** C^r $(r=0,1,\ldots,\infty,\omega)$ if E, B, F are †differentiable manifolds of class C^r, G is a †Lie group and a †transformation group of F of class C^r, and p and the coordinate functions are differentiable mappings of class C^r. Bundles of class C^0 are usual G-bundles, and those of class C^ω are **real analytic fiber bundles**. Similarly, **complex analytic fiber bundles** are defined by the notions of †complex manifolds, †complex Lie groups, and †holomorphic mappings. For example, the universal bundles $\xi(n-1,O(m))$ and $\eta(2n,U(m))$ (→ (2) in Section H) are real and complex analytic principal bundles, respectively, and the tangent bundle $\mathfrak{T}(M)$ of a C^{r+1} (or complex) manifold is a C^r (or complex analytic) vector bundle. The operations of the Whitney sum, etc., are defined analogously for these vector bundles.

The equivalence of C^r (complex analytic) bundles is defined by means of bundle mappings that are C^r-differentiable (holomorphic). Bundles of class C^r $(r\leqslant\infty)$ are classified by C^r mappings into a classifying space, in the same manner as for bundles of class C^0. Also, the connection of class C^r (→ 82 Connections) in C^r bundles is an important notion.

For complex analytic bundles, a similar classification has been obtained for restricted spaces by K. Kodaira, Serre, S. Nakano [11], and others. The classification of complex analytic bundles over a †Stein manifold is reduced to that of bundles of class C^0 (**Oka's principle** [12]), and similar results are valid for C^ω-manifolds [13]. The complex analytic (or holomorphic) connection does not necessarily exist, and M. F. Atiyah [14] found the condition for its existence and its relation with Chern classes.

References

[1] N. E. Steenrod, The topology of fiber bundles, Princeton Univ. Press, 1951.
[2] E. Stiefel, Richtungsfelder und Fernparallelismus in n-dimensionalen Mannigfaltigkeiten, Comment. Math. Helv., 8 (1936), 3–51.
[3] H. Whitney, Topological properties of differentiable manifolds, Bull. Amer. Math. Soc., 43 (1937), 785–805.
[4] S. S. Chern, Some new view-points in differential geometry in the large, Bull. Amer. Math. Soc., 52 (1946), 1–30.
[5] F. Hirzebruch, Topological methods in algebraic geometry, Springer, third edition, 1966.
[6] J. W. Milnor, Construction of universal bundles II, Ann. of Math., (2) 63 (1956), 430–436.
[7] J. C. Moore, Espaces classifiants, Séminaire H. Cartan, 12, no. 5 (1959–1960), Ecole Norm. Sup., Secrétariat Mathématique.
[8] M. E. Mahowald, On obstruction theory in orientable fiber bundles, Trans. Amer. Math. Soc., 110 (1964), 315–349.
[9] J. W. Milnor, Microbundles I, Topology, 3 (1964), Suppl. 1, 53–80.
[10] J. Kister, Microbundles are fiber bundles, Bull. Amer. Math. Soc., 69 (1963), 854–857.
[11] S. Nakano, On complex analytic vector bundles, J. Math. Soc. Japan, 7 (1955), 1–12.
[12] H. Grauert, Analytische Faserungen über holomorph-vollständigen Räumen, Math. Ann., 135 (1958), 263–273.

[13] K. Shiga, Some aspects of real-analytic manifolds and differentiable manifolds, J. Math. Soc. Japan, 16 (1964), 128–142.
[14] M. F. Atiyah, Complex analytic connections in fiber bundles, Trans. Amer. Math. Soc., 85 (1957), 181–207.
[15] D. Husemoller, Fiber bundles, McGraw-Hill, 1966.

156 (IX.19) Fiber Spaces

A. General Remarks

J.-P. Serre [1] generalized the concept of fiber bundles to that of fiber spaces by utilizing the covering homotopy property (→ Section D). He applied the theory of †spectral sequences, due to J. Leray, to the (cubic) †singular (co)homology groups of fiber spaces. These are quite useful for determining (co)homology structures and homotopy groups of topological spaces, and are now of fundamental importance in algebraic topology.

B. Definition

Let $p: E \to B$ be a continuous mapping of topological spaces, and let X be a topological space. Then we say that p has the **covering homotopy property** with respect to X if for any mapping $f: X \to E$ and †homotopy $g_t: X \to B$ with $p \circ f = g_0$, there is a homotopy $f_t: X \to E$ with $f_0 = f$ and $p \circ f_t = g_t$. We call (E,p,B) a **fiber space** if p has the covering homotopy property with respect to each cube $I^n = \{(x_1, \ldots, x_n) \mid 0 \leqslant x_i \leqslant 1\}$, $n = 0, 1, \ldots$ (then p has the covering homotopy property with respect to every †CW complex). Then E is called the **total space**, p the **projection**, B the **base space**, and $F_b = p^{-1}(b)$ the **fiber** over $b \in B$.

Let E, B, F be topological spaces and $p: E \to B$ a continuous mapping. We call (E,p,B,F) a **locally trivial fiber space** if for each $b \in B$, there exist an open neighborhood U of b and a homeomorphism $\varphi: U \times F \approx p^{-1}(U)$ with $p\varphi(b',y) = b'$ ($b' \in U, y \in F$). In this case, p has the covering homotopy property with respect to each †paracompact space, hence (E,p,B) is a fiber space. A †fiber bundle is clearly a locally trivial fiber space.

C. Path Spaces

Another important example of a fiber space is a path space. A **path** in a topological space X is a continuous mapping $w: I \to X$ ($I = [0,1]$). Given subsets A_0 and A_1 of X, the **path space** $\Omega(X; A_0, A_1)$ is the space of all paths $w: (I, 0, 1) \to (X, A_0, A_1)$ topologized by †compact-open topology. Define $p_\varepsilon: \Omega(X; A_0, A_1) \to A_\varepsilon$ by $p_\varepsilon(w) = w(\varepsilon)$ ($\varepsilon = 0, 1$). Then $(\Omega(X; A_0, A_1), p_\varepsilon, A_\varepsilon)$ is a fiber space; in fact, p_ε has the covering homotopy property with respect to every topological space. In particular, the total space of the fiber space $(\Omega(X; X, *), p_0, X)$ ($* \in X$) is †contractible, and the fiber $p_0^{-1}(*) = \Omega(X; *, *) = \Omega X$ over $*$ is the †loop space of X with base point $*$. For a continuous mapping $f: Y \to X$, consider the space $E_f = \{(y,w) \in Y \times \Omega(X; X, X) \mid f(y) = w(0)\}$ and the continuous mapping $p: E_f \to X$ defined by $p(y,w) = w(0)$. Then $Y \subset E_f$, and Y is a †deformation retract of E_f; furthermore, (E_f, p, X) is a fiber space with $f = p|Y$ (E_f is called the †mapping track of f).

D. Homotopy Groups of Fiber Spaces

For †homotopy groups of fiber spaces, the **Hurewicz-Steenrod isomorphism theorem** holds: Let (E,p,B) be a fiber space and $F = p^{-1}(*)$ the fiber over the base point $* \in B$. Then $p_*: \pi_n(E,F) \cong \pi_n(B)$ is an isomorphism for $n \geqslant 2$ and a bijection for $n = 1$. By this theorem, we have the **homotopy exact sequence** of a fiber space:

$$\ldots \to \pi_{n+1}(B) \xrightarrow{\Delta} \pi_n(F) \xrightarrow{i_*} \pi_n(Se) \xrightarrow{p_*} \pi_n(B) \to \ldots.$$

Furthermore, the more general exact sequence

$$\ldots \to \pi(Z; \Omega B)_0 \to \pi(Z; F)_0 \xrightarrow{i_*} \pi(Z; E)_0 \xrightarrow{p_*} \pi(Z; B)_0$$

is valid for each CW complex Z, where $\pi(Z;)_0$ is a †homotopy set with base point.

Example (1) A **cross section** of a fiber space (E,p,B) is a continuous mapping $f: B \to E$ with $p \circ f = 1$. If (E,p,B) has a †cross section or the fiber F is a †retract of E, then $\pi_n(E) \cong \pi_n(B) + \pi_n(F)$. If F is contractible in E, then $\pi_n(B) \cong \pi_n(E) + \pi_{n-1}(F)$ ($n \geqslant 2$).

Example (2) (E,p,B) is called an ***n*-connective fiber space** if B is †arcwise connected, E is †n-connected, and $p_*: \pi_i(E) \cong \pi_i(B)$ for $i > n$. For each arcwise connected space B and integer n, there is such a fiber space.

Example (3) For a CW complex X, there are topological spaces X_n and continuous mappings $f_n: X \to X_n$, $q_{n+1}: X_{n+1} \to X_n$ ($n = 0, 1, \ldots$) with the following four properties: (i) X_n ($0 \leqslant n \leqslant m$) is a point if X is m-connected; (ii) $f_{n*}: \pi_i(X) \cong \pi_i(X_n)$ ($i \leqslant n$); (iii) (X_n, q_n, X_{n-1}) is a fiber space, and its fiber is an †Eilenberg-MacLane space $K(\pi_n(X),n)$; (iv) $q_n \circ f_n$ is †homotopic to f_{n-1}. Such a system $\{X_n, f_n, q_n\}$ is called the **Postnikov system** of X and is in a sense considered a decomposition of X into Eilenberg-MacLane spaces.

E. Spectral Sequences of Fiber Spaces

The cohomological properties of fiber spaces are obtained mainly from the following results (which are valid similarly for homology except for properties of products). Assume that the base space B of a given fiber space (E,p,B) is †simply connected and the fiber $F=p^{-1}(*)$ is arcwise connected, and let R be a †principal ideal ring. Then there is a **spectral sequence** (of singular cohomology) **of the fiber space** (E,p,B) (with coefficients in R):

$\{E_r^{p,q}, d_r^{p,q}: E_r^{p,q} \to E_r^{p+r,q-r+1}\}$,

$r=0,1,\ldots,\infty$

(i.e., $E_r = \sum_{p,q} E_r^{p,q}$ is an †R-module, $d_r = \sum_{p,q} d_r^{p,q}$ is R-linear, $d_r \circ d_r = 0$, and $E_{r+1}^{p,q} = \operatorname{Ker} d_r^{p,q} / \operatorname{Im} d_r^{p-r,q+r-1}$, which means that $E_{r+1} = H(E_r)$), and the following five conditions hold: (i) $E_r^{p,q}=0$ for $p<0$ or $q<0$, $E_r^{p,q} = E_{r+1}^{p,q} = \ldots = E_\infty^{p,q}$ for $r > \max(p,q+1)$. (ii) E_r has a product for which $E_r^{p,q} \cdot E_r^{p',q'} \subset E_r^{p+p',q+q'}$ and $d_r(u \cdot v) = (d_r u) \cdot v + (-1)^{p+q} u \cdot d_r v$ ($u \in E_r^{p,q}$). Furthermore, the induced product in $H(E_r)$ coincides with the product in E_{r+1}. (iii) E_∞ is the †bigraded module associated with some filtration of the cohomology module $H^*(E;R)$; that is, $H^n(E;R) = D^{0,n} \supset D^{1,n-1} \supset \ldots \supset D^{n,0} \supset D^{n+1,-1} = 0$ and $E_\infty^{p,q} = D^{p,q} / D^{p+1,q-1}$. Furthermore, the †cup product $\smile$ in $H^*(E;R)$ satisfies $D^{p,q} \smile D^{p',q'} \subset D^{p+p',q+q'}$ and coincides with the given product in E_∞. (iv) $E_2^{p,q} = H^p(B; H^q(F;R))$, and the product in E_2 coincides with the cup product in $H^*(B; H^*(F;R))$. (v) The composition of $H^n(B;R) = E_2^{n,0} \to E_3^{n,0} \to \ldots \to E_{n+1}^{n,0} = E_\infty^{n,0} = D^{n,0} \subset H^n(E;R)$ is equal to p^*, and the composition of $H^n(E;R) = D^{0,n} \to E_\infty^{0,n} = E_{n+2}^{0,n} \subset \ldots \subset E_3^{0,n} \subset E_2^{0,n} = H^n(F;R)$ is equal to i^* ($i: F \subset E$), where each $\to$ is the projection onto the factor group. In the sequence

$$H^{n-1}(F;R) \xrightarrow{\partial^*} H^n(E,F;R) \xleftarrow{p^*} H^n(B;R),$$

we have $\partial^{*-1}(\operatorname{Im} p^*) = E_n^{0,n-1}$, $\operatorname{Coim} p^* = E_n^{n,0}$, and $d_n: E_n^{0,n-1} \to E_n^{n,0}$ is equal to the **transgression** $\tau^* = p^{*-1} \circ \partial^*: \partial^{*-1}(\operatorname{Im} p^*) \to \operatorname{Coim} p^*$. Each element of $\partial^{*-1}(\operatorname{Im} p^*)$ is called **transgressive**.

In each of the following five examples (where B and F are, as above, assumed to be simply connected and arcwise connected, respectively), condition (iv) gives rise to $E_2^{p,q} = H^p(B;R) \otimes H^q(F;R)$ by the †universal coefficient theorem.

Example (4) For the †Poincaré polynomial $P_E(t) = \sum_n b_n t^n$, $b_n = \dim_k H_n(E;k)$ (k is a commutative field), we have $P_E(t) = P_B(t) P_F(t) - (1+t)\varphi(t)$, where $\varphi(t)$ is a polynomial with nonnegative coefficients (Leray). In particular, for the †Euler characteristic $\chi(E) = P_E(-1)$, we have $\chi(E) = \chi(B)\chi(F)$. Also, if $i_*: H_n(F;k) \to H_n(E;k)$ is monomorphic for each $n \geqslant 0$, then $P_E(t) = P_B(t) P_F(t)$.

Example (5) Isomorphism theorem: If $H_n(B;R)=0$ $(0<n<r)$ and $H_n(F;R)=0$ $(0<n<s)$, then $p_*: H_n(E,F;R) \to H_n(B;R)$ is isomorphic for $0<n<r+s$ and epimorphic for $n=r+s$, and we have the following **homology exact sequence**:

$$\ldots \to H_n(F;R) \xrightarrow{i_*} H_n(E;R) \xrightarrow{p_*} H_n(B;R) \xrightarrow{\tau_*} H_{n-1}(F;R) \to \ldots, \quad n<r+s$$

(similarly for the cohomology).

Example (6) Assume that $H^n(F;R) \cong H^n(S^r;R)$ (S^r is the r-sphere, $r \geqslant 1$). Let $\tilde{E} = M_p$ be the †mapping cylinder of $p: E \to B$ and $\tilde{p}: \tilde{E} \to B$ be the continuous mapping defined by p. Then the **Thom-Gysin isomorphism** $\tilde{g}: H^{n-r-1}(B;R) \cong H^n(\tilde{E},E;R)$ $(n \geqslant 0)$ with $\tilde{g}(\alpha) = \tilde{p}^*(\alpha) \smile \tilde{g}(1)$ holds (→ 410 Topology of Differentiable Manifolds F). Also, we have the **Gysin exact sequence**:

$$\ldots \to H^n(B;R) \xrightarrow{p^*} H^n(E;R) \to H^{n-r}(B;R) \xrightarrow{g} H^{n+1}(B;R) \to \ldots,$$

where g satisfies $g(\alpha) = \alpha \smile \Omega = \Omega \smile \alpha$ ($\Omega = g(1) \in H^{r+1}(B;R)$). Here Ω is equal to the image of a generator of $H^r(F;R) = R$ by the transgression τ^*, and $2\Omega = 0$ if r is even. (These results hold also for $r=0$ and $R = \mathbf{Z}_2$.)

Example (7) Assume that $H^n(B;R) \cong H^n(S^r;R)$ $(r \geqslant 2)$. Then, we have $H^{n-r}(F;R) \cong H^n(E,F;R)$ $(n \geqslant 0)$ and the **Wang exact sequence**:

$$\ldots \to H^n(E;R) \xrightarrow{i^*} H^n(F;R) \xrightarrow{\theta} H^{n-r-1}(F;R) \to H^{n+1}(E;R) \to \ldots,$$

where θ satisfies $\theta(\alpha \smile \beta) = \theta(\alpha) \smile \beta + (-1)^{n(r-1)} \alpha \smile \theta(\beta)$ ($\alpha, \beta \in H^n(F;R)$).

Example (8) For a field k of odd characteristic, if $H^n(E;k)=0$ $(n>0)$ and the algebra $H^*(F;k)$ is generated by a finite number of elements of odd degree, then $H^*(F;k) \cong \wedge_k(x_1, \ldots, x_l)$ (the †exterior algebra) and $H^*(B;k) \cong k[y_1, \ldots, y_l]$, where $y_i = \tau^*(x_i)$ (A. Borel).

References

[1] J.-P. Serre, Homologie singulière des espaces fibrés, Ann. of Math., (2) 54 (1951), 425–505.

[2] S. T. Hu, Homotopy theory, Academic Press, 1959.

[3] A. Borel, Sur la cohomologie des espaces fibrés principaux et des espaces homogènes de groupes de Lie compacts, Ann. of Math., (2) 57 (1953), 115–207.

[4] A. Komatu, M. Nakaoka, and H. Toda, Isôkikagaku (Japanese; Topology), Kyôritu, 1957.

[5] A. Komatu, M. Nakaoka, and M. Sugawara, Isôkikagaku (Japanese; Topology) I, Iwanami, 1967.

157 (III.7) Fields

A. Definition

A set K having at least two elements is called a **field** if two operations, called †addition $(+)$ and †multiplication $(\cdot)$, are defined in K and satisfy the following three axioms.

(1) For any two elements a, b of K, the sum $a+b$ is defined; the associative law $(a+b)+c=a+(b+c)$ and the commutative law $a+b=b+a$ hold; and there exists for arbitrary a, b a unique element x such that $a+x=b$, that is, K is an †Abelian group with respect to the addition (the †identity element of this group is denoted by 0 and is called the **zero element** of K).

(2) For any two elements a, b of K, the product ab $(=a\cdot b)$ is defined; the associative law $(ab)c=a(bc)$ and the commutative law $ab=ba$ hold; and there exists for arbitrary a, b with $a\neq 0$ a unique element x such that $ax=b$, that is, the set K^* of all nonzero elements of K is an Abelian group with respect to the multiplication. K^* is called the **multiplicative group** of K, while the identity element of K^* is denoted by 1 and is called the **unity element (unit element** or **identity element**) of K.

(3) The distributive law $a(b+c)=ab+ac$ holds. In other words, a field is a †commutative ring whose nonzero elements form a group with respect to the multiplication.

A noncommutative ring whose nonzero elements form a group is called a **noncommutative field (skew field** or **s-field**). It should be noted that sometimes a field is defined as a ring whose nonzero elements form a group without assuming the commutativity of that group, and in this case our "field" defined before is called a commutative field. (The term "skew field" is sometimes used to mean either a commutative or a noncommutative field.) In this article we limit ourselves to commutative fields (for noncommutative fields → 31 Associative Algebras).

B. General Properties

Since a field K is a commutative ring, we have properties such as $a0=0a=0$, $(-a)b=a(-b)=-ab$ for elements a, b in K. If a †subring k of K is a field, we say that k is a **subfield** of K or K is an **overfield (extension field** or simply **extension**) of k. If a field K has no subfield other than K, K is called a **prime field**.

A mapping f of a field K into another field K' is called a (field) **homomorphism** if it is a ring homomorphism, i.e., if it satisfies $f(a+b)=f(a)+f(b)$, $f(ab)=f(a)f(b)$. Since a field is †simple as a ring, every (field) homomorphism is an injection unless it maps everything to zero. A homomorphism of K into K' is called an **isomorphism** if it is a bijection, and K and K' are called isomorphic if there exists an isomorphism of K onto K'. An isomorphism of K onto itself is called an **automorphism** of K.

If there is a natural number n such that the sum $n1=\overbrace{1+\ldots+1}^{n}$ of the unity element 1 is 0, then the minimum of such n is a prime number p, called the **characteristic** of K. On the other hand, if there is no natural number n such that $n1=0$, we say that the characteristic of K is 0.

C. Examples of Fields

The rational number field $\mathbf{Q}$ consisting of all rational numbers, the real number field $\mathbf{R}$ consisting of all real numbers, and the complex number field $\mathbf{C}$ consisting of all complex numbers, are all fields of characteristic 0. A subfield of the complex number field $\mathbf{C}$ is called a **number field**. The rational number field is a prime field, and every prime field of characteristic 0 is isomorphic to the rational number field. For the ring $\mathbf{Z}$ of all rational integers, the residue class ring modulo a prime number p is a field $\mathbf{Z}/p\mathbf{Z}=\{0,1,2,\ldots,p-1 \pmod{p}\}$ of characteristic p, called the **residue class field** for p. Thus $\mathbf{Z}/p\mathbf{Z}$ is a prime field, and every prime field of characteristic p is isomorphic to $\mathbf{Z}/p\mathbf{Z}$. If the number of the elements of a field K is finite, K is called a **finite field**. $\mathbf{Z}/p\mathbf{Z}$ is an example of a finite field.

D. Extensions of a Field

In order to express that K is an extension field of k, we often use the notation K/k. Subfields of K containing k are called **intermediate fields** of K/k. Consider two extensions K_1/k_1 and K_2/k_2, and let $\varphi: K_1 \to K_2$ be an isomorphism which induces an isomorphism $\psi: k_1 \to k_2$. Then we call φ an **extension** of ψ. Suppose that k_1, K_2 are given fields and K_2 contains a subfield k_2 isomorphic to k_1. Then there exist an extension field K_1 of k_1 and an isomorphism $\varphi: K_1 \to K_2$ which is an extension of the given isomorphism: $k_1 \to k_2$; to construct the field K_1 is often called to imbed k_1 into K_2. When K_1 and K_2 are extensions of k, an isomorphism: $K_1 \to K_2$ is called a **k-isomorphism** if it leaves every element of k invariant.

In an extension K/k, let S be a subset of K. The smallest intermediate field of K/k containing S is called the field obtained by **adjoining** S to k or the field **generated** by S over k, denoted by $k(S)$. The field $k(S)$ consists of those elements in K each of which is a rational expression in a finite number of elements of S with coefficients in k. An extension field $k(t)$ obtained by adjoining a single element t to k is called a **simple extension** of k, and in this case t is called a **primitive element** of the extension. The †rational function field $k(X)$ with coefficient field k is a simple extension of k with a primitive element X.

When subfields k_λ $(\lambda \in \Lambda)$ of a field K are given, the smallest subfield of K containing all these subfields exists and is called the **composite field** of the k_λ.

E. Algebraic and Transcendental Extensions

An element α of an extension field K of a field k is called an **algebraic element** over k if α is a †zero point of a nonzero polynomial, say, $f(X) = a_0 + a_1X + \ldots + a_nX^n$ with coefficients in k. If α is not algebraic over k, then α is called a **transcendental element** over k. An algebraic element α is always a root of an irreducible polynomial over k which is uniquely determined up to a constant factor $(\in k^*)$ and is called the **minimal polynomial** of α over k. K is called an **algebraic extension** of k if all elements of K are algebraic over k; otherwise we call K a **transcendental extension** of k. If K_1 is an algebraic extension of K and K is an algebraic extension of k, then K_1 is also an algebraic extension of k. In an arbitrary extension field K of k, the set of all algebraic elements over k forms an algebraic extension field of k. A simple extension $k(t)$ with a transcendental element t is isomorphic to the rational function field of one variable with coefficient field k. If t is an algebraic element over k, $k(t)$ is isomorphic to the †residue class field of the polynomial ring $k[X]$ modulo the minimal polynomial $f(X)$ of t over k.

F. Finite Extensions

An extension field K of a field k is called a **finite extension** if K has no infinite set of elements that are †linearly independent over k, i.e., if K is a finite-dimensional linear space over k. The dimension of the linear space over k is called the **degree** of K over k and is denoted by $(K:k)$ (or $[K:k]$). If K is a finite extension of k and L is a finite extension of K, then L is also a finite extension of k and $(L:K)(K:k) = (L:k)$. Every finite extension field of k is an algebraic extension of k and is obtained by adjoining a finite number of algebraic elements to k. Conversely, every field obtained by adjoining a finite number of algebraic elements to k is a finite extension of k. If $K = k(\alpha)$ with an algebraic element α, then $(K:k)$ is equal to the degree of the minimal polynomial of α over k, also called the **degree** of α over k. Every element of $k(\alpha)$ is expressed as a polynomial in α with coefficients in k. On the other hand, for any nonconstant polynomial $f(X)$ of $k[X]$ there exists a simple extension $k(\alpha)$ such that α is a root of $f(X)$.

G. Normal Extensions

An algebraic extension field K of a field k is called a **normal extension** of k if every irreducible polynomial of $k[X]$ which has a root in K can always be decomposed into a product of linear factors in $K[X]$. An extension field K of k is called a **splitting field** of a (nonconstant) polynomial $f(X) \in k[X]$ if $f(X)$ can be decomposed as a product of linear polynomials, i.e., $f(X) = (X - \alpha_1)\ldots(X - \alpha_2)\ldots(X - \alpha_n)$ $(\alpha_i \in K)$. A splitting field K of $f(X)$ $(f(X) \in k[X])$ is called a **minimal splitting field** of $f(X)$ if K is a splitting field of $f(x)$, but any proper subfield L of K $(K \supset L \supset k)$ is not a splitting field of $f(X)$. A minimal splitting field of $f(X)$ is obtained by adjoining all the zero points of $f(X)$. A finite extension field of k is a normal extension if and only if it is a minimal splitting field of a polynomial of $k[X]$. For any given (nonconstant) polynomial $f(X) \in k[X]$, there exists a minimal splitting field of $f(X)$, and all minimal splitting fields of $f(X)$ are k-isomorphic.

H. Separable and Inseparable Extensions

An algebraic element α over k is called a **separable element** or **inseparable element** over k according as the minimal polynomial of α over k is †separable or †inseparable. An algebraic extension K of k is called a **separable extension** of k if all the elements of K are separable over k; otherwise, K is called an **inseparable extension**. An element α is separable with respect to k if and only if the minimal polynomial of α over k has no double root in its splitting field. If α is inseparable, then k has nonzero characteristic p, and the minimal polynomial $f(X)$ of α can be decomposed as $f(X)=(X-\alpha_1)^{p^r}(X-\alpha_2)^{p^r}\ldots(X-\alpha_m)^{p^r}$, $r\geqslant 1$, where $\alpha_1,\alpha_2,\ldots,\alpha_m$ are distinct roots of $f(X)$ in its splitting field; between the degree n of $f(X)$ and the number m of distinct roots of $f(X)$, the relation $n=mp^r$ holds. In particular, if $\alpha^{p^r}\in k$ for some r, we call α a **purely inseparable element** over k. An algebraic extension of k is called **purely inseparable** if all elements of the field are purely inseparable over k. In an algebraic extension K of k the set of all separable elements forms an intermediate field K_0 of K/k. The field K_0 is called the **maximal separable extension** of k in K. If K is inseparable over k, i.e., if $K\neq K_0$, then the characteristic of k is $p\neq 0$, and K is purely inseparable over K_0. The degrees $d=[K_0:k]$ and $p^r=[K:K_0]$ are denoted by $[K:k]_s$ and $[K:k]_i$, respectively. A separable extension of a separable extension of k is also separable over k, and every finite separable extension of k is a simple extension.

If no inseparable irreducible polynomial in $k[X]$ exists, we call k a **perfect field**; otherwise, an **imperfect field**. Every field of characteristic 0 is a perfect field. A field of characteristic p ($\neq 0$) is perfect if and only if for each $a\in k$ the polynomial X^p-a has a root in k. Every algebraic extension of a perfect field is a separable extension and a perfect field. Any imperfect field has an inseparable, in fact purely inseparable, proper extension.

I. Algebraically Closed Fields

If every nonconstant polynomial of $k[X]$ can be decomposed into a product of linear polynomials of $k[X]$, or equivalently, if every irreducible polynomial of $k[X]$ is linear, k is called an **algebraically closed field**; k is algebraically closed if and only if k has no algebraic extension field other than k, and hence every algebraically closed field is perfect. For any given field k there exists an algebraically closed algebraic extension field of k unique up to k-isomorphisms (E. Steinitz); hence we call such a field the **algebraic closure** of k. To proceed further, suppose that we are given a field k and its extension K. If there is no algebraic element of K over k outside of k, i.e., if k is the intersection of K and the algebraic closure of k, then we say that k is **algebraically closed** in K. The complex number field is an algebraically closed field (C. F. Gauss's fundamental theorem of algebra; → 12 Algebraic Equations).

J. Conjugates

Let k be a field and K an algebraic extension of k. Two elements α, β of K are called **conjugate** over k if they are roots of the same irreducible polynomial of $k[X]$ (or equivalently, if the minimal polynomials of α and β with respect to k coincide); in this case we call the subfields $k(\alpha)$, $k(\beta)$ **conjugate fields** over k. The conjugate fields $k(\alpha)$ and $k(\beta)$ are k-isomorphic under an isomorphism σ such that $\sigma(\alpha)=\beta$. In particular, if K is a normal extension of k, the number of conjugate elements of an element α of K is the number of distinct roots of the minimal polynomial $f(X)$ of α, which is independent of the choice of a normal extension K containing k. The element α is separable if and only if the number of conjugate elements in K is the same as the degree of $f(X)$. On the other hand, $k(\alpha)$ is normal over k if and only if $k(\alpha)$ coincides with all its conjugate fields.

Let α be a separable algebraic element over k, and let $\alpha_1=\alpha,\alpha_2,\ldots,\alpha_n$ be conjugate elements of α over k. The product $A=\alpha_1\alpha_2\ldots\alpha_n$ and sum $B=\alpha_1+\alpha_2+\ldots+\alpha_n$ are elements of k. Indeed, if $f(X)=X^n+c_1X^{n-1}+\ldots+c_n$ is the minimal polynomial of α with respect to k, we have $A=(-1)^n c_n$, $B=-c_1$, and A and B are called the **norm** and the **trace** of α, respectively, denoted by $A=N(\alpha)$, $B=Tr(\alpha)$. Let K be a finite separable extension of degree n over k, and let α be an element of K. Then the degree m of the minimal polynomial of α is a divisor of n; that is, $n=mr$ with a positive integer r. We define the norm and the trace of α with respect to K/k by $N_{K/k}(\alpha)=N(\alpha)^r$, $Tr_{K/k}(\alpha)=rTr(\alpha)$, respectively. Then these quantities satisfy $N_{K/k}(\alpha\beta)=N_{K/k}(\alpha)N_{K/k}(\beta)$, $Tr_{K/k}(\alpha+\beta)=Tr_{K/k}(\alpha)+Tr_{K/k}(\beta)$ for $\alpha,\beta\in K$. (For the Galois theory of algebraic extensions → 177 Galois Theory.)

K. Transcendental Extensions

Let K be an extension of k and $u_1,\ldots,u_n$ be elements of K. An element v of K is said to be **algebraically dependent** on the elements

$u_1, u_2, \ldots, u_n$ if v is algebraic over the field $k(u_1, u_2, \ldots, u_n)$. A subset S of K is called **algebraically independent** over k if no $u \in S$ is algebraically dependent on a finite number of elements of S different from u itself; S is called a **transcendence basis** of K over k if S is algebraically independent and K is algebraic over $k(S)$. Furthermore, if K is separable over $k(S)$, then S is called a **separating transcendence basis** of K over k. There always exists an algebraically independent basis of K over k, and the †cardinal number of S depends only on K and k; this cardinal number is called the **transcendence degree** (or **degree of transcendency**) of K over k. (When S is an infinite set, we sometimes say that the transcendence degree is infinite.) In particular, if $K = k(S)$ with an algebraically independent S, K is called a **purely transcendental extension** of k.

An extension K of k is called a **separably generated extension** if every finitely generated intermediate field of K/k has a separating transcendence basis over k. If K itself has a separating transcendence basis over k, then K is separably generated, but not conversely.

A purely transcendental extension field of k having a finite transcendence degree n is also called a **rational function field in** n **variables** over k, and a finite extension of such a rational function field is called an **algebraic function field in** n **variables** over k.

Let K and L be extension fields of k, both contained in a common extension field. We say that K and L are **linearly disjoint** over k if every subset of K linearly independent over k is also linearly independent over L, or equivalently, if every subset of L linearly independent over k is also linearly independent over K. An algebraic function field $K = k(x_1, x_2, \ldots, x_n)$ over k (whose transcendence degree is $\leqslant n$) is called a **regular extension** of k if K and the algebraic closure $\bar{k}$ of k are linearly disjoint. In order that K be regular over k it is necessary and sufficient that k be algebraically closed in K and that K be separably generated over k.

L. Derivations

A mapping D of a field K into itself is called a **derivation** of K if it satisfies $D(a+b) = D(a) + D(b)$ and $D(ab) = aD(b) + bD(a)$ for all $a, b \in K$. The set of elements c of K for which $D(c) = 0$ is a subfield. If the characteristic of K is p ($\neq 0$), then $D(x^p) = 0$ for all $x \in K$. Let k be a subfield of K. A derivation D of K is called a **derivation over** k if $D(c) = 0$ for all $c \in k$; the totality of derivations over k is a †k-module. If $K = k(x_1, x_2, \ldots, x_n)$ is an algebraic function field over k, then the k-module of derivations over k has finite dimension s ($\leqslant n$) and we can choose s suitable elements $u_1, u_2, \ldots, u_s$ of K such that K is separably algebraic over $k(u_1, u_2, \ldots, u_s)$. Generally, the transcendence degree r of K over k does not exceed s, and K is separably generated over k if and only if $r = s$.

M. Finite Fields

Finite fields were first considered by E. Galois (1830), so they are also called **Galois fields**. There is no finite noncommutative field (**Wedderburn's theorem**, J. H. M. Wedderburn, *Trans. Amer. Math. Soc.*, 6 (1905)). A simple proof for this was given by E. Witt (*Abh. Math. Sem. Univ. Hamburg*, 8 (1931)). The characteristic of a finite field is a prime p, and the number of elements of the field is a power of p. Conversely, for any given prime number p and natural number α, there exists a finite field with p^α elements. Such a field is unique up to isomorphism, which we denote by $GF(p^\alpha)$ or $\mathbf{F}_q$ $(q = p^\alpha)$. For any positive integer m, $GF(p^{m\alpha})$ is an extension field of $GF(p^\alpha)$ of degree m and a †cyclic extension. Every element a of $GF(p^\alpha)$ satisfies $a^{p^\alpha} = a$; hence a has its pth root in $GF(p^\alpha)$. Therefore every finite field is perfect. The multiplicative group of $GF(p^\alpha)$ is a cyclic group of order $p^\alpha - 1$.

N. Ordered Fields

A field K is called an **ordered field** if there is given a †total order in K such that $a > b$ implies $a + c > b + c$ for all c and $a > b$, $c > 0$ implies $ac > bc$. The characteristic of an ordered field is always 0. An element a of K is called a **positive element** or a **negative element** according as $a > 0$ or $a < 0$. For an element a of K the **absolute value** of a, denoted by $|a|$, is a or $-a$ according as $a \geqslant 0$ or $a \leqslant 0$. If we define neighborhoods of a by the sets $\{x \mid a - \varepsilon < x < a + \varepsilon\}$ with positive elements ε, K becomes a †Hausdorff space. If, for any two positive elements a, b of K, there exists a natural number n such that $na > b$, then we call K an **Archimedean ordered field**. Two ordered fields are called **similarly isomorphic** if there exists an isomorphism between them under which positive elements are always mapped to positive elements. The rational number field and the real number field are examples of Archimedean ordered fields, while every Archimedean ordered field is similarly isomorphic to a subfield of the real number field. (For non-Archimedean ordered fields, see R. Baer, *S.-B. Heidelberger Akad. Wiss.* (1927).)

The concept of absolute values for numbers can also be extended to abstract fields (→ 425 Valuations).

O. Real Fields

A field k is called a **formally real field** (or simply **real field**) if -1 (1 is the unity element of k) cannot be expressed as a finite sum of squares of elements of k. The real number field is a model of formally real fields. More generally, every ordered field is a formally real field. A formally real field is called a **real closed field** if no proper algebraic extension of it is a formally real field. The real number field is a real closed field. The algebraic closure of a real closed field is obtained by adjoining a root of the polynomial X^2+1. If a is a nonzero element of a real closed field, then either a or $-a$ can be a square of an element of the field. Every real closed field can be made an ordered field in a unique way, namely, by defining squares of nonzero elements to be positive elements. Since it is known that every formally real field is a subfield of a real closed field, it follows that every formally real field is an ordered field and therefore is of characteristic 0. The notion of formally real fields was introduced by E. Artin (*Abh. Math. Sem. Univ. Hamburg*, 5 (1927)). By making use of the theory of formally real fields, Artin succeeded in solving affirmatively †Hilbert's 17th problem, which asked whether every positive definite rational expression (i.e., a rational expression with real coefficients that takes positive values for all real variables) can be expressed as a sum of squares of rational expressions. More precisely, it was shown by A. Pfister that every positive definite function in $\mathbf{R}(X_1, \ldots, X_n)$ is a sum of at most 2^n squares.

References

[1] K. Shoda, Tyûsyô daisûgaku (Japanese; Abstract algebra), Iwanami, 1932.
[2] M. Moriya, Daisûgaku (Japanese; Algebra) I, II, Asakura, 1949.
[3] K. Shoda and K. Asano, Daisûgaku (Japanese; Algebra) I, Iwanami, 1952.
[4] Y. Akizuki and M. Suzuki, Kôtô daisûgaku (Japanese; Higher algebra) I, Iwanami, 1952.
[5] S. Iyanaga and K. Kodaira, Gendai sûgaku gaisetu (Japanese; Introduction to modern mathematics) I, Iwanami, 1961.
[6] E. Steinitz, Algebraische Theorie der Körper, J. Reine Angew. Math., 137 (1910), 167–309.
[7] H. Hasse, Höhere Algebra I, II, Sammlung Göschen, Walter de Gruyter, 1926–1927.
[8] E. Steinitz and H. Hasse, Algebraische Theorie der Körper, Walter de Gruyter, 1930 (Chelsea, 1950).
[9] B. L. van der Waerden, Algebra I, Springer, seventh edition, 1966.
[10] A. A. Albert, Fundamental concepts of higher algebra, Univ. of Chicago Press, 1956.
[11] N. Bourbaki, Eléments de mathématique, Algèbre, ch. 5, Actualités Sci. Ind., 1102b, Hermann, second edition, 1959.
[12] N. Jacobson, Lectures in abstract algebra III, Van Nostrand, 1964.
[13] S. Lang, Algebra, Addison-Wesley, 1965.
[14] L. Rédei, Algebra I, Akademische Verlag., 1959.

158 (XIX.26) Field Theory

A. General Remarks

When a quantity $\psi(x)$, such as velocity, is defined at every point x in a certain region of a space, we say that a field of the quantity ψ is given. This general concept is used in many branches of science. Here we confine ourselves to some branches of physics, in particular to the quantum theory of fields, which describes †elementary particles.

B. History

The †theories of elasticity and †hydrodynamics (in particular, concerning †Euler's equation of motion) deal with displacement and velocity fields, respectively. However, a field in a vacuum (ether), which is quite different from a field in a space filled with matter, first became a subject of physics in †electromagnetism. M. Faraday (1837) introduced the electromagnetic field and discovered its fundamental laws, and J. C. Maxwell (1837) completed the mathematical formulation. On the basis of this formalism, Einstein (1905) established the theory of †relativity and later developed the general theory of relativity and of gravity.

Although †quantum theory originated from the problem of blackbody radiation, quantum theory of the electromagnetic field was first developed by P. A. M. Dirac (1927) after the development of †quantum mechanics. Along similar lines P. Jordan and E. P. Wigner (1927) quantized the matter wave (electron field), and W. Heisenberg and W. Pauli (1929) developed quantum theory of wave fields in general. Furthermore, Jordan and Pauli (1927), Pauli (1939), S. Tomonaga (1943), J. Schwinger (1948), and others reformulated the theory in a relativistically covariant manner. Quantum electrodynamics, dealing with an electromagnetic field interacting with electrons (and positrons), gave excellent agreement with experimental measurements when formulated in this way. Divergence difficulties inherent to quantized field theory were bypassed by the renormalization procedure of Tomonaga, Schwinger, R. P. Feynman, and F. J. Dyson (1947) (→ Section E; 379 *S*-Matrices B).

On the other hand, H. Yukawa (1934) applied the concept of the quantized field to the interpretation of nuclear force and predicted the existence of π-mesons. Many kinds of mesons have since been found, including π-mesons and μ-particles. The field theory of π-mesons can explain the qualitative features of the meson-nucleon system. Similar theories can be formulated for other types of unstable mesons that have been found in cosmic rays since 1949.

During the progress of meson theory, various types of fields were investigated, and a general theory of elementary particles was developed. Dirac (1936) proposed a general wave equation for elementary particles, and Pauli and M. Fierz (1939) proved the connection between spin and statistics (→ 142 Elementary Particles). Schwinger (1951) derived quantum mechanical equations of motion and commutation relations from a unified variational principle. For the cases where no interaction is present, a general theory of elementary particles consistent with the requirements of relativity and of quantum theory was established.

C. Field Equations

A field quantity at one space-time point is considered to be determined only by the field quantities at neighboring points (local action). Variation of a field quantity is governed by a †hyperbolic partial differential equation (†wave equation). (A static field, which is constant in time, obeys an †elliptic differential equation.) Given a field equation and suitable initial conditions, the field is determined in the whole space-time. A field equation can be regarded as a set of equations of a mechanical system with infinite degrees of freedom. Usually, field equations are derived from the †variational principle $\delta\int L\,dx=0$, where L represents the **Lagrangian density** $L(\psi,\psi^*,\partial\psi/\partial x,\partial\psi^*/\partial x,\ldots,x)$. (For relativistic invariance, L must be a 4-dimensional scalar and $dx=dx_1dx_2dx_3dt$.) From this Lagrangian, the **energy-momentum tensor** $T_{\mu\nu}$, the **4-dimensional current** s_μ of the field, other similar quantities, and their conservation laws can be derived. (If L does not contain the second or higher derivatives of the field, we have $T_{\mu\nu}=\Sigma_\alpha(\partial\psi_\alpha/\partial x_\mu)(\partial L/\partial(\partial\psi_\alpha/\partial x_\nu))+$ (complex conjugate) $-L\delta_{\mu\nu}$, and $s_\mu=$ $i\varepsilon\Sigma_\alpha((\partial L/\partial(\partial\psi_\alpha/\partial x_\mu))\psi_\alpha-$ (complex conjugate).)

D. Fields of Elementary Particles

Elementary particles are described by fields in 4-dimensional space-time. The theory of elementary particles must satisfy the requirement of special relativity. Thus the wave field ψ satisfies the wave equation $(-\partial^2/\partial x_0^2+\nabla^2-\kappa^2)\psi=0$ (where $x_0=ct$), and its transformation law is determined by an †irreducible representation of the †Lorentz group. On the other hand, an elementary particle possesses both a particle and a wave nature (→ 346 Quantum Mechanics). This means that the field ψ of an elementary particle cannot be a classical wave but must be a **quantized field**: ψ is an operator that operates on state vectors, and the wave equation is a relation among operators. The quantization must be done according to the †Bose statistics for integral spin particles and the †Fermi statistics for half-integral spins, respectively (→ 372 Second Quantization). As a result of quantization the system is particle-like (e.g., the total energy, charge, and current of the field can be identified with those of particles). The theory of the quantized field (†commutation relations, the †Heisenberg equations of motion, the †Schrödinger equation, etc.) can be derived from the variational principle for the transformation function:

$$\delta(\zeta_1',\sigma_1|\zeta_2',\sigma_2)$$
$$=(i/\hbar c)\left(\zeta_1',\sigma_1\middle|\delta\int_{\sigma_2}^{\sigma_1}L\,dx\middle|\zeta_2'',\sigma_2\right),$$

where σ is a 3-dimensional spacelike hypersurface, ζ_1 is a complete set of commuting dynamical variables constructed from the field variables on σ_1, and ζ_1' is an eigenvalue of ζ_1. A fundamental vector is determined by a set ζ' on σ.

Lagrangian densities and commutation relations for various fields are given below (commutators not listed here are all zero):

(1) (Pseudo)scalar field (spin 0). Example: π-meson.

$$L=c^2\sum_\mu\frac{\partial\psi^*}{\partial x_\mu}\frac{\partial\psi}{\partial x_\mu}+\kappa^2c^2\psi^*\psi$$

$$[\psi(x),\psi^*(x')]=(\hbar/i)\Delta_\kappa(x-x').$$

(2) (Pseudo)vector field (spin 1).

$$L=\frac{1}{2}c^2\sum_{\mu,\nu}\left(\frac{\partial\psi_\nu^*}{\partial x_\mu}-\frac{\partial\psi_\mu^*}{\partial x_\nu}\right)\left(\frac{\partial\psi_\nu}{\partial x_\mu}-\frac{\partial\psi_\mu}{\partial x_\nu}\right)+\kappa^2c^2\sum_\nu\psi_\nu^*\psi_\nu.$$

$$[\psi_\mu(x),\psi_\nu^*(x')]=\frac{\hbar}{i}\left(\delta_{\mu\nu}-\frac{1}{\kappa^2}\frac{\partial^2}{\partial x_\mu\partial x_\nu}\right)\times\Delta_\kappa(x-x').$$

Case $\kappa=0$. Example: photon.

$[\psi_\mu(x),\psi_\nu(x')]=(\hbar/i)\delta_{\mu\nu}\Delta_0(x-x')$.

(3) Spinor field (spin 1/2). Examples: electron, μ-particle, proton, neutron.

$L=i\hbar c\psi^*\left(\partial\psi/\partial x_0+\left(\vec{\alpha},\ \mathrm{grad}\,\psi\right)+i\kappa\beta\psi\right)$,

$[\psi_\rho(x),\psi_\sigma^*(x')]_+$

$=c\left(\delta_{\rho\sigma}-\left(\vec{\alpha}_{\rho\sigma},\ \mathrm{grad}\right)-i\kappa\beta_{\rho\sigma}\right)\Delta_\kappa(x-x')$.

Here Δ_κ on the right-hand side of these commutation relations is an †improper function satisfying the following relations:

$(\Box-\kappa^2)\Delta_\kappa=0$,

$\Delta_\kappa=0,\quad \partial\Delta_\kappa/\partial x_0=\delta(x)\quad$ for $\quad x_0=0$.

E. Interactions between Fields

Actual particles interact with each other and can be created or destroyed. The methods of handling these interactions are not yet established. To describe them, terms including products of two or more kinds of fields are added to the Lagrangian. As a result, the equations of motion become nonlinear in the fields. Usually, the interaction (after a suitable †contact transformation) is regarded as a small perturbation, and the solution is obtained by successive approximations. However, in this procedure, expectation values of physical quantities often diverge. For a particular type of interaction, finite results can be obtained after "renormalization." For other types, divergences must be "cut off," a prescription which is quite unsatisfactory. The present quantum field theory of pointlike particles breaks down in an infinitesimally small space-time region (corresponding to the high-frequency component of the field), and drastic reformulation of the theory is needed.

Heisenberg anticipated that future theory would accommodate a universal constant r_0 of the dimension of length besides $\hbar$ and c, and he developed the theory of the S-matrix as a framework of the new theory (→ 379 S-Matrices). At present, masses, spins, and charges of particles are determined only by experiments. In the future, it is expected that the mass spectrum of elementary particles will be derived from a more fundamental principle. New approaches in this direction, besides S-matrix theory, include Yukawa's nonlocal field theory and Heisenberg's nonlinear theory.

References

[1] Л. Д. Ландау, Е. М. Лифшиц (L. D. Landau and E. M. Lifšic), Теория поля (Теор. Физика, Т. II), физматгиз, fifth edition, 1967; English translation, The classical theory of fields, Cambridge Univ. Press, 1951.
[2] G. Wentzel, Einführung in die Quantentheorie der Wellenfelder, F. Deuticke, 1943; English translation, Quantum theory of fields, Interscience, 1949.
[3] W. Pauli, Relativistic field theories of elementary particles, Rev. Mod. Phys., 13 (1941), 203–232.
[4] H. Umezawa, Quantum field theory, North-Holland, 1956.
[5] H. Yukawa, Soryûsiron zyosetu (Japanese; Elements of the theory of elementary particles) I, Iwanami, 1948.
[6] H. Yukawa and M. Kobayashi (eds.), Soryûsiron (Japanese; Theory of elementary particles), Kyôritu, 1951.
[7] S. Tomonaga, N. Fukuda, H. Fukuda, and K. Sawada, Ba no ryôsiron (Japanese; Quantum theory of fields), Iwanami Coll. of Modern Phys., 1959.
[8] K. Nishijima, Fields and particles, Benjamin, 1969.

159 (XIII.17) Finite Differences

A. General Remarks

Let y be a function of a real variable x defined on a domain D. When for a fixed finite quantity Δx, the point $a+\Delta x$ is in D for any a in D, we define the **difference** $\Delta y(a)$ of y at a by $\Delta y(a)=y(a+\Delta x)-y(a)$; Δx is called the difference of x. Without loss of generality, we may take $\Delta x=1$, for otherwise there is a constant b such that $\Delta x'=1$ for the new independent variable $bx=x'$. We set $\Delta x=1$ unless otherwise mentioned. Hence the **difference quotient** $\Delta y(x)/\Delta x$ is equal to the difference Δy.

The **second difference** $\Delta^2y(x)=\Delta(\Delta y(x))$ is given by

$$\Delta^2y(x)=\Delta y(x+1)-\Delta y(x)$$
$$=y(x+2)-2y(x+1)+y(x).$$

Similarly, the **difference of the *n*th order** is defined by $\Delta^n y(x)=\Delta(\Delta^{n-1}y(x))$, and

$$\Delta^n y(x)=\sum_{k=0}^{n}(-1)^{n-k}\binom{n}{k}y(x+k).$$

Conversely, $y(x+n)$ is expressed by differences as

$$y(x+n)=\sum_{k=0}^{n}\binom{n}{k}\Delta^k y(x)$$

(→ 224 Interpolation).

B. Summation

Given a function $g(x)$ and Δx, a function $y(x)$ that satisfies $\Delta y(x)/\Delta x=g(x)$ is called a **sum** of $g(x)$. **Summation** of $g(x)$ is to find a sum of $g(x)$. Given a sum $y(x)$ of $g(x)$, an **indefinite sum** of $g(x)$, written as $\mathbf{S}g(x)\Delta x$, is given by $\mathbf{S}g(x)\Delta x=y(x)+c(x)$, where $c(x)$ is an arbitrary †periodic function of period Δx. In many cases, $c(x)$, which corresponds to an arbitrary constant in an indefinite integral, is omitted. For example, a sum of $g(x)=nx^{n-1}$ for $\Delta x=1$ is the nth-order †Bernoulli polynomial $B_n(x)$ for $n\neq 0$; a sum of x^{-1} is $\psi(x)$, given by $\psi(x)=d\log\Gamma(x)/dx$ (→ 179 Gamma Function).

When $-\Delta x\sum_{k=0}^{\infty}g(x+k\Delta x)$ or $\Delta x\sum_{k=1}^{\infty}g\cdot(x-k\Delta x)$ converges, both can be sums of $g(x)$. Since the requirement of convergence for these series was found to be too strict, the following requirement was given by N. E. Nörlund instead: Let x be a real variable and $g(x)$ be continuous for $x\geqslant b$. Define $\lambda(x)$ by $\lambda(x)=x^p(\log x)^q$ $(p\geqslant 1, q\geqslant 0)$. Then if for a positive η

$$F(x,\Delta x,\eta)=\int_a^\infty g(z)e^{-\eta\lambda(z)}dz \\ -\Delta x\sum_{k=0}^{\infty}g(x+k\Delta x)e^{-\eta\lambda(x+k\Delta x)}$$

is convergent for $a>b$, F satisfies $\Delta F(x,\Delta x,\eta)/\Delta x=g(x)\exp(-\eta\lambda(x))$. Accordingly, if $F(x,\Delta x,\eta)$ approaches a limit $F(x)$ as $\eta\to 0$, $F(x)$ is a solution of $\Delta F(x)/\Delta x=g(x)$. $F(x)$ is called the **principal solution** of $\Delta F(x)/\Delta x=g(x)$ and is denoted by $\mathbf{S}_a^x g(\xi)\Delta\xi$.

C. Difference Equations

Let $\Delta x=1$. An equation $F(x,y(x),\Delta y(x),\ldots,\Delta^n y(x))=0$ in x and differences of an unknown function $y(x)$ is called a **difference equation**. If the substitution $y=\varphi(x)$ satisfies the equation for x in some domain, $\varphi(x)$ is a **solution** of the equation. Because of the relation between $y(x),y(x+1),\ldots,y(x+n)$ and the difference of y at x, we can transform the given difference equation in the form $G(x,y(x),y(x+1),\ldots,y(x+n))=0$. This form appears more often in applications and is called the **standard form** of a difference equation.

If the equation is linear in $y(x),y(x+1),\ldots,y(x+n)$, namely, if it is given by

$$\sum_{i=0}^{n}p_i(x)y(x+i)=q(x),$$

the difference equation is said to be **linear**. When $q(x)\equiv 0$, it is **homogeneous**; otherwise, it is **inhomogeneous** (or **nonhomogeneous**).

D. Linear Difference Equations

Assume that $p_0(x),\ldots,p_n(x)$ are single-valued †analytic functions without poles and common zeros in some domain. Consider the linear difference equation

$$\sum_{i=0}^{n}p_i(x)y(x+i)=0. \tag{1}$$

If $\varphi_1(x),\varphi_2(x),\ldots,\varphi_m(x)$ are solutions of (1), then a linear combination $a_1(x)\varphi_1(x)+a_2(x)\varphi_2(x)+\ldots+a_m(x)\varphi_m(x)$ with arbitrary periodic functions $a_1(x),a_2(x),\ldots,a_m(x)$ of period 1 is also a solution of (1).

Let $\beta_1, \beta_2, \ldots$ be singular points of $p_1(x)$, $p_2(x),\ldots,p_n(x)$, $\alpha_1, \alpha_2,\ldots$ be the zeros of $p_0(x)$, and $\gamma_1, \gamma_2, \ldots$ be the zeros of $p_n(x+n)$. Then the set of **singular points** of the linear difference equation (1) is the set $\{\alpha_i,\beta_i,\gamma_i\}$.

A function $\varphi_m(x)$ is said to be **linearly dependent** on the functions $\varphi_1(x),\varphi_2(x),\ldots,\varphi_{m-1}(x)$ with respect to the difference equation (1) if $\varphi_m(x)=a_1(x)\varphi_1(x)+a_2(x)\varphi_2(x)+\ldots+a_{m-1}(x)\varphi_{m-1}(x)$, where $a_1(x),a_2(x),\ldots,a_{m-1}(x)$ are functions of period 1, every one of which takes a nonzero finite value at least at one point not congruent $(\mathrm{mod}\,\mathbf{Z})$ to any of the singular points, where $\mathbf{Z}$ is the additive group of integers.

A set of m functions is called linearly independent if none of the functions is dependent on the other $m-1$ functions. When a set of n solutions of equation (1) is linearly independent, it is a **fundamental system** for (1). Any solution of (1) can be expressed as a linear combination of n solutions of a fundamental system.

The determinant

$$\begin{vmatrix} \varphi_1(x) & \varphi_2(x) & \ldots & \varphi_n(x) \\ \varphi_1(x+1) & \varphi_2(x+1) & \ldots & \varphi_n(x+1) \\ \ldots & \ldots & & \ldots \\ \varphi_1(x+n-1) & \varphi_2(x+n-1) & \ldots & \varphi_n(x+n-1) \end{vmatrix}$$

formed from n functions $\varphi_1(x),\varphi_2(x),\ldots,\varphi_n(x)$ is called **Casorati's determinant** and is denoted by $D(\varphi_1(x),\varphi_2(x),\ldots,\varphi_n(x))$. A necessary and sufficient condition for a given set of n functions to be independent is that

Casorati's determinant is nonzero at every point except those which are congruent to singular points of (1). Casorati's determinant is used to determine whether a given set of solutions is fundamental or not.

Let $\psi(x)$ be a solution of a nonhomogeneous linear difference equation

$$P_x(y)=\sum_{i=0}^{n} p_i(x)y(x+i)=q(x). \qquad (2)$$

If $\varphi_1(x),\varphi_2(x),\ldots,\varphi_n(x)$ are n linearly independent solutions of (1), then an arbitrary solution of (2) is given by

$$y=a_1(x)\varphi_1(x)+a_2(x)\varphi_2(x)+\ldots$$
$$+a_n(x)\varphi_n(x)+\psi(x),$$

where $a_1(x),\ldots,a_n(x)$ are arbitrary periodic functions of period 1. Then the expression for y is called a **general solution** of (2). If we abbreviate Casorati's determinant of a fundamental system of solutions $\varphi_1(x),\varphi_2(x),\ldots,\varphi_n(x)$ of (1) by $D(x)$ and write $\mu_i(x)$ as the quotient of the cofactor of $\varphi_i(x+n)$ of $D(x+1)$ by $D(x+1)$, we have

$$\psi(x)=\sum_{i=1}^{n}\varphi_i(x)\mathop{\mathbf{S}}_{a}^{x} q(z)\mu_i(z)\Delta z,$$

assuming that the summation $\mathbf{S}$ on the right-hand side is known. This is the analog of Lagrange's †method of variation of constants in the theory of linear ordinary differential equations.

E. Linear Difference Equations with Constant Coefficients

If all the coefficients in

$$\sum_{i=0}^{n} p_i\, y(x+i)=0 \qquad p_0\neq 0,\quad p_n\neq 0, \qquad (3)$$

are constants, n linearly independent solutions are obtained easily. Indeed, if λ is a root of the algebraic equation $\sum_{i=0}^{n}p_i\lambda^i=0$, λ^x is a solution of (3). This algebraic equation is called the **characteristic equation** of (3). If it has n distinct roots, $\lambda_1,\lambda_2,\ldots,\lambda_n$ then λ_1^x, $\lambda_2^x,\ldots,\lambda_n^x$ are n linearly independent solutions. In general, if λ is an m-tuple root of the characteristic equation, then $\lambda^x, x\lambda^x$, $\ldots,x^{m-1}\lambda^x$ are solutions of (3). Accordingly, if λ_j is a root of multiplicity m_j ($\sum_{j=1}^{s}m_j=n$, $j=1,2,\ldots,s$), then $\lambda_j^x, x\lambda_j^x,\ldots,x^{m_j-1}\lambda_j^x$ ($j=1,\ldots,s$) constitute a set of n linearly independent solutions.

Even if all the p_i are real, the characteristic equation may have complex roots. In such a case real solutions are obtained as follows: When $\lambda=\mu+i\nu$ is a root of multiplicity m, $\bar{\lambda}=\mu-i\nu$ is also a root of the same multiplicity. If we write $\rho=\sqrt{\mu^2+\nu^2}$, $\tan\varphi=\nu/\mu$, then $\rho^x\cos\varphi x$, $\rho^x\sin\varphi x$, $x\rho^x\cos\varphi x$, $x\rho^x\sin\varphi x,\ldots$, $x^{m-1}\rho^x\cos\varphi x$, $x^{m-1}\rho^x\sin\varphi x$ are $2m$ independent real solutions.

Nonhomogeneous equations with constant coefficients can be generally solved by Lagrange's method with these solutions. However, when the nonhomogeneous term has a special form such as

$$\sum_{i=0}^{n} p_i y(x+i)=p(x)\lambda^x,$$

where $p(x)$ is a polynomial in x, λ is a root of multiplicity m of the characteristic equation, and we can use the method of undetermined coefficients. In this particular case, the substitution $(A_0+A_1x+\ldots+A_kx^k)x^m\lambda^x$ with undetermined coefficients $A_0,A_1,\ldots,A_k$ gives solutions.

F. Difference and Differential Equations

The †differential operator d/dx acts on the family of functions $\{x^m\mid m=0,\pm1,\ldots\}$ according to $dx^m/dx=mx^{m-1}$, just as the difference operator Δ acts on the family $\{x^{(m)}=\Gamma(x+1)/\Gamma(x-m+1)\mid m=0,\pm1,\ldots\}$ according to $\Delta x^{(m)}=mx^{(m-1)}$. Hence by using the **factorial series** $\sum a_m x^{(m)}$ and its similarity with the power series $\sum a_m x^m$, we may obtain some analogies with the theory of differential equations. For example, the †Frobenius method in the theory of †regular singular points can be applied for the system of difference equations

$$(z-1)\Delta_{-1}w_k(z)=\sum_{j=1}^{n} a_{jk}(z)w_j(z),$$

$$k=1,2,\ldots,n.$$

However, there are certain essential differences between functions defined as solutions of differential and difference equations. For example, **Hölder's theorem** states that no solution of the simple difference equation $y(x+1)-y(x)=x^{-1}$ satisfies any †algebraic differential equation. Consequently, the gamma function, which is related to a solution of the equation $\psi(x)=d\log\Gamma(x)/dx$, cannot be a solution of any algebraic differential equation.

For an arbitrary complex number q, an equation of the form $y(qx)=f(x,y(x))$ is called a **geometric difference equation**. For example, the ordinary difference equation (1) can be transformed into

$$\sum_{k=0}^{n} p_k(z)U(zq^k)=B(z) \qquad (1')$$

by the change of variable $z=q^x$. Although it is possible to transform an equation of the type (1′) into that of the type (1), there are theories specially developed for the type (1′), since the coefficients of the equation may become more complicated by such a transformation. (For the numerical solution of ordinary differential equations by difference equation approximation → 298 Numerical Solution of Ordinary Differential Equations.)

References

[1] T. Takahashi, Sabun hôteisiki (Japanese; Difference equations), Baihûkan, 1962.
[2] L. M. Milne-Thomson, The calculus of finite differences, Macmillan, 1951.
[3] H. Meschkowski, Differenzengleichungen, Vandenhoeck & Ruprecht, 1959.
[4] А. О. Гельфонд (A. O. Gel'fond), Исчисление конечных разностей, Гостехиздат, 1952; German translation, Differenzenrechnung, Deutscher Verlag der Wiss., 1958.
[5] W. A. Harris, Linear systems of difference equations, in J. P. LaSalle and J. B. Diaz (eds.), Contributions to differential equations, Interscience, 1963, vol. 1, p. 489–518.
[6] N. E. Nörlund, Vorlesungen über Differenzenrechnung, Springer, 1924.
[7] N. E. Nörlund, Leçons sur les équations linéaires aux différences finies, Gauthier-Villars, 1929.
[8] K. S. Miller, Linear difference equations, Benjamin, 1968.

160 (IV.4) Finite Groups

A. The Number of Finite Groups of a Given Order

A group is called a **finite group** if its order is finite (→ 193 Groups). Since the early years of the theory of finite groups, a major problem has been to find the number of distinct isomorphism classes of groups having a given order. It is almost impossible, however, to find a general solution to the problem unless the values of n are restricted to a (small) subset of the natural numbers. Let $f(n)$ denote the number of isomorphism classes of finite groups of order n. If p is a prime number, then $f(p)=1$ and any group of prime order is a cyclic group. If p is prime, then any group of order p^2 is an †Abelian group and $f(p^2)=2$. If p and q are distinct primes and $p>q$, then $f(pq)=2$ or 1 according as p is congruent to 1 modulo q or not. If $p\equiv 1 \pmod q$, there is a non-Abelian group of order pq as well as a cyclic group of order pq. For small n, the value of $f(n)$ is as follows:

n	8	12	16	18	20	24	27	28	30	32	60
$f(n)$	5	5	14	5	5	15	5	4	4	51	13

For any n, $f(n)\geqslant 1$. When p is prime, $f(p^m)$ is known for $m\leqslant 6$: $f(p^3)=5$, $f(p^4)=15$ if $p>2$. For $f(p^5)$ see O. Schreier, *Abh. Math. Sem. Univ. Hamburg*, 4 (1926). For $f(2^6)$ see [15]. Set $f(p^m)=p^l$ and $l=Am^3$. Then $A\to 2/27$ as $m\to\infty$ (G. Higman, *Proc. London Math. Soc.*, 10 (1960); C. C. Sims, Symposium on Group Theory, Harvard, 1963).

B. Fundamental Theorems on Finite Groups

Following are some of the fundamental theorems that are useful in studying finite groups:

(1) The order of any subgroup of a finite group G divides the order of G (J. L. Lagrange). The converse is not necessarily true. If a finite group G contains a subgroup of order n for any divisor n of the order of G, then G is a †solvable group. Furthermore, if G contains a unique subgroup of order n for each divisor n of the order of G, then G is a cyclic group.

Let p be a prime number. Let the order of a finite group G be $p^n m$, where m is not divisible by p. A subgroup of order p^n of G is called a p-**Sylow subgroup** (or simply **Sylow subgroup**) of G. The importance of this concept may be seen from the next theorem.

(2) A finite group contains a p-Sylow subgroup for any prime divisor p of the order of the group. Furthermore, p-Sylow subgroups are conjugate to each other. The number of distinct p-Sylow subgroups is congruent to 1 modulo p. In general, the number of distinct p-Sylow subgroups of G that contain a given subgroup whose order is a power of p is congruent to 1 modulo p (**Sylow's theorems**).

(3) A p-group is a †nilpotent group (a finite group is called a p-**group** if its order is a power of p). Thus any finite p-group G of order >1 contains a nonidentity element in the †center of G. Furthermore, any proper subgroup of G is different from its †normalizer. A paper by P. Hall, *Proc. London Math. Soc.*, (2) 36 (1933), is a classical and fundamental work on p-groups.

A group G of order 8 with 2 generators σ, τ and relations $\sigma^4=1$, $\tau\sigma\tau^{-1}=\sigma^{-1}$, $\sigma^2=\tau^2$ is called the **quaternion group**. This group is isomorphic to the multiplicative group consisting of $\{\pm 1, \pm i, \pm j, \pm k\}$ in the †quaternion field. The **generalized quaternion group** is a group of order 2^n with 2 generators σ, τ and relations $\sigma^{2^{n-1}}=1$, $\tau\sigma\tau^{-1}=\sigma^{-1}$, $\tau^2=\sigma^{2^{n-2}}$. A non-Abelian group all of whose subgroups are normal subgroups is called a **Hamilton group**. A Hamilton group is the direct product of a quaternion group, an Abelian group of odd order, and an Abelian group of exponent 2 (i.e., $\rho^2=1$ for each element ρ).

Let G be a finite group, and let $G_0=G\supset G_1\supset\ldots\supset G_r=1$ be a †composition series of G. The set of isomorphism classes G_{i-1}/G_i, $i=1,2,\ldots,r$, of simple groups G_{i-1}/G_i is uniquely determined (up to arrangement) by the †Jordan-Hölder theorem. Thus the two most fundamental problems of finite groups are (i) the study of the simple groups and (ii) the study of a group with a given set of composition factors. The first has been one of the leading problems of the theory, although it has stayed in a state of stagnancy until rather recently (→ Section I). As to the second problem, initial works by H. Wielandt and others are under way (particularly in the direction of various generalizations of Sylow's theorems). For the class of finite solvable groups, the first problem has a rather trivial solution; only the second problem is important, and even in this case the theory seems to leave something to be desired.

C. Finite Nilpotent Groups

A finite group is nilpotent if and only if it is the †direct product of its p-Sylow subgroups, where p ranges over all the prime divisors of the order. Any maximal subgroup of a nilpotent group is normal. The converse holds for finite groups; that is, a finite group is nilpotent if and only if all its maximal subgroups are normal.

D. Finite Solvable Groups

One of the most profound results on finite groups, an affirmative answer to the long-standing **Burnside conjecture**, is the **Feit-Thompson theorem** (*Pacific J. Math.*, 13 (1963)): A finite group of odd order is solvable. The index of a maximal subgroup of a finite solvable group is a power of a prime number (E. Galois). But the converse is not true. The unique simple group of order 168 has the property that all maximal subgroups are of prime power index. A finite solvable group contains a self-normalizing nilpotent subgroup (i.e., a nilpotent subgroup H such that $N_G(H)=H$), and any two such subgroups are conjugate (R. W. Carter, *Math. Z.*, 75 (1960); cf. W. Gaschütz, *Math. Z.*, 80 (1963), for a generalization). Such a subgroup is called a **Carter subgroup** and is an analog of a Cartan subalgebra of a Lie algebra. But unlike Cartan subalgebras, most simple groups do not contain any self-normalizing nilpotent subgroups. A finite solvable group of order mn (m,n are relatively prime) contains a subgroup of order m; two subgroups of order m are conjugate; if l is a divisor of m, then any subgroup of order l is contained in a subgroup of order m (P. Hall). The converse of the first part of this theorem is also true: A finite group is solvable if it contains a subgroup of order m for any decomposition of the order in the form mn, $(m,n)=1$ (P. Hall's solvability criterion). This generalizes the famous **Burnside's theorem** asserting the solvability of a group of order p^aq^b, where both p and q are prime numbers. If the sequence of quotient groups of a †principal series of a finite group G consists of cyclic groups, then the group G is called **supersolvable**. A finite group is supersolvable if and only if the index of any maximal subgroup is a prime number (B. Huppert, *Math. Z*, 60 (1954)). If p is the largest prime divisor of the order of a finite supersolvable group G, then a p-Sylow group of G is a normal subgroup.

E. Hall Subgroups

A subgroup is called a **Hall subgroup** if its order is relatively prime to its index (see the theorems of P. Hall on finite solvable groups). There is no general theorem known on the existence of a Hall subgroup. If a finite group G has a normal Hall subgroup N, then G contains a Hall subgroup H that is a complement of N in the sense that $G=NH$ and $N\cap H=1$; furthermore, any two complements are conjugate (**Schur-Zassenhaus theorem**). The analog of Hall's theorem on finite solvable groups fails for nonsolvable groups. But if a finite group, solvable or not, contains a nilpotent Hall subgroup H of order n, then any subgroup of an order dividing n is conjugate to a subgroup of H (Wielandt, *Math. Z.*, 60 (1954); cf. P. Hall, *Proc. London Math. Soc.*, 4 (1954), for a generalization). There are some generalizations of these results for maximal π-subgroups which may not be Hall subgroups.

F. π-Solvable Groups

Let π be a set of prime numbers. Denote the set of prime numbers not in π by π'. A finite group is called a **π-group** if all the prime divisors of the order belong to π. A finite group is called **π-solvable** if any composition factor is either a π'-group or a solvable π-group. If $\pi=\{p\}$ consists of a single prime number p, we use terms such as *p-group* or *p-solvable* (instead of *$\{p\}$-solvable*). Let G

be a π-solvable group. A series of subgroups $P_0 = 1 \subset N_0 \subsetneqq P_1 \subsetneqq N_1 \subsetneqq \ldots \subsetneqq P_l \subset N_l = G$ defined by the properties that P_i/N_{i-1} is a maximal normal π-group of G/N_{i-1} and N_i/P_i is a maximal normal π'-group of G/P_i is called the **π-series** of G, and the integer l is called the **π-length** of G. A solvable group is π-solvable for any set π of prime numbers. A π-solvable group contains a Hall subgroup which is a π-group and also a Hall subgroup which is a π'-group; an analog of Hall's theorem on finite solvable groups holds. Hall and Higman (*Proc. London Math. Soc.*, 7 (1956)) discovered deep relations between the p-length of a p-solvable group and invariants of its p-Sylow group. For example, the p-length is 1 if a p-Sylow group is Abelian.

G. Permutation Groups

The set of all permutations on a set Ω of n elements forms a group of order $n!$ whose structure depends only on n. This group is called the **symmetric group of degree** n, denoted by S_n. Any subgroup of S_n is a **permutation group of degree** n. When it is necessary to mention the set Ω on which permutations operate, S_n may be denoted as $S(\Omega)$, and a subgroup of $S(\Omega)$ is called a permutation group on Ω. The set of n elements on which S_n operates is usually assumed to be $\{1, 2, \ldots, n\}$, and an element σ of S_n is written as

$$\sigma = \begin{pmatrix} 1 & 2 & \ldots & n \\ 1' & 2' & \ldots & n' \end{pmatrix},$$

for example $\begin{pmatrix} 1 & 2 & 3 & 4 & 5 & 6 \\ 2 & 3 & 1 & 5 & 4 & 6 \end{pmatrix}$,

where i' is the image of i by σ: $i' = \sigma(i)$. The element σ may be written as $(\ldots)(abc\ldots z)(\ldots)$, which means that σ cyclically maps a into b, b into c, and so on, and finally z back into a. In the example, $\sigma = (1\ 2\ 3)(4\ 5)(6)$. It is customary to omit the cycle with only one letter in it, such as (6) in the example. With this convention, $\sigma = (1\ 2\ 3)(4\ 5)$ may be an element of S_n for any $n \geqslant 5$, leaving all the letters $i \geqslant 6$ invariant. A **cycle** of length l is an element σ of S_n which moves l letters $a_1, \ldots, a_l$ cyclically and leaves all the rest; i.e., $\sigma = (a_1, \ldots, a_l)$. Then an expression such as $\sigma = (1\ 2\ 3)(4\ 5)$ is the same as the product of two cycles (1 2 3) and (4 5). In general, any permutation can be expressed as the product of mutually disjoint cycles (two cycles $(a_1, \ldots, a_l)$ and $(b_1, \ldots, b_m)$ are said to be disjoint if $a_i \neq b_j$ for all i and j). Furthermore, the expression of the permutation as the product of mutually disjoint cycles is unique up to the order in which these cycles are written. A cycle of length 2 is called a **transposition**. Any permutation may be written as a product of transpositions. This expression is not unique, but the parity of the number of transpositions in the expression is determined by the permutation. A permutation is called **even** if it is the product of an even number of transpositions and **odd** otherwise. The symmetric group S_n contains the same number of even and odd permutations.

The totality of even permutations forms a normal subgroup of order $(n!)/2$ $(n \geqslant 2)$, called the **alternating group of degree** n and usually denoted by A_n. An even permutation is the product of cycles of length 3. The alternating group A_5 of degree 5 is the nonsolvable group of minimal order. This fact was known to Galois. If $n \neq 4$, then the alternating group A_n is a simple group and a unique proper normal subgroup of S_n. If $n = 4$, A_4 contains a normal noncyclic subgroup V of order 4. In this case A_4 and V are the only proper normal subgroups of S_4. A noncyclic group of order 4 is called a **(Klein's) four-group**. If $n \geqslant 5$, the symmetric group S_n is not a solvable group. This is the group-theoretic ground for the famous theorem, proved by Ruffini, Abel, and Galois, which asserts the impossibility of an algebraic solution of a general algebraic equation of degree more than four. If $n \leqslant 4$, S_n is solvable. The group S_4 has a composition series with composition factors of orders 2, 3, 2, 2, and S_4 is realized as the group of motions of 3-dimensional space which preserve an octahedron. Hence S_4 is called the **octahedral group**. Similarly, A_4 (A_5) is realized as the group of motions in the space which preserve a tetrahedron (icosahedron); thus A_4 is called the **tetrahedral group** and A_5 the **icosahedral group**.

These groups have been extensively studied in view of their geometric aspect. The group of motions of a plane that preserve a regular polygon is called a **dihedral group**. If a regular polygon has n sides, then the group has order $2n$. Sometimes Klein's four-group is included in the class of dihedral groups (for $n = 2$). The dihedral groups, octahedral group, etc., are called **regular polyhedral groups**. A finite subgroup of the group of motions in 3-dimensional space is either cyclic or one of the regular polyhedral groups. A dihedral group is generated by two elements of order 2. Conversely, a finite group generated by two elements of order 2 is a dihedral group. This simple fact has surprisingly many consequences in the theory of finite groups of even order ([19], ch. 9). A dihedral group of order $2n$ contains a cyclic normal subgroup of order n, and hence is solvable.

If $n \neq 6$, every automorphism of S_n is inner.

The order of the group of automorphisms of S_6 is twice the order of S_6. The index $(S_n:H)$ of a subgroup H of S_n is at least equal to n unless $H=A_n$. If $(S_n:H)=n$, then H is isomorphic to S_{n-1}. If $n\neq 6$, S_n contains a unique conjugate class of subgroups of index n. But S_6 contains two such classes, which are exchanged by an automorphism of S_6 (→ Section I).

H. Transitive Permutation Groups

A permutation group G on a set Ω is called a **transitive permutation group** if for any pair (a, b) of elements of Ω, there exists a permutation of G which sends a into b. Otherwise G is said to be **intransitive**. Let G be a transitive permutation group on a set Ω, and let a be an element of Ω. The totality of elements of G which leave a invariant forms a subgroup of G called the **stabilizer** of a (in G). The index of the stabilizer is equal to the number of elements of Ω, the degree of G. Thus the degree of a transitive permutation group G divides the order of G (a fundamental theorem).

The concept of orbits is important. Let G be a permutation group on a set Ω. A subset Γ of Ω is called an **orbit** of G if it is G-invariant and G acts transitively on Γ. In other words, a subset Γ of Ω is an orbit of G if the following two conditions are satisfied: (i) If $a\in\Gamma$ and $g\in G$, the image $g(a)$ also lies in Γ; and (ii) if a and b are two elements of Γ, there exists an element x of G such that $b=x(a)$. Thus each element x of G induces a permutation $\varphi_\Gamma(x)$ on Γ. The set of all these permutations $\varphi_\Gamma(x)$ $(x\in G)$ forms a permutation group on Γ, which may be denoted by $\varphi_\Gamma(G)$. Then $\varphi_\Gamma(G)$ is transitive on Γ, and φ_Γ is a homomorphism of G onto $\varphi_\Gamma(G)$. Thus the number of elements in an orbit Γ is a divisor of the order of G. It is clear that the set Ω on which G acts is the union of mutually disjoint orbits $\Gamma_1, \ldots, \Gamma_r$ of G. This implies that the degree of G is the sum of the numbers of elements in the orbits Γ_i. The resulting equation often contains nontrivial relations. If φ_i denotes the homomorphism φ_{Γ_i} defined before, then G is isomorphic to a subgroup of the direct product of the groups $\varphi_i(G)$, $i=1,2,\ldots,r$.

A transitive permutation group is called **regular** if the stabilizer of any letter is the identity subgroup $\{1\}$. A transitive permutation group is regular if and only if its order equals its degree. Any group can be realized as a regular permutation group (**Cayley's theorem**). A transitive permutation group which is Abelian is always regular.

Let G be a transitive permutation group on a set Ω. If the stabilizer of an element a of Ω is a maximal subgroup, G is called **primitive**, and otherwise **imprimitive**. A normal subgroup, which is $\neq\{1\}$, of a primitive permutation group is transitive. An imprimitive permutation group induces a decomposition of the set Ω into the union of mutually disjoint subsets $\Delta_1, \ldots, \Delta_s$ $(s>1)$ such that each Δ_i contains at least two elements, and if $x\in G$ maps an element a of Δ_i onto an element b of Δ_j, then x maps every element of Δ_i into $\Delta_j: x(\Delta_i)=\Delta_j$. The set $\{\Delta_1, \ldots, \Delta_s\}$ is called a system of imprimitivity. A subset Δ of Ω is called a **block** if $x(\Delta)\cap\Delta$ equals Δ or the empty set for all x in G. A block is called nontrivial if $\Delta\neq\Omega$ and Δ contains at least two elements. Each member of a system of imprimitivity is a nontrivial block. A transitive permutation group is primitive if and only if there is no nontrivial block.

A permutation group G on a set Ω is called k-**transitive** (or k-**ply transitive**, where k is a natural number) if for two arbitrary k-tuples $(a_1, \ldots, a_k)$ and $(b_1, \ldots, b_k)$ of distinct elements of Ω, there is an element of G which maps a_i into b_i for all $i=1,2,\ldots,k$. If $k\geqslant 2$, G is called **multiply transitive**. A doubly transitive permutation group is always primitive. The symmetric group S_n of degree n is n-transitive, while the alternating group A_n is $(n-2)$-transitive for $n\geqslant 3$. Conversely, an $(n-2)$-transitive permutation group on $\{1,2,\ldots,n\}$ is either S_n or A_n. The problem of finding all the multiply transitive permutation groups has not yet been solved. For multiply transitive permutation groups which are simple, see the list in Section I. If $k\geqslant 6$, no k-transitive groups are known at present except S_n and A_n. If the Schreier conjecture (→ Section I) is true, then there are no 7-transitive permutation groups except S_n and A_n (Wielandt, *Math. Z.*, 74 (1960); H. Nagao, *Nagoya J. Math.*, 27 (1964)).

Two 5-transitive permutation groups other than S_n and A_n are known: the groups M_{12} and M_{24} of degrees 12 and 24, respectively, discovered by E. L. Mathieu in 1864 and 1871. The stabilizer of a letter in M_{12} (M_{24}) is a 4-transitive permutation group of degree 11 (23), denoted by M_{11} (M_{23}). No 4-transitive permutation groups other than S_n, A_n, M_i $(i=11, 12, 23$, and $24)$ are known. The groups M_{12}, M_{11}, M_{24}, M_{23} and the stabilizer M_{22} of a letter in M_{23} are called **Mathieu groups**. They are simple groups which have quite exceptional properties. For Mathieu groups, see E. Witt, *Abh. Math. Sem. Univ. Hamburg*, 12 (1938). A k-transitive permutation group G on Ω of degree n and order $n(n-1)\ldots(n-k+1)$ has the property that no nonidentity element of G leaves k distinct letters of Ω invariant. If $k \geqslant 4$, such a group is one of the following: S_k, A_{k+2}, M_{12}, and M_{11} (C. Jordan). For $k=2$ and 3, see H. Zassenhaus, *Abh. Math. Sem. Univ. Hamburg*, 11 (1936).

A multiply transitive permutation group G contains a normal subgroup S such that S is a non-Abelian simple group and G is isomorphic to a subgroup of the group $\operatorname{Aut} S$ of the automorphisms of S, except when the degree n of G is a power of a prime number and G contains a regular normal subgroup of order n which is an †elementary Abelian group (W. S. Burnside). Furthermore, in these exceptional cases, G is at most 2-transitive if n is odd, while G is at most 3-transitive if n is even but more than 4. The symmetric group S_4 of degree 4 is the only 4-transitive group that contains a proper solvable normal subgroup.

A transitive extension of a permutation group H on Ω is defined as follows. Let ∞ be a new element not contained in Ω. A **transitive extension** G of H is a transitive permutation group on the set $\{\Omega, \infty\}$ in which the stabilizer of ∞ is the given permutation group H on Ω. Transitive extensions do not exist for some H. Suppose that a permutation group H admits a transitive extension G that is primitive. If H is simple, then G is also simple unless the degree of G is a power of a prime number. Constructing transitive extensions has been an effective method for constructing sporadic simple groups.

Permutation groups of prime degree have been studied extensively since the last century, partly because of their connection with algebraic equations of prime degree. Let p be a prime number. A transitive permutation group of degree p is either multiply transitive and nonsolvable or has a normal subgroup of order p with factor group isomorphic to a cyclic group of order dividing $p-1$ (Burnside). Choose two cycles x and y of length p in S_p. If y is not a power of x, then the subgroup $\langle x,y \rangle$ generated by x and y is a multiply transitive permutation group which is simple. The structure of $\langle x,y \rangle$ is not known in spite of its simple definition. More attention has been paid to groups of degree p, where p is a prime number such that $(p-1)/2 = q$ is another prime number. The problem is to decide if such nonsolvable groups contain the alternating group A_p. The Mathieu groups M_{11} and M_{23} are the only known exceptions for $p>7$. The search for additional exceptions has been aided by the development of high-speed computers. It is known that there is no exceptional group of degree $p = 2q+1$ for $23 < p \leqslant 4079$ (P. J. Nikolai and E. T. Parker, *Math. Tables Aids Comput.*, 12 (1958); see N. Ito, *Bull. Amer. Math. Soc.*, 69 (1963), for further results in this direction).

A **Frobenius group** is a nonregular transitive permutation group in which the identity is the only element leaving more than one letter invariant. A Frobenius group of degree n contains exactly $n-1$ elements which displace all the letters. These $n-1$ elements together with the identity form a regular normal subgroup of order n. This is a theorem of Frobenius; all the existing proofs depend on the theory of characters. The regular normal subgroup of a Frobenius group is nilpotent (J. G. Thompson, *Proc. Nat. Acad. Sci. US*, 45 (1959); [20, ch. 3]).

I. Finite Simple Groups

The class of simple groups of finite order is divided into four subclasses consisting of (1) cyclic groups of prime order, (2) alternating groups of degree $\geqslant 5$, (3) simple groups of Lie type, and (4) other simple groups.

The subclass (1) consists of cyclic groups of prime order p for any prime number p. An Abelian simple group belongs to this subclass. The subclass (4) consists of five Mathieu groups and sporadic simple groups. All sporadic groups are of recent discovery.

Simple groups of Lie type are analogs of simple Lie groups, and include the classical groups as well as the exceptional groups and the groups of twisted type.

Classical groups are divided into four types: †linear, †unitary, †symplectic, and †orthogonal (→ 63 Classical Groups). Let $q=p^r$ be a power of a prime number p. Consider a vector space V of dimension $n \geqslant 2$ over the field $\mathbf{F}_q$ of q elements, except in the unitary case where V is a vector space of dimension $n \geqslant 2$ over the field $\mathbf{F}_{q^2}$ of q^2 elements. Let f be a nondegenerate form on V which is †Hermitian in the unitary case (with respect to the automorphism of order 2 of $\mathbf{F}_{q^2}$ over $\mathbf{F}_q$), †skew symmetric bilinear in the symplectic case, and †quadratic in the orthogonal case. In the orthogonal case, the dimension of V is assumed $\geqslant 3$. Consider the group of all linear transformations of V (linear case) of determinant 1, or the group of all linear transformations of determinant 1 which leave the form f invariant (in other cases). In the orthogonal case, take the commutator subgroup. With each of these groups, the factor group of it by its center is a simple group with a few exceptions.

There are several notations to denote these groups. E. Artin's notation for simple groups, which is reasonably descriptive and simple, follows the name of the simple group, n and

q are as described in the preceding paragraph, g is the order of the simple group, and (a, b) denotes the greatest common divisor of two natural numbers a and b.

Linear simple group, $L_n(q)$

$$g=q^{n(n-1)/2}\prod_{i=2}^{n}(q^i-1)/d, \quad d=(n,q-1).$$

Unitary group, $U_n(q)$

$$g=q^{n(n-1)/2}\prod_{i=2}^{n}(q^i-(-1)^i)/d,$$
$$d=(n,q+1).$$

The structure of unitary groups does not depend on the form.

Symplectic group, $S_n(q)$, $n=2m$

$$g=q^{m^2}\prod_{i=1}^{m}(q^{2i}-1)/d, \qquad d=(2,q-1).$$

In the symplectic case, the dimension n of the space V must be even, so $n=2m$, and the structure does not depend on the form.

Orthogonal group in odd dimension, $n=2m+1$, $O_{2m+1}(q)$,

$$g=q^{m^2}\prod_{i=1}^{m}(q^{2i}-1)/d, \qquad d=(2,q-1).$$

The structure does not depend on the form in odd dimension.

Orthogonal groups in even dimension $n=2m$. There are two inequivalent forms, one with †index m (which is maximal) and the other with index $m-1$. The two orthogonal groups are denoted by $O_{2m}(\varepsilon, q)$, $\varepsilon=\pm 1$, where $\varepsilon=1$ if the form is of maximal index and -1 otherwise. Then,

$$g=q^{m(m-1)}(q^m-\varepsilon)\prod_{i=1}^{m-1}(q^{2i}-1)/d,$$
$$d=(4,q^m-\varepsilon).$$

The value of ε is determined by the form f: $\varepsilon=1$ if f is equivalent to $\Sigma_{i=1}^{m}x_{2i-1}x_{2i}$, and $\varepsilon=-1$ if $f\sim x_1^2+\gamma x_1x_2+x_2^2+\Sigma_{i=2}^{m}x_{2i-1}x_{2i}$, where the polynomial $t^2+\gamma t+1$ is irreducible over $\mathbf{F}_q$.

There are other ways to denote these groups. Let $X=X(*,*)$ be a group of nonsingular linear transformations of a vector space V. Two asterisks indicate two invariants, such as the dimension of V and the number of elements in the ground field. The notation SX stands for the subgroup of X consisting of linear transformations with determinant 1, and the notation PX stands for the factor group of the linear group X by its center. Thus PX is a subgroup of the group of all projectivities of the projective space formed by the linear subspaces of V. The following list is self-explanatory, except the last term in each row, which is the notation of L. E. Dickson [5]:

$$L_n(q)=PSL(n,q)=LF(n,q)$$
$$U_n(q)=PSU(n,q)=HO(n,q^2)$$
$$S_{2n}(q)=PSp(n,q)=A(2n,q).$$

(LF: linear fractional group; HO: hyperorthogonal group; A: Abelian linear group.) If f is a nondegenerate quadratic form, then the subgroup of $GL(n,q)$ consisting of all the elements leaving the form f invariant may be denoted by $O(n,q,f)$. Let $\Omega(n,q,f)$ denote the commutator subgroup of $O(n,q,f)$. Set $\varepsilon=1$ if f is of maximal index, and $\varepsilon=-1$ otherwise. Then

$$O_n(\varepsilon,q)=P\Omega(n,q,f).$$

Dickson's notation for orthogonal groups is complicated and seldom used.

Finite simple groups corresponding to Lie groups of some exceptional type were studied by Dickson early in this century, but C. Chevalley (*Tôhoku Math. J.*, (2) 7 (1955)) proved the existence, simplicity, and other properties of groups of any (exceptional) type over any field by a unified method. Simple Lie algebras over the field $\mathbf{C}$ of complex numbers are completely classified, and according to classification theory they are in one-to-one correspondence with the †Dynkin diagrams. Let L be a simple Lie algebra (over $\mathbf{C}$) corresponding to the †Dynkin diagram of type X (→ 247 Lie Algebras). Let $L=L_0+\Sigma L_\alpha$ be a Cartan decomposition of L, where α ranges over the †root system Δ of L. It is possible to choose a basis B of L with the following properties (†Chevalley's canonical basis): B consists of e_α $(\alpha\in\Delta)$ and a basis of L_0; the structure constants of L with respect to B are all rational integers; the automorphism $x_\alpha(\zeta)$ in the †adjoint group defined by

$$x_\alpha(\zeta)=\exp(\zeta\,\mathrm{ad}\,e_\alpha) \quad (\zeta\in\mathbf{C})$$

maps each element of B into a linear combination of elements of B with coefficients which are polynomials in ζ with integer coefficients. Thus the matrix $A_\alpha(\zeta)$ representing the transformation $x_\alpha(\zeta)$ with respect to B has coefficients which are polynomials in ζ with integer coefficients.

The elements of B span a Lie algebra $L_{\mathbf{Z}}$ over the ring $\mathbf{Z}$ of integers. Let F be a field and form $L_F=F\otimes_{\mathbf{Z}}L_{\mathbf{Z}}$. Then L_F is a Lie algebra over F, and the set B may be identified with a basis of L_F over F. Let t be any element of F, $A_\alpha(t)$ be the matrix obtained from $A_\alpha(\zeta)$ by replacing the complex variable ζ by the element t, and finally $x_\alpha(t)$ be the linear transformation of L_F represented by the matrix $A_\alpha(t)$ with respect to B. The group generated by the $x_\alpha(t)$ for each root α and each element t of F is called the **Chevalley**

group of type X over F. The commutator subgroups of the Chevalley groups are simple, with a few exceptions which will be stated after the complete list of simple groups of Lie type. Suppose that $X=A_n$, D_n, or E_6 (→ 248 Lie Groups S). Then the Dynkin diagram of type X has a nontrivial symmetry; let it be $\alpha \to \beta$. Suppose that the field F has an automorphism σ of the same order as the order of the symmetry of the Dynkin diagram. Let θ be the automorphism of the Chevalley group which sends $x_\alpha(t)$ to $x_\beta(t^\sigma)$. Let U (resp. V) be the subgroup of the Chevalley group generated by $x_\alpha(t)$ with $\alpha>0$, $t\in F$ ($x_\beta(t)$, $\beta<0$), and let U^1 (V^1) be the subgroup consisting of all the elements of U (V) which are left invariant by θ. The group generated by U^1 and V^1 is called the **group of twisted type**. If the order of σ is i, this group is said to be of twisted type ${}^iX$. In all but one case, the group of twisted type is simple (see R. Steinberg, *Pacific J. Math.*, 9 (1959)). The value i is 2 except when $X=D_4$. Since D_4 admits symmetries of orders 2 and 3, there are two twisted types. If $X=B_2$, G_2, or F_4, then the diagram has a symmetry. If the characteristic p of the ground field F is 2, 3, or 2 according as $X=B_2$, G_2, or F_4 and the field F has an automorphism σ such that $(t^\sigma)^\sigma=t^p$ for any $t\in F$, then a procedure similar to the one described before is applicable, and the group of twisted type X' is obtained (R. H. Ree, *Amer. J. Math.*, 83 (1961)). The group of twisted type is simple if the field F has more than three elements.

The following list contains all the simple groups of Lie type. For each classical group, we list the type followed by identification:

$A_n=L_{n+1}(q)$	$(n\geqslant 1)$
${}^2A_n=U_{n+1}(q)$	$(n\geqslant 1)$
$B_n=O_{2n+1}(q)$	$(n>1)$
$C_n=S_{2n}(q)$	$(n>2)$
$D_n=O_{2n}(1,q)$	$(n>3)$
${}^2D_n=O_{2n}(-1,q)$	$(n>3)$

For other groups the type of a group is followed by a customary name or notation, if any, and the order g:

B_2' **Suzuki group**, $Sz(q)$, $q=2^{2n+1}$, $g=q^2(q-1)(q^2+1)$

${}^3D_4$ $g=q^{12}(q^8+q^4+1)(q^6-1)(q^2-1)$

G_2 $g=q^6(q^6-1)(q^2-1)$

G_2' **Ree group**, $\mathrm{Re}(q)$, $q=3^{2n+1}$ $g=q^3(q^3+1)(q-1)$

F_4 $g=q^{24}(q^{12}-1)(q^8-1)(q^6-1)\times(q^2-1)$

F_4' $q=2^{2n+1}$ $g=q^{12}(q^6+1)(q^4-1)\times(q^3+1)(q-1)$

E_6 $dg=q^{36}(q^{12}-1)(q^9-1)(q^8-1)\times(q^6-1)(q^5-1)(q^2-1)$
$d=(3,q-1)$

${}^2E_6$ $dg=q^{36}(q^{12}-1)(q^9+1)(q^8-1)\times(q^6-1)(q^5+1)(q^2-1)$
$d=(3,q+1)$

E_7 $dg=q^{63}(q^{18}-1)(q^{14}-1)(q^{12}-1)\times(q^{10}-1)(q^8-1)(q^6-1)(q^2-1)$
$d=(2,q-1)$

E_8 $g=q^{120}(q^{30}-1)(q^{24}-1)(q^{20}-1)\times(q^{18}-1)(q^{14}-1)(q^{12}-1)\times(q^8-1)(q^2-1)$.

B_2': M. Suzuki, *Proc. Nat. Acad. Sci. US*, 46 (1960); G_2' and F_4': Ree, *Amer. J. Math.*, 83 (1961); G_2: Dickson, *Trans. Amer. Math. Soc.*, 2 (1901), *Math. Ann.*, 60 (1905); other Chevalley groups: Chevalley, *Tôhoku Math. J.*, (2) 7 (1955); twisted types: Steinberg, *Pacific J. Math.*, 9 (1959), J. Tits, *Séminaire Bourbaki* (1958), *Publ. Math. Inst. HES* (1959), D. Hertzig, *Amer. J. Math.*, 83 (1961), *Proc. Amer. Math. Soc.*, 12 (1961).

Nonsimple cases: $L_2(2)$, $L_2(3)$, $U_3(2)$, and $Sz(2)$ are solvable groups of orders 6, 12, 72, and 20, respectively. The groups $O_5(2)$, $G_2(2)$, $G_2'(3)$, and $F_4'(2)$ contain normal subgroups of indices 2, 2, 3, and 2, respectively. These normal subgroups are simple and identified as $L_2(9)$, $U_3(3)$, $L_2(8)$ in the first three cases. The normal subgroup of $F_4'(2)$ is not in the list of simple groups of Lie type and is quite exceptional (**Tits's simple group**, *Ann. Math.*, (2) 80 (1964)).

Isomorphisms between various simple groups: $L_2(q)=U_2(q)=S_2(q)=O_3(q)$; $O_5(q)=S_4(q)$; $O_4(1,q)=L_2(q)\times L_2(q)$; $O_4(-1,q)=L_2(q^2)$; $O_6(1,q)=L_4(q)$; $O_6(-1,q)=U_4(q)$; $O_{2n+1}(q)=S_{2n}(q)$ if q is a power of 2; $L_2(2)=S_3$; $L_2(3)=A_4$; $L_2(4)=L_2(5)=A_5$; $L_2(7)=L_3(2)$; $L_2(9)=A_6$; $L_4(2)=A_8$; $U_4(2)=S_4(3)$. If q is odd and $2n\geqslant 6$, then $S_{2n}(q)$ and $O_{2n+1}(q)$ have the same order but are not isomorphic. $L_3(4)$ and $L_4(2)$ have the same order but are not isomorphic. There is no other isomorphism or coincidence of orders among the known simple groups (Artin, *Comm. Pure Appl. Math.*, 8 (1955)).

The groups of the automorphisms of simple groups belonging to subclasses (1), (2), and (3) are known. For the simple groups of Lie type, see Steinberg, *Canad. J. Math.*, 10 (1960).

The following list contains all the simple groups known as of March 1972 which belong to class (4):

Five Mathieu groups (M_{11}, M_{12}, M_{22}, M_{23}, M_{24}), whose orders are

$7920=2^4\cdot 3^2\cdot 5\cdot 11$, $95{,}040=2^6\cdot 3^3\cdot 5\cdot 11$,
$443{,}520=2^7\cdot 3^2\cdot 5\cdot 7\cdot 11$,
$10{,}200{,}960=2^7\cdot 3^2\cdot 5\cdot 7\cdot 11\cdot 23$,
$244{,}823{,}040=2^{10}\cdot 3^3\cdot 5\cdot 7\cdot 11\cdot 23$.

Janko group (J. Z. Janko, *J. Algebra*, 3

(1966)). A subgroup of the Chevalley group $G_2(11)$; $g=175{,}560=2^3\cdot 3\cdot 5\cdot 7\cdot 11\cdot 19$.

The following groups have been discovered since August 1967. Each group is identified by the symbol $(x)_i$, indicating that it is the ith group discovered in the year $19x$. The Janko group is $(64)_1$. The list continues with the name or names of discoverers, the order of the group, and a brief description.

$(67)_1$: M. Hall, $g=604{,}800=2^7\cdot 3^3\cdot 5^2\cdot 7$, a transitive extension of $U_3(3)$ of degree 100.

$(67)_2$: D. G. Higman and Sims, $g=44{,}352{,}000=2^9\cdot 3^2\cdot 5^3\cdot 7\cdot 11$, a transitive extension of M_{22} of degree 100. It is a normal subgroup of index 2 in the group of automorphisms of a certain graph with 100 vertices.

$(67)_3$: Suzuki, $g=448{,}345{,}497{,}600=2^{13}\cdot 3^7\cdot 5^2\cdot 7\cdot 11\cdot 13$, a transitive extension of $G_2(4)$; defined from the automorphism group of a graph of 1782 vertices.

$(67)_4$: J. McLaughlin, $g=898{,}128{,}000=2^7\cdot 3^6\cdot 5^3\cdot 7\cdot 11$, a transitive extension of $U_4(3)$; defined from a graph of 275 vertices.

$(68)_1$: G. Higman and J. McKay, $g=50{,}232{,}960=2^7\cdot 3^5\cdot 5\cdot 17\cdot 19$, a transitive extension of the group which is obtained from $L_2(16)$ by adjoining the field automorphism of order 2. The existence was verified by using a computer.

$(68)_2$, $(68)_3$, $(68)_4$: J. H. Conway,

$$g=2^{21}\cdot 3^9\cdot 5^4\cdot 7^2\cdot 11\cdot 13\cdot 23$$
$$=4{,}157{,}776{,}806{,}543{,}360{,}000,$$
$$g=2^{18}\cdot 3^6\cdot 5^3\cdot 7\cdot 11\cdot 23,$$
$$g=2^{10}\cdot 3^7\cdot 5^3\cdot 7\cdot 11\cdot 23.$$

The big group is obtained from the automorphism group of a lattice in 24-dimensional space, and the two smaller ones are subgroups of it. The lattice was defined by J. Leech in connection with a problem of close packing of spheres in 24 dimensions (*Canad. J. Math.*, 19 (1967)).

$(68)_5$: B. Fischer, $g=2^{17}\cdot 3^9\cdot 5^2\cdot 7\cdot 11\cdot 13=70{,}321{,}751{,}654{,}400$, a transitive extension of $U_6(2)$ derived by means of a certain graph.

$(69)_1$: D. Held and others, $g=2^{10}\cdot 3^3\cdot 5^2\cdot 7^3\cdot 17=4{,}030{,}387{,}200$.

$(69)_2$: B. Fischer, $g=2^{18}\cdot 3^{13}\cdot 5^2\cdot 7\cdot 11\cdot 13\cdot 17\cdot 23=4{,}089{,}470{,}473{,}293{,}004{,}800$.

$(69)_3$: B. Fischer, $g=2^{22}\cdot 3^{16}\cdot 5^2\cdot 7^3\cdot 11\cdot 13\cdot 17\cdot 23\cdot 29=2{,}510{,}411{,}418{,}381{,}323{,}442{,}585{,}600$.

$(71)_1$: R. N. Lyons and C. C. Sims, $g=2^8\cdot 3^7\cdot 5^6\cdot 7\cdot 11\cdot 31\cdot 37\cdot 67$.

The existence of $(69)_1$ and $(71)_1$ was verified by using computers; $(69)_2$ and $(69)_3$ were derived by means of certain graphs. For a more detailed account of these new simple groups, see J. Tits, *Séminaire Bourbaki* (1970), No. 375, and the references [22, 23, 24.]

The possibility of the existence of the groups $(67)_1$ and $(68)_1$ was announced by Janko (1967). Like his group $(64)_1$, these groups are characterized by the structure of the centralizer of an element of order 2. The Conway group $(68)_2$ contains M_{24}, $(67)_2$, and $(67)_4$. The McLaughlin group $(67)_4$ contains M_{22}. The Hall-Janko group $(67)_1$ is a subgroup of $G_2(4)$.

Simple groups of order <1000 are A_5 ($g=60$), $L_2(7)$ (168), A_6 (360), $L_2(8)$ (504), and $L_2(11)$ (660). All simple groups of order $\leqslant$ 20,000 are known.

Among the known simple groups, the following multiply transitive permutation presentations are known: Alternating groups A_n (degree n), A_8 and A_7 (degree 15), A_6 (degree 10), A_5 (degree 6), Mathieu groups M_i (degree i), M_{11} (degree 12), $L_n(q)$ (degree $(q^n-1)/(q-1)$), $L_2(p)$ (degree p for $p=5,7,11$), $U_3(q)$ (degree $1+q^3$), $S_z(q)$ (degree $1+q^2$), $\mathrm{Re}(3^n)$ (degree $1+3^{3n}$), $S_{2n}(2)$ (degrees $2^{n-1}(2^n\pm 1)$), and the Higman-Sims group $(67)_2$ (degree 176). Among them, A_n (degree n, $n\geqslant 5$), M_i (degree i), M_{11} (degree 12), $L_2(2^m)$ (degree $1+2^m$), and $L_2(5)$ (degree 5) are triply transitive.

Let us state some theorems on non-Abelian finite simple groups. The order of non-Abelian simple groups is even (Feit and Thompson). The order of non-Abelian simple groups is divisible by at least three distinct prime numbers (Burnside). If a finite simple group G has the property that two distinct 2-Sylow subgroups have only the identity element in common, then $G=L_2(q)$, $Sz(q)$, or $U_3(q)$ for $q=2^n$ (Suzuki). If a 2-Sylow group of a simple group is dihedral, then $G=A_7$ or $L_2(q)$ for odd q (D. Gorenstein and J. H. Walter). Any simple group contains one of $L_3(3)$, $L_2(2^p)$, $L_2(3^p)$, $Sz(2^p)$ for some prime number p (Thompson).

There are several remarkable properties of known finite simple groups which have been conjectured to hold for arbitrary finite simple groups. One of the most famous is the **Schreier conjecture**, which asserts that the group of outer automorphisms of a simple group is solvable. This has been verified for all known cases except for some recently discovered simple groups. Another conjecture says that a finite simple group is generated by two elements. This has been verified also for almost all known groups. In many cases, there is a generating set of two elements, one of which has order 2. There is no counterexample known to disprove the universal validity of this property. Except for $Sz(q)$, the orders of known simple groups are divisible by 12.

J. Supplements (December 1975)

(1) Sporadic simple groups. Since March 1972, the following simple groups have been discovered:

$(72)_1$ A. Rudvalis: $g=2^{14}\cdot 3^3\cdot 5^3\cdot 7\cdot 13\cdot 29$,

$(73)_1$ M. O'Nan: $g=2^9\cdot 3^4\cdot 5\cdot 7^3\cdot 11\cdot 19\cdot 31$,

$(74)_1$ K. Harada: $g=2^{14}\cdot 3^6\cdot 5^6\cdot 7\cdot 11\cdot 19$,

$(74)_2$ J. G. Thompson: $g=2^{15}\cdot 3^{10}\cdot 5^3\cdot 7^2\cdot 13\cdot 19\cdot 31$.

The group $(72)_1$ is a transitive extension of Tits's simple group, i.e., a normal subgroup of $F_4'(2)$ of index 2. Concerning this group $(72)_1$, see the article of J. H. Conway and D. B. Wales, *J. Algebra*, 27 (1973). As for the groups $(73)_1$, $(74)_1$, and $(74)_2$, see *Proc. internat. symp. on theory of finite groups*, Sapporo, 1975, Japan Soc. for the Promotion of Sci., 1976. Three more possibilities have been announced. Their orders are

(a) $g=2^{46}\cdot 3^{20}\cdot 5^9\cdot 7^6\cdot 11^3\cdot 13^3\cdot 17\cdot 19\cdot 23\cdot 31\cdot 41\cdot 47\cdot 59\cdot 71$,

(b) $g=2^{41}\cdot 3^{13}\cdot 5^6\cdot 7^2\cdot 11\cdot 13\cdot 17\cdot 19\cdot 23\cdot 31\cdot 47$,

(c) $g=2^{21}\cdot 3^3\cdot 5\cdot 7\cdot 11^3\cdot 23\cdot 29\cdot 31\cdot 37\cdot 43$.

The existence of these groups has not yet been verified. The possibility of the first two groups (a) and (b) was suggested, together with $(74)_1$ and $(74)_2$, by B. Fischer. Z. Janko has announced the possibility of the third group (c).

(2) M. O'Nan generalized earlier results of H. Wielandt and H. Nagao (→ Section H), i.e., if the Schreier conjecture is true, there exists no 6-transitive group except S_n and A_n.

(3) We state some results on the problem of classifying finite simple groups with a given 2-Sylow subgroup. If a 2-Sylow subgroup of a simple group G is Abelian, $G=L_2(q)$ ($q\equiv 0,3,5 \bmod 8$) or G possesses an involution t whose centralizer is isomorphic to $\langle t\rangle\times L_2(q)$ ($q\equiv 3$ or $5 \bmod 8$, $q\geqslant 5$) (J. H. Walter, *Ann. of Math.*, (2) 89 (1969), H. Bender, *Math. Z.*, 111 (1970)). The latter group is called a group of Janko-Ree type (simply J-R type). Janko's group of order 175,600 and Ree's group $\mathrm{Re}(q)$ $=G_2'$ are the only known examples of groups of J-R type. If every 2-subgroup of a simple group G is generated by at most four elements, then G is isomorphic to one of the following groups: (i) Lie type of odd characteristic: $L_2(q)$, $L_3(q)$, $U_3(q)$, $G_2(q)$, $D_4^2(q)$, $L_4(q)$ ($q\not\equiv 1 \bmod 8$), $U_4(q)$ ($q\not\equiv 7 \bmod 8$), $L_5(q)$ ($q\equiv -1 \bmod 4$), $U_5(q)$ ($q\equiv 1 \bmod 4$); (ii) Lie type of even characteristic: $L_2(8)$, $L_2(16)$, $L_3(4)$, $U_3(4)$, $Sz(8)$; (iii) Alternating groups: A_7, A_8, A_9, A_{10}, A_{11}; (iv) Sporadic groups: Hall-Janko's group $(67)_1$, MacLaughlin's group $(67)_4$, Higman-MacKay's group $(68)_1$, Lyons-Sims's group $(71)_1$; and (v) Groups of J-R type (D. Gorenstein and K. Harada, *Mem. Amer. Math. Soc.*, no. 147 (1974)).

(4) Finally we mention a theorem of H. Bender (*J. Algebra*, 17 (1971)) which provides a fundamental tool in problems concerning the classification of finite simple groups. If a finite group has a proper subgroup H of even order with a property that $H^g\cap H$ is of odd order for any $g\in G-H$, then H is called a strongly embedded subgroup of G. Bender's theorem states that if a finite group G has a strongly embedded subgroup, then one of the following holds: (i) a 2-Sylow subgroup of G is cyclic or a generalized quaternion group; (ii) G possesses a normal series $G=G_0\supset G_1\supset G_2\supset 1$ such that G_0/G_1 and G_2 are of odd order and $G_1/G_2=L_2(2^n)$, $U_3(2^n)$, or $Sz(2^{2n-1})$ ($n\geqslant 2$). This theorem is a generalization of an earlier result of M. Suzuki (*Ann. of Math.*, (2) 79 (1964)), and the proof depends heavily on Suzuki's result.

References

[1] K. Shoda and K. Asano, Daisûgaku (Japanese; Algebra) I, Iwanami, 1952.
[2] Y. Akizuki and M. Suzuki, Kôtô daisûgaku (Japanese; Higher algebra) I, II, Iwanami, 1952, 1957.
[3] M. Osima, Gunron (Japanese; Theory of groups), Kyôritu, 1954.
[4] K. Asano and H. Nagao, Gunron (Japanese; Theory of groups), Iwanami, 1965.
[5] L. E. Dickson, Linear groups with an exposition of the Galois field theory, Teubner, 1901 (Dover, 1958).
[6] W. Burnside, Theory of groups of finite order, Cambridge Univ. Press, second edition, 1911.
[7] H. Zassenhaus, Lehrbuch der Gruppentheorie, Teubner, 1937; English translation, The theory of groups, Chelsea, 1958.
[8] A. Speiser, Die Theorie der Gruppen von endlicher Ordnung, Springer, third edition, 1937.
[9] W. Specht, Gruppentheorie, Springer, 1956.
[10] M. Suzuki, Structure of a group and the structure of its lattice of subgroups, Erg. Math., Springer, 1956.
[11] H. S. M. Coxeter and W. O. J. Moser, Generators and relations for discrete groups, Erg. Math., Springer, 1957.
[12] M. Hall, The theory of groups, Macmillan, 1959.
[13] C. W. Curtis and I. Reiner, Representation theory of finite groups and associative algebras, Interscience, 1962.
[14] W. R. Scott, Group theory, Prentice-Hall, 1964.

[15] M. Hall and J. K. Senior, The groups of order $2^n (n \leqslant 6)$, Macmillan, 1964.
[16] H. Wielandt, Finite permutation groups, Academic Press, 1964.
[17] J. Dieudonné, Sur les groupes classiques, Actualités Sci. Ind., Hermann, 1948.
[18] J. Dieudonné, La géométrie des groupes classiques, Erg. Math., Springer, 1955.
[19] D. Gorenstein, Finite groups, Harper & Row, 1968.
[20] D. Passman, Permutation groups, Benjamin, 1968.
[21] B. Huppert, Endliche Gruppen, Springer, 1967.
[22] R. Brauer and C. H. Sah (eds.), Theory of finite groups, a symposium, Benjamin, 1969.
[23] M. B. Powell and G. Higman, (eds.), Finite simple groups, Academic Press, 1971.
[24] W. Feit, The current situation in the theory of finite simple groups, Actes Congr. Intern. Math., 1970, Nice, Gauthier-Villars, vol. 1., p. 55–93.

161 (VII.9) Finsler Spaces

A. Definitions

Let $T(M)$ be the †tangent vector bundle of an n-dimensional †differentiable manifold M. An element of $T(M)$ is denoted by (x,y), where x is a point of M and y is a †tangent vector of M at x. Given a †local coordinate system $(x^1, \ldots, x^n)$ of M, we can obtain a local coordinate system of $T(M)$ by regarding $(x^1, \ldots, x^n, y^1, \ldots, y^n) = (x^i, y^j)$ as coordinates of the pair $(x,y) \in T(M)$, where $(x^1, \ldots, x^n)$ are coordinates of a point x of M and $y = \Sigma y^j \partial / \partial x^j$. A continuous real-valued function $L(x,y)$ defined on $T(M)$ is called a **Finsler metric** if the following conditions are satisfied: (i) $L(x,y)$ is differentiable at $y \neq 0$; (ii) $L(x, \lambda y) = |\lambda| L(x,y)$ for any element (x,y) of $T(M)$ and any real number λ; and (iii) if we put $g_{ij}(x,y) = (1/2)\partial^2 L(x,y)^2 / \partial y^i \partial y^j$, the symmetric matrix $(g_{ij}(x,y))$ is positive definite. A differentiable manifold with a Finsler metric is called a **Finsler space**. There exists a Finsler metric on a manifold M if and only if M is †paracompact. We call $F(x,y) = L(x,y)^2$ the **fundamental form** of the Finsler space. When $F(x,y)$ is a quadratic form of $(y^1, \ldots, y^n)$, $L(x,y)$ is a †Riemannian metric, and $F(x,y) = \Sigma_{i,j} g_{ij}(x) y^i y^j$. Therefore, a Finsler space is a Riemannian space if and only if g_{ij} does not depend on y. The matrix g_{ij} is also called a **fundamental tensor** of the Finsler space $(i,j = 1, \ldots, n)$.

Thus the notion of a Finsler metric is an extension of the notion of Riemannian metric. The study of differentiable manifolds utilizing such generalized metrics was considered by B. Riemann, but he stated that a Riemannian metric is more convenient for the purpose since "only nongeometrical results can be obtained" from using Finsler metrics [7]. P. Finsler initiated the systematic study of Finsler metrics and extended to a Finsler space many concepts and theorems valid in the classical theory of curves and surfaces [5].

B. The Finsler Metric

In a Finsler space, the arc length of a curve $x = x(t)$ $(a \leqslant t \leqslant b)$ is given by $\int_a^b L(x, dx/dt)\,dt$. Therefore, a †geodesic in a Finsler space is defined as a †stationary curve for the problem of †variation $\delta \int_a^b L(x, dx/dt)\,dt = 0$, and the differential equation of the geodesic is given by

$$\frac{d^2 x^i}{dt^2} + \sum_{j,k} \gamma_{jk}{}^i \left(x, \frac{dx}{dt} \right) \frac{dx^j}{dt} \frac{dx^k}{dt} = 0,$$

where $\gamma_{jk}{}^i(x,y)$ is the †Christoffel symbol of g_{ij}, i.e.,

$$\gamma_{jk}{}^i(x,y) = \frac{1}{2} \sum_a g^{ia} \left(\frac{\partial g_{ja}}{\partial x^k} + \frac{\partial g_{ak}}{\partial x^j} - \frac{\partial g_{jk}}{\partial x^a} \right),$$

where $(g^{ij}(x,y))$ is the inverse matrix of $(g_{ij}(x,y))$.

The distance between two points in a Finsler space is defined, as in a Riemannian space, as the infimum of the lengths of curves joining the two points. Many properties of Riemannian spaces as metric spaces can be extended to Finsler spaces. The topology defined by the Finsler metric coincides with the original topology of the manifold. A Finsler space M is said to be †complete if every Cauchy sequence of M as a metric space is convergent. The following three conditions are equivalent: (i) M is complete; (ii) each bounded closed subset of M is compact; (iii) each geodesic in M is infinitely extendable. In a complete Finsler space, any two points can be joined by the shortest geodesic.

A diffeomorphism φ of a Finsler space M preserves the distance between an arbitrary pair of points if and only if the transformation on $T(M)$ induced by φ preserves the Finsler metric $L(x,y)$. Such a transformation is called an †isometry of the Finsler space. In the †compact-open topology the set of all isometries of a Finsler space is a †Lie transformation group of dimension at most $n(n+1)/2$.

If a Finsler space admits the isometry group of dimension greater than $(n(n-1)+2)/2$, it is a Riemannian space of constant curvature.

C. The Theory of Connections

An important difference between a Finsler space and a Riemannian space relates to their properties with respect to the theory of †connections. In the case of a Riemannian space, the Christoffel symbols constructed from the fundamental tensor are exactly the coefficients of a connection, whereas in the case of a Finsler space, the Christoffel symbols $\gamma_{jk}{}^{i}(x,y)$ do not define a connection, for the fundamental tensor g_{ij} depends not only on the points of the space but also on the directions of tangent vectors at these points.

When we consider notions such as tensors, etc., in a Finsler space M, it is generally more convenient to take the whole tangent vector bundle $T(M)$ into consideration rather than restricting ourselves to the space M. For example, let P be the †tangent n-frame bundle over a Finsler space M and $Q=p^{-1}(P)$ be the †principal fiber bundle over $T(M)$ induced from P by the projection p of $T(M)$ onto M. We call the elements of fiber bundles associated with Q tensors. In this sense, the fundamental tensor g_{ij} in a Finsler space is the covariant tensor field of order 2. Therefore, it is natural to consider a connection in a Finsler space as a connection in the principal fiber bundle Q. The connection in a Finsler space defined by E. Cartan is exactly of this type [3]. Namely, he showed that by assigning to a connection in Q certain conditions related to the Finsler metric, we can determine uniquely a connection from the fundamental tensor so that the covariant differential of the fundamental tensor vanishes.

Cartan's introduction of the notion of connection produced a development in the theory of Finsler spaces that parallels the development in the theory of Riemannian spaces, and many important results have been obtained. O. Varga (1941) succeeded in obtaining a Cartan connection in a simpler way by using the notion of osculating Riemannian space. S. S. Chern (1943) studied general Euclidean connections that contain Cartan connections as a special case. Noticing that the tangent space of a Finsler space is a †normed linear space, H. Rund (1950) obtained many notions different from those of Cartan. However, as far as the theory of connections is concerned, the two theories do not seem to be essentially different. The theory of curvature in a Finsler space is more complicated than that in a Riemannian space because we have three curvature tensors in the Cartan connection. Using the fact that, in a local cross section of the tangent vector bundle of a Finsler space, a Riemannian metric can be introduced by the Finsler metric, L. Auslander (1955) [2] extended to Finsler spaces the results of J. L. Synge and S. B. Myers on the curvature and topology of Riemannian spaces. A. Lichnerowicz extended the †Gauss-Bonnet formula to Finsler spaces by considering an integral on the subbundle of the tangent vector bundle satisfying $L(x,y)=1$ [6]. M. H. Akbar-Zadeh studied †holonomy groups and transformation groups of Finsler spaces by using the theory of fiber bundles.

Connections of Finsler spaces have been investigated by many geometers, but most of them used methods considerably different from those of the modern theory of connections in principal fiber bundles. J. H. Taylor and Synge (1925) defined the covariant differential of a vector field along a curve. L. Berwald (1926) defined a connection from the point of view of the **general geometry of paths**. A curve on a manifold satisfying the differential equation

$$\frac{d^2x^i}{dt^2}+2G^i\left(x,\frac{dx}{dt}\right)=0$$

is called a **path**. The theory was originated by O. Veblen and T. Y. Thomas and generalized as above by J. Douglas. Characteristically, with respect to a Berwald connection, the covariant differential of the fundamental tensor does not vanish.

A Finsler space is a space endowed with a metric for line elements. As a dual concept, we have a **Cartan space**, which is endowed with a metric for **areal elements** [4]. A. Kawaguchi (1937) extended these notions further and studied a **space of line elements of higher order** (or **Kawaguchi space**).

References

[1] M. H. Akbar-Zadeh, Les espaces de Finsler et certaines de leurs généralisations, Ann. Sci. Ecole Norm. Sup., (3) 80 (1963), 1–79.

[2] L. Auslander, On curvature in Finsler geometry, Trans. Amer. Math. Soc., 79 (1955), 378–388.

[3] E. Cartan, Les espaces de Finsler, Actualités Sci. Ind., Hermann, 1934.

[4] E. Cartan, Les espaces métriques fondés sur la notion d'aire, Actualités Sci. Ind., Hermann, 1933.

[5] P. Finsler, Über Kurven und Flächen in allgemeinen Räumen, dissertation, Göttingen, 1918.
[6] A. Lichnérowicz, Quelques théorèmes de géométrie différentielle globale, Comment. Math. Helv., 22 (1949), 271–301.
[7] B. Riemann, Über die Hypothesen, welche der Geometrie zu Grunde liegen, Habilitationsschrift, 1854. (Gesammelte mathematische Werke, Teubner, 1876, p. 254–270; Dover, 1953.)
[8] H. C. Wang, On Finsler spaces with completely integrable equations of Killing, J. London Math. Soc., 22 (1947), 5–9.
[9] H. Rund, The differential geometry of Finsler spaces, Springer, 1959.

162 (IX.13) Fixed-Point Theorems

A. General Remarks

Given a mapping f of a space X into itself, a point x of X is called a **fixed point** of f if $f(x)=x$. When X is a topological space and f is a continuous mapping, we have various theorems concerning the fixed points of f.

B. X Is a Polyhedron

1. The Brouwer Fixed-Point Theorem. Let X be a †simplex and $f: X \rightarrow X$ a continuous mapping. Then f has a fixed point in X (*Math. Ann.*, 69 (1910), 71 (1912)).

2. X Is a Finite Polyhedron. Let $H_p(X)$ be the p-dimensional †homology group of a †finite polyhedron X (with integral coefficients), $T_p(X)$ the †torsion subgroup of $H_p(X)$, and $B_p(X)=H_p(X)/T_p(X)$. The continuous mapping $f: X \rightarrow X$ naturally induces a homomorphism f_* of the free **Z**-module $B_p(X)$ into itself. Let α_p be the †trace of f_* and $\Lambda_f = \sum_{p=0}^{n}(-1)^p \alpha_p$ $(n=\dim X)$. We call this integer Λ_f the **Lefschetz number** of f.

We have the **Lefschetz fixed-point theorem**: (i) Let f, g be continuous mappings sending X into itself. If f, g are †homotopic ($f \sim g$), then $\Lambda_f = \Lambda_g$. (ii) If $\Lambda_f \neq 0$, then f has at least one fixed point in X (*Trans. Amer. Math. Soc.*, 28 (1926)).

The condition $\Lambda_f \neq 0$ is, however, not necessary for the existence of a fixed point of f. The Brouwer fixed-point theorem is obtained immediately from (i) and (ii). In particular, if the mapping f is homotopic to the identity mapping 1_X, then α_p is the pth †Betti number of X, and Λ_f is equal to the †Euler characteristic $\chi(X)$ of X. Hence, in this case, if $\chi(X) \neq 0$, then f has a fixed point.

3. Lefschetz Number and Fixed-Point Indices. Suppose that $|K|$ is an n-dimensional homogeneous polyhedron (i.e., any simplex belonging to K and not a face of another simplex of K is of dimension n), and $f:|K| \rightarrow |K|$ is a continuous mapping. Then there exists a continuous mapping $g:|K| \rightarrow |K|$ homotopic to f and admitting only isolated fixed points $\{q_1, \ldots, q_r\}$, each of which is an inner point of an n-dimensional simplex of K. The †local degree λ_i of a mapping g at q_i is called the **fixed-point index** of g at q_i. Then $J_f = \sum_{i=1}^{r} \lambda_i$ does not depend on the choice of g and is equal to $(-1)^n \Lambda_f$.

4. Singularities of a Continuous Vector Field. Let X be an n-dimensional †differentiable manifold and F a †continuous vector field on X that assigns a tangent vector x_p to each point p of X. A point p is called a **singular point** of F if x_p is the zero vector. The vector field F induces in a natural manner a continuous mapping $f: X \rightarrow X$ that is homotopic to the identity mapping 1_X. Then a fixed point of f is a singular point of F, and vice versa. When such a singular point p is isolated, there exists a †neighborhood N of p that is homeomorphic to an n-dimensional open ball such that x_q is nonzero for every point q in N except for $q=p$. Let N' be the boundary of N. Then the †order of the point p with respect to N' is independent of the choice of N, and it is called the **index of the singular point** p. This index is equal to the fixed-point index λ_p of f at p. Hence, when X is compact, the sum of indices of (isolated) singular points of F is equal to $(-1)^n \chi(X)$. In particular, a compact manifold X admits a continuous vector field with no singular point if and only if $\chi(X)=0$ (**Hopf's theorem**, *Math. Ann.*, 96 (1927)).

5. Poincaré-Birkhoff Fixed-Point Theorem. In certain cases, a continuous mapping $f: X \rightarrow X$ of a finite polyhedron X into itself has fixed points even if $\Lambda_f = 0$. For example, let X be the annular space $\{(r,\theta) \mid \alpha \leqslant r \leqslant \beta\}$ ((r,θ) are the polar coordinates of points in a Euclidean plane) and $f: X \rightarrow X$ be a homeomorphism satisfying the following conditions: (i) there exist continuous functions $g(\theta)$, $h(\theta)$ such that $g(\theta) < \theta$, $h(\theta) > \theta$, $f(\alpha,\theta) = (\alpha, g(\theta))$, $f(\beta,\theta) = (\beta, h(\theta))$; (ii) there exists a continuous positive function $\rho(r,\theta)$ defined for $\alpha < r < \beta$ such that

$$\iint_X \rho(r,\theta)\, dr\, d\theta = \iint_X \rho(f(r,\theta))\, dr\, d\theta.$$

Then f has at least two fixed points. This theorem was conjectured in 1912 by Poincaré, who hoped to apply it to solve the †restricted three-body problem. The theorem was later proved by G. D. Birkhoff (*Trans. Amer. Math. Soc.*, 14 (1913)) and is called the **Poincaré-Birkhoff fixed-point theorem** or **the last theorem of Poincaré**.

6. Recently, M. F. Atiyah and R. Bott [5] extended the Lefschetz fixed-point theorem to include the case of elliptic complexes (→ 236 *K*-Theory), concerning compact differentiable manifolds and transversal differentiable mappings. This extension allows the application of the fixed-point theorem to the problems of various fields of study, such as the theory described in Section C.

C. Fixed-Point Theorems for Infinite-Dimensional Spaces

Birkhoff and O. D. Kellogg generalized Brouwer's fixed-point theorem to the case of function spaces (*Trans. Amer. Math. Soc.*, 23 (1922)). Their result was utilized to show the existence of solutions of certain differential equations and constituted a new method in the theory of functional equations.

J. P. Schauder obtained the following theorem: Let A be a closed convex subset of a Banach space, and assume that there exists a continuous mapping T sending A to a †countably compact subset $T(A)$ of A. Then T has fixed points (*Studia Math.*, 2 (1930)). This theorem is called the **Schauder fixed-point theorem**.

A. Tihonov generalized Brouwer's result and obtained the following **Tihonov fixed-point theorem** (*Math. Ann.*, 111 (1935)): Let R be a locally convex †topological linear space, A a compact convex subset of R, and T a continuous mapping sending A into itself. Then T has fixed points.

This theorem may be applied to the case where R is the space of continuous mappings sending an m-dimensional Euclidean space E^m into a k-dimensional Euclidean space E^k to show the existence of solutions of certain differential equations. For example, when $m=k=1$, consider the differential equation

$$dy/dx=f(x,y), \qquad y(x_0)=y_0.$$

We set $T(y)=y_0+\int_{x_0}^{x}f(t,y(t))\,dt$ to determine a continuous mapping $T:R\to R$. Then the fixed points of T are the solutions of the differential equation. Now we can apply the theorem of Tihonov to show the existence of solutions.

On the other hand, when we are given problems of functional analysis, Schauder's fixed-point theorem is usually more convenient to apply than Tihonov's theorem.

The following theorem, written in terms of functional analysis, is useful for applications: Let D be a subset of an n-dimensional Euclidean space, F the family of continuous functions defined on D, and $T:F\to F$ a mapping. Suppose that the following three conditions are satisfied: (i) For $f_1, f_2\in F$, $0<\lambda<1$ implies $\lambda f_1+(1-\lambda)f_2\in F$. (ii) If a series $\{f_k\}$ of functions in F converges uniformly in the wider sense to a function f, then $f\in F$; and furthermore, the series $\{Tf_k\}$ converges uniformly in the wider sense to Tf. (iii) The family $T(F)$ is a †normal family of functions on D. Then there exists a function $f\in F$ such that $Tf=f$.

Let R be a topological linear space and T a mapping assigning an arbitrary point x in R to a closed convex subset $T(x)$ of R. A point x of R is called a **fixed point** of T if $x\in T(x)$. The mapping T is called **semicontinuous** if the condition $x_n\to a, y_n\to b$ ($y_n\in T(x_n)$) implies that $b\in T(a)$. In particular, if K is a bounded closed convex subset of a finite-dimensional Euclidean space R and T a semicontinuous mapping sending points of K into convex subsets of K, then T admits fixed points (**Kakutani fixed-point theorem** *Duke Math. J.*, 8 (1941)). This result was further generalized to the case of locally convex topological linear spaces by Ky Fan (*Proc. Nat. Acad. Sci. US*, 38 (1952)).

References

[1] P. S. Aleksandrov and H. Hopf, Topologie I, Springer, 1935 (Chelsea, 1965).
[2] M. Hukuhara, Zyôbibun hôteisiki (Japanese; Ordinary differential equations), Iwanami, 1950.
[3] M. Nagumo, Syazôdo to sonzai teiri (Japanese; The degree of mappings and existence theorems), Kawade, 1948.
[4] M. Hukuhara and T. Sato, Bibunhôteisiki ron (Japanese; Theory of differential equations), Kyôritu, 1956–1957.
[5] M. F. Atiyah and R. Bott, A Lefschetz fixed point formula for elliptic complexes I, II, Ann. of Math., (2) 86 (1967), 374–407; (2) 88 (1968), 451–491.

163 (VI.2) Foundations of Geometry

A. Introduction

Geometry deals with figures. It depends, therefore, on our spatial intuition, but our intuition lacks objectivity. The Greeks originated the idea of developing geometry logically, based on explicitly formulated axioms, without resorting to intuition. From this intention resulted Euclid's *Elements*, which was long considered the perfect model of a logical system. As time passed, however, mathematicians came to notice its imperfections. Since the 19th century especially, with the awakening of a more rigorous critical spirit in science and philosophy, more systematic criticism of the *Elements* began to appear. Non-Euclidean geometry was formulated after reexamination of Euclid's axiom of parallels; but it was also discovered that even as a foundation of Euclidean geometry, Euclid's system of axioms was far from perfect. Various systems of axioms for Euclidean geometry were proposed by mathematicians in the latter half of the 19th century, among them one by Hilbert [1], which became the basis of far-reaching studies.

B. Hilbert's System of Axioms

Hilbert took as undefined elements **points** (denoted by A, B, C,...), **straight lines** (or simply **lines**, denoted by a, b, c,...), and **planes** (denoted by α, β, γ,...). Between these objects there exist incidence relations (expressed in phrases such as "A lies on a," "a passes through A," etc.); order relations ("B is between A and C"); congruence relations; and parallel relations. The relations are subject to the following five groups of axioms:

(I) **Incidence axioms**: (1) For two points A, B, there exists a line a through A and B. (2) If $A \neq B$, the line a through A, B is uniquely determined. We write $a = A \cup B$ and call a the **join** of A, B. (3) Every line contains at least two different points. There exist at least three points that do not lie on a line. (4) If A, B, C are points not on a line, there exists a plane α through A, B, C. (We also say that A, B, C lie on α.) For every plane α, there exists at least one point A on α. (5) If A, B, C are points not on a line, the plane α through A, B, C is uniquely determined. We write $\alpha = A \cup B \cup C$ and call α the **join** of A, B, C. (6) If A, B are two different points on a line a and if A, B lie on a plane α, then every point on a lies on α. (We say that a lies on α or α passes through a.) (7) If a point A lies on two planes α, β, there exists at least one other point B on α and β. (8) There exist at least four points not lying on a plane.

(II) **Ordering axioms**: (1) If B is between A and C, then A, B, C are three different points lying on a line; also, B is between C and A. (2) If A, C are two different points, then there exists a point B such that C is between A and B. (3) If B is between A and C, then A is not between B and C.

We define a **segment** as a set of two different points A, B, denoted as AB or BA, and we call A and B **ends** of this segment. The set of points between A, B is called the **interior** of AB, and the set of points of $A \cup B$ that are neither ends nor interior points of AB is called the **exterior** of AB.

(4) Let A, B, C be three points not lying on a line. If a line a on the plane $A \cup B \cup C$ does not pass through A, B, or C, but passes through a point of the interior of AB, then it also passes through a point of the interior of BC or CA (**Pasch's axiom**).

The following propositions are proved from the above axioms. Given n points A_1, A_2, $\ldots, A_n$ on a line ($n > 2$), we can rearrange them, if necessary, so that the point A_j is between A_i and A_k whenever we have $1 \leqslant i < j < k \leqslant n$. There are exactly two ways of arranging the points in this manner (**theorem of linear ordering**). Let O be a point on a line a, and let A, B be two points on the line different from O. Write $A \sim B$ when $A = B$ or O is not between A and B; write $A \nsim B$ otherwise. Then $\sim$ is an equivalence relation between points on the line different from O; from $A \nsim B$, $A \nsim C$ it follows that $B \sim C$. We say that A, B are on the same **side** or on different sides of O on a according to whether $A \sim B$ or $A \nsim B$. Two subsets a' and a'' of a defined by $a' = \{A' | A \sim A'\}$, $a'' = \{A'' | A \nsim A''\}$ are called **half-lines** or **rays** on a with O as the extremity (or starting from O). Denoting by a, for simplicity, the set of points on a, we have $a = a' \cup \{O\} \cup a''$ (disjoint union).

Using axiom II.4, we can also prove the following: Let a be a line on α, and let A, B be two points on α not lying on a. If $A = B$ or if the interior of the segment AB has no point in common with a, we say that A, B are on the same **side** of a on α, and write $A \sim B$. Otherwise, we say that A, B are on different sides of a on α and write $A \nsim B$. Then $\sim$ is an equivalence relation between points on α not lying on a, and from $A \nsim B$, $A \nsim C$ follows $B \sim C$. The subsets $\alpha' = \{A' | A \sim A'\}$, $\alpha'' = \{A'' | A \nsim A''\}$ of α are called **half-planes** on α **bounded** by a. Again, denoting by α, a the set of points on α and on a, respectively, we obtain $\alpha = \alpha' \cup a \cup \alpha''$ (disjoint union).

(III) **Congruence axioms**: Two segments AB, $A'B'$ can be in a relation of **congruence**, expressed symbolically as $AB \equiv A'B'$. (Segments AB and $A'B'$ are then said to be **congruent**. Since the segment AB is defined as the set $\{A, B\}$, the four relations $AB \equiv A'B'$, $BA \equiv A'B'$, $AB \equiv B'A'$, $BA \equiv B'A'$ are equivalent.) This relation is subject to the following three axioms: (1) Let A, B be two different points on a line a, and A_1 a point on a line a_1 (a_1 may or may not be equal to a). Let a_1' be a ray on a_1 starting from A_1. Then there exists a unique point B_1 on a_1' such that $AB \equiv A_1B_1$. (2) From $A_1B_1 \equiv AB$ and $A_2B_2 \equiv AB$ follows $A_1B_1 \equiv A_2B_2$. (Hence it follows that $\equiv$ is an equivalence relation between segments.) (3) Let A, B, C be three points such that B is between A and C, and let A_1, B_1, C_1 be three points such that B_1 is between A_1 and C_1. Then from $AB \equiv A_1B_1$, $BC \equiv B_1C_1$ follows $AC \equiv A_1C_1$.

Now let h, k be two different lines in a plane α and through a point O, and let h', k' be the rays on h, k starting from O. The set of two such rays h', k' is called an **angle** in α, denoted by $\angle(h', k')$ or $\angle(k', h')$. This angle is also denoted by $\angle AOB$, where A, B are points of h', k', respectively. The rays h', k' are called the **sides** and the point O is called the **vertex** of this angle. Then h' is a subset of a half-plane on α bounded by k, and k' is a subset of a half-plane on α bounded by h. The intersection of these two half-planes is called the **interior** of this angle, and the subset of $\alpha - O$ consisting of points belonging to neither the inside nor the sides of the angle is called the **exterior** of the angle. Between two angles $\angle(h', k')$, $\angle(h_1', k_1')$ there may exist the relation of congruence, again expressed by the symbol $\equiv$, as in the case of segments, and subject to the following two axioms: (4) Let $\angle(h', k')$ be an angle on a plane α and h_1 be a line on α_1 (α_1 may or may not be equal to α). Let O_1 be a point on h_1, h_1' a ray on α_1 starting from O_1, and α_1' a half-plane on α_1 bounded by h_1. Then there exists a unique ray k_1' starting from O_1 and lying in α_1' such that $\angle(h', k') \equiv \angle(h_1', k_1')$. Moreover, $\angle(h', k') \equiv \angle(h', k')$ always holds. (Hence it follows that $\equiv$ is an equivalence relation between angles.) (5) Let both A, B, C and A_1, B_1, C_1 be triples of points not lying on a line. Then from $AB \equiv A_1B_1$, $AC \equiv A_1C_1$, and $\angle BAC \equiv \angle B_1A_1C_1$, it follows that $\angle ABC \equiv \angle A_1B_1C_1$.

(IV) **Axiom of parallels**: Suppose that a, b are two different lines. Then it follows from axiom I.2 that if a and b share a point P, such a point is the unique point lying on both a and b. In this case we say that a, b **intersect** at P and write $a \cap b = P$. On the other hand, if a and b have no point in common and if a, b are on the same plane, we say that a, b are **parallel** and write $a // b$. If A, a are on a plane α and A is not on a, we can prove (utilizing the axioms I, II, and III) that there exists a line b passing through A in α such that $a // b$. The axiom of parallels postulates the uniqueness of such a b.

(V) **Axioms of continuity**: (1) Let AB, CD be two segments. Then there exist a finite number of points $A_1, A_2, \ldots, A_n$ on $A \cup B$ such that $CD \equiv AA_1 \equiv A_1A_2 \equiv \ldots \equiv A_{n-1}A_n$ and B is between A and A_n (**Archimedes' axiom**). (2) The set of points on a line a (again denoted for simplicity by a) is "maximal" in the following sense: It should satisfy axioms I.1–I.3, II.1–II.3, III.1, and V.1. If $\bar{a}$ is a set of points satisfying these axioms such that $\bar{a} \supset a$, then $\bar{a}$ should be $= a$ (**axiom of linear completeness**). Hence follows the **theorem of completeness**: the set of points, lines, and planes is maximal in the sense that it is not possible to add further points, lines, or planes to this set, so they still satisfy axioms I–IV and V.1.

C. Consistency

In formulating the above axioms and proving their consistency, Hilbert assumed the consistency of the theory of real numbers (→ 164 Foundations of Mathematics). To prove consistency, Hilbert constructed a model for the above axioms using the method of analytic geometry. He defined points as triples of real numbers (x_1, x_2, x_3), lines and planes as sets of points satisfying suitable systems of linear equations, and relations of ordering, congruence, and parallelism in the usual way. It is easy to verify that such a system satisfies all the axioms I–V. Thus the consistency of these axioms is reduced to the consistency of the theory of real numbers (→ 37 Axiom Systems).

A model for I–IV and V.1 can be obtained in the countable field $\mathbf{R}_0$ of all real †algebraic numbers instead of $\mathbf{R}$. Then $\mathbf{R}_0$ can be further restricted to its subfield $\mathbf{P}_0$ defined as follows: Let F be an arbitrary field. An †extension of F of the form $F(\sqrt{1+\lambda^2})$ with $\lambda \in F$ is called a **Pythagorean extension** of F, and F is said to be a **Pythagorean field** if any Pythagorean extension of F coincides with F (e.g. $\mathbf{R}_0$ and $\mathbf{R}$ are Pythagorean). It is easily verified that I–IV are satisfied in the "analytic geometry over any Pythagorean field." On the other hand, we can construct a minimal Pythagorean field containing a given field (the **Pythagorean closure** of the field) in the same way we construct the algebraic closure of a field. The field $\mathbf{P}_0$ is defined as the Pythagorean closure of the field $\mathbf{Q}$ of rational numbers.

D. Independence of Axioms

In Hilbert's system, the axioms I and II are used to formulate further axioms. On the other hand, it can be shown that each of the groups III, IV, and V is independent from other axioms.

The independence of IV is shown by the consistency of non-Euclidean geometry (→ 283 Non-Euclidean Geometry). The following model shows the independence of III.5: In the analytic model for I–V, we replace the definition of distance between two points (x_1,x_2,x_3), (y_1,y_2,y_3) by

$$\left((x_1-y_1+x_2-y_2)^2+(x_2-y_2)^2+(x_3-y_3)^2\right)^{1/2}.$$

Then III.5 does not hold, while all other axioms remain satisfied. The independence of V.2 is shown by the geometry over $\mathbf{R}_0$ or $\mathbf{P}_0$. The independence of V.1 follows from the existence of the non-Archimedean Pythagorean field: the Pythagorean closure of any non-Archimedean field (e.g., the field of rational functions of one variable over **Q** with a †non-Archimedean valuation) is such a field. A geometry in which V.1 does not hold is called a **non-Archimedean geometry**.

E. Categoricity of the System of Axioms and Relations between Axioms

The categoricity of the system of axioms I–V can be shown by introducing coordinates in the geometry with these axioms and representing it as †Euclidean geometry of three dimensions. Axiom group V is essential for the introduction of coordinates over **R**. Moreover, we have the following results:

(i) The geometry with the axioms I–IV can be represented as "Euclidean geometry" of three dimensions over a Pythagorean field, and the geometry with axioms I.1–I.3, II–IV can be represented as "Euclidean geometry" of two dimensions over a Pythagorean field.

(ii) The geometry with II, and a stronger axiom of parallels IV* (given a line a and a point A outside a, there exists one and only one line a' passing through A that is parallel to a), can be represented as an †affine geometry over a field K that is not necessarily commutative.

(iii) The field K is commutative if and only if the following holds: Suppose that in Fig. 1 $A'\cup B // A\cup B'$, $B'\cup C // B\cup C'$. Then it follows that $A'\cup C // A\cup C'$ (**Pascal's theorem**).

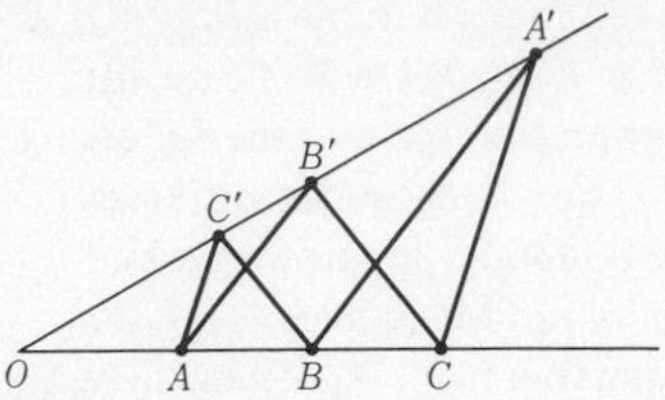

Fig. 1

(iv) The "two-dimensional geometry" with I.1–I.3, II, and IV* can be embedded in the "three-dimensional geometry" with axioms I, II, and IV* if and only if the following holds: Suppose that in Fig. 2 we have $A\cup B // A'\cup B'$, $B\cup C // B'\cup C'$. Then it follows that $C\cup A // C'\cup A'$ (**Desargues's theorem**).

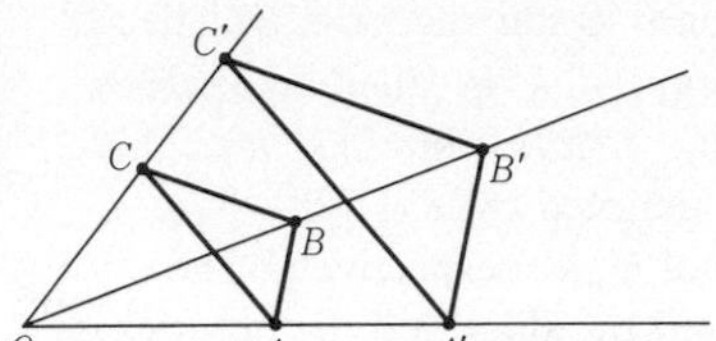

Fig. 2

(v) From I.1–I.3, II, IV*, and Pascal's theorem follows Desargues's theorem.

(vi) Desargues's theorem is independent of I.1–I.3, II, III.1–III.4, IV*, and V; that is, we can construct a **non-Desarguesian geometry** (a geometry in which Desargues's theorem does not hold) in which these axioms are satisfied.

Axioms I, II, and IV*, as well as the theorems of Pascal and Desargues, are propositions in affine geometry. Each has a corresponding proposition in †projective geometry, and the results concerning them can be transferred to the case of projective geometry (→ 340 Projective Geometry).

F. Polygons and Their Areas

Suppose that we are given a finite number of points A_i $(i=0,1,\ldots,r)$ in the geometry with axioms I and II. Then the set of segments (or, more precisely, the union of segments together with their interiors) A_iA_{i+1} $(i=0,1,\ldots,r-1)$ is called a **broken line** joining A_0 with A_r. In particular, if $A_0=A_r$, then this set is called a **polygon** with **vertices** A_i and **sides** A_iA_{i+1}. A polygon with r vertices is called an r-gon. (For $r=3,4,5,6$, r-gons are called **triangles, quadrangles, pentagons**, and **hexagons**, respectively.) A **plane polygon** is a polygon whose vertices all lie on a plane. A polygon is called **simple** if any three consecutive vertices do not lie on a line, and two sides A_iA_{i+1} and A_jA_{j+1} $(i \neq j)$ meet only when $j = i+1$ or $i = j+1$. In this article, we consider only simple plane polygons, and refer to them simply as polygons.

†Jordan's theorem implies that a polygon in the sense just defined divides the plane into two parts, its **interior** and its **exterior**. This special case of Jordan's theorem can be proved by I.1–I.3 and II only. A polygon P is divided into two polygons P_1, P_2 by a broken line joining two points on sides of P and lying in the interior of P (Fig. 3). In this case, we say that P is **decomposed** into P_1, P_2 and write $P = P_1 + P_2$. We may again decompose P_1, P_2 and thereby arrive at a decomposition of the form $P = P_1 + \ldots + P_k$. Axiom III is used to introduce the congruence relation $\equiv$ between polygons. Two polygons P, Q are called **decomposition-equal** if there exist decompositions $P = P_1 + \ldots + P_k$, $Q = Q_1 + \ldots + Q_k$ such that $P_1 \equiv Q_1, \ldots, P_k \equiv Q_k$. This is expressed by PzQ. We call P, Q **supplementation-equal** if there exist two polygons P', Q' such that $(P+P')z(Q+Q')$, $P'zQ'$. This will be expressed by PeQ. If we assume IV, we can use result (i) of Section E. Let K be the ground field of the geometry (K is Pythagorean, hence †ordered). The **area** of polygon P is defined as the positive element $m(P)$ of K assigned to P such that $m(P+Q) = m(P) + m(Q)$, and $m(P) = m(P')$ if $P \equiv P'$. From PzQ or PeQ, it follows that $m(P) = m(Q)$. Under these axioms, it is proved that $m(P) = m(Q)$ implies PeQ. If we also assume V.1, then $m(P) = m(Q)$ implies PzQ. Thus the theory of area of polygons can be constructed without assuming axiom V.2, though this result cannot be generalized to higher-dimensional cases. For the case of three dimensions, we can construct two solids that are not supplementation-equal [2].

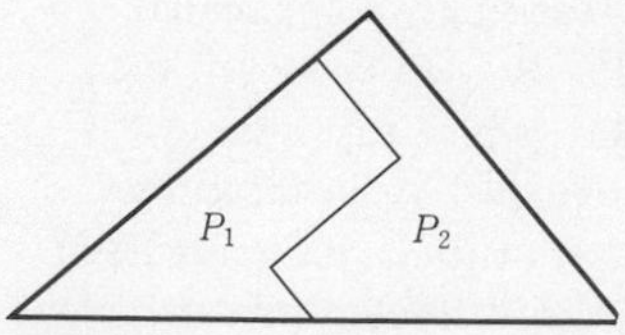

Fig. 3

G. Geometric Construction by Ruler and a Transferrer of Constant Lengths

The geometry with I–IV can be represented as 3-dimensional Euclidean geometry over a Pythagorean field. Conversely, all these axioms are valid in 3-dimensional Euclidean geometry over any Pythagorean field. Thus the minimal system of "quantities" whose existence is assured in geometry with these axioms is the field $\mathbf{P}_0$, the Pythagorean closure of $\mathbf{Q}$. Hilbert noticed that the existence of a geometric object under axioms I–IV can be expressed as its constructibility by **ruler** (i.e., an instrument to draw a straight line joining two points) and a **transferrer of constant lengths**. The latter, for a constant length x, is an instrument that permits finding the point X on the given ray AB such that $AX = x$. It is not possible to construct by ruler and transferrer all the points that can be constructed by means of ruler and compass (→ 183 Geometric Construction). However, it is possible to construct all the lengths λx, where λ is any element of $\mathbf{P}_0$. Hilbert conjectured that an element of $\mathbf{P}_0$ can be characterized as a totally positive algebraic number of degree 2^ν, $\nu \in \mathbf{N}$. This conjecture was proved by Artin [3].

H. Related Topics

While Hilbert's foundations are concerned with 3-dimensional Euclidean geometry, it is easy to generalize these results to the case of n-dimensional Euclidean geometry (→ 149 Euclidean Geometry). Also, for affine and projective geometries, there are well-organized systems of axioms (at least for the case of dimensions $\geqslant 3$). Hilbert [1, Appendix III] showed that plane †hyperbolic geometry can be constructed on a modified system of axioms, but for other non-Euclidean geometries (in particular, †elliptic geometries) there are no known systems of axioms as good as Hilbert's for the Euclidean case. On the other hand, Hilbert [1, Appendix IV] gives another method of constructing Euclidean geometry in characterizing the group of motions as the topological group with certain properties. G. Thomsen [4] rewrote Hilbert's system of axioms in group-theoretical language utilizing the fact that the group of motions is generated by symmetries with respect to points, lines, and planes. Finally, Hilbert's study of the foundations of geometry led him to research in the †foundations of mathematics.

References

[1] D. Hilbert, Grundlagen der Geometrie, Teubner, 1899, seventh edition, 1930.
[2] M. Dehn, Über den Rauminhalt, Math. Ann., 55 (1902), 465–478.
[3] E. Artin, Über die Zerlegung definiter Funktionen in Quadrate, Abh. Math. Sem.

Univ. Hamburg, 5 (1926), 100–115. (Collected papers, Addison-Wesley, 1965, p. 273–288.)
[4] G. Thomsen, Grundlagen der Elementargeometrie, Teubner, 1933.
[5] S. Iyanaga, Kikagaku kisoron (Japanese; Foundations of geometry), Iwanami Coll. of Math., 1935.

164 (I.1) Foundations of Mathematics

A. General Remarks

The notion of †set, introduced toward the end of the 19th century, has proved to be one of the most fundamental and useful ideas in mathematics. Nonetheless, it has given rise to well-known †paradoxes. Based on this notion, R. Dedekind developed the theories of natural numbers [4] and real numbers [5], defining the latter as "cuts" of the set of rational numbers. Thus, set theory served as a unifying principle of mathematics.

It has been noted, however, that some of the most commonly utilized arguments in set theory, which are at the same time the most useful in mathematics and belong almost to the basic framework of formal logic itself, resemble very much those which give rise to paradoxes. This fact has caused many critical mathematicians to question the very nature of mathematical reasoning. Thus, the new field, **foundations of mathematics**, came into being at the beginning of this century. This field was divided at its inception into different doctrines according to the views of its initiators: **logicism** by B. Russell, **intuitionism** by L. E. J. Brouwer, and **formalism** by D. Hilbert. In set theory, which was the origin of this controversy, it was pointed out that the "definition" of set as given by G. Cantor was too naive, and axiomatic treatments of this theory were proposed (→ 35 Axiomatic Set Theory).

B. Logicism

Russell asserted that mathematics is a branch of logic and that paradoxes come from neglecting the "types" of concepts. According to his opinion, mathematics deals formally with structures independently of their concrete meanings. Science of this character has been called logic from antiquity. According to him, logic is the youth of mathematics, and mathematics is the manhood of logic. To construct mathematics from this standpoint, asserted Russell, ordinary language is lengthy and inaccurate, and some proper system of symbols should be used instead. Thus, he tried to reconstruct mathematics using †symbolic logic.

Attempts to reorganize mathematics using logical symbols had formerly been made by G. Leibniz, who wrote *Dissertatio de arte combinatoria* in 1666, as well as by A. de Morgan, G. Boole, C. S. Peirce, E. Schröder, G. Frege, G. Peano, and others. Symbols used by the latter two authors resemble those of today. Russell studied these works and published his own theory in a monumental joint work with A. N. Whitehead: *Principia mathematica* (3 vols., 1st ed. 1910–1913, 2nd ed. 1925–1927), in which the theories of natural numbers and real numbers as well as analytic geometry are developed from the fundamental laws of logic.

If this work had been completely successful, it could have eliminated any possibility of the intrusion of paradoxes into mathematics. However, the authors were obliged to postulate an "unsatisfactory" axiom in order to construct mathematics. They introduced the notion of †type as follows: an object M defined as the set of all objects of a certain type belongs to a higher type than the types of the elements of M. This serves to eliminate certain paradoxes but brings about inconveniences such as the following. Suppose that we are trying to construct the theory of real numbers from that of rational numbers. Each real number can then be considered a †predicate about rational numbers. If this predicate contains only †quantifiers relating to variables running over all rational numbers, then the corresponding real number will be said to be **predicative**, otherwise **impredicative**. According to Russell, the latter should have a higher type than the former, which makes the theory of real numbers exceedingly complicated. To avoid this difficulty, Russell proposed the **axiom of reducibility**, which says that every predicate can be replaced by a predicative one. With this rather artificial axiom Russell himself expressed dissatisfaction. Russell also postulated the †axiom of infinity and the †axiom of choice, which are also problematic. After examining the philosophical background of the book, H. Weyl wrote about *Principia mathematica*, "Mathematics is no more based on logic than the utopia built by the logician." Nevertheless, logic as formulated in this book, as well as the theory of types as developed by F. P. Ramsay in the school of Russell and Whitehead, is still an important subject of mathematical logic.

C. Intuitionism

The intuitionist claims that mathematical objects or truths do not exist independently from mathematically thinking spirit or intuition, and that these objects or truths should be directly seized by mental or intuitional activity. The philosophical standpoints of such mathematicians as L. Kronecker and H. Poincaré in the 19th century or as E. Borel, H. Lebesgue, and N. N. Lusin at the turn of this century can be assimilated to intuitionism, but those of the latter three are often said to belong to **semi-intuitionism** or to **French empiricism**. Brouwer took a narrower standpoint, strongly antagonistic to Hilbert's formalism. Today the word "intuitionism" is generally interpreted in Brouwer's sense.

Brouwer sharply criticized the usual way of reasoning in mathematics and claimed that indiscriminate use of the **law of excluded middle** (or **tertium non datur**) $P \vee \neg P$ cannot be permitted. According to him, the proposition "Either there exists a natural number with a given property P, or else no such number exists" is to be regarded as proved only when an actual construction of a natural number with the property P is given or when the absurdity of the existence of such a natural number can be constructively proved. When neither of these two results can be shown, then one can say nothing about the truth of the above proposition. Thus, the usual method of proof, known as the method of **reductio ad absurdum**, i.e., of proving a proposition P by proving its double negation $\neg\neg P$, is not generally considered valid. It is a difficult but important problem of mathematical logic to determine which parts of usual mathematics can be reconstructed intuitionistically, though it does not seem easy to reconstruct any part of mathematics elegantly from this standpoint.

D. Formalism

To eliminate paradoxes, Hilbert tried to apply his axiomatic method. From Hilbert's standpoint, any part of mathematics is a deductive system based on its axioms. In the deductive development, however, "logic," including set theory and elementary number theory, is used. Paradoxes appear already in such logic. Hilbert's idea was to axiomatize such logic and to prove its consistency. Thus, one must first formalize the most elementary part of mathematics, including logic proper.

Hilbert proved the consistency of Euclidean geometry by assuming the consistency of the theory of real numbers. This is an example of a **relative consistency** proof, which reduces the consistency proof of one system to that of another. Such a proof can be meaningful only when the latter system can somehow be regarded as based on sounder ground than the former. To carry out the **consistency proof** of logic proper and set theory, one must reduce it to that of another system with sounder ground. For this purpose, Hilbert initiated **metamathematics** and the **finitary standpoint**.

The finitary standpoint recognizes as its foundation only those facts that can be expressed in a finite number of symbols and only those operations that can be actually executed in a finite number of steps. Essentially, it does not differ from the standpoint of intuitionism. The methods based on this standpoint are also called **constructive methods**.

Metamathematics is also called **proof theory**. Its subject of research is mathematical proof itself. Hilbert was the first to insist on its importance. The theory is indispensable for consistency proofs of mathematical systems, but it may be also used for other purposes. In fact, the same idea can be seen in the †duality principle of projective geometry, which dates from long before Hilbert's proclamation of formalism. This is not a theorem of projective geometry deduced from its axioms; rather, it is a proposition about the theorems in projective geometry, based on the type of axioms and proofs in this subject.

According to Hilbert's method, one must develop proof theory from the finitary standpoint with the aim of proving the consistency of axiomatized mathematics. For this purpose, one must formalize the mathematical theory in question by means of symbolic logic. A theory thus formalized is called a **formal system**.

E. Some Results of Formalist Theory

One of the most remarkable results hitherto obtained with Hilbert's method is the consistency proof of pure number theory by G. Gentzen [9]. This consistency proof covers the largest domain for which an explicit consistency proof has been obtained until today. However, the methods of formalist proof theory have proved to be most effective in studying the logical structure of mathematical theories and have led to various results on the consistency of formalized mathematical systems, on symbolic logic, and on axiomatic set theory. We give some examples.

1. Gödel's Incompleteness Theorem. K. Gödel [8] showed that if a system obtained by formalizing the theory of natural numbers is consistent, then this system contains a †logical formula A such that neither A nor its negation $\neg A$ can be proved within the system. He originally proved this under the assumption that the system is **ω-consistent**. This is a stronger condition for the system than simple consistency, but J. B. Rosser [15] succeeded in replacing this by the latter. This result shows the incompleteness not only of the usual theory of natural numbers but of any consistent theory (from the finitary standpoint) containing the theory of these numbers.

At the same time, Gödel also obtained the following important result: Let S be any consistent formal system containing the theory of natural numbers. Then it is impossible to prove the consistency of S by utilizing only arguments that can be formalized in S. This means that a consistency proof from the finitary standpoint of a formal system S inevitably necessitates some argument that cannot be formalized in S. This concerns, however, the definition of consistency.

2. Consistency Proofs for Pure Number Theory. G. Gentzen [9] called **pure number theory** the theory of natural numbers not depending on the free use of set theory (differing consequently from the usual theory of natural numbers based on †Peano axioms; → 290 Numbers) and proved its consistency. W. Ackermann [16] proved the consistency of a similar theory admitting the use of Hilbert's †ε-symbol. G. Takeuti [17] showed that Gentzen's result can be obtained as a corollary to his theorem extending †Gentzen's fundamental theorem on †predicate logic of the first order to a †theory of types of a certain kind.

According to the result of Gödel mentioned in (1), some reasoning outside pure number theory must be used to prove its consistency. In all consistency proofs of pure number theory mentioned above, †transfinite induction up to the first †ε-number ε_0 is used, but all the other reasoning used in these proofs can be presented in pure number theory. This shows that the legitimacy of transfinite induction up to ε_0 cannot be proved in this latter theory. A direct proof of this fact was given by Gentzen [15]. On the other hand, the legitimacy of transfinite induction up to an ordinal number $< \varepsilon_0$ can be proved within pure number theory.

Again, transfinite induction is not the only method by which to prove the consistency of pure number theory. Actually, Gödel [19] carried out the proof utilizing what he called computable functions of finite type on natural numbers.

By restricting pure number theory further, one obtains weaker theories of natural numbers whose consistency can be proved with finitary methods without recourse to such methods as transfinite induction up to ε_0. M. Presburger [20] proved the consistency of a theory in which only the addition of numbers is considered an operation. Ackermann [21], Von Neumann [22], J. Herbrand [23], and K. Ono [24] proved the consistency of theories in which some restrictions are placed on the use of the axiom of †mathematical induction.

On the other hand, K. Schütte [25] gave a consistency proof for number theory including what he called "infinite induction" from a stronger standpoint than Hilbert's finitary standpoint; he attempted to find a standpoint that makes such a proof possible.

3. The Consistency of Analysis. No definitive result has yet been obtained from the standpoint of formalism, though many attempts are being made, among which a recent one by C. Spector [26] should be mentioned.

4. Axiomatic Set Theory. There are different kinds of axiom systems (→ 35 Axiomatic Set Theory). To give a consistency proof for any of these systems is considered a very difficult problem today, but many interesting results are known concerning the relative consistency or independence of these axioms.

5. The Skolem-Löwenheim Theorem. The metamathematical Skolem-Löwenheim theorem states: Given a consistent system of axioms stated in the first-order predicate logic whose cardinality is at most countable, there always exists an †object domain consisting of countable objects satisfying all these axioms.

For example, axiomatic set theory is stated in predicate logic of the first order, and the cardinality of its axioms is countable. Thus there exists an object domain consisting of countable objects satisfying all these axioms, provided that they are consistent. Such a domain is called a **countable model** of axiomatic set theory. On the other hand, from the axioms of this theory one can prove that there exists a family of sets that is more than countable. This should also hold in a model of the theory, in which each object represents a set. This situation is known as the **Skolem paradox**.

This does not imply, however, the inconsistency of axiomatic set theory. In fact, the term "countable" is to be interpreted in its mathematical sense when one says "there exists a family of sets that is more than countable," while it should be interpreted in its metamathematical sense when one speaks of a countable model of the theory. It is the confusion of these two different interpretations that leads to the "paradox."

6. Skolem's Theorem on the Impossibility of Characterizing the System of Natural Numbers by Axioms. T. Skolem [29] proved that it is not possible to characterize the system of natural numbers by a countable system of axioms stated in the predicate logic of the first order. More precisely, given any consistent countable system of axioms satisfied by the system of natural numbers, there always exists another †linearly ordered system satisfying all these axioms and yet not isomorphic to the system of natural numbers as an ordered system.

Gödel's incompleteness theorem and the Skolem paradox, as well as this result, seem to indicate a certain limit to the effectiveness of the formalist method.

References

[1] S. Kuroda, Sûgaku kisoron (Japanese; Foundations of mathematics), Iwanami Coll. of Math., 1935.
[2] G. Takeuti, Sûgaku kisoron (Japanese; Foundations of mathematics), Kyôritu, 1957.
[3] S. C. Kleene, Introduction to metamathematics, Van Nostrand, 1952.
[4] R. Dedekind, Was sind und was sollen die Zahlen? F. Vieweg, 1888.
[5] R. Dedekind, Stetigkeit und irrationale Zahlen, F. Vieweg, 1872.
[6] B. Russell, Introduction to mathematical philosophy, Allen and Unwin, Macmillan, 1919.
[7] A. Heyting, Intuitionism, North-Holland, 1956.
[8] K. Gödel, Über formal unentscheidbare Sätze der Principia Mathematica und verwandter Systeme I, Monatsh. Math. Phys., 38 (1931), 173–198.
[9] G. Gentzen, Die Widerspruchsfreiheit der reinen Zahlentheorie, Math. Ann., 112 (1936), 493–565.
[10] D. Hilbert and P. Bernays, Grundlagen der Mathematik, Springer, I, second edition, 1968; II, 1970.
[11] A. Tarski, Logic, semantics, metamathematics, Clarendon Press, 1956.
[12] S. C. Kleene, Mathematical logic, John Wiley, 1967.
[13] J. R. Schoenfield, Mathematical logic, Addison-Wesley, 1967.
[14] O. Becker, Grundlagen der Mathematik in geschichtlicher Entwicklung, Karl Alber, 1954.
[15] J. B. Rosser, Extensions of some theorems of Gödel and Church, J. Symbolic Logic, 1 (1936), 87–91.
[16] W. Ackermann, Zur Widerspruchsfreiheit der Zahlentheorie, Math. Ann., 117 (1940), 162–194.
[17] G. Takeuti, On the fundamental conjecture of GLC I, J. Math. Soc. Japan, 7 (1955), 249–275.
[18] G. Gentzen, Beweisbarkeit und Unbeweisbarkeit von Anfangsfällen der transfiniten Induktion in der reinen Zahlentheorie, Math. Ann., 119 (1943), 140–161.
[19] K. Gödel, Über eine bisher noch nicht benützte Erweiterung des finiten Standpunktes, Dialectica, 12 (1958), 280–287.
[20] M. Presburger, Über die Vollständigkeit eines gewissen Systems der Arithmetik ganzer Zahlen, in welchem die Addition als einzige Operation hervortritt, C. R. du I Congrès des Math. des Pays Slaves, Warsaw (1929), 92–101, 395.
[21] W. Ackermann, Begründung des "tertium non datur" mittels der Hilbertschen Theorie der Widerspruchsfreiheit, Math. Ann., 93 (1924), 1–36.
[22] J. Von Neumann, Zur Hilbertschen Beweistheorie, Math. Z., 26 (1927), 1–46.
[23] J. Herbrand, Sur la non-contradiction de l'arithmétique, J. Reine Angew. Math., 166 (1931), 1–8.
[24] K. Ono, Logische Untersuchungen über die Grundlagen der Mathematik, J. Fac. Sci. Univ. Tokyo, 3 (1938), 329–389.
[25] K. Schütte, Beweistheoretische Erfassung der unendlichen Induktion in der Zahlentheorie, Math. Ann., 122 (1951), 369–389.
[26] C. Spector, Provably recursive functionals of analysis: a consistency proof of analysis by an extension of principles formulated in current intuitionistic mathematics, Recursive function theory, Amer. Math. Soc., Proc. Symposia in Pure Math., 5 (1962), 1–27.
[27] L. Löwenheim, Über Möglichkeiten in Relativkalkül, Math. Ann., 76 (1915), 447–470.
[28] T. Skolem, Einige Bemerkungen zur axiomatischen Begründung der Mengenlehre, 5 Congress der Skandinavischen Mathematiker, Helsingfors (1923), 217–232.
[29] T. Skolem, Über die Nichtcharakterisierbarkeit der Zahlenreihe mittels endlich oder abzählbar unendlich vieler Aussagen mit ausschliesslich Zahlenvariablen, Fund. Math., 23 (1934), 150–161.

165 (VI.26) Four-Color Problem

Are four colors necessary and sufficient to color a given geographical map on a sphere (or plane)? This is the famous **four-color problem**. To state the problem precisely, we must add the following two conditions: (i) Every country on a map is a †connected domain; a connected part of the sea is considered to be a country. (ii) Two countries sharing boundary lines must be colored differently. On the other hand, if two countries share only a finite number of points, then they may share the same color.

The conjecture was made by Francis Guthrie and communicated to De Morgan in 1852. A. Cayley called attention to the problem in 1878. Although the necessity is obvious, the sufficiency of the conjecture has not yet been proved. (For a report of an affirmative solution of this problem using a computer, see W. Haken and K. Appel, *Bull. Amer. Math. Soc.*, 82 (1976).)The problem for the case of maps where the number of countries $\leqslant 40$ has been solved (O. Ore and J. Stemple, 1968). It is easily proved that five colors are sufficient for coloring any map (P. J. Heawood, 1890).

On the other hand, seven colors are necessary and sufficient to color a map on a †torus. Generally, the problem of coloring maps on closed surfaces with †Euler characteristic $\chi<2$ is easier than coloring these on spheres ($\chi=2$). Let p_k be the least number of colors sufficient to color any map on a surface with Euler characteristic $\chi=k(<2)$. This number p_k is called the **chromatic number** of the surface. Then we have the Heawood inequality (1890)

$$p_k \leqslant \left[(7+\sqrt{49-24k}\,)/2\right], \qquad (1)$$

where $[\alpha]$ means the largest integer $a \leqslant \alpha$. J. W. T. Youngs and G. Ringel proved in 1968 that the equality holds in (1) except for Klein's bottle (the nonorientable surface with $\chi=0$). P. Franklin proved in 1934 that the chromatic number for Klein's bottle is 6.

References

[1] P. Franklin, The four-color problem, Scripta Math., 6 (1939), 149–156.
[2] D. Hilbert and S. Cohn-Vossen, Anschauliche Geometrie, Springer, 1932; English translation, Geometry and the imagination, Chelsea, 1952.
[3] O. Ore, The four-color problem, Academic Press, 1967.
[4] G. Ringel, Map color theorem, Springer, 1974.

166 (XX.23) Fourier, Jean Baptiste Joseph

Jean Baptiste Joseph Fourier (March 21, 1768–May 16, 1830) was born in Auxerre, France, the son of a tailor, and was orphaned at the age of eight. In 1790, he was appointed professor at the Ecole Polytechnique. Under Napoleon he went to Egypt as a soldier and worked with G. Monge as a cultural attaché for the French army. On his return to France, he was made governor of the department of Isère. With the downfall of Napoleon, he lost his position, however, he was later appointed to the French Academy of Science as a result of his research on the transmission of heat. In 1827, he was elected a member of the Académie Française.

His research on heat transmission was begun in 1800. In 1811, he presented a prize-winning solution to a problem put forth by the Academy of Science. He developed the †equation for heat transmission and obtained solutions under various †boundary conditions. Fourier also stated (without rigorous proof) that an arbitrary function could be represented by †trigonometric series (→ 167 Fourier Series), a statement that gave rise to many developments in analysis.

References

[1] J. B. J. Fourier, Oeuvres I, II, edited by G. Darboux, Gauthier-Villars, 1888–1890.
[2] T. Takagi, Kinsei sûgakusidan (Japanese; Topics from the history of mathematics of the 19th century), Kawade, 1943.

167 (X.22) Fourier Series

A. Introduction

The set of functions

$$1/\sqrt{2\pi},\ \cos x/\sqrt{\pi},\ \sin x/\sqrt{\pi},\ \ldots,$$
$$\cos kx/\sqrt{\pi},\ \sin kx/\sqrt{\pi},\ \ldots,$$

which is called the **trigonometric system**, is an †orthonormal system in $(-\pi,\pi)$ (→ 312 Orthogonal Functions). Let $f(x)$ be an element of $L_1(-\pi,\pi)$ (i.e., †Lebesgue integrable in $(-\pi,\pi)$). Throughout this article we assume that integrals are always †Lebesgue integrals. We put

$$a_k=\frac{1}{\pi}\int_{-\pi}^{\pi} f(t)\cos kt\,dt,$$

$$b_k=\frac{1}{\pi}\int_{-\pi}^{\pi} f(t)\sin kt\,dt, \quad k=0,1,\ldots \tag{1}$$

and call a_k, b_k the **Fourier coefficients** of f. The formal series

$$\frac{1}{2}a_0+\sum_{k=1}^{\infty}(a_k\cos kx+b_k\sin kx) \tag{2}$$

is called the **Fourier series** of f and is often denoted by $\mathfrak{S}(f)$. To indicate that a formal series $\mathfrak{S}(f)$, as above, is the Fourier series of a function f, we write

$$f(x)\sim\frac{1}{2}a_0+\sum_{k=1}^{\infty}(a_k\cos kx+b_k\sin kx).$$

The sign $\sim$ means that the numbers a_k, b_k are connected with f by the formula (1); it does not imply that the series is convergent, still less that it converges to f. Generally, **trigonometric series** are those of form (2), where the a_k, b_k are arbitrary real numbers. Since the trigonometric series have period 2π, we assume that the functions considered are extended for all real x by the condition of periodicity $f(x+2\pi)=f(x)$. To study the properties of the series $\mathfrak{S}(f)$ and the representation of f by $\mathfrak{S}(f)$ are major objects of the theory of Fourier series. Since $e^{ix}=\cos x+i\sin x$, if we set $2c_k=a_k-ib_k$, $c_{-k}=\bar{c}_k$ $(k=0,1,2,\ldots)$, we have

$$c_k=\frac{1}{2\pi}\int_{-\pi}^{\pi} f(t)e^{-ikt}\,dt, \qquad k=0,\pm1,\ldots.$$

Then $\mathfrak{S}(f)$ is represented by the **complex form** $\sum_{k=-\infty}^{\infty}c_ke^{ikx}$, and $\{e^{ikx}\}$ $(k=0,\pm1,\ldots)$ is an orthogonal system in $(-\pi,\pi)$. In this complex form, we take symmetric partial sums such as $\sum_{k=-n}^{n}c_ke^{ikx}$ $(n=1,2,\ldots)$.

Consider the power series $\frac{1}{2}a_0+\sum_{k=1}^{\infty}(a_k-ib_k)z^k$ on the unit circle $z=e^{ix}$ in the complex plane. Its real part is the trigonometric series (2), and the imaginary part (with vanishing constant term) is

$$\sum_{k=1}^{\infty}(a_k\sin kx-b_k\cos kx), \tag{3}$$

which is called the **conjugate series** (or **allied series**) of f and is denoted by $\tilde{\mathfrak{S}}(f)$. In complex form, the conjugate series is $-i\sum_{k=-\infty}^{\infty}(\operatorname{sgn}k)c_ke^{ikx}$.

If f and g belong to $L_1(-\pi,\pi)$ and

$$f(x)\sim\sum_{k=-\infty}^{\infty}c_ke^{ikx}, \quad g(x)\sim\sum_{k=-\infty}^{\infty}d_ke^{ikx},$$

then

$$\frac{1}{2\pi}\int_0^{2\pi}f(x-t)g(t)\,dt\sim\sum_{k=-\infty}^{\infty}c_kd_ke^{ikx}.$$

The function

$$f*g(x)=\frac{1}{2\pi}\int_0^{2\pi}f(x-t)g(t)\,dt$$

is called the **convolution** of f and g.

If f is †absolutely continuous, then the derivative $f'(x)$ satisfies

$$f'(x)\sim i\sum_{k=-\infty}^{\infty}kc_ke^{ikx}$$
$$=\sum_{k=1}^{\infty}k(b_k\cos kx-a_k\sin kx).$$

If $F(x)$ is an indefinite integral of f, then

$$F(x)-c_0x\sim C+\sum_{k=-\infty}^{\infty}{}'\frac{c_k}{ik}e^{ikx}$$
$$=C+\sum_{k=1}^{\infty}\frac{a_k\sin kx-b_k\cos kx}{k},$$

where C is a constant of integration and the symbol $'$ indicates that the term $k=0$ is omitted from the sum.

If $f\in L_1(-\pi,\pi)$, then the Fourier coefficients c_n converge to 0 as $n\to\infty$ (**Riemann-Lebesgue theorem**). If f satisfies the †Lipschitz condition of order α $(0<\alpha\leqslant1)$, then $c_n=O(n^{-\alpha})$, and if f is of †bounded variation, then $c_n=O(n^{-1})$. When $f\in L_2(-\pi,\pi)$,

$$\frac{1}{\pi}\int_0^{2\pi}|f(x)|^2dx=2\sum_{k=-\infty}^{\infty}|c_k|^2$$
$$=\frac{1}{2}a_0^2+\sum_{k=1}^{\infty}(a_k^2+b_k^2),$$

which is called the **Parseval equality**. If $\sum|c_k|^2<\infty$, then there exists a function $f\in L_2(-\pi,\pi)$ which has the c_k as its Fourier coefficients. This converse is implied by the †Riesz-Fischer theorem (→ Appendix A, Table 11.I).

B. Convergence Tests

The nth partial sums $s_n(x)=s_n(x;f)$ of the Fourier series $\mathfrak{S}(f)$ can be written in the following form:

$$s_n(x)=\frac{1}{\pi}\int_{-\pi}^{\pi}f(x+t)D_n(t)\,dt,$$

where

$$D_n(t)=\{\sin(n+1/2)t\}/2\sin(t/2).$$

The function $D_n(t)$ is called the **Dirichlet kernel**. For a fixed point x we set $\varphi_x(t)=f(x+t)+f(x-t)-2f(x)$; then

$$s_n(x)-f(x)=\frac{1}{\pi}\int_0^{\pi}\varphi_x(t)D_n(t)\,dt.$$

Hence if the integral on the right-hand side tends to zero as $n\to\infty$, $\lim_{n\to\infty}s_n(x)=f(x)$. If f vanishes in an interval $I=(a,b)$, then $\mathfrak{S}(f)$ converges uniformly in any interval $I'=(a+\varepsilon,b-\varepsilon)$ interior to I, and the sum of $\mathfrak{S}(f)$ is 0. This is called the **principle of localization**.

Here we give four convergence tests. (1) If f is of bounded variation, $\mathfrak{S}(f)$ converges at every point x to the value $\{f(x+0)+f(x-0)\}/2$. In addition, if f is continuous at every point of a closed interval I, $\mathfrak{S}(f)$ is uniformly convergent in I (**Jordan's test**). As a special case of this test, bounded functions having a finite number of maxima and minima and no more than a finite number of points of discontinuity have convergent Fourier series (**Dirichlet's test**). (2) If the integral $\int_0^\pi|\varphi x(t)|/t\,dt$ is finite, then $\mathfrak{S}(f)$ converges at x to $f(x)$ (**Dini's test**). (3) If

$$\int_0^h|\varphi_x(t)|\,dt=o(h),$$

$$\lim_{\eta\to 0}\int_\eta^\pi\frac{|\varphi_x(t)-\varphi_x(t+\eta)|}{t}\,dt=0,$$

then $\mathfrak{S}(f)$ converges at x to $f(x)$ (**Lebesgue's test**). Jordan's and Dini's tests are mutually independent and both are included in Lebesgue's test, which, although not as convenient in certain cases, is quite powerful. (4) If $f(x)$ is continuous in (a,b) and its modulus of continuity satisfies the condition $\omega(\delta)\cdot\log(1/\delta)\to 0$ as $\delta\to 0$ in this interval, then $\mathfrak{S}(f)$ converges uniformly in $(a+\varepsilon,b-\varepsilon)$ (**Dini-Lipschitz test**).

C. Summability

Let $s_n(x)$ be the nth partial sum of the Fourier series $\mathfrak{S}(f)$, and $\sigma_n(x)=\sigma_n(x;f)$ be the first arithmetic mean ((C, 1)-mean) of $s_n(x)$ (i.e., $\sigma_n(x)=(s_0(x)+s_1(x)+\ldots+s_n(x))/(n+1)$). Then we have

$$\sigma_n(x)-f(x)=\frac{1}{\pi}\int_0^\pi\varphi_x(t)K_n(t)\,dt,$$

where

$$K_n(t)=\frac{1}{2(n+1)}\left(\frac{\sin((n+1)t/2)}{\sin(t/2)}\right)^2.$$

The expression $K_n(t)$ is called the **Fejér kernel**, and the $\sigma_n(x)$ are often called **Fejér's means**. If the right and left limits $f(x\pm 0)$ exist, $\mathfrak{S}(f)$ is †(C, 1)-summable at the point x to the value $(f(x+0)+f(x-0))/2$. If f is continuous at every point of a closed interval I, $\mathfrak{S}(f)$ is uniformly (C, 1)-summable in I (**Fejér's theorem**, 1904). As we explain in Section H, there exist continuous functions whose Fourier series are divergent at some points. Thus the summability of $\mathfrak{S}(f)$ is more important than its convergence. Fejér's theorem remains true if we replace (C, 1)-summability by (C,α)-summability ($\alpha>0$). More generally, if $f\in L_1(-\pi,\pi)$, then $\mathfrak{S}(f)$ is (C,α)-summable for $\alpha>0$ to the value $f(x)$ almost everywhere (H. Lebesgue). Since (C,α)-summability ($\alpha>0$) implies †summability by Abel's method, the result of Fejér's theorem is valid for †A-summability. However, the direct study of A-summability is also important. Let $f(r,x)$ be Abel's mean of $\mathfrak{S}(f)$; that is,

$$f(r,x)=\frac{1}{2}a_0+\sum_{k=1}^\infty(a_k\cos kx+b_k\sin kx)r^k$$
$$=\frac{1}{\pi}\int_0^\pi f(x+t)P(r,t)\,dt,$$

where $P(r,t)=(1-r^2)/2(1-2r\cos t+r^2)$, $0\leqslant r<1$. We call $P(r,t)$ the **Poisson kernel**. The function $f(r,x)$ is †harmonic inside the unit circle and tends to $f(x)$ as $r\to 1$ almost everywhere. Hence $f(r,x)$ gives the solution of †Dirichlet's problem for the case of the unit circle.

D. Gibbs's Phenomenon

Let $f(x)$ be of bounded variation and not continuous at $0: f(0)=0, f(+0)=l>0, f(-0)=-l$. Then the partial sum $s_n(x)$ converges to $f(x)$ in the neighborhood of 0, but not uniformly. Moreover,

$$\lim_{n\to\infty}s_n\left(\frac{\pi}{n}\right)=lG,$$

$$G=\frac{2}{\pi}\int_0^\pi\frac{\sin t}{t}\,dt=1.1789\ldots.$$

Hence as x tends to 0 from above and n tends to ∞, the values $y=s_n(x)$ accumulate in the interval $[l,lG]$, while $s_n(-x)=-s_n(x)$ in the neighborhood of 0. This phenomenon is called **Gibbs's phenomenon**. If f is of bounded variation, then $\mathfrak{S}(f)$ exhibits Gibbs's phenomenon at every point of simple discontinuity of f. However, the (C, 1)-means of $\mathfrak{S}(f)$ do not exhibit this phenomenon.

E. Conjugate Functions

For any integrable $f\in L_1(-\pi,\pi)$, the integral

$$\tilde{f}(x)=\lim_{h\to 0}-\frac{1}{\pi}\int_h^\pi\left((f(x+t)-f(x-t))\Big/2\tan\frac{t}{2}\right)dt$$

exists almost everywhere. The function $\tilde{f}(x)$ is called the **conjugate function** of $f(x)$. The conjugate series $\tilde{\mathfrak{S}}(f)$ is (C,α)-summable ($\alpha>0$) to the value $\tilde{f}(x)$ at almost every point, and a fortiori summable by Abel's method. Even if $f\in L_1(-\pi,\pi)$, $\tilde{f}$ does not always belong to the class $L_1(-\pi,\pi)$. For example, $\sum_{n=2}^\infty\cos nx/\log n$ is the Fourier series of a function $f\in L_1(-\pi,\pi)$, but its conjugate series $\sum_{n=2}^\infty\sin nx/\log n$ is not the Fourier series of a function in $L_1(-\pi,\pi)$. However, if both f and $\tilde{f}$ are integrable, $\tilde{\mathfrak{S}}(f)=\mathfrak{S}(\tilde{f})$. If $f\in L_p$ $(p>1)$, then $\tilde{f}\in L_p$ and $\|\tilde{f}\|_p\leqslant A_p\|f\|_p$; also, $\tilde{\mathfrak{S}}(f)=\mathfrak{S}(\tilde{f})$. If $|f|\log^+|f|$ is integrable (such a function is said to belong to the **Zygmund class**), then $\tilde{f}$ is integrable and $\tilde{\mathfrak{S}}(f)=\mathfrak{S}(\tilde{f})$. Moreover, in this case there exist constants A and B such that

$$\int_0^{2\pi}|\tilde{f}|\,dx \leqslant A\int_0^{2\pi}|f|\log^+|f|\,dx+B.$$

If f is merely integrable, so is $|\tilde{f}|^p$ for any $0<p<1$, and $\|\tilde{f}\|_p \leqslant B_p\|f\|$ $(0<p<1)$. If $f\in \mathrm{Lip}\,\alpha$ $(0<\alpha<1)$, then $\tilde{f}\in\mathrm{Lip}\,\alpha$, but the theorem fails for $\alpha=0$ and $\alpha=1$. The conjugate function is important for convergence of partial sums of Fourier series.

F. Mean Convergence

The theorems on conjugate functions enable us to obtain some results for the †mean convergence of the partial sums s_n of $\mathfrak{S}(f)$. If $f\in L_p$ $(p>1)$, then $\|f-s_n\|_p\to 0$; if $f\in L_1$, then $\|f-s_n\|_p\to 0$, $\|\tilde{f}-\tilde{s}_n\|_p\to 0$ for every $0<p<1$. Also, if $|f|\log^+|f|\in L_1$, then $\|f-s_n\|\to 0$, $\|\tilde{f}-\tilde{s}_n\|\to 0$. As a corollary of this result, we obtain the following theorem, which is a generalization of the Parseval equality: If the Fourier coefficients of functions $f\in L_p$ and $g\in L_q$ $(1/p+1/q=1)$ are a_n, b_n and a_n', b_n', respectively, we have the Parseval formula

$$\frac{1}{\pi}\int_0^{2\pi} fg\,dx=\frac{1}{2}a_0a_0'+\sum_{n=1}^{\infty}(a_na_n'+b_nb_n'),$$

where this series is convergent.

G. Analytic Functions of the Class H_p

Let $p>0$. A complex function $\varphi(z)$ holomorphic for $|z|<1$ is said to belong to the class H_p (**Hardy class**) if there exists a constant M such that

$$\lim_{r\to 1}\frac{1}{2\pi}\int_0^{2\pi}|\varphi(re^{i\theta})|^p\,d\theta\leqslant M.$$

When $\varphi(z)\in H_p$, the nontangential limit $\varphi(e^{i\theta})=\lim_{z\to e^{i\theta}}\varphi(z)$ exists for almost all θ. We write this as $\varphi(e^{i\theta})=f(\theta)+i\tilde{f}(\theta)$, where $f(\theta)$, $\tilde{f}(\theta)$ belong to the class L_p. Also, $\tilde{f}(\theta)$ coincides with the conjugate function of $f(\theta)$ for $p\geqslant 1$. For $1<p<\infty$, H_p is isomorphic to L_p, but for $p=1$ and $p=\infty$, H_p and L_p are different classes. Using the theory of functions of H_p, we can discuss some properties of Fourier series. If $\varphi(e^{i\theta})=f(\theta)+i\tilde{f}(\theta)$ is of bounded variation, then $\varphi(e^{i\theta})$ is absolutely continuous and its Fourier series converges absolutely. We set

$$g(\theta)=\left(\int_0^1(1-r)|\varphi'(re^{i\theta})|^2\,dr\right)^{1/2},$$

$$g^*(\theta)=$$

$$\left(\int_0^1(1-r)\,dr\int_0^{2\pi}|\varphi'(re^{i(\theta+t)})|^2P(r,t)\,dt\right)^{1/2},$$

where $P(r,t)$ is the Poisson kernel. Then $g(\theta)\leqslant 2g^*(\theta)$, and there exist constants A_p, B, C, and A_μ such that

$$\int_0^{2\pi}|g(\theta)|^p\,d\theta\leqslant A_p\int_0^{2\pi}|\varphi(e^{i\theta})|^p\,d\theta,\quad p>0,$$

$$\int_0^{2\pi}|g^*(\theta)|^p\,d\theta\leqslant A_p\int_0^{2\pi}|\varphi(e^{i\theta})|^p\,d\theta,\quad p>1,$$

$$\int_0^{2\pi}|g^*(\theta)|\,d\theta\leqslant B\int_0^{2\pi}|\varphi|\log^+|\varphi|\,d\theta+C,$$

$$\int_0^{2\pi}|g^*(\theta)|^\mu\,d\theta\leqslant A_\mu\left(\int_0^{2\pi}|\varphi(e^{i\theta})|\,d\theta\right)^\mu,$$

$$0<\mu<1.$$

Denote by $s_n(\theta)$ and $\sigma_n(\theta)$, respectively, the partial sums and arithmetic means of the Fourier series of $\varphi(e^{i\theta})$, and set

$$\gamma(\theta)=\left(\sum_{n=1}^{\infty}\frac{|s_n(\theta)-\sigma_n(\theta)|^2}{n}\right)^{1/2}.$$

Then $0\neq A_1\leqslant g^*(\theta)/\gamma(\theta)\leqslant A_2\neq\infty$. From these relations, we can prove that if the indices n_k satisfy the conditions $\beta>n_{k+1}/n_k>\alpha>1$, $s_{n_k}(\theta)$ converges almost everywhere to $\varphi(e^{i\theta})$ for $\varphi(e^{i\theta})\in H_p$ $(1\leqslant p)$. If we set $\Delta_k(\theta)=\sum_{\nu=n_{k-1}+1}^{n_k}c_\nu e^{i\nu\theta}$, $\delta(\theta)=(\sum_{k=0}^{\infty}|\Delta_k(\theta)|^2)^{1/2}$, where $\varphi(e^{i\theta})\sim\sum_{\nu=0}^{\infty}c_\nu e^{i\nu\theta}$, then $\|\delta(\theta)\|_p\leqslant A_p\|\varphi\|_p$ $(p>1)$. If $\varphi(z)\in H_p$ $(0<p<1)$, then $\sum c_ne^{in\theta}$ is $(\mathrm{C},p^{-1}-1)$-summable to $\varphi(e^{i\theta})$ almost everywhere. These functions and relations were introduced mainly by J. E. Littlewood and R. E. A. C. Paley and were later generalized by A. Zygmund. There are more precise results by E. Stein [10], G. Sunouchi [11], S. Yano [12], and others.

H. Almost Everywhere Convergence and Divergence

P. du Bois Reymond (1876) first showed that there exists a continuous function whose Fourier series diverges at a point, but the problem whether the Fourier series of continuous functions converge almost everywhere (the so-called **du Bois Reymond problem**) remained unsolved for many years. At last in 1966, L. Carleson [13] proved that the Fourier series of a function belonging to L_2 converges almost everywhere; hence the du Bois Reymond problem was solved affirmatively. Using Carleson's method, R. A. Hunt [15] proved that

$$\int_0^{2\pi}\left(\sup_n|s_n(x)|\right)^p dx\leqslant A_p\int_0^{2\pi}|f(x)|^p\,dx,$$

$$1<p<\infty,$$

which implies that the Fourier series of $f\in L_p$ $(1<p<\infty)$ converges almost everywhere.

Hunt also proved that

$$\int_0^{2\pi}\left(\sup_n |s_n(x)|\right)dx$$

$$\leqslant A\int_0^{2\pi}|f(x)|(\log^+|f(x)|)^2\,dx + A.$$

Moreover, P. Sjölin proved that if

$$\int_0^{2\pi}|f|\cdot\log^+|f|\cdot\log^+\log^+|f|\,dx<\infty,$$

then the Fourier series of $f(x)$ converges almost everywhere.

On the other hand, A. N. Kolmogorov gave an integrable function with Fourier series diverging everywhere (more precisely, $\limsup_{n\to\infty}|s_{2^n}(x)|=\infty$ almost everywhere). Using this example, J. Marcinkiewicz showed that there is an $f\in L_1$ such that $s_n(x;f)$ oscillates finitely almost everywhere. Moreover, there exists an integrable function with integrable conjugate and almost everywhere diverging Fourier series (G. H. Hardy, W. W. Rogosinski, and Sunouchi). For any given †null set E, we can construct a continuous function whose Fourier series diverges on every $x\in E$.

I. Absolute Convergence

The convergence of the series (1) $\Sigma(|a_n|+|b_n|)$ implies the absolute convergence of the trigonometric series (2) $a_0/2+\Sigma(a_n\cos nx+b_n\sin nx)$. Conversely, if the series (2) converges absolutely in a set of positive measure, the series (1) converges (**Denjoy-Lusin theorem**). For the absolute convergence of Fourier series, we have the following tests: If $f\in\mathrm{Lip}\,\alpha$ $(\alpha>1/2)$, then $\mathfrak{S}(f)$ converges absolutely, but for $\alpha=1/2$, this is no longer true. If $f(x)$ is of bounded variation and belongs to $\mathrm{Lip}\,\alpha$ $(\alpha>0)$, $\mathfrak{S}(f)$ converges absolutely.

Suppose that the Fourier series of a function $f(x)$ is absolutely convergent and the value of $f(x)$ belongs to an interval (a,b). If $\varphi(z)$ is a function of a complex variable holomorphic at every point of the interval (a,b), the Fourier series of $\varphi\{f(x)\}$ converges absolutely (**Wiener-Lévy theorem**). As a corollary we obtain that if $\mathfrak{S}(f)$ converges absolutely and $f(x)\neq 0$, then $\mathfrak{S}(1/f)$ converges absolutely. The converse of the Wiener-Lévy theorem was proved by Y. Katznelson [19]. For a given $\varphi(x)$ defined in $[-1,1]$, if the Fourier series of $\varphi\{f(x)\}$ converges absolutely for every $f(x)$ with absolutely convergent Fourier series ($|f(x)|\leqslant 1$), then $\varphi(z)$ is holomorphic at every point of the interval $[-1,1]$.

Many problems concerning this topic still remain unsolved. In particular, the determination of the structure of the functions with absolutely convergent Fourier series has not been completed.

J. Sets of Uniqueness

If $a_n\cos nx+b_n\sin nx$ converges to 0 on a set of positive measure, then $a_n, b_n\to 0$ (**Cantor-Lebesgue theorem**). A point set $E\subset(0,2\pi)$ is called a **set of uniqueness** (or **U-set**) if every trigonometric series converging to 0 outside E vanishes identically. A set that is not a U-set is called a **set of multiplicity** (or **M-set**). G. Cantor showed that every finite set is a U-set, and W. H. Young showed that every denumerable set is a U-set. It is clear that any set E of positive measure is an M-set, but D. E. Menchoff showed that there are †perfect M-sets of measure 0. Moreover, N. K. Bari showed that there exist perfect sets of type U. However, the structure problem of sets of uniqueness has not yet been solved completely.

A set E is said to be of type H if there exists a sequence of positive integers $n_1<n_2<\ldots$ and an interval I such that for each $x\in E$, no point $\{n_kx\}_{k=1}^{\infty}$ is in I (mod 2π). H-sets are sets of uniqueness, a fact given by A. Rajchman. I. I. Pjateckiĭ-Šapiro generalized H-sets to $H^{(m)}$-sets [3].

Lacunary trigonometric series are series in which very few terms differ from zero. Such series may be written in the form

$$\sum_{k=1}^{\infty}(a_k\cos n_kx+b_k\sin n_kx)=\sum_{k=1}^{\infty}A_{n_k}(x).$$

S. Sidon established some of the characteristic properties of such series; he generalized them further and obtained the notion of Sidon sets (→ 194 Harmonic Analysis). We often define a lacunary series more specifically as a series for which the n_k satisfy Hadamard's gaps; that is, $n_{k+1}/n_k>q>1$. Then if $\Sigma_{k=1}^{\infty}(a_k^2+b_k^2)$ is finite, the series $\Sigma_{k=1}^{\infty}A_{n_k}(x)$ converges almost everywhere. Conversely, if $\Sigma_{k=1}^{\infty}A_{n_k}(x)$ is convergent in a set of positive measure, then $\Sigma_{k=1}^{\infty}(a_k^2+b_k^2)$ converges. This theorem is related to the Rademacher series and random Fourier series [4].

K. Multiple Fourier Series

Routine extensions to multiple Fourier series from the case of a single variable are easy, but significant results are difficult to obtain. Recently, however, there have been several important contributions in this field.

References

[1] A. Zygmund, Trigonometrical series, Warsaw, 1935.

[2] A. Zygmund, Trigonometric series I, II, Cambridge Univ. Press, second edition, 1959.
[3] G. H. Hardy and W. W. Rogosinski, Fourier series, Cambridge Univ. Press, 1950.
[4] Н. К. Бари (N. K. Bari), Тригонометрические ряды, физматгиз, 1958; English translation, A treatise on trigonometric series, Pergamon, 1964.
[5] J.-P. Kahane and R. Salem, Ensembles parfaits et séries trigonométriques, Actualités Sci. Ind., Hermann, 1963.
[6] K. K. Chen, Sankaku kyûsûron (Japanese; Theory of trigonometric series), Iwanami, 1930.
[7] G. Sunouchi, Hûrie kaiseki (Japanese; Fourier analysis), Kyôritu, 1956.
[8] T. Tsuchikura, Hûrie kaiseki (Japanese; Fourier analysis), Sibundô, 1964.
[9] Y. Katznelson, Sur les fonctions opérant sur l'algèbre des séries de Fourier absolument convergentes, C. R. Acad. Sci. Paris, 247 (1958), 404–406.
[10] E. M. Stein, A maximal function with applications to Fourier series, Ann. of Math., (2) 68 (1958), 584–603.
[11] G. Sunouchi, Theorems on power series of the class H^p, Tôhoku Math. J., (2) 8 (1956), 125–146.
[12] S. Yano, On a lemma of Marcinkiewicz and its applications to Fourier series, Tôhoku Math. J., (2) 11 (1959), 191–215.
[13] L. Carleson, On convergence and growth of partial sums of Fourier series, Acta Math., 116 (1966), 135–157.
[14] J.-P. Kahane and Y. Katznelson, Sur les ensembles de divergence des séries trigonométriques, Studia Math., 26 (1966), 305–306.
[15] R. A. Hunt, On the convergence of Fourier series, orthogonal expansions and their continuous analogues, Proc. Conf. Edwardsville, Ill. (1967), Southern Illinois Univ. Press, 1968, p. 235–255.
[16] P. Sjölin, An inequality of Paley and convergence a.e. of Walsh-Fourier series, Ark. Mat., 7 (1968), 551–570.
[17] R. Salem, Algebraic numbers and Fourier analysis, Heath and Co., 1963.
[18] J.-P. Kahane, Some random series of functions, Heath and Co., 1968.

168 (X.23) Fourier Transform

A. Fourier Integrals

In this article we assume that $f(x)$ is a complex-valued function defined on $\mathbf{R}=(-\infty,\infty)$ and †(Lebesgue) integrable on any finite interval. If the integral

$$\int_{-\infty}^{\infty} f(x)e^{-ixt}dx = \lim_{\substack{A\to-\infty\\ B\to\infty}} \int_A^B f(x)e^{-ixt}dx$$

exists, it is called the **trigonometric integral** or **Fourier integral**. We have a general result: If $f(x)\in L_1(-\infty,\infty)$, $K(x)$ is bounded on $(-\infty,\infty)$, and $\int_0^T K(x)dx=O(T)$ $(T\to\pm\infty)$, then $\int_{-\infty}^{\infty} f(x)K(xt)dx$ exists and

$$\lim_{t\to\pm\infty}\int_{-\infty}^{\infty} f(x)K(xt)dx=0.$$

In particular, it follows that if $f(x)\in L_1(-\infty,\infty)$, then $\int_{-\infty}^{\infty} f(x)e^{itx}dx$ exists and

$$\lim_{t\to\pm\infty}\int_{-\infty}^{\infty} f(x)e^{-itx}dx=0$$

(Riemann-Lebesgue theorem).

B. Fourier's Integral Theorems

Suppose that $f(x)$ is of †bounded variation in an interval including x, or more generally satisfies one of the other convergence tests for Fourier series (→ 167 Fourier Series B). Then **Fourier's single integral theorem**:

$$\frac{1}{2}(f(x+0)+f(x-0)) = \lim_{A\to\infty}\frac{1}{\pi}\int_{-\infty}^{\infty} f(t)\frac{\sin A(t-x)}{t-x}dt \tag{1}$$

holds if one of the following three conditions is satisfied: (1) $f(x)/(1+|x|)$ belongs to $L_1(-\infty,\infty)$; (2) $f(x)/x$ tends to zero monotonically as $x\to\pm\infty$; (3) $f(x)/x=g(x)\sin(px+q)$, where $g(x)$ tends to zero monotonically as $x\to\pm\infty$ (S. Izumi, 1934). The right-hand side of (1) is called **Dirichlet's integral**.

Let $f(x)$ be of bounded variation in an interval including x (or satisfy some other convergence test for Fourier series). Then **Fourier's double integral theorem**

$$\frac{1}{2}(f(x+0)+f(x-0)) = \frac{1}{\pi}\lim_{T\to\infty}\int_0^T dt\int_{-\infty}^{\infty} f(u)\cos t(u-x)du \tag{2}$$

holds if one of the following three conditions is satisfied: (4) $f(x)\in L_1(-\infty,\infty)$; (5) $f(x)/(1+|x|)\in L_1(-\infty,\infty)$, and $f(x)$ tends to zero monotonically as $x\to\pm\infty$; (6) $f(x)/(1+|x|)\in L_1(-\infty,\infty)$ and $f(x)=g(x)\sin(px+q)$, where $g(x)$ tends to zero monotonically as $x\to\pm\infty$. If $f(x)\in L_1(-\infty,\infty)$, the formula

$$f(x)=\lim_{A\to\infty}\frac{1}{\pi A}\int_{-\infty}^{\infty} f(t)\frac{\sin^2 A(t-x)}{A(t-x)^2}dt$$

holds almost everywhere, and in particular at any x where $f(x)$ is continuous. More generally, the formula

$$f(x)=\lim_{A\to\infty} A\int_{-\infty}^{\infty} f(t)K(A(t-x))dt = \lim_{A\to\infty}\int_{-\infty}^{\infty} f\left(x-\frac{t}{A}\right)K(t)dt \tag{3}$$

holds at any point x where $f(x+0)$ and $f(x-0)$ exist, if $K(t)\in L_1(-\infty,\infty)$, $\int_{-\infty}^{\infty}K(t)dt=1$, $|K(t)|<M$, $K(t)=o(t^{-1})$ as $|t|\to\infty$, and $f(x)\in L_1(-\infty,\infty)$, or if $K(t)\in L_1(-\infty,\infty)$, $\int_{-\infty}^{\infty}K(t)dt=1$, and $f(x)$ is bounded. Similarly, the formula

$$\lim_{A\to 0}\int_0^{\infty} f\left(\frac{x}{A}\right)K(x)dx=\mathfrak{M}\{f\}\int_0^{\infty}K(x)dx$$

holds if $\mathfrak{M}\{f\}=\lim_{T\to\infty}T^{-1}\int_0^T f(t)dt$ exists, $K(x)$ is differentiable, $|x^2K(x)|\leqslant C$ $(1\leqslant x)$, and $x^{-1}\int_0^x|f(t)|\,dt\leqslant D$, where C, D are constants (**Wiener's formula**).

C. Fourier Transforms (→ Appendix A, Table 11.II)

Let $f(x)\in L_1(0,\infty)$. Then

$$F(t)=\sqrt{\frac{2}{\pi}}\int_0^{\infty}f(u)\cos ut\,du$$

is called the **(Fourier) cosine transform** of $f(x)$. Under the same condition as for the validity of (2), the **inversion formula** of the cosine transform $f(x)=\sqrt{2/\pi}\int_0^{\infty}F(t)\cos xt\,dt$ holds, where we suppose that $f(x)=\frac{1}{2}(f(x+0)+f(x-0))$. If we define $f(-x)=f(x)$, then this is equivalent to the formula (2). Analogously,

$$G(t)=\sqrt{\frac{2}{\pi}}\int_0^{\infty}f(u)\sin ut\,du$$

is called the **(Fourier) sine transform** of $f(x)$. Under the same condition as for the validity of (2), we get the inversion formula $f(x)=\sqrt{2/\pi}\int_0^{\infty}G(t)\sin xt\,dt$. More generally, for any $f(x)\in L_1(-\infty,\infty)$,

$$F(t)=\frac{1}{\sqrt{2\pi}}\int_{-\infty}^{\infty}f(x)e^{-ixt}dx$$

is called the **Fourier transform** of $f(x)$. Under the same condition as for the validity of (2), the formula

$$f(x)=\frac{1}{\sqrt{2\pi}}\lim_{T\to\infty}\int_{-T}^{T}F(t)e^{ixt}dt$$

holds. The cosine transform and the sine transform coincide with the Fourier transform when $f(x)=f(-x)$ and $-f(x)=f(-x)$, respectively.

If for any $f(x)$, $F(t)\in L_q(-\infty,\infty)$ $(1\leqslant q<\infty)$,

$$\int_{-\infty}^{\infty}\left|\frac{1}{\sqrt{2\pi}}\int_{-T}^{T}f(x)e^{-ixt}dx-F(t)\right|^q dt\to 0\qquad (T\to\infty)$$

exists (i.e., $(1/\sqrt{2\pi})\int_{-T}^{T}f(t)e^{-ixt}dt$ converges to $F(x)$ as $T\to\infty$ in the mean of order q), then we say that $f(x)$ has the **Fourier transform** $F(t)$ **in** $L_q(-\infty,\infty)$. If $f(x)\in L_p(-\infty,\infty)$ $(1<p\leqslant 2)$, then $f(x)$ has the Fourier transform $F(t)$ in L_q $(1/p+1/q=1)$, and $F(t)$ has the Fourier transform $f(-x)$ in L_p (E. C. Titchmarsh). Moreover,

$$\int_{-\infty}^{\infty}|F(t)|^q dx\leqslant\frac{1}{(2\pi)^{(q/2)-1}}\left(\int_{-\infty}^{\infty}|f(x)|^p dx\right)^{1/(p-1)}.$$

If $f(x)$, $G(x)\in L_p(-\infty,\infty)$ $(1<p\leqslant 2)$ and their Fourier transforms in L_q are $F(t)$, $g(t)$, respectively, then the **Parseval equality**

$$\int_{-\infty}^{\infty}F(t)G(t)dt=\int_{-\infty}^{\infty}f(x)g(-x)dx$$

holds. The **Fourier reciprocity** holds, in the sense that

$$\int_0^x F(u)du=\frac{1}{\sqrt{2\pi}}\int_{-\infty}^{\infty}f(t)\frac{e^{-ixt}-1}{-it}dt,$$

$$\int_0^x f(t)dt=\frac{1}{\sqrt{2\pi}}\int_{-\infty}^{\infty}F(u)\frac{e^{ixu}-1}{iu}du.$$

The theory of Fourier transforms is valid for cosine and sine transforms. Specifically for $p=2$, if $f\in L_2(0,\infty)$, then $\sqrt{2/\pi}\int_0^T f(x)\cdot\cos xt\,dt$ †converges in the mean in $L_2(0,\infty)$ to the cosine transform $F(t)$ as $T\to\infty$, and reciprocally, the cosine transform of $F(t)$ in L_2 is $f(x)$. The transforms $f(x)$, $F(x)$ are connected by the formulas

$$\int_0^x F(u)du=\sqrt{\frac{2}{\pi}}\int_0^{\infty}f(t)\frac{\sin xt}{t}dt,$$

$$\int_0^x f(t)dt=\sqrt{\frac{2}{\pi}}\int_0^{\infty}F(u)\frac{\sin xu}{u}du,$$

$$\int_0^{\infty}|f(x)|^2dx=\int_0^{\infty}|F(t)|^2dt.$$

The theory of Fourier transforms was generalized as follows by G. N. Watson (*Proc. London Math. Soc.*, 35 (1933)). We suppose that $\chi(x)/x\in L_2(0,\infty)$ and

$$\int_0^{\infty}\frac{\chi(xu)\chi(yu)}{u^2}du=\min(x,y). \tag{4}$$

If $f(x)\in L_2(0,\infty)$, then there exists an $F(t)\in L_2(0,\infty)$ such that

$$\int_0^x F(t)dt=\int_0^{\infty}\frac{\chi(xu)f(u)}{u}du,$$

the reciprocal formula

$$\int_0^x f(u)du=\int_0^{\infty}\frac{\chi(xt)F(t)}{t}dt,$$

and the Parseval equality

$$\int_0^{\infty}|f(x)|^2dx=\int_0^{\infty}|F(t)|^2dt$$

hold. $F(t)$ is called the **Watson transform** of $f(x)$. For any $f\in L_2$, equality (4) is necessary for the existence of the Watson transform $F(t)$ for which the reciprocal formula holds. S. Bochner (1934) generalized this theory further to †unitary transformations in L_2 (→ 194 Harmonic Analysis).

D. Conjugate Functions

Corresponding to Fourier's double integral theorem (2), the integral

$$\lim_{\lambda\to\infty}\frac{1}{\pi}\int_0^{\lambda}dt\int_{-\infty}^{\infty}\sin t(u-x)f(u)du$$

is called the **conjugate Fourier integral** of the integral in the right side of (2). Formally, this is written $\lim_{\lambda\to\infty}t^{-1}\int_0^{\infty}(1-\cos\lambda t)t^{-1}(f(x+t)-f(x-t))\,dt$. If $f(x)$ is a sufficiently regular function, the part involving $\cos\lambda t$ tends to 0 as $\lambda\to\infty$. Now let

$$g(x)=\frac{1}{\pi}\lim_{\substack{A\to\infty\\ \varepsilon\to 0}}\int_{\varepsilon}^{A}\frac{f(x+t)-f(x-t)}{t}dt.$$

For any $f\in L_1\,(0,\infty)$, the integral exists almost everywhere, and $g(x)$ is called the **conjugate function** or **Hilbert transform** of $f(x)$. If $f\in L_p$ $(p>1)$, then $g(x)\in L_p$ also and we have

$$f(x)=-\frac{1}{\pi}\lim_{\substack{A\to\infty\\ \varepsilon\to 0}}\int_{\varepsilon}^{A}\frac{g(x+t)-g(x-t)}{t}dt$$

and $\int_{-\infty}^{\infty}|g(x)|^p dx\leqslant M_p\int_{-\infty}^{\infty}|f(x)|^p dx$, where M_p is a constant depending on p only. In particular,

$$\int_{-\infty}^{\infty}|f(x)|^2dx=\int_{-\infty}^{\infty}|g(x)|^2dx\quad\text{for } p=2.$$

E. Boundary Functions of Analytic Functions

Suppose that a complex-valued function $f(z)$ $(z=x+iy)$ is †holomorphic for $y>0$, $f(x+iy)$ converges as $y\to 0$ for almost all x to $f(x)$ (which is called the **boundary function**), and $f(x)\in L_p(-\infty,\infty)$ $(p\geqslant 1)$. Moreover, suppose that $f(z)$ is represented by †Cauchy's integral formula or †Poisson's integral of $f(x)$ on the real line. If either $f(x)\in L_p$ $(p\geqslant 1)$ and has $F(t)$ as its Fourier transform or $f(x)$ is an L_p-Fourier transform of $F(t)\in L_q$ $(q\geqslant 1)$, then a necessary and sufficient condition for the function $f(x)$ to be the boundary function of an analytic function is that $F(t)$ is 0 almost everywhere for $t>0$ (N. Wiener, R. E. A. C. Paley, E. Hille, J. D. Tamarkin).

F. Generalized Fourier Integrals

Let $|f(x)|/(1+|x|^k)\in L_1(-\infty,\infty)$ for a positive integer k and

$$L_k=L_k(t,x)=\begin{cases}\displaystyle\sum_{\nu=0}^{k-1}\frac{(-itx)^{\nu}}{\nu!} & \text{for } |x|\leqslant 1,\\ 0 \text{ for } |x|>1,\end{cases}$$

$$E_k(t)=\frac{1}{2\pi}\int_{-\infty}^{\infty}f(x)\frac{e^{-itx}-L_k}{(-ix)^k}dx.$$

The function $E_k(t)$ or one that differs from $E(k)$ by a polynomial of at most the kth degree is called the **kth transform** of $f(x)$. We write formally $f(x)=\int_{-\infty}^{\infty}e^{ixt}d^kE_k(t)$. Actually, if we give an appropriate meaning to the integral, this formula itself is valid (H. Hahn, *Wiener Berichte*, 134 (1925); S. Izumi, *Tôhoku Science Rep.*, 23 (1935)). (For the theory and applications of kth transforms → [2, ch. 6].)

G. Applications of Fourier Transforms

Suppose that $f(x)\in L_1(-\infty,\infty)$, $F(t)$ is its Fourier transform, and $f(x)=o(e^{-\theta(x)})$, where $\theta(x)$ is positive and increasing. If $\int_1^{\infty}\theta(x)x^{-2}dx=\infty$ and $F(t)$ vanishes identically in an interval, then $F(t)\equiv 0$ over $(-\infty,\infty)$. If $\int_1^{\infty}\theta(x)x^{-2}dx<\infty$, then there exists a function $f(x)$ such that $F(t)$ vanishes identically in an interval, but $F(t)$ does not vanish identically in $(-\infty,\infty)$ (Wiener, Paley, N. Levinson). These results are applicable to the theory of †quasi-analytic functions.

Let $f(x)\in L_1(-\infty,\infty)$. Then a necessary and sufficient condition for any function in $L_1(-\infty,\infty)$ to be approximated as closely as we wish by linear combinations of the translations $\sum_{k=1}^{N}a_kf(x+h_k)$ of $f(x)$ with respect to the L_1-norm is that the Fourier transform of $f(x)$ does not vanish at any real number. When $f(x)\in L_2(-\infty,\infty)$, a necessary and sufficient condition that it can be approximated as closely as we wish by $\sum_{k=1}^{N}a_kf(x+h_k)$ with respect to the L_2-norm is that the zeros of the Fourier transform of $f(x)$ have measure zero (N. Wiener). This result was used by Wiener to prove the **generalized Tauberian theorem**: Suppose that $g_1(x)\in L_1(-\infty,\infty)$ and its Fourier transform never vanishes. Moreover, let $g_2(x)\in L_1(-\infty,\infty)$ and $p(x)$ be bounded over $(-\infty,\infty)$. Then $\lim_{x\to\infty}\int_{-\infty}^{\infty}g_1(x-t)p(t)dt=A\int_{-\infty}^{\infty}g_1(t)dt$ implies that $\lim_{n\to\infty}\int_{-\infty}^{\infty}g_2(x-t)p(t)dt=A\int_{-\infty}^{\infty}g_2(t)dt$. Another type of Wiener theorem is concerned with †Stieltjes integrals. Suppose

$$\sum_{n=-\infty}^{\infty}\sup_{n\leqslant x<n+1}|g_1(x)|<\infty$$

(hence $g_1(x)\in L_1$) and the Fourier transform of $g_1(x)$ never vanishes. Moreover, let

$$\sum_{n=-\infty}^{\infty} \sup_{n\leqslant x<n+1} |g_2(x)|<\infty$$

and $\int_x^{x+1}|d\alpha(t)|$ be bounded. Then

$$\lim_{x\to\infty}\int_{-\infty}^{\infty} g_1(x-t)d\alpha(t)=A\int_{-\infty}^{\infty} g_1(t)dt$$

implies

$$\lim_{x\to\infty}\int_{-\infty}^{\infty} g_2(x-t)d\alpha(t)=A\int_{-\infty}^{\infty} g_2(t)dt.$$

From these general theorems, we can prove various †Tauberian theorems about the summation of series. Also, these results were applied to the proof of the †prime number theorem by S. Ikehara and E. Landau (Wiener [1]). In the general Tauberian theorem, the boundedness of $p(t)$ may be replaced by one-sided boundedness (H. R. Pitt, 1938). In fact, the first form of the theorem still holds if we replace the condition by the following one: g_1 and g_2 are continuous, $g_1(x)\geqslant 0$, the Fourier transform of $g_1(x)$ does not vanish, $g_2(x)$ satisfies the conditions of the second theorem, and $p(x)\geqslant C$. (Concerning Fourier transforms on the topological groups → 194 Harmonic Analysis; and concerning Fourier transforms of the distributions → 130 Distributions (Generalized Functions)).

References

[1] N. Wiener, The Fourier integral and certain of its applications, Cambridge Univ. Press, 1933.
[2] S. Bochner, Vorlesungen über Fouriersche Integrale, Akademische-Verlag, 1932; English translation, Lectures on Fourier integrals, Ann. Math. Studies, Princeton Univ. Press, 1959.
[3] E. C. Titchmarsh, Introduction to the theory of Fourier integrals, Clarendon Press, 1937.
[4] R. E. A. C. Paley and N. Wiener, Fourier transforms in the complex domain, Amer. Math. Soc. Colloq. Publ., 1934.
[5] S. Bochner and K. Chandrasekharan, Fourier transforms, Ann. Math. Studies, Princeton Univ. Press, 1949.
[6] M. J. Lighthill, Introduction to Fourier analysis and generalized functions, Cambridge Univ. Press, 1958.
[7] И. М. Гельфанд, Г. Е. Шилов (I. M. Gel'fand and G. E. Šilov), Обобщенные функция, физматгиз, 1958; English translation, Generalized functions I, Academic Press, 1964.
[8] S. Bochner, Harmonic analysis and the theory of probability, Univ. of California Press, 1955.
[9] R. R. Goldberg, Fourier transforms, Cambridge tracts, Cambridge Univ. Press, 1961.
[10] T. Carleman, L'intégrale de Fourier et questions qui s'y rattachent, Almquist & Wiksells, 1944.
Also, for formulas for Fourier transforms, → references to 222 Integral Transforms.

169 (IV.3) Free Groups

A. General Remarks

A group F is called a **free group** if it is the †free product (→ 193 Groups M) of †infinite cyclic groups $G_1, \ldots, G_n$ generated by $a_1, \ldots, a_n$, respectively. Then n is called the **rank** of F. A free product of †semigroups is defined similarly to that of groups, and the free product of infinite cyclic semigroups $G_i=\{1,a_i,a_i^2,\ldots\}$ $(i=1,\ldots,n)$ is called a **free semigroup** generated by n elements a_i $(i=1,\ldots,n)$.

If a group G is generated by subgroups H_i $(i=1,\ldots,n)$ isomorphic to G_i, then G is a homomorphic image of the free product of the groups G_i. A subgroup $\neq\{e\}$ of the free product F of groups G_i is itself the free product of a free group and several subgroups, each of which is conjugate in F to a subgroup of some G_j (A. G. Kuroš, 1934). Notably, a subgroup $\neq\{e\}$ of a free group is itself a free group (O. Schreier, *Abh. Math. Sem. Univ. Hamburg*, 5 (1927)). A subgroup of index j of a free group of rank n is a free group of rank $1+j(n-1)$ (Schreier).

Let F be the free group generated by n elements $a_1, \ldots, a_n$, and let G be a group generated by n elements $b_1, \ldots, b_n$. Then there is a homomorphism of F onto G. Let N be its kernel. If the class of a †word $w(a_1,\ldots,a_n)$ belongs to N, then we have $w(b_1,\ldots,b_n)=1$ We call $w(b_1,\ldots,b_n)=1$ a **relation** among the generators $b_1,\ldots,b_n$. If N is the minimal normal subgroup of F containing the classes of words $w_1(a_1,\ldots,a_n),\ldots,w_m(a_1,\ldots,a_n)$, then the relations $w_1(b_1,\ldots,b_n)=1,\ldots,w_m(b_1,\ldots,b_n)=1$ are called **defining relations** (or **fundamental relations**). If generators $a_1,\ldots,a_n$ and words $w_1(a_1,\ldots,a_n),\ldots,w_m(a_1,\ldots,a_n)$ are given, then there is a group generated by $a_1,\ldots,a_n$ with defining relations $w_1(a_1,\ldots,a_n)=1,\ldots,w_m(a_1,\ldots,a_n)=1$. In fact, let F be the free group generated by $a_1, \ldots, a_n$ and N the minimal normal subgroup containing the classes of words $w_1(a_1, \ldots, a_n), \ldots, w_m(a_1, \ldots, a_n)$. Then the factor group F/N is such a group. A free group is a group with an empty set of defining relations. In the preceding discussion, n and m are not necessarily finite. If both n and m are finite, then G is called **finitely presented**.

B. The Word Problem

If a finitely presented group G is given, then a general procedure has to be determined by which it can be decided, in a finite number of computational steps, whether a given word equals the identity element as an element of G. This is called the **word problem** (→ 193 Groups M). A solution to the word problem does not always exist (P. S. Novikov [9], 1955); in fact, there is a group with two generators and 32 defining relations for which the word problem cannot be solved (W. Boone [8]). However, it was shown by V. A. Tartakovskiĭ that the problem can be solved for a large class of groups. W. Magnus (1931) showed that it is solvable for any group with a single defining relation. The word problem is an example of decision problems (→ 100 Decision Problem). The word problem for groups is closely related to that for semigroups (A. M. Turing, 1937; E. L. Post, 1947; A. A. Markov, 1947). Similar problems for other algebraic systems can also be considered. The problem of determining a general procedure by which it can be decided, in a finite number of steps, whether two given words interpreted as elements of G can be transformed into each other by an (inner) automorphism of G is called the **transformation problem**.

Let F be a free group of rank n and $F = F_1 \supset \cdots \supset F_r \supset F_{r+1} \supset \cdots$ be the †lower central series of F. Then F_r / F_{r+1} is a †free Abelian group of rank $\mu_n(r) = (1/r)\Sigma_{d|r}\mu(r/d)n^d$, where μ is the †Möbius function (E. Witt). The intersection of all subgroups of F of finite index is the identity element.

C. The Burnside Problem

The original problem of Burnside is: If every element of a group G is of finite order (but not necessarily of bounded order) and G is finitely generated, is G a finite group? E. S. Golod [7] (1964) showed that this problem for p-groups has a negative solution. The following is the more usual form of the **Burnside problem**: If a group G is finitely generated and the orders of elements of G divide a given integer r, is G finite? Let F be a free group of rank n, N be the normal subgroup of F generated by all rth powers x^r of elements of F, and $B(r,n) = F/N$. Then the problem is the same as the question whether $B(r,n)$ is finite. For $r = 2,3,4,6$ the group is certainly finite (I. N. Sanov, M. Hall). The **restricted Burnside problem** is the question whether the orders of finite factor groups of $B(r,n)$ are bounded. It was solved affirmatively for r prime (A. I. Kostrikin [6], 1959).

A group generated by two generators x, y and satisfying the relations $x^u = y^v = (xy)^w = 1$ (where u, v, w are integers) is infinite if $1/u + 1/v + 1/w - 1 \leqslant 0$, and is of order g if $0 < 1/u + 1/v + 1/w - 1 = 2/g$.

There is also a finitely presented group which is isomorphic to its proper factor group (B. H. Neumann).

References

[1] S. Iyanaga, Ziyûgunron (Japanese; Theory of free groups), Series of math. lec. Osaka Univ., Iwanami, 1942.

[2] А. Г. Курош (A. G. Kuroš), Теория групп, Гостохиздат, second edition, 1953; English translation, The theory of groups I, II, Chelsea, 1960.

[3] M. Hall, The theory of groups, Macmillan, 1959.

[4] H. S. M. Coxeter and W. O. J. Moser, Generators and relations for discrete groups, Erg. Math., Springer, 1957.

[5] W. Magnus, A. Karrass, and D. Solitar, Combinatorial group theory, Interscience, 1966.

[6] А. И. Кострикин (A. I. Kostrikin), О проблеме Бернсайда, Izv. Akad. Nauk SSSR, 23 (1959), 1–34; English translation, The Burnside problem, Amer. Math. Soc. Transl., (2) 36 (1964), 63–100.

[7] Е. С. Голод (E. S. Golod), О Нильалгебрах и финитно-аппроксимируемых p-группах, Izv. Akad. Nauk SSSR, 28 (1964), 273–276; English translation, On nil-algebras and finitely approximable p-groups, Amer. Math. Soc. Transl., (2) 48 (1965), 103–106.

[8] W. W. Boone, The word problem, Ann. of Math., (2) 70 (1959), 207–265.

[9] П. С. Новиков (P. S. Novikov), Об алгорифмической неразрешимости проблемы тожества слов в теории групп, Trudy Mat. Inst. Steklov., 44 (1955), 1–143; English translation, On the algorithmic unsolvability of the word problem in group theory, Amer. Math. Soc. Transl., (2) 9 (1958), 1–122.

Also → references to 193 Groups.

170 (II.4) Functions

A. History

Leibniz first used the term *function* (Lat. *functio*) in 1694 to refer to certain line segments whose lengths depend on lines related to curves. Soon the term was used to refer to dependent quantities or expressions. In 1718, Johann Bernoulli used the notation φx, and by 1734 the modern functional notation $f(x)$ had been used by Clairaut and by Euler, who defined functions as analytic formulas constructed from variables and constants (1728) [1]. †Cauchy stated [2] (1821): "When there is a relation among many variables, which determines along with values of one of them the values of the others, we usually consider the others as expressed by the one. We then call the one an 'independent variable,' and the others 'dependent variables.'" †Dirichlet considered a function of $x \in [a,b]$ in his paper (1837) [3] concerning representations of "completely arbitrary functions" and stated that there was no need for the relation between y and x to be given by the same law throughout an interval, nor was it necessary that the relation be given by mathematical formulas. A function was simply a correspondence in which values of one variable determined values of another.

B. Functions

Today, the word "function" is used generally in mathematics in the same sense as a †mapping (→ 376 Sets C) or, which is the same thing, a †univalent correspondence (→ 354 Relations B). But this word is sometimes used in a wider sense, to mean a general (not necessarily univalent) correspondence, called a **many-valued** (or **multivalued**) **function**; in that case a univalent correspondence is called a **single-valued function**.

Specialists in each branch of mathematics have their respective ways of using the word. In analysis, values of a function are often considered real or complex numbers; such functions are called **real-valued functions** or **complex-valued functions**, respectively. Furthermore, if the domain of the function is also a set of real or complex numbers, then it is called a **real function** or a **complex function**, respectively (→ 141 Elementary Functions; 86 Continuous Functions; 200 Holomorphic Functions; and 24 Analytic Functions). If the domain of a real- or complex-valued function is contained in a †function space, the function is often called a **functional**; the †distribution is an example. In algebra we often fix a †field, †ring, etc., and consider functions whose domains and ranges are in such algebraic systems. Special names are given to functions having special properties, which can be defined according to the structures of the domain and the range. For example, when both domain and range of a function f are sets of real numbers, f is called an **even function** if $f(t)=f(-t)$, and an **odd function** if $f(t)=-f(-t)$. A function f which preserves the order relation between real numbers, i.e., such that $t_1<t_2$ implies $f(t_1)\leqslant f(t_2)$, is called a †monotone increasing function.

A mapping from a set I to a set F of functions, $\varphi: I \rightarrow F$, is called a **family of functions indexed by** I (or simply **family of functions**), and is denoted, using the form f_λ instead of $\varphi(\lambda)$, by $\{f_\lambda\}_{\lambda\in I}$ or $\{f_\lambda\}$ $(\lambda\in I)$. In particular, if I is the set of natural numbers, the family is called a **sequence of functions**.

C. Variables

A letter x, for which we may substitute a name of an element of a set X, is called a **variable**, and X is called the **domain** of the variable. An element of the domain of a variable x is called a **value** of x. In particular, if the domain is a set of real numbers or complex numbers, the variable is called a **real variable** or a **complex variable**, respectively. On the other hand, a letter which stands for a particular element is called a **constant**.

When the domain and range of a function f are X and Y, respectively, a variable x whose domain is X is called the **independent variable**, and a variable y whose domain is Y is called the **dependent variable**. Then we say y is a function of x, and write $y=f(x)$. When a concrete method is given by which we make a value of y correspond to each value of x, we say that y is an **explicit function** of x. When a function is determined only by a †binary relation such as $R(x,y)=0$, we say that y is an **implicit function** of x (→ 212 Implicit Functions).

Given functions f, g with an independent variable t, suppose that y is regarded as a function of x defined by relations $x=f(t)$, $y=g(t)$. Then we say that y is a function of x with the variable t as a **parameter**. A function whose range is a given set C with variable t as its independent variable is often called a **parametric representation** of C by t.

If the domain of a function f is contained

in a Cartesian product set $X_1 \times X_2 \times \ldots \times X_n$, the independent variable is denoted by $(x_1, x_2, \ldots, x_n)$, and f is often called a **function of** n **variables** or a **function of many variables** (when $n \geqslant 2$).

D. Families and Sequences

A function whose domain is a set I, $\varphi: I \rightarrow X$, is called a **family indexed by** I (or simply **family**), and I is called the **index set**. In the case $\varphi(\lambda) = x_\lambda$ $(\lambda \in I)$, the family is denoted by $\{x_\lambda\}_{\lambda \in I}$ or $\{x_\lambda\}(\lambda \in I)$. If the range X of a function φ is a set of points, a set of functions, a set of mappings, or a set of sets, then the family $\{x_\lambda\}_{\lambda \in I}$ is called a **family of points**, a **family of functions**, a **family of mappings**, or a **family of sets**, respectively. If the set I is a †directed set, the family is called a **directed family**. Generally, if J is a subset of I, the family $\{x_\lambda\}_{\lambda \in J}$ is called a **subfamily** of $\{x_\lambda\}_{\lambda \in I}$. In particular, if I is a finite or infinite set of natural numbers, the family indexed by I is called a **finite sequence** or **infinite sequence**, respectively. **Sequence** is a generic name for both, but in many cases it means an infinite sequence, and usually we have $I = \mathbf{N}$. Then the value corresponding to $n \in \mathbf{N}$ is called the n**th term** or, generally, a **term**. For convenience, the 0th term is often used as well. If each term of a sequence is a number, a point, a function, or a set, the sequence is called a **sequence of numbers**, a **sequence of points**, a **sequence of functions**, or a **sequence of sets**, respectively. A sequence is usually denoted by $\{a_n\}$. If it is necessary to show the domain of n explicitly, the sequence is denoted by $\{a_n\}_{n \in I}$. If J is a subset of I, a sequence $\{a_n\}_{n \in J}$ is called a **subsequence** of the sequence $\{a_n\}_{n \in I}$. And if $I = N$, the composite $\{a_{k_n}\}$ of $\{a_n\}$ and a sequence $\{k_n\}$ of natural numbers with $k_1 < k_2 < k_3 < \ldots$ is usually called a subsequence of $\{a_n\}$.

References

[1] L. Euler, Opera omnia ser. I: Opera mathematica VIII: Introductio in analysin infinitorum I, Teubner, 1922.

[2] A. L. Cauchy, Cours d'analyse de l'Ecole Royale Polytechnique pt. 1, 1821 (Oeuvres ser. 2, III, Gauthier-Villars, 1897).

[3] G. P. L. Dirichlet, Über die Darstellung ganz willkürlicher Funktion durch Sinus- und Cosinusreihen, Repertorium der Physik, Bd. 1 (1837), 152–174 (Werke, Bd. 1, G. Reimer, 1889, I, p. 133–160).

171 (X.5) Functions of Bounded Variation

A. Monotone Functions

A function (or mapping) f from an †ordered set X to another ordered set Y is called a **monotone increasing (monotone decreasing) function** if

$$x_1 < x_2 \quad \text{implies} \quad f(x_1) \leqslant f(x_2)$$

$$(x_1 < x_2 \quad \text{implies} \quad f(x_1) \geqslant f(x_2)). \tag{1}$$

A monotone increasing (decreasing) function is also called a **nondecreasing (nonincreasing)** function. In both cases, the function f is called simply a **monotone function**. If X and Y are †totally ordered sets and the inequality $<$ $(>)$ holds in (1) instead of $\leqslant$ $(\geqslant)$, then f is called a **strictly (monotone) increasing (strictly (monotone) decreasing)** function. In both cases, f is called simply a **strictly monotone function.**

In particular, when X and Y are subsets of the real line $\mathbf{R}$, a monotone function is continuous except for at most a countable number of points. Hence it is †Riemann integrable in a finite interval provided that it is bounded. A continuous real function $f(x)$ defined on an interval in $\mathbf{R}$ is †injective if and only if it is strictly monotone. In such a case, the range of the function $f(x)$ is also an interval, and the inverse function is also strictly monotone. Furthermore, a differentiable real function f defined on an interval is monotone if and only if its derivative f' is always $\geqslant 0$ (monotone increasing) or always $\leqslant 0$ (monotone decreasing). If $f' > 0$ (<0), f is strictly monotone increasing (decreasing).

B. Functions of Bounded Variation

Let $f(x)$ be a real bounded function defined on a closed interval $[a,b]$ in $\mathbf{R}$. Given a subdivision of the interval $a = x_0 < x_1 < x_2 < \ldots < x_n = b$, we denote the sum of positive differences $f(x_i) - f(x_{i-1})$ by P and the sum of negative differences $f(x_i) - f(x_{i-1})$ by $-N$. Then we easily obtain

$$P - N = f(b) - f(a),$$

$$P + N = \sum_i |f(x_i) - f(x_{i-1})|.$$

The suprema of P, N, and $P+N$ for all possible subdivisions of $[a,b]$ are called the **positive variation**, the **negative variation**, and the **total variation** of the function $f(x)$ in the interval $[a,b]$, respectively. If any of these three values is finite, then all three values are finite. In such a case, the function $f(x)$ is called a **function of bounded variation**. The positive and negative variations $\pi(t)$, $\nu(t)$ of the function $f(x)$ in the interval $[a,t]$ are monotone increasing functions with respect to t, and we have

$$f(x)-f(a)=\pi(x)-\nu(x) \tag{2}$$

if $f(x)$ is a function of bounded variation. A monotone function is a function of bounded variation, and the sum, the difference, or the product of two functions of bounded variation is also a function of bounded variation. Hence $f(x)$ is a function of bounded variation if and only if it is the difference of two monotone functions. The representation (2) (representing a function of bounded variation as the difference of two monotone functions) is called the **Jordan decomposition** of the function $f(x)$. A function of bounded variation is Riemann integrable, continuous except for at most a countable number of points, and differentiable †almost everywhere.

A continuous function is not necessarily a function of bounded variation (e.g., $x\sin\frac{1}{x}$). A discontinuous function may be a function of bounded variation (e.g., sgn(x)). However, an †absolutely continuous function, a differentiable function with bounded derivative, or a function satisfying the †Lipschitz condition is a function of bounded variation.

The notion of functions of bounded variation was introduced by C. Jordan in connection with the notion of the length of curves (→ 245 Length and Area). The notion of bounded variation has been extended to more general †set functions. Functions of bounded variation are frequently used in the †Stieltjes integral with a continuous integrand.

References

[1] T. Takagi, Kaiseki gairon (Japanese; A course of analysis), Iwanami, third edition, 1961.

[2] S. Saks, Theory of the integral, Warsaw, 1937.

[3] M. Tsuji, Zitu kansûron (Japanese; Theory of real functions), Maki, 1962.

172 (XIV.7) Functions of Confluent Type

A. Confluent Hypergeometric Functions

If some singularities of an ordinary differential equation of †Fuchsian type are confluent to each other, we obtain a **confluent differential equation** whose solutions are called **functions of confluent type**. The equations that appear frequently in practical problems are the **confluent hypergeometric differential equations**:

$$z\frac{d^2w}{dz^2}+(\gamma-z)\frac{dw}{dz}-\alpha w=0 \tag{1}$$

and related equations. Equation (1) corresponds to the †hypergeometric differential equation for which a †regular singular point coincides with the point at infinity and is an †irregular singular point of class 1. For (1), $z=0$ is a regular singular point, and a series solution (radius of convergence ∞) is given by

$$F(\alpha,\gamma;z)={}_1F_1(\alpha,\gamma;z)$$
$$=\sum_{n=0}^{\infty}\frac{\alpha(\alpha+1)\dots(\alpha+n-1)}{n!\gamma(\gamma+1)\dots(\gamma+n-1)}z^n \quad (|z|<\infty), \tag{2}$$

where γ is not equal to zero or a negative integer. The function ${}_1F_1$ in (2) is a †generalized hypergeometric function due to Barnes and is called a **hypergeometric function of confluent type** or **Kummer function**. If γ is not equal to a positive integer, the other solution of (1) independent of (2) is given by $z^{1-\gamma}F(1+\alpha-\gamma,\ 2-\gamma;\ z)$ (→ Appendix A, Table 19.I).

B. Whittaker Functions

Equation (1) with $w=e^{z/2}z^{-\gamma/2}W$, $\gamma-2\alpha=k$, $\gamma^2-2\gamma=4m^2-1$ reduces to **Whittaker's differential equation**

$$\frac{d^2W}{dz^2}+\left(-\frac{1}{4}+\frac{k}{z}+\frac{\frac{1}{4}-m^2}{z^2}\right)W=0. \tag{3}$$

If $2m$ is not equal to an integer, (3) has two series solutions for any finite z:

$$M_{k,m}(z)$$
$$=z^{(1/2)+m}e^{-z/2}F(\tfrac{1}{2}+m-k,1+2m;z),$$
$$M_{k,-m}(z)$$
$$=z^{(1/2)-m}e^{-z/2}F(\tfrac{1}{2}-m-k,\ 1-2m;z).$$

If $2m$ is an integer, since the functions $M_{k,m}$ and $M_{k,-m}$ are linearly dependent, E. T. Whittaker considered a solution of the form:

$$W_{k,m}(z)=-\frac{1}{2\pi i}\Gamma(k+\tfrac{1}{2}-m)z^ke^{-z/2}\times$$
$$\int_{\infty}^{(0+)}(-t)^{-k-(1/2)+m}\left(1+\frac{t}{z}\right)^{k-(1/2)+m}e^{-t}dt.$$

If $k-\frac{1}{2}-m$ is equal to a negative integer, this integral does not exist. The function

$$W_{k,m}(z)=\frac{z^ke^{-z/2}}{\Gamma(\frac{1}{2}-k+m)}\int_0^{\infty}t^{-k-(1/2)+m}$$
$$\times\left(1+\frac{t}{z}\right)^{k-(1/2)+m}e^{-t}dt$$

for Re $(k-\frac{1}{2}-m)\leqslant 0$ is defined for any m, k, and for any z except when z is a negative real number. We call $M_{k,m}$ and $W_{k,m}$ the **Whittaker functions.** †Bessel functions are particular cases of these functions, and the following relation is satisfied:

$$J_n(z)=\frac{z^{-1/2}}{2^{2n+(1/2)}i^{n+(1/2)}\Gamma(n+1)}M_{0,n}(2iz).$$

In Whittaker's differential equation, since $W_{-k,m}(-z)$ is also a solution and $W_{k,m}(z)/W_{-k,m}(-z)$ is not equal to a constant, $W_{k,m}(z)$ and $W_{-k,m}(-z)$ may be considered a pair of fundamental solutions (→ Appendix A, Table 19.II).

C. Parabolic Cylinder Functions

Putting $x=(\xi^2-\eta^2)/2$ and $y=\xi\eta$, the curves corresponding to $\xi=$constant and to $\eta=$ constant, respectively, constitute families of orthogonal parabolas. The curvilinear coordinates (ξ,η,z) in three dimensions are called **parabolic cylindrical coordinates**. By using parabolic coordinates, separating variables in Laplace's equation into the form $f(\xi)g(\eta)e^{i\alpha z}$, and making a simple transformation, we find that f and g satisfy a differential equation of the form

$$\frac{d^2F}{dz^2}+(n+\tfrac{1}{2}-\tfrac{1}{4}z^2)F=0. \qquad (4)$$

By means of the Whittaker function $W_{k,m}(z)$, a solution $D_n(z)$ of (4) is represented by

$$D_n(z)=2^{n/2+(1/4)}z^{-1/2}W_{n/2+(1/4),-1/4}(\tfrac{1}{2}z^2).$$

Equation (4) is called **Weber's differential equation** (or the **Weber-Hermite differential equation**), and $D_n(z)$ the **Weber function**. Another solution of (4) is $D_{-n-1}(iz)$ or $D_{-n-1}(-iz)$. The solutions of (4) are called **parabolic cylinder functions**. In particular, if n is equal to a nonnegative integer, then

$$H_n(z)=2^{-n/2}\exp(\tfrac{1}{2}z^2)D_n(\sqrt{2}\,z)$$

is the †Hermite polynomial of degree n. Solutions of differential equations for harmonic oscillators in quantum mechanics are of this form.

In general, suppose that three regular singular points are confluent to the point at infinity, and that they are reduced to an irregular singular point of class 2. Suppose further that there are no other singularities. Then differential equations of order 2 with these conditions are transformed into the form (4), whose solutions are represented by parabolic cylinder functions. Differential equations of the form (4) are reduced to confluent hypergeometric differential equations if z^2 is chosen as an independent variable (→ Appendix A, Table 20.III).

D. Indefinite Integrals of Elementary Functions

Since exponential and trigonometric functions can be represented by particular types of Kummer functions, their indefinite integrals that cannot be represented by elementary functions (e.g., †incomplete Γ-functions and the error function $\mathrm{Erf}\,z=\int_0^z\exp(-t^2)\,dt$) may be represented by Kummer or Whittaker functions. They are included in a family of †special functions of confluent type. The functions defined by

$$C(z)=\frac{1}{\sqrt{2\pi}}\int_0^z\frac{\cos t}{\sqrt{t}}\,dt,$$

$$S(z)=\frac{1}{\sqrt{2\pi}}\int_0^z\frac{\sin t}{\sqrt{t}}\,dt$$

are called **Fresnel integrals**, which are also represented in terms of the Whittaker function as

$$C(z)-iS(z)$$
$$=\frac{1-i}{2}\left(1-\frac{e^{-\pi i/8}}{\sqrt{\pi}}z^{1/4}e^{-z^2/2}W_{-1/4,1/4}(iz)\right).$$

Fresnel integrals first appeared in the theory of the diffraction of waves. More recently they have been applied to designing highways for high-speed automobiles. Furthermore, the functions

$$C(u)=\int_0^u\cos\left(\frac{\pi}{2}s^2\right)ds,$$

$$S(u)=\int_0^u\sin\left(\frac{\pi}{2}s^2\right)ds$$

(obtained by a change of variables $z=\pi u^2/2$) are also called Fresnel integrals. Numerical tables are available for them. The curves $x=C$ and $y=S$ with a parameter z or u are called **Cornu's spiral** (Fig. 1). The functions

$$\mathrm{Li}\,x=\int_0^x\frac{dt}{\log t},\qquad \mathrm{Ei}\,x=\int_{-\infty}^x\frac{e^t}{t}\,dt,$$

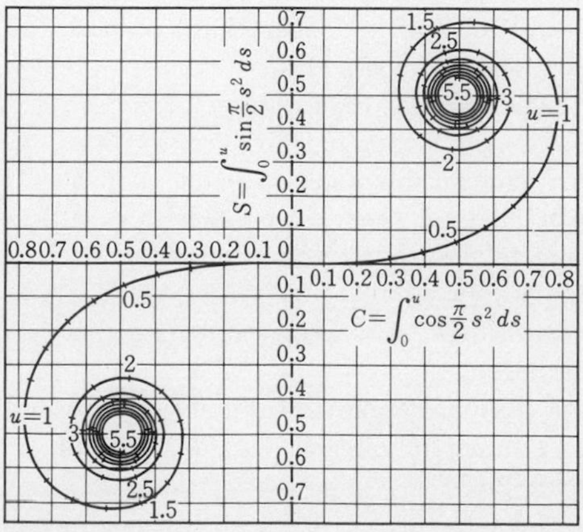

Fig. 1

where a †principal value must be taken at $t=0$ if $x>0$,

$$\mathrm{Si}\,x=\int_0^x \frac{\sin t}{t}\,dt, \quad \text{and} \quad \mathrm{Ci}\,x=-\int_x^\infty \frac{\cos t}{t}\,dt$$

are called the **logarithmic integral**, **exponential integral**, **sine integral**, and **cosine integral** (or **integral logarithm**, **integral exponent**, **integral sine**, and **integral cosine**), respectively. They satisfy the relations

$$\mathrm{Ei}\,x=\mathrm{Li}\,e^x, \quad \mathrm{Ei}\,ix=\mathrm{Ci}\,x+i\,\mathrm{Si}\,x+(\pi/2)i.$$

They have important applications, $\mathrm{Ei}\,x$ in quantum mechanics, $\mathrm{Si}\,x$ and $\mathrm{Ci}\,x$ in electronic engineering, and $\mathrm{Li}\,x$ in estimating the number of †primes less than x (→ 128 Distribution of Prime Numbers). $\mathrm{Li}\,x$ is also denoted by $\mathrm{li}\,x$ (→ Appendix A, Table 19.II).

E. Stokes's Equation

Consider a linear differential equation of the second order with five regular singular points including the point at infinity such that the difference of the characteristic indices at every singularity is equal to $1/2$. Such equations are called **generalized Lamé's differential equations**. F. Klein and M. Bôcher have shown that every linear differential equation that is commonly treated in mathematical physics is represented by a confluent type of generalized Lamé's equation. Among these equations, if all five singularities are confluent to the point at infinity, the resulting equation is called **Stokes's differential equation**, which is applied to the investigation of diffraction. This is reduced to †Bessel's differential equation of order $1/3$ by suitable transformations of the independent and dependent variables.

References

[1] H. Buchholz, The confluent hypergeometric function with special emphasis on its applications, Springer, 1969.
[2] L. J. Slater, Confluent hypergeometric functions, Cambridge Univ. Press, 1960.
For the logarithmic integral, etc.,
[3] N. Nielsen, Theorie der Integrallogarithmus und verwandter Transzendenten, Teubner, 1906.
[4] British Ass. Adv. Sci., Mathematical tables 1, London, 1931.
[5] Nat. Bur. Standards, Tables of sine, cosine and exponential integrals I, II, Washington, D. C., 1940.
Also → references to 381 Special Functions.

173 (XII.1) Function Spaces

A. General Remarks

It is a general method in modern analysis to consider a set X of mappings of a space Ω into another space A as a space (→ 376 Sets) and its elements (namely, mappings of Ω into A) as points of the space X, and to investigate them as geometric objects. In particular, it is important to consider the case where Ω is a †topological space or a †measure space and A is the real number field $\mathbf{R}$ or the complex number field $\mathbf{C}$, that is, the case of a set of real- or complex-valued functions defined on Ω. Spaces of such functions satisfying certain conditions, such as continuity, measurability, etc., are considered. Such spaces are generally called **function spaces**; they are †topological linear spaces (→ 39 Banach Spaces; 407 Topological Linear Spaces).

B. Examples of Function Spaces

In this article, some examples of important function spaces are given. Throughout this section, all functions are real- or complex-valued, and two functions that are equal to each other †almost everywhere on a measure space are identified.

(1) The Function Spaces $C(\Omega)$ and $C_0(\Omega)$. The totality of continuous functions $f(x)$ defined on a compact †Hausdorff space Ω is denoted by $C(\Omega)$. With $f(x)+g(x)$ and $\alpha f(x)$ for α a real or complex number (denoted by $f+g$ and αf, respectively), $C(\Omega)$ forms a †linear space. Furthermore, if the †norm of f is defined by $\|f\|=\sup_{x\in\Omega}|f(x)|$, called the **supremum norm** (or **uniform norm**) of f, then $C(\Omega)$ is †complete with respect to this norm. This means that $C(\Omega)$ is complete with respect to the metric ρ defined by $\rho(f,g)=\|f-g\|$, and accordingly, $C(\Omega)$ is a †Banach space. Then the condition that $\lim_{n\to\infty}\|f_n-f\|=0$ is equivalent to the condition that $\{f_n(x)\}$ converges to $f(x)$ uniformly on Ω as $n\to\infty$. Suppose that a subset R of $C(\Omega)$ satisfies the following three conditions: (1) R is a †ring over the complex number field with respect to the usual addition and multiplication of functions and contains the function identically equal to one. (2) For any two points x and y of Ω, there exists a function $f\in R$ satisfying $f(x)\neq f(y)$. (3) For any $f\in R$, there exists an $f^*\in R$ such that $f^*(x)=\overline{f(x)}$ on Ω. Then R is dense in $C(\Omega)$ with respect to the supremum norm (namely, in the sense of uniform convergence). This fact is called the **Weierstrass-Stone theorem** (or **Stone-Gel'fand theorem**). If Ω is a compact †uniform space, then in order for a subset E of $C(\Omega)$ to be †totally bounded (i.e., any sequence of functions in E contains a subsequence that converges uniformly on Ω), it is necessary and sufficient that E be †uniformly bounded and †equicontinuous (**Ascoli-Arzelà theorem**). When Ω is a topological space that is not necessarily compact, the totality of bounded continuous functions on Ω (denoted by $C(\Omega)$) is also a Banach space with respect to the supremum norm $\|f\|=\sup_{x\in\Omega}|f(x)|$. In particular, if Ω is locally compact, the totality of continuous functions with compact support is denoted by $C_0(\Omega)$, where the **support** (or **carrier**) of a function f is the †closure of the set $\{x \mid f(x)\neq 0\}$ in Ω and is usually denoted by supp$f$. Since $C_0(\Omega)$ is not complete, it is not a Banach space with respect to the supremum norm.

(2) The Function Space $L_p(\Omega)$ $(1\leqslant p<\infty)$. Let Ω be a space in which a †measure μ is defined. The totality of †measurable functions $f(x)$ on Ω such that $|f(x)|^p$ is integrable,

$$\int_\Omega |f(x)|^p\,d\mu(x)<\infty,$$

is denoted by $L_p(\Omega)$. In particular, if Ω is the interval (a,b), it may be denoted by $L_p(a,b)$. If $f, g\in L_p(\Omega)$, then $f+g\in L_p(\Omega)$ by the †Minkowski inequality. Accordingly, $L_p(\Omega)$ is a linear space. Furthermore, $L_p(\Omega)$ is a Banach space with respect to the norm

$$\|f\|=\|f\|_p=\left(\int_\Omega |f(x)|^p\,d\mu(x)\right)^{1/p}.$$

If $\lim_{n\to\infty}\|f_n-f\|_p=0$, we say that the sequence $\{f_n\}$ converges to f **in the mean of order** p (or **in the mean of power** p), and write $\text{l.i.m.}_{n\to\infty} f_n=f$. If $\{f_n\}$ converges to f in the mean of order 2, we simply say that $\{f_n\}$ converges to f in the mean. (The notation l.i.m. means the **limit in the mean** and is used mostly when $p=2$.) For any f, $g\in L_2(\Omega)$, $(f,g)=\int_\Omega f(x)\overline{g(x)}\,d\mu(x)$ is well defined, by the †Schwarz inequality, and has the properties of the †inner product. Hence, $L_2(\Omega)$ is a †Hilbert space.

The L_p spaces, $1<p<\infty$, are generalized in the following way. Let $\Phi(s)$ be a convex and nondecreasing function on $[0,\infty)$ satisfying $\Phi(0)=0$ and $\Phi(s)/s\to\infty$ as $s\to\infty$. Denote by $L_\Phi(\Omega)$ ($L_\Phi^*(\Omega)$) the set of all functions $f(x)$ such that $\Phi(|f(x)|)$ is integrable ($\Phi(k|f(x)|)$ is integrable for some $k>0$). $L_\Phi(\Omega)=L_\Phi^*(\Omega)$ if $\Phi(2s)\leqslant C\Phi(s)$. $L_\Phi^*(\Omega)$ is a Banach space, called the **Orlicz space**, under the norm

$$\|f\|=\inf\left\{\lambda>0\,\middle|\,\int\Phi(\lambda^{-1}|f(x)|)\,d\mu(x)\leqslant 1\right\}.$$

$L_p(\Omega)$, $1<p<\infty$, is the Orlicz space for $\Phi(s)=s^p$.

(3) The Function Space $M(\Omega)$. Let Ω and μ be as in (2). A measurable function $f(x)$ on Ω is said to be **essentially bounded** if there exists a positive number α such that $|f(x)|\leqslant\alpha$ almost everywhere on Ω. The infimum of such α is called the **essential supremum** of f, denoted by $\text{ess sup}_{x\in\Omega}|f(x)|$. The totality of essentially bounded measurable functions on Ω (denoted by $M(\Omega)$) is a Banach space with respect to the norm $\|f\|=\|f\|_\infty=\text{ess sup}_{x\in\Omega}|f(x)|$. If $\mu(\Omega)<\infty$, then $M(\Omega)\subset L_p(\Omega)$ for any $p\geqslant 1$, and $\|f\|_\infty=\lim_{p\to\infty}\|f\|_p$ for any $f\in M(\Omega)$. From this point of view, $M(\Omega)$ is also denoted by $L_\infty(\Omega)$ even when $\mu(\Omega)=\infty$. This is also the reason why the notation $\|\cdot\|_\infty$ is used for the norm in $M(\Omega)$.

(4) The Function Space $S(\Omega)$. Let μ be a measure in Ω such that $\mu(\Omega)<\infty$, and denote by $S(\Omega)$ the totality of measurable functions on Ω that take finite value almost everywhere. Then $\|f\|=\int_\Omega(|f(x)|/(1+|f(x)|))\,d\mu(x)$ for $f\in S(\Omega)$ has the properties of the †quasinorm, and $S(\Omega)$ is a †Fréchet space. We have $\lim_{n\to\infty}\|f_n-f\|=0$ if and only if

$$\lim_{n\to\infty}\mu(\{x \mid |f_n(x)-f(x)|>\varepsilon\})=0$$

for any positive number ε. Convergence of this type is called **convergence in measure** (or **asymptotic convergence**), and is the same notion as †convergence in probability of a sequence of †random variables (→ 339 Probability). In general, if $\{f_n\}\in L_p(\Omega)$ converges to $f\in L_p(\Omega)$ in the mean of order p, then $\{f_n\}$ converges to f asymptotically, but the converse is not true. If $\{f_n\}\in S(\Omega)$ converges to $f\in S(\Omega)$ almost everywhere, then $\{f_n\}$ converges to f asymptotically. Any sequence $\{f_n\}$ that converges to f asymptotically contains a subsequence $\{f_{n_\nu}\}$ that converges to f almost everywhere.

(5) The Sequence Spaces c, l_p, m, and s. The space $C(\Omega)$, where Ω is the compact space $\{1,1/2,1/3,\ldots,1/n,\ldots,0\}$, is denoted by (c) or simply c. The spaces $L_p(\Omega)$ (resp. $M(\Omega)$), where Ω is the space $\{1,2,3,\ldots,n,\ldots\}$, of which each point has unit mass, are denoted by (l_p) (resp. (m)) or simply l_p (resp. m). On the other hand, the space $S(\Omega)$, where $\Omega=\{1,2,\ldots,n,\ldots\}$, provided with a measure such that the point n has mass $1/2^n$, is denoted by (s) or s. Assume that the space $L_2(\Omega)$ mentioned in (2) is †separable and that $\{\varphi_n\}$ is a †complete orthonormal system in $L_2(\Omega)$. Then putting

$$\xi_n = \int_\Omega f(x)\overline{\varphi_n(x)}\,d\mu(x)$$

(†Fourier coefficient) for any $f \in L_2(\Omega)$, we have $\{\xi_n\} \in l_2$ and $\sum_{n=1}^{\infty} |\xi_n|^2 = \|f\|^2$. Conversely, for any $\{\xi_n\} \in l_2$, there exists an $f = \sum_{n=1}^{\infty} \xi_n \varphi_n \in L_2(\Omega)$ whose Fourier coefficients are the given ξ_n (**Riesz-Fischer theorem**). By means of this correspondence, separable spaces $L_2(\Omega)$ and l_2 are mutually isomorphic as Hilbert spaces.

(6) The Function Spaces $A(\Omega)$ and $A_p(\Omega)$ ($p \geqslant 1$). Let Ω be a subdomain of the complex plane **C**. The totality of functions bounded and continuous on the closure of Ω and †holomorphic in Ω (denoted by $A(\Omega)$) is a Banach space with respect to the norm $\|f\| = \sup_{z\in\Omega} |f(z)|$. For any $p \geqslant 1$, the totality of functions f holomorphic in Ω and satisfying $\int_\Omega |f(z)|^p\,dx\,dy < \infty$ ($z = x + iy$; $dx\,dy$ is a 2-dimensional Lebesgue measure), denoted by $A_p(\Omega)$, is also a Banach space with respect to the norm $\|f\|_p = (\int_\Omega |f(z)|^p\,dx\,dy)^{1/p}$. In particular, it is a Hilbert space when $p = 2$.

We introduce notation to describe some function spaces over domains in the n-dimensional Euclidean space $\mathbf{R}^n$. For an arbitrary point $x = (x_1, \ldots, x_n) \in \mathbf{R}^n$, we set $|x| = (x_1^2 + \ldots + x_n^2)^{1/2}$. For an arbitrary n-tuple $\alpha = (\alpha_1, \ldots, \alpha_n)$ of nonnegative integers, we set $|\alpha| = \alpha_1 + \ldots + \alpha_n$ and $D^\alpha = D_1^{\alpha_1} \ldots D_n^{\alpha_n}$ (where $D_j = \partial/\partial x_j$). We assume that Ω is a subdomain of $\mathbf{R}^n$ unless otherwise mentioned.

(7) The Function Spaces $C^l(\Omega)$, $C_0^l(\Omega)$ ($l = 0, 1, 2, \ldots, \infty$), $\mathfrak{D}(\Omega)$, and $\mathfrak{E}(\Omega)$. The totality of l-times continuously differentiable functions in Ω (namely, differentiable functions of †class C^l in Ω) is denoted by $C^l(\Omega)$. We say that a sequence $\{f_\nu\}$ of functions in $C^l(\Omega)$ converges to 0 in $C^l(\Omega)$ if $|D^\alpha f_\nu(x)|$ converges to 0 uniformly on every compact subset of Ω for every α satisfying $0 \leqslant |\alpha| \leqslant l$ ($0 \leqslant |\alpha| < \infty$ if $l = \infty$). The totality of functions in $C^l(\Omega)$ whose supports are compact subsets of Ω is denoted by $C_0^l(\Omega)$. We say that a sequence $\{f_\nu\}$ of functions in $C_0^l(\Omega)$ converges to 0 in $C_0^l(\Omega)$ if $\operatorname{supp} f_\nu$ ($\nu = 1, 2, \ldots$) is contained in a compact subset of Ω independent of ν and $\{f_\nu\}$ converges to 0 in $C^l(\Omega)$. $C_0^\infty(\Omega)$ and $C^\infty(\Omega)$ are also denoted by $\mathfrak{D}(\Omega)$ and $\mathfrak{E}(\Omega)$, respectively.

When Ω is the closure of a subdomain of $\mathbf{R}^n$, we denote by $C^l(\Omega)$ the totality of functions $f(x)$ on Ω such that for all α satisfying $0 \leqslant |\alpha| \leqslant l$, the $D^\alpha f(x)$ are bounded and uniformly continuous in the interior of Ω (and accordingly, they can be continuously extended up to the boundary of Ω). Convergence in $C^l(\Omega)$ is defined in the same way as mentioned above; that is, it is equivalent to convergence with respect to the norm

$$\|f\| = \sup\{|D^\alpha f(x)| \mid x \in \Omega,\ \ |\alpha| \leqslant l\}.$$

In particular, if Ω is compact, $C^0(\Omega)$ is the same as $C(\Omega)$ mentioned in (1).

(8) The Function Spaces $\mathfrak{D}_{L_p}(\Omega)$ ($1 \leqslant p \leqslant \infty$) and $\mathfrak{B}(\Omega)$. The totality of functions $f(x)$ in $C^\infty(\Omega)$ such that, for all α, $D^\alpha f(x)$ belongs to $L_p(\Omega)$ with respect to †Lebesgue measure is denoted by $\mathfrak{D}_{L_p}(\Omega)$. The neighborhood $V_{l,\varepsilon}$ of 0 in $\mathfrak{D}_{L_p}(\Omega)$ is defined to be the totality of functions $f(x)$ such that $\|D^\alpha f\|_p < \varepsilon$ for any α satisfying $|\alpha| \leqslant l$. In particular, $\mathfrak{D}_{L_\infty}(\Omega)$ is also denoted by $\mathfrak{B}(\Omega)$.

To represent each of the spaces mentioned in (7) and (8) for $\Omega = \mathbf{R}^n$, we usually omit $(\mathbf{R}^n)$; for example, $\mathfrak{D}(\mathbf{R}^n)$ and $\mathfrak{D}_{L_p}(\mathbf{R}^n)$ are denoted by $\mathfrak{D}$ and $\mathfrak{D}_{L_p}$, respectively.

(9) The Function Space $\mathfrak{S}$. A function $f(x)$ is called a **rapidly decreasing C^∞-function** if it belongs to $C^\infty(\mathbf{R}^n)$ and satisfies

$$\lim_{|x|\to\infty} |x|^k |D^\alpha f(x)| = 0$$

for any α and any integer $k > 0$. The totality of rapidly decreasing C^∞-functions is denoted by $\mathfrak{S}$. The neighborhood $V_{l,k,\varepsilon}$ of 0 in $\mathfrak{S}$ is defined to be the totality of functions $f(x)$ such that $(1 + |x|^2)^k |D^\alpha f(x)| < \varepsilon$ for any α satisfying $|\alpha| \leqslant l$.

(10) The Function Space $\mathcal{O}_M$. A function $f(x)$ is called a **slowly increasing C^∞-function** if it belongs to $C^\infty(\mathbf{R}^n)$ and if, for every α, $|D^\alpha f(x)|$ is dominated by a polynomial of $|x|$ (which may depend on α). (A function $f(x)$ that satisfies $|f(x)| \leqslant |P(x)|$ for a polynomial $P(x)$ of x is often called a **slowly increasing function**.) The set of all slowly increasing C^∞-functions is denoted by $\mathcal{O}_M$. We say that a sequence $\{f_\nu\}$ of functions in $\mathcal{O}_M$ converges to 0 in $\mathcal{O}_M$ if $\{\varphi(x) D^\alpha f_\nu(x)\}$ converges to 0 uniformly on $\mathbf{R}^n$ for any α and any $\varphi \in \mathfrak{S}$.

(11) Spaces of Distributions and $\mathcal{O}'_C$. The †dual spaces (→ 407 Topological Linear Spaces) of $\mathfrak{D}$, $\mathfrak{S}$, and $\mathfrak{B}$ are denoted by $\mathfrak{D}'$, $\mathfrak{S}'$, and $\mathfrak{B}'$, respectively. Then $\mathfrak{D}'$ is the set of all Schwartz distributions (→ 130 Distributions (Generalized Functions)). From the relation $\mathfrak{D} \subset \mathfrak{S} \subset \mathfrak{B}$, it follows that $\mathfrak{D}' \supset \mathfrak{S}' \supset \mathfrak{B}'$. A distribution T is called a †rapidly decreasing distribution if $(1+|x|)^k T \in \mathcal{B}'$ for any integer $k>0$. The totality of rapidly decreasing distributions is denoted by $\mathcal{O}'_C$. The condition $T \in \mathcal{O}'_C$ is equivalent to the condition that the †regularization $T * \varphi$ of T by any $\varphi \in \mathcal{D}$ belongs to $\mathcal{S}$. Accordingly, the distribution T_f defined by a function $f \in \mathcal{S}$ belongs to $\mathcal{O}'_C$.

(12) The Sobolev Space $W_p^l(\Omega)$. Consider functions $f(x)$ such that for all α satisfying $|\alpha| \leqslant l$, the derivatives $D^\alpha f(x)$ in the sense of distribution (→ 130 Distributions (Generalized Functions)) belong to $L_p(\Omega)$ with respect to Lebesgue measure in Ω. The totality of such functions (denoted by $W_p^l(\Omega)$) is a Banach space with respect to the norm

$$\|f\| = \|f\|_{W_p^l} = \left(\sum_{0 \leqslant |\alpha| \leqslant l} \int_\Omega |D^\alpha f(x)|^p \, dx \right)^{1/p}.$$

Clearly $W_p^0(\Omega) = L_p(\Omega)$. $W_2^l(\Omega)$ is a Hilbert space with respect to the inner product

$$(f,g) = \int_\Omega \sum_{0 \leqslant |\alpha| \leqslant l} D^\alpha f(x) \cdot \overline{D^\alpha g(x)} \, dx.$$

Sometimes $W_2^l(\Omega)$ is denoted by $H^l(\Omega)$. Also, $l > (n/p) + k$ implies $W_p^l(\Omega) \subset C^k(\Omega)$. More precisely, this implication holds if two functions that are equal to each other almost everywhere in Ω are identified (**Sobolev's embedding theorem**).

(13) The Hilbert Space $H_0^l(\Omega)$. The Hilbert space obtained as the †completion of $\mathcal{D}(\Omega)$ with respect to the norm

$$\|f\| = \left(\int_\Omega \sum_{0 \leqslant |\alpha| \leqslant l} |D^\alpha f(x)|^2 \, dx \right)^{1/2}$$

is denoted by $H_0^l(\Omega)$. We have $H_0^0(\Omega) = W_2^0(\Omega) = L_2(\Omega)$. However, if $l \geqslant 1$, we have $H_0^l(\Omega) \subset W_2^l(\Omega)$, and identity does not hold.

C. Dual Spaces

Denote the †norm of a †bounded linear functional Φ on a normed linear space by $\|\Phi\|$ (→ 39 Banach Spaces). Any bounded linear functional Φ on $C(\Omega)$ (Ω a compact Hausdorff space) is expressible by the †Stieltjes integral

$$\Phi(f) = \int_\Omega f(x) \, d\varphi(x), \quad f \in C(\Omega), \tag{1}$$

where φ is a (real- or complex-valued) †countably additive set function of bounded variation defined on the Borel sets in Ω. The totality of such φ is denoted by $BV(\Omega)$. Conversely, any $\varphi \in BV(\Omega)$ gives a bounded linear functional on $C(\Omega)$ defined by (1), and

$$\|\Phi\| = \text{the total variation of } \varphi \text{ over } \Omega. \tag{2}$$

Hence the dual space of $C(\Omega)$ is isomorphic to the Banach space obtained by defining the norm by (2) in the space $BV(\Omega)$.

Any bounded linear functional Φ on $L_1(\Omega)$ is expressible as

$$\Phi(f) = \int_\Omega f(x) \varphi(x) \, d\mu(x), \quad f \in L_1(\Omega), \tag{3}$$

with a suitable $\varphi \in M(\Omega)$; and $\|\Phi\| = \|\varphi\|_\infty$. Conversely, any $\varphi \in M(\Omega)$ defines a bounded linear functional on $L_1(\Omega)$ by means of (3). Accordingly, the dual space of $L_1(\Omega)$ is isomorphic to $M(\Omega)$.

The dual space of $L_p(\Omega)$ $(1 < p < \infty)$ is isomorphic to $L_q(\Omega)$, where q is the real number defined by $(1/p) + (1/q) = 1$ (accordingly, $1 < q < \infty$) and is called the **conjugate exponent** of p. Any bounded linear functional on $L_p(\Omega)$ is expressible by the formula in (3) (where $f \in L_p(\Omega)$) with $\varphi \in L_q(\Omega)$, and $\|\Phi\| = \|\varphi\|_q$.

The dual space of $M(\Omega)$ is isomorphic to the normed linear space of all (real- or complex-valued) finitely additive set functions φ defined on all measurable sets in Ω, of bounded variation over Ω, and absolutely continuous with respect to the measure μ given in Ω (i.e., $\mu(N) = 0$ implies $\varphi(N) = 0$), where $\|\varphi\|$ is defined as the total variation of φ over Ω. Any bounded linear functional Φ on $M(\Omega)$ is expressible by the formula in (1), where the precise meaning of the integral in the right-hand side is as follows: If both f and φ are real-valued, consider the division Δ of the closed interval $[-\|f\|, \|f\|]$:

$$-\|f\| = \alpha_0 < \alpha_1 < \ldots < \alpha_n = \|f\|, \tag{Δ}$$

and put $\varepsilon_\Delta = \sup_j(\alpha_j - \alpha_{j-1})$ and $s_\Delta = \sum_j \alpha_j \varphi(A_j)$, where $A_j = \{x \mid \alpha_{j-1} < f(x) \leqslant \alpha_j\}$. Then as $\varepsilon_\Delta \to 0$, s_Δ tends to a definite value, which we write as $\int_\Omega f \, d\varphi$. If f and φ are complex-valued, consider the real and imaginary parts of these functions, and apply the method just given to every combination of them.

No bounded linear functional on $S(\Omega)$ exists except the functional that is identically zero on $S(\Omega)$.

The sequence spaces c, l_p $(1 \leqslant p < \infty)$, m, and s are special cases of $C(\Omega)$, $L_p(\Omega)$, $M(\Omega)$, and $S(\Omega)$, respectively. Hence their dual spaces may be described explicitly. For example, the dual space of c (resp. l_1) is l_1 (resp. m), and if $1 < p < \infty$, the dual space of l_p is l_q (where $(1/p) + (1/q) = 1$).

From previous results of this section, we see that $L_p(\Omega)$ and l_p are †reflexive if $1 < p < \infty$; in particular, each of $L_2(\Omega)$ and l_2 has a dual space isomorphic to itself, while $C(\Omega)$, $L_1(\Omega)$, $M(\Omega)$, c, l_1, and m are not reflexive.

Since the dual space $\mathcal{D}'$ of $\mathcal{D}$ is the space of all Schwartz distributions, the dual spaces of $\mathcal{D}_{L_p}$ and $\mathcal{S}$ are (algebraically) linear subspaces of $\mathfrak{D}'$. Moreover, we have the following diagram, where $E \subset F$ means that E is contained in F and the topology in E is †stronger than the topology induced from F:

$$\begin{array}{ccccccccccccc} \mathfrak{D} & \subset & \mathfrak{S} & \subset & \mathfrak{D}_{L_p} & \subset & \mathfrak{D}_{L_q} & \subset & \mathfrak{B} & \subset & \mathcal{O}_M & \subset & \mathfrak{E} \\ \cap & & \cap & & \cap & & \cap & & \cap & & \cap & & \cap \\ \mathfrak{E}' & \subset & \mathcal{O}'_C & \subset & \mathfrak{D}'_{L_p} & \subset & \mathfrak{D}'_{L_q} & \subset & \mathfrak{B}' & \subset & \mathfrak{S}' & \subset & \mathfrak{D}' \end{array}$$

(here we assume that $1 \leqslant p \leqslant q < \infty$). This diagram shows the relations between function spaces over $\mathbf{R}^n$. However, if we omit $\mathfrak{S}$, $\mathfrak{S}'$, $\mathcal{O}_M$, and $\mathcal{O}'_C$, the relations hold between function spaces over an arbitrary subdomain Ω of $\mathbf{R}^n$.

D. Interpolation of Operators

Let (Ω, μ) and (Ω', ν) be σ-finite measure spaces. Denote by Simp(Ω) the space of μ-measurable simple functions on Ω with support of finite measure, and by Meas(Ω') the space of ν-measurable functions on Ω'. An operator T: Simp(Ω)$\rightarrow$Meas(Ω') is said to be of **(strong) type** (p,q), $p, q \in [1, \infty]$, if it satisfies

$$\|Tf\|_{L_q(\Omega')} \leqslant M \|f\|_{L_p(\Omega)}, \quad f \in \mathrm{Simp}(\Omega), \qquad (4)$$

where M is a constant independent of f.

Suppose that a linear operator T: Simp(Ω) $\rightarrow$Meas(Ω') is simultaneously of type (p_0, q_0) and (p_1, q_1). Then T is of type (p,q) for all p and q determined by

$$\frac{1}{p} = \frac{1-\theta}{p_0} + \frac{\theta}{p_1}, \qquad \frac{1}{q} = \frac{1-\theta}{q_0} + \frac{\theta}{q_1}, \qquad (5)$$

for $0 < \theta < 1$ and

$$\|Tf\|_{L_q(\Omega')} = M_0^{1-\theta} M_1^{\theta} \|f\|_{L_p(\Omega)}, \quad f \in \mathrm{Simp}(\Omega), \qquad (6)$$

where the M_i are the bounds for (p_i, q_i) (Riesz-Thorin theorem).

When $f \in$ Meas(Ω), the **distribution function** $\mu_f(s)$, the **rearrangement** $f^*(t)$, and the **average function** $f^{**}(t)$ are defined by

$$\mu_f(s) = \mu\{x \in \Omega \mid |f(x)| > s\}, \qquad s > 0,$$

$$f^*(t) = \inf\{s > 0 \mid \mu_f(s) \leqslant t\}, \qquad t > 0,$$

$$f^{**}(t) = \frac{1}{t}\int_0^t f^*(s)\,ds.$$

Denote by L_p^*, $1 \leqslant p \leqslant \infty$, the L_p space of functions on $(0, \infty)$ relative to the measure dt/t. Then the **Lorentz space** $L_{(p,q)}(X)$, for any $p, q \in [1, \infty]$, is defined to be the space of all $f \in$ Meas(X) such that

$$\|f\|^*_{(p,q)} = \|t^{1/p} f^*(t)\|_{L_q^*} < \infty.$$

$L_{(p,q)}(X)$ is a linear subspace of Meas(X). If $q_0 \leqslant q_1$, then the inclusion $L_{(p,q_0)}(X) \subset L_{(p,q_1)}(X)$ holds.

If $p = q$, $\|f\|^*_{(p,q)}$ coincides with the norm $\|f\|_{L_p(X)}$, hence $L_{(p,p)}(X) = L_p(X)$. Otherwise, $\|f\|^*_{(p,q)}$ does not necessarily satisfy the triangle inequality.

However, if $1 < p < \infty$, $L_{(p,q)}(X)$ turns out to be a Banach space under the norm

$$\|f\|_{L_{(p,q)}(X)} = \|t^{1/p} f^{**}(t)\|_{L_q^*}.$$

This norm is equivalent to $\|f\|^*_{(p,q)}$ in the sense that

$$\|f\|^*_{(p,q)} \leqslant \|f\|_{L_{(p,q)}(X)} \leqslant \frac{p}{p-1} \|f\|^*_{(p,q)}.$$

An operator T which maps a linear subspace of Meas(Ω) into Meas(Ω') is said to be **quasilinear** if $T(f+g)$ is uniquely defined whenever Tf and Tg are defined, and if

$$|T(f+g)(y)| \leqslant K(|Tf(y)| + |Tg(y)|),$$

where K is a constant independent of f and g. If $K = 1$, T is said to be **sublinear**.

Suppose that a quasilinear operator T satisfies

$$\|Tf\|^*_{(q_i, s_i)} \leqslant M_i \|f\|^*_{(p_i, r_i)}, \quad f \in L_{(p_i, r_i)}(\Omega), \quad i = 0, 1,$$

where $p_0 < p_1$ and $q_0 \neq q_1$. Then for every $0 < \theta < 1$ and $s \geqslant r$,

$$\|Tf\|^*_{(q,s)} \leqslant K_\theta M_0^{1-\theta} M_1^{\theta} \|f\|^*_{(p,r)}, \quad f \in L_{(p,r)}(\Omega),$$

where p and q are defined by (5) and $K_\theta = O(\theta^{-1} + (1-\theta)^{-1})$ (**Marcinkiewicz-Hunt theorem**).

An operator T is said to be of **weak type** (p,q) if there exists a constant M independent of $f \in L_p(\Omega)$ such that

$$\sup_{0 < s < \infty} s \cdot \nu_{Tf}(s)^{1/q} \leqslant M \|f\|_{L_p(\Omega)}, \quad q < \infty,$$

$$\|Tf\|_{L_\infty(\Omega')} \leqslant M \|f\|_{L_p(\Omega)}, \quad q = \infty \qquad (7)$$

(J. Marcinkiewicz), and of **restricted type** (p,q) (**restricted weak type** (p,q)) if (4) (resp. (7)) holds for the characteristic functions of sets of finite measure (E. M. Stein and G. Weiss). An operator T is of weak type (p,q) if and only if

$$\|Tf\|^*_{(q,\infty)} \leqslant M \|f\|_{L_p(X)}.$$

If $1 \leqslant p < \infty$ and $1 < q \leqslant \infty$, a sublinear operator T: Simp(Ω)$\rightarrow$Meas(Ω') is of restricted weak type (p,q) if and only if

$$\|Tf\|^*_{(q,\infty)} \leqslant M \|f\|^*_{(p,1)}, \qquad f \in \mathrm{Simp}(\Omega).$$

Hence we have the following theorem.

Suppose that a sublinear operator T: Simp(Ω)$\rightarrow$Meas(Ω') is simultaneously of restricted weak type (p_0, q_0) and (p_1, q_1), $p_i \leqslant q_i$, $p_0 < p_1$, $q_0 \neq q_1$. Then T is of strong type (p,q) for p, q defined by (5) (**Stein-Weiss theorem**).

These theorems have been successfully applied to prove the boundedness of (singular) integral operators such as the Fourier transform, and its generalization (A. P. Calderón and A. Zygmund) and the Hardy-Littlewood-Sobolev inequality.

References

[1] S. Banach, Théorie des opérations linéaires, Warsaw, 1932 (Chelsea, 1963).
[2] N. Dunford and J. T. Schwartz, Linear operators, Interscience, I, 1958; II, 1963; III, 1971.
[3] L. Schwartz, Théorie des distributions, Hermann, revised edition, 1966.
[4] A. Friedman, Generalized functions and partial differential equations, Prentice-Hall, 1963.
[5] С. Л. Соболев (S. L. Sobolev), Некоторые применения функционального анализа в математической физике, Leningrad, 1950; English translation, Applications of functional analysis in mathematical physics, Amer. Math. Soc. Transl. of Math. Monographs, vol. 7, 1963.
[6] K. Yosida, Functional analysis, Springer, 1965.
[7] A. Zygmund, Trigonometric series, Cambridge Univ. Press, second edition, 1959.
[8] R. A. Hunt, On $L(p,q)$ spaces, Enseignement Math., (2) 12 (1966), 249–276.

174 (X.33) Function-Theoretic Null Sets

A. General Remarks

By a **function-theoretic null set** we mean an exceptional set that appears in a theorem such as one asserting that a certain property holds with a "small exception." We give below some of the more important examples of exceptional sets. For simplicity, we shall limit ourselves to the n-dimensional Euclidean space $\mathbf{R}^n$ $(n \geqslant 2)$.

B. Sets of Harmonic Measure Zero

Denote by χ_E the characteristic function of a set E on the boundary ∂D of a bounded domain D in $\mathbf{R}^n$. We call the †hypofunction $\underline{H}_{\chi_E}$ and †hyperfunction $\overline{H}_{\chi_E}$ (→ 124 Dirichlet Problem) the **inner** and **outer harmonic measures** of E (with respect to D), respectively. When they coincide, the function is called the **harmonic measure** of E. A necessary and sufficient condition for E to be of inner harmonic measure zero is that $u \leqslant 0$ holds in D whenever a †subharmonic function u bounded above in D satisfies $\limsup u(P) \leqslant 0$ as P tends to any point of $\partial D - E$. This theorem implies the following uniqueness theorem: If h is bounded and harmonic in D, if E is a set of inner harmonic measure zero on ∂D, and if $h(P) \to 0$ as P tends to any point of $\partial D - E$, then $h \equiv 0$. A necessary and sufficient condition for E to be of outer harmonic measure zero is that there exists a positive †superharmonic function v in D such that $v(P) \to \infty$ as P tends to any point of E. (Concerning the existence of a limit for a subharmonic or †harmonic function at every boundary point except those on a set of harmonic measure zero, → 397 Subharmonic Functions; 195 Harmonic Functions.)

C. Sets of Capacity Zero

Although there are many kinds of capacity (→ 50 Capacity), here we shall consider only †logarithmic capacity and α-capacity $(\alpha > 0)$. Let K be a nonempty compact set in $\mathbf{R}^n$. Set $W(K) = \inf_\mu \int\int \overline{PQ}^{-\alpha} d\mu(P)\, d\mu(Q)$, where μ runs through the class of nonnegative †Radon measures of total mass 1 supported by K, and write $C_\alpha(K) = (W(K))^{-1/\alpha}$. Define $C_\alpha(\varnothing) = 0$ for an empty set $\varnothing$. For a general set $E \subset \mathbf{R}^n$, define the inner capacity by $\sup_{K \subset E} C_\alpha(K)$ and the outer capacity by the infimum of the inner capacity of an open set containing E. When the inner and outer capacities coincide, the common value is called the **α-capacity** (or **capacity of order** α) of E and is denoted by $C_\alpha(E)$. We denote the logarithmic capacity of E by $C_0(E)$. In order that $C_0(K) = 0$ $(n = 2)$ or the †Newtonian capacity $C_{n-2}(K) = 0$ $(n \geqslant 3)$ for a compact set K, it is necessary and sufficient that the harmonic measure of K with respect to $G - K$ vanishes for any bounded domain G containing K. Then K is removable for any harmonic function that is bounded or has a finite †Dirichlet integral in $G - K$. In general, K is said to be **removable** for a family F of functions if for any domain G containing K and $f \in F$ defined in $G - K$, there exists $g \in F$ defined in G such that $g = f$ in $G - K$. Let K be a compact set in $\mathbf{R}^2$ with $C_0(K) = 0$ and G be a domain containing K. Let f be a holomorphic function defined in $G - K$ for which every point of K is an †essential singularity. Then the set of †exceptional values for f at every point of K is of logarithmic capacity zero. If f is †meromorphic in $|z| < 1$ and

$$\int\int \frac{|f'|^2}{(1+|f|^2)^2}(1-|z|)^\alpha\, dx\, dy < \infty \qquad (0 \leqslant \alpha < 1),$$

then f has a finite limit in any angular domain with a vertex at every point of $|z| = 1$ except for those belonging to a set of α-capacity zero (logarithmic capacity zero if $\alpha = 0$).

D. Hausdorff Measure

Let $\alpha>0$ and a set E be given in $\mathbf{R}^n$. Denote by σ a covering of E by a countable number of balls with radii $d_1, d_2, \ldots$, all of which are smaller than $\varepsilon(>0)$. As $\varepsilon\to 0$, $\inf_\sigma \sum_k d_k^\alpha$ increases. The limit is called the **Hausdorff measure** of E of dimension α and is denoted by $\Lambda_\alpha(E)$. In order for a compact set K to be removable for the family of harmonic functions defined in a bounded domain and satisfying the †Hölder condition of order α, it is necessary and sufficient that $\Lambda_{n-2+\alpha}(K)=0$. Next, suppose that K is a compact set in a plane and the complement G of K with respect to the plane is connected. Set $\|f\|_q=(\iint_G|f|^q\,dx\,dy)^{1/q}$ for f holomorphic in G and q, $1\leqslant q<\infty$, and $\|f\|_\infty=\sup_G|f|$. Denote by H^q the family of $f\not\equiv 0$ with $\|f\|_q<\infty$. If p is defined by $1/p+1/q=1$, then $\Lambda_{2-p}(K)<\infty$ implies $H^q=\varnothing$ for q, $2<q<\infty$, and $\Lambda_1(K)=0$ implies $H^\infty=\varnothing$. Moreover, $H^2=\varnothing$ if and only if $C_0(K)=0$, and $H^q=\varnothing$ implies $C_{2-p}(K)=0$ for q, $2<q\leqslant\infty$ [2].

E. Null Sets Defined with Respect to Families of Functions

Conversely, L. V. Ahlfors and A. Beurling characterized the size of sets in a plane by means of families of functions [1]. Let D be a domain, and let f represent a holomorphic function in D. Fix a point z_0 in D. Set

$$\mathfrak{B}=\{f\mid |f|\leqslant 1\},$$

$$\mathfrak{D}=\left\{f\mid \iint_D |f'|^2\,dx\,dy\leqslant\pi\right\},$$

$$\mathfrak{E}=\{f\mid \text{the area of } R^c\geqslant\pi\},$$

where R^c is the complement of the †range R of $(f(z)-f(z_0))^{-1}$. Denote by $\mathfrak{S}\mathfrak{B}$, $\mathfrak{S}\mathfrak{D}$, $\mathfrak{S}\mathfrak{E}$ the families consisting of constants and †univalent functions in $\mathfrak{B}$, $\mathfrak{D}$, $\mathfrak{E}$, respectively. Use the notation $\mathfrak{F}$ to represent any one of these six families, and define $M_\mathfrak{F}=M_\mathfrak{F}(z_0;D)$ by $\sup\{|f'(z_0)|\mid f\in\mathfrak{F}\}$. Then $M_\mathfrak{B}=M_\mathfrak{E}\geqslant M_\mathfrak{D}=M_{\mathfrak{S}\mathfrak{E}}\geqslant M_{\mathfrak{S}\mathfrak{B}}=M_{\mathfrak{S}\mathfrak{D}}$, and $M_\mathfrak{F}(z_0;D)=0$ implies $M_\mathfrak{F}(z;D)=0$ for any $z\in D$.

Denote by $N_\mathfrak{F}$ the class of compact sets K such that the complement K^c of K is connected and $M_\mathfrak{F}(z;K^c)=0$. We call $K\in N_\mathfrak{F}$ a **null set of class** $N_\mathfrak{F}$. In order for K to be removable for $\mathfrak{B}$ or $\mathfrak{D}$, it is necessary and sufficient that $K\in N_\mathfrak{B}$ or $\in N_\mathfrak{D}$, respectively. We generally have

$$\{K\mid C_0(K)=0\}\underset{\neq}{\subset} N_\mathfrak{B}\underset{\neq}{\subset} N_\mathfrak{D}\underset{\neq}{\subset} N_{\mathfrak{S}\mathfrak{B}}.$$

If $\Lambda_1(K)=0$, then $K\in N_\mathfrak{B}$. In the case where K is a subset of an analytic arc A, $K\in N_\mathfrak{B}$ implies $\Lambda_1(K)=0$, $N_\mathfrak{D}$ is equal to $N_{\mathfrak{S}\mathfrak{B}}$, and K belongs to $N_\mathfrak{D}$ if and only if $C_0(A)=C_0(A-K)$. If an †analytic function has an essential singularity at every point of $K\in N_\mathfrak{B}$, then any compact subset of the set of exceptional values at every point of K belongs to $N_\mathfrak{B}$. A necessary and sufficient condition for $K\in N_\mathfrak{D}$ is either that the complement of any one-to-one †conformal mapping of K^c is of plane measure zero or that any †univalent analytic function in K^c is reduced to a †linear fractional function.

References

[1] L. V. Ahlfors and A. Beurling, Conformal invariants and function-theoretic null-sets, Acta Math., 83 (1950), 101–129.
[2] L. Carleson, Selected problems on exceptional sets, Van Nostrand, 1967.
[3] M. Ohtsuka, Kansûron tokuron (Japanese; Topics of the theory of functions), Kyôritu, 1957.
[4] K. Oikawa, On function-theoretic null sets (Japanese), Sûgaku, 7 (1955), 161–170.

175 (IX.9) Fundamental Group

A †continuous mapping f from the interval $I=\{t\mid 0\leqslant t\leqslant 1\}$ into a topological space Y is called a **path** connecting the **initial point** $f(0)$ and the **terminal point** $f(1)$. In particular, a path satisfying $f(0)=f(1)=y_0$ is called a **loop** (or **closed path**) with y_0 as the **base point**. For a path f, the **inverse path** $\bar f$ of f is defined by $\bar f(t)=f(1-t)$. When the terminal point of f and the initial point of g coincide, the path F defined by $F(t)=f(2t)$ for $0\leqslant t\leqslant 1/2$ and $F(t)=g(2t-1)$ for $1/2\leqslant t\leqslant 1$ is called the **product** (or **concatenation**) of f and g, and is denoted by $f\cdot g$. With $[f]$ standing for the equivalence class of a path f under the relation of †homotopy relative to 0, 1 ($\in I$) (i.e., by homotopy with fixed 0 and 1), the inverse $[f]^{-1}=[\bar f]$ and the product $[f]\cdot[g]=[f\cdot g]$ are defined. In particular, in the set of homotopy classes of loops with one common base point y_0, the product is always defined, and the set forms a group $\pi_1(Y,y_0)$. This group is called the **fundamental group** (or **Poincaré group**) (H. Poincaré, 1895) of Y (with respect to y_0). If Y is †arcwise connected, then $\pi_1(Y,y_0)\cong\pi_1(Y,y_1)$ for arbitrary points y_0, y_1, and the structure of the group is independent of the choice of the base point. This group is then denoted simply by $\pi_1(Y)$. A continuous mapping $\varphi:(Y,y_0)\to(Y',y_0')$ induces a homomorphism $\varphi_*:\pi_1(Y,y_0)\to\pi_1(Y',y_0')$ by sending

$[f]$ to $\varphi_*[f]=[\varphi\circ f]$, and $(\varphi'\circ\varphi)_*=\varphi'_*\circ\varphi_*$ holds for the composite $\varphi'\circ\varphi$ of mappings. Thus $\pi_1(Y)$ is a †topological invariant of Y. If $\pi_1(Y)$ consists of only one class (the class of the constant path), we say that Y is **simply connected**. For example, cells and spheres S^n $(n\geqslant 2)$ are simply connected. The famous †Poincaré conjecture states that a simply connected 3-dimensional compact †manifold is homeomorphic to the 3-dimensional sphere. We have **van Kampen's theorem**: Let P be a polyhedron, P_1 and P_2 be its subpolyhedra, $P_1\cap P_2$ be connected, and $P=P_1\cup P_2$. Then $\pi_1(P)$ is isomorphic to the group (†amalgamated product) obtained from the †free product of $\pi_1(P_1)$ and $\pi_1(P_2)$ by giving the relations that the images of each element of $\pi_1(P_1\cap P_2)$ in $\pi_1(P_1)$ and in $\pi_1(P_2)$ are equivalent. Also, the fundamental group of the †product spaces is the direct product of the fundamental groups of the spaces involved. Any group is the fundamental group of some †CW complex. The Abelization $\pi_1/[\pi_1,\pi_1]$ of the fundamental group $\pi_1=\pi_1(Y)$ (Y arcwise connected) is isomorphic to the 1-dimensional integral †homology group $H_1(Y)$. For example, the fundamental groups of a circle S^1 and a †torus T^n are an infinite cyclic group and a free Abelian group of rank n, respectively; the fundamental group of a 1-dimensional CW complex is a free group; and the fundamental group of an orientable 2-dimensional closed surface of †genus p is a group having $2p$ generators $\{a_1,\ldots,a_p,b_1,\ldots,b_p\}$ and a relation $\Pi_k a_k b_k a_k^{-1} b_k^{-1}=1$. If x_0 is a fixed point of the circle S^1, then the fundamental group can be defined as the set of the homotopy classes of continuous mappings f: $(S^1,x_0)\rightarrow(Y,y_0)$.

Extending the definition of the fundamental group by replacing I, S^1 with I^n, S^n, we obtain the n-dimensional homotopy group (→ 205 Homotopy Groups; 93 Covering Spaces).

References

[1] H. Seifert and W. Threlfall, Lehrbuch der Topologie, Teubner, 1934 (Chelsea, 1965).
[2] C. Chevalley, Theory of Lie groups I, Princeton Univ. Press, 1946.
[3] S. Lefschetz, Introduction to topology, Princeton Univ. Press, 1949.
[4] W. S. Massey, Algebraic topology, an introduction, Harcourt, Brace & World, 1967.

G

176 (XX.24) Galois, Evariste

Evariste Galois (October 25, 1811–May 31, 1832) was born in Bourg-la-Reine, a suburb of Paris. In 1828, while still in junior high school, he published a paper on periodic †continued fractions. Although he published four papers, his most important works were submitted to the French Academy of Science and either lost or rejected. He was unsuccessful in his attempt to enter the Ecole Polytechnique and instead entered the Ecole Normale Supérieure in 1829. Active in political affairs, he was expelled from school, imprisoned, and died in a duel soon after his release.

The night before the duel, he left his research outline and manuscripts to his friend, A. Chevalier. These were published by J. Liouville in *J. Math. Pures Appl.*, first series, 11 (1846). The contents include the concept of groups and what essentially became the Galois theory of algebraic equations. The manuscript also contained such expressions as "theory of ambiguity," which seems to indicate that Galois intended to study the theory of algebraic functions along the same line of thought.

References

[1] E. Galois, Oeuvres mathématiques, edited by E. Picard, Gauthier-Villars, 1897.
[2] F. Klein, Vorlesungen über die Entwicklung der Mathematik im 19. Jahrhundert I, Springer, 1926 (Chelsea, 1956).
[3] T. Takagi, Kinsei sûgakusidan (Japanese; Topics from the history of mathematics of the 19th century), Kawade, 1943.
[4] R. Bourgne and J.-P. Azra, Ecrits et mémoires mathématiques d'Evariste Galois, Gauthier-Villars, 1962.

177 (III.8) Galois Theory

A. History

After the discovery of solutions of algebraic equations of degrees 3 and 4 in the 16th century, efforts to solve equations of degree 5 remained unsuccessful. Early in the 19th century P. Ruffini and N. H. †Abel showed that an algebraic solution is impossible. Shortly afterward, E. †Galois established a general principle concerning the construction of roots of algebraic equations by radicals. The principle was described in terms of the structure of a certain permutation group (the Galois group) of the roots of the equation. Even in this original form of the theory (**Galois theory**), Galois not only completed the research started by J. L. Lagrange, P. Ruffini, and Abel, but also made an epochal discovery that opened the way to modern algebra. J. W. R. Dedekind (*Werke* III, 1894) interpreted this result as a duality theorem concerning the automorphism groups of a field. It was shown later that Galois theory plays an important role in the general theory of commutative fields established by E. Steinitz. In the 1920s, W. Krull generalized the idea of Dedekind, using the concept of topological algebraic systems, and obtained Galois theory of infinite algebraic extensions (*Math. Ann.*, 100 (1928)). Another line of development of this theory, also originated by Dedekind (*Werke* III, 1876/77), led to the Galois theory of rings, an object of active research by N. Jacobson (*Ann. of Math.*, 41 (1940)), T. Nakayama, and others since the 1940s.

B. Definition

Given a group G of †automorphisms of a given †field L, the subfield $F(G)=\{a\in L \mid a^\sigma=a, \sigma\in G\}$ is called the **invariant field** associated with G. Conversely, given a subfield K of L, the group consisting of all automorphisms of L leaving every element of K invariant is denoted by $G(L/K)$. An †algebraic extension L/K is called a **Galois extension** if there exists a group G of automorphisms of L such that $F(G)=K$; in this case, $G(L/K)$ is called the **Galois group** of L/K. The invariant field of $G(L/K)$ is K. Furthermore, if L/K is a finite extension, we always have $G=G(L/K)$. A necessary and sufficient condition for L/K to be a Galois extension is that it is both †separable and †normal. A Galois extension L/K is called an **Abelian extension** or a **cyclic extension** when $G(L/K)$ is †Abelian or †cyclic, respectively (→ 157 Fields).

C. Fundamental Theorem of Galois Theory

Let L/K be a finite Galois extension and G its Galois group. Then there exists a †dual lattice isomorphism between the set of intermediate fields of L/K and the set of subgroups of G, under which an intermediate field M of L/K corresponds to the subgroup $H=G(L/M)$; conversely, a subgroup H of G corresponds to $M=F(H)$. The degree of extension $[L:M]$ is equal to the order of the corresponding subgroup H (in particular, $[L:K]$ is the order of G), and $[M:K]$ coincides with the index $(G:H)$. If subfields M and M' are †conjugate over K, then the corresponding subgroups $G(L/M)$ and $G(L/M')$ are conjugate to each other in G, and vice versa. In particular, M/K is a Galois extension if and only if the subgroup H corresponding to M is a †normal subgroup of G, and in this case, the Galois group $G(M/K)$ is isomorphic to the factor group G/H.

D. Extensions of a Ground Field

Let L/K be a finite Galois extension, K'/K any extension, and L' the †composite field of L and K'. Then L'/K' is also a Galois extension, and its Galois group is isomorphic to $G(L/L\cap K')$ by the restriction mapping.

E. Normal Basis Theorem

Let L/K be a finite Galois extension with Galois group G. Then there exists an element u of L such that $\{u^\sigma \mid \sigma\in G\}$ forms a basis for L over K called a **normal basis**. If we denote by $K[G]$ the †group ring of G over K, a †$K[G]$-module structure can be introduced in L by the operation $\Sigma a_\sigma\sigma(x)=\Sigma a_\sigma x^\sigma$; the existence of a normal basis implies that L is isomorphic to $K[G]$ itself as a $K[G]$-module, or in other words, that the K-linear representation of G by means of L is equivalent to the †regular representation of G.

F. Examples of Galois Extensions

(1)†Cyclotomic Fields. Let m be a positive integer not divisible by the †characteristic of K; ζ a †primitive mth root of unity, and $L=K(\zeta)$. Then L/K is an Abelian extension, and its Galois group is isomorphic to a subgroup of the †reduced residue class group $(\mathbf{Z}/m\mathbf{Z})^*$; in particular, if $K=\mathbf{Q}$, the subgroup coincides with $(\mathbf{Z}/m\mathbf{Z})^*$, by the irreducibility of †cyclotomic polynomials. Hence the degree $[\mathbf{Q}(\zeta):\mathbf{Q}]$ is equal to $\varphi(m)$, where φ is †Euler's function.

(2) Finite Fields. A †finite field K has nonzero characteristic p, and the number q of elements of K is a power of p. Also, K is uniquely determined by q (up to isomorphism), hence it is denoted by $GF(q)$ or $\mathbf{F}_q$. Thus $GF(q^n)$ is the only extension of $GF(q)$ of degree n; moreover, it is a cyclic extension.

(3) Kummer Extensions. Assume that K contains a primitive mth root ζ of unity, and denote by K^* the multiplicative group of K. An extension L of K can be expressed in the form $L=K(\sqrt[m]{a_1},\ldots,\sqrt[m]{a_r})$ $(a_i\in K)$ if and only if L/K is an Abelian extension and all $\sigma\in G(L/K)$ satisfy $\sigma^m=1$; in this case L/K is called a **Kummer extension** of **exponent** m. There exists a one-to-one correspondence between Kummer extension L of exponent m over K and finite subgroups $H/(K^*)^m$ of the factor group $K^*/(K^*)^m$, given by the relations $H=L^m\cap K^*$, $L=K(\sqrt[m]{H})$. Moreover, there exists a canonical isomorphism between $H/(K^*)^m$ and the †character group of $G(L/K)$, so that $H/(K^*)^m$ is isomorphic to $G(L/K)$. Let $L=K(\theta)$ be a cyclic Kummer extension of degree m of K, and let σ be a generator of the Galois group $G(L/K)$. Then the **Lagrange resolvent** $(\zeta,\theta)=\theta+\zeta\theta^\sigma+\ldots+\zeta^{m-1}\theta^{\sigma^{m-1}}$ satisfies $(\zeta,\theta)^\sigma=\zeta^{-1}(\zeta,\theta)$, $(\zeta,\theta)^m\in K$, and θ and its conjugates can be expressed in terms of (ζ,θ). In particular, L is generated by (ζ,θ) over K.

(4) Artin-Schreier Extensions. Assume that K is of characteristic $p\neq0$. For any element a of an extension of K, we denote by $\mathcal{P}a$ the element a^p-a and by $(1/\mathcal{P})a$ a root of $\mathcal{P}X-a=0$. A finite extension L of K is of the form $L=K((1/\mathcal{P})a_1,\ldots,(1/\mathcal{P})a_r)$ $(a_i\in K)$ if and only if L/K is a Galois extension whose Galois group is an Abelian group of †exponent p; in this case, L/K is called an **Artin-Schreier extension**. There exists a one-to-one correspondence between Artin-Schreier extensions L over K and finite subgroups $H/\mathcal{P}K$ of the additive group $K/\mathcal{P}K$, given by the relations $H=\mathcal{P}L\cap K$, $L=K((1/\mathcal{P})H)$; moreover, $H/\mathcal{P}K$ is isomorphic to the character group $G(L/K)$ (therefore also to $G(L/K)$ itself). More generally, for Abelian extensions L of **exponent** p^n (i.e., Galois extensions whose Galois groups are Abelian groups of exponent p^n), we obtain similar descriptions by using the additive group of †Witt vectors of length n instead of K (→ 434 Witt Vectors).

G. Galois Group of an Equation

L/K is a finite Galois extension if and only if L is a †minimal splitting field of a †separable polynomial $f(X)$ in $K[X]$. In this case, we call $G(L/K)$ the **Galois group of the polynomial** $f(X)$ or **of the algebraic equation** $f(X)=0$. The equation $f(X)=0$ is called an **Abelian equation** or a **cyclic equation** if its Galois group is Abelian or cyclic, respectively, while $f(X)=0$ is called a **Galois equation** if L is generated by any root of $f(X)$ over K.

Generally, $G(L/K)$ can be †faithfully represented as a permutation group of roots of $f(X)=0$. If this group is †primitive, then $f(X)=0$ is called a **primitive equation**. The index of the group in the group of all permutations of roots is called the **affect** of the equation $f(X)=0$; if the affect is 1, the equation $f(X)=0$ is called **affectless**. Let $u_1, \ldots, u_n$ be †algebraically independent elements over K. Then for the polynomial $F_n(X)=X^n-u_1X^{n-1}+\ldots+(-1)^n u_n$ in $K(u_1, \ldots, u_n)[X]$, the equation $F_n(X)=0$ is called a **general equation** of degree n. The Galois group of $F_n(X)=0$ is isomorphic to the †symmetric group $\mathfrak{S}_n$ of degree n, and if K is not of characteristic 2, then the quadratic subfield corresponding to the †alternating group $\mathfrak{A}_n$ is the field $K(\sqrt{D})$ obtained by adjoining the quadratic root of the †discriminant D of $F_n(X)$.

H. Solvability of an Algebraic Equation

Assume that K is of characteristic 0, $f(X) \in K[X]$, and L is the minimal splitting field of $f(X)$. We say that the equation $f(X)=0$ is **solvable by radicals** if there is a chain of subfields $K=L_0 \subset L_1 \subset \ldots \subset L_r=L$ such that $L_i=L_{i-1}(\sqrt[n_i]{a})$ with some $a_i \in L_{i-1}$, and this is the case if and only if the Galois group of $f(X)$ is †solvable (Galois). In particular, Abelian equations are solvable by radicals. Cyclic equations are solved by using the Lagrange resolvent, and theoretically the general solvable equation can be solved by repeating this procedure. Since $\mathfrak{S}_n$ is solvable if and only if $n \leqslant 4$, it follows that a general equation of degree n is solvable only if $n=1$, 2, 3, 4 (Abel). Also, a polynomial is solvable by square roots if and only if the order of the Galois group is a power of 2. This fact enables us to answer some questions concerning geometric construction problems such as trisection of an angle or division of a circumference in equal parts (→ 183 Geometric Construction).

I. Infinite Galois Extensions

If a Galois extension L/K is infinite, then its Galois group is an infinite group. Let $\{M_\nu\}$ be the family of intermediate fields of L/K that are finite and normal over K, and put $H_\nu = G(L/M_\nu)$. Then by taking $\{H_\nu\}$ as a †base of a neighborhood system of the unity element, G becomes a †topological group. This topology is called the **Krull topology** (Krull, *Math. Ann.*, 100 (1928)). G is then isomorphic to the †projective limit of the family of finite groups $\{G/H\}$ and is †totally disconnected and †compact. There is a one-to-one correspondence between the set of intermediate fields of L/K and the set of closed subgroups of G given by the mapping (Galois group) ↔ (invariant field), and thus we have a generalization of Galois theory for finite extensions as described in Section C. Various theories, including the theory of Kummer extensions, can be generalized to the case of infinite extensions (→ 406 Topological Groups).

J. Galois Cohomology

Let L/K be a finite Galois extension and G its Galois group. Then both the additive group L and the multiplicative group L^* have G-module structures. The †cohomology groups of G with coefficient module L are 0 for all dimensions because of the existence of a normal basis (→ 202 Homological Algebra). As for the multiplicative group L^*, we have $\hat{H}^0(G,L^*) \cong K^*/\mathrm{N}(L^*)$ (N is the †norm $\mathrm{N}_{L/K}$), $H^1(G,L^*)=0$ (**Hilbert's theorem 90** or the **Hilbert-Speiser theorem**). In particular, if G is a cyclic group with generator σ, then every element a such that $\mathrm{N}(a)=1$ can be expressed in the form $a=b^{1-\sigma}$. $H^2(G,L^*)$ is isomorphic to the †Brauer group of †central simple algebras over K which have L as a †splitting field. In the case of number fields, a number of G-modules arise, such as †principal orders, †unit groups, †ideal groups, †idele groups, and so on, whose cohomological considerations are important (→ 7 Adeles and Ideles; 62 Class Field Theory). In many cases, we are more concerned with the †category of Galois extensions of K with K-isomorphisms between them than with a single extension L/K. In other words, we consider a †functor $L \rightarrow \mathfrak{F}(L)$ of the category of Galois extensions of K into the category of Abelian groups, and study the cohomology related to $G(L/K)$-module structures derived from $\mathfrak{F}(L)$. In the case of infinite algebraic extensions, we consider the inductive limit of cohomology of subfields of finite degrees, making use of continuous cocycles of Galois groups relative to the Krull topology [13, 14]. Galois theory gives a model of a successful theory, summarizing the essentials of the theory of separable algebraic extensions.

K. Inseparable Extensions

In the case of †inseparable extensions, †derivations often play roles which correspond to those played by automorphisms in Galois extensions, and it is possible to construct a theory similar to Galois theory by using them.

Let K be a field of characteristic $p \neq 0$ and L/K a finite †purely inseparable extension such that $L^p \subset K$. The set of all derivations of L/K forms a †restricted Lie algebra $D(L/K)$ over K and has the structure of a linear space over L whose dimension is equal to the degree of L/K. There exists a one-to-one and dual lattice-isomorphic correspondence between the set of intermediate fields M of L/K and the set of Lie subalgebras H of the restricted Lie algebra $D(L/K)$, given by the relations $H = D(L/M)$, $M = \{a \in L \mid d(a) = 0,\ d \in H\}$ (the constant field of H) (Jacobson). J. Dieudonné obtained a method of analyzing the subfields of a general purely inseparable extension by using the concept of semiderivations, while Jacobson succeeded in constructing a general Galois theory which includes both the cases of Galois extensions and purely inseparable extensions by making use of notions such as self-representations of fields and Galois composites. Their extension to noncommutative rings has also been studied.

L. Galois Theory of Rings

The theory of †centralizers in simple algebras can be interpreted as the theory of a certain Galois correspondence with respect to inner automorphism groups. Also, by using †crossed products, we can deduce from the theory of centralizers in simple algebras the Galois theory of commutative fields. On the other hand, Jacobson obtained a Galois theory of †division rings with respect to finite groups of outer automorphisms that is similar to the commutative case. Since then, many algebraists have proceeded with investigations that aim either at unifying these two theories by admitting inner automorphisms in the group of automorphisms, at extending the theory from division rings to general rings such as simple rings, †primitive rings, or †semiprimary rings, or at weakening the finiteness conditions. One principal method in these theories lies in considering first the endomorphism ring $\mathrm{Hom}_R(S, S)$ for an extension S/R and then the roles of endomorphisms (or derivations) in it [3,11] (→ 31 Associative Algebras).

M. Galois Algebras

As a generalization of the theory of Kummer extensions, H. Hasse (*J. Reine Angew. Math.*, 187 (1950)) introduced the concept of Galois algebras. If G is a finite group of automorphisms of a commutative algebra A over K, A has a $K[G]$-module structure ($K[G]$ is the group ring of G over K). If A is $K[G]$-isomorphic to $K[G]$ itself, we call A a **Galois algebra** with Galois group G.

Investigations have been carried out on various problems of Galois algebras, such as the construction problem, the embedding problem, and the problem of decomposing A corresponding to a decomposition of the regular representation of G, all of which are applicable to number-theoretic research. Although A is mostly the direct sum of fields, some attempts at generalization, including the study of noncommutative Galois algebras, have been made.

References

[1] K. Shoda and K. Asano, Daisûgaku I (Japanese; Algebra I), Iwanami, 1952.
[2] Y. Akizuku and M. Suzuki, Kôtô daisûgaku I (Japanese; Higher algebra I), Iwanami, 1952.
[3] T. Nakayama and G. Azumaya, Daisûgaku II (Japanese; Algebra II–Theory of rings), Iwanami, 1954.
[4] M. Moriya, Hôteisiki (Japanese; Algebraic equations), Sibundô, 1964.
[5] J. W. R. Dedekind, Über die Permutationen des Körpers aller algebraischen Zahlen, Gesammelte mathematische Werke, F. Viewig, 1930–1932, vol. 2, p. 272–291.
[6] E. Artin, Galois theory, Univ. of Notre Dame Press, second edition, 1948.
[7] N. Bourbaki, Eléments de mathématique, Algèbre, ch. 5, Actualités Sci. Ind. 1102b, Hermann, second edition, 1959.
[8] N. G. Cebotarev, Grundzüge der Galois'schen Theorie, Noordhoff, 1950.
[9] P. Wolf, Algebraische Theorie des Galoisschen Algebren, Deutscher Verlag der Wiss., 1956.
[10] B. L. van der Waerden, Algebra I, Springer, seventh edition, 1966.
[11] N. Jacobson, Structure of rings, Amer. Math. Soc. Colloq. Publ., 1956.
[12] L. Rédei, Algebra I, Akademische Verlag, 1959.
[13] J.-P. Serre, Corps locaux, Actualités Sci. Ind., Hermann, 1962.
[14] J.-P. Serre, Cohomologie galoisienne, Lectures notes in math. 5, Springer, 1964.
[15] N. Jacobson, Lectures in abstract algebra III, Van Nostrand, 1964.
[16] М. М. Постников (M. M. Postnikov), Теория Галуа, физматгиз, 1960; English translation, Foundations of Galois theory, Pergamon, 1962.
[17] S. Lang, Algebra, Addison-Wesley, 1965.

178 (XVIII.6) Game Theory

A. History

Game theory is now generally understood to be a specific mathematical approach to social phenomena, which has progressed under the direct or indirect influence of Von Neumann's pioneering work [6] and his book written in collaboration with O. Morgenstern [7]. One basis for the theory was the recognition of the similarities of the behavioral patterns of contending individuals and the strategies of players in various games such as chess and cards which involve bargaining, negotiations, coalition, and profit-sharing. The only fairly complete part of the theory deals with **zero-sum 2-person games** (→ Section B), which are the simplest. The general theory for **n-person games** (→ Section B) is far from complete, although various approaches have been proposed since the work of Von Neumann and Morgenstern [7]. Game theory has greatly influenced the development of mathematical theories concerning economics and other social sciences, and has encouraged application of modern topological and algebraic methods in these fields. Its principal lines of approach may be classified as the theory of noncooperative games and that of cooperative games, of which the latter seems to occupy a central position because of the abundance of unsolved problems.

B. Theory of n-Person Noncooperative Games

An **n-person game in normalized form** is formulated by specifying n pairs $\{K_i, X_i\}$ $(i=1, \ldots, n)$ of sets X_i $(i=1, \ldots, n)$ and real-valued functions $K_i(x_1, \ldots, x_n)$ $(i=1, \ldots, n)$ on the †Cartesian product $\prod_{i=1}^{n} X_i$. The x_i $(x_i \in X_i)$, X_i, and K_i are termed the **strategy**, **strategy space**, and **payoff functions** of the player i, respectively. When $\sum_{i=1}^{n} K_i(x_1, \ldots, x_n) = c$ is a constant, the game is called a **constant-sum game**, and in particular, when $c=0$, a **zero-sum game**. A **noncooperative game**, by definition, precludes any **coalition** among players in choosing their respective strategies; otherwise, the game is classified as a **cooperative game**.

The basic theory of noncooperative games is rather simple and is formulated on the basis of the concept of equilibrium point due to J. Nash [5]. An n-tuple $(\hat{x}_1, \ldots, \hat{x}_i, \ldots, \hat{x}_n)$ of strategies is called an **equilibrium point** if for each i,

$$K_i(\hat{x}_1, \ldots, \hat{x}_{i-1}, \hat{x}_i, \hat{x}_{i+1}, \ldots, \hat{x}_n) \geqslant K_i(\hat{x}_1, \ldots, \hat{x}_{i-1}, x_i, \hat{x}_{i+1}, \ldots, \hat{x}_n)$$

for all $x_i \in X_i$. The concept of equilibrium point was already present in the work of A. Cournot [15], a pioneer in the study of imperfect competition.

Von Neumann's **saddle-point theorem** [7] on zero-sum 2-person games is the simplest case of an existence theorem for equilibrium points, where X_1, X_2 are †simplexes in Euclidean spaces and K_1, K_2 are bilinear forms satisfying the identity $K_1(x_1,x_2)+K_2(x_1,x_2)=0$. In this case, an equilibrium point is a **saddle point**, which is defined to be a pair of strategies $(\hat{x}_1, \hat{x}_2)$ fulfilling

$$K_1(x_1,\hat{x}_2) \leqslant K_1(\hat{x}_1,\hat{x}_2) \leqslant K_1(\hat{x}_1,x_2), \qquad x_1 \in X_1,\ x_2 \in X_2$$

The saddle-point theorem is often referred to as the **minimax theorem**, since the existence of a saddle point is equivalent to the validity of the equation

$$\max_{x_1 \in X_1} \min_{x_2 \in X_2} K_1(x_1,x_2) = \min_{x_2 \in X_2} \max_{x_1 \in X_1} K_1(x_1,x_2).$$

A zero-sum 2-person game is said to be **strictly determined** if the minimax theorem holds for it. Various algorithms to locate saddle points (e.g., the †simplex method in linear programming; → 255 Linear Programming) have been devised.

Let M, N be two sets of strategies that may be chosen by two players. We assume that M, N are finite and consist of m, n strategies, respectively. Let $a(i,j)=a_{ij}$ $(i \in M, j \in N)$ be the first player's payoff function. Since $a(i,j)=a_{ij}$ does not always have a saddle point, we augment the spaces M, N to X, Y by letting them be the sets of †probability distributions $x=(x_1,x_2,\ldots,x_m)$ $(x_i \geqslant 0, \sum x_i = 1)$, $y=(y_1,y_2,\ldots,y_n)$ $(y_j \geqslant 0, \sum y_j = 1)$ on M, N, and letting $K(x,y)=\sum a_{ij}x_iy_j$ be the new payoff function of the first player (equal to his †mathematical expectation of payoff). Then for this augmented game, the function K has a saddle point. This is a probability-theoretic interpretation of Von Neumann's saddle-point theorem. A strategy in this augmented sense is called a **mixed strategy**. A special mixed strategy whose corresponding probability distribution is a 1-point distribution is called a **pure strategy**.

A typical existence theorem for equilibrium points of general games asserts the existence of an equilibrium for the case where the X_i are †compact †convex sets, the K_i are continuous, and for each i and any fixed $n-1$ elements $x_j \in X_j$ $(j \neq i)$, the set of x_i maximizing $K_i(x_1, \ldots, x_{i-1}, x_i, x_{i+1}, \ldots, x_n)$ is a convex set. The proof of this theorem is given by means of †Kakutani's fixed-point theorem [3] or its generalization to †locally convex linear topological space (→ 162 Fixed-Point Theorems).

C. Theory of n-Person Cooperative Games

The **characteristic function** of an n-person cooperative game is a real-valued set function $v(S)$, defined for all subsets S of the set of n players $N=\{1,2,\ldots,n\}$, satisfying (i) $v(\varnothing)=0$ and (ii) (superadditivity) $v(S\cup T)\geqslant v(S)+v(T)$ for $S,T\subset N$, $S\cap T=\varnothing$. If a game is given in normalized form, it is possible to construct a characteristic function from the given strategy spaces and payoff functions under appropriate conditions. Then $v(S)$ represents the total gain secured by the coalition of players belonging to S. A vector $\alpha=(\alpha_1,\ldots,\alpha_n)$, where α_i denotes a final allotment of the gain to player i, is termed an **imputation** if it satisfies (i) $\alpha_i\geqslant v(\{i\})$ and (ii) $\sum_{i=1}^n\alpha_i=v(N)$. An imputation may include not only gains paid off according to the rules of the game but also **side payments** made among players in compensation for the formation of coalitions. An imputation $\alpha=(\alpha_1,\ldots,\alpha_n)$ is said to **dominate** another imputation $\beta=(\beta_1,\ldots,\beta_n)$ if, for a subset $S\subset N$, (1) $v(S)\geqslant\sum_{i\in S}\alpha_i$ and (2) $\alpha_i>\beta_i$ for all $i\in S$. This concept of the dominance relation induces a nontransitive binary relation in the set of all imputations I. A subset of I which satisfies appropriate conditions (Von Neumann and Morgenstern's are notable) relative to the dominance relation is considered a **solution of the cooperative game**. However, for the "appropriate conditions," we have various condidates, none of which is fully satisfactory. In this respect, the theory of cooperative games is far from complete. Two examples of such solutions will be explained in the two following paragraphs.

A set P of imputations is termed a **Von Neumann-Morgenstern solution** if (i) no dominance relation exists between any two elements $\alpha,\beta\in P$ and (ii) any $\gamma\notin P$ is dominated by some α in P. On the other hand, the set C of all imputations that cannot be dominated by any imputation is called the **core** of the cooperative game. The core concept was formulated by D. B. Gillies and L. S. Shapley. The core is the set of solutions of linear inequalities $\sum_{i=1}^n\alpha_i\leqslant v(N)$, $\sum_{i\in S}\alpha_i\geqslant v(S)$ $(\forall S\subset N)$, and the nature of it is well known and can be studied by the techniques of linear programming.

From early research on Von Neumann-Morgenstern solutions it had been conjectured that they might exist for all games. However, the conjecture was settled in the negative, since W. F. Lucas [16] discovered a ten-person game for which no solution exists.

Another concept of the solution for a cooperative game is the Shapley value [9, vol. II, paper 17]. A **Shapley value** is a function $\eta(v)=(\eta_1(v),\ldots,\eta_n(v))$ sending arbitrary characteristic functions $v(S)$ and $w(S)$ on N into the n-dimensional Euclidean space satisfying the conditions (i) $\eta_{\pi_i}(\pi_v)=\eta_i(v)$; (ii) $\sum_{i\in N}\eta_i(v)=v(N)$; and (iii) $\eta(v+w)=\eta(v)+\eta(w)$, where π is any permutation of the elements in N and $\pi_v(S)=v(\pi^{-1}S)$. Conditions (i), (ii), and (iii) uniquely determine the Shapley value, which can be expressed as

$$\eta_i(v)=\sum_{i\in S}\frac{(s-1)!(n-s)!}{n!}(v(S)-v(S-\{i\})),$$

where s is the number of elements of S and the summation extends over all S containing i. For a fixed characteristic function v and the Shapley value y, the vector $y(v)$ is an imputation.

The contemporary progress of the theory of cooperative games centers around the study of the properties of various solution concepts, as exemplified above. The concepts of characteristic functions, imputations, dominance, and side payments presuppose the existence of a utility measure transferable among players. However, even in the case where no transferable utility exists, counterparts of these concepts have recently been proposed by R. J. Aumann and B. Peleg [17], and analogous theories have been thereby developed, with special regard to their applications in economic theory and other fields.

D. The Extensive Form of Games

If the specific structures of social games are taken into account, a game can be formulated in the **extensive form**. A social game can be regarded as a (partially) †ordered set of **moves** at each of which a specific player chooses among alternatives, with or without the aid of a chance mechanism such as a die or a roulette. This ordering structure is represented schematically by a **tree of the game**. A junction point (or vertex) of a tree of a game indicates a move and also tells which player is supposed to make the move. A path leading from the lowest junction point via consecutive junction points to an uppermost point (including no cycle) is called a **play** and represents a particular instance in which the game is played. Each play assigns specific gains to all players. A game in the extensive form can be transformed to its normalized form by introducing the concept of strategy. The theory of extensive games is due to Von Neumann and Morgenstern [7, ch. II] and H. W. Kuhn [9, vol. II, paper 11].

E. Applications to Mathematical Economics

Mathematical approaches in economics are classified into two fields, namely, pure theory, called **mathematical economics**, which concerns the qualitative study of economic phenomena, and **econometrics**, which concerns their statistical properties. Historically, mathematical economics originated with the work of A. Cournot (1801–1877) and was systematized by the theory of general equilibrium of L. Walras (1834–1910). However, the full development of the theory began only in the 1940s, and notably after World War II. From both the ideological and methodological viewpoints, the direct or indirect influence of game theory on modern mathematical economics is quite remarkable. A few typical problems in mathematical economics are illustrated below.

The main thesis of Walras's theory of general equilibrium is the analysis of a **competitive economy** that involves many goods, consumers (households), and producers (firms). There exists a set of appropriate prices (rates of exchange for goods) that ensures a **competitive equilibrium** situation where individual competitive behavior realizes maximum utility to each consumer, maximum profit to each producer, and equality of demand and supply for each good. This proposition, referred to as the **existence theorem of competitive equilibrium**, was conjectured by Walras and mathematically proved for the first time in the 1950s, half a century after its formulation. The proof is based on Kakutani's fixed-point theorem [3]. A competitive economy is similar to a noncooperative game situation.

Walras also developed the theory of **tâtonnement**, anticipating the convergence of the price adjustment process in which an excess of demand over supply of a good results in a rise in its price, or the reverse situation results in a fall in its price. This theory of tâtonnement has been formulated and studied as the global stability of the solution of a system of ordinary differential equations of n variables. Differential equations describing a tâtonnement process have certain specific properties originating from economic requirements, on which basis the convergence to a competitive equilibrium is discussed. The result of K. Arrow, H. Block, and L. Hurwicz [1] is important in this area.

There have recently been interesting game-theoretic approaches to the classical view of perfect competition, which presumes that the influence of an individual on the whole economy is very small or negligible in an economy involving a great many individuals. If a national economy is viewed as a cooperative game among individuals, competitive equilibrium allocations belong to the core. It has been proved by H. Scarf and G. Debreu that if the number of individuals increases toward infinity in a specific way, the core shrinks to the set of competitive equilibrium allocations. On the other hand, R. J. Aumann and K. Vind showed that the core coincides with the set of competitive equilibrium allocations for certain types of economy involving infinitely many individuals and formulated in terms of †measure-theoretic concepts. Moreover, the convergence of the Shapley value to a competitive equilibrium with an increasing number of individuals has also been discussed.

Interindustry input-output analysis and linear programming (→ 255 Linear Programming) are typical representatives of approaches that consider the static equilibrium and dynamic change in simple linear models of a national economy. Truly dynamic problems such as economic growth and capital accumulation are also dealt with by recent mathematical studies. From among these, two typical ones may be mentioned. The first is the **turnpike theorem** ([4] gives an excellent survey), which concerns the convergence of optimal paths of economic growth to a balanced growth path. The second is a variational study of maximization of the integral of utility over a whole period and is connected with the mathematical theory of optimal control, due to L. S. Pontrjagin and others [8].

References

[1] K. J. Arrow, H. D. Block, and L. Hurwicz, On the stability of the competitive equilibrium II, Econometrica, 27 (1959), 82–109.
[2] G. Debreu, Theory of value, John Wiley, 1959.
[3] S. Kakutani, A generalization of Brouwer's fixed point theorem, Duke Math. J., 8 (1941), 457–459.
[4] T. C. Koopmans, Economic growth at a maximal rate, Quart. J. Economics, 78 (1964), 355–394.
[5] J. F. Nash, Non-cooperative games, Ann. of Math., (2) 54 (1951), 286–295.
[6] J. Von Neumann, Zur Theorie der Gesellschaftsspiele, Math. Ann., 100 (1928), 295–320.
[7] J. Von Neumann and O. Morgenstern, Theory of games and economic behavior,

Princeton Univ. Press, 1944.
[8] Л. С. Понтрягин, В. Г. Болтанский, Р. В. Гамкрелидзе, Е. Ф. Мищенко (L. S. Pontrjagin, V. G. Boltjanskiĭ, R. V. Gamkrelidze, and E. F. Miščenko), Математическая теория оптимальиых процессов, физматгиз, 1961; English translation, The mathematical theory of optimal processes, Interscience, 1962.
[9] Contributions to the theory of games I, II, III, IV, Ann. Math. Studies, Princeton Univ. Press, 1950, 1953, 1957, 1959.
[10] M. Dresher, L. S. Shapley, and A. W. Tucker (eds.), Advances in game theory, Ann. Math. Studies, Princeton Univ. Press, 1964.
[11] R. D. Luce and H. Raiffa, Games and decisions, John Wiley, 1957.
[12] M. Shubik (ed.), Game theory and related approaches to social behavior, John Wiley, 1964.
[13] J. C. C. McKinsey, Introduction to the theory of games, McGraw-Hill, 1952.
[14] R. P. Isaacs, Differential games, John Wiley, 1965.
[15] A. A. Cournot, Recherches sur les principes mathématiques de la théorie des richesses, L. Hachette, 1838; English translation, Researches into the mathematical principles of the theory of wealth, Macmillan, 1929.
[16] W. F. Lucas, The proof that a game may not have a solution, Trans. Amer. Math. Soc., 137 (1969), 219–229.
[17] R. J. Aumann and B. Peleg, Von Neumann-Morgenstern solutions to cooperative games without side payments, Bull. Amer. Math. Soc., 66 (1960), 173–179.

179 (XIV.4) Gamma Function

A. The Gamma Function

The function $\Gamma(x)$ was defined by L. Euler (1729) as the infinite product

$$\frac{1}{x}\prod_{n=1}^{\infty}\left(1+\frac{1}{n}\right)^{x}\left(1+\frac{x}{n}\right)^{-1}.$$

Legendre later called it the **gamma function** or **Euler's integral of the second kind**. The latter name is based on the fact that for positive real x, we have

$$\Gamma(x)=\int_0^{\infty}e^{-t}t^{x-1}\,dt.$$

This function satisfies the **functional relation**

$$\Gamma(x+1)=x\Gamma(x),$$

and hence for positive integral x, we have $\Gamma(x+1)=x!$. C. F. Gauss denoted the function $\Gamma(x+1)$ by $\Pi(x)$ or $x!$, even when x is not a positive integer. The function $x!$ is also called the **factorial function**. The gamma function can also be defined as the solution of the functional relation $\Gamma(x+1)=x\Gamma(x)$ satisfying the conditions

$$\Gamma(1)=1,\ \lim_{n\to\infty}\frac{\Gamma(x+n)}{\Gamma(n)\,n^{x}}=1.$$

Furthermore, we have

$$\frac{1}{\Gamma(x)}=xe^{Cx}\prod_{n=1}^{\infty}\left(1+\frac{x}{n}\right)e^{-x/n}.$$

This expression is known as the **Weierstrass canonical form**. **Euler's constant,** defined by

$$C=\lim_{n\to\infty}\left(1+\frac{1}{2}+\ldots+\frac{1}{n}-\log n\right)$$

and approximated by
0.57721566490153286060651209008...,
is conjectured to be †transcendental, but as yet even its irrationality remains unproved.

The value of C was calculated by Adams (1878) to 260 decimal places, and recently to more than 7000 decimal places by using an electronic computer (W. A. Beyer and M. S. Waterman, *Math. Comp.*, 28 (1974)).

$\Gamma(x)$ is †holomorphic on the complex x-plane except at the points $x=0,\ -1,\ -2,\ldots,$ where it has simple †poles. When $\operatorname{Re}x>0$, we have Hankel's integral representation

$$\Gamma(x)=-\frac{1}{2i\sin\pi x}\int_C(-t)^{x-1}e^{-t}\,dt,\qquad x\neq\text{integer},$$

where the contour C lies in the complex plane cut along th. positive real axis, starting at ∞, going around the origin once counterclockwise, and ending at ∞ again.

Among various properties of this function (→ Appendix A, Table 17.I), the following two formulas are especially useful for numerical calculations: **Binet's formula**

$$\log\Gamma(x)=\left(x-\frac{1}{2}\right)\log x-x+\frac{\log 2\pi}{2}+\int_0^{\infty}\frac{\arctan(t/x)}{e^{2\pi t}-1}\,dt,\quad \operatorname{Re}x>0,$$

and the †asymptotic expansion formula that holds when $|\arg x|\leqslant(\pi/2)-\delta\ (\delta>0)$,

$$\log\Gamma(x)\sim\left(x-\frac{1}{2}\right)\log x-x+\frac{\log 2\pi}{2}+\sum_{n=1}^{\infty}\frac{(-1)^{n-1}B_{2n}}{2n(2n-1)x^{2n-1}}$$

(**Stirling's formula**), where the B_n are †Bernoulli numbers. This last formula can be rewritten as

$$\Gamma(x+1)=x!\sim x^{x}e^{-x}\sqrt{2\pi x}\,,$$

which is used for large positive integers x.

The integrals

$$\int_0^{\lambda} e^{-t}t^{x-1}dt, \qquad \int_{\lambda}^{\infty} e^{-t}t^{x-1}dt, \qquad \operatorname{Re} x>0$$

are known as the **incomplete gamma functions** and are used in statistics, the theory of molecular structure, and other fields.

B. Polygamma Functions

Functions derived from the gamma function are named and defined as follows: the **digamma function** (or **psi function**) $\psi(x)= d\log\Gamma(x)/dt$; the **trigamma function** $\psi'(x)$; the **tetragamma function** $\psi''(x)$; the **pentagamma function** $\psi'''(x)$, etc. These functions are called **polygamma functions**. In particular, $\psi(x)$ is the solution of the functional equation

$$\psi(x+1)-\psi(x)=1/x, \qquad \psi(1)=-C,$$
$$\lim_{n\to\infty}(\psi(x+n)-\psi(1+n))=0.$$

C. The Beta Function

Euler's integral of the first kind

$$B(x,y)=\int_0^1 t^{x-1}(1-t)^{y-1}dt, \qquad \operatorname{Re} x>0, \quad \operatorname{Re} y>0$$

is called the **beta function** and is an analytic function of two variables x, y. This function is related to the gamma function as follows:

$$B(x,y)=\frac{\Gamma(x)\Gamma(y)}{\Gamma(x+y)}.$$

If the upper limit 1 in the integral is replaced by α, the result is the **incomplete beta function** $B_{\alpha}(x,y)$.

References

[1] M. Hukuhara, Ganma kansu (Japanese; Gamma function), Kôbundo, 1951.
[2] W. Shibagaki, Ganma kansu no riron to ôyô (Japanese; Theory and applications of the gamma function), Iwanami, 1952.
[3] E. T. Whittaker and G. N. Watson, A course of modern analysis, Cambridge Univ. Press, fourth edition, 1958.
[4] K. Pearson, Tables of the incomplete Γ-function, Cambridge Univ. Press, second edition, 1968.
[5] E. Artin, The gamma function, Holt, Rinehart and Winston, 1964.
Also → references to 381 Special Functions.

180 (XX.25) Gauss, Carl Friedrich

Carl Friedrich Gauss (April 30, 1777–February 23, 1855) was born into a poor family in Braunschweig, Germany. From childhood, Gauss showed genius in mathematics. He gained the favor of Grand Duke Wilhelm Ferdinand and under his sponsorship attended the University of Göttingen. In 1797, on proving the †fundamental theorem of algebra, he received his doctorate from the University of Halle. From 1807 until his death, he was a professor and director of the Observatory at the University of Göttingen.

On March 30, 1796, he demonstrated that it is possible to draw a 17-sided †regular polygon with ruler and compass. The publication of his *Disquisitiones arithmeticae* in 1801 opened an entirely new era in number theory. In pure mathematics, he did excellent research on †non-Euclidean geometry, †hypergeometric series, the theory of functions of a complex variable, and the theory of †elliptic functions. In the field of applied mathematics he made outstanding contributions to astronomy, geodesy, and electromagnetism; he also studied the †method of least squares, the theory of surfaces (→ 114 Differential Geometry of Curves and Surfaces), and the theory of †potential. He considered the perfection of papers for publication of utmost importance; thus his published works are few relative to his amount of research. However, the scope of his work may be seen in his diary and letters, some of which are included in his complete works, comprising 12 volumes. He is generally considered the greatest mathematician of the first half of the 19th century.

References

[1] C. F. Gauss, Werke I–XII, Königlichen Gesellschaft der Wissenschaften, Göttingen, 1863–1933.
[2] C. F. Gauss, Disquisitiones arithmeticae; English translation, Yale Univ. Press, 1966; German translation, Untersuchungen über höhere Arithmetik, Springer, 1889 (Chelsea, 1965).
[3] F. Klein, Vorlesungen über die Entwicklung der Mathematik im 19. Jahrhundert I, II, Springer, 1926–1927 (Chelsea, 1956).
[4] T. Takagi, Kinsei sûgakusidan (Japanese; Topics from the history of mathematics of the 19th century), Kawade, 1943.

181 (VII.6) Gears

A. General Remarks

A gear is a solid centered on an axis of rotation, called the shaft, and is obtained by superimposing a concavoconvex form, the **tooth surface**, upon an original surface of rotation called the **pitch surface** of the gear. A pair of gears is used to transmit rotary motion continuously from one shaft to another by means of successively engaging tooth surfaces.

We restrict our treatment to cases where the tooth surfaces are in contact along a line and each gear turns with constant angular velocity (for gears that are in contact at a point or gears whose ratio of angular velocities is not constant, → [2]).

B. Conditions for Meshing

Let g_1 and g_2 be a pair of gears in space having two fixed straight lines l_1, l_2 as the axis of their respective gear shafts and rotating about the shafts with constant angular velocities $\boldsymbol{\omega}_1$, $\boldsymbol{\omega}_2$, respectively. We take a fixed †orthogonal frame $\Sigma=(O,\mathbf{i},\mathbf{j},\mathbf{k})$ in a 3-dimensional Euclidean space E^3 and assume that a pair of tooth surfaces Γ_1 and Γ_2 make contact along a curve C_t defined by $\mathbf{x}=\mathbf{x}(t,r)$ at a time t, where r is a parameter. The curve C_t is called a **line of contact** at the time t, and the surface that is the locus of lines of contact as t varies is called a **surface of contact**. Let λ denote 1 or 2. We fix a point O_λ on the axis l_λ and consider the moving orthogonal frame $\Sigma_\lambda(t)=(O_\lambda,\mathbf{i}_\lambda(t),\mathbf{j}_\lambda(t),\mathbf{k}_\lambda(t))$ fixed with the gear λ and rotating with t. We denote the line of contact C_t with respect to $\Sigma_\lambda(t)$ by $\mathbf{x}_\lambda= x_\lambda(t,r)\mathbf{i}_\lambda(t)+y_\lambda(t,r)\mathbf{j}_\lambda(t)+z_\lambda(t,r)\mathbf{k}_\lambda(t)$. Then we have $\mathbf{x}=\mathbf{x}_\lambda+\mathbf{e}_\lambda$, where $\overrightarrow{OO_\lambda}=\mathbf{e}_\lambda$. With respect to $\Sigma_\lambda(t_0)$ for any fixed t_0, the tooth surface Γ_λ is represented by $x=x_\lambda(t,r), y= y_\lambda(t,r), z=z_\lambda(t,r)$, where t and r are parameters. We put $\mathbf{x}_{\lambda t}=(\partial x_\lambda/\partial t)\mathbf{i}_\lambda+(\partial y_\lambda/\partial t)\mathbf{j}_\lambda+(\partial z_\lambda/\partial t)\mathbf{k}_\lambda$ and $\mathbf{x}_r=\partial\mathbf{x}/\partial r=\partial(\mathbf{x}_\lambda+\mathbf{e}_\lambda)/\partial r= (\partial x_\lambda/\partial r)\mathbf{i}_\lambda+(\partial y_\lambda/\partial r)\mathbf{j}_\lambda+(\partial z_\lambda/\partial r)\mathbf{k}_\lambda$. Then the †vector product $\mathbf{x}_{\lambda t}\times\mathbf{x}_r$ gives the direction of the normal to the tooth surface Γ_λ at the point $\mathbf{x}=\mathbf{x}(t,r)$ on C_t provided that $\mathbf{x}_{\lambda t}\times\mathbf{x}_r\neq 0$. If we denote by $\mathbf{v}_\lambda$ the velocity vector at $\mathbf{x}$ determined by the rotation of $\mathbf{x}$ about l_λ with the angular velocity $\boldsymbol{\omega}_\lambda$, then we have $\mathbf{v}_\lambda=\boldsymbol{\omega}_\lambda\times\mathbf{x}_\lambda$ and $\mathbf{x}_{\lambda t}=\mathbf{x}_t-\mathbf{v}_t$. The point $\mathbf{x}$ satisfying $\mathbf{x}_{\lambda t}\times\mathbf{x}_r=0$ is a singular point on the tooth surface Γ_λ. It is characterized as a point such that the normal to the surface of contact at the point is in the same plane as the gear axis l_λ [4]. We neglect such singular points and consider only general points on Γ_λ. Then the condition for the tooth surfaces Γ_1 and Γ_2 to be meshed at the common point $\mathbf{x}$ is given by $(\mathbf{x}_{2t}\times\mathbf{x}_r)\times(\mathbf{x}_{1t}\times\mathbf{x}_r)=0$, or equivalently, $[\mathbf{x}_{2t},\mathbf{x}_{1t},\mathbf{x}_r]=0$. If we put $\mathbf{w}=\mathbf{v}_2-\mathbf{v}_1$ ($\mathbf{w}$ is the relative velocity of the gear g_2 in relation to the gear g_1 and $\mathbf{w}=\mathbf{x}_{1t}-\mathbf{x}_{2t}$), then the condition for meshing can be expressed as $(\mathbf{x}_{\lambda t}\times\mathbf{x}_r)\cdot\mathbf{w}=0$. Hence if we denote by $\boldsymbol{\sigma}$ the unit normal vector at $\mathbf{x}$ on one of the two tooth surfaces Γ_1, Γ_2, the condition can also be expressed as $\boldsymbol{\sigma}\cdot\mathbf{w}=0$ or $\boldsymbol{\sigma}\cdot\mathbf{v}_2=\boldsymbol{\sigma}\cdot\mathbf{v}_1$. In particular, the condition for meshing holds for each point $\mathbf{x}$ satisfying the condition $\mathbf{w}\times\mathbf{x}_r=0$, because $\mathbf{x}_{1t}\times\mathbf{x}_r-\mathbf{x}_{2t}\times\mathbf{x}_r=\mathbf{w}\times\mathbf{x}_r$.

Now suppose that O_λ is the point where l_λ intersects the common perpendicular to l_1 and l_2, and the origin O is the point on the common perpendicular O_1O_2 such that $\mathbf{e}_2(\boldsymbol{\omega}\cdot\boldsymbol{\omega}_2)=\mathbf{e}_1(\boldsymbol{\omega}\cdot\boldsymbol{\omega}_1)$ $(\boldsymbol{\omega}=\boldsymbol{\omega}_2-\boldsymbol{\omega}_1)$. Then we obtain $\mathbf{w}=\boldsymbol{\omega}\times\mathbf{x}+h\boldsymbol{\omega}$ for the relative velocity $\mathbf{w}$, where $h=[\boldsymbol{\omega}_1,\boldsymbol{\omega}_2,\mathbf{e}]/\omega^2$, $\overrightarrow{O_1O_2}=\mathbf{e}$ [7]. Thus the relative motion of the gear g_2 in relation to the gear g_1 is a screw motion whose axis is a straight line perpendicular to the common perpendicular to both gear axes. The axis of the screw motion is called an **instantaneous axis**. The **reduced pitch** (i.e., the pitch divided by 2π) h is equal to 0 if and only if $\boldsymbol{\omega}_1\times\boldsymbol{\omega}_2=0$ (i.e., the gear axes are parallel) or $\mathbf{e}=0$ (i.e., the gear axes intersect at a point). In these cases the condition for meshing holds for any point on the instantaneous axis since $\mathbf{w}=0$ holds for such a point (for gears satisfying the condition $\mathbf{w}\times\mathbf{x}_r=0$ for any point on the contact line, → [7]).

We consider only cases where $\mathbf{w}\times\mathbf{x}_r\neq 0$. When $\mathbf{w}\times\mathbf{x}_r\neq 0$, we have the following alternative expression of the condition for meshing: $\mathbf{x}_{\lambda t}\times\mathbf{x}_r=-b_\lambda\mathbf{w}\times\mathbf{x}_r$, or equivalently, $\mathbf{x}_{\lambda t}=-b_\lambda\mathbf{w}+a\mathbf{x}_r$, $\mathbf{x}_t=b_2\mathbf{v}_1-b_1\mathbf{v}_2+a\mathbf{x}_r$, where $b_\lambda\neq 0$ by assumption and $b_2-b_1=1$.

C. Sliding Curves and Trajectories

Let $d_\lambda\mathbf{x}_\lambda$ and $d\mathbf{x}$ be infinitesimal vectors on the tooth surface Γ_λ and the surface of contact, respectively, directed to the meshing point $(t+dt,r+dr)$ from the meshing point (t,r). Then we have $d_\lambda\mathbf{x}_\lambda=\mathbf{x}_{\lambda t}dt+\mathbf{x}_r dr= -b_\lambda\mathbf{w}\,dt+(a\,dt+dr)\mathbf{x}_r$, $d\mathbf{x}=\mathbf{x}_t dt+\mathbf{x}_r dr=(b_2\mathbf{v}_1-b_1\mathbf{v}_2)dt+(a\,dt+dr)\mathbf{x}_r$. The curve defined by $a\,dt+dr=0$ on the tooth surface Γ_λ is called a **sliding curve** on Γ_λ, and the curve defined by $a\,dt+dr=0$ on the surface of contact is called a **trajectory**. In the equation of the surface of contact $\mathbf{x}=\mathbf{x}(t,r)$, we can replace the parameter r so that t-curves represent trajectories (i.e., a becomes 0). Then t-curves on the tooth surface Γ_λ are sliding curves, and we have $\mathbf{x}_{\lambda t}=-b_\lambda\mathbf{w}$ ($b_\lambda\neq 0$, $b_2-b_1=1$) along sliding curves. Accordingly, two sliding curves represented by t-curves corresponding to the same value of r touch at the common point $\mathbf{x}$, and the locus of the point of contact corresponding to each fixed value of r on the surface of contact is a trajectory.

Let ds_λ be the line element of the sliding

curve through the common point $\mathbf{x}$ on the tooth surface Γ_λ. Then $(ds_2 - ds_1)/ds_\lambda$ is called the **specific sliding** of the tooth surface Γ_λ at $\mathbf{x}$. If we take the positive sense of s_λ so that $ds_\lambda = -b_\lambda(\mathbf{w}^2)^{1/2}dt$, the specific sliding of the tooth surface Γ_λ at $\mathbf{x}$ is equal to $1/b_\lambda$.

The common unit tangent vector of the sliding curves on Γ_1 and Γ_2 at $\mathbf{x}$ can be expressed as $\mathbf{w}/(\mathbf{w}^2)^{1/2}$. Using this, we can find the curvatures of the sliding curves at $\mathbf{x}$. Moreover, we can see that the †normal curvature of the tooth surface Γ_λ at $\mathbf{x}$ with respect to the direction of the relative velocity is expressed as $1/\rho_\lambda = -\boldsymbol{\sigma}\cdot\mathbf{q}/b_\lambda\mathbf{w}^2 + [\omega, \mathbf{w}, \boldsymbol{\sigma}]/\mathbf{w}^2$, where $\mathbf{q} = \omega_2\times\mathbf{v}_1 - \omega_1\times\mathbf{v}_2$. Hence we have $1/\rho_2 - 1/\rho_1 = \boldsymbol{\sigma}\cdot\mathbf{q}/b_1b_2\mathbf{w}^2$. We call $1/\rho_2 - 1/\rho_1$ the relative curvature of the tooth surfaces at $\mathbf{x}$ with respect to the direction of the relative velocity. If we denote by θ the angle formed by the line of contact passing through $\mathbf{x}$ and the direction of the relative velocity at $\mathbf{x}$, then we have the formula

$$\frac{1}{R_2} - \frac{1}{R_1} = \operatorname{cosec}^2\theta\left(\frac{1}{\rho_2} - \frac{1}{\rho_1}\right) = \frac{\mathbf{x}_r^2(\boldsymbol{\sigma}\cdot\mathbf{q})}{b_1b_2(\mathbf{w}\times\mathbf{x}_r)^2} = \frac{\mathbf{x}_r^2(\boldsymbol{\sigma}\cdot\mathbf{q})}{(\mathbf{x}_{2t}\times\mathbf{x}_r)\cdot(\mathbf{x}_{1t}\times\mathbf{x}_r)},$$

where $1/R_\lambda$ is the normal curvature of Γ_λ with respect to the direction orthogonal to the line of contact. We call $1/R_2 - 1/R_1$ the **relative curvature of the tooth surfaces** at $\mathbf{x}$ with respect to the direction orthogonal to the line of contact. The last expression also holds for the exceptional case $\mathbf{w}\times\mathbf{x}_r = 0$ [3,5,7].

D. Critical Points for Meshing

Taking account of the fact that both gears are solid bodies, we can distinguish the gear that is a driver. We see that $\boldsymbol{\sigma}\cdot\mathbf{v}_2/R_2 - \boldsymbol{\sigma}\cdot\mathbf{v}_1/R_1 = (\boldsymbol{\sigma}\cdot\mathbf{v}_\lambda)(\boldsymbol{\sigma}\cdot\mathbf{q})\mathbf{x}_r^2/(\mathbf{x}_{2t}\times\mathbf{x}_r)\cdot(\mathbf{x}_{1t}\times\mathbf{x}_r) \gtrless 0$ according as the gear g_1 is the driver or not. Thus critical points for meshing that may occur on a line of contact by the meshing of tooth surfaces (excluding singular points) of a pair of gears are given as points satisfying the condition $\boldsymbol{\sigma}\cdot\mathbf{v}_1 = \boldsymbol{\sigma}\cdot\mathbf{v}_2 = 0$ or $\boldsymbol{\sigma}\cdot\mathbf{q} = 0$. The normal for tooth surfaces at the meshing point $\mathbf{x}$ satisfying the condition $\boldsymbol{\sigma}\cdot\mathbf{q} = 0$ is Wildhaber's **limit normal** [3,5,7].

E. Spur Gears and Bevel Gears

When the axes of a pair of gears are parallel (i.e., $\omega_1\times\omega_2 = 0$), we have $\mathbf{x}_t\cdot\omega_\lambda = 0$, $\mathbf{x}_{\lambda t}\cdot\omega_\lambda = 0$ for the tangent vector $\mathbf{x}_t$ of the trajectory at the meshing point $\mathbf{x}$ and the tangent vector $\mathbf{x}_{\lambda t}$ of the sliding curve on Γ_λ at $\mathbf{x}$. Therefore, both the trajectory and the sliding curves passing through $\mathbf{x}$ are on a plane orthogonal to the gear axes. When the axes of a pair of gears intersect (i.e., $\mathbf{e} = 0$), we have $\mathbf{x}_t\cdot\mathbf{x} = 0$, $\mathbf{x}_{\lambda t}\cdot\mathbf{x} = 0$ at the meshing point $\mathbf{x}$. Therefore, both the trajectory and the sliding curves passing through $\mathbf{x}$ are on a sphere whose center is at the point of intersection of the axes. These results illustrate why we may study meshing of gears with parallel or intersecting axes as problems on the meshing of curves on a plane or a sphere. In such simple problems each trajectory becomes a **path of contact**, and each sliding curve becomes a **tooth curve**. Moreover, since each point on each instantaneous axis satisfies the condition for meshing, the surface of contact is usually given so that it passes through each instantaneous axis of relative motion of the gear g_2 with respect to the gear g_1. The pitch surfaces are circular cylinders or circular cones whose axes are the axes of the gears and contain the instantaneous axes as their generators.

F. Skew Gears

Let C be the trajectory and C_λ be the sliding curve on Γ_λ passing through a meshing point $\mathbf{x}$. Also, let D_λ be the surface of revolution obtained by rotating the sliding curve C_λ about the gear axis l_λ. As time passes, D_1 and D_2 touch along the trajectory C, and the common normal at $\mathbf{x}$ is a straight line having the direction given by the vector $\mathbf{v}_2\times\mathbf{v}_1$ and intersecting both gear axes in general. We may consider that a pair of skew gears has surfaces of revolution D_1 and D_2 as pitch surfaces and sliding curves as **tooth traces**, the curves of intersection of tooth surfaces with their corresponding pitch surfaces. Let $\overline{D}_\lambda$ be a circular cone touching D_λ at the meshing point $\mathbf{x}$ and having the gear axis l_λ as its axis. A **hypoid gear** is a pair of gears such that its pitch surfaces are the cones $\overline{D}_1$ and $\overline{D}_2$ [3,4,6,8].

References

[1] G. B. Grant, A treatise on gear wheels, Philadelphia, twelfth edition, 1909.
[2] S. Watanabe, Haguruma hakeiron (Japanese; Theory of distortion of gears), Corona-sha, 1949.
[3] K. Takahashi, Study on the biting of the hypoid gears (Japanese), Inoue Printing, 1960.
[4] E. Stübler, Über hyperboloidische Verzahnung, Z. Angew. Math. Mech., 2 (1922), 429–446.

[5] M. Tanimura, On the gears of constant rotational ratio, 1st report (Japanese), Papers by Mechanical Engineering of Japan, 5 (1939), 184–190.
[6] E. Wildhaber, Basic relationship of hypoid gears I, II, Amer. Machinist, 90 (1946), 108–111, 131–134.
[7] S. Ogino, Study on the clutching of general gears, the gears of 0 specific sliding with non-intersecting axes (Japanese), Papers by Mechanical Engineering of Japan, 30 (1964), 379–388.
[8] J. Capelle, Théorie et calcul des engrenages hypoids, Dunod, 1949.

182 (XIV.2) Generating Functions

A. General Remarks

A power series $g(t)=\sum_{n=0}^{\infty}a_nt^n$ in t which converges in a certain neighborhood of $t=0$ defines the sequence of numbers $\{a_n\}$. The function $g(t)$ is called the **generating function** of the sequence. Similarly, the series $K(x,t)=\sum_{n=0}^{\infty}f_n(x)t^n$ which is convergent for x and t in a certain domain in (x,t)-space is called the **generating function** of the sequence of functions $\{f_n(x)\}$. Sometimes the function $\tilde{g}(t)=\sum_{n=0}^{\infty}(a_n/n!)t^n$ is called the **exponential generating function** of the sequence $\{a_n\}$. For example, the generating functions of the †binomial coefficients and †Legendre polynomials are $(1+t)^n$ and $(1-2tx+t^2)^{1/2}$, respectively. When a generating function of $\{a_n\}$ or $\{f_n(x)\}$ is given, we can obtain a_n or $f_n(x)$ by integral expressions; for example, in the latter case we have

$$f_n(x)=\frac{1}{2\pi i}\int_C\frac{K(x,t)}{t^n}dt,$$

where the contour C is a sufficiently small circle going counterclockwise around the origin. A generating function may be continued beyond the domain of convergence of the power series. Simple generating functions are known for many important orthogonal systems of functions (→ 312 Orthogonal Functions). Generating functions are widely used because they enable us to derive analytically the properties of sequences of numbers or functions. For a system of numbers or functions depending on a continuous parameter instead of the integral parameter n, we define the generating function in the form of a †Laplace or †Fourier transform. In particular, for a probability †distribution function $F(x)$, the exponential generating function $f(t)=\int_{-\infty}^{\infty}e^{-tx}dF(x)$ for the moments $\{a_n\}$ of $F(x)$ is called the **moment-generating function** of $F(x)$.

B. Bernoulli Polynomials

A system of polynomials

$$B_n(x)=B_0(0)x^n+\binom{n}{1}B_1(0)x^{n-1}+\binom{n}{2}B_2(0)x^{n-2}+\ldots+B_n(0)$$

is defined by the generating function

$$\frac{te^{xt}}{e^t-1}=\sum_{n=0}^{\infty}B_n(x)\frac{t^n}{n!}.$$

$B_n(x)$ is called the **Bernoulli polynomial** of degree n. Since $B_k(0)$ is the coefficient of

$$\frac{t^k}{k!}\quad\text{in}\quad\frac{t}{e^t-1}=\sum_{n=0}^{\infty}B_n(0)\frac{t^n}{n!},$$

we have $B_0(0)=1$, $B_1(0)=-1/2$, $B_2(0)=1/6,\ldots$; $B_{2n+1}(0)=0$ for $n\geqslant1$; and $(-1)^{n-1}B_{2n}(0)>0$ for $n\geqslant1$. The nth **Bernoulli number** B_n (→ Appendix B, Table 3) is defined as $|B_n(0)|$, or sometimes as $B_n(0)$ or B_{2n}. Bernoulli polynomials satisfy the relations

$$B_n(x+1)-B_n(x)=nx^{n-1},$$
$$dB_n(x)/dx=nB_{n-1}(x),$$

which are used in †interpolation. For example, a polynomial solution of the †difference equation $f(x+1)-f(x)=\sum_{n=0}^{m}a_nx^n$ is given by $f(x)=\sum_{n=0}^{m}a_nB_{n+1}(x)/(n+1)+$(arbitrary constant). In particular, we have $1^n+2^n+\ldots+p^n=(B_{n+1}(p+1)-B_{n+1}(1))/(n+1)$.

C. Euler Polynomials

A system of polynomials

$$E_n(x)=a_0x^n+\binom{n}{1}a_1x^{n-1}+\binom{n}{2}a_2x^{n-2}+\ldots+a_n$$

is defined by the generating function

$$\frac{2e^{xt}}{e^t+1}=\sum_{n=0}^{\infty}E_n(x)\frac{t^n}{n!}.$$

We call $E_n(x)$ the **Euler polynomial** of degree n. Here a_k is defined by $a_k=E_k(0)$ and

$$\frac{2}{e^t-1}=\sum_{n=0}^{\infty}a_n\frac{t^n}{n!},$$

so that we have $a_0=1$, $a_1=-1/2$, $a_3=1/4,\ldots$; $a_{2n}=0$ for $n\geqslant1$. The nth Euler number is sometimes a_n but more often is defined by

$$E_n=(-1)^n\sum_{\mu=0}^{n}2^{\mu}\binom{n}{\mu}a_{\mu},$$

i.e., by

$$\frac{2}{e^{x}+e^{-x}}=\operatorname{sech} x=\sum_{n=0}^{\infty}(-1)^{n}\frac{E_{n}}{n!}x^{n}$$

(→ Appendix B, Table 3). All the E_n are integers, $E_{2m+1}=0$ $(m=0,1,\ldots)$ and $E_{2m}>0$ $(m=0,1,\ldots)$; in the decimal expressions for E_n the last digit is 5 for E_{4m} $(m\geqslant 1)$ and 1 for E_{4m+2} $(m\geqslant 0)$. (Sometimes E_{2n} is denoted by E_n.) We have the relations

$$\sec x=\sum_{n=0}^{\infty}\frac{E_{n}x^{n}}{n!},$$

$$E_n(x)+E_n(x+1)=2x^n,$$

$$E_n(1-x)=(-1)^n E_n(x),$$

$$\frac{dE_n(x)}{dx}=nE_{n-1}(x),$$

and in particular,

$$-1^n+2^n-3^n+4^n-\ldots+(-1)^p p^n$$
$$=((-1)^p E_n(p+1)-E_n(1))/2.$$

References

[1] K. Hayashi, Sûti keisan (Japanese; Numerical computation), Iwanami, 1941.
[2] T. Ishizu, Tokusyu kansû ron (Japanese; Theory of special functions), Asakura, 1963.
[3] E. B. McBride, Obtaining generating functions, Springer, 1971.
Also → references to 381 Special Functions.

183 (VI.5) Geometric Construction

A. General Remarks

A **geometric construction problem** is a problem of drawing a figure satisfying given conditions using certain prescribed tools only a finite number of times. If the problem is solvable, then it is called a **possible construction problem**; if it is unsolvable, even though there exist figures satisfying the given conditions, then it is an **impossible construction problem**. Furthermore, if there does not exist a figure satisfying the given conditions, then we say that the problem is **inconsistent**.

Among problems of geometric construction, the oldest and the best known are those of constructing plane figures by means of **ruler** and **compass**. In this article, we call these problems simply problems of elementary geometric construction. The following are some of the more famous problems of this kind; the first four are possible construction problems:

(1) Suppose that we are given three straight lines l, m, n and three points P, Q, R in a plane. Draw a triangle ABC in such a way that vertices A, B, C lie on l, m, n and sides BC, CA, AB pass through P, Q, R (**Steiner's problem**).

(2) Suppose that we are given a circle O and three points P, Q, R not lying on O. Draw a triangle ABC inscribed in O in such a way that the sides BC, CA, AB pass through P, Q, R (**Cramer-Castillon problem**).

(3) Draw a circle tangent to all of three given circles (**Apollonius' problem**).

(4) Suppose we are given a triangle. Draw three circles inside this triangle in such a way that each is tangent to two sides of the triangle and any two of the circles are tangent to each other (**Malfatti's problem**).

(5) Let n be a natural number. For the division of the circumference of a circle into n equal parts (consequently, the construction of a †regular n-gon) to be a possible construction problem, it is necessary and sufficient that the representation of n as a product of prime numbers takes the form $n=2^{\lambda}p_1\ldots p_k$, where $\lambda\geqslant 0$, $p_1,\ldots,p_k$ are all different prime numbers of the form 2^h+1 (†Fermat number) (Gauss, 1801).

(6) The following are three famous impossible construction problems of Greek origin: (i) divide a given angle into three equal parts (**trisection of an angle**); (ii) construct a cube whose volume is double that of a given cube (**duplication of a cube** or the **Delos problem**); and (iii) construct a square whose area is that of a given circle (**quadrature of a circle**). P. L. Wantzel (1837) proved that problems (i) and (ii) are impossible; and C. L. F. Lindemann (1882) proved the impossibility of (iii) while proving that the number π is †transcendental.

B. Conditions for Constructibility

A problem of elementary geometric construction amounts to a problem of determining a certain number of points by drawing straight lines that pass through given pairs of points and circles having given points as centers and passing through given points. Let (a_1,b_1), (a_2,b_2), $\ldots,(a_n,b_n)$ be rectangular coordinates of given points, and let K be the smallest †number field containing the numbers $a_1,\ldots,b_n$. Straight lines that join given pairs of points and circles that have given points as centers and that pass through given points are represented by equations of the first or second degree with coefficients belonging to K. Consequently, the coordinates of points of intersection of these straight lines and circles belong to a quadratic extension

$K' = K(\sqrt{d})$ of K. Let A be the set of coordinates of the points that are to be determined. Then the problem is solvable if and only if any number α in A is contained in a field $L = K(\sqrt{d_1}, \sqrt{d_2}, \ldots, \sqrt{d_r})$, where $d_{i+1} \in K(\sqrt{d_1}, \ldots, \sqrt{d_i})$ $(i = 0, 1, \ldots, r-1)$. Thus L is a †normal extension field of K whose degree over K is a power of 2. Using this theorem we can prove the impossibility of trisection of an angle and duplication of a cube.

Since the 18th century, besides the problem of construction by ruler and compass, problems of construction by ruler alone or by compass alone have also been studied. We state here some of the more notable results: (1) If by drawing a straight line we mean the process of finding two different points on that line, then we can solve all the problems of elementary geometric construction by means of compass alone (G. Mohr, L. Mascheroni). (2) If by drawing a circle we mean the process of finding its center and a point on its circumference, and if a circle and its center are given, then we can solve any problem of elementary geometric construction by means of ruler alone. (3) It is not possible to find the center of a given circle by ruler alone (D. Hilbert). (4) It is impossible to bisect a given segment by ruler alone. (5) When two intersecting circles or concentric circles are given, we can find the centers of these circles by ruler alone. When nonintersecting and nonconcentric circles are given, it is not possible to find their centers by ruler alone (D. Cauer).

Cases have been considered in which the radius of a circle we can draw by compass or the length of a segment we can draw by ruler is required to satisfy certain conditions. Also, various considerations have been made concerning cases in which we can use tools other than ruler and compass. For example, it is known that although not all possible elementary geometric construction problems are solvable by ruler alone, all these problems are possible if we have either a pair of parallel rulers or two rulers making a rectangle or a fixed acute angle. If we use two rulers making a rectangle and a compass, then the trisection of an angle and the duplication of a cube are possible (L. Bieberbach). Also, when a conic section other than a circle is given, the trisection of an angle and the duplication of a cube become possible by ruler and compass (H. J. S. Smith and H. Kortum). By ruler and †transferrer of constant lengths, we can solve Malfatti's problem but not Apollonius' problem (Feldblum).

Even when a problem is possible, the method of construction may be rather complicated and impractical. In these cases, various methods of highly accurate approximate construction have been investigated.

References

[1] H. L. Lebesgue, Leçons sur les constructions géométriques, Gauthier-Villars, 1950.
[2] L. Bieberbach, Theorie der geometrischen Konstruktionen, Birkhäuser, 1952.

184 (XIX.17) Geometric Optics

A. General Remarks

Geometric optics is a mathematical theory of light rays. It is not concerned with the properties of light rays as waves (e.g., their wavelength and frequency), but studies their properties as pencils of rays that follow three laws: the law of rectilinear propagation, the law of reflection (i.e., angles of incidence and reflection on a smooth plane are equal (Euclid)), and the law of refraction (i.e., if θ and θ' are angles of incidence and refraction of a light ray refracted from a uniform medium to a second uniform medium and if n, n' are the refractive indices of the first and the second medium, respectively, then $n\sin\theta = n'\sin\theta'$ (R. W. Snell, Descartes)). These three laws follow from **Fermat's principle**, which states that the path of a light ray traveling from a point A' to A in a medium with refractive index $n(P)$ at P is such that the integral $\int_{A'}^{A} n(P)\,ds$ attains its extremal value, where ds is the line element along the path. This line integral is called the **optical distance** from A' to A. Therefore, Fermat's principle may be taken as a foundation of geometric optics and is, in a way, similar to the †variation principle in particle dynamics (**Maupertuis's principle**)

$$\delta\int\sqrt{2h-2U(P)}\;ds = 0,$$

which is satisfied by the path of a particle of unit mass having constant energy h passing through a field of †potential $U(P)$. The quantity $\sqrt{2h-2U}$ corresponds to the refractive index n.

In an optical system, express the position of a point on the path of a light ray by orthogonal coordinates (x, y, z), and define the †Lagrangian $L = n\sqrt{1+\dot{x}^2+\dot{y}^2}$ $(\dot{x} = dx/dz,$ $\dot{y} = dy/dz)$, **optical direction cosines** $p = \partial L/\partial\dot{x}$, $q = \partial L/\partial\dot{y}$, and the †Hamiltonian $H = \dot{x}p + \dot{y}q - L = -\sqrt{n^2-p^2-q^2}$. Then the †canonical equations of the path are obtained as in particle dynamics; x, y and p, q

are called †canonical variables. Utilizing a linear form $p\,dx + q\,dy - H\,dz = \omega_d$ and the variation principle, the variation of the integral along the light path may be written as $\omega_d(A) - \omega_d(A')$. Hence the latter integral is a function $S(A',A)$ of the points A', A. Then the optical direction cosines and the Hamiltonians of the system at A and A' are given by

$$\frac{\partial S}{\partial x'} = -p', \quad \frac{\partial S}{\partial y'} = -q', \quad \frac{\partial S}{\partial z'} = H',$$

$$\frac{\partial S}{\partial x} = p, \quad \frac{\partial S}{\partial y} = q, \quad \frac{\partial S}{\partial z} = -H.$$

Hence we obtain †Hamilton's equations

$$\left(\frac{\partial S}{\partial x'}\right)^2 + \left(\frac{\partial S}{\partial y'}\right)^2 + \left(\frac{\partial S}{\partial z'}\right)^2 = n'^2,$$

$$\left(\frac{\partial S}{\partial x}\right)^2 + \left(\frac{\partial S}{\partial y}\right)^2 + \left(\frac{\partial S}{\partial z}\right)^2 = n^2.$$

As a corollary of these relations we obtain **Malus's theorem** which states that a pencil of light rays perpendicular to a given surface at a given moment is also perpendicular to the surface after an arbitrary number of reflections and refractions.

Suppose that light rays travel from an object space into an image space through an optical apparatus. If any one of the rays starting from a point converges to a point of the image space and the mapping given by this correspondence is †bijective, then we say that this imaging is **perfect.** Examples of perfect imaging systems are realized by optical apparatus such as **Maxwell's fisheye** (having refractive index $n(r) = a/(b + r^2)$, where r denotes the distance from the center of the system) and **Luneburg's lens** ($n(r) = \sqrt{a - r^2}$).

Perfect imaging conserves optical distance, yields the relation $n(A')ds' = n(A)ds$, and gives a †conformal mapping, with the magnification inversely proportional to the refractive index.

B. Gauss Mappings

Consider an optical system with a symmetrical axis of rotation, its **optical axis**. A ray of light that is near the optical axis and has a small inclination to the axis is called a **paraxial ray**. A mapping realizable by paraxial rays, where the canonical variables x, y, p, q may be considered as infinitesimal variables whose squares are negligible, is called a **Gauss mapping**. When the positions of an object point and its image under a Gauss mapping are represented by homogeneous coordinates, the mapping is represented as a linear transformation, i.e., a †collineation, which maps a point to a point and a line to a line. A point in one space corresponding to the point at infinity in the other space is called a **focus**. If we take a focus as the origin of a coordinate system in each space and use the †homogeneous coordinates x_i such that $x = x_1/x_4$, $y = x_2/x_4$, $z = x_3/x_4$, then a Gauss mapping can be represented as $x_1 = x_1'$, $x_2 = x_2'$, $x_3 = fx_4'$, $f'x_4 = x_3'$. The ratio of x to x', i.e., the lateral magnification, is $x/x' = z/f = f'/z'$, where x' is the length of an object orthogonal to the axis and x is the length of its image. The distance f between a focus and a point where the lateral magnification is 1 (such a point is called a **principal point**) is called **focal length** in each space. The telescopic mapping, i.e., $x_1 = x_1'$, $x_2 = x_2'$, $x_3 = ax_3'$, $x_4 = bx_4'$, is also a Gauss mapping, in which the lateral magnification is constant.

C. Aberration

When a mapping is realized not only by paraxial rays but also by rays having larger inclinations, a departure from the Gauss mapping arises. This departure is generally called **aberration**. Suppose that a light ray that passes through the point (x',y',z') of a plane perpendicular to the optical axis and has optical direction cosines p', q' is transformed by the optical apparatus into a light ray that passes through the point (x,y,z) of a plane perpendicular to the optical axis and has optical direction cosines p, q there. Then by the variation principle, $p\,dx + q\,dy - p'\,dx - q'\,dy' = dW$ (dW is an †exact differential). Therefore, the transformation $(x',y',p',q') \to (x,y,p,q)$ is a †canonical transformation. The mapping may be described by

$$p = \frac{\partial W}{\partial x}, \qquad q = \frac{\partial W}{\partial y},$$

$$p' = -\frac{\partial W}{\partial x'}, \qquad q' = -\frac{\partial W}{\partial y'}$$

in terms of W, and can also be represented in terms of $V = W + p'x' + q'y'$ or $U = W + p'x' + q'y' - px - qy$. For a given optical system, one of these functions W, U, V (called a **characteristic function** or **eikonal**) may be used to estimate the aberration. By developing such a characteristic function in power series of canonical variables and observing its terms of less than the fifth power, we may single out five kinds of aberration: spherical aberration, curvature of image field, distortion, coma, and astigmatism in a rotationally symmetric optical system. To eliminate these aberrations an optical system must satisfy Abbe's sine condition (the elimination of spherical aberration and coma), Petzval's condition (the elimination of astigmatism and curvature of image field), and the tangent condition (the elimination of distortion).

The path of a charged particle in an electromagnetic field can be treated in the same way as the path of a light ray. Let ε represent the specific charge of the particle, h the energy, A_0 the electrostatic potential, and A_x, A_y, A_z vector potentials. Then the index of refraction is $\sqrt{2(h-\varepsilon A_0)}+\varepsilon(A_x dx/ds+A_y dy/ds+A_z dz/ds)$. In this case, the index of refraction shows the anisotropy caused by the existence of the magnetic field. The paths of paraxial rays are determined by a set of linear differential equations of the second order, and the Gauss mapping is realized as in geometric optics.

References

[1] C. Carathéodory, Geometrische Optik, Springer, 1937.
[2] M. Herzberger, Modern geometrical optics, Interscience, 1958.
[3] R. K. Luneburg, Mathematical theory of optics, Univ. of California Press, 1964.

185 (VI.1) Geometry

The Greek word for geometry, which means *measurement of the earth*, was used by the historian Herodotus, who wrote that in ancient Egypt people used geometry to restore their land after the inundation of the Nile. Thus the theoretical use of figures for practical purposes goes back to pre-Greek antiquity. Tradition holds that Thales of Miletus knew some properties of congruent triangles and used them for indirect measurement, and that the Pythagoreans had the idea of systematizing this knowledge by means of proofs (→ 28 Ancient Mathematics; 190 Greek Mathematics). †Euclid's *Elements* is an outgrowth of this idea. In this work, we can see the entire mathematical knowledge of the time presented as a logical system. It includes a chapter (Book V) on the theory of quantity (i.e., the theory of positive real numbers in present-day terminology) and chapters on the theory of integers (Books VII–IX), but for the most part, it treats figures in a plane or in space and presents number-theoretic facts in geometric language.

Geometry in today's usage means the branch of mathematics dealing with spatial figures. In ancient Greece, however, all of mathematics was regarded as geometry. In later times, the French word *géomètre* or the German word *Geometer* was sometimes used as a synonym for *mathematician*. In a fragment of his *Pensées*, B. Pascal speaks of the *esprit de géométrie* as opposed to the *esprit de finesse*. The former means simply the mathematical way of thinking.

Algebra was introduced into Europe from the Middle East toward the end of the Middle Ages and was further developed during the Renaissance. In the 17th and the 18th centuries, with the development of analysis, geometry achieved parity with algebra and analysis.

As R. Descartes pointed out, however, figures and numbers are closely related [1]. Geometric figures can be treated algebraically or analytically by means of †coordinates (the method of **analytic geometry**, so named by S. F. Lacroix [2]); conversely, algebraic or analytic facts can be expressed geometrically. Analytic geometry was developed in the 18th century, especially by L. Euler [3], who for the first time established a complete algebraic theory of †curves of the second order. Previously, these curves had been studied by Apollonius (262–200? B.C.) as †conic sections. The idea of Descartes was fundamental to the development of analysis in the 18th century. Toward the end of that century, analysis was again applied to geometry. For example, G. Monge's contribution [4] can be regarded as a forerunner of †differential geometry.

However, we cannot say that the analytic method is always the best manner of dealing with geometric problems. The method of treating figures directly without using coordinates is called **synthetic** (or **pure**) **geometry**. In the 17th century, a new field called †projective geometry was initiated synthetically by G. Desargues and B. Pascal. It was further developed in the 19th century by J.-V. Poncelet, L. N. Carnot, and others. In the same century, J. Steiner insisted on the importance of this field.

On the other hand, the †axiom of parallels in Euclid's *Elements* has been an object of criticism since ancient times. In the 19th century, by denying the a priori validity of Euclidean geometry, J. Bolyai and N. I. Lobačevskiĭ formulated non-Euclidean geometry, whose logical consistency was shown by models constructed in both Euclidean and projective geometry (→ 283 Non-Euclidean Geometry).

In analytic geometry, physical space and planes, as we know them, are represented as 3-dimensional or 2-dimensional Euclidean spaces E^3, E^2. It is easy to generalize these spaces to n-dimensional Euclidean space E^n. A "point" of E^n is an n-tuple of real numbers $(x_1, \ldots, x_n)$, and the distance between two points $(x_1, \ldots, x_n)$, $(y_1, \ldots, y_n)$ is $((y_1-x_1)^2+\ldots+(y_n-x_n)^2)^{1/2}$. The geometries of E^2,

E^3 are called **plane geometry** and **space** (or **solid**) **geometry**, respectively. The geometry of E^n is called n-**dimensional Euclidean geometry**. We obtain n-dimensional projective or non-Euclidean geometries similarly. F. Klein [6] proposed systematizing all these geometries in group-theoretic terms. He called a "space" a set S on which a group G operates and a "geometry" the study of properties of S invariant under the operations of G (→ 147 Erlangen Program).

B. Riemann [5] initiated another direction of geometric research when he investigated n-dimensional †manifolds and, in particular, †Riemannian manifolds and their geometries. Riemannian geometry is not always considered geometry as Klein saw it. It was a starting point for the broad field of modern differential geometry, that is, the geometry of †differentiable manifolds of various types (→ 112 Differential Geometry).

The reexamination of the system of axioms of Euclid's *Elements* led to Hilbert's †foundations of geometry and to the axiomatic tendency of present-day mathematics. The study of algebraic curves, which started with the study of conic sections, developed into the theory of algebraic manifolds, the algebraic geometry that is now so rapidly developing (→ 14 Algebraic Geometry). Another branch of geometry is topology, which has developed since the end of the 19th century. Its influence on the whole of mathematics today is considerable (→ 409 Topology; 117 Differential Topology). Geometry has now permeated all branches of mathematics, and it is sometimes difficult to distinguish it from algebra or analysis. The importance of geometric intuition, however, has not diminished from antiquity until today.

References

[1] R. Descartes, Géométrie, Paris, 1637 (Oeuvres, IV, 1901).
[2] S. F. Lacroix, Traité élémentaire de trigonométrie rectiligne et sphérique et d'application de l'algèbre à la géométrie, Bachelier, 1798–1799.
[3] L. Euler, Introductio in analysin infinitorum, Lausanne, 1748 (Opera omnia VIII, IX, 1922); French translation, Introduction à l'analyse infinitesimale I, II, Paris, 1835; German translation, Einleitung in die Analysis des Unendlichen, Springer, 1885.
[4] G. Monge, Application de l'analyse à la géométrie, fourth edition, Paris, 1809.
[5] B. Riemann, Über die Hypothese, welche der Geometrie zu Grunde liegen, Habilitationsschrift, 1854. (Gesammelte mathematische Werke, Teubner, 1876, p. 254–269; Dover, 1953.)
[6] F. Klein, Vergleichende Betrachtungen über neuere geometrische Forschungen, Das Erlanger Programm, 1872. (Gesammelte mathematicshe Abhandlungen I, Springer, 1921, p. 460–497.)

186 (VI.20) Geometry of Nets

A. General Remarks

Let G be a simply connected domain in a Euclidean plane E. Suppose that we are given n families of curves in G such that for each point $P \in G$, there exist lines l_i belonging to the ith family ($i = 1, \ldots, n$) containing P. Moreover, suppose that two distinct curves belonging to the same family never intersect, while two curves belonging to different families intersect at most at one point. In this case, the set of n families is called an n-**web** of curves in G.

As a simple example, we have the 3-web in E formed by three families of parallel lines in E, each line of which makes an angle of 60° with any line belonging to another family. We call such a web a **normal 3-web**. Suppose that we are given a 3-web of curves in G. Let P be a point in G and A a point lying on the curve of the first family passing through P. Suppose that the curve of the second family passing through A meets the curve of the third family passing through P at B (Fig. 1). Continuing the process illustrated in Fig. 1, we may be able to find the point H on the curve connecting P and A. If every point P in G has a neighborhood V in which we can draw figures like Fig. 1 for any choice of A ($\in V$) and A always coincides with H, then such a 3-web is called a **hexagonal web**. A 3-web of curves in E can be mapped topologically onto the normal 3-web if and only if it is a hexagonal web (W. Blaschke and G. Thomsen). A hexagonal web of straight lines consists of three families of tangents to an †algebraic curve in G (not necessarily irreducible) of the

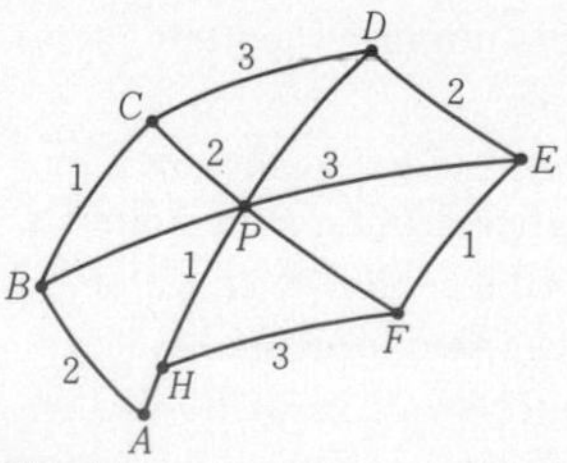

Fig. 1

third †class; the converse is also true (Graf-Sauer theorem).

B. Abel's Theorem

Suppose that we are given an n-web of curves in G and that the curves in the ith family can be represented as $u_i(x,y)=u_i=$ constant $(i=1,\ldots,n)$, where (x,y) are Cartesian coordinates of points in G. Then the u_i are called **canonical parameters** of the curves. In particular, if the web is a hexagonal web, then we can choose the u_i so that $u_1(x,y)+u_2(x,y)+u_3(x,y)=0$ for any point (x,y) in G (in this case we write $u_1+u_2+u_3=0$). Conversely, if canonical parameters of curves of a 3-web satisfy $u_1+u_2+u_3=0$, then the web is hexagonal. On the other hand, an n-web of straight lines coincides with the families of tangents to an algebraic curve of the nth class if and only if the curves of the web have canonical parameters u_i satisfying $u_1+\ldots+u_n=0$. This is an extension of the Graf-Sauer theorem and is the dual of †Abel's theorem in the theory of algebraic functions.

Suppose that the domain G has the structure of a differentiable manifold and each curve in the given families is differentiable. Two such systems of n-webs of curves in G_j $(j=1,2)$ are said to be differentiably equivalent if there exists a diffeomorphism φ sending G_1 onto G_2 and curves belonging to the ith family in G_1 onto curves belonging to the ith family in G_2 $(i=1,\ldots,n)$. The problem of finding a complete system of differential invariants that characterize equivalence classes of n-webs has been solved.

C. Abstract 3-Webs

An abstract 3-web can be defined axiomatically by taking points and lines as nondefined elements and assuming that lines are classified into three families so that the following two axioms are satisfied: (1) Through each point passes one and only one line of each family. (2) Any two lines belonging to different families have one and only one point in common. We call any two lines belonging to the same family *parallel*, and denote the relation by $//$. We define vectors AB and CD to be equal when $AB//CD$ and $AC//BD$ or when there is an EF such that $AB//EF$, $AE//BF$; $CD//EF$, $CE//DF$. Then we can construct a vector CD equal to a given vector AB so that it has an arbitrary given point C as initial point. If **Reidemeister's figure** (Fig. 2) is closed in such an abstract 3-web, equality of vectors is an equivalence relation, and addition may be defined among equivalence classes of vectors on a line as usual. Equivalence classes of vectors on a line constitute a group under this addition, and any two such groups for different lines are isomorphic. Conversely, for any given group, it is possible to construct an abstract 3-web such that the group constructed as outlined is isomorphic to the given group. **Thomsen's figure** T (Fig. 3) is closed if and only if its group is Abelian.

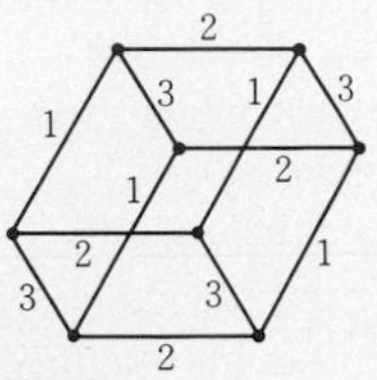

Fig. 2
R.

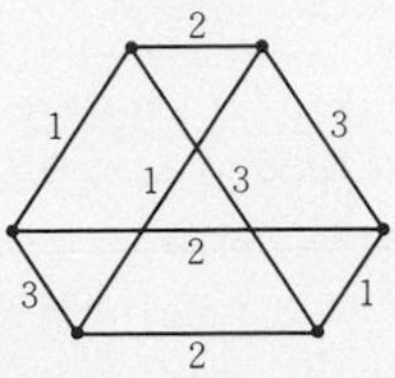

Fig. 3
T.

The n-webs of surfaces as well as the n-webs of curves in 3-dimensional Euclidean space have also been studied.

References

[1] W. Blaschke and G. Bol, Geometrie der Gewebe, Springer, 1938.
[2] K. Reidemeister, Vorlesungen über Grundlagen der Geometrie, Springer, 1930.
[3] G. Pickert, Projektive Ebenen, Springer, 1955.

187 (V.10) Geometry of Numbers

A. History

H. Minkowski introduced the notions of lattice and convex set in the †algebraic theory of numbers. He developed a simple yet powerful method of arithmetic investigation using these geometric notions to simplify the analytic theory of †Diophantine approximation, which had been developed by Dirichlet and C. Hermite. His theory, the **geometry of numbers**, has continued its development and contributed to various fields of mathematics (→ 85 Continued Fractions).

B. Lattices

Let E^n be an n-dimensional Euclidean space identified with the linear space $\mathbf{R}^n$. For a point P in E^n, we denote the corresponding vector in $\mathbf{R}^n$ by $v(P)={}^t(x_1,\dots,x_n)$. A subset Λ of E^n is called an n-dimensional **(homogeneous) lattice** if there exists a basis $\{v_1,\dots,v_n\}$ of $\mathbf{R}^n$ such that $\Lambda=\{P\in E \mid v(P)=\sum_{i=1}^n \lambda_i v_i, \lambda_i\in\mathbf{Z}\}$. The set of points $\{X_1,\dots,X_n\}$ such that $v(X_i)=v_i$ $(i=1,\dots,n)$ is called a **basis** of the lattice Λ. A typical example of a lattice is the point set corresponding to $\mathbf{Z}^n$ in $\mathbf{R}^n$. The †free module generated by v_i $(i=1,\dots,n)$ is denoted by Λ^* and is called the **lattice group** of Λ. We have $\Lambda^*=\{v\in\mathbf{R}^n \mid v=v(P), P\in\Lambda\}$. If $\{u_i,\dots,u_n\}$ is another basis of the free module Λ^*, then there exists an element (α_{ij}) of $GL(n,\mathbf{Z})$ (i.e., $\alpha_{ij}\in\mathbf{Z}$ and $|\det(\alpha_{ij})|=1$) such that $u_i=\sum_{j=1}^n\alpha_{ij}v_j$. Hence the quantity $|\det(v_1,\dots,v_n)|$ is independent of the choice of the basis $\{v_1,\dots,v_n\}$. We denote this quantity by $d(\Lambda)$ and call it the **determinant of the lattice**. We denote the minimum distance between the points belonging to Λ by $\delta(\Lambda)$.

A subset L of the space E^n is called an **inhomogeneous lattice** if there exists a homogeneous lattice Λ in E^n and a point P_0 in E^n such that $L=\{P\in E^n \mid v(P)-v(P_0)\in\Lambda^*\}$. Thus an inhomogeneous lattice is obtained from a homogeneous lattice by translation. In this article we restrict ourselves to the case of homogeneous lattices and henceforth omit the adjective "homogeneous."

Suppose we are given a sequence of lattices $\Lambda_1, \Lambda_2, \dots,$ in E^n with bases $\{X_i^{(1)}\}$, $\{X_i^{(2)}\},\dots$. If the sequence of points $X_i^{(\nu)}$ converges to X_i $(i=1,\dots,n)$ and the set $\{X_1,\dots,X_n\}$ forms a basis of a lattice Λ, we call Λ the **limit of the sequence** $\{\Lambda_\nu\}$; we also say that the sequence $\{\Lambda_\nu\}$ **converges** to the lattice Λ. In this case we have $d(\Lambda_\nu)\to d(\Lambda)$, $\delta(\Lambda_\nu)\to\delta(\Lambda)$. The notion of convergence of lattices gives rise to a topology of the space M_0 of all the lattices in E. A sequence $\{\Lambda_\nu\}$ of lattices is said to be **bounded** if there exist positive numbers c and c' such that $d(\Lambda_\nu)\leqslant c, \delta(\Lambda_\nu)\geqslant c'$ for all ν. A bounded sequence of lattices has a convergent subsequence.

Let S be a subset of the space E^n. A lattice Λ is called ***S*-admissible** if we have $\Lambda\cap S^i=\{O\}$, where S^i is the interior of S and O is the origin. We denote the set of S-admissible lattices by $A(S)$. Given a closed subset M of M_0, we put $\Delta(S\setminus M)=\inf_{\Lambda\in A(S)\cap M}d(\Lambda)$ if $A(S)\cap M$ is nonempty, while if $A(S)\cap M$ is empty, we put $\Delta(S\setminus M)=\infty$. When $M=M_0$, we write $\Delta(S\setminus M)=\Delta(S)$ and call it the **critical determinant** of S. Generally, a lattice Λ in M is said to be **critical in *M* with respect to *S*** if $\Lambda\in A(S)$ and $d(\Lambda)=\Delta(S\setminus M)$. Suppose that we have $0<\Delta(S\setminus M)<\infty$. Then for a lattice critical in M with respect to S to exist it is necessary and sufficient that there exists a bounded sequence $\{\Lambda_\nu\}$ such that $\Lambda_\nu\in M\cap A(S)$ and $d(\Lambda_\nu)\to\Delta(S\setminus M)$.

C. Successive Minima and Minkowski's Theorem

A subset S of the space E^n is called a **bounded star body** (symmetric with respect to the origin) if there exists a continuous function F defined on the space E satisfying the following four conditions: (i) $F(0)=0$; (ii) if $X\neq 0$, then $F(X)>0$; (iii) for an arbitrary real number t and a point X, we have $F(tX)=|t|F(X)$; (iv) $S=\{X \mid F(X)\leqslant 1\}$. A bounded closed †convex body that is symmetric with respect to the origin is a bounded star body. If we are given a star body S, the associated function F, and a lattice Λ, there exist a set of points $\{P_1,\dots,P_n\}$ in Λ and a set of positive numbers $\{p_1,\dots,p_n\}$ satisfying the following four conditions: (1) $v(P_1),\dots,v(P_n)$ are linearly independent; (2) $F(P_i)=p_i$ $(i=1,\dots,n)$; (3) $p_1\leqslant\dots\leqslant p_n$; (4) if P is a point in Λ and $v(P)$ is not contained in the subspace spanned by $\{v(P_1),\dots,v(P_{m-1})\}$, then $F(P)\geqslant p_m$. The set $\{p_1,\dots,p_n\}$ is uniquely determined by S and Λ. The numbers p_i are called the **successive minima** of S in Λ; the points P_i are the **successive minimum points** of S in Λ.

Minkowski's theorem: Let Λ be a lattice in a Euclidean space E^n and S a bounded subset of E^n. Then we have the following:

(I) If the volume $V(S)$ is larger than $d(\Lambda)$, then there exist points X_1 and X_2 in S such that $X_1\neq X_2$ and $v(X_1)-v(X_2)\in\Lambda^*$. Suppose, moreover, that S is convex and symmetric with respect to the origin. Then, if $V(S)>2^n d(\Lambda)$, there exists a point X in $S\cap\Lambda$ different from the origin. Hence we have $2^n\Delta(S)\geqslant V(S)$ $(n=\dim E^n)$.

(II) Let S be a bounded closed convex body that is symmetric with respect to the origin, and let $p_1,\dots,p_n$ be the successive minima of S in Λ. Then we have $p_1\dots p_n\cdot V(S)\leqslant 2^n d(\Lambda)$.

D. Minkowski-Hlawka Theorem

Suppose that we are given a subset S of the n-dimensional Euclidean space E^n such that the characteristic function $\chi(X)$ of S is †integrable in the sense of Riemann. Then we have the **Minkowski-Hlawka theorem**: (i) If $n\geqslant 2$ and S is open, then $\Delta(S)\leqslant V(S)$, and (ii) if, moreover, S is symmetric with respect to the origin, then $2\Delta(S)\leqslant V(S)$; (iii) if $n\geqslant 2$ and S is a star body, then $\zeta(n)\Delta(S)\leqslant V(S)$, where $\zeta(S)$ is the †Riemann zeta function, and (iv) if, moreover, S is symmetric with respect to the origin, then $2\zeta(n)\Delta(S) \leqslant V(S)$.

A proof for the theorem was given by E. Hlawka (1944); (iii) and (iv) were conjectured by Minkowski. C. L. Siegel obtained another proof (1945), and C. A. Rogers simplified the original proof by Hlawka (1947). There are results concerning the estimation of $\Delta(S)/V(S)$ for various subsets S.

E. Siegel's Mean Value Theorem

In an attempt to obtain a proof for the latter half of the Minkowski-Hlawka theorem, Minkowski observed the necessity of establishing the arithmetic theory of the linear transformation groups. Siegel was inspired by this observation and obtained the following theorem, **Siegel's mean value theorem**, which implies the Minkowski-Hlawka theorem: Let F be a †fundamental domain of the group $SL(n,\mathbf{R})$ with respect to a discrete subgroup $SL(n,\mathbf{Z})$. Let ω be the †invariant measure on $SL(n,\mathbf{R})$ such that $\int_F d\omega = 1$ (→ 225 Invariant Measures). Let f be a bounded Riemann integrable function with compact support defined on the space $\mathbf{R}^n$. Note that the lattice $\mathbf{Z}^n$ is stabilized by the subgroup $SL(n,\mathbf{Z})$. We have

$$\int_F \sum_{\substack{x\in \mathbf{Z}^n\\ x\neq 0}} f(g\cdot x)\,d\omega = \int_{\mathbf{R}^n} f(x)\,dx,$$

where the right-hand side of the equation is the usual Riemann integral of the function f. A. Weil considered this theorem in general perspective (*Summa Brasil. Math.*, 1 (1946)).

F. Diophantine Approximation

Minkowski initiated the notion of Diophantine approximation in reference to the problem of estimating the absolute value $|f(x)|$ of a given function f where x varies in $\mathbf{Z}$ or in a given ring of †algebraic integers. (A †Diophantine equation is an equation $f(x)=0$, where x varies in $\mathbf{Z}$.) Today Diophantine approximation (in the wider sense) refers to the investigation of the scheme of values $f(x)$, where x varies in a suitable ring of algebraic integers. The geometry of lattices is a powerful tool in this investigation. A typical problem in this field of study is that of approximating irrational numbers by rational numbers; here †continued fractions play an important role (→ Section G). For the problem of uniform distribution considered by H. Weyl, the analytic method, especially that of trigonometric series, is useful (→ Section H). **Dirichlet's drawer principle** (to put n objects in m drawers with $n>m$, it is necessary to put more than one object in at least one drawer) is one of the basic principles used in the theory of Diophantine approximation. Recently, the theory has been applied to the theory of †transcendental numbers and the theory of †Diophantine equations.

G. Approximation of Irrational Numbers by Rational Numbers

Given an irrational number θ, we have the problem of finding rational integers x (>0) and y such that $|\theta-y/x|$ can be written as ε/x, where ε is a small positive number. Suppose that we are given a positive integer N. Using Dirichlet's drawer principle we can show the existence of x $(\leqslant N)$ and y such that $|\theta-y/x|<1/xN$. Let $M(\theta)$ be the supremum of positive numbers M such that the inequality $|\theta-y/x|<1/Mx^2$ holds for infinitely many pairs of integers x, y. We have $1\leqslant M(\theta)$ $(\leqslant\infty)$. Two irrational numbers θ and θ' are said to be **equivalent** if there exists an element $(a_{ij})\in GL(2,\mathbf{Z})$ such that $\theta'=(a_{11}\theta+a_{12})/(a_{21}\theta+a_{22})$. In this case we have $M(\theta)=M(\theta')$.

If the irrational number θ satisfies the quadratic equation $a\theta^2+b\theta+c=0$ (a,b,c are rational integers), then we have $M(\theta)=k^{-1}\sqrt{b^2-4ac}$, where $k=\min\{ax^2+bxy+cy^2 \mid x,y\in\mathbf{Z}, x\neq 0, y\neq 0\}$. In general, for an irrational number θ of degree two, we have $M(\theta)\geqslant\sqrt{5}$. The equality $M(\theta)=\sqrt{5}$ holds if θ is equivalent to $\theta_1=(1+\sqrt{5})/2$. If θ is not equivalent to θ_1, then $M(\theta)\geqslant\sqrt{8}$; the equality holds if θ is equivalent to $\theta_2=1+\sqrt{2}$. Similarly, we have $\theta_3, \theta_4, \ldots$; and $M(\theta_n)\to 3$ $(n\to\infty)$. If $M(\theta)<3$, there exists a θ_n such that θ is equivalent to θ_n. The set of irrational numbers θ satisfying $M(\theta)=3$ is uncountable (A. A. Markov [16]). We have no information about $M(\theta)$ for the general algebraic irrational number θ. Let $\mu(\theta)$ be the supremum of real numbers μ such that the inequality $|\theta-y/x|<1/x^{\mu}$ holds for infinitely many pairs of integers x, y. Given a number $\kappa>2$, it can be shown that the †Lebesgue measure of the set of real numbers θ such that $\mu(\theta)\geqslant\kappa$ is zero. If θ is a real algebraic number of degree n, then $\mu(\theta)\leqslant n$ (J. Liouville). Concerning $\mu(\theta)$, results have been obtained by A. Thue, Siegel, A. O. Gel'fond, and F. J. Dyson. K. Roth recently proved that $\mu(\theta)=2$ (**Roth's theorem** [15]), which settled the problem of $\mu(\theta)$. Roth's theorem means that if κ is larger than 2, then there exist only a finite number of pairs x, y satisfying $|\theta-y/x|<1/x^{\kappa}$. This can be generalized to the case of the approximation of an element θ that is algebraic over an A-field k by an element of the field k (S. Lang [12]). (An A-field is either an algebraic number field of finite degree or an algebraic function field in one variable over a finite constant field.)

In 1970 W. M. Schmidt [23] obtained theorems on simultaneous approximation which generalize Roth's theorem. Thus, if $\alpha_1, \ldots, \alpha_n$ are real algebraic numbers such that 1, $\alpha_1, \ldots, \alpha_n$ are linearly independent over the field of rational numbers, then for every $\varepsilon > 0$ there are only finitely many positive integers q with

$$\|q\alpha_1\| \ldots \|q\alpha_n\| q^{1+\varepsilon} < 1,$$

where $\|\xi\|$ denotes the distance from a real number ξ to the nearest integer; in particular, we have

$$|\alpha_i - p_i/q| < q^{-(n+1)/n-\varepsilon}, \qquad i = 1, \ldots, n,$$

for only finitely many n-tuples of rationals $p_1/q, \ldots, p_n/q$. A dual to this result is as follows. Let $\alpha_1, \ldots, \alpha_n$, ε be as before. Then there are only finitely many n-tuples of nonzero integers $q_1, \ldots, q_n$ with

$$\|q_1\alpha_1 + \ldots + q_n\alpha_n\| \cdot |q_1 \ldots q_n|^{1+\varepsilon} < 1.$$

This last theorem can be used to prove that if α is an algebraic number, k a positive integer, and $\varepsilon > 0$, then there are only finitely many algebraic numbers ω of degree at most k such that $|\alpha - \omega| < H(\omega)^{-k-1-\varepsilon}$, where $H(\omega)$ denotes the height of ω. See also [19].

The work of Thue, Siegel, and Roth had the basic limitation of noneffectiveness. A. Baker (1968) succeeded in proving that for any algebraic number of degree $n \geqslant 3$ and any $\kappa > n$, there exists an effectively computable number $c = c(\theta, \kappa) > 0$ such that $|\theta - y/x| > cx^{-n}\exp(\log x)^{1/\kappa}$ for all integers x, y $(x > 0)$ [18]. This result is an immediate consequence of the following effective version of a classical theorem on binary Diophantine equations (Thue, 1909): Let $f = f(x,y)$ be an irreducible binary form of degree $n \geqslant 3$ with integer coefficients, and suppose that $\kappa > n$. Then for any positive integer m, all integer solutions x, y of the equation $f(x,y) = m$ satisfy $\max(|x|,|y|) < c\exp(\log m)^{\kappa}$, where $c > 0$ is an effectively computable number depending on n, κ, and the coefficients of f. Baker obtained this result by making use of his theorems which give effective estimates of moduli of linear forms in the logarithms of algebraic numbers with algebraic coefficients. A typical theorem reads as follows: Let $\alpha_1, \ldots, \alpha_n$ be nonzero algebraic numbers with $\log\alpha_1, \ldots, \log\alpha_n$ linearly independent over the rationals, and let $\beta_0, \ldots, \beta_n$ be algebraic numbers, not all 0, with degrees and heights at most d and H, respectively. Then for any $\kappa > n+1$, we have $|\beta_0 + \beta_1 \log\alpha_1 + \ldots + \beta_n \log\alpha_n| > c\exp(-(\log H)^{\kappa})$, where $c > 0$ is an effectively computable number depending only on n, κ, $\log\alpha_1, \ldots, \log\alpha_n$, and d [17, 22].

Results of this kind have many important applications in number theory. For instance, we obtain a generalization of the Gel'fond-Schneider theorem on transcendental numbers. Furthermore, the imaginary quadratic number fields of class number 1 can be completely determined on the basis of Baker's result. This was actually done by Baker (1966) and independently by H. M. Stark (1966) (→ 343 Quadratic Fields).

Refinements and generalizations of Thue's theorem on the finiteness of solutions of binary Diophantine equations have been obtained by Baker and his collaborators. (→ [20], and also [19]). Also, p- and $\mathfrak{p}$-adic analogs of Baker's results are known [21].

H. Uniform Distribution

Let θ be a real number, x a positive integer, and $[\theta x]$ the maximum integer not larger than θx. We write $(\theta x) = \theta x - [\theta x] = \theta x \pmod 1$. Jacobi showed that if θ is irrational, then the set $\{\theta x \pmod 1) | x \in \mathbf{N}\}$ is densely distributed in the interval (0, 1) ($\mathbf{N}$ is the set of positive integers). In general, let f be a real-valued function defined on $\mathbf{N}$. We say that $f(x)$ (mod 1) is **uniformly distributed** in the unit interval, or $f(x)$ is **uniformly distributed (mod** 1), if the following condition is satisfied: Let α, β be an arbitrary pair of real numbers such that $0 \leqslant \alpha < \beta \leqslant 1$, and let N be a given positive integer. Let $T(N)$ be the number of positive integers x such that $x \leqslant N$, $\alpha \leqslant (f(x)) < \beta$, where $(f(x)) = f(x) - [f(x)]$. Then $\lim_{N\to\infty} T(N)/N = \beta - \alpha$. In order for $f(x)$ (mod 1) to be uniformly distributed, it is necessary and sufficient that $\lim_{N\to\infty} N^{-1} \cdot \sum_{x=1}^{N} e^{2\pi i h f(x)} = 0$ for any nonzero integer h (**Weyl's principle**, 1914). Weyl proved that if θ is an irrational number, then $\theta x \pmod 1$ is uniformly distributed.

The following theorem, due to J. G. van der Corput, is often useful: Let f be a real-valued function defined on $\mathbf{N}$. Consider the function $f_h(x) = f(x+h) - f(x)$ for an arbitrary positive integer h. If $f_h(x)$ (mod 1) is uniformly distributed (mod 1) for all such h, then $f(x)$ (mod 1) is also uniformly distributed (mod 1).

Utilizing this theorem, it can be shown that if $f(x) = \theta_r x^r + \theta_{r-1} x^{r-1} + \ldots + \theta_0$, where at least one of the coefficients θ_i is irrational, then $f(x)$ (mod 1) is uniformly distributed.

The notion of uniform distribution of sequences of real numbers has an analog in compact Hausdorff spaces and in various topological groups. A systematic treatment of such generalized notions of uniform distribution can be found in [24].

References

[1] T. Takagi, Syotô seisûron kôgi (Japanese; Lectures on elementary theory of numbers), Kyôritu, 1931.
[2] T. Takagi, The geometry of lattices (Japanese), Sûgaku zatudan (Miscellaneous topics in mathematics), Kyôritu, 1935, ch. 1.
[3] M. Fujiwara, Daisûgaku (Japanese; Algebra) I, II, Utida-rôkakuho, 1928, 1929.
[4] H. Minkowski, Geometrie der Zahlen, Teubner, 1910 (Chelsea, 1953).
[5] H. Minkowski, Diophantische Approximationen, Teubner, second edition, 1927 (Chelsea, 1957).
[6] H. Minkowski, Gesammelte Abhandlungen I, II, Teubner, 1911 (Chelsea, 1967).
[7] J. F. Koksma, Diophantische Approximationen, Erg. Math., Springer, 1936 (Chelsea, 1950).
[8] H. Weyl, C. L. Siegel, and K. Mahler, Geometry of numbers, Mimeographed notes, Princeton, 1950.
[9] J. W. S. Cassels, An introduction to Diophantine approximation, Cambridge Univ. Press, 1957.
[10] J. W. S. Cassels, An introduction to the geometry of numbers, Springer, 1959.
[11] T. Schneider, Einführung in die transzendentalen Zahlen, Springer, 1957.
[12] S. Lang, Diophantine geometry, Interscience, 1962.
[13] C. L. Siegel, Über einige Anwendungen diophantischer Approximationen, Abh. Preuss. Akad. Wiss., no. 1 (1929). (Gesammelte Abhandlungen, Springer, 1966, vol. 1, p. 209–266.)
[14] C. L. Siegel, A mean value theorem in geometry of numbers, Ann. Math., (2) 46 (1945), 340–347. (Gesammelte Abhandlungen, Springer, 1966, vol. 3, p. 39–46.)
[15] K. F. Roth, Rational approximations to algebraic numbers, Mathematika, 2 (1955), 1–20.
[16] A. A. Markoff (A. A. Markov), Sur les formes quadratiques binaires indéfinies, Math. Ann., 15 (1879), 381–407.
[17] A. Baker, Linear forms in the logarithms of algebraic numbers I, II, III, IV, Mathematika, 13 (1966), 204–216; 14 (1967), 102–107; 14 (1967), 220–228; 15 (1968), 204–216.
[18] A. Baker, Contributions to the theory of Diophantine equations I, II, Philos. Trans. Roy. Soc. London, A 263 (1967–1968), 173–191, 193–208.
[19] A. Baker, Transcendental number theory, Cambridge Univ. Press, 1975.
[20] A. Baker and J. Coates, Integer points on curves of genus 1, Proc. Cambridge Philos. Soc., 67 (1970), 595–602.
[21] A. Brumer, On the units of algebraic number fields, Mathematika, 14 (1967), 121–124.
[22] Н. И. Фельдман (N. I. Fel'dman), О линейиой форме от логарифмов алгебраических чисел, Dokl. Akad. Nauk SSSR, 182 (1968), 1278–1279.
[23] W. M. Schmidt, Simultaneous approximation to algebraic numbers by rationals, Acta Math., 125 (1970), 189–201.
[24] L. Kuipers and H. Niederreiter, Uniform distribution of sequences, Interscience, 1974.

188 (I.9) Gödel Numbers

A. General Remarks

K. Gödel [1] devised the following method to prove his incompleteness theorem (→ Section B).

Let $\mathfrak{S}$ be a †formal system. In this article, we call its basic symbols, †terms, †formulas, and formal proofs the "constituents" of $\mathfrak{S}$. Let g be an †injection from the constituents of $\mathfrak{S}$ into the natural numbers satisfying the following two conditions: (1) Given a constituent C, we can compute the value $g(C)$ in a finite number of steps. (2) Given a natural number n, there exists a finitary procedure to find out whether there exists a constituent C of $\mathfrak{S}$ such that $g(C)=n$; furthermore, when such a C exists, it can actually be specified in a finite number of steps.

If such a mapping g is given for the system $\mathfrak{S}$, then the number $g(C)$ corresponding to a constituent C of $\mathfrak{S}$ is called a **Gödel number** of C (with respect to g).

B. An Example of Gödel Numbers

(1) Let $\alpha_0, \alpha_1, \dots$ be the †basic symbols of $\mathfrak{S}$. With each α_i we associate a distinct odd number q_i: $g(\alpha_i)=q_i$ $(i=0,1,\dots)$. (2) Let F be a constituent of $\mathfrak{S}$. If F is constructed from any other constituents $F_0, F_1, \dots, F_k$ of $\mathfrak{S}$ by a rule peculiar to $\mathfrak{S}$ (for convenience we write this $F=(F_0,F_1,\dots,F_k)$), and if, for each F_i, $g(F_i)$ is already defined, then we put $g(F)=\langle g(F_0), g(F_1), \dots, g(F_k)\rangle$, where $\langle a_0, a_1, \dots, a_k\rangle$ denotes the number $p_0^{a_0}p_1^{a_1}\dots p_k^{a_k}$ (p_i is the $(i+1)$st prime number). For example, suppose

that $\mathfrak{S}$ contains 0, $=$, v_j (variables), and $\neg$ (negation) among the basic symbols, and let their Gödel numbers be 7, 9, 11^{j+1}, and 13, respectively. Since the formula $\neg\ (0=v_j)$ can be analyzed in the form $(\neg, (0, =, v_j))$, its Gödel number is $\langle 13, \langle 7, 9, 11^{j+1} \rangle\rangle$. For details →[1, 2, 3].

By means of Gödel numbers a †metamathematical proposition about $\mathfrak{S}$ can be regarded as a number-theoretic proposition. This method is called the **arithmetization** of metamathematics. Consequently, if $\mathfrak{S}$ is a system obtained by formalizing a theory containing number theory in the classical †predicate calculus, then there exists a formula of $\mathfrak{S}$ that may be †interpreted as a metamathematical proposition about $\mathfrak{S}$ itself. Gödel found a †closed formula A that may be interpreted as expressing its own unprovability. Furthermore, he showed that although A holds with respect to the given interpretation, neither A nor $\neg A$ is provable in $\mathfrak{S}$. This result is called Gödel's **incompleteness theorem**. The method of arithmetization is important and useful in mathematical logic and foundations of mathematics. The notion of the Gödel number of a †general recursive function is one of its applications (→ 352 Recursive Functions).

References

[1] K. Gödel, Über formal unentscheidbare Sätze der Principia Mathematica und verwandter Systeme I, Monatsh. Math. Phys., 38 (1931), 173–198.

[2] S. C. Kleene, Introduction to metamathematics, Van Nostrand, 1952.

[3] E. Mendelson, Introduction to mathematical logic, Van Nostrand, 1964.

[4] J. B. Rosser, Extensions of some theorems of Gödel and Church, J. Symbolic Logic, 1 (1936), 87–91.

[5] S. C. Kleene, Mathematical logic, John Wiley, 1967.

[6] J. R. Schoenfield, Mathematical logic, Addison-Wesley, 1967.

189 (XV.15) Graphical Calculation

A. General Remarks

Graphical calculation is a method of computation with geometric constructions, using ruler, compass, and other common drawing tools. Sometimes simple auxiliary instruments are used. The most serious defect of graphical calculation is limitation of accuracy, but it is sometimes useful as a supplementary method in complicated numerical computations. In the wider sense, it includes graphical calculation in geometric optics and graphical mechanics (→ Section D; 282 Nomograms).

Some typical examples of practical graphical calculations in the field of algebraic computations are linear formulas of several variables by J. Massau (1887); graphical solution of systems of linear equations by F. J. van den Berg (1888); computation of the value of a polynomial by J. A. Segner (1761); and solution of an algebraic equation by Lill (1867).

In **graphical differentiation**, to draw a tangent to the graph of a function $y=f(x)$, more accurate results are generally obtained by finding the tangential point for a given direction than by drawing the tangent at a given point.

B. Graphical Integration

Massau (1878) gave a method of **graphical integration** which draws an approximate integration curve of a function $f(x)$ (Fig. 1). By this method, the original curve $y=f(x)$ is replaced by a step function, and the indefinite integral

$$F(x)=\int_a^x f(x)\,dx$$

is approximated by the line segments. Massau further gave a method of graphical integration of an ordinary differential equation $dy/dx = f(x,y)$ by which the integral curves can be drawn in a similar manner. Other methods are Czuber's method for a linear differential equation; Lord Kelvin's method (1892) for an ordinary differential equation of the second order; and graphical integration, which exactly follows the †Runge-Kutta method.

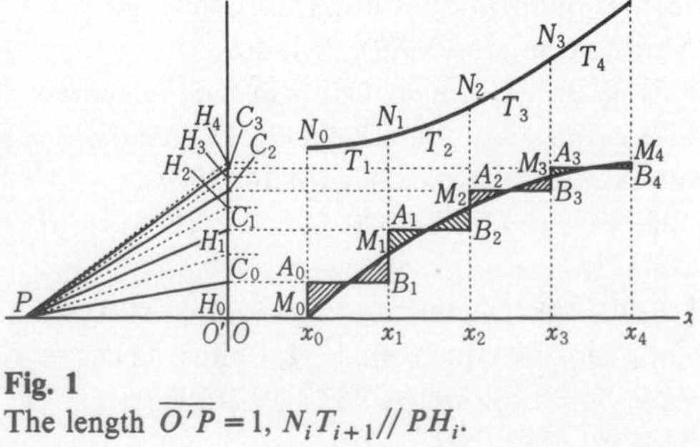

Fig. 1
The length $\overline{O'P}=1$, $N_iT_{i+1}//PH_i$.

C. Graphical Methods in Statistical Inference

Let X_1 and X_2 be two †random variables obeying mutually †independent †normal distributions, and let $m_1, m_2; \sigma_1^2, \sigma_2^2$ be the †population means and †population variances of these variables. Put $z = X_1/X_2$. In an approximation which neglects the probability of negative X_1, the variable

$$t = (zm_1 - m_2)/\sqrt{\sigma_1^2 z^2 + \sigma_2^2}$$

obeys a normal distribution whose mean is 0 and variance is 1. Hence on graph paper where the square roots of the horizontal and vertical scales represent the actual distances from the origin, the length of the orthogonal segment from the point $((m_1/2\sigma_1)^2, (m_2/2\sigma_2)^2)$ to the line connecting the origin and the point $((X_1/\sigma_1)^2, (X_2/\sigma_2)^2)$ (called the **split** of the ratio $(X_1/\sigma_1)^2 : (X_2/\sigma_2)^2$) is equal to $d = t/2$. On the other hand, let χ_1^2 and χ_2^2 be two mutually independent †χ^2-distributions of †degrees of freedom f_1 and f_2, respectively. Then $\sqrt{2}\,\chi_i - \sqrt{2\delta_i}$ is approximately a normal distribution whose mean is 0 and variance is 1. Therefore, if $F = (\chi_1^2/f_1)/(\chi_2^2/f_2)$ is a random variable obeying an †F distribution of degree of freedom (f_1, f_2), then the length of the orthogonal segments from the point $(f_1/2, f_2/2)$ to the split of the ratio $f_1F : f_2$ on the bisquare-root graph paper described before is equal to $d = t/2$. Hence distributions that can be reduced to the F distribution (for example, a †binomial distribution) can be handled graphically on such paper. Such graphical calculation is called the **graphical method of statistical inference**. Bisquare-root graph paper is sold under the name of **binomial probability paper** or **stochastic paper**. If $f_2 \to \infty$, the distance between the fixed point $(f_1/2, 0)$ and the moving point $(\chi_1^2/2, 0)$ on the horizontal axis is equal to d. Hence distributions reducible to the χ^2-distribution, such as †Poisson distributions, can also be handled graphically on the same graph paper.

D. Graphical Mechanics

Graphical mechanics is the graphical treatment of mechanical problems, especially problems of equilibrium. In this field the fundamental constructions are composition, resolution, and equilibrium of forces. The method is also applicable to problems in dynamics if we reduce them to problems of equilibrium by d'Alembert's principle. Composition or resolution of forces obeys the addition rules of

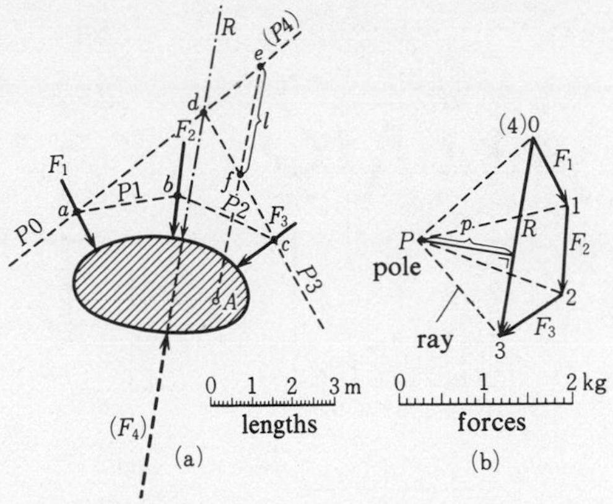

Fig. 2
The graphical method of constructing the composite and the line of action of three given forces F_1, F_2, and F_3. In the force polygon in (b), the vector $\overrightarrow{03}$ represents the composite R. On the link polygon in (a), the point d is a point on the line of action of the force. If the fourth force F_4 has the same magnitude as R with opposite direction, both polygons are closed, and the four forces F_1, F_2, F_3, and F_4 are in equilibrium.

vectors. The vector figure (see Fig. 2(b)) for the composition of forces 0123 is called the **force polygon**. To construct the position of the line of action of the composite R, it is convenient to construct the **link polygon**. As in Fig. 2(b), take an arbitrary point P and construct the **rays** $P0, P1, P2, P3$ with **pole** at P. The polygon $abcd$ in Fig. 2(a) is the link polygon, which is constructed such that each segment is parallel to one of the rays $P0$, $P1, P2, P3$. The construction begins by choosing an arbitrary point a on F_1 and determining successively the points b, c, and d. Then the line of action passes through the point d. For a given point A, we construct a line parallel to the vector R through A. Let $ef = l$(m) be the length of the line segment cut off by the two rays $P0$ and $P3$, which represent the composite R, and let p(kg) be the length of the orthogonal segment from the pole P to the composite R. Then the moment M around the point A is the product $M = lp$(kg-m).

In general, for equilibrium of the forces $\mathbf{F}_i$ $(i = 1, 2, \ldots, n)$, it is necessary and sufficient that the conditions $\Sigma_{i=1}^n \mathbf{F}_i = 0$, $\Sigma_{i=1}^n \mathbf{r}_i \times \mathbf{F}_i = 0$ are satisfied. Here $\times$ is the outer product of the vectors, and $\mathbf{r}_i$ is the position vector from a fixed point to the point of action of the force $\mathbf{F}_i$. If all forces lie in the same plane, this means that both the force polygon and the link polygon are closed. As an example, we show the moment at each point on a beam supported at both ends under vertical forces (Fig. 3). The moment is p times the ordinates l of the link polygon.

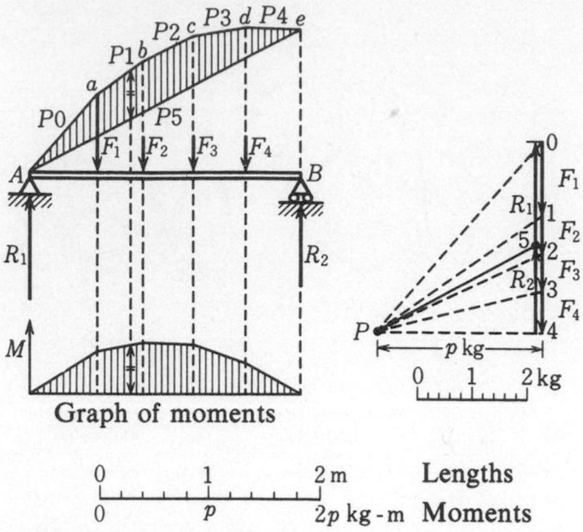

Fig. 3
The graphical determination of the supporting forces at the supports and the graph of moments when four vertical forces F_1, F_2, F_3, and F_4 work on a beam supported at the ends.

E. Graphical Mechanics for Solid Figures

The equilibrium condition of forces applied at a single point in space is the simultaneous closure of the force polygons projected on two arbitrary different coordinate planes. If the forces are applied to a rigid body, the equilibrium condition is that two force polygons in the image plane and three link polygons in each coordinate plane are closed simultaneously. For example, to resolve a force **P** acting at a point 0 in the space into three directions $(\mathbf{S}_1, \mathbf{S}_2, \mathbf{S}_3)$, we can construct two closed force polygons $a'b'c'd'$ and $a''b''c''d''$ (as shown in Fig. 4), where $'$ and $''$ are used to denote the quantities in the plane and the elevation, respectively.

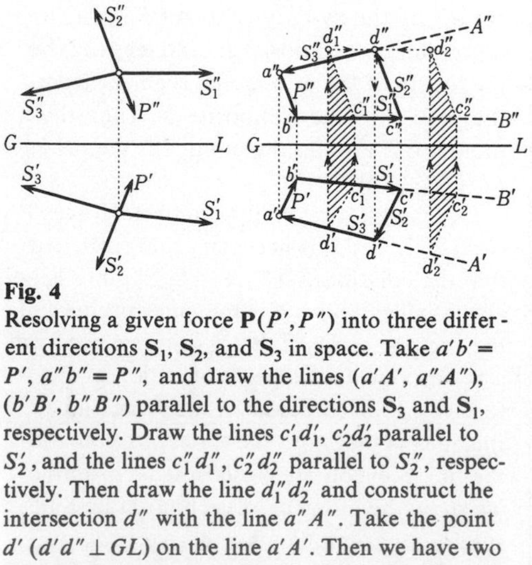

Fig. 4
Resolving a given force $\mathbf{P}(P', P'')$ into three different directions $\mathbf{S}_1$, $\mathbf{S}_2$, and $\mathbf{S}_3$ in space. Take $a'b' = P'$, $a''b'' = P''$, and draw the lines $(a'A', a''A'')$, $(b'B', b''B'')$ parallel to the directions $\mathbf{S}_3$ and $\mathbf{S}_1$, respectively. Draw the lines $c_1'd_1'$, $c_2'd_2'$ parallel to S_2', and the lines $c_1''d_1''$, $c_2''d_2''$ parallel to S_2'', respectively. Then draw the line $d_1''d_2''$ and construct the intersection d'' with the line $a''A''$. Take the point d' $(d'd'' \perp GL)$ on the line $a'A'$. Then we have two closed force polygons $a'b'c'd'$, $a''b''c''d''$. The sides of these polygons give the components of the force **P** in the given directions.

References

[1] C. Runge and F. A. Willers, Numerische und graphische Quadratur und Integration gewöhnlicher und partieller Differentialgleichungen, Enzykl. der Math., Bd. II, Teil 3, Heft 2 (1915), 47–176.
For the graphical method of statistical inference,
[2] M. Masuyama, Graphical method of statistical inference, Maruzen, 1954.
For graphical mechanics,
[3] M. Kuranishi and M. Miyagawa, Zusiki rikigaku (Japanese; Graphical mechanics), Kinbara, 1956.
[4] C. B. Biezeno and R. Grammel, Technische Dynamik, Springer, 1939.
[5] H. F. B. Müller-Breslau, Die graphische Statik der Baukonstruktionen, Kröner, I, sixth edition, 1927; II, pt. 1, fifth edition, 1922; II, pt. 2, second edition, 1925.

190 (XX.2) Greek Mathematics

It is generally believed that theoretical mathematics originated with the Greeks. The Greeks learned the arts of land surveying and commercial arithmetic from earlier civilizations. They developed theoretical mathematics themselves, toward the middle of the 4th century B.C. The creation of a mathematics that transcends practical purposes is one of the most remarkable events in the history of human culture, and one that had an immense impact on the development of all branches of science. We owe the reestablishment of important Greek mathematical texts and the reconstruction of the development of Greek mathematics to the historians of the 19th century; however, a certain amount of unfavorable criticism is now being made of their work.

The earliest known Greek mathematicians are Thales of Miletus (c. 639–546 B.C.) and Pythagoras of Samos (fl. 510? B.C.). Both were Ionians, but the latter went to what is now southern Italy and founded a semireligious school whose members called themselves Pythagoreans. Their motto was "Everything is number"; their studies were called *mathema* ("what is learned") and consisted of music, astronomy, geometry, and arithmetic (the subject group called the **quadrivium**, which formed the core of medieval and later higher education) "for the purification of soul." Their research delved into the theories of proportion (in relation to music) and †polygonal numbers (triangular numbers, square numbers, etc.), and more generally into the theory of numbers and geometric algebra. It is said that they knew of the irrationality of $\sqrt{2}$, though no evidence of this has been found. Even after the demise of the Pythagorean school, its followers continued to promote mathematics in collaboration with the Academy of Plato.

Another significant school was the Eleatic. Among its members Zeno (c. 490–c. 430 B.C.) is especially important. †Zeno's paradoxes are arguments leading to absurdity. Some see within them the origin of logical reasoning and, consequently, of theoretical mathematics [3]. It is chronologically difficult to attribute to Zeno the consideration of the continuum and irrational numbers, but we may find in him the impetus toward atomistic reasoning. The computation of the volume of pyramids (by dividing them into "atomistic" laminas) by Democritus (fl. 430? B.C.) and the atomistic calculation of the area of circles by Antiphon (fl. 430 B.C.) came shortly after the time of Zeno.

The middle decades of the 4th century B.C. are known as the Age of Pericles, the Golden Age of Athens. The †trisection of an angle, the †duplication of a cube, and the †quadrature of a circle, known at that time as the "**three big problems**" (→ 183 Geometric Construction), were studied by the Sophists. Hippias of Elis (fl. 420 B.C.), Hippocrates of Chios (fl. 430 B.C. in Athens), Archytas of Taras (c. 430–365 B.C.), Menaechmus (fl. 350 B.C.), and his brother Dinostratus (fl. 350 B.C.) solved these problems using conic sections and the **quadratrix** (a transcendental curve whose equation is $y = x\cot(\pi x/2)$).

By 400 B.C. Athens had lost its political influence, but it remained the center of Greek culture. It was during this time that Plato's Academy flourished, and Plato (427–347 B.C.) and his followers laid particular importance on mathematics. Archytas, Minaechmus, and Dinostratus belonged to or were closely associated with this school. During the first fifty years of the Academy, research in the following fields was pursued: methodology of mathematics or science in general (i.e., dialectics, analysis, synthesis); geometric reconstruction of Mesopotamian algebra; the theory of irrationals in relation to the geometrization of algebra (Theodorus of Cyrene (5th century B.C.), who was Plato's teacher, as well as Theaitetus of Athens (415?–369 B.C.) contributed to this study, and the general theory of proportion by Eudoxus of Cnidos (c. 408–c. 355 B.C.) also belongs to this field); the method of exhaustion (by Eudoxus); and studies of the "three big problems" and conic sections. It was this school in which the term *mathema* came to be used in its present mathematical sense rather than the sense of disciplines in general.

The conquests of Alexander the Great accelerated the already considerable cultural influence of Athens. Later, during the Ptolemaic period, the center of culture moved to Alexandria. The Mouseion at Alexandria, the combined library and university, is said to have possessed hundreds of thousands of volumes.

At Alexandria, Euclid (about 300 B.C.) compiled his *Elements*, which became a model for scientific works for centuries to come —Newton's *Principia* as well as Spinoza's *Ethics* are modeled on it. Most historians say that Euclid's method derives from Aristotle (384–322 B.C.), who, after studying at Plato's Academy, founded a new school, the Peripatetics, whose doctrines are at many points opposed to those of the Academy. Others, however, see the origin of Euclid's axiomatic method in the Eleatics [5]: we may find prototypes of some parts of the *Elements* in both Oenopides, who lived during the time of Zeno, and in Hippocrates.

The third century B.C. is the Golden Age of Greek mathematics. Archimedes of Syracuse (c. 282–212 B.C.) was the greatest mathematician, mechanic, and technician of antiquity. He did important work in the field of mathematics, studying the exact quadrature of the parabola. According to his *ephodos* (method), he found the result by mechanical experiments and then proved it by the method of "exhaustion." He also computed the value of π; studied spirals and other curves, spheres, and circular cylinders; contributed to the development of statics and optics and their application; and had a profound influence on later mathematicians. During the same period, Apollonius of Perga (fl. 210 B.C.) wrote *Konikon biblia* (Books on Conics) in eight books, of which the last has been lost. The geometric theory of †conic sections contained in this work is not much different from the one we know today; it had a great influence on 17th-century scientists especially. Other mathematicians of this period worth noting are Eratosthenes (c. 275–195 B.C.), who conceived the †sieve method of finding prime numbers and who measured the earth, and Hipparchus (fl. 150 B.C.), called the father of astronomy, who made a table of sines.

Hellenistic influence began to decline in the first century B.C., and the influence of Alexandria decreased. The Mouseion burned in 48 B.C., but was rebuilt. Among the mathematicians of this time, we may count Heron (fl. 60? A.D.); Menelaus (fl. 100 A.D.), who wrote *Sphaerica*; Theon of Smyrna (fl. 125 A.D.); Ptolemy (fl. 150 A.D.), the author of *Almagest*; Nicomachus (50?–150? A.D.), the author of *Arithmetike eisagoge*; Diophantus (fl. 250? A.D.), whose career is not fully known but who wrote *Arithmetika*, of which six of the original thirteen books remained to influence †Fermat; and Pappus (fl. 300 A.D.), the last creative mathematician in Greece, who left eight books of *Synagoge*, which influenced †Descartes and which still exist today.

The period following the fall of the Western Roman Empire was a difficult one for Greco-Egyptian science. The Mouseion was destroyed for the second time in 392 A.D. Theon of Alexandria (fl. 380) and his daughter, Hypatia (c. 370–418), were at that time working on commentaries on the classics. Among the few remaining works of the period is Proclus' (410–485) commentaries on the first book of Euclid's *Elements*. The Athenian Academy was closed in 529 by order of the Emperor Justinian; the last director was Simplicius, who commented on Aristotle. Soon afterwards, Alexandria fell into the hands of the Moors, and many scholars fled as refugees to Constantinople, the capital of the Eastern Empire.

References

[1] T. L. Heath, A history of Greek mathematics I, II, Clarendon Press, 1921.
[2] M. B. Cantor, Vorlesungen über Geschichte der Mathematik I, Teubner, third edition, 1907.
[3] B. L. van der Waerden, Zenon und die Grundlagenkrise der Griechischen Mathematik, Math. Ann., 117 (1940), 141–161.
[4] Á. Szabó, Anfänge des euklidischen Axiomensystems, Arch. History Exact Sci., 1 (1960), 37–106.
[5] J. L. Heiberg (ed.), Euclidis opera omnia I–XIII and suppl., Leipzig, 1883–1916.
[6] T. L. Heath, The thirteen books of Euclid's Elements I, II, III, Cambridge Univ. Press, 1908 (Dover, 1956).
[7] P. Ver Eecke (trans.), Les oeuvres complètes d'Archimède, Desclée de Brouwer & Cie, 1921 (Blanchard, 1961).
[8] P. Ver Eecke (trans.), Proclus Diadochus, les commentaires sur le premier livre des Eléments d'Euclide, Desclée de Brouwer & Cie, 1948 (Blanchard, 1959).
[9] P. Tannery, Le géométrie grecque, comment son histoire nous est parvenue et ce que nous en savons, Paris, 1887.
[10] B. L. van der Waerden, Science awakening, Noordhoff, 1954.
[11] M. Clagett, Greek science in antiquity, Abelard-Schuman, 1955 (Collier, 1963).
[12] O. Becker (ed.), Zur Geschichte der griechischen Mathematik, Darmstadt, 1965.
[13] Á. Szabó, Anfänge der griechischen Mathematik, Oldenbourg, 1969.
For the history of mathematics in general,
[14] R. C. Archibald, Outline of the history of mathematics, Mathematical Association of America, sixth edition, 1949.
[15] N. Bourbaki, Eléments d'histoire des mathématiques, Hermann, second edition, 1969.
[16] M. B. Cantor, Vorlesungen über Geschichte der Mathematik I–IV, Teubner, second edition, 1894–1908.
[17] E. Montucla, Histoire des mathématiques I–IV, Paris, new edition, 1799–1802.
[18] D. E. Smith, A source book in mathematics, McGraw-Hill, 1929 (Dover, 1959).
[19] D. J. Struik, A concise history of mathematics, Dover, third edition, 1967.

191 (XIII.30) Green's Functions

A. General Remarks

Green's functions are usually considered in connection with †boundary value problems for ordinary differential equations and also with †elliptic and †parabolic partial differential equations. For example, consider boundary value problems for the †Laplacian in 3-dimensional space: $L[u]=(\partial^2/\partial x_1^2+\partial^2/\partial x_2^2+\partial^2/\partial x_3^2)u$. Let D be a bounded domain with a smooth boundary S and the boundary condition B on S be either $u(x)=0$ $(x\in S)$ (the first kind) or $\partial u/\partial n+\beta u=0$ $(x\in S)$ (the third kind), where n is the outer normal of unit length, $\beta(x)\geqslant 0$, and $\beta(x)\not\equiv 0$. We say that the function $g(x_1,x_2,x_3;\xi_1,\xi_2,\xi_3)$ is the **Green's function** of L (or the partial differential equation $L[u]=0$) relative to the boundary condition B, when (i) $g(x,\xi)$ satisfies $L_x[g(x,\xi)]=0$ except for $x=\xi$; (ii) $g(x,\xi)=-1/4\pi r+\omega(x,\xi)$, where $r=(\sum_{i=1}^3(x_i-\xi_i)^2)^{1/2}$ and $\omega(x,\xi)$ is a regular function, i.e., of class C^ν for a suitable value ν; (iii) $g(x,\xi)$ satisfies the boundary condition B, i.e., $g(x,\xi)=0$, $x\in S$ (the first kind), or $(\partial/\partial n+\beta)g(x,\xi)=0$, $x\in S$ (the third kind). Conditions (i) and (ii) mean that $g(x,\xi)$ is a †fundamental solution of L, i.e., $L_x[g(x,\xi)]=\delta(x-\xi)$, where $\delta(x-\xi)$ is †Dirac's measure at the point $x=\xi$. Note that if $g(x,\xi)$ is a fundamental solution, then by adding any solution u

of the equation $L[u]=0$ to g, we obtain another fundamental solution $g+u$. Thus Green's function is the fundamental solution that satisfies the given boundary condition. To be more precise, in the boundary value problem, if the boundary condition is of the first kind, $g(x,\xi)$ can be obtained by adding to a fundamental solution $-1/4\pi r$ the solution $\omega(x,\xi)$ of the following Dirichlet problem: $\Delta_x\omega(x,\xi)=0$, $\omega(x,\xi)=1/4\pi r$ $(x\in S)$. We remark that there are slightly different definitions for Green's function. For example, there are cases where $g(x,\xi)$ is defined by $g(x,\xi)=1/4\pi r+\omega(x,\xi)$ or by $g(x,\xi)=1/r+\omega(x,\xi)$.

In the case in the previous paragraph, Green's functions satisfying the boundary conditions are uniquely determined. In general, if we are given a Green's function, then for any regular function $v(x)$ the function

$$u(x)=\int_D g(x,\xi)v(\xi)d\xi$$

represents the solution of $L[u]=v$ with the boundary condition B. More precisely, if $v(x)$ satisfies the †Hölder condition $|v(x)-v(x')|\leqslant L|x-x'|^\alpha$ $(0<\alpha\leqslant 1)$ $(L,\alpha$ positive constants), then $u(x)$ is of class C^2. Conversely, if $u(x)$ satisfies the equation $L[u]=v$ and the boundary condition B, it is represented by the formula for $u(x)$. This means that if we denote the operator that associates u to v by G, then G is the inverse operator of the Laplacian L with the boundary condition B, and Green's function is the †integral kernel of the operator G. Using this property, the boundary value problem relative to L can be reduced to a problem of †integral equations. For example, the differential equation with the boundary condition B containing the complex parameter λ, $L[u]+\lambda u=f$, is equivalent to the integral equation $u+\lambda G[u]=G[f]$, which is obtained by letting G act from the left on the above differential equation. In this way, the problem may be simplified.

In the case of general boundary value problems for higher-order elliptic operators, Green's functions are defined in the same way as before (→ 192 Green's Operator). The important case is when L and the boundary condition B define a †self-adjoint operator. In this case, Green's function is symmetric $(g(x,\xi)=g(\xi,x))$. To obtain Green's function is not easy in general. However, in some cases such functions can be obtained fairly easily.

B. Self-Adjoint Ordinary Differential Equations of the Second Order

Consider the operator $L[u]\equiv(p(x)u')'+q(x)u$ $(p(x)>0)$ defined in the interval $a\leqslant x\leqslant b$, with boundary conditions of the form $\alpha u'+\beta u=0$ at the two endpoints. Then Green's function $g(x,\xi)$ is defined in the following way: (i) For $x\neq\xi$, $L[g(x,\xi)]=0$; (ii) $[\partial g(x,\xi)/\partial x]_{x=\xi-0}^{x=\xi+0}=1/p(\xi)$; (iii) for ξ fixed, $g(x,\xi)$ satisfies the homogeneous boundary conditions at $x=a$ and $x=b$.

Conditions (i) and (ii) mean that $L[g(x,\xi)]=\delta(x-\xi)$. We can construct $g(x,\xi)$ in the following way: Let u_1 (u_2) be the solution of $L[u]=0$ satisfying the boundary condition at $x=a$ (at $x=b$). If u_1 and u_2 are linearly independent, we can satisfy $p(u_1'u_2-u_1u_2')=1$ by choosing the constants suitably. Then Green's function $g(x,\xi)$ is defined by $g(x,\xi)=u_1(x)u_2(\xi)$ for $x\leqslant\xi$, and $g(x,\xi)=u_1(\xi)u_2(x)$ for $\xi\leqslant x$. If u_1 and u_2 are linearly dependent, there exists no Green's function that satisfies conditions (i), (ii), and (iii). However, by modifying the definition, we can get a generalized Green's function playing a similar role [3]. This method can be applied to the case of ordinary differential equations of higher order.

C. The Laplace Operator

When the domain D is the n-dimensional sphere of radius a with center at the origin, Green's function of the Laplacian relative to the boundary condition $u=0$ is obtained in the following way. Let $E(r)$ be the following fundamental solution of the Laplacian: $E(r)=(2\pi)^{-1}\log r$ for $n=2$ and $E(r)=-((n-2)\cdot\omega_n r^{n-2})^{-1}$ for $n\geqslant 3$, where $\omega_n=2\pi^{n/2}\Gamma(n/2)$ is the $(n-1)$-dimensional surface area of the n-dimensional unit sphere. Then Green's function $g(x,\xi)$ is defined by

$$g(x,\xi)=E(r)-E(\rho r'/a),$$

where $\rho=(\sum_{i=1}^n\xi_i^2)^{1/2}$, $r'=(\sum_{i=1}^n(x_i-\xi_i')^2)^{1/2}$, $\xi_i'=(a/\rho)^2\xi_i$.

D. Helmholtz's Differential Equation

Let D be an exterior domain with a smooth boundary S in $\mathbf{R}^3$. In mathematical physics, the boundary value problem of finding a solution $u(x)$ of **Helmholtz's differential equation** $(\Delta+k^2)u(x)=f(x)$ $(k>0)$ satisfying $u(x)=0$ $(x\in S)$ is of particular interest. In this case, concerning the behavior of $u(x)$ at infinity, we usually assume the following **Sommerfeld's radiation condition**:

When $|x|\to+\infty$, $\quad u(x)=O(|x|^{-1})$,

$$\left(\frac{\partial}{\partial r}u-iku\right)\Big|_{|x|=r}=o(|x|^{-1}),$$

where $\partial/\partial r$ is the derivative along the radial direction. It is known that this condition ensures the uniqueness of the solution (**Rellich's uniqueness theorem**). We can construct Green's function $G(x,\xi)$ for any $k(>0)$ so that for smooth $f(x)$ with bounded †support,

$$u(x)=\int_D G(x,\xi)f(\xi)d\xi$$

represents the solution satisfying $u(x)=0$ $(x\in S)$ and Sommerfeld's radiation condition. Then, with

$$G(x,\xi)=-\frac{e^{ik|x-\xi|}}{4\pi|x-\xi|}+K_c(x,\xi),$$

where

$$|x-\xi|=\left(\sum_{i=1}^{3}(x_i-\xi_i)^2\right)^{1/2},$$

$K_c(x,\xi)$ can be obtained by solving an integral equation of †Fredholm type [4, 5]. In this case, there exists no Green's operator in L_2 space, and $G(x,\xi)$ may be considered a generalized Green's function [5].

E. Stokes's Differential Equation

Let D be a bounded domain in $\mathbf{R}^3$ with smooth boundary S, and consider **Stokes's differential equation** in D

$$\mu\Delta u_i=\frac{\partial p}{\partial x_i}-\rho X_i,\quad i=1,2,3,\qquad \sum_{j=1}^{3}\frac{\partial u_j}{\partial x_j}=0,$$

where μ, ρ are positive constants. In hydrodynamics, we consider the boundary value problem of finding solutions $(u_1(x),u_2(x),u_3(x),$ $p(x))$ of Stokes's equation satisfying the boundary condition $u_i(x)=0$, $x\in S$ $(i=1,2,3)$. In this case, **Green's tensors** $G_{ij}(x,\xi)$, $g_i(x,\xi)$ can be constructed, and for smooth functions $X_i(x)$ $(i=1,2,3)$ the unique solution of this boundary value problem is represented by

$$u_i(x)=\rho\sum_{j=1}^{3}\int_D G_{ij}(x,\xi)X_j(\xi)d\xi,$$

$$p(x)=\rho\sum_{j=1}^{3}\int_D g_j(x,\xi)X_j(\xi)d\xi$$

[6].

F. Parabolic Equations

Consider the following boundary value problem (the initial-boundary value problem):

$$L[u]\equiv\frac{\partial u}{\partial t}-c^2\frac{\partial^2 u}{\partial x^2}=f(x,t),$$

$$t>0,\quad a<x<b,$$

$$u(x,0)=\varphi(x),$$

where at $x=a$ and $x=b$, $u(x,t)$ satisfies some homogeneous boundary conditions. In this case, we can construct the function $g(x,t;\xi,\tau)$ $(t\geqslant\tau)$ satisfying the following conditions:
(i) $L[g]=0$ except for $x=\xi$, $t=\tau$;
(ii)

$$g(x,t;\xi,\tau)=\frac{\exp\left(-(x-\xi)^2/4c^2(t-\tau)\right)}{2c\sqrt{\pi(t-\tau)}}+(\text{regular function})$$

in a neighborhood of $x=\xi$, $t=\tau$; and (iii) $g(x,t;\xi,\tau)$ satisfies the given homogeneous boundary conditions at $x=a$ and $x=b$. Then

$$u(x,t)=\int_0^t\int_a^b g(x,t;\xi,\tau)f(\xi,\tau)d\xi\,d\tau+\int_a^b g(x,t;\xi,0)\varphi(\xi)d\xi$$

represents the solution of the problem stated in this section for regular functions f, g. The function $g(x,t;\xi,\tau)$ is called Green's function relative to the boundary value problem. Detailed consideration of such elementary cases is found in [7].

G. Kernel Functions

The kernel function is closely related to Green's function of Δ (the Laplacian) relative to the first boundary value problem in a domain in $\mathbf{R}^2$.

First we explain the general definitions of the kernel function. Let E be a general set, and let $\mathfrak{F}$ be a †linear space of complex-valued functions defined on E, endowed with the structure of a †Hilbert space by a suitable inner product (f,g). Suppose that we are given a function $K(x,y)$ defined on $E\times E$ satisfying the following conditions: (i) For any fixed y, $K(x,y)$ regarded as a function of x belongs to $\mathfrak{F}$; and (ii) for any $f(x)\in\mathfrak{F}$, $(f(x),K(x,y))_x=f(y)$. Then $K(x,y)$ is called a **kernel function** or **reproducing kernel**. The kernel function, if it exists, is unique and is †positive definite Hermitian; that is,

$$\sum_{j,k=1}^{n}K(y_j,y_k)\xi_j\bar{\xi}_k\geqslant 0. \tag{1}$$

Conversely, any positive definite function is a reproducing kernel of some Hilbert space. A necessary and sufficient condition for the existence of the kernel function is that for any $y\in E$, the linear functional: $f\to f(y)$ is bounded. In this case, the minimum of $\|f\|$ under the condition $f(y)=1$ $(f\in\mathfrak{F})$ is attained by the element $K(x,y)/K(y,y)$, and its value is $K(y,y)^{-1/2}$. When $\mathfrak{F}$ is a †separable space, then by an †orthonormal system $\{\varphi_\nu(x)\}$, we can represent $K(x,y)$ as

$$K(x,y)=\sum_{\nu=1}^{\infty}\varphi_\nu(x)\overline{\varphi_\nu(y)}. \tag{2}$$

As an example of kernel functions, the following case is of particular importance. Let E be an n-dimensional †complex manifold, φ and ψ be holomorphic †differential forms of degree n on E, and let the following inner product be given:

$$(\varphi,\psi)=\int_E \varphi\wedge\bar{\psi}.$$

Now $\mathfrak{F}$ is given by $\mathfrak{F}=\{\varphi \mid (\varphi,\varphi)<+\infty\}$, and the kernel function is called the **kernel differential**. When E is a domain in $\mathbf{C}^n$, regarding the coefficients of the differential form as functions, we call the kernel function **Bergman's kernel function**. Moreover, if E can be mapped onto a bounded domain by a one-to-one holomorphic mapping, then

$$ds^2=\sum\left(\partial^2 \log K(z,\bar{z})/\partial z_j \partial \bar{z}_k\right) dz_j d\bar{z}_k$$

is positive definite and gives a Kähler metric which is called the **Bergman metric**.

H. Kernel Functions for Domains in the Complex Plane

Let E be a domain D in the complex plane ($z=x+iy$). Let $K(z,\zeta)$ be Bergman's kernel function of D, and let $G(z,\zeta)$ be Green's function of Δ relative to the first boundary condition with pole at ζ. Then we have

$$K(z,\zeta)=-(2/\pi)\partial^2 G(z,\zeta)/\partial z \partial \bar{\zeta}. \tag{3}$$

Next, let $U(z,\zeta)$ be the kernel function of the Hilbert space consisting of the holomorphic differential forms whose integrals are single-valued, and let $N(z,\zeta)$ be **Neumann's function** of Δ, i.e., the function that is harmonic in $D-\{\zeta\}$, has the same singularity as G at ζ, and whose derivative in the normal direction $\partial N/\partial n$ is constant along the boundary. Then we have

$$U(z,\zeta)=(2/\pi)\partial^2 N(z,\zeta)/\partial z \partial \bar{\zeta}. \tag{4}$$

Now the kernel $H(z,\zeta)=(N(z,\zeta)-G(z,\zeta))/2\pi$ is the kernel function relative to the Hilbert space consisting of all real †harmonic functions whose integral mean value along the boundary Γ is 0 and having the inner product

$$(\varphi,\psi)=\iint_D\left(\frac{\partial\varphi}{\partial x}\frac{\partial\psi}{\partial x}+\frac{\partial\varphi}{\partial y}\frac{\partial\psi}{\partial y}\right)dx\,dy.$$

The kernel $H(z,\zeta)$ is called a **harmonic kernel function**.

Suppose that the boundary Γ is †piecewise smooth, and consider the space of all holomorphic functions in D that are continuous on the boundary of D. The inner product of such functions φ and ψ is given by $(\varphi,\psi)=\int_\Gamma \varphi\bar{\psi}\,ds$ (ds is the element of the arc length of Γ). Hence we have a Hilbert space. Then the kernel function relative to this Hilbert space is called **Szegö's kernel function**, which has a close relation with †bounded functions.

The kernel functions enable us to represent holomorphic mappings that map the domain D onto various canonical domains (→ 79 Conformal Mapping).

References

[1] R. Courant and D. Hilbert, Methods of mathematical physics, Interscience, I, 1953; II, 1962.
[2] T. Inui, Oyô henbibun hôteisiki ron (Japanese; Theory and applications of partial differential equations), Iwanami, 1951.
[3] K. Yosida, Sekibun hôteisiki ron (Japanese; Theory of integral equations), Iwanami, 1950; English translation, Lectures on differential and integral equations, Interscience, 1960.
[4] В. Д. Купрадзе (V. D. Kupradze), Граничные задачи теории колебаний и интегральные уравнения, Гостехиздат, 1950; German translation, Randwertaufgaben der Schwingungstheorie und Integralgleichungen, Deutscher Verlag der Wiss., 1956.
[5] S. Mizohata, Henbibun hôteisiki ron (Japanese; Theory of partial differential equations), Iwanami, 1965.
[6] F. K. G. Odqvist, Über die Randwertaufgaben der Hydrodynamik zäher Flüssigkeiten, Math. Z., 32 (1930), 329–375.
[7] E. E. Levi, Sull' equazione del calore, Ann. Mat. Pura Appl., (3) 14 (1908), 187–264.
For kernel functions,
[8] S. Bergman, The kernel function and conformal mapping, Amer. Math. Soc., Math. Surveys, 1950.
[9] S. Bergman and M. Schiffer, Kernel functions and elliptic differential equations in mathematical physics, Academic Press, 1953.
[10] N. Aronszajn, Theory of reproducing kernels, Trans. Amer. Math. Soc., 68 (1950), 337–404.
[11] H. Meschkowski, Hilbertsche Räume mit Kernfunktion, Springer, 1962.

192 (XIII.31) Green's Operator

A. General Remarks

Consider the first and the third boundary value problems for the elliptic equation

$$A[u]\equiv-\Delta u+\sum_{i=1}^{n}a_i(x)\frac{\partial u}{\partial x_i}+c(x)u=f(x)$$

(→ 318 Partial Differential Equations of

Elliptic Type). Let D be a bounded domain of $\mathbf{R}^n$ whose boundary S consists of a finite number of smooth hypersurfaces. By the †method of orthogonal projection, we take the domain $\mathfrak{D}(A)$ as follows: (i) $\{u(x)|u(x)\in H^2(D)$ and $u(x)=0$ for $x\in S\}$ or (ii) $\{u(x)|u(x)\in H^2(D)$ and $\partial u/\partial n+\beta(x)u=0$ for $x\in S\}$ according as we are considering the first or third boundary value problem, where $H^2(D)$ is the †Sobolev space (→ 173 Function Spaces). If the operator A is a one-to-one mapping from $\mathfrak{D}(A)$ onto the function space $L_2(D)$, we call the inverse operator A^{-1} **Green's operator** relative to the boundary condition, and we denote it by G. In general, the existence of A^{-1} is not guaranteed. However, if we take real t large enough, $G_t=(A+tI)^{-1}$ exists.

Consider the general case where λ is a complex parameter: $(\lambda I-A)[u]=f(x), f(x)\in L_2(D)$. Letting G_t act from the left, we have $(I-(\lambda+t)G_t)[u]=-G_tf$. Conversely, if $u\in L_2(D)$ is a solution, clearly $u(x)\in\mathfrak{D}(A)$, and $u(x)$ satisfies the first partial differential equation and the boundary condition. Since G_t is a †compact operator in $L_2(D)$, the †Riesz-Schauder theorem can be applied (→ 72 Compact Operators). In particular, if $\lambda+t$ is not an †eigenvalue of G_t, $u(x)=(\lambda I-A)^{-1}f=-(I-(\lambda+t)G_t)^{-1}G_tf$ represents a unique solution.

In the equations in the first paragraph of this section, if $a_i(x)=0$ and $c(x)$ and $\beta(x)$ are real, then G_t is a †self-adjoint operator in $L_2(D)$, and therefore the †Hilbert-Schmidt theorem can be applied. Namely, let $\{\lambda_i\}$ be the eigenvalues of A such that $A\omega_i(x)=\lambda_i\omega_i(x)$, where $\{\omega_i(x)\}$ is an †orthonormal system in $L_2(D)$. Then for any $f(x)\in L_2(D)$, $f(x)=\sum_{i=1}^{\infty}(f,\omega_i)\omega_i(x)$, where the right-hand side is taken in the sense of †mean convergence. Furthermore, for $f(x)\in\mathfrak{D}(A)$, we have the expansion $(Af)(x)=\sum_{i=1}^{\infty}\lambda_i(f,\omega_i)\omega_i(x)$ in the same sense.

When G_t is not self-adjoint, let G_t^* be the †adjoint operator of G_t in $L_2(D)$. Then G_t^* represents Green's operator relative to the equation

$$(A^*+tI)[v]=-\Delta v-\sum_{i=1}^{n}\frac{\partial}{\partial x_i}\left(\overline{a_i(x)}\,v\right)+\left(\overline{c(x)}+t\right)v=g(x),$$

corresponding to the boundary conditions (i) $v(x)=0$, $x\in S$ (first boundary condition) and (ii) $(\partial/\partial n)v+\beta'(x)v=0$, $x\in S$, where

$$\beta'(x)=\overline{\beta(x)}+\sum_{i=1}^{n}\overline{a_i(x)}\cos nx_i,$$

with n the outer normal (third boundary condition) [2].

B. Elliptic Equations of Higher Order

Green's operator can be defined for elliptic equations of higher order. Consider the equation

$$A(x,\partial/\partial x)u(x)=f(x),\quad x\in D;$$
$$B_j(x,\partial/\partial x)u(x)=0,\quad x\in S,\qquad j=1,2,\ldots,b(=m/2),\quad (1)$$

where A is an †elliptic operator of order m and the boundary operators $\{B_j\}$ satisfy: (i) At every point x of S, the normal direction is not †characteristic for any B_j; and (ii) the order m_j of B_j is less than m, and $m_j\neq m_k$ $(j\neq k)$. The domain $\mathfrak{D}(A)$ of A is defined by

$$\mathfrak{D}(A)=\{u(x)|u\in H^m(D)\text{ and }B_j(x,\partial/\partial x)u(x)=0\text{ for }x\in S,\ j=1,2,\ldots,b\}.$$

When A is a one-to-one mapping from $\mathfrak{D}(A)$ onto $L_2(D)$, the inverse $G=A^{-1}$ is called **Green's operator**.

This general boundary value problem, especially the existence theorem, was treated under some algebraic conditions on A and $\{B_j\}$ by M. Schechter [5] who showed that $G[n]\in H^m(D)$ if $u(x)\in L_2(D)$ and that $G[n]$ depends continuously on n. In particular, if $m>n/2$, then by †Sobolev's theorem, G is a continuous mapping from $L_2(D)$ into $C^0(\bar{D})$, and G is represented by an †integral operator of Hilbert-Schmidt type (L. Gårding [6]). Namely, for any $f(x)\in L_2(D)$,

$$(Gf)(x)=\int_D G(x,\xi)f(\xi)\,d\xi,$$

$$\int\int|G(x,\xi)|^2dx\,d\zeta<+\infty.$$

In general, the function $G(x,\xi)$, obtained by the **kernel representation** of Green's operator G, is called **Green's function**.

On the other hand, consider, for example, the †Dirichlet problem of Δ in $\mathbf{R}^3$. Then Green's function is defined in the following way (→ 191 Green's Functions): $G(x,\xi)=-(4\pi|x-\xi|)^{-1}+u(x,\xi)$, where $u(x,\xi)$ satisfies (i) $\Delta_x u(x,\xi)=0$, and (ii) $G(x,\xi)|_{x\in S}=0$. The function defined in this manner coincides with Green's function defined as a kernel representation of Green's operator [1, 2].

Suppose that in problem (1) the †elliptic operator $A(x,\partial/\partial x)$ is independent of x, and let $E(x)$ be a †fundamental solution, i.e., $E(x)$ is a distribution solution of $A(\partial/\partial x)E(x)=\delta(x)$ ($\delta(x)$ is †Dirac's δ-function; → 115 Differential Operators). Then $E(x)$ is a C^∞-

function away from the origin. Moreover, in a neighborhood of the origin the following estimates hold: For $|\alpha|<m$,

$$\left|\left(\frac{\partial}{\partial x}\right)^{\alpha}E(x)\right|\leqslant\begin{cases}c|x|^{m-n-|\alpha|}, & m-n-|\alpha|<0,\\ c\log|x|^{-1}, & m-n-|\alpha|=0,\\ c, & m-n-|\alpha|>0,\end{cases}$$

where c is some positive constant. When problem (1) has Green's operator G, we can say the following, using the fundamental solution: Green's function $G(x,\xi)$ exists and can be written as $G(x,\xi)=E(x-\xi)+u(x,\xi)$, where for any fixed $\xi\in D$, $u(x,\xi)$ satisfies (i) $A(\partial/\partial x)u(x,\xi)=0$ and (ii) $B_j(x,\partial/\partial x)G(x,\xi)=0$, $x\in S$, $j=1,2,\ldots,b$.

C. Hypoelliptic Operators

Let

$$A(x,\partial/\partial x)=\sum_{|\alpha|\leqslant m}a_\alpha(x)\left(\frac{\partial}{\partial x}\right)^{\alpha},$$

$$\left(\frac{\partial}{\partial x}\right)^{\alpha}=\frac{\partial^{|\alpha|}}{\partial x_1^{\alpha_1}\ldots\partial x_n^{\alpha_n}}$$

be a general partial differential operator with C^∞-coefficients. If the kernel $E(x,\xi)$ satisfies $A(x,\partial/\partial x)E(x,\xi)=\delta(x-\xi)$, that is, if we have the relation $\langle E(x,\xi),{}^tA(x,\partial/\partial x)\varphi(x)\rangle_x=\varphi(\xi)$ for any $\varphi(x)\in\mathfrak{D}$, then $E(x,\xi)$ is called a **fundamental solution** of A, where ${}^tA$ is the **transposed operator** of A:

$${}^tA(x,\partial/\partial x)v(x)$$
$$=\sum_{|\alpha|\leqslant m}(-1)^{|\alpha|}\left(\frac{\partial}{\partial x}\right)^{\alpha}(a_\alpha(x)v(x)).$$

Now, if there exists a fundamental solution $E'(x,\xi)$ of ${}^tA(x,\partial/\partial x)$ such that (i) $E'(x,\xi)$ defines a kernel that gives rise to two continuous mappings, one of which maps the space $\mathfrak{D}_\xi$ into $\mathfrak{E}_x$, and the other of which maps the space $\mathfrak{D}_x$ into $\mathfrak{E}_\xi$ (→ 173 Function Spaces), and (ii) for $x\neq\xi$, $E'(x,\xi)$ is a C^∞-function of (x,ξ); then any distribution $u(x)$ satisfying $A(x,\partial/\partial x)u(x)=g(x)$ is a C^∞-function, where $g(x)$ is of class C^∞. In general, an operator A with the property that any solution $u(x)$ of $A(x,\partial/\partial x)u(x)=g(x)$ is of class C^∞, where $g(x)$ is of class C^∞, is called **hypoelliptic**. Elliptic and parabolic operators are cboth hypoelliptic. L. Hörmander characterized the hypoelliptic differential operators with constant coefficients [8] (→ 115 Differential Operators).

A kernel $E(x,\xi)$ such that

$$A(x,\partial/\partial x)E(x,\xi)=\delta(x-\xi)+\varpi(x,\xi),$$

with $\varpi(x,\xi)$ a C^∞-function, is called a **parametrix** of A. To prove the hypoellipticity of A, it suffices to show the existence of a parametrix $E'(x,\xi)$ of the operator ${}^tA$ having the properties (i) and (ii) mentioned in the previous paragraph.

To explain the notion of the fundamental solution for the †evolution equations, suppose that we are given an evolution equation

$$L[u]=\frac{\partial^m}{\partial t^m}u(x,t)$$
$$+\sum_{j<m}a_{\alpha,j}(x,t)\left(\frac{\partial}{\partial x}\right)^{\alpha}\frac{\partial^j}{\partial t^j}u(x,t)=0,$$

$x\in\mathbf{R}^n$, $t_0\leqslant t\leqslant T$. A kernel $E(x,t;\xi,t_0)$ ($t_0\leqslant t\leqslant T$) is called a **fundamental solution** to the evolution equation if

$$L_{x,t}(E(x,t;\xi,t_0))=0,\quad t>t_0,$$

and

$$\lim_{t\to t_0+0}\frac{\partial^j}{\partial t^j}E(x,t;\xi,t_0)=\begin{cases}0, & 0\leqslant j\leqslant m-2,\\ \delta(x-\xi), & j=m-1.\end{cases}$$

References

[1] H. G. Garnir, Les problèmes aux limites de la physique mathématique, Birkhäuser, 1958.

[2] S. Mizohata, Henbibun hôteisiki ron (Japanese; Theory of partial differential equations), Iwanami, 1965.

[3] R. Courant and D. Hilbert, Methods of mathematical physics II, Interscience, 1962.

[4] L. Gårding, Dirichlet's problem for linear elliptic differential equations, Math. Scand., 1 (1953), 55–72.

[5] M. Schechter, General boundary value problems for elliptic partial differential equations, Comm. Pure Appl. Math., 12 (1959), 457–486.

[6] L. Gårding, Applications of the theory of direct integrals of Hilbert spaces to some integral and differential operators, Univ. of Maryland lecture series no. 11, 1954.

[7] L. Schwartz, Théorie des distributions, Hermann, second edition, 1966.

[8] L. Hörmander, On the theory of general partial differential operators, Acta Math., 94 (1955), 161–248.

193 (IV.1) Groups

A. Definition

Let G be a nonempty set. Suppose that for any elements a, b of G there exists a uniquely determined element c of G, which is called the **product** of a and b, written $c=ab$. We call

G a **group** or **multiplicative group** if (i) the **associative law** $a(bc)=(ab)c$ holds, and (ii) for any elements a, $b \in G$ there exist uniquely determined elements $x, y \in G$ satisfying $ax = b$, $ya = b$. Then the mapping $(a,b) \rightarrow ab$ is called **multiplication** in G. Condition (ii) is equivalent to the following two conditions: (iii) There exists an element a (called the **identity element** or **unit element** of G) such that $ae = ea = a$ for any element a of G; and (iv) for any element a of G there exists an element x such that $ax = xa = e$.

The element x in condition (iv) is called the **inverse** (or **inverse element**) of a, denoted by a^{-1}. The uniqueness of the identity element e and the inverse a^{-1} follows readily from the axioms. The identity element of a multiplicative group is sometimes denoted by 1. If $ab = ba$, then we say that a and b **commute**. The **commutative law**, $ab = ba$ for any elements a, $b \in G$, is not assumed in general. A group satisfying the commutative law is called an **Abelian group** (or **commutative group**) in honor of N. H. Abel, who made use of commutative groups in his study of the theory of equations. The product in a commutative group is often written in the form $a+b$, and in this case the mapping $(a,b) \rightarrow a+b$ is called **addition**. The element $a+b$ is called the **sum** of a and b, and G is called an **additive group**. In an additive group the identity element is usually denoted by 0 and called the **zero element**, and the inverse of a is denoted by $-a$ (→ 2 Abelian Groups; 275 Modules). To describe the †law of composition, we sometimes use notation different from multiplication or addition (→ 396 Structures).

B. Examples

A †linear space over a †field K is an additive group with respect to the usual addition of vectors (→ 256 Linear Spaces). A field is an additive group with respect to the addition, and the set of nonzero elements of a field forms a group with respect to the multiplication, which is called the **multiplicative group of the field** (→ 157 Fields).

All †invertible $n \times n$ matrices over a ring R form a group with respect to the usual multiplication of matrices. This group is called the †general linear group of degree n over R (→ 63 Classical Groups).

All one-to-one mappings from a set M onto itself (i.e., all **permutations** on M) form a group with respect to the composition defined by $(f \circ g)(x) = f(g(x))$ $(x \in M)$. (Sometimes the product $f \circ g$ is denoted by gf and $f(x)$ by xf. Then $x(gf) = (xg)f$.) The group of all permutations on M is called the **symmetric group** on M. A group G is called a **permutation group** (on M) if every element of G is a permutation on M. For instance, the general linear group of degree n over a field K may be regarded as a permutation group on the set of n-dimensional vectors, and it is also regarded as a permutation group on a †tensor space.

All †motions in a Euclidean space form a group with respect to the usual composition of motions. All invertible $n \times n$ matrices over K leaving a given †quadratic form invariant form a group with respect to the usual multiplication of matrices. This group is called the †orthogonal group belonging to the given quadratic form. If K is the †complex number field or the real number field, then these groups are Lie groups (→ 15 Algebraic Groups; 160 Finite Groups; 169 Free Groups; 247 Lie Groups; 406 Topological Groups).

C. Fundamental Concepts

If a group G consists of a finite number of elements, then G is called a **finite group**; otherwise, G is called an **infinite group**. The number of elements of G is called the **order** of G. A nonempty subset H of G is called a **subgroup** of G if H is a group with respect to the multiplication of the group G. Hence a nonempty subset H is a subgroup of G if and only if $a^{-1}b \in H$ for any a, $b \in H$. For a family $\{H_\lambda\}$ of subgroups of G, the intersection $\bigcap_\lambda H_\lambda$ is also a subgroup.

The associative law of multiplication says that elements a_1, a_2, a_3 of G determine the product $a_1a_2a_3$, which is the common value of $(a_1a_2)a_3$ and $a_1(a_2a_3)$. This law can be generalized to say that any ordered set of n elements $a_1, a_2, \ldots, a_n$ $(n > 2)$ of G determines their product $a_1a_2 \ldots a_n$ (**general associative law**). When $a_1 = a_2 = \ldots = a_n = a$, we denote the product $aa \ldots a$ by a^n. If we define a^{-n} for $n \geqslant 0$ by $a^0 = e$ and $a^{-n} = (a^n)^{-1}$, we then have $a^na^m = a^{n+m}$, $(a^n)^m = a^{nm}$ for any n, $m \in \mathbf{Z}$. If there exists a positive integer n such that $a^n = e$, then the smallest positive integer d with $a^d = e$ is called the **order** of the element a. If there is no such n, then a is called an element of **infinite order**. If a is of infinite order, then its powers a^0 $(=e)$, $a^{\pm 1}$, $a^{\pm 2}, \ldots$, are all unequal. If a is of order d, then the different powers of a are a^0 $(=e)$, $a, a^2, \ldots, a^{d-1}$. All the powers of a form a subgroup $\langle a \rangle$ of G, called a **cyclic subgroup**. The order of an element a is the same as the order of the subgroup $\langle a \rangle$. The group $\langle a \rangle$ itself is called a **cyclic group** and is an example of an Abelian group (→ 2 Abelian Groups).

Let S be a subset of a group G. Then the intersection of all subgroups of G containing S is called the subgroup **generated** by S and is denoted by $\langle S\rangle$. It is the smallest subgroup containing S, and if S is nonempty, $\langle S\rangle$ consists of all the elements of the form $a_1^{m_1}a_2^{m_2}\ldots a_r^{m_r}$ $(a_i\in S, m_i\in\mathbf{Z})$. If $\langle S\rangle=G$, the elements of S are called **generators** of G. When G has a finite set of generators, G is said to be **finitely generated**. When $S=\{a\}$, then $\langle S\rangle$ coincides with $\langle a\rangle$, and the element a is the generator of the cyclic group $\langle a\rangle$. Suppose that elements $a_1,\ldots,a_n$ of G satisfy an equation of the form $a_1^{m_1}a_2^{m_2}\ldots a_n^{m_n}=1$. This equation is then called a **relation** among the elements $a_1,\ldots,a_n$. If we have a system of generators and all relations among the generators, then they define a group (→ 169 Free Groups). It is, however, still an open problem to find a general procedure to decide whether the group determined by a given system of generators and the relations among them contains elements other than the identity (→ 169 Free Groups B).

For a subset S and an element x of a group G, the set of all elements $x^{-1}sx$ $(s\in S)$ is denoted by $x^{-1}Sx$ or S^x, and S and S^x are called **conjugate**. We have $(ab)^x=a^xb^x$, $(a^{-1})^x=(a^x)^{-1}$. If H is a subgroup, then H^x is also a subgroup. For a subset S, the set of all elements x satisfying $S^x=S$ forms a subgroup $N(S)$, called the **normalizer** of S. The set of all elements that commute with every element of S forms a subgroup $Z(S)$, called the **centralizer** of S. The centralizer Z of G is called the **center** of G. The set of all elements conjugate to a given element a of G is called a **conjugacy class**. A group G is the disjoint union of its conjugate classes.

Let H be a subgroup of a group G and x an element of G. The set of elements of the form hx $(h\in H)$ is denoted by Hx and is called a **right coset** of H. A **left coset** xH is defined similarly. G is the disjoint union of left (right) cosets of H. The cardinality of the set of left cosets of H equals that of the set of right cosets of H; it is called the **index** of the subgroup H and is denoted by $(G:H)$. Given two subgroups H and K of G, the set $HxK=\{hxk\mid h\in H, k\in K\}$ is called the **double coset** of H and K, and G is the disjoint union of different double cosets of H and K. If the left cosets of a subgroup H are also the right cosets, i.e., if $Hx=xH$ for every $x\in G$, then H is called a **normal subgroup** (or **invariant subgroup**) of G. An equivalent condition is that $H=H^x$ for all $x\in G$. The center of G is always a normal subgroup of G. If H is a normal subgroup of G, the set of all products of an element of Ha and an element of Hb coincides with Hab. Thus if we define the product of two cosets Ha and Hb to be Hab, then the set of cosets of H forms a group. This group is denoted by G/H and is called the **factor group** (or **quotient group**) of G modulo H. (When G is an additive group, G/H is also denoted by $G-H$ and is called the **difference group**.) The group G itself and $\{e\}$ are normal subgroups of G. If G has no normal subgroup other than these two, then G is called a **simple group**. A subgroup of finite index contains a normal subgroup of finite index. If H is a subgroup of finite index, then we can find a common complete system of representatives of the left cosets and the right cosets of H. If G is finitely generated, then so is any subgroup of G of finite index.

Let R be an †equivalence relation defined in a group G. If xRx' and yRy' always imply $(xy)R(x'y')$, then we say that R is **compatible** with the multiplication. The †quotient set G/R is a group with respect to the induced multiplication. This group is called the **quotient group** of G with respect to R. The equivalence class H containing e is a normal subgroup, and xRx' if and only if $x^{-1}x'\in H$, i.e., x and x' are contained in the same coset of H. Thus G/R coincides with G/H.

If G is a finite group of order n, then the order and the index of any subgroup of G, the order of any element of G, the cardinal number of any conjugacy class of G, and the number of different conjugate subgroups of any subgroup of G are all divisors of n.

D. Isomorphisms and Homomorphisms

If there is a one-to-one mapping $a\leftrightarrow a'$ of the elements of a group G onto those of a group G' and if $a\leftrightarrow a'$ and $b\leftrightarrow b'$ imply $ab\leftrightarrow ab'$, then we say that G and G' are **isomorphic** and write $G\cong G'$. If we put $a'=f(a)$, then $f: G\to G'$ is a †bijection satisfying $f(ab)=f(a)f(b)$ $(a,b\in G)$. More generally, if a mapping $f: G\to G'$ satisfies $f(ab)=f(a)f(b)$ for all $a,b\in G$, then f is called a **homomorphism** of G to G'. An †injective (†surjective) homomorphism is also called a †monomorphism (†epimorphism). If there is a surjective homomorphism $G\to G'$, then we say that G' is **homomorphic** to G. The composite of two homomorphisms is also a homomorphism. If a homomorphism $f: G\to G'$ is a bijection, then f is called an **isomorphism**. In this case f^{-1} is also an isomorphism, and we have $G\cong G'$.

For a subgroup H of a group G, the injective homomorphism $f: H\to G$ defined by $f(a)=a$ $(a\in H)$ is called the **canonical injection** (or **natural injection**). For a factor group G/R of G, the surjective homomorphism $f: G\to G/R$ such that $a\in f(a)$ $(a\in G)$ is called the **canonical surjection** (or **natural surjection**).

Let $f: G\to G'$ be a homomorphism. Then the **image** $f(G)$ of f is a subgroup of G, and the **kernel** $H=\{a\in G|f(a)=e'$ (the identity of $G')\}$ of f is a normal subgroup of G. The equivalence classes of the equivalence relation given by $f(x)=f(y)$ are just the cosets of H, and f induces an isomorphism $\bar{f}: G/H\rightarrow f(G)$. The latter proposition is called the **homomorphism theorem** of groups. This theorem is extended in the following way: For simplicity let $f: G\rightarrow G'$ be a surjective homomorphism. (i) If H' is a normal subgroup of G', then the inverse image $H=f^{-1}(H')$ is a normal subgroup of G, and f induces the isomorphism $\bar{f}: G/H\rightarrow G'/H'$. (ii) If H is a subgroup and N is a normal subgroup of G, then $HN=\{hn|h\in H, n\in N\}$ is a subgroup of G, and the canonical injection $H\rightarrow HN$ induces an isomorphism $H/H\cap N\rightarrow HN/N$. (iii) If H and N are two normal subgroups of G such that $H\supset N$, then the canonical surjection $G\rightarrow G/N$ induces an isomorphism $G/H\rightarrow (G/N)/(H/N)$. Propositions (i), (ii), and (iii) are called the **isomorphism theorems** of groups.

A homomorphism of G to itself is called an **endomorphism** of G, and an isomorphism of G to itself is called an **automorphism** of G. The set of automorphisms of G forms a group with respect to the composition of mappings, called the **group of automorphisms** of G. Given an element a of G, the mapping $x\rightarrow a^{-1}xa$ $(x\in G)$ yields an automorphism of G which is called an **inner automorphism** of G. The set of inner automorphisms of G forms a normal subgroup of the group of automorphisms of G, called the **group of inner automorphisms** of G, which is isomorphic to the factor group of G modulo its center. The factor group of the group of automorphisms of G modulo the group of inner automorphisms of G is called the **group of outer automorphisms** of G.

If a mapping $f: G\rightarrow G'$ from a group G to another group G' satisfies $f(ab)=f(b)f(a)$ $(a,b\in G)$, then f is called an **antihomomorphism**. A bijective antihomomorphism is called an **anti-isomorphism**. When $G=G'$, f is called an **anti-endomorphism** or **anti-automorphism** (e.g., $f: G\rightarrow G$ defined by $f(a)=a^{-1}$ is an anti-isomorphism).

E. Groups with Operator Domain

Let Ω be a set and G a group. Suppose that for each $\theta\in\Omega$ and $x\in G$, the product $\theta x\in G$ is defined and satisfies $\theta(xy)=(\theta x)(\theta y)$. Then Ω is called an **operator domain** of G, and G is called a group with operator domain Ω, or simply an **Ω-group**. (We sometimes write x^{θ} instead of θx.) The mapping $(\theta,x)\rightarrow\theta x$ from $\Omega\times G$ to G is called the †operation of Ω on G. If G is an Ω-group, then any element θ of Ω induces an endomorphism $\theta_G: x\rightarrow\theta x$ of G. Conversely, if we are given a mapping $\theta\rightarrow\theta_G$ of Ω to the set of endomorphisms of G, then we may regard G as an Ω-group. Any group may be regarded as an Ω-group with Ω equal to the empty set or to the set consisting of the identity automorphism of G. Thus the general theory of groups can be extended to the theory of groups with operator domain, and in some cases the latter can also be applied to the former (→ 2 Abelian Groups; 275 Modules).

A subgroup H of an Ω-group G is called an **Ω-subgroup** (or **admissible subgroup**) if $\theta x\in H$ for any $\theta\in\Omega$ and $x\in H$. In this case, H is also an Ω-group. If an equivalence relation R defined in G is compatible with the multiplication and also compatible with the operators, namely, if xRx' implies $(\theta x)R(\theta x')$ for any $\theta\in\Omega$, then the quotient group G/R is also an Ω-group. The equivalence class containing e is an **admissible normal subgroup**. Conversely, if H is an admissible normal subgroup, then the equivalence relation defined by H is compatible with the operators, and the factor group G/H is an Ω-group. A homomorphism $f: G\rightarrow G'$ of an Ω-group G to an Ω-group G' is called an **Ω-homomorphism (admissible homomorphism** or **operator homomorphism)** if $f(\theta x)=\theta f(x)$ for any $\theta\in\Omega$ and $x\in G$. If f is an isomorphism, then f is called an **Ω-isomorphism (admissible isomorphism** or **operator isomorphism)**. We have the homomorphism theorem and the isomorphism theorems of Ω-groups if we consider only admissible subgroups and admissible homomorphisms.

F. Sequences of Subgroups

Let $H_1, H_2, \ldots$ be an infinite sequence of (normal) subgroups of a group G. If $H_i \subsetneqq H_{i+1}$ $(i=1,2,\ldots)$, then the sequence is called an **ascending chain** of (normal) subgroups. If $H_i \supsetneqq H_{i+1}$ $(i=1,2,\ldots)$, then it is called a **descending chain** of (normal) subgroups. If there is no ascending (or descending) chain of (normal) subgroups of G, we say that G satisfies the **ascending** (or **descending**) **chain condition** for (normal) subgroups. These conditions are the same as the ascending (or descending) chain condition in the ordered set of all (normal) subgroups of G (→ 305 Ordering C). A group G satisfies the ascending chain condition for subgroups if and only if every subgroup of G is finitely generated. Also, for groups with operator domain we have similar results. The structure of Abelian groups satisfying the ascending (descending) chain condition is completely determined (→

2 Abelian Groups). It is not known whether there is an infinite group satisfying both the ascending and descending chain conditions for subgroups. A group satisfying the descending chain condition for subgroups has no element of infinite order, but the converse is not true. There is an infinite group which is finitely generated and has no element of infinite order (→ 169 Free Groups C).

G. Normal Chains

A finite sequence $G=G_0\supset G_1\supset G_2\supset\ldots\supset G_r=\{e\}$ of subgroups of a group G is called a **normal chain** if G_i is a normal subgroup of G_{i-1} for $i=1,2,\ldots,r$. We call r the **length** of the chain. The sequence G_0/G_1, G_1/G_2, $\ldots, G_{r-1}/G_r$ is called the **sequence of factor groups** of the normal chain. A normal chain $G=H_0\supset H_1\supset H_2\supset\ldots\supset H_s=\{e\}$ is called a **refinement** of the chain $G=G_0\supset G_1\supset\ldots\supset G_r=\{e\}$ if every G_i appears in this chain. Two normal chains with the same length are called isomorphic if there is a one-to-one correspondence between their sequences of factor groups such that corresponding factor groups are isomorphic. Any two normal chains have refinements which are isomorphic to each other (Schreier's refinement theorem). A normal chain is called a **composition series** (or **Jordan-Hölder sequence**) if it consists of different subgroups of G and in any proper refinement there appear two successive subgroups which are the same. The sequence of factor groups of a composition series is called a **composition factor series**, and the factor groups appearing in this series are called **composition factors**. Any composition factor is a simple group. As a direct consequence of the refinement theorem we see that if a group G has a composition series, then the composition factor series is unique up to isomorphism and the ordering of the factors. (This theorem is due to O. Hölder. C. Jordan proved that if G is a finite group, then the set of orders of composition factors is independent of the choice of composition series. Hence we call the theorem the **Jordan-Hölder theorem**.)

For an Ω-group G, if we consider only Ω-subgroups, we have definitions and theorems similar to those in this section. When we take the group of inner automorphisms of G as Ω, then a composition series of the Ω-group G is called a **principal series**. If we take the group of automorphisms of G as Ω, then a composition series is called a **characteristic series**. An infinite group G does not always have a composition series. Even if G has a composition series, G may have an infinite normal chain $G_1\subset G_2\subset\ldots\subset G$ such that each G_i is a normal subgroup of G_{i+1} and $\bigcup G_i=G$. In fact, there is a simple group which has such an infinite normal chain (P. Hall). Two groups which have isomorphic composition series are not necessarily isomorphic. A subgroup of a group G is called a **subnormal subgroup** of G if it may appear in some normal chain. The intersection of two subnormal subgroups is also subnormal, but their **join** (i.e., the subgroup generated by both of them) is not necessarily subnormal in an infinite group. The set of subgroups and the set of normal subgroups of a group form †lattices with respect to the inclusion relation (for the relationship between these lattices and the group structure → [13]).

H. Commutator Subgroups

Given two elements a and b of a group G, we call $a^{-1}b^{-1}ab=[a,b]$ the **commutator** of a and b. The subgroup C generated by all commutators in G is called the **commutator subgroup** (or **derived group**) of G. The subgroup C is a normal subgroup of G, and the factor group G/C is Abelian. On the other hand, if B is a normal subgroup of G and G/B is Abelian, then B contains C. For two subsets A, B of G, the subgroup generated by the commutators $[a,b]$ $(a\in A, b\in B)$ is called the **commutator group of** A **and** B and is denoted by $[A,B]$. If A and B are normal subgroups of G, then $C'=[A,B]$ is also a normal subgroup of G, and A/C' commutes with B/C' elementwise in the factor group G/C'. Furthermore, $[A,B]$ is the minimal normal subgroup with this property. The subgroup $[G,G]$ is the commutator subgroup of G.

If the commutator subgroup of G is Abelian, then G is called a **meta-Abelian group**. If a group G has a normal chain $G(=G_0)\supset G_1\supset G_2(=\{e\})$ of length 2 and the factor groups G/G_1, G_1/G_2 are Abelian, then G is meta-Abelian. Meta-Abelian groups are special cases of solvable groups discussed in Section I.

I. Solvable Groups

Suppose that we are given a series of subgroups G_i $(i=0,1,2,\ldots)$ of G such that $G=G_0$ and $[G_i,G_i]=G_{i+1}$. Then we have a normal chain $G=G_0\supset G_1\supset G_2\supset\ldots$. If $G_r=\{e\}$ for some r, then G is called a **solvable group**. For the normal chain G $(=G_0)\supset G_1\supset\ldots\supset G_r$ $(=\{e\})$ the factor groups G_i/G_{i+1} $(i=0, 1,\ldots,r-1)$ are all Abelian. A finite group G is solvable if and only if G has a composition series $G=H_0\supset H_1\supset H_2\supset\ldots\supset H_s=\{e\}$ such that the factor groups H_i/H_{i+1} $(i=0, 1,\ldots,s-1)$ are all of prime order. An †irreducible algebraic equation over a field of †characteristic 0 is solvable by radicals if and only if its Galois group is solvable (→ 177 Galois Theory).

J. Nilpotent Groups

The sequence of subgroups $G=G_0\supset G_1\supset G_2\supset\ldots$ defined inductively by setting $G_r=[G,G_{r-1}]$ $(r=1,2,\ldots)$ is called the **lower central series** of G. If $G_n=\{e\}$ for some n, then G is called a **nilpotent group**, and the least number n with $G_n=\{e\}$ is called the **class** of the nilpotent group G. A nilpotent group is solvable. Let Z_1 be the center of G, Z_2/Z_1 be the center of G/Z_1, and so on. Then we have a sequence of subgroups $Z_0=\{e\}\subset Z_1\subset Z_2\subset\ldots$, called the **upper central series** of G. A group G is nilpotent if and only if $Z_m=G$ for some m, and the least number m with $Z_m=G$ is the class of G. For the subgroups G_r and Z_r $(r=1,2,\ldots)$, we have $[G_{i-1},Z_i]=\{e\}$. If G is a †Lie group, then G is nilpotent if and only if the corresponding †Lie algebra $\mathfrak{g}$ is nilpotent, i.e., $\mathfrak{g}^n=0$.

K. Infinite Solvable Groups

The concepts of solvability and nilpotency are generalized in several ways for infinite groups. For instance, a group G is called a **generalized solvable group** if any homomorphic image of G which is unequal to $\{e\}$ contains an Abelian normal subgroup unequal to $\{e\}$, and G is called a **generalized nilpotent group** if any homomorphic image $(\neq\{e\})$ of G has center unequal to $\{e\}$. These definitions coincide with the previous ones for finite groups but not for infinite groups [12].

L. Direct Products

Let $G_1,\ldots,G_n$ be a finite number of groups. The set G of all elements $(x_1,\ldots,x_n)$ with $x_i\in G_i$ $(i=1,\ldots,n)$ is a group if we define the product of two elements $x=(x_1,\ldots,x_n)$ and $y=(y_1,\ldots,y_n)$ to be $xy=(x_1y_1,\ldots,x_ny_n)$. We call G the **direct product** of groups $G_1,\ldots,G_n$ and write $G=G_1\times\ldots\times G_n$. If e_j is the identity element of G_j, then $e=(e_1,\ldots,e_n)$ is the identity element of G. The mapping $(x_1,\ldots,x_n)\rightarrow x_i$ from G to G_i is a surjective homomorphism, called the **canonical surjection**. The subgroup $H_i=\{(e_1,\ldots,e_{i-1},x_i,e_{i+1},\ldots,e_n)\mid x_i\in G_i\}$ is isomorphic to G_i. The subgroups H_i $(i=1,\ldots,n)$ satisfy the following conditions: (i) H_i is a normal subgroup of G. (ii) H_i commutes with H_j elementwise if $i\neq j$. (iii) Any element of G can be written uniquely as the product of elements of $H_1,\ldots,H_n$. Conversely, if a group G has subgroups $H_1,\ldots,H_n$ satisfying these three conditions, then G is isomorphic to $H_1\times\ldots\times H_n$. In this case we also write $G=H_1\times\ldots\times H_n$, and we call this a **direct decomposition** of G. Each H_i is called a **direct factor** of G. Conditions (i), (ii), and (iii) are equivalent to condition (i), (ii′) $G=H_1H_2\ldots H_n$, and $H_1\ldots H_{i-1}\cap H_i=\{e\}$ $(i=2,\ldots,n)$.

A group G is called **indecomposable** if G cannot be decomposed into the direct product of two subgroups unequal to $\{e\}$, and **completely reducible** if G is the direct product of simple groups. If G satisfies the ascending or descending chain condition for normal subgroups, then G can be decomposed into the direct product of indecomposable groups. Such a decomposition is not unique in general, but if G has two direct product decompositions $G=G_1\times\ldots\times G_m=H_1\times\ldots\times H_n$, where G_i and H_j are indecomposable and not equal to $\{e\}$, then $m=n$ and the factors G_i are isomorphic to the factors H_j for some j; moreover, if G_1 corresponds, say, with H_1, then we have $G=H_1\times G_2\times\ldots\times G_m$. This fact was first stated by J. H. M. Wedderburn, and a complete proof of the theorem was given by R. Remak and O. Schmidt. Later W. Krull extended it to more general groups (with operator domain), and we call it the **Krull-Remak-Schmidt theorem**. O. Ore formulated it as a theorem on †modular lattices.

For an infinite number of groups G_λ $(\lambda\in\Lambda)$ we define the direct product $\prod_{\lambda\in\Lambda}G_\lambda$ of these groups similarly. The set of all elements $(\ldots,x_\lambda,\ldots)$ $(x_\lambda\in G_\lambda)$ such that almost all x_λ (i.e., all except a finite number of λ) are identity elements is a subgroup of the direct product, called the **direct sum** (or **restricted direct product**) of $\{G_\lambda\}$.

M. Free Products

Given a family of groups $\{G_\lambda\}_{\lambda\in\Lambda}$, we define the most general group G generated by these groups, called the free product of $\{G_\lambda\}$, together with canonical injections $f_\lambda: G_\lambda\rightarrow G$.

Let S be the disjoint union of the sets $\{G_\lambda\}_{\lambda\in\Lambda}$, and regard G_λ as a subset of S. A **word** is either void or a finite sequence $a_1, a_2,\ldots,a_n$ of elements of S, and we denote the set of all words by W. The product of two words w and w' is defined by connecting w

with w' so that the †associative law holds. We write $w \succ w'$ when two words w and w' satisfy one of the following two conditions: (i) The word w has successive members a, b which belong to the same group G_λ, and the word w' is obtained from w by replacing a, b by the product ab. (ii) Some member of w is an identity element, and w' is the word obtained from w by deleting this member. For two words w and w', we write $w \equiv w'$ if there is a finite sequence of words $w = w_0, w_1, \ldots, w_n = w'$ such that for each i $(1 \leqslant i \leqslant n)$, either $w_{i-1} \succ w_i$ or $w_i \succ w_{i-1}$. This relation is an equivalence relation and is compatible with the multiplication. Thus we may define a multiplication for the quotient set G of W by this equivalence relation, and then G is a group whose identity element is the equivalence class containing the void word. Any $x \in G_\lambda$ is regarded as a word, and we have an injective homomorphism $f_\lambda : G_\lambda \rightarrow G$ by assigning the corresponding class to each element of G_λ. The group G is called the **free product** of the system of groups $\{G_\lambda\}_{\lambda \in \Lambda}$, and f_λ is called the **canonical injection**. The free product G of $\{G_\lambda\}_{\lambda \in \Lambda}$ is characterized by the following universal property: Given a group G' and homomorphisms $f'_\lambda : G_\lambda \rightarrow G'$ $(\lambda \in \Lambda)$, we can find a unique homomorphism $g : G \rightarrow G'$ such that $g \circ f_\lambda = f'_\lambda$. The free product is the dual concept of direct product and is also called the †coproduct (→ 53 Categories and Functors). If each G_λ is an infinite cyclic group generated by a_λ, then the free product of the G_λ is the †free group generated by $\{a_\lambda\}$ (→ 169 Free Groups).

The concept of free product is generalized in the following way. Let H be a fixed group. We consider the family of pairs (G,j), where G is a group and $j : H \rightarrow G$ is an injective homomorphism. A homomorphism of pairs $f : (G,j) \rightarrow (G',j')$ is defined to be a group homomorphism $f : G \rightarrow G'$ such that $f \circ j = j'$. For a given family of pairs $\{(G_\lambda, j_\lambda)\}$, we have the **amalgamated product** (G,j) of the family and the canonical homomorphism $f_\lambda : (G_\lambda, j_\lambda) \rightarrow (G,j)$, which is characterized by the following universal property: Given a pair (G',j') and homomorphisms $f'_\lambda : (G_\lambda, j_\lambda) \rightarrow (G',j')$, we have a unique homomorphism $g : (G,j) \rightarrow (G',j')$ such that $g \circ f_\lambda = f'_\lambda$. If $H = \{e\}$, then the amalgamated product is the same as the free product. Now f_λ is an injection. If we regard G_λ as a subgroup of G, then G is generated by the subgroups G_λ and $G_\lambda \cap G_\mu = j_\lambda(H) = j_\mu(H)$ $(\lambda \neq \mu)$.

The notion of the amalgamated product is useful in constructing groups with interesting properties. For instance, we have a group whose nonidentity elements are all conjugate (B. H. Neumann and G. Higman), and a group generated by a finite number of elements such that its homomorphic image ($\neq \{e\}$) is always an infinite group (Higman) so that we have an infinite simple group generated by a finite number of elements.

N. Extensions

Let N and F be groups. A group G is called an **extension** of F by N if G has a normal subgroup $\bar{N}$ isomorphic to N and $G/\bar{N} \cong F$. The problem of finding all extensions was solved by Schreier (*Monatsh. Math. Phys.*, 34 (1926); *Abh. Math. Sem. Univ. Hamburg*, 4 (1928)). Suppose that (1) to each $\sigma \in F$ there corresponds an automorphism s_σ of N; (2) there exist elements $c_{\sigma,\tau}$ $(\sigma, \tau \in F)$ of N such that $s_\sigma(s_\tau(a)) = c_{\sigma,\tau}(s_{\sigma\tau}(a))c_{\sigma,\tau}^{-1}$ $(a \in N)$; and (3) $c_{\sigma,\tau}c_{\sigma\tau,\rho} = s_\sigma(c_{\tau,\rho})c_{\sigma,\tau\rho}$. Then the set G of all symbols as_σ $(a \in N, \sigma \in F)$ is an extension of F by N if we define multiplication by $as_\sigma \cdot bs_\tau = (as_\sigma(b)c_{\sigma,\tau})s_{\sigma\tau}$. In fact, the set of all elements $\bar{a} = ac_{1,1}^{-1}s_1$ $(a \in N)$ is a normal subgroup $\bar{N}$ of G such that $G/\bar{N} \cong F$. Any extension can be obtained in this way. A system $(s_\sigma, c_{\sigma,\tau})$ satisfying (1), (2), and (3) above is called a **factor set** belonging to F. Two factor sets $(s_\sigma, c_{\sigma,\tau})$ and $(t_\sigma, d_{\sigma,\tau})$ are said to be **associated** if there exist elements a_σ $(\sigma \in F)$ of N such that $t_\sigma(a) = s_\sigma(a_\sigma a a_\sigma^{-1})$ and $d_{\sigma,\tau} = a_\sigma(s_\sigma(a_\tau))c_{\sigma,\tau}a_{\sigma\tau}^{-1}$. In this case, two extensions determined by these factor sets are isomorphic. If $(s_\sigma, c_{\sigma,\tau})$ is associated with $(t_\sigma d_{\sigma,\tau})$ $(d_{\sigma,\tau} = 1$ for any $\sigma, \tau \in F)$, then we say that the corresponding extension is a **split extension**. In this case, the extension G contains a subgroup $\bar{F}$ isomorphic to F, and $G = \bar{F}\bar{N}$, $\bar{F} \cap \bar{N} = \{e\}$. We call such an extension a **semidirect product** of N and F.

If N is Abelian, then condition (2) is simply $s_\sigma(s_\tau(a)) = s_{\sigma\tau}(a)$, since the only inner automorphism of N is the identity mapping. The conditions of associated factor sets and split extension are also simplified. If N is contained in the center of G, then G is called a **central extension** of N.

O. Transfers

Let H be a subgroup of finite index in G and g_i $(i = 1, \ldots, h)$ be representatives of the right cosets of H. For $b \in Hg_i$ we write $g_i = \bar{b}$. Then for $x \in G$ an element $\tilde{x} = H'\prod_{i=1}^{h} g_i x (\overline{g_i x})^{-1}$ of H/H' is determined uniquely (independent of the choice of representatives), where H' is the commutator subgroup of H. The correspondence $G'x \rightarrow \tilde{x}$ yields a homomorphism of G/G' to H/H', which is called the **transfer** from G/G' to H/H'.

P. Generalizations

The concept of group can be generalized in several ways. A set S in which a multiplication $(a,b)\to ab$ satisfying $(ab)c=a(bc)$ (the associative law) is defined is called a **semigroup**. If S is a commutative semigroup in which $ax=bx$ implies $a=b$ (the **cancellation law**), then S can be embedded in a group G so that the multiplication in S is preserved in G and any $x\in G$ is the quotient of two elements of S: $x=a^{-1}b=ba^{-1}$ $(a,b\in S)$. Such a group G is determined uniquely by S. We call it the **group of quotients** of S.

The notion of semigroup is obtained by taking only associativity from the group axioms. On the other hand, if Q is a set with a law of composition $(a,b)\to ab$ which is not necessarily associative but satisfies the condition that any two among a, b, c in the equation $ab=c$ determine the third uniquely, then Q is called a **quasigroup**. A quasigroup with an identity element e such that $ea=ae=a$ for every element a is called a **loop**. For loops, we have an analog of the structure theory of groups (R. H. Bruck, *Trans. Amer. Math. Soc.*, 60 (1946)).

If we give up the possibility of forming products for all pairs of elements or the uniqueness of the product in the axioms for groups, then we have the following generalizations of groups. A set M with multiplication under which to any elements a, $b\in M$ there corresponds a nonempty subset ab of M is called a **hypergroupoid**. Moreover, if the associative law $(ab)c=a(bc)$ holds and for any elements a, $b\in M$ there exist $x,y\in M$ such that $b\in xa$, $b\in ay$, then M is called a **hypergroup**.

A set M is called a **mixed group** if (1) M can be partitioned into disjoint subsets M_0, M_1, M_2, ... ; (2) for $a\in M_0$, $b\in M_i$ $(i=0,1,2,\ldots)$, elements ab, $a\backslash b$ of M_i are defined such that $a(a\backslash b)=b$; (3) for b, $c\in M_i$, an element b/c of M_0 is defined such that $(b/c)\cdot c=b$; and (4) the associative law $(ab)c=a(bc)$ $(a,b\in M_0, c\in M)$ holds (A. Loewy, 1927).

A set M is called a **groupoid** if (1) M can be partitioned into disjoint subsets M_{ij} $(i,j=1,2,\ldots)$; (2) for $a\in M_{ij}$ and $b\in M_{jk}$, an element $ab\in M_{ik}$ is defined; (3) for $a\in M_{ij}$ and $b\in M_{ik}$, an element $a\backslash b\in M_{jk}$ is defined such that $a(a\backslash b)=b$; (4) for $a\in M_{ij}$ and $b\in M_{kj}$, an element $a/b\in M_{ik}$ is defined such that $(a/b)b=a$; and (5) for $a\in M_{ij}$, $b\in M_{jk}$, and $c\in M_{kl}$, the associative law $a(bc)=(ab)c$ holds (H. Brandt, 1926). These generalized concepts also have some practical applications (→ 2 Abelian Groups; 15 Algebraic Groups; 63 Classical Groups; 71 Compact Groups; 94 Crystallographic Groups; 126 Discontinuous Groups; 160 Finite Groups; 169 Free Groups; 241 Lattices; 248 Lie Groups; 275 Modules; 358 Representations; 405 Topological Abelian Groups; 406 Topological Groups; 423 Unitary Representations).

Q. History

The concept of the group was first introduced in the early 19th century, but its rudiments can be found in antiquity; in fact, it was virtually contained in the concept of motion or transformation used in ancient geometry. From the time it took explicit form in the late 19th century, it has played a fundamental role in all fields of mathematics.

In their study of algebraic equations in the late 18th century, J. L. †Lagrange, A. T. Vandermonde, and P. Ruffini saw the importance of the group of permutations of roots; using this idea N. H. Abel showed that a general equation of degree $\geqslant 5$ cannot be solved algebraically. A. L. †Cauchy studied the group of permutations of roots for its own interest, but a complete description of the relationship between groups and algebraic equations was first given by E. †Galois. C. Jordan developed a detailed exposition of the theory due to Abel and Galois in his *Traité des substitutions* (1870) [6]. Up to that time, a group meant a permutation group; the axiomatic definition of a group was given by A. Cayley (1854) and L. Kronecker (1870). F. †Klein emphasized the significance of group theory in geometry in his †Erlangen program (1872), and M. S. †Lie developed the theory of †Lie groups in the 1880s. In 1897, W. Burnside published his *Theory of groups* [8], whose second edition (1911) is one of the classics in group theory and is still valuable. Since 1896, G. Frobenius [7] and others have developed the theory of representation of groups by matrices (→ 358 Representations). By that time, the theory of finite groups had acquired all its essential features. Among the branches of abstract algebra, the theory of groups was the first to develop; it led to the progress of abstract algebra in the 1930s. Since the latter half of that decade, the theory of finite groups has been developed further; there has been increased interest in the theory, and many significant results have been obtained, especially since 1955 (→ 160 Finite Groups).

References

[1] K. Shoda, Tyûsyô daisûgaku (Japanese; Abstract algebra), Iwanami, 1932.
[2] K. Shoda and K. Asano, Daisûgaku (Jap-

anese; Algebra) I, Iwanami, 1952.
[3] Y. Akizuki and M. Suzuki, Kôtô daisûgaku (Japanese; Higher algebra), Iwanami, I, 1952; II, 1957.
[4] K. Asano and H. Nagao, Gunron (Japanese; Theory of groups), Iwanami, 1965.
[5] M. Osima, Gunron (Japanese; Theory of groups), Kyôritu, 1956.
[6] C. Jordan, Traité des substitutions et des équations algébriques, Gauthier-Villars, 1870 (Blanchard, 1957).
[7] G. Frobenius, Gesammelte Abhandlungen I, II, III, Springer, 1968.
[8] W. Burnside, Theory of groups of finite order, Cambridge Univ. Press, second edition, 1911.
[9] H. Weyl, The classical groups, Princeton Univ. Press, revised edition, 1946.
[10] B. L. van der Waerden, Algebra, Springer, I, seventh edition, 1966; II, fifth edition, 1967.
[11] B. L. van der Waerden, Gruppen von linearen Transformationen, Erg. Math., Springer, 1935 (Chelsea, 1948).
[12] А. Г. Курош (A. G. Kuroš), Теория групп, Гостехиздат, second edition, 1953; English translation, The theory of groups I, II, Chelsea, 1960.
[13] M. Suzuki, Structure of a group and the structure of its lattice of subgroups, Erg. Math., Springer, 1956.
[14] M. Hall, The theory of groups, Macmillan, 1959.
[15] H. Zassenhaus, Lehrbuch der Gruppentheorie I, Teubner, 1937; English translation, The theory of groups, Chelsea, 1958.
[16] A. Speiser, Die Theorie der Gruppen von endlicher Ordnung, Springer, third edition, 1937.
[17] W. Specht, Gruppentheorie, Springer, 1956.
[18] R. H. Bruck, A survey of binary systems, Erg. Math., Springer, 1958.
[19] A. H. Clifford and G. B. Preston, The algebraic theory of semigroups, Amer. Math. Soc. Math. Surveys, I, 1961; II, 1967.
[20] E. S. Liapin, Semigroups, Amer. Math. Soc. Transl. of Math. Monographs, 1963.

H

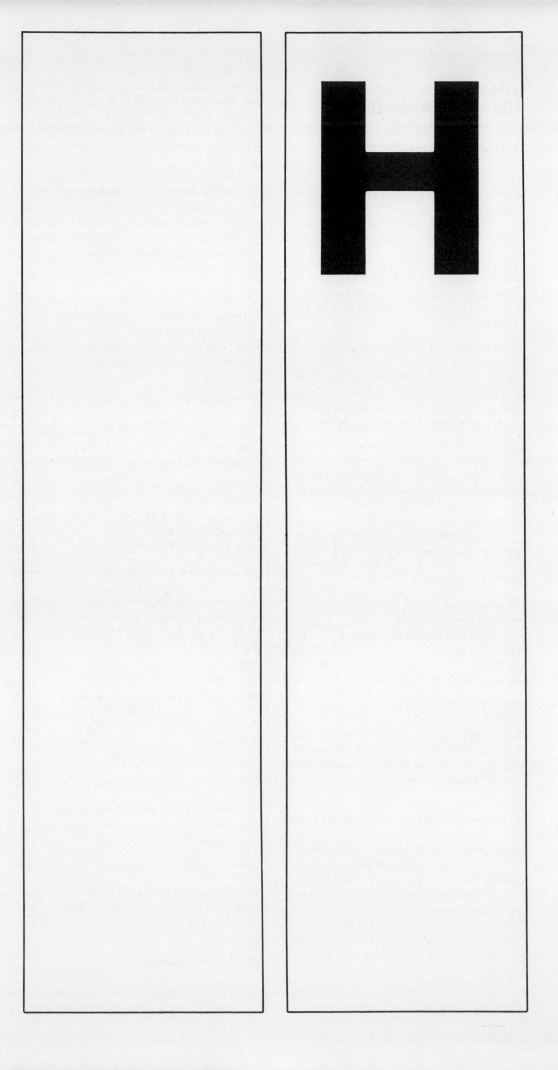

194 (X.24) Harmonic Analysis

A. Fourier Transforms

Let $f(x)$ be an element of the †function space $L_1(-\infty,\infty)$ and t a real number. Then the integral

$$\hat{f}(t)=(2\pi)^{-1/2}\int_{-\infty}^{\infty} f(x)e^{-itx}\,dx \qquad (1)$$

converges, and the function $\hat{f}(t)$ is continuous in $(-\infty,\infty)$. We call $\hat{f}$ the †Fourier transform of f. If $f(x)\in L_2(-\infty,\infty)$, then $f(x)\in L_1(-a,a)$ for any finite interval $(-a,a)$, and if we set

$$\hat{f}_a(t)=(2\pi)^{-1/2}\int_{-a}^{a} f(x)e^{-itx}\,dx,$$

then $\hat{f}_a$ †converges in the mean of order 2 as $a\to\infty$ to a function $\hat{f}$ in L_2. In this case, we define $\hat{f}$ to be the †Fourier transform of f $(\in L_2)$. Furthermore, in this case, if we set

$$f_a(x)=(2\pi)^{-1/2}\int_{-a}^{a}\hat{f}(t)e^{itx}\,dt,$$

then $\mathrm{l.i.m.}_{a\to\infty} f_a(x)=f(x)$ (**Plancherel theorem**). Moreover, we have †Parseval's equality: $\int_{-\infty}^{\infty}|f(x)|^2dx=\int_{-\infty}^{\infty}|\hat{f}(t)|^2dt$ (→ 168 Fourier Transform).

Suppose that $f(x)$ is periodic with period 2π and $f\in L_2(-\pi,\pi)$. We set

$$a_n=(2\pi)^{-1}\int_{-\pi}^{\pi} f(x)e^{-inx}\,dx$$

(†Fourier coefficients). The nth partial sum of the †Fourier series $s_n(x)=\sum_{\nu=-n}^{n}a_\nu e^{i\nu x}$ converges in the mean of order 2 to $f(x)$, and Parseval's equality

$$\frac{1}{2\pi}\int_{-\pi}^{\pi}|f(x)|^2dx=\sum_{n=-\infty}^{\infty}|a_n|^2$$

holds. On the other hand, if $\{a_n\}$ is a given sequence such that $\sum_{n=-\infty}^{\infty}|a_n|^2<\infty$, then $\sum_{\nu=-n}^{n}a_\nu e^{i\nu x}$ converges in the mean of order 2 to a function $f(x)$, the Fourier coefficients of $f(x)$ are $\{a_n\}$, and Parseval's equality holds (→ 167 Fourier Series).

B. Bochner's Theorem and Herglotz's Theorem

A complex-valued function $f(x)$ defined on $(-\infty,\infty)$ is said to be of **positive type** (or **positive definite**) if $\sum_{j,k=1}^{n} f(x_j-x_k)\xi_j\bar{\xi}_k\geqslant 0$ for any finite number of reals $x_1,x_2,\ldots,x_n$ and complex numbers $\xi_1,\xi_2,\ldots,\xi_n$. If $f(x)$ is measurable on $(-\infty,\infty)$ and of positive type, then there exists a †monotone increasing real-valued bounded function $\alpha(t)$ such that

$$f(x)=\int_{-\infty}^{\infty}e^{itx}\,d\alpha(t) \qquad (2)$$

for almost all x. If $\alpha(-\infty)=0$ and $\alpha(t)$ is right continuous, then $\alpha(t)$ is unique (**Bochner's theorem**). Conversely, if $\alpha(t)$ is nondecreasing and bounded and $f(x)$ is defined by (2), then $f(x)$ (called the **Fourier-Stieltjes transform** of $\alpha(t)$) is continuous and of positive type.

A sequence $\{a_n\}$ $(-\infty<n<\infty)$ is said to be **positive definite** (or of **positive type**) if $\sum_{j,k=1}^{n}a_{j-k}\xi_j\bar{\xi}_k\geqslant 0$ for finitely many arbitrarily chosen complex numbers $\xi_1,\xi_2,\ldots,\xi_n$. If $\{a_n\}$ is of positive type, then there exists a monotone increasing bounded function $\alpha(t)$ on $[-\pi,\pi]$ such that

$$a_n=\int_{-\pi}^{\pi}e^{int}\,d\alpha(t)$$

(**Herglotz's theorem**). Conversely, if $\alpha(t)$ is monotone increasing and bounded, and a_n is defined by the above integral, then the sequence $\{a_n\}$ is of positive type.

C. Poisson's Summation Formula

If $f(x)\in L_1(-\infty,\infty)$ is of †bounded variation and continuous, and if $\hat{f}(t)$ is its Fourier transform, then we have

$$\sqrt{a}\sum_{k=-\infty}^{\infty}f(ak)=\sqrt{b}\sum_{k=-\infty}^{\infty}\hat{f}(bk),$$

where $ab=2\pi$ $(a>0)$. This is called **Poisson's summation formula**.

D. Generalized Tauberian Theorems of Wiener

Suppose that we are given a function $f(x)\in L_1(-\infty,\infty)$ whose Fourier transform $\hat{f}(t)$ is never zero. Then the set of functions given by

$$h(x)=\int_{-\infty}^{\infty}f(x-y)\,g(y)\,dy,$$

where $g\in L_1(-\infty,\infty)$, is dense in $L_1(-\infty,\infty)$. Hence we can deduce the †generalized Tauberian theorem of Wiener: If the Fourier transform $\hat{k}_1(t)$ of $k_1(x)\in L_1(-\infty,\infty)$ does not vanish for any real t and

$$\lim_{x\to\infty}\int_{-\infty}^{\infty}k_1(x-y)f(y)\,dy=C\int_{-\infty}^{\infty}k_1(y)\,dy$$

for a function $f(x)$ that is bounded and measurable on $(-\infty,\infty)$, then for any $k_2\in L_1(-\infty,\infty)$,

$$\lim_{x\to\infty}\int_{-\infty}^{\infty}k_2(x-y)f(y)\,dy=C\int_{-\infty}^{\infty}k_2(y)\,dy$$

(→ 168 Fourier Transform G). Hence we can deduce †Tauberian theorems of the J. E. Littlewood type [3].

E. Harmonic Analysis

Let $\alpha(\lambda)$ be a complex-valued function on $(-\infty,\infty)$ that is of bounded variation and right continuous. If we have the expression

$$f(t)=\int_{-\infty}^{\infty} e^{i\lambda t}\,d\alpha(\lambda), \qquad (2')$$

then we say that the function $f(t)$ is represented by the superposition of harmonic oscillations $e^{i\lambda t}$. Conversely, when $f(t)$ is given, we have the problem of finding a function $\alpha(\lambda)$ as above such that $f(t)$ can be expressed in the form of (2′). When such a function $\alpha(\lambda)$ exists, it is also an important problem (in harmonic analysis) to find the frequency $\alpha(\lambda)-\alpha(\lambda-0)$ of the component of a proper oscillation. Concerning this problem, we have the following three theorems:

(I) A necessary and sufficient condition for $f(x)$ to be representable in the form (2′) is

$$\sup_{1\leqslant n}\int_{-\infty}^{\infty}\left|\int_{-\infty}^{\infty}\left(\frac{\sin(t/n)}{t/n}\right)^2 f(t)e^{-i\lambda t}\,dt\right|d\lambda<\infty.$$

(II) For a function $f(t)$ expressed in the form (2′), we have (i) for any λ_0,

$$\alpha(\lambda_0)-\alpha(\lambda_0-0)=\lim_{T\to\infty}\frac{1}{2T}\int_{-T}^{T} f(t)e^{-i\lambda_0 t}\,dt,$$

and (ii) if $\alpha(\lambda)$ is continuous at $\lambda=\lambda_0-\sigma$ and $\lambda=\lambda_0+\sigma$ $(\sigma>0)$, then

$$\alpha(\lambda_0+\sigma)-\alpha(\lambda_0-\sigma)$$
$$=\lim_{T\to\infty}\sqrt{\frac{2}{\pi}}\int_{-T}^{T}\frac{\sin\sigma t}{t}f(t)e^{-i\lambda_0 t}\,dt.$$

(III) In (2′), suppose that the discontinuity points of $\alpha(\lambda)$ are $\lambda_1,\lambda_2,\ldots$, and set $a_n=\alpha(\lambda_n)-\alpha(\lambda_n-0)$ $(n=1,2,\ldots)$; then

$$\lim_{S\to\infty}\frac{1}{S}\int_0^S f(t+s)\overline{f(s)}\,ds=\sum_{n=1}^{\infty}|a_n|^2e^{i\lambda_n t}.$$

F. The Paley-Wiener Theorem

In formula (1), if we change the variable from t to a complex variable $\zeta=t+i\sigma$, then we have

$$F(\zeta)=(2\pi)^{-1/2}\int_{-\infty}^{\infty} f(x)e^{-i\zeta x}\,dx, \qquad (3)$$

which is called the **Fourier-Laplace transform** of $f(x)$. In particular, if $f(x)$ has bounded †support, then $F(\zeta)$ is an †entire function. Concerning Fourier-Laplace transforms, Paley and Wiener proved the following theorems:

(I) A necessary and sufficient condition for an entire function $F(\zeta)$ to be the Fourier-Laplace transform of a †function of class C^∞ having its support in a finite interval $[-B,B]$ is that for any N, there exists a constant $C_N>0$ such that $|F(\zeta)|\leqslant C_N(1+|\zeta|)^{-N}e^{B|\sigma|}$ for all $\zeta=t+i\sigma$.

(II) If $g(t)\in L_2(0,\infty)$, then its one-sided †Laplace transform

$$f(z)=(2\pi)^{-1/2}\int_0^{\infty} g(t)e^{-tz}\,dt$$

satisfies: (i) $f(t)$ is holomorphic in the right half-plane $\mathrm{Re}\,z>0$, and (ii)

$$\sup_{x>0}\int_{-\infty}^{\infty}|f(x+iy)|^2\,dy<\infty.$$

Conversely, if $f(z)=f(x+iy)$ $(x>0)$ satisfies (i) and (ii), then the boundary function $f(iy)\in L_2(-\infty,\infty)$ exists and is such that

$$\lim_{x\downarrow 0}\int_{-\infty}^{\infty}|f(iy)-f(x+iy)|^2\,dy=0,$$

its Fourier transform in L_2

$$g(t)=\underset{N\to\infty}{\mathrm{l.i.m.}}\,(2\pi)^{1/2}\int_{-N}^{N} f(iy)e^{ity}\,dy$$

vanishes at almost all negative t, and $f(z)$ is the one-sided Laplace transform of $g(t)$.

G. Harmonic Analysis on Locally Compact Abelian Groups

The general theory of harmonic analysis on the real line was extended to a theory on locally compact Abelian groups by A. Weil, I. M. Gel′fand, D. A. Raĭkov, and others. The theory of normed rings was utilized for the development of the theory (→ 38 Banach Algebras). This theory is called **harmonic analysis on locally compact Abelian groups**.

H. Group Rings

Let $L_1=L_1(G)$ be the set of all integrable functions with respect to a †Haar measure on a locally compact Abelian group G. If we define the norm and multiplication in $L_1(G)$ by

$$\|f\|=\int_G|f(x)|\,dx,$$

$$f\cdot g(x)=\int_G f(xy^{-1})g(y)\,dy,$$

respectively, then L_1 has the structure of a commutative †Banach algebra. (We call $f\cdot g$ the **convolution** (or **composition product**) of f and g.) If the topology of G is not †discrete, $L_1(G)$ does not have a unity for multiplication. Hence, adjoining a formal unity $\mathbf{1}$ to $L_1(G)$, we set $R=\{\alpha\mathbf{1}+f \mid \alpha$ is a complex number, $f\in L_1(G)\}$, and

$$\|\alpha\mathbf{1}+f\|=|\alpha|+\|f\|,$$
$$(\alpha\mathbf{1}+f)+(\beta\mathbf{1}+g)=(\alpha+\beta)\mathbf{1}+(f+g),$$
$$(\alpha\mathbf{1}+f)\cdot(\beta\mathbf{1}+g)=(\alpha\beta)\mathbf{1}+(\beta f+\alpha g+f\cdot g).$$

Then R is a commutative Banach algebra with unity. When G is discrete, $R=L_1(G)$. The Banach algebra R is called the **group algebra** of G. Any group algebra R of G is †semisimple. R is algebraically isomorphic to a subalgebra of $C(\mathfrak{M})$, which is the associative algebra of all continuous functions on the compact Hausdorff space $\mathfrak{M}$ consisting of all maximal ideals in R. By this correspondence, if $\varphi \in R$ corresponds to $\varphi(M)$, which is a function on $\mathfrak{M}$, then $\sup_{M \in \mathfrak{M}} |\varphi(M)| \leqslant \|\varphi\|$. $L_1(G)$ belongs to $\mathfrak{M}$ if and only if the group G is not discrete.

I. Fourier Transforms

According as the group G is discrete or not, we set $\mathfrak{N} = \mathfrak{M}$ or $\mathfrak{N} = \mathfrak{M} - \{L_1(G)\}$. Then there exists a one-to-one correspondence between the elements $M \in \mathfrak{N}$ and the elements χ of the †character group $\hat{G}$ of G such that the following formulas are valid:

$$f(M) = \int_G \chi(x) f(x)\,dx, \quad f \in L_1, \tag{4}$$

$$\chi(y) = f_y(M)/f(M), \tag{5}$$

where $f_y(x) = f(xy^{-1})$ and f is a function such that $f(M) \neq 0$. This correspondence $M \leftrightarrow \chi$ gives a homeomorphism between the locally compact space $\mathfrak{N}$ and $\hat{G}$. Hence if we identify M with χ and set $f(M) = \hat{f}(\chi)$, then $\hat{f}$ is a continuous function on $\hat{G}$, called the **Fourier transform** of $f(x)$. Since $f \rightarrow f(M)$ is an algebraic isomorphism, $(fg)\hat{}(\chi) = \hat{f}(\chi)\hat{g}(\chi)$. If $\hat{f}(\chi) \equiv \hat{g}(\chi)$, then f is equal to g in $L_1(G)$. This is the **uniqueness theorem** of Fourier transforms. From it we can deduce the †maximal almost periodicity of locally compact Abelian groups.

If G is not discrete, then $L_1(G) \in \mathfrak{M}$, and $f(L_1(G)) = 0$ for $f \in L_1(G)$. Hence $\{\chi \,|\, |\hat{f}(\chi)| \geqslant \varepsilon\}$ is a compact subset of $\hat{G}$. This means that $\hat{f}$ is a continuous function vanishing at infinity on $\hat{G}$. (This is a generalization of the †Riemann-Lebesgue theorem concerning the cases $G = \mathbf{R}^1$ or $\mathbf{T}^1 = \mathbf{R}/\mathbf{Z}$.) Any continuous function $u(\chi)$ on $\hat{G}$ vanishing at infinity is approximated uniformly by $\hat{f}(\chi)$, which is the Fourier transform of $f \in L_1(G)$.

J. Positive Definite Functions

A function $\varphi(x)$ defined on G is said to be **positive definite** (or of **positive type**) if the inequality $\sum_{j,k=1}^{n} \varphi(x_j x_k^{-1}) \alpha_j \bar{\alpha}_k \geqslant 0$ holds for arbitrary elements $x_1, \ldots, x_n$ in G and arbitrary complex numbers $\alpha_1, \ldots, \alpha_n$. We denote by P_G the set of all positive definite functions on G. If $\varphi \in P_G$, then $\varphi(e) \geqslant 0$ (e is the identity of G), $|\varphi(x)| \leqslant \varphi(e)$, and $\overline{\varphi(x^{-1})} = \varphi(x)$. If G is locally compact, we further assume that $\varphi \in P_G$ is measurable with respect to the Haar measure of G. Then for any $f \in L_1(G)$,

$$\iint \varphi(xy^{-1}) f(y) \overline{f(y)}\, dx\, dy \geqslant 0.$$

Any $\varphi \in P_G$ is equal almost everywhere to a continuous $\varphi_1 \in P_G$. (Concerning positive definite functions on locally compact groups and their relation with unitary representations → 423 Unitary Representations B.)

K. Harmonic Analysis and the Duality Theorem

If G is a locally compact Abelian group, then a function $\varphi(x)$ on G belongs to P_G if and only if there exists a nonnegative measure $\mu(\hat{G}) < \infty$ on $\hat{G}$ such that $\varphi(x) = \int_{\hat{G}} \chi(x)\, d\mu(\chi)$. (When $G = \mathbf{R}^1$, this theorem is †Bochner's theorem, whereas when $G = \mathbf{Z}$, it is Herglotz's theorem.) Hence we can prove a spectral resolution of unitary representations $U(x) = \int_{\hat{G}} \chi(x)\, dE(\chi)$ (a generalization of †Stone's theorem). If $f \in L_1(G) \cap P_G$, then $\hat{f}(\chi) \geqslant 0$, $\hat{f} \in L_1(\hat{G})$, and the **inversion formula** of Fourier transforms $f(x) = \int_{\hat{G}} \overline{\chi(x)} \hat{f}(\chi)\, d\chi$ holds, provided that the Haar measure is suitably chosen. If $f \in L_1(G) \cap L_2(G)$, then $\hat{f} \in L_2(\hat{G})$ and **Parseval's equality** $\int_G |f(x)|^2 dx = \int_{\hat{G}} |\hat{f}(\chi)|^2 d\chi$ holds. If we set $\bar{U}f = \hat{f}$, $\bar{V}\hat{f} = f$, then $\bar{U}$ is extendible uniquely to an isometry of $L_2(G)$ onto $L_2(\hat{G})$ and $\bar{V}$ is extendible uniquely to its inverse transformation, respectively (**Plancherel's theorem** on locally compact Abelian groups). By the inversion formula and Plancherel's theorem, we can prove that the character group $\hat{\hat{G}}$ of $\hat{G}$ is isomorphic to G as a topological group. This is called the **Pontrjagin duality theorem** of locally compact Abelian groups (→ 405 Topological Abelian Groups). In particular, if G is compact, $\hat{G}$ is a discrete Abelian group. Then we can normalize the Haar measure of G and $\hat{G}$ so that the measure of G and the measure of each element of $\hat{G}$ are 1. Plancherel's theorem implies that the set of characters of G is a †complete †orthonormal set in $L_2(G)$.

L. Poisson's Summation Formula

Suppose that G is a locally compact Abelian group, H is its discrete subgroup, and G/H is compact. Then the †annihilator Γ of H is a discrete subgroup of $\hat{G}$. For any continuous function $f(x)$ on G, if $\sum_{y \in H} f(xy)$ is convergent absolutely and uniformly (hence $f \in L_1(G)$) and $\sum_{\xi \in \Gamma} \hat{f}(\xi)$ is convergent absolutely, then $\sum_{y \in H} f(y) = c \sum_{\xi \in \Gamma} \hat{f}(\xi)$, where c is a constant depending on the Haar measures of G and $\hat{G}$. This is called **Poisson's summation formula** on a locally compact Abelian group (a generalization of †Poisson's summation formula on $G = \mathbf{R}$).

M. Closed Ideals in $L_1(G)$

For a function f defined on G and any $g \in G$, the translation operator τ_g is defined by $\tau_g f(x) = f(x-g)$. A closed subspace of $L_1(G)$ is an ideal in $L_1(G)$ if and only if it is invariant under all translations (N. Wiener). A closed ideal I coincides with $L_1(G)$ if and only if the set of zeros of I, i.e., $Z(I) = \bigcap_{f \in I} \hat{f}^{-1}(0)$, is empty (Wiener's Tauberian theorem). A closed ideal I is maximal if and only if $Z(I)$ consists of a single point. If the dual $\hat{G}$ of G is discrete, then the closed ideals in $L_1(G)$ are completely characterized by the zeros; that is, spectral synthesis is possible, but this situation does not generally hold. P. Malliavin's theorem states that if $\hat{G}$ is not discrete, then there exists a set E in $\hat{G}$ and two different closed ideals I and J such that $Z(I) = Z(J) = E$. Such a set E is called a non-S-set. For example, if $\hat{G} = \mathbf{R}^3$, the unit sphere is a non-S-set (L. Schwartz).

N. Operating Functions

Denote by $A(\hat{G})$ the set of all Fourier transforms of the functions in $L_1(G)$. Let $\hat{f} \in A(\hat{G})$ and Φ an analytic function in a neighborhood of the range of $\hat{f}$. Furthermore, assume that $\Phi(0)=0$ if G is not discrete. Then there exists a $\hat{g} \in A(\hat{G})$ such that $\hat{g}(\gamma) = \Phi(\hat{f}(\gamma))$ for $\gamma \in \hat{G}$ (Wiener-Lévy theorem). The converse is also true. Let G be an infinite Abelian group and Φ a function on the interval $[-1,1]$. If $\Phi(\hat{f}) \in A(\hat{G})$ for every $\hat{f} \in A(\hat{G})$, then Φ is analytic in a neighborhood of the origin if G is compact, and analytic in a neighborhood of $[-1,1]$ if G is not compact [18, 19]. A function Φ for which the Wiener-Lévy theorem holds is called an **operating function**. Let E be a subset of $\hat{G}$, and I_E the set of all $\hat{f} \in A(\hat{G})$ such that $\hat{f}=0$ on E. Then $A(E) = A(\hat{G})/I_E$ is the quotient algebra. The set E is called a **set of analyticity** if every operating function on $A(E)$ is analytic. For a characterization of such a set $\rightharpoonup$ [13].

O. Measure Algebras

For a locally compact Abelian group G, let $M(G)$ be the set of all †regular bounded measures. For λ and μ in $M(G)$, the convolution $\lambda * \mu$ is defined by $(\lambda * \mu)(E) = \int_G \lambda(E-y) \cdot d\mu(y)$, where E is a Borel set of G. Then $M(G)$ is a semisimple commutative Banach algebra whose product is defined to be the convolution. The Fourier-Stieltjes transform of $\mu \in M(G)$ is defined by

$$\hat{\mu}(\gamma) = \int_G (x, \gamma)\, d\mu(x), \quad \gamma \in \hat{G}.$$

A continuous function on $\hat{G}$ is positive definite if and only if it is the Fourier-Stieltjes transform of a positive measure in $M(G)$ (Bochner's theorem). Assume that G is not discrete. A function on the interval $[-1,1]$ which operates on the Fourier-Stieltjes transforms of measures in $M(G)$ can be extended to an entire function (Helson, Kahane, Katznelson, and Rudin [19]). From this fact it follows that $M(G)$ is asymmetric and nonregular. Furthermore, there exists a measure $\mu \in M(G)$ such that $\hat{\mu}(\gamma) \geqslant 1$ but $1/\hat{\mu}$ is not a Fourier-Stieltjes transform of $M(G)$. (See also Wiener and Pitt [20], Šreĭder [21], Williamson [22], and Hewitt and Kakutani [23]; for the general description of measure algebra, see Rudin [13] and Hewitt and Ross [14].)

P. Idempotent Measures

A measure $\mu \in M(G)$ is called **idempotent** if $\mu * \mu = \mu$, that is, $\hat{\mu}(\gamma) = 0$ or 1 for all $\gamma \in \hat{G}$. Then $\hat{\mu}$ is the characteristic function of the set $\{\gamma \in \hat{G} \mid \hat{\mu}(\gamma) = 1\}$. The smallest ring of subsets of $\hat{G}$ that contains all open cosets in $\hat{G}$ is called the coset ring of $\hat{G}$. The characteristic function of a set E in $\hat{G}$ is the Fourier-Stieltjes transform of an idempotent measure in $M(G)$ if and only if E belongs to the coset ring of $\hat{G}$ (P. J. Cohen's theorem). A simple proof is given by Ito and Amemiya [28]. When G is the unit circle, the coset ring consists of sequences periodic except at a finite number of points, and for this case the theorem was obtained by H. Helson. Let $n_1, n_2, \ldots, n_k$ be distinct integers and $d\mu(x) = \sum_{j=1}^{k} e^{in_j x} \cdot dx$. Then μ is an idempotent measure on the unit circle. J. E. Littlewood conjectured that the norm of μ exceeds $c \log k$, where c is a positive constant not depending on the choice of $\{n_j\}$. A partial answer was given by P. J. Cohen [24] for compact connected Abelian groups and was improved by H. Davenport [25] and E. Hewitt and H. S. Zuckerman [26].

Q. Mappings of Group Algebras

Let G and H be two locally compact Abelian groups and φ a nontrivial homomorphism of $L_1(G)$ into $M(H)$. Associated with φ there is a mapping φ^* of a subset Y of $\hat{H}$ into $\hat{G}$ such that $\hat{\varphi}(f)(\gamma) = \hat{f}(\varphi^*(\gamma))$ for $\gamma \in Y$ and $=0$ for $\gamma \notin Y$, or symbolically $\hat{\varphi}(f) = \hat{f}(\varphi^*)$.

A continuous mapping α of Y into $\hat{G}$ is said to be **piecewise affine** if there exist a finite number of disjoint subsets $S_j, j=1,\ldots,n$, of the coset ring of $\hat{H}$ and mappings α_j such that (i) $Y=\bigcup_{j=1}^{n}S_j$; (ii) α_j is defined on an open coset K_j of $\hat{H}$, where $K_j\supset S_j$; (iii) $\alpha_j=\alpha$ on S_j; and (iv) $\alpha_j(\gamma+\gamma'-\gamma'')=\alpha_j(\gamma)+\alpha_j(\gamma')-\alpha_j(\gamma'')$ for all $\gamma,\gamma',\gamma''\in K_j, j=1,\ldots,n$. P. J. Cohen's theorem is: If φ is a homomorphism of $L_1(G)$ into $M(H)$, then Y belongs to the coset ring of $\hat{H}$ and φ^* is a piecewise affine mapping of Y into $\hat{G}$. Conversely, for any piecewise affine mapping α, there is a homomorphism φ of $L_1(G)$ into $M(H)$ such that $\varphi^*=\alpha$. Related theorems have been studied by A. Beurling, H. Helson, J.-P. Kahane, Z. L. Leibenson, and W. Rudin.

R. Exceptional Sets

Let G be a locally compact Abelian group. A subset E is said to be independent if $n_1x_1+\ldots+n_kx_k=0$, where the n_j are integers and $x_j\in E$ implies $n_jx_j=0, j=1,\ldots,k$. A set E in G is called a **Kronecker set** if for every continuous function φ on E of absolute value 1 and $\varepsilon>0$ there exists a $\gamma\in\hat{G}$ such that $|\varphi(x)-(x,\gamma)|<\varepsilon$, $x\in E$. Every Kronecker set is independent and of infinite order, but independent sets are not necessarily Kronecker sets. For a group G whose elements are of finite order p, a set E is called of type K_p if for every continuous function φ on E with values $\exp(2\pi ik/p)$, $k=0,\ldots,p-1$, there is a $\gamma\in\hat{G}$ such that $\varphi=\gamma$ on E. If E is a compact Kronecker set in G and μ is a measure with support in E, i.e., $\mu\in M(E)$, then $\|\mu\|=\|\hat{\mu}\|_\infty$. A compact set E is called a **Helson set** if there is a constant C such that $\|\mu\|\leqslant C\|\hat{\mu}\|_\infty$ for $\mu\in M(E)$. Every K_p set is also a Helson set. For a Helson set E, $C(E)=A(E)$. A discrete analog of a Helson set is a Sidon set. A subset F of a discrete group $\hat{G}$ is called a **Sidon set** if there is a constant C such that $\Sigma_{\gamma\in F}|a_\gamma|\leqslant C\sup_x|\Sigma_{\gamma\in F}a_\gamma(x,\gamma)|$ for every polynomial $\Sigma a_\gamma(x,\gamma)$. For example, a †lacunary sequence $\{n_k\}$, $n_{k+1}/n_k>q>1$, of integers is a Sidon set. These sets are deeply connected with harmonic analysis on groups; for example, see [13, 15].

S. Tensor Algebras and Group Algebras

Let X and Y be compact Hausdorff spaces. Denote by $V(X,Y)$ the projective tensor product $C(X)\hat{\otimes}C(Y)$ of continuous function spaces $C(X)$ and $C(Y)$. The norm of $\varphi(x,y)=\sum_{j=1}^{\infty}f_j(x)g_j(y)$ is defined by $\|\varphi\|=\inf\sum_{j=1}^{\infty}\|f_j\|_\infty\|g_j\|_\infty$, where the infimum is taken for all expressions of φ. If $\hat{G}$ is an infinite compact group, then there exist two subsets K_1 and K_2 such that (i) K_1 and K_2 are homeomorphic to the †Cantor ternary set; (ii) the representation $\gamma_1+\gamma_2$ is unique, where $\gamma_1\in K_1$ and $\gamma_2\in K_2$; (iii) $K_1\cap K_2=\varnothing$; and (iv) $K_1\cup K_2$ is a Kronecker set or a set of type K_p for some p. Varopoulos's theorem states that the algebra $V(K_1,K_2)$ is isomorphic to $A(K_1+K_2)$. By this theorem the problems of spectral synthesis and operating functions of group algebras are transformed into problems of tensor algebras. For a precise discussion → [27].

References

[1] E. C. Titchmarsh, Introduction to the theory of Fourier integrals, Clarendon Press, 1937.
[2] A. Zygmund, Trigonometric series I, Cambridge Univ. Press, second edition, 1959.
[3] N. Wiener, The Fourier integrals and certain of its applications, Cambridge Univ. Press, 1933.
[4] R. E. A. C. Paley and N. Wiener, Fourier transforms in the complex domain, Amer. Math. Soc. Colloq. Publ., 1934.
[5] K. Yosida, Functional analysis, Springer, 1965.
[6] М. А. Наймарк (M. A. Naĭmark), Нормированные кольца, Гостехиздат, 1956; English translation, Normed rings, Noordhoff, 1959.
[7] Д. А. Райков (D. A. Raĭkov), Гармонический анализ на коммитативных группах с мерой Хаара характеров, Trudy Mat. Inst. Steklov., 14 (1945), 1–86.
[8] L. H. Loomis, An introduction to abstract harmonic analysis, Van Nostrand, 1953.
[9] H. Cartan and R. Godement, Théorie de la dualité et analyse harmonique dans les groupes abéliens localement compacts, Ann. Sci. Ecole Norm. Sup., (3) 64 (1947), 79–99.
[10] R. Godement, Théorèmes taubériens et théorie spectrale, Ann. Sci. Ecole Norm. Sup., (3) 64 (1947), 118–138.
[11] A. Weil, L'intégration dans les groupes topologiques et ses applications, Actualités Sci. Ind., Hermann, 1940, second edition, 1951.
[12] T. Tannaka, Sôtaigenri (Japanese; Duality principles), Iwanami, 1951.
[13] W. Rudin, Fourier analysis on groups, Interscience, 1962.
[14] E. Hewitt and K. A. Ross, Abstract harmonic analysis, Springer, I, 1963; II, 1970.
[15] J.-P. Kahane and R. Salem, Ensembles parfaits et séries trigonométriques, Actualités Sci. Ind., Hermann, 1963.
[16] L. Ehrenpreis, Fourier analysis in several complex variables, John Wiley, 1970.

[17] Y. Katznelson, An introduction to harmonic analysis, John Wiley, 1968.
[18] Y. Katznelson, Sur le calcul symbolique dans quelques algèbres de Banach, Ann. Sci. Ecole Norm. Sup., (3) 76 (1959), 83–123.
[19] H. Helson, J.-P. Kahane, Y. Katznelson, and W. Rudin, The functions which operate on Fourier transforms, Acta Math., 102 (1959), 135–157.
[20] N. Wiener and H. R. Pitt, On absolutely convergent Fourier-Stieltjes transforms, Duke Math. J., 4 (1938), 420–436.
[21] Ю. А. Шрейдер (Yu. A. Šreĭder), Строение максимальных идеалов в кольцах мер со сверткой, Mat. Sb., (N.S.) 27 (1950), 297–318; English translation, The structure of maximal ideals in rings of measures with convolution, Amer. Math. Soc. Transl., 81 (1953).
[22] J. H. Williamson, A theorem on algebras of measures on topological groups, Proc. Edinburgh Math. Soc., 11 (1958–1959), 195–206.
[23] E. Hewitt and S. Kakutani, Some multiplicative linear functionals on $M(G)$, Ann. of Math., (2) 79 (1964), 489–505.
[24] P. J. Cohen, On a conjecture of Littlewood and idempotent measures, Amer. J. Math., 82 (1960), 191–212.
[25] H. Davenport, On a theorem of P. J. Cohen, Mathematika, 7 (1960), 93–97.
[26] E. Hewitt and H. S. Zuckerman, On a theorem of P. J. Cohen and H. Davenport, Proc. Amer. Math. Soc., 14 (1963), 847–855.
[27] N. T. Varopoulos, Tensor algebras and harmonic analysis, Acta Math., 119 (1967), 51–112.
[28] T. Ito and I. Amemiya, A simple proof of the theorem of P. J. Cohen, Bull. Amer. Math. Soc., 70 (1964), 774–776.

195 (X.29) Harmonic Functions

A. General Remarks

A real-valued function u of †class C^2 defined in a domain D in the n-dimensional Euclidean space $\mathbf{R}^n$ is called **harmonic** if

$$\Delta u(P)=\frac{\partial^2 u}{\partial x_1^2}+\ldots+\frac{\partial^2 u}{\partial x_n^2}=0 \qquad (P=(x_1,\ldots,x_n))$$

in D. The set of harmonic functions in D forms a †linear space over $\mathbf{R}$. Every harmonic function is at the same time †superharmonic and †subharmonic, and vice versa.

B. Invariance of Harmonicity

Harmonicity in $\mathbf{R}^2$ is invariant under any †conformal transformation. Namely, when there exists a conformal bijection sending a domain D in the xy-plane onto a domain D' in the $\xi\eta$-plane, every harmonic function $u(x,y)$ on D is transformed to a harmonic function of (ξ,η) on D'. In $\mathbf{R}^n$ for $n\geqslant 3$, harmonicity is not generally preserved under conformal transformations. However, harmonicity is preserved in the following special case: Let D be a domain in $\mathbf{R}^n$ $(n\geqslant 3)$, and consider the **inversion** $f: D\rightarrow D'$ defined by $f(x_1,\ldots,x_n)=(x_1',\ldots,x_n')=(a^2x_1/r^2,\ldots,a^2x_n/r^2)$, $r=(x_1^2+\ldots+x_n^2)^{1/2}$. Let $u(x_1,\ldots,x_n)$ be a harmonic function on D, and let $v(x_1',\ldots,x_n')$ be the function on D' obtained by applying the **Kelvin transformation** to u. Namely, $v(x_1',\ldots,x_n')=(a/r')^{n-2}u(a^2x_1'/r'^2,\ldots,a^2x_n'/r'^2)$, where $r'^2=x_1'^2+\ldots+x_n'^2$. Then the function v is harmonic on D'. A function u that is harmonic outside a compact set is called **regular at the point at infinity** if any Kelvin transform of u is harmonic in a neighborhood of the origin, in which case $u(P)\rightarrow 0$ as $\overline{OP}\rightarrow\infty$. Now let $T: x_k=x_k(x_1',\ldots,x_n')$, $1\leqslant k\leqslant n$, be a one-to-one analytic transformation of a domain D' onto another domain D. If there exists a positive function $\varphi(x_1',\ldots,x_n')$ in D' such that $\varphi(x_1',\ldots,x_n')u(x_1(x_1',\ldots,x_n'),\ldots,x_n(x_1',\ldots,x_n'))$ is harmonic for any harmonic function $u(x_1,\ldots,x_n)$ in D, then T is conformal. A conformal transformation as it is known in differential geometry is either (i) a †similarity transformation, (ii) an inversion with respect to a sphere or a plane, or (iii) a finite combination of transformations of types (i) and (ii).

C. Examples of Harmonic Functions

(1) If u is a polynomial in $x_1,\ldots,x_n$ and harmonic in $\mathbf{R}^n$, then the terms of degree k in u form a harmonic function for each $k\geqslant 0$. A harmonic homogeneous polynomial is said to be **spherical harmonic**. (2) $\log r$ in $\mathbf{R}^2$ and r^{2-n} in $\mathbf{R}^n$ $(n\geqslant 3)$ are harmonic except at $r=0$. (3) Every †logarithmic potential in $\mathbf{R}^2$ and every †Newtonian potential in $\mathbf{R}^n$ $(n\geqslant 3)$ is harmonic outside the †support of the measure. Conversely, any harmonic function defined on a domain D is represented in an arbitrary relatively compact domain D' in D as the sum of a logarithmic $(n=2)$ or Newtonian $(n\geqslant 3)$ potential of a measure on $\partial D'$ and the potential of a †double layer. (4) Both the real part u and the imaginary part v of an analytic function are harmonic. We call v a **conjugate harmonic function** of u. If u is harmonic on a †simply connected domain D,

then a conjugate v of u is given by

$$v(x,y)=\int_{(a,b)}^{(x,y)}\left(-\frac{\partial u}{\partial y}dx+\frac{\partial u}{\partial x}dy\right)+\text{constant},$$

where (a,b) is a fixed point in D and the path of integration is contained in D. When D is a †multiply connected domain, v may take multiple values in accordance with the †homology classes of the paths of integration.

D. Green's Formulas

Let D be a bounded domain whose boundary S consists of a finite number of closed surfaces that are piecewise of class C^1. Let u and v be harmonic in D, and suppose that u_x, u_y, v_x, v_y have finite limits at every boundary point. We call D the inside of S. Let n be a normal on S situated outside S. Then the relation

$$\int_S\left(u\frac{\partial v}{\partial n}-v\frac{\partial u}{\partial n}\right)d\sigma=0 \qquad (1)$$

follows immediately from †Gauss's formula, where $d\sigma$ is the †surface element on S. In particular, when v is identically equal to 1, formula (1) gives

$$\int_S\frac{\partial u}{\partial n}d\sigma=0. \qquad (2)$$

Both (1) and (2) are called **Green's formulas**. Conversely, u is harmonic in D if u is a function of class C^2 in D and at every point $P\in D$, there is a sequence $\{r_k\}$ decreasing to zero and such that $\int_{S(P,r_k)}(\partial u/\partial n)d\sigma=0$ $(k=1,2,\ldots)$, where $S(P,r_k)$ is the spherical surface with center at P and radius r_k. Another sufficient condition for u to be harmonic is that u is continuous and, at every point $P\in D$, there is an $r_P>0$ such that $\int_{S(P,r)}(\partial u/\partial n)d\sigma=0$ for every r, $0<r<r_P$.

E. Mean Value Theorems

We assume that u is a harmonic function, D the domain of definition of u, and S the boundary of D. The mean value of u on the surface or the interior of any ball in D is equal to the value of u at the center of the ball. Namely,

$$u(P)=\frac{1}{\tau_n r^n}\int_{B(P,r)}u\,d\tau=\frac{1}{\sigma_n r^{n-1}}\int_{S(P,r)}u\,d\sigma,$$

where τ_n and σ_n are the volume and surface area of a unit ball in $\mathbf{R}^n$, respectively, $B(P,r)$ is the open ball with center at P and radius r, and $d\tau$ is the volume element. These relations are called **mean value theorems**. Conversely, if v is continuous in D and at every point $P\in D$, there is a sequence $\{r_k\}$ decreasing to zero and such that the mean value of v on $B(P,r_k)$ or $S(P,r_k)$ is equal to $v(P)$ for each k, then v is harmonic in D. This result is called **Gauss's theorem**. From the mean value theorems the **maximum principle** for harmonic functions follows: Any nonconstant u assumes neither maximum nor minimum in D. In general, a value β is called the **boundary value** at $Q\in S$ of a function defined in D if the function converges to β as the variable in D tends to Q. If both u and v are harmonic in D and have the same finite boundary value at every point on S, then $u\equiv v$ in D. This is called the **uniqueness theorem**.

F. Boundary Value Problems

The **first boundary value problem** (or **Dirichlet problem**) is the problem of the determination of a harmonic function defined on D that assumes boundary values prescribed on S (→ 124 Dirichlet Problem). The **second boundary value problem** (or **Neumann problem**) is the problem of the determination of a harmonic function u whose normal derivative $\partial u/\partial n$ is equal to a function f prescribed on S. If formula (2) is satisfied by a solution u of the Neumann problem, then $\int_S f\,d\sigma=0$. The **third boundary value problem** is the problem of the determination of a harmonic function u on D that satisfies $\partial u/\partial n=hu+f$ on S, where h and f are functions prescribed on S. All these problems may be reduced to certain †Fredholm integral equations. There is also the boundary value problem of mixed type, in which the boundary values are prescribed in a part of S and the normal derivatives are prescribed on the rest.

G. The Poisson Integral

Let D be a bounded domain with smooth boundary S and u a function harmonic in D and continuous on $D\cup S$. Let $G(P,Q)$ be Green's function in D. Then (1) yields

$$u(P)=-\frac{1}{\sigma_n}\int_S u(Q)\frac{\partial G(P,Q)}{\partial n_Q}d\sigma(Q).$$

In particular, if $D=B(0,r)$, then

$$u(P)=\frac{r^2-\overline{OP}^2}{\sigma_n r}\int_{S(0,r)}\frac{u(Q)}{\overline{PQ}^n}d\sigma(Q).$$

Conversely, given an integrable function f on $S(0,r)$, we set

$$u(P)=\frac{r^2-\overline{OP}^2}{\sigma_n r}\int_{S(0,r)}\frac{f(Q)}{\overline{PQ}^n}d\sigma(Q).$$

Then $u(p)$ is harmonic in $B(0,r)$ and converges to $f(Q)$ as P tends to any point Q on

$S(0,r)$ where f is continuous. We call u a **Poisson integral**. Sometimes it is possible to represent a harmonic function u in $B(0,r)$ in the following form, which is more general than the Poisson integral:

$$u(P)=\frac{r^2-\overline{OP}^2}{\sigma_n r}\int_{S(0,r)}\frac{1}{\overline{PQ}^n}d\alpha(Q), \qquad (3)$$

where α is a †Radon measure of general sign on $S(0,r)$. In order for u to admit such a representation, it is necessary and sufficient that $\int_{S(0,r')}|u|d\sigma$ be a bounded function of r' for $0<r'<r$, or equivalently, that the †subharmonic function $|u|$ have a †harmonic majorant. Furthermore, if α is absolutely continuous, then the Poisson integral representation of u is possible, and vice versa. A necessary and sufficient condition for the function u to admit the Poisson integral representation is the existence of a positive convex function $\varphi(t)$ on $t>0$ such that $\varphi(t)/t\to\infty$ as $t\to\infty$ and $\varphi(|u|)$ has a harmonic majorant.

When D is a general domain in which Green's function exists, every positive harmonic function $u(P)$ is represented uniquely as the integral $\int K(P,Q)d\mu(Q)$, where $K(P,Q)$ is a †Martin kernel and μ is a Radon measure on the Martin boundary B such that μ vanishes on the complement of a certain essential part of B whose points are called minimal points. A similar integral representation appears in the theory of Markov processes (→ 261 Markov Chains H). The representation $\int K(P,Q)d\mu(Q)$ is a generalization of (3). In terms of a function similar to φ, we can give a necessary and sufficient condition for $u(P)$ to be represented in the form $\int K(P,Q)f(Q)d\nu(Q)$, which corresponds to the Poisson integral representation and in which ν is determined by $1=\int K(P,Q)d\nu(Q)$. This condition is equivalent to the condition that $u(P)$ be **quasibounded**, i.e., that there exist an increasing sequence of bounded harmonic functions that converges to u [7].

H. Expansion

Let $P_0=(x_1^0,\ldots,x_n^0)$ be a point in D, and denote the distance from P_0 to S_0 by r. Then u is expanded uniquely into a power series

$$\sum a_{k_1,\ldots,k_n}(x_1-x_1^0)^{k_1}\ldots(x_n-x_n^0)^{k_n}, \qquad k_1,\ldots,k_n\geqslant 0,$$

in $B(P_0,(\sqrt{2}-1)r)$. Thus u is (real) analytic in D. If the power series is written as $\sum_k h_k$ with spherical harmonics h_k of degree $k=1,2,\ldots,$ then this series converges over all of $B(P_0,r)$.

I. Sequences of Harmonic Functions

In this section, $\{u_m\}$ is a sequence of harmonic functions in D. First, if each u_m is continuous on $D\cup S$ and $\{u_m\}$ converges uniformly on S, then $\{u_m\}$ converges uniformly in D, and the limiting function u is harmonic in D. Moreover, $\partial^{k_1+\ldots+k_n}u_m/\partial x_1^{k_1}\ldots\partial x_n^{k_n}$ converges to $\partial^{k_1+\ldots+k_n}u/\partial x_1^{k_1}\ldots\partial x_n^{k_n}$ uniformly on any compact subset of D (**Harnack's first theorem**). Second, if $u_1\leqslant u_2\leqslant\ldots$ in D and there is a point of D at which $\{u_m\}$ is bounded, then $\{u_m\}$ converges uniformly on any compact subset of D (**Harnack's second theorem**). In the proof of this theorem (and in other connections too), **Harnack's lemma** is used: If u is positive harmonic in D, P_0 is a point of D, and K is a compact subset of D, then there exist constants c and c', depending only on P_0 and K, such that $0<cu(P_0)\leqslant u(P)\leqslant c'u(P_0)$ on K.

Any family of bounded harmonic functions is †normal. Boundedness may be replaced by boundedness from above or below by Harnack's lemma. If $\int_D|u_k-u_m|^p d\tau\to 0$ as $k,m\to\infty$ for $p>1$, then †Hölder's inequality implies that $\{u_m\}$ converges uniformly on any compact subset of D. It follows that if $\int_D|\mathrm{grad}(u_k-u_m)|^p d\tau\to 0$ as $k,m\to\infty$, then there exists a harmonic function u in D such that $\int_D|\mathrm{grad}(u_m-u)|^p d\tau\to 0$ as $m\to\infty$ and $u_m(P)-u_m(P_0)$ converges to $u(P)-u(P_0)$ uniformly on any compact subset of D, where P_0 is any point in D. Finally, if $\int_D|u_n|^p d\tau$ $(p>1)$ is bounded, then $\{u_n\}$ forms a normal family.

J. Level Surfaces and Orthogonal Trajectories

The set $\{P|u(P)=\text{constant}\}$ is called a **level surface** (**niveau** or **equipotential surface**). When a is given as the constant, the level surface is called the a-level surface. Assume that u is not a constant. A point where $\mathrm{grad}\,u$ vanishes is called **critical**. The set of critical points is locally of dimension $n-2$ and consists of at most countably many †real analytic manifolds of dimension $\leqslant n-2$ ($n=\dim D$, and a manifold of dimension 0 is understood to be a point). Any compact subset of D intersects only a finite number of such manifolds; we express this fact by saying that the manifolds do not cluster in D. Each of these manifolds is contained in a certain level surface. The complement of the critical points with respect to any level surface consists of real analytic manifolds of dimension $n-1$ that do not cluster in D.

For each noncritical point there exists an analytic curve passing through it such that grad u is tangent to the curve at each point on the curve. A maximal curve with this property is called an **orthogonal trajectory** (or **line of force**). Along every orthogonal trajectory, u increases strictly in one direction and hence decreases in the other, so that no orthogonal trajectory is a closed curve. There is exactly one orthogonal trajectory passing through any noncritical point. Therefore, no two orthogonal trajectories intersect, and no orthogonal trajectory terminates at a noncritical point. Moreover, the set of limit points of any orthogonal trajectory in each direction does not contain any noncritical point. When u is a Green's function $G(P,Q)$, every orthogonal trajectory is called a **Green line**, and a Green line that issues from the pole Q and along which u decreases to 0 is called **regular**. For any sufficiently large a, the a-level surface Σ_a is an analytic closed surface homeomorphic to a spherical surface. Let E be a family of orthogonal trajectories issuing from the pole. If the intersection A of E and a closed level surface Σ_a is an $(n-1)$- dimensional measurable set, then the †harmonic measure of A at Q with respect to the interior of Σ_a is called the **Green measure** of E. M. Brelot and G. Choquet proved that all orthogonal trajectories issuing from the pole except those belonging to a family of Green measure zero are regular. Consider a domain D bounded by two compact sets, and denote by u the harmonic measure of one compact set with respect to D. Assume that u is not a constant. Then u changes from 0 to 1 along all orthogonal trajectories except those belonging to a family that is small with respect to a measure similar to the Green measure (see "flux" defined in Section K).

K. Harmonic Flows

Denote by Σ_a the a-level surface for a general u, and by Σ_a^0 the complement of the set of critical points with respect to Σ_a. Let σ be an $(n-1)$-dimensional domain in Σ_a^0 such that the $(n-2)$-dimensional boundary of σ is piecewise of class C^2. Suppose that u assumes the value b $(>a)$ on each orthogonal trajectory passing through σ. Consider the union of orthogonal trajectories that pass through σ. The subset of this union on which u assumes values between a and b forms a set called a **regular tube**. The parts of the boundary corresponding to a and b are called the lower and upper bases of the tube; accordingly, σ is the lower base. The integral $\int(\partial u/\partial n)d\sigma$ on any section (i.e., the part of a level surface in the tube) is constant and is called the **flux** of the tube. The family of orthogonal trajectories passing through an $(n-1)$-dimensional domain (not necessarily bounded by a smooth boundary) in Σ_a^0 is called a **harmonic flow**, and a subfamily is called a harmonic subflow if its intersection A with Σ_a^0 is measurable (in the $(n-1)$-dimensional sense). The flux of a harmonic subflow is defined to be $\int_A(\partial u/\partial n)d\sigma$. Then the Green measure of a family E of Green lines issuing from the pole is equal to the flux of E divided by σ_n. We can compute the exact value of the †extremal length of any harmonic subflow.

L. Isolated Singularities

Let u be harmonic in an open ball except at the center O. It is expressed as the sum of a function $h(P)$ harmonic in the entire ball and $\sum_{m=0}^{\infty}H_m(P)/\overline{OP}^{2m+1}$, where H_m is a †spherical harmonic of degree m. If $\overline{OP}^{\alpha}u(P)\to 0$ as $P\to O$ for $\alpha>0$, then $u(P)$ is equal to $h(P)+c/\overline{OP}+\ldots+H_m(P)/\overline{OP}^{2m+1}$ with $m<\alpha-1$. In particular, if u is bounded in a neighborhood of O, then O is a **removable singularity** for u. If u is bounded above (below), then $u(P)=h(P)+c/\overline{OP}$, where $c\leqslant 0$ $(c\geqslant 0)$ (for the removability of a set of capacity zero, → 174 Function-Theoretic Null Sets).

If u is harmonic near the point at infinity, i.e., outside some closed ball, then

$$u(P)=\sum_{m=0}^{\infty}\frac{H_m(P)}{\overline{OP}^{\,2m+1}}+\sum_{m=0}^{\infty}U_m(P),$$

where the first sum is regular at the point at infinity and U_m is a spherical harmonic of degree m. If $\overline{OP}^{-\alpha}u(P)\to 0$ as $\overline{OP}\to\infty$ with $\alpha\geqslant 0$, then $U_m\equiv 0$ for all $m>\alpha$. If u is bounded above or below, then $U_m\equiv 0$ for all $m\geqslant 1$. If u is harmonic in $\mathbf{R}^n$ and $\overline{OP}^{-\alpha}u(P)\to 0$ as $\overline{OP}\to\infty$ with $\alpha\geqslant 0$, then u is a polynomial of degree m $(<\alpha)$. If u is harmonic and bounded above or below in $\mathbf{R}^n$, then u is constant. Brelot called a function u harmonic at the point at infinity if

$$u(P)=\text{constant}+\sum_{m=1}^{\infty}\frac{H_m(P)}{\overline{OP}^{\,2m+1}}$$

(note that $m\geqslant 1$) near the point at infinity [1].

M. Harmonic Continuation

If u vanishes in a subdomain of D, then $u\equiv 0$ in D. If u_k is harmonic in D_k $(k=1,2)$, $D_1\cap D_2\neq\varnothing$, and $u_1\equiv u_2$ in $D_1\cap D_2$, then u_1 and u_2 define a harmonic function in $D_1\cup D_2$. If the boundaries of mutually disjoint domains D_1 and D_2 have a surface S_0 of class C^1 in common, u_k is harmonic in D_k $(k=1,2)$,

$u_1 = u_2$ on S_0, and $\partial u_1/\partial n$ and $-\partial u_2/\partial n$ exist and coincide on S_0, then u_1 and u_2 define a harmonic function in the domain $D_1 \cup S_0 \cup D_2$. We express this fact by saying that one of u_1 and u_2 is a **harmonic continuation** of the other. It follows that $u \equiv 0$ in D if the boundary of D contains a surface S_0 of class C^1 and $u = \partial u/\partial n = 0$ on S_0. Consider the case $n = 2$. If the boundary of a Jordan domain contains an analytic arc C and u or $\partial u/\partial n$ vanishes on C, then a harmonic continuation of u into a certain domain beyond C is possible. If $n = 3$, however, nothing is known except in the case where S_0 is a part of a spherical surface or a plane and $u = 0$ or $\partial u/\partial n = 0$ on S_0.

Boundary values of u do not always exist, but in some special cases, u has limits if the variable is restricted to a part of the domain (→ 397 Subharmonic Functions). We note that a positive harmonic function in a ball has a finite limit at almost every boundary point Q if the variable is restricted to any angular domain with vertex at Q.

N. Green Spaces

As a generalization of †Riemann surfaces, Brelot and Choquet introduced **$\mathfrak{S}$-spaces** [3]. It is required that $\mathfrak{S}$ be a separable connected topological space and satisfy the following two conditions: (i) At each point P there exists a neighborhood V_P of P and a homeomorphism between V_P and an open set V'_P in the †Alexandrov compactification $\mathbf{R}^n \cup \{\infty\}$; (ii) if $A = V_{P_1} \cap V_{P_2} \neq \varnothing$ and A'_k is the part of V'_{P_k} that corresponds to A $(k = 1, 2)$, then the correspondence between A'_1 and A'_2 via A is a conformal (possibly with the sense of angles reversed) transformation when $n = 2$ and an †isometric transformation that keeps ∞ invariant when $n \geqslant 3$. If a †Green's function exists on $\mathfrak{S}$, then $\mathfrak{S}$ is called a **Green space**. Harmonic functions and the †Dirichlet problem on a Green space have been discussed from various points of view.

O. Biharmonic Functions

A function v is called **polyharmonic** if $\Delta^k v = 0$ $(k \geqslant 2)$ and **biharmonic** if $\Delta\Delta v = 0$; sometimes, polyharmonic functions are also called biharmonic. A biharmonic function in a plane domain D is written as $\mathrm{Re}(\bar{z}f(z) + g(z))$, where f and g are complex analytic in D (Goursat's representation). Biharmonic functions are used in the theory of elasticity and hydrodynamics.

For an axiomatic treatment of harmonic functions → 397 Subharmonic Functions.

References

[1] M. Brelot, Sur le rôle du point à l'infini dans la théorie des fonctions harmoniques, Ann. Sci. Ecole Norm. Sup., (3) 61 (1944), 301–332.

[2] M. Brelot, Eléments de la théorie classique du potentiel, Centre de Documentation Universitaire, Paris, third edition, 1965.

[3] M. Brelot and G. Choquet, Espaces et lignes de Green, Ann. Inst. Fourier, 3 (1951), 199–263.

[4] O. D. Kellogg, Foundations of potential theory, Springer, 1929.

[5] M. Nicolescu, Les fonctions polyharmoniques, Actualités Sci. Ind., Hermann, 1936.

[6] M. Ohtsuka, Extremal length of level surfaces and orthogonal trajectories, J. Sci. Hiroshima Univ., 28 (1964), 259–270.

[7] M. Parreau, Sur les moyennes des fonctions harmoniques et analytiques et la classification des surfaces de Riemann, Ann. Inst. Fourier, 3 (1952), 103–197.

196 (VII.13)
Harmonic Integrals

A. Introduction

†De Rham's theorem shows that the cohomology group with real coefficients of a †differentiable manifold of class C^∞ is isomorphic to the cohomology group of the cochain complex of †differential forms with respect to the exterior derivative d. Thus every element of the cohomology group can be represented by a class of †closed differential forms. Harmonic forms enable us to choose one definite differential form in each cohomology class. The theory of harmonic forms, called the theory of **harmonic integrals**, is modeled after the theory of holomorphic differentials and their integrals (Abelian integrals) in function theory [1, 7].

B. Definitions

Let X be an oriented n-dimensional differentiable manifold of class C^∞ with a †Riemannian metric ds^2 of class C^∞ (→ 108 Differentiable Manifolds). For every p-form φ on X we define an $(n-p)$-form $*\varphi$ on X as follows: First denote the †volume element of X by dv. If we choose a basis $\{\omega_1, \ldots, \omega_n\}$ of the space of 1-forms on an open set U of X such that $ds^2 = \sum_i \omega_i^2$ and $dv = \omega_1 \wedge \ldots \wedge \omega_n$, then φ can be expressed on U in the form

$\varphi=(1/p!)\Sigma\varphi_{i_1,\ldots,i_p}\omega_{i_1}\wedge\ldots\wedge\omega_{i_p}$. If we let $*\varphi=(1/(n-p)!)\Sigma(*\varphi)_{j_1,\ldots,j_{n-p}}\omega_{j_1}\wedge\ldots\wedge\omega_{j_{n-p}}$, where $(*\varphi)_{j_1\ldots j_{n-p}}=(1/p!)\Sigma\delta^{1\,2\ldots n}_{i_1\ldots i_p j_1\ldots j_{n-p}}\varphi_{i_1\ldots i_p}$, then $*\varphi$ is an $(n-p)$-form on U which does not depend on the choice of $(\omega_1,\ldots,\omega_n)$ and is determined only by φ. Since X is covered by open sets as above, $*$ defines a linear mapping that transforms p-forms to $(n-p)$-forms. If we let $ds^2=\Sigma g_{jk}\,dx^j\,dx^k$ in terms of the local coordinate system $(x^1,\ldots,x^n)$ and $\varphi=(1/p!)\varphi_{i_j\ldots i_p}dx^{i_1}\wedge\ldots\wedge dx^{i_p}$, then in the notation of tensor calculus, we have

$$*\varphi=(1/(n-p)!)\times(*\varphi)_{j_1\ldots j_{n-p}}dx^{j_1}\wedge\ldots\wedge dx^{j_{n-p}},$$

$$(*\varphi)_{j_1\ldots j_{n-p}}=\sqrt{g}\cdot\delta^{1\,2\ldots n}_{k_1\ldots k_p j_1\ldots j_{n-p}}\varphi^{k_1\ldots k_p}$$

$$(g=\det(g_{ij})).$$

For two p-forms φ and ψ, we define the inner product by $(\varphi,\psi)=\int_X\varphi\wedge*\psi$ if the right-hand side converges. In order for the inner product (φ,ψ) to be defined, it suffices that either φ or ψ has compact †support. Then (φ,ψ) is a symmetric, positive definite bilinear form.

If we let $\delta=(-1)^{np+n+1}*d*$ operate on p-forms, where d is the †exterior derivative, then d and δ are **adjoint** to each other with respect to the inner product. That is, if either φ or ψ has compact support, we have $(d\varphi,\psi)=(\varphi,\delta\psi)$ (†Stokes's theorem). We call $\Delta=d\delta+\delta d$ the **Laplace-Beltrami operator**, which is a self-adjoint †elliptic differential operator. These operators satisfy relations such as $**=(-1)^{p(n-p)}$, $dd=0$, $\delta\delta=0$, $*\Delta=\Delta*$, $*\delta=(-1)^p d*$, and $\delta*=(-1)^{n-p+1}*d$ (when they operate on p-forms).

A differential form φ is said to be **harmonic** if $d\varphi=0$ and $\delta\varphi=0$. Then $\Delta\varphi=0$. Since Δ is an elliptic operator, a †weak solution φ of the equation $\Delta\varphi=\mu$ is an ordinary solution of class C^∞ on the domain where μ is of class C^∞. Therefore, if φ is harmonic (as a weak solution), φ is of class C^∞

C. Harmonic Forms on Compact Manifolds

On a compact manifold X, any φ with $\Delta\varphi=0$ is harmonic, since $(\varphi,\Delta\varphi)=(d\varphi,d\varphi)+(\delta\varphi,\delta\varphi)$. Let $L_p(X)$ be the linear space of p-forms of class C^∞ on X, and denote by $\mathfrak{L}_p(X)$ the completion of $L_p(X)$ with respect to the inner product (φ,ψ). Then $\mathfrak{L}_p(X)$ is the Hilbert space of square integrable measurable p-forms. Then $\mathfrak{H}_p(X)=\{\varphi\in\mathfrak{L}_p(X)|\Delta\varphi=0$ (in the weak sense)$\}$ is a finite-dimensional subspace of $\mathfrak{L}_p(X)$ and is contained in $L_p(X)$, as we have seen before. Also, $\mathfrak{H}_p(X)$ is closed in $\mathfrak{L}_p(X)$, and the †projection operator H: $\mathfrak{L}_p(X)\to\mathfrak{H}_p(X)$ is an †integral operator with kernel of class C^∞. The orthogonal complement of $\mathfrak{H}_p(X)$ in $\mathfrak{L}_p(X)$ is mapped onto itself by Δ and has the inverse operator G of Δ, which is a continuous operator of the Hilbert space. By letting $G=0$ on $\mathfrak{H}_p(X)$, we can extend G to an operator from $\mathfrak{L}_p(X)$ to $\mathfrak{L}_p(X)$ that is called **Green's operator** and is also denoted by G, maps L_p into itself, commutes with d and δ, and satisfies $GH=HG=0$, $H+\Delta G=1$ (=identity mapping). Therefore, for $\varphi\in L_p(X)$ we have $\varphi=H\varphi+G\delta d\varphi+d\delta G\varphi$, which shows that H is †homotopic to the identity mapping of the †cochain complex $(\Sigma_p L_p(X),d)$. From this we infer that every cohomology class of de Rham cohomology contains a unique harmonic form that represents the cohomology class. However, since products of harmonic forms are not always harmonic, it is not appropriate to use harmonic forms to study the ring structure of cohomology. G is also an integral operator with kernel of class C^∞ outside the diagonal subset in $X\times X$.

D. Harmonic Forms on Noncompact Manifolds

If X is a noncompact manifold, let $L_p(X)$ be the space of p-forms of class C^∞ with compact support, and let $\mathfrak{L}_p(X)$ be its completion. Let $\mathfrak{B}_p(X)$ and $\mathfrak{B}^*_p(X)$ be the respective closures of $dL_{p-1}(X)$ and $\delta L_{p+1}(X)$ in $\mathfrak{L}_p(X)$, and let $\mathfrak{Z}_p(X)$ and $\mathfrak{Z}^*_p(X)$ be the respective orthogonal complements of $\mathfrak{B}_p(X)$ and $\mathfrak{B}^*_p(X)$ in $\mathfrak{L}_p(X)$. Then $\mathfrak{Z}_p(X)\cap\mathfrak{Z}^*_p(X)=\mathfrak{H}_p(X)$ is a subspace of the square integrable harmonic forms, and we have the direct sum decomposition $\mathfrak{L}_p(X)=\mathfrak{B}_p(X)+\mathfrak{B}^*_p(X)+\mathfrak{H}_p(X)$. In this decomposition any component of a form of class C^∞ is also of class C^∞.

If X is an open submanifold of another manifold Y, $\bar{X}$ is compact, and $\partial X=\bar{X}-X$ is a closed submanifold of Y, then the theory in this section is generalized potential theory with boundary condition $\varphi=0$ on ∂X. We sometimes treat decompositions of other Hilbert spaces that correspond to other boundary conditions.

E. Generalization to Complex Manifolds

If X is a complex manifold, we consider complex-valued differential forms (→ 74 Complex Manifolds). Then the space $L_p(X)$ of p-forms is the direct sum of the spaces $L_{r,s}(X)$ of forms of type (r,s), and the exterior derivative d has the expression $d=d'+d''$, where d' is of type $(1,0)$ (i.e., $L_{r,s}(X)\to L_{r+1,s}(X)$) and d'' is of type $(0,1)$. If we are given a holomorphic vector bundle E on X, we can define an operator d'' on differential forms with values in E, and we have the generalized †Dolbeault theorem. If X is compact and has a †Hermitian metric, we can define a Hermitian inner product on E as follows: There is an open covering $\{U_j\}$ such that over each U_j the vector bundle E is isomorphic to $U_j \times \mathbf{C}^q$. A point of E over U_j is represented by (x, ξ_j), where $x \in U_j$ and $\xi_j \in \mathbf{C}^q$. For $x \in U_j \cap U_k$ we have $(x, \xi_j) = (x, \xi_k)$ (the sides are the respective expressions over U_j and U_k) if and only if $\xi_j = g_{jk}(x)\xi_k$, where $g_{jk}(x)$ is a holomorphic mapping from $U_j \cap U_k$ to $GL(g, \mathbf{C})$ satisfying $g_{jk} g_{kl} = g_{jl}$ on $U_j \cap U_k \cap U_l$. A differential form φ with values in E is expressed as a family $\{\varphi_j\}$ of differential forms on U_j with values in $\mathbf{C}^q$ such that $\varphi_j(x) = g_{jk}(x)\varphi_k(x)$ on $U_j \cap U_k$. If we take a positive definite Hermitian matrix h_j whose components are C^∞-functions on U_j such that ${}^t g_{jk} h_j \bar{g}_{jk} = h_k$ on $U_j \cap U_k$, then $\{{}^t\xi_j h_j \bar{\xi}_j\}$ determines a Hermitian inner product on each fiber of E. We can also endow the space $L_{r,s}(E, X)$ (of forms of type (r,s) of class C^∞ with values in E) with a Hermitian inner product by setting $(\varphi, \psi) = \int_X \sum_{\alpha,\beta} h_{j\alpha\beta} \varphi_j^\alpha \wedge \overline{* \psi_j^\beta}$ for $\varphi, \psi \in L_{r,s}(E, X)$ (where the φ_j^α $(\alpha = 1, \ldots, q)$ are the components of φ_j). If we denote by $\mathfrak{d}$ the adjoint operator of d'' with respect to this inner product and let $A = d''\mathfrak{d} + \mathfrak{d}d''$, then A is a self-adjoint elliptic differential operator, and results similar to those for Δ mentioned above hold for A. For example, the space $\mathfrak{H}_{r,s}(E, X)$ of harmonic forms of type (r,s) is of finite dimension, and there is a continuous linear operator G on $\mathfrak{L}_{r,s}(E, X)$, the completion of $L_{r,s}(E, X)$, which satisfies $1 = H + AG$, $HG = GH = 0$, $d''G = Gd''$, and $\mathfrak{d}G = G\mathfrak{d}$. Here H denotes the projection $\mathfrak{L} \to \mathfrak{H}$, which is an integral operator with kernel of class C^∞. Also, G maps $L_{r,s}(E, X)$ into itself. Therefore, H is homotopic to the identity on the cochain complex $(\sum_s L_{r,s}(E, X), d'')$, and any element of †Dolbeault's cohomology groups (d''-cohomology groups) is represented by a unique harmonic form (→ 232 Kähler Manifolds).

F. Other Generalizations

Even if a manifold X is not of class C^∞, if X is a manifold of class C_1^1, we can develop the theory of harmonic forms [6]. We say that X is of class C_1^1 if it is of class C^1 and has a set of local coordinate systems whose transition functions have derivatives satisfying the †Lipschitz condition.

If X is a real analytic manifold with a real analytic Riemannian metric, then harmonic forms are also real analytic. Using this fact, we can real analytically embed a compact manifold with a real analytic Riemannian metric into a Euclidean space (P. Bidal and G. de Rham; this result is now included in the theorems of C. B. Morrey and H. Grauert).

We can consider the theory of harmonic forms with singularities [5, 8], a generalization of the theory of differential forms of the second and third kinds. Here the notion of †current is very useful.

G. Cohomology Vanishing Theorems

Since the operator Δ is closely related to the Riemannian metric, some metrics may admit no harmonic forms of certain degrees except zero. This is important since it means that the corresponding cohomology group of the manifold vanishes. The condition for this phenomenon to occur can be described in terms of the curvature of the metric. This study has its origin in S. Bochner's results [9].

Here is an example of a **vanishing theorem**: Let B be a holomorphic line bundle on a compact complex manifold X of dimension n. If the †Chern class of B is expressed by a real closed differential form of type (1, 1) as $\omega = \sqrt{-1} \sum a_{\alpha\beta} dz^\alpha \wedge d\bar{z}^\beta$, where the Hermitian matrix $(a_{\alpha\beta})$ is positive definite at every point of X, then $H^q(X, \Omega^p(B)) = 0$ for $p + q > n$. In this case, $ds^2 = 2\sum a_{\alpha\beta} dz^\alpha \wedge d\bar{z}^\beta$ is a †Hodge metric on X.

References

[1] Y. Akizuki, Tyôwa sekibunron (Japanese; Harmonic integrals), Iwanami, I, 1955; II, 1956.

[2] W. L. Bailey, The decomposition theorem for V-manifolds, Amer. J. Math., 78 (1956), 862–888.

[3] S. I. Goldberg, Curvature and homology, Academic Press, 1962.

[4] W. V. D. Hodge, The theory and applications of harmonic integrals, Cambridge Univ. Press, second edition, 1952.

[5] K. Kodaira, Harmonic fields in Riemannian manifolds (generalized potential theory), Ann. of Math., (2) 50 (1949), 587–665.

[6] C. B. Morrey and J. Eells, A variational method in the theory of harmonic integrals I, Ann. of Math., (2) 63 (1956), 91–128.

[7] G. de Rham, Variétés différentiables, Actualités Sci. Ind., Hermann, 1955.

[8] G. de Rham and K. Kodaira, Harmonic integrals, Lecture notes, Institute for Advanced Study, Princeton, 1950.

[9] K. Yano and S. Bochner, Curvature and Betti numbers, Ann. Math. Studies, Princeton Univ. Press, 1953.

197 (I.14) Hierarchies

The theory of †Borel sets (or †Baire functions) and †projective sets may be considered as an example of **hierarchy theory**. In particular, one has the notion of "classes" of such sets, and as the class gets higher their structure becomes essentially more complicated. This theory, also called **descriptive set theory**, has been investigated mostly by French empiricists (→ 164 Foundations of Mathematics). Utilizing the theory of †recursive functions, S. C. Kleene succeeded in establishing a theory of hierarchies that essentially contains classical descriptive set theory as an extreme case [4, 6, 7, 9]. Although research following a similar line had also been done by M. Davis, A. Mostovskiĭ, and others, it was Kleene who succeeded in bringing the theory to its present form. (For notations and terminology used in the following discussion → 352 Recursive Functions.)

Sets or functions are described by †predicates, which we classify as follows. Let $a, b, \ldots, a_1, a_2, \ldots, x, y, \ldots$ be variables ranging over the set $\mathbf{N}$ of natural numbers, and $\alpha, \beta, \ldots, \alpha_1, \alpha_2, \ldots, \xi, \eta, \ldots$ be variables ranging over the set $\mathbf{N}^{\mathbf{N}}$ of all †number-theoretic functions with one argument. Let $\psi_1, \ldots, \psi_l$ $(l \geqslant 0)$ be number-theoretic functions. A predicate $P(\alpha_1, \ldots, \alpha_m, a_1, \ldots, a_n)$ $(m, n \geqslant 0, m+n>0)$ with variables of two †types is called **analytic** in $\psi_1, \ldots, \psi_l$ $(l \geqslant 0)$ if it is expressible syntactically by applying a finite number of logical symbols: $\rightarrow, \vee, \wedge, \neg, \exists x, \forall x, \exists \xi, \forall \xi$, to †general recursive predicates in $\psi_1, \ldots, \psi_l$. In particular, when P is expressible without function quantifiers $\exists \xi, \forall \xi$, it is called **arithmetical** in $\psi_1, \ldots, \psi_l$ $(l \geqslant 0)$. When $l=0$, they are called simply analytic and arithmetical, respectively.

For brevity, consider the case $l=0$, and denote by $\mathfrak{a}$ a finite list of variables $(\alpha_1, \ldots, \alpha_m, a_1, \ldots, a_n)$. Every arithmetical predicate $P(\mathfrak{a})$ is expressible in a form contained in the following table (a):

(a) $R(\mathfrak{a})$;

$\exists x R(\mathfrak{a}, x), \quad \forall x \exists y R(\mathfrak{a}, x, y), \ldots,$

$\forall x R(\mathfrak{a}, x), \quad \exists x \forall y R(\mathfrak{a}, x, y), \ldots,$

where each R is †general recursive. In order to obtain such an expression we first transform the given predicate into its †prenex normal form and then contract successive quantifiers of the same kind by the formula

$$\exists x_1 \ldots \exists x_n A(x_1, \ldots, x_n) \Leftrightarrow \exists x A((x)_0, \ldots, (x)_{n-1}) \qquad (1)$$

and its "dual form." Each form in (a) (or the class of all predicates with that form) is denoted by Σ_k^0 or Π_k^0, where the suffix k refers to the number of quantifiers prefixed, and Σ or Π shows that the outermost quantifier is existential or universal, respectively. A predicate that is expressible in both forms Σ_k^0 and Π_k^0 (or the class of such predicates) is denoted by Δ_k^0. A predicate belongs to Δ_1^0 if and only if it is general recursive (an analog of †Souslin's theorem).

For $k \geqslant 1$, there exists in Σ_k^0 (or Π_k^0) an **enumerating predicate** that specifies every predicate in Σ_k^0 (Π_k^0). For example, for Π_2^0 and $m=n=1$, there is a †primitive recursive predicate $S(\alpha, z, a, x, y)$ such that, given a general recursive predicate $R(\alpha, a, x, y)$, we have a natural number e such that

$$\forall x \exists y R(\alpha, a, x, y) \Leftrightarrow \forall x \exists y S(\alpha, e, a, x, y)$$

(**enumeration theorem**). In this theorem, we can take $T_2^\alpha(z, a, x, y)$ (→ 352 Recursive Functions F) as $S(\alpha, z, a, x, y)$. For each $k \geqslant 0$, there exists a Σ_{k+1}^0 (Π_{k+1}^0) predicate that is not expressible in its dual form Π_{k+1}^0 (Σ_{k+1}^0) (hence, of course, in neither Σ_k^0 nor Π_k^0) (**hierarchy theorem**). Therefore, table (a) gives the classification of the arithmetical predicates in a hierarchy. This hierarchy is called the **arithmetical hierarchy**. For each $k \geqslant 1$, there exists a **complete** predicate with respect to Σ_k^0 (Π_k^0), that is, a Σ_k^0 (Π_k^0) predicate with only one variable such that any Σ_k^0 (Π_k^0) predicate is expressible by substituting a suitable general (or more strictly, primitive) recursive function for its variable (**theorem on complete form**). When $m=0$, all the general recursive predicates in Σ_k^0 exhaust Δ_{k+1}^0 (**Post's theorem**).

Concerning the function quantifiers, we have

$$\exists \alpha_1 \ldots \exists \alpha_m A(\alpha_1, \ldots, \alpha_m) \Leftrightarrow \exists \alpha A(\lambda t(\alpha(t))_0, \ldots, \lambda t(\alpha(t))_{m-1}) \qquad (2)$$

$$\exists x A(x) \Leftrightarrow \exists \alpha A(\alpha(0)), \qquad (3)$$

$$\forall x \exists \alpha A(x, \alpha) \Leftrightarrow \exists \alpha \forall x A(x, \lambda t \alpha(2^x \cdot 3^t)), \qquad (4)$$

and their dual forms. For any general recursive predicate R, there is a primitive recursive predicate S such that

$$\exists \alpha R(\alpha, \mathfrak{a}) \Leftrightarrow \exists \alpha \exists x S(2(x), \mathfrak{a}) \Leftrightarrow \exists y S(y, \mathfrak{a}) \qquad (5)$$

and its dual hold. Using these facts, we can classify the forms of all analytic predicates by the table (b):

(b) $A(\mathfrak{a})$;

$\forall \alpha \exists x R(\mathfrak{a}, \alpha, x), \quad \exists \alpha \forall \beta \exists x R(\mathfrak{a}, \alpha, \beta, x), \ldots,$

$\exists \alpha \forall x R(\mathfrak{a}, \alpha, x), \quad \forall \alpha \exists \beta \forall x R(\mathfrak{a}, \alpha, \beta, x), \ldots,$

where A is arithmetical and each R is general recursive. Similarly, denote by Σ_k^1, Π_k^1 each form of predicate in (b) (or the class of all predicates reducible to that form), where k is the number of function quantifiers prefixed; also, denote by Δ_k^1 the (class of) predicates expressible in both forms Σ_k^1 and Π_k^1. For Σ_k^1, Π_k^1 ($k \geqslant 1$), we have the enumeration theorem, the hierarchy theorem, and the theorem on complete form. The hierarchy given by table (b) is called the **analytic hierarchy**.

For $l>0$ (namely, when predicates are arithmetical or analytic in $\psi_1, \ldots, \psi_l$), we can †uniformly †relativize the above results with respect to $\psi_1, \ldots, \psi_l$. Now, let $\{\Sigma_k^{r,\psi_1,\ldots,\psi_l}, \Pi_k^{r,\psi_1,\ldots,\psi_l}\}_k$ ($r=0,1$) be the corresponding hierarchy relative to $\psi_1, \ldots, \psi_l$. Given a set C ($\subset \mathbf{N}^\mathbf{N}$) of functions with one argument, we can consider hierarchies of predicates which are arithmetical or analytic in a finite number of functions in C. Such a hierarchy is called a **C-arithmetical** or **C-analytic hierarchy** and denoted by $\{\Sigma_k^0[C], \Pi_k^0[C]\}_k$ or $\{\Sigma_k^1[C], \Pi_k^1[C]\}_k$, respectively. That is, when we regard $\Sigma_k^r[C]$ as a class of predicates (or sets) P, it is the family $\{P \mid P \in \Sigma_k^{r,\xi_1,\ldots,\xi_l}, \xi_1, \ldots, \xi_l \in C, l=0,1,2,\ldots\}$. These notations have been given by J. W. Addison [1, 2]. The $\mathbf{N}^\mathbf{N}$-arithmetical hierarchy and the $\mathbf{N}^\mathbf{N}$-analytic hierarchy for sets correspond respectively to the finite Borel hierarchy and the projective hierarchy in the †space of irrational numbers. Addison called the theory of those hierarchies **classical descriptive set theory**, and in contrast to this, the theory of arithmetical and analytic hierarchies for sets ($C=\varnothing$) **effective descriptive set theory** [1].

We now restrict our consideration to predicates for natural numbers (i.e., to the case $m=0$). Define the predicates L_k by $L_0(a) \Leftrightarrow a=a$, $L_{k+1}(a) \Leftrightarrow \exists x T_1^{L_k}(a,a,x)$. For each $k \geqslant 0$, $L_{k+1}(a)$ is a Σ_{k+1}^0 predicate which is of the highest †degree of recursive unsolvability among the Σ_{k+1}^0 predicates, and its degree is properly higher than that of $L_k(a)$. Thus L_k, $k=0,1,2,\ldots$, determine the **arithmetical hierarchy of degrees of recursive unsolvability**. Kleene has extended the series of L_k by using the system S_3 (→ 83 Constructive Ordinal Numbers) of notations for the constructive ordinal numbers as follows [6]: $H_1(a) \Leftrightarrow a=a$; for $y \in O$, $H_{2^y}(a) \Leftrightarrow \exists x T_1^{H_y}(a,a,x)$; for $3.5^y \in O$, $H_{3.5^y}(a) \Leftrightarrow H_{y_{(a)_1}}((a)_0)$, where $y_n = \{y\}(n_o)$. This H_y is defined for each $y \in O$, and it is of a properly higher degree than that of H_z when $z <_O y$. If $|y|=|z|$ ($|y|$ is the ordinal number represented by y), then H_y and H_z are of the same degree (C. Spector [11]). Thus a hierarchy of degrees is uniquely determined by constructive ordinal numbers. This hierarchy is called the **hyperarithmetical hierarchy of degrees of recursive unsolvability**. A function or predicate is said to be **hyperarithmetical** if it is recursive in H_y for some $y \in O$. These concepts and the results mentioned below can be relativized to any given functions or predicates.

A necessary (Kleene [6]) and sufficient (Kleene [7]) condition for a predicate to be hyperarithmetical is that it is expressible in both one-function quantifier forms Δ_1^1 (an effective version of Souslin's theorem). Denote by Hyp the set ($\subset \mathbf{N}^\mathbf{N}$) of all hyperarithmetical functions α. For an arithmetical predicate $A(\alpha,a)$, $\exists \alpha_{\alpha \in \mathrm{Hyp}} A(\alpha,a)$ is always a Π_1^1 predicate (Kleene [8]). Conversely, for any Π_1^1 predicate P, there is a general recursive predicate R such that $P(a) \Leftrightarrow \exists \alpha_{\alpha \in \mathrm{Hyp}} \forall x R(a,\alpha,x)$ (Spector [12]). As to †uniformization, for a Π_1^1 predicate $P(a,b)$, we have $\forall x \exists y P(x,y) \Rightarrow \exists \alpha_{\alpha \in \mathrm{Hyp}} \forall x P(x,\alpha(x))$ (G. Kreisel, 1962). Let $\mathbf{E}$ be an object of type 2 defined by: $\mathbf{E}(\alpha)=0$ if $\exists x(\alpha(x)=0)$, otherwise $\mathbf{E}(\alpha)=1$. A function $\varphi(a_1, \ldots, a_n)$ is hyperarithmetical if and only if it is general recursive in $\mathbf{E}$ (Kleene [9]). A predicate that is hyperarithmetical relative to Π_k^1 predicates ($k \geqslant 0$) is of Δ_{k+1}^1 (Kleene [7]), but the converse does not hold in general (Addison and Kleene, 1957).

Kleene extended his theory of hierarchy to the case of predicates of variables of any type by utilizing the theory of general recursive functions with variables of finite types $0,1,2,\ldots$ [9]. Let $\mathfrak{a}^t$ be a list of variables of types $\leqslant t$. We say a predicate $P(\mathfrak{a}^t)$ is of order r in completely defined functions $\psi_1, \ldots, \psi_l$ ($l \geqslant 0$) (for brevity, denote them by Ψ) if P is syntactically expressible in terms of variables of finite types, predicates that are general recursive in Ψ, and symbols of the †predicate calculus with quantification consisting only of variables of types $<r$. The predicates of order 0 in Ψ are exactly the general recursive ones in Ψ. When $t \geqslant 1$, and Ψ are functions of variables of type 0, a predicate $P(\mathfrak{a}^t)$ is of order 1 (of order 2) in Ψ if and only if P is arithmetical (analytic).

We have theorems similar to (2)–(4) and the following theorem and its dual for $r \geqslant 2$: For any given general recursive predicate $P(\mathfrak{a}^r, \sigma^r, \xi^{r-2})$, there is a primitive recursive predicate $R(\mathfrak{a}^r, \eta^{r-1}, \xi^{r-2})$ such that

$$\exists \sigma^r \forall \xi^{r-2} P(\mathfrak{a}^r, \sigma^r, \xi^{r-2}) \Leftrightarrow \exists \eta^{r-1} \forall \xi^{r-2} R(\mathfrak{a}^r, \eta^{r-1}, \xi^{r-2}). \qquad (6)$$

Using these equivalences, each predicate $P(\mathfrak{a}^t)$ of order $r+1$ ($r \geqslant 0$) is expressible in

one of the following forms:
(c) $B(\mathfrak{a})$;

$\forall \alpha^r \exists \xi^{r-1} R(\mathfrak{a}, \alpha^r, \xi^{r-1})$,

$\exists \alpha^r \forall \beta^r \exists \xi^{r-1} R(\mathfrak{a}, \alpha^r, \beta^r, \xi^{r-1}), \ldots,$

$\exists \alpha^r \forall \xi^{r-1} R(\mathfrak{a}, \alpha^r, \xi^{r-1})$,

$\forall \alpha^r \exists \beta^r \forall \xi^{r-1} R(\mathfrak{a}, \alpha^r, \beta^r, \xi^{r-1}), \ldots,$

where B is of order r and each R is general recursive. When $t = r+1$, table (c) gives the classification of the predicates of order $r+1$ into the hierarchy. In fact, for the predicates $P(\mathfrak{a}^{r+1})$ in each form, we have the enumeration theorem, the hierarchy theorem, and the theorem on complete form (Kleene [9]). D. A. Clarke [3] has published a detailed review on the general theory of hierarchies.

References

[1] J. W. Addison, Separation principles in the hierarchies of classical and effective descriptive set theory, Fund. Math., 46 (1958–1959), 123–135.
[2] J. W. Addison, Some consequences of the axiom of constructibility, Fund. Math., 46 (1958–1959), 337–357.
[3] D. A. Clarke, Hierarchies of predicates of finite types, Mem. Amer. Math. Soc., 1964.
[4] S. C. Kleene, Recursive predicates and quantifiers, Trans. Amer. Math. Soc., 53 (1943), 41–73.
[5] S. C. Kleene, Introduction to metamathematics, Van Nostrand, 1952.
[6] S. C. Kleene, Arithmetical predicates and function quantifiers, Trans. Amer. Math. Soc., 79 (1955), 312–340.
[7] S. C. Kleene, Hierarchies of number-theoretic predicates, Bull. Amer. Math. Soc., 61 (1955), 193–213.
[8] S. C. Kleene, Quantification of number-theoretic functions, Compositio Math., 14 (1959), 23–40.
[9] S. C. Kleene, Recursive functionals and quantifiers of finite types I, Trans. Amer. Math. Soc., 91 (1959), 1–52.
[10] S. C. Kleene, Recursive functionals and quantifiers of finite types II, Trans. Amer. Math. Soc., 108 (1963), 106–142.
[11] C. Spector, Recursive well-orderings, J. Symbolic Logic, 20 (1955), 151–163.
[12] C. Spector, Hyperarithmetical quantifiers, Fund. Math., 48 (1959–1960), 313–320.

198 (XX.26) Hilbert, David

David Hilbert (January 23, 1862–February 14, 1943) was born in Königsberg, Germany. He attended the University of Königsberg from 1882 to 1885, when he received his doctoral degree with a thesis on the theory of invariants. It was there that he established a lifelong friendship with H. Minkowski. In 1892 he became a professor at the University, and in 1895 he was appointed to a professorship at the University of Göttingen, a position he held until his death. He obtained his basic theorem on invariants between 1890 and 1893, and next began research on the †foundations of geometry and the theory of †algebraic number fields. Concerning the former, he published *Grundlagen der Geometrie* (first edition, 1899), in which he gave the complete axioms of Euclidean geometry and a logical examination of them. Concerning the latter, he systematized all the important known results of algebraic number theory in his monumental *Zahlbericht* (1897). In number theory, he enunciated his significant conjecture on †class field theory. At the international congress of mathematicians held in Paris in 1900, he put forth 23 problems as targets for mathematics of the 20th century (Table 1). Between 1904 and 1906 he conducted research on the †Dirichlet principle of †potential theory and on the direct method in the †calculus of variations. Around 1909 he established the foundations of the theory of †Hilbert spaces. After 1910 he was chiefly involved in research on the †foundations of mathematics, and he advocated the standpoint of †formalism. He is one of the greatest mathematicians of the first half of the 20th century.

Table 1. The 23 Problems of Hilbert

(1) To prove the continuum hypothesis (→ 35 Axiomatic Set Theory D).

(2) To investigate the consistency of the axioms of arithmetic (→ 164 Foundations of Mathematics E).

(3) To show that it is impossible to prove the following fact utilizing only congruence axioms: Two tetrahedrons having the same altitude and base area have the same volume. Solved (M. Dehn, 1900).

(4) To investigate geometries in which the line segment between any pair of points gives the shortest path between the pair (→ 163 Foundations of Geometry).

(5) To obtain the conditions under which a topological group has the structure of a Lie group (→ 406 Topological Groups M). Solved (A. M. Gleason; D. Montgomery and L. Zippin, 1952; H. Yamabe, 1953).

(6) To axiomatize those physical sciences in which mathematics plays an important role.

(7) To establish the transcendence of certain numbers (→ 414 Transcendental Numbers B). Solved (the transcendence of $2^{\sqrt{2}}$, which was one of the numbers put forth by Hilbert, was shown by A. Gel'fond, 1934, T. Schneider, 1935).

(8) To investigate problems concerning the distribution of prime numbers; in particular, to show the correctness of the Riemann hypothesis (→ 436 Zeta Functions). Unsolved.

(9) To establish a general law of reciprocity (→ 62 Class Field Theory A). Solved (T. Takagi, 1921, E. Artin, 1927).

(10) To establish effective methods to determine the solvability of Diophantine equations (→ 100 Decision Problem; 187 Geometry of Numbers). Solved affirmatively for equations of two unknowns (A. Baker, *Philos. Trans. Roy. Soc. London*, (A) 263, 1968); solved negatively for the general case (Ju. V. Matijasevič, 1970).

(11) To investigate the theory of quadratic forms over an arbitrary algebraic number field of finite degree (→ 344 Quadratic Forms).

(12) To construct class fields of algebraic number fields (→ 75 Complex Multiplication).

(13) To show the impossibility of the solution of the general algebraic equation of the seventh degree by compositions of continuous functions of two variables. Solved negatively. In general, V. I. Arnold proved that every real, continuous function $f(x_1,x_2,x_3)$ on $[0,1]$ can be represented in the form $\sum_{i=1}^{9} h_i(g_i(x_1,x_2),x_3)$, where h_i and g_i are real, continuous functions, and A. N. Kolmogorov proved that $f(x_1,x_2,x_3)$ can be represented in the form $\sum_{i=1}^{7} h_i(g_{i1}(x_1)+g_{i2}(x_2)+g_{i3}(x_3))$ where h_i and g_{ij} are real, continuous functions and g_{ij} can be chosen once for all independently of f (*Dokl. Akad. Nauk SSSR*, 114 (1957), *Amer. Math. Soc. Transl.*, 28 (1963)).

(14) Let k be a field, $x_1,\ldots,x_n$ be variables, and $f_i(x_1,\ldots,x_n)$ be given polynomials in $k[x_1,\ldots,x_n]$ $(i=1,\ldots,m)$. Furthermore, let R be the ring formed by rational functions $F(X_1,\ldots,X_m)$ in $k(X_1,\ldots,X_m)$ such that $F(f_1,\ldots,f_m)\in k[x_1,\ldots,x_n]$. The problem is to determine whether the ring R has a finite set of generators. Solved negatively (M. Nagata, *Amer. J. Math.*, 81 (1959)).

(15) To establish the foundations of algebraic geometry (→ 14 Algebraic Geometry). Solved (B. L. van der Waerden, 1938–1940, A. Weil, 1950, and others).

(16) To conduct topological studies of algebraic curves and surfaces.

(17) Let $f(x_1,\ldots,x_n)$ be a rational function with real coefficients that takes a positive value for any real n-tuple $(x_1,\ldots,x_n)$. The problem is to determine whether the function f can be written as the sum of squares of rational functions (→ 157 Fields O). Solved in the affirmative (E. Artin, 1927).

(18) To express Euclidean n-space as a disjoint union $\bigcup_\lambda P_\lambda$, where each P_λ is congruent to one of given polyhedra.

(19) To determine whether the solutions of regular problems in the calculus of variations are necessarily analytic (→ 318 Partial Differential Equations of Elliptic Type). Solved (S. Bernstein, I. G. Petrovskiĭ, and others).

(20) To investigate the general boundary value problem (→ 318 Partial Differential Equations of Elliptic Type; 124 Dirichlet Problem).

(21) To show that there always exists a linear differential equation of the Fuchsian class with given singular points and monodromic group (→ 253 Linear Ordinary Differential Equations (Global Theory)). Solved (H. Röhrl and others, 1957).

(22) To uniformize complex analytic functions by means of automorphic functions (→ 362 Riemann Surfaces). Solved for the case of one variable (P. Koebe, 1907).

(23) To develop the methodology of the calculus of variations (→ 47 Calculus of Variations).

References

[1] D. Hilbert, Gesammelte Abhandlungen I–III, Springer, 1932–1935 (Chelsea, 1967).
[2] D. Hilbert, Grundlagen der Geometrie, Teubner, seventh edition, 1930.
[3] D. Hilbert, Grundzüge der allgemeinen Theorie der linearen Integralgleichungen, Teubner, second edition, 1924 (Chelsea, 1953).
[4] D. Hilbert and W. Ackermann, Grundzüge der theoretischen Logik, Springer, third edition, 1949; English translation, Principles of mathematical logic, Chelsea, 1950.
[5] D. Hilbert and P. Bernays, Grundlagen der Mathematik, Springer, second edition, I, 1968; II, 1970.
[6] F. Klein, Vorlesungen über die Entwicklung der Mathematik im 19. Jahrhundert I, Springer, 1926 (Chelsea, 1956).
[7] H. Weyl, David Hilbert and his mathematical work, Bull. Amer. Math. Soc., 50 (1944), 612–654.
[8] C. Reid, Hilbert. With an appreciation of Hilbert's mathematical work by H. Weyl, Springer, 1970.

199 (XII.2) Hilbert Spaces

A. General Remarks

The theory of Hilbert spaces arose from problems in the theory of †integral equations. D. Hilbert noticed that a linear integral equation can be transformed into an infinite system of linear equations involving the †Fourier coefficients of the unknown function. He considered the linear space l_2 consisting of all infinite sequences of numbers $\{x_n\}$ for which $\sum_{n=1}^{\infty}|x_n|^2$ is finite, and for each pair of elements $x=\{x_n\}, y=\{y_n\}\in l_2$ defined their inner product as $(x,y)=\sum_{n=1}^{\infty}x_n\bar{y}_n$. The space l_2 may be regarded as an infinite-dimensional extension of the notion of a Euclidean space. F. Riesz considered the space of functions now termed L_2-space.

B. Definition

Abstracting from Euclidean, l_2, and L_2-spaces, J. Von Neumann gave the following definition of a Hilbert space: Let K be the field of complex or real numbers, the elements of which we denote by $\alpha, \beta, \ldots$. Let H be a †linear space over K, and to any pair of elements $x, y\in H$ let there correspond a number $(x,y)\in K$ satisfying the following five conditions: (i) $(x_1+x_2,y)=(x_1,y)+(x_2,y)$; (ii) $(\alpha x,y)=\alpha(x,y)$; (iii) $(x,y)=\overline{(y,x)}$; (iv) $(x,x)\geqslant 0$; and (v) $(x,x)=0 \Leftrightarrow x=0$. Then we call H a **pre-Hilbert space** and (x,y) the **inner product** of x and y.

With norm $\|x\|=\sqrt{(x,x)}$, H is a †normed linear space. If H is †complete with respect to the distance $\|x-y\|$ (i.e., $\|x_n-x_m\|\to 0$ implies the existence of $\lim x_n=x$), then we call H a **Hilbert space**. According as K is complex or real, we call H a **complex** or **real Hilbert space**. A Hilbert space is a †Banach space.

A normed linear space with norm $\|x\|$ can be made a Hilbert space, by defining in it an inner product (x,y) such that $\|x\|=\sqrt{(x,x)}$, if and only if for any x,y the equality $\|x+y\|^2+\|x-y\|^2=2(\|x\|^2+\|y\|^2)$ holds.

C. Orthonormal Sets

Two elements $x, y\in H$ are said to be mutually **orthogonal** if $(x,y)=0$. A subset Σ of H is called an **orthogonal set (system)** if $0\notin H$ and every distinct pair $x, y\in\Sigma$ is mutually orthogonal. If every element of an orthogonal set Σ is of norm 1, then Σ is called an **orthonormal set**. Any orthogonal set $\Sigma=\{x_i\}$ can be **normalized**; that is, from Σ we can make an orthonormal set $\{x_i/\|x_i\|\}$. A maximal orthonormal set is called a **complete orthonormal set**. All the complete orthonormal sets of a given H have the same cardinal number, called the **dimension** of H. Two Hilbert spaces of the same dimension are isomorphic.

If $\Sigma=\{x_i\}$ is a given orthonormal set, then for every $x\in H$, its **Fourier coefficients** (x,x_i) vanish for all but a countable number of i, and the **Bessel inequality** $\|x\|^2\geqslant\sum_i|(x,x_i)|^2$ holds. The following three statements are equivalent: (i) $\Sigma=\{x_i\}$ is complete; (ii) for every x **Parseval's equality** $\|x\|^2=\sum_i|(x,x_i)|^2$ holds; (iii) every x can be expanded in a **Fourier series** $x=\sum_i(x,x_i)x_i$.

D. Examples of Hilbert Spaces

The space l_2 (→ Section A) is a Hilbert space of dimension $\aleph_0$. The †function space L_2 on a measure space (X,μ) is a Hilbert space if the inner product of $f, g\in L_2$ is defined by $(f,g)=\int_X f\bar{g}\,d\mu$. In the case of a †Lebesgue measure in a Euclidean space, L_2 is of dimension $\aleph_0$, so that it is a Hilbert space isomorphic to l_2. Further examples of Hilbert spaces are $A_2(\Omega)$, $W_2^l(\Omega)$ $(=H^l(\Omega))$, and $H_0^l(\Omega)$ (→ 173 Function Spaces).

E. Closed Linear Subspaces and Projections

Let M be a **closed linear subspace** of a Hilbert space H, i.e., a linear subspace that is closed in the norm topology of H. For a given M the set of all $x\in H$ such that $(x,y)=0$ for every $y\in M$ forms a closed linear subspace M' called the **orthogonal complement** of M. The orthogonal complement of M' is M (that is, $M''=M$), and H is the direct sum of M and M' (that is, every $x\in H$ can be uniquely represented as $x=y+y', y\in M, y'\in M'$). The operator P_M that maps x to y is called the **projection** (or **projection operator**) to M. A bounded linear operator P is a projection if and only if it is idempotent ($P^2=P$) and self-adjoint ($(Px,y)=(x,Py)$ for any $x,y\in H$) (→ 251 Linear Operators).

F. Conjugate Spaces

A linear transformation from H to K is called a **linear functional**. The set H' of all continuous linear functionals f on H forms a Hilbert space with norm $\|f\|=\sup\{|f(x)|\mid\|x\|=1\}$. For every $f\in H'$ there exists a unique $y\in H$ such that $f(x)=(x,y)$ for all $x\in H$ (**Riesz's theorem**), and the correspondence $f\to y$ gives an †antilinear isometric linear transformation from H' onto H (for †linear operators on

Hilbert spaces → 72 Compact Operators; 135 Eigenvalue Problems; 251 Linear Operators).

References

[1] D. Hilbert, Grundzüge einer allgemeinen Theorie der linearen Integralgleichungen, Teubner, second edition, 1924 (Chelsea, 1953).
[2] E. Hellinger and O. Toeplitz, Integralgleichungen und Gleichungen mit unendlichvielen Unbekannten, Enzykl. Math., Teubner, 1928 (Chelsea, 1953).
[3] J. Von Neumann, Mathematische Grundlagen der Quantenmechanik, Springer, 1932.
[4] J. Von Neumann, Collected works II, III, Pergamon, 1961.
[5] M. H. Stone, Linear transformations in Hilbert space and their applications to analysis, Amer. Math. Soc. Colloq. Publ., 1932.
[6] B. Sz.-Nagy, Spektraldarstellung linearer Transformationen des Hilbertschen Raumes, Erg. Math., Springer, 1942.
[7] F. Riesz and B. Sz.-Nagy, Leçons d'analyse fonctionelle, Akademiai Kiadó, third edition, 1955; English translation, Functional analysis, Ungar, 1955.
[8] N. I. Akhiezer and I. M. Glazman, Theory of linear operators in Hilbert space, Ungar, I, 1961; II, 1963.
[9] K. Yosida, Functional analysis, Springer, 1965.
[10] K. Yosida, Hiruberuto kûkan (Japanese; Theory of Hilbert spaces), Kyôritu, 1955.
[11] P. R. Halmos, Introduction to Hilbert space and the theory of spectral multiplicity, Chelsea, second edition, 1957.
[12] H. Meschkowski, Hilbertsche Räume mit Kernfunktion, Springer, 1962.
[13] B. Sz.-Nagy and C. Foiaş, Analyse harmonique des opérateurs de l'espace de Hilbert, Masson, 1967.
[14] P. R. Halmos, A Hilbert space problem book, Van Nostrand, 1967.

200 (XI.1) Holomorphic Functions

A. Differentiation of Complex Functions

Let $f(z)$ be a †complex-valued function defined in an open set D in the †complex plane **C**. We say that $f(z)$ is **differentiable** at z if the limit

$$\lim_{h\to 0}(f(z+h)-f(z))/h=f'(z) \tag{1}$$

exists and is finite as the complex number h tends to zero. We call $f'(z)$ the **derivative** of $f(z)$ in D. This definition is a formal extension of the definition of differentiability of a function of a real variable to that of a complex variable (→ 109 Differential Calculus), but it is a much stronger condition than the differentiability of a real function, since $z+h$ in (1) may be an arbitrary point in a 2-dimensional neighborhood of z. Hence many results follow from it. In particular, if $f'(z)\neq 0$, letting $z+h$ and $z+k$ be two points in a neighborhood of z, we have by (1)

$$(f(z+h)-f(z))/(f(z+k)-f(z))\doteqdot h/k.$$

This implies that the image under f of the triangle with vertices at the points z, $z+h$, $z+k$ is approximately similar to the original triangle. Therefore, the mapping f is a †conformal mapping.

If a function $f(z)$ is differentiable at each point of an open set D, it is said to be **holomorphic** (or **regular**) in D, or $f(z)$ is a **holomorphic function** on D. (For the definition of holomorphy of a complex-valued function of several complex variables → 25 Analytic Functions of Several Complex Variables C.)

Let E be an arbitrary nonempty subset of **C**. We say that $f(z)$ is holomorphic on E if it is defined in an open set D containing E and is holomorphic on D. However, when E is a point p and $f(z)$ is differentiable at p, we sometimes say that $f(z)$ is holomorphic at p. Many results valid for differentiable real functions also hold for holomorphic functions. For instance, a function holomorphic at a point is continuous at the point. The class of all functions holomorphic in a region D is a †ring.

The following four conditions are equivalent for a function $f=u+iv$ defined on an open set D. (1) f is holomorphic in D. (2) $u=u(x,y)$ and $v=v(x,y)$ are †totally differentiable at each point $z=x+iy$ and satisfy the **Cauchy-Riemann differential equations**

$$\partial u/\partial x=\partial v/\partial y, \qquad \partial u/\partial y=-\partial v/\partial x.$$

(3) f is represented by a †power series $\sum_{n=0}^{\infty}c_n(z-a)^n$ in a neighborhood of each point a of D; that is, $f(z)$ is **analytic** in D (→ 24 Analytic Functions). (4) If a rectifiable †Jordan curve C and its interior are contained in D, $\int_C f(z)dz=0$. The proposition that (1) implies (4) is called **Cauchy's integral theorem**, and the proposition that (4) implies (1) is called **Morera's theorem**.

Suppose that $f=u+iv$ is holomorphic in D. Then the existence and continuity of the partial derivatives of u and v with respect to x and y implies total differentiability of u and v. However, in order to show that $f=u+iv$ is holomorphic in D, the assumption of the continuity of the partial derivatives of u and v can be weakened. Actually, we have the

Looman-Men'šov theorem: Suppose that u and v are continuous in D, $\partial u/\partial x$, $\partial u/\partial y$, $\partial v/\partial x$, and $\partial v/\partial y$ exist at every point of D except for at most a countable number of points, and the Cauchy-Riemann equations hold in D except for a set of 2-dimensional †measure zero; then $f=u+iv$ is holomorphic in D. D. E. Men'šov extended this theorem and obtained various conditions for holomorphy. For example, he proved the following theorem: If f is a topological mapping of D and f is conformal in D (i.e.,

$$\lim_{h\to 0}\arg(f(z+h)-f(z))/h$$

exists) except for at most a countable number of points, then f is holomorphic in D.

B. Cauchy's Integral Theorem

Cauchy's integral theorem may be stated as follows: If $f(z)$ is a holomorphic function in a †simply connected domain D, the equality $\int_C f(z)\,dz=0$ holds for every (rectifiable) closed curve C in D. In particular, the integral $\int_\alpha^\beta f(z)\,dz$ ($\alpha,\beta\in D$) is uniquely determined by α and β provided that its path of integration lies in D. In the proof of this theorem, Cauchy assumed the existence and continuity of the derivative $f'(z)$ in D. However, E. Goursat proved the theorem utilizing only the existence of $f'(z)$. Actually, by virtue of the integral formula (2) in this section, the existence of $f'(z)$ in D implies the continuity of $f'(z)$. This is sometimes called **Goursat's theorem**. Let $C, C_1, C_2, \ldots, C_n$ be rectifiable Jordan curves. Suppose that $C_1, C_2, \ldots, C_n$ are in the interior of C and that each pair of them are exterior to each other. If $f(z)$ is holomorphic in the region D bounded by these $n+1$ curves and continuous on $D\cup C\cup C_1\cup\ldots\cup C_n=\overline{D}$, then we have

$$\int_C f(z)\,dz=\int_{C_1}f(z)\,dz+\int_{C_2}f(z)\,dz+\ldots+\int_{C_n}f(z)\,dz.$$

Here the curvilinear integral is taken in the **positive direction** (i.e., we take the direction such that $\int_C(z-a)^{-1}dz=2\pi i$ for a point a in the interior of C). Henceforward, an integral along a closed curve is taken in the positive direction unless otherwise noted. Cauchy's integral theorem under the assumption of continuity of $f(z)$ on the boundary of D is sometimes called the **stronger form of Cauchy's integral theorem**.

Under the same assumptions as in the stronger form, we have **Cauchy's integral formula** for $z\in D$:

$$f(z)=\frac{1}{2\pi i}\int_C\frac{f(\zeta)}{\zeta-z}d\zeta-\frac{1}{2\pi i}\int_{C_1}\frac{f(\zeta)}{\zeta-z}d\zeta\ldots-\frac{1}{2\pi i}\int_{C_n}\frac{f(\zeta)}{\zeta-z}d\zeta. \tag{2}$$

This integral formula expresses the value of $f(z)$ at a point z in the domain D in terms of the behavior of f on the boundary of D. Conversely, if a function admits an integral formula such as (2), then it is holomorphic in D.

In particular, when $n=0$, the integral formula is as follows:

$$f(z)=\frac{1}{2\pi i}\int_C\frac{f(\zeta)}{\zeta-z}d\zeta. \tag{3}$$

Furthermore, if C is a circle $|z|=R$ (i.e., D is the disk $|z|<R$), we obtain **Poisson's integral formula**:

$$f(z)=\frac{1}{2\pi}\int_0^{2\pi}f(Re^{i\varphi})\frac{R^2-r^2}{R^2+r^2-2Rr\cos(\theta-\varphi)}d\varphi,$$

$$f(z)=\frac{1}{2\pi}\int_0^{2\pi}\mathrm{Re}f(Re^{i\varphi})\frac{Re^{i\varphi}+z}{Re^{i\varphi}-z}d\varphi+i\,\mathrm{Im}f(0),$$

$$z=re^{i\theta},\quad 0\leqslant r<R. \tag{3$'$}$$

The formula (3)′ is valid for a †harmonic function.

On the other hand, if D is the annulus $0<r<|z|<R$, **Villat's integral formula** holds:

$$f(z)=\frac{i\omega_1}{\pi^2}\int_0^{2\pi}(\mathrm{Re}\,f(Re^{i\varphi}))\zeta\left(\frac{\omega_1}{\pi i}\log\frac{z}{R}-\frac{\omega_1}{\pi}\varphi\right)d\varphi-\frac{i\omega_1}{\pi^2}\int_0^{2\pi}(\mathrm{Re}f(re^{i\varphi}))\zeta_3\left(\frac{\omega_1}{\pi i}\log\frac{z}{R}-\frac{\omega_1}{\pi}\varphi\right)d\varphi+i\,\mathrm{Im}\,f(0),$$

where $\zeta(u)$ is †Weierstrass's ζ-function, $\zeta_3(u)$ is defined by $\zeta_3(u)=\sigma_3'(u)/\sigma(u)$, and $\zeta(u)$ and $\zeta_3(u)$ have half-periods ω_1, ω_3 determined by

$$r/R=\exp(-\pi\omega_3/\omega_1 i),\quad \omega_1>0,\quad \omega_3/i>0.$$

The hypothesis of Morera's theorem can be weakened as follows: Let $f(z)$ be continuous in a domain D. If $\int_C f(z)\,dz=0$ for every rectangle C with sides parallel to the axes and whose interior consists of only points of D, then $f(z)$ is holomorphic in D. In the statement of this theorem, if we let C be an arbitrary circle, we get the same conclusion.

C. Zero Points

Let $f(z)$ be a holomorphic function not identically equal to zero. If $f(a)=0$, we call a a **zero point** of f. Every zero point of f is an isolated zero point, and there exists a unique positive integer k such that

$$f(z)=(z-a)^{k}g(z), \quad g(a)\neq 0. \qquad (4)$$

We call k the **order of the zero point** a and a a **zero point of the kth order**. The equality (4) implies that the †Taylor series of $f(z)$ at a begins with the term $c_k(z-a)^k$. Suppose that $f(z)=\gamma$ and a is a zero point of $f(z)-\gamma$ of the kth order. We call a a **γ-point of the kth order**.

For a function $f(z)$ defined in a neighborhood of the †point at infinity, we set $f(1/w)=g(w)$ ($f(\infty)=g(0)$) and call f holomorphic at ∞ if g is holomorphic at 0; f is said to be of **order** k at ∞ if g is of order k at 0.

D. Isolated Singularities

A point a in C is called a **singular point** (or **singularity**) of a complex function $f(z)$ if $f(z)$ is not holomorphic at a. If $f(z)$ is single-valued and holomorphic in $D=\{z\,|\,0<|z-a|<R\}$ (or, if $a=\infty$, in $D=\{z\,|\,R^{-1}<|z|<+\infty\}$) but not holomorphic in $D\cup\{a\}$, we call a an **isolated singularity** of $f(z)$. By utilizing the †local canonical parameter $t=z-a$ (or $t=1/z$ for $a=\infty$), $f(z)$ is expanded by the †Laurent series

$$f(z)=\sum_{n=-\infty}^{-1}c_nt^n+\sum_{n=0}^{\infty}c_nt^n. \qquad (5)$$

This is called the **Laurent expansion** of $f(z)$. The second term of (5) is an ordinary †power series, called the **holomorphic part** of $f(z)$. The first term is a power series of $1/t$ with no constant term, called the **singular part** of $f(z)$ at a or the **principal part** of the singularity (or of the Laurent expansion at a).

If we have $\lim_{t\to 0}tf(z)=0$, the Laurent expansion (5) of $f(z)$ lacks its singular part, and the limit of $f(z)$ exists as $t\to 0$ ($z\to a$) and is equal to c_0. If we set $f(a)=c_0$, then the function $f(z)$ is holomorphic in $D\cup\{a\}$. In this case, the point a is called a **removable singularity**. If $f(z)$ is bounded in a neighborhood of a singularity a, then a is removable (**Riemann's theorem**). Usually, we assume that the removable singularities of a function have already been removed in this way.

When a singular part of $f(z)$ at a exists and consists of a finite number of terms, the point a is called a **pole**; when it consists of an infinite number of terms, the point is called an **essential singularity**. If a is a pole, $f(z)$ is represented by the Laurent series $\sum_{n=-k}^{\infty}c_nt^n$ ($c_{-k}\neq 0$) and $f(z)\to\infty$ as $z\to a$. In this case, the index k is called the **order of the pole** a. Then a relation such as (4) holds, where the index k is replaced by $-k$. Hence the point a is sometimes called a **zero point of** the $-k$**th order**. If a is an essential singularity, then for an arbitrary number c there exists a sequence z_n converging to a such that $\lim_{n\to\infty}f(z_n)=c$ (the **Casorati-Weierstrass theorem** or simply **Weierstrass's theorem**). Related to Weierstrass's theorem, we have †Picard's theorem, which gives a detailed description of the behavior of a function around its singularities.

E. Residues

The coefficient c_{-1} of $(z-a)^{-1}$ in the Laurent expansion (5) of $f(z)$ is called the **residue** of $f(z)$ at a and is denoted by $\mathrm{Res}[f]_a$, $R(a;f)$, or $R(a)$ if we need not indicate f. We have

$$R(a)=c_{-1}=\frac{1}{2\pi i}\int_{|\zeta-a|=r}f(\zeta)\,d\zeta,$$

where the integral is taken in the positive direction along a path for $0<r<R$. If $f(z)$ is holomorphic at $z=a$, then $R(a)=0$. If $f(z)$ has a pole of the first order at a,

$$R(a)=\lim_{z\to a}(z-a)f(z).$$

The residue at the point at infinity is defined to be $-a_{-1}$, where a_{-1} is the coefficient with index -1 of the Laurent expansion of $f(z)$ at ∞: $f(z)=\sum_{n=-\infty}^{\infty}a_nz^n$, and we have

$$-a_{-1}=\frac{-1}{2\pi i}\int_{|\zeta|=r}f(\zeta)\,d\zeta, \quad R^{-1}<r<+\infty.$$

Thus the notion of the residue of $f(z)$ is actually related to the differential form $f(z)\,dz$ and not to $f(z)$ itself.

From the first formula in this section and the formula for $-a_{-1}$, the **residue theorem** follows (Cauchy, 1825): Let C be a rectifiable Jordan curve in the complex number plane. Let $a_1,\ldots,a_m$ be a finite number of points inside C, and let D be a domain containing C and its interior. If $f(z)$ is a function holomorphic in $D-\{a_1,\ldots,a_m\}$, we have

$$\frac{1}{2\pi i}\int_C f(z)\,dz=\sum_{n=1}^{m}R(a_n).$$

Furthermore, if $f(z)$ is holomorphic in the whole complex number plane (including the point at infinity) except for a finite number of poles, the sum of all residues is equal to zero.

F. Calculus of Residues

The **calculus of residues** is a field of calculus based on application of the notion of residues. For example, we have methods for the calculation of definite integrals. Actually, one of the reasons why Cauchy studied the theory of complex functions was that he believed that the theory would provide a unified method of computing definite integrals. For example, if $\varphi(z)$ is a rational function without poles on the real axis and with a zero point at infinity whose order is at least 2, then we have

$$\int_{-\infty}^{\infty}\varphi(x)\,dx=2\pi i\sum_{\operatorname{Im}\alpha>0}R(\alpha;\varphi(z)),\qquad(6)$$

$$\int_{-\infty}^{\infty}e^{ix}\varphi(x)\,dx=2\pi i\sum_{\operatorname{Im}\alpha>0}R(\alpha;e^{iz}\varphi(z)).\qquad(7)$$

Here the sums are taken over all the poles in the upper half-plane. Formula (7) is valid also for a rational function $\varphi(z)$ with a simple zero at infinity. If $\varphi(z)$ has simple poles at a_k $(k=1,\ldots,n)$ on the path of integration, then we take the principal values of the integrals at those poles and add $\pi i R(a)$ $(k=1,\ldots,n)$ to the terms on the right-hand side of (6) and (7). Sometimes we use the residue theorem to obtain the value of the sum of a series (e.g., the †Gaussian sum) by expressing it as an integral.

Let $f(z)$ be a single-valued function that is †meromorphic and not identically equal to zero in a domain D, and let $\varphi(z)$ be a function holomorphic in D. Draw a rectifiable Jordan curve C such that the interior of C is contained in D and $f(z)$ has neither zeros nor poles on C. Let $\alpha_1,\ldots,\alpha_N$ and $\beta_1,\ldots,\beta_P$ be the zeros and poles inside C, respectively (where each of them is repeated as often as its order). Then we have

$$\sum_{n=1}^{N}\varphi(\alpha_n)-\sum_{p=1}^{P}\varphi(\beta_p)=\frac{1}{2\pi i}\int_C\varphi(z)\frac{f'(z)}{f(z)}dz.$$

If $\varphi(z)=1$, we get

$$\frac{1}{2\pi}\int_C d\arg f(z)=N-P.$$

This is called the **argument principle**. Next, let $f(z)$ be a function meromorphic for $|z|<R\leqslant+\infty$, $f(0)\neq0,\neq\infty$, and set $\varphi(z)=\log z$. Take C as a closed curve consisting of the boundary of an annulus $0<\rho<|z|<r<R$ (where ρ is sufficiently small) and two sides of a suitable †crosscut joining a point of $|z|=\rho$ and $z=r$. Then we obtain **Jensen's formula**:

$$\log\left|\frac{\alpha_1,\alpha_2,\ldots,\alpha_N}{\beta_1,\beta_2,\ldots,\beta_P}\right|=\log|f(0)|+(N-P)\log r-\frac{1}{2\pi}\int_0^{2\pi}\log|f(re^{i\psi})|\,d\psi.$$

This formula can be utilized to prove **Rouché's theorem**: Let $f(z)$ and $g(z)$ be functions holomorphic in a domain D that contains a rectifiable Jordan curve C and its interior. Suppose that $f(z)+\lambda g(z)$ never vanishes on C for any λ with $0\leqslant\lambda\leqslant1$. Then the number of zeros of $f(z)$ in the interior of C is equal to that of $f(z)+g(z)$. If $|f(z)|>|g(z)|$ on C or $\arg f(z)-\arg g(z)\neq(2n+1)\pi$ (n is an integer), the hypothesis of Rouché's theorem is satisfied. This theorem is useful in proving the existence of a zero of a complex function (for example, a polynomial) and in finding its position.

For other properties of holomorphic functions → 24 Analytic Functions; 44 Bounded Functions; 413 Transcendental Entire Functions.

References

[1] J. Yoshikawa, Kansûron (Japanese; Theory of functions), Huzanbô, 1913.
[2] T. Takenouchi, Kansûron (Japanese; Theory of functions) I, II, Syôkabô, 1926, revised edition, 1966.
[3] S. Kakeya, Ippan kansûron (Japanese; General theory of functions), Iwanami, 1930.
[4] Y. Yosida, Kansûron (Japanese; Theory of functions), Iwanami, 1938, revised edition, 1965.
[5] M. Tsuji, Kansûron (Japanese; Theory of functions) I, II, Maki syoten, revised edition, 1968.
[6] S. Hitotumatu, Kansûron nyûmon (Japanese; Introduction to theory of functions), Baihûkan, 1956.
[7] Y. Kusunoki, Kaiseki kansû (Japanese; Analytic functions), Hirokawa, 1957.
[8] E. Borel, Leçons sur les fonctions monogènes uniformes d'une variable complexe, Gauthier-Villars, 1917.
[9] A. Hurwitz and R. Courant, Vorlesungen über allgemeine Funktionentheorie und elliptische Funktionen, Springer, fourth edition, 1964.
[10] L. Bieberbach, Lehrbuch der Funktionentheorie, Teubner, I, 1921; II, 1927 (Johnson Reprint Co., 1969).
[11] E. C. Titchmarsh, The theory of functions, Oxford Univ. Press, second edition, 1939.
[12] C. Carathéodory, Funktionentheorie I, II, Birkhäuser, 1950; English translation, Theory of functions, Chelsea, I, 1958; II, 1960.
[13] S. Saks and A. Zygmund, Analytic functions, Warsaw, 1952.
[14] L. V. Ahlfors, Complex analysis, McGraw-Hill, 1953, second edition, 1966.
[15] H. Behnke and F. Sommer, Theorie der analytischen Funktionen einer komplexen Veränderlichen, Springer, second edition, 1962.
[16] H. Kneser, Funktionentheorie, Vandenhoeck & Ruprecht, 1958.
[17] E. Hille, Analytic function theory, Ginn, I, 1959; II, 1962.
[18] H. P. Cartan, Théorie élémentaire des

fonctions analytiques d'une ou plusieurs variables complexes, Hermann, 1961; English translation, Elementary theory of analytic functions of one or several complex variables, Addison-Wesley, 1963.
[19] M. Heins, Complex function theory, Academic Press, 1968.
[20] Б. А. Фукс, В.И. Левин (B. A. Fuks and V. I. Levin), Функции комрлекского переменного и их приложения, Гостехиздат, I, II, 1951; English translation, Functions of a complex variable and some of their applications I, II, Pergamon, 1961.
[21] W. H. I. Fuchs, Topics in the theory of functions of one complex variable, Van Nostrand, 1967.
As to the Looman-Men'šov theorem,
[22] D. Menchoff (Men'šov), Les conditions de monogénéité, Actualités Sci. Ind., Hermann, 1936.
[23] D. Menchoff (Men'šov), Sur la généralisation des conditions de Cauchy-Riemann, Fund. Math., 25 (1935), 59–97.
[24] S. Saks, Theory of the integral, Warsaw, 1937, p. 188–201.
[25] K. Kunugi, Hukuso kansûron (Japanese; Theory of functions of a complex variable), Iwanami Coll. Modern Math. Sci., 1958.
For a new proof of Cauchy's integral theorem,
[26] E. Artin, On the theory of complex functions, Notre Dame mathematical lectures, 1944, p. 57–70.
[27] M. Tsuji, Hukuso hensû kansûron (Japanese; Theory of functions of a complex variable), Kyôritu, 1934, p. 310–316.
For Villat's integral formula,
[28] T. Sasaki, Tôkaku syazô no ôyô (Japanese; Applications of conformal mappings), Huzanbô, 1939, p. 256–261.
[29] Y. Komatu, Tôkaku syazôron (Japanese; Theory of conformal mapping) II, Kyôritu, 1949, p. 103–112, 351–362.
For residue calculus,
[30] E. L. Lindelöf, Le calcul des résidus et ses applications à la théorie des fonctions, Gauthier-Villars, 1905.

201 (IV.13) Homogeneous Spaces

A. General Remarks

Let M be a †differentiable manifold. If a †Lie group G acts †transitively on M as a †Lie transformation group, the manifold M is said to be a **homogeneous space** having G as its transformation group (→ 415 Transformation Groups). The †stabilizer (isotropy subgroup) H_x of G at a point x of M is a closed subgroup of G, and a one-to-one correspondence between G/H_x and M preserving the action of G is defined by associating the element sH_x ($s \in G$) of G/H_x with the point $s(x)$ of M. This correspondence is a †diffeomorphism between the manifold M and the quotient manifold G/H_x if the number of connected components of G is at most countable. Under this condition we may therefore identify a homogeneous space M with the quotient manifold G/H of a Lie group G by a closed Lie subgroup H (→ 248 Lie Groups). However, H is not uniquely determined by M, and it may be replaced by $H_{s(x)} = sH_xs^{-1}$ ($s \in G$). Each element h of the stabilizer H_x at a point x induces a linear transformation $\tilde{h}$ on the †tangent space V_x of M at the point x. The set $\tilde{H}_x$ of all $\tilde{h}$ is called the **linear isotropy group** at the point x.

If we represent the homogeneous space M as G/H, we obtain the canonical mapping $\pi: s \rightarrow sH$ of G onto M which we call the **projection** of G onto M. Let $\mathfrak{g}$ be the †Lie algebra of G and $\mathfrak{h}$ be the Lie subalgebra corresponding to the closed subgroup H. When we identify $\mathfrak{g}$ with the tangent space at the identity element e of G and $\mathfrak{h}$ with its subspace, the projection π induces a linear isomorphism of $\mathfrak{g}/\mathfrak{h}$ with the tangent space V_x of M at the point $x = \pi(e)$. The †adjoint representation of G gives rise to a linear representation $h \rightarrow \mathrm{Ad}(h)$ modulo $\mathfrak{h}$ of the group H on the linear space $\mathfrak{g}/\mathfrak{h}$. Through the linear isomorphism between $\mathfrak{g}/\mathfrak{h}$ and the tangent space V_x defined by the projection π, this representation of H is equivalent to the one which associates with h the linear transformation $\tilde{h}$ defined by h on the tangent space V_x.

The homogeneous space G/H is said to be **reductive** if there exists a linear subspace $\mathfrak{m}$ of $\mathfrak{g}$ such that $\mathfrak{g} = \mathfrak{h} + \mathfrak{m}$ (direct sum as linear spaces) and $(\mathrm{Ad}\,H)\mathfrak{m} \subset \mathfrak{m}$. H is said to be **reductive** in $\mathfrak{g}$ if the representation $h \rightarrow \mathrm{Ad}(h)$ of H in $\mathfrak{g}$ is †completely reducible.

If a †tensor field P on the homogeneous space $M = G/H$ is G-invariant (namely, invariant under the transformations defined by the elements of G), then the value of P at the point $x = \pi(e)$ is a †tensor over the tangent space V_x at x which is invariant under the linear isotropy group $\tilde{H}$. Conversely, such a tensor over V_x is uniquely extended to a G-invariant tensor field on M. If G/H is reductive, then G-invariant tensor fields over M are in one-to-one correspondence with $\tilde{H}$-invariant tensors over $\mathfrak{m}$. For instance, if H is compact, then H is reductive in $\mathfrak{g}$ and an $\tilde{H}$-invariant positive definite quadratic form on $\mathfrak{m}$ defines a G-invariant †Riemannian metric on G/H.

We say that the homogeneous space $M = G/H$ is a **Riemannian (linearly connected, complex Hermitian, Kähler) homogeneous space** if there exists on M a G-invariant Riemannian metric (†linear connection, †Hermitian metric, †Kähler metric). Concerning such homogeneous spaces, there are various results on their structures and geometric properties [1–5] (→ 401 Symmetric Riemannian Spaces; 411 Topology of Lie Groups and Homogeneous Spaces).

B. Examples

Stiefel Manifold. A **k-frame** ($1 \leqslant k \leqslant n$) in a real n-dimensional Euclidean vector space $\mathbf{R}^n$ is an ordered system consisting of k linearly independent vectors. If we regard the real †general linear group of degree n, $GL(n,\mathbf{R})$, as the regular linear transformation group of $\mathbf{R}^n$, $GL(n,\mathbf{R})$ acts transitively on the set $V'_{n,k}(\mathbf{R})$ of all k-frames in $\mathbf{R}^n$. Therefore, if H denotes the subgroup of $GL(n,\mathbf{R})$ consisting of the elements which leave fixed a given k-frame v_0^k, we may identify the set $V'_{n,k}$ and the quotient set $GL(n,\mathbf{R})/H$. Transferring the differentiable manifold structure of $GL(n,\mathbf{R})/H$ to $V'_{n,k}$ through this identification, $V'_{n,k}(\mathbf{R}) = GL(n,\mathbf{R})/H$ becomes a homogeneous space (the differentiable manifold structure of $V'_{n,k}$ is defined independently of the choice of v_0^k). The space $V'_{n,k}(\mathbf{R})$ is called the **(real) Stiefel manifold of k-frames** in $\mathbf{R}^n$.

A k-frame is called an **orthogonal k-frame** if the vectors belonging to the frame are of length 1 and are orthogonal to each other. The set $V_{n,k}(\mathbf{R})$ of all orthogonal k-frames is a submanifold of $V'_{n,k}(\mathbf{R})$. The †orthogonal group $O(n)$ acts transitively on $V_{n,k}(\mathbf{R})$, which is a homogeneous space represented as $V_{n,k}(\mathbf{R}) = O(n)/I_k \times O(n-k)$. The manifold $V_{n,1}(\mathbf{R})$ is actually the $(n-1)$-dimensional sphere. We call $V_{n,k}(\mathbf{R})$ the **(real) Stiefel manifold of orthogonal k-frames** (or simply **Stiefel manifold**). The **complex Stiefel manifold** $V_{n,k}(\mathbf{C}) = U(n)/I_k \times U(n-k)$ is defined analogously.

Grassmann Manifold. Let $M_{n,k}(\mathbf{R})$ ($1 \leqslant k \leqslant n$) be the set of all k-dimensional linear subspaces of $\mathbf{R}^n$. The group $O(n)$ acts transitively on $M_{n,k}(\mathbf{R})$, so that we may put $M_{n,k}(\mathbf{R}) = O(n)/O(k) \times O(n-k)$. Here $O(k)$ and $O(n-k)$ are identified with the subgroups of $O(n)$ consisting of all elements leaving fixed every point of a fixed $(n-k)$-dimensional subspace and of its orthogonal complement, respectively. In this way, $M_{n,k}(\mathbf{R})$ is a homogeneous space which we call the **(real) Grassmann manifold**. The †proper orthogonal group $SO(n)$ acts transitively on $M_{n,k}(\mathbf{R})$, and $M_{n,k}(\mathbf{R})$ may be represented as a homogeneous space having $SO(n)$ as its transformation group. It follows that $M_{n,k}(\mathbf{R})$ is connected. The homogeneous space $\tilde{M}_{n,k}(\mathbf{R}) = SO(n)/SO(k) \times SO(n-k)$ is called the **Grassmann manifold formed by oriented subspaces**. $M_{n,1}(\mathbf{R})$ and $\tilde{M}_{n,1}(\mathbf{R})$ may be identified with the $(n-1)$-dimensional real projective space and the $(n-1)$-dimensional sphere, respectively.

Applying the above process for real Grassmann manifolds to the complex Euclidean vector space $\mathbf{C}^n$ instead of $\mathbf{R}^n$, the set $M_{n,k}(\mathbf{C})$ of all k-dimensional linear subspaces in $\mathbf{C}^n$ is a homogeneous space with the †unitary group $U(n)$ of degree n as its transformation group, and is represented as $U(n)/U(k) \times U(n-k)$. This space is called the **complex Grassmann manifold**. The manifold $M_{n,k}(\mathbf{C})$ is a simply connected complex manifold and has a cellular decomposition as a †CW complex whose cells are †Schubert varieties (→ 58 Characteristic Classes E). On the other hand, $M_{n,k}(\mathbf{C})$ may be regarded as the set of all $(k-1)$-dimensional linear subspaces in the $(n-1)$-dimensional complex projective space. Then, using the †Plücker coordinates of these subspaces, $M_{n,k}(\mathbf{C})$ is realized as an †algebraic variety without singularity in the projective space of dimension $\binom{n}{k} - 1$ (→ 92 Coordinates B). Sometimes $M_{n,k}(\mathbf{R})$ is denoted by $G_{n,k}(\mathbf{R})$ or $G(n,k)$. In the same way, the homogeneous space represented as $Sp(n)/Sp(k) \times Sp(n-k)$ is called the †quaternion Grassmann manifold and is denoted by $M_{n,k}(\mathbf{H})$.

Flag Manifold. Let $k_1, \ldots, k_r$ be a sequence of integers such that $n > k_1 > \ldots > k_r > 0$, and let $F(k_1, \ldots, k_r)$ be the set of all monotone sequences $V_1 \supset \ldots \supset V_r$, where V_i ($i = 1, \ldots, r$) is a k_i-dimensional linear subspace in $\mathbf{R}^n$. For the two sequences $V_1 \supset \ldots \supset V_r$ and $V'_1 \supset \ldots \supset V'_r$ belonging to $F(k_1, \ldots, k_r)$, there exists an element $s \in GL(n,\mathbf{R})$ such that $s(V_i) = V'_i$ ($i = 1, \ldots, r$). Therefore, $F(k_1, \ldots, k_r)$ is a homogeneous space with $GL(n,\mathbf{R})$ as its transformation group, and is called the **proper flag manifold**. Since the unitary group $U(n)$ of degree n acts transitively on it, $F(k_1, \ldots, k_r)$ is also regarded as a homogeneous space admitting $U(n)$ as its transformation group. In this case, putting $F(k_1, \ldots, k_r) = U(n)/H$, H is isomorphic to the direct product $U(k_1 - k_2) \times U(k_2 - k_3) \times \ldots \times U(k_r)$. In the particular case where $r = n-1$, $k_i = n - i$, the homogeneous space is the quotient space of the compact Lie group $U(n)$ by a maximal †torus T. In general, the quotient space G/T of a compact connected Lie group G by a maximal torus of G is called a **flag manifold**. If G acts effectively on G/T, G is a †semisimple compact Lie group. The complex Lie group $G^{\mathbf{C}}$ is a Lie transformation group of †biregular transformations which acts transitively on the flag manifold G/T, a simply connected Kähler homogeneous space. Here $G^{\mathbf{C}}$ is a complex Lie group having G as a maximal compact subgroup. If B is a maximal †solvable Lie subgroup (†Borel subgroup) of $G^{\mathbf{C}}$, G/T is represented as $G^{\mathbf{C}}/B$.

References

[1] A. Borel and R. Remmert, Über kompakte homogene Kählersche Mannigfaltigkeiten, Math. Ann., 145 (1961–1962), 429–439.
[2] Y. Matsushima, Espaces homogènes de Stein des groupes de Lie complexes, Nagoya Math. J., 16 (1960), 205–218.
[3] K. Nomizu, Invariant affine connections on homogeneous spaces, Amer. J. Math., 76 (1954), 33–65.
[4] H. C. Wang, Closed manifolds with homogeneous complex structure, Amer. J. Math., 76 (1954), 1–32.
[5] Foundations and applications of differential geometry (Japanese), Report of seminar on math. Sûgaku sinkôkai, mimeographed note, 1956.

202 (III.26) Homological Algebra

A. General Remarks

Homological algebra is a new branch of mathematics that developed rapidly after World War II. The introduction of the theory was motivated by the observation that some algebraic ideas and mechanisms initiated in the development of †algebraic topology, in particular †homology theory, are powerful tools for treating various problems in algebra that had been treated separately from a unified viewpoint. One of its characteristic features lies in emphasizing, from the standpoint of categories and functors (→ 53 Categories and Functors), the functional structure of the objects to be investigated rather than their inner structure. This new theory turned out to have wide applications in other areas of mathematics, and the philosophy embodied in the theory has been influential in the general progress of mathematics. For general references → [2, 4, 6, 7, 8].

B. Homology and Cohomology of Complexes

We mainly consider general †Abelian categories $\mathcal{C}$. Consideration may, however, be restricted to the †category (Ab) of Abelian groups (whose †objects are Abelian groups and whose †morphisms are homomorphisms) or the †category ${}_R\mathfrak{M}$ of R-modules (→ 57 Chain Complexes).

A **(cochain) complex** C in an Abelian category $\mathcal{C}$ consists of objects $C^n \in \mathcal{C}$ ($n \in \mathbf{Z}$) and morphisms $d^n: C^n \rightarrow C^{n+1}$ subject to the condition that $d^{n+1} \circ d^n = 0$ ($n \in \mathbf{Z}$). The d^n are called the **differentiations** (or **boundary operators**). The nth **cohomology** $H^n(C)$ of C is defined by the †exact sequence $0 \rightarrow B^n(C) \rightarrow Z^n(C) \rightarrow H^n(C) \rightarrow 0$, where $B^n(C)$ and $Z^n(C)$ are objects representing $\operatorname{Im} d^{n-1}$ and $\operatorname{Ker} d^n$, respectively. The complex C is called **positive (negative)** if $C^n = 0$ for $n<0$ ($n>0$). We sometimes interchange positive superscripts and negative subscripts and write C_{-n} instead of C^n. Then the differentiations become $d_n: C_n \rightarrow C_{n-1}$, and C is then called a **chain complex**. The quotient of $\operatorname{Ker} d_n = Z_n$ by $\operatorname{Im} d_{n+1} = B_n$ is called the nth **homology** $H_n(C)$. Negative complexes are usually described in this manner. When C^n, Z^n, B^n, and H^n are sets, as in the category ${}_R\mathfrak{M}$ of R-modules, their elements are called **cochains, cocycles, coboundaries**, and **cohomology classes**, respectively. Similarly, in the group C_n of **chains**, residue classes of **cycles** ($\in Z_n$) modulo **boundaries** ($\in B_n$) are called **homology classes** ($\in H_n$).

A **morphism** (or **chain transformation**) $f: C \rightarrow C'$ is a †natural transformation of the complexes considered as †functors $\mathbf{Z} \rightarrow \mathcal{C}$; i.e., f is a family of morphisms $f^n: C^n \rightarrow C'^n$ ($n \in \mathbf{Z}$) satisfying $f^{n+1} \circ d^n = d^n \circ f^n$. It induces a morphism of cohomology $H^n(C) \rightarrow H^n(C')$. A **subcomplex** of C is an equivalence class of †monomorphisms $D \rightarrow C$, usually denoted by any representative D of the class. A **(chain) homotopy** between two chain transformations $f, g: C \rightarrow C'$ is a family of morphisms $h^n: C^n \rightarrow C'^{n-1}$ ($n \in \mathbf{Z}$) satisfying $f^n - g^n = h^{n+1} \circ d^n + d^{n-1} \circ h^n$. If there exists a homotopy between f and g, then f and g induce the same morphism of cohomology. A morphism $f: C \rightarrow C'$ is called a **(chain) equivalence** if there exists a morphism $f': C' \rightarrow C$ such that $f' \circ f$ and $f \circ f'$ are homotopic to the identities of C and C', respectively. In this case we have $H^n(C) \cong H^n(C')$. An exact sequence of complexes $0 \rightarrow C' \rightarrow C \rightarrow C'' \rightarrow 0$ gives rise to the **connecting morphisms** $H^n(C'') \rightarrow H^{n+1}(C')$ ($n \in \mathbf{Z}$), and the resulting sequence $\ldots \rightarrow H^{n-1}(C'') \rightarrow H^n(C') \rightarrow H^n(C) \rightarrow H^n(C'') \rightarrow H^{n+1}(C') \rightarrow \ldots$ is exact (the **exact cohomology sequence**), and similarly for homology instead of cohomology. An object $A \in \mathcal{C}$ defines a complex (also denoted by A) such that $A^0 = A$, $d^0 = 0$. A positive complex C together with a morphism $\varepsilon: A \rightarrow C$ is called a **complex over** A, and ε is the **augmentation**. A complex C over A is

acyclic $0 \to A \overset{\varepsilon}{\to} C^0 \to C^1 \to \ldots$ is exact. An acyclic positive complex over A is called a right **resolution** of A. Let $\{C,\varepsilon\}$, $\{C',\varepsilon'\}$ be complexes over A and A', respectively, and α a morphism $A \to A'$. Then a morphism $f: C \to C'$ satisfying $f \circ \varepsilon = \varepsilon' \circ \alpha$ is called a morphism over α. For a negative complex C, we define similarly augmentations $\varepsilon: C \to A$, acyclicity, left resolutions, etc.

A **bicomplex** (or **double complex**) C in $\mathcal{C}$ consists of objects $C^{p,q}$ $(p,q \in \mathbf{Z})$ and two differentiations $d_{\mathrm{I}}: C^{p,q} \to C^{p+1,q}$, $d_{\mathrm{II}}: C^{p,q} \to C^{p,q+1}$ subject to $d_{\mathrm{I}}^2 = d_{\mathrm{II}}^2 = 0$ and $d_{\mathrm{I}} d_{\mathrm{II}} = d_{\mathrm{II}} d_{\mathrm{I}}$ (sometimes replaced by anticommutativity, $d_{\mathrm{I}} d_{\mathrm{II}} + d_{\mathrm{II}} d_{\mathrm{I}} = 0$). Morphisms of bicomplexes are defined as for single complexes. A bicomplex C becomes a (single) complex if we put $C^n = \sum_{p+q=n} C^{p,q}$ (when the sum exists) and define the differentiation d to be $d_{\mathrm{I}} + (-1)^p d_{\mathrm{II}}$ on $C^{p,q}$. Then d is called the **total differentiation** and d_{I}, d_{II} the **partial differentiations**. On the other hand, $C_{\mathrm{I}}^q = \{C^{p,q} (p \in \mathbf{Z}), d_{\mathrm{I}}\}$ constitutes a complex for each q, whose cohomology $H^p(C_{\mathrm{I}}^q)$ is denoted by $H_{\mathrm{I}}^p(C^q)$. Then d_{II} induces morphisms $H_{\mathrm{I}}^p(C^q) \to H_{\mathrm{I}}^p(C^{q+1})$, so that we obtain a complex $H_{\mathrm{I}}^p(C)$. The cohomology of $H_{\mathrm{I}}^p(C)$ is denoted by $H_{\mathrm{II}}^q(H_{\mathrm{I}}^p(C))$. We define $H_{\mathrm{I}}^p(H_{\mathrm{II}}^q(C))$ similarly. The cohomology of C with respect to the total differentiation is denoted simply by $H^n(C)$. Similar constructions are applied to double chain complexes $\{C_{p,q}\}$ and further to **multiple complexes**, as we shall show in the case of bicomplexes.

Let T be a †bifunctor $\mathcal{C}_1 \times \mathcal{C}_2 \to \mathcal{C}'$ and C_i be complexes in $\mathcal{C}_i$ $(i=1,2)$. Then $T(C_1, C_2)$ is a bicomplex in $\mathcal{C}'$. For instance, $\mathrm{Hom}(C', C)$ is a positive (bipositive) complex if C (C') is a positive (negative) complex in $\mathcal{C}$. If C, C' are complexes in $\mathfrak{M}_R$, ${}_R\mathfrak{M}$, respectively, the †tensor product $C \otimes_R C'$ is a complex in (Ab) (the **product complex**). There is a canonical morphism $H_p(C) \otimes H_q(C') \to H_{p+q}(C \otimes C')$. If C_n and B_n are †flat for all $n \in \mathbf{Z}$, we have the following exact sequence (**Künneth's formula**):

$$0 \to \sum_{p+q=n} H_p(C) \otimes H_q(C') \to H_n(C \otimes C') \to$$

$$\sum_{p+q=n-1} \mathrm{Tor}_1(H_p(C), H_q(C')) \to 0$$

(for the definition of Tor → Section E). For $C' = A \in {}_R\mathfrak{M}$, Künneth's formula reduces to the exact sequence $0 \to H_n(C) \otimes A \to H_n(C \otimes A) \to \mathrm{Tor}_1(H_{n-1}(C), A) \to 0$ (**universal coefficient theorem**). The corresponding exact sequence for cohomology is

$$0 \to \mathrm{Ext}^1(H_{n-1}(C), A) \to H^n(C, A)$$
$$\to \mathrm{Hom}(H_n(C), A) \to 0$$

(→ 57 Chain Complexes; 203 Homology Groups).

C. Satellites and Derived Functors

Let $\mathcal{C}$ and $\mathcal{C}'$ be Abelian categories. All functors in this section are †additive. A †covariant functor $T: \mathcal{C} \to \mathcal{C}'$ is called **exact** if T maps every exact sequence in $\mathcal{C}$ to an exact sequence in $\mathcal{C}'$. T is called **half-exact**, **left exact**, or **right exact** if for every **short exact sequence** $0 \to A \to B \to C \to 0$, the sequence $T(A) \to T(B) \to T(C)$, $0 \to T(A) \to T(B) \to T(C)$, or $T(A) \to T(B) \to T(C) \to 0$, respectively, is exact. Similar definitions apply for †contravariant functors. The functor Hom: $\mathcal{C} \times \mathcal{C} \to$(Ab) (which defines the category $\mathcal{C}$) is left exact in both factors. An object P is **projective** if $h_P(\cdot) = \mathrm{Hom}(P, \cdot)$ is exact, while Q is **injective** if $h^Q(\cdot) = \mathrm{Hom}(\cdot, Q)$ is exact. If every object A admits an †epimorphism from a projective object $P \to A$ (resp. †monomorphism into an injective object $A \to Q$), $\mathcal{C}$ is said to have enough projectives (injectives). An object G is called a **generator** (**cogenerator**) if the natural mapping $\mathrm{Hom}(A,B) \to \mathrm{Hom}(h_G(A), h_G(B))$ $(\mathrm{Hom}(h^G(B), h^G(A)))$ is one-to-one.

An Abelian category $\mathcal{C}$ is called a **Grothendieck category** if (1) $\mathcal{C}$ has a generator, (2) †direct sums always exist, and (3) the identity $(\bigcup A_i) \cap B = \bigcup (A_i \cap B)$ holds for any object A, its †subobject B, and a †totally ordered family $\{A_i\}$ of subobjects. A Grothendieck category has enough injectives (R. Baer, 1940, for (Ab); A. Grothendieck, 1957, for general $\mathcal{C}$). A monomorphism into an injective object $f: A \to Q$ is called an **injective envelope** if $\mathrm{Im} f \cap \mathrm{Im} g \neq 0$ for any nonzero monomorphism $g: B \to Q$. Every object A in a Grothendieck category admits an injective envelope, which is unique up to isomorphism (B. Eckmann and A. Schopf, 1953, for ${}_R\mathfrak{M}$; B. Mitchell, 1960, for general $\mathcal{C}$).

We say that a covariant **∂-functor** $\mathcal{C} \to \mathcal{C}'$ is given if we have a sequence of covariant functors $T = \{T^i: \mathcal{C} \to \mathcal{C}'\}$ and the **connecting morphisms** $\partial: T^i(A'') \to T^{i+1}(A')$ for an arbitrary short exact sequence $0 \to A' \to A \to A'' \to 0$ satisfying the following conditions: (i) $\partial \circ T^i(f'') = T^{i+1}(f') \circ \partial$ for a morphism f of short exact sequences; and (ii) the sequence $\ldots \to T^{i-1}(A'') \overset{\partial}{\to} T^i(A') \to T^i(A) \to T^i(A'') \overset{\partial}{\to} T^{i+1}(A') \to \ldots$ constitutes a complex. $T = \{T^i\}$ is called a covariant **∂*-functor** if instead of ∂ there are given $\partial^*: T^i(A'') \to T^{i-1}(A')$ satisfying similar conditions (i*) and (ii*). By taking duals, we define the notion of contravariant ∂- and ∂*-functors. They are also called **connected sequences of functors**. A ∂- (∂*-)functor defined for $-\infty < i < +\infty$ is called a **cohomological functor** (**homological functor**) if the sequence in condition (ii) (resp. (ii*)) is always exact. A morphism of ∂-functors $f: S \to T$ consists of natural transformations $f^i: S^i \to T^i$ that commute with the connecting morphisms. A ∂-functor S defined for $a \leqslant i < b$ is called **universal** if for any ∂-functor T defined in the same interval and any natural transformation $\varphi: S^a \to T^a$, there exists one and only one morphism $f: S \to T$ such that $f^a = \varphi$. Let $F: \mathcal{C} \to \mathcal{C}'$ be a covariant functor and b any positive integer. A universal covariant ∂-functor S defined for $0 \leqslant i < b$ is called a **right satellite** of F if $S^0 = F$ (S is then denoted by $\{S^i F\}$). If such an S exists, then it is unique and satisfies $S^{i+1}(F) = S^1(S^i F)$. If $\mathcal{C}$ has enough injective objects, the right satellites always exist, and if F is left exact, then $\{S^i F\}$ is a cohomological functor. The universality of ∂^*-functors is defined by reversing the arrows; the satellites $\{S_i F\}$ are then written as $\{S^{-i} F\}$ and called the **left satellites.**

Let $\mathcal{C}$ be an Abelian category with enough injectives. An **injective resolution** of an object A is a right resolution $Q = \{Q^i\}$ such that all Q^i are injective. Every A admits an injective resolution, which is unique up to chain equivalence (H. Cartan, 1950). For a covariant functor $F: \mathcal{C} \to \mathcal{C}'$, the functor $A \to H^i(F(Q))$, called the ith **right derived functor** $R^i F$ of F, is independent of Q. $\{R^i F\}$ is a cohomological functor. By the universality of satellites, there exists a morphism of ∂-functors $\{S^i F\} \to \{R^i F\}$ which is an isomorphism if and only if F is left exact. The **left derived functors** $L_i F$ of a contravariant functor F are defined similarly and are isomorphic to the left satellites when F is right exact. If $\mathcal{C}$ has enough projectives (instead of injectives), we define left (right) derived functors of covariant (contravariant) functors via **projective resolutions.** For a multifunctor, we define **partial derived functors** as well as (total) derived functors of the functor viewed as a functor defined in the †product category. For instance, let $T(A,B): \mathcal{C}_1 \times \mathcal{C}_2 \to \mathcal{C}'$ be contravariant in A and covariant in B. When $\mathcal{C}_2$ has enough injectives, we obtain $R_2^i T(A,B) = H^i(T(A,Q))$ using an injective resolution Q of B. Suppose that T satisfies condition (i) $A \to T(A,B)$ is exact for any injective B. Then for a fixed injective B, $R_2^i T(A,B)$ is a cohomological functor in A. When A has a projective resolution P in $\mathcal{C}_1$, we obtain $R_1^i T(A,B) = H^i(T(P,B))$ as well as the equation for the total derived functor $R^i T(A,B) = H^i(T(P,Q))$. We say that a functor T is **right balanced** if it satisfies (i) and also (ii) $B \to T(A,B)$ is exact for any projective A. In this case, the three derived functors are isomorphic. The **left balanced** functors are defined similarly. When the right derived functors of the functor Hom (which defines the category) exist, they are denoted by $\mathrm{Ext}^i(A,B)$.

D. Spectral Sequences

In this section, we deal with cohomology in the category ${}_R\mathfrak{M}$ of R-modules. A similar theory for homology is obtained by modifying the theory in a natural way. Similar constructions are also possible for general Abelian categories [3, 10].

A **filtration** F of a module A is a family of submodules $\{F^p(A) \mid p \in \mathbf{Z}\}$ such that $F^p(A) \supset F^{p+1}(A)$. We say that the filtration F is **convergent from above** (or **exhaustive**) if $\bigcup_p F^p(A) = A$, and F is **bounded from below** (or **discrete**) if $F^p(A) = 0$ for some p. The †graded module $G(A) = \{G^p(A) = F^p(A)/F^{p+1}(A) \mid p \in \mathbf{Z}\}$ is said to be **associated** with A. A **morphism** of filtered modules $f: A \to A'$ is a module homomorphism such that $f(F^p(A)) \subset F^p(A')$. It induces a homomorphism of the graded modules $G(A) \to G(A')$. A filtration of a complex $C = \{C^n, d\}$ consists of subcomplexes $F^p(C) = \{F^p(C^n)\}$ such that $F^p(C) \supset F^{p+1}(C)$. We assume that the complex C satisfies $\bigcup_p F^p(C) = C$, and is bounded from below; i.e., for every n there exists some p such that $F^p(C^n) = 0$. In particular, if $F^0(C) = C$, $F^{p+1}(C^p) = 0$, the complex C is called **canonically bounded**. Writing $C^{p,q} = G^p(C^{p+q})$, we obtain a †bigraded module $\{C^{p,q}\}$, in which p, q, and $p+q$ are called the **filtration degree**, the **complementary degree**, and the **total degree**, respectively.

A **spectral sequence** $\{E_r\}$ with a graded module $D = \{D^n\}$ as its **limit** (denoted by $E_2^{p,q} \Rightarrow_p D^n$) consists of a family of doubly graded modules $E_r = \{E_r^{p,q} \mid p,q \in \mathbf{Z}\}$ ($r \geqslant 2$ or sometimes $r \geqslant 1$) and differentiations $d_r: E_r^{p,q} \to E_r^{p+r,q-r+1}$ ($p,q \in \mathbf{Z}$) of degree $(r, 1-r)$ satisfying $d_r^2 = 0$ and satisfying the following two conditions: (i) $H(E_r)$ (with respect to d_r) is isomorphic to E_{r+1} (hence there exists a sequence of graded submodules of $E_2: 0 = B_2 \subset B_3 \subset \ldots \subset Z_3 \subset Z_2 = E_2$ such that $Z_r / B_r \cong E_r$); and (ii) there are submodules Z_∞ and B_∞ such that $\bigcup_k B_k \subset B_\infty \subset Z_\infty \subset \bigcap_k Z_k$, and $E_\infty = Z_\infty / B_\infty$ is isomorphic to the doubly graded module associated with a certain filtration F of D (i.e., $E_\infty^{p,q} \cong G^p(D^{p+q})$). We assume that $Z_\infty = \bigcap_k Z_k$ and $B_\infty = \bigcup_k B_k$ (**weak convergence**). Suppose that F is convergent from above and bounded from below and that $Z_k(E_2^{p,q})$ is stationary for every p,q. Then $\{E_r\}$ is called **regular.** $\{E_r\}$ is **bounded from below** if for every n there exists a p_0 such that $E_2^{p,n-p} = 0$ for $p < p_0$. In particular, if $E_2^{p,q} = 0$ ($p < 0$, $q < 0$), then $\{E_r\}$ is called the **first quadrant** (or **cohomology spectral sequence**). In the latter case, the **edge homomorphisms** $E_2^{p,0} \to E_\infty^{p,0}$, $E_\infty^{0,q} \to E_2^{0,q}$ are defined through **base terms** $E_k^{p,0}$ and **fiber terms** $E_k^{0,q}$, respectively. A morphism of spectral sequences $f: \{E_r, D\} \to \{E_r', D'\}$ consists of $f_r: E_r \to E_r'$ of degree (0,0) and $f: D \to D'$ of degree 0 which preserve the mechanism of spectral sequences. When the spectral sequences are regular, a morphism f is an isomorphism if one of the f_r is an isomorphism. Addition is naturally introduced in the set of morphisms so that spectral sequences form an additive category. An additive functor from an Abelian category $\mathcal{C}$ to this category is called a **spectral functor**. A filtered complex $\{C,F\}$ gives rise to a spectral sequence $E_2^{p,q} \Rightarrow G(H(C))$ if we put $Z_r^p = \{a \in F^p(C) | da \in F^{p+r}(C)\}$, $B_r^p = dZ_r^{p-r}$, $E_r^p = Z_r^p/(Z_{r-1}^{p+1} + B_{r-1}^p)$, $E_r = \Sigma_p E_r^p$. A double complex $C = \{C^{p,q}, d_{\mathrm{I}}, d_{\mathrm{II}}\}$ admits two natural filtrations $F_{\mathrm{I}}: F_{\mathrm{I}}^p(C) = \Sigma_{s \geqslant p} \Sigma_q C^{s,q}$ and $F_{\mathrm{II}}: F_{\mathrm{II}}^q(C) = \Sigma_{t \geqslant q} \Sigma_p C^{p,t}$. By the procedure above, these filtrations give rise to spectral sequences $H_{\mathrm{I}}^p(H_{\mathrm{II}}^q(C)) \Rightarrow_p H^n(C)$ and $H_{\mathrm{II}}^q(H_{\mathrm{I}}^p(C)) \Rightarrow_q H^n(C)$, respectively. Comparison of these sequences yields many useful results. Let T be an additive covariant functor from an Abelian category $\mathcal{C}$ to ${}_R\mathfrak{M}$, C be a complex in $\mathcal{C}$, and $Q = \{Q^{p,q}\}$ be an injective resolution of C. The double complex Q gives rise to spectral sequences $H^p(R^qT(C)) \Rightarrow H(T(Q))$ and $R^pT(H^q(C)) \Rightarrow H(T(Q))$. The limit $H(T(Q))$ is independent of Q and is called the **hypercohomology** of T with respect to C [2, 10]. We can similarly define hypercohomology of multifunctors. The theory of spectral sequences was initiated by J. Leray (1946) and was given suitable algebraic formulations by J. L. Koszul (1950).

E. Categories of Modules

The category ${}_R\mathfrak{M}$ (resp. $\mathfrak{M}_R$) of left (right) R-modules over a †unitary ring R is not only an Abelian category but also a Grothendieck category (→ 275 Modules). The †full embedding theorem permits us to deduce many propositions about general Abelian categories from the consideration of ${}_R\mathfrak{M}$. An object P of ${}_R\mathfrak{M}$ is projective if and only if it is isomorphic to a direct summand of a †free module. Any projective module is the direct sum of countably generated projective modules (I. Kaplansky, 1958). Finitely generated projective modules P_1 and P_2 are said to be equivalent if there exist finitely generated free modules F_1, F_2 such that $P_1 \oplus F_1 \cong P_2 \oplus F_2$. The equivalence classes then form an Abelian group with respect to the direct sum construction called the **projective class group** of a ring R. The category of complex †vector bundles over a compact space X is equivalent to the category of projective modules over $\mathbf{C}(X)$, the ring of complex-valued continuous functions on X, and similarly for other types of spaces and bundles. Many investigations have been made involving the problem of whether every projective module over a polynomial ring is free (J.-P. Serre, 1955). It has been observed that "big" projective modules are often free: for example, nonfinitely generated projective modules over an †indecomposable weakly Noetherian ring are free (Y. Hinohara, 1963).

The nth right derived functor of $\mathrm{Hom}_R(A, B)$ is denoted by $\mathbf{Ext}_R^n(A,B)$ (→ 57 Chain Complexes I). This is a bifunctor ${}_R\mathfrak{M} \times {}_R\mathfrak{M} \to (\mathrm{Ab})$, contravariant in A and covariant in B. Ext_R^0 is isomorphic to and identified with Hom_R. An exact sequence $0 \to A' \to A \to A'' \to 0$ gives rise to the connecting homomorphisms $\Delta^n: \mathrm{Ext}_R^n(A',B) \to \mathrm{Ext}_R^{n+1}(A'',B)$, and the following sequence is exact: $\ldots \to \mathrm{Ext}_R^{n-1}(A',B) \xrightarrow{\Delta} \mathrm{Ext}_R^n(A'',B) \to \mathrm{Ext}_R^n(A,B) \to \mathrm{Ext}_R^n(A',B) \xrightarrow{\Delta} \mathrm{Ext}_R^{n+1}(A'',B) \to \ldots$ (the exact sequence of Ext). Similarly, an exact sequence $0 \to B' \to B \to B'' \to 0$ gives rise to $\Delta^n: \mathrm{Ext}_R^n(A,B'') \to \mathrm{Ext}_R^{n+1}(A,B')$ and to an exact sequence of Ext. An **extension** of A by B (or of B by A) is an exact sequence $(E): 0 \to B \to X \to A \to 0$. The set of equivalence classes of extensions of A by B is in one-to-one correspondence with $\mathrm{Ext}_R^1(A,B)$ by assigning to (E) its **characteristic class** $\chi_E = \Delta^0(1) \in \mathrm{Ext}_R^1(A,B)$, where 1 denotes the identity of $\mathrm{Hom}_R(B,B)$. In this correspondence, the sum of two extensions is obtained by a construction called **Baer's sum** of extensions. Similarly, $\mathrm{Ext}_R^n(A,B)$ is interpreted as the set of the equivalence classes of n-fold extensions $0 \to B \to X_{n-1} \to \ldots \to X_0 \to A \to 0$ (exact). This point of view permits us to establish a theory of Ext, etc., in more general (additive) categories lacking enough projectives or injectives (N. Yoneda, 1954, 1960).

The tensor product $A \otimes_R B$ is a right exact covariant bifunctor $\mathfrak{M}_R \otimes {}_R\mathfrak{M} \to (\mathrm{Ab})$. If the functor $t_P(\cdot) = \cdot \otimes P$ is exact, P is called a †flat module. A projective module is flat. In general, a flat module is the †inductive limit of finitely generated free modules (M. Lazard, 1964). A flat module P is called †faithfully flat if $P \neq \mathfrak{m}P$ for every maximal ideal $\mathfrak{m}$ of R. The functors $\otimes$ and Hom are related by †adjointness (→ 53 Categories and Functors). From this viewpoint $\otimes$ can be introduced in more general categories. Left-derived functors of $A \otimes_R B$ are denoted by $\mathbf{Tor}_n^R(A,B)$ and are called nth **torsion products** of A and B. $\mathrm{Tor}_1^R(A,B)$ is often denoted by $A *_R B$. The functor $\otimes_R$ is left balanced, hence Tor is calculated by using projective resolutions of A, B, or both A and B. We have $\mathrm{Tor}_0^R = \otimes_R$. An exact sequence $0 \to A' \to A \to A'' \to 0$ gives rise to $\Delta_n: \mathrm{Tor}_{n+1}^R(A'',B) \to \mathrm{Tor}_n^R(A',B)$ and the infinite exact sequence of Tor, and similarly for the second variables.

From the various relations between Hom and $\otimes$ follow the corresponding relations between their derived functors. When Λ and Γ are algebras over K and $\Omega=\Lambda\otimes\Gamma$, we can define the **external product** ($\top$**-product**), which is a mapping

$$\top:\mathrm{Tor}_p^{\Lambda}(A,B)\otimes\mathrm{Tor}_q^{\Gamma}(A',B')\to\mathrm{Tor}_{p+q}^{\Omega}(A\otimes A',B\otimes B').$$

In particular, if Λ and Γ are K-projective and $\mathrm{Tor}_n^k(A,A')=0$ $(n>0)$, then we can define the **wedge product** ($\vee$**-product**) $\vee:\mathrm{Ext}_{\Lambda}^p(A,B)\otimes\mathrm{Ext}_{\Gamma}^q(A',B')\to\mathrm{Ext}_{\Omega}^{p+q}(A\otimes A',B\otimes B')$. The latter is described in terms of the composition of module extensions. When $K=\Lambda=\Gamma=\Omega$, the $\top$-product reduces to the **internal product**, called the $\pitchfork$**-product**. If Λ is a †Hopf algebra over K, the †comultiplication $\Lambda\to\Lambda\otimes\Lambda$ induces $\mathrm{Ext}_{\Lambda\otimes\Lambda}\to\mathrm{Ext}_{\Lambda}$. This, combined with the $\vee$-product, yields the **cup product** ($\smile$**-product**) $\smile:\mathrm{Ext}_{\Lambda}^p(A,B)\otimes\mathrm{Ext}_{\Lambda}^q(A',B')\to\mathrm{Ext}_{\Lambda}^{p+q}(A\otimes A',B\otimes B')$. We define similarly $\perp$**-product**, $\wedge$**-product**, $\uplus$**-product**, and $\frown$**-product (cap product)** [2]. Let Λ, Γ, and Σ be algebras over K, with Λ K-projective; let $A\in\mathfrak{M}_{\Lambda\otimes\Gamma}$, $B\in{}_{\Lambda}\mathfrak{M}_{\Sigma}$, $C\in\mathfrak{M}_{\Gamma\otimes\Sigma}$, and assume $\mathrm{Tor}_n^{\Lambda}(A,B)=0$ $(n>0)$. The natural isomorphism $\mathrm{Hom}_{\Lambda\otimes\Gamma}(A,\mathrm{Hom}_{\Sigma}(B,C))\cong\mathrm{Hom}_{\Gamma\otimes\Sigma}(A\otimes_{\Lambda}B,C)$ then yields a spectral sequence $\mathrm{Ext}_{\Lambda\otimes\Gamma}^p(A,\mathrm{Ext}_{\Sigma}^q(B,C))\Rightarrow_p\mathrm{Ext}_{\Gamma\otimes\Sigma}^n(A\otimes_{\Lambda}B,C)$ by the double complex argument in Section D.

The **homological dimension** $\mathrm{h\,dim}_R A$, $\mathrm{dh}_R A$, or **projective dimension** $\mathrm{proj\,dim}_R A$ of $A\in{}_R\mathfrak{M}$ is the supremum ($\leqslant\infty$) of n such that $\mathrm{Ext}_R^n(A,B)\neq0$ for some B. The relation $\mathrm{h\,dim}_R A\leqslant0$ means that A is projective. The **injective dimension** $\mathrm{inj\,dim}_R B$ of $B\in{}_R\mathfrak{M}$ is defined similarly by means of the functor $\mathrm{Ext}_R^n(\cdot,B)$, and the **weak dimension** $\mathrm{w\,dim}_R C$ of $C\in{}_R\mathfrak{M}$ by the functor $\mathrm{Tor}_n^R(\cdot,C)$. The common value $\sup\{\mathrm{proj\,dim}_R A\mid A\in{}_R\mathfrak{M}\}=\sup\{\mathrm{inj\,dim}_R B\mid B\in{}_R\mathfrak{M}\}$ is called the **left global dimension** $\mathrm{l\ gl\ dim}\,R$ of R. It is identical to the supremum of homological dimensions of †cyclic modules (M. Auslander, 1955). The **right global dimension** $\mathrm{r\ gl\ dim}\,R$ is defined similarly. The common value $\sup\{\mathrm{w\,dim}_R A\mid A\in\mathfrak{M}_R\}=\sup\{\mathrm{w\,dim}_R C\mid C\in{}_R\mathfrak{M}\}$ is called the **weak global dimension** $\mathrm{w\ gl\ dim}\,R$ of R. We have $\mathrm{w\ gl\ dim}\,R\leqslant\mathrm{l\ gl\ dim}\,R$, $\mathrm{r\ gl\ dim}\,R$. The equality may fail to hold (Kaplansky, 1958). If R is †Noetherian, the three global dimensions coincide (Auslander, 1955) and are called simply the **global dimension** of R: $\mathrm{gl\,dim}\,R$. The condition $\mathrm{l\ gl\ dim}\,R=0$ (or $\mathrm{r\ gl\ dim}\,R=0$) holds if and only if R is an †Artinian semisimple ring, while $\mathrm{w\ gl\ dim}\,R=0$ if and only if R is a †regular ring in the sense of J. Von Neumann (M. Harada, 1956).

A ring R is called **left (right) hereditary** if $\mathrm{l\ gl\ dim}\,R\leqslant1$ ($\mathrm{r\ gl\ dim}\,R\leqslant1$), and **left (right) semihereditary** if every finitely generated left (right) ideal is projective. A left and right (semi)hereditary ring is called a **(semi)hereditary ring**. Since projectivity and †invertibility of an ideal of a (commutative) integral domain R are equivalent, R is hereditary if and only if R is a †Dedekind ring. In this case, the projective class group of R reduces to the †ideal class group. An integral domain R is semihereditary if and only if $\mathrm{w\ gl\ dim}\,R\leqslant1$ (A. Hattori, 1957), and in that case R is called a **Prüfer ring**. A †maximal order over a Dedekind ring is hereditary. A commutative semihereditary ring R is characterized by the property that flatness of R-modules is equivalent to torsion-freeness (S. Endo, 1961). A Noetherian ring R is left self-injective if and only if R is †quasi-Frobenius (M. Ikeda, 1952), and the global dimension of a quasi-Frobenius ring is 0 or ∞ (S. Eilenberg and T. Nakayama, 1955). A polynomial ring $R=K[X_1,\ldots,X_n]$ over a commutative ring K satisfies $\mathrm{gl\ dim}\,R=\mathrm{gl\ dim}\,K+n$. When K is a field, this is a reformulation of Hilbert's theory of †syzygy sequences (→ 364 Rings of Polynomials). In this sense, the study of the global dimension of rings and categories is sometimes called **syzygy theory** (Eilenberg, 1956). The homological algebra of commutative Noetherian rings has been studied extensively and is useful in algebraic geometry. Since $\mathrm{gl\ dim}\,R=\sup_{\mathfrak{m}}\mathrm{gl\ dim}\,R_{\mathfrak{m}}$ ($R_{\mathfrak{m}}$ is the †ring of quotients relative to $\mathfrak{m}$), with $\mathfrak{m}$ running over the maximal ideals of R, the problem of determining $\mathrm{gl\ dim}\,R$ reduces to the case of †local rings. A finitely generated flat module over a local ring R is free. If K denotes the residue field $R/\mathfrak{m}$, where $\mathfrak{m}$ is the maximal ideal of the local ring R, $\mathrm{Tor}^R(K,K)$ has the structure of a Hopf algebra (E. F. Assmus Jr., 1959). Detailed results concerning the **Betti numbers** $\dim\mathrm{Tor}_i^R(K,K)$ of R have been obtained (J. Tate, 1957, et al.). In particular, R is †regular if and only if $\mathrm{gl\ dim}\,R<\infty$ (Serre, 1955). A local ring R is called a **Gorenstein ring** if the injective dimension of R-module R is finite. This is a notion intermediate between regular rings and †Macaulay rings (→ 281 Noetherian Rings).

Consideration of a ring R in relation to a subring S leads to **relative homological algebra**. Foundations for this theory were established by Hochschild (1956). An exact sequence of R-modules that †splits as a sequence of S-modules is called an **(R,S)-exact sequence**. An R-module P is called an **(R,S)-projective module** if $\mathrm{Hom}_R(P,\cdot)$ maps any (R,S)-exact sequence to an exact sequence. **(R,S)-injective modules** are defined similarly. Based on these notions, $\mathrm{Ext}_{(R,S)}$ and

$\mathrm{Tor}^{(R,S)}$ are defined as the **relative derived functors** of Hom_R and $\otimes_R$, respectively. We also have a relative theory from a different viewpoint (S. Takasu, 1957). Relative theory is extended to general categories from various viewpoints [8].

F. Cohomology Theory for Associative Algebras

Let Λ be an †algebra over a commutative ring K and A a †two-sided Λ-module. Let C^n be the module of all n-linear mappings of Λ into A called ***n*-cochains** ($C^0 = A$). Define the coboundary operator $\delta^n : C^n \to C^{n+1}$ by $(\delta^n f)(\lambda_1, \ldots, \lambda_{n+1}) = \lambda_1 f(\lambda_2, \ldots, \lambda_{n+1}) + \sum_{i=1}^{n}(-1)^i f(\lambda_1, \ldots, \lambda_i\lambda_{i+1}, \ldots, \lambda_{n+1}) + (-1)^{n+1} f(\lambda_1, \ldots, \lambda_n)\lambda_{n+1}$.

We thus obtain a complex whose cohomology is denoted by $H^n(\Lambda, A)$ and is called the nth **Hochschild's cohomology group** of Λ relative to the **coefficient module** A (Hochschild, 1945). A cochain f is called **normalized** if $f(\lambda_1, \ldots, \lambda_n) = 0$ whenever one of the λ_i is 1. We obtain the same cohomology group $H^n(\Lambda, A)$ from the subcomplex of normalized cochains. $\{H^n(\Lambda, \cdot)\}$ is a cohomological functor from the category ${}_\Lambda\mathfrak{M}_\Lambda$ of two-sided Λ-modules to the category ${}_K\mathfrak{M}$. Using the **enveloping algebra** $\Lambda^e = \Lambda \otimes_K \Lambda^0$, where Λ^0 is an anti-isomorphic copy of Λ, ${}_\Lambda\mathfrak{M}_\Lambda$ may be identified with ${}_{\Lambda^e}\mathfrak{M}$ (and $\mathfrak{M}_{\Lambda^e}$). If Λ is K-projective, $\{H^n(\Lambda, \cdot)\}$ is isomorphic to $\{\mathrm{Ext}^n_{\Lambda^e}(\Lambda, \cdot)\}$. (In [2], $H^n(\Lambda, A)$ is defined as $\mathrm{Ext}^n_{\Lambda^e}(\Lambda, A)$ in general.) We have $H^0(\Lambda, A) = \{a \in A \mid \lambda a = a\lambda, \forall \lambda \in \Lambda\}$. We call 1-cocycles **derivations** (or **crossed homomorphisms**) of Λ in A and 1-coboundaries **inner derivations**. Thus $H^1(\Lambda, A)$ is the derivation class group and is related to the †ramification. When K is a field, $H^1(\Lambda, \cdot) = 0$ if and only if Λ is a †separable algebra. In general, an algebra Λ over a commutative ring K is called a **separable algebra** if Λ is Λ^e-projective, i.e., if $\mathrm{Ext}^1_{\Lambda^e}(\Lambda, \cdot) = 0$ (Auslander and O. Goldman, 1960). This is a generalization of the notion of †maximally central algebras (Nakayama and G. Azumaya, 1948). We have a one-to-one correspondence of $H^2(\Lambda, A)$ to the family of **algebra extensions** of Λ with kernel A (i.e, K-algebras Γ containing A as a two-sided ideal such that $\Gamma/A = \Lambda$) satisfying $A^2 = 0$. Any extension of an algebra Λ over a field K such that $H^2(\Lambda, \cdot) = 0$ splits over a nilpotent kernel (J. H. C. Whitehead and G. Hochschild). This holds in particular for a separable algebra, and we obtain the †Wedderburn-Mal'cev theorem. There are some interpretations of $H^3(\Lambda, A)$ in terms of extensions.

The supremum ($\leqslant \infty$) of n such that $H^n(\Lambda, A) \neq 0$ for some A is called the **cohomological dimension** of Λ and written $\dim \Lambda$. For a finite-dimensional algebra Λ over a field K, $\dim \Lambda < \infty$ if and only if Λ/N is separable and $\mathrm{gl\,dim}\,\Lambda < \infty$, where N is the †radical of Λ (N. Ikeda, H. Nagao, and Nakayama, 1954).

The homology groups $H_n(\Lambda, A)$ of Λ relative to a coefficient module A are defined similarly. If Λ is K-projective, $\{H_n(\Lambda, \cdot)\}$ is isomorphic to $\{\mathrm{Tor}^{\Lambda^e}_n(\cdot, \Lambda)\}$.

G. Cohomology of Groups

The pair consisting of an algebra Λ over K and an algebra homomorphism $\varepsilon : \Lambda \to K$ is called a **supplemented algebra** [2] (or **augmented algebra** [8]), of which ε is the **augmentation**. The †group algebra $\mathbf{Z}[G]$ of a group G over the ring of rational integers is a supplemented algebra, in which the augmentation is defined by $\varepsilon(x) = 1$ ($x \in G$). The category of left G-modules is identified with the category of left $\mathbf{Z}[G]$-modules. For a finite group G, a finitely generated projective G-module is not necessarily free (D. S. Rim, 1959) and is isomorphic to the direct sum of a free module and a left ideal of $\mathbf{Z}[G]$. It follows that the projective class group of $\mathbf{Z}[G]$ is a finite group (R. G. Swan, 1960). The **cohomology groups** and **homology groups** of G relative to $A \in {}_G\mathfrak{M}$ (Eilenberg and S. MacLane, 1943) are defined by $H^n(G, A) = \mathrm{Ext}^n_{\mathbf{Z}[G]}(\mathbf{Z}, A)$ and $H_n(G, A) = \mathrm{Tor}^{\mathbf{Z}[G]}_n(\mathbf{Z}, A)$, respectively. Their concrete description is given usually via the **$\mathbf{Z}[G]$-standard resolution** of $\mathbf{Z}$.

1. Homogeneous Formulation. The group of **homogeneous *n*-chains** is the free Abelian group with basis $G \times \ldots \times G$ ($n+1$ times), on which G operates by $x(x_0, \ldots, x_n) = (xx_0, \ldots, xx_n)$, and the boundary operator is defined by $d(x_0, \ldots, x_n) = \sum_{i=0}^{n}(-1)^i(x_0, \ldots, \hat{x}_i, \ldots, x_n)$.

2. Nonhomogeneous Formulation. The group of **nonhomogeneous *n*-chains** is the $\mathbf{Z}[G]$-free module with basis $G \times \ldots \times G$ (n times), and the boundary operator is defined by $d(x_1, \ldots, x_n) = x_1(x_2, \ldots, x_n) + \sum_{i=1}^{n-1}(-1)^i(x_1, \ldots, x_i x_{i+1}, \ldots, x_n) + (-1)^n(x_1, \ldots, x_{n-1})$. A nonhomogeneous 2-cocycle is sometimes called a **factor set**. $H^0(G, A)$ is the submodule A^G of A consisting of the G-invariant elements, while $H_0(G, A)$ is the largest residue class module A_G of A on which G acts trivially. Given two groups G and K, an exact sequence of group homomorphisms $1 \to K \to E \to G \to 1$ is called a **group extension** of G over the kernel K. When K is Abelian, the extension canonically induces a G-module structure on

K, and the deviation of K from being a semidirect factor of E is measured by a factor set. The group $H^2(G,A)$ is thus in one-to-one correspondence with the set of equivalence classes of the †group extensions of G over A which induce the originally given G-module structure on A (→ 193 Groups N). This point of view is essential in the proof of the †Schur-Zassenhaus theorem (→ 160 Finite Groups). $H^3(G,A)$ is interpreted as the set of obstructions for extensions (Eilenberg and MacLane, 1947). For a †free group F, $H^n(F,A)=0$ ($n>1$). If a group G is represented as a factor group F/R of a free group F, we have a group extension $1 \to K \to E \to G \to 1$, where $K=R/[R,R]$ and $E=F/[R,R]$. Let $\xi \in H^2(G,K)$ correspond to this extension. Then for any G-module A, the cup product $\chi \to \chi \smile \xi$ followed by the pairing $\mathrm{Hom}(K,A) \otimes K \to A$ provides isomorphisms $H^n(G,\mathrm{Hom}(K,A)) \cong H^{n+2}(G,A)$ ($n>0$) (the **cup product reduction theorem** of Eilenberg and MacLane, 1947); similarly, we have the reduction theorem for the homology. The $\mathbf{Z}$-algebra $H(G,\mathbf{Z})=\sum_{n=0}^{\infty} H^n(G,\mathbf{Z})$ under the multiplication defined by the cup product is finitely generated if G is a finite group (B. B. Venkov, 1959; L. Evens, 1961).

The following are mappings relative to a subgroup H. (1) The inner automorphism by $x \in G$ induces an isomorphism of $H^n(H,A)$ and $H^n(xHx^{-1},A)$ which reduces to the identity of $H^n(G,A)$ if $H=G$. Hence if H is a normal subgroup, $H^n(H,A)$ has the structure of a G/H-module, and similarly for $H_n(H,A)$.

(2) When H is a normal subgroup of G, the mapping $(x_1,\ldots,x_n) \to (x_1H,\ldots,x_nH)$ of nonhomogeneous chains induces the **inflation** (or **lift**) $\mathrm{Inf}: H^n(G/H,A^H) \to H^n(G,A)$ and the **deflation** $\mathrm{Def}: H_n(G,A) \to H_n(G/H,A_H)$.

(3) The embedding of nonhomogeneous chains induces the **restriction** $\mathrm{Res}: H^n(G,A) \to H^n(H,A)$ and the **injection** Inj (or **corestriction** Cor)$: H_n(H,A) \to H_n(G,A)$. The theory of †induced representation gives another construction of these mappings; that is, if we put $\iota^G(A)=\mathrm{Hom}_{\mathbf{Z}[H]}(\mathbf{Z}[G],A)$, Res is obtained by the isomorphism $H^n(G,\iota^G(A)) \cong H^n(H,A)$ combined with the homomorphism induced by $A \to \iota^G(A)$; while if we put $\iota_G(A)=\mathbf{Z}[G] \otimes_{\mathbf{Z}[H]} A$, Inj is obtained by the isomorphism $H_n(H,A) \cong H_n(G,\iota_G(A))$ followed by the homomorphism induced by $\iota_G(A) \to A$.

(4) If $(G:H)<\infty$, we have $\iota^G(A) \cong \iota_G(A)$. The composition of $H^n(H,A) \to H^n(G,\iota_G(A)) \to H^n(G,A)$ defines on the cohomology groups the $\mathrm{Inj}: H^n(H,A) \to H^n(G,A)$, while the composition $H_n(G,A) \to H_n(G,\iota^G(A)) \to H_n(H,A)$ gives the $\mathrm{Res}: H_n(G,A) \to H_n(H,A)$. In particular, $\mathrm{Res}: H_1(G,\mathbf{Z}) \to H_1(H,\mathbf{Z})$ coincides with the †transfer $G/[G,G] \to H/[H,H]$.

(5) Let H be a normal subgroup of G. Consider the additive relation ρ (the correspondence) between $h \in Z^n(H,A)^G$ and $f \in Z^{n+1}(G/H,A^H)$ determined by $\rho(h,f)$ if and only if there exists a $g \in C^n(G,A)$ such that $h=\mathrm{Res}\, g$ and $\mathrm{Inf} f=\delta g$. If the relation induces a homomorphism $H^n(H,A)^G \to H^{n+1}(G/H,A^H)$, then it is called the **transgression**. If $H^i(H,A)=0$ ($0<i<n$), the sequence $0 \to H^n(G/H,A^H) \to H^n(G,A) \to H^n(H,A)^G \to H^{n+1}(G/H,A^H) \to H^{n+1}(G,A)$ composed of inflation, restriction, and transgression mappings is exact (Hochschild and Serre, 1953) and is called the **fundamental exact sequence**. This exact sequence may be derived from a certain spectral sequence $H^n(G/H,H^q(H,A)) \Rightarrow_p H^n(G,A)$ (R. C. Lyndon, 1948; Hochschild and Serre, 1953).

The relative (co)homology theory relative to a subgroup (I. T. Adamson, 1954) may be dealt with in terms of the relative Ext and the relative Tor (Hochschild, 1956). Many results in the absolute case are generalized to the relative case: for example, the fundamental exact sequence (Nakayama and Hattori, 1958). The relative theory is further generalized to the cohomology theory of †permutation representations of G (E. Snapper, 1964).

H. Non-Abelian Cohomology

For a non-Abelian G-group A, the cohomology "set" $H^1(G,A)$ (and $H^0(G,A)$) is defined as in the Abelian case by means of the nonhomogeneous cochains [e.g., 11]. Some efforts are being made toward the construction of a more general non-Abelian theory.

I. Finite Groups

Let G be a finite group and A a G-module. Define the norm $N: A \to A$ by $N(a)=\sum_{x \in G} xa$, and denote $\mathrm{Ker}\, N$ by ${}_NA$. The kernel of the augmentation $\varepsilon: \mathbf{Z}[G] \to \mathbf{Z}$ is denoted by I. Put $\hat{H}^n(G,A)=H^n(G,A)$ ($n>0$), $\hat{H}^0(G,A)=A^G/NA$, $\hat{H}^{-1}(G,A)={}_NA/IA$, and $\hat{H}^{-n}(G,A)=H_{n-1}(G,A)$ ($n>1$). Then $\{\hat{H}^n(G,\cdot)\}$ forms a cohomological functor (E. Artin and J. T. Tate) and can be described as the set of cohomology groups concerning a certain complex called a **complete free resolution** of $\mathbf{Z}$. (Similar arguments are valid more generally for quasi-Frobenius rings (Nakayama, 1957), and a theory of this kind is called **complete cohomology theory**.) We have $\hat{H}^n(G,A)=0$ ($n \in \mathbf{Z}$) if and only if $\mathrm{h\,dim}_{\mathbf{Z}[G]} A \leqslant 1$ (Nakayama, 1957). If A satisfies the conditions (i) $\hat{H}^1(G_p,A)=0$ for any Sylow p-subgroup G_p of G, and (ii) there exists a $\xi \in \hat{H}^2(G,A)$ such that $\mathrm{Res}\,\xi \in \hat{H}^2(G_p,A)$ has the same order as G_p and generates all of $\hat{H}^2(G_p,A)$, then the homomorphisms $\hat{H}^n(H,B)\to\hat{H}^{n+2}(H,A\otimes B)$ $(n\in\mathbf{Z})$ defined by the cup product with $\operatorname{Res}\xi$ are isomorphisms for every subgroup H and every G-module B such that $\operatorname{Tor}(A,B)=0$ (Nakayama, 1957; for $B=\mathbf{Z}$, Tate, 1952). If G is cyclic, the mappings $\hat{H}^n(A)\to\hat{H}^{n+2}(A)$ $(n\in\mathbf{Z})$ defined by the cup product with a generator of $\hat{H}^2(\mathbf{Z})$ are isomorphisms. (The notation is abbreviated by omitting G.) If the orders of $\hat{H}^0(A)$ and $\hat{H}^1(A)$ are finite, their ratio is called the **Herbrand quotient** $h(A)$ of A. If $0\to A'\to A\to A''\to 0$ is exact, then $h(A)=h(A')h(A'')$. If A is finite, then $h(A)=1$. By combining these two facts we obtain **Herbrand's lemma**: If A' is a sub-G-module of A of finite index and $h(A')$ exists, then $h(A)$ also exists and $h(A)=h(A')$. The periodicity $\hat{H}^n(A)=\hat{H}^{n+p}(A)$ $(n\in\mathbf{Z}, A\in{}_G\mathfrak{M})$ holds if and only if every Sylow subgroup is cyclic or a †generalized quaternion group (Artin and Tate; [2]).

Let L/K be a finite †Galois extension with the †Galois group G. The cohomology groups of various types of G-modules related to L/K are called the **Galois cohomology** groups (→ 177 Galois Theory). Using **continuous cocycles**, a cohomology theory (**Tate cohomology**) is developed for infinite Galois extensions as well [11, 12]. By means of Galois cohomology (→ 62 Class Field Theory), the cohomology theory of finite groups and of †totally disconnected compact groups (which are †profinite groups) plays an important role in class field theory and its related branches.

J. Cohomology Theory of Lie Algebras

Let $\mathfrak{g}$ be a †Lie algebra over a commutative ring K, and assume that $\mathfrak{g}$ is K-free. The †enveloping algebra $U=U(\mathfrak{g})$ is a †supplemented algebra over K. For a $\mathfrak{g}$-module (= U-module) A, $\operatorname{Ext}^n_U(K,A)$ and $\operatorname{Tor}^U_n(K,A)$ are called the **cohomology groups** $H^n(\mathfrak{g},A)$ and **homology groups** $H_n(\mathfrak{g},A)$, respectively, of $\mathfrak{g}$ relative to the coefficient module A. They are usually described by means of the U-free resolution $U\otimes\bigwedge_K(\mathfrak{g})$ of K (called the **standard complex** of $\mathfrak{g}$) constructed by C. Chevalley and Eilenberg (1948), where $\bigwedge_K(\mathfrak{g})$ is the exterior algebra of the K-module $\mathfrak{g}$ and (denoting $1\otimes(x_1\wedge\ldots\wedge x_n)$ by $(x_1,\ldots,x_n)$) the differentiation is given by

$$d(x_1,\ldots,x_n)=$$
$$\sum_{i=1}^{n}(-1)^{i+1}x_i(x_1,\ldots,\hat{x}_i,\ldots,x_n)+$$
$$\sum_{1\leqslant i<j\leqslant n}(-1)^{i+j}([x_i,x_j],x_1,\ldots,\hat{x}_i,\ldots,\hat{x}_j,\ldots,x_n).$$

For $n>[\mathfrak{g}:K]$, $H^n(\mathfrak{g},A)=H_n(\mathfrak{g},A)=0$. If $\mathfrak{g}$ is a †semisimple Lie algebra over a field K of characteristic 0, we have $H^1(\mathfrak{g},A)=0$, $H^2(\mathfrak{g},A)=0$, while $H^3(\mathfrak{g},K)\neq 0$. $H^1(\mathfrak{g},A)=0$ is equivalent to Weyl's theorem, which asserts the complete reducibility of finite-dimensional representations (→ 247 Lie Algebras E). $H^2(\mathfrak{g},A)$ and $H^3(\mathfrak{g},A)$ are interpreted by means of Lie algebra extensions as in the cohomology of groups. The theorem on †Levi decomposition is derived from $H^2(\mathfrak{g},A)=0$. Chevalley and Eilenberg constructed this cohomology by algebraization of the cohomology of compact †Lie groups. They also introduced the notion of relative cohomology groups $H^n(\mathfrak{g},\mathfrak{h},A)$ relative to a Lie subalgebra $\mathfrak{h}$ of $\mathfrak{g}$, which correspond to the cohomology of homogeneous spaces. $H^n(\mathfrak{g},\mathfrak{h},A)$ does not always coincide with $\operatorname{Ext}^n_{U(\mathfrak{g}),U(\mathfrak{h})}(K,A)$ (Hochschild, 1956), but does so in an important case where K is a field of characteristic 0 and $\mathfrak{h}$ is †reductive in $\mathfrak{g}$ (→ 247 Lie Algebras).

For †transformation spaces of †linear algebraic groups G over a field K, the **rational cohomology groups** are introduced using the notion of **rational injectivity** (Hochschild, 1961). In particular, if G is a †unipotent algebraic group over a field K of characteristic 0, then $H(G,A)$ is isomorphic to $H(\mathfrak{g},A)$, where $\mathfrak{g}$ is the Lie algebra of G. There is also a relative theory.

K. Amitsur Cohomology

Let R be a commutative ring, and F a covariant functor from the category $\mathcal{C}_R$ of commutative R-algebras to the category of Abelian groups. For $S\in\mathcal{C}_R$ and $n=0,1,2,\ldots$, we write $S^{(n)}=S\otimes\ldots\otimes S$ (n-fold tensor product over R). Let $\varepsilon_i:S^{(n+1)}\to S^{(n+2)}$ $(i=0,1,\ldots,n+1)$ be $\mathcal{C}_R$-morphisms defined by $\varepsilon_i(x_0\otimes\ldots\otimes x_n)=x_0\otimes\ldots\otimes x_{i-1}\otimes 1\otimes x_i\otimes\ldots\otimes x_n$. Defining $d^n:F(S^{(n+1)})\to F(S^{(n+2)})$ by $d^n=\sum_{i=0}^{n+1}(-1)^iF(\varepsilon_i)$, we obtain a cochain complex $\{F(S^{(n+1)}),d^n\}$. This complex and its cohomology groups are called the **Amitsur complex** and the **Amitsur cohomology groups**, and are usually denoted by $C(S/R,F)$ and $H^n(S/R,F)$ respectively.

If S/R is a finite Galois extension with Galois group G, then the group $H^n(S/R,U)$ of the unit group functor U is naturally isomorphic to $H^n(G,U(S))$. If S/R is a finite purely inseparable extension, then $H^n(S/R,U)=0$ for $n\geqslant 3$. The group $H^2(S/R,U)$ is related to the †Brauer group $B(S/R)$ (→ 31 Associative Algebras K).

References

[1] Séminaire H. Cartan, 1950–1951, Paris, 1951.

[2] H. P. Cartan and S. Eilenberg, Homological algebra, Princeton Univ. Press, 1956.
[3] A. Grothendieck, Sur quelques points d'algèbre homologique, Tôhoku Math. J., (2) 9 (1957), 119–221.
[4] T. Nakayama and A. Hattori, Homorozî daisûgaku (Japanese; Homological algebra), Kyôritu, 1957.
[5] R. Godement, Topologie algébrique et théorie des faisceaux, Actualités Sci. Ind., Hermann, 1958.
[6] D. G. Northcott, An introduction to homological algebra, Cambridge Univ. Press, 1960.
[7] S. Iyanaga and K. Kodaira, Gendai sûgaku gaisetu I (Japanese; Introduction to modern mathematics I), Iwanami, 1961.
[8] S. MacLane, Homology, Springer, 1963.
[9] P. Freyd, Abelian categories, Harper, 1964.
[10] A. Grothendieck (and J. Dieudonné), Eléments de géometrie algébrique I, II, III, Publ. Math. Inst. HES, 1960–1963.
[11] J.-P. Serre, Cohomologie galoisienne, Lecture notes in math. 5, Springer, 1964.
[12] S. Lang, Rapport sur la cohomologie des groupes, Benjamin, 1966.
[13] E. Weiss, Cohomology of groups, Academic Press, 1969.
[14] H. Bass, Algebraic K-theory, Benjamin, 1968.

203 (IX.4) Homology Groups

A. History

Let γ_1, γ_1' be two oriented paths on a †torus T (Fig. 1), and let Δ be the domain bounded by γ_1 and γ_1' and supplied with the orientation shown. Then the oriented boundary of Δ may be written as $\gamma_1 - \gamma_1'$, where $-\gamma_1'$ is the path γ_1' with its orientation reversed. In such a case we say that γ_1 and γ_1' are homologous on T and write $\gamma_1 \sim \gamma_1'$. Generally, let $C = \Sigma c_i \gamma_i$ and $C' = \Sigma c_i' \gamma_i$ be linear combinations with integral coefficients of finite closed paths γ_i on T. If there exists a domain Δ in T such that $C - C'$ is an integral multiple of the boundary of Δ, then we write $C \sim C'$. When T is the †Riemann surface of an †elliptic function and $w\,dz$ is a †differential of the first kind on T, we set $\int_C w\,dz = \Sigma c_i \int_{\gamma_i} w\,dz$. If $C \sim C'$, then we have $\int_C w\,dz = \int_{C'} w\,dz$.

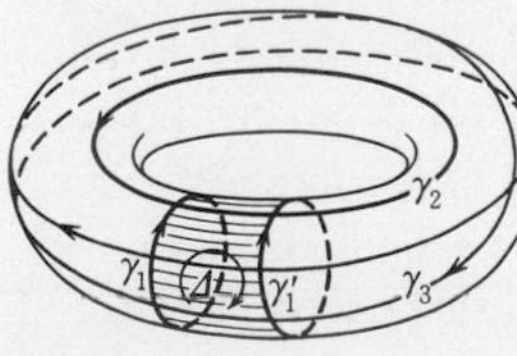

Fig. 1

In the first example (Fig. 1), it can be shown that any closed path on T is homologous to a linear combination of γ_1 and γ_2 with integral coefficients (for example, $\gamma_3 \sim 2\gamma_2$). If we classify all the closed paths on T by the homology relation $\sim$, then the equivalence classes constitute an †Abelian group. This group is called the 1-dimensional homology group on T. Thus we see that this group is a †free Abelian group with two generators, the equivalence classes of γ_1 and γ_2. Generalizing just such a consideration, E. Betti (1870) defined the notion of the r-dimensional homology group $H_r(M^n)$ $(0 \leqslant r \leqslant n)$ for any n-dimensional manifold M^n. The definition given by Betti was made precise by H. Poincaré (1895), who introduced the notions of †simplicial complexes and cycles, thereby also introducing algebraic or combinatorial methods into topology. J. W. Alexander proved the †topological invariance of the homology group $H_r(K)$ of a simplicial complex K. Namely, he showed that if the polyhedra $|K|$ and $|K'|$ of given simplicial complexes K and K' are homeomorphic, then their homology groups $H_r(K)$ and $H_r(K')$ of each dimension are isomorphic (**invariance theorem**, *Trans. Amer. Math. Soc.*, 16 (1915)). Thus it was shown that homology groups are algebraic objects that reflect topological properties of a polyhedron. Since then, the notion of homology groups has undergone great development.

We can choose as coefficient groups various groups other than the additive group of integers. Notions such as those dealing with †cohomology rings and †cohomology operators can be utilized to investigate further geometrical properties of complexes. Furthermore, homology groups have been defined on larger classes of spaces such as CW complexes and compact spaces.

B. Homology Groups of a Complex

We consider permutations of $n+1$ vertices $a_0, \ldots, a_n$ of an n-dimensional †simplex Δ. When two ordered $(n+1)$-tuples of the vertices are transformed to each other by an even number of transpositions of the vertices, they are said to be equivalent. The set of all such $(n+1)$-tuples is divided into two equivalence classes. We assign the positive sign for one class and the negative sign for the other.

An ordered $(n+1)$-tuple of the vertices $(a_0, \ldots, a_n)$ with its sign is called an **oriented simplex**. If an $(n+1)$-tuple $(a_0, \ldots, a_n)$ is contained in the positive equivalence class, then we write $x^n = (a_0, \ldots, a_n)$; otherwise we write $-x^n = (a_0, \ldots, a_n)$. We define the **incidence number** $[x^n : x^{n-1}]$ between two oriented simplexes x^n and x^{n-1} as follows: If x^{n-1} is not a face simplex of x^n, then $[x^n : x^{n-1}]=0$; if $x^n=(a_0, \ldots, a_n)$ and $x^{n-1}=(a_0, \ldots, \hat{a}_i, \ldots, a_n)$ (this means that the vertex a_i is omitted from x^n), then $[x^n : x^{n-1}]=(-1)^i$.

Let K be a finite †Euclidean simplicial complex; namely, K is a set of a finite number of simplexes such that: (i) if $\Delta \in K$ and Δ' is a face of Δ, then $\Delta' \in K$; (ii) if $\Delta_1, \Delta_2 \in K$ and $\Delta_1 \cap \Delta_2 \neq \varnothing$, then $\Delta_1 \cap \Delta_2$ is a face of both Δ_1 and Δ_2. Let $\{x_1^r, \ldots, x_{\alpha_r}^r\}$ $(r=0, 1, \ldots, n)$ be oriented r-simplexes (Fig. 2) of K. An **integral r-chain** of K is a function mapping the set of r-simplexes $\{x_i^r\}$ into the integers $\{g_i\}$ and is denoted by a linear combination

$$C^r = g_1 x_1^r + \ldots + g_{\alpha_r} x_{\alpha_r}^r. \qquad (1)$$

Now let $g(-x^r)=(-g)x^r$; for $C_1^r=\sum_i g_i x_i^r$ and $C_2^r=\sum_i h_i x_i^r$, $C_1^r+C_2^r=\sum_i (g_i+h_i)x_i^r$; and for an integer m, $mC^r=\sum_i (mg_i)x_i^r$. If all g_i are 0, the chain is called a zero chain and is denoted by 0. For a chain x^r, the **boundary** of x^r is defined as the $(r-1)$-chain $\partial x^r = \sum_j [x^r : x_j^{r-1}] x_j^{r-1}$. That is, the boundary of $x^r=(a_0, a_1, \ldots, a_r)$ is $\partial x^r = \sum_{j=0}^r (-1)^j (a_0, \ldots, a_{j-1}, a_{j+1}, \ldots, a_r)$. Generally, for an r-chain $C^r=\sum_i g_i x_i^r$, the $(r-1)$-chain $\partial C^r = \sum_i g_i (\partial x_i^r) = \sum_j (\sum_i g_i [x_i^r : x_j^{r-1}]) x_j^{r-1}$ is called the **boundary** of C^r.

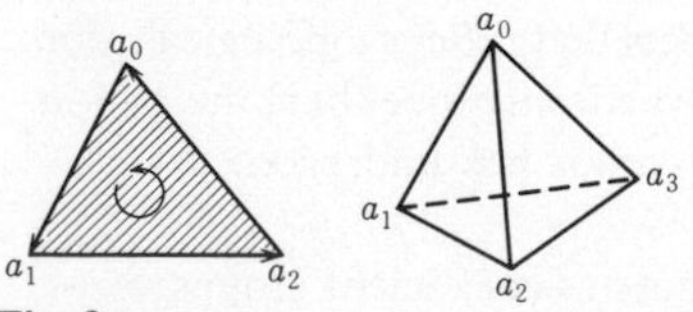

Fig. 2

A chain Z^r satisfying $\partial Z^r=0$ is called an **r-cycle**. From the definition of ∂, we have $\partial(\partial x^r)=0$; hence, in general, $\partial\partial C^r=0$. Therefore, a boundary is necessarily a cycle and is called a **bounding cycle**. If an r-cycle Z^r is bounding (i.e., there exists an $(r+1)$-chain C^{r+1} such that $\partial C^{r+1}=Z^r$), then Z^r is said to be **homologous** to 0, and we write $Z^r \sim 0$. If two cycles Z_1^r, Z_2^r satisfy $Z_1^r - Z_2^r \sim 0$, then they are said to be **homologous** and we write $Z_1^r \sim Z_2^r$. This relation is clearly an equivalence relation. The definitions in this section so far are valid for any finite simplicial complex K.

Next we consider a †locally finite simplicial complex K. A chain is a function that maps the set of oriented r-simplexes $\{x_i^r\}$ into the integers $\{g_i\}$ and is zero on all but a finite number of r-simplexes. It is written as a linear combination $C^r=\sum_i g_i x_i^r$ (finite sum), called an **integral r-chain**. The set of chains constitutes a free Abelian group (we set $\sum_i g_i x_i^r + \sum_i g_i' x_i^r = \sum_i (g_i+g_i') x_i^r$). The notions of boundary, cycle, bounding cycle, and the relation of being homologous are defined as in the case of finite simplicial complexes.

C. Homology Groups

The set of integral r-chains of a simplicial complex K constitutes a free Abelian group $C_r(K)$ with generators $\{x_j^r\}$. We call $C_r(K)$ the (integral) **r-chain group** of K. The mapping $\partial_r : C_r(K) \to C_{r-1}(K)$ sending each r-chain C_r to its boundary $\partial_r C_r$ is a homomorphism of the chain groups. We call ∂ the **boundary operator**, and denote by $Z_r(K)$, $B_r(K)$ the set of r-cycles and the set of r-bounding cycles, respectively, which are subgroups of $C_r(K)$. We call $Z_r(K)$ and $B_r(K)$ the **group of r-cycles** and the **group of r-bounding cycles** of K. Using ∂_r we have $Z_r(K)=\operatorname{Ker}\partial_r=\{C^r \in C_r(K) | \partial_r C^r=0\}$ and $B_r(K)=\operatorname{Im}\partial_{r+1}=\{\partial_{r+1}C^{r+1} | C^{r+1} \in C_{r+1}(K)\}$, and $B_r(K)$ is a subgroup of $Z_r(K)$. The factor group $H_r(K)=Z_r(K)/B_r(K)$ is called the (integral) **r-homology group** or **r-Betti group** of K. An element of $H_r(K)$ considered as an equivalence class of $Z_r(K)$ is called a **homology class**.

Given a †simplicial mapping $f: K \to K'$ from a simplicial complex K into a simplicial complex K', we can define a homomorphism $f_\# : C_r(K) \to C_r(K')$ as follows: For an oriented simplex $(a_0, \ldots, a_r)$ of K, we define $f_\#(a_0, \ldots, a_r)$ as $(f(a_0), \ldots, f(a_r))$ if $f(a_0), \ldots, f(a_r)$ are all different vertices of K', and as 0 in other cases. We have $\partial \circ f_\# = f_\# \circ \partial$ and hence $f_\#$ maps $Z_r(K)$ into $Z_r(K')$ and $B_r(K)$ into $B_r(K')$. Consequently, we can define the **induced homomorphism** $f_* : H_r(K) \to H_r(K')$.

D. Betti Numbers and Torsion Coefficients

Let K be a finite simplicial complex. Then $C_r(K)$ is a finitely generated free Abelian group. Hence $H_r(K)$ is also a finitely generated Abelian group and is decomposed into the direct sum of a free Abelian subgroup $\mathfrak{B}_r(K)$ and a finite Abelian group $T_r(K)$. The †rank of $\mathfrak{B}_r(K)$, denoted by $p_r(K)$, is called the r-dimensional **Betti number** of K. (We sometimes call $\mathfrak{B}_r(K)$ the **r-Betti group** of K.) We call $T_r(K)$ the **r-torsion group** of K, and its †invariants t_i^r the system of r-dimensional **torsion coefficients**. For example: (1) For an n-dimensional sphere S^n, we have $H_n(S^n) \cong \mathbf{Z}$ (additive group of integers), $H_r(S^n)=0$ $(0<r<n)$, $H_0(S^n) \cong \mathbf{Z}$. (2) For a †projective plane $\mathbf{P}^2$, we have $H_r(\mathbf{P}^2)=0$ $(r \geqslant 3)$, $H_2(\mathbf{P}^2)=0$, $H_1(\mathbf{P}^2) \cong \mathbf{Z}_2$ (cyclic group of order 2), that is, Betti numbers of positive dimensions are 0, $H_0(\mathbf{P}^2) \cong \mathbf{Z}$. The 1-dimensional torsion coefficient is 2, and the others are 0. (3) For a †torus T^2, we have $H_r(T^2)=0$ $(r \geqslant 3)$, $H_2(T^2) \cong \mathbf{Z}$, $H_1(T^2) \cong \mathbf{Z}+\mathbf{Z}$ (direct sum), $H_0(T^2) \cong \mathbf{Z}$, and the torsion coefficients are all 0.

As a basis of $C_r(K)$ $(r=0,1,\dots)$ we can take the basis $\{a_i^r, b_j^r, c_k^r, d_l^r, e_m^r\}$ satisfying the following relations, called the **canonical** (or **standard) basis** of $C_r(K)$. The numbers of $\{e_m^r\}$, $\{d_l^r\}$ are respectively equal to the numbers of $\{a_i^{r-1}\}$, $\{b_j^{r-1}\}$, and $\partial a_i^r=0$, $\partial b_j^r=0$, $\partial c_k^r=0$, $\partial d_l^r=t_l^{r-1}b_l^{r-1}$, $\partial e_m^r=a_m^{r-1}$. Thus the number of $\{c_k^r\}$ is equal to the r-dimensional Betti number $p_r(K)$, and the t_l^r are the r-dimensional torsion coefficients. A polynomial $P(t,K)=\sum_r p_r(K)t^r$ with variable t is called the **Poincaré polynomial** of K.

E. Homology Groups of a Polyhedron

We consider a †polyhedron $|K|$ of a finite simplicial complex K and put $H_r(|K|)=H_r(K)$. By the invariance theorem, $H_r(|K|)$ is independent of the choice of the †simplicial decomposition of $|K|$ and is a topological invariant of $|K|$. This group is called the **homology group** of the polyhedron $|K|$. The Betti number $p_r(|K|)=p_r(K)$ is also a topological invariant of $|K|$. For a finite Euclidean cellular complex K, we can define the homology group $H_r(|K|)$ of the polyhedron $|K|$ by making use of a proper simplicial subdivision of K.

F. The Euler Characteristic

Let α_0, α_1, α_2, respectively, be the number of vertices, edges, and faces of a 2-dimensional finite polyhedron homeomorphic to S^2; then we have $\alpha_0-\alpha_1+\alpha_2=2$. This relation, **Euler's theorem on polyhedra**, was the first important result of topology (L. Euler, 1752). It is said that this result was previously known to R. Descartes. The theorem can be generalized to the case of an n-dimensional Euclidean cellular complex K (Poincaré). Namely, if we denote by α_r the number of †r-cells of K and by $p_r(|K|)$ the r-dimensional Betti number, then

$$\sum_{r=0}^{n}(-1)^r\alpha_r=\sum_{r=0}^{n}(-1)^r p_r(|K|)=P(-1:|K|)$$

(**Euler-Poincaré formula**). The right-hand term of this formula is a †topological invariant of the polyhedron $|K|$; hence the left-hand side, denoted by $\chi(K)$, is independent of the choice of cellular decompositions of $|K|$. This is also written $\chi(|K|)$ and is called the **Euler characteristic** (or **Euler-Poincaré characteristic**) of K. The Euler characteristic $\chi(K)$ is important because it expresses some interesting geometric properties of K. When M is a †differentiable manifold, $\chi(M)=0$ is a necessary and sufficient condition for the existence of a continuous †vector field on M without a singular point. The notion of †Lefschetz number (→ 162 Fixed-Point Theorems) is a generalization of χ.

G. Homology Groups of Chain Complexes

The definition of the r-chain group $C_r(K)$ $(r=0,1,\dots)$ of a simplicial complex K can be generalized to the case of general †chain complexes. Namely, we consider modules M_r $(r=0,\pm1,\pm2,\dots)$, called **r-chain groups**, and homomorphisms $\partial_r: M_r\to M_{r-1}$ $(r=0,\pm1,\pm2,\dots)$, called **boundary operators**, which satisfy $\partial_r\circ\partial_{r+1}=0$:

$$\dots\to M_{r+1}\xrightarrow{\partial_{r+1}} M_r\xrightarrow{\partial_r} M_{r-1}\xrightarrow{\partial_{r-1}} M_{r-2}\to\dots.$$

This sequence (M_r,∂_r) is called a **chain complex** M (→ 57 Chain Complexes). Here $Z_r=\operatorname{Ker}\partial_r$, $B_r=\operatorname{Im}\partial_{r+1}$ are submodules of M_r, and $Z_r\supset B_r$. The factor groups $H_r=Z_r/B_r$ $(r=0,\pm1,\pm2,\dots)$ are called **homology groups** of the chain complex M. Given two chain complexes $M=(M_r,\partial_r)$ and $N=(N_r,\partial_r')$, if there exist homomorphisms $f_r: M_r\to N_r$ such that $\partial_{r+1}\circ f_{r+1}=f_r\circ\partial_{r+1}$ for all r, then $\{f_r\}$ is called the **chain mapping** from M to N, and f_r maps $Z_r(M)$ and $B_r(M)$ into $Z_r(N)$ and $B_r(N)$, respectively; hence f_r induces a homomorphism $f_*: H_r(M)\to H_r(N)$.

H. Relative Homology Groups

Let L be a †subcomplex of a simplicial complex K. We consider chain groups $C_r(K)$ and $C_r(L)$. The group $C_r(L)$ is a subgroup of $C_r(K)$, and the boundary operator ∂_r maps $C_r(K)$ and $C_r(L)$ into $C_{r-1}(K)$ and $C_{r-1}(L)$, respectively. Hence the factor groups $C_r(K,L)=C_r(K)/C_r(L)$ admit homomorphisms $\partial_r: C_r(K,L)\to C_{r-1}(K,L)$ satisfying $\partial\circ\partial=0$. Thus we obtain a new chain complex $(C_r(K,L),\partial_r)$, and $C_r(K,L)$ is a free Abelian group generated by the $\{x_i^r\}$, which are r-simplexes of K not included in L. The boundary operator ∂_r is defined by $\partial_r x_i^r\equiv\sum_j[x_i^r:x_j^{r-1}]x_j^{r-1}$ $(\operatorname{mod} C_{r-1}(L))$. The homology group $H_r(K,L)$ of this chain complex is called the **relative r-homology group** (or simply **r-homology group**) of $K \bmod L$. In this case r-cycles and r-bounding cycles are called **r-relative cycles** of (K,L) (or $\operatorname{mod} L$) and **r-relative bounding cycles** of (K,L) (or $\operatorname{mod} L$), respectively.

Let L, L' be subcomplexes of simplicial complexes K, K', respectively, and $f: K\to K'$ be a simplicial mapping such that $f(L)\subset L'$, written $f:(K,L)\to(K',L')$. Then f induces homomorphisms $C_r(K)\to C_r(K')$ and $C_r(L)\to C_r(L')$. Hence it gives rise to the homomorphism $f_\#: C_r(K,L)\to C_r(K',L')$ satisfying $\partial\circ f_\#=f_\#\circ\partial$ and the homomorphism $f_*: H_r(K,L)\to H_r(K',L')$.

I. Homology Groups with General Coefficient Groups

Let G be any Abelian group, for example, the additive group of real numbers $\mathbf{R}$, the additive group of rational numbers $\mathbf{Q}$, $\mathbf{Z}_m(=\mathbf{Z}/m\mathbf{Z})$, $\mathbf{R}/\mathbf{Z}$, the additive group of a field k, etc. The r-chain group with coefficients in G of a simplicial complex K is the †tensor product $C_r(K)\otimes G$ (over $\mathbf{Z}$) of the integral r-chain group $C_r(K)$ and G and is written $C_r(K;G)$. The boundary operator $\partial_r: C_r(K)\rightarrow C_{r-1}(K)$ gives rise to the operator $\partial_r\otimes 1_G: C_r(K;G)\rightarrow C_{r-1}(K;G)$, which is also denoted by ∂_r. Thus we obtain a chain complex $(C_r(K;G),\partial_r)$. Its homology groups, written $H_r(K;G)$, are called the **homology groups with coefficients in** G of K or the **homology groups with coefficient group** G of K. Similarly, using $C_r(K,L)$ we can define the **relative homology groups with coefficient group** $G: H_r(K,L;G)$. If G is a module over a †unitary ring R, then these homology groups are also R-modules.

The structure of these groups is determined by the integral homology groups as follows:

$$H_r(K,L;G)\cong H_r(K,L)\otimes G + \operatorname{Tor}(H_{r-1}(K,L),G).$$

This formula is called the **universal coefficient theorem** on homology groups (→ 57 Chain Complexes). If the coefficient group is the additive group of a field k of †characteristic 0, then $H_r(K,L;k)\cong H_r(K,L)\otimes k$, $p_r(K)=\dim_k H_r(K;k)$. Hence the Euler characteristic is expressed as $\chi(K)=\Sigma_r(-1)^r\dim_k H_r(K;k)$. Let $K\times K'$ be the †Cartesian product of simplicial complexes K and K'. Then the homology group $H_r(K\times K';G)$ is decomposed into a direct sum as follows:

$$H_r(K\times K';G)\cong \sum_{p+q=r} H_p(K)\otimes H_q(K';G) + \sum_{p+q=r-1}\operatorname{Tor}(H_p(K),H_q(K';G)).$$

This formula is called the **Künneth formula**. In particular, if G is the additive group of a field k, we have

$$H_r(K\times K';k)\cong \sum_{p+q=r} H_p(K;k)\otimes H_q(K';k).$$

Let f be a simplicial mapping $(K,L)\rightarrow(K',L')$; then the homomorphisms $f_{\#}\otimes 1_G: C_r(K,L)\otimes G\rightarrow C_r(K',L')\otimes G$ induce the homomorphisms $f_*: H_r(K,L;G)\rightarrow H_r(K',L';G)$. Given subcomplexes L, M of a simplicial complex K, the monomorphism $L\rightarrow K$ (inclusion mapping) induces $(L,L\cap M)\rightarrow(L\cup M,M)$. This mapping induces isomorphisms $H_r(L,L\cap M;G)\cong H_r(L\cup M,M;G)$, called the **excision isomorphisms**.

For a subcomplex L of K, inclusion mappings $i:L\rightarrow K$ and $j:(K,\varnothing)\rightarrow(K,L)$, respectively, induce homomorphisms $i_*: H_r(L;G)\rightarrow H_r(K;G)$ and $j_*: H_r(K;G)\rightarrow H_r(K,L;G)$. Furthermore, we can define a homomorphism $\partial_*: H_r(K,L;G)\rightarrow H_{r-1}(L;G)$ utilizing the mapping that sends a cycle $z\in C_r(K,L)\otimes G$ to a cycle $i^{-1}\circ\partial\circ j^{-1}(z)\in C_{r-1}(L)\otimes G$. ∂_* is called the **boundary homomorphism** (or **connecting homomorphism**). Thus we have the following **homology exact sequence**:

$$(K,L):\ldots\rightarrow H_r(L;G)\xrightarrow{i_*} H_r(K;G)\xrightarrow{j_*} H_r(K,L;G)\xrightarrow{\partial_*} H_{r-1}(L;G)\rightarrow\ldots$$

(i.e., $\operatorname{Im} i_*=\operatorname{Ker} j_*$, $\operatorname{Im} j_*=\operatorname{Ker}\partial_*$, $\operatorname{Im}\partial_*=\operatorname{Ker} i_*$).

J. Cohomology Groups

Given an Abelian group G, using the r-chain group $C_r(K)$ of a simplicial complex K we can define the r-**cochain group** $C^r(K;G)=\operatorname{Hom}(C_r(K),G)$ with coefficient group G and **coboundary operator** $\delta=\operatorname{Hom}(\partial;1)$: $C^r(K;G)\rightarrow C^{r+1}(K;G)$ (i.e., for $f^r\in C^r(K,G)$, $(\delta f^r)(x^{r+1})=f^r(\partial x^{r+1})=\Sigma_i[x^{r+1}:x_i^r]\cdot f^r(x_i^r)$). From these notions we can define the **cochain complex** $(C^r(K;G),\delta)$ with coefficients in G: $\delta\circ\delta=0$, $\ldots\rightarrow C^{r-1}(K;G)\rightarrow C^r(K;G)\rightarrow C^{r+1}(K;G)\rightarrow\ldots$. We call the element $f^r\in C^r(K,G)$ an r-**cochain**, the $f^r\in C^r(K;G)$ satisfying $\delta f^r=0$ an r-**cocycle**, and f^r, which can be written as $f^r=\delta f^{r-1}$ for some $f^{r-1}\in C^{r-1}(K;G)$, an r-**coboundary**. From these notions we can define the **group of** r-**cocycles** $Z^r(K;G)=\operatorname{Ker}\delta$, the **group of** r-**coboundaries** $B^r(K;G)=\operatorname{Im}\delta$, and the r-**cohomology group** $H^r(K;G)=Z^r(K;G)/B^r(K;G)$. In particular, when $G=\mathbf{Z}$, we write $H^r(K)$ instead of $H^r(K;\mathbf{Z})$.

Similarly, the **relative** r-**cohomology group** $H^r(K,L;G)$ of $K \bmod L$ is defined using the relative r-chain group $C_r(K,L)$. If the homology group $H_r(K,L)$ is finitely generated and is decomposed into the direct sum of the free Abelian group $\mathfrak{B}_r(K,L)$ and the finite group $T_r(K,L)$ (torsion group), then the group $H^r(K,L)$ is also finitely generated and can be similarly decomposed into the direct sum of the free Abelian subgroup $\mathfrak{B}^r(K,L)$ and the torsion group (finite group) $T^r(K,L)$. We have $\mathfrak{B}_r(K,L)\cong\mathfrak{B}^r(K,L)$, $T_{r-1}(K,L)\cong T^r(K,L)$, and the **universal coefficient theorem** on cohomology groups:

$$H^r(K,L;G)\cong\operatorname{Hom}(H_r(K,L),G) + \operatorname{Ext}(H_{r-1}(K,L),G).$$

In particular, if G is the additive group of a field k of characteristic 0, then $H^r(K,L;G)\cong\operatorname{Hom}(H_r(K,L),k)$.

A simplicial mapping $f:(K,L)\rightarrow(K',L')$ induces homomorphisms $f^*: H^r(K',L';G)\rightarrow$

$H^r(K,L;G)$. Specifically, utilizing the mappings $i:L\rightarrow K, j:(K,\varnothing)\rightarrow(K,L)$, and ∂ defined as in the previous section, we obtain the **cohomology exact sequence**:

$$\ldots\rightarrow H^{r-1}(L;G)\xrightarrow{\partial^*}H^r(K,L;G)$$

$$\xrightarrow{j^*}H^r(K;G)\xrightarrow{i^*}H^r(L;G)\rightarrow\ldots.$$

We call ∂^* the **coboundary homomorphism**.

If the coefficient group is the additive group of a commutative ring R, then for two cochains $f^r\in\mathrm{Hom}(C_r(K),R)$, $f^s\in\mathrm{Hom}(S_s(K),R)$, a third cochain (called the †cup product) $f^r\smile f^s\in\mathrm{Hom}(C_{r+s}(K),R)$ is defined as follows: $(f^r\smile f^s)(a_0,\ldots,a_{r+s})=f^r(a_0,\ldots,a_r)f^s(a_r,\ldots,a_{r+s})$, where $(a_0,\ldots,a_{r+s})$ is an oriented $(r+s)$-simplex of K. This product gives rise to the cup product of the cohomology groups and thus supplies the set $H^*(K;R)=\Sigma_r H^r(K;R)$ with the structure of a ring (→ 68 Cohomology Rings).

K. Cohomology of Semisimplicial Complexes

Let K be an †s.s. (semisimplicial) complex. Namely, K is a family of triples (K_r,∂_i,s_i), where the K_r are given sets and r runs over the set of nonnegative integers and $\partial_i:K_r\rightarrow K_{r-1}$, $s_i:K_{r-1}\rightarrow K_r$ $(0\leqslant i\leqslant r)$ are mappings satisfying certain conditions (→ 73 Complexes). Let $C_r(K)$ be the free Abelian group generated by elements of K_r. The boundary operator ∂ is defined as $\partial\sigma_\lambda=\Sigma_{i=0}^r\partial_i\sigma_\lambda$ for every element σ_λ of K_r. Generally for $C^r=\Sigma g_\lambda\sigma_\lambda$ we set $\partial C^r=\Sigma g_\lambda\partial\sigma_\lambda$. Then we have $\partial:C_r(K)\rightarrow C_{r-1}(K)$, $\partial\circ\partial=0$; hence a chain complex $C_*(K)=(C_r(K),\partial)$ is defined. A subgroup $D_r(K)$ of $C_r(K)$ generated by the set of †degenerate r-simplexes of K_r gives rise to a subchain complex $D_*(K)=(D_r(K),\partial)$ of $C_*(K)$. $C_*(K)$ is called the chain complex of an s.s. complex K, and the factor chain complex $C_*(K)/D_*(K)$ is called the **normalization** of $C_*(K)$. The homology groups induced from $C_*(K)$ are isomorphic to that of the normalized chain complex. These homology groups are called simply the **homology groups** of an s.s. complex K. An †s.s. mapping $f:K\rightarrow L$ induces a mapping from the normalized chain complex of K to that of L and a homomorphism $f_*:H_r(K)\rightarrow H_r(L)$. The chain complex $C_*(K\times L)$ of the Cartesian product of two s.s. complexes K and L determines the homology groups $H_r(K\times L)$, which are isomorphic to the homology groups determined by the chain complex $C_*(K)\otimes C_*(L)$. An †ordered complex $O(K)$ defined for a simplicial complex K is an s.s. complex, and the chain complex constructed from $O(K)$ as above is called the **ordered chain complex** of K. In contrast to this, the chain complex of K defined earlier is called the **oriented chain complex**. These two chain complexes are different, but they both define isomorphic homology groups $H_r(K)$.

L. Singular Homology and Cohomology Groups of a Topological Space

Singular homology theory and cohomology theory were defined by S. Lefschetz (*Bull. Amer. Math. Soc.*, 39 (1933)) and S. Eilenberg (*Ann. of Math.*, 45 (1944)).

For a topological space X, a chain complex $S(X)=\Sigma C_r(S(X))$, called the **singular chain complex**, is determined from the s.s. complex made up of the set of †singular simplexes of X (→ 73 Complexes). For a pair (X,A) consisting of a topological space X and a subspace A of X, $S(A)$ is a chain subcomplex of $S(X)$. For any Abelian group G, the (co)-homology group

$$H_r(X,A;G)=H_r(S(X)/S(A);G)$$

$$(H^r(X,A;G)=H^r(S(X)/S(A);G))$$

is called the r-dimensional **singular (co)homology group** with coefficients in G of (X,A). A continuous mapping $f:(X,A)\rightarrow(Y,B)$ determines a chain mapping that assigns to an r-singular simplex $\sigma:\Delta(r)\rightarrow X$ of X an r-singular simplex $f\circ\sigma:\Delta(r)\rightarrow Y$ (composition of mappings) of Y. Hence f gives rise to the **induced homomorphisms** $f_*:H_r(X,A;G)\rightarrow H_r(Y,B;G)$, $f^*:H^r(Y,B;G)\rightarrow H^r(X,A;G)$.

A function that assigns to any pair (X,A) of topological spaces (we always assume that $X\supset A$) a group $H_r(X,A;G)$ $(H^r(X,A;G))$ and assigns to a continuous mapping f a homomorphism f_* (f^*) is a †covariant (contravariant) functor from the †category consisting of pairs of topological spaces and continuous mappings to the category of all Abelian groups and homomorphisms (→ 53 Categories and Functors).

For a pair of spaces (X,A), a connecting homomorphism $\partial_*:H_r(X,A;G)\rightarrow H_{r-1}(A;G)$, called the **homology boundary homomorphism**, is defined using the exact sequence of chain complexes $0\rightarrow S(A)\otimes G\rightarrow S(X)\otimes G\rightarrow(S(X)/S(A))\otimes G\rightarrow 0$. This homomorphism is natural with respect to continuous mappings: namely, for a continuous mapping $f:(X,A)\rightarrow(Y,B)$, we have $\partial_*\circ f_*=(f|A)_*\circ\partial_*$ ($f|A$ is the restriction of $f:A\rightarrow B$). Similarly, a connecting homomorphism $\delta^*:H^{r-1}(A;G)\rightarrow H^r(X,A;G)$, called the **cohomology coboundary homomorphism**, is defined using the exact sequence of cochain complexes $0\rightarrow\mathrm{Hom}(S(X)/S(A),G)\rightarrow\mathrm{Hom}(S(X),G)\rightarrow\mathrm{Hom}(S(A),G)\rightarrow 0$.

M. Fundamental Properties of Singular Homology and Cohomology Groups

(I) **Homotopy theorem**: If $f \sim f'$ (†homotopic): $(X,A) \rightarrow (Y,B)$, then $f_* = f'_* : H_r(X,A;G) \rightarrow H_r(Y,B;G)$. (II) **Exactness theorem**: The exact sequences of homology groups explained in Section I hold if the pair (K,L) of simplicial complexes is replaced by the pair (X,A) of topological spaces. (III) **Excision isomorphism theorem**: If U satisfies $\bar{U} \subset \operatorname{Int} A$, then we have the isomorphism $i_* : H_r(X-U, A-U;G) \cong H_r(X,A;G)$, where i_* is induced by the inclusion mapping. (IV) For a point P, $H_0(P;G)=G$, $H_r(P;G)=0$ $(r \neq 0)$.

If X is †arcwise connected, then $H_0(X)=\mathbf{Z}$. Given a triple (X,A,B) $(X \supset A \supset B)$, condition (II) implies the validity of the **homology exact sequence**:

$$\ldots \rightarrow H_r(A,B;G) \xrightarrow{i_*} H_r(X,B;G)$$

$$\xrightarrow{j_*} H_r(X,A;G) \xrightarrow{\partial_*} H_{r-1}(A,B;G) \rightarrow \ldots .$$

The homology groups of two pairs of topological spaces that are †homotopically equivalent to each other are isomorphic by (I), and the homology groups of a space that is †contractible to a point are isomorphic to the homology groups of a point.

For subcomplexes X_1, X_2 of a †CW complex X, the excision isomorphism i_*: $H_r(X_1, X_1 \cap X_2) \cong H_r(X_1 \cup X_2, X_2)$ is valid by (III). If (X,A) has the †homotopy extension property (for example, if X is a CW complex and A is a subcomplex of X), then the †standard mapping $h:(X,A) \rightarrow (X/A, *)$, where X/A is given by †contracting A to a point $*$, induces $h_* : H_r(X,A;G) \cong \tilde{H}_r(X/A;G)$. (Here $\tilde{H}_r(X;G)$, the **reduced homology group** of X, is isomorphic to $H_r(X;G)$ $(r>0)$, and $H_0(X;G)=G+\tilde{H}_0(X;G)$.)

For a standard mapping $h:(CX,X) \rightarrow (SX,*)$, where CX is the †cone of X and SX is the †suspension of X, the composite $\sigma_* : h_* \circ \partial_*^{-1} : \tilde{H}_r(X;G) \rightarrow \tilde{H}_{r+1}(SX;G)$ is an isomorphism called the **suspension isomorphism**. From this we can easily obtain $\tilde{H}_r(S^n)=0$ $(r \neq 0)$, $H_n(S^n)=\mathbf{Z}$ for an n-sphere S^n. Suppose that X_1, X_2 are subsets of a topological space X and that the inclusion mapping induces an isomorphism $H_*(X_1, X_1 \cap X_2) \cong H_*(X_1 \cup X_2, X_2)$. In this case the triple (X, X_1, X_2) is called **proper** and we have the **Mayer-Vietoris exact sequence**: $\ldots \rightarrow H_r(X_1 \cap X_2;G) \rightarrow H_r(X_1;G)+H_r(X_2:G) \rightarrow H_r(X_1 \cup X_2;G) \rightarrow H_{r-1}(X_1 \cap X_2;G) \rightarrow \ldots$.

Concerning (singular) cohomology groups, we can prove theorems dual to those mentioned so far in this section. The Künneth theorem for a product space $X \times Y$ and the universal coefficient theorem hold.

N. Cellular Homology and Cohomology Groups

Let X be a CW complex, A a subcomplex of X, and X^r the r-†skeleton of X (i.e., the subcomplex consisting of all †cells of dimension less than or equal to r), and put $\bar{X}^r = A \cup X^r$. Set $C_r(X,A;G)=H_r(\bar{X}^r, \bar{X}^{r-1};G)$, and define the boundary homomorphism $\partial : C_r(X,A;G) \rightarrow C_{r-1}(X,A;G)$ to be the homomorphism $\partial_* : H_r(\bar{X}^r, \bar{X}^{r-1};G) \rightarrow H_{r-1}(\bar{X}^{r-1}, \bar{X}^{r-2};G)$ determined by the exact sequence associated with the triple $(\bar{X}^r, \bar{X}^{r-1}, \bar{X}^{r-2})$. Then we have $\partial \circ \partial = 0$ and obtain a chain complex $C_*(X,A;G)=\Sigma C_r(X,A;G)$. The homology groups $H_r(C_*(X,A;G))$ are called the **cellular homology groups** of (X,A). The **cellular cohomology groups** $H^r(C^*(X,A;G))$ are defined similarly. These groups are isomorphic to $H_r(X,A;G)$ and $H^r(X,A;G)$, respectively. Here we have $C_*(X,A;G)=C_*(X,A) \otimes G$ and $C^*(X,A;G)=\operatorname{Hom}(C_*(X,A),G)$. The r-chain group $C_r(X,A)=H_r(\bar{X}^r, \bar{X}^{r-1})$ of $C_*(X,A)$ is the free Abelian group generated by all r-cells of X except those belonging to A, and is useful for calculating homology groups of CW complexes. In particular, if $|K|$ is the polyhedron determined by a simplicial complex K, the (co)homology group of K coincides with the cellular (co)homology group of the CW complex $|K|$, which again coincides with the singular (co)homology group of $|K|$. Hence we obtain the invariance theorem mentioned in Section E.

O. Axioms of Homology Theory

S. Eilenberg and N. E. Steenrod selected some fundamental properties as axioms to characterize homology theory [7]. Namely, they defined homology theory as follows: For any integer $r \geqslant 0$ and any pair (X,A) of topological spaces, Abelian groups $H_r(X,A)$ and homomorphisms $\partial_* : H_r(X,A) \rightarrow H_{r-1}(A)$ $(=H_{r-1}(A,\varnothing))$ are given. To any continuous mapping $f:(X,A) \rightarrow (Y,B)$ homomorphisms $f_* : H_r(X,A) \rightarrow H_r(Y,B)$ are assigned which satisfy the following axioms: (I) $(g \circ f)_* = g_* \circ f_*$, $1_* = 1$, $\partial_* \circ f_* = (f|A)_* \circ \partial_*$. (II) **Homotopy axiom**: If $f \sim f'$, then $f_* = f'_*$. (III) **Exactness axiom**: The sequence $\ldots \rightarrow H_r(A) \xrightarrow[i_*]{} H_r(X) \xrightarrow[j_*]{} H_r(X,A) \xrightarrow[\partial_*]{} H_{r-1}(A) \rightarrow \ldots$ is exact. (IV) **Excision axiom**: If U is open and $\bar{U} \subset \operatorname{Int} A$, then $i_* : H_r(X-U, A-U) \cong H_r(X,A)$. (V) **Dimension axiom**: For a point space P, we have $H_r(P)=0$ $(r \neq 0)$, $G = H_0(P)$. The group $H_0(P)$ is called the coefficient group of this homology theory. For a pair consisting of a finite CW complex X and a subcomplex A of X, $H_r(X,A)$ is determined uniquely by its coefficient group and coincides with the singular homology group. The cohomology theory is constructed similarly by dual conditions.

P. Homology and Cohomology Groups with Local Coefficients

Suppose that we are given Abelian groups G_x associated with points x of a topological space X and isomorphisms $l^*: G_{l(0)} \cong G_{l(1)}$ for any †path l in X. Assume further that if two paths l and l' are homotopic with endpoints fixed, then $l^* = l'^*$. Furthermore, if l and m are paths such that $l(1) = m(0)$, and $l \circ m$ is the composite path, then we assume that $(l \circ m)^* = m^* \circ l^*$. Under these conditions, the set $\{G_x\}$ is called the **local system of groups** over X. For example, the †homotopy groups $\{\pi_n(X,x)\}$ are a local system of groups. Let X be a polyhedron $|K|$, y_i a fixed point of each oriented r-simplex $|x_i^r|$ of $|K|$, and g_i an element of G_{y_i}. Let $C_r(K; \{G_x\})$ be the Abelian group formed by the set of all finite linear forms $\Sigma_i g_i x_i^r$, and define the homomorphism $\partial: C_r(K; \{G_x\}) \to C_{r-1}(K; \{G_x\})$ by $\partial \Sigma_i g_i x_i^r = \Sigma_j (\Sigma_i [x_i^r : x_j^{r-1}] l_{ij}^*(g_i)) x_j^{r-1}$, where l_{ij} is a path from $y_i \in |x_i^r|$ to $y_j \in |x_j^{r-1}|$ in $|x_i^r|$. Then we have $\partial \circ \partial = 0$. Hence we get the chain complex $C(K; \{G_x\}) = \Sigma C_r(K; \{G_x\})$ and the homology groups, written $H_r(K; \{G_x\})$. These groups are called the **homology groups with local coefficients in** $\{G_x\}$ (or **homology groups with the local system of groups** $\{G_x\}$). Similarly, the set of all functions f^r that assign to every oriented simplex x_i^r an element $g_i \in G_{y_i}$ (written $C^r(K; \{G\})$) constitutes the cochain group with the homomorphism $\delta: C^r(K; \{G_x\}) \to C^{r+1}(K; \{G_x\})$ defined by $(\delta f^r)(x_j^{r+1}) = \Sigma_i [x_j^{r+1} : x_i^r] l_{ij}^{*-1}(f^r(x_i^r))$. From these relations we can define the **cohomology groups with the local system of groups** $\{G_x\}$. These definitions can be extended to the case of singular cohomology groups. In particular, if the local system of groups $\{G_x\}$ is **trivial** (for example, if X is †simply connected), then the homology groups with the local system of groups $\{G_x\}$ coincide with the usual homology groups with coefficient group G_x. Generally, however, the $H_r(K; \{G_x\})$ do not coincide with the usual homology groups. This theory of homology with local systems of groups is useful for †obstruction theory and †cohomology spectral sequences of fiber spaces. The theory has been generalized to homology groups with sheaves as their coefficients (→ 377 Sheaves).

Q. Local Homology Groups

For a point x of a topological space X, the homology group $H_r(X, X - x; G)$ is called the **local homology group** of x. We consider the †star complex $St(x)$ of a vertex x in a polyhedron $|K|$ and its subcomplex $B(x)$ consisting of all simplexes that are disjoint with x. Then we have $H_r(|K|, |K| - x; G) = H_r(St(x), B(x); G)$.

R. Various Homology Groups

Many mathematicians endeavored to extend the homology theory of polyhedra to general topological spaces before the discovery of singular homology theory. Their works served to clarify the relations between combinatorial and set-theoretic methods in topology (→ 409 Topology). L. Vietoris defined the **Vietoris homology group** (*Math. Ann.*, 97 (1927)) as follows: Let X be a †compact metric space, and call every point of X a vertex. A set of finite points whose distances are smaller than $\varepsilon > 0$ is defined to be a simplex. We get a simplicial complex $K(\varepsilon)$, called an **ε-complex**. For a decreasing sequence ε_n ($\varepsilon_n \to 0$), we consider ε_n-complexes and denote a cycle in $K(\varepsilon_n)$ by z_n^r. An r-cycle is defined to be a sequence $z^r = \{z_n^r\}$ such that z_n^r and z_{n+1}^r are homologous in $C_r(K(\varepsilon_n))$ (we write $z_n^r \overset{\varepsilon_n}{\sim} z_{n+1}^r$). An r-cycle $\{z_n^r\}$ is called a bounding cycle if there exists a decreasing sequence η_n ($\eta_n \to 0$) such that $z_n^r \overset{\eta_n}{\sim} 0$. Following a routine process, we define the Vietoris homology. Vietoris's homology theory developed into the **Kolmogorov-Alexander homology groups** for †locally compact spaces (A. N. Kolmogorov, *C. R. Acad. Sci. Paris*, 202 (1936); Alexander, *Ann. of Math.*, (2) 39 (1938)) and the **Kolmogorov-Spanier cohomology groups** (H. Spanier, *Ann. of Math.*, (2) 49 (1948)).

A set C^r of functions ψ^r, which assign to every point of the $(r+1)$-fold Cartesian product X^{r+1} of X an element of an Abelian group G, forms an Abelian group in a natural manner. A homomorphism $\delta: C^r \to C^{r+1}$, defined by $(\delta\psi^r)(x_0, \ldots, x_{r+1}) = \Sigma_i (-1)^i \psi^r(x_0, \ldots, \hat{x}_i, \ldots, x_{r+1})$, satisfies $\delta \circ \delta = 0$. We call $\psi^r \in C^r$ locally zero when the values of ψ^r are always 0 in a neighborhood of the diagonal $\{(x, \ldots, x) | x \in X\}$ of X^{r+1}. The set of such chains ψ^r forms a subgroup C_0^r of C^r, and $\delta(C_0^r) \subset C_0^{r+1}$. Thus we get a cochain complex $\Sigma C^r / C_0^r$; its cohomology groups are the Kolmogorov-Spanier groups.

Let λ, μ be †open coverings of a topological space X. If λ is a †refinement of μ, denoted by $\lambda > \mu$, then we get a †simplicial mapping $\pi_\mu^\lambda: N(\lambda) \to N(\mu)$, where $N(\lambda)$ is the nerve of the open covering λ, by assigning to every element O_i of λ an element of μ containing O_i. P. Aleksandrov defined the homology group for a compact metric space X (*Ann.*

of Math., (2) 30 (1928)) as follows: He considered the directed set of all finite open †coverings and its †cofinal subset Λ_0 such that for each pair $\lambda, \mu \in \Lambda_0$ $(\lambda > \mu)$, the mapping π_μ^λ is uniquely determined. Defining a chain complex $C(\Lambda_0) = \varprojlim \{C(N(\lambda)), \pi_{\mu\#}^\lambda\}$ to be the limit group of the sequence of the chain complexes $\{C(N(\lambda)), \pi_{\mu\#}^\lambda\}$, he obtained the homology groups called the **projective homology groups** of X. The definition of these groups was extended to any compact space by E. Čech (*Fund. Math.*, 19 (1932)) and to general topological spaces by C. H. Dowker (*Ann. of Math.*, (2) 49 (1948)).

S. Čech Homology and Cohomology Groups

Let Λ be the directed set of all open coverings of a topological space X and A be a subset of X. For each open covering $\lambda \in \Lambda$, $\lambda \cap A = \{U \cap A \mid U \in \lambda\}$ is an open covering of A, and the nerve $N(\lambda \cap A)$ is considered as a subcomplex of $N(\lambda)$. Then the (co)homology groups of simplicial complexes $H_r(N(\lambda), N(\lambda \cap A); G)$, $H^r(N(\lambda), N(\lambda \cap A); G)$ are defined for any coefficient group G. The mapping $\pi_\mu^\lambda : (N(\lambda), N(\lambda \cap A)) \rightarrow (N(\mu), N(\mu \cap A))$ is not uniquely determined for $\lambda, \mu \in \Lambda$ $(\lambda > \mu)$, but the induced homomorphisms

$$\pi_{\mu *}^\lambda : H_r(N(\lambda), N(\lambda \cap A); G) \rightarrow H_r(N(\mu), N(\mu \cap A); G),$$

$$\pi_\mu^{\lambda *} : H^r(N(\mu), N(\mu \cap A); G) \rightarrow H^r(N(\lambda), N(\lambda \cap A); G)$$

are uniquely determined. The limit group $\check{H}_r(X, A; G) = \varprojlim \{H_r(N(\lambda), N(\lambda \cap A); G), \pi_{\mu *}^\lambda\}$ $(\check{H}^r(X, A; G) = \varinjlim \{H^r(N(\lambda), N(\lambda \cap A); G), \pi_\mu^{\lambda *}\})$ is called the **Čech homology (cohomology) group** of (X, A). When (X, A) is a pair of compact spaces, we consider the directed set of all finite open coverings (whereas if (X, A) is a pair of †paracompact spaces, we consider the directed set of all locally finite open coverings) and apply a process similar to that in this section to obtain a (co)homology group that is actually isomorphic to the Čech (co)homology group of (X, A).

Given a continuous mapping $f : (X, A) \rightarrow (Y, B)$ and an open covering λ' of Y, we consider the open covering $f^{-1}\lambda'$ of X, which is defined to be $\{f^{-1}(U) \mid U \in \lambda'\}$. The mapping f gives rise to the simplicial mapping $f_\# : N(f^{-1}\lambda') \rightarrow N(\lambda')$, which induces the homomorphisms $f_* : \check{H}_r(X, A; G) \rightarrow \check{H}_r(Y, B; G)$, $f^* : \check{H}^r(Y, B; G) \rightarrow \check{H}^r(X, A; G)$. Also utilizing the mapping $\partial_* : H_r(N(\lambda), N(\lambda \cap A); G) \rightarrow H_{r-1}(N(\lambda \cap A); G)$, we can define the boundary homomorphisms $\partial_* : \check{H}_r(X, A; G) \rightarrow \check{H}_{r-1}(A; G)$; the homomorphisms $\delta^* : \check{H}^r(A, G) \rightarrow \check{H}^{r+1}(X, A; G)$ are obtained similarly.

The Čech (co)homology groups have similar properties as the singular (co)homology group except for the exactness property of homology. Concerning property (II) of homology (→ Section M), the sequence is in general only of order 2; i.e., the composite of the two consecutive homomorphisms is trivial. However, if (X, A) is a pair of compact spaces, or if the coefficient group is a finite group or a field, then the exactness property holds.

Furthermore, concerning the Čech homology groups, the **continuity theorem** is valid: $H_r(\varprojlim X, A) \cong \varinjlim H_r(X, A)$. Here (X, A) is a sequence of pairs of compact spaces $(X_\lambda, A_\lambda)_{\lambda \in \Lambda}$ (Λ is a directed set) supplied with continuous mappings $\pi_\mu^\lambda : (X_\lambda, A_\lambda) \rightarrow (X_\mu, A_\mu)$ for $\lambda > \mu$ such that $\pi_\lambda^\lambda = 1$, $\pi_\nu^\lambda = \pi_\nu^\mu \circ \pi_\mu^\lambda$ $(\lambda > \mu > \nu)$; $\varprojlim(X, A)$ is the †projective limit of the sequence. Similar relations hold concerning the cohomology groups.

The axioms of homology theory with conditions (I)–(V) (of which (III) is modified to state that the sequence in question is of order 2), together with the continuity theorem, characterize the Čech homology groups. For compact spaces the Čech homology coincides with the Kolmogorov-Spanier homology, and for finite CW complexes the Čech homology groups coincide with the singular homology groups. However, even for compact spaces X, $\check{H}(X)$ does not necessarily coincide with $H(X)$.

T. De Rham Cohomology Groups

A †differentiable manifold X is †triangulable, hence it is a polyhedron. We can, however, utilize the differentiable structure of X to define cohomology groups of X instead of using the combinatorial structure of X (G. de Rham, 1931). More precisely, we can define the homology groups and cohomology ring of X with real number coefficients using the †differential forms of class C^∞ on M (→ 108 Differentiable Manifolds R).

U. Homology and Cohomology Groups with Coefficients in Topological Groups

The coefficient group G of homology groups may be a †topological Abelian group. In this case the chain group C_r is also a topological group, and the homology groups are defined to be the topological groups $H_r = Z_r / \overline{B}_r$ ($\overline{B}_r$ is the †closure of B_r). If G is a compact Abelian group, then for a finite simplicial complex K, $H_r(K; G)$, $H^r(K; G)$ are also compact Abelian groups. The character group G^* of G with respect to $\mathbf{R}_1=\mathbf{R}/\mathbf{Z}$ is a discrete Abelian group; $H_r(K;G)$ and $H^r(K;G^*)$ are an orthogonal group pair by taking as the inner product the †Kronecker product, and they are †character groups of each other.

V. Homology and Cohomology Groups of the Second Kind

Assume that a simplicial complex K is infinite and †locally finite. A formal (infinite) sum $\Sigma_i g_i x_i^r$, where the $x_i^r \in K$ are r-dimensional oriented simplexes and $g_i \in G$ (the coefficient group), is called an **infinite chain** over G. Instead of this chain, the usual chain is called a **finite chain**. The Abelian group $C_r^{\mathrm{II}}(K;G)$ consisting of the infinite chains is a chain complex if we define the boundary ∂ as in the case of finite chains. The homology groups $H_r^{\mathrm{II}}(K;G)$ thus obtained are called the **homology groups of the second kind**. A function f^r that is defined on all x_i^r, takes the values $f^r(x_i^r)$ in G, and is 0 except for a finite number of x_i^r is called a **finite cochain**. The set of finite cochains $C_{\mathrm{II}}^r(K;G)$ is a cochain complex because δf^r is also a finite cochain. The cohomology groups $H_{\mathrm{II}}^r(K;G)$ thus defined are called the **cohomology groups of the second kind**.

Similar notions with respect to the singular (co)homology groups of a topological space X can be defined. We consider an infinite chain $C^r=\Sigma g_i\sigma_i^r$, where the σ_i^r are the singular simplexes of X and the g_i are such that the family $\{\operatorname{supp}\sigma_i^r \mid g_i\neq 0\}$ of compact sets is locally finite. Using these chains we define the **singular (co)homology group of the second kind** $H_r^{\mathrm{II}}(X;G)$. For a compact space X this group coincides with the usual homology group of X. If X is a polyhedron $|K|$, these groups are isomorphic to $H_r^{\mathrm{II}}(K;G)$, $H_{\mathrm{II}}^r(K;G)$. In particular, if X is a finite polyhedron and A is a subcomplex of X, $H_r(X,A;G)$ and $H^r(X,A;G)$ are isomorphic to $H_r^{\mathrm{II}}(X-A;G)$ and $H_{\mathrm{II}}^r(X-A;G)$, respectively.

W. Generalized Homology Theory

Let X be a finite CW complex and A a subcomplex of X. **Generalized homology groups** for CW complexes are defined as follows: For any integer $n\geqslant 0$ and any pair of CW complexes (X,A) $(A\subset X)$, a module $H_n(X,A)$ and a homomorphism $\partial: H_n(X,A)\rightarrow H_{n-1}(A)$ are given. For a cellular mapping $f:(X,A)\rightarrow(Y,B)$, we assign homomorphisms $f_*: H_n(X,A)\rightarrow H_n(Y,B)$. We choose as a system of axioms the usual system of axioms of homology theory (I)–(IV) except the dimension axiom. The cohomology axioms are given as the "dual" of (I)–(IV). Various examples of **generalized cohomology groups** are known, of which †K-theory is most famous and important (→ 236 K-Theory).

References

[1] A. Komatu, M. Nakaoka, and M. Sugawara, Isôkikagaku (Japanese; Topology) I, Iwanami, 1967.
[2] P. S. Aleksandrov and H. Hopf, Topologie I, Springer, 1935 (Chelsea, 1965).
[3] H. Seifert and W. Threlfall, Lehrbuch der Topologie, Teubner, 1934 (Chelsea, 1965).
[4] S. Lefschetz, Algebraic topology, Amer. Math. Soc. Colloq. Publ., 1942.
[5] S. Lefschetz, Introduction to topology, Princeton Univ. Press, 1949.
[6] H. Cartan, Séminaire de topologie algébrique, Ecole Norm. Sup., 1948–1949.
[7] S. Eilenberg and N. E. Steenrod, Foundations of algebraic topology, Princeton Univ. Press, 1952.
[8] P. J. Hilton and S. Wylie, Homology theory, Cambridge Univ. Press, 1960.
[9] G. W. Whitehead, Generalized homology theories, Trans. Amer. Math. Soc., 102 (1962), 227–283.
[10] E. H. Spanier, Algebraic topology, McGraw-Hill, 1966.

204 (IX.8) Homotopy

A. General Remarks

If a family $f_t: X\rightarrow Y$ $(t\in I=\{t \mid 0\leqslant t\leqslant 1\})$ of †continuous mappings from a †topological space X into a topological space Y is also continuous with respect to t, that is, if the mapping F from the product space $X\times I$ into Y defined by $F(x,t)=f_t(x)$ $(x\in X, t\in I)$ is continuous, then $\{f_t\}$ or F is called a **homotopy**. In this case, f_0 and f_1 are said to be **homotopic**. This relation between f_0 and f_1 is indicated by $f_0\simeq f_1: X\rightarrow Y$, or simply $f_0\simeq f_1$, and is called the relation of homotopy. It is an †equivalence relation, and the equivalence class $[f]$ of a mapping $f: X\rightarrow Y$ under $\simeq$ is called the **homotopy class** (or **mapping class**) of f. The set of all homotopy classes of mappings of X into Y is called the **homotopy set** and denoted by $\pi(X;Y)$ or $[X,Y]$. A function γ of continuous mappings $f\in Y^X$ is called a **homotopy invariant** if $f\simeq g$ implies $\gamma(f)=\gamma(g)$. When X consists of a point $*$ we write $\pi(*;Y)=\pi_0(Y)$. If all continuous mappings in Y^X are homotopic to each other, we write

$\pi(X, Y)=0$; $\pi_0(Y)=0$ means that Y is †arcwise connected.

These concepts are generalized as follows: Let A_i and B_i $(i=1,2,\ldots)$ be subspaces of X and Y, respectively, and denote by $Y^X(A_1, A_2, \ldots; B_1, B_2, \ldots)$ the set of continuous mappings $f \in Y^X$ satisfying $f(A_i) \subset B_i$. If a homotopy $\{f_t\}$ is such that $f_t \in Y^X(A_i; B_i)$, then $\{f_t\}$ is called a **restricted homotopy** with respect to A_i, B_i or a homotopy from a system of spaces $(X, A_1, A_2, \ldots)$ into a system of spaces $(Y, B_1, B_2, \ldots)$. The notation $f_0 \simeq f_1 : (X, A_1, A_2, \ldots) \rightarrow (Y, B_1, B_2, \ldots)$ and the homotopy set $\pi(X, A_1, A_2, \ldots; Y, B_1, B_2, \ldots)$ are defined accordingly.

For the composite $g \circ f \in Z^X(A_i; C_i)$ of $f \in Y^X(A_i; B_i)$ and $g \in Z^Y(B_i; C_i)$, $f \simeq f'$ and $g \simeq g'$ imply $g \circ f \simeq g' \circ f'$. Thus the **composite** $\beta \circ \alpha = [g \circ f] \in \pi(X, A_i; Z, C_i)$ of $[f] = \alpha \in \pi(A, A_i; Y, B_i)$ and $[g] = \beta \in \pi(Y, B_i; Z, C_i)$ is defined. By putting $g_*[f] = [g \circ f] = f^*[g]$ we induce two mappings,

$$g_* : \pi(X, A_i; Y, B_i) \rightarrow \pi(X, A_i; Z, C_i),$$

$$f^* : \pi(Y, B_i; Z, C_i) \rightarrow \pi(X, A_i; Z, C_i).$$

Then $f \simeq f'$ implies $f^* = f'^*$ and $g \simeq g'$ implies $g_* = g'_*$. Also $(g \circ f)_* = g_* \circ f_*$, $(g \circ f)^* = f^* \circ g^*$, and $h^* \circ g_* = g_* \circ h^*$, where $h \in X^W(D_i; A_i)$.

The **category of pointed topological spaces** is defined to be the †category in which each object X, which is a topological space, has a point fixed as a **base point** and each mapping $X \rightarrow Y$ carries the base point of X to the base point of Y. In this category, we define a homotopy set, denoted by $\pi(X; Y)_0$ or $[X; Y]_0$, as follows: Denoting the base points by $*$, we have $\pi(X, A_i, *; Y, B_i, *) = \pi(X, A_i; Y, B_i)_0$. A continuous mapping f homotopic to the constant mapping $X \rightarrow * \in Y$ is said to be **homotopic to zero** (or **null-homotopic**). This is indicated by $f \simeq 0$, and $\pi(X; Y)_0 = 0$ means that all continuous mappings are homotopic to zero. Let S^0 be a set of two points; then $\pi(X; S^0)_0 = 0$ means that X is arcwise connected.

Suppose that a homotopy $\{f_t\}$ $(f_t : X \rightarrow Y)$ is such that the restriction of f_t to a subspace A of X is stationary, that is, $f_t(a) = f_0(a)$ $(a \in A, t \in I)$. Then f_0, f_1 are said to be **homotopic relative to** A, indicated by $f_0 \simeq f_1$ (rel. A). If a homotopy $\{f_t\}$ $(f_t : X \rightarrow Y)$ is such that each f_t is a homeomorphism into Y, then $\{f_t\}$ is called an **isotopy** and f_0 is called **isotopic** to f_1 (→ 234 Knot Theory). In contrast to these specific homotopies, the usual homotopy is sometimes called a **free homotopy**.

Research done by L. E. Brouwer, H. Hopf, W. Hurewicz, K. Borsuk, L. S. Pontrjagin, and S. Eilenberg has contributed to the theory of homotopy, an important field of topology still in the process of development.

B. Mapping Spaces

Denote by Y^X the set of all continuous mappings $f : X \rightarrow Y$, and supply it with the †compact-open topology. The topological space Y^X is called a **mapping space**. In particular we denote $Y^I(0,1; *, *)$ $(* \in Y)$ by $\Omega(Y) = \Omega(Y, *)$ and call it the **space of closed paths** (or **loop space**) of Y. Two points f, g of Y^X are connected by a †path in Y^X if and only if $f \simeq g : X \rightarrow Y$. Thus $\pi_0(Y^X) = \pi(X; Y)$ and $\pi_0(Y^X(A_i; B_i)) = \pi(X, A_i; Y, B_i)$.

C. Retracts

Let A be a subspace of a topological space X. If there exists an $f \in A^X$ such that the restriction $f|A$ is the identity mapping of A, then A is called a **retract** of X, and f a **retraction**. If A is a retract of X, any continuous mapping of A into any topological space can be extended to a continuous mapping of X. If A is a retract of some neighborhood $U(A)$, A is called a **neighborhood retract** (or **NR**) of X. If for any †embedding of a metric space A in a closed subspace A_0 of any metric space X, A_0 is a retract (neighborhood retract) of X, then A is called an **absolute retract** or **AR** (**absolute neighborhood retract** or **ANR**). For example, an n-dimensional simplex or an n-dimensional Euclidean space is an AR. If a retraction f is homotopic to the identity mapping of X (resp. $U(A)$), we call A a **deformation retract** (**neighborhood deformation retract**) of X. Moreover, if $f \simeq 1_X$ (rel. A), then A is called a **strong deformation retract**. In particular, if a point x_0 is a (strong) deformation retract of X, we say that X is **contractible** to the point x_0. For example, any †polyhedron P and any compact n-dimensional †topological manifold are ANRs; any polyhedron P_0 contained in P is a strong deformation retract of some neighborhood in P. X is called **locally contractible** if each point x of X has a contractible neighborhood U of x.

D. The Extension Property

Let X, Y be topological spaces, $A \subset X$, f_0, $f_1 \in Y^X$, and $\{g_t : A \rightarrow Y\}$ a homotopy such that $g_i = f_i|A$ $(i=0,1)$. We can extend $\{g_t\}$ to a homotopy $\{f_t\}$ of X if and only if the mapping $F : (X \times 0) \cup (A \times I) \cup (X \times 1) \rightarrow Y$ defined by $F(x, i) = f_i(x)$, $F(a, t) = g_t(a)$ can be extended to a continuous mapping sending $X \times I$ into Y. Therefore, the problem whether $f_0 \simeq f_1$ can be reduced to a problem whether a continuous mapping defined on a subspace can be extended to the whole space. If for any homotopy $\{g_t: A \to Y\}$ and any continuous mapping $f_0; X \to Y$ into any topological space Y satisfying $f_0|A = g_0$, there exists a homotopy $\{f_t: X \to Y\}$ satisfying $f_t|A = g_t$, then we say that (X,A) has the **homotopy extension property**. This occurs if and only if $(X \times 0) \cup (A \times I)$ is a retract of $X \times I$. A pair (X,A) of ANRs, where A is closed in X, and a pair (P, P_0) with P a †CW complex and P_0 a subcomplex of P have this property. Given a continuous mapping $h: B \to A$ of a subspace B of a topological space Y into a topological space A, we identify $b \in B$ with $h(b) \in A$ in the †direct sum $A \cup Y$ and obtain the †identification space denoted by $A \cup_h Y$, which is called an **attaching space** under h. If (Y,B) has the homotopy extension property, then $(Y \times X, B \times X)$ and $(A \cup_h Y, A)$ also have the same property. When A consists of a point $*$, we write $* \cup_h Y = Y/B$ and call the space Y/B a **space smashing (shrinking** or **pinching) B to a point**. If $Y = B \times I$, $B = B \times 0$, then we call $A \cup_h (B \times I)$ a **mapping cylinder** of h, $(A \cup_h (B \times I))/(B \times 1)$ a **mapping cone** of h, and the mapping cylinder (mapping cone) of $h: B \to *$ the **cone** over B (**suspension** of B).

E. Homotopy Type

For systems (X, A_i), (Y, B_i) of topological spaces, if there exist $f \in Y^X(A_i; B_i)$, $g \in X^Y(B_i; A_i)$ such that $g \circ f$, $f \circ g$ are homotopic to the identity mappings of (X, A_i), (Y, B_i), respectively, then we say that (X, A_i) and (Y, B_i) have the same **homotopy type** or are **homotopy equivalent**. Such mappings f and g are called **homotopy equivalences**. For a homotopy equivalence f, the induced mappings f_* and f^* are bijective. Therefore, in homotopy theory, systems of spaces having the same homotopy type are considered equivalent. If A is a deformation retract of X, then A and X have the same homotopy type, and the injection of A into X and the retraction of X onto A are homotopy equivalences. A contractible space has the same homotopy type as a point. Spaces having the same homotopy type have isomorphic †homotopy groups and †(co)homology groups. Since the mapping cylinder $Z_f = Y \cup_f (X \times I)$ of $f \in Y^X$ contains Y as its deformation retract, it has the same homotopy type as Y. By this homotopy equivalence, f can be replaced by the injection of $X \times 1$ into Z_f. If to each topological space there corresponds a value (which may be some element of $\mathbf{R}$ or some algebraic structure) and the values are the same for homotopy equivalent spaces, then the value is called a **homotopy type invariant**. A homotopy type invariant is a †topological invariant; for example, $\pi(X; Y)$ is a homotopy type invariant of X. If a continuous mapping $f: X \to Y$ induces isomorphisms of the homotopy groups of each †arcwise connected component, then f is called a **weak homotopy equivalence**. Conversely, if X and Y are CW complexes, then a weak homotopy equivalence is a homotopy equivalence (J. H. C. Whitehead).

Now we consider the category of pointed topological spaces. The **reduced join** of topological spaces A, B is the space obtained from $A \times B$ by smashing its subspace $A \vee B = (A \times *) \cup (* \times B)$ to a point, and is denoted by $A \wedge B$. We call $A \wedge S^1$ the **(reduced) suspension** of A and denote it by SA. Repeating the suspension n times, we have the **n-fold reduced suspension** of A. We call $CA = A \wedge I$ ($I = [0,1]$) the **reduced cone** of A (I has the base point 1). For a continuous mapping $f: X \to Y$, the space obtained by identifying each point $(x, 0)$ of the base (X) of CX with $f(x) \in Y$ is called the **reduced mapping cone** and is denoted by $C_f = Y \cup_f CX$. The **reduced join** of mappings $f: Y \to X$ and $f': Y' \to X'$ is the mapping $f \wedge f': Y \wedge Y' \to X \wedge X'$ induced from the product mapping $f \times f': Y \times Y' \to X \times X'$. The reduced join of $f: Y \to X$ and $1: S^1 \to S^1$ (identity mapping) is written as $Sf = f \wedge 1$ and is called the **suspension** of f.

F. Puppe Exact Sequences

For $f: X \to Y$ and $g: Y \to Z$, we have $g \circ f \simeq 0$ if and only if g can be extended to a continuous mapping from C_f into Z. In other words, the sequence

$$\pi(C_f; Z)_0 \xrightarrow{i^*} \pi(Y; Z)_0 \xrightarrow{f^*} \pi(X; Z)_0$$

is exact (i.e., $\operatorname{Im} i^* = \operatorname{Ker} f^* = f^{*-1}(0)$, where $i: Y \to C_f$ is the canonical inclusion and 0 is the class of the constant mapping). The inclusion $i: Y \to C_f$ gives rise to the reduced mapping cone C_i. We also have the canonical inclusion $i': C_f \to C_i$. Adding the term $\pi(C_i : Z)_0 \xrightarrow{i'^*}$ to the left-hand side of the sequence above, we have a new exact sequence. Continuing this process, we obtain an exact sequence of infinite length. If X, Y satisfy a suitable condition (e.g., X, Y are CW complexes), then C_i has the same homotopy type as the reduced suspension SX of X; i'^* is equivalent to $p^*: \pi(SX; Z)_0 \to \pi(C_f; Z)_0$ induced by a mapping $p: C_f \to SX$ smashing Y to a point; and furthermore C_p has the same homotopy type as SY and the inclusion $i_0: SX \to C_p$ is equivalent to the suspension $Sf: SX \to SY$ of f. Thus the following **Puppe**

exact sequence is obtained:

$$\ldots \xrightarrow{Sp^*} \pi(SC_f; Z)_0 \xrightarrow{Si^*} \pi(SY; Z)_0 \xrightarrow{Sf^*} \pi(SX; Z)_0$$

$$\xrightarrow{p^*} \pi(C_f; Z)_0 \xrightarrow{i^*} \pi(Y; Z)_0 \xrightarrow{f^*} \pi(X; Z)_0.$$

In this exact sequence, if Y is a CW complex, X is a subcomplex of Y, and f is the inclusion $i: X \rightarrow Y$, then $C_i = C_f = Y \cup C_X$ is homotopy equivalent to the space $C_i / CX = Y/X$ obtained by smashing CX to a point, and an exact sequence of the following type is obtained:

$$\ldots \xrightarrow{Si^*} \pi(SX; Z)_0 \xrightarrow{\delta^*} \pi(Y/X; Z)_0 \xrightarrow{p^*} \pi(Y; Z)_0$$

$$\xrightarrow{i^*} \pi(X; Z)_0.$$

For a continuous mapping $f: X \rightarrow Y$, consider the subspace $E_f = \{(x, \varphi) | f(x) = \varphi(0)\}$ of the product space $X \times Y^I$. By identifying X with $\{(x, \varphi_x) | \varphi_x(I) = x\}$, we can regard X as a deformation retract of E_f. By putting $p_1(x, \varphi) = \varphi(1)$, we obtain a †fiber space (E_f, p_1, Y). The fiber $T_f = p_1^{-1}(*)$ is called a **mapping track** of f. Using the †covering homotopy property, we see that the sequence

$$\pi(W; T_f)_0 \xrightarrow{p_*} \pi(W; X)_0 \xrightarrow{f_*} \pi(W; Y)_0$$

is exact, where $p(x, \varphi) = x$. This sequence is also extended infinitely to the left as

$$\ldots \longrightarrow \pi(W; \Omega X)_0 \xrightarrow{(\Omega f)_*} \pi(W; \Omega Y)_0$$

$$\xrightarrow{i_*} \pi(W; T_f)_0 \xrightarrow{p_*},$$

where i is the inclusion of the loop space $\Omega(Y)$ into T_f and $\Omega f: \Omega X \rightarrow \Omega Y$ is the correspondence of the loops induced from f.

G. Homotopy Sets That Form Groups

If $X = SX'$ or $Y = \Omega Y'$ (or, generally, if Y is a †homotopy associative †H-space having a †homotopy inverse), then $\pi(X; Y)_0$ forms a group. In the general case the product of the loops induces the product of $\pi(X; \Omega Y')_0$, and SX is obtained from $X \times I$ by smashing the subset $(X \times \dot{I}) \cup (* \times I)$ to a point (where $\dot{I} = \{0, 1\}$). We represent a point of SX by (x, t) $(x \in X, t \in I)$ and define the mapping $\Omega_0 g: X \rightarrow \Omega Y$ for each $g: SX \rightarrow Y$ by $\Omega_0 g(x)(t) = g(x, t)$. Hence an isomorphism $\Omega_0: \pi(SX; Y)_0 \cong \pi(X; \Omega Y)_0$ is obtained. Each of the following pairs of homomorphisms is equivalent: $f_*: \pi(SX; Y)_0 \rightarrow \pi(SX; Y')_0$ and $\Omega f_*: \pi(X; \Omega Y)_0 \rightarrow \pi(X; \Omega Y')_0$; and $h^*: \pi(X'; \Omega Y)_0 \rightarrow \pi(X; \Omega Y)_0$ and $Sh^*: \pi(SX'; Y)_0 \rightarrow \pi(SX; Y)_0$.

If S^n is an n-dimensional sphere, then $\pi_n(X) = \pi(S^n; X)_0$ is the n-dimensional †homotopy group. Let $\pi^n(X) = \pi(X; S^n)_0$. If X is a CW complex of dimension less than $2n-1$, $\pi^n(X)$ is the †cohomotopy group isomorphic to $\pi(X; \Omega S^{n+1})_0$ (→ 205 Homotopy Groups). Let K_n be an †Eilenberg-MacLane space of type (Π, n). Then we have $K_n = \Omega K_{n+1}$, and if (X, A) is a pair of CW complexes, then $\pi(X/A; K_n)_0$ coincides with the cohomology group $H^n(X, A; \Pi)$. For the †classifying space $B_{\mathbf{O}}$ $(B_{\mathbf{U}})$ of the †infinite orthogonal group $\mathbf{O}$ (†infinite unitary group $\mathbf{U}$), $\pi(X/A; B_{\mathbf{O}})$ $(\pi(X/A; B_{\mathbf{U}}))$ may be considered the KO-group $KO(X, A)$ (K-group $K(X, A)$) (→ 236 K-Theory).

H. Hopf's Classification Theorem

With each homotopy class of a continuous mapping f of an n-dimensional †polyhedron K^n into an n-dimensional sphere S^n, we associate the homomorphism $f_*: H_n(K^n; \mathbf{R}_1) \rightarrow H_n(S^n; \mathbf{R}_1) \cong \mathbf{R}_1$ $(\mathbf{R}_1 = \mathbf{R}/\mathbf{Z})$ of the n-dimensional homology groups induced by f. We thus obtain a bijective relation $\pi(K^n; S^n) \rightarrow \mathrm{Hom}(H_n(K^n; \mathbf{R}_1), \mathbf{R}_1)$, called **Hopf's classification theorem** (Hopf, *Comment. Math. Helv.*, 5 (1933)). This theorem is generalized as follows: First we replace $\mathrm{Hom}(H_n(K^n; \mathbf{R}_1), \mathbf{R}_1)$ by the n-dimensional integral cohomology group $H^n(K^n; \mathbf{Z})$ (H. Whitney), S^n by an †$(n-1)$-connected †n-simple topological space Y (Hurewicz), and K^n by a pair (K, K_0) of CW complexes with $\dim(K - K_0) \leqslant n$. Then we obtain a one-to-one correspondence $\pi(K, K_0; Y, *) \rightarrow H^n(K, K_0; \pi_n(Y))$. If u_n is the element (called the fundamental class of Y) of $H^n(Y; \pi_n(Y)) = \mathrm{Hom}(H_n(Y), \pi_n(Y))$ given by the †Hurewicz isomorphism $H_n(Y) \rightarrow \pi_n(Y)$, then $[f] \rightarrow f^*(u_n)$ gives this one-to-one correspondence.

Let K_0 be a subpolyhedron of a polyhedron K. Then a necessary and sufficient condition for a continuous mapping $f: K_0 \rightarrow S^n$ to be extended to $K_0 \cup K^{n+1}$ (K^{n+1} is the $(n+1)$-†skeleton of K) is that $f_*(\mathrm{Ker}\, i_*) = 0$ $(f_*: H_n(K_0; \mathbf{R}_1) \rightarrow H_n(S^n; \mathbf{R}_1)$, $i_*: H_n(K_0; \mathbf{R}_1) \rightarrow H_n(K; \mathbf{R}_1))$ (**Hopf's extension theorem**, which corresponds to Hopf's classification theorem). In general, in the study of the classification problem and extension problem for mappings of a polyhedron into a topological space, the theory of obstructions is useful (→ 300 Obstructions).

I. Essential Mappings

A mapping f from a compact space into an n-dimensional sphere S^n is called **essential** if any mapping g homotopic to f satisfies $g(X) = S^n$. A mapping is inessential if and only if it is homotopic to the constant mapping.

References

[1] P. S. Aleksandrov and H. Hopf, Topologie I, Springer, 1935 (Chelsea, 1965).
[2] P. J. Hilton, An introduction to homotopy theory, Cambridge Univ. Press, 1953.
[3] A. Komatu, M. Nakaoka, and H. Toda, Isôkikagaku (Japanese; Topology), Kyôritu, 1957.
[4] S. T. Hu, Homotopy theory, Academic Press, 1959.
[5] A. Komatu, M. Nakaoka, and M. Sugawara, Isôkikagaku (Japanese; Topology) I, Iwanami, 1967.
[6] D. Puppe, Homotopiengen und ihre induzierten Abbildungen I, Math. Z., 69 (1958), 299–344.
[7] J. F. Adams, Stable homotopy theory, Lecture notes in math. 3, Springer, third edition, 1969.

205 (IX.15) Homotopy Groups

A. General Remarks

Given a topological space X, we utilize the concept of homotopy to define the †fundamental group, homotopy groups, and cohomotopy groups of X. These groups, together with (co)homology groups, are useful tools in topology. Since W. Hurewicz defined the notion of homotopy groups in 1935, their theory has made rapid progress and now plays an important role in topology.

B. Homotopy Groups

Let X be a topological space with a base point $*$, $I^n=\{t=(t_1,t_2,\ldots,t_n)\,|\,0\leqslant t_1,t_2,\ldots,t_n\leqslant 1\}$ be the unit n-cube, and $\dot{I}^n$ its boundary. Denote by $\Omega^n(X,*)=X^{I^n}(\dot{I}^n,*)$ the †mapping space consisting of all continuous mappings $f:(I^n,\dot{I}^n)\rightarrow(X,*)$ (in particular $\Omega^1(X,*)$ is the †loop space), and denote by $\pi_n(X,*)$ or simply $\pi_n(X)$ the set of arcwise connected components of $\Omega^n(X,*)$, i.e., the †homotopy classes $[f]$. Using the notation of †homotopy sets, we have $\pi_n(X,*)=\pi(I^n,\dot{I}^n;X,*)$. If we choose the constant mapping as the base point $*$ of $\Omega^n(X,*)$, then $\Omega^m(\Omega^n(X,*),*)=\Omega^{m+n}(X,*)$. Thus $\pi_m(\Omega^n(X,*),*)=\pi_{m+n}(X,*)$. Since π_1 is the fundamental group, $\pi_n(X,*)=\pi_1(\Omega^{n-1}(X),*)$ is also a group, called the n-dimensional **homotopy group** of X with base point $*$. "Multiplication" in homotopy groups is defined as follows: Given $f_1,f_2\in\Omega^n(X,*)$ we define $f_1+f_2\in\Omega^n(X,*)$ by

$$(f_1+f_2)(t)=\begin{cases} f_1(2t_1,t_2,\ldots,t_n), & 0\leqslant t_1\leqslant\frac{1}{2},\\ f_2(2t_1-1,t_2,\ldots,t_n), & \frac{1}{2}\leqslant t_1\leqslant 1\end{cases}$$

(Fig. 1). Then the product or sum of $[f_1]$ and $[f_2]$ is given by $[f_1+f_2]$. The identity is the class of the constant mapping (denoted by 0), and the inverse of $[f]$ is $[\bar{f}]$, represented by $\bar{f}(t)=f(1-t_1,t_2,\ldots,t_n)$. The space $\Omega^n(X,*)$ is an †H-space, where multiplication is given by the correspondence $(f_1,f_2)\rightarrow f_1+f_2$. Since the fundamental group of an H-space is commutative, $\pi_n(X,*)$ is an Abelian group for $n\geqslant 2$.

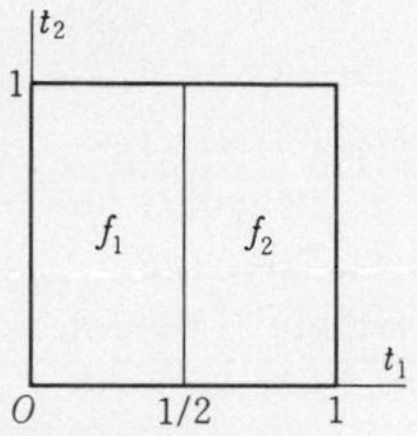

Fig. 1

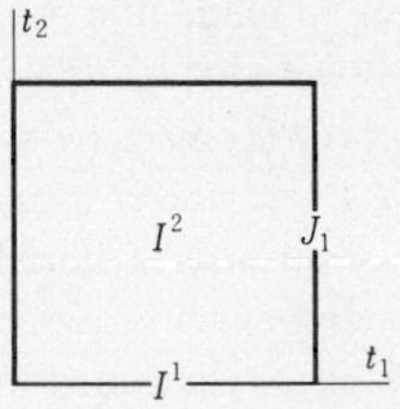

Fig. 2

Let $S^n=\{t=(t_1,\ldots,t_{n+1})\,|\,\Sigma t_i^2=1\}$ be the n-sphere, and take $*=(1,0,\ldots,0)$ as its base point. Suppose that we are given a continuous mapping $\psi_n:(I^n,\dot{I}^n)\rightarrow(S^n,*)$ such that $\psi_n: I^n-\dot{I}^n\rightarrow S^n-*$ is homeomorphic. Then the correspondence $\psi_n^*:\pi(S^n;X)_0\rightarrow\pi_n(X,*)$ determined by $\psi_n^*[g]=[g\circ\psi_n]$ is bijective. Thus we can identify the homotopy group $\pi_n(X,*)$ with $\pi(S^n;X)_0$.

C. Relative Homotopy Groups

Suppose that we are given a topological space X and a subspace A of X sharing the same base point $*$. Identify I^{n-1} with the face $t_n=0$ of I^n, and let J^{n-1} be the closure of $\dot{I}^n-I^{n-1}$ (Fig. 2). Denote by $\pi_n(X,A,*)$ the set of homotopy classes of continuous mappings $f:(I^n,\dot{I}^n,J^{n-1})\rightarrow(X,A,*)$. Let $\Omega^n(X,A,*)$ be the mapping space consisting of such mappings f, and let $\pi_n(X,A,*)=\pi_0(\Omega^n(X,A,*))$. Since $\Omega^m(\Omega^n(X,A,*),*)$ is homeomorphic to $\Omega^{m+n}(X,A,*)$, we have $\pi_m(\Omega^n(X,A,*),*)\cong\pi_{m+n}(X,A,*)$. Thus $\pi_n(X,A,*)$ is a group for $n\geqslant 2$ and an Abelian group for $n\geqslant 3$. This group is called the n-dimensional **relative homotopy group** of (X,A) with respect to the base point $*$, or simply the n-dimensional homotopy group of (X,A). In the same manner as in Section B multiplication in this group can be defined using f_1+f_2. Since $\Omega^n(X,*,*)$ and $\Omega^n(X,*)$ are identical, we have $\pi_n(X,*,*)=\pi_n(X,*)$. Hence homotopy groups are special cases of relative homotopy groups.

Let $g:(X,A,*)\to(Y,B,*)$ be a continuous mapping. Then a correspondence $g_*:\pi_n(X,A,*)\to\pi_n(Y,B,*)$ is obtained by $g_*[f]=[g\circ f]$, with g_* a homomorphism of homotopy groups for $n\geqslant 2$ and for $n=1$, $A=*$. We call g_* the **homomorphism induced by** g. Let $E^n=\{t=(t_1,\ldots,t_n)\,|\,\Sigma t_i^2=1\}$ be the unit n-cell with boundary S^{n-1}. Utilizing a suitable relative homeomorphism $\psi_n':(I^n,J^{n-1})\to(E^n,*)$, $\psi_n'(\dot I^n)=S^{n-1}$, we obtain a one-to-one correspondence $\psi_n'^*:\pi(E^n,S^{n-1};X,A)_0\to\pi_n(X,A,*)$, and $\Omega^n(X,A,*)$ is homeomorphic (via ψ_n') to the mapping space $X^{E^n}(S^{n-1},*;A,*)$.

D. Homotopy Exact Sequences

Given an element $\alpha=[f]\in\pi_n(X,A,*)$, and letting $\partial\alpha=[f\,|\,I^{n-1}]\in\pi_{n-1}(A,*)$, we obtain a homomorphism $(n\geqslant 2)$ $\partial:\pi_n(X,A,*)\to\pi_{n-1}(A,*)$, which is called the **boundary homomorphism**. Furthermore, we have the following exact sequence involving homomorphisms i_*,j_* induced by two inclusions $i:(A,*)\to(X,*)$, $j:(X,*,*)\to(X,A,*)$:

$$\ldots\xrightarrow{\partial}\pi_n(A,*)\xrightarrow{i_*}\pi_n(X,*)\xrightarrow{j_*}\pi_n(X,A,*)$$

$$\xrightarrow{\partial}\ldots\to\pi_1(X,*)\xrightarrow{j_*}\pi_1(X,A,*)\xrightarrow{\partial}\pi_0(A)$$

$$\xrightarrow{i_*}\pi_0(X).$$

This sequence is called the **homotopy exact sequence** of the pair (X,A). A system of topological spaces $X\supset A\supset B\ni *$ is called a **triple**. In this homotopy exact sequence, if we replace $(A,*)$, $(X,*)$ by $(A,B,*)$, $(X,B,*)$, respectively, we obtain an exact sequence, called the **homotopy exact sequence of the triple** (X,A,B).

The homotopy group $\pi_n(A\times B)$ of the product space is isomorphic to the direct sum $\pi_n(A)+\pi_n(B)$, and the projections $p(p'):A\times B\to A(B)$ of the product space induce the projections from $\pi_n(A\times B)$ onto the direct summands $\pi_n(A)$, $\pi_n(B)$. This is a special case of the †Hurewicz isomorphism theorem in fiber spaces (→ 156 Fiber Spaces). Setting $A\vee B=(A\times *)\cup(*\times B)$, we obtain a direct sum decomposition $\pi_n(A\vee B)\cong\pi_n(A)+\pi_n(B)+\pi_{n+1}(A\times B,A\vee B)$. Next we consider a fixed pair (X,A) and move the base point $*$ to investigate its effect on the elements of the homotopy group. Suppose that we are given a path $h:I\to A$ with terminal point $*=h(1)$ and an element $\alpha\in\pi_n(X,A,*)$ $(\alpha=[f],\ f:(I^n,\dot I^n,J^{n-1})\to(X,A,*))$. By the homotopy extension property, we can construct a homotopy $f_\theta:(I^n,\dot I^n)\to(X,A)$ satisfying $f_\theta(J^{n-1})=h(\theta)$ and $f_1=f$. Then the homotopy class $[f_0]$ of f_0 with respect to the base point $*'=h(0)$ is determined only by α and the †homotopy class ω of the path h. We denote the homotopy class $[f_0]$ by $\alpha^\omega\in\pi_n(X,A,*')$. The correspondence $\alpha\to\alpha^\omega$ is a group isomorphism, and $(\alpha^\omega)^{\omega'}=\alpha^{\omega\omega'}$. Thus if A is arcwise connected, $\pi_n(X,A,*)$ is isomorphic to $\pi_n(X,A,*')$. Hence, in this case, we may simply write $\pi_n(X,A)$ instead of $\pi_n(X,A,*)$. When $*=*'$, the correspondence $\alpha\to\alpha^\omega$ determines the action of the group $\pi_1(A,*)$ on $\pi_n(X,A,*)$. Given an element $\alpha\in\pi_n(X,*)$ and a class ω of paths in X, we define $\alpha^\omega\in\pi_n(X,*')$ as for relative homotopy. Specifically, if $\omega\in\pi_1(X,*)$, then $\alpha^\omega-\alpha$ coincides with the †Whitehead product $[\omega,\alpha]$ (when $n=1$, we have $\alpha^\omega\cdot\alpha^{-1}=[\omega,\alpha]=\omega\alpha\omega^{-1}\alpha^{-1}$).

A pair (X,A) consisting of a topological space X and an arcwise connected subspace A of X is said to be ***n*-simple** if the operation of $\pi_1(A)$ on $\pi_n(X,A)$ is trivial. Similarly, an arcwise connected space X is called ***n*-simple** if the operation of $\pi_1(X)$ on $\pi_n(X)$ is trivial. For example, a pair (X,A) consisting of an H-space X and a sub-H-space A is **simple**, i.e., n-simple for each n. If a topological space X satisfies $\pi_i(X)=0$ $(0\leqslant i\leqslant n)$, then X is said to be ***n*-connected**. 0-connectedness coincides with arcwise connectedness and 1-connectedness means †simple connectedness. S^n is $(n-1)$-connected. A pair (X,A) is said to be ***n*-connected** if $\pi_0(A)=\pi_0(X)=\pi_i(X,A)=0$ $(1\leqslant i\leqslant n)$, and (E^n,S^{n-1}) is $(n-1)$-connected.

E. Homotopy Groups of Triads

Let $(X;A,B,*)$ be a system, called a **triad**, of a topological space X and its subspaces A, B satisfying $A\cap B\ni *$ (base point). Let $\pi_n(X;A,B,*)=\pi_{n-1}(\Omega^1(X,B),\Omega^1(A,A\cap B),*)$ $(n\geqslant 2)$; $\pi_n(X;A,B,*)$ is a group for $n\geqslant 3$ and an Abelian group for $n\geqslant 4$. We call $\pi_n(X;A,B,*)$ the **homotopy group of the triad**. From the homotopy exact sequence of the pair, we obtain the following **homotopy exact sequence of the triad**:

$$\ldots\xrightarrow{\partial}\pi_j(A,A\cap B,*)\xrightarrow{i_*}\pi_j(X,B,*)\xrightarrow{j_*}$$

$$\pi_j(X;A,B,*)\xrightarrow{\partial}\pi_{j-1}(A,A\cap B,*)\xrightarrow{i_*}\ldots.$$

Assume for simplicity that $A\cap B$ is simply connected, $X=\operatorname{Int}A\cup\operatorname{Int}B$ ($\operatorname{Int}A$ is the †interior of A), $(A,A\cap B)$ is m-connected, and $(B,A\cap B)$ is n-connected. Then $(X;A,B)$ is $(m+n)$-connected, i.e., $\pi_j(X;A,B,*)=0$ $(2\leqslant j\leqslant m+n)$ **(Blakers-Massey theorem)**.

Furthermore, in this case we have a replica of the †excision axiom for $j<m+n$; that is, we have the isomorphism $i_*:\pi_j(A,A\cap B,*)\cong\pi_j(X,B,*)$ induced by the inclusion $i:(A,A\cap B)\to(X,B)$. On the other hand,

$\pi_{m+n+1}(X;A,B,*)$ is isomorphic to $\pi_{m+1}(A, A\cap B,*)\otimes\pi_{n+1}(B,A\cap B,*)$. This shows that the excision axiom does not always hold for homotopy groups, an important difference from homology theory. However, if we replace the excision axiom by the Hurewicz-Steenrod isomorphism theorem, which is valid for fiber spaces (→ 156 Fiber Spaces), then we can construct homotopy theory axiomatically in the same manner as homology theory (→ 203 Homology Groups).

F. The Hurewicz Isomorphism Theorem

The **Hurewicz homomorphism** τ of $\pi_n(X,A)$ into the n-dimensional integral homology group $H_n(X,A)$ is defined by $\tau([f])=f_*(\varepsilon_n)$ (where ε_n is a generator of $H_n(I^n,\dot{I}^n)$). Then we have the **Hurewicz isomorphism theorem**: Suppose that the pair (X,A) is n-simple (e.g., $A=*$) and $(n-1)$-connected. Then we have $H_i(X,A)=0$ $(i<n)$ and the isomorphism $\tau:\pi_n(X,A)\cong H_n(X,A)$ (for $n=1$ → 175 Fundamental Group). Let X, Y be simply connected topological spaces, and let $f:X\to Y$ be a continuous mapping. Then the following two conditions are equivalent: (1) $f_*:\pi_i(X)\to\pi_i(Y)$ is injective for $i<n$ and surjective for $i\leqslant n$. (2) $f_*:H_i(X)\to H_i(Y)$ is injective for $i<n$ and surjective for $i\leqslant n$ (**Whitehead's theorem**).

J.-P. Serre generalized these theorems as follows: A family $\mathcal{C}$ of Abelian groups satisfying condition (i) is called a **class of Abelian groups**: (i) If a sequence $F\to G\to H$ of Abelian groups is exact and $F, H\in\mathcal{C}$, then $G\in\mathcal{C}$. Furthermore, we consider the following conditions: (ii) The tensor product $G\otimes F$ of an arbitrary Abelian group F with an element $G\in\mathcal{C}$ also belongs to $\mathcal{C}$. (ii′) If both $F, G\in\mathcal{C}$, then $F\otimes G$, $\mathrm{Tor}(F,G)\in\mathcal{C}$. (iii) If $G\in\mathcal{C}$, then its homology group $H_i(G)\in\mathcal{C}$ $(i>0)$. Condition (ii′) is implied by (ii). A homomorphism $f:F\to G$ is called $\mathcal{C}$-injective if $\mathrm{Ker}\, f\in\mathcal{C}$, $\mathcal{C}$-surjective if $\mathrm{Coker}\, f=G/\mathrm{Im}\, f\in\mathcal{C}$, and a $\mathcal{C}$-isomorphism if f is $\mathcal{C}$-injective and $\mathcal{C}$-surjective. Two Abelian groups G and G' are called $\mathcal{C}$-isomorphic if there exist $\mathcal{C}$-isomorphisms $f:F\to G$ and $f':F\to G'$. In particular, if the class $\mathcal{C}_0$ consists of only the trivial group 0, then concepts such as $\mathcal{C}_0$-isomorphism coincide with the usual concepts of isomorphism, and so on. Let $\mathcal{C}_p$ be the class of finite Abelian groups whose orders are relatively prime to a fixed prime number p. Here, instead of the terms $\mathcal{C}_p$-isomorphism and so on, we use the terms **mod p isomorphism** and so on. Let $\mathfrak{D}$ be a class of finitely generated Abelian groups. Then $\mathcal{C}_p$ satisfies conditions (ii) and (iii), and $\mathfrak{D}$ satisfies (ii′) and (iii). We have the following **generalized Hurewicz theorem**: (A) Suppose that a class $\mathcal{C}$ satisfies (ii) and (iii) and we are given a 2-connected pair (X,A) of simply connected spaces X, A. If $\pi_i(X,A)\in\mathcal{C}$ $(i<n)$, then $H_i(X,A)\in\mathcal{C}$, and $\tau:\pi_n(X,A)\to H_n(X,A)$ is a $\mathcal{C}$-isomorphism. (B) Suppose that $A=*$, $\mathcal{C}$ satisfies conditions (ii′) and (iii), and X is simply connected. Then an assertion similar to (A) holds. In particular, a simply connected space X having finitely generated homology groups (e.g., a simply connected finite polyhedron) has finitely generated homotopy groups. As a corollary to theorem (A), we obtain a **generalized Whitehead theorem**. In particular, applying the theorem to the class $\mathfrak{D}\cap\mathcal{C}_p$, we obtain the following frequently used theorem: Suppose that we are given simply connected spaces X, Y whose homology groups are finitely generated and $f:X\to Y$ satisfies $f_*\pi_2(X)=\pi_2(Y)$. Then the following two conditons are equivalent: (1) $f_*:\pi_i(X)\to\pi_i(Y)$ is a mod p isomorphism for $i<n$ and a mod p surjection for $i=n$. (2) $f_*:H_i(X,\mathbf{Z}_p)\to H_i(Y,\mathbf{Z}_p)$ is an isomorphism for $i<n$ and a surjection for $i=n$ (where $\mathbf{Z}_p=\mathbf{Z}/p\mathbf{Z}$). The theory above, which makes use of the notion of class $\mathcal{C}$, is an example of **Serre's $\mathcal{C}$-theory**. Concepts such as †spectral sequences for fiber spaces and n-connected fiber spaces are important tools in Serre's $\mathcal{C}$-theory (→ 156 Fiber Spaces).

To calculate homotopy groups, we use notions such as exact sequences, fiber spaces, (co)homology groups of n-connected fiber spaces, and †Postnikov systems. Given an arbitrary group (more generally, a Postnikov system), there exists a CW complex having the given group (system) as its homotopy group (Postnikov system) (**realization theorem of homotopy groups**). For an arbitrary arcwise connected topological space X there exist topological spaces (X,n) and continuous mappings $p_n:(X,n+1)\to(X,n)$ $(n=1,2,\ldots)$ satisfying the following two conditions: (i) $((X,n+1),p_n,(X,n))$ is a fiber space whose fiber is an †Eilenberg-MacLane space. (ii) $(X,1)=X$, and $((X,n+1),p_1\circ\ldots\circ p_n,X)$ is an n-connected fiber space. The method of obtaining the homotopy group $\pi_n(X)\cong H_n((X,n))$ by computing (co)homology groups of (X,n) is called a **killing method**.

G. Homotopy Groups of Spheres

The spheres S^n and their homotopy groups are basic objects in homotopy theory. Although much research has been done concerning these objects, there are still open problems.

S^n is $(n-1)$-connected: $\pi_i(S^n)=0$ $(i<n)$. The fact that $\pi_n(S^n)\cong\mathbf{Z}$ (infinite cyclic

group) was obtained from the †Brouwer mapping theorem. Also, $\pi_i(S^1)=0$ $(i>1)$ follows from the fact that the universal covering space of S^1 is contractible. Suppose that we are given a continuous mapping $f: S^{2n-1}\rightarrow S^n$. We approximate it by a †simplicial mapping φ. Then the inverse image $\varphi^{-1}(*)$ of a point $*$ in the interior of an n-simplex of S^n is an $(n-1)$-dimensional †pseudomanifold which is orientable by means of a suitable generator $\varepsilon\in H_{n-1}(\varphi^{-1}(*))$. The boundary isomorphism $\partial: H_n(S^{2n-1},\varphi^{-1}(*))\cong H_{n-1}(\varphi^{-1}(*))$ and the homomorphism $\varphi_*: H_n(S^{2n-1},\varphi^{-1}(*))\rightarrow H_n(S^n,*)$ give rise to an integer $\gamma(\varphi)$ determined by the relation $\varphi_*\partial^{-1}(\varepsilon)=\gamma(\varphi)\varepsilon_n$ (ε_n is an orientation of S^n). This integer is independent of the choice of φ, so we may set $\gamma(f)=\gamma(\varphi)$. Then $f\simeq g$ implies that $\gamma(f)=\gamma(g)$. We call $\gamma(f)$ the **Hopf invariant** of f. H. Hopf defined γ and showed $\gamma:\pi_3(S^2)\cong\mathbf{Z}$ (1931); $\gamma(\pi_{2n-1}(S^n))=0$ for odd n; $\gamma(\pi_{2n-1}(S^n))\supset 2\mathbf{Z}$ for even n; and $\gamma(\pi_{2n-1}(S^n))=\mathbf{Z}$ for $n=4,8$ (1935). H. Freudenthal defined a homomorphism $E:\pi_i(S^n)\rightarrow\pi_{i+1}(S^{n+1})$, $E[f]=[Sf]$, and proved the **Freudenthal theorem**: (1) E is an isomorphism for $i<2n-1$; (2) E is a surjection for $i=2n-1$; and (3) the image of E coincides with the kernel of γ for $i=2n$. Furthermore he obtained $\pi_{n+1}(S^n)\cong\mathbf{Z}_2$ $(n\geqslant 3)$ (1937). For $n=2,4,8$, a mapping $f: S^{2n-1}\rightarrow S^n$ (†Hopf mapping) such that $\gamma(f)=1$ (given by Hopf) is the projection of a †fiber bundle S^{2n-1} over the base space S^n, and the correspondence $(\alpha,\beta)\rightarrow E\alpha+f_*\beta$ gives an isomorphism $\pi_{i-1}(S^{n-1})+\pi_i(S^{2n-1})$ (direct sum) $\cong\pi_i(S^n)$. Hence we obtain $\pi_4(S^2)=\mathbf{Z}_2$. It was shown by G. W. Whitehead and L. S. Pontrjagin that $\pi_{n+2}(S^n)$ $(n\geqslant 3)$ is isomorphic to $\mathbf{Z}_2$ (1949). Whitehead also defined a **generalized Hopf homomorphism** $H:\pi_i(S^n)\rightarrow\pi_i(S^{2n-1})$ for a range of $i<3n-3$, and this restriction on the dimension was removed by P. J. Hilton and I. M. James. Using H, many nontrivial results concerning $\pi_i(S^n)$ have been obtained. Serre obtained the following (1951–1953): $\pi_i(S^n)$ is finite except when $i=n$ or $i=4m-1$ and $n=2m$. Furthermore, $\pi_{4m-1}(S^{2m})$ is the direct sum of $\mathbf{Z}$ and a finite group. Let p be an odd prime and n be even. Then $\pi_i(S^n)$ is $\mathcal{C}_p$-isomorphic to $\pi_{i-1}(S^{n-1})+\pi_i(S^{2n-1})$. Let n be odd. Then $\pi_{n+k}(S^n)\in\mathcal{C}_p$ $(k<2p-3)$, and $\pi_{n+2p-3}(S^n)$ is $\mathcal{C}_p$-isomorphic to $\mathbf{Z}_p$. Serre and H. Toda determined $\pi_{n+k}(S^n)$ for $k=3,4,5$, and Serre further determined it for $k=6,7,8$. Utilizing the reduced product space of S^n, James gave the sequence

$$\ldots\rightarrow\pi_i(S^n)\xrightarrow{E}\pi_{i+1}(S^{n+1})\xrightarrow{H}\pi_{i+1}(S^{2n+1})$$

$$\rightarrow\pi_{i-1}(S^n)\rightarrow\ldots$$

and showed that it is an exact sequence if n is odd and an exact sequence mod 2 if n is even (1953). Using this exact sequence and the †secondary composition, Toda determined $\pi_{n+k}(S^n)$ for $k\leqslant 19$ [6] (→ Appendix A, Table 6.V).

By the Freudenthal theorem (1), the $\pi_{n+k}(S^n)$ $(n>k+1)$ for a fixed k are isomorphic to each other. We call $\pi_{n+k}(S^n)$ $(n>k+1)$ the **stable homotopy group of the k-stem** of the sphere and denote it by G_k. For $k=0,1,2,\ldots,15,\ldots$, $G_k\cong\mathbf{Z},\mathbf{Z}_2,\mathbf{Z}_2,\mathbf{Z}_{24},0,0,\mathbf{Z}_2,\mathbf{Z}_{240},\mathbf{Z}_2+\mathbf{Z}_2,\mathbf{Z}_2+\mathbf{Z}_2+\mathbf{Z}_2,\mathbf{Z}_6,\mathbf{Z}_{504},0,\mathbf{Z}_3,\mathbf{Z}_2+\mathbf{Z}_2,\mathbf{Z}_{480}+\mathbf{Z}_2,\ldots$. For the computation of G_k, the notion of n-connected fiber spaces is important. By utilizing the Adams spectral sequence, we can show that G_k is closely related to the cohomology of the †Steenrod algebra. The p-component of G_k was determined for $k<2p^2(p-1)-3$ (Toda). Let $\pi_i(S^n:p)$ be the p-component of $\pi_i(S^n)$. To survey this group for the nonstable case $(i\geqslant 2n-1)$, we utilize Serre's mod p direct sum decomposition (for n even), and we have the following two exact sequences for the case of odd n:

$$\ldots\rightarrow\pi_i(S^n)\xrightarrow{E^2}\pi_{i+2}(S^{n+2})$$

$$\rightarrow\pi_i(\Omega^2(S^{n+2}),S^n)\xrightarrow{\partial}\ldots,$$

$$\ldots\rightarrow\pi_{i+3}(S^{pn+p+1}:p)\xrightarrow{\Delta}\pi_{i+1}(S^{pn+p-1}:p)$$

$$\rightarrow\pi_i(\Omega^2(S^{n+2}),S^n:p)$$

$$\rightarrow\pi_{i+2}(S^{pn+p+1}:p)\xrightarrow{\Delta}\ldots,$$

where $E^2=E\circ E$ and $\Delta E^2(\alpha)=p\alpha$ (→ Appendix A, Table 6.V).

H. Homotopy Groups of Classical Groups

Consider the classical group $U(n,\Lambda)$, which is either the orthogonal group $O(n)$ $(\Lambda=\mathbf{R})$; the unitary group $U(n)$ $(\Lambda=\mathbf{C})$; or the symplectic group $Sp(n)$ $(\Lambda=\mathbf{H})$. The **infinite classical group** $U(\infty,\Lambda)$ is defined to be the inductive limit group of $\{U(n,\Lambda)\,|\,n=1,2,\ldots\}$ with respect to the natural injection $U(n,\Lambda)\subset U(n+1,\Lambda)$. We call $U(\infty,\Lambda)$ the **infinite orthogonal group**, **infinite unitary group**, and **infinite symplectic group** for $\Lambda=\mathbf{R}$, $\mathbf{C}$, and $\mathbf{H}$, respectively. The dimensions of the cells of $U(\infty,\Lambda)-U(n,\Lambda)$ are $\lambda(n+1)-1$, where $\lambda=\dim_{\mathbf{R}}\Lambda$ $(=1\ (\Lambda=\mathbf{R}),\ =2\ (\Lambda=\mathbf{C}),\ =4\ (\Lambda=\mathbf{H}))$. It follows that $\pi_k(U(n,\Lambda))$ is isomorphic to $\pi_k(U(\infty,\Lambda))$ for $k<\lambda(n+1)-2$, which is called the kth **stable homotopy group of the classical group**. Let $\mathbf{O}=U(\infty,\mathbf{R})$, $\mathbf{U}=U(\infty,\mathbf{C})$, $\mathbf{Sp}=U(\infty,\mathbf{H})$. The homotopy groups of the classical groups are periodic $(k\geqslant 0)$:

$$\pi_k(\mathbf{U}) \cong \pi_{k+2}(\mathbf{U}) \cong \mathbf{Z}, \quad k \text{ odd},$$
$$\cong 0, \quad k \text{ even},$$
$$\pi_k(\mathbf{O}) \cong \pi_{k+4}(\mathbf{Sp}) \cong \pi_{k+8}(\mathbf{O}),$$
$$\cong \mathbf{Z}, \quad k \equiv 3,7 \pmod 8,$$
$$\cong \mathbf{Z}_2, \quad k \equiv 0,1 \pmod 8,$$
$$\cong 0, \quad k \not\equiv 0,1,3,7 \pmod 8.$$

This is called the **Bott periodicity theorem**. The relations are deduced from †weak homotopy equivalences $\mathbf{U} \to \Omega(B_{\mathbf{U}})$, $B_{\mathbf{U}} \times \mathbf{Z} \to \Omega(\mathbf{U})$, $B_{\mathbf{O}} \times \mathbf{Z} \to \Omega(\mathbf{U}/\mathbf{O})$, $\mathbf{U}/\mathbf{O} \to \Omega(\mathbf{Sp}/\mathbf{U})$, $\mathbf{Sp}/\mathbf{U} \to \Omega(\mathbf{Sp})$, $\mathbf{Sp} \to \Omega(B_{\mathbf{Sp}})$, $B_{\mathbf{Sp}} \times \mathbf{Z} \to \Omega(\mathbf{U}/\mathbf{Sp})$, $\mathbf{U}/\mathbf{Sp} \to \Omega(\mathbf{O}/\mathbf{U})$, $\mathbf{O}/\mathbf{U} \to \Omega(\mathbf{O})$. This result is applied to nonstable cases; for example, $\pi_{2n}(U(n))$ is a cyclic group of order $n!$ (→ Appendix A, Table 6.V). The 2-dimensional homotopy group $\pi_2(G)$ of any Lie group is trivial.

Let $\alpha \in \pi_k(O(n))$, where $\alpha = [f]$, $f: S^k \to O(n)$. We define $\bar{f}: S^k \times S^{n-1} \to S^{n-1}$ by $\bar{f}(x,y) = f(x) \cdot y$ and identify S^{k+n} with the boundary $(E^{k+1} \times S^{n-1}) \cup (S^k \times E^n)$ of $E^{k+1} \times E^n$. We extend $\bar{f}$ to $f: S^{k+n} \to S^n$ so that it maps $E^{k+1} \times S^{n-1}$, $S^k \times E^n$ into the upper and lower hemisphere of S^n ($S^{n-1} =$ the equator), respectively. Let $J(\alpha) \in \pi_{n+k}(S^n)$ be the class of the mapping thus obtained. This homomorphism $J: \pi_k(O(n)) \to \pi_{n+k}(S^n)$ is called a ***J*-homomorphism** of Hopf and Whitehead. For the stable case, $J: \pi_k(\mathbf{O}) \to G_k$ is injective for $k \equiv 0, 1 \pmod 8$, and the order of the image of J is the denominator of $B_{2t}/4t$ (B_{2t} is a †Bernoulli number) or its double for $k = 4t-1$ (J. F. Adams).

I. Cohomotopy Groups

K. Borsuk defined a sum of mapping classes of X into S^n (1936), which was named Borsuk's **cohomotopy group** by E. Spanier. Spanier also studied the duality of the cohomotopy group with the homotopy group and its relations to the usual cohomology groups. A cohomotopy group of (X,A) is defined to be $\pi^n(X,A) = \pi(X,A; S^n, *)$, which forms a group if $\dim X/A < 2n-1$. A mapping $F: X/A \to S^n \times S^n$ given by $F(x) = (f(x), g(x))$ with $f, g: X/A \to S^n$ is homotopic to a mapping into $S^n \vee S^n$. If we compose F with a folding mapping of $S^n \vee S^n$ onto S^n, we obtain a mapping that represents the sum $[f]+[g]$ (→ 204 Homotopy).

References

[1] N. E. Steenrod, The topology of fiber bundles, Princeton Univ. Press, 1951.
[2] Y. Kawada and K. Oguchi, Isôkikagaku (Japanese; Topology), Asakura, 1967.
[3] P. J. Hilton, An introduction to homotopy theory, Cambridge Univ. Press, 1953.
[4] A. Komatu, M. Nakaoka, and H. Toda, Isôkikagaku (Japanese; Topology), Kyôritu, 1957.
[5] S. T. Hu, Homotopy theory, Academic Press, 1959.
[6] H. Toda, Composition methods in homotopy groups of spheres, Princeton Univ. Press, 1962.
[7] H. Hopf, Über die Abbildungen von Sphären auf Sphären niedriger Dimension, Fund. Math., 25 (1935), 427–440.
[8] W. Hurewicz, Beiträge zur Topologie der Deformationen I–IV, Proc. Acad. Amsterdam, 38 (1935), 112–119, 521–528; 39 (1936), 117–125, 215–224.
[9] H. Freudenthal, Über die Klassen der Sphärenabbildungen, Compositio Math., 5 (1937), 299–314.
[10] G. W. Whitehead, A generalization of the Hopf invariant, Ann. of Math., (2) 51 (1950), 192–237.
[11] A. L. Blakers and W. S. Massey, The homotopy groups of a triad II, Ann. of Math., (2) 55 (1952), 192–201.
[12] J.-P. Serre, Groupes d'homotopie et classes des groupes abéliens, Ann. of Math., (2) 58 (1953), 258–294.
[13] R. Bott, The stable homotopy of the classical groups, Ann. of Math., (2) 70 (1959), 313–337.

206 (IX.16) Homotopy Operations

A. General Remarks

Let X, Y, X', Y' be topological spaces. If to each continuous mapping $f \in Y^X$ there corresponds a †homotopy class $\Phi(f) \in \pi(X'; Y')$ that is a †homotopy invariant of f (satisfying a certain naturality condition), then Φ is called a **homotopy operation** (→ 204 Homotopy). More generally, we may consider the case where Φ is a mapping from $\pi(X_1; Y_1) \times \ldots \times \pi(X_r; Y_r)$ into $\pi(X'; Y')$. The **naturality** of Φ is defined as follows: Consider the †category $\mathcal{C}$ of topological spaces (or its subcategory). Let $Y = Y'$ be an arbitrary †object of $\mathcal{C}$, and fix X and X'. In this case, the naturality of $\Phi_Y: \pi(X; Y) \to \pi(X'; Y)$ is defined to be the commutativity of the following diagram:

$$\begin{array}{ccc} \pi(X;Y) & \xrightarrow{\Phi_Y} & \pi(X';Y) \\ \downarrow g_* & & \downarrow g_* \\ \pi(X;Z) & \xrightarrow{\Phi_Z} & \pi(X';Z) \end{array}$$

i.e., $g_* \circ \Phi_Y = \Phi_Z \circ g_*$ for an arbitrary †morphism (i.e., continuous mapping) $g: Y \rightarrow Z$ of the category $\mathcal{C}$. Similarly, when objects Y, Y' of the category $\mathcal{C}$ are fixed and $X = X'$ is an arbitrary object of $\mathcal{C}$, to say that a homotopy operation $\Phi_X: \pi(X; Y) \rightarrow \pi(X; Y')$ is **natural** means that $h^* \circ \Phi_X = \Phi_W \circ h^*$ for an arbitrary morphism $h: W \rightarrow X$.

We have the following theorem: In the category of topological spaces and continuous mappings, the homotopy operations $\Phi_Y: \pi(X; Y) \rightarrow \pi(X'; Y)$ and the elements of $\pi(X'; X)$ are in one-to-one correspondence. The correspondence is obtained by associating a homotopy operation $\Phi(\beta) = \beta \circ \alpha$ ($\beta \in \pi(X; Y)$) with each $\alpha \in \pi(X'; X)$. Similarly, the homotopy operations $\Phi_X: \pi(X; Y) \rightarrow \pi(X; Y')$ and the elements of $\pi(Y; Y')$ are in one-to-one correspondence. This theorem holds also for the case involving several variables if we consider $\pi(X'; X_1 \vee X_2 \vee \ldots)$ or $\pi(Y_1 \times Y_2 \times \ldots; Y')$ instead of $\pi(X'; Y)$ or $\pi(X; Y')$. The theorem remains valid if we replace the spaces X, Y by systems of spaces.

B. Homotopy Operations in Homotopy Groups

(1) If X, X' are spheres S^n, S^p with base points and Y, Y' are topological spaces with base points, a homotopy operation $\Phi_Y: \pi_n(Y) \rightarrow \pi_p(Y)$ is said to be of type (n,p). By the theorem in Section A, the homotopy operations of type (n,p) are in one-to-one correspondence with the elements of the †homotopy group of the sphere $\pi_p(S^n)$.

(2) As an example of the 2-variable homotopy operations $\Phi_Y: \pi_m(Y) \times \pi_n(Y) \rightarrow \pi_p(Y)$ of type $(m,n;p)$ we have the Whitehead product defined as follows: Suppose that $\alpha \in \pi_m(Y)$, $\beta \in \pi_n(Y)$ are elements represented by $f: (I^m, \dot{I}^m) \rightarrow (Y, *)$ and $g: (I^n, \dot{I}^n) \rightarrow (Y, *)$, respectively. Define a continuous mapping F from the boundary $\dot{I}^{m+n} = (I^m \times \dot{I}^n) \cup (\dot{I}^m \times I^n)$ of $I^{m+n} = I^m \times I^n$ into Y by $F(x,y) = f(x)$ for $(x,y) \in I^m \times \dot{I}^n$ and $F(x,y) = g(y)$ for $(x,y) \in \dot{I}^m \times I^n$. Since $\dot{I}^{m+n}$ is homeomorphic to S^{m+n-1}, we may identify them. The homotopy class represented by F is an element of $\pi_{m+n-1}(Y)$ determined by α and β, denoted by $[\alpha, \beta]$ and called the **Whitehead product** of α and β (J. H. C. Whitehead, *Ann. of Math.*, 42 (1941)). The Whitehead product is a homotopy operation of type $(m,n;m+n-1)$. Let $\psi_m: (I^m, \dot{I}^m) \rightarrow (S^m, *)$ be a mapping that smashes $\dot{I}^m$ to a point. The product of ψ_m and ψ_n defines a mapping $\psi_{m,n}: S^{m+n-1} \rightarrow S^m \vee S^n = (S^m \times *) \cup (* \times S^n)$ (→ 204 Homotopy E). Let $\iota \in \pi_m(S^m \vee S^n)$, $\iota' \in \pi_n(S^m \vee S^n)$ be the homotopy classes of the natural inclusions of S^m, S^n into $S^m \vee S^n$; then the homotopy class of $\psi_{m,n}$ is $[\iota, \iota']$. G. W. Whitehead showed that a direct sum decomposition $\pi_p(S^m \vee S^n) = \iota_* \pi_p(S^m) + \iota'_* \pi_p(S^n) + [\iota, \iota']_* \pi_p(S^{m+n-1})$ ($\iota_*, \iota'_*, [\iota, \iota']_*$ are injective) holds for $1 < p < m + n + \min(m,n) - 3$. Furthermore, P. J. Hilton showed that for general $p > 1$, $\pi_p(S^m \vee S^n)$ is the direct sum of the images of injections ι_*, ι'_*, $[\iota, \iota']_*$, $[[\iota, \iota'], \iota]_*$, $[[\iota, \iota'], \iota']_*$, etc. The homotopy operations of type $(m,n;p)$ are in one-to-one correspondence with the elements of $\pi_p(S^m \vee S^n)$; hence such operations can be constructed by means of composition and the Whitehead product. The last proposition is also valid for homotopy operations of type $(m_1, \ldots, m_r; p)$. The Whitehead product $[\alpha, \beta]$ ($\alpha \in \pi_m(X)$, $\beta \in \pi_n(X)$) is distributive with respect to α (resp. β) for $m > 1$ ($n > 1$), and we have $[\beta, \alpha] = (-1)^{mn} [\alpha, \beta]$ and $f_*[\alpha, \beta] = [f_* \alpha, f_* \beta]$ for $f: X \rightarrow Y$. Moreover, for $\gamma \in \pi_r(X)$ the **Jacobi identity** holds: $(-1)^{mr}[[\alpha, \beta], \gamma] + (-1)^{mn}[[\beta, \gamma], \alpha] + (-1)^{rn}[[\gamma, \alpha], \beta] = 0$ (M. Nakaoka and H. Toda; H. Uehara and W. S. Massey; Hilton).

C. Suspensions and Generalized Hopf Invariants

We denote by $\alpha \wedge \beta \in \pi(X \wedge X'; Y \wedge Y')_0$ the class of the †reduced join of f, g, where f represents $\alpha \in \pi(X; Y)_0$ and g represents $\beta \in \pi(X'; Y')_0$. We call $\alpha \wedge \beta$ the **reduced join** of α and β. In particular, if $Y = Y' = S^1$, β is the identity mapping of S^1, and α is represented by f, then $\alpha \wedge \beta$ is called the **suspension** of α and is denoted by $S\alpha$. $S\alpha$ is the class of the †suspension Sf of f and belongs to $\pi(SX; SY)_0$, where SX indicates the †reduced suspension of X. The suspension $S\alpha$ is often denoted by $E\alpha$ in reference to the German term *Einhängung*. The identity mapping 1 of SY gives rise to an injection $i = \Omega_0 1$ sending Y into the †loop space $\Omega(SY)$ determined by the formula $i(y)(t) = (y,t)$. Then we have

$$i_* = \Omega_0 \circ S: \pi(X; Y)_0 \rightarrow \pi(SX; SY)_0 \xrightarrow{\approx} \pi(X; \Omega SY)_0,$$

and S and i_* are equivalent. Let Y_k be the identifying space $Y^k/\sim$, where Y^k is the product space $Y \times \ldots \times Y$ of k copies of Y and $\sim$ is the equivalence relation determined by

$$(*, y_1, y_2, \ldots, y_{k-1}) \sim (y_1, *, y_2, \ldots, y_{k-1}) \sim \ldots \sim (y_1, \ldots, y_{k-1}, *).$$

Denote by $Y_\infty = \bigcup_k Y_k$ the limit space with respect to the injection $Y_{k-1} \rightarrow Y_k$ given by $(y_1, \ldots, y_{k-1}) \rightarrow (y_1, \ldots, y_{k-1}, *)$ and call it the **reduced product space** of Y. Let Y be a †CW complex of 0-†section $*$. The mapping $i: Y = Y_1 \rightarrow SY$ can then be extended to

$\bar{i}: Y_\infty \to SY$, where $\bar{i}$ is a †weak homotopy equivalence. If X is also a CW complex, then $\Omega_0^{-1} \circ \bar{i}_*: \pi(X; Y_\infty)_0 \to \pi(SX; SY)_0$ is bijective. By smashing the subset Y of Y_2, we have $Y \wedge Y = Y_2 / Y$. This smashing mapping can be extended to $h: Y_\infty \to (Y \wedge Y)_\infty$ (I. M. James). Utilizing $h_*: \pi(X; Y_\infty)_0 \to \pi(X; (Y \wedge Y)_\infty)_0$ and the bijection $\Omega_0^{-1} \circ \bar{i}_*$, we obtain a correspondence $H: \pi(SX; SY)_0 \to \pi(SX; S(Y \wedge Y))_0$. We call $H(\alpha)$ the **generalized Hopf invariant** of α. When $X = S^{2n-2}$, $Y = S^{n-1}$, H is equivalent to the †Hopf invariant $\gamma: \pi_{2n-1}(S^n) \to \mathbf{Z}$. In general, we have $H \circ S = 0$, and the exactness of $\xrightarrow{S} \xrightarrow{H}$ holds under various conditions (→ 205 Homotopy Groups). Furthermore, we have $S(\alpha \circ \beta) = S\alpha \circ S\beta$ and $H(\alpha \circ S\beta) = H\alpha \circ S\beta$. Also, $H(S\alpha \circ \beta) = S(\alpha \wedge \alpha) \circ H\beta$. Under the condition $i < 3n-3$, we have $(\alpha_1 + \alpha_2) \circ \beta = \alpha_1 \circ \beta + \alpha_2 \circ \beta + [\alpha_1, \alpha_2] \circ H(\beta)$ (G. W. Whitehead). Thus the composition $\alpha \circ \beta$ is not always left distributive but is always right distributive, and $\alpha \circ \beta$ is left distributive if $\beta = S\beta'$. The composition is defined over the †stable homotopy groups G_r of spheres: $\alpha \circ \beta \in G_{p+q}$ ($\alpha \in G_p$, $\beta \in G_q$). It is distributive and satisfies $\beta \circ \alpha = (-1)^{pq} \alpha \circ \beta$.

When Y and Y' are †Eilenberg-MacLane spaces, B_{O}, and B_{U}, we have †cohomology operations on cohomology groups $H^n(\ ;\Pi)$, KO groups, and K groups, respectively. As typical examples there are †Steenrod square operations $Sq^i: H^n(X; \mathbf{Z}_2) \to H^{n+i}(X; \mathbf{Z}_2)$, †Steenrod pth power operations $\mathscr{P}^i: H^n(X; \mathbf{Z}_p) \to H^{n+2i(p-1)}(X; \mathbf{Z}_p)$, †Chern characters $ch^n: K(X) \to H^{2n}(X; \mathbf{Q})$ ($\mathbf{Q}$: rational field), †Adams operations $\Phi_i: KO(X) \to KO(X)$ ($K(X) \to K(X)$). They are all homomorphisms (→ 67 Cohomology Operations; 236 K-Theory).

D. Secondary Compositions

Suppose that $\alpha \circ \beta = 0$, $\beta \circ \gamma = 0$ for $\gamma \in \pi(W; X)_0$, $\beta \in \pi(X; Y)_0$, $\alpha \in \pi(Y; Z)_0$. In the commutative diagram of †Puppe exact sequences

$$\begin{array}{ccccccc} \xrightarrow{S\gamma^*} \pi(SW; Y)_0 & \xrightarrow{p^*} & \pi(C_\gamma; Y)_0 & \xrightarrow{i^*} & \pi(X; Y)_0 & \xrightarrow{\gamma^*} \\ \downarrow \alpha_* & & \downarrow \alpha_* & & \downarrow \alpha_* & \\ \xrightarrow{S\gamma^*} \pi(SW; Z)_0 & \xrightarrow{p^*} & \pi(C_\gamma; Z)_0 & \xrightarrow{i^*} & \pi(X; Z)_0 & \xrightarrow{\gamma^*} \end{array}$$

the set of elements $\bar{\beta}$ in $\pi(SW; Z)$ such that $p^*(\bar{\beta}) \in \alpha_* i^{*-1}(\beta)$ is denoted by $\{\alpha, \beta, \gamma\}$ and is called a **secondary composition** or **Toda bracket**. If θ, η are elements of $\pi(SW; Y)_0$, $\pi(SX; Z)_0$, respectively, then we have $\{\alpha, \beta, \gamma\} + \alpha_* \theta = \{\alpha, \beta, \gamma\}$, $\{\alpha, \beta, \gamma\} + S\gamma^* \eta = \{\alpha, \beta, \gamma\}$. Hence we may consider the set $\{\alpha, \beta, \gamma\}$ to be a residue class modulo a submodule generated by $\alpha_* \pi(SW; Y)_0$ and $S\gamma^* \pi(SX; Z)_0$.

The secondary composition $\{\alpha, \beta, \gamma\}$ has the following properties: (i) $\{\alpha, \beta, \gamma\}$ is linear with respect to α, β, γ (if the sum is defined); (ii) $\alpha \circ \{\beta, \gamma, \delta\} = \{\alpha, \beta, \gamma\} \circ (-S\delta)$; (iii) $S\{\alpha, \beta, \gamma\} \equiv -\{S\alpha, S\beta, S\gamma\}$; (iv) $\alpha \circ \{\beta, \gamma, \delta\} \equiv \{\alpha \circ \beta, \gamma, \delta\}$, $\{\alpha \circ \beta, \gamma, \delta\} \equiv \{\alpha, \beta \circ \gamma, \delta\}, \ldots$; (v) $\{\{\alpha, \beta, \gamma\}, S\delta, S\varepsilon\} + \{\alpha, \{\beta, \gamma, \delta\}, S\varepsilon\} + \{\alpha, \beta, \{\gamma, \delta, \varepsilon\}\} \equiv 0$. Suppose that the spaces X, Y, Z, W are spheres. Then by (iii) the secondary composition $\{\alpha, \beta, \gamma\} \in G_{p+q+r+1} / (\alpha \circ G_{q+r+1} + \gamma \circ G_{p+q+1})$ is defined in the stable homotopy groups $G_r = \lim_{n \to \infty} \pi_{n+r}(S^n)$ of spheres. From this we obtain (vi) $\{\gamma, \beta, \alpha\} = (-1)^{pq+qr+rp+1} \{\alpha, \beta, \gamma\}$ and (vii) $(-1)^{pr} \cdot \{\alpha, \beta, \gamma\} + (-1)^{qp} \{\beta, \gamma, \alpha\} + (-1)^{rq} \{\gamma, \alpha, \beta\} \equiv 0$.

E. Functional Operations

Let Φ be an operation corresponding to α and γ be the class of f. We put $\Phi_f(\beta) = \{\alpha, \beta, \gamma\}$ and call Φ_f a **functional Φ-operation**. When Φ is a cohomology operation, Φ_f is called a **functional cohomology operation**. Then $\Phi_f(\beta)$ is defined for β satisfying $f^*(\beta) = \Phi(\beta) = 0$, and $\Phi_f(\beta)$ is determined modulo $\operatorname{Im} Sf^* + \operatorname{Im} \Phi$. For $f: S^{n+k} \to S^n$, $k = 2i(p-1) - 1$, we denote by $H_p(f) \cdot \varepsilon_{n+k+1} \in H^{n+k+1}(S^{n+k+1}; \mathbf{Z}_p)$ the image of a generator ε_n of $H^n(S^n; \mathbf{Z}_p)$ under the functional $\mathscr{P}^i_f$ operation. Then the **Hopf invariant modulo** p (or **mod** p **Hopf invariant**) $H_p: \pi_{n+k}(S^n) \to \mathbf{Z}_p$ is obtained (we use Sq^{2i} for $p = 2$). The following statements are equivalent: (i) The mod 2 Hopf invariant is not trivial ($H_2 \neq 0$); (ii) there exists a mapping: $S^{2k+1} \to S^{k+1}$ of Hopf invariant 1; (iii) S^k is an H-space; (iv) the Whitehead product $[\iota, \iota]$ of a generator ι of $\pi_k(S^k)$ vanishes. Also, $H_2 \neq 0$ if and only if $k = 2, 4, 8$ (J. Adams), and for an odd prime p, $H_p \neq 0$ if and only if $k = 2p - 3$ (A. L. Liulevicius; N. Shimada and T. Yamanoshita).

References

[1] P. J. Hilton, An introduction to homotopy theory, Cambridge Univ. Press, 1953.

[2] A. Komatu, M. Nakaoka, and H. Toda, Isôkikagaku (Japanese; Topology), Kyôritu, 1957.

[3] S. T. Hu, Homotopy theory, Academic Press, 1959.

[4] H. Toda, Composition methods in homotopy groups of spheres, Princeton Univ. Press, 1962.

[5] E. H. Spanier, Secondary operations on mappings and cohomology, Ann. of Math., (2) 75 (1962), 260–282.

[6] H. Hopf, Über die Abbildungen von Sphären auf Sphären niedriger Dimension, Fund. Math., 25 (1935), 427–440.
[7] H. Freudenthal, Über die Klassen der Sphärenabbildungen, Compositio Math., 5 (1937), 299–314.
[8] G. W. Whitehead, A generalization of the Hopf invariant, Ann. of Math., (2) 51 (1950), 192–237.

207 (IX.7) Hopf Algebras

A. General Remarks

The notion of Hopf algebras arose from the study of homology and cohomology of Lie groups or, more generally, H-spaces. It was introduced by H. Hopf [1], whose basic structure theorem was generalized and applied to several problems by A. Borel [2]. Hopf algebras are now used as standard tools in algebraic topology.

B. Graded Algebras

A †graded module $A=\Sigma_{n\geqslant 0}A_n$ over a field k is said to be of **finite type** when each A_n is finite-dimensional. A is **connected** when an isomorphism $\eta:k\cong A_0$ is given. The †tensor product of two graded modules A and B is a graded module with $A\otimes B=\Sigma_n(A\otimes B)_n$, $(A\otimes B)_n=\Sigma_p A_p\otimes B_{n-p}$. We call $A^*=\Sigma A_n^*$ (where A_n^* is the †dual module of A_n) the **dual graded module** of A. When A and B are of finite type, $A\otimes B$ and A^* are also of finite type, and we have $(A\otimes B)^*=A^*\otimes B^*$ and $A^{**}=A$. When A and B are connected, A^* and $A\otimes B$ are also connected.

Let A be a graded module. If there exists a degree-preserving linear mapping $\varphi:A\otimes A\to A$, we call (A,φ) a **graded algebra**, whereas if there exists a degree-preserving linear mapping $\psi:A\to A\otimes A$, we call (A,ψ) a **graded coalgebra**. In this article we call a graded algebra or a graded coalgebra simply an algebra or a coalgebra. We call φ a **multiplication**, and ψ a **comultiplication** (or **diagonal mapping**). Usually we write $\varphi(a\otimes b)=ab$ (the **product** of a, $b\in A$), and call $\psi(a)$ the **coproduct** of a. Multiplication and comultiplication are dual operations. If A is of finite type and (A,φ) is an algebra, then (A^*,φ^*) (where φ^* is the dual mapping of φ) is a coalgebra, and vice versa. A multiplication φ is called **associative (commutative)** if $\varphi(1\otimes\varphi)=\varphi(\varphi\otimes 1)$ $(\varphi\circ T=\varphi)$, where $T:A\otimes A\to A\otimes A$ is the mapping defined by $T(a\otimes b)=(-1)^{pq}b\otimes a$ for $a\in A_p$ and $b\in A_q$. Associativity and commutativity of a comultiplication are defined dually. Let (A,ψ) be a connected coalgebra of finite type, and identify k with A_0 via η. Then the algebra (A^*,ψ^*) has the unity of k as unity if and only if ψ satisfies: $\psi(1)=1\otimes 1$ and $\psi(x)=1\otimes x+x\otimes 1+\Sigma_i x_i'\otimes x_i''$ $(0<\deg x_i'<\deg x)$ for $\deg x>0$. In this case we say that ψ has the unity of k as **counity**. For algebras (A,φ) and (B,φ'), if $\varphi''=(\varphi\otimes\varphi')\circ(1\otimes T\otimes 1):A\otimes B\otimes A\otimes B\to A\otimes B$, then $(A\otimes B,\varphi'')$ is also an algebra, which we denote by $(A\otimes B,\varphi'')=(A,\varphi)\otimes(B,\varphi')$. The tensor product of coalgebras is defined as the dual notion of $(A,\varphi)\otimes(B,\varphi')$.

C. Hopf Algebras

For simplicity we assume that graded modules are defined over a field, connected, and of finite type. Let a graded module A be equipped with a multiplication φ and a comultiplication ψ. If φ and ψ have the unity of k as unity and $\psi:(A,\varphi)\to(A,\varphi)\otimes(A,\varphi)$ is an algebra homomorphism, then we call (A,φ,ψ) a **Hopf algebra**. The last condition for a Hopf algebra is satisfied if and only if $\varphi:(A,\psi)\otimes(A,\psi)\to(A,\psi)$ is a homomorphism of coalgebras. The dual (A^*,ψ^*,φ^*) is also a Hopf algebra, called the **dual Hopf algebra** of (A,φ,ψ).

D. H-spaces

Let X be a topological space. The †cohomology group $H^*(X)$ (†homology group $H_*(X)$) considered over a field k has a multiplication d^* (comultiplication d_*), which is induced by the diagonal mapping $d:X\to X\times X$ and becomes a commutative and associative algebra (coalgebra). The groups $H^*(X)$ and $H_*(X)$ are dual to each other (→ 68 Cohomology Rings C). When X is equipped with a base point x_0 and a base point-preserving continuous mapping $h:X\times X\to X$ such that $h\circ\iota_i\simeq 1_X$ (†homotopic) for $i=1$ and 2 (where $\iota_1(x)=(x,x_0)$ and $\iota_2(x)=(x_0,x)$), we call (X,h) an H-**space**, h a **multiplication**, and x_0 a **homotopy identity** of X. Then h induces, through a †Künneth isomorphism, a comultiplication $h^*:H^*(X)\to H^*(X)\otimes H^*(X)$ (**Hopf comultiplication**) and a multiplication $h_*:H_*(X)\otimes H_*(X)\to H_*(X)$ (**Pontrjagin multiplication**). Then $h^*(\alpha)$ $(\alpha\in H_*(X))$ is called the **Hopf coproduct** of α, and $h_*(\beta\otimes\gamma)$ $(\beta,\gamma\in H_*(X))$ is called the **Pontrjagin product** of β and γ. When X is †arcwise connected and $H_*(X)$ is of finite type, $(H^*(X),d^*,h^*)$ and $(H_*(X),h_*,d_*)$ are Hopf algebras dual to each other. In particular, when h is **homotopy associative**, i.e., $h\circ(h\times 1_x)\simeq h\circ(1_x\times h)$ (ho-

motopy commutative, i.e., $h \simeq h \circ T$, where $T(x_1, x_2) = (x_2, x_1)$ for $x_i \in X$), then h^* and h_* are associative (commutative). †Topological groups and †loop spaces are homotopy associative H-spaces. If a continuous mapping $g: X \to X$ satisfies $h \circ (1_X \times g) \simeq h \circ (g \times 1_X) \simeq c$ (constant mapping $X \to \{x_0\}$), then g is called a **homotopy inverse** for X, h.

Suppose that a Hopf algebra A is defined over a field k of characteristic p and equipped with associative and commutative multiplication, and A is generated by a single element $a \in A_n$. Then A is a †polynomial ring $k[a]$ (n is even when $p \neq 2$) or a †quotient ring $k[a]/(a^2)$ (n is odd when $p \neq 2$) or $k[a]/(a^{p^f})$ (only when $p \neq 0$; n is even when $p \neq 2$). These are called **elementary Hopf algebras**. Every Hopf algebra over a †perfect field k with associative and commutative multiplication is isomorphic (as an algebra) to a tensor product of elementary Hopf algebras (Borel's theorem) [2]. In particular, the cohomology algebra over a field of characteristic 0 of a †compact connected Lie group is isomorphic to a †Grassmann algebra generated by elements of odd degrees [1].

E. Steenrod Algebras

The †Steenrod algebra $\mathcal{A}_p$ over $\mathbf{Z}_p$ is generated by †Steenrod operations Sq^i ($p=2$), $\mathcal{P}^i$ ($p>2$), and the †Bockstein operation Δ_p ($p>2$), with composition of operations defined as multiplication. Then $\mathcal{A}_p$ is a connected associative graded algebra of finite type (not commutative). Defining a comultiplication ψ of $\mathcal{A}_p$ by $\psi(Sq^n) = \sum Sq^i \otimes Sq^{n-i}$, $\psi(\mathcal{P}^n) = \sum \mathcal{P}^i \otimes \mathcal{P}^{n-i}$, and $\psi(\Delta_p) = 1 \otimes \Delta_p + \Delta_p \otimes 1$, $\mathcal{A}_p$ becomes a Hopf algebra with an associative and commutative comultiplication. Thus its dual $\mathcal{A}_p^*$ is a Hopf algebra with an associative and commutative multiplication, and we can apply Borel's theorem to $\mathcal{A}_p^*$ in order to investigate the structure of $\mathcal{A}_p$ [3].

Let (A, φ, ψ) be a Hopf algebra with associative multiplication and comultiplication. Putting $c(1) = 1$ and $c(a) = -a - \sum a_i' \cdot c(a_i'')$ for $\deg a > 0$ (where $\psi(a) = 1 \otimes a + a \otimes 1 + \sum a_i' \otimes a_i''$), we obtain a linear mapping $c: A \to A$ satisfying $c\varphi = \varphi(c \otimes c)T$. We call c the **conjugation mapping** of A. When the multiplication or comultiplication is commutative, we obtain the relation $c^2 = 1$, and c is a bijection. The conjugation mapping is utilized in studying Steenrod algebras [3, 4].

References

[1] H. Hopf, Über die Topologie der Gruppen-Mannigfaltigkeiten und ihre Verallgemeinerungen, Ann. of Math., (2) 42 (1941), 22–52.

[2] A. Borel, Sur la cohomologie des espaces fibrés principaux et des espaces homogènes des groupes de Lie compacts, Ann. of Math., (2) 57 (1953), 115–207.

[3] J. W. Milnor, The Steenrod algebra and its dual, Ann. of Math., (2) 67 (1958), 150–171.

[4] J. W. Milnor and J. C. Moore, On the structure of Hopf algebras, Ann. of Math., (2) 81 (1965), 211–264.

208 (XIX.12) Hydrodynamics

A. General Remarks

Gases and liquids are easily deformed and share many kinetic properties. They are examples of fluids. By definition, a **fluid** is a continuous substance having the property that when it is not moving, any part of the substance separated from the rest by a surface exerts an outward force that is perpendicular to the given surface.

Hydrodynamics (or **fluid dynamics**) concerns the equilibrium and motion of gases and liquids without considering their molecular structure. In particular, the branch of the theory concerning the equilibrium of fluid is called **hydrostatics**, and **hydrodynamics** sometimes means the branch concerning the motion of fluid.

There are two methods of describing the motion of a fluid. One regards a fluid as a system of an infinite number of particles and discusses the motion of each particle as a function of time. This is **Lagrange's method**. For example, suppose that a fluid particle with the coordinates $(x,y,z) = (a,b,c)$ at the moment $t=0$ has coordinates $x = f_1(a,b,c,t)$, $y = f_2(a,b,c,t)$, $z = f_3(a,b,c,t)$ at an arbitrary time t. Then the motion of the fluid is perfectly determined by the functions f_1, f_2, and f_3.

The other is **Euler's method**, which discusses the values of the velocity $\mathbf{v}(u,v,w)$, the density ρ, the pressure p, etc., of the fluid at arbitrary times and positions. From this standpoint each quantity of the fluid is regarded as a function of a space-time point (x,y,z,t).

The rate at which any physical quantity F varies while moving with the fluid particle is the **Lagrangian derivative** DF/Dt, which is related to the ordinary partial derivatives by

$$\frac{DF}{Dt} = \frac{\partial F}{\partial t} + u\frac{\partial F}{\partial x} + v\frac{\partial F}{\partial y} + w\frac{\partial F}{\partial z}.$$

The three components (u,v,w) and the two state quantities (p,ρ) (in general, other state quantities, for example, the temperature T and the †entropy S, are assumed to be determined by equations of state such as $T=T(p,\rho)$, $S=S(p,\rho)$) are determined by five $(=1+3+1)$ relations derived from the conservation laws of mass, momentum, and energy, namely, the **equation of continuity**, which corresponds to the conservation of mass,

$$\partial\rho/\partial t+\operatorname{div}(\rho\mathbf{v})=0; \tag{1}$$

the **equation of motion**, which corresponds to the conservation of momentum,

$$\partial(\rho\mathbf{v})/\partial t+\operatorname{div}(\rho\mathbf{v}\otimes\mathbf{v}-\mathbf{p})=\rho\mathbf{K}, \tag{2}$$

where $\mathbf{K}$ is the external force per unit mass, $\mathbf{p}$ is the stress tensor, and $\otimes$ denotes the †tensor product, while †divergence is applied to each row vector, and by virtue of (1), (2) may be expressed component-wise as

$$\rho\frac{Du}{Dt}=\frac{\partial p_{xx}}{\partial x}+\frac{\partial p_{xy}}{\partial y}+\frac{\partial p_{xz}}{\partial z}+\rho K_x,$$

$$\rho\frac{Dv}{Dt}=\frac{\partial p_{yx}}{\partial x}+\frac{\partial p_{yy}}{\partial y}+\frac{\partial p_{yz}}{\partial z}+\rho K_y, \tag{2'}$$

$$\rho\frac{Dw}{Dt}=\frac{\partial p_{zx}}{\partial x}+\frac{\partial p_{zy}}{\partial y}+\frac{\partial p_{zz}}{\partial z}+\rho K_z;$$

and the **energy equation**, which corresponds to the conservation of energy,

$$\partial(\rho v^2/2+\rho E)/\partial t + \operatorname{div}\big(\rho v(v^2/2+E)-\mathbf{v}\cdot\mathbf{p}+\mathbf{h}\big)=0, \tag{3}$$

or the **equation of entropy production**, which is another expression of (3),

$$\rho T\,DS/Dt=-\operatorname{div}\mathbf{h}+Q, \tag{3'}$$

where E is the internal energy per unit mass, Q the heat generated per unit time and volume, and $\mathbf{h}$ the heat flux. Here $\mathbf{K}$, p_{ik}, $\mathbf{h}$, and Q or their relations with other quantities (e.g., $\mathbf{h}=-\kappa\operatorname{grad}T$, where κ is the thermal conductivity) are assumed to be known.

B. Perfect Fluids

When there is a velocity gradient in the flow, a tangential stress appears which tends to make the velocity uniform, so that $\mathbf{p}$ is not a diagonal tensor ($-p\,\delta_{ik}$, i.e., pressure). This property is called fluid **viscosity**. Generally, Q and $\mathbf{h}$ do not vanish in this case. However, in order to simplify the problem we consider a nonviscous (sometimes also adiabatic) fluid, which is called a **perfect fluid** and is a good approximation in the large of actual fluids. The motion of a perfect fluid is determined by **Euler's equation of motion**

$$\rho\,D\mathbf{v}/Dt=-\operatorname{grad}p+\rho\mathbf{K}, \tag{4}$$

which is obtained from (1) and (2) by replacing p_{ik} by the pressure only, and also by the thermodynamic relation $DS/Dt=0$ obtained from (3) by putting $Q=0$ and $\mathbf{h}=0$ or its integral $S=0$ in **homentropic flow**, which is governed by the **adiabatic law** $p\propto\rho^{\gamma}$. In particular, for a liquid, the density variation can be neglected. Putting $\rho=$ constant in (1), we have

$$\operatorname{div}\mathbf{v}=0, \tag{5}$$

which, in conjunction with (2), determines four unknowns (u,v,w,p) as functions of (x,y,z,t).

A fluid of constant density is called an **incompressible fluid**, and one of variable density a **compressible fluid**. Even though it might seem natural to consider gases as examples of compressible fluids, they can be treated as incompressible fluids if the speed of the flow of the gas $q=|\mathbf{v}|$ is small compared with the velocity $c=\sqrt{dp/d\rho}$ of sound propagating in the gas. We call $q/c=M$ the **Mach number**.

The vector $\omega(\xi,\eta,\zeta)$, which is derived from the velocity vector $\mathbf{v}$ as $\omega=\operatorname{rot}\mathbf{v}$, is called the **vorticity**. A small part of the fluid rotates with angular velocity $\omega/2$. If $\omega=0$, the flow is called **irrotational**, otherwise **rotational**. The curves $dx:dy:dz=u:v:w$ and $dx:dy:dz=\xi:\eta:\zeta$ are called, respectively, **stream lines** and **vortex lines**. The line integral $\oint_C v_s\,ds$ along a closed circuit C is called the **circulation** around C.

In irrotational flow, the velocity is expressed as $\mathbf{v}=\operatorname{grad}\Phi$, where Φ is called a **velocity potential**. When the external force K has a potential Ω ($K=-\operatorname{grad}\Omega$) and p is a definite function of ρ, we have the **pressure equation**

$$\frac{\partial\Phi}{\partial t}+\frac{1}{2}q^2+\int\frac{dp}{\rho}+\Omega=\text{constant},$$

which is valid everywhere in the flow. In a steady flow,

$$\frac{1}{2}q^2+\int\frac{dp}{\rho}+\Omega=\text{constant} \tag{6}$$

is valid along each stream line; this is called the **Bernoulli theorem**. These two equations correspond to †energy integrals of the equation of motion. Furthermore, corresponding to the conservation of †angular momentum, we have **Helmholtz's vorticity theorem**: When $\mathbf{K}=-\operatorname{grad}\Omega$ and $p=f(\rho)$, vorticity is neither created nor annihilated in the fluid.

For the irrotational motion of an incompressible fluid, †Laplace's equation $\Delta\Phi=0$ is derived from (1). Hence the problem reduces to the determination of a †harmonic function Φ under appropriate boundary conditions (e.g., for a fixed wall, normal velocity $v_n=\partial\Phi/\partial n=0$). For the 2-dimensional problem a **stream function** Ψ is introduced to satisfy (1) by the relation $u=\partial\Psi/\partial y$, $v=$

$-\partial\Psi/\partial x$. Since the †Cauchy-Riemann equations $\partial\Phi/\partial x = \partial\Psi/\partial y$, $\partial\Phi/\partial y = -\partial\Psi/\partial x$ are valid in this case, $f = \Phi + i\Psi$ is an †analytic function of $z = x + iy$. Therefore, the theory of 2-dimensional irrotational motion is essentially equivalent to the theory of complex †analytic functions, and the theory of †conformal mapping is a powerful method in the first theory.

For irrotational steady flow of a compressible fluid in which $\Omega = 0$, c is determined from (6) as a function of q. Then (1) and (4) yield a †nonlinear partial differential equation for Φ:

$$\left(1-\frac{u^2}{c^2}\right)\frac{\partial^2\Phi}{\partial x^2}+\left(1-\frac{v^2}{c^2}\right)\frac{\partial^2\Phi}{\partial y^2}+\left(1-\frac{w^2}{c^2}\right)\frac{\partial^2\Phi}{\partial z^2}-2\frac{vw}{c^2}\frac{\partial^2\Phi}{\partial y\partial z}-2\frac{wu}{c^2}\frac{\partial^2\Phi}{\partial z\partial x}-2\frac{uv}{c^2}\frac{\partial^2\Phi}{\partial x\partial y}=0. \tag{7}$$

This equation is †elliptic or †hyperbolic (→ 321 Partial Differential Equations of Mixed Type) according as M is less than 1 (**subsonic**) or greater than 1 (**supersonic**).

For 2-dimensional flow, we can introduce a stream function Ψ from (1) by $u = \partial\Phi/\partial x = (1/\rho)(\partial\Psi/\partial y)$, $v = \partial\Phi/\partial y = -(1/\rho)(\partial\Psi/\partial x)$. By utilizing the idea of †Legendre transformation, this system of nonlinear equations for Φ and Ψ may be reduced to a system of linear equations in the **hodograph plane** (q,θ):

$$\frac{\partial\Phi}{\partial q}=q\frac{d}{dq}\left(\frac{1}{\rho q}\right)\frac{\partial\Psi}{\partial\theta}=-\frac{1-M^2}{\rho q}\frac{\partial\Psi}{\partial\theta},$$

$$\frac{\partial\Phi}{\partial\theta}=\frac{q}{\rho}\frac{\partial\Psi}{\partial q}$$

($d(\rho q)/dq = \rho(1-M^2)$), where the independent variables q and θ are the magnitude and the inclination of the velocity, respectively. The treatment of 2-dimensional compressible flow on the basis of this system is called the **hodograph method**. For a flow of small M, there is a method of successive approximation (**M^2-expansion method**) which starts from Laplace's equation, neglecting the term of $O(M^2)$ in (7). For uniform flow (velocity U in the x-direction) past a thin wing or slender body where v and w are small, we have **thin wing theory** or **slender body theory**, whose first approximation is

$$(1-M^2)\frac{\partial^2\Phi}{\partial x^2}+\frac{\partial^2\Phi}{\partial y^2}+\frac{\partial^2\Phi}{\partial z^2}=0. \tag{8}$$

If $M<1$ or $M>1$ (although not too large) a linearization (**Prandtl-Glauert approximation**) is possible by replacing M by the Mach number at infinity $M_\infty = U/C_\infty$. For $M>1$, (8) has a †characteristic surface, which is the **Mach cone** whose central axis makes an angle $\arcsin c/q = \arcsin 1/M$ with the flow. This can be interpreted also as an envelope produced by spherical sound waves with velocity c from a source drifting with velocity q. For $M \sim 1$, we put $\Phi = c_* x + \varphi$ (c_* is the fluid velocity when $q = c$). Then for an adiabatic gas, (8) is approximated by the following partial differential equation of elliptic type:

$$\frac{\partial^2\varphi}{\partial y^2}+\frac{\partial^2\varphi}{\partial z^2}=\frac{\gamma+1}{c_*}\frac{\partial\varphi}{\partial x}\frac{\partial^2\varphi}{\partial x^2}. \tag{9}$$

Such a flow in which both domains $M \gtrless 1$ coexist is called the **transonic flow**, and exact solutions by the hodograph method are known. However, continuous deceleration from $M>1$ to $M<1$ is generally apt to be unstable or impossible, and the appearance of a **shock wave**, i.e., a discontinuous surface of state quantities, is not unusual. This belongs to the †weak solution of (1), (2), (3) for a perfect fluid. In particular, in the coordinate system fixed to the surface, its integrated form can be obtained as follows: $[\rho v_n]=0$, $[p\delta_{in}+\rho v_i v_n]=0$, $[q^2/2+E+p/\rho]=0$ ([ ] is the jump of the quantity at the surface, and n is the normal component). Supplemented by the entropy increase, these formulas give relations between the fluid velocity and the state variable at the front and back of the shock. In an ideal gas they are called the **Rankine-Hugoniot relation**. Entropy is not uniform behind a curved shock, and the flow is not irrotational. For a weak shock starting from the tip of a pointed slender body, however, the discontinuity is small and approaches the †characteristic surface of (7), i.e., the **Mach wave** (compressible wave, in this case). Rarefactive Mach waves are found in the supersonic flow of acceleration around a convex surface. Such waves contribute to the drag on an obstacle placed in supersonic flow.

C. Viscous Fluids

A body moving uniformly in a fluid at rest (with velocity less than that of sound) suffers no drag as long as the viscosity of the fluid is negligible and the flow is continuous (**d'Alembert's paradox**). Hence we must take the viscosity into account in order to discuss the creation and annihilation of vortices, the generation and structure of shock waves, and the drag acting on obstacles. For this purpose, we extend **Newton's law** that frictional stress is proportional to the velocity gradient and assume that the stress tensor **p** is a linear function of the rate-of-strain tensor:

$$p_{xx} = -p + \mu e_{xx} - \frac{2}{3}\mu \operatorname{div} \mathbf{v}$$
$$= -p + 2\mu \frac{\partial u}{\partial x} - \frac{2}{3}\mu \operatorname{div} \mathbf{v}, \ldots,$$
$$p_{yz} = \mu\left(\frac{\partial w}{\partial y} + \frac{\partial v}{\partial z}\right).$$

The proportional constant μ is called the **coefficient of viscosity**. When a fluid satisfies this assumption it is called a **Newtonian fluid**. Otherwise, it is called a **non-Newtonian fluid**. Except for a few cases, such as colloid solutions, fluids may be regarded as Newtonian.

If we take the viscosity into account, the equation of motion of an incompressible fluid becomes

$$\rho D\mathbf{v}/Dt = \rho \mathbf{K} - \operatorname{grad} p + \mu \Delta \mathbf{v}. \qquad (10)$$

This is called the **Navier-Stokes equation**. A nondimensional quantity $R = \rho UL/\mu$ formed from representative length L, velocity U, density ρ, and viscosity μ of a flow is called the **Reynolds number**. In order that two flows with geometrically similar boundaries share similar kinetic properties, their Reynolds numbers must be equal. This is called the **Reynolds law of similarity**.

For small R, we can approximate the equation of motion (10) by replacing the acceleration $D\mathbf{v}/Dt$ by $\partial \mathbf{v}/\partial t$ (**Stokes approximation**) or by $\partial \mathbf{v}/\partial t + U\partial \mathbf{v}/\partial x$ (**Oseen approximation**) for a body placed in the uniform flow of velocity U in the x-direction.

For large R, the flow may be regarded as that of a perfect fluid, since we can neglect $\mu \Delta \mathbf{v}$ as long as the velocity gradient is not too large. In the vicinity of a fixed wall, however, the velocity gradient becomes large, because in a very thin layer the velocity decreases rapidly from the value U of a perfect fluid to zero at the wall. This layer is called the **boundary layer**. For the boundary layer, **Prandtl's boundary layer equation**

$$\frac{\partial u}{\partial t} + u\frac{\partial u}{\partial x} + v\frac{\partial u}{\partial y} = \frac{\partial U}{\partial t} + U\frac{\partial U}{\partial x} + \frac{\mu}{\rho}\frac{\partial^2 u}{\partial y^2},$$
$$\frac{\partial u}{\partial x} + \frac{\partial v}{\partial y} = 0 \qquad (11)$$

is valid, where x and y are the coordinates parallel and perpendicular to the wall, respectively, and U is the velocity outside the boundary layer.

If $\partial U/\partial x < 0$, it sometimes happens that the boundary layer separates from the surface of the body. In this case a vortex is generated in the flow, as large vorticities in the boundary layer are carried into the flow. For a body without separation of the boundary layer, the d'Alembert paradox holds and is no longer a "paradox," and the drag is small. Such bodies are called **streamlined**.

In compressible flow the new problem arises of the interaction of shock waves with the boundary layer. A rapid increase in pressure due to the shock wave formed on the surface of a body invalidates the assumption of a boundary layer and causes its separation. If the Mach number becomes sufficiently large ($M \gtrsim 5$, **hypersonic flow**), the bow shock approaches the body and interferes with the boundary layer. The generation of heat at the boundary layer (e.g., viscous dissipation in Q) requires the consideration of heat transfer as well as viscosity. In this manner, we are obliged to treat a complete system of equations which take into account the energy equation (3) as well as the temperature dependence of κ and μ.

D. Laws of Similarity

For such complicated systems, dimensional analysis is often useful (→ 120 Dimensional Analysis). As laws of similarity, we may consider not only those such as the Reynolds law but also others for bodies which transform similarly by †affine transformations. Corresponding to equation (8), the **Prandtl-Glauert law of similarity** for subsonic flow is famous: The pressure coefficient (nondimensional pressure change) for a thin wing of chord (i.e., the width in the direction of flow) 1, length L, and thickness τ is

$$C_p(L,\tau) = \lambda C_{p_0}\left(\sqrt{1-M_\infty^2}\, L, \tau/\sqrt{1-M_\infty^2}\, \lambda\right),$$

where λ is an arbitrary constant and C_{p_0} is C_p for a body of scaled length and thickness placed in an incompressible flow. Corresponding to (9), an extension of the famous **von Kármán transonic similarity** is possible:

$$C_p(L,\tau) = \tau^{2/3}(\gamma+1)^{-1/3}$$
$$\times f\left(\sqrt{|1-M_\infty^2|}\, L, (\gamma+1)\tau/|1-M_\infty^2|^{3/2}\right).$$

E. Turbulence

For low Reynolds numbers, the flow generally has smooth streamlines. For high Reynolds numbers, however, extremely irregular motion in space and time appears. The former is called **laminar flow**, and the latter **turbulent flow** (→ 418 Turbulent Flow). The transition from laminar to turbulent flow is considered to be due to the instability of the laminar flow, and this transition has been studied by the method of small oscillations. Regarding the internal structure of turbulence, statistical theories originated by T. von Kármán and G. I. Taylor (*Proc. Roy. Soc. London*, 151 (1935)) and A. N. Kolmogorov (*Dokl. Akad.*

Nauk. SSSR, 30 (1941)) are of central importance.

F. Mathematical Problems

These branches of fluid mechanics give rise to various kinds of mathematical problems. Here we review only the recent outstanding advances—the theory on the Navier-Stokes equation (10) for viscous incompressible fluids.

The problem for steady flow in a domain G of space is reduced to a boundary value problem consisting of (5), (10), and the boundary condition

$$\mathbf{v}|_{\partial G} = \boldsymbol{\beta}, \tag{12}$$

where $\mathbf{v}=\mathbf{v}(\mathbf{x})$ and $p=p(\mathbf{x})$ are the unknowns. If G is not bounded (for the external boundary value problem), the boundary condition at infinity

$$\mathbf{v}(\mathbf{x})\to\mathbf{U}_0, \quad |\mathbf{x}|\to\infty$$

is imposed ($\mathbf{x}$ is the position vector). It should also be noted that $D\mathbf{v}/Dt=(\mathbf{v}\cdot\nabla)\,\mathbf{v}$ in (10). Mathematical studies on the existence and properties of solutions for this boundary value problem were intiated by J. Leray and others [6]. Recently, their methods and results were improved or completed by R. Finn, H. Fujita, and others [8, 10]. For a review, Finn's report [9] may be consulted. The problem of unsteady flow in G is reduced to an initial and boundary value problem for unknowns $\mathbf{v}=\mathbf{v}(t,\mathbf{x})$ and $p=p(t,\mathbf{x})$ consisting of (5), (10), (12), and the initial condition

$$\mathbf{v}|_{t=0}=\boldsymbol{\alpha}. \tag{13}$$

The case in which G is the entire space was treated by Leray [6, 7]. In the general case where the boundary exists, many kinds of weak solutions were obtained by E. Hopf [13], A. A. Kiselev and O. A. Ladyženskaja [14], and others. Then the existence of regular solutions was established by S. Itô [17] and P. E. Sobolevskiĭ [18]. Furthermore, these results were improved by T. Kato, H. Fujita, and others [11, 12]. The uniqueness of the solution is also proved in all the cited papers except Hopf's.

For initial and boundary value problems, the dimension of the space m and the existence of a solution in an infinite time interval $0\leqslant t<\infty$ are intimately related. In fact, of the cases with $m=2, 3$, which are physically interesting, the existence of a solution in an infinite time interval was established by Ladyženskaja and others [15, 19, 21] for $m=2$. For $m=3$, however, no existence proof in an infinite time interval has yet been obtained under general assumptions.

For a general survey of compressible fluids → [20].

References

[1] H. Lamb, Hydrodynamics, Cambridge Univ. Press, sixth edition, 1932.

[2] S. Goldstein, Modern developments in fluid dynamics, Clarendon Press, 1938.

[3] S. Tomotika, Ryûtai rikigaku (Japanese; Hydrodynamics), Kyôritu, 1940.

[4] L. Howarth, Modern developments in fluid dynamics, High speed flow I, II, Oxford Univ. Press, 1953.

[5] I. Imai, Ryûtai rikigaku (Japanese; Hydrodynamics), Iwanami, 1970.

[6] J. Leray, Etude de diverses équations intégrales non linéaires et de quelques problèmes que pose l'hydrodynamique, J. Math. Pures Appl., (9) 12 (1933), 1–82.

[7] J. Leray, Sur le mouvement d'un liquid visqueux emplissant l'espace, Acta Math., 63 (1934), 193–248.

[8] R. Finn, On steady-state solutions of the Navier-Stokes partial differential equations, Arch. Rational Mech. Anal., 3 (1959), 381–396.

[9] R. Finn, Stationary solutions of the Navier-Stokes equation, Amer. Math. Soc. Proc. Symp. Appl. Math., 17 (1965), 121–153.

[10] H. Fujita, On the existence and regularity of the steady-state solutions of the Navier-Stokes equation, J. Fac. Sci. Univ. Tokyo, 9 (1961), 59–102.

[11] H. Fujita and T. Kato, On the Navier-Stokes initial value problem I, Arch. Rational Mech. Anal., 16 (1964), 269–315.

[12] T. Kato and H. Fujita, On the nonstationary Navier-Stokes system, Rend. Sem. Mat. Univ. Padova, 32 (1962), 243–260.

[13] E. Hopf, Über die Anfangswertaufgabe für die hydrodynamischen Grundgleichungen, Math. Nachr., 4 (1950–1951), 213–231.

[14] А. А. Киселев, О. А. Лалыженская (A. A. Kiselev and O. A. Ladyženskaja), О существовании и единственности решения нестационарной задачи для вязкой несжимаемой жидкости, Izv. Akad. Nauk SSSR, 21 (1957), 655–680.

[15] O. A. Ladyženskaja, Solution "in the large" of the nonstationary boundary value problem for the Navier-Stokes system with two space variables, Comm. Pure Appl. Math., 12 (1959), 427–433.

[16] О. А. Ладыженская (O. A. Ladyženskaja), Математические вопросы динамики вязкой несжимаемой жидкости, Физматгиз, 1961; English translation, The mathematical theory of viscous incompressible flows, Gordon and Breach, 1963.

[17] S. Itô, The existence and uniqueness of a regular solution of the nonstationary Navier-Stokes equation, J. Fac. Sci. Univ. Tokyo, 9 (1961), 103–140.

[18] П. Е. Соболевский (P. E. Sobolevskiĭ), О гладкости обобщенных решений уравнений Навье-Стокса, Dokl. Akad. Nauk SSSR, 131 (1960), 758–760.
[19] П. Е. Соболевский (P. E. Sobolevskiĭ), Нестационарных уравнениях гидродинамики вязкой жидкости, Dokl. Akad. Nauk SSSR, 128 (1959), 45–48.
[20] L. Bers, Mathematical aspects of subsonic and transonic gas dynamics, John Wiley, 1958.

209 (XIV.5) Hypergeometric Functions

A. Hypergeometric Functions

The †power series

$$F(\alpha,\beta,\gamma;z) = \frac{\Gamma(\gamma)}{\Gamma(\alpha)\Gamma(\beta)}\sum_{n=1}^{\infty}\frac{\Gamma(\alpha+n)\Gamma(\beta+n)}{n!\Gamma(\gamma+n)}z^n$$

in the complex variable z is called the **hypergeometric series** (or **Gauss's series**).

It is convergent for any α, β, and γ if $|z|<1$, and is convergent for $\mathrm{Re}(\alpha+\beta-\gamma)<0$ if $|z|=1$. If $z=1$, its sum is equal to $\Gamma(\gamma)\Gamma(\gamma-\alpha-\beta)/\Gamma(\gamma-\alpha)\Gamma(\gamma-\beta)$ (except when γ is a nonpositive integer). The **hypergeometric functions** are obtained as analytic continuations of the functions determined by hypergeometric series which are single-valued analytic functions defined on the domain obtained from the complex plane by deleting a line connecting branch points $z=0$ and $z=\infty$ (→ Appendix A, Table 18.I).

A hypergeometric function is a solution of the differential equation

$$z(1-z)\frac{d^2w}{dz^2}+(\gamma-(\alpha+\beta+1)z)\frac{dw}{dz}-\alpha\beta w=0, \tag{1}$$

which is called the **hypergeometric differential equation** (or **Gaussian differential equation**). This equation is a differential equation of †Fuchsian type with †regular singular points at 0, 1, and ∞, whose solutions are expressed, in terms of the †P-function of Riemann, by

$$w=P\left\{\begin{matrix}0 & \infty & 1 & \\ 0 & \alpha & 0 & z \\ 1-\gamma & \beta & \gamma-\alpha-\beta & \end{matrix}\right\}.$$

If any one of the values of γ, $\gamma-\alpha-\beta$, or $\alpha-\beta$ is integral, there exists a series containing $\log z$, representing a solution of the differential equation (1) in the neighborhood of the corresponding singular points. When none of the γ, $\gamma-\alpha-\beta$, or $\alpha-\beta$ values is integral, since the linear transformations $z'=z$, $z'=1/z$, $z'=1-z$, $z'=z/(z-1)$, $z'=(z-1)/z$, $z'=1/(1-z)$ permute singular points, there exist 24 particular solutions around the singular points. The latter fact was first proved by E. E. Kummer (1836).

There exist various curves C for which the integral

$$w=\int_C u^{\alpha-1}(1-u)^{\gamma-\alpha-1}(1-zu)^{-\beta}du$$

is a solution of (1). Among them we can take the segment [0, 1], when $\mathrm{Re}\,\alpha>0$, $\mathrm{Re}(\gamma-\alpha)>0$. Then the corresponding solution is holomorphic in the interior of the unit circle, and

$$F(\alpha,\beta,\gamma;z)=\frac{\Gamma(\gamma)}{\Gamma(\alpha)\Gamma(\gamma-\alpha)}\times\int_C u^{\alpha-1}(1-u)^{\gamma-\alpha-1}(1-zu)^{-\beta}du.$$

Since the integrand has branch points at 0, 1, and $1/z$, we have the following expression when γ is not an integer:

$$F(\alpha,\beta,\gamma;z) = \frac{1}{(1-e^{2\pi i(\gamma-\alpha)})(1-e^{2\pi i\alpha})}\frac{\Gamma(\gamma)}{\Gamma(\alpha)\Gamma(\gamma-\alpha)}\times\oint^{(1+,0+,1-,0-)}u^{\alpha}(1-u)^{\gamma-\alpha-1}(1-zu)^{-\beta}du,$$

where $\mathrm{Re}\,\alpha>0$, $\mathrm{Re}(\gamma-\alpha)>0$; whereas if γ is an integer, then

$$F(\alpha,\beta,\gamma;z)=\frac{1}{(1-e^{-2\pi i\alpha})}\frac{\Gamma(\gamma)}{\Gamma(\alpha)\Gamma(\gamma-\alpha)}\times\oint^{(1+,0+)}u^{\alpha}(1-u)^{\gamma-\alpha-1}(1-zu)^{-\beta}du,$$

where the contour in the first expression encircles successively each of 1, 0, 1, and 0 once with indicated directions. These expressions may be adopted as a definition of the hypergeometric functions for the general value of z. Other integral expressions are also known (→ Appendix A, Table 18.I).

B. The Ladder Method

A linear ordinary differential equation of the second order having three regular singular points on the complex sphere is easily transformed into an equation of the form (1). To solve such an equation with a parameter, it is often useful to decompose, in two different ways, the main part of the equation into two factors of the first order, and find a recurrence formula involving the parameter, as we shall see in the following example. This method is called the **ladder method** or **factorization method**.

For example, †Legendre's differential equation

$$L_n[w]\equiv(1-z^2)((1-z^2)w')'+n(n+1)w=0$$

is decomposed as follows:

$$L_n = S_n \cdot T_n + n^2 = T_{n+1} \cdot S_{n+1} + (n+1)^2,$$

$$T_n = (1-z^2)\frac{d}{dz} + nz, \quad S_n = (1-z^2)\frac{d}{dz} - nz.$$

If w_n is a solution of $L_n[w]=0$, then multiplying both sides of $S_n \cdot T_n[w_n] + n^2 w_n = 0$ by T_n, we find that $T_n \cdot S_n(T_n[w_n]) + n^2(T_n[w_n]) = 0$, that is, $T_n[w_n]$ is a solution of $L_{n-1}[w]=0$. Similarly, we see that $S_{n+1}[w_n]$ is a solution of $L_{n+1}[w]=0$. In this sense, S_n and T_n are called, respectively, the **step-up operator** (or **up-ladder**) and the **step-down operator** (or **down-ladder**) with respect to the parameter n.

The above relation constitutes a recurrence formula for Legendre functions (→ Appendix A, Table 18.II).

C. Extensions of Hypergeometric Functions

J. Thomas (1870) proposed the series

$$1 + \sum_{n=1}^{\infty} \frac{(\alpha_1)_n (\alpha_2)_n \dots (\alpha_h)_n}{(\beta_1)_n (\beta_2)_n \dots (\beta_h)_n} z^n,$$

$$(\lambda)_n = \lambda(\lambda+1)\dots(\lambda+n-1)$$

as an extension of the hypergeometric series. The sum of this series satisfies the hth-order differential equation

$$(1-z)\frac{d^h w}{dt^h} + (A_1 - B_1 z)\frac{d^{h-1} w}{dt^{h-1}}$$

$$+ (A_2 - B_2 z)\frac{d^{h-2} w}{dt^{h-2}} + \dots + (A_h - B_h z)w = 0,$$

$$t = \log z.$$

When $h=2$ and $\beta_1 = 1$, it reduces to the ordinary hypergeometric series. The notation

$$_pF_q(\alpha_1, \alpha_2, \dots, \alpha_p; \beta_1, \beta_2, \dots, \beta_q; z)$$

$$= \sum_{n=0}^{\infty} \frac{(\alpha_1)_n (\alpha_2)_n \dots (\alpha_p)_n}{n!(\beta_1)_n (\beta_2)_n \dots (\beta_q)_n} z^n, \qquad (2)$$

which is due to L. Pochhammer and modified by E. W. Barnes, is used to denote the extended hypergeometric series, and the function defined by (2) is often called **Barnes's extended hypergeometric function**. For example, Gauss's series in this notation is $_2F_1(\alpha, \beta, \gamma; z)$.

Corresponding to Barnes's integral expression for hypergeometric functions, it is known that the integral

$$W(z) = \frac{1}{2\pi i}\int_{c-i\infty}^{c+i\infty} K(\zeta)H(\zeta)z^{-\zeta}\,d\zeta,$$

where

$$K(\zeta) = K(\zeta+1)$$

and

$$H(\zeta)$$

$$= \frac{\Gamma(\zeta+\alpha_1)\Gamma(\zeta+\alpha_2)\dots\Gamma(\zeta+\alpha_h)}{\Gamma(\zeta+1+\beta_1)\Gamma(\zeta+1+\beta_2)\dots\Gamma(\zeta+1+\beta_h)},$$

is a solution of the hth-order differential equation at the beginning of this section. The hypergeometric function expressed by the definite integral

$$\int \zeta^a (\zeta-1)^b (\zeta-z)^c\,d\zeta$$

has an obvious formal extension

$$\int (\zeta-a_1)^{b_1}(\zeta-a_2)^{b_2}\dots(\zeta-a_m)^{b_m}(\zeta-z)^c\,d\zeta.$$

On the other hand, the equation

$$\sum_{\nu=0}^{m} \varphi_\nu(z)\frac{d^\nu w}{dz^\nu} = 0,$$

where

$$\varphi_\nu(z) = \frac{(-1)^{m-1-\nu}}{(h+m-2)\dots(h+1)h}$$

$$\times\left(\binom{h+m-\nu-2}{m-\nu-1}P_1^{(m-1-\nu)}(z)\right.$$

$$\left. + \binom{h+m-\nu-2}{m-\nu}P_0^{(m-\nu)}(z)\right),$$

$$P_0(z) = (z-a_1)(z-a_2)\dots(z-a_m),$$

$$P_1(z) = P_0(z)\left(\frac{\beta_1}{z-a_1} + \frac{\beta_2}{z-a_2} + \dots + \frac{\beta_m}{z-a_m}\right),$$

called the **Tissot-Pochhammer differential equation**, has a solution

$$w(z) = \int_C (\zeta-a_1)^{\beta_1-1}(\zeta-a_2)^{\beta_2-1}\dots$$

$$\times(\zeta-a_m)^{\beta_m-1}(\zeta-z)^{h+m-2}\,d\zeta.$$

Pochhammer (1870) called this function an extended hypergeometric function.

As another extension of Gauss's series, H. E. Heine (1846) introduced Heine's series

$$\varphi(a,b,c;q;z) = 1 + \frac{(1-q^a)(1-q^b)}{(1-q)(1-q^c)}q^z$$

$$+ \frac{(1-q^a)(1-q^{a+1})(1-q^b)(1-q^{b+1})}{(1-q)(1-q^2)(1-q^c)(1-q^{c+1})}q^{2z} + \dots.$$

Setting $q = 1+\varepsilon$, $z = (1/\varepsilon)\log x$, and letting $\varepsilon \to 0$, we obtain Gauss's series as the limit of Heine's series.

P. Appell (1880) formally extended Gauss's series to the case of two variables and defined four kinds of functions [3]. They are called **Appell's hypergeometric functions of two variables** (→ Appendix A, Table 18.I.5). C. E. Picard (1881) showed that they are expressed by integrals of the form

$$\int_0^1 u^\alpha (1-u)^\beta (1-xu)^\gamma (1-yu)^\delta\,du.$$

D. Hypergeometric Functions with Matrix Argument

For symmetric matrices Z of degree m, C. S. Herz defined **hypergeometric functions with matrix argument** as follows [5]: Denoting by

$\operatorname{etr} Z$ the exponential $\exp(\operatorname{tr} Z)$ of the †trace of Z, let

$${}_0F_0(Z)=\operatorname{etr} Z,$$

$${}_{p+1}F_q(\alpha_1,\ldots,\alpha_p;\beta_1,\ldots,\beta_q;\gamma;Z)$$

$$=\frac{1}{\Gamma_m(\gamma)}\int_{\Lambda>0}\operatorname{etr}(-\Lambda)\,{}_pF_q(\alpha_1,\ldots,\alpha_p;\beta_1,\ldots,\beta_q;\Lambda Z)(\det\Lambda)^{\gamma-p}d\lambda_{11}d\lambda_{22}\ldots d\lambda_{mm}, \quad (3)$$

$${}_pF_{q+1}(\alpha_1,\ldots,\alpha_p;\beta_1,\ldots,\beta_q;\gamma;\Lambda)$$

$$=\frac{\Gamma_m(\gamma)}{(2\pi i)^{m(m+1)/2}}\int_{\operatorname{Re}Z=X_0>0}\operatorname{etr}Z\,{}_pF_q(\alpha_1,\ldots,\alpha_p;\beta_1,\ldots,\beta_q;\Lambda Z^{-1})(\det Z)^{-\gamma}dz_{11}dz_{22}\ldots dz_{mm}, \quad (4)$$

where

$$\Lambda=(\lambda_{ij})_{i,j=1,\ldots,m},$$

$$Z=((1+\delta_{ij})z_{ij}/2)_{i,j=1,\ldots,m},$$

$$\Gamma_m(\gamma)=\pi^{m(m-1)/4}\Gamma(\gamma)\Gamma(\gamma-1/2)\ldots\times\Gamma(\gamma-(m-1)/2),$$

and $\Lambda>0$ means that Λ is †positive definite. The integral (3) converges for $-Z>0$ if $\operatorname{Re}\gamma>(m-1)/2$. If $\operatorname{Re}\gamma$ is sufficiently large, then for suitably chosen X_0, (4) converges in a domain of the space of Λ and represents an analytic function of its argument. In particular, we have

$${}_1F_0(\alpha;Z)=(\det(E-Z))^{-\alpha}.$$

Based on this definition, many special functions and formulas are extended to the case of a matrix argument. For example,

$$A_\delta(Z)={}_0F_1(\delta+(m+1)/2;-Z)/\Gamma_m(\delta+(m+1)/2) \quad (5)$$

is an extension of the †Bessel function, and this reduces to

$$(t/2)^{-\delta}J_\delta(t)=A_\delta((t/2)^2)$$

when $m=1$. Formula (5) is applied to the †noncentral Wishart distribution in mathematical statistics.

References

[1] F. Klein, Vorlesungen über die hypergeometrische Funktionen, Springer, 1933.
[2] M. Hukuhara, Zyôbibun hôteisiki no kaihô II, Senkei no bu (Japanese; Methods of solution of ordinary differential equations pt. II. Linear differential equations), Iwanami, 1941.
Also → references to 381 Special Functions.
For hypergeometric functions of several variables,
[3] P. Appell, Sur les fonctions hypergéométriques de plusieurs variables, Mémor. Sci. Math., Gauthier-Villars, 1925.
For applications of the ladder method,
[4] L. Infeld and T. E. Hull, The factorization method, Rev. Mod. Phys., 23 (1951), 21–68.
For the case of a matrix variable,
[5] C. S. Herz, Bessel functions of matrix argument, Ann. of Math., (2) 61 (1955), 474–523.
Also →
[6] L. J. Slater, Generalized hypergeometric functions, Cambridge Univ. Press, 1966.

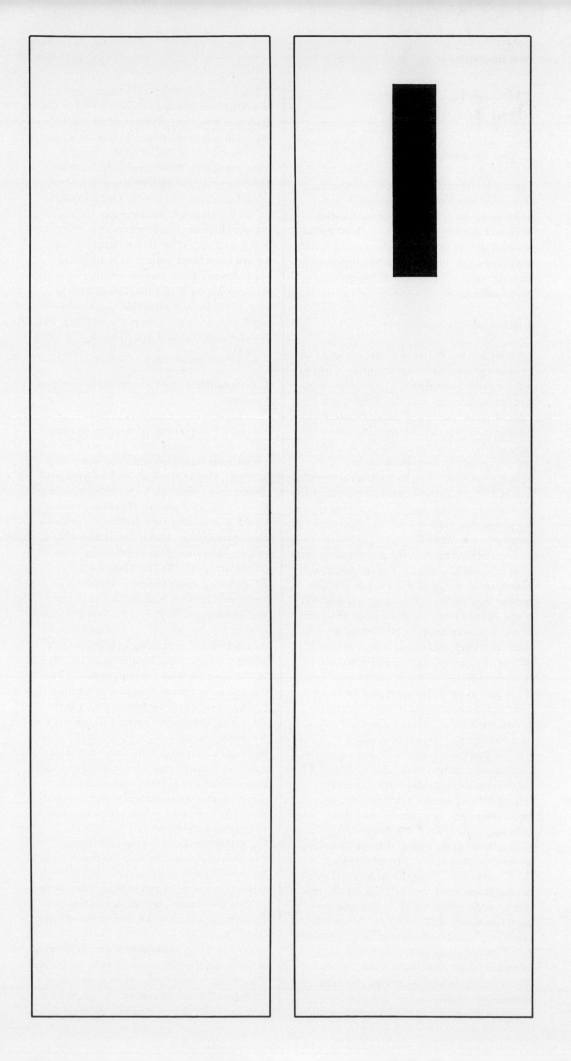

210 (XI.13) Ideal Boundaries

A. Ideal Boundaries

For a given Hausdorff space R, a †compact Hausdorff space R^* that contains R as its dense subspace is called a **compactification** of R, and $\Delta = R^* - R$ is called an **ideal boundary** of R. In the present article, we deal mainly with properties (in particular function-theoretic properties) of ideal boundaries of †Riemann surfaces R.

B. Harmonic Boundaries

By R we mean a Riemann surface. The set Γ of points p^* in Δ such that $\liminf_{R \ni p \to p^*} P(p) = 0$ for every †potential P, i.e., for every positive †superharmonic function P for which the class of nonnegative †harmonic functions smaller than P consists of only the constant function 0, is a compact subset of R^*. The set Γ is called the **harmonic boundary** of R with respect to R^*. For an arbitrary compact subset K in $\Delta - \Gamma$, there exists a finite-valued potential P_K with $\lim_{R \ni p \to p^*} P_K(p) = \infty$ ($p^* \in K$). From this, various kinds of †maximum principles are derived.

There are infinitely many compactifications of R unless R is compact. For two compactifications R_i^* ($i = 1, 2$) of R, we say that R_1^* is **greater than** R_2^* or, equivalently, **lies over** R_2^*, if the identity mapping of R can be extended to a continuous mapping of R_1^* onto R_2^*. In order that deep function-theoretic studies of R^* may be carried out, various conditions must be imposed on R^*. A compactification R^* is said to be of **Stoïlow type** if for every †connected open subset G^* in R^* whose boundary in R^* is contained in R, $G^* - \Delta$ is also connected. Next suppose that $R \in O_G$ (→ 362 Riemann Surfaces E). For a given real-valued function f on Δ, let $\overline{\mathfrak{U}}_f^{R,R^*}$ ($\underline{\mathfrak{U}}_f^{R,R^*}$) be the class of †superharmonic functions s bounded from below (†subharmonic functions s bounded from above) such that $\liminf_{R \ni p \to p^*} s(p) \geqslant f(p^*)$ ($\limsup_{R \ni p \to p^*} s(p) \leqslant f(p^*)$) for every $p^* \in \Delta$. If these classes are nonempty, then $\overline{H}_f^{R,R^*}(p) = \inf\{s(p) \mid s \in \overline{\mathfrak{U}}_f^{R,R^*}\}$ and $\underline{H}_f^{R,R^*}(p) = \sup\{s(p) \mid s \in \underline{\mathfrak{U}}_f^{R,R^*}\}$ are harmonic on R, and $\underline{H}_f^{R,R^*} \leqslant \overline{H}_f^{R,R^*}$. In particular, if $\underline{H}_f^{R,R^*} \equiv \overline{H}_f^{R,R^*}$, then the common function is denoted by H_f^{R,R^*}, and the function f is said to be **resolutive** with respect to R^*. A compactification such that every bounded continuous function on Δ is resolutive is called a **resolutive compactification**. In such a case, a point p^* in Δ is said to be **regular** with respect to the †Dirichlet problem if $\lim_{R \ni p \to p^*} H_f^{R,R^*}(p) = f(p^*)$ for every bounded continuous function f on Δ (→ 124 Dirichlet Problem). The set Δ_r of regular points in Δ is contained in Γ. If R^* is a resolutive compactification, then there exists a unique positive †Borel measure μ_p such that $H_f^{R,R^*}(p) = \int_\Delta f(p^*) d\mu_p(p^*)$ for every bounded continuous function f on Δ. This measure is called the **harmonic measure** with respect to $p \in R$. There exists a function $P(p,p^*)$ on $R \times \Delta$ with $d\mu_p(p^*) = P(p,p^*) d\mu_o(p^*)$ for an arbitrary fixed point o in R satisfying the following three conditions: (i) $P(p,p^*)$ is harmonic on R as a function of p; (ii) $P(p,p^*)$ is Borel measurable as a function of p^*; (iii) $k(o,p)^{-1} \leqslant P(p,p^*) \leqslant k(o,p)$, with the Harnack constant $k(o,p)$ of $\{o,p\}$ relative to R [5].

C. Compactifications Determined by Function Families

A family F of real-valued continuous functions on R admitting infinite values is called a **separating family** on R if there exists an f in F such that $f(p) \neq f(q)$ for two given distinct points p and q in R. A compactification R^* is called an **F-compactification**, denoted by R_F^*, if every function in F can be continuously extended to R^* and the family of extended functions again constitutes a separating family on R^*. The correspondence $\varphi: F \to R_F^*$ defines a single-valued mapping of all separating families F on R onto all F-compactifications of R. If $F_1 \supset F_2$, then $\varphi(F_1)$ lies over $\varphi(F_2)$. For any R^*, $\varphi^{-1}(R^*)$ contains infinitely many separating families, among which the separating families constituting †associative algebras are important [3]. The following are typical examples of compactifications determined by function families:

(1) The **Aleksandrov compactification** is the $\mathfrak{A}$-compactification $R_{\mathfrak{A}}^*$ with the family $\mathfrak{A}$ of bounded continuous functions on R with compact support. It is the smallest compactification of R and is often used in function theory in discussing Dirichlet problems for relatively noncompact subregions in reference to relative boundaries.

(2) The **Stone-Čech compactification** is the $\mathfrak{C}$-compactification $R_{\mathfrak{C}}^*$ with the family $\mathfrak{C}$ of bounded continuous functions on R. It is the largest compactification of R. It is rarely used in function theory, but an example of its powerful application is found in the works of M. Nakai [4].

(3) The **Kerékjártó-Stoïlow compactification** is the $\mathfrak{S}$-compactification $R_{\mathfrak{S}}^*$ with the family $\mathfrak{S}$ of bounded continuous functions f on R such that there exist compact sets K_f with the property that the f are constants on each

connected component of $R-K_f$. This is the smallest compactification of Stoïlow type. Many applications of this compactification can be found in function theory, among which the investigation done by M. Ohtsuka on the Dirichlet problem and the theory of conformal mappings is typical.

(4) The **Royden compactification** is the $\mathfrak{R}$-compactification $R^*_{\mathfrak{R}}$ with the family $\mathfrak{R}$ of bounded C^∞ functions f on R with finite Dirichlet integrals $\iint_R df \wedge {}^* df$. It was introduced by H. L. Royden and further developed by S. Mori, M. Ôta, Y. Kusunoki, Nakai, and others. This compactification has been effectively used in the study of HD-functions and the classification problem of Riemann surfaces (→ 362 Riemann Surfaces).

(5) The **Wiener compactification** is the $\mathfrak{W}$-compactification $R^*_{\mathfrak{W}}$ with the family $\mathfrak{W}$ of bounded continuous functions f on R such that $\{H_f^{G_n}\}$ converges to a unique harmonic function independent of the choice of exhaustions $\{G_n\}$ of an arbitrary fixed subregion $G \in O_G$ consisting of relatively compact subregions of G. It is the largest resolutive compactification, and compactifications smaller than $R^*_{\mathfrak{W}}$ are always resolutive. This compactification was studied by Mori, K. Hayashi, Kusunoki, C. Constantinescu and A. Cornea, and others, and is useful for the study of HB-functions and the classification of Riemann surfaces.

(6) The **Martin compactification** is the $\mathfrak{M}$-compactification $R^*_{\mathfrak{M}}$ with the family $\mathfrak{M}$ of bounded continuous functions f on R such that there exist relatively compact regions R_f with the property that $f = H_{f^*}^{R-R_f, R^*_{\mathfrak{A}}-R_f} / H_{1^*}^{R-R_f, R^*_{\mathfrak{A}}-R_f}$ on $R-R_f$. Here f^* coincides with f on R_f and equals 0 on $R^*_{\mathfrak{A}}-R$, and 1^* is similarly defined. The set $R^*_{\mathfrak{M}}-R$ is called the **Martin boundary** of R. If †Green's function g exists on R, then the function $m(p,q)=g(p,q)/g(o,q)$ for an arbitrary fixed $o \in R$ can be extended continuously to $R \times R^*_{\mathfrak{M}}$, which is called the **Martin kernel**. By the metric $d_{\mathfrak{M}}(q,r) = \sup_{p \in R_0} |m(p,q)/(1+m(p,q)) - m(p,r)/(1+m(p,r))|$ with a parametric disk R_0 in R, $R^*_{\mathfrak{M}}$ is †metrizable. This compactification was introduced by R. S. Martin, and many applications of it to the study of HP-functions, potential theory, and cluster sets were obtained by M. H. Heins, Z. Kuramochi, J. L. Doob, Constantinescu and Cornea, and others.

(7) For a function f on R, $(R)\partial f/\partial n = 0$ means that there exists a relatively compact subregion R_f such that f is of class C^∞ on R outside R_f and the Dirichlet integral of f over $R-R_f$ is not greater than those of functions on $R-R_f$ that coincide with f on the boundary of R_f. The **Kuramochi compactification** is the $\mathfrak{K}$-compactification $R^*_{\mathfrak{K}}$ with the family $\mathfrak{K}$ of bounded continuous functions f on R satisfying $(R)\partial f/\partial n = 0$. The continuous function $k(p,q)$ on R which vanishes in a fixed parametric disk R_0 in R and is harmonic in $R-\overline{R}_0$ except for a positive †logarithmic singularity at a point q can be extended continuously to $R \times R^*_{\mathfrak{K}}$, which is called the **Kuramochi kernel**. By this kernel, $R^*_{\mathfrak{K}}$ is metrizable, as in the case of Martin compactification. This compactification was introduced by Kuramochi, and its important applications to the study of HD-functions, potential theory, and cluster sets were made by Kuramochi, Constantinescu and Cornea, and others.

Among examples (1)–(7), no boundary point in (2), (4), or (5) satisfies the †first countability axiom, while the others are all metrizable. In (4) and (5), $\Delta_r = \Gamma$. Fig. 1 shows the relationship among the seven examples. Here $A \to B$ means that A lies over B, and $A \not\rightleftarrows B$ means that in general neither $A \to B$ nor $B \to A$.

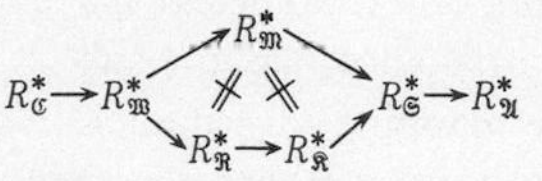

Fig. 1

References

[1] C. Constantinescu and A. Cornea, Ideale Ränder Riemannscher Flächen, Springer, 1963.

[2] Z. Kuramochi, On the boundary of Riemann surfaces (Japanese), Sûgaku, 16 (1964), 80–94.

[3] M. Nakai, On function algebras on Riemann surfaces (Japanese), Sûgaku, 13 (1962), 130–140.

[4] M. Nakai, On Evans potential, Proc. Japan Acad., 38 (1962), 624–629.

[5] M. Nakai, Radon-Nikodym densities between harmonic measures on the ideal boundary of an open Riemann surface, Nagoya Math. J., 27 (1966), 71–76.

[6] F. Maeda, M. Ohtsuka, et al., Kuramochi boundaries of Riemann surfaces, Lecture notes in math. 58, Springer, 1968.

211 (IX.25) Immersion and Embedding

A. General Remarks

If there exists a homeomorphic mapping f from a topological space V into a topological space W, then f is called the **embedding** (more precisely, **topological embedding**) of V into W, and V is called **embeddable** into W. If there exists a continuous mapping f from V into W such that for some neighborhood U_p of any point p of V the restricted mapping $f|U_p$ is a homeomorphic mapping from U_p into W, then f is called the **immersion** (more precisely, **topological immersion**) of V into W. The space W is usually the Euclidean space $\mathbf{R}^N$, the projective space $\mathbf{P}^N\mathbf{R}$, or a manifold.

In the above, we have considered the †category of topological spaces and continuous mappings. The notion of embedding (immersion) can be defined similarly for other categories, such as those of simplicial complexes and simplicial mappings, differentiable manifolds and differentiable mappings, Riemannian manifolds and isometric mappings, and analytic manifolds and analytic mappings. The notions thus defined are called, respectively, **combinatorial** or **simplicial embeddings (immersions), differentiable embeddings (immersions), isometric embeddings (immersions)**, and **analytic embeddings (immersions)**.

If f is an embedding of a topological space V into W and V is homeomorphic to a subspace of W, then the mapping f is called a **regular embedding**. If V is †compact, then the embedding f is necessarily regular. For example, an n-dimensional local Euclidean simplicial complex K^n is (simplicially) immersed into $\mathbf{R}^{2n}$ and regularly embedded into $\mathbf{R}^{2n+1}$. Generally, $\mathbf{R}^{2n}$ (resp. $\mathbf{R}^{2n+1}$) is the lowest-dimensional Euclidean space into which K^n can be immersed (embedded) (W. T. Wu and others) (→ 73 Complexes; Appendix A, Table 6.VII).

B. Immersions and Embeddings of Differentiable Manifolds

An **immersion** of an n-dimensional †C^∞-differentiable manifold (without boundary) M^n in an m-dimensional C^∞-differentiable manifold X^m is a C^∞-differentiable mapping $f: M^n \rightarrow X^m$ such that f is †regular at each point of M^n. In Sections B and C, embedding always means regular embedding. Some of the first results in the theory of embeddings and immersions were given by H. Whitney [1] (1936). He proved by "general position" arguments that M^n can always be immersed in the $2n$-dimensional Euclidean space $\mathbf{R}^{2n}$ and M^n can always be embedded in $\mathbf{R}^{2n+1}$.

Two immersions $f_0, f_1: M^n \rightarrow X^m$ are said to be **regularly homotopic** if there exists a †homotopy $f_t: M^n \rightarrow X^m$, $0 \leqslant t \leqslant 1$, such that f_t is an immersion for each t and the induced mapping (†differential of f_t) $df_t: T(M^n) \rightarrow T(X^m)$ on the tangent spaces naturally gives rise to a continuous mapping on $T(M^n) \times I$. Two embeddings are **isotopic** if they are regularly homotopic and the homotopy mapping f_t is an embedding for each t. Given M^n and X^m, a fundamental problem of embedding (immersion) theory is to classify the embeddings (immersions) of M^n in X^m according to their isotopy (regular homotopy) classes. The following results are also due to Whitney [1]: Any two immersions of M^n in X^m that are homotopic are regularly homotopic if $m \geqslant 2n+2$, and any two embeddings of M^n in X^m that are homotopic are isotopic if $m \geqslant 2n+3$. The range $m \geqslant 2n+3$ is called the **stable range** of embeddings.

In [2, 3] (1944) Whitney improved his classical theorems and showed that M^n can always be immersed in $\mathbf{R}^{2n-1}$ for $n > 1$ and M^n can always be embedded in $\mathbf{R}^{2n}$. The methods used in his proof have played an important role in the subsequent development of the theory. Classification of immersions of the n-sphere S^n in $\mathbf{R}^m$ was determined by Smale [4]. The following theorems due to Hirsch [5] are fundamental to immersion theory: The regular homotopy classes of immersions of M^n in X^m are in one-to-one correspondence with the homotopy classes of †cross sections of the bundle associated with the †bundle of n-frames of M^n whose fiber is the bundle of n-frames of X^m. If M^n is immersible in $\mathbf{R}_+^{m+r}$, where $m > n$ with r linearly independent fields of †normal vectors, then M^n is immersible in $\mathbf{R}^m$. In particular, if M^n is a †π-manifold, M^n is immersible in $\mathbf{R}^{n+1}$ (→ 410 Topology of Differentiable Manifolds).

Let $\mathcal{E}(M^n)$ be the set of isomorphism classes of real †vector bundles over M^n, and consider $\theta: \mathcal{E}(M^n) \rightarrow KO(M^n)$ (→ 236 K-Theory). An element $\xi \in KO(M^n)$ is said to be **positive** if ξ is in the image of θ. If $\xi_0 \in \widetilde{KO}(M^n)$, the **geometric dimension** of ξ_0, written $g(\xi_0)$, is the least integer k such that $\xi_0 + k$ is positive. Then Hirsch's theorem [6] can be expressed as follows: M^n is immersible in $\mathbf{R}^{n+k}$ $(k > 0)$ if and only if $g(n - \tau(M^n)) \leqslant k$, where $\tau(M^n)$ is the tangent bundle of M^n.

Haefliger [7] obtained the following important result: Let M^n be compact and †$(k-1)$-connected, and let X^m be k-connected. Then any continuous mapping of M^n in X^m is homotopic to an embedding if $2k < n$, $m \geqslant 2n - k + 1$, and any two homotopic embeddings of M^n in X^m are isotopic if $2k < n+1$, $m \geqslant 2n - k + 2$. Thus if $m > 3(n+1)/2$, any two embeddings of S^n in $\mathbf{R}^m$ are isotopic. The range $m > 3(n+1)/2$ is called the **metastable range**. Haefliger [8] further classified the embeddings of S^{4n-1} in $\mathbf{R}^{6n}$ and showed the existence of embeddings of S^{4n-1} in $\mathbf{R}^{6n}$ that are not isomorphic to the natural one. More complete results for the classification of embeddings of S^n in S^m were obtained by Levine [9].

We list some recent results about embeddings and immersions. If M^n is noncompact, M^n can always be embedded in $\mathbf{R}^{2n-1}$; if M^n is a noncompact π-manifold, M^n can always be immersed in $\mathbf{R}^n$; if M^n is compact and orientable and $n>4$, M^n can always be embedded in $\mathbf{R}^{2n-1}$.

C. Nonembedding and Nonimmersion Theorems

We denote the †total Stiefel-Whitney class of M^n by $w(M^n)$ and the †total Pontrjagin class of M^n by $p(M^n)$ (→ 58 Characteristic Classes). Then $(w(M^n))^{-1}$ $(\in H^*(M^n;\mathbf{Z}_2))$ and $(p(M^n))^{-1}$ $(\in H^*(M^n;\mathbf{Z}))$ can be written as $\bar{w}(M^n)=\Sigma\bar{w}_i(M^n)$ $(\bar{w}_i\in H^i(M^n;\mathbf{Z}_2))$ and $\bar{p}(M^n)=\Sigma\bar{p}_i(M^n)$ $(\bar{p}_i\in H^{4i}(M^n;\mathbf{Z}))$. Then the property of characteristic classes for the †Whitney sum implies the following theorem: If M^n can be immersed in $\mathbf{R}^{n+k}$, then $\bar{w}_i(M^n)=0$ for $i>k$ and $\bar{p}_i(M^n)=0$ for $i>[k/2]$. Furthermore, if M^n can be embedded in $\mathbf{R}^{n+k}$, then $\bar{w}_k(M^n)=0$. As an application, these results yield the nonembedding (nonimmersion) theorem for projective spaces (→ Appendix A, Table 6.VII). Sharper theorems were obtained subsequently. In particular, Atiyah [6] proved the following: Let λ^i $(i=0,1,\dots)$ be †exterior power operations (→ 236 K-Theory), and let γ^i be the operations defined by the formal power series $\Sigma_{i=0}^{\infty}\gamma^i t^i=(\Sigma_{i=0}^{\infty}\lambda^i t^i)(1-t)^{-1}$. Then $\gamma^i(n-\tau(M))=0$ for $i>k$ $(i\geqslant k)$ if M^n can be immersed (embedded) in $\mathbf{R}^{n+k}$. Furthermore, we have an interesting result for the differentiable case. For any positive integer q, there exists a differentiable manifold M^n such that M^n is immersible in $\mathbf{R}^k$ but not embeddable in $\mathbf{R}^{k+q}$.

D. Embeddings of Piecewise Linear Manifolds

An embedding of an n-dimensional †piecewise linear manifold M^n in an m-dimensional piecewise linear manifold X^m is a †piecewise linear mapping $f: M^n \rightarrow X^m$ such that f is a homeomorphism onto $f(M^n)$, $f(M^n)$ is a †subcomplex of X^m, and $f^{-1}: f(M^n)\rightarrow M^n$ is piecewise linear. Two embeddings $f_0, f_1: M^n \rightarrow X^m$ are isotopic if there exists a homotopy $F: M^n\times[0,1]\rightarrow X^m\times[0,1]$ such that for $0\leqslant t\leqslant 1$, $F(M^n\times t)\subset X^m\times t$, $F|(M^n\times t): M^n\times t\rightarrow X^m\times t$ is an embedding, and $F|M\times 0=f_0$, $F|M\times 1=f_1$. It is obvious that an n-dimensional simplicial complex can be piecewise linearly embedded in $\mathbf{R}^{2n+1}$. The following is Wu's theorem: Any n-dimensional piecewise linear manifold can be embedded in $\mathbf{R}^{2n}$. Among many significant results obtained in the 1960s, we mention the following theorem of Zeeman [10]: If $m-n\geqslant 3$, any two embeddings of S^n in S^m are isotopic; hence S^n is unknotted in S^m.

References

[1] H. Whitney, Differentiable manifolds, Ann. of Math., (2) 37 (1936), 645–680.
[2] H. Whitney, The self-intersections of a smooth n-manifold in $2n$-space, Ann. of Math., (2) 45 (1944), 220–246.
[3] H. Whitney, The singularities of a smooth n-manifold in $(2n-1)$-space, Ann. of Math., (2) 45 (1944), 247–293.
[4] S. Smale, The classification of immersions of spheres in Euclidean spaces, Ann. of Math., (2) 69 (1959), 327–344.
[5] M. W. Hirsch, Immersions of manifolds, Trans. Amer. Math. Soc., 93 (1959), 242–276.
[6] M. F. Atiyah, Immersions and embeddings of manifolds, Topology, 1 (1962), 125–132.
[7] A. Haefliger, Plongements différentiables de variétés dans variétés, Comment. Math. Helv., 36 (1961), 47–82.
[8] A. Haefliger, Knotted $(4k-1)$-spheres in $6k$-space, Ann. of Math., (2) 75 (1962), 452–466.
[9] J. Levine, A classification of differentiable knots, Ann. of Math., (2) 82 (1965), 15–50.
[10] E. C. Zeeman, Unknotting combinatorial balls, Ann. of Math., (2) 78 (1963), 501–526.

212 (X.7) Implicit Functions

A. General Remarks

Historically, a function y of x was called an implicit function of x if there was given a functional relation $f(x,y)=0$ between x and y, but no explicit representation of y in terms of x (→ 170 Functions). Today, however, the notion of implicit function is rigorously defined as follows: Suppose that a function $f(x_1,\dots,x_n,y)$ is of †class C^1 in a domain G in the real $(n+1)$-dimensional Euclidean space $\mathbf{R}^{n+1}$ and that $f(x_1^0,\dots,x_n^0,y^0)=0$, $f_y(x_1^0,\dots,x_n^0,y^0)\neq 0$ at a point $(x_1^0,\dots,x_n^0,y^0)$ in G. Then there is a unique function $g(x_1,\dots,x_n)$ of †class C^1 in a neighborhood of the point $(x_1^0,\dots,x_n^0)$ that satisfies $f(x_1,\dots,x_n,g(x_1,\dots,x_n))=0$, $y^0=g(x_1^0,\dots,x_n^0)$ (**implicit function theorem**). The function g is called the **implicit function** determined by $f=0$. The derivative of g is given by the relation

$\partial g/\partial x_j = -(\partial f/\partial x_j)/(\partial f/\partial y),$

where $y = g(x_1, \ldots, x_n)$. If the function f is of class C^r $(1 \leqslant r \leqslant \infty$ or $r = \omega)$, then the function g is also of class C^r. In particular, when $n = 1$, letting x_1 be x, we have $dg/dx = -f_x/f_y$.

B. Jacobian Matrices and Jacobian Determinants

A mapping u from a domain G in $\mathbf{R}^n$ into $\mathbf{R}^m$

$u(x) = (u_1(x_1, \ldots, x_n), \ldots, u_m(x_1, \ldots, x_n)),$

$x = (x_1, \ldots, x_n),$

is called a **mapping of class** C^r if each component $u_1, \ldots, u_m$ is of class C^r $(0 \leqslant r \leqslant \infty$ or $r = \omega)$ in G. Given a mapping u of class C^1 from G into $\mathbf{R}^m$, we consider the following matrix, which gives rise to the differential du_x of the mapping u (→ 108 Differentiable Manifolds I):

$$\partial(u)/\partial(x) = (\partial u_j/\partial x_k)_{1 \leqslant j \leqslant m, 1 \leqslant k \leqslant n}. \qquad (1)$$

This matrix is called the **Jacobian matrix** of the mapping u at x. If there is another mapping v of class C^1 from a domain containing the †range U of u into $\mathbf{R}^l$, then we have the law of composition:

$$(\partial(v)/\partial(u))(\partial(u)/\partial(x)) = \partial(v)/\partial(x).$$

When $n = m$, the †determinant of the matrix (1) is called the **Jacobian determinant** (or simply **Jacobian**), and is denoted by $D(u)/D(x)$, $D(u_1, \ldots, u_n)/D(x_1, \ldots, x_n)$, or

$$\frac{D(u_1, \ldots, u_n)}{D(x_1, \ldots, x_n)}.$$

Sometimes the notation ∂ is used instead of D, but in the present article we distinguish the matrix and the determinant, and use ∂ for the matrix and D for the determinant.

If $m = n$ and $D(u)/D(x)$ never vanishes at any point of the domain G, then u is called a **regular** (or **nonsingular**) **mapping of class** C^1. If the Jacobian $D(u)/D(x)$ is 0 at x, we say that u is **singular** at x. A mapping that is singular at every point in a set $S \subset G$ is said to be **degenerate** on S. For a regular mapping u, the sign of the Jacobian is constant in a connected domain G. If it is positive, the mapping u preserves the orientation of the coordinate system at each point in G, while if it is negative, the mapping changes the orientation. A point where u is degenerate is called a **critical point** of the mapping u, and its image under u is called a **critical value**. In general, the image of the mapping is "folded" along the set of critical points. The set of critical values of a mapping u of class C^1 (sending a domain in $\mathbf{R}^n$ into $\mathbf{R}^n$) is of Lebesgue measure 0 in $\mathbf{R}^n$ (**Sard's theorem**). If u is a regular mapping, then each point in the domain G of u has a neighborhood V such that the restriction of u on V is a †topological mapping. Its inverse mapping $x(u)$ is also a regular mapping of class C^1 and satisfies the relation

$$\frac{D(u)}{D(x)}\frac{D(x)}{D(u)} = 1$$

(**inverse mapping theorem**). If u is of class C^r $(1 \leqslant r \leqslant \infty$ or $r = \omega)$, then so is its inverse mapping.

C. Functional Relations

A function $F(u_1, \ldots, u_n)$ defined on a domain B in $\mathbf{R}^n$ is called a **function with scattered zeros** if F has a zero point (i.e., there exists a point u for which $F(u) = 0$), but any open subset of B contains a point u such that $F(u) \neq 0$. An †analytic function has scattered zeros. Let $u(x)$ be a mapping from a domain G in $\mathbf{R}^n$ into $B \subset \mathbf{R}^n$. Suppose that there exists a function $F(u)$ defined in B, of class C^r, with scattered zeros, and such that $F(u(x)) = 0$ for any x in G. Then we say that the components $u_1, \ldots, u_n$ of the mapping u have a **functional relation of class** C^r or are **functionally dependent of class** C^r. In such a case, we sometimes say simply that $u_1, \ldots, u_n$ are **functionally dependent** or that they have a **functional relation**. If the components $u_1, \ldots, u_n$ of a mapping u of class C^1 are functionally dependent of class C^0, then the Jacobian $D(u)/D(x)$ of u must vanish. Conversely, if the Jacobian $D(u)/D(x)$ of a mapping u of class C^1 is identically 0 in the domain G, the components $u_1, \ldots, u_n$ are functionally dependent of class C^∞ on every compact set in G (**Knopp-Schmidt theorem**). It is easy to show from the implicit function theorem that they are functionally dependent in a neighborhood of each point. However, it is rather difficult to show global functional dependence for the nonanalytic case, and the theorem was first proved rigorously by K. Knopp and R. Schmidt in 1926 [1].

D. Implicit Functions Determined by Systems of Functions

Suppose that the †rank of the Jacobian matrix of (1) is $r < m$ everywhere in G. Suppose that $u(x)$ is a mapping of class C^1 from a domain G in $\mathbf{R}^n$ into $\mathbf{R}^m$, the Jacobian determinant $D(u_1, \ldots, u_r)/D(x_1, \ldots, x_r)$ never vanishes in G, and further $D(u_1, \ldots, u_r, u_\rho)/D(x_1, \ldots, x_r, x_\sigma)$ is identically 0 in G for each ρ, σ with $r < \rho \leqslant m$, $r < \sigma \leqslant n$. Then the values

$u_{r+1}(x), \ldots, u_m(x)$ are determined by the values $u_1(x), \ldots, u_r(x)$, and each u_ρ is represented as a function of class C^1 of $u_1, \ldots, u_r$.

Let $u(x)$ be a mapping of class C^1 from a domain G in $\mathbf{R}^n$ into $\mathbf{R}^m$ and V the †inverse image of a point u^0. To study the properties of the set V, we assume, for simplicity, that u^0 is the origin. Suppose that the rank of the matrix $\partial(u)/\partial(x)$ is r for every point x in G, and each of $u_{r+1}, \ldots, u_m$ is functionally dependent on $u_1, \ldots, u_r$. Then each u_ρ $(r<\rho\leqslant m)$ is a function u_ρ $(u_1, \ldots, u_r)$ of $u_1, \ldots, u_r$. The set V is empty if there is a ρ such that $u_\rho(0, \ldots, 0)\neq 0$. On the other hand, if $u_\rho(0, \ldots, 0)=0$ for all ρ $(r<\rho\leqslant m)$, then V is the set of common zero points of the functions $u_1(x), \ldots, u_r(x)$. Therefore, to study the set V, we may assume that $r=m\leqslant n$. If $m=n$, V consists of isolated points only. If $m<n$, then changing the order of the variables $x_1, \ldots, x_n$ if necessary, we may assume that $D(u_1, \ldots, u_m)/D(x_1, \ldots, x_m)\neq 0$ at a point (x^0) in V. In this case, there is a unique function $\xi_\mu(x_{m+1}, \ldots, x_n)$ of class C^1 $(1\leqslant\mu\leqslant m)$ in a neighborhood of (x_i^0) satisfying the following two conditions: (i) $x_\mu^0=\xi_\mu(x_{m+1}^0, \ldots, x_n^0)$; (ii) if the point $(x_{m+1}, \ldots, x_n)$ is in a neighborhood of $(x_{m+1}^0, \ldots, x_n^0)$, then the point

$$(\xi_1(x_{m+1}, \ldots, x_n), \ldots, \xi_m(x_{m+1}, \ldots, x_n), x_{m+1}, \ldots, x_n)\in V.$$

Each function ξ_μ is called an **implicit function** of $x_{m+1}, \ldots, x_n$ determined by the relations $u_1=\ldots=u_m=0$. The †total derivatives of the ξ_μ are determined from the following system of linear equations:

$$\sum_{k=1}^{m}\frac{\partial u_j}{\partial x_k}d\xi_k+\sum_{l=m+1}^{n}\frac{\partial u_j}{\partial x_l}dx_l=0,$$

$$j=1, \ldots, m.$$

E. Linear Relations

Suppose that $u(x)$ is a mapping of class C^{m-1} from $\mathbf{R}^1$ into $\mathbf{R}^m$. Its components $(u_1, \ldots, u_m)$ are functions of class C^{m-1}. Then the determinant

$$\begin{vmatrix} u_1 & u_2 & \ldots & u_m \\ u_1' & u_2' & \ldots & u_m' \\ \ldots & \ldots & \ldots & \ldots \\ u_1^{(m-1)} & u_2^{(m-1)} & \ldots & u_m^{(m-1)} \end{vmatrix}$$

is called the **Wronskian determinant** (or simply **Wronskian**) of the functions $u_1, \ldots, u_m$ and is denoted by $W(u_1, u_2, \ldots, u_m)$. If the functions $u_1, \ldots, u_m$ are †linearly dependent, i.e., if there exist constants c_j not all zero satisfying $\sum_{j=1}^m c_j u_j(x)=0$ identically, then the Wronskian vanishes identically. Therefore, if $W(u_1, \ldots, u_m)\neq 0$, then the functions $u_1, \ldots, u_m$ are linearly independent. Conversely, if $W(u_1, \ldots, u_m)=0$ identically, and further if there is at least one nonvanishing Wronskian for $u_1, \ldots, u_{i-1}, u_{i+1}, \ldots, u_m$ $(1\leqslant i\leqslant m)$, then the functions $u_1, \ldots, u_m$ are linearly dependent. The necessity of the additional condition is shown by the following example: $u_1=x^3$ and $u_2=|x|^3$ are of class C^1 in the interval $[-1,1]$ and linearly independent, but satisfy $W(x^3,|x|^3)=0$ identically. However, the additional condition is unnecessary if the functions $u_1, \ldots, u_m$ are analytic. Similar theorems are valid in a domain in a complex plane.

Furthermore, if $u(x)$ is a continuous mapping from an interval $[a,b]$ in $\mathbf{R}^1$ into $\mathbf{R}^m$, the determinant

$$G(u_1, \ldots, u_m)=\begin{vmatrix} (1,1) & \ldots & (1,m) \\ (2,1) & \ldots & (2,m) \\ \ldots & \ldots & \ldots \\ (m,1) & \ldots & (m,m) \end{vmatrix},$$

$$(j,k)=\int_a^b u_j(x)u_k(x)\,dx$$

is called the **Gramian determinant** (or simply **Gramian**). The Gramian is the †discriminant of the quadratic form

$$\int_a^b\left(\sum_{j=1}^m \xi_j u_j(x)\right)^2dx$$

of $\xi_1, \ldots, \xi_m$ and is equal to

$$\frac{1}{m!}\int_a^b\cdots\int_a^b\left(\det\left(u_j(x_k)\right)\right)^2dx_1\ldots dx_m.$$

We always have $G(u_1, \ldots, u_m)\geqslant 0$, and $G(u_1, \ldots, u_m)=0$ if and only if $u_1, \ldots, u_m$ are linearly dependent. The Gramian is defined if the functions $u_1, \ldots, u_m$ are †square integrable in the sense of Lebesgue. In that case, the condition $G(u_1, \ldots, u_m)=0$ holds if and only if $u_1, \ldots, u_m$ are linearly dependent †almost everywhere, i.e., there are constants $c_1, \ldots, c_m$ not all zero such that the relation $c_1u_1(x)+\ldots+c_mu_m(x)=0$ holds except on a set of Lebesgue measure 0.

References

[1] K. Knopp and R. Schmidt, Funktionaldeterminanten und Abhängigkeit von Funktionen, Math. Z., 25 (1926), 373–381.
[2] G. Doetsch, Die Funktionaldeterminante als Deformationsmass einer Abbildung und als Kriterium der Abhängigkeit von Funktionen, Math. Ann., 99 (1928), 590–601.
[3] T. Takagi, Daisûgaku kôgi (Japanese; Lectures on algebra), Kyôritu, revised edition, 1965.
[4] T. Takagi, Kaiseki gairon (Japanese; A course of analysis), Iwanami, third edition, 1961.

[5] M. Fujiwara, Sûgaku kaiseki I; Bibunsekibun gaku (Japanese; Mathematical analysis pt. 1. Differential and integral calculus) II, Utida-rôkakuho, 1939.
[6] S. Hitotumatu, Kaisekigaku zyosetu (Japanese; Elements of analysis) II, Syôkabô, 1963.
[7] W. Rudin, Principles of mathematical analysis, McGraw-Hill, second edition, 1964.
[8] H. Cartan, Calcul différentiel, Hermann, 1967.

213 (XX.5) Indian Mathematics

India was one of the earliest civilizations, but because it has no precise chronological record of ancient times, it is said to possess no history. Indian mathematics seems to have developed under the influence of the cult of Brahma, as did the calendar. It may also have some relation to the mathematics of the Near East and China, but this is difficult to trace. The word *ganita* (computation) appears in early religious writings; after the beginning of the Christian Era, it was classified into *pâti-ganita* (arithmetic), *bija-ganita* (algebra), and *krestra-ganita* (geometry), thus showing some systematization. The Buddhists (notably Nagarjuna) had a kind of logic, but it had no relation to mathematics. Unlike the Greeks, the Indians had no demonstrational geometry, but they had symbolic algebra and a position system of numeration.

Indian geometry was computational: Ayra-Bhatta (c. 476–c. 550) computed the value of π as 3.1416; Brahmagupta (598–c. 660) had a formula to compute the area of quadrangles inscribed in a circle; and Bhâskara (1114–1185) gave a proof of the Pythagorean theorem. In trigonometry, Arya-Bhatta made a table of sines of angles between 0° and 90° for every 3.75° interval. The name "sine" is related to the Sanskrit *jya*, which referred to half of the chord of the double arc.

The Indians had a remarkable system of algebra. At the beginning they had no operational symbols and described in words the rules for solving equations. Brahmagupta worked on the †Pell equation $ax^2+1=y^2$. Bhâskara knew that a quadratic equation may have two roots that can be positive and negative, but did not adopt the negative root in such cases. Bhâskara also introduced algebraic symbols.

The symbol 0 was used in India from about 200 B.C. to denote the void place in the position system of numeration; 0 as a number is found in a book by Bakhshâli published in the 3rd century A.D. The number 0 is defined as $a-a=0$ in our notation, and the rules $a\pm 0=a$, $0\times a=0$, $\sqrt{0}=0$, $0\div a=0$ are mentioned. Brahmagupta prohibited division by 0 in arithmetic, but in algebra he called the "quantity" $a\div 0$ *taccheda*. Bhâskara called it *khahara* and made it play a similar role to infinity. Some historians assert that the Indians had the ideas of infinity and infinitesimal. Some explain the origin of the Indian position system of numeration by the fact that the names of numbers differed according to their positions. The Indian numeration system was exported to Europe through Arabia and had great influence on the development of mathematics.

References

[1] M. B. Cantor, Vorlesungen über Geschichte der Mathematik I, Teubner, third edition, 1907.
[2] G. Sarton, Introduction to the history of science I, From Homer to Omar Khayyam Carnegie Institute of Washington, 1927.
[3] B. Datta and A. N. Singh, History of Hindu mathematics I, II, Lahore, 1935–1938 (Asia Publ. House, 1962).

214 (II.26) Inductive Limit and Projective Limit

A. General Remarks

Inductive and projective limits can be defined over any †preordered set I and in any †category. We first explain the definition of these limits in the special case where I is a †directed set and the category is that of sets, of groups, or of topological spaces. The simplest case is when I is the ordered set **N** of the natural numbers.

B. The Limit of Sets

Let I be a directed set. Suppose that we are given a set X_i for each $i\in I$ and a mapping $\varphi_{ji}:X_i\rightarrow X_j$ for each pair (i,j) of elements of I with $i\leqslant j$, such that $\varphi_{ii}=1_{X_i}$ (the identity mapping on X_i) and $\varphi_{ki}=\varphi_{kj}\circ\varphi_{ji}$ $(i\leqslant j\leqslant k)$. Then we denote the system by (X_i,φ_{ji}) and call it an **inductive system** (or **direct system**) of sets over I. Let S be the †direct sum $\coprod_i X_i$ of the sets X_i $(i\in I)$, and define an equivalence relation in S as follows: $x\in X_i$ and

$y \in X_j$ are equivalent if and only if there exists a $k \in I$ such that $i \leqslant k, j \leqslant k$, $\varphi_{ki}(x) = \varphi_{kj}(y)$. Let D be the †quotient set of S by this equivalence relation, and let $f_i : X_i \to D$ $(i \in I)$ be the canonical mappings. Then we have I(1) $f_j \circ \varphi_{ji} = f_i$ $(i \leqslant j)$; I(2) for any set X, and for any system of mappings $g_i : X_i \to X$ $(i \in I)$ satisfying $g_j \circ \varphi_{ji} = g_i$ $(i \leqslant j)$, there exists a unique mapping $f : D \to X$ such that $f \circ f_i = g_i$ $(i \in I)$. We call (D, f_i) the **inductive limit** (or **direct limit**) of the inductive system (X_i, φ_{ji}) over I, and denote it by $\varinjlim X_i$ or $\operatorname{ind\,lim} X_i$ (more precisely, by $\varinjlim_{i \in I} X_i$ or $\operatorname{ind\,lim}_{i \in I} X_i$).

Suppose, dually, that we are given a set X_i for each $i \in I$ and a mapping $\psi_{ij} : X_j \to X_i$ for each $i \leqslant j$, such that $\psi_{ii} = 1_{X_i}$ and $\psi_{ik} = \psi_{ij} \circ \psi_{jk}$ $(i \leqslant j \leqslant k)$. Then we denote the system by (X_i, ψ_{ij}) and call it a **projective system** (or **inverse system**) of sets over I. Let P be the subset of the Cartesian product $\prod X_i$ defined by $P = \{(x_i) \mid \psi_{ij}(x_j) = x_i \ (i \leqslant j)\}$, and let $p_i : P \to X_i$ be the canonical mapping. Then we have P(1) $\psi_{ij} \circ p_j = p_i$ $(i \leqslant j)$; P(2) for any set X, and for any system of mappings $q_i : X \to X_i$ satisfying $\psi_{ij} \circ q_j = q_i$ $(i \leqslant j)$, there exists a unique mapping $p : X \to P$ such that $p_i \circ p = q_i$ $(i \in I)$. We call (P, p_i) the **projective limit** (or **inverse limit**) of the projective system (X_i, ψ_{ij}) over I and denote it by $\varprojlim X_i$ or $\operatorname{proj\,lim} X_i$.

Note that we may replace I by any †cofinal subset of I without changing the limits.

C. The Limit of Groups and of Topological Spaces

If, in the notation of Section B, X_i is a group and φ_{ji} (ψ_{ij}) is a homomorphism, then we say that (X_i, φ_{ji}) $((X_i, \psi_{ij}))$ is an **inductive (projective) system of groups**. The inductive limit (as a set) $D = \varinjlim X_i$ has the structure of a group for which the canonical mappings f_i are homomorphisms. With this group structure, D is called the **inductive limit (group)** of the inductive system of groups. It satisfies properties I(1) and I(2) with group X and homomorphisms g_i and f. Similarly, the projective limit (as a set) $P = \varprojlim X_i$ has a unique group structure such that each $p_i : P \to X_i$ is a homomorphism, namely, that of a subgroup of the direct product group $\prod X_i$. The group P is called the **projective limit (group)** of the projective system of groups. When each X_i is a module over a fixed ring A, we get entirely similar results by considering A-homomorphisms instead of group homomorphisms.

Next, let X_i be a topological space and φ_{ji} and ψ_{ij} be continuous mappings. Then (X_i, φ_{ji}) $((X_i, \psi_{ij}))$ is called an **inductive (projective) system of topological spaces**. If we introduce in $D = \varinjlim X_i$ the topology of a quotient space of the †topological direct sum of the spaces X_i $(i \in I)$, then the f_i are continuous, and I(1) and I(2) hold with sets and mappings replaced by topological spaces and continuous mappings. Similarly, if we view $P = \varprojlim X_i$ as a subspace of the †product space $\prod X_i$, then the p_i are continuous and P(1) and P(2) hold with the same modification as before. The spaces D and P are called the **inductive limit (space)** and the **projective limit (space)** of the system of topological spaces, respectively. The projective limit of Hausdorff (compact) spaces is also Hausdorff (compact).

Furthermore, if the X_i $(i \in I)$ form a topological group and φ_{ji}, ψ_{ij} are continuous homomorphisms, then $\varinjlim X_i$ and $\varprojlim X_i$ are topological groups, and properties I(1), I(2), P(1), and P(2) are satisfied for topological groups and continuous homomorphisms (→ 406 Topological Groups). In particular, projective limits of finite groups are †totally disconnected compact groups and are called **profinite groups**; they occur, e.g., as the ring of †p-adic integers and as the †Galois group of an infinite †Galois extension. Conversely, the †germs of continuous functions at a point x in a topological space X, and other kinds of germs (→ 377 Sheaves), are important examples of inductive limit of groups.

D. Limits in a Category

Let I be a preordered set and $\mathcal{C}$ a category. If we are given an object X_i of a category $\mathcal{C}$ for each $i \in I$ and a †morphism $\varphi_{ji} : X_i \to X_j$ of $\mathcal{C}$ for each pair (i, j) of elements of I with $i \leqslant j$, and if the conditions $\varphi_{ii} = 1_{X_i}$, $\varphi_{ki} = \varphi_{kj} \circ \varphi_{ji}$ $(i \leqslant j \leqslant k)$ are satisfied, then we call the system (X_i, φ_{ji}) an **inductive system** over I in the category $\mathcal{C}$. A **projective system** in $\mathcal{C}$ is defined dually: It is an inductive system in the †dual category $\mathcal{C}^\circ$. If we view I as a category (→ 53 Categories and Functors B), then an inductive (projective) system in $\mathcal{C}$ over the index set I is a †covariant (†contravariant) functor from I to $\mathcal{C}$. Now if an object $D \in \mathcal{C}$ and morphisms $f_i : X_i \to D$ $(i \in I)$ satisfy conditions I(1) and I(2) with the modification that X is an object and g_i, f are morphisms in $\mathcal{C}$, then the system (D, f_i) is called the **inductive limit** of (X_i, φ_{ji}) and is denoted by $\varinjlim X_i$ Similarly, if an object $P \in \mathcal{C}$ and morphisms $p_i : P \to X_i$ $(i \in I)$ satisfy P(1) and P(2) with a similar modification, then (P, p_i) is called the **projective limit** of (X_i, ψ_{ij}) and is written $\varprojlim X_i$. By I(2) and P(2), these limits are unique if they exist.

In the categories of sets, of groups, of modules, and of topological spaces, inductive and projective limits always exist. Note that, if the ordering of I is such that $i \leqslant j$ implies

$i=j$, i.e., if there is no ordering between two distinct elements of I, then the inductive (projective) limit is the †direct sum (†direct product) (→ 53 Categories and Functors E).

Let (X_i, φ_{ji}), (X', φ'_{ji}) be two inductive systems over the same index set I, and let $\varphi_i: X_i \to X'_i$ $(i \in I)$ be morphisms satisfying $\varphi'_{ji} \circ \varphi_i = \varphi_j \circ \varphi_{ji}$ $(i \leqslant j)$. Then the system (φ_i) is called a **morphism** between the inductive systems. Such a morphism is a †natural transformation between the inductive systems viewed as functors $I \to \mathcal{C}$. If $\varinjlim X_i$ and $\varinjlim X'_i$ exist, then (φ_i) induces a morphism $\varinjlim \varphi_i : \varinjlim X_i \to \varinjlim X'_i$ in a natural way, and similarly for projective limits.

For the more abstract notion of limit of a functor → [6]. For the theory of procategories → [5].

References

[1] S. Iyanaga and K. Kodaira, Gendai sûgaku gaisetu (Japanese; Introduction to modern mathematics) I, Iwanami, 1961.
[2] S. Lefschetz, Algebraic topology, Amer. Math. Soc. Colloq. Publ., 1942.
[3] S. Eilenberg and N. Steenrod, Foundations of algebraic topology, Princeton Univ. Press, 1952.
[4] H. P. Cartan and S. Eilenberg, Homological algebra, Princeton Univ. Press, 1956.
[5] A. Grothendieck, Technique de descente et théorèmes d'existence en géométrie algébrique II. Le théorème d'existence en théorie formelle des modules, Sém. Bourbaki, Exposé 195, 1959/1960 (Benjamin, 1966).
[6] M. Artin, Grothendieck topology, Lecture notes, Harvard University, 1962.

215 (X.3) Inequalities

A. General Remarks

In this article we consider various properties of inequalities between real numbers. An inequality that holds for every real number (e.g., $x^2 \geqslant 0$) is called an **absolute inequality**; otherwise it is called a **conditional inequality**. When we are given a conditional inequality (e.g., $x(x-1)<0$), the set of reals that satisfy it is called the **solution** of the inequality.

B. The Solution of a Conditional Inequality

Suppose that a conditional inequality is given by $f(x)>0$ (or $f(x) \geqslant 0$), where f is a continuous function defined for every real. If the equation $f(x)=0$ has no solution, then we have either $f(x)>0$ or $f(x)<0$ for all x. On the other hand, if α and β $(\alpha<\beta)$ are adjacent roots of the equation $f(x)=0$, the sign of $f(x)$ is unchanged in the open interval (α, β). Therefore, the solution of the given inequality depends essentially on the solution of the equation $f(x)=0$. If inequalities involve two variables x, y and are given by $f(x,y)>0$, $g(x,y)>0$ for continuous functions f and g, the solution is, in general, a domain in the xy-plane bounded by the curves $f(x,y)=0$ and $g(x,y)=0$. Similar results hold for the case of inequalities involving more than two variables.

C. Famous Absolute Inequalities

(1) Inequalities concerning **means** (or **averages**): Suppose that we are given an n-tuple $a=(a_1, \ldots, a_n)$, $a_\nu \geqslant 0$. We set

$$M_r = M_r(a) = \left(\frac{1}{n} \sum_{\nu=1}^{n} a_\nu^r \right)^{1/r}.$$

If at least one a_ν is 0 and $r<0$, we put $M_r=0$. In particular, we put

$$A = M_1,$$

$$G = \lim_{r \to 0} M_r = \left(\prod_{\nu=1}^{n} a_\nu \right)^{1/n},$$

$$H = M_{-1};$$

these are called the **arithmetic mean**, **geometric mean**, and **harmonic mean** of a_ν $(\nu=1, \ldots, n)$, respectively. Except when either all a_ν are identical, or some a_ν is 0 and $r \leqslant 0$, the function M_r increases †strictly monotonically as r increases, and $M_r \to \min a_\nu$ $(r \to -\infty)$, $M_r \to \max a_\nu$ $(r \to +\infty)$. Therefore, we always have $\min a_\nu \leqslant M_r \leqslant \max a_\nu$. In particular, we have $H<G<A$ if the a_ν are all positive and not all equal.

Let $p(x)$ (>0), $f(x)$ $(\geqslant 0)$ be †integrable functions on a †measurable set E. We put

$$M_r(f) = \left(\int_E p f^r \, dx \Big/ \int_E p \, dx \right)^{1/r}, \quad r \neq 0.$$

Furthermore, if $M_r(f)$ is strictly positive for some $r>0$, we put

$$M_0(f) = \lim_{r \to +0} M_r(f)$$
$$= \exp\left(\int_E p \log f \, dx \Big/ \int_E p \, dx \right).$$

We call $M_r(f)$ the **mean of degree** r of the function $f(x)$ with respect to the **weight function** $p(x)$. It has properties similar to those of $M_r(a)$. In particular, when the weight function $p=1$, the means $M_1(f)$, $M_0(f)$, $M_{-1}(f)$ are called the **arithmetic mean**, **geometric mean**, and **harmonic mean** of f, respectively.

(2) The Hölder inequality: Suppose that $p \neq 0, 1$ and $(p-1)(q-1)=1$; that is, $1/p+1/q=1$, and $a_\nu > 0, b_\nu > 0$. Then, in general, we have the **Hölder inequality**:

$$\sum_r a_\nu b_\nu \gtrless \Big(\sum_\nu a_\nu^p\Big)^{1/p} \Big(\sum_\nu b_\nu^q\Big)^{1/q}, \quad p \lessgtr 1,$$

where the inequality signs in the first inequality are taken in accordance with $p<1$ or $p>1$. The summation may be infinite if the sums are convergent. The inequality sign is replaced by the equality sign if and only if there exist constant factors λ and μ such that $\lambda a_\nu^p = \mu b_\nu^q$ for all ν. The Hölder inequality for $p=q=2$ is called the **Cauchy inequality** (or **Cauchy-Schwarz inequality**).

For two measurable positive functions $f(x)$, $g(x)$, we have the **Hölder integral inequality**:

$$\int_E fg\,dx \gtrless \Big(\int_E f^p\,dx\Big)^{1/p} \Big(\int_E g^q\,dx\Big)^{1/q}, \quad p \lessgtr 1,$$

except when there exist two constant factors λ and μ such that $\lambda f^p = \mu g^q$ holds †almost everywhere. The above inequality is replaced by equality if and only if we are in the latter, exceptional case. The case where $p=q=2$ is called the **Schwarz inequality** (or **Bunjakovskiĭ inequality**).

(3) The Minkowski inequality: Suppose that $p \neq 0, 1$ and $a_\nu > 0, b_\nu > 0$. Then we have the **Minkowski inequality**:

$$\Big(\sum_\nu (a_\nu + b_\nu)^p\Big)^{1/p} \gtrless \Big(\sum_\nu a_\nu^p\Big)^{1/p} + \Big(\sum_\nu b_\nu^p\Big)^{1/p}, \quad p \lessgtr 1,$$

except when $\{a_\nu\}$ and $\{b_\nu\}$ are proportional. The inequality is replaced by equality if and only if we are in the exceptional case.

The corresponding integral inequality for positive functions $f(x)$, $g(x)$ is

$$\Big(\int_E (f+g)^p\,dx\Big)^{1/p} \gtrless \Big(\int_E f^p\,dx\Big)^{1/p} + \Big(\int_E g^p\,dx\Big)^{1/p}, \quad p \lessgtr 1,$$

except when $f(x)/g(x)$ is constant almost everywhere. The inequality is replaced by equality if and only if we are in the latter, exceptional case.

(For other famous inequalities → Appendix A, Table 8; for related topics → 90 Convex Functions; for inequalities concerning convex functions → 255 Linear Programming; for the linear inequalities → 91 Convex Sets.)

References

[1] G. H. Hardy, J. E. Littlewood, and G. Pólya, Inequalities, Cambridge Univ. Press, revised edition, 1952.
[2] V. I. Levin and S. B. Stečkin, Inequalities, Amer. Math. Soc. Transl., (2) 14 (1960), 1–29 (English translation of the appendix to the Russian translation of [1]).
[3] E. F. Beckenbach and R. Bellman, Inequalities, Erg. Math., Springer, second revised printing, 1965.
[4] E. F. Beckenbach and R. Bellman, An introduction to inequalities, Random House, 1961.
[5] N. D. Kazarinoff, Geometric inequalities, Random House, 1961.
[6] N. D. Kazarinoff, Analytic inequalities, Holt, Rinehart and Winston, 1961.
[7] W. Walter, Differential- und Integral-Ungleichungen, Springer, 1964.
[8] O. Shisha (ed.), Inequalities, Academic Press, I, 1967; II, 1970.

216 (XVIII.10) Information Theory

A. General Remarks

The mathematical theory of information transmission in communication systems, first developed by C. E. Shannon [1] and now called **information theory**, is one of the most important fields of mathematical science. It combines various methods, including those of probability, statistics, functional analysis, Fourier analysis, and some parts of algebra. A system of information transmission has, to begin with, an **information source** and a **channel** through which information is conveyed. Information supplied by the source is codified by a **coding** process before it is fed into the channel. **Decoding** is therefore needed to transmit the information from the channel to the receiver (→ 66 Coding Theory). Sometimes, because of **noise** in the communication system, the final information obtained by the receiver is erroneous. In this case, the channel is called a **noisy channel**; otherwise it is called a **noiseless channel**.

B. Entropy

In information theory, the **amount of information** is estimated by means of **entropy**, to be defined below. The information source, consisting of a finite set A and a †probability distribution p over A, is denoted by the pair $[A,p]$ and may be regarded as a finite †probability space. For instance, $A=\{\alpha_1, \ldots, \alpha_N\}$ may be an alphabet and p the †probability distribution on A giving the rate $p_i = p(\alpha_i)$ of appearance of $\alpha_i \in A$ in the data ($p_i \geqslant 0$, $\sum p_i = 1$); this information source is sometimes denoted by $(\alpha_1,\ldots,\alpha_N;p_1,\ldots,p_N)$. In the information source $[A,p]$, the amount of information supplied by each $\alpha_k \in A$ is defined to be $I(\alpha_k) = -\log p(\alpha_k)$, which is called **self-information**. Furthermore, the †mean value of self-information (i.e., $H(A) = -\sum_{k=1}^{N} p(\alpha_k)\log p(\alpha_k)$) is called the **entropy** of the source $[A,p]$. To give a numerical value to entropy, we usually choose 2 as the base of the logarithm, and the unit of such values is called a **bit**. When we choose 10 or e as the base, the corresponding unit is called a **decit** or **nat**.

C. Characterizations of Entropy

Several characterizations of the entropy $H(\cdot)$ as a function of N variables p_i $(i=1,\ldots,N,$ $p_i \geqslant 0, \sum_{i=1}^{N} p_i = 1)$ with values in $\{\lambda \mid 0 \leqslant \lambda \leqslant +\infty\}$ have been given by Shannon, A. Ja. Hinčin, A. D. Faddeev, and others [13]. Among such characterizations, the following are important: (i) The function $f(p) = H(p, 1-p)$ is continuous on $0 \leqslant p \leqslant 1$, and $f(p_0) > 0$ for at least one $0 < p_0 < 1$. (ii) For any permutation $(p_1',\ldots,p_N')$ of $(p_1,\ldots,p_N)$, $H(p_1,\ldots,p_N) = H(p_1',\ldots,p_N')$. (iii) For any $p_N = q + r > 0$ with $q \geqslant 0$ and $r \geqslant 0$, $H(p_1, \ldots,p_{N-1},q,r) = H(p_1,\ldots,p_N) + p_N H(q/p_N, r/p_N)$.

Furthermore, the following propositions hold: (I) $-\sum_{i=1}^{N} p(\alpha_i)\log p(\alpha_i) \leqslant \log N$. The equality holds if and only if $p(\alpha_i) = 1/N$ $(i = 1,\ldots,N)$. (II) Given two information sources $[A,p]$ and $[B,p]$ which are subspaces of a probability space (X,p), we may define the †joint probability distribution $p(\alpha_i,\beta_j)$ of $\alpha_i \in A$ and $\beta_j \in B$. Then we have

$$H(A,B) \leqslant H(A) + H(B),$$

where $H(A,B) = -\sum p(\alpha_i,\beta_j)\log p(\alpha_i,\beta_j)$. The equality holds if and only if $p(\alpha_i,\beta_j) = p(\alpha_i)p(\beta_j)$ for all i,j. Denote the †conditional probability by $p(\alpha_i \mid \beta_j)$. Then $H(A \mid \beta_j) = -\sum_i p(\alpha_i \mid \beta_j)\log p(\alpha_i \mid \beta_j)$ is called the conditional entropy of A with respect to the occurrence of $\beta_j \in B$, and

$$H(A \mid B) = \sum_j p(\beta_j) H(A \mid \beta_j)$$

is called the **conditional entropy** of A given B. (III) Using this notation, we have $H(A,B) = H(A) + H(B \mid A) = H(B) + H(A \mid B)$ and $H(A \mid B) \leqslant H(A)$, where the equality holds if and only if $p(\alpha_i,\beta_j) = p(\alpha_i)p(\beta_j)$ for all i, j.

D. Information Sources

Given a finite set $A = \{\alpha_1^0,\ldots,\alpha_N^0\}$ which is part of an information source, we often consider the sets $A^n = A \times \ldots \times A$ or $A^{\mathbf{Z}} = \prod_{k=-\infty}^{\infty} A_k$ $(A_k = A, k = 0, \pm 1, \pm 2, \ldots)$, where each $a_n \in A^n$ and $a \in A^{\mathbf{Z}}$ are expressed as $a_n = (\alpha_1,\ldots,\alpha_n)$ $(\alpha_k \in A)$ and $a = (\ldots,\alpha_{-2}, \alpha_{-1},\alpha_0,\alpha_1,\ldots)$ $(\alpha_k \in A)$. Let $\mathfrak{F}_A$ be the σ-algebra generated by all †cylinder sets in $A^{\mathbf{Z}}$. Given a †probability measure P over $\mathfrak{F}_A$, an information source $[A^{\mathbf{Z}},P]$ is defined. When P is invariant under the †shift transformation T on $A^{\mathbf{Z}}$: $P(E) = P(TE)$ $(E \in \mathfrak{F}_A)$, $[A^{\mathbf{Z}},P]$ is said to be a **stationary information source**. In particular, if $P(E) = 0$ or 1 whenever $TE = E \in \mathfrak{F}_A$, then $[A^{\mathbf{Z}},P]$ is said to be **ergodic** or an **ergodic information source**. In a stationary information source $[A^{\mathbf{Z}},P]$, let $a_n = (\alpha_1,\ldots,\alpha_n) \in A^n$ correspond to each $a = (\ldots,\alpha_{-1},\alpha_0, \alpha_1,\ldots) \in A^{\mathbf{Z}}$. Then an information source $[A^n,P_n]$ can be naturally induced. Let $H(A^n)$ be the entropy of $[A^n,P_n]$. Then $\lim_{n\to\infty} H(A^n)/n$ exists. We denote it by $H(A^{\mathbf{Z}},P)$ and call it the **mean entropy**. When $[A^{\mathbf{Z}},P]$ is an ergodic information source, $-\log P_n(a_n)/n$ (where $-\log P_n(a_n)$ $(a_n \in A^n, a \in A^{\mathbf{Z}})$ is the self-information of $[A^n,P_n]$) converges in probability to $H(A^{\mathbf{Z}},P)$ as $n \to \infty$. This is **McMillan's theorem** [13].

E. Channels

The simplest channels are the noiseless ones, for which there is a one-to-one correspondence between input and output and no loss of information in transmission through the channel. (In noisy channels, loss of information always occurs.) Given a message $a = (\ldots,\alpha_{-1},\alpha_0,\alpha_1,\ldots)$ belonging to the source $[A^{\mathbf{Z}},P]$, if each $\alpha_k \in A$ is successively transmitted through the channel, then the receiver will get a corresponding sequence $b = (\ldots, \beta_{-1},\beta_0,\beta_1,\ldots)$ of elements β_j contained in a certain finite set B. Let $\mathfrak{F}_B$ be the σ-algebra generated by all cylinder sets in $B^{\mathbf{Z}}$. Any $a \in A^{\mathbf{Z}}$ determines a conditional probability $\nu(S \mid a)$ $(S \in \mathfrak{F}_B)$ which may be regarded as the †transition probability from $A^{\mathbf{Z}}$ to $B^{\mathbf{Z}}$. The channel characterized by $B^{\mathbf{Z}}$ and ν is denoted by $[A^{\mathbf{Z}},\nu,B^{\mathbf{Z}}]$.

(1) Measurability. If for any set $S \in B^{\mathbf{Z}}$, $\nu(S \mid \cdot)$ is a †measurable function on $(A^{\mathbf{Z}},\mathfrak{F}_A)$, then the channel $[A^{\mathbf{Z}},\nu,B^{\mathbf{Z}}]$ is said to be **measurable**. (2) Stationary channels. If for any $a \in A^{\mathbf{Z}}$ and $S \in \mathfrak{F}_B$, $\nu(TS \mid Ta) = \nu(S \mid a)$, then the channel is said to be **stationary**. (3) Nonanticipating channels. Suppose that for $S \in \mathfrak{F}_B$ and a fixed index t, the message $(\ldots,\beta_{-1},\beta_0,\beta_1,\ldots) \in S$ is equivalent to the message $(\ldots,\beta'_{-1},\beta'_0,\beta'_1,\ldots) \in S$ when $\beta_j' = \beta_j$, $j \leqslant t$. If in addition, $\nu(S \mid a) = \nu(S \mid a')$ whenever $a_i' = a_i$, $i \leqslant t$, the channel is said to be **nonanticipating**. (4) Finite memory. If there exists a fixed integer $m > 0$ such that given

any cylinder set $c_{n,t}=[\beta_n,\beta_{n+1},\dots,\beta_t]\in\mathfrak{F}_B$, the equality $\nu(c_{n,t}|a)=\nu(c_{n,t}|a')$ holds for any pair $a=(\dots,\alpha_{-1},\alpha_0,\alpha_1,\dots)$, $a'=(\dots,\alpha'_{-1},\alpha'_0,\alpha'_1,\dots)\in A^{\mathbf{Z}}$ with $\alpha_k=\alpha'_k$ $(n-m\leqslant k\leqslant t)$, then the channel is said to have **finite memory**, where $[\beta_n,\dots,\beta_t]$ is the cylinder set of all $b\in B^{\mathbf{Z}}$ whose kth coordinate equals β_k $(n\leqslant k\leqslant t)$. The minimum integer $m>0$ satisfying such a requirement is called the **length of memory**. The transition probability $\nu([\beta_n,\dots,\beta_t]|[\alpha_{n-m},\dots,\alpha_t])$ is induced by such a memory. (5) M-dependency. Suppose we are given a positive integer M. If, for any integers n, r, s, t with $n\leqslant r<s\leqslant t$,

$$\nu(c_{n,r}\cap c_{s,t}|a)=\nu(c_{n,r}|a)\cdot\nu(c_{s,t}|a)$$

whenever $s-r>M$, then the channel is said to be **M-dependent** (Takano, [13]). (6) Memoryless channels. If for every pair of integers n, r $(n<r)$, $\nu([\beta_n,\dots,\beta_r]|[\alpha_n,\dots,\alpha_r])=\prod_{k=1}^{r}\nu([\beta_k]|[\alpha_k])$, then the channel is called a **memoryless channel**.

F. Capacities of Channels

Given an information source $[A^{\mathbf{Z}},P]$ and channel $[A^{\mathbf{Z}},\nu,B^{\mathbf{Z}}]$, two probability measures P_{AB} on $(A\times B,\mathfrak{F}_A\otimes\mathfrak{F}_B)$ and P_B on $(B,\mathfrak{F}_B)$ are determined by

$$P_{AB}(E\times F)=\int_E\nu(F|a)P(da)$$

and $P_B(F)=P_{AB}(A\times F)$, $E\in\mathfrak{F}_A$, $F\in\mathfrak{F}_B$. We call P the **input measure**, P_{AB} the **compound measure**, and P_B the **output measure**. If $[A^{\mathbf{Z}},P]$ is ergodic (an ergodic input measure) and $[A^{\mathbf{Z}},\nu,B^{\mathbf{Z}}]$ has finite memory, then $[B^{\mathbf{Z}},P_B]$ is ergodic (an ergodic output measure). In general, when P and ν are stationary, P_{AB} and P_B are also stationary and the mean entropies $H(A^{\mathbf{Z}},P)$, $H(B^{\mathbf{Z}},P_B)$, $H(A^{\mathbf{Z}}\times B^{\mathbf{Z}},P_{AB})$ are defined. The **transmission rate** of $[A^{\mathbf{Z}},\nu,B^{\mathbf{Z}}]$ with respect to the input measure P is defined to be $R(P)=H(A^{\mathbf{Z}},P)+H(B^{\mathbf{Z}},P_B)-H(A^{\mathbf{Z}}\times B^{\mathbf{Z}},P_{AB})\geqslant 0$ and may be regarded as the mean amount of information carried by each element of $A^{\mathbf{Z}}$. When the channel has finite memory and is M-dependent, $C_e=(C_s=)\sup R(P)$ exists and is called the **ergodic (stationary) transmission capacity** of the channel, where sup is taken over the set of all possible ergodic (stationary) input measures P. Since the equality $C_e=C_s$ has been established by L. Carlson, I. P. Čaregradskiĭ, and A. Feinstein [13], either is called the **transmission capacity** of the channel. L. Breiman later proved that $C_e=R(P)$ for some ergodic input measure P [13]; that is, the capacity C_e is achieved by an ergodic input.

G. The Fundamental Theorem of Information Theory

The **fundamental theorem** was first known as Shannon's theorem for memoryless channels [i]. A precise proof was given by A. Feinstein [2]. The theorem claims that given a stationary memoryless channel with capacity $C_s>0$ and arbitrary data taken from an ergodic information source, then for any $\varepsilon>0$ and R $(0<R<C_s)$ there exists a codified message $(\alpha_1,\dots,\alpha_n)$ with a sufficiently large length n which conveys the given message with transmission rate R and probability of decoding errors λ_n less than ε.

The **weak converse theorem** of the fundamental theorem states: If, in the fundamental theorem, $R>C_e$, then the probability λ_n does not converge to 0 as n tends to infinity (J. Wolfowitz [2], D. Blackwell, L. Breiman, and A. J. Thomasian [14]). In this case, $\lambda_n\to 1$ $(n\to\infty)$ does not always hold. If the conclusion of the weak converse is replaced by $\lambda_n\to 1$, then the result is called the **strong converse theorem**. A necessary and sufficient condition for the strong converse theorem to hold has been given (Wolfowitz [2], K. Yosihara [15]). The fundamental theorem was extended by Hinčin to stationary, nonanticipating, and finite memory channels.

H. The Coding Problem

The fundamental theorem implies the existence of ideal codes under suitable conditions. Various investigations have sought to obtain practical methods of producing ideal coding processes. For noiseless channels the problem has been completely solved, and there are several coding methods, for example, the coding methods of Shannon, R. M. Fano, and D. A. Huffman, and the equal length coding of K. Kunisawa, N. Honda, and S. Ikeno. For noisy channels, there are several investigations by R. W. Hamming, Z. Kiyasu, D. E. Muller, and M. J. E. Golay (Kunisawa [12], R. B. Ash [16], W. W. Peterson [17]). In particular, for two-way symmetric channels, by means of parity test a coding theorem similar to the fundamental theorem has been established [16].

I. The Continuous Cases

A. N. Kolmogorov and his colleagues studied the case of a continuous information source and discussed Shannon's theory from a generalized viewpoint. Let ξ, η,... be †random variables with values in †measurable spaces

$(X,\mathfrak{X})$, $(Y,\mathfrak{Y})$, ..., respectively. Let P_ξ and P_η be probability distributions for ξ and η, and $P_{\xi\eta}$ be the joint distribution of ξ, η. Then $I(\xi,\eta)=\sup\sum_{i,j}P_{\xi,\eta}(E_i\times F_j)\log[P_{\xi\eta}(E_i\times F_j)/P_\xi(E_i)P_\eta(F_j)]$ is defined to be the amount of information of ξ (resp. η) when η (resp. ξ) is chosen, where the sup is taken over the set of all measurable partitions of X (resp. Y). The **entropy** of ξ is defined to be $I(\xi)=I(\xi,\xi)$. The notions of integral representations and information densities of $I(\xi,\xi)$ and $I(\xi)$ were also introduced by I. M. Gel'fand and A. M. Jaglom, A. Pérez, and others [5], and various fundamental properties of $I(\ ,\ )$ have been studied [18]. Using these, a mathematical model of continuous channels was introduced by Kolmogorov, and an exact description of the channels was discussed by R. L. Dobrušin [5]. Moreover, the concepts of **exactness of reproduction** and **information stability** between message and channel were introduced, and Shannon's coding theorem was discussed for these generalized cases [5,8]. Kolmogorov and his school introduced the concept of entropy for †flows over †dynamical systems, and many valuable results have been obtained by them (V. A. Rohlin [6], Ja. G. Sinai [7]).

J. ε-Entropy

Let A be a †totally bounded subset of a †metric space (X,ρ), and let $\varepsilon>0$ be given. A finite family $U_1,\ldots,U_n$ of subsets of X is called an **ε-covering** of A if the diameter $d(U_k)\leqslant 2\varepsilon$ for each U_k and $A\subset\bigcup_{k=1}^n U_k$. A finite set of points $x_1,\ldots,x_n\in X$ is called an †ε-net for A if for each $a\in A$ there exists at least one point x_k of the ε-net with $\rho(x,x_k)\leqslant\varepsilon$. A finite set of points $a_1,\ldots,a_n\in A$ is called **ε-distinguishable** if $\rho(a_i,a_j)>\varepsilon$ for all i and j with $i\neq j$. For a given $\varepsilon>0$, the minimal values of the numbers of sets in ε-coverings of A and of the numbers of elements in ε-nets for A are denoted by $N_\varepsilon(A)$ and $N_\varepsilon^X(X)$, respectively, and the maximal value of the numbers of ε-distinguishable points of A is denoted by $M_\varepsilon(A)$. We set $H_\varepsilon(A)=\log N_\varepsilon(A)$, $H_\varepsilon^X(A)=\log N_\varepsilon^X(A)$, and $C_\varepsilon(A)=\log M_\varepsilon(A)$, and call them the **ε-entropy of** A, the **ε-entropy of** A **with respect to** X, and the **ε-capacity** of A, respectively. They satisfy

$$C_{2\varepsilon}(A)\leqslant H_\varepsilon(A)\leqslant H_\varepsilon^X(A)\leqslant C_\varepsilon(A).$$

These values are adopted in computing the exactness of reproductions, discriminations, and codings of messages. These notions are used in approximation problems in various function spaces and were also used to solve †Hilbert's thirteenth problem (Kolmogorov and V. M. Tihomirov [13], G. G. Lorentz [21]).

Besides the notion of entropy, there are several other concepts of amount of information, such as †K-L information for discrimination of populations, given by S. Kullback and R. A. Leibler [9]; negentropy related to physical thermodynamics, given by L. Brillouin [10]; and the information concept, in J. Von Neumann's theory of measurements (M. Nakamura and H. Umegaki [11]).

References

[1] C. E. Shannon, A mathematical theory of communication, Bell System Tech. J., 27 (1948), 379–423, 623–656.

[2] A. Feinstein, Foundations of information theory, McGraw-Hill, 1958.

[3] A. Feinstein, On the coding theorem and its converse for finite-memory channels, Inform. Control, 2 (1959), 25–44.

[4] A. N. Kolmogorov, Theory of transmission of information, Acad. Rep. Populare Romine An Romino-Soviet, ser. Mat. Fiz. (3) 13 (1959), no. 1 (28), 5–33.

[5] Р. Л. Добрушин (R. L. Dobrušin), Общая Формулировка основной теорему Шеннова в теории информации, Uspehi Mat. Nauk, 14 (1959), no. 6, 3–104; English translation, General formulation of Shannon's main theorem in information theory, Amer. Math. Soc. Transl., (2) 33, 323–438.

[6] В. А. Рохлин (V. A. Rohlin), Новый прогресс в теории преобразований с инвариантной мерой, Uspehi Mat. Nauk, 15 (1960), no. 4, 3–26.

[7] Я. Г. Синай (Ja. G. Sinai), Вероятностные идеи в эргодической теории, Proc. Intern. Congr. Math., 1962, Stockholm, Almqvist & Wiksells, p. 540–559.

[8] М. С. Пинскер (M. S. Pinsker), Информация и информационная устойчивость случайных величин и процессов, Acad. Sci. SSSR, 1960; English translation, Information and information stability of random variables and processes, Holden-Day, 1964.

[9] S. Kullback, Information theory and statistics, John Wiley, 1957.

[10] L. Brillouin, Science and information theory, Academic Press, 1956.

[11] M. Nakamura and H. Umegaki, On Von Neumann's theory of measurements in quantum statistics, Math. Japonicae, 7 (1962), 151–157.

[12] K. Kunisawa, Zyôhôriron (Japanese; Information theory), Kyôritu, 1960.

[13] K. Kunisawa and H. Umegaki (eds.), Zyôhôriron no sinpo—Entoropî gainen no hatten (Japanese; Progress of information

theory—The development of the notion of entropy), Iwanami, 1963.
[14] D. Blackwell, L. Breiman, and A. J. Thomasian, Proof of Shannon's transmission theorem for finite-state indecomposable channels, Ann. Math. Statist., 29 (1958), 1209–1220.
[15] K. Yosihara, Simple proof of the strong converse theorems in some channels, Kôdai Math. Sem. Rep., 16 (1964), 213–222.
[16] R. B. Ash, Information theory, Interscience, 1965.
[17] W. W. Peterson, Error-correcting codes, John Wiley, 1961.
[18] R. M. Fano, Transmission of information, MIT Press, 1961.
[19] R. G. Gallager, Information theory and reliable communication, John Wiley, 1968.
[20] F. Jelinek, Probabilistic information theory, McGraw-Hill, 1968.
[21] G. G. Lorentz, Approximation of functions, Holt, 1966.

217 (XVII.16) Insurance Mathematics

A. General Remarks

Insurance is a system in which a large number of people contribute a small precalculated amount of money (called a **premium**) to fill the economic need that arises when a person meets adversity. The amount of economic need filled by this system is called the **amount of insurance** (or **amount insured**). The **insurer** mediates the system. **Actuarial mathematics** is the branch of applied mathematics that studies the mathematical basis of insurance, one of the first cases in which mathematics was successfully applied to a social question. Actuarial mathematics may be divided into two branches according to its application. The first includes the calculation of various values of each individual policy, such as premiums or reserves. The second is mainly connected with management of an insurance business, and includes the study of reinsurance systems, of the maximum amount of insurance, of the contingency fund, or the analysis of profits. There is only one basic principle in actuarial mathematics. Called the **principle of equivalence**, it determines the premium and reserve in each year so that the present value of future premium income of the insurer is equal to the present value of future benefits, for each policy.

The basic factors of actuarial calculations are (1) probabilities of contingencies, (2) an expected rate of interest in the future (often referred to as the **assumed rate of interest**), and (3) cost of administration of the system. Premiums are calculated using these factors and the principle of equivalence. The following is an example of the classical method of calculation for a life insurance policy with the use of "commutation symbols," which is an old device for the convenience of calculations.

We write P for the net premium (in which the cost of administration mentioned before is disregarded), P' for the gross premium, T_t for the amount of death benefits payable in the tth year after the policy is issued, E_t for the amount of survival benefits payable at the beginning of the tth year, n for the period for which the insurance is effective, and m for the period for which premiums are to be paid. Let α, β, and γ stand for three positive constants determining the initial expenses $=\alpha$ $(T_1$ or $E_1)$, the premium collection expenses $=\beta P'$, and the general expenses for maintenance $=\gamma(T_t$ or $E_t)$. The factor that comes into consideration next is a **model of human death and survival** (measurement). Assume that l_x is the number of lives attaining age x, and write q_x for the probability that a life of x years will end within one year. Then d_x, the number of lives ending within one year out of l_x, is $l_x q_x$, and l_{x+1}, the number of lives remaining after one year at age $x+1$, is l_x-d_x $=l_x(1-q_x)$. The commutation symbols commonly employed are defined as follows: Writing $v=1/(1+i)$ (where i is the assumed rate of interest),

$$D_x=l_x v^x, \qquad C_x=d_x v^{x+1},$$

$$N_x=\sum_{t=0}^{n} D_{x+t}, \qquad M_x=\sum_{t=0}^{n} C_{x+t}.$$

For a policy issued at an insured person's age x, the present value of the insurer's future income may be expressed as $P'(N_x-N_{x+m})/D_x$, and the present value of his future payments may be expressed as

$$(1/D_x)\left(\sum_{t=1}^{n} T_t C_{x+t-1}-\sum_{t=1}^{n+1} E_t D_{x+t-1}\right.$$
$$+\alpha(T_1 \text{ or } E_1)D_x+\gamma\sum_{t=0}^{n}(T_t \text{ or } E_t)D_{x+t-1}$$
$$\left.+\beta P'(N_x-N_{x+m})\right).$$

By assuming that the present value of the future income is equal to the present value of the future payments, the value of the gross premium P' is obtained. (The P' obtained from the assumption $\alpha=\beta=\gamma=0$ is denoted by P and is called the **net premium**. The difference $P'-P$ is called the **loading**.) For a policy in which benefits are payable on disability or contingencies other than death, we have only to obtain a model of contingencies and apply a similar calculation.

B. Liability Reserve

During the term of an insurance contract, it often happens that the present value of the future income is less than the present value of the future payments. If this is the case, the difference is to be held by the insurer as the **liability reserve**. The source of this fund is the past premium income plus interest. The net premium reserve, which disregards expenses, is calculated as

$$V_t=(1/D_{x+t})\times\left(\sum_{r=t+1}^{n} T_r C_{x+r-1}-\sum_{r=t+1}^{n+1} E_r D_{x+r-1}-P(N_{x+t}-N_{x+m})\right).$$

Between the net premium P and the net premium reserve V, we have the following relation:

$$P=(V_t-V_{t-1})+(T_t-V_t)vq_{x+t}+T_t.$$

The first term of the right-hand side of this formula is called the **savings premium**, since it is the amount left out of the premium income of the tth year and added to the reserve. The second term is called the **cost of insurance** or **risk premium**, and is applied to cover the difference between the amount of insurance and that of the existing reserve in case the contingency of death arises. The third term is applied to the payment of the survival benefits (or annuities in case of an **annuity contract**). If T_t-V_t, the amount of risk insured by the insurer, is positive for all values of t during the period of insurance, the policy is called **death insurance**. On the other hand, if the value of T_t-V_t is negative for all values of t, the policy is called **survival insurance**. If the value of T_t-V_t varies between positive and negative according to the different values of t, the policy is called **mixed insurance**. If T_t is always equal to V_t, the policy constitutes mere savings. Most of the insurance policies issued today are one or another type of death insurance, while life annuity policies are a type of survival insurance. For a long time, studies have been made of the effect on premiums and reserves of changes in the three basic factors (1), (2), and (3) in Section A.

C. Risk Theory

Risk Theory occupies a special position in the field of actuarial mathematics. Actuarial mathematics was first born from the theory of probabilities and developed parallel to it. Since a new branch of the theory of probability based on measure theory arose from the contributions made by A. N. Kolmogorov and other mathematicians (→ 339 Probability), revised approaches have inevitably been made to actuarial mathematics. An outstanding example is risk theory.

Risk theory may be divided into two branches. One is called **classical risk theory** (or **individual risk theory**), in which the profit or loss that may result during a certain term of an insurance contract is regarded as a †random variable. Since the insurer's profit equals the sum of these random variables over all the individual contracts, various probability functions can be obtained by applying the theory of probability. The second, called **collective risk theory**, pays no attention to each individual contract, but studies changes in the insurer's balance as a whole with the lapse of time. The basis of collective risk theory was given by F. Lundberg, H. Cramér, and other mathematicians.

Assume that we have an insurer who issues no policies other than death insurance and makes no expenditures except the policy claims. Write dP for the risk premium income during the period of time dt, $d\pi$ for the probability that the contingency will arise during dt, and $p(z)dz$ for the conditional probability that the amount of insurance will fall between z and $z+dz$ at the time the contingency arises. If we take the average amount at risk as the unit amount, we have $\int_0^\infty zp(z)dz=1$. Hence the expectation of claims paid during dt will be $dt\int_0^\infty zp(z)dz=d\pi$. By the principle of equivalence, $dP=d\pi$. If we take the value of risk premium income P as the value of time for simplicity, the insurer's profit during the period of time dP is expressed as follows: (i) The probability that the contingency will not arise equals $1-dp$, and the profit that then arises equals dP; (ii) the probability that the contingency will arise equals dP, the conditional probability that the amount of insurance then will be between z and $z+dz$ equals $p(z)dz$, and the profit that arises equals $dP-z$. Its †characteristic function is expressed as follows:

$$e^{it\,dP}(1-dP)+dP\int_0^\infty e^{it(dP-z)}p(z)\,dz.$$

Since the insurer's profit during the time interval $(0,P)$ is the value obtained by integrating the profit that arises in each period of time dP, its characteristic function is expressed as follows:

$$\exp\left(P\int_0^\infty(e^{-itz}-1+itz)p(z)\,dz\right).$$

Assume that an insurer starts business with the initial contingency reserve. The probability that the insurer's fund, which equals u at the start and increases or decreases by future profits or losses, will become negative in the future is called the **ruin probability**. If we write this as $\psi(u)$ and if the amount at risk is always positive, we have

$$\psi(u)=\alpha_u e^{-Ru}, \quad 0<\alpha_u<1,$$

where R is a constant determined by the relation $\int_0^\infty e^{Rz}p(z)\,dz=1+(1+\lambda)R$, and λ is the rate of safety margin added to the risk premium. When this method is applied to a practical problem, the ruin probability is generally made fairly large by assuming $\alpha_u=1$.

References

[1] C. W. Jordan, Society of Actuaries' textbook on life contingencies, Society of Actuaries, 1952.
[2] W. Saxer, Versicherungsmathematik, Springer, I, 1955; II, 1958.
[3] E. Zwinggi, Versicherungsmathematik, Birkhäuser, 1945.
[4] P. F. Hooker and L. H. Longley-Cook, Life and other contingencies I, Cambridge Univ. Press, 1953.
[5] H. Cramér, Collective risk theory, Nordiska bokhandeln, 1955.
[6] T. Morita, Hokensûgaku (Japanese; Insurance mathematics), Seimeihoken bunka kenkyûzyo, 1963.

218 (X.10) Integral Calculus

A. The Riemann Integral

Let $f(x)$ be a bounded real-valued function defined on an interval $[a,b]$. We shall divide this interval $I=[a,b]$ into subintervals $I_i=[x_{i-1},x_i]$ $(i=1,\ldots,n)$ by a finite number of points x_i $(a=x_0<x_1<\ldots<x_n=b)$. This division into subintervals is uniquely determined by the set $D=\{x_i\}$, called the **partition** of I. We set $M_i=\sup_{x\in I_i}f(x)$, $m_i=\inf_{x\in I_i}f(x)$, and put $\bar{\sigma}(D)=\sum_{i=1}^n M_i(x_i-x_{i-1})$, $\underline{\sigma}(D)=\sum_{i=1}^n m_i(x_i-x_{i-1})$. Considering all possible partitions D of I, we set $\overline{\int_a^b} f(x)\,dx=\inf_D\bar{\sigma}(D)$, $\underline{\int_a^b} f(x)\,dx=\sup_D\underline{\sigma}(D)$, which are called the **Riemann upper integral** and **Riemann lower integral** of f, respectively. If they coincide, then the common value is called the **Riemann integral** of f on $[a,b]$ and is denoted by $\int_a^b f(x)\,dx$. In this case, we say that f is **Riemann integrable** (or simply **integrable**) on $[a,b]$ and call f the **integrand**; a and b are called the **lower limit** and the **upper limit**, respectively. In this case, by **integrating** f from a to b we mean the process of obtaining the value $\int_a^b f(x)\,dx$.

Darboux's theorem: For each $\varepsilon>0$ there exists a positive δ such that the inequalities

$$\left|\bar{\sigma}(D)-\overline{\int_a^b} f(x)\,dx\right|<\varepsilon,$$

$$\left|\underline{\sigma}(D)-\underline{\int_a^b} f(x)\,dx\right|<\varepsilon$$

hold for any partition D with $\delta(D)=\max_i(x_i-x_{i-1})<\delta$. In other words, we have

$$\lim_{\delta(D)\to 0}\bar{\sigma}(D)=\overline{\int_a^b} f(x)\,dx,$$

$$\lim_{\delta(D)\to 0}\underline{\sigma}(D)=\underline{\int_a^b} f(x)\,dx.$$

From Darboux's theorem it follows that a necessary and sufficient condition for $f(x)$ to be integrable on $[a,b]$ is that for each positive ε there exists a positive δ such that $\delta(D)<\delta$ implies $\bar{\sigma}(D)-\underline{\sigma}(D)=\sum_{i=1}^n(M_i-m_i)(x_i-x_{i-1})<\varepsilon$.

We call M_i-m_i the **oscillation** of f on I_i and $\bar{\sigma}(D)$ and $\underline{\sigma}(D)$ the **Darboux sums**. Obviously, if f is integrable on $[a,b]$, then for each positive ε there exists a positive δ such that the following inequality holds for all $D=\{x_j\}$ with $\delta(D)<\delta$ and for an arbitrary ξ_j from each I_j $(j=1,\ldots,n)$:

$$\left|\sum_{j=1}^n f(\xi_j)(x_j-x_{j-1})-\int_a^b f(x)\,dx\right|<\varepsilon.$$

The sum $\sum_{j=1}^n f(\xi_j)(x_j-x_{j-1})$ is often called a **Riemann sum** (or **sum of products**). A function that is continuous on $[a,b]$, or bounded and continuous except for a finite number of points in the interval, is integrable. Furthermore, a bounded function that is continuous on $[a,b]$ except for an infinite number of points x_λ is integrable if for an arbitrary positive number ε there exist a finite number of intervals I_i of which the total length is less than ε and if the set $\{x_\lambda\}$ of exceptional points is contained in $\bigcup I_i$. Generally, a necessary and sufficient condition for a bounded function defined on $[a,b]$ to be integrable is that the set of points where the function is not continuous is of †measure 0 (in the sense of Lebesgue). A function that is either †monotonic on $[a,b]$ (and consequently bounded) or of †bounded variation is integrable. A function that is integrable on $[a,b]$ is integrable on any subinterval of $[a,b]$, the integrand being the restriction of the given function to this subinterval.

B. Basic Properties of Integrals

Let **I** be the set of all functions integrable on $[a,b]$. If $f, g \in \mathbf{I}$, then for any numbers α, β we have $\alpha f + \beta g \in \mathbf{I}$, $fg \in \mathbf{I}$, $\min\{f,g\} \in \mathbf{I}$, $\max\{f,g\} \in \mathbf{I}$, and $f/g \in \mathbf{I}$ provided that there exists a positive constant A such that the inequality $|g| > A$ holds. Furthermore, if $f \in \mathbf{I}$, then $|f| \in \mathbf{I}$; and if $f_n \in \mathbf{I}\,(n=1,2,\ldots)$ and f_n converges †uniformly to f, then $f \in \mathbf{I}$. Corresponding to these properties, the following formulas hold:

(1) Linearity:

$$\int_a^b (\alpha f(x) + \beta g(x))\,dx$$

$$= \alpha \int_a^b f(x)\,dx + \beta \int_a^b f(x)\,dx,$$

where α, β are constants.

(2) Monotonicity: If $f(x) \geqslant 0$, then

$$\int_a^b f(x)\,dx \geqslant 0.$$

If, further, f is continuous at a point $x_0 \in [a,b]$ and $f(x_0) > 0$, then $\int_a^b f(x)\,dx > 0$.

(3) Additivity with respect to intervals: If a, b, and c are points belonging to an interval on which f is integrable and $a < c < b$, then

$$\int_a^b f(x)\,dx = \int_a^c f(x)\,dx + \int_c^b f(x)\,dx.$$

Adopting the conventions that $\int_a^a f(x)\,dx = 0$ and $\int_b^a f(x)\,dx = -\int_a^b f(x)\,dx$, the additivity formula holds independent of the order of a, b, and c.

It follows from (2) that $|\int_a^b f(x)\,dx| \leqslant \int_a^b |f(x)|\,dx$ if $a < b$. Further, if $f_n(x)$ converges to $f(x)$ uniformly on $[a,b]$, then $\lim_{n\to\infty} \int_a^b f_n(x)\,dx = \int_a^b f(x)\,dx$.

Replacing $f_n(x)$ by partial sums of a series, we obtain the following theorem: Let $\Sigma a_n(x)$ be a series in which each term $a_n(x)$ is integrable on an interval $[a,b]$. If the series converges uniformly on $[a,b]$, then the sum $s(x)$ is integrable on $[a,b]$, and the series is **termwise integrable**, that is,

$$\int_a^b s(x)\,dx = \sum_{n=1}^{\infty} \int_a^b a_n(x)\,dx.$$

Also, the series $\Sigma_{n=1}^{\infty} \int_a^x a_n(t)\,dt$ converges uniformly on $[a,b]$ to the integral $\int_a^x s(t)\,dt$. Assume that $\Sigma a_n(x)$ is convergent but not uniformly convergent. If all $a_n(x)$, together with $s(x) = \Sigma a_n(x)$, are integrable and there is a constant M independent of n such that $|s_n(x)| \leqslant M$ $(x \in [a,b])$ for all n, where $s_n(x)$ are partial sums, then the series is termwise integrable (C. Arzelà).

The first mean value theorem: If $f(x)$ is continuous on $[a,b]$ and $\varphi(x)$ is integrable and of constant sign on $[a,b]$, then there exists θ $(0 < \theta < 1)$ such that

$$\int_a^b f(x)\varphi(x)\,dx = f(a + \theta(b-a)) \int_a^b \varphi(x)\,dx.$$

When $\varphi(x) = 1$, we have

$$\int_a^b f(x)\,dx = f(a + \theta(b-a))(b-a).$$

The second mean value theorem: If $f(x)$ is a positive, monotone decreasing function defined on $[a,b]$ and $\varphi(x)$ is an integrable function, then there exists η $(a < \eta \leqslant b)$ such that

$$\int_a^b f(x)\varphi(x)\,dx = f(a+0) \int_a^\eta \varphi(x)\,dx.$$

In the hypothesis of the second mean value theorem, if $f(x)$ is assumed to be monotonic but not necessarily positive, then there exists η $(a < \eta < b)$ such that

$$\int_a^b f(x)\varphi(x)\,dx$$

$$= f(a+0) \int_a^\eta \varphi(x)\,dx + f(b-0) \int_\eta^b \varphi(x)\,dx.$$

Although in the usual statement of this theorem, $a < \eta < b$ is replaced by $a \leqslant \eta \leqslant b$, this form was shown to be valid by H. Okamura (*Sûgaku*, 1 (1947)).

In the case $f(x) \geqslant 0$ on $[a,b]$, we consider the figure F bounded by the graph of $f(x)$, the x-axis, and the lines $x = a$ and $x = b$. Then $\bar{\sigma}(D)$ and $\underline{\sigma}(D)$ are areas of polygons of which one encloses F and the other is enclosed by F, as shown in Fig. 1. Hence it can be shown that the integrability of $f(x)$ in the sense of Riemann is equivalent to the measurability of F in the sense of Jordan. The Riemann integral $\int_a^b f(x)\,dx$ is the area of F with respect to its †Jordan measure.

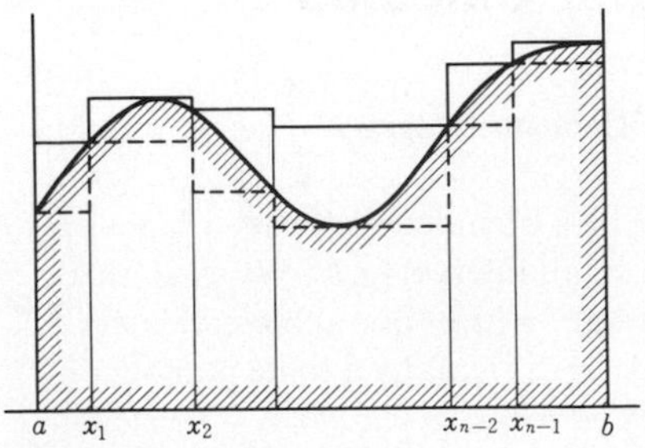

Fig. 1

C. Relations between Differentiation and Integration

Suppose that $f(x)$ is integrable on an interval I. We fix a point a of I and consider the integration $F(x) = \int_a^x f(t)\,dt$, where x varies in I. The function $F(x)$ is called the **indefinite integral** of $f(x)$. In contrast with this, the integral on a fixed interval, as considered in the previous sections, is often called the **defi-**

nite integral. The indefinite integral $F(x)$ is continuous on the interval I and of bounded variation. If $f(x)$ is continuous at a point x_0 in I, then $F(x)$ is differentiable at x_0 and $F'(x_0)=f(x_0)$. In general, if a function $G(x)$ satisfies $G'(x)=f(x)$ everywhere in I, then $G(x)$ is called a **primitive function** of $f(x)$. If $f(x)$ is continuous, the indefinite integral of $f(x)$ is one of the primitive functions of $f(x)$. Furthermore, if a function $G(x)$ is a primitive function of $f(x)$, then any other primitive function may be written in the form $G(x)+C$, where C is a constant, called an **integral constant**. For a continuous function $f(x)$ on $[a,b]$ and any one of its primitive functions $G(x)$, we have

$$\int_a^b f(x)\,dx = G(b)-G(a)=[G(x)]_a^b$$

(fundamental theorem of calculus) (→ Appendix A, Table 9). From the differentiation formulas we obtain the following integration formulas:

Integration by parts: If $f(x)$ and $g(x)$ have continuous derivatives on $[a,b]$, then

$$\int_a^b f(x)g'(x)\,dx = [f(x)g(x)]_a^b - \int_a^b f'(x)g(x)\,dx.$$

More generally, if $f(x)$ and $g(x)$ are integrable on $[a,b]$, then

$$\int_a^b g(x)\left(\int_a^x f(t)\,dt\right)dx = \int_a^b f(x)\,dx\int_a^b g(x)\,dx - \int_a^b f(x)\left(\int_a^x g(t)\,dt\right)dx.$$

Change of variables: If $f(x)$ is integrable on $[a,b]$ and $x=\varphi(t)$ and $\varphi'(t)$ are continuous on $[\alpha,\beta]$, where $a=\varphi(\alpha)$, $b=\varphi(\beta)$ $(a\leqslant\varphi(t)\leqslant b)$, then

$$\int_a^b f(x)\,dx=\int_\alpha^\beta f(\varphi(t))\varphi'(t)\,dt.$$

D. Improper Integrals

The concept of the integral may be generalized to the case where the integrand or the interval on which integration is accomplished is not bounded. Assume $f(x)$ is not bounded on $[a,b)$ but is bounded and integrable on any interval $[a,b-\varepsilon]$ $(\subset[a,b))$. If $\int_a^{b-\varepsilon}f(x)\,dx$ has a finite limit for $\varepsilon\to0$, the limit is denoted by $\int_a^b f(x)\,dx$ and is called the **improper Riemann integral** (or simply **improper integral**) of $f(x)$ on $[a,b)$. For example, if $f(x)$ is continuous on $[a,b)$ and $f(x)=O((b-x)^\alpha)$ for some α $(0>\alpha>-1)$, where O is the †notation of Landau, then the improper integral $\int_a^b f(x)\,dx$ exists. On the other hand, if f is integrable on $[a+\varepsilon,b]$ for each $\varepsilon>0$ but not bounded in any neighborhood of a, we can define the integral on $(a,b]$ in the same way. If f is not bounded in any neighborhood of a or b and if there exists a point c $(a<c<b)$ for which the improper integrals $\int_a^c f(x)\,dx$ and $\int_c^b f(x)\,dx$ exist, then we define $\int_a^b f(x)\,dx = \int_a^c f(x)\,dx+\int_c^b f(x)\,dx$, which is independent of the choice of the point c. Furthermore, assume that $f(x)$ is not bounded in any neighborhood of each point c_j $(j=1,\ldots,n)$ $(c_1<\ldots<c_n)$. Then we define

$$\int_a^b f(x)\,dx=\int_a^{c_1}f(x)\,dx+\int_{c_1}^{c_2}f(x)\,dx+\ldots+\int_{c_{n-1}}^{c_n}f(x)\,dx+\int_{c_n}^b f(x)\,dx,$$

provided that all improper integrals

$$\int_a^{c_1}f(x)\,dx,\ldots,\int_{c_n}^b f(x)\,dx$$

exist. Suppose that $f(x)$ is defined on $[a,b]$ and bounded outside any neighborhood of $c\in(a,b)$ but not bounded in either $[c-\varepsilon,c]$ or $[c,c+\varepsilon]$ for any $\varepsilon>0$. It may well happen that, although neither $\lim_{\varepsilon\to0}\int_a^{c-\varepsilon}f(x)\,dx$ nor $\lim_{\varepsilon'\to0}\int_{c+\varepsilon'}^b f(x)\,dx$ exists (accordingly, the improper integral $\int_a^b f(x)\,dx$ does not exist), if we put $\varepsilon=\varepsilon'$, the limit

$$\lim_{\varepsilon\to0}\left(\int_a^{c-\varepsilon}f(x)\,dx+\int_{c+\varepsilon}^b f(x)\,dx\right)$$

does exist. This limit is called **Cauchy's principal value** and is denoted by p.v.$\int_a^b f(x)\,dx$ (v.p. in French). For example, p.v.$\int_{-1}^1(dx/x)$ $=\lim_{\varepsilon\to0}(\int_{-1}^{-\varepsilon}(1/x)\,dx+\int_\varepsilon^1(1/x)\,dx)=0$.

E. Integrals on Infinite Intervals

Suppose that we are given a formal function $f(x)$ defined on an infinite interval $[a,\infty)$ and integrable on any finite interval $[a,b]$. If $\lim_{b\to\infty}\int_a^b f(x)\,dx$ exists and is finite, then this limit is called the **improper integral** of f on $[a,\infty)$ and is denoted by $\int_a^\infty f(x)\,dx$. We define similarly $\int_{-\infty}^b f(x)\,dx=\lim_{a\to-\infty}\int_a^b f(x)\,dx$, where f is defined on $(-\infty,b]$ and integrable on any interval $[a,b]$. Furthermore, $\int_{-\infty}^\infty f(x)\,dx$ is, by definition, $\int_{-\infty}^c f(x)\,dx+\int_c^\infty f(x)\,dx$, which is independent of the choice of c. Suppose that $f(x)$ is integrable on $[a,b]$ for a fixed a and an arbitrary b larger than a. If $f(x)=O(x^\alpha)$ for some $\alpha<-1$, then $\int_a^\infty f(x)\,dx$ exists. Generally, for α, β such that $-\infty\leqslant\alpha<\infty$ and $-\infty<\beta\leqslant\infty$, if the improper integral $\int_\alpha^\beta f(x)\,dx$ exists, we say that the integral is **convergent**; otherwise, it is **divergent**. Improper integrals also satisfy the three basic properties of integrals (1), (2), and

(3) (→ Section B). However, the existence of an improper integral of a function f on an interval I does not imply the existence of the improper integral of $|f|$ on the same interval I. For example, let f be a function determined by $f(0)=0$, $f(x)=(1/x)\sin(1/x)$ for $0<x\leqslant\pi$. Then $\int_0^\pi f(x)\,dx$ exists, but $\int_0^\pi|f(x)|\,dx$ does not. On the other hand, if the improper integral of $|f(x)|$ exists, then the improper integral of $f(x)$ exists and we have

$$\left|\int_\alpha^\beta f(x)\,dx\right|\leqslant\int_\alpha^\beta|f(x)|\,dx,$$

where $-\infty\leqslant\alpha<\beta\leqslant+\infty$. In this case, we say that f is **absolutely integrable** on the interval $[\alpha,\beta]$. Assume now that $f(x)$ is defined on $(-\infty,\infty)$ and integrable on any finite interval. If $\lim_{\alpha\to\infty}\int_{-\alpha}^{\alpha}f(x)\,dx$ exists, then it is called **Cauchy's principal value** of the integral of f in $(-\infty,\infty)$.

If $f(x)$ is a monotone decreasing, positive, and continuous function defined on $[k,\infty)$ (where k is an integer), then according as $\sum_{\nu=k}^\infty f(\nu)$ converges or diverges, so does $\int_k^\infty f(x)\,dx$.

Suppose that a series $\sum_{n=1}^\infty f_n(x)$, where all the functions $f_n(x)$ $(n=1,2,\ldots)$ are defined on an infinite interval $[a,\infty)$, satisfies $\int_a^b(\sum_{n=1}^\infty f_n(x))\,dx=\sum_{n=1}^\infty\int_a^b f_n(x)\,dx$ for arbitrary $b>a$. Then, according as $\sum_{n=1}^\infty\int_a^\infty f_n(x)\,dx$ converges or diverges, so does $\int_a^b(\sum_{n=1}^\infty f_n(x))\,dx$. When they converge, the following equality holds: $\int_a^\infty\sum_{n=1}^\infty f_n(x)\,dx=\sum_{n=1}^\infty\int_a^\infty f_n(x)\,dx$. In this theorem, if $\int_a^\infty\sum_{n=1}^\infty|f_n(x)|\,dx$ or $\sum_{n=1}^\infty\int_a^\infty|f_n(x)|\,dx$ converges, then the same conclusion as above will follow even when the $f_n(x)$ are not necessarily positive (→ Appendix A, Table 9).

F. Multiple Integrals

Suppose that $f(x,y)$ is a function defined and bounded on an interval $I=\{(x,y)\,|\,a\leqslant x\leqslant b,\ c\leqslant y\leqslant d\}$ in the xy-plane. Partitions $\{x_j\}$ and $\{y_k\}$ of $[a,b]$ and $[c,d]$ with $a=x_0<x_1<\ldots<x_m=b$ and $c=y_0<y_1<\ldots<y_n=d$ determine a "partition," denoted by D, of I into subintervals of the form $I_{jk}=\{(x,y)\,|\,x_{j-1}\leqslant x\leqslant x_j,\ y_{k-1}\leqslant y\leqslant y_k\}$ $(j=1,\ldots,m;\ k=1,\ldots,n)$. Writing

$$M_{jk}=\sup_{(x,y)\in I_{jk}}f(x,y),$$

$$m_{jk}=\inf_{(x,y)\in I_{jk}}f(x,y),$$

we set

$$\bar\sigma(D)=\sum_{j=1}^m\sum_{k=1}^n M_{jk}(x_j-x_{j-1})(y_k-y_{k-1}),$$

$$\underline{\sigma}(D)=\sum_{j=1}^m\sum_{k=1}^n m_{jk}(x_j-x_{j-1})(y_k-y_{k-1}).$$

Then we obtain $\inf_D\bar\sigma(D)\geqslant\sup_D\underline{\sigma}(D)$. If $\inf_D\bar\sigma(D)=\sup_D\underline{\sigma}(D)$, then $f(x,y)$ is called integrable on I, and the common value is called the **double integral** of f on I and is denoted by $\iint_I f(x,y)\,dx\,dy$. Analogously, we may define **n-tuple integrals** (or **multiple integrals**) and the integrability of functions of n variables.

Let K be a bounded set in the xy-plane and I an interval containing K. Let $\varphi(x,y)$ be the characteristic function of K defined on I, that is, φ is determined by

$$\varphi(x,y)=1\quad\text{for}\quad(x,y)\in K,$$
$$\varphi(x,y)=0\quad\text{for}\quad(x,y)\in I-K.$$

Replacing $f(x,y)$ by this $\varphi(x,y)$, we consider $\inf_D\bar\sigma(D)$ ($\sup_D\underline{\sigma}(D)$). These values can be shown to be independent of the choice of such an interval I and are called the **outer area (inner area)** of K, respectively. When these two values coincide, K is said to be **of definite area**, and the common value is called the **area of** K. A necessary and sufficient condition for K to be of definite area is that the outer area of the †boundary of K should be zero. Now consider a bounded function defined on a set K of definite area. Then, taking an interval I containing K, define an extension $\varphi(x,y)$ of $f(x,y)$ as follows:

$$\varphi(x,y)=f(x,y)\quad\text{for}\quad(x,y)\in K,$$
$$\varphi(x,y)=0\quad\text{for}\quad(x,y)\in I-K.$$

If $\varphi(x,y)$ is integrable on I, then $f(x,y)$ is called **integrable** on K, and the integral of f on K is defined by $\iint_K f(x,y)\,dx\,dy=\iint_I\varphi(x,y)\,dx\,dy$, which is independent of the special choice of I. The set K is called the **domain of integration**. Since K is of definite area, the set of boundary points of K at which $\varphi(x,y)$ is not continuous can be contained in a union of intervals whose total area may be made smaller than any preassigned positive number. Consequently, a function bounded on K and continuous at each †interior point of K is integrable on K. Like integrals of functions of a single variable, multiple integrals satisfy the three basic properties of integrals (→ Section B).

G. Multiple Integrals and Iterated Integrals

Suppose that we are given a function $f(x,y)$ that is continuous on an interval $I=\{(x,y)\,|\,a\leqslant x\leqslant b,\ c\leqslant y\leqslant d\}$. Then, for a fixed y in $[c,d]$, the function $f(x,y)$, regarded as a function of x, may be integrated with respect to x on the interval $[a,b]$, and the integral thus obtained is a continuous function of y. The integral of the function defined on $[c,d]$, namely, $\int_c^d(\int_a^b f(x,y)\,dx)\,dy$, is called the **iterated integral** (or **repeated integral**) of

$f(x,y)$ and is often written as $\int_c^d dy \int_a^b f(x,y)\,dx$. The following formula gives a representation of a double integral by iterated ones:

$$\iint_I f(x,y)\,dx\,dy = \int_c^d dy \int_a^b f(x,y)\,dx$$
$$= \int_a^b dx \int_c^d f(x,y)\,dy.$$

More generally, $\varphi_1(x)$ and $\varphi_2(x)$ being continuous on $[a,b]$ and $\varphi_1(x) \leqslant \varphi_2(x)$, consider the following subset $K=\{(x,y)|a \leqslant x \leqslant b, \varphi_1(x) \leqslant y \leqslant \varphi_2(x)\}$ of the xy-plane. Suppose further that $f(x,y)$ is continuous on K. Then the following equality holds:

$$\iint_K f(x,y)\,dx\,dy = \int_a^b dx \int_{\varphi_1(x)}^{\varphi_2(x)} f(x,y)\,dy.$$

In the case of unbounded integrands or unbounded domains of integration, we can still define integrals under suitable restrictions. For instance, assume the following two properties: (1) There exists a sequence $\{K_n\}$ of sets, each of which is of definite area, satisfying $K_1 \subset K_2 \subset \ldots$ and $K = \bigcup_{n=1}^{\infty} K_n$. (2) $f(x,y)$ is bounded and integrable on each K_n $(n=1,2,\ldots)$. If a finite limit $\lim_{n\to\infty} \iint_{K_n} f(x,y)\,dx\,dy$ exists and is independent of the choice of $\{K_n\}$, then $f(x,y)$ is called integrable on K. This limit is called the integral of $f(x,y)$ on K and is denoted by $\iint_K f(x,y)\,dx\,dy$:

$$\lim_{n\to\infty} \iint_{K_n} f(x,y)\,dx\,dy = \iint_K f(x,y)\,dx\,dy.$$

When the integral thus defined exists, we say that the integral is convergent. If a finite limit $\lim_{n\to\infty} \iint_{K_n} |f(x,y)|\,dx\,dy$ exists for some sequence $\{K_n\}$ with property (1), then f is integrable on K. Let $f(x,y)$ be continuous and nonnegative on $K=\{(x,y)|\alpha<x<\beta, \gamma<y<\delta\}$, where $-\infty \leqslant \alpha < \beta \leqslant \infty$, $-\infty \leqslant \gamma < \delta \leqslant \infty$. Furthermore, let $f(x,y)$ be integrable on K, and assume that the improper integral $F(x) = \int_\gamma^\delta f(x,y)\,dy = \lim_{c\downarrow\gamma, d\uparrow\delta} \int_c^d f(x,y)\,dy$ exists and converges uniformly with respect to x as $c\downarrow\gamma$, $d\uparrow\delta$. Then $\int_\alpha^\beta F(x)\,dx$ is well defined, and we have

$$\iint_K f(x,y)\,dx\,dy = \int_\alpha^\beta dx \int_\gamma^\delta f(x,y)\,dy.$$

In particular, if $\alpha=a$, $\gamma=b$, $\beta=\delta=\infty$, then we have

$$\int_a^\infty \int_b^\infty f(x,y)\,dx\,dy = \int_a^\infty dx \int_b^\infty f(x,y)\,dy.$$

H. Interchanging the Order of Differentiation and Integration

If both $f(x,y)$ and $\partial f(x,y)/\partial y$ are continuous on an interval $\{(x,y)|a \leqslant x \leqslant b, y_0-\eta \leqslant y \leqslant y_0+\eta\}$, then we may interchange the order of differentiation and integration as follows:

$$\frac{d}{dy}\int_a^b f(x,y)\,dx = \int_a^b \frac{\partial f(x,y)}{\partial y}\,dx$$

for $y=y_0$.

Assume further that this equality holds for every $b(>a)$, the improper integral

$$\int_a^\infty f(x,y)\,dx = \lim_{b\to\infty} \int_a^b f(x,y)\,dx$$

converges, and the improper integral

$$\int_a^\infty \frac{\partial f(x,y)}{\partial y}\,dx = \lim_{b\to\infty}\int_a^b \frac{\partial f(x,y)}{\partial y}\,dx$$

converges as $b\to\infty$ uniformly for y with $|y-y_0|<\eta$. Then

$$\frac{d}{dy}\int_a^\infty f(x,y)\,dx = \int_a^\infty \frac{\partial f(x,y)}{\partial y}\,dx$$

for $y=y_0$. Several other similar theorems are known. Though the previous theorems are written in terms of two variables, analogous theorems hold for n variables.

I. Change of Variables in Multiple Integrals

Let G be a bounded domain of definite area in an n-dimensional Euclidean space $\mathbf{R}^n(x)$. Assume that a mapping $x\to y(x) = (y_1(x_1,\ldots,x_n),\ldots,y_n(x_1,\ldots,x_n))$ is of class C^1 from an open set containing the closure $\bar{G}$ of G into an n-dimensional Euclidean space $\mathbf{R}^n(y)$. We denote the image of G under this mapping by B. If $f(y_1,\ldots,y_n)$ is continuous on B, then the following formula on change of variables holds:

$$\int\ldots\int_B f(y_1,\ldots,y_n)\,dy_1\ldots dy_n =$$
$$\int\ldots\int_G g(x_1,\ldots,x_n)\left|\frac{D(y_1,\ldots,y_n)}{D(x_1,\ldots,x_n)}\right| dx_1\ldots dx_n,$$

where $g(x_1,\ldots,x_n) = f(y_1(x_1,\ldots,x_n),\ldots, y_n(x_1,\ldots,x_n))$ and $D(y_1,\ldots,y_n)/D(x_1,\ldots,x_n)$ is the †Jacobian determinant of the mapping $y(x)$. This formula is usually utilized in the case where $y_1,\ldots,y_n$ are †functionally independent, though otherwise both sides vanish and the formula still holds. For improper integrals, a similar formula will hold under suitable restrictions, for example, the absolute convergence of the integrals.

(For related topics, → 97 Curvilinear Integrals and Surface Integrals; 243 Lebesgue Integral; and 270 Measure Theory.)

References

[1] T. Takagi, Kaiseki gairon (Japanese; A course of analysis), Iwanami, third edition, 1961.
[2] M. Fujiwara, Sûgaku kaiseki I; Bibun-

sekibun gaku (Japanese; Mathematical analysis pt. 1. Differential and integral calculus), Utida-rôkakuho, I, 1934; II, 1939.
[3] S. Kametani, Syotô kaisekigaku (Japanese; Elementary analysis), Iwanami, I, 1953; II, 1958.
[4] S. Hitotumatu, Kaisekikaku zyosetu (Japanese; Elements of analysis), Syôkabô, I, 1962; II, 1963.
[5] T. M. Apostol, Mathematical analysis, Addison-Wesley, 1957.
[6] N. Bourbaki, Eléments de mathématique, Fonctions d'une variable réelle, Actualités Sci. Ind., 1074b, 1132a, Hermann, second edition, 1958, 1961.
[7] R. C. Buck, Advanced calculus, McGraw-Hill, second edition, 1965.
[8] R. Courant, Differential and integral calculus I, II, Nordemann, 1938.
[9] G. H. Hardy, A course of pure mathematics, Cambridge Univ. Press, seventh edition, 1938.
[10] E. Hille, Analysis I, II, Blaisdell, 1964–1966.
[11] W. Kaplan, Advanced calculus, Addison-Wesley, 1952.
[12] E. Landau, Einführung in die Differentialrechnung und Integralrechnung, Noordhoff, 1934; English translation, Differential and integral calculus, Chelsea, 1965.
[13] J. M. H. Olmsted, Advanced calculus, Appleton-Century-Crofts, 1961.
[14] A. Ostrowski, Vorlesungen über Differential- und Integralrechnung I, II, III, Birkhäuser, second edition, 1960–1961.
[15] M. H. Protter and C. B. Morrey, Modern mathematical analysis, Addison-Wesley, 1964.
[16] W. Rudin, Principles of mathematical analysis, McGraw-Hill, second edition, 1964.
[17] V. I. Smirnov, A course of higher mathematics I. Elementary calculus, II. Advanced calculus, Addison-Wesley, 1964.
[18] A. E. Taylor, Advanced calculus, Ginn, 1955.

219 (XIII.32) Integral Equations

A. General Remarks

Equations including the integrals of unknown functions are called **integral equations**. The most studied ones are the **linear integral equations**, i.e., linear in unknown functions.

Let D be a domain of n-dimensional Euclidean space and $f(x)$ and $K(x,y)$ be functions defined for $x=(x_1,x_2,\ldots,x_n)\in D$, $y=(y_1,y_2,\ldots,y_n)\in D$. **Integral equations of Fredholm type** (or **Fredholm integral equations**) are those of the following forms:

$$\int_D K(x,y)\varphi(y)\,dy=f(x), \tag{1}$$

$$\varphi(x)-\int_D K(x,y)\varphi(y)\,dy=f(x), \tag{2}$$

$$A(x)\varphi(x)-\int_D K(x,y)\varphi(y)\,dy=f(x), \tag{3}$$

where $\varphi(x)$ is an unknown function and $\int_D dy$ means the n-tuple integral $\int\ldots\int_D dy_1\ldots dy_n$. Equations of the forms (1), (2), and (3) are called equations of the **first**, **second**, and **third kind**, respectively. Equations of the second kind have been investigated in great detail. Equations of the third kind, in many cases, can be reduced formally to those of the second kind. The function $K(x,y)$ is called a **kernel** of the integral equation.

Integral equations of Volterra type (or **Volterra integral equations**) are those of the following forms:

$$\int_a^x K(x,y)\varphi(y)\,dy=f(x), \tag{1'}$$

$$\varphi(x)-\int_a^x K(x,y)\varphi(y)\,dy=f(x), \tag{2'}$$

$$A(x)\varphi(x)-\int_a^x K(x,y)\varphi(y)\,dy=f(x), \tag{3'}$$

where $\varphi(x)$ is an unknown function. Equations of the forms (1′), (2′), and (3′) are also called equations of the **first**, **second**, and **third kind**, respectively. Integral equations of Volterra type can be regarded as integral equations of Fredholm type having kernels equal to 0 for $x<y$, but these two types of equations are usually treated separately, since they have considerably different characters.

The kernels in equations (1)–(3) and (1′)–(3′) are frequently written in the form $\lambda K(x,y)$ with a parameter λ, in particular when the equations are related to eigenvalue problems, which will be explained in Section F.

The theory of integral equations was originated in 1823 by N. H. Abel, who investigated the relationship between time and the path of a falling body in the field of gravitation. Let $\varphi(t)$ be a quantity varying with time, which is connected by some law with its value in some time interval of the past or the future. Then the law of variation of $\varphi(t)$ can be described mathematically by an integral equation. The situation is the same even when the variable t is not time but a coordinate of the space. In this way, various problems in physics can be reduced to the solution of integral equations [16].

B. Relation to Differential Equations

Many problems in differential equations can be reduced to problems relating to integral equations. Such reduction often makes the problems easier to handle and clarifies the nature of the solutions. For example, consider the problem of finding a solution of the ordinary second-order linear differential equation $d^2y/dx^2+\lambda y=0$ with the boundary condition $y(0)=y(1)=0$. Let $d^2y/dx^2=u(x)$. If we integrate the equation twice, change the order of integration, and make use of the boundary condition, then we have

$$y=\int_0^x(x-\xi)u(\xi)d\xi-\int_0^1 x(1-\xi)u(\xi)d\xi,$$

from which we see that the given differential equation can be written in the form

$$u=\lambda\int_0^1 x(1-\xi)u(\xi)d\xi-\lambda\int_0^x(x-\xi)u(\xi)d\xi.$$

Decomposing the first integral in the right-hand member into the sum of an integral over $(0,x)$ and one over $(x,1)$, and combining the integral over $(0,x)$ with the second integral in the right-hand member, we obtain a Fredholm integral equation of the first kind as follows:

$$u=\lambda\int_0^1 G(x,\xi)u(\xi)d\xi,$$

$$G=\begin{cases}\xi(1-x), & 0\leqslant\xi\leqslant x,\\ x(1-\xi), & x\leqslant\xi\leqslant 1.\end{cases}$$

Clearly, the solution of this integral equation is equivalent to that of the original differential equation. The function G is called †Green's function of the boundary value condition $y(0)=y(1)=0$. Differential equations of higher orders can be treated analogously (→ 309 Ordinary Differential Equations (Boundary Value Problems); 135 Eigenvalue Problems). †Initial value problems of linear ordinary differential equations can be reduced to the solution of Volterra integral equations in a similar way.

As another example, consider the †Dirichlet problem on a plane, i.e., the problem of finding a function u satisfying the conditions (i) u is †harmonic in the interior of the region D bounded by a closed curve $C\,(\xi=\varphi(s),\ \eta=\psi(s),\ 0\leqslant s\leqslant l)$; (ii) $u(x,y)\to F(s)$ uniformly with respect to (x_0,y_0) as (x,y) approaches (x_0,y_0) from the inside of D, where (x_0,y_0) is an arbitrary point on C and $F(s)$ is a continuous function given on C. Put $f(s)=F(s)/\pi$ and

$$K(s;t)=\frac{(\psi(s)-\psi(t))\varphi'(t)-(\varphi(s)-\varphi(t))\psi'(t)}{\pi\big((\varphi(s)-\varphi(t))^2+(\psi(s)-\psi(t))^2\big)}.$$

Then it is known that a solution u of the problem can be given in the form

$$u(x,y)=\int_0^l\mu(s)\frac{\partial}{\partial n}\log\frac{1}{r}\,ds,$$

where $r^2=(\varphi(s)-x)^2+(\psi(s)-y)^2$, n is the inner normal of C, and $\mu(s)$ is a continuous solution of the following Fredholm integral equation of the second kind:

$$\mu(s)=f(s)-\int_0^l K(s;t)\,\mu(t)\,dt.$$

We can treat the †Neumann problem similarly, i.e., the problem in which condition (ii) is replaced by (ii′) $(\partial u/\partial n)(x,y)\to F(s)$ uniformly with respect to (x_0,y_0) as (x,y) approaches (x_0,y_0) from the inside of D. In the Neumann problem, put $f(s)=F(s)/\pi$ and

$$L(s;t)=\frac{(\psi(s)-\psi(t))\varphi'(s)-(\varphi(s)-\varphi(t))\psi'(s)}{\pi\big((\varphi(s)-\varphi(t))^2+(\psi(s)-\psi(t))^2\big)}.$$

Then we have a solution u in the form

$$u(x,y)=-\int_0^l\mu(s)\log\frac{1}{r}\,ds+c,$$

where $\mu(s)$ is a solution of the following Fredholm integral equation of the second kind:

$$\mu(s)=f(s)-\int_0^l L(s;t)\,\mu(t)\,dt.$$

A solution of this integral equation, however, exists when and only when $\int_0^l F(s)\,ds=0$. In the expression of a solution $u(x,y)$ of the Neumann problem, c is an arbitrary additive constant, up to which a solution of the problem is determined uniquely [2, ch. 5]. We can also treat partial differential equations of †elliptic type in an analogous way.

C. Integral Equations with Continuous Kernel

We describe some results for integral equations with m-dimensional independent variables, i.e., equations in which D is an m-dimensional closed domain. We assume that $K(x,y)$ and $f(x)$ are continuous in Sections D–H.

D. The Method of Successive Iteration

Among methods of solving Fredholm integral equations of the second kind, the simplest is the **method of successive iteration**, sometimes called the **method of successive approximation**. In the method of successive iteration, we rewrite (2) in the form

$$\varphi(x)=f(x)+\int_D K(x,y)\varphi(y)\,dy,$$

and replace the function $\varphi(y)$ in the right-hand member by the function

$$f(y)+\int_D K(y,z)\varphi(z)\,dz.$$

If we repeat the process successively, then we have

$$\varphi(x)=f(x)+\sum_{i=1}^{n}\int_D K_i(x,y)f(y)\,dy$$
$$+\int_D K_{n+1}(x,y)\varphi(y)\,dy,$$

where

$$K_1(x,y)=K(x,y),$$

$$K_i(x,y)=\int_D K_{i-1}(x,s)K(s,y)\,ds.$$

The functions $K_i(x,y)$ are called the **iterated kernels**. Assume that $\sum_{n=1}^{\infty}K_n(x,y)$ converges uniformly. Then, putting

$$R(x,y)=\sum_{n=1}^{\infty}K_n(x,y), \tag{4}$$

we obtain a solution of (2) in the form

$$\varphi(x)=f(x)+\int_D R(x,y)f(y)\,dy. \tag{5}$$

The series (4) is called a **Neumann series**.

For a given kernel $K(x,y)$, a function $R(x,y)$ satisfying

$$K(x,y)-R(x,y)+\int_D K(x,s)R(s,y)\,ds=0$$

and

$$K(x,y)-R(x,y)+\int_D R(x,s)K(s,y)\,ds=0$$

is called a **resolvent** of $K(x,y)$ (in some cases $-R(x,y)$ instead is called a resolvent). If a resolvent of $K(x,y)$ exists, the solution of (2) can be given uniquely by (5). If a Neumann series converges uniformly, (4) gives a resolvent of $K(x,y)$.

If we apply a similar process to Volterra integral equations of the second kind, then we have the iterated kernels defined by

$$K_{i+1}(x,y)=\int_y^x K_i(x,s)K(s,y)\,ds \qquad (i=1,2,\ldots).$$

For these iterated kernels, a Neumann series defined by (4) always converges uniformly.

E. Fredholm's Method

Let D be a bounded closed domain and $K(x,y)$ a continuous kernel. A Neumann series (4) converges uniformly and gives a resolvent if $|K(x,y)|$ or the region D is sufficiently small, but otherwise it does not necessarily converge. E. I. Fredholm gave a method of constructing a resolvent for a more general case. Write a kernel in the form $\lambda K(x,y)$, and put

$$K\begin{pmatrix}x_1,\ldots,x_n\\ y_1,\ldots,y_n\end{pmatrix}=\begin{vmatrix}K(x_1,y_1) & \ldots & K(x_1,y_n)\\ & \ldots & \\ K(x_n,y_1) & \ldots & K(x_n,y_n)\end{vmatrix}.$$

Define $D(\lambda)$ and $D(x,y;\lambda)$ by

$$D(\lambda)=1+\sum_{n=1}^{\infty}\frac{(-\lambda)^n}{n!}\int_D\cdots\int_D K\begin{pmatrix}s_1,\ldots,s_n\\ s_1,\ldots,s_n\end{pmatrix}ds_1\ldots ds_n$$

and

$$D(x,y;\lambda)=K(x,y)+\sum_{n=1}^{\infty}\frac{(-\lambda)^n}{n!}\int_D\cdots\int_D K\begin{pmatrix}x,s_1,\ldots,s_n\\ y,s_1,\ldots,s_n\end{pmatrix}ds_1\ldots ds_n.$$

The series in these two equations both converge uniformly and hence define †entire functions of λ. The functions $D(\lambda)$ and $D(x,y;\lambda)$ are called **Fredholm's determinant** and **Fredholm's first minor** of the kernel $K(x,y)$, respectively. For small $|\lambda|$, we have

$$\frac{D'(\lambda)}{D(\lambda)}=\sum_{n=1}^{\infty}\lambda^{n-1}\int_D K_n(s,s)\,ds,$$

where the $K_n(x,y)$ are iterated kernels corresponding to $K(x,y)$. Now if $D(\lambda)\neq 0$, a resolvent $\lambda R(x,y;\lambda)$ of the kernel $\lambda K(x,y)$ can be given by

$$\frac{D(x,y;\lambda)}{D(\lambda)}=R(x,y;\lambda).$$

If $D(\lambda)=0$, some extension of the method in this section is needed. Fredholm [2, ch. 4] introduced for this purpose **Fredholm's rth minor**

$$D\begin{pmatrix}x_1,\ldots,x_r\\ & ;\lambda\\ y_1,\ldots,y_r\end{pmatrix}$$

defined by

$$D\begin{pmatrix}x_1,\ldots,x_r\\ & ;\lambda\\ y_1,\ldots,y_r\end{pmatrix}=K\begin{pmatrix}x_1,\ldots,x_r\\ y_1,\ldots,y_r\end{pmatrix}+\sum_{n=1}^{\infty}\frac{(-\lambda)^n}{n!}$$
$$\times\int_D\cdots\int_D K\begin{pmatrix}x_1,\ldots,x_r,s_1,\ldots,s_n\\ y_1,\ldots,y_r,s_1,\ldots,s_n\end{pmatrix}ds_1\ldots ds_n.$$

F. Eigenvalue Problems and Fredholm's Alternative Theorem

Consider a **homogeneous integral equation** of the second kind

$$\varphi(x)-\lambda\int_D K(x,y)\varphi(y)\,dy=0, \tag{6}$$

where D is a bounded closed domain and

$K(x,y)$ is continuous in $D \times D$. When (6) has a nontrivial solution $\varphi(x)$ for some λ, then λ is called an **eigenvalue** corresponding to the kernel $K(x,y)$, and the corresponding nontrivial solution $\varphi(x)$ is called an **eigenfunction** corresponding to the kernel $K(x,y)$. If $D(\lambda) \neq 0$, then (6) has no nontrivial solution, from which it follows that eigenvalues must be zero points of the entire function $D(\lambda)$. For an arbitrary eigenvalue λ, there is a set of linearly independent eigenfunctions corresponding to λ such that any eigenfunction corresponding to λ can be written as a linear combination of the eigenfunctions belonging to the set under consideration. Such a set of linearly independent eigenfunctions corresponding to an eigenvalue λ is called a **fundamental system** corresponding to the eigenvalue λ. The number of elements of the fundamental system is called the **index** of the eigenvalue λ. The index of an eigenvalue is always finite. The homogeneous integral equation of the form

$$\psi(y)-\lambda\int_D K(x,y)\psi(x)\,dx=0 \qquad (6')$$

is called an **associated** (or **transposed**) **integral equation** of (6). The associated equation has the same eigenvalues as the original equation; moreover, the index of a common eigenvalue is the same for both equations. For any eigenvalue λ, the order of the zero point λ of the entire function $D(\lambda)$ is called the **multiplicity** of the eigenvalue λ. If an eigenvalue λ is a pole of $R(x,y;\lambda)$, then we have $p+1 \geqslant r+q$, where r is the order of the pole, p is the multiplicity of λ, and q is the index of λ. In particular, if λ is a simple pole of $R(x,y,\lambda)$, we have $p=q$, namely, the multiplicity is equal to the index. An example with this particular property is the integral equation with a symmetric kernel to be discussed in Section G [6]. For the set of eigenvalues there is no finite accumulation point even if there are infinitely many eigenvalues.

If λ is not an eigenvalue, the inhomogeneous equation

$$\varphi(x)-\lambda\int_D K(x,y)\varphi(y)\,dy=f(x) \qquad (7)$$

can be solved uniquely for any continuous function $f(x)$. In this case we have $D(\lambda)\neq 0$, and the resolvent $R(x,y;\lambda)$ of the kernel $\lambda K(x,y)$ exists. If λ is an eigenvalue, we have $D(\lambda)=0$, and equation (7) has a solution if and only if

$$\int_D \psi(x)f(x)\,dx=0$$

for all solutions $\psi(y)$ of (6′) (linearly independent solutions $\psi(y)$ are finite in number). The last statement is called **Fredholm's alternative theorem** (→ 72 Compact Operators D).

A kernel of the type

$$K(x,y)=\sum_{j=1}^{n} X_j(x)Y_j(y)$$

is called a **separated kernel**, **degenerate kernel**, or **Pincherle-Goursat kernel**. For such a kernel, we have $D(\lambda)=\det(\delta_{jk}-\lambda\int_a^b X_j(t)Y_k(t)\,dt)$, and hence we can easily obtain eigenvalues and eigenfunctions. A nondegenerate kernel can be studied using the results obtained for separated kernels, since we can regard a kernel of the general form as the limit of a sequence of separated kernels.

G. Symmetric Kernels

A kernel $K(x,y)$ is called a **symmetric kernel** if it is real and $K(x,y)=K(y,x)$. Let D be a bounded closed domain and $K(x,y)$ be a continuous symmetric kernel. In this case the associated equation (6′) clearly coincides with the original equation (6).

Corresponding to any nontrivial symmetric kernel $K(x,y)$, there exist at least one eigenvalue and one eigenfunction. The eigenvalues are all real, and the eigenfunctions corresponding to distinct eigenvalues are mutually orthogonal. If we orthonormalize the eigenfunctions belonging to all fundamental systems and number them according to the order of increasing absolute values of the corresponding eigenvalues, then we have an orthonormal system $\{\varphi_i(x)\}$, called a **complete orthonormal system of fundamental functions** or simply a **complete orthogonal** (or **orthonormal**) **system**. If we number the eigenvalues taking their multiplicities into account and according to the order of increasing absolute values, then we have the equality

$$\sum_{i=1}^{\infty}\frac{1}{\lambda_i^2}=\int_D\int_D K^2(x,y)\,dx\,dy.$$

Corresponding to an iterated kernel $K_m(x,y)$, we have the eigenvalues $\{\lambda_i^m\}$, and we can choose the corresponding orthonormal system so that it coincides with the one corresponding to $K(x,y)$. Eigenvalues and eigenfunctions corresponding to an iterated kernel can be obtained in the following way: Put $\int_D K_n(s,s)\,ds=u_n$; then the following limit exists:

$$\lim_{n\to\infty} u_{2n}/u_{2n+2}=\lambda^2<+\infty,$$

which gives an eigenvalue of the iterated kernel $K_2(x,y)$. A function $\varphi(x,y)$ defined by

$$\lim_{n\to\infty}\lambda^{2n}K_{2n}(x,y)=\varphi(x,y)$$

(uniformly convergent) gives the corresponding eigenfunction $\varphi(x,c)$ for any constant c satisfying $\varphi(x,c)\not\equiv 0$.

Let λ^n be an eigenvalue corresponding to

an iterated kernel $K_n(x,y)$ and $\varphi(x)$ be a corresponding eigenfunction. Consider the functions $\psi_j(x)$ $(j=0,1,\ldots,n-1)$ defined by

$$\psi_j(x)=\varphi(x)+\sum_{k=1}^{n-1}\varepsilon^{kj}\lambda^k\int_D K_k(x,y)\varphi(y)\,dy \qquad (j=0,1,2,\ldots,n-1),$$

where ε is one of the nth primitive roots of 1. Then for at least one value of j, $\varepsilon^j\lambda$ is an eigenvalue corresponding to the kernel $K(x,y)$, and ψ_j is a corresponding eigenfunction. This relationship of eigenvalues and eigenfunctions corresponding to an iterated kernel with those corresponding to the original kernel is valid even for kernels that are not necessarily symmetric [1].

Let a kernel $K^{(n)}(x,y)$ be defined by

$$K^{(n)}(x,y)=K(x,y)-\sum_{i=1}^{n}\frac{\varphi_i(x)\varphi_i(y)}{\lambda_i},$$

where the λ_i $(i=1,2,\ldots,n)$ are the eigenvalues corresponding to the kernel $K(x,y)$ and the $\varphi_i(x)$ $(i=1,\ldots,n)$ are the corresponding orthonormalized eigenfunctions. Then eigenvalues and eigenfunctions corresponding to $K^{(n)}(x,y)$ are those corresponding to $K(x,y)$, with $\lambda_1,\ldots,\lambda_n$ and $\varphi_1(x),\ldots,\varphi_n(x)$ excluded.

Let $\varphi(x)$ be any function that satisfies $\int_D(\varphi(x))^2dx=1$. Then the integral

$$J=\int_D\int_D K_2(x,y)\varphi(x)\varphi(y)\,dx\,dy$$

assumes the maximum value when $\varphi(x)$ is an eigenfunction corresponding to $K_2(x,y)$ with the smallest eigenvalue λ_1^2. Let the eigenvalues λ_n of $K(x,y)$ be numbered in order of increasing absolute values, so that $|\lambda_n|\leqslant|\lambda_{n+1}|$. Let $\varphi(x)$ be any function satisfying

$$\int_D\psi_i(x)\varphi(x)\,dx=0 \quad (i=1,2,\ldots,n),$$

$$\int_D(\varphi(x))^2dx=1$$

for given functions $\psi_i(x)$ $(i=1,\ldots,n)$. Then the maximum value of the integral

$$J=\int_D\int_D K_2(x,y)\varphi(x)\varphi(y)\,dx\,dy$$

is least when the set of all linear combinations of $\{\psi_1(x),\ldots,\psi_n(x)\}$ coincides with the set of all linear combinations of $\{\varphi_1(x),\ldots,\varphi_n(x)\}$, and the minimum is attained by some eigenfunction $\varphi(x)$ corresponding to $K_2(x,y)$ with the eigenvalue λ_{n+1}^2. The results in this paragraph show that we can obtain eigenvalues by solving a variational problem concerned with the integral J.

H. Expansion Theorems

Let $K(x,y)$ be a continuous symmetric kernel and $h(x)$ be a function square integrable on a bounded closed domain D. Then a function $f(x)$ such that

$$f(x)=\int_D K(x,y)h(y)\,dy$$

can be expanded in the form

$$f(x)=\sum_{n=1}^{\infty}c_n\varphi_n(x),$$

where $\{\varphi_i(x)\}$ is a complete orthonormal system of fundamental functions corresponding to $K(x,y)$ and

$$c_n=\int_D f(x)\varphi_n(x)\,dx \quad (n=1,2,\ldots).$$

The series in the expansion of $f(x)$ converges uniformly. These facts are the content of the **Hilbert-Schmidt expansion theorem**. By using this theorem, for a λ that is not an eigenvalue, we can obtain a solution $\varphi(x)$ of the Fredholm integral equation (7) with a symmetric kernel in the form:

$$\varphi(x)=f(x)+\lambda\sum_{i=1}^{\infty}\frac{\varphi_i(x)}{\lambda_i-\lambda}\int_D f(x)\varphi_i(x)\,dx.$$

For $m\geqslant2$, iterated kernels can be expanded in the form

$$K_m(x,y)=\sum_{i=1}^{\infty}\frac{\varphi_i(x)\varphi_i(y)}{\lambda_i^m}$$

(uniformly convergent).

If $\lambda R(x,y;\lambda)$ is a resolvent of a symmetric kernel $\lambda K(x,y)$, then $R(x,y;\lambda)$ can be expanded as follows:

$$R(x,y;\lambda)=K(x,y)+\lambda\sum_{i=1}^{\infty}\frac{\varphi_i(x)\varphi_i(y)}{\lambda_i(\lambda_i-\lambda)}.$$

If a symmetric kernel $K(x,y)$ satisfies the inequality

$$\int_D\int_D K(x,y)\varphi(x)\varphi(y)\,dx\,dy\geqslant0$$

for all $\varphi(x)$, it is called a **positive (semidefinite) kernel**. If in this inequality the equality holds only for $\varphi(x)\equiv0$, $K(x,y)$ is called a **positive definite kernel**. For a positive definite kernel, eigenvalues are all positive, and the kernel can be expanded in the form

$$K(x,y)=\sum_{i=1}^{\infty}\frac{\varphi_i(x)\varphi_i(y)}{\lambda_i}$$

(uniformly convergent).

This result is called **Mercer's theorem**.

When a real continuous kernel $K(x,y)$ is not symmetric, we consider two positive kernels $\hat{K}'(x,y)$ and $\hat{K}''(x,y)$ defined by

$$\int_D K(x,s)K(y,s)\,ds=\hat{K}'(x,y)$$

and

$$\int_D K(s,x)K(s,y)\,ds=\hat{K}''(x,y).$$

The eigenvalues corresponding to these kernels are the same, and they are all positive. Let λ_i^2 $(i=1,2,\ldots)$ be these eigenvalues and $\{\varphi_i(x)\}$ and $\{\psi_i(x)\}$ be the corresponding complete orthonormal systems corresponding to $\hat{K}'$ and $\hat{K}''$, respectively. Then we have

$$\lambda_i \int_D K(y,x)\varphi_i(y)\,dy = \psi_i(x),$$

$$\lambda_i \int_D K(x,y)\psi_i(y)\,dy = \varphi_i(x).$$

Let $f(x)$ be an arbitrary function such that

$$f(x) = \int_D K(x,y)h(y)\,dy,$$

where $h(x)$ is a function square integrable on D. The function $f(x)$ can be then expanded in the form

$$f(x) = \sum_{i=1}^{\infty} c_i\varphi_i(x), \tag{8}$$

where

$$c_i = \int_D f(x)\varphi_i(x)\,dx \quad (i=1,2,\ldots).$$

The series in the expansion (8) converges uniformly.

The Fredholm integral equation (1) of the first kind with a general kernel (i.e., a kernel that is not necessarily symmetric) has a square integrable solution $\varphi(x)$ if and only if $f(x)$ has a uniformly convergent expansion (8) and $\sum_{i=1}^{\infty}(c_i\lambda_i)^2 < +\infty$. When this condition is satisfied, equation (1) has a solution given by $\sum_{n=1}^{\infty} c_n\lambda_n\varphi_n(x)$ that converges in the sense of †mean convergence [6].

It should be noted that the theory concerning symmetric kernels can be extended to complex-valued **Hermitian kernels**, i.e., kernels such that $K(x,y) = \overline{K(y,x)}$. Also, we can establish the theory in Section G and this section, concerning Fredholm integral equations with continuous kernels, by methods of functional analysis that treat $\int_D K(x,y)\varphi(y)\,dy$ as a †compact operator in the space of continuous functions (→72 Compact Operators).

I. Kernels of Hilbert-Schmidt Type

Kernels of Hilbert-Schmidt type are kernels which are square integrable in the sense of Lebesgue over $D \times D$, where D is an arbitrary domain. Most of the results mentioned in the previous section concerning integral equations with kernels continuous on bounded domains are valid also for equations with kernels of Hilbert-Schmidt type, because every operator mentioned in the previous section is also a compact operator in the space concerned, i.e., the space L_2 [4] (→ 72 Compact Operators).

J. Singular Kernels

For general kernels that are not necessarily continuous, the theory in the previous sections does not apply properly, but when an iterated kernel $K_m(x,y)$ has a resolvent $R_m(x,y)$, we can find a resolvent $R(x,y)$ of $K(x,y)$ in the following form:

$$R(x,y) = R_m(x,y) + H_m(x,y) + \int_D R_m(x,s)H_m(s,y)\,ds,$$

where $H_m(x,y) = \sum_{i=1}^{m-1} K^i(x,y)$. When a kernel is of the form $\lambda K(x,y)$, the relationship between eigenvalues and eigenfunctions corresponding to an iterated kernel and those corresponding to the original kernel, which was stated in Sections G and H, is still valid for general kernels. If a kernel $K(x,y)$ is continuous for $x \neq y$ and has a singularity of the form $|x-y|^{-\alpha}$ $(0<\alpha<1)$ on $x=y$, the iterated kernels $K_m(x,y)$ are continuous provided that $(1-\alpha)m \geqslant 1$. †Green's functions of partial differential equations of elliptic type have this property.

A kernel that is not square integrable is called a **singular kernel**. An integral equation whose domain of definition is unbounded or whose kernel is singular is called a **singular integral equation**. Singular integral equations have some particular properties that are not seen in ordinary integral equations, i.e., integral equations with kernels continuous in a bounded closed domain. For example, consider the identity:

$$\sqrt{\frac{2}{\pi}} \int_0^{\infty} \sin xy \left(\sqrt{\frac{\pi}{2}}\, e^{-\alpha y} + \frac{y}{\alpha^2 + y^2} \right) dy$$

$$= \sqrt{\frac{\pi}{2}}\, e^{-\alpha x} + \frac{x}{\alpha^2 + x^2},$$

where α is an arbitrary real number. This equality shows that for the continuous kernel $\sqrt{2/\pi}\,\sin xy$, unity is an eigenvalue and $\sqrt{\pi/2}\,e^{-\alpha x} + x/(\alpha^2 + x^2)$ is a corresponding eigenfunction. Since α is arbitrary, the index of the eigenvalue 1 is evidently infinity. As another example, observe the equality

$$\int_{-\infty}^{\infty} e^{-|x-y|} e^{-i\alpha y}\,dy = \frac{2}{1+\alpha^2} e^{-i\alpha x},$$

where α is an arbitrary real number. From this equality, we see that for the continuous kernel $e^{-|x-y|}$ defined on $(-\infty,\infty)$, any number $\lambda = (1+\alpha^2)/2$ greater than or equal to $1/2$ is an eigenvalue. In this example, the **spectrum**, i.e., the set of eigenvalues, is a continuum. Such a spectrum is called a **continuous spectrum**.

In applications, an important role is played

by integral equations with **kernels of Carleman type**:

$K(x,y)=G(x,y)/(y-x)$,

where $G(x,y)$ is a bounded function. In integral equations with such kernels, the integral means the †Cauchy principal value. Such integral equations are studied by the use of †Hilbert transformations whether or not domains of definition are bounded [8]. (For singular integral operators → 251 Linear Operators K; also → 135 Eigenvalue Problems.)

K. Systems of Integral Equations

A system of Fredholm integral equations of the second kind can always be reduced to a single equation. In fact, as is seen easily, a system of integral equations

$$\varphi_i(x)-\lambda\sum_j\int_0^1 K_{ij}(x,y)\varphi_j(y)\,dy=f_i(x) \qquad (i=1,2,\ldots,n)$$

can be reduced to a single equation

$$\Phi(x)-\lambda\int_0^n K(x,y)\Phi(y)\,dy=F(x) \qquad (0\leqslant x\leqslant n),$$

where

$$\Phi(x)=\varphi_i(x-i+1),\quad F(x)=f_i(x-i+1),$$
$$K(x,y)=K_{ij}(x-i+1,y-j+1)$$
$$(i-1\leqslant x,j-1\leqslant y\leqslant j;\ i,j=1,2,\ldots,n).$$

A system of Volterra integral equations of the second kind can be reduced to a single equation by eliminating the unknown functions successively.

L. Integral Equations of Volterra Type

Consider a Volterra integral equation of the first kind

$$\int_a^x K(x,y)\varphi(y)\,dy=f(x)$$

such that $K(x,x)\neq 0$ and $K_x(x,y)$ and $f'(x)$ are continuous. If we differentiate both sides of the equation, then we have a Volterra integral equation of the second kind:

$$\varphi(x)+\int_a^x\frac{K_x(x,y)}{K(x,x)}\varphi(y)\,dy=\frac{f'(x)}{K(x,x)}.$$

Abel's integral equation of general form is

$$\int_0^x\frac{G(x,y)}{(x-y)^\alpha}\varphi(y)\,dy=f(x)\quad(0<\alpha<1).\quad(9)$$

If G, G_x, and f' are continuous and $G(x,x)\neq 0$, equation (9) can be reduced to the equation

$$\int_0^u H(u,y)\varphi(y)\,dy=\int_0^u f(x)(u-x)^{\alpha-1}\,dx,$$

where

$$H(u,y)=\int_y^u\frac{G(x,y)\,dx}{(u-x)^{1-\alpha}(x-y)^\alpha}.$$

Since $H(u,u)=(\pi/\sin\alpha\pi)G(u,u)\neq 0$, it follows that

$$\varphi(u)+\int_0^u\frac{H_u(u,y)}{H(u,u)}\varphi(y)\,dy=g(u),\qquad(9a)$$

where

$$g(u)=H(u,u)^{-1}\frac{d}{du}\int_0^u f(x)(u-x)^{\alpha-1}\,dx$$
$$=H(u,u)^{-1}\left(u^{\alpha-1}f(0)+\int_0^u(u-x)^{\alpha-1}f'(x)\,dx\right).$$

Clearly (9a) is a Volterra integral equation of the second kind. **Abel's problem** (→ Section A) was to find the path of a falling body for a given time of descent. The problem then can be reduced to the solution of equation (9) with $G(x,y)\equiv 1$ and $\alpha=1/2$. When $G(x,y)\equiv 1$, we can solve equation (9) explicitly to get

$$\varphi(x)=\frac{\sin\alpha\pi}{\pi}\left(x^{\alpha-1}f(0)+\int_0^x(x-t)^{\alpha-1}f'(t)\,dt\right).$$

M. Nonlinear Integral Equations

When a **nonlinear integral equation** includes a parameter λ, it may happen that the parameter has a **bifurcation point**, i.e., a value λ_0 such that the number of real solutions is changed when λ varies through λ_0 taking real values. For example, consider the equation

$$\varphi(x)-\lambda\int_0^1\varphi^2(y)\,dy=1.$$

This equation has real solutions $\varphi(x)=(1\pm\sqrt{1-4\lambda})/2\lambda$ for $\lambda\leqslant 1/4$ but no real solution for $\lambda>1/4$. Hence $\lambda_0=1/4$ is a bifurcation point.

Among nonlinear integral equations, **Hammerstein's integral equation** has been studied in detail [8, 9]. It is an equation of the form

$$\varphi(x)+\int_D K(x,y)f(y,\varphi(y))\,dy=0.\qquad(10)$$

If $K(x,y)$ and $f(y,0)$ are square integrable and $f(y,u)$ satisfies a †Lipschitz condition in u with a sufficiently small coefficient, then the integral equation (10) can be solved by successive approximations. If $K(x,y)$ is a square integrable positive kernel and $\int_D|K(x,y)|^2dy$ is bounded, then we can prove the existence and uniqueness of a solution of (10) under a condition on $f(y,u)$ weaker than a Lipschitz condition. We can prove similar results for equation (10) with a nonsymmetric kernel when $K(x,y)$ is **continuous in the mean**, that is,

$$\lim_{x'\to x}\int_D |K(x',y)-K(x,y)|^2 dy=0,$$

$$\lim_{y'\to y}\int_D |K(x,y')-K(x,y)|^2 dx=0.$$

A nonlinear Volterra integral equation of the form

$$\varphi(x)=f(x)+\int_a^x F(x,y;\varphi(y))\,dy$$

can be solved by successive approximations if $F(x,y;u)$ and $f(x)$ are square integrable, $F(x,y;u)$ satisfies a Lipschitz condition $|F(x,y;u')-F(x,y;u'')|\leqslant k(x,y)|u'-u''|$ with some square integrable function $k(x,y)$, and $\int_a^x F(x,y;f(y))\,dy$ is majorized by some square integrable function of x. When $F(x,y;u)$ and $f(x)$ are continuous, we can obtain theorems on the existence and uniqueness of continuous solutions and †comparison theorems similar to those for initial value problems in ordinary differential equations [12].

N. Numerical Solution

In the **numerical solution of integral equations**, we assume thoroughout that the functions appearing are all continuous and the solution of every equation is unique. Methods of numerical solution can be divided roughly into two classes. Methods of one class apply numerically the analytical solution described in the preceding sections, and methods of the other obtain a solution by transforming the problem to one that is numerically solvable.

(1) A method based on numerical quadrature. Consider the integral equation

$$\int_a^b F(x,y,\varphi(x),\varphi(y))\,dy=0.$$

Let $a=x_1<x_2<\ldots<x_n=b$ be points on the interval $[a,b]$ and $\varphi_1,\varphi_2,\ldots,\varphi_n$ be the values of $\varphi(x)$ at $x_1,x_2,\ldots,x_n$. By the use of numerical quadrature, we then have the following system of equations in φ_i:

$$\sum_{j=1}^n a_j F(x_i,x_j,\varphi_i,\varphi_j)=0 \quad (i=1,2,\ldots,n).$$

The method corresponds to solving ordinary differential equations by their difference equation analogs. Hence the errors of solutions obtained by this method can be analyzed similarly to those in the numerical solution of ordinary differential equations (→ 298 Numerical Solution of Ordinary Differential Equations). If the given integral equation is a Fredholm equation of the second kind, then we have a system of linear equations in φ_i. If we apply quadrature formulas to the integral appearing in the integral equation using the values of φ_i obtained, then we have a formula by which the solution can be evaluated directly, that is, without using an interpolation formula.

(2) A method utilizing recurrence formulas. Let d_n and $d_n(x,y)$ be the respective coefficients of λ^n in the expansions of Fredholm's determinant $D(\lambda)$ and Fredholm's first minor $D(x,y;\lambda)$. They satisfy the recurrence formulas:

$$d_n(x,y)=d_nK(x,y)+\int_a^b K(x,s)\,d_{n-1}(s,y)\,ds,$$

$$d_{n+1}=-\frac{1}{n+1}\int_a^b d_n(s,s)\,ds,$$

$$d_0=1,\quad d_0(x,y)=K(x,y).$$

By the use of these formulas, we can compute d_n and $d_n(x,y)$ successively and hence evaluate $D(\lambda)$ and $D(x,y;\lambda)$ approximately, and by means of these recurrence formulas we can readily obtain a solution of a Fredholm equation of the second kind.

(3) A method utilizing approximate kernels. If we replace a kernel by an approximate one in a Fredholm integral equation of the second kind, then we have an integral equation that has a solution approximately equal to the solution of the original equation. Hence if we can find an approximate kernel for which an integral equation can be solved numerically or analytically, then we can find an approximation to the desired solution by solving the modified equation. For such solutions, a method of error estimation was given by F. G. Tricomi [17].

(4) An iterative method. Consider the integral equation

$$\varphi(x)=\int_a^b F(x,y,\varphi(x),\varphi(y))\,dy.$$

Let $\varphi_0(x)$ be an adequate function, and define $\varphi_n(x)$ successively by

$$\varphi_{n+1}(x)=\int_a^b F(x,y,\varphi_n(x),\varphi_n(y))\,dy.$$

If the sequence $\{\varphi_n(x)\}$ converges, then the limit $\lim_{n\to\infty}\varphi_n(x)=\varphi(x)$ is a solution of the given equation, and hence we can obtain an approximation to a solution by calculating $\varphi_n(x)$ for some finite n. This method can be used effectively for Fredholm integral equations of the second kind with a parameter λ, provided that the absolute value of λ is smaller than the least absolute value of the eigenvalues.

(5) Variational method. If some conditions are fulfilled, an integral equation of the form

$$G(x,\varphi(x))+\int_a^b F(x,y,\varphi(x),\varphi(y))\,dy=0$$

can be regarded as an †Euler-Lagrange equation for a variational problem

$$J[u]=\int_a^b\int_a^b E(x,y,u(x),u(y))\,dx\,dy + \int_a^b H(x,u(x))\,dx=\text{extremal.} \quad (11)$$

In this case, we can find a solution of the given integral equation numerically by solving the variational problem (11) numerically [18].

(6) Enskog's method. Suppose that $\{\varphi_n(x)\}$ is a complete orthonormal system for the Fredholm integral equation (7). If we put $\psi_n(x)=\varphi_n(x)-\lambda\int_a^b K(y,x)\varphi_n(y)\,dy$, then from (7) we have

$$\int_a^b \varphi(x)\psi_n(x)\,dx=\int_a^b f(x)\varphi_n(x)\,dx, \quad (12)$$

and furthermore we see that $\{\psi_n(x)\}$ can be orthonormalized to yield a complete orthonormal system $\{\chi_n(x)\}$. The equality (12) then shows that the Fourier coefficients of a solution $\varphi(x)$ with respect to the system $\{\chi_n(x)\}$ can be obtained readily from the Fourier coefficients of $f(x)$ with respect to the system $\{\varphi_n(x)\}$. This method of obtaining a solution is called **Enskog's method**.

For Volterra integral equations, methods (1) and (4) can be used effectively. We usually transform equations of the first kind into equations of the second kind by differentiation and then apply the above numerical methods. This is done for the sake of securing stability of the numerical methods.

References

[1] G. Vivanti and F. Schwank, Elemente der Theorie der Lineare Integralgleichungen, Helwingsche Verlagsbuchhandlung, 1929.
[2] T. Takenouchi, Sekibun hôteisiki ron (Japanese; Theory of integral equations), Kyôritu, 1934.
[3] E. Hellinger and O. Toeplitz, Integralgleichungen und Gleichungen mit unendlichvielen Unbekannten, Teubner, 1928.
[4] K. Yosida, Sekibun hôteisiki ron (Japanese; Theory of integral equations), Iwanami, 1950; English translation, Lectures on differential and integral equations, Interscience, 1960.
[5] L. Lichtenstein, Vorlesungen über einige Klassen nichtlinearer Integralgleichungen und Integro-differentialgleichungen, Springer, 1931.
[6] T. Sato, Sekibun hôteisiki to Gurîn kansû (Japanese; Integral equations and Green functions), Hakuyô syoin, 1944.
[7] M. Hukuhara, Sekibun hôteisiki (Japanese; Integral equations), Kyôritu, 1955.
[8] F. G. Tricomi, Integral equations, Interscience, 1957.
[9] H. Schaefer, Neue Existenzsatze in der Theorie nichtlinearer Integralgleichungen, Akademie-Verlag, 1955.
[10] М. А. Красносельский (M. A. Krasnosel'skiĭ), Топологические методы в теории нелинейных интегральных уравнений, Гостехиздат, 1956.
[11] Н. И. Мусхелишвили (N. I. Musheli-švili), Сингулярные интегральные уравнения, Гостехиздат, 1946; English translation, Singular integral equations, Noordhoff, 1953.
[12] T. Satô, Sur l'équation intégrale non linéaire de Volterra, Compositio Math., 11 (1953), 271–290.
[13] F. Smithies, Integral equations, Cambridge Univ. Press, 1958.
[14] С. Г. Михлин (S. G. Mihlin), Многомерные сингулярные интегралы н интегральные уравнения, Физматгиз, 1962; English translation, Multidimensional singular integrals and integral equations, Pergamon, 1965.
[15] Л. Я. Канторович, В. И. Крылов (L. V. Kantorovitch and V. I. Krylov), Приближенные метолы высшего анализа, изд. 3, Гостехздат, 1949; English translation, Approximate methods of higher analysis, Interscience, 1958.
For numerical solution,
[16] K. Hidaka, Ôyô sekibun hôteisiki ron (Japanese; Applications of the theory of integral equations), Kawade, 1943.
[17] H. Bückner, Die praktische Behandlung von Integral-gleichungen, Springer, 1952.
[18] L. Collatz, The numerical treatment of differential equations, Springer, third edition, 1960.

220 (IV.22) Integral Geometry

A. General Remarks

Integral geometry, in the broad sense, is the branch of geometry concerned with integrals on manifolds, but the problems considered in **integral geometry** are of a more limited nature. If a †Lie group G acts on a †differentiable manifold M as a †Lie transformation group, G also acts on various figures on M, by which we mean geometric objects such as †submanifolds of M, †tangent r-frame bundles on M, etc. Let $\mathfrak{F}$ be a set of such figures on M invariant under G (i.e., $gF \in \mathfrak{F}$ for $g \in G$, $F \in \mathfrak{F}$). Consider the following problems: (i) to know whether any G-invariant †measure μ on $\mathfrak{F}$ exists, and how to determine μ if it exists; (ii) to find the integral $\int \varphi(F)\, d\mu(F)$ of functions φ on $\mathfrak{F}$ with respect to the measure μ.

The term **integral geometry** was introduced by W. Blaschke, who considered the special case of problem (ii) in which $\varphi(F)$ is a function representing geometric properties of F and the integral is to be evaluated by means of the geometric invariants concerning $\mathfrak{F}$ [1]. Problems of so-called **geometric probability** (such as the problem of Buffon's needle) belong to this category. The measure μ is called the **kinetic measure** (or **kinetic density**), and $d\mu(F)$ is also denoted by dF. If $\mathfrak{F}$ has the structure of an n-dimensional differentiable manifold and the measure μ is given by a †volume element ω (i.e., a positive †differential form of degree n), we denote ω also by dF. Problem (i) is simple: If G acts †transitively on $\mathfrak{F}$, then $\mathfrak{F} = G/H$, where H is the †isotropy subgroup of G. In this case $\mathfrak{F}$ has the structure of a differentiable manifold, and if a G-invariant †(Radon) measure exists, it is unique up to a multiplicative constant. A condition for the existence of a G-invariant measure μ can be given by means of †Haar measures of G and H (→ 225 Invariant Measures). We now consider some examples.

B. Crofton's Formula

Let $G(p,\theta)$ be a straight line defined by the equation $x_1 \cos\theta + x_2 \sin\theta = p$ with respect to orthogonal coordinates in a Euclidean plane. Let $n(p,\theta)$ be the number of intersections of $G(p,\theta)$ with a curve C of length L. Then we have **Crofton's formula**,

$$\int n(p,\theta)\, dp\, d\theta = 2L, \tag{1}$$

where $dp\, d\theta$ is the †exterior product of the differential forms dp, $d\theta$ of degree 1, and the integral is extended over $p \in (-\infty, \infty)$ and $\theta \in [0, 2\pi]$.

C. Poincaré's Formula and the Principal Formula of Integral Geometry

The kinetic density dF of a figure F congruent (with the same orientation) to a fixed figure in a Euclidean plane is defined as follows: Let R be an orthogonal frame attached to F, (x_1, x_2) be the coordinates of the origin of R with respect to a fixed orthogonal frame R_0, and θ be the angle between the first axis of R and the first axis of R_0. If we put $dF = dx_1\, dx_2\, d\theta$ (exterior product), dF has the following invariance properties: (i) dF is not changed by displacements of F; (ii) dF is not changed if instead of R we take another orthogonal frame R' attached to F.

Let two plane curves C_1, C_2 of length L_1, L_2, respectively, be given, and suppose that C_1 is fixed and C_2 is mobile. If the number of intersections of C_2 in an arbitrary position with C_1 is finite and equal to n, then the integral of n extended over all possible positions of C_2 is given by

$$\int n\, dC_2 = 4L_1 L_2 \tag{2}$$

(**Poincaré's formula**). L. A. Santaló applied this result to give a solution of the †isoperimetric problem (*Abh. Math. Sem. Univ. Hamburg*, 10 (1935)).

Let C_1, C_2 be two plane †Jordan curves of length L_1, L_2, respectively, and let S_1, S_2 be the areas of the domains bounded by C_1, C_2, respectively. Suppose that C_1 is fixed and C_2 mobile, and let χ be the number of connected domains common to the domains bounded by C_1, C_2 for C_2 in an arbitrary position. Then the integral of χ extended over all possible positions of C_2 intersecting C_1 is given by

$$\int \chi\, dC_2 = L_1 L_2 + 2\pi(S_1 + S_2) \tag{3}$$

(Blaschke [1]). This is the **principal formula of integral geometry**. Many formulas can be derived from it as special cases or limiting cases.

D. Generalization to Dimension n

The kinetic density of subspaces of dimension k in a Euclidean space and in a spherical space of dimension n was given by W. Blaschke, while the generalization of the principal formula (3) to a Euclidean space of dimension n was given by S. S. Chern, applying the methods of E. Cartan.

Let $(e_1, \ldots, e_n)$ be a positively oriented orthonormal frame with vertex A. The infinitesimal relative displacements are then given by

$$dA = \sum_{i=1}^{n} \omega_i e_i, \quad de_i = \sum_{j=1}^{n} \omega_{ij} e_j,$$

where $\omega_i = (dA, e_i)$, $\omega_{ij} = (de_i, e_j) = -\omega_{ji}$ are differential forms of degree 1 in the orthogonal coordinates of A and the $n(n-1)/2$ variables that determine $e_1, \ldots, e_n$. For various positions of a figure, we take an orthogonal frame $(A, e_1, \ldots, e_n)$ fixed to this figure, and form the †exterior product

$$dK = \bigwedge_i \omega_i \bigwedge_{i<j} \omega_{ij}$$

of all the ω_i and ω_{ij} $(i<j)$. This has the invariance properties (i) and (ii) of Section C and is, by definition, the kinetic density of the figure in an n-dimensional Euclidean space. Moreover, the kinetic density of dE of k-dimensional subspaces E can be obtained by considering the orthogonal frames such that the vertex A and $e_1, \ldots, e_k$ lie on E, and forming the exterior product of the corresponding ω_α, $\omega_{\alpha\lambda}$ $(\alpha = 1, \ldots, k; \lambda = k+1, \ldots, n)$,

$$dE = \bigwedge \omega_\alpha \bigwedge \omega_{\alpha\lambda}.$$

Let Σ be a compact orientable hypersurface of class C^2 in an n-dimensional Euclidean space, and let k_α $(\alpha = 1, \ldots, n-1)$ be the †principal curvatures at a point on Σ. Denote by S_i the †elementary symmetric form of degree i in k_α $(i = 1, \ldots, n-1)$, and put $S_0 = 1$. Then consider the integrals over Σ

$$M_i = \int_\Sigma S_i \, dS \Big/ \binom{n-1}{i}, \quad i = 0, 1, \ldots, n-1, \tag{4}$$

where dS denotes the surface element of Σ. Let D_1, D_2 be the domains bounded by two compact orientable hypersurfaces Σ_1, Σ_2 of class C^2 with volume V_1, V_2, respectively, and let $M_i^{(1)}$, $M_i^{(2)}$ be the integrals M_i defined by (4) for Σ_1, Σ_2. If Σ_1 is fixed, Σ_2 is mobile, and the †Euler-Poincaré characteristic χ of $D_1 \cap D_2$ is finite, then the generalization of (3) has the following form:

$$\int \chi \, d\Sigma_2 = I_1 \ldots I_{n-1} \left(M_{n-1}^{(1)} V_2 + M_{n-1}^{(2)} V_1 + \frac{1}{n} \sum_{k=0}^{n-2} M_k^{(1)} M_{n-2-k}^{(2)} \right) \tag{5}$$

(**Chern's formula**), where $d\Sigma_2$ is the kinetic density of Σ_2 and I_k $(k = 1, \ldots, n-1)$ is the area of the unit sphere in a Euclidean space of dimension $k+1$, with the integral extended over all positions of Σ_2 intersecting Σ_1.

Let dE be the kinetic density of the subspaces E of dimension k intersecting a compact orientable hypersurface Σ of class C^2, and let χ be the Euler characteristic of the intersection of E with the domain bounded by Σ. The integral $\int \chi \, dE$ extended over all hyperplanes of dimension k intersecting Σ is proportional to M_k relative to the hypersurface Σ defined by (4). This fact generalizes (1). (For the results in this section → [5].)

E. Other Generalizations

For 2-dimensional spaces of constant curvature, Santaló derived formulas analogous to those in a Euclidean plane (1942–1943) and thus solved the isoperimetric problem in these spaces. In 1952, he derived a formula corresponding to (5) in n-dimensional spaces of constant curvature, following Chern's method. He investigated further integral geometry in affine, projective, and Hermitian spaces.

Chern and others obtained the results of the previous sections by Cartan's method of general moving frames and studied integral geometry in the setting of the geometry of Lie transformation groups in the sense of F. Klein (→ 147 Erlangen Program).

F. Radon Transforms

Let $\mathfrak{F}$ be the set of hyperplanes $\xi(\omega, p) = \{x \in \mathbf{R}^n \mid (x, \omega) = p\}$ in the Euclidean space $\mathbf{R}^n$, where $\omega = (\lambda_1, \ldots, \lambda_n)$ is a unit vector, $(x, \omega) = \sum x_i \lambda_i$, and p is real. For a function f defined in $\mathbf{R}^n$, define $\hat{f}(\xi) = \hat{f}(\omega, p)$ on $\mathfrak{F}$ by

$$\hat{f}(\xi) = \int_\xi f(x) \, d_\xi x, \tag{6}$$

where $d_\xi x$ is the †volume element on the hyperplane ξ such that $d_\xi x \wedge \sum \lambda_i dx_i = dx$, with dx the volume element of $\mathbf{R}^n$. Then $\hat{f}$ is called the **Radon transform** of f. For example, if f is the †characteristic function of a bounded domain V, the value $\hat{f}(\xi)$ of the Radon transform $\hat{f}$ of f at $\xi \in \mathfrak{F}$ is the volume of the section of V by ξ. Now the group G of †motions of $\mathbf{R}^n$ (the †connected component of the identity of the group of isometries) acts on $\mathfrak{F}$ transitively. For every $x \in \mathbf{R}^n$, the †isotropy subgroup G_x of G with respect to x acts transitively on the set $\check{x} = \{\xi \in \mathfrak{F} \mid x \in \xi\}$. Since G_x is compact, there exists a unique normalized G_x-invariant measure μ on $\check{x}$ such that $\mu(\check{x}) = 1$. For a function g on $\mathfrak{F}$, the **conjugate Radon transform** $\check{g}$ can now be defined by

$$\check{g}(x) = \int_{x \in \xi} g(\xi) \, d\mu(\xi) \tag{7}$$

as a function on $\mathbf{R}^n$.

The determination of $\check{g}$ belongs to problem (ii) of integral geometry mentioned in Section A. In particular, it is important to determine $\check{g} = (\hat{f})^\vee$ for $g = \hat{f}$ and to find the relation between $(\hat{f})^\vee$ and f. These problems were solved by J. Radon for $n = 2, 3$ and by F. John in the general case. The results can be formulated as follows:

In the case of odd n, let $\mathfrak{S}$ be the space of †rapidly decreasing C^∞-functions (→ 173 Function Spaces). Let $\mathfrak{S}^*(\mathfrak{F})$ be the set of $g(\omega, p) \in \mathfrak{S}(\mathfrak{F})$ such that $\int_{-\infty}^{\infty} g(\omega, p) p^k \, dp = 0$ for every natural number k and every ω. For every $f \in \mathfrak{S}(\mathbf{R}^n)$ and every $g \in \mathfrak{S}^*(\mathfrak{F})$, we then have

$$f = c\Delta^{(n-1)/2}((\hat{f})^\vee), \quad g = cL^{(n-1)/2}((\check{g})^\wedge),$$

where Δ is the †Laplacian in $\mathbf{R}^n$ and $Lg(\omega, p)$

$= d^2g(\omega,p)/dp^2$, $c=\Gamma(n/2)^{-1}(2\pi i)^{1-n}\pi^{n/2}$.

In the case of even n, for every $f\in\mathcal{S}(\mathbf{R}^n)$, $g\in\mathcal{S}^*(\mathfrak{F})$,

$$f=c_1J_1((\hat{f})^{\vee}),\quad g=c_2J_2((\check{g})^{\wedge}),$$

where

$$J_1(f)(x)=\int_{\mathbf{R}^n}f(y)|x-y|^{1-2n}dy,$$

$$J_2(g)(\omega,p)=\int_{\mathbf{R}}g(\omega,p)|p-q|^{-n}dq.$$

These integrals are in general divergent, and they must be interpreted as regularizations defined by analytic continuation with respect to the powers of $|x-y|$ or $|p-q|$ (S. Helgason [6]).

A formula corresponding to †Plancherel's theorem for the Fourier transform is valid for the Radon transform (I. M. Gel'fand and others [7]).

G. Horospheres

The theory of the Radon transform is also important in noncompact †symmetric Riemannian spaces M. The connected component G of the identity in the group of isometries of M is isomorphic to the †adjoint group $\mathrm{Ad}\,G$ and can be considered as a linear group. Maximal †unipotent subgroups of G are conjugate to each other. If N is such a subgroup, we call the †orbits on M of gNg^{-1} **horospheres** on M for $g\in G$. These correspond to the hyperplanes in $\mathbf{R}^n$. If M is the complex upper half-plane with the †hyperbolic non-Euclidean metric, the horospheres are precisely the circles tangent to the real axis and the straight lines parallel to the real axis.

The group G acts transitively on the set $\mathfrak{F}$ of horospheres on M, and we have $\mathfrak{F}=G/M_0N$. Here $G=KAN$ is an †Iwasawa decomposition of G, and M_0 is the †centralizer of A in K. For a horosphere $\xi\in\mathfrak{F}$, let $d_\xi x$ be the volume element on ξ with respect to the †Riemannian metric on ξ induced by the Riemannian metric on M, and define the Radon transform $\hat{f}$ of a function f on M by (6) as a function on $\mathfrak{F}$. For every $x\in M$, there exists a unique normalized measure on $\check{x}=\{\xi\in\mathfrak{F}\,|\,x\in\xi\}$ invariant under the (compact) isotropy subgroup G_x of G at x ($\mu(\check{x})=1$). The conjugate Radon transform $\check{g}$ of a function g on $\mathfrak{F}$ is defined by formula (7) by means of this measure μ. Then there exists an integrodifferential operator $\wedge$ such that if $\wedge^*$ is the adjoint operator, we have the inversion formula $f=(\wedge\wedge^*\hat{f})^{\vee}$ and **Plancherel's theorem**:

$$\int_M|f(x)|^2dx=\int_{\mathfrak{F}}|\wedge\hat{f}\,(\xi)|^2d\xi,$$

where dx, $d\xi$ are G-invariant measures on M, $\mathfrak{F}$, respectively, and f is an arbitrary C^∞-function with compact support. If the †Cartan subgroups of G are conjugate to each other, $\wedge$ is a differential operator; the inversion formula can then be written in the form $f=L((\hat{f})^{\vee})$ with some differential operator L on M [6].

Horospheres and Radon transforms can be defined not only for symmetric Riemannian spaces G/K, but also for various †homogeneous spaces G/H of noncompact semisimple Lie groups G. The Radon transform $f\to\hat{f}$ maps a function f on G/H into a function on the space of horospheres on G/H. If a †unitary representation U of G is realized in a function space over G/H, the Radon transform $f\to\hat{f}$ helps to clarify the properties of U by going over to the function space on $\mathfrak{F}$. In several examples, I. M. Gel'fand and others have by this method decomposed U explicitly into direct integrals of irreducible representations [7, 8].

H. Another Generalization

Integral geometry can also be investigated in spaces admitting no displacement. Let $F(x_1,x_2,y_1,y_2)$ be positive and homogeneous of degree 1 with respect to y_1,y_2, and consider a †stationary curve of the integral $\int F(x_1,x_2,dx_1/dt,dx_2/dt)\,dt$. Let $p_i=\partial F/\partial y_i$ ($y_i=dx_i/dt$) along this curve. Poincaré found out that for a 2-parameter set of stationary curves, $dx_1dp_1+dx_2dp_2$ is not changed by any displacement of the line element (x_i,p_i) along a stationary curve. Blaschke took this as the kinetic density of the stationary curve and proved a formula containing formula (1) as a special case. On the other hand, Santaló introduced kinetic density on 2-dimensional surfaces and proved a generalization of formula (2).

References

[1] W. Blaschke, Vorlesungen über Integralgeometrie, Teubner, I, 1936; II, 1937.
[2] S. S. Chern, On integral geometry in Klein spaces, Ann. of Math., (2) 43 (1942), 178–189.
[3] S. S. Chern, On the kinematic formula in the Euclidean space of n dimensions, Amer. J. Math., 74 (1952), 227–236.
[4] L. A. Santaló, Introduction to integral geometry, Hermann, 1953.
[5] M. Kurita, Sekibunkikagaku (Japanese; Integral geometry), Kyôritu, 1956.
[6] S. Helgason, A duality in integral geometry; some generalizations of the Radon transform, Bull. Amer. Math. Soc., 70 (1964), 435–446.

[7] И. М. Гельфанд, М. И. Граев, Н. Я. Виленкин, (I. M. Gel'fand, M. I. Graev, and N. Ja. Vilenkin), Обобщенные функции, вып. 5, Интегральная геометрия и связанные с ней вопросы теории представлений, Физматгиз, 1962; English translation, Generalized functions V, Integral geometry and representation theory, Academic Press, 1966.
[8] И. М. Гельфанд, М. И. Граев (I. M. Gel'fand and M. I. Graev), Геометрия однородных пространств, представления групп в однородных пространствах и связнные с ними вопросы интегральной геометрии, I, Trudy Moskov. Mat. Obšč., 8 (1959), 321–390; English translation, Geometry of homogeneous spaces, representations of groups in homogeneous spaces and related questions of integral geometry, Amer. Math. Soc. Transl., (2) 37 (1964), 351–429.
[9] R. Deltheil, Probabilités géométriques, Gauthier-Villars, 1926.
[10] K. Yano, Integral formulas in Riemannian geometry, Marcel Dekker, 1970.

221 (XIII.16) Integral Invariants

A. General Remarks

Let us view a †system of differential equations $dx_i/dt = X_i(x_1, \ldots, x_n, t)$ $(i=1,2,\ldots,n)$ as defining the motion of a point whose coordinates are $(x_1, \ldots, x_n)$ in the n-dimensional space $\mathbf{R}^n$ at time t. Let K be a †p-dimensional surface $(1 \leqslant p \leqslant n)$ in $\mathbf{R}^n$, and let K_t be the set occupied at the instant t by the points which occupy K for $t=0$. If the integral

$$\int_{K_t} F(x_1, \ldots, x_n, t)\, dw, \tag{1}$$

where dw is the surface element of K_t, does not depend on t for any p-dimensional surface K, then $\int F(x_1, \ldots, x_n, t)\, dw$ is called an **integral invariant of degree** p of the original system of differential equations. In particular, a necessary and sufficient condition that an integral $\int M(x,t)\, dx_1 \ldots dx_n$ be an integral invariant of degree n is $\partial M/\partial t + \sum_{i=1}^n \partial(MX_i)/\partial x_i = 0$, and in this case $M(x,t)$ is called the **last multiple**. Furthermore, a necessary and sufficient condition that an integral $\int \sum_{i=1}^n M_i(x,t)\, dx_i$ be an integral invariant of degree 1 is $\partial M_i/\partial t + \sum_{j=1}^n ((\partial M_i/\partial x_j) X_j + (\partial X_i/\partial x_j) M_j) = 0$. If the integral (1) does not depend on t for any closed p-dimensional surface K, then $\int F(x_1, \ldots, x_n, t)\, dw$ is called a **relative integral invariant** of degree p. Corresponding to this terminology, an integral invariant is sometimes called an **absolute integral invariant**.

If $\int \sum_{i=1}^n M_i\, dx_i$ is a relative integral invariant of degree 1, then by †Stokes's theorem $\int \sum_{i,j} (\partial M_i/\partial x_j - \partial M_j/\partial x_i)\, dx_i\, dx_j$ is an absolute integral invariant of degree 2. In general, we can similarly construct an absolute integral invariant of degree $p+1$ from a relative integral invariant of degree p.

For a system of equations of even order $dp_i/dt = P_i(p_1, \ldots, p_m, q_1, \ldots, q_m, t)$, $dq_i/dt = Q_i(p_1, \ldots, p_m, q_1, \ldots, q_m, t)$ $(i=1, \ldots, m)$ to be a †Hamiltonian system, that is, for the existence of the function $H = H(p_1, \ldots, p_m, q_1, \ldots, q_m, t)$ satisfying $P_i = -\partial H/\partial q_i$ and $Q_i = \partial H/\partial p_i$, it is necessary and sufficient that the integral $\int \sum_{i=1}^n p_i\, dq_i$ be a relative integral invariant of degree 1. If a system of differential equations is a Hamiltonian system, then $\int \ldots \int dp_1 \ldots dp_m\, dq_1 \ldots dq_m$ is an integral invariant, that is, if we write $p_i = p_i(t, p^0, q^0)$, $q_i = q_i(t, p^0, q^0)$ $(i=1, \ldots, m)$ for the solution of the Hamiltonian system that passes through the point (p^0, q^0) at time $t=0$, then we have the functional determinant equation

$$\left| \frac{D(p_1, \ldots, p_m, q_1, \ldots, q_m)}{D(p_1^0, \ldots, p_m^0, q_1^0, \ldots, q_m^0)} \right| = 1.$$

In other words, a $2m$-dimensional figure in $(p_1, \ldots, p_m, q_1, \ldots, q_m)$ space may change its form according to the motion of points, but its volume remains unaltered (**Liouville's theorem**). This fact is of importance in applications to †statistical mechanics.

Poincaré developed the theory of integral invariants and applied the theory to the †three-body problem and the problem of †stability.

B. Cartan's Extension

Poincaré treated the position $(x_1, \ldots, x_n)$ and the time t separately. E. Cartan extended Poincaré's theory by unifying the treatment of position and time. Consider the solution curve of a system of differential equations $dx_i/dt = X_i$ $(i=1, \ldots, n)$ through each point of a p-dimensional surface K in n-dimensional space $(x_1, \ldots, x_n)$, and let K_i $(i=1,2)$ be a p-dimensional surface intersected by the solution curve at only one point. Let $\int F\, dw$ be an integral invariant of degree p. If we denote by Φ the †exterior differential form obtained by substituting $dx_1 - X_1\, dt, \ldots, dx_n - X_n\, dt$ for $dx_1, \ldots, dx_n$, respectively, in the exterior differential form

$$F\, dw = \sum_{i_1 < \ldots < i_p} f_{i_1, \ldots, i_p}(x_1, \ldots, x_n, t)\, dx_{i_1} \wedge \ldots \wedge dx_{i_p},$$

then $\int_{K_1} \Phi = \int_{K_2} \Phi$. Cartan called $\int \Phi$ an **integral invariant of degree** p. The relative integral invariant can be extended in a similar way.

We may consider the solution curves of a Hamiltonian system $dp_i/dt = -\partial H/\partial q_i$, $dq_i/dt = \partial H/\partial p_i$ $(i = 1, \ldots, n)$ through all points of a closed curve C in $(2m+1)$-dimensional space $(p_1, \ldots, p_m, q_1, \ldots, q_m, t)$ together with the tube consisting of these solution curves. For any closed curve C_1 lying on and enclosing the tube, we have $\int_C \sum_{i=1}^m p_i dq_i - dH = \int_{C_1} \sum_{i=1}^m p_i dq_i - dH$, that is, $\sum_{i=1}^m p_i dq_i - dH$ is a relative integral invariant, in Cartan's terminology. Conversely, if a system of differential equations of order $2m$, $dp_i/dt = P_i(p,q,t)$, $dq_i/dt = Q_i(p,q,t)$ $(i = 1, 2, \ldots, m)$, possesses the relative integral invariant $\int \sum_{i=1}^m p_i dq_i - H\,dt$ for some $H = H(p,q,t)$, then $P_i = -\partial H/\partial q_i$, $Q_i = \partial H/\partial p_i$. Cartan called the differential form of degree 1, $\omega = \sum_{i=1}^m p_i dq_i - dH$, the **momentum-energy tensor**. If the curve C lies on t = constant, then $\int \omega$ is a relative integral invariant, in Poincaré's terminology. If the curve C is a solution curve of the system of differential equations, then $\int_{t_0}^{t_1} \omega$ is a Hamilton action integral. Moreover, the integral invariant for the continuous transformation group can be defined.

References

[1] E. Cartan, Leçons sur les invariants intégraux, Hermann, 1922.
[2] H. Poincaré, Les méthodes nouvelles de la mécanique céleste III, Gauthier-Villars, 1899.

222 (X.27) Integral Transforms

A. General Remarks

Given (real- or complex-valued) functions $f(y)$ and $K(x,y)$ such that their product is †integrable as a function of y in the interval $[a,b]$, we set

$$g(x) = \int_a^b K(x,y) f(y)\,dy. \qquad (1)$$

This transformation of f to g is called the **integral transform** with the **kernel** $K(x,y)$. Now fix the kernel $K(x,y)$. When the correspondence $f \to g$ given by (1) from a set of functions f to a set of functions g is bijective, we may consider the **inverse transform** $g \to f$. The formula that describes the inverse transform $g \to f$ in terms of an integral transform is called the **reciprocal formula**. The kernel $K(x,y)$ can often be written as $k(x-y)$, $k(xy)$, etc., where $k(t)$ is an integrable function. Table 1 contains integral transforms that are important in applications.

Table 1

Kernel	Interval	Name
e^{ixy}	$(-\infty, \infty)$	Fourier transform
$\cos xy$	$(0, \infty)$	Fourier cosine transform
$\sin xy$	$(0, \infty)$	Fourier sine transform
e^{-xy}	$(0, \infty)$	Laplace transform
$\sqrt{xy}\, J_\nu(xy)$	$(0, \infty)$	Hankel transform
$1/(x-y)$	$(-\infty, \infty)$	Hilbert transform
x^{y-1}	$(0, \infty)$	Mellin transform
$(x+y)^{-\rho}$	$(0, \infty)$	Stieltjes transform
$e^{-(x-y)^2}$	$(-\infty, \infty)$	Gauss transform

In the Hankel transform, J is the †Bessel function. In the Hilbert transform, the †principal value is to be taken in the integral. In the Stieltjes transform, ρ is assumed positive. (For further details on two of these transforms → 168 Fourier Transform, 239 Laplace Transform.)

Since the explanations for the Fourier transform and Laplace transform are given in the corresponding articles, we deal here only with generalized Fourier, Hilbert, Mellin, and Stieltjes transforms.

B. Generalized Fourier Transform

Suppose that the kernels of the transform (1) and its inverse transform are both of the form $k(xy)$. Hence

$$f(x) = \int_a^b k(xy) g(y)\,dy. \qquad (2)$$

In this case, we call the integral transform (1) the **generalized Fourier transform** of symmetric type, or the **Watson transform**, and $k(t)$ the **Fourier kernel** of (2). The functions $\sqrt{2/\pi}\,\cos t$, $\sqrt{2/\pi}\,\sin t$, and $\sqrt{t}\,J_\nu(t)$ defined in the interval $(0,\infty)$ are examples of such Fourier kernels (→ 168 Fourier Transform). The last kernel gives rise to the **Hankel transform**. Suppose, in general, that a function $K(1/2+it)$ satisfies $K(1/2+it)K(1/2-it) = 1$, $|K(1/2+it)| = 1$. Then the function $K(1/2+it)/(1/2-it) = k(t)$ belongs to $L_2(-\infty,\infty)$. We set

$$k_1(x) = \frac{x}{2\pi} \operatorname*{l.i.m.}_{T\to\infty} \int_{-T}^{T} k(t) x^{-1/2-it}\,dt$$

(the limit is the †mean convergence of order 2). Then for a function $f(x) \in L_2(0,\infty)$,

$$g(x) = \frac{d}{dx} \int_0^\infty k_1(xt) f(t) \frac{dt}{t}$$

exists almost everywhere, and $g(x) \in L_2(0,\infty)$.

In this case, we have the **inversion formula**

$$f(t)=\frac{d}{dt}\int_0^{\infty}k_1(xt)\,g(x)\frac{dx}{x}$$

and the **Parseval equality**

$$\int_0^{\infty}(f(t))^2\,dt=\int_0^{\infty}(q(x))^2\,dx.$$

If a function $f(x)$ is invariant under a generalized Fourier transform, then $f(x)$ is called a **self-reciprocal function**. Such a function $f(x)$ is a solution of the homogeneous integral equation

$$f(y)=\int_a^b k(xy)f(x)\,dx.$$

The function $x^{-1/2}$ is an example of a function that is self-reciprocal with respect to the Fourier cosine transform. Using a function that is self-reciprocal with respect to the Hankel transform, we can derive the **lattice-point formula** of number theory: Let $r(n)$ be the number of possible ways in which a nonnegative integer n can be represented as the sum of two square numbers. Set

$$\bar{P}(x)=\sum_{0\leqslant n\leqslant x}{}' \; r(n)-\pi x,$$

where Σ' means that if x is an integer, we take $r(n)/2$ instead of $r(n)$. Then $f(x)=x^{-3/2}\cdot(\bar{P}(x^2/2\pi)-1)$ is self-reciprocal with respect to the Hankel transform with $\nu=2$. Utilizing this, G. H. Hardy proved (1925)

$$\bar{P}(x)=\sqrt{x}\sum_{n=1}^{\infty}\frac{r(n)}{\sqrt{n}}J_1(2\pi\sqrt{nx}).$$

A. Z. Walfisz (1926) and A. Oppenheim (1927) generalized this formula and obtained a formula concerning the number of ways in which n can be represented as the sum of p square numbers (→ 240 Lattice-Point Problems).

C. Mellin Transform

If $f(x)x^{k-1}\in L_1(0,\infty)$, then

$$F(s)=\int_0^{\infty}f(x)x^{s-1}\,dx, \qquad s=k+it$$

is called the **Mellin transform** of f. If $f(x)$ is of †bounded variation in a neighborhood of x, then the **inversion formula**

$$\frac{f(x+0)+f(x-0)}{2}$$
$$=\frac{1}{2\pi i}\lim_{T\to\infty}\int_{k-iT}^{k+iT}F(s)x^{-s}\,ds$$

holds. If $f(x)x^{k-1/2}\in L_2(0,\infty)$, then $\int_{1/A}^{A}f(x)x^{s-1}\,dx$ $(s=k+it)$ †converges in the mean of order 2 to a function $F(s)$ for a fixed k and $-\infty<t<\infty$, and the **Parseval equality**

$$\int_0^{\infty}|f(x)|^2x^{2k-1}\,dx=\frac{1}{2\pi}\int_{-\infty}^{\infty}|F(k+it)|^2\,dt$$

holds. $F(s)$ is also called the **Mellin transform** of $f(x)$. If $f(x)x^{k-1/2}$, $g(x)x^{1/2-k}\in L_2(0,\infty)$, and $F(s)$, $G(s)$ are the Mellin transforms of $f(x)$, $g(x)$, respectively, then

$$\int_0^{\infty}f(x)g(x)\,dx$$
$$=\frac{1}{2\pi i}\int_{k-i\infty}^{k+i\infty}F(s)G(1-s)\,ds.$$

The theory of the Mellin transform in the function space L_p is analogous to the theory of the Fourier transform [1, ch. 4].

D. Stieltjes Transform

For a function $\alpha(t)$ of bounded variation,

$$f(s)=\int_0^{\infty}\frac{d\alpha(t)}{(s+t)^{\rho}}, \qquad \rho>0$$

is called the **Stieltjes transform** of $\alpha(t)$. Usually we assume that $\rho=1$. If the Laplace transform is applied twice in succession to $\alpha(t)$, we obtain formally the Stieltjes transform of $\alpha(t)$. The Stieltjes transform has been studied systematically in connection with the theory of the Laplace transform by D. V. Widder, R. P. Boas, and others.

Assume that $\rho=1$. Let D be the domain obtained from the complex plane by removing the negative part of the real axis. If the Stieltjes transform converges at a point $s=s_0\in D$, then it converges uniformly on any compact set in D. The inversion formula is:

$$\lim_{\eta\to+0}\frac{1}{2\pi i}\int_0^t(f(-\sigma-i\eta)-f(-\sigma+i\eta))\,d\sigma$$
$$=(\alpha(t+0)+\alpha(t-0)-(\alpha(+0)+\alpha(-0)))/2, \qquad t>0.$$

If $\alpha(t)=\int_0^t\varphi(u)\,du$ and $\varphi(t\pm0)$ exist, then

$$\lim_{\eta\to+0}\frac{1}{2\pi i}(f(-t-i\eta)-f(-t+i\eta))$$
$$=(\varphi(t+0)+\varphi(t-0))/2, \qquad t>0.$$

E. Hilbert Transform

Let $\varphi(z)=U(x,y)+iV(x,y)$ $(z=x+iy)$ be holomorphic in the upper half-plane and $f(x)=U(x,0)$, $g(x)=-V(x,0)$ be the respective boundary values on the real axis. Then if $f, g\in L_1(-\infty,\infty)$,

$$g(x)=\frac{1}{\pi}\,\text{p.v.}\int_{-\infty}^{\infty}\frac{f(x+t)}{t}\,dt,$$

$$f(x)=-\frac{1}{\pi}\,\text{p.v.}\int_{-\infty}^{\infty}\frac{g(x+t)}{t}\,dt.$$

Here p.v. means Cauchy's †principal value, that is,

$$\text{p.v.}\int_{-\infty}^{\infty}F(t)\,dt=$$
$$=\lim_{A\to\infty,\,\varepsilon\to0}\left(\int_{-A}^{-\varepsilon}F(t)\,dt+\int_{\varepsilon}^{A}F(t)\,dt\right).$$

We call g the **Hilbert transform** of f. If $f\in L_2(-\infty,\infty)$, the **inversion formula** and the

Parseval equality hold. More precisely, for any $f \in L_2(-\infty, \infty)$, these relations hold almost everywhere, $g \in L_2(-\infty, \infty)$, and the L_2-norms of f and g are identical (→ 168 Fourier Transform). The importance of Hilbert transforms lies in the fact that they establish relations between the real and imaginary parts of an analytic function (→ 127 Dispersion Relations).

References

[1] E. C. Titchmarsh, Introduction to the theory of Fourier integrals, Oxford Univ. Press, 1937.
[2] I. I. Hirschman and D. V. Widder, The convolution transform, Princeton Univ. Press, 1955.
For formulas of integral transforms,
[3] A. Erdélyi (ed.), Tables of integral transformations I, II, McGraw-Hill, 1954.
[4] G. A. Campbell and R. M. Foster, Fourier integrals for practical applications, Bell Telephone Lab., 1931, revised edition, Van Nostrand, 1948.
[5] S. Moriguti, K. Udagawa, and S. Hitotumatu, Sûgaku kôsiki II (Japanese; Mathematical formulas II. Series and Fourier analysis), Iwanami, 1957.
[6] F. Oberhettinger, Tabellen zur Fourier Transformation, Springer, 1957.

223 (XIII.33) Integrodifferential Equations

A. General Remarks

A †functional equation involving an operator T of the form

$$f(t, x'(t), x(t), (Tx)(t)) = 0, \quad t_0 \leqslant t \leqslant t_1, \tag{1}$$

$$x(t_0) = x_0, \tag{2}$$

reduces to a †differential equation, †differential-difference equation, †integral equation, integrodifferential equation, or other functional equation when the operator T is given a special form [1,6]. In particular, by letting

$$f(t,x,y,z) = x - g(t,y) - z, \tag{3}$$

$$(Tx)(t) = \int_{t_0}^{\varphi(t)} K(t,s,x(s))\, ds, \tag{4}$$

we obtain an **integrodifferential equation**

$$x'(t) = g(t,x(t)) + \int_{t_0}^{\varphi(t)} K(t,s,x(s))\, ds. \tag{5}$$

If $\varphi(t) \equiv$ constant or $\varphi(t) \equiv t$, (5) is said to be an **integrodifferential equation of Fredholm type** or of **Volterra type**, respectively.

B. The Initial Value Problem

Let I be an interval $t_0 \leqslant t \leqslant t_1$, I_0 the interval $t_0 < t \leqslant t_1$, $C(I)$ a set of continuous functions on I, T an operator such that $Tx \in C(I_0)$ if $x \in C(I)$, $\mathfrak{F}$ a family of all such T, $\mathfrak{F}_+$ a subset of $\mathfrak{F}$ consisting of all T in $\mathfrak{F}$ for which $Tx \leqslant Ty$ holds at $t = s$ whenever x, y are functions in $C(I)$ satisfying $x(t) < y(t)$ for $t_0 \leqslant t < s$ for some $s \in I_0$, and Z a set of continuous functions which are differentiable on I_0. Let M be the class of all functions $f(t,x,y,z)$ defined for $t \in I_0$, $|x|$, $|y|$, $|z| < \infty$ and satisfying $f(t,x_1,y,z_1) \geqslant f(t,x_2,y,z_2)$ $(x_1 \geqslant x_2,\ z_1 \leqslant z_2)$.

Suppose that in (1) $f(t,x,y,z)$ is defined for $t \in I_0$, $|x|$, $|y|$, $|z| < \infty$, and $T \in \mathfrak{F}$. Suppose further that for some $\gamma > 0$ and two solutions x_1 and x_2 of (1) with (2) belonging to Z, there always exist a function $\omega \in M$ and an operator $\Omega \in \mathfrak{F}_+$ such that the inequality

$$\omega\big(\bar{t}, x_2' - x_1', x_2 - x_1, \Omega(x_2 - x_1)\big) \leqslant 0$$

holds for every $\bar{t} \in I_0$ such that $x_2(\bar{t}) - x_1(\bar{t}) = \gamma$ and $x_2(t) - x_1(t) < \gamma$ for $t_0 < t < \bar{t}$. Furthermore, suppose that there exists a function $\rho \in Z$ satisfying the following inequalities:

$$0 \leqslant \rho \leqslant \gamma, \quad t \in I_0;$$

$$\omega(t, \rho', \rho, \Omega\rho) > 0, \quad t \in I_0;$$

$$\rho(t_0 + 0) > x_2(t_0 + 0) - x_1(t_0 + 0).$$

Then equation (1) with (2) has at most one solution $x \in Z$. This result can be established by obtaining an estimate for the difference $|x_1(t) - x_2(t)|$, where $x_i(t)$ is a solution of (1) with the condition $x(0) = \eta_i$ $(i = 1, 2)$.

For the particular case of integrodifferential equations of Fredholm type, suppose that the following conditions are satisfied:

$$g(t,y_1) - g(t,y_2) \leqslant L(t)(y_1 - y_2), \quad t \in I_0;$$

$$\int_0^{t_0} \big(K(t,s,w_1(s)) - K(t,s,w_2(s))\big)\, ds$$

$$\leqslant N(t) \int_0^t M(s)(w_1(s) - w_2(s))\, ds, \quad t \in I;$$

$$\int_0^t sM(s)\, ds < \infty, \quad t \in I_0;$$

$$N(t) + tL(t) \leqslant 1 + t^2 M(t), \quad t \in I_0;$$

where $y_1 \geqslant y_2$, $L \in C(I_0)$, w_1, w_2, $M, N \in C(I)$, $w_1 \geqslant w_2$, $M \geqslant 0$, and $N \geqslant 0$. Then it is possible to obtain more practical expressions for ω, Ω, and ρ:

$$\omega(t,x,y,z) = x - L(t)y - N(t)z, \quad t \in I_0,$$

$$\Omega w = \int_0^t M(s) w(s)\, ds,$$

$$\rho(t) = \beta t(1+t) \exp \int_0^t s(1+s) M(s)\, ds,$$

where $\beta > 0$ is sufficiently small.

C. Another Problem

A problem analogous to the †boundary value or †eigenvalue problems of linear ordinary differential equations is to find a solution of the linear integrodifferential equation

$$(pu')' - qu + \lambda\left(\rho u + \int_G k(x,y)u(y)\,dy\right) = 0$$

with the boundary value $u=0$. This equation can be derived from the problem of minimizing the functional $D[\varphi]$ under the condition that $H[\varphi]$ is constant, where

$$D[\varphi] = \int_G p\varphi'^2\,dx + \int_G q\varphi^2\,dx,$$

$$H[\varphi] = \int_G \rho\varphi^2\,dx + \int_G\int_G k(x,y)\varphi(x)\varphi(y)\,dx\,dy.$$

The orthogonality condition for this boundary value problem is given by

$$\int_G \rho u_i(x)u_j(x)\,dx + \int_G\int_G k(x,y)u_i(x)u_j(y)\,dx\,dy = \begin{cases} 1, & i=j, \\ 0, & i\neq j, \end{cases}$$

[2].

Integrodifferential equations are closely related to problems of mathematical physics and engineering, and there are many investigations of such equations in the study of the equilibrium of rotating fluids [3], Prandtl's integrodifferential equation for aircraft wings in 3-dimensional space [4], the dynamics of reactors, and so on. In the second example, the circulation of the airflow $\Gamma(y)$ with constant velocity V around the profile is determined by the equation

$$\alpha(y) = \frac{\Gamma(y)}{\pi k(y)t(y)V} + \frac{1}{\pi V}\,\text{p.v.}\int_{-b/2}^{b/2} \frac{d\Gamma(y')}{y-y'},$$

called **Prandtl's integrodifferential equation**, where b is the wingspan, y and y' are variables whose range is $[-b/2, b/2]$ (y is assumed fixed), t the length of the chord, α an angle of incidence from the point with buoyancy 0, $2\pi k$ the slope of the curve defined from buoyancy by the angle of incidence, and p.v. †Cauchy's principal value.

As a problem having applications in the theory of †stochastic processes, the existence of solutions that have finite limits as $t\to\infty$ has been investigated for the **Wiener-Hopf integrodifferential equation**

$$\prod_{k=1}^{n}(D+\lambda_k)f(x) = \lambda_1\ldots\lambda_n\int_0^\infty f(x+t)\,dH(t), \quad x\geqslant 0, \quad D=d/dx. \quad (6)$$

For this problem, by means of the method of †semigroups of operators, equation (6) can be extended to

$$\prod_{k=1}^{n}(A+\lambda_k)f(x) = \lambda_1\ldots\lambda_n\int_0^\infty (T_t f)(x)\,dH(t), \quad x\in I,$$

where A is the †infinitesimal generator of a semigroup of operators $\{T_t\}$ and it has been shown that analogous results to those for (6) can be obtained for this more general equation [5].

References

[1] K. Nickel, Fehlerabschätzungs- und Eindeutigkeitssätze für Integro-Differentialgleichungen, Arch. Rational Mech. Anal., 8 (1961), 159–180.

[2] R. Courant and D. Hilbert, Methods of mathematical physics I, Interscience, 1953.

[3] L. Lichtenstein, Vorlesungen über einige Klassen nichtlinearer Integralgleichungen und Integro-Differentialgleichungen, Springer, 1931.

[4] W. F. Durand, Aerodynamic theory II, Springer, 1935.

[5] S. Karlin and G. Szegö, On certain differential-integral equations, Math. Z., 72 (1959–1960), 205–228.

[6] M. N. Oğuztöreli, Time-lag control systems, Academic Press, 1966.

[7] Н. И. Мусхелишвили (N. I. Mushelišvili), Сингулярные интегральные уравнения, Гостехиздат, 1946; English translation, Singular integral equations, Noordhoff, 1953.

[8] T. L. Saaty, Modern nonlinear equations, McGraw-Hill, 1967.

224 (XV.2) Interpolation

A. Lagrange's Interpolation

Assume that the values of a real function $f(x)$ with some regularity property (e.g., differentiability up to a certain order) are given at each of $n+1$ distinct real values x_i ($i=0,1,\ldots,n$). The method of finding the values $f(x)$ at x ($\neq x_i$) by using these values $f_0=f(x_0), f_1=f(x_1),\ldots,f_n=f(x_n)$ is called **interpolation**. To obtain such values, we usually use a polynomial of degree n that coincides with $f(x)$ at the points $x_0, x_1, \ldots, x_n$. Such a polynomial is called an **interpolation polynomial**, and the method of using such a polynomial is called **Lagrange's interpolation**.

If we let $\Pi(x)=\Pi_{i=0}^{n}(x-x_i)$ and $l_i(x)=\Pi(x)/((x-x_i)\Pi'(x_i))$, Lagrange's interpolation polynomial can then be expressed in the following form:

$$L(x)=\sum_{i=0}^{n} l_i(x)f_i.$$

For the given values of x, the value of $L(x)$ may be used as an approximation of $f(x)$. Sometimes, by **interpolation** we mean the method of finding $f(x)$ for x lying between the maximum and minimum of x_i. When x is otherwise situated, such a method is called **extrapolation**. The method of finding x satisfying $f(x)=f$ for a given value is called **inverse interpolation.** If $f(x)$ is $(n+1)$-times differentiable, the deviation of $L(x)$ from $f(x)$ is given by $f^{(n+1)}(\xi)\Pi(x)/(n+1)!$, where ξ lies between the maximum and minimum of x_i.

The **Aitken interpolation scheme** is useful to find the value of the interpolation polynomial at x by the successive application of a simple process. Define the expressions

$$I_{0i}(x)=\frac{1}{x_i-x_0}\begin{vmatrix} f_0 & x_0-x \\ f_i & x_i-x \end{vmatrix}, \qquad i=1,2,\ldots,n,$$

and successively

$$I_{01\ldots ki}=\frac{1}{x_i-x_k}\begin{vmatrix} I_{01\ldots k}(x) & x_k-x \\ I_{01\ldots k-1i}(x) & x_i-x \end{vmatrix}, \qquad i=k+1,\ldots,n.$$

Then continue successive evaluation of $I_{01}(x), I_{012}(x), \ldots$ until the value coincides with the last one to the desired degree of accuracy. In this case x_i are not necessarily assumed to be arranged in monotonic order. It is better to arrange them in order of their distance from x rather than in ascending or descending order. Also, $I_{01\ldots n}(x)$ coincides with $L(x)$. It should be remarked that interpolation polynomials do not always converge to $f(x)$ as n increases to infinity.

B. Interpolation by Finite Differences

In practical use, when the points of interpolation are equally spaced, interpolation by using finite differences is more useful. Suppose that for $x_i=x_0+ih$, the values $f_0, f_1, \ldots, f_n$ of the function $f(x)$ are known. The differences $\Delta f_i=f_{i+1}-f_i$ are then called **(finite) differences** of first order. Furthermore, we define the differences of order $k+1$ inductively by $\Delta^{k+1}f_i=\Delta^k f_{i+1}-\Delta^k f_i$. If we use the shift operator E defined by $Ef_i=f_{i+1}$, the difference operator Δ is represented as $\Delta=E-1$. Sometimes the following operators may be used also: the **backward difference** $\nabla=1-E^{-1}$ (Δ is called **forward difference** in contrast to ∇), and the **central difference** $\delta=E^{1/2}-E^{-1/2}$. Here, the operator representing the **average operation** $\mu=(E^{1/2}+E^{-1/2})/2$ in contrast to δ may also be used. Table 1 shows the relations between them using the notation D for the **differentiation operator** ($Df(x)=df(x)/dx$).

A table in which differences are arranged next to values of functions is called a **difference table**. Table 2 shows three difference tables; although the notations are different, the values in the corresponding places in the three tables are the same. If $f(x)$ is a polynomial of degree k, then $\Delta f(x)$ is a polynomial of degree $k-1$, $\Delta^k f(x)$ is a constant, and $\Delta^{k+1}f(x)$ is zero. Therefore, looking at the difference table, we can find the degree of an interpolation polynomial that can satisfactorily approximate $f(x)$. And since any error in the values $f_0, f_1, \ldots, f_n$ is multiplied by binomial coefficients corresponding to the place in the difference table, the errors in the tabulated values are easily found.

The following are interpolation formulas using Table 2. Suppose that we want to interpolate the value at $x=x_0+ph$. Considering $f_p=E^pf_0=(1+\Delta)^pf_0$, we have **Newton's forward interpolation formula:**

$$f_p=f_0+p\Delta f_0+\frac{p(p-1)}{2}\Delta^2 f_0+\ldots.$$

And from the formula $f_p=E^pf_0=(1-\nabla)^{-p}f_0$, **Newton's backward interpolation formula**

Table 1

	E	Δ	δ	∇	hD
E	E	$1+\Delta$	$1+\frac{\delta^2}{2}+\delta\mu$	$\frac{1}{1-\nabla}$	e^{hD}
Δ	$E-1$	Δ	$\delta\mu+\frac{\delta^2}{2}$	$\frac{\nabla}{1-\nabla}$	$e^{hD}-1$
δ	$E^{1/2}-E^{-1/2}$	$\frac{\Delta}{(1+\Delta)^{1/2}}$	δ	$\frac{\nabla}{(1-\nabla)^{1/2}}$	$2\sinh(hD/2)$
∇	$1-E^{-1}$	$\frac{\Delta}{1+\Delta}$	$\delta\mu-\frac{\delta^2}{2}$	∇	$1-\frac{1}{e^{hD}}$
hD	$\log E$	$\log(1+\Delta)$	$2\,\mathrm{arc\,sinh}(\delta/2)$	$-\log(1-\nabla)$	hD
μ	$\frac{E^{1/2}+E^{-1/2}}{2}$	$\frac{1+\Delta/2}{(1+\Delta)^{1/2}}$	μ	$\frac{1-\nabla/2}{(1-\nabla)^{1/2}}$	$\cosh(hD/2)$

$\mu=(1+\delta^2/4)^{1/2}$

Table 2

f_{-2}			f_{-2}		
	Δf_{-2}			∇f_{-1}	
f_{-1}		$\Delta^2 f_{-2}$	f_{-1}		$\nabla^2 f_0$
	Δf_{-1}			∇f_0	
f_0		$\Delta^2 f_{-1}$ ...	f_0		$\nabla^2 f_1$ ...
	Δf_0			∇f_1	
f_1		$\Delta^2 f_0$	f_1		$\nabla^2 f_2$
	Δf_1			∇f_2	
f_2			f_2		

f_{-2}		
	$\delta f_{-3/2}$	
f_{-1}		$\delta^2 f_{-1}$
	$\delta f_{-1/2}$	
f_0		$\delta^2 f_0$ ...
	$\delta f_{1/2}$	
f_1		$\delta^2 f_1$
	$\delta f_{3/2}$	
f_2		

follows:

$$f_p = f_0 + p\nabla f_0 + \frac{p(p+1)}{2}\nabla^2 f_0 + \ldots.$$

These formulas are important as foundations of the theory, but they are seldom used practically. The first formula may also be obtained as follows: we may express f_p in the form $f_p = a_0 + a_1 x^{(1)} + a_2 x^{(2)} + \ldots$, where $x^{(k)}$ is the factorial polynomial $x(x-1)\ldots(x-k+1)$. The coefficients a_k are determined from the values of Δf_p at $p=0$. The most useful formula is **Everett's interpolation formula**:

$$f_p = q f_0 + E_2\delta^2 f_0 + E_4\delta^4 f_0 + \ldots + p f_1 + F_2\delta^2 f_1 + F_4\delta^4 f_1 + \ldots,$$

where $q = 1-p$,

$$E_2 = (q-1)q(q+1)/3!,$$

$$E_4 = (q-2)(q-1)q(q+1)(q+2)/5!, \ldots,$$

and

$$F_2 = (p-1)p(p+1)/3!,$$

$$F_4 = (p-2)(p-1)p(p+1)(p+2)/5!, \ldots.$$

In addition to these formulas, there are several useful ones by Bessel, Gauss, Stirling, and others, which are essentially the same.

C. Interpolation by Divided Differences

For points located at unequal intervals, we may use **divided differences** (or **difference quotients**)

$$f_{ij} = \frac{f_i - f_j}{x_i - x_j}, \quad f_{ijk} = \frac{f_{ij} - f_{jk}}{x_i - x_k}, \ldots.$$

These differences have symmetry with respect to the suffixes. For example, $f_{01\ldots k}$ can be expressed as

$$f_{01\ldots k} = \sum_{i=0}^{k} \frac{f_i}{\Pi'(x_i)}, \quad \text{where}$$

$$\Pi(x) = \prod_{i=0}^{k} (x - x_i).$$

The difference of a polynomial of degree n is one of degree $n-1$, and the higher the difference, the lower the degree. With divided differences, the value $f(x)$ can be expressed as follows:

$$f(x) = f_0 + (x-x_0)f_{01} + (x-x_0)(x-x_1)f_{012} + \ldots + (x-x_0)(x-x_1)\ldots(x-x_n)f_{01\ldots nx}.$$

The formula without the last term is a polynomial of degree n with values equal to f_i at $x = x_i$ $(i = 0, 1, \ldots, n)$. We can use this expression for interpolation. Interpolation of functions with two or more variables can be handled similarly.

References

[1] C. Lanczos, Linear differential operators, Van Nostrand, 1961.

[2] W. E. Milne, Numerical calculus, Princeton Univ. Press, 1949.

[3] National Physical Laboratory (ed.), Modern computing methods, second edition, Notes on Applied Science, no. 16, Philosophical Library, 1961.

[4] J. Walsh (ed.), Numerical analysis, an introduction, Academic Press, 1967.

[5] P. J. Davis, Interpolation and approximation, Blaisdell, 1963.

[6] Y. Ishida, Hokan keisû hyô (Japanese; Tables of coefficients of interpolation formulas), Baihûkan, 1953.

[7] J. Yamanouchi, S. Moriguti, and S. Hitotumatu (eds.), Densi keisanki no tameno sûti keisan hô (Japanese; Numerical methods for computer programmers), Baihûkan, 1965, vol. 1, ch. 3.

225 (IV.20) Invariant Measures

A. Definitions

Let μ be a †measure defined on a †completely additive family $\mathfrak{B}$ of subsets of a set X, and let G be a †transformation group acting on X from the left (right), such that $sA \in \mathfrak{B}$ $(As \in \mathfrak{B})$ for $A \in \mathfrak{B}$, $s \in G$. For $s \in G$, define the measures $\gamma(s)\mu$ and $\delta(s)\mu$ on $\mathfrak{B}$ by $(\gamma(s)\mu)(sA) = \mu(A)$ and $(\delta(s)\mu)(As) = \mu(A)$. If $\gamma(s)\mu = \mu$ $(\delta(s)\mu = \mu)$ for all $s \in G$, μ is called a **left- (right-) invariant measure** with respect to G (or **left (right)** G**-invariant measure**).

We consider the case where G is a †topological group, X is a †locally compact Hausdorff space, and G is a †topological transformation group of X. We further suppose that $\mathfrak{B}$ is the smallest †σ-additive family containing the family $\mathfrak{C}$ of all compact subsets of X, and that $\mu(K)<\infty$ for all $K\in\mathfrak{C}$ (→ 270 Measure Theory). Let $C_0(X)$ be the space of all real-valued continuous functions with compact †support defined on X. For example, if X is an †oriented †Riemannian manifold and ω is the †volume element associated with the Riemannian metric on X, there exists a unique measure μ on X such that

$$\int_X f(x)\,d\mu(x)=\int_X f\omega \qquad (1)$$

for every $f(x)\in C_0(X)$. This measure μ is invariant under the group G of †isometries of X. In the case of a nonorientable X, a G-invariant measure can also be defined from the †Riemannian metric.

We consider G-invariant measures on †homogeneous spaces X of a locally compact Hausdorff topological group (abbreviated to locally compact group) G.

B. Haar Measures

Most fundamental is the case in which G is locally compact and $X=G$, with sx (resp. xs) defined by the group multiplication law. In this case, a nonzero G-invariant measure on G is called a **left- (right-) invariant Haar measure** on G. On every locally compact group, there exists a left- (right-) invariant Haar measure, which is unique up to a positive multiplicative constant **(Haar's theorem)**. For example, Haar measures on the additive group $\mathbf{R}$ of real numbers and the additive group $\mathbf{R}^n$ are the usual †Lebesgue measures. A Haar measure μ on the multiplicative group $\mathbf{R}_+^*$ of positive real numbers is given by

$$\int_0^\infty f(x)\,d\mu(x)=\int_0^\infty f(x)\,dx/x.$$

For an n-dimensional †Lie group G, a left-invariant Haar measure μ is defined by formula (1) with a left-invariant †differential form ω of degree n.

A Haar measure μ on a locally compact group G is †regular in the following sense. If $\mathfrak{D}$ is the set of all open subsets of G, then for every $A\in\mathfrak{B}$, we have

$$\mu(A)=\sup\{\mu(K)\mid K\in\mathfrak{C},\ K\subset A\}$$
$$=\inf\{\mu(U)\mid U\in\mathfrak{B}\cap\mathfrak{D},\ A\subset U\}.$$

For $U\in\mathfrak{B}\cap\mathfrak{D}$ $(U\neq\varnothing)$, we have $\mu(U)>0$, and $\mu(A)<\infty$ for a compact A (→ 270 Measure Theory H). The measure $\mu(s)$ of one point s is >0 if and only if G is †discrete. The †outer measure $\mu^*(G)$ of G is finite if and only if G is compact.

C. Modular Functions

Let μ be a left-invariant Haar measure on a locally compact group G. Since $\delta(s)\mu$ is also a left-invariant Haar measure, there exists a positive real number $\Delta(s)$ such that $\delta(s)\mu=\Delta(s)\mu$, by virtue of the uniqueness of the left-invariant Haar measure. The function $\Delta=\Delta_G$ on G is called the **modular function** of G. For an †integrable function on G with respect to μ, we have

$$\int_G f(xs)\,d\mu(x)=\Delta(s)^{-1}\int_G f(x)\,d\mu(x),$$

$$\int_G f(x^{-1})\Delta(x)^{-1}\,d\mu(x)=\int_G f(x)\,d\mu(x).$$

If ν is a right-invariant Haar measure on G, we have the following formulas:

$$\int_G f(sx)\,d\nu(x)=\Delta(s)\int_G f(x)\,d\nu(x),$$

$$\int_G f(x^{-1})\Delta(x)\,d\nu(x)=\int_G f(x)\,d\nu(x).$$

Moreover, $\Delta^{-1}\mu$ is a right-invariant Haar measure, while $\Delta\nu$ is a left-invariant Haar measure.

The modular function Δ of G is a continuous homomorphism of G into the multiplicative group $\mathbf{R}_+^*$ of positive real numbers. If the modular function Δ of G is equal to the constant 1, i.e., if a left-invariant Haar measure is also right-invariant, G is said to be **unimodular**. G is unimodular if G is compact, commutative, or discrete. If G is a †Lie group, we have $\Delta(s)=|\det \mathrm{Ad}(s)^{-1}|$, where $s\to\mathrm{Ad}(s)$ is the †adjoint representation. In particular, G is unimodular if G is a †semisimple Lie group, a connected †nilpotent Lie group, or a Lie group for which Ad G is compact. However, the group $T(n;\mathbf{R})$ of right triangular matrices $(n>1)$ is not unimodular.

D. Product Measures

Let $\{G_\alpha\}_{\alpha\in A}$ be a family of locally compact groups, and let μ_α be a left-invariant Haar measure on G_α for every $\alpha\in A$. Suppose that there exists a finite subset B of A such that G_α is compact and $\mu_\alpha(G_\alpha)=1$ for $\alpha\in A-B$. The product measure $\mu=\prod_{\alpha\in A}\mu_\alpha$ is then a left-invariant Haar measure on the †Cartesian product $G=\prod_{\alpha\in A}G_\alpha$, which is also a locally compact group. Moreover, if Δ_α is the module of G_α, then $\Delta_G(x)=\prod_{\alpha\in A}\Delta_{G_\alpha}(x_\alpha)$ for $x=(x_\alpha)_{\alpha\in A}$.

E. The Product Formula

Let H, L be two closed subgroups of a locally compact group G, and suppose that $\Omega=HL$

contains a neighborhood V of e in G. This means that Ω is an open subset of G. If we put $D=\{(s,s)|s\in H\cap L\}$, then the mapping $(s,t)\rightarrow st^{-1}$ of $H\times L$ into Ω induces a one-to-one continuous mapping φ of the quotient space $H\times L/D$ onto Ω. Suppose that φ is a homeomorphism. This is the case, for example, if G is †paracompact. Furthermore, if, $H\cap L$ is compact, we have the **product formula**:

$$\int_{\Omega} f(\omega)\,d\mu(\omega)$$

$$=a\iint_{H\times L} f(hl)\Delta_G(l)\Delta_L(l)^{-1}\,d\mu_H(h)\,d\mu_L(l),$$

when μ, μ_H, μ_L denote left-invariant Haar measures on G, H, L, respectively, and $a>0$ is a constant independent of f.

F. Weil Measures

If A is a measurable subset with respect to a left-invariant Haar measure μ and $\mu(A)>0$, then $A^{-1}A=\{s^{-1}t|s,t\in A\}$ is a neighborhood of the identity element of G, and such subsets form a base for the neighborhood system of the identity. This shows that the topology of a locally compact group is determined by its Haar measure. Conversely, we shall consider the definiton of a topology in an abstract group G with a measure μ.

Let μ be a †σ-finite measure defined on a †σ-additive family $\mathfrak{B}$ in G, such that $sA\in\mathfrak{B}$ for $A\in\mathfrak{B}$ and $s\in G$. μ is called a **Weil measure** if it satisfies the following two conditions: (W1) $\mu(sA)=\mu(A)$; (W2) if $f(x)$ is $\mathfrak{B}$-measurable, then $f(xy^{-1})$ is $\mathfrak{B}\times\mathfrak{B}$- measurable.

If a Weil measure $\mu\neq 0$ exists in a group G, then $\{A^{-1}A\,|\,\mu(A)>0\}$ forms a base for the neighborhood system of the identity element of a topology, which makes G a locally †totally bounded topological group. If for every $s\in G$ there exists an $A\in\mathfrak{B}$ such that $\mu(A\cap sA)<\mu(A)<\infty$, $\mu(A)>0$, then G is a Hausdorff space. In this case the †completion $\bar{G}$ of G is a locally compact group, and for a suitable left-invariant Haar measure $\bar{\mu}$ on $\bar{G}$, we have $A=\bar{A}\cap G\in\mathfrak{B}$ and $\mu(A)=\bar{\mu}(\bar{A})$ for every $\bar{A}\in\bar{\mathfrak{B}}$ (the completely additive family of $\bar{G}$).

G. Relative Invariant Measures

Let G be a transformation group acting on a set X from the left. A measure μ on X is said to be a **relative invariant measure** with respect to G if for every $s\in G$, $\gamma(s)\,\mu$ is proportional to μ, i.e., $\gamma(s)\,\mu=\chi(s)\cdot\mu$ $(\chi(s)\in\mathbf{R}^*_+)$. If $\mu\neq 0$, $\chi(s)$ is uniquely determined by s, and $s\rightarrow\chi(s)$ is a continuous homomorphism from G into the multiplicative group $\mathbf{R}^*_+$ of positive real numbers. We call χ the **multiplicator** of the relative invariant measure.

We now consider relative invariant measures with respect to a locally compact group G on the †quotient space G/H of G by a closed subgroup H. Let μ, β be left-invariant Haar measures on G, H, respectively, and let $x\rightarrow x^*$ be the canonical mapping of G onto G/H. For any measure λ on G/H, there exists a unique measure $\lambda^{\#}$ on G satisfying the following condition:

$$\int_{G/H}\left(\int_H f(xh)\,d\beta(h)\right)d\lambda(x^*)$$

$$=\int_G f(x)\,d\lambda^{\#}(x)$$

for every continuous function f with compact support on G. For every $h\in H$, we have $\delta(h)\lambda^{\#}=\Delta_H(h)\lambda^{\#}$. Conversely, for a measure ν on G such that $\delta(h)\nu=\Delta_H(h)\nu$ for every $h\in H$, there exists a unique measure λ on G/H such that $\lambda^{\#}=\nu$. This measure λ is called the **quotient measure** of ν by β and is denoted by $\lambda=\nu/\beta$. For a continuous homomorphism χ of G into the multiplicative group $\mathbf{R}^*_+$ of positive real numbers, a necessary and sufficient condition for the existence of a not identically zero, relative invariant measure on G/H with the multiplicator χ is that $\chi(h)=\Delta_H(h)/\Delta_G(h)$ for every $h\in H$. If this condition is satisfied, the relative invariant measure on G/H with multiplicator χ is unique up to a multiplicative constant and is given by the quotient measure $\nu=(\chi\mu)/\beta$ of $\chi\mu$ by β. In particular, for the existence of a G-invariant measure on G/H, it is necessary and sufficient that the modular functions Δ_G and Δ_H coincide on H. Hence if G, H are unimodular, there exists an invariant measure on G/H.

H. Weyl's Integral Formula

Let G be a compact connected †semisimple Lie group and H a †Cartan subgroup (maximal torus) of G. Then a Haar measure μ on G can be expressed by means of a Haar measure β on H and a G-invariant measure λ on G/H. If μ, β, λ are all normalized to be of total measure 1, we have the following formula for every continuous function f on G:

$$\int_G f(g)\,d\mu(g)$$

$$=\frac{1}{w}\int_H\int_{G/H} f(ghg^{-1})J(h)\,d\lambda(g^*)\,d\beta(h)$$

(Weyl's integral formula), where w is the order

of the †Weyl group of G and $J(h)$ is given by

$$J(\exp X)=\left|\prod_{\alpha\in P}\left(e^{\alpha(X)/2}-e^{-\alpha(X)/2}\right)\right|^2,$$

with P the set of all †positive roots α of G with respect to H and X an arbitrary element of the †Lie algebra of H. For an element h of H, the element X with $h=\exp X$ is not unique, but the function J is a single-valued function. A similar formula is valid on a †symmetric Riemannian manifold. Weyl's integral formula can also be generalized to the case of noncompact semisimple Lie groups. However, it is then necessary to replace the right-hand side by a sum extended over a system of representatives of mutually nonconjugate Cartan subgroups.

I. Quasi-Invariant Measures

If a group G acts on a set X from the left, a measure μ on X is said to be a **quasi-invariant measure** with respect to G if the measures $\gamma(s)\mu$ and μ are equivalent for every $s\in G$. Here two measures λ, μ defined on $\mathfrak{B}$ are **equivalent** if $\lambda=g\mu$ for some measurable function $g(x)$ which is >0 almost everywhere with respect to μ and μ-integrable on every $A\in\mathfrak{B}$ ($\mu(A)<\infty$).

We now consider quasi-invariant measures with respect to a locally compact group G on a quotient space G/H of G by a closed subgroup H. There are always quasi-invariant measures on G/H with respect to G, and they are all mutually equivalent. They can be constructed as follows. There exists a positive continuous function ρ on G such that $\rho(gh)=\Delta_H(h)\Delta_G(h)^{-1}\rho(g)$ for $g\in G$, $h\in H$. Then the quotient measure $\lambda=(\rho\mu)/\beta$ is a nonzero quasi-invariant measure on G/H with respect to G if μ, β are Haar measures on G, H, respectively. If G is a Lie group, we can take a function ρ of †class C^∞. If X is an infinite-dimensional †locally convex topological vector space over $\mathbf{R}$, there exists no †σ-finite Borel measure on X that is quasi-invariant with respect to translations by the elements of X [8].

References

[1] N. Bourbaki, Eléments de mathématique, Intégration, ch. 7, 8, Actualités Sci. Ind., 1306, Hermann, 1963.
[2] A. Weil, L'intégration dans les groupes topologiques et ses applications, Actualités Sci. Ind., Hermann, 1940.
[3] P. R. Halmos, Measure theory, Van Nostrand, 1950.
[4] S. Helgason, Differential geometry and symmetric spaces, Academic Press, 1962.
[5] L. H. Loomis, An introduction to abstract harmonic analysis, Van Nostrand, 1953.
[6] K. Kunugi, Zitukansûron oyobi sekibunron (Japanese; Theory of functions of real variables and theory of integrations), Kyôritu, 1957.
[7] Y. Kawada, Sekibunron (Japanese; Theory of integrals), Kyôritu, 1959.
[8] Y. Umemura, Measures on infinite dimensional vector spaces, Publ. Res. Inst. Math. Sci., (A) 1 (1965), 1–47.
[9] A. Haar, Der Massbegriff in der Theorie der kontinuierlichen Gruppen, Ann. of Math., 34 (1933), 147–169.
[10] J. Von Neumann, The uniqueness of Haar's measure, Mat. Sbornik, 1 (1936), 721–734.
[11] H. Cartan, Sur la mesure de Haar, C. R. Acad. Sci. Paris, 211 (1940), 759–762.

226 (IV.21) Invariants and Covariants

A. The General Case

Let R be a †commutative ring. We say that a group G **acts** on R if (i) each element σ of G defines an †automorphism $f\to\sigma f$ of R, and (ii) $\sigma(\tau f)=(\sigma\tau)f$ for any $\sigma,\tau\in G, f\in R$. In this case, an element f of R is said to be **G-invariant** (or simply **invariant**) if $\sigma f=f$ for any $\sigma\in G$. An element f is called **(G-)semi-invariant** if for each σ in G there is an invariant $a(\sigma)$ such that $\sigma f=a(\sigma)f$, that is, if f is invariant up to an invariant multiplier depending on σ. A semi-invariant may also be called a **relative invariant**, and an invariant may be called an **absolute invariant**. The correspondence $\sigma\to a(\sigma)\bmod(0:f)$ ($(0:f)=\{x\in R\mid xf=0\}$) is a †representation of degree 1 of G, and $a(\sigma)$ is called the **multiplier** of the semi-invariant f, or the **character** defined by f.

B. Invariants of Matrix Groups

Let K be a commutative ring with a unity element. Then we can consider a **matrix group** (or **matric group**) G over K, i.e., a subgroup of the group of $n\times n$ invertible matrices over K; this latter group is called the **general linear group** of degree n over K and is denoted by $GL(n,K)$ (→ 63 Classical Groups). Assume that R is a commutative ring generated by $x_1,\ldots,x_n$ over K and an action of the group G on R is defined such that ${}^t(\sigma x_1,\ldots,\sigma x_n)=\sigma\,{}^t(x_1,\ldots,x_n)$ (^{t} means the †transpose of a matrix). In this case, we say that G acts on

R as a matrix group. If K is a field, then the smallest †algebraic group $\bar{G}$ (in $GL(n,K)$) containing G acts on R as a matrix group, and an element f of R is $\bar{G}$-invariant if and only if f is G-invariant, and similarly for semi-invariants. These results can be generalized to the case where K is not a field.

A †homomorphism ρ of a matrix group G ($\subset GL(n,K)$) into $GL(m,K)$ is called a **rational representation** of G if there exist rational functions φ_{kl} ($1 \leqslant k,l \leqslant m$) in n^2 variables x_{ij} ($1 \leqslant i,j \leqslant n$) with coefficients in K such that $\rho((\sigma_{ij}))=(\varphi_{kl}(\sigma_{ij}))$ for all $(\sigma_{ij}) \in G$. Assume that ρ is a rational representation of a matrix group G and $\rho(G)$ acts on a ring R as a matrix group. Then we have an action of G on R defined by $\sigma f=(\rho\sigma)f$ ($\sigma \in G, f \in R$), called the **rational action** defined by ρ. If the following condition is satisfied, then the action is called **semireductive** (or **geometrically reductive**): If $N=f_1K+\ldots+f_rK$ is a G-†admissible module ($f_1,\ldots,f_r \in R$) and if f_0 mod N ($f_0 \in R$) is G-invariant, then there is a †homogeneous form h in $f_0,\ldots,f_r$ of positive degree with coefficients in K such that h is †monic in f_0 and is G-invariant. This action is called **reductive** (or **linearly reductive**) if h can always be chosen to be a linear form.

(1) Rational actions of a matrix group G in each of the following three cases are all reductive. (i) K is either the real number field or the complex number field and G is a dense subset of a †Lie group ($\subset GL(n,K)$) which is either †semisimple or †compact. (ii) K is a field of †characteristic 0, and letting $\bar{G}$ be the smallest algebraic group containing G, the †radical of $\bar{G}$ is a †torus group (→ 15 Algebraic Groups). (iii) K is a field of characteristic $p \neq 0$, and $\bar{G}$ (as in (ii)) contains a torus group T of finite †index that is relatively prime to p.

(2) Any action of a finite group is a semireductive action. If we omit the condition that the characteristic of K is 0 in (ii), then rational actions of G are semireductive. This was known as Mumford's conjecture and was proved by W. J. Haboush recently.

(3) Assume that φ is a G-†admissible homomorphism of the ring R onto another ring R'. We denote the sets of G-invariants in R and R' by $I_G(R)$ and $I_G(R')$, respectively.

If the actions of G are reductive, then (i) $\varphi(I_G(R))=I_G(R')$; (ii) $h_i \in I_G(R)$ implies $(\Sigma_i h_i R) \cap I_G(R)=\Sigma_i h_i I_G(R)$; and (iii) if K is †Noetherian, then $I_G(R)$ is finitely generated over K.

If rational actions of G are semireductive, then (i) for each element a of $I_G(R')$, there is a natural number t such that $a^t \in \varphi(I_G(R))$, and hence $I_G(R')$ is †integral over $\varphi(I_G(R))$; (ii) if $h_i \in I_G(R)$ and $f \in (\Sigma_i h_i R) \cap I_G(R)$, then a suitable power f^t of f is in $\Sigma_i h_i I_G(R)$; and (iii) if K is a †pseudogeometric ring (in particular, if K is a field), then $I_G(R)$ is finitely generated over K [3].

When $I_G(R)$ is generated by $f_1,\ldots,f_s$ over K, $f_1,\ldots,f_s$ are called **basic invariants**.

C. Polynomial Rings

Let $\rho_1,\ldots,\rho_u$ be matrix representations of a group G over a commutative ring K of respective degrees $n_1,\ldots,n_u$. Let $x_j^{(i)}$ ($1 \leqslant i \leqslant u, 1 \leqslant j \leqslant n_i$) be Σn_i †algebraically independent elements over K. Then we define an action of G on the †polynomial ring $K[x_1^{(1)},\ldots,x_{n_s}^{(s)}]$ by ${}^t(\sigma x_1^{(i)},\ldots,\sigma x_{n_i}^{(i)})=\rho_i(\sigma)\,{}^t(x_1^{(i)},\ldots,x_{n_i}^{(i)})$. In this case, a (relative) invariant is the sum of (relative) invariants that are homogeneous in each $(x_1^{(i)},\ldots,x_{n_i}^{(i)})$. (Because of this fact, in some literature, a (relative) invariant means a (relative) invariant that is homogeneous in each of $(x_1^{(i)},\ldots,x_{n_i}^{(i)})$.) On the existence of basic invariants, the following theorem is known (besides the one on (semi)reductive actions): Assume that K is a field of characteristic 0, G is dense under the †Zariski topology in an algebraic linear group $\bar{G}$ such that the †unipotent part $(\bar{G})_u$ of the radical of $\bar{G}$ is at most 1-dimensional (these conditions hold if G is a 1-dimensional Lie group), and all of the ρ_i are rational representations; then basic invariants exist (R. Weitzenböck).

Furthermore, if G is a matrix group and each ρ_i is either the †identity mapping or the contragredient mapping $A \to {}^tA^{-1}$, then the invariants are called **vector invariants**. If K is a field of characteristic zero, the basic invariants and a basis for the ideal of algebraic relations of the basic invariants are explicitly given in several cases [1,2].

D. Classical Terminology

The classical theory of invariants considers the following objects. Let K be a field of characteristic zero (e.g., the real number field or the complex number field), and let G be $GL(n,K)$. Consider a homogeneous form F of degree d in n variables $\xi_1,\ldots,\xi_n$ with coefficients in K: $F=\Sigma c_{i_1\ldots i_n} m_{i_1\ldots i_n}$ ($\Sigma i_\alpha=d$, $m_{i_1\ldots i_n}=(d!/\Pi(i_\alpha!))\xi_1^{i_1}\ldots\xi_n^{i_n}$). For each $\sigma \in G$, we define $\sigma\xi_i$ by $(\sigma\xi_1,\ldots,\sigma\xi_n)=(\xi_1,\ldots,\xi_n)\sigma^{-1}$ and then $(\sigma c)_{i_1\ldots i_n}$ by $F=\Sigma(\sigma c)_{i_1\ldots i_n}(\sigma m_{i_1\ldots i_n})$. Then the transformation

$$[\sigma]_d:(c_{d0\ldots0},\ldots,c_{i_1\ldots i_n},\ldots,c_{0\ldots0d})$$

$$\to((\sigma c)_{d0\ldots0},\ldots,(\sigma c)_{i_1\ldots i_n},\ldots,(\sigma c)_{0\ldots0d})$$

is a †linear transformation of a ${}_{d+n-1}C_{n-1}$-dimensional †affine space. Let us denote the matrix of the linear transformation by the same symbol $[\sigma]_d$:

$$^t(\ldots,(\sigma c)_{i_1\ldots i_n},\ldots)=[\sigma]_d{}^t(\ldots,c_{i_1\ldots i_n},\ldots).$$

Then $\varphi_d:\sigma\to[\sigma]_d$ is a rational representation of G.

Now fix an F such that the coefficients $c_{i_1\ldots i_n}$ are independent variables, and consider the action of G on the polynomial ring $R=K[\ldots,c_{i_1\ldots i_n},\ldots]$ defined by the rational representation φ_d. If g is a relative invariant, then $\sigma g=a(\sigma)g$ with a rational representation $\sigma\to a(\sigma)$. Hence there is an integer w such that $a(\sigma)=(\det\sigma)^w$. Then g is called an **invariant of weight** w (note that g is an $SL(n,K)$-invariant); g is an absolute invariant if and only if $w=0$. The group G acts naturally on the ring $R[\xi_1,\ldots,\xi_n]$ also. Then relative invariants in this case are called **covariants**. The **weight** of a covariant and the **absolute covariant** are defined as in the case of invariants. When we want to refer to n and d, we call the invariants (covariants) **invariants (covariants) of** n**-ary forms of degree** d: binary for $n=2$; ternary for $n=3$; linear forms for $d=1$; quadratic forms for $d=2$, etc.

For example, (1) if $n=2$, $d=2$, then the †discriminant $D=c_{02}c_{20}-c_{11}^2$ is an invariant of weight 2.

(2) Assume that $d=2$ (n arbitrary), and let u_{ij} be the coefficient of $\xi_i\xi_j$ in F, or, more precisely, $u_{ij}=u_{ji}=c_{\alpha_1\ldots\alpha_n}$ where (i) if $i=j$, then $\alpha_i=2$ and the other α_k are zero and (ii) if $i\neq j$, then $\alpha_i=\alpha_j=1$ and the other α_k are zero. In this case, $D=\det(u_{ij})$ is an invariant of weight 2.

(3) If $n=2$ and $d=4$, then $g_2=c_{40}c_{04}-4c_{13}c_{31}+3c_{22}^2$, $g_3=c_{04}c_{22}c_{40}-c_{04}c_{31}^2-c_{40}c_{13}^2+2c_{13}c_{22}c_{31}-c_{22}^3$ are invariants of respective weights 4, 6. The discriminant is expressed as $2^3(g_2^3-27g_3^2)$ and is an invariant of weight 12.

(4) For $d=1$, if we denote the coefficient of ξ_i in F by c_i, then $\sum_i c_i\xi_i$ is an absolute covariant and is also a vector invariant.

(5) For arbitrary n and d, $\det(\partial^2F/\partial\xi_i\partial\xi_j)$ is a covariant of weight 2 and is termed the **Hessian**.

Instead of one form F, we may take a finite number of homogeneous forms $F_1,\ldots,F_r$ of degree $d_1,\ldots,d_r$ such that the coefficients $c^{(i)}_{\alpha_1\ldots\alpha_n}$ are algebraically independent. Then we consider an action of $GL(n,K)$ on the polynomial ring $K[\ldots,c^{(i)}_{\alpha_1\ldots\alpha_n},\ldots][\xi_1,\ldots,\xi_n]$ given by $\varphi_{\alpha_i}:\sigma\to[\sigma]_{d_i}$ on the coefficients of F_i and by $\sigma\to{}^t\sigma^{-1}$ on ξ_j, as in the case of one form. Invariants in this case are called **covariants**, and covariants containing no ξ_i are called **invariants**. Weight, absolute covariants, and absolute invariants are defined similarly. The forms $F_1,\ldots,F_r$ are called **ground forms**, and the covariant is called a **covariant with ground forms** $F_1,\ldots,F_r$.

(6) If $r=n$, then the †Jacobian $\det(\partial F_i/\partial\xi_j)$ is a covariant of weight 1.

E. Multiple Covariants

Consider the situation described in Section C for the case of polynomials, and assume that K is a field of characteristic zero, $G=GL(n,K)$, and each ρ_i is either a φ_d (d arbitrary) or the contragredient mapping κ. Then these invariants are called **multiple covariants**, and **weights** and **absolute multiple covariants** are defined as before. Let s be the number of ρ_i equal to κ. Then the invariants and covariants of the preceding section correspond to the cases $s=0$ and $s=1$, respectively. Now assume that $\rho_i=\kappa$ if and only if $i=1,\ldots,s$. **Gram's theorem** states: for each $\alpha=1,\ldots,s$, let H_α be a polynomial in $x_j^{(i)}$ ($i>s$) homogeneous in $x_1^{(i)},\ldots,x_{n_i}^{(i)}$ for each i, and assume that the set V of common zeros of $H_1,\ldots,H_s$ (in the affine space of dimension $\sum_{i>s}n_i$) is G-stable. Then there exist a finite number of absolute multiple covariants $c_1,\ldots,c_t$ such that V is the set of $(\ldots,a_j^{(i)},\ldots)$ ($i>s$) satisfying the condition that $(\ldots,a_m^{(\alpha)},\ldots,a_j^{(i)},\ldots)$ is a zero point of $c_1,\ldots,c_t$ for any $a_m^{(\alpha)}$ with $\alpha\leqslant s$.

Since every rational action of $GL(n,K)$ is reductive, the set I of absolute multiple covariants (in a fixed polynomial ring over K) is a finitely generated ring over K. Furthermore, the set of multiple covariants of a given weight is a finitely generated I-module.

If we omit the assumption that K is of characteristic zero, it is difficult to define φ_d; this difficulty may be avoided by considering transformations of coefficients $a_{i_1\ldots i_n}$ of $F=\sum a_{i_1\ldots i_n}\xi_1^{i_1}\ldots\xi_n^{i_n}$. Although we can give similar definitions in that case, the theory does not proceed similarly because, for example, rational actions of $GL(n,K)$ are not reductive.

F. Invariants of Lie Groups

Let G be a Lie group (hence K must be either the real number field or the complex number field) and ρ be a †differentiable representation of G such that $\rho(G)\subset GL(n,K)$, and assume that an action of G on $K[x_1,\ldots,x_n]$ is defined by ρ. Then, by means of an infinitesimal transformation X_a corresponding to G, an invariant (or semi-invariant) is characterized as an element f satisfying the condition $X_a f = 0$ $(\forall X_a)$ (or $X_a f = \alpha_a f$, $\alpha_a \in K (\forall X_a)$). For instance, if $G = GL(2, K)$, then f is an invariant if $X_a f = 0$ for only two infinitesimal transformations that correspond to $\begin{pmatrix} 1 & t \\ 0 & 1 \end{pmatrix}$ and $\begin{pmatrix} 1 & 0 \\ t & 1 \end{pmatrix}$, respectively. Similarly, for each Lie group G, there exist a finite number of X_a such that $X_a f = 0$ for these X_a characterizes f as an invariant.

G. History

In connection with geometry, the theory of invariants, especially that of binary forms, was first studied by A. Cayley (*J. Reine Angew. Math.*, 30 (1846)). The theory was further developed by J. J. Sylvester, R. F. A. Clebsch, P. Gordan, and others. Since the theory was originated for applications to projective spaces, homogeneous semi-invariants were important; this is why semi-invariants were called invariants in the classical theory. On the other hand, in the theory of binary quadratic forms invariants of discontinuous groups were studied from the viewpoint of the theory of numbers. It was D. Hilbert who introduced clearly the notion of invariants for general groups. He proved the existence of basic invariants in the classical case, making use of the †Hilbert basis theorem. Hilbert's 14th problem (→ 198 Hilbert) is related to this result, but its answer is negative, that is, even in the case of a polynomial ring over the real number field or the complex number field, there are groups acting on the ring without basic invariants (M. Nagata). Though the theory of invariants has not been studied effectively for a long time, it is again under active study because of its importance in algebraic geometry.

References

[1] R. Weitzenböck, Invariantentheorie, Noordhoff, 1923.
[2] I. Schur and H. Grunsky, Vorlesungen über Invariantentheorie, Springer, 1968.
[3] H. Weyl, Classical groups, Princeton Univ. Press, revised edition, 1946.
[4] M. Nagata and T. Miyata, Remarks on matric groups, J. Math. Kyoto Univ., 4 (1965), 381–384.
[5] D. Mumford, Geometric invariant theory, Springer, 1965.
[6] J. Fogarty, Invariant theory, Benjamin, 1969.
[7] J. A. Dieudonné and J. B. Carrell, Invariant theory, old and new, Advances in Math., 4 (1970), 1–80.

227 (XVIII.16) Inventory Control

A mathematical formulation of **inventory control** theory must be based upon an inventory structure in which the following items are considered: (i) costs and revenue, (ii) demand, and (iii) deliveries. Item (i) involves 1. order and production costs, 2. storage costs, 3. discount rates, 4. penalties, 5. revenues (under the assumption that price and demand are not controlled by the firm), 6. costs of changing the production rate, 7. salvage costs, and so on. For item (ii), there are several different situations due to the combination of predictability and stability conditions of demand for the commodity. Current theories are concerned with two particular situations. In the first, the †probability distribution of demand is known to the firm. In the second, the date of occurrence of orders is a †random variable with a known probability distribution, while the amount of demand is a known constant. A †minimax principle is applied when some of these probability distribution functions are unknown to the firm.

Problems in (iii) are due to delays after an order for inventory is placed or a decision is made to produce for an uncertain demand. In some models we assume that all orders and deliveries take place at a sequence of equally spaced time points, while in other cases we assume that demands in successive periods are †independent random variables. Let x_t be the initial stock level in a period, y_t the stock level after ordering, z_t the amount ordered for inventory, and ξ_t the demand.

The revenue equals price r multiplied by deliveries $\xi_t - \max(0, \xi_t - y_t)$. The ordering cost is $c(z_t)$, the penalty cost is $p(\xi_t - y_t)$ when $\xi_t - y_t > 0$ and zero when $\xi_t - y_t \leqslant 0$, the storage cost is assumed to depend upon the inventory y_t and is designated as $h(y_t)$, the salvage cost is $v(y_t - \xi_t)$, and the cost associated with changing the rate of production is $G(z_t - z_{t-1})$. Then the profit π_t is given by

$$\pi_t = r(\xi_t - \max(0, \xi_t - y_t)) - c(z_t) - p(\xi_t - y_t) - h(y_t) - v(y_t - \xi_t) - G(z_t - z_{t-1}). \quad (1)$$

The †optimal policy is given as follows. (a) If the demand ξ_t is known, then we choose z_t to maximize π_t in (1); of course z_{t-1} is given, and y_t is determined by $z_t = y_t - x_t$. (b) If demand is considered to be a †random variable, then z_t is chosen to maximize the expected value of π_t. Since in any case ξ_t is independent of the firm's control, the term $r\xi_t$ can be ignored when comparing policies. The term $r\max(0, \xi_t - y_t)$ can then be absorbed in the penalty function. The problem in the static model can then be restated as that of minimizing the loss

$$L_t(z_t|x_t)=c(z_t)+p(\xi_t-y_t)+h(y_t)$$
$$+v(y_t-\xi_t)+G(z_t-z_{t-1}), \qquad (2)$$

or, if ξ_t is a random variable, of maximizing the expected value in (2). If we consider T time periods, the discounted stream of losses is

$$\lambda(z|x_1)=\sum_{t=1}^{T} a^{t-1}L_t(z_t|x_t)+V(y_T-\xi_T), \quad (3)$$

where a is a discount factor, $0<a<1$, V is salvage cost discounted to the initial period, and z is the sequence of z_t in (3). Furthermore, in an infinite sequence of time periods, the total discounted loss is given by

$$\lambda(z|x_1)=\sum_{t=1}^{\infty} a^{t-1}L_t(z_t|x_t).$$

In general it is not easy to give an optimal strategy in a constructive way by appealing to an analysis of these mathematical models. Instead, there are three ways of handling these models in specific situations. The first is concerned with conventional calculation formulas useful when various parameters in the models are known. Examples are the formulas giving economical lot sizes for the deterministic and stochastic cases. The second uses a functional equation valid for an optimal policy and solves it by a †dynamic programming method due to R. Bellman.

The third method is application of †mathematical programming techniques in which nonnegativity and boundary conditions play essential roles. Detailed discussions were given by the Stanford University group. Several well-known inventory strategies, which are actually used in practice, can be deduced from these general considerations. For instance, the total expected cost incurred from use of an (s,S) policy (or **two-bin system**) satisfies a solvable renewal equation. This policy is implemented as follows: One orders or produces only if the present stock level falls below the given value s. When ordering is done, the stock is increased to the second value S. For each assigned specific inventory policy, the resultant fluctuating inventory level is a stochastic phenomenon, governed primarily by statistical aspects of demand. The operating characteristics of an (s,S) policy are determined in this way.

References

[1] K. J. Arrow, S. Karlin, and H. E. Scarf, Studies in the mathematical theory of inventory and production, Stanford Univ. Press, 1958.

[2] H. E. Scarf, D. M. Gilford, and M. W. Shelly (eds.), Multistage inventory models and techniques, Standord Univ. Press, 1963.

[3] T. M. Whitin, The theory of inventory management, Princeton Univ. Press, 1953.

[4] T. Yokoyama and Y. Hukuba, Zaiko kanri (Japanese: Inventory control), Kyôritu, 1959.

[5] T. Kitagawa, Inventory control (Japanese), Gendai tôkeigaku daiziten (Encyclopedia of modern statistics), Tôyôkeizai sinpôsya, 1962.

[6] G. Hadley and T. M. Whitin, Analysis of inventory systems, Prentice-Hall, 1963.

228 (X. 36) Isoperimetric Problems

A. The Classical Isoperimetric Problem

Two curves are called **isoperimetric** if their perimeters are equal. The term *curve* is used here to mean a †Jordan curve. The classical **isoperimetric problem** is to find, among all curves J with a given perimeter L, the curve enclosing the maximum area. This problem is also called the **special isoperimetric problem** or **Dido's problem**. Its solution is a circle. The analogous problem in 3-dimensional space has a sphere as its solution; that is, among all closed surfaces with a given surface area, the sphere has the maximum volume. The following variational problem can be regarded as a generalization of the classical isoperimetric problem: To find the curve $C:y=f(x)$ that gives the maximum value of the functional $\int_C F(x,y,y')dx$ under the subsidiary condition $\int_C G(x,y,y')dx=$ constant. This is sometimes called the **generalized isoperimetric problem**. The classical isoperimetric problem can be solved by variational methods. It can also be solved by using inequalities among quantities in the figure. For example, the inequality

$$L^2-4\pi F\geqslant 0 \qquad (1)$$

between the area F and the perimeter L of a Jordan curve J solves the problem, since the equality holds only for a circle. For refinements of (1) there are further inequalities due to T. Bonnesen (1921):

$$L^2-4\pi F\geqslant(L-2\pi r)^2,$$
$$L^2-4\pi F\geqslant(2\pi R-L)^2,$$
$$L^2-4\pi F\geqslant \pi^2(R-r)^2,$$

where r is the radius of the largest circle inscribed in the curve J and R the radius of the smallest circumscribed circle. These inequalities can also be used to solve the isoperimetric problem. Moreover, we have the following inequality for curves on the sphere with radius a:

$$L^2-4\pi F+F^2a^{-2}\geqslant 8\pi a^2\sin\frac{R-r}{4a(1+2\pi)}$$

(F. Bernstein, 1905). For curves on the surface of negative constant curvature $-1/a^2$, we have

$$L^2-4\pi F-F^2a^{-2}$$

$$\geqslant\frac{1}{4}a^2(4\pi+Fa^{-2})\left(\tanh\frac{R}{2a}-\tanh\frac{r}{2a}\right)^2$$

(L. A. Santaló [2]). From these inequalities, we see that the circle remains the solution to the isoperimetric problem in each of the non-Euclidean planes.

The corresponding problem in the 3-dimensional case is more difficult. Without going into detail, we offer the following example: For an †ovaloid with surface area S and volume V,

$$S^3-36\pi V^2\geqslant 0,$$

where the equality holds only for the sphere.

B. Isoperimetric Inequalities on Eigenvalues

In recent years the concept of **isoperimetric inequality** has been extended to include all inequalities connecting two or more geometric or physical quantities depending on the shape and size of a figure. For example, it includes inequalities on †eigenvalues of partial differential equations under given boundary conditions.

Lord Rayleigh conjectured in 1877 that, in the equation $\Delta u+\lambda u=0$ for a vibrating membrane on a region D of fixed area F (with $u=0$ on the boundary), the first eigenvalue λ_1 is least when D is a circle. This conjecture is true. In fact, in 1923 G. Faber and E. Krahn proved independently that

$$\lambda_1\geqslant(\pi/F)j, \tag{2}$$

and that the equality in (2) holds if and only if the domain D is a circle, where $j=2.4048\ldots$ is the first positive zero of the †Bessel function $J_0(x)$. For the second eigenvalue λ_2 of the same problem, the circle does not give the minimum value. I. S. Hong (1954) gave the inequality

$$\lambda_2\geqslant(2\pi/F)j^2$$

and showed that λ_2 approaches its greatest lower bound, $(2\pi/F)j^2$, as the shape of the domain approaches a figure consisting of two equal tangent circles, each having area $F/2$ [4].

Many other results were found with regard to isoperimetric inequalities on eigenvalues of partial differential equations, for example, relating eigenvalues for a membrane under other boundary conditions, such as $\partial u/\partial x=0$, and for other types of equations, such as $\Delta\Delta u-\lambda u=0$.

A method devised by J. Steiner and called **symmetrization** has been a powerful means of discovering isoperimetric inequalities. **Steiner's symmetrization** with respect to the line l changes the plane domain P into another plane domain Q characterized as follows: Q is symmetric with respect to l, any straight line g perpendicular to l that intersects one of the domains P or Q also intersects the other, both intersections have the same length, and the intersection of Q is a segment bisected by l. This operation can be extended to a space of higher dimension by replacing l with a hyperplane. Steiner's symmetrization preserves area (or volume) and diminishes perimeter (or surface area). Steiner first used these properties to solve the classical isoperimetric problem in 1838. In 1945, G. Pólya and G. Szegö found that the electrostatic capacity of a solid is diminished by Steiner's symmetrization. The concepts developed in their papers made possible a systematic technique of dealing with many isoperimetric inequalities and estimations of mathematical and physical quantities.

References

For the classical isoperimetric problem,
[1] W. Blaschke, Kreis und Kugel, Verlag von Veit, 1916 (Chelsea, 1949).
[2] L. A. Santaló, La desigualdad isoperimetrica sobre superificies de curvatura constante negativa, Rev. Univ. Tucumán, (A) 3 (1942), 243–259.
For isoperimetric inequalities on eigenvalues,
[3] G. Pólya and G. Szegö, Isoperimetric inequalities in mathematical physics, Princeton Univ. Press, 1951.
[4] I. Hong, On an inequality concerning the eigenvalue problem of a membrane, Kôdai Math. Sem. Rep. (1954), 113–114.
[5] T. Kubo, Symmetrization and its applications (Japanese), Sûgaku, 9 (1957), 45–55.

J

229 (XX.27) Jacobi, Carl Gustav Jacob

Carl Gustav Jacob Jacobi (December 10, 1804–February 18, 1851) was born into a wealthy banking family in Potsdam, Germany. He was well educated at home and was highly cultured in many areas. He entered the University of Berlin, studied mathematics largely on his own through Euler's texts, and obtained the doctorate in 1825. The following year he became a private lecturer at the University of Königsberg, and in 1831 a professor. For the next seventeen years he worked vigorously in Königsberg, where his influence was considerable. Toward the end of his life, his health failed; moreover, he lost his property and met with general misfortune because of the political situation of the time. He made no further contributions after 1843; he died of smallpox at 47.

Because he had an intense personality, there were times when he invited the animosity of people; however, he did have early contact with †Abel, and in his later years he enjoyed the friendship of †Dirichlet. Jacobi's mathematical works lacked formal completeness and rigor, but were very original and contributed to many fields. The †Hamilton-Jacobi equation in dynamics and the †Jacobian determinant of a differentiable mapping are well-known products of his ideas, but even more noteworthy are his contributions to the theory of elliptic functions [2].

References

[1] K. G. J. Jacobi's gesammelte Werke I–VII, edited by C. W. Borchardt and K. Weierstrass, G. Reimer, 1881–1891 (Chelsea, second edition, 1969).
[2] G. J. Jacobi, Fundamenta nova theoriae functionum ellipticarum, Bornträger, 1829.
[3] F. Klein, Vorlesungen über die Entwicklung der Mathematik im 19. Jahrhundert I, II, Springer, 1926, 1927 (Chelsea, 1956).
[4] T. Takagi, Kinsei sûgakusidan (Japanese; Topics from the history of mathematics of the 19th century), Kawade, 1943.

230 (XX.7) Japanese Mathematics (Wasan)

Before the introduction of Western mathematics, indigenous Japanese mathematics developed along its own characteristic lines. In Japanese, this form of mathematics is called **wasan**. The first development took place in the 8th century A.D. during the Nara era under the influence of the Tang dynasty in China. After a period of decline came another period of development from the 13th to the 17th century. During this second wave, Chinese mathematical books such as *Suanhsueh chimeng* and *Suanfa tangtsung* were imported, along with the abacus, or *soroban* as it is called in Japanese, and calculating rods, *sangi* in Japanese, with which Chinese mathematicians performed algebraic operations. Japanese mathematicians absorbed these methods and invented and developed their own written algebra, called *endan-zyutu* or *tenzan*.

The fundamental concepts of *wasan* are attributed to Seki Takakazu, or Seki Kowa (Seki is the surname; likewise for other Japanese names in this article), Takebe Katahiro, and Kurusima Yosihiro. Its main developments stem from Azima Naonobu and Wada Yasusi (or Wada Nei), among others. *Wasan* scholars obtained interesting and significant results, but they pursued mathematics as an art in the Japanese manner rather than as a science in the Western sense. *Wasan* had no philosophical background as did the Greek tradition, nor had it an intimate relation with the natural sciences. Thus it lacked the character of systematized science and dissolved after the introduction of Western mathematics into the school system by the Meizi government (1867–1912).

Among *wasan* works of the earlier period, *Zinkôki* by Yosida Mituyosi (1598–1672), the first edition of which appeared in 1627, contributed much to popularize the *soroban* and to arouse general interest in mathematics.

Seki Takakazu (1642?–1708) was born in Huzioka in the Gunma prefecture. Some historians say he learned mathematics from Takahara Yositane, while others say he was completely self-taught. His achievements include the following: (1) the invention of *endan-zyutu*, or written algebra; (2) the discovery of determinants; (3) the solution of numerical equations by a method similar to †Horner's; (4) the invention of an iteration method to solve equations, similar to Newton's; (5) the introduction of derivatives and of discriminants of polynomials; (6) the discovery of conditions for the existence of positive and of negative roots of polynomials; (7) a method of finding maxima and minima; (8) the transformation theory of algebraic equations; (9) continued fractions; (10) the solution of some Diophantine equations; (11) the introduction of †Bernoulli numbers; (12)

the study of regular polygons; (13) the calculation of π and the volume of the sphere; (14) †Newton's interpolation formula; (15) some properties of ellipses; (16) the study of the †spirals of Archimedes; (17) the discovery of the Pappus-Guldin theorem; (18) the study of magic squares; and (19) the theoretical study of some questions of mathematical recreations (called *mamako-date*, *metuke-zi*, etc.).

Takebe Takahiro (1664–1739) was a disciple of Seki. He is the author of the book *Enri tetuzyutu*. (*Enri*, or circle theory, is one of the favorite subjects of *wasan* scholars.) It contains the formula:

$$\left(\frac{1}{2}\arcsin x\right)^2 = \frac{x^2}{2} + \frac{2^2 \cdot x^4}{4} + \frac{2^2 \cdot 4^2 \cdot x^6}{6} + \dots .$$

He also obtained other formulas of trigonometry and some approximation formulas, by means of which he compiled trigonometric tables to 11 decimal places. In collaboration with Seki he wrote the 20 volumes of *Taisei sankei* and *Hukyû tetuzyutu*, the latter elucidating the methodology of the Seki school. It contains a value of π to 42 decimal places.

Kurusima Yosihiro (?–1757) was an original scholar influenced by Nakane Genkei (1662–1733), a disciple of Takebe. He generalized an approximation formula for sines obtained by Takebe, treated problems of maxima and minima involving trigonometric functions, improved the theory of determinants and the theory of equations, obtained a formula for $S_p = 1^p + \dots + n^p$ without using Bernoulli numbers, and found a relation between a, b, c, n by eliminating x from

$$x + (x+c) + \dots + (x + (n-1)c) = a,$$

$$x^k + (x+c)^k + \dots + (x + (n-1)c)^k = b.$$

Furthermore, he studied †Euler's function $\varphi(n)$ before Euler and obtained the †Laplace expansion theorem for determinants before Laplace. He is said to have contributed to *Hôen sankei*, an important work on *enri* written by Matunaga Yosisuke (c. 1694–1744) in which a value of π is given up to the 50th decimal place. The Seki school, continued under Takebe, Nakane, Kurusima, and Matunaga, became a center of *wasan*. Scholars of this school lived mainly in Edo (the ancient name for Tokyo). Yamazi Nusizumi (or Syuzyû) (1704–1772) studied *wasan* with Nakane, Kurusima, and Matunaga. Arima Yoriyuki (1712–83), one of his disciples, for the first time made public the teachings of this school in the book *Syûki sanpô*. The practice of dedicating to Shinto shrines or Buddhist temples tablets engraved with solved mathematical problems became popular during this period.

Azima Naonobu (1739–98) was a disciple of Yamazi. He improved *enri*, simplified its theory, and amplified its applications. He treated problems of finding volumes involving double integrals, discovered the binomial theorem for exponent $1/n$, compiled a table of logarithms to 14 decimal places, and treated Diophantine problems. No trace of demonstrative geometry is found in *wasan*, but Azima and his school did treat geometric problems, such as †Malfatti's problem, dealing with several circles tangent to each other.

Wada Yasusi (or Wada Nei, 1787–1840) studied *wasan* with Kusaka Makoto (1764–1839), a disciple of Azima. He made tables containing more than 100 definite integrals, including, for example,

$$\int_0^1 x^p (1-x)^q \, dx, \quad \int_0^1 x^p (1-x^2)^q \, dx.$$

However, there is no evidence that *wasan* scholars, even in this period, knew of the fundamental theorem of infinitesimal calculus.

Apart from the Seki school, there were Tanaka Yosizane (1651–1719), a contemporary of Seki, and his disciple Iseki Tomotoki (c. 1690). Tanaka is said to have ranked with Seki in his work on determinants and magic squares, but most of his writings have been lost. Iseki wrote a text on determinants called *Sanpô hakki* (1690), the first of its kind in the history of mathematics. A little later, Takuma-ryû (Takuma school) was formed in Osaka. Inô Tadataka (1745–1821), famous for making the first precise map of Japan, had studied *wasan* with Takahasi Yositoki (1764–1804), who belonged to this school. Aida Yasuaki (1747–1817), a contemporary of Azima, founded *Saizyô-ryû*, or the "superlative school," and rivaled Huzita Sadasuke (1734–1807) of the Seki school.

Toward the end of the Tokugawa era, the study of geometric problems became popular among *wasan* scholars, such as Hasegawa Kan (1782–1838), Utida Gokan (1805–82), and his disciple Hôdôzi Zen (1820–68), who used the method of †inversion. Hasegawa wrote *Sanpô sinsyo*, a popular work containing an explanation of the methods of *enri*.

The influence of Western mathematics is hardly recognizable outside astronomy, calendar making, and the compilation of logarithmic tables. However, some *wasan* scholars took a more positive attitude and began studying Western mathematics towards the close of the Tokugawa period, thus helping to lay the foundations for the development of mathematics in Japan in the new era.

Following the Meizi restoration, Kikuchi Dairoku (1855–1917), Hayashi Tsuruichi

(1873–1935), Fujiwara Matsusaburo (1861–1933), and recent scholars contributed much to preserve *wasan* literature and to clarify its content, but their undertaking has not yet been completed.

References

[1] T. Endo, Zôsyû Nippon sûgakusi (Japanese; History of Japanese mathematics, revised edition), Kôseisya, new edition, 1960.
[2] Hayashi hakushi wasan kenkyû syûroku (Japanese; Collection of Dr. T. Hayashi's papers on the history of Japanese mathematics), Kaiseikan, 1937.
[3] M. Fujiwara, Meizi-zen Nippon sûgakusi (Japanese; History of mathematics in Japan before the Meizi era), I–V, under the auspices of the Japan Academy. Iwanami, 1954–1960.
[4] H. Kato, Wasan no kenkyû, gyôretusiki oyobi enri (Japanese; Studies on the history of Japanese mathematics, Determinants and theory of circles), Kaiseikan, 1944.
[5] H. Kato, Wasan no kenkyû, zaturon (Japanese; Studies on the history of Japanese mathematics, Miscelleneous) I–III, Japan Society for the Promotion of Science, 1954–1956.
[6] H. Kato, Wasan no kenkyû, hôteisikirion (Japanese; Studies on the history of Japanese mathematics, Algebraic equations), Japan Society for the Promotion of Science, 1957.
[7] H. Kato, Wasan no kenkyû, seisûron (Japanese; Studies on the history of Japanese mathematics, Number theory), Japan Society for the Promotion of Science, 1946.
[8] K. Ogura, Nippon no sûgaku (Japanese; A short history of Japanese mathematics), Iwanami, 1940.
[9] Y. Mikami, Tôzai sûgakusi (Japanese; History of mathematics in Orient and Occident), Kyôritu, 1928.
[10] Y. Mikami, The development of mathematics in China and Japan, Teubner, 1913 (Chelsea, 1961).

231 (III.23) Jordan Algebras

A. Definitions

Let A be a †linear space over a field K. If there is given a †bilinear mapping (multiplication) $A \times A \rightarrow A$, the space A is called a **distributive algebra** (or **nonassociative algebra**) over K. In particular, if this multiplication satisfies the associative law, A is called an **associative algebra** over K (→ 31 Associative Algebras). We assume that A is a distributive algebra over K of finite †dimension over K. Denote by $E(A)$ the associative algebra of all K-†endomorphisms of the K-linear space A. With every $a \in A$ we associate $R_a \in E(A)$, $L_a \in E(A)$ by $R_a(x) = xa$, $L_a(x) = ax$, where x belongs to A. The †subalgebra of $E(A)$ generated by the R_a, L_a $(a \in A)$ and the †identity mapping of A is called the **enveloping algebra** of A. The left, right, and two-sided †ideals of A are defined as in the case of associative algebras. An element c of A is said to be in the **center** of A if (i) $ac = ca$ and (ii) $a(bc) = (ab)c$, $a(cb) = (ac)b$, and $c(ab) = (ca)b$ for every a, b in A. We denote the product of two elements a, b in A by $a \cdot b$ in order to distinguish it from multiplication in the case of associative algebras. Denote $a \cdot a$ by $a^{\cdot 2}$ and put $A^{\cdot 2} = \{a_1 \cdot a_2 | a_1, a_2 \in A\}$. Define $A^{(n)}$ successively by $A^{(0)} = A$, $A^{(1)} = A^{\cdot 2}, \ldots, A^{(k+1)} = (A^{(k)})^{\cdot 2}$. Then A is called a **solvable algebra** if $A^{(n)} = 0$ for some n and a **nilalgebra** if every element of A is †nilpotent. A distributive algebra A is called a **Jordan algebra** if the following two conditions are satisfied for every a, u in A: (i) $a \cdot u = u \cdot a$ and (ii) $a^{\cdot 2} \cdot (u \cdot a) = (a^{\cdot 2} \cdot u) \cdot a$. A distributive algebra A is called an **alternative algebra** if the following two conditions are satisfied for every a, u in A: (i) $a \cdot u^{\cdot 2} = (a \cdot u) \cdot u$ and (ii) $u^{\cdot 2} \cdot a = u \cdot (u \cdot a)$. An alternative algebra A is called an **alternative field** if $L_a(A) = A = R_a(A)$ for every a in A such that $a \neq 0$. Generalizing these algebras we obtain the notion of a **power associative algebra**. A distributive algebra A is a power associative algebra if every element of A generates an associative subalgebra. Jordan algebras and alternative algebras are power associative.

We will consider Jordan algebras A over a field K. We assume that the †characteristic of K is not 2 and that A is of finite dimension over K. If A and B are Jordan algebras, a linear mapping $\sigma : A \rightarrow B$ is called a **Jordan homomorphism** if (i) $(a^{\cdot 2})^\sigma = (a^\sigma)^{\cdot 2}$ for every a in A and (ii) $(a \cdot b \cdot a)^\sigma = a^\sigma \cdot b^\sigma \cdot a^\sigma$ for every a, b in A (where we denote by a^σ the image of $a \in A$ by the mapping σ). If B does not contain a †zero divisor, then $(a \cdot b)^\sigma = a^\sigma \cdot b^\sigma$ or $(a \cdot b)^\sigma = b^\sigma \cdot a^\sigma$.

Let A be an associative algebra. Define a new multiplication $\cdot$ in A by $a \cdot b = (ab + ba)/2$. We then have a Jordan algebra A^+. A subalgebra of the Jordan algebra A^+ is called a **special Jordan algebra**. Let $K[x_1, \ldots, x_n]$ be the noncommutative free ring in the indeterminates $x_1, \ldots, x_n$ (that is, $K[x_1, \ldots, x_n]$ is the associative algebra over K that has as its K-bases the free semigroup with identity element 1 over the free generators $x_1, \ldots, x_n$). The subalgebra $K[x_1, \ldots, x_n]^+$ generated by 1 and the x_i is called the **free special Jordan algebra** of n generators and is denoted by $J_0^{(n)}$. A Jordan algebra A is **special** if and only if there is an isomorphism from A onto B_0^+, where B is some associative algebra. A Jordan algebra that is not special is called **exceptional**. All homomorphic images of $J_0^{(2)}$ are special. Denote by $\mathfrak{K}$ the ideal of $J_0^{(3)}$ generated by $x^{\cdot 2} - y^{\cdot 2}$ (note that $J_0^{(3)} \subsetneq K[x,y,z]$). Then $J_0^{(3)}/\mathfrak{K}$ is exceptional. A Jordan algebra is special if it contains the unity element and is generated by two elements. If A is an alternative algebra, the associated Jordan algebra A^+ is special.

Condition (ii) for a Jordan algebra A is equivalent to $[R_a, R_{a^{\cdot 2}}] = 0$. In A, we have $[R_a, R_{b \cdot c}] + [R_b, R_{c \cdot a}] + [R_c, R_{a \cdot b}] = 0$; and $[[R_c, R_a], R_b] = R_{[a, b \cdot c]}$. Here we put $[S, T] = ST - TS$, $[a, b \cdot c] = (a \cdot b) \cdot c - a \cdot (b \cdot c)$. Such an equation in A is called an **identity** in a Jordan algebra.

B. Structure of Jordan Algebras

A Jordan algebra A has a unique largest solvable ideal N, which contains all nilpotent ideals of A and is called the **radical** of A. If $N = 0$, A is called **semisimple**. The quotient A/N is always semisimple. A semisimple Jordan algebra A contains the unity element and can be decomposed into a direct sum $A = A_1 \oplus \ldots \oplus A_r$ of minimal ideals A_i. Each A_i is a †simple algebra. In particular, if K is of characteristic 0, there is a semisimple subalgebra S of A such that $A = S \oplus N$. Let e be an idempotent element of A, and let $\lambda \in K$. Put $A_e(\lambda) = \{x \mid x \in A, e \cdot x = \lambda x\}$. Then we have

$$A = A_e(1) \oplus A_e(1/2) \oplus A_e(0).$$

This decomposition of A is called the **Peirce decomposition** of A relative to e. Suppose that the unity element 1 is expressed as a sum of the mutually orthogonal idempotents e_i. Then, putting $A_{i,i} = A_{e_i}(1)$, $A_{i,j} = A_{e_i}(1/2) \cap A_{e_j}(1/2)$, we have $A = \sum_{i \leqslant j} \oplus A_{i,j}$. The $A_{i,j}$ are called **Peirce spaces**. Furthermore, suppose that for every i, $A_{i,i}$ is of the form $A_{i,i} = K \cdot e_i + N_i$, with N_i a nilpotent ideal of $A_{i,i}$. In this case, A is called a **reduced algebra** and the number of the e_i is called the **degree** of A.

Let D be an alternative algebra with the unity element 1 and with an †involution $^-$. Furthermore, suppose that there is a †quadratic form $Q(X)$ such that $x \cdot \bar{x} = \bar{x} \cdot x = Q(x) \cdot 1$ $(x \in D)$ and that $f(X, Y) = (Q(X+Y) - Q(X) - Q(Y))/2$ is a †nondegenerate bilinear form. Then D is called a **composition algebra**.

Reduced Jordan algebras A are classified into the following three types:

(1) $A = K \cdot 1$.

(2) There exist idempotents e_1, e_2 such that $A = K \cdot 1 \oplus K \cdot (e_1 - e_2) \oplus A_{1,2}$, where $A_{1,2} \neq 0$; and furthermore, the multiplication of A is determined as follows: Let $x = \alpha(e_1 - e_2) + a_{1,2}$, $y = \beta(e_1 - e_2) + b_{1,2}$ be in A with $\alpha, \beta \in K$, $a_{1,2} \in A_{1,2}$, $b_{1,2} \in A_{1,2}$. Then $x \cdot y = (\alpha\beta + f(a_{1,2}, b_{1,2})) \cdot 1$, where f is a nondegenerate symmetric bilinear form.

(3) Let D be a composition algebra with the involution $^-$, and let D_n be the †total matrix algebra over D of degree n. Let $r = \operatorname{diag}\{r_1, \ldots, r_n\}$ $(r_i \neq 0)$ be given in D_n. Define $J: D_n \to D_n$ by $a^J = r \cdot \bar{a} \cdot r^{-1}$. Then A is of the form $A = \{x \in D_n \mid x = x^J\}$.

Let A be a special, reduced Jordan algebra such that $A_{i,i} = Ke_i$. If A is of degree 2, then A is a †Clifford algebra. If A is of degree $\geqslant 3$, then A is classified into five types.

Let A be a simple Jordan algebra over a field K. Then there is an extension field P of K of finite degree such that A_P $(= A \otimes_K P)$ is isomorphic to one of the following five types: (i) P_n^+, where P_n is the total matrix algebra of degree n over P; (ii) the subalgebra of P_n^+ consisting of all symmetric matrices in P_n^+; (iii) $\{x \in P_{2m}^+ \mid x^J = x\}$, where

$$x^J = q^{-1}\, {}^t x q, \qquad q = \begin{pmatrix} 0 & I_m \\ -I_m & 0 \end{pmatrix},$$

with ${}^t x$ the transpose of x and I_m the unit matrix of degree m; (iv) an algebra generated by the generators $s_0, s_1, \ldots, s_n$ together with the defining relations $s_0 \cdot s_i = s_i$, $s_i^{\cdot 2} = s_0$, $s_i \cdot s_j = 0$ $(i \neq j)$; (v) $\mathcal{H}_3^+$, where $\mathcal{H}_3$ is the algebra of all 3×3 Hermitian matrices over a †Cayley algebra.

C. Representations of Jordan Algebras

A **representation** S of a Jordan algebra A on a K-linear space M is a K-linear mapping $a \to S_a$ from A into the associative algebra $E(M)$ of all K-endomorphisms of M such that (i) $[S_a, S_{b \cdot c}] + [S_b, S_{c \cdot a}] + [S_c, S_{a \cdot b}] = 0$ and (ii) $S_a S_b S_c + S_c S_b S_a + S_{(a \cdot c) \cdot b} = S_a S_{b \cdot c} + S_b S_{a \cdot c} + S_c S_{a \cdot b}$ (for all a, b, c in A). A K-linear space M is called a **Jordan module** of A if there are given bilinear mappings $M \times A \to M$ (denoted by $(x,a) \to x \cdot a$), $A \times M \to M$ (denoted by $(a,x) \to a \cdot x$) such that for every $x \in M$ and every a, b, $c \in A$, (i) $x \cdot a = a \cdot x$; (ii) $(x \cdot a) \cdot (b \cdot c) + (x \cdot b) \cdot (a \cdot c) + (x \cdot c) \cdot (a \cdot b) = (x \cdot (b \cdot c)) \cdot a + (x \cdot (a \cdot c)) \cdot b + (x \cdot (a \cdot b)) \cdot c$; and (iii) $x \cdot a \cdot b \cdot c + x \cdot c \cdot b \cdot a + x \cdot a \cdot c \cdot b = (x \cdot c) \cdot (a \cdot b) + (x \cdot a) \cdot (b \cdot c) + (x \cdot b) \cdot (a \cdot c)$. As usual, there is a natural bijection between the representations of A and the Jordan modules of A. A **special representation** of a Jordan algebra A is a homomorphism $A \to E^+$, where E is an associative algebra. Among the special representations of A, there exists a unique universal one in the following sense. There exists a special representation $S: A \to U^+$ with the following property: For every special representation $\sigma: A \to E^+$ there exists a unique homomorphism $\eta: U^+ \to E^+$ such that $\sigma = \eta S$. The pair (U, S) is uniquely determined. Furthermore, if A is n-dimensional over K, U is of dimension $\binom{2n+1}{n}$ over K. The pair (U, S) is called the **special universal enveloping algebra** of A. A is special if and only if $S: A \to U^+$ is injective.

A Jordan algebra A has only a finite number of inequivalent †irreducible Jordan modules. Suppose that the base field K is of characteristic 0. Let S be a representation of A. If N is the radical of A, $S(N)$ is contained in the radical of the associative algebra $S(A)^*$ generated by $S(A)$. If A is semisimple, so is $S(A)^*$. In general, the radical R of $S(A)^*$ is an ideal of $S(A)^*$ generated by $S(N)$. Furthermore, for every semisimple subalgebra T of A such that $A = T \oplus N$, we have $S(A)^* = S(T)^* \oplus R$.

References

[1] A. A. Albert, Non-associative algebras I, II, Ann. of Math., (2) 43 (1942), 685–707.
[2] A. A. Albert, A structure theory for Jordan algebras, Ann. of Math., (2) 48 (1947), 546–567.
[3] N. Jacobson, General representation theory of Jordan algebras, Trans. Amer. Math. Soc., 70 (1951), 509–530.
[4] F. D. Jacobson and N. Jacobson, Classification and representation of semi-simple Jordan algebras, Trans. Amer. Math. Soc., 65 (1949), 141–169.
[5] N. Jacobson and C. E. Rickart, Jordan homomorphisms of rings, Trans. Amer. Math. Soc., 69 (1950), 479–502.
[6] N. Jacobson, Structure and representations of Jordan algebras, Amer. Math. Soc. Colloq. Publ., 1968.
[7] H. Braun and M. Koecher, Jordan-Algebren, Springer, 1966.

K

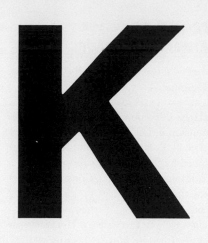

232 (VII.12) Kähler Manifolds

A. Definitions

Let X be a †complex manifold of complex dimension n. Let J denote the †tensor field of type (1, 1) of †almost complex structure induced by the complex structure of X (→ 74 Complex Manifolds). Considering J as a linear transformation of vector fields on X, we have $J^2=-1$. A †Riemannian metric g of class C^∞ on X is called a **Hermitian metric** on X if $g(x,y)=g(Jx,Jy)$ for any two vector fields x,y on X. On a †paracompact complex manifold X there always exists a Hermitian metric. If we put $\Omega(x,y)=g(Jx,y)$ for a Hermitian metric g, then Ω is an †alternating covariant tensor field of order 2, and hence a †differential form of degree 2. We call Ω the **fundamental form** associated with the Hermitian metric g. If the †exterior derivative of Ω vanishes, i.e., $d\Omega=0$, the given Hermitian metric g on X is called a **Kähler metric** on X, and X is called a **Kähler manifold**. With a †holomorphic local coordinate system $(z^1, \ldots, z^n)$ on a coordinate neighborhood U in X, we can write $g=\sum_{\alpha,\beta=1}^{n} g_{\alpha\beta}\,dz^\alpha\,d\bar{z}^\beta$ in U, where the matrix $(g_{\alpha\beta})$ is an $n\times n$ †positive definite Hermitian matrix. Then the fundamental form can be expressed as

$$\Omega=(\sqrt{-1}/2)\sum g_{\alpha\beta}\,dz^\alpha\wedge d\bar{z}^\beta.$$

If a complex manifold X with a Hermitian metric g is a Kähler manifold, then X has the following properties. (1) For every point p of X there exists a real-valued function ψ of class C^∞ on a suitable neighborhood of p such that $g_{\alpha\beta}=\partial^2\psi/\partial z^\alpha\partial\bar{z}^\beta$. (2) For every point p there exists a holomorphic local coordinate system at p whose real and imaginary parts form a †geodesic coordinate system at p in the weak sense. (We say a coordinate system $(x_1,\ldots,x_n)$ is a **geodesic coordinate system at** p **in the weak sense** if $[\nabla_{\partial/\partial x^i}(\partial/\partial x^j)]_p=0$, $i,j=1,\ldots,n$ (→ 403 Tensor Calculus). Each of these properties characterizes a Kähler metric. The Kähler metric was introduced by E. Kähler with property (1) as the definition. W. V. D. Hodge applied it to the theory of harmonic integrals [4].

B. Harmonic Forms on Compact Kähler Manifolds (→ 196 Harmonic Integrals)

We now consider complex-valued differential forms on a compact complex manifold X of complex dimension n endowed with a Hermitian metric. The operators d and $*$ on real-valued differential forms can be uniquely extended to complex linear operators, and the inner product of real differential forms can be extended to a Hermitian inner product of the form $(\varphi,\psi)=\int_X\varphi\wedge\overline{*\psi}$. Then the †adjoint operator δ of d is also complex linear, and the decomposition $d=d'+d''$ gives rise to $\delta=\delta'+\delta''$, where δ' and δ'' are the adjoint operators of d' and d'', respectively. We also have $\delta'=(-1)*d''*$ and $\delta''=(-1)*d'*$. Define the operator L by $L\varphi=\Omega\wedge\varphi$, and denote by Λ the adjoint operator of L. Then for p-forms, we have $\Lambda=(-1)^p*L*$. We say that φ is **primitive** when $\Lambda\varphi=0$. Some important properties of L and Λ follow: (3) $\Lambda\varphi^p=0$ if and only if $L^q\varphi=0$ $(q=\max(n-p+1,0))$. (4) If $\Lambda\varphi=0$ and φ is of type (r,s), then for all $0\leqslant q\leqslant n-p$,

$$*L^q\varphi=\{(-1)^{p(p+1)/2}\times(\sqrt{-1})^{r-s}q!/(n-p-q)!\}L^{n-p-q}\varphi$$

(where $p=r+s$). (5) A p-form φ can be uniquely decomposed in the form $\varphi=\varphi_0+L\varphi_1+\ldots+L^r\varphi_r$ with $r=[p/2]$ and φ_i primitive (Hodge [4], Weil [9]). Properties (3)–(5) are shared by all Hermitian metrics. When the metric is Kählerian, we further have the following:

$$Ld-dL=0,\quad \Lambda d'-d'\Lambda=\sqrt{-1}\,\delta'',$$
$$\Lambda d''-d''\Lambda=-\sqrt{-1}\,\delta'.$$

From these we also obtain:

$$\Delta L=L\Delta,\qquad \Delta\Lambda=\Lambda\Delta,$$
$$d'\delta''+\delta''d'=0,\qquad d''\delta'+\delta'd''=0,$$
$$\Delta=2(d'\delta'+\delta'd')=2(d''\delta''+\delta''d''),$$

where Δ is the †Laplace-Beltrami operator $\Delta=d\delta+\delta d$. †Green's operator G commutes with d', d'', δ', and δ''. Thus when the metric is Kählerian, we have: (6) Let $L_p(X)=\sum_{r+s=p}L_{r,s}(X)$ be the decomposition of the space of p-forms into the spaces $L_{r,s}(X)$ of forms of type (r,s). Then the space $H_p(X)$ of harmonic p-forms is the direct sum $H_p(X)=\sum_{r+s=p}H_{r,s}(X)$ of the spaces $H_{r,s}(X)$ of harmonic forms of type (r,s). (7) If we let $A=d''\delta''+\delta''d''$, then $\Delta\varphi=0$ is equivalent to $A\varphi=0$. Denote the projection to the space of harmonic forms by H. Then using the formula $1=H+\Delta G$ (→ 196 Harmonic Integrals), we can infer that not only in the †cochain complex $(\sum_p L_p(X),d)$ but also in the cochain complexes $(\sum_s L_{r,s}(X),d'')$ and $(\sum_r L_{r,s}(X),d')$, H is homotopic to the identity mapping and the respective cohomology groups of degree s and degree r of the last two complexes are both isomorphic to $H_{r,s}(X)$. (8) For a harmonic p-form φ, the forms φ_i in the decomposition (5) are harmonic. (9) The exterior powers Ω^r $(r=0,1,\ldots,n)$ of the fundamental form Ω are harmonic. (10) Holomorphic differential forms are harmonic.

C. Further Properties of Compact Kähler Manifolds

The p-dimensional †de Rham cohomology group with complex coefficients of a compact Kähler manifold X is canonically isomorphic to the direct sum of the †Dolbeault cohomology groups of type (r,s), where $r+s=p$ (→ 74 Complex Manifolds). Let b_p and $h^{r,s}$, respectively, denote the pth Betti number of X and $\dim_{\mathbf{C}} H^s(X,\Omega^r)$. Then it follows that $b_p=\sum_{r+s=p}h^{r,s}$. If φ is harmonic, then $\bar{\varphi}$ is also harmonic, so we have $h^{r,s}=h^{s,r}$. Therefore, if p is odd, b_p is even. This is a generalization of the fact that the first Betti number of a closed Riemann surface is even. From (9) we have $b_p \geqslant 1$ if p is even. Furthermore, a linear combination of irreducible analytic subsets of X of (complex) dimension r with positive real coefficients is a cycle of degree $2r$ that is never homologous to zero on X, since the integral of Ω^r on the cycle is never zero.

For a compact Kähler manifold X, we can define a †complex torus $\mathfrak{A}$ and a holomorphic mapping $\lambda: X \rightarrow \mathfrak{A}$ such that (i) $\mathfrak{A}$ is generated by $\lambda(X)$ as a group; (ii) any holomorphic mapping $\mu: X \rightarrow T$ can be decomposed as $\mu=\alpha\circ\lambda+c$, where T is another complex torus, α is a complex analytic homomorphism from $\mathfrak{A}$ to T_0, and c is a point of T. $\mathfrak{A}$ is called the **Albanese variety** of X. The set $\mathfrak{P}$ of complex analytic isomorphism classes of †complex line bundles that are trivial as topological bundles has a natural complex structure with a canonically associated structure of a family of vector bundles (→ 74 Complex Manifolds G). With this structure $\mathfrak{P}$ is, in fact, a complex torus and is called the **Picard variety** of X. Then $\mathfrak{P}$ and $\mathfrak{A}$ are constructed using $H^1(X,\mathbf{R})$ and $H^{2n-1}(X,\mathbf{R})$, respectively, and are dual complex tori. If X is a Hodge manifold (→ Section D), $\mathfrak{P}$ and $\mathfrak{A}$ are Abelian varieties (→ 3 Abelian Varieties) [10].

A small deformation of a compact Kähler manifold is also Kählerian (a Kähler metric can be taken to be of class C^∞ with respect to parameters) [6]. But a limit (in the sense of deformation) of a Kähler manifold is not always Kählerian. An example was given by Hironaka [3].

D. Hodge Manifolds

One of the most important examples of Kähler manifolds is an †algebraic variety in a projective space. If we let $(\zeta_0,\dots,\zeta_N)$ be a homogeneous coordinate system in the N-dimensional projective space $\mathbf{P}^N(\mathbf{C})$, then the subset $\zeta_k\neq 0$ has its holomorphic coordinate system $(z^1,\dots,z^N)$, where $z^1=\zeta_0/\zeta_k,\dots,z^k=\zeta_{k-1}/\zeta_k$, $z^{k+1}=\zeta_{k+1}/\zeta_k,\dots,z^N=\zeta^N/\zeta^k$. If we let $\psi_k=(1/2\pi)\log(1+\sum_j|z^j|^2)$ and $g_{\alpha\beta}=\partial^2\psi/\partial z^\alpha\partial\bar{z}^\beta$, then $ds^2=\sum g_{\alpha\beta}\,dz^\alpha\,d\bar{z}^\beta$ is independent of k and determines a Kähler metric on $\mathbf{P}^N(\mathbf{C})$, which is called a **standard Kähler metric** of $\mathbf{P}^N(\mathbf{C})$. In this case, we have the following results: (11) Let $\mathfrak{H}$ be a hyperplane of $\mathbf{P}^N$. Then the integral $\int_Z\Omega$ of the fundamental form Ω on a 2-cycle Z is equal to the †Kronecker index $\mathrm{KI}(Z,\mathfrak{H})$ of the †intersection of Z and $\mathfrak{H}$. Therefore, (12) an integral (a period) of Ω on a 2-cycle with integral coefficients is an integer. In other words, Ω corresponds to the cohomology class in $H^2(X,\mathbf{R})$ of a cocycle with integral coefficients. Property (12) holds for the induced Kähler metric on an analytic submanifold X of $\mathbf{P}^N(\mathbf{C})$ (which is algebraic by †Chow's theorem). Property (11) also holds if we replace $\mathfrak{H}$ by the intersection Y of X and $\mathfrak{H}$. Property (8) can be thought of as an expression (in terms of harmonic forms) of Lefschetz's theorem [17,4] on the topology of projective algebraic varieties.

Generally, if a compact complex manifold X admits a Kähler metric with property (12), we say that the metric is a **Hodge metric** and X is a **Hodge manifold. Kodaira's theorem** asserts that a Hodge manifold has a biholomorphic embedding into a projective space [5]. Cohomology vanishing theorems and the properties of †monoidal transformations of complex manifolds are used to prove this theorem.

On a closed Riemann surface $\mathfrak{R}$, i.e., a 1-dimensional compact complex manifold, any Hermitian metric is Kählerian. Moreover, a metric with total volume 1 is a Hodge metric. This proves that $\mathfrak{R}$ is isomorphic to an algebraic curve in a projective space. This is a proof (using Kodaira's theorem) of the existence of nonconstant meromorphic functions on $\mathfrak{R}$. The condition on the †Riemann matrix for a complex torus $T=\mathbf{C}^n/D$ (D is a discrete subgroup of rank $2n$) to be a projective algebraic variety is that T admits a Hodge metric.

E. Examples and Other Properties

Concerning differential geometry on compact Kähler manifolds, the analytic transformation group and the †isometric transformation group are also studied. For an example, let X be a compact Kähler manifold of complex dimension n. Then the Lie group of isometric transformations on X is of dimension $\leqslant n^2+$

$2n$, and the equality holds if and only if $X = \mathbf{P}^n(\mathbf{C})$ [8].

A nonalgebraic complex torus is the most important example of a compact Kähler manifold that is not algebraic. A complex torus is not an †abstract algebraic variety (in the sense of Weil) if it is not a Hodge manifold.

An example of a noncompact Kähler manifold is a bounded domain in $\mathbf{C}^n$ with the Kähler metric

$$ds^2 = \sum_{\alpha,\beta} (\partial^2 \log K(z,\bar{z}) / \partial z^\alpha \partial \bar{z}^\beta) dz^\alpha d\bar{z}^\beta,$$

where $K(z,\bar{\zeta})$ is †Bergman's kernel function. This metric is significant because of its invariance under the analytic automorphisms of the domain. More generally, a †Stein manifold admits a complete Kähler metric [2].

Hironaka's example shows that there is a non-Kähler compact complex manifold that is an abstract algebraic variety (in the sense of Weil). The Hopf manifold is another example of a non-Kähler compact complex manifold, where the **Hopf manifold** is the quotient space of $W = \mathbf{C}^2 - \{(0,0)\}$ by the transformation group generated by the analytic automorphism $g: W \ni (z,w) \rightarrow (2z, 2w) \in W$.

References

[1] Y. Akuzuki, Tyôwa sekibunron (Japanese; Harmonic integrals), Iwanami, I, 1955; II, 1956.
[2] H. Grauert, Charakterisierung der Holomorphiegebiete durch die vollständige Kählersche Metrik, Math. Ann., 131 (1956), 38–75.
[3] H. Hironaka, An example of a non-Kählerian complex-analytic deformation of Kählerian complex structures, Ann. of Math., (2) 75 (1962), 190–208.
[4] W. V. D. Hodge, The theory and applications of harmonic integrals, Cambridge Univ. Press, second edition, 1952.
[5] K. Kodaira, On Kähler varieties of restricted type (An intrinsic characterization of algebraic varieties), Ann. of Math., (2) 60 (1954), 28–48.
[6] K. Kodaira and D. C. Spencer, On deformations of complex analytic structures III, Ann. of Math., (2) 71 (1960), 43–76.
[7] S. Lefschetz, L'analysis situs et la géométrie algébrique, Gauthier-Villars, 1924.
[8] A. Lichnerowicz, Géométrie des groupes de transformations, Dunod, 1958.
[9] A. Weil, Introduction à l'étude des variétés kähleriennes, Actualités Sci. Ind., Hermann, 1958.
[10] A. Weil, On Picard varieties, Amer. J. Math., 74 (1952), 865–894.

233 (XX.28) Klein, Felix

Felix Klein (May 25, 1849–June 22, 1925) was one of the leading mathematicians in Germany in the latter half of the 19th century. Born in Düsseldorf and graduated from the University of Bonn, Klein went to study in Paris. In 1872, he became a professor at the University of Erlangen, and in 1886 attained a chair at the University of Göttingen, where he was employed until the end of his life. His accomplishments cover all aspects of mathematics, but his main field was geometry. In his inaugural lecture at the University of Erlangen, he presented a bird's-eye view of all the then known fields of geometry from the standpoint of group theory, which is referred to as the †Erlangen program (→ 147 Erlangen Program). In it he stated that both Euclidean and non-Euclidean geometry are included in †projective geometry. Klein said in his last lectures [4] that he spent the greatest part of his energies in the field of †automorphic functions. These last lectures are important as historical material on the mathematics of the 19th century. Klein was a leader of reforms of mathematical education in Germany [3].

References

[1] F. Klein, Gesammelte mathematische Abhandlungen I–III, Springer, 1921–1923.
[2] R. Fricke and F. Klein, Vorlesungen über die Theorie der automorphen Funktionen, Teubner, I, 1897; II, 1901.
[3] F. Klein, Elementarmathematik von höheren Standpunkte aus, Springer, I, 1924; II, 1925; III, 1928.
[4] F. Klein, Vorlesungen über die Entwicklung der Mathematik im 19. Jahrhundert I, II, Springer, 1926–1927 (Chelsea, 1956).
[5] F. Klein, Vorlesungen über nicht-Euklidische Geometrie, Springer, 1928 (Chelsea, 1960).
[6] F. Klein, Vorlesungen über höhere Geometrie, Springer, third edition, 1926 (Chelsea, 1957).

234 (IX.11) Knot Theory

A. General Remarks

The knot problem is a special case of the **placement problem**, stated as follows: Given two homeomorphic topological spaces X_1 and X_2 and their respective homeomorphic subsets A_1 and A_2, is there a homeomorphism $f: X_1 \rightarrow X_2$ such that $f(A_1)=A_2$? A simple closed curve in a Euclidean 3-space $\mathbf{R}^3$ (or in a 3-sphere S^3 obtained from $\mathbf{R}^3$ by one-point compactification, sometimes used in preference to $\mathbf{R}^3$) is called a **knot**. If two knots K_1 and K_2 can be mapped from one to the other by a homeomorphism of $\mathbf{R}^3$ (or of S^3), they are called **equivalent**. All knots are classified into **knot types** by this equivalence.

Let (x,y,z) be Cartesian coordinates in $\mathbf{R}^3$. Knots equivalent to $x^2+y^2=1$, $z=0$ and their knot type are called **trivial** or **unknotted**; knots of other types are called **knotted**. Knots equivalent to those given by polygons in $\mathbf{R}^3$ and their knot types are called **tame**; other knots are called **wild**. Given a polygonal knot K in $\mathbf{R}^3$, there exists a plane such that the orthogonal projection π on it has the following two properties: (1) The image πK has no multiple points other than a finite number of double points. (2) The projections of the vertices of K are not double points of πK. Then πK is called a **regular knot projection** of K. Let $z=0$ be the plane in question. Of the two points of K corresponding to a double point of πK, the one with the greater z-coordinate is called the **overcrossing point**; the one with the smaller z-coordinate is called the **undercrossing point**. Suppose that overcrossing points and undercrossing points appear alternately when we move along K in a fixed direction; then K is called **alternating**.

Knots K_1 and K_2 in $\mathbf{R}^3$ are said to be of the same **isotopy type** if there is an †isotopy $\{h_t\}$ $(0 \leqslant t \leqslant 1)$ of $\mathbf{R}^3$ such that $h_t: \mathbf{R}^3 \rightarrow \mathbf{R}^3$ are orientation-preserving homeomorphisms with the identity h_0 and such that $h_1(K_1)=K_2$. Knots K_1 and K_2 of $\mathbf{R}^3$ are of the same isotopy type if and only if K_1 is mapped onto K_2 by an orientation-preserving homeomorphism of $\mathbf{R}^3$. Thus knots of the same isotopy type are equivalent. A knot that can be mapped onto itself by an orientation-reversing homeomorphism of $\mathbf{R}^3$ is called **amphicheiral**. A knot K is called **invertible** if K is mapped onto itself by an orientation-preserving homeomorphism h such that $h|K$ reverses the orientation of K.

A tame knot K is always the boundary of an orientable surface in $\mathbf{R}^3$ (H. Seifert, *Math. Ann.*, 110 (1934)). The minimum of the †genera of surfaces having K as their boundary is called the **genus** of K. Using Dehn's lemma and the †sphere theorem, C. D. Papakyriakopoulos showed that a tame knot K is unknotted if and only if $\pi_i(S^3-K)=0$ $(i \geqslant 2)$ and $\pi_1(S^3-K)=\mathbf{Z}$ (*Proc. Nat. Acad. Sci. US*, 66 (1957)). The so-called **Alexander-Briggs classification** of all tame knot types with at most 9 double points is given in [1]. This classification was accomplished by using the invariants of knots introduced below (→ Appendix A, Table 7).

Suppose that we are given knot types represented by knots K_1, K_2 placed so that K_1 does not intersect with a surface having K_2 as its boundary. Then we can choose small arcs P_1P_2, Q_1Q_2 on K_1, K_2, respectively, and curves C_1, C_2 connecting P_1Q_1, P_2Q_2, respectively, so that the curve C passing through $P_1P_2Q_2Q_1P_1$ successively is closed and simple. We then delete the arcs P_1P_2, Q_1Q_2 and obtain a knot that is called the product of K_1 and K_2 (Fig. 1). The knot type of the product of K_1 and K_2 is uniquely determined by the knot types of K_1, K_2 and is called the product of respective knot types. The set of knot types has the structure of a †semigroup with respect to this product. Any knot type is decomposed uniquely into finite products (composition) of the prime knot types, which cannot be decomposed into products of unknotted knot types (H. Schubert, S.-B. Heidelberger, *Akad. Wiss.*, 3 (1949)).

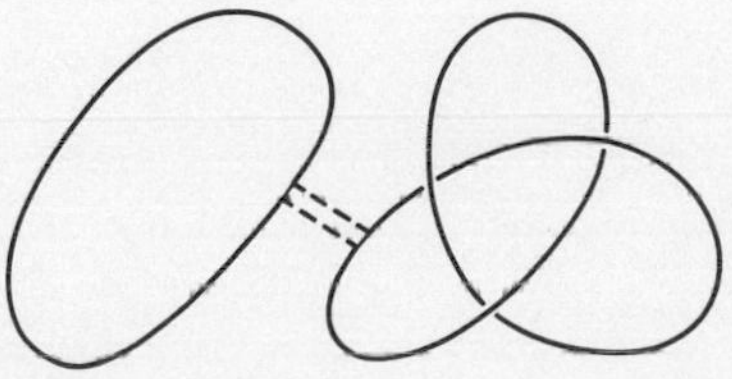

Fig. 1

Until about 1930 the theory of knots had been studied chiefly by J. W. Alexander in the United States and by K. Reidemeister, H. Seifert, and others in Germany. Little progress was made, however, until a new development of the theory was introduced by R. H. Fox [5] and his school in the United States. In Japan, significant contributions have been made by T. Homma, S. Kinoshita, K. Murasugi, and others [6].

B. Knot Groups

If K is a knot, the †fundamental group $G=\pi_1(\mathbf{R}^3-K)$ of $\mathbf{R}^3-K$ is called the **knot group** of K. Consider a regular knot K given by a polygon and its regular projection π, specifically a projection on a plane, having no multiple points other than double points. If the projection $\pi(K)$ has n double points $\pi(d_i)$, then K is divided into n arcs z_i by n undercrossing points. We consider a free group F generated by n letters $x_1, \ldots, x_n$. To each double point $\pi(d_i)$ we associate a word r_i in the following manner. Suppose that the point d_i separates the arcs z_λ, $z_{\lambda+1}$ and that the arc z_μ overcrosses the composite arc $z_\lambda z_{\lambda+1}$. We fix a positive orientation of K and the plane of projection π. We draw a small oriented circle around $\pi(d_i)$ meeting the arcs $\pi(z_{\lambda+1})$, $\pi(z_\mu)$, $\pi(z_\lambda)$, $\pi(z_\mu)$ in that order. Now, the word r_i is given by $r_i = x_{\lambda+1}^{-1} x_\mu^e x_\lambda x_\mu^{-e}$, where $e = +1$ or -1 according as the signature of the rotation of the angle less than π sending the oriented tangent (in the plane of projection) to $\pi(z_\mu)$ at $\pi(d_i)$ onto the oriented tangent to $\pi(z_\lambda)$ at $\pi(d_i)$ is positive or negative (Fig. 2). Let $G = (x_1, \ldots, x_n; r_1, \ldots, r_n)$ be the factor group F/N, where N is the minimum normal subgroup of F containing $r_1, \ldots, r_n$. Then the group G is isomorphic with the knot group of K. The word r_i corresponds to the relation $1 = x_{\lambda+1}^{-1} x_\mu^e x_\lambda x_\mu^{-e}$, and any one of such relations is a consequence of the rest. Hence G may be written as $(x_1, x_2, \ldots, x_n; r_1, \ldots, r_{n-1})$, which is usually known as the **Wirtinger presentation** of G. If G' is the †commutator subgroup of G, then G/G' is an infinite cyclic group **Z**. A knot group is an invariant of a knot type, but two knots with isomorphic knot groups are not necessarily equivalent.

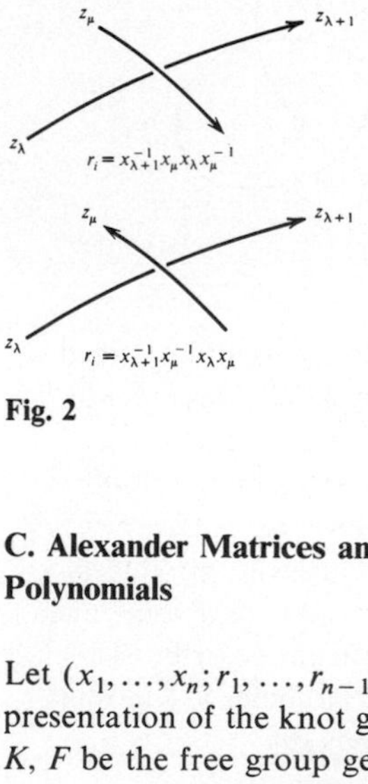

Fig. 2

C. Alexander Matrices and Alexander Polynomials

Let $(x_1, \ldots, x_n; r_1, \ldots, r_{n-1})$ be a Wirtinger presentation of the knot group G of a knot K, F be the free group generated by $x_1, x_2, \ldots, x_n$, and $\varphi: F \rightarrow G$, $\psi: G \rightarrow H = G/G'$ be canonical homomorphisms. Then φ and ψ can be extended to homomorphisms between group rings with integral coefficients $\varphi: \mathbf{Z}[F] \rightarrow \mathbf{Z}[G]$ and $\psi: \mathbf{Z}[G] \rightarrow \mathbf{Z}[H]$. For any word r in F, we have the **free derivative** $\partial r/\partial x_i$ for $i = 1, \ldots, n$ satisfying the following:

$$\partial x_i/\partial x_j = \delta_{ij}, \quad \partial x_i^{-1}/\partial x_i = -x_i^{-1},$$

$$\partial(rs)/\partial x_i = \partial r/\partial x_i + r \cdot (\partial s/\partial x_i)$$

(Fox [6]). Now we have the **Alexander matrix** (a_{ij}) $(i = 1, \ldots, m; j = 1, \ldots, n)$ of the knot K defined by $a_{ij} = \psi \circ \varphi\,(\partial r_i/\partial x_j)$. The ideal of $\mathbf{Z}[H]$ generated by the $n-1$ minors of the matrix is a principal ideal, called the **Alexander ideal** of K. Since $H \cong \mathbf{Z}$, any element of $\mathbf{Z}[H]$ is a finite sum of elements of the form mt^n $(m \in \mathbf{Z}, n = 0, 1, \ldots)$. The generator $\Delta(t)$ of the Alexander (principal) ideal is called the **Alexander polynomial** of the knot K. The †elementary divisors of the Alexander matrix of K, including $\Delta(t)$, are invariants of the knot type of K. The divisor $\Delta(t)$ is uniquely determined by K up to the factor $\pm t^\lambda$ and has the following properties: (i) $\Delta(1) = 1$; (ii) $\Delta(1/t) = t^\lambda \Delta(t)$. Conversely, any polynomial with integral coefficients satisfying (i) and (ii) is the Alexander polynomial of some knot. For example, the knot group G of the **trefoil knot** (or **cloverleaf knot**) K (Fig. 3) is presented by $G = (a, b; aba(bab)^{-1})$. The Alexander matrix of K is $(1 - t + t^2, -1 + t - t^2)$, and the Alexander polynomial of K is $\Delta(t) = 1 - t + t^2$. Two trefoil knots that are mirror images of each other have the same knot group, but they are not isotopic (M. Dehn, 1914).

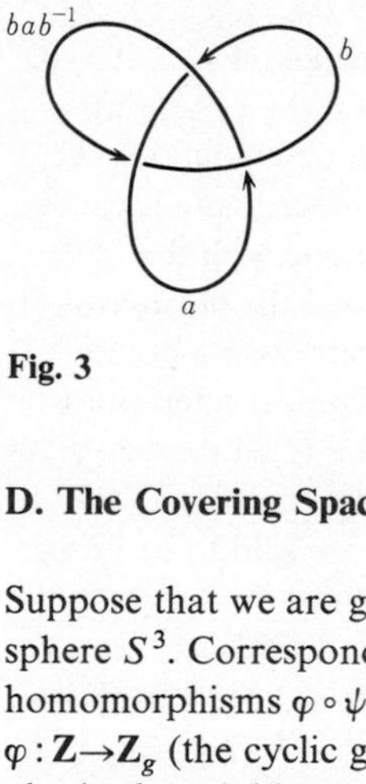

Fig. 3

D. The Covering Space of a Knot

Suppose that we are given a knot in the sphere S^3. Corresponding to the kernel of homomorphisms $\varphi \circ \psi$, where $\psi: G \rightarrow \mathbf{Z}$ and $\varphi: \mathbf{Z} \rightarrow \mathbf{Z}_g$ (the cyclic group of order g), we obtain the g-fold covering space $\Sigma_g - \Lambda_g$ of $S^3 - K$. By adjoining K to $\Sigma_g - \Lambda_g$ suitably, we obtain a covering space Σ_g of S^3 with branch points on K. Then the 1-dimensional homology group $H_1(\Sigma_g - \Lambda_g)$ is isomorphic to the direct sum of $H_1(\Sigma_g)$ and **Z**, and the order θ of $H_1(\Sigma_g)$ is given by $\theta = \prod_{j=0}^{g-1} \Delta(\omega^j)$, where ω denotes the primitive gth root of unity. If $H_1(\Sigma_g)$ is infinite, then the †Betti number of Σ_g is the number of roots of $\Delta(t) = 0$; if $H_1(\Sigma_g)$ is finite, then the quotient group by the commutator group of $\pi_1(\Sigma_g - \Lambda_g)$ is isomorphic to **Z**. By considering the fundamental group of $\Sigma_g - \Lambda_g$, we can define the Alexander polynomial $\tilde{\Delta}(\tau)$ for $\Sigma_g - \Lambda_g$, and we have $\tilde{\Delta}(\tau) = \prod_{j=0}^{g-1} \Delta(\omega^j \tau)$, where $\tau = t^g$.

A more detailed account of the covering spaces was given by H. Seifert as follows (*Math. Ann.*, 110 (1934)): Suppose that K is a knot of genus h and F is an orientable surface of genus h having K as its boundary.

Let $a_1, a_2, \ldots, a_{2h-1}, a_{2h}$ be †normal sections of F. We deform F as illustrated in Fig. 4 to obtain a disc with $2h$ strips along normal sections attached to it. By a further manipulation to undo the twist of strips, we obtain a figure called a **Seifert surface** (the right-hand figure in Fig. 4). Let v_{ij} be the number of times the jth strip (along a_j) passes under the ith strip (where the orientation of a_i, a_j is taken account of so that if the jth strip passes the ith strip from "left to right" x_{ij} times and from "right to left" y_{ij} times, then $v_{ij} = x_{ij} - y_{ij}$). We now have the $2h \times 2h$ matrix $V = (v_{ij})$. Further, let $I(a_i, a_j)$ be the †intersection number of a_i and a_j on F and define the $2h \times 2h$ matrix $I^* = (I(a_i, a_j))$. Let $\Gamma = VI^*$, and let $F(t) = I + (t-1)\Gamma$, where I denotes the unit matrix. Then the Alexander polynomial $\Delta(t)$ of K is $\Delta(t) = \det F(t)$.

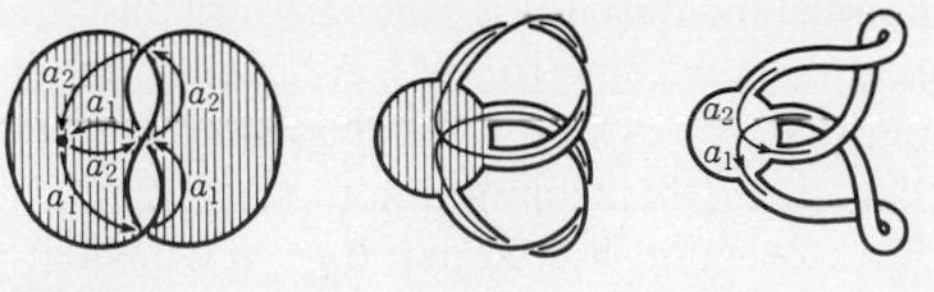

Fig. 4

If we put $F_g = \Gamma^g - (\Gamma - I)^g$, then the $2h \times 2h$ matrix F_g is the relation matrix of $H_1(\Sigma_g)$, and Σ_g is a compact, orientable 3-manifold. If a, b are 1-dimensional cycles of Σ_g representing †torsion elements of $H_1(\Sigma_g)$, then there is a 2-dimensional chain A with the boundary ma, where m is an integer. In this case $L(a,b) \equiv (1/m)\, I(A,b) \pmod 1$ will be called the **self-linking coefficient** of a and b, and $L(a,b)$ is uniquely determined by homology classes α, β of a, b. Hence we call $L(a,b)$ the self-linking coefficient of α, β and denote it by $L(\alpha, \beta)$.

The 1-dimensional torsion group of $H_1(\Sigma_g)$ can be decomposed into a direct sum of primary cyclic groups, i.e., cyclic groups each of which has order equal to a power of a prime. We fix a prime power p^c $(p \neq 2)$. Let $g_1, g_2, \ldots, g_r$ be generators of these primary cyclic groups of order p^c. We set $A = p^c(L(g_i, g_j))$. Then the †quadratic residue symbol $((\det A)/p)$ is an invariant of the knot type and is called the **Seifert invariant** of K.

A regular projection of a knot K gives rise to a system of domains $X_0, X_1, \ldots, X_n$ (as indicated by oblique lines in Fig. 5), where X_0 denotes the unbounded domain. Let $e_{ij} = \Sigma \eta(C)$, where C is a double point of the intersection of the closures $\bar{X}_i$ and $\bar{X}_j$ and $\eta(C)$ is $+1$ or -1 as indicated (Fig. 6). We set $\hat{f}(x_0, x_1, \ldots, x_n) = \Sigma_{i<j} e_{ij}(x_i - x_j)^2$ and $f(x_1, x_2, \ldots, x_n) = \hat{f}(0, x_1, \ldots, x_n)$. These are called the quadratic forms associated with the domains $X_0, \ldots, X_n$, such as in Fig. 5 (L. Goeritz, 1933). If we set $a_{ij} = -e_{ij}$ $(i \neq j)$ and $a_{ii} = \Sigma_{j \neq i} e_{ij}$, then $\hat{f}(x_0, x_1, \ldots, x_n) = \Sigma_{i,j=0}^{n} a_{ij} x_i x_j$. The matrix $A = (a_{ij})$ is the relation matrix of the 1-dimensional homology group $H_1(\Sigma_2)$ of the covering space Σ_2 of K. $A^{-1} \pmod 1$ is represented as the matrix of self-linking numbers and is useful in determining the amphicheirality of K.

All the known invariants of knot types that can be calculated by projection are obtained by means of the matrix Γ.

Fig. 5

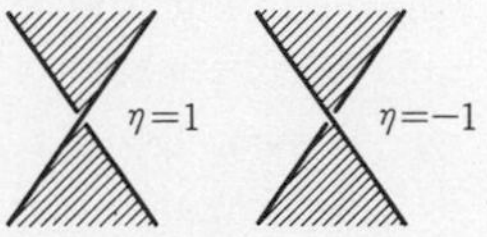

Fig. 6

E. Braids

The theory of **braids** has been used as a tool to investigate the theory of knots. An illustration of a braid of fourth order is given in Fig. 7. In general, we consider a rectangle P, called the **frame of the braid**, and two sets of n points $A_1, A_2, \ldots, A_n$; $B_1, B_2, \ldots, B_n$ placed at equal distances on a pair of opposite sides g_1 and g_2, respectively. Suppose that $A_i \to B_i$ is a one-to-one correspondence between these sets of points. The frame of the braid P is assumed to be placed in the space $\mathbf{R}^3$. We now take simple broken lines l_i in $\mathbf{R}^3$ connecting A_i and B_{k_i} such that l_i, l_j are disjoint for $i \neq j$, the projection l_i' of l_i on the plane of P lies inside P, and any straight line parallel to g_1 and g_2 and lying between them intersects l_i at a single point for each i. Finally, we assume that l_i' and l_j' for $i \neq j$ intersect at finite points and that they are all in different levels. Such a configuration is called a braid of nth order.

Two braids $\mathfrak{z}_1$ and $\mathfrak{z}_2$ are said to be **isotopic**, $\mathfrak{z}_1 \approx \mathfrak{z}_2$, if there is a homeomorphism of $\mathbf{R}^3$ onto itself mapping $\mathfrak{z}_1$ onto $\mathfrak{z}_2$ such that it is an identity outside a sufficiently large sphere including $\mathfrak{z}_1$ and $\mathfrak{z}_2$. This isotopy relation is an equivalence relation among braids.

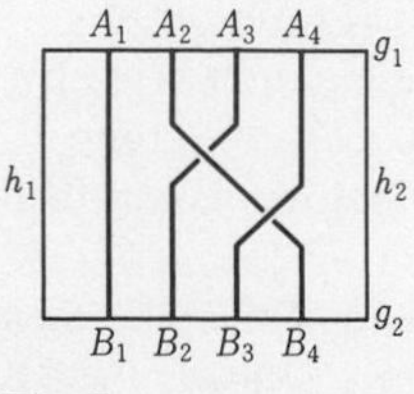

Fig. 7

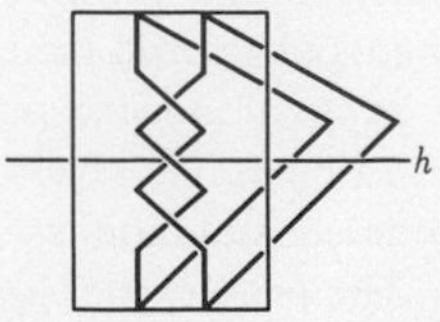

Fig. 8

Suppose that we are given a braid such as in Fig. 7. If we connect A_i and B_i by Λ-shaped broken lines passing outside the frame P (Fig. 8) and remove the frame, then we obtain a **closed braid**. In general, a closed braid consists of several polygons (knots); conversely, any knot is equivalent to a closed braid.

The product $\mathfrak{z}_1\mathfrak{z}_2$ of two braids $\mathfrak{z}_1$ and $\mathfrak{z}_2$ is defined as the braid obtained by connecting $\mathfrak{z}_1$ and $\mathfrak{z}_2$ as shown in Fig. 9. If $\mathfrak{z}_1 \approx \mathfrak{z}_1'$ and $\mathfrak{z}_2 \approx \mathfrak{z}_2'$, then $\mathfrak{z}_1\mathfrak{z}_2 \approx \mathfrak{z}_1'\mathfrak{z}_2'$. Hence we can define the product $[\mathfrak{z}_1][\mathfrak{z}_2]=[\mathfrak{z}_1,\mathfrak{z}_2]$ of equivalence classes of braids of nth order. Thus the totality of $[\mathfrak{z}]$ forms a group $\mathfrak{Z}_n$ called the **braid group** of order n. Thus $\mathfrak{Z}_n$ is generated by the equivalence classes σ_i $(i=1,2,\ldots,n-1)$ of braids as shown in Fig. 10(a); note σ_i^{-1} shown in Fig. 10(b). The †fundamental relations between $\{\sigma_i\}$ are $S_{ji}=1$, where $S_{ji}=\sigma_j^{-1}\sigma_i^{-1}\sigma_j\sigma_i$ or $\sigma_j^{-1}\sigma_i^{-1}\sigma_j^{-1}\sigma_i\sigma_j\sigma_i$ according as $|j-i|\geqslant 2$ or $|i-j|=1$.

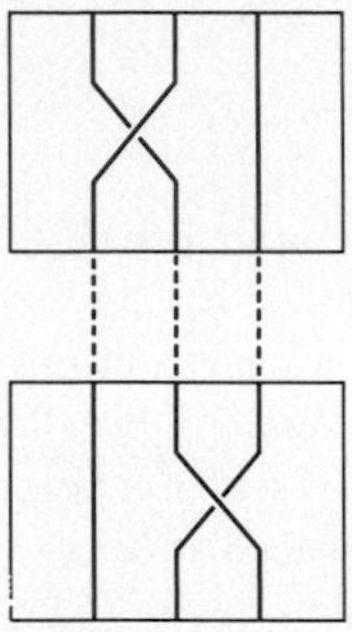

Fig. 9

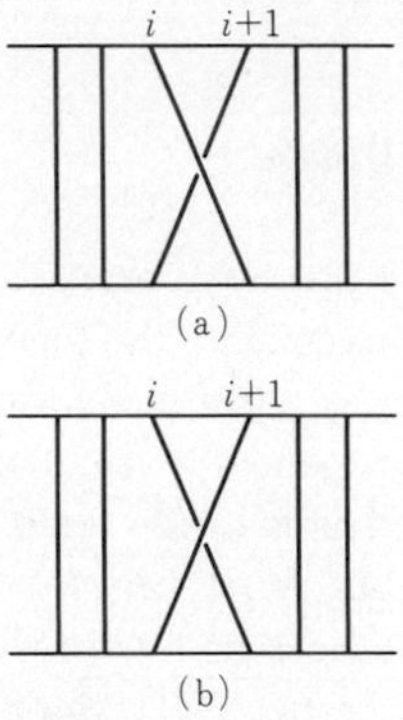

Fig. 10

Suppose that we are given braids $\mathfrak{z}_1$, $\mathfrak{z}_2$ represented as the products of σ_i. Then $\mathfrak{z}_1$ and $\mathfrak{z}_2$ are equivalent if and only if the two products represent the same element in $\mathfrak{Z}_n$. The problem of deciding whether $\mathfrak{z}_1$ and $\mathfrak{z}_2$ are equivalent reduces to the †word problem in $\mathfrak{Z}_n$. On the other hand, the problem of deciding whether or not two closed braids are equivalent reduces to the †transformation problem in $\mathfrak{Z}_n$ (→ 169 Free Groups).

The theory of braids was initiated by E. Artin, who also gave a solution to the word problem in $\mathfrak{Z}_n$ (*Abh. Math. Sem. Univ. Hamburg* (1926)). The theory was developed by Artin (*Ann. of Math.*, 48 (1947)) and F. Bohnenblust (*Ann. of Math.*, 48 (1947)). Principal results obtained by the latter are the determination of the normal forms of elements of $\mathfrak{Z}_n$ and the representation of $\mathfrak{Z}_n$ by a group of automorphisms of a free group utilizing purely algebraic and free group-theoretic considerations.

The problems of links and of the placement of graphs in $\mathbf{R}^3$ are similarly treated.

F. Higher-Dimensional Knots

The problem of knots, that is, the problem of placement of simple closed curves in $\mathbf{R}^3$, is extended to the problem of placement of q-dimensional spheres in p-space $\mathbf{R}^p$ or in a p-dimensional sphere S^p.

In this section, the explanation is restricted to the case of †combinatorial manifolds and †piecewise linear homeomorphisms between them. A similar theory can be developed for other categories.

Let Δ^n denote an n-dimensional simplex. We assume that Δ^n is a face of Δ^{n+1}. If S^q is a subcomplex of S^p, then (S^p,S^q) $(p>q)$ is called a (p,q) **sphere pair**. If a q-dimensional cell B^q is a subcomplex of B^p and if the boundary $\dot{B}^q$ of B^q is contained in $\dot{B}^p$, then (B^p,B^q) is called a (p,q) **ball pair**. Two pairs $X=(X^p,X^q)$ and $Y=(Y^p,Y^q)$ are said to be homeomorphic if there is a homeomorphism $h:X^p\rightarrow Y^p$ such that $h(X^q)=Y^q$. We classify (p,q) ball pairs and (p,q) sphere pairs into equivalence classes via homeomorphisms. A (p,q) ball pair (B^p,B^q) is said to be **unknotted** (or **flat**) if it is homeomorphic to the normal pair $\Gamma^{p,q}=(\Delta^p,\Delta^q)$. A (p,q) sphere pair (S^p,S^q) is said to be **unknotted** (or **flat**) if it is homeomorphic to $(\partial\Delta^{p+1},\partial\Delta^{q+1})$. E. C. Zeeman showed that if $p-q\geqslant 3$, then the (p,q) ball pair and the (p,q) sphere pair are both unknotted (*Ann. of Math.*, (2) 78 (1963)). Similar results have been obtained by T. Stallings (*Ann. of Math.*, (2) 77 (1963)).

References

[1] K. Reidemeister, Knotentheorie, Erg. Math., Springer, 1932 (Chelsea, 1948).
[2] H. Seifert and W. Threlfall, Lehrbuch der Topologie, Teubner, 1934 (Chelsea, 1965).
[3] S. Iyanaga, Ziyûgunron (Japanese; Theory of free groups), Series of math. lectures, Osaka Univ., Iwanami, 1942.
[4] H. Seifert and W. Threlfall, Old and new results on knots, Canad. J. Math., 2 (1950), 1–15.
[5] R. H. Fox, Recent development of knot theory at Princeton, Proc. Intern. Congr. Math.,1950, Cambridge, Mass., Amer. Math. Soc., II, 453–457.
[6] R. H. Crowell and R. H. Fox, Introduction to knot theory, Ginn, 1963.
[7] M. K. Fort, Jr. (ed.), Topology of 3-manifolds and related topics, Prentice-Hall, 1962.
[8] H. Terasaka, Theory of knots (Japanese), Sûgaku, 12 (1960), 1–20.

235 (XX.29) Kronecker, Leopold

Leopold Kronecker (December 7, 1823–December 29, 1891) was born in Liegnitz near Breslau in Germany (now Legnica in Poland).

He entered the University of Berlin in 1841 but studied at various universities throughout the country, finally studying under E. E. Kummer at Breslau. In 1845, he received his doctorate with a thesis on units of †algebraic number fields. Then he succeeded to his uncle's business in the management of banks and farms, which kept him away from publishing mathematical papers for eight years. In 1853, he published a paper on algebraic solution of equations, containing the assertions that Abelian extensions of the rational number field are contained in cyclotomic fields and that Abelian extensions of imaginary quadratic fields can be obtained using †complex multiplication. The latter is called "Kronecker's dream in his youth." It remained a conjecture until it was solved by means of †class field theory. He gave lectures at the University of Berlin, first in his capacity of academician, then as a professor in succession to his teacher Kummer from 1883. His lectures covered many fields of algebra and analysis. He also did pioneer work in topology. His statement that mathematics as a whole should be based solely on the intuition of natural numbers (→ 164 Foundations of Mathematics) often brought on disputes with his colleague K. †Weierstrass. His rejection of the bold reasoning of †set theory produced anxieties for G. †Cantor. His famous statement "Natural numbers were made by God; the rest is the work of man," may be contrasted to the liberal statements of Cantor and R. Dedekind.

References

[1] Leopold Kronecker's Werke I–V, edited by K. Hensel, Teubner, 1895–1931 (Chelsea, 1968).
[2] H. Weber, Leopold Kronecker, Jber. D.M.V., 2 (1891–1892), 5–31.

236 (IX.22) *K*-Theory

A. General Remarks

K-theory was introduced by M. F. Atiyah and F. Hirzebruch, after the original idea was suggested by A. Grothendieck. The †Bott periodicity theorem is essential for the development of the theory. There are important applications of *K*-theory to †differential topology, such as the Riemann-Roch theorems for differentiable manifolds (due to Atiyah and Hirzebruch) [8,9], the solution of the vector field problem on spheres (due to J. Adams) [1], and applications to immersion and embedding problems (→ 410 Topology of Differentiable Manifolds; 211 Immersion and Embedding).

B. Construction of $K_\Lambda(X)$

Let the basic field Λ be the real number field **R**, the complex number field **C**, or the quaternion field **H**. Let Λ' be the †center of Λ, and X be a finite †CW complex. Then $\mathcal{E}_\Lambda(X)$ denotes the set of all isomorphism classes of Λ-†vector bundles over X and is a commutative semigroup under the †Whitney sum $\xi\oplus\eta$. Let $F_\Lambda(X)$ be the free Abelian group generated by the set $\mathcal{E}_\Lambda(X)$, and let $Q_\Lambda(X)$ be the subgroup of $F_\Lambda(X)$ generated by the elements of the form $\xi\oplus\eta-\xi-\eta$. Then the Abelian group $K_\Lambda(X)$ is defined as the quotient group $K_\Lambda(X)=F_\Lambda(X)/Q_\Lambda(X)$.

From this we obtain the canonical mapping $\theta:\mathcal{E}_\Lambda(X)\subset F_\Lambda(X)\to K_\Lambda(X)$, which is a homomorphism of the semigroups. Moreover, the pair $(K_\Lambda(X),\theta)$ is universal in the following sense: Given an Abelian group G and a semigroup homomorphism $g:\mathcal{E}_\Lambda(X)\to G$, there exists a unique group homomorphism h: $K_\Lambda(X)\to G$ such that $g=h\circ\theta$. We call h the extension of g.

Let $f:X\to Y$ be a continuous mapping from X into another finite CW complex Y. Then for $\eta\in\mathcal{E}_\Lambda(Y)$, the †induced bundle $f^*(\eta)\in\mathcal{E}_\Lambda(Y)$ is defined. Since the mapping f^*: $\mathcal{E}_\Lambda(Y)\to\mathcal{E}_\Lambda(X)$ is a semigroup homomorphism, it induces a homomorphism $K_\Lambda(f)$: $K_\Lambda(Y)\to K_\Lambda(X)$, which is the extension of $\theta\circ f^*$, so that $K_\Lambda(f)\circ\theta=\theta\circ f^*$. Usually, $K_\Lambda(f)$ is also denoted by f^*. Thus K_Λ is a †contravariant functor. According to the case as $\Lambda=\mathbf{R}$, **C**, or **H**, the notations **KO**, **K**, or **KSP** are often used for K_Λ.

If $X=\{x_0\}$, the semigroup homomorphism $\mathcal{E}_\Lambda(x_0)\ni\xi\to\dim_\Lambda\xi\in\mathbf{Z}$ induces an isomorphism $K_\Lambda(x_0)\cong\mathbf{Z}$. If X is a CW complex with base point x_0, then the reduced group $\tilde{K}_\Lambda(X)$ is defined to be the kernel of i^*, where $i:x_0\to X$ is the inclusion. Then we have the canonical splitting $K_\Lambda(X)\cong\mathbf{Z}\oplus\tilde{K}_\Lambda(X)$, $\tilde{K}_\Lambda$ is a functor defined on the category of pointed CW complexes.

Two isomorphism classes ξ and η of $\mathcal{E}_\Lambda(X)$ are said to be **stably equivalent** if there exist trivial bundles θ_1 and θ_2 such that $\xi\oplus\theta_1=\eta\oplus\theta_2$. An equivalence class with respect to this relation is called a **stable vector bundle**. If X is connected, then the set of all stable Λ-vector bundles can be identified with $\tilde{K}_\Lambda(X)$.

When $\Lambda=\mathbf{R}$ or **C**, the †tensor product of vector bundles induces a ring structure on $K_\Lambda(X)$, and f^* becomes a ring homomorphism.

The complexification of real vector bundles $\xi \rightarrow \xi \otimes_{\mathbf{R}} \mathbf{C} = i(\xi)$ is a semigroup homomorphism $\mathcal{E}_{\mathbf{R}}(X) \rightarrow \mathcal{E}_{\mathbf{C}}(X)$ and induces a ring homomorphism $i : KO(X) \rightarrow K(X)$ such that $i \circ \theta = \theta \circ i$. If $\xi \in \mathcal{E}_{\mathbf{H}}(X)$, then ξ can be viewed as a complex vector bundle under the scalar restriction of basic field, which we shall denote by $\rho(\xi) \in \mathcal{E}_{\mathbf{C}}(X)$. The mapping ρ: $\mathcal{E}_{\mathbf{H}}(X) \rightarrow \mathcal{E}_{\mathbf{C}}(X)$ induces a homomorphism $\rho : KSP(X) \rightarrow K(X)$. Similarly, the scalar restriction from **C** to **R** induces a homomorphism $r : K(X) \rightarrow KO(X)$. All these are †natural transformations.

Let ξ be a complex vector bundle. We can formally write the †Chern class (→ 58 Characteristic Classes) $c(\xi)$ of ξ as

$$c(\xi) = \prod (1 + x_i).$$

Then the **Chern character** $ch(\xi) \in H^*(X; \mathbf{Q})$ is defined by

$$ch(\xi) = \sum \exp x_i,$$

where **Q** is the field of rational numbers. The mapping $ch : \mathcal{E}_{\mathbf{C}}(X) \rightarrow H^*(X; \mathbf{Q})$ is extended to a ring homomorphism $ch : K(X) \rightarrow H^*(X; \mathbf{Q})$. We denote by the same notation ch the ring homomorphisms $ch \circ i : KO(X) \rightarrow H^*(X; \mathbf{Q})$ and $ch \circ \rho : KSP(X) \rightarrow H^*(X; \mathbf{Q})$. These are natural transformations from the functor K_Λ to the functor $H^*(\ ;\mathbf{Q})$.

C. Cohomology Theory

$O_\Lambda(n)$ denotes $O(n)$, $U(n)$, or $Sp(n)$ according as the basic field Λ is **R**, **C**, or **H**. Let O_Λ be the †inductive limit group with respect to the usual inclusion $O_\Lambda(n) \subset O_\Lambda(n+1)$. Provided with the weak topology, the group O_Λ becomes a †CW group. The set of all equivalence classes of stable Λ-vector bundles corresponds bijectively to the set of †principal O_Λ-bundles. Let B_Λ be the †classifying space for the group O_Λ, X, Y be finite CW complexes with base points, $[X, Y]$ be the set of all †homotopy classes of mapping from X to Y, and $[X, Y]_0$ be the set of all †homotopy classes in the †category of pointed topological spaces. Then by the classification theorem of fiber bundles (→ 155 Fiber Bundles), we have $K_\Lambda(X) = [X, \mathbf{Z} \times B_\Lambda]$ and $\tilde{K}_\Lambda(X) = [X, \mathbf{Z} \times B_\Lambda]_0$. The space B_Λ has the structure of a weak †H-space, so that the induced group structure of the homotopy set $[X, \mathbf{Z} \times B_\Lambda]$ coincides with that of $K_\Lambda(X)$. If f is a continuous mapping, then the induced homomorphism f^* of the homotopy set $[X, \mathbf{Z} \times B_\Lambda]$ coincides with that of $K_\Lambda(X)$.

For a finite CW pair (X,A), we put $K_\Lambda^{-n}(X,A) = \tilde{K}_\Lambda(S^n(X/A))$, $n = 0, 1, 2, \ldots$, where X/A is the space obtained from X by collapsing A to a point that becomes the base point of X/A, and S^n denotes the †n-fold reduced suspension. This gives rise to a cohomology theory (indexed by nonpositive integers) (→ 204 Homotopy; 203 Homology Groups).

The tensor product of vector bundles induces the following pairing, called the **cross product**:

$$K_\Lambda^{-m}(X,A) \otimes K_\Lambda^{-n}(Y,B) \rightarrow K_{\Lambda'}^{-(m+n)}(X \times Y, X \times B \cup A \times Y).$$

When $\Lambda = \mathbf{R}$ or **C**, the **cup product**

$$K_\Lambda^{-m}(X) \otimes K_\Lambda^{-n}(X) \rightarrow K_\Lambda^{-(m+n)}(X),$$

$$K_\Lambda^{-m}(X) \otimes K_\Lambda^{-n}(X,A) \rightarrow K_\Lambda^{-(m+n)}(X,A)$$

is defined as the composite of the cross product and the induced homomorphism Δ^*, where $\Delta : X \rightarrow X \times X$ is the diagonal mapping. The complexification $i : KO^{-n}(X,A) \rightarrow K^{-n}(X,A)$ preserves cup product. The composite of

$$ch : K_\Lambda^{-n}(X,A) \rightarrow \tilde{H}^*(S^n(X/A); \mathbf{Q})$$

and the †suspension isomorphism

$$\tilde{H}^*(S^n(X/A); \mathbf{Q}) \rightarrow H^*(X,A; \mathbf{Q})$$

is denoted by

$$ch : K_\Lambda^{-n}(X,A) \rightarrow H^*(X,A; \mathbf{Q}).$$

The homomorphism ch preserves cup product when $\Lambda = \mathbf{R}$ or **C**.

D. Bott Periodicity

Let ξ_Λ be the †canonical Λ-line bundle over the Λ-projective line. The elements

$$g_{\mathbf{C}} = \theta(\xi_{\mathbf{C}}) - 1 \in \tilde{K}(S^2) = K^{-2}(x),$$

$$g_{\mathbf{H}} = \theta(\xi_{\mathbf{H}}) - 1 \in \widetilde{KSP}(S^4) = KSP^{-4}(x),$$

and

$$g_{\mathbf{R}} = g_{\mathbf{H}} \times g_{\mathbf{H}} \text{(cross product)} \in \widetilde{KO}(S^8) = KO^{-8}(x)$$

are called **Bott generators**.

The **Bott periodicity theorem** [13] is as follows: (1) $\tilde{K}(S^2)$, $\widetilde{KSP}(S^4)$, and $\widetilde{KO}(S^8)$ are infinite cyclic groups generated by $g_{\mathbf{C}}$, $g_{\mathbf{H}}$, and $g_{\mathbf{R}}$, respectively. Moreover, $ch(g_{\mathbf{C}}) = \sigma^2$, $ch(g_{\mathbf{H}}) = \sigma^4$, and $ch(g_{\mathbf{R}}) = \sigma^8$, where $\sigma^n \in H^n(S^n; \mathbf{Z})$ is a generator. (2) The cross products

$$K^{-n}(X,A) \otimes K^{-2}(x) \rightarrow K^{-(n+2)}(X,A),$$

$$KSP^{-n}(X,A) \otimes KSP^{-4}(x) \rightarrow KO^{-(n+4)}(X,A),$$

$$KO^{-n}(X,A) \otimes KO^{-8}(x) \rightarrow KO^{-(n+8)}(X,A)$$

are isomorphisms.

The isomorphisms

$$K^{-n}(X,A) \cong K^{-(n+2)}(X,A),$$

$$KSP^{-n}(X,A) \cong KO^{-(n+4)}(X,A),$$

and

$$KO^{-n}(X,A) \cong KO^{-(n+8)}(X,A),$$

defined by $a \to a \times g_\Lambda$, are called **Bott isomorphisms**. They commute with f^*, the coboundary operator δ, and ch. Identifying K^{-n} with $K^{-(n+2)}$, KSP^{-n} with $KO^{-(n+4)}$, and KO^{-n} with $KO^{-(n+8)}$, we get periodic cohomology theories $K^* = \sum_{n \in \mathbf{Z}_2} K^n$ and $KO^* = \sum_{n \in \mathbf{Z}_8} KO^n$, which are multiplicative cohomology theories, and ch is a multiplication-preserving homomorphism into $H^*(\ ;\mathbf{Q})$. The cohomology of a point $K_\Lambda^{-n}(x) = \tilde{K}_\Lambda(S^n) = \pi_n(B_\Lambda) = \pi_{n-1}(O_\Lambda)$ is given by Bott [14] (→ Appendix A, Table 6.IV).

E. Cohomology Operations

We assume that Λ is $\mathbf{R}$ or $\mathbf{C}$. The exterior powers λ^q are basic operations in $K_\Lambda(X)$. For $\xi \in \mathcal{E}_\Lambda(X)$, the pth †exterior power of ξ, $\lambda^p(\xi) \in \mathcal{E}_\Lambda(X)$, has the following properties: $\lambda^0(\xi) = 1$, $\lambda^1(\xi) = \xi$, and $\lambda^p(\xi \oplus \eta) = \sum_{q+r=p} \lambda^q(\xi) \otimes \lambda^r(\xi)$. Let G be the multiplicative group consisting of the formal power series $\in K_\Lambda(X)\{t\}$ whose constant term is 1. The assignment $\xi \to \lambda_t(\xi) = \sum \lambda^q(\xi) t^q$ gives rise to the homomorphism $\mathcal{E}_\Lambda(X) \to G$. Let $\lambda_t : K_\Lambda(X) \to G$ be its extension. The operation $\lambda^q : K_\Lambda(X) \to K_\Lambda(X)$ is defined by $\lambda_t(x) = \sum_{0 \leqslant q} \lambda^q(x) t^q$.

An important series of operations ψ^k, called the **Adams operations**, is derived from the exterior powers. Put

$$\psi_{-t}(x) = -t \frac{d\lambda_t(x)}{dt} \frac{1}{\lambda_t(x)} \in K_\Lambda(X)\{t\},$$

and define $\psi^k : K_\Lambda(X) \to K_\Lambda(X)$ by $\psi_t(x) = \sum_{0 < k} \psi^k(x) t^k$. When $\Lambda = \mathbf{C}$, define ψ^{-1} as the extension of $\xi \to \bar{\xi}$, where $\bar{\xi}$ is the †conjugate complex vector bundle of ξ. The operation ψ^k is a ring homomorphism preserving 1, and the relation $\psi^k \circ \psi^l = \psi^{kl}$ holds. If $\xi \in \mathcal{E}_\Lambda(X)$ is a line bundle, then $\psi^k(\theta(\xi)) = (\theta(\xi))^k$. If $x \in K_\Lambda(X)$ and $ch(x) = \sum_n ch_n(x)$, where $ch_n(x) \in H^{2n}(X;\mathbf{Q})$, then

$$ch(\psi^k(x)) = \sum_n k^n ch_n(x).$$

The operations λ^q and ψ^k commute with the complexification $i : KO(X) \to K(X)$. If $\beta_{\mathbf{C}} : \tilde{K}(X) \to \tilde{K}(S^2X)$ and $\beta_{\mathbf{R}} : \widetilde{KO}(X) \to \widetilde{KO}(S^8X)$ are Bott isomorphisms, then $\psi^k \circ \beta_{\mathbf{C}} = k\beta_{\mathbf{C}} \circ \psi^k$ and $\psi^k \circ \beta_{\mathbf{R}} = k^4 \beta_{\mathbf{R}} \circ \psi^k$ [1].

F. Thom-Gysin Isomorphisms

A real oriented vector bundle ξ over a finite CW complex X is called a **spin bundle** if $w_2(\xi) = 0$, where $w_2(\xi)$ is the second †Stiefel-Whitney class of the bundle ξ. The bundle ξ is called a c_1-**bundle** if there is given a cohomology class $c_1(\xi) \in H^2(X;\mathbf{Z})$ such that $c_1(\xi) \equiv w_2(\xi) \bmod 2$. Let X^ξ be the †Thom complex of the vector bundle ξ. The group $\tilde{K}_\Lambda^*(X^\xi)$ has the structure of a $K_\Lambda^*(X)$-module. If we assume that ξ is a c_1-bundle when $\Lambda = \mathbf{C}$ and a spin bundle when $\Lambda = \mathbf{R}$, then there exists a canonical $K_\Lambda^*(X)$-module isomorphism $\varphi : K_\Lambda^n(X) \to \tilde{K}_\Lambda^{n+\dim\xi}(X^\xi)$ such that $ch\varphi(1) = \varphi'((\hat{\mathfrak{A}}(\xi)\exp(c_1(\xi)/2))^{-1})$ when $\Lambda = \mathbf{C}$ and $ch\varphi(1) = \varphi'(\hat{\mathfrak{A}}(\xi)^{-1})$ when $\Lambda = \mathbf{R}$ [2]. Here $\varphi' : H^*(X;\mathbf{Q}) \to \tilde{H}^*(X^\xi;\mathbf{Q})$ is the usual †Thom-Gysin isomorphism and $\hat{\mathfrak{A}}(\xi)$ is the $\hat{\mathfrak{A}}$-**characteristic class** of the bundle ξ defined as follows: Write the †Pontrjagin class $p(\xi)$ of ξ formally as $p(\xi) = \prod(1 + x_i^2)$; then the class $\hat{\mathfrak{A}}(\xi)$ is given by

$$\hat{\mathfrak{A}}(\xi) = \prod(x_i/2)/(\sinh(x_i/2)).$$

If ξ is a complex vector bundle, then its first †Chern class $c_1(\xi)$ gives a c_1-bundle structure to $r(\xi) \in \mathcal{E}_{\mathbf{R}}(X)$. In this case, the class $\mathfrak{T}(\xi) = \hat{\mathfrak{A}}(\xi)\exp(c_1/2)$ is the **Todd characteristic class** of the complex vector bundle ξ.

G. Riemann-Roch Theorems for Differentiable Manifolds

Let M and N be connected closed differentiable manifolds. A continuous mapping $f : M \to N$ is called a **spin mapping** if $w_1(M) = f^* w_1(N)$ and $w_2(M) = f^* w_2(M)$. If $w_1(M) = f^* w_1(N)$ and there is given a class $c_1 \in H^2(M;\mathbf{Z})$ such that $w_2(M) - f^* w_2(N) \equiv c_1 (\operatorname{mod} 2)$, f is called a c_1-**mapping**. If we assume that f is a c_1-mapping when $\Lambda = \mathbf{C}$ and a spin mapping when $\Lambda = \mathbf{R}$, then there is a canonical homomorphism $f_! : K_\Lambda^n(M) \to K_\Lambda^{n+\dim N - \dim M}(N)$ such that

$$f_!(f^*(x) \cdot y) = x \cdot f_!(y)$$

and

$$ch(f_!(y))\hat{\mathfrak{A}}(N) = f_!(ch(y)\hat{\mathfrak{A}}(M)\exp(c_1/2))$$

for $y \in K_\Lambda^*(M)$ and $x \in K_\Lambda^*(N)$. This is the **Riemann-Roch theorem for differentiable manifolds** [8] (→ 361 Riemann-Roch Theorems). In the second formula, if $\Lambda = \mathbf{R}$ we set $c_1 = 0$, and the homomorphism $f_! : H^*(M;\mathbf{Q}) \to H^*(N;\mathbf{Q})$ on the right-hand side is the usual †Gysin homomorphism. The homomorphism $f_!$ depends only on the homotopy class of f and has the usual functorial properties $1_! = 1$ and $(f \circ g)_! = f_! \circ g_!$.

H. The Atiyah-Singer Index Theorem

Let X be an n-dimensional compact differentiable manifold of class C^∞ without boundary (→ 108 Differentiable Manifolds). As we shall see later, any †elliptic differential operator (or, more generally, any elliptic complex) d on X has analytic index $\operatorname{ind}_a(d)$ and topological index $\operatorname{ind}_t(d)$, the latter of which is deeply related to *K*-theory. The **Atiyah-Singer index theorem** asserts that $\mathrm{ind}_a(d)=\mathrm{ind}_t(d)$ [11, 12]. We shall describe the details of the definitions and the theorem.

Let E and F be complex vector bundles of class C^∞ over X with $\dim E=s$ and $\dim F=t$ (→ 155 Fiber Bundles). Let $\Gamma(E)$ and $\Gamma(F)$ be the linear spaces over $\mathbf{C}$ consisting of C^∞-cross sections of E and F, respectively. A linear mapping d from $\Gamma(E)$ to $\Gamma(F)$ is called a **differential operator of the *k*th order** if d is locally expressed by some differential operator of the kth order. This means that if we choose a local coordinate neighborhood U of X and trivializations of E and F on U such that $E|U\cong U\times\mathbf{C}^s$ and $F|U\cong U\times\mathbf{C}^t$, then d is a differential operator of the kth order from $C^\infty(U,\mathbf{C}^s)$ to $C^\infty(U,\mathbf{C}^t)$. Thus d is locally expressed by the matrix form $(\sum_{|\alpha|\leqslant k}a_\alpha^{(i,j)}(x)D^\alpha)$ $(i=1,\ldots,t;j=1,\ldots,s)$, each component of which is a differential operator (→ 115 Differential Operators). Using this expression we define the symbol $\sigma(d)$ of d as follows: Let $T^*(X)$ be the cotangent bundle of X. Given any $\eta_x\in T_x^*(X)$, put $\sigma(d)(\eta_x)=(\sum_{|\alpha|=k}a_\alpha^{(i,j)}(x)\eta_x^\alpha)$, where η_x^α stands for $\eta_1^{\alpha_1}\ldots\eta_n^{\alpha_n}$ for any multi-index $\alpha=(\alpha_1,\ldots,\alpha_n)$ and for a local coordinate expression $\eta_x=(\eta_x^1,\ldots,\eta_x^n)$. We call $\sigma(d)(\eta_x)$ the symbol of d. Now a differential operator d is called **elliptic** if for each $\eta_x\neq 0$ the symbol $\sigma(d)(\eta_x)$ gives an isomorphism from E_x onto F_x. For the elliptic differential operator d, we have $\dim\mathrm{Ker}\,d<\infty$ and $\dim\mathrm{Coker}\,d<\infty$ [7]. The **analytic index** $\mathrm{ind}_a(d)$ is defined to be the integer $\dim\mathrm{Ker}\,d-\dim\mathrm{Coker}\,d$, and it has the characteristic property that $\mathrm{ind}_a(d)$ is invariant under deformation of d.

More generally, an elliptic complex $\mathcal{E}$ on X and the analytic index $\mathrm{ind}_a(\mathcal{E})$ can be defined as follows: Given a finite number of smooth complex vector bundles $\{E_i\}_{i=1,\ldots,l}$ on X and differential operators $d_i:\Gamma(E_i)\to\Gamma(E_{i+1})$, we call $\mathcal{E}=\{E_i,d_i\}_{i=1,\ldots,l}$ an **elliptic complex** on X if the following two conditions are satisfied: (i) $d_{i+1}\circ d_i=0$; (ii) for any $\eta_x\in T_x^*(X)$, $\eta_x\neq 0$, the symbol sequence $\to E_{i,x}\xrightarrow{\sigma(d_i)(\eta_x)}E_{i+1,x}\to$ is exact. For an elliptic complex $\mathcal{E}$, we have $\dim H^i(\mathcal{E})=\dim(\mathrm{Ker}\,d_i/\mathrm{Im}\,d_{i-1})<\infty$ [7, 9, 16]. The integer $\sum(-1)^i\dim H^i(\mathcal{E})$ is called the **analytic index** of $\mathcal{E}$. An elliptic complex with the form $0\to\Gamma(E)\xrightarrow{d}\Gamma(F)\to 0$ is an elliptic operator. An important example of an elliptic complex arises from de Rham theory (→ 108 Differentiable Manifolds Q): Take $E_i=\Lambda^iT^*(X)$ and the exterior differentiations as differential operators. The elliptic complex thus obtained is the **de Rham complex**.

The topological index $\mathrm{ind}_t(\mathcal{E})$ of the elliptic complex $\mathcal{E}$ is introduced in the following way: From the theory of difference bundles [4] we can deduce that by virtue of the exactness for $\eta_x\neq 0$, the symbol sequence $\to E_{i,x}\xrightarrow{\sigma(d_i)(\eta_x)}E_{i+1,x}\to$ determines a definite element $[\sigma(d)]$ of $K(X^\tau)$, where X^τ is the †Thom complex associated with the cotangent bundle of X. Embed X in some Euclidean space $\mathbf{R}^N$. Then the mapping $j:T^*(X)\to T^*(\mathbf{R}^N)\cong\mathbf{R}^{2N}$ canonically induces the homomorphism $j_!:K(X^\tau)\to K((\mathbf{R}^N)^\tau)\cong K(S^{2N})\cong\mathbf{Z}$, and $j_!$ is obtained from some modified construction of the Thom homomorphism. We set $\mathrm{ind}_t(\mathcal{E})=j_![\sigma(\mathcal{E})]$ and call this the **topological index** of $\mathcal{E}$. We have

$$\mathrm{ind}_t(\mathcal{E})=ch([\sigma(d)])\mathfrak{T}(X)[X^\tau],$$

where $ch([\sigma(d)])$ $(\in H^*(X^\tau;\mathbf{Q}))$ is the Chern character of $\sigma(d)$, $\mathfrak{T}(X)$ $(\in H^*(X;\mathbf{Q}))$ is the Todd class of $\mathfrak{T}(X)\otimes\mathbf{C}$, and $[X^\tau]$ is the fundamental cycle of X^τ [12].

The Atiyah-Singer index theorem, in general form, now asserts that $\mathrm{ind}_a(\mathcal{E})=\mathrm{ind}_t(\mathcal{E})$. For the de Rham complex E, it follows from the definition that $\mathrm{ind}_t(E)$ is equal to the Euler characteristic of X. Let X be a compact complex analytic manifold, and W be a complex analytic vector bundle on X. Applying the theorem to the Dolbeault complex with value in $W\ldots\to A^{0,i}(W)\xrightarrow{d''}A^{0,i+1}(W)\to\ldots$ (→ 74 Complex Manifolds), we can conclude that Hirzebruch's formulation of Riemann-Roch theorem (→ 361 Riemann-Roch Theorems B) holds not only for projective algebraic manifolds but also for compact complex manifolds. Moreover, from the index theorem we can deduce the Hirzebruch index theorem (→ 58 Characteristic Classes G) and various integrability theorems [12].

The Atiyah-Singer index theorem can be extended to compact manifolds with boundary, where it is related to the elliptic boundary value problem [5]. Another generalization of the theorem was obtained for the case where a compact Lie group operates on manifolds, vector bundles, and elliptic operators in a reasonable way [12]. In this case, the index theorem is formulated in the framework of equivariant *K*-theory [10] and has an intimate relation with the †Lefschetz fixed-point theorem. Actually, this enables us to apply the index theorem to the study of transformation groups on manifolds.

I. Algebraic *K*-theory

Algebraic *K*-theory is a branch of algebra concerned mainly with a series of Abelian group valued functors K_n of rings (and, more generally, of certain categories), which have certain features of generalized homology theory. It originated in the *K*-group construction used in Grothendieck's work on the Riemann-Roch theorem. The theory was initiated in the early sixties by H. Bass, who introduced K_1 and extensively studied K_0 and K_1 in collaboration with other researchers [20, 21, 22, 23]. Then K_2 was introduced by J. Milnor [26], and higher K-theories were constructed by D. G. Quillen and others from various viewpoints ([27], I). There is also a K-theory with respect to Hermitian structure ([27], III). Algebraic K-theory is intimately related to various other branches of mathematics such as topology, algebraic geometry, and arithmetic.

The **Grothendieck group** $K_0(A)$ of a ring A is the Abelian group generated by the set of isomorphism classes $[P]$ of finitely generated projective A-modules subject to the relation $[P \oplus P'] = [P] + [P']$ for every pair of projective modules P, P'. If A is finitely generated as a $\mathbf{Z}$-algebra, then $K_0(A)$ is a finitely generated group. The assignment $n \rightarrow n[A]$ defines a homomorphism $\mathbf{Z} \rightarrow K_0(A)$, whose cokernel is the †projective class group of A. If A is commutative, a similar construction for the category of rank 1 projective A-modules with respect to the tensor product $\otimes$ leads to the **Picard group** $Pic(A)$. We then have an epimorphism $K_0(A) \rightarrow Pic(A)$ defined by $P \rightarrow \Lambda^r P$ where P is of rank r [21]. If X is a compact Hausdorff space, the topological $K(X)$ is isomorphic to the algebraic $K_0(A)$, where A is the ring of complex-valued continuous functions on X.

The **Whitehead group** $K_1(A)$ is defined as follows. Let $GL(A)$ be the direct limit of the sequence $\ldots \rightarrow GL_n(A) \xrightarrow{i_n} GL_{n+1}(A) \rightarrow \ldots$ where $i_n(X) = \begin{pmatrix} X & 0 \\ 0 & 1 \end{pmatrix}$. Let $E_n(A)$ be the subgroup of $GL_n(A)$ generated by all elementary matrices $1 + ae_{ij}$ $(i \neq j, a \in A)$, where the e_{ij} are matrix units. The limit $E(A)$ of $E_n(A)$ coincides with the commutator subgroup of $GL(A)$. Now define $K_1(A) = GL(A)/E(A)$ [20, 23]. For $A = \mathbf{Z}\pi$, the integral group algebra of a group π, the cokernel of the natural homomorphism $\pm \pi \rightarrow K_1(A)$ is denoted by $Wh(\pi)$. The torsion invariant of J. H. C. Whitehead is defined in $Wh(\pi)$ [25]. If A is commutative, we put $SK_1(A) = SL(A)/E(A)$ where $SL(A) = \varinjlim SL_n(A)$. By the determinant homomorphism, we have $K_1(A) \cong SK_1(A) \oplus U(A)$, where $U(A)$ is the group of units of A. $SK_1(A) = 0$ when A is a field or a local ring A. A deeper result states that $SK_1(A) = 0$ for the ring of integers of an algebraic number field (Bass, Milnor, and Serre [22]). This and related results have been applied to investigate $SK_1(\mathbf{Z}\pi)$ for a finite group π.

We define $K_2(A) = H_2(E(A), Z)$ (the Schur multiplier of $E(A)$). This yields a universal central extension of $E(A)$, defined by $0 \rightarrow K_2(A) \rightarrow St(A) \rightarrow E(A) \rightarrow 0$, where $St(A) = \varinjlim St_n(A)$ and $St_n(A)$ $(n \geqslant 3)$ is the **Steinberg group** generated by $x_{ij}(a)$ $(a \in A; i, j = 1, \ldots, n, i \neq j)$ subject to the relations (i) $x_{ij}(a)x_{ij}(b) = x_{ij}(a+b)$, and (ii) the commutator $(x_{ij}(a), x_{kl}(b))$ equals $x_{il}(ab)$ for $j = k$, $i \neq l$; and equals 1 for $j \neq k$, $i \neq l$.

Let F be a field. A bimultiplicative mapping $s: F^* \times F^* \rightarrow C$ (C an Abelian group) satisfying $s(x, 1-x) = 1$ $(x \neq 0, 1)$ is called a **(Steinberg) symbol** on F. There exists a universal symbol $F^* \times F^* \rightarrow K_2(F)$ which, followed by homomorphisms $K_2(F) \rightarrow C$, yields all C-valued symbols on F. Since certain Steinberg symbols such as the †Hilbert symbol are important in arithmetic, the group $K_2(F)$ of a global or local field F is intimately related to the arithmetic of F. If F is an algebraic number field and R its ring of integers, we have an exact sequence $0 \rightarrow K_2(R) \rightarrow K_2(F) \rightarrow \coprod_{\mathfrak{p}} (R/\mathfrak{p})^* \rightarrow 0$ (the last term being the direct sum over all prime ideals $\mathfrak{p}$ of R), and $K_2(R)$ is a finite group.

References

[1] J. F. Adams, Vector fields on spheres, Ann. of Math., (2) 75 (1962), 603–632.
[2] J. F. Adams, On the groups $J(X)$ I, II, III, IV, Topology, 2 (1963), 181–195; 3 (1965), 137–171, 193–222; 5 (1966), 21–71.
[3] M. F. Atiyah, Algebraic topology and elliptic operators, Comm. Pure Appl. Math., 20 (1967), 237–249.
[4] M. F. Atiyah, K-theory, Benjamin, 1967.
[5] M. F. Atiyah and R. Bott, The index problem for manifolds with boundary, Bombay Colloquium on Differential Analysis, Oxford Univ. Press (1964), 175–186.
[6] M. F. Atiyah, R. Bott, and A. Shapiro (Šapiro), Clifford modules, Topology, 3 (1964), Suppl. 1, 3–38.
[7] M. F. Atiyah and R. Bott, A Lefschetz fixed point formula for elliptic complexes I, Ann. of Math., (2) 86 (1967), 374–407.
[8] M. F. Atiyah and F. Hirzebruch, Riemann-Roch theorems for differentiable manifolds, Bull. Amer. Math. Soc., 65 (1959), 276–281.
[9] M. F. Atiyah and F. Hirzebruch, Vector bundles and homogeneous spaces, Amer. Math. Soc., Proc. Symposia in Pure Math. (1961), 7–38.
[10] M. F. Atiyah and G. B. Segal, Seminar on equivariant K-theory, Lecture notes, Oxford Univ., 1965.
[11] M. F. Atiyah and I. M. Singer, The index of elliptic operators on compact manifolds, Bull. Amer. Math. Soc., 69 (1963), 422–433.

[12] M. F. Atiyah and I. M. Singer, The index of elliptic operators I, II, III, Ann. of Math., (2) 87 (1968), 484–604.
[13] R. Bott, Quelques remarques sur les théorèmes de périodicité, Bull. Soc. Math. France, 87 (1959), 293–310.
[14] R. Bott, The stable homotopy of the classical groups, Ann. of Math., (2) 70 (1959), 313–337.
[15] R. Bott, Lectures on $K(X)$, Benjamin, 1969.
[16] Séminaire H. Cartan, Ecole Norm. Sup., 1963–1964.
[17] P. E. Conner and E. E. Floyd, The relation of cobordism to K-theories, Lecture notes in math. 28, Springer, 1966.
[18] E. Dyer, Cohomology theories, Benjamin, 1969.
[19] R. Palais, Seminar on the Atiyah-Singer index theorem, Ann. Math. Studies, Princeton Univ. Press, 1965.
[20] H. Bass, K-theory and stable algebra, Publ. Math. Inst. HES, 22 (1964), 5–60.
[21] H. Bass and M. P. Murthy, Grothendieck groups and Picard groups of abelian group rings, Ann. of Math., (2) 86 (1967), 16–73.
[22] H. Bass, J. Milnor, and J.-P. Serre, Solution of the congruence subgroup problem for $SL_n (n \geqslant 3)$ and $Sp_{2n} (n \geqslant 2)$, Publ. Math. Inst. HES, 33 (1968), 59–137.
[23] H. Bass, Algebraic K-theory, Benjamin, 1968.
[24] R. Swan, K-theory of finite groups and orders, Lecture notes in math. 149, Springer, 1970.
[25] J. Milnor, Whitehead torsion, Bull. Amer. Math. Soc., 72 (1966), 358–426.
[26] J. Milnor, Introduction to algebraic K-theory, Ann. Math. Studies, Princeton Univ. Press, 1971.
[27] Algebraic K-theory, Battelle Inst. Conf. 1972, I, II, III, Lecture notes in math. 341, 342, 343, Springer, 1973.

L

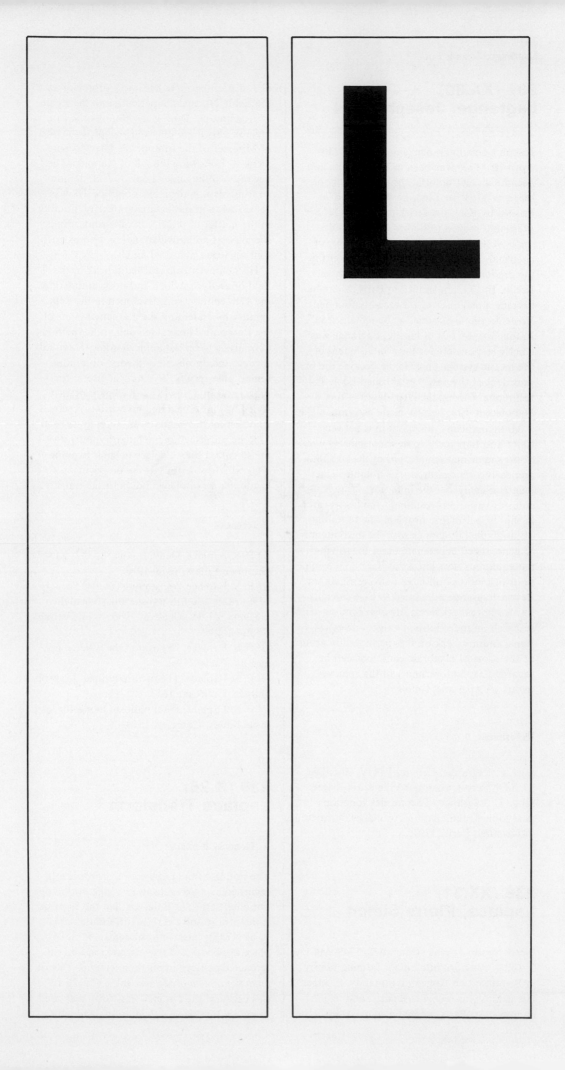

237 (XX.30) Lagrange, Joseph Louis

Joseph Louis Lagrange (January 25, 1736–April 10, 1813) was born in Turin, Italy, and became an instructor at the military school there in 1753. In 1766, he was invited to Prussia by King Frederick the Great (1712–1786) and moved to Berlin, where he filled the post (formerly occupied by L. Euler) of chairman of the mathematics department in the graduate division of the University of Berlin. In 1787, he moved to Paris, where he became a professor at the recently founded Ecole Normale Supérieure; he remained in France for the rest of his life. Lagrange was chiefly responsible for the establishment of the metric system. In 1795, he became the first president of the newly established Ecole Polytechnique. During the later stages of the Napoleonic Era, he was made a count.

Mathematically, his position is between Euler and Laplace, and he is considered one of the major mathematicians of the late 18th and early 19th centuries. His notable achievements in analysis—the initiation of the †calculus of variations resulting from his research in the †isoperimetric problem, the founding of †analytical dynamics with the introduction of generalized coordinates, and the solving of the equations now known as †Lagrange's equations of motion—all have a strong algebraic flavor. Lagrange attempted to base calculus on †formal power series. He also conducted research on the solution of algebraic equations, and his work on the †permutation group of the roots of algebraic equations can be regarded as the forerunner of the achievements of Abel and Galois.

References

[1] J. L. Lagrange, Oeuvres I–XIV, edited by M. J. A. Serret, Gauthier-Villars, 1867–1892.
[2] J. L. Lagrange, Théorie des fonctions analytiques, contenant les principes du calcul différentiel, Paris, 1797.

238 (XX.31) Laplace, Pierre Simon

Pierre Simon Laplace (March 23, 1749–March 5, 1827) was born into a poor farming family in Beaumont en Auge in Normandy, France. His genius was recognized early, and in 1767 he moved to Paris, where he enjoyed the favor of J. d'Alembert. He became a professor at the Ecole Normale Supérieure and the Ecole Polytechnique. During the Napoleonic Era, he took part in politics and accepted the post of Minister of the Interior; he later became a count. Following the fall of Napoleon, he became a duke under Louis XVIII, demonstrating his nonideological politics. He was also lacking in professional integrity, notably in the matter of priority in obtaining results; social position notwithstanding, he was not well regarded and died lonely.

His achievements reached a peak in the field of analysis, which had been initiated in the 17th century and developed in the 18th century by Euler and the mathematicians of the Bernoulli family. He applied the methods of analysis to †celestial mechanics, †potential theory, and †probability theory, obtaining remarkable results. Without the use of formulas and in a flowing and elegant literary style [3, 5], he elucidated his various results. Concerning the origin of the solar system, in 1796 he published the nebular hypothesis—the so-called **Kant-Laplace nebular hypothesis**—which is famous as the predecessor of the theory of the evolution of the universe.

References

[1] P. S. Laplace, Oeuvres complètes I–XIV, Gauthier-Villars, 1878–1912.
[2] P. S. Laplace, Mécanique céleste, Paris, 1798–1825; English translation, Celestial mechanics I–IV, Chelsea, 1966; V (in French), Chelsea, 1969.
[3] P. S. Laplace, Exposition du système du monde, Paris, 1796.
[4] P. S. Laplace, Théorie analytique des probabilités, Courcier, 1812.
[5] P. S. Laplace, Essai philosophique sur les probabilités, Courcier, 1814.

239 (X.26) Laplace Transform

A. General Remarks

The notion of the Laplace transform can be regarded as a generalization of the notion of Dirichlet series. L. Euler applied the Laplace transform to solve certain differential equations (1737); later, independently, P. S. Laplace applied it to solve differential and difference equations in his famous book *Théorie analytique des probabilités*, vol. 1 (1812). In this century, the Laplace transform was used to justify Heaviside's †operational calculus, and the notion has become an important tool in applied mathematics.

Let $\alpha(t)$ be a function of †bounded variation in the interval $0 \leqslant t \leqslant R$ for every positive R. If

$$L(s)=\int_0^{\infty} e^{-st}\,d\alpha(t)=\lim_{R\to\infty}\int_0^R e^{-st}\,d\alpha(t)$$

converges for some complex number s_0, then it converges for all s satisfying $\mathrm{Re}\,s>\mathrm{Re}\,s_0$. We call $L(s)$ the **Laplace-Stieltjes transform** of $\alpha(t)$. If $\alpha(t)=\int_0^t\varphi(u)\,du$ (where $\varphi(u)$ is †Lebesgue integrable in the interval $0\leqslant t\leqslant R$ for every R), then we call

$$L(s)=\int_0^{\infty} e^{-st}\varphi(t)\,dt$$

the **Laplace transform** of $\varphi(t)$ (→ Appendix A, Table 12.I).

B. Regions of Convergence

Given a Laplace transform $L(s)$ of $\alpha(t)$, there exists a real number (or $\pm\infty$) σ_c such that the maximal region of convergence of $L(s)$ is the set of all s such that $\mathrm{Re}\,s>\sigma_c$. In extreme cases, when the integral never converges, we write $\sigma_c=+\infty$, and when it converges everywhere, we write $\sigma_c=-\infty$. The number σ_c is called the **abscissa of convergence** of $L(s)$, and the line $\mathrm{Re}\,s=\sigma_c$ the **axis of convergence** of $L(s)$. A formula to determine the abscissa of convergence in terms of $\alpha(t)$ is known. If $k=\limsup_{t\to\infty}(\log|\alpha(t)|)/t \neq 0$, then $\sigma_c=k$; if $k=0$ and $\alpha(t)$ does not converge as t tends to infinity, then $\sigma_c=0$. If $L(s)$ has a nonnegative abscissa σ_c, then $\sigma_c=\limsup_{t\to 0}(\log|\alpha(t)|)/t$, and if $\sigma_c<0$, then $\sigma_c=\limsup_{t\to\infty}(\log|\alpha(\infty)-\alpha(t)|)/t$ (E. Landau, S. Pincherle). Generally,

$$\limsup_{t\to\infty}\frac{\log|\alpha(t)-\alpha([t])|}{t}=\sigma_c,$$

where [ ] is the †Gauss symbol (K. Kurosu, T. Kojima, T. Ugaeri, and K. Knopp).

If $\int_0^{\infty}e^{-st}|d\alpha(t)|<\infty$, then the Laplace-Stieltjes integral $L(s)=\int_0^{\infty}e^{-st}\,d\alpha(t)$ is said to be **absolutely convergent**. There exists a real number σ_a such that $L(s)$ converges absolutely for $\mathrm{Re}\,s>\sigma_a$ and does not converge absolutely for $\mathrm{Re}\,s<\sigma_a$. We call σ_a the **abscissa of absolute convergence** of $L(s)$. There exists a real number σ_u such that $L(s)$ converges uniformly for $\mathrm{Re}\,s\geqslant\sigma_u+\varepsilon$ (for every $\varepsilon>0$) and fails to do so for $\mathrm{Re}\,s\geqslant\sigma_u-\varepsilon$. We call σ_u the **abscissa of uniform convergence** of $L(s)$. It is clear that $\sigma_c\leqslant\sigma_u\leqslant\sigma_a$. Formulas determining σ_u and σ_a are analogous to the Dirichlet series formulas given by Kojima and M. Kunieda (→ 125 Dirichlet Series B) [1].

C. Regularity

In the region of convergence $\mathrm{Re}\,s>\sigma_c$, $L(s)=\int_0^{\infty}e^{-st}\,d\alpha(t)$ is †holomorphic and we have $L^{(k)}(s)=\int_0^{\infty}e^{-st}(-t)^k\,d\alpha(t)$ in $\mathrm{Re}\,s>\sigma_c$. If $\alpha(t)$ is monotonic, then the real point $s=\sigma_c$ on the axis of convergence is a singular point of $L(s)$. However, there may not be any singular point on the axis of convergence in general. The **abscissa of regularity** is the infimum of all σ such that $L(s)$ is holomorphic in $\mathrm{Re}\,s>\sigma$. Also, $L(\sigma+i\tau)=O(|\tau|)$ uniformly in $\sigma_c+\delta\leqslant\sigma<\infty$ for every positive δ as $|\tau|\to\infty$. Any analytic function that is holomorphic at ∞ may be represented by a Laplace transform. To be precise, let $f(s)=f(\infty)+\sum_{n=0}^{\infty}(a_n-n!)/s^{n+1}$ ($|s|>c$). Then the function $\varphi(t)=\sum_{n=0}^{\infty}a_nt^n$ is entire, $f(s)=f(\infty)+\int_0^{\infty}e^{-st}\varphi(t)\,dt$, and $\sigma_c>c$.

D. Inversion Formulas

We say that a function $\alpha(t)$ that is of †bounded variation in any positive interval is normalized if $\alpha(0)=\alpha(+0)=0$ and $\alpha(t)=(\alpha(t+0)+\alpha(t-0))/2$. A normalized function $\alpha(t)$ ($t>0$) is uniquely determined by its Laplace transform. Moreover, the **inversion formula** determining $\alpha(t)$ in terms of $L(s)$ is known. That is, if $L(s)=\int_0^{\infty}e^{-st}\,d\alpha(t)$, then for $c>\max(\sigma_c,0)$,

$$\lim_{T\to\infty}\frac{1}{2\pi i}\int_{c-iT}^{c+iT}\frac{L(s)}{s}e^{st}\,ds=\begin{cases}\alpha(t), & t>0,\\ \alpha(+0)/2, & t=0,\\ 0, & t<0.\end{cases}$$

The integral on the left-hand side is often called the **Bromwich integral**. Suppose that $\alpha(t)=\int_0^t\varphi(u)\,du$, $L(s)$ converges absolutely on $\mathrm{Re}\,s=c$, and $\varphi(u)$ is of bounded variation in a neighborhood of $u=t$ ($t\geqslant 0$). Then we have

$$\lim_{T\to\infty}\frac{1}{2\pi i}\int_{c-iT}^{c+iT}L(s)e^{st}\,ds=\begin{cases}(\varphi(t+0)+\varphi(t-0))/2, & t>0,\\ \varphi(+0)/2, & t=0,\\ 0, & t<0.\end{cases}$$

There is another form of the inversion formula by E. L. Post and D. V. Widder. Namely, set

$$L_{k,t}[f(x)]=(-1)^kf^{(k)}(k/t)(k/t)^{k+1}$$

for a C^{∞}-function $f(x)$ ($t>0$, k is a positive integer). Then

$$\lim_{k\to\infty}\int_0^t L_{k,u}[L(x)]\,du=\alpha(t)-\alpha(+0).$$

If $\alpha(t)=\int_0^t\varphi(u)\,du$, then for almost all $t\,(>0)$,

$$\lim_{k\to\infty} L_{k,t}[L(x)]=\varphi(t)$$

(→ Appendix A, Table 12.I).

E. The Representation Theorem

If $f(x)$ is of class C^∞ on (a,b) and $(-1)^k f^{(k)}(x)\geqslant 0$ for $k=1,2,,\ldots$, then $f(x)$ is called **completely monotonic** in (a,b). Moreover, if $f(x)$ is continuous on $[a,b]$, then $f(x)$ is called completely monotonic on $[a,b]$. A necessary and sufficient condition for a function $f(x)$ to be completely monotonic in $0\leqslant x<\infty$ is that $f(x)=\int_0^\infty e^{-xt}\,d\alpha(t)$, where $\alpha(t)$ is bounded and increasing and the integral converges for $0\leqslant x<\infty$ (**Bernstein's theorem**). A necessary and sufficient condition for $f(x)$ to be representable in the form $\int_0^\infty e^{-xt}\varphi(t)\,dt$, where $\varphi(t)\in L_p(0,\infty)$ $(p>1)$, is that (i) $f(x)$ has derivatives of all orders in $0<x<\infty$, (ii) $f(x)$ vanishes at infinity, and (iii) there exists a constant M such that $\int_0^\infty|L_{k,t}[f(x)]|^p\,dt<M$ for $k=1,2,3,\ldots$. In the representation $f(x)=\int_0^\infty e^{-xt}\varphi(t)\,dt$, a necessary and sufficient condition for $\varphi(t)$ to be bounded in $0<t<\infty$ is that $f(x)$ is of class C^∞ in $0<x<\infty$ and there exists a constant M such that $|L_{k,t}[f(x)]|<M$ and $|xf(x)|<M$ for $0<x<\infty$. In order that $\varphi(t)\in L_1$ $(0\leqslant t<\infty)$, it is necessary and sufficient that (i) $f(x)$ is of class C^∞ in $0<x<\infty$, (ii) $f(x)$ vanishes at infinity, and (iii) $\int_0^\infty|L_{k,t}[f(x)]|\,dt<\infty$ and

$$\lim_{\substack{j\to\infty\\k\to\infty}}\int_0^\infty|L_{k,t}[f(x)]-L_{j,t}[f(x)]|\,dt=0$$

(Widder).

F. Operations on Laplace Transforms

Let the Laplace transform of $f(x)$ be $L(s,f(t))=\int_0^\infty e^{-st}f(t)\,dt$. It is important in operational calculus to know the formula for the Laplace transform of φf, where φ is an operation. We mention here some important formulas:

$$L(s,f(at-b))=\frac{1}{a}\exp\left(-\frac{b}{a}s\right)L\left(\frac{s}{a},f(t)\right)$$

$$(f(at-b)=0 \quad \text{if} \quad at<b,\quad a>0,\quad b\geqslant 0);$$

$$L\left(s,\int_0^t f(t)\,dt\right)=\frac{1}{s}L(s,f(t))$$

$$(\operatorname{Re}s>\max(0,\sigma_c));$$

and

$$L(s,f'(t))=sL(s,f)-f(0),$$

provided that $L(s,f'(t))$ converges at $s\,(>0)$ and $f(t)\to f(0)$ as $t\to+0$. Generally speaking, if $f(+0),\ldots,f^{(k-1)}(+0)$ exist and the Laplace transforms $L(s,f^{(k)}(t))$ converge at $s\,(>0)$, then

$$L(s,f^{(k)}(t))=s^kL(s,f(t))-f(+0)s^{k-1}$$
$$-f'(+0)s^{k-2}-\ldots-f^{(k-1)}(+0).$$

Furthermore, given functions f_1 and f_2, if $L(s,f_1)$ and $L(s,f_2)$ are both convergent and if one of them converges absolutely or $L(s,f_1*f_2)$ converges, then

$$L(s,f_1*f_2)=L(s,f_1)L(s,f_2).$$

G. Asymptotic Properties of the Laplace Transform

If $L(s)=\int_0^\infty e^{-st}\,d\alpha(t)$ $(s>0)$, then for any $c\geqslant 0$ and any constant A, we have

$$\limsup_{s\to+0}|s^cL(s)-A|$$
$$\leqslant\limsup_{t\to\infty}|\alpha(t)t^{-c}\Gamma(c+1)-A|,$$

$$\limsup_{s\to\infty}|s^cL(s)-A|$$
$$\leqslant\limsup_{t\to+0}|\alpha(t)t^{-c}\Gamma(c+1)-A|.$$

In particular, if we set $c=0$ and assume that $\alpha(t)\to A$ as $t\to\infty$, then $f(s)\to A$ as $s\to+0$; and if we assume $\alpha(t)\sim At^c/\Gamma(c+1)$ as $t\to\infty$ (or $t\to+0$), then $f(s)\sim As^{-r}$ as $s\to+0$ (or $s\to\infty$). These results are called **Abelian theorems**, because if we take $\alpha(t)$ appropriately and change variables, we get †Abel's continuity theorem on power series: If Σa_n converges to s, then Σa_nx^n tends to s as x tends to $1-0$. More generally, we have the following theorem: If $\int_0^\infty e^{-st}\,d\alpha(t)=L(s)\to A$ as $s\to+0$, then $\lim_{t\to\infty}\alpha(t)=A$ if and only if $\beta(t)=\int_0^t u\,d\alpha(u)=o(t)$ $(t\to\infty)$.

H. The Bilateral Laplace Transform

If $\alpha(t)$ is of bounded variation in every finite interval, and if for some s

$$\lim_{R\to\infty}\int_0^R e^{-st}\,d\alpha(t),\qquad \lim_{R'\to\infty}\int_{-R'}^0 e^{-st}\,d\alpha(t)$$

exist, we set $L(s)=\int_{-\infty}^\infty e^{-st}\,d\alpha(t)$.

If $L(s)$ converges at $s_1=\sigma_1+i\tau_1$, $s_2=\sigma_2+i\tau_2$, then $L(s)$ converges in the vertical strip $\sigma_1<\operatorname{Re}s<\sigma_2$. If $L(s)$ converges in the strip $\sigma_c'<\operatorname{Re}s<\sigma_c''$ and diverges for $\operatorname{Re}s>\sigma_c''$ and $\operatorname{Re}s<\sigma_c'$, then each of the numbers σ_c' and σ_c'' is called an **abscissa of convergence** of $L(s)$. If

$$\limsup_{t\to\infty}\frac{\log|\alpha(t)|}{t}=k\neq 0,$$

$$\liminf_{t\to-\infty}\frac{\log|\alpha(t)|}{t}=l\neq 0$$

with $k<1$, then k and l are the the abscissas of convergence. If $\alpha(t)$ is a normalized function of bounded variation in every finite interval and $L(s)$ converges in the strip $k<\operatorname{Re}s<l$, then for all t

$$\lim_{T\to\infty}\frac{1}{2\pi i}\int_{c-iT}^{c+iT}\frac{L(s)}{s}e^{st}\,ds = \begin{cases}\alpha(t)-\alpha(-\infty), & c>0, \quad k<c<l,\\ \alpha(t)-\alpha(\infty), & c<0, \quad k<c<l.\end{cases}$$

Suppose that $\alpha(t)=\int_0^t\varphi(u)\,du$, $L(s)$ converges absolutely on the line $\operatorname{Re}s=c$, and $\varphi(t)$ is of bounded variation in some neighborhood of $t=t_0$. Then

$$\lim_{T\to\infty}\frac{1}{2\pi i}\int_{c-iT}^{c+iT}L(s)e^{st_0}\,ds = (\varphi(t_0+0)+\varphi(t_0-0))/2.$$

There are other formulas analogous to those for the ordinary Laplace transform.

References

[1] D. V. Widder, Laplace transform, Princeton Univ. Press, 1941.
[2] G. Doetsch, Theorie und Anwendung der Laplace-Transformation, Springer, 1937.
[3] B. Van der Pol and H. Bremmer, Operational calculus based on the two-sided Laplace integral, Cambridge Univ. Press, second edition, 1964.

240 (V.7) Lattice-Point Problems

A. General Remarks

Suppose that we are given in a Euclidean plane a closed †Jordan arc C of length L that bounds a †region of area F. We denote by A the number of †lattice points on the arc C or in the region bounded by C. In many cases it can be verified that $A=F+O(L)$ (O is the †notation of Landau). Specifically, if we take a circle whose center is the origin and whose radius is $\sqrt{x}$, then $A(x)=\pi x+O(\sqrt{x})$. Next, consider the closed region defined by $uv\leqslant x$, $u\geqslant 1$, and $v\geqslant 1$ on the uv-plane. Let $D(x)$ be the number of lattice points lying in this closed region. Then $D(x)=x\log x+(2C-1)x+O(\sqrt{x})$, where C is the †Euler constant. To observe another aspect of the problem of estimating the number of lattice points in a given region, consider the series

$$\sum_{\substack{m,n=-\infty\\(m,n)\neq(0,0)}}^{+\infty}(m^2+n^2)^{-s}=f(s),$$

$$\sum_{n=1}^{\infty}(n^{-s})^2=g(s).$$

Then we have $f(s)=\sum_{n=1}^{\infty}r(n)n^{-s}$, $g(s)=\sum_{n=1}^{\infty}d(n)n^{-s}$, where $r(n)$ is the number of integral solutions (u,v) of the equation $u^2+v^2=n$ and $d(n)$ is the number of positive integral solutions (u,v) of the equation $uv=n$. Thus the problem of estimating $A(x)$ is identical to that of estimating $H(x)=\sum_{n\leqslant x}r(n)$; this is called **Gauss's circle problem**. We also have $D(x)=\sum_{n\leqslant x}d(n)$, and the problem of estimating the latter is called **Dirichlet's divisor problem**.

We set $P(x)=A(x)-\pi x$ and $\Delta(x)=D(x)-(x\log x+(2C-1)x)$. W. Sierpiński (1906) showed that $P(x)=O(x^{1/3})$, and G. Voronoi (1903) showed that $\Delta(x)=O(x^{1/3}\log x)$. There are further investigations concerning the estimations of $P(x)$ and $\Delta(x)$. J. G. van der Corput and E. C. Titchmarsh devised methods to estimate more general †trigonometric sums. For instance, let $f(x)$ be a real-valued function of †class C^k ($k\geqslant 3$). If $0<\lambda\leqslant f^{(k)}(x)\leqslant h\lambda$ or $0<\lambda\leqslant -f^{(k)}(x)\leqslant h\lambda$ in the interval $a\leqslant x\leqslant b$ (with $b-a\geqslant 1$), then

$$\sum_{a\leqslant n\leqslant b}\exp(2\pi i f(x)) = O\left(h^{2^{2-k}}(b-a)\lambda^{(2^k-2)^{-1}} + (b-a)^{1-2^{2-k}}\lambda^{-(2^k-2)^{-1}}\right).$$

As of 1968, we have an estimate slightly better than $O(x^{13/40+\varepsilon})$ for $P(x)$ and $\Delta(x)$ obtained by L. K. Hua (1942), C. J. Cheng (1963), and W. L. Yin (1959). It is conjectured that $P(x)$ and $\Delta(x)$ are $O(x^{1/4+\varepsilon})$ where ε is an arbitrary positive number. G. H. Hardy (1916) and A. E. Ingham (1941) showed that

$$\limsup_{x\to\infty}\frac{P(x)}{x^{1/4}}=\infty, \quad \liminf_{x\to\infty}\frac{P(x)}{x^{1/4}\log^{1/4}x}<0,$$

$$\limsup_{x\to\infty}\frac{\Delta(x)}{x^{1/4}\log^{1/4}x}>0, \quad \liminf_{x\to\infty}\frac{\Delta(x)}{x^{1/4}}=-\infty.$$

H. Cramér (1926) showed that

$$\frac{1}{x}\int_1^x|P(y)|\,dy=O(x^{1/4}),$$

$$\frac{1}{x}\int_1^x|\Delta(y)|\,dy=O(x^{1/4}).$$

G. Voronoi (1904) proved that if x is positive, then

$$\sum_{n\leqslant x}{}' r(n)=\pi x+\sqrt{x}\sum_{n=1}^{\infty}\frac{r(n)}{\sqrt{n}}J_1(2\pi\sqrt{nx}),$$

$$\sum_{n\leqslant x}{}' d(n)=x\log x+(2C-1)x+1/4 + \sqrt{x}\sum_{n=1}^{\infty}\frac{d(n)}{\sqrt{n}}F(2\pi\sqrt{nx}),$$

where Σ' means that when x is an integer m, the mth term $d(m)$ is replaced by $(1/2)d(m)$. Here $J_1(x)$ is the †Bessel function of the first kind, and $F(x)=(2/\pi)\int_0^{\infty}\cos(xu)\sin(x/u)\,du$.

There are many proofs for these expansion formulas. E. Landau's proof (1920) of the estimation of $\Sigma_{n\leqslant x}r(n)$ is interesting from the geometric point of view, and W. Rogosinski's proof (1922) of the estimation of $\Sigma'_{n\leqslant x}d(n)$ uses real analytic methods in an ingenious manner. A. Oppenheim (1926) generalized these problems.

B. Other Extensions

Let $a_{\mu\nu}$ be rational, $a_{\mu\nu}=a_{\nu\mu}$, and $Q(u_1,\ldots,u_n)=\Sigma^n_{\mu,\nu=1}a_{\mu\nu}u_\mu u_\nu$ be a †positive definite quadratic form with †discriminant D. As an extension of Gauss's circle problem, it is natural to consider the number of lattice points $(m_1,\ldots,m_n)$ satisfying $Q(m_1,\ldots,m_n)\leqslant x$. In connection with the †Epstein zeta function, this problem was extended to that of estimating the sum

$$F(x)=\sum_{Q(m_1,\ldots,m_n)\leqslant x}\exp(2\pi i(\alpha_1 m_1+\ldots+\alpha_n m_n)).$$

Namely, the weight $\exp(2\pi i(\alpha_1 m_1+\ldots+\alpha_n m_n))$ is placed at each lattice point. We now define δ such that $\delta=1$ if $\alpha_1,\ldots,\alpha_n$ are all integers and $\delta=0$ otherwise. Landau (1915) obtained three exquisite proofs of

$$F(x)=\delta\frac{\pi^{n/2}}{\sqrt{D}\,\Gamma(n/2+1)}x^{n/2}+O(x^{n(n-1)/2(n+1)}).$$

I. M. Vinogradov (1960) obtained a deeper result for the special case $n=3$: $\Sigma_{u^2+v^2+w^2\leqslant x}1=(4/3)\pi x^{3/2}+O(x^{19/28+\varepsilon})$.

Four times the †Dedekind zeta function of the Gaussian field $\mathbf{Q}(\sqrt{-1})$ is equal to $\Sigma^{\infty}_{n=1}r(n)n^{-s}$. Hence Gauss's circle problem can be extended to that of estimating $H(x)$ for the Dedekind zeta function. The generalized divisor problem, including Gauss's circle problem and Dirichlet's divisor problem, was the principal theme of Landau's research after 1912. We now consider the case where the Dirichlet series $\Sigma^{\infty}_{n=1}F(n)n^{-s}$ is a finite product of Dedekind zeta functions. With a slight modification the following result is valid for the product of †Hecke L-functions: Let k_j $(1\leqslant j\leqslant\tau)$ be an algebraic number field of degree n_j, $\zeta_j(s)$ be the †Dedekind zeta function of k_j, and ρ_j be the residue at the pole $s=1$ of $\zeta_j(s)$. Further, we assume that

$$n_1+n_2+\ldots+n_\tau=N,$$

$$\zeta_1(s)\zeta_2(s)\ldots\zeta_\tau(s)=\sum_{n=1}^{\infty}F(n)n^{-s},$$

$$H(x)=\sum_{n\leqslant x}F(n).$$

Then

$$H(x)=x(a_1\log^{\tau-1}x+\ldots+a_{\tau-1}\log x+a_\tau)+O(x^{(N-1)/(N+1)}\log^{\tau-1}x),$$

$$a_1=\rho_1\rho_2\ldots\rho_\tau/(\tau-1)!,$$

and the remainder O-term of the right-hand side cannot be replaced by $O(x^{\theta})$ for $\theta<1/2-1/2N$. There are some algebraic results (Z. Suetuna (1929), H. Hasse and Suetuna (1931)) concerning the estimation of $H(x)$. In particular, if in the definition of $F(n)$, the $\zeta_j(s)$ are all equal to the Riemann zeta function, then we obtain $\zeta(s)^k=\Sigma^{\infty}_{n=1}d_k(n)n^{-s}$, where $d_k(n)$ is the number of ways of expressing n as a product of k factors. In this special case the remainder term can be replaced by $O(x^{c+\varepsilon})$ where $c=\max(1/2,(k-1)/(k+2))$ $(k\geqslant 3)$ (G. H. Hardy and J. E. Littlewood, 1922). An appropriate application of the Artin L-function was obtained by Suetuna (1925), who extended to algebraic number fields of finite degree the result obtained by Landau (1912) —which states that the number of positive integers not larger than x that can be expressed as the sum of two squares is approximately equal to $ax/\sqrt{\log x}$, where a is a positive constant.

References

[1] G. H. Hardy and J. E. Littlewood, The approximate functional equation in the theory of the zeta-function, with application to the divisor-problems of Dirichlet and Piltz, Proc. London Math. Soc., (2) 21 (1923), 39–74.

[2] L. K. Hua, Die Abschätzung von Exponentialsummen und ihre Anwendung in der Zahlentheorie, Enzykl. Math., Bd. I, 2, Heft 13, Teil I, 1959.

[3] E. G. H. Landau, Zur analytischen Zahlentheorie der definiten quadratischen Formen, S.-B. Preuss. Akad. Wiss. (1915), 458–476.

[4] E. G. H. Landau, Einführung in die elementare und analytische Theorie der algebraischen Zahlen und der Ideale, Teubner, 1918, second edition, 1927.

[5] E. G. H. Landau, Vorlesungen über Zahlentheorie II, Hirzel, 1927 (Chelsea, 1969).

[6] E. C. Titchmarsh, The theory of the Riemann zeta-function, Clarendon Press, 1951.

[7] Z. Suetuna, Kaisekiteki seisûron (Japanese; Analytic theory of numbers), Iwanami, 1950.

[8] E. G. H. Landau, Ausgewählte Abhandlungen zur Gitterpunktlehre (edited by A. Walfisz), Deutscher Verlag der Wiss., 1962.

241 (II.15) Lattices

A. Definition

When x and y are elements of an †ordered set L, the †supremum and †infimum of $\{x,y\}$, whenever they exist, are called the **join** and **meet** of x, y and denoted by $x \cup y$ and $x \cap y$, respectively. L is called a **lattice** (or **lattice-ordered set**) when every pair of its elements has a join and a meet. The following three laws hold in any lattice L: (i) $x \cup y = y \cup x$, $x \cap y = y \cap x$ (**commutative law**); (ii) $x \cup (y \cup z) = (x \cup y) \cup z$, $x \cap (y \cap z) = (x \cap y) \cap z$ (**associative law**); and (iii) $x \cup (y \cap x) = (x \cup y) \cap x = x$ (**absorption law**). Conversely, if in a set L two operations $\cup, \cap$ are given that satisfy (i)–(iii), then the conditions $x \cup y = y$ and $x \cap y = x$ are equivalent and define an †ordering $x \leqslant y$ in L with respect to which L becomes a lattice. The supremum and infimum of $\{x,y\}$ in this lattice coincide with the elements $x \cup y$ and $x \cap y$, respectively. Accordingly, a lattice can also be defined as an †algebraic system with operations $\cup$, $\cap$ satisfying laws (i)–(iii). The **idempotent law** $x \cup x = x \cap x = x$ holds in any lattice.

An ordered set L is called an **upper semilattice** if each pair of elements x, y always has a join (supremum) $x \cup y$, and a **lower semilattice** if each pair of elements x, y always has a meet (infimum) $x \cap y$.

B. Examples

The set $\mathfrak{P}(S)$ of all subsets of a given set S is a complete and distributive lattice with respect to the inclusion relation (→ Sections D, E). The set of all †normal subgroups of a given †group is a complete and modular lattice (→ Section F) with respect to the inclusion relation. This remains true if the normal subgroups are replaced by the †admissible subgroups with respect to a given †operator domain. This applies in particular to the case of the set of all †ideals in a given commutative †ring. The set of all subspaces of a given †projective space is a modular lattice.

C. Further Definitions

A mapping f of a lattice L into a lattice L' that satisfies the conditions $f(x \cup y) = f(x) \cup f(y)$ and $f(x \cap y) = f(x) \cap f(y)$ is called a **lattice homomorphism** (or simply **homomorphism**). A †bijective lattice homomorphism f is called a **lattice isomorphism** (or simply **isomorphism**); its inverse mapping is also a lattice isomorphism. When such an f exists, the lattices L and L' are said to be **isomorphic**. More generally, a mapping f between ordered sets is said to be order-preserving when it satisfies the condition: $x \leqslant y$ implies $f(x) \leqslant f(y)$. Any lattice homomorphism is order-preserving, but the converse is not always true; however, an order-preserving bijection is an isomorphism.

If the ordering in a lattice L is replaced by the †dual ordering, then the join and the meet are interchanged, and a new lattice L' is obtained. This new lattice is called the **dual lattice** for L.

A mapping f of a lattice L into a lattice L' satisfying the conditions $f(x \cap y) = f(x) \cup f(y)$ and $f(x \cup y) = f(x) \cap f(y)$ is called a **dual homomorphism** (or **antihomomorphism**). Moreover, when f is a bijection, f is called a **dual isomorphism** (or **anti-isomorphism**), and we say that L and L' are **dually isomorphic** (or **anti-isomorphic**) to each other.

When a lattice L' is a subset of a lattice L and the canonical injection $L' \rightarrow L$ is a lattice homomorphism, L' is called a **sublattice** of L. If a subset L' of a lattice L satisfies the condition that $x, y \in L'$ implies $x \cup y$, $x \cap y \in L'$, then two operations $\cup$, $\cap$ can be induced in L' so that L' becomes a sublattice. For example, when a, b are given elements of a lattice L, the set of elements x satisfying $a \leqslant x \leqslant b$ is a sublattice, denoted by $[a,b]$ and called an **interval** of L. When the quotient set L/R of a lattice L by an equivalence relation R in L is also a lattice and the canonical surjection $L \rightarrow L/R$ is a homomorphism, then L/R is called a **quotient lattice** of L. If an equivalence relation R in a lattice L satisfies the condition that $x \equiv x', y \equiv y' \pmod R$ implies $x \cup y \equiv x' \cup y'$, $x \cap y \equiv x' \cap y' \pmod R$, then two operations $\cup$, $\cap$ can be induced in L/R so that L/R becomes a quotient lattice. The Cartesian product $L = \prod_{i \in I} L_i$ of a family $\{L_i\}_{i \in I}$ of lattices becomes a lattice if the operations $\cup$, $\cap$ are defined by $(x_i) \cup (y_i) = (x_i \cup y_i)$, $(x_i) \cap (y_i) = (x_i \cap y_i)$. This lattice is called the **direct product** of lattices $\{L_i\}_{i \in I}$.

D. Complete Lattices

An ordered set L is called a **complete lattice** if every nonempty subset of L has a supremum and an infimum in L, and a **σ-complete lattice** if every nonempty countable subset has a supremum and an infimum. Naturally, such ordered sets are lattices. And a sort of converse holds: Any lattice is a sublattice of some complete lattice. An ordered set L is said to be **conditionally complete** when every subset †bounded from above (below) has a supremum (infimum) in L, and **conditionally σ-complete** when every countable subset bounded from above (below) has a supremum (infimum). For any ordered set L there exist a complete lattice $\bar{L}$ and an order-preserving injection $f: L \rightarrow \bar{L}$ satisfying the condition that each element $\xi \in \bar{L}$ is the supremum and infimum of the images $f(X)$ and $f(Y)$, respectively, for some sets $X, Y \subset L$. This condition is equivalent to the condition that for any complete lattice $\bar{L}'$ and order-preserving injection $f': L \rightarrow \bar{L}'$, there exists an order-preserving injection $\varphi: \bar{L} \rightarrow \bar{L}'$ for which $\varphi \circ f = f'$. Hence $(\bar{L}, f)$ is unique up to lattice isomorphisms. $\bar{L}$ is called the **completion** of the ordered set L. For example, the set of real numbers supplemented by $+\infty$ and $-\infty$ is the completion of the set of rational numbers.

E. Distributive Lattices

A lattice L is said to be **distributive** when it satisfies the following equivalent conditions (**distributive laws**) for $x, y, z \in L$: (i) $x \cup (y \cap z) = (x \cup y) \cap (x \cup z)$; (ii) $x \cap (y \cup z) = (x \cap y) \cup (x \cap z)$; and (iii) $(x \cup y) \cap (y \cup z) \cap (z \cup x) = (x \cap y) \cup (y \cap z) \cup (z \cap x)$. The dual lattices, sublattices, quotient lattices, and direct products of distributive lattices are distributive. The set $\mathfrak{P}(S)$ of subsets of a given set S is a distributive lattice, and each of its sublattices is called a **lattice of sets** in S. A distributive lattice is isomorphic to a certain lattice of sets. A homomorphism from a distributive lattice L into $\mathfrak{P}(S)$ is called a **representation** of L in S.

A lattice L is said to be **complemented** when a greatest element I and a least element 0 exist in L and for every element x, there exists an element x' satisfying $x \cup x' = I$, $x \cap x' = 0$. Such an x' is called a **complement** of x. A lattice that is distributive and complemented is called a **Boolean lattice** (or †**Boolean algebra**). In a Boolean lattice, every element has a unique complement. The lattice $\mathfrak{P}(S)$ of all subsets of a given set S is a Boolean lattice, in which S is the greatest element and $\varnothing$ the least element. A sublattice of $\mathfrak{P}(S)$ that contains the complement of each of its elements is also a Boolean lattice and so called a **Boolean lattice of sets**. Any Boolean lattice can be represented isomorphically by some Boolean lattice of sets (→ 43 Boolean Algebra).

An element a of a lattice is called a **neutral element** if for any pair of elements x and y, the sublattice generated by a, x, y is distributive. When a is neutral and has a complement, a is called a **central element**. The set of all central elements of a lattice L is called the **center** of L.

F. Modular Lattices

A lattice L is said to be **modular** if the following condition (**modular law**) is satisfied: $x \leqslant z$ implies $x \cup (y \cap z) = (x \cup y) \cap z$. A distributive lattice is always modular. The dual lattices, sublattices, quotient lattices, and direct product of modular lattices are also modular.

The †Jordan-Hölder theorem and the refinement theorem of O. Schreier on normal subgroups (→ 195 Groups) are generalized as follows to the case of any modular lattice: A pair of elements x, y in a lattice L satisfying $x \geqslant y$ is called a **quotient** and is denoted by x/y. In particular, when $x > y$ and there exists no element z such that $x > z > y$, then x/y is called a **prime quotient**, x is said to be **prime over** y, and y is said to be **prime under** x. A sequence $C: x_0, x_1, \ldots, x_k$ of elements of L satisfying the conditions $x_{i-1} \geqslant x_i$ $(1 \leqslant i \leqslant k)$ is called a **descending chain**, and k is called its **length**. Each x_{i-1}/x_i is called a quotient determined by C. When each of these quotients is prime, C is called a **composition series**. A descending chain $D: y_0, y_1, \ldots, y_l$ is called a **refinement** of a descending chain $C: x_0, x_1, \ldots, x_k$ when $x_0 = y_0$, $x_k = y_l$ and each x_i is equal to some y_j. We now define an equivalence relation between descending chains. First, a relation $x/y \approx x'/y'$ between quotients x/y and x'/y' is defined to mean that either the condition $x = x' \cup y$, $y' = x' \cap y$ or the condition $x' = x \cup y'$, $y = x \cap y'$ holds. Then x/y and x'/y' are called equivalent if there exists a finite number of quotients $q_0, q_1, \ldots, q_r$ satisfying the conditions $x/y = q_0$, $x'/y' = q_r$ and $q_{i-1} \approx q_i (1 \leqslant i \leqslant r)$. Descending chains C and C' are said to be equivalent if they have the same length and the set of quotients determined by C is mapped bijectively to the set of quotients determined by C' so that the quotient and its image are equivalent. Now let L be a modular lattice. If two quotients x/y and x'/y' in L are equivalent, the intervals $[y, x]$ and $[y', x']$, considered as lattices, are isomorphic (**Dedekind's principle**). If two descending chains $C: x_0, x_1, \ldots, x_k$ and $C': y_0, y_1, \ldots, y_l$ have the same ends $x_0 = y_0$ and $x_k = y_l$, then there exist a refinement of C and a refinement of C' which are equivalent. In particular, any two composition series connecting the same elements are equivalent (→ 87 Continuous Geometry).

In a modular lattice L with a least element 0, if there exists a composition series connecting 0 and a given element a, then all such composition series have a common length k, denoted by $d(a)$ and called the **height** of the element a. If no such composition series exists, the height is defined to be $d(a) = \infty$. If two elements a, b of L satisfy $d(a \cup b) < \infty$, then $d(a \cup b) + d(a \cap b) = d(a) + d(b)$. This fact is called the **dimension theorem** of modular lattices. If L has a greatest element I, $d(I)$ is called the **height** of the lattice L.

In a complemented modular lattice L, an element which is prime over the least element 0 is called an **atomic element**. A complemented modular lattice L is said to be **irreducible** if any two atomic elements have a common complement.

G. Lattice-Ordered Groups

An ordered set G in which a group operation is defined is called an **ordered group** when $x \leqslant y$ implies $xz \leqslant yz$ and $zx \leqslant zy$ for all x, y, z in G. Moreover, if it is a †totally ordered set, the ordered group G is called a **totally ordered group**. If G is a lattice, the condition for an ordered group is equivalent to the condition that $(x \cup y)z = xz \cup yz$, $(x \cap y)z = xz \cap yz$, $z(x \cup y) = zx \cup zy$, and $z(x \cap y) = zx \cap zy$. In this case, G is called a **lattice-ordered group**. A lattice-ordered group is a distributive lattice and has neither a greatest nor a least element. If $\{x_i\}$ has a supremum in a lattice-ordered group, then we have $(\sup_i x_i) \cap y = \sup_i (x_i \cap y)$ and its dual (**complete distributive law**).

The lattice-theoretic structure of a lattice-ordered group was clarified by P. Lorentzen, A. H. Clifford, T. Nakayama. In particular, a lattice-ordered commutative group is isomorphic (as a lattice-ordered group) to some subgroup of a direct product of totally ordered groups. A lattice-ordered group has no element of finite order other than the identity element. Conversely, a commutative group which has no element of finite order other than the identity element can be made a lattice-ordered group with respect to some total ordering. Any free group can also be made a lattice-ordered group with respect to some total ordering. K. Iwasawa (1948) and others have done further research on totally ordered groups.

An element x ($\neq e$) of a lattice-ordered group G is called **positive** (**negative**) when $x \geqslant e$ ($x \leqslant e$), where e is the identity. G is called an **Archimedean lattice-ordered group** when the following condition is satisfied: if, for some y, $x^n \leqslant y$ for all natural numbers n, then $x \leqslant e$. An Archimedean lattice-ordered group is isomorphic to some subgroup of a complete lattice-ordered group. Conversely, a complete lattice-ordered group is Archimedean; moreover, it is commutative and isomorphic to the direct product of some †lattice-ordered linear spaces and copies of the lattice-ordered group of rational integers (Iwasawa, 1943). In particular, any totally ordered Archimedean lattice-ordered group is isomorphic to a subgroup of the lattice-ordered group of all real numbers.

If the †minimal condition holds for the set of all positive elements in a lattice-ordered group, then the group is commutative, and each of its elements can be decomposed uniquely into a product of powers of elements that are prime over the identity element. The set of all †fractional ideals of an algebraic number field is a typical example of such a lattice-ordered group. For further reference → 304 Ordered Linear Spaces, 87 Continuous Geometry.

References

[1] T. Nakayama, Sokuron (Japanese; Lattice theory) I, Iwanami, 1944.
[2] T. Iwamura, Sokuron (Japanese; Lattice theory), Kyôritu, 1966.
[3] S. Iyanaga and K. Kodaira, Gendai sûgaku gaisetu (Japanese; Introduction to modern mathematics) I, Iwanami, 1961.
[4] M. Suzuki, Gun to soku, §4 Sokugun (Japanese; Groupes and lattices, §4 Lattice-ordered groups), Gendai no sûgaku, Kyôritu, 1950.
[5] G. Birkhoff, Lattice theory, Amer. Math. Soc. Colloq. Publ., 1940, revised edition, 1948.
[6] P. Dubreil and M. L. Dubreil-Jacotin, Leçons d'algèbre moderne, Dunod, second edition, 1961; English translation, Lectures on modern algebra, Oliver & Boyd, 1967.

242 (XX.32) Lebesgue, Henri Léon

Henri Lebesgue (June 28, 1875–July 26, 1941) was one of the most influential French analysts of this century and is known as the inventor of the †Lebesgue integral. With deep insight based on intuitive geometric conceptions, he was able to initiate a new era in analysis. He began his research on Lebesgue integrals in his doctoral thesis [1] with reflections on length and area. Not only was this theory the start of modern integration; it was also a turning point in the theory of †Fourier series and †potential theory. His introduction of the †Lebesgue number for compact sets and the definition of compactness are also important contributions. He referred to mathematical entities that can be individually "named" as "effective entities," and, although he supported †French empiricism, he took a more idealistic standpoint than E. Borel or R. Baire [2, 4]. He taught for many years at the University of Paris and at the Collège de France.

References

[1] H. Lebesgue, Intégrale, longueur, aire, Ann. Mat. Pura Appl., (3) 7 (1902), 231–359.
[2] H. Lebesgue, Sur les fonctions représentables analytiquement, J. Math. Pures Appl., (6) 1 (1905), 139–216.
[3] H. Lebesgue, Leçons sur l'intégration et la recherche des fonctions primitives, Gauthier-Villars, 1904, second edition, 1928.
[4] Cinq lettres sur la théorie des ensembles, in E. Borel, Leçons sur la théorie des fonctions, Gauthier-Villars, fourth edition, 1950, note IV, p. 153–156.

243 (X.13) Lebesgue Integral

A. General Remarks

Let X be an arbitrary set. If a †completely additive class $\mathfrak{B}$ of subsets of X and a †measure μ defined on $\mathfrak{B}$ are given, then we say that a †measure space $(X,\mathfrak{B},\mu)$ is defined. For example, the Euclidean space $\mathbf{R}^n$, the class $\mathfrak{M}_n$ of all †Lebesgue measurable sets in $\mathbf{R}_n$, and the †Lebesgue measure m_n on $\mathfrak{M}_n$ form the measure space $(\mathbf{R}^n, \mathfrak{M}_n, m_n)$ (→ 270 Measure Theory). We consider only $\mathfrak{B}$-measurable sets and $\mathfrak{B}$-measurable functions, so a $\mathfrak{B}$-measurable set is called simply a set and a $\mathfrak{B}$-measurable function simply a function.

The integral $\int_E f(x)\,d\mu(x)$ on a set $E\subset X$ of a real-valued function $f(x)$ (we write simply $\int_E f d\mu$ or $\int_E f$) can be defined in steps as follows. (1) Let $f(x)\geqslant 0$ be a **simple function**, that is, a function whose range is a finite set $\{a_i\}$ $(i=1,2,\ldots,n)$, not containing $\pm\infty$. If $f(x)=a_i$ for $x\in E_i$, where $E=\bigcup_{j=1}^{n}E_j$, $E_j\cap E_k=\varnothing$ $(j\neq k)$, then we put $\int_E f=\sum_{a_j\neq 0}a_j\mu(E_j)$. (The value of the integral is a real number or $+\infty$. For operations concerning $\pm\infty$, → 270 Measure Theory.) (2) For an arbitrary $f(x)\geqslant 0$, we define $\int_E f$ as the †supremum of $\int_E g$, where the supremum is taken for all simple functions g such that $0\leqslant g\leqslant f$. (3) In general, letting $f(x)=f^+(x)-f^-(x)$, where $f^+(x)=\max\{f(x),0\}$, $f^-(x)=\max\{-f(x),0\}$, we define $\int_E f=\int_E f^+-\int_E f^-$ if at least one of $\int_E f^+$ and $\int_E f^-$ is finite, and say that f **has an integral on** E. In particular, if $\int_E f$ is finite, then we say f is **integrable** on E.

If the given measure space is $(\mathbf{R}^n,\mathfrak{M}_n,m_n)$, the integral defined in this section is called the **Lebesgue integral** (or simply **L-integral**), and the function that is integrable in this case is said to be **Lebesgue integrable**. If $n=1$ the integral is often written as $\int_E f(x)\,dx$, and if E is the interval $[a,b]$, as $\int_a^b f(x)\,dx$. In the general case of an arbitrary measure space, the integral is called the Lebesgue integral in the general sense, but some authors call it also simply the Lebesgue integral.

Let $\mu(E)<\infty$ and $f(x)$ be bounded. Let $\Delta: E=\bigcup_{j=1}^{n}E_j$, $E_j\cap E_k=\varnothing$ $(j\neq k)$ be a partition of E, and put $S_\Delta=\sum_{j=1}^{n}\sup\{f(E_j)\}\cdot\mu(E_j)$ and $s_\Delta=\sum_{j=1}^{n}\inf\{f(E_j)\}\cdot\mu(E_j)$. Then both $\inf S_\Delta$ and $\sup s_\Delta$ coincide with $\int_E f$, where inf and sup are taken over all partitions of E. Put

$$\delta(\Delta)=\max\big(\sup\{f(E_i)\}-\inf\{f(E_i)\}\big).$$

Then for every sequence of partitions $\{\Delta_n\}$ such that $\delta(\Delta_n)\to 0$, both S_{Δ_n} and s_{Δ_n} tend to $\int_E f$. This is the original definition given by Lebesgue. In a †σ-finite measure space, we can define the integral of a function $f(x)\geqslant 0$ starting from this definition as

$$\int_E f=\lim_{n\to\infty}\int_{E_n} f,$$

where $E_n=\{x\,|\,f(x)\leqslant n,\ x\in X_n\}$ and $\{X_n\}$ is an arbitrary sequence of sets such that $X_n\uparrow X$ and $\mu(X_n)<\infty$.

B. Properties of Integrals

(1) The set of all functions integrable on E forms a †linear space over $\mathbf{R}$, that is, if f and g are integrable on E, then for any real α, β $\alpha f+\beta g$ is also, and

$$\int_E(\alpha f+\beta g)\,d\mu=\alpha\int_E f\,d\mu+\beta\int_E g\,d\mu.$$

(2) The integrability of f, of both f^+ and f^-, and of $|f|$ are mutually equivalent. (3) If $g\leqslant f$ on E, then $\int_E g\leqslant\int_E f$. In particular, if $m\leqslant f(x)\leqslant l$ on E, then $m\mu(E)\leqslant\int_E f\,d\mu\leqslant l\mu(E)$ (**mean value theorem**). The †second mean value theorem may be expressed in the same form as for the Riemann integral (→ 218 Integral Calculus). (4) If $f(x)$ is integrable on E (has an integral on E), then it is integrable (has an integral) on every subset of E. (5) If $E=\bigcup_{n=1}^{\infty}E_n$, $E_j\cap E_k=\varnothing$ $(j\neq k)$, and $\int_E f\,d\mu$ exists, then the series $\sum_{n=1}^{\infty}\int_{E_n} f\,d\mu$ converges and is equal to $\int_E f\,d\mu$ (**complete additivity of the integral**). (6) If $f(x)\geqslant 0$ and $\int_E f\,d\mu=0$, then $f(x)=0$ †almost everywhere on E. (7) Modification of the values of $f(x)$ on a †null

set influences neither the integrability nor the value of the integral. Consequently, if the function f is not defined on a null set, we may define the value of f on this set arbitrarily so that the integral has meaning. (8) If $f(x)$ is integrable on E, $E \supset E_n$, and $\mu(E_n) \to 0$, then $\lim_{n\to\infty} \int_{E_n} f d\mu = 0$.

C. Relation to the Riemann Integral

Integrability of $f(x)$ on $[a,b] \subset \mathbf{R}$ in the sense of Riemann implies integrability in the sense of Lebesgue, and the values of the two integrals coincide. The converse is not true. For example, the function equal to 0 for irrational x and 1 for rational x, which is called the **Dirichlet function**, is Lebesgue integrable but not Riemann integrable. In fact, in order that a bounded function be Riemann integrable on an interval E it is necessary and sufficient that the function be continuous †almost everywhere. On the other hand, the existence of the †improper Riemann integral does not imply that of the Lebesgue integral. For example, $\int_0^\infty (\sin x/x)dx = \pi/2$ is not a Lebesgue integral because, while $f(x)$ and $|f(x)|$ should be Lebesgue integrable at the same time, we have $\int_0^\infty (|\sin x|/x)dx = \infty$. The theory of the Denjoy integral gives insight into this situation (→ 103 Denjoy Integrals).

For the relation between $\lim_{n\to\infty} \int_E f_n$ and $\int_E (\lim_{n\to\infty} f_n)$, it may be said that Lebesgue's theory settled a problem to which the theory of the Riemann integral could not give a satisfactory answer. We have the following propositions: (1) If $f_n(x) \leqslant f_{n+1}(x)$ on E, and there exists a $\varphi(x)$ such that $f_n(x) \geqslant \varphi(x)$ and $\int_E \varphi > -\infty$ (for example, if $f_n(x) \geqslant 0$), then

$$\lim_{n\to\infty} \int_E f_n = \int_E \Big(\lim_{n\to\infty} f_n \Big).$$

(2) If $\lim_{n\to\infty} f_n(x)$ exists almost everywhere on E, and there exists a $\varphi(x)$ such that $|f_n(x)| \leqslant \varphi(x)$ and $\int_E \varphi < +\infty$ (for example, if $\mu(E) < \infty$ and $|f_n(x)| < M$), then

$$\lim_{n\to\infty} \int_E f_n = \int_E \Big(\lim_{n\to\infty} f_n \Big)$$

(**Lebesgue's convergence theorem**). (3) If there exists a $\varphi(x)$ such that $f_n(x) \geqslant \varphi(x)$ on E and $\int_E \varphi > -\infty$, then

$$\liminf_{n\to\infty} \int_E f_n \geqslant \int_E \Big(\liminf_{n\to\infty} f_n \Big)$$

(**Fatou's theorem**).

D. Indefinite Integrals

If we put $F(e) = \int_e f$ for each measurable subset e of E, with $f(x)$ integrable on E, $F(e)$ is a †μ-absolutely continuous completely additive †set function (properties of integrals (4), (5), (8)). We call $F(e)$ the **indefinite integral** of $f(x)$. In the special case where X is the set $\mathbf{R}$ of real numbers and $f(x)$ is integrable on the interval $[a,b]$, the function $F(x) = \int_a^x f(t)dt$ defined for $x \in [a,b]$ is also called the indefinite integral of $f(x)$. The $F(x)$ so defined is an †absolutely continuous function. Conversely, if a function $F(x)$ is absolutely continuous on $[a,b]$, then it is differentiable almost everywhere on $[a,b]$, and we have $F(x) - F(a) = \int_a^x F'(t)dt$. (For the relationship between differentiation and integration in the case of $\mathbf{R}^n$, or more generally in the case of an arbitrary measure space, → 375 Set Functions.)

E. Fubini's Theorem

Let $(X, \mathfrak{B}_1, \mu_1)$ and $(Y, \mathfrak{B}_2, \mu_2)$ be two σ-finite measure spaces and $(X \times Y, \mathfrak{B}, \mu)$ be their †direct product measure space. Assume that $f(x,y)$ is $\mathfrak{B}$-measurable and integrable (has an integral) on $X \times Y$. Then for almost all fixed $y \in Y$, $f(x,y)$ considered as a function of x is $\mathfrak{B}_1$-measurable and integrable (has an integral), and $\int_X f(x,y) d\mu_1(x)$ is a $\mathfrak{B}_2$-measurable function of y. Moreover, in this case we have

$$\int_{X\times Y} f(x,y)\, d\mu(x,y)$$

$$= \int_Y \Big(\int_X f(x,y)\, d\mu_1(x) \Big) d\mu_2(y)$$

(**Fubini's theorem**). The integral on the left-hand side of this equation is called a **multiple integral**, while that on the right-hand side is called an **iterated** (or **repeated**) **integral**.

Even if an iterated integral exists, the corresponding multiple integral need not always exist. For example, let $f(x,y)$ be defined as $(x^2-y^2)/(x^2+y^2)^2$ on $(0,1) \times (0,1)$, and otherwise 0. Then $\int_{\mathbf{R}^2} f^+ = \int_{\mathbf{R}^2} f^- = \infty$, so that $\int_{\mathbf{R}^2} f$ does not exist, but

$$\int_{-\infty}^{\infty} \Big(\int_{-\infty}^{\infty} f(x,y)\, dx \Big) dy = -\frac{\pi}{2},$$

$$\int_{-\infty}^{\infty} \Big(\int_{-\infty}^{\infty} f(x,y)\, dy \Big) dx = \frac{\pi}{4}.$$

By Fubini's theorem, if f is a nonnegative function defined on a $\mathfrak{B}$-measurable subset of a σ-finite measure space $(X, \mathfrak{B}, \mu)$, the $\mathfrak{B}$-measurability of f is equivalent to the measurability of the **ordinate set** $E_f = \{(x,y) | 0 \leqslant y \leqslant f(x), x \in E\}$ considered as a subset of the measure space $(X \times \mathbf{R}, \mathfrak{B}', \mu')$, which is the direct product of $(X, \mathfrak{B}, \mu)$ and $(\mathbf{R}, \mathfrak{M}_1, m_1)$. In this case we have $\int_E f d\mu = \mu'(E_f)$, which may serve as a definition of the Lebesgue integral. When $(X, \mathfrak{B}, \mu)$ coincides with

$(\mathbf{R}, \mathfrak{M}_1, m_1)$, the †Jordan measure (area) of the ordinate set coincides with the Riemann integral.

References

[1] P. R. Halmos, Measure theory, Van Nostrand, 1950.
[2] N. Bourbaki, Eléments de mathématique, Intégration, Actualités Sci. Ind., 1175a, 1244b, 1281, 1306, 1343, Hermann, second edition, 1965, second edition, 1967, 1959, 1963, 1969.
Also → references to 270 Measure Theory.

244 (XX.33) Leibniz, Gottfried Wilhelm

Baron Gottfried Wilhelm von Leibniz (July 1, 1646–November 4, 1716) was born the son of a professor and grew up to be a genius with encyclopedic knowledge. He took part in politics and touched all scholarly fields, contributing creatively to modern technology as well. His posthumously published works cover theology, philosophy, mathematics, the natural sciences, history, and technology, and are classified into 41 fields. A complete edition of his works has yet to be published. *Ars combinatoria*, written upon his graduation from Altdorf in 1666, was a scheme to systematize the various fields using mathematics as a model. During his stay in Paris (1672–1676), when not involved in politics, he studied the works of †Descartes and †Pascal, as suggested by C. Huygens. He discovered the †fundamental theorem of differential and integral calculus and set up a basis for calculus with the introduction of an ingenious system of notation. After 1676, he worked on historical compilations under the Duke of Hanover.

He worked not only on the synthesis of modern mechanistic philospphy and medieval theological philosophy, but also on the reconciliation of Protestantism and Catholicism. With his monadism he attempted to unify the old and new philosophy. In addition he worked on plans for a world academy for the development of learning and on the unification of all knowledge. This was to be accomplished using, for example, universal symbolism and universal linguistics. Under his influence, the Berlin Academy was established in 1700. After his death, his conceptions of †symbolic logic and †computers were realized.

References

[1] C. I. Gerhardt (ed.), Leibnizens mathematische Schriften I–VII, H. W. Schmidt, 1849–1863.
[2] C. I. Gerhardt (ed.), G. W. Leibniz, Philosophische Schriften I–VII, Berlin, 1875–1890.
[3] T. Shimomura, Leibniz (Japanese), Kôbundô, 1938.

245 (X.15) Length and Area

A. Length of a Curve

A continuous mapping C sending each point u of an interval $I=\{u \mid a \leqslant u \leqslant b\}$ to the point $p=p(u)=(x_1(u), \ldots, x_k(u))$ of the k-dimensional Euclidean space $\mathbf{R}^k$ $(k \geqslant 2)$, or the image $C(I)$, is called a †continuous arc. We sometimes denote the image $C(I)$ by C, or $C: p=p(u)$. The supremum of the length of a polygonal curve inscribed in C is called the **length** of C (in the sense of Jordan) and is denoted by $l(C)$. Namely, it is equal to $\sup_\delta \sum_{i=1}^m |p(u_i)-p(u_{i-1})|$, where δ is a partition of $I: a=u_0<u_1 \ldots <u_m=b$. Let $C: p=p(u)$, and $C_n: p=p_n(u)$, $n=1,2, \ldots (u \in I)$ be continuous arcs. If $p_n(u) \rightarrow p(u)$ on I, then $l(C) \leqslant \liminf_n l(C_n)$. This property is called **lower semicontinuity** of length. Given two continuous arcs $C: p=p(u)$ $(u \in I)$ and C_1: $p=q(v)$ $(v \in I_1)$, if for any $\varepsilon>0$, there exists a homeomorphism $u=h_\varepsilon(v)$ of I_1 onto I such that $|p(h_\varepsilon(v))-q(v)|<\varepsilon$ on I_1, then C and C_1 are called **equivalent** in the sense of Fréchet. Equivalent continuous arcs have the same length. A continuous image of an open interval or a circle is called a †curve. The length of a curve is defined to be the supremum of the length of a continuous arc contained in the curve, and the notion of the equivalence of curves is defined the same way as that for continuous arcs. The set of all curves are equivalent to a given curve is called a **Fréchet curve**, and its length is defined uniquely.

Suppose that a continuous arc C is expressed by $(x_1(u), \ldots, x_k(u))$, $u \in I$. Then the length $l(C)$ is finite if and only if every $x_i(u)$ is of †bounded variation. When $l(C)$ is finite, C is called **rectifiable**. In this case each $\partial x_i / \partial u$ exists †almost everywhere on I, and the inequality

$$l(C) \geqslant \int_a^b \left(\sum_{i=1}^k \left(\frac{\partial x_i}{\partial u} \right)^2 \right)^{1/2} du \qquad (1)$$

holds. The equality holds if and only if each $x_i(u)$ is absolutely continuous. Among the continuous arcs equivalent to C, there exists a unique continuous arc C_1 such that $C_1: q=$

$q(s)$ $(0 \leqslant s \leqslant l(C))$ and the length of every subarc $q = q(s)$ $(0 \leqslant s \leqslant s'(\leqslant l(C)))$ is equal to s'. C_1 is called the **representation in terms of arc length** of C. For C_1, the equality holds in (1). A similar argument is valid for any curve. When every subarc of C is rectifiable, C is called **locally rectifiable**. If Λ_1 is the 1-dimensional †Hausdorff measure in $\mathbf{R}^k$ and $n(p)$ is the number of points on I corresponding to $p \in \mathbf{R}^k$, then $l(C) = \int n(p)\,d\Lambda_1(p)$ (→ e.g., M. Ohtsuka, *Nagoya Math. J.*, 3 (1951), 125–126).

B. Surface Area

In this section, we deal with area of †surfaces in $\mathbf{R}^3$, using [1] as the main reference. In contrast to the situation for curves, the area of a polyhedral surface P inscribed in a given surface does not necessarily tend to a fixed value as P approximates the surface. In a letter of 1880, H. A. Schwarz gave the following example: Approximate a circular cylinder with height h and radius r by a sequence $\{P_n\}$ of inscribed polyhedral surfaces, each P_n consisting of similar triangles of height b_n and base length a_n. If b_n/a_n^2 is suitably chosen, then the surface area of P_n tends to an arbitrary value not smaller than $2\pi rh$ (Fig. 1).

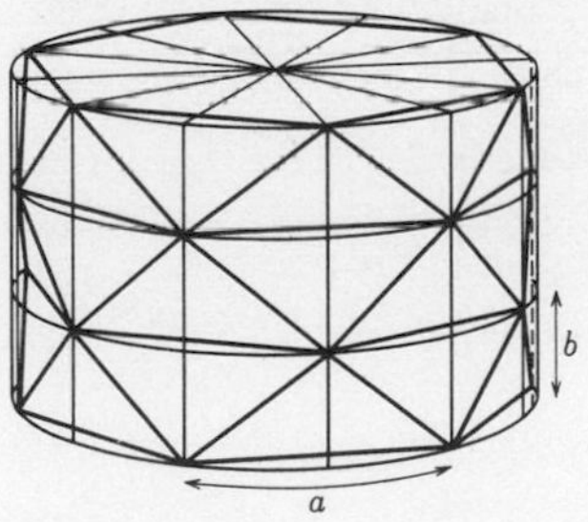

Fig. 1

C. Lebesgue Area

Suppose that we are given a plane domain A and a continuous mapping T of A into $\mathbf{R}^3$. The pair (T,A) is called a surface. Let $d(T, T', B) = \sup_{w \in B} |T(w) - T'(w)|$ for surfaces (T,A), (T',A') and a set $B \subset A \cap A'$. Let (T,A), (T_1,A_1), (T_2,A_2), ... be given. If $A_n \uparrow A$ and $d(T,T_n,A_n) \rightarrow 0$, then (T_n,A_n) (or simply T_n) is said to converge to (T,A) (or T), and the convergence is expressed by $T_n \rightarrow T$. In particular, if A consists of a finite number of triangles and T is linear on each triangle, i.e., the image of A under T consists of triangles, then the notation (P,F) is used for (T,A), and the area of $T(A)$ is denoted by $a(P,F)$. Given a surface (T,A), denote the totality of sequences $\{(P_n,F_n)\}$ converging to (T,A) by Φ, and call $\inf_\Phi \liminf_n a(P_n,F_n)$ the **Lebesgue area** of (T,A). This area is denoted by $L(T,A)$. By definition there exists a sequence $\{(P_n,F_n)\}$ converging to (T,A) such that $a(P_n,F_n) \rightarrow L(T,A)$. Like length, Lebesgue area has the lower semicontinuity property. Namely, $T_n \rightarrow T$ implies $L(T,A) \leqslant \liminf_n L(T_n,A_n)$. When A is a Jordan domain, the same value $L(T,A)$ is obtained if Φ is replaced by the set Φ^* of all sequences $\{(P_n,F_n)\}$ such that $F_n \uparrow A$ and $P_n(w) \rightarrow T(w)$.

Let $f(x,y)$ be a continuous function defined on $0 \leqslant x \leqslant 1$, $0 \leqslant y \leqslant 1$. Regard $f(x,y)$ as a function of y (resp. x) for a fixed x (y), and denote it by $f_x(y)$ ($f_y(x)$) and its †total variation by $V(x)$ ($V_1(y)$). When $\int_0^1 V(x)\,dx + \int_0^1 V_1(y)\,dy < \infty$, $f(x,y)$ is said to be of **bounded variation in the sense of Tonelli**. Furthermore, if $f_x(y)$ and $f_y(x)$ are †absolutely continuous for almost every x and y, respectively, then $f(x,y)$ is said to be **absolutely continuous in the sense of Tonelli**. Similar definitions are also given in the case where f is defined in a general domain. Suppose that a surface (T,A) is expressed by a set of three functions $x = x(u,v)$, $y = y(u,v)$, $z = z(u,v)$ all of which are absolutely continuous in the sense of Tonelli, and that the partial derivatives $x_u, x_v, \ldots, z_v$ are square integrable. Then $L(T,A) = \int\int_A J\,du\,dv < \infty$ (as was shown by C. B. Morrey), where $J = (J_1^2 + J_2^2 + J_3^2)^{1/2}$ with †functional determinants J_1, J_2, J_3 of the transformations $(u,v) \rightarrow (y,z)$, (z,x), (x,y).

D. The Geöcze Problem

The **Geöcze problem** is the problem of determining whether $L(T,A)$ coincides with the area obtained by using (instead of Φ) the set of all sequences $\{(P_n,F_n)\}$ such that (P_n,F_n) converges to (T,A) and each (P_n,F_n) is inscribed in (T,A). The answer is affirmative when A is a Jordan domain with $L(T,A) < \infty$. Let T be a surface expressed by a function $z = F(x,y)$ $(0 \leqslant x \leqslant 1, 0 \leqslant y \leqslant 1)$ that is absolutely continuous in the sense of Tonelli, and let $\{(P_n,F_n)\}$ be as before. If the ratio of the length of the largest side and the smallest height of each triangle in (P_n,F_n) is uniformly bounded, then $a(P_n,F_n)$ tends to $L(T,A)$ [1, p. 74].

E. Geöcze Area

Consider a surface (T,A). Let E_1, E_2, E_3 be coordinate planes in $\mathbf{R}^3$, and denote by T_i $(i = 1,2,3)$ the composition of the mapping T and the projection of $\mathbf{R}^3$ onto E_i. Let $\partial\pi$ be the positively oriented boundary of a polygonal domain π in A and C_i be the oriented image of $\partial\pi$ by T_i. Then the †order $O(z;C_i)$

of z with respect to C_i is a measurable function of z. Set $v_i(T;\pi)=v_i=\int\int_{E_i}|O(z;C_i)|\,dx\,dy$ $(z=x+iy)$ and $v(T;\pi)=v=(v_1^2+v_2^2+v_3^2)^{1/2}$. The quantity

$$V(T;A)=\sup_S \sum_{\pi\in S} v(T;\pi) \qquad (2)$$

is called the **Geöcze area** of (T,A), where S is a finite collection of polygonal domains in A such that no two of them overlap. If $V(T,A)<\infty$, u_i is defined by $\int\int_{E_i} O(z;C_i)\,dx\,dy$, and $U(T,A)$ is defined as in (2) by means of u_i, then $U(T,A)=V(T,A)$. The inequalities $V(T_i,A)\leqslant V(T,A)\leqslant V(T_1,A)+V(T_2,A)+V(T_3,A)$ hold trivially.

F. Peano Area

Consider (T,A) and π as in the previous section. Let τ be the projection of $\mathbf{R}^3$ onto a plane E, and denote by C' the image of the boundary of π under the composite mapping $\tau\circ T$. Set $v(T,\pi,E)=\int\int_E|O(z;C')|\,d\sigma$, where $d\sigma$ is the surface element on E, and set $\psi(T,\pi)=\sup_E v(T,\pi,E)$. Define

$$P(T,A)=\sup_S \sum_{\pi\in S}\psi(T,\pi)$$

as in (2). This is called the **Peano area** of (T,A). H. Okamura defined area by integrating the †mapping degree instead of $|O(z;C')|$. L, V, P all coincide, and hence L and V are invariant under any orthogonal transformation of $\mathbf{R}^3$.

G. Other Definitions of Area

As in the definition of Peano area, consider π, τ, E, and denote the Lebesgue measure of $\tau\circ T(\pi)$ by $m(\pi,E)$. If we set $\mu(\pi)=\sup_E m(\pi,E)$, then we may define the area of (T,A) by $\sup_E\Sigma_{\pi\in S}\mu(\pi)$. If $v(\pi)=(m^2(\pi,E_1)+m^2(\pi,E_2)+m^2(\pi,E_3))^{1/2}$ is used instead of $\mu(\pi)$, then the **Banach area** of (T,A) is obtained, and if $\int\int|O(z;C_i)|\,dx\,dy$ is used instead of $m(\pi,E_i)$, then the Geöcze area is obtained. Let us define various kinds of area for an arbitrary †Borel set X in $\mathbf{R}^3$. Divide $\mathbf{R}^3$ into meshes $M_1, M_2, \ldots$ which are half-open cubes with diameter of equal length d, and denote by $m_i^{(j)}$ $(i=1,2,3)$ the †Lebesgue measure of the projection of $M_j\cap X$ onto the ith coordinate plane. The limit of $\Sigma_j((m_1^{(j)})^2+(m_2^{(j)})^2+(m_3^{(j)})^2)^{1/2}$ as $d\to 0$ is called the **Janzen area** of X. Denote by m_j the supremum with respect to the set of planes E in $\mathbf{R}^3$ of the Lebesgue measure of the projection of $M_j\cap X$ to E. Then $\lim\Sigma_j m_j$ as $d\to 0$ is called the **Gross area** of X. C. Carathéodory covered X by a countable number of convex sets $K_1, K_2, \ldots$ each of whose diameters is less than $\delta>0$, denoted by m_j' the supremum of the Lebesgue area of the projection of K_j into E, and adopted $\lim\Sigma_j m_j'$ as $\delta\to 0$ as his definition of area of X. If the K_j are limited to be spheres, then Carathéodory's area divided by $\pi/4$ is identical with the †Hausdorff measure $\Lambda_2(X)$. (For other definitions and mutual relations → [4].)

We give a measure-theoretic definition of an area for a surface (T,A) as follows. Denote by $n(p)$ the number of points in A corresponding to a point p in $\mathbf{R}^3$, and call $n(p)$ the **multiplicity function** of the mapping T. The integral $\Lambda(T,A)=(\pi/4)\int n(p)\,d\Lambda_2(p)$ may be taken as a definition of the area. However, a different definition of the multiplicity function is needed for the integral to be equal to $L(T,A)$ [3, 6]. In fact, if $x=\varphi(u)$, $y=\psi(u)$ $(0\leqslant u\leqslant 1)$ represent a †Peano curve filling the square $0\leqslant x\leqslant 1$, $0\leqslant y\leqslant 1$, then the Lebesgue area of the surface (T,A) defined by $A=\{0<u<1, 0<v<1\}$ and $T: x=\varphi(u)$, $y=\psi(u)$, $z=0$ is zero, but $\Lambda(T,A)\geqslant 1$.

H. Mappings of Bounded Variation

Let T be a mapping of a domain A in the w-plane into the z-plane, and define π, $O=O(z;C)$, and S as in Section E. Set

$$O^+(z;C)=(|O|+O)/2,$$
$$O^-(z;C)=(|O|-O)/2,$$
$$v(T,\pi)=\int\int|O(z;C)|\,dx\,dy,$$
$$v^{\pm}(T,\pi)=\int\int O^{\pm}(z;C)\,dx\,dy$$

(the same signs correspond to each other),

$$V(T,A)=\sup_S\Sigma_{\pi\in S}v(T,\pi),$$
$$V^{\pm}(T,A)=\sup_S\Sigma_{\pi\in S}v^{\pm}(T,\pi),$$
$$N(z;T,A)=\sup_S\Sigma_{\pi\in S}|O(z;C)|,$$

and

$$N^{\pm}(z;T,A)=\sup_S\Sigma_{\pi\in S}O^{\pm}(z;C).$$

Then $N, N^{\pm}$ are lower semicontinuous in the z-plane. The integrals $W(T,A)=\int\int N\,dx\,dy$, $W^+(T,A)=\int\int N^+\,dx\,dy$, and $W^-(T,A)=\int\int N^-\,dx\,dy$ are called the **total variation**, **positive variation**, and **negative variation** of T, respectively, and the equalities $W=W^++W^-$, $V=W$, and $V^{\pm}=W^{\pm}$ hold. When $W(T,A)<\infty$, T is said to be of **bounded variation**. A related notion is defined as follows: T is **absolutely continuous** if the following two conditions hold. (1) For any given $\varepsilon>0$ there exists a $\delta>0$ such that $\Sigma_{\pi\in S}v(T,\pi)\leqslant\varepsilon$ whenever the sum of the areas of $\pi \in S$ is $\leqslant \delta$. (2) For any polygonal domain π_0 such that $\pi_0 \cup \partial\pi_0 \subset A$ and any polygonal subdivision S of π_0, $V(T,\pi_0) = \Sigma_{\pi \in S} V(T,\pi)$. If the area of A is finite and T is absolutely continuous, then T is of bounded variation.

Let T be a continuous mapping of bounded variation of a domain A in the w-plane into the z-plane. The derivatives $V'(w)$, $V'_+(w)$, $V'_-(w)$ of the set functions $V(T,A)$, $V^+(T,A)$, $V^-(T,A)$ exist †almost everywhere (a.e.) in A and are finite. The difference $J(W) = V'_+(w) - V'_-(w)$ is called the **generalized Jacobian**, and the relation $J(w) = V'(w)$ holds a.e. If $x(u,v), y(u,v)$ are differentiable a.e., then $J(w)$ coincides with the ordinary †functional determinant a.e. Next, let T be a continuous mapping of A into $\mathbf{R}^3$ with $V(T,A) < \infty$, and denote by $J_i(w)$ the generalized Jacobian of (T_i,A). Then $J(w) = (J_1^2(w) + J_2^2(w) + J_3^2(w))^{1/2}$ is called the generalized Jacobian of (T,A). The relation $J(w) = \pm V'(w)$ holds a.e. in A. Therefore,

$$V(T,A) \geqslant \iint_A J(w)\,du\,dv \tag{3}$$

is valid. The equality holds if and only if each (T_i,A) is absolutely continuous.

I. Fréchet Distance

Let (T_1,A_1) and (T_2,A_2) be surfaces, and assume that the set H of homeomorphisms between A_1 and A_2 is nonempty. Define the **Fréchet distance** between two surfaces by

$$\|T_1, T_2\| = \inf_{h \in H} \sup_{w \in A_1} |T_1(w) - T_2(h(w))|.$$

It satisfies the three axioms of distance (→ 273 Metric Spaces). When $\|T_1,T_2\| = 0$, T_1 and T_2 are called **equivalent** (in the sense of Fréchet). A set of all equivalent surfaces is called a **Fréchet surface**. Equivalent surfaces have equal Lebesgue areas; hence Lebesgue area is well defined for any Fréchet surface. Given a surface (T,A) with $L(T,A) < \infty$, there exists a pair (T_1,A_1), equivalent to (T,A) such that the functional determinants $J_i(w)$ exist for (T_1,A_1) and

$$V(T_1,A_1) = \iint_{A_1} J(w)\,du\,dv,$$

where $J^2(w) = \Sigma_{i=1}^3 J_i^2(w)$, as before. The problem of finding such a pair (T_1,A_1) is called the **representation problem**. Moreover, we can choose (T_1,A_1) to be a **generalized conformal mapping** in the following sense: Express (T_1,A_1) by $x = x(u,v)$, $y = y(u,v)$, $z = z(u,v)$. Then $x_u, x_v, \ldots, z_v$ exist a.e. in A_1 and are square integrable, $x_u^2 + y_u^2 + z_u^2 = x_v^2 + y_v^2 + z_v^2$, and $x_u x_v + y_u y_v + z_u z_v = 0$ a.e. in A_1.

References

[1] L. Cesari, Surface area, Ann. Math. Studies, Princeton Univ. Press, 1956.
[2] L. Cesari, Recent results in surface area theory, Amer. Math. Monthly, 66 (1959), 173–192.
[3] H. Federer, Measure and area, Bull. Amer. Math. Soc., 58 (1952), 306–378.
[4] G. Nöbeling, Über die Flächenmasse im Euklidischen Raum, Math. Ann., 118 (1943), 687–701.
[5] H. Okamura, On the surface integral and Gauss-Green's theorem, Mem. Coll. Sci. Univ. Kyôto, (A. Math.) 26 (1950), 5–14.
[6] T. Radó, Length and area, Amer. Math. Soc. Colloq. Publ., 1948.
[7] T. Radó, Lebesgue area and Hausdorff measure, Fund. Math., 44 (1957), 198–237.
[8] S. Saks, Theory of the integral, Warsaw, 1937.

246 (X.X.34)
Lie, Marius Sophus

Marius Sophus Lie (December 17, 1842–February 18, 1899), a Norwegian mathematician, is famous as the founder of †Lie groups. In 1869–1870, while conducting research with F. Klein on †sphere geometry, he developed the concept of continuous groups. This discovery was a stepping stone for Klein to complete his ideas for the †Erlangen program. In 1872, Lie became a professor at the University of Christiania (now Oslo), where he was employed until his death.

The continuous groups that Lie dealt with are today called the †Lie transformation group germs. With the free use of geometric concepts and analytic methods (especially the theory of †differential equations) he was able to develop his theory, and with its application, he was able to build the foundations of geometry. He also applied it to the theory of differential equations. The significance of his work was not duly acknowledged until after his death. Early in the 20th century, E. †Cartan and H. †Weyl were able to complete the theory of Lie groups, and by the middle of the century, the characteristics of the Lie group as a †topological group were clarified.

References

[1] F. Engel and P. Heegaard (eds.), Sophus Lie, Gesammelte Abhandlungen I–VI, Teubner, 1922–1937.

[2] S. Lie and F. Engel, Theorie der Transformationsgruppen I–III, Teubner, 1888–1893 (Chelsea, second edition, 1970).
[3] S. Lie and G. Scheffers, Vorlesungen über continuierliche Gruppen, Teubner, 1893 (Chelsea, second edition, 1971).

247 (IV.10) Lie Algebras

A. Basic Concepts

Let K be a †commutative ring with unity. A set $\mathfrak{g}$ is called a **Lie algebra** over K if the following four conditions are satisfied: (i) $\mathfrak{g}$ is a †left K-module, where we assume that the unity of K acts on $\mathfrak{g}$ as the identity operator. (ii) There is given a K-bilinear mapping (called the **bracket product**) $(X, Y)\rightarrow[X, Y]$ from $\mathfrak{g}\times\mathfrak{g}$ into $\mathfrak{g}$:

$$\left[\sum \alpha_i X_i, \sum \beta_j Y_j\right]=\sum \alpha_i\beta_j[X_i, Y_j]$$

for all α_i, β_j in K and X_i, Y_j in $\mathfrak{g}$. (iii) $[X,X]=0$ for every X in $\mathfrak{g}$. (Hence, $[X, Y]=-[Y,X]$ for every X, Y in $\mathfrak{g}$ (**alternating law**).) (iv) $[X,[Y,Z]]+[Y,[Z,X]]+[Z,[X,Y]]=0$ for every X, Y, Z in $\mathfrak{g}$ (**Jacobi identity**). In particular, if $K=\mathbf{C}$ (the complex number field) or $K=\mathbf{R}$ (the real number field), $\mathfrak{g}$ is called a **complex Lie algebra** or **real Lie algebra**, respectively.

For example, let $\mathfrak{A}$ be an †associative algebra over K. Putting $[X, Y]=XY-YX$, we can supply $\mathfrak{A}$ with the structure of a Lie algebra over K, which is called the Lie algebra associated with $\mathfrak{A}$. In particular, if $\mathfrak{A}$ is the †total matrix algebra K_n of degree n over K, then the Lie algebra associated with K_n is called the **general linear Lie algebra** of degree n over K and is denoted by $\mathfrak{gl}(n,K)$.

Let $\mathfrak{g}$ be a Lie algebra over K and $\mathfrak{a}$, $\mathfrak{b}$ be †K-submodules of $\mathfrak{g}$. The subset of $\mathfrak{g}$ consisting of elements of the form $\Sigma[A,B]$ (finite sum) with $A\in\mathfrak{a}$, $B\in\mathfrak{b}$ is denoted by $[\mathfrak{a},\mathfrak{b}]$, which is a K-submodule of $\mathfrak{g}$. A K-submodule $\mathfrak{a}$ of $\mathfrak{g}$ is called a **Lie subalgebra** of $\mathfrak{g}$ if $[\mathfrak{a},\mathfrak{a}]\subset\mathfrak{a}$. A subalgebra $\mathfrak{a}$ of $\mathfrak{g}$ is called an **ideal** of $\mathfrak{g}$ if $[\mathfrak{a},\mathfrak{g}]\subset\mathfrak{a}$ (this condition is equivalent to $[\mathfrak{g},\mathfrak{a}]\subset\mathfrak{a}$). If $\mathfrak{a}$ is a subalgebra of $\mathfrak{g}$, the restriction of the bracket product of $\mathfrak{g}$ on $\mathfrak{a}$ makes $\mathfrak{a}$ a Lie algebra over K. If $\mathfrak{a}$ is an ideal of $\mathfrak{g}$, the †quotient K-module $\mathfrak{g}/\mathfrak{a}$ is a Lie algebra over K relative to the following bracket product: $[X+\mathfrak{a}, Y+\mathfrak{a}]=[X, Y]+\mathfrak{a}$. This Lie algebra $\mathfrak{g}/\mathfrak{a}$ is called the **quotient Lie algebra** of $\mathfrak{g}$ modulo $\mathfrak{a}$.

Let $\mathfrak{g}_1$, $\mathfrak{g}_2$ be Lie algebras over K. A mapping $f:\mathfrak{g}_1\rightarrow\mathfrak{g}_2$ is called a **homomorphism** of $\mathfrak{g}_1$ into $\mathfrak{g}_2$ if f is K-linear and $f([X, Y])=[f(X),f(Y)]$ for every X, Y in $\mathfrak{g}_1$. A bijective homomorphism is called an **isomorphism**. Then $\mathfrak{g}_1$ is said to be **isomorphic** to $\mathfrak{g}_2$ if there exists an isomorphism from $\mathfrak{g}_1$ onto $\mathfrak{g}_2$, and we write $\mathfrak{g}_1\cong\mathfrak{g}_2$. If $f:\mathfrak{g}_1\rightarrow\mathfrak{g}_2$ is a homomorphism, then $f(\mathfrak{g}_1)$ is a subalgebra of $\mathfrak{g}_2$, and the kernel $\mathfrak{a}=f^{-1}(0)$ of f is an ideal of $\mathfrak{g}_1$. Furthermore, the homomorphism f induces an isomorphism $\bar{f}:\mathfrak{g}_1/\mathfrak{a}\rightarrow f(\mathfrak{g}_1)$ (**homomorphism theorem**).

The **direct sum** $\mathfrak{g}_1+\mathfrak{g}_2$ of two Lie algebras $\mathfrak{g}_1$, $\mathfrak{g}_2$ over K is defined as in the case of associative algebras. Then $\mathfrak{g}_1$, $\mathfrak{g}_2$ are ideals of $\mathfrak{g}_1+\mathfrak{g}_2$.

The set $A(\mathfrak{g})$ of all automorphisms of a Lie algebra $\mathfrak{g}$ is a subgroup of the general linear group $GL(\mathfrak{g})$. $A(\mathfrak{g})$ is called the (full) **automorphism group** of $\mathfrak{g}$.

B. Representations

Let $\mathfrak{g}$ be a Lie algebra over K, and let V be a K-module. Denote by $\mathfrak{L}(V)$ the associative algebra consisting of all K-linear mappings from V into V. Denote by $\mathfrak{gl}(V)$ the Lie algebra associated with $\mathfrak{L}(V)$. (Note that if V has a basis consisting of m elements over K, then $\mathfrak{gl}(V)\cong\mathfrak{gl}(m,K)$.) A homomorphism $\rho:\mathfrak{g}\rightarrow\mathfrak{gl}(V)$ is called a **representation** (more precisely, **linear representation**) of $\mathfrak{g}$ over V, and V is called the **representation space** of ρ. If V is a †free K-module of rank m, then m is called the **degree** of the representation ρ. We also use (ρ, V) instead of ρ to mention the representation space explicitly. The concepts concerning the representations such as †equivalence, †irreducibility, or complete reducibility are similar to the corresponding concepts found in the representation theory of associative algebras (→ 358 Representations). In particular, by taking $V=\mathfrak{g}$ and putting $\rho(X)Y=[X, Y]$ $(X, Y\in\mathfrak{g})$, we obtain a representation of $\mathfrak{g}$, which is called the **adjoint representation** of $\mathfrak{g}$, and $\rho(X)$ is denoted by $\mathrm{ad}(X)$. Then $\mathrm{ad}(\mathfrak{g})=\{\mathrm{ad}(X)|X\in\mathfrak{g}\}$ is a subalgebra of $\mathfrak{gl}(\mathfrak{g})$ and is called the **adjoint Lie algebra** of $\mathfrak{g}$.

Let (ρ, V) be a representation of $\mathfrak{g}$. Then there is associated with this representation a †symmetric bilinear form $B_\rho:\mathfrak{g}\times\mathfrak{g}\rightarrow K$ given by $B_\rho(X, Y)=\mathrm{tr}\,\rho(X)\rho(Y)$, where B_ρ satisfies the following invariance property: $B_\rho([X,Z], Y)=B_\rho(X,[Z, Y])$. In particular, if $\rho=\mathrm{ad}$, then we write B instead of B_ρ, and B is called the **Killing form** of $\mathfrak{g}$.

Let $\mathfrak{g}$ be a Lie algebra over $\mathbf{R}$ of dimension n, and let $\mathrm{ad}:\mathfrak{g}\rightarrow\mathfrak{gl}(\mathfrak{g})$ be the adjoint representation of $\mathfrak{g}$. Put $\det(tI-\mathrm{ad}(X))=\sum_{j=0}^{n}t^jP_j(X)$ for every element $X\in\mathfrak{g}$. Then the $P_j(X)$ are polynomial functions on $\mathfrak{g}$, and $P_n=1$. Let l be the least integer such that $P_l\neq0$. Then

$l = {}^{\dagger}$rank $\mathfrak{g}$. An element X of $\mathfrak{g}$ is called **regular (singular)** if $P_l(X) \neq 0$ ($P_l(X) = 0$). The subset $\mathfrak{g}'$ of $\mathfrak{g}$ consisting of all regular elements of $\mathfrak{g}$ is open and dense in $\mathfrak{g}$. The subset $\mathfrak{g} - \mathfrak{g}'$ of $\mathfrak{g}$ consisting of all singular elements of $\mathfrak{g}$ is of measure zero with respect to the †Lebesgue measure of $\mathfrak{g}$, which is obtained uniquely up to positive scalar multiples by means of a linear isomorphism of $\mathfrak{g}$ onto $\mathbf{R}^n$ using any basis of $\mathfrak{g}$ over $\mathbf{R}$.

Now suppose that $\mathfrak{g}$ is reductive (→ Section G). Then an element $X \in \mathfrak{g}$ is regular if and only if the centralizer $\mathfrak{z}_X = \{ Y \in \mathfrak{g} \mid \mathrm{ad}(X) Y = 0\}$ of X is a Cartan subalgebra (→ Section I) of $\mathfrak{g}$. Furthermore, if $X \in \mathfrak{g}$ is regular, then $\mathrm{ad}(X)$ is a †semisimple linear endomorphism of $\mathfrak{g}$.

C. Structure of Lie Algebras

Suppose that $\mathfrak{a}$, $\mathfrak{b}$ are ideals of a Lie algebra $\mathfrak{g}$. Then $[\mathfrak{a}, \mathfrak{b}]$ is also an ideal of $\mathfrak{g}$. In particular, $\mathfrak{g}$ has the following ideals: $\mathfrak{g}' = [\mathfrak{g}, \mathfrak{g}]$, $\mathfrak{g}'' = [\mathfrak{g}', \mathfrak{g}'], \ldots, \mathfrak{g}^{(i+1)} = [\mathfrak{g}^{(i)}, \mathfrak{g}^{(i)}], \ldots$. Furthermore, we have $\mathfrak{g} \supset \mathfrak{g}' \supset \mathfrak{g}'' \supset \ldots$. This series is called the **derived series** of $\mathfrak{g}$, and $\mathfrak{g}'$ is called the **derived algebra** of $\mathfrak{g}$. The Lie algebra $\mathfrak{g}$ is said to be **Abelian** if $\mathfrak{g}' = 0$, and **solvable** if $\mathfrak{g}^{(k)} = 0$ for some k. Now put $\mathfrak{g}^1 = \mathfrak{g}$, $\mathfrak{g}^2 = [\mathfrak{g}, \mathfrak{g}^1]$, $\mathfrak{g}^3 = [\mathfrak{g}, \mathfrak{g}^2], \ldots, \mathfrak{g}^{i+1} = [\mathfrak{g}, \mathfrak{g}^i], \ldots$. Then $\mathfrak{g}^1, \mathfrak{g}^2, \ldots$ are all ideals of $\mathfrak{g}$, and we have $\mathfrak{g}^1 \supset \mathfrak{g}^2 \supset \mathfrak{g}^3 \supset \ldots$. This series is called the **descending central series** of $\mathfrak{g}$, and $\mathfrak{g}$ is said to be **nilpotent** if $\mathfrak{g}^k = 0$ for some k. An ideal $\mathfrak{a}$ of $\mathfrak{g}$ is called **Abelian (solvable, nilpotent)** if the subalgebra $\mathfrak{a}$ is Abelian (solvable, nilpotent).

Put $\mathfrak{z} = \{ A \in \mathfrak{g} \mid [X, A] = 0$ for every X in $\mathfrak{g}\}$. Then $\mathfrak{z}$ is an Abelian ideal of $\mathfrak{g}$, called the **center** of $\mathfrak{g}$, and is the kernel of the adjoint representation of $\mathfrak{g}$. Define the ideals $\mathfrak{z}_1, \mathfrak{z}_2, \ldots$ of $\mathfrak{g}$ as follows: $\mathfrak{z}_1$ is the center of $\mathfrak{g}$, $\mathfrak{z}_2/\mathfrak{z}_1$ is the center of $\mathfrak{g}/\mathfrak{z}_1, \ldots, \mathfrak{z}_{i+1}/\mathfrak{z}_i$ is the center of $\mathfrak{g}/\mathfrak{z}_i, \ldots$. Then we have $0 \subset \mathfrak{z}_1 \subset \mathfrak{z}_2 \subset \ldots$. This series is called the **ascending central series** of $\mathfrak{g}$, and $\mathfrak{g}$ is nilpotent if and only if $\mathfrak{z}_k = \mathfrak{g}$ for some k.

We assume that K is a field of characteristic 0 and Lie algebras over K are of finite dimension. Let $X_1, \ldots, X_n$ be a basis of $\mathfrak{g}$ over K. Then the n^3 elements c_{ij}^k in K defined by $[X_i, X_j] = \sum c_{ij}^k X_k$ are called the **structural constants** of $\mathfrak{g}$ relative to the basis (X_i).

D. Radicals and Largest Nilpotent Ideals

The union $\mathfrak{r}$ of all solvable ideals of $\mathfrak{g}$ is also a solvable ideal of $\mathfrak{g}$, which is called the **radical** of $\mathfrak{g}$. The union $\mathfrak{n}$ of all nilpotent ideals of $\mathfrak{g}$ is also a nilpotent ideal of $\mathfrak{g}$, which is called the **largest nilpotent ideal** of $\mathfrak{g}$. The ideal $\mathfrak{s} = [\mathfrak{r}, \mathfrak{g}]$ is called the **nilpotent radical** of $\mathfrak{g}$. We have $\mathfrak{g} \supset \mathfrak{r} \supset \mathfrak{n} \supset \mathfrak{s}$.

E. Semisimplicity

A Lie algebra is called **semisimple** if its radical is 0. A semisimple Lie algebra $\mathfrak{g}$ over K is called **simple** if $\mathfrak{g}$ has no ideals other than $\mathfrak{g}$ and 0. If $\mathfrak{r}$ is the radical of a Lie algebra K, then $\mathfrak{g}/\mathfrak{r}$ is semisimple. Every semisimple Lie algebra is the direct sum of simple Lie algebras.

For example, put $\mathfrak{t}(n, K) = \{ A = (a_{ij}) \in \mathfrak{gl}(n, K) \mid a_{ij} = 0$ for every $i < j\}$ and $\mathfrak{n}(n, K) = \{ A = (a_{ij}) \in \mathfrak{t}(n, K) \mid a_{11} = a_{22} = \ldots = a_{nn} = 0\}$. Note that $\mathfrak{t}(n, K)$ is the set of all lower triangular matrices and $\mathfrak{n}(n, K)$ is the set of nilpotent lower triangular matrices. Then $\mathfrak{t}(n, K)$ is a solvable subalgebra of $\mathfrak{gl}(n, K)$, and $\mathfrak{n}(n, K)$ is a nilpotent subalgebra of $\mathfrak{gl}(n, K)$. Put $\mathfrak{sl}(n, K) = \{ A \in \mathfrak{gl}(n, K) \mid \mathrm{tr} A = 0\}$. Then $\mathfrak{sl}(n, K)$ is an ideal of $\mathfrak{gl}(n, K)$. For $n \geqslant 2$, $\mathfrak{sl}(n, K)$ is a simple Lie algebra.

F. Theorems

The following theorems are fundamental in the theory of Lie algebras:

(1) **Engel's Theorem** (valid even if K is of positive characteristic): Let V be a finite-dimensional vector space over a field K such that $V \neq \{0\}$. Let $\mathfrak{g}$ be a subalgebra of $\mathfrak{gl}(V)$ consisting of nilpotent elements. Then there is a nonzero element v in V such that $Xv = 0$ for every X in $\mathfrak{g}$. (Thus by choosing a suitable basis of V and identifying $\mathfrak{gl}(V)$ with $\mathfrak{gl}(n, K)$, we have $\mathfrak{g} \subset \mathfrak{n}(n, K)$, where $n = \dim V$.)

(2) **Lie's Theorem**: Let (ρ, V) be an irreducible representation of a solvable Lie algebra $\mathfrak{g}$. Then $\rho(\mathfrak{g})$ is Abelian. In particular, if K is algebraically closed, then $\dim V = 1$. (Thus for every representation (ρ, V) of a solvable Lie algebra $\mathfrak{g}$ over an †algebraically closed field K, we have $\rho(\mathfrak{g}) \subset \mathfrak{t}(n, K)$ by choosing a suitable basis of V.)

(3) **Cartan's Criterion of Solvability**: Let $\mathfrak{g}$ be a subalgebra of $\mathfrak{gl}(n, K)$. Then $\mathfrak{g}$ is solvable if and only if $\mathrm{tr}\, XY = 0$ for every $X \in \mathfrak{g}$ and $Y \in [\mathfrak{g}, \mathfrak{g}]$.

(4) **Cartan's Criterion of Semisimplicity**: A Lie algebra $\mathfrak{g}$ is semisimple if and only if the Killing form B of $\mathfrak{g}$ is †nondegenerate (i.e., $B(X, \mathfrak{g}) = 0$, $X \in \mathfrak{g}$ implies $X = 0$).

(5) **Weyl's Theorem**: Every representation (of finite degree) of a semisimple Lie algebra is completely reducible.

(6) **Levi Decomposition**: Let $\mathfrak{r}$ be the radical of a Lie algebra $\mathfrak{g}$. Then there is a semisimple subalgebra $\mathfrak{s}$ of $\mathfrak{g}$ such that $\mathfrak{g} = \mathfrak{r} + \mathfrak{s}$, $\mathfrak{r} \cap \mathfrak{s} = 0$.

Furthermore, such a subalgebra $\mathfrak{s}$ is unique up to automorphisms of $\mathfrak{g}$ (A. I. Mal'cev).

(7) **Ado's Theorem** (originally proved only for the case of characteristic 0 for K; the case of positive characteristic was proved by K. Iwasawa): Let $\mathfrak{g}$ be a finite-dimensional Lie algebra over a field K. Then there exists a representation (ρ, V) of $\mathfrak{g}$ of finite degree such that $\mathfrak{g} \cong \rho(\mathfrak{g})$.

G. Reductive Lie Algebras

A Lie algebra $\mathfrak{g}$ is called **reductive** if the radical $\mathfrak{r}$ of $\mathfrak{g}$ coincides with the center $\mathfrak{z}$ of $\mathfrak{g}$. The following four conditions for a Lie algebra $\mathfrak{g}$ are mutually equivalent: (i) $\mathfrak{g}$ is reductive; (ii) the nilpotent radical $\mathfrak{s}$ of $\mathfrak{g}$ is 0; (iii) the adjoint representation of $\mathfrak{g}$ is completely reducible; and (iv) the derived algebra $[\mathfrak{g},\mathfrak{g}]$ of $\mathfrak{g}$ is semisimple and $\mathfrak{g} = \mathfrak{z} + [\mathfrak{g},\mathfrak{g}]$ (direct sum), where $\mathfrak{z}$ is the center of $\mathfrak{g}$.

A representation (ρ, V) of a reductive Lie algebra $\mathfrak{g}$ is completely reducible if and only if $\rho(X)$ is diagonalizable for every X in $\mathfrak{z}$. For example, the Lie algebra $\mathfrak{gl}(n,K)$ is reductive.

H. Derivations

A linear mapping $\delta: \mathfrak{g} \to \mathfrak{g}$ is called a **derivation** of the Lie algebra $\mathfrak{g}$ if $\delta([X,Y]) = [\delta(X), Y] + [X, \delta(Y)]$ for every X, Y in $\mathfrak{g}$. The set $\mathfrak{D}(\mathfrak{g})$ of all derivations of $\mathfrak{g}$ is a subalgebra of $\mathfrak{gl}(\mathfrak{g})$, and $\mathfrak{D}(\mathfrak{g})$ is called the **Lie algebra of derivations** of $\mathfrak{g}$. The adjoint Lie algebra $\mathrm{ad}(\mathfrak{g})$ is an ideal of $\mathfrak{D}(\mathfrak{g})$, and elements of $\mathrm{ad}(\mathfrak{g})$ are called **inner derivations** of $\mathfrak{g}$. If $\mathfrak{g}$ is semisimple, then $\mathfrak{D}(\mathfrak{g}) = \mathrm{ad}(\mathfrak{g}) \cong \mathfrak{g}$.

Now suppose that $K = \mathbf{R}$ ($K = \mathbf{C}$). Then the group $A(\mathfrak{g})$ of automorphisms of $\mathfrak{g}$ is a Lie group (complex Lie group), and the Lie algebra of $A(\mathfrak{g})$ is given by $\mathfrak{D}(\mathfrak{g})$. Let $\delta \in \mathfrak{D}(\mathfrak{g})$ $(\subset \mathfrak{gl}(\mathfrak{g}))$. Then $\exp\delta$ $(\in GL(\mathfrak{g}))$ is in $A(\mathfrak{g})$. The connected subgroup $I(\mathfrak{g})$ of $A(\mathfrak{g})$ generated by $\{\exp\delta \mid \delta \in \mathrm{ad}(\mathfrak{g})\}$ is a †Lie subgroup of $A(\mathfrak{g})$. Furthermore, $I(\mathfrak{g})$ is a normal subgroup of $A(\mathfrak{g})$, called the **group of inner automorphisms** of $\mathfrak{g}$ or the **adjoint group** of $\mathfrak{g}$. Thus $\mathrm{ad}(\mathfrak{g})$ is the Lie algebra associated with $I(\mathfrak{g})$. The quotient group $A(\mathfrak{g})/I(\mathfrak{g})$ is called the **group of outer automorphisms** of $\mathfrak{g}$. If $\mathfrak{g}$ is semisimple, then $I(\mathfrak{g})$ coincides with the identity component of $A(\mathfrak{g})$.

I. Cartan Subalgebras

A subalgebra $\mathfrak{h}$ of a Lie algebra $\mathfrak{g}$ over K is called a **Cartan subalgebra** of $\mathfrak{g}$ if (i) $\mathfrak{h}$ is nilpotent and (ii) the normalizer $\mathfrak{n}$ of $\mathfrak{h}$ in $\mathfrak{g}$ (i.e., $\mathfrak{n} = \{X \in \mathfrak{g} \mid [X,\mathfrak{h}] \subset \mathfrak{h}\}$) coincides with $\mathfrak{h}$. If K is algebraically closed, then for every two Cartan subalgebras $\mathfrak{h}_1$, $\mathfrak{h}_2$ of $\mathfrak{g}$ there exists an automorphism σ of $\mathfrak{g}$ such that $\sigma(\mathfrak{h}_1) = \mathfrak{h}_2$. Furthermore, for such a σ we can take an automorphism of the form $\sigma = \exp(\mathrm{ad}(A_1)) \ldots \exp(\mathrm{ad}(A_r))$ $(A_1, \ldots, A_r \in \mathfrak{g})$, where all the $\mathrm{ad}(A_i)$ are nilpotent.

J. Universal Enveloping Algebras

Let $\mathfrak{g}$ be a Lie algebra over a field K. Regarding $\mathfrak{g}$ as a vector space over K, let $T(\mathfrak{g})$ be the †tensor algebra over $\mathfrak{g}$. Let J be the †two-sided ideal of $T(\mathfrak{g})$ generated by all elements of the form $X \otimes Y - Y \otimes X - [X,Y]$ $(X, Y \in \mathfrak{g})$. The quotient associative algebra $U(\mathfrak{g}) = T(\mathfrak{g})/J$ is called the **universal enveloping algebra** of $\mathfrak{g}$. The composite of the natural mappings $\mathfrak{g} \to T(\mathfrak{g}) \to U(\mathfrak{g})$ is an injection $\mathfrak{g} \to U(\mathfrak{g})$, and we identify $\mathfrak{g}$ with a linear subspace of $U(\mathfrak{g})$ by this mapping. Then we have $[X,Y] = XY - YX$ $(X, Y \in \mathfrak{g})$ in $U(\mathfrak{g})$. The algebra $U(\mathfrak{g})$ has no †zero divisors. In particular, if $\mathfrak{g}$ is the Lie algebra of a connected †Lie group G, then $U(\mathfrak{g})$ is isomorphic to the associative algebra of all †left-invariant differential operators on G. For every subalgebra $\mathfrak{h}$ of $\mathfrak{g}$, the universal enveloping algebra $U(\mathfrak{h})$ of $\mathfrak{h}$ is isomorphic to the subalgebra of the associative algebra $U(\mathfrak{g})$ generated by 1 and $\mathfrak{h}$. If $\mathfrak{g}$ is the direct sum of two Lie algebras $\mathfrak{g}_1$, $\mathfrak{g}_2$, then $U(\mathfrak{g})$ is isomorphic to the †tensor product $U(\mathfrak{g}_1) \otimes_K U(\mathfrak{g}_2)$. Let $\mathfrak{a}$ be an ideal of $\mathfrak{g}$, and let $\mathfrak{A}$ be the two-sided ideal of $U(\mathfrak{g})$ generated by $\mathfrak{a}$. Then we have $U(\mathfrak{g})/\mathfrak{A} \cong U(\mathfrak{g}/\mathfrak{a})$. Now put $U_0 = K \cdot 1$. Define a linear subspace U_i of $U(\mathfrak{g})$ by

$$U_i = K \cdot 1 + \mathfrak{g} + \mathfrak{g} \cdot \mathfrak{g} + \ldots + \underbrace{\mathfrak{g} \ldots \mathfrak{g}}_{i}.$$

Then we have $U_0 \subset U_1 \subset \ldots$, $U_i U_j \subset U_{i+j}$, $\bigcup_i U_i = U(\mathfrak{g})$. Thus $\{U_i\}$ defines a †filtration of $U(\mathfrak{g})$. Denote by $G = G^0 + G^1 + G^2 + \ldots$ $(G^0 = U_0, G^i = U_i / U_{i-1})$ the †graded ring associated with this filtration. Then we have $\mathfrak{g} = G^1 \subset G$. Let $X_1, \ldots, X_n$ be a basis of $\mathfrak{g}$, and let $S = K[Y_1, \ldots, Y_n]$ be a †polynomial ring on K in n indeterminates $Y_1, \ldots, Y_n$. Then there exists a unique algebra homomorphism $\omega: S \to G$ such that $\omega(1) = 1$, $\omega(Y_i) = X_i$ $(i = 1, \ldots, n)$. Furthermore, ω is bijective, and the ith homogeneous component S^i is mapped by ω onto G^i. Thus the set of monomials $\{X_1^{i_1} X_2^{i_2} \ldots X_n^{i_n}\}$ $(i_1 \geqslant 0, \ldots, i_n \geqslant 0)$ forms a basis of $U(\mathfrak{g})$ over K (the **Poincaré-Birkhoff-Witt theorem**).

Every representation (ρ, V) of $\mathfrak{g}$ over K can be extended to a unique representation (ρ', V) of $U(\mathfrak{g})$. Furthermore, ρ is irreducible (completely reducible) if and only if ρ' is irreducible (completely reducible). Given two representations ρ_1, ρ_2 of $\mathfrak{g}$, ρ_1 is equivalent to ρ_2 if and only if ρ_1' is equivalent to ρ_2'.

Now suppose that $\mathfrak{g}$ is semisimple, and let $X_1, \ldots, X_n$ be a basis of $\mathfrak{g}$. Using the Killing form B of $\mathfrak{g}$, put $g_{ij} = B(X_i, X_j)$. Denote the inverse matrix of (g_{ij}) by (g^{ij}). Define $c \in U(\mathfrak{g})$ by $c = \sum_{ij} g^{ij} X_i X_j$. The element c, called the **Casimir element** of the Lie algebra, is independent of the choice of the basis (X_i), is a well-defined element of $U(\mathfrak{g})$, and belongs to the center of $U(\mathfrak{g})$. For every absolutely irreducible representation ρ of $U(\mathfrak{g})$, $\rho(c)$ is a scalar operator and $\operatorname{tr}\rho(c)$ is a positive rational number.

K. Complex Semisimple Lie Algebras

We assume that $K = \mathbf{C}$, although there is no essential change if we assume that K is an algebraically closed field of characteristic 0.

A subalgebra $\mathfrak{h}$ of a complex semisimple Lie algebra $\mathfrak{g}$ is a Cartan subalgebra of $\mathfrak{g}$ if and only if $\mathfrak{h}$ is a maximal Abelian subalgebra of $\mathfrak{g}$ such that $\operatorname{ad}(H)$ is diagonalizable for every H in $\mathfrak{h}$. We fix a Cartan subalgebra $\mathfrak{h}$; $\dim \mathfrak{h}$ is called the **rank** of $\mathfrak{g}$, and we denote the linear space consisting of all $\mathbf{C}$-valued forms on $\mathfrak{h}$ by $\mathfrak{h}^*$. For every α in $\mathfrak{h}^*$, let

$$\mathfrak{g}_\alpha = \{X \in \mathfrak{g} \mid \operatorname{ad}(H)X = \alpha(H)X \quad \text{for all } H \text{ in } \mathfrak{h}\}.$$

Then $\mathfrak{g}_\alpha$ is a linear subspace of $\mathfrak{g}$, and $\mathfrak{g}_0 = \mathfrak{h}$. Define a subset Δ of $\mathfrak{h}^*$ by

$$\Delta = \{\alpha \in \mathfrak{h}^* \mid \alpha \neq 0,\ \mathfrak{g}_\alpha \neq \{0\}\}.$$

Then Δ is a finite set. Elements of Δ are called **roots** of $\mathfrak{g}$ relative to $\mathfrak{h}$, and Δ is called the **root system** of $\mathfrak{g}$ relative to $\mathfrak{h}$. For every root α, $\mathfrak{g}_\alpha$ is of dimension one, and $\mathfrak{g}$ is decomposed into a direct sum of linear subspaces as follows:

$$\mathfrak{g} = \mathfrak{h} + \sum_{\alpha \in \Delta} \mathfrak{g}_\alpha.$$

For each root α, $\mathfrak{g}_\alpha$ is called the **root subspace** corresponding to α.

The restriction $B_{\mathfrak{h}}$ of the Killing form B of $\mathfrak{g}$ on $\mathfrak{h}$ is nondegenerate. Hence for every λ in $\mathfrak{h}^*$ there exists a unique element H_λ in $\mathfrak{h}$ such that $\lambda(H) = B(H_\lambda, H)$ for all H in $\mathfrak{h}$. Thus we get a linear bijection $\mathfrak{h}^* \to \mathfrak{h}$ defined by $\lambda \to H_\lambda$. Via this bijection, $B_{\mathfrak{h}}$ gives rise to a symmetric bilinear form $(\lambda, \mu) = (H_\lambda, H_\mu)$ $(\lambda, \mu \in \mathfrak{h}^*)$ on $\mathfrak{h}^*$. Denote by $\mathfrak{h}_{\mathbf{R}}^*$ the real linear subspace of $\mathfrak{h}^*$ spanned by Δ. Then the inner product (λ, μ) defined on $\mathfrak{h}^*$ is positive definite on $\mathfrak{h}_{\mathbf{R}}^*$. Hence with respect to this inner product, $\mathfrak{h}_{\mathbf{R}}^*$ is an l-dimensional Euclidean space, where $l = \dim \mathfrak{h}$. The root system Δ is a finite subset of the Euclidean space $\mathfrak{h}_{\mathbf{R}}^*$.

L. Properties of Root Systems

(i) $\alpha \in \Delta$ implies $-\alpha \in \Delta$. Furthermore, among the scalar multiples of α, only $\pm\alpha$ belong to Δ. (ii) Let $\alpha, \beta \in \Delta$. Then $2(\alpha, \beta)/(\alpha, \alpha)$ is a rational integer. (iii) Let $\alpha, \beta \in \Delta$ and $\beta \neq \pm\alpha$. Then there exist unique nonnegative integers j, i such that $\{\beta + \nu\alpha \mid \nu \in \mathbf{Z}\} \cap \Delta = \{\beta - j\alpha, \beta - (j-1)\alpha, \ldots, \beta - \alpha, \beta, \beta + \alpha, \ldots, \beta + i\alpha\}$. Furthermore, $j - i = 2(\alpha, \beta)/(\alpha, \alpha)$, $i + j \leqslant 3$. The set $\{\beta + \mathbf{Z}\alpha\} \cap \Delta$ is called the **α-string** of β.

Now let $\alpha \in \Delta$. Denote by w_α the †reflection mapping of $\mathfrak{h}_{\mathbf{R}}^*$ with respect to the hyperplane $P_\alpha = \{x \in \mathfrak{h}_{\mathbf{R}}^* \mid (\alpha, x) = 0\}$, which is orthogonal to α. Then we have $w_\alpha(\beta) = \beta - (\alpha^*, \beta)\alpha \in \Delta$ for every β in Δ, where $\alpha^* = 2\alpha/(\alpha, \alpha)$. Thus we have $w_\alpha(\Delta) = \Delta$ for every α in Δ. (iv) Let $\alpha, \beta \in \Delta$ and $\beta \neq \pm\alpha$. Then the angle θ between α and β is one of the following: 30°, 45°, 60°, 90°, 120°, 135°, 150°. Suppose, moreover, that $0 \leqslant \theta \leqslant 90°$ and $(\alpha, \alpha) \leqslant (\beta, \beta)$. Then we have the following criteria: $\theta = 30° \Leftrightarrow 3(\alpha, \alpha) = (\beta, \beta)$; $\theta = 45° \Leftrightarrow 2(\alpha, \alpha) = (\beta, \beta)$; $\theta = 60° \Leftrightarrow (\alpha, \alpha) = (\beta, \beta)$. (v) Let $\alpha, \beta, \alpha + \beta \in \Delta$. Then $[\mathfrak{g}_\alpha, \mathfrak{g}_\beta] = \mathfrak{g}_{\alpha+\beta}$.

Conversely, suppose that a finite subset Δ of a finite-dimensional Euclidean space E satisfies conditions (i) and (ii) together with a part of (iii): $w_\alpha(\Delta) = \Delta$ for every α in Δ. Then Δ is a root system of some complex semisimple Lie algebra.

M. Lexicographic Linear Ordering in $\mathfrak{h}_{\mathbf{R}}^*$

Let $\lambda_1, \ldots, \lambda_l$ be a basis of $\mathfrak{h}_{\mathbf{R}}^*$ over $\mathbf{R}$. Define a linear ordering $\lambda > \mu$ on $\mathfrak{h}_{\mathbf{R}}^*$ as follows: If $\lambda = \sum \xi_i \lambda_i$, $\mu = \sum \eta_i \lambda_i$ (ξ_i, η_i all in $\mathbf{R}$), then $\lambda > \mu$ if and only if there exists an index s $(1 \leqslant s \leqslant l)$ such that $\xi_i = \eta_i$ for $i = 1, \ldots, s-1$ and $\xi_s > \eta_s$. This linear ordering is called the **lexicographic linear ordering** of $\mathfrak{h}_{\mathbf{R}}^*$ associated with the basis (λ_i). Relative to this linear ordering, a root α is called a **positive (negative) root** if $\alpha > 0$ $(\alpha < 0)$. We denote the subset of Δ consisting of all positive (negative) roots by Δ^+ (Δ^-). A subset S of Δ coincides with Δ^+ for some lexicographic linear ordering of $\mathfrak{h}_{\mathbf{R}}^*$ if and only if the following conditions are satisfied: $\Delta = S \cup (-S)$; $S \cap (-S) = \varnothing$; $\alpha, \beta \in S$, $\alpha + \beta \in \Delta$ imply $\alpha + \beta \in S$. A positive root $\alpha \in \Delta^+$ is called a **simple root** if α cannot be expressed as the sum of two positive roots.

N. Fundamental Root Systems

Let Π be a subset of Δ consisting of l roots $\alpha_1, \ldots, \alpha_l$. Then Π is called a **fundamental root system** of Δ if the following conditions are satisfied: (i) Every element α of Δ is expressed uniquely as an integral linear combination of the α_i ($\alpha=\Sigma m_i\alpha_i$); and (ii) in this expression, $m_1, \ldots, m_l$ are either all $\geqslant 0$ or all $\leqslant 0$. For any lexicographic linear ordering of $\mathfrak{h}_{\mathbf{R}}^*$, the set of all simple roots forms a fundamental root system of Δ. Moreover, every fundamental root system of Δ is obtained in this manner. Let $\Pi=\{\alpha_1, \ldots, \alpha_l\}$ be a fundamental root system of Δ. Then $\Sigma\mathfrak{g}_{\alpha_i}+\Sigma\mathfrak{g}_{-\alpha_i}$ generates $\mathfrak{g}$. The Lie algebra $\mathfrak{g}$ is not simple if and only if Π admits an orthogonal partition, i.e., $\Pi=\Pi_1\cup\Pi_2$, $\Pi_1\neq\varnothing$, $\Pi_2\neq\varnothing$, $\Pi_1\cap\Pi_2=\varnothing$, and $(\alpha,\beta)=0$ for every $\alpha\in\Pi_1$ and $\beta\in\Pi_2$. The l^2 integers $a_{ij}=-2(\alpha_i,\alpha_j)/(\alpha_j,\alpha_j)$ ($1\leqslant i, j\leqslant l$) are called the **Cartan integers** of $\mathfrak{g}$ relative to the fundamental root system Π. Then we have $a_{ii}=-2$, $a_{ij}\geqslant 0$ for $i\neq j$.

O. Borel Subalgebras and Parabolic Subalgebras

Let $\mathfrak{b}=\mathfrak{h}+\Sigma_{\alpha>0}\mathfrak{g}_\alpha$. Then $\mathfrak{b}$ is a maximal solvable subalgebra of $\mathfrak{g}$. The group $I(\mathfrak{g})$ acts transitively on the set of all maximal solvable subalgebras of $\mathfrak{g}$. A maximal solvable subalgebra of $\mathfrak{g}$ is called a **Borel subalgebra** of $\mathfrak{g}$. A subalgebra of $\mathfrak{g}$ is called a **parabolic subalgebra** if it contains a Borel subalgebra of $\mathfrak{g}$. Now let Φ be any subset of a given fundamental root system $\Pi=\{\alpha_1, \ldots, \alpha_l\}$. Denote by $\Delta^-(\Phi)$ the set of all negative roots $\alpha=\Sigma n_i\alpha_i$ such that $n_j=0$ for all α_j in Φ. Then $\mathfrak{p}_\Phi=\mathfrak{b}+\Sigma_{\alpha\in\Delta^-(\Phi)}\mathfrak{g}_\alpha$ is a parabolic subalgebra. Thus we get 2^l parabolic subalgebras $\{\mathfrak{p}_\Phi \mid \Phi\subset\Pi\}$. Every parabolic subalgebra is conjugate under $I(\mathfrak{g})$ to one and only one of the parabolic subalgebras $\{\mathfrak{p}_\Phi \mid \Phi\subset\Pi\}$.

P. Weyl's Canonical Basis

Let $H_1, \ldots, H_l$ be a basis of $\mathfrak{h}$, and let E_α be a basis of $\mathfrak{g}_\alpha$ for each root α. Then we have a basis $\{H_i, E_\alpha\}$ of $\mathfrak{g}$. Such a basis is called **Weyl's canonical basis** if the following three conditions are satisfied: (i) $\alpha(H_j)\in\mathbf{R}$ ($j=1, \ldots, l$) for every $\alpha\in\Delta$; (ii) the Killing form B of $\mathfrak{g}$ satisfies $B(E_\alpha, E_{-\alpha})=-1$ for every $\alpha\in\Delta$; and (iii) if α, β, $\alpha+\beta\in\Delta$ and $[E_\alpha, E_\beta]=N_{\alpha,\beta}E_{\alpha+\beta}$ ($N_{\alpha,\beta}\in\mathbf{C}$), then $N_{\alpha,\beta}$ is in $\mathbf{R}$ and $N_{\alpha,\beta}=N_{-\alpha,-\beta}$. The Lie algebra $\mathfrak{g}$ always has Weyl's canonical basis. For such a basis $\{H_i, E_\alpha\}$, the linear space

$$\mathfrak{g}_u=\sum\mathbf{R}\sqrt{-1}\,H_j+\sum\mathbf{R}(E_\alpha+E_{-\alpha})$$
$$+\sum\mathbf{R}\left(\sqrt{-1}\,(E_\alpha-E_{-\alpha})\right)$$

is a semisimple Lie algebra over $\mathbf{R}$. The Killing form of $\mathfrak{g}_u$ is negative definite. Every connected Lie group whose Lie algebra is $\mathfrak{g}_u$ is always compact. Furthermore, $\mathfrak{g}=\mathfrak{g}_u+\sqrt{-1}\,\mathfrak{g}_u$, $\mathfrak{g}_u\cap\sqrt{-1}\,\mathfrak{g}_u=0$. Thus $\mathfrak{g}$ is isomorphic to the Lie algebra $\mathfrak{g}_u^{\mathbf{C}}=\mathbf{C}\otimes_{\mathbf{R}}\mathfrak{g}_u$ over $\mathbf{C}$ obtained from $\mathfrak{g}_u$ by extending the basic field $\mathbf{R}$ to $\mathbf{C}$, and $\mathfrak{g}_u$ is called the **unitary restriction** of $\mathfrak{g}$ relative to Weyl's canonical basis $\{H_i, E_\alpha\}$.

A Lie algebra $\mathfrak{a}$ over $\mathbf{R}$ is called a **real form** of $\mathfrak{g}$ if $\mathfrak{a}^{\mathbf{C}}=\mathbf{C}\otimes_{\mathbf{R}}\mathfrak{a}$ is isomorphic to $\mathfrak{g}$. When this is the case, $\mathfrak{g}$ is called a **complex form** (or the **complexification**) of $\mathfrak{a}$. Note that a real form $\mathfrak{a}$ of $\mathfrak{g}$ can be regarded as a real subalgebra of $\mathfrak{g}$ such that $\mathfrak{g}=\mathfrak{a}+\sqrt{-1}\,\mathfrak{a}$, $\mathfrak{a}\cap\sqrt{-1}\,\mathfrak{a}=0$. A real Lie algebra $\mathfrak{a}$ is called a **compact real Lie algebra** if its Killing form is negative definite. A real Lie algebra is compact if and only if it is semisimple and is the Lie algebra of some compact Lie group. The Lie algebra $\mathfrak{a}$ of a compact Lie group A is the direct sum of its center $\mathfrak{z}$ and some compact Lie algebra; hence $\mathfrak{a}$ is reductive.

A compact real form $\mathfrak{g}_u$ of a complex semisimple Lie algebra $\mathfrak{g}$ is called a **compact form** of $\mathfrak{g}$. The group $I(\mathfrak{g})$ of all inner automorphisms of $\mathfrak{g}$ acts transitively on the set of all compact real forms of $\mathfrak{g}$ (regarding the real forms of $\mathfrak{g}$ as real subalgebras of $\mathfrak{g}$).

Q. Chevalley's Canonical Basis

A complex semisimple Lie algebra $\mathfrak{g}$ always has a basis $\{H_i, E_\alpha\}$ (consisting of a basis $H_1, \ldots, H_l$ of $\mathfrak{h}$ together with a basis E_α of $\mathfrak{g}_\alpha$ for each root α) such that: (i) $\alpha(H_i)\in\mathbf{Z}$ for every $\alpha\in\Delta$ and $i=1, \ldots, l$; (ii) $B(E_\alpha, E_{-\alpha})=2/(\alpha,\alpha)$ for every $\alpha\in\Delta$; and (iii) if α, β, $\alpha+\beta\in\Delta$ and $[E_\alpha, E_\beta]=N_{\alpha,\beta}E_{\alpha+\beta}$ ($N_{\alpha,\beta}\in\mathbf{C}$), then $N_{\alpha,\beta}\in\mathbf{Z}$ and $N_{\alpha,\beta}=-N_{-\alpha,-\beta}$. Such a basis $\{H_i, E_\alpha\}$ is called **Chevalley's canonical basis**. When we take this basis, the structural constants of $\mathfrak{g}$ relative to $\{H_i, E_\alpha\}$ are all integers. Thus $\mathfrak{g}_{\mathbf{Z}}=\Sigma\mathbf{Z}H_i+\Sigma\mathbf{Z}E_\alpha$ is a Lie algebra over $\mathbf{Z}$. Furthermore, $\mathfrak{g}_{\mathbf{R}}=\Sigma\mathbf{R}H_i+\Sigma\mathbf{R}E_\alpha$ is a real form of $\mathfrak{g}$ with the property that there exists a Cartan subalgebra $\mathfrak{h}_{\mathbf{R}}$ of $\mathfrak{g}_{\mathbf{R}}$ such that for each element H in $\mathfrak{h}_{\mathbf{R}}$, all the eigenvalues of $\mathrm{ad}(H)$ on $\mathfrak{g}_{\mathbf{R}}$ are contained in $\mathbf{R}$. (In fact, we may take $\Sigma\mathbf{R}H_i$ as $\mathfrak{h}_{\mathbf{R}}$.) Such a real form of $\mathfrak{g}$ is called a **normal real form**. The group $I(\mathfrak{g})$ acts transitively on the set of all normal real forms of $\mathfrak{g}$.

R. Weyl Groups

The reflections w_α ($\alpha\in\Delta$) of the Euclidean space $\mathfrak{h}_{\mathbf{R}}^*$ generate a subgroup W of the group of all †congruent transformations of $\mathfrak{h}_{\mathbf{R}}^*$. W is called the **Weyl group** of $\mathfrak{g}$ relative to $\mathfrak{h}$ and is represented faithfully as a †permutation group over the finite set Δ. Hence W is a finite group. Let $\Pi=\{\alpha_1, \ldots, \alpha_l\}$ be a fundamental root system of Δ. Then W is generated by

$w_{\alpha_1}, \ldots, w_{\alpha_l}$. The root system Δ coincides with the set $\{w(\alpha) \mid w \in W, \alpha \in \Pi\}$. Let $\mathfrak{F}$ be the set of all fundamental root systems of Δ. Then W acts on $\mathfrak{F}$ and is †simply transitive on $\mathfrak{F}$. If $\mathfrak{g}$ is simple, two roots α, β are conjugate under W if and only if $(\alpha,\alpha)=(\beta,\beta)$. Now let Γ be the complement in $\mathfrak{h}_{\mathbf{R}}^*$ of the union of all the hyperplanes P_α $(\alpha \in \Delta)$ orthogonal to α. Then Γ is a W-stable open subset of $\mathfrak{h}_{\mathbf{R}}^*$. A connected component of Γ is called a **Weyl chamber**. W acts on the set $\mathfrak{F}_0$ of all Weyl chambers and is simply transitive on $\mathfrak{F}_0$. Let $\Pi=\{\alpha_1, \ldots, \alpha_l\}$ be a fundamental root system. Then the set $\{x \in \mathfrak{h}_{\mathbf{R}}^* \mid (x, \alpha_i)>0 \text{ for } i=1, \ldots, l\}$ is a Weyl chamber, called the **positive Weyl chamber** associated with Π. Now fix any lexicographic linear ordering of $\mathfrak{h}_{\mathbf{R}}^*$ that has Π as the set of simple roots. For $w \in W$, put $\Delta_w^+=\{\alpha \in \Delta^+ \mid w(\alpha) \in \Delta^-\}$. Denote the cardinality of Δ_w^+ by $n(w)$. Then $n(w)=0 \Leftrightarrow w=1$. Furthermore, w can be expressed as a product of $n(w)$ factors $w=w_{\alpha_i} \ldots w_{\alpha_j}$, where each factor is taken from $\{w_{\alpha_1}, \ldots, w_{\alpha_l}\}$ admitting repetitions. In fact, $n(w)$ is the minimum length of the expression of w as a product $w=w_{\alpha_i} \ldots w_{\alpha_j}$ $(i, j=1, \ldots, l)$. With respect to the generators $w_{\alpha_1}, \ldots, w_{\alpha_l}$, W has the following system of †defining relations:

$$w_{\alpha_i}^2=1, \quad 1 \leqslant i \leqslant l,$$
$$(w_{\alpha_i} w_{\alpha_j})^{m_{ij}}=1, \quad 1 \leqslant i<j \leqslant l,$$

where m_{ij} is the order of $w_{\alpha_i} w_{\alpha_j}$. Thus if θ_{ij} is the angle between α_i and α_j, we have $m_{ij}=\pi/(\pi-\theta_{ij})$.

Denote by T the set of all linear transformations σ of $\mathfrak{h}_{\mathbf{R}}^*$ such that $\sigma(\Delta)=\Delta$. Then T is also the subgroup of all congruent transformations of $\mathfrak{h}_{\mathbf{R}}^*$; furthermore, T is a finite group, and W is a normal subgroup of T. Let Π be a fundamental root system of Δ and put $P=\{\alpha \in T \mid \sigma(\Pi)=\Pi\}$. Then P is a subgroup of T, and we have a semidirect product $T=P \cdot W$. Elements of P are called **particular transformations** relative to Π. The group $P \cong T/W$ is isomorphic to the group $A(\mathfrak{g})/I(\mathfrak{g})$ of outer automorphisms of $\mathfrak{g}$.

S. Classification of Complex Simple Lie Algebras

Let $\mathfrak{h}$ be a Cartan subalgebra of a complex semisimple Lie algebra $\mathfrak{g}$, and let $\Pi=\{\alpha_1, \ldots, \alpha_l\}$ be a fundamental root system relative to $\mathfrak{h}$. We associate with Π the diagram (†1-dimensional complex) indicated in Fig. 1. This diagram is called the **Dynkin diagram** of $\mathfrak{g}$ (also called the **Schläfli diagram** or **Coxeter diagram**). It is constructed as follows: with each α_i there is associated a vertex (denoted by a small white circle). These l vertices are connected by several segments as follows. Let θ_{ij} be the angle between α_i and α_j. (i) If $\theta_{ij}=150°$, α_i and α_j are connected by three oriented segments as in (1) of Fig. 1, where the orientation means $(\alpha_i, \alpha_i)>(\alpha_j, \alpha_j)$. (ii) If $\theta_{ij}=135°$, α_i and α_j are connected by two oriented segments as in (2) of Fig. 1. (iii) If $\theta_{ij}=120°$, α_i and α_j are connected by a non-oriented single segment as in (3) of Fig. 1. (iv) If $\theta_{ij}=90°$, α_i and α_j are not connected.

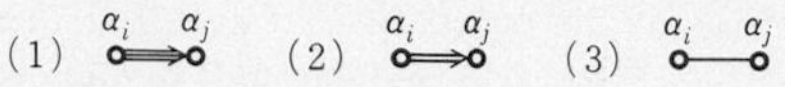

Fig. 1
A Dynkin diagram.

The Dynkin diagram of $\mathfrak{g}$ is independent of the choice of $\mathfrak{h}$, Π. Furthermore, two complex semisimple Lie algebras are isomorphic if and only if they have the same Dynkin diagram. A complex semisimple Lie algebra is simple if and only if its Dynkin diagram is connected.

Fig. 2 gives all possible Dynkin diagrams associated with complex semisimple simple Lie algebras. There are seven categories. (The index l in A_l means the rank of $\mathfrak{g}$.) Among these A_l $(l \geqslant 1)$, B_l $(l \geqslant 2)$, C_l $(l \geqslant 3)$, D_l $(l \geqslant 4)$ are called **classical complex simple Lie algebras**. E_l $(l=6,7,8)$, F_4, and G_2 are called **exceptional complex simple algebras**. Note that $A_1 \cong B_1 \cong C_1$, $B_2 \cong C_2$, $A_3 \cong D_3$, $D_2=A_1+A_1$. A_l (resp. B_l, C_l, D_l) is the Lie algebra of the complex Lie group $SL(l+1, \mathbf{C})$ $(SO(2l+1, \mathbf{C})$, $Sp(l, \mathbf{C})$, $SO(2l, \mathbf{C}))$.

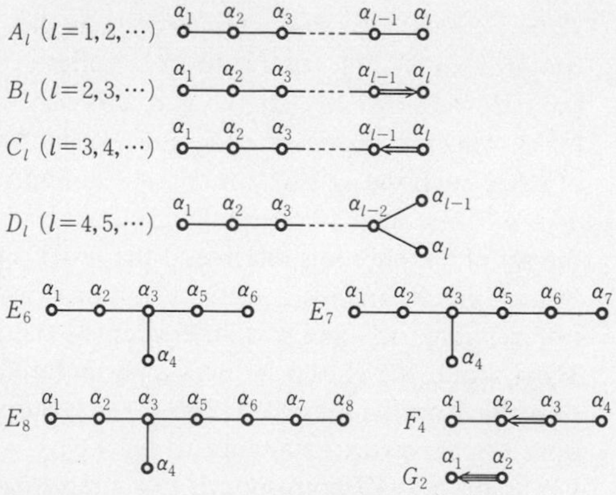

Fig. 2
Dynkin diagrams of simple Lie algebras. Note that $\dim \mathfrak{g}$ is as follows. A_l:l^2+2l; B_l:$2l^2+l$; C_l: $2l^2+l$; D_l:$2l^2-l$; E_6:78; E_7:133; E_8:248; F_4:52; G_2: 14.

T. Classification of Real Simple Lie Algebras

We refer the reader to [4] and the references at the end of [4] for the classification of simple Lie algebras over a general field k; in particular, for $k=\mathbf{R}$. (→ Appendix A, Tables 5.I, 5.II), the algebras of which are closely related to the classification of irreducible symmetric Riemannian manifolds (→ 401

Symmetric Riemannian Spaces).

In particular, since compact semisimple real Lie algebras $\mathfrak{g}$ are in one-to-one correspondence (up to isomorphism) with complex semisimple Lie algebras $\mathfrak{g}^{\mathbf{C}}$ obtained as the complexification of $\mathfrak{g}$, the classification of compact real simple Lie algebras reduces to the classification of complex simple Lie algebras. Hence they are also represented by the same Dynkin diagrams. Compact real simple Lie algebras of the types A_l, B_l, C_l, and D_l are called **classical compact real simple Lie algebras**. They are the Lie algebras of the compact Lie groups $SU(l+1)$, $SO(2l+1)$, $Sp(l)$, and $SO(2l)$, respectively. Compact real simple Lie algebras of the type E_l $(l=6,7,8)$, F_4, and G_2 are called **exceptional compact real simple Lie algebras**.

U. Satake Diagrams of Real Semisimple Lie Algebras

Let $\mathfrak{g}$ be a real semisimple Lie algebra, $\mathfrak{k}$ be the subalgebra associated with a †maximal compact subgroup of the †adjoint group of $\mathfrak{g}$, and $\mathfrak{p}$ be the orthogonal complement of $\mathfrak{k}$ in $\mathfrak{g}$ relative to the Killing form of $\mathfrak{g}$. Let $\mathfrak{a}$ be a maximal Abelian subalgebra contained in $\mathfrak{p}$, and let $\mathfrak{h}$ be a Cartan subalgebra of $\mathfrak{g}$ containing $\mathfrak{a}$. Denote by $\mathfrak{g}^{\mathbf{C}}$, $\mathfrak{h}^{\mathbf{C}}$ the complexifications of $\mathfrak{g}$, $\mathfrak{h}$, respectively. Let σ be the †semilinear †automorphism of $\mathfrak{g}^{\mathbf{C}}$ defined by $\sigma(X+\sqrt{-1}\ Y)=X-\sqrt{-1}\ Y$ $(X, Y\in\mathfrak{g})$. Then we have $\sigma(\mathfrak{h}^{\mathbf{C}})=\mathfrak{h}^{\mathbf{C}}$. Thus σ acts on the root system Δ of $\mathfrak{g}^{\mathbf{C}}$ relative to $\mathfrak{h}^{\mathbf{C}}$ as follows: $(\sigma\alpha)(H)=\overline{\alpha(\sigma H)}$ $(\alpha\in\Delta)$. Thus σ acts on $(\mathfrak{h}^{\mathbf{C}})^*_{\mathbf{R}}$. There is a lexicographic linear ordering of $(\mathfrak{h}^{\mathbf{C}})^*_{\mathbf{R}}$ such that $\alpha\in\Delta^+$ and $\sigma\alpha\neq-\alpha$ imply $\sigma\alpha\in\Delta^+$. Fix such an ordering and let Π be the set of simple roots relative to the ordering. Put $\sigma=pw$, where p is a particular transformation relative to Π and w is an element in the Weyl group W. Then p induces a permutation of order 2 on the set $\{\alpha\in\Pi\mid\sigma\alpha\neq-\alpha\}$. Suppose that a vertex belonging to the Dynkin diagram of Π corresponds to a simple root α such that $\sigma\alpha=-\alpha$. Then replace the vertex by a small black circle. Also, if two vertices are mapped to each other by p, then connect the two vertices by an arc with two arrows on the end. The diagram thus obtained is called the **Satake diagram** of $\mathfrak{g}$. The Satake diagram of $\mathfrak{g}$ is independent of the choice of $\mathfrak{k}$, $\mathfrak{a}$, $\mathfrak{h}$ and of the ordering of $(\mathfrak{h}^{\mathbf{C}})^*_{\mathbf{R}}$. Two real semisimple Lie algebras are isomorphic if and only if they have the same Satake diagram, and $\mathfrak{g}$ is simple if and only if its Satake diagram is connected. Thus real simple Lie algebras are classified by their Satake diagrams [17].

V. Iwasawa Decomposition of Real Semisimple Lie Algebras

Let $\mathfrak{g}$ be a real semisimple Lie algebra. Take $\mathfrak{k}$, $\mathfrak{a}$, $\mathfrak{h}$ and the ordering of $(\mathfrak{h}^{\mathbf{C}})^*_{\mathbf{R}}$ as in the construction of Satake diagrams (→ Section U). Let $\mathfrak{n}$ be the intersection of $\mathfrak{g}$ with the subspace $\Sigma(\mathfrak{g}^{\mathbf{C}})_\alpha$, where the sum is taken over $\alpha\in\Delta^+$ such that $\sigma\alpha\neq-\alpha$. Then $\mathfrak{n}$ is a nilpotent subalgebra of $\mathfrak{g}$, and we have a decomposition of $\mathfrak{g}$ into the direct sum of linear spaces as follows: $\mathfrak{g}=\mathfrak{k}+\mathfrak{a}+\mathfrak{n}$. This decomposition is called an **Iwasawa decomposition** of $\mathfrak{g}$. Iwasawa decompositions are unique in the following sense: Let $\mathfrak{g}=\mathfrak{k}'+\mathfrak{a}'+\mathfrak{n}'$ be another Iwasawa decomposition; then there exists an inner automorphism A of $\mathfrak{g}$ such that $A\mathfrak{k}=\mathfrak{k}'$, $A\mathfrak{a}=\mathfrak{a}'$, $A\mathfrak{n}=\mathfrak{n}'$.

For the cohomology theory of Lie algebras and Lie algebras over fields of characteristic $p>0$, in particular the theory of **restricted Lie algebras**, → [4]. For the relationship between Lie algebras and the theory of finite groups (e.g., Chevalley's simple groups, †Burnside problems), → [8] and the references therein.

W. Representations

Let $\mathfrak{g}$ be a complex semisimple Lie algebra and $\mathfrak{h}$ be a Cartan subalgebra of $\mathfrak{g}$. We fix $\mathfrak{h}$ and a lexicographic linear ordering on $\mathfrak{h}^*_{\mathbf{R}}$. Let $\Pi=\{\alpha_1,\ldots,\alpha_l\}$ be the set of simple roots. Since every representation of $\mathfrak{g}$ is completely reducible, we restrict ourselves to the explanation of irreducible representations. Let (ρ, V) be a representation of $\mathfrak{g}$. For each $\lambda\in\mathfrak{h}^*$, put $V_\lambda=\{v\in V\mid\rho(H)v=\lambda(H)v$ for all $H\in\mathfrak{h}\}$. Then V_λ is a linear subspace of V, and λ is called a **weight** of the representation ρ (relative to $\mathfrak{h}$) if $V_\lambda\neq\{0\}$; then $\dim V_\lambda$ is called the **multiplicity** of the weight λ. The set of all weights of ρ is a finite W-stable subset of $\mathfrak{h}^*_{\mathbf{R}}$. Denote this set by $\{\lambda_1,\ldots,\lambda_r\}$. Then V is decomposed into a direct sum as follows: $V=V_{\lambda_1}+\ldots+V_{\lambda_r}$. The maximum element among $\{\lambda_1,\ldots,\lambda_r\}$ with respect to the given ordering of $\mathfrak{h}^*_{\mathbf{R}}$ is called the **highest weight** of ρ.

The following two theorems are basic in determining irreducible representations.

(1) **Cartan's theorem**: Let Λ_1, Λ_2 be the highest weights of irreducible representations ρ_1, ρ_2 of $\mathfrak{g}$, respectively. Then ρ_1 is equivalent to ρ_2 if and only if $\Lambda_1=\Lambda_2$.

(2) **Cartan-Weyl theorem**: Let $\lambda\in\mathfrak{h}^*_{\mathbf{R}}$. Then there exists an irreducible representation ρ of $\mathfrak{g}$ which has λ as its highest weight if and only if (i) $2(\lambda,\alpha)/(\alpha,\alpha)\in\mathbf{Z}$ for every root $\alpha\in\Delta$ (such an element $\lambda\in\mathfrak{h}^*_{\mathbf{R}}$ is called an

integral form on $\mathfrak{h}$); and (ii) $w(\lambda) \leqslant \lambda$ for every $w \in W$ (such an element $\lambda \in \mathfrak{h}_{\mathbf{R}}^*$ is called **dominant**).

These theorems lead to the concept of the system of fundamental representations. Put $\alpha_i^* = 2\alpha_i/(\alpha_i, \alpha_i)$, and let $\Lambda_1, \ldots, \Lambda_l$ be the basis of $\mathfrak{h}_{\mathbf{R}}^*$ dual to $\alpha_1^*, \ldots, \alpha_l^*$ ($(\Lambda_i, \alpha_j^*) = \delta_{ij}$). Then the †free Abelian group $\Sigma \mathbf{Z}\Lambda_i$ coincides with the module P of all integral forms. An element $\Sigma m_i \Lambda_i$ ($m_i \in \mathbf{Z}$) is dominant if and only if $m_1 \geqslant 0, \ldots, m_l \geqslant 0$. Denote by P^+ the semigroup in P consisting of all dominant elements in P. For $j = 1, \ldots, l$, let (ρ_j, V_j) be the irreducible representation of $\mathfrak{g}$ which has Λ_j as its highest weight. The system $\{\rho_1, \ldots, \rho_l\}$ is called the **fundamental system of irreducible representations** associated with Π. The irreducible representation that has $\Lambda = \Sigma m_i \Lambda_i \in P^+$ as its highest weight is constructed as follows: Put $V_j^m = V_j \otimes \ldots \otimes V_j$ (mth tensor power of V_j) and $\tilde{V} = V_1^{m_1} \otimes \ldots \otimes V_l^{m_l}$. Then $\tilde{V}$ can be regarded as a representation space of $\mathfrak{g}$ in a natural manner. Let V be the smallest $\mathfrak{g}$-stable subspace of $\tilde{V}$ containing $\tilde{V}_\Lambda = (V_1)_{\Lambda_1}^{m_1} \otimes \ldots \otimes (V_l)_{\Lambda_l}^{m_l}$. Then V gives an irreducible representation of $\mathfrak{g}$ with highest weight Λ. Thus by decomposing $V_1^{m_1} \otimes \ldots \otimes V_l^{m_l}$, we get all irreducible representations of $\mathfrak{g}$. This is why $\{\rho_1, \ldots, \rho_l\}$ is called the fundamental system of irreducible representations.

X. Relation with Representations of Compact Lie Groups

Let G be a compact, connected, semisimple Lie group. Then every Cartan subalgebra $\mathfrak{h}$ of the Lie algebra $\mathfrak{g}$ of G is Abelian. We call $\dim \mathfrak{h}$ the **rank** of $\mathfrak{g}$ or of G. Let H be the connected Lie subgroup of G associated with $\mathfrak{h}$. Then H is a **maximal torus (toroidal subgroup)** of G. Furthermore, every maximal torus of G is conjugate to H in G. Also, every element of G is conjugate to an element of H in G. Let $N = N(H)$ be the †normalizer of H in G. Then every $\sigma \in N$ induces an automorphism $\mathrm{Ad}(\sigma)$ of $\mathfrak{g}$, which induces an automorphism (denoted by the same symbol $\mathrm{Ad}(\sigma)$) of the complexification $\mathfrak{g}^{\mathbf{C}}$ of $\mathfrak{g}$. We have $\mathrm{Ad}(\sigma)(\mathfrak{h}^{\mathbf{C}}) = \mathfrak{h}^{\mathbf{C}}$, and $\mathrm{Ad}(\sigma)$ preserves the root system of $\mathfrak{h}^{\mathbf{C}}$. Furthermore, the restriction of $\mathrm{Ad}(\sigma)$ on $(\mathfrak{h}^{\mathbf{C}})_{\mathbf{R}}^*$ is an element w_σ of the Weyl group W of $\mathfrak{g}^{\mathbf{C}}$ relative to $\mathfrak{h}^{\mathbf{C}}$. Via this mapping $\sigma \rightarrow w_\sigma$, we have $N/H \cong W$. Thus we identify N/H with W.

Since $(\mathfrak{h}^{\mathbf{C}})_{\mathbf{R}} = \sqrt{-1}\,\mathfrak{h}$, we can define a lexicographic linear ordering using a basis of $\mathfrak{h}$. Fix such an ordering. Now every representation ρ of G over a complex vector space V induces a representation $d\rho$ of $\mathfrak{g}$ over V, where $d\rho$ is the differential of ρ (→ 248 Lie Groups). Then we have a representation $d\rho$ of $\mathfrak{g}^{\mathbf{C}}$ over V. We also call a weight λ relative to $\mathfrak{h}^{\mathbf{C}}$ of the representation $d\rho$ of $\mathfrak{g}^{\mathbf{C}}$ a weight of the representation ρ of G relative to H. Representations ρ_1, ρ_2 of G are equivalent if and only if representations $d\rho_1, d\rho_2$ of $\mathfrak{g}^{\mathbf{C}}$ are equivalent. Denote by Λ_1, Λ_2 the highest weights of $d\rho_1, d\rho_2$, respectively. Then ρ_1 is equivalent to ρ_2 if and only if $\Lambda_1 = \Lambda_2$.

The module P of all integral forms in $(\mathfrak{h}^{\mathbf{C}})_{\mathbf{R}}^*$ coincides with the set of all elements in $(\mathfrak{h}^{\mathbf{C}})_{\mathbf{R}}^*$ that are weights (relative to $\mathfrak{h}^{\mathbf{C}}$) of some representation of $\mathfrak{g}^{\mathbf{C}}$. Denote by P_G the subset of P consisting of all elements in P that are weights (relative to H) of some representation of G. Then P_G is a submodule of P such that $[P : P_G] < \infty$. Put $P_G^+ = P_G \cap P^+$. Then the mapping (representation ρ)→(the highest weight Λ of ρ) induces a bijective mapping from the set of all classes of irreducible representations of G onto P_G^+.

The exponential mapping $\exp : \mathfrak{h} \rightarrow H$ is a surjective homomorphism from $\mathfrak{h}$ to H. The kernel Γ_G of this homomorphism is a †lattice group in $\mathfrak{h}$ of rank $l (= \dim \mathfrak{h})$; a basis $H_1, \ldots, H_l$ of Γ_G over $\mathbf{Z}$ is also a basis of $\mathfrak{h}$ over $\mathbf{R}$. Now define linear forms $\lambda_1, \ldots, \lambda_l$ on $\mathfrak{h}$ by $(\lambda_i, H_j) = \delta_{ij}$. Then we have

$$P_G = 2\pi\sqrt{-1}\,\sum \mathbf{Z}\lambda_j.$$

In other words, an element $\lambda \in (\mathfrak{h}^{\mathbf{C}})_{\mathbf{R}}^*$ is in P_G if and only if $\lambda(H) \in 2\pi\sqrt{-1}\,\mathbf{Z}$ for every element H in Γ_G. This characterization of P_G also characterizes $P_G^+ = P_G \cap P^+$. In particular, G is simply connected $\Leftrightarrow P = P_G \Leftrightarrow \Gamma_G = 2\pi\sqrt{-1}\,\Sigma \mathbf{Z}\alpha_i^*$, where $\alpha_1, \ldots, \alpha_l$ are simple roots and $\alpha_i^* = 2\alpha_i/(\alpha_i, \alpha_i)$ for $i = 1, \ldots, l$. We have also: $G = I(\mathfrak{g})$ = the group of inner automorphisms of $\mathfrak{g} \Leftrightarrow P_G = \Sigma \mathbf{Z}\alpha_i \Leftrightarrow \Gamma_G = 2\pi\sqrt{-1}\,\Sigma \mathbf{Z}\varepsilon_i$, where $\varepsilon_1, \ldots, \varepsilon_l$ are elements of $(\mathfrak{g}^{\mathbf{C}})_{\mathbf{R}}^*$ defined by $(\alpha_i, \varepsilon_j) = \delta_{ij}$. In general, we have $P \supset P_G \supset \Sigma \mathbf{Z}\alpha_i, 2\pi\sqrt{-1}\,\Sigma \mathbf{Z}\alpha_i^* \subset \Gamma_G \subset 2\pi\sqrt{-1}\,\mathbf{Z}\varepsilon_i$. Furthermore, the †fundamental group $\pi_1(G)$ of G is given by $P/P_G \cong \Gamma_G/(2\pi\sqrt{-1}\,\Sigma \mathbf{Z}\alpha_i^*)$. Also, the kernel of the adjoint representation $G \rightarrow I(G) = I(\mathfrak{g})$ is given by $P_G/\Sigma \mathbf{Z}\alpha_i \cong (2\pi\sqrt{-1}\,\Sigma \mathbf{Z}\varepsilon_i)/\Gamma_G$.

Y. Invariant Measure on G

Keeping the definitions in this section the same as in the previous section, every root α defines a representation $h \rightarrow \chi_\alpha(h)$ of the group H of dimension 1 over $(\mathfrak{g}^{\mathbf{C}})_\alpha$ as follows: Let E_α be a basis of $(\mathfrak{g}^{\mathbf{C}})_\alpha$. Then $\mathrm{Ad}(h)E_\alpha = \chi_\alpha(h)E_\alpha$. We have $\chi_\alpha(h) = e^{\alpha(X)}$ if $h = \exp X$ ($X \in \mathfrak{h}$), i.e., $\chi_\alpha \circ \exp = e^\alpha$. We let e^α stand for $\chi_\alpha : e^\alpha(h) = \chi_\alpha(h)$, $h \in H$. Now let dg, dh, dm be the †invariant measures on G, H, $M =$

G/H, respectively, normalized by

$$\int_G dg = 1, \quad \int_H dh = 1, \quad \int_M dm = 1.$$

Then for every continuous function f on G, we have the following formula:

$$\int_G f(g)\,dg = \frac{1}{w}\int_H \left(\int_M f(m,h)\,dm\right)\Omega(h)\,dh, \tag{1}$$

where w is the order of the Weyl group W and $f(m,h)$ is a function on $M \times H$ defined by $f(m,h)=f(ghg^{-1})$, $m=gH$. Note that $f(m,h)$ is well defined. Finally, $\Omega(h)$ is a function on H defined by

$$\Omega(h) = \prod_{\alpha\in\Delta}\left(e^{\alpha(X)/2}-e^{-\alpha(X)/2}\right)$$

for $h=\exp X$ ($X \in \mathfrak{h}$), and $\Omega(h)$ is called the **density** on H. Denote by $D(h)$ the same product as $\Omega(h)$, letting α range over Δ^+. Then we have $\Omega(h)=D(h)\overline{D(h)}=|D(h)|^2 \geqslant 0$. In particular, if f is a †class function on G (i.e., $f(xyx^{-1})=f(y)$ for every $x, y \in G$), $f(m,h)=f(h)$. Hence (1) is simplified into the following integral formula for class functions:

$$\int_G f(g)\,dg = \frac{1}{w}\int_H f(h)\Omega(h)\,dh. \tag{2}$$

Z. Weyl's Character Formula

Keeping the definitions in this section the same as those in the previous section, let (ρ, V) be an irreducible representation of G, and let χ_ρ be the character of ρ ($\chi_\rho(g)=\operatorname{tr}\rho(g)$). Let Λ be the highest weight of ρ. Then since Λ determines ρ up to equivalence, χ_ρ must be determined by Λ. In fact, χ_ρ is given via Λ by **Weyl's character formula** (3). (Note that $G=\bigcup gHg^{-1}$. Hence χ_ρ is determined by its restriction on H.) Now for $h \in H$, we have

$$\chi_\rho(h) = \frac{\xi_{\Lambda+\delta}(h)}{\xi_\delta(h)}, \tag{3}$$

where ξ_λ ($\lambda \in P$) is the following alternating sum:

$$\xi_\lambda(h) = \sum_{w\in W}\det(w)e^{(w(\lambda))(X)}, \quad h=\exp X. \tag{4}$$

Finally, δ in (3) is given by

$$\delta = \frac{1}{2}\sum_{\alpha\in\Delta^+}\alpha,$$

that is, δ is the half-sum of the positive roots. In particular, we have $\xi_\delta(h)=D(h)$. Denote by $m_\Lambda(\lambda)$ the multiplicity $\dim V_\lambda$ of a weight λ of ρ. Then we have $\chi_\rho(h)=\Sigma_\lambda m_\Lambda(\lambda)e^\lambda(h)$. Furthermore, we have $m_\Lambda(\Lambda)=1$, $m_\Lambda(\lambda)=m_\Lambda(w(\lambda))$ ($w \in W$), and $m_\Lambda(\lambda)$ is given by **Kostant's formula**:

$$m_\Lambda(\lambda) = \sum_{w\in W}\det(w)P(w(\Lambda+\delta)-(\lambda+\delta)), \tag{5}$$

where for each $\mu \in P$, $P(\mu)$ is the number of ways μ can be expressed as a sum of positive roots. Thus $P(\mu)$ is the number of nonnegative integral solutions $\{k_\alpha\}$ of $\mu=\Sigma_{\alpha\in\Delta^+}k_\alpha\alpha$.

Now suppose that (ρ_1, V_1), (ρ_2, V_2) are irreducible representations of G. Their tensor product $(\rho_1\otimes\rho_2, V_1\otimes V_2)$ is decomposed into a direct sum of irreducible constituents: $\rho_1\otimes\rho_2=\Sigma m(\mu)\rho_\mu$, where ρ_μ is the irreducible representation of G which has μ as its highest weight and $m(\mu)$ is the multiplicity of ρ_μ in $\rho_1\otimes\rho_2$. Then $m(\mu)$ is given by **Steinberg's formula**:

$$m(\mu) = \sum_{w\in W}\sum_{w'\in W}\det(ww')P(w(\Lambda_1+\delta) + w'(\Lambda_2+\delta)-(\mu+2\delta)), \tag{6}$$

where $\rho_1=\rho_{\Lambda_1}$, $\rho_2=\rho_{\Lambda_2}$. The partition function $P(\mu)$ that appears in (5) and (6) satisfies the following recursive formula (B. Kostant):

$$P(\mu) = -\sum_{w\in W, w\neq 1}\det(w)P(\mu-(\delta-w(\delta))). \tag{7}$$

References

[1] Séminaire "Sophus Lie," Ecole Norm. Sup., 1954–1955.
[2] Y. Matsushima, Rîkan ron (Japanese; Theory of Lie algebras), Kyôritu, 1957.
[3] N. Bourbaki, Eléments de mathématique, Groupe et algèbres de Lie., Actualités Sci. Ind., Hermann, ch. 1–3, 1960; ch. 4–6, 1968.
[4] N. Jacobson, Lie algebras, Interscience, 1962.
[5] S. Helgason, Differential geometry and symmetric spaces, Academic Press, 1962.
[6] E. Cartan, Oeuvres complètes pt. I, Gauthier-Villars, 1952.
[7] C. Chevalley, Theory of Lie groups I, Princeton Univ. Press, 1946; Théorie des groupes de Lie, Actualités Sci. Ind., Hermann, II, 1951; III, 1954.
[8] I. Kaplansky, Lie algebras, Lectures on modern mathematics, T. L. Saaty (ed.), John Wiley, 1963, vol. 1, p. 115–132.
[9] H. Weyl, Theorie der Darstellung kontinuierlicher halb-einfacher Gruppen durch linearen Transformationen I, II, III, Math. Z., 23 (1925), 271–309; 24 (1926), 328–376; 377–395 (Gesammelte Abhandlungen II, Springer, 1968, 543–647).
[10] H. Weyl, The classical groups, Princeton Univ. Press, revised edition, 1946.

[11] I. D. Ado, Über die Darstellung von Lieschen Gruppen durch linearen Substitutionen, Bull. Soc. Physico-Math. Kazan, (3) 7 (1934–1935), 3–43.
[12] F. R. Gantmacher, On the classification of real simple Lie groups Mat. Sb., 5 (47) (1939), 217–250.
[13] Е. Б. Дынкин (E. B. Dynkin), Структура полупростых алгебр, Uspehi Mat. Nauk (N.S.), 2, no. 4 (1947), 59–127; English translation, The structure of semi-simple Lie algebras, Amer. Math. Soc. Transl., 17 (1950).
[14] R. Bott, Homogeneous vector bundles, Ann. of Math., (2) 66 (1957), 203–248.
[15] I. Satake, On representations and compactifications of symmetric Riemannian spaces, Ann. of Math., (2) 71 (1960), 77–110.
[16] H. Freudenthal, Oktaven, Ausnahmegruppen und Oktavengeometrie, Mathematisch Instituut der Rijksuniversiteitte, Utrecht, mimeographed note, 1951.
[17] S. Araki, On root systems and an infinitesimal classification of irreducible symmetric spaces, J. Math. Osaka City Univ., 13 (1962), 1–34.
[18] J.-P. Serre, Lie algebras and Lie groups, Benjamin, 1965.
[19] J.-P. Serre, Algèbres de Lie semi-simples complexes, Benjamin, 1966.
Also → references for [4], [5].

248 (IV.9) Lie Groups

A. Definitions

A set G is called a **Lie group** if there is given on G a structure satisfying the following three axioms: (i) G is a group; (ii) G is a †paracompact, †real analytic manifold (G need not be connected); and (iii) the mapping $G \times G \rightarrow G$ defined by $(x,y) \rightarrow xy^{-1}$ is †real analytic (→ 108 Differentiable Manifolds).

For simplicity, real analyticity is denoted by C^ω: for example, C^ω-functions, C^ω-mappings. If we replace real analyticity by †complex analyticity in axioms (ii) and (iii), then we have the axioms (i), (ii′), and (iii′) of a **complex Lie group**. We consider the real analytic case, since the complex analytic case can be dealt with similarly.

Every element σ of G defines a mapping $G \rightarrow G$ given by $x \rightarrow \sigma x$ ($x \rightarrow x\sigma$), denoted by L_σ (R_σ) and called the **left (right) translation** of G by σ. L_σ, R_σ are automorphisms of G as a C^ω-manifold. Therefore, given a †vector field X on G, a †differential form ω on G, or, in general, a †tensor field T on G, we can apply L_σ and R_σ in a natural manner to the given tensor field and obtain a tensor field $L_\sigma T$, $R_\sigma T$ on G. A tensor field T on G is called **left (right) invariant** if $L_\sigma T = T$ ($R_\sigma T = T$) for every $\sigma \in G$. A tensor field on G is of class C^ω if it is left or right invariant.

B. Lie Algebras of Lie Groups

Let G be a Lie group. Then the set $\mathfrak{X}(G)$ of all C^ω-vector fields on G has the structure of a vector space over the real number field $\mathbf{R}$. Furthermore, with respect to the bracket operation $[X, Y] = XY - YX$ ($X, Y \in \mathfrak{X}(G)$), $\mathfrak{X}(G)$ forms a †Lie algebra over $\mathbf{R}$. Denote by $\mathfrak{g}$ the subset of $\mathfrak{X}(G)$ consisting of all left invariant vector fields on G. Then $\mathfrak{g}$ is a †subalgebra of the Lie algebra $\mathfrak{X}(G)$. Thus $\mathfrak{g}$ is also a Lie algebra over $\mathbf{R}$. This Lie algebra $\mathfrak{g}$ is called the **Lie algebra of the Lie group** G. The linear mapping from $\mathfrak{g}$ into the †tangent space $T_e(G)$ of G at the identity element e given by $X \rightarrow X_e$ is bijective. Hence $\dim \mathfrak{g} = \dim G$. The Lie algebra $\mathfrak{g}$ is often identified with $T_e(G)$ via this bijection.

C. Simply Connected Covering Lie Groups

For any finite-dimensional Lie algebra $\mathfrak{g}$ over $\mathbf{R}$, there exists a †connected Lie group G that has $\mathfrak{g}$ as its Lie algebra. Such Lie groups are all †locally isomorphic. Among these groups, there exists a †simply connected one that is unique up to isomorphism. This group is called the **simply connected covering Lie group** of the Lie algebra $\mathfrak{g}$.

D. Lie Subgroups

A subgroup H of a Lie group G is called a **Lie subgroup** if (i) H has the structure of a Lie group and (ii) the inclusion mapping $\varphi: H \rightarrow G$ is a C^ω-mapping, and the †differential $d\varphi$ is injective at every point of H (i.e., as a manifold, H is a †submanifold of G). Moreover, if H is a connected manifold, it is called a **connected Lie subgroup**. For a subgroup H of a Lie group G, there exists at most one structure of a Lie group on H that makes H a connected Lie subgroup. (If H is not assumed to be connected, then this uniqueness statement is not generally true.) A subgroup H of a Lie group G is a connected Lie subgroup if and only if H is arcwise connected (M. Kuranishi and H. Yamabe).

Let H be a Lie subgroup of a Lie group G. Then the tangent space $T_e(H)$ is a subspace of $T_e(G)$. Under the isomorphism $T_e(G) \cong \mathfrak{g}$ mentioned in Section B, the subspace $T_e(H)$ corresponds to $\mathfrak{h}$, which is a †subalgebra of the Lie algebra $\mathfrak{g}$. We call $\mathfrak{h}$ the **subalgebra of $\mathfrak{g}$ associated with the Lie subgroup** H, and $\mathfrak{h}$ can be identified with the Lie algebra of H in a natural manner. By this mapping $H \to \mathfrak{h}$, we have a bijection from the set of all connected Lie subgroups of G onto the set of all subalgebras of $\mathfrak{g}$. For example, the connected Lie subgroup G' of G corresponding to the †derived algebra $\mathfrak{g}'$ of $\mathfrak{g}$ is the †commutator subgroup of G. The Lie algebra of a normal subgroup of G is an †ideal of $\mathfrak{g}$. Conversely, if the Lie algebra $\mathfrak{h}$ of a connected Lie subgroup H of a connected Lie group G is an ideal of $\mathfrak{g}$, then H is a normal subgroup of G.

A connected Lie group G is called **semisimple, simple, solvable, nilpotent**, or **Abelian** (or **commutative**), respectively, whenever the Lie algebra $\mathfrak{g}$ is (→ 247 Lie Algebras). This definition of solvability, nilpotency, and commutativity agrees with the corresponding definition in which G is regarded as an abstract group.

E. Closed Subgroups

The topology of a Lie subgroup H (which we call the **inner topology** of H) as a submanifold of a Lie group G need not coincide with the relative topology of H regarded as a subspace of a topological space G. The inner topology of H coincides with the relative topology of H if and only if H is closed in G. Conversely, for every closed subgroup H of G, H has the structure of a Lie subgroup such that the inner topology and relative topology coincide (E. Cartan). Moreover, such a structure of a Lie subgroup on H is unique. We always regard closed subgroups of a Lie group as Lie subgroups in this sense. Also, we denote the Lie algebras of Lie groups $G, H, \ldots$ by the corresponding lower-case German letters $\mathfrak{g}$, $\mathfrak{h}, \ldots$.

F. Homogeneous Spaces

Let G be a Lie group, and let H be a closed subgroup of G. Then the †quotient topological space (coset space) $M = G/H$ of the topological group G modulo H has the structure of a C^ω-manifold such that the canonical mapping $\pi: G \to M$ and the action of G on $M: G \times M \to M$ (defined by $(g, xH) \to gxH$) are both C^ω-mappings. Furthermore, such a C^ω-manifold structure on M is unique. The C^ω-manifold M thus obtained is called the **homogeneous space** of G over H (→ 201 Homogeneous Spaces). Put $\pi(e) = p$. Then $\mathfrak{h}$ is the kernel of the differential mapping $T_e(G) = \mathfrak{g} \to T_p(M)$. Thus we can identify the tangent space $T_p(M)$ of $M = G/H$ at p with the quotient linear space $\mathfrak{g}/\mathfrak{h}$ via this differential mapping.

G. Quotient Lie Groups

Suppose that H is a closed, normal subgroup of a Lie group G. Then G/H has the structure of a quotient group together with the structure of a manifold as a homogeneous space. Now G/H is a Lie group with respect to these two structures. The Lie group G/H is called the **quotient Lie group** of G over H. In this case, $\mathfrak{h}$ is an †ideal of $\mathfrak{g}$, and the quotient Lie algebra $\mathfrak{g}/\mathfrak{h}$ is isomorphic to the Lie algebra of the Lie group G/H.

H. Direct Product of Lie Groups

Let G_1, G_2 be Lie groups. Then the direct product $G_1 \times G_2$ satisfies axioms (i), (ii), and (iii) of a Lie group (→ Section A) in a natural manner. The Lie group $G_1 \times G_2$ thus obtained is called the **direct product** of the Lie groups G_1, G_2. The Lie algebra of $G_1 \times G_2$ can be identified with the direct sum of $\mathfrak{g}_1, \mathfrak{g}_2$.

I. Cartan Subgroups

A subgroup H of a group G is called a **Cartan subgroup** of G if H is a maximal nilpotent subgroup of G and moreover, for every subgroup H_1 of H of finite index in H, $[N(H_1): H_1] < \infty$, where $N(H_1) = \{g \in G \mid gH_1g^{-1} = H_1\}$ is the †normalizer of H_1 in G. A connected Lie group G always has a Cartan subgroup. Furthermore, every Cartan subgroup H of G is closed; hence H is a Lie subgroup. The Lie algebra $\mathfrak{h}$ of H is a †Cartan subalgebra of the Lie algebra $\mathfrak{g}$. This mapping $H \to \mathfrak{h}$ is a bijection from the set of all Cartan subgroups of G onto the set of all Cartan subalgebras of $\mathfrak{g}$ [7].

J. Borel Subgroups

Let G be a complex connected semisimple Lie group. A maximal connected solvable Lie subgroup of G is called a **Borel subgroup** of G. Any two Borel subgroups are conjugate in G. A subgroup P of G is called a **parabolic subgroup** if P contains a Borel subgroup of G. Every parabolic subgroup is a connected closed Lie subgroup of G.

K. Simple Examples

Let V be a finite-dimensional vector space over **R**. Let $\mathfrak{L}(V)$ be the †associative algebra of linear endomorphisms of V. The †general linear group $GL(V)=\{x\in\mathfrak{L}(V)\mid \det x\neq 0\}$ over V is a Lie group. The Lie algebra of $GL(V)$ is denoted by $\mathfrak{gl}(V)$. We can identify $\mathfrak{gl}(V)$ with $\mathfrak{L}(V)$, and we have $[X,Y]=XY-YX$ for all $X,Y\in\mathfrak{gl}(V)$. If $\dim V=n$, $GL(V)$ may be identified with the group $GL(n,\mathbf{R})$ of all real nonsingular $n\times n$ matrices. Then $\mathfrak{gl}(V)$ may be identified with the Lie algebra $\mathfrak{gl}(n,\mathbf{R})$ of all $n\times n$ real matrices.

The following are examples of (closed) Lie subgroups of the general linear group: the †special linear group $SL(n,\mathbf{R})$, whose Lie algebra is $\{X\in\mathfrak{gl}(n,\mathbf{R})\mid \operatorname{tr}X=0\}$; the †unitary group $U(n)$, whose Lie algebra is $\{X\in\mathfrak{gl}(n,\mathbf{C})\mid {}^t X+\bar{X}=0\}$; the †orthogonal group $O(n)$, whose Lie algebra is $\{X\in\mathfrak{gl}(n,\mathbf{R})\mid {}^t X+X=0\}$; and the †symplectic group $Sp(n)$, whose Lie algebra is $\{X\in\mathfrak{gl}(2n,\mathbf{C})\mid JX+{}^t XJ=0, {}^t X+\bar{X}=0\}$, where

$$J=\begin{pmatrix}0 & I\\ -I & 0\end{pmatrix}$$

and I is the $n\times n$ unit matrix.

Examples of complex Lie groups are: the complex †general linear group $GL(n,\mathbf{C})$, whose Lie algebra is $\mathfrak{gl}(n,\mathbf{C})$; the complex †orthogonal group $O(n,\mathbf{C})$, whose Lie algebra is $\{X\in\mathfrak{gl}(n,\mathbf{C})\mid {}^t X+X=0\}$; and the complex †symplectic group $Sp(n,\mathbf{C})$, whose Lie algebra is $\{X\in\mathfrak{gl}(2n,\mathbf{C})\mid JX+{}^t XJ=0\}$, where J is the matrix given before.

L. Compact Simple Lie Groups

If a connected Lie group G has a †compact real Lie algebra, then G is compact. Thus for each compact real simple Lie algebra $\mathfrak{g}$, the simply connected covering Lie group G is compact, and the center of G is a finite group. The connected compact Lie groups $SU(l+1)$, $SO(2l+1)$, $Sp(l)$, $SO(2l)$ have, respectively, the compact real simple Lie algebras A_l $(l\geqslant 1)$, B_l $(l\geqslant 2)$, C_l $(l\geqslant 3)$, D_l $(l\geqslant 4)$ as their Lie algebras; these groups are called **classical compact simple Lie groups**. $Sp(n)$ $(n\geqslant 2)$ and $SU(n)$ $(n\geqslant 2)$ are simply connected, but $SO(n)$ $(n=3$ or $n\geqslant 5)$ is not, and has the fundamental group of order 2. The universal covering group of $SO(n)$ (i.e., the simply connected, connected Lie group that has the same Lie algebra as $SO(n)$) is $Spin(n)$. The connected compact Lie groups that have E_l $(l=6,7,8)$, F_4, or G_2 as their Lie algebras are called **exceptional compact simple Lie groups**. The simply connected covering Lie groups of E_8, F_4, G_2 have the identity group $\{e\}$ as their center. Hence they coincide with their †adjoint groups. The simply connected covering Lie groups E_6, E_7 have as their centers the groups of orders 3 and 2, respectively. Hence their adjoint groups are not simply connected.

M. Complex Simple Lie Groups

For a complex Lie group G, the Lie algebra $\mathfrak{g}$ is defined similarly to the case of real Lie groups. Then $\mathfrak{g}$ is a Lie algebra over the complex number field **C** and is called the **complex Lie algebra of the complex Lie group** G. The complex connected Lie groups $SL(l+1,\mathbf{C})$, $SO(2l+1,\mathbf{C})$, $Sp(l,\mathbf{C})$, $SO(2l,\mathbf{C})$, which have the †classical complex simple Lie algebras A_l, B_l, C_l, D_l as their Lie algebras, respectively, are called **classical complex simple Lie groups**. Complex connected Lie groups that have †exceptional complex simple Lie algebras E_l $(l=6,7,8)$, F_4, G_2 as their Lie algebras are called **exceptional complex simple Lie groups**.

N. Homomorphisms

A mapping $\varphi: G_1\to G_2$ from a Lie group G_1 into a Lie group G_2 is called an **analytic homomorphism** (**C^ω-homomorphism**) if (i) φ is a homomorphism between groups and (ii) φ is a C^ω-mapping between manifolds. Moreover, if φ is bijective and φ^{-1} is also a C^ω-mapping, then φ is called an **analytic isomorphism** (**C^ω-isomorphism**). Two Lie groups G_1, G_2 are said to be **isomorphic** as Lie groups if there exists a C^ω-isomorphism from G_1 onto G_2; we denote this situation by $G_1\cong G_2$. Now let $\varphi: G_1\to G_2$ be a C^ω-homomorphism. Then the †differential $(d\varphi)_{e_1}: T_{e_1}(G_1)\to T_{e_2}(G_2)$ induces a Lie algebra homomorphism $d\varphi:\mathfrak{g}_1\to\mathfrak{g}_2$. The kernel of $d\varphi$ is the Lie algebra of the kernel of φ. Suppose that G_1 is connected and simply connected. Then for every Lie group G_2 and every Lie algebra homomorphism $\psi:\mathfrak{g}_1\to\mathfrak{g}_2$, there exists a unique C^ω-homomorphism $\varphi: G_1\to G_2$ such that $\psi=d\varphi$. We can replace condition (ii) in the definition of C^ω-homomorphism by the following weaker one: (ii′) φ is a continuous mapping. In particular, for a given topological group G, there exists at most one structure of a Lie group on G preserving the group structure and the topology. A topological group G admits a structure of a Lie group (preserving the group structure and the topology) if and only if G is a †locally Euclidean topological group (→ 406 Topological Groups N).

O. Representations

Let G be a Lie group, and let V be a finite-dimensional vector space over **C** (**R**). Then a continuous (hence C^ω-) homomorphism $\rho: G \to GL(V)$ is called a **complex (real) representation** of G. The linear space V is called the **representation space** of ρ, and $\dim V$ is called the **degree** of ρ. To be more precise, a representation is denoted by (ρ, V) instead of ρ. The function $\chi_\rho: G \to \mathbf{C}$ defined by $\chi_\rho(g) = \operatorname{tr}\rho(g)$ $(g \in G)$ is called the **character** of ρ (→ 358 Representations). The representation (ρ, V) of G gives rise to a representation $(d\rho, V)$ of $\mathfrak{g}$. Suppose G is connected. Then two representations (ρ_i, V_i) $(i=1,2)$ of G are †equivalent if and only if the representations $d\rho_1$ and $d\rho_2$ of $\mathfrak{g}$ are equivalent. For the †direct sum of representations and the contragredient representation, we have $d(\rho_1 + \rho_2) = d\rho_1 + d\rho_2$, $d({}^t\rho^{-1}) = -{}^t(d\rho)$. For the †tensor product of representations, we have $(d(\rho_1 \otimes \rho_2))(X) = (d\rho_1)(X) \otimes 1_{V_2} + 1_{V_1} \otimes (d\rho_2)(X)$. (For representations of compact Lie groups → 247 Lie Algebras W.)

P. Adjoint Representations

An element σ of a Lie group G defines an analytic automorphism $\varphi_\sigma: x \to \sigma x \sigma^{-1}$ of G. The †differential $\operatorname{Ad}(\sigma)$ of φ_σ is an automorphism of the Lie algebra $\mathfrak{g}$. Since $\operatorname{Ad}(\sigma) \in GL(\mathfrak{g})$, we have a representation $\sigma \to \operatorname{Ad}(\sigma)$ of G whose representation space is $\mathfrak{g}$. This is called the **adjoint representation** of G, and the image $\operatorname{Ad}(G)$ of G is called the **adjoint group** of G. $\operatorname{Ad}(G)$ is a Lie subgroup of $GL(\mathfrak{g})$. The †center Z of G is a closed subgroup of G, and $G/Z \cong \operatorname{Ad}(G)$. The analytic homomorphism $\sigma \to \operatorname{Ad}(\sigma)$ from G into $GL(\mathfrak{g})$ gives rise to a Lie algebra homomorphism $\mathfrak{g} \to \mathfrak{gl}(\mathfrak{g})$ by taking the differential of $\sigma \to \operatorname{Ad}(\sigma)$. Denote this Lie algebra homomorphism by $X \to \operatorname{ad}(X)$ $(X \in \mathfrak{g})$. Then $\operatorname{ad}(X)Y = [X, Y]$ for every X, Y in $\mathfrak{g}$. Thus $\operatorname{ad}: \mathfrak{g} \to \mathfrak{gl}(\mathfrak{g})$ coincides with the †adjoint representation of the Lie algebra $\mathfrak{g}$. If G is connected, then the Lie subalgebra $\operatorname{ad}(\mathfrak{g})$ (the †adjoint Lie algebra of $\mathfrak{g}$) of $\mathfrak{gl}(\mathfrak{g})$ is associated with the connected Lie subgroup $\operatorname{Ad}(G)$ of $GL(\mathfrak{g})$. Thus the adjoint group $\operatorname{Ad}(G)$ coincides with the group $I(\mathfrak{g})$ of †inner automorphisms of $\mathfrak{g}$ (the †adjoint group of $\mathfrak{g}$).

If G is connected and semisimple, $\mathfrak{g}$ is also semisimple. Furthermore, we have $\mathfrak{D}(\mathfrak{g}) = \operatorname{ad}(\mathfrak{g})$. Hence the adjoint group $\operatorname{Ad}(G) = I(\mathfrak{g})$ of G coincides with the connected component of the †automorphism group $A(\mathfrak{g})$ of $\mathfrak{g}$ containing the identity element. If G is compact or semisimple, then every representation of G is completely reducible.

Let G be a connected Lie group of dimension n, $\mathfrak{g}$ be the †Lie algebra of G, and $\operatorname{Ad}: G \to GL(\mathfrak{g})$ be the adjoint representation of G. Put $\det((t+1)I - \operatorname{Ad}(x)) = \sum_{j=0}^{n} t^j D_j(x)$ for every $x \in G$. Then each D_j is an analytic function on G, and $D_n = 1$. Let l be the least integer such that $D_l \neq 0$. Then $l = \operatorname{rank} G = \operatorname{rank} \mathfrak{g}$. An element x of G is called **regular (singular)** if $D_l(x) \neq 0$ $(D_l(x) = 0)$. The subset G^* of G consisting of all regular elements of G is open and dense in G. The subset $G - G^*$ of G consisting of all singular elements of G is of measure zero with respect to the left-invariant †Haar measure of G.

Now suppose that $\mathfrak{g}$ is †reductive. Then an element $x \in G$ is regular if and only if the centralizer $\mathfrak{z}_x = \{X \in \mathfrak{g} \mid \operatorname{Ad}(x)X = 0\}$ is a †Cartan subalgebra of $\mathfrak{g}$. Furthermore, if $x \in G$ is regular, then $\operatorname{Ad}(x)$ is a semisimple linear endomorphism of $\mathfrak{g}$.

Q. Canonical Coordinates

With each element X of the Lie algebra $\mathfrak{g}$ of the Lie group G there is associated a **one-parameter subgroup** of G, i.e., a continuous homomorphism $t \to \varphi(t)$ from the additive group $\mathbf{R}$ of real numbers into G, such that $d\varphi(X_0) = X$, where $X_0 = d/dt$ is the basis of the Lie algebra of $\mathbf{R}$. Furthermore, the continuous homomorphism φ is unique. Putting $\varphi(1) = \exp X$, we define the **exponential mapping** $\exp: \mathfrak{g} \to G$. Then we have $\varphi(t) = \exp tX$ for every $t \in \mathbf{R}$. In particular, for the case $G = GL(n, \mathbf{C})$, $\mathfrak{g} = \mathfrak{gl}(n, \mathbf{C})$, we have $\exp X = \sum X^m/m!$ for every $X \in \mathfrak{g}$. Thus $\exp X$ coincides with the usual exponential mapping of matrices (→ 269 Matrices). The mapping $\exp: \mathfrak{g} \to G$ is a C^ω-mapping whose differential at $X = 0$ is a bijection from $T_0(\mathfrak{g}) = \mathfrak{g}$ onto $T_e(G)$. Thus for each basis $X_1, \ldots, X_n$ of $\mathfrak{g}$, there exists a positive real number ε with the following property: $\{\exp(\sum x_i X_i) \mid |x_i| < \varepsilon\ (i = 1, \ldots, n)\}$ is an open neighborhood of the identity element e in G on which $\sigma = \exp(\sum x_i X_i) \to (x_1, \ldots, x_n)$ $(|x_i| < \varepsilon, i = 1, \ldots, n)$ is a †local coordinate system. These local coordinates are called the **canonical coordinates of the first kind** associated with the basis (X_i) of $\mathfrak{g}$. Similarly, we have a local coordinate system $\tau = (\exp x_1 X_1) \ldots (\exp x_n X_n) \to (x_1, \ldots, x_n)$ in a neighborhood of e; these $x_1, \ldots, x_n$ are called the **canonical coordinates of the second kind** associated with the basis (X_i) of $\mathfrak{g}$.

Let $\varphi: G_1 \to G_2$ be a continuous homomorphism from a Lie group G_1 into a Lie group G_2. Then $\varphi(\exp X) = \exp(d\varphi(X))$ for every $X \in \mathfrak{g}_1$. The Lie subalgebra $\mathfrak{h}$ associated with a connected Lie subgroup H of G is characterized using the exponential mapping as follows: $\mathfrak{h} = \{X \in \mathfrak{g} \mid \exp tX \in H \text{ for all } t \in \mathbf{R}\}$.

R. Multiplication Functions

Fix a basis $X_1, \ldots, X_n$ of the Lie algebra $\mathfrak{g}$ of a Lie group G. Let $(u_1, \ldots, u_n)$ be the canonical coordinates of the first kind associated with the basis (X_i). With respect to this coordinate system, let (x_i), (y_i), (z_i) be the coordinates of the elements σ, τ, $\sigma\tau$, respectively. Then each z_i is a real analytic function in $(x_1, \ldots, x_n; y_1, \ldots, y_n)$. The †Taylor expansion of z_i at $x_1 = x_2 = \ldots = y_1 = y_2 = \ldots = 0$ is given as follows: Put $S = \Sigma x_i X_i$, $T = \Sigma y_i X_i$. Then

$$z_i = \sum_{k,l=0}^{\infty} \frac{1}{k!l!}(S^k T^l u_i)_0$$

$$= x_i + y_i + (STu_i)_0 + (\text{terms of degree} \geqslant 3).$$

Furthermore, let (w_i) be the coordinates of $\tau^{-1}\sigma^{-1}\tau\sigma$. Then we have

$$w_i = ([T, S]u_i)_0 + (\text{terms of degree} \geqslant 3).$$

Let $\mathfrak{g}^*$ be the set of all **R**-valued linear forms on $\mathfrak{g}$, that is, $\mathfrak{g}^*$ is the dual vector space of $\mathfrak{g}$. We can identify the elements of $\mathfrak{g}^*$ with the left-invariant †differential forms of degree 1 on G. These forms are called **Maurer-Cartan differential forms**. Let $\omega_1, \ldots, \omega_n$ be a basis of $\mathfrak{g}^*$ which is dual to the basis $X_1, \ldots, X_n$ of $\mathfrak{g}$: $\omega_i(X_j) = \delta_{ij}$ $(1 \leqslant i, j \leqslant n)$. The †exterior derivative $d\omega_k$ of ω_k is a left-invariant differential form of degree 2 on G and is expressed as a linear combination of the †exterior products $\omega_i \wedge \omega_j$ as follows:

$$d\omega_k = -\frac{1}{2}\sum_{i,j} c_{ij}^k \omega_i \wedge \omega_j, \tag{1}$$

where (c_{ij}^k) are the †structural constants of $\mathfrak{g}$ with respect to the basis (X_i), i.e., $[X_i, X_j] = \Sigma_k c_{ij}^k X_k$.

Using the canonical coordinates of the first kind (u_i) associated with the basis (X_i), put $\omega_k = \omega_k(u, du) = \Sigma_j A_{kj}(u)\, du_j$. Then the matrix $\mathfrak{A} = (A_{kj}(u))$ is given by

$$\mathfrak{A} = I + \frac{\mathfrak{X}}{2!} + \frac{\mathfrak{X}^2}{3!} + \ldots, \tag{2}$$

where $\mathfrak{X} = (c_{ij}(u))$, $c_{ij}(u) = \Sigma_k c_{jk}^i u_k$, i.e., $-\mathfrak{X}$ is the matrix of $\mathrm{ad}(\Sigma u_k X_k)$ relative to the basis (X_i). Thus, in particular, if G is a nilpotent group, $\mathfrak{X}$ is a †nilpotent matrix.

Once the functions $A_{kj}(u)$ are known, then the functions $z_i = f_i(x_1, \ldots, x_n; y_1, \ldots, y_n)$ describing the multiplication in G (note that $\sigma \leftrightarrow (x_i)$, $\tau \leftrightarrow (y_i)$, $\sigma\tau \leftrightarrow (z_i)$) are obtained as follows. By the left-invariance of the ω_i, we have the following **Maurer-Cartan system of differential equations**:

$$\omega_i(z, dz) = \omega_i(y, dy), \quad 1 \leqslant i \leqslant n. \tag{3}$$

Regarding $x_1, \ldots, x_n$ as parameters, put $z_i = \varphi_i(y_1, \ldots, y_n)$. Then (3) is equivalent to

$$\sum_j A_{ij}(z) \frac{\partial \varphi_j}{\partial y_k} = A_{ik}(y), \quad 1 \leqslant i, k \leqslant n. \tag{3'}$$

Now (3′) is †completely integrable, by (1). By solving the system (3′) of differential equations under the initial conditions

$$\varphi_j(0, \ldots, 0) = x_j, \quad 1 \leqslant j \leqslant n, \tag{4}$$

we obtain the multiplication functions $z_i = f_i(x_1, \ldots, x_n; y_1, \ldots, y_n)$ [1].

S. Maximal Compact Subgroups

Any compact subgroup of a connected Lie group G is contained in a maximal compact subgroup of G. Any two maximal compact subgroups of G are conjugate in G. Let K be a maximal compact subgroup of G. Then K is connected and G is homeomorphic to the direct product of K with a Euclidean space $\mathbf{R}^m$ **(Cartan-Mal'cev-Iwasawa theorem)**.

T. Iwasawa Decomposition

Let G be a connected semisimple Lie group. Suppose that the center of G is a finite group. Let $\mathfrak{g} = \mathfrak{k} + \mathfrak{a} + \mathfrak{n}$ be an †Iwasawa decomposition of the Lie algebra $\mathfrak{g}$ of G (→ 247 Lie Algebras). Then the connected Lie subgroups K, A, N associated with $\mathfrak{k}$, $\mathfrak{a}$, $\mathfrak{n}$, respectively, are all closed subgroups of G. Furthermore, K is a maximal compact subgroup of G, A is isomorphic to the additive group $\mathbf{R}^s$ for a suitable s, and N is a simply connected nilpotent Lie group. The mapping $K \times A \times N \to G$ given by $(k, a, n) \to kan$ is bijective and is an isomorphism of analytic manifolds. The decomposition $G = KAN$ is called an **Iwasawa decomposition** of G. An Iwasawa decomposition is unique in the following sense: Let $G = K'A'N'$ be another Iwasawa decomposition. Then there exists an element g in G such that $gKg^{-1} = K'$, $gAg^{-1} = A'$, $gNg^{-1} = N'$ [10, 3].

U. Complexification of Compact Lie Groups

Let G be a compact Lie group. Denote by $C(G)$ the commutative associative algebra of all **C**-valued continuous functions defined on G relative to the usual multiplication in $C(G)$. Then for each $\sigma \in G$, L_σ (R_σ) acts on $C(G)$ by $L_\sigma f = f \circ L_\sigma$ $(R_\sigma f = f \circ R_\sigma)$. Now put $\mathfrak{o}(G) = \{f \in C(G) \mid \dim \Sigma_{\sigma \in G} \mathbf{C} L_\sigma f < \infty\}$. Then $\mathfrak{o}(G)$ is a subalgebra of $C(G)$, called the **representative ring** of G. Elements of $\mathfrak{o}(G)$ are called **representative functions** of G because for an element $f \in C(G)$, $f \in \mathfrak{o}(G)$ if and only if there exists a continuous matrix representation $\sigma \to (d_{ij}(\sigma))$ of G such that f is a $\mathbf{C}$-linear combination of the d_{ij}. For each $\sigma \in G$, L_σ and R_σ preserve $\mathfrak{o}(G)$. With respect to the uniform norm $\|f\|_\infty = \max_{\sigma \in G}|f(\sigma)|$ of $C(G)$, $\mathfrak{o}(G)$ is everywhere dense in $C(G)$ (†**Peter-Weyl theory**). This implies the existence of a faithful (=injective) representation $\rho: G \to GL(n, \mathbf{C})$ of G. Furthermore, $\mathfrak{o}(G)$ is a finitely generated algebra over $\mathbf{C}$. Thus there exists a representation $\sigma \to (d_{ij}(\sigma))$ such that $\mathfrak{o}(G)$ is generated over $\mathbf{C}$ by the d_{ij}. Such a representation is called a **generating representation**.

Denote the group of all automorphisms of the algebra $\mathfrak{o}(G)$ by A. Let $\tilde{G}$ be the centralizer in A of the subgroup $\{L_\sigma | \sigma \in G\}$ of A. Then we have a bijective mapping $\alpha \to \alpha'$ defined by $\alpha'(f) = (\alpha(f))(e)$ $(f \in \mathfrak{o}(G))$ from the group $\tilde{G}$ onto the set $\mathrm{Hom}_{\mathbf{C}}(\mathfrak{o}(G), \mathbf{C})$, which is an affine variety. Thus every generating representation $\rho: \sigma \to (d_{ij}(\sigma))_{1 \leqslant i,j \leqslant n}$ defines a faithful matrix representation $\tilde{\rho}$ of $\tilde{G}$ by $\tilde{\rho}(\alpha) = (\alpha'(d_{ij}))_{1 \leqslant i,j \leqslant n}$ using the bijection $\alpha \to \alpha'$ from $\tilde{G}$ onto $\mathrm{Hom}_{\mathbf{C}}(\mathfrak{o}(G), \mathbf{C})$. Furthermore, the image $\tilde{\rho}(\tilde{G})$ is an †algebraic subgroup of $GL(n, \mathbf{C})$ given by $\{\beta(d_{ij})_{1 \leqslant i,j \leqslant n} | \beta \in \mathrm{Hom}_{\mathbf{C}}(\mathfrak{o}(G), \mathbf{C})\}$. Thus $\tilde{G}$ has the structure of a †linear algebraic group, which is actually independent of the choice of a generating representation. (Hence $\tilde{G}$ also has the structure of a complex Lie group.) Now $\sigma \to R_\sigma$ defines an injective continuous homomorphism from G into $\tilde{G}$, and we may regard G as a subgroup of $\tilde{G}$. Then for $\alpha \in \tilde{G}$, α is in G if and only if $\alpha(\bar{f}) = \overline{\alpha(f)}$ for every f in $\mathfrak{o}(G)$ (**Tannaka's duality theorem**) (→ 71 Compact Groups). G is a maximal compact subgroup of $\tilde{G}$, and every maximal compact subgroup of $\tilde{G}$ is conjugate to G in $\tilde{G}$. The complex Lie algebra $\tilde{\mathfrak{g}}$ of $\tilde{G}$ is isomorphic to the complexification $\mathfrak{g}^{\mathbf{C}} = \mathbf{C} \otimes_{\mathbf{R}} \mathfrak{g}$ of the Lie algebra $\mathfrak{g}$ of G. $\tilde{G}$ is homeomorphic to the direct product of G with a Euclidean space $\mathbf{R}^N$, where $N = \dim G$. For a complex analytic function φ defined on $\tilde{G}$, $\varphi = 0 \Leftrightarrow \varphi|_G = 0$. In particular, $\tilde{G}$ is the closure of G relative to the †Zariski topology of $\tilde{G}$. The group $\tilde{G}$ is called the **Chevalley complexification** of G, which we denote by $G^{\mathbf{C}}$. Let G_1, G_2 be compact Lie groups, and let $\varphi: G_1 \to G_2$ be a continuous homomorphism. Then φ can be extended uniquely to a rational homomorphism $\varphi^{\mathbf{C}}: G_1^{\mathbf{C}} \to G_2^{\mathbf{C}}$. In particular, every complex representation (ρ, V) can be extended uniquely to a †rational representation $(\rho^{\mathbf{C}}, V)$ of $G^{\mathbf{C}}$. Every complex analytic representation $\tilde{\rho}$ of $G^{\mathbf{C}}$ is a rational representation, and $\tilde{\rho} = \rho^{\mathbf{C}}$, where $\rho = \tilde{\rho}|_G$. Thus $\tilde{\rho}$ is completely reducible. Also, we have a bijection from the classes of irreducible continuous representations of G onto the classes of irreducible complex analytic representations of $G^{\mathbf{C}}$. For a closed subgroup H of G, the complexification $H^{\mathbf{C}}$ of H coincides with the closure of H in $G^{\mathbf{C}}$ relative to the Zariski topology of $G^{\mathbf{C}}$.

If (ρ, V) is a generating representation of G, then $G^{\mathbf{C}} \cong \rho^{\mathbf{C}}(G^{\mathbf{C}}) \subset GL(V)$. Furthermore, $\rho^{\mathbf{C}}(G^{\mathbf{C}})$ is an algebraic subgroup of $GL(V)$ that is completely reducible on V. Conversely, let F be an algebraic subgroup of $GL(V)$ that is completely reducible on V. Then there exists a compact Lie group G such that $G^{\mathbf{C}} \cong F$ (as algebraic groups). If an algebraic subgroup F of $GL(n, \mathbf{C})$ satisfies ${}^t\bar{F} = F$, then $F \cong G^{\mathbf{C}}$, where $G = F \cap U(n)$. Now let F be a connected semisimple complex Lie group. Then $F \cong G^{\mathbf{C}}$ for every maximal compact subgroup G of F. In particular, F has a faithful representation.

For example, let $G \cong U(n)$, $O(n)$, $SO(n)$, $Sp(n)$, respectively. Then $G^{\mathbf{C}} \cong GL(n, \mathbf{C})$, $O(n, \mathbf{C})$, $SO(n, \mathbf{C})$, $Sp(n, \mathbf{C})$, respectively [1].

V. History

In the late 19th century Lie considered Lie groups (then called continuous groups) for the first time. His motivation was to treat the various geometries from a group-theoretic point of view (→ 147 Erlangen Program) and to investigate the relationship between differential equations and the group of transformations preserving their solutions. Lie groups were studied locally, and the notion of Lie algebras was introduced. It was observed that the properties of Lie groups are remarkably reflected by the properties of Lie algebras. Even in this early stage, the notions of solvable and semisimple Lie algebras were introduced together with proofs of basic properties. The Galois theory of linear differential equations (by E. Vessiot and others) is contained in Lie's theory. Early in the 20th century, the theory of infinite-dimensional Lie groups was studied by E. Cartan. After him, however, there was a long lull until it was taken up again by M. Kuranishi and others in the 1950s. We restrict ourselves to the consideration of finite-dimensional Lie groups.

From 1900–1930, E Cartan and H. Weyl obtained a complete classification of the semisimple Lie algebras and determined their representations and characters. They also devised useful methods of investigating the structure of these algebras, and did pioneering work on global structures of the underlying manifolds of Lie groups. After them (1930–1950), these results were systematized and refined by C. Chevalley, Harish-Chandra, and others. In the same period, K. Iwasawa [10] clarified Cartan's idea, showing that the only Lie groups that are topologically important are compact Lie groups. He also obtained the Iwasawa decomposition, which has become a basic tool in the study of semisimple Lie groups. At the same time, Iwasawa contributed to Hilbert's fifth problem, which seeks to characterize Lie groups among topological groups. This problem was solved by A. M. Gleason, D. Montgomery, L. Zippin, and H. Yamabe in 1952. Since 1950, the topological properties of Lie groups have attracted considerable attention. The methods of Cartan, who postulated de Rham's theory, and H. Hopf, who used the properties of groups extensively, were succeeded by systematic applications of the general theory of algebraic topology. In particular, the topological theory of †fiber bundles was applied to homogeneous spaces G/H by A. Borel and others. Thus homology groups of Lie groups were completely determined, and homotopy groups of Lie groups were determined to a considerable extent. At the present time, the following fields, rather than Lie groups themselves, are objects of extensive study: structure and analysis of homogeneous spaces, algebraic groups, infinite-dimensional unitary representations, and finite or discrete subgroups (→ 15 Algebraic Groups, 112 Differential Geometry, 411 Topology of Lie Groups and Homogeneous Spaces, 423 Unitary Representations).

References

[1] C. Chevalley, Theory of Lie groups I, Princeton Univ. Press, 1946.
[2] Л. С. Понтрягин (L. S. Pontrjagin), Непрерывные группы, Гостехиздат, 1954; English translation, Topological groups, first edition, Princeton Univ. Press, 1939, second edition, Gordon and Breach, 1966.
[3] S. Helgason, Differential geometry and symmetric spaces, Academic Press, 1962.
[4] E. Cartan, Sur la structure des groupes de transformations finis et continus, Thèse, Paris, Nony, 1894. (Oeuvres complètes, Gauthier-Villars, 1952, pt. I, vol. 1, p. 137–287.)
[5] H. Weyl, The structure and representation of continuous groups, Lecture notes, Institute for Advanced Study, Princeton, 1935.
[6] H. Weyl, The classical groups, Princeton Univ. Press, revised edition, 1946.
[7] S. Lie, Theorie der Transformationsgruppen, Teubner, I, 1888; II, 1890; III, 1893 (Chelsea, 1970).
[8] F. Peter and H. Weyl, Die Vollständigkeit der primitiven Darstellungen einer geschlossenen kontinuierlichen Gruppe, Math. Ann., 97 (1927), 737–755. (Weyl, Gesammelte Abhandlungen, Springer, 1968, vol. 3, p. 58–75.)
[9] T. Tannaka, Sôtuigenri (Japanese; Duality principles), Iwanami, 1951.
[10] K. Iwasawa, On some types of topological groups, Ann. of Math., (2) 50 (1949), 507–558.
[11] G. D. Mostow, A new proof of E. Cartan's theorem on the topology of semi-simple groups, Bull. Amer. Math. Soc., 55 (1949), 969–980.
[12] C. Chevalley, On the topological structure of solvable groups, Ann. of Math., (2) 42 (1941), 668–675.
[13] D. Montgomery and L. Zippin, Topological transformation groups, Interscience, 1955.
[14] A. Borel, Topology of Lie groups and characteristic classes, Bull. Amer. Math. Soc., 61 (1955), 397–432.
[15] Y. Matsushima, Tayôtai nyûmon (Japanese; Introduction to the theory of manifolds), Syôkabô, 1965; English translation, Differentiable manifolds, Marcel Dekker, 1972.
[16] N. Iwahori, Rî-gunron (Japanese; Theory of Lie groups) I, II, Iwanami Coll. of Modern Math. Sci., 1957.
[17] G. P. Hochschild, The structure of Lie groups, Holden-Day, 1965.
[18] P. M. Cohn, Lie groups, Cambridge Univ. Press, 1957.
[19] J.-P. Serre, Lie algebras and Lie groups, Benjamin, 1965.
[20] H. Freudenthal, Linear Lie groups, Academic Press, 1969.

249 (XVI.4) Limit Theorems in Probability Theory

A. General Remarks

In †probability theory, theorems concerning †convergence in distribution, †convergence in probability, and †almost certain convergence of sequences of †random variables are subsumed under the generic term of **limit theorems**. When the †probability distributions of a sequence of random variables converge to the distribution F (→ 131 Distributions (of Random Variables)), then F is called the **limit distribution**.

B. Convergence in Distribution of Sums of Independent Random Variables

A sequence of random variables $\{X_{nk}\}$ $(1 \leqslant k \leqslant k_n, n \geqslant 1)$ $(k_n \to \infty)$ is said to be **infinitesimal** if $\max_{1 \leqslant k \leqslant k_n} P(|X_{nk}| \geqslant \varepsilon) \to 0$ $(n \to \infty)$ for every $\varepsilon > 0$. When $X_{n1}, \ldots, X_{nk_n}$ are †independent for every n and $\{X_{nk}\}$ $(1 \leqslant k \leqslant k_n,\ n \geqslant 1)$ is infinitesimal, the set of all limit distributions for the sums $S_n = X_{n1} + \ldots + X_{nk_n} - A_n$ (the A_n are suitable constants) coincides with the class of †infinitely divisible distributions (→ 131 Distributions (of Random Variables)). Suppose that the †characteristic function f of an infinitely divisible distribution F is given by †Lévy's canonical form

$$\log f(t) = i\gamma t - \frac{\sigma^2 t^2}{2}$$

$$+ \int_{-\infty}^{0-} \left(e^{itx} - 1 - \frac{itx}{1+x^2} \right) dM(x)$$

$$+ \int_{0+}^{\infty} \left(e^{itx} - 1 - \frac{itx}{1+x^2} \right) dN(x)$$

and the distribution function of X_{nk} is $F_{nk}(x)$. Then a necessary and sufficient condition for the distribution of S_n to converge to F is that (i) for every †continuity point x of $M(x)$ and $N(x)$,

$$\sum_{k=1}^{k_n} F_{nk}(x) \to M(x) \ (x < 0),$$

$$\sum_{k=1}^{k_n} (F_{nk}(x) - 1) \to N(x) \ (x > 0),$$

and (ii)

$$\lim_{\varepsilon \to 0} \limsup_{n \to \infty} \sum_{k=1}^{k_n} \left(\int_{|x|<\varepsilon} x^2 dF_{nk}(x) - \left(\int_{|x|<\varepsilon} x dF_{nk}(x) \right)^2 \right)$$

$$= \lim_{\varepsilon \to 0} \liminf_{n \to \infty} \sum_{k=1}^{k_n} \left(\int_{|x|<\varepsilon} x^2 dF_{nk}(x) - \left(\int_{|x|<\varepsilon} x dF_{nk}(x) \right)^2 \right)$$

$$= \sigma^2.$$

Applying these results to special distribution functions, we obtain various kinds of limit distributions.

(1) The Central Limit Theorem. In the central limit theorem, the limit distribution is a †normal distribution. A necessary and sufficient condition that the distributions of the sums $S_n = B_n^{-1}(X_1 + \ldots + X_n) - A_n$ $(B_n \to \infty)$ of a sequence of independent random variables $\{X_n\}$ converge to the †standard normal distribution $N(0,1)$ and that $\{B_n^{-1} X_k\}$ $(1 \leqslant k \leqslant n,\ n \geqslant 1)$ becomes infinitesimal is that

$$\lim_{n \to \infty} \sum_{k=1}^{n} \int_{|x| > \varepsilon B_n} dF_n(x) = 0 \text{ and}$$

$$\lim_{n \to \infty} B_n^{-2} \sum_{k=1}^{n} \left(\int_{|x| < \varepsilon B_n} x^2 dF_k(x) - \left(\int_{|x| < \varepsilon B_n} x dF_k(x) \right)^2 \right) = 1,$$

where $F_k(x)$ is the distribution function of X_k. When the †variance of each X_k is finite, we put

$$B_n^2 = \sum_{k=1}^{n} V(X_k), \quad A_n = B_n^{-1} \sum_{k=1}^{n} E(X_k),$$

where $V(X)$, $E(X)$ stand for the variance and mean of X, respectively. Then the necessary and sufficient condition is replaced by the **Lindeberg condition**:

$$\lim_{n \to \infty} B_n^{-2} \sum_{k=1}^{n} \int_{|x| > \varepsilon B_n} x^2 dF_k(x + E(X_k)) = 0$$

$$\text{for every } \varepsilon > 0.$$

In particular, if the X_n have the same distribution with finite variance, the corresponding Lindeberg condition is satisfied, and the central limit theorem holds. When the variables give outcomes of independent †Bernoulli trials, the proposition is reduced to the classical theorem of Gauss and Laplace. Moreover, if X_n has the finite †absolute moment $m_{2+\delta}^{(n)} = E(|X_n|^{2+\delta})$ of order $2+\delta$ and $E(X_n) = 0$, then the Lindeberg condition is implied by the **Ljapunov condition**

$$B_n^{-(2+\delta)} \sum_{k=1}^{n} m_{2+\delta}^{(k)} \to 0 \quad (n \to \infty).$$

(2) The Law of Small Numbers. In order that the distributions of the sums $S_n = X_{n1} + \ldots + X_{nk_n}$ of infinitesimal independent random variables converge to the †Poisson distribution $P(\lambda)$ with mean λ, it is necessary and sufficient that for every $0 < \varepsilon < 1$,

$$\lim_{n \to \infty} \sum_{k=1}^{k_n} \int_{R_\varepsilon} dF_{nk}(x) = 0,$$

$$R_\varepsilon = \{ x \,\big|\, |x-1| \geqslant \varepsilon, \quad |x| \geqslant \varepsilon \};$$

$$\lim_{n \to \infty} \sum_{k=1}^{k_n} \int_{|x-1|<\varepsilon} dF_{nk}(x) = \lambda;$$

$$\lim_{n \to \infty} \sum_{k=1}^{k_n} \int_{|x|<\varepsilon} x\, dF_{nk}(x) = 0; \text{ and}$$

$$\lim_{n \to \infty} \sum_{k=1}^{k_n} \left(\int_{|x|<\varepsilon} x^2 dF_{nk}(x) - \left(\int_{|x|<\varepsilon} x\, dF_{nk}(x) \right)^2 \right) = 0.$$

From this proposition follows the classical **law of small numbers**: If the probability of success p_k for the kth outcome of independent trials satisfies $kp_k = \lambda$ with constant λ independent of k, then the total number S_n of successes up to the nth trial converges in distribution to the Poisson distribution with mean λ.

(3) The Law of Large Numbers. In the law of large numbers the limit distribution is the †unit distribution. Given a sequence of independent random variables X_n with distribution function $F_n(x)$ with mean a_n, a necessary and sufficient condition that $n^{-1}\Sigma_{k=1}^{n}(X_k - a_k)$ converges to 0 in probability is that

$$\lim_{n\to\infty}\sum_{k=1}^{n}\int_{|x|>n} dF_k(x+a_k)=0,$$

$$\lim_{n\to\infty}\frac{1}{n}\sum_{k=1}^{n}\int_{|x|<n} x\,dF_k(x+a_k)=0, \text{ and}$$

$$\lim_{n\to\infty}\frac{1}{n^2}\sum_{k=1}^{n}\left(\int_{|x|<n} x^2\,dF_k(x+a_k) - \left(\int_{|x|<n} x\,dF_k(x+a_k)\right)^2\right)=0.$$

In particular, this is the case when (i) X_n has the finite variance $V(X_n)$ and $n^{-2}\Sigma_{k=1}^{n}V(X_k)\to 0$; or (ii) the X_n, $n\geqslant 1$, obey the same distribution with finite mean.

(4) Convergence to †Quasistable Distributions. The set of limit distributions for the sums $S_n = B_n^{-1}(X_1+\ldots+X_n)-A_n$ of identically distributed independent random variables $\{X_n\}$ (with suitable constants A_n, B_n) coincides with the set of quasistable distributions (→ 131 Distributions (of Random Variables)). If $G(x)$ is the distribution function of the X_i, in order for the limit distribution to be normal it is necessary and sufficient that

$$\lim_{K\to\infty} K^2\int_{|x|>K} dG(x)\Big/\int_{|x|<K} x^2 dG(x)=0.$$

In order for the limit distribution to be quasistable with index α $(0<\alpha<2)$ it is necessary and sufficient that

$$\lim_{x\to\infty}\frac{G(-x)}{1-G(x)}=\frac{c_1}{c_2},\quad c_1+c_2>0;\text{ and}$$

$$\lim_{x\to\infty}\frac{1-G(x)+G(-x)}{1-G(ax)+G(-ax)}=a^{\alpha}\quad\text{for every } a>0.$$

In this case the characteristic function of the limit distribution is given by Lévy's canonical form with $M(x)=c_1|x|^{-\alpha}$, $N(x)=-c_2|x|^{-\alpha}$, $\sigma=0$.

(5) Refinement of Central Limit Theorem. Let $\{X_n\}$ be a sequence of identically distributed independent random variables with $E(X_i)=0$, $\sigma^2=V(X_i)$, $E(|X_i|^3)<\infty$. Let $\Phi_n(x)$ be the distribution function of $S_n=(X_1+\ldots+X_n)/\sigma\sqrt{n}$ and $\Phi(x)$ the distribution function of the normal distribution $N(0,1)$. Then we have

$$\Phi_n(x)-\Phi(x) = \frac{\exp(-x^2/2)}{\sqrt{2\pi n}}(Q(x)+R(x))+o\left(\frac{1}{\sqrt{n}}\right)$$

uniformly in x, where $Q(x)=(1-x^2)E(X_1^3)/6\sigma^3$; $R(x)$ is identically zero when X_i does not obey a †lattice distribution; but $R(x)=d\sigma^{-1}R_1((x+a_n)\sigma\sqrt{n}\,d^{-1})$, $R_1(x)=[x]-x+1/2$, and $a_n=-\sqrt{n}\,a\sigma^{-1}$ when X_i obeys the lattice distribution with †maximal span d: $P(X_i\in\{a+kd\mid k=0,\pm 1,\ldots\})=1$. More accurate asymptotic expansions for $\Phi_n(x)$ are derived when the X_i have absolute moments of higher orders [1]. Asymptotic expressions are also obtained when the X_i are not identically distributed [2] or the limit distribution is †stable [3].

(6) The Local Limit Theorem. The local limit theorem is concerned with the density function of the limit distribution. Let $\{X_n\}$ be a sequence of identically distributed lattice variables with finite variances as in (5). Let

$$S_n=(\sigma\sqrt{n})^{-1}\sum_{k=1}^{n}(X_k-E(X_k)),$$

$$s_{n,j}=(\sigma\sqrt{n})^{-1}d(j-nE(X_1)).$$

Then uniformly in j,

$$\sigma\sqrt{n}\,P(S_n=s_{nj})-(2\pi)^{-1/2}\exp(-s_{nj}^2/2)\to 0\quad(n\to\infty).$$

The classical **de Moivre-Laplace theorem** for Bernoulli trials is a special case of this theorem. When X_i has a density function, a corresponding limit theorem holds. Local limit theorems are derived also when (i) higher moments exist, (ii) the limit distribution is stable, or (iii) the component variables are not identically distributed [2, 3].

In connection with limit distributions, many papers have been devoted to large deviations, that is, to the asymptotic behavior of $P(S_n>a_n)$, where S_n is a sum of independent random variables and a_n tends to infinity with n [4].

C. The Law of Large Numbers and Its Refinements

A sequence of random variables $\{X_n\}$ is said to obey the **strong law of large numbers** if $n^{-1}\Sigma_{k=1}^{n}(X_k-a_k)$ tends to zero with probability 1 as $n \to \infty$, when constants a_k are properly chosen. When the component variables of the sequence are independent, a useful sufficient condition for validity of the strong law of large numbers is

$$\sum_{n=1}^{\infty} \frac{1}{n^2} E(X_n - E(X_n))^2 < \infty.$$

According to Birkhoff's special ergodic theorem, it is necessary and sufficient that the mean of a component variable exist for a sequence of identically distributed independent variables to obey the strong law of large numbers.

Suppose that $\{X_n\}$ is a sequence of independent random variables with $E(X_n)=0$, $\sigma_n^2 = E(X_n^2) < \infty$, $b_n^2 = \sigma_1^2 + \ldots + \sigma_n^2 \to \infty$, and $P(|X_n| < \lambda_n b_n, n=1,2,\ldots)=1$ for a decreasing sequence λ_n tending to 0 as $n \to \infty$. If $\lambda_n = O(1/\varphi^3(b_n^2))$ for a certain increasing continuous function φ, the probability that $X_1 + \ldots + X_n > b_n \varphi(b_n^2)$ for infinitely many n equals 0 or 1 according as the integral $\int_1^{\infty} \frac{1}{t} \varphi(t) e^{-\varphi^2(t)/2} dt$ converges or diverges [5]. In particular, when

$$\varphi(t) = (2\log_{(2)} t + 3\log_{(3)} t + 2\log_{(4)} t + \ldots + 2\log_{(k-1)} t + (2+\varepsilon)\log_{(k)} t)^{1/2},$$

where $\log_{(2)} t = \log\log t$, etc., the relevant probability is equal to 0 or 1 according as ε is positive or negative. If we take $\varphi(t) = 2\log_{(2)} t$, we are led to **Hinčin's law of the iterated logarithm**: If

$$|X_n| = o\left(b_n / \sqrt{\log_{(2)} b_n^2}\right),$$

then

$$P\left(\limsup_{n\to\infty} \frac{X_1 + \ldots + X_n}{\sqrt{2b_n^2 \log_{(2)} b_n^2}} = 1\right) = 1.$$

D. Functionals of Sums of Independent Variables

(1) Consider a **recurrent event** which occurs at successive time periods $\tau_1, \tau_1 + \tau_2, \ldots$, where $\{\tau_n\}$ is a †sequence of nonnegative independent random variables with the same distribution. Let F be the distribution function of τ_1, $m = E(\tau_1)$, $\sigma^2 = V(\tau_1) < \infty$, and N_n be the number of realizations up to time n. Then

$$\lim_{n\to\infty} P\left(N_n \geqslant \frac{n}{m} - \frac{\sigma\sqrt{n}}{m^{3/2}} x\right) = \frac{1}{\sqrt{2\pi}} \int_{-\infty}^{x} e^{-u^2/2} du.$$

When $\sigma^2 = \infty$, a necessary and sufficient condition that suitably normalized N_n converges in distribution is that for every $a > 0$ the relation $(1 - F(x))/(1 - F(ax)) \to a^{\alpha}$ $(x \to \infty)$, $0 \leqslant \alpha < 2$, holds. When $0 \leqslant \alpha < 1$, $P(N_n(1 - F(n)) < x) \to \Psi_\alpha(x)$. When $1 < \alpha < 2$, $P((N_n - nm^{-1})m^{1+1/\alpha} b_n^{-1} < x) \to \Psi_\alpha(x)$ for b_n such that $1 - F(b_n) \sim n^{-1}$, where $\Psi_0(x) = 1 - e^{-x}$ $(x > 0)$, $\Psi_\alpha(x) = 1 - \Phi_\alpha(x^{-1/\alpha})$ $(x > 0)$ for $0 < \alpha < 1$, $\Psi_\alpha(x) = 1 - \Phi_\alpha(-x)$ for $1 < \alpha < 2$, and Φ_α is the quasistable distribution function whose characteristic function is

$$\exp(-|t|^{\alpha}(\cos 2^{-1}\pi\alpha - i \operatorname{sgn} t \sin 2^{-1}\pi\alpha)\Gamma(1-\alpha)).$$

In particular, we have

$$\Psi_{1/2}(\sqrt{2/\pi}\, x) = \sqrt{2/\pi} \int_0^x \exp(-u^2/2)\, du.$$

Moreover, the strong law of large numbers and the law of the iterated logarithm connected with N_n are obtained [7].

(2) Let $\{X_n\}$ be a sequence of independent identically distributed random variables taking integer values with the same distribution, and put $S_n = X_1 + \ldots + X_n$. If the maximum span of the X_i is d and $0 < E(X_i) \leqslant +\infty$, then

$$E\left(\sum_{k=1}^{\infty} \chi_{\{nd\}}(S_k)\right) \to d/E(X_i) \qquad (n \to \infty),$$

$$\to 0 \qquad (n \to -\infty);$$

$$t^{-1} E\left(\sum_{k=1}^{\infty} \chi_{(0,t)}(S_k)\right) \to 1/E(X_i) \qquad (t \to \infty),$$

where χ_E is the †indicator function of the set E, and $1/E(X_i)$ is defined as zero when $E(X_i) = \infty$. If the limit distribution of the normalized sums S_n of $\{X_n\}$ is stable with index β $(1 \leqslant \beta \leqslant 2)$, X_i is symmetric for $\beta = 1$, and $E(X_i) = 0$ for $\beta > 1$, then the S_n return to 0 recurrently. For these cases, setting $\alpha = 1 - \beta^{-1}$, the conditions in (1) imposed on the distribution of the recurrence time τ_1 are satisfied, and the limit distribution for $N_n = \chi_{\{0\}}(S_1) + \ldots + \chi_{\{0\}}(S_n)$ is determined. In this connection, we can show that a similar result is valid when $\chi_{\{0\}}$ is replaced by a member of a considerably wider class of functions, and that the limit distribution is determined as well for M_n which is the number of realizations of the event $(S_k \geqslant 0, S_{k+1} \leqslant 0)$, $1 \leqslant k \leqslant n$. Similar theorems are established for the case when X_i obeys a nonlattice distribution [8, 9, 10].

(3) Let $\{X_n\}$ be a sequence of independent random variables with the same distribution, and put $S_n = X_1 + \ldots + X_n$, $L_n = \chi_{(0,\infty)}(S_1) + \ldots + \chi_{(0,\infty)}(S_n)$. If $n^{-1}E(L_n) \to \alpha$ $(0 \leqslant \alpha \leqslant 1)$, then $P(L_n \leqslant nx) \to G_\alpha(x)$, where G_α is a distribution function such that

$$G_0(0+) - G_0(0-) = 1,$$

$$G_\alpha(x) = \pi^{-1} \sin \pi\alpha \int_0^x u^{\alpha-1}(1-u)^{-\alpha} du$$

for $0 < \alpha < 1$, and $G_1(1+0) - G_1(1-0) = 1$. When $\alpha = 1/2$, the corresponding limit distribution is given by $G_{1/2}(x)=2\pi^{-1}\arcsin\sqrt{x}$ (the **arc sine law**). In (2) L_n, N_n, and M_n are considered †additive functionals of the †Markov process S_n. When $E(X_i)=0$ and $E(X_i^2)=1$, we have the limiting relation

$$\lim_{n\to\infty} P\Big(\max_{1\leqslant k\leqslant n} S_k < x\sqrt{n}\Big) = \begin{cases} 0, & x\leqslant 0 \\ \sqrt{\dfrac{2}{\pi}}\displaystyle\int_0^x \exp\Big(-\frac{u^2}{2}\Big)du, & x>0, \end{cases}$$

$$\lim_{n\to\infty} P\Big(\max_{1\leqslant k\leqslant n} |S_k| < x\sqrt{n}\Big) = \frac{4}{\pi}\sum_{m=0}^{\infty}\frac{(-1)^m}{2m+1}\exp\Big(-\frac{(2m+1)^2\pi^2}{8x^2}\Big), \quad x\geqslant 0.$$

Furthermore, the limit distributions of $n^{-1}(S_1^2+\ldots+S_n^2)$ and $n^{-3/2}(|S_1|+\ldots+|S_n|)$ are also determined explicitly [12]. When X_1 is nonnegative, let ν_x be the number of S_n $(n=1,2,\ldots)$ not exceeding x, and write $Y_x=S_{\nu_x+1}-x$, $Y_x'=x-S_{\nu_x}$. Then a fundamental theorem of †renewal theory guarantees the existence of the limit distributions of Y_x, Y_x' as $x\to\infty$ [3].

E. Convergence in Distribution of Sums of Dependent Variables

Sufficient conditions for the asymptotic normality of the distributions of sums of dependent variables were studied by S. N. Bernstein, P. Lévy, and others [18]. Important in those investigations is the replacement of the concept of ordinary independence by that of asymptotic independence. Suppose that there is a sequence $\{X_n\}$ with $E(X_n)=0$, and finite variance B_n of $S_n=X_1+\ldots+X_n$. Let $\mathfrak{B}_n$ be the †σ-algebra determined by $X_1,\ldots,X_n$ and assume that there exist real sequences $\{\xi_n\}$, $\{\eta_n\}$, $\{\zeta_n\}$ such that $|E(X_n|\mathfrak{B}_{n-1})|\leqslant\xi_n$, $|E(X_n^2|\mathfrak{B}_{n-1})-E(X_n^2)|\leqslant\eta_n$, $|E(|X_n|^3|\mathfrak{B}_{n-1})|\leqslant\zeta_n$, and $(\xi_1+\ldots+\xi_n)/\sqrt{B_n}\to 0$, $(\eta_1+\ldots+\eta_n)/B_n\to 0$, $(\zeta_1+\ldots+\zeta_n)/B_n^3\to 0$, as $n\to\infty$. Then $S_n/\sqrt{B_n}$ is asymptotically distributed according to the standard normal law.

Extensive studies have been devoted to limit theorems for †Markov chains [18, 19]. Suppose that to every n there corresponds a real Markov chain $X_1^{(n)},\ldots,X_n^{(n)}$ and state spaces $\Omega^{(n)}$ for $X_j^{(n)}$ $(1\leqslant j\leqslant n)$ endowed with σ-algebra $\mathfrak{B}(\Omega_j^{(n)})$. Let $P_j^{(n)}(x,A)$ $(x\in\Omega_j^{(n)}$, $A\in\mathfrak{B}(\Omega_{j+1}^{(n)}))$ be the †transition probability, and put $S_n=X_1^{(n)}+\ldots+X_n^{(n)}$. L. Dobrušin [18] proved that in this case the probability law of $(S_n-E(S_n))/\sqrt{(V(S_n))}$ converges to the standard normal law when $|X_j^{(n)}|\leqslant c<\infty$, $V(X_j(n))\geqslant d>0$ for all n and j, $\alpha^{(n)}n^{1/3}\to\infty$ as $n\to\infty$, where c, d are positive constants and

$$\alpha^{(n)}=\min_{1\leqslant j\leqslant n-1}\alpha_j^{(n)},$$

and

$$\alpha_j^{(n)}=1-\sup|P_j^{(n)}(x,A)-P_j^{(n)}(y,A)|,$$

with the supremum taken for $x,y\in\Omega_j^{(n)}$, $A\in\mathfrak{B}(\Omega_{j+1}^{(n)})$. As to possible limit distributions, Dobrušin [18] obtained the following results: If $0<c\leqslant V(X_j^{(n)})\leqslant C<\infty$ for all n and j, $\alpha^{(n)}n^{1/3}\to\infty$ as $n\to\infty$, and $(S_n-E(S_n))/\sqrt{(V(S_n))}$ converges in distribution, then the limit distribution must be infinitely divisible. Let $\{X_n\}$ be a stationary Markov chain with its kth transition probability $P^{(k)}(x,A)$, and suppose that

$$\sup_{x,y,A}|P^{(k)}(x,A)-P^{(k)}(y,A)|<1$$

for some k. Then for a real-valued measurable function f over the state space and constants A_n, $B_n>0$, if $S_n=B_n^{-1}(f(X_1)+\ldots+f(X_n))-A_n$ converges in distribution, then the limit distribution is quasistable [19]. Dobrušin completely determined the limit distributions for the sums of a Markov chain with two states. Among them appear probability distributions that are not infinitely divisible.

F. Convergence in Distribution for Stochastic Processes

(1) General Theory. Let T be a finite or infinite time interval, and let $(\Omega,\mathfrak{B},P)$ and S be a probability space and topological state space, respectively. Given an S-valued stochastic process $X=(X(t,\omega), t\in T, \omega\in\Omega)$, we denote by $\mathfrak{B}(S^T)$ the σ-algebra generated by the †cylinder sets and denote by P_X and $P_X^{t_1\ldots t_k}$, respectively, the probability measures over $\mathfrak{B}(S^T)$ induced by the processes X and $(X(t_1,\omega),\ldots,X(t_k,\omega))$. Suppose that there is a sequence of stochastic processes X_n $(n=1,2,\ldots)$ whose †sample paths are contained in a subset E of S^T, and we can introduce a topology τ on E such that if $\mathfrak{B}_n$ $(n=1,2,\ldots)$ is the P_{X_n}-completion of $\mathfrak{B}(S^T)$, the topological σ-algebra $\mathfrak{B}_E^\tau$ on E becomes a subfamily of $\mathfrak{B}_n\cap E$ $(n=1,2,\ldots)$. If there exists a probability measure P_0 over $(E,\mathfrak{B}_E^\tau)$ such that

$$\lim_{n\to\infty}\int_E f(y)\,dP_{X_n}=\int_E f(y)\,dP_0$$

for every bounded τ-continuous real function f on E, then the probability law of $\{X_n\}$ is said to converge to P_0.

Let (E,ρ) be a †complete metric separable space and P' the †Borel probability measure on $\Omega'=[0,1]$. If a sequence of E-valued random variables $\{X_n(\omega), 1\leqslant n<\infty\}$ has the probability distribution P_0 over the σ-algebra of Borel sets of (E,ρ) as its limit distribution, then it can be shown that there is a sequence of E-valued random variables $\{\xi_n(\omega')\}$ $(\omega'\in\Omega';$ $0\leqslant n<\infty)$ such that the probability distribution of $\xi_n(\omega')$ coincides with that of $X_n(\omega)$ $(1\leqslant n<\infty)$, $\xi_0(\omega')$ has the probability distribution P_0, and $P'(\xi_n(\omega')\to\xi_0(\omega'))=1$.

Consider a sequence of real-valued stochastic processes $\{X_n(t,\omega)\}$. If (i) there is a probability measure P_0 on $\mathfrak{B}(\mathbf{R}^T)$ such that $P_{X_n}^{t_1\ldots t_k}$ converges to $P_0^{t_1\ldots t_k}$ for every choice of $(t_1,\ldots,t_k)$ and (ii)

$$\lim_{h\downarrow 0}\limsup_{n\to\infty}\sup_{|s-t|\leqslant h} P(|X_n(s)-X_n(t)|>\varepsilon)=0$$

for every $\varepsilon>0$, then there exists a sequence of real-valued processes $\{\xi_n(t,\omega')\}$ $(\omega'\in\Omega';$ $0\leqslant n<\infty)$ such that for every $t\in T$, $\xi_n(t)$ converges in probability to $\xi_0(t)$, ξ_n and X_n $(1\leqslant n<\infty)$ induce the same probability measure over $\mathfrak{B}(\mathbf{R}^T)$, and ξ_0 is continuous in probability with the probability distribution P_0.

(2) Convergence in Distribution of Stochastic Processes Whose Sample Functions Have †Discontinuities of Only the First Kind. Ju. V. Prohorov and A. V. Skorohod carried out a systematic study of convergence in distribution for those processes whose sample functions are (S,ρ)-valued and have discontinuities of only the first kind, where (S,ρ) is a complete metric separable space. Let D be the set of functions $x(t)$ from $T=[0,1]$ to S having discontinuities of only the first kind, with $x(1-0)=x(1)$, and introduce a metric ρ_D by $\rho_D(x,y)=\inf_\lambda\{\sup_{t\in T}|t-\lambda(t)|+\sup_{t\in T}\rho(x(t),y(\lambda(t)))\}$, where the infimum is taken for all homeomorphisms λ on T. Then (i) in the notation of (1), $\mathfrak{B}_D^{\rho_D}=\mathfrak{B}(S^T)\cap D$; and (ii) (D,ρ_D) is †separable but not necessarily †complete. When almost all sample functions of X are elements of D, we write $X\in D$. Given $X_n\in D$ $(n\geqslant 0)$, a necessary and sufficient condition for P_{X_n} to converge to P_{X_0} over $\mathfrak{B}_D^{\rho_D}$ is that (i) $P_{X_n}^{t_1\ldots t_k}$ converges to $P_{X_0}^{t_1\ldots t_k}$ for any $(t_1,\ldots,t_n)\subset N$, where N is a dense subset of $[0,1]$ containing 0, 1; and that (ii) $\lim_{h\downarrow 0}\limsup_{n\to\infty}\rho(\Delta_D(h,X_n)>\delta)=0$ for every $\delta>0$, where $\Delta_D(h,x)=\sup\{\min(\rho(x(t_1),x(t)),\rho(x(t_2),x(t)))\,|\,t-h<t_1<t<t_2<t+h\}$.

If C is a subspace of D consisting of all continuous functions, ρ_D is equivalent to $\rho_C(x,y)=\max\{\rho(x(t),y(t))\,|\,0\leqslant t\leqslant 1\}$, and $\Delta_C(h,x)=\max\{\rho(x(s),x(t))\,|\,|s-t|\leqslant h\}$ corresponds to Δ_D. For a sequence of processes $X_n\in C$, a necessary and sufficient condition for P_{X_n} to converge to P_{X_0} over $\mathfrak{B}_C^{\rho_C}$ is that (i) $P_{X_n}^{t_1\ldots t_k}$ converges to $P_{X_0}^{t_1\ldots t_k}$ for any $(t_1,\ldots,t_k)\subset N$ (N is a dense set as in the case of D); and that (ii) $\lim_{h\downarrow 0}\limsup_{n\to\infty}P_{X_n}(\Delta_C(h,X_n)>\delta)=0$ for every $\delta>0$. Under these conditions, there exists a sequence $\xi_n(t,\omega')\in D$ $(\in C)$ $(t\in T,\omega'\in\Omega')$ for $0\leqslant n<\infty$, such that $P_{\xi_n}=P_{X_n}$ for $0\leqslant n<\infty$ and $P'(\rho_D(\xi_n,\xi_0)\to 0)=1$ $(P'(\rho_C(\xi_n,\xi_0)\to 0)=1)$.

The following theorem, due to Prohorov, is useful for practical applications. Let $\{X_\alpha(t)\,|\,0\leqslant t<1,\alpha\in A\}$ be a family of real-valued processes that satisfy

$$E(|X_\alpha(t)-X_\alpha(s)|^a)\leqslant C|t-s|^{1+b},\quad \alpha\in A,$$

where a, b, c are positive constants independent of α. If the family of distributions generated by $\{X_\alpha(0)\,|\,\alpha\in A\}$ is †totally bounded, then $X_\alpha(t)\in C$ $(\alpha\in A)$ and $\{P_{X_\alpha}\}$ is a totally bounded set of probability distributions over C.

(3) Convergence in Distribution of Markov Processes. Consider a Markov process $\{X(t),$ $0\leqslant t\leqslant 1\}$ with complete metric separable space (S,ρ) as its state space and with topological σ-algebra $\mathfrak{B}_\rho$. Suppose that its transition probabilities $P(s,x;t,A)$, $0\leqslant s<t\leqslant 1$, $A\in\mathfrak{B}_\rho$ are measurable in $x\in S$. Define $V^\varepsilon(x)=\{y\,|\,\rho(x,y)\geqslant\varepsilon\}$ and introduce two kinds of conditions: (D) for every $\varepsilon>0$ and $0\leqslant t\leqslant 1$, $\lim_{h\downarrow 0}\sup_{x\in S,t_1,t_2}P(t_1,x;t_2,V^\varepsilon(x))=0$, where t_1, t_2 are subject to the conditions $t\leqslant t_1<t_2\leqslant t+h$, $t-h\leqslant t_1<t_2<t$, $1-h\leqslant t_1<t_2\leqslant 1$; (C) for every $\varepsilon>0$, $\sup_{x\in S,t-s\leqslant h}P(s,x;t,V^\varepsilon(x))\leqslant h\Psi_\varepsilon(h)$, where $\Psi_\varepsilon(h)$ is such that $\Psi_\varepsilon(h)\downarrow 0$ as $h\downarrow 0$ for every $\varepsilon>0$. According as $\{X(t),0\leqslant t\leqslant 1\}$ satisfies conditions (D) or (C), there is a process $X'(t)$ which is equivalent (→ 395 Stochastic Processes) to $X(t)$ belonging respectively to (D) or (C). Convergence in distribution of Markov processes is based on this fact. Suppose that a sequence of Markov processes $\{X_n(t)\,|\,0\leqslant t\leqslant 1\}$, $0\leqslant n<\infty$, satisfies (i) the condition (D) (condition (C) with Ψ_ε independent of n), (ii) with N as in (2), $P_{X_n}^{t_1\ldots t_k}$ converges to $P_{X_0}^{t_1\ldots t_k}$ for every $(t_1,\ldots,t_k)\subset N$. Then P_{X_n} converges to P_{X_0} over $\mathfrak{B}_D^{\rho_D}$ (over $\mathfrak{B}_C^{\rho_C}$).

(4) Convergence of Generators and Convergence in Distribution for Markov Processes. Consider a †temporally homogeneous Markov process $X=\{X(t)\,|\,0\leqslant t\leqslant 1\}$ with the state space $\mathbf{R}^m$. Define C to be the space of bounded continuous functions on $\mathbf{R}^m$ and $D^{(2)}$ to be those subspaces of C whose elements have uniformly continuous second derivatives. Suppose that the †transition operator T_t $(0\leqslant t<\infty)$ of X maps C into itself and is †weakly continuous in t. Assume furthermore that the following conditions are satisfied: (i) The †Green's operator of order α of T_t satisfies $G_\alpha D^{(2)}\subset D^{(2)}$. (ii) The domain $\mathfrak{D}(\tilde{I})$ of the †weak generator $\tilde{I}$ of T_t contains

$D^{(2)}$, and for $\varphi \in D^{(2)}$ we have

$$\tilde{I}\varphi(x) = \frac{1}{2}\sum_{i,j=1}^{m} b_{ij}(x)\frac{\partial^2\varphi(x)}{\partial x^i \partial x^j} + \sum_{i=1}^{m} a_i(x)\frac{\partial\varphi(x)}{\partial x^i} + \int_{\mathbf{R}^m}\left[\varphi(y)-\varphi(x) - \sum_{i=1}^{m}\frac{\partial\varphi(x)}{\partial x^i}\frac{y^i-x^i}{1+|y-x|^2}\right]\Pi(x,dy).$$

In this equality $x=(x^1,\ldots,x^m)\in\mathbf{R}^m$; the $a^i(x)$ are bounded; the $b_{ij}(x)$ are subject to the condition that there exist $c_1, c_2>0$, independent of x, such that

$$c_1\sum_i \xi_i^2 \leqslant \sum_{i,j} b_{ij}(x)\xi_i\xi_j \leqslant c_2\sum_i \xi_i^2$$

for every $(\xi_1,\ldots,\xi_m)\in\mathbf{R}^m$; and $\Pi(x,dy)$ $(x\in\mathbf{R}^m)$ is a measure on $\mathbf{R}^m$ subject to the conditions that for every $r>0$

$$\sup_x \Pi(x, V^r(x)) < \infty,$$

$$\sup_x \Pi(x, V^r(x)) \to 0 \quad (r\to\infty),$$

$$\sup_x \int_{|y-x|\leqslant 1} |x-y|^2\Pi(x,dy) < \infty,$$

and for a nonnegative $\varphi\in C$ that vanishes around x_0,

$$\int_{\mathbf{R}^m}\varphi(y)\Pi(x_n,dy) \to \int_{\mathbf{R}^m}\varphi(y)\Pi(x_0,dy) \quad (x_n\to x_0).$$

Suppose that there is a sequence of Markov processes $\{X_n(t)\}$ $(0\leqslant n<\infty)$ each of which satisfies conditions (i), (ii). Furthermore, assume that the sequence satisfies the three conditions (a)

$$\lim_{n\to\infty}\int_{\mathbf{R}^m}\psi(y-x)\Pi^{(n)}(x,dy) = \int_{\mathbf{R}^m}\psi(y-x)\Pi^{(0)}(x,dy)$$

uniformly in x, where $\psi\in C$ and $\psi(x)=0$ around $x=0$, (b) for $\varphi\in D^{(2)}$ the expression

$$\sum_{i,j=1}^{m} b_{ij}^{(n)}(x)\frac{\partial^2\varphi(x)}{\partial x^i\partial x^j} + \int_{\mathbf{R}^m}\sum_{i,j=1}^{m}\frac{\partial^2\varphi(x)}{\partial x^i\partial x^j}(y^i-x^i)(y^j-x^j) \times \alpha_\varepsilon(y-x)\Pi^{(n)}(x,dy)$$

converges uniformly in x to the same expression with $n=0$, where $\alpha_\varepsilon\in D^{(2)}$, $0\leqslant\alpha_\varepsilon\leqslant 1$, $\alpha_\varepsilon(x)=1$ $(|x|\leqslant\varepsilon)$, $=0$ $(|x|\geqslant 2\varepsilon)$, and for $\varphi\in D^{(2)}$,

$$\lim_{n\to\infty}\sum_{i,j=1}^{m} a_i^{(n)}(x)\frac{\partial\varphi(x)}{\partial x^i} = \sum_{i=1}^{m} a_i^{(0)}(x)\frac{\partial\varphi(x)}{\partial x^i}$$

uniformly in x. Then if the distribution of $X_n(0)$ converges to that of $X_0(0)$ as $n\to\infty$, P_{X_n} converges to P_{X_0}.

Given a sequence of Markov chains $X^{(n)}=\{X_{n0},X_{n1},\ldots,X_{nn}\}$, $1\leqslant n<\infty$, define a sequence of Markov processes $\{X_n(t)|0\leqslant t\leqslant 1\}$ by writing $X_n(t)=X_{nk}$ for $k(n+1)^{-1}\leqslant t<(k+1)(n+1)^{-1}$, $X_n(1)=X_{nn}$. Then the convergence in distribution of $X^{(n)}$ is reduced to that of $\{X_n(t)|0\leqslant t\leqslant 1\}$, and we obtain results similar to those in this section so far. The temporally inhomogeneous case is reduced to the temporally homogeneous case [26].

G. Convergence of Empirical Distributions

Let $\{X_k(\omega)|1\leqslant k\leqslant n\}$, $\{Y_j(\omega)|1\leqslant j\leqslant m\}$ be independent random samples from a population with population distribution F. Let $N_n(x,\omega)$ be the number of k $(k=1,\ldots,n)$ such that $X_k\leqslant x$. Then $F_n(x,\omega)=n^{-1}N_n(x,\omega)$ is a distribution function in x called an **empirical distribution function**. It is easily shown that $H_n^+=\sup_x|F_n(x,\omega)-F(x)|\to 0$ $(n\to\infty)$ with probability 1. Let $G_m(x,\omega)$ be the empirical distribution function constructed from the Y_j, and put

$$H_n = \sup_x(F_n(x,\omega)-F(x)),$$

$$H^+(n,m) = \sup_x|F_n(x,\omega)-G_m(x,\omega)|,$$

$$H(n,m) = \sup_x(F_n(x,\omega)-G_m(x,\omega)).$$

If F is continuous, then according to **Kolmogorov's theorem**, as $n\to\infty$ we have

$$P(\sqrt{n}\,H_n^+<x)\to K(x), \text{ where } K(x) = \sum_{k=-\infty}^{\infty}(-1)^k\exp(-2k^2x^2) \quad (x>0);$$
$$=0 \quad (x\leqslant 0).$$

On the other hand, **Smirnov's theorem** asserts that (i)

$$P(\sqrt{n}\,H_n<x)=0 \quad (x\leqslant 0);$$
$$=1-(1-n^{-1/2}x)^n - x\sqrt{n}\sum_{k=r+1}^{n-1} n^{-n}\binom{n}{k} \times(k-x\sqrt{n})^k(n-k+x\sqrt{n})^{n-k-1}$$
$$(0<x\leqslant\sqrt{n},\ r=[x\sqrt{n}]);$$
$$=1 \quad (x>\sqrt{n}),$$

and, as $n\to\infty$,

$$P(\sqrt{n}\,H_n<x)\to L(x),$$

$$L(x)=1-\exp(-2x^2) \quad (x>0),$$
$$=0 \quad (x\leqslant 0);$$

(ii) when $n\to\infty$, $mn^{-1}\to r$, r a constant,

$$P((nm)^{1/2}(n+m)^{-1/2}H^+(n,m)<x)\to K(x),$$

$$P((nm)^{1/2}(n+m)^{-1/2}H(n,m)<x)\to L(x).$$

In the sense of convergence in distribution for Markov processes these results are interpreted as follows: $F(X_k)$ is uniformly distributed over $[0,1]$. Let $\nu_n(t)$ be the number of k $(k=1,2,\ldots,n)$ such that $F(X_k)\leqslant t$, and put $X_n(t)=\sqrt{n}\,(n^{-1}\nu_n(t)-t)$ $(0<t\leqslant 1)$, $X_n(0)=0$. Then $E(X_n(t))=0$, $E(X_n(t)X_m(t))=\min(s,t)-st$, $H_n^+=\sup_t|X_n(t)|$, and $H_n(t)=\sup_t X_n(t)$. Let $\{X(t)\,|\,0\leqslant t\leqslant 1\}$ be a †Gaussian process with $E(X(t))=0$, $E(X(s)X(t))=\min(s,t)-st$, and put $H^+=\sup_t|X(t)|$, $H=\sup_t X(t)$. Then P_{X_n} converges to P_X over $\mathfrak{B}^{\rho}_{C}$. Therefore

$$P(\sqrt{n}\,H_n^+\geqslant x)\to P(H^+\geqslant x),\quad n\to\infty,$$

$$P(\sqrt{n}\,H_n\geqslant x)\to P(H\geqslant x),\quad n\to\infty.$$

Similarly, if we denote the number of j such that $G(Y_j)\leqslant t$ by $\mu_m(t)$, then $H^+(n,m)=\sup_t|n^{-1}\nu_n(t)-m^{-1}\mu_m(t)|$, and therefore

$$P\left((nm)^{1/2}(n+m)^{-1/2}H^+(n,m)\geqslant x\right)\to$$
$$P\left((1+r)^{-1/2}\sup_t|X(t)-\sqrt{r}\,Y(t)|\geqslant x\right)$$

when $n\to\infty$, $mn^{-1}\to r$, where $\{Y(t)\,|\,0\leqslant t\leqslant 1\}$ is independent of $\{X(t)\,|\,0\leqslant t\leqslant 1\}$ and distributed according to the same law as the latter. The exact as well as asymptotic expressions for the distributions H_n^+, $H^+(n,m)$ are also obtained [32, 33].

References

[1] Б. В. Гнеденко, А. Н. Колмогоров (B. V. Gnedenko and A. N. Kolmogorov), Предельные распределения для сумм независимых случайных величин, Гостехиздат, 1949; English translation, Limit distributions for sums of independent random variables, Addison-Wesley, 1954.

[2] В. В. Петров (V. V. Petrov), Асимптотические разложения для распределений сумм независимых случайных величин, Teor. Verojatnost. i Primenen., 4 (1959), 220–224; English translation, Asymptotic expansions for distributions of sums of independent random variables, Theory of Prob. Appl., 4 (1959), 208–211.

[3] H. Cramér, On asymptotic expansions for sums of independent random variables with a limiting stable distribution, Sankhyā, (A) 25 (1963), 13–24.

[4] Ju. V. Linnik, On the probability of large deviations for the sums of independent variables, Proc. 4th Berkeley Symp. on Math. Stat. and Prob., II, Univ. of California Press, 1961, p. 289–306.

[5] W. Feller, The general form of the so-called law of the iterated logarithm, Trans. Amer. Math. Soc., 54 (1943), 373–402.

[6] K. L. Chung, The strong law of large numbers, Proc. 2nd Berkeley Symp. on Math. Stat. and Prob., Univ. of California Press, 1951, p. 341–352.

[7] W. Feller, Fluctuation theory of recurrent events, Trans. Amer. Math. Soc., 67 (1949), 98–119.

[8] G. Maruyama, Fourier analytic treatment of some problems on the sums of random variables, Natural Sci. Report, Ochanomizu Univ., 6 (1955), 7–24.

[9] N. Ikeda, Fluctuation of sums of independent random variables, Mem. Fac. Sci. Kyūsyū Univ., (A) 10 (1956), 15–28.

[10] G. Kallianpur and H. Robbins, The sequence of sums of independent random variables, Duke Math. J., 21 (1954), 285–307.

[11] F. Spitzer, A combinatorial lemma and its application to probability theory, Trans. Amer. Math. Soc., 82 (1956), 323–339.

[12] P. Erdös and M. Kac, On certain limit theorems of the theory of probability, Bull. Amer. Math. Soc., 52 (1946), 292–302.

[13] Е. Б. Дынкин (E. B. Dynkin), Некоторые предельные теоремы для сумм независимых случайных величин с бесконечными математическими ожиданиями, Izv. Akad. Nauk SSSR, 19 (1955), 247–266.

[14] S. N. Bernstein, Sur le théorème limite du calcul des probabilités, Math. Ann., 85 (1922), 237–241.

[15] S. N. Bernstein, Sur l'extension du théorème limite du calcul des probabilités aux sommes de quantités dépendantes, Math. Ann., 97 (1927), 1–59.

[16] P. Lévy, Théorie de l'addition des variables aléatoires, Gauthier-Villars, 1937.

[17] M. Loève, On almost sure convergence, Proc. 2nd Berkeley Symp. on Math. Stat. and Prob., Univ. of California Press, 1951, p. 279–303.

[18] Р. Л. Добрушин (R. L. Dobrušin), Центральная предельная теорема для неоднородных цепей Маркова I, II, Teor. Verojatnost. i Primenen., 1 (1956), 72–89, 365–425; English translation, Central limit theorem for nonstationary Markov chains I, II, Theory of Prob. Appl., 1 (1956), 65–80, 329–383.

[19] С. В. Нагаев (S. V. Nagaev), Некоторые предельные теоремы для однородных цепей Маркова, Teor. Verojatnost. i Primenen., 2 (1957), 389–416; English translation, Some limit theorems for stationary Markov chains, Theory of Prob. Appl., 2 (1957), 378–406.

[20] Р. Л. Добрушин (R. L. Dobrušin), Предельные теоремы для цепи Маркова из двух состояний, Izv. Akad. Nauk SSSR, 17 (1953), 291–330.

[21] А. Н. Колмогоров (A. N. Kolmogorov), Локальная предельная теорема для классических цепей Маркова, Izv. Akad. Nauk SSSR, 13 (1949), 281–300.

[22] С. Х. Сираждинов (S. H. Siraždinov), Предельные теоремы для однородных

цепей Маркова, Ташкент, 1955.
[23] M. D. Donsker, An invariance principle for certain probability limit theorems, Mem. Amer. Math. Soc. (1951), 1–12.
[24] Ю. В. Прохоров (Ju. V. Prohorov), Сходимость случайных процессов и предельные теоремы теории вероятностей, Teor. Verojatnost. i Primenen., 1 (1956), 177–238; English translation, Convergence of random processes and limit theorems in probability theory, Theory of Prob. Appl., 1 (1956), 155–214.
[25] А. В. Скороход (A. V. Skorohod), Предельные теоремы для случайных процессов, Teor. Verojatnost. i Primenen., 1 (1956), 289–319; English translation, Limit theorems for stochastic processes, Theory of Prob. Appl., 1 (1956), 261–290.
[26] А. В. Скороход (A. V. Skorohod), Предельные теоремы для процессов Маркова, Teor. Verojatnost. i Primenen., 3 (1958), 217–264; English translation, Limit theorems for Markov processes, Theory of Prob. Appl., 3 (1958), 202–246.
[27] А. В. Скороход (A. V. Skorohod), Исследования по теории случайных процессов, Киев, 1961.
[28] A. N. Kolmogorov, Sulla determinazione empirica di una legge di distribuzione, Giorn. Inst. d. Ital. Attuari, 4 (1933), 83–91.
[29] Н. В. Смирнов (N. V. Smirnov), Приближение случайных величин по эмпирическим данным, Uspehi Mat. Nauk, 10 (1944), 179–206.
[30] W. Feller, On the Kolmogorov-Smirnov limit theorems for empirical distributions, Ann. Math. Statist., 19 (1948), 177–189.
[31] J. L. Doob, Heuristic approach to the Kolmogorov-Smirnov theorems, Ann. Math. Statist., 20 (1949), 393–403.
[32] В. С. Королюк (V. S. Koroljuk), О расхождении эмпирических распределений для случая двух независимых выборок, Izv. Akad. Nauk SSSR, 19 (1955), 81–96.
[33] В. С. Королюк (V. S. Koroljuk), Асимптотические разложения для критериев согласия А. Н. Колмогорова и Н. В. Смирнова, Izv. Akad. Nauk SSSR, 19 (1955), 103–124.
[34] P. Billingsley, Convergence of probability measures, John Wiley, 1968.

250 (III.4) Linear Equations

A. General Remarks

A **linear equation** is an †algebraic equation of the first degree. Let K be a †field, $f_i = a_{i1}X_1 + \ldots + a_{in}X_n$ $(i=1,\ldots,m)$ be m †linear forms: $K^n \rightarrow K$, and $b_1, \ldots, b_m$ be m given elements of K. Then a set of n elements $x_1, \ldots, x_n$ of K, or an n-tuple $\mathbf{x} = (x_1, \ldots, x_n) \in K^n$, satisfying the **system of m linear equations**

$$a_{i1}x_1 + \ldots + a_{in}x_n = b_n, \quad i = 1, \ldots, m, \tag{1}$$

is called a **solution** of equations (1). In particular, a system with $b_1 = \ldots = b_m = 0$, i.e.,

$$a_{i1}x_1 + \ldots + a_{in}x_n = 0, \quad i = 1, \ldots, m, \tag{2}$$

is called a **system of linear homogeneous equations**. In the theory of linear equations, the field K need not be commutative. In those cases where K is noncommutative, we have to distinguish between right multiplication and left multiplication. We make this distinction by adding the word "right" or "left" in parentheses whenever it is necessary to do so.

If $\mathbf{x}_1, \ldots, \mathbf{x}_t$ are solutions of the system (2), then any of their (right) linear combinations $\sum_i \mathbf{x}_i c_i$ $(c_i \in K)$ is also a solution of (2), so that the solutions of (2) form a †(right) linear space L over K. L is the kernel of the (left) linear mapping $K^n \rightarrow K^m$ given by $\mathbf{x} \rightarrow (f_1(\mathbf{x}), \ldots, f_m(\mathbf{x}))$. Let $s = \dim L$. If $s = 0$, the system (2) has only $\mathbf{x} = 0$ as solution, called the **trivial solution**. If $s > 0$, L has a basis $\{\mathbf{x}_1, \ldots, \mathbf{x}_s\}$. Then $\mathbf{x}_1, \ldots, \mathbf{x}_s$ are (right) linearly independent solutions of (2), and every solution of (2) is a (right) linear combination of them. We say that $\mathbf{x}_1, \ldots, \mathbf{x}_s$ form a **system of fundamental solutions** of (2). Let r denote the maximum number of (left) linearly independent forms in the set $\{f_1, f_2, \ldots, f_m\}$. Then we have $s + r = n$, so that the system (2) has a nontrivial solution if and only if $r < n$, and the number of linearly independent fundamental solutions is $n - r$. Accordingly, if the number of equations is less than the number of unknown quantities, there always exists a nontrivial solution.

Suppose now that the system (1) has a solution $\mathbf{x}_0$. Then every solution $\mathbf{x}$ of (1) can be written $\mathbf{x}_0 + \mathbf{y}$ for a solution $\mathbf{y}$ of (2); and conversely, for any solution $\mathbf{y}$ of (2), $\mathbf{x}_0 + \mathbf{y}$ is a solution of (1). Furthermore, in order for (1) to have a solution, it is necessary and sufficient that $\sum_i c_i b_i = 0$ whenever $\sum_i c_i f_i = 0$ $(c_i \in K)$, i.e., that the following two matrices have the same †rank:

$$A = \begin{pmatrix} a_{11} & \cdots & a_{1n} \\ & \cdots & \\ a_{m1} & \cdots & a_{mn} \end{pmatrix},$$

$$\tilde{A} = \begin{pmatrix} a_{11} & \cdots & a_{1n} & b_1 \\ & \cdots & & \\ a_{m1} & \cdots & a_{mn} & b_m \end{pmatrix}.$$

In particular, equation (1) has a unique solution if and only if A and $\tilde{A}$ have the same rank n. If $m = n$ (i.e., A is a square matrix), (1) has a unique solution if and only if A has the inverse A^{-1}, and then the solution is given by ${}^t\mathbf{x}=A^{-1}\,{}^t\mathbf{b}$, where $\mathbf{b}=(b_1,b_2,\ldots,b_n)$.

We have also the following result. Let K' be an †extension field of K. If a system of linear equations in K has a solution in K', it has a solution contained in K. In particular, if (2) in K has a nontrivial solution in K', it has a nontrivial solution already in K, and any system of fundamental solutions in K is itself a system of fundamental solutions in K'.

B. Cramér's Rule

Suppose now that K is commutative. Then we have an explicit formula for solving equation (1) by means of †determinants.

Consider first the case $m=n$. Let $\Delta=|A|$, the determinant of the matrix A. If $\Delta\neq 0$, then (1) has a unique solution, given by $x_k=\Delta_k/\Delta$ $(k=1,\ldots,n)$, where Δ_k is the determinant obtained from Δ by replacing its kth column, $a_{1k},a_{2k},\ldots,a_{nk}$ by $b_1,b_2,\ldots,b_n$. This is called **Cramér's rule**. To consider the general case, let r be the common rank of A and $\tilde{A}$. We may assume that $|a_{ik}|\neq 0$ $(i,k=1,\ldots,r)$, appropriately changing the order of the equations and unknowns. Then by Cramér's rule we can solve the first r equations for $x_1,\ldots,x_r$, assigning arbitrary values to $x_{r+1},\ldots,x_n$. In order for equation (2) to have a nontrivial solution, it is necessary and sufficient that the rank of A is less than n, and hence that $|A|=0$ if A is a square matrix. For geometric applications of the theory of linear equations → 9 Affine Geometry, and for numerical solution → 297 Numerical Solution of Linear Equations.

References

[1] T. Takagi, Daisûgaku kôgi (Japanese; Lectures on algebra), Kyôritu, revised edition, 1948.
[2] K. Asano, Senkei daisûgaku teiyô (Japanese; A manual of linear algebra), Kyôritu, 1948.
[3] S. Lang, Linear algebra, Addison-Wesley, 1966.
Also → references to 107 Determinants; 256 Linear Spaces; 269 Matrices; 297 Numerical Solution of Linear Equations.

251 (XII.8) Linear Operators

A. General Remarks

In †functional analysis, when we talk about an operator or a mapping T from one space to another, it is important to specify not only the domain $\mathfrak{D}(T)$ and range $\mathfrak{R}(T)$ of T, but also the spaces $\mathfrak{X}$ and $\mathfrak{Y}$ of which $\mathfrak{D}(T)$ and $\mathfrak{R}(T)$, respectively, are regarded to be subsets. Thus a mapping T from a subset $\mathfrak{D}(T)$ of a (real or complex) linear space $\mathfrak{X}$ to a linear space $\mathfrak{Y}$ is called an **operator** from $\mathfrak{X}$ to $\mathfrak{Y}$. The image of $x\in\mathfrak{X}$ under T is customarily denoted by Tx. An operator T from $\mathfrak{X}$ to $\mathfrak{Y}$ is said to be a **linear operator (linear transformation** or **additive operator)** if (i) $\mathfrak{D}(T)$ is a †linear subspace of $\mathfrak{X}$ and (ii) $T(\alpha_1x_1+\alpha_2x_2)=\alpha_1Tx_1+\alpha_2Tx_2$ for all $x_1,x_2\in\mathfrak{D}(T)$ and all scalars α_1, α_2. The set of all linear continuous operators T from $\mathfrak{X}$ to $\mathfrak{Y}$ with $\mathfrak{D}(T)=\mathfrak{X}$ is denoted by $B(\mathfrak{X},\mathfrak{Y})$. We denote $B(\mathfrak{X},\mathfrak{X})$ by $B(\mathfrak{X})$.

For simplicity we suppose throughout the present article that $\mathfrak{X}$ and $\mathfrak{Y}$ are †Banach spaces. Some of the statements given remain true in more general situations (→ 407 Topological Linear Spaces). More information about operators in $B(\mathfrak{X},\mathfrak{Y})$ will be found elsewhere (→ 39 Banach Spaces). Examples are grouped together at the end of the article.

B. Linear Operations on Operators

When T_1 and T_2 are operators from $\mathfrak{X}$ to $\mathfrak{Y}$, the sum T_1+T_2 is the operator defined by $\mathfrak{D}(T_1+T_2)=\mathfrak{D}(T_1)\cap\mathfrak{D}(T_2)$ and $(T_1+T_2)x=T_1x+T_2x$, $x\in\mathfrak{D}(T_1+T_2)$. When T_1 is from $\mathfrak{X}$ to $\mathfrak{Y}$ and T_2 is from $\mathfrak{Y}$ to $\mathfrak{Z}$, the product T_2T_1 is the operator defined by $\mathfrak{D}(T_2T_1)=\{x\in\mathfrak{D}(T_1)\,|\,T_1x\in\mathfrak{D}(T_2)\}$ and $T_2T_1x=T_2(T_1x)$, $x\in\mathfrak{D}(T_2T_1)$. The product αT of a scalar α and an operator T is defined in a similar and obvious way. These operators are linear whenever T_1 and T_2 are linear. For an operator T from $\mathfrak{X}$ to $\mathfrak{Y}$ the subset $\Gamma(T)=\{\{x,Tx\}\,|\,x\in\mathfrak{D}(T)\}$ of the product space $\mathfrak{X}\times\mathfrak{Y}$ is said to be the **graph** of T. If $\Gamma(T_1)\subset\Gamma(T_2)$, the operator T_2 is said to be an **extension** of T_1 (we write $T_1\subset T_2$). If a linear operator T from $\mathfrak{X}$ to $\mathfrak{Y}$ satisfies the condition that $x\neq 0$ implies $Tx\neq 0$, then T has the **inverse operator** T^{-1}, which is a linear operator satisfying $\mathfrak{D}(T^{-1})=\mathfrak{R}(T)$ and $\mathfrak{R}(T^{-1})=\mathfrak{D}(T)$.

C. Convergence of Operators

The following three topologies are most frequently used in the linear space $B(\mathfrak{X},\mathfrak{Y})$, which becomes a †locally convex linear topological space under each of them: (1) the **uniform operator topology**, (2) the **strong operator topology**, and (3) the **weak operator topology**. These topologies are determined by the respective basic systems of neighborhoods of 0 consisting of all sets of type (1) $\{T\,|\,\|T\|<\varepsilon\}$, (2) $\{T\,|\,\|Tx\|<\varepsilon,\ x\in X\}$, and (3) $\{T\,|\,|f(Tx)|<\varepsilon,\ x\in X, f\in Y'\}$, where ε varies over all positive numbers, X over all finite subsets of $\mathfrak{X}$, and Y' over all finite subsets of $\mathfrak{Y}'$, the †dual space of $\mathfrak{Y}$. The uniform operator topology is thus the metric topology determined by the †norm in $B(\mathfrak{X},\mathfrak{Y})$ (→ 39 Banach Spaces C). Convergence of T_n to T with respect to one of these topologies is referred to as (1) **uniform convergence**, (2) **strong convergence**, or (3) **weak convergence**. We have such convergence if and only if (1) $\|T_n - T\| \to 0$, (2) $\|(T_n - T)x\| \to 0$ for every $x \in \mathfrak{X}$, or (3) $|f(T_n x - Tx)| \to 0$ for every $x \in \mathfrak{X}$ and $f \in \mathfrak{Y}'$.

D. Closed Operators

Closed operators play an important role when we deal, as is frequent in applications, with operators that are not necessarily continuous. An operator T from $\mathfrak{X}$ to $\mathfrak{Y}$ is said to be a **closed operator** if the graph of T is closed in $\mathfrak{X} \times \mathfrak{Y}$, or equivalently, if $x_n \in \mathfrak{D}(T)$, $x_n \to x$, and $Tx_n \to y$ imply $x \in \mathfrak{D}(T)$ and $Tx = y$. If T is continuous and $\mathfrak{D}(T)$ is closed, then T is closed. Conversely, if a linear operator T is closed and $\mathfrak{D}(T)$ is closed, then T is necessarily continuous (the **closed graph theorem**). An operator T from $\mathfrak{X}$ to $\mathfrak{Y}$ is said to be a **closable operator** if T has a closed extension, or equivalently if $x_n \in \mathfrak{D}(T)$, $x_n \to 0$, and $Tx_n \to y$ imply $y = 0$. When T is a linear operator from $\mathfrak{X}$ to $\mathfrak{Y}$ with $\mathfrak{D}(T)$ dense in $\mathfrak{X}$, the †dual operator (or **adjoint operator**) T' of T is defined to be an operator from $\mathfrak{Y}'$ to $\mathfrak{X}'$ determined by the relations $\mathfrak{D}(T') = \{f \in \mathfrak{Y}' \mid$ there exists $g \in \mathfrak{X}'$ such that $g(x) = f(Tx)$ for every $x \in \mathfrak{D}(T)\}$ and $T'f = g$, $f \in \mathfrak{D}(T')$ (→ 39 Banach Spaces). T' is always closed.

E. Operators between Hilbert Spaces

Throughout this section we suppose that $\mathfrak{X}$ and $\mathfrak{Y}$ are complex †Hilbert spaces. Let T be a densely defined linear operator from $\mathfrak{X}$ to $\mathfrak{Y}$. Instead of the dual T', it is sometimes convenient to use the operator T^* from $\mathfrak{Y}$ to $\mathfrak{X}$ determined by the relation $(x, T^*y) = (Tx, y)$, $x \in \mathfrak{D}(T)$. The operator T^* is called the **adjoint operator** (or **Hilbert space adjoint**) of T. By means of an antilinear isomorphism $\pi_{\mathfrak{X}}$ from $\mathfrak{X}$ onto $\mathfrak{X}'$ given by $(\pi_{\mathfrak{X}} x)(u) = (u, x)$ (†Riesz's theorem), T^* is related to T' by $T^* = \pi_{\mathfrak{X}}^{-1} T' \pi_{\mathfrak{Y}}$. The correspondence $T \to T'$ is linear, while the correspondence $T \to T^*$ is antilinear. $\mathfrak{D}(T^*)$ is dense in $\mathfrak{Y}$ if and only if T has a closed extension, and in this case the smallest closed extension $\bar{T}$ of T coincides with $T^{**} = (T^*)^*$. We call $\bar{T}$ the **closure** of T.

If a densely defined linear operator T in $\mathfrak{X}$ (i.e., T is from $\mathfrak{X}$ to $\mathfrak{X}$) satisfies $T \subset T^*$, then T is said to be a **symmetric operator** (or **Hermitian operator**). If $T = T^*$, then T is said to be a **self-adjoint operator**. A symmetric operator is always closable. A symmetric operator T is said to be essentially self-adjoint if the closure (i.e., the smallest closed extension) of T is self-adjoint. The operator $T \in B(\mathfrak{X},\mathfrak{Y})$ is said to be **partially isometric** if there exists a closed subspace $\mathfrak{M}$ of $\mathfrak{X}$ such that T is isometric on $\mathfrak{M}$ (i.e., $\|Tx\| = \|x\|$, $x \in \mathfrak{M}$) and zero on the †orthogonal complement of $\mathfrak{M}$. The closed subspace $\mathfrak{M}$ ($T\mathfrak{M}$) is called the **initial (final) set** of T. An operator $T \in B(\mathfrak{X},\mathfrak{Y})$ is partially isometric if and only if T^*T and TT^* are (orthogonal) †projections. The ranges of these projections are the initial and final sets of T, respectively. In particular, if $\mathfrak{M} = \mathfrak{X}$, then T is said to be an **isometric operator**. If $\mathfrak{X} = \mathfrak{Y} = \mathfrak{M} = \mathfrak{N}$, then T is said to be a **unitary operator**, and $T \in B(\mathfrak{X})$ is unitary if and only if $T^* = T^{-1}$. The set of self-adjoint (or unitary) operators in $\mathfrak{X}$ forms a subclass of the class of all **normal operators** T in $\mathfrak{X}$, and normal operators are characterized by the relation $T^*T = TT^*$. The structure of normal operators, especially that of self-adjoint operators, has been studied in detail by means of spectral analysis (→ 135 Eigenvalue Problems). Let T be a densely defined closed operator from $\mathfrak{X}$ to $\mathfrak{Y}$. Then there exist a nonnegative self-adjoint operator P in $\mathfrak{X}$ (here "nonnegative" means $(Px, x) \geqslant 0$, $x \in \mathfrak{D}(P)$) and a partially isometric operator W with initial set $\overline{\mathfrak{R}(P)} = \overline{\mathfrak{R}(T^*)}$ such that $T = WP$. The operators P and W are determined uniquely by these requirements. This is called the **canonical decomposition** of T.

A closed subspace $\mathfrak{M}$ of $\mathfrak{X}$ is said to be an **invariant subspace** of an operator $T \in B(\mathfrak{X})$ if T maps $\mathfrak{M}$ into $\mathfrak{M}$. Problems concerning invariant subspaces in a Hilbert space have been studied by many authors (see Helson [14]). For the shift operator in L_2-space (or the so-called Hardy space H_2) over the unit circle, a theorem of A. Beurling gives a function-theoretic characterization of all the invariant subspaces. The question whether an arbitrary bounded linear operator in a separable infinite-dimensional Hilbert space has a nontrivial invariant subspace remains open.

F. Resolvents and Spectra

Let T be a closed operator in a complex Banach space $\mathfrak{X}$, and I be the identity in $\mathfrak{X}$. The set $\rho(T)$ of all complex numbers λ such that $\lambda I - T$ has an inverse in $B(\mathfrak{X})$ is called the **resolvent set** of T, and the complement $\sigma(T)$ of $\rho(T)$ is called the **spectrum** of T. For $\lambda \in \rho(T)$ the operator $R(\lambda; T) = (\lambda I - T)^{-1} \in B(\mathfrak{X})$ (or sometimes $-R(\lambda; T)$) is called the

resolvent of T. If $\lambda_0 \in \rho(T)$, then $\{\lambda \mid |\lambda-\lambda_0| < \|R(\lambda_0; T)\|^{-1}\} \subset \rho(T)$. Hence $\rho(T)$ is an open, and $\sigma(T)$ a closed, set. The closedness of T implies that $\sigma(T) \neq \varnothing$. However, $\rho(T)$ may be empty (example (2) in Section K). The operator $R(\lambda; T)$ is a †holomorphic function (→ Section G) of λ in $\rho(T)$ and satisfies the (first) **resolvent equation**

$$R(\lambda_1; T) - R(\lambda_2; T)$$
$$= (\lambda_2 - \lambda_1) R(\lambda_1; T) R(\lambda_2; T)$$

for every $\lambda_1, \lambda_2 \in \rho(T)$. For every $T \in B(\mathfrak{X})$ the limit $r(T) = \lim \|T^n\|^{1/n} \leqslant \|T\|$, $n \to \infty$, exists, is called the †spectral radius of T, and satisfies $\{\lambda \mid |\lambda| > r(T)\} \subset \rho(T)$, $\{\lambda \mid |\lambda| = r(T)\} \cap \sigma(T) \neq \varnothing$. Concerning the dual operator, we have $\rho(T') = \rho(T)$ and $R(\lambda; T') = R(\lambda; T)'$. If $\mathfrak{X}$ is a Hilbert space, then $\rho(T^*) = \overline{\rho(T)}$ and $R(\lambda; T^*) = R(\bar{\lambda}; T)^*$, where $\bar{\ }$ stands for the complex conjugate.

G. Differential and Integral Calculus of Vector-Valued Functions

Calculus involving vector-valued functions $x(t)$ from a set Ω to a linear space X is an effective tool in operator theory. Some aspects of it will be discussed here, restricting $\mathfrak{X}$ to a Banach space. First, it is necessary to specify the topology for $\mathfrak{X}$. Usually, either the norm or the †weak topology is used. When $\mathfrak{X}$ is a Banach space consisting of operators, however, the †strong (or †weak) operator topology is also used. Differentiability, integrability, etc., with respect to one topology imply the same with respect to a weaker topology. Converse statements sometimes hold, possibly under additional conditions. (For example, an $\mathfrak{X}$-valued function on a finite measure space is †Bochner integrable if and only if it is †Gel'fand-Pettis integrable and almost separably valued.) Of course, the stronger the topology is, the more applicable the concept is.

When Ω is an interval, strong (weak) differentiability of $x(t)$ is defined as differentiability in terms of convergence in the normed (weak) topology. If $\Omega = [a, b]$ and $x(t)$ is strongly continuous (i.e., $\|x(t') - x(t)\| \to 0$ as $t' \to t$), the Riemann integral of $x(t)$ can be defined in a standard way, using convergence in the norm,

$$\int_a^b x(t)\,dt = \lim \sum x(t_i')(t_{i+1} - t_i).$$

The †fundamental theorem of calculus (i.e., strong differentiability of the indefinite integral) remains true. The Riemann integral is already quite useful in operator theory (→ 373 Semigroups of Operators). Like the Lebesgue integral of an $\mathfrak{X}$-valued function $x(t)$ defined on a †measure space Ω, the †Gel'fand-Pettis integral and the †Bochner integral (which are, in some sense, weak and strong integrals, respectively) are used most frequently (for more detailed discussions → 4 Abstract Integrals).

Let Ω be a domain in the complex plane, and $\mathfrak{X}$ a complex Banach space. If $f(x(t))$ is †holomorphic in Ω for every $f \in \mathfrak{X}'$, then there exists an $\mathfrak{X}$-valued function y on Ω such that

$$\|\delta^{-1}(x(t+\delta) - x(t)) - y(t)\| \to 0 \text{ as } \delta \to 0.$$

(When $\mathfrak{X} = B(\mathfrak{Y}, \mathfrak{Z})$, it suffices to assume that $g(x(t)y)$ is holomorphic for every $y \in \mathfrak{Y}$ and $g \in \mathfrak{Z}'$.) In short, there is no difference between "strong" and "weak" in analyticity. †Cauchy's integral theorem remains true for an $\mathfrak{X}$-valued holomorphic function $x(t)$, and the †Laurent expansion

$$x(t) = \sum_{n=-\infty}^{\infty} a_n (t - t_0)^n,$$

$$a_n = \frac{1}{2\pi i} \int_C x(t)(t - t_0)^{-n-1}\,dt,$$

is valid with the integral taken in the Riemannian sense. In this way results parallel to those in the theory of functions of a complex variable can be discussed. For instance, the †maximum (modulus) principle and certain theorems derived from it remain true (→ 200 Holomorphic Functions, 44 Bounded Functions).

Vector-valued holomorphic functions can also be defined in higher-dimensional spaces. Namely, let $x(t)$ be an $\mathfrak{X}$-valued function defined on a domain in a †complex (or †real analytic) manifold Ω. Suppose that for any $t \in \Omega$ there exists a neighborhood U of t such that $x(t)$ can be represented in U by a strongly convergent power series around t of holomorphic (real analytic) local coordinates. Then $x(t)$ is said to be an **$\mathfrak{X}$-valued holomorphic (real analytic) function** on Ω. This definition does not depend on the choice of the local coordinates.

When the domain of definition Ω of $f: \Omega \to \mathfrak{X}$ is an open subset in a complex Banach space $\mathfrak{Y}$ (with $\mathfrak{X}$ a complex space), we can still discuss derivatives of f. Suppose that for every $t \in \Omega$ and every $h \in \mathfrak{Y}$ the limit $\delta x(t; h) = \text{s-lim}_{\delta \to 0}(x(t + \delta h) - x(t))/\delta$ exists. Then $x(t)$ is said to be **Gâteaux differentiable** in Ω, and $\delta x(t; h)$ is called the **Gâteaux differential** of x at t. The function $\delta x(t; h)$ is linear in h. If there exists a $\delta x(t) \in B(\mathfrak{Y}, \mathfrak{X})$, $t \in \Omega$, such that $\lim_{\|h\| \to 0} \|x(t+h) - x(t) - \delta x(t) h\| / \|h\| = 0$, then $x(t)$ is said to be **Fréchet differentiable** at t, and $\delta x(t)$ is called the **Fréchet derivative** of x at t. Then $x(t)$ is Fréchet differentiable in Ω if and only if it is Gâteaux differentiable and $\delta x(t; h)$ is bounded on $\{\|h\| = 1\}$ for each $t \in \Omega$. In that case, $\delta x(t; h)$ can be written as

$\delta x(t;h)=\delta x(t)h$, and is called the **Fréchet differential** of x. When $\mathfrak{X}$ and $\mathfrak{Y}$ are real Banach spaces, linearity in h of $\delta x(t;h)$ is required as a part of the definition.

H. Operational Calculus

In operator theory, the term **operational calculus** generally indicates a way of defining "functions" $f(T)$ of an operator T so that a kind of algebraic isomorphism is established between a set of complex-valued functions f and the corresponding set of operators $f(T)$. The functions and operators that must be taken into consideration depend upon the nature of the problems to be solved, and accordingly there are several versions of operational calculus. We describe two typical ones. (1) Let T be a self-adjoint operator in a complex Hilbert space $\mathfrak{X}$ with the †spectral resolution $T=\int_{-\infty}^{\infty}\lambda dE(\lambda)$, and let f be a complex-valued Borel measurable function on $\mathbf{R}$. Then the operator $f(T)$ is uniquely determined by the following relations (→ 135 Eigenvalue Problems):

$$\mathfrak{D}(f(T))=\left\{x\,\middle|\,\int_{-\infty}^{\infty}|f(\lambda)|^2d(E(\lambda)x,x)<\infty\right\};$$

$$(f(T)x,y)=\int_{-\infty}^{\infty}f(\lambda)d(E(\lambda)x,y),$$

$$x\in\mathfrak{D}(f(T)),\quad y\in\mathfrak{H}.$$

Then $f(T)$ is normal and the correspondence $f\rightarrow f(T)$ satisfies the following relations: (i) $(\alpha f+\beta g)(T)\supset\alpha f(T)+\beta g(T)$; (ii) $(fg)(T)\supset f(T)g(T)$; and (iii) $f(T)^*=\bar{f}(T)$. If g is a bounded function, the inclusions in (i) and (ii) can be replaced by equalities. (2) Let $\mathfrak{X}$ be a complex Banach space, $T\in B(\mathfrak{X})$, and $\mathfrak{F}(T)$ the set of all functions holomorphic in a neighborhood of $\sigma(T)$. We define an operator $f(T)\in B(\mathfrak{X})$, $f\in\mathfrak{F}(T)$, by

$$f(T)=\frac{1}{2\pi i}\int_C f(t)R(t;T)\,dt, \qquad (*)$$

where C is a closed curve consisting of a finite number of rectifiable Jordan arcs, contains $\sigma(T)$ in its interior, and lies with its interior completely in the domain in which f is holomorphic. In this case relations (i) and (ii) hold with equality in place of inclusion. Instead of (iii) we have (iv) $\mathfrak{F}(T)=\mathfrak{F}(T')$ and $f(T')=f(T)'$. The integral appearing in $(*)$ is sometimes called the **Dunford integral**. In both situations described in (1) and (2) the **spectral mapping theorem** $\sigma(f(T))=f(\sigma(T))$ holds. When these two ways of defining $f(T)$ are possible, the resulting operators coincide.

Another kind of operational calculus can be constructed, for example, when T is the generator of a certain semigroup of operators (→ 373 Semigroups of Operators).

I. Isolated Singularities of the Resolvent

Let T be a densely defined closed operator in a complex Banach space $\mathfrak{X}$ and λ_0 an isolated point of $\sigma(T)$. Take a sufficiently small circle C around λ_0 and put

$$E=\frac{1}{2\pi i}\int_C R(\lambda;T)\,d\lambda,$$

which is a projection in $\mathfrak{X}$. Then the Laurent expansion around λ_0 of the resolvent is given by

$$R(\lambda;T)=\sum_{n=-\infty}^{\infty}A_n(\lambda-\lambda_0)^n,$$

with $A_n=(-1)^{n+1}(\lambda_0I-T)^{-(n+1)}E$ for $n<0$. When the dimension ν of the range of E is finite, λ_0 is a †pole of $R(\lambda;T)$ with order not exceeding ν, and λ_0 is an eigenvalue of T with multiplicity not exceeding ν. Furthermore, E is then a projection onto the †principal subspace belonging to the eigenvalue λ_0.

J. Extension of Symmetric Operators

In applications we frequently encounter the problem of finding self-adjoint extensions of a given symmetric operator, Let T be a closed symmetric operator in a Hilbert space $\mathfrak{X}$. Then $T\pm iI$ is one-to-one, and its range $\mathfrak{R}_\pm$ is a closed subspace of $\mathfrak{X}$. The operator $V_T=(T-iI)(T+iI)^{-1}$ from $\mathfrak{R}_+$ onto $\mathfrak{R}_-$ is isometric, and $(I-V_T)\mathfrak{R}_+$ is dense in $\mathfrak{X}$. We call V_T the **Cayley transform** of T. Conversely, let V be an isometric operator from a closed subspace $\mathfrak{M}$ of $\mathfrak{X}$ onto another closed subspace $\mathfrak{N}$ such that $(I-V)\mathfrak{M}$ is dense in $\mathfrak{X}$. Then the operator $T=i(I+V)(I-V)^{-1}$ is a closed symmetric operator satisfying $V_T=V$. Thus, the correspondence $T\rightarrow V_T$ is one-to-one onto; $T\subset S$ if and only if $V_T\subset V_S$, and T is self-adjoint if and only if V_T is unitary. The dimension $n_\pm$ of the subspace $\mathfrak{X}\ominus\mathfrak{R}_\pm=\{x\,|\,T^*x=\pm ix\}$ is called the **deficiency index** of T. Denoting the †residual spectrum of T by $\sigma_r(T)$ and putting $\Pi_\pm=\{\lambda|\mathrm{Im}\lambda\gtrless 0\}$, we have the following propositions: (i) According as $n_+>0$ or $n_+=0$, $\Pi_+\subset\sigma_r(T)$ or $\Pi_+\subset\rho(T)$ (similarly for n_- in place of n_+); (ii) T has a self-adjoint extension if and only if $n_+=n_-$; and (iii) T is self-adjoint if and only if $n_+=n_-=0$, or, equivalently, if and only if $\Pi_\pm\subset\rho(T)$. When T is symmetric but not closed, the previous arguments may be applied to the closure $\bar{T}$ of T. The deficiency index of $\bar{T}$ is also called the deficiency index of T. These arguments can be performed similarly with λ and $\bar{\lambda}$ $(\mathrm{Im}\lambda\neq 0)$ in place of i and $-i$.

We now describe two more concrete criteria for T to have self-adjoint extensions. (1) Semibounded operators: If there exists a real number γ such that $(Tx,x)\geqslant\gamma\|x\|^2$ for every

$x \in \mathfrak{D}(T)$, then T has a self-adjoint extension satisfying a similar inequality with the same constant γ. (2) Real operators: If T commutes with an involution in $\mathfrak{X}$, namely, if there exists an antilinear mapping J from $\mathfrak{X}$ onto $\mathfrak{X}$ such that $(Jx,Jy)=(y,x)$, $J^2=I$, and $JT=TJ$, then T has a self-adjoint extension (→ example (2) in Section K).

K. Examples of Linear Operators

(1) Integral Operators. Let E_j, $j=1,2$, be linear spaces consisting of functions defined on measure spaces Ω_j with measures μ_j. Let $k(t,s)$ be a measurable function on $\Omega_2 \times \Omega_1$ and define $\mathfrak{D}(\mathfrak{K})$ to be the set of all $x \in E_1$ such that $(Kx)(t)=\int_{\Omega_1} k(t,s)x(s)d\mu_1(s)$ belongs to E_2, where the integral is assumed to be absolutely convergent. The mapping that assigns Kx to each $x \in \mathfrak{D}(\mathfrak{K})$ determines a linear operator K from E_1 to E_2 with domain $\mathfrak{D}(K)=\mathfrak{D}(\mathfrak{K})$. K is called an **integral operator**, and $k(t,s)$ the **kernel** (or **integral kernel**) (of K). As an example, let $E_j=L_p(\Omega_j)$, $1 \leqslant p \leqslant \infty$, and suppose there exists an $M>0$ such that

$$\int_{\Omega_1} |k(t,s)|d\mu_1(s) \leqslant M,$$

$$\int_{\Omega_2} |k(t,s)|d\mu_2(t) \leqslant M.$$

Then $K \in B(E_1,E_2)$ with $\|K\| \leqslant M$ (→ 72 Compact Operators B).

(2) Differential Operators. For $\mathfrak{X}=L^2(0,1)$, let $\mathfrak{D}_0$ be the set of all $x \in C^2(0,1)$ with compact support in $(0,1)$ and $\mathfrak{D}_1$ the set of all $x \in C^1(0,1)$ such that $x'(t)$ is absolutely continuous in $(0,1)$ with $x'' \in \mathfrak{X}$. Then the operators T_j, $j=0,1$, determined by $(T_jx)(t)=-x''(t)$, $x \in \mathfrak{D}_j$, are linear in $\mathfrak{X}$. $T_0^*=T_1$, so that T_0 is a symmetric operator. Furthermore, T_0 is a real operator with respect to the involution $x \to \bar{x}$. Since two linearly independent solutions of $(T_1-\lambda I)x=0$ both belong to $\mathfrak{X}$, the deficiency indices $n_\pm$ of T_0 are 2. (Note that $\rho(T_1)=\varnothing$.) Self-adjoint extensions of T_0 are obtained by restricting the domains of T_1 by boundary conditions (→ 115 Differential Operators).

(3) †Fourier Transforms. For every $x \in L_p(\mathbf{R}^n)$, $1 \leqslant p \leqslant 2$,

$$(Ux)(t) = \lim_{m\to\infty} (2\pi)^{-n/2} \int_{|t|\leqslant m} \exp(-its)x(s)\,ds$$

converges in the norm of $L_q(\mathbf{R}^n)$, $p^{-1}+q^{-1}=1$. The operator U thus defined belongs to $B(L_p,L_q)$. When $p=q=2$, U is a unitary operator in L^2 [5, 8].

(4) Singular Integral Operators. Let $k(t)$ be a measurable bounded function on $\mathbf{R}^n$. Suppose that k is homogeneous of degree 0 and that the integral of k over the unit sphere vanishes. Then for every $x \in L_p(\mathbf{R}^n)$, $1<p<\infty$,

$$(Tx)(t) = \lim_{\varepsilon\downarrow 0} \int_{|t-s|>\varepsilon} \frac{k(t-s)}{|t-s|^n} x(s)\,ds$$

converges in the norm of $L_p(\mathbf{R}^n)$. The operator T thus defined belongs to $B(L_p)$. Such operators are called Calderón-Zygmund singular integral operators and form a subclass of a more general class of operators called †pseudodifferential operators. The †Hilbert transform is an example of a singular integral operator. They provide a powerful means of studying †unique continuation theorems for solutions of elliptic partial differential equations [10, 11] (→ 318 Partial Differential Equations of Elliptic Type).

References

[1] N. Dunford and J. T. Schwartz, Linear operators, Interscience, I, 1958; II, 1963.
[2] E. Hille and R. S. Phillips, Functional analysis and semigroups, Amer. Math. Soc. Colloq. Publ., second edition, 1957.
[3] F. Riesz and B. Sz. Nagy, Leçons d'analyse fonctionelle, Akadémiai Kiadó, third edition, 1955; English translation, Functional analysis, Ungar, 1955.
[4] M. H. Stone, Linear transformations in Hilbert space and their applications to analysis, Amer. Math. Soc. Colloq. Publ., 1932.
[5] G. Sunouchi, Hûrie kaiseki (Japanese; Fourier analysis), Kyôritu, 1956.
[6] K. Yosida, Isôkaiseki (Japanese; Functional analysis), Iwanami, 1951.
[7] K. Yosida, Functional analysis, Springer, 1965.
[8] E. C. Titchmarsh, Introduction to the theory of Fourier integrals, Clarendon Press, 1937.
[9] A. P. Calderón and A. Zygmund, Singular integral operators and differential equations, Amer. J. Math., 79 (1957), 901–921.
[10] A. P. Calderón, Uniqueness in the Cauchy problem for partial differential equations, Amer. J. Math., 80 (1958), 16–36.
[11] S. Mizohata, Unicité du prolongement des solutions pour quelques opérateurs différentiels paraboliques, Mem. Coll. Sci. Univ. Kyôto, 31 (1958), 219–239.
[12] Л. В. Канторович, Г. Р. Акилов (L. V. Kantorovič and G. R. Akilov), Функциональный анализ в нормированных пространствах, Физматгиз, 1965; English translation, Functional analysis in normed spaces, Pergamon, 1964.

[13] И. Ц. Гохберг, М. Г. Крейн (I. C. Gohberg and M. G. Kreĭn), Введение в теорию линейных несамосопряженных операторов, Наука, 1965; English translation, Introduction to the theory of linear non-selfadjoint operators, Amer. Math. Soc. Transl. of Math. Monographs, 1969, vol. 18.
[14] H. Helson, Lectures on invariant theory, Academic Press, 1964.
Also → references to 199 Hilbert Spaces.

252 (XIII.7) Linear Ordinary Differential Equations

A. General Remarks

Let $p_1(x), \ldots, p_n(x), q(x)$ be known functions of a real (or complex) variable x. An ordinary differential equation

$$y^{(n)}+p_1(x)y^{(n-1)}+\ldots+p_n(x)y=q(x) \qquad (1)$$

containing an unknown function y and its derivatives $y', y'', \ldots, y^{(n)}$ of order up to n is called a **linear ordinary differential equation** of the nth **order**. In particular, a linear differential equation

$$y^{(n)}+p_1(x)y^{(n-1)}+\ldots+p_n(x)y=0 \qquad (1')$$

with $q(x)\equiv 0$ is said to be **homogeneous**. If $q(x)\not\equiv 0$, (1) is said to be **inhomogeneous**. Every †singular point of a solution of (1) is a point of discontinuity (or singular point) for at least one of the coefficients $p_k(x), q(x)$ (→ 254 Linear Ordinary Differential Equations (Singular Points)). Namely, the **unique existence theorem for solutions** holds:

Let the coefficients $p_k(x)$, $q(x)$ be continuous in a real domain D. Then for every point x_0 in D and every n-tuple of numbers $\eta, \eta', \ldots, \eta^{(n-1)}$, there exists one and only one solution $y(x)$ of (1) satisfying the initial conditions

$$y(x_0)=\eta, y'(x_0)=\eta', \quad \ldots, y^{(n-1)}(x_0)=\eta^{(n-1)} \qquad (2)$$

and such that $y(x), y'(x), \ldots, y^{(n)}(x)$ are all continuous in D. If D is a complex domain (not containing the point at infinity) and the complex functions $p_k(x), q(x)$ are †holomorphic in D, then there exists one and only one complex-valued solution $y(x)$ of (1) satisfying (2) and such that $y(x)$ is holomorphic in D.

B. Fundamental Systems of Solutions

The totality of solutions of a homogeneous linear ordinary differential equation forms a †linear space (over the real or complex field). That is, any linear combination $y(x)=\sum_{i=1}^m C_i y_i(x)$ of the solutions $y_1, y_2, \ldots, y_m$ of (1′), where the C_i are †arbitrary constants, is also a solution of (1′). This is called the **principle of superposition**. More than $n+1$ solutions of (1′) are always linearly dependent, that is, if $m \geqslant n+1$, we can find m constants $C_1, C_2, \ldots, C_m$, not all equal to zero, such that $\sum_{i=1}^m C_i y_i(x)=0$. Equation (1′) has n linearly independent solutions. For instance, the n solutions $y_1, y_2, \ldots, y_n$ defined by the initial conditions

$$\begin{gathered} y_1(x_0)=1, y_1'(x_0)=0, \ \ldots, y_1^{(n-1)}(x_0)=0, \\ y_2(x_0)=0, y_2'(x_0)=1, \ \ldots, y_2^{(n-1)}(x_0)=0, \\ \ldots \\ y_n(x_0)=0, y_n'(x_0)=0, \ \ldots, y_n^{(n-1)}(x_0)=1 \end{gathered} \qquad (3)$$

are linearly independent. Such a system of n linearly independent solutions $y_1, \ldots, y_n$ of (1′) is called a **fundamental system of solutions** of (1′). In terms of a fundamental system of solutions $y_1, \ldots, y_n$, each solution y of (1′) is represented uniquely in the form $y(x)=\sum_{i=1}^n C_i y_i(x)$.

C. Liouville's Formula

In order for n solutions $y_1, y_2, \ldots, y_n$ of (1′) to be linearly independent, it is necessary and sufficient that the †Wronskian determinant $W(y_1, y_2, \ldots, y_n)\neq 0$ in D. Furthermore, the coefficients $p_k(x)$ can be represented in terms of an arbitrary fundamental system of solutions $y_1, y_2, \ldots, y_n$ since the coefficient of $y^{(n-k)}$ in the expansion of

$$\frac{(-1)^n W(y, y_1(x), y_2(x), \ldots, y_n(x))}{W(y_1(x), y_2(x), \ldots, y_n(x))} \qquad (4)$$

is identically equal to $p_k(x)$ in D. Using this equality for $p_1(x)$, we have **Liouville's formula**:

$$W(y_1(x), \ldots, y_n(x)) = W(y_1(x_0), \ldots, y_n(x_0)) \exp\left(-\int_{x_0}^x p_1(t)\,dt\right). \qquad (5)$$

D. Lagrange's Method of Variation of Constants

The difference of two solutions of the inhomogeneous equation (1) is a solution of the homogeneous equation (1′). Consequently, the †general solution of (1) can be represented as the sum of a †particular solution of (1) and the †general solution of (1′). Since a particular solution of (1) can be obtained from an arbitrary fundamental system of solutions $y_1, y_2, \ldots, y_n$ of (1′), (1) can be solved if a fundamental system of solutions for (1′) is known. In fact, if we consider $C_1, C_2, \ldots, C_n$

in the representation $y=\Sigma_{i=1}^{n}C_iy_i(x)$, not as constants, but as functions of x, and determine them by the conditions

$$y_1(x)C_1'(x)+y_2(x)C_2'(x)+\ldots+y_n(x)C_n'(x)=0,$$
$$y_1'(x)C_1'(x)+y_2'(x)C_2'(x)+\ldots+y_n'(x)C_n'(x)=0,$$
$$\ldots$$
$$y_1^{(n-1)}(x)C_1'(x)+y_2^{(n-1)}(x)C_2'(x)+\ldots+y_n^{(n-1)}(x)C_n'(x)=q(x), \quad (6)$$

then $y(x)=\Sigma_{i=1}^{n}C_i(x)y_i(x)$ is a solution of (1). This is always possible because from (6) we obtain

$$\frac{dC_i}{dx}=\frac{q(x)W_i(x)}{W(y_1(x),\ldots,y_n(x))}, \quad (7)$$

where $W_i(x)$ is the †cofactor of $y_i^{(n-1)}(x)$ in the determinant $W(y_1(x),\ldots,y_n(x))$. This method is called **Lagrange's method of variation of constants** (or **variation of parameters**).

E. Linear Ordinary Differential Equations with Constant Coefficients

A linear ordinary differential equation

$$y^{(n)}+a_1y^{(n-1)}+\ldots+a_{n-1}y'+a_ny=0 \quad (8)$$

with constant coefficients a_i has $y=\exp rx$ as a solution if r is a root of the algebraic equation

$$f(r)=r^n+a_1r^{n-1}+\ldots+a_{n-1}r+a_n=0, \quad (9)$$

called the **characteristic equation** of (8). Let $r_1,r_2,\ldots,r_m$ be the distinct roots of (9), and suppose that the root r_i has †multiplicity μ_i $(i=1,2,\ldots,m)$. Then the set of functions

$$e^{r_1x},xe^{r_1x},\ldots,x^{\mu_1-1}e^{r_1x};\ldots; \quad (10)$$
$$e^{r_mx},xe^{r_mx},\ldots,x^{\mu_m-1}e^{r_mx}$$

is a fundamental system of solutions of (8).

F. D'Alembert's Method of Reduction of Order

Let $y_1(x)$ be a solution, not identically equal to zero, of the homogeneous equation (1′). By substituting $y=y_1z$ into (1′), we see that z' satisfies a linear differential equation of order $n-1$. This method is called **d'Alembert's method of reduction of order**. Since linear ordinary differential equations of the first order can be integrated by quadrature (→ Appendix A, Table 14.I), a homogeneous linear ordinary differential equation of the second order can be integrated completely if one solution of the equation that does not identically vanish is known.

So far we have outlined a general theory of solutions of (1) in the domain where solutions are continuous or holomorphic, but in order to have thorough knowledge of all the solutions, we have to examine their behavior also in the neighborhood of a singular point (which is a discontinuity point or a singular point for at least one of the coefficients) (→ 253 Linear Ordinary Differential Equations (Global Theory), 254 Linear Ordinary Differential Equations (Singular Points)). Also, †boundary value problems are important as well as †initial value problems described before, especially for second-order equations in connection with mathematical physics [4]. For these → 309 Ordinary Differential Equations (Boundary Value Problems); 135 Eigenvalue Problems.

G. Systems of Linear Ordinary Differential Equations of the First Order

Let the $f_{ij}(x)$ be known functions. A **system of linear differential equations of the first order**

$$dy_1/dx=f_{11}(x)y_1+\ldots+f_{1n}(x)y_n+g_1(x),$$
$$dy_2/dx=f_{21}(x)y_1+\ldots+f_{2n}(x)y_n+g_2(x),$$
$$\ldots$$
$$dy_n/dx=f_{n1}(x)y_1+\ldots+f_{nn}(x)y_n+g_n(x) \quad (11)$$

with n unknowns $y_1,y_2,\ldots,y_n$ contains (1) as a special case, since (1) is transformed into (11) by setting $y=y_1$, $y'=y_2$, $\ldots$, $y^{(n-1)}=y_n$. A system (11) with $g_i(x)\equiv 0$, i.e., a system

$$dy_1/dx=f_{11}(x)y_1+\ldots+f_{1n}(x)y_n,$$
$$\ldots$$
$$dy_n/dx=f_{n1}(x)y_1+\ldots+f_{nn}(x)y_n \quad (11')$$

is said to be **homogeneous**, while (11) is said to be **inhomogeneous**. A point at which at least one solution $(y_1,y_2,\ldots,y_n)$ is discontinuous (singular) is a discontinuity point (singular point) for at least one of the coefficients $f_{ij}(x),g_i(x)$. Suppose that the $f_{ij}(x)$, $g_i(x)$ are continuous in a real domain D. Then the **unique existence theorem of solutions** holds:

To every point x_0 in D and every initial condition

$$y_1(x_0)=b_1,y_2(x_0)=b_2,\ \ldots,y_n(x_0)=b_n,$$

there corresponds one and only one solution that is continuous in D. If D is a complex domain (not containing the point at infinity) and the f_{ij}, g_i are holomorphic in D, then the solution $y_i(x)$ is holomorphic in D.

H. Fundamental Systems of Solutions

If $m\geqslant n+1$, m solutions $(y_{1i},y_{2i},\ldots,y_{ni})$ $(i=1,2,\ldots,m)$ of (11′) are linearly dependent, i.e., we can find constants $C_1,C_2,\ldots,C_m$, not all equal to zero, such that $\sum_{i=1}^{m} C_i\, y_{ki}(x)=0$ $(k=1,2,\ldots,n)$. System (11′) has n linearly independent solutions. To see this, we have only to choose the initial conditions so that

$$y_{ik}(x_0)=1,\quad i=k,$$
$$=0,\quad i\neq k.$$

Such a system of n linearly independent solutions $(y_{1i}, y_{2i}, \ldots, y_{ni})$ $(i=1,2,\ldots,n)$ is called a **fundamental system of solutions** of (11′). In terms of this fundamental system, any solution $(y_1,\ldots,y_n)$ of (11′) is represented uniquely in the form $y_k(x)=\sum_{j=1}^{n} C_j y_{kj}(x)$ $(k=1,2,\ldots,n)$.

The linear independence of n solutions $(y_{11},\ldots,y_{1n}),\ldots,(y_{n1},\ldots,y_{nn})$ is equivalent to the condition that the determinant

$$\Delta(x)=\begin{vmatrix} y_{11}(x) & y_{12}(x) & \ldots & y_{1n}(x) \\ y_{21}(x) & y_{22}(x) & \ldots & y_{2n}(x) \\ & & \ldots & \\ y_{n1}(x) & y_{n2}(x) & \ldots & y_{nn}(x) \end{vmatrix}$$

does not vanish in D. Corresponding to Liouville's formula (5), we have

$$\Delta(x)=\Delta(x_0)\exp\left(\sum_{i=1}^{n}\int_{x_0}^{x} f_{ii}(t)\,dt\right).$$

I. Method of Variation of Constants

The general solution $(Y_1, Y_2, \ldots, Y_n)$ of the inhomogeneous equation (11) is given as the sum of the general solution $(y_1, y_2, \ldots, y_n)$ of (11′) and one particular solution $(Y_{10}, Y_{20}, \ldots, Y_{n0})$ of (11), i.e., in the form

$$(y_1+Y_{10}, y_2+Y_{20},\ldots, y_n+Y_{n0}).$$

To obtain the particular solution $(Y_{10}, Y_{20}, \ldots, Y_{n0})$, we take any fundamental system of solutions

$$y_1=\varphi_{1k}(x), y_2=\varphi_{2k}(x),\ldots, y_n=\varphi_{nk}(x),\qquad k=1,2,\ldots,n$$

of (11′) and consider the constants u_k in the linear combination

$$y_i=\sum_{k=1}^{n}\varphi_{ik}(x)u_k,\qquad i=1,2,\ldots,n \tag{12}$$

as functions of x. Substituting (12) into (11), we obtain a system of differential equations

$$\sum_{k=1}^{n}\varphi_{ik}(x)u_k'(x)=g_i(x),\qquad i=1,2,\ldots,n \tag{13}$$

with unknowns u_k. Since the y_i form a fundamental system, the determinant of the matrix with elements $\varphi_{ik}(x)$ does not vanish. Hence (13) can be solved in the form

$$u_k'(x)=G_k(x),\qquad k=1,2,\ldots,n,$$

and the $u_k(x)$ can be obtained by quadrature. Consequently, it follows that one particular solution can be given, in terms of the $u_k(x)$, in the form (12). This method is also called the **method of variation of constants.**

J. Systems of Linear Ordinary Differential Equations with Constant Coefficients

If the coefficients f in (11′) are all constants, the general solution has the form

$$y_j=\sum_{k=1}^{m}P_{jk}(x)e^{\lambda_k x},\qquad j=1,2,\ldots,n.$$

Here $\lambda_1,\lambda_2,\ldots,\lambda_m$ are the distinct roots of the **characteristic equation**

$$\begin{vmatrix} f_{11}-\lambda & f_{12} & \ldots & f_{1n} \\ f_{21} & f_{22}-\lambda & \ldots & f_{2n} \\ & & \ldots & \\ f_{n1} & f_{n2} & \ldots & f_{nn}-\lambda \end{vmatrix}=0,$$

and if λ_k has multiplicity e_k $(\sum_{j=1}^{m}e_j=n)$, $P_{jk}(x)$ is a polynomial of degree at most e_k-1 which contains e_k arbitrary constants.

Suppose that the coefficients $f_{jk}(x)$ in (11′) are all periodic functions having the same period ω. Then there exists a linear transformation $y_j=\sum q_{jk}(x)z_k$ in which the q_{jk} are periodic with period ω, such that the original equation is reduced to $dz_j/dx=\sum c_{jk}z_k$, where the c_{jk} are constants. Hence if we can find such a linear transformation, we can integrate the original equation.

K. Adjoint Differential Equations

Consider a linear homogeneous ordinary differential equation

$$F(y)=p_0y^{(n)}+p_1y^{(n-1)}+\ldots+p_ny=0.$$

Integration by parts of $\int\bar{z}F(y)\,dx$ gives

$$\bar{z}F(y)-y\,\overline{G(z)}=d[R(y,x)]/dx,$$

$$R(y,z)=\sum_{k=1}^{n}\sum_{h=0}^{k-1}(-1)^h y^{(k-h-1)}(p_{n-k}\bar{z})^{(h)},$$

$$G(z)=(-1)^n\left((\bar{p}_0z)^{(n)}-(\bar{p}_1z)^{(n-1)}+\ldots+(-1)^n\bar{p}_nz\right). \tag{14}$$

We call $G(z)=0$ the †adjoint differential equation of $F(y)=0$. It turns out that the adjoint differential equation of the adjoint equation of $F(y)=0$ coincides with $F(y)=0$. If y is a solution of $F(y)=0$, then the solution z of the adjoint differential equation satisfies the $(n-1)$st-order differential equation $R(y,z)=$ constant. When $G(y)=F(y)$, $F(y)=0$ is called a †self-adjoint differential equation. In the case of second-order equations with real coefficients, its general form is

$$F(y)=d(p\,dy/dx)/dx+qy=0. \tag{15}$$

For systems of differential equations, the †adjoint system of (11′) is defined by

$$dz_j/dx = -\bar{f}_{1j}z_1 - \bar{f}_{2j}z_2 - \ldots - \bar{f}_{nj}z_n, \quad j=1,2,\ldots,n. \qquad (16)$$

Conversely, (11′) is the adjoint system to (16). If $(y_1, y_2, \ldots, y_n)$, $(z_1, z_2, \ldots, z_n)$ are solutions of (11′), (16), respectively, then we have $\sum_{j=1}^{n} y_j\bar{z}_j = \text{constant}$. The system (11′) is called a †self-adjoint system of differential equations if (11′) coincides with (16), i.e., if $f_{jk}(x) = -\bar{f}_{kj}(x)$ (→ 8 Adjoint Differential Equations).

L. Laplace and Euler Transforms

When the coefficients $p_i(x)$ $(i=1,2,\ldots,n)$ in (1′) are rational functions, it often happens that we can find a solution in the form

$$y(x) = \int_a^b v(t)e^{xt}\,dt. \qquad (17)$$

Namely, we can often find a suitable function $v(t)$ so that the †Laplace transform (17) of $v(t)$ is a solution of (1′). Similarly, it is often possible to find a solution of (1′) as the †Euler transform

$$y(x) = \int_a^b v(t)(1-x)^{\rho-1}\,dt \qquad (18)$$

of some suitable $v(t)$. These transforms are used for the integral representation of special functions.

M. Linear Ordinary Differential Equations and Special Functions

A number of transcendental functions, such as †hypergeometric functions, †Bessel functions, †Legendre functions, etc., and Hermite polynomials, Laguerre polynomials, Jacobi polynomials, etc., are defined by linear ordinary differential equations of the second order (→ 381 Special Functions).

References

[1] M. Fujiwara, Zyôbibun hôteisiki ron (Japanese; Theory of ordinary differential equations), Iwanami, 1930.
[2] M. Hukuhara, Zyôbibun hôteisiki (Japanese; Ordinary differential equations), Iwanami, 1950.
[3] T. Yoshiye, Syotô zyôbibun hôteisiki (Japanese; Elementary theory of ordinary differential equations), Syôkabô, second edition, 1945.
[4] K. Yosida, Sekibun hôteisiki ron (Japanese; Theory of integral equations), Iwanami, 1950; English translation, Lectures on differential and integral equations, Interscience, 1960.
[5] M. Hukuhara, Zyôbibun hôteisiki no kaihô II, Senkei no bu (Japanese; Methods of solution of ordinary differential equations II. Linear differential equations), Iwanami, 1941.
[6] Y. Komatu, Zyôbibun hôteisiki ron (Japanese; Theory of ordinary differential equations), Hirokawa, 1965.
[7] L. Bieberbach, Theorie der gewöhnlichen Differentialgleichungen, Springer, second edition, 1965.
[8] E. Picard, Traité d'analyse III, Gauthier-Villars, 1896.
[9] G. Sansone, Equazioni differenziali nel campo reale I, II, Zanichelli, second edition, 1948–1949.
[10] K. O. Friedrichs, Lectures on advanced ordinary differential equations, Gordon and Breach, 1965.
[11] E. Hille, Lectures on ordinary differential equations, Addison-Wesley, 1968.

253 (XIII.9) Linear Ordinary Differential Equations (Global Theory)

A. General Remarks

Let there be given a †linear differential equation of the nth order

$$y^{(n)} + p_1(x)y^{(n-1)} + \ldots + p_n(x)y = 0, \qquad (1)$$

or a †system of linear differential equations

$$\mathbf{y}' = A(x)\mathbf{y}, \qquad (2)$$

which is the vector-matrix expression of

$$y_j' = \sum_{k=1}^{n} a_{jk}(x)y_k, \qquad j=1,\ldots,n, \qquad (2')$$

where the coefficients $p_k(x)$, $a_{jk}(x)$ are complex analytic functions of x in a certain complex domain D. The solutions of (1) or (2) are known to be holomorphic when the coefficients are all holomorphic. However, at a †singular point of at least one of the coefficients, a †branch point of the solution usually appears. Thus a solution of (1) or (2) is, in general, a †multivalent analytic function of x. The object of global theory is the function-theoretic study of this function—that is, determination of its †Riemann surface and investigation of its behavior on the Riemann surface.

At a holomorphic point of the coefficients, a †Taylor expansion of the solution can be obtained easily. If the equation is of †Fuchsian type, an explicit expression of the solution can be obtained even at the singular point. More precisely, the solution can be expressed explicitly by a certain series convergent within a circle around the singular point not containing another singular point in its interior (→ 254 Linear Ordinary Differential Equations (Singular Points)). In the presence of an †irregular singular point, instead of a convergent expression, we can construct an †asymptotic expansion valid within a certain sector whose vertex is situated at the singular point. Once such expressions have been obtained, the remaining task is to find the relations connecting those locally valid expressions, which are called **connection formulas**. There lies the main and most difficult part of global theory.

B. Monodromy Groups

Suppose that the coefficients of equation (2) are defined on a certain Riemann surface $\mathfrak{F}$ with singular points $a_1, a_2, \ldots$. By deleting $a_1, a_2, \ldots$ from $\mathfrak{F}$, another Riemann surface $\mathfrak{F}'$ is obtained. Choose a point x on $\mathfrak{F}'$, and let $Y(x)$ be any fixed branch of a †fundamental system of solutions of (2). Also, let Γ be a circuit on $\mathfrak{F}'$ starting from x. By an analytic continuation along Γ, another branch of $Y(x)$ is obtained, which we denote by $Y(x\Gamma)$. It is known that in this case these two branches are connected by the relation $Y(x\Gamma) = Y(x)C_\Gamma$, where C_Γ is an $n \times n$ constant matrix, and also that if Γ_1 and Γ_2 are †homotopic circuits, $C_{\Gamma_1} = C_{\Gamma_2}$. So the branch $Y(x\Gamma)$ and the matrix C_Γ are determined by the †homotopy class γ to which Γ belongs. Thus we can write $Y(x\gamma)$ or C_γ instead of $Y(x\Gamma)$ or C_Γ.

Now let G be the †fundamental group of $\mathfrak{F}'$. Since $Y(x\gamma_2\gamma_1) = Y(x\gamma_2)C_{\gamma_1} = Y(x)C_{\gamma_2}C_{\gamma_1}$ for any $\gamma_1, \gamma_2 \in G$, the correspondence $\gamma \to C_\gamma$ defines a representation of G. The group $g = \{C_\gamma \mid \gamma \in G\}$, which is naturally homomorphic to G, is called the **monodromy group** of equation (2).

For equation (1), we can also define the monodromy group of the equation by $g = \{C_\gamma \mid \gamma \in G\}$, where C_γ is a matrix such that $(y_1(x\gamma), \ldots, y_n(x\gamma)) = (y_1(x), \ldots, y_n(x))C_\gamma$, where $y_1(x), \ldots, y_n(x)$ are linearly independent solutions of (1).

If the equation is of Fuchsian type, the global problem can be regarded as solved when the monodromy group of the equation has been completely determined.

If $\mathfrak{F}$ is a complex sphere and the equation is of Fuchsian type, the number of singular points of the coefficients is of course finite. Let $a_1, \ldots, a_m$ be those singular points, and γ_k be the homotopy class of $\mathfrak{F}'$ determined by a closed curve Γ_k which encloses only one singular point a_k. Then the monodromy group g is generated by the matrices $C_{\gamma_1}, \ldots, C_{\gamma_m}$. Obviously $C_{\gamma_1}, \ldots, C_{\gamma_m}$ are not necessarily independent. At least one relation $C_{\gamma_1} \ldots C_{\gamma_m} = I$ (a unit matrix) always holds. In this case, †Jordan canonical forms of C_{γ_k} are all determined from the convergent expression for the fundamental system of solutions valid around a_k constructed by the famous †Frobenius method. However, the calculation of C_{γ_k} itself is generally impossible.

If $n=2$, $m=3$, and the coefficients are all rational functions of x, equation (1) or (2) is completely determined if we fix the roots of †indicial equations at every a_k, as long as the equation is of Fuchsian type. Therefore, the monodromy group is determined by the values of the roots of indicial equations. Since these roots are calculated purely algebraically, the monodromy group is determined by algebraic procedure in this case [12].

Let $n=2$, $m=3$ in equation (1). Denote by a, b, c the three singular points and by λ, λ'; μ, μ'; ν, ν' the roots of indicial equations at a, b, c, respectively. As was mentioned in the previous paragraph, the equation is determined uniquely by these nine quantities. Hence they also determine a family of functions consisting of all the solutions of the equation. This family is usually written as

$$P\begin{Bmatrix} a & b & c & \\ \lambda & \mu & \nu & x \\ \lambda' & \mu' & \nu' & \end{Bmatrix}$$

and is called the ***P*-function of Riemann** [1, 6, 7]. A simple transformation of variables reduces it to the totality of solutions of Gauss's †hypergeometric differential equation

$$x(1-x)y'' + (\gamma - (\alpha+\beta+1)x)y' - \alpha\beta y = 0. \tag{3}$$

Any solution of (3) is expressed by a **hypergeometric integral**

$$\int_C t^{\alpha-\gamma}(t-1)^{\gamma-\beta-1}(t-x)^{-\alpha}\,dt,$$

where C is a suitably chosen path of integration [12] (→ 209 Hypergeometric Functions).

Calculation of the monodromy group of the equation (1) or (2) is still an unsolved problem except for the case $n=2$, $m=3$ and a few other particular cases, e.g., $n<2$ or $m<3$.

C. Equations with an Irregular Singular Point

In the presence of an irregular singular point, complete knowledge of the monodromy group is still insufficient for the solution of the global problem. It is only the structure of the Riemann surface that is known from the monodromy group, and the behavior of the solution on the Riemann surface still remains to be studied. At an irregular singular point, the solution can be expressed only by an †asymptotic series valid within a certain sector, and the same solution possesses completely different expressions in different sectors. This is called †Stokes's phenomenon (→ 254 Linear Ordinary Differential Equations (Singular Points)). Thus, to complete the global theory, connection formulas between different asymptotic expressions must be established.

For a second-order linear equation with two singular points, one of which is regular and the other irregular of the first rank, the problem is completely solved. In this case, the equation can be reduced to a †confluent hypergeometric differential equation [3, 7] (→ 172 Functions of Confluent Type). The problem is also partly solved for a linear equation of higher order with two singular points, one of which is regular and the other irregular [8, 11, 17].

If two singular points are both irregular, even the monodromy group cannot be calculated in general. For such a case, G. D. Birkhoff proposed a method of reducing one of the singular points to a regular one. He showed that this procedure is possible under certain assumptions on the monodromy matrix [2, 18].

D. Riemann's Problem

As a noteworthy result for equation (1) of Fuchsian type with algebraic coefficients, Poincaré's theory deserves special mention. According to his theory, a solution of (1) can be uniformized in the form $y=f(z)$, $x=g(z)$, where f and g are single-valued analytic functions of z. Although it is known generally that any analytic function admits such uniformization (→ 362 Riemann Surfaces), Poincaré's theory affords a more explicit and efficient uniformizing construction. As uniformizing parameter z, we may take a ratio of two independent solutions of a certain linear differential equation of the second order which is determined from (1), and f and g are, in general, †Fuchsian functions, i.e., †automorphic functions for a certain †Fuchsian group, save for a few exceptional cases in which they are rational or †elliptic functions.

Brief mention should be made of **Riemann's problem** as a problem closely related to the global theory of linear differential equations. This problem was taken up by †Hilbert in his famous Paris lecture as the 21st problem, and hence is often called the Riemann-Hilbert problem. The problem can be stated as follows: Suppose that we are given a Riemann surface $\mathfrak{F}$, points $a_1, a_2, \ldots$ on $\mathfrak{F}$, and a group g of $n\times n$ matrices homomorphic to the fundamental group of $\mathfrak{F}-\{a_1, a_2, \ldots\}$. Then find an equation of the form (2) such that (i) the coefficient $A(x)$ is single-valued and †meromorphic on $\mathfrak{F}$; (ii) singular points are all regular and situated at $a_1, a_2, \ldots$; and (iii) the monodromy group of the equation coincides with g if a fundamental system of solutions is suitably chosen. Extensive research was done by many mathematicians, and finally H. Röhrl succeeded in solving the problem [5, 9, 10, 14].

The monodromy group of a linear differential equation is generally influenced by the change of position of the singular points a_1, $a_2, \ldots$. Therefore, to keep the monodromy group invariant, the coefficients of the equation must be suitable functions of $a_1, a_2, \ldots$. Then what kind of functions should they be? This is the problem proposed by L. Schlesinger. He treated the case of equation (2), where $A(x)$ is of the form

$$A(x)=\sum_{j=1}^{m-1}\frac{A_j}{x-a_j}\qquad (A_j\text{: constant matrices}),$$

with singular points $a_1, a_2, \ldots, a_{m-1}, \infty$. According to Schlesinger, in order that the monodromy group remain invariant, A_j must depend on $a_1, \ldots, a_{m-1}$ in such a way that

$$\frac{\partial A_j}{\partial a_k}=\frac{A_kA_j-A_jA_k}{a_j-a_k},\quad j\neq k,$$

$$\sum_{j=1}^{m-1}\frac{\partial A_j}{\partial a_j}=0 \tag{4}$$

[15, 16].

The system (4) was found to be in close connection with the theory of nonlinear differential equations without †movable branch points. For example, take the case $n=2$, $m=4$, where $a_1=0$, $a_2=1$, $a_3=t$. Then the A_j are functions of one variable t, and (4) is a system of ordinary differential equations. In this case, (4) can be reduced to a nonlinear equation of the second order without movable branch points. Such equations were thoroughly studied by P. Painlevé, and were found to be reducible to one of the six normal forms established by him (→ 284 Nonlinear Ordinary Differential Equations (Global Theory)). In our case, the reduced equation belongs to type (VI) of those normal forms. R. Garnier investigated the system (4) in detail to solve Riemann's problem on a complex sphere [4].

References

[1] L. Bieberbach, Theorie der gewöhnlichen Differentialgleichungen, Springer, second edition, 1965.
[2] G. D. Birkhoff, Equivalent singular points of ordinary linear differential equations, Math. Ann., 74 (1913), 134–139.
[3] E. A. Coddington and N. Levinson, Theory of ordinary differential equations, McGraw-Hill, 1955.
[4], R. Garnier, Solution du problème de Riemann pour les systèmes différentielles linéaires du second ordre, Ann. Sci. Ecole Norm. Sup., 43 (1926), 177–307.
[5] E. Hilb, Lineare Differentialgleichungen im komplexen Gebiet, Enzykl. Math., Bd. II, 2, Heft 4 (1913), 471–562.
[6] M. Hukuhara, Zyôbibun hôteisiki no kaihô II, Senkei no bu (Japanese; Methods of solution of ordinary differential equations pt. II. Linear differential equations), Iwanami, 1941.
[7] M. Hukuhara, Zyôbibun hôteisiki (Japanese; Ordinary differential equations), Iwanami, 1951.
[8] R. E. Langer, The solutions of the differential equation $v''' - \lambda^2 z v' + 3\mu\lambda^2 v = 0$, Duke Math. J., 22 (1955), 525–541.
[9] I. A. Lappo-Danilevskiĭ, Mémoires sur la théorie des systèmes des équations différentielles linéaires, Chelsea, 1953.
[10] Н. И. Мусхелишвили (N. I. Muskhelišvili), Сингулярные интегральные уравнения, Гостехиздат, 1946; English translation, Singular integral equations, Noordhoff, 1953.
[11] K. Okubo, A global representation of a fundamental set of solutions and a Stokes phenomenon for a system of linear ordinary differential equations, J. Math. Soc. Japan, 15 (1963), 268–288.
[12] C. E. Picard, Traité d'analyse III, Gauthier-Villars, 1896.
[13] H. Poincaré, Sur le groupe des équations linéaires, Acta Math., 4 (1884), 201–311.
[14] H. Röhrl, Das Riemann-Hilbertsche Problem der Theorie der linearen Differentialgleichungen, Math. Ann., 133 (1957), 1–25.
[15] L. Schlesinger, Handbuch der Theorie der linearen Differentialgleichungen I, II_1, II_2, Teubner, 1895–1898.
[16] L. Schlesinger, Über eine Klasse von Differentialsystemen beliebiger Ordnung mit festen kritischen Punkten, J. Reine Angew. Math., 141 (1912), 96–145.
[17] H. L. Turrittin, Stokes multipliers for asymptotic solutions of a certain differential equation, Trans. Amer. Math. Soc., 68 (1950), 304–329.
[18] H. L. Turrittin, Reduction of ordinary differential equations to the Birkhoff canonical form, Trans. Amer. Math. Soc., 107 (1963), 485–507.

254 (XIII.8) Linear Ordinary Differential Equations (Singular Points)

A. General Remarks

Consider a system of n †linear ordinary differential equations

$$\frac{dy}{dx} = A(x)\mathbf{y}, \tag{1}$$

where the independent variable x belongs to a complex domain D (or a †Riemann surface D), $\mathbf{y}$ is a complex n-dimensional column vector $(y_1(x), y_2(x), \ldots, y_n(x))$, and the $n \times n$ matrix $A(x)$ has complex †analytic functions as elements. A †singular point $x = a$ of $A(x)$ is a **singular point** of the system (1). In particular, if $t = 0$ is a singular point of $A(x)$ after the change of variable $x = 1/t$, then $x = \infty$ is a singular point of the system. Hence, to describe the properties of singular points of (1), we may start with the case when $x = 0$ is a singular point of the system (1), applying a suitable change of variable if necessary.

B. Regular Singular Points

If $A(x)$ is single-valued and holomorphic in $0 < |x| < R$, then any solution of (1) is holomorphic in $0 < |x| < R$, but not necessarily single-valued. If all the solutions have $x = 0$ as a holomorphic point or as a pole, then $x = 0$ is an **apparent singular point** of (1). When $\mathbf{y}_1(x), \mathbf{y}_2(x), \ldots, \mathbf{y}_n(x)$ constitute a †fundamental system of solutions, the matrix formed by $Y(x) = (\mathbf{y}_1(x), \ldots, \mathbf{y}_n(x))$ undergoes a linear transformation $Y(xe^{2\pi i}) = Y(x)M$ when the independent variable x is transformed to $xe^{2\pi i}$. The constant matrix M is the **monodromy matrix** (or **circuit matrix**) of $Y(x)$ at $x = 0$. If we take a constant matrix S such that $M = e^{2\pi i S}$, then there is a single-valued holomorphic matrix $P(x)$ whose elements are single-valued holomorphic functions in $0 < |x| < R$, such that $Y(x)$ has the form $Y(x) = P(x)x^S$.

If for a solution $\mathbf{y}(x)$ of (1) and an arbitrary sector Σ there is a positive number r such that $\lim_{x \to 0} |x|^r \mathbf{y}(x) = 0$, $x = 0$ is a **regular singular point** of the solution $\mathbf{y}(x)$; and if there is no such number, $x = 0$ is an **irregular singular point** of $y(x)$. If $x = 0$ is a regular singular point of all the solutions of (1), it is a **regular singular point** of the system (1). If some of the solutions have $x=0$ as an irregular singular point, then it is an **irregular singular point** of the system (1). A necessary and sufficient condition for $x=0$ to be a regular singular point of the system is that an arbitrary fundamental matrix solution $Y(x)$ of (1) in the form $Y(x)=P(x)x^M$ as described above has $x=0$ as a pole of $P(x)$. In this case, with a suitable choice of the constant matrix S, the point 0 is a regular point of $P(x)$ with $\det P(0)\neq 0$.

More specifically, let us consider a system

$$x^r(d\mathbf{y}/dx)=A(x)\mathbf{y}, \tag{2}$$

where $A(x)$ is holomorphic at $x=0$ and $A(0)\neq 0$. Then the point $x=0$ is a pole of order r of the coefficient. The algebraic equation $\det(A(0)-\rho I)=0$ of degree n is called the **indicial equation** of (2) at $x=0$.

If $r=1$, $x=0$ is a regular singular point of the system (2). We denote the roots of the indicial equation by $\rho_1,\rho_2,\ldots,\rho_n$ and assume that the roots $\rho_1,\rho_2,\ldots,\rho_p$ are chosen such that the $\rho_i-\rho_j$ are all integers for $i,j=1,\ldots,p$, and that $\operatorname{Re}\rho_1\leqslant\operatorname{Re}\rho_2\leqslant\ldots\leqslant\operatorname{Re}\rho_p$. Then the system (2) has a set of solutions

$$\mathbf{y}_j=x^{\rho_j}\mathbf{p}_j(x,\log x), \qquad j=1,\ldots,p,$$

where $\mathbf{p}_j(x,\lambda)$ is a polynomial vector of order at most $p-j$ with single-valued holomorphic functions of x as coefficients that do not vanish simultaneously at $x=0$. In particular, if the difference $\rho_i-\rho_j$ $(i\neq j)$ is never equal to an integer, then there is a set of n solutions of the form $\mathbf{y}_j=x^{\rho_j}\mathbf{p}_j(x)$.

C. Asymptotic Expansions

When $r>1$, the system (2) may be transformed by means of the †formal power series

$$\mathbf{y}=p(x)\mathbf{z}=\left(\sum p_k x^{k/h}\right)\mathbf{z} \tag{3}$$

into

$$\xi^s(d\mathbf{z}/d\xi)=(\Lambda(\xi)+J\xi^{s-1})\mathbf{z}, \qquad \xi=x^{1/h}, \tag{4}$$

where h is a suitable positive integer, $\Lambda(\xi)$ is a diagonal matrix with $\rho_j(\xi)=\rho_{j0}+\rho_{j1}\xi+\ldots+\rho_{js-1}\xi^{s-1}$ as the jth diagonal element, and J is a constant matrix in †Jordan canonical form. In particular, ρ_{j0} is equal to a root ρ_j of the indicial equation of (2). Formally, we have $\det P(x)\neq 0$. Also, when $s=0$, we have $\rho_j(\xi)=0$ $(j=1,\ldots,n)$. By substituting a solution of (4) into (3), we obtain a set of formal solutions of (2). However, the formal power series $P(x)$ does not always converge. By introducing the notion of †asymptotic expansions, H. Poincaré proved under a very restrictive hypothesis that these formal solutions represent asymptotically actual solutions in some sector. Contributions in this direction had been made, notably by W. J. Trjitzinski and J. Malmquist, but a decisive result was obtained by M. Hukuhara, whose method is also applicable to the study of regular singular points.

A fundamental matrix solution $\Phi_1(x)$ formed by n solutions which are expressed asymptotically by formal solutions in a sector D_1, and another matrix solution $\Phi_2(x)$ of the same nature expressed by the same formal solutions but in a different sector D_2, are, in general, not the same matrix solution. This is the so-called **Stokes phenomenon**. The elements of the matrix C such that $\Phi_1(x)=\Phi_2(x)C$ are called the **Stokes multipliers**, and the problem of determining these multipliers is called the **connection problem**. G. D. Birkhoff proved that when the monodromy matrix at zero of the system (2) can be diagonalized, there is a nonsingular matrix $P(x)$ ($\det P(0)\neq 0$) such that the linear transformation $\mathbf{y}=P(x)\mathbf{z}$ transforms (2) into

$$x^r\frac{d\mathbf{z}}{dx}=\left(\sum_{k=0}^{r-1}B_kx^k\right)\mathbf{z}.$$

D. Differential Equations of Fuchsian Type

Consider a single nth-order differential equation

$$y^{(n)}+p_1(x)y^{(n-1)}+\ldots+p_n(x)y=0. \tag{5}$$

A necessary and sufficient condition that $x=0$ is a regular singular point of (5) is that every $p_k(x)$ has a pole of order at most k at $x=0$. Consequently, we can write (5) in the form

$$x^ny^{(n)}+P_1(x)x^{n-1}y^{(n-1)}+\ldots+P_n(x)y=0, \tag{6}$$

where each $P_k(x)$ is holomorphic at $x=0$. Equation (6) has a set of n linearly independent solutions of the form $y=x^{\rho_k}P_k(x,\log x)$, where the ρ_k are n roots of the indicial equation of (6): $\rho(\rho-1)\ldots(\rho-n+1)+P_1(0)\rho\cdot(\rho-1)\ldots(\rho-n+2)+\ldots+P_{n-1}(0)\rho+P_n(0)=0$, and $P_k(x,\lambda)$ is a polynomial in λ of degree at most equal to the number of roots of the indicial equation that are congruent to ρ_k (modulo integers) with coefficients holomorphic functions of x. In particular, no solution contains the logarithmic term when no pair of roots of the indicial equation is congruent modulo integers. (A pair (a,b) is congruent modulo integers when $a-b$ is an integer.) Furthermore, if the real part of ρ_k is largest among the real parts of the roots ρ_j that are congruent to ρ_k modulo integers, the logarithmic term is absent. Frobenius's method is convenient for finding these solutions (→ Appendix A, Table 14).

Equation (5) is called an **equation of Fuchsian type** when $p_1(x), \ldots, p_n(x)$ are rational and there are no singular points, finite or infinite, other than regular singular points. If (5) is of Fuchsian type with regular singular points $a_1, a_2, \ldots, a_{m+1} = \infty$ and $\rho_{j1}, \ldots, \rho_{jn}$ ($j = 1, \ldots, m+1$) are the roots of the indicial equation, the **Fuchsian relation**

$$\sum_{j=1}^{m+1} \sum_{k=1}^{n} \rho_{jk} = \frac{(m-1)n(n-1)}{2}$$

holds.

E. Irregular Singular Points

If at least one of the coefficients $p_j(x)$ of equation (5) has a pole of order more than $j+1$, then $x=0$ is an irregular singular point of (5). Let m_j be the order of the pole of $p_j(x)$ at $x=0$, with the convention that $m_j = \infty$ when $p_j(x) = 0$; and assume that $m_0 = 0$, for simplicity. Let A_ν ($\nu = 0, 1, \ldots, n$) be the points with coordinates (ν, m_ν) in the Euclidean plane and Π the subset of the plane consisting of points (a,b) not lower than the points contained in the minimal convex polygon containing all the A_ν. This polygon is called the **Newton diagram** of equation (5). Let (ν, r_ν) be the intersection of the straight line $x = \nu$ with Π, and consider the nonincreasing sequence of numbers $\{\sigma_j\}$, $\sigma_\nu = r_\nu - r_{\nu-1}$ ($r_0 = 0$). We assume μ to be an integer such that $\sigma_1 > \sigma_2 > \ldots > \sigma_\mu > 1 \geqslant \sigma_{\mu+1} > \ldots > \sigma_n$.

To the first μ sections of the diagram Π with slopes $\sigma_j > 1$ there correspond μ formal solutions of the following form which are formally linearly independent:

$$y = \exp(\Lambda_j(x)) x^{\lambda_j} P_j(x, \log x),$$

where the $\Lambda_j(x)$ are polynomials in fractional powers of x^{-1} and the $P_j(x, \lambda)$ are polynomials in λ with coefficients given by formal series in fractional powers of x. Associated with the sections of Π corresponding to those $n-\mu$ numbers $\sigma_j \leqslant 1$, there are $n-\mu$ linearly independent formal solutions of similar form but without the exponential term. The set of these formal solutions represents a system of fundamental solutions in a certain sector in the x-plane.

The existence of holomorphic solutions at an irregular singular point, namely the existence of convergent formal solutions, was first proved by O. Perron for (5), and generalizations to systems of equations were given by F. Lettenmeyer [3] and by Hukuhara and M. Iwano [8].

In particular, concerning the second-order equation $y'' + p(x)y' + q(x)y = 0$, $x = \infty$ is an irregular singular point if $r = \max(p, q/2) > -1$, where p and q are the orders of the poles at $x=0$ of $p(x)$ and $q(x)$, respectively. The irregular singular point at ∞, in this case, is called an irregular singular point of **rank** $r+1$. In this terminology, a regular singular point is a singular point of rank 0.

F. Singularities with Respect to a Parameter

An analogous theory has been obtained for a system of first-order linear differential equations with a small complex parameter ε:

$$\varepsilon^h \frac{dy}{dx} = A(x, \varepsilon)\mathbf{y},$$

$$A(x,\varepsilon) \cong \sum_{k=0}^{\infty} A_k(x)\varepsilon^k, \tag{7}$$

where h is a positive integer, the $n \times n$ matrices $A_k(x)$ ($k = 0, 1, \ldots$) are single-valued holomorphic functions of x in a neighborhood D of the origin, and the †asymptotic expansion is valid when $\varepsilon \to 0$ in a sector Σ. When all eigenvalues of the matrix $A_0(0)$ are distinct, there is a matrix of formal solutions of the form $P(x,\varepsilon)e^{Q(x,\varepsilon)}$, where $P(x,\varepsilon)$ is a formal series of the form $A(x,\varepsilon)$, $Q(x,\varepsilon)$ is a diagonal matrix with polynomials in $1/\varepsilon$ of degree h as diagonal elements, and all the coefficients are single-valued holomorphic functions of x on D^* ($\subset D$). In particular, the coefficient of ε^{-h} in the jth diagonal element of $Q(x,\varepsilon)$ is $\int_0^x \mu_j(t)\,dt$, where $\mu_j(x)$ is an eigenvalue of the matrix $A_0(x)$. If we take a certain subsector Σ^* of Σ, the matrix $P(x,\varepsilon)e^{Q(x,\varepsilon)}$ represents an actual matrix solution in Σ^*. When there is no †turning point (→ Section G), it is always possible, even if there are multiple eigenvalues in $A_0(0)$, to construct formal solutions that are asymptotic representations of some solutions in some sector. Similar theories were developed for cases where more than two parameters appear.

G. Turning Points

Consider the point $x=0$ in the system (7), and set $n=2$, $h=1$, and $A_0(x) = \begin{pmatrix} 0 & 1 \\ x & 0 \end{pmatrix}$. Then $A_0(x)$ has a multiple root when $x=0$ and distinct roots when $x \neq 0$. As in this example, when the Jordan canonical form of the leading matrix $A_0(x)$ has different structure for $x=0$ and for $x \neq 0$, the coefficients of the formal power series in ε are not single-valued holomorphic functions of x, and have worse regularities as the order becomes higher. Consequently, in the neighborhood of $x=0$, we cannot construct actual solutions which can be represented asymptotically by these formal solutions. Such a solution point is called a

turning point (or **transition point**) of the system (7).

If there is a nonsingular formal transformation $y = T(x,\varepsilon)z$; $T(x,\varepsilon) = \sum T_k(x)\varepsilon^k$ ($\det T_0(0) \neq 0$) having similar analytic properties to those of $A(x,\varepsilon)$, and if the transformed system has a well-known form, it is possible to give analytic meaning to the formal transformation $T(x,\varepsilon)$. Then the transformed system is called a **related differential equation** of (7); in the above example, $\varepsilon(dz/dx) = A_0(x)z$ is a related differential equation that has well-known solutions expressed by †Bessel functions. What is meant by well-known here is that the behavior of all solutions is known in the entire complex plane for a fixed ε. However, it is not easy to find a suitable related differential equation for an arbitrarily given system.

References

[1] M. Fukuhara (Hukuhara), Zyôbibun hôteisiki (Japanese; Ordinary differential equations), Iwanami, 1950.

[2] E. A. Coddington and N. Levinson, Theory of ordinary differential equations, McGraw-Hill, 1955.

[3] P. Hartman, Ordinary differential equations, John Wiley, 1964.

[4] W. Wasow, Asymptotic expansions for ordinary differential equations, Interscience, 1965.

[5] L. Bieberbach, Theorie der gewöhnlichen Differentialgleichungen, Springer, second edition, 1965.

[6] H. L. Turrittin, Reduction of ordinary differential equations to the Birkhoff canonical form, Trans. Amer. Math. Soc., 107 (1963), 485–507.

[7] K. Okubo, A global representation of a fundamental set of solutions and a Stokes phenomenon for a system of linear ordinary differential equations, J. Math. Soc. Japan, 15 (1963), 268–288.

[8] M. Fukuhara (Hukuhara) and M. Iwano, Etude de la convergence des solutions formelles d'un système différentielle ordinaire linéaire, Funkcial. Ekvac., 2 (1959), 1–18.

[9] J. Moser, The order of a singularity in Fuchs' theory, Math. Z., 72 (1959–1960), 379–398.

255 (XVIII.2) Linear Programming

A. General Remarks

Linear Programming (often abbreviated as **LP**) is a method of finding extremal values of certain linear functions (**objective functions**) satisfying conditions expressed by systems of linear equalities, inequalities, or both. Usually the variables of such functions are assumed to be nonnegative. In its theoretical treatment for establishing the existence of extremals or giving criteria for extremal values, the method has both algebraic and topological aspects.

In its classical form LP was expressed in terms of †convex sets; in particular, the †duality of convex cones and systems of linear inequalities played an important role, while in its modern presentation, the method makes use of duality theorems in LP and the saddle value condition for the Lagrangian form, which applies also to nonlinear cases. Generally, the spaces appearing in the theory are †linear topological spaces, for example, $\mathbf{R}^n$.

B. History

The history of LP goes back to G. Monge (1781) and Fourier (1823). Concerning †polyhedral convex cones, †convex polyhedra in $\mathbf{R}^n$, and the algebraic theory of systems of linear inequalities, we have classical results due to P. Gordon (1873), J. Farkas [6], E. Stiemke [19], H. Weyl [21], etc., and a later refinement due to A. W. Tucker [15, paper 1]. For general convex sets in $\mathbf{R}^n$, the results of C. Carathéodory [2] and A. Haar [9] are classical. The theory of F. Riesz (1911) for representation in terms of the †Stieltjes integral has been followed by many representation theorems in modern analysis. Here we cite only G. Choquet [3] for a review of the development of the theory.

P. C. Rosenbloom [18] succeeded in obtaining an extension of LP theory in †function spaces, especially the †sequence space (l). The Kuhn-Tucker theory [14] on the correspondence between the maximal problem and the saddle value problem was generalized notably by L. Hurwicz [10] and K. Isii [11] to the case of linear topological spaces.

The application of LP to economics was made possible through the works of J. Von Neumann, especially his †game theory and balanced linear growth model [20]. These and the interindustrial input-output analysis of W. Leontief [16] led to the works assembled in 1951 by T. C. Koopmans and others [13]. Concerning practical computation and applications to industry, there are isolated and long-neglected works by L. V. Kantorovič [12]; however, the main part of the methods was developed in the United States, especially after the discovery of the simplex method by G. B. Dantzig [4A, ch. XXI] and his followers.

We assume that coefficients of functions

are real numbers unless otherwise stated; X, Y are n-dimensional column vectors with coordinates X_i, Y_i; $'$ means the transpose of a matrix (or of a vector); and * means the †dual space or its element, or a conjugate linear operator. Three partial ordering relationships $\leqq$, $\leq$, $<$ for vectors must be distinguished: $X \leqq Y$ and $X < Y$ mean $X_i \leqslant Y_i$ and $X_i < Y_i$ $(i=1,2,\ldots,n)$, respectively, and $X \leq Y$ means $X \leqq Y$ but $X \neq Y$.

C. Linear Inequalities and Convex Sets

Let K be a †closed convex set K in $\mathbf{R}^n$. It has the following basic properties: Any point not belonging to K can be separated from it by a hyperplane (**separation theorem**). When K is a †polyhedral convex cone the subsets $\{X \mid AX \geqq 0, X \in K\}$ and $\{X \mid X = BY, Y \geqq 0, X \in K\}$ become identical for appropriate choice of matrices A and B. When K is a polyhedral convex cone and K^* is its conjugate cone $\{Y \mid Y'X \geqq 0, \forall X \in C\}$, we have the **duality** $K = K^{**}$. When K is a bounded †convex polyhedron, then K can be defined either as the bounded domain surrounded by a finite set of hyperplanes or as a convex set spanned by a finite set of points. An unbounded convex polyhedron is obtained by the †addition of a bounded convex polyhedron to a polyhedral convex cone (→ 91 Convex Sets).

The following theorems may be deduced from these basic properties of a closed convex set. **Minkowski-Farkas theorem**: Given an equation $AX = B$, where B is an element of $\mathbf{R}^n$, a necessary and sufficient condition for a solution $X \geqq 0$ to exist is that $U'B \geqq 0$ holds for any vector U such that $U'A \geqq 0$. **Stiemke theorem**: For a matrix A one of the following two alternatives holds: (i) $AX=0$, $X>0$ have a solution; (ii) $U'A \geq 0$ has a solution. **Tucker's theorem on complementary slackness**: For any matrix A, the two systems of linear inequalities (i) $AX=0$, $X \geqq 0$ and (ii) $U'A \geqq 0$ have solutions X, U satisfying $A'U + X > 0$.

D. Linear Programming in Finite-Dimensional Spaces

Suppose that A is an $m \times n$ matrix, $B \in \mathbf{R}^m$, and $P \in \mathbf{R}^{n*}$. Problem P: To find $X \in \mathbf{R}^n$ for which $x_0 = P'X$ is maximal under the conditions (i) $AX = B$ and (ii) $X \geqq 0$. We call any X satisfying (i) a **solution** and satisfying (i) and (ii) a **feasible solution**, and we denote the set of all feasible solutions by $\mathfrak{F}$. Then $\mathfrak{F} \neq \varnothing$ if and only if there exists no $V \in \mathbf{R}^{m*}$ such that (1) $V'A \geqq 0$, $V'B < 0$, and a necessary and sufficient condition on $X_0 \in \mathfrak{F}$ for $P'X$ to attain a maximal value at $X = X_0$ is the existence of a $U = U_0 \in \mathbf{R}^{m*}$ satisfying (2) $U_0'A \geqq P'$ and (3) $(U_0'A - P')X_0 = 0$.

The vector U_0 is called the **shadow price**, and condition (2) is called the **simplex criterion**. Conditions (i), (ii), (2), and (3) imply that for $X \geqq 0$,

$$\Phi(X,U) = P'X - U'AX + U'B \tag{4}$$

has a †saddle point at (X_0, U_0), and by rewriting (4) as

$$\Phi(X,U) = P'X - U'(AX - B) = (P' - U'A)X + U'B,$$

we obtain the duality theorem stated below (Problem P and Problem D, which follows, are called **dual** to each other.)

Problem D: To find $U \in \mathbf{R}^{m*}$ that minimizes $U'B$ and $U'A \geqq P'$.

The **duality theorem of linear programming** states: If problem P (resp. D) has a feasible solution and the objective function is upper (lower) bounded, then both problems have extremal solutions and the two optimal values (i.e., the maximal value in problem P and the minimal value in problem D) coincide. In this case the shadow price vector U_0 gives an **optimal solution** for problem D. If problem P (D) has a feasible solution, then the dual problem D (P) is lower (upper) bounded.

Kuhn and Tucker showed that the relation between the feasible solution, letting the objective function attain its maximal value, and the saddle point of (4) holds also for the nonlinear case [14]. Let $f(X)$ be a differentiable function of $X \in \mathbf{R}^n$, and let $A(X)$ be a differentiable mapping from $\mathbf{R}^n$ into $\mathbf{R}^m$. Suppose that the given function $f(X)$ attains a maximal value for $X = X_0$, and that $A(X_0) \geqq 0$, $X_0 \geqq 0$. Furthermore, consider a form

$$\Phi(X,U) = f(X) + U'A(X), \tag{5}$$

and set

$$\Phi_X^{0\prime} = (\partial\Phi/\partial X_1, \partial\Phi/\partial X_2, \ldots, \partial\Phi/\partial X_n)_{X=X_0},$$

etc. Then, under certain conditions, there exists some $U_0 \in \mathbf{R}^{m*}$ that satisfies (i) $\Phi_X^0 \leqq 0$, $\Phi_X^{0\prime} X_0 = 0$ $(X_0 \geqq 0)$, and (ii) $\Phi_U^0 \geqq 0$, $\Phi_U^{0\prime} U_0 = 0$ $(U_0 \geqq 0)$.

On the other hand, if any pair (X_0, U_0) satisfies, in addition to (i) and (ii), (iii) $\Phi(X, U_0) \leqq \Phi(X_0, U_0) + \Phi_X^{0\prime}(X - X_0)$ $(\forall X \geqq 0)$ and (iv) $\Phi(X_0, U) \geqq \Phi(X_0, U_0) + \Phi_U^{0\prime}(U - U_0)$ $(\forall U \geqq 0)$, then $f(X)$ attains a maximal value at $X = X_0$.

If each component of $A(X)$ and $f(X)$ is †concave and differentiable, $f(X)$ attains a maximal value when $X = X_0$ if and only if there exists $U_0 \geqq 0$ such that (X_0, U_0) is a saddle point of Φ. When $F(X)$ is a differentiable mapping from $\mathbf{R}^n$ into $\mathbf{R}^k$, and the problem is extended to finding an **efficient point**, i.e., a point X_0 where $F(X)$ attains its maximal point with respect to the partial ordering $\leqq$ in $\mathbf{R}^k$, we have the following general theorem: For X_0 there exists some $V_0 \in \mathbf{R}^{k^*}$, $V_0 > 0$, such that (i) and (ii) hold for $\Phi(X, U) = V_0' F(X) + U' A(X)$.

E. Generalizations and Applications

Linear programming in the sequence space $(l) = \{X = \{x_j\} \mid \sum_{j=1}^{\infty} |x_j| < \infty\}$ was treated by P. C. Rosenbloom [18]. In this case, the requirements for variables X are given in terms of †linear functionals $\lambda_i \in (l)^* = (m)$ $(i = 1, 2, \ldots, k)$, for example, as (i) equalities $\lambda_i(X) = c_i$ or (i') inequalities $\lambda_i(X) \leqslant c_i$, and (ii) $x_j \geqslant 0$ $(\forall j)$. The space (l) is supplied with the †weak topology as the conjugate space of $(c_0) = \{X = \{x_j\} \mid \lim_{j \to \infty} x_j = 0\}$. As before, we let $\mathfrak{F}$ be the set of all X satisfying (i) (or (i')) and (ii). Let $\lambda(X)$ be a †weakly upper semicontinuous functional, and suppose that the $\lambda_i(x)$ are †weakly lower semicontinuous, $\mathfrak{F} \neq \varnothing$, and $\lambda(X)$ is upper bounded. Then the maximal value of $\lambda(X)$ is attained at an extreme point of $\mathfrak{F}$, and furthermore, if the space is completely regular, then the solution is unique. If $\mathfrak{F}$ is bounded, it is the **convex hull** of its extreme points, i.e., the smallest closed convex set containing its extreme points. By applying the theory in this section to the family of functions that can be expressed as $f(x) = \sum_{j=1}^{\infty} a_j \varphi_j(x)$ in terms of a given system of functions $\varphi_j(x)$ $(j = 1, 2, \ldots)$ on $\mathbf{R}$, S. N. Bernšteĭn's approximation theory of function systems, the theory of absolutely monotonic functions, and several inequalities in the theory of functions of a complex variable may be treated in a unified fashion. If the theory is further extended to the case of †finitely additive measures defined by means of linear functionals on the Banach space of bounded functions (A. D. Aleksandrov, 1940–1943), we may treat the extremal problems of linear functionals on the function space of all $f(x)$ that can be expressed as $f(x) = \int_S K(x, s)\, d\mu(s)$, and apply it to obtain the interpolation formula of nonnegative †harmonic functions, results of Carathéodory and Fejér on its Fourier coefficients, an analogy of Harnack's theorem for the †heat conduction equation, and so on.

The extension of the Kuhn-Tucker theory to linear topological spaces is due to L. Hurwicz [10]. Let $\mathfrak{X}$ be a linear space, $\mathfrak{Y}$, $\mathfrak{Z}$ be linear topological spaces, P_Y, P_Z the nonnegativity cones of $\mathfrak{Y}$, $\mathfrak{Z}$, respectively, which are closed convex cones containing inner points, D a convex set in $\mathfrak{X}$, and F, G concave mappings (→ 90 Convex Functions) from D into $\mathfrak{Y}$, $\mathfrak{Z}$, respectively, such that $G(D)$ contains an inner point of P_Z. If $F(X)$ attains its maximal point when $X = X_0$ and X_0 satisfies $G(X_0) \geqq 0$, $X_0 \in D$, then there exist $Y_0^* \geqq 0$, $Z_0^* \geqq 0$ such that $\Phi(X, Z^*) = Y_0^*(F(X)) + Z_0^*(G(X))$ has a saddle point at (X_0, Z_0^*). The condition that P_Y and P_Z have inner points may be weakened to cover the cases of (l_p), (L_p), (s), (S) (Hurwicz and Uzawa [1, paper 5]).

In accordance with the presentation given by Hurwicz, we assume that F and A are linear mappings from D into $\mathfrak{Y}$ and $\mathfrak{Z}$, $\mathfrak{Y} = \mathbf{R}$, $\mathfrak{Z}$ is †locally convex, and P_X, P_Z are closed convex cones in $\mathfrak{X}$ and $\mathfrak{Z}$, respectively, and we consider Problem L: To maximize $F(X) = X^*(X)$ under the condition

$$G(X) = A(X) - B \geqq 0, \qquad X \geqq 0, \qquad B \in \mathfrak{Z}.$$

Put $\Phi(X, Z^*) = X^*(X) + Z^*(A(X) - B)$. If we define a linear mapping T from $\mathfrak{W} = \mathfrak{X} \times \mathbf{R}$ into $\mathfrak{V} = \mathfrak{Z} \times \mathfrak{W}$ by $T((X, \rho)) = (A(X) - \rho B, (X, \rho))$, where $\rho \in \mathbf{R}$, then under the condition that the image under T^* of the nonnegativity cone of $\mathfrak{V}^*$ is a †regularly convex set in $\mathfrak{W}^*$, X_0 is a solution of problem L if and only if there exists a $Z_0^* \in \mathfrak{Z}^*$ such that (X_0, Z_0^*) is a nonnegative saddle point of Φ.

When the space is a Banach space and the †Fréchet differential is utilized, there exists a similar correspondence between feasible maximal solutions of a nonlinear programming problem (i.e., the solutions where the given objective function attains its maximal values) and weak saddle points expressed in terms of differential coefficients, where the requirement for the variable is assumed to be not stronger than that of Kuhn and Tucker.

Under conditions expressed in simpler terms, Isii proved the coincidence of the supremum of the objective function and $\inf_{Z^*} \sup_{X \in D}$ of the †Lagrangian form, gave conditions for the supremum and the infimum to be attained, and developed a theory that generalizes the †Čebyšev inequality and the †Cramér-Rao inequality [11].

F. Practical Computational Methods

Among practical numerical algorithms for solving linear programming problems of finite dimension, the standard one is the **two-phase simplex method**, which is based on Dantzig's simplex method. Among other notable methods are the **dual simplex algorithm**, the **composite simplex algorithm**, and the **primal-dual algorithm**, which are all variants of a common basic operation—the †pivot †elimination method (P. Wolfe and L. Cutler [7, paper 23]).

Consider the $m+1$ equations in $n+1$ unknowns obtained from the m equation in $AX = B$ and $x_0 - P'X = 0$. Suppose that these $m+1$ equations may be written as $MT = N$, the rank of M is $m+1$, and $n \geqslant m$. Then we can choose $m+1$ linearly independent column vectors of M and reduce the equation to one that involves only $m+1$ unknowns x_0, $x_{i_1}, \ldots, x_{i_m}$. This reduced equation is called a **basic** (or **canonical**) **form**, the set of variables $x_0, x_{i_1}, \ldots, x_{i_m}$ is called a **basis**, and any solution with all the unknowns other than these $m+1$ set equal to 0 is called a **basic solution**. A **feasible basic solution** is abbreviated as f.b.s. in this section. When a pivot elimination procedure is applied to a basic form, we obtain a new basic form whose basis differs from the former by just one variable. The **simplex method** gives a rule of successive pivot choice so that in a finite number of steps the optimal solution, if any, is achieved. The numerical table of basic forms successively obtained by pivot elimination is called a **simplex tableau**.

In the two-phase simplex method three actions are taken successively: (0) finding a basic solution, (i) finding a feasible basic solution, (ii) finding an optimal solution. At each stage it is automatically determined whether an appropriate solution exists. In phase (ii), at each step an f.b.s. X and $U'A-P'$ derived from U satisfying (3) with respect to X are obtained, and the signs of the components of $U'A-P'$ are tested.

If the simplex criterion (2) is not satisfied, the rule gives a pivot choice so that the f.b.s. in the next step is an extreme point of $\mathfrak{F}$ situated next to X, with a higher value of the objective.

The problem related to the finiteness of the process in case of degeneracy is dealt with by utilizing lexicographic ordering in the pivot choice.

In phase (i) an effective algorithm is furnished by applying the rules of phase (ii) for an auxiliary objective function—the sum of all basic variables whose values are negative in phase (i). The algorithm automatically provides the vector V in (1) when $\mathfrak{F}=\varphi$.

Pivot elimination plays a central role also in **parametric analysis**, where B, P, or both depend linearly on a parameter t, and corresponding maximal solutions X_t, x_{0t}, U_t are analyzed. As a function of B and P, $\max x_0$ is concave in B and convex in P.

There are other methods which use operations besides pivot elimination as their main procedures.

Generalized linear programming, based on the decomposition principle of Dantzig and Wolfe [5], may serve as a bridge connecting extended infinite-dimensional linear programming to practical computations. The problem of **two-stage linear programming under uncertainty** can also be treated by applying the decomposition principle to the dual angular system.

G. Integer Programming

Linear programming where some or all of the variables are required to take integer values is called **integer programming**. Several methods of solution were proposed in 1959–1960 by R. E. Gomory [7]. The main idea calls for first finding an optimal solution in the usual sense. If the obtained solution does not satisfy the integer condition, then we find a stronger inequality condition that is satisfied by any feasible integer solution but not by the obtained one, so that the latter noninteger solution is eliminated.

Suppose that we are given a linear equation

$$\sum_{j=1}^{n} a_j x_j = a_0 \tag{6}$$

and a positive integer λ. Let $a_j=[a_j/\lambda]\lambda+\rho_j$ $(0\leqslant\rho_j<\lambda;\ j=0,1,\ldots,n)$, and consider $z\equiv(\sum_{j=1}^{n}\rho_j x_j-\rho_0)/\lambda$. For any nonnegative integer solution $\{x_j\}$ of (6), z is a nonnegative integer, and hence by requiring $z\geqslant 0$ we may eliminate the solutions with $\rho_0>0$. Now suppose that a row of a basic form is given by

$$x_i+\sum_{j=1}^{n} a_j x_j = a_0, \tag{7}$$

where a_0 is not an integer. We may follow the procedure in this paragraph by taking $\lambda=1$, and add the condition $z\geqslant 0$, to eliminate the basic solution with $x_i=a_0$. The procedure terminates in a finite number of steps if all coefficients and variables are integers.

The vast area of applications that would be opened by the appearance of a satisfactorily efficient algorithm for integer programming problems was indicated in part by Dantzig [4].

References

[1] K. J. Arrow, L. Hurwicz, and H. Uzawa (eds.), Studies in linear and non-linear programming, Stanford Univ. Press, 1958.
[2] C. Carathéodory, Über den Variabilitätsbereich der Fourierschen Konstanten von positiven harmonischen Funktionen, Rend. Circ. Mat. Palermo, 32 (1911), 193–217.
[3] G. Choquet, Existence et unicité des représentations intégrales au moyen des points extrémaux dans les cônes convexes, Sem. Bourbaki, Exposé 139, 1956/1957 (Benjamin, 1966).
[4] G. B. Dantzig, On the significance of solving linear programming problems with

some integer variables, Econometrica, 28 (1960), 30–44.
[4A] G. B. Dantzig, Linear programming and extensions, Princeton Univ. Press, 1963.
[5] G. B. Dantzig and P. Wolfe, Decomposition principle for linear programs, Operations Res., 8 (1960), 101–111.
[6] J. Farkas, Über die Theorie der einfachen Ungleichungen, J. Reine Angew. Math., 124 (1902), 1–27.
[7] R. L. Graves and P. Wolfe (eds.), Recent advances in mathematical programming, McGraw-Hill, 1963.
[8] S. I. Gass, Linear programming: methods and applications, McGraw-Hill, second edition, 1964.
[9] A. Haar, Über lineare Ungleichungen, Acta Sci. Math. Szeged., 2 (1924), 1–14.
[10] L. Hurwicz, Programming in linear spaces; in Studies in linear and non-linear programming, edited by K. J. Arrow, L. Hurwicz, and H. Uzawa, Stanford Univ. Press, 1958, p. 38–102.
[11] K. Isii, Inequalities of the types of Chebyshev and Cramér-Rao and mathematical programming, Ann. Inst. Stat. Math., 16 (1964), 277–293.
[12] Л. В. Канторович (L. V. Kantorovič), Математические методы организации и планирования производства, Ленинград (1904); 1–64; English translation, Mathematical methods of organization and planning production, Management Sci., 6 (1960), 366–422.
[13] T. C. Koopmans (ed.), Activity analysis of production and allocation, John Wiley, 1951.
[14] H. W. Kuhn and A. W. Tucker, Nonlinear programming, Proc. 2nd Berkeley Symp. Math. Stat. Prob., Univ. of California Press, (1951), p. 481–492.
[15] H. W. Kuhn and A. W. Tucker (eds.), Linear inequalities and related systems, Ann. Math. Studies, Princeton Univ. Press, 1956.
[16] W. Leontief, The structure of American economy 1919–1939, Oxford Univ. Press, second edition, 1951.
[17] F. Nikaido, Keizaigaku no tameno senkei sûgaku (Japanese; Linear algebra for econometrics), Baihûkan, 1961.
[18] P. C. Rosenbloom, Quelques classes de problèmes extrémaux, Bull. Soc. Math. France, 79 (1951), 1–58; 80 (1952), 183–215.
[19] E. Stiemke, Über positive Lösungen homogener linearer Gleichungen, Math. Ann., 76 (1914–1915), 340–342.
[20] J. Von Neumann, Über ein ökonomisches Gleichungssystem und eine Verallgemeinerung des Brouwerschen Fixpunktsatzes, Ergebnisse eines Math. Kolloquiums, 8 (1937), 73–83; English translation, A model of general economic equilibrium, Rev. Economic Studies, 13 (1945–1946), 1–9 (Collected works, Pergamon, 1963, Vol. 6, p. 29–37).
[21] H. Weyl, Elementare Theorie der konvexen Polyeder, Comment. Math. Helv., 7 (1934–1935), 290–306.

256 (III.9) Linear Spaces

A. Definition

Suppose that we are given a set L and a †field K satisfying the following two requirements: (i) Given an arbitrary pair (a,b) of elements in L, there exists a unique element $a+b$ (called the **sum** of a,b) in L; (ii) given an arbitrary element α in K and an arbitrary element a in L, there exists a unique element αa (called the **scalar multiple** of a by α) in L. The set L is called a **linear space over** K (or **vector space over** K) if the following eight conditions are satisfied: (i) $(a+b)+c=a+(b+c)$; (ii) there exists an element $0\in L$, called the **zero element** of L, such that $a+0=0+a=a$ for all $a\in L$; (iii) For any $a\in L$, there exists an element $x=-a\in L$ satisfying $a+x=x+a=0$; (iv) $a+b=b+a$; (v) $\alpha(a+b)=\alpha a+\alpha b$; (vi) $(\alpha\beta)a=\alpha(\beta a)$; (vii) $(\alpha+\beta)a=\alpha a+\beta a$; (viii) $1a=a$ (where 1 is the †unity element of K). An element of K is called a **scalar**, and an element of L is called a **vector**. K is called the **field of scalars** (**basic field** or **ground field**) of the linear space L.

In the definition of linear spaces, K can be noncommutative. L is also called a **left linear space** over K since the scalars act on L from the left ($a\rightarrow\alpha a, a\in L, \alpha\in K$). A **right linear space** is similarly defined. Actually, a left (right) linear space is a unitary left (right) K-module. If K is commutative, it is not necessary to specify left or right, since we can identify αa and $a\alpha$. In this article we consider only linear spaces over commutative fields. A similar theory can be established for linear spaces over noncommutative fields (→ 275 Modules).

If K is the field of real numbers $\mathbf{R}$ or the field of complex numbers $\mathbf{C}$, a linear space over K is called a **real linear space** or **complex linear space**, respectively. In the following discussion, we fix a field K, and by a linear space we mean a linear space over K.

Examples (1) Geometric vectors: In a †Euclidean space or, more generally, an †affine space, the set of vectors $\overrightarrow{PQ}$ associated with points P, Q in the space forms a linear space.

(2) n-tuples in K: K^n denotes the set of all sequences $(\alpha_1, \ldots, \alpha_n)$ of n elements in a field K. Defining two operations by $(\alpha_1, \ldots, \alpha_n) + (\beta_1, \ldots, \beta_n) = (\alpha_1 + \beta_1, \ldots, \alpha_n + \beta_n)$, $\lambda(\alpha_1, \ldots, \alpha_n) = (\lambda\alpha_1, \ldots, \lambda\alpha_n)$ $(\lambda \in K)$, the set K^n forms a linear space over K. An element of K^n is called an **n-tuple** in K, and α_i is called the ith **component** of $(\alpha_1, \ldots, \alpha_n)$. In general, the **inner product** of $a = (\alpha_1, \ldots, \alpha_n)$, $b = (\beta_1, \ldots, \beta_n)$ is defined by $(a,b) = \sum_{i=1}^n \alpha_i \beta_i$. However, when $K = \mathbf{C}$ (complex number field), we usually define (a,b) to be $\sum_{i=1}^n \alpha_i \bar{\beta}_i$.

(3) Sequences in K: All infinite sequences in a field K form a linear space over K under the operations defined in example (2).

(4) K-valued functions: Given a nonempty set I and a field K, let K^I be the set of all K-valued functions defined on I. Defining two operations by $(f+g)(x) = f(x) + g(x)$, $(\lambda f)(x) = \lambda f(x)$ $(x \in I, \lambda \in K)$, the set K^I forms a linear space over K. In particular, if we put $I = \{1, \ldots, n\}$, then the space K^I can be identified with the space of n-tuples given in example (2), and if we put $I = \mathbf{N}$ (natural numbers), then we obtain the space given in example (3). Let K be the field $\mathbf{R}$ of real numbers and I an interval in $\mathbf{R}$. The set $C(I)$ of all continuous functions on I, the set $D(I)$ of all differentiable functions on I, and the set $A(I)$ of all †real analytic functions on I are all linear spaces contained in the space $\mathbf{R}^I$.

(5) Polynomials in $K: K[X_1, \ldots, X_n]$ denotes the set of all polynomials of n variables with coefficients in a field K. This forms a linear space under the usual operations.

B. Linear Mappings

Let L, M be linear spaces over a field K. A mapping φ from L to M is called a **linear mapping** or **linear operator** if φ satisfies the following two conditions; (i) $\varphi(a+b) = \varphi(a) + \varphi(b)$; and (ii) $\varphi(\lambda a) = \lambda\varphi(a)$ $(a, b \in L,\ \lambda \in K)$. Namely, a linear mapping is a K-homomorphism between K-modules (→ 275 Modules). Regarding K as a linear space, a linear mapping $L \rightarrow K$ is called a **linear form**. A linear mapping from L to L is called a **linear transformation** of L. The identity mapping of L is a linear transformation. Given linear spaces L, M, N, and linear mappings $\varphi: L \rightarrow M$ and $\psi: M \rightarrow N$, the composite $\psi \circ \varphi: L \rightarrow N$ is also a linear mapping. If a linear mapping $\varphi: L \rightarrow M$ is †bijective, then the inverse mapping $\varphi^{-1}: M \rightarrow L$ is also a linear mapping. Such a mapping φ is called an **isomorphism**, and we write $L \cong M$ if there exists an isomorphism $L \rightarrow M$. A linear transformation $L \rightarrow L$ is called **regular** (or **nonsingular**) if it is an isomorphism.

Examples (1) Let L be the linear space formed by all geometric vectors in a Euclidean space (affine space) E. Then a †motion (†affine transformation) of E induces a linear transformation $L \rightarrow L$.

(2) Let (α_{ij}) be an $m \times n$ †matrix in K. Assigning $(\eta_1, \ldots, \eta_m)$ to $(\xi_1, \ldots, \xi_n)$, where $\eta_i = \sum_{j=1}^n \alpha_{ij}\xi_j$ $(1 \leqslant i \leqslant m)$, we have a linear mapping $K^n \rightarrow K^m$.

(3) Assigning the †derivative f' to a real-valued differentiable function f on an interval I, we have a linear mapping $D(I) \rightarrow \mathbf{R}^I$.

C. Linear Combinations

Let L be a linear space over a field K. An element of L of the form $\alpha_1 a_1 + \ldots + \alpha_n a_n$ $(\alpha_i \in K,\ a_i \in L)$ is called a **linear combination** of $a_1, \ldots, a_n$. A sequence $a_1, \ldots, a_n$ of elements in L is called **linearly dependent** if there exists a sequence $\alpha_1, \ldots, \alpha_n$ of elements in K such that not all the α_i are equal to 0 and $\alpha_1 a_1 + \ldots + \alpha_n a_n = 0$. A sequence of elements in L is called **linearly independent** if it is not linearly dependent. Suppose that there exists a linearly independent sequence of n elements in a linear space L, and no sequence of $n+1$ elements in L is linearly independent. Then n is called the **dimension** of L and is denoted by $\dim L$. If there exists such a number n, L is said to be **finite-dimensional**. Otherwise, L is said to be **infinite-dimensional**. In an infinite-dimensional linear space, there exist linearly independent sequences of elements having arbitrary length. The linear space K^n of n-tuples in K is of dimension n.

A sequence $(a_1, \ldots, a_n)$ of elements in a linear space L is called a basis if every element a of L is uniquely written in the form $a = \alpha_1 a_1 + \ldots + \alpha_n a_n$ $(\alpha_i \in K,\ i = 1, \ldots, n)$. This means that the linear mapping $K^n \rightarrow L$ assigning $\alpha_1 a_1 + \ldots + \alpha_n a_n \in L$ to $(\alpha_1, \ldots, \alpha_n) \in K^n$ is bijective and hence an isomorphism. The condition that $(a_1, \ldots, a_n)$ is a basis of L is equivalent to any two of the following three conditions: (i) $(a_1, \ldots, a_n)$ is linearly independent; (ii) every element of L is a linear combination of $a_1, \ldots, a_n$; (iii) L is of dimension n. It follows that the length n of a basis $(a_1, \ldots, a_n)$ is equal to the dimension and hence is independent of the choice of basis. In the expression $a = \sum \alpha_i a_i$, α_i is called the **ith component** (or **ith coordinate**) of the element a relative to the basis $(a_1, \ldots, a_n)$.

D. Spaces of Linear Mappings (Finite-Dimensional Case)

Let L, M, and N be finite-dinemsional linear spaces over a field K. The set $\mathrm{Hom}_K(L,M)$ of all linear mappings $L \to M$ is a linear space under the operations defined by $(\varphi+\varphi')(a)=\varphi(a)+\varphi'(a)$, $(\lambda\varphi)(a)=\lambda\varphi(a)$ $(a\in L, \lambda\in K)$.

Let $(a_1,\ldots,a_l)$, $(b_1,\ldots,b_m)$ be bases of L, M, respectively. Then any linear mapping $\varphi: L\to M$ can be represented by an $m\times l$ matrix (α_{ij}) determined by $\varphi(a_j)=\sum_{i=1}^m b_i\alpha_{ij}$ $(1\leqslant j\leqslant l)$. This assignment $\varphi\to(\alpha_{ij})$ gives an isomorphism from the linear space $\mathrm{Hom}_K(L,M)$ to the linear space of all $m\times l$ matrices (→ 269 Matrices). In addition, let $(c_1,\ldots,c_n)$ be a basis of N, and let the $n\times m$ matrix (β_{ki}) represent a linear mapping ψ: $M\to N$. Then the composite mapping $\psi\circ\varphi$: $L\to N$ is represented by the product $(\beta_{ki})(\alpha_{ij})$ of the matrices (β_{ki}) and (α_{ij}). The set of all linear transformations of a linear space N of dimension n forms an †associative algebra over K which is isomorphic to the †total matrix algebra $M_n(K)$ of degree n under the correspondence $\varphi\to(\alpha_{ij})$. Its †invertible elements are regular linear transformations, and they form a group which is denoted by $GL(N)$ and called the †**general linear group** on N. This corresponds to the group $GL(n,K)$ formed by all $n\times n$ †invertible matrices under the isomorphism $\varphi\to(\alpha_{ij})$.

E. Infinite-Dimensional Linear Spaces

In this section, we consider only the algebraic aspects of infinite-dimensional linear spaces (for the topological aspects → 405 Topological Abelian Groups L; 407 Topological Linear Spaces). Let $\{a_\lambda\}_{\lambda\in\Lambda}$ be a family of elements in a linear space L. A linear combination of the family is an element of L in the form $\sum_{\lambda\in\Lambda}\alpha_\lambda a_\lambda$ ($\alpha_\lambda\in K$, where $\alpha_\lambda=0$ except for a finite number of λ). The family $\{a_\lambda\}_{\lambda\in\Lambda}$ is called **linearly independent** if no linear combination $\sum_{\lambda\in\Lambda}\alpha_\lambda a_\lambda$ is equal to 0 unless all the coefficients α_λ are equal to 0. The family $\{a_\lambda\}_{\lambda\in\Lambda}$ is called a **basis** of L if every element of L is uniquely written in the form $\sum_{\lambda\in\Lambda}\alpha_\lambda a_\lambda$. These notions are generalizations of those defined for finite Λ. In general, Λ is an infinite set. Any linear space L has a basis (→ 36 Axiom of Choice and Equivalents C). The cardinality of a basis is determined by L. Two linear spaces are isomorphic if and only if their bases have the same cardinality.

F. Subspaces and Quotient Spaces

Let L be a linear space over a field K. A nonempty subset N of L forms a linear space over K under the induced operations if the following two conditions hold: (i) $a,b\in N$ implies $a+b\in N$; and (ii) $\lambda\in K$, $a\in N$ imply $\lambda a\in N$. In this case, the subset N is called a **linear subspace** of L (or simply **subspace** of L). The canonical mapping $\varphi: N\to L$ defined by $\varphi(a)=a$ $(a\in N)$ is an injective linear mapping.

Let S be a nonempty subset of L. The set of all linear combinations of elements in S forms the smallest subspace of L containing S; this space is called the subspace **generated** (or **spanned**) by S. For subspaces N, N' of L, the intersection $N\cap N'$ and the **sum** $N+N'=\{a+a' \mid a\in N, a'\in N'\}$ are both subspaces. Similar propositions hold for an arbitrary number of subspaces. If N, N' are of finite dimension, the equality $\dim N+\dim N'=\dim(N\cap N')+\dim(N+N')$ holds. We say that L is decomposed into the **direct sum** of N, N' if every element of L can be uniquely written in the form $a+a'$ $(a\in N, a'\in N')$. This is the case if and only if L is generated by N and N' and $N\cap N'=\{0\}$. In this case, N' is called a **complementary subspace** of N. Any subspace has a complementary subspace. For direct products and sums of linear spaces → 275 Modules F.

An equivalence relation R in a linear space L is said to be **compatible** with the operations in L if the following two conditions hold: (i) $R(a,a')$ and $R(b,b')$ imply $R(a+b,a'+b')$; (ii) $R(a,a')$ implies $R(\lambda a,\lambda a')$ $(\lambda\in K)$. Then the †quotient set L/R, namely, the set of all equivalence classes, forms a linear space over K under the induced operations; this is called the **quotient linear space** (or simply **quotient space**) of L with respect to R. The canonical mapping $\varphi: L\to L/R$, given by $a\in\varphi(a)$ $(a\in L)$, is a surjective linear mapping. The equivalence class N containing 0 forms a subspace of L, and the equivalence class containing $a\in L$ is the coset $a+N=\{a+b \mid b\in N\}$. We have $R(a,a')$ if and only if $a-a'\in N$. Conversely, for any subspace N, there exists an equivalence relation R compatible with the operations determined by $R(a,a')$ if and only if $a-a'\in N$. The quotient linear space L/R thus obtained is denoted by L/N and called the **quotient (linear) space** of L by N. If L/N is finite-dimensional, its dimension is called the **codimension** of N relative to L and is denoted by codim N.

Let $\varphi: L\to M$ be a linear mapping of linear spaces. Its **image** $\varphi(L)$ is a subspace of M, and the **kernel** $N=\{a\in L \mid \varphi(a)=0\}$ is a subspace of L. The mapping φ induces an isomorphism $\bar{\varphi}: L/N\to\varphi(L)$ (→ 275 Modules E). If L is finite-dimensional, $\dim L-\dim N=\dim\varphi(L)$. The dimension of the image of φ is called the **rank** of φ, and the dimension of the kernel of φ is called the **nullity** of φ. The rank and nullity of an $m \times n$ matrix (α_{ij}) (→ 269 Matrices) are the respective rank and nullity of the linear mapping $K^n \to K^m$ represented by the matrix (α_{ij}) (→ Section B, example (2)).

G. Dual Spaces

Let L be a linear space over a field K. The set $\mathrm{Hom}_K(L,K)$ of all linear forms on L is a linear space, denoted by L^* and called the **dual (linear) space** of L. The space L^* is the †dual module of L as a K-module (→ 275 Modules). For elements a of L and a^* of L^*, we denote the element $a^*(a)$ by $\langle a, a^* \rangle$ and call it the **inner product** of a and a^*. For a linear mapping $\varphi: L \to M$, we define a linear mapping ${}^t\varphi: M^* \to L^*$ by $({}^t\varphi)(b^*) = b^* \circ \varphi$ $(b^* \in M^*)$. The mapping ${}^t\varphi$ is called the **dual mapping (transposed mapping** or **transpose)** of φ, and is determined by the relation $\langle a, {}^t\varphi(b^*) \rangle = \langle \varphi(a), b^* \rangle$ $(a \in L,\ b^* \in M^*)$. We have ${}^t(\varphi_1 + \varphi_2) = {}^t\varphi_1 + {}^t\varphi_2$, ${}^t(\psi \circ \varphi) = {}^t\varphi \circ {}^t\psi$, ${}^t1_L = 1_{L^*}$. If φ is †surjective, then ${}^t\varphi$ is †injective, and if φ is injective, then ${}^t\varphi$ is surjective. If φ is bijective, then ${}^t\varphi$ is also bijective. The rank of ${}^t\varphi$ coincides with the rank of φ if the rank of φ is finite. For an isomorphism $\varphi: L \to M$, the inverse mapping ${}^t\varphi^{-1} = \check{\varphi}: L^* \to M^*$ of ${}^t\varphi$ is called the **contragredient** of φ. We have $(\psi \circ \varphi)^{\check{}} = \check{\psi} \circ \check{\varphi}$.

Given a subspace N of a linear space L, the subspace $\{a^* \in L^* \mid \langle a, a^* \rangle = 0\ (a \in N)\}$ of L^* is denoted by $N^\perp$ and is called the subspace (of L^*) **orthogonal** to N. Then we have the canonical isomorphisms $(L/N)^* \cong N^\perp$, $N^* \cong L^*/N^\perp$. Similarly, given a subspace N' of L^*, we obtain the subspace $N'^\perp$ of L orthogonal to N': $N'^\perp = \{a \in L \mid \langle a, a^* \rangle = 0\ (a^* \in N')\}$. Thus we have a one-to-one correspondence between the finite-codimensional subspaces of L and the finite-dimensional subspaces of L^* by assigning $N^\perp$ to N and $N'^\perp$ to N'. The codimension of N is equal to the dimension of $N^\perp$. If L is finite-dimensional, we have a canonical isomorphism $L \cong (L^*)^*$ and a one-to-one correspondence $N \to N' = N^\perp$ between the set $\{N\}$ of all subspaces of L and the set $\{N'\}$ of all subspaces of L^*. These properties of correspondence between the subspaces of L and L^* are called **duality properties** of the linear space.

Let $(e_1, \ldots, e_n)$ be a basis of a linear space L. Then the system of elements $(e_1^*, \ldots, e_n^*)$ in L^* defined by the relation $\langle e_j, e_i^* \rangle = 0$ $(i \neq j)$, $\langle e_i, e_i^* \rangle = 1$ forms a basis of L^*, called the **dual basis** of $(e_1, \ldots, e_n)$. Thus a finite-dimensional linear space L can be identified with its dual space utilizing the isomorphism given by assigning each element of the dual basis to the corresponding element of the basis in a natural manner.

H. Multilinear Mappings

Let L, M, N be linear spaces over a field K and f be a mapping from the Cartesian product $M \times N$ to L. Suppose that for any fixed $b \in N$, the mapping $M \to L$ assigning $f(x,b) \in L$ to $x \in M$ is linear, and for any fixed $a \in M$, the mapping $N \to L$ assigning $f(a,y) \in L$ to $y \in N$ is also linear. Then f is called a **bilinear mapping** from $M \times N$ to L. The set of all bilinear mappings from $M \times N$ to L forms a linear space under the operations $(f+g)(x,y) = f(x,y) + g(x,y)$, $(\lambda f)(x,y) = \lambda f(x,y)$ $(\lambda \in K)$; this space is denoted by $\mathfrak{L}(M,N;L)$. In general, for linear spaces $M_1, \ldots, M_n$, a mapping $f: M_1 \times \ldots \times M_n \to L$ is called a **multilinear mapping** if it is linear in each variable. The set of all multilinear mappings from $M_1 \times \ldots \times M_n$ to L forms a linear space, denoted by $\mathfrak{L}(M_1, \ldots, M_n; L)$. If $L = K$, a bilinear mapping and a multilinear mapping are called a **bilinear form** and **multilinear form**, respectively.

Suppose, in particular, that $M_1 = \ldots = M_n = M$. A multilinear mapping $f: M_1 \times \ldots \times M_n \to L$ is called **symmetric** if $f(x_{\sigma(1)}, \ldots, x_{\sigma(n)}) = f(x_1, \ldots, x_n)$ $(x_i \in M)$ for any permutation σ of $\{1, \ldots, n\}$. Also, f is called **alternating** if $f(x_1, \ldots, x_i, \ldots, x_j, \ldots, x_n) = 0$ for $x_i = x_j$, $i \neq j$. In this case, $f(x_{\sigma(1)}, \ldots, x_{\sigma(n)}) = \mathrm{sgn}\,\sigma \cdot f(x_1, \ldots, x_n)$ for any permutation σ ($\mathrm{sgn}\,\sigma$ is $+1$ if σ is †even and -1 if σ is †odd). On the other hand, f is called **skew-symmetric** (or **antisymmetric**) if it satisfies this equality. Therefore, if the characteristic of the field K is different from 2, a skew-symmetric mapping is alternating.

Let M, N be linear spaces over a field K and Φ be a bilinear form on $M \times N$. The mappings $d_\Phi: N \to M^*$, $s_\Phi: M \to N^*$ defined by $\Phi(x,y) = (d_\Phi(y))(x) = (s_\Phi(x))(y)$ $(x \in M,\ y \in N)$ are linear mappings. If M, N are finite-dimensional, then d_Φ and s_Φ have the same rank, called the **rank** of Φ. Let $(x_1, \ldots, x_m)$, $(y_1, \ldots, y_n)$ be bases of M, N and $(x_1^*, \ldots, x_m^*)$, $(y_1^*, \ldots, y_n^*)$ be their dual bases. Then we have $d_\Phi(y_j) = \sum_{i=1}^m x_i^* \Phi(x_i, y_j)$, $s_\Phi(x_i) = \sum_{j=1}^n \Phi(x_i, y_j) y_j^*$. The matrix $(\Phi(x_i, y_j))$ is called the **matrix of a bilinear form** Φ relative to the given bases, and its rank coincides with the rank of Φ. If d_Φ, s_Φ are both injective, they are also bijective, and in this case Φ is said to be **nondegenerate**. Then each of d_Φ and s_Φ can be regarded as the transpose of the other, and M, N can be identified with N^*, M^*, respectively. In particular, if Φ is a nondegenerate bilinear form on $M \times M$, we have an isomorphism from M to its dual space M^*; identifying M with M^* by this isomorphism, M is said to be **self-dual**.

Let M be a linear space over a field K. A mapping $Q: M \to K$ is called a **quadratic form** on M if the following two conditions hold: (i) $Q(\alpha x) = \alpha^2 Q(x)$ $(\alpha \in K, x \in M)$; and (ii) the mapping $\Phi: M \times M \to K$ defined by $\Phi(x,y) = Q(x+y) - Q(x) - Q(y)$ $(x,y \in M)$ is a bilinear form on $M \times M$. In this case, Φ is called the **bilinear form associated with the quadratic form** Q, and it can be shown to be symmetric. We have $\Phi(x,x) = 2Q(x)$ $(x \in M)$ and $Q(x) = (1/2)\Phi(x,x)$ if the characteristic of $K \neq 2$. In general, for any bilinear form $f: M \times M \to K$, the mapping $Q: M \to K$ defined by $Q(x) = f(x,x)$ is a quadratic form. If $(x_1, \ldots, x_n)$ is a basis of M, a quadratic form Q is expressed as follows: $Q(\Sigma_{i=1}^n \xi_i x_i) = \Sigma_{\{i,j\}} \alpha_{ij} \xi_i \xi_j$ (the sum over all unordered pairs $\{i,j\}$), where $\alpha_{ii} = Q(x_i)$, $\alpha_{ij} = \Phi(x_i, x_j)$ $(i \neq j)$ (→ 344 Quadratic Forms). A **metric vector space** is a linear space M supplied with a nondegenerate quadratic form Q on M, and is denoted by (M, Q). The bilinear form Φ associated with Q gives an **inner product** $\Phi(x,y)$ $(x,y \in M)$.

I. Tensor Products

Let M, N be linear spaces over a field K. The **tensor product** $M \otimes N$ of M, N is defined as follows and can be used to "linearize" bilinear mappings from $M \times N$ to any linear space. Let F be the linear space generated by $M \times N$ and R be the subspace of F generated by all elements of the forms $(x+x',y)-(x,y)-(x',y)$, $(x,y+y')-(x,y)-(x,y')$, $(\alpha x,y)-\alpha(x,y)$, $(x,\alpha y)-\alpha(x,y)$ $(x,x' \in M, y,y' \in N, \alpha \in K)$. Then the quotient space F/R is denoted by $M \otimes N$, and the canonical projection $F \to M \otimes N$ is denoted by ψ. Given an element $(x,y) \in M \times N$, we denote the image $\psi((x,y))$ by $x \otimes y$. The bilinear mapping $M \times N \to M \otimes N$ assigning $x \otimes y$ to (x,y) is called the **canonical bilinear mapping**. We have, by definition, $(x+x') \otimes y = x \otimes y + x' \otimes y$, $x \otimes (y+y') = x \otimes y + x \otimes y'$, $(\alpha x) \otimes y = \alpha(x \otimes y) = x \otimes (\alpha y)$ $(\alpha \in K)$. To emphasize the basic field K, we sometimes write $M \otimes_K N$ instead of $M \otimes N$.

The tensor product can be characterized by the property that for any linear space L and any bilinear mapping $f: M \times N \to L$, there exists a unique linear mapping $\varphi: M \otimes N \to L$ satisfying $f(x,y) = \varphi(x \otimes y)$. Thus assigning the bilinear mapping $f: M \times N \to L$ defined by $f(x,y) = \varphi(x \otimes y)$ to a linear mapping $\varphi: M \otimes N \to L$, we obtain an isomorphism $\operatorname{Hom}(M \otimes N, L) \cong \mathfrak{L}(M,N;L)$. Every element of $M \otimes N$ can be expressed as a finite sum of elements of the form $x \otimes y$ $(x \in M, y \in N)$. If $\{x_i\}_{i \in I}$, $\{y_j\}_{j \in J}$ are bases of M, N, respectively, then the family $\{x_i \otimes y_j\}_{i \in I, j \in J}$ forms a basis of $M \otimes N$. Hence if M and N are of finite dimension, $\dim(M \otimes N) = \dim M \dim N$.

Let $M_1, M_2, \ldots$ be linear spaces over a field K. We have a unique isomorphism $M_1 \otimes M_2 \to M_2 \otimes M_1$ that assigns $x_2 \otimes x_1$ to $x_1 \otimes x_2$ $(x_i \in M_i)$. We also have a unique isomorphism $(M_1 \otimes M_2) \otimes M_3 \to M_1 \otimes (M_2 \otimes M_3)$ that assigns $x_1 \otimes (x_2 \otimes x_3)$ to $(x_1 \otimes x_2) \otimes x_3$ $(x_i \in M_i)$; hence we may identify $(M_1 \otimes M_2) \otimes M_3$ and $M_1 \otimes (M_2 \otimes M_3)$, and we denote them simply by $M_1 \otimes M_2 \otimes M_3$. In general, assigning $x_1 \otimes \ldots \otimes x_n$ to $(x_1, \ldots, x_n)$, we obtain the canonical multilinear mapping $M_1 \times \ldots \times M_n \to M_1 \otimes \ldots \otimes M_n$. As before, given any linear space L, we have the natural isomorphism $\operatorname{Hom}(M_1 \otimes \ldots \otimes M_n, L) \cong \mathfrak{L}(M_1, \ldots, M_n; L)$. Conversely, given linear spaces $M_1, \ldots, M_n$, the space $M_1 \otimes \ldots \otimes M_n$ can be characterized as a linear space N with a given multilinear mapping $\psi: M_1 \times \ldots \times M_n \to N$ such that (i) N is generated by the image $\psi(M_1 \times \ldots \times M_n)$; and (ii) for any multilinear mapping $f: M_1 \times \ldots \times M_n \to L$, there exists a unique linear mapping $f': N \to L$ satisfying $f = f' \circ \psi$. The tensor product $M_1 \otimes \ldots \otimes M_n$ is sometimes written as $\bigotimes_{i=1}^n M_i$, and an element $x_1 \otimes \ldots \otimes x_n$ is written as $\bigotimes_{i=1}^n x_i$.

Given linear mappings $f_i: M_i \to M_i'$ $(1 \leqslant i \leqslant n)$, there exists a unique linear mapping $f: M_1 \otimes \ldots \otimes M_n \to M_1' \otimes \ldots \otimes M_n'$ satisfying $f(x_1 \otimes \ldots \otimes x_n) = f_1(x_1) \otimes \ldots \otimes f_n(x_n)$; we denote the mapping f by $f_1 \otimes \ldots \otimes f_n$ or $\bigotimes_{i=1}^n f_i$ and call it the **tensor product** of the f_i $(1 \leqslant i \leqslant n)$. The assignment $(f_1, \ldots, f_n) \to f_1 \otimes \ldots \otimes f_n$ gives an isomorphism $\bigotimes_{i=1}^n \operatorname{Hom}(M_i, M_i') \to \operatorname{Hom}(\bigotimes_{i=1}^n M_i, \bigotimes_{i=1}^n M_i')$ if the M_i are finite-dimensional. If in particular $M_1' = \ldots = M_n' = K$, we have an isomorphism $\bigotimes_{i=1}^n M_i^* \to (\bigotimes_{i=1}^n M_i)^*$ under the identification $\bigotimes_{i=1}^n M_i' = K$ given by the assignment $x_1' \otimes \ldots \otimes x_n' \to x_1' \ldots x_n'$. Explicitly, the isomorphism $f: \bigotimes_{i=1}^n M_i^* \to (\bigotimes_{i=1}^n M_i)^*$ is determined by $f(\bigotimes_{i=1}^n x_i^*)(\bigotimes_{i=1}^n x_i) = \prod_{i=1}^n \langle x_i, x_i^* \rangle$ $(x_i^* \in M_i^*, x_i \in M_i)$.

J. Tensors

Let $E^{(\lambda)}$ $(1 \leqslant \lambda \leqslant k)$ be linear spaces over a field K. If $E^{(1)} = \ldots = E^{(k)} = E$, then $\bigotimes_{\lambda=1}^k E^{(\lambda)}$ is written $\bigotimes^k E$ and called the **tensor space of degree** k of E ($\bigotimes^0 E$ denotes K). Also, $(\bigotimes^p E) \otimes (\bigotimes^q E^*)$ is written $T_q^p(E)$, where E^* is the dual space of E. We have $T_0^p(E) = \bigotimes^p E$, $T_q^0(E) = \bigotimes^q E^*$, and $T_0^0(E) = K$. $T_q^p(E)$ is called the **tensor space of type** (p,q) of E, and each of its elements is called a **tensor of type** (p,q). In particular, a tensor of type $(p,0)$ is called a **contravariant tensor of degree** p, and a tensor of type $(0,q)$ is called a **covariant tensor of degree** q. A tensor of type $(0,0)$ is a **scalar**. An element of $T_0^1(E)=E$ is called a **contravariant vector**, and an element of $T_1^0(E)=E^*$ is called a **covariant vector**. If $p\neq 0$, $q\neq 0$, a tensor of type (p,q) is called a **mixed tensor**.

Let $(e_1,\dots,e_n)$ be a basis of E and $(f^1,\dots,f^n)$ be the basis of E^* dual to $(e_1,\dots,e_n)$. Then the tensors $e_{i_1}\otimes\dots\otimes e_{i_p}\otimes f^{j_1}\otimes\dots\otimes f^{j_q}$ $(i_\lambda,j_\mu=1,\dots,n;\ \lambda=1,\dots,p;\ \mu=1,\dots,q)$ form a basis of $T_q^p(E)$. Therefore, any tensor of type (p,q) can be written uniquely in the form

$$t=\sum \xi^{i_1\dots i_p}_{j_1\dots j_q}e_{i_1}\otimes\dots\otimes e_{i_p}\otimes f^{j_1}\otimes\dots\otimes f^{j_q}.$$

Also, $\xi^{i_1\dots i_p}_{j_1\dots j_q}$ is called the **component** of t relative to the basis $(e_1,\dots,e_n)$, the index i_λ is called a **contravariant index**, and the index j_μ is called a **covariant index**.

Let $(\bar e_1,\dots,\bar e_n)$ be another basis of E and $(\bar f^1,\dots,\bar f^n)$ be its dual basis. Suppose that we have

$$\bar e_i=\sum_{j=1}^n \alpha_i^j e_j,\quad \bar f^i=\sum_{j=1}^n \beta_j^i f^j.$$

Then we have

$$\sum_{k=1}^n \beta_k^i\alpha_j^k=\delta_j^i$$

and the transformation formula

$$\bar\xi^{i_1\dots i_p}_{j_1\dots j_q}=\sum_{k_1,\dots,k_p}\sum_{l_1,\dots,l_q}\beta^{i_1}_{k_1}\dots\beta^{i_p}_{k_p}\times\alpha^{l_1}_{j_1}\dots\alpha^{l_q}_{j_q}\xi^{k_1\dots k_p}_{l_1\dots l_q},$$

where the $\xi^{k_1\dots k_p}_{l_1\dots l_q}$ are the components of t relative to $(e_1,\dots,e_n)$ and the $\bar\xi^{i_1\dots i_p}_{j_1\dots j_q}$ are the components of t relative to $(\bar e_1,\dots,\bar e_n)$.

In the tensor calculus, an index appearing after the symbol Σ is called a **dummy index** if it appears in both the upper and the lower positions. For example, in the expression $\sum_{i=1}^n\xi_iy^i$, the index i is a dummy. As a convention, we sometimes omit the symbol $\sum_{i=1}^n$ for a dummy index i; for example, by ξ_iy^i we mean the sum $\sum_{i=1}^n\xi_iy^i$. This convention is called **Einstein's convention**. Using it, we write the previous transformation formula as

$$\bar\xi^{i_1\dots i_p}_{j_1\dots j_q}=\beta^{i_1}_{k_1}\dots\beta^{i_p}_{k_p}\alpha^{l_1}_{j_1}\dots\alpha^{l_q}_{j_q}\xi^{k_1\dots k_p}_{l_1\dots l_q}.$$

We have a nondegenerate bilinear form Φ on $T_q^p(E)\times T_p^q(E)$ determined by

$$\Phi\left(\bigotimes_{i=1}^p x_i\otimes\bigotimes_{j=1}^q y_j^*,\ \bigotimes_{j=1}^q y_j\otimes\bigotimes_{i=1}^p x_i^*\right)=\prod_{i=1}^p\langle x_i,x_i^*\rangle\prod_{j=1}^q\langle y_j,y_j^*\rangle.$$

Thus the space $T_q^p(E)$ can be identified with the dual space of $T_p^q(E)$, and vice versa (→ Section H). In this identification, the basis $(e_{i_1}\otimes\dots\otimes e_{i_p}\otimes f^{j_1}\otimes\dots\otimes f^{j_q})$ of $T_q^p(E)$ and the basis $(e_{j_1}\otimes\dots\otimes e_{j_q}\otimes f^{i_1}\otimes\dots\otimes f^{i_p})$ of $T_p^q(E)$ are dual to each other. In addition, combining the natural isomorphism $T_p^q(E)^*\cong\mathfrak{L}(\Pi^qE,\Pi^pE^*;K)$ with the duality $T_p^q(E)^*=T_q^p(E)$, we have a natural isomorphism $T_q^p(E)\to\mathfrak{L}(\Pi^qE,\Pi^pE^*;K)$. Explicitly, identifying an element $t\in T_q^p(E)$ with the multilinear form $\Pi^qE\times\Pi^pE^*\to K$ corresponding to it under the natural isomorphism, we have

$$t(x_1,\dots,x_q,y_1^*,\dots,y_p^*)=\zeta^{i_1\dots i_p}_{j_1\dots j_q}\xi_1^{j_1}\dots\xi_q^{j_q}\eta^1_{i_1}\dots\eta^p_{i_p},$$

where $\{\xi_\lambda^j\}$ is the component of $x_\lambda\in E=T_0^1(E)$, $\{\eta_i^\mu\}$ is the component of $y_\mu^*\in E^*=T_1^0(E)$, and $\{\zeta^{i_1\dots i_p}_{j_1\dots j_q}\}$ is the component of $t\in T_q^p(E)$. By the natural isomorphisms $T_0^p(E)\cong\mathfrak{L}(\Pi^pE^*,K)$, $T_p^0(E)=\mathfrak{L}(\Pi^pE,K)$, a contravariant tensor of degree p and a covariant tensor of degree p can be identified with a multilinear p-form on E^* and on E, respectively.

K. Tensor Algebras

There exists a unique bilinear mapping $T_q^p(E)\times T_s^r(E)\to T_{q+s}^{p+r}(E)$ which assigns the element $x_1\otimes\dots\otimes x_p\otimes y_1\otimes\dots\otimes y_r\otimes x_1^*\otimes\dots\otimes x_q^*\otimes y_1^*\otimes\dots\otimes y_s^*$ to the pair $(x_1\otimes\dots\otimes x_p\otimes x_1^*\otimes\dots\otimes x_q^*, y_1\otimes\dots\otimes y_r\otimes y_1^*\otimes\dots\otimes y_s^*)$; we denote the element assigned to the latter by $t\otimes u$, where $t=x_1\otimes\dots\otimes x_p\otimes x_1^*\otimes\dots\otimes x_q^*$, $u=y_1\otimes\dots\otimes y_r\otimes y_1^*\otimes\dots\otimes y_s^*$, and call it the **product** of t and u. If the components of t, u, $t\otimes u$ are

$$\{\xi^{i_1\dots i_p}_{j_1\dots j_q}\},\quad\{\eta^{k_1\dots k_r}_{l_1\dots l_s}\},\quad\{\zeta^{a_1\dots a_{p+r}}_{b_1\dots b_{q+s}}\},$$

then we have

$$\zeta^{i_1\dots i_pk_1\dots k_r}_{j_1\dots j_ql_1\dots l_s}=\xi^{i_1\dots i_p}_{j_1\dots j_q}\eta^{k_1\dots k_r}_{l_1\dots l_s}.$$

Let $T(E)$ be the direct sum of $T_q^p(E)$ $(p,q=0,1,2,\dots)$. Then $T(E)$ is an associative algebra over K whose product is a natural extension of the product $\otimes$. We call $T(E)$ the **tensor algebra** on E. The direct sum of $T_0^p(E)$ $(p=0,1,2,\dots)$ forms a subalgebra of $T(E)$, also called the **(contravariant) tensor algebra**.

L. Contractions

The **contraction** of $T_q^p(E)$ relative to the kth contravariant index and the lth covariant index is by definition the linear mapping $C_l^k: T_q^p(E)\to T_{q-1}^{p-1}(E)$ determined by assigning $\langle x_k,x_l^*\rangle x_1\otimes\dots\otimes x_{k-1}\otimes x_{k+1}\otimes\dots\otimes x_p\otimes x_1^*\otimes\dots\otimes x_{l-1}^*\otimes x_{l+1}^*\dots\otimes x_q^*$ to $x_1\otimes\dots\otimes x_p\otimes x_1^*\otimes\dots\otimes x_q^*$, where $\langle x_k,x_l^*\rangle$ is the inner product of x_k, x_l^*. For a tensor t of type (p,q), the tensor $C_l^k(t)$ of type $(p-1,q-1)$ is called the **contracted tensor** of t. If the components of t are $\xi^{i_1\dots i_p}_{j_1\dots j_q}$, the components of $C_l^k(t)$ are given by

$$\eta^{i_1\dots i_{p-1}}_{j_1\dots j_{q-1}}=\sum_{s=1}^{n}\xi^{i_1\dots i_{k-1}s i_k\dots i_{p-1}}_{j_1\dots j_{l-1}s j_l\dots j_{q-1}}.$$

M. Tensor Representations

For a linear mapping $f: E\rightarrow F$, the tensor product $f\otimes\dots\otimes f: T^p_0(E)=\otimes^pE\rightarrow\otimes^pF=T^p_0(F)$ is denoted by f^p. The f^p $(p=0,1,2,\dots)$ give an algebra homomorphism $\Sigma^{\infty}_{p=0}T^p_0(E)\rightarrow\Sigma^{\infty}_{p=0}T^p_0(F)$. Next, let f be an isomorphism and $\check{f}={}^tf^{-1}$ be its contragredient. Then f_q denotes the tensor product $\check{f}\otimes\dots\otimes\check{f}: T^0_q(E)=\otimes^qE^*\rightarrow\otimes^qF^*=T^0_q(F)$, and f^p_q the tensor product $f^p\otimes f_q: T^p_q(E)\rightarrow T^p_q(F)$. The mapping f^p_q is an isomorphism, and the system $\{f^p_q\}$ $(p,q=0,1,2,\dots)$ gives rise to an algebra isomorphism $T(E)\rightarrow T(F)$. In particular, if f is a nonsingular linear transformation of E, then f^p_q is a nonsingular linear transformation of the linear space $T^p_q(E)$, and the assignment $f\rightarrow f^p_q$ gives a group homomorphism $GL(E)\rightarrow GL(T^p_q(E))$; this homomorphism is called a **tensor representation** of the group $GL(E)$.

N. Symmetric and Alternating Tensors

A contravariant tensor of degree p is called **symmetric (alternating)** if the corresponding multilinear p-form, under the natural isomorphism $T^p_0(E)\cong\mathfrak{L}(\Pi^pE^*,K)$, is symmetric (alternating). A covariant tensor is also called **symmetric (alternating)** if the corresponding multilinear form is symmetric (alternating). A **skew-symmetric** (or **antisymmetric**) tensor is defined similarly. We reformulate these definitions under the assumption that the field K is not of characteristic 2. Let $\mathfrak{S}_p$ be the group of all permutations of $1,\dots,p$ (the †symmetric group of degree p). For any $\sigma\in\mathfrak{S}_p$, we have a unique linear transformation $T^p_0(E)\rightarrow T^p_0(E)$ assigning $x_{\sigma^{-1}(1)}\otimes\dots\otimes x_{\sigma^{-1}(p)}$ to $x_1\otimes\dots\otimes x_p$. This transformation is nonsingular and is also denoted by σ. Similarly, we have a unique nonsingular linear transformation of $T^0_p(E)$, also denoted by σ. An element $t\in T^p_0(E)$ (or $\in T^0_p(E)$) is symmetric if and only if $\sigma t=t$ for all $\sigma\in\mathfrak{S}_p$, while t is alternating if and only if $\sigma t=(\operatorname{sgn}\sigma)t$ for all $\sigma\in\mathfrak{S}_p$. Let the $\xi^{i_1\dots i_p}$ (or the $\xi_{i_1\dots i_p}$) be the components of t. Then t is symmetric (alternating) if and only if the components are symmetric (alternating) relative to permutations of the indices $i_1,\dots,i_p$. The linear transformation $S_p=\Sigma_{\sigma\in\mathfrak{S}_p}\sigma$ of $T^p_0(E)$ or $T^0_p(E)$ is called the **symmetrizer**, and $A_p=\Sigma_{\sigma\in\mathfrak{S}_p}(\operatorname{sgn}\sigma)\sigma$ is called the **alternizer**. For any t, S_pt is a symmetric tensor, and A_pt is an alternating tensor.

The subspace of $T^p_0(E)$ consisting of all symmetric (or alternating) tensors is invariant under the transformation $R: GL(E)\rightarrow GL(T^p_0(E))$, the tensor representation where $\varphi(t)=R(\varphi)(t)$ for $\varphi\in GL(E)$ and $t\in T^p_0(E)$.

O. Exterior Product

For simplicity, we assume that the basic field K is of characteristic 0. We denote by N_p the kernel of the alternizer $A_p: T^p_0(E)\rightarrow T^p_0(E)$, namely, the subspace consisting of all t satisfying $A_pt=0$, and by $\bigwedge^pE$ the quotient space $T^p_0(E)/N_p$. The image of $x_1\otimes\dots\otimes x_p$ $(x_i\in E)$ under the natural mapping $T^p_0(E)\rightarrow\bigwedge^pE$ is denoted by $x_1\wedge\dots\wedge x_p$ and is called the **exterior product** of $x_1,\dots,x_p$. The linear space $\bigwedge^pE$ is called the **p-fold exterior power** of E. We have

$$x_1\wedge\dots\wedge(x_i+x_i')\wedge\dots\wedge x_p$$
$$=x_1\wedge\dots\wedge x_i\wedge\dots\wedge x_p+x_1\wedge\dots\wedge x_i'\wedge\dots\wedge x_p,$$
$$x_1\wedge\dots\wedge(\alpha x_i)\wedge\dots\wedge x_p$$
$$=\alpha(x_1\wedge\dots\wedge x_i\wedge\dots\wedge x_p),\qquad \alpha\in K,$$

and for every $\sigma\in\mathfrak{S}_p$,

$$x_{\sigma(1)}\wedge\dots\wedge x_{\sigma(p)}=(\operatorname{sgn}\sigma)x_1\wedge\dots\wedge x_p.$$

A_p induces a natural isomorphism $\bigwedge^pE\cong\mathfrak{A}^p$, where $\mathfrak{A}^p$ consists of all contravariant alternating tensors of degree p. Thus an element of $\bigwedge^pE$ may be identified with a contravariant alternating tensor of degree p. Then we have $A_p(x_1\otimes\dots\otimes x_p)=x_1\wedge\dots\wedge x_p$. Similarly, $\bigwedge^pE^*$ is identified with the linear space consisting of all covariant alternating tensors of degree p. An element of $\bigwedge^pE$ is sometimes called a **p-vector**, and an element of $\bigwedge^pE^*$ is called a **p-covector** (→ 92 Coordinates B). If $(e_1,\dots,e_n)$ is a basis of E, then the $e_{i_1}\wedge\dots\wedge e_{i_p}$ $(i_1<i_2<\dots<i_p)$ form a basis of $\bigwedge^pE$, and any element of $\bigwedge^pE$ is written in the form $t=\Sigma_{i_1<\dots<i_p}\alpha^{i_1\dots i_p}e_{i_1}\wedge\dots\wedge e_{i_p}$ or $t=(1/p!)\Sigma_{i_1,\dots,i_p}\alpha^{i_1\dots i_p}e_{i_1}\wedge\dots\wedge e_{i_p}$. In the latter form, $\alpha^{i_1\dots i_p}$ is alternating relative to permutations of $i_1,\dots,i_p$, and it is the component of t. The dimension of $\bigwedge^pE$ is equal to $\binom{n}{p}$, and $\bigwedge^pE=0$ if $p>n$. If $(f^1,\dots,f^n)$ is the dual basis of $(e_1,\dots,e_n)$, the inner product of an element $t=\Sigma_{i_1<\dots<i_p}\alpha^{i_1\dots i_p}e_{i_1}\wedge\dots\wedge e_{i_p}\in\bigwedge^pE$ and an element $s=\Sigma_{i_1<\dots<i_p}\beta_{i_1\dots i_p}\cdot f^{i_1}\wedge\dots\wedge f^{i_p}\in\bigwedge^pE^*$ is defined by

$$\langle s,t\rangle=\sum_{i_1<\dots<i_p}\alpha^{i_1\dots i_p}\beta_{i_1\dots i_p}.$$

Then we have $\langle x_1\wedge\dots\wedge x_p, y_1\wedge\dots\wedge y_p\rangle=\det(\langle x_i,y_j\rangle)$, where $\langle x_i,y_j\rangle$ is the inner product of $x_i\in E$ and $y_j\in E^*$. By this inner product, we can identify $\bigwedge^pE^*$ with the dual space of $\bigwedge^pE$.

The tensor product $\otimes^p E \times \otimes^q E \to \otimes^{p+q} E$ induces naturally a mapping $\Phi: \otimes^p E/N_p \times \otimes^q E/N_q \to \otimes^{p+q} E/N_{p+q}$. Using this bilinear mapping $\Phi: \bigwedge^p E \times \bigwedge^q E \to \bigwedge^{p+q} E$, we define the **exterior product** $t \wedge s$ of an element $t \in \bigwedge^p E$ and an element $s \in \bigwedge^q E$ by $t \wedge s = \Phi(t,s)$. Then $t \wedge s$ is an element of $\bigwedge^{p+q} E$, and we have $t \wedge s = (-1)^{pq} s \wedge t$, $(x_1 \wedge \ldots \wedge x_p) \wedge (x_{p+1} \wedge \ldots \wedge x_{p+q}) = x_1 \wedge \ldots \wedge x_{p+q}$.

We denote by $\bigwedge E$ the direct sum of $\bigwedge^p E$ $(p = 0, 1, 2, \ldots, n)$, and define the product of two elements $x = \sum_{p=0}^n x^p$, $y = \sum_{p=0}^n y^p$ $(x^p, y^p \in \bigwedge^p E)$ by $x \wedge y = \sum_{p+q=0}^n x^p \wedge y^q$. Then the product $\wedge$ satisfies the associative law. We call $\bigwedge E$ the **exterior algebra** (or **Grassmann algebra**) of the linear space E. If E is of dimension n, $\bigwedge E$ is of dimension 2^n. If $(e_1, \ldots, e_n)$ is a basis of E relative to K, $\bigwedge E$ is sometimes written as $\bigwedge_K(e_1, \ldots, e_n)$. The exterior algebra $\bigwedge E^*$ of the dual space E^* is similarly defined and can be considered as the dual space of $\bigwedge E$.

P. Linear Transformations

Let L be an n-dimensional linear space over a field K and φ be a linear transformation of L. An **eigenvector** (or **proper vector**) of φ is a vector $a \in L$ such that there exists a scalar $\alpha \in K$ satisfying $\varphi(a) = \alpha a$. If $a \neq 0$, α is determined by a and is called an **eigenvalue** (or **proper value**) of φ with respect to the eigenvector a. Let F be the square matrix representing φ relative to a basis of L. Then the †characteristic polynomial $\chi(X)$ of F does not depend on the choice of basis. We call $\chi(X)$ the **characteristic polynomial** of φ. The scalar $\alpha \in K$ is an eigenvalue of φ (with respect to some nonvanishing eigenvector) if and only if α is a root of the equation $\chi(X) = 0$. An eigenvalue of φ is sometimes called a **characteristic root** of φ. For an eigenvalue α of φ, the subspace $N_\alpha = \{a \in L \mid \varphi(a) = \alpha a\}$ of L is called the eigenspace belonging to α. Furthermore, the space $N'_\alpha = \{a \in L \mid (\varphi - \alpha 1_L)^k(a) = 0$ for some $k > 0\}$ is a subspace of L containing N_α and is sometimes called an **eigenspace in a weaker sense**. If all roots of $\chi(X) = 0$ are in K, L is decomposed into the direct sum of $N'_{\alpha_1}, \ldots, N'_{\alpha_s}$, where $\alpha_1, \ldots, \alpha_s$ are the distinct roots of the equation $\chi(X) = 0$. The dimension of N'_{α_i} is equal to the multiplicity of the root α_i in the equation $\chi(X) = 0$.

Any polynomial $p(X)$ satisfying $p(\varphi) = 0$ is divisible by a monic polynomial $\mu(X)$ uniquely determined by φ. We call $\mu(X)$ the **minimal polynomial** of φ. An eigenvalue of φ coincides with a root of the equation $\mu(X) = 0$. The characteristic polynomial $\chi(X)$ is divisible by $\mu(X)$, and we have $\chi(\varphi) = 0$ (W. R. Hamilton and A. Cayley). In detail, we note that L has the structure of a module over the polynomial ring $K[X]$ determined by $p(X) \cdot a = (p(\varphi))(a)$ $(p(X) \in K[X], a \in L)$. There exist monic polynomials $q_1(X), \ldots, q_n(X)$ satisfying the following two conditions: (i) $q_{i+1}(X)$ is divisible by $q_i(X)$ $(1 \leqslant i \leqslant n-1)$; and (ii) L is decomposed into the direct sum of some †monomial submodules $L_1, \ldots, L_n$ such that $q_i(X)$ is the minimal polynomial of the restriction of φ to L_i $(1 \leqslant i \leqslant n)$. Such $q_i(X)$ $(1 \leqslant i \leqslant n)$ are uniquely determined by φ, and they are the †elementary divisors of $XI - F$, where F is a matrix corresponding to φ relative to a basis (→ 269 Matrices E, 70 Commutative Rings K). Furthermore, $q_1(X) \ldots q_n(X) = \chi(X)$, $q_n(X) = \mu(X)$.

A linear transformation φ of L is called **semisimple** if L has the structure of a †semisimple $K[X]$-module determined by φ. Hence φ is semisimple if and only if the minimal polynomial $\mu(X)$ of φ has no square factor different from constants in $K[X]$. In particular, the condition that all roots of $\mu(X) = 0$ are in K and simple is sufficient for semisimplicity of φ. This condition is equivalent to the condition that φ is represented by a †diagonal matrix relative to some basis of L. Then L is decomposed into the direct sum of the eigenspaces N_α, and φ is said to be **diagonalizable**.

If K is a †perfect field, any linear transformation φ of L is represented as the sum of a semisimple linear transformation φ_s and a nilpotent linear transformation φ_n: $\varphi = \varphi_s + \varphi_n$ (**Jordan decomposition**). Also, φ_s and φ_n commute with each other, and they are uniquely determined by φ. We call φ_s and φ_n the **semisimple** and **nilpotent component** of φ, respectively. Furthermore, φ_s and φ_n can be represented as polynomials of φ without a constant term. The transformation φ is nonsingular if and only if φ_s is nonsingular. A nonsingular linear transformation φ is called **unipotent** if φ_s is equal to the identity transformation 1_L. Any nonsingular linear transformation φ is uniquely represented as a product of a semisimple linear transformation and a unipotent linear transformation, which are commutative: $\varphi = \varphi_s \varphi_u$ (**multiplicative Jordan decomposition**). Here φ_s is the semisimple component and $\varphi_u = 1_L + \varphi_s^{-1} \varphi_n$ is unipotent (φ_u is called the **unipotent component** of φ; → 15 Algebraic Groups).

Q. Semilinear Mappings

Let L be a linear space over a field K and L' be a linear space over a field K'. A pair (φ, ρ) consisting of a mapping $\varphi: L \to L'$ and a mapping $\rho: K \to K'$ is called a **semilinear mapping** if the following four conditions hold (for convenience here we write the scalars to the right of the vectors): (i) $\varphi(a+b)=\varphi(a)+\varphi(b)$; (ii) $\varphi(a\lambda)=\varphi(a)\rho(\lambda)$; (iii) $\rho(\alpha+\beta)=\rho(\alpha)+\rho(\beta)$; and (iv) $\rho(\alpha\beta)=\rho(\alpha)\rho(\beta)$ $(a,b\in L;\lambda,\alpha,\beta\in K)$. In this case, φ is sometimes said to be semilinear relative to ρ (→ 275 Modules L). Conditions (iii) and (iv) mean that ρ is a field homomorphism. If $K=K'$ and ρ is the identity, then the semilinear mapping $\varphi:L\to L'$ is a linear mapping. If $L=L'$, $K=K'$ (and ρ is an automorphism), φ is called a **semilinear transformation** relative to ρ.

For semilinear mappings $(\varphi,\rho):L\to L'$, $K\to K'$ and $(\varphi',\rho'):L'\to L''$, $K'\to K''$, where L'' is a linear space over K'', the composite $(\varphi'\circ\varphi,\rho'\circ\rho)$ is also a semilinear mapping. If a basis $(e_1,\ldots,e_n)$ for L over K and a basis $(e'_1,\ldots,e'_{n'})$ for L' over K' are given, a semilinear mapping $(\varphi,\rho):L\to L'$, $K\to K'$ determines a matrix (α_{ij}) by the relation $\varphi(e_j)=\sum_{i=1}^{n'}e'_i\alpha_{ij}$ $(1\leqslant j\leqslant n)$. Conversely, a homomorphism ρ and an $n'\times n$ matrix $A=(\alpha_{ij})$ determine a semilinear mapping φ by this relation. Hence for fixed bases, a semilinear mapping is represented by a pair (A,ρ), where A is an $n'\times n$ matrix. If a semilinear mapping φ' relative to ρ' is represented by (A',ρ'), the composite $(\varphi'\circ\varphi,\rho'\circ\rho)$ is represented by $(A'A^{\rho},\rho'\circ\rho)$, where A^{ρ} is the matrix $(\rho(\alpha_{ij}))$.

Let $\varphi:L\to L$ be a semilinear transformation relative to an automorphism $\rho:K\to K$. Suppose that φ is represented by (A,ρ) relative to a basis $(e_1,\ldots,e_n)$ for L and by (B,ρ) relative to another basis $(f_1,\ldots,f_n)$. If we define a matrix $P=(p_{ij})$ by the relation $f_j=\sum_{i=1}^{n}e_ip_{ij}$ $(1\leqslant j\leqslant n)$, we have $B=P^{-1}AP^{\rho}$. Two pairs (A,ρ), (B,ρ) having the relation $B=P^{-1}AP^{\rho}$ are said to be **similar**.

R. Sesquilinear Forms

Let K be a field (not necessarily commutative) and J be its †antiautomorphism. For left linear spaces M, N over K, a mapping $\Phi:M\times N\to K$ is called a (right) **sesquilinear form** relative to J if the following four conditions are satisfied: (i) $\Phi(x+x',y)=\Phi(x,y)+\Phi(x',y)$; (ii) $\Phi(x,y+y')=\Phi(x,y)+\Phi(x,y')$; (iii) $\Phi(\alpha x,y)=\alpha\Phi(x,y)$; and (iv) $\Phi(x,\alpha y)=\Phi(x,y)\alpha^{J}$ $(x,x'\in M;y,y'\in N;\alpha\in K)$. If J is the identity automorphism, then K is necessarily commutative and Φ is a bilinear form (→ Section H). As an example of K and J, we may take K as the field of complex numbers and J as complex conjugation. In general, for a left linear space E over K, we denote by E^{J} the right linear space with the scalar multiplication $x\lambda=\lambda^{J^{-1}}x$ $(x\in E,\lambda\in K)$. Then condition (iv) becomes (iv′) $\Phi(x,y\alpha)=\Phi(x,y)\alpha$; and if K is commutative, Φ is a bilinear form on $M\times N^{J}$. For a sesquilinear form Φ on $M\times N$, we have the linear mappings $d_{\Phi}:N^{J}\to M^{*}$, $s_{\Phi}:M^{J^{-1}}\to N^{*}$ defined by the relation $\Phi(x,y)=\langle x,d_{\Phi}(y)\rangle=\langle y,s_{\Phi}(x)\rangle^{J}$ $(x\in M,y\in N)$. If M, N are finite-dimensional, d_{Φ}, s_{Φ} have the same rank, which is called the **rank** of Φ. We assume that all linear spaces are finite-dimensional.

Let $(x_1,\ldots,x_m)$, $(y_1,\ldots,y_n)$ be bases of M, N and $(x_1^{*},\ldots,x_m^{*})$, $(y_1^{*},\ldots,y_n^{*})$ be their dual bases. Then we have $d_{\Phi}(y_j)=\sum_{i=1}^{m}x_i^{*}\Phi(x_i,y_j)$, $s_{\Phi}(x_i)=\sum_{j=1}^{n}y_j^{*}\Phi(x_i,y_j)^{J^{-1}}$. The matrix $(\Phi(x_i,y_j))$ is called the **matrix of the sesquilinear form** Φ relative to the given bases; its rank is equal to the rank of Φ. If d_{Φ}, s_{Φ} are both injective (therefore bijective), Φ is said to be **nondegenerate**. Let $\Phi':M'\times N'\to K$ be another sesquilinear form relative to J. Then for any linear mapping $u:M\to M'$, there exists a unique linear mapping $u^{*}:N'\to N$ such that $\Phi'(u(x),y')=\Phi(x,u^{*}(y'))$ $(x\in M,y'\in N')$; this is called the **left-adjoint** linear mapping of u. Similarly, for any linear mapping $v:N\to N'$, there exists a unique linear mapping $v^{*}:M'\to M$ such that $\Phi'(x',v(y))=\Phi(v^{*}(x'),y)$ $(x'\in M',y\in N)$; this is called the **right-adjoint** linear mapping of v. We have $u^{*}=d_{\Phi}^{-1}\circ{}^{t}u\circ d_{\Phi'}$, $v^{*}=s_{\Phi}^{-1}\circ{}^{t}v\circ s_{\Phi'}$. In particular, let u, v be isomorphisms. Then we have $\Phi(x,y)=\Phi'(u(x),v(y))$ $(x\in M,y\in N)$ if and only if $u^{-1}=v^{*}$, $v^{-1}=u^{*}$.

A sesquilinear form on $M\times M$ is called simply a **sesquilinear form** on M. Let J be an †involution (namely $J=J^{-1}$), and write $\lambda^{J}=\bar{\lambda}$ $(\lambda\in K)$. If condition (v) $\Phi(x,y)=\overline{\Phi(y,x)}$ $(x,y\in M)$ holds, Φ is called a **Hermitian form** on M. On the other hand, if the condition (v′) $\Phi(x,y)=-\overline{\Phi(y,x)}$ holds, Φ is called an **anti-Hermitian form** (or **skew-Hermitian form**) on M. In particular, if $J=1_K$, then a Hermitian form (anti-Hermitian form) is a symmetric bilinear form (antisymmetric bilinear form). A linear space M supplied with a nondegenerate Hermitian form Φ is called a **Hermitian linear space**, and $\Phi(x,y)$ is called the **Hermitian inner product** (or simply **inner product**) of $x,y\in M$.

References

[1] S. Iyanaga and K. Kodaira, Gendai sûgaku gaisetu I (Japanese; Introduction to modern mathematics I), Iwanami, 1961.
[2] K. Asano, Senkei daisûgaku teiyô (Japanese; A manual of linear algebra), Kyôritu, 1948.
[3] Y. Akizuki and M. Suzuki, Kôtô daisûgaku II (Japanese; Higher algebra II), Iwanami, 1957.
[4] S. Iyanaga and M. Sugiura, Ôyôsûgakusya no tameno daisûgaku (Japanese; Algebra for

applied mathematicians), Iwanami, 1960.
[5] O. Schreier and E. Sperner, Einführung in die analytische Geometrie und Algebra I, II, Vandenhoeck, fourth edition, 1959; English translation, Introduction to modern algebra and matrix theory, Chelsea, 1959.
[6] N. Bourbaki, Eléments de mathématique, Algèbre, ch. 2, 3, 7, 8, 9, Actualités Sci. Ind. 1236b, 1044, 1179a, 1261a, 1272a, Hermann, 1958–1964.
[7] И. М. Гельфанд (I. M. Gel'fand), Лекции по линейной алгебре, Гостехиздат, 1951; English translation, Lectures on linear algebra, Interscience, 1961.
[8] N. Jacobson, Lectures in abstract algebra II, Van Nostrand, 1953.
[9] C. Chevalley, Fundamental concepts of algebra, Academic Press, 1956.
[10] W. H. Graeub, Lineare Algebra, Springer, 1958.
[11] P. R. Halmos, Finite-dimensional vector spaces, Van Nostrand, second edition, 1958.
[12] R. Godement, Cours d'algèbre, Hermann, 1963.
[13] Д. А. Райков (D. A. Raĭkov), Векторные пространства, Физматгиз, 1962; English translation, Vector spaces, Noordhoff, 1965.
[14] S. Lang, Algebra, Addison-Wesley, 1965.

257 (V.17) Local Fields

A. General Remarks

A field k that is †complete with respect to a †discrete valuation is called a **local field** if its field of residue classes is finite. (Real and complex number fields are sometimes also called local fields; these, however, are not considered in this article.) A local field k is isomorphic either to the †completion with respect to a †$\mathfrak{p}$-adic valuation determined by a prime ideal $\mathfrak{p}$ of a number field of finite degree or to the field of †formal power series over a finite field. In the former case, k is called a **$\mathfrak{p}$-adic number field**. We let $\mathfrak{o}$ stand for the †valuation ring of k, $\mathfrak{p}$ stand for the †valuation ideal of k, p stand for the characteristic of the field $\mathfrak{o}/\mathfrak{p}$ of residue classes, and $N(\mathfrak{p})$ stand for the number of the elements of $\mathfrak{o}/\mathfrak{p}$. An additive valuation of k whose set of values coincides with the set of all rational integers is denoted by $\operatorname{ord}\alpha$ ($\alpha\in k$); here we understand $\operatorname{ord}0=\infty$. The †regular (multiplicative) valuation of k is defined by $|\alpha|=(N(\mathfrak{p}))^{-\operatorname{ord}\alpha}$ (→ 425 Valuations).

B. Construction of Local Fields

A $\mathfrak{p}$-adic number field k is an extension of finite degree of the p-adic field $\mathbf{Q}_p$. If $n=[k:\mathbf{Q}_p]=ef$, where e is the †ramification index of $k/\mathbf{Q}_p$ and f is the †relative degree of $k/\mathbf{Q}_p$ (→Section D), then there exists one and only one field F such that $k\supset F\supset\mathbf{Q}_p$, $[k:F]=e$, $[F:\mathbf{Q}_p]=f$, and $F/\mathbf{Q}_p$ is †unramified. The field κ of residue classes of F is isomorphic to $GF(p^f)$, and F is uniquely determined by κ by means of Witt vectors (→ 434 Witt Vectors). Every residue class ($\neq\bar{0}$) of $\mathfrak{o}_F/\mathfrak{p}_F\cong\kappa$ contains one and only one mth root of unity (m is a divisor of p^f-1), and F is obtained by adjoining to $\mathbf{Q}_p$ a primitive (p^f-1)th root. Then k is a †completely ramified extension of F and is obtained by adjoining to F a root of an †Eisenstein polynomial (→ 334 Polynomials F).

C. The Topology of k

Taking $\mathfrak{p}^m$ ($m=0,1,2,\ldots$) as a †base for a neighborhood system of 0, k becomes a †locally compact †totally disconnected †topological field, and $\mathfrak{p}^m$ ($m=0,1,2,\ldots$) are compact subgroups of the additive group k. The multiplicative group $k^\times$ of nonzero elements of k is a locally compact †Abelian group, and the $\mathfrak{u}^{(m)}=\{\alpha\in\mathfrak{o}\mid\alpha\equiv1\ (\operatorname{mod}\mathfrak{p}^m)\}$ ($m=0,1,2,\ldots$) form a base for the neighborhood system of 1. The †character group in the sense of Pontrjagin of the additive group k is isomorphic to k. This isomorphism is obtained by the following natural correspondence: For a $\mathfrak{p}$-adic field k, denote by φ the composition of the natural mapping of $\mathbf{Q}_p$ onto $\mathbf{Q}_p/\mathbf{Z}_p$ ($\subset\mathbf{Q}/\mathbf{Z}$) and the †trace Tr from k to $\mathbf{Q}_p$, and put $\chi_x(y)=\exp(2\pi\sqrt{-1}\,\varphi(xy))$ ($y\in k$); for the field k of the formal power series over a finite field κ, put $\chi_x(y)=\psi(xy)$ ($y\in k$) with $\psi(\alpha)=\exp(2\pi\sqrt{-1}\,\operatorname{Tr}(\operatorname{Res}\alpha)/p)$ ($\alpha\in k$), where $\operatorname{Res}\alpha$ is the residue of $\alpha\in k$ and Tr is the trace from κ to $\mathbf{Z}/p\mathbf{Z}$. Then in both cases χ_x is a character of k, and $x\to\chi_x$ gives an isomorphism between k and the character group.

D. Ramification Theory

A valuation v of k has a unique †prolongation to an extension K of finite degree over k (we denote the prolongation also by v). K is complete under the valuation and is therefore a local field. Denoting by $\mathfrak{k}$ and κ the field of residue classes of K and k, respectively, we call $[\mathfrak{k}:\kappa]=f$ the **relative degree** of K/k, and $e=[v(K^\times):v(k^\times)]$ the **ramification index** of K/k. Then we have the equality $[K:k]=ef$.

If $e=1$, we call K/k an **unramified extension**.

An unramified extension K/k is †normal, and its †Galois group is a cyclic group generated by the Frobenius automorphism, i.e., the element σ of the Galois group of K/k such that $\alpha^{\sigma} \equiv \alpha^{N(\mathfrak{p})} \pmod{\mathfrak{p}}$ for any element α in the valuation ring of K. For a given natural number f, there exists one and only one unramified extension of degree f over k in an algebraic closure of k.

Let K/k be a normal extension of finite degree with Galois group G, let Π be a †prime element of K, that is, a generator of the †valuation ideal $\mathfrak{P}$ of K, and put $V^{(i)} = \{\sigma \in G \mid \Pi^{\sigma} \equiv \Pi \pmod{\mathfrak{P}^{i+1}}\}$. Then $V^{(i)}$ is independent of the choice of Π. We call $V^{(0)}$ the **inertia group** and $V^{(i)}$ the ith **ramification group**. Then $V^{(i)}$ is normal in G, $[G:V^{(0)}]=f$, $[V^{(0)}:1]=e$, and $V^{(1)}$ is the p-Sylow subgroup of $V^{(0)}$. Furthermore, $G/V^{(0)}$ and $V^{(0)}/V^{(1)}$ are cyclic, and $V^{(i)}/V^{(i+1)}$ $(i=1,2,\ldots)$ is an Abelian group of type $(p,p,\ldots,p)$. Ramification theory for Abelian extensions is described in Section F.

For $x \in k$, the series $\exp(x) = \sum_{n=0}^{\infty} x^n/n!$ (resp. $\log(1+x) = \sum_{n=1}^{\infty} (-1)^{n-1} x/n$) converges for $\operatorname{ord} x > e/(p-1)$ $(\operatorname{ord} x > 0)$. The additive group $\mathfrak{p}^m$ and the multiplicative group $\mathfrak{u}^{(m)}$ $(m > e/(p-1))$ are isomorphic as topological groups under the mappings $x \rightarrow y = \exp(x)$, $y \rightarrow x = \log(y)$. If we fix an element $\pi \in k$ with $\operatorname{ord}\pi = 1$, then an arbitrary element $x \in k$ with $\operatorname{ord} x = r$ is uniquely expressed in the form $x = \pi^r \zeta \alpha$, $\zeta^{p^f-1} = 1$, $\alpha \in \mathfrak{u}^{(1)}$. The group $\mathfrak{u}^{(1)}$ is a multiplicative group on which $\mathbf{Z}_p$ operates, and the structure of $\mathfrak{u}^{(1)}$ as $\mathbf{Z}_p$-group can be determined explicitly (→ [2], ch. II).

E. Cohomology

For a normal extension K/k with Galois group G, we may consider the †cohomology groups $H^r(G, K^{\times})$ $(r=1,2,\ldots)$ of G operating on the multiplicative group $K^{\times}$. In particular, the 2-cohomology group $H^2(G, K^{\times})$ is important in local class field theory and theory of algebras over k.

If C is the separable algebraic closure of k, i.e., the †maximal separable field over k of the †algebraic closure of k, and Γ is the Galois group of C/k, we can consider the 2-cohomology group $H^2(\Gamma, C^{\times})$. Here we take as cocycles only those mappings $f(\sigma,\tau)$ of $\Gamma \times \Gamma$ into $C^{\times}$ that are continuous with respect to the †Krull topology of Γ and the discrete topology of $C^{\times}$. A fundamental theorem about the structure of $H^2(\Gamma, C^{\times})$ states that the cocycles of $H^2(\Gamma, C^{\times})$ that split in an extension K/k of degree n are exactly those cocycles that split in the unramified extension of degree n over k. Here a cocycle is said to **split** in K if it belongs to the †kernel of the homomorphism $H^2(\Gamma, C^{\times}) \xrightarrow{\text{res}} H^2(H, C^{\times})$, where H is the subgroup of Γ corresponding to K, and res is the mapping obtained by restricting σ, τ of $f(\sigma,\tau)$ to the elements of H.

Let K/k be a normal extension of degree n with the Galois group G, and let H be the subgroup of Γ corresponding to K. Then the fundamental theorem combined with the exact sequence

$$0 \rightarrow H^2(G, K^{\times}) \rightarrow H^2(\Gamma, C^{\times}) \rightarrow H^2(H, C^{\times}) \qquad (1)$$

known in the theory of Galois cohomology (→ 202 Homological Algebra I; 62 Class Field Theory H), yields $H^2(G, K^{\times}) \cong \mathbf{Z}/n\mathbf{Z}$ (cyclic group of order n), and $H^2(\Gamma, C^{\times}) \cong \mathbf{Q}/\mathbf{Z}$, where $\mathbf{Q}$ is the additive group of rational numbers and $\mathbf{Z}$ is the additive group of rational integers. These isomorphisms can be given in canonical form as follows: Denote the unramified extension of degree n by K_n/k, and the Frobenius automorphism of K_n/k by σ. Then an element a of $H^2(G_n, K_n^{\times})$ (where G_n is the Galois group of K_n/k) is represented by the cocycle

$$f(\sigma^i, \sigma^j) = a^{[(i+j)/n]-[i/n]-[j/n]}, \quad i,j \in \mathbf{Z}$$

with $a \in k$, and conversely, every $a \in k$ determines an element c of $H^2(G_n, K_n^{\times})$ in this manner. Under these conditions, the correspondence between c and a gives rise to an isomorphism $H^2(G_n, K_n^{\times}) \cong k^{\times}/N_{K_n/k}(K_n^{\times})$. Next define an element $\operatorname{inv} c$ of $\mathbf{Q}/\mathbf{Z}$ by $\operatorname{inv} c \equiv (\operatorname{ord} a)/n \pmod{\mathbf{Z}}$. Then since $c \in H^2(\Gamma, C^{\times})$ splits in an unramified extension of degree n by the fundamental theorem, the exact sequence (1) determines in a natural way an element c' of $H^2(G_n, K_n^{\times})$ corresponding to c. Putting $\operatorname{inv} c = \operatorname{inv} c'$, we can show that $H^2(\Gamma, C^{\times}) \ni c \rightarrow \operatorname{inv} c$ gives an isomorphism $H^2(\Gamma, C^{\times}) \cong \mathbf{Q}/\mathbf{Z}$. We call $\operatorname{inv} c$ the **invariant** of $c \in H^2(\Gamma, C^{\times})$. The invariant of an element of the cohomology group $H^2(G, K^{\times})$ is defined to be the invariant of the corresponding element of $H^2(\Gamma, C^{\times})$, which is determined by the exact sequence (1). Mapping an element of $H^2(G, K)$ to its invariant, we obtain an isomorphism $H^2(G, K^{\times}) \cong \mathbf{Z}/n\mathbf{Z}$.

F. Local Class Field Theory

Let K/k be a normal extension of degree n with the Galois group G, let $f(\sigma,\tau)$ be a cocycle representing the element of $H^2(G, K^{\times})$

with the invariant $1/n$, and put

$$\left(\frac{K/k}{\sigma}\right)=\prod_{\tau\in G} f(\tau,\sigma)^{-1}.$$

Then $\sigma\rightarrow\left(\frac{K/k}{\sigma}\right)$ gives an isomorphism between G/G' (where G' is the commutator subgroup of G) and $k^{\times}/N_{K/k}(K^{\times})$. It follows from this that $[k^{\times}:N_{E/k}(E^{\times})]\leqslant[E:k]$ for any extension E/k of finite degree, and the equality holds if and only if E/k is Abelian. The inverse mapping of $G\ni\sigma\rightarrow\left(\frac{K/k}{\sigma}\right)\in k^{\times}/N_{K/k}(K^{\times})$ for an Abelian extension K/k is written as $k^{\times}\ni\alpha\rightarrow(\alpha,K/k)\in G$, and $(\alpha,K/k)$ is called the **norm-residue symbol**. If L/k is Abelian and K/k is a subfield of L, then the restriction of $(\alpha,L/k)$ to K coincides with $(\alpha,K/k)$. Let k_a be the maximal Abelian extension of k, i.e., the union of all Abelian extensions of finite degree over k. Then for any $\alpha\in k^{\times}$, an element (α,k) of the Galois group $G(k_a/k)$ of k_a/k is uniquely determined by $(\alpha,k)(\gamma)=(\alpha,k(\gamma)/k)(\gamma)$, $\gamma\in k_a$. The mapping $k^{\times}\ni\alpha\rightarrow(\alpha,k)\in G(k_a/k)$ is a one-to-one continuous homomorphism, and the image is †dense in $G(k_a/k)$. It has also been proved that there exists one and only one Abelian extension K/k with $N_{K/k}(K^{\times})=A$ for any given closed subgroup A of finite index of $k^{\times}$. Therefore, closed subgroups A of finite index of $k^{\times}$ are in one-to-one correspondence with finite Abelian extensions K of k through the relation $A=N_{K/k}(K^{\times})$, and in this case A is called the subgroup of $k^{\times}$ corresponding to K/k.

Let K/k be an Abelian extension of finite degree, A be the corresponding subgroup of $k^{\times}$, and $V^{(i)}$ $(i=0,1,2,\ldots)$ be ramification groups of K/k. Furthermore, define constants $v_1,v_2,\ldots,v_r$ by

$$V^{(0)}=V^{(1)}=\ldots=V^{(v_1)}\underset{\neq}{\supset}V^{(v_1+1)}=\ldots$$
$$=V^{(v_2)}\underset{\neq}{\supset}\ldots\underset{\neq}{\supset}V^{(v_{r-1}+1)}=\ldots=V^{(v_r)}$$
$$\underset{\neq}{\supset}V^{(v_r+1)}=(1),$$

denote the order of $V^{(v_i+1)}$ by n_i $(i=1,2,\ldots,r)$, and put $u_\rho=v_0+(n_0/n_0)(v_1-v_0)+\ldots+(n_{\rho-1}/n_0)(v_\rho-v_{\rho-1})$, $\rho=1,2,\ldots,r$ (here we understand $v_0=-1$, $n_0=[V^{(0)}:1]$). Then $u_1,\ldots,u_r$ are rational integers, and we have

$$A\mathfrak{u}^{(0)}=\ldots=A\mathfrak{u}^{(u_1)}\underset{\neq}{\supset}A\mathfrak{u}^{(u_1+1)}=\ldots$$
$$=A\mathfrak{u}^{(u_2)}\underset{\neq}{\supset}\ldots\underset{\neq}{\supset}A\mathfrak{u}^{(u_r+1)}=A$$

(H. Hasse). If m is the smallest integer with $A\supseteq\mathfrak{u}^{(m)}$, then $\mathfrak{p}^m$ is called the **conductor** of K/k. The above results of Hasse show that $m=u_r+1$. On the other hand, it is known that the correspondence between $A\mathfrak{u}^{(u_\rho)}$ and $V^{(v_\rho)}$ $(\rho=1,2,\ldots,r)$ is given by the norm-residue symbol $(\alpha,K/k)$ (→ 62 Class Field Theory).

G. Theory of Algebras

By the general theory of †crossed products of algebras, the structure of the †Brauer group formed by the classes of †normal simple algebras over a local k is directly obtained from results concerning cohomology (→ Section F). Namely, a †normal simple algebra over k splits over a separable extension of degree n if and only if it splits over the unramified extension of degree n, and the Brauer group of k is isomorphic to $\mathbf{Q}/\mathbf{Z}$. Furthermore, the †exponent (the order as an element of the Brauer group) of a normal simple algebra $\mathfrak{A}$ over k coincides with the †Schur index, and if $[\mathfrak{A}:k]=n^2$, then $\mathfrak{A}$ is expressed as a crossed product with respect to any normal extension of degree n over k. The invariant of the factor set (2-cocycle) that appears in this crossed product expression is called the **invariant** of $\mathfrak{A}$ (→ 31 Associative Algebras; for the properties of $\mathfrak{A}$ as a topological ring and as a topological group of the group of invertible elements of $\mathfrak{A}$ → 7 Adeles and Ideles).

H. Explicit Formulas

Let K/k be an Abelian extension of finite degree. When we have an explicit formula for the norm-residue symbol $(\alpha,K/k)$, we say that we have an explicit reciprocity law.

Let k be a $\mathfrak{p}$-adic number field containing a primitive mth root ζ_m of unity, and let β be an element in $k^{\times}$. Let p be a prime number contained in $\mathfrak{p}$. Since the Kummer extension $K=k(\sqrt[m]{\beta})$ over k is Abelian, the Hilbert norm-residue symbol $(\alpha,\beta)_m$ is defined by $(\alpha,K/k)(\sqrt[m]{\beta})=(\alpha,\beta)_m\sqrt[m]{\beta}$, where $(\alpha,\beta)_m$ is an mth root of unity. In this case, the problem of obtaining an explicit reciprocity law is solved if we can express the symbol $(\alpha,\beta)_m$ in terms of α, β and suitable parameters depending on the ground field k.

In particular, if $p=2$, $m=2$, we have a simple formula due to Hasse, $(\alpha,\beta)_2=(-1)^{\mathrm{Tr}((\alpha-1)(\beta-1)/4)}$, where α, β are two units in k satisfying $\alpha\equiv\beta\equiv1\ (\mathrm{mod}\ 2)$ and Tr is the trace from k to $\mathbf{Q}_p$. Similar formulas for the complementary laws are also known [12].

On the other hand, if $m=p$ is an odd prime and $k=\mathbf{Q}_p(\zeta_p)$, we have the following formulas for a prime element $\lambda_p=1-\zeta_p$ and two units α, β satisfying $\alpha\equiv1\ (\mathrm{mod}\ \mathfrak{p}^2)$, $\beta\equiv1$

(mod $\mathfrak{p}$):

$$(\alpha,\beta)_p=\zeta_p^{(1/p)\mathrm{Tr}(\zeta_p\log\alpha\, D\log\beta)}, \tag{2}$$

$$(\zeta_p,\beta)_p=\zeta_p^{(1/p)\mathrm{Tr}(\log\beta)}, \tag{3}$$

$$(\lambda_p,\beta)_p=\zeta_p^{-(1/p)\mathrm{Tr}((\zeta_p/\lambda_p)\log\beta)}, \tag{4}$$

where $D\log\beta=(1/\beta)\Sigma_{i=1}^{\infty} ib_i\lambda_p^{i-1}$, while the b_i are determined by the λ_p-expansion $\beta=\Sigma_{i=0}^{\infty} b_i\lambda_p^i$, $b_i\in\mathbf{Z}_p$ [5]. Furthermore, we have an explicit Kummer-Hilbert formula deduced from (2) in terms of Kummer's logarithmic differential quotients [10, 12]. Concerning the complementary laws (3), (4), the following Artin-Hasse formulas are known for $k=\mathbf{Q}_p(\zeta_{p^n})$ and $\beta\equiv 1$ (mod $\mathfrak{p}$):

$$(\zeta_{p^n},\beta)_{p^n}=\zeta_{p^n}^{(-1/p^n)\mathrm{Tr}(\log\beta)}, \tag{3'}$$

$$(\lambda_{p^n},\beta)_{p^n}=\zeta_{p^n}^{-(1/n)\mathrm{Tr}((\zeta_{p^n}/\lambda_{p^n})\log\beta)}, \tag{4'}$$

where $\lambda_{p^n}=1-\zeta_{p^n}$ [11]. Utilizing these formulas, K. Iwasawa obtained a formula for $(\alpha,\beta)_{p^n}$ that is a natural generalization of (1) [15].

Concerning $(\alpha,\beta)_{p^n}$, we have **Šafarevič's reciprocity law** [13]. To explain it, let k_0 be the inertia field in k, i.e., the maximal subfield in k that is unramified over $\mathbf{Q}_p$. For an arbitrary integer a in k_0, we consider the **Artin-Hasse function** $E(a,x)=\exp(-L(a,x))$, where $L(a,x)=\Sigma_{i=0}^{\infty}(a^{\mathbf{p}^i}/p^i)x^{p^i}$ with the †Frobenius automorphism $\mathbf{p}$ of $k_0/\mathbf{Q}_p$. We choose a prime element $\tilde{\pi}$ in k_0 such that $\zeta_{p^n}=E(1,\tilde{\pi})$ and an integer $\bar{a}$ in a suitable unramified extension field of k_0 such that $\bar{a}^{\mathbf{p}}-\bar{a}=a$ for a given integer a in k_0. Given any p^n-primary element x in k (i.e., $k(x^{1/p^n})$ is unramified over k), there exists an integer a in k_0 such that $x\approx E(a)=E(p^n\bar{a},\tilde{\pi})$, where $x\approx y$ $(x,y\in k^\times)$ means $x=y\cdot u$ for an element u in $k^{\times p^n}$. Furthermore, if δ is an element of $k^\times$, we have the following canonical decomposition formula:

$$\delta\approx\pi^{d^*}E(d)\prod_{\substack{1\leqslant j<e_0p\\(j,p)=1}}E(d_j,\pi^j),$$

where π is a prime element in k, $d^*\in\mathbf{Z}$, d, d_j are integers in k_0, and $e_0=e/(p-1)$ with the ramification index e of k. The explicit formula due to Šafarevič and Hasse is expressed as follows:

$$(\alpha,\beta)_{p^n}=\zeta_{p^n}^{\mathrm{Tr}_0(a^*b-ab^*+c)},$$

where Tr_0 is the trace from k_0 to $\mathbf{Q}_p$ and a^*, b^* are determined by the canonical decompositions of α, β as stated above. The element c in k_0 is determined by

$$\prod_{\substack{1\leqslant i,j<e_0p\\(i,p)=(j,p)=1}}E(ja_ib_i,\pi^{i+j})$$

$$=\gamma\approx E(c)\prod_{\substack{u\leqslant 1<e_0p\\(j,p)=1}}E(c_j,\pi^j)$$

for odd p, and by

$$(-1)^{a^*b^*}\prod_{\substack{1\leqslant i,j<2e_0\\(i,2)=(j,2)=1}}E(ja_ib_j,\pi^{i+j})$$

$$\times\prod_{\mu,\nu=1}^{\infty}E((i2^{\mu-1}+j2^{\nu-1})a_i^{\mathbf{p}^\mu}b_j^{\mathbf{p}^\nu},\pi^{i2^\mu+j2^\nu})$$

$$=\gamma\approx E(c)\prod_{\substack{1\leqslant j<2e_0\\(j,2)=1}}E(c_j,\pi^j)$$

for $p=2$. Several formulas for the case where k is a function field are also known [1, 5].

Let k be a general local field. When K is an extension field of k obtained by adjoining π^n-division points $\{\lambda\}$ of the Lubin-Tate formal group defined over the integer ring of k, the extension K/k is totally ramified and Abelian, and we have an explicit formula: $(u^{-1},K/k)\cdot\lambda=[u]_F(\lambda)$, where u is a unit in k and $[u]_F$ is an endomorphism of F corresponding to the unit u. In particular, this formula implies the cyclotomic reciprocity law [9, 14].

The problem of obtaining an explicit reciprocity law arises in the problem of obtaining the reciprocity law for power residues from the law of reciprocity in global class field theory. The problem is closely connected with T. Kubota's recent works (for example [16]), which clarify the analytic meaning of the reciprocity law in algebraic number fields.

References

[1] E. Artin, Algebraic numbers and algebraic functions, Lecture notes, Princeton Univ., 1950–1951 (Gordon and Breach, 1967).
[2] H. Hasse, Zahlentheorie, Akademie-Verlag, 1949.
[3] H. Hasse, Normenresttheorie galoisscher Zahlkörper mit Anwendungen auf Führer und Diskriminante abelscher Zahlkörper, J. Fac. Sci. Univ. Tokyo, 2 (1934), 477–498.
[4] M. Deuring, Algebren, Erg. Math., Springer, second edition, 1968.
[5] E. Artin and J. Tate, Class field theory, Lecture notes, Princeton Univ., 1951 (Benjamin, 1967).
[6] J.-P. Serre, Corps locaux, Actualités Sci. Ind., Hermann, 1962.
[7] O. F. G. Schilling, The theory of valuations, Amer. Math. Soc. Math. Surveys, 1950.
[8] T. Takagi, Daisûteki seisûron (Japanese; Algebraic theory of numbers), Iwanami, second edition, 1971.
[9] Y. Kawada, Daisûteki seisûron (Japanese; Theory of algebraic numbers), Kyôritu, 1957.
[10] T. Takagi, On the law of reciprocity in the cyclotomic corpus, Proc. Phys.-Math. Soc. Japan, 4 (1922), 173–182.

[11] E. Artin and H. Hasse, Die beiden Ergänzungssätze zum Reziprozitätsgesetz der l^n-ten Potenzreste im Körper der l^n-ten Einheitswurzeln, Abh. Math. Sem. Univ. Hamburg, 6 (1928), 146–162.
[12] H. Hasse, Bericht über neuere Untersuchungen und Probleme aus der Theorie der algebraischen Zahlkörper II, Jber. Deutsch. Math. Verein. (1930) (Physica Verlag, 1965).
[13] И. Р. Шафаревич (I. R. Šafarevič), Общий закон в заимности, Mat. Sb., 26 (68) (1950), 113–146; English translation, A general reciprocity law, Amer. Math. Soc. Transl., 4 (1956), 73–106.
[14] J. Lubin and J. Tate, Formal complex multiplication in local fields, Ann. of Math., (2) 81 (1965), 380–387.
[15] K. Iwasawa, On explicit formulas for the norm residue symbol, J. Math. Soc. Japan, 20 (1968), 151–165.
[16] T. Kubota, Ein arithmetischer Satz über eine Matrizengruppe, J. Reine Angew. Math., 222 (1966), 55–57.
[17] J. W. S. Cassels and A. Frölich (eds.), Algebraic number theory, Academic Press, 1967.

M

258 (XIX.13) Magnetohydrodynamics

Magnetohydrodynamics (also called **hydromagnetics** or **magnetofluid dynamics**) concerns the motion of an electrically conductive fluid in the presence of a magnetic field. An electromotive force (e.m.f.) induced by the motion of the fluid in the magnetic field perturbs the original magnetic field by itself creating induced electric currents. On the other hand, an electromagnetic force due to the magnetic field deforms the original motion. Many important and interesting phenomena result from such interactions of the magnetic field and the motion of the fluid.

In ordinary magnetohydrodynamics we assume that (i) the fluid is continuous, (ii) electric conductivity σ is not negligible, and (iii) fluid velocity is small compared with the light velocity, i.e., $\max(L^2/(c^2T^2), U^2/c^2) \ll \min(1, R_m)$, where L is a representative length, T a representative time, U a representative velocity, and $R_m = \sigma\mu UL$ (μ is the magnetic permeability) is a nondimensional number called the **magnetic Reynolds number**. In this case we may neglect in the †Maxwell equations the displacement and convective currents compared with the conductive current $\mathbf{J}$, and write

$$\operatorname{div}\mathbf{B}=0, \qquad \operatorname{rot}\mathbf{H}=\mathbf{J},$$
$$\operatorname{rot}\mathbf{E}=-\partial\mathbf{B}/\partial t, \tag{1}$$
$$\rho_e=\operatorname{div}\varepsilon\mathbf{E}, \tag{2}$$

where ε is the dielectric constant, and

$$\mathbf{B}=\mu\mathbf{H}, \qquad \mathbf{J}=\sigma(\mathbf{E}+\mathbf{v}\times\mathbf{B}). \tag{3}$$

Here the latter equation, **Ohm's law** for a moving medium, is valid when the effects of temperature gradient, the Hall effect, etc., are small. The motion of fluids (→ 208 Hydrodynamics) is governed by the †equation of continuity

$$\partial\rho/\partial t+\operatorname{div}\rho\mathbf{v}=0, \tag{4}$$

the †equation of motion

$$\partial(\rho\mathbf{v})/\partial t=-\operatorname{div}(\rho\mathbf{v}\otimes\mathbf{v}-\mathbf{P}-\mathbf{T}) \tag{5}$$

($\mathbf{P}$ is the mechanical stress tensor, $\mathbf{T}$ is the Maxwell stress tensor $T_{ij}=\mu(H_iH_j-\frac{1}{2}H^2\delta_{ij})$) or

$$\rho D\mathbf{v}/Dt=\operatorname{div}\mathbf{P}+\mathbf{K}, \quad \mathbf{K}=\mathbf{J}\times\mathbf{B}, \tag{5'}$$

the †equation of state, and the †equation of energy. (From assumption (iii), the force $\rho_e\mathbf{E}$ on the electric charge ρ_e can be neglected compared with the force $\mathbf{J}\times\mathbf{B}$ on the electric current, and (2) can be separated from the other equations merely to determine ρ_e.)

When μ and σ are uniform, we can eliminate $\mathbf{E}$ and $\mathbf{J}$ from (1) and Ohm's law to obtain the **induction equation**

$$\partial\mathbf{B}/\partial t=\operatorname{rot}(\mathbf{v}\times\mathbf{B})+\lambda\Delta\mathbf{B}$$
$$\Delta=\operatorname{grad}\operatorname{div}-\operatorname{rot}\operatorname{rot}, \quad \lambda=1/(\mu\sigma). \tag{6}$$

This is of the same form as the equation $\partial\omega/\partial t=\operatorname{rot}(\mathbf{v}\times\omega)+\nu\Delta\omega$ (ν is the kinematic viscosity) for the †vorticity ω of an ordinary †incompressible viscous fluid. We call $\lambda=1/(\mu\sigma)$ the **magnetic viscosity**, and the ratio of the first term (the convection term) to the second one (the diffusion term) in the right-hand side of (6) is the magnetic Reynolds number $R_m=UL/\lambda=\mu\sigma UL$. $R_m=\infty$ corresponds to the †perfect fluid as $\sigma\to\infty$ or $L\to\infty$. In this case, the magnetic flux moves with the fluid as if both were frozen together, as in the †Helmholtz theorem about vorticity. The existence of a transversal wave of velocity $\alpha=\sqrt{\mu H^2/\rho}$ along magnetic lines of force in the fluid, owing to the tension μH^2 ((5) except for the magnetic pressure $-\frac{1}{2}\mu H^2\delta_{ij}$) of the magnetic flux frozen to the fluid, was noted for the first time by H. Alfvén (1943), and this wave is called the **Alfvén wave**. In a compressible perfect fluid ($R_m=\infty$, $P_{ij}=-p\delta_{ij}$, where p is the pressure), (1)–(5) reduce to a system of †hyperbolic partial differential equations, yielding as †characteristic surfaces in addition to the pure Alfvén wave two kinds of magneto-sound waves of phase velocities

$$a_{\pm}=\left(\frac{1}{\sqrt{2}}\right)\left(a^2+\alpha^2\pm\sqrt{(a^2+\alpha^2)-4a^2\alpha^2\cos^2\theta}\right)^{1/2}$$

(θ is the angle between the magnetic field and the wave normal, a is the velocity of sound interfering with the Alfvén wave). We call a_+ and a_-, respectively, the **fast wave** and the **slow wave**. **Hydromagnetic dynamo theory** explains the generation and maintenance of the magnetic field inside the earth on the basis of (6). Applications are also made to cosmic problems and MHD generation of electricity. Mercury, liquid sodium, etc., can be used to verify the theoretical results. Applications are made to the real plasma used in thermonuclear fusion to the extent that the hydrodynamic treatment is valid as a first approximation.

References

[1] T. G. Cowling, Magnetohydrodynamics, Interscience, 1957.

[2] H. Alfvén and C. G. Falthammar, Cosmical electrodynamics, Clarendon Press, second edition, 1963.

[3] T. Kihara, Henbibun hôteisiki no ôyô (Japanese; Applications of partial differential

equations), ch. 8. Dynamics of plasma, Iwanami Coll. of Modern Appl. Math., 1958.
[4] I. Imai and A. Sakurai, Denzi ryûtai rikigaku (Japanese; Magnetohydrodynamics), Iwanami Coll. of Modern Phys., 1959.
[5] T. Kihara and Y. Mizuno, Purazuma no buturigaku (Japanese; Physics of plasma), Iwanami Coll. of Modern Phys., 1959.
[6] Progress of theoretical physics, suppl. no. 24, 1962.

259 (IX.3) Manifolds

A. General Remarks

The rudimentary concept of n-dimensional manifolds can already be seen in J. Lagrange's dynamics. In the middle of the 19th century n-dimensional †Euclidean space was known as a continuum of n real parameters (A. Cayley; H. Grassmann, 1844, 1861; L. Schläfli, 1852). The notion of general n-dimensional manifolds was introduced by B. Riemann as a result of his differential geometric observations (1854). He considered an n-dimensional manifold to be a set formed by a 1-parameter family of $(n-1)$-dimensional manifolds, just as a surface is formed by the motion of a curve. Analytical studies of topological structures of manifolds and their local properties were initiated and developed by Riemann, E. Betti, H. Poincaré, and others. To avoid the difficulties and disadvantages of analytical methods, Poincaré restricted his consideration to those topological spaces X that are †connected, †triangulable, and such that each point of X has a neighborhood homeomorphic to an n-dimensional Euclidean space. We often refer to such spaces as **Poincaré manifolds**; Poincaré called them n-dimensional manifolds. His research led to later development of combinatorial topology (→ 73 Complexes; 203 Homology Groups).

B. Topological Manifolds

Let H^n be the half-space $\{(x_1, \ldots, x_n) \in \mathbf{R}^n \mid x_1 \geqslant 0\}$ of the n-dimensional Euclidean space $\mathbf{R}^n$. An n-dimensional **topological manifold** M is by definition a †Hausdorff space in which each point p has a neighborhood $U(p)$ homeomorphic to $\mathbf{R}^n$ or H^n. The **boundary** ∂M of M is the set of all points p of M whose neighborhoods $U(p)$ are homeomorphic to H^n. The **interior** of M is the complement $M_0 = M - \partial M$ of the boundary. A manifold M is called a **manifold without boundary** if $\partial M = \varnothing$; otherwise, M is called a **manifold with boundary**. The boundary of an n-dimensional manifold is an $(n-1)$-dimensional manifold. A manifold without boundary is called a **closed** or **open manifold** according as it is compact or not. There exist connected manifolds that are not †paracompact; among them, the 1-dimensional ones are called **long lines**. We consider connected paracompact manifolds. Such a manifold M has a †countable open base and is †metrizable.

The boundary $B = \partial M$ of a manifold M can always be **collared** in M, that is, B has an open neighborhood homeomorphic to $B \times [0,1)$ with B corresponding to $B \times 0$ (M. Brown). A **submanifold** of M is a closed subset N of M satisfying the following two conditions: (i) N itself is a manifold with relative topology; (ii) $\partial N = \partial M \cap N$. An **open submanifold** of M is simply a connected open subset of M. Let N be an $(n-1)$-dimensional submanifold of an n-dimensional manifold M. If for each point p of N there exists a neighborhood $U(p)$ in M and a homeomorphism $f_p: U(p) \to \mathbf{R}^n$ such that $f_p(U(p) \cap N) = \mathbf{R}^{n-1} \subset \mathbf{R}^n$, then N is said to be **locally flat** in M.

Let M_0 be the interior of an n-dimensional manifold M. Then M_0 is **locally Euclidean**, that is, each point p of M_0 has a neighborhood homeomorphic to $\mathbf{R}^n$. Thus the †local homology groups $H_i(p) = H_i(M_0, M_0 - \{p\})$ at p with integral coefficients vanish except for $i = n$, and $H_n(p) \cong \mathbf{Z} =$ the group of integers. If we associate the local homology group $H_n(p)$ with each point p of M_0, then we obtain a system $G(\mathbf{Z}) = \{H_n(p) \mid p \in M_0\}$ of †local coefficient groups. If this system is †simple, namely, if $G(\mathbf{Z})$ is isomorphic to $M_0 \times \mathbf{Z}$, then M is said to be **orientable**; otherwise, it is **nonorientable**. The 2-dimensional sphere is orientable, while the 2-dimensional real †projective space P^2 is nonorientable. When M is orientable, there are two possible canonical †cross sections $s_i: M_0 \to G(\mathbf{Z})$ ($i = 1, 2$; $s_1(p) = -s_2(p)$ is a generator of $H_n(p)$). The choice of one s determines an **orientation** of M, and $-s$ determines its **opposite orientation**; the pair (M, s) is called an **oriented manifold**. Thus the orientation s of M determines the **local orientation** $s(p)$ at each point p (of M_0).

For the †singular homology groups of an oriented connected compact n-dimensional manifold M, we have $H_n(M, \partial M; \mathbf{Z}) \cong \mathbf{Z}$, and $H_n(M, \partial M; \mathbf{Z}_2) \cong \mathbf{Z}_2$ for nonorientable M ($\mathbf{Z}_2 = \mathbf{Z}/2\mathbf{Z}$). The †canonical generator m of $H_n(M, \partial M; \mathbf{Z})$ (or $H_n(M, \partial M; \mathbf{Z}_2)$) is called the **fundamental homology class** (or **fundamen-**

tal class) of $(M, \partial M)$. For an oriented manifold M, the fundamental class m corresponds to the local orientation $s(p)$ under the †excision isomorphism $H_n(M, \partial M) \cong H_n(M_0, M_0 - \{p\})$.

For an oriented compact n-dimensional manifold M, we have the **Poincaré (-Lefschetz) duality theorem** giving relations between †singular homology and †cohomology groups with arbitrary coefficients;

$$D: \quad H^i(M) \cong H_{n-i}(M, \partial M),$$
$$H^i(M, \partial M) \cong H_{n-i}(M).$$

These isomorphisms are given by the †cap product $Du = u \frown m$ with the fundamental class m. For a nonorientable manifold, the theorem holds with coefficient $\mathbf{Z}_2$ instead of $\mathbf{Z}$. The classical †intersections introduce multiplications into the direct sums $H_*(M) = \Sigma_i H_i(M)$ and $H_*(M, \partial M) = \Sigma_i H_i(M, \partial M)$ with ring coefficients. Then $H_*(M)$ and $H_*(M, \partial M)$ form rings, which are called **homology rings**. By the duality D, the homology rings $H_*(M)$ and $H_*(M, \partial M)$ are isomorphic to the cohomology rings $H^*(M, \partial M)$ and $H^*(M)$, respectively. Let M be an orientable closed n-dimensional manifold, b_r its rth †Betti number, and T_r its rth †torsion group. Then from the duality D follows $b_r = b_{n-r}$, $T_r = T_{n-r-1}$.

Let F be a closed set in the n-sphere S^n. Then we have the **Alexander-Pontrjagin duality**: $\tilde{H}^i_C(F) = \tilde{H}_{n-i-1}(S^n - F)$ $(i \geqslant 0)$, where the left-hand term denotes the †reduced †Čech cohomology and the right-hand term the reduced singular homology, both with the same arbitrary fixed coefficient group. A special case of this theorem is the †Jordan curve theorem: Any Jordan curve in a plane separates the plane into two domains.

C. Pseudomanifolds

Here we explain notions of triangulated spaces and polyhedra and some combinatorial treatment of manifolds. A paracompact †polyhedron $M = |K|$ is called an n-dimensional **pseudomanifold** if the triangulating complex K satisfies the following three conditions: (i) Every simplex of K is either an n-simplex or a face of an n-simplex; (ii) each $(n-1)$-simplex is a face of at most two n-simplexes; (iii) for any two n-simplexes σ, τ of a connected component of K, there exists a finite sequence of n-simplexes $\sigma = \sigma_0, \sigma_1, \ldots, \sigma_s = \tau$ such that σ_i and σ_{i+1} have an $(n-1)$-face in common. The triangulation K of a pseudomanifold $M = |K|$ is necessarily locally finite.

Consider the set S of $(n-1)$-simplexes of K each of which is a face of only one n-simplex. Then the set of all $\sigma \in S$ together with their faces forms a subcomplex L of K. The polyhedron $|L|$ of L is called the **boundary** of the pseudomanifold $M = |K|$ and is denoted by ∂M. The boundary of a pseudomanifold is not necessarily a pseudomanifold.

Let $|K|$ be an n-dimensional polyhedron (or †cell complex). A point p in $|K|$ is called a **regular point** of $|K|$ if it has an open neighborhood in $|K|$ homeomorphic to $\mathbf{R}^n$ or H^n; otherwise, p is called a **singular point**. A pseudomanifold without singular points is a topological manifold. An n-dimensional polyhedron $|K|$ is a pseudomanifold if and only if the dimension of the set of all singular points in $|K|$ is less than $n-1$.

Let τ_1, τ_2 be oriented n-simplexes of an n-dimensional pseudomanifold $M = |K|$. Suppose that they have an $(n-1)$-face σ in common. If the †incidence numbers of these simplexes with σ satisfy the relation $[\tau_1, \sigma] = -[\tau_2, \sigma]$, then τ_1 and τ_2 are said to be **coherently oriented**. We call $M = |K|$ with all its n-simplexes oriented an **oriented** pseudomanifold if any two n-simplexes τ_1 and τ_2 are coherently oriented whenever they have an $(n-1)$-face in common. For an oriented pseudomanifold, the formal sum of all its oriented n-simplexes forms an integral cycle of M $(\operatorname{mod} \partial M)$, which is called the **fundamental cycle** of $(M, \partial M)$. Its homology class, or fundamental class, generates the homology group $H_n(M, \partial M; \mathbf{Z}) \cong \mathbf{Z}$. When $M = |K|$ is nonorientable, the fundamental class with coefficient $\mathbf{Z}_2$ (instead of $\mathbf{Z}$) is similarly defined. A pseudomanifold is orientable if and only if the topological manifold consisting of all its regular points is orientable.

D. Combinatorial Manifolds and Homology Manifolds

Let $M = |K|$ be a paracompact polyhedron. If the †star of each vertex p of the triangulation K is combinatorially equivalent to an n-simplex, $M = |K|$ is called a **combinatorial manifold**. This is necessarily both a topological manifold and a pseudomanifold. If for each point p of a paracompact polyhedron $M = |K|$ the local homology groups are $H_i(p) = 0$ $(i \neq n)$ and $H_n(p) = \mathbf{Z}$ or 0, we call $M = |K|$ an n-dimensional **homology manifold**. It is also a pseudomanifold. The set ∂M of points p of $M = |K|$ with $H_n(p) = 0$, called the **boundary** of M, coincides with the boundary of $M = |K|$ as a pseudomanifold. If the boundary ∂M of an n-dimensional homology manifold M is an $(n-1)$-dimensional homology manifold, M is said to have a **regular boundary**.

For an n-dimensional homology manifold $M=|K|$ with regular boundary, the **dual subdivision** of the triangulation K is defined as follows: First, given simplexes σ, τ of K, $\sigma<\tau$ means that σ is a face of τ. By definition, a vertex of the †barycentric subdivision K' is the barycenter of a simplex of K. Every k-simplex τ of K' corresponds bijectively to a sequence of simplexes $(\sigma_0,\sigma_1,\ldots,\sigma_k)$ of K with $\sigma_0<\sigma_1<\ldots<\sigma_k$ in such a way that the barycenters of σ_i constitute the set of vertices of τ. Now for an r-simplex σ^r of K, consider all sequences $(\sigma^r,\sigma^{r+1},\sigma^{r+2},\ldots,\sigma^n)$ with $\sigma^r<\sigma^{r+1}<\cdots<\sigma^n$. Then the **dual cell** of σ^r is defined to be the $(n-r)$-dimensional subcomplex y^{n-r} of K' that consists of the $(n-r)$-simplexes of K' corresponding to the sequences $(\sigma^r,\sigma^{r+1},\ldots,\sigma^n)$ and of their faces. This polyhedron $|y^{n-r}|$ is an orientable pseudomanifold contained in the star $|St(\sigma^r)|$ in K of σ^r, and its boundary $\partial|y^{n-r}|$ has the same homology groups as the $(n-r-1)$-sphere if σ^r is not contained in the boundary of $|K|$. For an r-simplex σ^r of K and its dual cell $y^{n-r}=D\sigma^r$, we have $\partial|y^{n-r}|\cap|\sigma^r|=|y^{n-r}|\cap\partial|\sigma^r|=\varnothing$, and $|y^{n-r}|\cap|\sigma^r|$ consists of one point, the barycenter of σ^r. The boundary of a dual cell consists of several dual cells, and $D\tau<D\sigma$ for $\sigma<\tau$. In particular, the dual cell $y^n=D\sigma^0$ of a 0-simplex σ^0 is the star $St'(\sigma^0)$ in K' of σ^0, and the dual cell $y^0=D\sigma^n$ of an n-simplex σ^n of K is the barycenter of σ^n. The set of all these dual cells gives a cellular subdivision of the homology manifold M, which is called the **dual cellular subdivision** of the triangulation of M. This is used in the proof of the Poincaré duality theorem for homology manifolds and gives a geometric interpretation for the theorem.

Let $\{U_j\}$ be a system of coordinate neighborhoods of an n-dimensional topological manifold M^n and $h_i: U_i\to\mathbf{R}^n$ be a homeomorphism of U_i onto an open set in $\mathbf{R}^n$. If the coordinate transformation $h_i\circ h_j^{-1}: h_j(U_i\cap U_j)\to h_i(U_i\cap U_j)$ is **piecewise linear** (namely, for suitable subdivisions of $h_j(U_i\cap U_j)$ and $h_i(U_i\cap U_j)$ with respect to the standard triangulation of $\mathbf{R}^n$, $h_i\circ h_j^{-1}$ can be considered as a simplicial mapping), then M^n is called a **piecewise linear manifold** (or **PL-manifold**). A PL-manifold can be triangulated (see J. Stallings, Lectures on polyhedral topology, Lecture notes at Tata Institute, 1968, and J. F. P. Hudson, Piecewise linear topology, Mathematical lecture notes, Univ. of Chicago, 1966–1967). Thus the notion of PL-manifold is equivalent to that of combinatorial manifold. E. Zeeman introduced the concept of polymanifold analogous to combinatorial manifold [9, p. 57–64]. (For more general concepts of manifolds and their duality theorems → [8].)

A combinatorial manifold is a homology manifold. However, the converse does not hold except for dimensions 1, 2, and 3. Topological manifolds of dimension $\leqslant 3$ are triangulable and can be considered as combinatorial manifolds; moreover, all triangulations are combinatorially equivalent (T. Radó, *Abh. Math. Sem. Univ. Hamburg*, 11 (1935); C. Papakyriakopoulos and E. Moise, *Ann. of Math.*, (2) 56 (1952); R. Bing, *Ann. of Math.*, (2) 69 (1959)). However, it was shown in 1969 by Kirby, Siebenmann, and Wall that there exist topological manifolds of dimension 6 that do not admit any combinatorial structure, and that the †Hauptvermutung does not hold for combinatorial manifolds of dimension 5 (R. Kirby and L. Siebenmann, *Bull. Amer. Math. Soc.*, 75 (1969)).

Any †C^r-manifold M has a †C^r-triangulation K, and $M=|K|$ is a combinatorial manifold. Two C^r-triangulations of a C^r-manifold are combinatorially equivalent. Conversely, for the †smoothing problem of combinatorial manifolds, the following results have been obtained: Let M be an n-dimensional submanifold of the $(n+k)$-dimensional Euclidean space $\mathbf{R}^{n+k}$, $p\in M$, and let ν be a k-plane through p. If there exists a neighborhood U_p of p in M such that for any two points x, y in U_p the angle between ν and $\overrightarrow{xy}$ is greater than a constant $\varepsilon>0$, then ν is said to be **transverse** to M at p. Let $\nu: M\to G_{n,k}$ be a continuous mapping of M into the †Grassmann manifold. If for each point p of M the plane through p parallel to $\nu(p)$ is transverse to M at p, then ν is called a **transverse field** on M. An n-dimensional combinatorial manifold $M=|K|$ is **smoothable** (i.e., admits †smoothing) if and only if there exists a †piecewise linear embedding $f: M\to\mathbf{R}^{n+k}$ such that the image $f(M)$ has a transverse field on itself. Moreover, if two transverse fields $\nu_1,\nu_2: f(M)\to G_{n,k}$ are homotopic to each other, they define the same differentiable structures on M (S. Cairns, *Ann. of Math.*, (2) 41 (1940); J. H. C. Whitehead, *Ann. of Math.*, (2) 73 (1961)). J. Milnor [11] gave another criterion for smoothing in terms of PL-microbundles.

A †simply connected 3-dimensional closed manifold is homeomorphic to a 3-sphere—this is the famous **Poincaré conjecture**, which is still open in spite of much effort by mathematicians. The conjecture is generalized for dimension n as follows: Any homotopy n-sphere is homeomorphic to an n-sphere (**gen-**

eralized Poincaré conjecture), where homotopy n-spheres are combinatorial manifolds that have the same homotopy type as the n-sphere. This has been affirmatively answered for $n \geqslant 5$ (see J. Stallings, *Bull. Amer. Math. Soc.*, 66 (1960); E. Zeeman [9, p. 198–204]; S. Smale, *Ann. of Math.*, (2) 74 (1961)).

E. 3-Manifolds

Much research on 3-manifolds has been done, and we have, among others, the following results:

(1) **Sphere theorem**: Let M be an orientable 3-manifold with $\pi_2(M) \neq 0$. Then there exists a piecewise linear embedding f of a 2-sphere into M such that f is not homotopic to 0 in M (see C. Papakyriakopoulos, *Ann. of Math.*, (2) 66 (1957); J. H. C. Whitehead, *Bull. Amer. Math. Soc.*, 64 (1958)).

(2) **Dehn's lemma**: Let D be a †2-cell with singularities in a 3-manifold M, with its boundary a simple polygon C. If some neighborhood U of C in D has no singularities, then there exists a 2-cell in M without singularities whose boundary is C. In 1910, M. Dehn asserted this lemma, but his proof was incomplete. In 1957, Papakyriakopoulos and T. Homma proved the lemma independently (*Ann. of Math.*, (2) 66 (1957); *Yokohama Math. J.*, 5 (1957); J. H. C. Whitehead and A. Shapiro, *Bull. Amer. Math. Soc.*, 64 (1958)).

(3) **Loop theorem**: Let N be the boundary of a 3-manifold M and U be an open set in a component of N. If there is a closed curve in U which is homotopic to 0 in M but not in N, then there exists a simple closed curve in U homotopic to 0 in M but not in N (cf. C. Papakyriakopoulos, *Proc. London Math. Soc.*, 7 (1957); J. Stallings, *Ann. of Math.*, (2) 72 (1960)).

(4) **Unique decomposition theorem** for a 3-manifold (J. Milnor, *Amer. J. Math.*, 82 (1962)): We assume that all manifolds are connected, oriented, and triangulated 3-manifolds without boundaries and that all homeomorphisms are piecewise linear. Two manifolds M, M' are said to be isomorphic ($M \approx M'$) if there exists an orientation-preserving homeomorphism of M onto M'. Removing an open 3-cell from each of two 3-manifolds M, M' and identifying the boundaries of these removed cells, we obtain a 3-manifold $M \# M'$, called the **connected sum** of M and M'. A manifold that is not isomorphic to the 3-sphere S^3 is called **nontrivial**. A nontrivial manifold P is called **prime** if P cannot be decomposed as $P = M_1 \# M_2$ with M_1 and M_2 both nontrivial. A manifold M is called **irreducible** if every 2-sphere in M bounds a 3-cell. Then from the sphere theorem the following results can be deduced: If a compact 3-manifold M is nontrivial, then M is isomorphic to a connected sum $P_1 \# P_2 \# \dots \# P_k$ of prime manifolds P_i ($i = 1, 2, \dots, k$), where $P_1, P_2, \dots, P_k$ are determined uniquely up to order. Every irreducible 3-manifold M has $\pi_2(M) = 0$. Conversely, if a 3-manifold M has $\pi_2(M) = 0$, then M is irreducible provided that the Poincaré conjecture is correct. (For further results on 3-manifolds → [9].)

F. Wild Spaces

Nonclosed 3-manifolds behave very differently from closed 3-manifolds. For example, there exists an open 3-manifold U that has the same †homotopy type as an open 3-cell but is not homeomorphic to it (M. Newman and J. Whitehead, *Quart. J. Math.*, 8 (1937)). The construction of this and similar examples generally involves infinite processes. Such examples, which do not have the usual properties we expect, are commonly termed **pathological** (or **wild**) **spaces**. We give some typical examples.

L. Antoine (*J. Math. Pures Appl.*, 8 (1921)) initiated research on wild spaces. Thereafter, Alexander (*Proc. Nat. Acad. Sci. US*, 10 (1924)) constructed the so-called Alexander horned sphere as a counterexample for the †Schönflies problem with $n = 2$. The Alexander horned sphere is a topological 2-sphere S^2 in the Euclidean 3-space $\mathbf{R}^3$ for which the bounded component of $\mathbf{R}^3 - S^2$ is not simply connected (Fig. 1). R. Fox and E. Artin (*Ann. of Math.*, (2) 49 (1948)) gave an analogous example applying †knot theory (Fig. 2). These results were further enriched by Bing (*Ann. of Math.*, (2) 69 (1959)), B. Ball (*Ann. of Math.*, (2) 69 (1959)), and others. For the Newman-Whitehead manifold U mentioned in this section, the product $U \times \mathbf{R}^1$ is found to be homeomorphic to the Euclidean 4-space $\mathbf{R}^4$. R. Moore [10] proved that the quotient space of the Euclidean 2-space $\mathbf{R}^2$ by any †upper semicontinuous decomposition consisting of continua not separating $\mathbf{R}^2$ is homeomorphic to $\mathbf{R}^2$. On the other hand, Bing (*Ann. of Math.*, (2) 70 (1959)) gave an example of a similar upper semicontinuous decomposition of $\mathbf{R}^3$ for which the quotient space B is not homeomorphic to $\mathbf{R}^3$ and is not even a manifold. This quotient space has the property $B \times \mathbf{R}^1 \approx \mathbf{R}^4$. Bing (*Fund. Math.*, 47 (1957)) also proved that at most countably many copies of a wild closed surface can simultaneously be embedded in $\mathbf{R}^3$. On the other hand, Stallings (*Ann. of Math.*, (2) 72 (1960)) proved that there exists a wild disk of which a set of copies with the power of the continuum can simultaneously be embedded in $\mathbf{R}^3$ [9].

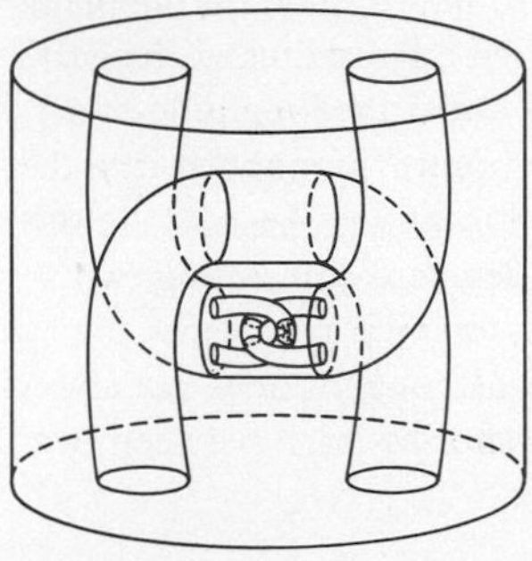

Fig. 1

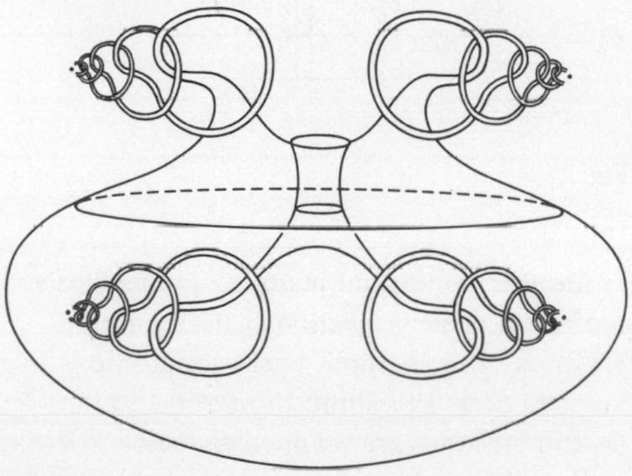

Fig. 2

References

[1] F. Enriques, Prinzipien der Geometrie, Enzykl. Math., III, 1A (1907), 1–129.
[2] H. Weyl, Die Idee der Riemannsche Fläche, Teubner, 1913, third edition, 1955; English translation, The concept of a Riemann surface, Addison-Wesley, 1964.
[3] S. Lefschetz, Topology, Amer. Math. Soc. Colloq. Publ., 1931.
[4] S. Lefschetz, Algebraic topology, Amer. Math. Soc. Colloq. Publ., 1942.
[5] S. Lefshetz, Introduction to topology, Princeton Univ. Press, 1949.
[6] H. Seifert and W. Threlfall, Lehrbuch der Topologie, Teubner, 1934 (Chelsea, 1965).
[7] S. S. Cairns, Triangulated manifolds and differentiable manifolds (R. L. Wilder and W. L. Ayres, Lectures in topology, Univ. of Michigan, 1941), 143–157.
[8] R. L. Wilder, Topology of manifolds, Amer. Math. Soc. Colloq. Publ., 1949.
[9] M. K. Fort, Jr. (ed.), Topology of 3-manifolds and related topics, Prentice-Hall, 1962.
[10] R. L. Moore, Foundations of point set theory, Amer. Math. Soc. Colloq. Publ., 1962.
[11] J. Milnor, Microbundles and differentiable structures, Lecture notes, Princeton Univ., 1961.
[12] D. Sullivan, On the Hauptvermutung for manifolds, Bull. Amer. Math. Soc., 73 (1967), 598–600.

260 (VI.21) Map Projections

A. General Remarks

Map projections are methods of constructing latitudes and longitudes on maps of the earth, which is assumed to be a sphere. Since a sphere cannot be developed on a plane, it is impossible to avoid some degree of distortion when representing figures on a sphere in a plane. Therefore, we use various types of map projections, the choice depending on which type of distortion will least affect the intended purpose of a particular map.

In constructing a map, it is desirable to preserve the proportions of at least one of the quantities: length, angle, or area. We can retain these proportions only under certain conditions; for example, the proportions of length may be retained for loxodromic curves (e.g., latitudes) (→ Section D) or parts of great circles passing through a fixed point. The proportions of either angles or areas can be preserved, but not both. There are, though, map projections whose straight lines correspond to the great circles on the globe or to loxodromic curves.

B. Perspective Projections

With **perspective projections**, we are given the point of vision and the projecting surface (plane or other developable surface, such as a cylinder or cone) on which we draw the perspective image of the sphere, just as we do in †descriptive geometry. A typical example of a projection of this type is the **stereographic projection**. In this projection, the point of vision S is on the sphere, and the projecting surface is the plane E passing through the center O of the sphere and perpendicular to the line OS (Fig. 1). The plane E divides the sphere into two half-spheres. One sphere contains S and the other does not. The latter half-sphere is projected on the plane E as in Fig. 1. This projection is †conformal, and all latitudes and longitudes are represented by circular arcs. Generally, a rectangular coordinate system can be introduced on the plane E so that the coordinates (x,y) of a point P' which is the projection of P satisfy the following formulas:

$$x=\frac{r\sin\lambda\cos\varphi}{1+\sin\beta\sin\varphi+\cos\beta\cos\varphi\cos\lambda},$$

$$y=\frac{r(\cos\beta\sin\varphi-\sin\beta\cos\varphi\cos\lambda)}{1+\sin\beta\sin\varphi+\cos\beta\cos\varphi\cos\lambda},$$

where r is the radius of the earth, φ the latitude of P, λ the difference of the longitudes of P and S, and β the latitude of the point S.

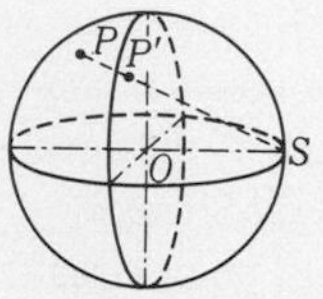

Fig. 1

Another example of a perspective projection is the **central projection** (or **gnomonic projection**), where the point of vision is the center of the sphere and the projecting surface is a tangent plane of the earth. Under this mapping, any great circle on the sphere is mapped to a straight line. The **polyhedral projection**, commonly used to draw a variety of basic maps, is sometimes described as a central projection for small regions of a sphere, but it is actually a polyconic projection as will next be described.

C. Polyconic Projections

There are methods of drawing latitudes and longitudes that preserve the actual relations as exactly as possible without utilizing a point of vision and a fixed projecting plane. For example, we take a series of strips of circular cones tangential to latitudes so that on each of the strips we can map a region of the earth between two sufficiently close latitudes. By unrolling the strips, we obtain a map. This method is called a **polyconic projection**. Let r be the radius of the earth. Then to the latitude φ there corresponds a circular arc of radius $r\cot\varphi$. Two points on the latitude φ whose longitudes differ by λ will be mapped to points on the arc separated by a distance $r\cot\varphi\lambda$ along the arc. For maps of small regions, this projection offers a fairly accurate method of preserving distances and angles.

D. Conformal Mappings

Mercator's projection, introduced by G. Mercator in 1569, is still widely used for charts (marine maps). It is a conformal mapping that maps latitudes and longitudes into mutually orthogonal straight lines and each **loxodromic curve** (a curve cutting all meridian lines at a constant angle, also called an **equidirection line**) into a straight line (Fig. 2). We introduce a rectangular coordinate system on the plane E of the map. Let r be the radius of the earth, and φ the latitude and λ the longitude of a point P. Then the point P is mapped into the point $x=r\lambda, y=r\log\tan(\pi/4+\varphi/2)$ in the plane E. The line segment connecting two points on a Mercator map represents the loxodromic curve, although it does not always correspond to the shortest path between two end points.

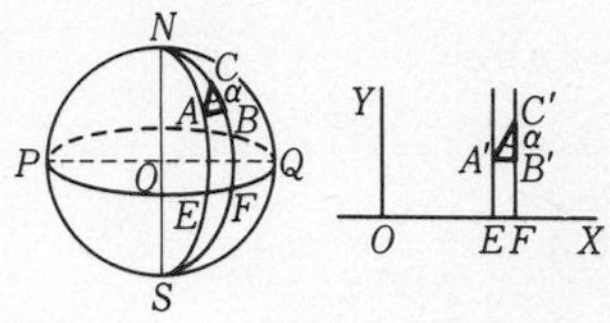

Fig. 2

Another conformal mapping is **Lambert's conformal conic projection** (J. H. Lambert, 1772). By this method, every longitude is mapped to a ray issuing from a fixed point O, and latitudes are mapped to concentric circles whose center is the point O. This projection preserves fairly exactly not only the relations of angles but also the relations of lengths, and it is useful in drawing maps of regions of higher latitude that spread widely from east to west. There are two techniques used in Lambert conformal conic projection: a double projection with two basic latitudes, and a single projection with one basic latitude, the former being the more accurate. Denote the two basic latitudes by φ_1 and φ_2, the radii of the concentric circles corresponding to φ_1 and φ_2 by ρ_1 and ρ_2, and the angle between the rays corresponding to the central meridian and the meridian with longitude λ by θ. We take a point P with longitude λ and latitude φ. Let ρ be the radius of the circle corresponding to φ. Then we have

$$\rho_1=r\cos\varphi_1/\mu,\quad \rho_2=r\cos\varphi_2/\mu,$$
$$\rho_1=c\tan^{\mu}\chi_1/2,\quad \rho_2=c\tan^{\mu}\chi_2/2,$$

where $\chi_i=\pi/2-\varphi_i$ $(i=1,2)$, c is a constant, and

$$\mu=\frac{\log\cos\varphi_1-\log\cos\varphi_2}{\log\tan\chi_1/2-\log\tan\chi_2/2}.$$

Further, we have

$$\rho=c\tan^{\mu}\chi/2,\quad \theta=\mu\lambda,\quad \text{where } \chi=\pi/2-\varphi.$$

E. Projections That Preserve Area

Let r be the radius of the earth, and φ the latitude and λ the longitude of a point to be projected to (x,y). Then we have the following three formulas, which give typical projections that preserve area:

(1) $x = rk\lambda, \quad y = (r/k)\sin\varphi,$

where k is a constant (M. Eckert, 1906).

(2) $x = ((2\sqrt{2})/\pi)r\lambda\cos g(\varphi),$

$y = \sqrt{2}\, r\sin g(\varphi),$

where the function $g(\varphi)$ is determined by the relation

$2g(\varphi) + \sin 2g(\varphi) = \pi\sin\varphi$

(K. B. Mollweide, 1805).

(3) $x = ((2\sqrt{2})/\pi)r\lambda\sin(\pi/4 - \varphi/2),$

$y = r\sqrt{\pi}\,(1 - \sqrt{2}\,\sin(\pi/4 - \varphi/2))$

(Collingmon, 1865).

References

[1] T. W. Birch, Maps, topographical and statistical, Clarendon Press, 1949.
[2] E. J. Raisz, General cartography, McGraw-Hill, second edition, 1948.
[3] W. Chamberlin (ed.), The round earth on flat paper, map projections used by cartographers, National Geographic Society, 1950.

261 (XVI.7) Markov Chains

A. General Remarks

A random process X_t $(t \geqslant 0)$ is one governed by probability laws. One of the most important is the process whose probability law of X_t under the condition $X_{s_1} = a_1, \ldots, X_{s_n} = a_n$ $(s_1 < s_2 < \ldots < s_n < t)$ coincides with that under the condition $X_{s_n} = a_n$. This is called the †Markov property, and a process with this property is called a †Markov process. In particular, if the process takes place in a countable set S, it is called a **Markov chain**.

We consider only †temporally homogeneous Markov chains, which are described as †stochastic processes in the following way. Let S be a countable set (called the †state space of the movement) and T be $\{0,1,2,\ldots\}$ or $[0,\infty)$ (called the **time parameter space**). A family $(X_t, P_x)_{t\in T, x\in S}$ is called a **Markov chain** if P_x, the †probability law of the process X_t starting at x, is subject to

$$P_x(X_{s_1+t_n} = y_1, \ldots, X_{s_m+t_n} = y_m \mid X_{t_1} = x_1, \ldots, X_{t_n} = x_n) = P_{x_n}(X_{s_1} = y_1, \ldots, X_{s_m} = y_m), \qquad (1)$$

for every $t_j, s_k \in T$ $(j = 1, \ldots, n; k = 1, \ldots, m)$ such that $t_1 < t_2 < \ldots < t_n, 0 < s_1 < s_2 \ldots < s_m$. Then the function defined by

$$p_t(x,y) = P_x(X_t = y), \quad t \in T, \quad x,y \in S$$

satisfies

$$0 \leqslant p_t(x,y) \leqslant 1, \qquad (2)$$

$$\sum_{y\in S} p_t(x,y) = 1, \qquad (3)$$

$$p_{t+s}(x,y) = \sum_{z\in S} p_t(x,z)p_s(z,y) \qquad (4)$$

by (1) and the general properties of probability laws. We call $p_t(x,y)$ $(t \in T, x,y \in S)$ the **transition probability** (or **transition function**) of the Markov chain, and the relation (4) is called the **Chapman-Kolmogorov equation**. Furthermore, the matrix $p_t = (p_t(x,y))$ is called the **transition matrix**.

Conversely, for a given $p_t(x,y)$ with the properties (2), (3), and (4), we can construct a Markov chain that satisfies

$$P_x(X_{t_1} = x_1, \ldots, X_{t_n} = x_n) = p_{t_1}(x,x_1)p_{t_2-t_1}(x_1,x_2)\ldots p_{t_n-t_{n-1}}(x_{n-1},x_n)$$

by using Kolmogorov's extension theorem. Such a Markov chain is essentially unique. If we are given a **stochastic matrix** $p = (p_{x,y})$, i.e., a matrix satisfying $0 \leqslant p_{x,y} \leqslant 1$ and $\sum_{y\in S} p_{x,y} = 1$, the components $p_n(x,y)$ of the iterated matrix p^n satisfy (2), (3), (4), and hence there exists a Markov chain with discrete parameter having $p_n(x,y)$ as its transition probability.

Let $\mu_x = P(X_0 = x)$ be the initial distribution over S at time 0. Then the distribution $\mu_y^{(t)} = P(X_t = y)$ at time t is obtained from $\mu_y^{(t)} = \sum_{x\in S}\mu_x p_t(x,y)$. In particular, if $\mu_y^{(t)}$ is independent of t, i.e.,

$$\mu_x = \sum_{x\in S} \mu_x p_t(x,y), \qquad (5)$$

then μ is called the **invariant distribution** of the Markov chain. An arbitrary real nonnegative solution of (5) is often called an **invariant measure**.

When S is the set of all d-dimensional lattice points and $p_{x,y} = \pi_{x-y}$, the associated Markov chain is called a **(general) random walk**. In particular, if $\pi_x = 2^{-n}$ for every neighboring point x of the origin and $=0$ otherwise, it is called a **standard random walk** (or simply **random walk**).

Now, when the equality sign in (3) is replaced by $<$, we consider the space S^* obtained by adjoining ∂ (†death point) to S, and define

$$p_t^*(x,y)=p_t(x,y), \quad x,y\in S;$$

$$p_t^*(x,\partial)=1-\sum_{y\in S} p_t(x,y), \quad x\in S;$$

$$p_t^*(\partial,\partial)=1; \quad p_t^*(\partial,x)=0.$$

Then we can associate a Markov chain (X_t, P_x^*) on S^* with $\{p_t^*\}$. If X_t is ∂, the particle of the process is regarded as extinct at the random time $\zeta=\inf\{t\mid X_t=\partial\}$, called the **lifetime** (or **killing time**). In this case, the process X_t restricted to t smaller than ζ is also called a Markov chain on S with the transition probabilities $\{p_t(x,y)\}$. Then the conditions $\Sigma_{y\in S}p_t(x,y)=1$ and $P_x(\zeta=\infty)=1$ are equivalent, and the chain is called **conservative** if $P_x(\zeta=\infty)=1$ for every x.

A Markov process with discrete parameter is often called a **Markov chain** whether the state space is countable or not.

In this section we have restricted ourselves to the †temporally homogeneous case. In the temporally inhomogeneous case, we have to consider the probability law $P_{x,t}$ of the path starting from $x\in S$ at time t, instead of P_x. Equation (1) becomes

$$P_{x,t}(X_{s_1}=y_1,\ldots,X_{s_m}=y_m\mid X_{t_1}=x_1,\ldots,X_{t_n}=x_n)$$

$$=P_{x_n,t_n}(X_{s_1}=y_1,\ldots,X_{s_m}=y_m),$$

$$t_1<\ldots<t_n<s_1<\ldots<s_n.$$

For the rest of this article, we consider only the homogeneous case.

B. Markov Chains with Discrete Parameter

Let $\mathfrak{X}=(X_n,P_x)$ be a conservative Markov chain. For a subset A of the state space S, $\sigma_A=\min\{n\geqslant 1\mid X_n\in A\}$ $(\min\varnothing=+\infty)$ is called the **hitting time** for the set A. If $P_x(\sigma_y<\infty)>0$ $(=0)$, we write $x\to y$ $(x\not\to y)$. Then $x\to y$ is equivalent to the existence of $n\geqslant 1$ with $P_n(x,y)>0$. When $x\to y$ and $y\to x$, we write $x\leftrightarrow y$. The set of all x for which there exists $y\neq x$ with $x\to y$ and $y\not\to x$ is denoted by F and is called the **dissipative part** of S. For elements of $S-F$, the relation $\leftrightarrow$ is an equivalence relation. Each equivalence class E_α is called an **ergodic class**. When $F=\varnothing$ and S consists of a single class, the chain $\mathfrak{X}$ is called **ergodic** (or **irreducible**). For an ergodic class E, the greatest common divisor d of $\{n>1\mid P_n(x,x)>0\}$ for $x\in E$ does not depend on the choice of $x\in E$ and is called the **period** of the class E. Set $G_n=\{y\in E\mid P_n(x,y)>0\}$ $(n=1,2,\ldots,d)$ for a fixed $x\in E$. Then we have a decomposition of $E: E=\bigcup G_n$, $G_n\cap G_m=\varnothing$ $(n\neq m)$, $|G_n|=d^{-1}|E|$ ($|\ |$ denotes the number of points in the set), and $\Sigma_{z\in G_{m+1}}P(y,z)=1$ $(y\in G_m)$. Each G_n is called a **cyclic part**. The decomposition may depend on the choice of x but is unique up to ordering.

The point x is called **recurrent** or **nonrecurrent (transient)** according as $P_x(\sigma_x<\infty)=1$ or <1. A necessary and sufficient condition for x be be recurrent is $\Sigma_{n=0}^{\infty}P_n(x,x)=\infty$. The probability $P_x(X_n=x$ occurs infinitely often) is 1 or 0 according as x is recurrent or nonrecurrent. A chain is called **recurrent** or **nonrecurrent** according as every point in S is recurrent or nonrecurrent. A recurrent point x is called **positive recurrent** or **null recurrent** according as $m_x\equiv E_x(\sigma_x)$ is finite or infinite. Let E be an ergodic class. If there exists a positive recurrent element in E, then all elements are also positive recurrent, and the class E is called **positive recurrent**. We can define null recurrence or nonrecurrence of an ergodic class similarly. For a finite state space, every Markov chain has at least one ergodic class, and all ergodic classes are positive recurrent.

C. Limit Theorems for Markov Chains

The properties of a recurrent chain reduce to those of an ergodic recurrent chain, since a recurrent chain decomposes into ergodic classes. We assume that $\mathfrak{X}=(X_t,P_x)$ is an ergodic recurrent chain. Concerning the transition function we have the following limit theorems: Let d be the period of X and $S=\bigcup_{r=1}^{d}G_r$ be the decomposition into cyclic parts. If $x\in G_r$, then $\lim_{n\to\infty}P_{nd+r}(x,y)=dm_y^{-1}$ $(y\in G_r)$, $=0$ $(y\notin G_r)$. For every $x,y\in S$, $\lim N^{-1}\Sigma_{n=1}^{N}P_n(x,y)=m_y^{-1}$. Furthermore, $r_{x,y}=\lim_{N\to\infty}(\Sigma_{n=1}^{N}P_n(y,y)/\Sigma_{n=1}^{N}P_n(x,x))$ exists and is finite. Then $r_{x,y}$ is an invariant measure as a function of y, and every nonnegative invariant measure is a constant multiple of it. The chain is positive recurrent if and only if $\Sigma_y r_{x,y}<\infty$, and then we have $r_{x,y}=m_x m_y^{-1}$. Consequently there exists an invariant measure μ_x such that $0<\Sigma_{x\in S}|\mu_x|<\infty$ if and only if the chain is positive recurrent, and in this case μ_x is a constant multiple of m_x^{-1}.

Let μ be a nonnegative invariant measure, and set $I_\mu(f)=\Sigma_{x\in S}\mu_x f(x)$. For f, g such that $I_\mu(|f|)<\infty$, $0<I_\mu(|g|)<\infty$, and $I_\mu(g)\neq 0$, we have

$$P_x\left(\lim_{N\to\infty}\left(\sum_{n=1}^{N}f(X_n)\Big/\sum_{n=1}^{N}g(X_n)\right)=I_\mu(f)/I_\mu(g)\right)=1.$$

The †law of large numbers, the †central limit theorem, and the †law of the iterated logarithm hold for the asymptotic behavior of the sum $\Sigma_{n=1}^{N} f(x_n)$ as $N \to \infty$.

D. Potential Theory for Markov Chains

For a given Markov chain,

$$G(x,y) = \sum_{n=0}^{\infty} P_n(x,y)$$

is well defined, including the value ∞. If $G(x,y)$ is not identically ∞, we can define a (generalized) †potential with kernel $G(x,y)$ (→ 262 Markov Processes; 46 Brownian Motion; 335 Potential Theory). For a real function φ over S, the function $G\varphi(x) = \Sigma_{y\in S} G(x,y)\varphi(y)$ is called the **potential** with charge φ if the sum exists. Even if $G\varphi$ does not exist, the infinite sum $\Sigma_{n=0}^{\infty} P_n\varphi$ may exist. In this case this sum is also called the potential with charge φ. A real function f $(-\infty < f \leqslant +\infty)$ over S is defined to be **superharmonic** (or **superregular**) if $Pf \leqslant f$. Here P is the operator associated with the kernel $P_1(x,y)$, that is, $Pf(x) = \Sigma_{y\in S} P_1(x,y)f(y)$. Furthermore, if $f \geqslant 0$ and $f \geqslant Pf$, f is called **excessive**, and if $-\infty < f < +\infty$ and $f = Pf$, f is called **harmonic**. If $Pf \leqslant f$ at a point x, f is called **superharmonic at** x, etc. For an ergodic recurrent chain, every nonnegative superharmonic function is constant.

1. For a nonrecurrent chain we may consider the potential $f = G\varphi$ for every function φ with finite support. G satisfies $(P-I)G = -I$ and $\lim_{n\to\infty} P_n G = 0$, where I is the unit matrix. Consequently, the operator $P-I$ corresponds to the †Laplacian Δ of †Newtonian potential theory, and the equation $(P-I)f = 0$ corresponds to the Laplace equation $\Delta f = 0$. If the limit $w = \lim_{n\to\infty} P_n f$ exists for a function f on S, f can be expressed uniquely as the sum of the potential $G\varphi$ $(\varphi = f - Pf)$ and the harmonic function w. This decomposition is called the **Riesz decomposition**, following the terminology in Newtonian potential theory (→ 335 Potential Theory). Let E be a finite subset of S and $\sigma_E^* = \min\{n \geqslant 0 \mid X_n \in E\}$. Then $f(x) = P_x(\sigma_E^* < \infty)$ is a (unique) potential which is harmonic at $x \in E^c$ and takes the value 1 on E. This is called the **equilibrium potential** of the set E, and its total charge $C(E) = \Sigma_{x\in S} f(x) - Pf(x)$ is called the **capacity** of E. The †maximum principle and the †balayage principle (→ 335 Potential Theory) are valid in this potential theory.

2. For a recurrent chain, we cannot define the potential kernel as in 1., since $G(x,x) = \infty$. However, we can define a kernel analogous to the †logarithmic potential. We assume that the chain is ergodic. When the limit

$$A(x,y) = \lim_{n\to\infty} \left(G_n(x,x)\,\mu_y\mu_x^{-1} - G_n(x,y)\right),$$

$$G_n(x,y) = \sum_{k=0}^{n} P_k(x,y),$$

exists for a nonnegative invariant measure μ, the chain is called (right) **normal**. A chain is normal if and only if

$$\lambda_E(\cdot) = \lim_{n\to\infty} \sum_{z\in S} P_n(x,z) P_z(\sigma_E \in \cdot)$$

exists. If $\lambda_E(\cdot)$ exists, it is independent of x. Every positive recurrent chain is normal. Let φ be a function with finite support. The potential f of φ exists if and only if $\Sigma_{x\in S} \mu_x \cdot \varphi(x) = 0$, and $f = -A\varphi$ in this case. $A\varphi$ satisfies $(P-I)A\varphi = \varphi$ and $\lim_{n\to\infty} P_n A\varphi = 0$. The potential f is a bounded potential of a function with finite support E if and only if f is bounded and harmonic in E^c and satisfies $\Sigma_x \lambda_E(x) f(x) = 0$.

E. Random Walks

Consider a random walk defined on the set s of all lattice points in a d-dimensional Euclidean space. Let $S^+ = \{x \mid 0 \to x\}$, $\bar{S} = \{z \mid z = x - y,\ x,y \in S^+\}$. F. Spitzer obtained the following results for the random walk with $S = \bar{S}$. The random walk is recurrent if the following conditions are satisfied: (i) $d = 1$, $\Sigma |x| P(0,x) < \infty$ ($|x|$ is the distance between 0 and x), and $m \equiv \Sigma x P(0,x) = 0$; (ii) $d = 2$, $m = 0$, and $\sigma \equiv \Sigma |x|^2 P(0,x) < \infty$. When $d \geqslant 3$, the random walk is always nonrecurrent. The measure $\mu_x \equiv 1$ is invariant whether the random walk is recurrent or not.

Every recurrent random walk is right normal, and the potential kernel A satisfies $(P - I)A = I$. Several interesting results are known on the uniqueness of the kernel A such that $(P-I)A = I$ [8].

For the case $m = 0$, called the **symmetric** random walk, there are a number of results similar to those of †Brownian motion, including: (i) When $d = 1$ and $0 < \sigma < \infty$,

$$\lim_{n\to\infty} P_0\left(\max |X_k| \leqslant \sigma\sqrt{n}\,x\right) = 1 - F(x^{-2}) \quad (x > 0),$$

where

$$F(x) = 1 - \frac{4}{\pi} \sum_{k=0}^{\infty} \frac{(-1)^k}{2k+1} \exp\left(-\frac{\pi^2}{8}(2k+1)^2 x\right);$$

(ii) the **arc sine law**: Let T_n be the number of $k \leqslant n$ for which X_k becomes > 0. When $d = 1$ and $0 < \sigma < \infty$,

$$\lim_{n\to\infty} P_0(T_n \leqslant nx) = \frac{2}{\pi} \arcsin \sqrt{x} \quad (0 \leqslant x \leqslant 1);$$

(iii) the **Wiener test**: The set E is called **recurrent** if $P_x(\sigma_E < \infty) = 1$ holds for every x.

When $d=3$ and $\sigma<\infty$, a set E is recurrent if and only if $\sum_{n=1}^{\infty}C(E_n)2^{-n}=\infty$, where $E_n=E\cap\{x\mid 2^n<|x|<2^{n+1}\}$.

F. Markov Chains with Continuous Time Parameter

Suppose that the transition probability $p_t(x,y)$ is measurable in t. Then $p_t(x,y)$ is uniformly continuous at the complement of any neighborhood of $t=0$, and there exists $m_{x,y}=\lim_{t\uparrow\infty}p_t(x,y)$ for which $m_{x,y}=\sum_{z\in S}m_{x,z}\cdot p_t(z,y)=\sum_{z\in S}p_t(x,z)m_{z,y}$. If $p_t(x,y)=0$ for all $x\in S$ and $t>0$, y is called a **fictitious state.** Let F be the set of all fictitious states of S. Then the restriction of $p_t(x,y)$ to $S-F$ forms a transition probability on $S-F$. If $F=\varnothing$ and the function family $\mathbf{p}=\{p_t(\cdot,y)\mid t>0,y\in S\}$ separates points of S, the transition probability $p_t(x,y)$ is called **standard.** Then $p_t(x,y)$ is standard if and only if $\lim_{t\downarrow 0}p_t(x,y)=\delta_{x,y}$. The case where $F=\varnothing$ and $\mathbf{p}$ does not separate points of S may be reduced to the standard case by a suitable identification of states. We assume that $p_t(x,y)$ is standard. Then $\lim_{t\downarrow 0}t^{-1}(p_t(x,y)-\delta_{x,y})=q_{x,y}$ exists and satisfies $0\leqslant q_{x,y}<\infty$ if $x\neq y$.

Set $q_x=-q_{x,x}$ ($\geqslant 0$). We call x a **stable state** if $q_x<\infty$ and an **instantaneous state** if $q_x=\infty$. If $q_x<\infty$, $p_t'(x,y)$ (the derivative with respect to t) exists and is continuous in $t>0$. When every point of S is stable, $\pi(x,y)$ defined by

$$\pi(x,y)=\begin{cases} q_{x,y}q_x^{-1} & (x\neq y)\\ 0 & (x=y)\end{cases}$$

satisfies

$$0\leqslant\pi(x,y)\leqslant 1,\quad \pi(x,x)=0,$$
$$\sum_{y\in S}\pi(x,y)\leqslant 1. \tag{6}$$

From the †Kolmogorov-Chapman equation we can formally derive **Kolmogorov's backward equation**

$$p_t'(x,y)=-q_xp_t(x,y)+q_x\sum_{z\in S}\pi(x,z)p_t(z,y) \tag{7}$$

and its dual, **Kolomogorov's forward equation**

$$p_t'(x,y)=-q_yp(x,y)+q_y\sum_{z\in S}p_t(x,z)\pi(z,y). \tag{8}$$

Strictly speaking, these equations hold only under suitable conditicns: For instance, a conservative transition probability $p_t(x,y)$ satisfies (7) if and only if $\sum_{y\in S}\pi(x,y)=1$. If there exist $\xi_x>0$ ($x\in S$) such that $\sum_{x\in S}\xi_x p_t(x,y)\leqslant\xi_y$ ($\forall t>0$), then $p_t(x,y)$ satisfies (8) if and only if $\sum_{y\in S}\xi_yq_y\pi(y,x)=\xi_x$. Conversely, given $0\leqslant q_x<\infty$ and π satisfying (6), there exist in general many solutions of (7) and (8) with the initial condition $\lim_{t\downarrow 0}p_t(x,y)=\delta_{x,y}$. Among them, the minimal solution $p_t^0(x,y)$ exists. The chain with $p_t^0(x,y)$ as its transition probability is called the **minimal chain** associated with $\{q,\pi\}$.

For the path of a Markov chain with standard transition probability, there exists a measurable modification X_t, but in general there exists neither a right continuous modification nor a modification having the strong Markov property for all stopping times. (For detailed properties of paths → [7, 1].) Set $\tau_0=0$, $\tau_n=\inf\{t>\tau_{n-1}|X_t\neq X_{\tau_{n-1}}\}$ ($n\geqslant 1$), and $\tau_\infty=\lim_{n\to\infty}\tau_n$. If each point x of S is stable, τ_1 is subject to the exponential distribution $P_x(\tau_1>t)=e^{-q_xt}$, so that $E_x(\tau_1)=q_x^{-1}$, $P_x(X_{\tau_1}=y)=\pi(x,y)$ and τ_1 and X_{τ_1} are independent with respect to P_x-measure. Furthermore $p_t^0(x,y)=P_x(X_t=y,\tau_\infty>t)$ is the minimal solution. For the minimal chain there exists a right continuous modification with left-hand limits, which has the strong Markov property.

For a finite Markov chain with a standard transition probability, all states are stable and both (7) and (8) are fulfilled. Furthermore, the transition probability satisfying (7) and (8) is the unique minimal solution.

G. Birth and Death Processes

A chain is called a **birth and death process** if $S=\{0,1,2,\dots\}$, $q_0=0$, $0<q_n<\infty$ ($n\geqslant 1$), and $\pi(n,m)=0$ ($m\neq n+1,n-1$). We call $\sigma_0=\inf\{t>0|X_t=0\}$ and τ_∞ the **extinction time** and **explosion time**, respectively. In particular, if $\pi(n,n+1)=1$ ($\forall n\geqslant 1$), the Markov chain X is called a **birth process**, and if $\pi(n,n-1)=1$ ($\forall n\geqslant 1$), X is called a **death process.** The birth and death process that satisfies $\pi(n,n+1)>0$ and $\pi(n,n-1)>0$ enjoys a number of properties similar to those of a 1-dimensional †diffusion process.

Set $q_n=\lambda_n+\mu_n$, $\pi(n,n+1)=\lambda_n(\lambda_n+\mu_n)^{-1}$, and $\pi(n,n-1)=\mu_n(\lambda_n+\mu_n)^{-1}$ for $n\geqslant 1$. The sequence x_n, called the **natural scale**, is defined as follows: $x_1=\mu_1^{-1}$, $x_2=\mu_1^{-1}+\lambda_1^{-1}$, $x_n=\mu_1^{-1}+\lambda_1^{-1}+\ldots+(\mu_2\ldots\mu_{n-1})(\lambda_1\ldots\lambda_{n-1})^{-1}$ ($n\geqslant 3$), $x_\infty=\lim_{n\to\infty}x_n$. The measure m with the mass $m_n=\lambda_1\ldots\lambda_{n-1}(\mu_2\ldots\mu_n)^{-1}$ ($n\geqslant 2$) and $=1$ ($n=1$) at the point x_n is called the **canonical measure**. Then $p_t(x_i,x_k)=p_t(i,k)m_k^{-1}$ is a transition probability on $E=\{x_1,x_2,\dots\}$, and $f(x_i,t)=p_t(x_i,x_k)$ satisfies a differential-difference equation

$$\partial f/\partial t=D_mf^+, \tag{9}$$

which is equivalent to (7). Here, $f^+(x_n)=(f(x_{n+1})-f(x_n))(x_{n+1}-x_n)^{-1}$ and $D_mg(x_n)=$

$(g(x_n)-g(x_{n-1}))m_n^{-1}$. The operator $f \to D_m \cdot f^+$ is similar to Feller's expression of the infinitesimal generators of 1-dimensional diffusion processes.

Every birth and death process is obtained from †Brownian motion by time change. Furthermore, x_∞ is regarded as a boundary point, and is classified as a natural, exit, entrance, or regular boundary point. Every birth and death process is determined by (9) and the boundary condition at x_∞ (→ 119 Diffusion Processes).

H. Boundary Value Problems for Markov Chains

To discuss the behavior of the path of a Markov chain beyond the time τ_∞, we have to introduce a suitable boundary of the state space S. Among several conceivable boundaries, the Martin boundary, which is most frequently utilized, is explained later in this section. The name comes from its similarity to the †Martin boundary in the theory of harmonic functions. Most results stated in this section are also valid for the discrete time parameter case.

Let $\mathfrak{X}=(X_t,P_x)$ be the minimal, nonrecurrent chain associated with $\{g,\pi\}$ for which τ_∞ equals the lifetime ζ. Harmonic functions, etc., are defined as in the discrete time parameter case (with p replaced by π). Let $\gamma\equiv\gamma(x)$ be a measure such that $0<\gamma G(y)\equiv\sum_x\gamma(x)\cdot G(x,y)<\infty$, where $G(x,y)=\int_0^\infty p_t(x,y)\,dt$. Let ρ_1 be the metric in the one-point compactification of the state space S equipped with the discrete topology. Set $K(x,y)=G(x,y)/\gamma G(y)$, and define

$$\rho_2(y,y')=\sum_{n=1}^{\infty}\frac{1}{2^n}\frac{|K(x_n,y)-K(x_n,y')|}{1+|K(x_n,y)-K(x_n,y')|},$$

$\{x_n\}=S$.

The set ∂S of all points adjoined to S by the completion of S relative to $\rho=\rho_1+\rho_2$ is called the **Martin boundary** of S. $M=S\cup\partial S$ is a compact separable space. By the definition of ρ, $K(x,\xi)=\lim_{y\to\xi}K(x,y)$ exists, is continuous in ξ, and is superharmonic in x. A nonnegative superharmonic function u is called **minimal** if every superharmonic function v such that $0\leqslant v\leqslant u$ is a constant multiple of u. The set $\partial S_1=\{\xi\mid K(\cdot,\xi)$ is minimal harmonic$\}$, called the **essential part** of ∂S, is an †F_σ-set. Then every γ-integrable nonnegative superharmonic function u is represented by $u(x)=\int K(x,\xi)\,\mu(d\xi)$ by means of a unique measure μ on $M_1=S\cup\partial S_1$, called the **canonical measure** of u. In particular, if u is harmonic, μ is concentrated in ∂S. A number of representation problems in analysis, such as the Hausdorff †moment problem, may be considered representation problems of suitable Markov chains. Let u be a γ-integrable nonnegative superharmonic function and (X_t,P_x^u) the Markov chain (called the **u-chain**) having $p_t^u(x,y)=u(x)^{-1}u(y)p_t(x,y)$ $(0/0=0)$ as its transition probability. Then $X_{\zeta-}=\lim_{t\uparrow\zeta}X_t$ exists and

$$P_x^u(X_{\zeta-}\in B)=u(x)^{-1}\int_B K(x,\xi)\,\mu(d\xi),$$

where μ is the canonical measure of u.

A measure ν on S is called a **superharmonic measure (harmonic measure)** if $\nu q(\pi-I)\leqslant 0$ $(=0)$, that is, $\sum_{y\in S}\nu_y q_y(\pi_{yx}-\delta_{yx})\leqslant 0$ $(=0)$. Fix a function $g\geqslant 0$ such that $0<Gg<\infty$ (→ Section D), and set $K^*(x,y)=G(x,y)Gg(x)^{-1}$. Define the metric ρ_2^* similar to ρ_2, using the function family $\{K^*(\cdot,y)\}$ $(y\in S)$. The set adjoined to S by the completion relative to $\rho^*=\rho_1+\rho_2^*$ is called the **dual Martin boundary**. Extend K^* to $S\cup\partial S^*$ and denote by ∂S_1^* the set of all $\eta\in\partial S^*$ such that $K^*(\eta,\cdot)$ is a minimal superharmonic measure. Then every superharmonic measure ν with $\int\nu(dx)g(x)<\infty$ is represented uniquely by $\nu=\int\mu(d\eta)K^*(\eta,\cdot)$ in terms of a measure μ on $S\cup\partial S_1^*$.

Let $\xi\in M_1$ and denote the $K(\cdot,\xi)$-chain by $(X_t,P_x^{*,\xi})$. Then $P_x^\xi(\zeta<\infty)=0$ or $=1$. We call ξ a **passive boundary point** in the first case and an **exit boundary point** in the second. On the other hand, denote by $(X_t,P_x^{*,\eta})$ the chain having $P^{*,\eta}(x,y)=K^*(\eta,x)^{-1}K^*(\eta,y)p_t(y,x)$ as its transition probability. Then $P_x^{*,\eta}(\zeta<\infty)=0$ or $=1$. We call η a **dual passive boundary point** in the first case and an **entrance boundary point** in the second.

The Feller Representation of †Green's Operator. Fix $\{q,\pi\}$ and let $\mathfrak{X}=(X_t,P_x)$ be the minimal chain associated with $\{q,\pi\}$. Let $(\partial S)_{\mathrm{ex}}$ and $(\partial S^*)_{\mathrm{en}}$ be the sets of exit and entrance boundary points, respectively. If $P_x(X_{\zeta-}\in(\partial S)_{\mathrm{ex}})=0$, (7) has no solution other than the minimal solution. But if $P_x(X_{\zeta-}\in(\partial S)_{\mathrm{ex}})>0$ for some $x\in S$, (7) has infinitely many solutions. Furthermore, there are infinitely many solutions satisfying both (7) and (8).

Now let $K_\alpha(x,y)=K(x,y)E_x(e^{-\alpha\zeta})$, $K_\alpha^*(\eta,x)=K^*(\eta,x)\,E_x^{*,\eta}(e^{-\alpha\zeta})$, and $p_t(x,y)$ be the transition probability of X. Suppose that $p_t(x,y)$ satisfies (7), (8), and $\int_0^\infty p_t(x,x)\,dt<\infty$. Then there exists a unique family of measures $M_\alpha(\xi,\cdot)$ $(\xi\in(\partial S)_{\mathrm{en}})$ over (∂S^*) for which the **Feller representation**

$$G_\alpha(x,y)=G_\alpha^0(x,y)$$
$$+\iint K_\alpha(x,\xi)M_\alpha(\xi,d\eta)K^*(\eta,y)(d\xi)\quad(10)$$

holds, where, $G_\alpha^0(x,y)=\int_0^\infty e^{-\alpha t}p_t^0(x,y)\,dt$ and μ is the canonical measure of the function identically equal to 1. The problem of finding all the transition probabilities satisfying (7) and (8) is equivalent to that of finding all $M_\alpha(\xi,d\eta)$ whose Green's operator is given by the right-hand member of (10). Since M_α is a function on the boundary, the problem of finding all M_α corresponds to solving (7) and (8) with a suitable boundary condition. There are many open problems in this connection [4, 5].

References

[1] K. L. Chung, Markov chains with stationary transition probabilities, Springer, 1960.
[2] J. L. Doob, Stochastic processes, John Wiley, 1953.
[2A] J. L. Doob, Discrete potential theory and boundaries, J. Math. Mech., 8 (1959), 433–458.
[3] W. Feller, An introduction to probability theory and its applications, John Wiley, 1950.
[4] W. Feller, On boundaries and lateral conditions for the Kolmogorov differential equations, Ann. of Math., (2) 65 (1957), 527–570.
[4A] W. Feller, The birth and death processes as diffusion processes, J. Math. Pures Appl., (9) 38 (1959), 301–345.
[5] G. A. Hunt, Markov chains and Martin boundaries, Illinois J. Math., 4 (1960), 313–340.
[6] J. G. Kemeny and J. L. Snell, Potentials for denumerable Markov chains, J. Math. Anal. Appl., 3 (1961), 196–260.
[7] P. Lévy, Systèmes markoviens et stationaires, Cas dénombrable, Ann. Sci. Ecole Norm. Sup., (3) 68 (1951), 327–381.
[8] F. Spitzer, Principles of random walk, Van Nostrand, 1964.
[9] G. Maruyama, Kakurituron (Japanese; Theory of probability), Kyôritu, 1957.
[10] T. Watanabe, On the theory of Martin boundaries induced by countable Markov processes, Mem. Coll. Sci. Univ. Kyōto, (A) 33 (1960–1961), 39–108.

262 (XVI.6) Markov Processes

A. General Remarks

Let $\{X_t\}_{t\in T}$ be a †stochastic process defined on the †probability space $(\Omega,\mathfrak{B},P)$. The †state space S of X_t is the set of real numbers $\mathbf{R}$ or N-dimensional Euclidian space $\mathbf{R}^N$. In general, S may be a †locally compact Hausdorff space satisfying the second countability axiom. T is an interval $[0,\infty)$ or a set $\{0,1,2,\dots\}$. (T may also be any interval in the real line or a set $\{\dots,-2,1,0,1,2,\dots\}$.) We call this $\{X_t\}_{t\in T}$ a **Markov process** if, for any choice of points $s_1<s_2<\dots<s_n<t$ in T, the †conditional probability law of X_t relative to $X_{s_1}, X_{s_2},\dots,X_{s_n}$ is equal to the conditional probability law of X_t relative to X_{s_n}. Namely, for any A in $\mathfrak{B}(S)$,

(1) $$P(X_t\in A\,|\,X_{s_1}=x_{s_1},\dots,X_{s_n}=x_{s_n}) = P(X_t\in A\,|\,X_{s_n}=x_{s_n})$$

holds with probability one, where $\mathfrak{B}(S)$ is the least †σ-algebra that contains all open sets of S.

For a Markov process $\{X_t\}_{t\in T}$, there exists a function $P(s,x,t,A)$ of s, $t\in T$ $(s\leqslant t)$, $x\in S$, and $A\in\mathfrak{B}(S)$ which has the following properties:

(2) For fixed s and t, $P(s,x,t,A)$ is a †probability measure in A and is $\mathfrak{B}(S)$-measurable in x.

(2′) $P(s,x,s,A)=1$ if $x\in A$,
$=0$ if $x\notin A$.

(3) $P(X_t\in A\,|\,X_s=x_s)=P(s,x_s,t,A)$ with probability 1.

(4) The **Chapman-Kolmogorov equality**

$$P(s,x,u,A)=\int_S P(s,x,t,dy)P(t,y,u,A) \qquad (s<t<u)$$

holds except on a set whose measure is 0 relative to the distribution of X_s. The function $P(s,x,t,A)$ is called **transition probability** of $\{X_t\}_{t\in T}$.

Conversely, for a given function $P(s,x,t,A)$ satisfying (2), (2′), and (4), and for a given probability distribution μ on $\mathfrak{B}(S)$, there exists a Markov process $\{X_t\}_{t\in T}$ such that its transition probability is $P(s,x,t,A)$ and such that the distribution of X_0 coincides with μ. This process is uniquely determined up to equivalence of stochastic processes (→ 395 Stochastic Processes). The distribution of X_0 is called the **initial distribution** of the Markov process $\{X_t\}_{t\in T}$.

If the transition probability depends only on the difference of s and t, that is, if there exists a function $P(t,x,A)$ of t, x, and A such that $P(s,x,t,A)=P(t-s,x,A)$, then the Markov process is called **temporally homogeneous**. If S is an additive group and there exists a function $P(s,t,A)$ of s, t, and A such that $P(s,x,t,A)=P(s,t,A-x)$, where $A-x=$

$\{y-x \mid y \in A\}$, then the Markov process is called **spatially homogeneous**. When $S=\mathbf{R}^N$, a spatially homogeneous Markov process is an †additive process. A Markov process whose state space S is countable is called a †Markov chain (→ 261 Markov Chains). A Markov process whose sample †path is continuous with probability 1 (→ 395 Stochastic Processes) is called a †diffusion process (→ 119 Diffusion Processes). For a given initial distribution, the †finite-dimensional distribution of a Markov process is uniquely determined by its transition probability. In this sense, the transition probability is a characteristic quantity of the Markov process. Consider the case $T=[0,\infty)$ and $S=\mathbf{R}^N$ and assume that the transition probability has a †density $p(s,x,t,y)$ with respect to Lebesgue measure and satisfies certain analytic conditions which assure the continuity of the sample path, etc. Then A. N. Kolmogorov proved that $p(s,x,t,y)$ satisfies the †Fokker-Planck partial differential equation. Conversely, he raised the problem of finding conditions for the existence and uniqueness of the transition probability satisfying the given Fokker-Planck equation [21] (→119 Diffusion Processes). W. Feller extended this equation to an †integrodifferential equation of a certain type and solved the problem partially by classical analytic methods [11]. Also, S. N. Bernstein and P. Lévy made probabilistic approaches to the problem by †stochastic differential equations, and K. Itô made this approach more precise [1, 24, 17].

The problem is also related to †semigroup theory. Let $B(S)$ and $C(S)$ be the sets of all bounded measurable functions and of all bounded continuous functions on S, respectively. If S is not compact, let $C_\infty(S)$ be the set of all functions in $C(S)$ which converge to 0 at the point at infinity in the †one-point compactification of S. Using a transition probability $P(t,x,A)$ of a temporally homogeneous Markov process, set

$$(5)\quad T_t f(x)=\int_S P(t,x,dy)f(y)$$

for a function f in $B(S)$. Then T_t is an operator on $B(S)$, and the family of operators $\{T_t\}$ satisfies the following conditions:

$$(6)\quad T_s T_t = T_{s+t},$$

T_t is a linear †positive operator, and $\|T_t\| \leqslant 1$. Sometimes, a subspace of $B(S)$ such as $C(S)$ or $C_\infty(S)$ is invariant under T_t; then T_t may be regarded as an operator on the subspace. Conversely, assume that S is compact. Then for a given family of operators on $C(S)$ satisfying (6), there corresponds a unique transition probability that is related to the family by (5) (→ 173 Function Spaces C). If S is not compact, the same results hold for the family of operators on $C_\infty(S)$. From this point of view, the relation between the Fokker-Planck equation and the transition probability of the Markov process is better clarified in the scheme of Hille-Yosida semigroup theory (→ 373 Semigroups of Operators).

In the case of 1-dimensional diffusion processes W. Feller determined all the possible boundary conditions and gave the intrinsic representation of the second-order differential operator from the standpoint of semigroup theory (→ 119 Diffusion Processes) to characterize completely the analytic structure of 1-dimensional diffusion processes [12, 13]. Feller's work was followed by the sample function approaches of E. B. Dynkin [7], K. Itô and H. P. McKean [18], D. B. Ray [30], and others. In a sense, research on the 1-dimensional diffusion process was completed analytically and probabilistically. In succession, G. A. Hunt [16] established the general theory of the relation between Markov processes and †potentials, and W. Feller [14, 15] investigated the boundary value problem for Markov processes. These approaches clarified the relation between probabilistic properties of Markov processes and analytic entities in †potential theory, boundary value problems, and semigroup theory. They also established a method for precise treatment of properties such as continuity of the sample path and the strong Markov property (→ Section B). Thus the theory of Markov processes has revealed new aspects of probability theory and analysis. A profound study of the structure of Brownian motion and the additive process by P. Lévy [23, 24] and a systematic method for rigorous treatment of the sample path by J. L. Doob [4] had prepared for this development.

We now present a formulation for the temporally homogeneous Markov process with continuous time parameter [8] and state the main properties of the processes.

B. Definitions and Fundamental Properties

Adjoin an isolated point ∂ to S, set $\bar{S}=S\cup\{\partial\}$, and let $\mathfrak{B}(\bar{S})$ be the σ-algebra that consists of all the Borel sets in $\bar{S}$. Let $\tilde{W}$ be the set of all right continuous functions $w(t)$ whose †discontinuities are at most of the first kind and such that $w(t)=\partial$ for $t \geqslant s$ if $w(s)=\partial$. By convention, we shall set $w(\infty)=\partial$. Let $\zeta(w)$ be the minimum of the t-values such that $w(t)=\partial$. For $w \in \tilde{W}$, w_t^- and w_t^+ in $\tilde{W}$

are defined as $w_t^-(s)=w(\min(t,s))$ and $w_t^+(s)=w(s+t)$, respectively. Let W be a subset of $\tilde{W}$ that is closed under the operation $w\to w_t^+$ and $w\to w_t^-$, and $\mathfrak{B}=\mathfrak{B}(W)$ be the σ-algebra generated by the sets $\{w\in W\,|\,w(t)\in A\}$ $(A\in\mathfrak{B}(\bar{S}))$. We shall often write $X_t(w)$ for $w(t)$. The subclass of $\mathfrak{B}$ that consists of sets represented by $\{w\in W\,|\,w_t^-\in B\}$ $(B\in\mathfrak{B})$ is denoted by $\mathfrak{B}_t$. Suppose that the family of probability measures $\{P_x\}$ $(x\in\bar{S})$ on (W,B) satisfies the following conditions:

(7) For a fixed B in $\mathfrak{B}$, $P_x(B)$ is $\mathfrak{B}(\bar{S})$-measurable in x.

(8) $P_x(X_0(w)=x)=1$ for $x\in\bar{S}$.

(9) The **Markov property**

$$P_x(w_t^+\in B\,|\,\mathfrak{B}_t)=P_{x_t(w)}(B)$$

holds with P_x-measure 1 for $B\in\mathfrak{B}$.

Then the triple $\mathfrak{M}=(X_t,W,P_x\,|\,x\in\bar{S})$ is called a **Markov process**. This is a mathematical model of the random motion of a particle moving in S whose †probability law is independent of its past history once the present position of the particle is known (→ 261 Markov Chains). We shall call S the **state space** of $\mathfrak{M}$, W the **path space**, and the element w in W the **path**, respectively. P_x represents the probability law of a particle that starts from x. In view of the interpretation that the particle vanishes if it reaches ∂, we call $\zeta(w)$ the **lifetime** (or **terminal time**) and ∂ the **terminal point**.

Now, for any $s_1<s_2<\ldots<s_n$ and t,

$$P_x(X_{s_n+t}\in A\,|\,X_{s_1},\ldots,X_{s_n})$$
$$=P_x(X_{s_n+t}\in A\,|\,X_{s_n})=P(X_t\in A)$$

holds with P_x-measure 1 for any x. This equality means that $\{X_t\}_{t\geqslant 0}$ is a Markov process on S with initial distribution δ_x in the sense mentioned before, where δ_x is a unit measure at x. The transition probability of this process is given by $P(t,x,A)=P_x(X_t\in A)$, and

(4′) $$P(s+t,x,A)=\int_{\bar{S}}P(s,x,dy)P(t,y,A)$$

holds without exception. We can extend a real-valued function on S to a unique function on $\bar{S}$ by defining its value at ∂ to be 0. Therefore $B(S)$, $C(S)$, $C_\infty(S)$, etc., may also be regarded as †function spaces on $\bar{S}$. For f in $B(S)$, let

$$T_tf(x)=E_x(f(X_t))=\int_S P(t,x,dy)f(y),$$

where $E_x(\ )$ represents the integral with respect to P_x. Then $\{T_t\}$ $(t\geqslant 0)$ is a family of operators on $B(S)$ that satisfies (6). Let G_α be the †Laplace transform of $\{T_t\}$

$$G_\alpha f(x)=E_x\left(\int_0^\infty e^{-\alpha t}f(X_t)\,dt\right)$$
$$=\int_0^\infty e^{-\alpha t}T_tf(x)\,dt\quad(\alpha>0).$$

Then G_α is a linear †positive operator on $B(S)$, $\|G_\alpha\|\leqslant 1/\alpha$, and the **resolvent equation**

(10) $$G_\alpha-G_\beta+(\alpha-\beta)G_\alpha G_\beta=0$$

holds. Also $\{T_t\}$ and G_α are called the **semigroup** and **Green's operator** (or **resolvent operator**) of $\mathfrak{M}$, respectively. If $C(S)$ is invariant under G_α $(\alpha>0)$, then $\mathfrak{R}=G_\alpha\{C(S)\}$ is independent of α, and $G_\alpha^{-1}\{0\}=\{0\}$. Therefore $\mathfrak{G}=\alpha I-G_\alpha^{-1}$ can be defined on $\mathfrak{R}$ and is independent of α. $\mathfrak{G}$ is called the generator of M. In this case $\{T_t\}$ can be regarded as a semigroup on $C(S)$. If it is †strongly continuous at $t=0$, then $\mathfrak{G}$ coincides with the generator in the sense of Hille and Yosida (→ 373 Semigroups of Operators). For example, replacing $C(S)$ by $C_\infty(S)$, we have similar results. It is always necessary to specify the subspace on which G_α operates. Sometimes a Markov process $\mathfrak{M}$ is called a **Feller process** if the semigroup of $\mathfrak{M}$ is strongly continuous on $C_\infty(S)$. (If S is compact, $C_\infty(S)$ is replaced by $C(S)$.)

About the existence and the uniqueness of Markov processes, the following results are known. (1) A Markov process is uniquely determined if its transition probability and its path space W are given. However, for a given transition probability the choice of W is arbitrary to some extent. (2) For the compact state space S, if $P(t,x,A)$ $(x\in\bar{S},A\in B(\bar{S}))$ satisfies the conditions (2), (2′), (4), and

(11) $$\lim_{h\downarrow 0}\ \sup_{t\leqslant h,x\in S}P(t,x,U^\varepsilon(x)^c)=0\qquad(\varepsilon>0),$$

where $U^\varepsilon(x)$ is the ε-neighborhood of x, then there exists a Markov process with transition probability $P(t,x,A)$. Furthermore, in this case, we can take $W=\tilde{W}$. If $P(t,x,A)$ satisfies the conditions (2), (2′), (4′), and

(11′) $$\sup_{t\leqslant h,x\in S}P(t,x,U^\varepsilon(x)^c)\leqslant h\psi_\varepsilon(h),$$

where $\psi_\varepsilon(h)$ is monotone in h and $\psi_\varepsilon(h)\to 0$ $(h\to 0)$, then there corresponds a Markov process with continuous paths, namely a diffusion process. For the noncompact state space S, by properly restricting the behavior of the process at infinity, we obtain similar results under conditions similar to (11) and (11′) [10]. (3) If S is compact (noncompact), for a given semigroup $\{T_t\}$ on $C(S)$ $(C_\infty(S))$ that satisfies (6) and is strongly continuous at $t=0$, there exists a transition probability satisfying (11) that corresponds to $\{T_t\}$ by (5). Therefore, there exists a Markov process whose semigroup is an extension of $\{T_t\}$ to $B(S)$. On the condition of the continuity of the path, L. V. Seregin obtained a precise result [31] which can be applied to a general stochastic process [6, 29, 30].

A Markov process $\mathfrak{M}$ is called **conservative** if $P(t,x,S)=1$ holds for every x in S and every t. A point x in S is called a **recurrent point** if for every neighborhood U of x and every $t>0$, the path starting from x returns to U after time t with probability 1. Under some conditions, x is recurrent if and only if $\int_0^\infty p(t,x,U)\,dt=\infty$ for any neighborhood U of x. $\mathfrak{M}$ is called **recurrent** if all points in S are recurrent. Otherwise, it is called **transient**. $\mathfrak{M}$ is of course conservative if it is recurrent. There are other instances in which $\{X_t\}$ is called recurrent, if a particle starting from any point in S reaches any neighborhood of any other point in S in finite time with probability 1. In this case, under suitable conditions for regularity, the †mixing property holds, whence follows the †ergodic property (→ 146 Ergodic Theory). If the transition probability of $\{X_t\}$ has an †invariant measure with total mass 1, †Birkhoff's individual ergodic theorem holds [4].

For a $[0,\infty]$-valued $\mathfrak{B}$-measurable function σ on W, set $w_\sigma^-(t)=w(\min(t,\sigma(w)))$ and $w_\sigma^+(t)=w(t+\sigma(w))$. Let $\mathfrak{B}_\sigma$ be a family of all sets represented by $\{w\in W\,|\,w_\sigma^-\in B\}$ $(B\in\mathfrak{B})$ and $\mathfrak{B}_{\sigma+}=\bigcap_n B_{\sigma+(1/n)}$. If $\{w\,|\,\sigma(w)<t\}\in\mathfrak{B}_t$ for all t, σ is called a **Markov time**. If for every Markov time σ and every x in S

$$(9')\quad P_x(w_\sigma^+\in B\,|\,\mathfrak{B}_{\sigma+})=P_{w(\sigma)}(B),\quad B\in\mathfrak{B}$$

holds with P_x-measure 1, then it is said that $\mathfrak{M}$ has the **strong Markov property**, and such an $\mathfrak{M}$ is called a **strong Markov process**. Since a constant time is also a Markov time, the strong Markov property requires stronger conditions than the Markov property. A sufficient condition for a Markov process $\mathfrak{M}$ to have the strong Markov property is that $C(S)$ be invariant under Green's operator of $\mathfrak{M}$. For a set A of $\bar{S}$, let

$$\sigma_A=\inf\{t\,|\,t>0,\ \ x_t\in A\}\ \text{if such a } t \text{ exists},$$
$$=\infty\ \text{if } x_t\notin A \text{ for all } t>0.$$

Then σ_A is called the **hitting time** for A. (Sometimes the condition $t>0$ in the definition of σ_A is replaced by $t\geqslant 0$.) This σ_A is a Markov time for any open set A. If W consists of the paths that are continuous in $[0,\zeta(w))$, then σ_A is also a Markov time for any closed set A. Under suitable conditions, $P_x(X_{\sigma_A}\in E)$ as a function of E is a measure on $(S,\mathfrak{B}(S))$ whose support is $\bar{A}$. This measure is called a **hitting measure** from x to A and is written as $H_A(x,\cdot)$. For a subset A, $\tau_A=\sigma_{A^c}$ is called an exit time from A. If $\mathfrak{M}$ is a strong Markov process, the exit time τ_a from a point a is subject to the exponential distribution

$$P_a(\tau_a>t)=e^{-\lambda(a)t},\quad \text{where}\quad 0\leqslant\lambda(a)\leqslant\infty.$$

In particular, a is called an **instantaneous state** if $\lambda(a)=\infty$ and a **trap** if $\lambda(a)=0$. For a strong Markov process $\mathfrak{M}$ with the generator $\mathfrak{G}$, let σ be a Markov time such that $E_x(\sigma)<\infty$. Then we have **Dynkin's formula**:

$$f(x)=-E_x\left(\int_0^\sigma \mathfrak{G}f(X_t)\,dt\right)+E_x(f(X_\sigma))$$

and **Dynkin's representation of generator**

$$\mathfrak{G}f(x)=\lim_{U\downarrow\{x\}}\frac{E_x(f(X_{\tau_U}))-f(x)}{E_x(\tau_U)}\quad \text{if } x \text{ is not a trap},$$
$$=0\ \text{if } x \text{ is a trap},$$

where f is in the domain of $\mathfrak{G}$ and U is an open neighborhood of x. In particular, if S is compact and $C(S)$ is invariant under G_α, then $\mathfrak{G}$ is determined by this formula.

For a strong Markov process $\mathfrak{M}$, $P_x(B)=0$ or 1 if B is in $\mathfrak{B}_{0+}$. This is called **Blumenthal's zero-one law** [10, 18].

C. Potential Theory Corresponding to Markov Processes

There is a natural correspondence between †Newtonian potential and Brownian motion in $\mathbf{R}^N$ $(N\geqslant 3)$ (→ 46 Brownian Motion). Let $p(t,x,y)$ be the density function of the transition probability of Brownian motion relative to Lebesgue measure. Then the Newtonian kernel $(1/2)\pi^{-N/2}\Gamma(N/2-1)|x-y|^{2-N}$ is equal to $\int_0^\infty p(t,x,y)\,dt$. In Brownian motion, the hitting time σ_A of an †analytic set A is a Markov time. Consider the event where the path starting from x hits the set A. Then the †capacity of A is positive or not according as the probability of this event is positive for every x or not, and x is a †regular point of A or not according as $P_x(\sigma_A=0)=1$ or 0. The capacity itself can also be defined in terms of probability theory. Let D be a domain in $\mathbf{R}^N$. Then for a continuous function f on the boundary ∂D of D, the †Perron-Brelot solution of the †Dirichlet problem with boundary function f can be expressed as

$$E_x(f(X_{\sigma_{\partial D}}))=\int_{\partial D} f(y)H_{\partial D}(x,dy).$$

Therefore the hitting measure coincides with the †harmonic measure. This solution becomes an ordinary one if and only if every point in ∂D is †regular for D^c. If f is a †superharmonic function on $\mathbf{R}^N$, then it is known that $f(X_t(w))$ is continuous in t with probability one [5]. Keeping these facts in mind, we can establish a potential theory corresponding to the general Markov process [16, 34]. For example, the †Riesz potential corresponds to the †symmetric stable process. To describe a general theory, let us first introduce some notation. Let μ be a measure on $(\bar{S}, \mathfrak{B}(\bar{S}))$, and write $\mathfrak{B}(\bar{S})_\mu$ for the completion of $\mathfrak{B}(\bar{S})$ by μ. Write $\overline{\mathfrak{B}(\bar{S})}$ for the intersection of all $(\mathfrak{B}(\bar{S}))_\mu$, where μ runs over all probability measures on $\mathfrak{B}(\bar{S})$. Let P_μ be the measure on $(W, \mathfrak{B})$ defined by $P_\mu(\Lambda) = \int_{\bar{S}} \mu(dx) P_x(\Lambda)$; and let $\bar{\mathfrak{B}}_t = \bigcap_\mu (\mathfrak{B}_t)_\mu$, $\bar{\mathfrak{B}}_\sigma = \bigcap_\mu (\mathfrak{B}_\sigma)_\mu$, and $\bar{\mathfrak{B}} = \bigcap_\mu (\mathfrak{B})_\mu$, where $(\mathfrak{B}_t)_\mu$, $(\mathfrak{B}_\sigma)_\mu$, and $(\mathfrak{B})_\mu$ are the completions of $\mathfrak{B}_t$, $\mathfrak{B}_\sigma$, and $\mathfrak{B}$ by P_μ, respectively, and $\bar{\mathfrak{B}}_t$, $\bar{\mathfrak{B}}_\sigma$, and $\bar{\mathfrak{B}}$ are all σ-algebras. The definition of the Markov time and strong Markov property can be modified by using $\bar{\mathfrak{B}}_t$, $\bar{\mathfrak{B}}_\sigma$, and $\bar{\mathfrak{B}}$ instead of $\mathfrak{B}_t$, $\mathfrak{B}_\sigma$, and $\mathfrak{B}$. A Markov process $\mathfrak{M}$ is said to have **left quasicontinuity** if the following condition is satisfied: If $\{\sigma_n\}$ is any increasing sequence of Markov times and $\sigma = \lim \sigma_n$, then for $\sigma < \infty$, $\lim X_{\sigma_n} = X_\sigma$ holds except for a set of P_μ-measure 0. A Markov process $\mathfrak{M}$ is called a **Hunt process** if it has the strong Markov property in the modified sense and has left quasicontinuity. If the semigroup $\{T_t\}$ on $B(S)$ can be regarded as a Hille-Yosida semigroup $C_\infty(S)$ ($C(S)$ if S is compact), then a Hunt process corresponds to this semigroup.

Let $\mathfrak{M}$ be a Hunt process. If

$$Uf(x) = E_x\left(\int_0^\infty f(X_t)\,dt\right) = \int_0^\infty T_t f(x)\,dt$$

is well-defined for an extended real-valued function f on S, Uf is called the **potential** of f, and U is called a **potential operator**. Let f, $g \geq 0$, and suppose that Uf and Ug are well-defined. If $Uf \leq Ug$ holds on the support of f, then $Uf \leq Ug$ holds everywhere. Then U is said to satisfy the **maximum principle**. Conversely, the following fact can be proved: Let $C_0(S)$ be the set of all functions in $C(S)$ with compact support and U be a †positive linear operator on $C_0(S)$ whose image is a dense set in $C_\infty(S)$. If U satisfies the maximum principle and if there exists a sequence $\{f_n\}$ of functions in $C_0(S)$ such that $Uf_n \uparrow 1$, then there exists a Hunt process whose potential operator is U.

For a Markov process $\mathfrak{M}$, a nonnegative $\overline{\mathfrak{B}(\bar{S})}$-measurable function f is called **α-excessive** ($\alpha \geq 0$) if $e^{-\alpha t} T_t f \leq f$ for any t and $e^{-\alpha t} T_t f(x) \to f(x)$ ($t \to 0$). A 0-excessive function is called simply **excessive**. It should be noted that the operator T_t on $B(S)$ has a natural extension to nonnegative $\overline{\mathfrak{B}(\bar{S})}$-measurable functions. If f is nonnegative and Uf is well-defined, then Uf is excessive. For a Hunt process $\mathfrak{M}$, any excessive function can be uniquely decomposed into the sum of two excessive functions, $f = g + h$, where h satisfies the relation $h(x) = E_x(h(x_{\sigma_{G^c}}))$ for every relatively compact open set G, and $\lim_{t \to \zeta} g(X_t(w)) = 0$ for the path w such that $\lim_{t \to \zeta} X_t(w) =$ infinity (in one-point compactification). This decomposition is called the **Riesz decomposition** of f. In the case of Brownian motion, excessive functions are superharmonic functions, and this decomposition is the †Riesz decomposition in classical potential theory. Also, h is a maximal harmonic function dominated by f. For a Hunt process, if f is excessive, $f(X_t(w))$ is a right continuous function of t with discontinuities of at most the first kind. Regular points of an analytic set for a general Hunt process can be defined as in the case of Brownian motion. Natural topology or fine topology can be introduced in the state space S on this basis. All excessive functions are continuous according to this topology. If the potential U satisfies certain conditions, capacity and other concepts can be defined as in the case of Brownian motion. The Newtonian kernel is symmetric, but the potential kernels appearing in the general theory are not necessarily so. The †Dirichlet problem, Martin's representation of harmonic functions (→ 195 Harmonic Functions G), and other problems are studied from the probabilistic standpoint, and many facts analogous to classical ones hold in these cases (→ 261 Markov Chains).

We have so far mainly discussed potential theory for transient Markov processes. However, we can also establish potential theory for recurrent Markov processes following the model of the †logarithmic potential in 2-dimensional Brownian motion. Most results obtained in this connection up to now are confined to the case of Markov chains [20, 10, 16, 18].

D. Additive and Multiplicative Functionals

The notion of additive functionals was first introduced in relation to the study of time change of Markov processes, and in particular, to the study of the †local time of Brownian motion. Later it was studied in relation to potential theory and †martingale theory [26, 31]. Additive functionals play an important role in the study of Markov processes. For a Markov process $\mathfrak{M}$, a function $\varphi = \varphi_t(w)$ of t and w is called a (right continuous and homogeneous) **additive functional** if the following conditions are satisfied: (1) $-\infty < \varphi_t(w) \leq \infty$; (2) $\varphi_t(w)$ is right continuous in

t $(t<\zeta)$, $\varphi_{\zeta-}$ exists, and $\varphi_t=\varphi_{\zeta-}$ for $t \geqslant \zeta$; (3) $\varphi_t(w)$ is $\mathfrak{B}_t$-measurable in w for fixed t; and (4) for any t and s,

(12) $\varphi_{s+t}(w)=\varphi_s(w)+\varphi_t(w_s^+)$.

We call φ an **almost additive functional**, if it satisfies (1), (2), (3) and satisfies (12) except for a set of P_μ-measure 0 for any fixed s, t, and μ. Two (almost) additive functionals are **equivalent** if they are equal except on a set of P_μ-measure 0 for any t and μ, and φ is called nonnegative if there exists a nonnegative (almost) additive functional which is equivalent to φ. The concept of continuous (almost) additive functionals can be defined similarly. The function $\alpha=\alpha_t(w)$ of t and w is called a (right continuous and homogeneous) **multiplicative functional** if the following conditions are satisfied: (1′) $0 \leqslant \alpha_t(w) < \infty$; (2′) $\alpha_t(w)$ is †right continuous in t $(t<\zeta)$, the †limit on the left $\alpha_{\zeta-}$ exists, and $\alpha_t=\alpha_{\zeta-}$ for $t \geqslant \zeta$; (3′) $\alpha_t(w)$ is $\mathfrak{B}_t$-measurable in w for fixed t; (4′) for any t and s,

(12′) $\alpha_{s+t}(w)=\alpha_s(w)\alpha_t(w_s^+)$.

There exists an one-to-one correspondence between the set of all additive functionals and the set of all multiplicative functionals under the following relation:

(13) $\alpha_t(w)=\exp(-\varphi_t(w))$,

$\varphi_t(w)=-\log\alpha_t(w)$.

Almost multiplicative functionals, equivalence between such functionals, their continuity, and their nonincreasing property are defined as in the case of additive functionals. The multiplicative functional α is continuous (nonincreasing) if and only if the additive functional which corresponds to α by (13) is continuous (nonnegative).

A nonnegative additive functional is uniquely decomposed into the sum of three additive functionals:

$$\varphi_t(w)=\varphi_t^{(1)}(w)+\varphi_t^{(2)}(w)+\varphi_t^{(3)}(w),$$

where $\varphi^{(1)}$ is continuous, and $\varphi^{(2)}$ and $\varphi^{(3)}$ increase only by jumps. Moreover, $\varphi^{(2)}$ and the path of the process $\mathfrak{M}$ have no jump in common, while $\varphi^{(3)}$ increases only at points of discontinuity and is identically zero if and only if the path of $\mathfrak{M}$ is continuous. Even though the path is continuous, $\varphi^{(2)}=0$ does not necessarily hold. In the case of Brownian motion, $\varphi^{(2)}\equiv\varphi^{(3)}\equiv 0$ holds for every nonnegative φ, and therefore φ is always continuous.

For a nonnegative functional φ,

(14) $u_\alpha(x)=E_x\left(\int_0^\infty e^{-\alpha t}\,d\varphi_t\right)$

is α-excessive if it is finite. Let $\mathfrak{M}$ be a Hunt process and $W=\tilde{W}$. A measure μ_0 on $(S, \mathfrak{B}(S))$ is called a **reference measure** if the following condition is satisfied: If any α-excessive function is equal to 0 except on a set of μ_0-measure 0, then it is equal to 0 everywhere. Assume there exists a reference measure for $\mathfrak{M}$. Then an α-excessive function u_α can be expressed in the form (14) by a nonnegative additive functional φ if and only if for every increasing sequence $\{\sigma_n\}$ of Markov times such that $\sigma_n\uparrow\zeta$,

$$E_x(e^{-\alpha\sigma_n}u_\alpha(X_{\sigma_n}))\to 0$$

holds. Moreover, in this case, it is possible to choose such φ with the condition $\varphi^{(3)}\equiv 0$. The correspondence between φ with $\varphi^{(3)}\equiv 0$ and u_α satisfying the above condition is one-to-one. It is possible to choose a continuous φ corresponding to u_α if and only if for every increasing sequence $\{\sigma_n\}$ of Markov times, $E_x(e^{-\alpha\sigma_n}u_\alpha(X_{\sigma_n}))\to E_x(e^{-\alpha\sigma}u_\alpha(X_\sigma))$ holds, where $\sigma=\lim\sigma_n$. In particular, u_α may correspond to a continuous φ if the following two conditions are satisfied: (1) u_α is bounded and $e^{-\alpha t}T_t u_\alpha(x)\to 0$ $(t\to\infty)$, (2) $e^{-\alpha t}T_t u_\alpha(x)\to u_\alpha(x)$ $(t\to 0)$ uniformly in x.

In the previous discussion, mainly nonnegative additive functionals have been considered. However, there are also some works about additive functionals φ such that φ_t is a †supermartingale relative to $\mathfrak{B}_t$. In particular, the φ with mean 0 ($E_x(\varphi_t)\equiv 0$) is important in applications to the transformation of Markov processes. Additive functionals are also applied to the probabilistic extension of the †Dirichlet integral and to the extension of †Lévy measure.

Replacing (12) or (12′) by the "associative law,"

$\varphi_t^s(w)+\varphi_u^t(w)=\varphi_u^s(w)$ or

$\alpha_t^s(w)\alpha_u^t(w)=\alpha_u^s(w)$ $(s\leqslant t\leqslant u)$,

we can also define temporally inhomogeneous additive functionals or multiplicative functionals [10, 18, 26, 32, 33].

E. Transformation of Markov Processes

There are several methods by which a given Markov process can be transformed to a new one. Here we shall mention some important transformations. (For the transformation by a †stochastic integral → 46 Brownian Motion; 6 Additive Processes; and 263 Martingales.)

Transformation by a Multiplicative Functional. For a Markov process $\mathfrak{M}$, let α be a multiplicative functional such that $E_x(\alpha_t) \leqslant 1$ and $P_x(\alpha_0 = 1) = 1$. Set

$P^\alpha(t,x,\Gamma) = E_x(\alpha_t \chi_\Gamma(X_t))$ $(\Gamma \in \mathfrak{B}(S))$ and

$P^\alpha(t,x,\{\partial\}) = 1 - P(t,x,S)$.

Then $P^\alpha(t,x,E)$ is a transition probability on S and corresponds to a Markov process $\mathfrak{M}^\alpha = (S, W^\alpha, P_x, x \in \bar{S})$. We call $\mathfrak{M}^\alpha$ a **transformation** of $\mathfrak{M}$ **by a multiplicative functional** α. It is possible to choose $W^\alpha = \tilde{W}$. If $\mathfrak{M}$ is a strong Markov process, so is $\mathfrak{M}^\alpha$. For an almost multiplicative functional α, this assertion also holds if α satisfies the following condition: For any Markov time σ and $\mathfrak{B}$-measurable τ such that $\sigma \leqslant \tau$, $\alpha_\tau(w) = \alpha_\sigma(w)\alpha_{\tau-\sigma}(w_\sigma^+)$ holds except on a set of P_ν-measure 0, where ν is any bounded measure on S. Conversely, let $\mathfrak{M}$ and $\mathfrak{M}'$ be two Markov processes with the same path space $\tilde{W}$ on the same state space S. If the probability law P_x' of $\mathfrak{M}'$ is absolutely continuous relative to P_x of $\mathfrak{M}$, then $\mathfrak{M}'$ is a transformation of $\mathfrak{M}$ by a certain almost multiplicative functional of $\mathfrak{M}$. This transformation includes killing, transformation by drift, and superharmonic transformation as special cases.

(1) Killing. Transformation by a multiplicative functional α is called **killing** if $0 \leqslant \alpha_t \leqslant 1$ holds. In fact, $\mathfrak{M}^\alpha$ can be constructed as follows: A particle going along a path w of $\mathfrak{M}$ is "killed" (jumped to ∂) in such a way that its surviving probability up to time t is $\alpha_t(w)$. For a nonnegative bounded continuous function $C(x)$ on S, let

$$\alpha_t(w) = \exp\left(-\int_0^t c(X_s(w))\,ds\right).$$

Then α satisfies the condition given previously. If the semigroup $\{T_t\}$ of $\mathfrak{M}$ is regarded as a set of operators on $C(S)$ and has the generator $\mathfrak{G}$, then the semigroup $\{T_t\}$ of $\mathfrak{M}^\alpha$ is also regarded as a semigroup of operators on $C(S)$, its generator $\mathfrak{G}^\alpha$ has the same domain as $\mathfrak{G}$, and $\mathfrak{G}^\alpha = \mathfrak{G} - c$ [19].

(2) Transformation by drift. Let φ be a continuous additive functional such that $E_x(\varphi_t) = 0$ and $E_x(\varphi_t^2) < \infty$, and let ψ be a nonnegative continuous additive functional determined by the relation $E_x(\varphi_t^2) = E_x(\psi_t)$ for every t and every x. Then $\alpha_t(w) = \exp(\varphi_t - \psi_t/2)$ is a multiplicative functional. The transformation determined by α is called a **transformation by drift**. Let $\mathfrak{M}$ be N-dimensional Brownian motion and $b_1, \ldots, b_N$ be bounded measurable functions on S. Setting

$$\varphi_t = \int_0^t \sum_{i=1}^N b_i(X_t)\,dX_i(t)$$

([†]stochastic integral)

and

$$\psi_t = \int_0^t \sum_{i=1}^N b_i^2(X_t)\,dt$$

in the above formula, we obtain a transformation by drift. Moreover, if $b_1, \ldots, b_N$ are in $C_\infty(S)$, then the semigroup of $\mathfrak{M}^\alpha$ maps $C_\infty(S)$ into $C_\infty(S)$, and for a bounded function f with bounded continuous derivatives up to the second order, f is in the domain of $\mathfrak{G}^\alpha$ and $\mathfrak{G}^\alpha f = (1/2)\Delta f + \sum_{i=1}^N b_i(\partial f/\partial x_i)$ [9, 27].

(3) Superharmonic transformation. Let u be an excessive function of $\mathfrak{M}$ and $A = \{x \mid 0 < u(x) < \infty\}$. Set

$$\alpha_t(w) = u(X_t(w))/u(X_0(w)) \quad \text{if} \quad X_0(w) \in A$$
$$= 0 \quad \text{if} \quad X_0(w) \notin A.$$

Then α_t is a multiplicative functional. The transformation defined by α, first introduced by Doob, is called a **superharmonic transformation**. The transition probability $P(t,x,\Gamma)$ of $\mathfrak{M}^\alpha$ is equal to $u(x)^{-1}\int_\Gamma P(t,x,dy)u(y)$ if $x \in A$, 0 if $x \notin A$ and $t > 0$, and $\delta_x(\Gamma)$ if $x \notin A$ and $t = 0$, for Γ in $\mathfrak{B}(S)$. In particular, if $\mathfrak{M}$ is a Hunt process, $C(S)$ is invariant under the Green operator of $\mathfrak{M}$, and u is a continuous function such that $0 < c \leqslant u \leqslant k < \infty$, then $C(S)$ is also invariant under the Green operator of $\mathfrak{M}^\alpha$ and $\mathfrak{G}^\alpha f = u^{-1}\mathfrak{G}(uf)$, where the domain of $\mathfrak{G}^\alpha$ is the set of f for which uf is in the domain of $\mathfrak{G}$.

We understand the term **time change** in a broad sense, including the following two important special cases.

(4) Time change by an additive functional. Let $\mathfrak{M}$ be a Hunt process and φ be a nonnegative continuous additive functional such that $P_x(\varphi_0 = 0) = 1$. Set $S^* = \{x \mid P_x(\varphi_t(w) > 0$ for every $t > 0) = 1\}$ and assume that S^* is locally compact. Let $\tilde{W}^*$ be the set of all right continuous functions on S^*. They have discontinuities of at most the first kind, and $\mathfrak{B}^*$ is the σ-algebra on $\tilde{W}^*$ generated by all [†]Borel cylinder sets. Set $P_x^*(B) = P_x(X_{\varphi_t^{-1}(w)}(w) \in B)$ $(B \in \mathfrak{B}^*)$, where φ_t^{-1} is a right continuous inverse function of φ_t. Then $\mathfrak{M}^* = \{S^*, \tilde{W}^*, P_x^*, x \in \bar{S}^*\}$ is a Hunt process on S^*, and it is said that M^* is obtained by time change from $\mathfrak{M}$ by φ. Roughly speaking, $\mathfrak{M}^*$ may be considered as a Markov process with path $X_t^*(w) = X_{\varphi_t^{-1}(w)}(w)$. The Green's operator of $\mathfrak{M}^*$ is given by $E_x(\int_0^\infty e^{-\lambda\varphi_t} f(X_t)\,d\varphi_t)$. Suppose, in particular, that the Green's operator of $\mathfrak{M}$ maps $C(S)$ into $C(S)$ and that $a(x)$ is a continuous function on S such that $0 < c \leqslant a(x) \leqslant k < \infty$, and set $\varphi_t(w) = \int_0^t a(X_s(w))\,ds$. Then $S^* = S$, $\tilde{W}^* = \tilde{W}$, and $\mathfrak{B}^* = \mathfrak{B}$, and the Green's operator of $\mathfrak{M}^*$ also maps $C(S)$ into $C(S)$. The domain of the generator $\mathfrak{G}^*$ of $\mathfrak{M}^*$

coincides with that of $\mathfrak{G}$ of $\mathfrak{M}$ and $\mathfrak{G}^*f = a^{-1}\mathfrak{G}f$. Let $\mathfrak{M}$ and $\mathfrak{M}'$ be Markov processes with the same state space and the same path space $\tilde{W}$. If they have the same hitting probabilities $\{H_k(x,\cdot)\}$, then each one of them can be transformed from the other by time change by a strictly increasing additive functional. The converse is also true [2].

(5) Subordination. This concept, first introduced by Bochner, was extended as follows: Let $e^{-\psi(\lambda)}$ be the †Laplace transform of an †infinitely divisible distribution with support $[0,\infty)$, and let $F_t(\cdot)$ be the distribution with Laplace transform $e^{-t\psi(\lambda)}$. Let $\{T_t\}$ be a Hille-Yosida semigroup on a certain †Banach space, and set $T_t^{\psi} = \int_0^{\infty} T_s F_t(ds)$ (†Bochner integral). Then $\{T_t^{\psi}\}$ is also a semigroup and is called the **subordination** of $\{T_t\}$ by ψ. If $\mathfrak{G}$ is a generator of $\{T_t\}$, then $-\psi(-\mathfrak{G})$ is a generator of $\{T_t^{\psi}\}$.

In particular, we may assume $\{T_t\}$ to be a nonnegative semigroup on $C(S)$ $(C_\infty(S))$ such that $T_t 1 = 1$, and $\{X_t\}$ to be a Markov process corresponding to the semigroup T_t. Let $\{\Psi(t)\}$ be an additive process that is independent of $\{X_t\}$ and satisfies $E(e^{-\lambda\Psi(t)}) = e^{-t\psi(\lambda)}$. Set $Y_t(\omega) = X_{\Psi(t,\omega)}(\omega)$; then $\{Y_t\}$ is a Markov process corresponding to the semigroup $\{T_t^{\psi}\}$. The operation by which we obtain $\{Y_t\}$ from $\{X_t\}$ by using $\{\Psi(t)\}$ is also called **subordination**. In particular, if $\{\Psi(t)\}$ is a one-sided stable process of the αth order (→ 6 Additive Processes), this operation is called **subordination of the αth order**. If $\{X_t\}$ is an additive process, then the process obtained from it by subordination is also an additive process. The subordination of the αth order of Brownian motion gives a †symmetric stable process of the 2αth order. Let $\{\psi_1(t)\}$ and $\{\psi_2(t)\}$ be independent of $\{X_t\}$. Then the superposition of two subordinations of $\{X_t\}$ by $\psi_1(t)$ and $\psi_2(t)$ coincides with subordination by $\{\Psi_1(\Psi_2(t,\omega),\omega)\}$ [3, 28].

(6) Reversed processes. Let $\{X_t\}_{t\in T}$ be a Markov process on $(\Omega,\mathfrak{B},P)$ and $X_t^* = X_{-t}$ for $t \in T^* = \{t \mid -t \in T\}$. Then $\{X_t^*\}_{t\in T^*}$ is a Markov process and is called a **reversed process** of $\{X_t\}$. If the state space of S is countable, then the transition probability $P(s,x,t,y)$ of $\{X_t\}$ and $P^*(s,x,t,y)$ of $\{X_t^*\}$ satisfy the following condition: $P^*(s,x,t,y) = Q(-t,y)P(-t,y,-s,x)Q(-s,x)^{-1}$, where $Q(t,x) = P(X_t(\omega)=x)$ and we assume $Q(t,x) \neq 0$.

For a given temporally homogeneous Markov process $\{X_t\}$ with state space S, let $\{T_t\}$ be the semigroup corresponding to $\{X_t\}$. Then a †σ-finite measure m on $(S,\mathfrak{B}(S))$ is called a **subinvariant measure** (or **excessive measure**) if the inequality $\int_S T_t f(x)m(dx) \leqslant \int_S f(x)m(dx)$ holds for every nonnegative function f with compact support. The measure m is called an **invariant measure** if the equality holds instead. Let $\{X_t^*\}$ be another Markov process whose state space is S and whose semigroup is $\{T_t^*\}$. If for some subinvariant measure m of $\{X_t\}$ and for every pair of f, g in $C_0(S)$ the equality $\int_S T_t f(x)g(x)m(dx) = \int_S f(x)T_t^* g(x)m(dx)$ holds, then $\{X_t^*\}$ is called the **adjoint process** of $\{X_t\}$. In fact, $\{X_t\}$ is the adjoint process of $\{X_t^*\}$ under the previous condition, since m is also a subinvariant measure of $\{X_t^*\}$.

The concept of reversed process is related to that of adjoint process. For example, if $\{X_t\}$ has an invariant probability measure m and if the distribution of X_0 is m, then the distribution of X_t is also m for every t, and the reversed process of $\{X_t\}$ coincides with the adjoint process [22].

References

[1] S. Bernstein, Equations différentielles stochastiques, Actualités Sci. Ind., Hermann, 1938, p. 5–31.

[2] R. M. Blumenthal, R. K. Getoor, and H. P. McKean Jr., Markov processes with identical hitting distributions, Illinois J. Math., 6 (1962), 402–420.

[3] S. Bochner, Harmonic analysis and the theory of probability, Univ. of California Press, 1955.

[4] J. L. Doob, Stochastic processes, John Wiley, 1953.

[5] J. L. Doob, Semimartingales and subharmonic functions, Trans. Amer. Math. Soc., 77 (1954), 86–121.

[6] Е. Б. Дынкин (E. B. Dynkin), Критерин непрерывности и отсутсвия разрывов второго рода для траекторий Марковского случайного процесса, Izv. Acad. Nauk SSSR, 16 (1952), 563–572.

[7] Е. Б. Дынкин (E. B. Dynkin), Марковские процессы и полугруппы операторов, Teor. Verojatnost. i Primenen., 1 (1956), 25–37; English translation, Infinitesimal operators of Markov processes, Theory of Prob. Appl., 1 (1956), 34–54.

[8] Е. Б. Дынкин (E. B. Dynkin), Основания теории Марковских процессов, Физматгиз, 1959; English translation, The theory of Markov processes, Pergamon, 1960.

[9] E. B. Dynkin, Transformations of Markov processes connected with additive functionals, Proc. 4th Berkeley Symp. Math. Stat. Prob., Univ. of California Press (1961), vol. II, p. 117–142.

[10] Е. Б. Дынкин (E. B. Dynkin), Марковские процессы, Физматгиз, 1963; English translation, Markov processes I, II, Springer, 1965.
[11] W. Feller, Zur Theorie der stochastichen Prozesse, Math. Ann., 113 (1936), 113–160.
[12] W. Feller, The parabolic differential equations and the associated semi-groups of transformations, Ann. of Math., (2) 55 (1952), 468–519.
[13] W. Feller, On second order differential operators, Ann. of Math., (2) 61 (1955), 90-105.
[14] W. Feller, Boundaries induced by non-negative matrices, Trans. Amer. Math. Soc., 83 (1956), 19–54.
[15] W. Feller, On boundaries and lateral conditions for the Kolmogorov differential equations, Ann. of Math., (2) 65 (1957), 527–570.
[16] G. A. Hunt, Markov processes and potentials, Illinois J. Math., 1 (1957), 44–93, 316–369; 2 (1958), 151–213.
[17] K. Itô, On stochastic differential equations, Mem. Amer. Math. Soc., 1951.
[18] K. Itô and H. P. McKean, Jr., Diffusion processes and their sample paths, Springer, 1965.
[19] M. Kac, On some connections between probability theory and differential and integral equations, Proc. 2nd Berkeley Symp. Math. Stat. Prob., Univ. of California Press (1951), p. 189–215.
[20] J. G. Kemeny and J. L. Snell, Potentials for denumerable Markov chains, J. Math. Anal. Appl., 3 (1961), 196–260.
[21] A. N. Kolmogorov, Über die analytischen Methoden in der Wahrscheinlichkeitsrechnung, Math. Ann., 104 (1931), 415–458.
[22] A. N. Kolmogorov, Zur Theorie der Markoffschen Ketten, Math. Ann., 112 (1935–1936), 155–160.
[23] P. Lévy, Théorie de l'addition des variables aléatoires, Gauthier-Villars, 1937.
[24] P. Lévy, Processus stochastiques et mouvement brownien, Gauthier-Villars, 1948.
[25] G. Maruyama and H. Tanaka, Ergodic property of N-dimensional recurrent Markov processes, Mem. Fac. Sci. Kyushu Univ., (A) 13 (1959), 157–172.
[26] P.-A. Meyer, Fonctionelles multiplicatives et additives de Markov, Ann. Inst. Fourier, 12 (1962), 125–230.
[27] M. Motoo, Diffusion process corresponding to $(1/2)\Sigma\partial^2/\partial x^{i2}+\Sigma b^i(x)\partial/\partial x^i$, Ann. Inst. Stat. Math., 12 (1960), 37–61.
[28] E. Nelson, A functional calculus using singular Laplace integrals, Trans. Amer. Math. Soc., 88 (1958), 400–413.
[29] Ю. В. Прохоров (Ju. V. Prohorov), Сходимость случайных процессов и предельные теоремы теории вероятностей, Teor. Verojatnost. i Primenen., 1 (1956), 177–238; English translation, Convergence of random processes and limit theorems in probability theory, Theory of Prob. Appl., 1 (1956), 157–214.
[30] D. Ray, Stationary Markov processes with continuous paths, Trans. Amer. Math. Soc., 82 (1956), 452–493.
[31] Л. В. Серегин (L. V. Seregin), Условия непрерывности вероятностных процессов, Teor. Verojatnost. i Primenen., 6 (1961), 3–30; English translation, Continuity conditions for stochastic processes, Theory of Prob. Appl., 6 (1961), 1–26.
[32] H. Tanaka, Note on continuous additive functionals of the 1-dimensional Brownian path, Z. Wahrscheinlichkeitstheorie, 1 (1963), 251–257.
[33] В. А. Волконский (V. A. Volkonskiĭ), Аддитивные функционалы от Марковских процессов, Trudy Moskov. Obšč., 9 (1960), 143–189.
[34] R. M. Blumenthal and R. K. Getoor, Markov processes and potential theory, Academic Press, 1968.

263 (XVI.12) Martingales

A. General Remarks

Let $(\Omega, \mathfrak{B}, P)$ be a †probability space and T a time parameter set. For each $t \in T$, let $\mathfrak{F}_t$ be a †σ-algebra such that $\mathfrak{F}_s \subset \mathfrak{F}_t \subset \mathfrak{B}$ $(s<t)$. Without loss of generality we may assume that the probability space $(\Omega, \mathfrak{B}, P)$ is †complete and each $\mathfrak{F}_t$ contains all measurable subsets of Ω with P-measure zero. A real-valued †stochastic process $\{X_t\}_{t\in T}$ on $(\Omega, \mathfrak{B}, P)$ (which is also denoted by $(X_t, t\in T)$) is called a **martingale** with respect to $\mathfrak{F}_t$ provided that (i) X_t is $\mathfrak{F}_t$-measurable and $E(|X_t|)<\infty$; and (ii) if $s<t$, then

$$E(X_t \mid \mathfrak{F}_s) = X_s \quad \text{(a.s.)} \tag{1}$$

where (a.s.) means †almost surely, which will be omitted when there is no room for confusion. If the equality in (1) is replaced by the inequality $\leqslant$ ($\geqslant$), $\{X_t\}_{t\in T}$ is called a **supermartingale** (**submartingale**). For the case of martingales the values of X_t may be complex numbers. We write martingale as (M) and submartingale as (SM), for short. Since the definition of (SM) depends on a family of σ-algebras $\{\mathfrak{F}_t\}_{t\in T}$, it is sometimes convenient to denote an (SM) with respect to $\mathfrak{F}_t$ by $(X_t, \mathfrak{F}_t, t\in T)$. If $(X_t, \mathfrak{F}_t, t\in T)$ is an (SM), then $(X_t, \mathfrak{B}_t, t\in T)$ is also an (SM), where $\mathfrak{B}_t = \mathfrak{B}(X_u, u\in T\cap(-\infty, t])$. When the family of σ-algebras involved in the definition of an (SM) $\{X_t\}_{t\in T}$ is not explicitly mentioned, we understand that $\{X_t\}_{t\in T}$ is an (SM) with respect to $\mathfrak{B}_t$. This convention is used for an (M) also. The term martingale is due to J. Ville. P. Lévy had already made use of the concept in his work, and J. L. Doob introduced the concept of (SM) in addition to that of (M) and developed a systematic theory of martingales. These concepts are widely used in the theory of †stochastic processes.

As can be observed from the definition, super- and submartingales have some properties similar to those of super- and subharmonic functions. We state some simple properties of (SM): (1) For any given (SM) X_t, $m(t) = E(X_t)$ is an increasing function of t, and X_t is an (M) if and only if $m(t) =$ constant. (2) Let X_t and Y_t be (SM). Then $aX_t + bY_t$ $(a, b\geqslant 0)$ and $\sup(X_t, Y_t)$ are (SM). (3) Let X_t be an (SM) and $f(x)$ an increasing †convex function defined in $(-\infty, \infty)$. For any given $t_0\in T$, if $E(|f(X_{t_0})|)<\infty$, then $(f(X_t), t\in(-\infty, t_0]\cap T)$ is an (SM) and furthermore, when X_t is an (M), $(f(X_t), t\in (-\infty, t_0]\cap T)$ is an (SM) even if $f(x)$ is not increasing. In particular, $X_t^+ = \sup(X_t, 0)$ is an (SM) if X_t is an (SM), and $|X_t|$ is an (SM) if X_t is an (M). (4) If $a, b\in T$ and $a<b$, then $E(|X_t|)\leqslant 2E(|X_b|) - E(X_a)$. (5) If $X_t\geqslant 0$ $(t\in T)$ and $t_1\in T$, then the family of †random variables $\{X_t, t\in(-\infty, t_1]\cap T\}$ is uniformly integrable (see (6)). (6) If $t_n\in T$ and $t_n\downarrow$, then $\{X_{t_n}\}$ is uniformly integrable if and only if $\lim_{n\to\infty} E(X_{t_n}) > -\infty$. Here a family of random variables $\{X_t\}_{t\in T}$ is said to be **uniformly integrable** if we have

$$\lim_{n\to\infty} \sup_t \int_{A_{n,t}} |X_t(\omega)|\, dP(\omega) = 0, \tag{2}$$

where $A_{n,t} = \{\omega \mid |X_t(\omega)| > n\}$.

Example 1. For any sequence of random variables $Y_1, Y_2, \ldots,$ if the relations

$$E(Y_{n+1} | Y_1, \ldots, Y_n) \geqslant 0, \quad n = 1, 2, \ldots \tag{3}$$

hold, then

$$X_n = \sum_{\nu=1}^{n} Y_\nu$$

is an (SM). If the inequality sign in (3) is replaced by the equality sign, then X_n is an (M). In particular, if $\{Y_n\}$ is a sequence of independent random variables such that $E(Y_n) = 0$, then

$$X_n = \sum_{\nu=1}^{n} Y_\nu$$

is an (M).

B. Convergence Theorems

The following results on the †sample sequences of an (SM) are very useful. Let $(X_j, 1\leqslant j\leqslant n)$ be an (SM). Then

$$P\Big(\max_{1\leqslant j\leqslant n} |X_j| > \lambda\Big) \leqslant 2E(|X_n|) - E(X_1) \qquad (\lambda > 0).$$

For any interval $I = [a, b]$ and any sequence $\{x_j\}$ $(1\leqslant j\leqslant n)$ of real numbers, define the **upcrossing number** $N(I)$ of I by $\{x_j\}$ as follows: When $x_j < a$ or $x_j > b$ for all j, $N(I) = 0$; otherwise, $N(I)$ is the number of j such that the sequence $x_{j-1} < a$ and $x_j > b$ $(j = 2, \ldots, n)$. If $N(I)$ is the upcrossing number of I by a sample sequence $\{X_j(\omega)\}$ $(1\leqslant j\leqslant n)$, then $E(N(I)) \leqslant (E(|x_n| + |a|))/(b-a)$ (**Doob's inequality**).

Using these inequalities, we have the following **convergence theorems**: (i) Let $(X_n, 1\leqslant n<\infty)$ be an (SM). (a) If $\sup_n E(|x_n|)<\infty$, then $\lim_{n\to\infty} X_n = X_\infty$ exists with probability 1 and $E(|X_\infty|)<\infty$. (b) Furthermore, if $\{X_n, 1\leqslant n<\infty\}$ is uniformly integrable, then $\lim_{n\to\infty} X_n = X_\infty$ exists with probability 1 by (a), and $(X_n, 1\leqslant n\leqslant\infty)$ is also an (SM). (c) If X_n is an (SM) such that $\{E(|X_n|)\}$ is bounded, then $\lim_{n\to\infty} X_n$ exists, and if $(X_n, 1\leqslant n\leqslant \infty)$ is an (SM), then $\lim_{n\to\infty} E(X_n)\leqslant E(X_\infty)$, where the equality holds if and only if $\{X_n, 1\leqslant n<\infty\}$ is uniformly integrable. (ii) If $(X_n, -\infty<n\leqslant -1)$ is an (SM), then $\lim_{n\to-\infty} X_n = X_{-\infty}$ exists and $-\infty\leqslant X_{-\infty}<\infty$. Furthermore, if $E(X_{-\infty})>-\infty$, then $-\infty<X_{-\infty}<\infty$ and $(X_n, -\infty\leqslant n\leqslant -1)$ is a uniformly integrable (SM). (iii) Let $(X_1, X_2, \ldots, Z)$ be an (SM). (a) $\lim_{n\to\infty} X_n = X_\infty$ exists and $\lim_{n\to\infty} E(X_n)\leqslant E(X_\infty)\leqslant E(Z)$. (b) $\lim_{n\to\infty} E(X_n) = E(X_\infty)$ if and only if $\{X_n, 1\leqslant n<\infty\}$ is uniformly integrable, and in this case $(X_1, X_2, \ldots, X_\infty, Z)$ is an (SM). (iv) Let $\{\mathfrak{F}_n\}$ $(-\infty<n<\infty$ and $\mathfrak{F}_n\subset\mathfrak{F}_{n+1}\subset\mathfrak{B})$ be a sequence of σ-algebras on $(\Omega, \mathfrak{B}, P)$. Put $\mathfrak{F}_{-\infty} = \bigcap_n \mathfrak{F}_n$ and $\mathfrak{F}_{+\infty} = \bigvee_n \mathfrak{F}_n$ (the smallest σ-algebra containing all $\mathfrak{F}_n$). If Z is a random variable with $E(|Z|)<\infty$, then $\lim_{n\to\pm\infty} E(Z|\mathfrak{F}_n) = E(Z|\mathfrak{F}_{\pm\infty})$ (a.s.). We mention some applications of these convergence theorems.

Example 2. Let $(\Omega, \mathfrak{B}, P)$ be a probability space and $\{\pi_n\}$ $(n = 1, 2, \ldots)$ a sequence of partitions of Ω into $\mathfrak{B}$-measurable sets with positive P-measure such that for each n, π_{n+1}

is finer than π_n. Let π_n be $\{M_1^{(n)}, M_2^{(n)}, \dots\}$, denote the smallest σ-algebra containing $\{M_j^{(n)}\}_{j=1,2,\dots}$ by $\mathfrak{F}_n$ for each n, and set $\mathfrak{F}_\infty = \bigvee_n \mathfrak{F}_n$. For a given †completely additive set function φ on $(\Omega, \mathfrak{F}_\infty)$, if we define $X_n(\omega)$ by $X_n(\omega) = \varphi(M_j^{(n)})/P(M_j^{(n)})$ for $\omega \in M_j^{(n)}$, $j = 1, 2, \dots$, then $(X_n, \mathfrak{F}_n, 1 \leqslant n < \infty)$ is an (M) and $\lim_{n\to\infty} X_n = X_\infty$ exists. If $\hat{P}$ is the restriction of P to $\mathfrak{F}_\infty$, then φ is †absolutely continuous with respect to $\hat{P}$ if and only if $\{X_n, 1 \leqslant n < \infty\}$ is uniformly integrable, and in this case, $X_\infty = d\varphi/d\hat{P}$ with $\hat{P}$-measure 1. If φ is †singular with respect to $\hat{P}$, then $X_\infty = 0$ with $\hat{P}$-measure 1.

Example 3. Let $X_1, X_2, \dots$ be any sequence of random variables and Z an integrable random variable that is measurable with respect to $\mathfrak{B}(X_1, X_2, \dots)$. Then $\lim_{n\to\infty} E(Z \mid X_1, X_2, \dots, X_n) = Z$ (a.s.). In particular, if $X_1, X_2, \dots$ are independent and Z is $\mathfrak{B}(X_n, X_{n+1}, \dots)$-measurable for every n, then Z is equal to a constant almost surely. This is the so-called †Kolmogorov zero-one law.

Example 4. If $\{X_t\}$ $(0 \leqslant t < \infty)$ is a †Brownian motion, then both X_t and $X_t^2 - t$ are (M). Conversely, if a stochastic process with continuous sample functions is such that X_t and $X_t^2 - t$ are (M), then it is a Brownian motion.

C. Sample Functions

Let $(X_t, \mathfrak{F}_t, t \in T)$ be an (SM). Here the parameter set T may be an arbitrary subset of $(-\infty, \infty)$. However, we can always find an interval $I \supset T$ and for each $t \in I$ a σ-algebra $\tilde{\mathfrak{F}}_t$ and a random variable $\tilde{X}_t$ such that $(\tilde{X}_t, \tilde{\mathfrak{F}}_t, t \in I)$ is an (SM) and $P(\tilde{X}_t = X_t) = 1$, $\tilde{\mathfrak{F}}_t = \mathfrak{F}_t$ for every $t \in T$. Therefore we may assume without loss of generality that the parameter set T is an interval. Furthermore, we assume that the stochastic process $\{X_t\}_{t\in T}$ is †separable. Using inequalities and convergence theorems for the sample sequences of (SM) with discrete parameters, we obtain the following properties of the †sample functions of (SM) with continuous parameters: (i) The sample function of an (SM) $\{X_t\}$ is bounded on every finite interval $[a,b] \subset T$ with probability 1. (ii) Let T_0 be the interior of T. Then $P(X_{t+0}$ and X_{t-0} exist for all $t \in T_0) = 1$, and for each $t \in T_0$, $\lim_{s\uparrow t} E(X_s) \leqslant E(X_{t-0}) \leqslant E(X_t) \leqslant E(X_{t+0}) \leqslant \lim_{s\downarrow t} E(X_s)$. (iii) Let D be the set of fixed discontinuity points of $\{X_t\}$ (t is called a †fixed discontinuity point of $\{X_t\}$ if $P(X_{t-0} = X_t = X_{t+0}) \neq 1$). Then D is an at most countable set.

We assume for simplicity that the parameter set is $\mathbf{R}^+ = [0, \infty)$ and the sample functions of $\{X_t\}$ are right continuous with probability 1. In this case, if $(X_t, \mathfrak{F}_t, t \in \mathbf{R}^+)$ is an (SM), then $(X_t, \mathfrak{F}_{t+}, t \in \mathbf{R}^+)$ is an (SM), where $\mathfrak{F}_{t+} = \bigcap_{s>t} \mathfrak{F}_s$. Therefore we may assume that $\mathfrak{F}_t = \mathfrak{F}_{t+}$ for all $t \in \mathbf{R}^+$. Denote by $\mathfrak{T}$ the collection of all stopping times with respect to $\mathfrak{F}_t$ (a random variable $\tau(\omega)$ is called a †stopping time with respect to $\mathfrak{F}_t$ provided that $P(0 \leqslant \tau \leqslant \infty) = 1$ and $\{\tau \leqslant t\} \in \mathfrak{F}_t$ for all $t > 0$). For any $\tau \in \mathfrak{T}$, the set of all measurable subsets A of Ω such that $A \cap \{\tau \leqslant t\} \in \mathfrak{F}_t$ for all $t > 0$ is a σ-algebra on Ω, which is denoted by $\mathfrak{F}_\tau$. If $\tau, \sigma \in \mathfrak{T}$ and $\tau \leqslant \sigma$, then $\mathfrak{F}_\tau \subset \mathfrak{F}_\sigma$, and if $P(\tau < \infty) = 1$, the random variable X_τ is $\mathfrak{F}_\tau$-measurable. Let A be an interval and $\{\tau_\alpha\}_{\alpha\in A}$ a family of stopping times such that $\tau_\alpha \leqslant \tau_\beta < \infty$ whenever $\alpha < \beta$. Put $X_\alpha^* = X_{\tau_\alpha}$ and $\mathfrak{F}_\alpha^* = \mathfrak{F}_{\tau_\alpha}$. Then $(X_\alpha^*, \mathfrak{F}_\alpha^*, \alpha \in A)$ is called the stochastic process obtained by an **optional sampling** from $(X_t, \mathfrak{F}_t, t \in T)$. Suppose that one of the following conditions is satisfied: (1) $X_t \leqslant 0$ (a.s.) for all $t \in \mathbf{R}^+$; (2) $\{X_t, t \in \mathbf{R}^+\}$ is uniformly integrable; (3) for each $\alpha \in A$, τ_α is bounded with probability 1. Then $(X_\alpha^*, \mathfrak{F}_\alpha^*, \alpha \in A)$ is also an (SM). If (2) or (3) is satisfied, $E(X_\alpha^*) \leqslant \sup E(X_t)$.

D. Decompositions of Submartingales (→[3,4])

If an (SM) $(X_t, \mathfrak{F}_t, t \in \mathbf{R}^+)$ is uniformly integrable, $\lim_{t\to\infty} X_t = X_\infty$ exists and $(X_t, \mathfrak{F}_t, t \in [0, \infty])$ is an (SM) for which $\mathfrak{F}_\infty = \bigvee_t \mathfrak{F}_t$. In addition, when the family $\{X_\tau \mid \tau \in \mathfrak{T}\}$ is uniformly integrable, it is said to belong to class (D). If for each a $(0 < a < \infty)$ the family $\{X_\tau \mid \tau \in \mathfrak{T}$ and $\tau \leqslant a\}$ is uniformly integrable, the family is said to belong locally to class (D).

If an (SM) X_t is uniformly integrable, X_t is decomposed as

$$X_t = E(X_\infty \mid \mathfrak{F}_t) - (E(X_\infty \mid \mathfrak{F}_t) - X_t), \tag{4}$$

and if we take appropriate †versions of conditional expectations, $E(X_\infty \mid \mathfrak{F}_t)$ becomes a right continuous martingale. In the decomposition (3), $M_t = E(X_\infty \mid \mathfrak{F}_t)$ is an (M) and $Z_t = E(X_\infty \mid \mathfrak{F}_t) - X_t$ is a potential, i.e., a nonnegative right continuous supermartingale with $\lim_{t\to\infty} Z_t = 0$ (a.s.). The decomposition (4) is called the **Riesz decomposition** of the (SM) X_t; the names "potential" and "Riesz decomposition" come from †potential theory in view of the obvious similarity.

A stochastic process $(A_t, t \in \mathbf{R}^+)$ on $(\Omega, \mathfrak{B}, P)$ is called a (right continuous) **increasing process** provided that (i) A_t is $\mathfrak{F}_t$-measurable, and (ii) with probability 1, the sample function is a right continuous and increasing function with $A_0 = 0$. If $E(A_\infty) < \infty$, where

$A_\infty = \lim_{t\to\infty} A_t$, the stochastic process is said to be **integrable**. We have the following relations between increasing processes and (SM): (i) A potential X_t is decomposed into

$$X_t = E(A_\infty | \mathfrak{F}_t) - A_t \qquad (5)$$

(**Doob-Meyer decomposition**) by a suitably chosen integrable increasing process A_t if and only if X_t belongs to class (D). (ii) In (i), A_t can be chosen as a continuous process if and only if for any sequence $\tau_n \in \mathfrak{T}$ such that $\tau_n \uparrow \tau$, $\lim_{n\to\infty} E(X_{\tau_n}) = E(X_\tau)$. (iii) An (SM) X_t is decomposed into $X_t = X_t' + A_t$, where X_t' is an (M) and A_t is an increasing process, if and only if X_t belongs locally to class (D). If we consider only natural increasing processes, these decompositions are unique. Here an increasing process A_t is said to be **natural** if for every bounded right continuous martingale Y_t and every constant a $(0 < a < \infty)$,

$$E\Big(\sum_{t \leqslant a} (A_t - A_{t-})(Y_t - Y_{t-})\Big) = 0,$$

where

$$\sum_{t \leqslant a} (A_t - A_{t-})(Y_t - Y_{t-})$$

is the limit of

$$S_n = \sum_{\substack{|Y_t - Y_{t-0}| \geqslant 1/n \\ t \leqslant a}} (A_t - A_{t-0})(Y_t - Y_{t-0})$$

in the sense of †L_1. These decompositions are used in studying †additive functionals for Markov processes, †stochastic integrals, and so on.

References

[1] J. L. Doob, Stochastic processes, John Wiley, 1953.
[2] J. L. Doob, Semimartingales and subharmonic functions, Trans. Amer. Math. Soc., 77 (1954), 86–121.
[3] P. A. Meyer, A decomposition theorem for supermartingales, Illinois J. Math., 6 (1962), 193–205.
[4] P. A. Meyer, Decomposition of supermartingales: the uniqueness theorem, Illinois J. Math., 7 (1963), 1–17.
[5] P. A. Meyer, Probability and potentials, Blaisdell, 1966.
[6] D. L. Burkholder, Martingale transform, Ann. Math. Stat., 37 (1966), 1494–1504.
[7] J. Neveu, Martingales à temps discret, Masson et Cie, 1972.
[8] A. M. Garsia, Martingale inequalities, Benjamin, 1973.

264 (XVIII.1) Mathematical Programming

Mathematical programming in its narrow sense means the mathematical theory of programming, including †linear programming, †quadratic programming, †nonlinear programming (→ 288 Nonlinear Programming), and †dynamic programming, as well as **stochastic programming**. However, mathematical programming is sometimes understood to be equivalent to mathematics for programming, in which †queuing theory, †game theory, and †inventory control theory (→ 227 Inventory Control) are included. The stimulus and foundation for the development of mathematics of programming were given by †operations research and the theory of economic planning (→ 134 Econometrics).

Since the 1940s scientists have realized the necessity of establishing an adequate mathematical model in order to discuss the properties of the decision process in controlling a system. When no human means are available to change the structure of a system, we recognize such a structure to be natural and dismiss it as being beyond the applicability of a program. On the other hand, when we apply a program to some occurrence, we assume that there exists a certain regularity to the effect of such an action.

The variables appearing in a model are classified into the following types: (1) data or **external variables** (or **planning coefficients**), (2) **target variables** (or **object variables**), (3) **strategic variables** (or **planning variables**). The type of model is determined by the relations existing among these variables. The **equations of structure** describe the static side of the relations, whereas the **equations of motion** describe the relations in reference to a time change. In actual programming, the variables may take values only within certain domains —they must satisfy certain **boundary conditions**. Such boundary conditions are frequently given by a set of inequalities. It is in the interior of the domain defined by such a set of inequalities that the equations of structure or the equations of motion are applicable.

For instance, we may consider a typical problem of linear programming as follows. Given n-dimensional row vectors x and c, an $m \times n$ real matrix A, and an m-dimensional column vector b, find the maximum of the inner product $Z = (c, x)$ under the condition that $x \geqslant 0$ and $Ax \leqslant b$. Here x is a strategic variable, A, b, c are external variables, and Z is the object variable. The conditions $x \geqslant 0$ and $Ax \leqslant b$ define the boundary for the admissible domain of x.

Programming is concerned with actions to be realized in the future, and consequently has an intrinsic connection with prediction. It is to be noted that in most cases we have no complete information regarding data, objectives, or structure. Hence some theory of decision making with uncertain information is required. For this purpose the following three typical models have been proposed:

(1) Deterministic models. All variables and parameters in both the equations of structure and boundary conditions are deterministic (e.g., linear programming).

(2) Stochastic models. Some variables and parameters in the equations of structure or boundary conditions are stochastic (e.g., queuing theory, statistical control, etc.).

(3) Game-theoretic models. In some situations the problem of planning may be formulated as a game involving several participants. When action has to be based on incomplete information and the situation cannot be assumed to have a probabilistic character, there is some degree of realism in assuming the existence of a competing opponent. There is a good possibility that application of game theory can provide a methodology for dealing with the uncertainties associated with planning.

Furthermore, a more detailed analysis of the process of planning makes it clear that a participant in a game is not concerned merely with a choice of optimal actions from a prescribed set. He also acts as an explorer in the sense that he obtains new information through his experiences and thereby is able to improve his course of action.

In this manner the theory of planning is connected with the theory of adaptation, learning, and self-organization. It should be noted also that in the actual execution of programming, †simulation techniques are sometimes used to avoid the difficulties of finding analytical formulas and obtaining numerical solutions based on complicated mathematical models. Corresponding to the three types of models, there are three types of simulation: (1) deterministic, (2) probabilistic (†Monte Carlo), and (3) game-theoretic (e.g., business games).

References

[1] N. V. Reinfeld and W. R. Vogel, Mathematical programming, Prentice-Hall, 1958.
[2] S. Karlin, Mathematical methods and theory in games, programming and economics I, II, Addison-Wesley, 1959.
[3] T. Kitagawa, Mathematical programming (Japanese), Gendai tôkeigaku daiziten (Encyclopedia of modern statistics), Tôyôkeizai-sinpôsya, 1962.
[4] K. J. Arrow, L. Hurwicz, and H. Uzawa (eds.), Studies in linear and non-linear programming, Stanford Univ. Press, 1958.
[5] H. Hancock, Theory of maxima and minima, Dover, 1960.
[6] R. T. Rockafellar, Convex analysis, Princeton Univ. Press, 1970.
[7] R. L. Graves and P. Wolfe (eds.), Recent advances in mathematical programming, McGraw-Hill, 1963.

265 (XX.9) Mathematics in the 17th Century

The 17th century abounds in remarkable events in the history of science: the work on mechanics by Galileo (1564–1642); the discovery of analytic geometry by R. †Descartes (1596–1650); the early research in the theory of probability by P. de†Fermat (1601–1665) and B. †Pascal (1623–1662); the discovery of †mathematical induction by Pascal; the discovery of **infinitesimal calculus** (i.e., †differential calculus and †integral calculus) by I. †Newton (1642–1727) and G. W. †Leibniz (1646–1716). Compared with these events, the results of mathematical research from the Middle Ages to the 16th century seem minute. These results nonetheless exist, and historians of mathematics are now reevaluating them, particularly those of the 15th and the 16th centuries.

Before Galileo, Tycho Brahe (1546–1601) kept precise records of astronomical observations. J. Kepler (1571–1630), motivated by a mystic faith in the "harmony of the universe," studied Brahe's record and discovered three laws on the motion of planets. He also treated a question of cubature in his paper on the form and volume of the wine barrel (1615). His contemporaries J. Napier (1550–1617) and J. Bürgi (1552–1632) discovered logarithms, which helped astronomers tremendously in their calculations. Napier used the concept of velocity in his introduction of logarithms; thus analysis began to germinate. Galileo founded the modern approach to the concepts of velocity and acceleration in his *Dialogue on two new sciences* (1638). Using a self-made telescope, he discovered four of Jupiter's moons and observed sunspots. His espousal of the heliocentric theory of Copernicus (1473–1543) led to his denunciation before the Inquisition, which ordered him to refrain from holding or defending the theory. This is the most famous episode of his life, but his most significant contribution to science lies in his foundations of theoretical mechanics, which he freed from the Aristotelian tradition, thereby opening the way for Newton. F. Cavalieri (1598–1647), a disciple of Galileo, applied the notion of *indivisibilis* (originating in scholastic philosophy) to questions of quadrature in his *Geometry of the indivisible* (1635). This idea influenced Pascal and J. Wallis (1616–1703).

Descartes established the method of analytic geometry in his *Geometry* (1637), published as an appendix to his *Discourse on method*. The use of †coordinates can be traced back to Apollonius of Perga; Fermat used them occasionally, but Descartes made the first clear formulation of the method of representing general figures by means of equations, an essential step beyond Greek geometry. He also surpassed F. †Viète (1540–1603) by abolishing the restriction that quantities represented by letters should be of one dimension.

Contemporary to Descartes, Fermat made remarkable contributions to †number theory and Pascal to †projective geometry, and through their correspondence there started the theory of probability; both also made precursory contributions to analysis. Fermat treated questions on maxima and minima of functions and tangents of curves; Pascal solved some questions on tangents, centers of gravity, quadrature, and cubature concerning †cycloids. Pascal also contributed to hydrostatics, made positive use of the idea of the point of infinity in projective geometry, and clearly formulated the principle of mathematical induction in his theory of arithmetic triangles, the so-called †Pascal's triangles. (Freudenthal [4] established that the first discovery of the principle of mathematical induction is due to Pascal; the exact date of the discovery was studied by Hara [5].)

In England, Wallis and I. Barrow (1630–1677) preceded Newton. Wallis solved questions concerning quadrature and cubature (by bold use of the methods established by Cavalieri), infinite series, and interpolation. Barrow was Newton's teacher. He came close to the fundamental theorem of calculus, and Newton certainly owed some ideas which led to his discovery to Barrow's suggestions. Newton completed his method of fluxion, corresponding to our differential calculus, toward 1669–1671, but his paper on this method was published only after his death (1736). In his main work, *Principia mathematica philosophiae naturalis* (1687), he used this method and its converse, without naming them, to solve the †two-body problem. The work begins with three laws of mechanics and covers the motion of the moon and hydromechanics. Leibniz discovered infinitesimal calculus slightly later than Newton but independently. He invented convenient new symbols that gave great impetus to the development of calculus: the symbols dx and $\int$ are due to him. Leibniz was in Paris in 1672–1677, where he made the acquaintance of Father Arnauld (1612–1694) of Port Royal (the monastery to which Pascal belonged) and the Dutch physicist C. Huygens (1629–1695). Through their suggestions, he studied the work of Descartes and Pascal. Leibniz's first papers on calculus were published in 1684 in the scientific journal *Acta Eruditorum*, which he also edited. The methods of calculus he initiated were transmitted to mathematicians of the †Bernoulli family and then to L. †Euler, who developed them into the wide field of analysis.

Thus the mathematics of the 17th century went clearly beyond Greek mathematics. The importance of numbers over figures was recognized, and mathematicians were no longer hesitant to use infinity. Moreover, people became aware of the importance of experimental methods in science. The position of mathematics as an important method of natural science was established; mathematics became a rational basis of scientific research.

It was also in this century that a peculiar kind of mathematics was developed in Japan by T. Seki (1642?–1708). However, it lacked the Greek tradition of viewing logical foundations as being important, and cannot be compared with Western mathematics (→ 230 Japanese Mathematics (*Wasan*)).

References

[1] M. B. Cantor, Vorlesungen über Geschichte der Mathematik, Teubner, II, 1892; III, 1898.

[2] P. L. Boutroux, L'idéal scientifique des mathématiciens dans l'antiquité et les temps moderns, Presses Universitaires de France, new edition, 1955.

[3] D. T. Whiteside, Patterns of mathematical thought in the later seventeenth century, Arch. History Exact Sci., 1 (1961), 179–388.

[4] H. Freudenthal, Zur Geschichte der vollständigen Induktion, Arch. Internat. Histoire Sci., 6, no. 22 (1953), 17–37.

[5] K. Hara, Pascal et l'induction mathématique, Rev. Histoire Sci. Appl., 15 (1962), 287–302.

266 (XX.10) Mathematics in the 18th Century

During the Age of Enlightenment, mathematical analysis developed steadily after its initiation in the preceding century. It found numerous applications in theoretical physics and contributed to the growth of rationalistic thought.

The central figures in mathematics during the late 17th and early 18th century were I. †Newton (1642–1727) and G. W. †Leibniz (1646–1716). C. Maclaurin (1698–1746) of Scotland followed Newton, but no mathematician of Newton's stature appeared in Great Britain. An unfortunate dispute over the priority of discovery of infinitesimal calculus arose between Newton and Leibniz, after which their followers came into conflict. This prevented the members of the English school from giving up their inconvenient notation system, which hindered their progress in calculus.

On the Continent, Leibniz was succeeded by the mathematicians of the †Bernoulli family and by L. †Euler (1707–1783), who brought about brilliant developments in calculus and its applications. They solved various kinds of †differential equations and invented the †calculus of variations. F. †Viète used the term *analysis* in the sense of algebra as a heuristic method; the same term meant "infinitesimal calculus" in Newton's usage. It was during this century that analysis secured a position as a branch of mathematics independent from algebra and geometry. †Analytical dynamics, initiated by Euler, was further developed by J. L. †Lagrange (1736–1813) and P. S. de †Laplace (1749–1827). Laplace, in systematizing †celestial mechanics and the †theory of probability, showed what a powerful instrument analysis was. A. M. Legendre (1752–1833) investigated †elliptic integrals and opened the way for C. F. †Gauss and other mathematicians of the next century.

The growth of the Ecole Polytechnique, established during the time of the French Revolution, contributed to the brilliant progress of French mathematics. Lagrange, Laplace, and Legendre were all active in Paris during this period, as were S. D. Poisson (1781–1840) and J. B. J. †Fourier (1768–1830), both of whom made major contributions to analysis, and G. Monge (1746–1818), L. Carnot (1753–1823), and J. V. Poncelet (1788–1867). A problem proposed by Fourier in his theory of heat propagation gave rise to an important question of analysis, one that later formed the basis of †harmonic analysis. Fourier and Poisson aimed at clarifying the laws of nature, while Monge, Carnot, and Poncelet developed †projective geometry and †descriptive geometry for their purely geometric interest. Monge also did precursory work on †differential geometry.

The mathematics of this century left many remarkable results in geometry and analysis and their applications; however, it inherited its methods from the preceding century and lacked critical spirit. Mathematicians were more interested in obtaining new results than in reflecting upon the rigor of their methods. Reexamination and reestablishment of the foundations of mathematics were left to the next century.

References

[1] M. B. Cantor, Vorlesungen über Geschichte der Mathematik III, Teubner, 1898.
[2] D. J. Struik, A concise history of mathematics, Dover, 1948, third edition, 1967.

267 (XX.11) Mathematics in the 19th Century

The 19th century was a critical period in the history of mathematics. When the century began, the memory of the French Revolution was still fresh, and World War I followed closely upon the turn of the 20th century. During this time mathematics made enormous progress and left a tremendous inheritance to the present century. Increased personal liberty released people from traditions and allowed culture to spread to wider classes of society, producing a greater reservoir of talent. Research activities were intensified in the universities, and many specialists collaborated or competed with each other. The century may be divided into three periods: the first 20 years, during which many new fields of mathematics arose; the next 30 years, a period of further development; and the latter half of the century, when these fields attained maturity.

In 1801, *Disquisitiones arithmeticae* by the young C. F. †Gauss (1777–1855) appeared. It contained a systematized theory of numbers, ushering in a new era of mathematics. In France, many mathematicians studied at the Ecole Polytechnique, established during the French Revolution. Among them, A. L. †Cauchy (1789–1857) was one of the most prominent. He gave the exact definitions of limit and convergence, thus giving solid foundations to calculus some 150 years after its discovery. N. H. †Abel (1802–1829) and C. G. J. †Jacobi (1804–1851) studied †elliptic functions during the same time, producing results sensational to their contemporaries. Gauss gave a rigorous proof of the existence of roots of algebraic equations in the field of complex numbers; Abel proved the algebraic nonsolvability of algebraic equations of degree $\geqslant 5$; and E. †Galois (1811–1832) created his theory of algebraic equations, which began a new phase in algebra. J. V. Poncelet (1788–1867), another graduate of the Ecole Polytechnique, developed †projective geometry along the lines pursued by G. Monge (1746–1818). His research was continued in Germany by A. F. Möbius (1790–1868), J. Steiner (1796–1863), and J. Plücker (1801–1868). Steiner investigated, in particular, †algebraic curves and surfaces by synthetic methods; Plücker introduced projective coordinates, thus enlarging the usage of analytic methods in geometry. These brilliant results in the first period of the 19th century were all achieved by young mathematicians, most of them still in their twenties.

The new geometry was developed in the period 1830–1840 by K. G. C. von Staudt (1798–1867) in Germany and M. Chasles (1793–1880) in France. In the 1840s, the theory of †invariants was taken up in connection with geometry; outstanding in this domain were the English mathematicians A. Cayley (1821–1895) and J. J. Sylvester (1814–1897). P. G. L. †Dirichlet (1805–1859) endeavored to simplify Gauss's number theory and introduced the †Dirichlet series in his computation of †class numbers of binary †quadratic forms. He also initiated the theory of †trigonometric series by giving a rigorous proof to a theorem on expansion in †Fourier series, introduced by J. B. J. †Fourier (1768–1830) in his theory of heat propagation. Another notable event was the independent and almost simultaneous discovery of †non-Euclidean geometry by J. Bolyai (1802–1860) and N. I. Lobačevskiĭ (1793–1856), which aroused philosophical interest since it changed the character of axioms. The invention of †quaternions by W. R. Hamilton (1805–1865), publication of *Ausdehungslehre* (theory of extensions) by H. G. Grassmann (1809–1877), and development of the algebra of logic by G. Boole (1815–1864) also occurred during this period, but these notions did not win deep comprehension or sympathy until later.

During the latter half of the century, G. F. B. †Riemann (1826–1866) and K. T. W. †Weierstrass (1815–1897) were prominent. Both have had great influence on the mathematics of the 20th century, the former by his brilliant and abundant production, the latter by his mature and critical spirit. Riemann lived only 40 years, and published, in rapid succession, his epoch-making ideas on the theory of functions of a complex variable, †Abelian functions, trigonometric series, †foundations of geometry, †distribution of primes, and †zeta functions. Weierstrass was already 49 when, after teaching in a country *gymnasium* (secondary or college preparatory school), he became a professor at the University of Berlin. The theory of the functions of a complex variable, initiated by Cauchy in the 1820s, had to wait for the contribution of these men to attain completion in the form of the theory of †elliptic functions. Riemann's influence is also considerable in algebraic geometry and the theory of †differential equations. Weierstrass reformed the †calculus of variations. His critical approach gave rise to pathological functions, such as continuous nowhere differentiable functions (→ 109 Differential Calculus) and Peano space-filling curves (→ 325 Peano Curves), and to real analysis, based largely on the set theory of G. †Cantor (1848–1918).

Concerning the foundations of mathematics, Cantor, M. C. Méray, and J. W. R. Dedekind (1831–1916) established the theory of irrational numbers. Dedekind and G. Peano (1858–1932) developed the theory of natural numbers; their results brought about the "arithmetization" of mathematics and led to the research in the foundations of mathematics of the present century.

Cayley and F. †Klein (1849–1925) interpreted non-Euclidean geometry by means of metrics introduced in projective geometry. Toward the end of the century, D. †Hilbert (1862–1943) examined the roles of axioms of congruence, continuity, and parallelism in Euclidean geometry, thus initiating the study of axiomatic systems in general.

The theory of †groups, in particular finite groups, was developed around 1870 by C. Jordan (1838–1922), G. Frobenius (1849–1917), and W. S. Burnside (1852–1927). M. S. †Lie (1842–1899) applied infinitesimal transformations to differential equations, and Klein applied the groups of linear transformations to geometry. The discovery of †automorphic functions by Klein and H. †Poincaré (1854–1912) was another brilliant application of the theory of groups. In the algebraic theory of numbers, originated by Gauss, E. E. †Kummer (1810–1893) developed the idea of "ideal numbers" (→ 16 Algebraic Number

Fields); Dedekind then established the theory of †ideals. L. †Kronecker (1823–1891), an admirer of Abel's work, studied algebraic equations and discovered that every †Abelian extension of the rational number field is contained in a †cyclotomic field; he believed that relations of a similar kind would hold between †modular equations of elliptic functions with †complex multiplications and Abelian extensions of †imaginary quadratic fields, and enunciated a famous conjecture known as his "dream in his youth."

Finally, we note the appearance of another important mathematician, S. V. Kovalevskaja (1850–1891). After studying with Weierstrass, in 1884 she was invited by G. M. Mittag-Leffler (1846–1927) to teach at the University of Stockholm, where she remained until her death.

Toward the end of the 19th century, the subjects of mathematical research became highly differentiated. Branches were further ramified into more specialized branches, while unexpected relations were found between previously unconnected fields. The situation became so complicated that it was difficult to view mathematics as a whole. It was in these circumstances that in 1898, at the suggestion of F. Meyer and under the sponsorship of the Academies of Göttingen, Berlin, and Vienna, a project was initiated to compile an encyclopedia of the mathematical sciences. The *Enzyklopädie der mathematischen Wissenschaften* was completed in 20 years; it provided a useful overview of the mathematics of the 19th century.

Toward the end of the century, the International Congress of Mathematicians (ICM) was established to foster communication among mathematicians from all parts of the world. Before World War I broke out, the ICM met in Zürich (1896), Paris (1900), Heidelberg (1904), Rome (1908), and Cambridge, Mass. (1912). During this period, mathematical societies were formed in many countries, e.g., the London Mathematical Society (1865), the Société Mathématique de France (1872), the American Mathematical Society (1888), the Deutsche Mathematiker Vereinigung (1907), and the Mathematical Society of Tokyo (1877), which later became the Physico-Mathematical Society of Japan and subdivided in 1946. The present Mathematical Society of Japan evolved from this division.

Five years after the 1872 reform of the Japanese educational system, the University of Tokyo was established, and D. Kikuchi (1855–1917) and R. Fujisawa (1861–1933) taught at the Department of Mathematics during its early years. Under their influence, Japanese research in European-style mathematics (based on Greek traditions) began. The Universities of Kyoto and Tôhoku were established in 1897 and 1911, respectively. From the beginning of the 20th century, original results were obtained and published in the *Proceedings of the Physico-Mathematical Society of Japan* and in the journals of the faculties of science of these universities. In 1911, the *Tôhoku Mathematical Journal* was founded by T. Hayashi (1873–1935). In 1920, a paper on †class field theory by T. Takagi (1875–1960) was published in the *Journal of the College of Science of the University of Tokyo*. Thus the position of Japanese mathematics gradually came to be established.

References

[1] F. Klein, Vorlesungen über die Entwicklung der Mathematik im 19. Jahrhundert, Springer, I, 1926; II, 1927 (Chelsea, 1956).
[2] T. Takagi, Kinsei sûgakusidan (Japanese; Topics from the history of mathematics of the 19th century), Kawade, 1943 (Kyôritu, 1970).
[3] A. Kobori, Sûgakusi (Japanese; History of mathematics), Asakura, 1955.
[4] D. J. Struik, A concise history of mathematics, Dover, 1948, third edition, 1967.
[5] N. Bourbaki, Les éléments d'histoire des mathématiques, Hermann, second edition, 1969.
[6] Enzyklopädie der mathematischen Wissenschaften mit Einschluss ihrer Anwendungen, Teubner, 1898–1934.
[7] Encyclopédie des sciences mathématiques pures et appliquées (French translation of [6]), Paris, 1904–1914.

268 (XIV.10) Mathieu Functions

A. Mathieu's Differential Equation

The 2-dimensional †Helmholtz equation $(\Delta+k^2)\Psi=0$ $(\Delta=\partial^2/\partial x^2+\partial^2/\partial y^2)$, separated in †elliptic coordinates ξ, η given by $x=c\cosh\xi\cos\eta$, $y=c\sinh\xi\sin\eta$, has a solution of the form $\Psi=X(\xi)Y(\eta)$ whose factors $X(\xi)$, $Y(\eta)$ satisfy

$$d^2u/dz^2+(a-2q\cos 2z)u=0, \qquad (1)$$

$$d^2u/dz^2-(a-2q\cosh 2z)u=0, \qquad (2)$$

respectively, where a is an arbitrary constant and $q=k^2c^2/4$. By the substitution $z\to\pm iz$, (1) becomes (2). (1) and its solutions are known as **Mathieu's differential equation** and the **Mathieu functions**, and (2) and its solutions are called the **modified Mathieu differential equation** and the **modified Mathieu functions.**

B. Hill's Differential Equation

Hill's differential equation is a linear ordinary differential equation of the second order

$$d^2u/dx^2+F(x)u=0 \tag{3}$$

with $F(x+2\pi)=F(x)$. It is named after G. W. Hill, who investigated it in his study of lunar motion. This equation includes Mathieu's differential equation and †Lamé's differential equation as particular cases, and by suitable transformations, †Legendre's differential equation and the †confluent hypergeometric differential equations as well.

While $F(x)$ is periodic, solutions of (3) are not necessarily so. There always exists, however, a particular solution that is **quasiperiodic** in the sense that

$$u(x+2\pi)=\sigma u(x),\quad \sigma=\text{constant} \tag{4}$$

(**Floquet's theorem**).That is, the differential equation (3) has a solution of the form

$$u(x)=e^{\mu x}\varphi(x), \tag{5}$$

where $\varphi(x+2\pi)=\varphi(x)$ and μ (defined by $\sigma=e^{2\pi\mu}$) is called the **characteristic exponent**.

Being a particular case of Hill's equation, (1) has a general solution of the form

$$u(z)=Ae^{\mu z}\varphi(z)+Be^{-\mu z}\varphi(-z), \tag{6}$$

where $\varphi(z+\pi)=\varphi(z)$. For those values of a (called **eigenvalues**) that make the characteristic exponent μ equal to 0 or i, $u(z)$ has period π or 2π and is called a **Mathieu function of the first kind**, also called an **elliptic cylinder function** when employed in problems of diffraction by an elliptic cylinder. Sometimes it is called simply the Mathieu function, and other solutions of (1) are referred to as **general Mathieu functions.**

The †Fourier series expansion of $F(x)$, i.e.,

$$F(x)=\sum_{m=-\infty}^{\infty} a_m e^{imx}, \tag{7}$$

suggests, in conjunction with Floquet's theorem, a solution of (3) in the form

$$u=e^{\mu x}\sum_{n=-\infty}^{\infty} b_n e^{inx}. \tag{8}$$

Substituting (7) and (8) into (3) and comparing coefficients of $e^{(\mu+in)x}$, we have infinitely many linear equations

$$(\mu+in)^2 b_n+\sum_{m=-\infty}^{\infty} a_m b_{n-m}=0,$$
$$n=\ldots,-2,-1,0,1,2,\ldots. \tag{9}$$

By eliminating the b_n in (9), we also obtain an infinite determinantal equation called **Hill's determinantal equation,**

$$\Delta(\mu)=|B_{rs}|=0, \tag{10}$$

where the elements B_{rs} of **Hill's determinant** $\Delta(\mu)$ are such that

$$B_{rs}=\begin{cases}1 & \text{if } r=s,\\ \dfrac{a_{r-s}}{(\mu+ir)^2+a_0} & \text{if } r\neq s.\end{cases}$$

Here an **infinite determinant** $D=|B_{mn}|(m,n=-\infty,\ldots,\infty)$ is defined as the limit, if it exists, of $D_m=\det(B_{ij})$ $(i,j=-m,\ldots,m)$ as $m\to\infty$. The formula (10) can be reduced to a simpler form

$$\sin^2\pi i\mu=\Delta(0)\sin^2\pi\sqrt{a_0}\ . \tag{11}$$

This determines a characteristic exponent μ, which in turn determines the b_n in (9) and a solution (8). This procedure is called **Hill's method of solution.**

Applying Hill's method of solution to (1), we have an equation for the characteristic exponent μ,

$$\sin^2(\pi/2)i\mu=\Delta(0)\sin^2(\pi/2)\sqrt{a}\ , \tag{12}$$

where the infinite determinant $\Delta(0)=|B_{mn}|$ has elements such that

$$B_{mn}=\begin{cases}1 & \text{if } m=n\\ \dfrac{-2q}{a-m^2} & \text{if } n=m\pm 1\\ 0 & \text{otherwise } (m,n=\ldots,-1,0,1,\ldots).\end{cases}$$

When $q\to 0$, we have

$$\Delta(0)$$
$$=1+2(2q)^2\frac{\pi}{8\sqrt{a}\,(1-a)}\cot\frac{\pi\sqrt{a}}{2}+O(q^4). \tag{13}$$

Thus if $q=0$, we have $\Delta(0)=1$, and $a=4n^2$, $(2n+1)^2$ correspond to $\mu=0$, i in (12).

C. Mathieu Functions of the First Kind

Mathieu functions of the first kind are further classified into the four following types:

$$ce_{2n}(z,q)=\sum A_r^{(2n)}\cos 2rz,\quad (a_{2n}), \tag{14.1}$$

$$se_{2n+1}(z,q)=\sum B_{2r+1}^{(2n+1)}\sin(2r+1)z,\quad (b_{2n+1}), \tag{14.2}$$

$$ce_{2n+1}(z,q)=\sum A_{2r+1}^{(2n+1)}\cos(2r+1)z,\quad (a_{2n+1}), \tag{14.3}$$

$$se_{2n+2}(z,q)=\sum B_{2r+2}^{(2n+2)}\sin(2r+2)z, \qquad (b_{2n+2}) \qquad (14.4)$$

$(n,r=0,1,2,\ldots)$, where the final terms in parentheses are eigenvalues, ordered by $a_{2n}<b_{2n+1}<a_{2n+1}<b_{2n+2}$ for a given q, and increasing with n. Each of these series converges absolutely and uniformly for all finite z and has n zeros in $0<z<\pi/2$. In addition, orthonormality relations

$$\int_0^{2\pi} ce_m(x)se_n(x)\,dx=0,$$

$$\int_0^{2\pi} ce_m(x)ce_n(x)\,dx=\int_0^{2\pi} se_m(x)se_n(x)\,dx=\pi\delta_{mn}$$

hold. When $q\to 0$, $ce_0(z)\to 1/\sqrt{2}$, $ce_m(z)\to \cos mz$, and $se_m(z)\to \sin mz$.

For small q we assume that the quantities involved have power series expansions in q, e.g.,

$$a=m^2+\alpha q+\beta q^2+\ldots,$$

$$ce_m(z)=\cos mz+qF_1(z)+q^2F_2(z)+\ldots,$$

and we substitute them in (1) to determine successively $\alpha,\beta,\ldots,F_1(z),F_2(z),\ldots$ (**Mathieu's method**). For larger q, (14) is substituted in (1) to give recurrence formulas for the coefficients, e.g., for $ce_{2n}(z)$,

$$-aA_0^{(2n)}+qA_2^{(2n)}=0,$$

$$2qA_0^{(2n)}+(4-a)A_2^{(2n)}+qA_4^{(2n)}=0,$$

$$qA_{2r-2}^{(2n)}+(4r^2-a)A_{2r}^{(2n)}+qA_{2r+2}^{(2n)}=0, \quad r>1. \qquad (15)$$

After we eliminate the $A_{2r}^{(2n)}$, the formulas lead to an equation for the eigenvalues a_{2n}

$$\begin{vmatrix} a & -q & 0 & 0 & 0 & \ldots \\ -2q & a-4 & -q & 0 & 0 & \ldots \\ 0 & -q & a-16 & -q & 0 & \ldots \\ & & \ldots & & & \end{vmatrix}=0, \qquad (16)$$

or, equivalently, to a †continued fraction

$$a=\frac{2q^2}{a-4}-\frac{q^2}{a-16}-\frac{q^2}{a-36}-\cdots. \qquad (17)$$

Given q, we can find the a_{2n} from (17) and determine the $A_{2r}^{(2n)}$ for each a_{2n} from (15) (**Ince-Goldstein method**).

D. Mathieu Functions of the Second Kind and Modified Mathieu Functions

There exists only one (half-)periodic solution of (1) corresponding to each (half-)periodic eigenvalue (→ Section E). Therefore other solutions corresponding to the same a_m or b_m and independent of $ce_m(z,q)$ or $se_m(z,q)$ are nonperiodic. They are called the **Mathieu functions of the second kind** and are denoted by $fe_m(z,q)$ or $ge_m(z,q)$.

By the substitution $z\to iz$ in (14) we have formulas for the **modified Mathieu functions of the first kind**,

$$Ce_m(z,q)=ce_m(iz,q),$$

$$Se_m(z,q)=-ise_m(iz,q). \qquad (18)$$

When $q\to 0$, then $Ce_0(z)\to 1/\sqrt{2}$, $Ce_m(z)\to \cosh mz$, $Se_m(z)\to\sinh mz$. Similarly, by the substitution $z\to iz$ we obtain **modified Mathieu functions of the second kind** from Mathieu functions of the second kind. In addition, we introduce **modified Mathieu functions of the third kind** as those linear combinations of modified Mathieu functions of the first and second kinds that have the asymptotic form $e^{-y}y^{-1/2}$ $(y=q^{1/2}e^z)$ as $z\to\infty$. In addition to the Fourier expansion (7), expansion of the Mathieu functions in terms of †Bessel functions is possible, e.g., after taking $q=h^2$,

$$Ce_{2n}(z,q)=\sum A_{2r}\cosh 2rz \qquad (19.1)$$

$$=(A_0)^{-1}ce_{2n}(\pi/2,q)\times\sum(-1)^rA_{2r}J_{2r}(2h\cosh z) \qquad (19.2)$$

$$=(A_0)^{-1}ce_{2n}(0,q)\times\sum A_{2r}J_{2r}(2h\sinh z) \qquad (19.3)$$

$$=(A_0)^{-2}ce_{2n}(0,q)ce_{2n}(\pi/2,q)\times\sum(-1)^rA_{2r}J_r(he^{-z})J_r(he^z). \qquad (19.4)$$

These series converge absolutely and uniformly for all finite z. Replacing the J on the right-hand sides of (19) by $N_{2r}(2h\cosh z)$, $N_{2r}(2h\sinh z)$, $J_r(he^{-z})N_r(he^z)$, respectively, in an obvious way we obtain infinite series for a function that again satisfies (2) and is denoted by $Fey_{2n}(z,q)$. In a similar manner other modified Mathieu functions of the second kind $Gey_{2n+1}(z,q)$, $Fey_{2n+1}(z,q)$, $Gey_{2n+2}(z,q)$ can be obtained. These are more convenient for practical applications than $fe_m(iz,q)$ and $ge_m(iz,q)$ since they converge more rapidly.

The equations

$$d^2u/dz^2+(a+2q\cos 2z)u=0, \qquad (20)$$

$$d^2u/dz^2-(a+2q\cosh 2z)u=0 \qquad (21)$$

obtained from (1) and (2) by the substitution $q\to -q$ are the results of separating $(\Delta-k^2)\varphi=0$. In general, if $f(z,q)$ is a solution of (1), then $f(\pi/2-z,q)$ is a solution of (20). Thus the formulas

$$ce_{2n}(z,-q)=(-1)^nce_{2n}(\pi/2-z,q), \qquad (22.1)$$

$$ce_{2n+1}(z,-q)=(-1)^nse_{2n+1}(\pi/2-z,q), \qquad (22.2)$$

$$se_{2n+1}(z,-q)=(-1)^nce_{2n+1}(\pi/2-z,q), \qquad (22.3)$$

$$se_{2n+2}(z,-q)=(-1)^n se_{2n+2}(\pi/2-z,q) \tag{22.4}$$

can be adopted as definitions of $ce_m(z,q)$, $se_m(z,q)$ for $q<0$ (**Ince's definition**). Accordingly, an expansion of Ce holds in terms of modified Bessel functions I_m in place of the J_m in (19), which in turn becomes a solution of (21) if we replace I_m by $(-1)^m K_m/\pi$:

$$Fek_{2n}(z,-q)=(-1)^n(\pi A_0)^{-1}ce_{2n}(\pi/2,q) \times \sum A_{2r}K_{2r}(2h\sinh z). \tag{23}$$

In a similar manner, $Fek_{2n+1}(z,-q)$, $Gek_{2n+1}(z,-q)$, $Gek_{2n+2}(z,-q)$ can be defined. They decrease exponentially as $z\to\infty$ and hence are precisely the modified Mathieu functions of the third kind.

E. Stability

Let $u_1(x), u_2(x)$ be a fundamental system of solutions of (3) such that $u_1(0)=1$, $u_1'(0)=0$; $u_2(0)=0$, $u_2'(0)=1$; then σ, μ in (4), (5) are given by

$$\sigma=e^{2\pi\mu}, \qquad 2\pi\mu=\operatorname{arccosh}A,$$
$$2A=u_1(2\pi)+u_2'(2\pi). \tag{24}$$

There are two values of μ satisfying (24); let them be μ_1 and μ_2 ($\mu_2=-\mu_1$). If $F(x)$ is a real function, then $u_1(x)$, $u_2(x)$ are also real functions and A is a real number. It may be seen from (24) that $A<-1$, $-1<A<0$, $0<A<1$, $1<A$ correspond to $\pm 2\pi\mu=$ $\operatorname{arccosh}|A|+\pi i$, $i(\arccos|A|+\pi)$, $i\arccos A$, $\operatorname{arccosh}A$, respectively. Consequently, when $|A|<1$, the general solution of (3) neither diverges nor vanishes as x tends to infinity. Such solutions are called **stable solutions** of Hill's equation, or sometimes **Hill's functions**. When $|A|>1$, either $e^{\mu_1 x}$ or $e^{\mu_2 x}$ tends to infinity with x. Such solutions are called **unstable solutions**. If $A=1$ (-1), we have $\mu=0$ $(i/2)$ and a solution of the form of $u=\varphi(x)$ $(e^{ix/2}\varphi(x))$ with the period 2π (4π) which is called a **periodic (half-periodic) solution**.

When we apply the Mathieu functions to physical and engineering sciences, such as the theory of oscillation and quantum mechanics, it is convenient to modify (3) in the form

$$d^2u/dx^2+(\lambda+\gamma\Phi(x))u=0 \tag{25}$$

involving parameters λ, γ. Then with γ fixed, there exist countably many values of λ (called eigenvalues) corresponding to periodic or half-periodic solutions of (25). Let them be $\lambda_0\leqslant\lambda_1\leqslant\ldots$ or $\bar\lambda_1\leqslant\bar\lambda_2\leqslant\ldots$, respectively; then we have

$$\lambda_0<\bar\lambda_1\leqslant\bar\lambda_2<\lambda_1\leqslant\lambda_2<\ldots<\bar\lambda_{2k-1}\leqslant\bar\lambda_{2k} <\lambda_{2k-1}\leqslant\lambda_{2k}<\ldots$$

as shown in Fig. 1, where the values of λ on the intervals drawn with solid and broken lines correspond to stable and unstable solutions, respectively. This assertion is known as **Haupt's theorem**.

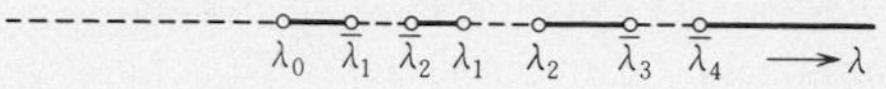

Fig. 1

When both λ and γ vary, the $\lambda\gamma$-plane is divided into regions corresponding to stable or unstable solutions according as the characteristic exponent μ is purely imaginary or not. When $\Phi(x)=2\cos x$ in equation (25), we obtain the Mathieu equation. In this case the $\lambda\gamma$-plane is divided as shown in Fig. 2, where shaded (unshaded) regions correspond to stable (unstable) solutions and boundary curves give eigenvalues corresponding to periodic or half-periodic solutions. In another particular case when $\Phi(x)$ is a step function,

$$\Phi(x)=\begin{cases}1, & 0<x<\pi,\\ -1, & \pi<x<2\pi,\end{cases}$$

Hill's equation is readily integrated in terms of trigonometric functions, and the division in the stable and unstable regions of the $\lambda\gamma$-plane is shown in Fig. 3.

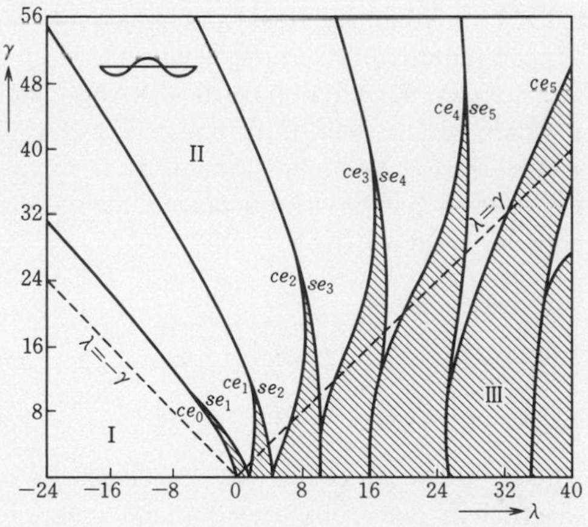

Fig. 2

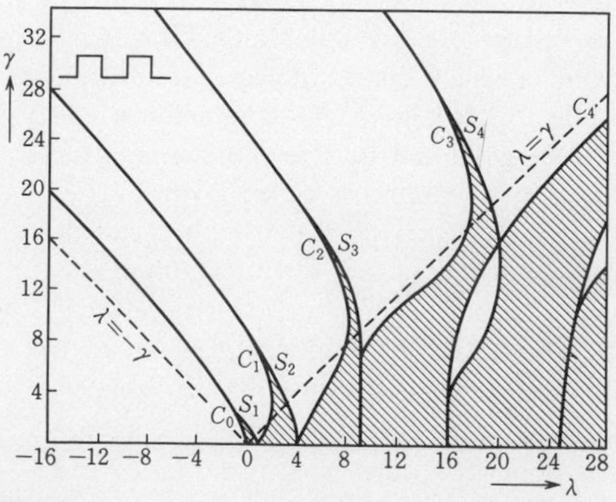

Fig. 3

References

[1] E. T. Whittaker and G. N. Watson, A course of modern analysis, Cambridge Univ. Press, fourth edition, 1958.
[2] M. J. O. Strutt, Lamésche-, Mathieusche-, und verwandte Funktionen in Physik und Technik, Erg. Math., Springer, 1932 (Chelsea, 1967).
[3] N. W. McLachlan, Theory and application of Mathieu functions, Clarendon Press, 1947.
[4] J. Meixner and F. W. Schäfke, Mathieusche Funktionen und Sphäroid-funktionen, Springer, 1954.
[5] M. Kotani and H. Hashimoto, Tokusyu kansû (Japanese; Special functions), Iwanami Coll. of Modern Appl. Math., 1958.
[6] K. Terazawa (ed.), Sizen kagakusya no tameno sûgaku gairon (ôyô hen) (Japanese; Introduction to mathematics for natural scientists pt. II. Applications), Iwanami, 1960.

269 (III.2) Matrices

A. General Remarks

Let K be a †ring or a †field (which may be noncommutative). As examples of such K, we may take the real number field **R** and the complex number field **C**. By a **matrix** in K, we mean an array of mn elements a_{ik} ($i = 1, \ldots, m; k = 1, \ldots, n$) of K arranged in a rectangular form as follows:

$$\begin{bmatrix} a_{11} & a_{12} & \cdots & a_{1n} \\ a_{21} & a_{22} & \cdots & a_{2n} \\ & & \cdots & \\ a_{m1} & a_{m2} & \cdots & a_{mn} \end{bmatrix}.$$

The element a_{ik} is called its (i,k)**-component** (**coefficient, entry,** or **element**). (Sometimes, instead of using parentheses, we use $\| \;\|$ or [ ].) More precisely, this matrix is said to be an m **by** n **matrix** ($m \times n$ **matrix** or **matrix of** (m,n)**-type**). In particular, an $n \times n$ matrix is called a **square matrix** of **degree** (or **order**) n, while a matrix in general is sometimes called a **rectangular matrix**. Each horizontal n-tuple in an $m \times n$ matrix is called a **row** of the matrix, and each vertical m-tuple is called a **column** of the matrix. We often abbreviate the notation for the matrix given previously by writing (a_{ik}) or simply A. A square matrix is called a **diagonal matrix** if all its components are zero except possibly for diagonal components a_{ii}, that is, if $a_{ik} = 0$ for $i \neq k$. If the components a_{ii} of a diagonal matrix are all equal, it is called a **scalar matrix**. An $n \times n$ matrix whose (i,k)-component is equal to δ_{ik} is called the **unit matrix** (or **identity matrix**) of degree n, where δ_{ik} is the **Kronecker delta**, which is defined by $\delta_{ii} = 1$ and $\delta_{ij} = 0$ for $i \neq j$. We denote this matrix by I_n, or simply by I if there is no need to specify the n.

In particular, a $1 \times n$ matrix $(a_1, a_2, \ldots, a_n)$ in K is called a **row vector** of dimension n over K, and an $m \times 1$ matrix

$$\begin{bmatrix} b_1 \\ \cdot \\ \cdot \\ \cdot \\ b_m \end{bmatrix}$$

is called a **column vector** of dimension m. The m rows and n columns of an $m \times n$ matrix A are called row vectors and column vectors of A.

B. Operations on Matrices

Two matrices $A = (a_{ik})$ and $B = (b_{ik})$ are said to be equal if and only if they have the same type and $a_{ik} = b_{ik}$ ($i = 1, \ldots, m; k = 1, \ldots, n$). If both A and B are $m \times n$ matrices, we define the **sum** of A and B by $A + B = (a_{ik} + b_{ik})$. The product of two matrices A and B is defined by $AB = (c_{ik}), c_{ik} = \sum_j a_{ij} b_{jk}$, provided that the number of columns of A is equal to the number of rows of B. We further define the (left and right) multiplication of a matrix A by an element a of K by $aA = (aa_{ik})$ and $Aa = (a_{ik}a)$. The set of all matrices in K of the same type forms a †K-module. The multiplication of matrices satisfies the associative law and the left and right distributive laws with respect to addition. Thus the set of all $n \times n$ matrices in K forms a ring, which is called the **total matrix algebra** (or **full matrix algebra**) of degree n over K; it is usually denoted by $M_n(K)$ or K_n. If K has the †unity element 1, then I_n is the unity element of $M_n(K)$. The matrix whose components are all 0 is called the **zero matrix** and is denoted by the same symbol 0. Suppose that K has the unity element 1. Let E_{ik} be the matrix whose (i,k)-component is 1 and whose other components are all 0. Then every matrix $A = (a_{ik})$ in $M_n(K)$ can be expressed uniquely as $A = \sum a_{ik} E_{ik}$, a linear combination of E_{ik}. The matrix E_{ik} is called a **matrix unit**. We have $E_{ij} E_{kl} = 0$ if $j \neq k$, $E_{ik} E_{kl} = E_{il}$, and $aE_{ik} = E_{ik} a$ for all $a \in K$.

Let A be a square matrix in K. If there exists a matrix A^{-1} such that $AA^{-1} = A^{-1}A =$

I, then A^{-1} is called the **inverse matrix** of A, and A is called a **regular matrix (nonsingular matrix** or **invertible matrix)**. The inverse A^{-1} is unique if it exists. In the case where K is commutative, A is regular if and only if its †determinant $|A|$ is a †regular element of K; in particular, in the case where K is a field, A is regular if and only if $|A| \neq 0$.

Let $A=(a_{ik})$ be an $m \times n$ matrix. Then the $n \times m$ matrix (b_{ik}) such that $b_{ik}=a_{ki}$ for all i and k is called the **transpose** of A, and is usually denoted by ${}^tA$. Taking the transpose of a matrix amounts to changing rows into columns and vice versa. In the case where K is commutative, $AB=C$ implies ${}^tC={}^tB\,{}^tA$. A square matrix such that ${}^tA=A$ is called a **symmetric matrix**, and one such that ${}^tA=-A$ is called an **alternating (skew-symmetric** or **antisymmetric) matrix**. A square matrix $A=(a_{ik})$ is called an **upper (lower) triangular matrix** if $a_{ik}=0$ for $i>k$ $(i<k)$.

C. The Kronecker Product of Matrices

We assume that the ring K is commutative. Let A be an $m \times n$ matrix (a_{ik}), let B be an $r \times s$ matrix (b_{jl}) in K, and write $c_{\lambda,\mu}=a_{ik}b_{jl}$ by means of indexes $\lambda=(i,j)$ and $\mu=(k,l)$. The **Kronecker product** of A and B, usually denoted by $A \otimes B$, is defined as the $mr \times ns$ matrix $C=(c_{\lambda,\mu})$. By an appropriate arrangement of λ and μ, it can be expressed as

$$\begin{pmatrix} a_{11}B & a_{12}B & \dots & a_{1n}B \\ a_{21}B & a_{22}B & \dots & a_{2n}B \\ & & \dots & \\ a_{m1}B & a_{m2}B & \dots & a_{mn}B \end{pmatrix},$$

$$\text{or}\quad \begin{pmatrix} b_{11}A & b_{12}A & \dots & b_{1s}A \\ b_{21}A & b_{22}A & \dots & b_{2s}A \\ & & \dots & \\ b_{r1}A & b_{r2}A & \dots & b_{rs}A \end{pmatrix}.$$

We have the following formulas:

$$A \otimes (B_1+B_2)=A \otimes B_1 + A \otimes B_2,$$
$$(A_1+A_2) \otimes B = A_1 \otimes B + A_2 \otimes B,$$
$$c(A \otimes B)=(cA) \otimes B = A \otimes (cB),$$
$$(A_1 \otimes B_1)(A_2 \otimes B_2)=(A_1A_2) \otimes (B_1B_2),$$

provided that the sums and products can be defined.

D. The Rank of a Matrix

Let A be an $m \times n$ matrix (a_{ik}) in a field K. If there exists a nonzero †minor of A of degree r, and if all minors of degree $\geqslant r+1$ are equal to 0, then the number r is called the **rank** of A. Denoting the rank of a matrix A by $\rho(A)$, we have $\rho(PAQ) \leqslant \rho(A)$ for any matrices P, Q for which the product PAQ exists; the equality holds if P and Q are regular (square) matrices. The rank r of A is equal to the maximum number of †linearly independent row vectors of A, and it is also equal to the maximum number of linearly independent column vectors of A. The number $n-\rho(A)$ is called the **(column) nullity** of the matrix A. It is equal to the dimension of the linear space consisting of solutions of homogeneous linear equations $A\mathfrak{x}=0$, that is, the number of fundamental solutions of these equations. (The **row nullity** $m-\rho(A)$ of A is equal to the dimension of the linear space consisting of solutions of $\mathfrak{x}A=0$.) Even if K is not commutative, we can define the rank of a matrix A as the maximum number of left (right) linearly independent row (column) vectors of A (→ 256 Linear Spaces F).

E. Elementary Divisors

Let $\mathfrak{o}$ be a †principal ideal domain (e.g., the ring $\mathbf{Z}$ of rational integers, or the †polynomial ring $K[x]$ over a field K), and let A be a matrix in $\mathfrak{o}$. Then, multiplying appropriate invertible matrices from the left and right, we can transform A into a diagonal matrix of the following form:

$$\begin{pmatrix} e_1 & & & & & & 0 \\ & e_2 & & & & & \\ & & \ddots & & & & \\ & & & e_r & & & \\ & & & & 0 & & \\ & & & & & \ddots & \\ 0 & & & & & & 0 \end{pmatrix},$$

where r denotes the rank of A, each $e_i \neq 0$, and each e_{i+1} is divisible by e_i. The components $e_1, e_2, \dots, e_r$, called the **elementary divisors** of A, are determined by A, and they are unique up to regular element factors. This fact is generalized to some extent for the case where $\mathfrak{o}$ is noncommutative (→ 256 Linear Spaces Q). Using the concept of determinant, the elementary divisors can also be defined as follows. Let d_k denote the greatest common divisor of all minors of degree k of A. The elements $e_i=d_i/d_{i-1}$ $(i=1,\dots,r;\ d_0=1)$ of $\mathfrak{o}$ are the elementary divisors of A. The numbers d_i $(1 \leqslant i \leqslant r)$ are called the **determinant factors** of A. Let p be one of prime divisors of e_r, and suppose that e_i is divisible by p^{ρ_i}

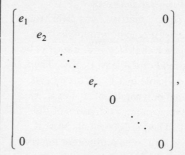

but not divisible by p^{ρ_i+1}. Then p^{ρ_i} ($i=1,\ldots,r$) are called the **simple elementary divisors** with base p of A (→ 281 Noetherian Rings).

F. Characteristic Polynomials and Eigenvalues

Let $A=(a_{ik})$ be a square matrix of degree n in a field K. The determinant $F(x)=|xI-A|$ is a polynomial in x with coefficients in K, called the **characteristic polynomial** of A. The algebraic equation $F(x)=0$ is called the **characteristic equation** of A, and its roots $\lambda_1, \lambda_2, \ldots, \lambda_n$ are called the **eigenvalues (proper values** or **characteristic roots)** of A. The †determinant of A, $\det A$, is equal to $\prod_{i=1}^n \lambda_i$, the product of the eigenvalues. The sum of eigenvalues, $\sum_{i=1}^n \lambda_i = \sum_{i=1}^n a_{ii}$, is called the **trace** (G., **Spur**) or **diagonal sum** of A, and is denoted by $\operatorname{tr} A$ or $\operatorname{Sp}(A)$. In case K is †algebraically closed, there exists a nonzero vector $\mathfrak{x}$ which satisfies the equations $A\mathfrak{x}=\lambda\mathfrak{x}$ if and only if λ is an eigenvalue of A. This solution $\mathfrak{x}$ is called an **eigenvector (proper vector** or **characteristic solution)** belonging to the eigenvalue λ. In particular, if all the a_{ik} are real and A is symmetric, then all eigenvalues of A are real, and the characteristic equation $F(x)=0$ of A is called a **secular equation**. Every square matrix A satisfies its characteristic equation, i.e., $F(A)=0$. This is called the **Hamilton-Cayley theorem**, and it is useful in numerical calculation of inverse matrices.

Let A be a square matrix. As is clear from above, there exist monic polynomials $f(x)$ ($\neq 0$) in $K[x]$ such that $f(A)=0$. Let $\varphi(x)$ be such a polynomial of the least degree. Then every $f(x)$ is divisible by $\varphi(x)$. This $\varphi(x)$ is called a **minimal polynomial** of A. Let $e_1(x), \ldots, e_n(x)$ be the elementary divisors of the matrix $xI-A$. Then $e_n(x)=\varphi(x)$. Now if $F(x)=x^n$, A is said to be a **nilpotent matrix**, and if $F(x)=(x-1)^n$, A is said to be a **unipotent matrix** (→ 256 Linear Spaces B).

G. Canonical Form

Two square matrices A and B in a field K are said to be **similar** to each other if we can write $B=P^{-1}AP$ with a regular matrix P. The matrices A and B are similar if and only if $xI-A$ and $xI-B$ have the same elementary divisors. Now suppose that K is an algebraically closed field (e.g., the field of complex numbers), and let λ be an eigenvalue of a matrix A. Furthermore, let $(x-\lambda)^{\rho_1}, \ldots, (x-\lambda)^{\rho_s}$ be the simple elementary divisors $\neq 1$ with base $x-\lambda$ of $xI-A$ and form the following matrix:

$$A_\lambda = \begin{bmatrix} P_1 & & & & 0 \\ & P_2 & & & \\ & & \cdot & & \\ & & & \cdot & \\ 0 & & & & P_s \end{bmatrix},$$

$$\text{where } P_i = \begin{bmatrix} \lambda & & & & 0 \\ 1 & \cdot & & & \\ & \cdot & \cdot & & \\ & & \cdot & \cdot & \\ 0 & & & 1 & \lambda \end{bmatrix}.$$

(Note that P_i is a $\rho_i \times \rho_i$ matrix. If $\rho_i = 1$, then P_i is composed of a single element λ.) If $\lambda_1, \ldots, \lambda_t$ are all the distinct eigenvalues of A, then A is similar to the block-diagonal matrix in which A_{λ_i} ($i=1,\ldots,t$) are arranged along the diagonal. This is called **Jordan's canonical form** of A. It is a diagonal matrix if and only if the minimal polynomial of A has no multiple roots. If this is the case, A is said to be **semisimple** (→ 256 Linear Spaces B).

H. The Exponential Function of a Matrix

Let $A=(a_{ij}^{(\nu)})$ ($\nu=0,1,2,\ldots$) be square matrices in the complex number field $\mathbf{C}$. The series $A_0+A_1+\ldots+A_\nu+\ldots$ is said to be convergent if the series of components $a_{ij}^{(0)}+a_{ij}^{(1)}+\ldots+a_{ij}^{(\nu)}+\ldots$ is convergent for each (i,j). For every square matrix A, the following series is convergent:

$$I+A+\frac{1}{2!}A^2+\ldots+\frac{1}{\nu!}A^\nu+\ldots.$$

We denote it by $\exp A$. Then $\exp({}^tA) = {}^t(\exp A)$; $\exp(-A)=(\exp A)^{-1}$; and if $AB=BA$, $\exp(A+B)=\exp A \exp B$. Moreover, $\det(\exp A)=\exp(\operatorname{tr} A)$. Furthermore, if we set $F(t)=\exp(tA)$, then $dF(t)/dt = AF(t)$ (→ 415 Transformation Groups).

I. Normal Matrices, Unitary Matrices, and Hermitian Matrices

Let $A=(a_{ik})$ be a square matrix in the complex number field $\mathbf{C}$. Then the **adjoint matrix** A^* of A is the conjugate transpose ${}^t\bar{A}=(\bar{a}_{ki})$, where $\bar{a}$ is the complex conjugate of a. If $AA^*=A^*A$, then A is said to be a **normal matrix**. A matrix U such that $U^*U=I$, that is, $U^{-1}=U^*$, is normal. Following G. Frobenius, we call it a **unitary matrix** (in the following discussion, we shall call it simply a u-matrix). A matrix H such that $H^*=H$ is also normal. It is called a **Hermitian matrix** (which we shall refer to as simply an h-matrix). An h-matrix P such that $P^2 = P$ is called a **projection matrix**.

The set of all u-matrices forms a †group with respect to matrix multiplication. If A is a normal matrix and U is a u-matrix, U^*AU is a normal matrix. If H is an h-matrix, Q^*HQ is also an h-matrix for any matrix Q. Every normal matrix A can be transformed into a diagonal matrix by a suitable u-matrix (i.e., there exists a u-matrix U such that $U^*AU = U^{-1}AU$ is a diagonal matrix), and conversely, every matrix with this property is a normal matrix. Thus, in particular, any h-matrix or u-matrix can be transformed into a diagonal matrix by a suitable u-matrix. Moreover, every real symmetric matrix, which is naturally an h-matrix, can be transformed into a diagonal matrix by an orthogonal matrix; this will be defined later. If $A_1, \ldots, A_m$ are mutually commutative normal matrices, we can transform them into diagonal matrices by the same u-matrix, that is, there exists a u-matrix U such that $U^{-1}A_iU$ is of diagonal form for all i. All eigenvalues of an h-matrix are real, and all eigenvalues of a u-matrix have absolute value 1.

If all eigenvalues of an h-matrix H are positive (positive or zero), then H is said to be **positive definite** (**positive semidefinite**). For an h-matrix H, $\exp H$ is a positive definite h-matrix, and conversely, every positive definite h-matrix can be expressed as $\exp H$ by a unique h-matrix H. Furthermore, every regular matrix A can be uniquely expressed as $A = UH$ (or $A = H'U'$) by a u-matrix U (or U') and a positive definite h-matrix H (or H'), and A is normal if and only if $UH = HU$.

A square matrix A such that $A^* = -A$ is called a **skew-Hermitian matrix** (simply **skew h-matrix** or **anti-Hermitian matrix**). All eigenvalues of a skew h-matrix are purely imaginary numbers. If A is a skew h-matrix, $\exp A$ is a u-matrix. Conversely, if a u-matrix U lies in a sufficiently small neighborhood of the identity matrix I (i.e., if all elements of $U - I$ have sufficiently small absolute values), U can be uniquely expressed in the form $\exp A$ with a skew h-matrix A.

J. Orthogonal Matrices

An h-matrix whose components are all real is necessarily symmetric. A u-matrix whose components are all real, that is, a real matrix R such that $R^{-1} = {}^tR$, is called an **orthogonal matrix**. The totality of orthogonal matrices forms a group with respect to matrix multiplication. If S is a real symmetric matrix, there exists an orthogonal matrix T for which $T^{-1}ST$ is a diagonal matrix; that is, S can be transformed into a diagonal form by T. Every orthogonal matrix R can be transformed by an orthogonal matrix T into the following diagonal form:

$$T^{-1}RT = \begin{bmatrix} 1 & & & & & & & & & \\ & \ddots & & & & & & & & \\ & & 1 & & & & & & & \\ & & & -1 & & & & & & \\ & & & & \ddots & & & & & \\ & & & & & -1 & & & & \\ & & & & & & P_1 & & & \\ & & & & & & & \ddots & & \\ & & & & & & & & & P_t \end{bmatrix},$$

$$P_j = \begin{pmatrix} \cos\theta_j & \sin\theta_j \\ -\sin\theta_j & \cos\theta_j \end{pmatrix} \quad \text{for } j = 1, \ldots, t.$$

The determinant $|R|$ of an orthogonal matrix R is either 1 or -1. If $|R| = 1$, then R is called a **proper orthogonal matrix**. If A is a real alternating matrix, then $\exp A$ is a proper orthogonal matrix; conversely, any proper orthogonal matrix in a sufficiently small neighborhood of the identity matrix I can be expressed uniquely in this form. Moreover, there exists a one-to-one correspondence between real alternating matrices A and proper orthogonal matrices R without eigenvalues equal to -1, given by $R = (I - A)(I + A)^{-1}$ and $A = (I - R)(I + R)^{-1}$. This is called a **Cayley transformation**.

A (complex) square matrix T with the property ${}^tT = T^{-1}$ is called a **complex orthogonal matrix**. This matrix can be uniquely expressed as $T = R\exp(iA)$, where R is an orthogonal matrix, A is a real alternating matrix, and $i^2 = -1$.

K. Infinite Matrices

By an **infinite matrix** in a (noncommutative) field K, we mean an array of elements of K with infinite numbers of rows and columns as follows:

$$\begin{pmatrix} \cdots & \cdots & \cdots & \cdots \\ \cdots & a_{\sigma\tau} & \cdots & \cdots \\ \cdots & \cdots & \cdots & \cdots \end{pmatrix},$$

where σ, τ are indexes denoting the row and column for the element $a_{\sigma\tau}$ and each index ranges over an infinite set Γ. Equality, addition, and multiplication by an element of K of infinite matrices are defined in the same manner as for ordinary matrices. Generally, however, multiplication of infinite matrices cannot be defined. If the elements of each row (column) of an infinite matrix are zero except for a finite number of them, then it is called a **row (column) finite matrix**. For row (column) finite matrices $A=(a_{\sigma\tau})$ and $B=(b_{\sigma\tau})$, the product $AB=(c_{\sigma\tau})$ is defined by $c_{\sigma\tau}=\sum_\nu a_{\sigma\nu}b_{\nu\tau}$ for all $\sigma,\tau\in\Gamma$. By this definition the totality of such infinite matrices forms a ring. Now let $\mathfrak{M}$ be a †linear space of infinite dimension over K, that is, an infinite-dimensional †right K-module. Then $\mathfrak{M}$ has a basis $\{u_\tau\}$ consisting of linearly independent elements u_τ, and every element of $\mathfrak{M}$ can be expressed as a linear combination of a finite number of u_τ. Let A be a †linear transformation of $\mathfrak{M}$ expressed as $Au_\tau=\sum_\sigma u_\sigma a_{\sigma\tau}$ ($a_{\sigma\tau}\in K$). Then the coefficients $a_{\sigma\tau}$ for a fixed τ are zero except for a finite number of σ. Thus to A there corresponds a column finite matrix $(a_{\sigma\tau})$, and under this correspondence the ring of linear transformations of $\mathfrak{M}$ is isomorphic to the ring of column finite matrices. In the case where K is a topological field, these rings can be further discussed in detail.

Now let K be the complex number field and Γ the set of natural numbers, and consider an infinite matrix (a_{ik}). It is called a **bounded matrix** if the following inequality holds for arbitrary x_i and y_k:

$$\left|\sum_{i,k=1}^{m,n} a_{ik}x_iy_k\right|$$
$$\leqslant M\left(\sum_{i=1}^{m}|x_i|^2\right)^{1/2}\left(\sum_{k=1}^{n}|y_k|^2\right)^{1/2},$$

where M is a constant. The set of all bounded matrices forms a ring. If $\varphi_1,\varphi_2,\ldots$ is a complete orthonormal system of a †Hilbert space $\mathfrak{H}$, then for any continuous linear operator A we have $A\varphi_k=\sum_{i=1}^{\infty}\varphi_i a_{ik}$, and A corresponds to a bounded matrix (a_{ik}). By this correspondence we have a ring isomorphism between the ring of continuous linear operators of $\mathfrak{H}$ and the ring of bounded matrices (→ 199 Hilbert Spaces).

References

[1] M. Fujiwara, Gyôretu oyobi gyôretusiki (Japanese; Matrices and determinants), Iwanami, revised edition, 1961.
[2] K. Asano, Senkei daisûgaku teiyô (Japanese; A manual of linear algebra), Kyôritu, 1948.
[3] H. Tôyama, Gyôreturon (Japanese; Theory of matrices), Kyôritu, 1951.
[4] S. Furuya, Gyôretu to gyôretusiki (Japanese; Matrices and determinants), Baihûkan, 1957.
[5] I. Satake, Gyôretu to gyôretusiki (Japanese; Matrices and determinants), Syôkabô, 1958.
[6] O. Schreier and E. Sperner, Einführung in die analytische Geometrie und Algebra I, II, Teubner, 1931–1935; English translation, Introduction to modern algebra and matrix theory, Chelsea, 1959.
[7] C. C. MacDuffee, Theory of matrices, Springer, 1933 (Chelsea, 1946).
[8] J. H. M. Wedderburn, Lectures on matrices, Amer. Math. Soc. Colloq. Publ., 1934.
[9] N. Bourbaki, Eléments de mathématique, Algèbre, ch. 2, Actualités Sci. Ind. 1236b, Hermann, third edition, 1962.
[10] Ф. Р. Гантмахер (F. R. Gantmaher), Теория матриц, 1953; English translation, The theory of matrices I, II, Chelsea, 1959.
[11] А. Г. Курош (A. G. Kuroš), Курс высшей алгебры, 1955; English translation, Lectures on general algebra, Chelsea, 1970.
[12] А. И. Мальцев (A. I. Mal'cev), Основы линейной алгебры, 1956.
[13] R. Godement, Cours d'algèbre, Hermann, 1963.
[14] S. Lang, Linear algebra, Addison-Wesley, 1966.
[15] S. MacLane and G. Birkhoff, Algebra, Macmillan, 1967.
[16] S. Lang, Algebra, Addison-Wesley, 1965.

270 (X.12) Measure Theory

A. Rings of Sets

A collection $\mathfrak{B}$ of subsets of a fixed set X ($\neq\varnothing$) is called a **ring of sets** (more precisely, a **ring of sets in the space** X) if for an arbitrary pair A,B of elements in $\mathfrak{B}$, $A\cup B\in\mathfrak{B}$ and $A\cap B^c\in\mathfrak{B}$. A nonempty ring of sets $\mathfrak{B}$ is called a **field of sets** (or **finitely additive class**) if the complement of every element of $\mathfrak{B}$ also belongs to $\mathfrak{B}$. A ring of sets $\mathfrak{B}$ is called **completely additive** (**totally additive** or **countably additive**) if the union of a countable number of elements of $\mathfrak{B}$ always belongs to $\mathfrak{B}$. A completely additive field of sets is called a **σ-field** or a **completely additive class**. (A completely additive class is also called a **σ-algebra, countably additive class, σ-additive class, additive class, Borel class**, or **Borel field**. Some authors distinguish among these terms, but in general, they are used synonymously.) A pair $(X, \mathfrak{B})$, where X is a set and $\mathfrak{B}$ a completely additive class of subsets of X, is sometimes called a **measurable space**. A finitely additive class (resp. completely additive class) $\mathfrak{B}$ can also be defined as a collection of subsets of X satisfying (1), (2), (3) (resp. (3′)): (1) $\varnothing \in \mathfrak{B}$; (2) $A \in \mathfrak{B} \Rightarrow A^c \in \mathfrak{B}$; (3) $A, B \in \mathfrak{B} \Rightarrow A \cup B \in \mathfrak{B}$; (3′) $A_n \in \mathfrak{B}\,(n = 1, 2, \ldots) \Rightarrow \bigcup_{n=1}^{\infty} A_n \in \mathfrak{B}$. When $\mathfrak{B}$ is a completely additive class, a set E belonging to $\mathfrak{B}$ is called **$\mathfrak{B}$-measurable**.

For a sequence $\{A_n\}$ $(n \in \mathbf{N})$ of subsets of X, the **superior limit** (or **limit superior**) and the **inferior limit** (or **limit inferior**) are defined by $\limsup_{n\to\infty} A_n = \bigcap_{m=1}^{\infty}(\bigcup_{n=m}^{\infty} A_n)$ and $\liminf_{n\to\infty} A_n = \bigcup_{m=1}^{\infty}(\bigcap_{n=m}^{\infty} A_n)$, respectively. The symbols $\overline{\lim}$ and $\underline{\lim}$ may be used in place of lim sup and lim inf, respectively. The inferior limit is always contained in the superior limit, and when they coincide, they define the **limit** of $\{A_n\}$, which is denoted by $\lim_{n\to\infty} A_n$ (→ 89 Convergence). A **monotone sequence of sets**, i.e., a sequence $\{A_n\}$ of sets satisfying either $A_1 \subset A_2 \subset \ldots \subset A_n \subset A_{n+1} \subset \ldots$ (monotone increasing) or $A_1 \supset A_2 \supset \ldots \supset A_n \supset A_{n+1} \supset \ldots$ (monotone decreasing) has a limit, which equals $\bigcup_{n=1}^{\infty} A_n$ if $\{A_n\}$ is monotone increasing and $\bigcap_{n=1}^{\infty} A_n$ if $\{A_n\}$ is monotone decreasing. In a completely additive class $\mathfrak{A}$, both $\liminf_{n\to\infty} A_n$ and $\limsup_{n\to\infty} A_n$ belong to $\mathfrak{A}$ if $A_n \in \mathfrak{A}$ $(n \in \mathbf{N})$, and in particular, the limit of $\{A_n\}$ belongs to $\mathfrak{A}$, if it exists. Conversely, a finitely additive class $\mathfrak{A}$ is completely additive if $\lim_{n\to\infty} A_n \in \mathfrak{A}$ for every monotone increasing sequence $\{A_n\}$ of sets $A_n \in \mathfrak{A}$.

B. Construction of Completely Additive Classes

When a collection $\mathfrak{M}$ of subsets of X is given, there exists a smallest completely additive class $\mathfrak{B}(\mathfrak{M})$ containing $\mathfrak{M}$. Namely, (1) $\mathfrak{B}(\mathfrak{M})$ is a completely additive class and $\mathfrak{B}(\mathfrak{M}) \supset \mathfrak{M}$; (2) if a completely additive class $\mathfrak{B}$ contains $\mathfrak{M}$, then $\mathfrak{B} \supset \mathfrak{B}(\mathfrak{M})$. We call $\mathfrak{B}(\mathfrak{M})$ the completely additive class **generated** by $\mathfrak{M}$. Given a class of sets $\mathfrak{A}$, let $\mathfrak{A}_\sigma$ $(\mathfrak{A}_\delta)$ denote the collection of all sets that can be represented as the union (intersection) of a countable number of sets belonging to $\mathfrak{A}$. If for a class of sets $\mathfrak{M}$ we let $\mathfrak{M}_0$ be the collection of all sets of the form

$$\bigcup_{k=1}^{l} (A_{k1} \cap \ldots \cap A_{km_k} \cap B_{k1}^c \cap \ldots \cap B_{kn_k}^c),$$

where $A_{kj}, B_{kj} \in \mathfrak{M}$ and $l, m_1, \ldots, m_l, n_1, \ldots, n_l \in \mathbf{N}$, then $\mathfrak{M}_0$ is a finitely additive class. For an †ordinal number ξ we define the class $\mathfrak{M}_\xi$ as follows: If $\mathfrak{M}_\eta$ has already been defined for every ordinal number η such that $\eta < \xi$, we set $\mathfrak{M}_\xi = (\bigcup_{\eta<\xi} \mathfrak{M}_\eta)_\sigma$ if $\xi = 2\zeta$, and $\mathfrak{M}_\xi = (\bigcup_{\eta<\xi} \mathfrak{M}_\eta)_\delta$ if $\xi = 2\zeta + 1$. If ω_1 is the first uncountable ordinal number, then $\bigcup_{\xi<\omega_1} \mathfrak{M}_\xi$ is the completely additive class $\mathfrak{B}(\mathfrak{M})$ generated by $\mathfrak{M}$.

C. Borel Sets

In the n-dimensional Euclidean space $\mathbf{R}^n$, the sets of the form $\{x \mid a_k \leqslant x_k < b_k,\ k = 1, 2, \ldots, n\}$ $(a_k \leqslant b_k\ (k = 1, 2, \ldots, n))$ are called **half-open intervals**. The completely additive class generated by the class of all half-open intervals is called the **class of Borel sets** in $\mathbf{R}^n$, and a set belonging to this class is called a **Borel set**. Open sets and closed sets are Borel sets. Since any half-open interval can be represented as the union of a countable number of closed sets or the intersection of a countable number of open sets, the class of Borel sets in $\mathbf{R}^n$ can also be defined as the completely additive class generated by the class of all closed sets or the class of all open sets in $\mathbf{R}^n$.

More generally, in a topological space X, the completely additive class generated by the class of all closed sets is called the **Borel field**, and sets belonging to it are called **Borel sets** (or **$\mathfrak{B}$-measurable sets**). The completely additive class generated by the class of all open sets coincides with the Borel field. If the †characteristic function of a set is a †Baire function, then the set is a Borel set. Such a set is called a **Borel set in the strict sense** (or a **Baire set**). The union (intersection) of at most a countable number of closed (open) sets is called a **F_σ set** (**G_δ set**). Both F_σ sets and G_δ sets are Borel sets. Denote by $\mathfrak{F}_0$ the class of all closed subsets of X and by $\mathfrak{G}_0$ the class of all open subsets, and define classes $\mathfrak{F}_\xi$ and $\mathfrak{G}_\xi$ for an ordinal ξ as we defined $\mathfrak{M}_\xi$ in Section B. Then by construction, every set belonging to either $\mathfrak{F}_\xi$ or $\mathfrak{G}_\xi$ is a Borel set. If $\omega_1 < \xi$, then $\mathfrak{F}_\xi = \mathfrak{F}_{\omega_1}, \mathfrak{G}_\xi = \mathfrak{G}_{\omega_1}$. In particular, in a completely normal topological space (for example, a metric space), $\mathfrak{F}_\alpha \subset \mathfrak{G}_\beta$, $\mathfrak{G}_\alpha \subset \mathfrak{F}_\beta$ if $\alpha < \beta$, and both $\bigcup_{\xi<\omega_1} \mathfrak{F}_\xi$ and $\bigcup_{\xi<\omega_1} \mathfrak{G}_\xi$ coincide with the Borel field. In this case, an arbitrary Borel set B is a Baire set, and thus can be classified according to the †Baire class of the characteristic function of B. It can also be classified according to the smallest ordinal number ξ with respect to which $B \in \mathfrak{F}_\xi$ (or $B \in \mathfrak{G}_\xi$).

D. Measure

A real-valued set function m defined on a finitely additive class $\mathfrak{M}$ in a space X is called a **finitely additive measure** (or **Jordan measure**) if it satisfies the following two conditions: (I) $0 \leqslant m(E) \leqslant \infty$, $m(\varnothing)=0$; (II_0) $A, B \in \mathfrak{M}$, $A \cap B=\varnothing$ imply $m(A \cup B)=m(A)+m(B)$. We call m a **bounded measure** if $m(X)<\infty$, and a **σ-finite measure** if there exists a sequence $\{X_n\}$ such that $X_1 \subset X_2 \subset \ldots$, $X=\bigcup_{n=1}^{\infty} X_n$, $m(X_n)<\infty$, and for an arbitrary $A \in \mathfrak{M}$, $m(A)=\lim_{n \rightarrow \infty} m(A \cap X_n)$.

A set function μ defined on a completely additive class $\mathfrak{B}$ in a space X is called a **measure (completely additive measure** or **σ-additive measure)** on $\mathfrak{B}$ (or on X) if it satisfies the following two conditions: (I) $0 \leqslant \mu(E) \leqslant \infty$, $\mu(\varnothing)=0$; (II) $E_n \in \mathfrak{B}$ $(n=1,2, \ldots)$, $E_j \cap E_k=\varnothing$ $(j \neq k)$ imply $\mu(\bigcup_{n=1}^{\infty} E_n)=\sum_{n=1}^{\infty} \mu(E_n)$ (**complete additivity**). The triple $(X, \mathfrak{B}, \mu)$ or the pair (X, μ), where X is a space, $\mathfrak{B}$ a completely additive class of subsets of X, and μ a measure, is called a **measure space**. We call $(X, \mathfrak{B}, \mu)$ **bounded** if $\mu(X)<\infty$, and **σ-finite** if there exists a sequence $\{X_n\}$ satisfying $\mu(X_n)<\infty$ and $\bigcup_{n=1}^{\infty} X_n=X$. The simplest bounded measure is given by defining $m(A)=1$ if $a \in A$ and $m(A)=0$ if $a \notin A$, where a is a fixed point of X. Such a measure is called the Dirac **δ-measure**. A measure m with $m(X)=1$ is called a †probability measure (→ 339 Probability).

Equalities appearing in (II_0) and (II) include the possibility that both sides equal ∞. Addition and multiplication involving $\pm\infty$ are carried out as follows (a denotes a real number): $(\pm\infty)+a=a+(\pm\infty)=\pm\infty$; $(\pm\infty)-a=\pm\infty$, $a-(\pm\infty)=\mp\infty$; $(\pm\infty)+(\pm\infty)=\pm\infty$; $(\pm\infty)-(\mp\infty)=\pm\infty$; $a\cdot(\pm\infty)=(\pm\infty)\cdot a=\pm\infty$ if $a>0$; $a\cdot(\pm\infty)=(\pm\infty)\cdot a=\mp\infty$ if $a<0$; $(+\infty)\cdot(+\infty)=+\infty$. (Sometimes it is further agreed to put $0\cdot(\pm\infty)=(\pm\infty)\cdot 0=0$.)

A $\mathfrak{B}$-measurable set in a measure space $(X, \mathfrak{B}, \mu)$ is also called **μ-measurable** (or simply **measurable**). For a sequence $\{A_n\}$ of μ-measurable sets the following conditions hold: (i) $\mu(\liminf_{n \rightarrow \infty} A_n) \leqslant \liminf_{n \rightarrow \infty} \mu(A_n)$; (ii) if $\mu(\bigcup_{n=n_0}^{\infty} A_n)<+\infty$ for some n_0, then $\mu(\limsup_{n \rightarrow \infty} A_n) \geqslant \limsup_{n \rightarrow \infty} \mu(A_n)$; and (iii) if $\lim_{n \rightarrow \infty} A_n$ exists and the hypothesis of (ii) holds, then $\mu(\lim_{n \rightarrow \infty} A_n)=\lim_{n \rightarrow \infty} \mu(A_n)$. For a finitely additive measure μ on a completely additive class $\mathfrak{B}$ to be a completely additive measure, it is necessary and sufficient that $\mu(\lim_{n \rightarrow \infty} A_n)=\lim_{n \rightarrow \infty} \mu(A_n)$ holds for an arbitrary monotone increasing sequence $\{A_n\}$ of sets in $\mathfrak{B}$.

If every subset of a set E satisfying $\mu(E)=0$ belongs to $\mathfrak{B}$, then we say that $(X, \mathfrak{B}, \mu)$ **is complete**, or more simply, μ is complete. **If** $(X, \mathfrak{B}, \mu)$ is not complete, the complete measure space obtained by adjoining to $(X, \mathfrak{B}, \mu)$ every subset of a set E with $\mu(E)=0$ is called the **completion** of $(X, \mathfrak{B}, \mu)$. A set E satisfying $\mu(E)=0$ is called a **null set**. When the set of all points for which a property **P** fails to hold is a null set, then we say that **P** holds **almost everywhere (a.e.** or **at almost all points)**. Sometimes we use the expression "almost **P**" to describe the same situation.

E. Construction of Measures

Suppose that a completely additive class $\mathfrak{A}$ in the space X satisfies the property that if $A \in \mathfrak{A}$ and $B \subset A$ then $B \in \mathfrak{A}$. A set function $\mu^*(A)$ defined on $\mathfrak{A}$ is called a **Carathéodory outer measure** (or **outer measure**) if it has the following three properties: (i) $0 \leqslant \mu^*(A) \leqslant \infty$, $\mu^*(\varnothing)=0$; (ii) $A \subset B \Rightarrow \mu^*(A) \leqslant \mu^*(B)$; (iii) $\mu^*(\bigcup_{n=1}^{\infty} A_n) \leqslant \sum_{n=1}^{\infty} \mu^*(A_n)$. Because of (iii), the inequality $\mu^*(B) \leqslant \mu^*(B \cap A)+\mu^*(B \cap A^c)$ always holds, whereas if the equality holds for every $B \in \mathfrak{A}$, A is called **measurable with respect to** μ^*. It follows from this definition that a set A satisfying $\mu^*(A)=0$ is always measurable with respect to μ^*. The class $\mathfrak{B}$ of all sets measurable with respect to μ^* is a completely additive class, and if we define $\mu(A)=\mu^*(A)$ for A belonging to $\mathfrak{B}$, then $\mu(A)$ gives a complete measure on $\mathfrak{B}$. This measure is called the **Carathéodory measure** induced by μ^* or the **generalized Lebesgue measure**.

A finitely additive measure m defined on a finitely additive class $\mathfrak{M}$ is called **completely additive** when $A_n \in \mathfrak{M}$, $A_j \cap A_k=\varnothing$ $(j \neq k)$, $\bigcup_{n=1}^{\infty} A_n \in \mathfrak{M}$ imply $m(\bigcup_{n=1}^{\infty} A_n)=\sum_{n=1}^{\infty} m(A_n)$. If by means of such an m we define $\mu^*(A)$ for an arbitrary $A \subset X$ to be the infimum of all possible values $\sum_{n=1}^{\infty} m(A_n)$, where $A \subset \bigcup_{n=1}^{\infty} A_n$ $(A_n \in \mathfrak{M})$, then μ^* gives a Carathéodory outer measure. From this μ^* a measure μ on a completely additive class $\mathfrak{B}$ is induced as described in this section, and we have $\mathfrak{B} \supset \mathfrak{M}$ and $\mu(A)=m(A)$ for $A \in \mathfrak{M}$. Therefore, m can be extended to a measure μ on $\mathfrak{B}(\mathfrak{M})$ (**E. Hopf's extension theorem**). In particular, if m is σ-finite or bounded, then this extension is unique.

For a Carathéodory outer measure μ^* defined on a completely additive class $\mathfrak{A}$, let $\mathfrak{B}$ denote the collection of all sets measurable with respect to μ^*, and μ the measure μ^* induces. If $\mu^*(A)=\inf\{\mu(B) \mid B \in \mathfrak{B}, A \subset B\}$ for every $A \in \mathfrak{A}$, then the outer measure μ^* or the measure μ is called **regular**. In contrast to the outer measure, for a regular measure we can define the **inner measure** of a set $A \in \mathfrak{A}$ by $\mu_*(A) = \sup\{\mu(B) \mid B \in \mathfrak{B}, A \supset B\}$. If $A \in \mathfrak{A}$, $B \in \mathfrak{B}$, $\mu(B) < +\infty$, and $A \subset B$, then $\mu_*(A) = \mu(B) - \mu^*(B - A)$. Also, a set A with $\mu^*(A) < \infty$ is measurable if and only if $\mu^*(A) = \mu_*(A)$. For an arbitrary $A \in \mathfrak{A}$ there exist measurable sets B and C such that $A \subset B$, $\mu^*(A) = \mu(B)$, $C \subset A, \mu_*(A) = \mu(C)$. We call B the **measurable cover** of A, and C the **measurable kernel** of A.

F. Lebesgue Measure

In the n-dimensional Euclidean space $\mathbf{R}^n$, the volume of a half-open interval $I = \{x \mid a_k \leqslant x_k < b_k, k = 1, 2, \ldots, n\}$ $(-\infty \leqslant a_k < b_k \leqslant \infty)$ is defined by $m(I) = \prod_{k=1}^{n}(b_k - a_k)$. Let $\mathfrak{M}_0$ be the collection of all sets that can be represented as the finite union of disjoint half-open intervals, and for such expression $A = \bigcup_{j=1}^{r} I_j$ in $\mathfrak{M}_0$ define $m(A) = \sum_{j=1}^{r} m(I_j)$. Let $\mathfrak{M} = \mathfrak{M}_0 \cup \{\varnothing\}$ and $m(\varnothing) = 0$. Then m gives a finitely additive measure on $\mathfrak{M}$ that is completely additive. Therefore, m determines an outer measure μ^*, which in turn determines a measure μ. This μ^* is called the **Lebesgue outer measure** (or simply **outer measure**). This outer measure is the one that Carathéodory first considered. Sets that are measurable with respect to μ^* are said to be **Lebesgue measurable** (or simply **measurable**), and the measure μ is called the **Lebesgue measure** (or simply **measure**). Every interval is measurable, and its measure coincides with its volume. Open sets, closed sets, and Borel sets are all measurable. More generally, suppose that an outer measure μ^* defined on a metric space X with a metric d satisfies the condition: If $d(A, B) > 0$, then $\mu^*(A \cup B) = \mu^*(A) + \mu^*(B)$. Then every closed subset of X is μ^*-measurable, and therefore so is every Borel subset. Here $d(A, B) = \inf\{d(a, b) \mid a \in A, b \in B\}$ denotes the distance between the two sets A and B. A measure μ defined on the class of all Borel subsets of a topological space X is called a **Borel measure**. It is customary to assume that a Borel measure satisfies the condition $\mu(E) < \infty$ for every compact subset E. The Lebesgue measure on $\mathbf{R}^n$ satisfies this condition. The cardinality of the set of all Lebesgue measurable subsets of $\mathbf{R}^n$ is $2^{\mathfrak{c}}$, while the cardinality of the class of Borel sets is $\mathfrak{c}$ (here $\mathfrak{c}$ is the cardinal number of the †continuum, i.e., the cardinal number of $\mathbf{R}$). Therefore, there exists a Lebesgue measurable set that is not a Borel set.

Historically, for a bounded subset A of $\mathbf{R}^n$, C. Jordan defined $\overline{m}(A)$ to be the infimum of all possible values $m(B)$, where $B \in \mathfrak{M}$ and $B \supset A$, and $\underline{m}(A)$ to be the supremum of all possible values $m(B)$, where $B \in \mathfrak{M}$ and $B \subset A$. He called $\overline{m}(A)$ the **outer volume** of A and $\underline{m}(A)$ the **inner volume** of A (in the case of $\mathbf{R}^2$, the **outer area** and **inner area**, respectively). When $\overline{m}(A) = \underline{m}(A)$, A is called **Jordan measurable**, and this common value is defined to be the **Jordan measure** (**Jordan content**) of A. Jordan measure is only finitely additive, and was found to be unsatisfactory in many respects. It was Lebesgue who modified this notion and introduced completely additive measures. Jordan measurable sets are always Lebesgue measurable.

The Lebesgue measure in $\mathbf{R}^n$ is regular. Therefore it can be defined alternatively as follows: For a set A with Lebesgue outer measure $\mu^*(A) < \infty$, the value $\mu^*(G) - \mu^*(G - A)$ for an open set G of finite outer measure containing A is independent of G. Define the **Lebesgue inner measure** $\mu_*(A)$ of A to be this number. Then $\mu_*(A) \leqslant \mu^*(A)$ for any A. Call a set A **Lebesgue measurable** if the equality holds, and denote the common value by $\mu(A)$. If $\mu^*(A) = \infty$, then call A measurable if $A \cap G$ is measurable for every bounded open set G, and assign $\mu(A) = \infty$. In $\mathbf{R}^n$, we can choose for any set A a measurable cover that is a G_δ set and a measurable kernel that is an F_σ set. Therefore, any measurable set can be written as the difference of a G_δ set and a null set or the union of an F_σ set and a null set.

Using the †axiom of choice, a set that is not Lebesgue measurable can be constructed (G. Vitali). For example, a set obtained by choosing exactly one element from each coset of the additive group of all rationals in the additive group of the reals is not measurable [3, p. 67–70].

If two subsets A and B of $\mathbf{R}^n$ are congruent, i.e., if one can be mapped onto the other by means of a Euclidean motion, then A is measurable if and only if B is, and if measurable, they have the same Lebesgue measure. In other words, Lebesgue measure can be characterized as †Haar measure for the group of Euclidean motions.

G. Product Measure

When two measures μ_X and μ_Y are defined on completely additive classes $\mathfrak{B}_X$ and $\mathfrak{B}_Y$ of X and Y, respectively, an element C belonging to the smallest finitely additive class $\mathfrak{R}$ in the Cartesian product that contains $\{A \times B \mid A \in \mathfrak{B}_X, B \in \mathfrak{B}_Y\}$ can be represented as a finite disjoint union $C = \bigcup_{j=1}^{n}(A_j \times B_j)$ $(A_j \in \mathfrak{B}_X, B_j \in \mathfrak{B}_Y)$. If we define $\nu(C) = \sum_{j=1}^{n} \mu_X(A_j) \cdot \mu_Y(B_j)$ (here we agree to put $0 \cdot \infty = 0$), then this value is independent of the way the set

C is represented, and ν defines a completely additive measure on $\mathfrak{R}$. By extending ν by means of Hopf's extension theorem, we obtain a (complete) measure space, called the **(complete) product measure space** obtained from the measure spaces $(X, \mathfrak{B}_X, \mu_X)$ and $(Y, \mathfrak{B}_Y, \mu_Y)$. The measure obtained in this way on the space $X \times Y$ is called the **product measure** of μ_X and μ_Y and is denoted by $\mu_X \times \mu_Y$. If we denote by $\mathfrak{M}_p$ the class of all Lebesgue measurable subsets of the p-dimensional Euclidean space $\mathbf{R}^p$ and by m_p the p-dimensional Lebesgue measure, then the (complete) product measure space of $(\mathbf{R}^p, \mathfrak{M}_p, m_p)$ and $(\mathbf{R}^q, \mathfrak{M}_q, m_q)$ is $(\mathbf{R}^{(p+q)}, \mathfrak{M}_{p+q}, m_{p+q})$. The product measure space of any finite number of measure spaces $(X_i, \mathfrak{B}_i, \mu_i)$ $(i=1, \ldots, n)$ is defined similarly.

Let X_λ $(\lambda \in \Lambda)$ be spaces with an index set of arbitrary cardinality. For the product space $X = \prod_{\lambda \in \Lambda} X_\lambda$, an **$n$-cylinder set**, or simply a **cylinder set**, is a set of the form $A \times \prod_{\lambda \notin \{\lambda_1, \ldots, \lambda_n\}} X_\lambda$ $(A \subset X_{\lambda_1} \times \ldots \times X_{\lambda_n})$. If a finitely additive class $\mathfrak{A}_\lambda$ is given for each X_λ, the class of all subsets that can be represented as the union of a finite number of cylinder sets of the form $A_1 \times A_2 \times \ldots \times A_n \times \prod_{\lambda \notin \{\lambda_1, \ldots, \lambda_n\}} X_\lambda$ $(A_j \in \mathfrak{A}_{\lambda_j}, j=1,2, \ldots, n)$ is a finitely additive class in the space X. When each $\mathfrak{A}_\lambda$ is completely additive, the completely additive class $\mathfrak{A}$ generated by this finitely additive class is called the **product** of the completely additive classes $\mathfrak{A}_\lambda$ and is denoted by $\prod_{\lambda \in \Lambda} \mathfrak{A}_\lambda$. When a measure space $(X_\lambda, \mathfrak{B}_\lambda, \mu_\lambda)$ with $\mu_\lambda(X_\lambda) = 1$ is given for $\lambda \in \Lambda$, a measure μ can be defined in the following way on the completely additive class $\mathfrak{B} = \prod_{\lambda \in \Lambda} \mathfrak{B}_\lambda$ in the product space X: To begin with, for a cylinder set of the form $A_1 \times \ldots \times A_n \times X'$ (here $A_j \in \mathfrak{B}_{\lambda_j}$ and $X' = \prod_{\lambda \notin \{\lambda_1, \ldots, \lambda_n\}} X_\lambda$), we define $\mu(A_1 \times \ldots \times A_n \times X') = \mu_{\lambda_1}(A_1) \ldots \mu_{\lambda_n}(A_n)$. If we extend μ to the finitely additive class $\mathfrak{C}$ consisting of all sets that can be represented as the finite union of such cylinder sets, then this extension gives a completely additive measure on $\mathfrak{C}$, and therefore, by Hopf's extension theorem, there exists a unique extension to a measure μ on $\mathfrak{B}$, and μ satisfies $\mu(X) = 1$. We denote this μ by $\mu = \prod_{\lambda \in \Lambda} \mu_\lambda$.

H. Radon Measure

Let X be a †locally compact Hausdorff space, and let $C_0(X)$ be the real linear space of all real-valued continuous functions f on X having †compact support (i.e., the closure of the set $\{x \mid f(x) \neq 0\}$ is compact). A (real) †linear functional φ defined on $C_0(X)$ is called a **positive Radon measure** if $\varphi(f) \geqslant 0$ whenever $f \geqslant 0$. For such a functional φ there corresponds a regular measure μ with respect to which every Borel subset of X is $\mathfrak{B}$-measurable and for which $\varphi(f) = \int_X f d\mu$ holds for every $f \in C_0(X)$. Here a positive Radon measure μ is called **regular** if $\mu(B) = \inf\{\mu(G) \mid G \supset B, G \text{ open}\}$ for every $B \in \mathfrak{B}$. If X is σ-compact (i.e., X can be represented as the countable union of compact sets), then the property $\varphi(f) = \int_X f d\mu$ for all $f \in C_0(X)$ defines the measure μ uniquely on the class of Borel sets of X. A linear functional on $C_0(X)$ that can be written as the difference of two positive Radon measures is called a **Radon measure**. For a linear functional φ defined on $C_0(X)$ to be a Radon measure, it is necessary and sufficient that for an arbitrary $f \in C_0(X)$, the set $\{\varphi(g) \mid |g| \leqslant |f|, g \in C_0(X)\}$ is bounded. Equivalently, it is necessary and sufficient that the restriction of φ to the subspace of all functions in $C_0(X)$ having their support in a fixed compact subset of X must be continuous with respect to the topology of uniform convergence. Therefore, if X is compact, an arbitrary continuous linear functional on $C(X)$ is a Radon measure. L. Schwartz investigated Radon measures on spaces that are not locally compact [15].

I. Measurable Functions

When a completely additive class $\mathfrak{B}$ on a space X is given, a function f defined on a $\mathfrak{B}$-measurable set E and taking real (and possibly $\pm\infty$) values is called a **$\mathfrak{B}$-measurable function** on E if for an arbitrary real number α, the set $\{x \mid f(x) > \alpha\}$ is $\mathfrak{B}$-measurable. The condition $f > \alpha$ may be replaced by $f \geqslant \alpha$, $f < \alpha$, $f \leqslant \alpha$. A function f may also be defined to be $\mathfrak{B}$-measurable if the †inverse image under f of any Borel set is $\mathfrak{B}$-measurable. When f and g are $\mathfrak{B}$-measurable, so are $af + bg$ (a, b constants), $f - g$, f/g, $\max(f,g)$, $\min(f,g)$, $|f|^p$ (p a constant), whenever they are well defined. The superior and inferior limits of a sequence of $\mathfrak{B}$-measurable functions are also $\mathfrak{B}$-measurable. In a complete product space of two σ-finite measure spaces, a function $f(x,y)$ may fail to be measurable as a function of two variables even if it is measurable with respect to each of the variables x and y separately. For example, let $\mathfrak{F}$ be the class of all closed subsets F of $\mathbf{R}^2$ satisfying $m_2(F) > 0$, $F \subset [0,1] \times [0,1]$. Using the fact that the cardinality of $\mathfrak{F}$ does not exceed the cardinal number of the continuum, we can prove by transfinite induction that for every $F_\xi \in \mathfrak{F}$ we can pick two points $z_\xi = (x_\xi, y_\xi)$, $z'_\xi = (x'_\xi, y'_\xi)$ in such a way that if $F_\xi \neq F_\eta$, then $x_\xi, x'_\xi, x_\eta, x'_\eta$ are all distinct and $y_\xi, y'_\xi, y_\eta, y'_\eta$ are all distinct. Furthermore, we can prove that the set E consisting of all such z_ξ is not measurable. Therefore, denoting the characteristic function of the set E by $f(x,y)$, $f(x,y)$ is not measurable; but if we fix x (resp. y), then as a function of y (resp. x), $f(x,y)$ is measurable, since $f(x,\cdot)$ (resp. $f(\cdot,y)$) is always 0 except possibly at one point. If $\mathfrak{B}$ is the class of all Lebesgue measurable sets or the class of all Borel sets, then a $\mathfrak{B}$-measurable function is called a **Lebesgue measurable function** (or simply **measurable function**) or a **Borel measurable function**, respectively. The composite of two measurable functions may not be measurable. In particular, for composites of Lebesgue measurable functions and Borel measurable functions in $\mathbf{R}^n$, the following statements are valid:

$$B\circ B=B, \quad L\circ B=L, \quad B\circ L=\times, \quad L\circ L=\times,$$

where B denotes Borel measurability, L denotes Lebesgue measurability, and $\times$ means that the resultant may not be measurable.

In Euclidean space, the class of all Borel measurable functions coincides with the class of all Baire functions, and an arbitrary Lebesgue measurable function is equal almost everywhere to a Baire function of at most the second class. For a function f that is finite almost everywhere on a Lebesgue measurable set E to be **Lebesgue measurable**, it is necessary and sufficient that for an arbitrary $\varepsilon>0$ we can find a closed subset F such that $m(E-F)<\varepsilon$ and f is continuous on F (**Luzin's theorem**).

If a sequence $\{f_n\}$ of $\mathfrak{B}$-measurable functions on a measure space $(X,\mathfrak{B},\mu)$ converges almost everywhere to f on a set E with $\mu(E)<\infty$, then for an arbitrary $\varepsilon>0$ we can find a set F $(F\subset E, F\in\mathfrak{B})$ such that $\mu(E-F)<\varepsilon$ and f_n converges uniformly on F. If $X=\mathbf{R}^n$ and $\mathfrak{B}$ is either the class of Borel sets or the class of Lebesgue measurable sets, then the set F can be chosen to be a closed set (**Egorov's theorem**).

For a finite measurable function $f(x)$ defined on the real line, there exists a sequence $\{h_n\}$ such that $\lim_{n\to\infty}f(x+h_n)=f(x)$ almost everywhere (H. Auerbach).

The functional equation $f(x+y)=f(x)+f(y)$ has infinitely many nonmeasurable solutions (G. Hamel; → 380 Special Functional Equations).

For related topics, → 243 Lebesgue Integral, 375 Set Functions, 225 Invariant Measures.

References

[1] S. Saks, Theory of the integral, Warsaw, 1937.

[2] F. Hausdorff, Mengenlehre, Teubner, second edition, 1927; English translation, Set theory, Chelsea, 1962.

[3] P. R. Halmos, Measure theory, Van Nostrand, 1950.

[4] C. Carathéodory, Vorlesungen über reelle Funktionen, Teubner, 1918 (Chelsea, 1968).

[5] H. L. Royden, Real analysis, Macmillan, second edition, 1968.

[6] T. Inagaki, Tensyûgôron (Japanese; Theory of point sets), Iwanami, 1949.

[7] H. Nakano, Sokudo ron (Japanese; Measure theory), Syôkabô, I, 1947; II, 1948; III, 1950.

[8] Y. Kawada, Sekibunron (Japanese; Theory of integrals), Kyôritu, 1959.

[9] T. Takagi, Kaiseki gairon (Japanese; A course of analysis), Iwanami, third edition, 1961.

[10] M. Tsuji, Zitu kansûron (Japanese; Theory of real functions), Maki, 1962.

[11] S. Ito, Rubegu sekibun nyûmon (Japanese; Introduction to Lebesgue integrals), Syôkabô, 1963.

[12] Y. Yosida, Rubegu sekibun nyûmon (Japanese; Introduction to the theory of Lebesgue integrals), Baihûkan, 1965.

[13] Y. Kawada and Y. Mimura, Gendai sûgaku gaisetu (Japanese; Introduction to modern mathematics) II, Iwanami, 1965.

For the history of Lebesgue integrals,

[14] S. Iyanaga, Gendai sûgaku no kiso gainen (Japanese; Fundamental concepts of modern mathematics), Kôbundô, 1944, ch. 6, sec. 3-V.

[15] L. Schwartz, Probabilités cylindriques et applications radonifiantes, J. Fac. Sci. Univ. Tokyo, IA, 18 (1971), 139–286.

[16] N. Bourbaki, Elements de mathématique, Intégration, Actualités Sci. Ind., 1175a, 1244b, 1281, 1306, 1343, Herman, second edition, 1965, 1967, 1959, 1963, 1969.

271 (XI.8) Meromorphic Functions

A. General Remarks

A single-valued †analytic function in a domain D in the complex plane $\mathbf{C}$ is called **meromorphic** in D if it has no singularities other than †poles. A function that is meromorphic in the whole complex plane including the point at infinity is a rational function (**Liouville's theorem**). Specifically, if a function is meromorphic in the domain $\mathbf{C}$, then the function is called simply a **meromorphic function**, and a meromorphic function that is not a rational function is called a **transcen-**

dental meromorphic function. A meromorphic function $f(z)$ can be represented as a quotient of two †entire functions. Let $\{z_k\}$ $(k=1,2,\ldots)$ be poles of $f(z)$, and let $f_k(z)=a_{n_k}^{(k)}/(z-z_k)^{n_k}+\ldots+a_1^{(k)}/(z-z_k)$ denote the †singular parts of $f(z)$ at z_k $(k=1,2,\ldots)$. Then $f(z)$ can also be written in the form

$$f(z)=g(z)+\sum_{k=1}^{\infty}(f_k(z)-p_k(z)),$$

where $g(z)$ is an entire function and the $p_k(z)$ $(k=1,2,\ldots)$ are rational entire functions **(Weierstrass's theorem)**. Assume that a sequence $\{z_k\}$ $(k=1,2,\ldots)$ converges only to the point at infinity and that $f_k(1/(z-z_k))$ $(k=1,2,\ldots)$ are rational entire functions of $1/(z-z_k)$ which have no constant terms. Then there exists a meromorphic function of z with $f_k(1/(z-z_k))$ as its †singular part at z_k **(Mittag-Leffler's theorem)**.

B. Nevanlinna Theory

The theory of meromorphic functions can be considered an extension of the theory of entire functions. In particular, value distribution theory, originating in Picard's theorem, was studied by many people, and in 1925 R. Nevanlinna published a systematic theory unifying the results obtained until then. This is called **Nevanlinna theory**.

We let $f(z)$ denote a meromorphic function in $|z|<R\leqslant+\infty$, and when we say that $f(z)$ takes on a value, the value may be ∞. For a value α, $n(r,\alpha)$ denotes the number of α-**points** of $f(z)$, i.e., points z with $f(z)=\alpha$, in $|z|\leqslant r<R$, where each α-point is counted with its multiplicity. We set

$$N(r,\alpha)=\int_0^r\frac{n(t,\alpha)-n(0,\alpha)}{t}dt+n(0,\alpha)\log r,$$

$$m(r,\alpha)=\frac{1}{2\pi}\int_0^{2\pi}\log^+\left|\frac{1}{f(re^{i\theta})-\alpha}\right|d\theta$$

if $\alpha\neq\infty$, and

$$N(r,\infty)=\int_0^r\frac{n(t,\infty)-n(0,\infty)}{t}dt$$
$$+n(0,\infty)\log r,$$

$$m(r,\infty)=\frac{1}{2\pi}\int_0^{2\pi}\log^+|f(re^{i\theta})|\,d\theta$$

if $\alpha=\infty$, where $\log^+a=\max(\log a,0)$ for $a>0$. The functions N and m are called the **counting function** and **proximity function** of $f(z)$, respectively, and $T(r)=m(r,\infty)+N(r,\infty)$ is the **order function** (or **characteristic function**) of $f(z)$. $T(r)$ is an increasing function of r and a †convex function of $\log r$, and is useful for expressing $f(z)$ as an infinite product, etc.

The following relation holds among $T(r)$, $m(r,\alpha)$, and $N(r,\alpha)$ for any α:

$$T(r)=m(r,\alpha)+N(r,\alpha)+O(1),\qquad(1)$$

where $O(1)$ is †Landau's symbol (**Nevanlinna's first fundamental theorem**). By this theorem, if a bounded remainder is disregarded, then $m(r,\alpha)+N(r,\alpha)$ is equal to $T(r)$ for all α. This equality thus demonstrates a beautifully balanced distribution of α-points.

We see that $N(r,\alpha)$ is in a sense the mean value of the number of α-points in $|z|\leqslant r$, and $m(r,\alpha)$ is the mean proximity to α of the value $f(z)$ on $|z|=r$. If the term $\log^+$ in the definition of the proximity function is replaced by the logarithm of the reciprocal of the chordal distance between $f(re^{i\theta})$ and α on the complex sphere, then the remainder term in (1) is eliminated. Hence the definition of the proximity function is sometimes given in this form.

C. The Order of Meromorphic Functions

For an entire function $f(z)$, the equality

$$\limsup_{r\to\infty}\frac{\log T(r)}{\log r}=\limsup_{r\to\infty}\frac{\log\log M(r)}{\log r}$$

holds, where $M(r)=\max_{|z|=r}|f(z)|$. Since the right-hand side is the order of $f(z)$ (→ 413 Transcendental Entire Functions), we define the **order** ρ of a meromorphic function $f(z)$ by

$$\limsup_{r\to\infty}\frac{\log T(r)}{\log r}=\rho.$$

The **order** of a meromorphic function in $|z|<R$ is also defined by

$$\limsup_{r\to R}\frac{\log T(r)}{\log(1/(R-r))}=\rho.$$

D. Meromorphic Functions on a Disk

The order function $T(r)$ is bounded if and only if $f(z)$ can be represented as the quotient of two bounded holomorphic functions $h_1(z)$, $h_2(z)$ in $|z|<R$ (Nevanlinna). If $T(r)$ is bounded, $\lim_{r\to R}f(re^{i\theta})$ exists and is finite for every θ, $0\leqslant\theta<2\pi$, except possibly for a set with †linear measure zero (P. Fatou and Nevanlinna). Among functions $f(z)$ such that $\lim_{r\to R}T(r)=\infty$, those satisfying

$$\limsup_{r\to R}\frac{T(r)}{\log(1/(R-r))}=\infty$$

have properties similar to those of transcendental meromorphic functions.

E. Meromorphic Functions in the Whole Finite Plane

Any meromorphic function such that

$$\limsup_{r\to R}\frac{T(r)}{\log r}<K$$

is a rational function. If $f(z)$ is a meromorphic function of order ρ and $\{r_j(\alpha)\}, r_j(\alpha)\leqslant r_{j+1}(\alpha)$ $(j=1,2,\ldots)$ is the set of absolute values of α-points, then $\sum_{j=1}^{\infty}(1/r_j(\alpha))^{\rho+\varepsilon}$ converges for any α. Furthermore,
$f(z)=z^k e^{P(z)}$

$$\times\lim_{r\to\infty}\prod_{\substack{|a_\mu|<r\\|b_\nu|<r}}\frac{\left(1-\dfrac{z}{a_\mu}\right)\exp\left(\dfrac{z}{a_\mu}+\ldots+\dfrac{z^p}{pa_\mu^p}\right)}{\left(1-\dfrac{z}{b_\nu}\right)\exp\left(\dfrac{z}{b_\nu}+\ldots+\dfrac{z^p}{pb_\nu^p}\right)},$$

where p is the smallest integer satisfying $p+1>\rho$, the a_μ and b_ν are the zeros and poles of $f(z)$, respectively, k is an integer, and $P(z)$ is a polynomial of degree at most p (**Hadamard's theorem**).

Let $\alpha_1,\ldots,\alpha_q$ $(q\geqslant 3)$ be distinct values. Then for any meromorphic function in $|z|<R\leqslant\infty$,

$$(q-2)T(r)<\sum_{j=1}^{q}N(r,\alpha_j)-N_1(r)+D(r),\qquad 0\leqslant r<R.$$

Here

$$N_1(r)=\int_0^r\frac{n_1(t)-n_1(0)}{t}dt+n_1(0)\log r,$$

$n_1(r)$ is the number of †multiple points in $|z|\leqslant r$ (a multiple point of order k is counted $k-1$ times), and $D(r)$ is the remainder such that if $R=\infty$, then $D(r)<K(\log T(r)+\log r)$ for some K except possibly for values of r belonging to the union of a countable number of intervals with finite total length, and if $R<\infty$, then $D(r)<K(\log T(r)+\log(1/(R-r)))$ except possibly for the union of a countable number of intervals $\{I_j\}$ with $\sum_j\int_{I_j}d(1/(R-r))<\infty$ (**Nevanlinna's second fundamental theorem**).

Several theorems on value distribution of meromorphic functions can be obtained directly from this theorem. For instance, if $f(z)$ is a transcendental meromorphic function, the equation $f(z)=\alpha$ has an infinite number of roots for every value α except for at most two values called **Picard's exceptional values (Picard's theorem)**. For a meromorphic function of order ρ, $\lim_{r\to\infty}\sum_{r_j\leqslant r}(r_j(\alpha))^{-\lambda}$ $(\lambda<\rho)$ diverges for every value α except for at most two values (**Borel's theorem**). A value α for which the series converges is called a **Borel exceptional value**. We call $\delta(\alpha)=1-\limsup_{r\to\infty}N(r,\alpha)/T(r)$ the **defect** of f. It always satisfies $0\leqslant\delta(\alpha)\leqslant 1$, and the values with $\delta(\alpha)>0$ are called **Nevanlinna's exceptional values**. The number of values α (may be ∞) with $\delta(\alpha)>0$ is at most countable for any meromorphic function $f(z)$, and $\sum_{i=1}^{\infty}\delta(\alpha_i)\leqslant 2$. There are many studies concerning the values α with $\delta(\alpha)=0$.

F. Julia Directions

Among functions that have an essential singularity at the point at infinity and are meromorphic in the whole plane, there are some that possess no †Julia directions. These functions, called **Julia exceptional functions**, are of order 0. A necessary and sufficient condition for $f(z)$ to be a Julia exceptional function is that $f(z)$ can be written in the form $z^m\prod_\mu(1-z/a_\mu)/\prod_\nu(1-z/b_\nu)$ (A. Ostrowski, 1925). Concerning zeros a_μ and poles b_ν of $f(z)$, the following three properties are obtained using the theory of †normal families due to P. Montel: (1) There are constants K_1, K_2, and K_3 independent of r such that $|n(r,\infty)-n(r,0)|<K_1$, $n(2r,\infty)-n(r,\infty)<K_2$, and $n(2r,0)-n(r,0)<K_3$. (2) There are constants K_4 and K_5 such that for any p and q,

$$|a_p|^m\prod_{|a_\mu|<|a_p|}\left|\frac{a_p}{a_\mu}\right|\Big/\prod_{|b_\nu|<|a_p|}\left|\frac{a_p}{b_\nu}\right|<K_4,$$

$$|b_q|^{-m}\prod_{|b_\nu|<|b_q|}\left|\frac{b_q}{b_\nu}\right|\Big/\prod_{|a_\mu|<|b_q|}\left|\frac{b_q}{a_\mu}\right|<K_5.$$

(3) There exists an $\varepsilon>0$ satisfying $|a_p/b_q-1|\geqslant\varepsilon>0$ for any p and q.

G. Valiron gave a precise form of Julia directions that corresponds to Borel's theorem (*Acta Math.*, 52 (1928), [6]). Namely, if the order ρ of a meromorphic function $f(z)$ in $|z|<\infty$ is positive and finite, then there exists a direction J defined by $\arg z=\alpha$ such that the zeros $z_\nu(a,\Delta)$ of $f(z)-a$ in any angular domain $\Delta:|\arg z-\alpha|<\delta$ containing J have the property $\sum_\nu|z_\nu(a,\Delta)|^{-(\rho-\varepsilon)}=\infty$ for any $\varepsilon>0$ except for at most two values of a. The direction J is called a **Borel direction**.

G. Relations between Two or More Meromorphic Functions

If entire functions $f_1(z), f_2(z), f_3(z)$ with no zeros satisfy $c_1f_1(z)+c_2f_2(z)+c_3f_3(z)\equiv 0$, then the quotient of any two of them is a constant (E. Borel). If two meromorphic functions $f_1(z)$, $f_2(z)$ have the same α_j-points for five distinct values α_j $(j=1,\ldots,5)$ (where multiplicity is not taken into account), then they coincide everywhere. If the †Riemann surface of the †algebraic function $w(z)$ defined by a polynomial $P(z,w)=0$ of z, w is of †genus >1, it is impossible to find meromorphic functions $z=f(\zeta)$, $w=g(\zeta)$ that satisfy $P(f(\zeta), g(\zeta))=0$ (†uniformization by meromorphic functions).

H. Asymptotic Values

If a meromorphic function $f(z)\to\alpha$ as $z\to\infty$ along a curve C, the value α and the curve C are called an **asymptotic value** and **asymptotic path**, respectively. Each (Picard's) exceptional value of $f(z)$ is an asymptotic value. For meromorphic functions, no simple relation is known between their order and the number of their asymptotic values. There exists a meromorphic function of order 0 with an infinite number of asymptotic finite values. Some results analogous to those for entire functions are obtained for meromorphic functions by applying the theory of normal families. F. Marty established a systematic theory of normal families of meromorphic functions by using spherical distance.

I. Inverse Functions

Generally, the inverse function of a meromorphic function $w=f(z)$ is infinitely multiple-valued. Let $P(w,w_0)$ be a function element of the inverse function with center at w_0, and let C be an arbitrary curve starting at w_0 and with ω its terminal point. For any domain S containing C, $P(w,w_0)$ can be continued analytically in S up to a point arbitrarily near ω (**Iversen's theorem**). Continue the function element analytically along each half-line starting at its center. Then the set of arguments of half-lines along which the analytic continuation meets a singularity at a finite point is of zero linear measure (**Gross's theorem**).

By considering the inverse image of the suitably cut Riemann surface of the inverse function, the z-plane can be divided into fundamental domains such that each domain is the inverse image of the whole w-plane (with suitable slits removed) and has a boundary each point of which is †accessible from the inside of the domain, and the boundary curves of fundamental domains cluster nowhere in the plane.

For a meromorphic function $f(z)$, the set of functions $z'=\varphi(z)$ defined by $f(z')=f(z)$ (i.e., transformations between points that give $f(z)$ the same value) has the property of a †hypergroup. If $\varphi(z)$ is single-valued, then it is a linear entire function, and if it is finitely multiple-valued, then it is an algebraic function. The †cluster set of the inverse function at a transcendental singularity consists of only one point, ∞, that is, it is an †ordinary singularity. To an analytic continuation along a curve that determines a transcendental singularity of the inverse function there corresponds a curve in the z-plane terminating at ∞. This curve is an asymptotic path of $f(z)$. Namely, the value $f(z)$ tends to the coordinate of the transcendental singularity as $z\to\infty$ along this path. Each asymptotic value of a transcendental meromorphic function $w=f(z)$ corresponds to a transcendental singularity of its inverse function $z=\varphi(w)$, and if we consider two asymptotic paths to be the same if they correspond to the same singularity, then there exists a one-to-one correspondence between the set of asymptotic paths and the set of transcendental singularities of the inverse function. The inverse function of any meromorphic function of order ρ has at most 2ρ †direct transcendental singularities if $\rho\geqslant 1/2$ and at most 1 such singularity if $\rho<1/2$ (L. V. Ahlfors).

J. Theory of Covering Surfaces

Any simply connected open Riemann surface can be mapped conformally onto a disk or the whole finite plane. The former case occurs when the Riemann surface is †hyperbolic, and the latter when the Riemann surface is †parabolic. The problem of determining the type of Riemann surface from its structure as a covering surface of the sphere was studied first by A. Speiser. Since then many results have been obtained. For instance, a simply connected open Riemann surface that has only algebraic branch points within a finite distance is parabolic if the number $n(\rho)$ of branch points in the subdomain consisting of points that can be joined with a given point by a curve of length $\leqslant\rho$ satisfies $\int_0^\infty d\rho/\rho n(\rho)=\infty$ (Ahlfors). Furthermore, Ahlfors established the theory of covering surfaces by a metricotopological method, and in applying it obtained Nevanlinna theory and many other results on meromorphic functions.

Let F_r denote the covering surface of the Riemann sphere F_0 with radius $1/2$; F_r is the image of $|z|\leqslant r$ under a meromorphic function $w=f(z)$. The area of F_r divided by π, where π is the area of F_0, is given by

$$A(r)=\frac{1}{\pi}\int\int_{|z|<r}\frac{|f'(z)|^2}{(1+|f(z)|^2)^2}\rho\,d\rho\,d\theta,$$

$z=\rho e^{i\theta}$

and is called the **mean number of sheets** of F_r. The length of the boundary of F_r is given by

$$L(r)=\int_{|z|=r}\frac{|f'(z)|}{1+|f(z)|^2}|dz|.$$

The relation

$$T(r)=\int\frac{A(r)}{r}dr+O(1)$$

holds (T. Shimizu, Ahlfors).

Consider the Riemann surface of the inverse function of a meromorphic function $w=f(z)$ in $|z|<R\leqslant+\infty$. It has a countable number of components Q_ν over a domain on the w-plane. Let Δ_ν denote the inverse image of Q_ν on the z-plane. If Δ_ν together with its boundary is contained in $|z|<R$, the component Q_ν is called an **island**, and otherwise, a **peninsula**.

Let D be a simply connected domain of the w-plane, $n(r,D)$ be the sum of the sheet numbers of the islands of F_r over D, and $m(r,D)$ be the sum of the areas of the peninsulas of F_r over D divided by the area of D. Then

$$m(r,D)+n(r,D)=A(r)+O(L(r)).$$

Let D_j $(j=1,\ldots,q)$ $(q\geqslant3)$ be disjoint simply connected domains on the w-plane. Then

$$\sum_{j=1}^{q}n(r,D_j)-\sum_{j=1}^{q}n_1(r,D_j)>(q-2)A(r)-O(L(r)),$$

where $n_1(D)$ is the sum of the orders of branch points of all islands of F_r over D.

For a meromorphic function $f(z)$, $L(r)<A(r)^{1/2+\varepsilon}$, where r satisfies $0\leqslant r<\infty$ except for $r\in\bigcup_j I_j$ for some intervals I_j. Hence in the case where D_j is a point α_j, this inequality yields

$$\sum_{j=1}^{q}n(r,\alpha_j)-\sum_{j=1}^{q}n_1(r,\alpha_j)>(q-2)A(r)-O\left(A(r)^{1/2+\varepsilon}\right)$$

with some exceptional intervals of values of r. This latter important inequality corresponds to Nevanlinna's second fundamental theorem. Let D_j $(j=1,\ldots,q)$ $(q\geqslant3)$ be disjoint simply connected domains on the w-plane. If every simply connected island over D_j has at least μ_j sheets, then $\sum_{j=1}^{q}(1-(1/\mu_j))\leqslant2$ (**disk theorem**). It follows from this theorem that given three disjoint disks D_j on the Riemann sphere, there is at least one D_j that has an infinite number of islands over it, and also that given five D_j, there exists at least one D_j that has a 1-sheeted island over it (**Ahlfors's five-disk theorem**). This theorem corresponds to †Bloch's theorem for entire functions. These theorems can also be obtained for meromorphic functions on a disk. Ahlfors established a more important theory by introducing a differential metric.

K. History

The value distribution theory of meromorphic functions had its inception with the classical Picard theorem. It first appeared as the value distribution theory of entire functions and was developed into a concise field by the Nevanlinna theory and the Ahlfors theory of covering surfaces. In recent years, emphasis has also been placed on the study of meromorphic functions on open Riemann surfaces (→ 362 Riemann Surfaces). The value distribution of a set of several meromorphic functions was studied first by A. Bloch and developed into the study of **meromorphic curves** by Ahlfors, and H. and J. Weyl [4]. The behavior of meromorphic functions in neighborhoods of general singularities has also been studied. An example of results in that field is the theory of †cluster sets.

References

[1] K. Noshiro, Saikin no kansûron (Japanese; Modern theory of functions), Series of math. lec. Osaka Univ., Iwanami, 1941.
[2] T. Shimizu, Bankin kansûron (Japanese; Modern theory of functions), Iwanami Coll. of Math., 1935.
[3] R. H. Nevanlinna, Eindeutige analytische Funktionen, Springer, 1935, revised edition, 1953; English translation, Analytic functions, Springer, 1970.
[4] H. Weyl and J. Weyl, Meromorphic functions and analytic curves, Princeton Univ. Press, 1943.
[5] K. Noshiro, Kindai kansûron (Japanese; Modern theory of functions), Iwanami, 1954.
[6] G. Valiron, Directions de Borel des fonctions méromorphes, Mémor. Sci. Math., Gauthier-Villars, 1938.
[7] W. K. Hayman, Meromorphic functions, Clarendon Press, 1964.
Also → references to 129 Distribution of Values of Functions of a Complex Variable.

272 (XV.13) Method of Steepest Descent

The method of steepest descent originated with B. Riemann. Later, P. Debye successfully applied the method to obtain the †asymptotic representation of a function given by

$$f(z)=\int_C g(t)e^{zh(t)}\,dt, \tag{1}$$

where z is a complex variable with large $|z|$, $g(t)$ and $h(t)$ are †analytic functions, and C is a contour in the complex plane [1]. The method is sometimes called the **saddle-point method**. Since by the †Laplace transform we can get an integral representation of the form (1) of the general solution of an ordinary differential equation whose coefficients are linear functions of z, we can get its asymptotic representation by the present method. In particular, this method is useful for obtaining asymptotic representations of special functions such as †Bessel functions. The method is also used systematically in †statistical dynamics to obtain asymptotic estimates for systems with a large number of degrees of freedom.

In the expression (1), a point t_0 where $h'(t)$ vanishes is called a **saddle point** (or **col**) of the function $e^{zh(t)}$. In the neighborhood of such a point t_0, let

$$h(t)=h(t_0)+(1/2)h''(t_0)(t-t_0)^2+\ldots$$

be the Taylor expansion of the function $h(t)$. When t moves along the line

$$\arg(t-t_0)=z/2-(1/2)\arg(zh''(t_0)),$$

the value $|e^{zh(t)}|$ decreases from its maximum value $|e^{zh(t_0)}|$ at t_0 more rapidly than when t moves along a line in any other direction, because $zh''(t_0)(t-t_0)^2\leqslant 0$. Furthermore, the imaginary part $\operatorname{Im} zh(t)$ is constant along the line. This direction is called the **direction of steepest descent** of the function $e^{zh(t)}$ at the point t_0. When $|z|$ is sufficiently large, we may deform the contour C in such a way that it passes through the saddle point t_0 along the direction of steepest descent. Then the absolute value of the integrand on C is largest when t is in the neighborhood of the saddle point t_0 and decreases rapidly outside the neighborhood. Therefore, the integral (1) along the deformed contour C is approximated nicely by the value of the integral in a small neighborhood of the saddle point t_0. The method of steepest descent is a means for obtaining the principal terms of the integral (1) for large $|z|$.

References

[1] P. Debye, Näherungsformeln für die Zylinderfunktionen für grosse Werte des Arguments und unbeschränkt veränderliche Werte des Index, Math. Ann., 67 (1909), 535–558.

[2] В. И. Смирнов (V. I. Smirnov), Курс высшей математики III, Физматгиз, 1960; English translation, A course of higher mathematics III, pt. 2. Complex variables, Special functions. Pergamon, 1964.
Also → references to 32 Asymptotic Series.

273 (II.19) Metric Spaces

A. General Remarks

The distance between two points $x=(x_1,\ldots,x_n)$ and $y=(y_1,\ldots,y_n)$ in the n-dimensional †Euclidean space $\mathbf{R}^n$ is defined by $\rho(x,y)=\sqrt{(y_1-x_1)^2+\ldots+(y_n-x_n)^2}$. The function $\rho(x,y)$ is nonnegative for every pair (x,y) and has the following properties: (i) $\rho(x,y)=0$ if and only if $x=y$; (ii) $\rho(x,y)=\rho(y,x)$; and (iii) $\rho(x,z)\leqslant\rho(x,y)+\rho(y,z)$ for any three points x, y, z. Property (iii) is called the **triangle inequality**.

B. Definition

Abstracting the notion of distance from Euclidean spaces, M. Fréchet defined metric spaces [1] (1906). A **metric** on a set X is a nonnegative function ρ on $X\times X$ that satisfies (i), (ii), and (iii) of Section A, and a **metric space** (X,ρ), or simply X, is a set X provided with a metric ρ. The members of X are called points, ρ is also called a **distance function**, and $\rho(x,y)$ is called the **distance** from x to y. The distance function is sometimes denoted by $d(x,y)$ or $\operatorname{dis}(x,y)$. If (i) is replaced by its weaker form (i′) $\rho(x,x)=0$, the function ρ is called a **pseudometric** (or **pseudodistance function**), and X is called a **pseudometric space**.

The spaces in (1)–(6) are examples of metric spaces:

(1) The n-dimensional Euclidean space $\mathbf{R}^n$, in particular the real number system $\mathbf{R}$ with $\rho_0(x,y)=|x-y|$. (2) The †function space $L_p(\Omega)$. (3) The †function space $C(\Omega)$. (4) The †sequence space s, i.e., the space $\mathbf{R}^{\mathbf{N}}$ of all sequences of real numbers with metric $\rho(x,y)=\sum_{n=1}^{\infty}2^{-n}|x_n-y_n|/(1+|x_n-y_n|)$, where $x=(x_1,x_2,\ldots)$ and $y=(y_1,y_2,\ldots)$. (5) The †sequence space m, i.e., the space of all bounded sequences of real numbers with metric $\rho(x,y)=\sup_n|x_n-y_n|$ for $x=(x_1,x_2,\ldots)$ and $y=(y_1,y_2,\ldots)$. (6) A **Baire zero-dimensional space** $(\Omega^{\mathbf{N}},\rho)$, where Ω is a set and the distance $\rho(x,y)$ between $x=(x_1,x_2,\ldots)$ and

$y=(y_1,y_2,\dots)$ is equal to the reciprocal of the minimum of n such that $x_n \neq y_n$. (7) For any set X, define ρ by setting $\rho(x,x)=0$ and $\rho(x,y)=1$ when $x \neq y$. Then (X,ρ) is a metric space, called a **discrete metric space**. (8) For any set X, define ρ by setting $\rho(x,y)=0$ for any members x and y. Then ρ is a pseudometric, and the resulting space X is called an **indiscrete pseudometric space**.

For a subset M of a metric space X, $\sup\{\rho(x,y) \mid x,y \in M\}$ is called the **diameter** of M (denoted by $d(M)$), and M is said to be **bounded** if its diameter is finite (including $M=\varnothing$). For two subsets A, B of X, $\inf\{\rho(x,y) \mid x \in A, y \in B\}$ is called the **distance** between A and B, denoted by $\rho(A,B)$. We have $\rho(A,B)=\rho(B,A)$. When a family $\mathfrak{M}=\{M_\lambda \mid \lambda \in \Lambda\}$ of subsets of X is a covering of X, i.e., $X=\bigcup_\lambda M_\lambda$, the supremum of the diameters $d(M_\lambda)$ of M_λ in $\mathfrak{M}$, $\sup\{d(M_\lambda) \mid \lambda \in \Lambda\}$, is called the **mesh of the covering** $\mathfrak{M}$. For a positive number ε, a covering whose mesh is less than ε is called an **ε-covering**. A metric space X is called **totally bounded** (or **precompact**) (F. Hausdorff, 1927) if for each positive number ε there exists a finite ε-covering of X. A subset X_1 of a metric space X becomes a metric space if we define its metric ρ_1 by setting $\rho_1(x,y)=\rho(x,y)$ for $x,y \in X_1$, where ρ is the metric of X. The space (X_1,ρ_1) is called a **metric subspace** of (X,ρ). A subset of X is called **totally bounded** (or **precompact**) if it is totally bounded as a metric subspace. Any totally bounded subset is bounded. Conversely, in the Euclidean space $\mathbf{R}^n$ any bounded subset is totally bounded.

A bijection f from a metric space (X_1,ρ_1) onto a metric space (X_2,ρ_2) is called an **isometric mapping** if f preserves the metric, i.e., $\rho_2(f(x),f(y))=\rho_1(x,y)$ for any points $x,y \in X_1$; and X_1 and X_2 are called **isometric** if there is an isometric mapping from X_1 onto X_2.

Let X be a metric space with metric ρ, and let f be a bijection from a set Y onto X. Then the function $\rho'(y_1,y_2)=\rho(f(y_1),f(y_2))$ ($y_1, y_2 \in Y$) is a distance function on Y, and with this metric the set Y becomes a metric space called the **metric space induced by** f; f is an isometric mapping from (X,ρ) to (Y,ρ').

For a finite number of metric spaces $(X_1,\rho_1),\dots,(X_n,\rho_n)$, we can define a metric ρ on their Cartesian product $X=X_1\times\dots\times X_n$ by setting

$$\rho(x,y)=\sqrt{\rho_1(x_1,y_1)^2+\dots+\rho_n(x_n,y_n)^2}$$

for two points $x=(x_1,\dots,x_n), y=(y_1,\dots,y_n)$ of X. Thus we obtain a metric space (X,ρ), called the **product metric space** of $(X_1,\rho_1),\dots,(X_n,\rho_n)$. The n-dimensional Euclidean space $\mathbf{R}^n$ is the product metric space of n copies of the real line $(\mathbf{R},\rho_0)$.

C. Topology for Metric Spaces

For a point x of a metric space (X,ρ) and any positive number ε, the set $U_\varepsilon(x)$ of all points y such that $\rho(x,y)<\varepsilon$ is called the **ε-neighborhood** (or **ε-sphere**) of x. We can introduce a topology for X by taking the family of all ε-neighborhoods as a †base for the neighborhood system (→ 408 Topological Spaces). Then the following five propositions hold, any one of which can be used to define the same topology: (i) A subset O is †open if and only if for any point x in O there is a positive number ε such that the ε-neighborhood of x is contained in O. (ii) A subset F is †closed if and only if any point whose every ε-neighborhood contains at least one point of F is contained in F. (iii) A subset U is a neighborhood of a point x if and only if U contains some ε-neighborhood of x. (iv) A point x is an †interior point of a subset A if and only if some ε-neighborhood of x is contained in A; the interior A^i of A is the set of all such points. (v) A point x is a †boundary point of a subset A if every ε-neighborhood of x contains at least one point of A; the closure $\overline{A}$ of A is the set of all such points, and $x \in \overline{A}$ if and only if $\rho(x,A)=0$.

Every metric space X satisfies the †first countability axiom. A metric space X is a †Hausdorff space and, more specifically, a †normal space. Any subset of X is also a metric space; therefore, it is normal. Hence X is †completely normal; it is also †paracompact.

In the same way, we can define a topology for each pseudometric space that satisfies the first countability axiom, but a pseudometric space is not necessarily Hausdorff.

D. Convergence of Sequences

A sequence $\{x_n\}$ of points in a metric space is said to **converge** to a point x (written $\lim_{n\to\infty} x_n = x$) if $\rho(x_n,x)$ tends to zero as $n\to\infty$. The point x is called the **limit** of $\{x_n\}$. This convergence is equivalent to convergence with respect to the topology defined in Section C (→ 89 Convergence). As the first countability axiom is satisfied, we may define the topology by means of convergent sequences of points: the closure $\overline{A}$ of a subset A is the set of all limits of sequences of points in A.

E. Separable Metric Spaces

For a metric space X the following three conditions are equivalent: (i) There exists a countable family $\mathfrak{O}_0$ of open sets of X such that each open set of X is the union of members of $\mathfrak{O}_0$ (†second countability axiom). (ii) X is †separable, that is, X has a countable subset that is †dense in X. (iii) Every open covering of X has a countable subcovering (†Lindelöf property). A metric space with any of these properties is called a **separable metric space**. The sequence space s is separable. Any separable metric space can be isometrically embedded in the sequence space m, i.e., is isometric to a subspace of m (→ 173 Function Spaces B).

F. Compact Metric Spaces

For a metric space X, the following five conditions are equivalent: (i) X is †compact, that is, every open covering of X has a finite subcovering. (ii) X is †countably compact, that is, every countable open covering of X has a finite subcovering. (iii) X is †sequentially compact, that is, any sequence of points in X has a convergent subsequence. (iv) Every nested family $F_1 \supset F_2 \supset \ldots$ of nonempty closed sets of X has a nonempty intersection. (v) Every infinite subset M of X has an accumulation point x, i.e., $x \in \overline{M - \{x\}}$. A metric space satisfying any of these conditions is called a **compact metric space** (M. Fréchet [1]). Every real-valued continuous function defined on a compact metric space has a maximum and a minimum. Every compact metric space is totally bounded, and every totally bounded metric space is separable. Consequently, every compact metric space is separable.

Let $\mathfrak{U} = \{U_\lambda\}$ be an open covering of a compact metric space X. There exists a positive number δ such that every set with $d(A) < \delta$ is contained in some U_λ. The number δ is called the **Lebesgue number** of the open covering $\mathfrak{U}$.

A subset A of a metric space is said to be **compact** if it is compact as a metric subspace, and A is said to be **relatively compact** if its closure is compact. Bounded closed sets in $\mathbf{R}^n$, in particular closed intervals of real numbers, are compact. For these sets, conditions (i), (iv), and (v) are called the **Heine-Borel theorem** (or **Borel-Lebesgue theorem**), **Cantor's intersection theorem**, and the **Bolzano-Weierstrass theorem**, respectively.

G. Product Spaces of Metric Spaces

Let $(X_1, \rho_1), \ldots, (X_n, \rho_n)$ be metric spaces. Then the Cartesian product $X = X_1 \times \ldots \times X_n$ has distance functions

$$\bar{\rho}_p(x,y) = \{\rho_1(x_1,y_1)^p + \ldots + \rho_n(x_n,y_n)^p\}^{1/p}, \qquad p \geqslant 1$$

and

$$\bar{\rho}_\infty(x,y) = \max\{\rho_1(x_1,y_1), \ldots, \rho_n(x_n,y_n)\},$$

where $x = (x_1, \ldots, x_n)$ and $y = (y_1, \ldots, y_n)$. The topology of X induced by each one of these metrics coincides with the product topology. In particular, for the n-dimensional Euclidean space $\mathbf{R}^n$, the metrics $\bar{\rho}_p$ $(p \geqslant 1)$ and $\bar{\rho}_\infty$ define the same topology.

Let $(X_1, \rho_1), \ldots, (X_n, \rho_n), \ldots$ be a countable number of metric spaces. If we define a metric ρ on the Cartesian product $X = \prod_{n=1}^{\infty} X_n$ by

$$\rho(x,y) = \sum_{n=1}^{\infty} \frac{1}{2^n} \frac{\rho_n(x_n,y_n)}{1 + \rho_n(x_n,y_n)},$$

where $x = (x_1, x_2, \ldots)$ and $y = (y_1, y_2, \ldots)$, then the topology defined by ρ is identical with the product topology. For the Cartesian product of an uncountable number of metric spaces, we cannot construct a metric ρ such that the topology induced by ρ agrees with the product topology in general.

H. Uniformity of Metric Spaces

Every metric space X is a †uniform space, for which we may take a countable number of subsets $\{(x,y) \mid \rho(x,y) < 2^n\}$, $n = 1, 2, \ldots$ of $X \times X$ as a base of †uniformity (→ 422 Uniform Spaces).

I. Uniform Continuity

A mapping f from a metric space (X, ρ) into a metric space (Y, σ) is †continuous if for any point x in X and any positive number ε there is a positive number δ such that $f(U_\delta(x)) \subset V_\varepsilon(f(x))$, where $U_\delta(x)$ is a δ-neighborhood for ρ and $V_\varepsilon(y)$ is an ε-neighborhood for σ; that is, $\rho(x, x') < \delta$ implies $\sigma(f(x), f(x')) < \varepsilon$. In this case, we must generally choose δ depending on x and ε. In the special case where we can choose δ depending only on ε, independently of x, we call f **uniformly continuous** in X. (The notion of uniform continuity may be generalized to uniform spaces.) Not every continuous mapping is necessarily uniformly continuous, but every continuous mapping from a compact metric space into a metric space must be uniformly continuous.

J. Complete Metric Spaces

A sequence $\{x_n\}$ of points in a metric space

(X,ρ) is called a **fundamental sequence** (or **Cauchy sequence**) if $\rho(x_n, x_m)\rightarrow 0$ as $n,m\rightarrow\infty$. Every convergent sequence is a fundamental sequence, but the converse is not always true. A metric space is called **complete** if every fundamental sequence in the space converges to some point of the space (M. Fréchet [1]). The metric spaces introduced in examples (1) through (5) of Section B are complete. (In example (3) we must assume that the space Ω is a compact Hausdorff space.) Every compact metric space is complete. Conversely, a metric space is compact if and only if it is complete and totally bounded.

For a metric space X, we can construct a complete metric space Y such that there is an isometric mapping φ from X onto a dense subset X_1 of Y (F. Hausdorff, 1914). Such a pair (Y,φ) is called the **completion** of X. If X has two completions (Y_1,φ_1) and (Y_2,φ_2), then there is an isometric mapping f from Y_1 onto Y_2 with $\varphi_2=f\circ\varphi_1$. In this sense the completion of X is unique. By identifying X with $\varphi(X)$ when (Y,φ) is the completion of X, any metric space can be regarded as a dense subset of a complete metric space. For example, the completion of the rational number system **Q** is the real number system **R**.

Baire-Hausdorff theorem: In a complete metric space every set of the †first category is a †boundary set. That is, every set that can be expressed as the union of a countable number of sets whose closures have no interior point has no interior point. In other words, if the union $\bigcup_{n=1}^{\infty} F_n$ of closed sets $F_1, F_2, \ldots$ of X has an interior point, then at least one of the F_n must have an interior point.

K. The Metrization Problem

A topological space X is called **metrizable** if we can introduce a suitable metric for X which induces a topology identical to the original one. A †T_1-space satisfying the second countability axiom is metrizable if and only if it is †regular (**Uryson-Tihonov theorem**; P. S. Uryson, *Math. Ann.*, 94 (1925), A. Tihonov, *Math. Ann.*, 95 (1925)). However, a metric space does not necessarily satisfy the second countability axiom. Therefore, the Uryson-Tihonov theorem does not provide the necessary and sufficient conditions for metrizability. Following are some of the necessary and sufficient conditions for a topological space X to be metrizable:

(1) There exists a nonnegative real-valued function d on $X\times X$ satisfying the first two axioms given in Section A and the following condition: There exists a real-valued function $\varphi(w)$ that converges to zero as $w\rightarrow 0$ such that, for any three points x,y,z and any positive number ε, $d(x,y)<\varphi(\varepsilon)$ and $d(y,z)<\varphi(\varepsilon)$ imply $d(x,z)<\varepsilon$ (E. W. Chittenden, *Trans. Amer. Math. Soc.*, 18 (1917)).

(2) X is a T_1-space that has a countable number of open coverings $\mathfrak{M}_1, \mathfrak{M}_2, \ldots$ satisfying the following two conditions: (i) If $U_1, U_2 \in\mathfrak{M}_{n+1}$ have a common point, there is a set U with $U\supset U_1\cup U_2$, $U\in\mathfrak{M}_n$; (ii) for any point x in X, if U_n is any member of $\mathfrak{M}_n$ containing x, the family $\{U_n\}_{n=1,2,\ldots}$ is a base for the neighborhood system of x (P. S. Aleksandrov and Uryson, *C. R. Acad. Sci., Paris*, 177 (1923), and N. Aronszajn).

(3) X is a T_1-space, and every point in X has a countable base for the neighborhood system $\{U_n(x)\}$ and a sequence $\{S_n(x)\}$ of neighborhoods with the following properties: (i) $S_n(x)$ and $S_n(y)$ have no common points if $y\notin U_n(x)$; (ii) $S_n(y)\subset U_n(x)$ for $y\in U_n(x)$ (J. Nagata, *J. Inst. Polytech. Osaka City Univ.*, 8 (1957)).

(4) S is regular and has a †σ-locally finite base (Nagata, *J. Inst. Polytech. Osaka City Univ.*, 1 (1950); Ju. M. Smirnov, *Uspehi Mat. Nauk*, 6 (1951)).

L. Quotient Spaces of Metric Spaces

Let a surjection f from a metric space X onto a topological space Y be continuous and closed. Then Y is metrizable if and only if $f^{-1}(y)$ is compact for each y in Y (A. H. Stone, K. Morita and S. Hanai). Any metric space can be expressed as a quotient space with respect to a continuous closed mapping from a subspace of a certain †Baire zero-dimensional space (K. Morita).

Reference

[1] M. Fréchet, Sur quelques points du calcul fonctionnel, Rend. Circ. Mat. Palermo, 22 (1906), 1–74.
Also → references to 408 Topological Spaces.

274 (I.7) Model Theory

A. Language

Every mathematical theory has an appropriate language. To determine a language for a theory means to determine a language for the related mathematical system. Such a language consists of the following symbols (the actual symbols given below are only examples of one notational system).

(1) Symbols that express logical concepts (†logical symbols): $\forall, \exists, \neg, \wedge, \vee, \rightarrow$;

(2) †free variables: $a_0, a_1, a_2, \ldots$;

(3) †bound variables: $x_0, x_1, x_2, \ldots$;

(4) symbols that denote individual objects (individual constants): $c_0, c_1, c_2, \ldots, c_\alpha, \ldots$;

(5) †function symbols: $f_0, f_1, f_2, \ldots, f_\alpha, \ldots$;

(6) †predicate symbols: $P_0, P_1, P_2, \ldots, P_\alpha, \ldots$.

The †cardinalities of the sets of symbols in (4), (5), and (6) are arbitrary, except that there must be at least one predicate symbol. It is assumed that each set of symbols is †well ordered. Also, it is understood that to each f_j in (5) there corresponds a positive integer i_j, while to each P_j there corresponds a nonnegative integer (these integers are called the number of arguments of f_j and P_j, respectively).

In practice, other kinds of languages are also dealt with. One example is a system with infinitely long expressions, which permits †transfinite ordinals for the numbers of arguments of f_j and P_j, and which has the extended concepts $\bigwedge_{\alpha<\beta}$, $\bigvee_{\alpha<\beta}$, $\exists x_1 \ldots \exists x_\alpha \ldots$ for $\alpha<\beta$, and $\forall x_1 \forall x_2 \ldots \forall x_\alpha \ldots$ for $\alpha<\beta$, where α and β are transfinite ordinals. Another language includes variables of higher †types as well as $\forall$ and $\exists$ over those variables. Free and bound variables may not be distinguished typographically. In that case a variable not bound by $\forall$ or $\exists$ in a †formula is called **free**. (The notion of a formula is defined later.) To simplify this discussion, however, we restrict ourselves to the †first-order predicate calculus with a typographic distinction between free and bound variables. We also assume that there are only a countable number of variables, and hence we use natural numbers as subscripts.

Set $L_1=\{$logical symbols$\}$, $L_2=\{a_0, a_1, a_2, \ldots\}$, $L_3=\{x_0, x_1, x_2, \ldots\}$, $L_4=\{c_0, c_1, c_2, \ldots\}$, $L_5=\{f_0, f_1, f_2, \ldots\}$, $L_6=\{P_0, P_1, P_2, \ldots\}$, $L=\langle L_1, L_2, L_3, L_4, L_5, L_6\rangle$. To determine a language L is to specify such a list L. Since $\forall, \exists, \neg, \wedge, \vee, \rightarrow$ are normally used for L_1, $a_0, a_1, a_2, \ldots$ for L_2, and $x_0, x_1, x_2, \ldots$ for L_3, we may assume that these are fixed, and hence to determine a language is to determine $\langle L_4, L_5, L_6\rangle$. We take $\langle L_1, L_2, L_3\rangle$ just described and assume an arbitrary but fixed $L=\langle L_4, L_5, L_6\rangle$. First we define the notions term of L (or L-term) and formula of L (or L-formula).

Definition of the **terms** of L (L-terms): (1) Each free variable a_j is an L-term. (2) Each individual constant c_j of L is an L-term. (3) If f_j is a function symbol of L, i_j is the number of arguments of f_j, and each of $t_1, \ldots, t_{i_j}$ is an L-term, then $f_j(t_1, \ldots, t_{i_j})$ is also an L-term. (4) The L-terms are only those constructed by (1)–(3).

A term that does not contain a free variable is called a **closed term**.

Definition of the **formulas** of L (or L-formulas): (1) Let P_j be a predicate symbol of L and i_j be the corresponding natural number. If each of $t_1, \ldots, t_{i_j}$ is an L-term, then $P_j(t_1, \ldots, t_{i_j})$ is an L-formula. This type of formula is called a **prime formula** (or **atomic formula**). (2) If A and B are L-formulas, then each of $\neg(A)$, $(A)\wedge(B)$, $(A)\vee(B)$, and $(A)\rightarrow(B)$ is an L-formula. (3) Let F be an L-formula and x_i be a bound variable that does not occur in F. Then an expression obtained by putting ( ) around F, replacing some occurrences in F of a free variable, say a_j, by x_i, and prefixing $\forall x_i$ or $\exists x_i$ is an L-formula. (4) The L-formulas are only those constructed by (1)–(3).

A formula that has no occurrence of a free variable is called a **closed formula**. The parentheses used in the formation of a formula may be omitted if no ambiguity arises.

B. Structures

Let L be a specific language as described in the previous section. The $\mathfrak{M}=[M:\rho;\sigma;\tau]$ defined by (1)–(4) below is called a **structure** for L (or L-structure).

(1) M is a set. (M is called the **universe** of $\mathfrak{M}$.)

(2) ρ is a mapping from L_4 into M.

(3) Let $L_5^i=\{f_j \mid$ the number of arguments of f_j is $i\}$. Then $L_5=L_5^1 \cup L_5^2 \cup \ldots \cup L_5^i \cup \ldots$ provides a partition of L_5. Let $\mathfrak{F}_i$ be the set of all mappings from $M^i=M\times\ldots\times M$ (i times) into M and σ_i be a mapping from L_5^i into $\mathfrak{F}_i$. We define σ for an arbitrary f of L_5 by $\sigma(f)=\sigma_i(f)$, where i is the number of arguments of f. Then σ is obviously a mapping from L_5 into $\bigcup_{i=1}^{\infty}\mathfrak{F}_i$.

(4) Decompose L_6 into $L_6^0 \cup L_6^1 \cup \ldots \cup L_6^i \cup \ldots$ as in (3). Let P_i be the set of all subsets of M^i and τ_i be a mapping from L_6^i into P_i, where P_0 is the set $\{M, \varnothing\}$ ($\varnothing$ is the empty set). Then τ is defined for every i and for an arbitrary P of L_6^i by $\tau(P)=\tau_i(P)$.

If we denote $\rho(c)$ by $\bar{c}$, $\sigma(f)$ by $\bar{f}$, and $\tau(P)$ by $\bar{P}$, then we may understand that ρ is represented by $\bar{c}_0, \bar{c}_1, \ldots$, σ is represented by $\bar{f}_0, \bar{f}_1, \ldots$, and τ is represented by $\bar{P}_0, \bar{P}_1, \ldots$. Therefore $\mathfrak{M}$ is normally expressed as

$$\mathfrak{M}=\left[M:\bar{c}_0, \bar{c}_1, \ldots; \bar{f}_0, \bar{f}_1, \ldots; \bar{P}_0, \bar{P}_1, \ldots\right].$$

C. Satisfiability

We fix not only a language L but also a structure $\mathfrak{M}$ for L. Then the property that an L-formula is **satisfiable** is defined by the following procedure:

Let $\mathfrak{m}, \mathfrak{n}, \ldots$ stand for †sequences of the elements of M, say $(m_0, m_1, \ldots)$, $(n_0, n_1, \ldots)$, $\ldots$, called $\mathfrak{M}$-sequences. We write $\mathfrak{m} \overset{i}{=} \mathfrak{n}$ to indicate that each entry of $\mathfrak{m}$ except the ith one is equal to the corresponding entry of $\mathfrak{n}$. Using these concepts, the value of an L-term at an $\mathfrak{M}$-sequence $\mathfrak{m}$, denoted by $t[\mathfrak{m}]$, is defined as follows:

(1) If t is a free variable a_j, then $t[\mathfrak{m}] = m_j$.

(2) If t is an individual constant c_j, then $t[\mathfrak{m}] = \bar{c}_j$.

(3) If t is of the form $f_j(t_1, \ldots, t_i)$, then $t[\mathfrak{m}] = \bar{f}_j(t_1[\mathfrak{m}], \ldots, t_i[\mathfrak{m}])$. If t is an L-term, then evidently $t[\mathfrak{m}]$ is an element of M.

Based on this definition of $t[\mathfrak{m}]$, the relation A is satisfiable by $\mathfrak{m}$ in $\mathfrak{M}$, denoted by $\mathfrak{M}, \mathfrak{m} \models A$, is defined for an arbitrary L-formula A and an arbitrary $\mathfrak{M}$-sequence $\mathfrak{m}$ as follows:

(1) $\mathfrak{M}, \mathfrak{m} \models P_j(t_1, \ldots, t_i) \Leftrightarrow \langle t_i[\mathfrak{m}], \ldots, t_i[\mathfrak{m}] \rangle \in \bar{P}_j$.

(2) $\mathfrak{M}, \mathfrak{m} \models \neg B \Leftrightarrow \mathfrak{M}, \mathfrak{m} \models B$ is false.

(3) $\mathfrak{M}, \mathfrak{m} \models B \wedge C \Leftrightarrow \mathfrak{M}, \mathfrak{m} \models B$ and $\mathfrak{M}, \mathfrak{m} \models C$.

(4) $\mathfrak{M}, \mathfrak{m} \models B \vee C \Leftrightarrow \mathfrak{M}, \mathfrak{m} \models B$ or $\mathfrak{M}, \mathfrak{m} \models C$.

(5) $\mathfrak{M}, \mathfrak{m} \models B \rightarrow C \Leftrightarrow \mathfrak{M}, \mathfrak{m} \models B$ implies $\mathfrak{M}, \mathfrak{m} \models C$.

(6) $\mathfrak{M}, \mathfrak{m} \models \forall x_j F(x_j) \Leftrightarrow \mathfrak{M}, \mathfrak{n} \models F(a_i)$ for an arbitrary $\mathfrak{n}$ that satisfies $\mathfrak{m} \overset{i}{=} \mathfrak{n}$, where a_i has the least index among the free variables that do not occur in $F(x_j)$.

(7) $\mathfrak{M}, \mathfrak{m} \models \exists x_j F(x_j) \Leftrightarrow$there exists an $\mathfrak{n}$ such that $\mathfrak{n} \overset{i}{=} \mathfrak{m}$ and $\mathfrak{M}, \mathfrak{n} \models F(a_i)$, where a_i satisfies the same condition as in (6).

Following are some consequences of this definition.

(1) For an arbitrary L-formula A and an arbitrary $\mathfrak{M}$-sequence $\mathfrak{m}$, exactly one of $\mathfrak{M}, \mathfrak{m} \models A$ and $\mathfrak{M}, \mathfrak{m} \models \neg A$ holds.

(2) Let $a_{j_1}, \ldots, a_{j_i}$ include all free variables that occur in A, and let $\mathfrak{m}$ and $\mathfrak{n}$ be $\mathfrak{M}$-sequences for which $m_{j_1} = n_{j_1}, \ldots, m_{j_i} = n_{j_i}$. Then $\mathfrak{M}, \mathfrak{m} \models A$ and $\mathfrak{M}, \mathfrak{n} \models A$ are equivalent.

(3) If A is a closed formula, then for an arbitrary pair of sequences $\mathfrak{m}$ and $\mathfrak{n}$, $\mathfrak{M}, \mathfrak{m} \models A$ and $\mathfrak{M}, \mathfrak{n} \models A$ are equivalent. Therefore, for a closed formula A, we may express the statement "for some (or, equivalently, for all) $\mathfrak{M}$-sequence $\mathfrak{m}$, $\mathfrak{M}, \mathfrak{m} \models A$ holds" by $\mathfrak{M} \models A$.

(4) Let a_i be an arbitrary variable that does not occur in $\forall x F(x)$ or $\exists x F(x)$. Then $\mathfrak{M}, \mathfrak{m} \models \forall x F(x)$ is equivalent to $\mathfrak{M}, \mathfrak{n} \models F(a_i)$ for an arbitrary $\mathfrak{n}$ such that $\mathfrak{n} \overset{i}{=} \mathfrak{m}$. Likewise, $\mathfrak{M}, \mathfrak{m} \models \exists x F(x)$ is equivalent to the statement that there is an $\mathfrak{n}$ such that $\mathfrak{n} \overset{i}{=} \mathfrak{m}$ and $\mathfrak{M}, \mathfrak{n} \models F(a_i)$.

D. Models

Here again we fix a language L. Let A be a closed L-formula and $\mathfrak{M}$ an L-structure. If $\mathfrak{M} \models A$, then $\mathfrak{M}$ is called a **model** of A. Furthermore, if $\Gamma = \{A_0, A_1, \ldots\}$ is an arbitrary set of closed formulas and $\mathfrak{M} \models A_i$ for all A_i in Γ, then the structure $\mathfrak{M}$ is called a **model** of Γ.

(1) **Consistency**. Consider a logical system whose language is L. If there is a model of the set of all provable closed formulas of the system, then the system is †consistent. In particular, the †first-order predicate calculus is consistent.

(2) **Completeness**. A logical system is said to be **complete** if every closed formula that is satisfied in every structure is provable in the system. In particular, the first-order predicate calculus is complete.

K. Gödel proved (2). Later L. Henkin gave an alternative proof whose essential idea contributed to proving the following proposition: If a set Γ of closed L-formulas is consistent, then there is a model of Γ. Henkin also introduced a (nonstandard) second-order semantics, relative to which the †second-order predicate calculus is complete. This can be shown by extending Henkin's technique (for the first order) to the second-order language.

(3) Here we extend the language slightly by adding the second-order free predicate variables $\alpha_1^n, \alpha_2^n, \ldots, \alpha_i^n, \ldots$ $(n = 1, 2, \ldots)$ and the second-order bound predicate variables $\varphi_1^n, \varphi_2^n, \ldots, \varphi_i^n, \ldots$ $(n = 1, 2, \ldots)$, where n indicates the number of arguments of a variable. Otherwise the definition of the language is the same as for the case of the first-order predicate calculus. For simplicity, however, we assume that there are no individual constants, function symbols, or predicate symbols.

The structure is defined as follows: Put $\mathfrak{M} = [M : S_1, S_2, \ldots, S_n, \ldots]$, where M is a set and S_n is a set of subsets of $M \times \ldots \times M$ (n times). An $\mathfrak{M}$-sequence $\mathfrak{m}$ is defined as before, and $\mathfrak{s}_n$ denotes $(s_1^n, s_2^n, \ldots, s_i^n, \ldots)$, where each s_i^n is a member of S_n. The concept of satisfiability is defined as follows:

$\mathfrak{M}, (\mathfrak{m}, \mathfrak{s}_1, \ldots, \mathfrak{s}_n, \ldots) \models \alpha_j^n(x_{i_1}, \ldots, x_{i_n}) \Leftrightarrow (m_{i_1}, \ldots, m_{i_n}) \in s_j^n$.

$\mathfrak{M}, (\mathfrak{m}, \mathfrak{s}_1, \ldots, \mathfrak{s}_n, \ldots) \models \forall \varphi_k^n A(\varphi_k^n) \Leftrightarrow$for an arbitrary $\mathfrak{s}_n'$ for which $\mathfrak{s}_n' \overset{j}{=} \mathfrak{s}_n$, $\mathfrak{M}, (\mathfrak{m}, \mathfrak{s}_1, \ldots, \mathfrak{s}_n', \ldots) \models A(\alpha_j^n)$, where α_j^n has the smallest index among the free predicate variables that do not occur in $A(\varphi_k^n)$.

$\mathfrak{M},(\mathfrak{m},\mathfrak{s}_1,\ldots,\mathfrak{s}_n,\ldots)\models \exists\varphi_k^n A(\varphi_k^n)\Leftrightarrow$there exists an $\mathfrak{s}_n'$ such that $\mathfrak{s}_n' \overset{j}{=} \mathfrak{s}_n$ and $\mathfrak{M},(\mathfrak{m},\mathfrak{s}_1,\ldots,\mathfrak{s}_n,\ldots)\models A(\alpha_j^n)$, where α_j^n satisfies the same condition as in the previous clause.

Satisfiability for other cases is defined as for first-order predicate calculus. A structure $\mathfrak{M}$ is called **normal** if all axioms of the second-order predicate calculus are true in $\mathfrak{M}$.

The completeness of the second-order predicate calculus: Every closed formula that is satisfiable in all normal structures is provable in the second-order predicate calculus.

(4) Let the cardinality of L_4 be τ, and Γ be an arbitrary set of closed L-formulas. If Γ has a model, then Γ has a model of cardinality max $(\tau,\aleph_0)$. This follows from Henkin's method. Historically, however, it was first proved by Th. Skolem and L. Löwenheim for a special case, and was later generalized by A. I. Mal'cev and A. Robinson.

(5) The following results are all due to A. Tarski and R. L. Vaught.

Definition 1. Two L-structures $\mathfrak{M}$ and $\mathfrak{N}$ are said to be **elementarily (arithmetically) equivalent** if for an arbitrary closed L-formula $A,\mathfrak{M}\models A \Leftrightarrow \mathfrak{N}\models A$.

Definition 2. Let

$\mathfrak{M}=[M:q_0,q_1,\ldots;g_0,g_1,\ldots;Q_0,Q_1,\ldots]$

and

$\mathfrak{N}=[N:r_0,r_1,\ldots;h_0,h_1,\ldots;R_0,R_1,\ldots]$

be two structures. $\mathfrak{M}$ is an **elementary extension** of $\mathfrak{N}$ if the following two conditions are satisfied: (i) $M\supset N$; $q_j=r_j$ $(j=0,1,\ldots)$; the restriction of g_j to N is identical to h_j $(j=0,1,\ldots)$; the restriction of Q_j to N is identical to R_j $(j=0,1,\ldots)$. (If this condition holds, then $\mathfrak{M}$ is said to be an extension of $\mathfrak{N}$.) (ii) For an arbitrary L-formula A and an arbitrary $\mathfrak{N}$-sequence $\mathfrak{n}$, if $\mathfrak{N},\mathfrak{n}\models A$ then $\mathfrak{M},\mathfrak{n}\models A$.

Theorem 1. Let $\mathfrak{M}$ be an extension of $\mathfrak{N}$. A necessary and sufficient condition for $\mathfrak{M}$ to be an elementary extension of $\mathfrak{N}$ is that for an arbitrary L-formula of the form $\exists xF(x)$ and an arbitrary $\mathfrak{N}$-sequence $\mathfrak{n}$, if $\mathfrak{M},\mathfrak{n}\models \exists xF(x)$, then there is some element n of N such that for the $\mathfrak{M}$-sequence $\mathfrak{m}$ for which $\mathfrak{m}\overset{i}{=}\mathfrak{n}$ and $m_i=n$, $\mathfrak{M},\mathfrak{m}\models F(a_i)$, where a_i is an arbitrary free variable that does not occur in $F(x)$.

Theorem 2. Here we place a condition on L that each set of symbols be at most countable and arranged in the ω-type (→ 306 Ordinal Numbers). Let the cardinality of the universe M of $\mathfrak{M}$ be an infinite cardinal $\mathfrak{a}$, M' be a subset of M of cardinality $\mathfrak{c}$, and $\mathfrak{b}$ be an infinite cardinal that satisfies $\mathfrak{c}\leqslant\mathfrak{b}\leqslant\mathfrak{a}$. Then there exists an L-structure $\mathfrak{N}$ whose universe N has cardinality $\mathfrak{b}$ and such that $M'\subset N$ and $\mathfrak{M}$ is an elementary extension of $\mathfrak{N}$.

Theorem 3. Suppose that L satisfies the same condition as in Theorem 2. Let the cardinality of the universe M of $\mathfrak{M}$ be $\mathfrak{a}$ ($\mathfrak{a}$ is an infinite cardinal) and $\mathfrak{b}$ be a cardinal for which $\mathfrak{a}\leqslant\mathfrak{b}$. Then there exists an L-structure $\mathfrak{N}$ that is a proper elementary extension of $\mathfrak{M}$ and whose universe has cardinality $\mathfrak{b}$.

E. Ultraproducts

Assume that for a set of L-structures Σ and a set of indices I, there is a mapping θ from I onto Σ. If α is a member of I, $\mathfrak{M}$ is a member of Σ, and $\theta(\alpha)=\mathfrak{M}$, then $\mathfrak{M}$ may be denoted by $\mathfrak{M}^\alpha$. It should be noted that there may be more than one α corresponding to the same structure. If D is a †maximal filter of I and $\mathfrak{M}^\alpha$ is expressed as

$\mathfrak{M}^\alpha=\left[M^\alpha:\bar{c}^\alpha,\ldots:\bar{f}^\alpha,\ldots:\bar{P}^\alpha,\ldots\right]$,

then $\Pi_{\alpha\in I}M^\alpha$ is defined by

$$\prod_{\alpha\in I} M^\alpha=\{\varphi\,|\,\varphi \text{ is a mapping from } I \text{ into } \bigcup_{\alpha\in I} M^\alpha, \text{ where } \varphi(\alpha)\in M^\alpha\}.$$

For any two elements φ and ψ of $\Pi_{\alpha\in I}M^\alpha$, $\varphi\overset{D}{=}\psi$ is defined by

$\varphi\overset{D}{=}\psi\Leftrightarrow\{\alpha\,|\,\varphi(\alpha)=\psi(\alpha)\}\in D.$

Then $\varphi\overset{D}{=}\psi$ is an equivalence relation between the elements of $\Pi_{\alpha\in I}M^\alpha$. Furthermore, the set $\Pi_{\alpha\in I}M^\alpha$ partitioned by $\overset{D}{=}$ is expressed by $\Pi_{\alpha\in I}M^\alpha/D$, and each element m of $\Pi_{\alpha\in I}M^\alpha/D$ is expressed by $m=[\varphi]$, where φ is a representing element of m.

Next we define an operator that produces a new structure from Σ. Put $M=\Pi_{\alpha\in I}M^\alpha/D$. For an individual constant c of L, let $\bar{c}=[\varphi]$, where $\varphi(\alpha)=\bar{c}^\alpha$ for every α. For an n-ary function f of L and arbitrary elements $m_1=[\varphi_1],\ldots,m_n=[\varphi_n]$ of M, define $\bar{f}(m_1,\ldots,m_n)=[\psi]$, where $\psi(\alpha)=\bar{f}^\alpha(\varphi_1(\alpha),\ldots,\varphi_n(\alpha))$ for every α. For an n-ary predicate P of L, define $\langle m_1,\ldots,m_n\rangle\in\bar{P}$ by

$\langle m_1,\ldots,m_n\rangle\in\bar{P}$

$\Leftrightarrow\{\alpha\,|\,\langle\varphi_1(\alpha),\ldots,\varphi_n(\alpha)\rangle\in P^\alpha\}\in D.$

According to these definitions, put

$\mathfrak{M}=\left[M:\bar{c},\ldots,\bar{f},\ldots,\bar{P},\ldots\right]$, and denote it by $\prod_{\alpha \in I} \mathfrak{M}^\alpha / D$, called the **ultraproduct** of $\{\mathfrak{M}^\alpha\}_{\alpha \in I}$ (with respect to D). $\mathfrak{M}$ is an L-structure.

The fundamental theorem of ultraproducts: Let $\mathfrak{M} = \prod_{\alpha \in I} \mathfrak{M}^\alpha / D$ be the ultraproduct of $\{\mathfrak{M}^\alpha\}_{\alpha \in I}$, $\mathfrak{m} = (m_1, m_2, \ldots)$ be an $\mathfrak{M}$-sequence, φ_i be a representing element of m_i, and A be an arbitrary formula. Then $\mathfrak{M}, \mathfrak{m} \models A \Leftrightarrow \{\alpha \mid \mathfrak{M}^\alpha, (\varphi_1(\alpha), \varphi_2(\alpha), \ldots) \models A\} \in D$.

In the case where all the structures $\mathfrak{M}^\alpha$ coincide with the single structure $\mathfrak{N}$, the ultraproduct of $\{\mathfrak{M}^\alpha\}_{\alpha \in I}$ (with respect to D) may be written $\mathfrak{N}^I / D$ and called the **ultrapower** of $\mathfrak{N}$ (with respect to D).

Let

$$\mathfrak{M} = [M; q_0, q_1, \ldots; g_0, g_1, \ldots; Q_0, Q_1, \ldots]$$

and

$$\mathfrak{N} = [N; r_0, r_1, \ldots; h_0, h_1, \ldots; R_0, R_1, \ldots]$$

be two structures. $\mathfrak{M}$ and $\mathfrak{N}$ are said to be **isomorphic** if there is a bijection f from M to N such that the following three conditions hold: (i) $f(q_0) = r_0, f(q_1) = r_1, \ldots$. (ii) The sequences $g_0, g_1, \ldots$ and $h_0, h_1, \ldots$ are of the same type and $f(g_j(a_1, \ldots, a_n)) = h_j(f(a_1), \ldots, f(a_n))$ holds for every n-tuple $a_1, \ldots, a_n$ in M. (iii) The sequences $Q_0, Q_1, \ldots$ and $R_0, R_1, \ldots$ are of the same type and $R_j = \{\langle f(a_1), \ldots, f(a_n) \rangle \mid \langle a_1, \ldots, a_n \rangle \in Q_j\}$.

Let j be the function from N to N^I / D defined by $j(a) = [\varphi_a]$ for each $a \in N$, where φ_a is the constant function from I to N such that $\varphi_a(\alpha) = a$ for each $\alpha \in I$. Let $\mathfrak{M}$ be the substructure of $\mathfrak{N}^I / D$ whose universe is the range of j. Then j is an isomorphism of $\mathfrak{N}$ to $\mathfrak{M}$. In the following we identify a and $j(a)$ for each $a \in N$. Then $\mathfrak{N}$ is an elementary substructure of $\mathfrak{N}^I / D$ by the fundamental theorem of ultraproducts.

If $\mathfrak{M}$ and $\mathfrak{N}$ are isomorphic, then $\mathfrak{M}$ and $\mathfrak{N}$ are elementarily equivalent. By using this fact and the fundamental theorem of ultraproducts, we have the following result. Let $\mathfrak{M}$ and $\mathfrak{N}$ be two structures. If there is a nonempty set I and a maximal filter D on I such that $\mathfrak{M}^I / D$ and $\mathfrak{N}^I / D$ are isomorphic, then $\mathfrak{M}$ and $\mathfrak{N}$ are elementarily equivalent. H. J. Keisler proved the converse of this proposition by using the G.C.H. (generalized continuum hypothesis), and later S. Shelah proved it without the G.C.H. **Keisler-Shelah isomorphism theorem**: Let $\mathfrak{M}$ and $\mathfrak{N}$ be two structures. Then $\mathfrak{M}$ and $\mathfrak{N}$ are elementarily equivalent if and only if there is a nonempty set I and a maximal filter D on I such that $\mathfrak{M}^I / D$ and $\mathfrak{N}^I / D$ are isomorphic.

The ultraproduct operation has various applications in number theory, algebraic geometry, and analysis. Here we give an example due to J. Ax and S. Kochen. Let P be the set of prime numbers. Let $\mathbf{Q}_p$ and $Z_p((t))$ be the field of p-adic numbers and the field of formal power series over $Z_p = \{0, 1, \ldots, p-1\}$ for each p in P, respectively. **Ax-Kochen isomorphism theorem**: Suppose that D is a nonprincipal maximal filter on P. Then $\prod_{p \in P} \mathbf{Q}_p / D$ and $\prod_{p \in P} Z_p((t)) / D$ are isomorphic.

As an immediate consequence of this theorem, we have the following partial solution of Artin's conjecture on Diophantine equations. **Theorem**: For each positive integer d, there exists a finite set Y of primes such that every homogeneous polynomial $f(t_1, \ldots, t_n)$ of degree d over $\mathbf{Q}_p$, with $n > d^2$, has a nontrivial zero in $\mathbf{Q}_p$ for every $p \notin Y$ (→ 122 Diophantine Equations).

We give another example in nonstandard analysis. A. Robinson developed the general theory of nonstandard analysis in [3]. Here we explain a theorem due to A. R. Bernstein. Let X be a nonempty set and $U(X)$ be the smallest transitive set (i.e., $a \in b$ and $b \in U(X)$ imply $a \in U(X)$) which has X as a member and is closed under the following operations: pairing, union, power set, and subset operation (i.e., $a \in U(X)$ and $b \subset a$ imply $b \in U(X)$). Let L be the first-order predicate logic with equality whose set of nonlogical constants consists of a binary predicate symbol $\in$ and individual constant symbols c_a for $a \in U(X)$ (→ 400 Symbolic Logic F). Then the first-order structure $\mathfrak{N} = [U(X), a\ (a \in U(X)); E]$ is an L-structure, where E is the $\in$ relation on the set $U(X)$. Let $\mathfrak{M} = [U(X)^I_D : a\ (a \in U(X)); E^I_D]$ be the ultrapower $\mathfrak{N}^I_D$ of $\mathfrak{N}$ with respect to a nonprincipal ultrafilter D on a set I. For each $a \in U(X)$, let a^* be the set of all elements $[\varphi]$ in $U(X)^I_D$ such that $\{i \in I \mid \varphi(i) \in a\} \in D$. Then a is a proper subset of a^* if a is infinite. Since $\mathfrak{N}$ and $\mathfrak{M}$ are elementarily equivalent, these two sets a and a^* have common first-order properties in the following sense: for each formula $\Phi(x)$ in L,

$$(\forall b \in a)(\mathfrak{N} \models \Phi[b]) \Leftrightarrow (\forall b \in a^*)(\mathfrak{M} \models \Phi[b]).$$

From this it follows that if r is a relation on a set a in $\mathfrak{N}$, then r^* is a relation on the set a^*; and if f is a mapping from a to b in $\mathfrak{N}$, then f^* is a mapping from a^* to b^*. Hence a^* is a mathematical object which greatly resembles a. By using this type of resemblance between a and a^* we have the following result.

Let H be a Hilbert space over the complex number field $\mathbf{C}$ such that $\dim(H) = \omega$ and let T be a bounded linear operator on H. Let $X = H \cup \mathbf{C}$ and consider the first-order structure $\mathfrak{M}$ as above. Since $\mathbf{R}$ (the set of all real numbers) and $\mathbf{N}$ (the set of all natural numbers) are infinite sets which belong to $U(X)$,

$\mathbf{R}^*$ and $\mathbf{N}^*$ have elements which do not belong to $\mathbf{R}$ and $\mathbf{N}$, respectively. Such elements are called **nonstandard real numbers** and **nonstandard natural numbers**, respectively. By the fundamental theorem of ultraproducts we can conclude that there are many nonstandard real numbers α such that $0 < {}^*\alpha < {}^*a$ in $\mathbf{R}^*$ for any $a \in \mathbf{R}$. Such a nonstandard real number α is called an **infinitesimal real number**. Since the norm operator $\| \quad \|$ is a mapping from H to $\mathbf{R}$, $\| \quad \|^*$ is a mapping from H^* to $\mathbf{R}^*$. If $\|x\|^*$ $(x \in H^*)$ is infinitesimal, then x is said to be **infinitesimal** in H^*. Let S be the set of all linear subspaces of H. For a linear subspace K of H^* that is contained in S^* let K° be the set of elements $x \in H$ such that $x - x_0$ is infinitesimal for some x_0 in K. Then K° is a closed linear subspace of H. Let $\mathfrak{e} = \{e_i\}_{i \in \mathbf{N}}$ be an orthonormal basis of the Hilbert space H; $\mathfrak{e}$ can be considered as a mapping from $\mathbf{N}$ to H, and hence $\mathfrak{e}^* = \{e_j\}_{j \in \mathbf{N}^*}$ is a mapping from $\mathbf{N}^*$ to H^*. For each $j \in \mathbf{N}^*$, let H_j be the linear subspace of H^* spanned by $\{e_k \mid k \leqslant j\}$. For a given bounded linear operator T on H, T^* is a linear operator on H^*. We define $T_j = P_j T^* P_j$, where P_j is the projection from H^* to H_j. Since $\dim(H_j) = j$ is a (nonstandard) natural number, there exists a tower $J_{0j} \subset J_{1j} \subset \ldots \subset J_{jj} = H_j$ of closed, T_j-invariant linear subspaces of H_j such that $\dim(J_{kj}) = k$ $(k \leqslant j)$. Then $J_{kj}{}^\circ$ is a closed, T-invariant linear subspace of H. If there is a polynomial $p(x)$ such that $p(T)$ is a compact operator, then we get a nonstandard natural number j such that $J_{kj}{}^\circ$ is a proper subspace of H for some $k \leqslant j$. This gives the following result which is an affirmative solution of a problem of K. Smith and P. R. Halmos. **Theorem** (Bernstein [8]): Let T be a bounded linear operator on an infinite-dimensional Hilbert space H over the complex numbers and let $p(x) \neq 0$ be a polynomial with complex coefficients such that $p(T)$ is compact. Then T leaves invariant at least one closed linear subspace of H other than H or $\{0\}$.

F. Categoricity in Powers

Let Γ be a set of closed formulas in a first-order language L which has a designated binary predicate symbol P_0. In the following, we assume that the interpretation $\bar{P}_0$ of P_0 by $\mathfrak{M}$ is the equality relation on the universe of $\mathfrak{M}$ for every L-structure $\mathfrak{M}$. Γ is said to be categorical if all the models of Γ are isomorphic. By Theorem 3 in Section D, any Γ having a model of infinite cardinality is not categorical. Hence, there exists no interesting Γ which is categorical. Therefore, we consider the weaker notion of **categoricity in powers**. Let κ be an infinite cardinal and $n(\Gamma, \kappa)$ be the number of nonisomorphic models of Γ of cardinality κ. Then Γ is said to be categorical in κ if $n(\Gamma, \kappa) = 1$, i.e., if all the models of Γ of cardinality κ are isomorphic. There exist many interesting Γ's which are categorical in κ for some κ. For example, the set of axioms of algebraically closed fields of characteristic 0 is categorical in $\aleph_1$, and the set of axioms of dense linear orderings without endpoints is categorical in $\aleph_0$. With respect to this notion, J. Łoś conjectured that if Γ is categorical in κ for some $\kappa > \overline{\overline{L}}$ (the cardinality of L), then Γ is categorical in κ for all $\kappa > \overline{\overline{L}}$. This conjecture was solved affirmatively by M. Morley in the case $\overline{\overline{L}} = \aleph_0$, and later by S. Shelah in the general case. Theorem (1): Let Γ be a set of closed formulas in L. Then Γ is categorical in κ for some $\kappa > \overline{\overline{L}}$ if and only if Γ is categorical in κ for all $\kappa > \overline{\overline{L}}$. Also we have the following interesting theorem, due to J. T. Baldwin and A. H. Lachlan. Theorem (2): Let Γ be a set of closed formulas in L such that $\overline{\overline{L}} = \aleph_0$. If Γ is categorical in $\aleph_1$, then $n(\Gamma, \aleph_0) = 1$ or $\aleph_0$.

References

[1] A. Robinson, Introduction to model theory and to the metamathematics of algebra, North-Holland, 1963.
[2] S. C. Kleene, Mathematical logic, John Wiley, 1967.
[3] A. Robinson, Non-standard analysis, North-Holland, 1966.
[4] M. Machover and J. Hirschfeld, Lectures on non-standard analysis, Lecture notes in math. 94, Springer, 1969.
[5] A. Tarski and R. L. Vaught, Arithmetical extensions of relational systems, Compositio Math., 13 (1958), 81–102.
[6] C. C. Chang and H. J. Keisler, Model theory, North-Holland, 1973.
[7] G. E. Sacks, Saturated model theory, Benjamin, 1972.
[8] A. R. Bernstein, Non-standard analysis, in Studies in model theory 8, M. D. Morley ed., Math. Association of America, 1973, p. 35–58.

275 (III.24) Modules

A. General Remarks

In this article, we consider mainly modules with operator domain (Section C), in particular modules over a †ring. Modules over a field are linear spaces (→ 256 Linear Spaces).

Modules over a commutative ring are important in algebraic geometry (→ 18 Algebraic Varieties, 70 Commutative Rings, 281 Noetherian Rings). The theory of modules over a †group ring can be identified with the theory of linear representations of a group (→ 358 Representations). Modules without operator domain may be regarded as modules over the ring **Z** of rational integers, and the theory of finitely generated Abelian groups can be generalized to the theory of modules over a †principal ideal domain (→ 53 Categories and Functors, 202 Homological Algebra).

B. Modules

A **module** (without operator domain) is a †commutative group M whose law of composition is written additively: $a+b=b+a$ $(a,b\in M)$; the †identity element is denoted by 0, and the inverse element of a by $-a$. Every subgroup H of M is a normal subgroup. For any $a\in M$, the left and right cosets of H containing the element a are identical: $H+a=a+H$ (→ 193 Groups A).

In the set N^M of all mappings from a set M to a module N, we define an addition by the sums of values: $(f+g)(x)=f(x)+g(x)$. Then N^M forms a module. The set $\mathrm{Hom}(M,N)$ of all homomorphisms from a module M to a module N forms a subgroup of the module N^M, called the **module of homomorphisms** from M to N. The composite of homomorphisms is a homomorphism. Hence the set $\mathrm{Hom}(M,M)=\mathfrak{E}(M)$ of all endomorphisms of M forms a †rings with respect to the addition and the multiplication defined by composition; this is called the **endomorphism ring** of M. The †unity element of $\mathfrak{E}(M)$ is the identity mapping of M, and the †invertible elements of $\mathfrak{E}(M)$ are the automorphisms of M.

Let $\{x_\lambda\}_{\lambda\in\Lambda}$ be a family of elements in a module M. The sum $\sum_{\lambda\in\Lambda}x_\lambda$ is well defined if $x_\lambda=0$ $(\lambda\in\Lambda)$ except for a finite number of λ. For any family $\{N_\lambda\}_{\lambda\in\Lambda}$ of subsets of M, $\sum_{\lambda\in\Lambda}N_\lambda$ denotes the set of all elements of the form $\sum_{\lambda\in\Lambda}x_\lambda$ $(x_\lambda\in N_\lambda)$, where $x_\lambda=0$ except for a finite number of λ. If all the N_λ are subgroups of M, then $N=\sum_{\lambda\in\Lambda}N_\lambda$ is also a subgroup, called the **sum** of $\{N_\lambda\}_{\lambda\in\Lambda}$. If every element of N can be written uniquely in the form $\sum_{\lambda\in\Lambda}x_\lambda(x_\lambda\in N_\lambda)$, N is called the **direct sum** of $\{N_\lambda\}_{\lambda\in\Lambda}$. When the N_λ are subgroups, this is equivalent to the condition that $N_\lambda\cap\sum_{\lambda\neq\mu\in\Lambda}N_\mu=\{0\}$ $(\lambda\in\Lambda)$.

C. Modules with Operator Domain

Suppose that we are given a set A and a module M. If with each pair of elements $a\in A$ and $x\in M$ there is associated a unique element $ax\in M$ satisfying the condition (1) $a(x+y)=ax+ay$ $(a\in A;x,y\in M)$, we say that A is an **operator domain** of M and M is a **module with operator domain** A (**module over** A or A**-module**) (→ 193 Groups E). The mapping $A\times M\rightarrow M$ given by $(a,x)\rightarrow ax$ is called the †**operation** of A on M. Any $a\in A$ induces an endomorphism $a_M:x\rightarrow ax$ of M as a module (not as an A-module). To give the structure of an A-module to a module M is simply to give a mapping $A\rightarrow\mathfrak{E}(M)$ $(a\rightarrow a_M)$.

If N is a subgroup of an A-module M such that $ax\in N$ for any $a\in A$ and $x\in N$, then N forms an A-module, called a **sub-**A**-module** (or **allowed submodule**) of M. If $\{N_\lambda\}_{\lambda\in\Lambda}$ is a family of sub-A-modules of an A-module M, then the intersection $\bigcap_{\lambda\in\Lambda}N_\lambda$ and the sum $\sum_{\lambda\in\Lambda}N_\lambda$ are both sub-A-modules of M.

Let R be an †equivalence relation in an A-module M such that if $a\in A$ and $R(x,y)$, then $R(ax,ay)$. Then R is said to be **compatible with the operation** of A. In this case, an operation of A is induced on the quotient set M/R. Moreover, if R is compatible with the addition, namely, $R(x,x')$ and $R(y,y')$ imply $R(x+y,x'+y')$, then M/R forms an A-module, called a **quotient** A**-module** of M. The equivalence class N containing 0 is a sub-A-module of M, and M/R coincides with M/N.

D. Modules over a Group or a Ring

If the structure of a group (with the operation written multiplicatively) is given to the operator domain A of a module M, we always assume (in addition to condition (1) in Section C) that the following two conditions are satisfied: (2) $(ab)x=a(bx)$; (3) $1x=x$ $(a,b\in A,$ $x\in M)$.

If the structure of a ring is given to A, we always assume (besides conditions (1) and (2)) that a further condition holds: (4) $(a+b)x=ax+bx$ $(a,b\in A,\ x\in M)$. This means that the mapping $A\rightarrow\mathfrak{E}(M)$ $(a\rightarrow a_M)$ is a †ring homomorphism. If the ring A has unity element 1, and $1_M=$identity mapping (namely, condition (3) holds), then the A-module M is called **unitary**. We consider only unitary A-modules. Any module M is regarded as a **Z**-module, where **Z** is the ring of rational integers. M is regarded also as an $\mathfrak{E}(M)$-module.

When M is a module over a ring A, an element of A is called a **scalar**, A is called the **ring of scalars (basic ring** or **ground ring)**, and the operation $A\times M\rightarrow M$ is called the **scalar multiplication**. The elements ax $(a\in A)$ are called **scalar multiples** of x, and the totality of these elements is denoted by Ax. Let $\{x_\lambda\}_{\lambda\in\Lambda}$ be a family of elements in M. An element of the form $\Sigma_{\lambda \in \Lambda} a_\lambda x_\lambda$, where the a_λ are elements of A and equal to 0 except for a finite number of λ, is called a linear combination of $\{x_\lambda\}_{\lambda \in \Lambda}$. The set N of linear combinations of $\{x_\lambda\}_{\lambda \in \Lambda}$ is the smallest sub-A-module of M containing all the x_λ ($\lambda \in \Lambda$) and equal to the sum $\Sigma_{\lambda \in \Lambda} A x_\lambda$. The A-module N is said to be **generated** by $\{x_\lambda\}_{\lambda \in \Lambda}$, and $\{x_\lambda\}_{\lambda \in \Lambda}$ is called a **system of generators** of N. A module having a finite system of generators is said to be **finitely generated (of finite type** or simply **finite**). The module Ax generated by a single element x is called **monomial**. If A is a †field (which may be noncommutative), an A-module is a linear space over A (→ 256 Linear Spaces).

Let a be an element of an A-module M. If there exists an element λ of A that is not a zero divisor and that satisfies $\lambda a = 0$, then a is called a **torsion element** of M. We say that M is a **torsion A-module** if every element of M is a torsion element, and M is **torsion free** if M has no torsion element other than 0. An element a of M is called **divisible** if it can be written in the form λb ($b \in M$) for an arbitrary $\lambda \in A$ that is not a zero divisor. M is called a **divisible A-module** if every element of M is divisible.

Strictly speaking, the A-modules we have considered so far are called **left A-modules.** If we use the notation xa ($a \in A$, $x \in M$) instead of ax and modify conditions (1)–(4) accordingly (in particular, condition (2) becomes $x(ab) = (xa)b$), then M is called a **right A-module**. If we take a group or ring A° anti-isomorphic to A, then a left A-module can be naturally identified with a right A°-module. For a commutative group or ring A, we can disregard the distinction between left and right A-modules.

Let A and B be groups or rings. Sometimes we consider the structures of an A-module and B-module simultaneously in the same module M. If the operations of A and B commute with each other, namely, $a(bx) = b(ax)$ ($a \in A$, $b \in B$, $x \in M$), it is convenient to put one of the operations to the right. Suppose that M has the structure of a left A-module and a right B-module satisfying condition (5) $(ax)b = a(xb)$. Then M is called an **A-B-bimodule**. If G is a group and K is a commutative ring, considering the G-K-bimodules is equivalent to considering the left $K[G]$-modules, where $K[G]$ is the †group ring.

E. Operator Homomorphisms

If a module homomorphism $f: M \to N$ between A-modules M and N satisfies $f(ax) = af(x)$ ($a \in A$, $x \in M$), it is called an **A-homomorphism (operator homomorphism** or **allowed homomorphism**). If A is a ring, f is also called an **A-linear mapping**. Regarding A as an A-module, an A-linear mapping $M \to A$ is called a **linear form** on M. The composite of A-homomorphisms is an A-homomorphism.

Let $f: M \to L$ be an A-homomorphism between A-modules. The sub-A-module $\mathrm{Im} f = f(M)$ of L is called the **image** of f, and the sub-A-module $\mathrm{Ker} f = \{x \mid x \in M, f(x) = 0\}$ of M is called the **kernel** of f. $\mathrm{Coim} f = M/\mathrm{Ker} f$ is called the **coimage** of f, and $\mathrm{Coker} f = L/\mathrm{Im} f$ the **cokernel** of f. The binary relation $R(x,y)$ on M defined by $f(x) = f(y)$ ($x, y \in M$) coincides with the equivalence relation defined by $x - y \in N = \mathrm{Ker} f$ ($x, y \in M$), and the mapping f induces an A-isomorphism $\bar{f}: M/N \to f(M)$. This is the **homomorphism theorem** for modules (→ 193 Groups E).

A sequence of A-homomorphisms of A-modules M_n ($n \in \mathbf{Z}$)

$$\cdots \to M_{n-1} \xrightarrow{f_{n-1}} M_n \xrightarrow{f_n} M_{n+1} \to \cdots$$

is called an **exact sequence** if $\mathrm{Im} f_{n-1} = \mathrm{Ker} f_n$ for all n. The A-module $\{0\}$ is denoted simply by 0. Exactness of $0 \to N \xrightarrow{f} M$ or $M \xrightarrow{g} L \to 0$ means that the mapping $f: N \to M$ is injective or the mapping $g: M \to L$ is surjective, respectively. In an †inductive (†projective) system $\{M_\lambda, f_{\lambda\mu}\}$ of A-modules, in which every $f_{\lambda\mu}: M_\lambda \to M_\mu$ is an A-homomorphism, the limit $M = \varinjlim M_\lambda$ $\left(\varprojlim M_\lambda\right)$ is also an A-module. If $0 \to L_\lambda \to M_\lambda \to N_\lambda \to 0$ is an exact sequence for every λ and

$$\begin{array}{ccccc} L_\lambda & \to & M_\lambda & \to & N_\lambda \\ \downarrow & & \downarrow & & \downarrow \\ L_\mu & \to & M_\mu & \to & N_\mu \end{array}$$

is a †commutative diagram for $\lambda < \mu$, then $0 \to \varinjlim L_\lambda \to \varinjlim M_\lambda \to \varinjlim N_\lambda \to 0$ is an exact sequence. However, this does not necessarily hold for the projective limit.

The set of all A-homomorphisms from an A-module M to an A-module N, denoted by $\mathrm{Hom}_A(M,N)$ and called the **module of A-homomorphisms**, is a subgroup of the module $\mathrm{Hom}(M,N)$. The set $\mathrm{Hom}_A(M,M) = \mathfrak{E}_A(M)$ of all A-endomorphisms of an A-module M forms a †subring of the ring $\mathfrak{E}(M)$ and coincides with the set of all elements commuting with any a_M ($a \in A$). $GL(M)$ is the group consisting of all †invertible elements in $\mathfrak{E}_A(M)$. If A is a commutative ring, $\mathrm{Hom}_A(M,N)$ is made into an A-module by defining $(af)(x) = af(x)$, namely, $af = a_N \circ f$, and in particular $\mathfrak{E}_A(M)$ is an †associative algebra over A. If M is an A-B-bimodule, $\mathrm{Hom}_A(M,N)$ forms a left B-module by $(bf)(x) = f(xb)$. If N is an A-B-bimodule, $\mathrm{Hom}_A(M,N)$ forms a right B-module by $(fb)(x) = f(x)b$.

F. Direct Products and Direct Sums

In the Cartesian product $P=\prod_{\lambda\in\Lambda}M_\lambda$ of a family $\{M_\lambda\}_{\lambda\in\Lambda}$ of A-modules, we define addition and an operation of A as follows: $\{x_\lambda\}+\{y_\lambda\}=\{x_\lambda+y_\lambda\}$, $a\{x_\lambda\}=\{ax_\lambda\}$. Then P forms an A-module, called the **direct product of modules** $\{M_\lambda\}_{\lambda\in\Lambda}$ and denoted by $\prod_{\lambda\in\Lambda}M_\lambda$. The **canonical surjection** assigning x_λ to $\{x_\lambda\}$ is denoted by $p_\lambda: P\rightarrow M_\lambda$. Suppose that an A-module M and A-homomorphisms $f_\lambda: M\rightarrow M_\lambda (\lambda\in\Lambda)$ are given. Then there exists a unique A-homomorphism $f: M\rightarrow P$ such that $p_\lambda\circ f=f_\lambda$ $(\lambda\in\Lambda)$; f is given by $f(x)=\{f_\lambda(x)\}$.

In the direct product $\prod_{\lambda\in\Lambda}M_\lambda$ of a family $\{M_\lambda\}_{\lambda\in\Lambda}$ of A-modules, the set S of all elements whose components x_λ are equal to 0 except for a finite number of λ is denoted by $\sum_{\lambda\in\Lambda}M_\lambda$ (or $\coprod_{\lambda\in\Lambda}M_\lambda, \bigotimes_{\lambda\in\Lambda}M_\lambda$) and called the **direct sum** of modules $\{M_\lambda\}_{\lambda\in\Lambda}$. The **canonical injection** assigning $\{\ldots,0,x_\lambda,0,\ldots\}\in S$ to $x_\lambda\in M_\lambda$ is denoted by $j_\lambda: M_\lambda\rightarrow S$. Suppose that an A-module M and A-homomorphisms $f_\lambda: M_\lambda\rightarrow M$ are given. Then there exists a unique A-homomorphism $f: S\rightarrow M$ such that $f\circ j_\lambda=f_\lambda$ $(\lambda\in\Lambda)$; f is given by $f(\{x_\lambda\})=\sum_{\lambda\in\Lambda}f_\lambda(x_\lambda)$. When M is an A-module and $\{N_\lambda\}_{\lambda\in\Lambda}$ is a family of sub-A-modules of M, the A-homomorphism $f:\sum_{\lambda\in\Lambda}N_\lambda\rightarrow M$ defined by $f(\{x_\lambda\})=\sum_{\lambda\in\Lambda}x_\lambda$ is an A-isomorphism if and only if M is the direct sum of $\{N_\lambda\}$ as a module (without operator domain).

If $M_\lambda=M$ for all $\lambda\in\Lambda$, $\prod_{\lambda\in\Lambda}M_\lambda$ and $\sum_{\lambda\in\Lambda}M_\lambda$ are denoted by M^Λ and $M^{(\Lambda)}$, respectively. M^Λ can be regarded as the set of all mappings from Λ to M. The direct product and direct sum of A-modules $M_1,\ldots,M_n$, written $M_1\times\ldots\times M_n$ and $M_1+\ldots+M_n$, respectively, can be identified with each other and written M^n if $M_i=M$ $(1\leqslant i\leqslant n)$.

G. Free Modules

Let A be a ring. A family $\{x_\lambda\}_{\lambda\in\Lambda}$ of elements in an A-module M is called **linearly independent** if $\sum_{\lambda\in\Lambda}a_\lambda x_\lambda=0$ $(a_\lambda\in A)$ implies $a_\lambda=0$ for all $\lambda\in\Lambda$. This is equivalent to saying that the mapping $A^{(\Lambda)}\rightarrow M$ that assigns $\sum_{\lambda\in\Lambda}a_\lambda x_\lambda\in M$ to $\{a_\lambda\}$ is injective. A linearly independent family $\{x_\lambda\}_{\lambda\in\Lambda}$ generating M is called a **basis** of M. A family $\{x_\lambda\}_{\lambda\in\Lambda}$ is a basis if and only if every element of M can be written uniquely in the form $\sum_{\lambda\in\Lambda}a_\lambda x_\lambda$ $(a_\lambda\in A)$.

An A-module that has a basis is called a **free module** over A. If A is a field (which may be noncommutative), every A-module is a free module (→ 256 Linear Spaces). The †cardinality of a basis of a free module M over A depends only on M if A is a field (which may be noncommutative) or a commutative ring; this number is called the **rank** (or **dimension**) of M. Any submodule of a free module over a †principal ideal domain is a free module.

H. Simple Modules and Semisimple Modules

An A-module M is called **simple** if $M\neq 0$ and M has no sub-A-modules except M and 0. If M and N are simple A-modules, any A-homomorphism from M to N is an isomorphism or the **zero homomorphism** (i.e., the homomorphism sending every element of M to 0) (**Schur's lemma**). If an A-module M is the sum of a family $\{M_\lambda\}_{\lambda\in\Lambda}$ of simple submodules, M is the direct sum of a suitable subfamily $\{M_{\lambda'}\}_{\lambda'\in\Lambda'}$ $(\Lambda'\subset\Lambda)$. In this case, M is called **semisimple** (or **completely reducible**).

Let an A-module M be decomposed into the direct sum of sub-A-modules N and N'. Then N' is called a **complementary submodule** of N. An A-module M is semisimple if and only if every sub-A-module of M has a complementary submodule. Let A be a ring. Then the A-module M is semisimple if and only if every A-module is semisimple. In this case A is called a †semisimple ring (→ 363 Rings G). Every simple module over a semisimple ring A is A-isomorphic to a †minimal left ideal of A regarded as an A-module.

I. Chain Conditions

The set of all sub-A-modules of an A-module M forms an †ordered set under the inclusion relation. An A-module is called a **Noetherian module** if the ordered set satisfies the †maximal condition, and an **Artinian module** if it satisfies the †minimal condition (→ 305 Ordering C).

Let N be a sub-A-module of an A-module M. If M is Noetherian (Artinian), N and M/N are both Noetherian (Artinian). The converse also holds. A ring A is called a †left Noetherian ring (†left Artinian ring) if A is Noetherian (Artinian) as a left A-module, and similarly for right Noetherian and Artinian rings. Every finitely generated module over a Noetherian (Artinian) ring is Noetherian (Artinian). Over an arbitrary ring A, a module M is Noetherian if and only if every sub-A-module of M is finitely generated.

A finite sequence $\{M_i\}_{0\leqslant i\leqslant r}$ of sub-A-modules of an A-module M is called a †Jordan-Hölder sequence if $M=M_0$, $M_i\supset M_{i+1}$, $M_r=\{0\}$, and the M_i/M_{i+1} $(0\leqslant i<r)$ are simple. If such a sequence exists, M is said to be **of finite length**. The number r, called the **length**

of M, depends only on M. The quotient modules M_i/M_{i+1} $(0 \leqslant i < r)$ are uniquely determined by M up to A-isomorphism and permutation of the indices (C. Jordan and O. Hölder). An A-module M is of finite length if and only if M is Noetherian and Artinian. A semisimple A-module is of finite length if and only if it is finitely generated.

An A-module M is called **indecomposable** if M cannot be decomposed into the direct sum of two sub-A-modules different from M and $\{0\}$. Any A-module of finite length can be decomposed into the direct sum of a finite sequence $N_1, \ldots, N_n$ of indecomposable sub-A-modules different from $\{0\}$. The direct summands N_i $(1 \leqslant i \leqslant n)$ are unique up to A-isomorphism and permutation of the indices (W. Krull, R. Remak, and O. Schmidt).

J. Tensor Products

Let A be a ring. Given a right A-module M and a left A-module N, we construct a module $M \otimes_A N$ (called the **tensor product** of M and N) and a canonical mapping $M \times N \rightarrow M \otimes_A N$ as follows. Let F be a free $\mathbf{Z}$-module (free Abelian additive group) generated by $M \times N$, and R be the subgroup generated by the elements of the forms $(x+x',y)-(x,y)-(x',y)$, $(x,y+y')-(x,y)-(x,y')$, $(xa,y)-(x,ay)$ $(x,x' \in M, y,y' \in N, a \in A)$. We define $M \otimes_A N = F/R$, denote by $x \otimes y$ the element of $M \otimes_A N$ containing $(x,y) \in F$, and define a mapping $M \times N \rightarrow M \otimes_A N$ by the assignment $(x,y) \rightarrow x \otimes y$. Then we have $(x_1+x_2)\otimes y = x_1 \otimes y + x_2 \otimes y$, $x \otimes (y_1+y_2) = x \otimes y_1 + x \otimes y_2$, and $(xa)\otimes y = x \otimes (ay)$. Any element of $M \otimes_A N$ is written in the form $\sum x_i \otimes y_i$ $(x_i \in M, y_i \in N)$.

The tensor product $M \otimes_A N$ of M and N and the canonical mapping $M \times N \rightarrow M \otimes_A N$ can be characterized as follows: For a module L, a mapping $f: M \times N \rightarrow L$ is called **biadditive** if the conditions $f(x+x',y) = f(x,y) + f(x',y)$, $f(x,y+y') = f(x,y) + f(x,y')$ hold. A biadditive mapping f satisfying the condition $f(xa,y) = f(x,ay)$ is called an **A-balanced mapping**. Then we have (i) the canonical mapping $M \times N \rightarrow M \otimes_A N$ is A-balanced; and (ii) for any module L and any A-balanced mapping $f: M \times N \rightarrow L$, there exists a unique homomorphism $\varphi: M \otimes_A N \rightarrow L$ such that $f(x,y) = \varphi(x \otimes y)$ $(x \in M, y \in N)$.

A right (left) A-module can be regarded as a left (right) A°-module, where A° is the ring anti-isomorphic to A. In this sense, we have $M \otimes_A N \cong N \otimes_{A^\circ} M$.

Let A be a commutative ring. For A-modules M, N, and L, a mapping $f: M \times N \rightarrow L$ is called a **bilinear mapping** if f is biadditive and satisfies $f(ax,y) = f(x,ay) = af(x,y)$ $(a \in A, x \in M, y \in N)$. The set $\mathfrak{L}(M,N;L)$ of all bilinear mappings $M \times N \rightarrow L$ forms a sub-A-module of the A-module $L^{M \times N}$. A bilinear mapping $M \times N \rightarrow A$ is called a **bilinear form** on $M \times N$. The tensor product $M \otimes_A N$ becomes an A-module if we define $a(x \otimes y) = (ax) \otimes y$ $(= x \otimes (ay))$, and the canonical mapping $M \times N \rightarrow M \otimes_A N$ is bilinear. For any A-module L and bilinear mapping $f: M \times N \rightarrow L$, there exists a unique A-linear mapping $\varphi: M \otimes_A N \rightarrow L$ satisfying $f(x,y) = \varphi(x \otimes y)$. By this correspondence $f \leftrightarrow \varphi$, we get an A-isomorphism $\mathfrak{L}(M,N;L) \cong \mathrm{Hom}_A(M \otimes_A N, L)$. If A is a field, $M \otimes_A N$ coincides with the tensor product $M \otimes N$ as a linear space (→ 256 Linear Spaces H, I).

In general, let M be a B-A-bimodule and N be a left A-module. Then $M \otimes_A N$ becomes a left B-module if we define $b(x \otimes y) = (bx) \otimes y$. Let N be an A-B-bimodule and M be a right A-module. Then $M \otimes_A N$ becomes a right B-module if we define $(x \otimes y)b = x \otimes (yb)$. In particular, we have $A \otimes_A N \cong N$, $M \otimes_A A \cong M$.

Let M, M' be right A-modules and N, N' be left A-modules. For A-homomorphisms $f: M \rightarrow M'$ and $g: N \rightarrow N'$, there exists a unique homomorphism $h: M \otimes_A N \rightarrow M' \otimes_A N'$ satisfying $h(x \otimes y) = f(x) \otimes g(y)$; h is called the **tensor product** of f, g and is denoted by $f \otimes g$. We give here some simple examples (also → Section L).

Examples. (1) Let M, N be free modules (linear spaces, for example) over a commutative ring A. If $\{x_i\}_{i \in I}$ and $\{y_j\}_{j \in J}$ are bases of M and N, respectively, $M \otimes_A N$ is also a free module with a basis $\{x_i \otimes y_j\}_{i \in I, j \in J}$. If the dimensions $\dim M$, $\dim N$ are finite, $\dim M \otimes_A N = \dim M \dim N$.

(2) For an †ideal $\mathfrak{m}$ of a commutative ring A, the †quotient ring $M = A/\mathfrak{m}$ is regarded as an A-module, and we have $M \otimes_A N \cong N/\mathfrak{m}N$. For instance, $(\mathbf{Z}/m\mathbf{Z}) \otimes_{\mathbf{Z}} (\mathbf{Z}/n\mathbf{Z}) \cong \mathbf{Z}/(m,n)\mathbf{Z}$, where (m,n) denotes the greatest common divisor of m and n.

K. Hom and ⊗

We continue to consider modules over a ring A. Concerning the direct sum and product, we have

$$\mathrm{Hom}_A\Big(\sum_\lambda M_\lambda, \prod_\mu N_\mu\Big) \cong \prod_{\lambda,\mu} \mathrm{Hom}_A(M_\lambda, N_\mu)$$

and

$$\Big(\sum_\lambda M_\lambda\Big) \otimes_A \Big(\sum_\mu N_\mu\Big) \cong \sum_{\lambda,\mu} (M_\lambda \otimes_A N_\mu).$$

Concerning projective and inductive limits

we have

$$\mathrm{Hom}_A\left(\varinjlim M_\lambda, \varprojlim N_\mu\right) \cong \varprojlim \mathrm{Hom}_A(M_\lambda, N_\mu)$$

and

$$\left(\varinjlim M_\lambda\right) \otimes_A \left(\varinjlim N_\mu\right) = \varinjlim (M_\lambda \otimes_A N_\mu).$$

An A-homomorphism $f: M \to M'$ induces a homomorphism $\mathrm{Hom}_A(M', N) \to \mathrm{Hom}_A(M, N)$ by the assignment $g \to g \circ f$. An exact sequence $M' \to M \to M'' \to 0$ gives rise to the following exact sequence:

$$0 \to \mathrm{Hom}_A(M'', N) \to \mathrm{Hom}_A(M, N) \to \mathrm{Hom}_A(M', N). \quad (1)$$

An A-homomorphism $f: N \to N'$ induces a homomorphism $\mathrm{Hom}_A(M, N) \to \mathrm{Hom}_A(M, N')$ by the assignment $g \to f \circ g$, and an exact sequence $0 \to N' \to N \to N''$ gives rise to the following exact sequence:

$$0 \to \mathrm{Hom}_A(M, N') \to \mathrm{Hom}_A(M, N) \to \mathrm{Hom}_A(M, N'') \quad (2)$$

Let M be a right A-module and N', N, N'' be left A-modules. An A-homomorphism $f: N \to N'$ induces the homomorphism $1_M \otimes f: M \otimes_A N \to M \otimes_A N'$, and an exact sequence $N' \to N \to N'' \to 0$ gives rise to the following exact sequence:

$$M \otimes_A N' \to M \otimes_A N \to M \otimes_A N'' \to 0. \quad (3)$$

Exchanging left and right, we obtain similar results (→ 53 Categories and Functors B; 57 Chain Complexes D, I; 202 Homological Algebra).

Let Q be an A-module. If for any exact sequence of A-homomorphisms of A-modules

$$0 \to M' \xrightarrow{\varphi} M \xrightarrow{\psi} M'' \to 0, \quad (4)$$

the induced sequence

$$0 \to \mathrm{Hom}_A(M'', Q) \to \mathrm{Hom}_A(M, Q) \to \mathrm{Hom}_A(M', Q) \to 0 \quad (5)$$

is exact, then Q is called an **injective A-module**. This is equivalent to the condition that if M' is a sub-A-module of an A-module M, then any A-homomorphism $M' \to Q$ can be extended to an A-homomorphism $M \to Q$. Any A-module is a sub-A-module of some injective A-module, and any injective A-module is a divisible A-module. If A is a †Dedekind ring, any divisible A-module is an injective A-module.

Let P be an A-module. If for any exact sequence (4), the induced sequence

$$0 \to \mathrm{Hom}_A(P, M') \to \mathrm{Hom}_A(P, M) \to \mathrm{Hom}_A(P, M'') \to 0 \quad (6)$$

is exact, then P is called a **projective A-module**. This is equivalent to the condition that for any surjective A-homomorphism $g: M \to M''$ and any A-homomorphism $f: P \to M''$, there exists an A-homomorphism $h: P \to M$ satisfying $g \circ h = f$. Any A-module is a quotient A-module of some projective A-module. A projective A-module has no torsion element. A free A-module is a projective A-module. In general, an A-module is a projective A-module if and only if it is a direct summand of a free A-module.

Let R be a right A-module. If for any exact sequence (4), the induced sequence

$$0 \to R \otimes_A M' \to R \otimes_A M \to R \otimes_A M'' \to 0 \quad (7)$$

is exact, then R is called a **flat A-module**. Any projective A-module is a flat A-module. A flat A-module R is called **faithfully flat** if $R \otimes_A M = \{0\}$ implies $M = \{0\}$. A flat right A-module R is faithfully flat if and only if $R \neq R\mathfrak{A}$ for any left ideal $\mathfrak{A}$ $(\neq A)$ of A. Let A be a †principal ideal domain. Then an A-module R is flat if and only if R has no torsion element, and R is faithfully flat if and only if R has no torsion element and $R \neq Rp$ for any †prime element p of A. We have the following important examples:

(1) For a commutative ring A and its multiplicatively closed subset S, the †ring of quotients A_S is flat as an A-module. However, A_S is not faithfully flat. For instance, the field of rational numbers $\mathbf{Q}$ is not faithfully flat as a $\mathbf{Z}$-module.

(2) Let A be a †semilocal ring and $\bar{A}$ be its completion. Then $\bar{A}$ is faithfully flat as an A-module (→ 281 Noetherian Rings; also [3, 10]).

In the exact sequence (4), if $\mathrm{Im}\,\varphi = \mathrm{Ker}\,\psi$ is a direct summand of the A-module M, we say that (4) **splits**. Then (5), (6), and (7) are exact for any A-modules Q, P, R. The exact sequence (4) splits if M' is injective or M'' is projective.

By ${}_AM$, M_A, and ${}_AM_B$, we mean that M is a left A-module, a right A-module, and an A-B-bimodule, respectively. As already stated, ${}_AM_B$ and ${}_AN$ imply ${}_B(\mathrm{Hom}_A(M,N))$, and ${}_AM$ and ${}_AN_B$ imply $(\mathrm{Hom}_A(M,N))_B$. Similarly ${}_BM_A$ and N_A imply $(\mathrm{Hom}_A(M,N))_B$, and M_A and ${}_BN_A$ imply ${}_B(\mathrm{Hom}_A(M,N))$. Furthermore, ${}_BM_A$ and ${}_AN$ imply ${}_B(M \otimes_A N)$, and M_A and ${}_AN_B$ imply $(M \otimes_A N)_B$.

If ${}_BL_A$, ${}_AM$, and ${}_BN$, then we have

$$\mathrm{Hom}_A(M, \mathrm{Hom}_B(L, N)) \cong \mathrm{Hom}_B(L \otimes_A M, N). \quad (8)$$

Similarly, if ${}_AM_B$, L_A, and N_B, then we have

$$\mathrm{Hom}_A(L, \mathrm{Hom}_B(M, N)) \cong \mathrm{Hom}_B(L \otimes_A M, N). \quad (8')$$

If B is a commutative ring, (8) and (8′) are B-isomorphisms. Furthermore, if L_A, ${}_AM_B$, and ${}_BN$, then we have

$$(L\otimes_A M)\otimes_B N\cong L\otimes_A (M\otimes_B N). \qquad (9)$$

We denote by M^* the set $\mathrm{Hom}_A(M,A)$ of all linear forms on an A-module M. Then ${}_AM$ implies M^*_A, and M_A implies ${}_AM^*$; the A-module M^* is called the **dual module** of M. A as a left A-module is dual to A as a right A-module, and vice versa. For a family of A-modules $\{M_\lambda\}_{\lambda\in\Lambda}$, we have a canonical correspondence $(\Sigma_{\lambda\in\Lambda}M_\lambda)^*\cong\Pi_{\lambda\in\Lambda}M^*_\lambda$. From this, we have a canonical isomorphism $(M^*)^*\cong M$ for any finitely generated projective A-module M. Many facts concerning this †duality are similar to those valid for linear spaces (→ 256 Linear Spaces G).

Let A be a commutative ring. Considering the case $A=B=N$ in (8) and (8′), we have the canonical A-isomorphisms

$$\mathrm{Hom}_A(M,L^*)\cong\mathrm{Hom}_A(L,M^*)$$
$$\cong(L\otimes_A M)^*=\mathfrak{L}(L,M;A),$$

namely, any bilinear form on $L\times M$ is represented by a linear mapping $M\to L^*$ or $L\to M^*$.

L. Extension and Restriction of a Basic Ring

Fix a ring homomorphism $\rho:A\to B$. We regard B as a B-A-bimodule by defining a right operation of A on B by $b\cdot a=b\rho(a)$ $(a\in A, b\in B)$. This bimodule is denoted by B_ρ.

For every left A-module M, we construct the left B-module $\rho^*(M)=B_\rho\otimes_A M$, which is called the **scalar extension** of M by ρ. Every A-homomorphism of A-modules $f:M\to M'$ induces the B-homomorphism $\rho^*(f)=1_B\otimes f$: $\rho^*(M)\to\rho^*(M')$.

For every left B-module N, we construct the left A-module $\rho_*(N)=\mathrm{Hom}_B(B_\rho,N)$, which is called the **scalar restriction** (or **scalar change**) of N by ρ. By the assignment $h\to h(1)$, we have a module isomorphism $\mathrm{Hom}_B(B_\rho,N)=N$. We identify $\rho_*(N)$ with N under this isomorphism. The operation of A on N is then given as follows: $a\cdot y=\rho(a)y$ $(a\in A, y\in N)$. If A is a subring of B and ρ is the canonical injection, then the operation of A on $\rho_*(N)$ is the restriction of the operation of B on N. Every B-homomorphism of B-modules $f:N\to N'$ induces the A-homomorphism $\rho_*(f):\rho_*(N)\to\rho_*(N')$. For any left A-module M and left B-module N, an A-linear mapping $f:M\to\rho_*(N)=N$ is called a **semilinear mapping** with respect to ρ. This means that f is an additive homomorphism satisfying $f(ax)=\rho(a)f(x)$ $(a\in A, x\in M)$.

The extension and the restriction of a basic ring are related by the isomorphism $\mathrm{Hom}_A(M,\rho_*(N))\cong\mathrm{Hom}_B(\rho^*(M),N)$ for an A-module M and B-modules N, where an element α of the left-hand side and an element β of the right-hand side are associated by the relation $\alpha(x)=\beta(1\otimes x)$ $(x\in M)$ (→ 53 Categories and Functors).

Let A and B be commutative rings. Then for A-modules M and M', we have the canonical B-linear mapping $\rho^*:B\otimes_A\mathrm{Hom}_A(M,M')\to\mathrm{Hom}_B(B\otimes_A M, B\otimes_A M')$, which is a B-isomorphism if M is a finitely generated free (or more generally, projective) module. Using the notation ρ^*, we have $\rho^*(\mathrm{Hom}_A(M,M'))\cong\mathrm{Hom}_B(\rho^*(M),\rho^*(M'))$.

We now give some examples where the basic rings are noncommutative. Let G be a group and H its subgroup. Let ρ denote the homomorphism of group rings $K[H]\to K[G]$ induced by the canonical injection $H\to G$, where K is a commutative ring. For any $K[H]$-module M, $\rho^*(M)$ is called the **induced module** of M. The representation of G associated with $\rho^*(M)$ is the †induced representation of the representation of H associated with M. Next, we fix a group G and consider a homomorphism $\rho:K[G]\to\bar{K}[G]$ induced by a homomorphism of commutative rings $\sigma:K\to\bar{K}$. If $K=\bar{K}$ and σ is an automorphism, then the representation associated with the "scalar extension" $\rho^*(M)$ of a $K[G]$-module M is the †conjugate representation to the representation associated with M. If $\bar{K}=K/\mathfrak{A}$ ($\mathfrak{A}$ is an ideal of K) and σ is the canonical surjection, then the representation over $\bar{K}$ associated with the scalar extension $\rho^*(M)$ of a $K[G]$-module M is the **reduction modulo** $\mathfrak{A}$ of the representation over K associated with M, and $\rho^*(M)$ is canonically isomorphic to $M/\mathfrak{A}M$. The †localization and the †completion can also be treated under the formulation of scalar extension (→ 70 Commutative Rings G, 281 Noetherian Rings B).

References

[1] T. Nakayama and A. Hattori, Homorozî daisûgaku (Japanese; Homological algebra), Kyôritu, 1957.
[2] S. Iyanaga and K. Kodaira, Gendai sûgaku gaisetu I (Japanese; Introduction to modern mathematics I), Iwanami, 1961.
[3] M. Nagata, Local rings, Interscience, 1962.
[4] Y. Akizuki and M. Suzuki, Kôtô daisûgaku (Japanese; Higher algebra), Iwanami, I, 1952; II, 1957.
[5] H. P. Cartan and S. Eilenberg, Homological algebra, Princeton Univ. Press, 1956.
[6] D. G. Northcott, An introduction to homological algebra, Cambridge Univ. Press, 1960.
[7] S. MacLane, Homology, Springer, 1963.
[8] C. Chevalley, Fundamental concepts of algebra, Academic Press, 1956.

[9] N. Bourbaki, Eléments de mathématique, Algèbre, ch. 2, Actualités Sci. Ind, 1236b, Hermann, third edition, 1962.
[10] N. Bourbaki, Eléments de mathématique, Algèbre commutative, ch. 1, 2, Actualités Sci. Ind., 1290a, Hermann, 1961.
[11] C. W. Curtis and I. Reiner, Representation theory of finite groups and associative algebras, Interscience, 1962.
[12] R. Godement, Cours d'algèbre, Hermann, 1963.
[13] S. Lang, Algebra, Addison-Wesley, 1965.
[14] S. T. Hu, Elements of modern algebra, Holden-Day, 1965.

276 (XIII.25) Monge-Ampère Equations

A. Monge-Ampère Equations

A **Monge-Ampère differential equation** is a second-order partial differential equation of the form

$$Hr+2Ks+Lt+M+N(rt-s^2)=0, \qquad (1)$$

where H, K, L, M, and N are functions of x, y, z, p, and q, and r, s, t, p, and q represent the partial derivatives

$$r=\frac{\partial^2 z}{\partial x^2}, \qquad s=\frac{\partial^2 z}{\partial x \partial y}, \qquad t=\frac{\partial^2 z}{\partial y^2},$$

$$p=\frac{\partial z}{\partial x}, \qquad q=\frac{\partial z}{\partial y}.$$

The characteristic manifolds are integrals of a system of differential equations defined as follows:

Case (i) $N \neq 0$.

$$N\,dp+L\,dx+\lambda_1\,dy=0,$$
$$N\,dq+\lambda_2\,dx+H\,dy=0,$$
$$dz-p\,dx-q\,dy=0, \qquad (2)$$

$$N\,dp+L\,dx+\lambda_2\,dy=0,$$
$$N\,dq+\lambda_1\,dx+H\,dy=0,$$
$$dz-p\,dx-q\,dy=0, \qquad (3)$$

where λ_1 and λ_2 are the two roots of the equation $\lambda^2+2K\lambda+HL-MN=0$.

Case (ii) $N=0$, $H \neq 0$.

$$dy=\lambda_1\,dx, \quad H\,dp+H\lambda_2\,dq+M\,dx=0,$$
$$dz-p\,dx-q\,dy=0, \qquad (4)$$

$$dy=\lambda_2\,dx, \quad H\,dp+H\lambda_1\,dq+M\,dx=0,$$
$$dz-p\,dx-q\,dy=0, \qquad (5)$$

where λ_1 and λ_2 are the two roots of the equation $H\lambda^2-2K\lambda+L=0$.

Case (iii) $N=0$, $H=0, L \neq 0$.

$$dx=0, \quad M\,dy+2K\,dp+L\,dq=0,$$
$$dz-p\,dx-q\,dy=0, \qquad (6)$$

$$2K\,dy-L\,dx=0, \quad M\,dy+L\,dq=0,$$
$$dz-p\,dx-q\,dy=0. \qquad (7)$$

Case (iv) $N=0$, $H=0$, $L=0$.

$$dx=0, \quad 2K\,dp+M\,dy=0,$$
$$dz-p\,dx-q\,dy=0, \qquad (8)$$

$$dy=0, \quad 2K\,dq+M\,dx=0,$$
$$dz-p\,dx-q\,dy=0. \qquad (9)$$

A manifold $x(\lambda), y(\lambda), z(\lambda), p(\lambda), q(\lambda)$ that satisfies the system (2), (3) of differential equations for case (i), (4), (5) for case (ii), (6), (7) for case (iii), or (8), (9) for case (iv) is a characteristic manifold of equation (1).

The following result is known concerning Monge-Ampère equations: The union of surface elements of an integral surface of (1) is generated in two ways by characteristic manifolds depending on one parameter, and vice versa.

B. Intermediate Integrals

If a relation $dV(x, y, z, p, q)=0$ is a consequence of a system of differential equations of characteristic manifolds, $V(x,y,z,p,q)=c$ (c an arbitrary constant) is called an **integral** of the system of differential equations. (i) If $V=c$ is an integral of a system of differential equations of characteristic manifolds, the solution of $V=c$ considered as a partial differential equation of the first order is a solution of (1). Conversely, if every solution of $V=c$ (excepting †singular ones) satisfies equation (1), $V=c$ is an integral of a system of differential equations of characteristic manifolds. (ii) If a system of differential equations of characteristic manifolds has two integrals $u=c$, $v=c$, every solution of (1) satisfies a partial differential equation of the first order $\varphi(u,v)=0$, where φ is a suitable function of u, v, that is, equation (1) is equivalent to $\varphi(u,v)=0$ with an arbitrary function φ. The relation $\varphi(u,v)=0$ is called an **intermediate integral** of (1). Sometimes an integral of a system of differential equations of characteristic manifolds is also called an intermediate integral. If each of the two systems of differential equations defining the characteristic manifolds has an intermediate integral, then the two intermediate integrals $\varphi(u,v)=0$ and $\psi(u,v)=0$ form a †complete system of partial differential equations of the first order. Integrating this complete system, the †general solution of equation (1) is obtained.

References

[1] E. Goursat, Cours d'analyse mathématique III, Gauthier-Villars, fourth edition, 1927.

[2] E. Goursat, Leçons sur l'intégration des équations aux dérivées partielles du second ordre à deux variables indépendantes I, Hermann, 1896.
[3] M. Hukuhara, Henbibun hôteisiki ron (Japanese; Theory of partial differential equations), Iwanami Coll. of Math., 1935.

277 (XVII.5) Multivariate Analysis

A. General Remarks

Multivariate analysis consists of methods of statistical analysis of **multivariate data**, characterized as consisting of several observations on each of a set of objects, or, mathematically, represented by a collection of points in a finite-dimensional Euclidean space $\mathbf{R}^p$. Multivariate techniques may be classified into two groups, direct extensions of univariate techniques to the multivariate case and multivariate techniques proper.

B. The Multivariate Linear Model

The multivariate linear model is the immediate extension of the univariate linear model. Suppose that $\mathbf{X}=(\mathbf{X}^{(1)}\ldots\mathbf{X}^{(n)})$ denotes the $p\times n$ matrix of n observations of p-dimensional data. Suppose that it can be expressed as

$$\mathbf{X}=BZ+\mathbf{U}, \qquad (1)$$

where B is a $p\times m$ matrix of unknown parameters, Z is a known $m\times n$ matrix of independent variables, and $\mathbf{U}$ is a $p\times n$ matrix of errors. We assume (i) that the †expectations of the elements of $\mathbf{U}$ are zero, that is, the $m\times n$ matrix $E(\mathbf{U})=0$, and call the relation (1) a **multivariate linear model**. We usually assume further (ii) that the column vectors $\mathbf{U}^{(i)}$, $i=1,\ldots,n$, of $\mathbf{U}$ are independent and identically distributed, and (iii) that $\mathbf{U}^{(i)}$ is distributed according to a multivariate †normal distribution with †variance-covariance matrix Σ. Analogous to the univariate case, the †least squares estimator $\hat{\mathbf{B}}$ of B is defined to be the $p\times m$ matrix that minimizes

$$\mathrm{tr}(\mathbf{X}-BZ)(\mathbf{X}-BZ)'$$

and is given explicitly by

$$\hat{\mathbf{B}}=\mathbf{X}Z'(ZZ')^{-1} \quad \text{when} \quad |ZZ'|\neq 0.$$

Here the symbol $'$ means the †transpose of a matrix. Also, an †unbiased estimator of Σ is given by

$$\hat{\Sigma}=\mathbf{Q}_e/(n-m), \qquad \mathbf{Q}_e=\mathbf{X}\mathbf{X}'-\hat{\mathbf{B}}Z\mathbf{X}'.$$

$\hat{\mathbf{B}}$ is an †unbiased estimator of B under assumption (i) and the †best linear unbiased estimator under (i) and (ii), while $\hat{\Sigma}$ is unbiased when (i) and (ii) are assumed. Under the assumptions (i)–(iii), $\hat{\mathbf{B}}$ and $\hat{\Sigma}$ form a set of †complete †sufficient statistics; hence they are †uniformly minimum variance unbiased estimators. Also under (i)–(iii), elements of $\hat{\mathbf{B}}$ are normally distributed, and their variance-covariance can be expressed by

$$\Sigma\otimes M \quad (\otimes \text{ denotes the †Kronecker product}),$$

where $M=(ZZ')^{-1}$. Applying †Cochran's theorem for the multivariate case, $\mathbf{Q}_e$ is shown to be distributed according to a †Wishart distribution with $n-m$ degrees of freedom.

To test the hypothesis $B=B_0$ under (i)–(iii), we put $\mathbf{Q}_B=(\hat{\mathbf{B}}-B_0)ZZ'(\hat{\mathbf{B}}-B_0)'$ and have $(\mathbf{X}-B_0Z)(\mathbf{X}-B_0Z)'=\mathbf{Q}_B+\mathbf{Q}_e$, where $\mathbf{Q}_B$ and $\mathbf{Q}_e$ are independently distributed. The distribution of $\mathbf{Q}_B$ is a Wishart distribution with m degrees of freedom when the hypothesis is true, and a †noncentral Wishart distribution when $B\neq B_0$. Based on this fact, several procedures have been proposed. If we require the invariance of procedures with respect to linear transformations of the coordinates of p-dimensional vectors, the roots $\lambda_1,\ldots,\lambda_p$ of the †characteristic equation $|\mathbf{Q}_B-\lambda\mathbf{Q}_e|=0$ form a †maximal invariant statistic; hence the testing procedures should be defined in terms of these roots (→ 387 Statistic J). Also, the consideration of †power leads to procedures that reject the hypothesis when these roots are large. Commonly used test statistics are (1) the †likelihood ratio test $W=|\mathbf{Q}_e|/|\mathbf{Q}_B+\mathbf{Q}_e|=\Pi_i(1+\lambda_i)^{-1}$ (S. S. Wilks); (2) tr $\mathbf{Q}_e^{-1}\mathbf{Q}_B=\Sigma\lambda_i$ (D. N. Lawley); (3) $\max\lambda_i$ (S. N. Roy). Small sample distributions of these statistics are complicated, but when $n\to\infty$, $-(n-m)\cdot\log W$ and $n\,\mathrm{tr}\,\mathbf{Q}_e^{-1}\mathbf{Q}_B$ are asymptotically distributed according to a †chi-square distribution with pm degrees of freedom under the hypothesis. As a special case, if $m=1$ there exists only one nonzero λ, and the procedures in this paragraph all coincide and are equivalent to one based on $T^2=M^{-1}(\hat{\mathbf{B}}-\beta_0)'\hat{\Sigma}^{-1}(\hat{\mathbf{B}}-\beta_0)$. It is known that under the hypothesis, $(n-p)T^2/(n-1)p$ is distributed according to an †F-distribution with $(p,n-p)$ degrees of freedom. Also, when $p=2$ (resp. $m=2$), $(n-m-1)(1-\sqrt{W})/m\sqrt{W}$ (resp. $(n-1-p)(1-\sqrt{W})/p\sqrt{W}$) is distributed according to an F-distribution with degree of freedom $(2m, 2(n-m-1))$ (resp. $(2p,2(n-1-p))$). Simultaneous †confidence regions of B can be derived from the testing procedures in this paragraph, that is,

$$\mathrm{tr}\mathbf{Q}_e^{-1}(B-\hat{\mathbf{B}})ZZ'(B-\hat{\mathbf{B}})'<c.$$

Moreover, when the matrix B is decomposed as $B=(B_1 \vdots B_2)$, where B_1 and B_2 are a $p\times q$ matrix and a $p\times(m-1)$ matrix, respectively, and the hypothesis to be tested is of the form $B_1=0$, the test procedures can be obtained as follows: Decompose Z as

$$Z=\begin{pmatrix} Z_1 \\ Z_2 \end{pmatrix},$$

where Z_1 is a $q\times n$ matrix and Z_2 is an $(m-q)\times n$ matrix, and put

$$\hat{\mathbf{B}}_2^*=\mathbf{X}Z_2'(Z_2Z_2')^{-1},$$

$$\mathbf{Q}^*=\mathbf{XX}'-\hat{\mathbf{B}}_2^*Z_2\mathbf{X}', \quad \mathbf{Q}_{B_1}=\mathbf{Q}^*-\mathbf{Q}_e.$$

Then $\mathbf{Q}_{B_1}$ and $\mathbf{Q}_e$ are independent, $\mathbf{Q}_{B_1}$ is distributed according to a Wishart distribution with q degrees of freedom when the hypothesis is true, and we can apply the procedures in the previous paragraph, simply replacing $\mathbf{Q}_B$ by $\mathbf{Q}_{B_1}$.

Such a procedure is called **multivariate analysis of variance** (or **MANOVA**, for short). Various standard situations can be treated in this way (after some linear transformation of variables, if necessary). Some examples are (1) $\mathbf{X}=(\mathbf{X}^{(1)}\ldots\mathbf{X}^{(n)})$, where the $\mathbf{X}^{(i)}$ $(i=1,\ldots,n)$ are distributed independently according to a p-dimensional normal distribution $N(\boldsymbol{\mu},\Sigma)$. We can express $\mathbf{X}$ as $\mathbf{X}=\boldsymbol{\mu}\mathbf{1}'+\mathbf{U}$, and the estimators are given by $\hat{\boldsymbol{\mu}}=\overline{\mathbf{X}}=\mathbf{X1}/n$, $\hat{\boldsymbol{\Sigma}}=(\mathbf{X}-\overline{\mathbf{X}}\mathbf{1}')(\mathbf{X}-\overline{\mathbf{X}}\mathbf{1}')'/(n-1)$. In this case, we obtain a test for the hypothesis $\boldsymbol{\mu}=\boldsymbol{\mu}_0$ based on **Hotelling's** T^2 statistic, i.e., the test with †critical region

$$T^2=n(\overline{\mathbf{X}}-\boldsymbol{\mu}_0)'\hat{\boldsymbol{\Sigma}}^{-1}(\overline{\mathbf{X}}-\boldsymbol{\mu}_0)>c.$$

(2) Suppose that $p\times n_i$ matrices $\mathbf{X}_i$, $i=1,\ldots,k$, are samples of size n_i from p-dimensional normal distributions $N(\boldsymbol{\mu}_i,\Sigma)$ with common variance-covariance matrix Σ. The tests for the hypothesis $\boldsymbol{\mu}_1=\ldots=\boldsymbol{\mu}_k$ are obtained from the following observation: Let $\mathbf{Q}_e=\Sigma_i(\mathbf{X}_i-\overline{\mathbf{X}}_i\mathbf{1}')(\mathbf{X}_i-\overline{\mathbf{X}}_i\mathbf{1}')'$, where $\overline{\mathbf{X}}_i=\mathbf{X}_i\mathbf{1}/n_i$, $\mathbf{Q}_\mu=\Sigma n_i(\overline{\mathbf{X}}_i-\overline{\mathbf{X}})(\overline{\mathbf{X}}_i-\overline{\mathbf{X}})'$, $\overline{\mathbf{X}}=\Sigma n_i\overline{\mathbf{X}}_i/\Sigma n_i$. Then $\Sigma(\mathbf{X}_i-\overline{\mathbf{X}}\mathbf{1}')(\mathbf{X}_i-\overline{\mathbf{X}}\mathbf{1}')'=\mathbf{Q}_\mu+\mathbf{Q}_e$. We call $\mathbf{Q}_e$ the **matrix of the sum of squares within classes**, and $\mathbf{Q}_\mu$ the **matrix of the sum of squares between classes**. The latter is distributed according to a Wishart distribution when the hypothesis is true. (3) Suppose that $\mathbf{X}_{ij}$ are p-dimensional vectors, and that

$$\mathbf{X}_{ij}=\boldsymbol{\mu}+\boldsymbol{\alpha}_i+\boldsymbol{\beta}_j+\mathbf{U}_{ij}, \qquad i=1,\ldots,m, \quad j=1,\ldots,n,$$

where $\boldsymbol{\mu}$, $\boldsymbol{\alpha}_i$, $\boldsymbol{\beta}_j$ are p-dimensional constant vectors such that $\Sigma\boldsymbol{\alpha}_i=0$ and $\Sigma\boldsymbol{\beta}_j=0$, and the $\mathbf{U}_{ij}$ are independently distributed according to a p-dimensional normal distribution $N(\mathbf{0},\Sigma)$. We set

$$\mathbf{Q}_\alpha=n\sum(\overline{\mathbf{X}}_i-\overline{\mathbf{X}})(\overline{\mathbf{X}}_i-\overline{\mathbf{X}})',$$

$$\mathbf{Q}_\beta=m\sum(\overline{\mathbf{X}}_j-\overline{\mathbf{X}})(\overline{\mathbf{X}}_j-\overline{\mathbf{X}})',$$

$$\mathbf{Q}_e=\sum\sum(\mathbf{X}_{ij}-\overline{\mathbf{X}}_i-\overline{\mathbf{X}}_j+\overline{\mathbf{X}})\times(\mathbf{X}_{ij}-\overline{\mathbf{X}}_i-\overline{\mathbf{X}}_j+\overline{\mathbf{X}})',$$

where $\overline{\mathbf{X}}_i=\sum_j\mathbf{X}_{ij}/n$, $\overline{\mathbf{X}}_j=\sum_i\mathbf{X}_{ij}/m$,

$$\overline{\mathbf{X}}=\sum\sum\mathbf{X}_{ij}/mn.$$

Then we have $\Sigma\Sigma(\mathbf{X}_{ij}-\overline{\mathbf{X}})(\mathbf{X}_{ij}-\overline{\mathbf{X}})'=\mathbf{Q}_\alpha+\mathbf{Q}_\beta+\mathbf{Q}_e$, and $\mathbf{Q}_\alpha$, $\mathbf{Q}_\beta$, $\mathbf{Q}_e$ are distributed independently according to (noncentral) Wishart distributions with degrees of freedom $m-1$, $n-1$, and $(n-1)(m-1)$, respectively. The tests for the hypothesis $\boldsymbol{\alpha}_i=0$ $(i=1,\ldots,m)$ or $\boldsymbol{\beta}_j=0$ $(j=1,\ldots,n)$ are obtained from these matrices.

C. Correlation among Variables

The problems that are proper to multivariate analysis generally deal with the structure of the interrelations between the components of p-dimensional observations. If $\mathbf{X}$ is a p-dimensional random vector, then the most common method used to represent the relations between the components of $\mathbf{X}$ is to compute the **variance-covariance matrix**

$$\boldsymbol{\Sigma}=\{\sigma_{ij}\}=E(\mathbf{X}-\boldsymbol{\mu})(\mathbf{X}-\boldsymbol{\mu})',$$

where $\boldsymbol{\mu}=E(\mathbf{X})$, or the correlation matrix $P=(\rho_{ij})=(\sigma_{ij}/\sqrt{\sigma_{ii}\sigma_{jj}})$. Similarly, when $\mathbf{X}=(\mathbf{X}^{(1)}\ldots\mathbf{X}^{(n)})$ (the $\mathbf{X}^{(i)}$ are p-dimensional random vectors), we obtain $p\times p$ matrices such as

$$\mathbf{S}=(S_{ij})=(\mathbf{X}-\overline{\mathbf{X}}\mathbf{1}')(\mathbf{X}-\overline{\mathbf{X}}\mathbf{1}')'/n$$

and

$$\mathbf{R}=(S_{ij}/\sqrt{S_{ii}S_{jj}}),$$

which are called the **sample variance-covariance matrix** and the **sample correlation matrix**. If we denote cofactors of P by P_{ij}, the **multiple correlation coefficient** of the ith coordinate X_i of $\mathbf{X}$ and $X_1,\ldots,X_p$ (excluding X_i) is defined by

$$\rho_{i\cdot 1\ldots(i)\ldots p}=\sqrt{1-|P|/P_{ii}}.$$

Also, given $X_1,\ldots,X_p$ (excluding X_i and X_j), the **partial correlation coefficient** of X_i and X_j is defined by

$$\rho_{ij\cdot 1\ldots(i)\cdot(j)\ldots p}=-P_{ij}/\sqrt{P_{ii}P_{jj}}.$$

$\rho_{i\cdot 1\ldots(i)\ldots p}$ is equal to the correlation coefficient of X_i and the †linear regression function of X_i on $X_1,\ldots,X_p$ (excluding X_i). Also, $\rho_{ij\cdot 1\ldots(i)\cdot(j)\ldots p}$ is equal to the correlation coefficient of $X_i - \hat{X}_i$ and $X_j - \hat{X}_j$, where $\hat{X}_i$ and $\hat{X}_j$ are linear regression functions of X_i and X_j on $X_1, \ldots, X_p$ (X_i and X_j excluded). Similarly, the **sample multiple correlation coefficient** and the **sample partial correlation coefficient** are defined in terms of the sample correlation matrix. That is,

$$R_{i\cdot 1 \ldots (i) \ldots p} = \sqrt{1 - |\mathbf{R}|/R_{ii}}$$

and

$$R_{ij\cdot 1 \ldots (i) \ldots (j) \ldots p} = -R_{ij}/\sqrt{R_{ii}R_{jj}}\ ,$$

where R_{ij} is the cofactor of $\mathbf{R}$. When $\mathbf{X} = (\mathbf{X}^{(1)} \ldots \mathbf{X}^{(n)})$ is a sample of size n from a multivariate normal population, the sampling distributions of $R_{i\cdot 1 \ldots (i) \ldots p}$ and $R_{ij\cdot 1 \ldots (i) \ldots (j) \ldots p}$ are known (→ 368 Sampling Distributions).

The determinant of the variance-covariance matrix $|\Sigma|$ or $|\mathbf{S}|$, called the **(sample) generalized variance**, is a measure of the dispersion of a p-dimensional distribution. The distance of two distributions with mean vectors $\boldsymbol{\mu}_1$ and $\boldsymbol{\mu}_2$, respectively, and with common variance Σ is often expressed by

$$\delta = (\boldsymbol{\mu}_1 - \boldsymbol{\mu}_2)' \Sigma^{-1} (\boldsymbol{\mu}_1 - \boldsymbol{\mu}_2),$$

which is called the **Mahalanobis generalized distance**.

When the data consists of $(p+q)$-dimensional vectors $\begin{pmatrix} \mathbf{X} \\ \mathbf{Y} \end{pmatrix}$ with $q < p$, the interrelation of $\mathbf{X}$ and $\mathbf{Y}$ as a whole can be expressed in the following way: Let the variance-covariance matrix be partitioned as

$$\Sigma = \begin{pmatrix} \Sigma_{\mathbf{XX}} & \Sigma_{\mathbf{YX}} \\ \Sigma_{\mathbf{XY}} & \Sigma_{\mathbf{YY}} \end{pmatrix}$$

and the nonzero roots of the equation $|\rho\Sigma_{\mathbf{YY}} - \Sigma_{\mathbf{YX}}\Sigma_{\mathbf{XX}}^{-1}\Sigma_{\mathbf{XY}}| = 0$ be $\rho_1, \ldots, \rho_q$. Then $\rho_1^{1/2}, \ldots, \rho_q^{1/2}$, called the **canonical correlation coefficients**, are the maximal invariant statistics with respect to linear transformation of $\mathbf{X}$ and $\mathbf{Y}$. Also, if we denote the characteristic vector corresponding to a root ρ_i by $\boldsymbol{\eta}_i$, i.e.,

$$(\rho_i \Sigma_{\mathbf{YY}})\boldsymbol{\eta}_i = (\Sigma_{\mathbf{XY}}\Sigma_{\mathbf{YY}}^{-1}\Sigma_{\mathbf{XY}})\boldsymbol{\eta}_i,$$

the linear statistics $\boldsymbol{\eta}_i'\mathbf{Y}$ and $\boldsymbol{\eta}_i'\Sigma_{\mathbf{YX}}\Sigma_{\mathbf{XX}}^{-1}\mathbf{X}$ are called the **canonical variates**.

D. Principal Components

An important problem in multivariate analysis is to express the variations of many variables by a small number of indices. **Principal component analysis** is a technique of dealing with this problem. Let $\mathbf{X}$ be a $p \times n$ matrix of p-dimensional data. Suppose that $\mathbf{T}$ is the $r \times n$ matrix of the indices. We fix r and determine $\mathbf{T}$ so that the sum of squares of the multiple correlation coefficients of each of the components of $\mathbf{X}$ on the components of $\mathbf{T}$ is maximized. Then $\mathbf{T}$ is given by $\mathbf{T} = A\mathbf{X}$, where A is an $r \times p$ matrix formed by the r characteristic vectors of the correlation matrix of $\mathbf{X}$ corresponding to the r largest roots, and the sum of squares of multiple correlation coefficients is equal to the sum of the r characteristic roots. The matrix $\mathbf{T}$ is called the **principal component**.

When we assume normality, the characteristic roots of the sample correlation matrix are the †maximum likelihood estimators of the characteristic roots of the population correlation matrix, and the sampling distribution can be obtained. A hypothesis relevant to principal component analysis is, for example, that the smallest $p-r$ roots of the correlation matrix are equal, which can be tested by the statistic

$$R_{p-r} = |\mathbf{R}|/(\lambda_1 \ldots \lambda_r)\big((p - \lambda_1 - \ldots - \lambda_r)/(p-r)\big)^{p-r},$$

where $\lambda_1, \ldots, \lambda_r$ are the r largest roots of R. Under the hypothesis, $-c\log R_{p-r}$ (c a constant) is asymptotically distributed according to a chi-square distribution when $n \to \infty$.

Variations of principal component analysis can be obtained by taking the characteristic vectors of the variance-covariance matrix, or of a multiple of it by some weight matrix.

E. Factor Analysis

Factor analysis is closely related to principal component analysis. We assume a model

$$\mathbf{X} = BF + \mathbf{U},$$

where B and F are unknown $p \times r$ and $r \times n$ matrices of constants ($p > r$) and $\mathbf{U}$ is a $p \times n$ matrix of independent errors. F is called the matrix of **factor scores** and B the matrix of **factor loadings**. We assume $FF' = nI$. If $E(\mathbf{UU}') = n\Phi$ is known, then by applying the least squares principle, we can determine B and F so as to minimize the trace of $(\mathbf{X} - BF)'\Phi^{-1}(\mathbf{X} - BF)$. Then B is obtained by taking the r characteristic vectors of $\Phi^{-1}\mathbf{XX}'$ corresponding to the r largest characteristic roots. When Φ is diagonal but unknown, we can solve the simultaneous equation for B and Φ, whose solutions are the matrix $\hat{\mathbf{B}}$ with columns equal to characteristic vectors of $\hat{\mathbf{\Phi}}^{-1}\mathbf{XX}'$ and the diagonal matrix $\hat{\mathbf{\Phi}}$ with elements equal to the diagonal elements of $\mathbf{XX}'/n - \hat{\mathbf{B}}\hat{\mathbf{B}}'$.

If we assume that $\mathbf{U}$ is normal, the procedure in the preceding paragraph for the case when Φ is known is equivalent to the maximum likelihood method. When Φ is unknown, we may further assume that the columns in F are also multivariate normal vectors distributed independently of $\mathbf{U}$, which implies that the columns of $\mathbf{X}$ are also normal vectors with the variance-covariance matrix $\Sigma = BB' + \Phi$. B and Φ are estimated from the sample variance-covariance matrix, and the solution of the simultaneous equation for B and Φ gives the maximum likelihood estimator. However, there is a so-called **identification** problem, to determine whether for given Σ and r the decomposition $\Sigma = BB' + \Phi$ is unique, which is not yet completely settled.

F. Canonical Correlation Analysis

Canonical correlation coefficients and canonical variates can also be computed from the sample. They have various descriptive implications. Suppose that $\boldsymbol{\eta}_1'\mathbf{Y}$ and $\boldsymbol{\xi}_1'\mathbf{X}$ are the first canonical variates corresponding to the largest canonical correlation $\rho_1^{1/2}$. Then $\rho_1^{1/2}$ is the largest possible correlation between a linear function of $\mathbf{X}$ and a linear function of $\mathbf{Y}$, and is actually equal to the correlation of $\boldsymbol{\eta}_1'\mathbf{Y}$ and $\boldsymbol{\xi}_1'\mathbf{X}$. Similarly, the second canonical correlation is equal to the largest possible correlation between linear functions in $\mathbf{X}$ and in $\mathbf{Y}$ which are orthogonal to $\boldsymbol{\xi}_1'\mathbf{X}$ and $\boldsymbol{\eta}_1'\mathbf{Y}$, respectively, and so forth.

As a second interpretation of the canonical variates, we consider the linear regression model

$$\mathbf{Y} = B\mathbf{X} + \mathbf{U},$$

where $\mathbf{Y}$, B, $\mathbf{U}$ are $q \times n$, $q \times p$, $q \times n$ matrices, respectively, and $\operatorname{rank} B = r < q$. Then there exists an $r \times q$ matrix A such that $\mathbf{Y} = C\mathbf{T} + \mathbf{U}$, where $\mathbf{T} = A\mathbf{X}$. If $E(\mathbf{U}\mathbf{U}') = n\Sigma$ is known, least squares considerations lead to minimizing $\operatorname{tr}(\mathbf{Y} - B\mathbf{X})'\Sigma^{-1}(\mathbf{Y} - B\mathbf{X})$ with the condition $\operatorname{rank} B = r$, and if Σ is replaced by $\hat{\Sigma} = n^{-1}(\mathbf{S}_{\mathbf{YY}} - \mathbf{S}_{\mathbf{YX}}\mathbf{S}_{\mathbf{XX}}^{-1}\mathbf{S}_{\mathbf{XY}})$, then the resulting A consists of the r characteristic vectors corresponding to the r largest roots of the matrix $\mathbf{S}_{\mathbf{YY}}^{-1}\mathbf{S}_{\mathbf{YX}}\mathbf{S}_{\mathbf{XX}}^{-1}\mathbf{S}_{\mathbf{XY}}$ or the roots of the equation $|\mathbf{S}_{\mathbf{YY}} - \rho\mathbf{S}_{\mathbf{YX}}\mathbf{S}_{\mathbf{XX}}^{-1}\mathbf{S}_{\mathbf{XY}}| = 0$. Hence $\mathbf{T} = A\mathbf{X}$ and $\mathbf{Z} = C'\mathbf{Y}$ are equal to the matrices of canonical variates. If we assume that $\mathbf{U}$ is normal, this procedure is equivalent to the maximum likelihood method. It should be remarked that although the model here is not symmetric in $\mathbf{X}$ and $\mathbf{Y}$, the results are symmetric in $\mathbf{X}$ and $\mathbf{Y}$, and therefore they will be the same if $\mathbf{X}$ and $\mathbf{Y}$ interchange their roles in this model.

G. Linear Discriminants and Classificatory Problems

Let $p \times n_i$ matrices $\mathbf{X}_i$ $(i = 1,, \ldots, k)$ be the set of observations for k distinct populations with a common variance-covariance matrix. We shall determine a vector $\mathbf{a}$ such that $\mathbf{T}_i = \mathbf{a}'\mathbf{X}_i$ reveals the differences of the k populations as much as possible, or, more precisely, so that the ratio of the sum of squares between classes of $\mathbf{T}$ to the sum of squares within classes is maximized. If the matrices of the sums of squares between and within classes are $\mathbf{Q}_b$ and $\mathbf{Q}_w$, respectively, the ratio is equal to

$$l = \mathbf{a}'\mathbf{Q}_b\mathbf{a} / \mathbf{a}'\mathbf{Q}_w\mathbf{a},$$

which is maximized when $\mathbf{a}$ is equal to the characteristic vector of $\mathbf{Q}_w^{-1}\mathbf{Q}_b$ corresponding to the largest root. The linear function $t = \mathbf{a}'\mathbf{x}$ is called the **linear discriminant function**. When $k = 2$, $\mathbf{a}$ is given by $\mathbf{a} = \mathbf{Q}_w^{-1}(\overline{\mathbf{X}}_1 - \overline{\mathbf{X}}_2)$, where $\overline{\mathbf{X}}_1$ and $\overline{\mathbf{X}}_2$ are sample mean vectors. When $k > 2$, we let A be the matrix formed by the r characteristic vectors corresponding to the first r largest characteristic roots of $\mathbf{Q}_w^{-1}\mathbf{Q}_b$, and set $\mathbf{T}_i = A\mathbf{X}_i$. From this we can construct the r-dimensional discriminant function. These functions can be used to locate the k populations in r-dimensional space, and also to decide to which population a new observation belongs. For the latter problem we can also construct k quadratic functions $s_i = (\mathbf{X} - \overline{\mathbf{X}}_i)'\mathbf{Q}_w^{-1}(\mathbf{X} - \overline{\mathbf{X}}_i)$, where $\overline{\mathbf{X}}_i$ is the sample mean vector of the ith population, and $\mathbf{X}$ a new observation. Then we may decide that $\mathbf{X}$ belongs to the population corresponding to the minimum s_i. Such a method is called a **classificatory procedure.**

H. Other Problems

The sampling distributions associated with the procedures discussed in this article are usually very complicated, and often only asymptotic properties are known (→ 368 Sampling Distributions). Also, the entire discussion is heavily dependent on the normality assumption, and although some nonparametric procedures for MANOVA have been discussed recently, purely multivariate techniques independent of the normality assumption are yet to be found.

References

[1] T. W. Anderson, An introduction to multivariate statistical analysis, John Wiley, 1958.
[2] M. G. Kendall, A course in multivariate analysis, Hafner, 1957.
[3] D. N. Lawley and A. E. Maxwell, Factor analysis as a statistical method, Butterworths, 1963.
[4] C. R. Rao, Advanced statistical methods in biometric research, John Wiley, 1952.